BURGER'S MEDICINAL CHEMISTRY, DRUG DISCOVERY AND DEVELOPMENT

BURGER'S MEDICINAL CHEMISTRY, DRUG DISCOVERY AND DEVELOPMENT

BURGER'S MEDICINAL CHEMISTRY, DRUG DISCOVERY AND DEVELOPMENT

Seventh Edition

Volume 2: Discovery Lead Molecules

Edited by

Donald J. Abraham
Virginia Commonwealth University

David P. Rotella
Wyeth Research

Burger's Medicinal Chemistry, Drug Discovery and Development
is available Online in full color at
http:/mrw.interscience.wiley.com/emrw/9780471266945/home/

A JOHN WILEY & SONS, INC., PUBLICATION

Published by John Wiley & Sons, Inc., Hoboken, New Jersey
Published simultaneously in Canada

Library of Congress Cataloging-in-Publication Data:

Abraham, Donald J., 1936-
Burger's medicinal chemistry, drug discovery, and development/Donald J. Abraham, David P. Rotella. – 7th ed.
p. ; cm.
Other title: Medicinal chemistry, drug discovery, and development
Rev. ed. of: Burger's medicinal chemistry and drug discovery. 6th ed. / edited by Donald J. Abraham. c2003.
Includes bibliographical references and index.
ISBN 978-0-470-27815-4 (cloth)
1. Pharmaceutical chemistry. 2. Drug development. I. Rotella, David P. II. Burger, Alfred, 1905-2000. III. Burger's medicinal chemistry and drug discovery. IV. Title. V. Title: Medicinal chemistry, drug discovery, and development.
[DNLM: 1. Chemistry, Pharmaceutical–methods. 2. Biopharmaceutics–methods. 3. Drug Compounding–methods. QV 744 A105b 2010]
RS403.B8 2010
615'.19–dc22 2010010779

Printed in Singapore

10 9 8 7 6 5 4 3 2 1

CONTENTS

PREFACE

The seventh edition of Burger's Medicinal Chemistry resulted from a collaboration established between John Wiley & Sons, the editorial board, authors, and coeditors over the last 3 years. The editorial board for the seventh edition provided important advice to the editors on topics and contributors. Wiley staff effectively handled the complex tasks of manuscript production and editing and effectively tracked the process from beginning to end. Authors provided well-written, comprehensive summaries of their topics and responded to editorial requests in a timely manner. This edition, with 8 volumes and 116 chapters, like the previous editions, is a reflection of the expanding complexity of medicinal chemistry and associated disciplines. Separate volumes have been added on anti-infectives, cancer, and the process of drug development. In addition, the coeditors elected to expand coverage of cardiovascular and metabolic disorders, aspects of CNS-related medicinal chemistry, and computational drug discovery. This provided the opportunity to delve into many subjects in greater detail and resulted in specific chapters on important subjects such as biologics and protein drug discovery, HIV, new diabetes drug targets, amyloid-based targets for treatment of Alzheimer's disease, high-throughput and other screening methods, and the key role played by metabolism and other pharmacokinetic properties in drug development.

The following individuals merit special thanks for their contributions to this complex endeavor: Surlan Alexander of John Wiley & Sons for her organizational skills and attention to detail, Sanchari Sil of Thomson Digital for processing the galley proofs, Jonathan Mason of Lundbeck, Andrea Mozzarelli of the University of Parma, Alex Tropsha of the University of North Carolina, John Block of Oregon State University, Paul Reider of Princeton University, William (Rick) Ewing of Bristol-Myers Squibb, William Hagmann of Merck, John Primeau and Rob Bradbury of AstraZeneca, Bryan Norman of Eli Lilly, Al Robichaud of Wyeth, and John Lowe for their input on topics and potential authors. The many reviewers for these chapters deserve special thanks for the constructive comments they provided to authors. Finally, we must express gratitude to our lovely, devoted wives, Nancy and Mary Beth, for their tolerance as we spent time with this task, rather than with them.

As coeditors, we sincerely hope that this edition meets the high expectations of the scientific community. We assembled this edition with the guiding vision of its namesake in mind and would like to dedicate it to Professor H.C. Brown and Professor Donald T. Witiak. Don collaborated with Dr. Witiak in the early days of his research in sickle cell drug discovery. Professor Witiak was Dave's doctoral advisor at Ohio State University and provided essential guidance to a young scientist. Professor Brown, whose love for chemistry infected all organic graduate students at Purdue University, arranged for Don to become a medicinal chemist by securing a postdoctoral position for him with Professor Alfred Burger.

It has been a real pleasure to work with all concerned to assemble an outstanding and up-to-date edition in this series.

Donald J. Abraham
David P. Rotella

March 2010

CONTRIBUTORS

Adriano D. Andricopulo, Universidade de São Paulo, São Carlos-SP, Brazil
Valerie S. Bernan, Wyeth Research, Pearl River, NY
Julian Blagg, Haddow Laboratories, Sutton, Surrey, UK
Frank E. Blaney, GlaxoSmithKline plc, Harlow, Essex, UK
John H. Block, Oregon State University, Corvallis, OR
Nicholas Bodor, University of Florida, Gainesville, FL
Peter Buchwald, University of Miami, Miami, FL
Guy T. Carter, Wyeth Research, Princeton, NJ; Wyeth Research, Pearl River, NY
Richard J. Clark, The University of Queensland, Brisbane, Australia
Pietro Cozzini, University of Parma, Parma, Italy; Italian National Institute of Biostructures and Biosystems, Rome, Italy
Gordon M. Cragg, NCI-Frederick, Frederick, MD
David J. Craik, The University of Queensland, Brisbane, Australia
Li Di, Wyeth Research, Princeton, NJ
Alexander Dömling, University of Pittsburgh, Pittsburgh, PA
Emir Duzic, Discovery Research Cephalon, Inc., West Chester, PA
Sean Ekins, Collaborations in Chemistry, Jenkintown, PA; University of Maryland, Baltimore, MD; University of Medicine and Dentistry, Piscataway, NJ; Collaborative Drug Discovery, Burlingame, CA
Steven L. Garland, GlaxoSmithKline plc, Harlow, Essex, UK
Holger Gohlke, Heinrich-Heine-University, Düsseldorf, Germany
Rafael V.C. Guido, Universidade de São Paulo, São Carlos-SP, Brazil
J. Dana Honeycutt, Accelrys, San Diego, CA
Edgar Jacoby, Novartis Institutes for BioMedical Research, Basel, Switzerland
Glen Eugene Kellogg, Virginia Commonwealth University, Richmond, VA
Edward H. Kerns, Wyeth Research, Princeton, NJ
Frank E. Koehn, Wyeth Research, Pearl River, NY
Hannes Kopitz, Heinrich-Heine-University, Düsseldorf, Germany
Andrei Leitão, University of New Mexico, Albuquerque, NM
Michael J. Marino, Discovery Research Cephalon, Inc., West Chester, PA
James T. Metz, Abbott Laboratories, Abbott Park, IL
Carlos Montanari, Universidade de São Paulo, São Carlos-SP, Brazil
Ingo Muegge, Boehringer Ingelheim Pharmaceuticals, Inc., Ridgefield, CT
Adrian J. Mulholland, University of Bristol, Bristol, UK
David J. Newman, NCI-Frederick, Frederick, MD
Scott Oloff, Boehringer Ingelheim Pharmaceuticals, Inc., Ridgefield, CT
Igor Polikarpov, Universidade de São Paulo, São Carlos-SP, Brazil
Iosif-Daniel Rosca, University of Bucharest, Bucharest, Romania; Concordia University, Montreal, Canada
Katherine E. Shaw, University of Bristol, Bristol, UK
Phillippa A. Smith, The University of Queensland, Brisbane, Australia
Francesca Spyrakis, University of Parma, Parma, Italy; Italian National Institute of Biostructures and Biosystems, Rome, Italy
Peter W. Swaan, University of Maryland, Baltimore, MD
Daniela B.B. Trivella, Universidade de São Paulo, São Carlos-SP, Brazil
Jean-Maurice Vergnaud, University of Saint-Etienne, Saint-Chamond, France
Michael Williams, Discovery Research Cephalon, Inc., West Chester, PA
Christopher J. Woods, University of Bristol, Bristol, UK

VIRTUAL SCREENING

INGO MUEGGE
SCOTT OLOFF
Boehringer Ingelheim
Pharmaceuticals,
Inc., Ridgefield, CT

1. INTRODUCTION

Since the 6th edition of *Burger's Medicinal Chemistry and Drug Discovery* went to print six years ago, the field of virtual screening (VS) has expanded dramatically in scope as well as in breadth of application. Figure 1 attests eloquently to the increased use of VS in medicinal chemistry-related research and drug discovery. In 2008 alone, there are more than 404 articles in journals published by the American Chemical Society that contain the phrase "virtual screening"; 50 of them maintain this phrase in its title (compared to only 6 in 2002). These numbers amount to a >2.5-fold increase in yearly publications related to VS over these 6 years. New areas of interest surfaced during the past few years including a resurgence of fragment-based screening techniques often guided by VS, the virtual screening of trillions of combinatorial compounds, a growing interest in the *in silico* screening of synthetically accessible compounds (also often know as *de novo* design), the increased use of machine learning techniques, the emphasis on scaffold hopping, and combining ligand-based and structure-based approaches in synergy to name a few. In addition to highlighting some of these new developments in VS, we have revised the topics presented in the last edition and include more recent examples for VS successes.

Virtual screening, sometimes also called *in silico* screening, is an established branch of medicinal chemistry that represents a fast and cost-effective tool for computationally screening virtual compound databases in search for novel drug leads. The roots for virtual screening go back to structure-based drug design and molecular modeling. In the 1970s, researchers hoped to find novel drugs designed rationally using a fast growing number of diverse protein structures being solved by X-ray crystallography [1,2] or nuclear magnetic resonance spectroscopy (NMR) [3]. However, only very few drugs resulted from those early efforts. Examples include captopril as an angiotensin-converting enzyme inhibitor [4] and metothrexate as a dihydrofolate reductase inhibitor [5]. The reasons for this somewhat disappointing drug yield lie in the low resolution of the protein structures as well as limitations in compute power and methods. Researchers have often tried to *de novo* design the final drug candidate on a computer screen. The compounds suggested have often been difficult to synthesize; initial failure in exhibiting potency has often resulted in the termination of structure-based projects. At the end of the 1980s, rational drug design techniques became somewhat discredited due to the high failure rate in drug discovery projects.

In the 1990s, drastic changes occurred in the way drugs are discovered in the pharmaceutical industry. High-throughput synthesis [6,7] and screening techniques [8] changed the lead identification process that is now governed not only by large numbers of compounds processed but also by fast prosecution of many putative drug targets in parallel. The characterization of the human genome has resulted in a large number of novel putative drug targets. Improved screening techniques make it also possible to look at the entire gene families, at orphan targets, or at otherwise uncharacterized putative drug targets. In this environment of data explosion, rational design techniques experienced a comeback [9]. While the growing number of solved protein structures at high resolution makes it possible to embark on structure-based design for many drug targets, virtual screening—the computational counterpart to high-throughput screening—has become a particularly successful computational tool for lead finding in drug discovery.

Proprietary screening collections typically hold approximately 10^6 compounds. This is only a tiny fraction of the conceivable drug-like chemical space that is estimated to hold more than 10^{60} compounds [10,11]. Even if this space is reduced to compounds that are comparably easy to access synthetically it is still an astronomical number. Virtual screening attempts to suggest which compounds

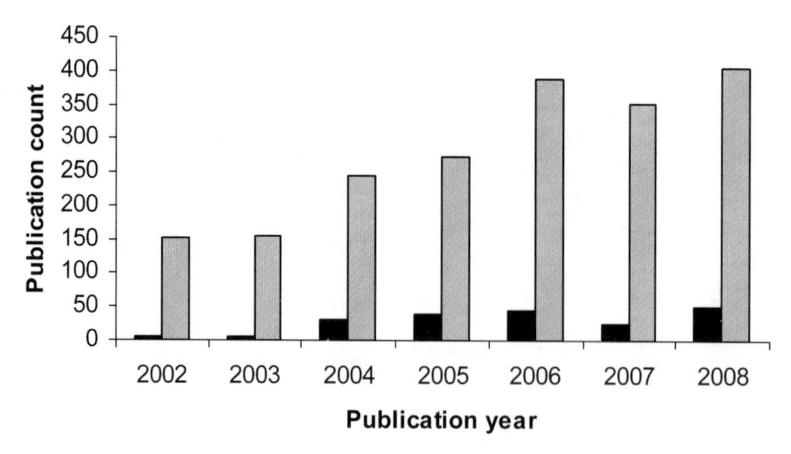

Figure 1. Number of ACS publications with the phrase "virtual screening" in the text (gray patterned bars) or in the title (black bars).

should actually be synthesized or tested against a drug target of interest. Large virtual libraries of up to 10^{12} individual compounds or 10^{14} combinatorial chemistry compounds or 10^{18}*de novo* compounds reassembled from fragments [12] can be screened today using a cascade of various screening tools to reduce the chemical space.

This chapter describes the different concepts and tools used today for virtual screening. They reach from the assessment of the overall "drug-likeness" of a small organic molecule to its ability to specifically bind to a given drug target. Ligand-based as well as structure-based approaches are described and illustrated with specific examples. The ability of virtual screening to assist in scaffold hopping (identifying isofunctional molecular structures with significantly different molecular backbones [13]) is also discussed. Additional topics include the triaging of VS hit sets, the use of machine learning techniques, and finding synergies between different VS approaches. For additional information, the interested reader is referred to a selection of recent books and reviews on the subject of virtual screening [14–25].

2. LIGAND-BASED VIRTUAL SCREENING

Casting a wide net, ligand-based virtual screening (LBVS) encompasses all techniques to computationally screen for novel compounds using ligand information only without including any information of the target structure. LBVS ranges from compound similarity methods for which only a single template molecule needs to be known to very sophisticated pharmacophore elucidation and machine learning methods using a multitude of information on multiple compounds. The interested reader is referred to recent reviews on the topic of similarity searching [21,26], pharmacophore elucidation methods [27,28], and machine learning [29,30] for virtual screening.

2.1. Compound Similarity

Compound similarity as basis for virtual screening relies on the assumption that similar compounds exhibit similar biological activities. The degree to which this is true varies substantially, however. Martin et al. showed from the analysis of a series of HTS experiments at Abbott Laboratories that only approximately 30% of compounds with Daylight fingerprint Tanimoto similarities >0.85 show activity against the same drug target [31]. Compound similarity is generally measured in either structural terms or property terms. Topological fingerprints such as Daylight fingerprints [32] or atom pair descriptors [33], circular fingerprints such as SciTegic fingerprints [34], and structural keys such as ISIS keys [35] encode structural features of the molecules. MolconnZ descriptors [36] present a mixture of molecular connectivity and property description, graph theoretical description, and topological description. Properties such as lipophilicity (calculated $\log P$), molecular weight, the number of rotatable bonds, hydrogen-bond acceptors and donors, pK_a values etc. form a group of property descriptors that can link compounds that are structurally quite different. Using such molecular descriptors is quite popular for virtual screening purposes. Especially 2D fingerprints have performed very well in virtual screening [37–41]. However, 3D descriptors have gained in popularity [42–44]. Jenkins et al. showed that 3D pharmacophore feature descriptors called FEPOPS can outperform standard 2D descriptors in a virtual screening protocol [43]. Similarly, Good et al. have demonstrated applying a variety of 2D and 3D descriptors that 3D pharmacophore fingerprints are better suited than 2D descriptors for scaffold hopping [45]. 3D descriptors used in the context of similarity searching often capture pharmacophore features such as hydrophobic or aromatic moieties, hydrogen-bond (HB) acceptors or donors, and both negative and positive ionizable groups. 3D pharmacophore fingerprints store the information of pairs, triplets, or quartets of these pharmacophore features for multiple conformations in the form of binned distance ranges [46–53]. 3D pharmacophore fingerprints are used more often now as descriptors for QSAR as well as for the design of compound libraries [52]. The usefulness of 3D pharmacophore fingerprints in finding new chemotypes through virtual screening has recently been demonstrated [54]. The use of fingerprints to describe chemotypes and biological effects are described in Chapter 12 of Volume 1.

2.1.1. Example 1: Virtual Screening for 15-Lipoxygenase Inhibitors

Weinstein et al. reported the identification of a new 15-lipoxygenase inhibitor from the virtual screen of a corporate database of compounds [55]. Starting from the literature compound PD-146176 as a template, 4-point pharmacophore fingerprints [48] were generated for the template compound as well as for all database compounds and then compared. Figure 2 shows the template compound as well as the resulting virtual screening hit. The figure illustrates how structurally different the virtual screening hit is from the template molecule. Some resemblance of the indole moiety is noticeable. This example illustrates the scaffold hopping abilities of 3D pharmacophore fingerprints.

2.2. Pharmacophore Screening

Pharmacophore models are built more often based on small-molecule information. They are a preferred filter tool for virtual screening [56]. Feature-based pharmacophore eluci-

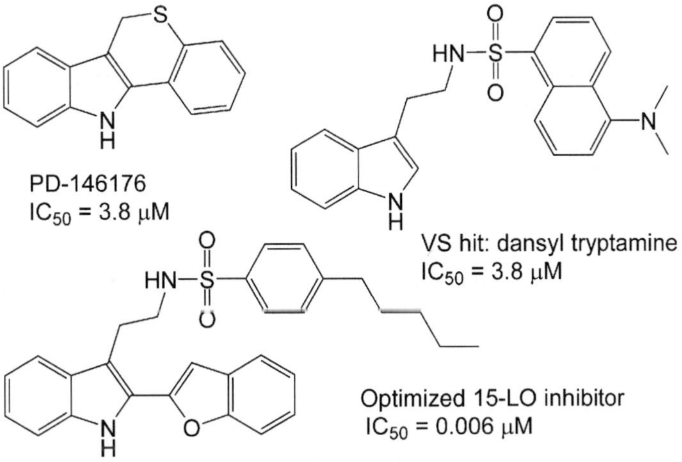

PD-146176
IC_{50} = 3.8 μM

VS hit: dansyl tryptamine
IC_{50} = 3.8 μM

Optimized 15-LO inhibitor
IC_{50} = 0.006 μM

Figure 2. Discovery of a new class of 15-lipoxygenase inhibitors through virtual screening using a 3D pharmacophore fingerprint similarity technique.

dation algorithms such as HipHop [57] and other automated pharmacophore query builders [58] allow for the fast generation of pharmacophore hypotheses. Today sophisticated tool such as LigandScout [59] are available that extract pharmacophore models from one or more protein–ligand crystal structures making pharmacophore modeling a hybrid between ligand-based and structure-based modelings. Traditionally, however, pharmacophore modeling has been most often applied to the target class of GPCRs [60] where until very recently structural information [61–65] of the target receptors was unavailable. Pharmacophore filters are generally much faster than molecular docking approaches and can, therefore, greatly reduce the number of compounds screened further using more expensive docking applications. Today, pharmacophore features are guiding most 3D virtual screening approaches including docking in the form of constraints or prefilters. Most time consuming is still the generation of 3D conformations for small molecules—a task that can only be solved for a small number of molecules when precalculation and storage is necessary.

An interesting aspect of pharmacophores in virtual screening is 3D pharmacophore diversity. While the diversity concept for virtual compounds in general is not applicable because of the enormity of the chemical space, diversity in pharmacophore space is a feasible concept. Virtual libraries can therefore be optimized to cover a wide pharmacophore space [66].

2.2.1. Introduction to Pharmacophores In 1894, Emil Fischer proposed the "lock and key" hypothesis to characterize the binding of compounds to proteins [67]. This can be considered the first attempt to explain binding of small molecules to a biological target. Proteins recognize substrates through specific interactions, forming the pharmacophore. Inhibitors intended to block substrate binding to the protein, should also capture these interactions. The first definition of the pharmacophore formulated by Paul Ehrlich was "a molecular framework that carries (phoros) the essential features responsible for a drug's (pharmacon) biological activity" [68]. This definition was slightly modified by Peter Gund to "a set of structural features in a molecule that is recognized at a receptor site and is responsible for that molecule's biological activity" [69]. An example is shown in Fig. 3. An X-ray structure of CDK2 complexed with the adenine-derived inhibitor H717 [70–72] was solved. Interactions that are essential to substrate and inhibitor binding to the enzyme will form the pharmacophore that should be captured by inhibitors binding the same way H717 does. As shown in Fig. 3, the inhibitor binds to the hinge region (Phe82 and Leu83) through two hydrogen bonds, to a hydrophobic region through the cyclopentyl group, and to Asp145 and Asn132 through hydrogen bonds. The pharmacophore that reflects these interactions has a hydrogen-bond donor and a hydrogen-bond acceptor pair that ensures binding to the hinge region, a hydrophobic group that corresponds to the cyclopentyl binding site, and a hydrogen-bond donor that ensures binding to Asp145 and/or Asn132. Pharmacophore hypotheses can be generated using structural information from ligands or from the protein active site itself. Pharmacophore-based drug design is further elaborated in Chapter 11 of Volume 1.

2.2.2. Databases of Organic Compounds It is much more practical to virtually screen databases of compounds that have already been synthesized than to operate in a completely virtual chemical space. If no information about possible lead structures is available, if in-house efforts such as HTS were not successful, and if cost and time are an issue, virtual screening of vendor databases and the subsequent purchase of hits is the method of choice for the generation of novel lead chemical matter.

There is a wealth of databases of available chemical matter including academic sources such as the ZINC database [73] and the National Cancer Institute Database [74,75] as well as commercial vendor databases such as eMolecules [76] that holds approximately 10 million commercially available compounds, Chemspider [77] that is publicly available holding approximately 20 million compounds, and the Available Chemicals Directory screening collection [78]. Some vendors maintain

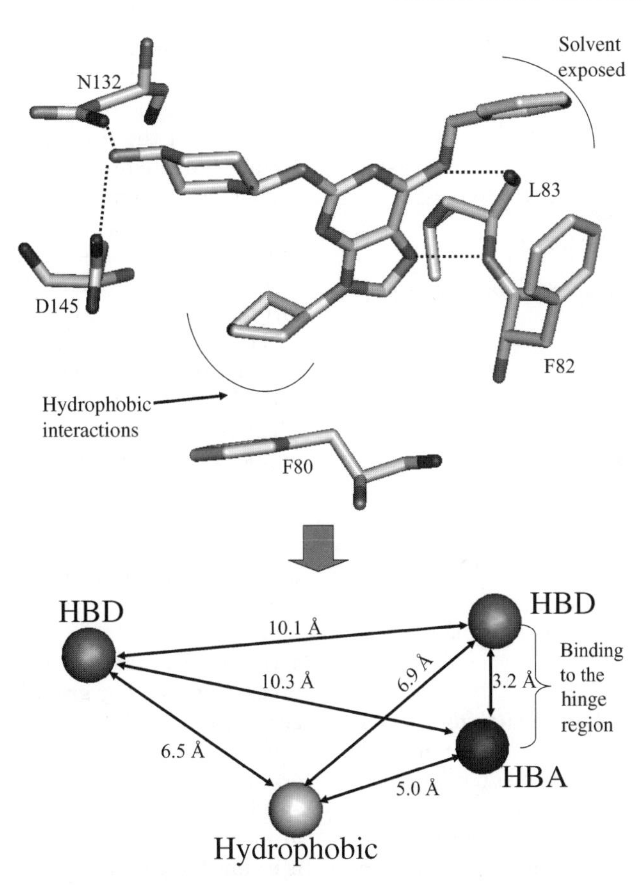

Figure 3. Pharmacophore derived based on the interactions between human cyclin-dependent kinase 2 and the adenine-derived inhibitor H717 as observed in the X-ray structure of the complex (PDB ID 1g5s). Dashed lines highlight hydrogen-bonding interactions. HBD: hydrogen-bond donor, HBA: hydrogen-bond acceptor, and the hinge region is linking the N- and C-terminal domains of a kinase.

databases of virtual compounds that can be custom synthesized. Enamine's REAL database of close to 10 million compounds is an example [79]. In addition, many vendors of chemicals also provide searchable databases with 2D-structure and property information of their compounds. Sometimes, compounds are coded in linear representations such as the SMILES [80,81] notation. The SMILES codes obtained using CACTVS and Daylight programs for 4-benzyl pyridine and R-cocaine are shown in Fig. 4.

Managing compound collections from different sources is not trivial. Chemical databases frequently contain incorrect structures. Careful treatment and curation of compound collections is therefore important to assure optimal virtual screening outcomes. This includes the appropriate assignment of charges, chirality, and tautomerization states, filtering for duplicates, and unwanted structures.

The primary source of 3D experimental structures of organic molecules is the Cambridge Structural Database [82]. Alternatively, 2D databases of organic compounds can be converted into 3D databases using several software programs [83]. Each program starts with generating a crude structure that is subsequently optimized using a force field. To mention the most commonly used programs, CONCORD [84] applies rules derived from experimental structures and a univariate strain function for building an initial structure; CORINA [85] generates an initial structure using a standard set of bond lengths, angles, and dihedrals and rules for cyclic systems.

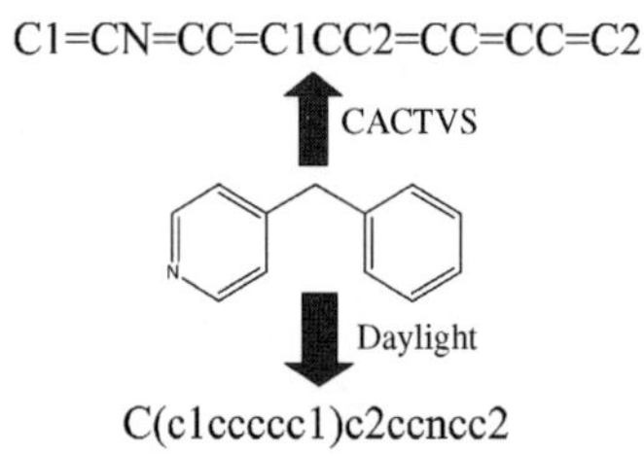

Figure 4. Examples of SMILES notations for two compounds obtained using CACTVS and Daylight.

2.2.3. 2D Pharmacophore Searching Different strategies are pursued to search a 2D database. An exact structure search is applied to find out if a compound is present in the database. Substructure searches are used to find larger molecules with the query imbedded. Superstructure searches are used to find smaller molecules that are embedded in the query. Similarity searches are used to find compounds that are structurally related to the query. Some searching software combines similarity with substructure or super structure searches. Flexible match searches are used for identifying compounds that differ from the query structure in user specified ways. In addition, isomer, tautomer, parent molecule searches may be done to find in a database isomers, tautomers or parent molecules of the query.

2.2.4. Ligand-Based 3D Pharmacophore Generation Ligand-based pharmacophores are typically used when the crystallographic, solution structure, or modeled structure of a protein cannot be obtained. When a set of active compounds is known and it is hypothesized that all compounds bind in a similar way to the protein, then common groups should interact with the same protein residues. Thus, a pharmacophore capturing these common features should be able to identify from a database novel compounds that bind to the same site of the protein as the known compounds do. The process of deriving a pharmacophore, called pharmacophore mapping, consists of three steps: (1) identifying common binding elements that are responsible for biological activity, (2) generating potential conformations that active compounds may adopt, and (3) measuring distances between binding pharmacophore elements in each conformation. To build a pharmacophore based on a set of active compounds, two methods are usually applied. One method generates a set of minimum energy conformations for each ligand and searches for common structural features. Another method considers all possible conformations of each ligand to evaluate shared orientations of common functional groups. Analyzing many low energy conformers of active compounds can suggest a range of distances between key groups that will take into account the flexibility of the ligands and of the protein. This task can be performed manually or automatically.

Pharmacophore generation through conformational analysis and manual alignment is a time-consuming task especially when the list of active ligands is large and the elements of the pharmacophore model are not obvious. There are several software products and algorithms such as HipHop [86], HypoGen [87], PHASE [88], MOE [89], and older tools such as Disco [90], Gasp [91], Flo [92], APEX [93] that can automatically generate potential pharmacophores from a list of known inhibitors. A collection of views on pharmacophore elucidation can be found in Chapter 11 of Volume 1 as well as in the book by Güner [94]. The programs use algorithms that identify common pharmacophore features in the training set molecules that are ranked with a scoring function. Common pharmacophore features include: hydrogen-bond donors, acceptors, negative and positive charges or ionizable centers, and surface accessible hydrophobic regions that can be aliphatic, aromatic, or nonspecific. Most of the programs incorporate ligand flexibility when generating pharmacophores since compounds may not bind to the protein in the minimum energy conformation.

2.2.5. Example 2: Virtual Screening for Dopamine Transporter Inhibitors The dopamine transporter (DAT) is a 12 transmembrane helix protein that plays a critical role in terminating dopamine neurotransmission by taking up dopamine released into the synapse. DAT is involved in several diseases such as drug addiction and attention deficit disorder [95]. DAT inhibitors share one or more common 3D pharmacophore models [96–98]. A pharmacophore model was derived based on two known potent DAT inhibitors R-cocaine and WIN-35065-2 (Fig. 5) [98]. The common binding elements of these compounds are a ring nitrogen that may be substituted, a carbonyl oxygen, and an aromatic ring that can be defined by the position of its center. A systematic conformational search followed by conformer clustering was performed to obtain all possible conformations these compounds could assume when bound to DAT. The resulting 3D pharmacophore model (Fig. 5) was used to search the NCI 3D-database [75] of 206,876 compounds using the program Chem-X [99]. A total of 4094 compounds, 2% of the database, were identified as hits. After further reduction using molecular weight, structural novelty, simplicity, diversity, and hydrogen-bond acceptor nitrogen filters, 70 compounds were selected for testing in biochemical assays.

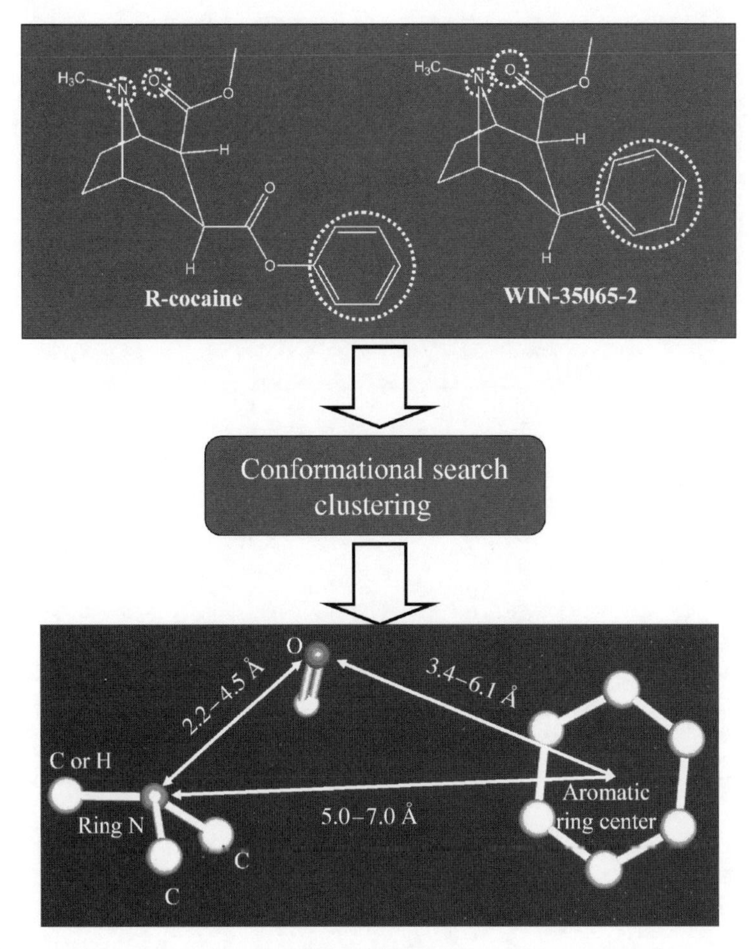

Figure 5. Pharmacophore proposed for identifying DAT inhibitors. The pharmacophore was obtained based on two known DAT inhibitors, R-cocaine and WIN-35065-2. Distance ranges between pharmacophore points were obtained through systematic search of all possible conformations that the two compounds may adopt when bound to DAT. (This figure is available in full color at http://mrw.interscience.wiley.com/emrw/9780471266945/home.)

Forty-four compounds displayed more than 20% inhibition at 10 μM in a [^{3}H]mazindol binding assay. Figure 6 shows selected DAT inhibitors found in the virtual screen.

2.2.6. Receptor-Based 3D Pharmacophore Generation If the 3D structure of a receptor is known a pharmacophore model can be derived directly from the receptor active site. The presence of a cocrystallized ligand in the active site can greatly help with the identification of the pharmacophore. The program LigandScout generates pharmacophore hypotheses in a fully automated way from protein–ligand complexes [59]. In the absence of a ligand costructure, biochemical data can be used for identifying key residues that are important for substrate and/or inhibitor binding. Most ligands bind to proteins through nonbonded interactions such as hydrogen-bond and hydrophobic interactions. Programs such as LUDI [100] or GRID [101] can use the structure of the protein to generate interaction sites or grids to characterize favorable positions that ligand atoms should occupy. Since proteins are not rigid, Carlson et al. [102] proposed using molecular dynamics simulation for generating a set of diverse protein conformations to include protein flexibility in the pharmacophore development. In this case,

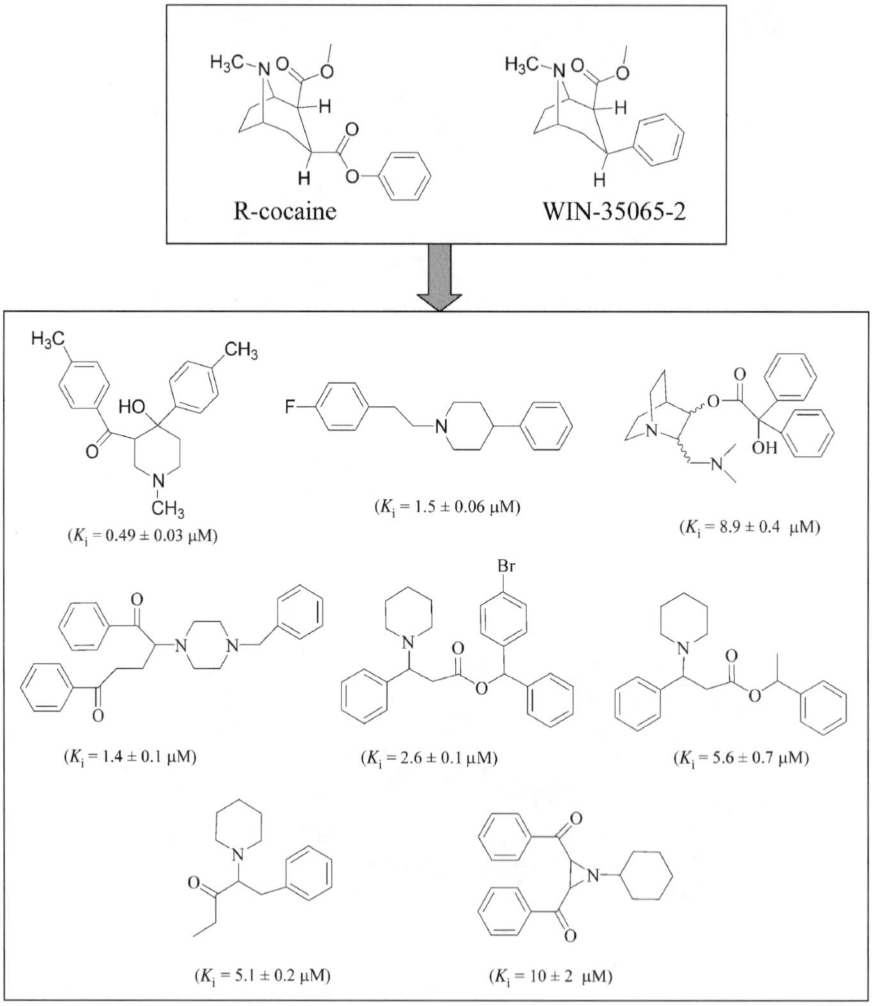

Figure 6. Selected DAT inhibitors identified from the NCI database through virtual screening.

distance ranges between pharmacophores are obtained by examining several conformations of the protein. This technique is similar to the one used for the generation of flexible pharmacophores based on active compounds when several conformations of compound(s) are considered for pharmacophore mapping.

2.2.7. Example 3: Virtual Screening for Novel PPAR Ligands To illustrate how receptor-derived pharmacophore models are used in virtual screening we discuss here a recent example of identifying novel PPAR ligands [103]. A series of LigandScout models were generated based on agonist-PPAR complex structures from the PDB. In addition, ligand-based HipHop models were generated. The pharmacophore models were evaluated using 357 PPAR ligands and 12,775 PPAR decoys. The most selective PPARα agonist model was obtained from LigandScout applied to the PPARα complex structure with PDB ID 1k7l. The best PPARδ agonist model was generated likewise using the PDB complex 1gwx. The best PPARγ agonist model was obtained from a HipHop model. More than one million compounds were screened in Catalyst [104] using all three PPAR-agonist models. A total of 14,311 virtual screening hits were obtained. Filtering by physicochemical properties as found in 321 PPAR agonists retained 5898 hits. 3D shape screening, electrostatic similarity to known PPAR agonists, diversity clustering, visual inspection, and elimination of compounds with reported PPAR activity led to 21 compounds of which 10 were purchased and tested. Several novel PPAR agonists with varying PPAR selectivity were discovered (Fig. 7).

2.2.8. 3D Pharmacophore-Based Virtual Screening Techniques Pharmacophore-based virtual screening is the process of matching atoms and/or functional groups and the geometric relations between them to the pharmacophoric query. Once a pharmacophore model is generated, virtual screening of a database against such a query model becomes straightforward. Programs that perform pharmacophore-based searches include Catalyst [104], UNITY [105], ROCS [106], PHASE, and MOE. There are also some Web-based applications that can perform pharmacophore searches [107,108]. Usually pharmacophore-based searches are done in two steps. (1) The software checks if the compound contains the atom types and/or functional groups prescribed by the pharmacophore. (2) It checks if the spatial arrangement of these elements matches the query. Since most small molecules are flexible, multiple conformations have to be checked. These conformations can be either precomputed and stored (such as in MOE and Catalyst) or calculated on the fly. Precalculating conformations will speed up the virtual screen tremendously. Therefore, this is the method of choice for databases of existing compounds that are screened repeatedly. On the other hand, for larger virtual libraries the data handling requirements become too complex. Therefore, generating conformations on the fly and discarding them after the pharmacophore query has been checked remains an important option for pharmacophore screening programs. Generating a representative ensemble of conformations is essential for the success of a virtual screen using a 3D pharmacophore query.

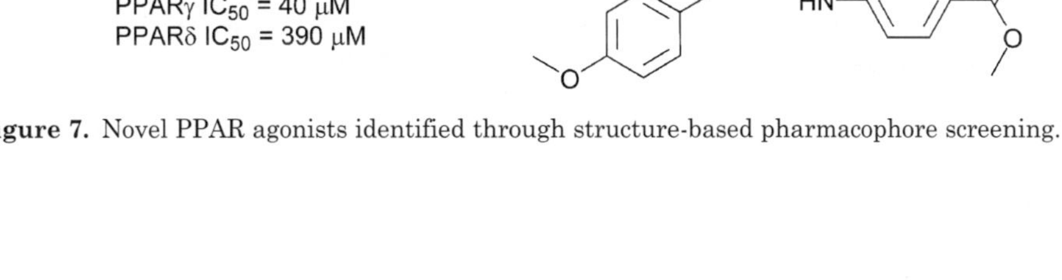

PPARα IC$_{50}$ = 1.0 μM
PPARγ IC$_{50}$ = 40 μM
PPARδ IC$_{50}$ = 390 μM

PPARγ IC$_{50}$ = 13 μM
PPARδ IC$_{50}$ = 7.2 μM

Figure 7. Novel PPAR agonists identified through structure-based pharmacophore screening.

Many ligands do not adopt a minimum energy conformation as the binding pose [109–111]. Therefore, continued research is necessary to represent a conformational ensemble of a ligand including higher energy conformations with a limited number of conformations.

2.3. Application of Machine Learning for Virtual Screening

2.3.1. QSAR Model Generation and Validation

Machine learning is a form of ligand-based VS that builds predictive quantitative structure–activity relationship (QSAR) models that are based on available assay data. Compounds are described with various molecular descriptors to provide numerical representations of a compound's properties. There are numerous types of 2D and 3D descriptors that are commonly used but will not be discussed in detail here [112–117]. Particular descriptors and machine learning methods are chosen because they are believed to be linked or directly correlated to the property being modeled. The process for model generation, validation, and application to virtual screening is shown in Fig. 8 (figure adapted from a recent review by Tropsha and Golbraikh [118]). Many different algorithms and computer software are available for the purpose of predictive modeling; they are based on linear as well as nonlinear methods. In all approaches, descriptors serve as independent variables, and biological activities as dependent variables.

To establish the ability of a QSAR model to predict biological activities the model needs to be thoroughly validated. This is arguably the most important part of QSAR model development [119,120]. Most QSAR modeling methods implement a leave-one-out (LOO) (or leave-some-out) cross-validation procedure. The outcome of this procedure is a cross-validated R^2 (q^2), which does not always guarantee the predictive ability of the model [119,121] illustrating the necessity for thorough model validation. A widely used approach to establish the model's robustness is called *y*-randomization (randomization of the dependent variable, that is, biological activities). *Y*-randomization consists of repeating the calculation procedure with randomized activities and subsequent probability assessment of the results. Often, it is used along with the LOO cross-validation to ensure the model is not based on a chance correlation.

2.3.2. Machine Learning Algorithms Commonly Used in QSAR Models Collections of validated QSAR models can be useful tools for VS

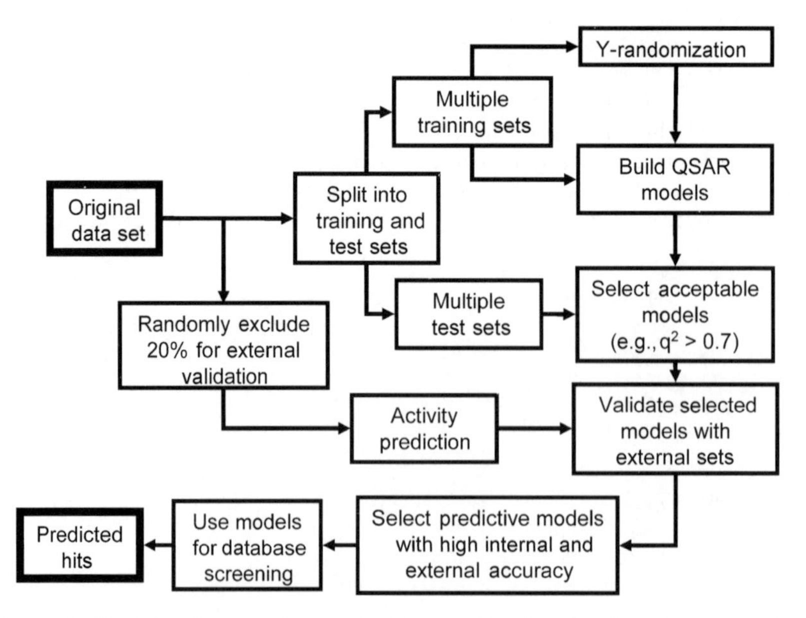

Figure 8. Workflow for generating and validating QSAR models for virtual screening

with the aid of a model applicability domain. Here, we include a short synopsis of the most prominent machine learning techniques in addition to highlighting their applications in VS. Several recent success stories have been reported in the literature and a few examples of popular machine learning techniques are listed in Table 1. Each method has its own advantages and disadvantages that should be understood to select the best approach for a particular virtual screen.

The first approach listed in the table, a self-organizing map (SOM), is quite simple and makes it easy to visualize groups of compounds with the same biological properties. A data set of compounds is mapped on to a 2D grid such that most similar compounds are grouped together. Compounds in the vicinity of compounds with a desired biological property are considered potential hits in a virtual screen. As a disadvantage the SOM approach has a high false-positive rate. SOMs have been used recently to identify several purinergic receptor agonists [122].

Binary QSAR uses all compounds in the training set to predict the biological property of test compounds in a virtual screen rather than the most similar training compounds only. While not providing an image of the training data, the binary QSAR approach is fast and like SOMs it works well if the training data are highly similar to the test compounds being screened. Compounds that are significantly different from the training set are not expected to be predicted accurately and are commonly missed in a virtual screen [123].

While SOMs and binary QSAR techniques do not require a prebuilt model to perform a virtual screen most other machine learning methods do. These techniques form correlations between training data and descriptors that describe training compounds to predict a biological property for a virtual set of compounds. The Bayesian classifier requires a prebuilt model. It distinguishes itself from binary QSAR through identifying specific descriptors that best separate compounds with a desired biological property from others. The search for pertinent descriptors is called variable selection and helps the model eliminate descriptors that are not relevant to the current problem and cloud the separation of one biological property class from another. While the Bayesian classifier algorithm has not performed as well as more sophisticated approaches [124], it does handle large training sets very well.

Decision trees (or forests) incorporate the simplest form of variable selection and can be considered as a set of Boolean functions. Descriptors that capture molecular features of the training compounds are systematically added to a decision tree model one at a time until training compounds with different biological properties are adequately separated. This approach allows for easy determination of the chemical features most relevant to the target biological property. The resulting rules including specific property ranges from the data partitioning can then be used to design new molecules. Comparison studies have shown that decision tree methods slightly outperform the Bayesian classifier; however, other more advanced approaches show even higher enrichment rates in a virtual screen [124].

Mathematical functions are often used to correlate descriptor values with a biological property. Multiple linear regression establishes a linear correlation between dependent and independent variables. A popular extension, called partial least squares (PLS), helps simplify the model optimization to allow for large training sets. Variable selection techniques are commonly added on top of the PLS algorithm to optimize the choice of descriptors. This approach has shown success in enriching a virtual database for various GPCR ligands [125]. The obvious drawback is that there is not always a linear correlation between the property modeled and the descriptors describing the data set. The artificial neural network (ANN) and support vector machine (SVM) approaches try to overcome this problem by building nonlinear models. Therefore, ANN and SVM have become popular tools for model building and virtual screening. In a side by side comparison, both ANN and SVM showed similar enrichment rates for several virtual screens [124].

The k-nearest neighbors (kNN) approach does not require the use of a mathematical function to split one property class from another. This behavior can be very useful if the

Table 1. Overview of Popular Regression and Learning Algorithms Used for Virtual Screening

	Classification	Regression	Variable Selection	Explanatory	Virtual Screening
SOM	Yes, useful for visualizing global data trends	No	Various techniques show success extracting pertinent dependent variables by grouping compounds in the same target class together	Yes, when the pertinent dependent variables are optimized	Identified purinergic receptor antagonists from a virtual combinatorial library [122]
Binary QSAR	Yes	No	No	No	Showed superior enrichment rates when compared to Bayesian Classifiers and PLS [123]
Bayesian classifier	Yes	Yes	Descriptors are weighted based on how well each divides the training data	Yes if the significance of each descriptor can be extracted	Performed poorly compared to SVM, kNN, ANN, and Decision Trees [124]
Decision trees	Yes	No	Descriptors that best divide one class from another are used to separate the data	Variables used in the tree(s) suggest activity dependency	Slightly outperformed a Bayesian classifier in a comparison study [124]
PLS variants	Yes	Yes	Variable selection techniques are commonly added above PLS model building	Yes, when a variable selection technique is incorporated	Ligands for various GPCR targets were successfully enriched from a test database [125]
ANN	Yes	Yes	Performed internally	No	Comparable enrichment rates in a direct comparison to SVM and kNN [124]
SVM	Yes	Yes	Performed internally	Yes, if the weights of each descriptor are explicitly solved.	Identified previously characterized Dopamine D1 Inhibitors and suggested new hits [127]
kNN	Yes	Yes	Commonly a genetic algorithm or simulated annealing is used	Descriptors selected by multiple models imply relevance to the target property	Identified several Anticonvulsant compounds that were experimentally confirmed [126]

problem is complex. Compounds in a virtual screen are predicted based on known activities of the most similar training compounds. Similarity between compounds is calculated using only a small set of descriptors that are optimized during the model building (this optimization can be quite slow for large data sets). When used in a virtual screen, kNN shows database enrichment rates similar to those obtained with both ANN and SVM [124]. For an in-depth view on QSAR the interested reader is encouraged to review Chapter 13 in Volume 1.

2.3.3. The Applicability Domain Concept

With all predictive models, the question of their predictive reach is central to their success. Therefore, the concept of quantitatively assessing the applicability of a predictive model needs to be established. Applicability domain techniques have been applied successfully for instance for kNN [126] and SVM [127]. However, limitations lie in the fact that a very narrow applicability domain only identifies potential compound hits that are highly similar to the training set compounds. Such hits could have possibly been found using simple similarity approaches as well.

Technically, a QSAR model can predict the target property for any compound for which chemical descriptors have been calculated. Since the training set models are developed by interpolating activities, an applicability domain (i.e., similarity threshold) is typically introduced to avoid making predictions for compounds that differ substantially from the training set molecules [120]. In general, there are four different approaches used to determine whether a test compound is similar to the training set within a multivariate descriptor space: range, distance, geometrical shapes, and probability density. Range approaches use the min/max descriptor values within the training set to define an M-dimensional rectangle surrounding the training set. This approach works well only if the training set is evenly distributed. Distance approaches define a distance cutoff to the training set as a boundary often resulting in an M-dimensional sphere that surrounds the training set. Geometrical shapes such as a convex hull are difficult to compute in large descriptor spaces and cannot account for unpopulated descriptor regions within the convex hull. Probability density approaches are gaining in popularity because, in theory, they can account for empty descriptor space within a training set and are not dependent upon any one training set reference point.

Thus far, the most successful applicability domain approaches measure the similarity to the training set via distance and/or the number of training set nearest neighbors within a predefined cutoff. Merck researchers found these approaches to be independent of the machine learning approach being used [128]. For this reason, we will use the remainder of this section do discuss a traditional distance based applicability domain.

To measure distance similarity, each compound is represented by a point in the M-dimensional descriptor space (where M is the total number of descriptors in the descriptor pharmacophore) with the coordinates X_{i1}, X_{i2}, . . ., X_{iM}, where X_is are the values of individual descriptors. The molecular similarity between any two molecules is characterized by a Euclidean distance (sometimes weighted) between their representative points. In the weighted case, the contribution of each descriptor to the overall similarity is weighted based upon correlations found within the training set model. The contribution of each descriptor to the pairwise distance is weighted by a descriptor's relevance to the model, w_k. This provides a more accurate representation of what compounds in the training set the model believes to be most similar. The Euclidean distance $d_{i,j}$ between two points i and j (which correspond to compounds i and j) in M-dimensional space can be calculated as follows:

$$d_{ij} = \sqrt{\sum_{k=1}^{M} w_k \times (X_{ik} - X_{jk})^2} \qquad (1)$$

Compounds with the smallest distance between one another are considered to have the highest similarity. In parallel, the unweighted Euclidean or Tanimoto distances can be used as a measure of overall similarity. The distances (similarity) of compounds in the training set are then compiled to produce an applicability domain threshold, $D_{Threshold}$, calculated as follows:

$$D_{Threshold} = Z \times D_{Train} \quad (2)$$

Here, $D_{Thershold}$ is a distance threshold for a test compound, D_{Train} is the minimum distance required to predict all compounds in the training set in a leave-one-out fashion, and Z is an arbitrary parameter to control the radii of the applicability domain. The default value of this parameter is 1.0 that would classify each training compound within the applicability domain if it was individually removed and placed into a test set. Thus, if the pairwise distance exceeds $D_{Threshold}$ for an external compound to all training compounds, the prediction is considered unreliable. This assumes the compounds are evenly distributed in the training set descriptor space. A modified vector Z' can also be used to account for changes in the data density. For example, one may wish to decrease the applicability domain in portions of the descriptor space where training compounds are very sparse. More insight can be gained from Chapter 13 of Volume 1.

2.3.4. Example 4: Enrichment of an HTS Collection for Kinase Inhibitors When large databases of compounds are screened, often only small portions of the hit set can be ordered for testing. Therefore, hits need to be prioritized. This provides a means to select the top X% of hits chosen for testing or additional filtering. In this example, the kScore algorithm, a hybrid of kNN and SVM, was used to develop several QSAR models of kinase inhibition to enrich a large HTS collection [129]. The training set contained approximately 10,200 compounds of which approximately 17% inhibited a specific kinase. The remaining training set compounds are known inhibitors of other kinases, but showed no activity against the kinase in question.

The validated models were generated and used to predict approximately 775,000 HTS compounds of which approximately 3700 were found to be experimentally active kinase inhibitors corresponding to an approximately 0.5% hit rate. Each kScore model assigned a float activity prediction to the HTS compounds corresponding to their likelihood of being active. These predictions were averaged for each database compound and sorted based on the predicted likelihood an HTS compound is active. This ranking allows for the selection of the top X% of a database to enrich a large percentage of active compounds within a smaller compound collection.

In Fig. 9, the database enrichment plot for kScore models developed with Ghose and Crippen descriptors [130] is shown with and without the use of an applicability domain. The figure demonstrates that the addition of an applicability domain (AD) can significantly improve the results of a virtual screen.

3. STRUCTURE-BASED VIRTUAL SCREENING

In direct analogy to high-throughput screening, docking, and scoring techniques are applied to computationally screen a database of a million compounds and sometimes more against a specific target protein. Computational methods that predict the three-dimensional structure of a protein–ligand complex are often referred to as molecular docking approaches [131]. Protein structures can be employed to dock ligands into the binding site of the protein and to study their interactions [132]. For virtual screening, the crucial task is the fast and reliable ranking of a database of putative protein–ligand complexes according to their binding affinities. Docking programs used today (Table 2) can facilitate this task in less than 1 min per processor and molecule. Energy functions that evaluate the binding free energy between protein and ligand sometimes employ rather heuristic terms. Therefore, those functions are more broadly referred to as scoring functions.

3.1. Protein Structures and Protein Flexibility

A 3D-protein structure of the receptor at atomic resolution is necessary to start a protein–ligand docking experiment. Homology models [133] and pseudo-receptor models [134,135] represent alternatives. The chances of a successful virtual screen very much depend on the

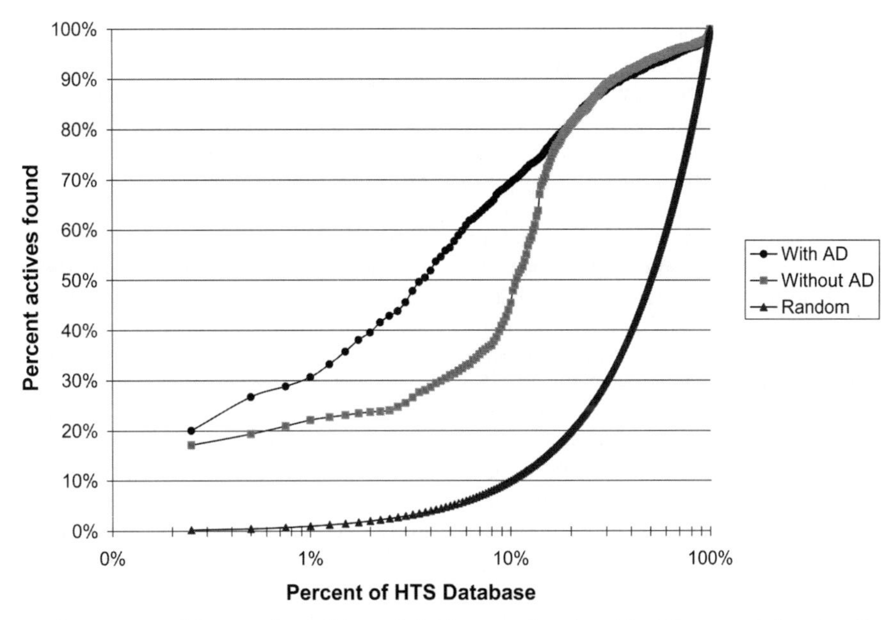

Figure 9. Database enrichment plot of kinase inhibitors rank-ordered by kScore models using Ghose and Crippen descriptors.

Table 2. Selection of Available Protein–Ligand Docking Software for Structure-Based Virtual Screening.

Docking Program	Docking/Sampling Method	Scoring Method
Glide (www.schrodinger.com)	Rigid protein; multiple conformation rigid docking; grid based energy evaluation	Empirical scoring including penalty term for unformed hydrogen bonds; force field scoring
DOCK (www.cmpharm.ucsf.edu/kuntz/dock.html)	Rigid protein; flexible ligand docking (incremental construction)	Force field scoring; chemical scoring, contact scoring
FlexX (www.biosolveit.com)	Rigid protein; flexible ligand docking (incremental construction)	Empirical scoring intertwined with sampling
DockVision (www.dockvision.com)	Monte Carlo, genetic algorithm	Various force fields
DockIT (www.daylight.com/meetings/emug00/Dixon)	Ligand conformations generated inside binding site spheres using distance geometry	PLP, PMF
FRED (www.eyesopen.com/fred.html)	Exhaustive sampling; rigid protein, multiple conformation rigid docking	ChemScore, PLP, ScreenScore, and Gaussian shape scoring
LigandFit (www.accelrys.com)	Monte Carlo	LIGSCORE, PLP, PMF, LUDI
eHITS (www.simbiosys.com)	Exhaustive sampling of all possible binding modes	Customizable
Gold (www.ccdc.cam.ac.uk/prods/gold/)	Genetic algorithm	Soft core vdW potential and hydrogen bond potentials

For a Comprehensive List, Please See the Review of Moitessier et al. [171]

quality of the available structure as well as the flexibility and adaptability of the ligand binding site [136–138]. Optimally, the crystal structure resolution should be at least 2.5 Å [139]. Small changes in structure can drastically alter the outcome of a computational docking experiment [140].

To account for induced fitting effects upon protein–ligand binding in molecular docking experiments different strategies have been employed. Explicit side-chain flexibility can be invoked in AutoDock [141,142] and GOLD for a limited number of residues [143]. FlexE allows for the use of alternative composite protein structures derived from an ensemble of protein structures [144]. A series of promising approaches have been explored, but have not found their way into widely used commercial applications yet. They include among others the use of normal modes [138], elastic network theory [145], constrained geometry [146], constructing receptor ensembles using rotamer libraries [147], multiple crystal structures [148], alanine scanning [149,150], and cross-docking [151].

While taking induced fitting into account is undoubtedly useful in exploring the relevant binding modes of small molecules in protein binding sites it is less obvious how beneficial it actually is for virtual screening applications. While the chances increase to dock the desired active ligands, increased protein flexibility also increases the potential for false positives. Therefore, induced fitting plays a limited role in today's structure-based virtual screening setups. Protein flexibility is also the subject of Chapter 19 in Volume 2.

3.2. Scoring for Virtual Screening

Structure-based virtual screening methods such as high-throughput ligand docking into a receptor site rely on fast scoring functions for ranking large databases of compounds according to their predicted complementarity to a target protein of interest [132,152–155]. No scoring function attempts to rigorously capture all relevant contributions to a ligand–protein binding free energy (even sophisticated and time-consuming methods such as free energy perturbation have difficulties doing so [156]). Each scoring function design, therefore, focuses on terms deemed relevant by the particular author. The result is a wealth of different scoring functions being promoted by different researchers focusing the demonstration of usefulness almost exclusively on empirical data comparisons rather than the conceptual improvement of the underlying methods. This fact has led to an explosion of literature that compares the merits of different scoring functions in the context of virtual screening; [157–172] the most recent and comprehensive review was provided by Moitessier et al. [171]. Before we go into discussing the various contributions to scoring function evaluations, we start with some general remarks about scoring functions.

Scoring functions need to be able to enrich active compounds in the upper ranks of a virtual screening hit list. The question is what is a reasonable enrichment? The answer obviously depends on the number of actives and inactives in the database, as well as the number of compounds that can ultimately be tested in a biological assay. The number of compounds that can be tested is typically in the area of 10^2–10^3 compounds depending on the assay throughput and compound availability (if a virtual screen is performed on vendor compounds, only a limited number of compounds can be purchased because of budget constraints). If one considers a typical virtual screening setup with 1 million individual compounds being screened *in silico* and one assumes to have 100 actives among them there is only an approximately 4.7% chance to find at least one active compound among 500 compounds tested. If a scoring function is able to choose an active compound with a 10-fold higher chance than a false positive (inactive compound with high score) the chances of finding at least one active among the top-ranked 500 virtual screening hits is still below 40%. It takes an enrichment of 100-fold to almost guarantee (99%) at least one hit under the given hypothetical scenario. The situation is even worse considering that more and more often screening (including virtual screening) collections are loaded with combinatorial compounds that are highly structurally correlated. If a combinatorial library is enriched with actives, one may experience a higher confirmed hit rate of a virtual screen; however, all hits will be of the same chemotype and, therefore, the gain in information may be limited. On the other hand, in the more likely case that the library does not contain actives, it dilutes the database that contains actives and most likely contributes overproportionally to the number of false positives thereby making it harder to find true actives in a virtual screen. This means that in the imperfect world of scoring functions not only the target and the actual choice of the scoring function make a significant difference in the performance of the virtual screen but also the composition of the database of compounds that is screened. There are indeed quite drastic differences observed in enrichment rates of the same actives in virtual screening hit lists depending on the decoys being used as demonstrated for instance for CDK2 and SRC kinase virtual screening studies involving decoys from different databases [173].

With few exceptions scoring functions rarely exhibit enrichment rates high enough to be reliable primary decision makers for selecting virtual screening hits. Scoring functions exhibit various scoring artifacts that can be studied in detail and used to improve scoring function design [174], but nevertheless limit their robustness. Therefore, it is vital to use all other information one can assemble about the target protein or the known active ligands, especially the interaction to known ligands to filter virtual screening hits suggested by scoring functions. Especially, information about specific pharmacophore interactions or the preference for certain binding subpockets [175] needs to be translated into relevant filters or constraints for virtual screening/scoring runs. For instance, preferences for hinge binders in kinases [173] or metal ion interactions in metalloproteinases [159] can be defined. The point is illustrated in a recent report in the literature that the application of the same scoring function with pharmacophore-enhanced docking/scoring increases the enrichment of actives in virtual screening hits for BACE1 [176] when applying FlexX [177] and FlexX-Pharm in comparison [177,178]. Also, one should keep in mind, that the experienced chemist is most often the best scoring function. Once a hit list is reduced to less than 1000 compounds, visual inspection along with diversity considerations should be the final selector for compounds to be tested.

Following these introductory remarks on scoring functions, we will review the current state of scoring functions as well as efforts to increase the effectiveness of scoring through consensus combination of these functions.

3.2.1. Force Field Scoring Using nonbonded terms of force fields has been the first choice for scoring functions in early docking programs such as DOCK [179]. When the concept of protein–ligand docking was born, nobody really thought of using this tool in the context of virtual screening. The goal was rather to establish binding hypotheses for known or putative ligands and learn from the resulting models. Force field-based methods with their emphasis on eletrostatics and sterics are a reasonable choice for finding good binding modes because they will favor the formation of hydrogen bonds and salt bridges and avoid van der Waals clashes. However, for binding free energies, hydrogen bonds may not add much to the overall binding energy because of competing desolvation of the ligand. Also, in some cases solvation effects alone are the major driver of ligand potency [180]. Therefore, force field-based methods have had less importance in recent developments of scoring functions aiming at improving the ranking of a large ligand database according to their putative binding free energy against a target protein of interest.

Notwithstanding the above remarks, force field-based scoring functions can be quite successful in virtual screening. Especially, if soft Lennard-Jones potentials are used attenuating repulsive terms (as for instance in the commercial software package GOLD) [139] the hit rates of virtual screens can improve as shown for virtual screens against T4 lysozyme and aldose reductase [181]. To kinetically access the binding pocket, Vieth et al. has also concluded in an analysis of several parameters to be optimized for force field scoring that a soft vdW potential is needed [182]. Also, the use of containment potentials is useful as additional constraints forcing a ligand to dock in a certain region [183]. In addition to including soft vdW potentials, considerable effort has been put into including solvation effects into docking [184–188]. For virtual screening purposes, however, these methods have been less used so far partly because they tend to be slower.

3.2.2. Empirical Scoring Regression methods have been in vogue for almost as long as computational chemistry has been around [189]. As a natural extension in the area of structure-based methods, regression-based functions have been derived using properties of protein–-ligand interactions such as hydrogen bonds, buried surface area, metal contacts, and lipophilic contacts to name a few. These functions are then trained based on known binding affinities of ligands for which the 3D structures of protein–ligand complexes are known at atomic resolution [190–194]. The popular empirical scoring function ChemScore has been designed by Eldridge et al. [192,195]. To illustrate the concept of empirical scoring, the ChemScore function is outlined here in some detail.

$$\Delta G_{\mathrm{binding}} = \Delta G_0 + \Delta G_{\mathrm{hbond}} \sum_{iI} g_1(\Delta r) g_2(\Delta\alpha) + \Delta G_{\mathrm{metal}} \sum_{a\mathrm{M}} f(r_{a\mathrm{M}}) + \Delta G_{\mathrm{lipo}} \sum_{i\mathrm{L}} f(r_{i\mathrm{L}}) + \Delta G_{\mathrm{rot}} H_{\mathrm{rot}} \quad (3)$$

where the various ΔG coefficients are optimized using multiple linear regression. $\sum_{iI} g_1 g_2$ is a hydrogen-bond term calculated for all ligand atoms i and protein atoms I.

$$g_1(\Delta r) = \begin{cases} 1 & \text{if} \quad \Delta r \leq 0.25\,\text{Å} \\ \dfrac{1-(\Delta r - 0.25)}{0.4} & \text{if} \quad 0.25\,\text{Å} < \Delta r \leq 0.65\,\text{Å} \\ 0 & \text{if} \quad \Delta r > 0.65\,\text{Å} \end{cases} \quad (4)$$

$$g_2(\Delta\alpha) = \begin{cases} 1 & \text{if} \quad \Delta\alpha \leq 30^{\circ} \\ \dfrac{1-(\Delta\alpha - 30)}{50} & \text{if} \quad 30 < \Delta\alpha \leq 80^{\circ} \\ 0 & \text{if} \quad \Delta\alpha > 80^{\circ} \end{cases} \quad (5)$$

Ionic and nonionic hydrogen bonds are combined into one term. The deviation from the ideal hydrogen-bond length of 1.85 Å (H–O/N) is given by Δr. The deviation from the ideal angle of 180° is given by $\Delta\alpha$. The lipophilic $\sum_{i\mathrm{L}} f(r_{i\mathrm{L}})$ and metal $\sum_{a\mathrm{M}} f(r_{a\mathrm{M}})$ terms are calculated as simple contact terms. The flexibility penalty of ligands for frozen rotatable bonds is calculated as

$$H_{\mathrm{rot}} = 1 + \left(\frac{1-1}{N_{\mathrm{rot}}}\right) \sum_r \frac{P_{\mathrm{nl}}(r) + P'_{\mathrm{nl}}(r)}{2} \quad (6)$$

where N_{rot} is the number of frozen rotatable bonds and $P_{n1}(r)$ and $P'_{\mathrm{nl}}(r)$ are the percentages of nonlipophilic heavy atoms on either side of the rotatable bond. ChemScore has been trained on 82 protein–ligand complexes yielding $\Delta G_0 = -5.48$ kJ/mol, $\Delta G_{\mathrm{hbond}} = -3.34$ kJ/mol, $\Delta G_{\mathrm{metal}} = -6.03$ kJ/mol, $\Delta G_{\mathrm{lipo}} = 0.117$ kJ/mol, $\Delta G_{\mathrm{rot}} = 2.56$ kJ/mol with a standard error of 8.68 kJ/mol. As with all regression-type functions, the predictability of ChemScore becomes less reliable as the function extrapolates further from protein–ligand complexes in the training set. Lead Finder is the the most recent example of an improved empirical scoring function [196]. Lead Finder trained the same function differently for the three purposes of optimized docking pose prediction, free energy estimation, and virtual screening.

3.2.3. Knowledge-Based Scoring Statistical preferences for protein–ligand atom pair interactions can be derived from large databases of protein–ligand complexes assuming that the deposited structures represent assemblies of high affinity and structural accuracy. For virtual screening purposes a series of knowledge-based scoring functions were developed including SmoG [197], PMF scoring [198], DurgScore [199], and BLEEP [200]. Knowledge-based scoring functions have been extensively reviewed by Gohlke and Klebe [201]. Therefore, we focus here on explaining the concept of knowledge-based screening using PMF scoring as example.

The PMF scoring function [198,202–204] has recently been updated exploiting the exponential increase in publicly available crystal structures of protein–ligand complexes. The new version of PMF, called PMF04, has been derived from 7152 protein–ligand complexes from the PDB. Compared to the original PMF99 score (697 complexes) this is approximately a 10-fold increase. This increase in the statistical knowledge based for PMF scoring has resulted in new potentials being introduced for metal ions. Also, while similarly to SmoG halogen-containing potentials have mostly been ignored in PMF99, for many halogen atom-containing protein–ligand atom pairs PMF04 now contains statistically significant potentials.

Here is a brief outline of the PMF scoring function. The PMF score is defined as a sum over all protein–ligand atom pair interaction free energies $A_{ij}(r)$ at distance r,

$$\text{PMF\_score} = \sum_{\substack{kl \\ r < r^{ij}_{\text{cutoff}}}} A_{ij}(r) \tag{7}$$

where kl is a ligand–protein atom pair of type ij. r^{ij}_{cutoff} is the distance at which atom pair interactions are truncated. The $A_{ij}(r)$ are calculated as

$$A_{ij}(r) = -k_{\text{B}} T \ln \left[f^{j}_{\text{Vol\_corr}}(r) \frac{\rho^{ij}_{\text{seg}}(r)}{\rho^{ij}_{\text{bulk}}} \right] \tag{8}$$

where k_{B} is the Boltzmann factor, T is the absolute temperature, and $f^{j}_{\text{Vol\_corr}}(r)$ is a ligand volume correction factor. $\rho^{ij}_{\text{seg}}(r)$is the number density of atom pairs of type ij at a certain atom pair distance r. $\rho^{ij}_{\text{bulk}}$is the number density of a ligand–protein atom pair of type ij in a reference sphere with a radius of 12 Å. For docking purposes, the PMF score adds a vdW term to account for short ranged interaction [205]. Knowledge-based potentials have been proven useful in virtual screening scenarios against FK506 binding protein [205], aldose reductase [206], and a series of other targets [207].

Figure 10 shows some of the newly derived PMF04 potentials and how they compare to PMF99. Figure 11 shows for a set of 170 protein–ligand complexes assembled by Wang et al. [208] how PMF99 and PMF04 compare. While there is no substantial overall improvement in correlating PMF scores to experimental binding affinities the figure shows how some outliers have been fixed in PMF04. As an example, elastase complexed with FTAP has been highlighted (7est). Because of the statistically insufficient potentials for fluorine atom types one of trifluoromethyl groups of FTAP has scored very badly in PMF99. The combined score of the three fluoro atoms has been 45.0 (Fig. 12). In PMF04 this problem has been rectified because of the better potentials for halogens. The combined score of the three fluoro atoms is now -0.1. The same improvement in scoring can be observed for metal binding protein–ligand complexes (data not shown). For other, statistically more established atom types, the differences in PMF99 and PMF04 scores are minimal (Fig. 12).

3.2.4. Hybrid and Alternative Scoring Methods

There is a wealth of other scoring functions being introduced ranging from steric fit, surface, and pharmacophore complementarity [209,210] to empirically motivated scores such as PLP (piecewise linear potential) [211,212]. An interesting example of a hybrid scoring method has been introduced by Feher et al. called BHB [213,214]. BHB stands for buriedness, hydrogen bonding, and binding energy. The first term assesses how well a ligand occupies a binding pocket. Interestingly, the authors focus on the buriedness of the receptor site rather than that of the ligand as suggested by others [213,215]. The hydrogen-bonding term along with a term for steric fits based on vdW potentials has been taken from the GOLD docking function [139,216]. A binding energy is elaborately calculated using a thermodynamic cycle approach subtracting the energies of the unbound rigid solvated receptor and ligand from that of the complex structure. Using a series of different protein–ligand complexes including androgen receptor, estrogen receptor, neuraminidase, thymidine kinase, inosin monophosphate dehydrogenase, CDK3, and

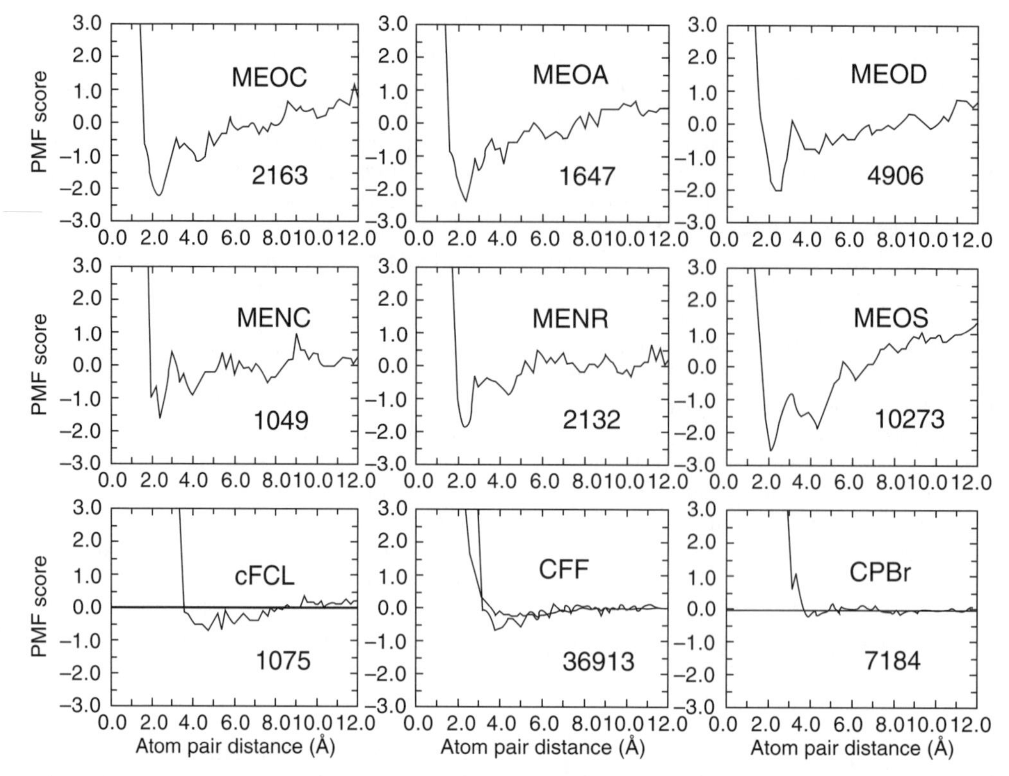

Figure 10. PMF04 (bold) and PMF99 potentials for a selected set of protein–ligand atom pairs [213]. The four-letter code refers to the atom pair types, where the first two letters indicate the protein atom types and the last one or two letters indicate the ligand atom type (ME: metal, OC: negatively charged oxygen, OA: oxygen as HB acceptor, OD: oxygen as HB donor, NC: positively charged nitrogen, NR: ring nitrogen, OS: other oxygen (e.g., phosphate), CF: aliphatic carbon, cF: aromatic carbon, CL: chlorine, F: fluorine, and Br: bromine). The number of atom pair occurrences in the set of protein–ligand complexes used to derive the potentials is given for each pair potential exemplified in the figure. A complete list of protein and ligand atom types for PMF99 and PMF04 as well as complete access to the potentials has been given in a recent publication [213].

p38 MAP kinase Feher et al. [214] derive the following generalized function:

$$F_{\text{score}} = -0.05\text{BE} + 20(\text{RB} + \text{RB}^3) + \left(\frac{10}{n}\right)\sum_{i}^{n}\text{HB}_i + 0.5(\text{H}_{\text{ext}} + \text{vdW}_{\text{ext}}) \quad (9)$$

where BE is the calculated binding energy, RB is the receptor buriedness, HB_i is an indicator for the ith specific hydrogen bond (1 if exist, 0 otherwise), and n is the total number of hydrogen-bonding residues considered for the receptor. H_{ext} and vdW_{ext} are terms describing van der Waals interaction and hydrogen bonding as defined in GOLD. In a virtual screening setting involving the Glide decoys as negative controls [217], the authors report drastically improved enrichments of actives in the virtual screening hit lists for all the targets mentioned.

3.2.5. Some Comments on Using Enrichment Factors

Before assessing the performance of different scoring functions it is useful to add some remarks on how they are compared. The performance of virtual screening methods/scoring functions is often measured in enrichment factors. A typical enrichment factor measure is defined as

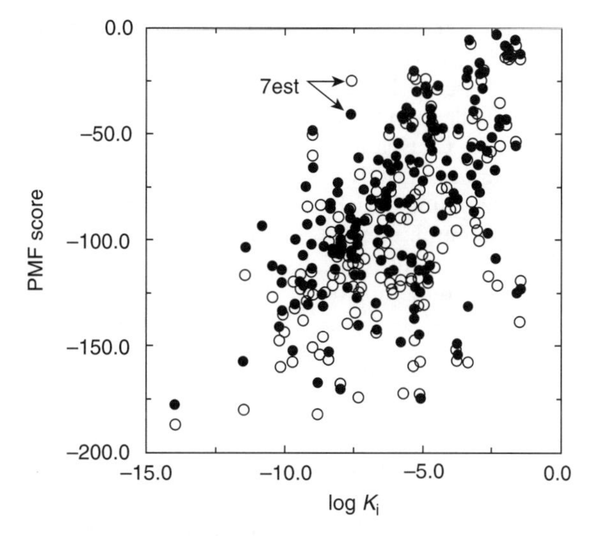

Figure 11. PMF04 (filled circles) and PMF99 (open circles) scores compared to the experimental binding affinities for a set of 170 protein–ligand complexes taken from the training set of the Score function [208].

$$\mathrm{EF} = \frac{\%\mathrm{actives\_retrieved}}{\%\mathrm{compounds\_sampled}} \qquad (10)$$

A random enrichment would result in EF = 1. The higher the enrichment factor the better the scoring function performs. This enrichment factor is ideal for comparing different scoring functions or virtual screening scenarios for the same set (or at least the same number) of actives and inactives independent of the target. However, even for the same target, if literature data are compared employing different numbers of actives and inactives comparing enrichment factors becomes cumbersome even if the same number of compounds is being sampled. To illustrate this point let us consider Case 1 with 1000 compounds and 10 actives compared to Case 2 with 1000 compounds and 100 actives among them. Let us assume for Case 1, 50% of the actives (5/10) have been sampled among the

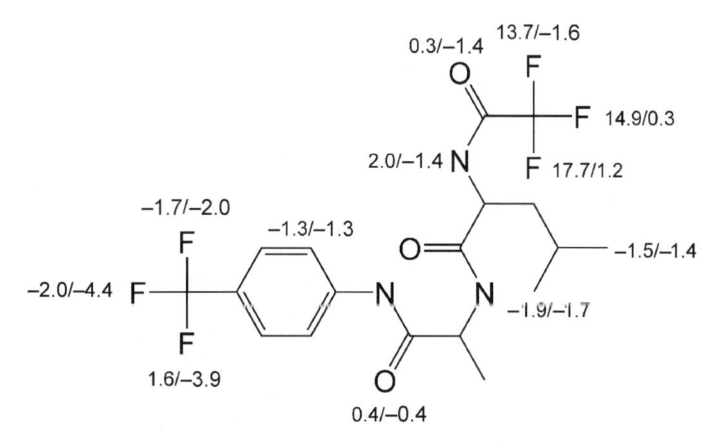

Figure 12. Trifluoroacetyl-L-leucyl-L-alanyl-*P*-trifluoromethylphenylanilide (TFAP) ligand of the serine protease elastase. The PMF99/PMF04 contributions of selected atoms are shown as calculated from the elastase/TFAP complex taken from the PDB ID 7est.

top 2% of the compounds and 100% (20/100) have been sampled in Case 2. The enrichment factor for Case 1 would be better (EF = 25) than that of Case 2 (EF = 10). The reason is that the maximum enrichment that can be achieved depends on the number of actives and inactives. While the maximum enrichment in Case 1 is 50, the maximum enrichment in Case 2 is only 10. One way of avoiding the problems described above is to define an enrichment factor as

$$\mathrm{EF} = \frac{\%\mathrm{actives\_retrieved}}{\%\mathrm{inactives\_retrieved}} \tag{11}$$

In this case, the enrichment factor can become infinite.

The term "%actives_retrieved" is often called recovery. A relevant question for virtual screening purposes is often how much of a database needs to be screened to retain a certain recovery. An additional problem with Equation 10 for calculating enrichment factors at a certain recovery percentage is that this measure only captures the performance of the worst active within the desired recovery. It would make no difference if ligands scored 1st, 2nd, and 50th, or if they scored 48th, 49th, and 50th. However, it would be much more desirable to have higher ranking actives among the recovered as in the first case. Therefore, for full recovery an alternative enrichment function has been suggested [205] as

$$\mathrm{EF}' = \frac{50\%}{\%\mathrm{APR}(\mathrm{actives\_retrieved})} \tag{12}$$

where %APR(actives_retrieved) is the average percentile rank of the retrieved actives. Equation 12 can be extended to observation points of less than full recovery as [218]

$$\mathrm{EF}'' = \frac{50\%}{\%\mathrm{APR}(\mathrm{actives\_retrieved})}\,\%\mathrm{actives\_retrieved} \tag{13}$$

Enrichment factors EF′ and EF″ consider the rank of each retrieved hit. As a consequence, enrichment factors can be larger than those obtained from Equation 10.

3.2.6. Comparing, Consolidating, and Finding Consensus After most scoring functions have conceptually been developed in the 1990s, recently more effort has been put on validating various scoring functions in the context of docking and scoring tools. One of the early comparative studies of docking/scoring protocols has interestingly been done on the more challenging task of fragment docking using Abbott's SAR from an NMR fragment library. Comparing DOCK/force field scoring and DOCK/PMF scoring 3200 fragments (K_i values between 0.06 and 2 mM) were virtually screened against the KF605 binding protein reaching an enrichment of up to EF′=3 (Eq. 12) [205]. While an enrichment factor of three would be considered very low for fully enumerated compounds, it is still typical for weakly binding fragments.

An important observation made by several authors is that different targets require different scoring schemes. Comparing DOCK, FlexX [177], and GOLD docking programs along with ChemScore, DOCK, FlexX, Fresno [219,220], GOLD, PMF, and Score [220] scoring functions, Bissantz et al. [221] evaluate virtual screens against thymidine kinase (TK) and estrogen receptor (ER). Mixing 10 ligands with 990 decoys, the authors found that FlexX docking and PMF scores correctly identify the 10 TK ligands but fail for the estrogen case. Similarly, Stahl and Rarey report varying performances of docking/scoring combinations for different targets [207]. For instance, PMF performed well for neuraminidase but poorly for gelatinase A and thrombin. Yet another study by Kontoyianni et al. [222] further supports the finding of varying performances for four docking methods, 10 scoring functions, and 6 drug targets with a few front runners being combinations between LigandFit, FlexX, and DOCK as docking engines combined with Ligscore, FlexX, GOLD, and PMF as scoring functions. Without trying to be exhaustive, a few other studies by Perola et al. [163], Schulz-Gasch and Stahl [164], Muegge and Enyedy [173], and Warren et al. [167] are mentioned here solidifying further the general observation of greatly varying performances of docking/scoring experiments. There are two major consequences from these observations:

(i) For optimum virtual screening purposes a validation phase needs to be introduced using known active and inactive ligands to fine-tune a virtual screening protocol before it is applied in a production run.
(ii) Consensus approaches need to be considered to average out the volatile performances of single scoring functions.

As pointed out above, because of the varying performance of scoring functions, methods have been deployed that allow focusing on scoring functions that are better (or worse) than others by consensus. In general, consensus scoring is rarely able to improve scoring beyond the performance of the single best scoring function used [223]. However, because one does not know up-front which of the scoring function performs best for a given target, consensus scoring allows for an averaging effect that often (but not always) is only marginally worse in performance than the best individual scoring function. The first comprehensive consensus scoring study has been presented by Charifson et al. [224]. Using 11 scoring functions (LUDI [100], ChemScore, Score, PLP, Merck force field [225], DOCK energy score, DOCK chemical score, Flog [226], strain energy, Poisson Boltzmann [227], buried lipophilic surface area [228], DOCK contact score, and volume overlap [210]), two docking programs (DOCK, Gambler), and three protein targets (p38 MAP kinase, inosine monophosphate dehydrogenase, and HIV-1 protease) the authors showed a significant reduction of false positives in the virtual screening hit sets. Other authors report success with consensus scoring also suggesting general hybrid scores applicable to any target such as ScreenScore [207]. Consensus scoring tools are available in commercial software packages such as Sybyl [229] and Fred [190].

Often scoring functions exhibit artifactual dependencies on ligand properties such as compound size, prevalence of hydrogen-bonding features, or flexibility. To normalize binding scores of compounds, Vigers and Rizzi have suggested a form of consensus scoring coming not only from different scoring functions applied to the same protein–ligand complex but also from a series of different protein targets a given ligand is exposed to [230]. Using a set of arbitrary active sites, a reference score for a given ligand is calculated leading to a simple statistical correction for a calculated ligand score against a target of interest. This correction has been called MASC (multiple active site correction) and is calculated as:

$$S'_{ij} = \frac{(S_{ij} - \mu_i)}{\sigma_i} \tag{14}$$

where S'_{ij} is the modified score for compound i in the active site j and S'_{ij} is the original score. μ_i and σ_i are mean and standard deviations of the scores for compound i across all active sites j. If a statistically significant higher score is obtained for the active site of interest, the compound is more likely to be a true active. The MASC procedure requires a substantial initial investment in CPU time due to the need to dock the ligands of interest against not only 1 but 10 or more targets to calculate a reference score. However, especially for databases used repeatedly such as an in-house screening collection this up-front expenditure is reasonable. Vigers and Rizzi report significant improvements in virtual screening of p38 MAP kinase and Ptp-1b using their MASC scoring method in the context of FlexX and Gold scores.

3.2.7. Example 5: Identification of a Novel Liver X Receptor (LXR) Modulator

LXRa and LXRb are oxysterol-activated nuclear receptors that are involved in controlling the expression of various genes associated with cholesterol, glucose, and fatty acid metabolism. LXR agonists are therefore of potential pharmacological interest as antiatherosclerotic, anti-inflammatory, and antidiabetic agents [231]. Combining Glide virtual screening with gene profiling using an ArrayPlate mRNA assay resulted in the identification of a novel LXR modulator (Fig. 13) [232]. To validate the Glide docking method, 127 representative LXR agonists from 6 different chemical classes were selected from the literature. These LXR ligands were

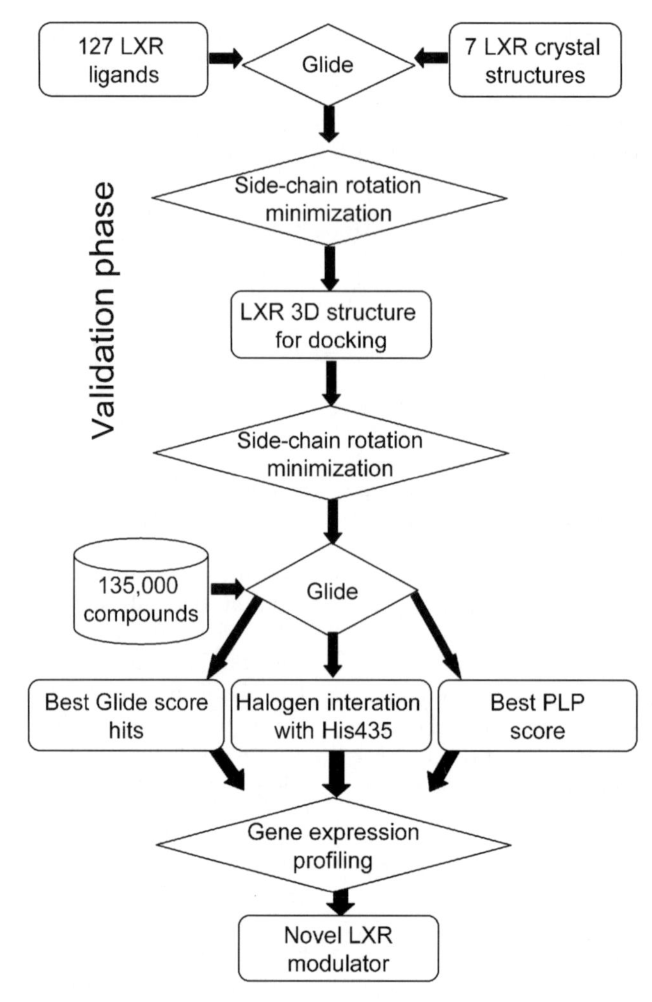

Figure 13. Structure-based virtual screening flowchart used by Cheng et al. [232] to identify a novel LXR modulator.

docked against seven LXR crystal structures from the PDB. The PDB structure 1P8D recovered the most LXR ligands (27 of 127) through Glide docking. To increase the recovery rate further, the hydrophobic pocket of the ligand binding site was widened through rotating side chains of Phe340 and Ile253. Minimization of such an altered protein structure in presence of the LXR agonist GW3965 (Fig. 14) resulted in the final protein structure used in the virtual screen of 135,000 compounds from an internal collection. Applying a cutoff of −9.0 using the GlideSP scoring function 10,407 compounds were retrieved. Ligands clashing with the protein matrix to more than 20% were removed. This reduced the hit rate to 3263. Compounds were prioritized further by requiring a close contact to His435 that was believed to be a required interaction for LXR agonists. A total of 647 compounds were thus selected. Compounds prioritized by the PLP1 score [233] were added. In addition, 247 compounds with halogen atoms interacting with His435 such as in GW3965 were added to the hit list. A total of 1308 compounds were ultimately selected for high-throughput gene expression profiling. Gene profiles of one virtual screening hit (Fig. 14) were found to be extremely similar to those of other LXR agonists such as

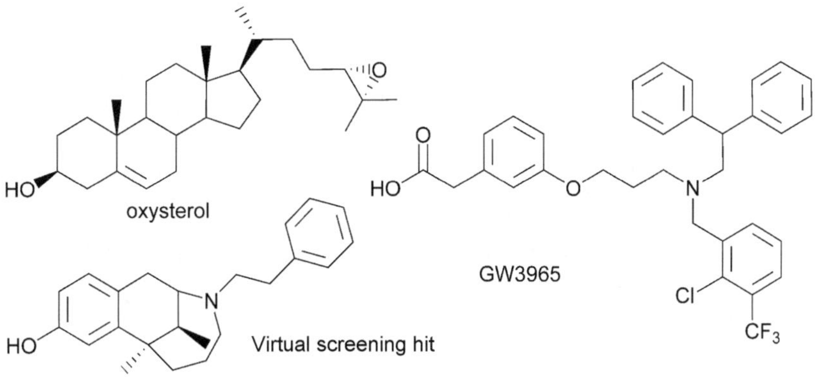

Figure 14. LXR ligands.

GW3965. The hit was subsequently confirmed as full LXR agonist by a cofactor recruitment assay and relevant functional assays.

3.3. Fragment-Based Virtual Screening

Fragment-based drug discovery methods have experienced a renaissance in recent years [234]. Fragments are mostly identified using high-concentration screening methods with NMR follow-up and crystallographic binding mode elucidation. However, virtual screening is sometimes used to narrow the set of fragments chosen for fragment screening. Although fragment virtual screening methods have historically produced low enrichment rates [205], there are some success stories reported in the literature.

A virtual screen for S1-binding fragments of dipeptidyl peptidase IV was reported by Rummey et al. [235]. DPP IV is an important diabetes target. DPP IV inhibitors prevent the rapid degradation of GLP-1 that has beneficial effects on glucose homeostasis. Using a pharmacophore-constrained docking approach, the S1 pocket of DPP IV was targeted. Ten thousand primary aliphatic amines (positively charged group in the S2 pocket was required) were docked using FlexX [236]. The formation of at least two hydrogen bonds was required for a hit in the S1 pocket. With such specific constraints fragment docking identified a series of known and unknown anchor fragments in the S1 pocket. Similar examples of DPP IV inhibitors obtained through fragment-based screening were described by Makara [237].

A second example illustrates again the necessity to guide fragment based *in silico* screening through specific constraints to be successful. Indolin-2-one analogs were discovered by *in silico* fragment-based virtual screening of DNA gyrase [238]. DNA gyrase is a bacterial enzyme targeted for discovering antibacterial agents. A new core fragment to sit in the DNA gyrase active-site pocket was desired. The crystal structure with PDB ID 1ei1 of the N-terminal fragment of DNA gyrase B was used. LUDI identified key interactions in the active site. The MDL-ACD and MDL-SCD databases of commercially available compounds were searched to find a novel fragment with MW < 180 Da to fill the active site and engage it in all relevant interactions. LUDI was used in combination with compound similarity to known DNA gyrase binders to subsequently identify indolin-2-ones with up to 60 μM activity.

The interested reader can learn more on the topic of fragment-based drug design from Chapter 10 in Volume 2.

4. PRE- AND POSTPROCESSING OF VIRTUAL SCREENING DATA

The preprocessing of databases subjected to virtual screens as well as the postprocessing of hits needs some attention. The choice of compound databases to screen has been discussed elsewhere in different contexts already. Hence, only a few words about preparing a database for a virtual screen. Generating 3D coordinates if necessary and enumerating stereocenters of small molecules can be done in a straightforward fashion by programs such as CORINA [85]. Generating the appropriate ionization states of titrable groups is often more challenging. Commercial software to predict ionization states at different pH values with varying degree of accuracy is available from software vendors such as Schrodinger, Chemaxon, and ACS Labs. Tautomer enumeration is often ignored due to the absence of commercial software to reliably determine the tautomer with the highest population among many possible tautomers. As mentioned in Section 2, sometimes ligand conformations need to be precalculated. Also, various fingerprint representations can be precalculated and stored for databases that are recurrently screened.

4.1. Hit Triaging

Smart hit triaging is an essential step in a virtual screen. Unfortunately, there is no suitable commercial software available that can powerfully assist in hit triaging. Virtual screening filter cascades rely on the power of multiple exclusion criteria to reduce the number of virtual screening hits to a manageable size. If target-specific criteria such as pharmacophore matches, shape complementarity, and similarity to known hits are not discriminative enough, target-independent criteria are typically used. These include properties such as general drug-likeness [239], perceived selectivity, ligand efficiency [234,240,241], and diversity [242].

4.1.1. Traffic Lights Virtual screening often produces as many hits or more than high-throughput screening does. If we set aside for a moment the fact that for virtual screening we are not sure of the true activity due to false positives, both approaches face a similar problem of hit triaging. An example of how hits are triaged using traffic lights has been given by Lobell at al. [243]. The authors highlight the fact that leads are often very similar to hits [244] that ADME-related liabilities are hard to address during lead optimization and that more hit classes are generated. They propose to use ADMET (absorption, distribution, metabolism, excretion, toxicology) traffic lights and oral PhysChem scores to triage hit lists. The traffic lights center around solubility, octanol-water partitioning, polar surface area, rotatable bonds, and molecular weight descriptors. The interested reader may consult Chapter 2 in Volume 2 for further reading on the topic of *in silico* ADMET prediction.

4.1.2. Ligand Efficiency When evaluating hits identified in a screen, they must be judged not only for their experimental potency but also for the potential to develop the hit into a drug. Often this lead optimization process results in an increased molecular weight to improve potency and target selectivity [245,246]. Unfortunately, increased molecular weight often compromises critical drug properties such as solubility or permeability that are necessary for bioavailability [247,248]. Therefore, it is better to optimize a small, less potent molecule than a much larger molecule that already contains liabilities that are difficult to remove. To relate potency to compound size, Andrews [249] and then later Kuntz [250] defined a measure of binding energy and demonstrated there is a correlation between potency and the number of nonhydrogen atoms in a ligand. This concept was used to define the term ligand efficiency by Hopkins et al. [251] whereby the binding energy is divided by the number of nonhydrogen atoms, shown in Equation 15.

$$\text{Ligand efficiency} = \frac{-RT\ln K_\text{d}}{\text{Nonhydrogen atoms}} \qquad (15)$$

Ligand efficiency has gained wide use in the pharmaceutical industry for HTS hit selection [252]. Increased ligand efficiency increases the likelihood that a hit can be developed into a clinical candidate [253].

The minimum ligand efficiency viewed as acceptable for a drug-like compound is approximately 0.3. This number is calculated from the hypothetical case of a 10 nM optimized compound with approximately 36 heavy atoms (the maximum number of atoms to able to achieve a MW < 500) [254]. The maximum ligand efficiency appears to plateau as the number of heavy atoms increases [255].

This is thought to be the consequence of enthalpic effects from structural changes in the target protein that ultimately occur as a larger molecule binds. The number of contact points with the target protein increases with the ligand size making it challenging to optimize each contact simultaneously [256]. Entropic effects should also play a significant role in ligand efficiency but as of yet no technique has shown an accurate estimation of ligand efficiency loss due to entropy.

4.1.3. HTS Triaging High-throughput screening (HTS) is arguably the major technique used today for the discovery of new chemical matter [257,258]. Often the lead-like matter identified by an HTS is clouded with nonspecific inhibitors that do not directly bind to the target protein in a one-to-one fashion. Instead, compounds may absorb and emit light at the wavelengths tested [259] or may aggregate to form a colloidal particle that directly interacts with the target protein [260,261]. These phenomena have led to the creation of structural filters [262] and experimental techniques [263–265] that identify such potential aggregators. Likewise, there are also nonspecific activators characterized that may elicit activity through a similar mechanism [266]. These filters derived from HTS experiences can be applied directly as promiscuity filters for virtual screening hit lists.

Within the class of HTS hits that direct binding to the target protein in a one-to-one fashion, there are structural elements that allow binding to a large range of targets. These compounds are commonly referred to as promiscuous binders or privileged structures [267–271]. They contain specific chemical groups that interact with a portion of the target protein that is conserved throughout a protein family. There can also be strong cross-reactivity between completely unrelated protein families if the pharmacophoric elements necessary for binding are similar [272]. These classes of compounds commonly must be filtered out if the main contributor to binding is also the cause of the selectivity liabilities. To identify these compounds, the known assay data are associated with each HTS hit to flag probable liabilities that would have to be overcome if the compound was pursued [273]. There are several databases available to serve this purpose including PubChem [244,274,275], ChemIDplus [276], and DiscoveryGate [277]. These resources can be used to identify an exact chemical match to HTS or virtual screening hits or highly similar compounds that highlight potential activities.

5. SCAFFOLD HOPPING

The ultimate goal of a virtual screening experiment is to identify promising hits of new chemical matter. It is often irrelevant what the actual hit rate of a virtual screen is. Rather the novelty, patentability, drug-likeness, synthetic accessibility, and optimizability of a hit class are of importance. Therefore, it is essential to understand the ability of a virtual screening approach to identify new chemotypes. For structure-based virtual screening methods the ability to scaffold hop is usually taken for granted because there is no ligand (other than the one being docked) involved in docking experiments. While the assumption of scaffold-independent docking preferences should be debated at some point, it is not the central subject here and is therefore not further discussed. We would rather like to say some words about ligand-based scaffold hopping opportunities in the context of compound similarity and virtual screening. In particular, we would like to illustrate a few simple concepts such as exhaustive scaffold enumeration and scaffold hopping using similarity descriptors. The more wide-ranging concept of extending virtual screening to virtually accessible chemicals, an approach often referred to as *de novo* design [278], will not be discussed here.

5.1. Example 6: Scaffold Enumeration for PDE5 Inhibitors

For any given scaffold, there are a number of analogs that may produce a similar clinical endpoint. In particular, for a specific ring system the necessary topology of the ring can be maintained by modifying the heteroatoms and bond characteristics to preserve the shape, critical H-bonding characteristics, or

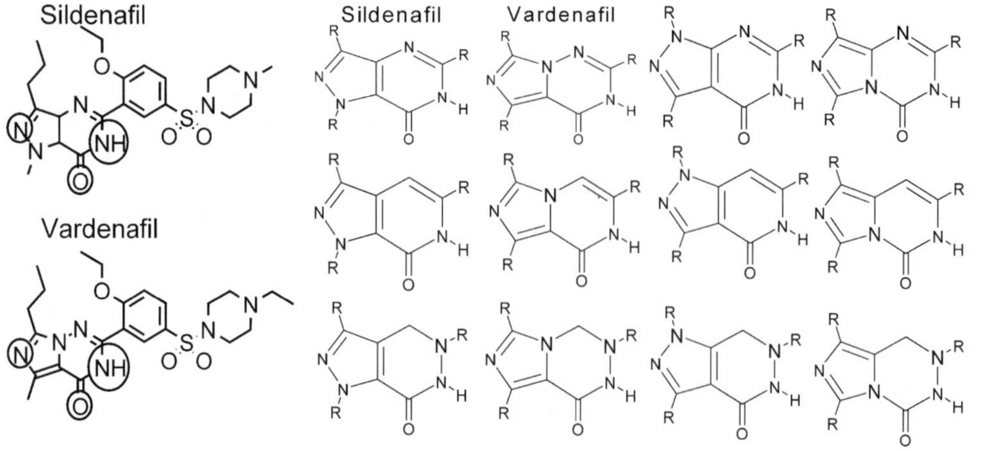

Figure 15. Example analogs of sildenafil and vardenafil using a scaffold enumeration approach that maintains the critical pharmacophores circled on both drugs.

attachment points [279]. An instructive example of modifying the shape, attachment, and pharmacophoric properties in a molecule yet scaffold hop to a new chemotype with similar biological activity is illustrated in Fig. 15. The discovery of PDE5 inhibitor vardenafil appears to be a direct result of a scaffold hop starting from sildenafil. Vardenafil was designed by simply shifting the heteroatom placement in the core. Such subtle modifications can sometime even improve the bioavailability, solubility, selectivity, etc. without dramatically modifying the clinical efficacy. In the case of vardenafil, the shift in heteroatoms improved the selectivity over PDE6, which reduced the risk of altered color perception, a rare side effect that sometimes occurs with sildenafil. In Fig. 15, the atoms circled in the core scaffold are critical pharmacophoric elements for potency. By holding these atoms constant and maintaining the 3 attachment points, there are only 12 scaffolds that would likely produce a similar efficacy.

To calculate the possible analogs for a specific core the algorithm is quite simple. First, one must define the atom types to be considered; an example list is shown in Fig. 16. Second, one must define a SMILES string that represents a core to start from. Then one simply loops through every atom in the core and sequentially replaces each atom with an atom from the atom type list. This procedure produces a large number of SMILES codes that may have valence problems, may be

```
Atom enumeration algorithm.
Start with a SMILES string of your ring system (e.g., C1CCCCC1)
While SUM(SMILES_String[Atom_Type]) < #Atom_Types*#Atoms LOOP
        SMILES_String[1]++;
        IF SMILES_String[1] > #Atom_Types THEN
                FOR "i" in 1 through the #Atoms LOOP
                        IF SMILES_String[i] > #Atom_Types THEN
                                SMILES_String[i+1]++;
                                SMILES_String[i]=1;
        Filter atom sequence against internal filters or write to output
END WHILE LOOP
Postfilter output using external applications if necessary
```

```
Atom types:
1→C
2→C=
3→C(=O)
4→N
5→N=
6→O
7→S
```

Figure 16. Pseudo-code for an atom enumeration algorithm.

unstable, or are duplicates. A valence filter check or a substructure filter eliminates SMILES strings with known stability problems. Other problematic SMILES strings are removed through postfiltering. If there are specific atom restrictions, such as those circled in Fig. 15, then an additional substructure filter can identify the relevant scaffolds. A pseudo-code of the algorithm is shown in Fig. 16.

The scaffold enumeration approach works well not only for rings but also for complete molecules. A critical detail is to restrict the search by requiring specific pharmacophores, atoms, or groups at different locations. In the example of sildenafil, without restricting the three atoms necessary for potency and stability filters, this search would have yielded over a thousand possibilities. An example set of stability filters that can be applied is shown in Fig. 17.

5.2. Scaffold Hopping Through Similarity Descriptors

As described in Section 2, compound similarity descriptors have different abilities to identify compounds with new scaffolds yet similar biological activity. Molecular descriptors such as atom pairs [41], 3D pharmacophore fingerprints [54], molecular field descriptors [280], and FeatureTrees [281] have been shown to facilitate scaffold hopping. Recently, a systematic study of the scaffold hopping ability of different compound similarity descriptors was conducted that involved seven drug targets of different classes (p38 kinase, CDK2, thrombin, neuraminidase, estrogen receptor, HIV-1 protease, COX2) as well as several popular descriptors (atom pairs, Daylight fingerprints, 3D pharmacophore fingerprints, SciTegic fingerprints) [18,54]. Not surprisingly, it was concluded that 3D descriptors as well as circular fingerprints were found to facilitate scaffold hopping in more cases than topological fingerprints did (Daylight fingerprints). What was more surprising was the finding that ligand similarity was better able to identify new chemotypes than structure-based virtual screening was, a phenomenon that was observed before [41] but hitherto not studied systematically.

6. VIRTUAL SCREENING OF COMBINATORIAL LIBRARIES

Virtual screening success obviously relies on accessing a hit molecule synthetically for biological testing. The number of available compounds (either in-house or commercially) is limited to approximately 10 million. On the other hand, synthesizing individual hit molecules (for instance as a result of a *de novo* design campaign) is typically limited to few analogs and may involve significant chemistry efforts. Finding an alternative between the practically boundless *in silico* world of molecules and the quite limited synthetic options is therefore crucial for realizing the full potential of virtual screening.

The concept of extending virtual screens to the area of chemically accessible combinatorial libraries has several advantages.

1. Combinatorial libraries of previously established chemistry can be synthesized rather quickly.
2. Combinatorial libraries compensate for the limited accuracy in many virtual screens by the power of large numbers of analogs thereby introducing an

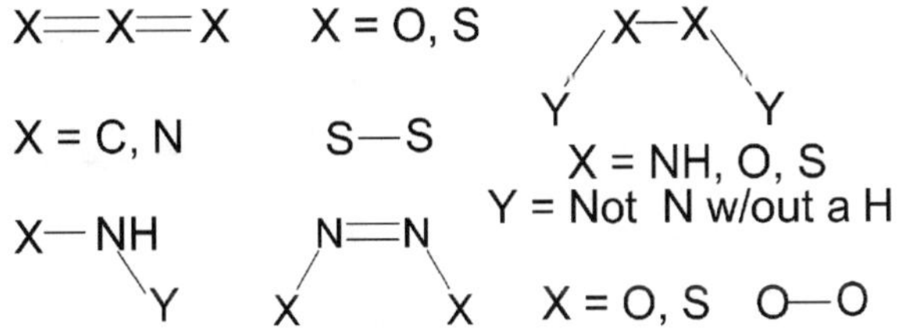

Figure 17. An example of compound stability filters that can be used to filter the compound lists produced by the atom enumeration algorithm.

element of serendipity into the drug discovery process.

3. Virtual screening methods applied to combinatorial libraries allow for dimensionality reduction techniques to be deployed that render the storage and processing of individual compounds unnecessary.

Large combinatorial library collections for the purpose of virtual screening have been described in the literature by Nikitin et al. [282] based on published combinatorial chemistry collected from approximately 200 individual publications. Boehm et al. described the use of large sets of combinatorial libraries for applications within the pharmaceutical industry [283]. Typically, combinatorial library collections include a few hundred libraries with up to 10 trillion compounds. A few 3D methods have been described to search in such a large molecular spaces. Fragmentation of the molecules is used to reduce the dimensionality of the search space. Topomer descriptors are an example of 3D compound fragmentation used for this purpose [42]. Nikitin et al. used their collection of reactions to feed into a 3D *de novo* design program. For more insight into the design of combinatorial libraries, the reader may consult Chapter 8 in Volume 1.

Very fast searching of large combinatorial libraries can only be done using 2D similarity descriptors. FeatureTrees-FS [12] an extension of the FeatureTrees [284] algorithm has been applied by Boehm et al. to search through 358 libraries containing $>10^{12}$ compounds [283]. The FeatureTree approach is a 2D compound similarity method that represents a small molecule by a tree structure where the nodes represent small building blocks of the compound. These nodes are either ring systems or acyclic atoms that are connected according to the molecular graph (Fig. 18). The nodes are labeled with information related to the properties of the building blocks they represent. Shape and size of a building block as well as chemical descriptors such as pharmacophore character profiles are linearly combined to characterize each node. The FeatureTrees-FS extension of the Ftree approach relies on dynamic programming to perform a similarity search of combinatorial libraries of up to 10^{18} compounds in a few minutes searching within the entire combinatorial space without enumeration.

Combining reaction rules applicable to combinatorial chemistry is a concept that can be deployed in *de novo* design programs. Recent attempts in retrosynthetic *de novo* design programming go beyond the use of simple RECAP [285] rules for combining fragments to fully enumerated molecules. Recently, SimBioSys announced plans to combine their retrosynthtic chemistry package ARChem with their eHITS technology to build a new retrosynthetic *de novo* design tool in 2009 [286].

6.1. Privileged Structures

Privileged structures are structural types of small molecules that are able to bind with high affinity to multiple classes of receptors [287]. An enrichment of libraries with privileged structures may increase the chance of finding active compounds. Examples of privileged structures include benzazepine analogs found to be effective ligands for an enzyme that cleaves the peptide angiotensin I while others are effective CCK-A receptor ligands. Cyproheptadine derivatives were found to have per-

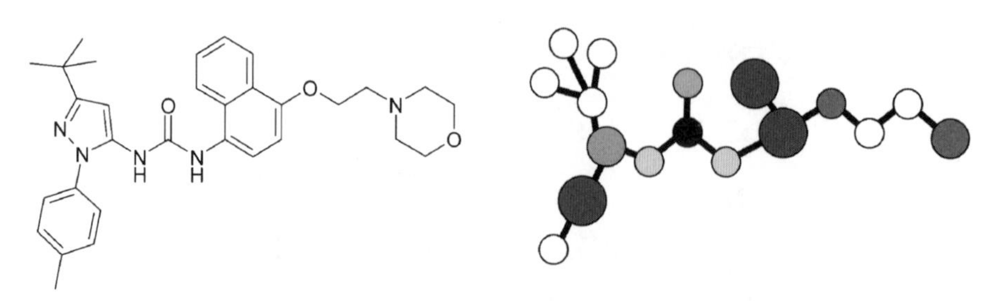

Figure 18. FeatureTree representation of a small molecule.

ipheral anticholinergic, antiserotonin, antihistaminic, and orexigenic activity. Hydroxamate and benzamidine derivatives have been shown to be privileged structures for metalloproteases and serine proteases, respectively. For the class of 7-transmembrane G-protein-coupled receptors, a large number of privileged structures has been found including, for example, diphenylmethane, diazepine, benzazepine, biphenyltetrazole, spiropiperidine, indole, and benzylpiperidine [48]. Some ubiquitously privileged structures have been identified [288]. They include carboxylic acids, biphenyls, diphenylmethane, and to a lesser extent, naphthyl, phenyl, cyclohexyl, bibenzyl, benzimidazole, and quinoline.

6.2. Targeting Protein Families

As alluded to above already, it is as important to screen the most appropriate compound collection, as it is to choose the best virtual screening method. In addition to enriching screening collections with privileged structures, the choice of databases shaped for specific protein classes is also very useful. It can be expected that the number of hits in a targeted compound collection is higher than that of a general compound collection. At the same time, the number of false positives will also be possibly higher. It is therefore observed that virtual screening enrichment rates for targeted libraries and general drug-like libraries

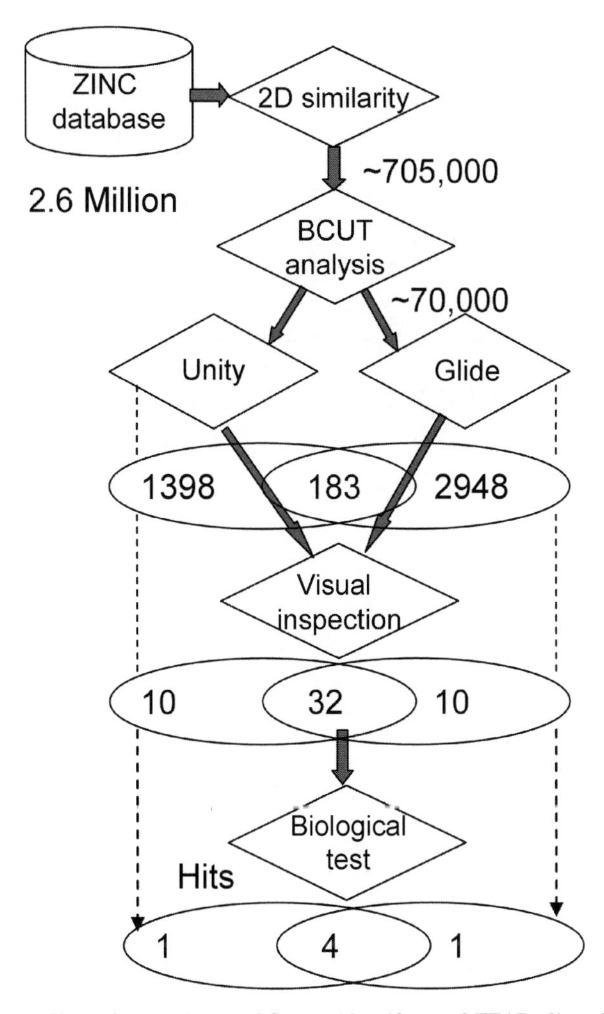

Figure 19. Virtual screening workflow to identify novel FFAR1 ligands [290].

Table 3. Recently Pursued Targets Using Synergistic Virtual Screening Strategies

Target Class	Target	VS Approaches Used	OUTCOME	Reference
Kinases	YpkA	Machine learning and multiple conformational virtual screening	First reported single digit micromolar inhibitors	Hu et al., 2007 [294]
	TMPKmt	Pharmacophore searching, docking, structure interaction fingerprints	Three hits showed MIC of 3.12 μg/mL	Kumar et al., 2008 [295]
	JAK3	Docking, similarity searching	Three hits with nanomolar activity	Chen et al., 2008 [296]
Phosphatases	CDC25	Several docking programs (FRED, Surflex, LigandFit)	13 μM hit	Montes et al., 2008 [297]
G-protein-coupled receptors	CCR2	pharmacophore fingerprints, HSQSAR, docking	Two putative CCR2 ligands	Nair and Sobhia, 2008 [298]
	FFAR1	2D similarity, 3D pharmacophore searching, docking	15 agonists, partial agonists, and antagonists	Tikhonova et al., 2008 [290]
	CCR5	2D pharmacophore similarity, docking, clustering	Agonist found that promoted receptor internalization	Kellenberger et al., 2007 [299]
Proteases	CatepsinS	Ligand-based and structure-based pharmacophore searching	7 hits in low micromolar range	Markt et al., 2008 [300]
	SARS CoV protease	Docking, 3D-QSAR, pharmacophore	25 inhibitors >3 μM	Tsai et al., 2006 [301]
Ion channels	$K_v1.5$	Protein-derived pharmacophores, FeatureTrees, 2D-similarity	5 inhibitor classes <10 μM	Pirard et al., 2005 [302]
Nuclear receptors	PPAR	Ligand-based and structure-based pharmacophore searching	Novel PPAR agonists with varying PPAR selectivity	Markt et al., 2008 [103]
Other enzymes	Chorismate mutase	Ligand-based pharmacophore models, docking	6 μM inhibitors	Agrawal et al., 2007 [303]
	DPPIV	Pharmacophore, docking	51 inhibitors found	Ward et al., 2005 [304]
	COX-2	machine learning, pharmacophore descriptor	new inhibitors found	Franke et al., 2005 [305]
protein-RNA interactions	HIV-1 reverse transcriptase	3D-QSAR, docking	2 potent inhibitors from SPECS	Zhang et al., 2006 [306]
transporters	SHGB	2D-QSAR, docking	ligands found	Cherkasov et al., 2005 [307]
others	HRV coat protein	structure-based pharmacophore model, docking, PCA	6 structures with antirhinoviral activity	Steindl et al., 2005 [308]
	spermidine synthase	3D pharmacophore filtering, docking	seven novel binders	Jacobsson et al., 2008 [309]

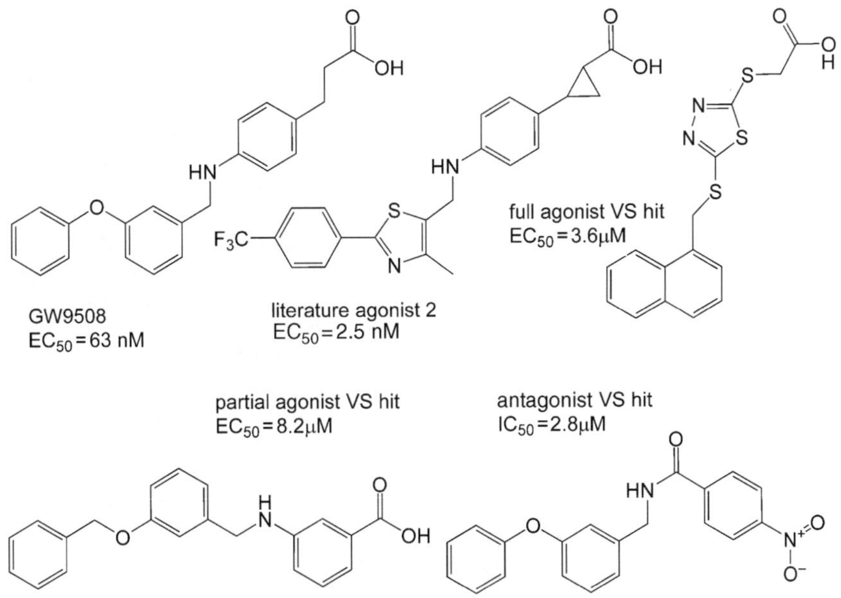

GW9508
$EC_{50} = 63$ nM

literature agonist 2
$EC_{50} = 2.5$ nM

full agonist VS hit
$EC_{50} = 3.6\mu M$

partial agonist VS hit
$EC_{50} = 8.2\mu M$

antagonist VS hit
$IC_{50} = 2.8\mu M$

Figure 20. FFAR1 ligands. GW9508 and the literature agonist 2 are full agonists [293]. The other three compounds depict representative virtual screening hits [290].

is often similar [289]. However, targeted libraries have the potential of offering structurally richer hit sets.

7. SYNERGIES

Early on virtual screening has been considered a cascade of filters involving many different components considering 1D to 3D properties according to computational expense [11]. Nonetheless, in practical applications one primary filter often prevailed. For instance, if target structures at atomic resolution were available structure-based virtual screening was often the primary virtual screening tool [22]. Alternatively, for target classes such as GPCRs mostly ligand-based methods were preferred. After some debate about the strengths of ligand-based and structure-based virtual screening approaches [41,54], it is now common practice to combine different virtual screening approaches in synergy. Table 3 illustrates recent virtual screening campaigns involving different drug targets and synergetic virtual screening protocols.

7.1. Example 7: Free Fatty Acid Receptor 1 Ligands

A particularly instructive example of using virtual screening synergies has been presented recently by Tikhonova et al. [290]. Agonists, partial agonists, and an antagonist of free fatty acid receptor (FFAR) 1 were identified through a combination of 2D similarity searching of the ZINC database [291], BCUT diversity analysis, 3D pharmacophore searching, receptor modeling involving receptor mutagenesis data [292], molecular docking, and pharmacological testing. Due to its perceived role in regulating insulin and glucagon secretion FFAR1 is considered a promising target for treating type 2 diabetes. Figure 19 illustrates the workflow of the virtual screening procedure. Figure 20 shows the compounds used for the ligand-based part of the virtual screen as well as the hits identified. It is particularly interesting to note that the docking into the FFAR1 homology model that was generated using a ground-state rhodopsin structure in presence of GW9508 only yielded agonist and partial agonist hits. The only antagonist hit was obtained through the pharmacophore search. This example shows not only how different virtual screening methods can work together to discover structurally novel hits but also how the different strengths contribute to an overall increased richness of the hits' properties.

ACKNOWLEDGMENT

I.M. would like to thank Dr. Istvan Enyedy for his work on the previous version of this book article that appeared in the 6th edition, Vol. 1 of *Burger's Medicinal Chemistry and Drug Discovery*.

REFERENCES

1. Abraham DJ. The potential role of single crystal X-ray diffraction in medicinal chemistry. Intra-Science Chem Rep 1974;8:1–9.
2. Perutz M. Protein Structure. New Approaches to Disease and Therapy. New York: W. H. Freeman and Company; 1992.
3. Fesik SW. NMR Studies of molecular complexes as a tool in drug design. J Med Chem 1991;34:2937–2945.
4. Cushman DW, Cheung HS, Sabo EF, Ondetti MA. Design of potent competitive inhibitors of agiotensin-converting enzyme. Carboxyalkanoyl and mercaptoalkanoyl amino acids. Biochemistry 1977;16:5484–5491.
5. Kuyper LF, Roth B, Baccanari DP, Ferone R, Bedell CR, CHampness JN, Stammers DK, Dann JG, Norrington FE, Blaker DJ, Goodford PJ. Receptor-based design of dihydrofolate reductase inhibitors: comparison of crystallographically determined enzyme binding with enzyme affinity in a series of carboxy-substituted trimethoprim analogues. J Med Chem 1982;25:1120–1122.
6. Gallop MA, Barrett RW, Dower WJ, Fodor SPA, Gordon AM. Applications of combinatorial technologies to drug discovery. 1. Background and peptide combinatorial libraries. J Med Chem 1994;37:1233–1251.
7. Gordon EM, Barrett RW, Dower WJ, Fodor SPA, Gallop MA. Applications of combinatorial technologies to drug discovery 2. Combinatorial organic synthesis, library screening strategies, and future directions. J Med Chem 1994;37:1385–1401.
8. Lutz MW, Menius JA, Choi TD, Laskody RG, Domanico PL, Goetz AS, Saussy DL. Experimental-design for high throughput screening. Drug Discov Today 1996;1:277–286.
9. Böhm HJ, Klebe G. What can we learn from molecular recognition in protein–ligand complexes for the design of new drugs. Angew Chem Int Ed Engl 1996;35:2589–2614.
10. Martin YC. Challenges and prospects for computational aids to molecular diversity. Perspect Drug Discov Des 1997;7/8:159–172.
11. Walters WP, Stahl MT, Murcko MA. Virtual screening: an overview. Drug Discov Today 1998;3:160–178.
12. Rarey M, Stahl M. Similarity searching in large combinatorial chemistry spaces. J Comput Aided Mol Des 2001;15:497–520.
13. Schneider G, Neidhart W, Giller T, Schmid G. "Scaffold-hopping" by topological pharmacophore search: a contribution to virtual screening. Angew Chem Int Ed Engl 1999;38: 2894–2896.
14. Klebe G. Virtual ligand screening: strategies, perspectives and limitations. Drug Discov Today 2006;11:580–594.
15. Lyne PD. Structure-based virtual screening: an overview. Drug Discov Today 2002;7: 1047–1055.
16. Muegge I, Enyedy I, Virtual screening. In: Abraham D, editors. Burger's Medicinal Chemistry. 6th ed. New York: Wiley; 2003. p 243–280.
17. Muegge I, Oloff S. Advances in virtual screening. Drug Discov Today Technol 2006;3: 405–411.
18. Muegge I. Synergies of virtual screening approaches. Mini Rev Med Chem 2008;8:927–933.
19. Oprea TI, Matter H. Integrating virtual screening in lead discovery. Curr Opin Chem Biol 2004;8:349–358.
20. Waszkowycz B. Towards improving compound selection in structure-based virtual screening. Drug Discov Today 2008;13:219–226.
21. Willet P. Similarity-based virtual screening using 2D fingerprints. Drug Discov Today 2006;11:1046–1053.
22. Good A. Structure-based virtual screening protocols. Curr Opin Drug Discov Dev 2001;4: 301–307.
23. Good AC, Mason JS, Three dimensional structure database searches. In: Lipkowitz KB, Boyd DB, editors. Reviews in Computational Chemistry. Vol.7. New York: VCH; 1995. p 67–117.
24. Klebe G. Virtual Screening: An Alternative or Complement to High Throughput Screening? Leiden: Kluwer/Escom; 2000.

25. Varnek A, Tropsha A, Chemoinformatics Approaches to Virtual Screening. Cambridge, UK: RSC Publishing; 2008.
26. Eckert H, Bajorath J. Molecular similarity analysis in virtual screening: foundataions, limitations and novel approaches. Drug Discov Today 2007;12:225–233.
27. Mason JS, Good AC, Martin EJ. 3-D pharmacophores in drug discovery. Curr Pharm Des 2001;7:567–597.
28. Sun H. Pharmacophore-based virtual screening. Curr Med Chem 2008;15:1018–1024.
29. Vert JP, Jacob L. Machine learning for *in silico* virtual screening and chemical genomics: new strategies. Comb Chem High Throughput Screen 2008;11:677–685.
30. Li H, Yap CW, Ung CY, Xue Y, Li ZR, Han LY, Lin HH, Chen YZ. Machine learning approaches for predicting compounds that interact with therapeutic and ADMET related proteins. J Pharm Sci 2007;96:2838–2860.
31. Martin YC, Kofron JL, Traphagen LM. Do structurally similar molecules have similar biological activity? J Med Chem 2002;45: 4350–4358.
32. Daylight Chemical Information Systems, Inc. Mission Viejo, CA; www.daylight.com.
33. Carhart RE, Smith DH, Venkataraghavan R. Atom pairs as molecular features in structure–activity studies: definition and application. J Chem Inf Comput Sci 1985;25:64–73.
34. Pipeline Pilot Basic Chemistry Component Collection, SciTegic Inc., San Diego, CA. 2008.
35. MDL Information Systems, Inc., San Leandro, CA.
36. Kier LB, Hall LH, Molecular Connectivity in Chemistry and Drug Research. New York: Academic Press; 1976.
37. Hert J, Willett P, Wilton DJ, Acklin P, Azzaoui K, Jacoby E, Schuffenhauer A. Comparison of fingerprint-based methods for virtual screening using multiple bioactive reference structures. J Chem Inf Comput Sci 2004;44: 1177–1185.
38. Hert J, Willett P, Wilton DJ, Acklin P, Azzaoui K, Jacoby E, Schuffenhauer A. Comparison of topological descriptors for similarity-based virtual screening using multiple bioactive reference structures. Org Biomol Chem 2004;2: 3256–3266.
39. Matter H, Potter T. Comparing 3D pharmacophore triplets and 2D fingerprints for selecting diverse compound subsets. J Chem Inf Comput Sci 1999; 1211–1225.
40. Matter H, Schwab W, Barbier D, Billen G, Haase B, Neises B, Schudok M, Thorwart W, Schreuder H, Brachvogel V, Lonze P, Weithmann KU. Quantitative structure–activity relationship of human neutrophil collagenase (MMP-8) inhibitors using comparative molecular field analysis and X-ray structure analysis. J Med Chem 1999;42:1908–1920.
41. Sheridan RP, Kearsley SK. Why do we need so many chemical similarity search methods? Drug Discov Today 2002;7:903–911.
42. Andrews KM, Cramer RD. Toward general methods of targeted library design: topomer shape similarity searching with diverse structures as queries. J Med Chem 2000;43: 1723–1740.
43. Jenkins JL, Glick M, Davies JW. A 3D similarity method for scaffold hopping from known drugs or natural ligands to new chemotypes. J Med Chem 2004;47:6144–6159.
44. Makara GM. Measuring molecular similarity and diversity: total pharmacophore diversity. J Med Chem 2001;44:3563–3571.
45. Good AC, Hermsmeier MA, Hindle SA. Measuring CAMD technique performance: a virtual screening case study in the design of validation experiments. J Comput Aided Mol Des 2004;18:529–536.
46. Good AC, Cho SJ, Mason JS. Descriptors you can count on? Normalized and filtered pharmacophore descriptors for virtual screening. J Comput Aided Mol Des 2004;18:523–527.
47. Mason JS, Cheney DL. Library design and virtual screening using multiple 4-point pharmacophore fingerprints. Pac Symp Biocomput 2000;5:576–587.
48. Mason JS, Morize I, Menard PR, Cheney DL, Hulme C, Labaudiniere RF. New 4-point pharmacophore method for molecular similarity and diversity applications: overview of the method and applications, including a novel approach to the design of combinatorial libraries containing privileged substructures. J Med Chem 1999;42:3251–3264.
49. Mason JS, Cheney DL. Ligand–receptor 3-D similarity studies using multiple 4-point pharmacophores. Pac Symp Biocomput 1999;4: 456–467.
50. Deng Z, Chuaqui C, Singh J. Structural interaction fingerprint (SIFt): a novel method for analyzing three-dimensional protein–ligand binding interactions. J Med Chem 2004;47: 337–344.
51. McGregor MJ, Muskal SM. Pharmacophore fingerprinting. 1. Application to QSAR and

focused library design. J Chem Inf Comput Sci 1999;39:569–574.

52. McGregor MJ, Muskal SM. Pharmacophore fingerprinting. 2. Application to primary library design. J Chem Inf Comput Sci 2000; 40:117–125.
53. Abrahamian E, Fox PC, Naerum L, Christensen IT, Thogersen H, Clark RD. Efficient generation, storage, and manipulation of fully flexible pharmacophore multiplets and their use in 3-D similarity searching. J Chem Inf Comput Sci 2005;43:458–468.
54. Zhang Q, Muegge I. Scaffold hopping through virtual screening using 2D and 3D similarity descriptors: ranking, voting, and consensus scoring. J Med Chem 2006;49:1536–1548.
55. Weinstein DS, Liu W, Gu Z, Langevin C, Ngu K, Fadnis L, Combs DW, Sitkoff D, Ahmad S, Zhuang S, Chen X, Wang F-L, Loughny DA, Atwal KS, Zahler R, Macor JE, Madsen CS, Murugesan N. Tryptamine and homotryptamine-based sulfonamides as potent and selective inhibitors of 15-lipoxygenase. Bioorg Med Chem Lett 2005;15:1435–1440.
56. Van Drie JH, Weininger D, Martin YC. Aladdin: an integrated tool for computer-assisted molecular design and pharmacophore recognition from geometric, steric and substructure searching of three-dimensional molecular structures. J Comput Aided Mol Des 1989;3:225–251.
57. Clement OO, Mehl AT, HipHop: pharmacophores based on multiple common-feature alignments. In: Güner OF, editors. Pharmacophore Perception, Development, and Use in Drug Design. La Jolla CA: International University Line; 2000. p 71–84.
58. Jia L, Zou J, So SS, Sun H. Automated pharmacophore query optimization with genetic algorithms: a case study using the MC4R system. J Chem Inf Model 2007;47:1545–1552.
59. Wolber G, Langer T. LigandScout: 3-D pharmacophores derived from protein-bound ligands and their use as virtual screening filters. J Chem Inf Model 2005;45:160–169.
60. Jacob L, Hoffmann B, Stoven V, Vert JP. Virtual screening of GPCRs: an *in silico* chemogenomics approach. BMC Bioinformatics 2008;9:363
61. Jaakola VP, Griffith MT, Hanson MA, Cherezov V, Chien EY, Lane JR, Ijzerman AP, Stevens RC. The 2.6 Å crystal structure of a human A2A adenosine receptor bound to an antagonist. Science 2008;322:1211–1217.
62. Scheerer P, Park JH, Hildebrand PW, Kim YJ, Krauss N, Choe HW, Hofmann KP, Ernst OP. Crystal structure of opsin in its G-protein-interacting conformation. Nature 2008;455: 497–502.
63. Warne T, Serrano-Vega MJ, Baker JG, Moukhametzianov R, Edwards PC, Henderson R, Leslie AG, Tate CG, Schertler GF. Structure of a beta1-adrenergic G-protein-coupled receptor. Nature 2008;454:486–491.
64. Rasmussen SG, Choi HJ, Rosenbaum DM, Kobilka TS, Thian FS, Edwards PC, Burghammer M, Ratnala VR, Sanishvili R, Fischetti RF, Schertler GF, Weis WI, Kobilka BK. Crystal structure of the human beta2 adrenergic G-protein-coupled receptor. Nature 2007;450: 383–387.
65. Palczewski K, Kumasaka T, Hori T, Behnke CA, Motoshima H, Fox BA, Le TI, Teller DC, Okada T, Stenkamp RE, Yamamoto M, Miyano M. Crystal structure of rhodopsin: a G protein-coupled receptor. Science 2000;289:739–745.
66. Leach AR, Green DV, Hann MM, Judd DB, Good AC. Where are the GaPs? A rational approach to monomer acquisition and selection. J Chem Inf Comput Sci 2000;40: 1262–1269.
67. Fischer E. Einfluss der Konfiguration auf die Wirkung der Enzyme. Ber Dtsch Chem Ges 1894;27:2985–2993.
68. Ehrlich P. Uber den jetzigen stand der chemotherapie. Ber Dtsch Chem Ges 1909;42: 17–47.
69. Gund P. Three-dimensional pharmacophoric pattern searching. Prog Mol Subcell Biol 1977;5:117–143.
70. Berman HM, Westbrook J, Feng Z, Gilliland G, Baht TN, Weissig H, Shindyalov IN, Bourne PE. The Protein Data Bank. Nucleic Acids Res 2000;28:235–242.
71. Bernstein FC, Koetzle TF, Williams G, Meyer EF Jr, Brice MD, Rodgers JR, Kennard O, Shimanouchi T, Tasumi M. The Protein Data Bank: a computer based archival file for macromolecular structures. J Mol Biol 1977;112: 535–542.
72. Dreyer MK, Borcherding DR, Dumont JA, Peet NP, Tsay JT, Wright PS, Bitonti AJ, Shen J, Kim S-H. Crystal structure of human cyclin-dependent kinase 2 in complex with the adenine-derived inhibitor H717. J Med Chem 2001;44:524–530.
73. Irwin JJ, Shoichet BK. ZINC: a free database of commercially available compounds for virtual screening. J Chem Inf Model 2005;45:177–182.
74. Milne GWA, Miller JA. The NCI Drug Information System Parts 1-6. J Chem Inf Comput Sci 1986;26:154–159.

75. Milne GWA, Nicklaus MC, Driscoll JS, Wang S, Zaharevitz DW. The NCI Drug Information System 3D Database. J Chem Inf Comput Sci 1994;34:1219–1224.
76. eMolecules; www.emolecules.com. 2008.
77. ChemSpider; www.chemspider.com. 2009.
78. Available Chemicals Directory is available from MDL Information Systems, Inc., San Leandro, CA, and contains specialty bulk chemicals from commercial sources; www.mdli.com.
79. Enamine; www.enamine.net. 2009.
80. Weininger D. SMILES, a chemical language and information system. 1. Introduction to methodology and encoding rules. J Chem Inf Comput Sci 1988;28:31–36.
81. Weininger D, Weininger A, Weininger JL. Algorithm for generation of unique SMILES notation. J Chem Inf Comput Sci 1989;29: 97–101.
82. Allen FH, Bellard S, Brice MD, Cartwright BA, Doubleday A, Higgs H, Hummelink-Peters T, Kenard O, Motherwell WDS, Rodgers JR, Watson DG. The Cambridge Crystallographic Data Center: computer-based search, retrieval, analysis and display of information. Acta Cryst 1979;B35:2331–2339.
83. DesJarlais RL, Generation and use of three-dimensional databases for drug-discovery. In: Charifson PS, editors. Practical Application of Computer-Aided Drug Design. New York: Marcel Dekker, Inc; 1997. p 73–104.
84. Rusinko A III, Skell JM, Balducci R, McGarity CM, Pearlman RS,CONCORD, University of Texas, Austin, TX and Tripos Associates, St. Louis, MO. 1988.
85. Gasteiger J, Rudolph C, Sadowski J. Automatic generation of 3D-atomic coordinates for organic molecules. Tetrahedron Comput Methodol 1990;3:537–547.
86. Barnum D, Greene J, Smellie A, Sprague P. Identification of common functional configurations among molecules. J Chem Inf Comput Sci 1996;36:563–571.
87. Sprague PW. Automated chemical hypothesis generation and database searching with Catalyst. Perspectives in Drug Discovery and Design 1995;3:21–33.
88. Dixon SL, Smondyrev AM, Knoll EH, Rao SN, Shaw DE, Friesner RA. PHASE: a new engine for pharmacophore perception, 3D QSAR model development, and 3D database screening: 1. Methodology and preliminary results. J Comput Aided Mol Des 2006;20:647–671.
89. CCG; http://www.chemcomp.com/journal/ph4.htm. 2008.
90. Martin YC, Bures MG, Danaher EA, DeLazzer J, Lico I, Pavlik PA. A fast new approach to pharmacophore mapping and its application to dopaminergic and benzodiazepine agonists. J Comput Aided Mol Des 1993;7:83–102.
91. Beusen D,Alignment of angiotensin II receptor antagonists using GASP, Tripos Technical Notes 1996:1.(4).
92. Flo96, Thistlesoft, Colebrook, CT. 1996.
93. Golender V, Vesterman B, Eliyahu O, Kardash A, Kletzin M, Shokhen M, Vorpagel E, 1995. Knowledge-engineering approach to drug design and its implementation in the Apex-3D expert system. In: Sanz F, Giraldo J, Manaut F, editors. QSAR and Molecular Modeling: Concepts, Computational Tools and Biological Applications: Proceedings of the 10th European Symposium on Structure-Activity Relationships, QSAR and Molecular Modeling: 1994 September 4–9; Barcelona, Spain; J. R. Prous Science Publishers; 1995. p 246–251.
94. OF.Pharmacophore Perception, Development, and use in Drug Design. La Jolla, CA: International University Line; 2000.
95. Millichap JG. Drugs in management of minimal brain dysfunction. Ann NY Acad Sci 1973;205:321–334.
96. Enyedy IJ, Zaman WA, Sakamuri S, Kozikowski AP, Johnson KM, Wang S. Pharmacophore-based discovery of 3, 4-disubstituted pyrrolidines as a novel class of monoamine transporter inhibitors that function as weak cocaine antagonists. Bioorg Med Chem Lett 2001; 11:1113–1118.
97. Froimowitz M, Patrick KS, Cody V. Conformational analysis of methylphenidate and its structural relationship to other dopamine re-uptake blockers such as CFT. Pharm Res 1995;12:1430–1434.
98. Wang S, Sakamuri S, Enyedy IJ, Kozikowski AP, Deschaux O, Bandyopadhyay BC, Tella SR, Zaman WA, Johnson KM. Discovery of a novel dopamine transporter inhibitor, 4-hydroxy-1-methyl-4-(4-methylphenyl)-3-piperidyl 4-methylphenyl ketone, as a potential cocaine antagonist through 3D-database pharmacophore searching. Molecular modeling, structure–activity relationships, and behavioral pharmacological studies. J Med Chem 2000;43:351–360.
99. Chem-X, version 96; Oxford Molecular Group, Inc., Hunt Valley, MD (company exists no longer). 2001.

100. Böhm H-J. LUDI: Rule-based automatic design of new substituents for enzyme inhibitor leads. J Comput Aided Mol Des 1992;6: 593–606.
101. Goodford PJ. A computational procedure for determining energetically favorable binding sites on biologically important macromolecules. J Med Chem 1985;28:849–857.
102. Carlson HA, Masukawa KM, Rubins K, Bushman FD, Jorgensen WL, Lins RD, Briggs JM, McCammon JA. Developing a dynamic pharmacophore model for HIV-1 integrase. J Med Chem 2000;43:2100–2114.
103. Markt P, Petersen RK, Flindt EN, Kristiansen K, Kirchmair J, Spitzer G, Distinto S, Schuster D, Wolber G, Laggner C, Langer T. Discovery of novel PPAR ligands by a virtual screening approach based on pharmacophore modeling, 3D shape, and electrostatic similarity screening. J Med Chem 2008;51:6303–6317.
104. Catalyst, version 4.11; Accelrys, Inc., San Diego, CA. 2006.
105. UNITY, version 6.6; Tripos Associates, St. Louis, MO; www.tripos.com.
106. Rush TS III, Grant JA, Mosyak L, Nicholls A. A shape-based 3-D scaffold hopping method and its application to a bacterial protein–protein interaction. J Med Chem 2005;48:1489–1495.
107. Fang X, Wang S. A web-based 3D-database pharmacophore searching tool for drug discovery. J Chem Inf Comput Sci 2002;42:192–198.
108. Ihlenfeldt W-D, Voigt JH, Bienfait B, Oellien F, Nicklaus MC. Enhanced CACTVS browser of the open NCI database. J Chem Inf Comput Sci 2002;42:46–57.
109. Nicklaus MC, Wang S, Driscoll JS, Milne GWA. Conformational changes of small molecules binding to proteins. Bioorg Med Chem 1995;3:411–428.
110. Perola E, Charifson PS. Conformational analysis of drug-like molecules bound to proteins: an extensive study of ligand reorganization upon binding. J Med Chem 2004;47: 2499–2510.
111. Hao MH, Haq O, Muegge I. Torsion angle preference and energetics of small-molecule ligands bound to proteins. J Chem Inf Model 2007;47:2242–2252.
112. EduSoft. Molconn-Z, version 4.09 is available from EduSoft; http://www.edusoft-lc.com/molconn/. 2006.
113. Cruciani G, Pastor M, Guba W. VolSurf: a new tool for the pharmacokinetic optimization of lead compounds. Eur J Pharm Sci 2000;11 (Suppl 2): S29–S39
114. Carhart RE, Smith DH, Venkataraghavan R. Atom pairs as molecular features in structure–activity studies: definition and applications. J Chem Inf Comput Sci 1985;25:64–73.
115. Cruciani G, Crivori P, carrupt PA, Testa B. Molecular fields in quantitative structure–permeation relationships: the VolSurf approach. J Mol Struct 2000;503:17–30.
116. Todeschini R, Consonni V, Handbook of Molecular Descriptors. Weinheim: Wiley-VCH; 2000.
117. Golbraikh A, Bonchev D, Tropsha A. Novel chirality descriptors derived from molecular topology. J Chem Inf Comput Sci 2001;41: 147–158.
118. Tropsha A, Golbraikh A. Predictive QSAR modeling workflow, model applicability domains, and virtual screening. Curr Pharm Des 2007;13:3494–3504.
119. Golbraikh A, Tropsha A. Beware of q2! J Mol Graph Model 2002;20:269–276.
120. Golbraikh A, Gramatica P, Gombar VK. The importance of being earnest: Validation is the absolute essential for successful application and interpretation of QSPR models. QSAR Comb Sci 2003;22:69–77.
121. Kubinyi H, Hamprecht FA, Mietzner T. Three-dimensional quantitative similarity–activity relationships (3D QSiAR) from SEAL similarity matrices. J Med Chem 1998;41:2553–2564.
122. Schneider G, Nettekoven M. Ligand-based combinatorial design of selective purinergic receptor (A2A) antagonists using self-organizing maps. J Comb Chem 2003;5:233–237.
123. Prathipati P, Saxena AK. Evaluation of binary QSAR models derived from LUDI and MOE scoring functions for structure based virtual screening. J Chem Inf Model 2006;46:39–51.
124. Plewczynski D, Spieser SA, Koch U. Assessing different classification methods for virtual screening. J Chem Inf Model 2006;46: 1098–1106.
125. Evers A, Hessler G, Matter H, Klabunde T. Virtual screening of biogenic amine-binding G-protein coupled receptors: comparative evaluation of protein- and ligand-based virtual screening protocols. J Med Chem 2005; 48:5448–5465.
126. Shen M, Beguin C, Golbraikh A, Stables JP, Kohn H, Tropsha A. Application of predictive QSAR models to database mining: identification and experimental validation of novel anticonvulsant compounds. J Med Chem 2004; 47:2356–2364.

127. Oloff S, Mailman RB, Tropsha A. Application of validated QSAR models of D1 dopaminergic antagonists for database mining. J Med Chem 2005;48:7322–7332.
128. Sheridan RP, Feuston BP, Maiorov VN, Kearsley SK. Similarity to molecules in the training set is a good discriminator for prediction accuracy in QSAR. J Chem Inf Comput Sci 2004; 44:1912–1928.
129. Oloff S, Muegge I. kScore: a novel machine learning approach that is not dependent on the data structure of the training set. J Comput Aided Mol Des 2007;21:87–95.
130. Ghose AK, Crippen GM. Atomic physicochemical parameters for three-dimensional-structure-directed quantitative structure–activity relationships. 2. Modeling dispersive and hydrophobic interactions. J Chem Inf Comput Sci 1987;27:21–35.
131. Blaney JM, Dixon JS. A good ligand is hard to find: automatic docking methods. Perspect Drug Disc Des 1993;15:301–319.
132. Muegge I, Rarey M. Small Molecule Docking and Scoring. In: Boyd DB, Lipkowitz KB, editors. Reviews in Computational Chemistry. Vol. 17. New York: Wiley-VCH; 2001. p 1–60.
133. Blundell TL, Sibanda BL, Sternberg MJE, Thornton JM. Knowledge-based prediction of protein structures and the design of novel molecules. Nature 1987;326:347–352.
134. Vedani A, Zbinden P, Snyder JP, Greenidge PA. Pseudoreceptor modeling: the construction of three-dimensional receptor surrogates. J Am Chem Soc 1995;117:4987–4994.
135. Tanrikulu Y, Schneider G. Pseudoreceptor models in drug design: bridging ligand- and receptor-based virtual screening. Nat Rev Drug Discov 2008;7:667–677.
136. Teague SJ. Implications of protein flexibility for drug discovery. Nat Rev Drug Discov 2003;2:527–541.
137. Cozzini P, Kellogg GE, Spyrakis F, Abraham DJ, Costantino G, Emerson A, Fanelli F, Gohlke H, Kuhn LA, Morris GM, Orozco M, Pertinhez TA, Rizzi M, Sotriffer CA. Target flexibility: an emerging consideration in drug discovery and design. J Med Chem 2008;51: 6237–6255.
138. Cavasotto CN, Kovacs JA, Abagyan RA. Representing receptor flexibility in ligand docking through relevant normal modes. J Am Chem Soc 2005;127:9632–9640.
139. Jones G, Willett P, Glen RC, Leach AR. Development and validation of a genetic algorithm for flexible docking. J Mol Biol 1997;267: 727–748.
140. Muegge I. The effect of small changes in protein structure on predicted binding modes of known inhibitors of influenza virus neuraminidase: PMF-scoring in DOCK4. Med Chem Res 1999;9:490–500.
141. Goodsell DS, Olson AJ. Automated Docking of Substrates to proteins by Simulated Annealiong. Proteins 1990;8:195–202.
142. Huey R, Morris GM, Olson AJ, Goodsell DS. A semiempirical free energy force field with charge-based desolvation. J Comput Chem 2007;28:1145–1152.
143. Verdonk ML, Cole JC, Hartshorn MJ, Murray CW, Taylor RD. Improved protein–ligand docking using GOLD. Proteins 2003;52: 609–623.
144. Claussen H, Buning C, Rarey M, Lengauer T. FlexE: efficient molecular docking considering protein structure variations. J Mol Biol 2001; 308:377–395.
145. Ahmed A, Gohlke H. Multiscale modeling of macromolecular conformational changes combining concepts from rigidity and elastic network theory. Proteins 2006;63: 1038–1051.
146. Zavodszky MI, Lei M, Thorpe MF, Day AR, Kuhn LA. Modeling correlated main-chain motions in proteins for flexible molecular recognition. Proteins 2004;57:243–261.
147. Nabuurs SB, Wagener M, de VJ. A flexible approach to induced fit docking. J Med Chem 2007;50:6507–6518.
148. Bottegoni G, Kufareva I, Totrov M, Abagyan R. Four-dimensional docking: a fast and accurate account of discrete receptor flexibility in ligand docking. J Med Chem 2009;52:397–406.
149. Totrov M, Abagyan R. Flexible ligand docking to multiple receptor conformations: a practical alternative. Curr Opin Struct Biol 2008;18: 178–184.
150. Bottegoni G, Kufareva I, Totrov M, Abagyan R. A new method for ligand docking to flexible receptors by dual alanine scanning and refinement (SCARE). J Comput Aided Mol Des 2008;22:311–325.
151. Zentgraf M, Fokkens J, Sotriffer CA. Addressing protein flexibility and ligand selectivity by "*in situ* cross-docking". ChemMedChem 2006;1:1355–1359.
152. Tame JRH. Scoring functions: a view from the bench. J Comput Aided Mol Des 1999;13: 99–108.

153. Böhm HJ, Stahl M. Rapid empirical scoring functions in virtual screening applications. Med Chem Res 1999;9:445–462.

154. Good A. Structure-based virtual screening protocols. Curr Opin Drug Discov Devel 2001;4: 301–307.

155. Böhm H.-J. Stahl M. The use of scoring functions in drug discovery applicatong. In: Lipkowitz KB, Boyd DB, editors. Reviews in Computational Chemistry. Vol.18.Hoboken, NJ: Wiley-VCH; 2002. p 41–87

156. Kollman P. Free energy calculations-applications to chemical and biological phenomena. Chem Rev 1993;7:2395–2417.

157. Bissantz C, Folkers G, Rognan D. Protein-based virtual screening of chemical databases. 1. Evaluation of different docking/scoring combinations. J Med Chem 2000;43:4759–4767.

158. Ferrara P, Gohlke H, Price DJ, Klebe G, Brooks CL III. Assessing scoring functions for protein–ligand interactions. J Med Chem 2004;47:3032–3047.

159. Ha S, Andreani R, Robbins A, Muegge I. Evaluation of docking/scoring approaches: a comparative study based on MMP-3 inhibitors. J Comput Aided Mol Des 2000;14:435–448.

160. Kellenberger E, Rodrigo J, Muller P, Rognan D. Comparative evaluation of eight docking tools for docking and virtual screening accuracy. Proteins 2004;57:225–242.

161. Knegtel RMA, Wagener M. Efficacy and selectivity in flexible database docking. Proteins 1999;37:334–345.

162. Kontoyianni M, Sokol GS, McClellan LM. Evaluation of library ranking efficacy in virtual screening. J Comput Chem 2005;26:11–22.

163. Perola E, Walters WP, Charifson PS. A detailed comparison of current docking and scoring methods on systems of pharmaceutical relevance. Proteins 2004;56:235–249.

164. Schulz-Gasch T, Stahl M. Binding site characteristics in structure-based virtual screening: evaluation of current docking tools. J Mol Model 2003;9:47–57.

165. Stahl M, Rarey M. Detailed analysis of scoring functions for virtual screening. J Med Chem 2001;44:1035–1042.

166. Kontoyianni M, Madhav P, Suchanek E, Seibel W. Theoretical and practical considerations in virtual screening: a beaten field?. Curr Med Chem 2008;15:107–116.

167. Warren GL, Andrews CW, Capelli AM, Clarke B, LaLonde J, Lambert MH, Lindvall M, Nevins N, Semus SF, Senger S, Tedesco G, Wall ID, Woolven JM, Peishoff CE, Head MS. A critical assessment of docking programs and scoring functions. J Med Chem 2006;49: 5912–5931.

168. Wang R, Lu Y, Wang S. Comparative evaluation of 11 scoring functions for molecular docking. J Med Chem 2003;46:2287–2303.

169. Kontoyianni M, McClellan LM, Sokol GS. Evaluation of docking performance: comparative data on docking algorithms. J Med Chem 2004; 47:558–565.

170. Liebeschuetz JW. Evaluating docking programs: keeping the playing field level. J Comput Aided Mol Des 2008;22:229–238.

171. Moitessier N, Englebienne P, Lee D, Lawandi J, Corbeil CR. Towards the development of universal, fast and highly accurate docking/scoring methods: a long way to go. Br J Pharmacol 2008;153(Suppl 1): S7–S26

172. Holloway MK, McGaughey GB, Coburn CA, Stachel SJ, Jones KG, Stanton EL, Gregro AR, Lai MT, Crouthamel MC, Pietrak BL, Munshi SK. Evaluating scoring functions for docking and designing beta-secretase inhibitors. Bioorg Med Chem Lett 2007;17:823–827.

173. Muegge I, Enyedy I. Virtual screening of kinase targets. Curr Med Chem 2004;11: 691–705.

174. Graves AP, Brenk R, Shoichet BK. Decoys for docking. J Med Chem 2005;48:3714–3728.

175. Jansen JM, Martin EJ. Target-biased scoring approaches and expert systems in structure-based virtual screening. Curr Opin Chem Biol 2004;8:359–364.

176. Polgar T, Keseru GM. Virtual screening for beta-secretase (BACE1) inhibitors reveals the importance of protonation states at Asp32 and Asp228. J Med Chem 2005;48:3749–3755.

177. Rarey M, Kramer B, Lengauer T, Klebe G. A fast flexible docking method using an incremental construction algorithm. J Mol Biol 1996;261: 470–489.

178. Hindle SA, Rarey M, Buning C, Lengaue T. Flexible docking under pharmacophore type constraints. J Comput Aided Mol Des 2002; 16:129–149.

179. Blaney JD, Dixon JS. Distance geometry in molecular modeling. In: Lipkowitz KB, Boyd DB, editors. Reviews in Computational Chemistry. New York: Wiley-VCH; 1994. p 299–335

180. Nauchatel V, Villaverde MC, Sussman F. Solvent accessibility as a predictive tool for the free energy of inhibitor binding to the HIV-1 protease. Protein Sci 1995;4:1356–1364.

181. Ferrari AM, Wei BQ, Costantino L, Shoichet BK. Soft docking and multiple receptor con-

formations in virtual screening. J Med Chem 2004;47:5076–5084.

182. Vieth M, Hirst JD, Kolinski A, Brooks CL III. Assessing energy functions for flexible docking. J Comput Chem 1998;14:1612–1622.
183. Liu M, Wang SM. MCDOCK: a Monte Carlo simulation approach to the molecular docking problem. J Comput Aided Mol Des 1999;13:435–451.
184. Majeux N, Scarsi M, Apostolakis J, Ehrhardt C, Caflisch A. Exhaustive docking of molecular fragments with electrostatic solvation. Proteins 1999;37:88–105.
185. Qui D, Shenkin PS, Hollinger EP, Still WC. The GB/SA contunuum model for salvation. A fast analytical method for the calculation of approximate Born radii. J Phys Chem 1997; 101:3005–3014.
186. Shoichet BK, Leach AR, Kuntz ID. Ligand solvation in molecular docking. Proteins 1999;34:4–16.
187. Wang J, Wang W, Huo S, Lee M, Kollman PA. Solvation model based on weighted solvent accessible surface area. J Phys Chem B 2001;105:5055–5067.
188. Zou X, Sun Y, Kuntz ID. Inclusion of solvation in ligand binding free energy calculations using the generalized-Born model. J Am Chem Soc 1999;121:8033–8043.
189. Hansch C, Anderson SM. The structure–activity relationship in barbiturates and its similarity to that in other narcotics. J Med Chem 1967;10:745–753.
190. OpenEye; http://www.eyesopen.com/fred.html.
191. Böhm H-J. The development of a simple empirical scoring function to estimate the binding constant for a protein–ligand complex of known three-dimensional structure. J Comput Aided Mol Design 1994;8: 243–256.
192. Eldridge MD, Murray CW, Auton TR, Paolini GV, Mee RP. Empirical scoring functions. I. The development of a fast empirical scoring function to estimate the binding affinity of ligands in receptor complexes. J Comput Aided Mol Des 1997;11:425–445.
193. Head RD, Smythe ML, Oprea TL, Waller CL, Green SM, Marshall GM. VALIDATE. a new method for the receptor-based prediction of binding affinites of novel ligands. J Am Chem Soc 1996;118:3959–3969.
194. Jain AN. Scoring noncovalent protein–ligand interactions: a continuous differentiable function tuned to compute binding affinities. J Comput Aided Mol Des 1996;10:427–440.
195. Murray CW, Auton TR, Eldridge MD. Empirical scoring functions. II. The testing of an empirical scoring function for the prediction of ligand–receptor binding affinities and the use of Bayesian regression to improve the quality of the model. J Comput Aided Mol Des 1998;12:503–519.
196. Stroganov OV, Novikov FN, Stroylov VS, Kulkov V, Chilov GG. Lead finder: an approach to improve accuracy of protein–ligand docking, binding energy estimation, and virtual screening. J Chem Inf Model 2008;48 (12): 2371–2385.
197. DeWitte RS, Shakhnovich EI. SMoG: *de novo* design method based on simple, fast, and accurate free energy estimates. 1. Methodology and supporting evidence. J Am Chem Soc 1996;118:11733–11744.
198. Muegge I, Martin YC. A general and fast scoring function for protein–ligand interactions: a simplified potential approach. J Med Chem 1999;42:791–804.
199. Gohlke H, Hendlich M, Klebe G. Knowledge-based scoring function to predict protein–ligand interactions. J Mol Biol 2000;295: 337–356.
200. Mitchell JBO, Laskowski RA, Alex A, Thornton JM. BLEEP-potential of mean force describing protein–ligand interactions. I. Generating potential. J Comput Chem 1999;20: 1165–1176.
201. Gohlke H, Klebe G. Statistical potentials and scoring functions applied to protein–ligand binding. Curr Opin Struct Biol 2001;11: 231–235.
202. Muegge I. A knowledge-based scoring function for protein–ligand interactions: probing the reference state. Perspect Drug Disc Des 2000;20:99–114.
203. Muegge I. Effect of ligand volume correction on PMF-scoring. J Comput Chem 2001;22: 418–425.
204. Muegge I. PMF scoring revisited. J Med Chem 2006;49:5895–5902.
205. Muegge I, Martin YC, Hajduk PJ, Fesik SW. Evaluation of PMF scoring in docking weak ligands to the FK506 binding protein. J Med Chem 1999;42:2498–2503.
206. Kraemer O, Hazemann I, Podjarny AD, Klebe G. Virtual screening for inhibitors of human aldose reductase. Proteins 2004;55:814–823.
207. Stahl M, Rarey M. Detailed analysis of scoring functions for virtual screening. J Med Chem 2001;44:1035–1042.

208. Wang R, Liu L, Lai L, Tang Y. SCORE: a new empirical method for estimating the binding affinity of a protein–ligand complex. J Mol Model 1998;4:379–394.
209. Bohacek RS, McMartin C. Definition and display of steric, hydrophobic, and hydrogen-bonding properties of ligand binding sites in proteins using Lee and Richards accessible surface: validation of a high-resolution graphical tool for drug design. J Med Chem 1992; 35:1671–1684.
210. Stouch TR, Jurs PC. A simple method for the representation, quantification, and comparison of the volumes and shapes of chemical compounds. J Chem Inf Comput Sci 1986;26: 4–12.
211. Gehlhaar DK, Verkhivker GM, Rejto PA, Sherman CJ, Fogel DB, Fogel LJ, Freer ST. Molecular recognition of the inhibitor AG-1343 by HIV-1 protease: conformationally flexible docking by evolutionary programming. Chem Biol 1995;2:317–324.
212. Gehlhaar DK, Bouzida D, Rejto PA. Reduced dimensionality in ligand–protein structure prediction: covalent inhibitors of serine proteases and design of site-directed combinatorial libraries. In: Parrill L. Reddy RM. editors. ACS Symposium Series 719. Washington, DC: American Chemical Society; 1999. p 292–311
213. Muegge I. PMF scoring revisited. J Med Chem 2006;49:5895–5902.
214. Feher M, Deretey E, Roy S. BHB: a simple knowledge-based scoring function to improve the efficiency of database screening. J Chem Inf Comput Sci 2003;43:1316–1327.
215. Stahl M, Böhm HJ. Development of filter functions for protein–ligand docking. J Mol Graph Model 1998;16:121–132.
216. Jones G, Willett P. Docking small-molecule ligands into active sites. Curr Opin Biotechnol 1995;6:652–656.
217. Halgren TA, Murphy RB, Friesner RA, Beard HS, Frye LL, Pollard WT, Banks JL. Glide: a new approach for rapid, accurate docking and scoring, 2. Enrichment factors in database screening. J Med Chem 2004;47:1750–1759.
218. Schrodinger, LLC; technical documentation; http://www.schrodinger.com/. 2005.
219. Rognan D, Lauemoller SL, Holm A, Buus S, Tschinke V. Predicting binding affinities of protein ligands from three-dimensional models: application to peptide binding to class I major histocompatibility proteins. J Med Chem 1999;42:4650–4658.
220. Wang RX, Liu L, Lai LH, Tang YQ. SCORE: a new empirical method for estimating the binding affinity of a protein–ligand complex. J Mol Model 1998;4:379–394.
221. Bissantz C, Folkers G, Rognan D. Protein-based virtual screening of chemical databases, 1. Evaluation of different docking/scoring combinations. J Med Chem 2000;43:4759–4767.
222. Kontoyianni M, Sokol GS, McClellan LM. Evaluation of library ranking efficacy in virtual screening. J Comput Chem 2005;26:11–22.
223. Verdonk ML, Berdini V, Hartshorn MJ, Mooij WT, Murray CW, Taylor RD, Watson P. Virtual screening using protein–ligand docking: avoiding artificial enrichment. J Chem Inf Comput Sci 2004;44:793–806.
224. Charifson PS, Corkery JJ, Murcko MA, Walters WP. Consensus scoring: a method for obtaining improved hit rates from docking databases of three-dimensional structures into proteins. J Med Chem 1999;42:5100–5109.
225. Halgren TA. Merck molecular force field. II. MMFF94 van der Waals and electrostatic parameters for intermolecular interactions. J Comput Chem 1996;17:520–552.
226. Miller MD, Kearsley SK, Underwood DJ, Sheridan RP. FLOG: a system to select "quasi-flexible" ligands complementary to a receptor of known three-dimensional structure. J Comp Aided Mol Des 1994;8:153–174.
227. Honig B, Nicholls A. Classical electrostatics in biology and chemistry. Science 1995;268: 1144–1149.
228. Flower DR. SERF: a program for accessible surface area calculations. J Mol Graph Model 1998;15:238–244.
229. Tripos; www.tripos.com.
230. Vigers GP, Rizzi JP. Multiple active site corrections for docking and virtual screening. J Med Chem 2004;47:80–89.
231. Bradley MN, Hong C, Chen M, Joseph SB, Wilpitz DC, Wang X, Lusis AJ, Collins A, Hseuh WA, Collins JL, Tangirala RK, Tontonoz P. Ligand activation of LXR beta reverses atherosclerosis and cellular cholesterol overload in mice lacking LXR alpha and apoE. J Clin Invest 2007;117:2337–2346.
232. Cheng JF, Zapf J, Takedomi K, Fukushima C, Ogiku T, Zhang SH, Yang G, Sakurai N, Barbosa M, Jack R, Xu K. Combination of virtual screening and high throughput gene profiling for identification of novel liver X receptor modulators. J Med Chem 2008;51: 2057–2061.
233. Gehlhaar DK, Verkhivker GM, Rejto PA, Fogel DB, Fogel LJ, Freer ST. Docking confor-

mationally flexible small molecules into a protein binding site through evolutionary programming. In: McDonnell JR, Reynolds RG, Fogel DB, editors. Proceedings of the Fourth Annual Conference on Evolutionary Programming. MIT PRESS; 1995. p 615–627

234. Carr RA, Congreve M, Murray CW, Rees DC. Fragment-based lead discovery: leads by design. Drug Discov Today 2005;10:987–992.
235. Rummey C, Nordhoff S, Thiemann M, Metz G. *In silico* fragment-based discovery of DPP-IV S1 pocket binders. Bioorg Med Chem Lett 2006;16:1405–1409.
236. Rarey M, Kramer B, Lengauer T, Klebe G. A fast flexible docking method using an incremental construction algorithm. J Mol Biol 1996;261:470–489.
237. Makara GM. On sampling of fragment space. J Med Chem 2007;50:3214–3221.
238. Oblak M, Grdadolnik SG, Kotnik M, Jerala R, Filipic M, Solmajer T. *In silico* fragment-based discovery of indolin-2-one analogues as potent DNA gyrase inhibitors. Bioorg Med Chem Lett 2005;15:5207–5210.
239. Muegge I. Selection criteria for drug-like compounds. Med Res Rev 2003;23:302–321.
240. Cele AZ, Metz JT. Ligand efficiency indices as guideposts for drug discovery. Drug Discov Today 2005;10:464–469.
241. Hopkins AL, Groom CR, Alex A. Ligand efficiency: a useful metric for lead selection. Drug Discov Today 2004;9:430–431.
242. Martin YC. Diverse Viewpoints on computational aspects of molecular diversity. J Comb Chem 2001;3:1–20.
243. Lobell M, Hendrix M, Hinzen B, Keldenich J, Meier H, Schmeck C, Schohe-Loop R, Wunberg T, Hillisch A. *In silico* ADMET traffic lights as a tool for the prioritization of HTS hits. ChemMedChem 2006;1:1229–1236.
244. Proudfoot JR. Drugs, leads, and drug-likeness: an analysis of some recently launched drugs. Bioorg Med Chem Lett 2002;12:1647–1650.
245. Oprea TI, Davis AM, Teague SJ, Leeson PD. Is there a difference between leads and drugs? A historical perspective. J Chem Inf Comput Sci 2001;41:1308–1315.
246. Teague SJ, Davis AM, Leeson PD, Oprea T. The design of leadlike combinatorial libraries. Angew Chem Int Ed Engl 1999;38: 3743–3748.
247. Lipinski CA. Drug-like properties and the causes of poor solubility and poor permeability. J Pharmacol Toxicol Methods 2000;44: 235–249.
248. Lipinski A. Lead- and drug-like compounds: the rule-of-five revolution. Drug Discov Today Technol 2004;1:337–341.
249. Andrews PR, Craik DJ, Martin JL. Functional group contributions to drug–receptor interactions. J Med Chem 1984;27: 1648–1657.
250. Kuntz ID, Chen K, Sharp KA, Kollman PA. The maximal affinity of ligands. Proc Natl Acad Sci U S A 8-31- 1999;96:9997–10002.
251. Hopkins AL, Groom CR, Alex A. Ligand efficiency: a useful metric for lead selection. Drug Discov Today 2004;9:430–431.
252. Bleicher KH, Bohm HJ, Muller K, Alanine AI. Hit and lead generation: beyond high-throughput screening. Nat Rev Drug Discov 2003;2: 369–378.
253. Wunberg T, Hendrix M, Hillisch A, Lobell M, Meier H, Schmeck C, Wild H, Hinzen B. Improving the hit-to-lead process: data-driven assessment of drug-like and lead-like screening hits. Drug Discov Today 2006;11: 175–180.
254. Carr RA, Congreve M, Murray CW, Rees DC. Fragment-based lead discovery: leads by design. Drug Discov Today 2005;10: 987–992.
255. Reynolds CH, Bembenek SD, Tounge BA. The role of molecular size in ligand efficiency. Bioorg Med Chem Lett 2007;17:4258–4261.
256. Reynolds CH, Tounge BA, Bembenek SD. Ligand binding efficiency: trends, physical basis, and implications. J Med Chem 2008;51: 2432–2438.
257. Posner BA. High-throughput screening-driven lead discovery: meeting the challenges of finding new therapeutics. Curr Opin Drug Discov Devel 2005;8:487–494.
258. Pereira DA, Williams JA. Origin and evolution of high throughput screening. Br J Pharmacol 2007;152:53–61.
259. Walters WP, Namchuk M. Designing screens: how to make your hits a hit. Nat Rev Drug Discov 2003;2:259–266.
260. McGovern SL, Caselli E, Grigorieff N, Shoichet BK. A common mechanism underlying promiscuous inhibitors from virtual and high-throughput screening. J Med Chem 2002;45: 1712–1722.
261. McGovern SL, Helfand BT, Feng B, Shoichet BK. A specific mechanism of nonspecific inhibition. J Med Chem 2003;46:4265–4272.

262. Seidler J, McGovern SL, Doman TN, Shoichet BK. Identification and prediction of promiscuous aggregating inhibitors among known drugs. J Med Chem 2003;46: 4477–4486.
263. Feng BY, Shoichet BK. A detergent-based assay for the detection of promiscuous inhibitors. Nat Protoc 2006;1:550–553.
264. Feng BY, Shelat A, Doman TN, Guy RK, Shoichet BK. High-throughput assays for promiscuous inhibitors. Nat Chem Biol 2005;1: 146–148.
265. Ryan AJ, Gray NM, Lowe PN, Chung CW. Effect of detergent on "promiscuous" inhibitors. J Med Chem 2003;46:3448–3451.
266. Goode DR, Totten RK, Heeres JT, Hergenrother PJ. Identification of promiscuous small molecule activators in high-throughput enzyme activation screens. J Med Chem 2008; 51:2346–2349.
267. Kubinyi H. Chemogenomics in drug discovery. Ernst Schering Res Found Workshop 2006;58: 1–19.
268. Bywater RP. Privileged structures in GPCRs. Ernst Schering Found Symp Proc Vol. 2. 2006; 75–91.
269. Costantino L, Barlocco D. Privileged structures as leads in medicinal chemistry. Curr Med Chem 2006;13:65–85.
270. DeSimone RW, Currie KS, Mitchell SA, Darrow JW, Pippin DA. Privileged structures: applications in drug discovery. Comb Chem High Throughput Screen 2004;7:473–494.
271. Muller G. Medicinal chemistry of target family-directed masterkeys. Drug Discov Today 2003;8:681–691.
272. McGovern SL, Shoichet BK. Kinase inhibitors: not just for kinases anymore. J Med Chem 2003;46:1478–1483.
273. Zhou Y, Zhou B, Chen K, Yan SF, King FJ, Jiang S, Winzeler EA. Large-scale annotation of small-molecule libraries using public databases. J Chem Inf Model 2007;47: 1386–1394.
274. Wheeler DL, Barrett T, Benson DA, Bryant SH, Canese K, Chetvernin V, Church DM, Dicuccio M, Edgar R, Federhen S, Feolo M, Geer LY, Helmberg W, Kapustin Y, Khovayko O, Landsman D, Lipman DJ, Madden TL, Maglott DR, Miller V, Ostell J, Pruitt KD, Schuler GD, Shumway M, Sequeira E, Sherry ST, Sirotkin K, Souvorov A, Starchenko G, Tatusov RL, Tatusova TA, Wagner L, Yaschenko E. Database resources of the National Center for Biotechnology Information. Nucleic Acids Res 2008;36:D13–D21
275. PubChem; http://pubchem.ncbi.nlm.nih.gov/.
276. ChemIDplus; http://chem.sis.nlm.nih.gov/chemidplus/.
277. DiscoveryGate; www.discoverygate.com.
278. Schneider G, Hartenfeller M, Reutlinger M, Tanrikulu Y, Proschak E, Schneider P. Voyages to the (un)known: adaptive design of bioactive compounds. Trends Biotechnol 2009; 27:18–26.
279. Ertl P, Jelfs S, Muhlbacher J, Schuffenhauer A, Selzer P. Quest for the rings. In silico exploration of ring universe to identify novel bioactive heteroaromatic scaffolds. J Med Chem 2006;49:4568–4573.
280. Cheeseright TJ, Mackey MD, Melville JL, Vinter JG. FieldScreen: virtual screening using molecular fields. Application to the DUD data set. J Chem Inf Model 2008;48 (11): 2108–2117.
281. Evers A, Klebe G. Successful virtual screening for a submicromolar antagonist of the neurokinin-1 receptor based on a ligand-supported homology model. J Med Chem 2004; 47:5381–5392.
282. Nikitin S, Zaitseva N, Demina O, Solovieva V, Mazin E, Mikhalev S, Smolov M, Rubinov A, Vlasov P, Lepikhin D, Khachko D, Fokin V, Queen C, Zosimov V. A very large diversity space of synthetically accessible compounds for use with drug design programs. J Comput Aided Mol Des 2005;19:47–63.
283. Boehm M, Wu TY, Claussen H, Lemmen C. Similarity searching and scaffold hopping in synthetically accessible combinatorial chemistry spaces. J Med Chem 2008;51: 2468–2480.
284. Rarey M, Dixon JS. Feature trees: a new molecular similarity measure based on tree matching. J Comput Aided Mol Des 1998;12: 471–490.
285. Lewell XQ, Judd DB, Watson SP, Hann MM. RECAP-retrosynthetic combinatorial analysis procedure: a powerful new technique for identifying privileged molecular fragments with useful applications in combinatorial chemistry. J Chem Inf Comput Sci 1998;38: 511–522.
286. www.simbiosys.com. 2008.
287. Evans BE, Rittle KE, Bock MG, DiPardo RM, Freidinger RM, Whitter WL, Lundell GF, Veber DF, Anderson PS, Chang RSL, Lotti VJ,

Cerino DJ, Chen TB, Kling PJ, Kunkel KA, Springer JP, Hirshfield J. Methods for drug discovery: development of potent, selective, orally effective cholecystokinin antagonists. J Med Chem 1988;31:2235–2246.

288. Hajduk PJ, Bures M, Praestgaard J, Fesik SW. Privileged molecules for protein binding identified from NMR-based screening. J Med Chem 2000;43:3443–3447.

289. Muegge I, Enyedy IJ. Virtual screening for kinase targets. Curr Med Chem 2004;11: 693–707.

290. Tikhonova IG, Sum CS, Neumann S, Engel S, Raaka BM, Costanzi S, Gershengorn MC. Discovery of novel agonists and antagonists of the free fatty acid receptor 1 (FFAR1) using virtual screening. J Med Chem 2008;51:625–633.

291. Irwin JJ, Shoichet BK. ZINC: a free database of commercially available compounds for virtual screening. J Chem Inf Model 2005;45: 177–182.

292. Tikhonova IG, Sum CS, Neumann S, Thomas CJ, Raaka BM, Costanzi S, Gershengorn MC. Bidirectional, iterative approach to the structural delineation of the functional "chemoprint" in GPR40 for agonist recognition. J Med Chem 2007;50:2981–2989.

293. Garrido DM, Corbett DF, Dwornik KA, Goetz AS, Littleton TR, McKeown SC, Mills WY, Smalley TL Jr, Briscoe CP, Peat AJ. Synthesis and activity of small molecule GPR40 agonists. Bioorg Med Chem Lett 2006;16:1840–1845.

294. Hu X, Prehna G, Stebbins CE. Targeting plague virulence factors: a combined machine learning method and multiple conformational virtual screening for the discovery of Yersinia protein kinase A inhibitors. J Med Chem 2007;50:3980–3983.

295. Kumar A, Chaturvedi V, Bhatnagar S, Sinha S, Siddiqi MI. Knowledge based identification of potent antitubercular compounds using structure based virtual screening and structure interaction fingerprints. J Chem Inf Model 2009;49:35–42.

296. Chen X, Wilson LJ, Malaviya R, Argentieri RL, Yang SM. Virtual screening to successfully identify novel Janus kinase 3 inhibitors: a sequential focused screening approach. J Med Chem 2008;51:7015–7019.

297. Montes M, Braud E, Miteva MA, Goddard ML, Mondesert O, Kolb S, Brun MP, Ducommun B, Garbay C, Villoutreix BO. Receptor-based virtual ligand screening for the identification of novel CDC25 phosphatase inhibitors. J Chem Inf Model 2008;48:157–165.

298. Nair PC, Sobhia ME. Fingerprint directed scaffold hopping for identification of CCR2 antagonists. J Chem Inf Model 2008;48: 1891–1902.

299. Kellenberger E, Springael JY, Parmentier M, Hachet-Haas M, Galzi JL, Rognan D. Identification of nonpeptide CCR5 receptor agonists by structure-based virtual screening. J Med Chem 2007;50:1294–1303.

300. Markt P, McGoohan C, Walker B, Kirchmair J, Feldmann C, De MG, Spitzer G, Distinto S, Schuster D, Wolber G, Laggner C, Langer T. Discovery of novel cathepsin S inhibitors by pharmacophore-based virtual high-throughput screening. J Chem Inf Model 2008;48: 1693–1705.

301. Tsai KC, Chen SY, Liang PH, Lu IL, Mahindroo N, Hsieh HP, Chao YS, Liu L, Liu D, Lien W, Lin TH, Wu SY. Discovery of a novel family of SARS-CoV protease inhibitors by virtual screening and 3D-QSAR studies. J Med Chem 2006;49:3485–3495.

302. Pirard B, Brendel J, Peukert S. The discovery of K_v1, 5 blockers as a case study for the application of virtual screening approaches. J Chem Inf Model 2005;45:477–485.

303. Agrawal H, Kumar A, Bal NC, Siddiqi MI, Arora A. Ligand based virtual screening and biological evaluation of inhibitors of chorismate mutase (Rv1885c) from *Mycobacterium tuberculosis* H37Rv. Bioorg Med Chem Lett 2007;17:3053–3058.

304. Ward RA, Perkins TD, Stafford J. Structure-based virtual screening for low molecular weight chemical starting points for dipeptidyl peptidase IV inhibitors. J Med Chem 2005;48: 6991–6996.

305. Franke L, Byvatov E, Werz O, Steinhilber D, Schneider P, Schneider G. Extraction and visualization of potential pharmacophore points using support vector machines: application to ligand-based virtual screening for COX-2 inhibitors. J Med Chem 2005;48:6997–7004.

306. Zhang Z, Zheng M, Du L, Shen J, Luo X, Zhu W, Jiang H. Towards discovering dual functional inhibitors against both wild type and K103N mutant HIV-1 reverse transcriptases: molecular docking and QSAR studies on 4, 1-benzoxazepinone analogues. J Comput Aided Mol Des 2006;20:281–293.

307. Cherkasov A, Shi Z, Fallahi M, Hammond G. Successful *in silico* discovery of novel non-steroidal ligands for human sex hormone binding globulin (SHBG). J Med Chem 2005;48: 3203–3213.

308. Steindl T, Crump CE, Hayden FG, Langer T. Pharmacophore modeling, docking, and principal component analysis based clustering: combined computer-assisted approaches to identify new inhibitors of the human thinovirus coat protein. J Med Chem 2005;48: 6250–6260.

309. Jacobsson M, Garedal M, Schultz J, Karlen A. Identification of *Plasmodium falciparum* spermidine synthase active site binders through structure-based virtual screening. J Med Chem 2008;51:2777–2786.

ADMET *IN VITRO* PROFILING: UTILITY AND APPLICATIONS IN LEAD DISCOVERY

Edward H. Kerns
Li Di
Guy T. Carter
Chemical Sciences, Chemical Technologies, Wyeth Research, Princeton, NJ

1. INTRODUCTION TO UTILITY AND APPLICATIONS OF ADMET PROFILING IN LEAD DISCOVERY

ADMET properties are widely acknowledged as an integral part of medicinal chemistry. Fundamentally, ADMET properties affect the exposure (concentration and duration) of the dosed compound at the therapeutic target and its safety [1–6]. Achieving quality exposure and safety are major goals for lead discovery. ADMET properties are among the many requirements for advancing a compound to clinical development (e.g., target binding, *in vivo* efficacy, novelty, and selectivity). Thus, it can be a challenge for medicinal chemists to prioritize ADMET properties among the many criteria. To assist with prioritization, Section 2 focuses on the highly valuable opportunities for using ADMET properties in positively impacting drug discovery projects.

ADME is an acronym that has been used in pharmacokinetics (PK) for decades. It represents absorption (A), distribution (D), metabolism (M), and excretion (E), the essential processes that determine the PK of a compound. In recent years, the term toxicity (T) has been added in order to reflect the importance of safety. To improve the ADMET performance of discovery leads, the underlying physicochemical, metabolic, and safety properties of compounds are studied. Section 3 discusses ADMET properties for practical application.

Reliable data for ADMET properties is a crucial need. Assays for use in measuring properties during drug discovery are introduced in Section 4. The purpose is to acquaint medicinal chemists with the assays so that the data can be best interpreted and applied. It is important to understand what the data reliably indicate, for optimal application, and what they do not indicate, so that the data are not overinterpreted.

Overall, this chapter provides an introduction to ADMET properties; how they affect biological assays, *in vivo* PK, and safety in drug discovery; assays used to measure properties; and many practical insights and strategies for applying this information for success in lead discovery. Medicinal chemists are encouraged to expand their knowledge of ADMET on an ongoing basis by reading the referenced books and papers. Of particular utility are the practical strategies for ADMET property optimization [1] and case studies on structure–property relationships [1–6].

2. OPPORTUNITIES FOR ADMET *IN VITRO* DATA TO ENHANCE LEAD DISCOVERY

Many approaches have been successful in applying ADMET properties to positively impact lead discovery. These opportunities are discussed below, along with examples.

ADMET properties are used in studies leading up to first-in-human clinical studies (Phase I) to predict human pharmacokinetics. Models have been developed to combine several properties (e.g., microsomal stability and plasma protein binding) to quantitatively predict human PK clearance. This article focuses on utility and applications in lead discovery support. A key message is that *in vitro* ADMET data guide diagnosing and improving PK parameters. For example, we know prospectively that a low *in vitro* microsomal stability assay half-life indicates that clearance will likely be high and that structural modifications may improve microsomal stability and reduce PK clearance. We also know that if high PK clearance is observed, metabolic stability may be a contributing cause.

It is also important to remember that *in vivo* PK parameters are the result of multiple properties operating together. For example, clearance is affected by plasma protein binding, tissue binding, Phase I and Phase II metabolism, biliary extraction, renal extraction, transporters, and plasma enzyme reactions. One property may dominate or multiple properties may be involved. For example, clearance is often improved by increasing the metabolic stability, but in some cases clearance may be influenced by a high rate of efflux at the bile canaliculus.

2.1. Develop, Optimize, and Interpret Biological Assays

ADMET properties typically focus on *in vivo* dosing and pharmacokinetics, however, properties also affect *in vitro* biological assays. By applying property *in vitro* ADMET data to biology, there is an opportunity to assist project team biologists in optimizing bioassays and interpreting biological data. This insures that structure–activity relationships (SARs) are purely the result of binding and not of different physicochemical properties of analogs in the assay.

Low solubility of compounds can lead to precipitation, which produces a lower than planned concentration in the bioassay. The adverse result is lower exposure of compound molecules to the biological target. Thus, a higher IC_{50} is produced, making the compound appear less active. Solubility limitations can occur in the bioassay media, as well as during the dilution protocol for the compounds prior to the incubation [7]. These are significant issues for discovery teams and studies have shown that many compounds investigated in drug discovery have low solubility [8,9]. It has been reported by several researchers that approximately 30% of compounds in screening libraries have aqueous solubility less than 10 μM. Precipitation of the compounds may be unobserved by visual inspection and be unrecognized in the biological laboratory, owing to the low amount of material, small volume of the well in the 96- or 384-well plate (200–20 uL), and distortions from the plate. Table 1 illustrates the differences in IC_{50} that can be introduced in bioassays by low solubility.

The opportunity exists, during assay development in early discovery for a new therapeutic target, to include a dilution protocol and bioassay media that insure that low solubility compounds remain fully solubilized for the full time period of the assay. Di discussed how *in vitro* assays can be enhanced for increased solubility of compounds in the assay media or in the dilution protocol [3]. This can be included as part of the assay development process and validated using standard low solubility compounds with HPLC detection [7]. It is also important to perform the compound dilution in DMSO, rather than in aqueous buffer to reduce precipitation during the dilution protocol [7]. An assay developed to work with low solubility compounds will operate effectively to produce reliable and consistent data during the next several years when the assay is being routinely applied to a wide range of compounds of varying solubilities. This is also more efficient than routinely measuring, during full-scale use of the bioassay, the actual concentrations of thousands of project test compounds in the assay media and calculating the actual concentration tested. Low solubility can also affect NMR- and X-ray binding studies that have become so important for drug discovery [7].

Permeability affects cell-based assays in which the compound must penetrate the cell's lipid bilayer membrane to reach an intracellular target. Low passive diffusion permeability may have a negative effect. (Transporters may effect the intercellular concentration if the cell line expresses a transporter for which the compound is a substrate.) This insight can

Table 1. Example Data from a Bioassay Using the Nominal Assay Concentration for IC_{50} Values Versus Using Measured Buffer Concentration Values

Measured IC_{50}(uM)	Nominal IC_{50}(uM)
0.044	7.2
0.11	31.7
0.12	12.7
0.18	7.2
0.38	9.6
1.1	80.9
1.5	33.5
2.1	61.4
2.5	33.5
2.6	61.4
11.7	19.4
15.7	33.5

Adapted from Ref. [9].

Table 2. The Effect Permeability Can Have on a Cell-Based Bioassays Is Illustrated. Only Compounds That Had Good Enzyme Activity and Passive Diffusion Permeability Were Active in the Cell Assay

Analog	Enzyme Assay K_i (μM)	Permeability (PAMPA 10^{-6} cm/s)	Cell Assay IC_{50} (μM)
1	0.007	4.9	10.5
2	0.02	1.0	22.1
3	0.01	0.02	0
4	0.05	0.1	0
5	3.5	14.3	0
6	17	6.6	0
7	4.3	0.01	0

Adapted from Ref. [10].

assist interpretation of cell-based assay activity results. In Table 2, only compounds having both good enzyme activity and passive diffusion permeability were active in the cell-based model [10]. Medicinal chemists have the opportunity to use property data to evaluate cell-based bioassay data for the potentially negative affects of low permeability by using tools such as topical polar surface area (TPSA), log *D*, and passive diffusion from the high-throughput PAMPA method (see section on "Permeability Assays").

Chemical stability can also affect bioassay results. If the compound is unstable in the stock solution, dilution solution, or bioassay medium, the compound will appear less active and have a higher IC_{50}. Early diagnosis of this limitation provides an opportunity for improvement of the unstable substructure(s) in the chemical series. This can save valuable leads for which the chemical instability may be correctable by structure modification. When unusual or unexplained bioassay results are produced, testing selected compounds for their stability in the bioassay protocol and media can be beneficial in problem solving, especially if a substructure in the molecule appears to be potentially unstable.

It would be unfortunate to deprioritize a compound or chemical series owing to apparently low activity when the actual cause is a limiting property. By diagnosing the lack of activity to a particular property using *in vitro* property assays, there is an opportunity to improve the lead series and save valuable leads.

2.2. Early Review of Hit ADMET Properties for Lead Selection

Compounds for consideration early in a discovery project arise from many sources. These include "hits" from screening of fragment libraries, high-throughput screening (HTS) of large diverse libraries, virtual screening using computational techniques, purchased libraries, compound sets from alliance partner companies or academic groups, natural products, and ligands from the scientific literature. For these "hits," it is useful to obtain property information early in the evaluation process as soon as the compound is found to be active.

Early access to property data for "hits" provides the opportunity to evaluate the compound to

- continue to consider it, owing to a favorable property profile,
- drop the compound from further consideration,
- attempt to improve the performance of the compound series, and
- select it as a "template" for the full-scale lead optimization process.

Waiting until later to evaluate properties can lead to wasted time or irretrievable compounds.

The template (lead) locks in much of the structure of the final NCE arising from the optimization process [11], along with its properties. Thus, early property insights assist with insuring the project team has quality leads on which to base the remainder of the project, or has the opportunity to make some early structure modifications to improve problematic substructures and investigate whether a useful "template" can be generated.

2.3. Profile Properties and Activity Together for a Comprehensive Assessment

Newly synthesized compounds in the lead optimization process can be assayed for their ADMET properties at the same time as their biological activity and selectivity [12]. By having the entire activity and property profile of the compounds, the project team has the opportunity to review all of the data together in order to enable the comprehensive decision making. In addition, access to the property data enables the project team to review if structure modifications that were made to improve target binding had the unfortunate effect of deteriorating the ADMET properties, or vice versa. Without a comprehensive assessment, teams have focused on target binding IC_{50} for compound selection or structural optimization. This approach moved the lead series into structural space that was later abandoned, owing to inadequate properties for absorption, distribution, metabolism or safety. Pursuing such dead ends results in lost opportunities to invest time and resources in exploring more promising leads by using the full data profile.

2.4. Diagnose PK Liabilities

PK parameters (e.g., $t_{1/2}$, CL, V_d, AUC, and F) are commonly used in drug discovery as criteria to rank-order compounds, advance them to clinical development, and decide on dosing regimes. It is useful to recognize that PK parameters are determined by the underlying ADMET properties. For example, clearance (CL) is affected by metabolic stability, plasma stability, transporters, and others. Thus, if an inadequate PK parameter is observed from an *in vivo* dosing experiment, it is opportune to review all of the underlying properties that might contribute, in order to diagnose the causative property/properties. This focuses attention on a specific property that could be improved to improve the PK parameter [13,14].

2.5. Guide Structure Modification for Lead Optimization

A powerful opportunity for applying ADMET property data is to use it as a guide for structure modifications for the purpose of optimizing the performance of the compound at barriers in the body that limit drug exposure. Poor PK performance triggers the need for action and structure modifications are guided by data from *in vitro* ADMET property assays.

For improving the chemical series, ADMET data that provide simple discrete guidance for accessible structure modifications is useful. For example, if discrete data from PAMPA indicates that the passive diffusion permeability needs improvement, discrete structural modifications that increase the $\log P$ (i.e., lipophilicity), decrease the hydrogen bonding (i.e., H-bond donors, H-bond acceptors, and TPSA), or decrease the molecular size (i.e., molecular weight) are often successful. If the microsomal stability assay data are low, a Phase I metabolic liability is indicated and may be improved by discrete structure modification to improve the stability of the chemical series. Successful and proven structural modification options for improving properties have been tabulated and medicinal chemistry scientific literature case studies have been described elsewhere [15–17]. It is important to keep in mind that a single structure modification intended to improve one property can also change other properties. Therefore, it is a good idea to assay the new compound for all the properties to check if other properties have been affected.

2.6. Select Compounds and Plan for *In Vivo* Experiments

In vivo experiments are typically expensive and time consuming. Consequently, only a limited number can be performed each month for a discovery team. These include both *in vivo* pharmacology and PK studies. Thus, the compounds advancing to *in vivo* studies are carefully selected. ADMET properties contribute valuable insights to the ensemble of data used for selection of compounds and for planning *in vivo* experiments. ADMET data provide the opportunity to select the most promising compounds and the ones whose property profile suggests the possibility of good PK [5].

2.7. Optimize *In Vivo* Dosing to Enhance Exposure for Pharmacology and PK Studies

Drug dosing and therapeutic target exposure are affected by ADMET properties. By knowing the properties in advance, the project team has the opportunity to optimize *in vivo* dosing studies for PK and efficacy [18]. Following are a few examples:

- If the compound of interest has low aqueous solubility, a vehicle can be investigated that will enhance solubility in order to enhance oral absorption or enable IV dosing, resulting in increased exposure of the compound.
- If the compound is found to have low metabolic stability or poor properties for intestinal absorption, initial dosing for *in vivo* efficacy might best be performed using intravenous (IV) dosing, which avoids intestinal absorption and first pass metabolism so that sufficient exposure can be produced to check for the desired pharmacology and show proof of concept.
- Formulations can be screened for the compound of interest to identify a vehicle that maximizes solubility of the compound. Absorption is enhanced by the compound being fully in solution versus in suspension or solid dosage form.

3. ADMET PROPERTIES

3.1. General Structural Properties

Several descriptors are used that can be considered structural property descriptors. They affect all aspects of ADMET and are widely used.

3.1.1. Lipophilicity Lipophilicity is the tendency of a compound to partition into a lipid environment versus an aqueous environment. Thus, lipophilicity is determined by the compound's structure and the characteristics of the aqueous and lipid phases.

Lipophilicity has been extensively studied by A. Leo and C. Hansch [19,20] and van de Waterbeemd [21]. Their work has developed correlations to a large number of pharmacological effects, including permeability, solubility, protein binding, metabolism, and volume of distribution. This correlation is owing to the numerous lipid-aqueous partitioning occurrences in pharmacology, such as lipid bilayer membrane partitioning or binding to a protein. Lipophilicity of a compound is enhanced by nonpolar substructures in the molecule and reduced by polar, hydrogen bonding, and charged substructures, as well as by increasing molecular volume (correlated to molecular weight) [22].

A common approach to assessing lipophilicity is to measure the partitioning of a compound between octanol and aqueous phases. Octanol contains a nonpolar aliphatic chain and a hydrogen-bonding alcohol, in a similar manner to a lipid. When the preponderance of a compound's molecules are neutral, owing to the aqueous phase pH in relation to the pK_a, the equilibrium concentration of the compound in the octanol compared to the aqueous phase is termed "log P." When a portion of the molecules is ionized, owing to the aqueous phase pH in relation to the pK_a (the charged molecules are much more soluble in the aqueous phase), the equilibrium concentration of the compound in the octanol compared to the aqueous phase is termed "log D."

In general, intestinal absorption and metabolic stability are optimum for compounds having a log D in the range of 1–3 [23]. Below this range, solubility increases, permeability decreases, and renal elimination increases. Above this range, solubility decreases, permeability increases, and hepatic elimination increases (metabolism and biliary extraction). Volume of distribution generally increases with lipophilicity.

3.1.2. Hydrogen Bonding Hydrogen bonds, in which a hydrogen atom covalently bonded to an electronegative atom (e.g., N and O) is also electrostatically bonded to an electronegative atom usually of a different molecule, have a significant effect on absorption because they affect both solubility and permeability. Hydrogen bonds to water molecules in aqueous solution increase solubility. However, they must be broken for the compound to permeate across the lipid bilayer membrane by passive diffusion. If the number of hydrogen bonds gets too high they significantly reduce intestinal absorption for the compound. The importance of hydrogen bonding has been incorporated into the rule of five from Lipinski and colleagues [24] and bioavailability rules from Veber and colleagues [25]. Increasing hydrogen-bond donors (HBDs) and hydrogen-bond acceptors (HBAs) are correlated with decreasing absorption as they approach and exceed the recommended upper limits of 5 HBD and 10 HBA (Veber suggests 12 total hydrogen bonds). Topical polar surface area [26] is highly influenced by hydrogen bonding. TPSA provides guidance for medicinal chemists, with 140 Å^2 being the recommended upper limit for compounds to have acceptable bioavailability [25]. Hydrogen bonding is even more restricted at the blood–brain barrier, where less than eight total hydrogen bonds and less than 60 Å^2 TPSA are recommended [27].

3.1.3. pK_a The ionizability of a compound has a major effect on its solubility because the charged form has increased solubility, and on its permeability by passive diffusion because only the neutral form permeates the lipid bilayer. When the pH of the solution is the same as the pK_a, the population of molecules in solution is half ionized and half neutral. As the pH decreases below the pK_a, an increasing fraction of the molecules in solution are in the ionized form for bases and in the neutral form for acids. The opposite occurs as the pH increases above the pK_a. Basicity increases as pK_a increases, as indicated by the examples: aniline (pK_a 4.9), pyridine (pK_a 5.2), and piperazine (pK_a 9.8). Acidity increases as pK_a decreases, as indicated by the examples: phenol (pK_a 10), acetic acid (pK_a 4.8), and formic acid (pK_a 3.8).

Structure modification can change the pK_a of the chemical series and thus "tune" the ionizability in order to improve the absorption. If an increase in solubility is desired, the structure can be modified to increase the pK_a of a base or decrease the pK_a of an acid. The Henderson–Hasselbach equation indicates that for an acid,

$$\mathrm{HA} = \mathrm{H}^+ + \mathrm{A}^-$$

$$\frac{[\mathrm{HA}]}{[\mathrm{A}^-]} = 10^{(\mathrm{p}K_a - \mathrm{pH})}$$

For example, benzoic acid (pK_a 4.8) at pH 7.4 has $[\mathrm{HA}]/[\mathrm{A}^-] = 10^{-2.6}$ (highly ionized). For a base,

$$\mathrm{BH}^+ = \mathrm{H}^+ + \mathrm{B}$$

$$\frac{[\mathrm{BH}^+]}{[\mathrm{B}]} = 10^{(\mathrm{p}K_a - \mathrm{pH})}$$

For example, piperazine (pK_a 9.8) at pH 6.5 has $[\mathrm{BH}^+]/[\mathrm{B}] = 10^{3.3}$ (highly ionized).

3.2. Properties That Affect Absorption

Once-per-day oral dosing is preferable for patient-administered drugs. This is most favorable for patient convenience and compliance and efficient manufacturing, storage, and distribution to patients. In order to enable oral dosing the compound should be well absorbed from the intestine.

Absorption is primarily controlled by solubility and permeability in the intestine. Also, if a compound is unstable in the intestine, its absorption will be reduced. Lipophilicity, hydrogen bonding and pK_a are major determinants of solubility and permeability, thus it is also important for medicinal chemists to understand their effects. Transporters (influx and efflux) can also impact oral absorption.

3.2.1. Solubility Equilibrium solubility is the concentration of a compound at equilibrium with excess solid under specific chemical and physical solution conditions. Thus, aqueous solubility is determined by the structure of the compound, crystal lattice energy and by the solution specifics [28].

The neutral compound has an "intrinsic solubility" (S_0) that is affected by several physicochemical structural properties. S_0 has been shown by Yalkowsky and colleagues [29] to increases as the compound's lipophilicity decreases and as crystal energy increases. (As crystal energy increases, the crystal is less stable, has lower crystal packing, has lower melting point, and is more soluble.) Log $P_{\text{octanol-water}}$ is used as an indicator of lipophilicity and melting point as an indicator of the crystal lattice energy. Abraham and Le [30] showed that intrinsic solubility increases as hydrogen bonding in solution increases and as molecular volume (correlated to molecular weight) and hydrogen bonding in the crystal decrease. Thus, solubility generally increases as structural modifications reduce lipophilicity, molecular weight or crystal packing. Solubility also increases with hydrogen-bonding capacity, ionizability, or polarity (Fig. 1).

When the compound becomes ionized, the solubility increases. This is indicated by the Henderson–Hasselbach equation, which can be expressed as

$$\text{Acids}: S = S_0(1 + 10^{\text{pH}-\text{p}K_\text{a}})$$

$$\text{Bases}: S = S_0(1 + 10^{\text{p}K_\text{a}-\text{pH}})$$

The logarithmic effect on solubility of the difference between the pK_a and the solution pH results from the greater solubility of the ion than the neutral molecule. The solution pH causes a portion of the population of molecules in aqueous solution to be ionized. When the aqueous solution pH equals the pK_a, 50% of the molecules in solution are ionized and 50% are neutral. Structural modifications that increase the pK_a of a base or decrease the pK_a of an acid increase the solubility of the resulting analog compound.

Solution conditions, such as pH, cosolvent(s) (e.g., DMSO, ethanol), additives (e.g., albumin, lipid, surfactant, and bile salts), ionic strength, time, and temperature affect the solubility of the compound. Formulation provides an environment in which a compound is more soluble in the dosing solution, or has a higher dissolution rate compared to the crystalline solid, in order to increase the number of compound molecules in solution *in vivo*. Formulation involves a range of strategies, including salt form, particle size reduction, cosolvents, wetting agents, complexation, and others.

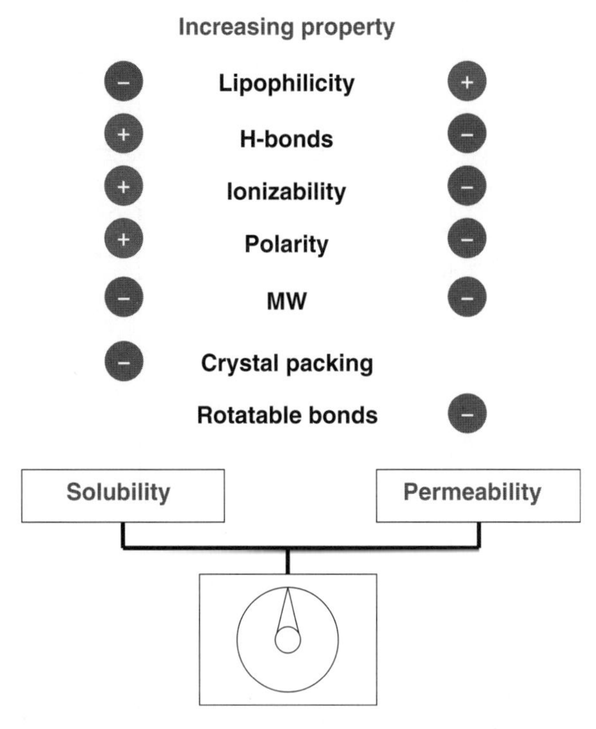

Figure 1. Structural properties affect solubility and permeability in different directions. (This figure is available in full color at http://mrw.interscience.wiley.com/emrw/9780471266945/home.)

Solubility has a major role in intestinal absorption. Individual molecules of the dosed drug must be in solution at the surface of the intestinal epithelial cells for absorption to occur. Solid material is not absorbed. This undissolved material results in reduced PK AUC (exposure) and C_{max} *in vivo*.

Structural modifications can increase solubility. Formulation techniques increase the dissolution rate and solubility to obtain more molecules in solution. The solubility of molecules in solution is determined by the Henderson–Hasselbach equation. Thus, precipitation can occur with time, even if molecules are initially dissolved by assistance from the salt form or formulation. While the high initial concentration exists, a high-concentration gradient exists across the intestinal epithelium to drive passive diffusion permeability for increased absorption. This is described by Fick's law of passive diffusion [28], which indicates the rate of compound flux across a membrane is directly related to the concentration difference (gradient). Low solubility can result in absorption of only a fraction of the dose and cause lower *in vivo* AUC (exposure), lower bioavailability, or nonlinear pharmacokinetics.

In drug discovery there is a great opportunity to increase solubility using structure modifications. If this cannot be achieved without major effects on other discovery priorities (e.g., decreased activity or selectivity), then formulation can be used to enhance solubility and dissolution rate. This may allow observation of the pharmacology in proof-of-concept studies that is vital for drug discovery. It would be unfortunate for solubility to limit this important milestone.

Lipinski and colleagues provided guidance for the solubility of drug discovery compounds [24]. The solubilities of over 87% of commercial drugs were ≥65 μg/mL and only 7% were ≤20 μg/mL. Curatolo [31] and Lipinski [32] also reported that a compound with a clinical dose of 1 mg/mL and average permeability would require a solubility of 52 μg/mL for full-dose absorption. Thus, a useful goal for drug discovery compounds is greater than 60 μg/mL solubility.

An example of solubility improvement is shown in Fig. 2 [33]. Solubility assays showed

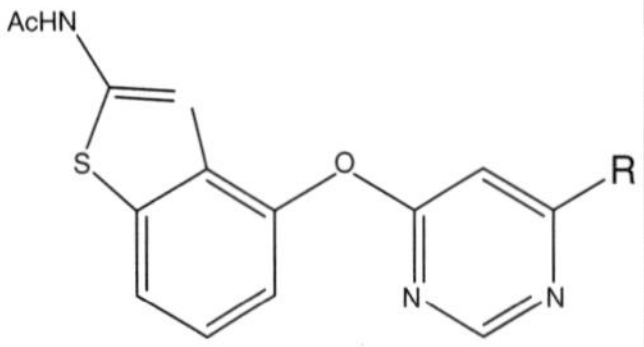

R	IC_{50} (nM)	Solubility (μg/mL) pH 2	pH 7.4	SIF	CL	F%
AMG 517 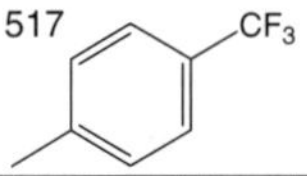 (CF_3)	0.5	<1	<1	6.6	0.2	32
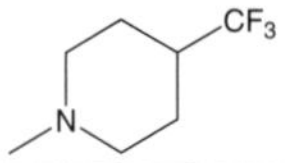 N (CF_3)	19	9				
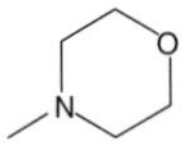 N (O)	>4000	32				
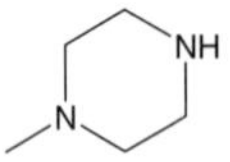 N (NH)	>4000	173				
AMG 628 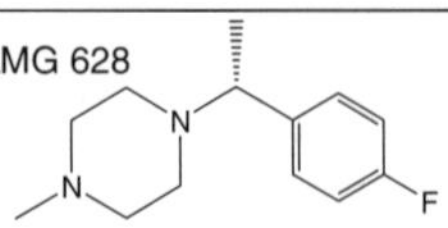 (N, N, F)	5	≥200	7	64	0.8	51

Figure 2. Example of structure modification to introduce an ionizable group to increase solubility to yield increased bioavailability [33]. SIF is simulated intestinal fluid (USP), CL is clearance, *F*% is percent bioavailability. (This figure is available in full color at http://mrw.interscience.wiley.com/emrw/9780471266945/home.)

that the initial lead had low solubility. Subsequent analogs with an ionizable amine produced greatly improved solubility, which led to improved absorption and bioavailability, while maintaining an IC_{50} that was advantageous for pharmacological activity.

3.2.2. Permeability Permeability is the rate at which a compound moves through a cell membrane barrier. It is affected by the structure of the compound and the characteristics of the membrane barrier. There are four permeability mechanisms that are of greatest importance for drug discovery: passive diffusion, efflux, active uptake and paracellular.

Passive Diffusion In passive diffusion [34], the molecule in aqueous solution sheds the water molecules and moves through the highly nonpolar region of the lipid bilayer membrane that envelops the cells of the membrane barrier (e.g., epithelial cells of the small intestine). Thus, the rate of passive diffusion permeation is affected by the lipophilicity of the compound. A balance of lipophilicity of the compound between favoring permeability (higher $\log P$) and solubility (lower $\log P$) is favorable for optimum absorption. The aliphatic chains of the phospholipids that form the bilayer are tightly packed, thus increasing molecular volume (correlated to molecular weight) reduces permeability. The neutral molecule is the species that moves through the membrane by passive diffusion, therefore, ionizability, which is controlled by the pK_a, can affect permeability. Other structural features that reduce permeability are increasing rotatable bonds, hydrogen bonds, topical polar surface area, and polarity (Fig. 1). Chemical modifications that affect one of these structural properties can affect permeability. Solution conditions, especially pH, that increase the concentration of neutral molecules in solution can increase passive diffusion permeability. However, structural features and solution conditions that improve permeability can reduce solubility. Artursson and colleagues [35] estimated that for 95% of commercial drugs, passive diffusion is the predominant mechanism of absorption. Therefore, it is of primary importance to medicinal chemistry. Formulation is not an effective and safe approach to increasing the permeability of a compound, thus, it can only be improved by structure modification. An increase in permeability can improve the PK AUC, C_{max}, t_{max}, and F of a dosed compound. Passive diffusion permeability is not generally saturable in the intestine. Passive diffusion permeability can affect *in vitro* drug discovery assays if the therapeutic target is inside the cell (see Section 2.1).

An example of structural effects on passive diffusion permeability is shown in Fig. 3 [36]. Permeability decreased with increasing hydrogen-bonding capacity.

Transporters Efflux is the expulsion of compound molecules from the cell. This is performed by membrane transporters that use an energy source (e.g., ATP) to move molecules across membranes, even against the concentration gradient. Many efflux transporters are found in living systems and they function *in vivo* to reduce intestinal absorption, oppose entrance into sensitive tissues (e.g., blood–brain barrier), or increase elimination at various membranes (e.g., hepatic bile canaliculus). P-glycoprotein (Pgp) is a major efflux transporter found throughout these barrier tissues. Efflux by Pgp is a major problem for many drug discovery projects in achieving sufficient exposure for a lead compound series. Recommendations for structural features that reduce efflux by Pgp include the following: reduce hydrogen-bond acceptors ($\leq$4 nitrogen + oxygen atoms), reduce molecular weight (<400), reduce lipophilicity, and reduce basicity of bases. Other efflux transporters (e.g., MRP2, BCRP) have also been reported to affect permeability for a smaller number of compounds. Pgp can be saturated if the concentration is high enough. This has been observed as nonlinear PK in which Pgp limits absorption up to a certain dosing level, above which the absorption greatly increases owing to saturation of the Pgp efflux capacity in the intestine. Inhibitors of Pgp have also been coadministered with discovery compounds to enhance permeability for proof-of-concept studies, but this has not proven to be an effective clinical dosing approach because of the protective nature of Pgp against admitting toxic compounds. Caco-2 cells express some of the efflux transporters and other assays (e.g., MDR1-MDCKII) have been

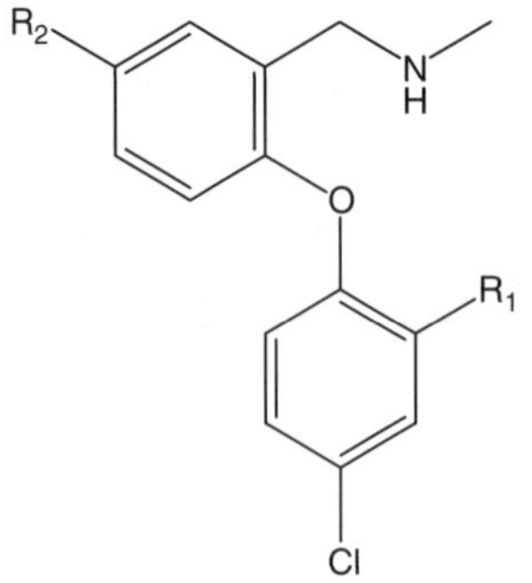

R_1	R_2	PAMPA (10^{-6} cm/s)	5-HT IC_{50} (nm)
OMe	$CONH_2$	11	12
OMe	CONHMe	14	13
OMe	$CONMe_2$	15	8
Me	$CONMe_2$	19	17
Me	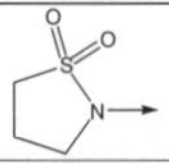	2.7	10
Me	$NHSO_2Me$	2.2	16

Figure 3. Example of structural effects on passive diffusion permeability as measured by PAMPA [36].

developed to assay specific transporters (see Section 4.1.4).

An example of Pgp efflux improvement is shown in Fig. 4 [37]. The Pgp efflux of a series of biphenylaminocyclopropane carboxamide based bradykinin B1 receptor antagonists was reduced by replacement of the trifluoropropionamide in the lead structure with polyhaloacetamides. The reduction in efflux appears to be owing to electron withdrawal from the hydrogen-bond acceptors.

Uptake transporters [38] can carry molecules of some compounds across cellular membranes. These have the natural function of increasing permeability of vital nutrients (e.g., amino acids, glucose, and nucleosides), which are too polar to permeate by passive diffusion. Affinity and transport by an uptake transporter is often discovered by serendipity, but some drug and prodrug structures have been designed to take advantage of uptake transporters to increase permeability. Examples of drugs that are transported include cyclosporins (peptide transporters) and statins (carboxylic acid transporter). PEPT1 has been useful for enhancing absorption in some drug discovery programs, owing to its high capacity compared to other transporters. Uptake transport is enhanced by structural similarity to the natural substrate (e.g., addition of a valine group may enhance transport by PEPT1 [39]). Assays for specific uptake transporters have been developed (see section on "Permeability Assays").

Transporters can also affect metabolism, because they can transport drug molecules into and out of hepatocytes. Transporters can also affect excretion because they are present in hepatocytes at the bile canaliculus for biliary extraction and in the kidney nephrons for active secretion.

Paracellular Permeability Paracellular permeation is the movement of molecules through the junctions between cells. In the intestine, these junctions are small (8 Å), polar, and comprise only 0.3% of the surface area. Thus, compounds that permeate paracellularly must be small (MW < 180), polar and charged. Less than 5% of drugs have appreciable paracellular permeability and most drug discovery targets require larger molecules for sufficient activity.

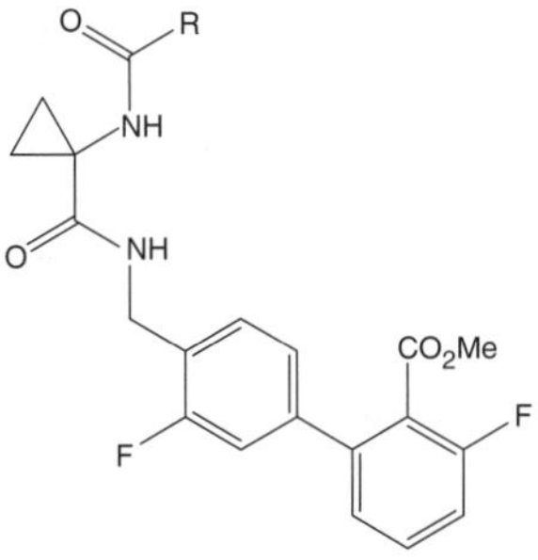

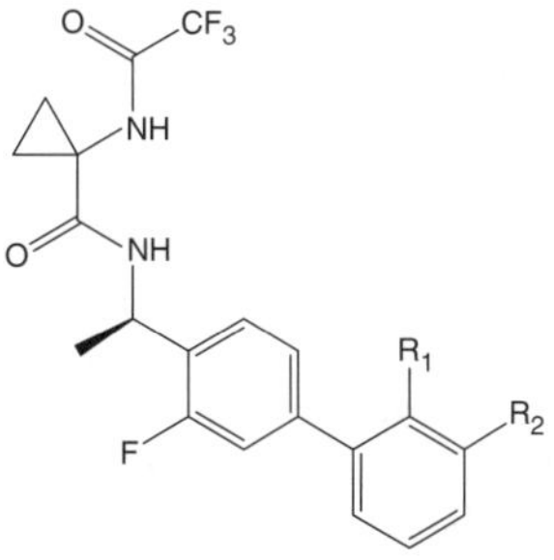

R	Efflux Ratio	hBK$_1$ K_i (nm)
CH_2CF_3	18.4	0.81
CH_3	6.3	11.6
CHF_2	3.4	13.5
CF_3	4.1	1.47
CF_2CF_3	2.2	2.95
$CHCl_2$	2.8	0.54

R_1	R_2	Efflux Ratio	hBK$_1$ K_i (nm)
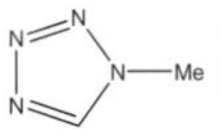 N N N—Me N	F	15.5	62.5
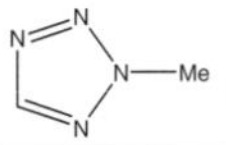 N N N—Me N	F	4	0.6
CO_2Me	F	8.6	0.13
CO_2Me	H	2.2	10.3
COEt	Cl	1.6	1.95
CO_2Me	Cl	1.9	0.44

Figure 4. Example of structure modification to improve Pgp efflux and yield increased blood–brain barrier penetration. Compound **13b** was selected for further study [37].

3.2.3. Chemical and Gastrointestinal Stability

There are many potential causes of compound decomposition in drug discovery as compounds encounter reactive conditions in the laboratory, *in vitro* experiments, during storage, and *in vivo* [40]. These can degrade the compound and reduce exposure of the parent compound *in vivo*, modify the activity profile of the compound, or generate toxic compounds that affect biological results. In the laboratory, compounds may be unstable when exposed to light, oxygen, or water in the atmosphere.

In *in vitro* assays, compounds are exposed to bioassay media pH, reactive components (e.g., DTT), hydrolytic enzymes, and elevated temperature that may cause decomposition.

After oral dosing the compound is exposed to pHs from 1 to 8 in the stomach, intestine and colon. These pHs can cause chemical reactions that decompose the compound prior to absorption. Instability can also occur in the intestine, owing to hydrolytic enzymes. These enzymes have the normal function of hydrolyzing macromolecules (protein, lipid, and carbohydrates) into monomers or dimmers for absorption. If a drug molecule binds to one of these enzymes and has a hydrolysable group, it can undergo decomposition. Recognizing GI instability early provides the opportunity to counteract the reactions via structure modification to save the valuable pharmacophore series.

3.3. Properties That Affect Metabolism

For drug discovery purposes, medicinal chemists can readily assess the effects of metabolism. Metabolic enzymes exist in many tissues, but attention to metabolism in the liver and intestine are most beneficial for improving the pharmacokinetic properties of a chemical series.

Drug molecules that permeate into the epithelial cells of the intestine are exposed to metabolic enzymes, especially cytochrome P450 (CYP) 3A4 isozyme. If the compound binds to and is reactive at 3A4, metabolism can occur to some extent at this barrier even before a molecule can permeate into the bloodstream. All absorbed molecules are collected into the hepatic portal vein and are routed to the liver, where they can permeate into hepatocytes and be exposed to an ensemble of metabolizing enzymes. Different metabolic enzymes have different binding specificities and reactivities for different compounds. Thus, drug molecules are exposed to diverse metabolic challenges in the intestine and liver, termed "first pass metabolism" before they enter general systemic circulation and can reach the target therapeutic tissue. Metabolism generally reduces the compound's activity, although some metabolites are more active than the parent compound. Metabolism also increases the rate of excretion of molecules from the body because it increases the aqueous solubility for easier extraction in the kidney or adds conjugating species (e.g., glucuronic acid) that can be bound by transporters at the bile canaliculus or in the kidney for enhanced excretion.

3.3.1. Metabolic Stability With oral dosing, first pass metabolism starts in the intestinal epithelial cells and continues, to a greater extent, in the liver. With intravenous dosing or after an orally dosed drug enters systemic circulation, metabolism continues to occur as a portion of the systemic blood flow circulates through the liver. Drug discovery scientists are primarily focused on optimizing the metabolic stability of compounds in the liver, although some metabolism occurs in other tissues.

Molecules are exposed to an array of metabolizing enzymes. The enzymes are typically categorized into Phase I, which catalyze smaller structure modifications, such as hydroxylation, and Phase II, which catalyze additions of larger polar structures, such as glucuronic acid. Examples of some major metabolizing enzymes are listed in Table 3.

These enzymes encompass a wide diversity of binding specificities. Each of the types of metabolizing enzymes listed in Table 3 has a set of isozymes. For example, there are many isozymes of CYP enzymes. In humans, the major CYP isozymes are 3A4, 2D6, 2C9, and 1A2. CYP isozyme 3A4 constitutes 28% of the CYP protein in the liver and metabolizes 50% of commercial drugs. Isozyme 2D6 constitutes only 2% of the CYP protein but metabolizes 30% of drugs. 2C9 and 1A2 constitute 18% and 13% of CYP protein and metabolize 10% and 4% of drugs, respectively. There is some diversity among human individuals in their expression levels of different isozymes. For example, some Caucasians do not express 2D6, thus, a drug that is primarily cleared by 2D6 has different pharmacokinetics (increased exposure for the same dose) in individuals lacking or underexpressing the isozyme.

It is beneficial that a drug candidate be metabolized by more than one isozyme. In this way, if one metabolism pathway is reduced by different expression levels or drug–drug interaction, there will be other pathways that can clear the drug. Drug candidates approaching selection for drug development are often studied to carefully determine the specific enzyme types and isozymes by which they are metabolized. This experiment is called metabolic (or CYP) "phenotyping."

Different isozymes have different binding and reactivity specificities [41]. Some favor planar amines and amides, others favor substrates with carboxylic acids, oxygen-rich regions and an aromatic ring or lipophilic group. CYP 2D6 favors amines with a hydrophobic region, hence its importance in metabolizing many basic drugs.

Different species have different isozymes. For example, the CYP 3A family isozymes are differently expressed in humans and rats. This is one of the reasons that metabolic stability measurements in one species may not translate directly to another species.

Metabolic stability affects the PK parameters clearance, half-life, area under the curve, and bioavailability. Correlation of experimental *in vivo* PK parameters with *in vitro* metabolic stability measurements are fruitful in indicating how PK can be improved, which is a major goal of drug discovery.

Structures can be modified to improve metabolic stability [15]. This usually involves one of the following: (1) addition of a group to block metabolism at a site, (2) electronic effects to withdraw electrons and make the site less reactive, or (3) steric effects to make the site less accessible to attack. An example of a sequence of structure modifications to improve metabolic stability is shown in Fig. 5 [42]. A single structure modification may cause a significant reduction in activity or shifting of

Table 3. Major Metabolizing Enzymes

Metabolizing Enzymes	Reaction	Examples	Location in Hepatocyte
Phase I			
Cytochrome P450 (CYP)	Aliphatic or aromatic oxidation	Hydroxylation; epoxidation; dehydration; *S*-, *N*-, or *O*-dealkylation; *N*-hydroxylation	ER
Alcohol dehydrogenase	Alcohol to aldehyde	Alcohol to aldehyde	Cytosol
Aldehyde dehydrogenase	Aldehyde to carboxylic acid	Aldehyde to carboxylic acid	Cytosol, mitochondria
Monoamine oxidase	Oxidative deamination	Amine to carbonyl	Mitochondria
Diamine oxidase	Oxidative deamination		Mitochondria
Flavin-mono-oxygenases (FMO)	*N*- or *S*-oxidation	Amine to *N*-oxide, sulfide to sulfoxide or sulfone	ER
Reductases	Reduction	Nitro to amine, azo to amines, and ketone to alcohol	ER, cytosol
Esterases	Hydrolysis of ester		ER, cytosol
Phase II			
UDP-glucuronosyltransferases (UGT)	Glucuronidation	Addition to phenol, aliphatic hydroxyl, carboxylic acid, and others	ER
Sulfotransferases	Sulfonation	Addition to phenol	Cytosol
N-Acetyltransferases	Acetylation	Addition to aniline, amine, and *N*-hydroxyl	Cytosol
Amino acid addition	Amino acid conjugation	Addition of glycine, taurine, and glutamine to carboxylic acid	Kupffer cells
Glutathione-*S*-transferase	Glutathione addition	Reaction of −SH of glutathione with reactive metabolite	Cytosol
Methyltransferase, catechol *O*-methyl transferase	Methylation	Methylation of amine or catechol	

ER: endoplasmic reticulum.
Adapted from Ref. [1].

metabolism of the compound to another metabolic enzyme ("metabolic switching"); therefore, structure modifications at multiple sites in the molecule may be necessary to achieve stability and maintain activity.

3.3.2. Plasma Stability When compound molecules enter the bloodstream they are exposed to hydrolytic enzymes in the bloodstream. These have diverse specificities and are different from the enzymes in the intestine.

3.4. Properties that Affect Distribution

Compound molecules that are circulating in the bloodstream are carried through the tissues of the body. Molecules partition into the tissues and reach equilibrium between the blood and the tissue. Physiological characteristics affect distribution to tissues. Blood flow to organs differs, with tissues such as brain and heart receiving higher blood flow than skin, bone, and adipose tissue. Tissues that are more lipophilic tend to accumulate a higher concentration of lipophilic compounds, and tissues that are more acidic (e.g., muscle) tend to accumulate more acidic compounds. Other ADMET properties affecting distribution into the tissues are described below.

3.4.1. Plasma Protein Binding There are proteins in the blood that serve as carriers of endogenous compounds to tissues. These

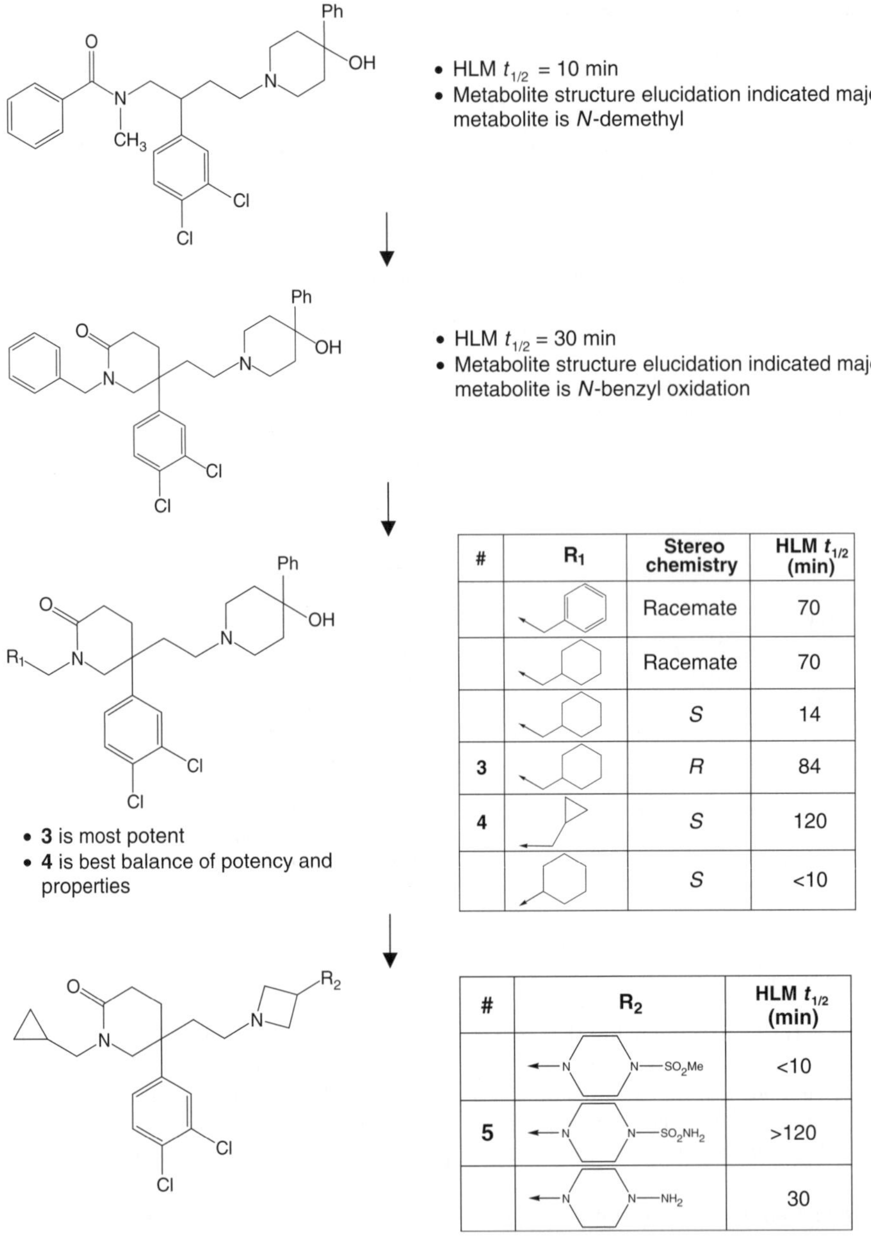

- HLM $t_{1/2}$ = 10 min
- Metabolite structure elucidation indicated major metabolite is *N*-demethyl

- HLM $t_{1/2}$ = 30 min
- Metabolite structure elucidation indicated major metabolite is *N*-benzyl oxidation

- **3** is most potent
- **4** is best balance of potency and properties

#	R_1	Stereo chemistry	HLM $t_{1/2}$ (min)
		Racemate	70
		Racemate	70
		S	14
3		*R*	84
4		*S*	120
		S	<10

#	R_2	HLM $t_{1/2}$ (min)
	N N—SO_2Me	<10
5	N N—SO_2NH_2	>120
	N N—NH_2	30

Figure 5. Example sequence of structure modifications for metabolic stability optimization. HLM is human liver microsomes. Compound **5** was preferred [42].

include serum albumin, α_1-acid glycoprotein and lipoproteins. For some drug compounds, greater than 99% of the molecules in the blood are protein bound, while binding of some compounds is less than 40%. Only the "free drug" in blood ("unbound") can permeate through the capillary to reach the tissue. Typically, with higher protein binding there is lower distribution into tissues (lower PK volume of distribution). Plasma protein binding also reduces excretion and metabolism by reducing the penetration of compounds into the liver and kidney.

3.4.2. Tissue Binding Once a molecule enters the tissue, it can bind nonspecifically to proteins or lipids in the tissue. This reduces the free drug concentration in the tissue, which is available to interact with the therapeutic target. It also increases the total concentration of drug (free drug plus bound drug) in the tissue (higher PK volume of distribution). The free drug in tissue is the species that produces the therapeutic effect. At equilibrium, the concentration of free drug in tissue equals free drug in blood, however, the total concentration is not the same, depending on whether the compound has a relatively higher affinity for the plasma proteins (low distribution) or higher affinity for the tissue proteins and lipids.

3.4.3. Distribution Barriers at Some Organs Some tissues have barrier characteristics that affect the penetration of compounds into them [43,44]. For example, the blood–brain barrier has a preponderance of phospholipids with negatively charged head groups that enhance permeation of molecules with basic moieties compared to other tissues. The membrane also has reduced permeability for compounds with hydrogen bonding, TPSA, and molecular weight (i.e., size). Moreover, there is a high expression level of Pgp; thus, efflux substrates for Pgp are reduced in their permeation rate. These factors reduce the distribution of molecules to tissues with such barriers (e.g., brain, testes, and placenta).

3.5. Properties that Affect Excretion

The two primary organs of clearance of compounds are the liver and the kidney. These both work to excrete drug and metabolite molecules from the body.

The liver excretes some molecules of the compound and its metabolites into the bile, which are eliminated via the intestine into the feces. These molecules reach the bile via the bile canaliculus, a capillary that is formed by the hepatocytes at their junctions. The canalicular membrane is a lipid bilayer, thus, molecules can reach the bile by passive diffusion. Moreover, the canalicular membrane has several transporters (e.g., Pgp, MRP2, BCRP, and BSEP) that actively transport molecules from the hepatocyte cytoplasm to the bile and has been termed biliary excretion or extraction. Structural features that enhance binding to these transporters increase the rate of biliary excretion.

The kidney excretes compounds into the urine. Compound and metabolite molecules reach the urine via the nephron. In this device, molecules are removed from the blood by filtration through leaky pours (paracellular permeation) and passive diffusion in the glomerulus and by active secretion by specific transporters from the proximal tubule to an adjacent blood capillary. Thus, the kidney serves to excrete compounds that are hydrophilic or are substrates for these transporters.

3.6. Properties That Affect Toxicity

Safety is a primary requirement for any new drug. High safety standards have resulted in a high level of attrition of clinical candidates during preclinical development through Phase III. For this reason, toxicity effects are being increasingly assessed during drug discovery and compounds are being optimized to reduce or eliminate the potential toxic mechanisms to reduce attrition and wasted development resources. *In vitro* and *in vivo* toxicity assays, for use during drug discovery, are being increasingly developed and implemented. The confidence of medicinal chemists and toxicologists in this approach will rely on the development of good correlations to *in vivo* toxicity endpoints. There are a large number of toxicity mechanisms. A common strategy in drug discovery is to screen for the major toxic mechanisms in order to have the maximum impact for the resources expended.

Toxicity is an unwanted deleterious effect that arises from the reversible binding of the compound to (1) the therapeutic target to produce an unforeseen negative side effect ("target mediated") or (2) a different biomolecule target in the body ("off target"). Toxicity can also arise from unwanted irreversible (covalent or quai-irreversible) binding of the compound or a reactive metabolite to a biomolecule in the body. Oxidative stress toxicity can occur when a compound or metabolite depletes the natural reductive trapping capacity of the cell (e.g., glutathione) and allows reactions with macromolecules.

Toxicity is a pharmacological process with negative results. It normally follows a typical dose–response relationship of increasing response with increasing concentration. All compounds produce toxic responses, thus, the goal is to achieve wide separation between the beneficial response (e.g., IC_{50}, ED_{50}) and the toxic response (e.g., lethal dose LD_{50}). This separation is often termed therapeutic index or window. Toxicity responses can be thought of as selectivity targets in which medicinal chemists seek high ratios of IC_{50} values. For toxicity, these ratios may be in terms of a ratio of IC_{50} for binding to the toxicity target divided by IC_{50} for binding to the therapeutic target, or the ratio may be the maximum unbound concentration of the drug in the bloodstream (C_{max}) after dosing at the maximum anticipated dose divided by the IC_{50} or K_i for binding to the toxicity target.

Irreversible reactions with endogenous biomolecules can be caused by *reactive metabolites*. A metabolic reaction, often oxidation, produces a reactive intermediate or product, which can react with a macromolecule to deactivate it or an immune response can be generated that is deleterious to the tissue. A list of some substructures that can produce a reactive metabolite is found in Table 4. These substructures are commonly termed "toxic alerts" and compounds containing them should be checked for toxicity in association with metabolizing materials. Troglitazone was withdrawn from the market owing to hepatic toxicity caused by a reactive metabolite.

3.6.1. Common Types of Toxicity and The Causative Mechanisms

Cytotoxicity A compound may interfere with one of the thousands of enzymes that catalyze reactions in cells, or a vital receptor or transporter. This can be by reversible inhibition or agonism or by irreversible reaction that depletes an enzyme. This can result in disruption in normal biochemistry and be toxic to cells.

Genotoxicity and Mutagenicity Interference of a compound or metabolite with the DNA sequence or disruption of the chromosomes can alter the functioning of the cell in the current generation and can be passed to later generations via mitosis. Mutagenicity can lead to carcinogenicity [45].

Teratogenicity A teratogen is a compound that causes changes in the morphology or health of a developing embryo in the uterus. Thalidomide is a well-known example. One current theory is that the compound causes oxidative stress that downregulates the Wnt and Akt survival pathways, causing apoptosis in embryonic limb development [46].

Table 4. Partial List of Substructures that Form Reactive Metabolites

Substructure	Reactive Metabolite
Aromatic amine	Nitroso, quinone-imine
Hydroxyl amine	Nitroso
Aromatic nitro	Nitroso
Nitroso	Nitroso, diazonium ions
Alkyl halide	Acylhalide
Polycyclic aromatic	Epoxide
α,β-Unsaturated aldehyde	Michael acceptor
Carboxylic acid	Acyl glucuronide
Bromo aromatic	Epoxide
Thiophene	S-oxide, epoxide
Hydrazine	Diazene, diazonium, carbenium ion
Hydroquinones	*p*-Benzoquinone
Azo	Nitrenium
Furans	α,β-Unsaturated dicarbonyl
Nitrosamine	Carbenium ion

Adapted from Ref. [1].

Drug–Drug Interactions When a drug goes through clinical development, its normal PK parameters are determined and the dose regimen is established in order to produce drug exposure levels in the patient that cause the desired therapeutic effect without approaching toxic levels. In practice, a patient may take two or more drugs at the same time for different treatments. It is possible that one drug may change the PK parameters of another codosed drug. This is called drug–drug interaction (DDI). One mechanism of DDI is when a drug inhibits a metabolic enzyme that is important for the clearance of a codosed drug, causing its AUC and C_{max} to be higher than expected. A second DDI mechanism is when one drug inhibits a transporter, which affects a second drug's absorption, distribution, or excretion, such as decreasing excretion in the liver or kidney. A third type of DDI is when a drug induces increased levels of a metabolic enzyme, which increases the clearance of a second drug, causing its AUC and C_{max} to be lower than expected and, thus, reducing the beneficial pharmacological action.

For CYP inhibition a therapeutic index of $C_{max}/K_i > 1$ indicates likely CYP inhibition *in vivo* and $C_{max}/K_i < 0.1$ indicates CYP inhibition is not likely *in vivo* [47]. Compounds or their metabolites can also bind to CYP enzymes irreversibly via a covalent reaction or quasi-irreversible interaction and this is termed "mechanism-dependent inhibition" (MDI). In this case, the CYP molecule is permanently deactivated.

Arrhythmia One safety concern is heart arrhythmia caused by disruption of normal cardiac action potential. A known cause of this is binding of compound molecules inside the hERG potassium ion channels in cardiac cells and, thus, blocking the normal flow of potassium ions [48]. This lengthens the cardiac action potential and can trigger Torsades de Pointe arrhythmia, which can be fatal. A therapeutic index of hERG $IC_{50}/C_{max} > 30$ is recommended.

4. ADMET PROPERTY ASSESSMENT

Medicinal chemists use various tools to assess the properties of compounds. "Rules" provide simple yet powerful guidance using only the structure. *In silico* (software) tools are convenient and rapid, but are only predictions. *In vitro* assays provide data for confident decision making, such as structure modifications to improve ADMET properties and guide compound selection for expensive *in vivo* ADMET and biology experiments. A series of reviews on *in vitro* ADMET property assays in drug discovery provide current information for medicinal chemists [49].

4.1. Rules for ADMET Property Assessment

Sets of rules have been developed based on statistical correlation of structural features of compounds to desirable PK parameters. Rules rapidly provide an alert to potential compound liabilities based on structural features that are easy to determine. They guide medicinal chemists away from structural elements that limit the chemical series' ADMET performance. Rules are actually guidelines and are not meant to be strict cutoff values. They are very useful as guides to important structural characteristics and values that are at increasing risk of causing poor ADMET performance. Below are some rules that have proven to be very beneficial to medicinal chemists. In addition, rule sets have been developed for other purposes, such as blood–brain barrier [50] and fragment library compounds [51].

4.1.1. Rule of Five (Lipinski's Rules)

A highly cited paper [22] evaluated the structural characteristics of compounds that had acceptable absorption in humans. Rules that encompassed 90% of compounds with good PK were as follows:

Poor absorption or permeation is more likely when

- H-bond donors > 5 (sum of OHs and NHs),
- MW > 500,
- $C\log P > 5$,
- H-bond acceptors > 10 (sum of Ns and Os), and
- substrates for biological transporters are exceptions to the rule.

As discussed above, hydrogen bonds enhance water solubility and reduce partitioning into lipids, as is necessary for permeability. Increasing molecular weight reduces water solubility and reduces movement through the lipid bilayer membrane. Increasing $\log P$ reduces solubility. These rules highlight deficiencies in structures that can reduce absorption and permeation.

4.1.2. Bioavailability Rules (Veber's Rules) Another study [38] used computational approaches to examine structural characteristics of compounds having good rat oral bioavailability:

There is a high probability of good oral bioavailability in rats when

- rotatable bonds ≤ 10,
- TPSA $\leq 140\,\text{Å}^2$, and
- H-bond total (donors + acceptors) ≤ 12

4.1.3. Rules for Leads Entering the Optimization Phase The lead optimization process usually adds substructures that improve target binding (e.g., lipophilicity and H-bonds) but reduce their solubility and permeability. Therefore, it is useful to build screening libraries and select leads for optimization that have more restricted property ranges than the Lipinski and Veber rules.

One study [52] used computational approaches to suggest the following guidelines for lead-like compounds that can be successfully modified to produce drug-like compounds:

- MW ≤ 460
- $-4 \leq \log P \leq 4.2$
- Rotatable bonds ≤ 10
- Rings ≤ 4
- Hydrogen-bond donors ≤ 5
- Hydrogen-bond acceptors ≤ 9

The structural "template" of the lead is well preserved though lead optimization to produce the NCE [11], thus it makes sense to assure that the lead has good properties and allows the opportunity for structure modifications without making the compound nondrug-like.

4.2. *In Silico* Tools for ADMET Property Assessment

Many software tools have been developed to predict properties. It is important to keep in mind that they are predictions and not data. These are comprised of algorithms developed by a pharmaceutical company, software company or academic research laboratory using their own database of measurements or data obtained from the scientific literature. Training sets of compounds and measured data are used to develop the algorithm and separate validation sets are used to evaluate the quality of the predictions.

Tools have been developed to predict all of the properties listed above, as well as modeling the absorption and pharmacokinetics of compounds, based on their structures. Lists of available *in silico* tools have been compiled [1]. *In silico* tools are best used to predict the properties of large compound sets (e.g., compound libraries), when material is not available for assay, or predicting properties in advance when planning library synthesis. *In silico* models are also available for modeling absorption and PK based on the structure or data from *in vitro* assays.

In silico tools can vary greatly in their quality of prediction. It is always beneficial to compare the actual measured values from a trusted laboratory to the software predictions prior to purchase.

In silico tools for toxicity can evaluate potential toxic alerts. QSAR, statistical, and expert approaches have been developed to predict toxicities based on structure.

4.3. *In Vitro* Assays for ADMET Property Assessment

Assays provide measured data for ADMET properties. The purpose is to assess the properties of a compound that contribute to *in vivo* PK or toxicity. Measurements have greater certainty for discovery project decision making than rules or software. However, it is important that the assays be carefully developed and validated. Control compounds should be coassayed with each set of compounds insure quality. Assay validation should include correlation to actual *in vivo* data when possible. The assay conditions should be close to the *in vivo* chemical environments encountered by the compound in order to assure relevant data. Medicinal chemists should also become familiar with the assays used to generate the data they are using for key decisions, so that the data can be used properly and not over- or underinterpreted. Some assays measure one discrete variable (e.g., PAMPA measures passive diffusion permeability), while the data from other assays are the result of multiple variables (e.g., Caco-2 is the composite of passive diffusion, efflux, active uptake, and paracellular permeation mechanisms). Discrete variable assays are most useful for guiding discrete structural modifications to optimize a property, while multiple variable assays are most useful as models of *in vivo* processes (e.g., Caco-2 predicts intestinal absorption). Some *in vitro* assays are performed in vials. Microtitre plates with 96 or more wells allow higher throughput via laboratory robots.

Assays can maximize throughput by using robotics and generic assay conditions. These "standard assays" are useful for screening a large number of compounds for a set of structure–property relationships. Alternatively, custom assay conditions can be used that allow investigation of specific research questions.

4.3.1. Physicochemical Property Assays

Physicochemical property assays have been reviewed [53,54].

Lipophilicity Assays The definitive approach for lipophilicity measurement is equilibrium partitioning between aqueous and octanol phases. The compound concentration in each phase is measured using HPLC or LC/MS and the $\log_{10}$ (concentration in octanol/concentration in aqueous) is calculated. For log P, a pH is used at which the compound is completely in the neutral form. For log D, the pH is specified because it affects the level of ionization and solubility in octanol versus aqueous. An estimate of lipophilicity can be obtained from the HPLC retention time, in which a nonpolar stationary bonded phase (e.g., C_{18}) is the lipophilic phase and the mobile phase is the aqueous phase [55,56]. *In silico* tools for lipophilicity are highly developed and it is common during drug discovery to calculate lipophilicity, which allows available resources to be used for other assays.

pK_a Assays The pK_a is determined by titration with equivalents of acid or base [57]. The pH at the inflection point is the pK_a. Commercial instruments are commonly used in the industry for automated assays for pK_a (Sirius Analytical) and have been compiled for other properties [1]. Software is often used to estimate the pK_a values of a compound during drug discovery.

Solubility Assays Equilibrium aqueous solubility assays are performed by adding solid compound to an aqueous matrix and allowing it to reach thermodynamic equilibrium with shaking for a long period (e.g., 3 days) to promote interaction of the solvent and the solid and allow the solid to reach the most stable polymorphic crystal form. This is most useful for development research on dosage forms for oral administration, but the requirements (time, expense, and large amount of compound) permit only a few compounds to be assayed in this manner during discovery.

Discovery experiments for solubility often reduce these requirements by shortening the incubation time and/or predissolving the compound in organic solvent [58]. Thus, true equilibrium conditions are not established and the experiment is termed "kinetic" solubility. For kinetic assays, there may not be excess compound present, predissolution in organic solvent destroys the crystal structure, and there may not be sufficient time to reach equilibrium. Nevertheless, kinetic solubility assays are useful for assaying a large number of compounds for structure–solubility relationships (i.e., lead optimization). Moreover, most discovery experiments (e.g., bioassays, *in vivo* PK and pharmacology) are not conducted under equilibrium conditions, thus the kinetic data are relevant. In a kinetic solubility assay a small volume of a DMSO stock solution of a compound is added to an aqueous buffer in the well of a 96-well plate. After an incubation period, the solution is filtered to remove undissolved material and the concentration of compound in solution is measured using a UV plate reader, HPLC or LC/MS and is taken as the solubility [59,60]. Alternatively, the compound can be serially diluted across several wells in a plate and the concentration of the well having the highest concentration without undissolved material is taken as the solubility [61]. This assay has been termed the "nephelometric" method.

The solubility assay format permits variations that allow custom assays to answer specific questions. Examples of these include testing the solubility of compounds in the assay media and using the protocol of a bioassay [7], or screening a series of compounds in the discovery dosing vehicle to select compound with highest solubility for animal dosing experiments [18].

Permeability Assays Several assays have been developed to measure compound permeability through lipid membranes. The parallel artificial membrane permeability assay (PAMPA) [62–64] uses a lipid membrane supported by the pores of a filter membrane held as a "sandwich" between the two aqueous compartments. The molecules from the donor compartment dissolve into the artificial membrane, diffuse across the membrane to the opposite interface, and then dissolve into the aqueous acceptor compartment. The 96-well format facilitates automation for high throughput, as well as measurement of concentrations in the compartments using a UV plate reader. Variations on PAMPA permit customization to various physiological membranes (e.g., BBB [65,66], skin [67]) and solution matrices barriers (e.g., various pHs). PAMPA is conducted for 1–18 h and assesses passive diffusion permeability.

A model of intestinal absorption is provided by the Caco-2 assay [68–70]. This assay uses a human colon carcinoma cell line that develops the morphology, some of the transporter expression, and tight junctions between the cells that are characteristic of intestinal epithelial cells. The assay is conduced for 1–2 h and is a composite of passive diffusion, active uptake and efflux. The assay can be conducted with the compound starting in the apical (A) compartment (top, intestine-facing side) to measure the compound appearance in the basolateral (B) compartment (bottom, bloodstream-facing side) or vice versa. The experiment will also indicate if a compound is an efflux substrate by the ratio of permeability in the B to A direction ($P_{eff,B>A}$) divided by permeability in the A to B direction ($P_{eff,A>B}$), which is termed "efflux ratio" (ER). If ER > 3, the compound is likely an efflux transporter substrate. Evidence for the identity of the particular efflux transporter involved is obtained by coincubating with an inhibitor for a specific transporter to observe whether the P_{eff} changes.

Other cell lines are used to address other transporter-related questions [71]. Cell lines have been transfected with specific transporter genes to provide an assay for a specific transporter. For example, transfection of the human MDR1 gene into MDCKII cells provided a relatively specific assay for Pgp efflux. One use of this cell line is to study the involvement of Pgp efflux at the BBB in reducing brain exposure of compounds.

Plasma Protein Binding Assays Binding to proteins in blood restricts the penetration of compound molecules into tissues. This can affect exposure to the therapeutic target and to selectivity targets, as well as exposure to organs that metabolize and excrete the compound. Plasma protein binding is commonly measured by incubating the test compound with plasma in a compartment that is separated from buffer by a dialysis membrane [72]. The unbound drug permeates through the membrane and establishes equilibrium with the buffer, thus, the unbound drug concentration is equal to the concentration on the buffer-only side of the membrane.

Tissue binding is performed in a similar manner. Tissue homogenate is used instead of the plasma. This is very useful in correlations to pharmacodynamics.

4.3.2. Metabolic Property Assays Metabolic instability is one of the most consistent liabilities of lead series early in drug discovery, thus, many metabolic stability assays are performed. Teams can often improve metabolic stability using structure modifications, but they first need assay data.

Metabolic Stability Assays Metabolic stability assays are performed either with intact hepatocytes in culture or using subcellular fractions of liver hepatocytes. These fractions are obtained by tissue homogenization followed by differential centrifugation. The S9 fraction is the supernatant from centrifugation at 9000 g for 20 min. Further centrifugation of S9 at 100,000 g for 60 min separates it into cytosol and microsomes, each containing metabolizing enzymes. Microsomes are comprised of the membranes of the endoplasmic reticulum with metabolizing enzymes (CYPs on the outside and UGTs on the inside).

Microsomes are the most frequently used reagent for metabolic stability assays [73]. They are incubated with cofactor NADPH, other cofactors and the test compound to assess Phase I metabolic stability. For Phase II glucuronidation they are incubated with cofactor UDPGA, alamethicin (to open pores in the microsomes for compound access to UGTs on the inside) and other cofactors. S9 fraction typically has a quarter of the activity of microsomes because it consists of microsomes plus cytosol. Vendors sell cellular fractions from various tissues and many species. It has been shown that if a compound is unstable in microsomes, it has a high risk of high PK clearance *in vivo* [74]. Different fractions contain different enzyme sets. Recombinantly expressed metabolic enzymes are individually commercially available. Hepatocytes may be used fresh or obtained frozen in liquid nitrogen (cryopreserved). They are more expensive and less active than microsomes and quantitation must be linked to the number of cells.

The differences in metabolic enzymes from species to species makes it important to assay metabolic stability with the microsomes of key species, including the efficacy species used by the project team. In many cases, compounds are more stable in human microsomes than in rat or mouse, but the reverse also occurs.

It is common to incubate test compounds at 1 μM concentration. This is consistent with tissue concentration levels reached after oral dosing, but is not so high as to saturate the metabolizing enzymes. LC/MS/MS is usually used to measure the concentration of the test compound that remains unmetabolized at the end of the incubation. It is sufficient in drug discovery to measure the concentration of test compound remaining at only one time point, because a plot of log percent remaining versus time is linear for first-order kinetics [75]. Additional measurement points not only add to the precision of the assay but also add considerably to the resources needed for the assay. Microsomal "half-life" is calculated from the slope of the plot of log percent remaining versus linear time, according to first-order kinetics approaches. It is better for discovery teams to convert stability to $t_{1/2}$ because it can be linearly compared from compound to compound, while percent remaining is not linear. Fifteen minutes of incubation allows full coverage, using a single sampling time point, of highly unstable compounds ($t_{1/2}$ up to 15 min) through moderately stable compounds, which is the area of greatest concern for drug discovery [75]. LC/MS/MS is necessary for quantitation, owing to the low concentration of test compound and compounds from the microsome and cofactor matrix that would otherwise interfere with quantitation. It is important to check the activity of each batch of metabolizing enzyme that will be used in the assay, because activity can vary from batch to batch.

Enzymatic Stability Assays Enzymes in the intestine and blood may catalyze the decomposition of leads if they have a substructure that is susceptible to hydrolysis. It is valuable to incubate compounds with plasma and gastrointestinal fluids if this may occur.

An optimized assay for plasma stability has been reported [76]. It involves diluting with buffer to stabilize the pH. The assay can tolerate up to 2.5% DMSO.

The stability of compounds in simulated gastric fluid (SGF) and simulated intestinal fluid (SIF) is also useful [77]. Enzymes in GI fluids can decompose compounds. Standard recipes for SGF and SIF have been described in the U.S. Pharmacopeia.

Metabolite Identification Once it is learned that a compound is predicted to be unstable *in vitro*, it is often useful to identify the metabolites in order to understand the reaction and attempt to counteract it by structure modification [78].

Metabolites can be rapidly profiled using LC/MS/MS techniques. A microsomal incubation solution is injected onto the instrument and the metabolite HPLC peaks are analyzed on-the-fly for molecular weight and MS/MS product ions (fragments), from which the structure of the metabolite is deduced. When the MS/MS fragmentation does not provide sufficient detail for regiospecific structure elucidation, NMR analysis of an isolated incubation fraction is necessary. It is increasingly common in drug discovery for the structures of metabolites to be identified to aid structure optimization [1] to improve metabolic stability (see Section 3.2.1). This also provides the opportunity to synthesize major metabolites during the concerted optimization synthesis stage in order to test their pharmacological activity and safety.

4.3.3. Toxicity Property Tools Improved *in vitro* assays for toxicity are continually emerging and being refined. Concomitantly, there are studies being performed to correlate the *in vitro* data to *in vivo* toxicity endpoints to achieve the most reliable approaches and greatest acceptance within drug discovery. There is a strong opportunity for efficiency by solving safety issues during drug discovery rather than investing resources and time in a compound or chemical series that is later found to be toxic during development.

Cytotoxicity Assays Many biochemical mechanisms can be affected by a compound to cause cell damage or death. Cytotoxicity assays have been developed to assess major cellular functions that may be affected. Hepatocytes are often used in order to include potentially toxic metabolites in addition to the compounds themselves. The neutral red (NR) assay detects the normal accumulation of this red dye into lysosomes and the red intensity is reduced if lysosome function is not healthy. MTT is normally metabolized by mitochondria to purple formazan and the purple intensity is reduced if the mitochondria are not healthy. Lactate dehydrogenase (LDH) is a plentiful enzyme in the cytoplasm and its release into the test media can be measured if the unhealthy cells lyse. Other *in vitro* assays incude ATP depletion, cell growth rate, caspase apoptosis pathyway activation, and glutathione depletion (oxidative stress). *In vitro* assays provide rapid assessment of interference by test compounds with normal cell biochemistry.

Genotoxicity and Mutagenicity Assays Genotoxicity can be assayed through the tendency of the compound to cause changes in DNA sequence or disruption of chromosomes [45]. The micronucleus assay microscopically examines for chromosome breaks that result in partial chromosomes surrounded by membranes after compound incubation and cell mitosis. Abnormal chromosomes and DNA fragments can also be observed by microscopic examination and by the "comet assay." The Ames and Mouse TK assays examine for DNA frame shifts or deletions. Activation of the GADD45a gene has also been used as a biomarker for stress to the DNA by the test compound and a kit (Greenscreen) is commercially available.

Reactive Metabolites Assays The detection of reactive metabolites [79] is often first attempted *in vitro* using glutathione trapping. Glutathione is coincubated with the test compound and metabolizing system (e.g., microsomes) and the presence of a glutathione metabolite is detected using specific ions on the LC/MS/MS. Detection of glutathione adducts does not guarantee that the test compound forms significant amounts of reactive metabolites, but can alert the team to the need for more detailed studies.

Teratogenicity Assays Damage to fetuses [80] has traditionally been monitored by dosing pregnant rodents and examining the embryos after a period. The development of zebrafish embryos is an emerging *in vitro* assay for detecting teratogenicity potential.

Drug–Drug Interaction Assays CYP inhibition is the most commonly studied DDI during drug discovery. Standard assays have been developed to determine whether the test compound reduces the rate of metabolism of a known substrate compound [81–83]. The substrate and test compound are coincubated with microsomes or rhCYP enzymes and the rate of production of the substrate metabolite is determined using LC/MS/MS, fluorescence or luminescence. Incubations are similar to the metabolic stability assay. Test kits of fluorogenic and luminogenic substrates and recombinantly produced human CYP isozymes are commercially available for high throughput plate reader-based assays. Otherwise, drugs that are metabolized by specific isozymes are used with human microsomes, which contain a mixture of natural isozymes at their native concentrations. Increased throughput has been obtained by combining multiple specific drug substrates into a "cocktail", coincubating them with microsomes, and using LC/MS/MS to specifically quantitate each substrate metabolite using different LC/MS/MS detection conditions. Alternatively, a mixture of rhCYP isozymes can be used with a mixture of specific probe substrates to provide the optimal enzyme initial rate conditions for the best kinetic conditions for determination of IC_{50} [83].

Arrhythmia Assays Blocking of the hERG potassium channel in heart cells can lead to arrhythmia [48]. It is assayed *in vitro* using cells (e.g., CHO) transfected with hERG gene, so that they express the potassium ion channel. hERG blocking compounds are detected using membrane sensitive dyes, competitive radioligand binding, and using rubidium as a surrogate for potassium. These assays are useful for screening, but consensus is that the high throughput patch-clamp assay is the best approach. Cells expressing the hERG channel are plated into wells of a device that contains small holes on the bottom, the cells settle to the bottom of the well and a negative pressure is applied to break the cell membrane at the hole to create an electrical circuit between an electrode in the aqueous media and one below the hole. This allows the current through the ion channels to be measured and if changes in current occur when with test compound dosing indicates a hERG-channel blocker.

5. CONCLUSIONS

A wide range of *in vitro* assays are available for assessing ADMET properties for drug discovery leads. These enable a wide array of important opportunities for medicinal chemists: decision making for lead selection, diagnosis of the causes of PK shortcomings, property optimization through structure modification, problem solving, accurate bioassay SAR, selection of the best compounds for *in vivo* studies, optimizing *in vivo* exposure for proof of concept, and achievement of advancement criteria to clinical development. A successful and efficient approach is to integrate ADMET property research into the drug discovery process in parallel with activity and selectivity research. It is imperative for medicinal chemists to modify structures during the discovery optimization phase to solve ADMET property issues. An ensemble of structure modification strategies to successfully improve ADMET properties is available [1].

REFERENCES

1. Kerns EH, Di L. Drug-Like Properties: Concepts, Structure Design and Methods: from ADME to Toxicity Optimization. Oxford: Academic Press; 2008.
2. van de Waterbeemd H, Smith DA, Beaumont K, Walker DK. Property-based design: optimization of drug absorption and pharmacokinetics. J Med Chem 2001;44:1313–1332.
3. Smith DA, van de Waterbeemd H, Walker DK. Pharmacokinetics and Metabolism in Drug Design. Weinheim: Wiley-VCH; 2001.
4. Borchardt RT, Kerns EH, Hageman MJ, Thakker HR, Stevens JL, editors. Optimizing the Drug-like Properties of Leads in Drug Discovery. Arlington, VA: AAPS Press; 2006.
5. Di L, Kerns EH. Profiling drug-like properties in discovery research. Curr Opin Chem Biol 2003;7:402–408.
6. Testa B, Krämer SD, Wunderli-Allenspach H, Folkers G, editors. Pharmacokinetic Profiling in Drug Research. Zurich: Verlag Helvetica Chimica Acta; 2006.
7. Di L, Kerns EH. Biological assay challenges from compound solubility: strategies for bioassay optimization. Drug Discov Today 2006;11 (9–10): 446–451.
8. Lipinski CA. Solubility in water and DMSO: issues and potential solutions. Biotechnology: pharmaceutical aspects. 2004;1:93–125.
9. Popa-Burke IG, Issakova O, Arroway JD, Bernasconi P, Chen M, Coudurier L, Galasinski S, Jadhav AP, Janzen WP, Lagasca D, Liu D, Lewis RS, Mohney RP, Sepetov N, Sparkman DA, Hodge CN. Streamlined system for purifying and quantifying a diverse library of compounds and the effect of compound concentration measurements on the accurate interpretation of biological assay results. Anal Chem 2004;76:7278–7287.
10. Kerns EH, Di L. Drug-Like Properties: Concepts, Structure Design and Methods: from ADME to Toxicity Optimization. Oxford: Academic Press; 2008. p 91.
11. Proudfoot JR. Drugs, leads, and drug-likeness: an analysis of some recently launched drugs. Bioorg Med Chem Lett 2002;12:1647–1650.
12. Di L, Kerns EH. Application of pharmaceutical profiling assays for optimization of drug-like properties. Curr Opin Drug Discov Dev 2005;8 (4): 495–504.
13. Gan LSL, Eads C, Niederer T, Bridgers A, Yanni S, Hsyu PH, Pritchard FJ, Thakker D. Use of Caco-2 cells as an *in vitro* intestinal absorption

and metabolism model. Drug Dev Ind Pharmacy 1994;20(4):615–31.

14. Gan LSL, Thakker DR. Application of Caco-2 model in the design and development of orally active drugs: elucidation of biochemical and physicochemical barriers posed by the intestinal epithelium. Adv Drug Deliv Rev. 1997; 2377–98.
15. Kerns EH, Di L. Drug-Like Properties: Concepts, Structure Design and Methods: from ADME to Toxicity Optimization. Oxford: Academic Press; 2008. p 70–78, 92–97, 114–115, 119, 131–134, 146–156, 172–174, 180–183, 193–194 203–204, 213, 222.
16. Nassar AF, Kamel AM, Clarimont C. Improving the decision-making process in the structural modification of drug candidates: enhancing metabolic stability. Drug Discov Today 2004;9 (23):1020–1028.
17. Nassar AF, Kamel AM, Clarimont C. Improving the decision-making process in the structural modification of drug candidates: reducing toxicity. Drug Discov Today 2004;9(23):1055–1064.
18. Di L, 2006. Impacts of solubility in drug discovery from exploratory to candidate selection: applications and methodologies. Improving Solubility Conference, London, UK.
19. Hansch C, Leo A, Hoekman D. Exploring QSAR. Fundamentals and Applications in Chemistry and Biology, Vol. 1. Hydrophobic, Electronic and Steric Constants, Vol. 2. New York: Oxford University Press; 1995.
20. Hansch C, Leo A, Mekapati SB, Kurup A. QSAR and ADME. Bioorg Med Chem 2004;12: 3391–3400.
21. van de Waterbeemd H, Testa B, edirors. Drug Bioavailability. Weinheim: Wiley-VCH; 2003.
22. Abraham MH, Chadha HS, Leitao RAE, Mitchell RC, Lambert WJ, Kaliszan R, Nasal A, Haber P. Determination of solute lipophilicity, as log *P*(octanol) and log *P*(alkane) using poly(styrene-divinylbenzene) and immobilized artificial membrane stationary phases in reversed-phase high-performance liquid chromatography. J Chrom A 1997;766:35–47.
23. Kerns EH, Di L. Drug-Like Properties: Concepts, Structure Design and Methods: from ADME to Toxicity Optimization. Oxford: Academic Pres; 2008. p 46.
24. Lipinski CA, Lombardo F, Dominy BW, Feeney PJ. Experimental and computational approaches to estimate solubility and permeability in drug discovery and development settings. Adv Drug Delivery Rev 1997;23:3–25.
25. Veber DF, Johnson SR, Cheng HY, Smith BR, Ward KW, Kopple KD. Molecular properties that influence the oral bioavailability of drug candidates. J Med Chem 2002;45:2615–2623.
26. Stenberg P, Bergström CA, Luthman K, Artursson P. Theoretical predictions of drug absorption in drug discovery and development. Clin Pharmacokinet 2002;41(11): 877–899.
27. Clark DE. *In silico* prediction of blood–brain barrier permeation. Drug Discov Today 2003; 8:927–933.
28. Kerns EH, Di L, Carter GT. *In vitro* solubility assays in drug discovery. CurrDrug Metab 2008;9(8):879–885.
29. Yalkowsky S, Banerjee S. Aqueous Solubility: Methods of Estimation for Organic Compounds. New York: Marcel Dekker; 1992.
30. Abraham M, Le J. The correlation and prediction of the solubility of compounds in water using an amended solvation energy relationship. J Pharm Sci 1999;88(9):868–880.
31. Curatolo W. Physical chemical properties of oral drug candidates in the discovery and exploratory development settings. Pharm Sci Technol Today 1998;1:387–393.
32. Lipinski CA. Drug-like properties and the causes of poor solubility and poor permeability. J Pharm Tox Methods 2000;44:235–249.
33. Wang HL, Katon J, Balan C, Bannon AW, Bernard C, Doherty EM, Dominguez C, Gavva NR, Gore V, Ma V, Nishimura N, Surapaneni S, Tang P, Tamir R, Thiel O, Treanor JJS, Norman MH. Novel vanilloid receptor-1 antagonists. 3. The identification of a second-generation clinical candidate with improved physicochemical and pharmacokinetic properties. J Med Chem 2007;50(15):3528–3539.
34. Stenberg P, Bergström CA, Luthman K, Artursson P. Theoretical predictions of drug absorption in drug discovery and development. Clin Pharmacokinet 2002;41(11):877–899.
35. Artursson P, 2002. Prediction of drug absorption: Caco-2 and beyond. PAMPA 2002 Conference, San Francisco, California.
36. Chan LMS, Lowes S, Hirst BH. The ABCs of drug transport in intestine and liver: efflux proteins limiting drug absorption and bioavailability. Eur J Pharm Sci 2004;21:25–51.
37. Kuduk SD, Di Marco CN, Chang RK, Wood MR, Schirripa KM, Kim JJ, Wai JMC, DiPardo RM, Murphy KL, Ransom RW, Harrell CM, Reiss DR, Holahan MA, Cook J, Hess JF, Sain N, Urban MO, Tang C, Prueksaritanont T, Pettibone DJ, Bock MG. Development of orally bioavailable and CNS penetrant biphenylaminocyclopropane carboxamide bradykinin B_1 receptor antagonists. J Med Chem 2007;50:272–282.

38. Yang CY, Dantzig AH, Pidgeon C. Intestinal peptide transport systems and oral drug availability. Pharm Res 1999;16(9):1331–1343.
39. Rubio-Aliaga I, Daniel H. Mammalian peptide transporters as targets for drug delivery. Trends Pharmacol Sci 2002;23(9):434–440.
40. Kerns EH, Di L. Chemical stability. In: John BT, David JT, editors. Comprehensive Medicinal Chemistry II. Oxford: Elsevier; 2006. p 489–507.
41. Kerns EH, Di L. Drug-Like Properties: Concepts, Structure Design and Methods: from ADME to Toxicity Optimization. Oxford: Academic Press; 2008. p 162–165.
42. MacKenzie AR, Marchington AP, Middleton DS, Newman SD, Jones BC. Structure–activity relationships of 1-alkyl-5-(3,4-dichloro-phenyl)-5 {2-[3-substitututed)-1-azetidinyl]ethyl}-2-piperidones. 1. Selective antagonists of the neurokinin-2 receptor. J Med Chem 2002;45: 5365–5377.
43. Hammarlund-Udenaes M, Fridén M, Syvänen S, Gupta A. On the rate and extent of drug delivery to the brain. Pharm Res 2008;25 (8):1737–1750.
44. Di L, Kerns EH, Carter GT. Strategies to assess blood–brain barrier penetration. Expert Opin Drug Discov 2008;3(6):677–687.
45. Knobloch J, Rüther U. Shedding light on an old mystery: thalidomide suppresses survival pathways to induce limb defects. Cell Cycle 2008;7 (9):1122–1128.
46. Custer LL, Sweder K. The role of genetic toxicology in drug discovery and optimization. Curr Drug Metab 2008;9(8):978–985.
47. Bjornsson TD, Callaghan JT, Einolf HJ, Fischer V, Gan L, Grimm S, Kao J, King SP, Miwa G, Ni L, Kumar G, Mcloed J, Obach RS, Roberts S, Roe A, Shah A, Snikers F, Sullivan JT, Tweedie D, Vega JM, Walsh J, WRIGHT SA. On the conduct of *in vitro* and *in vivo* drug–drug interaction studies: a pharmaceutical research and manufacturers of America (PhRMA) perspective. Drug Metab Dispos. 2003;31(7):815–832.
48. Bowlby MR, Peri R, Zhang H, Dunlop J. herg (kcnh2 or kv11.1) k + channels: screening for cardiac arrhythmia risk. Curr Drug Metab 2008;9(8):965–970.
49. Kerns, EH. In vitro ADME/Tox profiling for drug discovery. Curr Drug Metab 2008;9(8):845–994.
50. Pardridge WM. CNS drug design based on principles of blood–brain barrier transport. J Neurochem 1998;70:1781–1792.
51. Congreve M, Carr R, Murray C, Jhoti H.A 'rule of three' for fragment-based lead discovery?. Drug Discov Today 2003;8:876–877.
52. Hann MM, Oprea TI. Pursuing the leadlikeness concept in pharmaceutical research. Curr Opin Chem Biol 2004;8:255–263.
53. Kerns EH. High throughput physicochemical profiling for drug discovery. J Pharm Sci 2001;90:1838–1858.
54. Kerns EH, Di L. Physicochemical profiling: overview of the screens. Drug Discov Today Technol 2004;1:343–348.
55. Lombardo F, Shalaeva MY, Tupper KA, Gao F, Abraham MH. ElogPoct: a tool for lipophilicity determination in drug discovery. J Med Chem 2000;43:2922–2928.
56. Kerns EH, Di L, Petusky S, Kleintop T, McConnell O, Carter G, Pharmaceutical profiling method for lipophilicity and Integrity using LC/MS. J Chromatogr B 2003;791:381–388.
57. Box KJ, Comer JEA. Using measured p*K*a, log *P* and solubility to investigate supersaturation and predict BCS class. Curr Drug Metab 2008;9(8): 869–878.
58. Kerns EH, Di L, Carter GT. *In vitro* solubility assays in drug discovery. Curr Drug Metab 2008;9(8):879–885.
59. Avdeef A. High throughput measurements of solubility profiles. In: Bernard T, Han van de W, Gerd F, Richard G, eds., Pharmacokinetics Optimization in Drug Research. Zurich: Verlag Helvetica Chimica Acta; 2001. p 305–326.
60. Goodwin JJ. Rationale and benefit of using high throughput solubility screens in drug discovery. Drug Discov Today Technol 2006;3(1):67–71.
61. Bevan C, Lloyd RS. A high-throughput screening method for the determination of aqueous drug solubility using laser nephelometry in microtiter plates. Anal Chem 2000;72:1781–1787.
62. Kansy M, Senner F, Gubernator K. Physicochemical high throughput screening: parallel artificial membrane permeability assay in the description of passive absorption processes. J Med Chem 1998;41:1007–1010.
63. Faller B. Artificial membrane assays to assess permeability. Curr Drug Metab 2008;9(8): 886–892.
64. Avdeef A, Bendels S, Di L, Faller B, Kansy M, Sugano K, Yamauchi Y. Parallel artificial membrane permeability assay (PAMPA)-critical factors for better predictions of absorption. J Pharm Sci 2007;96(11):2893–2909.
65. Di L, Kerns EH, McConnell OJ, Carter G. High throughput artificial membrane permeability assay for blood–brain barrier. Eur J Med Chem 2003;38:223–232.

66. Di L, Kerns EH, Bezar IF, Petusky SL, Huang Y. Comparison of blood–brain barrier permeability assays: in situ brain perfusion, MDR1-MDCKII and PAMPA-BBB. J Pharm Sci 2009; 98(6): 1980–1991. DOI: 10. 1002/jps.21580.
67. Ottaviani G, Martel S, Carrupt PA. Parallel artificial membrane permeability assay: a new membrane for the fast prediction of passive human skin permeability. J Med Chem 2006;49(13):3948–54.
68. Hildago IJ. Assessing the absorption of new pharmaceuticals. Curr Topics Med Chem 2001;1: 385–401.
69. Balimane PV, Han YH, Chong S, Current industrial practices of assessing permeability and P-glycoprotein interaction. AAPS J 2006;8(1): E1–13.
70. Press B, Di Grandi D. Permeability for intestinal absorption: Caco-2 assay and related issues. Curr Drug Metab 2008;9(8):893–900.
71. Jin H, Di L. Permeability: *in vitro* assays for assessing drug transporter activity. Curr Drug Metab 2008;9(8):911–920.
72. Banker MJ, Clark TH. Plasma/serum protein binding determinations. Curr Drug Metab 2008;9(8):854–859.
73. Laine R. Metabolic stability: Main enzymes involved and best tools to assess it. Curr. Drug Metab. 2008;9(8):921–927.
74. Di L, Kerns EH, Ma XJ, Huang Y, Carter GT. Applications of high throughput microsomal stability assay in drug discovery. Comb Chem High Through Screen 2008;11(6):469–476.
75. Di L, Kerns EH, Gao N, Li SQ, Huang Y, Bourassa JL, Huryn DM. Experimental design on single-time-point high-throughput microsomal stability assay. J Pharm Sci 2004;93: 1537–1544.
76. Di L, Kerns EH, Chen H, Petusky SL. Development and application of an automated solution stability assay for drug discovery. J Biomolec Screen 2006;11(1):40–47.
77. Di L, Kerns EH. Solution stability: plasma, gastrointestinal, bioassay. Curr Drug Metab 2008;9(8):860–868.
78. Nassar AEF, Talaat RE. Strategies for dealing with metabolite elucidation in drug discovery and development. Drug Discov Today. 2004;9: 317–325.
79. Prakash C, Sharma R, Gleave M, Nedderman A. *In vitro* screening techniques for reactive metabolites for minimizing bioactivation potential in drug discovery. Curr Drug Metab 2008;9 (8):952–964.
80. Augustine-Rauch KA. Predictive teratology: teratogenic risk-hazard identification partnered in the discovery process. Curr Drug Metab 2008;9(8):971–977.
81. Walsky RL, Boldt SE. *In vitro* cytochrome P450 inhibition and induction. Curr Drug Metab 2008;9(8):928–939.
82. Obach RS, Walsky RL, Venkatakrishnan K, Gaman EA, Houston JB, Tremaine LM. The utility of *in vitro* cytochrome P450 inhibition data in the prediction of drug–drug interactions. J Pharmacol Exp Ther 2006;316:336–348.
83. Di L, Kerns EH, Li SQ, Carter GT. Comparison of cytochrome P450 inhibition assays for drug discovery using human liver microsomes with LC-MS, rhCYP450 isozymes with fluorescence, and double cocktail with LC-MS. Int. J Pharmaceut 2007;335(1–2):1–11.

RETROMETABOLISM-BASED DRUG DESIGN AND TARGETING

NICHOLAS BODOR[1]
PETER BUCHWALD[2]
[1] Center for Drug Discovery, University of Florida, Gainesville, FL
[2] Diabetes Research Institute and Molecular and Cellular Pharmacology, Miller School of Medicine, University of Miami, Miami, FL

1. INTRODUCTION

Whereas considerable medical progress, mainly driven by drug research, took place during the last century [1–4], there is increasing evidence for a slowdown as the number of new chemical entities (NCEs) launched per year has essentially stagnated during the past 30 years despite exponentially increasing R&D expenditures [5–8]. Medicinal chemistry has certainly witnessed significant changes due to advancements in the elucidation of the molecular–biochemical mechanisms of drug action and due to other developments such as combinatorial chemistry and high-throughput screening (HTS). Nevertheless, rational drug design, which would allow the development of effective pharmaceutical agents with minimal side effects on as rational a basis as possible, is still an elusive goal. In fact, the situation seems to have worsened as a result of increasing regulation and the increased complexity of drug research, which has arguably hampered true innovation. By all indications, not only are most relatively recently launched drugs derived by modifications of known drug structures or published lead structures but they also are even closely related to their original leads and new technologies such as combinatorial chemistry or HTS seem to have had no significant impact [9]. It even has been suggested that some of the new technologies may be generating bigger haystacks as opposed to more needles [10]. It is a telling fact that two ideas formulated quite a while ago by two of the most important figures in medicinal chemistry still hold to a surprising degree. Successful drug discovery still relies on the same four Gs stated (in German) by Paul Ehrlich: "Glück, Geduld, Geschick und Geld" (that is luck, patience, skill, and money), and Sir James Black's advice that "the most fruitful basis for the discovery of a new drug is to start with an old drug" still holds true [11,12]. Retrometabolic drug design (RMDD) approaches provide general drug design strategies very much along the lines of the latter as they usually start from a known lead structure and focus on designing safer, less toxic, and intrinsically better targeted drugs through either a soft drug or a chemical delivery system designs.

A main reason behind the low success rate of most drug design processes is that while focusing on increasing pharmacological potency, they often tend to ignore the related side effect and toxicity aspects. Many new therapeutic agents designed to bind to a specific receptor or found to have high activity ultimately have to be discarded when unacceptable toxicity or unavoidable side effects are encountered in later stages of the development. However, this should not be surprising. Many side effects are usually closely related to the intrinsic receptor affinity responsible for the desired activity. Additionally, even if, in most cases, the desired response should be localized to some organ or cell, the corresponding receptors and/or other members of the same receptor family are in fact distributed throughout the body. It is worth noting that most small-molecule drugs are quite active: a recent study of all binding-affinity-related endpoints of all identifiable drug-efficacy target pairs (i.e., IC_{50}, EC_{50}, ED_{50}, K_i, K_d, and pA_2) found the median affinity for current small-molecule drugs to be approximately 20 nM [11]. Furthermore, many drugs show clinically relevant polypharmacology (i.e., they are "dirty" drugs in the sense that they bind to multiple target proteins and their clinical effects might be mediated through modulation of a set of protein targets) [11]. In addition, closely related members of the gene family show significant drug promiscuity, and, as a result of the generally similar function of these proteins, give rise to complex, composite clinical pharmacology. Hence, any sufficiently potent drug is likely to generate an array of, often undesired, biological responses. Finally, for most drugs, metabolism generates multiple metabolites that can have a qualitatively or quantitatively different type of biological activity, including enhanced toxicity, which further complicates the picture.

Consequently, drug design should focus not on increasing activity alone, but on increasing the therapeutic index (TI), which reflects the degree of selectivity or margin of safety. TI is usually defined as the ratio between the median toxic dose (TD_{50}) and the median effective dose (ED_{50}) (or some other corresponding ratio):

$$\mathrm{TI} = \frac{\mathrm{TD}_{50}}{\mathrm{ED}_{50}} \tag{1}$$

To fully estimate the potential for toxicity, one must also take into account that metabolic conversion of a drug (D) can generate multiple metabolites including (i) analog metabolites (D_1, D_2, and so on) that have structures and activities similar to the original drug but have different pharmacokinetic properties, (ii) other metabolites (M_1, M_2, and so on) including inactive ones (M_i), and (iii) potential reactive intermediates (I^*_1, I^*_2, and so on) that can be responsible for various kinds of cell damage by forming toxic species (Fig. 1). All these compounds are present simultaneously and at varying concentrations as they tend to have different pharmacokinetic profiles, hence, the overall toxicity (T) of any drug is in fact a combination of the intrinsic toxicity–selectivity of the original drug, $T_i(\mathrm{D})$, and of all the toxicities due to the various metabolic products that are formed:

$$\begin{aligned} T(\mathrm{D}) = {} & T_i(\mathrm{D}) + T(\mathrm{D}_1, \mathrm{D}_2, \ldots) \\ & + T(\mathrm{M}_1, \mathrm{M}_2, \ldots) + T(\mathrm{I}^*_1, \mathrm{I}^*_2, \ldots) \end{aligned} \tag{2}$$

Accordingly, targeting and metabolism considerations should be an integral part of any drug design process to ensure that the desired NCE is designed with targeting and a preferred metabolic route in mind. The new molecule should be designed with site specificity and corresponding selectivity considerations already incorporated into its molecular structure. Site-specific delivery and site-specific action, if achievable and if sufficient for efficacy, might alleviate the undesired effects due to intrinsic toxicity. By designing and predicting the major metabolic pathways, the formation of undesired toxic, active, or high-energy intermediates might be avoidable. Surprisingly, the importance of early integration of metabolism, pharmacokinetic, and general physicochemical considerations in the drug design process has been clearly recognized in industrial settings only during the mid-1990s [13,14] despite numerous publications describing these concepts and methodologies in the early 1980s [15–21].

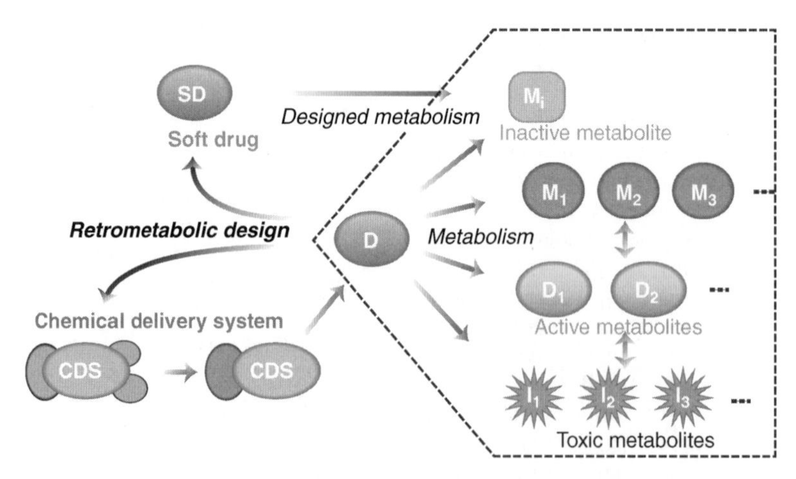

Figure 1. The RMDD loop, including chemical delivery system design and soft drug design. A schematic representation of possible metabolic pathways for drugs (D) in general is also included (dashed box; see text for details).

2. RETROMETABOLIC DRUG DESIGN

2.1. Principles

RMDD approaches represent systematic methodologies that thoroughly integrate structure–activity and structure–metabolism relationships and are aimed to design safe, locally active compounds with an improved therapeutic index [22–26]. They include two distinct methods (Fig. 1). One approach is the design of soft drugs (SDs) [15–21,24,27,28]. Soft drugs are new, active therapeutic agents, often isosteric–isoelectronic analogs of a lead compound, with a chemical structure specifically designed to allow predictable metabolism into inactive metabolites after exerting their desired therapeutic effect(s). The other approach is the design of chemical delivery systems (CDSs) [24,29–36]. CDSs are biologically inert molecules intended to enhance drug delivery to a particular organ or site and requiring several conversion steps before releasing the active drug.

Both approaches involve chemical modifications of the molecular structure, and both require enzymatic reactions to fulfill drug targeting; however, the principles of SD and CDS design are distinctly different. Whereas CDSs are inactive as administered and sequential enzymatic reactions provide the differential distribution and ultimately release the active drug, SDs are active as administered and are designed to be easily metabolized into inactive species. Assuming an ideal situation, with a CDS, the drug is present at the site and nowhere else in the body because enzymatic processes produce the drug only at the site, whereas with a SD, the drug is present at the site and nowhere else in the body because enzymatic processes destroy the drug at those sites. Whereas CDSs are designed to achieve drug targeting at a selected organ or site, SDs are designed to afford a differential distribution that can be regarded as reversed targeting.

2.2. Terminology

The *retrometabolic* designation has been introduced for these drug design approaches to emphasize that metabolic pathways are designed going backward compared to the actual metabolic processes in a manner somewhat similar to E.J. Corey's *retrosynthetic* analysis [37] in which synthetic pathways are designed going backward compared to actual synthetic laboratory operations.

2.2.1. Soft Drug Versus Hard Drug The *soft drug* terminology was originally introduced in 1976 [15,38–40] to contrast Ariëns's theoretical drug design concept of nonmetabolizable, *hard drugs*. Hard drugs do not undergo any metabolism and, hence, avoid the problems caused by reactive intermediates or active metabolites. For existing drugs, this is rarely the case; however, certain strongly lipophobic drugs, such as enalaprilat, lisinopril, cromolyn, and bisphophonates (e.g., alendronate), are essentially not metabolized *in vivo*, and they can be regarded as hard drug examples [13].

2.2.2. Soft Drug Versus Prodrug Because *soft drugs* and *prodrugs* are still often confused, the difference between these two concepts should also be clarified [41]. Prodrugs are pharmacologically inactive compounds that result from transient chemical modifications of biologically active species [42–50]. This concept was introduced by A. Albert in 1958 [51], and the rationale for prodrug design is that the structural requirements needed to elicit a desired pharmacological action and those needed to provide optimal delivery to the targeted receptor sites may not be the same. Hence, a chemical change is introduced to improve some deficient physicochemical property, such as membrane permeability or water solubility, or to overcome some other problem, such as rapid elimination, bad taste, or a formulation difficulty. After administration, the prodrug, by virtue of its improved characteristics, is more systemically and/or locally available than the parent drug. However, before exerting its biological effect, the prodrug must undergo chemical or biochemical conversion to the active form. Therefore, in theoretical terms, the prodrug and the soft drug concepts are opposite to each other. Whereas prodrugs are, ideally, inactive by design and are converted by a predictable mechanism to the active drug, soft drugs are active *per se* and are designed to undergo a predictable and controllable metabolic deactivation.

Prodrugs and soft drugs tend to be confused mainly because both are designed to undergo predictable metabolic changes and both tend to rely primarily on enzymatic hydrolysis for these designed-in metabolic changes [52], which, however, as just mentioned, are activation for prodrugs, but deactivation for soft drugs. According to one estimate [48], approximately 5–8% of the marketed drugs are prodrugs; approximately half of them are activated by hydrolysis and approximately a quarter are bioprecursors activated by a biosynthetic reaction. A more recent and more careful analysis, found an even larger proportion of prodrugs: 16% of the small-molecule drugs (192 out of the 1024 unique small molecules identified from the >21,000 FDA-approved products) [11]. Well-known drugs that are in fact prodrugs are, for example, acyclovir, enalapril, omeprazole, lovastatin, simvastatin, and ximelagatran; most of them, however, are accidental prodrugs as they were not intentionally designed as such, but the proportion of designed prodrugs reaching FDA approval is increasing [50].

Unfortunately, the introduction of the *antedrug* terminology by H. J. Lee in 1982 [53] only created additional confusion. The definition of an "antedrug" [53–56] is essentially the same as that of a soft drug [15,38–40], and it was introduced later. The Latin *ante*- prefix, which is very similar in meaning to the Greek *pro*- prefix (e.g., prior to, precedent), implies the conceptual opposite of a soft drug, that is, an inactive agent that has to be activated by metabolism. Hence, the "antedrug" terminology should be dropped to avoid further confusion since, as the authors themselves recognized it [55], prodrugs and antedrugs are "two diametrical approaches in designing safer drugs."

2.2.3. Chemical Delivery System Versus Prodrug The *CDS* concept evolved from the *prodrug* concept, but became essentially different by the introduction of targetor moieties and by the employment of multistep activation. Prodrug approaches may often be hampered by problems due to poor selectivity, poor retention, and the possibility for reactive metabolites, which may often end up decreasing the TI of drugs masked as prodrugs. Because of the introduction of additional complexity (chemical and/or enzymatic stability issues, the possibility of species-dependent metabolism, possible toxicity of the promoiety, etc.), prodrug approaches are worth considering whenever pharmacokinetically problematic moieties are an essential part of the molecule and are required for the desired biological activity [48], but, in general, they should be used only after careful consideration of all alternatives [47] even when used to improved oral bioavailability, one of the most popular uses of this strategy. CDSs were developed starting in the early 1980s to address such challenges [24,29–35]. In addition to functional moieties, which are also contained by prodrugs to provide protected or enhanced overall delivery, CDSs also contain targetor moieties responsible for targeting, site-specificity, and lock-in. CDSs are designed to undergo sequential metabolic conversions, disengaging the modifier functions and finally the targetor, after this moiety fulfills its site- or organ-targeting role.

3. SOFT DRUGS

As mentioned, soft drugs are new, active therapeutic agents, often isosteric–isoelectronic analogs of a lead compound, with a chemical structure specifically designed to allow predictable metabolism into inactive metabolites after exerting their desired therapeutic effect [15–21,24,27,28,57]. Accordingly, SDs are specifically designed by incorporating into their molecular structure specific structural elements, in addition to those required for activity, that are responsible for an optimized deactivation and detoxification route. In general, the desired activity is local, and the SD is applied or administered at or near the site of action. Therefore, in most cases, SDs produce pharmacological activity locally, but their distribution away from the site results in a prompt metabolic deactivation that prevents any kind of undesired pharmacological activity or toxicity—Fig. 2 provides a schematic illustration.

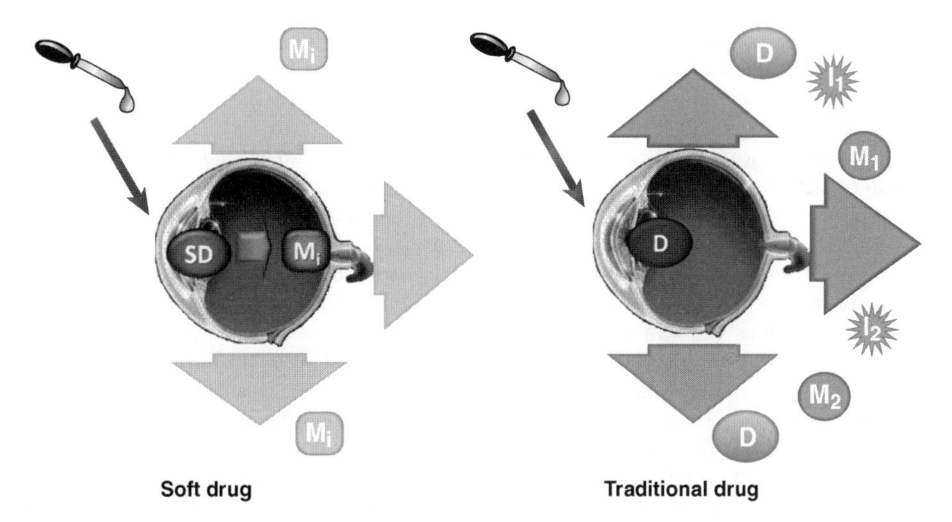

Figure 2. Illustration of the difference between soft and traditional drugs for ocular drugs. For soft drugs, the designed-in metabolism, which generates an inactive metabolite M_i, rapidly deactivates any fraction that reaches the systemic circulation; hence, the local effect is accompanied by no or just minimal side effect.

In SD design, the goal is not to avoid metabolism, but rather to control and direct it. Inclusion of a metabolically sensitive moiety into the drug molecule makes possible the design and prediction of the major metabolic pathway and makes possible the avoidance of the formation of undesired toxic, active, or high-energy intermediates. If possible, inactivation should take place as the result of a single, low-energy, and high-capacity step that yields inactive species subject to rapid elimination. Most critical metabolic pathways are mediated by oxygenases, which is not surprising since, as Adrian Albert already noticed, "an organism's normal reaction to a foreign substance is to burn it up as food" [58]. An analysis of the top 200 drugs found metabolism as the listed clearance mechanism for approximately 75% of them, and it was predominantly oxidative: of the drugs cleared via metabolism, approximately 75% are metabolized by cytochrome P450 (CYP) enzymes, mainly CYP3A (46% of all CYP-mediated ones), CYP2C9 (16%), CYP2C19 (12%), CYP2D6 (12%), and CYP1A (9%) [59,60]. Because oxygenases exhibit not only interspecies but also interindividual variability, and are subject to inhibition and induction [61], and because the rates of hepatic monooxygenase reactions are at least two orders of magnitude lower than the slowest of the other enzymatic reactions [62], it is usually desirable to avoid oxidative pathways as well as these slow, easily saturable oxidases. Accordingly, the design of soft drugs should be based on moieties inactivated by hydrolytic enzymes. Rapid metabolism can be more reliably carried out by ubiquitously distributed esterases. Not relying exclusively on metabolism or clearance by organs such as liver or kidney is an additional advantage because blood flow and enzyme activity in these organs can be seriously impaired in critically ill patients.

3.1. Enzymatic Hydrolysis

Most carboxylic ester-containing chemicals are very efficiently hydrolyzed into the respective free acids by a class of enzyme designated as carboxylic ester hydrolases (EC 3.1.1). Unfortunately, because these widely occurring enzymes exhibit broad and overlapping substrate specificity toward esters and amides, and because, in many cases, their exact physiological role remains unclear, their classification is difficult [63–69]. According to an older but still used classification system [70], the more important subclasses include carboxylesterase, EC 3.1.1.1 (carboxylic-ester hydrolase, ali-esterase, B-esterase, monobutyrase, cocaine esterase, and so on); arylesterase, EC 3.1.1.2 (A-esterase, paraoxonase); acetylcholinesterase, EC 3.1.1.7 (choline esterase I); and cholinesterase, EC 3.1.1.8 (choline esterase II, pseudocholinesterase, butyryl-choline esterase, benzoylcholinesterase, and so on). A possible solution to the existing confusion might come from accumulation of adequate sequence information and emergence of a novel classification system based on this information [69]. According to the phylogenetic alignment of genes from several species, mammalian carboxylesterase isozymes could be classified into four major classes, CES-1–CES-4 [69]. There are two major human carboxylesterases, hCE-1 (CES1A1) and hCE-2 (CES2) [69,71], and a more recently reported third isoenzyme, CES3 [72]. There seems to be some separation in their activity with hCE-1 preferentially hydrolyzing substrates with relatively smaller alcohol parts as compared to the acyl part (e.g., cocaine methyl ester, delapril, meperidine) and hCE-2 preferentially hydrolyzing substrates with relatively bulkier alcohol parts as compared to the acyl part (e.g., cocaine benzoyl ester, heroin, irinotecan) [71,73]. Both hCE-1 and hCE-2 are highly expressed in the liver, and hCE-1 seems more important for clearance through the kidney, whereas hCE-2 seems more important for the clearance of orally administered drugs through the small intestine and colon [71].

Esterases belong to a superfamily of α/β-fold proteins, which contain a similar core of β-sheets connected by α-helices. Esterases have widely varying sequence identities, but share similarities in secondary and tertiary structure and are related by divergent evolution [74,75]. They also seem to rely on a similar mechanism centered on a catalytic triad of which the most common set is Ser-His-Glu (or Asp) and that seems to lie at the base of a deep catalytic gorge. Experimental and modeling data seem to indicate that the active site serine residue lies at the bottom of an approximately 25 Å deep gorge that is negatively charged and relatively narrow (3–4 Å diameter) [76,77]. This can explain why negatively charged (e.g., carboxylic acid-substituted) compounds are poor substrates, which is expected because carboxylic acids are products of hydrolysis and should be repelled from the active site. It might also explain why, in certain cases, relatively small increases in size can completely limit access to the active site and, hence, hydrolysis. On the other hand, the active site of hCE-1 seems to contain both a small rigid pocket, which accommodates the ester moiety, and a large flexible region, which can accommodate structurally diverse side chains resulting in its wide substrate specificity [78,79].

Esterase activity not only depends on the substrate but also shows strong interspecies, interindividual, and interorgan variability [63,69,80–83]. Aliphatic esters (e.g., clevidipine, esmolol, isocarbacyclin methyl ester TEI-9090, remifentanil, and soft cannabinoid analogs) tend to be metabolized much faster by rodents (rats and guinea pigs) than by humans [80,81,84–89]. Aromatic esters, however, might show an opposite trend, as exemplified by flestolol [90] and by nicotinate esters [91]. Therefore, compared to the usual problems related to the extrapolation of animal test results to humans [92,93], preclinical evaluation of ester-based SDs might be even more challenging and animal data less predictive of human clinical trial results. Contrary to the two most frequently used animal models—rodents, which, in general, tend to hydrolyze faster than humans, and dogs, which, in general, tend to hydrolyze slower than humans (at least for aliphatic esters) [81,89], monkeys might serve as better models for the hydrolytic metabolism of humans [80,94].

Despite their metabolic sensitivity, ester structures are of sufficient chemical stability to provide the shelf life necessary for marketable drugs. One of the earliest therapeutic agents that made use of the advantages attainable by introduction of an ester moiety into the structure was etomidate (Amidate™) (**1**, Fig. 3). This is a unique short-acting nonbarbiturate hypnotic agent discovered in 1964 [95]. It is eliminated by ester hydrolysis in plasma and liver [96]. Etomidate is a potent intravenous (i.v.) hypnotic agent with a very rapid onset of action. However, its acid metabolite is inactive, and the duration of hypnosis after etomidate administration can be very short (<5 min) [97]. Therefore, the TI of etomidate (18.0–32.0) is considerably larger than that of other hypnotic agents, such as thiopental (2.5–4.3) or methohexital (4.9–11.7) [97].

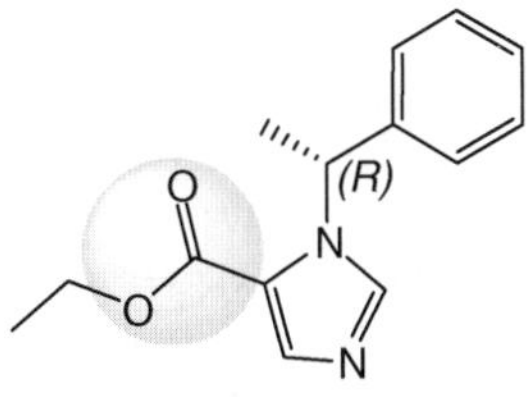

1
Etomidate

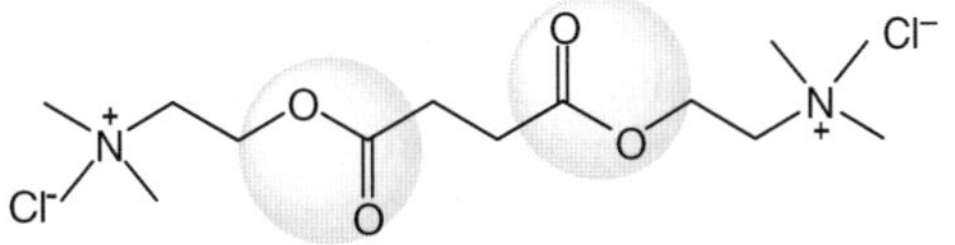

2
Succinylcholine (chloride)

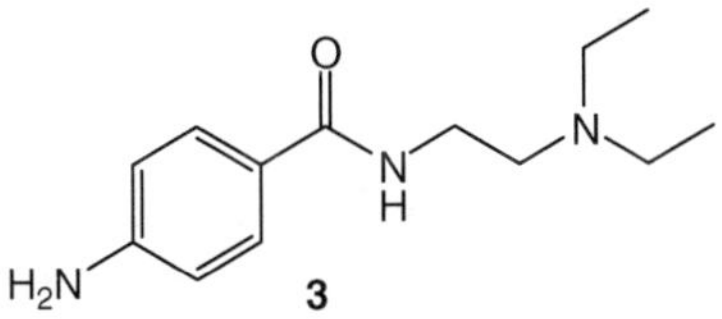

3
Procainamide ($t_{1/2}$=~3h)

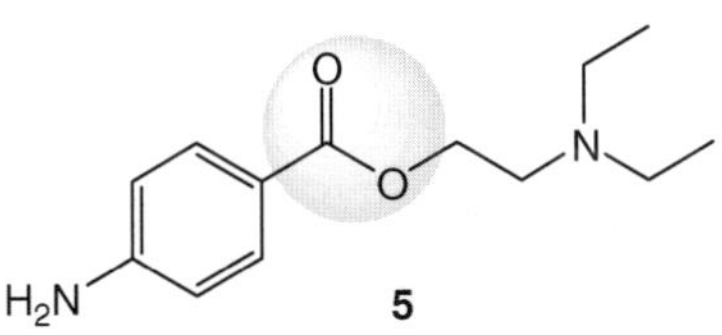

5
Procaine ($t_{1/2}$=~1min)

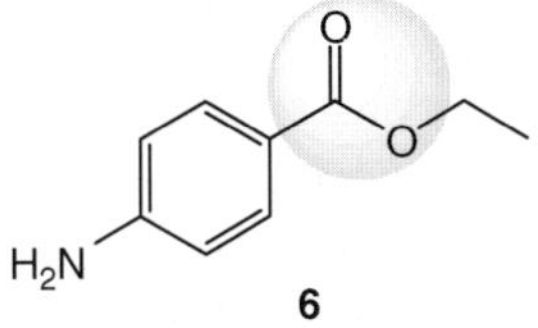

6
Benzocaine

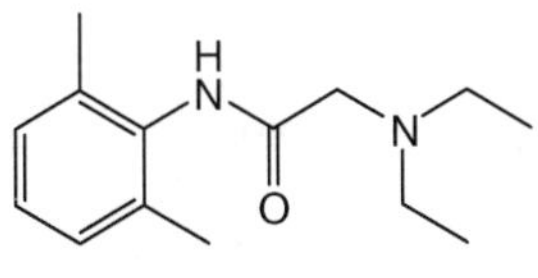

4
Lidocaine

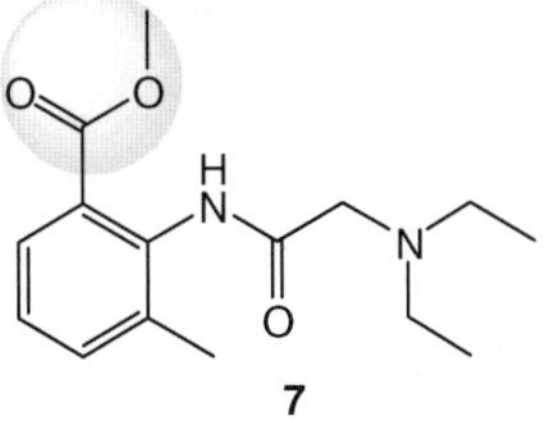

7
Tolycaine

Figure 3. Etomidate (**1**), a short-acting hypnotic agent, and succinylcholine (**2**), a short-acting depolarizing neuromuscular agent, were early examples of compounds with short-acting and safe character provided by esterase-mediated hydrolysis. As another example, ester-containing local anesthetics such as procaine (**5**) or benzocaine (**6**) typically have shorter duration of action then their corresponding amide-containing analogs, such as procainamide (**3**) or lidocaine (**4**), because of their susceptibility to hydrolytic degradation. (This figure is available in full color at http://mrw.interscience.wiley.com/emrw/9780471266945/home.)

Short-acting ester-containing neuromuscular drugs, such as succinylcholine (Anectine™, **2**, Fig. 3) and mivacurium chloride (Mivacron™) [98] designed to undergo hydrolysis by human plasma cholinesterase, also exploit similar principles to ensure fast and spontaneous recovery upon cessation of administration. Hence, their durations of action are only 6–8 and 12–18 min, respectively, and the corresponding elimination routes remain functional even in renal failure. Local anesthetics provide another example. They can be considered as being composed of three main building blocks: a lipophilic (usually aromatic) group and an ionizable (usually tertiary amine) group connected by an alkyl/alylene chain that incorporates either an amide or an ester linkage (Fig. 3) [99]. Because of their susceptibility to hydrolytic degradation, ester local anesthetics such as procaine (**5**), benzocaine (**6**), or cocaine, typically have a shorter duration of action then amide local anesthetics, such as lidocaine (**4**), bupivacaine, or prilocaine. For example, procaine (**5**,

Novocain™, originally derived from "novus" and "caine" from cocaine) has a very short half-life of approximately 1 min as it is hydrolyzed to *para*-aminobenzoic acid (PABA, which may cause allergic reactions and inhibit the action of sulfonamides), whereas the analogous procainamide (**3**) has a considerably longer half-life of approximately 3 h. On a related note, tolycaine (**7**, Baycaine™), which has also been used as a local anesthetic, can be considered a soft analog of lidocaine (**4**) with a methyl ester in **7** replacing the methyl functionality in **4** (as will be discussed later).

3.2. Soft Drug Approaches

The soft drug concept was first introduced in 1976 [15] and reiterated in 1980 [38–40]. Since then, a total of five major SD approaches have been identified [16,17,20,21,24,25,27]:

1. *Inactive Metabolite-Based Soft Drugs*: active compounds designed starting from a known (or hypothetical) inactive metabolite of an existing drug by converting this metabolite into an isosteric/isoelectronic analog of the original drug in such a way as to allow a facile, one-step controllable metabolic conversion after the desired therapeutic role has been achieved back to the very inactive metabolite the design started from.
2. *Soft Analogs*: close structural analogs of known active drugs that have a specific metabolically sensitive moiety built into their structure to allow a facile, one-step controllable deactivation and detoxication after the desired therapeutic role has been achieved.
3. *Active Metabolite-Based Soft Drugs*: metabolic products of a drug resulting from oxidative conversions that retain significant activity of the same kind as the parent drug. If activity and pharmacokinetic considerations allow it, the drug of choice should be the metabolite at the highest oxidation state that still retains activity.
4. *Activated Soft Drugs*: a somewhat separate class derived from nontoxic chemical compounds activated by introduction of a specific group that provides pharmacological activity. During expression of activity, the inactive starting molecule is regenerated.
5. *Prosoft Drugs*: inactive prodrugs (chemical delivery forms) of a soft drug of any of the above classes including endogenous soft molecules. They are converted enzymatically into the active soft drug, which is subsequently enzymatically deactivated.

Among these approaches, the *inactive metabolite-based* and the *soft analog* approaches have been the most useful and successful strategies for designing safe and selective drugs; they overlap somewhat and are not always clearly distinguishable. Both of these approaches focus on designing compounds that have a moiety that is susceptible to metabolic, preferentially hydrolytic, degradation built into their structure. This allows a one-step controllable decomposition into inactive, nontoxic moieties as soon as possible after the desired role is achieved and avoids other types of metabolic routes. Of course, judicious combination of *de novo* (e.g., receptor-based) design principles with general soft drug design principles can also result in *de novo* soft drugs (see, for example, the soft cytokine inhibitors described in Section 3.4.8).

3.3. Inactive Metabolite-Based Soft Drugs

This is a very versatile method for developing new and safe drugs, and, as of today, the most successful SD design strategy. The design process starts from a known or a designed inactive metabolite of a drug used as lead compound. Starting from the structure of this inactive metabolite, novel structures are designed that are isosteric and/or isoelectronic analogs of the drug from which the lead inactive metabolite was derived (*isosteric/isoelectronic analogy*). These new structures are designed in such a way as to yield the starting inactive metabolite in a single metabolic step (*metabolic inactivation*) without any other metabolic conversions (*predictable metabolism*). The specific binding and transport properties as well as the metabolic degradation rates of the new SDs can be controlled by structural modifications (*controlled metabolism*). The entire corresponding design process has been described in detail [28].

How freely one can make structural modifications while designing the new structures depends on how involved are the restrictive pharmacophore regions in the formation of the inactive metabolite. Evidently, if they are not involved, there is more freedom for structural modifications, and one can deviate considerably from the requirement of the isosteric/isoelectronic analogy. Inactive-metabolite-based SDs can also be obtained by starting not from an actual isolated and identified inactive metabolite, but from a designed useful inactive metabolite (a hypothetical inactive metabolite). Obviously, the inactivity of this metabolite will have to be confirmed in a later stage of the developmental process. Nevertheless, the design of such ester-based SDs is greatly enhanced by the fact that the introduction of a carboxylic acid moiety can often fundamentally change the biological activity, especially in smaller molecules. Very often, the initial biological activity is destroyed and the toxicity of the parent compound is greatly reduced [100]. Compare, for example, the toxic aniline (LD_{50} of 0.44 g/kg p.o., rats) with the nontoxic, former antirickettsial *p*-aminobenzoic acid ($LD_{50} > 6.0$ g/kg p.o., rats), the sympathomimetic phenethylamine (LD_{50} of 0.47 g/kg s.c., mouse) with the inactive phenylalanine, the hypnotic *tert*-pentyl alcohol (2-methyl-2-butanol) (LD_{50} of 1.0 g/kg p.o., rats) with the inactive 2,2-dimethylbutyric acid, or even the toxic antiseptic phenol (LD_{50} of 0.53 g/kg p.o., rats) with the nontoxic antiinflammatory salicylic acid. Hence, such acids can often serve as a starting point for hypothetical inactive-metabolite-based SD design approaches if their esterification can restore the original biological activity. Obviously, this is not a generally valid rule as in many structures, especially in larger ones, pharmacological activity is maintained despite the presence of carboxylic groups (e.g., β-lactam antibiotics, antiinflammatory arylacetic acids, or prostaglandins). To illustrate the general inactive metabolite-based SD design principles and their specific applications, a number of drug classes will be reviewed starting with those that already resulted in marketed drugs.

3.3.1. Soft β-Blockers β-Adrenergic blockers provide many examples, because in this class inactive metabolite-based SDs can be obtained by introducing the hydrolytically sensitive functionality at a flexible pharmacophore region; therefore, there is considerable freedom for structural modifications. Consequently, transport and metabolism properties can be controlled more easily. By blocking the cardiac β-adrenergic receptors, β-blockers protect the heart from the oxygen-wasting effect of sympathetic inotropism; an effect that is utilized prophylactically in angina pectoris to prevent myocardial stress that could trigger an ischemic attack [101]. They are also used to lower the cardiac rate and to protect the failing heart against excessive sympathetic drive, to lower elevated blood pressure, and to manage elevated intraocular pressure in glaucoma following topical application to the eye. Since the introduction of propranolol in the late 1960s, many other agents became commercially available including, for example, (in a more or less chronological order) alprenolol, timolol, metoprolol, nadolol, penbutolol, betaxolol, bopindolol, esmolol, and nebivolol [101].

Figure 4 compares the metabolism of the well-known β-blocker metoprolol (**8**) with that of the corresponding soft drugs (**13**) designed starting from one of its inactive metabolites (**12**). Metoprolol is extensively metabolized by the hepatic monooxygenase system both at the more restrictive β-amino alcohol pharmacophore region (pathway A resulting in **9**) and at the more flexible pharmacophore region *para* to the phenol ring (pathways B and C resulting in **10** and **11**, respectively) [102–104]. Two of these metabolites, α-hydroxymetoprolol (**10**) and *O*-demethylmetoprolol (**11**), have selective β_1-blocker activity, but are five to ten times less potent than metoprolol itself [102]. The main metabolites detected are, however, the acids **9** and **12**, and they are inactive in the sense that they are devoid of β-adrenoceptor activity or toxicity (LD_{50} in mice is greater than 500 mg/kg i.v.) [102]. Hence, the phenylacetic acid derivative **12** can be used as

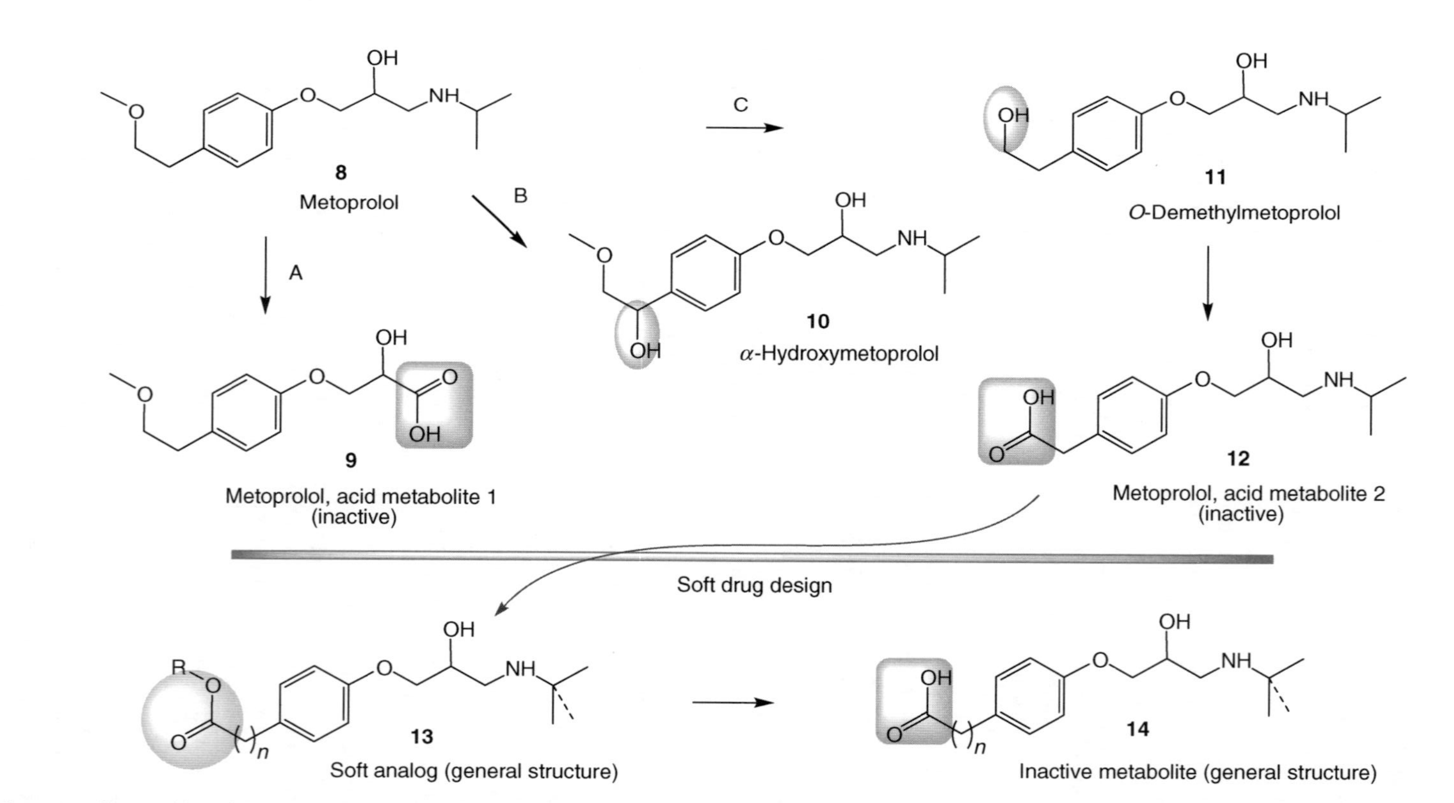

Figure 4. Comparison of the metabolism of metoprolol (**8**) with that of the corresponding soft drugs (**13**) designed starting from one of its inactive acid metabolites (**12**). The dashed line in structures **13** and **14** denotes the possibility of having either isopropyl- or *tert*-butyl-substituted amines. (This figure is available in full color at http://mrw.interscience.wiley.com/emrw/9780471266945/home.)

starting point for an inactive metabolite approach. By esterification of **12** and by introduction of some additional flexibility in the design (e.g., $n=0$ or 2 and not just $n=1$), a number of soft β-blockers structures (**13**) can be obtained with different receptor binding, transport, rate of cleavage, and metabolic properties.

Adaprolol SDs are ideally suited for ophthalmic applications (Fig. 2), which is especially important as most ophthalmic drugs were not designed and developed specifically for the treatment of eye diseases [105]. Despite the apparent easy accessibility of the eye for topical treatments, various defense mechanisms make it difficult to achieve an effective, and if possible localized, concentration within the eye. The cornea is the main biological barrier to drug penetration, and it is very effective as the corneal epithelium has tight annular junctions. In addition, topically applied drugs are rapidly eliminated (washed away by the tear flow) from the precorneal area that anyway can hold only a relatively limited volume. Various approaches have been tried to circumvent this problem of low ocular delivery and potential for substantial systemic side effects [106]; SDs are particularly well suited. β-Adrenergic blockers were found to be beneficial in reducing the elevated intraocular pressure (IOP) associated with glaucoma, and they are now a major class of drugs used in glaucoma patients (e.g., timolol, betaxolol, carteolol). However, β-blockers not only retain their cardiovascular activity when applied topically to the eye and are responsible for significant cardiovascular side-effects but even major respiratory events have been reported in association with the ophthalmic use of various β-blockers [107–109].

If membrane transport, which tends to be lipophilicity related, and relative stability are important to achieve pharmacological activity (i.e., they are needed for good corneal permeability), then the R group of **13** should be relatively lipophilic and impart ester stability. From the various soft β-blockers developed in our laboratory ($n=1$), adaprolol, the adamantane ethyl derivative (**16**; Fig. 5), was selected as a candidate for a topical antiglaucoma agent [110–114]. Adaprolol produces prolonged and significant reduction of IOP, but it hydrolyzes relatively rapidly [111,112]. Therefore, local activity can be separated from undesired systemic cardiovascular/pulmonary activity, a characteristic much sought after in the search for antiglaucoma therapy [115]. Following unilateral ocular treatment with adaprolol, no effects are produced in the contralateral eye because of systemic inactivation.

Several clinical studies of adaprolol maleate have already been completed, and no severe or clinically significant medical events have been reported. A double masked comparison of adaprolol and timolol performed on ocular hypertensive patients (IOP >21 mmHg, $n=67$) demonstrated that intraocular pressure was significantly reduced throughout the study in all treatment groups. Adaprolol reduced IOP by approximately 20%, while timolol reduced IOP by 25–30% [27,105]. In patients older than 70 years, the IOP-reducing effects of 0.2% adaprolol and 0.5% timolol were statistically indistinguishable after 10 days of application (Fig. 6) [27,105]. On the other hand, timolol reduced the systolic blood pressure with statistical significance, while neither of the adaprolol concentrations tested demonstrated such change (Fig. 6). Timolol also showed a trend, though, not statistically significant, to reduce the heart rate, while pulse was conserved in both adaprolol treatment groups. Hence, adaprolol has a safer cardiovascular profile than timolol, especially in the population elder than 70 years.

If systemic ultrashort-action (USA) is the objective, then R groups that make **13** susceptible toward rapid hydrolysis should be used. With such agents that have short half-lives ($\sim$15 min), steady-state plasma concentrations and readily adjustable effects can be rapidly achieved on i.v. administration and infusion. Also, drug effects can be rapidly eliminated by termination of the infusion. For example, a number of methyl–thiomethyl and related esters (**13**; $n=1$, $R=CH_2SCH_3$, CH_2SOCH_3, $CH_2SO_2CH_3$) were found as ultrashort acting [116]. When injected intravenously, these compounds hydrolyzed much faster than simple alkyl esters. Various other soft β-blockers within this family of general structures (**13**; $n=0$–3), have been developed

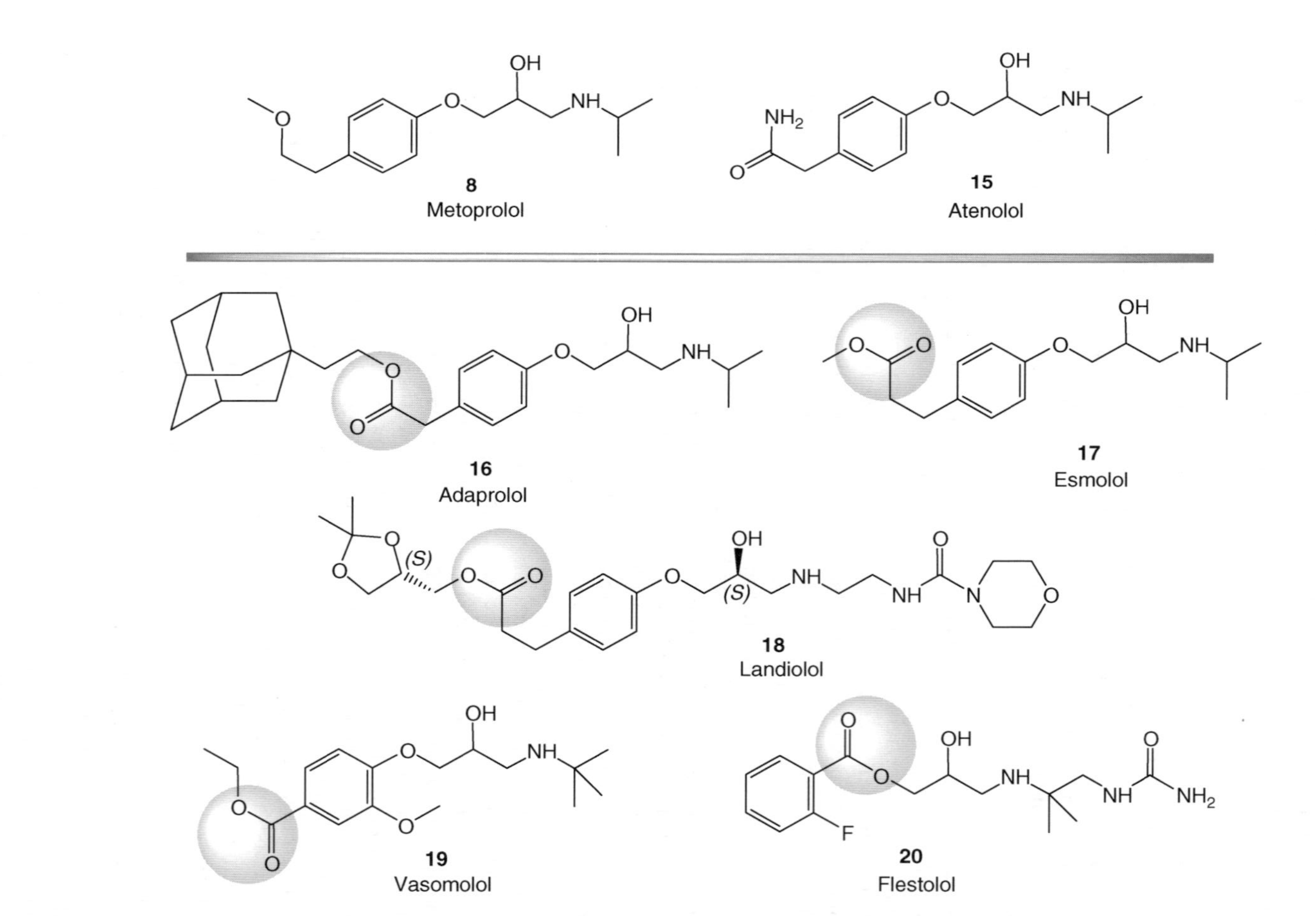

Figure 5. Representative ester-containing β-blocker structures (**16–20**) compared to metoprolol (**8**) and atenolol (**15**). (This figure is available in full color at http://mrw.interscience.wiley.com/emrw/9780471266945/home.)

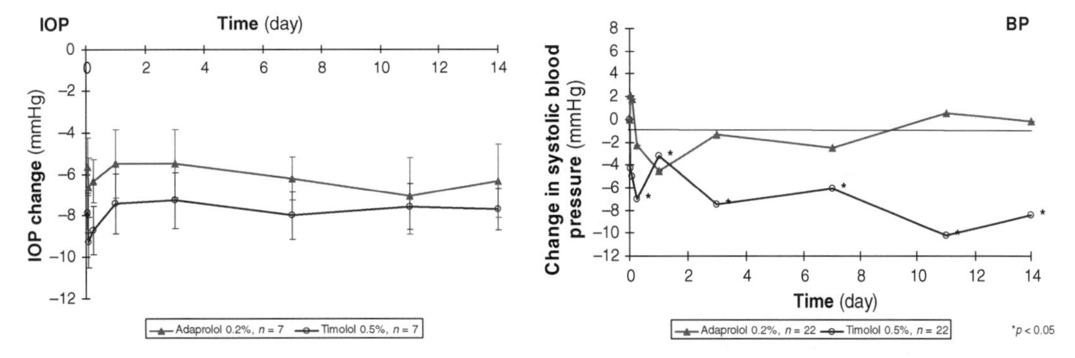

Figure 6. Comparison of the changes caused by administration of adaprolol (0.2%) and timolol (0.5%) in the IOP and systolic blood pressure (BP) of ocular hypertensive patients. IOP data shown is for patients older than 70 years. (This figure is available in full color at http://mrw.interscience.wiley.com/emrw/9780471266945/home.)

and tested in different laboratories. A number of these designs have proven quite successful; they will be briefly reviewed in the following subchapters.

Esmolol Esmolol (Brevibloc™, **17**) is an ultra-short-acting β-blocker also designed to rely on rapid metabolism by serum esterases [117,118]. By the late 1970s, it was already known that insertion of an ester moiety between the aromatic ring and the β-amino alcohol side chain [119] or at the more remote *para* position [120] might not affect β-blocking activity significantly. It was also shown that the acid metabolites **14** ($n = 0, 1$) are devoid of β-adrenoceptor activity [102,120]. Somewhat later, following a systematic search of different β-blocker series that contained ester moieties inserted at different positions, esmolol was selected as the best candidate for development [117,121–123].

The duration of action for the compounds of the general structure **13** in these series, decreased as n increased from 0 to 2, probably because of increasing degradation by hydrolysis with decreasing steric hindrance. Esmolol ($n = 2$, $R = CH_3$; **17**, Fig. 5) was the fourth β-blocker approved by the FDA (1986) for i.v. clinical use, but it was different from the previous three (propranolol, metoprolol, labetalol) because its pharmacological effects dissipate within 15–20 min after stopping the drug [124]. The elimination half-life of esmolol (5–15 min) is indeed considerably shorter than that of propranolol (2–4 h) [124]. Its acid metabolite formed by hydrolysis of the ester group (**14**, $n = 2$) has a relatively long half-life (3.7 h), but it is indeed inactive. It has a β-adrenoceptor antagonist potency approximately 1500-fold lower than esmolol, and it is unlikely to exert any clinically significant effects during esmolol administration [124].

The *in vitro* human blood half-life of esmolol is approximately 25–27 min [84,85]. As a nice confirmation of the SD design principles, the presence of other ester-containing drugs, such as acetylcholine, succinylcholine, procaine, or chloroprocaine, have been shown to have no effect on this hydrolytic half-life, and consequently no metabolic interactions are to be expected [84].

Landiolol Landiolol (ONO-1101, Onoact™, **18**) is a more recently developed USA β-blocker with improved cardioselectivity [125–128]. Modification of the R ester group of esmolol did not afford compounds superior to esmolol in β-blocking activity, duration of action, or cardioselectivity. However, additional modifications resulted in landiolol, which, compared to esmolol, has a nine times increased *in vivo* β-blocking activity and an eight times increased *in vitro* β_1/β_2 cardioselectivity (255 versus 32) [125]. Landiolol (**18**) contains a morpholinocarbonylamino moiety and has *S*-configured hydroxyl and ester functions. It has proved to be a potent β-antagonist with effects that are quickly removed by washout [126–128]. Landiolol was found to have an elimination half-life shorter than any other β-blocker (2.3–4.0 min, *in vivo* human) [128]. It is widely used in Japan, and clinical use has confirmed that it is valuable as a bridge toward starting oral β-adrenergic receptor blockers and as an antiarrhythmic agent. Because of its increased β$_1$-selectivity and its rapid onset and offset of action, it is also suitable for intensive care unit (ICU) patients [129].

Vasomolol, Flestolol, and Other Structures Vasomolol (**19**) was developed along quite similar design principles. It is a vanilloid-type USA β$_1$-adrenoceptor antagonist that has vasorelaxant activity and is devoid of intrinsic sympathomimetic activity [130]. As with other similar agents, vasomolol infusion was characterized by a rapid onset of action. Steady state of β-blockade was attained within 10 min after initial infusion, and a rapid recovery from blockade occurred after discontinuation of the infusion [130].

Because relatively early in the study of β-blockers, it was found that insertion of an ester moiety between the aromatic ring and the β-amino alcohol side chain preserves reasonable β-blocking activity [119,123], a variety of such structures have also been synthesized and investigated. Half-lives in blood and liver suggested that ester hydrolysis is the major pathway for the inactivation of these [(arylcarbonyl)-oxy]propanolamines. A bulky 2-CH_3 substituent prevented the hydrolysis of the ester, but the 2-F substituent, which offered minimum steric hindrance but maximum electron-withdrawing effect, was a promising aromatic substituent for short action [123]. Flestolol (**20**, ACC-9089) a compound with such a 2-F substituent was selected for further toxicological evaluation and clinical study. All clinical findings were consistent with the SD hypothesis: flestolol was safe and effective, had a short elimination half-life of approximately 7 min and, thus, a rapid recovery from β-block after termination of infusion, and was cleared mainly by extrahepatic routes [131,132]. Interestingly, an investigation of the metabolism of flestolol and other esters found polymorphic rates of ester hydrolysis in New Zealand white rabbit blood and cornea [90]. Approximately 30% of the animals studied were found as "slow" metabolizing ($t_{1/2} = 17$ min, *in vitro*, blood) and approximately 70% were found as "fast" metabolizing ($t_{1/2} < 1$ min). No such bimodal distribution of esterase activity was found in blood from rats, dogs, and humans or in the aqueous humor and iris–ciliary body complex of rabbits [90].

Another β-antagonist, which has a structure similar to the general structure **13** ($n = 2$) but contains a reversed ester in its *para* positioned side chain (L-653,328) [133], was also claimed to be a soft drug [115]. The ocular instillation of 2% of this drug to human volunteers resulted in a reduction in IOP; however, this was less than that elicited by 0.5% timolol. Nevertheless, in contrast to timolol, there was no evidence of systemic β-adrenoceptor blockade [134]. In addition, cumulative concentrations of L-653,328 up to 4% did not cause bronchoconstriction in asthmatic patients [135]. Despite these, L-653,328 cannot be considered a true soft drug because its hydrolytic cleavage releases an active alcohol, L-652,698. In fact, L-653,328 was originally designed as the acetate ester prodrug of this active alcohol, and both the ester L-653,328 and the alcohol L-652,698 have modest β-receptor blocking activity. The K_i value for displacement of ^{125}I-iodocyanopindolol binding to β_1-binding sites in membrane fractions of rabbit left ventricle is somewhat smaller for the ester: 3.1 versus 5.7 μM, and the more lipophilic ester causes somewhat better IOP lowering [133]. As it turns out, this case represents neither an ideal prodrug nor an ideal SD design. The lack of systemic effects is attributed to the rapid oxidation of the alcohol in the systemic circulation into inactive carboxylic acid metabolites [134].

Soft Bufuralol Analogs Bufuralol (**21**) is a potent, nonselective β-adrenoceptor antagonist with β$_2$ partial agonist properties. Its effectiveness in the treatment of essential hypertension is probably due to a favorable balance of β-blockade and β$_2$-agonist-mediated vasodilation. Bufuralol undergoes complex metabolism in humans, including stepwise oxidation toward an acid metabolite (**24**) through the corresponding hydroxy (**22**) and keto (**23**) intermediates (Fig. 7). These intermediates are still active [136,137] and have different (interestingly, longer) elimination half-lives [138]. Not only does a differential metabolism of the two enantiomers occur but also differences due to genetic polymorphism are encountered [139].

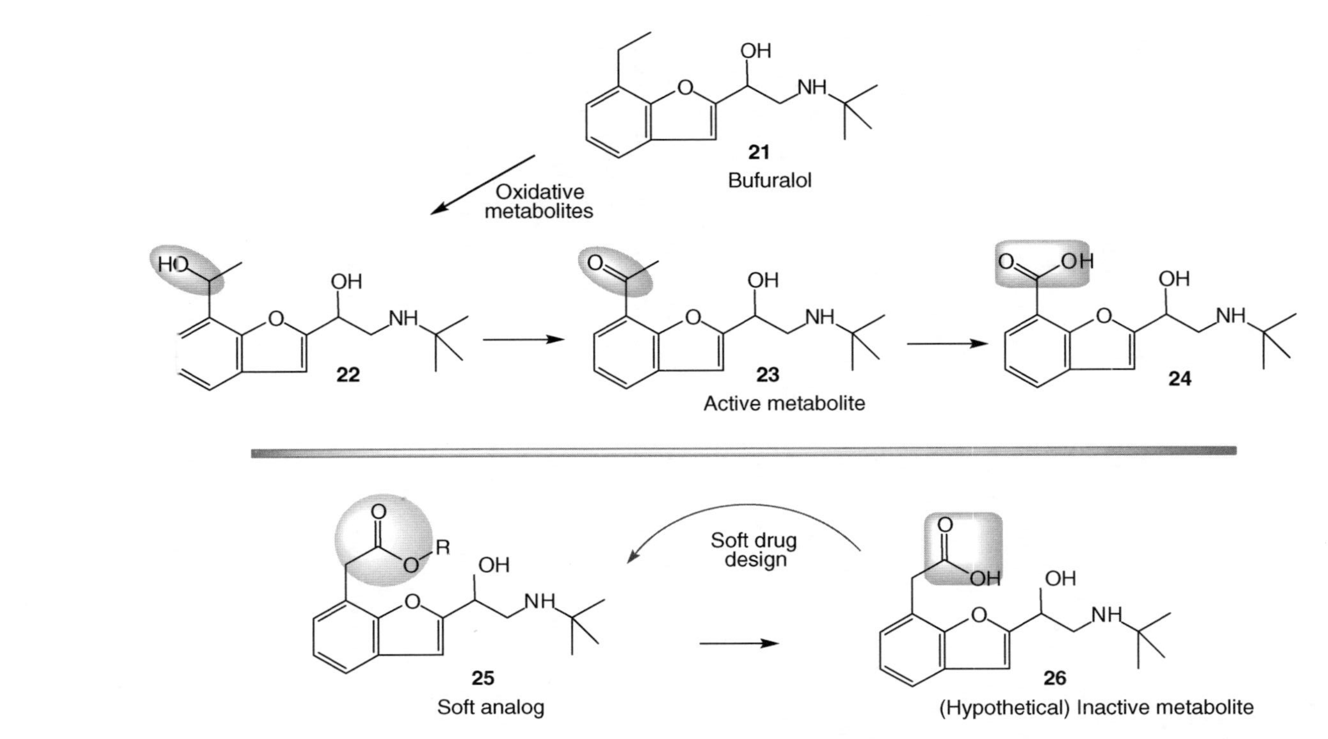

Figure 7. The metabolism of bufuralol (**21**) produces oxidative, active metabolites (**22**, **23**) that lead to a final inactive acid metabolite (**24**). Starting from a designed (hypothetical) inactive metabolite (**26**), a series of inactive metabolite-based soft compounds (**25**) were designed. (This figure is available in full color at http://mrw.interscience.wiley.com/emrw/9780471266945/home.)

A soft drug approach may help avoid these problems, and a number of ester-containing SD candidates (**25**) were synthesized and then tested for β-antagonist activity by recording ECG and intra-arterial blood pressure in rats [140]. This is an example of a hypothetical inactive metabolite-based approach, as the retrometabolic design starts not from an actual, major metabolite, but from a hypothetical one (**26**). Nevertheless, this metabolite was confirmed to be inactive during the study: it did not decrease the heart rate significantly. In the isoproterenol-induced tachycardia model, bufuralol (1 mg/kg, i.v.) diminished the heart rate for at least 2 h, whereas, the effects of equimolar SDs lasted for only 10–30 min (Fig. 8). The effects of the four most active compounds (**25**, R = methyl, ethyl, isopropyl, and *tert*-butyl) on resting heart rate and mean arterial pressure were also evaluated in comparison to esmolol following infusion (Fig. 8). These new SDs produced effects that were similar to that produced by esmolol both in magnitude and time-course, and the corresponding infusion rates were 10-fold smaller.

3.3.2. Soft Opioid Analgetics Several short-acting 4-anilidopiperidine opioids, such as fentanyl (**27a**), sufentanil (**27b**), or alfentanil (Fig. 9), have been introduced into anesthetic practice because they showed advantages over morphine. They do not cause significant histamine release and have shorter durations of action than morphine. Therefore, they can provide greater cardiovascular stability and less persistent postoperative respiratory depression. However, their terminal half-lives in humans are still longer than desired; for example, even the shortest-acting alfentanil has $t_{1/2} \approx 70$–90 min. This can result in drug accumulation and prolonged durations of action after multiple bolus injections or infusion. In addition, hepatic dysfunction may result in prolonged retention because the elimination of these compounds relies on hepatic metabolism [141].

A hypothetical inactive metabolite-based SD approach proved useful in solving these problems. A first attempt to incorporate ester and carbonate moieties into structure **27** (Fig. 9) at the R_2 side chain ($R_2 = -CH_2O_2CR'$, $-CH_2O_2COR'$) produced potent analgetics, but durations of action were still longer than desired [142]. However, another design based on esterification of the hypothetical inactive metabolite **29** yielded remifentanil (**28a**, Ultiva™), a unique ultrashort-acting opioid analgesic [86,143–145].

Even if there is no evidence for the opioid analgetics represented by **27** (Fig. 9) to metabolize into the acids **29** [141], structures of type **28** can represent possible soft, short-acting compounds susceptible to hydrolytic inactivation. During the synthesis and pharmacological evaluation of a number of such compounds [86], it was established that the carfentanil (**27c**) piperidine ($R_2 = CO_2CH_3$) provided more potent analgetics than the fentanyl nucleus ($R_2 = H$). A separation of two methylene units ($n = 2$) between the piperidine nitrogen and the ester function was found as optimal for added potency and decreased duration of action [86]. Durations of actions, as measured *in vivo* by a classic rat tail with-

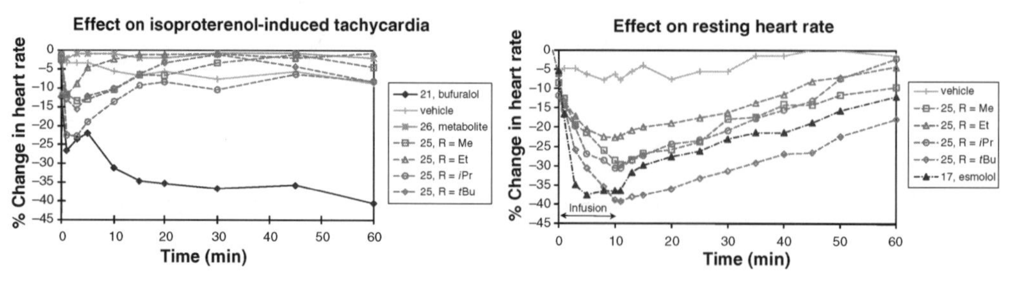

Figure 8. Effects of various bufuralol-related soft drugs (**25**) on isoproterenol-induced tachycardia (i.v. bolus of 3.8 μmol/kg; vehicle 10% DMSO in 30% hydroxypropyl β-cyclodextrin) and resting heart rate (i.v. infusion, 2 μmol/kg/min, R = Et: 4 μmol/kg/min, esmolol: 20 μmol/kg/min; vehicle 0.9% NaCl) in rats. Data represent the mean of at least three animals; error bars were omitted for better visibility. (This figure is available in full color at http://mrw.interscience.wiley.com/emrw/9780471266945/home.)

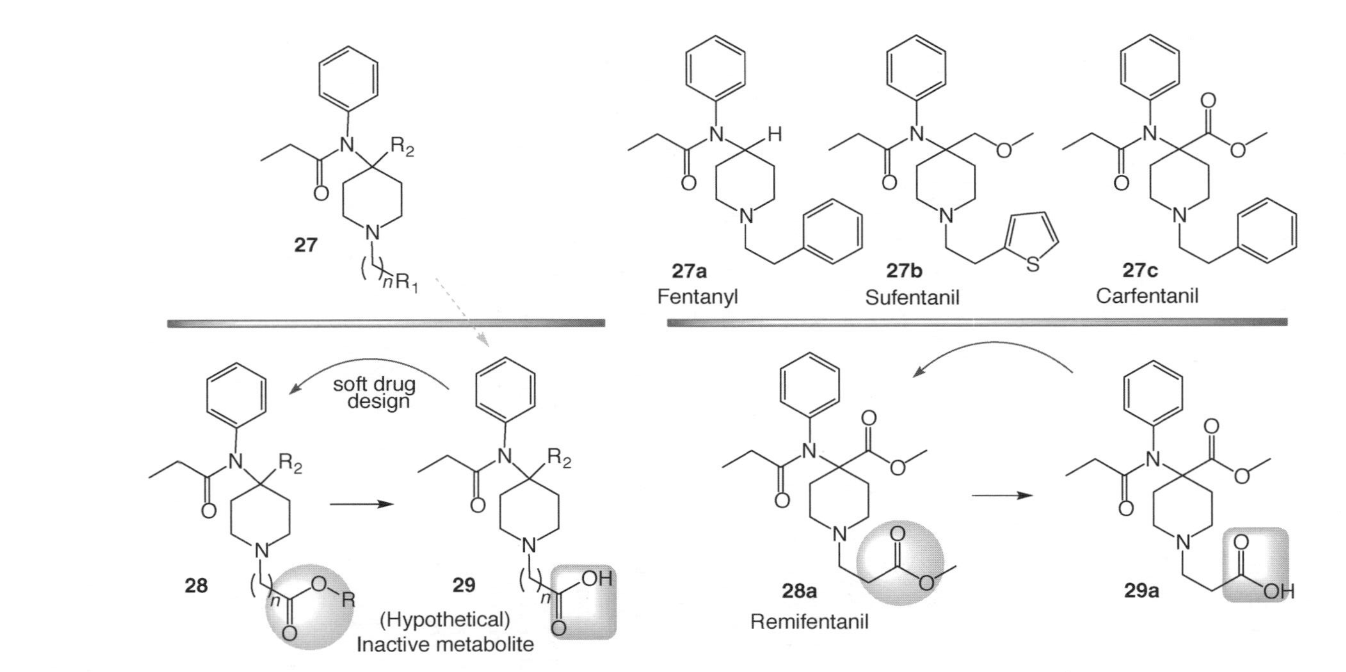

Figure 9. Design of soft opioid analgetics (**28**) based on the hypothetical inactive metabolite (**29**) and the structure of the SD remifentanil (**28a**) compared to that of other opioid analgetics such as fentanyl (**27a**) or carfentanil (**27c**). (This figure is available in full color at http://mrw.interscience.wiley.com/emrw/9780471266945/home.)

drawal assay [146], ranged from extremely short to long (5–85 min) depending upon the substitution of the alkyl portion *R* of the ester. One of these compounds, remifentanil (**28a**, Fig. 9), was approved by the FDA in 1996 for clinical use as an USA opioid analgetic (Ultiva) during general anesthesia and monitored anesthesia care. Remifentanil has a half-life of 37 min in human whole blood (*in vitro*) and is nearly quantitatively converted to the corresponding acid (**29a**). Furthermore, the carboxylic acid (**29a**) was indeed found to be approximately 1000 times less potent in the *in vitro* guinea pig ileum assay and 350 times less potent in the *in vivo* rat tail withdrawal reflex model than its parent drug **28a**. Remifentanil is roughly equivalent in potency to fentanyl in the guinea pig ileum assay (EC_{50}s of 2.4 and 1.8 nM, respectively), and its effect can be antagonized by naloxone, an opiate antagonist [143]. In 24 patients undergoing elective inpatient surgery, its terminal half-life ranged from 10 to 21 min, while that of its major metabolite (**29a**) ranged from 88 to 137 min [145].

In the case of remifentanil it was proved again that the possibility of drug interactions could be minimized by building metabolic considerations into the structure, as predicted by the basic principles of SD design. Clearance, volume of distribution, and terminal half-life data indicated that coadministration of esmolol had no significant ($p < 0.05$) effect on the pharmacokinetics (or pharmacodynamics) of remifentanil in rats despite both drugs being metabolized by nonspecific esterases [147,148].

3.3.3. Soft Corticosteroids Antiinflammatory corticosteroids represent one of the most active fields for SD design. Corticosteroids exert profound biologic effects in almost every organ; hence, they are one of the most widely used drug classes in a bewildering range of clinical diseases—mainly for their anti-inflammatory and immunosuppressive effects [149,150]. Unfortunately, systemic side effects, which typically include myopathy, osteoporosis, hypertension, insulin resistance, weight gain, fat redistribution, increased IOP, growth inhibition, and others in a dose-dependent manner [151–156], seriously limit their application. Even if corticosteroids are applied only topically (e.g., lung, nasal mucosa, gastrointestinal tract, eye, or skin), significant portions still reach the general circulatory system. Consequently, resulting systemic side effects, together with local side effects, such as oral candidiasis or dysphonia, often limit their use. Because most of these potential adverse effects usually arise only following long-term treatment, clinical studies of corticosteroids in the development phase monitor suppression of the hypothalamic–pituitary–adrenal (HPA) axis as a surrogate marker. Corticosteroids are also subject to different oxidative and/or reductive metabolic conversions, and formation of various steroidal metabolites can lead to undesirably complex situations as illustrated by the metabolism of hydrocortisone (**30**) (Fig. 10). A considerable number of attempts were aimed to improve this situation, and SD approaches are particularly well suited for this purpose (Fig. 2).

There is a frequent misconception regarding SDs, and in particularly soft steroids, that has to be clarified. Often, the soft nature is associated with fast hydrolytic degradation, but this is not necessarily so. If hydrolysis is too rapid, then only weak activity may be obtained. The desired increase of the TI can be achieved only if the drug is sufficiently stable to reach the receptor sites at the target organ and to produce its desired effect, but the free, nonprotein-bound drug undergoes facile hydrolysis to avoid unwanted, systemic side effects. In order to successfully separate the desired local activity from systemic toxicity, an adequate balance between intrinsic activity, solubility/lipophilicity, tissue distribution, protein binding, and rate of metabolic deactivation has to be achieved. In the case of slow, sustained release to the general circulatory system from the delivery site, even a relatively slow hydrolysis could result in a very low, almost steady state systemic concentration.

Cortienic Acid-Based Soft Steroids: First Generation (Loteprednol Etabonate and Analogs) Loteprednol etabonate (**41**, LE Figs. 11–13) is an active corticosteroid that lacks serious side effects and that received final FDA approval in 1998 as the active ingredient of two ophthalmic preparations, Lotemax™ and

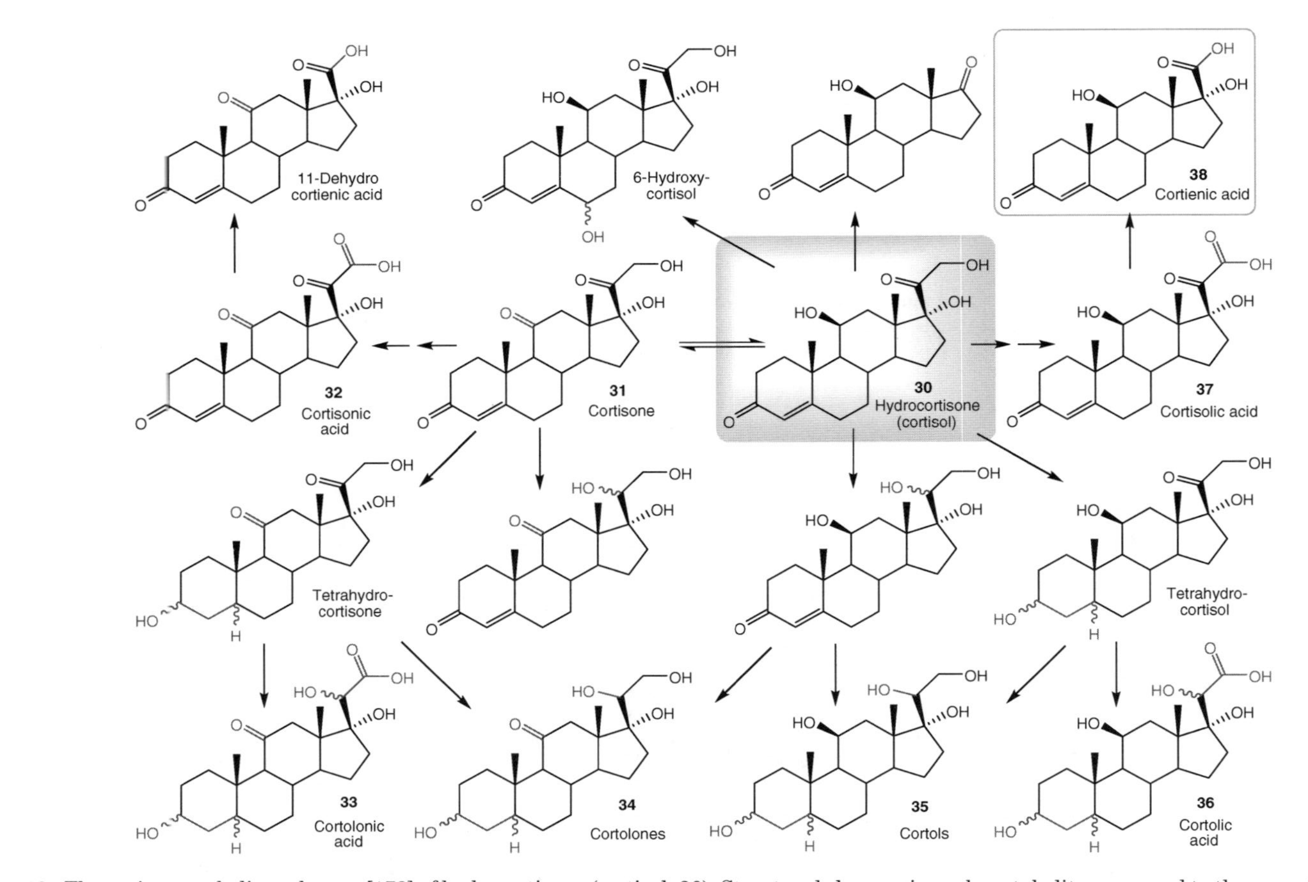

Figure 10. The major metabolic pathways [172] of hydrocortisone (cortisol; **30**). Structural changes in each metabolite compared to the parent hydrocortisone are marked in blue. (This figure is available in full color at http://mrw.interscience.wiley.com/emrw/9780471266945/home.)

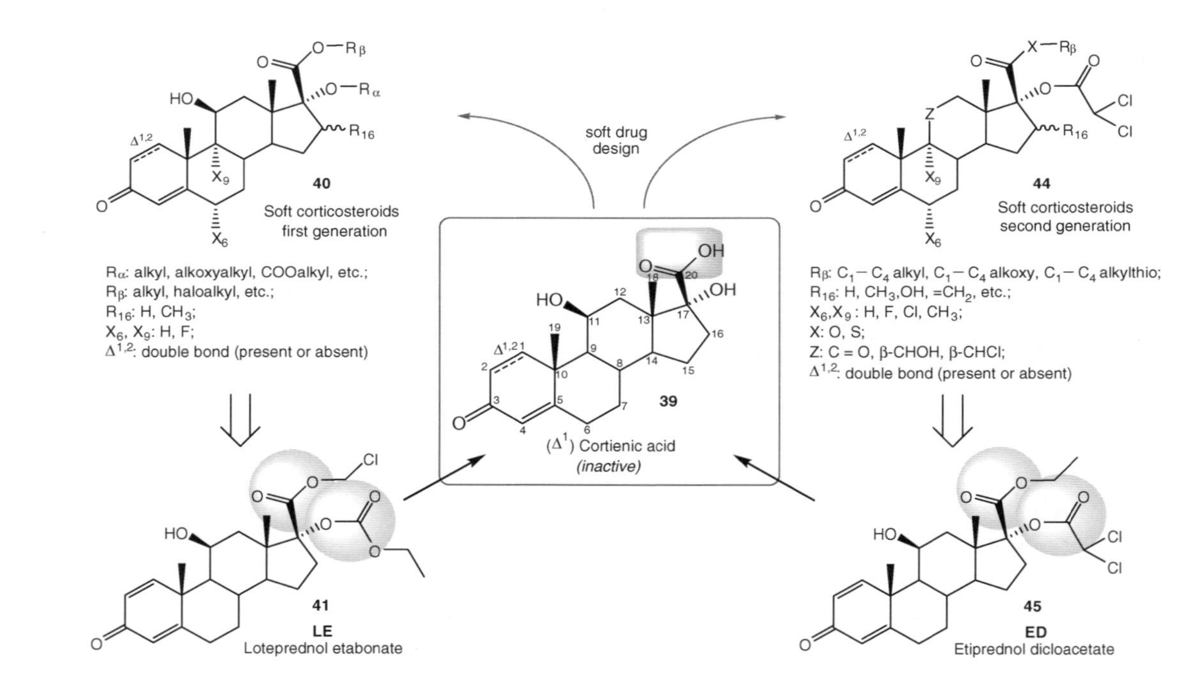

Figure 11. The inactive metabolite (**39**) based design of first- (**40**) and second-generation (**44**) cortienic acid-based soft steroids, and their selected representative compounds loteprednol etabonate (**41**) and etiprednol dicloacetate (**45**), respectively. Structure **39** includes the traditional numbering of the steroid framework. (This figure is available in full color at http://mrw.interscience.wiley.com/emrw/9780471266945/home.)

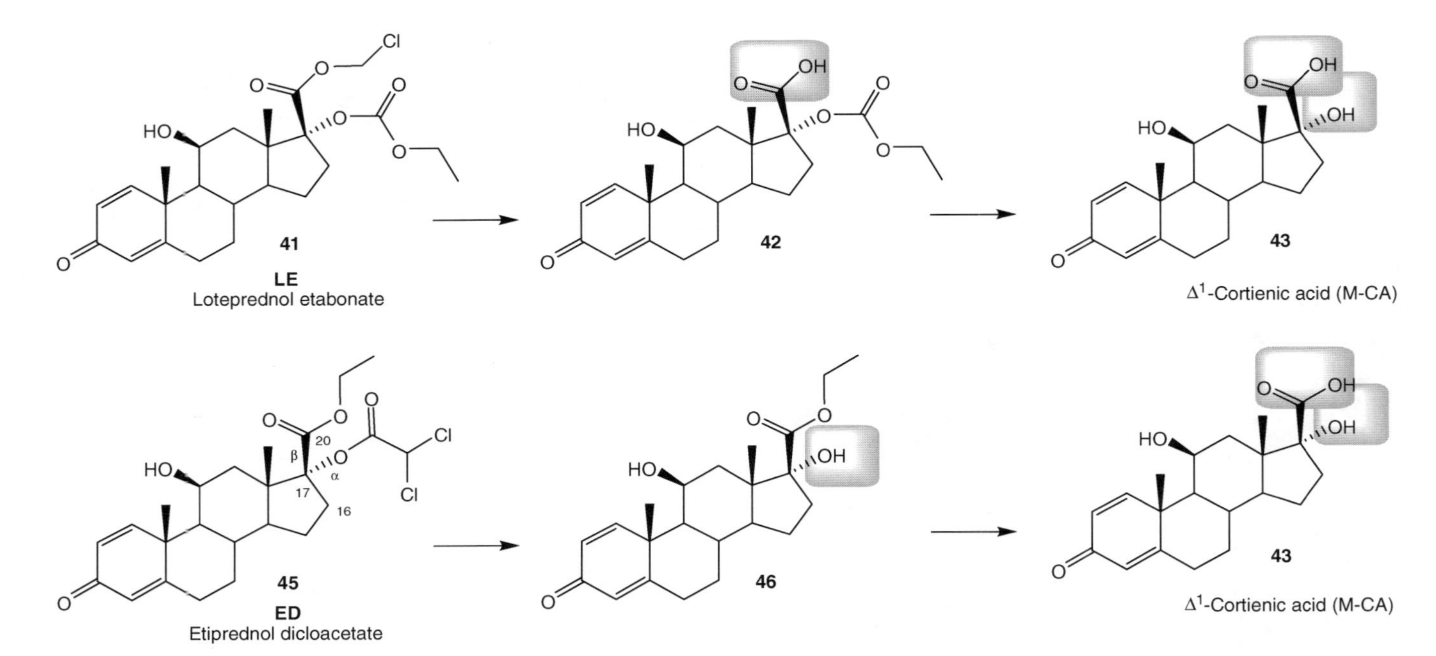

Figure 12. The main inactivating metabolic pathways of loteprednol etabonate (**41**) and etiprednol dicloacetate (**45**) both ultimately leading to the formation of Δ^1-cortienic acid (**43**). (This figure is available in full color at http://mrw.interscience.wiley.com/emrw/9780471266945/home.)

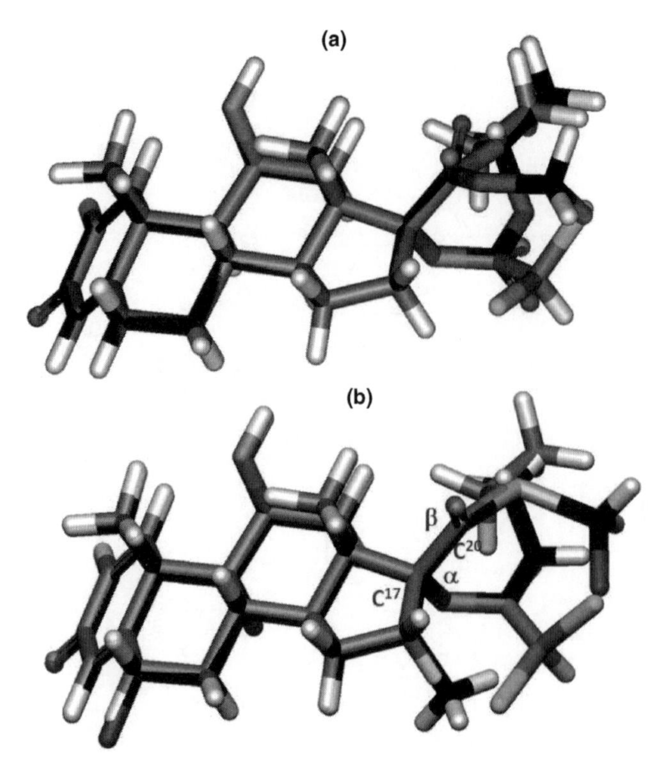

Figure 13. Overlapping pharmacophore structures of etiprednol dicloacetate (**45**) (in lighter colors) with loteprednol etabonate (**41**) (a) and fluticasone propionate (**59**) (b). Both views are from the β side, from slightly above the steroid ring system. (See color insert.)

Alrex™ [157–160]. Currently, it is the only corticosteroid approved by the FDA for use in all inflammatory and allergy-related ophthalmic disorders, including inflammation after cataract surgery, uveitis, allergic conjunctivitis, and giant papillary conjunctivitis. It was later also approved as part of a combination ophthalmic suspension with an anti-infective agent (Zylet®, loteprednol etabonate 0.5%/tobramycin 0.3%). LE resulted from a classic inactive metabolite-based SD approach [28,159,161–171].

As mentioned, hydrocortisone (**30**) undergoes a variety of oxidative and reductive metabolic conversions (Fig. 10) [172]. Oxidation of its dihydroxyacetone side chain leads to formation of cortienic acid (**38**) through a 21-aldehyde (21-dehydrocortisol) and a 21-acid (cortisolic acid, **37**). Cortienic acid is an ideal lead for the inactive metabolite approach because it lacks corticosteroid activity and is a major metabolite excreted in human urine. To obtain active compounds, the important pharmacophores found in the 17α and 17β side chains had to be restored (Fig. 11). Suitable isosteric/isoelectronic substitution of the α-hydroxy and β-carboxy substituents with esters or other types of functions should restore the original corticosteroid activity and also incorporate hydrolytic features to help avoid accumulation of toxic levels. More than 120 of these first-generation soft steroids (**40**) that resulted from modifications of the 17β carboxyl function and the 17α hydroxy function together with other changes intended to enhance corticosteroid activity [e.g., introduction of Δ^1, fluorination at 6α (X_6) and/or 9α (X_9), methylation at 16α or 16β (R_3)] have been synthesized.

The first soft analogs of this kind were synthesized during the late 1970s soon after the introduction of the SD concept [15], followed by a systematic synthetic study performed in collaboration with Otsuka

Pharmaceutical Company (Japan) in 1980–1981 [17,161,173]. A haloester in the 17β position and a novel carbonate [163] or ether [174] substitution in the 17α-position were found as critical functions for activity. Incorporation of 17α carbonates or ethers was preferred over 17α esters to enhance stability and to prevent formation of mixed anhydrides that might be produced by reaction of a 17α ester with a 17β acid functionality. Such mixed anhydrides were assumed toxic and probably cataractogenic. A variety of 17β esters were synthesized. Because this position is an important pharmacophore that is sensitive to small modifications, the freedom of choice was relatively limited. For example, although chloromethyl or fluoromethyl esters showed very good activity, the chloroethyl or α-chloroethylidene derivatives demonstrated very weak activity. Simple alkyl esters also proved virtually inactive. Consequently, the 17β chloromethyl ester was held constant and 17α-carbonates with different substituents on the steroid skeleton were varied for further investigation. LE, and some of the other soft steroids, provided a significant improvement of the TI determined as the ratio between the antiinflammatory activity and the thymus involution activity [159,175,176]. Furthermore, binding studies using rat lung cytosolic corticosteroid receptors showed that some of the compounds approach and even exceed the binding affinity of the most potent corticosteroids known [177,178].

LE (**41**) was selected for development based on various considerations including the TI, availability, synthesis, and "softness" (the rate and easiness of metabolic deactivation). Early studies in rabbits [164,167] and rats [168] demonstrated that, consistent with its design, **41** is indeed active, is metabolized into its predicted metabolites (**42**, **43**) (Fig. 12), and these metabolites are inactive [163]. The PK profile of LE indicated that, when absorbed systemically, it is rapidly transformed to the inactive metabolite **42** and eliminated from the body mainly through the bile and urine [168,169,171]. It did not affect the IOP in rabbits [167] an observation confirmed later in various human studies (Fig. 14) [179]. Consistent with the soft nature of this steroid, systemic levels or effects cannot be detected even after chronic ocular administration [180].

LE has been shown to be a safe and clinically effective treatment for contact lens-associated giant papillary conjunctivitis (GPC), seasonal allergic conjunctivitis, postoperative inflammation, uveitis, and dry eye syndrome [157,158,160]. A recent retrospective study confirmed that, in agreement with its design principles, even long-term use (>12 months) of LE caused no reported adverse effects [181]. LE in combination with tobramycin was also found to be effective in the treatment of ocular inflammation associated with blepharokeratoconjunctivitis [182], while significantly less likely to produce ele-

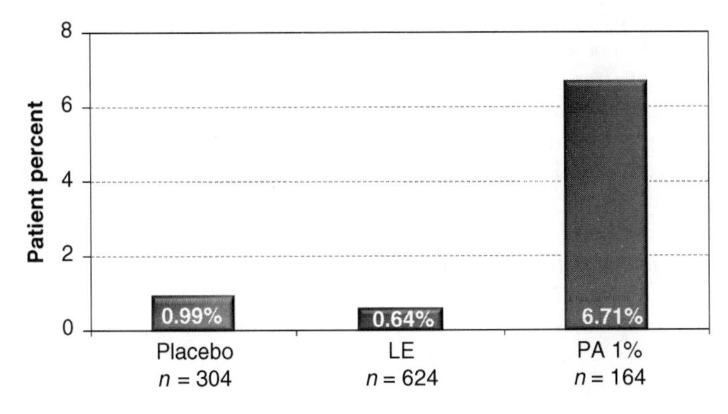

Figure 14. Pooled data showing the percent of patients with IOP elevation greater than 10 mmHg among patients not wearing contact lenses and treated for more than 28 days. The number of patients within each group was as follows: placebo, $n = 304$; loteprednol etabonate (LE; 0.2% and 0.5%), $n = 624$; prednisolone acetate (PA; 1.0%), $n = 164$ [179]. (This figure is available in full color at http://mrw.interscience.wiley.com/emrw/9780471266945/home.)

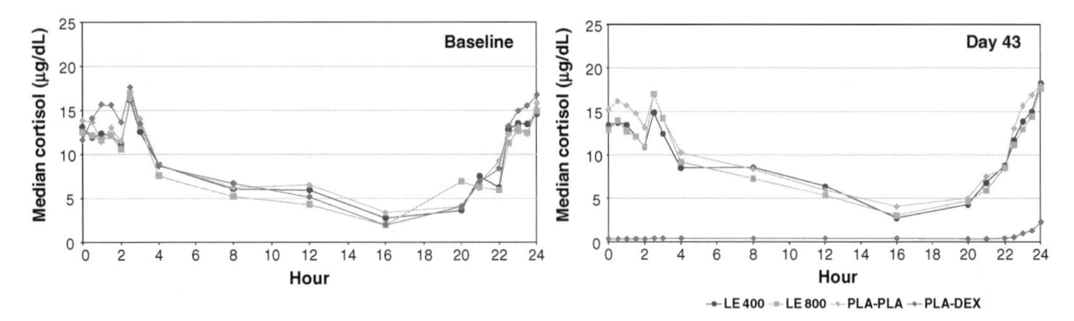

Figure 15. Median cortisol profiles obtained in a randomized, double-blind, placebo- and active-controlled, single center trial with four parallel groups designed to investigate the effects of glucocorticoid nasal sprays on the HPA axis function, as well as safety, general and local tolerability, and PK in patients ($n = 80$) with perennial allergic rhinitis (PAR) in a 6-week study. (See color insert.)

vations in IOP then the dexamethasone/tobramycin combination [183].

Based on promising results from animal studies [159,170,171], LE is also being developed for treatment of asthma, rhinitis, colitis, and dermatological problems; pertaining results have been reviewed recently [184]. It might also be a promising agent for localized immunosuppression, for example, in pancreatic islet transplantations using a biohybrid device as a bioartificial pancreas [185,186]. LE nasal spray for treatment of allergic rhinitis has already been evaluated in a number of clinical studies, and they indicate that LE nasal spray (400 and 800 μg once daily) could be a safe and well-tolerated treatment for up to 2 weeks for allergic rhinitis (Fig. 15) [187,188]. A large ($n = 165$) environmental exposure unit (EEU) study in patients with seasonal allergic rhinitis (SAR) also confirmed that LE 400 μg once daily is effective and superior to placebo: after 14 days of treatment, patient in this LE group had significantly lower total nasal symptom scores (TNSS) than those receiving placebo ($p = 0.007$) [189].

Cortienic Acid-Based Soft Steroids: Second Generation (Etiprednol Dicloacetate and Analogs) More recently, a new, second generation of soft steroids with a unique 17α-dichloroester substituent has been identified (**44**, Fig. 11) [184,190–193]. This is a unique design: no known corticosteroid contains halogen substituents at the 17α position. Nevertheless, the pharmacophore portions of these second-generation soft steroids, including the halogen atoms at 17α, can be positioned so as to provide excellent overlap with those of the traditional corticosteroids (Fig. 13) [177]. Dichlorinated substituents seem required for activity and sufficiently soft nature, and two justifications seem likely. First, with dichlorinated substituents, one of the Cl atoms will necessarily point in the direction needed for pharmacophore overlap, but with monochlorinated substituents, steric hindrance will force the lone Cl atom to point away from this desired direction. Second, whereas dichloro substituents increase the second-order rate constant k_{cat}/K_M of enzymatic hydrolysis in acetate esters by a factor of approximately 20 compared to the unsubstituted ester, monochloro substituents do not cause any change [194].

Contrary to first-generation soft steroids, in this class, hydrolysis primarily cleaves not the 17-β positioned, but the 17α-positioned ester (Fig. 12). Nevertheless, the corresponding metabolites are also inactive. From this series, etiprednol dicloacetate (ED, **45**, Figs. 11 and 12) was selected for development. ED has shown better receptor binding affinity than LE, and was proven as, or even more effective, than budesonide (BUD) in various asthma models (Fig. 16). In agreement with its soft nature, ED was found as having low toxicity in animal models and in human clinical trials [190–192,195]. The transrepressing and transactivating activity of ED and BUD were compared by measuring their inhibition in IL-1β production of a stimulated human monocyte cell line and by evaluating glucocorticoid-

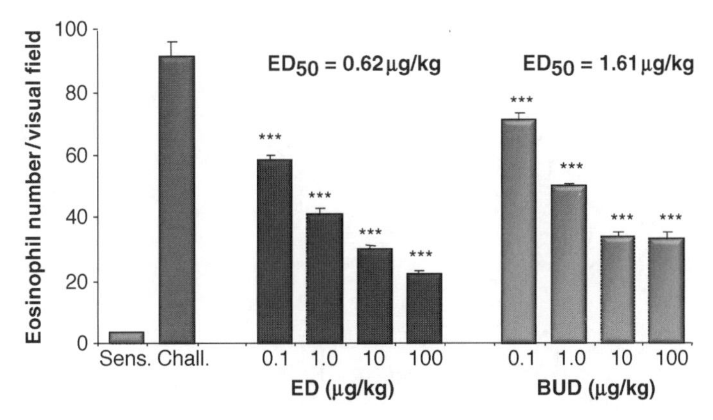

Figure 16. Effect of ED and BUD on antigen-induced airway eosinophil infiltration into the bronchoalveolar tissues of Brown Norway rats (peribronchial eosinophil numbers shown; $n = 4$–25 per treatment group). Sensitized rats were treated intratracheally with different doses (0.1, 1, 10, and 100 μg/kg) of the drugs and 2 h later were challenged with ovalbumin aerosol. The level of significance (Mann–Whitney U test) in case of each drug-treated group compared with vehicle treated challenged controls was $p < 0.001$. (This figure is available in full color at http://mrw.interscience.wiley.com/emrw/9780471266945/home.)

induced increase in the activity of tyrosine-amino-transferase of a rat hepatoma cell line, respectively [192]. ED was found to be a dissociated glucocorticoid, that is, to possess reduced transactivating activity with a preserved transrepressing activity. Transactivation is mediated by binding of the hormone-activated receptor to a defined DNA sequence, called glucocorticoid response element (GRE). This process may account for some of the unwanted effects of glucocorticoids via the increase in expression of genes involved in gluconeogenesis and development of arterial or ocular tensions. Transrepression, which seems to be the main mechanism by which glucocorticoids suppress inflammation, may be the result of binding to negative GREs, but it occurs mainly by interaction with transcription factors (AP-1 and NF-κB), which control the gene of many inflammatory mediators from IL-1β to RANTES. Hence, the dissociation of transactivating and transrepressing activity seen for ED is a likely advantage that may further help in separating the beneficial anti-inflammatory activity from the undesired side effects, and is in line with the development of dissociated steroids, one of the novel mechanistic approaches pursued in development of new inhaled corticosteroids [196,197].

Fluocortin Butyl Fluocortin butyl (**47**, Fig. 17) is an anti-inflammatory steroid obtained in one of the early approaches aimed at integrating ester moieties into steroid structures. Metabolism studies on fluocortolone revealed a number of oxidative and reductive metabolites in human urine [198] including fluocortolone-21-acid, an inactive metabolite. Synthesis and pharmacological evaluation of its different ester derivatives yielded fluocortin butyl (**47**, Vaspit®, Novoderm®, Varlane®), the butyl ester of a C-21 carboxy steroid [199–202]. The ester is an anti-inflammatory agent of rather weak activity, and any portion absorbed systemically following topical application is hydrolyzed into inactive species.

The widespread use of this steroid has been hindered by its low intrinsic activity. The glucocorticoid receptor affinity and the topical anti-inflammatory potency of fluocortin butyl (**47**) are several folds lower than those of dexamethasone [200]. Fluocortin butyl ameliorated allergic rhinitis at daily doses of 2–8 mg divided into two to four daily inhalations [203,204], but it did not protect against bronchial obstruction in bronchial provocation tests even at 8 mg doses divided into four daily inhalations, in contrast to a 10-fold lower dose of beclomethasone dipropionate (**61**) [205].

Itrocinonide A soft steroid series containing 17β methyl-carbonate ester moieties susceptible toward hydrolysis was also developed during the 1980s [206]. Since unsubstituted 17β esters tend not to be very susceptible

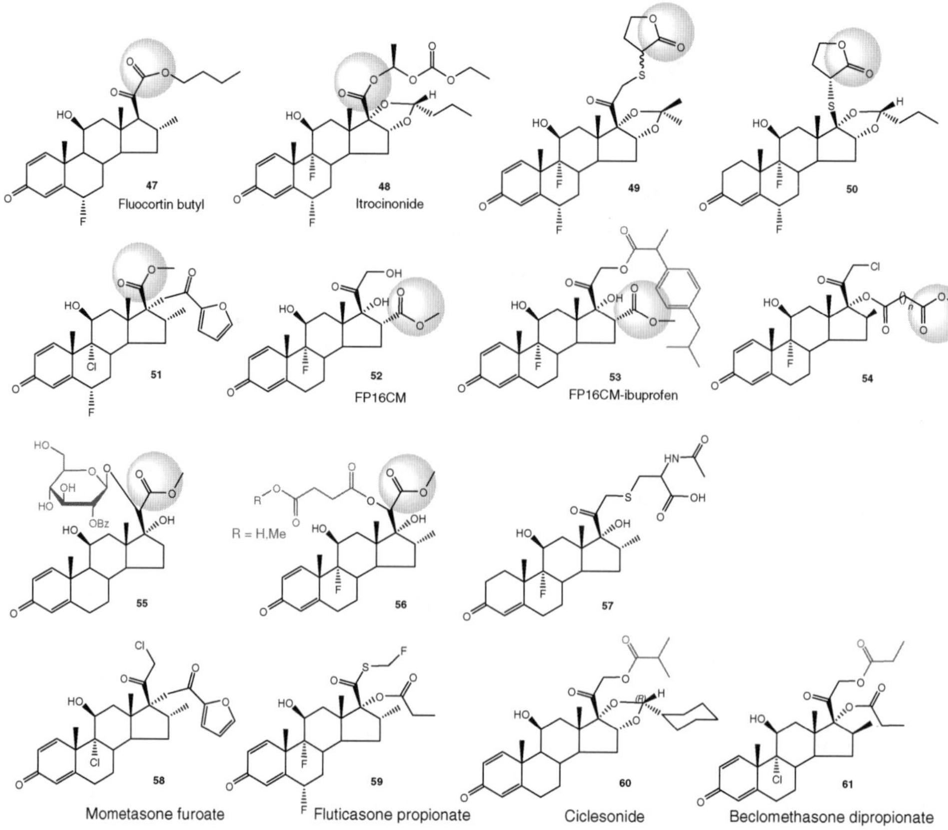

Figure 17. Corticosteroid structures discussed in the text. (This figure is available in full color at http://mrw.interscience.wiley.com/emrw/9780471266945/home.)

toward hydrolysis due to their steric hindrance, in these structures, the metabolically sensitive spot has been moved farther away along this side chain by introduction of a carbonate moiety (**48**, Fig. 17). Receptor affinities varied significantly with the substituents and with the stereochemistry of the chiral center at the 17β ester. The selected double-fluorinated compound, itrocinonide (**48**), had a receptor affinity similar to budenoside and a sufficiently rapid rate of *in vitro* hydrolysis ($t_{1/2} = 30$ min in human blood at 37°C) [206]. The *in vivo* potency of itrocinonide was less than that of budesonide, but, in agreement with SD design principles, the ratio between its anti-inflammatory efficacy in airways/lung and its systemic steroid activity (i.e., thymus involution or plasma cortisol suppression) was much better than the corresponding ratio for budesonide. It also had very good systemic tolerance in human volunteers and asthmatics. Owing to its short plasma half-live (~30 min), which is approximatelyt one-fifth of that of budenoside and fluocortin butyl, itrocinonide lacked measurable systemic glucocorticoid activity. In patients with asthma or seasonal rhinitis, itrocinonide administered as a dry powder formulation did exert some antiasthmatic and antirhinitic efficacy, but these effects were not sufficient enough to compete with the efficacy of current inhaled steroids [206].

Glucocorticoid γ-Lactones A group at GlaxoSmithKline has explored a series of various γ-lactone derivatives including 21-thio derivatives of fluocinolone acetonide with γ-lactones and cyclic carbonates (e.g., **49**) [207] and sulfur-linked γ-lactones incorporated at the 17β-position (e.g., **50**) [208]. For these compounds, human serum paraoxonase (E.C. 3.1.8.1) was claimed to be the metabolizing enzyme [207], which is of interest because this enzyme has a much lower activity in lung tissue than in plasma, and thus it can provide improved site-specific activity for inhaled compounds. Contrary to the corresponding esters, 21-thio-linked lactones were stable in human lung S9 preparation ($t_{1/2} > 480$ min for **49**), but rapidly hydrolyzed in human plasma ($t_{1/2} < 1$ min) [207]. A higher metabolic stability of such structures in lung tissues than in corresponding plasma has been independently confirmed [206]. The rate of hydrolysis was also rapid in plasma for the 17-linked lactones possessing a sulfur in the α-position of the butyrolactone group ($t_{1/2} < 5$ min), whereas C-linked lactones were stable [208]. Among the compounds of this series, **50** showed promising topical antiinflammatory activity in the rat ear edema model and much lower systemic effects than budesonide in the thymus involution test. Nevertheless, development of these series for asthma has been discontinued, most likely because too rapid enzymatic inactivation at the desired site of action resulted in reduced anti-inflammatory efficacy [209].

Soft 17-Furoate Androstadienes A SD design approach aiming toward a safe, inhaled corticosteroid has also been recently reported by a group at Novartis [210]. The starting point of this design was essentially a 6α-F substituted mometasone furoate (**58**) analog with a 17β acid inactive metabolite. Hence, various 17α furoate ester substituents were explored while maintaining the 17β methyl esters substituent as the intended soft metabolic deactivation site. The obtained structures showed good (1–10 nM) affinity toward the glucocorticoid receptor. *In vivo* screening of the synthesized analogs in a rodent model identified **51** (Fig. 17) as a promising lead with minimal oral absorption as well as superior efficacy and duration of action and similar intratracheal (i.t.) side effect profile compared to budesonide [210]. Compound **51** showed complete inhibition of eosinophilia 24 h after 1 mg/kg dosing without any significant oral side effects. By comparing the structure of **51** to that of mometasone furoate (**58**), it is obvious that it can be considered as an intended soft analog of **58**. It has to be mentioned, however, that even if these compounds were intended as soft drugs, they seem to be unusually stable in plasma showing no significant extrahepatic metabolism. Compound **51** was found not to degrade in rat or human plasma over 30 min at 37°C, indicating that the C-21 methyl ester, which is sterically strongly hindered in these structures, is stable to plasma esterases under these conditions.

"Antedrug" Steroids Various, mostly prednisolone-based ester derivatives, were synthesized and investigated in a series of attempts designated as "antedrug" designs [53,54, 211–219]. They were aimed to improve the local-to-systemic activity ratio of anti-inflammatory steroids and may be considered as SD designs based on hypothetical inactive metabolites. Studied compounds include ester derivatives of steroid 21-oic acids [53], a number of 16α-carboxylate analogs (e.g., **52**) [54,211–213,216,220], 6-carboxylate analogs [214], and (16α,17α-*d*) isoxazolines derivatives [215,217,218]. Some of these compounds were found to have relatively low activity, similar to that of hydrocortisone or prednisolone, and they also achieved some, 8w?>but not very significant, improvement in the local-to-systemic activity ratio. Relative binding affinities (RBA considering $RBA_{Dex} = 100$) of two such 9α-fluorinated steroids for the cytosolic glucocorticoid receptor are 11 and 4 for FP16CM (**52**) and its 21-acetate derivative FP16CMAc, respectively [54,178]. For comparison, loteprednol etabonate (**41**), a nonfluorinated soft steroid, has a RBA of around 150 [163,178]. As an additional possible attempt to increase topical potency by a different approach, the 17α-dehydro steroid FDP16CM was conjugated through an ester bond (**53**) to nonsteroidal antiinflammatory drugs such as ibuprofen [55], an approach that could be qualified as essentially a prosoft drug. Topical potency was not increased compared to the nonconjugated steroid (because of decreased receptor binding); however, the local-to-systemic activity ratio was still improved compared to prednisone.

Other Corticosteroid Designs Other groups also made attempts to separate local and systemic effects by integrating moieties susceptible to rapid, nonhepatic metabolism within the corticosteroid structure. One of the more successful attempts explored 17α-(alkoxycarbonyl)alkanoates analogs (**54**) of clobetasol propionate [221]. Again, this can be considered as a hypothetical inactive metabolite-based approach, and a corresponding metabolite (**54**, $n=2$, R = H) has indeed been shown to be inactive. Esters that were susceptible to rapid hydrolysis exhibited good separation of topical antiinflammatory to systemic activity. The study also indicated the existence of an optimal volume for the 17α side chain. For example, the methyl succinate derivative (**54**, $n=2$, R = methyl) showed as potent topical antiinflammatory activity as clobetasol propionate, but a dramatically reduced thymolytic activity. Therefore, the corresponding TI was increased more than 130-fold compared to clobetasol propionate. It has to be mentioned, however, that for this compound (**54**) as well as for the glucocorticoid γ-lactones (**49**, **50**), active compounds may be formed from the inactive metabolite; for example, in the case of **54**, by chemical cleavage of the succinate ester to the active clobetasol resulting in a possible active → inactive → active sequence, which should not be the case in a good SD design.

Another effort involved the design, synthesis, and testing of a colon-targeted prosoft drug (**55**) for possible oral treatment of ulcerative colitis [222]. These C-20 oxyprednisolonate 21-esters contain glucopyranosyl ethers to render the pro-SDs hydrophilic and thus poorly absorbable in the small intestine. Removal of the glucopyranosyl ethers releases the corresponding active soft drug. This process is mediated by colonic bacteria within the colonic lumen as demonstrated *in vivo* after administration in the jejunum of guinea pigs. In the systemic circulation, degradation of the C-21 esters rapidly releases the inactive acid metabolites. Interestingly, the half-lives in guinea pig plasma for the two different ester stereoisomers were quite different, being 2.6 and 166.8 min for the 20*R*-dihydroprednisolonate and 20*S*-dihydroprednisolonate, respectively. Somewhat later, steroid-17-yl methyl glycolates with succinyl group at C20 derived from prednisolone and dexamethasone (**56**) were also investigated by the same group [223]. In fact, this is again a prosoft drug-type approach, as first an active compound, the 21-ester, is released, and then this is further metabolized into an inactive metabolite.

In another separate study, three series of compounds were synthesized in which sulfur-containing amino acids were incorporated into the steroidal structure at the 21 position [224]. The rationale for this, which the authors considered as being more-or-less along the principles of SD design, was that sulfur-containing compounds have shown, generally, a good cutaneous distribution as well as relative rapid biotransformation and fast elimination with the oxidized metabolites being inactive in most cases. However, drugs relying on such oxidative metabolism cannot be considered true SDs (see also the next paragraph). The selected most promising compound of this series was **57**, for which *in vivo* results showed a local activity approximately 10 times less and a systemic activity approximately 970 times less than that of dexamethasone.

Finally, before closing this section, it should be mentioned that some other steroid drugs such as fluticasone propionate (FP, **59**), tipredane [225,226], or butixocort 21-propionate [227,228] are often and erroneously called soft drugs (see, for example, Refs. [209,229,230]). These drugs are indeed metabolized rapidly, but by oxidative mechanisms in the liver [226,228,231]. Thiol ester corticosteroids such as FP have been shown to be metabolized in the liver by oxidative cleavage of the thiol ester bond and not by hydrolysis in the plasma [231]. Consequently, even if FP itself lacks oral activity because high hepatic first-pass metabolism to the corresponding (inactive) 17-carboxylic acid, it has systemic effects if given subcutaneously [232]. In a few cases, even severe systemic side-effects have been seen with inhaled or intranasal FP especially when coadministered with CYP3A4 inhibitors [233–236], and since 2004, the combination of inhaled fluticasone and ritonavir is no longer recommended by GlaxoSmithKline, because of the risk of Cushing's syndrome, unless the benefits overcome the risks [236]. FP was found to have a terminal half-life of ~8 h in 12 healthy male subjects after inhaled administration of 500, 1000, and 2000 μg of drug using a metered-dose inhaler (MDI). In these subjects, it produced dose-related cortisol suppression; the highest administered dose of FP resulted in cortisol concentrations that were lower than the limit of detection [237]. The slow elimination of FP led to accumulation during repeated dosing. This accumulation may explain the marked decrease in plasma cortisol seen during treatment with FP within the clinical dose range [238]. Furthermore, it is a highly lipophilic steroid and it shows increased terminal half-lives after inhalation, which usually is an indication of slow, rate-limiting absorption ("flip-flop pharmacokinetics") [239]. In fact, FP has been shown to exhibit significantly steeper dose-related systemic bioactivity as measured by cortisol suppression than beclomethasone dipropionate (BDP; **61**), budesonide, or triamcinolone acetate [152]. Ciclesonide (CIC, **60**) [240,241] sometimes is also erroneously classified as a soft drug (see, for example, [48,209]). CIC, just as BDP (**61**), is an inactive ester prodrug that has to be has to be metabolically transformed into its active metabolite (CIC_{am}) to exert activity at the GR. In fact, CIC is slightly more lipophilic, but otherwise very similar to BDP: both are inactive 21-esters activated by hydrolysis, both have active metabolites that are 17α monoesters, and both have about similar binding affinity values [178].

3.3.4. Soft Calcitriol (1α,25-Dihydroxyvitamin D_3) Analogs The naturally occurring hormone calcitriol (1α,25-dihydroxyvitamin D_3; **62**, Fig. 18) has antiproliferative and cell differentiation activities that are of potential therapeutic use if sufficiently separable from the undesirable calcemic activity. Analogs such as maxacalcitol (**63**) have been developed along these lines for treatment of various diseases such as psoriasis, secondary hyperparathyroidism, and osteoporosis, but they still require careful administration due to potential toxicity. Introduction of a 16 double bond in such structures was explored by a group at Chugai Pharmaceutical Company to accelerate the oxidative metabolism in liver,

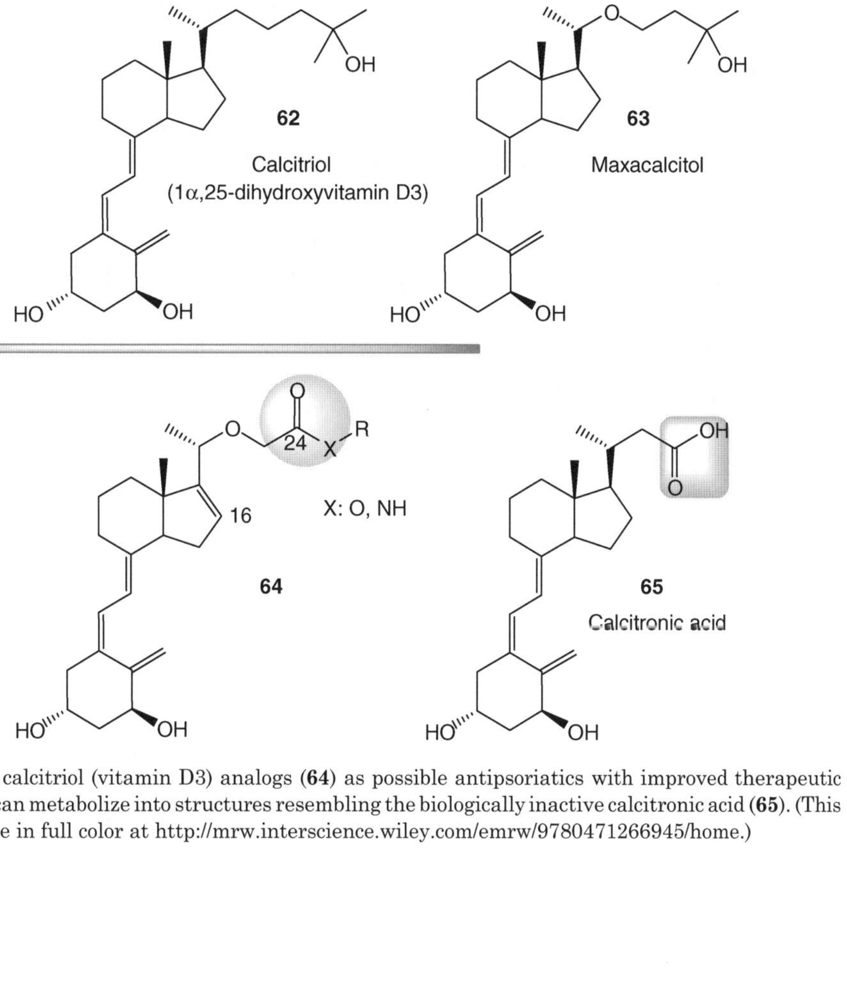

Figure 18. Soft calcitriol (vitamin D3) analogs (**64**) as possible antipsoriatics with improved therapeutic index since they can metabolize into structures resembling the biologically inactive calcitronic acid (**65**). (This figure is available in full color at http://mrw.interscience.wiley.com/emrw/9780471266945/home.)

presumed to be essential for the reduction of calcemic activity [242]. Later, in what qualifies as a true SD approach to develop antipsoriatics with improved TI, ester or amide moieties were inserted at the 24 position (**64**) to impart metabolic sensitivity with the knowledge that calcitriol's calcitronic acid metabolite (**65**) is biologically inactive [243]. This series (**64**) has been shown to have some extremely potent agonists of the vitamin D receptor with low calcemic action, and biological evaluations identified two structures as extremely potent, low-calcemic vitamin D_3 analogs, with concentration in rat skin comparable to maxacalcitol. These compounds (**64**; $X=O$, $R=C(CH_3)(C_2H_5)_2$, and $X=N$, $R=CH_2CF_2CF_3$) are being further evaluated for their possible clinical application as antipsoriatic agents.

3.3.5. Soft Estrogens The three major naturally occurring estrogens, the primary female sex hormone, are estradiol (**66**), estriol (**67**), and estrone (**68**) (Fig. 19), and they represent another important group of steroids in which SD approaches can provide new therapeutic agents with a beneficial separation of local and systemic effects. Menopause-related estrogen depletion is more than likely associated with a variety of symptoms ranging from vasomotor complaints to cognitive deficits. Estrogen administration (hormone therapy, HT) is known to alleviate most of these symptoms, but because of an association with increased risk for cancer, stroke, and other metabolic diseases, such therapies are either not recommended for or avoided by many women [244,245]. This is especially true now that two large-scale parallel randomized, double-blind, placebo-

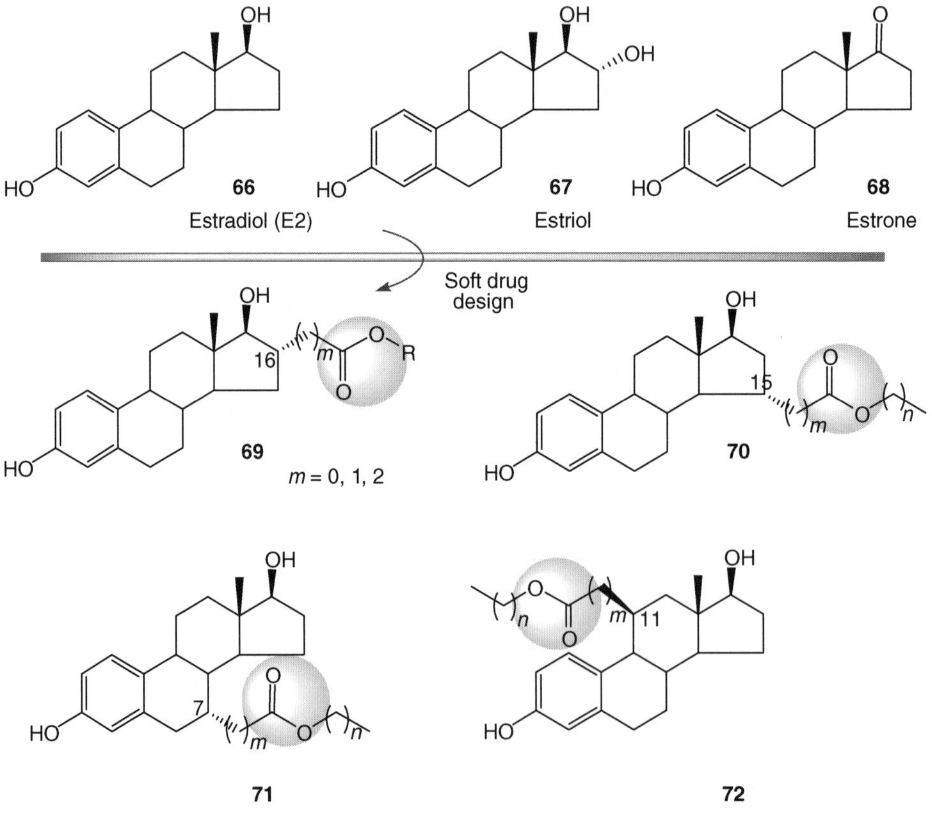

Figure 19. Estrogens (**66–68**), the primary female sex hormone (including estradiol, estriol, and estrone), and soft estrogens (**69–72**) designed for the treatment of vaginal dyspareunia. (This figure is available in full color at http://mrw.interscience.wiley.com/emrw/9780471266945/home.)

controlled clinical trials undertaken within the Women's Health Initiative (WHI) to determine whether conjugated equine estrogen (CEE) alone (in women with prior hysterectomy) or in combination with progestin (medroxy-progesterone acetate) reduces cardiovascular events in postmenopausal women had to be halted early because of unfavorable outcomes (increased risk for invasive breast cancer, stroke, and coronary heart disease) [246,247] (see also Section 4.1.6 for further discussion).

Vaginal dyspareunia is a common disease affecting a large proportion of menopausal women (approximately 40% within 10 years of the onset of menopause), and topical application of estrogen has been used for treatment. Locally active soft estrogens with reduced systemic activity may provide a therapeutic alternative. Estradiol (E_2; **66**), the most potent human estrogen, provides a good starting point for a SD design. Along these lines, a series of estradiol-16α-carboxylic acid esters (**69**, Fig. 19) were synthesized and examined first [248]. Whereas, none of the acids (**69**, R = H, $m = 0, 1, 2$) showed significant estrogen receptor binding, the esters did. For them, receptor binding decreased with increasing m (Fig. 19) or branching of the alcohol portion (R = isopropyl, neopentyl), but not with increasing length of the alcohol chain (R = methyl–butyl). The rate of hydrolysis in rat hepatic microsomes increased with increasing chain length (methyl to butyl) and was especially high for fluorinated alcohol chains (e.g., R = CH_2CHF_2). Three of the most promising compounds (**69**, $m = 0$, R = CH_3, CH_2CH_3, CH_2CH_2F) were also tested for systemic and local action in rodent *in vivo* models. All of them, especially the fluoroethyl ester, showed good separation of the local and the systemic estrogenic action.

In a follow-up study, additional carboxylic acid ester derivatives of estradiol have been synthesized with substitutions at the 7α-, 11β-, and 15α-positions (**70–72**) [249]. Whereas, again all short-chain carboxylic acids were devoid of hormonal activity (as required by the inactive metabolite principle), some of the esters (e.g., formate esters, $m = 0$, at 7α and 15α) showed good estrogenic activity. The position of the ester moiety relative to the steroid nucleus seemed important as lengthening of the chain from formate to acetate dramaticaly decreased hormonal activity. In general, lengthening of the alcohol moiety (from methyl to butyl) had only a small effect on receptor binding, but tended to increase esterase activity. The more promising candidates were also tested *in vivo* for local and systemic estrogen activity by using vaginal assay in mice and uterotrophic assay in rats, respectively, and the methyl and ethyl esters of estradiol 15α-formate (**70**, $m = 0$, $n = 0, 1$) showed high local and low systemic activity; hence, they are considered as the most promising soft estrogens and are being developed for clinical applications. Strangely, it turned out that among the 11-substituted compounds (**72**), increasing the ester substituent with a single methylene unit converted the compound from an estrogen to an antiestrogen: whereas the methyl ester (**72**, $m = 1$, $n = 0$) has high ER affinity and high estrogenic potency, the ethyl ester (**72**, $m = 1$, $n = 1$) had even higher ER affinity, but little or no estrogenic activity [250]. Thus, this small modification resulted in an unusual, steroidal selective estrogen receptor modulator (SERM).

3.3.6. Soft β_2-Agonists β_2-Agonists represent an important class of drugs in the therapy of asthma because of their β_2-receptor-mediated bronchodilating activity [251]. Compounds such as terbutaline (**73**, Fig. 20), fenoterol, or salbutamol are chemical analogs of epinephrine and are β_2-selective agents. These agents, including the longer acting formoterol or salmeterol, are most frequently taken by aerosol. The majority of the drug administered this way is swallowed, and only approximately 10–25% of the dose reaches the lung directly. Therefore, there is a great potential to produce unwanted side effects such as tachycardia or skeletal muscle tremor. Again, a SD approach might yield viable solutions. Incorporation of a metabolically labile ester group into such structures has been attempted both in the nitrogen substituent [252] and on the aryl system [253] (Fig. 20).

The activity of compounds **75** ($n = 0$ or 1, R = CH_3) or **76** (R = CH_3) surpassed that of terbutaline (**73**) or isoprenaline. The corresponding carboxylic acids (**75** or **76**, R = H) are essentially devoid of β_2-agonist activity

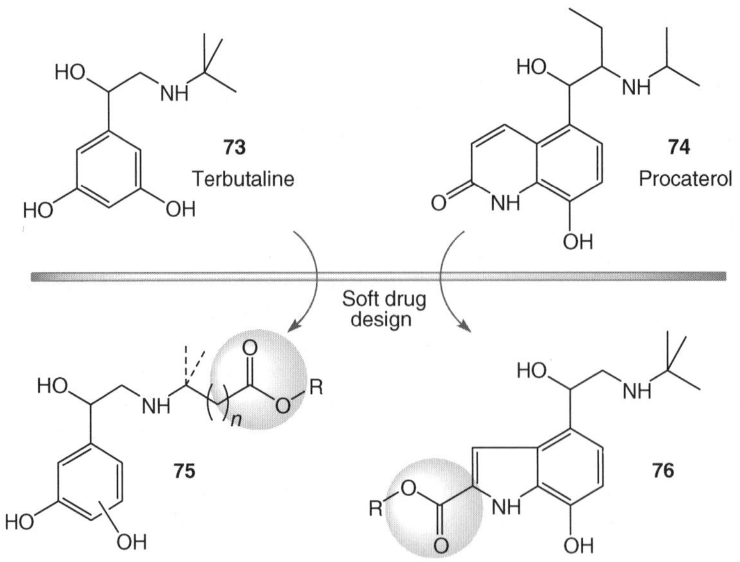

Figure 20. Soft β_2-agonists (**75**, **76**) designed on the basis of lead structures such as terbutaline (**73**) or procaterol (**74**). (This figure is available in full color at http://mrw.interscience.wiley.com/emrw/9780471266945/home.)

(they are 1–4 orders of magnitude less potent); therefore, their use as inactive metabolites in the design of more potent esters is justified (Fig. 20) [252,253]. Compound **76** with R = CH_3 (ZK 90.055) was selected for further pharmacological and toxicological evaluation. Consistent with the design, it rapidly hydrolyzed in the presence of guinea pig liver homogenate and showed good *in vivo* broncho-spasmolytic activity when given by inhalation. Meanwhile, it was almost inactive on oral administration and had no effect on the heart rate of guinea pigs following inhalation of up to 10-fold the dose that was active in the bronchospasm experiment [253].

Soft β-agonist structures were also investigated as possible soft antipsoriatic agents [254]. Both soft drugs and prosoft drugs were synthesized as models of topical antipsoriatic β-adrenergic agonists. The structure of the SDs was similar to **75** (Fig. 20, $n = 1$, R = CH_3, CH_2CH_3), and the pro-SDs were obtained by esterification of the phenolic functions. In the presence of porcine liver carboxyesterase, the pivaloyl ester groups of the prodrug underwent rapid hydrolysis ($t_{1/2} = 8.4$ min) to release the soft drug, which then also underwent hydrolysis ($t_{1/2} = 456$ min) to the inactive carboxylate anion [254]. The SD was a full β-agonist on the guinea pig tracheal preparation producing a maximal response similar to that achieved with isoprenaline. The pro-SD produced only slowly developing responses at high concentrations (>10 μM) and had better transport properties across a silicone membrane.

3.3.7. Soft Psychostimulants Methylphenidate (**77**, Fig. 21), a methyl ester-containing piperidine derivative that is structurally related to amphetamine (**79**, Fig. 21), is a mild CNS stimulant with more prominent effects on mental than on motor activities that has been in clinical use for almost 50 years. More recently, it has become widely used as the most prescribed drug in child and adolescent psychiatry (Ritalin™, Concerta™) for the treatment of attention deficit–hyperactivity disorder (ADHD). Its mechanism of action is not entirely elucidated, but there is mounting evidence for a dopaminergic basis of its action: its therapeutic effect in ADHD-treatment seems to be elicited primarily through an inhibition of the presynaptic dopamine transporter [255,256]. Methylphenidate (**77**) has two chiral centers and, therefore, a total of four isomers: *erythro-RS-d-*, *erythro-SR-l-*, *threo-SS-l-*, and *threo-RR-d*-methylphenidate

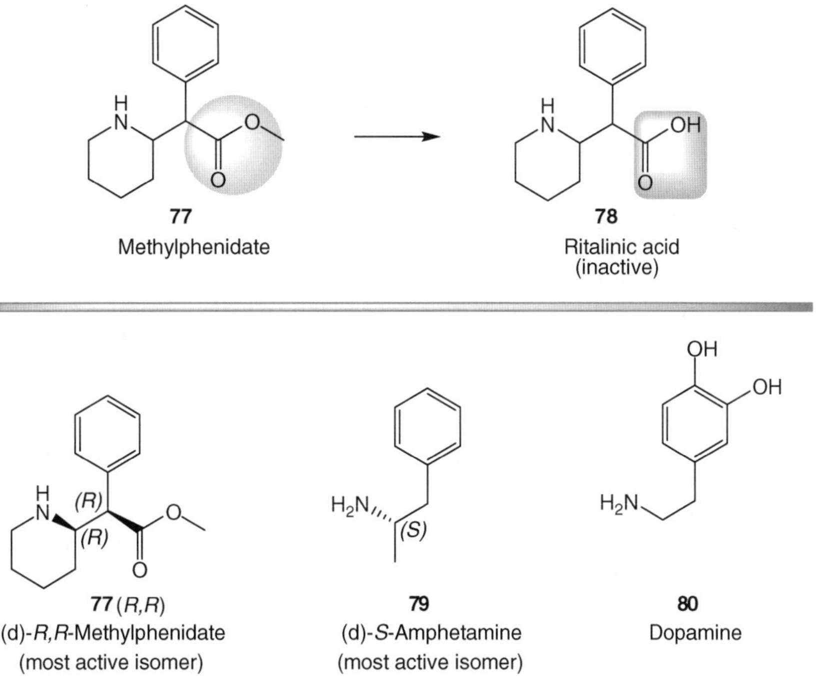

Figure 21. Methylphenidate (**77**) as an example of metabolically degradable soft psychostimulant as it is rapidly and hydrolytically cleaved into the inactive ritalinic acid (**78**). The structures of amphetamine (**79**) and dopamine (**80**) are shown for comparison. (This figure is available in full color at http://mrw.interscience.wiley.com/emrw/9780471266945/home.)

(Fig. 21). Early formulations contained all four isomers, but later the *erythro* isomers have been removed because of their association with some adverse effects. Most of the effect seems to reside with the *d*-enantiomer, and an enantiopure formulation has been recently introduced on the market, but most marketed formulations are 50:50 mixtures of the *threo-SS-l-* and the *threo-RR-d*-methylphenidate isomers.

Methlyphenidate is rapidly hydrolyzed [257] into an inactive [258], acidic metabolite (ritalinic acid; **78**); therefore, it can be considered a soft drug, even if it was not designed as such. This certainly is the main reason behind its safety that makes possible its widespread pediatric use. Nevertheless, methylphenidate is a schedule II drug, just as amphetamines; that is, it is considered a medication of high abuse potential. Plasma concentrations of the ritalinic acid metabolite **78** are much higher than those of the parent **77** (e.g., mean AUC values were 23 ± 4 times greater [257]), and 60–80% of methylphenidate is eliminated in urine as its acidic metabolite [256]. The hydrolysis of **77** is enantioselective with the active *d* isomer being apparently less susceptible to hydrolytic degradation [259]; consequently, the *d*-enantiomer has a longer half-life and is present in higher concentrations. Methylphenidate is rapidly distributed (t_{max} approximately 2 h after oral administration) and because of its fast, mainly hydrolytic, metabolism, it is also rapidly cleared: its elimination half-life is around 2–3 h, which allows no day-to-day accumulation [255,256]. In fact, its short half-life is one of its main problems, because immediate-release formulations require relatively frequent administration to maintain effectiveness, an inconvenience especially in pediatric populations. Extended-release formulations seem to provide good solutions, and they also make possible a sort of ideal SD therapeutic approach: slow release/infusion to maintain a safe pharmacological effect that rapidly disappears when administration stops. Because with methylphenidate the greatest behavioral effect in ADHD seems to occur when there is a rising blood concentration (a so-called gradient effect), most newer extended release formulations try to achieve not a concentration plateau, but prolonged or multiple rising phases [256].

3.3.8. Soft Insecticides/Pesticides The very same concepts used for soft drug design can be extended to the design of less toxic commercial chemical substances, provided that adequate structure–activity relationship (SAR) and structure–metabolism relationship (SMR) data of analogous substances can be gathered (*soft chemical* design). The following two examples are not actual designs based on such principles, but observations made in hindsight. Nevertheless, they illustrate the possibilities inherent to such approaches. Also, they provide examples for the design of environmentally safe, nontoxic chemicals (green chemistry) [260].

Chlorobenzilate One instance in which these principles have been used (unintentionally) for the design of nonpharmaceutical products is chlorobenzilate (Acaraben™, Folbex™, **88**), an ethyl ester-containing analog of dicofol (**82**) and dichlorodiphenyltrichloroethane (DDT, chlorophenothane, **81**) (Fig. 22). DDT was the first chemical that revolutionized pest control, and it was also used to control typhus and malaria. Although already synthesized in 1874 by Zeidler, its insecticidal properties were discovered by Paul Müller only in 1939 [261,262]. It was widely used by the German and later by the US army during World War II (WWII). It became extensively used as a pesticide in the United States, but it was banned in 1972 for all but essential public health use and a few minor uses. The decision was prompted by the prospect of ecological imbalance from continued use of DDT, by the development of resistant strains of insects, and by suspicions that it causes a variety of

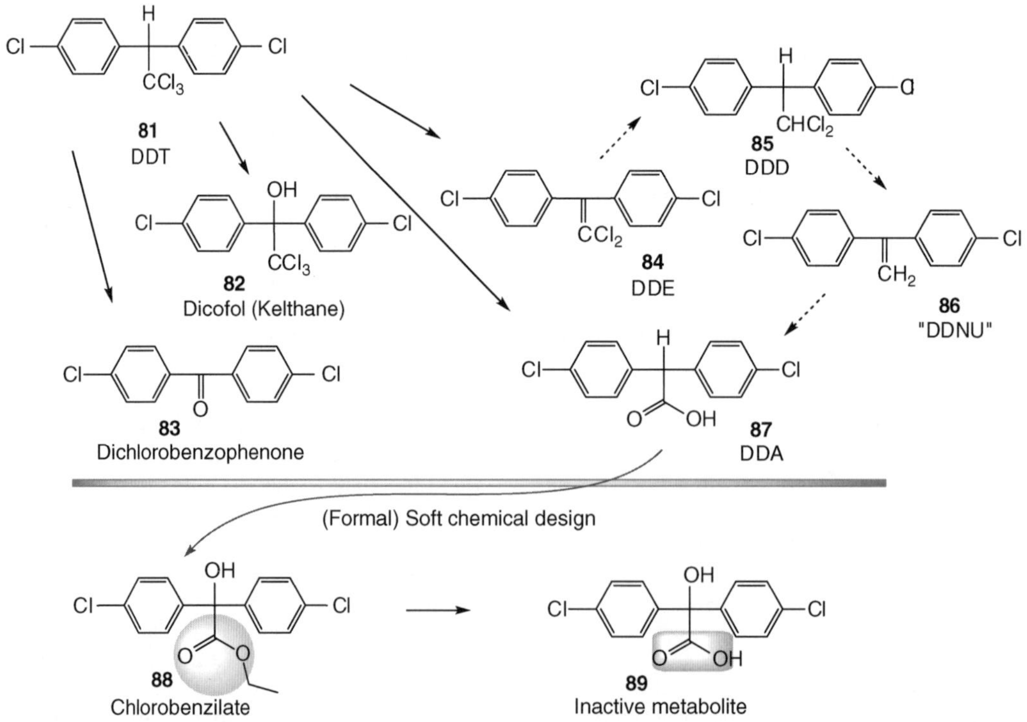

Figure 22. Chlorobenzilate (**88**) can be regarded as a soft chemical obtained using an inactive metabolite-based approach based on the metabolism of DDT (**81**). (This figure is available in full color at http://mrw.interscience.wiley.com/emrw/9780471266945/home.)

health problems, including cancer. DDT undergoes complex *in vivo* metabolism, including oxidation (**82**, **83**), iterative dehydrohalogenation/reduction cycles (**84**–**86**), and hydrolysis (**87**) (Fig. 22) [263]. The acid metabolite **87** (DDA) is of low toxicity, can be excreted as a water-soluble species, and is, indeed, a major metabolite detected in feces and urine. Therefore, it is an ideal lead compound for a formal inactive metabolite approach (Fig. 22).

Not surprisingly, the corresponding ethyl ester, ethyl-4,4′-dichlorobenzilate (chlorobenzilate, **88**), is also active as a pesticide, but has much lower carcinogenicity than DDT (**81**) or dicofol (kelthane, **82**). For example, the carcinogen concentration determined in mice is 6000 mg/kg for chlorobenzilate compared to 10 and 264 mg/kg for DDT and dicofol, respectively [264,265]. Similarly, the oral median lethal dose (LD_{50}) for female rats is 1220 mg/kg for chlorobenzilate compared to 118 and 1000 mg/kg for DDT and dicofol, respectively [266]. The ethyl ester moiety apparently can replace the trichloromethyl group of DDT to restoring pesticidal activity. However, because in exposed subjects, the labile ethyl ester group enables rapid metabolism to the free, nontoxic carboxylic acid **89**, chlorobenzilate is considerably less toxic than DDT.

Malathion Malathion (Derbac-M™, Malaspray™, **92**, Fig. 23) is an excellent example to illustrate an additional, not yet sufficiently explored aspect of the design of soft chemicals. As mentioned, it is desirable to design soft chemicals deactivated by carboxylesterases. For soft chemicals intended to be used as pesticides, in addition to the usual advantages of SD design, the differential distribution of these enzymes between vertebrates and insects may also provide selectivity based on metabolism. An elegant example is provided by malathion (**92**), a widely used organophosphate insecticide (Fig. 23). Malathion is detoxified through a variety of metabolic pathways, one of the most prominent ones being the hydrolysis of one of its two ethyl carboxylester groups leading to **94**. The carboxylesterase that hydrolysis and thereby detoxifies malathion is widely distributed in mammals, but only sporadically in insects, where in some rare cases is responsible for insecticide resistance [267,268]. In the meantime, insects seem to possess a very active oxidative enzyme system that transforms malathion (**92**) into malaoxon (**93**), a much more active cholinesterase inhibitor (Fig. 23). Probably, all insects and all vertebrates possess both an esterase and an NADPH-dependent oxidase system, but the balance of action of these two systems varies from one organism to another and provides selectivity of action. Similar mechanism may provide considerable selectivity for other soft chemicals to be designed and may result in safer, *soft* insecticides, for example, in the parathion family. These compounds are not susceptible to such deactivation mechanisms and, consequently, have unacceptably high mammalian toxicities. For example, acute oral LD_{50} in male rats are 2 and 5 mg/kg for phorate (**90**) and parathion (**91**), respectively, compared to 1400 mg/kg for malathion (**92**).

3.3.9. Soft Anticholinergics: Inactive Metabolite-Based Approach Soft anticholinergics provide a good illustration for the flexibility and potential of the general soft drug design concept. Our work in this area resulted in two entirely different classes of soft anticholinergics. *Inactive metabolite-based* classes (**102**, **104**, Fig. 24), which will be discussed here, were obtained by using methylatropine (**96**) [269–274], *N*-methylscopolamine (**97**) [275,276], or glycopyrrolate (**99**) [277] as lead. The *soft analog* class (**106**), which will be discussed in the following section, contains soft quaternary analogs [40,278].

Muscarinic receptor antagonists inhibit the effects of acetylcholine by blocking its binding to muscarinic cholinergic receptors at neuroeffector sites on smooth muscle, cardiac muscle, and gland cells, as well as in peripherial ganglia and in the central nervous system (CNS); therefore, they are used or are of therapeutic interest for a variety of applications including treatment of asthma/COPD, prevention of motion sickness, mydriasis/cycloplegia, Alzheimer's and Parkinson's disease, and disorders of intestinal motility, cardiac function, and urinary bladder function [279,280]. Muscarinic antagonists include the naturally occurring alkaloids of the belladonna plants such as atropine and

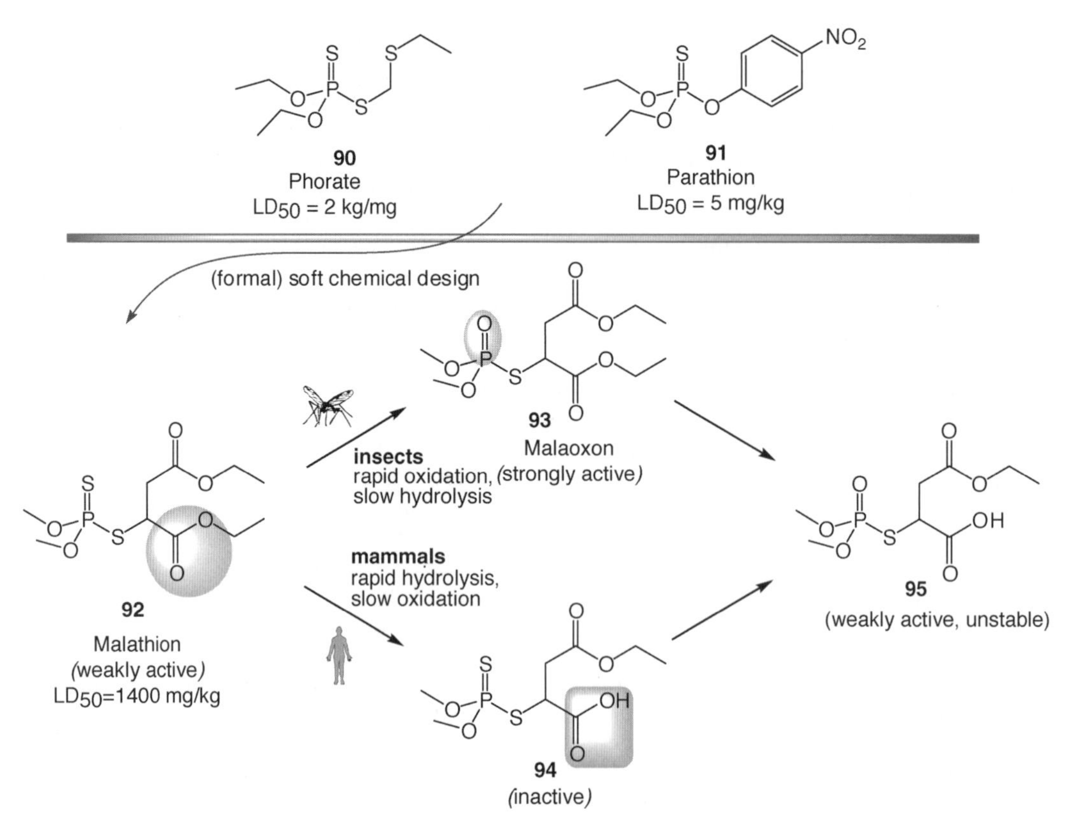

Figure 23. Malathion (**92**), its oxidative activation to malaoxon (**93**), a much more active cholinesterase inhibitor, and its deactivation by carboxylesterases. Oxidative activation is the dominant pathway in insect, but hydrolytic deactivation is the dominant pathway in mammals. Because of its hydrolytic deactivation in mammals, malathion (**92**) is much less toxic than other organophosphates such as phorate (**90**) or parathion (**91**) as illustrated by the corresponding LD_{50} values. (This figure is available in full color at http://mrw.interscience.wiley.com/emrw/9780471266945/home.)

scopolamine. Quaternary derivatives are usually more active, and they cannot cross the blood–brain barrier (BBB) and reach the CNS (usually an advantage because they are less likely to cause CNS-related side effects), but they also tend to be more poorly absorbed, which might cause problems in ensuring adequate bioavailability. Obviously, subtype selectivity, if achievable, can provide increased therapeutic advantage [281]. Because of their ability to inhibit local antisecretory activity, anticholinergics have even been explored as antiperspirants for excessive sweating [282–286].

There is renewed interest in these agents due to their applicability as inhaled bronchodilators for the treatment of bronchospasm associated with COPD and other diseases (tiotropium (**98**), ipratropium) [287]. However, in many cases, their use is still limited by the possibility of a number of side effects, such as cardiac arrhythmias, tachycardia, dry mouth, difficulty in urination, constipation, photophobia, irritability, restlessness, disorientation, dementia, and hallucinations [280,288]. Even topically applied anticholinergics can cause unwanted side effects [289–297] because they are absorbed into the systemic circulatory system and are eliminated only relatively slowly. A locally active soft drug may again represent a viable solution.

To obtain such compounds that have high local, but practically no systemic activity, different series of soft anticholinergics based on methylatropine, N-methylscopolamine, glycopyrrolate, or propantheline were designed in our laboratories [298,299]. Several new molecules synthesized were found to be potent anticholinergics both *in vitro* and *in vivo*, but in contrast to their hard analogs they had no systemic anticholinergic activity following topical administration. For example, methylatropine- or methylscopolamine-derived phenylmalonic acids (**102**, $n=0$) [269–272, 274,275] and phenylsuccinic acids (**102**, $n=1$) [273,276] served as useful hypothetical inactive metabolites for the design of soft anticholinergics. This *inactive metabolite-based* approach exploits the idea that the benzylic hydroxy function in methylatropine (**96**) or methylscopolamine (**97**) could be oxidized to the carboxylic acids **103** ($n=0$), which are hypothetical [288,300,301] inactive metabolites. Esterification of this carboxy function to afford soft drug series **102** may restore activity and meanwhile ensure facile, hydrolytic deactivation.

Soft anticholinergic esters of this kind (**102**) showed good intrinsic activity as indicated by the pA_2 value of 7.85 of tematropium (PM*TR*.Et, **102a**) compared with 8.29 for atropine. However, *in vivo* activities were much shorter than for the "hard" atropine. Accordingly, when equipotent mydriatic concentrations of atropine and tematropium (**102a**) were compared following ocular administration, the same maximal mydriasis was obtained, but the area under the curve (mydriasis versus time) for the soft compound was only 11–19% of that for atropine [269,271]. This is consistent with the facile hydrolytic deactivation of the soft drug. Similarly, the cardiovascular activity of compound **102a** showed ultrashort duration. The effect of **102a** on the heart rate and its ability to antagonize the cholinergic cardiac depressant action induced by acetylcholine injection or by electrical vagus stimulation was determined in comparison with atropine (sulfate) and methylatropine (nitrate). A dose of 1 mg/kg of atropine or methylatropine could completely abolish the bradycardia induced by acetylcholine injection or by electrical vagus stimulation for more than 2 h following i.v. injection. On the other hand, similar doses of tematropium exerted antimuscarinic activity for only 1–3 min following i.v. injection. Even a 10-fold increase in its dose to 10 mg/kg did not lead to any significant prolongation of the duration of anticholinergic activity.

As a further variation, the corresponding ester analogs of methylscopolamine were also investigated [275]. For example, the ethyl ester was shorter acting than even tropicamide, and, consistent with the SD approach, the untreated eye did not show any mydriasis, as opposed to administration of methscopolamine. Other structures such as the cyclopentyl derivative **102c** (PcPM*TR*.Me, PCMS-2) [274] or different phenylsuccinic analogs of methylatropine [273] and methylscopolamine [276] were also investigated. Compound **102c** was equipotent to atropine in protecting against carbachol-induced bradycardia in rats, but its duration of action was again significantly shorter (15–30 min versus more than 2 h) [274]. Similar ester analogs (PcPM*GP*.Me, **102d**) of glycopyrrolate (**99**) were also explored [277].

For soft anticholinergics, the inactive metabolite-based approach can also yield a different class of compounds in which the hydrolytically labile ester-containing side chain is attached to the quaternary nitrogen head (**104**). A number of such compounds derived from methylatropine (e.g., PcPA*TR*_NA.Me **104a**, **104b**) [302,303] or glycopyrrolate (e.g., PcPOA*GP*_NA.Et **104c**) [304–309] have been recently synthesized and tested. Several of them showed promising and stereospecific *in vitro* receptor-binding and *in vivo* mydriatic activities as well as some degree of and M_3/M_2 muscarinic-receptor subtype-selectivity. Furthermore, compounds such as **104d** (PcHC*TR*_NA.Me) have also been explored in an attempt to obtain soft anticholinergics with muscarinic receptor subtype selectivity [310]. Hence, a lead compound (LG50643) was selected that has been shown to be a potent and selective antagonist for the M_3 receptor subtype [311], and it was derivatized following the same procedure to obtain soft compounds such as **104d**. Receptor binding studies on cloned muscarinic receptors indicated that these soft anticholinergics have reasonable activity (pK_i values of 7.5–8.9) and that two of them show muscarinic receptor subtype selectivity (M_3/M_2) [310]. Consistent with their soft nature, these compounds were short acting and were rapidly eliminated from plasma.

3.4. Soft Analogs

Compounds classified as soft analogs are close structural analogs of known active drugs (lead compounds), but they have a moiety that is susceptible to metabolic, preferentially hydrolytic, degradation built into their structure. The built-in metabolism should be the major, and preferentially, the only metabolic route for drug deactivation (*predictable metabolism*), and the rate of the predictable metabolism should be controllable by structural modifications (*controlled metabolism*). The predicted metabolism should not require enzymatic processes leading to highly reactive intermediates, and the products resulting from the metabolism should be nontoxic and have no significant biological or other activities (*metabolic inactivation*). Finally, the metabolically weak spot should be located within the molecule so that the overall physical, physicochemical, steric, and complementary properties of the soft analog are very close to those of the lead compound (*isosteric/isoelectronic analogy*).

3.4.1. Soft Anticholinergics: Soft Quaternary Analogs As mentioned, in addition to the inactive metabolite-based classes of soft anticholinergics, an entirely different class containing soft quaternary analogs has also been explored. Structural differences between "hard" and "soft" anticholinergics of this kind are relatively small, but nonetheless profound as illustrated on the right side of Fig. 24. *Soft analog* anticholinergic structures were obtained by shortening the bridge of two or three carbon atoms separating the quaternary head and ester function of traditional, "hard" anticholinergics illustrated by the generalized structure **101** ($k = 3$) to just one carbon separation as shown in structure **106**. This allows facile hydrolytic deactivation via a short lived intermediate to the corresponding acid (**107**), tertiary amine, and aldehyde, all inactive as anticholinergics, as shown in Fig. 24.

At the time of the design, a separation of at least two carbon atoms between the ester oxygen and the quaternary nitrogen of such anticholinergic structures was thought to be critical for effective receptor binding. Nevertheless, several compounds of type **106** were found to be at least as potent as atropine [40]. For example, **106a** (SQA.PcP-*DMP*) was equipotent with atropine in various anticholinergic tests, but it was very short-acting after i.v. injection. Therefore, when applied topically to humans, it produced high local antisecretory activity but no systemic toxicity. In a more recent study, similarly designed soft analogs using propantheline (**100**) as lead have also been investigated as potential antiperspirants/antiulcerative agent [278].

Recently, we were able to put together a retrospective, comprehensive quantitative structure–activity relationship (QSAR) study for all quaternary soft anticholinergics ($n = 76$) discussed here from these two distinctly different classes, which were designed on the basis of the soft analog and the inactive metabolite approaches, respectively [299]. Modeled activity data included pA_2 values (the negative logarithm of the molar concentration of the antagonist that produces a twofold right-shift in the concentration–response curve of the agonist) measured using the *in vitro* guinea pig ileum assay and receptor binding pK_i values measured with cloned muscarinic receptors (M_3 subtype). Because of the clear biphasic (bilinear) nature of the activity data when all structures were considered as a function of molecular size (volume), a nonlinear model had to be used (Fig. 25), and the recently introduced linearized biexponential (LinBiExp) model (Eq. 3) [312,313] proved very adequate.

$$y = f(x) = \eta \times \ln[\mathrm{e}^{\alpha_1(x-\xi_c)/\eta} + \mathrm{e}^{\alpha_2(x-\xi_c)/\eta}] + \chi + \sum_{\varphi} \delta_\varphi I_\varphi \qquad (3)$$

This QSAR study indicated that for quaternary anticholinergics, molecular size seems a major determinant of activity, and best activity is achieved with ligands not considerably smaller or larger than the known highly active anticholinergics such as **96–100**, which seem to be close to the ideal ligand size at these receptors. In agreement with SD design principles, acid metabolites were indeed essentially inactive: their activities being around two orders of magnitude less than those of the corresponding esters. The impor-

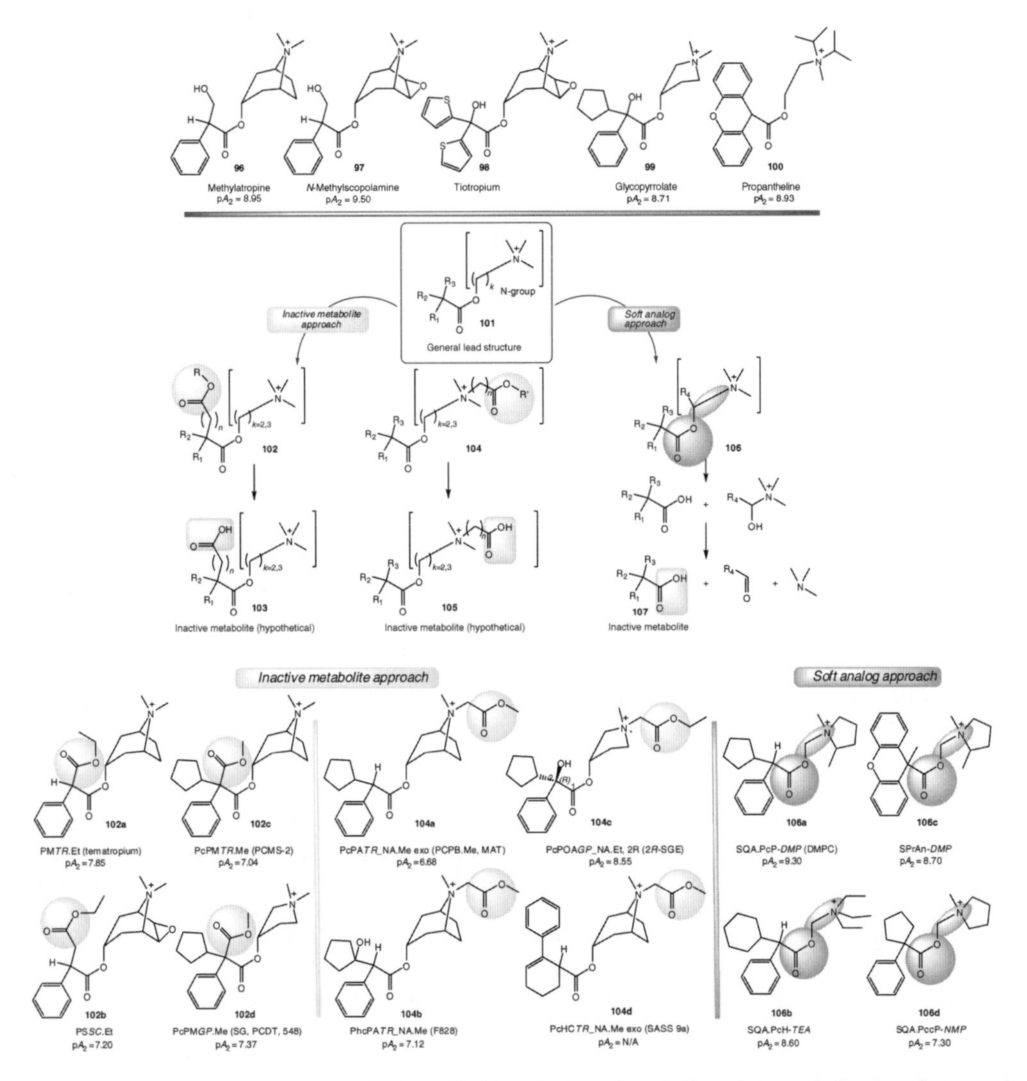

Figure 24. Design and metabolism of soft anticholinergics based on the inactive metabolite-based approach (**102**, **104**; substitutions at two different positions) and the soft analog approach (**106**) together with a set of representative classic (**96–100**) and soft (**102a–d**, **104a–d**, **106a–d**) quaternary anticholinergic structures. All soft structures can be traced back to a common general lead (**101**), and they are all inactivated in a single hydrolytic step. pA_2 values shown are for *in vitro* anticholinergic activity determined by guinea pig ileum assay with carbachol as agonist. (This figure is available in full color at http://mrw.interscience.wiley.com/emrw/9780471266945/home.)

tance of stereospecificity at muscarinic receptors was also confirmed on this series of compounds: 2*R* analogs are considerably more active than the corresponding isomeric mixtures. The effect of other substitutions (φ), such as inclusion of a scopolamine-type oxygen or a cycloalkyl ring, *endo*/*exo* isomerism, or distancing of the metabolically labile ester from the substitution center (phenylsuccinate versus phenylmalonate esters), could also be quantified through the corresponding δ_φ terms (Fig. 25) [299].

3.4.2. Soft Antimicrobials

Cetylpyridinium Analogs: Soft Quaternary Salts

The very first soft analogs designed were "soft quaternary salts" represented by the general-

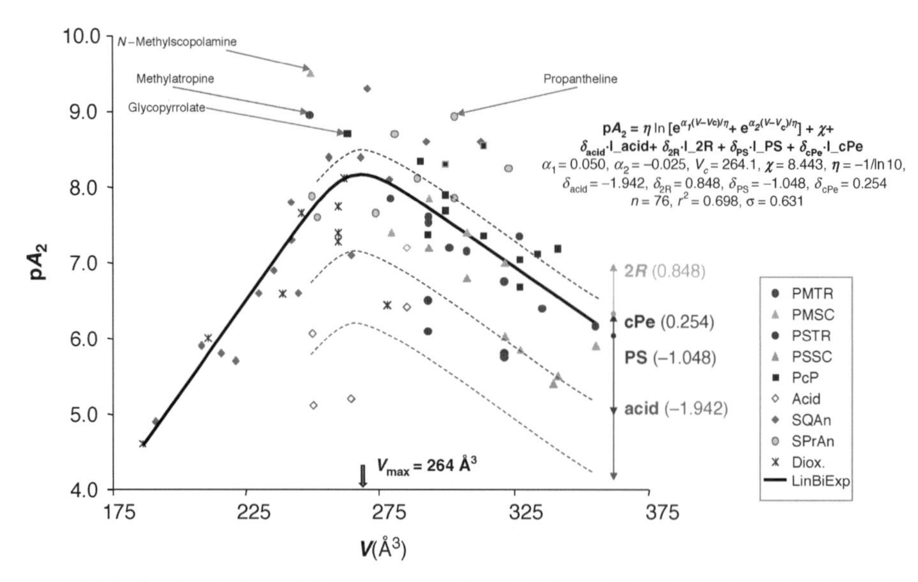

Figure 25. QSAR for the pA_2 data of all quaternary soft anticholinergics ($n = 76$) using molecular volume V as a main descriptor and the bilinear LinBiExp model with additional structural descriptors [I_acid for the presence of a carboxylic acid ($-COOH$), I_2R for enantiomerically pure $2R$ isomers, I_PS for succinic analogs where the carboxylic ester is one position away from the substitution center ($R_3 = -CH_2COOR$ versus $R_3 = -COOR$ in the malonic series), and I_cPe for the presence of a cyclopentyl substitution at the 2-position (as in glycopyrrolate)]. (See color insert.)

ized structure **111** intended for antimicrobial use [15,314]. Similar to the previously described anticholinergics structures, these substances undergo a facile hydrolytic cleavage process via a very short-lived intermediate (**113**) to deactivate and form an acid (**114**), an amine (**115**), and an aldehyde (**116**), as shown in Fig. 26. This mechanism was initially designed to develop a prodrug of aspirin, the synthesis of which, however, was unsuccessful [41].

The simplest example of useful true soft analogs (Fig. 26) is provided by the isosteric analogs (**109**, **110**) of cetylpyridinium chloride (**108**). Cetylpyridinium, a known "hard" quaternary antimicrobial agent, needs several oxidative (generally β-oxidation) steps to lose its surface-active, antimicrobial properties. The quaternary salts represented by **108** and **109** are very similar: both contain side chains that are essentially 16 atoms in length. Their physicochemical properties are also very similar. For example, their critical micelle concentrations (CMC) determined by a molecular light scattering method are 1.3×10^{-4} and 1.7×10^{-4} M, respectively [38]. *Hard* and *soft* compounds possess comparable antimicrobial activity as measured by their contact germicidal efficiency, but *soft* compounds undergo facile hydrolytic cleavage, leading to their deactivation. Because of this, the *soft* **110** is approximately 40 times less toxic than the *hard* **108**: the corresponding oral LD_{50} values for white Swiss male mice are 4110 and 108 mg/kg, respectively.

Environmental-Friendly Quaternary Analogs More recently, another set of similar cetylpyridinim chloride of benzalkonium chloride analogs have been synthesized and investigated, and the corresponding structure–activity relationships have been explored in additional details [315–317]. Similar to **109** or **110**, these soft antimicrobials also consisted of long alkyl chains connected to polar quaternary ammonium head groups (pyridinium or trialkyl ammonium) via chemically labile spacer groups, but more flexibility was allowed; that is, various polar quaternary head groups and spacer distances were explored. Building blocks were selected preferentially from natural

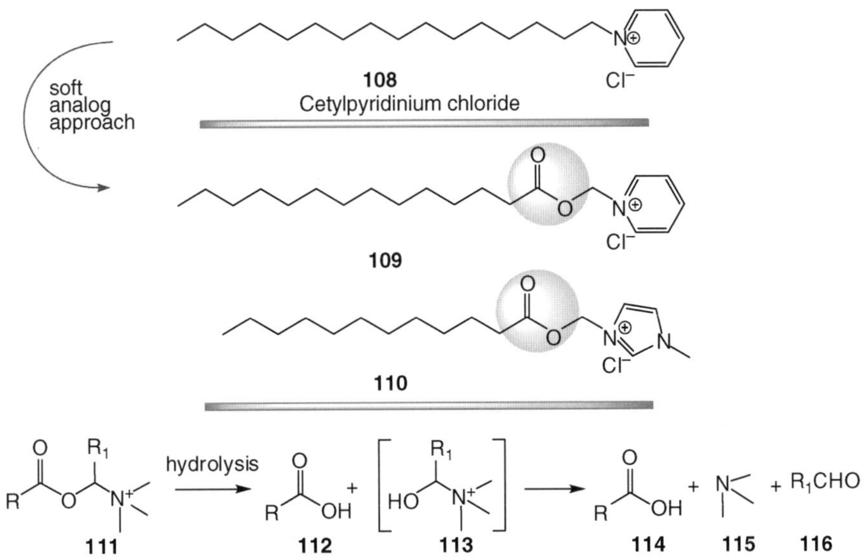

soft analog approach

108
Cetylpyridinium chloride

109

110

111 hydrolysis 112 + 113 → 114 + 115 + R_1CHO 116

Figure 26. Cetylpyridinium chloride (**108**) and soft analog antimicrobials (**109**, **110**). The general hydrolytic deactivation mechanism of soft quaternary salts (**111**) via a very short-lived intermediate (**113**) to an acid (**114**), an amine (**115**), and an aldehyde (**116**) is also shown. (This figure is available in full color at http://mrw.interscience.wiley.com/emrw/9780471266945/home.)

compounds, partly based on marine lipids, such as fatty acids, alcohols, or amides. The more active compounds had minimum inhibitory concentration (MIC) as low as 1 μg/mL and viral reduction greater than 6.7 log units. Best activities were achieved with alkyl chain-length of 12–18 carbon atoms, an observation long-known for such antimicrobial compounds [318,319], and also with smaller quaternary heads and inactivation half-lives larger than 3 h at 60°C [316]. Before closing this chapter, one should mention that a set of ester-containing soft bisquaternary ammonium salts with two long aliphatic chains derived from bis-(2-dimethylaminoethyl) glutarate have also been explored by a different group, and several of them were found to have adequate activity [320].

Long-Chain Esters of Betaine and Choline Another, unrelated effort was directed toward the development of long-chain esters (**117**, **120**) of betaine (**122**) or choline (**119**) as soft antimicrobial agents (Fig. 27) [321–325]. These compounds are ester-containing structural analogs of amphiphilic quaternary ammonium compounds, which, similar to cetylpyridinium (**108**), are surface active substances known for their membrane-disruptive and antimicrobial activities. Contrary to the design of **111**, however, hydrolysis here is not followed by additional, fast degradation, but results in well-investigated and common compounds, such as choline (**119**), betaine (**122**), and fatty acids (**118**) (Fig. 27). For example, the alkanoylcholines were found active against gram-negative and gram-positive bacteria, as well as yeasts [322]. Activity increased with increasing chain-length and was similar to that of the stable, *hard* quaternary ammonium compounds of similar length such as hexadecyltrimethylammonium bromide. Considerable differences in the binding affinity of compounds with different hydrocarbon chains at different concentrations to *Candida albicans* were observed, and they seemed related to the CMC of the compounds [323].

L-Carnitine Esters Another class of *soft* broad-spectrum antimicrobials devoted to curing dermatological infections was designed based on quaternary ammonium L-carnitine esters (**123**) (Fig. 27) [326]. The series, particularly members characterized by alkyl chains with a total of 16–18 carbons, showed good activity against a wide range of bacteria, yeasts, and fungi. They also showed low *in vitro* cytotoxicity and good *in vivo* dermal tolerance.

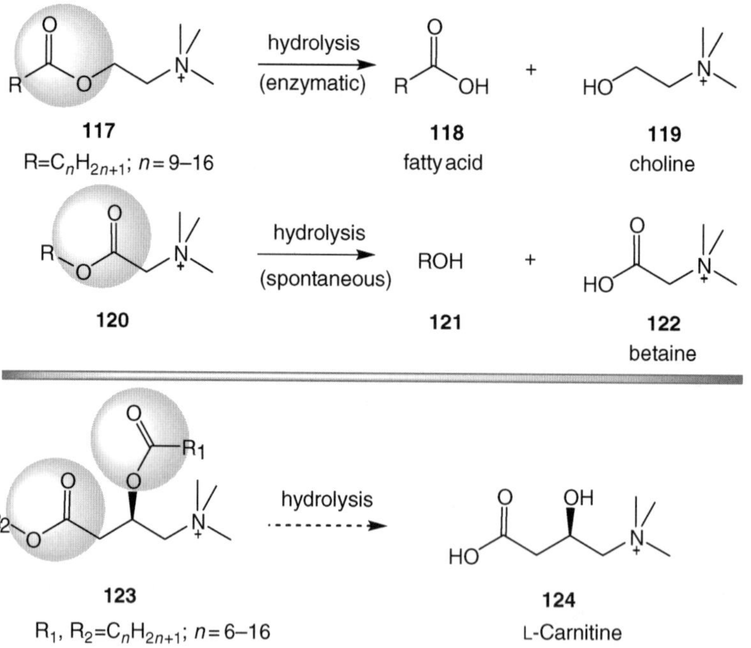

hydrolysis (enzymatic)

117 $R{=}C_nH_{2n+1}$; $n = 9–16$ | 118 fatty acid | 119 choline

hydrolysis (spontaneous)

120 | 121 ROH | 122 betaine

hydrolysis

123 R_1, $R_2{=}C_nH_{2n+1}$; $n = 6–16$ | 124 L-Carnitine

Figure 27. Soft quaternary ammonium antimicrobials designed as long-chain ester of choline (**119**), betaine (**122**), or L-carnitine (**124**). (This figure is available in full color at http://mrw.interscience.wiley.com/emrw/9780471266945/home.)

However, the decomposition of these compounds was not analyzed in detail. Judicious esterification (R_2) of the carboxy function is required for antimicrobial activity, but a free hydroxy group at position 3 of the carnitine skeleton does not annihilate activity. From the SD design point of view, it is also important to note that the common constituent of all these compounds, L-carnitine (**124**), has no pharmacological effects for doses of up to 15 g/day.

3.4.3. Soft Antiarrhythmic Agents

ACC-9358 Soft Analogs ACC-9358 (**125**, Fig. 28) is an orally active class Ic antiarrhythmic agent that underwent clinical trials and for which a number of soft analogs (**126**) were synthesized and tested [327]. Replacement of the formanilide function of ACC-9385 with alkyl esters resulted in compounds with similar antiarrhythmic activity. Esters attached directly to the aromatic ring of the bis(aminomethyl)phenol moiety (**126**, $n = 0$) were resistant to hydrolysis in human blood, but distancing the ester from the aromatic ring by one, two, and three methylene units (**126**, $n = 1, 2, 3$; $R = -CH_2CH(CH_3)_2$) afforded *soft* compounds with human blood half-lives of 8.7, 25.9, and 2.0 min, respectively [327]. As in most other cases, branching on the carbon attached to the oxygen atom of the alkoxy functionality tended to inhibit ester hydrolysis. The antiarrhythmic activity, as measured *in vitro* in the guinea pig right atrium, of a number of acid metabolites (**126**, $n = 0, 1, 2$; R = H) was indeed significantly less than that of the corresponding ester SDs.

In a follow-up study, additional esters **126** derived from aromatic and heterocyclic alcohols were investigated to improve the lipophilic character and enhance the *in vivo* potency, biodistribution, and duration profile [328]. A number of them showed consistent ability to convert acetylstrophanthidin-induced arrhythmias in guinea pig right atria to normal sinus rhythm with an ED_{50} of less than 10 μg/mL. Based on their shorter half-life in human blood, esters **126a** ($t_{1/2} = 3.5$ min) and **126b** ($t_{1/2} = 7.1$ min) were selected for *in vivo* evaluation. They both demonstrated greater potency than lidocaine in the 24 h Harris dog

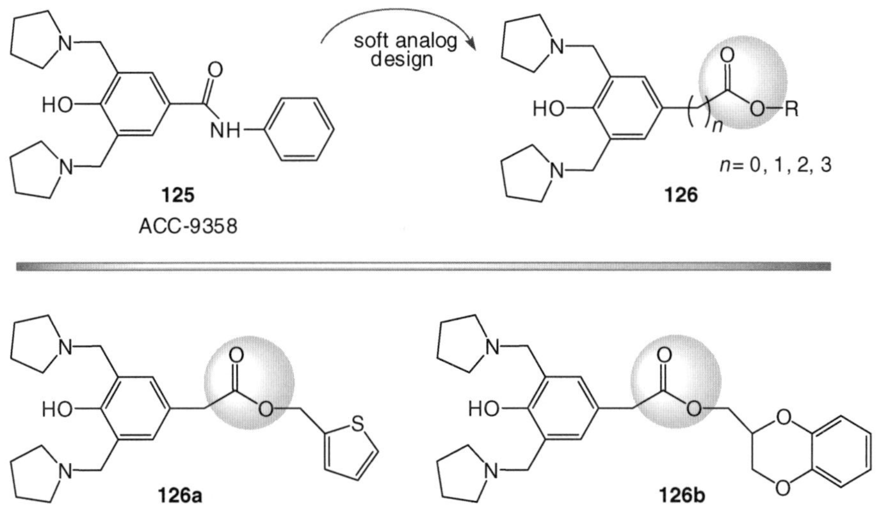

Figure 28. Soft analogs (**126**) of ACC-9358 (**125**), an orally active class Ic antiarrhythmic agent that underwent clinical evaluation. Soft analogs such as **126a** and **126b** showed good activity and sufficiently short half-lives in human blood. (This figure is available in full color at http://mrw.interscience.wiley.com/emrw/9780471266945/home.)

model and equal potency to lidocaine in the oubaine-intoxicated dog model [328]. In addition, the lipophilicities of these compounds were lower than that of lidocaine suggesting a lower ability to penetrate the blood–brain barrier and, thus, lower CNS liability. Considering all these observations, **126b** was chosen as a potential development candidate as it possessed the most desirable pharmacological and pharmacokinetic profile. Some analogs where the ester moiety was attached to the second aromatic ring of ACC-9385 [327] and some monoaminomethylene appended analogs were also explored [328].

Amiodarone Soft Analogs Among currently available antiarrhythmic agents, amiodarone (**127**), a structural analog of thyroid hormone, has electrophysiological effects that most closely resemble those of an ideal drug. However, amiodarone is highly lipophilic, is eliminated extremely slowly, and has unusually complex pharmacokinetic properties, frequent side effects, and clinically significant interactions with many commonly used drugs [329]. Consequently, despite its high efficacy, amiodarone is used only for life-threatening ventricular antiarrhythmias that are refractory to other drugs.

An active soft analog may solve many of these problems, and because amiodarone has a butyl side chain, its structure is well suited for such a design (Fig. 29). A number of possible soft analogs (ATI-2000 series, **128**) were synthesized and tested for activity and duration of action [330–333]. For example, the electrophysiological effects of the first investigated analog, the methyl ester ATI-2001 (**128**, $R = CH_3$), were found to be even greater than those of amiodarone in guinea pig isolated heart, and, in agreement with the SD design principles, they were more readily reversible. At equimolar concentration (1 μM), the soft analog caused significantly greater slowing of heart rate, depression of atrioventricular and intraventricular conduction, and prolongation of ventricular repolarization than amiodarone. However, unlike amiodarone, its effects were significantly reversed during washout of the drug.

Because the half-life of ATI-2001 in human plasma was found to be only 12 min, which may be too short to allow long-term management of cardiac arrhythmias, esters with longer or more branched side chains were also examined. These modifications were found to markedly alter the magnitude and time-course of the induced electrophysiological effects, and the *sec*-butyl and isopropyl esters were considered to merit further investigation [332]. In agreement with the principles of soft drug design, the

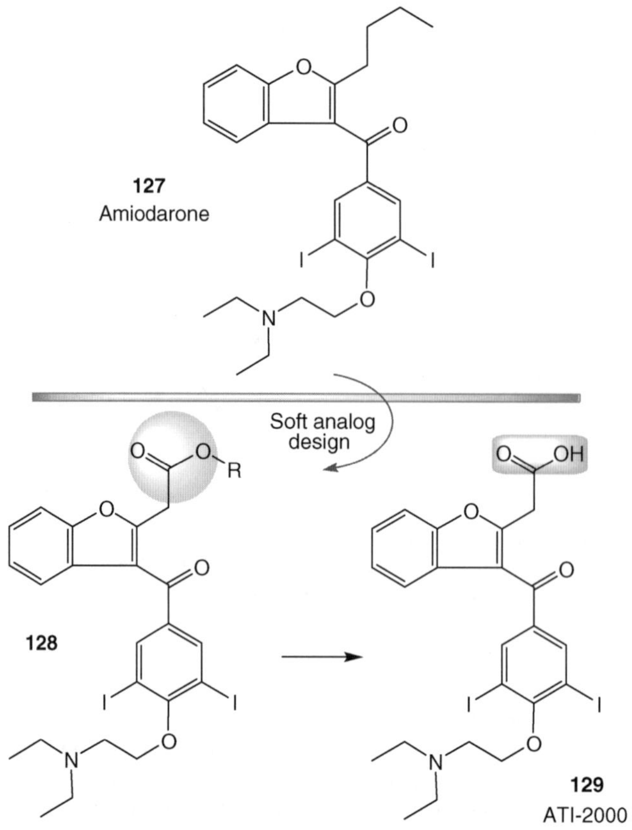

127 Amiodarone

Soft analog design

128

129 ATI-2000

128a R = Me (ATI-2001)
128b R = Et (ATI-2010)
128c R = *i*Pr (ATI-2064)
128d R = *i*Bu
128e R = *s*Bu (**ATI-2042**)
128f R = *n*Pe (ATI-2054)
128g R = cyclohexyl
128h R = (R)menthyl
128i R = (*R*)endobornyl
128j R = methyladamantane

Figure 29. Amiodarone (**127**) and possible soft analogs (ATI-2000 series, **128**), in which the butyl side chain at position 2 of the benzofurane moiety of amiodarone is replaced with an ester-containing side chain to allow facile hydrolytic degradation toward an inactive acid (**129**). (This figure is available in full color at http://mrw.interscience.wiley.com/emrw/9780471266945/home.)

common acid metabolite (**129**) was found to have no electrophysiological activity [332]. Taken together, these findings suggest that such SDs may prove to be a valuable addition to current antiarrhythmic therapy, but further studies are needed. ATI-2042 (budiodarone), the *sec*-butyl analog (**128e**) is currently in Phase II clinical trials for atrial fibrillation [333].

These compounds represent a possible example of an orally active SD. Even if the structures contain an ester moiety to allow enzymatic hydrolysis, it is possible to maintain activity for ester-containing drugs after oral administration. Indeed, many ester-containing drugs are orally administered. A study of the butyl ester prodrug of indomethacin in rats also showed that hydrolysis of the ester bond is mainly carried out in the circulatory system and the bond is barely hydrolyzed in the intestinal tract [334].

The above examples illustrate the generally applicable isosteric-type soft analog design, where an ester or a reversed ester function replaces two neighboring methylene groups. In some of these cases, when sufficient structural variability is introduced, the distinction between a soft analog and an (real or hypothetical) inactive metabolite-based

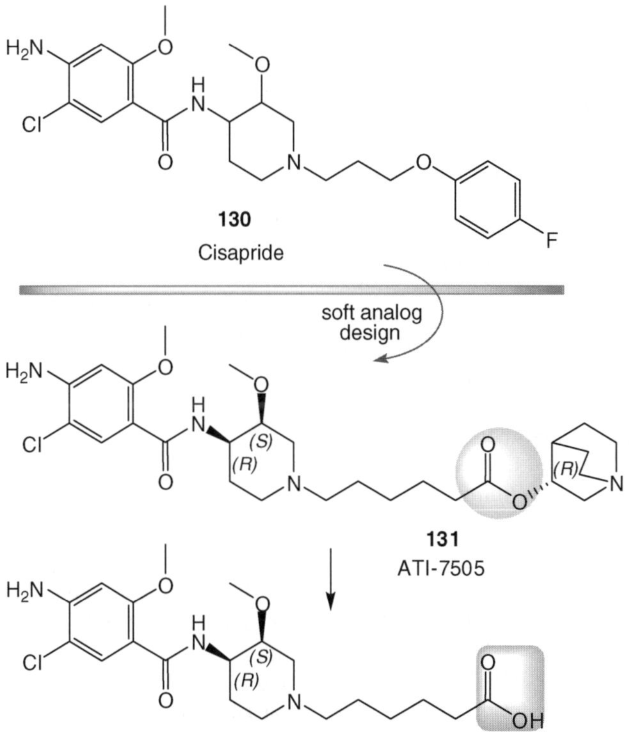

Figure 30. Cisapride (130) and its soft analog ATI-7505 (131) designed to avoid CYP450-mediated metabolism and the QT prolongation side effects of cisapride. (This figure is available in full color at http://mrw.interscience.wiley.com/emrw/9780471266945/home.)

design may become somewhat blurred. For example, larger esters can no longer be regarded as strict structural analogs, but they can be regarded as esters of the (hypothetical) inactive acid metabolite **129**.

3.4.4. Soft Serotonin Receptor Agonists Cisapride (**130**, Fig. 30), a serotonin 5-HT4 receptor agonist, is a parasympathomimetic that increases muscle tone in the esophageal sphincter used in gastroesophageal reflux disease (GERD) and in severe gastroparesis (e.g., in diabetic patients). In 2000, it has been withdrawn from the US market due to drug-related proarrhythmic events (QT prolongation) mainly occurring in patients taking other medications that are known CYP450 3A4 inhibitors, for example, erythromycin, fluconazole, and amiodarone. ATI-7505 (**131**; Fig. 30), an investigational 5-HT4 receptor agonist, was designed as a soft analog to have similar gastroprokinetic activity without the cardiac adverse effects by avoiding the CYP450 metabolism [335,336].

A randomized, parallel-group, double-blind, placebo-controlled study in healthy volunteers designed to evaluate the effects of 9-day treatment on scintigraphic GI and colonic transit found that ATI-7505 (10 or 20 mg, tid) accelerated gastric emptying, increased colonic transit, and accelerated ascending colon emptying [336]. This compound (**131**) has now also successfully completed Phase II clinical trials for the treatment of multiple gastrointestinal disorders including GERD and functional dyspepsia. In agreement with its SD design principles, a completed thorough QT (TQT) study designed to test the cardiac safety of ATI-7505 confirmed its safety profile by finding no significant effect on the QT interval at either therapeutic or supratherapeutic doses.

The group working on the development of ATI-2042 (**128e**) and ATI-7505 (**131**), is also working on other possible SD designs including compounds such as

- Soft warfarin analogs as oral anticoagulants [337]. The most promising compound of this series, tecarfarin (ATI-5923), is currently in Phase II/III clinical trials for the treatment of patients who are at risk for the formation of blood clots, such as those with atrial fibrillation or those at risk of venous thromboembolism. Consistent with its design focusing on esterase-mediated metabolism, a recent clinical trial confirmed that whereas fluconazole, a CYP450 inhibitor, increases the plasma concentrations of both (*R*)- and (*S*)-warfarin, it does not increases that of ATI-5923.
- Soft alosentron analogs as 5-HT3 receptor antagonists for the treatment of irritable bowel syndrome (IBS) [338].
- Short-acting soft hypnotic barbiturates [339].
- Soft fluvoxamine analogs as possible selective serotonin reuptake inhibitors (SSRIs) for the treatment of depression [340].
- Soft mibefradil analogs as calcium-channel blockers [341], and others.

3.4.5. Soft Angiotensin Converting Enzyme (ACE) Inhibitors ACE inhibitors, such as captopril (**133**) or enalapril, are widely accepted vasodilators in chronic heart failure. However, their use in acute conditions is restricted owing to the prolonged duration of their effect. Again, a *soft analog* approach may provide an active, ultrashort-acting ACE inhibitor that may represent a viable solution. In work based on these ideas, a number of captopril analogs (**134**, Fig. 31) with the proline amide bond of **133** replaced by esters susceptible to hydrolytic cleavage were investigated [342]. Whereas no captopril hydrolysis was observed in human blood, a number of soft analogs, especially those with thioalkyl substituents, were degraded in a sufficiently fast manner. Potency, as measured on purified rabbit lung ACE, could be further increased with larger substituents, but the hydrolysis of these compounds became unacceptably slow. Soft analog **135** (FPL 66564) was selected as a potential drug candidate because it showed the required balance of ACE inhibitory potency (5.7 nM) and degradation rate (human blood $t_{1/2} = 14$ min). As required by general SD design principles, its hydrolytic products (**136**,

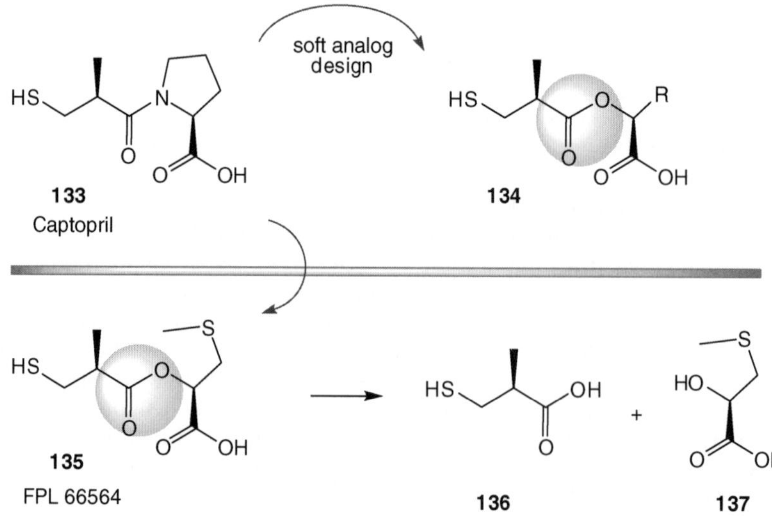

HS O N O OH
133
Captopril
soft analog design
HS O O R O OH
134
HS O O S O OH
135
FPL 66564
HS OH O + HO S O OH
136 137

Figure 31. Design and metabolism of ultrashort-acting ACE inhibitors (**134**) based on soft analogs of captopril (**133**). Hydrolytic cleavage of these compounds results in small, hydrophilic metabolites (e.g., **136**, **137**) that have no ACE inhibitory activity. (This figure is available in full color at http://mrw.interscience.wiley.com/emrw/9780471266945/home.)

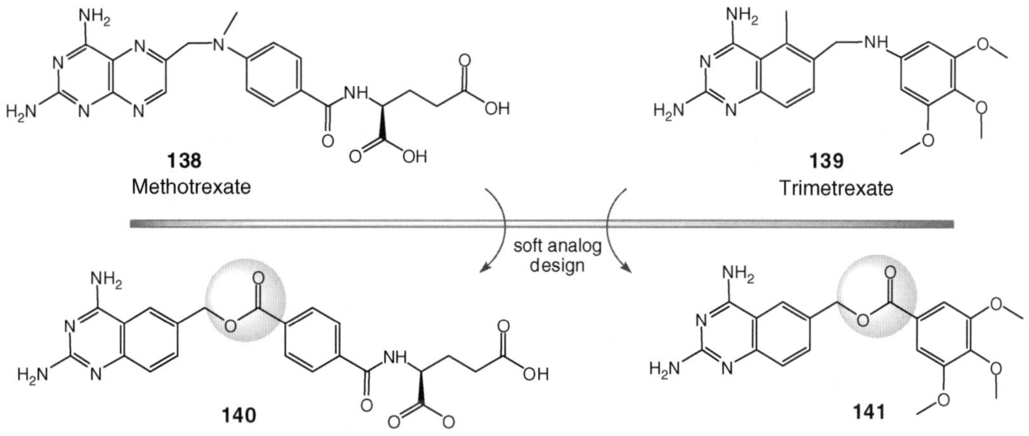

Figure 32. Design of soft analogs (**140, 141**) of methotrexate (**138**) and trimetrexate (**139**) as possible DHFR inhibitors. (This figure is available in full color at http://mrw.interscience.wiley.com/emrw/9780471266945/home.)

137) are without ACE inhibitor activity and, being small hydrophilic molecules, should not present any clearance problems. The *in vivo* testing of compound **135** showed a dose-dependent inhibition of angiotensin I pressor response in the anesthetized rat, but the effect rapidly dissipated to baseline levels on termination of the i.v. infusion [342].

3.4.6. Soft Dihydrofolate Reductase Inhibitors

A series of esters were explored as possible dihydrofolate reductase (DHFR) inhibitors that are susceptible toward hydrolytic degradation [229,343–347]. DHFR is involved in the reduction of dihydrofolate into tetrahydrofolate, and reduced folates are important cofactors in the biosynthesis of nucleic acids and amino acids. Hence, DHFR inhibitors can limit cellular growth. Consequently, classical DHFR inhibitors such as methotrexate (**138**, Fig. 32) or nonclassical DHFR inhibitors such as trimetrexate (**139**) have shown antineoplastic or antiprotozoal activity, and they could be useful in the treatment of inflammatory bowel disease (IBD), rheumatoid arthritis, psoriasis, and asthma. A number of compounds comprising a bicyclic aromatic unit connected by an ester-containing bridge to another aromatic ring (e.g., **140**, **141**) have been synthesized and investigated as possible therapeutic agents against *Pneumocystis carinii* pneumonia (PCP) and IBD with increased safety. Substitution of the methyleneamino bridge common to antifolates with an ester-containing bridge (Fig. 32) retained DHFR-inhibitory activity. The best ester-containing inhibitors were approximately 10 times less potent inhibitors than trimetrexate and piritrexim in the DHFR assay, but provided slightly better pcDHFR selectivity index, which was defined as the ratio of IC_{50} (rat liver DHFR) to IC_{50} (*P. carinii* DHFR). Furthermore, the hydrolytic metabolites were all poor inhibitors. *In vitro* hydrolysis using human and rat tissues or available esterases were relatively slow for most of the esters. Human and rat liver fractions were more active than human duodenal mucosa and human blood leukocytes in hydrolyzing the compounds. Contrary to **140**, effective *in vivo* hydrolysis and a favorable pharmacokinetic profile was found for **141** [344]. Finally, **140** exhibited good anti-inflammatory activity in a colitis model in mice [345], but showed unsatisfactory results in a rat arthritic model [343]. However, it showed some favorable *in vivo* effects at 15 mg/kg/day in a mouse colitis model [346].

3.4.7. Soft Calcineurin Inhibitors (Soft Immunosuppressants)

Compounds such as cyclosporine A (**142**) and tacrolimus (FK506; **144**) (Fig. 33) that block the signal transduction in T cells by inhibiting the phosphatase calcineurin A are widely used to treat and prevent organ transplant rejection [348]. They also have considerable therapeutic potential in the treatment of autoimmune diseases, such as asthma, psoriasis, atopic dermatitis, and rheumatoid arthritis. Unfortunately, long-term systemic exposure to such agents causes serious side effects such as nephrotoxicity, increased blood pressure, neurotoxicity, and hypertrichosis [349]. Again, a SD approach might prevent this problem. Because of the possibility of local delivery by inhalation, soft calcineurin inhibitors are of particular interest for asthma, a chronic inflammatory disease in which the airways become hyperreactive and constrict easily after a variety of diverse stimuli. The disease is characterized by a pulmonary inflammatory cell infiltrate, consisting mainly of eosinophils, neutrophils, and mononuclear cells, particularly activated T lymphocytes, and T cells seem to be key effector cells in the inflammatory pathology associated with asthmatic airways.

Soft Cyclosporine Analogs Cyclosporine (ciclosporin, cyclosporin A; Sandimmune™, Neoral™, Gengraf™ **142**) is a calcineurin inhibitor cyclic undecapeptide that is widely used as immunosuppressant in postallogeneic organ transplant to avoid organ rejection. A topical emulsion formulation (Restasis™) is available for the treatment of keratoconjunctivitis sicca (dry eye). Cyclosporine showed promising results in the treatment of asthma even when given orally [350] and especially when given by inhalation [351]. However, because its most serious side effect, nephrotoxi-

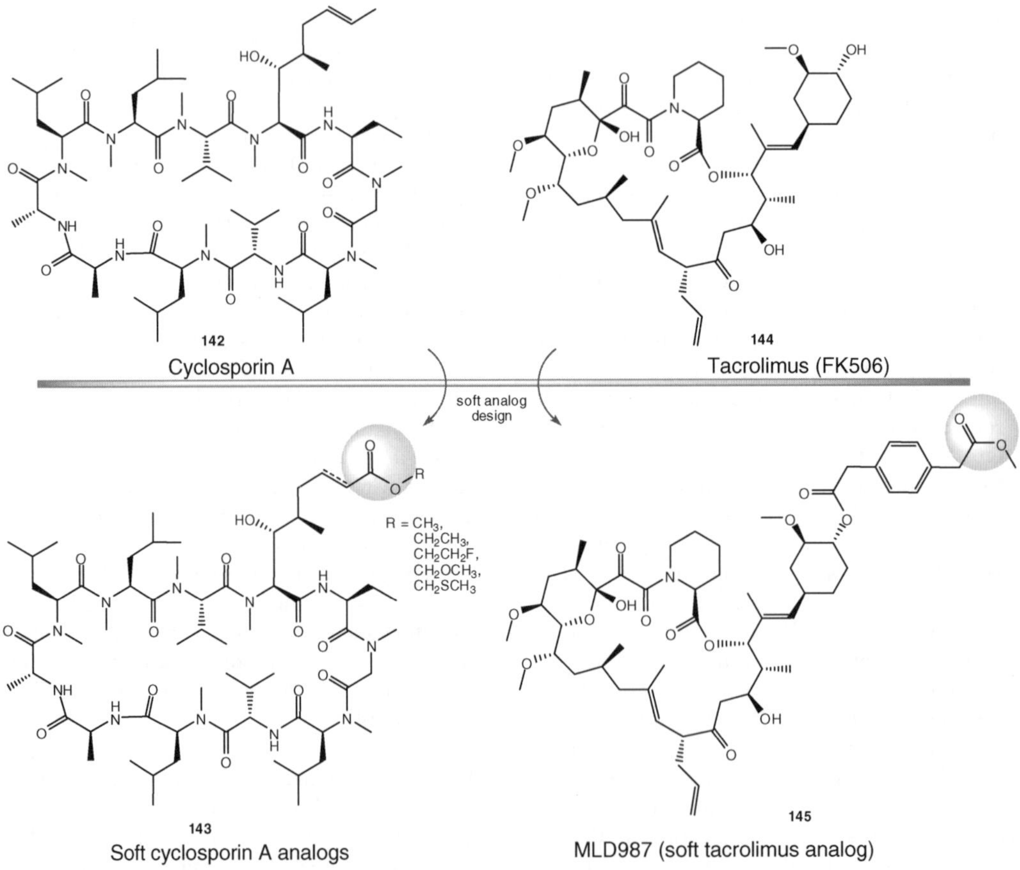

142 Cyclosporin A

144 Tacrolimus (FK506)

soft analog design

143 Soft cyclosporin A analogs

145 MLD987 (soft tacrolimus analog)

Figure 33. Soft immunosuppressants designed as soft analogs of the calcineurin inhibitors cyclosporine A (**142**) and tacrolimus (FK506, **144**) as possible inhalation therapies for asthma. (This figure is available in full color at http://mrw.interscience.wiley.com/emrw/9780471266945/home.)

city, seems to be mechanism-based (i.e., caused by calcineurin inhibition), a SD approach may be particularly promising to identify a locally active, but systemically nontoxic compound. Two series of drugs with an ester moiety inserted into the side chain of the main cyclosporine ring have been synthesized and tested (**143**, saturated and unsaturated side chain; Fig. 33) [352]. This unusual amino acid side chain is known to be essential (but not sufficient) for activity, and it is also the most accessible portion of the molecule for synthetic modifications. The olefin cross metathesis reaction with a ruthenium catalyst provided an efficient one-step synthetic transformation route from cyclosporine A to the soft analogs **143** [352]. A number of simple ester derivatives of both the saturated and the unsaturated side chain (**143**) were synthesized and tested for calcineurin inhibitory activity together with the corresponding acid metabolites (**143**, R = H) as well (Fig. 33). Some of the unsaturated ester analogs showed activity similar to cyclosporine with IC_{50}s in the 500–700 nM range, while the saturated ester analogs were less active. Both acid metabolites (**143**, R = H) showed no calcineurin inhibition activity ($IC_{50} > 5\,\mu M$), and a preliminary toxicology study confirmed the nontoxic nature of the unsaturated acid derivative. Hence, this approach may yield useful SD analogs of cyclosporine with considerable therapeutic potential for local activity with minimal systemic side effects.

Soft Tacrolimus Analogs Along the same conceptual lines as for cyclosporine, a soft analog of tacrolimus (FK506; Prograf™, **144**), another widely used calcineurin inhibitor, has also been explored as a potential candidate for the inhalation therapy of asthma [353]. The corresponding carboxylic ester incorporating compound, MLD987 (**145**) (Fig. 33) retained the potent immunosuppressant activity and inhibited the activation, proliferation, and release of cytokines from T cells with IC_{50} values in the low nanomolar range. In a rat model of allergic asthma, MLD987 reduced the influx of leukocytes into bronchoalveolar lavage fluid samples obtained from antigen-challenged animals when given into the airways by intratracheal administration (ED_{50} of 1 mg/kg) or by inhalation (ED_{50} of 0.4 mg/kg) while having an appreciably weaker activity when given orally or intravenously. The corresponding acid metabolite was less active in all assays. PK evaluations of **145** in rat and in rhesus monkey confirmed its low oral (2–4%) and pulmonary (7%, monkey) bioavailability. MLD987 administered i.t. in actively sensitized BN rats challenged 3 h previously with ovalbumin dose dependently reduced the response to adenosine at doses that are not anti-inflammatory (hence, might reflect direct suppression of mast cell activation) [354]. These findings are consistent with a local action making it likely to have an improved TI compared to other selective T-cell inhibitors [353].

3.4.8. Soft Cytokine Inhibitors In another SD approach toward possible asthma medications, a series of triazinylphenylalkyl-thiazolecarboxylic acid esters that are inhibitors of the production of interleukin (IL)-5, a primary eosinophil-activating and proinflammatory cytokine, have been synthesized and evaluated by a group at Johnson & Johnson Pharmaceutical Research [355]. Since the influx of leukocytes (eosinophils, lymphocytes, and monocytes) into the airways and their production of proinflammatory cytokines contribute to the severity of allergic asthma, a localized cytokine inhibitor is of considerable therapeutic interest, and inhaled SDs could provide the needed lung-specificity. These structures, represented here by **147** (Fig. 34) are loose structural analogs of R146225 (**146**), a novel, orally active inhibitor of IL-5 biosynthesis identified earlier [356], whose development was halted due to teratogenic effects. In fact, this could be considered an example of *de novo* SD design as compounds were screened for a novel target and close isosteric analogy was not a main guiding principle as the design did not start from well-known existing drugs.

Most esters were rather resistant to metabolic conversion when incubated with lung S9 fraction, but several were readily metabolized when incubated with liver S9 fraction. The combination of metabolic and activity data, suggested the hydroxypropyl ester (**147**, R = $(CH_2)_3OH$) as the most promising candidate. It showed good metabolic stability ($t_{1/2} > 240$ min) in human lung S9 fraction, but rapid

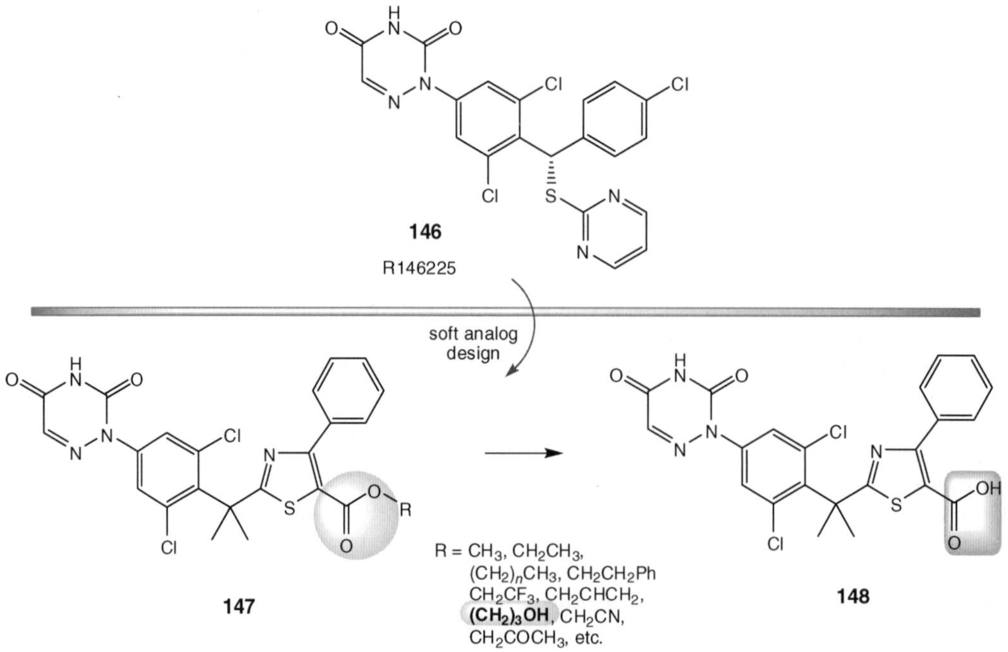

146
R146225

soft analog design

147

R = CH_3, CH_2CH_3, $(CH_2)_nCH_3$, CH_2CH_2Ph CH_2CF_3, CH_2CHCH_2, $(CH_2)_3OH$, CH_2CN, CH_2COCH_3, etc.

148

Figure 34. Soft cytokine inhibitors (**147**) designed as lung-specific SDs inhibiting the production of IL-5. (This figure is available in full color at http://mrw.interscience.wiley.com/emrw/9780471266945/home.)

($t_{1/2} = 15$ min) conversion into the pharmacologically inactive carboxylic acid **148** in human liver preparations. In stimulated human whole blood cultures, it reduced not only the production of IL-5 (IC_{50} of 78 nM) but also the biosynthesis of the monocyte chemotactic proteins MCP-1 (IC_{50} of 220 nM), MCP-2 (580 nM), and MCP-3 (80 nM). Its intratracheal administration to allergic sheep (6 mg/animal), either before (−4 h) or after (+1.5 h) pulmonary allergen challenge, completely abrogated the late-phase airway response and reduced the bronchial hyperreactivity to inhaled carbachol while causing only short-lived and minimal plasma exposure, hence, making it a promising candidate for the topical treatment of allergic disorders.

3.4.9. Soft Matrix Metalloproteinase (MMP) Inhibitors A series of phosphonamide-based inhibitors showed potent (micromolar) inhibitory activity against the shedding of epidermal growth factors, amphiregulin (AR), and heparin-binding EGF-like growth factor (HB-EGF), which are suspected to participate in the development of psoriasis. However, because they also inhibited matrix metalloproteinases (MMPs), soft analogs **149** (Fig. 35) that are susceptible to hydrolytic inactivation were investigated to avoid adverse effects such as the musculoskeletal syndrome of MMP inhibitors [357]. Activity was configuration sensitive: only the (*R*,*R*) configuration shown in **149** showed potent inhibition. In this series, hydrolysis of a phosphate ester and not that of a carboxylic ester was expected to provide metabolic inactivation. The P-OEt ester bond of the ethyl ester turned out to be stable in human plasma, but di- and trifluoroethyl esters (**149b**, **149c**) hydrolyzed rapidly ($t_{1/2}$ of 10 and <1 min, respectively) and resulted in inactive metabolites (**151**, **152**) via a spontaneously decomposing phosphonic acid intermediate (**150**). It seems that presence of an electron-withdrawing substituent that can cause an increase in the positive charge on the *P* atom is needed to facilitate enzymatic hydrolysis of these compounds. Because topical application of these fluoro-esters significantly suppressed tetradecanoylphorbol acetate (TPA)-induced epidermal hyperplasia in murine skin, a model of psoriasis, these compounds could have considerable therapeutic potential as antipsoriatics.

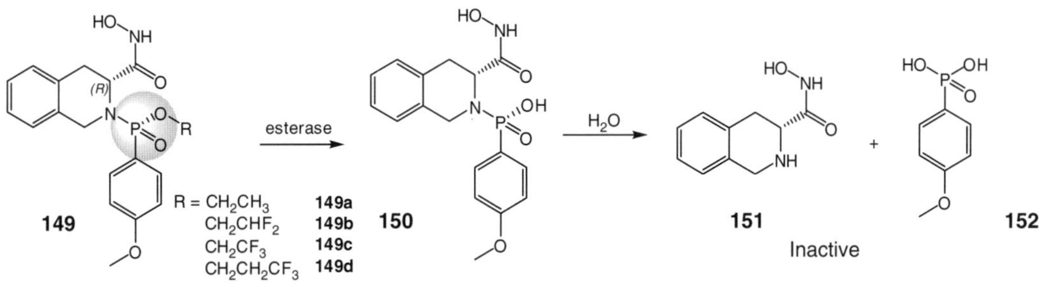

Figure 35. Phosphonamide-based soft metalloproteinase inhibitors (149) investigated as possible antipsoriatic drugs.

3.4.10. Soft Cannabinoids The most important pharmacologically active member of the cannabinoid family is Δ^9-tetrahydrocannabinol (THC, **153**, Fig. 36), the main active constituent of marijuana, a psychoactive agent used for thousands of years. Cannabinoids have many potential therapeutic benefits, one of the more interesting ones being the reduction of the intraocular pressure [358]. They are known to produce significant and dose-dependent IOP-lowering even if applied topically [358,359]. However, despite apparently doing this by a different mechanism than other antiglaucoma drugs [360], their therapeutic potential is diminished because this effect could not be separated from strong CNS and cardiovascular effects.

Soft SP-1 Analogs As in other cases discussed here, a topically applied, SD susceptible to metabolic inactivation may afford separation between desired local effect and unwanted systemic side effects. Consequently, new soft cannabinoids such as **156** were synthesized and evaluated [361,362]. They are structural analogs of SP-1 (**154**), a nitrogen-containing cannabinoid derivative that has been shown earlier to have IOP-lowering activity [359,360,363]. In agreement with the SAR hypothesis used for the design, all the compounds that were successfully synthesized by using a Pechmann condensation had IOP-lowering activity, but the common acid metabolite of the soft analogs (**157**) was inactive. Activities were evaluated in a number of *in vivo* experiments following both i.v. and topical administration in rabbits. The results obtained were somewhat equivocal due to the extremely low aqueous solubility of the compounds tested, but they were in agreement with a soft drug that is active and is rapidly inactivated. For example, when administered i.v. in rabbits, the ethyl-substituted soft analog **156** ($R = C_2H_5$, $R' = C_6H_5$) produced IOP-lowering that was parallel in both eyes, lasted for only 15 min, and had a maximum of $18 \pm 3\%$ ($p < 0.005$). When dissolved in emulphor (EL-719 PF618; PEG 40 castor oil) and administered topically, the same soft analog produced an IOP-lowering that lasted longer and was significant ($p < 0.05$) at $t = 1$ h. In a first evaluation [362], an indirect response E_{max} PK/PD model [364] connecting plasma concentration and IOP-lowering effect explained well the experimental results obtained in rabbits following i.v. administration [362].

Soft Biphenylic Analogs In another, entirely separate effort, structurally simple biphenylic derivatives such as **155**, which have been found to be good cannabinoid receptor CB-ligands and potent CB-agonists, were used as leads to design metabolically labile soft cannabinoids by incorporation of ester moieties into their long, aliphatic side chain [365]. Some of the ester analogs such as **158** and **159** showed promising binding activity at both the CB1 and CB2 receptors with K_i values in the low micromolar range—less than the lead, which had activity in the mid-nanomolar range, but, most importantly, the corresponding acid metabolites were completely devoid of any affinity for either receptor subtype. Hence, these soft cannabinoids can also become promising local analgesic or antiglaucoma agents devoid of psychotropic side effects.

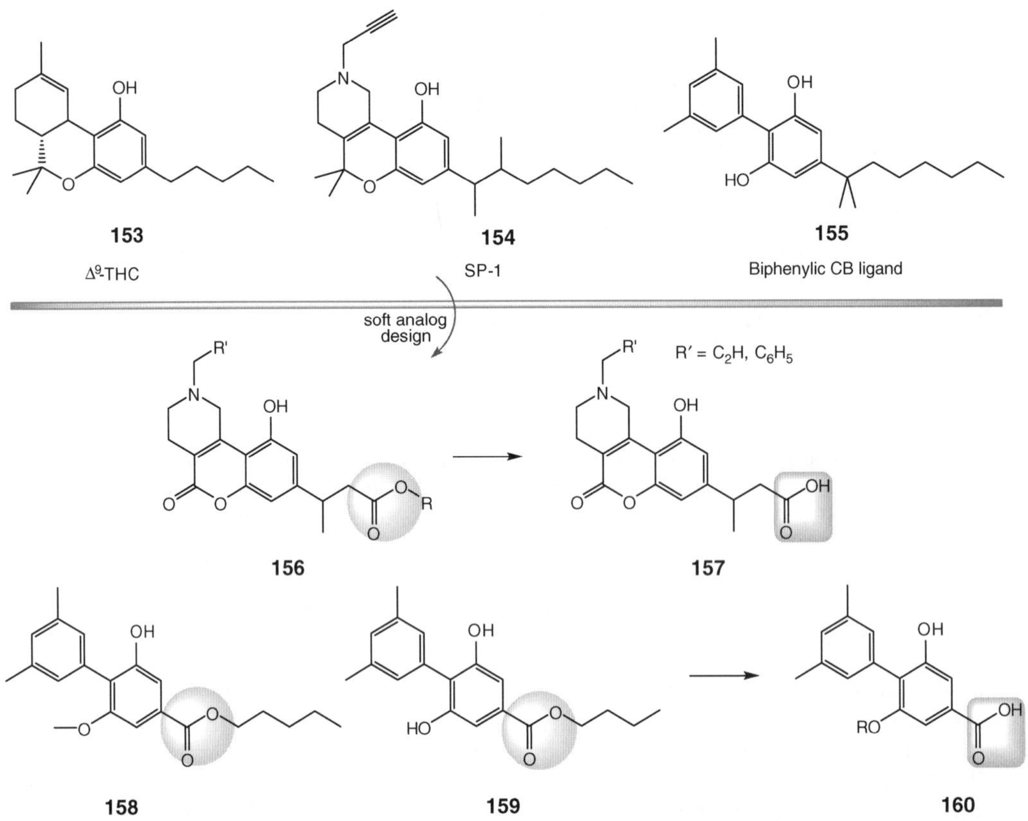

Figure 36. Structures of Δ^9-THC (**153**) and two other cannabinoids (SP-1, **154**, and the biphenylic **155**) that served as leads for two different soft cannabinoid designs represented by **156** [362] and **158/159** [365], respectively. (This figure is available in full color at http://mrw.interscience.wiley.com/emrw/9780471266945/home.)

3.4.11. Soft Benzodiazepine Analogs Since their introduction in the 1960s, benzodiazepines, a class of psychoactive drugs with hypnotic, sedative, anxiolytic, anticonvulsant, muscle relaxant, and amnesic properties, have been widely used to provide anxiolysis and sedation in numerous clinical settings. According to our current knowledge [366], benzodiazepines act at a specific site on the γ-aminobutyric acid type A ($GABA_A$) receptor, which is referred to as the benzodiazepine receptor, and enhance the action of the inhibitory neurotransmitter GABA via positive allosteric modulation of the receptor–chloride channel complex. A set of soft benzodiazepines were developed originally by GlaxoSmithKline on the basis of the remifentanil experience, and were intended as i.v. agent with a predictable fast-onset, short duration of action, and rapid recovery profile [367,368]. They are soft analogs of diazepam (**161**), bromazepam (**162**), or midazolam (**163**), and several ester-containing side chains at the 3-position were explored. A number of them maintained high binding affinity at the benzodiazepine receptor (<50 nM), showed good (>100-fold) separation of activity between the (active) ester and the (inactive) acid, and were able to cause loss of the righting reflex (LRR) in rats, an effect associated with benzodiazepine full agonism. Structures such as CNS 7259X (**164**) [367,369] and CNS 7056 (**166**) [368,370] (Fig. 37) were selected for development, which has been taken over by CeNeS and later PAION.

Selectivity studies confirmed that CNS 7056 (**166**) is highly selective for the $GABA_A$ receptor and that its acid metabolite CNS

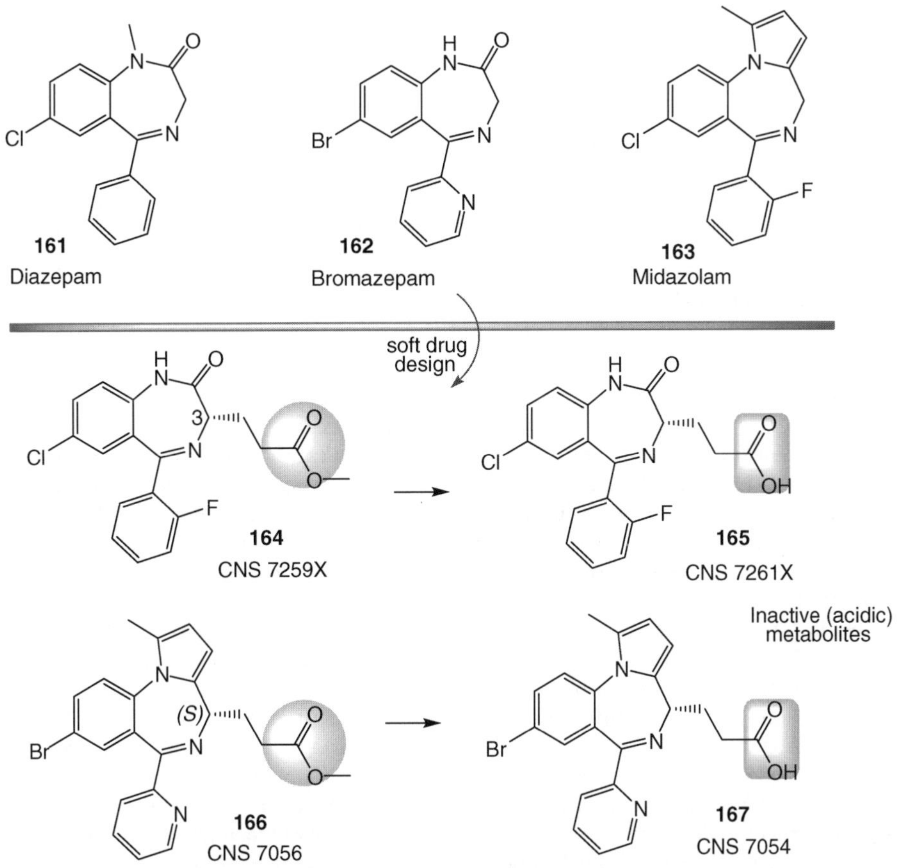

Figure 37. Soft benzodiazepines such as CNS 7259X (164) and CNS 7056 (166) developed as ultrashort-acting soft analogs of classical benzodiazepines such as diazepam (161) or midazolam (163), respectively. (This figure is available in full color at http://mrw.interscience.wiley.com/emrw/9780471266945/home.)

7054 (**167**) has no measurable affinity for any off-target site tested. In competition assays to displace for [^{3}H]flunitrazepam binding in tissue homogenates from human, rat, and pig brain, the acid metabolite **167** indeed showed a 3–400-fold reduced activity compared to the parent drug **166** having, for example, pK_i values of 4.91 versus 7.53 in human brain homogenate [370]. CNS 7056 induced rapid sedation in rats and mice, as measured by the LRR test, and while the onset time was similar to that observed with midazolam, the duration of LRR was short (<10 min)—clearly shorter than that induced by midazolam. In the meantime, the carboxylic acid metabolite **167** produced no LRR even after i.v. doses up to 100 mg/kg [370]. Escalating doses of CNS 7056 were also studied in sheep and were found to produce sedation for 9–25 min without excessive respiratory or cardiovascular depression [371]. CNS 7056 has recently completed successfully a Phase I study as i.v. sedative/anesthetic showing promising results and no safety concerns. Consequently, because of its expected quick on/off sedative profile and thus anticipated faster recovery, CNS 7056 is being developed for patients undergoing painful out- and inpatient procedures such as colonoscopy since fast recovery after sedation is in the interest of both the treating physician and the patient.

3.4.12. Soft Anesthetics As already discussed, etomidate (Amidate) (**1**, Fig. 3), a short

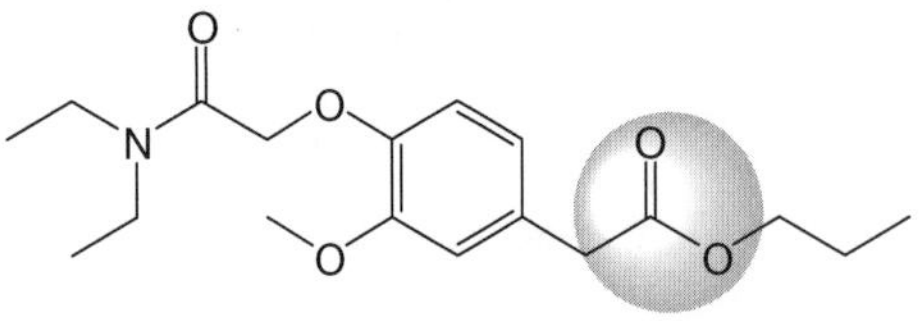

168

Propanidid

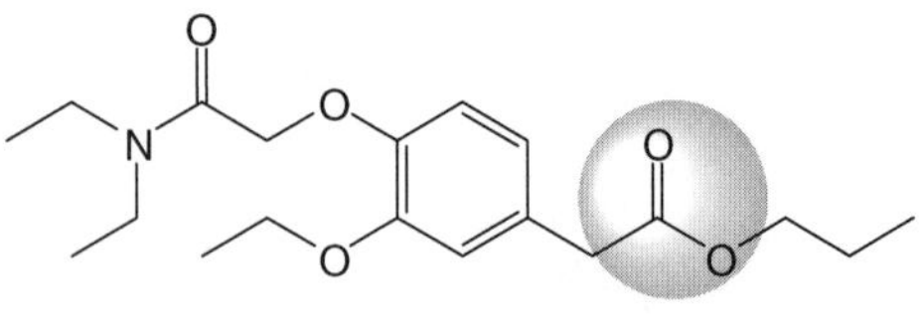

169

AZD3043 (TD-4756, THRX-918661) / (proposed structure)

Figure 38. Propanidid (**168**) and AZD3043 (TD-4756, **169**) as possible short-acting soft anesthetics incorporating a metabolically labile ester moiety.

acting i.v. anesthetic agent, can be considered a SD since it is rapidly deactivated by hepatic and plasma esterases [96]. Propanidid (Epontol™) (**168**) (Fig. 38) is an ultra short-acting phenylacetate general anesthetic originally introduced by Bayer. It contains an ester and is rapidly inactivated by esterases. Anaphylactic reactions caused it to be withdrawn, but it is possible that these reactions were in fact caused by the Cremophor solubilizing agent, a polyethoxylated castor oil, used in its formulation. AZD3043 (**169**), which is being developed by AstraZeneca (originally, TD-4756/THRX-918661 by Theravance), also seems to be a soft anesthetic [369] that is a $GABA_A$ allosteric modulator. Its structure has not been officially disclosed yet, but has been suggested [372] to be a close propanidid analog. It has been shown to produce rapid hypnosis following i.v. bolus administration or infusion in rats, mice, guinea pigs, ferrets, dogs, cats, pigs, and mini-pigs and to rapidly hydrolyze into its acid metabolite both *in vitro* (e.g., $t_{1/2}$ ~10 min in human blood) and *in vivo* [372–374].

3.4.13. Soft Ca^{2+} Channel Blockers

Perhexiline Analogs Perhexiline is a Ca^{2+} channel blocker that is effective in the treatment of angina pectoris, the most common symptom of chronic ischemic heart disease, but is of limited use due to side effects such as hepatotoxicity, weight loss, and peripheral neuropathy. As these undesirable effects are related to the slow metabolism and accumulation of perhexiline, soft analogs may represent a conceivable alternative. A class of analogs with an amide moiety inserted as a possible enzymatically labile center was investigated for perhexiline [375–377]. Most of these newly designed compounds proved to be more active calcium antagonists on depolarized pig coronary artery than perhexiline and produced concentration-related coronary dilation in the perfused guinea pig heart. Some of them also provided promising *in vivo* results. However, metabolic degradation of amide moieties may not be fast enough to provide adequate metabolic lability and may take place only with the involvement of cytochrome P450.

Clevidipine Clevidipine (Cleviprex™) (**170**, Fig. 39) is an ultrashort-acting, soft calcium channel blocker for i.v. use in the reduction and control of blood pressure in cardiac surgical procedures when oral therapy is not feasible or not desirable [88,378–381]. It has been recently approved by the FDA for intravenous use in lowering high blood pressure on the basis of several large-scale clinical trials [382,383]. It is a dihydropyridine-type calcium channel antagonist, structurally related to felodipine (**173**), but with an additional, less hindered ester moiety introduced into its structure through an acyloxyalkyl-type substitution. Contrary to the two, hindered esters already present in the felodipine structure, this ester is susceptible to enzymatic degradation and leads to an acidic, inactive metabolite (**172**) through an unstable intermediate (**171**) (Fig. 39) [88]. Such acyloxyalkyl-type double esters have been extensively used in prodrug design to trigger sufficiently fast, substitution controllable degradations [43]. This design resulted in sufficiently fast degradation (e.g., *in vitro* human blood $t_{1/2}$ of 5.8 min) and formation of an equivalent amount of the corresponding pri-

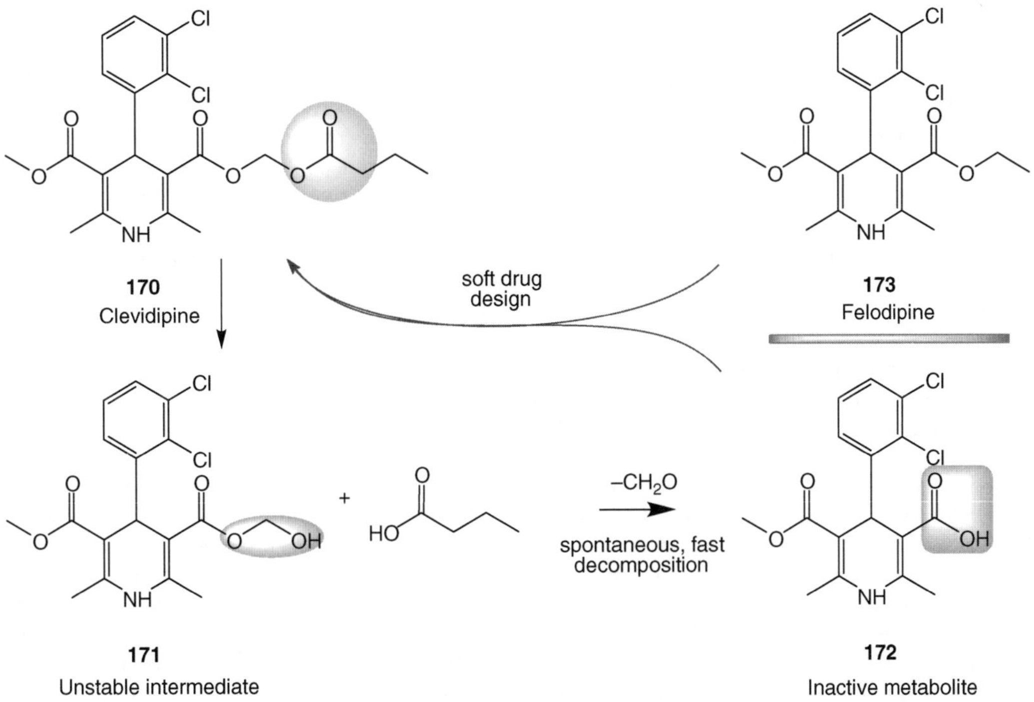

Figure 39. Metabolic pathway of clevidipine (**170**), a soft calcium channel antagonist structurally related to felodipine (**173**), but with an acyloxyalkyl type substitution to ensure hydrolytic sensitivity. (This figure is available in full color at http://mrw.interscience.wiley.com/emrw/9780471266945/home.)

mary metabolite, which has been shown to be devoid of any vasodilating effect both in animal and human experiments [88,378]. Clevidipine is a high clearance drug with a small volume of distribution, which results in very short half-lives. Recent large-scale clinical trials suggest that clevidipine offers a safe and effective alternative to older agents for treating perioperative or other acute severe hypertension [382,383].

Clevidipine has two possible enantiomers, and hydrolysis half-lives were only approximately 10% different for the *R* and *S* isomers. A slight stereoselectivity in the extensive plasma binding of clevidipine (free fractions of 0.43% and 0.32% for *S*- and *R*-clevidipine, respectively) might be the reason for this difference [88,380]. Hence, from a pharmacokinetic point of view, there seems to be no advantage in using a single enantiomer instead of a racemic mixture. The hydrolytic half-life of clevidipine was also investigated in blood from pseudocholinesterase-deficient volunteers, because the homozygous atypical allele for the corresponding gene has been reported to occur with a frequency of approximately 1 in 3500 for Caucasians. In the small number of subjects studied, *in vitro* half-life increased by only approximately 50% [88].

3.5. Active Metabolite-Based Soft Drugs

Active metabolite-based SDs are metabolic products of a drug resulting from oxidative conversions that retain significant activity of the same kind as the parent drug [16,19–21]. Most drugs undergo stepwise metabolic degradation to yield intermediates and structural analogs (D_1, D_2, and so on) that often have similar activity as the original drug molecule (D) (Fig. 1). These general oxidative metabolic transformations put a burden on the saturable and slow oxidative enzyme system, and result in compounds that have different selectivity, binding, distribution, and elimination properties. Therefore, at any given time, a mixture of active components is present in continuously changing relative concentration. This can result in complex, almost uncontrollable situations that make safe and effective dosing almost impossible.

In agreement with basic SD design principles, judicious selection of an active metabolite can yield a potent drug that will undergo a one-step deactivation process, as it is already at the highest oxidation state. For example, if sequential oxidative metabolic conversions take place, such as the quite common hydroxyalkyl → oxo → carboxy sequence in which the terminal carboxy function is generally the inactive form, the best choice for a drug would be the oxidative metabolite just one step before deactivation. Despite numerous publications suggesting the utility of active metabolites for drug design purposes [16,19–21], the idea was long neglected and only recently began to generate interest. Nevertheless, there are examples of active metabolites used as a source of new drug candidates because of better safety profiles. Examples of well-known active metabolites, mostly generated by oxidative and dealkylating processes, include, for example [384], oxyphenbutazone, the active *p*-hydroxy metabolite of phenylbutazone; oxazepam, the common active metabolite of chlordiazepoxide, halazepam, chlorazepate, and diazepam; pravastatin, the hydroxylated structure derived from mevastatin (compactin); fexofenadine, the active oxidative (acid) metabolite or terfenadine; cetirizine, the active oxidative (acid) metabolite of hydroxyzine, a first-generation H_1-receptor antagonist; or desloratadin, the active metabolite of loratadine.

Bufuralol (**21**, Fig. 7), the nonselective β-adrenoceptor antagonist with β_2 partial agonist properties discussed earlier, provides a good illustration for the principles of active metabolite-based SD design. As mentioned in Section 3.3.1.5, bufuralol undergoes complex metabolism in humans, including stepwise oxidation toward an acid metabolite (**24**) through the corresponding hydroxy (**22**) and keto (**23**) intermediates (Fig. 7). These intermediates are still active: the β-antagonist ED_{50} for inhibition of tachycardia in rats are 169, 46/284, and 203 μg/kg for bufuralol (**21**), the two stereoisomers of **22**, and **23**, respectively [136,137]. They also have different and long elimination half-lives: biological half-lives are 4, 7, and 12 h for bufuralol (**21**), **22**, and **23**, respectively [138]. Differential metabolism of the two enantiomers and differences due to genetic polymorphism also occur [139]. According to the principle of active metabolite design, the ketone **23**, the active metabolite in the highest oxidative state, should be the drug of choice [20]. This compound still retains most of the activity, but is deactivated by one-step oxidation to structure **24**.

3.6. Activated Soft Drugs

Activated soft compounds are not analogs of known drugs, but are derived from nontoxic chemical compounds activated by introduction of a group that provides pharmacological activity. During expression of activity, the inactive starting molecule is regenerated as a result of a hydrolytic process. An example of activated soft compound is provided by *N*-chloramine antimicrobials. During an effort to identify locally active antimicrobial agents of low toxicity, *N*-chloramines based on amino acids, amino alcohol esters, and related compounds were developed. These compounds (Fig. 40), particularly those derived from α-di-substituted amino acid esters and amides [385,386], serve as a source of positive chlorine (Cl^+), which was assumed to be primarily responsible for antimicrobial activity. However, when the chlorine is lost, before or after penetration through microbial cell walls, the nontoxic initial amine is regenerated (Fig. 40).

The antimicrobial activity of chloramines was known before, but chemical instability and high reactivity limited their widespread use. After establishing the mechanism of their decomposition [385], it was realized that stable *N*-chloramines that have much lower "chlorine potential" and are much less corrosive could be obtained if the α-carbons lack hydrogen [385–388]. Structures **174–176** in Fig. 40 illustrate some of these low chlorine potential, soft chloramines. Structure **174** proved to be a particularly effective bactericide in a laboratory water treatment plant [389]. It is exceptionally stable in water

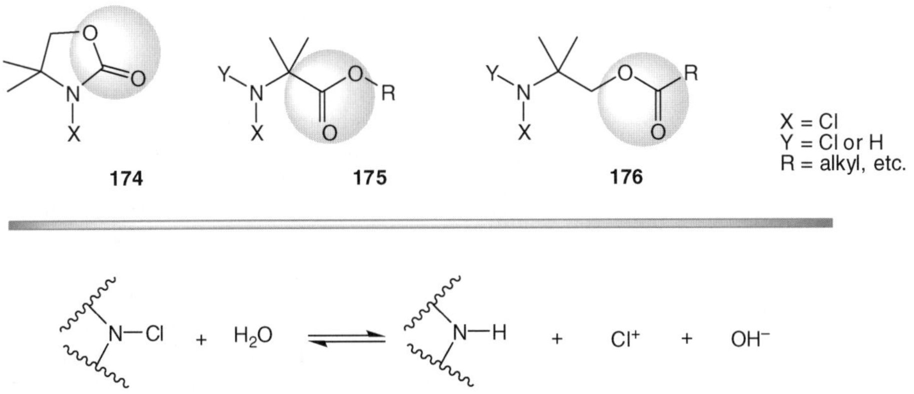

174 175 176

X = Cl
Y = Cl or H
R = alkyl, etc.

N–Cl + H_2O ⇌ N–H + Cl^+ + OH^-

Figure 40. General structures of representative low chlorine potential, soft *N*-chloramine antimicrobials (**174–176**) and the mechanism of their decomposition. (This figure is available in full color at http://mrw.interscience.wiley.com/emrw/9780471266945/home.)

and in dry storage [390,391]; it is nontoxic in chicken's drinking water at 200 mg/L, and it detoxifies aflatoxin [392]. The mechanism of action of these compounds involves inhibition of bacterial growth by inhibiting DNA, RNA, and protein synthesis [393]. When the chlorine interacts with –SH containing enzymes, the original precursor is regenerated. Thus, the predictable and controllable metabolism of soft chloramines gives a definite advantage over hard, lipophilic, aromatic antimicrobials containing C–Cl bonds.

Another example of activated soft compound is provided by the class of soft alkylating agents prepared during the development of soft quaternary salts [39]. These compounds are α-halo esters and are relatively mild alkylating agents. Their low alkylating potential should allow transport to tumor cells without indiscriminate alkylation. In addition, being activated esters, the circulating free part can be deactivated by esterases. These should result in a better separation of desired activity and unwanted toxicity. One of these compounds, chloromethyl hexanoate, was found to have anticancer activity in the P388 lymphocytic leukemia test [39].

3.7. Prosoft Drugs

As their name implies, prosoft drugs are inactive prodrugs (chemical delivery forms) of a soft drug of any of the above classes including endogenous soft molecules. They are converted enzymatically into the active SD, which is subsequently enzymatically deactivated. Soft drugs, as any other drug, can be the subject of prodrug design resulting in prosoft drugs, but, as results from the SD definition, one cannot conceive a "soft-prodrug." Some examples of pro-SDs have already been discussed; for example, the colon-targeted prosoft drug (**55**) for possible oral treatment of ulcerative colitis [222] and possible prosoft antipsoriatic β-adrenergic agonists obtained by esterification of **75** (Section 3.3.6) [254].

Derivatives of natural hormones and other biologically active agents such as neurotransmitters have well-developed mechanisms for their disposition and, therefore, can be considered as *natural soft drugs* (e.g., hydrocortisone and dopamine). Because their metabolism is usually fast and their transport is specific, they can become useful as drugs only if sustained, local, or site-specific delivery is developed for them. Hence, compounds designed for such purposes can also be considered pro-SDs.

A possible example for sustained chemical release at the site of application for hydrocortisone (**30**), a natural SD, is shown in Fig. 41. The 4,5-unsaturated 3-ketone group, being essential for binding and activity, is a good target for modification. Spirothiazolidine derivatives (**177**) were selected [394] because they should be subject to the biological cleavage of the imine formed after spontaneous cleavage of the carbon-sulfur bond [395].

Spirothiazolidines of hydrocortisone acetate were approximately three to four times more potent than hydrocortisone derivatives when tested for topical anti-inflammatory activity. Meanwhile, they delivered significantly less hydrocortisone transdermally than the unmodified hydrocortisone or hydrocortisone 21-acetate. These results are consistent with a local tissue binding through a disulfide bridge as suggested in Fig. 41. The opening of the spirothiazolidine ring of **177** as shown in structure **178** allows trapping of the steroid with a disulfide bridging (**179**) at the site of application. Hydrolysis releases the active component (**30**) only locally [396]. A similar behavior of the thiazolidine of progesterone confirmed this local binding in the skin [394].

3.8. Computer-Aided Design

Owing to the considerable flexibility of retrometabolic drug design, for certain lead compounds a large number of possible soft structural analogs can be designed by applying the various "soft transformation rules." Finding the best drug candidate among them may prove tedious and difficult. Fortunately, computer methods developed to calculate various molecular properties, such as molecular volume, surface area, charge distribution, polarizability, aqueous solubility, partition and permeability coefficient, and even hydrolytic lability, make possible more quantitative design [6,397–409]. The capabilities of quantitative design have been further advanced by developing expert systems that combine the various structure-generating rules of SD design with the developed predictive software to provide a ranking order based on isosteric/

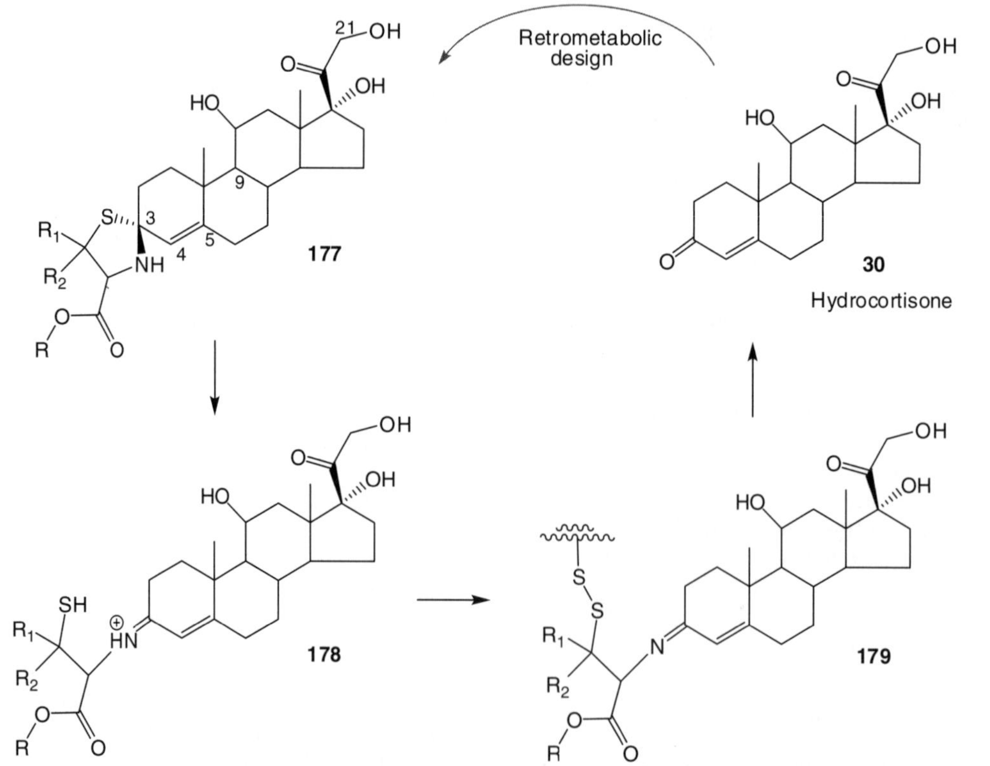

Figure 41. Hydrocortisone (**30**) can be regarded as a natural SD, and a spirothiazolidine derivative (**177**) serves as a pro-SD for controlled release. Opening of the thiazolidine ring (**178**) is followed by disulfide bridging (**179**) to trap the steroid at the site of application.

isoelectronic analogy [27,409–412]. The architecture and the interface of a recent version of the corresponding computer program are shown in Fig. 42.

The overall approach is general in nature and can be used starting with essentially any lead. The system can provide full libraries of possible new "soft" molecular structures and an analogy-based ranking of these candidates making possible more thorough and more quantitative design. Because candidates that are unlikely to have reasonable activity can be eliminated ahead of synthesis and experimental testing, considerable time-, labor-, and financial savings can be achieved.

3.8.1. Structure Generation The expert system designed to aid the generation and ranking of novel SD candidates has been described in the literature in detail [27,409–412]. As an important part, it contains the rules to transform certain substructures of the lead compound according to the principles of retrometabolic design and generate full virtual libraries of candidate compounds. For the SD approach, the two most successful strategies, the inactive metabolite-based and the soft analog strategies, were implemented. Following the rules of inactive metabolite-based design, it generates common oxidative metabolites of the lead compound and then functionalizes them to generate new SDs. For example, it finds $-CH_3$, $-CH_2OH$, or other alkyl groups in the lead and replaces them by $-COOH$ or $-(CH_2)_n-COOH$ groups, and then, it generates new soft drugs by functionalizing these "oxidized" metabolites to generate active SDs (i.e., $-CH_3$/$-CH_2OH \rightarrow -COOH \rightarrow -COOR$ transformations). Following the rules of soft analog design, the program generates SD structures by looking for neighboring methylene/hydroxymethylene groups ($-CH_2-CH_2-$) and replacing them with corresponding esters $-O-CO-$, reversed esters $-CO-O-$, or other isosteric/isoelectronic functions (i.e., $-CH_2-CH_2-$/$-CH_2-O- \rightarrow -COO-$/$-OCO-$ transformations).

3.8.2. Candidate Ranking Because in certain cases a large number of analog structures may be designed, it is desirable to have some

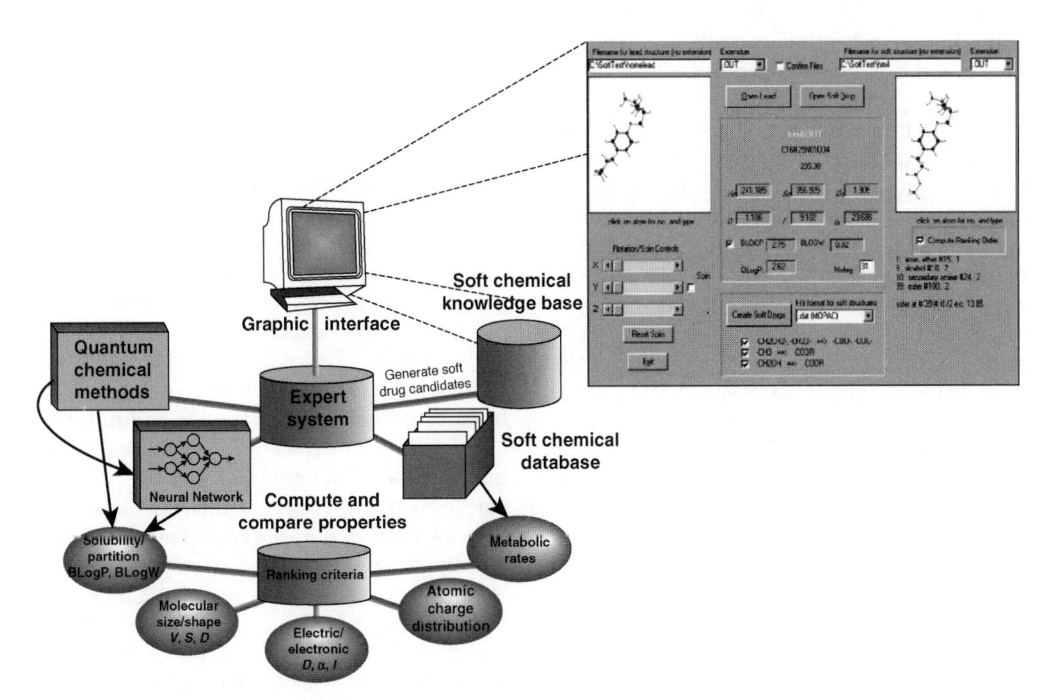

Figure 42. The conceptual architecture and the graphical user interface (GUI) of the computer-aided soft drug design program.

prediction regarding their activities and/or toxicities. The main underlying assumption of the present system is that structures that are closer steric and electronic analogs of the original lead are also more promising drug candidates. Accordingly, to quantify this analogy, a ranking-order is calculated on the basis of the closeness of calculated properties (descriptors) to those of the lead compound. The present expert system provides a ranking order using fully optimized geometries obtained from advanced semi-empirical AM1 calculations [413] for all the compounds that are of interest.

Because most new structures are close structural analogs (often exact isomers), it is important to include as many and as relevant properties as possible. At present, four parameter categories are used with equal weights to describe isosteric-isoelectronic relations: (i) molecular size/shape descriptors $\mathbb{S}$ (V, volume; S, surface area; and O, ovality); (ii) electronic properties $\mathcal{E}$ (D, dipole moment; α, average polarizability; and I, ionization potential); (iii) predicted solubility/partition properties $\mathcal{P}$ (log W and log $P_{o/w}$); (iv) and atomic charge distribution $\mathbb{Q}$ on the unchanged portions. For computable descriptors, size/shape-, electrostatic-, and H-bond-related descriptors consistently seem to be the most important factors describing physicochemical properties [414]. Each of these parameters can be obtained from the optimization output, and they all play important roles in determining binding and transport properties; they should give a relatively good description of the isosteric/isoelectronic analogy. Because all these properties are measured in different units and vary over different ranges, ranking factors (RF) were introduced for each property ($RF_{\mathbb{S}}$, $RF_{\mathcal{E}}$, $RF_{\mathcal{P}}$, and $RF_{\mathbb{Q}}$), and a weighted average is used for the final ranking [27,409–412]. Following synthesis and experimental testing of a few best candidates, further refinements are possible; for example, experimental results might suggest different weighting for the properties used in ranking. Illustrative examples were also provided in details in the reviews mentioned [27,409–412].

3.8.3. Hydrolytic Lability Hydrolytic lability is also of considerable relevance for a ranking of SD candidates as it is a measure of their "softness" (metabolic stability); therefore, we undertook a quantitative structure-metabolism relationship (QSMR) study to identify a structure-based QSMR equation that is not limited to congener series [405]. *In vitro* human blood data was used because it represented the data of interest available in the largest number over the widest range of structures under comparable experimental conditions. An equation accounting for 80% of the variability of the *in vitro* human blood log $t_{1/2}$ of 7 different drug classes containing a total of 67 compounds was obtained by introducing the inaccessible solid angle Ω_h as a measure of steric hindrance, and the current version of the computer-aided SD design expert system [409,412] uses this equation to estimate log $t_{1/2}$. The final equation used to estimate log $t_{1/2}$, in addition to $\Omega_h^{O=}$, includes the AM1-calculated charge on the carbonyl carbon ($q_{C=}$) and a calculated log octanol–water partition coefficient ($Q\,\text{Log}P$) [401,404,407] as parameters:

$$\log t_{1/2} = -3.805 + 0.172\,\Omega_h^{O=} - 0.146 q_{C=} + 0.112\,Q\,\text{Log}\,P$$
$$n = 67, \quad r = 0.899, \quad \sigma = 0.356, \quad F = 88.1 \tag{4}$$

In general agreement with previous results, this QSMR suggests steric effects as having the most important influence on the rate of enzymatic hydrolysis. Lipophilicity, as measured by $Q\,\text{Log}\,P$ and some of the electronic parameters, such as the charge on the carbonyl C ($q_{C=}$), also proved informative, but to a much lesser degree. Half-lives were found to increase with increasing steric hindrance around the ester moiety as measured by the inaccessible solid angle Ω_h. An important novelty was the finding that the rate of metabolism as measured by log $t_{1/2}$ seems to be more strongly correlated with the steric hindrance of the carbonyl sp^2 oxygen ($\Omega_h^{O=}$: $r^2 = 0.58$, $n = 79$) than with that of the carbonyl sp^2 carbon as measured by the inaccessible solid angle ($\Omega_h^{C=}$: $r^2 = 0.29$). This provides additional evidence for the important, possibly even rate-determining role played by hydrogen bonding at this oxygen atom in the mechanism of this enzymatic reaction.

Because most likely a number of different enzymes are involved in the hydrolysis of xenobiotics, it is unrealistic to expect accurate predictions of hydrolytic half-lives for arbitrary structures, and in a few cases, the model fails to fit compounds that are not hydrolyzed at all. Nevertheless, it is useful in distinguishing among compounds whose hydrolysis is fast, medium, or slow on the basis of structural information alone.

3.8.4. Illustrations

Esmolol as Homo-Metoprolol Analog A formal computer-aided SD design that uses homo-metoprolol (**180**) as lead compound provides a good illustration of the process of computerized structure-generation and candidate ranking. Starting with structure **180**, four different SD analogs are generated by oxidizing the methoxymethyl function to its corresponding carboxylic acid type metabolites and converting them into various esters (R methyl or ethyl) (**181a**, **181b**), respectively by replacing neighboring methylene groups with –O–CO– or –CO–O– functions (**181c**, **181d**), as shown in Fig. 43. Properties are then calculated based on AM1 optimized structures. Table 1 summarizes the analogy ranking in order of decreasing overall analogy together with the half-lives estimated using the present program (Fig. 42). Charge

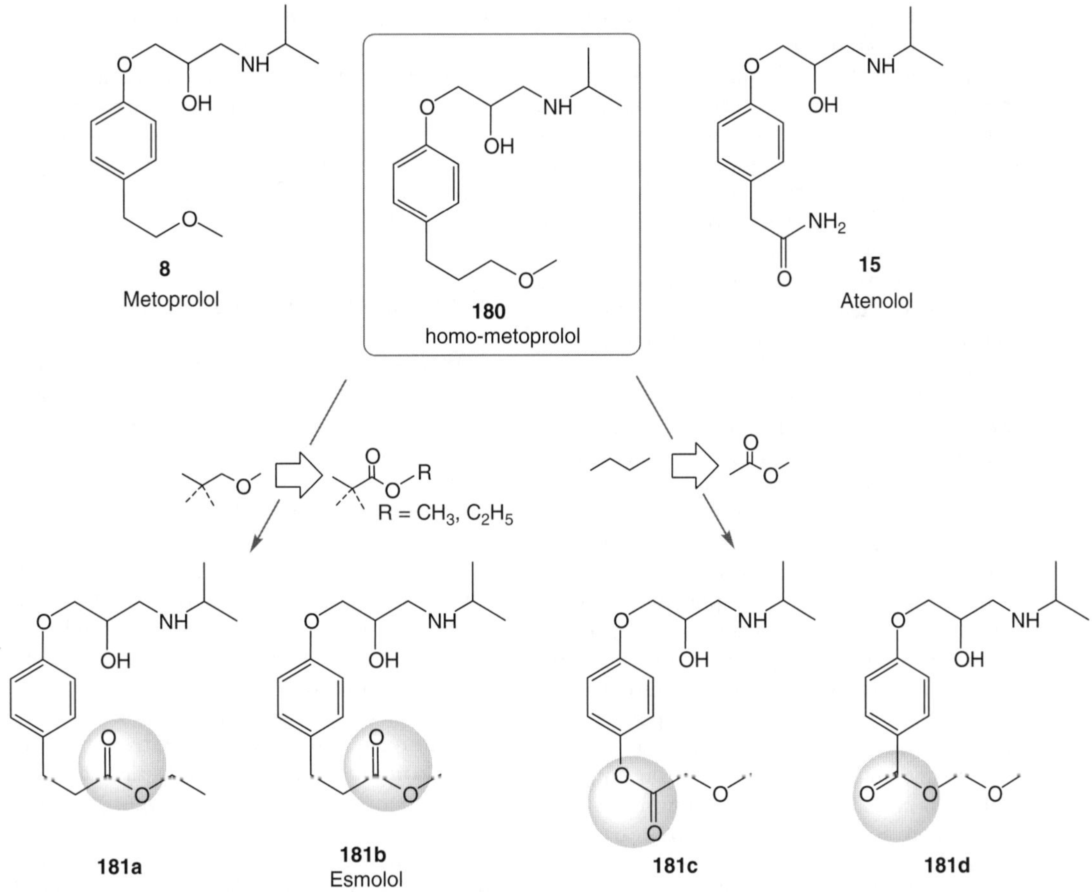

Figure 43. Illustration of the formal SD design process that uses homo-metoprolol (**180**) as lead to generate four soft drug analogs including esmolol (**181b**). (This figure is available in full color at http://mrw.interscience.wiley.com/emrw/9780471266945/home.)

Table 1. Ranking Order and Predicted Hydrolytic Half-Lives for Homo-Metoprolol (180) Analogs 181 (Fig. 43)

Compound	$RF_{\mathcal{S}}$*	$RF_{\mathcal{E}}$	$RF_{\mathcal{P}}$	$RF_{\mathcal{Q}}$	RF	Predicted $t_{1/2}$ (min)
181b	1.00	3.30	2.12	1.00	1.85	13.9
181c	13.12	2.59	9.94	18.99	11.16	21.3
181a	36.94	5.37	4.56	1.47	12.09	14.7
181d	17.11	16.18	13.88	15.89	15.76	1.8

*RF were computed as described in the references mentioned. Smaller values indicate better isosteric/isoelectronic analogy.

distributions were compared for atoms in the aromatic ring and the β-amino alcohol side chain (including hydrogens). The obtained results clearly show **181b** as an outstanding best analog in practically all categories. This is, of course, a hindsight design, but the nice aspect of it is that structure **181b** corresponds to esmolol (Brevibloc®), an ultrashort-acting β-blocker discussed earlier (**17**, Fig. 5).

Illustration: Virtual Soft Analog Library for Alprostadil Alprostadil (prostaglandin E1, PGE-1) (**182**, Fig. 44) is a potent vasodilator and a strong inhibitor of platelet aggregation. It is used in the treatment of erectile dysfunction, but needs localized delivery (as urethral suppository or injection). The metabolism of prostaglandins is often too fast or too slow to produce the desired activity with sufficient selectivity. Since most of the transformations are oxidative, by replacing them with hydrolytic processes, more controllable activity and targeting could be achieved. Alprostadil has a relatively lipophilic character ($\log P = 3.2$). Since it has two long side chains, it provides a good illustration for the possibility of generating relatively large libraries of soft analogs; here, 16 different SD analogs (**183a–183p**) obtained by replacing neighboring methylene groups in the two side chains with –O–CO– or –CO–O– functions are shown (Fig. 44). Quantitative differentiation among these structures could be particularly helpful as one certainly would not want to synthesize and test all of them; ranking calculations with the present system have been performed. Since these structures are relatively less rigid, results (particularly dipole moments and properties influenced by them) are considerably conformation-dependant; therefore, the ranking order should be treated with caution, especially because all generated compounds are exact structural isomers. Nevertheless, some differentiation could be made. Charge distributions were compared on the five-membered ring, the terminal acid function, and the side chain that was not altered. Charge on the ring seem less perturbed if the carbonyl and not the oxygen part of the introduced ester moiety is closer (**f** versus **a**, **g** versus **b**). Substitutions in the side chain with acidic termini seem to give better analogy; structures **c**, **g** (and **a** if charge distribution is disregarded) have scores that are particularly good. The second most preferred compound, structure **183g**, is part of a group of compounds patented by ONO Pharmaceuticals [415].

4. CHEMICAL DELIVERY SYSTEMS

CDS approaches provide novel, systematic methodologies for targeting active biological molecules to specific target sites or organs based on predictable enzymatic activation [24,33,35]. CDSs are inactive chemical derivatives of a drug obtained by one or more chemical modifications. They provide site-specific or site-enhanced delivery through sequential, multistep enzymatic and/or chemical transformations. The newly attached moieties are monomolecular units that are in general smaller or of similar size as the original molecule. Hence, the chemical delivery system term is used here in a stricter sense: they do not include systems in which the drug is covalently attached to large "carrier" molecules.

Bioremovable moieties used in CDS approaches can be classified into two general categories: targetor (T) moieties are responsible for targeting, site-specificity, and lock-in whereas modifier functions ($F_1 \ldots F_n$) serve as lipophilizers, protect certain functions, or fine-

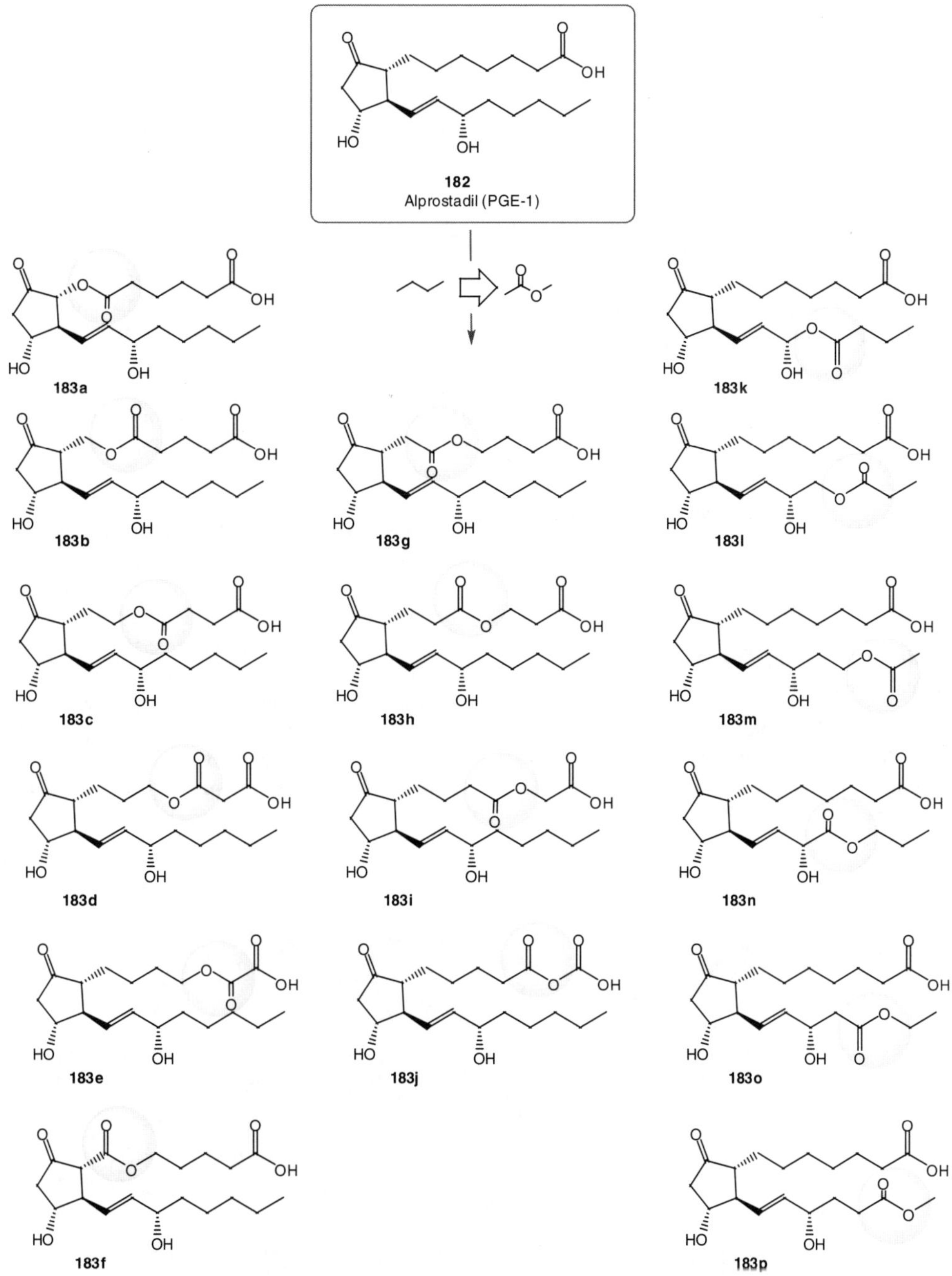

Figure 44. Virtual library of soft analogs (**183a–p**) generated for alprostadil (PGE-1; **182**). (This figure is available in full color at http://mrw.interscience.wiley.com/emrw/9780471266945/home.)

tune the necessary molecular properties to prevent premature, unwanted metabolic conversions. CDSs are designed to undergo sequential metabolic conversions, first disengaging the modifier functions and then the targetor moiety, after it fulfilled its site- or organ-targeting role. These transformations provide targeting (differential distribution) of the drug either by exploiting site-specific transport properties, such as those provided by the presence of a blood–brain barrier, or by recognizing specific enzymes found primarily, exclusively, or at higher activity at the site of action.

As of today, three major CDS classes have been identified:

1. *Brain-Targeting (Enzymatic Physical–Chemical-Based) CDSs*: exploit site-specific traffic properties by sequential metabolic conversions that result in considerably altered transport properties.
2. *Eye-Targeting (Site-Specific Enzyme-Activated) CDSs*: exploit specific enzymes found primarily, exclusively, or at higher activity at the site of action.
3. *Receptor-Based Transient Anchor-Type CDSs*: provide enhanced selectivity and activity through transient, reversible binding at the receptor.

4.1. Brain-Targeting (Enzymatic Physical–Chemical-Based) CDSs

4.1.1. Design Principles Brain-targeting chemical delivery systems represent the most developed CDS class. To obtain such a CDS, the drug (D) is chemically modified to introduce the protective modifier function(s) (F_i) and the targetor (T) moiety. Upon administration, the resulting CDS is distributed throughout the body. Predictable enzymatic reactions convert the original CDS by removing some of the protective functions and modifying the T moiety, leading to a precursor form (T^+-D), which is still inactive, but has significantly altered physicochemical properties (Fig. 45). Whereas these intermediates are continuously eliminated from the "rest of the body," they accumulate at the targeted site, for which efflux/influx processes are altered due to the presence of some specific membrane or other distribution barrier. Consequently, release of the active drug essentially occurs only at the site of action.

For example, the blood–brain barrier can be regarded as a biological membrane that is in general permeable to lipophilic compounds, but not to hydrophilic molecules (Fig. 46). In most cases, these transport criteria apply to both sides of the barrier. The BBB is a unique membranous barrier that tightly segregates the brain from the circulating blood [416,417]. Capillaries of the vertebrate brain and spinal cord lack the small pores that allow rapid

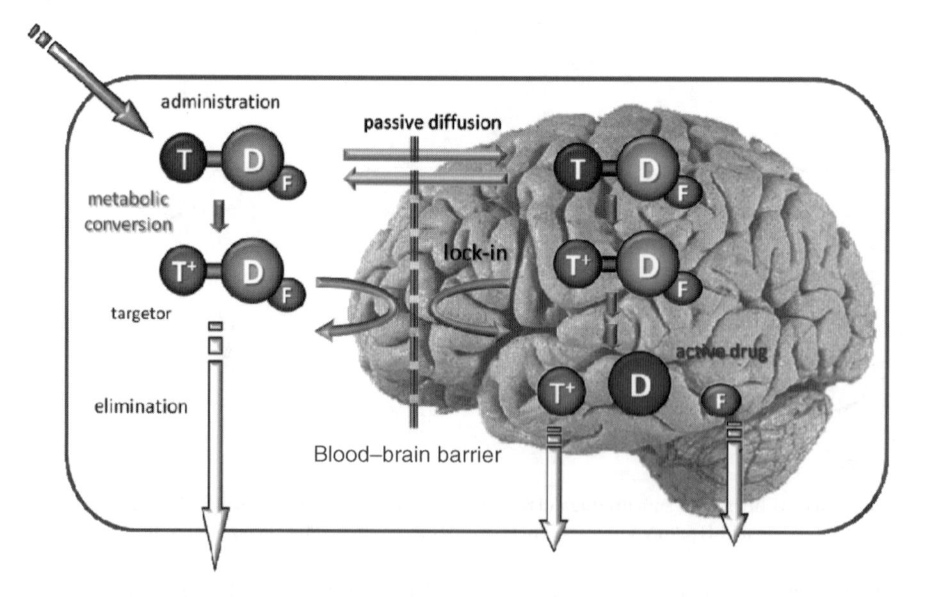

Figure 45. Schematic representation of the sequential metabolism employed by brain-targeting CDSs to allow targeted and sustained delivery of the active drug D. (This figure is available in full color at http://mrw.interscience.wiley.com/emrw/9780471266945/home.)

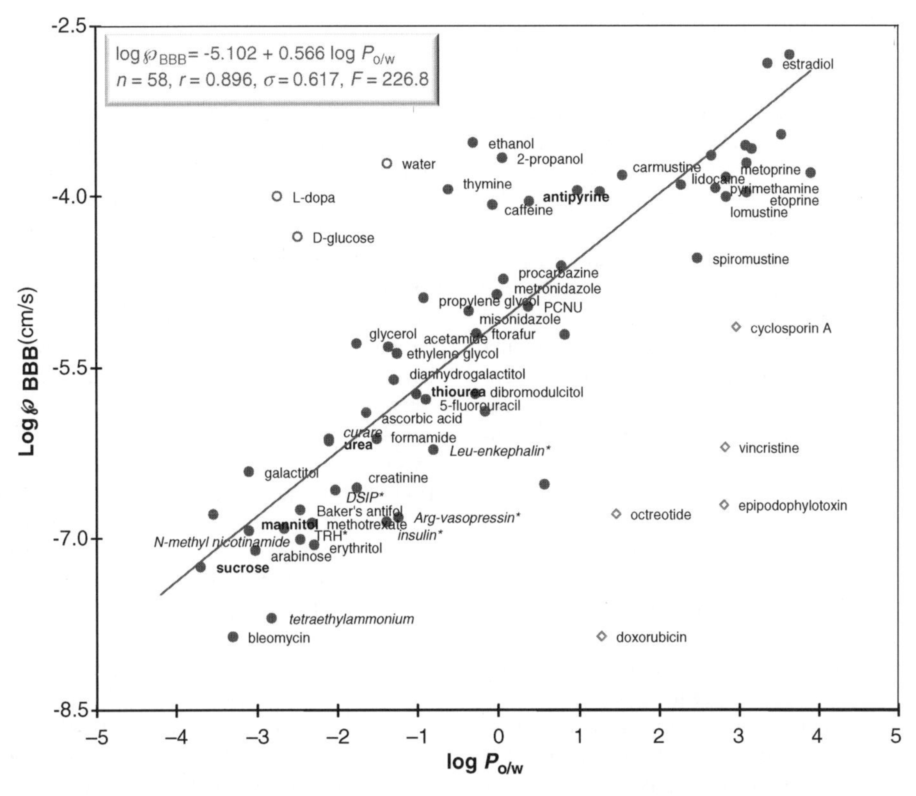

Figure 46. *In vivo* log permeability coefficient of rat brain capillaries (log $\wp_{BBB}$) as a function of log octanol/water partition coefficient (log $P_{o/w}$), the most commonly used measure of lipophilicity. For compounds in italic, the log distribution coefficient measured at physiological pH was used (log $D_{o/w}$). Values denoted with a star are for guinea pig. The strong deviants below the line that are denoted with diamonds are known substrates for P-glycoprotein, a multidrug transporter that actively removes them from the brain. (This figure is available in full color at http://mrw.interscience.wiley.com/emrw/9780471266945/home.)

movement of solutes from circulation into other organs; these capillaries are lined with a layer of special endothelial cells that lack fenestrations and are sealed with tight junctions. Tight epithelium, similar in nature to this barrier, can also be found in other organs (skin, bladder, colon, and lung) [418]. It is now well established that these endothelial cells, together with perivascular elements such as astrocytes and pericytes, constitute the blood--brain barrier. In brain capillaries, intercellular cleft, pinocytosis, and fenestrae are virtually nonexistent; exchange must pass transcellularly. Therefore, only lipid soluble solutes that can freely diffuse through the capillary endothelial membrane may passively cross the BBB. In general, such exchange is overshadowed by other nonspecific exchanges. Despite the estimated total length of 650 km and total surface area of 12 m^2 of capillaries in human brain covered by only ~1 mL of total capillary endothelial cell volume, this barrier is very efficient and makes the brain practically inaccessible for lipid insoluble compounds, such as polar molecules and small ions. The BBB also has an additional, enzymatic aspect: solutes crossing the cell membrane are subsequently exposed to degrading enzymes present in large numbers inside the endothelial cells that contain large densities of mitochondria, metabolically highly active organelles. Furthermore, active transport can significantly alter both inward and outward transports for compounds that are substrates of the corresponding transporters.

Not surprisingly, practically all drugs currently used to treat brain disorders are lipid soluble and can readily cross the BBB following oral administration. Brain delivery of other therapeutics requires some strategy to overcome the BBB [419]. Delivery of compounds such as neuropeptides or oligonucleotides is further complicated by their metabolic lability [420]. Various more or less sophisticated brain-targeted drug delivery approaches have been explored to overcome these difficulties including, for example, intracerebral delivery, intracerebroventricular delivery, intranasal delivery, BBB disruption, nanoparticles, receptor-mediated transport (vector-mediated transport or "chimeric" peptides), cell-penetrating peptides, and prodrugs [419]. Simple attempts, such as transient osmotic opening of the BBB [421,422], shunts, or biodegradable implants [423,424], are seriously limited by their invasive nature and their many toxic side effects. Glycosylation, cationization, or polymer conjugation have also been attempted ([420] and references therein). Some more recent attempts, such as those based on cell-penetrating peptides [425–427] or brain transport vectors [428], focus on exploiting natural transporters. They might provide solutions, but there are still many unresolved issues. Another possible strategy is to smuggle compounds across disguised as lipophilic precursors. Heroin, a diacyl derivative of morphine, is a notorious example that crosses the BBB approximately 100 times more easily than its parent drug just by being more lipophilic making it one of the most addictive and socially harmful drugs [429]. Codeine, which also is a derivative of morphine and has one hydroxyl group methylated, also has a close to 10-fold increased permeability. Along similar lines, prodrug approaches can be used to improve the brain uptake of drugs [430], but simple prodrugs suffer from important limitations such as nonselectivity in target-site delivery. As mentioned, the CDS concept evolved from the *prodrug* concept, but became essentially different by the introduction of targetor moieties and by the employment of multistep activation.

The CDS approach exploits the idea that if a lipophilic compound enters the brain and is converted there into a hydrophilic molecule, it will become "locked-in": it will no longer be able to come out (Fig. 45). With such a system, targeting is assisted because the same conversion, as it takes place in the rest of the body, accelerates peripheral elimination and further contributes to brain targeting. In principle, many targetor moieties are possible for a general system of this kind [431–440], but the one based on the 1,4-dihydrotrigonelline ↔ trigonelline (coffearine) system, in which the lipophilic 1,4-dihydro form (T) is converted *in vivo* to the hydrophilic quaternary form (T^+) (Fig. 47), proved the most useful. This conversion takes place easily everywhere in the body, because it is closely related to the ubiquitous NAD(P)H ↔ $NAD(P)^+$ coenzyme system (**186** ↔ **187**) associated with numerous oxidoreductases and cellular respiration [441,442]. Because oxidation takes place with direct hydride transfer [443] and without generating highly active or reactive radical intermediates, it provides a nontoxic targetor system [444]. Furthermore, it was shown [445] that the trigonellinate ion formed after cleavage of the CDS undergoes rapid elimination from the brain, most likely by involvement of an active transport mechanism that eliminates small organic ions; therefore, the T^+ moiety formed during the final release of the active drug D from the charged T^+-D form will not accumulate within the brain [445,446].

Whereas, the charged T^+-D form is locked behind the BBB into the brain, it is easily eliminated from the body as a result of the acquired positive charge, which enhances water solubility. After a relatively short time, the delivered drug D (as the inactive, locked-in T^+-D) is present essentially only in the brain, providing sustained and brain-specific release of the active drug. It has to be emphasized again that the system not only achieves delivery to the brain but also achieves preferential delivery that means brain targeting. This should allow smaller doses and reduce peripheral side effects. Furthermore, because the "lock-in" mechanism works against the concentration gradient, the system also provides more prolonged effects. Consequently, such CDSs can be used not only to deliver compounds that otherwise have no access to the brain but also to retain lipophilic compounds within the brain, as it has indeed been achieved, for example, with a variety of steroid hormones.

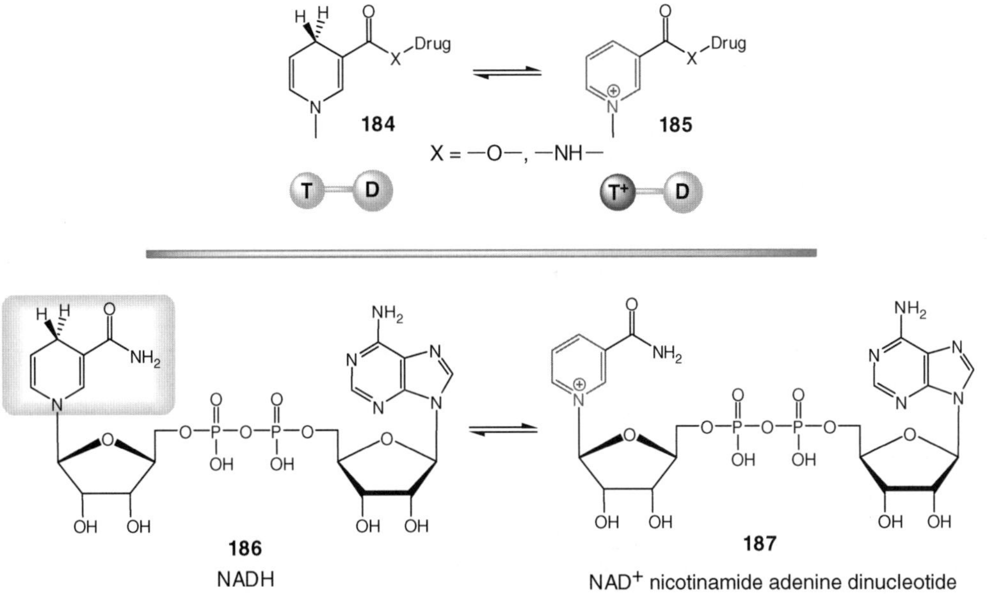

Figure 47. The trigonellinate ↔ 1,4-dihydrotrigonellinate redox system used in brain-targeting CDSs exploits the analogy with the ubiquitous NAD(P)H ↔ NAD(P)$^+$ coenzyme system (**186** ↔ **187**) to convert the lipophilic 1,4-dihydro form (T, **184**) into the hydrophilic quaternary form (T$^+$, **185**). (This figure is available in full color at http://mrw.interscience.wiley.com/emrw/9780471266945/home.)

The CDS approach has been explored with a wide variety of drug classes—biogenic amines: phenylethylamine [29,447–450], tryptamine [451,452]; steroids: testosterone [437,453–455], progestins [456], progesterone [457], ethinyl estradiol [458], dexamethasone [459,460], hydrocortisone [437], alfaxalone [461], estradiol [462–484]; anti-infective agents: penicillins [485–488], sulfonamides [489]; antivirals: acyclovir [33,490], trifluorothymidine [491,492], ribavirin [493–495], Ara-A [493], deoxyuridines [496–498], 2′-F-5-methylarabinosyluracil [499]; antiretrovirals: zidovudine (AZT) [500–514], 2′,3′-dideoxythymidine [446], 2′,3′-dideoxycytidine [515,516], ganciclovir [517], cytosinyl-oxathiolane [518]; anticancer agents: lomustine (CCNU) [519,520], HECNU [520], chlorambucil [521], bis(chloroethyl)amino [522]; neurotransmitters: dopamine [523–527], GABA [528,529]; nerve growth factor (NGF) inducers: catechol derivatives [530,531]; anticonvulsants: GABA [529], phenytoin [532], valproate [533], stiripentol [534], the calcium channel antagonist felodipine [535]; monoamine oxidase (MAO) inhibitors: tranylcypromine [536,537]; antidementia drugs (cholinesterase inhibitors): tacrine [538,539]; antioxidants: LY231617 [540]; chelating agents: ligands for technetium complexes [541]; nonsteroidal anti-inflammatory drugs (NSAIDs): indomethacin [542], naproxen [542], ibuprofen, diclofenac, tolmetin, and others [543]; anesthetics: propofol [544]; nucleosides: adenosine [545]; HIV protease inhibitors: the peptidomimetic KNI-279 [546]; and peptides: tryptophan [547,548], Leu-enkephalin analog [549,550], thyrotropin-releasing hormone (TRH) analogs [551–557], kyotorphin analogs [34,558], and *S*-adenosyl-L-methionine (SAM) [559]. In addition to the trigonelline targetor system, a system relying on the ring-closure reaction of *cis*-2-formylaminoethenylthio derivatives to quaternary thiazolium derivatives has also been explored for the brain delivery of DOPA (L-3,4-dihydroxyphenylalanine) [432], NMDA and AMPA receptor antagonists [436], and α-tocopherol analogs as free radical scavengers [436]. Selected brain-targeting CDS examples are summarized in the following chapters together with representative physicochemical properties, metabolic pathways, and pharmacological data.

Table 2. STI, SEF, TEF of the Brain-Targeting CDSs with Available *In Vivo* Data

Compound Species, Reference	Dose (μmol/kg)	Time (h)	D only AUC (μg × h/mL) Brain	D only AUC (μg × h/mL) Blood	D only STI	CDS AUC (μg × h/mL) Brain	CDS AUC (μg × h/mL) Blood	CDS STI	SEF Equation 6	TEF Equation 7
AZT										
Mouse [513] [a]	187	12	1.21	26.64	0.045	11.28	25.38	0.444	9.32	9.79
Rat [505]	130	6	0.23	26.33	0.009	7.09	9.67	0.733	31.51	85.84
Rabbit [502]	64	3	11.22	37.92	0.296	26.98	34.07	0.792	2.40	2.68
AZDU										
Mouse [513] [a]	197	12	2.09	25.83	0.081	11.43	25.79	0.443	5.47	5.48
Ganciclovir										
Rat [517] [b]	80	6	0.66	10.46	0.063	3.61	1.42	2.542	5.44	40.05
Benzylpenicillin, S = Me										
Rat [487]	60	1	0.20	6.05	0.033	0.74	4.56	0.162	3.70	4.91
Dog [488] [c]	30	4	0.16	24.52	0.007	5.07	11.75	0.431	31.69	66.13
Benzylpenicillin, S = Et										
Rat [487]	60	1	0.20	6.05	0.033	2.12	2.95	0.719	10.60	21.74
Dog [488] [c]	30	4	0.16	24.52	0.007	0.48	0.61	0.787	3.00	120.59
Dexamethasone										
Rat [459]	20	6	1.36	19.48	0.070	1.69	8.97	0.188	1.24	2.70

[a] Brain and serum concentrations monitored.
[b] Brain and plasma concentrations monitored.
[c] CSF and blood concentrations monitored.

4.1.2. Site-Targeting Index and Targeting Enhancement Factors

Whereas therapeutic agents are usually intended to exert their action at some selected site, in reality, usually only a small fraction of the systemically administered dose reaches the intended target site (e.g., organ or cell). Drug targeting or drug delivery systems (including most prodrugs) are specifically designed to improve this situation. To assess the site-targeting effectiveness of drugs or drug delivery systems in general from a pharmacokinetic perspective, one can define a *site-targeting index* (STI) [25,419] as the ratio between the area under the curve (AUC) for the concentration of the drug itself at the targeted site and that at a systemic site, for example, blood or plasma:

$$\mathrm{STI} = \frac{\mathrm{AUC}_{\mathrm{target}}}{\mathrm{AUC}_{\mathrm{blood}}} \qquad (5)$$

This index gives an accurate measure on how effectively the active therapeutic agent is actually delivered to its intended site of action. For example, as data in Table 2 indicate, some drugs (e.g., AZT and benzylpenicillin) are very ineffective in penetrating the BBB and reaching brain tissues; the AUCs of their brain concentrations are less than 1% of the corresponding blood values.

To assess the effectiveness of a drug delivery system compared to the drug itself, one can define a *site-exposure enhancement factor* (SEF) to measure the change of the AUC of the active drug at the target site after administration of the delivery system compared to administration of the drug alone:

$$\mathrm{SEF} = \frac{\mathrm{AUC}_{\mathrm{Target}}^{\mathrm{Delivery\ system}}}{\mathrm{AUC}_{\mathrm{Target}}^{\mathrm{Drug\ alone}}} \qquad (6)$$

Of course, this definition assumes equivalent doses. For slightly different doses, linearity between dose and AUC is a reasonable assumption, and the ratio of dose-normalized AUCs (AUC/dose) can be used in the above definition. Because a good delivery system not only increases exposure to the active agent at the target site but also decreases the corresponding systemic exposure, a *targeting enhancement factor* (TEF) [25,419] can be introduced that measures the relative improvement in the STI produced by administration of the delivery system compared to administration of the drug itself:

$$\mathrm{TEF} = \frac{\mathrm{STI}^{\mathrm{Delivery\ system}}}{\mathrm{STI}^{\mathrm{Drug\ alone}}} \tag{7}$$

TEF as defined above is the most rigorous PK measure that can be used to quantify the improvement in targeting produced by a delivery system. It compares not just concentrations, but concentrations along a time period, and it compares actual, active drug concentrations both at target and systemic sites. If (i) the targeting or delivery ("carrier") moiety produces no toxicity of its own, (ii) the therapeutic effect is linearly related to $\mathrm{AUC}_{\mathrm{target}}$, and (iii) side effects are linearly related to $\mathrm{AUC}_{\mathrm{blood}}$ (assumptions that are mostly satisfied), then TEF as defined in Equation 7 also represents the enhancement produced in the TI defined by Equation 1: $\mathrm{TI}^{\mathrm{Delivery\ system}}/\mathrm{TI}^{\mathrm{Drug\ alone}}$. TEF, as defined here [25,419], represents the same ratio as a drug-targeting index used in a theoretical PK treatment [560].

Table 2 presents such AUC-based SEF and TEF values for brain-targeting CDSs in different species with available *in vivo* experimental data. In a number of cases, CDSs produced very substantial increase of the targeting effectiveness, often more than an order of magnitude.

4.1.3. Zidovudine (AZT)-CDS One of the many devastating consequences of the infection by human immunodeficiency virus type 1 (HIV-1), the causative agent of AIDS, is an encephalopathy with subsequent dementia. To adequately treat AIDS dementia, antiviral agents must reach the CNS in therapeutically relevant concentrations, but many potentially useful drugs cannot penetrate into brain tissue. Zidovudine (azidothymidine, AZT; **188**, Fig. 48) was the first drug approved for the treatment of AIDS. This modified riboside has been shown to be useful in improving the

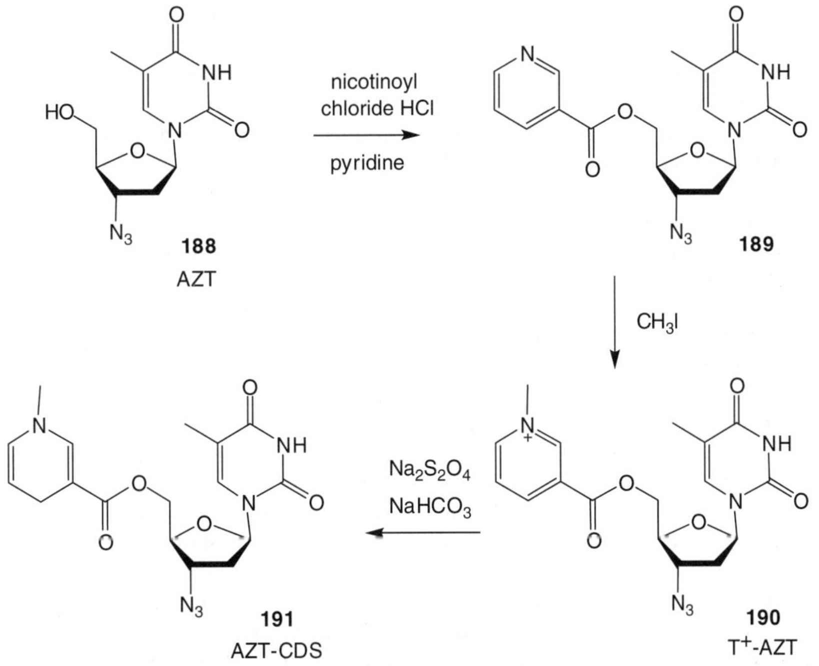

Figure 48. Synthesis of AZT-CDS (**191**) [501]. Because AZT (**188**) contains a primary alcohol at the 5'-position, attachment of the targetor through an ester bond can be relatively easily carried out, and it illustrates one of the simplest possible synthetic routes for brain-targeting CDSs.

neuropsychiatric course of AIDS encephalopathy in a few patients, but the doses required for this improvement usually precipitated severe anemia.

Following the increasing occurrence of AIDS in the 1980s, AZT-CDS (**191**) was investigated in a number of laboratories to enhance the access of AZT to brain tissues especially due to the relatively easy synthetic accessibility of AZT-CDS (Fig. 48) [500–514]. AZT-CDS is a crystalline solid, which is stable at room temperature for several months when protected from light and moisture. It is relatively stable in pH 7.4 phosphate buffer, but rapidly oxidizes in enzymatic media. In addition, T^+-AZT (**190**), the depot form of AZT, was shown to gradually release the parent compound.

The ability of AZT-CDS to provide increased brain AZT levels has been demonstrated in a number of different *in vivo* animal models. The corresponding AUC data for mice [513], rats [505], and rabbits [502] are presented in Table 2. In all species, AZT-CDS provided substantially increased brain-targeting of AZT. For example, the targeting enhancement factor of the AUC-based site-targeting index was 86 in rats. It is interesting to note that the relatively large difference in the TEF values for different species is mainly due to the variability of the brain-STI of AZT itself; AZT-CDS gave very similar values in all three species.

In rabbits, the brain/blood ratio never exceeded 1.0 after AZT administration, indicating that the drug is always in higher concentration in the blood than in the brain, but it approached 3.0 at 1 h postdosing with AZT-CDS. Similarly, in dogs, AZT levels were 46% lower in blood, but 175–330% higher in brain after AZT-CDS administration compared to parent drug administration [506]. Furthermore, *in vitro* experiments found AZT-CDS not only more effective in inhibiting HIV replication than AZT itself, presumably due to hydrolysis, but also less toxic to host lymphocytes [505,508,509]. In conclusion, a number of different studies found an improved delivery to the brain and a decreased potential for dose-limiting side effects suggesting that AZT-CDS could be useful in the treatment of AIDS encephalopathy and the related dementia.

4.1.4. Ganciclovir-CDS Enhanced brain delivery with a CDS approach has also been achieved for ganciclovir (**192**, Fig. 49), another antiviral agent that may be useful in treating human encephalitic cytomegalovirus (CMV) infection [517]. CMV infections are common and usually benign; however, human CMV occurs in 94% of all patients suffering with AIDS, and it has been implicated as a deadly cofactor. While positive results were achieved in the treatment of the associated retinitis, the treatment of encephalitic cytomegalic disease is far less successful with ganciclovir. A brain-targeted CDS (**193**, Fig. 49) was, therefore, investigated, and improved organ selectivity was demonstrated. As shown in Fig. 49 and Table 2, an approximately fivefold increase in the relative brain exposure to ganciclovir and a simultaneous approximately sevenfold decrease in blood exposure was achieved with the CDS in rats.

4.1.5. Benzylpenicillin-CDS The poor CNS penetration of β-lactam antibiotics, including penicillins and cephalosporins, often makes the treatment of various bacterial infections localized in brain and cerebrospinal fluid difficult. Furthermore, because these antibiotics readily return from the CNS back to the blood (active transport may be involved), simple prodrug approaches did not provide real solutions. Consequently, a number of CDSs have been synthesized and investigated [485–488]. Because in these structures the targetor moiety had to be coupled to an acid moiety, a spacer function had to be inserted between the targetor and the drug. This rendered additional flexibility to the approach, and several possibilities were investigated. Benzylpenicillin-CDS was one of the first cases where the influence of such functions on the stability and delivery properties of CDSs could be investigated. Systematic studies of the effects of various modifications have been performed since then [433,550].

Intravenous administration of benzylpenicillin-CDS with an ethylene 1,2-diol spacer in rats gave an AUC-based brain-exposure enhancement factor for benzylpenicillin of around 11 (Table 2) [487]. From the several coupling possibilities investigated, diesters of methylene diol and ethylene 1,2-diol proved

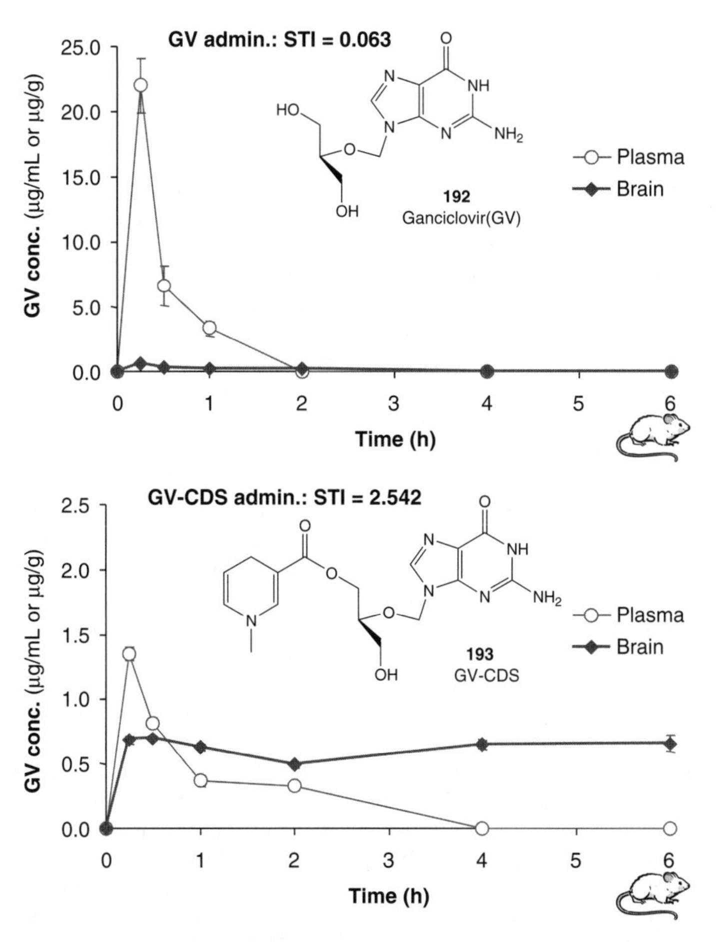

Figure 49. Ganciclovir (GV, **192**) concentrations in brain and plasma as a function of time after an i.v. dose of 80 µmol/kg of GV (**192**) or GV-CDS (**193**). Data are mean ± SEM in rats. (This figure is available in full color at http://mrw.interscience.wiley.com/emrw/9780471266945/home.)

worthy of further investigation, but ester-amide combinations did not prove successful. Consequently, the diesters have also been investigated in rabbits and dogs [488]. Both in rabbits and in dogs, i.v. administration of equimolar doses of CDSs provided benzylpenicillin levels in brain and cerebrospinal fluid (CSF) that were substantially higher and more prolonged than in the case of parent drug administration. In this study, the diester of methylene diol provided more significant increase in benzylpenicillin brain exposure (SEF = 32); the diester of ethylene 1,2-diol gave better targeting, but released lower drug concentrations due to formation of an inactive hydroxyethyl ester. Because dogs were not sacrificed, only blood and CSF data were collected (Table 2). These indicated better CSF penetration in dogs than in rabbits.

Although parenteral injection of benzylpenicillins may sometimes cause undesired side effects, such as seizures, the high CNS levels obtained in dogs and rabbits after CDS administration were not accompanied by any toxic side effects [488]. Similarly, although the delivered compound exhibited toxicity in the rotorod test at 0.5 h following a 300 mg/kg i.p. dose, felodipine-CDS displayed no toxicity at either 0.5 h or 4 h after a 100 mg/kg i.p. dose or 0.5 h for a 300 mg/kg i.p. dose [535]. Some toxicity, which is not surprising considering the toxicity of the delivered compound itself, was observed at 4 h for the 300 mg/kg CDS dose.

4.1.6. Estradiol-CDS Estradiol-CDS (Estredox®) is in the most advanced developmental stage among all CDSs (Phase I/II clinical trials). Estrogens, as discussed before (**66–68**, Fig. 19), are endogenous hormones that produce numerous physiological actions and are among the most commonly prescribed drugs in the United States, mainly for hormone therapy (HT) in postmenopausal women and as a component of oral contraceptives [561]. They are lipophilic steroids that are not impeded in their entry to the CNS; hence, they can readily penetrate the BBB and achieve high central levels after peripheral administration. However, estrogens are poorly retained within the brain. Therefore, to maintain therapeutically significant concentrations, frequent or sustained (e.g., transdermal) doses have to be administered. The constant peripheral exposure to estrogens, however, has been related to a number of pathological conditions including cancer, hypertension, and altered metabolism [244,245,562–564]. In light of the results of the recent large scale parallel randomized, double-blind, placebo-controlled clinical trials undertaken within the WHI that were halted early because of unfavorable outcomes (increased risk for invasive breast cancer, stroke, and coronary heart disease) [246,247], and the fact that there seems to be an approximately 30% increase in the risk of breast cancer caused by taking estrogen, which is suspected to rise to 50% if taken for more than 10 years [565], currently there is little justification for systemic HT. In industrialized nations, the average woman spends around a third of her life in the postmenopausal stage as menopause occurs at an average age of 51 [566]; hence, whether to take HT or not is an important decision. As the CNS is the target site for many estrogenic actions [567,568], brain-targeted delivery may provide safer and more effective agents in many cases. With the recent unraveling of the many roles estrogens play in males [569,570], therapeutic applications are not restricted to females only; there could be number of possible important therapeutic applications in males as well. Furthermore, CEEs mainly used for HT, are obtained from pregnant mares and contain a combination of at least 10 estrogenic compounds (mainly estrone and equiline) as their sulfate esters [571,572] that, contrary to E_2, hardly can cross the BBB [573] and are less likely to have cognitive effects [574].

Estrogen CDSs could be useful in reducing the secretion of luteinizing hormone-releasing hormone (LHRH) and, hence, in reducing the secretion of luteinizing hormone (LH) and gonadal steroids. As such, they could be used to achieve contraception and to reduce the growth of peripheral steroid-dependent tumors, such as those of the breast, uterus, and prostate, and to treat endometriosis. They also could be useful in male sexual dysfunction (MSD) and female sexual dysfunction (FSD) [569,575], and in the treatment of menopausal vasomotor symptoms ("hot flashes") [576]. Other potential uses are in neuroprotection [577], in the reduction of body weight, or in the treatment of Parkinson's disease [578], depression, and various types of dementia, including Alzheimer's disease (AD) [244,481,579,580]. AD, which still has no specific cure, results in progressively worsening symptoms that range from memory loss to declining cognitive ability. It affects an estimated 10% of the population over 65 years of age and almost 50% of that over 85 years of age [581]. Findings on the effect of estrogen in AD are somewhat conflicting [582,583]. Nevertheless, estrogen produces many effects [567,568] that may delay the onset of AD, prevent AD, or improve the quality of life in AD, and therapeutic effects may require sustained and sufficiently high estradiol levels [584,585]. A detailed review of possible therapeutic applications for brain-targeted estrogens has been published recently [484].

Estradiol (E_2, **66**, Fig. 50) is the most potent natural estrogenic steroid. It contains two hydroxy functions: one in the phenolic 3-position and one in the 17-position. With these synthetic handles, three possible CDSs can be designed attaching the targetor at the 17-, at the 3-, or at both positions. Attachment at either position, but especially at the 17-position, should greatly decrease the pharmacological activity of E_2, because these esters are known not to interact with estrogen receptors [586].

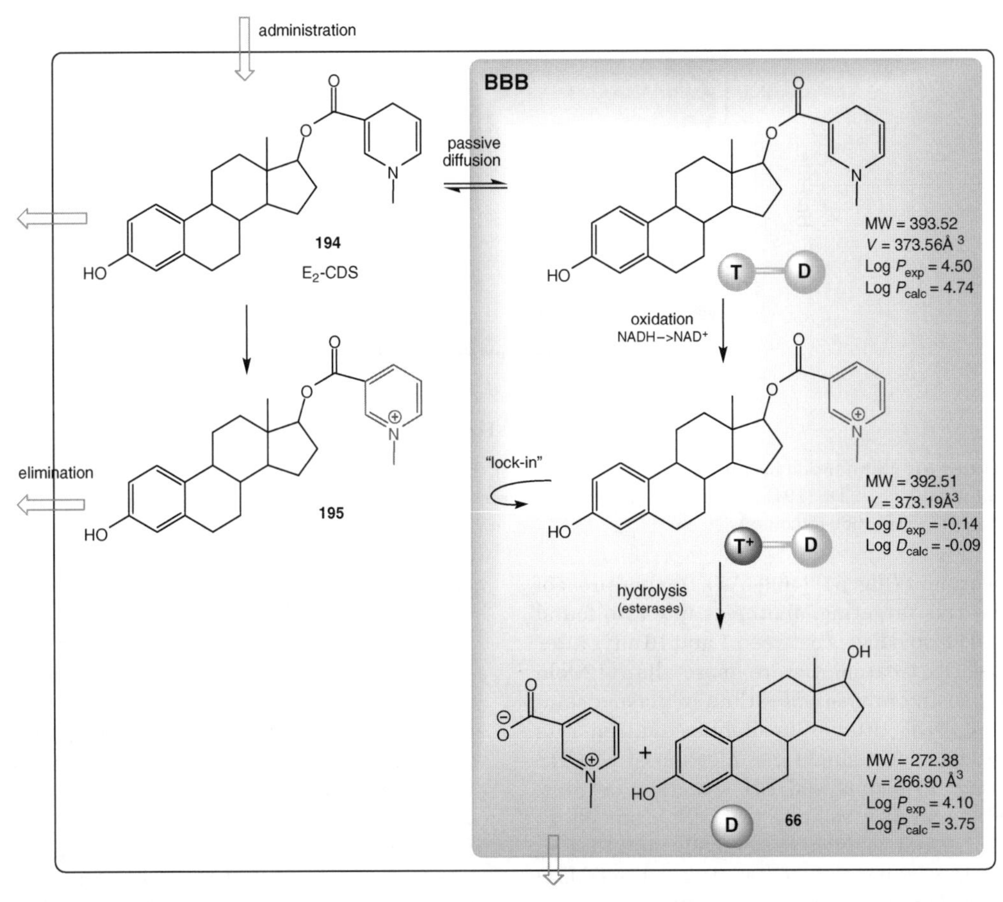

Figure 50. *Lock-in* mechanism for estradiol-CDS (E_2-CDS; **194**). Experimental and calculated logarithms of the octanol–water partition (log *P*) and distribution (log *D*) coefficients are shown to illustrate the significant changes in partition properties. The lipophilic CDS (log $P > 4$) can easily cross the BBB, while the hydrophilic intermediate (log $D < 0$) is no longer able to come out, providing a sustained release of the active estradiol (**66**). (This figure is available in full color at http://mrw.interscience.wiley.com/emrw/9780471266945/home.)

Since its first synthesis in 1986 [462], E_2-CDS (**194**) has been investigated in several models [463–473,475–482,484]. *In vitro* studies with rat organ homogenates as the test matrix indicated half-lives of 156.6, 29.9, and 29.2 min for E_2-CDS (T at the 17-position) in plasma, liver, and brain homogenates, respectively [462]. Thus, E_2-CDS is converted to the corresponding quaternary form (T^+-E_2) (**195**) faster in the tissue homogenates than in plasma. This is consistent with the hypothesis of a membrane-bound enzyme, such as the members of the NADH transhydrogenase family, acting as oxidative catalyst. These studies also indicated a very slow production of E_2 from T^+-E_2, suggesting a possible slow and sustained release of estradiol from brain deposits of T^+-E_2.

To detect doses of E_2-CDS (**194**), T^+-E_2 (**195**), and E_2 (**66**) of physiological significance, a selective and sensitive method was needed. Therefore a precolumn-enriching HPLC system was used [469] that allowed accurate detection in plasma samples and organ homogenates with limits of 10, 20, and 50 ng/mL or ng/g for T^+-E_2, E_2-CDS, and E_2, respectively. This study proved that in rats, E_2 released from the T^+-E_2 intermediate formed after i.v. E_2-CDS administration has an elimination half-life of more than 200 h and brain E_2-levels are elevated four to five times longer after administration than after simple estradiol

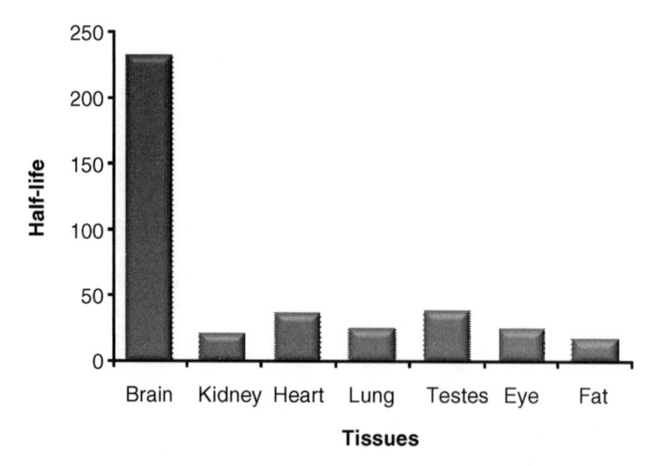

Figure 51. Elimination half-lives in various tissues for the T^+–E_2 (**195**) formed after i.v. administration of 38.1 μmol/kg E_2-CDS (**194**) in rats. (This figure is available in full color at http://mrw.interscience.wiley.com/emrw/9780471266945/home.)

treatment (Fig. 51) [469]. As a further proof of effective targeting, another study also found that steroid levels between 1 and 16 days after E_2-CDS treatment were more than 12-fold higher in brain samples than in plasma samples [473]. Studies in orchidectomized rats proved that a single i.v. injection of E_2-CDS (3 mg/kg) suppressed LH secretion by 88%, 86%, and 66% relative to DMSO controls at 12, 18, and 24 days, respectively, and that E_2 levels were not elevated relative to the DMSO control at any sampling time [465]. A single i. v. administration of doses as low as 0.5 mg/kg to ovariectomized rats induced prolonged (3–6 weeks) pharmacological effects as measured by LH suppression [463,465,473], reduced rate of weight gain [466,471–473], or, in castrated male rats, reestablishment of copulatory behavior [464]. A large number of other very encouraging results have been obtained in various animal models; most of them have been reviewed previously [33,340,475, 476,478,484].

E_2-CDS also was shown to provide very encouraging neuroprotective effects. In ovariectomized rats, pretreatment with E_2-CDS decreased the mortality caused by middle cerebral artery occlusion (MCAO) from 65% to 16% [479]. E_2-CDS reduced the area of ischemia by 45–90% or 31% even if administered after MCAO by 40 or 90 min, respectively. Another study provided evidence that treatment with E_2-CDS can protect cholinergic neurons in the medial septum from lesion-induced degeneration [480].

More recently, a dramatic effect on the sexual behavior of female rats has also been confirmed for E_2-CDS (Fig. 52) [482]. Considering the complexity of FSD [575,587], this was a somewhat surprising, but quite possibly a significant discovery. Because FSD is more complex than MSD, such a strong response caused by administration of an estrogenic compound alone was not expected. The dose and time dependency of the effects of E_2-CDS on the reestablishment of the copulatory behavior and on the plasma LH and E_2 levels in castrated male and female rats have been examined. E_2-CDS reestablished the copulatory behavior from day 7 to day 28 in orchidectomized males after single i.v. doses of 0.3 and 3.0 mg/kg. LH suppression started on day 1 and lasted until day 7 and 28, respectively. In ovariectomized female rats, sexual activity was monitored through the lordosis behavior (lordosis being the vertebral dorsiflexion performed by female quadrupeds in response to adequate stimuli from a reproductively competent male). Animals were treated daily i.v. for 5 days with E_2-CDS, estradiol benzoate (E_2-Bz), or vehicle, respectively. At a dose of only 0.01 mg/kg, E_2-CDS already normalized sexual behavior (Fig. 52). At the 0.03 mg/kg dose level, the effect of E_2-Bz was 10 days shorter than that of E_2-CDS. LH suppression lasted for up to 18 days at the 0.03 mg/kg dose,

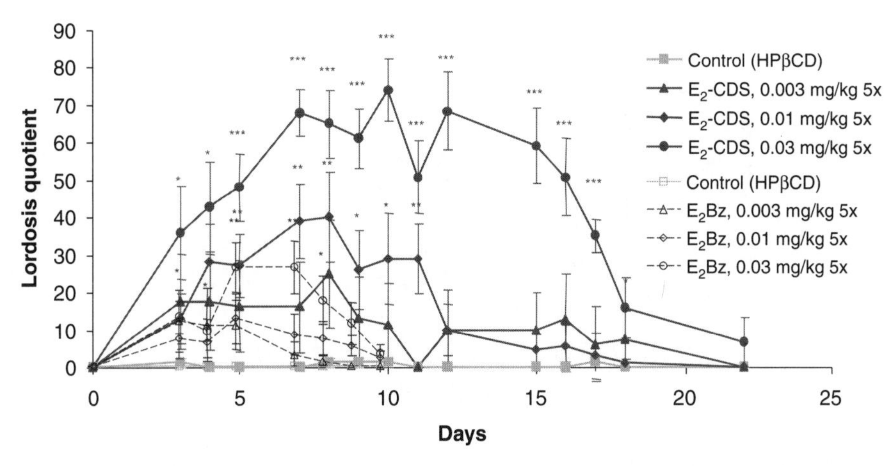

Figure 52. Effect of different doses of E_2-CDS (**194**) (closed blue symbols and solid line) and of the isolipophilic E_2-benzoate (E_2Bz, open red symbols and dashed line) on the lordosis behavior of castrated female rats. Lordosis quotient calculated as $100 \times$ number of lordosis/10 mounts. Data are mean $\pm$ SE for 8–12 animals per group. E_2-CDS was administered for 5 days i.v. dissolved in 27% HPβCD solution. $^{*}p<0.05$, $^{**}p<0.01$, $^{***}p<0.001$ using Mann–Whitney U test. (This figure is available in full color at http://mrw.interscience.wiley.com/emrw/9780471266945/home.)

and for up to 10 days at the 0.01 and 0.003 mg/kg doses. On the other hand, there was no significant decrease in LH levels after E_2-Bz treatment. The low plasma E_2 levels indicated fast rate of peripheral elimination in both males and females. These results confirm again the potential of E_2-CDS to cause significantly higher and more prolonged effects than E_2-Bz, which was used as an isolipophilic control.

For E_2-CDS, four human Phase I/II clinical trials have been completed already [33,470, 475,476,478,484,588–591]. The first study was designed as a rising dose safety study in postmenopausal women, where plasma LH and follicular stimulating hormone (FSH) could be readily assayed as indicators of pharmacological action. No adverse effects that could be attributed to E_2 were reported, and all hematological and clinical chemistry values were unaffected by administration of the drug. Results could be clustered in three dose-groups: 10–40 μg ($n=3$) showing minimal changes, 80–640 μg ($n=5$) showing threshold effects, and 1280 μg ($n=2$) showing substantial and sustained decrease in plasma LH levels (Fig. 53a). The mean peak decreases obtained for these groups (11%, 34%, and 50%, respectively) are comparable to those observed after one month of dosing with E_2 transdermal patches (28–36%). While LH was dampened in a clinically meaningful way through 96 h after E_2-CDS treatment, LH values returned to control values within 24 h after E_2 dosing.

The aim of a second clinical study was to increase the number of patients exposed to E_2-CDS, to collect further safety information in humans, to establish the feasibility of buccal dosing in humans, and to demonstrate the central action of E_2-CDS relative to a commercially available oral and parenteral E_2 preparations. A buccal formulation was considered so as to avoid the irreversible acid-catalyzed water addition at the 5,6-double bond of the dihydropyridine targetor moiety, which is unavoidable following traditional oral administration and would render the CDS ineffective for brain targeting. For buccal dosing, a comparison with biological effects suggested significant bioavailability. Both i.v. and buccal dosing demonstrated a dose-related suppression of circulating LH levels. A comparison of AUCs for LH suppression and plasma E_2 levels for E_2-CDS and Progynova® demonstrated that for a certain blood level of E_2, E_2-CDS is significantly more active than Progynova for both buccal and i.v. administration. As the comparison of i.v. E_2-CDS and E_2 indicates, while they both exert similar effects on LH suppression (ratio of 0.7), E_2-CDS elevates plasma E_2 levels to only a very small

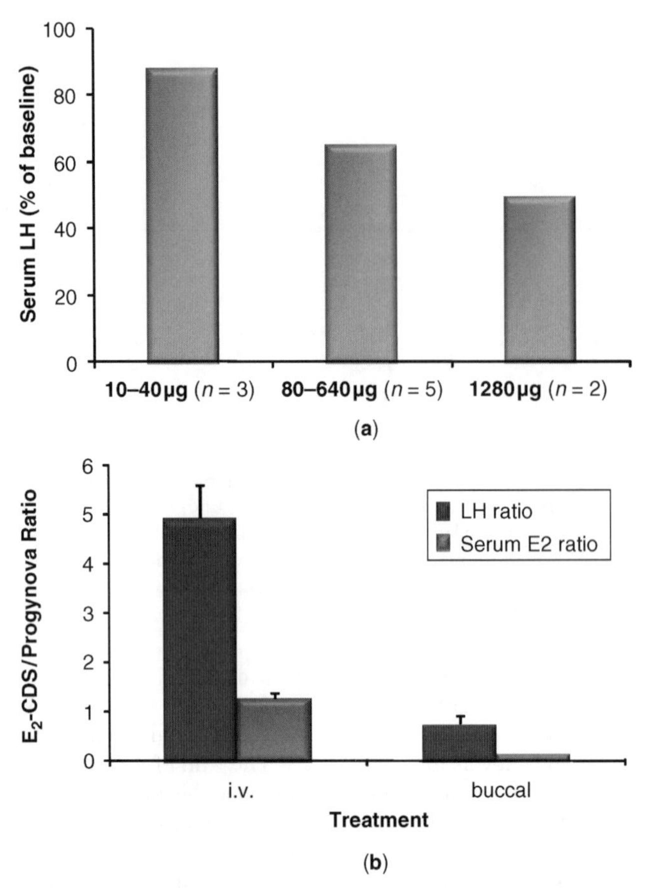

Figure 53. (a) Effect of various grouped doses of E_2-CDS (**194**) on mean LH suppression relative to baseline values in postmenopausal human volunteers. (b) The effect of i.v. or buccal E_2-CDS relative to oral Progynova (estradiol valerate) on LH suppression or E_2 serum elevation in a study of 18 postmenopausal volunteers. Data are collected from all dosing groups and are corrected for baseline and dose. E_2-CDS was significantly more active as measured by the LH suppression, yet had less of a tendency to elevate serum E_2 levels. (This figure is available in full color at http://mrw.interscience.wiley.com/emrw/9780471266945/home.)

fraction (0.044) of that produced by E_2 administration (Fig. 53b).

More recently, a pilot Phase II study for postmenopausal symptoms using circulating LH and FSH as biomarker was completed with a newly developed buccal formulation tablet in 12 healthy postmenopausal female volunteers whom were not on ET [484,590]. Following a single E_2-CDS administration, sustained plasma E_2 levels were maintained for up to 3 days confirming release of active E_2 from the intermediate "locked in" the brain. In the meantime, peak plasma E_2 levels were considerably lower than those expected to be produced by administration of an equivalent E_2 dose, and they were within or just slightly above the normal pre-menopausal fluctuating range of 20–250 pg/mL. AUC data also showed nice dose-response. Both LH and FSH blood levels were suppressed in a statistically significant manner as indicated by concentration and $AUC_{24\,h}$ data [484,590] confirming the potential of E_2-CDS to deliver brain-targeted estradiol producing CNS-related activities and to provide, in the same time, prolonged, sustained release resulting in safe, but sufficiently elevated peripheral hormone levels.

A Phase Ib/II study has also been completed with this buccal tablet in twelve healthy postmenopausal female subjects not on estrogen or hormone therapy (ET/HT) [484,591]. Buccal E_2-CDS (2.86 mg) ad-

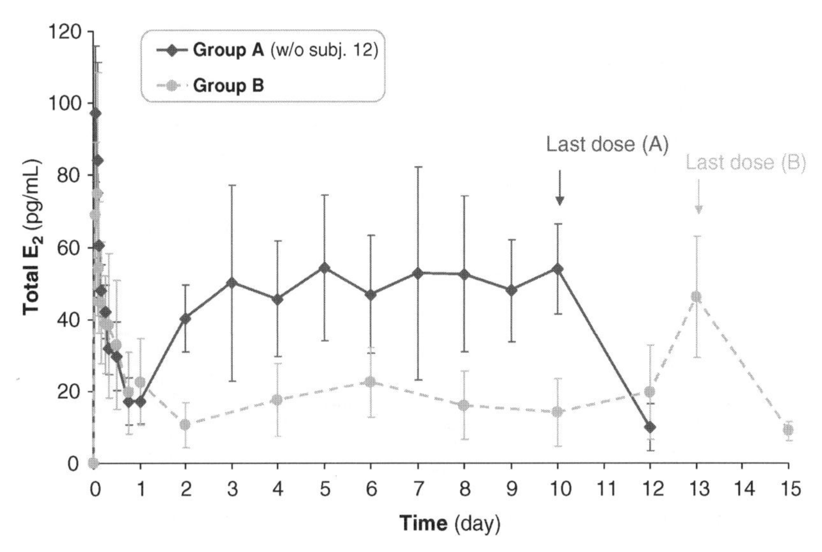

Figure 54. Baseline-adjusted total estradiol (E_2) concentrations for the whole study period following administration of 2.86 mg E_2-CDS as a buccal tablet once daily (Group A) or once every other day (Group B). Data are mean $\pm$ SD for 6 subjects in Group B and 5 subjects in Group A, where data of one subject, who had approximately 10-fold higher E_2 levels than the other subjects, were not included in this graphic. (This figure is available in full color at http://mrw.interscience.wiley.com/emrw/9780471266945/home.)

ministered daily (Group A) or every other day (Group B) was well tolerated. Serum E_2 levels were higher for the daily regimen (Group A), and differences were significant (Fig. 54) [484,591]. Both dosing-regimens caused significant LH- and FSH-suppression providing evidence of CNS penetration. Effects on LH and FSH were observed at low serum levels of estradiol. FSH-response was still evident three days after the last dose of E_2-CDS, when serum E_2 levels were near baseline. Furthermore, the $2OHE_1/16OHE_1$ ratio was increased by E_2-CDS indicating a potential low risk for breast cancer. In summary, clinical evaluations suggest a potent central effect with only marginal elevations in systemic estrogen levels; therefore, E_2-CDS could become a useful and safe therapy for menopausal symptoms, sexual disease, or for estrogen-dependent cognitive effects.

4.1.7. Cyclodextrin Complexes The same physicochemical characteristics that allow successful chemical delivery also complicate the development of acceptable pharmaceutical formulations. Increased lipophilicity allows not only partition into deep brain compartments but also confers poor aqueous solubility. The oxidative lability, which is needed for the lock-in mechanism, and the hydrolytic instability, which releases the modifier functions or the active drug, combine to limit the shelf life of the CDS. Cyclodextrins may provide a possible solution, and there are already a number of marketed pharmaceutical products containing cyclodextrins [592]. Cyclodextrin are torus-shaped oligosaccharides that contain various numbers of α-1,4-linked glucose units (6, 7, and 8 for α-, β-, and γ-cyclodextrin, respectively). The number of units determines the size of a cone-like cavity into which various compounds can include and form stable complexes (Fig. 55) [593–596]. Formation of the host-guest inclusion complex generally involves only the simple spatial entrapment of the guest molecule without formation of covalent bonds [597].

The corresponding inclusion complex with 2-hydroxypropyl-β-cyclodextrin (HPβCD) solved the formulation problems of E_2-CDS [598]. This modified cyclodextrin was selected based on its low toxicity observed using various administration routes and based on the fact that alkylation or hydroxyalkylation of the glucose oligomer can disrupt hydrogen bonding and provide increased

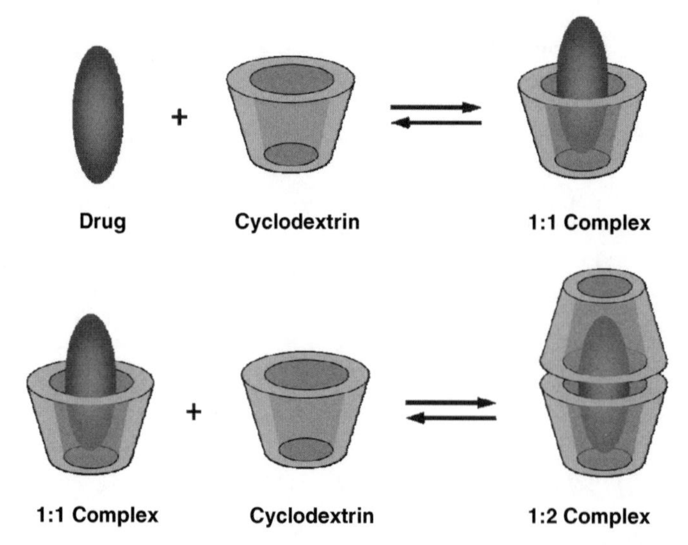

Figure 55. Schematic illustration of drug cyclodextrin 1:1 and 1:2 inclusion complex formations. (This figure is available in full color at http://mrw.interscience.wiley.com/emrw/9780471266945/home.)

water solubility for the compound and for its inclusion complexes as well [599–602]. Indeed, the aqueous solubility of E_2-CDS was enhanced approximately 250,000-fold in a 40% (w/v) HPβCD solution (from 65.8 ng/mL to 16.36 mg/mL). The phase solubility diagram indicated that a 1:1 complex forms at low HPβCD concentration but a 1:2 complex occurs at higher HPβCD concentrations. The stability of E_2-CDS was also significantly increased allowing formulation in acceptable form. The rate of ferricyanide-mediated oxidation, a good indicator of oxidative stability, was decreased approximately 10-fold, and shelf life was increased approximately 4-fold, as indicated by t_{90} and t_{50} values (the time required for 10% and 50% of the active principle to degrade) in a temperature range of 23–80°C [598]. Similarly promising results were obtained for other CDSs as well; for example, testosterone-CDS [455], lomustine-CDS [520], and benzylpenicillin-CDS [603]. For the latter, aqueous solubility was enhanced approximately 70,000-fold in a 20% (w/v) HPβCD solution (from 50–70 ng/mL to 4.2 mg/mL), and stability was also increased.

4.1.8. Molecular Packaging More recently, the CDS approach has been extended to achieve successful brain deliveries of a Leu-enkephalin analog [549,550], TRH analogs [551–557], and kyotorphin analogs [34,558]. Neuropeptides, peptides that act on the brain or spinal cord, represent the largest class of transmitter substance and have considerable therapeutic potential [34,604,605]. The number of identified endogenous peptides is growing exponentially; recently, almost 600 sequences for neuropeptides and related peptides have been listed [606]. However, delivery of peptides through the BBB is an even more complex problem than delivery of other drugs because they can be rapidly inactivated by ubiquitous peptidases [420,607–610]. For peptides, it is usually quite difficult to build up and to maintain effective blood levels in general, because they are very efficiently metabolized and eliminated—not a surprising observation if one considers that their name comes from the Greek πεπτοζ (*peptos*) meaning "digestible." Therefore, for a successful delivery, three issues need to be solved simultaneously: (i) enhancement of the passive transport to increase CNS access (e.g., by increasing the lipophilicity), (ii) assurance of the enzymatic stability to prevent premature degradation, and (iii) exploitation of the lock-in mechanism to provide targeting.

The solution developed by us is a complex molecular packaging strategy, in which the peptide unit is part of a bulky molecule

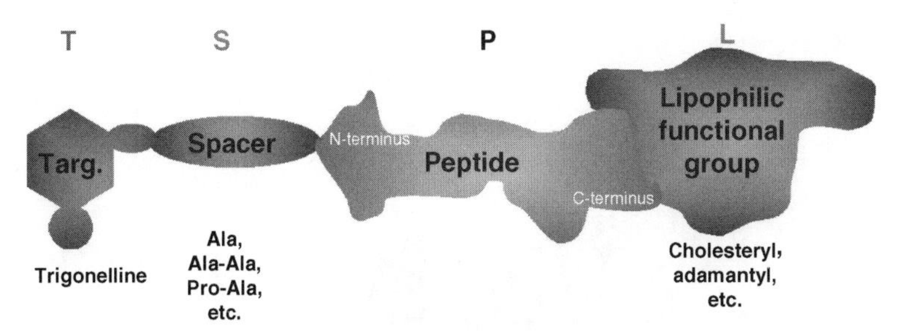

Figure 56. Schematic representation of the molecular packaging strategy used for brain delivery of neuropeptides. (This figure is available in full color at http://mrw.interscience.wiley.com/emrw/9780471266945/home.)

dominated by lipophilic modifying groups that direct BBB penetration and prevent recognition by peptidases [611]. Such a brain-targeted packaged peptide delivery system contains the following major components: the redox targetor (T); a spacer function (S), consisting of strategically used amino acids to ensure timely removal of the charged targetor from the peptide; the peptide itself (P); and a bulky lipophilic moiety (L) attached through an ester bond or sometimes through a C-terminal adjuster (A) at the carboxyl terminal to enhance lipid solubility and to disguise the peptide nature of the molecule (Fig. 56). To achieve delivery and sustained activity with such complex systems, it is very important that the designated enzymatic reactions take place in a specific sequence. On delivery, the first step must be the conversion of the targetor to allow for lock-in. This must be followed by removal of the L function to form a direct precursor of the peptide that is still attached to the charged targetor. Subsequent cleavage of the targetor–spacer moiety finally releases the active peptide.

Leu-Enkephalin Analog Since the discovery of endogenous peptides with opiate activity [612], many potential roles have been identified for these substances [613,614]. Possible therapeutic applications could extend from treatment of opiate dependence to numerous other CNS-mediated dysfunctions. Opioid peptides are implicated in regulating neuroendocrine activity [615–617], motor behavior [618], seizure threshold [619,620], feeding and drinking [621], mental disorders [613,622], cognitive functions [613,623], cardiovascular functions [624,625], gastrointestinal functions [613,626], and sexual behavior and development [614]. However, analgesia remains their best known role, and it is commonly used to evaluate endogenous opioid peptide activity. Whereas native opioid peptides could alter the pain threshold after intracerebroventricular (i.c.v.) dosing, they were ineffective after systemic injection. It was reasonable, therefore, to attempt a brain-targeted CDS approach.

The first successful delivery and sustained release was achieved for Tyr-D-Ala-Gly-Phe-D-Leu (DADLE), an analog of leucine-enkephalin, a naturally occurring linear pentapeptide (Tyr-Gly-Gly-Phe-Leu) that binds to opioid receptors (Fig. 57) [549]. In rat brain tissue samples collected 15 min after i.v. CDS administration, electrospray ionization mass spectrometry clearly showed the presence of the locked-in T^+-D form at an estimated concentration of 600 pmol/g of tissue. The same ion was absent from the sample collected from the control animals treated with the vehicle solution only. To optimize this delivery strategy, an effective synthetic route was established for peptide CDSs, and the role of the spacer and the lipophilic functions was investigated [550]. Four different CDSs were synthesized by a segment-coupling method. Their i.v. injection produced a significant and long-lasting (>5 h) response in rats monitored by classic [146] tail-flick latency test. Compared with the delivered peptide itself, the packaged peptide and the intermediates formed during the sequential metabolism had weak affinity to opioid receptors. The antinociceptive effect was naloxone-reversible and methylnaloxonium-irreversible. Because quaternary derivatives such as methylnaloxonium are unable to cross the BBB, these

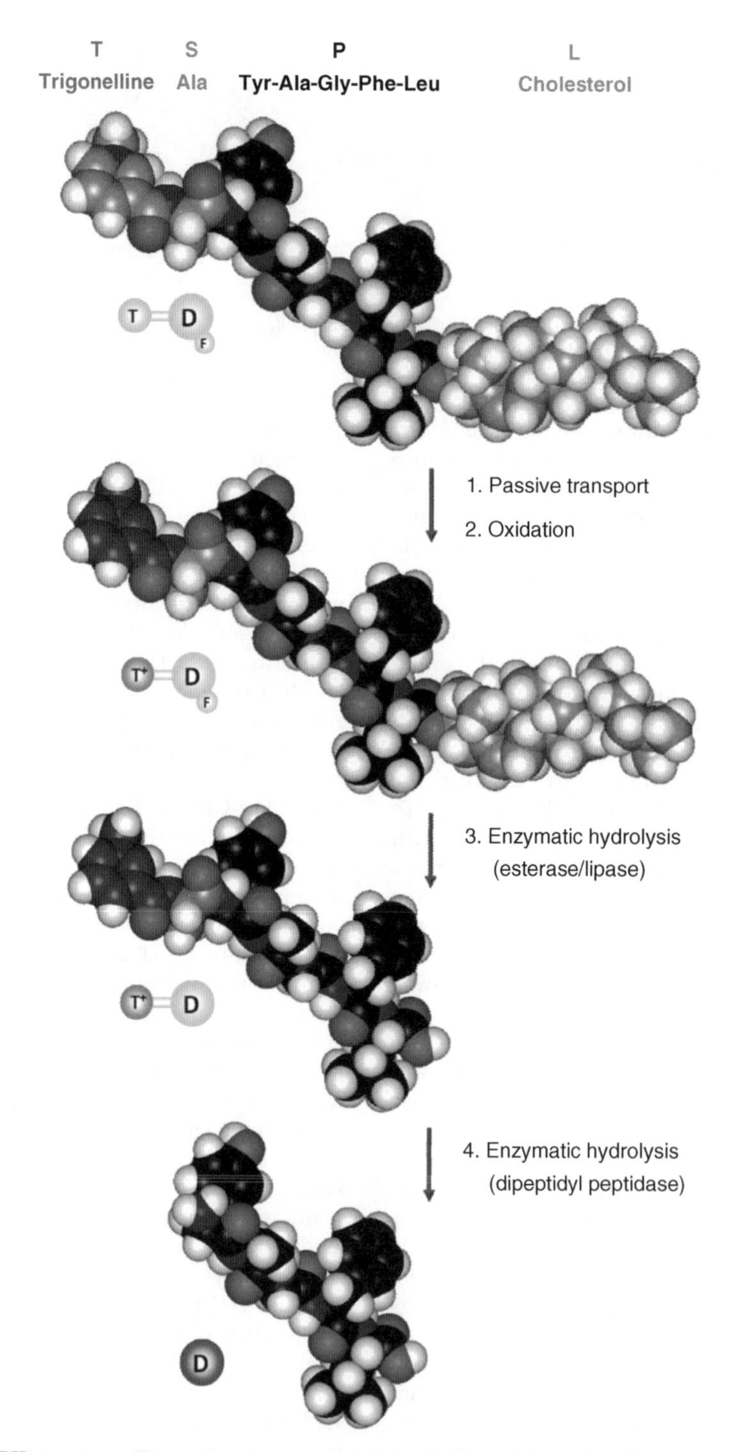

Figure 57. CPK structures illustrating the sequential metabolism of the molecular package used for brain delivery of a leucine enkephalin analog. (This figure is available in full color at http://mrw.interscience.wiley.com/emrw/9780471266945/home.)

techniques [550,627] are used to prove that central opiate receptors are responsible for mediating the induced analgesia. It could be concluded, therefore, that the peptide CDS successfully delivered, retained, and released the peptide in the brain.

The efficacy of the CDS package was strongly influenced by modifications of the spacer (S) and lipophilic (L) components proving that they have important roles determining molecular properties and the timing of the metabolic sequence. The bulkier cholesteryl group used as L showed a better efficacy than the smaller 1-adamantaneethyl, but the most important factor for manipulating the rate of peptide release and the pharmacological activity turned out to be the spacer (S) function: proline as spacer produced more potent analgesia than alanine.

TRH Analogs A similar strategy was used [551–555] for the CNS delivery of a TRH(Glp-His-Pro) analog (Glp-Leu-Pro) in which histidine (His^2) was replaced with leucine (Leu^2) to dissociate CNS effects from thyrotropin-releasing activity [628]. Such compounds can increase extracellular acetylcholine (ACh) levels, accelerate ACh turnover, improve memory and learning, and reverse the reduction in high affinity choline uptake induced by lesions of the medial septal cholinergic neurons [629,630]. Therefore, as reviewed in the literature [631,632], these peptides are potential agents for treating motor neuron diseases [633], spinal cord trauma [634], or neurodegenerative disorders such as Alzheimer's disease [635] and may also have a potential cytoprotective role [636].

Because the peptide that has to be delivered has no free -NH_2 and -COOH termini, a precursor sequence (Gln-Leu-Pro-Gly) that will ultimately produce the final TRH analog was packaged for delivery. Therefore, two additional steps had to be included in the metabolic sequence: one where the C-terminal adjuster glycine is cleaved by peptidyl glycine α-amidating monooxygenase (PAM) to form the ending prolinamide, and one where the N-terminal pyroglutamyl is formed from glutamine by glutaminyl cyclase [551]. In summary, the following sequential biotransformation has to take place after delivery to the brain: first, lock-in by oxidation of the dihydrotrigonellyl (T) to the corresponding pyridinium salt, then removal of cholesterol, oxidation of glycine by peptidyl glycine α-amidating monoxygenase to prolineamide, cleavage of the targetor–spacer combination, and, finally, cyclization of glutamine by glutaminyl cyclase to pyroglutamate.

Selection of a suitable spacer proved also important for the efficacy of TRH-CDSs as measured by the decrease in the barbiturate-induced sleeping time in mice, an interesting and well-documented effect of this neuropeptide [637,638]. After i.v. administration of 15 μmol/kg, the Leu^2-TRH analog itself produced only a slight decrease of $17 \pm 7\%$ compared to the vehicle control. Equimolar doses of the CDS compounds with Pro, Pro-Pro, and Pro-Ala spacers produced statistically significant ($p < 0.05$) decreases of $47 \pm 6\%$, $51 \pm 7\%$, and $56 \pm 4\%$, respectively [553]. A Nva^2-TRH analog, in which norvaline replaced leucine, was also used in such an approach with a Pro-Pro spacer, and very similar results were obtained with a dose of 20.7 μmol/kg in Swiss-Webster mice [555]. Treatment with a Leu^2-TRH–CDS also significantly improved memory-related behavior in a passive avoidance paradigm in rats bearing bilateral fimbrial lesions without altering thyroid function [552].

Because the amide precursor was found to be susceptible to deamination by TRH-deaminase, a side reaction process competitive with the designed-in cleavage of the spacer-Gln peptide bond, another analog of TRH was also examined [554]. The Pro-Gly C-terminal was replaced by pipecolic acid (Pip) to obtain a TRH analog that is active in the carboxylate form and, hence, to eliminate the need for the formation of the amide. The CDS was prepared by a 5 + 1 segment coupling approach, and the cholesteryl ester of pipecolic acid was prepared using either Fmoc (fluorenylmethyloxycarbonyl) or Boc (*tert*-butyloxycarbonyl) as protection. As before, the pharmacological activity was assessed by determining the antagonism of barbiturate-induced sleeping time in mice, and the pipecolic acid analog was found to be even somewhat more effective than the previous package.

Kyotorphin Analogs Finally, a kyotorphin analog (Tyr-Lys) that has activity similar to kyotorphin itself [639] was also successfully delivered [34,558]. Kyotorphin (Tyr-Arg) is an endogenous neuropeptide that exhibits analgesic action through the release of endogenous enkephalin, and its analgesic activity is approximately four times larger than that of Met-enkephalin [640]. It also has other effects that may prove therapeutically useful, such as seizure protective effects [641] or effects on nitric oxide synthase (NOS) [642–644]. Because this peptide contains a free amine residue, and because preliminary studies indicated that such ionizable functions might prevent successful delivery, this free amine functionality was additionally "packaged" by attachment of a Boc group (**196**, Fig. 58). It was assumed that this attachment is bioconvertible, and the lock-in mechanism will allow time for its enzymatic removal.

Because during the chemical synthesis of this CDS package, α- and ε-amine functions had to be differentially functionalized, combined liquid phase Fmoc and Boc chemistry was employed. With this synthetic method at hand, a number of various spacers were explored for this delivery system (proline, proline–alanine, and proline–proline), and the double proline (**196**, Fig. 58) provided the best pharmacological effect amongst them. The corresponding CDS showed good activity on the rat tail-flick latency test even with a single proline spacer. Activity was already significant at a 3 μmol/kg (1.0 mg/kg) dose and leveled off approximately 22 μmol/kg. This represents an improvement of approximately two orders of magnitude compared to the ~200 mg/kg dose necessary to observe activity when the peptide itself is given intravenously. Several intermediates and building blocks were also studied, but only administration of the whole molecular package produced significant pharmaceutical response confirming that only the designed metabolic sequence as a whole is effective in delivering peptides across the BBB.

Brain-Targeted Redox Analogs Kyotorphin analogs also served for the design and evaluation of a novel and conceptually different method that can provide brain-targeted activity for peptides containing amino acids with basic side chains (lysine, arginine) [558]. Within this approach, the targetor T moiety, which by conversion to its charged T^+ form is responsible for the lock-in mechanism, is not attached to the peptide from outside, but it is integrated within novel *redox* amino acids building blocks that replace the original basic amino acid of the active peptide. Consequently, isosteric/isoelectronic analogy between the original and the redox side chain and not cleavage of the redox targetor is expected to provide activity. Figure 58 compares the structure of the classical brain-targeted CDS (**196**) with the structure of the brain-targeted redox analog (BTRA) (**197**) used for the present kyotorphin analog. The ε-amine function of lysine could be directly replaced by the nicotinamide function using the Zincke procedure to form pyridinium salts [548,558,645–647]. As shown in Fig. 59, both the CDS and the BTRA produced significant and prolonged analgesic effect in rats, and both could be reversed by naloxone demonstrating again that central opiate receptors must be responsible for mediating the induced analgesia. Such novel amino acids may provide both effective replacement and targeting for other lysine or arginine containing peptides.

Somewhat later, in a copycat design, a BTRA approach has also been attempted for TRH by replacing its His^2 residue with a redox analog [556,557] as this His is believed to be essential for endocrine, TSH-releasing activity, but not for CNS effects [648]. These analogs lost endocrine activity as measured by the binding affinity to receptors from rat brain labeled by (^{3}H)(3-Me-His^2)TRH, but some of them seemed to maintain analeptic activity as measured by the decrease in the barbiturate-induced sleeping time in mice. An increase in the chain length separating the redox moiety from the amino acid backbone resulted in decreased potency.

In conclusion, the molecular packaging of peptides is a rational drug design approach that achieved the first documented noninvasive brain delivery of these important biomolecules in pharmacologically significant amounts. Because drug delivery is the weakest link of biotechnology's peptide drug industry [649], this approach may represent an important step toward future generations of

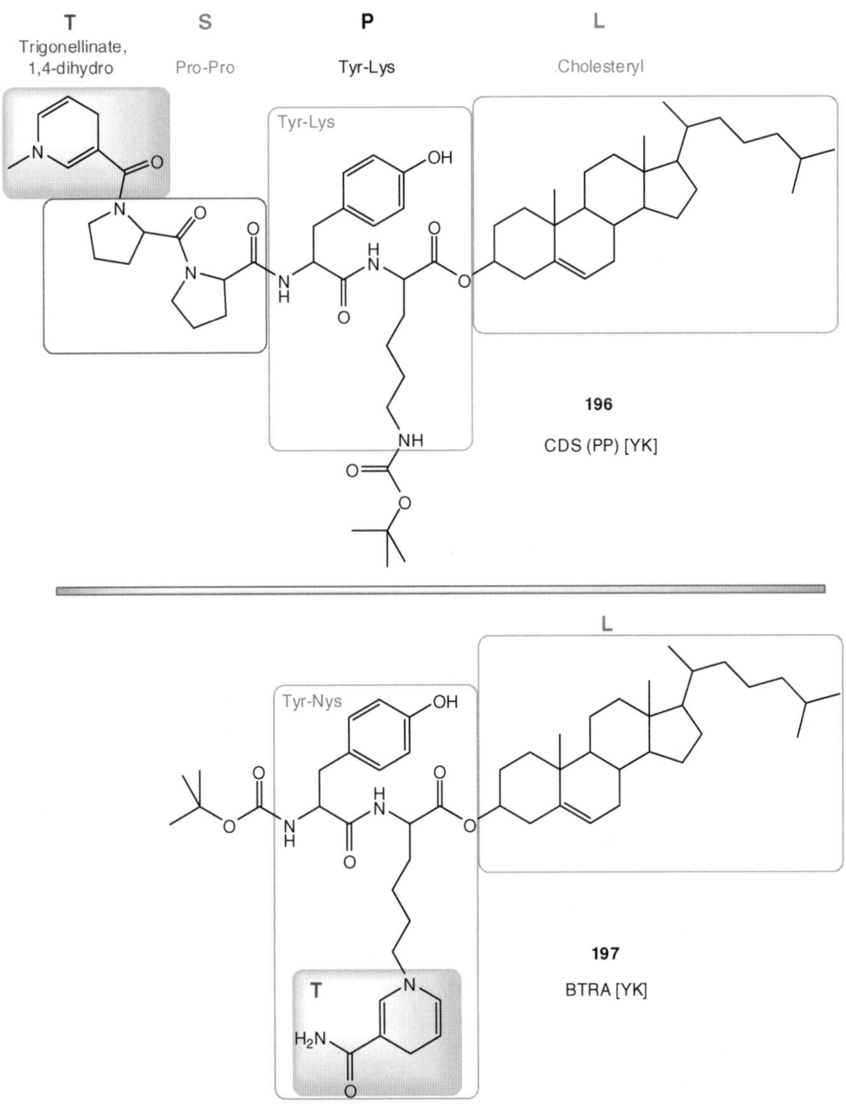

Figure 58. Delivery of kyotorphin analogs was achieved using both a chemical delivery system (CDS; **196**) and a brain-targeted redox analog (BTRA; **197**) approach. (This figure is available in full color at http://mrw.interscience.wiley.com/emrw/9780471266945/home.)

high efficiency neuropharmaceuticals obtained from biologically active peptides.

4.2. Eye-Targeting (Site-Specific Enzyme-Activated) CDS

4.2.1. Design Principles These CDSs exploit enzymatic conversions that take place primarily, exclusively, or at higher activity at the site of action as a result of differential distribution of certain enzymes. If such enzymes can be found, their use in a strategically designed system can lead to high site specificity. The targetor moiety and the eventual protective function(s) are introduced by chemical modifications that also involve important pharmacophore functionalities. Following administration, the resulting CDS is distributed throughout the body. It is continuously eliminated from the "rest of the body" without

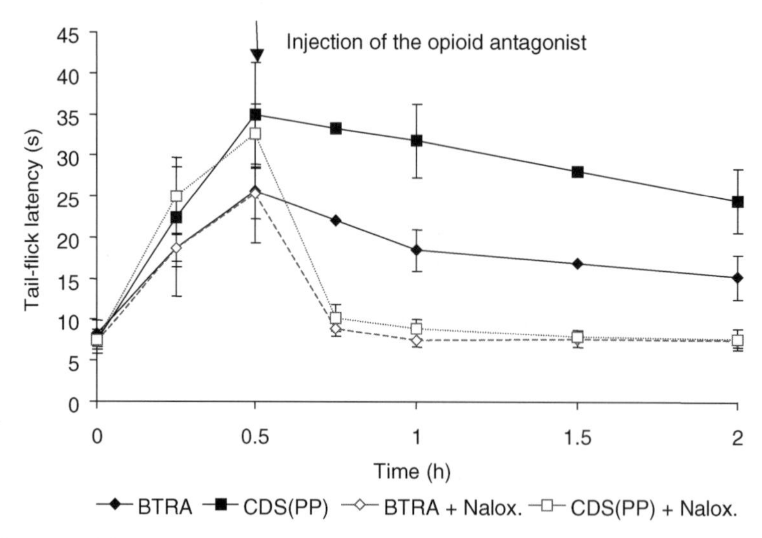

Figure 59. Reversal of the analgesia produced by 22.3 μmol/kg YK-CDS (**196**) and BTRA (**197**) administered i.v. by naloxone (6.1 μmol/kg, s.c.) administered 30 min later. Data represents mean ± SE of six rats for each group. (This figure is available in full color at http://mrw.interscience.wiley.com/emrw/9780471266945/home.)

producing any pharmacological effects. However, at the targeted site, where the necessary enzymes are found primarily, exclusively, or at higher activity, predictable enzymatic reactions convert the original CDS to the active drug (Fig. 60).

Certainly, the system depends upon the existence of such enzymes. Successful site- and stereospecific delivery of IOP-reducing β-adrenergic blocking agents to the eye was achieved with such an approach [105,176, 650–655] based on a general metabolic process that appears to apply to all lipophilic ketone precursors of β-amino alcohols. Contrary to the free catechol-containing adrenalone, lipophilic esters of adrenalone were effectively reduced within the eye to yield epinephrine via the corresponding esters. However, oral administration did not result in any formation of epinephrine [656]. Hence, the ketone function may act as a targetor that can be reduced to produce the β-amino alcohol pharmacophore necessary for activity. Because the approach could be extended to other β-agonists [657,658], the possibility arose to produce ophthalmically useful β-adrenergic antagonists within the eye from the corresponding ketones. However, because ketones of the phenol ether-type β-blockers decompose in aqueous solutions to form the corresponding phenols, they are not good bioprecursors to produce β-amino alcohols. Hence, they were converted to oximes in order to stabilize them assuming that they will undergo a facile enzymatic hydrolysis to the bioreducible ketone. Although the oximes are much more stable than the ketones, their aqueous stability not always provided an acceptable shelf life (e.g., even at pH 4.5, which can be considered the lowest acceptable pH-limit for ophthalmic vehicle solutions and where the oxime is more stable, $t_{90\%}$ at room temperature of a 1% w/v solution is only 44 ± 5 days for alprenoxime (**201**) [653]). Additional stabilization was obtained, however, by using methoximes, methylethers of the corresponding oximes. For example, for the methoxime analog of alprenolol, no significant decomposition was observed during storage at room temperature within one year at pH values around 6.5 [653].

To summarize, in these CDSs, a β-amino oxime or alkyloxime function (**198**, Fig. 61) replaces the corresponding β-amino alcohol pharmacophore part of the original molecules (**200**). These oxime or alkyloxime derivatives exist in alternative *Z* (syn) or *E* (anti) configuration. They are enzymatically hydrolyzed within the eye by enzymes located in the iris-ciliary body, and subsequently, reductive enzymes also located in the iris-ciliary body

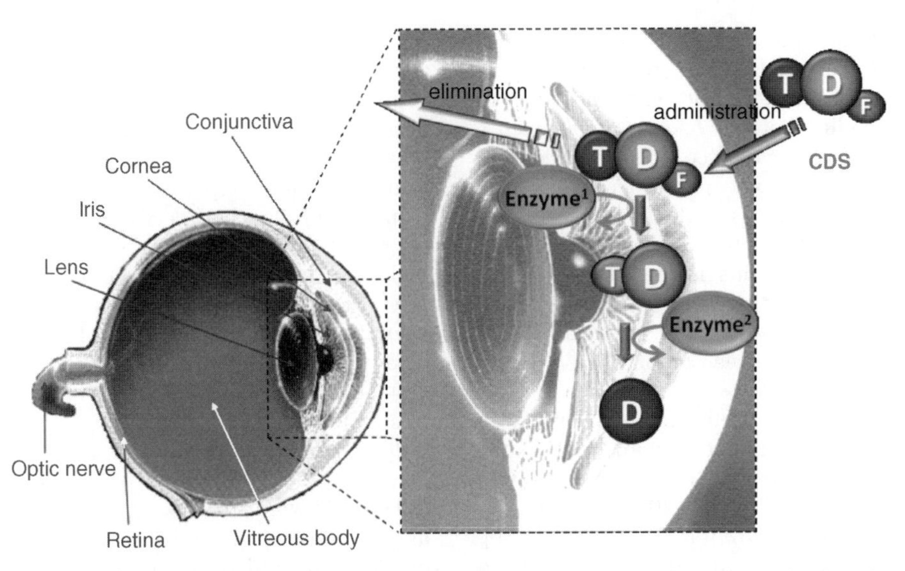

Figure 60. Schematic representation of the processes that provide site-specific targeting for the oxime-type CDS approach used for eye-targeted delivery of active β-blocker drugs (D). (This figure is available in full color at http://mrw.interscience.wiley.com/emrw/9780471266945/home.)

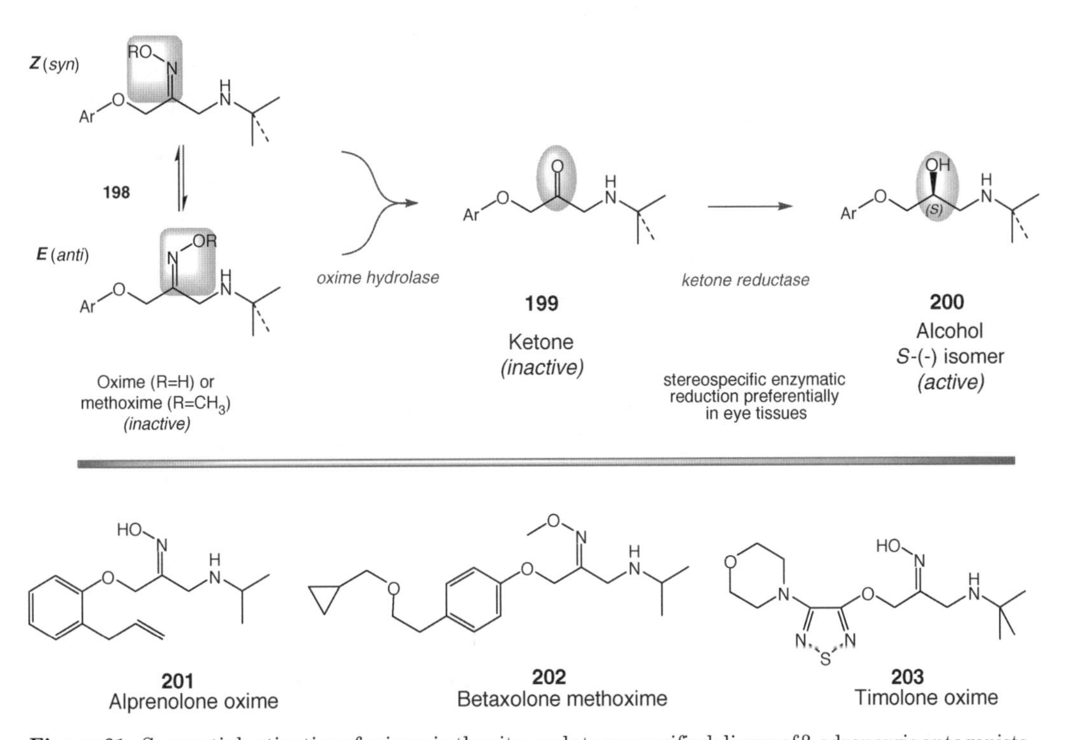

Figure 61. Sequential activation of oximes in the site- and stereospecific delivery of β-adrenergic antagonists to the eye. The original oximes or methoximes (**198**) and the intermediate ketones (**199**) are inactive; they are enzymatically converted into the active S-(−) β-adrenergic blocker alcohols (**200**) in a site- and stereospecific manner. (This figure is available in full color at http://mrw.interscience.wiley.com/emrw/9780471266945/home.)

produce only the active S-(−) stereoisomer of the β-blocker [32]. Besides these β-amino oximes, the biotransformation of oximes into the corresponding ketones has been shown in a number of other cases too [659–661]. For aryl β-amino alcohol-type β-adrenergic agonists and antagonists, most of the activity is known to reside with the (−)-stereoisomer [662–664], possibly because this isomer allows better interaction of all three important functionalities (aromatic, amino, and β-hydroxyl moieties) with the β-adrenoceptor [665].

4.2.2. Oxime and Methoxime Analogs of β-Blockers A variety of oxime and methoxime analogs of alprenolol (**201**), betaxolol (**202**), propranolol, and timolol (**203**) were synthesized and tested [32,105,650–655,666–669]. They were all shown to undergo the predicted specific activation within, and only within, the eye. Highest concentrations of the active β-antagonists were observed in the iris-ciliary body. When applied topically, both the oximes and the methoximes showed higher IOP-reducing activities in rabbits than the corresponding alcohols. On the other hand, topical or even i.v. administration did not produce any cardiovascular effect neither in normal rabbits, rats, and dogs nor after inducing tachycardia. Following i.v. injection, the oxime disappeared rapidly from the blood, and at no time could the corresponding alcohol be detected [650]. This indicates that the required enzymatic hydrolysis–reduction activation sequence, which occurs in the eye, does not occur in the systemic circulation. Therefore, the oxime derivatives are void of any cardiovascular activity, a major drawback of classical antiglaucoma agents.

Oxime or methoxime derivatives can be relatively simply prepared from the original β-blocker alcohols by oxidation of the secondary alcohol using activated DMSO (Pfitzner–-Moffat oxidation) [670] followed by coupling of the formed ketone with either hydroxylamine or methoxyamine in the same reaction medium. The Z (syn)–E (anti) isomerization in buffer is usually relatively slow, and is somewhat faster in alkaline than in acidic buffer. Equilibrium is pH-dependant; for example, for betaxoxime equilibrium is reached within two weeks and the final ratios are Z/E 46:54 at pH 6.5 and 48:52 at pH 7.4, respectively [655]. Stability studies were performed for a number of analogs in buffer solutions at different temperatures. Methoximes are usually more stable than the corresponding oxime derivatives; at room temperature, they have t_{90}s (time within which 10% of the active drug is degraded) in the range of a few years. In biological fluids, the Z/E isomer equilibration is much faster and indicates involvement of enzymatic catalysis. For example, for betaxolone oxime and methoxime, isomerization is 300–500 times faster in biological fluids than in simple buffer [655].

The oxime and methoxime analogs of alprenolol (**201**) [653,666] and betaxolol (**202**) [655] were found to produce significant and long-lasting reduction in the IOP of rabbits following uni- or bilateral administration (Fig. 62). In most cases, the novel analogs produced more pronounced and longer-lasting effects than the parent compounds and were also less irritating. On the other hand, the i.v. bolus injection of alprenoxime led to only insignificant transient bradycardia, and no activity was found after oral or topical administration in rats and rabbits [666]. Alprenolol in a similar dose produced sustained and significant bradycardia for more than 30 min. A study in dogs [669] confirmed that no significant cardiac electrophysiologic parameters are altered after systemic treatment with alprenoxime or its methyl ester analog even at doses that far exceed that effective in reducing IOP (no monitored parameter was affected more than 6%). In the meantime, administration of alprenolol itself exerted profound effects on cardiac function consistent with β-blocking activity and resulting in changes of 30–140%. Alprenoxime also did not alter isoproterenol induced tachycardia in dogs [669]. In similar manner and in contrast to betaxolol itself, the oxime or methoxime analogs of betaxolol had no effect on isoproterenol induced tachycardia in Sprague-Dawley rats at doses up to 20 μmol/kg [655]. Acute oral and i.v. toxicity studies have been completed in mice and rats for betaxoxime, the oxime derivative of betaxolol selected for proof of concept studies.

For alprenoxime (**201**), an open single dose escalating tolerance study in 14 male

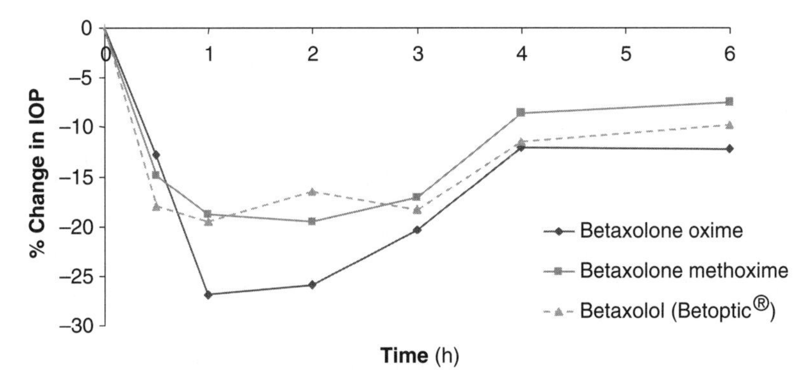

Figure 62. IOP reducing activities of betaxolol (Betoptic®) and its oxime and methoxime analogs in New Zealand albino rabbits. Doses of 100 μL solutions formulated in saline with 0.01% EDTA and 0.01% benzalkonium chloride (pH 6.0) were administered in one eye (1.84–1.86 mg). (This figure is available in full color at http://mrw.interscience.wiley.com/emrw/9780471266945/home.)

volunteers and a Phase I study to evaluate safety and tolerance of multiple doses in 14 normal male volunteers have been completed. These studies found single-drop instillation of alprenoxime ophthalmic solution up to 1.0% administered up to two times daily for 14 days to be well tolerated in normal male individuals. No clinically significant medical events were observed. Alprenoxime had no apparent clinically significant effect on the subject's hematology, biochemistry, or urine-analysis values. The data obtained from systemic monitoring showed that neither the cardiovascular nor the respiratory systems were significantly affected by the administration of alprenoxime at the doses given in this study. There were no occurrences of symptoms of pain and foreign body sensation by subjects of any treatment group. No evidence of lowered IOP was observed in these groups of volunteers. In the multiple dose study, the extent of systemic β-adrenoceptor blockade was evaluated by determining the effect of treatment on the dose of intravenous isoproterenol-induced tachycardia. No evidence of systemic β-blocking actions was evident in these studies. Further human development of alprenoxime was abandoned because it did not provide an acceptable shelf life in aqueous solutions.

Betaxoxime, the oxime derivative of betaxolol, has been selected for proof of concept studies, and has been synthesized and prepared as ophthalmic solution under good manufacturing practice (GMP) conditions. A Phase I single-center, open, single-group study designed to evaluate the tolerability, safety, and efficacy of two concentrations (0.25 and 0.50%) of betaxoxime administered into one eye of ocular hypertensive subjects compared to placebo (vehicle) administered into the contralateral eye of the same subjects has been completed recently [105]. Betaxoxime eye drops were well tolerated at both concentrations applied; they did not irritate the eye and did not cause unwanted ocular or systemic (e.g., cardiovascular) side effects. Statistically significant IOP-reductions in the treated eye were found for both doses, whereas no consistent statistically significant reductions were found in the control eye (Fig. 63). There was no significant change in the heart rate, and there were no consistent, statistically significant changes in the systolic or diastolic blood pressures. Results indicate betaxoxime to be a safe and well-tolerated eye drop with eye-targeted activity and promising IOP-lowering effects.

In conclusion, oxime or methoxime derivatives showed significant IOP-lowering activity, but even their i.v. administration did not produce the active β-blockers metabolically. Therefore, they are void of any cardiovascular activity, a major drawback of classical antiglaucoma agents. Methoxime derivatives provide sufficient stabilization for acceptable shelf life. Because of the advantages provided by this unique, eye-targeting profile, oxime-based eye-targeting CDSs could replace the β-blockers currently used for ophthalmic applications.

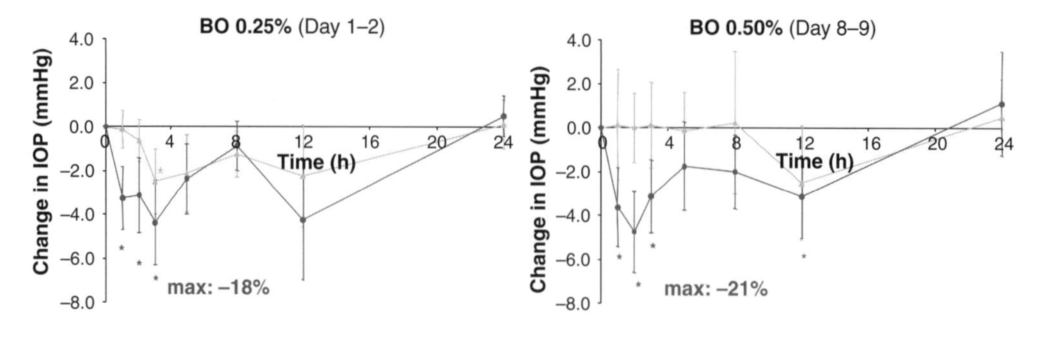

Figure 63. Effects of betaxoxime (BO) on the IOP of four ocular hypertensive patients in a Phase I/II single-center, open-label, single group study shown as change in IOP versus baseline [105]. Statistically significant reductions in the treated (L) eye (asterisk denotes $p < 0.05$; paired t-test versus predose at 0 h) were observed at the following time points: BO 0.25%: 1, 2, and 3 h; BO 0.50%: 1, 2, 3, and 12 h; no consistent statistically significant reductions were observed in the control (R) eye. (This figure is available in full color at http://mrw.interscience.wiley.com/emrw/9780471266945/home.)

4.3. Receptor-Based Transient Anchor-Type CDS

This CDS class is based on formation of a reversible covalent bond between the delivered entity and some part of the targeted receptor site to enhance selectivity and activity. The T targetor moiety of these CDSs is expected to undergo a modification that allows transient anchoring of the active agent. This class is less well developed than the previous ones, but it may become of increasing importance with the accumulation of receptor-related structural information.

The possibilities of a receptor-based CDS for naloxone, a pure opiate antagonist, were explored because opiate addiction is a very serious problem and long-lasting antagonists are highly desirable. The presence of an essential –SH group near the binding site in opioid receptors was long suspected [671], as agonist binding capacity is destroyed by –SH reagents such as N-ethylmaleimide. Mutagenesis and other studies [672,673] are beginning to elucidate the mechanism of this inhibition of binding at opioid receptors, receptors that are members of the G-protein-coupled receptor (GPCR) family. There is also evidence that Leu-enkephalin or even morphine analogs containing activated sulfhydryl groups can form mixed disulfide linkages with opioid receptors. The modified agonist caused receptor activation that persisted following extensive washing, was naloxone-reversible, and returned after naloxone was washed away [674]. Chlornaltrexamine, an irreversible alkylating affinity label on the opiate receptor, can maintain its antagonist effect for an astonishing period of 2–3 days, proving that covalent binding can increase the duration of action at these receptors.

Our investigation focused on a more reversible, spirothiazolidine-based system. Spirothiazolidine derivatives were selected because they should be subject to biological cleavage of the imine formed after spontaneous cleavage of the carbon-sulfur bond [395], and because they have been explored earlier in delivery systems for controlled-release endogenous agents such as hydrocortisone or progesterone (Fig. 41) [394,396]. As modifications at the 6-keto position of naloxone are possible without loss of activity, the spirothiazolidine ring was attached at this site in a manner similar to that done for hydrocortisone (Fig. 41). Opening of this spirothiazolidine ring allows oxidative anchoring of the delivered agent to nearby –SH groups through disulfide bridging. The blocking of the receptor is reversible; endogenous –SH compounds presumably will regenerate the –SH group of the receptor. If an adequate conformation can be obtained, stronger and longer binding is expected at the opiate receptors. Indeed, IC_{50}s of naloxone-6-spirothiazolidine for these receptors indicated increased affinities [24]. The increase was especially significant for the κ-receptors known to be involved, among others, in sedative analgesia (IC_{50}s of 3.9 and 60.0 nM for naloxone-CDS and naloxone, respectively). The intrinsic activity for guinea pig ileum of naloxone-CDS also showed an approximately 25-fold increase compared with that of naloxone. Because other GPCRs (e.g., TRH, D_1 and D_2 dopamine, vasopressin, follicle-stimulating hormone, and cannabinoid receptors) are also sensitive to sulfhydryl reagents [673], there is considerable potential for this CDS class if modifications of the corresponding ligands at adequate sites can be performed.

In addition, as the earlier evaluations of spirothiazolidine-based systems indicated an enhanced deposition to lung tissue, the possibility to use similar mechanisms in developing other lung tissue-targeting CDSs seemed worth exploring. Lipoic acid, a nontoxic coenzyme for acyl transfer and redox reactions, was investigated as targetor moiety for selective delivery to lung tissue because of its ability to also form disulfide linkages [675]. The corresponding CDSs for chlorambucil, an antineoplastic agent, and cromolyn, a bischromone used in antiasthmatic prophylaxis, were built via an ester linkage. *In vitro* kinetic and *in vivo* pharmacokinetic studies showed that the respective CDSs were sufficiently stable in buffer and biological media, hydrolyzed rapidly into the respective active parent drugs, and, relative to the underivatized parent compounds, they significantly enhanced delivery and retention of the active drug in lung tissue. For example, following administration to rats, approximately 20% of the administered dose was found in lung with cromolyn-CDS; with parent drug administration, only approximately 0.2% was found for the same period of up to 30 min. Compared with the administration of parent drug, chlorambucil-CDS also produced an over 20-fold increase in the amount of chlorambucil delivered to rabbit lung tissue after 30 min following administration. Increased lipophilicity in these systems is also likely to have a beneficial effect on the bioavailability of these highly hydrophilic compounds. These CDSs are designed to anchor the drug entity in lung tissue and provide a sustained release of active compound over a protracted period of time.

5. CONCLUSIONS

The present chapter attempted to systematize and summarize several novel metabolism-based drug design approaches collected under the umbrella of retrometabolic drug design and targeting. These approaches incorporate two major classes: soft drugs and chemical delivery systems. They achieve their drug-targeting roles in opposite ways, but they have in common the basic concept of designed metabolism to control drug action and targeting. These approaches are general in nature, can be applied to essentially all drug classes, and are based on specific design rules whose applications can be enhanced by specific computer programs. To illustrate the concepts, a number of specific examples from a variety of fields were described in detail including already marketed drugs that resulted from the successful application of such design principles.

REFERENCES

1. Le Fanu J. The Rise and Fall of Modern Medicine. New York: Carrol & Graf; 1999.
2. Drews J. Drug discovery: a historical perspective. Science 2000;287:1960–1964.
3. Corey EJ, Czakó B, Kürti L. Molecules and Medicine Hoboken, NJ: Wiley; 2007.
4. Nicolaou KC, Montagnon T. Molecules That Changed the World. Weinheim: Wiley-VCH; 2008.
5. Reuben BG. The consumption and production of pharmaceuticals. In: Wermuth CG, editor. The Practice of Medicinal Chemistry. London: Academic Press; 2008. p 894–921.
6. Buchwald P, Bodor N. Computer-aided drug design: the role of quantitative structure–property, structure–activity, and structure–metabolism relationships (QSPR, QSAR, QSMR). Drugs Future 2002;27:577–588.
7. Tufts Center for the Study of Drug Development. Outlook 2007. Boston, MA: Tufts Center for the Study of Drug Development; 2007; p 1–6.

8. Yildirim MA, Goh KI, Cusick ME, Barabasi AL, Vidal M. Drug-target network. Nat Biotechnol 2007;25:1119–1126.
9. Proudfoot JR. Drugs, leads, and drug-likeness: an analysis of some recently launched drugs. Bioorg Med Chem Lett 2002;12:1647–1650.
10. Horrobin DF. New drug discovery techniques are not so revolutionary. BMJ 2001;322:239.
11. Overington JP, Al-Lazikani B, Hopkins AL. How many drug targets are there? Nat Rev Drug Discov 2006;5:993–996.
12. Lipinski C, Hopkins A. Navigating chemical space for biology and medicine. Nature 2004;432:855–861.
13. Lin JH, Lu AY. Role of pharmacokinetics and metabolism in drug discovery and development. Pharmacol Rev 1997;49:403–449.
14. van de Waterbeemd H, Smith DA, Beaumont K, Walker DK. Property-based design: optimization of drug absorption and pharmacokinetics. J Med Chem 2001;44:1313–1333.
15. Bodor N. Novel approaches for the design of membrane transport properties of drugs. In: Roche EB, editor. Design of Biopharmaceutical Properties through Prodrugs and Analogs. Washington, DC: Academy of Pharmaceutical Sciences; 1977. p 98–135.
16. Bodor N. Soft drugs: strategies for design of safer drugs. In: Buisman JAK, editor. Strategy in Drug Research. Proceedings of the 2nd IUPAC-IUPHAR Symposium on Research, Noordwijkerhout, The Netherlands, Amsterdam: Elsevier; 1982. p 137–164.
17. Bodor N. Designing safer drugs based on the soft drug approach. Trends Pharmacol Sci 1982;3:53–56.
18. Bodor N. Novel approaches to the design of safer drugs: soft drugs and site-specific chemical delivery systems. Adv Drug Res 1984;13: 255–331.
19. Bodor N. Soft drugs: principles and methods for the design of safe drugs. Med Res Rev 1984;3:449–469.
20. Bodor N. The soft drug approach. Chemtech 1984;14(1):28–38.
21. Bodor N. Soft drugs. In: Dulbecco R, editor. Encyclopedia of Human Biology. San Diego, CA: Academic Press; 1991. p 101–117.
22. Bodor N. Drug targeting and retrometabolic drug design approaches. Adv Drug Deliv Rev 1994;14:157–166.
23. Bodor N. Retrometabolic approaches to drug targeting. In: Rapaka R, editor. Membranes and Barriers: Targeted Drug Delivery. Vol. 154. NIDAResearch Monograph Series. Rockville, IN: NIDA; 1995; p 1–27.
24. Bodor N, Buchwald P. Drug targeting via retrometabolic approaches. Pharmacol Ther 1997; 76:1–27.
25. Bodor N, Buchwald P. Retrometabolism-based drug design and targeting. In: Abraham DJ, editor. Burger's Medicinal Chemistry and Drug Discovery. Vol. 2. Drug Discovery and Drug Development, 6th ed. New York: Wiley; 2003. p 533–608.
26. Bodor N, Buchwald P. Retrometabolic drug design: principles and recent developments. Pure Appl Chem 2008;80:1669–1682.
27. Bodor N, Buchwald P. Soft drug design: general principles and recent applications. Med Res Rev 2000;20:58–101.
28. Bodor N. Inactive metabolite approach to soft drug design. US Patent 6,610,675. 2003.
29. Bodor N, Farag HH, Brewster ME. Site-specific, sustained release of drugs to the brain. Science 1981;214:1370–1372.
30. Bodor N. Targeting of drugs to the brain. Method Enzymol 1985;112:381–396.
31. Bodor N. Redox drug delivery systems for targeting drugs to the brain. Ann NY Acad Sci 1987;507:289–306.
32. Bodor N, Prokai L. Site- and stereospecific ocular drug delivery by sequential enzymatic bioactivation. Pharm Res 1990;7:723–725.
33. Bodor N, Brewster ME. Chemical delivery systems. In: Juliano RL, editor. Targeted Drug Delivery. Vol. 100. Handbook of Experimental Pharmacology. Berlin: Springer; 1991. p. 231–284.
34. Bodor N, Buchwald P. All in the mind. Chem Br 1998;34(1):36–40.
35. Bodor N, Buchwald P. Recent advances in the brain targeting of neuropharmaceuticals by chemical delivery systems. Adv Drug Deliv Rev 1999;36:229–254.
36. Bodor N, Buchwald P. Targeting of neuropharmaceuticals by chemical delivery systems. In: Dermietzel R, Spray DC, Needergard M, editors. Blood–Brain Interfaces: From Ontogeny to Artificial Barriers. Weinheim: Wiley-VCH; 2006. p 463–500.
37. Corey EJ. Retrosynthetic thinking: essentials and examples. Chem Soc Rev 1988; 17: 111–133.
38. Bodor N, Kaminski JJ, Selk S. Soft drugs. 1. Labile quaternary ammonium salts as soft antimicrobials. J Med Chem 1980;23:469–474.

39. Bodor N, Kaminski JJ. Soft drugs. 2. Soft alkylating compounds as potential antitumor agents. J Med Chem 1980;23:566–569.
40. Bodor N, Woods R, Raper C, Kearney P, Kaminski J. Soft drugs. 3. A new class of anticholinergic agents. J Med Chem 1980;23: 474–480.
41. Bodor N. Prodrugs versus soft drugs. In: Bundgaard H, editor. Design of Prodrugs. Amsterdam: Elsevier; 1985. p 333–354.
42. Stella VJ. Prodrugs: an overview and definition. In: Higuchi T, Stella VJ, editors. Prodrugs as Novel Drug Delivery Systems. Washington, DC: American Chemical Society; 1975. p 1–115.
43. Bundgaard H, editor. Design of Prodrugs. Amsterdam: Elsevier Science; 1985.
44. Bodor N, Kaminski JJ. Prodrugs and site-specific chemical delivery systems. Annu Rep Med Chem 1987;22:303–313.
45. Balant LP, Doelker E. Metabolic considerations in prodrug design. In: Wolff ME, editor. Burger's Medicinal Chemistry and Drug Discovery. New York: Wiley-Interscience; 1995. p 949–982.
46. Stella VJ. Themed issue. Low molecular weight prodrugs. Adv Drug Deliv Rev 1996;19:111–330.
47. Beaumont K, Webster R, Gardner I, Dack K. Design of ester prodrugs to enhance oral absorption of poorly permeable compounds: challenges to the discovery scientist. Curr Drug Metab 2003;4:461–485.
48. Ettmayer P, Amidon GL, Clement B, Testa B. Lessons learned from marketed and investigational prodrugs. J Med Chem 2004;47: 2393–2404.
49. Wermuth CG, Gaignault J-C, Marchandeau C. Designing prodrugs and bioprecursors. In: Wermuth CG, editor. The Practice of Medicinal Chemistry. London: Academic Press; 2008. p 721–746.
50. Rautio J, Kumpulainen H, Heimbach T, Oliyai R, Oh D, Järvinen T, Savolainen J. Prodrugs: design and clinical applications. Nat Rev Drug Discov 2008;7:255–270.
51. Albert A. Chemical aspects of selective toxicity. Nature 1958;182:421–427.
52. Liederer BM, Borchardt RT. Enzymes involved in the bioconversion of ester-based prodrugs. J Pharm Sci 2006;95:1177–1195.
53. Lee HJ, Soliman MRI. Anti-inflammatory steroids without pituitary-adrenal suppresion. Science 1982;215:989–991.
54. Heiman AS, Ko D-H, Chen M, Lee HJ. New steroidal anti-inflammatory antedrugs: methyl 3,20-dioxo-9α-fluoro-11β,17α,21-trihydroxy-1,4-pregnadiene-16α-carboxylate and methyl 21-acetyloxy-3,20-dioxo-11β,17α-dihydroxy-9α-fluoro-1,4-pregnadiene-16α-carboxylate. Steroids 1997;62:491–499.
55. Lee HJ, Cooperwood JS, You Z, Ko DH. Prodrug and antedrug: two diametrical approaches in designing safer drugs. Arch Pharm Res 2002;25:111–136.
56. Khan MO, Park KK, Lee HJ. Antedrugs: an approach to safer drugs. Curr Med Chem 2005;12:2227–2239.
57. Bodor N, Buchwald P. Designing safer (soft) drugs by avoiding the formation of toxic and oxidative metabolites. Mol Biotechnol 2004;26: 123–132.
58. Albert A. Selective Toxicity. The Physico–Chemical Basis of Therapy. 7th ed. London: Chapman and Hall; 1985.
59. Williams JA, Hyland R, Jones BC, Smith DA, Hurst S, Goosen TC, Peterkin V, Koup JR, Ball SE. Drug–drug interactions for UDP-glucuronosyltransferase substrates: a pharmacokinetic explanation for typically observed low exposure (AUCi/AUC) ratios. Drug Metab Dispos 2004;32:1201–1208.
60. Wienkers LC, Heath TG. Predicting *in vivo* drug interactions from *in vitro* drug discovery data. Nature Rev Drug Discov 2005;4:825–833.
61. Gillette JR. Effects of induction of cytochrome P-450 enzymes on the concentration of foreign compounds and their metabolites and on the toxicological effects of these compounds. Drug Metab Rev 1979;10:59–87.
62. Mannering GJ. Hepatic cytochrome P-450-linked drug-metabolizing systems. In: Testa B, Jenner P, editors. Concepts in Drug Metabolism Part B. New York: Marcel Dekker, Inc.; 1981; p 53–166.
63. Augustinsson K-B. Multiple forms of esterase in vertebrate blood plasma. Ann NY Acad Sci 1961;94:844–860.
64. Krisch K. Carboxylic ester hydrolases. In: Boyer PD, editor. The Enzymes. New York: Academic Press; 1971. p 43–69.
65. Heymann E. Hydrolysis of carboxylic esters and amides. In: Jakoby WB, Bend JR, Caldwell J, editors. Metabolic Basis of Detoxication: Metabolism of Functional Groups (Biochemical Pharmacology and Toxicology). New York: Academic Press; 1982. p 229–245.

66. Walker CH, Mackness MI. Esterases: problems of identification and classification. Biochem Pharmacol 1983;32:3265–3569.
67. Williams FW. Clinical significance of esterases in man. Clin Pharmacokin 1985;10:392–403.
68. Leinweber F-J. Possible physiological roles of carboxylic ester hydrolases. Drug Metab Rev 1987;18:379–439.
69. Satoh T, Hosokawa M. The mammalian carboxylesterases: from molecules to functions. Annu Rev Pharmacol Toxicol 1998;38: 257–288.
70. International Union of Biochemistry, Nomenclature Committee. Enzyme Nomenclature 1984. Recommendations of the Nomenclature Committee of the International Union of Biochemistry on the Nomenclature and Classification of Enzyme-Catalyzed Reactions. Orlando: Academic Press; 1984.
71. Satoh T, Taylor P, Bosron WF, Sanghani SP, Hosokawa M, La Du BN. Current progress on esterases: from molecular structure to function. Drug Metab Dispos 2002;30:488–493.
72. Sanghani SP, Quinney SK, Fredenburg TB, Davis WI, Murry DJ, Bosron WF. Hydrolysis of irinotecan and its oxidative metabolites, 7-ethyl-10-[4-*N*-(5-aminopentanoic acid)-1-piperidino] carbonyloxycamptothecin and 7-ethyl-10-[4-(1-piperidino)-1-amino]-carbonyloxycamptothecin, by human carboxylesterases CES1A1, CES2, and a newly expressed carboxylesterase isoenzyme, CES3. Drug Metab Dispos 2004;32:505–511.
73. Imai T, Taketani M, Shii M, Hosokawa M, Chiba K. Substrate specificity of carboxylesterase isozymes and their contribution to hydrolase activity in human liver and small intestine. Drug Metab Dispos 2006;34: 1734–1741.
74. Ollis DL, Cheah E, Cygler M, Dijkstra B, Frolow F, Franken SM, Harel M, Remington SJ, Silman I, Schrag J, Sussman JL, Verschueren KHG, Goldman A. The alpha/beta hydrolase fold. Protein Eng 1992;5:197–211.
75. Oakeshott JG, Claudianos C, Russell RJ, Robin GC. Carboxyl/cholinesterases: a case study of the evolution of a successful multigene family. Bioessays 1999;21:1031–1042.
76. Wadkins RM, Morton CL, Weeks JK, Oliver L, Wierdl M, Danks MK, Potter PM. Structural constraints affect the metabolism of 7-ethyl-10-[4-(1-piperidino)-1-piperidino]carbonyloxycamptothecin (CPT-11) by carboxylesterases. Mol Pharmacol 2001;60:355–362.
77. Bencharit S, Morton CL, Howard-Williams EL, Danks MK, Potter PM, Redinbo MR. Structural insights into CPT-11 activation by mammalian carboxylesterases. Nat Struct Biol 2002;9:337–342.
78. Bencharit S, Morton CL, Xue Y, Potter PM, Redinbo MR. Structural basis of heroin and cocaine metabolism by a promiscuous human drug-processing enzyme. Nat Struct Biol 2003;10:349–356.
79. Redinbo MR, Bencharit S, Potter PM. Human carboxylesterase 1: from drug metabolism to drug discovery. Biochem Soc Trans 2003;31:620–624.
80. Cook CS, Karabatsos PJ, Schoenhard GL, Karim A. Species dependent esterase activities for hydrolysis of an anti-HIV prodrug glycovir and bioavalability of active SC-48334. Pharm Res 1995;12:1158–1164.
81. Buchwald P. Structure-metabolism relationships: steric effects and the enzymatic hydrolysis of carboxylic esters. Mini Rev Med Chem 2001;1:101–111.
82. Buchwald P, Bodor N. Physicochemical aspects of the enzymatic hydrolysis of carboxylic esters. Pharmazie 2002;57:87–93.
83. Taketani M, Shii M, Ohura K, Ninomiya S, Imai T. Carboxylesterase in the liver and small intestine of experimental animals and human. Life Sci 2007;81:924–932.
84. Quon CY, Stampfli HF. Biochemical properties of blood esmolol esterase. Drug Metab Dispos 1985;13:420–424.
85. Quon CY, Mai K, Patil G, Stampfli HF. Species differences in the stereoselective hydrolysis of esmolol by blood esterases. Drug Metab Dispos 1988;16:425–428.
86. Feldman PL, James MK, Brackeen MF, Bilotta JM, Schuster SV, Lahey AP, Lutz MW, Johnson MR, Leighton HJ. Design, synthesis, and pharmacological evaluation of ultrashort- to long-acting opioid analgetics. J Med Chem 1991;34:2202–2208.
87. Minagawa T, Kohno Y, Suwa T, Tsuji A. Species differences in hydrolysis of isocarbacyclin methyl ester (TEI-9090) by blood esterases. Biochem Pharmacol 1995;49: 1361–1365.
88. Ericsson H, Tholander B, Regårdh CG. *In vitro* hydrolysis rate and protein binding of clevidipine, a new ultrashort-acting calcium antagonist metabolised by esterases, in different animal species and man. Eur J Pharm Sci 1999;8:29–37.

89. Pao LH, Hsiong CH, Hu OY, Wang JJ, Ho ST. *In vitro* and *in vivo* evaluation of the metabolism and pharmacokinetics of sebacoyl dinalbuphine. Drug Metab Dispos 2005;33: 395–402.
90. Stampfli HF, Quon CY. Polymorphic metabolism of flestolol and other ester containing compounds by a carboxylesterase in New Zealand white rabbit blood and cornea. Res Commun Mol Pathol Pharmacol 1995;88:87–97.
91. Durrer A, Wernly-Chung GN, Boss G, Testa B. Enzymic hydrolysis of nicotinate esters: comparison between plasma and liver catalysis. Xenobiotica 1992;22:273–282.
92. Chappell WR, Mordenti J. Extrapolation of toxicological and pharmacological data from animals to humans. Adv Drug Res 1991;20: 1–116.
93. Campbell DB. Extrapolation from animals to man. The integration of pharmacokinetics and pharmacodynamics. Ann NY Acad Sci 1996;801:116–135.
94. Prueksaritanont T, Gorham LM, Breslin MJ, Hutchinson JH, Hartman GD, Vyas KP, Baillie TA. *In vitro* and *in vivo* evaluations of the metabolism, pharmacokinetics, and bioavailability of ester prodrugs of L-767,679, a potent fibrinogen receptor antagonist: an approach for the selection of a prodrug candidate. Drug Metab Dispos 1997;25:978–984.
95. Godefroi EF, Janssen PAJ, van der Eycken CAM, van Heertum AHMT, Niemegeers CJE. DL-1-(1-Arylalkyl)imidazole-5-carboxylate esters. A novel type of hypnotic agents. J Med Chem 1965;8:220–223.
96. Lewi PJ, Heykants JJP, Janssen PAJ. Intravenous pharmacokinetic profile in rats of etomidate, a short-acting hypnotic drug. Arch Int Pharmacodyn Ther 1976;220:72–85.
97. Janssen PAJ, Niemegeers CJE, Schellekens KHL, Lenaerts FM. Etomidate, *R*-(+)-ethyl-1-(α-methyl-benzyl)imidazole-5-carboxylate (R 16659), a potent, short-acting and relatively atoxic intravenous hypnotic agent in rats. Arzneim Forsch 1971;21:1234–1243.
98. Savarese JJ, Ali HH, Basta SJ, Embree PB, Scott RPF, Sunder N, Weakly JN, Wastila WB, El-Sayad HA. The clinical neuromuscular pharmacology of mivacurium chloride (BW B1090U). A short-acting nondepolarizing ester neuromuscular blocking drug. Anesthesiology 1988;68:723–732.
99. Nogrady T, Weaver DF. Medicinal Chemistry. A Molecular and Biochemical Approach. 3rd ed. New York: Oxford University Press; 2005.
100. Bazzini P, Wermuth CG. Substituent groups. In: Wermuth CG, editor. The Practice of Medicinal Chemistry. London: Academic Press. 2008. p 431–463.
101. Lüllmann H, Hein L, Mohr K, Bieger D. Color Atlas of Pharmacology. 3rd ed. Stuttgart: Thieme; 2005.
102. Borg KO, Carlsson E, Hoffmann K-J, Jönsson T-E, Thorin H, Wallin B. Metabolism of metoprolol-(^{3}H) in man, the dog and the rat. Acta Pharmacol Toxicol 1975;36 (Suppl V):125–135.
103. Regårdh C-G, Johnsson G. Clinical pharmacokinetics of metoprolol. Clin Pharmacokin 1980;5:557–569.
104. Benfield P, Clissold SP, Brogden RN. Metoprolol. An updated review of its pharmacodynamic and pharmacokinetic properties, and therapeutic efficacy, in hypertension, ischaemic heart disease and related cardiovascular disorders. Drugs 1986;31:376–429.
105. Bodor N, Buchwald P. Ophthalmic drug design based on the metabolic activity of the eye: soft drugs and chemical delivery systems. AAPS J 2005;7:E820–E833.
106. Sasaki H, Yamamura K, Mukai T, Nishida K, Nakamura J, Nakashima M, Ichikawa M. Enhancement of ocular drug penetration. Crit Rev Ther Drug Carrier Syst 1999;16: 85–146.
107. Schoene RB, Martin TR, Charan NB, French CL. Timolol-induced bronchospasm in asthmatic bronchitis. JAMA 1981;245:1460–1461.
108. Schoene RB, Abuan T, Ward RL, Beasley CH. Effects of topical betaxolol, timolol, and placebo on pulmonary function in asthmatic bronchitis. Am J Ophthalmol 1984;97:86–92.
109. Richards R, Tattersfield AE. Bronchial β-adrenoceptor blockade following eyedrops of timolol and its isomer L-714,465 in normal subjects. Br J Clin Pharmacol 1985;20:459–462.
110. Bodor N, Oshiro Y, Loftsson T, Katovich M, Caldwell W. Soft drugs. 6. The application of the inactive metabolite approach for design of soft β-blockers. Pharm Res 1984;3:120–125.
111. Bodor N, El-Koussi A, Kano M, Khalifa MM. Soft drugs. 7. β-Blockers for systemic and ophthalmic use. J Med Chem 1988;31: 1651–1656.
112. Bodor N, El-Koussi A. Novel 'soft' β-blockers as potential safe antiglaucoma agents. Curr Eye Res 1988;7:369–374.
113. Polgar P, Bodor N. Cardiac electrophysiologic effects of adaprolol maleate, a new β-blocker, in closed chest dogs. Life Sci 1991;48: 1519–1528.

114. Bodor N, El-Koussi A, Zuobi K, Kovacs P. Synthesis and pharmacological activity of adaprolol enantiomers: a new soft drug for treating glaucoma. J Ocul Pharmacol Ther 1996;12:115–122.

115. Sugrue MF. New approaches to antiglaucoma therapy. J Med Chem 1997;40:2793–2809.

116. Yang H-S, Wu W-M, Bodor N. Soft drugs. XX. Design, synthesis and evaluation of ultra-short acting beta-blockers. Pharm Res 1995;12:329–336.

117. Erhardt PW, Woo CM, Anderson WG, Gorczynski RJ. Ultra-short-acting β-adrenergic receptor blocking agents. 2. (Arlyoxy)propanolamines containing esters on the aryl function. J Med Chem 1982;25:1408–1412.

118. Erhardt PW. A prodrug and a soft drug. In: Erhardt PW, editor. Drug Metabolism. Databases and High Throughput Testing During Drug Design and Development. Oxford: Blackwell Science; 1999. p 62–69.

119. Tatsuno H, Goto K, Shigenobu K, Kasuya Y, Obase H, Yamada Y, Kudo S. Synthesis and adrenergic β-blocking activity of some 1,3-benzodioxole derivatives. J Med Chem 1977;20:394–397.

120. O'Donnell JP, Parekh S, Borgman RJ, Gorczynski RJ. Synthesis and pharmacology of potential β-blockers. J Pharm Sci 1979;68: 1236–1238.

121. Erhardt PW, Woo CM, Gorczynski RJ, Anderson WG. Ultra-short-acting β-adrenergic receptor blocking agents. 1. (Arlyoxy)propanolamines containing esters in the nitrogen substituent. J Med Chem 1982;25: 1402–1407.

122. Erhardt PW, Woo CM, Matier WL, Gorczynski RJ, Anderson WG. Ultra-short-acting β-adrenergic receptor blocking agents. 3. Ethylenediamine derivatives of (arlyoxy)propanolamines having esters on the aryl function. J Med Chem 1983;26:1109–1112.

123. Kam S-T, Matier WL, Mai KX, Barcelon-Yang C, Borgman RJ, O'Donnell JP, Stampfli HF, Sum CY, Anderson WG, Gorczynski RJ, Lee RJ. [(Arylcarbonyl)oxy]propanolamines. 1. Novel β-blockers with ultrashort duration of action. J Med Chem 1984;27:1007–1016.

124. Benfield P, Sorkin EM. Esmolol. A preliminary review of its pharmacodynamic and pharmacokinetic properties, and therapeutic efficacy. Drugs 1987;33:392–412.

125. Iguchi S, Iwamura H, Nishizaki M, Hayashi A, Senokuchi K, Kobayashi K, Sakaki K, Hachiya K, Ichioka Y, Kawamura M. Development of a highly cardioselective ultra short-acting β-blocker, ONO-1101. Chem Pharm Bull 1992; 40:1462–1469.

126. Muraki K, Nakagawa H, Nagano N, Henmi S, Kawasumi H, Nakanishi T, Imaizumi K, Tokuno T, Atsuki K, Imaizumi Y, Watanabe M. Effects of ONO-1101, a novel beta-antagonist, on action potential and membrane currents in cardiac muscle. J Pharmacol Exp Ther 1996; 278:555–563.

127. Kitamura A, Sakamoto A, Inoue T, Ogawa R. Efficacy of an ultrashort-acting β-adrenoreceptor blocker (ONO-1101) in attenuating cardiovascular responses to endotracheal intubation. Eur J Clin Pharmacol 1997;51:467–471.

128. Atarashi H, Kuruma A, Yashima M, Saitoh H, Ino T, Endoh Y, Hayakawa H. Pharmacokinetics of landiolol hydrochloride, a new ultra-short-acting beta-blocker, in patients with cardiac arrhythmias. Clin Pharmacol Ther 2000;68:143–150.

129. Yoshida Y, Terajima K, Sato C, Akada S, Miyagi Y, Hongo T, Takeda S, Tanaka K, Sakamoto A. Clinical role and efficacy of landiolol in the intensive care unit. J Anesth 2008;22:64–69.

130. Lin Y-T, Wu B-N, Wu J-R, Lo Y-C, Chen L-C, Chen I-J. Vasomolol: an ultra short-acting and vasorelaxant vanilloid type β_1-adrenoreceptor antagonist. J Cardiovasc Pharmacol 1996;28: 149–157.

131. Achari R, Drissel D, Hulse JD, Bell V, Turlapaty P, Laddu A, Matier WL. Pharmacokinetics and pharmacodynamics of flestolol, a new short-acting, beta-adrenergic receptor antagonist. J Clin Pharmacol 1987;27:60–64.

132. Strom J, Josephson M, Frishman WH, Singh B, Heilbrunn S, Osterle S, Turlapaty P, Viray R, Coe J, Bell V. Hemodynamic effects of flestolol, a titratable short-acting intravenous beta-adrenergic receptor blocker. J Clin Pharmacol 1988;28:276–282.

133. Sugrue MF, Gautheron P, Grove J, Mallorga P, Viader M-P, Baldwin JJ, Ponticello GS, Varga SL. L-653,328: an ocular hypotensive agent with modest beta receptor blocking activity. Invest Ophthalmol Vis Sci 1988;29:776–784.

134. Bauer K, Brunner-Ferber F, Distlerath LM, Lippa EA, Binkowitz B, Till P, Kaik G. Assessment of systemic effects of different ophthalmic β-blockers in healthy volunteers. Clin Pharmacol Ther 1991;49:658–664.

135. Bauer KG, Brunner-Ferber F, Distlerath LM, Lippa EA, Binkowitz B, Till P, Kaik G. Assessment of bronchial effects following topical administration of butylamino-phenoxypropanol-

acetate, an oculoselective β-adrenoceptor blocker in asthmatic subjects. Br J Clin Pharmacol 1992;34:122–129.

136. Hamilton TC, Chapman V. Intrinsic sympathomimetic activity of β-adrenoceptor blocking drugs at cardiac and vascular β-adrenoceptors. Life Sci 1978;23:813–820.

137. Machin PJ, Hurst DN, Osbond JM. β-Adrenoceptor activity of the stereoisomers of the bufuralol alcohol and ketone metabolites. J Med Chem 1985;28:1648–1651.

138. Francis RJ, East PB, McLaren SJ, Larman J. Determination of bufuralol and its metabolites in plasma by mass fragmentography and by gas chromatography with electron capture detection. Biomed Mass Spectrom 1976;3: 281–285.

139. Dayer P, Leemann T, Kupfer A, Kronbach T, Meyer UA. Stereo- and regioselectivity of hepatic oxidation in man-effect of the debrisoquine/sparteine phenotype on bufuralol hydroxylation. Eur J Clin Pharmacol 1986;31:313–318.

140. Hwang S-K, Juhasz A, Yoon S-H, Bodor N. Soft drugs 22. Design, synthesis, and evaluation of soft bufuralol analogues. J Med Chem 2000;43:1525–1532.

141. Meuldermans W, Hendrickx J, Lauwers W, Hurkmans R, Swysen E, Thijssen J, Timmerman P, Woestenborghs R, Heykants J. Excretion and biotransformation of alfentanil and sufentanil in rats and dogs. Drug Metab Dispos 1987;15:905–913.

142. Colapret JA, Diamantidis G, Spencer HK, Spaulding TC, Rudo FG. Synthesis and pharmacological evaluation of 4,4-disubstituted piperidines. J Med Chem 1989;32:968–974.

143. James MK, Feldman PL, Schuster SV, Bilotta JM, Brackeen MF, Leighton HJ. Opioid receptor activity of GI 87084B, a novel ultra-short acting analgesic, in isolated tissues. J Pharmacol. Exp Ther 1991;259:712–718.

144. Egan TD, Lemmens HJM, Fiset P, Hermann DJ, Muir KT, Stanski DR, Shafer SL. The pharmacokinetics of the new short-acting opioid remifentanil (GI87084B) in healthy adult male volunteers. Anesthesiology 1993;79: 881–892.

145. Westmoreland CL, Hoke JF, Sebel PS, Hug CC, Muir KT. Pharmacokinetics of remifentanil (GI87084B) and its major metabolite (GI90291) in patients undergoing elective inpatient surgery. Anesthesiology 1993;79: 893–903.

146. D'Amour FE, Smith DL. A method for determining loss of pain sensation. J Pharmacol Exp Ther 1941;72:74–79.

147. Haidar SH, Moreton JE, Liang Z, Hoke JF, Muir KT, Eddington ND. Evaluating a possible pharmacokinetic interaction between remifentanil and esmolol in the rat. J Pharm Sci 1997;86:1278–1282.

148. Haidar SH, Moreton JE, Liang Z, Hoke JF, Muir KT, Eddington ND. The pharmacokinetics and electroencephalogram response of remifentanil alone and in combination with esmolol in the rat. Pharm Res 1997;14: 1817–1823.

149. Schimmer BP, Parker KL. Adrenocorticotropic hormone; adrenocortical steroids and their synthetic analogs; inhibitors of the synthesis and actions of adrenocortical hormones. In: Hardman JG, Limbird LE, editors. Goodman & Gilman's The Pharmacological Basis of Therapeutics. New York: McGraw-Hill; 1996. p 1459–1485.

150. Avery MA, Woolfrey JR. Anti-inflammatory steroids. In: Abraham DJ, editor. Burger's Medicinal Chemistry and Drug Discovery. Vol. 3. Cardiovascular Agents and Endocrines. 6th ed. New York: Wiley; 2003. p 747–853.

151. Buchman AL. Side effects of corticosteroid therapy. J Clin Gastroenterol 2001;33: 289–294.

152. Lipworth BJ. Systemic adverse effects of inhaled corticosteroid therapy: A systematic review and meta-analysis. Arch Intern Med 1999;159:941–955.

153. Allen DB. Systemic effects of intranasal steroids: an endocrinologist's perspective. J Allergy Clin Immunol 2000;106 (Suppl):S179–S190.

154. Perry RJ, Findlay CA, Donaldson MDC. Cushing's syndrome, growth impairment, and occult adrenal suppression associated with intranasal steroids. Arch Dis Child 2002;87: 45–48.

155. Israel E, Banerjee TR, Fitzmaurice GM, Kotlov TV, LaHive K, LeBoff MS. Effects of inhaled glucocorticoids on bone density in premenopausal women. N Engl J Med 2001;345: 941–947.

156. Rosen J, Miner JN. The search for safer glucocorticoid receptor ligands. Endocr Rev 2005; 26:452–464.

157. Noble S, Goa KL. Loteprednol etabonate. Clinical potential in the management of ocular inflammation. BioDrugs 1998;10:329–339.

158. Howes JF. Loteprednol etabonate: a review of ophthalmic clinical studies. Pharmazie 2000;55:178–183.

159. Bodor N, Buchwald P. Design and development of a soft corticosteroid, loteprednol etabonate. In: Schleimer RP, O'Byrne PM, Szefler SJ,

Brattsand R, editors. Inhaled Steroids in Asthma. Optimizing Effects in the Airways. Lung Biology in Health and Disease. Vol. 163. New York: Marcel Dekker; 2002. p 541–564.

160. Pavesio CE, Decory HH. Treatment of ocular inflammatory conditions with loteprednol etabonate. Br J Ophthalmol 2008;92:455–459.

161. Bodor N.Stéroïds doux exerçant une activité anti-inflammatoire. (Steroids having antiinflammatory activity.) Belgian Patent BE889,563. (Internat. Classif. C07J/A61K). 1981.

162. Bodor N, Varga M. Effect of a novel soft steroid on the wound healing of rabbit cornea. Exp Eye Res 1990;50:183–187.

163. Druzgala P, Hochhaus G, Bodor N. Soft drugs. 10. Blanching activity and receptor binding affinity of a new type of glucocorticoid: loteprednol etabonate. J Steroid Biochem 1991;38:149–154.

164. Druzgala P, Wu W-M, Bodor N. Ocular absorption and distribution of loteprednol etabonate, a soft steroid, in rabbit eyes. Curr Eye Res 1991;10:933–937.

165. Alberth M, Wu W-M, Winwood D, Bodor N. Lipophilicity, solubility and permeability of loteprednol etabonate: a novel, soft anti-inflammatory corticosteroid. J Biopharm Sci 1991;2:115–125.

166. Bodor NS, Kiss-Buris ST, Buris L. Novel soft steroids: effects on cell growth *in vitro* and on wound healing in the mouse. Steroids 1991;56:434–439.

167. Bodor N, Bodor N, Wu W-M. A comparison of intraocular pressure elevating activity of loteprednol etabonate and dexamethasone in rabbits. Curr Eye Res 1992;11:525–530.

168. Bodor N, Loftsson T, Wu W-M. Metabolism, distribution, and transdermal permeability of a soft corticosteroid, loteprednol etabonate. Pharm Res 1992;9:1275–1278.

169. Hochhaus G, Chen L-S, Ratka A, Druzgala P, Howes J, Bodor N, Derendorf H. Pharmacokinetic characterization and tissue distribution of the new glucocorticoid soft drug loteprednol etabonate in rats and dogs. J Pharm Sci 1992;81:1210–1215.

170. Bodor N, Murakami T, Wu W-M. Soft drugs. 18. Oral and rectal delivery of loteprednol etabonate, a novel soft corticosteroid, in rats-for safer treatment of gastrointestinal inflammation. Pharm Res 1995;12:869–874.

171. Bodor N, Wu W-M, Murakami T, Engel S. Soft drugs. 19. Pharmacokinetics, metabolism and excretion of a novel soft corticosteroid, loteprednol etabonate, in rats. Pharm Res 1995;12:875–879.

172. Monder C, Bradlow HL. Cortoic acids: explorations at the frontier of corticosteroid metabolism. Recent Prog Horm Res 1980;36:345–400.

173. Bodor N. Soft drugs: strategies for design of safer drugs. In: Briot M, Cautreels W, Roncucci R, editors. Metabolisme et Conception Medicaments: Quo Vadis? Proceedings of Symposium at Montpellier, France, November 26–27, 1981, 1983. CLIN MIDY, Montpellier. p 217–251.

174. Druzgala P, Bodor N. Regioselective O-alkylation of cortienic acid and synthesis of a new class of glucocorticoids containing a 17α-alkoxy, a 17α-(1′-alkoxyethyloxy), a 17α-alkoxymethyloxy, or a 17α-methylthiomethyloxy function. Steroids 1991;56:490–494.

175. Bodor N. The application of soft drug approaches to the design of safer corticosteroids. In: Christophers E, Kligman AM, Schöpf E, Stoughton RB, editors. Topical Corticosteroid Therapy: A Novel Approach to Safer Drugs. New York: Raven Press Ltd.; 1988. p 13–25.

176. Bodor N. Designing safer ophthalmic drugs. In: van der Goot H, Domány G, Pallos L, Timmerman H, editors. Trends in Medicinal Chemistry '88 Proceeding of the Xth International Symposium on Medicinal Chemistry. Amsterdam: Elsevier; 1989. p 145–164.

177. Buchwald P, Bodor N. Soft glucocorticoid design: structural elements and physicochemical parameters determining receptor-binding affinity. Pharmazie 2004;59:396–404.

178. Buchwald P. Glucocorticoid receptor binding: a biphasic dependence on molecular size as revealed by the bilinear LinBiExp model. Steroids 2008;73:193–208.

179. Novack GD, Howes J, Crockett RS, Sherwood MB. Change in intraocular pressure during long-term use of loteprednol etabonate. J Glaucoma 1998;7:266–269.

180. Howes J, Novack GD. Failure to detect systemic levels and effects of loteprednol etabonate and its metabolite, PJ-91, following chronic ocular administration. J Ocul Pharmacol Ther 1998;14:153–158.

181. Ilyas H, Slonim CB, Braswell GR, Favetta JR, Schulman M. Long-term safety of loteprednol etabonate 0.2% in the treatment of seasonal and perennial allergic conjunctivitis. Eye Contact Lens 2004;30:10–13.

182. White EM, Macy JI, Bateman KM, Comstock TL. Comparison of the safety and efficacy of loteprednol 0.5%/tobramycin 0.3% with dexa-

methasone 0.1%/tobramycin 0.3% in the treatment of blepharokeratoconjunctivitis. Curr Med Res Opin 2008;24:287–296.

183. Holland EJ, Bartlett JD, Paterno MR, Usner DW, Comstock TL. Effects of loteprednol/tobramycin versus dexamethasone/tobramycin on intraocular pressure in healthy volunteers. Cornea 2008;27:50–55.

184. Bodor N, Buchwald P. Corticosteroid design for the treatment of asthma: structural insights and the therapeutic potential of soft corticosteroids. Curr Pharm Des 2006; 12:3241–3260.

185. Bocca N, Pileggi A, Molano RD, Marzorati S, Wu W, Bodor N, Ricordi C, Buchwald P. Soft corticosteroids for local immunosuppression: exploring the possibility for the use of loteprednol etabonate in islet transplantation. Pharmazie 2008;63:226–232.

186. Bocca N, Buchwald P, Molano RD, Marzorati S, Stabler C, Pileggi A, Ricordi C. Biohybrid devices and local immunomodulation as opposed to systemic immunosuppression. In: Hallé JP, de Vos P, Rosenberg L, editors. The Bioartificial Pancreas and Other Biohybrid Therapies. Transworld Research Network; 2009. p 309–328.

187. Geldmacher H, Buckendahl A, Badorrek P, Nguyen D, LaVallee N, Petzold U, Hermann R, Hohlfeld JM, Krug N. A pilot study to assess the efficacy of loteprednol etabonate nasal spray as a treatment for allergic rhinitis in an environmental exposure unit (EEU). Allergy 2002;57(Suppl 73):234.

188. Hermann R, Locher M, Siebert-Weigel M, LaVallee N, Derendorf H, Hochhaus G. Intranasal loteprednol etabonate in healthy male subjects: pharmacokinetics and effects on endogenous cortisol. J Clin Pharmacol 2004;44:510–519.

189. Krug N, Hohlfeld JM, Geldmacher H, Larbig M, Heermann R, LaVallee N, Nguyen DT, Petzold U, Hermann R. Effect of loteprednol etabonate nasal spray suspension on seasonal allergic rhinitis assessed by allergen challenge in an environmental exposure unit. Allergy 2005;60:354–359.

190. Bodor N.Androstene derivatives. US Patent 5,981,517. 1999.

191. Kurucz I, Németh K, Mészáros S, Török K, Nagy Z, Zubovics Z, Horváth K, Bodor N. Anti-inflammatory effect and soft properties of etiprednol dicloacetate (BNP-166), a new, anti-asthmatic steroid. Pharmazie 2004;59: 412–416.

192. Kurucz I, Tóth S, Németh K, Török K, Csillik-Perczel V, Pataki A, Salamon C, Nagy Z, Székely JI, Horváth K, Bodor N. Potency and specificity of the pharmacological action of a new, antiasthmatic, topically administered soft steroid, etiprednol dicloacetate (BNP-166). J Pharmacol Exp Ther 2003;307:83–92.

193. Csanádi Á, Horváth G, Szekeres T, Haskó T, Ila L, Ivanics J, Patthy M, Salát J, Seres G, Pallagi I, Tóth G, Szederkényi F, Kónya A, Tegdes A, Bodor N, Zubovics Z. Etiprednol dicloacetate, a new soft glucocorticoid drug candidate. Development of chemistry. Pharmazie 2004;59:349–359.

194. Barton P, Laws AP, Page MI. Structure–activity relationships in the esterase-catalysed hydrolysis and transesterification of esters and lactones. J Chem Soc Perkin Trans 2 1994; 2021–2029.

195. Miklós A, Magyar Z, Kiss É, Novák I, Grósz M, Nyitray M, Dereszlay I, Czégeni E, Druga A, Howes J, Bodor N. 28-Day oral toxicity study with soft corticosteroid BNP-166 in rats and dogs, followed by a 14-day recovery period. Pharmazie 2002;57:142–146.

196. Jaffuel D, Demoly P, Gougat C, Balaguer P, Mautino G, Godard P, Bousquet J, Mathieu M. Transcriptional potencies of inhaled glucocorticoids. Am J Respir Crit Care Med 2000;162:57–63.

197. Bhalay G, Sandham DA. Recent advances in corticosteroids for the treatment of asthma. Curr Opin Investig Drugs 2002;3:1149–1156.

198. Gerhards E, Nieuweboer B, Schulz G, Gibian H, Berger D, Hecker W. Stoffwechsel von 6α-Fluor-16α-methyl-pregna-1,4-diene-11β,21-diol-3,20-dion (Fluocortolon) beim Menschen. Acta Endocr 1971;68:98–126.

199. Laurent H, Gerhards E, Wiechert R. New biologically active pregnan-21-oic acid esters. J Steroid Biochem 1975;6:185–192.

200. Kapp JF, Koch H, Töpert M, Kessler H-J, Gerhards E. Unterschungen zur Pharmakologie von 6α-Fluor-11β-hydroxy-16α-methyl-3,20-dioxo-1,4-pregnadiene-21-säure- butylester (Fluocortin-butylester). Arzneim Forsch 1977;27:2191–2202.

201. Kapp JF, Gliwitzki B, Josefiuk P, Weishaupt W. Dermale und systemische Nebenwirkungen von Fluocortin-butylester (FCB). Arzneim Forsch 1977;27:2206–2213.

202. Reckers R. Klinische Prüfungen von Fluocortin-butylester im doppelblinden Halbseitenvergleich gegenüber Fluocortolon und Hydrocortisonacetat. Arzneim Forsch 1977;27: 2240–2244.

203. Hartley TF, Lieberman PL, Meltzer EO, Noyes JN, Pearlman DS, Tinkelman DG. Efficacy

and tolerance of fluocortin butyl administered twice daily in adult patients with perennial rhinitis. J Allergy Clin Immunol 1985;75: 501–507.

204. Orgel HA, Meltzer EO, Bierman CW, Bronsky E, Connell JT, Lieberman PL, Nathan R, Pearlman DS, Pence HL, Slavin RG, et al. Intranasal fluocortin butyl in patients with perennial rhinitis: a 12-month efficacy and safety study including nasal biopsy. J Allergy Clin Immunol 1991;88:257–264.

205. Burge PS, Efthimiou J, Turner-Warwick M, Nelmes PT. Double-blind trials of inhaled beclomethasone diproprionate and fluocortin butyl ester in allergen-induced immediate and late asthmatic reactions. Clin Allergy 1982;12: 523–531.

206. Thalén A, Andersson PH, Andersson PT, Axelsson B, Edsbäcker S, Brattsand R. Prospects for developing inhaled steroids with extrahepatic metabolism: "soft steroids". In: Schleimer RP, O'Byrne PM, Szefler SJ, Brattsand R, editors. Inhaled Steroids in Asthma. Optimizing Effects in the Airways. Lung Biology in Health and Disease. Vol. 163. NewYork: Marcel Dekker; Vol. 2002. p 521–537.

207. Biggadike K, Angell RM, Burgess CM, Farrell RM, Hancock AP, Harker AJ, Irving WR, Ioannou C, Procopiou PA, Shaw RE, Solanke YE, Singh OMP, Snowden MA, Stubbs RJ, Walton S, Weston HE. Selective plasma hydrolysis of glucocorticoid γ-lactones and cyclic carbonates by the enzyme paraoxonase: an ideal plasma inactivation mechanism. J Med Chem 2000;43:19–21.

208. Procopiou PA, Biggadike K, English AF, Farrell RM, Hagger GN, Hancock AP, Haase MV, Irving WR, Sareen M, Snowden MA, Solanke YE, Tralau-Stewart CJ, Walton SE, Wood JA. Novel glucocorticoid antedrugs possessing a 17β-(γ-lactone) ring. J Med Chem 2001;44: 602–612.

209. Belvisi MG, Hele DJ. Soft steroids: a new approach to the treatment of inflammatory airways diseases. Pulm Pharmacol Ther 2003;16:321–325.

210. Sandham DA, Barker L, Beattie D, Beer D, Bidlake L, Bentley D, Butler KD, Craig S, Farr D, Ffoulkes-Jones C, Fozard JR, Haberthuer S, Howes C, Hynx D, Jeffers S, Keller TH, Kirkham PA, Maas JC, Mazzoni L, Nicholls A, Pilgrim GE, Schaebulin E, Spooner GM, Stringer R, Tranter P, Turner KL, Tweed MF, Walker C, Watson SJ, Cuenoud BM. Synthesis and biological properties of novel glucocorticoid androstene C-17 furoate esters. Bioorg Med Chem 2004;12:5213–5224.

211. Taraporewala IB, Kim HP, Heiman AS, Lee HJ. A novel class of local antiinflammatory steroids. 1st communication: Analogues of methyl 11β,17α,21-trihydroxy-3,20-dioxo-pregna-1,4-diene-16α-carboxylate. Arzneim Forsch/Drug Res 1989;39(II):21–25.

212. Heiman AS, Kim HP, Taraporewala IB, Lee HJ. A novel class of local antiinflammatory steroids. 2nd communication: Pharmacological studies of methyl 11β,17α,21-trihydroxy-3,20-dioxo-pregna-1,4-diene-16α-carboxylate and methyl 11β,21-dihydroxy-3,20-dioxo-pregna-1,4-diene-16α-carboxylate. Arzneim Forsch/Drug Res 1989;39(II):262–267.

213. McLean HM, Khalil MA, Heiman AS, Lee HJ. Novel fluorinated antiinflammatory steroid with reduced side effects: methyl 9α-fluoroprednisolone-16-carboxylate. J Pharm Sci 1994;83:476–479.

214. Heiman AS, Hong D, Lee HJ. Receptor binding affinity and antiproliferative activity of new antiinflammatory antedrugs: 6-methoxycarbonyl prednisolone and its derivatives. Steroids 1994;59:324–329.

215. Kwon T, Heiman AS, Oriaku ET, Yoon K, Lee HJ. New steroidal antiinflammatory antedrugs: Steroidal [16α,17α-*d*]-3′-carbethoxyisoxazolines. J Med Chem 1995;38:1048–1051.

216. Yoon K-J, Khalil MA, Kwon T, Choi S-J, Lee HJ. Steroidal anti-inflammatory antedrugs: synthesis and pharmacological evaluation of 16α-alkoxycarbonyl-17-deoxyprednisolone derivatives. Steroids 1995;60:445–451 (515–521).

217. Khalil MA, Maponya MK, Ko D-H, You Z, Oriaku ET, Lee HJ. New anti-inflammatory steroids: [16α,17α-*d*] isoxazoline derivatives of prednisolone and 9α-fluoroprednisolone. Med Chem Res 1996;6:52–60.

218. Ko D-H, Maponya MF, Khalil MA, Oriaku ET, You Z, Lee HJ. New anti-inflammatory steroids: [16α,17α-*d*]-3′-hydroxy-iminoformyl isoxazoline derivatives of prednisolone and 9α-fluoroprednisolone. Med Chem Res 1997;7:313–324.

219. Lee HJ, Ko D-H. Antedrug: a novel approach to the discovery of safer anti-inflammatory steroids. Soc Biomed Res Symp 1997;7:33–40.

220. Park K-K, Ko D-H, You Z, Heiman AS, Lee HJ. Synthesis and pharmacological evaluations of new steroidal anti-inflammatory antedrugs: 9α-Fluoro-11β,17α,21-trihydroxy-3,20-dioxo-pregna-1,4-diene-16α-carboxylate (FP16CM) and its derivatives. Steroids 2006;71:83–89.

221. Ueno H, Maruyama A, Miyake M, Nakao E, Nakao K, Umezu K, Nitta I. Synthesis and evaluation of antiinflammatory activities of a series of corticosteroids 17α-esters containing a functional group. J Med Chem 1991;34: 2468–2473.
222. Kimura T, Yamaguchi T, Usuki K, Kurosaki Y, Nakayama T, Fujiwara Y, Matsuda Y, Unno K, Suzuki T. Colonic mucosa-specific "pro-antedrugs" for oral treatment of ulcerative colitis: design, synthesis and fate of methyl 20-glucopyranosyloxyprednisolonates. J Control Release 1994;30:125–135.
223. Suzuki T, Sato E, Tada H, Tojima Y. Examination of local anti-inflammatory activities of new steroids, hemisuccinyl methyl glycolates. Biol Pharm Bull 1999;22:816–821.
224. Milioni C, Jung L, Koch B. Synthesis of new local anti-inflammatory thiosteroids based on antedrug concept. Eur J Med Chem 1991;26: 947–951.
225. Lutsky BN, Millonig RC, Wojnar RJ, Free CA, Devlin RG, Varma RK, Karanewsky DS. Androstene-17-thioketals. 2nd communication: pharmacological profiles of tipredane and (11 beta, 17 alpha)-17-(ethylthio)-9 alpha-fluoro-17-[2-(fluoroethyl)thio]-11 beta-hydroxy-androsta-1,4-dien-3-one, structurally novel 20-thiasteroids possessing potent and selective topical antiinflammatory activity. Arzneimittelforschung 1986;36: 1787–1795.
226. Cooper AE, Gray AJ, Collington J, Seddon H, Beattie I, Logan CJ. Excretion and metabolism of tipredane, a novel glucocorticoid, in the rat, mouse, monkey, and human. Drug Metab Dispos 1996;24:1071–1080.
227. Moodley I, Grouhel A, Lelievre V, Junien JL. Anti-inflammatory properties of tixocortol 17-butyrate,21-propionate (JO 1222), a novel, locally acting corticosteroid. J Lipid Mediat 1991;3:51–70.
228. Chanoine F, Grenot C, Heidmann P, Junien JL. Pharmacokinetics of butixocort 21-propionate, budesonide, and beclomethasone dipropionate in the rat after intratracheal, intravenous, and oral treatments. Drug Metab Dispos 1991;19:546–553.
229. Graffner-Nordberg M, Sjödin K, Tunek A, Hallberg A. Synthesis and enzymatic hydrolysis of esters, constituting simple models of soft drugs. Chem Pharm Bull 1998;46: 591–601.
230. Barnes PJ. Therapeutic strategies for allergic diseases. Nature 1999;402 (Suppl): B31–B38.
231. Ong JTH, Poulsen BJ, Akers WA, Scholtz JR, Genter FC, Kertesz DJ. Intrinsic potencies of novel thiol ester corticosteroids RS-85095 and RS-21314 as compared with clobetasol 17-propionate and fluocinonide. Arch Dermatol 1989;125:1662–1665.
232. Phillipps GH. Structure–activity relationships of topically active steroids: the selection of fluticasone propionate. Respir Med 1990;84 (Suppl A):19–23.
233. Chen F, Kearney T, Robinson S, Daley-Yates PT, Waldron S, Churchill DR. Cushing's syndrome and severe adrenal suppression in patients treated with ritonavir and inhaled nasal fluticasone. Sex Transm Infect 1999;75:274.
234. Gupta SK, Dubé MP. Exogenous Cushing syndrome mimicking human immunodeficiency virus lipodystrophy. Clin Infect Dis 2002;35: E69–E71.
235. Samaras K, Pett S, Gowers A, McMurchie M, Cooper DA. Iatrogenic Cushing's syndrome with osteoporosis and secondary adrenal failure in human immunodeficiency virus-infected patients receiving inhaled corticosteroids and ritonavir-boosted protease inhibitors: six cases. J Clin Endocrinol Metab 2005;90:4394–4398.
236. Pessanha TM, Campos JM, Barros AC, Pone MV, Garrido JR, Pone SM. Iatrogenic Cushing's syndrome in a adolescent with AIDSs on ritonavir and inhaled fluticasone. Case report and literature review. AIDS 2007;21:529–532.
237. Rohatagi S, Bye A, Falcoz C, Mackie AE, Meibohm B, Möllmann H, Derendorf H. Dynamic modeling of cortisol reduction after inhaled administration of fluticasone propionate. J Clin Pharmacol 1996;36:938–941.
238. Thorsson L, Dahlström K, Edsbäcker S, Källén A, Paulson J, Wirén J-E. Pharmacokinetics and systemic effects of inhaled fluticasone propionate in healthy subjects. Br J Clin Pharmacol 1997;43:155–161.
239. Derendorf H, Hochhaus G, Meibohm B, Möllmann H, Barth J. Pharmacokinetics and pharmacodynamics of inhaled corticosteroids. J Allergy Clin Immunol 1998;101:S440–S446.
240. Dent G. Ciclesonide (Byk Gulden). Curr Opin Investig Drugs 2002;3:78–83.
241. Derendorf H. Pharmacokinetic and pharmacodynamic properties of inhaled ciclesonide. J Clin Pharmacol 2007;47:782–789.
242. Shimizu K, Kawase A, Haneishi T, Kato Y, Kobayashi T, Sekiguchi N, Yamamoto T, Ishigai M, Tokuda K, Matsushita T, Shimaoka S,

Morikawa K. Novel vitamin D3 antipsoriatic antedrugs: 16-En-22-oxa-1alpha,25-(OH)2D3 analogs. Bioorg Med Chem Lett 2006;14: 1838–1850.

243. Shimizu K, Kawase A, Haneishi T, Kato Y, Kinoshita K, Ohmori M, Furuta Y, Emura T, Kato N, Mitsui T, Yamaguchi K, Morita K, Sekiguchi N, Yamamoto T, Matsushita T, Shimaoka S, Sugita A, Morikawa K. Design and evaluation of new antipsoriatic antedrug candidates having 16-en-22-oxa-vitamin D3 structures. Bioorg Med Chem Lett 2006;16: 3323–3329.
244. Lobo RA. Benefits and risks of estrogen replacement therapy. Am J Obstet Gynecol 1995;173:982–989.
245. Beral V, Banks E, Reeves G, Appleby P. Use of HRT and the subsequent risk of cancer. J Epidemiol Biostat 1999;4:191–210.
246. Rossouw JE, Anderson GL, Prentice RL, LaCroix AZ, Kooperberg C, Stefanick ML, Jackson RD, Beresford SAA, Howard BV, Johnson KC, Kotchen JM, Ockene J. Risks and benefits of estrogen plus progestin in healthy postmenopausal women: Principal results from the Women's Health Initiative randomized controlled trial. JAMA 2002;288:321–333.
247. Hays J, Ockene JK, Brunner RL, Kotchen JM, Manson JE, Patterson RE, Aragaki AK, Shumaker SA, Brzyski RG, LaCroix AZ, Granek IA, Valanis BG. Effects of estrogen plus progestin on health-related quality of life. N Engl J Med 2003;348:1839–1854.
248. Labaree DC, Reynolds TY, Hochberg RB. Estradiol-16α-carboxylic acid esters as locally active estrogens. J Med Chem 2001;44: 1802–1814.
249. Labaree DC, Zhang JX, Harris HA, O'Connor C, Reynolds TY, Hochberg RB. Synthesis and evaluation of B-, C-, and D-ring-substituted estradiol carboxylic acid esters as locally active estrogens. J Med Chem 2003;46:1886–1904.
250. Zhang JX, Labaree DC, Mor G, Hochberg RB. Estrogen to antiestrogen with a single methylene group resulting in an unusual steroidal selective estrogen receptor modulator. J Clin Endocrinol Metab 2004;89: 3527–3535.
251. Waldeck B. β-Adrenoceptor agonists after terbutaline. Pharmacol Toxicol 1995;77(Suppl III):25–29.
252. Albrecht R, Loge O. β_2-Agonists containing metabolically labile groups. I. The influence of ester groups in the nitrogen substituent. Eur J Med Chem 1985;20:51–55.
253. Albrecht R, Heindl J, Loge O. β_2-Agonists containing metabolically labile groups. II. The influence of ester groups in the aryl system. Eur J Med Chem 1985;20:57–60.
254. Gill HS, Freeman S, Irwin WJ, Wilson KA. Soft β-adrenergic agonists for the topical treatment of psoriasis. Eur J Med Chem 1996;31: 847–859.
255. Kimko HC, Cross JT, Abernethy DR. Pharmacokinetics and clinical effectiveness of methylphenidate. Clin Pharmacokinet 1999;37: 457–470.
256. Markowitz JS, Straughn AB, Patrick KS. Advances in the pharmacotherapy of attention-deficit-hyperactivity disorder: focus on methylphenidate formulations. Pharmacotherapy 2003;23:1281–1299.
257. Markowitz JS, DeVane CL, Boulton DW, Nahas Z, Risch SC, Diamond F, Patrick KS. Ethylphenidate formation in human subjects after the administration of a single dose of methylphenidate and ethanol. Drug Metab Dispos 2000;28:620–624.
258. Patrick KS, Kilts CD, Breese GR. Synthesis and pharmacology of hydroxylated metabolites of methylphenidate. J Med Chem 1981;24:1237–1240.
259. Sun Z, Murry DJ, Sanghani SP, Davis WI, Kedishvili NY, Zou Q, Hurley TD, Bosron WF. Methylphenidate is stereoselectively hydrolyzed by human carboxylesterase CES1A1. J Pharmacol Exp Ther 2004;310:469–476.
260. Bodor N. Design of biologically safer chemicals. Chemtech 1995;25:(10):22–32.
261. Müller P, editor. DDT. The Insecticide Dichlorodiphenyltrichloroethane and Its Significance. Basel; Birkhäuser Verlag; 1955.
262. Ihde AJ. The Development of Modern Chemistry. New York: Dover; 1984.
263. O'Brien RD. Insecticides. Action and Metabolism. New York: Academic Press; 1967.
264. Kashyap SK, Nigam SK, Karnik AB, Gupta RC, Chatterjee SK. Carcinogenicity of DDT (dichlorodiphenyltrichloroethane) in pure inbred Swiss mice. Int J Cancer 1979;19: 725–729.
265. IARC IARC Monograph Series 1983; 30: 73–101.
266. Gaines TB. Acute toxicity of pesticides. Toxicol Appl Pharmacol 1969;14:515–534.
267. Hassall KA. The Biochemistry and Uses of Pesticides. 2nd ed. London: Macmillan; 1990.
268. Hodgson E, Kuhr RJ, editors. Safer Insecticides. Development and Use. New York: Marcel Dekker; 1990.

269. Hammer RH, Amin K, Gunes ZE, Brouillette G, Bodor N. Novel soft anticholinergic agents. Drug Des Deliv 1988;2:207–219.
270. Bodor N, El-Koussi A, Hammer R. Soft drugs. 9. The ultra-short duration of the muscarinolytic activity of a new short-acting "soft" mydriatic agent. J Biopharm Sci 1990;1:215–223.
271. Hammer RH, Wu W-M, Sastry JS, Bodor N. Short acting soft mydriatics. Curr Eye Res 1991;10:565–570.
272. Kumar GN, Hammer RH, Bodor NS. Soft drugs. 16. Design, evaluation and transdermal penetration of novel soft anticholinergics based on methatropine. Bioorg Med Chem 1993;1:327–332.
273. Hammer RH, Gunes E, Kumar GN, Wu W-M, Srinivasan V, Bodor NS. Soft drugs. 14. Synthesis and anticholinergic activity of soft phenylsuccinic analogs of methatropine. Bioorg Med Chem 1993;1:183–187.
274. Juhász A, Huang F, Ji F, Buchwald P, Wu W-M, Bodor N. Design and evaluation of new soft anticholinergic agents. Drug Develop Res 1998;43:117–127.
275. Kumar GN, Hammer RH, Bodor N. Soft drugs. 12. Design, synthesis, stability and evaluation of soft anticholinergics. Drug Des Discov 1993;10:11–21.
276. Kumar GN, Hammer RH, Bodor N. Soft drugs. 13. Design and evaluation of phenylsuccinic analogs of scopolamine as soft anticholinergics. Drug Des Discov 1993;10:1–9.
277. Ji F, Huang F, Juhasz A, Wu W, Bodor N. Design, synthesis, and pharmacological evaluation of soft glycopyrrolate and its analog. Pharmazie 2000;55:187–191.
278. Brouillette G, Kawamura M, Kumar GN, Bodor N. Soft drugs. 21. Design and evaluation of soft analogs of propantheline. J Pharm Sci 1996;85:619–623.
279. Eglen RM, Watson N. Selective muscarinic receptor agonists and antagonists. Pharmacol Toxicol 1996;78:59–68.
280. Brown JH, Taylor P. Muscarinic receptor agonists and antagonists. In: Hardman JG, Limbird LE, editors. Goodman & Gilman's The Pharmacological Basis of Therapeutics. New York: McGraw-Hill; 1996. p 141–160.
281. Böhme TM, Keim C, Kreutzmann K, Linder M, Dingermann T, Dannhardt G, Mutschler E, Lambrecht G. Structure–activity relationships of dimethindene derivatives as new M_2-selective muscarinic receptor antagonists. J Med Chem 2003;46:856–867.
282. Shelley WB, Horvath PN. Comparative study on the effect of anticholinergic compounds on sweating. J Invest Dermatol 1951;16:267–274.
283. MacMillan FSK, Reller HH, Synder FH. The antiperspirant action of topically applied anticholinergics. J Invest Dermatol 1964;43:363–377.
284. Stoughton RB, Chiu F, Fritsch W, Nurse D. Topical suppression of eccrine sweat delivery with a new anticholinegic agent. J Invest Dermatol 1964;42:151–155.
285. Lasser AE. Results of a double-blind clinical study of a new anticholinergic agent as an axillary antiperspirant. IMJ Ill. Med J 1967;131:314–317.
286. Dolianitis C, Scarff CE, Kelly J, Sinclair R. Iontophoresis with glycopyrrolate for the treatment of palmoplantar hyperhidrosis. Australas J Dermatol 2004;45:208–212.
287. Tashkin DP, Cooper CB. The role of long-acting bronchodilators in the management of stable COPD. Chest 2004;125:249–259.
288. Ali-Melkillä T, Kanto J, Iisalo E. Pharmacokinetics and related pharmacodynamics of anticholinergic drugs. Acta Anaesthesiol Scand 1993;37:633–642.
289. Osterholm RK, Camoriano JK. Transdermal scopolamine psychosis. JAMA 1982;247:3081.
290. Birkhimer LJ, Jacobson PA, Olson J, Goyette DM. Ocular scopolamine-induced psychosis. J Fam Pract 1984;18:464–469.
291. Hamborg-Petersen B, Nielsen MM, Thordal C. Toxic effect of scopolamine eye drops in children. Acta Ophthalmol 1984;62:485–488.
292. Merli GJ, Weitz H, Martin JH, McClay EF, Adler AG, Fellin FM, Libonati M. Cardiac dysrhythmias associated with ophthalmic atropine. Arch Intern Med 1986;146:45–47.
293. Fraunfelder FT. Mydriatics and cycloplegics. In: Fraunfelder FT, Meyer SM, editors. Drug-Induced Ocular Side Effects and Drug Interactions. New York: Lea & Febiger; 1989. p 445–458.
294. Khurana AK, Ahluwalia BK, Rajan C, Vohra AK. Acute psychosis associated with topical cyclopentolate hydrochloride. Am J Ophthalmol 1988;105:91.
295. Wright BD. Exacerbation of akinetic seizures by atropine eye drops. Br J Ophthalmol 1992;76:179–180.
296. Diamond JP. Systemic adverse effects of topical ophthalmic agents. Implications for older patients. Drugs Aging 1997;11:352–360.

297. Jayamanne DG, Ray-Chaudhuri N, Wariyar R, Cottrell DG. Haemodynamic responses to subconjunctival mydriatic agents (Mydricaine) used for maintenance of perioperative mydriasis in patients undergoing vitreoretinal surgery. Eye 1998;12:792–794.
298. Kumar GN, Bodor N. Soft anticholinergics. Curr Med Chem 1996;3:23–36.
299. Buchwald P, Bodor N. Soft quaternary anticholinergics: comprehensive QSAR with a linearized biexponential (LinBiExp) model. J Med Chem 2006;49:883–891.
300. Gabourel JD, Gosselin RE. The mechanism of atropine detoxification in mice and rats. Arch Int Pharmacodyn 1958;115:416–431.
301. Gosselin RE, Gabourel JD, Wills JH. The fate of atropine in man. Clin Pharmacol Ther 1960;1:597–603.
302. Huang F, Buchwald P, Browne CE, Farag HH, Wu W-M, Ji F, Hochhaus G, Bodor N. Receptor binding studies of soft anticholinergic agents. AAPS PharmSci 2001;3(4):E30 http://www.pharmsci.org.
303. Huang F, Wu W-M, Ji F, Juhász A, Bodor N. Design, pharmacokinetic, and pharmacodynamic evaluation of soft anticholinergics based on tropyl α-phenylcyclopentylacetate. Pharmazie 2002;57:115–121.
304. Ji F, Wu W-M, Bodor N. Studies on soft glycopyrrolate analog, SG-1. Pharmazie 2002;57:138–141.
305. Ji F, Wu W, Dai X, Mori N, Wu J, Buchwald P, Bodor N. Synthesis and pharmacological effects of new, *N*-substituted soft anticholinergics based on glycopyrrolate. J Pharm Pharmacol 2005;57:1427–1436.
306. Wu W-M, Buchwald P, Mori N, Ji F, Wu J, Bodor N. Pharmacokinetic and pharmacodynamic evaluations of the zwitterionic metabolite of a new series of *N*-substituted soft anticholinergics. Pharm Res 2005;22:2035–2044.
307. Mori N, Buchwald P, Wu W-M, Ji F, Hochhaus G, Bodor N. Pharmacological effects of some newly developed soft anticholinergics and a receptor-binding QSAR study. Pharmazie 2006;61:148–153.
308. Tóth-Sarudy E, Tóth G, Pallagi I, Seres G, Vitális B, Tapfer M, Perczel V, Kurucz I, Bodor N, Zubovics Z. Preparation and biological effects of pure stereoisomeric novel soft anticholinergics. Pharmazie 2006;61:90–96.
309. Wu WM, Wu J, Mori N, Buchwald P, Bodor N. Stereoisomers of *N*-substituted soft anticholinergics and their zwitterionic metabolite based on glycopyrrolate: syntheses and pharmacological evaluations. Pharmazie 2008;63:200–209.
310. Huang F, Browne CE, Wu W-M, Juhász A, Ji F, Bodor N. Design, pharmacokinetic, and pharmacodynamic evaluation of a new class of soft anticholinergics. Pharm Res 2003;20:1681–1689.
311. D'Agostino G, Renzetti AR, Zonta F, Subissi A. Selectivity of LG50643 for postjunctional muscarinic-receptor subtype in the guinea-pig trachea. J Pharm Pharmacol 1994;46:332–336.
312. Buchwald P. General linearized biexponential model for QSAR data showing bilinear-type distribution. J Pharm Sci 2005;94:2355–2379.
313. Buchwald P. A general bilinear model to describe growth or decline time-profiles. Math Biosc 2007;205:108–136.
314. Bodor N.1-Hydrocarbonoyloxymethyl-3-carbamoyl or 3-carboethoxy-pyridinium salts. US Patent 3,998,815. 1976.
315. Thorsteinsson T, Loftsson T, Masson M. Soft antibacterial agents. Curr Med Chem 2003;10:1129–1136.
316. Thorsteinsson T, Másson M, Kristinsson KG, Hjálmarsdóttir MA, Hilmarsson H, Loftsson T. Soft antimicrobial agents: synthesis and activity of labile environmentally friendly long chain quaternary ammonium compounds. J Med Chem 2003;46:4173–4181.
317. Loftsson T, Thorsteinsson T, Másson M. Marine lipids as building blocks for soft quaternary ammonium compounds and their antibacterial activity. Pharmazie 2004;59:360–364.
318. Shelton RS, van Campen MG, Tilford CH, Lang HC, Nisonger L, Bandelin FJ, Rubenkoenig HL. Quaternary ammonium salts as germicides. I. Non-acylated quaternary ammonium salts derived from aliphatic amines. J Am Chem Soc 1946;68:753–755.
319. Jono K, Takayama T, Kuno M, Higashide E. Effect of alkyl chain length of benzalkonium chloride on the bactericidal activity and binding to organic materials. Chem Pharm Bull 1986;34:4215–4224.
320. Pavlikova-Moricka M, Lacko I, Devinsky F, Masarova L, Milynarcik D. Quantitative relationships between structure and antimicrobial activity of new "soft" bisquaternary ammonium salts. Folia Microbiol 1994;39:176–180.
321. Lindstedt M, Allenmark S, Thompson RA, Edebo L. Antimicrobial activity of betaine esters, quaternary ammonium amphiphiles which spontaneously hydrolyze into nontoxic

components. Antimicrob Agents Chemother 1990;34:1949–1954.

322. Ahlström B, Chelminska-Bertilsson M, Thompson RA, Edebo L. Long-chain alkanoylcholines, a new category of soft antimicrobial agents that are enzymatically degradable. Antimicrob Agents Chemother 1995;39:50–55.
323. Ahlström B, Chelminska-Bertilsson M, Thompson RA, Edebo L. Submicellar complexes may initiate the fungicidal effects of cationic amphiphilic compounds on *Candida albicans*. Antimicrob Agents Chemother 1997; 41:544–550.
324. Ahlström B, Edebo L. Hydrolysis of the soft amphiphilic antimicrobial agent tetradecyl betainate is retarded after binding to and killing *Salmonella typhimurium*. Microbiology 1998;144:2497–2504.
325. Ahlström B, Thompson RA, Edebo L. The effect of hydrocarbon chain length, pH, and temperature on the binding and bactericidal effect of amphiphilic betaine esters on Salm*onella typhimurium*. APMIS 1999;107:318–324.
326. Calvani M, Critelli L, Gallo G, Giorgi F, Gramiccioli G, Santaniello M, Scafetta N, Tinti MO, De Angelis F. L-Carnitine esters as "soft", broad-spectrum antimicrobial amphiphiles. J Med Chem 1998;41:2227–2233.
327. Stout DM, Black LA, Barcelon-Yang C, Matier WL, Brown BS, Quon CY, Stampfli HF. Ester derivatives of 2,6-bis(1-pyrrolidinylmethyl)-4-benzamidophenol as short-acting antiarrhythmic agents. 1. J Med Chem 1989;32: 1910–1913.
328. Chorvat RJ, Black LA, Ranade VV, Barcelon-Yang C, Stout DM, Brown BS, Stampfli HF, Quon CY. Mono- and bis(aminomethyl)phenylacetic acid esters as short-acting antiarrhythmic agents. 2. J Med Chem 1993;36: 2494–2498.
329. Vrobel TR, Miller PE, Mostow ND, Rakita L. A general overview of amiodarone toxicity: its prevention, detection, and management. Prog Cardiovasc Dis 1989;31:393–426.
330. Raatikainen MJP, Napolitano CA, Druzgala P, Dennis DM. Electrophysiological effects of a novel, short-acting and potent ester derivative of amiodarone, ATI-2001, in guinea pig isolated heart. J Pharmacol Exp Ther 1996;277: 1454–1463.
331. Raatikainen MJP, Morey TE, Druzgala P, Milner P, Gonzalez MD, Dennis DM. Potent and reversible effects of ATI-2001 on atrial and atrioventricular nodal electrophysiological properties in guinea pig isolated perfused heart. J Pharmacol Exp Ther 2000;295: 779–785.
332. Morey TE, Seubert CN, Raatikainen MJP, Martynyuk AE, Druzgala P, Milner P, Gonzalez MD, Dennis DM. Structure–activity relationships and electrophysiological effects of short-acting amiodarone homologs in guinea pig isolated heart. J Pharmacol Exp Ther 2001;297:260–266.
333. Arya A, Silberbauer J, Teichman SL, Milner P, Sulke N, Camm AJ. A preliminary assessment of the effects of ATI-2042 in subjects with paroxysmal atrial fibrillation using implanted pacemaker methodology. Europace 2009;11:458–464.
334. Ogiso T, Iwaki M, Tanino T, Nagai T, Ueda Y, Muraoka O, Tanabe G. Pharmacokinetics of indomethacin ester prodrugs: Gastrointestinal and hepatic toxicity and the hydrolytic capacity of various tissues in rats. Biol Pharm Bull 1996;19:1178–1183.
335. Druzgala P, Milner PG, Pfister J, Becker C. Materials and methods for the treatment of gastroesophageal reflux disease. US Patent 6,552,046. 2003.
336. Camilleri M, Vazquez-Roque MI, Burton D, Ford T, McKinzie S, Zinsmeister AR, Druzgala P. Pharmacodynamic effects of a novel prokinetic 5-HT_4 receptor agonist, ATI-7505, in humans. Neurogastroenterol Motil 2007;19: 30–38.
337. Druzgala P, Zhang X, Pfister J. Materials and methods for treating coagulation disorders. US Patent 7,145,020. 2006.
338. Zhang X, Pfister JR, Becker C, Druzgala P.5HT-3 Receptor antogonists and methods of use. US Patent 7,125,886. 2006.
339. Druzgala P, Milner PG. Ultrashort acting hypnotic barbiturates. US Patent 7,041,673,. 2006.
340. Druzgala P. Materials and methods for the treatment of depression. US Patent 6,809, 116. 2004.
341. Druzgala P, Milner PG, Pfister J, Zhang X. Materials and methods for the treatment of hypertension and angina. US Patent 7,265, 142. 2007
342. Baxter AJG, Carr RD, Eyley SC, Fraser-Rae L, Hallam C, Harper ST, Hurved PA, King SJ, Meghani P. (*R*)-2-(3-Mercapto-2(*S*)-methyl-1-oxo-propoxy)-3-(methylthio)propanoic acid, the first ultra-short-acting angiotensin converting enzyme inhibitor. J Med Chem 1992;35:3718–3720.

343. Graffner-Nordberg M, Marelius J, Ohlsson S, Persson Å, Swedberg G, Andersson P, Andersson SE, Åqvist J, Hallberg A. Computational predictions of binding affinities to dihydrofolate reductase: synthesis and biological evaluation of methotrexate analogues. J Med Chem 2000;43:3852–3861.

344. Graffner-Nordberg M, Kolmodin K, Åqvist J, Queener SF, Hallberg A. Design, synthesis, computational prediction, and biological evaluation of ester soft drugs as inhibitors of dihydrofolate reductase from *Pneumocystis carinii*. J Med Chem 2001;44:2391–2402.

345. Graffner Nordberg M. Approaches to soft drug analogues of dihydrofolate reductase inhibitors. Design and synthesis. Ph.D. Thesis, Uppsala University, Uppsala, Sweden, 2001.

346. Graffner-Nordberg M, Fyfe M, Brattsand R, Mellgard B, Hallberg A. Design and synthesis of dihydrofolate reductase inhibitors encompassing a bridging ester group. Evaluation in a mouse colitis model. J Med Chem 2003;46: 3455–3462.

347. Graffner-Nordberg M, Kolmodin K, Aqvist J, Queener SF, Hallberg A. Design, synthesis, and computational affinity prediction of ester soft drugs as inhibitors of dihydrofolate reductase from *Pneumocystis carinii*. Eur J Pharm Sci 2004;22:43–54.

348. Halloran PF. Immunosuppressive drugs for kidney transplantation. N Engl J Med 2004; 351:2715–2729.

349. Henry ML. Cyclosporine and tacrolimus (FK506): a comparison of efficacy and safety profiles. Clin Transplant 1999;13:209–220.

350. Alexander AG, Barnes NC, Kay AB, Corrigan CJ. Clinical response to cyclosporin in chronic severe asthma is associated with reduction in serum soluble interleukin-2 receptor concentrations. Eur Respir J 1995;8: 574–578.

351. Rohatagi S, Calic F, Harding N, Ozoux ML, Bouriot JP, Kirkesseli S, DeLeij L, Jensen BK. Pharmacokinetics, pharmacodynamics, and safety of inhaled cyclosporin A (ADI628) after single and repeated administration in healthy male and female subjects and asthmatic patients. J Clin Pharmacol 2000;40:1211–1226.

352. Lazarova T, Chen JS, Hamann B, Kang JM, Homuth-Trombino D, Han F, Hoffmann E, McClure C, Eckstein J, Or YS. Synthesis and biological evaluation of novel cyclosporin A analogues: potential soft drugs for the treatment of autoimmune diseases. J Med Chem 2003;46:674–676.

353. Hersperger R, Buchheit K-H, Cammisuli S, Enz A, Lohse O, Ponelle M, Schuler W, Schweitzer A, Walker C, Zehender H, Zenke G, Zimmerlin AG, Zollinger M, Mazzoni L, Fozard JR. A locally active antiinflammatory macrolide (MLD987) for inhalation therapy of asthma. J Med Chem 2004;47:4950–4957.

354. Tigani B, Hannon JP, Schaeublin E, Mazzoni L, Fozard JR. Effects of immunomodulators on airways hyperresponsiveness to adenosine induced in actively sensitised Brown Norway rats by exposure to allergen. Naunyn Schmiedebergs Arch Pharmacol 2003;368:17–25.

355. Freyne EJ, Lacrampe JF, Deroose F, Boeckx GM, Willems M, Embrechts W, Coesemans E, Willems JJ, Fortin JM, Ligney Y, Dillen LL, Cools WF, Goossens J, Corens D, De Groot A, Van Wauwe JP. Synthesis and biological evaluation of 1,2,4-triazinylphenylalkylthiazolecarboxylic acid esters as cytokine-inhibiting antedrugs with strong bronchodilating effects in an animal model of asthma. J Med Chem 2005;48:2167–2175.

356. Van Wauwe J, Aerts F, Cools M, Deroose F, Freyne E, Goossens J, Hermans B, Lacrampe J, Van Genechten H, Van Gerven F, Van Nyen G. Identification of R146225 as a novel, orally active inhibitor of interleukin-5 biosynthesis. J Pharmacol Exp Ther 2000;295:655–661.

357. Sawa M, Tsukamoto T, Kiyoi T, Kurokawa K, Nakajima F, Nakada Y, Yokota K, Inoue Y, Kondo H, Yoshino K. New strategy for antedrug application: development of metalloproteinase inhibitors as antipsoriatic drugs. J Med Chem 2002;45:930–936.

358. Razdan RK, Howes JF. Drugs related to tetrahydrocannabinol. Med Res Rev 1983;3: 119–146.

359. Green K, Kim K. Acute dose response of intraocular pressure to topical and oral cannabinoids. Proc Soc Exp Biol Med 1977;154:228–231.

360. Green K, Bigger JF, Kim K, Bowman K. Cannabinoid action on the eye as mediated through the central nervous system and local adrenergic activity. Exp Eye Res 1977;24:189–196.

361. Buchwald A.Soft cannabinoid analogues as potential anti-glaucoma agents. Ph.D. Thesis, University of Florida, Gainesville; 2001.

362. Buchwald A, Derendorf H, Ji F, Nagaraja NV, Wu W-M, Bodor N. Soft cannabinoid analogues as potential anti-glaucoma agents. Pharmazie 2002;57:108–114.

363. Pars HG, Granchelli FE, Razdan RK, Keller JK, Teiger DG, Rosenberg FJ, Harris LS. Drugs derived from cannabinoids. 1. Nitrogen

analogs, benzopyranopyridines and benzopyranopyrroles. J Med Chem 1976;19:445–454.

364. Meibohm B, Derendorf H. Basic concepts of pharmacokinetic/pharmacodynamic (PK/PD) modelling. Int J Clin Pharmacol Ther 1997; 35:401–413.
365. Minutolo F, Cascio MG, Carboni I, Bisogno T, Prota G, Bertini S, Digiacomo M, Bifulco M, Di Marzo V, Macchia M. Metabolically labile cannabinoid esters: A 'soft drug' approach for the development of cannabinoid-based therapeutic drugs. Bioorg Med Chem Lett 2007;17: 4878–4881.
366. Sigel E, Buhr A. The benzodiazepine binding site of $GABA_A$ receptors. Trends Pharmacol Sci 1997;18:425–429.
367. Stafford JA, Pacofsky GJ, Cox RF, Cowan JR, Dorsey GF Jr, Gonzales SS, Jung DK, Koszalka GW, McIntyre MS, Tidwell JH, Wiard RP, Feldman PL. Identification and structure–activity studies of novel ultrashort-acting benzodiazepine receptor agonists. Bioorg Med Chem Lett 2002;12:3215–3218.
368. Pacofsky GJ, Stafford JA, Cox RF, Cowan JR, Dorsey GF Jr, Gonzales SS, Kaldor I, Koszalka GW, Lovell GG, McIntyre MS, Tidwell JH, Todd D, Whitesell G, Wiard RP, Feldman PL. Relating the structure, activity, and physical properties of ultrashort-acting benzodiazepine receptor agonists. Bioorg Med Chem Lett 2002;12:3219–3222.
369. Kilpatrick GJ, Tilbrook GS. Drug development in anaesthesia: industrial perspective. Curr Opin Anaesthesiol 2006;19:385–389.
370. Kilpatrick GJ, McIntyre MS, Cox RF, Stafford JA, Pacofsky GJ, Lovell GG, Wiard RP, Feldman PL, Collins H, Waszczak BL, Tilbrook GS. CNS 7056: a novel ultra-short-acting benzodiazepine. Anesthesiology 2007;107:60–66.
371. Upton R, Martinez A, Grant C. A dose escalation study in sheep of the effects of the benzodiazepine CNS 7056 on sedation, the EEG and the respiratory and cardiovascular systems. Br J Pharmacol 2008;155:52–61.
372. Sneyd JR. Recent advances in intravenous anaesthesia. Br J Anaesth 2004;93:725–736.
373. Jenkins T, Beattie D, Jaw Tsai S, Amagasu S, Halladay J, Vanapalli S, Kern R, Shaw JP, Egan T, Shafer S. THRX-918661, a novel, pharmacokinetically responsive sedative/hypnotic agent. Anaesthesia 2004;59:100.
374. Beattie D, Jenkins T, McCullough J, Thibodeaux H, Renner T, Bolton J, Cook D, Steffensen S, Egan T, Shafer S. The *in vivo* activity of THRX-918661, a novel, pharmacokinetically responsive sedative/hypnotic agent. Anaesthesia 2004;59:101.
375. Marciniak G, Decolin D, Leclerc G, Decker N, Schwartz J. Synthesis and pharmacological properties of "soft drug" derivatives related to perhexiline. J Med Chem 1988;31:2289–2296.
376. Decker N, Grima M, Velly J, Marciniak G, Leclerc G, Schwartz J. Soft drug derivatives related to perhexiline. Part I. *In vitro* studies. Arzneim Forsch/Drug Res 1988;38(II):905–908.
377. Decker N, Grima M, Velly J, Marciniak G, Leclerc G, Schwartz J. Soft drug derivatives related to perhexiline. Part II. *In vivo* studies. Arzneim Forsch/Drug Res 1988;38(II): 1110–1114.
378. Ericsson H, Tholander B, Björkman JA, Nordlander M, Regårdh CG. Pharmacokinetics of new calcium channel antagonist clevidipine in the rat, rabbit, and dog and pharmacokinetic/pharmacodynamic relationship in anesthetized dogs. Drug Metab Dispos 1999;27: 558–564.
379. Kieler-Jensen N, Jolin-Mellgard A, Nordlander M, Ricksten SE. Coronary and systemic hemodynamic effects of clevidipine, an ultra-short-acting calcium antagonist, for treatment of hypertension after coronary artery surgery. Acta Anaesthesiol Scand 2000;44:186–193.
380. Ericsson H, Schwieler J, Lindmark BO, Lofdahl P, Thulin T, Regårdh CG. Enantioselective pharmacokinetics of the enantiomers of clevidipine following intravenous infusion of the racemate in essential hypertensive patients. Chirality 2001;13:130–134.
381. Nordlander M, Sjoquist PO, Ericsson H, Ryden L. Pharmacodynamic, pharmacokinetic and clinical effects of clevidipine, an ultrashort-acting calcium antagonist for rapid blood pressure control. Cardiovasc Drug Rev 2004;22: 227–250.
382. Aronson S, Dyke CM, Stierer KA, Levy JH, Cheung AT, Lumb PD, Kereiakes DJ, Newman MF. The ECLIPSE trials: comparative studies of clevidipine to nitroglycerin, sodium nitroprusside, and nicardipine for acute hypertension treatment in cardiac surgery patients. Anesth Analg 2008;107:1110–1121.
383. Singla N, Warltier DC, Gandhi SD, Lumb PD, Sladen RN, Aronson S, Newman MF, Corwin HL. Treatment of acute postoperative hypertension in cardiac surgery patients: an efficacy study of clevidipine assessing its postoperative antihypertensive effect in cardiac surgery-2 (ESCAPE-2), a randomized, double-blind, placebo-controlled trial. Anesth Analg 2008;107:59–67.

384. Fura A. Role of pharmacologically active metabolites in drug discovery and development. Drug Discov Today 2006;11:133–142.
385. Kaminski JJ, Bodor N, Higuchi T. N-Halo derivatives. III. Stabilization of nitrogen-chlorine bond in *N*-chloroamino acid derivatives. J Pharm Sci 1976;65:553–557.
386. Kaminski JJ, Bodor N, Higuchi T. N-Halo derivatives. IV. Synthesis of low chlorine potential soft *N*-chloramine systems. J Pharm Sci 1976;65:1733–1737.
387. Kaminski JJ, Huycke MM, Selk SH, Bodor N, Higuchi T. N-Halo derivatives. V. Comparative antimicrobial activity of soft *N*-chloramine systems. J Pharm Sci 1976;65:1737–1742.
388. Kosugi M, Kaminski JJ, Selk SH, Pitman IH, Bodor N, Higuchi T. N-Halo derivatives. VI. Microbiological and chemical evaluations of 3-chloro-2-oxazolidinones. J Pharm Sci 1976;65: 1743–1746.
389. Burkett HD, Faison JH, Kohl HH, Wheatley WB, Worley SD, Bodor N. A novel chloramine compound for water disinfection. Water Res Bull 1981;17:874–879.
390. Worley SD, Wheatley WB, Kohl HH, Burkett HD, Faison JH, Van Hoose JA, Bodor N. A novel bactericidal agent for treatment of water. In: Jolley RL, Brungs WA, Cotruvo JA, Cumming RB, Mattice JS, Jacobs VA, editors. Water Chlorination. Environmental Impact and Health Effects. Ann Arbor, MI: Ann Arbor Science Publishers; 1983. p 1105–1113.
391. Worley SD, Wheatley WB, Kohl HH, Burkett HD, Van Hoose JA, Bodor N. A new water disinfectant: a comparative study. Ind Eng Chem Prod Res Dev 1983;22:716–718.
392. Mora EC, Kohl HH, Wheatley WB, Worley SD, Faison JH, Burkett HD, Bodor N. Properties of a new chloramine disinfectant and detoxicant. Poultry Sci 1982;61:1968–1971.
393. Kohl HH, Wheatley WB, Worley SD, Bodor N. Antimicrobial activity of *N*-chloramine compounds. J Pharm Sci 1980;69:1292–1295.
394. Bodor N, Sloan KB. Soft drugs. 5. Thiazolidine-type derivatives of progesterone and testosterone. J Pharm Sci 1982;71:514–520.
395. Schubert WM, Motoyama Y. An example of SN1 cleavage of a sulfide. J Am Chem Soc 1965;87:5507–5508.
396. Bodor N, Sloan KB, Little RJ, Selk SH, Caldwell L. Soft drugs. 4. 3-Spirothiazolidines of hydrocortisone and its derivatives. Int J Pharm 1982;10:307–321.
397. Bodor N, Gabanyi Z, Wong C-K. A new method for the estimation of partition coefficient. J Am Chem Soc 1989;111:3783–3786.
398. Bodor N, Harget A, Huang M-J. Neural network studies. 1. Estimation of the aqueous solubility of organic compounds. J Am Chem Soc 1991;113:9480–9483.
399. Bodor N, Huang M-J. An extended version of a novel method for the estimation of partition coefficients. J Pharm Sci 1992;81:272–281.
400. Bodor N, Huang M-J. A new method for the estimation of the aqueous solubility of organic compounds. J Pharm Sci 1992;81: 954–960.
401. Bodor N, Buchwald P. Molecular size based approach to estimate partition properties for organic solutes. J Phys Chem B 1997;101: 3404–3412.
402. Buchwald P, Bodor N. Octanol–water partition of nonzwitterionic peptides: Predictive power of a molecular size-based model. Proteins 1998;30:86–99.
403. Buchwald P, Bodor N. Molecular size-based model to describe simple organic liquids. J Phys Chem B 1998;102:5715–5726.
404. Buchwald P, Bodor N. Octanol–water partition: Searching for predictive models. Curr Med Chem 1998;5:353–380.
405. Buchwald P, Bodor N. Quantitative structure-metabolism relationships: steric and nonsteric effects in the enzymatic hydrolysis of noncongener carboxylic esters. J Med Chem 1999;42: 5160–5168.
406. Buchwald P. Modeling liquid properties, solvation, and hydrophobicity: a molecular size-based perspective. Perspect Drug Disc Des 2000;19:19–45.
407. Buchwald P, Bodor N. Simple model for non-associative organic liquids and water. J Am Chem Soc 2000;122:10671–10679.
408. Buchwald P, Bodor N. A simple, predictive, structure-based skin permeability model. J Pharm Pharmacol 2001;53:1087–1098.
409. Buchwald P. Computer-aided retrometabolic drug design: soft drugs. Expert Opin Drug Discov 2007;2:923–933.
410. Bodor N, Buchwald P, Huang M-J. Computer-assisted design of new drugs based on retrometabolic concepts. SAR QSAR Environ Res 1998;8:41–92.
411. Bodor N, Buchwald P, Huang M-J. The role of computational techniques in retrometabolic drug design strategies. In: Leszczynski J, editor. Computational Molecular Biology. Theo-

retical and Computational Chemistry. Vol. 8. Amsterdam: Elsevier; 1999. p 569–618.

412. Buchwald P, Bodor N. Structure-based estimation of enzymatic hydrolysis rates and its application in computer-aided retrometabolic drug design. Pharmazie 2000;55:210–217.
413. Dewar MJS, Zoebisch EG, Healy EF, Stewart JJP. AM1: a new general purpose quantum mechanical molecular model. J Am Chem Soc 1985;107:3902–3909.
414. Brüstle M, Beck B, Schindler T, King W, Mitchell T, Clark T. Descriptors, physical properties, and drug-likeness. J Med Chem 2002;45: 3345–3355.
415. Konishi Y, Kawamura M.5-Hetero-6-oxo-PGE-derivatives. US Patent 5,164,412. 1992.
416. Goldstein GW, Betz AL. The blood–brain barrier. Sci Am 1986;255(3):74–83.
417. Bradbury MWB, editor. Physiology and Pharmacology of the Blood–Brain Barrier. Vol. 103. Berlin: Springer; 1992.
418. Crone C. The blood–brain barrier: a modified tight epithelium. In: Suckling AJ, Rumsby MJ, Bradbury MWB, editors. The Blood–Brain Barrier in Health and Disease. Chichester: Ellis Horwood; 1986. p 17–40.
419. Bodor N, Buchwald P. Brain-targeted delivery: experiences to date. Am J Drug Deliv 2003;1: 13–26.
420. Witt KA, Gillespie TJ, Huber JD, Egleton RD, Davis TP. Peptide drug modifications to enhance bioavailability and blood–brain barrier permeability. Peptides 2001;22:2329–2343.
421. Kroll RA, Neuwelt EA. Outwitting the blood--brain barrier for therapeutic purposes: osmotic opening and other means. Neurosurgery 1998;42:1083–1099.
422. Rapoport SI. Osmotic opening of the blood–brain barrier: principles, mechanism, and therapeutic applications. Cell Mol Neurobiol 2000;20:217–230.
423. Langer R. Polymer implants for drug delivery in the brain. J Control Release 1991;16:53–60.
424. Benoit JP, Faisant N, Venier-Julienne MC, Menei P. Development of microspheres for neurological disorders: from basics to clinical applications. J Control Release 2000;65: 285–296.
425. Lindgren M, Hällbrink M, Prochiantz A, Langel Ü. Cell-penetrating peptides. Trends Pharmacol Sci 2000;21:99–103.
426. Rousselle C, Clair P, Lefauconnier J-M, Kaczorek M, Scherrmann J-M, Temsamani J. New advances in the transport of doxorubicin through the blood–brain barrier by a peptide vector-mediated strategy. Mol Pharmacol 2000;57:679–686.
427. Bodor N, Tóth-Sarudy É, Holm T, Pallagi I, Vass E, Buchwald P, Langel Ü. Novel, cell-penetrating molecular transporters with flexible backbones and permanently charged side chains. J Pharm Pharmacol 2007;59: 1065–1076.
428. Pardridge WM. Drug and gene targeting to the brain with molecular Trojan Horses. Nature Rev Drug Disc 2002;1:131–139.
429. Nutt D, King LA, Saulsbury W, Blakemore C. Development of a rational scale to assess the harm of drugs of potential misuse. Lancet 2007;369:1047–1053.
430. Anderson BD. Prodrugs for improved CNS delivery. Adv Drug Deliv Rev 1996;19: 171–202.
431. Bodor N, Brewster ME. Problems of delivery of drugs to the brain. Pharmacol Ther 1983;19: 337–386.
432. Ishikura T, Senou T, Ishihara H, Kato T, Ito T. Drug delivery to the brain. DOPA prodrugs based on a ring-closure reaction to quaternary thiazolium compounds. Int J Pharm 1995;116: 51–63.
433. Pop E. Optimization of the properties of brain specific chemical delivery systems by structural modifications. Curr Med Chem 1997;4:279–294.
434. Somogyi G, Nishitani S, Nomi D, Buchwald P, Prokai L, Bodor N. Targeted drug delivery to the brain via phosphonate derivatives. I. Design, synthesis, and evaluation of an anionic chemical delivery system for testosterone. Int J Pharm 1998;166:15–26.
435. Somogyi G, Buchwald P, Nomi D, Prokai L, Bodor N. Targeted drug delivery to the brain via phosphonate derivatives. II. Anionic chemical delivery system for zidovudine (AZT). Int J Pharm 1998;166:27–35.
436. Yoshikawa T, Sakaeda T, Sugawara T, Hirano K, Stella VJ. A novel chemical delivery system for brain targeting. Adv Drug Deliv Rev 1999;36:255–275.
437. Bodor N, Farag HH, Barros MDC, Wu W-M, Buchwald P. *In vitro* and *in vivo* evaluations of dihydroquinoline- and dihydroisoquinoline-based targetor moieties for brain-specific chemical delivery systems. J Drug Targeting 2002;10:63–71.
438. Mahmoud S, Aboul-Fadl T, Sheha M, Farag H, Mouhamed AM. 1,2-dihydroisoquinoline-*N*-acetic acid derivatives as new carriers for specific brain delivery I: synthesis and estimation of oxidation kinetics using multivariate cali-

bration method. Arch Pharm 2003;336: 573–584.

439. Sheha M, Al-Tayeb A, El-Sherief H, Farag H. New carrier for specific delivery of drugs to the brain. Bioorg Med Chem Lett 2003;11: 1865–1872.

440. Patteux C, Foucout L, Bohn P, Dupas G, Leprince J, Tonon MC, Dehouck B, Marsais F, Papamicaël C, Levacher V. Solid phase synthesis of a redox delivery system with the aim of targeting peptides into the brain. Org Biomol Chem 2006;4:817–825.

441. Rydström J, Hoek JB, Ernster L. The nicotinamide nucleotide transhydrogenases. In: Boyer PD, editor. The Enzymes. Vol. 13. New York: Academic Press; 1976.

442. Hoek JB, Rydström J. Physiological roles of nicotinamide nucleotide transhydrogenase. Biochem J 1988;254:1–10.

443. Bodor N, Brewster ME, Kaminski JJ. Reactivity of biologically important reduced pyridines. Part III. Energetics and mechanism of hydride transfer between 1-methyl-1,4-dihydronicotinamide and the 1-methylnicotinamide cation, a theoretical study. J Mol Struct (Theochem) 1990;206:315–334.

444. Brewster ME, Estes KS, Perchalski R, Bodor N. A dihydropyridine conjugate which generates high and sustained levels of the corresponding pyridinium salt in the brain does not exhibit neurotoxicity in cynomolgus monkeys. Neurosci Lett 1988;87:277–282.

445. Bodor N, Roller RG, Selk SJ. Elimination of a quaternary pyridinium salt delivered as its dihydropyridine derivative from brain of mice. J Pharm Sci 1978;67:685–687.

446. Palomino E, Kessel D, Horwitz JP. A dihydropyridine carrier system for sustained delivery of 1′,3′-dideoxynucleosides to the brain. J Med Chem 1989;32:622–625.

447. Bodor N, Farag HH. Improved delivery through biological membranes. 11. A redox chemical drug-delivery system and its use for brain specific delivery of phenylethylamine. J Med Chem 1983;26:313–318.

448. Bodor N, Abdelalim AM. Improved delivery through biological membranes. XIX. Novel redox carriers for brain-specific chemical delivery systems. J Pharm Sci 1985;74:241–245.

449. Tedjamulia ML, Srivastava PC, Knapp FF Jr. Evaluation of the brain-specific delivery of radioiodinated (iodophenyl)alkyl-substituted amines coupled to a dihydropyridine carrier. J Med Chem 1985;28:1574–1580.

450. Mahmoud S, Sheha M, Aboul-Fadl T, Farag H. 1, 2-dihydroisoquinoline-*N*-acetic acid derivatives as new carriers for brain-specific delivery II: delivery of phenethylamine as model drug. Arch Pharm 2003;336:258–263.

451. Bodor N, Nakamura T, Brewster ME. Improved delivery through biological membranes. XXIII. Synthesis, distribution, and neurochemical effects of a tryptamine chemical delivery system. Drug Des Deliv 1986;1:51–64.

452. Bodor N, Farag HH, Polgar P. A tryptamine analog with high affinity to the heart tissues is a potential antiarrhytmic agent. J Pharm Pharmacol 2001;53:889–894.

453. Bodor N, Farag HH. Improved delivery through biological membranes. XIV. Brain-specific, sustained delivery of testosterone using a redox chemical delivery system. J Pharm Sci 1984;73:385–389.

454. Bodor N, Abdelalim AM. Improved delivery through biological membranes. XX. Nicotinamide–dihydronicotinamide based ester-linked redox carrier systems. J Pharm Sci 1986;75: 29–35.

455. Anderson WR, Simpkins JW, Brewster ME, Bodor N. Brain-enhanced delivery of testosterone using a chemical delivery system complexed with 2-hydroxypropyl-β-cyclodextrin. Drug Des Del 1988;2:287–298.

456. Brewster ME, Estes KS, Bodor N. Improved delivery through biological membranes. XXIV. Synthesis, *in vitro* studies, and *in vivo* characterization of brain-specific and sustained progestin delivery systems. Pharm Res 1986;3:278–285.

457. Brewster ME, Deyrup M, Czako K, Bodor N. Extension of a redox-based chemical delivery system to α,β-unsaturated ketones. J Med Chem 1990;33:2063–2065.

458. Brewster ME, Estes KS, Bodor N. Use of a chemical redox system for delivery of drugs to the brain: ethinyl estradiol. Ann NY Acad Sci 1988;529:298–300.

459. Anderson WR, Simpkins JW, Brewster ME, Bodor N. Evidence for prolonged suppression of stress-induced release of adrenocortopic hormone and corticosterone with a brain-enhanced dexamethasone-redox delivery system. Neuroendocrinology 1989;50:9–16.

460. Siegal T, Soti F, Biegon A, Pop E, Brewster ME. Effect of a chemical delivery system for dexamethasone (Dex-CDS) on peritumoral edema in an experimental brain tumor model. Pharm Res 1997;14:672–675.

461. Pop E, Brewster ME, Prókai-Tátrai K, Bodor N. Preparation of redox derivatives of 3α-hydroxy-5α-pregnane-11,20-dione. Org Prep Proced Int 1994;26:379–382.

462. Bodor N, McCornack J, Brewster ME. Improved delivery through biological membranes. XXII. Synthesis and distribution of brain-selective estrogen delivery systems. Int J Pharm 1987;35:47–59.

463. Simpkins JW, McCornack J, Estes KS, Brewster ME, Shek E, Bodor N. Sustained brain-specific delivery of estradiol causes long-term suppression of luteinizing hormone secretion. J Med Chem 1986;29:1809–1812.

464. Anderson WR, Simpkins JW, Brewster ME, Bodor N. Evidence for the reestablishment of copulatory behavior in castrated male rats with a brain-enhanced estradiol-chemical delivery system. Pharmacol Biochem Behav 1987;27:265–271.

465. Estes KS, Brewster ME, Simpkins JW, Bodor N. A novel redox system for CNS-directed delivery of estradiol causes sustained LH suppression in castrate rats. Life Sci 1987;40:1327–1334.

466. Estes KS, Brewster ME, Bodor NS. A redox system for brain targeted estrogen delivery causes chronic body weight decrease in rats. Life Sci 1988;42:1077–1084.

467. Anderson WR, Simpkins JW, Brewster ME, Bodor N. Effects of a brain-enhanced chemical delivery system for estradiol on body weight and serum hormones in middle-aged male rats. Endocr Res 1988;14:131–148.

468. Brewster ME, Estes KS, Bodor N. Improved delivery through biological membranes. 32. Synthesis and biological activity of brain-targeted delivery systems for various estradiol derivatives. J Med Chem 1988;31:244–249.

469. Mullersman G, Derendorf H, Brewster ME, Estes KS, Bodor N. High performance liquid chromatographic assay of a central nervous system (CNS)-directed estradiol chemical delivery system and its application after intravenous administration in rats. Pharm Res 1988;5:172–177.

470. Howes J, Bodor N, Brewster ME, Estes K, Eve M. A pilot study with PR-63 in postmenopausal volunteers. J Clin Pharmacol 1988;28:951 Abstract 181.

471. Simpkins JW, Anderson WR, Dawson R Jr, Seth E, Brewster M, Estes KS, Bodor N. Chronic weight loss in lean and obese rats with a brain-enhanced chemical delivery system for estradiol. Physiol Behav 1988;44:573–580.

472. Simpkins JW, Anderson WR, Dawson R Jr, Bodor N. Effects of a brain-enhanced chemical delivery system for estradiol on body weight and food intake in intact and ovariectomized rats. Pharm Res 1989;6:592–600.

473. Sarkar DK, Friedman SJ, Yen SSC, Frautschy SA. Chronic inhibition of hypothalamic-pituitary-ovarian axis and body weight gain by brain-directed delivery of estradiol-17β in female rats. Neuroendocrinology 1989;50: 204–210.

474. Millard WJ, Romano TM, Bodor N, Simpkins JW. Growth hormone (GH) secretory dynamics in animals administered estradiol utilizing a chemical delivery system. Pharm Res 1990;7:1011–1018.

475. Brewster ME, Simpkins JW, Bodor N. Brain-targeted delivery of estrogens. Rev Neurosci 1990;2:241–285.

476. Estes KS, Dewland PM, Brewster ME, Derendorf H, Bodor N. A redox-based chemical delivery system (CDS) applied to estradiol. Pharm Ztg Wiss 1991;136:153–158.

477. Brewster ME, Bartruff MSM, Anderson WR, Druzgala PJ, Bodor N, Pop E. Effect of molecular manipulation on the estrogenic activity of a brain-targeting estradiol chemical delivery system. J Med Chem 1994;37:4237–4244.

478. Estes KS, Brewster ME, Bodor N. Evaluation of an estradiol chemical delivery system (CDS) designed to provide enhanced and sustained hormone levels in the brain. Adv Drug Deliv Rev 1994;14:167–175.

479. Simpkins JW, Rajakumar G, Zhang Y-Q, Simpkins CE, Greenwald D, Yu CJ, Bodor N, Day AL. Estrogens may reduce mortality and ischemic damage caused by middle cerebral artery occlusion in the female rat. J Neurosurg 1997;87:724–730.

480. Rabbani O, Panickar KS, Rajakumar G, King MA, Bodor N, Meyer EM, Simpkins JW. 17 β-estradiol attenuates fimbrial lesion-induced decline of ChAT-immunoreactive neurons in the rat medial septum. Exp Neurol 1997;146: 179–186.

481. Bodor N, Buchwald P. Barriers to remember: brain-targeting chemical delivery systems and Alzheimer's disease. Drug Discov Today 2002;7:766–774.

482. Tapfer MK, Sebestyen L, Kurucz I, Horvath K, Szelenyi I, Bodor N. New evidence for the selective, long-lasting central effects of the brain-targeted estradiol, Estredox. Pharmacol Biochem Behav 2004;77:423–429.

483. Sziráki I, Horváth K, Bodor N. Comparative evaluation of Estredox, a brain-targeted estra-

diol delivery system versus traditional estrogen replacement therapy. Pharmazie 2006;61: 140–143.

484. Bodor N, Buchwald P. Brain-targeted delivery of estradiol: therapeutic potential and results obtained with a chemical delivery system approach. Am J Drug Deliv 2006;4:161–175.

485. Pop E, Wu W-M, Shek E, Bodor N. Brain-specific chemical delivery systems for β-lactam antibiotics. Synthesis and properties of some dihydropyridine and dihydroisoquinoline derivatives of benzylpenicillin. J Med Chem 1989;32:1774–1781.

486. Pop E, Wu W-M, Bodor N. Chemical delivery systems for some penicillinase-resistant semisynthetic penicillins. J Med Chem 1989;32: 1789–1795.

487. Wu W-M, Pop E, Shek E, Bodor N. Brain-specific chemical delivery systems for β-lactam antibiotics. *In vitro* and *in vivo* studies of some dihydropyridine and dihydroisoquinoline derivatives of benzylpenicillin in rats. J Med Chem 1989;32:1782–1788.

488. Wu W-M, Pop E, Shek E, Clemmons R, Bodor N. Brain and CSF specific chemical delivery systems for β-lactam antibiotics. Study of two dihydropyridine derivatives of benzylpenicillin in rabbits and dogs. Drug Des Delivery 1990;7:33–43.

489. Brewster ME, Deyrup M, Seyda K, Bodor N. Synthesis, characterization and *in vitro* evaluation of various sulfonamide chemical delivery systems. Int J Pharm 1991;68:215–229.

490. Venkatraghavan V, Shek E, Perchalski R, Bodor N. Brain-specific chemical delivery systems for acyclovir. Pharmacologist 1986;28:145 Abstract.

491. Rand K, Bodor N, El-Koussi A, Raad I, Miyake A, Houck H, Gildersleeve N. Potential treatment of herpes simplex virus encephalitis by brain-specific delivery of trifluorothymidine using a dihydropyridine-pyridinium salt redox delivery system. J Med Virol 1986;20:1–8.

492. El-Koussi A, Bodor N. Improved delivery through biological membranes. XXV. Enhanced and sustained delivery of trifluorothymidine to the brain using a dihydropyridine-pyridinium salt type redox delivery system. Drug Des Deliv 1987;1:275–283.

493. Canonico PG, Kende M, Gabrielsen B. Carrier-mediated delivery of antiviral agents. Adv Virus Res 1988;35:271–312.

494. Bhagrath M, Sidwell R, Czako K, Seyda K, Anderson W, Bodor N, Brewster ME. Improved delivery through biological membranes. Synthesis, characterization and antiviral activity of a series of ribavirin chemical delivery systems: 5′ and carboxamide derivatives. Antiviral Chem Chemother 1991;2: 265–286.

495. Deyrup M, Sidwell R, Little R, Druzgala P, Bodor N, Brewster ME. Improved delivery through biological membranes. Synthesis and antiviral activity of a series of ribavirin chemical delivery systems: 2′ and 3′ derivatives. Antiviral Chem Chemother 1991;2:337–355.

496. Morin KW, Wiebe LI, Knaus EE. Synthesis of brain-targeted 1-(2-deoxy-2-fluoro-β-D-ribofuranosyl)-(*E*)-5-(2-iodovinyl)uracil coupled to a dihydropyridine-pyridinium salt redox chemical-delivery system. Carbohydr Res 1993;249:109–116.

497. Morin KW, Knaus EE, Wiebe LI. Site-directed design and synthesis of a brain-targeted radioiodinated nucleoside for diagnosis of herpes simplex encephalitis. J Label Compds Radiopharm 1994;35:205–207.

498. Balzarini J, Morin KW, Knaus EE, Wiebe LI, De Clercq E. Novel (*E*)-5-(2-iodovinyl)-2′-deoxyuridine derivatives as potential cytostatic agents against herpes-simplex virus thymidine kinase gene transfected tumors. Gene Therapy 1995;2:317–322.

499. Pop E, Anderson W, Vlasak J, Brewster ME, Bodor N. Enhanced brain delivery of 2′-fluoro-5-methylarabinosyluracil. Int J Pharm 1992; 84:39–48.

500. Brewster ME, Little R, Venkatraghavan V, Bodor N. Brain-enhanced delivery of antiviral agents. Antiviral Res 1988;9:127. Abstract.

501. Little R, Bailey D, Brewster ME, Estes KS, Clemmons RM, Saab A, Bodor N. Improved delivery through biological membranes. XXXIII. Brain enhanced delivery of azidothymidine (AZT). J Biopharm Sci 1990;1:1–18.

502. Brewster ME, Anderson W, Bodor N. Brain, blood, and cerebrospinal fluid distribution of a zidovudine chemical delivery system in rabbits. J Pharm Sci 1991;80:843–845.

503. Pop E, Brewster ME, Anderson WR, Bodor N. Biodistribution of azidothymidine mediated by a highly lipophilic chemical delivery system. Med Chem Res 1992;2:457–466.

504. Pop E, Liu ZZ, Vlasak J, Anderson W, Brewster ME, Bodor N. A dihydroisoquinoline targetor-based acid resistant chemical delivery system of azidothymidine (AZT). Drug Del 1993;1: 143–149.

505. Mizrachi Y, Rubinstein A, Harish Z, Biegon A, Anderson WR, Brewster ME. Improved deliv-

ery and *in vitro* activity of zidovudine through the use of a redox chemical delivery system. AIDS 1995;9:153–158.

506. Brewster ME, Anderson WR, Webb AI, Pablo LM, Meinsma D, Moreno D, Derendorf H, Bodor N, Pop E. Evaluation of a brain-targeting zidovudine chemical delivery system in dogs. Antimicrob Agents Chemother 1997;41:122–128.
507. Torrence PT, Kinjo J, Lesiak K, Balzarini J, DeClerq E. AIDS dementia: synthesis and properties of a derivative of 3′-azido-3′-deoxythimidine (AZT) that may become "locked" in the central nervous system. FEBS Lett 1988;234:135–140.
508. Gogu SR, Aggarwal SK, Rangan SRS, Agrawal KC. A prodrug of zidovudine with enhanced efficacy against human immunodeficiency virus. Biochem Biophys Res Commun 1989;160: 656–661.
509. Aggarwal SK, Gogu SR, Rangan SRS, Agrawal KC. Synthesis and biological evaluation of prodrugs of zidovudine. J Med Chem 1990; 33:1505–1510.
510. Lupia RH, Ferencz N, Aggarwal SK, Agrawal KC, Lertora JJL. Plasma (P), cerebrospinal fluid (C) and brain (B) pharmacokinetics of two novel prodrugs of zidovudine (Z). Clin Res 1990;38:15A. Abstract.
511. Lupia RH, Ferencz N, Lertora JJL, Aggarwal SK, George WJ, Agrawal KC. Comparative pharmacokinetics of two prodrugs of zidovudine in rabbits: enhanced levels of zidovudine in brain tissue. Antimicrob Agents Chemother 1993;37:818–824.
512. Gallo J, Boubinot F, Doshi D, Etse J, Bhandti V, Schinazi R, Chu CK. Evaluation of brain targeting of anti-HIV nucleosides delivered via dihydropyridine prodrugs. Pharm Res 1989;6: S161. Abstract.
513. Chu CK, Bhadti VS, Doshi KJ, Etse JT, Gallo JM, Boudinot FD, Schinazi RF. Brain targeting of anti-HIV nucleosides: synthesis and *in vitro* and *in vivo* studies of dihydropyridine derivatives of 3′-azido-2′,3′-dideoxyuridine and 3′-azido-3′deoxythimidine. J Med Chem 1990;33:2188–2192.
514. Gallo JM, Etse JT, Doshi KJ, Boudinot FD, Chu CK. Hybrid pharmacokinetic models to describe anti-HIV nucleoside brain disposition following parent and prodrug administration in mice. Pharm Res 1991;8:247–253.
515. Palomino E, Kessel D, Horwitz JP. A dihydropyridine carrier system for delivery of 2′,3′-dideoxycytidine (DDC) to the brain. Nucleosides Nucleotides 1992;11:1639–1649.
516. Torrence PF, Kinjo J, Khamnei S, Greig NH. Synthesis and pharmacokinetics of a dihydropyridine chemical delivery system for the antiimmunodeficiency virus agent dideoxycytidine. J Med Chem 1993;36:529–537.
517. Brewster ME, Raghavan K, Pop E, Bodor N. Enhanced delivery of ganciclovir to the brain through the use of redox targeting. Antimicro Agents Chemother 1994;38:817–823.
518. Camplo M, Charvet Faury AS, Borel C, Turin F, Hantz O, Trabaud C, Niddam V, Mourier N, Graciet JC, Chermann JC, Kraus JL. Synthesis and antiviral activity of *N*-4′-dihydropyridinyl and dihydroquinolinylcarbonyl-2-hydroxymethyl-5-[cytosin-1′-yl]-1,3-oxathiolane derivatives against human immunodeficiency virus and duck hepatitis B virus. Eur J Med Chem 1996;31:539–546.
519. Raghavan K, Shek E, Bodor N. Improved delivery through biological membranes. XXX. Synthesis and biological aspects of a 1,4-dihydropyridine based chemical delivery system for brain-sustained delivery of hydroxy CCNU. Anticancer Drug Des 1987;2:25–36.
520. Raghavan K, Loftsson T, Brewster ME, Bodor N. Improved delivery through biological membranes. XLV. Synthesis, physical-chemical evaluation, and brain uptake studies of 2-chloroethyl nitrosourea delivery system. Pharm Res 1992;9:743–749.
521. Bodor N, Venkatraghavan V, Winwood D, Estes K, Brewster ME. Improved delivery through biological membranes. XLI. Brain-enhanced delivery of chlorambucil. Int J Pharm 1989;53:195–208.
522. El-Sherbeny MA, Al-Salem HS, Sultan MA, Radwan MA, Farag HA, El-Subbagh HI. Synthesis, in *vitro* and *in vivo* evaluation of a delivery system for targeting anticancer drugs to the brain. Arch Pharm 2003;336:445–455.
523. Bodor N, Simpkins JW. Redox delivery system for brain-specific, sustained release of dopamine. Science 1983;221:65–67.
524. Bodor N, Farag HH. Improved delivery through biological membranes. 13. Brain-specific delivery of dopamine with a dihydropyridine-pyridinium salt type redox delivery system. J Med Chem 1983;26:528–534.
525. Simpkins JW, Bodor N, Enz A. Direct evidence for brain-specific release of dopamine from a redox delivery system. J Pharm Sci 1985;74: 1033–1036.

526. Omar FA, Farag HH, Bodor N. Synthesis and evaluation of a redox chemical delivery system for brain-enhanced dopamine containing an activated carbamate-type ester. J Drug Targeting 1994;2:309–316.

527. Carelli V, Liberatore F, Scipione L, Impicciatore M, Barocelli E, Cardellini M, Giorgioni G. New systems for the specific delivery and sustained release of dopamine to the brain. J Control Release 1996;42:209–216.

528. Anderson WR, Simpkins JW, Woodard PA, Winwood D, Stern WC, Bodor N. Anxiolytic activity of a brain delivery system for GABA. Psychopharmacol 1987;92:157–163.

529. Woodard PA, Winwood D, Brewster ME, Estes KS, Bodor N. Improved delivery through biological membranes. XXI. Brain-targeted anti-convulsive agents. Drug Des Deliv 1990;6:15–28.

530. Kourounakis A, Bodor N, Simpkins J. Synthesis and evaluation of a brain-targeted catechol derivative as a potential NGF-inducer. Int J Pharm 1996;141:239–250.

531. Kourounakis A, Bodor N, Simpkins J. Synthesis and evaluation of a brain-targeted chemical delivery system for the neurotrophomodulator 4-methylcatechol. J Pharm Pharmacol 1997;49:1–9.

532. Shek E, Murakami T, Nath C, Pop E, Bodor NS. Improved anticonvulsant activity of phenytoin by a redox brain delivery system. III. Brain uptake and pharmacological effects. J Pharm Sci 1989;78:837–843.

533. Pop E, Bodor N. Chemical systems for delivery of antiepileptic drugs to the central nervous system. Epilepsy Res 1992;13:1–16.

534. Boddy AV, Zhang K, Lepage F, Tombret F, Slatter JG, Baillie TA, Levy RH. *In vitro* and *in vivo* investigations of dihydropyridine-based chemical delivery systems for anticonvulsants. Pharm Res 1991;8:690–697.

535. Yiu SH, Knaus EE. Synthesis, biological evaluation, calcium channel antagonist activity, and anticonvulsant activity of felodipine coupled to a dihydropyridine-pyridinium salt redox chemical delivery system. J Med Chem 1996;39:4576–4582.

536. Pop E, Prókai-Tátrai K, Anderson W, Lin J-L, Brewster ME, Bodor N. Application of a brain-specific chemical delivery system approach to tranylcypromine. Eur J Pharmacol 1990;183:1909–1909.

537. Prókai-Tátrai K, Pop E, Anderson W, Lin J-L, Brewster ME, Bodor N. Redox derivatives of tranylcypromine: synthesis, properties, and monoamine oxidase inhibitor activity of some chemical delivery systems. J Pharm Sci 1991;80:255–261.

538. Brewster ME, Robledo-Luiggi C, Miyakeb A, Pop E, Bodor N. Brain-enhanced delivery of anti-dementia drugs. In: Meyer EM, Simpkins JW, Yamamoto J, editors. Novel Approaches to the Treatment of Alzheimer's Disease. Advances in Behavioral Biology. Vol. 36.New York: Plenum; 1989. p 173–183.

539. Pop E, Prókai-Tátrai K, Scott JD, Brewster ME, Bodor N. Application of a brain-targeting chemical delivery system to 9-amino-1,2,3,4-tetrahydroacridine. Pharm Res 1990;7: 658–664.

540. Pop E, Soti F, Anderson WR, Panetta JA, Estes KS, Bodor NS, Brewster ME. Redox targeting of LY231617, an antioxidant with potential use in the treatment of brain damage. Int J Pharm 1996;140:33–44.

541. Bailey D, Perchalski R, Bhagrath M, Shek E, Winwood D, Bodor N. Synthesis of pyridinium salt derivatives of 1,1′-[[1-{aminomethyl}-1,2-ethandediyl]diimino]-bis[2-methyl-2-propanethiol]: potential brain specific Tc-99m complexes. J Biopharm Sci 1991;2:205–218.

542. Phelan MJ, Bodor N. Improved delivery through biological membranes. XXXVII. Synthesis and stability of novel redox derivatives of naproxen and indomethacin. Pharm Res 1989;6:667–676.

543. Perioli L, Ambrogi V, Bernardini C, Grandolini G, Ricci M, Giovagnoli S, Rossi C. Potential prodrugs of non-steroidal anti-inflammatory agents for targeted drug delivery to the CNS. Eur J Med Chem 2004;39:715–727.

544. Pop E, Anderson W, Prókai-Tátrai K, Vlasak J, Brewster ME, Bodor N. Synthesis and preliminary pharmacological evaluation of some chemical delivery systems of 2,6-diisopropylphenol (propofol). Med Chem Res 1992;2: 16–21.

545. Anderson W, Pop E, Lee S-K, Bodor N, Brewster M. Brain-targeting chemical delivery systems for adenosine depresses locomotor behavior in rats. Med Chem Res 1991;1:74–79.

546. Sheha MM, el-Koussi NA, Farag HH. Brain delivery of HIV protease inhibitors. Arch Pharm 2003;336:47–52.

547. Pop E, Anderson W, Prókai-Tátrai K, Brewster ME, Fregly M, Bodor N. Antihypertensive activity of redox derivatives of tryptophan. J Med Chem 1990;33:2216–2221.

548. Pop E, Prókai-Tátrai K, Brewster ME, Bodor N. Analogs of tryptophan. Org Prep Proced Int 1994;26:687–690.

549. Bodor N, Prokai L, Wu W-M, Farag HH, Jonnalagadda S, Kawamura M, Simpkins J. A strategy for delivering peptides into the central nervous system by sequential metabolism. Science 1992;257:1698–1700.

550. Prokai-Tatrai K, Prokai L, Bodor N. Brain-targeted delivery of a leucine-enkephalin analogue by retrometabolic design. J Med Chem 1996;39:4775–4782.

551. Prokai L, Ouyang X-D, Wu W-M, Bodor N. Chemical delivery system to transport a pyroglutamyl peptide to the central nervous system. J Am Chem Soc 1994;116: 2643–2644.

552. Prokai L, Ouyang X, Prokai-Tatrai K, Simpkins JW, Bodor N. Synthesis and behavioral evaluation of a chemical brain-targeting system for a thyrotropin-releasing hormone analogue. Eur J Med Chem 1998;33: 879–886.

553. Prokai L, Prokai-Tatrai K, Ouyang X, Kim H-S, Wu W-M, Zharikova A, Bodor N. Metabolism-based brain-targeting system for a thyrotropin-releasing hormone analogue. J Med Chem 1999;42:4563–4571.

554. Yoon S-H, Wu J, Wu W-M, Prokai L, Bodor N. Brain-targeted chemical delivery of [Leu^2, Pip^3]-TRH: synthesis and biological evaluation. Bioorg Med Chem 2000;9:1059–1063.

555. Wu J, Yoon S-H, Wu W-M, Bodor N. Synthesis and biological evaluation of a brain-targeted chemical delivery system of [Nva^2]-TRH. J Pharm Pharmacol 2002;54:945–950.

556. Prokai-Tatrai K, Perjési P, Zharikova AD, Li X, Prokai L. Design, synthesis, and biological evaluation of novel, centrally-acting thyrotropin-releasing hormone analogues. Bioorg Med Chem Lett 2002;12:2171–2174.

557. Prokai L, Prokai-Tatrai K, Zharikova AD, Nguyen V, Perjesi P, Stevens SM Jr. Centrally acting and metabolically stable thyrotropin-releasing hormone analogues by replacement of histidine with substituted pyridinium. J Med Chem 2004;47:6025–6033.

558. Chen P, Bodor N, Wu W-M, Prokai L. Strategies to target kyotorphin analogues to the brain. J Med Chem 1998;41:3773–3781.

559. Prokai L, Prokai-Tatrai K, Bodor N. Targeting drugs to the brain by redox chemical delivery systems. Med Res Rev 2000;20:367–416.

560. Hunt CA, MacGregor RD, Siegel RA. Engineering targeted *in vivo* drug delivery. I. The physiological and physicochemical principles governing opportunities and limitations. Pharm Res 1986;3:333–344.

561. Williams CL, Stancel GM. Estrogens and progestins. In: Hardman JG, Limbird LE, editors. Goodman & Gilman's The Pharmacological Basis of Therapeutics. New York: McGraw-Hill; 1996. p 1411–1440.

562. Kaplan NM. Cardiovascular complications of oral contraceptives. Annu Rev Med 1978;29: 31–40.

563. Fotherby K. Oral contraceptives, lipids and cardiovascular disease. Contraception 1985; 31:367–394.

564. Yager JD, Liehr JG. Molecular mechanism of estrogen carcinogenesis. Annu Rev Pharmacol Toxicol 1996;36:203–232.

565. Aldridge S. Magic Molecules: How Drugs Work. Cambridge: Cambridge University Press; 1998.

566. Greendale GA, Lee NP, Arriola ER. The menopause. Lancet 1999;353:571–580.

567. McEwen BS, Alves SE. Estrogen actions in the central nervous system. Endocr Rev 1999;20: 279–307.

568. Sherwin BB. Estrogen and cognitive functioning in women. Endocr Rev 2003;24:133–151.

569. Sharpe RM. The roles of oestrogen in the male. Trends Endocrinol Metabol 1998;9:371–377.

570. Faustini-Fustini M, Rochira V, Carani C. Oestrogen deficiency in men: where are we today? Eur J Endocrinol 1999;140:111–129.

571. Bhavnani BR. Pharmacokinetics and pharmacodynamics of conjugated equine estrogens: chemistry and metabolism. Proc Soc Exp Biol Med 1998;217:6–16.

572. Bhavnani BR. Estrogens and menopause: pharmacology of conjugated equine estrogens and their potential role in the prevention of neurodegenerative diseases such as Alzheimer's. J Steroid Biochem Mol Biol 2003;85:473–482.

573. Steingold KA, Cefalu W, Pardridge W, Judd HL, Chaudhuri G. Enhanced hepatic extraction of estrogens used for replacement therapy. J Clin Endocrinol Metab 1986;62:761–766.

574. Sherwin BB. Surgical menopause, estrogen, and cognitive function in women: what do the findings tell us? Ann NY Acad Sci 2005;1052:3–10.

575. Shepherd JE. Therapeutic options in female sexual dysfunction. J Am Pharm Assoc 2002;42:479–488.

576. Upton GV. Therapeutic considerations in the management of the climacteric. J Reprod Med 1984;29:71–79.

577. Behl C, Manthey D. Neuroprotective activities of estrogen: an update. J Neurocytol 2000;29: 351–358.

578. Leranth C, Roth RH, Elswoth JD, Naftolin F, Horvath TL, Redmond DE Jr. Estrogen is essential for maintaining nigrostriatal dopamine neurons in primates: implications for Parkinson's disease and memory. J Neurosci 2000;20:8604–8609.

579. Maggi A, Perez J. Role of female gonadal hormones in the CNS: clinical and experimental aspects. Life Sci 1985;37:893–906.

580. Yaffe K, Sawaya G, Lieberburg I, Grady D. Estrogen therapy in postmenopausal women: effects on cognitive function and dementia. JAMA 1998;279:688–695.

581. Gopinath L. Outsmarting Alzheimer's disease. Chem Br 1998;34(5):38–40.

582. Haskell SG, Richardson ED, Horwitz RI. The effect of estrogen replacement therapy on cognitive function in women: a critical review of the literature. J Clin Epidemiol 1997;50: 1249–1264.

583. Henderson VW. The epidemiology of estrogen replacement therapy and Alzheimer's disease. Neurology 1997;48:S27–S35.

584. Asthana S, Craft S, Baker LD, Raskind MA, Birnbaum RS, Lofgreen CP, Veith RC, Plymate SR. Cognitive and neuroendocrine response to transdermal estrogen in postmenopausal women with Alzheimer's disease: results of a placebo-controlled, double-blind, pilot study. Psychoneuroendocrinology 1999;24: 657–677.

585. Asthana S, Baker LD, Craft S, Stanczyk FZ, Veith RC, Raskind MA, Plymate SR. High-dose estradiol improves cognition for women with AD: results of a randomized study. Neurology 2001;57:605–612.

586. Janocko L, Lamer JM, Hochberg RB. The interaction of C-17 esters of estradiol with the estrogen receptor. Endocrinology 1984;114: 1180–1186.

587. Laumann EO, Paik A, Rosen RC. Sexual dysfunction in the United States: prevalence and predictors. JAMA 1999;281:537–544.

588. Howes J, Brewster M, Harris A, Griffith W, Garty N. Buccal and parenteral clinical evaluation of a brain-targeting estradiol chemical delivery system (E2-CDS). Clin Pharmacol Ther 1995;57:172.

589. Brewster ME, Howes J, Griffith W, Garty N, Bodor N, Anderson WR, Pop E. Intravenous and buccal 2-hydroxypropyl-β-cyclodextrin formulations of E2-CDS—Phase I clinical trials. In: Proceedings of the 8th International Cyclodextrins Symposium; 1996 March 30–April 2; Paris: Editions de Sante; 1996.

590. Juhász A, Howes J, Mantelle L, Halabi A, Bodor N. Evaluation of a buccal delivery formulation of E_2CDS in postmenopausal women. Presented at the Proceedings of the North American Menopause Society, 14th Annual Meeting; 2003 September 17–19; Miami, FL; 2003.

591. Juhász A, Howes J, Halabi A, Bodor N. Evaluation of the safety, and the effects on hormone levels of two different administration regimens of a buccal delivery tablet of estradiol chemical delivery system (E_2CDS-Estredox™) in postmenopausal women. Presented at the Proceedings of the North American Menopause Society, 15th Annual Meeting; 2004 October 6–9; Washington, DC; 2004.

592. Loftsson T, Brewster ME, Másson M. Role of cyclodextrins in improving oral drug delivery. Am J Drug Deliv 2004;2:261–275.

593. Saenger W. Cyclodextrin inclusion compounds in research and industry. Angew Chem Int Ed Engl 1980;19:344–362.

594. Szejtli J. Cyclodextrins and Their Inclusion Complexes. Budapest: Akadémiai Kiadó; 1982.

595. Pagington JS. β-Cyclodextrin: the success of molecular inclusion. Chem Br 1987;23: 455–458.

596. Szejtli J. Introduction and general overview of cyclodextrin chemistry. Chem Rev 1998;98: 1743–1753.

597. Buchwald P. Complexation thermodynamics of cyclodextrins in the framework of a molecular size-based model for nonassociative organic liquids that includes a modified hydration-shell hydrogen-bond model for water. J Phys Chem B 2002;106:6864–6870.

598. Brewster ME, Estes KE, Loftsson T, Perchalski R, Derendorf H, Mullersman G, Bodor N. Improved delivery through biological membranes. XXXI. Solubilization and stabilization of an estradiol chemical delivery system by modified β-cyclodextrins. J Pharm Sci 1988;77:981–985.

599. Pitha J, Pitha J. Amorphous water-soluble derivatives of cyclodextrins: nontoxic dissolution enhancing excipients. J Pharm Sci 1985;74:987–990.

600. Pitha J, Milecki J, Fales H, Pannell L, Uekama K. Hydroxypropyl-β-cyclodextrin: preparation and characterization, effects on solubility of drugs. Int J Pharm 1986;29:73–82.

601. Yoshida A, Arima H, Uekama K, Pitha J. Pharmaceutical evaluation of hydroxyalkyl ethers of β-cyclodextrins. Int J Pharm 1988;46:217–222.

602. Brewster ME, Estes KS, Bodor N. An intravenous toxicity study of 2-hydroxypropyl-β-cyclodextrin, a useful drug solubilizer, in rats and monkeys. Int J Pharm 1990;59:231–243.

603. Pop E, Loftsson T, Bodor N. Solubilization and stabilization of a benzylpenicillin chemical delivery system by 2-hydroxypropyl-β-cyclodextrin. Pharm Res 1991;8:1044–1049.

604. Kastin AJ, Ehrensing RH, Banks WA, Zadina JE. Possible therapeutic implications of the effects of some peptides on the brain. In: de Kloet ER, Wiegant VM, de Wied D, editors. Neuropeptides and Brain Function. Progress in Brain Research. Vol. 72. Amsterdam: Elsevier; 1987. p 223–234.

605. Nemeroff CBE. Neuropeptides in Psychiatric and Neurological Disorders. Baltimore: John Hopkins University Press; 1988.

606. Hoyle CHV. Neuropeptides. Essential Data. Chichester: Wiley; 1996.

607. Banks WA, Kastin AJ. Exchange of peptides between the circulation and the nervous system: Role of the blood–brain barrier. Adv Exp Med Biol 1990;274:59–69.

608. Brownlees J, Williams CH. Peptidases, peptides, and the mammalian blood–brain barrier. J Neurochem 1993;60:793–803.

609. Ermisch A, Brust P, Kretzschmar R, Rühle H-J. Peptides and blood–brain barrier transport. Physiol Rev 1993;73:489–527.

610. Oliyai R, Stella VJ. Prodrugs of peptides and proteins for improved formulation and delivery. Annu Rev Pharmacol Toxicol 1993;33:521–544.

611. Bodor N, Prokai L. Molecular packaging: peptide delivery to the central nervous system by sequential metabolism. In: Taylor M, Amidon G, editors. Peptide-Based Drug Design: Controlling Transport and Metabolism. Washington, DC: American Chemical Society; 1995. p 317–337.

612. Hughes J, Smith TV, Kosterlitz HW, Fothergill L, Morgan BA, Morris HR. Identification of two related pentapeptides from the brain with potent opiate agonist activity. Nature 1975;258:577–579.

613. Olson GA, Olson RD, Kastin AJ. Endogenous opiates: 1984. Peptides 1985;6:769–791.

614. Olson GA, Olson RD, Kastin AJ. Endogenous opiates: 1991. Peptides 1992;13:1247–1287.

615. Bicknell RJ. Endogenous opioid peptides and hypothalamic neuroendocrine neurones. J Endocr 1985;107:437–446.

616. Millan MJ, Herz A. The endocrinology of the opioids. Int Rev Neurobiol 1985;26:1–83.

617. Yen SSC, Quigley ME, Reid RL, Ropert JF, Cetel NS. Neuroendocrinology of opioid peptides and their role in the control of gonadotropin and prolactin secretion. Am J Obstet Gynecol 1985;152:485–493.

618. Sandyk R. The endogenous opioid system in neurological disorders of the basal ganglia. Life Sci 1985;37:1655–1663.

619. Frenk H. Pro- and anticonvulsant actions of morphine and the endogenous opioids: involvement and interactions of multiple opiate and non-opiate systems. Brain Res Rev. 1983;6: 197–210.

620. Tortella FC, Long JB, Holaday JW. Endogenous opioid systems: physiological role in the self-limitation of seizures. Brain Res 1985;332:174–178.

621. Baile CA, McLaughlin CL, Della-Fera MA. Role of cholecystokinin and opioid peptides in control of food intake. Physiol Rev 1986;66: 172–234.

622. Schmauss C, Emrich HM. Dopamine and the actions of opiates: a reevaluation of the dopamine hypothesis of schyzophrenia with special consideration of the role of endogenous opioid in the pathogenesis of schizophrenia. Biol Psychiatry 1985;20:1211–1231.

623. Izquierdo I, Netto CA. Role of β-endorphin in behavioral regulation. Ann NY Acad Sci 1985;444:162–177.

624. Holaday JW. Cardiovascular effects of endogenous opiate systems. Annu Rev Pharmacol Toxicol 1983;23:541–594.

625. Johnson MW, Mitch WE, Wilcox CS. The cardiovascular actions of morphine and the endogenous opioid peptides. Prog Cardiovasc Dis 1985;27:435–450.

626. Porreca F, Burks TF. The spinal cord as a site of opioid effects on gastrointestinal transit in the mouse. J Pharmacol Exp Ther 1983;227: 22–27.

627. Kastin AJ, Pearson MA, Banks WA. EEG evidence that morphine and an enkephalin analog cross the blood–brain barrier. Pharmacol Biochem Behav 1991;40:771–774.

628. Szirtes T, Kisfaludy L, Pálosi É, Szporny L. Synthesis of thyrotropin-releasing hormone analogues. 1. Complete dissociation of central nervous system effects from thyrotropin-releasing activity. J Med Chem 1984;27: 741–745.

629. Santori EM, Schmidt DE. Effects of MK-771, a TRH analog, on pentobarbital-induced alterations of cholinergic parameters in discrete regions of rat-brain. Regul Pept 1980;1:69–74.

630. Itoh Y, Ogasawara T, Mushiroi T, Yamazaki A, Ukai Y, Kimura K. Effect of NS-3, a thyrotropin-releasing hormone analog, on *in vivo* acetylcholine-release in rat-brain: regional differences and its sites of action. J Pharmacol Exp Ther 1994;271:884–890.

631. Kelly JA. Thyrotropin-releasing hormone: basis and potential for its therapeutic use. Essays Biochem 1995;30:133–149.

632. Bennett GW, Ballard TM, Watson CD, Fone KC. Effect of neuropeptides on cognitive function. Exp Gerontol 1997;32:451–469.

633. Yarbrough GG. Thyrotropin releasing hormone and CNS cholinergic neurons. Life Sci 1983;33:111–118.

634. Faden AI, Vink R, McIntosh TK. Thyrotropin-releasing-hormone and central nervous system trauma. Ann NY Acad Sci 1989;553: 380–384.

635. Mellow AH, Sunderland T, Cohen RM, Lawlor BA, Hill JL, Newhouse PA, Cohen MR, Murphy DL. Acute effects of high-dose thyrotropin-releasing hormone infusions in Alzheimer's disease. Psychopharmacology 1989;98:403– 407.

636. Horita A, Carino MA, Zabawska J, Lai H. TRH analog MK-771 reverses neurochemical and learning deficits in medial septal-lesioned rats. Peptides 1989;10:121–124.

637. Horita A, Carino MA, Smith JR. Effects of TRH on the central nervous system of the rabbit. Pharmacol Biochem Behav 1976;5(Suppl 1):111–116.

638. Horita A, Carino MA, Lai H. Analeptic activity produced by TRH microinjection into basal forebrain area of the rat. Federation Proc 1986;45:795. Abstract.

639. Rolka K, Oslisok E, Krupa J, Kruszinsky M, Baran L, Przegalinska E, Kupryszewski G. Opiate-like peptides. 4. Kyotorphin and its analogues: synthesis and analgesic activity. Pol J Pharmacol Pharm 1983;35:473–480.

640. Takagi H, Shiomi H, Ueda H, Amano H. A novel analgesic dipeptide from bovine brain is a possible Met-enkephalin releaser. Nature 1979;282:410–412.

641. Godlevsky LS, Shandra AA, Mikhaleva II, Vastyanov RS, Mazarati AM. Seizure-protecting effects of kyotorphin and related peptides in an animal model of epilepsy. Brain Res Bull 1995;37:223–226.

642. Arima T, Kitamura Y, Nishiya T, Takagi H, Nomura Y. Kyotorphin (L-tyrosyl-L-arginine) as a possible substrate for inducible nitric oxide synthase in rat glial cells. Neurosci Lett 1996;212:1–4.

643. Arima T, Kitamura Y, Nishiya T, Taniguchi T, Takagi H, Nomura Y. Effects of (L-tyrosyl-L-arginine) on [^{3}H]N^G-nitro-L-arginine binding to neuronal nitric oxide synthase in rat brain. Neurochem Int 1997;30:605–611.

644. Summy-Long JY, Bui V, Gestl S, Koehler-Stec E, Liu H, Terrell ML, Kadekaro M. Effects of central injection of kyotorphin and L-arginine on oxytocin and vasopressin release and blood pressure in conscious rats. Brain Res Bull 1998;45:395–403.

645. Zincke T. Über Dinitrophenylpyridiniumchlorid und dessen Umwandlungsproducte. Ann Chem 1903;330:361–374.

646. Lettré H, Haede W, Ruhbaum E. Zur Darstellung von Derivaten des Nicotinsäureamids. Ann Chem 1953;579:123–132.

647. Génisson Y, Marazano C, Mehmandoust M, Gnecco D, Das BC. Zincke reaction with chiral primary amines: a practical entry to pyridinium salts of interest in asymmetric-synthesis. Synlett 1992; 431–434.

648. Rivier J, Vale W, Monahan M, Ling N, Burgus R. Synthetic thyrotropin-releasing factor analogs. 3. Effect of replacement or modification of histidine residue on biological activity. J Med Chem 1972;15:479–482.

649. Wallace BM, Lasker JS. Stand and deliver: Getting peptide drugs into the body. Science 1993;260:912–913.

650. Bodor N, El-Koussi A, Kano M, Nakamuro T. Improved delivery through biological membranes. 26. Design, synthesis, and pharmacological activity of a novel chemical delivery system for β-adrenergic blocking agents. J Med Chem 1988;31:100–106.

651. El-Koussi A, Bodor N. Formation of propanolol in the iris-ciliary body from its propranolol ketoxime precursor: a potential antiglaucoma drug. Int J Pharm 1989;53:189–194.

652. Bodor N. Retrometabolic drug design concepts in ophthalmic target-specific drug delivery. Adv Drug Deliv Rev 1995;16:21–38.

653. Prokai L, Wu W-M, Somogyi G, Bodor N. Ocular delivery of the β-adrenergic antagonist alprenolol by sequential bioactivation of its methoxime analog. J Med Chem 1995;38: 2018–2020.

654. Bodor N, Farag HH, Somogyi G, Wu W-M, Barros MDC, Prokai L. Ocular-specific delivery of timolol by sequential bioactivation of its

oxime and methoxime analogs. J Ocul Pharmacol 1997;13:389–403.

655. Farag HH, Wu W-M, Barros MDC, Somogyi G, Prokai L, Bodor N. Ocular-specific chemical delivery system of betaxolol for safe local treatment of glaucoma. Drug Des Discov 1997;15:117–130.

656. Bodor N, Visor G. Formation of adrenaline in the iris-ciliary body from adrenalone diesters. Exp Eye Res 1984;38:621–626.

657. Reddy IK, Bodor N. *In vitro* evaluation of a controlled-release site-specific diisovaleryl tert-butalone chemical delivery system for the eye. J Pharm Sci 1994;83:450–453.

658. Reddy IK, Vaithiyalingam SR, Khan MA, Bodor NS. Design, *in vitro* stability, and ocular hypotensive activity of *t*-butalone chemical delivery systems. J Pharm Sci 2001;90: 1026–1033.

659. Tatsumi K, Ishigai M. Oxime-metabolizing activity of liver aldehyde oxidase. Arch Biochem Biophys 1987;253:413–418.

660. Parmar D, Burka LT. Metabolism and disposition of cyclohexanone oxime in male F-344 rats. Drug Metab Dispos 1991;19:1101–1107.

661. Kumpulainen H, Mähönen N, Laitinen ML, Jaurakkajärvi M, Raunio H, Juvonen RO, Vepsäläinen J, Järvinen T, Rautio J. Evaluation of hydroxyimine as cytochrome P450-selective prodrug structure. J Med Chem 2006;49:1207–1211.

662. Nathanson JA. Stereospecificity of beta adrenergic antagonists: R-enantiomers show increased selectivity for beta-2 receptors in ciliary process. J Pharmacol Exp Ther 1988;245: 94–101.

663. Mehvar R, Brocks DR. Stereospecific pharmacokinetics and pharmacodynamics of beta-adrenergic blockers in humans. J Pharm Pharm Sci 2001;4:185–200.

664. Sharif NA, Xu SX, Crider JY, McLaughlin M, Davis TL. Levobetaxolol (Betaxon) and other beta-adrenergic antagonists: preclinical pharmacology, IOP-lowering activity and sites of action in human eyes. J Ocul Pharmacol Ther 2001;17:305–317.

665. Nandel FS, Dhaliwal RK, Singh B. Modeling, design, chiral aspects and role of *para*-substituents in aryloxypropranolamine based beta-blockers. Indian J Biochem Biophys 1999;36: 29–35.

666. Bodor N, El-Koussi A. Improved delivery through biological membranes. LVI. Pharmacological evaluation of alprenoxime: a new potential antiglaucoma agent. Pharm Res 1991;8:1389–1395.

667. Simay A, Prokai L, Bodor N. Oxidation of aryloxy-β-amino alcohols with activated dimethylsulfoxide. A novel C–N oxidation facilitated by neighboring group effect. Tetrahedron 1989;45:4091–4102.

668. Simay A, Bodor N. Site- and stereospecific drug delivery to the eye. In: Sarel S, Mechoulam R, Agranat I, editors. Trends in Medicinal Chemistry '90, Oxford: Blackwell Scientific Publications; 1992. p 361–368.

669. Polgar P, Bodor N. Minimal cardiac electrophysiological activity of alprenoxime, a site-activated ocular β-blocker, in dogs. Life Sci 1995;56:1207–1213.

670. Pfitzner KE, Moffat JG. Sulfoxide-carbodiimide reactions. I. A facile oxidation of alcohols. J Am Chem Soc 1965;87:5661–5670.

671. Smith JR, Simon EJ. Selective protection of stereospecific enkephalin and opiate binding against inactivation by *N*-ethylmaleimide: evidence for two classes of opiate receptors. Proc Natl Acad Sci USA 1980;77:281–284.

672. Joseph DB, Bidlak JM. The κ-opioid receptor expressed on the mouse lymphoma cell line R1.1 contains a sulfhydryl group at the binding site. Eur J Pharmacol 1996;267:1–6.

673. Shahrestanifar M, Wang WW, Howells RD. Studies on inhibition of μ and δ opioid receptor binding by dithiothreitol and *N*-ethylmaleimide. J Biol Chem 1996;271:5505–5512.

674. Kanematsu K, Naito R, Shimohigashi Y, Ohno M, Ogasawara T, Kurono M, Yagi K. Design and synthesis of an opiod receptor probe: mode of binding of S-activated (−)-6β-sulfhydryldihydromorphine with the SH group in the μ-opioid receptor. Chem Pharm Bull 1990;38:1438–1440.

675. Saah M, Wu W-M, Eberst K, Marvanyos E, Bodor N. Design, synthesis, and pharmacokinetic evaluation of a chemical delivery system for drug targeting to lung tissue. J Pharm Sci 1996;85:496–504.

NEW STRATEGIES FOR NATURAL PRODUCTS LEAD GENERATION

Guy T. Carter
Valerie S. Bernan
Frank E. Koehn
Chemical Sciences, Wyeth Research, Pearl River, NY

1. INTRODUCTION

Searching for natural products with significant biological activities has been a fruitful approach to the development of new medicines and other useful products for many decades. The historical aspects of this process have been exceptionally well covered in several treatises and the reader is referred to the earlier chapters in this series, particularly that of Buss and Waigh [1]. The purpose of the present chapter is to critically review newer aspects of natural product screening prospectively with a goal of illuminating the path ahead. Our focus is directed toward sources and processes that will provide the initial link to new chemistry and/or new biology, through the investigation of secondary metabolites. We have emphasized microbial sources in this chapter, primarily because of the innovation derived from research in this area.

2. SOURCES OF NATURAL PRODUCTS FOR SCREENING

Although highly significant natural products derived from plants and other higher organisms have been discovered, studied and even commercialized during the past hundred years, our focus in this chapter is on the microbial world. Cultivable microorganisms have been recognized as a critical source of active therapeutic compounds with original chemical scaffolds. The usual definition of natural products emphasizes that they are chemicals obtained from living organisms. These compounds are mainly divided into two classes: primary metabolites that are globally shared by most living species and secondary metabolites that are of lower molecular weight (<3000 amu), and have not been associated with primary metabolic processes. Secondary metabolites encompass extremely diversified chemical structures that have shown efficacy in an array of therapeutic applications, ranging from antibacterial, antifungal, and antiprotozoan effects to antitumor, immunosuppressive, and antiatherosclerotic applications.

Owing to their capacity to adapt and colonize various ecosystems, microorganisms represent the most important part of the earth's biodiversity. Their adaptive ability, even in the most inhospitable environments such as the deep-sea hydrothermal vents, the Dead Sea, the Antarctic, alkaline lakes, and volcanic soils, also offers the potential for translation into an impressive brand of new chemical entities.

Biodiversity is a concept usually referring to the number of species in a given ecosystem or geographic area. The most straightforward definition is "variation of life at all levels of biological organization." A second definition holds that biodiversity is a measure of the relative diversity among organisms present in different ecosystems. "Diversity" in this definition includes diversity within a species and among species, and comparative diversity among ecosystems. A third definition that is often used by ecologists is the "totality of genes, species, and ecosystems of a region." An advantage of this definition is that it describes most circumstances and presents a unified view of the traditional three levels at which biodiversity has been identified. If the gene is the fundamental unit of biodiversity, according to E.O. Wilson [2] real biodiversity is genetic diversity of organisms. This includes processes such as mutations, gene exchanges, and genome dynamics that occur at the DNA level and generate evolution. The study of natural products is multidisciplinary and encompasses the three levels of biodiversity. However, for this chapter, biodiversity will be defined as concerning those biological species that are studied specifically for their value as producers of small organic compounds, the secondary metabolites.

The number of bacterial strains that have been isolated is negligible compared to the estimated number of existing bacterial species [3] that is in the range of 10^7–10^9. These estimates have been derived through molecular ecology studies, which have demonstrated that more than 99% of environmental bacteria are not cultivated with classically used culture methods in microbiology. Amman [4] has estimated that less than 0.3% of soil bacteria can be cultivated in well-known media, and this fraction decreases to 0.00001% for water-associated microorganisms. As for fungi, approximately 70,000 fungal species have been taxonomically described, but the number of species existing in the environment is estimated to be 1.5×10^6 [5]. It is clear that the potential for the discovery of new microorganisms is immense.

The discovery of penicillin from a *Penicillium* sp. by Alexander Fleming in 1928 introduced microbial sources for the discovery of biologically and therapeutically relevant natural products. This precedent continued during the "golden age" of the discovery of microbial natural products in the 1940s and 1950s, when many antibiotics were discovered (see Fig. 1). The incredible structural diversity and potent biological activity derived from microbial sources remained powerful forces driving the pharmaceutical industry during the ensuing 30 years. Among the estimated 250,000 bioactive natural products known today, more than 22,000 are derived from microbial sources [6]. Since essentially all identified microbial bioactive compounds have been isolated from cultivable strains, this cultivable fraction represents only a small fraction of the total number of existing microbes, as noted above. Therefore, an enormous reservoir of untapped active metabolites remains to be investigated and we have apparently only scraped the surface of this chemical universe. While plants and marine macroorganisms are prolific sources of bioactive metabolites, with enormous chemical diversity, they will not be highlighted in this chapter. Emphasis will be placed on microbial sources, mainly actinomycetes and fungi,

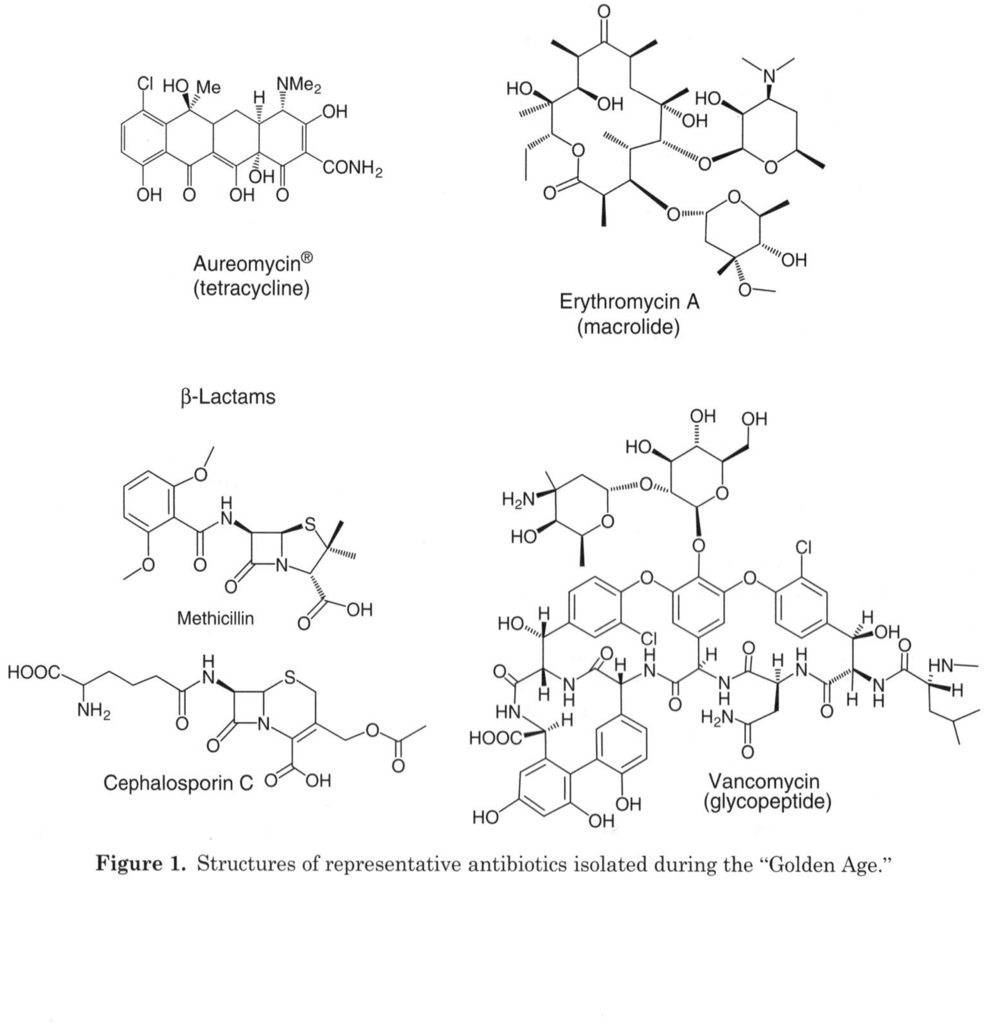

Figure 1. Structures of representative antibiotics isolated during the "Golden Age."

which provide a renewable source of material for the isolation of bioactive compounds.

2.1. Microbial Sources

One of the fundamental questions one has to address is what relevant microorganisms are the most productive and which phyla should be focused on to improve the chances of success in finding new metabolites. The distribution of secondary metabolite pathways does not appear to be uniform in all of the 50 prokaryotic phyla described presently, but appears to be concentrated in five groups; Actinobacteria, Firmicutes, Bacteriodetes, Cyanobacteria, and Proteobacteria. Interestingly, those five phyla contain more than 90% of all cultivated and published species [7].

Filamentous fungi and actinomycetes have been traditionally the focus of many industrial and academic groups for the past 60 years. The number of metabolites with biological activity reported from these two groups is astonishing, and accounts for approximately 60% of all the bioactive microbial metabolites reported to date. It is not unreasonable therefore to argue that since these microorganisms have been so extensively studied, the chances of discovering novel metabolites would be rare and not worth the time and effort. However, new species and even major taxa of fungi and actinomycetes are being found, opening new windows of opportunity and providing evidence that there is much to be discovered. Moreover, these species are producing new structural classes with interesting biological activities. In fact, mathematical models [8] suggest that the number of antibiotics still to be discovered from actinomycetes could be above 10^5, suggesting that the number of compounds characterized to date would represent less than 5% of the total. The concept of this untapped potential is supported by the results from the analysis of the first complete genome sequences from the actinomycetes (see genomic section). For example, the analysis of the genomic sequences of two *Streptomyces* sp., *S. coelicolor* [9] and *S. avermitilis* [10] revealed the presence of more than 20 gene clusters encoding for the synthesis of polyketides (PKS) or nonribosomal peptides (NRPS), the two most important biosynthetic classes of microbial secondary products. This number is far greater than the number of compounds isolated from these strains. This suggests that the potential to exploit the metabolic machinery of these organisms is vast.

Interestingly, secondary metabolic pathways are not evenly distributed across these two large microbial groups. Approximately 70% of all the metabolites described from Actinobacteria are produced by species of one single genus, *Streptomyces*, the rest being distributed across other families, mainly Micromonospraceae, Pseudonocardiaceae, and Streptosporangineae. Likewise, there are certain fungal genera that are more productive than others such as *Penicillium*, *Aspergillus*, *Trichoderma*, and *Fusarium*.

Recently, other microbial taxa that are known to produce bioactive secondary metabolites have been explored and include the cyanobacteria and myxobacteria. These two taxa, similar to the actinomycetes, harbor large genomes comprised of significant portions of PKS and NRPS genes. In fact, the cyanobaterium, *Nostoc punctiforme* contains up to 15 classes of these genes [11]. More than 600 bioactive metabolites have been described from cyanobacteria and some compounds, such as the dolastatins (see Fig. 2), are in clinical trial [12]. *Sorangium*, a myxobacterial genus, contains at least 10 gene clusters for PKS and NRPS that encode the biosynthesis of secondary metabolites. It is not surprising that more than 80 basic scaffolds and 450 structural variants have been described from various strains. Significantly, some myxobacterial compounds are in clinical trails, including derivatives of the compound epothilone [13], isolated from the fermentations of *Sorangium cellulsum*.

2.1.1. Terrestrial and Marine Habitats Historically, soil-derived actinomycetes and fungi dominated early natural products screening. In fact, all microbially derived drugs are produced by soil microorganisms. Many useful products were derived from this intensively exploited ecological niche, including the clinically important antibiotics tetracycline, erythromycin, vancomycin, and β-lactams (see Fig. 1). However, it is well known that there are many other ecological niches and

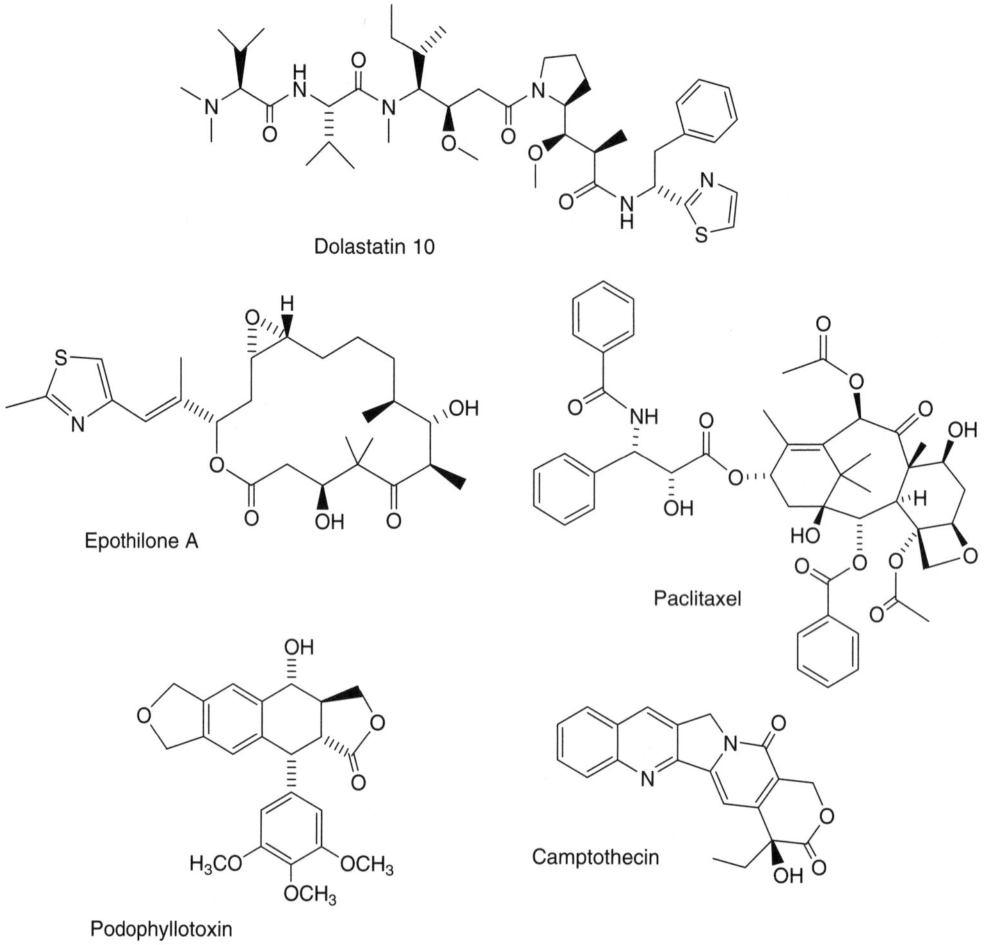

Dolastatin 10

Epothilone A

Paclitaxel

Podophyllotoxin

Camptothecin

Figure 2. Natural product structures for compounds covered in Section 2.1.

types of substrates harboring abundant and diverse microbial communities. Bacterial diversity has not been systematically explored and major environmental habitats have yet to be sampled for natural product discovery. A few of these habitats have been explored since the late 1980s, and have been a source of productive microbes. A good example is the plant-associated or endophytic actinomycetes and fungi. Twenty years ago, endophytic fungi and actinomycetes first received attention as a mechanism to protect their hosts against insect–pest pathogens. More recently, it became clear that these cohabitating microbes could confer other important characteristics to plants, such as resistance to stress conditions, alteration of physiological properties, and the production of phytohormones and other bioactive compounds. Endophytes form symptomless infections for all or part of their life cycle, within their host plant. They are usually host-specific and may even be tissue specific. Endophytes have been isolated from the roots, stems, and leaves of a large diversity of plants, including grasses, orchids, shrubs, and trees. They represent a significant contribution to diversity, since it has been estimated that as many as six unique species can inhabit a single species of vascular plant in the temperate environment. It has been postulated that endophytes may be able to synthesize the same metabolites that are produced by the host plant. The classic example is paclitaxel (see Fig. 2). Paclitaxel was originally identified from the plant *Taxus brevifolia*, but has also been reported to be produced by the endophytes *Taxomyces* and *Pestalotiopsis* sp., isolated from yew species [14]. Other examples include the anticancer compounds podophyllotoxin [15] and camptothecin [16] (see Fig. 2).

The marine environment is a major habitat that has not been efficiently explored for microbial diversity. The oceans of the world cover approximately 70% of the earth's surface, which represents better than 95% of the biosphere. This enormous area encompasses many unusual and diverse habitats that have created niches for the evolution of diverse life forms. Indeed, analysis using a variety of molecular biology tools has demonstrated that the marine environment contains a large number of rare or novel species that have not been investigated for the production of biologically active secondary metabolites. The advent of marine invertebrate sources, such as molluscs, tunicates, bryozoans, and sponges, for the discovery of natural products has produced a plethora of chemically unique metabolites. Some of the most interesting of these bioactive compounds harbor highly cytotoxic and anticancer properties. Owing to this type of bioactivity, the US National Cancer Institute has funded several clinical trials for bryostatin, ecteinascidin, aplidine, discodermolide, and several others (see Figs 3 and 5). One major issue, however, is the reproducible availability of these compounds. Most cannot be supplied economically by isolation from nature, aquaculture, or by chemical synthesis in sufficient amounts for continued use in therapy. In contrast, marine microorganisms offer a viable option of providing an adequate, reliable, and renewable supply.

The study of the metabolites produced by marine actinomycetes was pioneered by researchers at the Institute of Microbial Chemistry in Tokyo in the early 1970s. One of the first compounds described by Okami [17] was a benzanthraquinone antibiotic isolated from the actinomycete *Chainia purpurgensa* SS-228 collected in a mud sample from Sagami Bay, Japan. Most interestingly, this bacterium only produced the bioactive compound when it was cultivated in the presence of "Kobu Cha" (the brown seaweed *Laminaria*) with the addition of 3% NaCl. This observation demonstrated that marine actinomycetes have nutritional requirements corresponding to nutrients in their natural habititats. This study was also one of the first to report that bacteria from the order Actinomycetales could be isolated from the marine environment. Actinomycetes were originally thought to have entered the marine environment via rivers or runoff or to have existed as spores of terrestrial species. However, work by Moran [18] using a 16S rRNA genus-specific probe, demonstrated that

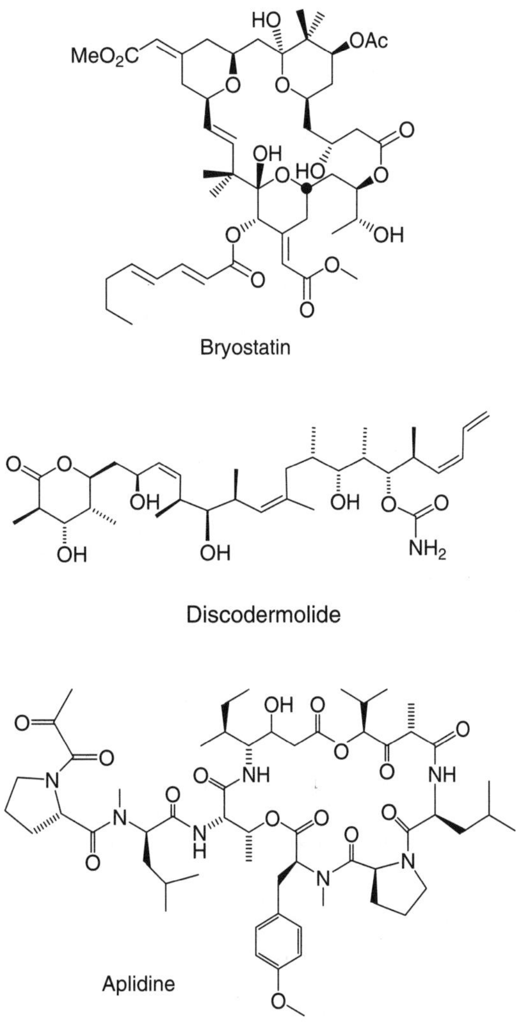

Bryostatin

Discodermolide

Aplidine

Figure 3. Selected compounds derived from marine invertebrates that have been evaluated in the clinic.

Streptomyces occurred as indigenous populations, and that populations increased in relative abundance in response to the availability of certain nutrients. The first formal taxonomic description of a marine actinomycete was *Rhodococcus marinonascens* in 1984, by Helmke and Weyland [19]. Since then, at least three additional marine actinomycete genera have been taxonomically described. These taxa include the *Salinispora* [20], *Serinicoccus* [21], and *Salinibacterium* [22]. Additional new genera, that have yet to be formally described, include "Marinispora" and "Solwaraspora" whose culture extracts have shown high levels of biological activity. Various *Salinispora* sp. produce novel metabolites with antibiotic and antitumor properties, such as the salinosporamids, sporolides, the arenicolides, the saliniketals, lomaiviticins, and others [23] (see Fig 4). In fact, DNA sequencing of the genome of *Salinispora tropica* revealed that approximately 10% of the genome encoded for secondary metabolites [24]. Intriguingly, secondary metabolite production in the *Salinispora* appears to be species specific. This contradicts the archetype that actinomycete secondary metabolite production is strain specific and may indicate that chemotaxonomy may exist for actinomycetes, as in the fungal world. This also suggests that the isolation of identical strains from geographically diverse locations will be less profitable for natural product discovery, but more studies are required to confirm this conclusion.

Numerous examples exist in which structurally related or identical compounds have been reported from taxonomically distinct marine invertebrates or from animals collected in certain geographical locations. The microbial-like structures of compounds isolated from marine sponges and ascidians lend support to the hypothesis that many marine invertebrate natural products may be of microbial origin (see Fig. 5). Furthermore, approximately two-thirds of the marine animal-derived antitumor compounds that are currently in preclinical or clinical trial can be structurally classified as complex polykektides or nonribosomal peptides [12]. One of the most striking examples is the isolation of the endiyne antitumor antibiotic namenamicin from the didemnid ascidian *Polysycraton lithostrotum* [25]. Members of this class of compounds were first isolated in the 1980s from terrestrial actinomycetes in the genera *Micromonospora* and *Actinomadura*, which produce calicheamicin and esperimicin, respectively. Additional examples of microbial-like chemistry include the isoqinoline alkaloids such as ecteinascidin and renieramycin. These compounds have been isolated from the ascidian *Ecteinascidia tubinata* and the sponge, *Reiniera* sp., with the terrestrial counterpart saframycin B, isolated from *Streptomyces lavendulae* [26].

Marine sessile invertebrates are considered a particularly rich source of novel metabolites and associated microorganisms. Sponges have provided more natural products with unprecedented molecular structure and bioactivities than any other phylum of marine invertebrates. Remarkably, sponges may have up to 40% of their cellular volume occupied by associated bacteria. The surfaces, tissues, and internal spaces of such invertebrates provide a variety of marine microhabitats that are conducive to the growth of diverse microorganisms. However, this abundance of microorgansisms poses the challenge of distinguishing between specific sponge symbionts and transient seawater microorganisms. Despite the numerous associations documented between sponges and microorganisms, most invertebrate-associated microbes remain uncultured. There is almost no taxonomic overlap between bacteria detected by culture-independent molecular studies and those cultivated from sponges. Cultivation of these organisms remains a major challenge, and new methods of isolation that include a consortium of microorganisms in combination with growth conditions that mimic the *in situ* environment must be developed. In the 1970s and 1980s, descriptions of invertebrate-associated microorganisms relied on microscopy and cell sorting to demonstrate their presence. Recently, molecular phylogenic surveys, which include mass spectral imaging, gene probing techniques such as catalyzed reporter deposition (CARD)-FISH, and *in situ* hybridization have proven the existence of these uncultivated microorganisms [27]. One early example of the existence of a true symbiont is the discovery and preliminary

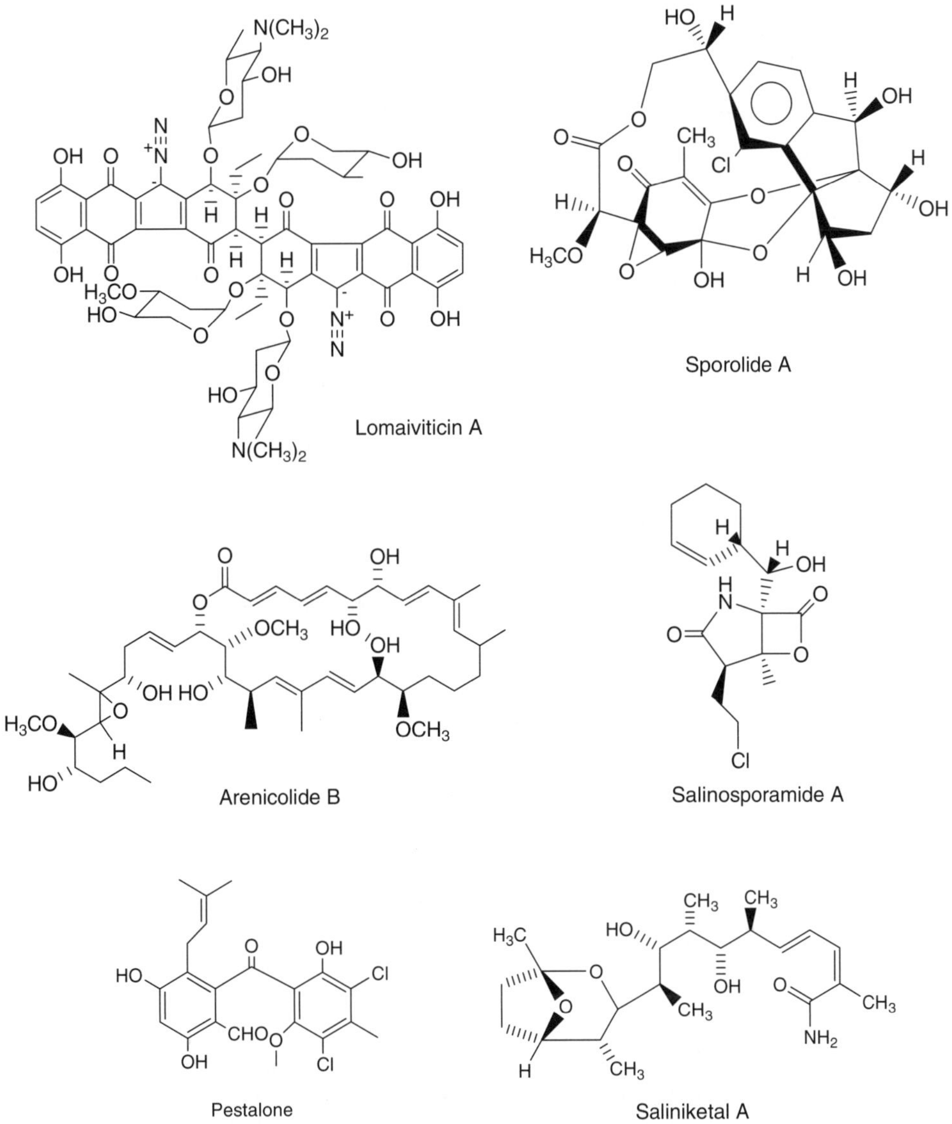

Lomaiviticin A

Sporolide A

Arenicolide B

Salinosporamide A

Pestalone

Saliniketal A

Figure 4. Examples of compounds isolated from new species of marine bacteria.

characterization of a marine archeon that inhabits the tissues of a temperate sponge. The microorganism, *Cenarchaeum symbiosum* (phylum: Crenarchaeota) inhabits a single species of *Axinella* [28]. *In situ* hybridization studies revealed the high abundance of a single, crenarchaeal phylotype in every specimen of actively dividing laboratory-maintained sponges over a long period of time, strongly suggesting a true symbiont was maintained in the invertebrate.

A recent advance is found in the work carried out by Schmidt and coworkers [29] who demonstrated that rare invertebrate-derived natural products can be produced by isolating and expressing genes from a cyanobacterial symbiont. Several cytotoxic peptides have been isolated from the ascidian,

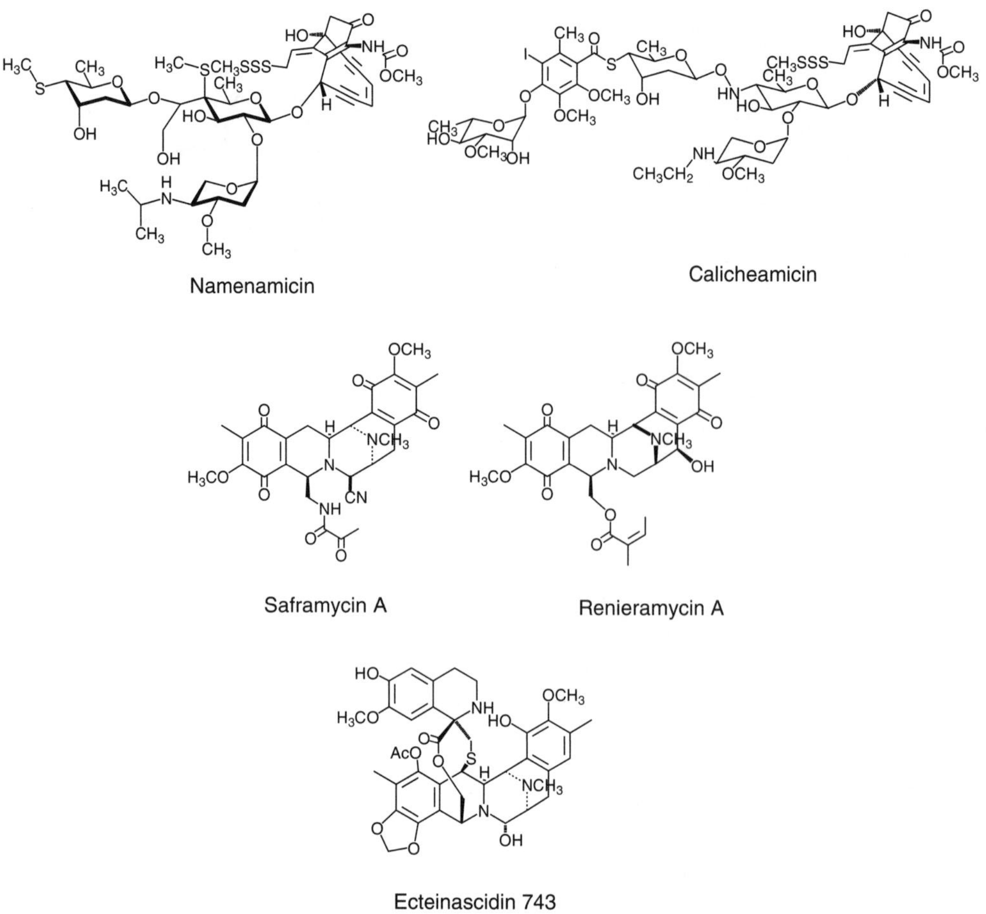

Namenamicin

Calicheamicin

Saframycin A

Renieramycin A

Ecteinascidin 743

Figure 5. Ecteinascidin and namenamicin and their structurally related counterparts.

Lissoclinum patella, which harbors the unculturable cyanobacterial symbionts of the genus *Prochloron*. Schmidt as well as Long and coworkers [30] independently demonstrated that the true source of these peptides is the unculturable cyanobacteria. In Schmidt's work, the biosynthetic genes were discovered *in silico* in the sequenced genome of *Prochloron didemni* and then functionally expressed in *Escherichia coli*. Long, working retrospectively, examined random clones of an *E. coli* library prepared from the symbiont's DNA for the presence of the biosynthetic pathway. This was the first account of expressing symbiont genes responsible for the biosynthesis of the natural products originally derived from a marine invertebrate.

Another example of utilizing unculturable symbionts for cloning the biosynthetic pathway was carried out by Haygood and coworkers. Using the bryozoan *Bugula neritina*, Haygood [31] investigated whether the biosynthetic source of the potent bryostatins is a marine microorganism. Bryostatins are an important family of cytotoxic PKC-activating macrolides that are currently in clinical trials [12]. An important advantage of this bryozoan is that the microbial community of *B. neritina* consists of a single bacterial symbiont located in the larvae and just a few other bacteria, in contrast to the complex microbial populations of most sponges. 16S-rDNA analysis and *in situ* hybridization revealed that the symbiont is a new genus of γ-proteobacterium and has been named "*Candidatus* Endobugula sertula." The symbiont is not viable in culture, but since bryostatin is a complex polykektide, presumably a PKS pathway would be responsible for its synthesis. By using degenerate primers, they were able to clone a component of the polyketide synthase, the ketosynthase (KS, see Fig. 11) domain and demonstrate that this was derived from the larvae. In the larvae, this gene was localized by *in situ* hybridization to "*Candidatus* Endobugula Sertula." Haygood [32] went on to isolate the *bryA* gene of the corresponding PKS cluster using metagenomic DNA of *B. neritina*. They proposed that *bryA* is involved in the assembly of the first four polykektide units of bryostatin. Interestingly, *bryA* is another example of a PKS gene belonging to the trans-AT type. Rescreening the metagenomic libraries revealed no additional clusters of the correct size that would encode bryostatin, suggesting the correct gene has been identified.

2.2. Techniques to Harvest New Biodiversity

Attempts have been made to mine the secondary metabolite reservoir present in as yet uncultured bacterial species by several approaches. One is by improving *in vitro* culture conditions through development of new artificial media or by developing new techniques such as colony encapsulation in microdroplets. Other methods have included the use of oligotrophic isolation media and the addition of cell signaling agents, such as cyclic AMP and homoserine lactones, resulting in additional microbial isolates [33]. Lewis and coworkers [34] recently described a diffusion chamber-based approach as an alternative method that allows microorganisms to grow in their natural environment, providing them access to natural substrates. In this case, an inoculum is sandwiched between two semipermeable membranes of the chamber, which is then returned to the source environment. The chamber allows for a free exchange of chemicals with the external milieu by diffusion while restricting the movement of cells. This strategy resulted in the isolation of species rarely cultivated by conventional methods. Since bioactive compounds may be produced in nature to provide a competitive advantage, it is possible that the pathways responsible for the biosynthesis are regulated by factors elicited by another associated microbe. This was demonstrated by Fenical and Clardy [35] when they reported the isolation of a new chlorinated benzophenone antibiotic, pestalone (see Fig. 4) from a mixed fermentation extract. Pestalone was only produced when a marine fungus, an unidentified species of *Pestalotia*, was cofermented with an unidentified, Gram-negative unicellular bacterium. Another strategy encompasses the exhaustive analysis of sequences of bacterial genomes to gradually improve our knowledge of the physiology of bacteria, and help define culture media better suited to *in vitro* culture [36]. Despite such advances, these methods are slow and cannot resolve all of the issues associated with microbial cultures, especially those involved in mutual growth regulation and feedback between microorganisms. The noncultivables largely remain recalcitrant to these methods. Therefore, in the absence of any cultivation process, molecular genetic techniques have begun to exploit the hidden biosynthetic pathways. This enormous reservoir of noncultivable microorganisms had remained untapped until the recent development and combination of molecular ecology and molecular biology, which gave rise to a new approach called metagenomics [37]. Metagenomics was developed in the early 1990s and is a result of pluridisciplinary collaboration between molecular biology, recombinant expression, microbiology, and high-throughput screening. The metagenomic approach is still in its infancy and is technically challenging but does offer new opportunities in discovering the hidden potential of the unculturable.

2.2.1. Metagenomics The discovery of novel bioactive compounds through metagenomic cloning and expression is an alternative for exploiting uncultivable microorganisms for natural product discovery. Unique structures with various bioactivites have been elucidated for terragines [38], violacein [39], indirubin [40], and turbomycins [41] (see Fig. 6). This method utilizes the extraction of total DNA from the environmental sample, which is digested, ligated and cloned in a surrogate or heterologous host. The resulting library of

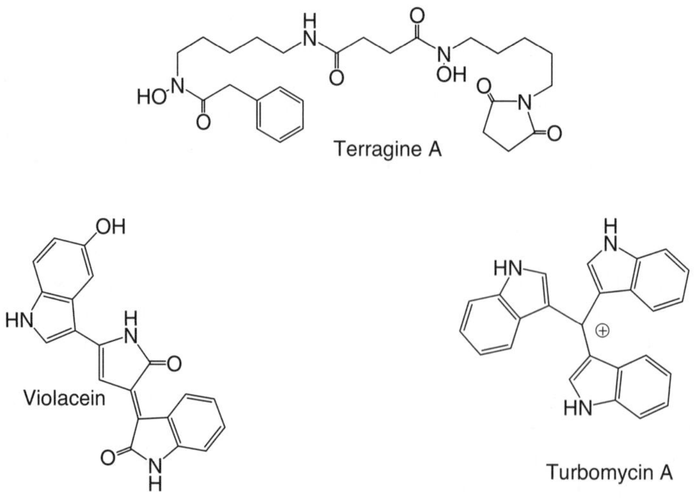

Figure 6. Examples of compounds isolated from metagenomic approaches.

DNA is called a metagenome, as it represents the overall genome of the sample. This approach relies on the findings that the genes involved in the biosynthesis of secondary metabolites are usually contiguously clustered in the microbial genome, and are capable of being expressed in a heterolgous host. The two most common heterologous hosts utilized are *E. coli* and *Streptomyces lividans*. One of the earliest reported successes of this technique was the discovery of novel *N*-hydroxyamides called terragine A and related compounds from a soil metagenomic library expressed in *S. lividans* [38]. *N*-Acyl-tyrosine derivatives with biological activity were also found in a soil metagenomic library by Brady and Clardy [39], while indirubin and related compounds were isolated by scientists at Aventis [40]. Other examples include the triarylcations, turbomycin A [41], and B and palmitoyl putrescine that were isolated from environmental DNA and found to exhibit broad-spectrum antibiotic activity against Gram-negative and Gram-positive microorganisms. However, most of the interest has been focused on the discovery of new polyketide biosynthetic pathways, showing promising biological activities such as antitumor effects.

A related approach using slow-growing fungi involves isolating cosmid-sized DNA from individual fungal colonies and cloning them into *Aspergillus nidulas* [42]. The transgenic strains are then fermented and analyzed for secondary metabolites not produced by the host. Many of the slow-growing fungi are not amenable to cultivation in sufficient mass for the isolation of secondary metabolites, but it is possible to obtain sufficient biomass for DNA extraction from individual colonies or directly from fruiting bodies or lichens. This approach may be a viable option to exploit the metabolic potential of fungi resistant to current cultivation techniques.

3. NATURAL PRODUCTS LIBRARIES

3.1. Library Complexity: Mixtures to Single Components

The success of any screening campaign is dependent on the nature and quality of the screening library. Natural product libraries may be composed of crude or complete extracts (tens to hundreds of compounds per sample), semipurified mixtures (roughly 3–10 compounds per sample) or pure natural products. Crude extract libraries are generally made by liquid- or solid-phase extraction of the fermentation broth or plant material. They require only modest sample preparation but demand the most resources for the identification and isolation of the bioactive constituents. In crude natural products libraries, highly polar or highly lipophilic components of the crude extract may interfere with the functioning of the assay, causing false positives or false negatives. In addition, one or more rounds of chemical purification and biological assay may be necessary to identify and isolate the active component(s) from the extract. This requires the continued availability of assay resources to support the isolation and purification along with additional time to resolve the hit and furnish pure compound for further biological evaluation.

Prefractionated libraries can be an effective strategy to alleviate interferences encountered with crude libraries, and may also shorten the time needed to identify the active principle. There are many variations of this approach and each offers advantages of expediency or purity gained at the cost of up-front partial purification [43].

Samples produced by the prefractionation approach are simpler mixtures and the final resolution of active components requires fewer purification steps. Interferences are reduced due to the fact that extremely polar and extremely nonpolar components are separated from the bulk of the library samples [44]. Moreover, the relative concentration of minor components is increased over that in the crude, thereby enhancing the opportunity to uncover novel biologically active metabolites. The advantages of the prefractionation approach need to be balanced against the resource investment necessary to select, prepare, characterize, and maintain such a partially purified natural product library. Since it creates several samples from a single extract, prefractionation increases the size (and cost) of the library for a given number of extracts. Given the often-substantial costs associated with assay reagents, especially against high-value targets, it is essential that redundancy in screening libraries be minimized. This entails analytical characterization of the natural product library contents in the form of HPLC–MS or other techniques, to assure a minimum of redundancy and a maximum of chemical diversity.

Purified natural product libraries offer the advantage that the hit detection process is similar to that of synthetic single component libraries, and the robustness of the hit identification process depends primarily on the purity and chemical integrity of the library itself. Purified libraries offer considerable advantage in the detection of quality hits and in moving forward immediately, since the bioactive principle requires no isolation from a mixture. The downside to this approach is that even with modern methods of separation and automation, substantial resources are required to prepare pure natural product libraries and trace components will not be fully captured in pure form [45]. However, the increasing migration of the industry toward precision ultrahigh-throughput screening coupled with the overall accelerated pace of drug discovery have prompted a move toward highly processed or even pure natural product libraries [46]. Advances in automated separations and sample processing have made purified natural products libraries much more achievable [47]. Besides familiar separation methods such as reversed phase HPLC, newer applications of countercurrent and super critical fluid extraction and chromatography are finding use in natural products library construction [48,49]. In any library strategy, it is important to consider two points: (1) it is essential to accurately characterize the content of the library to minimize duplication and maximize the chemical diversity and integrity [50] and (2) maintaining the purity and integrity of a sizable pure compound library is an often-underestimated technical challenge [51]. This is especially true for pure natural products libraries that contain complex unidentified molecules.

3.2. Attributes of Natural Products

It is now well established that natural products have chemical diversity and biochemical specificity, which clearly differentiate them from synthetic and combinatorial compounds and that make them favorable lead structures for drug discovery [52]. Natural products contain greater chirality and steric complexity than either synthetic drugs or combinatorial libraries, and while synthetic molecules contain significantly higher numbers of nitrogen, sulfur, and halogen containing groups, natural products bear higher numbers of oxygen atoms [53,54]. Natural products can be differentiated from trade drugs or other synthetic molecular libraries on the basis of scaffold architecture and pharmacophoric properties [55]. They differ significantly from synthetic drugs and combinatorial libraries in the ratio of aromatic ring atoms to total heavy atoms (lower), number of solvated hydrogen bond donors and acceptors (higher), and by greater molecular rigidity [56], and they have a broader distribution of molecular properties such as molecular weight, log P and diversity of ring systems. Indeed, Grabowski [57] reported more than 100 scafffolds in a limited set of natural products that were not contained in any other compound set.

It is often said that natural products lack suitable drug-like properties or are structurally too complex for efficient postscreen hit-to-lead development. Detailed analysis, however, shows that high-quality natural product libraries compare quite favorably in terms of drug-like properties. Feher and Schmidt [54] examined representative combinatorial, synthetic, and natural product compound libraries on the basis of molecular diversity and "drug-likedness" properties such as molecular weight, number of chiral centers, molecular flexibility as measured by number of rotatable bonds and ring topology, distribution of heavy atoms, and Linpiski-type descriptors. In the overall picture for drug-like properties, Schneider and Lee [55] determined that the fraction of natural product structures with two or more "rule-of-five" violations is equal to that of trade drugs, approximately 10%. By applying suitable selection criteria, it is possible to construct diverse pure natural product libraries containing a high proportion of molecules with good drug-like properties [58]. Indeed, it is now deemed important to incorporate a certain degree of "natural product-likedness" into any screening library [59]. If properly characterized and constructed, a purified natural products library need not be large to be an effective screening source of bioactive molecules [60].

3.3. Characterization of Natural Products Libraries by Mass Spectrometry

In any type of natural products library crude, prefractionated or pure, it is highly advantageous to analytically characterize or profile the library content to maximize the chemical diversity, efficiency of screening and identification of hits. The first important criterion is whether the library contains metabolites of potential interest, and secondly to identify undesirable compounds. In the case of microbial products, whether or not a culture produces secondary metabolites is primarily determined by the genetic capacity of the organism coupled with response of the microbe to the fermentation conditions. Most microbial cultures do not ordinarily produce secondary metabolites. It is thus important to select organisms and culture conditions that result in secondary metabolite production as well as processing methods that retain secondary metabolites. Secondly, candidate extracts for library inclusion should show the presence of metabolites that enhance the existing diversity of the library, at levels sufficient to give a detectable assay response at micromolar activity.

HPLC can be used to compare and rank chromatographic profiles of natural products samples. Initial approaches were based on transforming the HPLC chromatogram of an extract into a retention time-peak area matrix, emphasizing the most abundant peaks. This approach can be difficult to apply to complex chromatograms with widely different amounts of metabolites, particularly if low-abundance peaks of uncommon compounds are desired. Higgs and coworkers [61] used mean natural log of the area under the entire HPLC–ELSD chromatogram to determine the presence of secondary metabolites, and then later found that direct infusion electrospray mass spectrometry of the mixture gave comparable utility for assessing actinomycete secondary metabolism. Automated pattern recognition routines such as factor analysis, principal components analysis and cluster analysis can be applied to chromatograms and mass spectra to evaluate culture samples. Misalignment of peaks due to drifts in compound retention between runs can be overcome either by the use of alignment algorithms [62] or by the inclusion of internal standards. A typical HPLC–ELSD–MS chromatogram of a natural products sample contains a wealth of information: chromatographic peak retention time,

UV–VIS absorbance spectrum, mass abundance, molecular mass, and molecular formula if sufficient mass accuracy is obtained. All of these data can be incorporated into a database and statistically leveraged to gain an accurate profile of the library sample. Wolf and Siems [63] have described an LCMS-based peak library of microbial and plant natural products based on electrospray mass spectra, referenced retention time, and normalized peak area. The library can be used for different data mining strategies: such as dereplication of previously isolated natural products, clustering/ranking of extracts, as a selection tool for the focused isolation of bioactive natural products, and to search for alternative sources of a target natural product. Natural products libraries can also be effectively profiled and the source organisms can be classified using a metabolomics-based analysis of HPLC–MS data [64]. Overall, advances in computer and instrument capabilities have made the automated analysis and comparison of chromatographic extract profiles a valuable tool in increasing the chemodiversity of collections, media development and even investigation of taxonomic relationships between organisms [65].

For the identification of common or undesired compounds, the molecular ion and fragment product masses contained in HPLC–MS spectral libraries facilitate rapid assessment of HTS hits and library components [66,67]. Along with this, multichannel HPLC–MS technology has increased the sample throughput and has made it possible to assay the chemical integrity and content of sizable natural product libraries, such as a 36,000 sample fractionated plant library [68]. Often intramurally developed databases are most effective, since they are tailored to meet the particular needs of the specific laboratory. This was demonstrated by Fredenhagen and coworkers [69] who used an ESI–HPLC–MS system interfaced with an electrospray (ESI) source and ion trap mass spectrometer to develop an extensive MS/MS dereplication spectral library.

Nonspecific compounds along with pan-inhibitors, can be problematic in the prioritization of hits from a crude or fractionated natural products library. Compounds of this type are found in all source types, be they microbial, plant, and/or marine samples. Such microbial products include nonspecific inhibitors such as funalenone, fonsecin, diepoxins, equisetin, niphymicin, and others (Fig. 7). Potent pan-kinase inhibitors such as staurosporine, radicicol, cercosporamide, and others are often produced in actinomycete and fungal fermentations [70] (Fig. 7). The presence of these compounds and their associated bioactivity in a library sample can often mask the bioactivity of other minor metabolites, and they are usually of diminished interest due to their known activity. On occasion, however, a common metabolite can serve as a valuable medicinal chemistry lead for new targets. For example, potent, selective inhibitors of estrogen receptor β, have been derived from the common plant metabolite genistein [71]. Finally, it must be recognized that media components employed in the fermentation of microbial cultures may contain small molecules, such as saponins or flavonoids that will interfere with certain bioassays and therefore the presence of such constituents should be noted.

3.4. Characterization of Natural Products Libraries by NMR

NMR spectroscopy, by virtue of its rich chemical information, versatility and high dynamic range is an unsurpassed tool in the investigation of natural product mixtures [72]. NMR spectra can be analyzed with pattern recognition techniques such as principal components analysis (PCA) to characterize natural products library samples. After originating in the field of metabolomics of biofluids [73], microcoil NMR methods are now being applied to plant natural product libraries [74,75]; and also to marine natural product extracts in micro-titer plates [76]. Automated microcoil NMR can effectively characterize natural products screening library components without the necessity of reisolation in larger quantities. At 600 MHz with microcoil probes, one- and two-dimensional spectra are readily measured on 50–200 μg of sample [77], compound amounts readily obtained in a typical library primary fractionation [78]. Due to the high resolution and dynamic range of NMR methods, individual components in complex mixtures of limited quantity can be

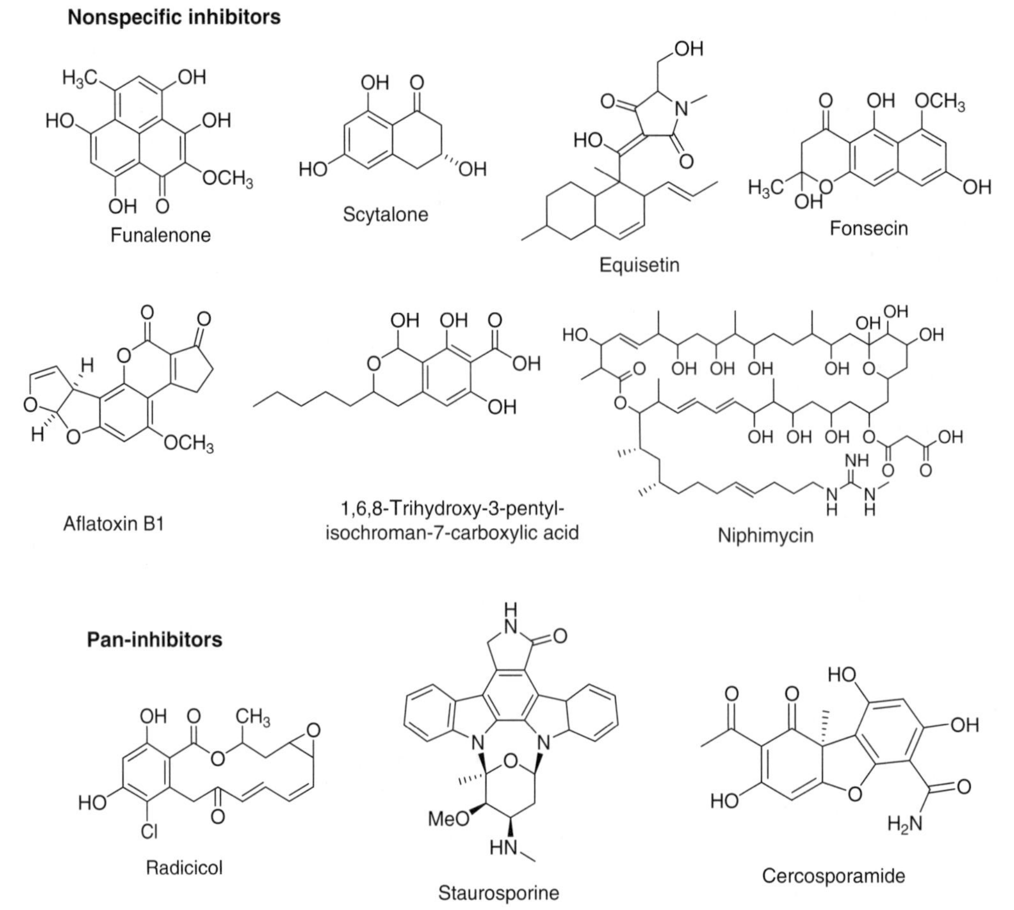

Nonspecific inhibitors

Funalenone

Scytalone

Equisetin

Fonsecin

Aflatoxin B1

1,6,8-Trihydroxy-3-pentyl-isochroman-7-carboxylic acid

Niphimycin

Pan-inhibitors

Radicicol

Staurosporine

Cercosporamide

Figure 7. Examples of common nonspecific and pan-inhibitors frequently identified in high-throughput screening of microbial natural products.

extensively characterized by NMR, even without separation—for example, crude spider venoms [79].

Recent advances in NMR, mass spectrometry, HPLC and SFC technology have made hyphenated LC–NMR, and LC–NMR–MS practical options for screening [80] and analysis of complex natural products mixtures, particularly those from plants [81,82]. For high-quality spectra of distinct extract components, chromatographic peaks are trapped in a sample handling device during the chromatographic run and later selectively delivered to the NMR probe for measurement. In-line peak storage units are used to reduce peak diffusion compared with earlier stopped-flow or on-flow methods, and recently include on-line automated solid-phase extraction (SPE) and peak trapping improve sensitivity by concentrating the sample in the NMR probe while reducing the need for background solvent suppression [83]. Chromatographic hyphenation techniques combined with cryogenic flow probes and capillary probe technologies have been effectively employed to natural products extracts and libraries for compound profiling, dereplication, and structural characterization [84–86]. Using LC–UV–SPE–NMR–MS it is now possible to perform automated analysis of natural products extracts where the individual components are present in 10–50 μg amounts [87,88].

4. SCREENING PLATFORMS FOR NATURAL PRODUCTS LIBRARIES

Many of our current therapeutic agents are derived from the screening of natural product extract libraries [89]. In early efforts the difficulty of isolating purified target proteins directly from tissues necessitated that natural product screening be performed using cellular *in vitro* or whole animal systems. The throughput of these assays was limited, and so it was necessary to screen crude or prefractionated natural products extracts to achieve sufficient diversity exposure. These screens were phenotypic in nature, and relatively effective when applied to crude or partially purified natural products libraries, particularly with respect to detecting antimicrobial compounds. Using these approaches, many important therapeutic agents currently in use were discovered, often without detailed knowledge of the specific target or mechanism of action.

4.1. High-Throughput Screening

In the past two decades, advances in biotechnology have made it possible to obtain purified protein targets in quantities sufficient to conduct large-scale biochemical assays to directly detect catalytic inhibition or target binding [90]. Advances in laboratory automation and miniaturization, and data acquisition have greatly enhanced assay throughput, the end result being a vastly expanded portfolio and screening capacity for protein targets. Developments such as these have placed demanding requirements on chemical screening library integrity and compound physical properties—requirements that pose a challenge to natural products extracts.

It is important to draw a distinction between three general types of assay platforms used to screen natural products. Targeted biochemical assays utilize purified target proteins in conjunction with a biochemical detection system to measure modulation of activity. An example here would be the case of protein kinases. HTS kinase assays can be based on fluorescence intensity, time resolved fluorescence or fluorescence polarization [91]. The second general platform is cell-based systems incorporating reporter assays engineered for specific targets [92]. Cell-based screening platforms have been developed for the major categories of drug targets—enzymes [93], GPCRs [94], and ion channels [95]. Cell-based screening led to the isolation of the neuromedin-U receptor agonist icariin (see Fig. 8) [96]. A fluorescent reporter assay engineered into a cell-based system was used to screen a natural products library of 3523 compounds for modulators of the multidrug transporter protein component ABCG-2 [97], identifying eupatin along with several other moderately potent compounds. Marion and coworkers [98] employed a whole cell fluorescence polarization based assay using human PI3 kinase α expressed in SF9 insect cells to screen a marine extract library and isolated liphagal, a selective inhibitor of PI3 kinase α. Structures for these compounds are shown in Fig. 8. It may turn out that cell-based systems will yield improved effectiveness with natural product libraries, given the cellular compatibility of natural products themselves.

Complications can arise in using assays of either type for screening natural products libraries. First, to detect competitive substrate inhibitors, it is best to screen at a compound concentration close to the K_m of the enzyme, which is typically in the micromolar range. This poses a challenge with natural products libraries where the concentrations of individual components in a sample are often undetermined, and may vary by three orders of magnitude or more. Consequently, at any given assay dose, the levels of trace components may not be high enough for detection while the highly abundant inactive components may exhibit inhibition via nonspecific binding, perturbation of the assay pH, or other physical properties. Alternatively, the amount of active compounds present may be far in excess, leading to dose independent nonspecific inhibition.

In fluorescence-based assays, difficulties are encountered when natural product samples contain compounds that either emit or absorb radiation at excitation or emission wavelengths of the fluorophore, or when insoluble components cause light scattering [99]. This liability is exhibited by synthetic screening libraries also, but the issue is exacerbated with natural products since the presence of

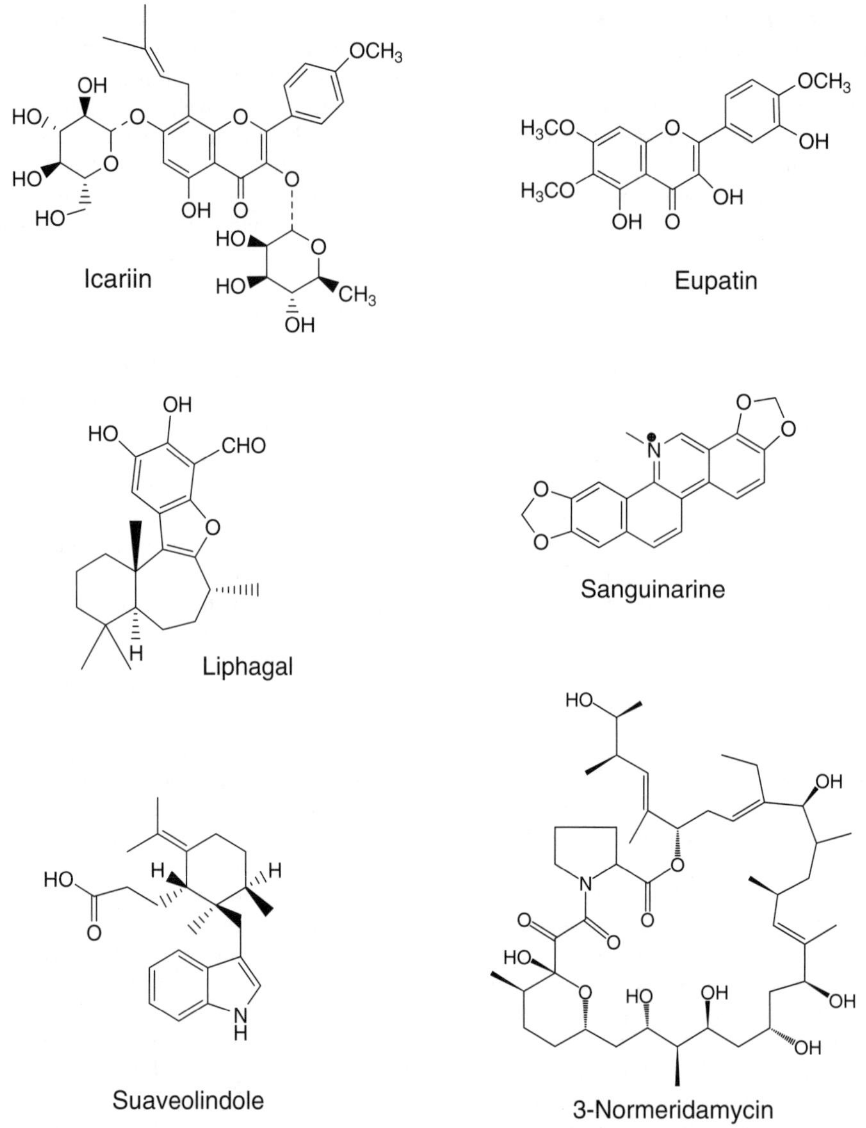

Icariin Eupatin

Liphagal Sanguinarine

Suaveolindole 3-Normeridamycin

Figure 8. Recently isolated bioactive natural products from high-throughput screening.

interfering compounds may not be fully characterized. Fluorescence interferences can be reduced in kinase, protease, and phosphatase HTS assays by increasing the fluorophore concentration in the assay, by using red-shifted wavelength dyes [100], or by the technique of lifetime discriminated polarization. Further, a very effective strategy that helps to alleviate these kinds of interferences as well as shorten the time needed to isolate the active principle is the use of purified or prefractionated screening samples from the original crude extract [44].

Obstacles in screening natural products libraries can be addressed through approaches such as quantitative high-throughput screening (QHTS), a titration-based approach that can be effective in screening libraries of varied composition. QHTS can furnish concentration–response curves in the primary screen, and relies on assay robustness, automation, and miniaturization in the form of a 1536 well format. The approach was demonstrated on a 60,000-member library screened against pyruvate kinase [101]. Further work in this laboratory has described the applicability of QHTS to identify three classes of nonsugar inhibitors of glucocerebrosidase, the lysosomal enzyme deficient in Gaucher disease [102] as a first stage in identifying chemical chaperones of the misfolded mutant glucocerebrosidase. Extension of the technique has been demonstrated to include inhibition assays of protein–protein interactions by Simeonov [103], who used an interleaved QHTS dual-flourophore assay to identify inhibitors of the BRCT-phosphopeptide interaction. A red-shifted dye was utilized to effectively reduce false positives and allow for the identification of reliable natural product and synthetic hits from a diverse 75,552-member library. The QHTS approach has been extended to include a whole-cell cAMP assay to identify small-molecule agonists the thyroid stimulating hormone (TSH, thyrotropin) receptor [104]. It has been used in conjunction with a cell-based imaging assay that measures derepression of a silenced green fluorescent protein (GFP) to identify epigenetic modulators of gene expression [105].

In the 1990s, HTS-based lead generation evolved away from the use of natural products libraries, and instead focused on synthetic compound libraries. The results of this approach have been disappointing, particularly with regard to anti-infectives. Payne et al [106] leveraged bacterial genomics information to design and execute 70 HTS antibacterial screens of which 67 were against molecular targets. The screening library was exclusively synthetic and yielded very little in terms of lead generation. The investigators attribute their lack of success in finding suitable anti-infective leads in significant part to the limited chemical diversity of their synthetic screening library. They proposed that screening natural products libraries with whole-cell phenotypic assays would enjoy greater success.

4.1.1. Analysis of Natural Products HTS Hits

When large natural products libraries are subjected to ultrahigh-throughput screening campaigns, new challenges emerge in the analysis and prioritization of hits. These screening campaigns typically consist of primary and confirmation phases followed by additional assays to identify "false positives" and determine inhibition IC_{50} values for statistically significant hits [107]. A primary screen hit rate of 1% for a natural products library consisting of 50,000–100,000 samples yields approximately 500–1000 hits. These primary hits are often reassayed to confirm the initial assay results. Depending on the statistical cutoff value and the noise characteristics of the primary assay, 30–75% of the primary hits will confirm in the confirmation assay, giving anywhere between 150 and 750 confirmed hits. If the natural products library is composed of multicomponent samples, then a follow-up process of bioassay-guided fractionation is necessary to isolate the active principle(s). If the natural products library is composed of pure compounds, then the confirmed hits can immediately be identified and prioritized according to hit-to-lead potential. The selection criteria are similar to those for synthetic hits including purity, integrity, molecular complexity, potency, intellectual property constraints, and other drug-like or lead-like properties such as molecular weight, solubility, membrane permeability, and stability [108]. In addition, availability of compound supply frequently becomes an important factor.

Targeted and cell-based high-throughput screening continues to yield natural products with novel structures, biological activity and drug potential. These include novel agents such as suaveolindole, discovered using natural products library screening and high-throughput microchemistry [109], sanguinarine a selective inhibitor of mitogen-activated protein kinase phophatase-1 of MKP-1 [110], and normeridamycin [111], a neuroprotective polyketide discovered by screening a natural products library for FKBP12 ligands (Fig. 8).

4.2. Phenotypic Screening "Back to the Future"

The challenges encountered with targeted screens coupled with advances in chemical biology have inspired a "back to the future" approach, which employs whole cell-based phenotypic screens coupled with high-content analysis and target identification as a means of more fully leveraging the chemical diversity and biological friendliness of natural products. This strategy embodies the concept that natural product molecules themselves are biased via natural selection to be predisposed for biological activity. Therefore, exposing the natural product library to a maximum number of cellular targets should be an effective means of screening, in principle an inverse approach to single target biochemical screening. Following detection of the desired cellular response by phenotypic characterization or high-content imaging analysis and identification of the bioactive component, the task remains to identify the target and mechanism responsible for the drug-induced phenotype.

Yeast systems, due to their eukaryotic nature and advanced molecular genetic amenability have become effective tools in this approach, particularly when applied to natural products discovery. One effective approach has been antisense-based screening strategies, which can be used to sensitize a microorganism and selectively detect inhibitors against a particular cellular target of interest. The use of antisense RNA technology for screening natural products libraries by cell-based assay using engineered strains that underexpress key targets has been effective in finding novel antibacterial leads. Investigators at Merck [112] screened a library of 250,000 microbial extracts using a whole-cell antisense differential sensitivity assay to discover platensimycin (see Fig. 9), one of the new classes of potent and selective inhibitors of β-ketoacyl- (acyl-carrier-protein synthase I/II) (FabF/B), a key enzyme in the biosynthetic pathway of fatty acids in bacteria. Additional screening using the same whole-cell approach led to the isolation of phomallenic acids A–C, [113], and platencin, a dual FabH and FabF inhibitor with broad-spectrum antibiotic activity [114] (Fig. 9). Recently, the Merck group [114] used a strain of *Staphylococcus aureus* that generates antisense RNA against SecA, part of the protein secretion machinery of bacteria, to screen an 115,000-member natural products library for inhibitors of SecA ATP-ase. From this effort the group isolated pannomycin [115], a novel *cis*-decalin metabolite as the active constituent from a culture of *Geomyces pannorum* (Fig. 9).

Phenotypic antifungal screening was used to isolate novel cell-wall active lipopeptides from the fungus *Pochonia bulbilosa* [116]. Mechanism of action detection of fermentation-based NP libraries by chemical genetic profiling is becoming an effective strategy for natural products-based drug discovery. For example, Lum [117] used genome-wide pool of tagged heterozygotes to assess the cellular effects of compounds in *Saccharomyces cerevisiae*. Specifically, lanosterol synthase in the sterol biosynthetic pathway was identified as a target of the antianginal drug molsidomin. Jiang [118] established a cell-based assay for high-throughput drug screening to identify several D1-like receptor agonists from natural products plant extracts. Ruocco and coworkers [119] reported a cell-based high-throughput assay to screen natural product libraries for noncytotoxic inhibitors of mitogen-activated AP-1 activity. The AP-1 actives were validated with inhibitors of kinases located upstream of AP-1 and with known natural product inhibitors of AP-1 (nordihydroguaiaretic acid and curcumin). The assay was able to identify a number of chemically diverse compounds with unknown mechanisms of action from natural products libraries. Diterpenoid (example in Fig. 9) and galactolipid inducers of apoptosis were isolated from the soft coral *Xenia elongate* and marine diatom *Phaeodactylum tricornutum* using a novel high-throughput cell-based screen for apoptosis-inducing potential anticancer compounds [120,121].

5. VISIBLE ON THE HORIZON

As the trends we have noted in the sections above continue to mature, it is possible to foresee how new technologies and scientific knowledge will combine to provide new directions in natural product screening. Some of the most intriguing prospects are noted below, where there is some preliminary research suggesting great potential reward.

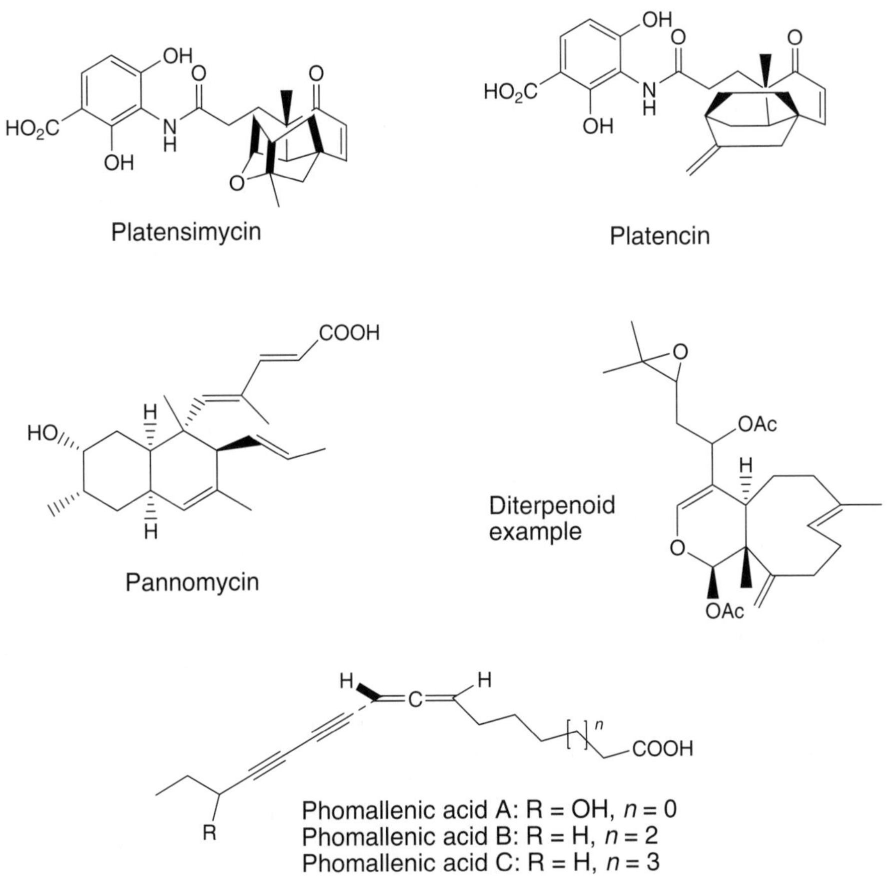

Figure 9. Recently isolated natural products from screening by whole cell phenotypic and yeast antisense assays.

5.1. Genome Mining: Prospecting in Bacterial Genomes

Enhancements to DNA sequencing technologies for the human genome project as well as subsequent developments, have expedited the compilation of whole genome sequences for bacteria. Sequencing resources have been primarily devoted to the analysis of pathogenic organisms in an effort to understand their biochemistry, and potentially to define new targets for antibiotic action. More recently, however, these tools have been applied to actinomycetes, the family of bacteria known for the prolific production of antibiotics and other useful secondary metabolites. Several examples have now been published of whole genome sequences of actinomycetes that produce commercially important products. One of the earliest of these was *S. avermitilis*, the organism that produces the antiparasitic agent avermectin [10]. This organism has been extensively investigated by the Merck Company, and it has been widely distributed for scientific experimentation, particularly in biosynthetic research. Thus, it was a stunning surprise when the genome sequence revealed pathways for at least 30 secondary metabolites, constituting 6.6% of the total genome. Only a handful of these products had ever been observed in cultures of *S. avermitilis*. Such cryptic pathways appear to be common, having been observed in several of the fully annotated genome sequences published to date [24,122]. The implication of these findings is that there is a wealth of previously undetected secondary metabolites yet to be discovered, and that these can be detected through genomic analysis.

The process of identifying the genes for these biosynthetic pathways is greatly facilitated by their tendency to be clustered on the bacterial chromosome. Various processes have been employed to localize these clustered genes. Certain signature sequences are sought for use as probes for homologous sequences within a new gene cluster. Thus, once a key segment of the putative pathway is recognized, such as polyketide synthetase or nonribosomal peptide synthetase genes, then the flanking regions can be mined for sequences potentially relating to tailoring or regulatory functions. Furthermore, owing to the colinearity principal for PKS and NRPS systems, the molecular structures of the encoded secondary metabolites can be predicted from the sequence data. An example of this approach for delineating new structures is shown in Fig. 10, for ECO-02301 [123], an antifungal agent derived from *S. aizunensis*.

Analysis of the sequence of this cluster revealed a long PKS string comprised of the 26 modules indicated in Fig. 10. Within each of these modules the functional domains were further defined as shown: acyl carrier protein (ACP) is the tethering unit to which the growing polyketide chain is transferred from module to module; ketosynthase (KS) is the polypeptide region which catalyzes the condensation between the growing chain and the newly presented acyl unit, which is in turn selected and delivered by the acyl transferase (AT) subunit. Reductive processing steps specified for each module are denoted as keto reductase (KR), dehydratase (DH), and enoyl reductase (ER) functions. These reductive steps determine the oxidation state of each newly introduced keto unit in the PKS, as indicated in Fig. 11. This arrangement of functions allows one to predict the structure of the growing carbon chain as it traverses the catalytic machinery. For ECO-02301 most of the ATs encode for malonyl CoA units, except for those in modules 12, 19, and 20 that specify methyl malonyl CoA, and hence result in methyl branching in the chain. Additional details revealed in the analysis of the sequences for the DHs in modules 6 and 11 and the KR in module 12 predicted that their active sites were substituted by amino acids that would render these nonfunctional. Therefore, in those instances the reductive step was not observed. The thioesterase function (TE) is responsible for cleaving the thioester tether that holds the nascent product on the enzymatic backbone. Identification of this TE function in module 26 effectively signifies the end of the chain and the partial structure of the resultant polyketide product is shown. By probing the sequences of genes flanking the PKS it was possible to determine the nature of the other structural moieties incorporated into the final product, which was fully elucidated by isolation and NMR analysis.

The ability to predict the essential chemical features and gross structure of ECO-02301 is an excellent example of the power of genomics as a screening tool. Knowledge of the potential structure of such products enables decisions to be made regarding the priority of follow-up experimental work for the production and isolation of the compounds. In this case, the prediction included a number of unusual features, including a long linear chain, and two nitrogenous units derived from arginine and glycine that suggested this would be an interesting compound. This research was greatly facilitated by the relative ease with which *S. aizunensis* was induced to produce ECO-02031. Often the ability to turn on the pathway's expression is problematic. Little is known regarding the exact environmental or physiological cues that are required to induce expression and therefore these experiments become highly empirical.

5.2. Surveying Genomes for Biosynthetic Pathways

An approach to finding the biosynthetic machinery for selected classes of metabolites is to "scan" genomes of potential producers for a sequence corresponding to a particular characteristic protein. For such methods, specific DNA sequences must be defined that characterize the encoding of a particular protein family. Once these sequences have been identified, probes can be constructed that contain the key DNA sequences for comparisons with genomic libraries derived from the test strains of interest. One advantage to this approach is

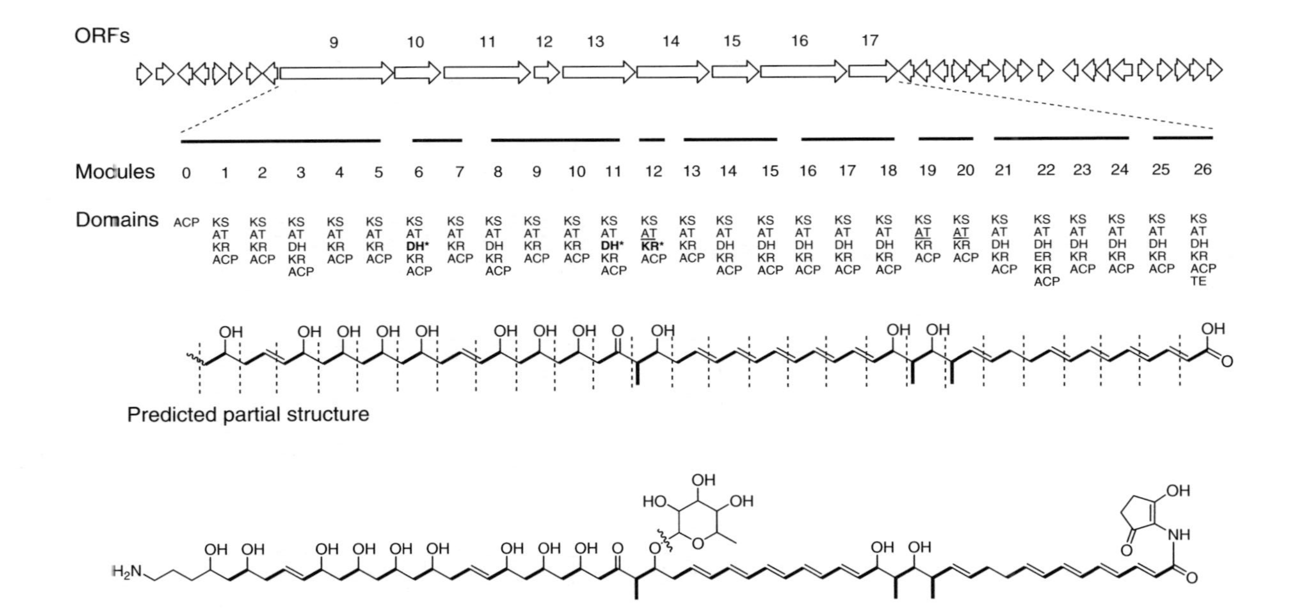

Figure 10. Gene cluster for ECO-02301 biosynthesis [123]. Open reading frames (ORFs) 9 through 17 encode the type I polyketide synthetase and the domains of the modules from these ORFs are shown. ACP: acyl carrier protein, KS: ketosynthase, AT: acyl transferase, KR = ketoreductase, DH: dehydratase, ER: enoylreductase, TE: thioesterase. (Reprinted with permission of the American Chemical Society.)

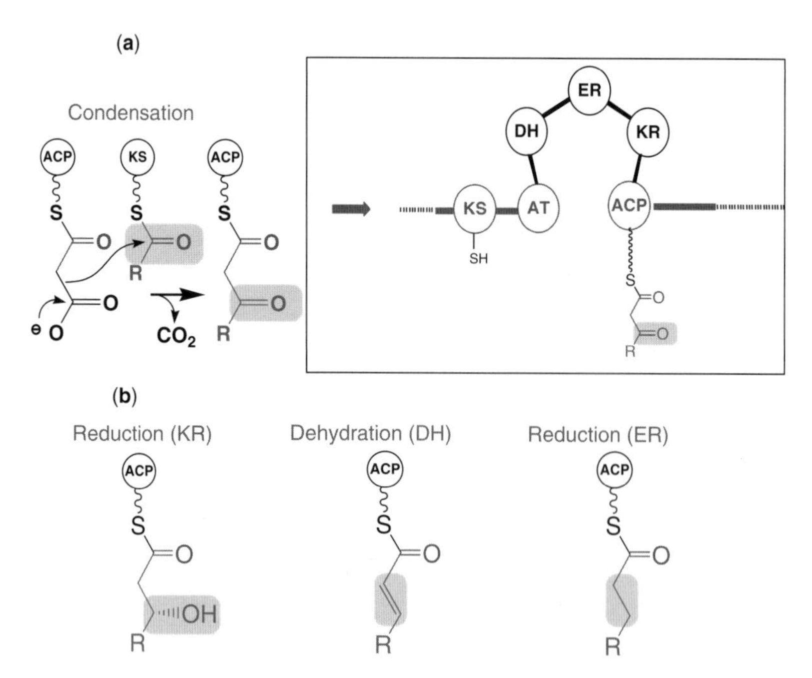

Figure 11. (a) Depiction of a single PKS module. The growing chain held by the ACP is condensed with the next acyl precursor as selected by the AT through the action of the KS, with the concomitant release of CO_2. (b) The resulting keto group is then optionally processed by the domains of the reductive loop, resulting in hydroxyl-bearing, olefinic, or saturated chains. (See color insert.)

that the knowledge of the whole genome sequence of the source organism is not required. Smaller fragments constituting the whole genome are ordinarily used for this screening process. These fragments are captured on smaller vectors that can be probed with the sequences of interest to determine regions of high homology. Those fragments that show strong correlation to the designated sequences are then further analyzed through DNA sequencing to determine the nature of the encoded product.

This approach is widely applied to identify sequences encoding specific proteins of interest; however, there are a few reports where the intent was focused on finding whole biosynthetic clusters. An early example termed "genome scanning" was published by workers [124] at Ecopia Biosciences, in which they utilized the sequences of five unique genes found in all known biosynthetic pathways for enediyne-containing antibiotics, such as calicheamicin (see Fig. 5) to identify related cassettes in other actinomycetes. The results of these experiments were astonishing. Fully 15% of randomly selected soil-derived actinomycetes appeared to possess the biosynthetic capability to produce an enediyne. These compounds had previously been regarded as quite rare occurrences in Nature. This is a particularly dramatic example of screening for a class of natural products by genomics alone.

A more recent example [125] employed a halogenase enzyme as the target for such screening. Halogens are often found as substituents in secondary metabolites, and the genes that encode the halogenating enzymes are often found in close proximity to the genes encoding the whole biosynthetic pathway. Localization of a halogenase therefore would provide a good chance to identify a secondary metabolite pathway. In this case, a genetic probe was designed to target conserved sequences in $FADH_2$-dependent halogenase genes. These probes were used to search for homologous sequences in 550 randomly selected actinomycetes. Approximately 20% of the strains examined showed a positive indication of halogenase genes. Furthermore, when flanking regions were examined, linkage to full secondary metabolite encoding pathways was established in several cases, which validated the utility of this straightforward approach.

Such examples of the utility of genomic information for the detection of secondary metabolite pathways are illustrative of the potential of this technology. It is clear that as our ability to translate genomic data into functional activity grows, so will the applications in this area. Progress will also be closely linked to continued developments in sequencing technologies that will drive down the cost of obtaining whole genome sequences, and thereby open new vistas to explore.

5.3. Virtual Screening of Natural Products

Computational methods, first developed for libraries of small synthetic compounds, are now beginning to be being employed to model interactions of natural products with macromolecular targets. The most straightforward of these methods aim to simulate the interactions of the metabolites with the binding sites of enzymes or receptors. The full three-dimensional structures of both the target macromolecule and the natural product are the preferred starting points from which to model binding interactions. Furthermore, an understanding of the key binding interactions between a small-molecule and key residues or pockets within the binding site is needed. Once again, the least ambiguous way to generate this information is through a cocrystal structure of the macromolecule with its small-molecule ligand in place. One such example [126] employed the binding of the alkaloid galanthamine (see Fig. 12) with acetylcholine esterase to create a pharmacophore model. Screening a 3D structure library of natural products yielded the coumarin derivatives scopolin and scopoletin as strong positive hits. Interestingly, both coumarins showed significant biological effects *in vivo*, which the authors attribute to their inhibitory effects on acetylcholinesterase.

The reader is referred to reviews by Rollinger [127] and Shen and Jiang [128] for additional background and examples. While it is clear that this type of approach has promise, the scope is limited by the ability of such computational methods to effectively model truly complex natural products.

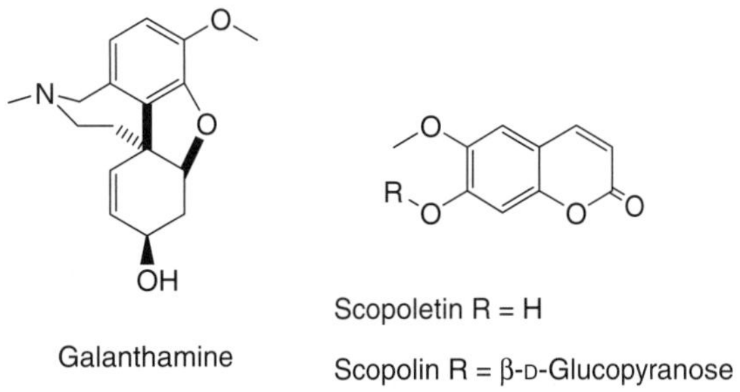

Galanthamine

Scopoletin R = H

Scopolin R = β-D-Glucopyranose

Figure 12. Galanthamine and compounds derived from virtual screening.

6. CONCLUSION

Natural products research is rich in its historical contribution to drug discovery. It can be argued that no other source of chemical diversity can compare with the depth and breadth of therapeutic agents that have their origins in the secondary metabolites produced by living organisms. These contributions have come via numerous strategies; some direct and focused, some by serendipitous routes, and many by screening. The rational approach to lead discovery, using natural products as biologically validated starting points, continues to be an effective means of developing drug candidates. Today's drug discovery environment, however, with its increasing reliance on automated high-throughput screening of chemical libraries and rapid hit-to-lead development, does not favor traditional natural products screen-based approaches. This is because the enhanced chemical diversity offered by natural products libraries is offset by the often-lengthy (by contemporary standards) periods needed for screening, hit validation and dereplication. If screening of natural products is to continue as an important engine for lead generation, it will be necessary to reduce the time and resources needed to assay natural products libraries and to identify and isolate the important bioactive components. Many of the necessary economies can be found in the construction of improved natural products libraries themselves, and in the means by which they are screened.

Encouragingly, it is becoming increasingly apparent that the rich chemical diversity obtainable from natural products is not nearly exhausted, as some have speculated. New microbial sources, many of which have their origins in the marine environment, offer the very real benefit of an inherent means of production. The ready availability of compounds addresses the supply issue that has often plagued the advancement of natural products in development programs.

As the facility with which we are able to generate and interpret genomic information continues to grow, new opportunities to apply this knowledge will be developed. As we apply these technologies and develop them even further, we can look forward to continued and even increased impact of natural products for the discovery of new medicines and other high-value products.

REFERENCES

1. Buss AD, Waigh RD. Natural products as leads for new pharmaceuticals. In: Wolfe ME, editor. Burger's Medicinal Chemistry and Drug Discovery. 5th ed. Vol. 1, *Principles and Practice*. New York: John Wiley & Sons; 1995. p 983–1033.
2. Wilson EO. The Diversity of Life. New York: W. W. Norton & Company; 1999. p 35–59.
3. Schloss PD, Handelsman J. Status of the microbial census. Microbiol Mol Biol Rev 2004;68:686–691.
4. Amann RI, Ludwig W, Schleifer KH. Phylogenetic identification and *in situ* detection of individual microbial cells without cultivation. Microbiol Rev 1995;59:143–169.
5. Hawksworth DL. The fungal dimension of biodiversity, magnitude, significance and conservation. Mycol Res 1991;95:641–655.
6. Berdy J. Bioactive microbial metabolites. J Antibiot 2005;58:1–26.
7. Keller M, Zengler K. Tapping into microbial diversity. Nat Rev Microbiol 2004;2:141–150.
8. Watve MG, Tickoo R, Jog MM, Bhole BD. How many antibiotics are produced by the genus *Streptomyces*? Arch Microbiol 2001;176: 386–390.
9. Bentley SD, Chater KF, Cerdeno-Tarraga A-M, Challis GL, Thomson NR, James KD, Harris DE, Quail MA, Kieser H, Harper D, Bateman A, Brown S, Chandra G, Chen CW, Collins M, Cronin A, Fraser A, Goble A, Hidalgo J, Hornsby T, Howarth S, Huang C-H, Kieser T, Larke L, Murphy L, Oliver K, O'Neil S, Rabbinowitsch E, Rajandream M-A, Rutherford K, Rutter S, Seeger K, Saunders D, Sharp S, Squares R, Squares S, Taylor K, Warren T, Wietzorrek A, Woodward J, Barrell BG, Parkhill J, Hopwood DA. Complete genome sequence of the model actinomycete *Streptomyces coelicolor* A3(2). Nature 2002;417 (6885): 141–147.
10. Ikeda H, Ishikawa J, Hanamoto A, Shinose M, Kikuchi H, Shiba T, Sakaki Y, Hattori M, Omura S. Complete genome sequence and comparative analysis of the industrial microorganism *Streptomyces avermitilis*. Nat Biotechnol 2003;21(5): 526–531.
11. Sielaff H, Christiansen G, Schwecke T. Natural products from cyanobacteria: exploiting a new source for drug discovery. IDrugs 2006;9:119–127.
12. Newman DJ, Cragg GM. Natural products as sources of new drugs over the last 25 years. J Nat Prod 2007;70:461–477.
13. Gerth K, Pradella S, Perlova O, Beyer S, Muler R. Myxobacteria: proficient producers of novel natural products with various biological activities—past and future biotechnological aspects with the focus on the genus *Sorangium*. J Biotechnol 2003;106:233–253.
14. Strobel GA, Daisy BH, Castillo U, Harper J. Natural products from rainforest endophytes. In: Zhang L, Demain A, editors. Natural Products Drug Discovery and Therapeutic Medicine. Totowa, NJ: Humana Press; 2005. p 329–352.
15. Eyberger AL, Dondapati R, Porter JR. Endophyte fungal isolates from *Podophyllum peltatum* produce podophyllotoxin. J Nat Prod 2006;69:1121–1124.
16. Puri SC, Handa G, Bhat BA, Gupta VK, Amna T, Verma N, Anand R, Dhar KL, Qazi GN. Separation of 9-methoxycamptothecin and camptothecin from *Nothapodytes foetida* by semipreparative HPLC. J Chromatogr Sci 2005;43(7): 348–350.
17. Okazaki T, Kitahara T, Okami Y. Studies on marine microorganisms. IV. A new antibiotic SS-228Y produced by *Chainia* isolated from shallow sea mud. J Antibiot 1975;28: 176–184.
18. Moran MA, Rutherford LT, Hodson RE. Evidence for indigenous *Streptomyces* populations in a marine environment determined with a

16S rRNA probe. Appl Environ Microbiol 1995;61:3695–3700.

19. Helmke E, Weyland H. *Rhodococcus marinonascens* sp. nov., an actinomycete from the sea. Int J Syst Bacteriol 1984;34:127–138.
20. Maldonado LA, Fenical W, Jensen PR, Kauffman CA, Mincer TJ, Ward AC, Bull AT, Goodfellow M. *Salinispora arenicola* gen. nov., sp. nov. and *Salinispora tropica* sp. nov., obligate marine actinomycetes belonging to the family Micromonosporaceae. Int J Syst Evol Microbiol 2005;55:1759–1766.
21. Yi H, Schumann P, Sohn K, Chun J. *Serinicoccus marinus* gen. nov. sp. nov., a novel actinomycte with L-ornithine and L-serine in the peptidoglycan. Int J Syst Evol Microbiol 2004;54:1585–1589.
22. Han SK, Nedashkovzkaya OI, Mikhailov VV, Kim SB, Bae KS. *Salinibacterium amurkyense* gen. nov., a novel genus of the family Microbacteriaceae from the marine environement. Int J Syst Evol Microbiol 2003;53:2061–2066.
23. Fenical W, Jensen PR. Developing a new resource for drug discovery: marine actinomyete bacteria. Nat Chem Biol 2006;2(12): 666–673.
24. Udwary DW, Zeigler L, Asolkar RN, Singan V, Lapidus A, Fenical W, Jensen PR, Moore BS. Genome sequencing reveals complex secondary metabolome in the marine actinomycete *Salinispora tropica*. Proc Natl Acad Sci USA 2007;104(25): 10376–10381.
25. McDonald LA, Capson TL, Krishnamurthy G, Ding WD, Ellestad GA, Bernan VS, Maiese WM, Lassota P, Discafani C, Kramer RA, Ireland CM. Namenamicin, a new enediyne antitumor antibiotic from the marine ascidian *Polysyncraton lithostrotum*. J Am Chem Soc 1996;118:10898.
26. Moore BS. Biosynthesis of marine natural products: microorganisms and macroalgae. Nat Prod Rep 1999;16:653–674.
27. Simmons TL, Coates RC, Clark BR, Engene N, Gonzalez D, Esquenazi E, Dorrestein PC, Gerwick WH. Biosynthetic origin of natural products isolated from marine microorganism–invertebrate assemblages. Proc Natl Acad Sci USA 2008;105(12): 4587–4594.
28. Preston CM, Wu KY, Molinski TF, Delong EF. A psychrophilic crenarchaeon inhabits a marine sponge: *Cenarchaeum symbiosum* gen. nov. sp. nov. Proc Natl Acad Sci USA 1996;93: 6241–6246.
29. Schmidt EW, Nelson JT, Rasko DA, Sudek S, Eisen JA, Haygood MG, Ravel J. Patellamide A and C biosynthesis by a microcin-like pathway in *Prochloron didemni*, the cyanobacterial symbiont of *Lissoclinum patella*. Proc Natl Acad Sci USA 2005;102:7315–7320.
30. Long PF, Dunlap WC, Battershill CN, Jaspars M. Shotgun cloning and heterologous expression of the patellamide gene cluster as a strategy to achieving sustained metabolite production. ChemBioChem 2005;6:1760–1765.
31. Davidson SK, Allen SW, Lim GE, Anderson CM, Haygood MG. Evidence for the biosynthesis of bryostatins by the bacterial symbiont "*Candidatus* Endobugula sertula" of the bryozoan *Bugula neritina*. Appl Environ Microbiol 2001;67:4531–4537.
32. Sudek S, Lopanik NB, Waggoner LE, Hildebrand M, Anderson C, Liu H, Patel A, Sherman DH, Haygood MG. Identification of the putative bryostatin polyketide synthase gene cluster from "*Candidatus* Endobugula sertula", the uncultivated microbial symbiont of the marine bryozoan *Bugula neritina*. J Nat Prod 2007;70:67–74.
33. Zhang L. Integrated approaches for discovering novel drugs from microbial natural products. In: Zhang L, Demain A, editors. Natural Products Drug Discovery and Therapeutic Medicine. Totowa, NJ: Humana Press; 2005. p 33–55.
34. Kaeberlein T, Lewis K, Epstein SS. Isolating 'uncultivable 'microorganisms in pure culture in a simulated natural environment. Science 2002;296:1127–1129.
35. Cueto M, Jensen PR, Kauffman C, Fenical W, Lobkovsky E, Clardy J. Pestalone, a new antibiotic produced by a marine fungus in response to bacterial challenge. J Nat Prod 2001;64: 1444–1446.
36. Piel J. Bacterial symbionts: prospects for the sustainable production of invertebrate-derived pharmaceuticals. Curr Med Chem 2006;13: 39–50.
37. Lefevre F, Robe P, Jarrin C, Ginolhac A, Zago C, Auriol D, Vogel TM, Simonet P, Nalin R. Drugs from hidden bugs: their discovery via untapped resources. Res Microbiol 2008;159: 153–161.
38. Wang GYS, Graziani E, Waters B, Pan W, Li X, McDermott J, Meurer G, Saxena G, Andersen RJ, Davies J. Novel natural products from soil DNA libraries in a streptomycete host. Org Lett 2000;2:2401–2404.
39. Brady SF, Chao CJ, Handelsman J, Clardy J. Cloning and heterologous expression of a natural product biosynthetic cluster from eDNA. Org Lett 2001;3:1981–1984.

40. MacNeil IA, Tiong CL, Minor C, August PR, Grossman TH, Loiacono KA, Lynch BA, Phillips T, Narula S, Sundaramoorthi R, Tyler A, Aldredge T, Long H, Gilman M, Holt D, Osburne MS. Expression and isolation of antimicrobial small molecules from soil DNA libraries. J Mol Microbiol Biotechnol 2001;3: 301–308.
41. Gillespie DE, Brady SF, Bettermann AD, Cianciotto NP, Liles MR, Rondon MR, Clardy J, Goodman RM, Handelsman J. Isolation of antibiotics turbomycin A and B from a metagenomic library of soil microbial DNA. Appl Environ Microbiol 2002;68: 4301–4306.
42. An Z, Harris GH, Zink D, Giacobbe R, Lu P, Sangari R, Svetnik V, Gunter B, Liaw A, Masurekar PS, Liesch J, Gould S, Strohl W, Bills G. Expression of cosmid-size DNA of slow-growing fungi in *Aspergillus nidulans* for secondary metabolite screening. In: An Z, editor. Handbook of Industrial Mycology. New York: Marcel Dekker; 2005. p 167–186.
43. Eldridge GR, Vervoort HC, Lee CM, Cremin PA, Williams CT, Hart SM, Goering MG, O'Neil-Johnson M, Zeng L. High-throughput method for the production and analysis of large natural product libraries for drug discovery. Anal Chem 2002;74(16): 3963–3971.
44. Wagenaar MM. Pre-fractionated microbial samples: the second generation natural products library at Wyeth. Molecules 2008;13(6): 1406–1426.
45. Bindseil KU, Jakupovic J, Wolf D, Lavayre J, Leboul J, van der Pyl D. Pure compound libraries; a new perspective for natural product based drug discovery. Drug Discov Today 2001;6(16): 840–847.
46. Koehn FE, Carter GT. The evolving role of natural products in drug discovery. Nat Rev Drug Discov 2005;4(3): 206–220.
47. Abel U, Koch C, Speitling M, Hansske FG. Modern methods to produce natural-product libraries. Curr Opin Chem Biol 2002;6(4): 453–458.
48. Barrow RA. Isolation of microbial natural products. In: Methods in Biotechnology. Vol. 20. Series: *Natural Products Isolation*. 2nd ed. 2005;20:391–414.
49. Henry MC, Yonker CR. Supercritical fluid chromatography, pressurized liquid extraction, and supercritical fluid extraction. Anal Chem 2006;78(12): 3909–3915.
50. Harrigan GG, Goetz GH. Chemical and biological integrity in natural products screening. Comb Chem High Throughput Screening 2005; 8(6):529–534.
51. Yan B, Fang L, Irving M, Zhang S, Boldi AM, Woolard F, Johnson CR, Kshirsagar T, Figliozzi GM, Krueger CA, Collins N. Quality control in combinatorial chemistry: determination of the quantity, purity, and quantitative purity of compounds in combinatorial libraries. J Comb Chem 2003;5(5): 547–559.
52. Newman DJ, Cragg GM, Kingston DGI. Natural products as pharmaceuticals and sources for lead structures. Practice of Medicinal Chemistry. 2nd ed. 2003; 91–109.
53. Henkel T, Brunne RM, Muller H, Reichel F. Statistical investigation into the structural complementarity of natural products and synthetic compounds. Angew Chem Int Ed Engl 1999;38(5): 643–647.
54. Feher M, Schmidt JM. Property distributions: differences between drugs, natural products, and molecules from combinatorial chemistry. J Chem Inf Comput Sci 2003;43(1): 218–227.
55. Lee M-L, Schneider G. Scaffold architecture and pharmacophoric properties of natural products and trade drugs: application in the design of natural product-based combinatorial libraries. J Comb Chem 2001;3(3): 284–289.
56. Stahura FL, Godden JW, Xue L, Bajorath J. Distinguishing between natural products and synthetic molecules by descriptor Shannon entropy analysis and binary QSAR calculations. J Chem Inf Comput Sci 2000;40(5): 1245–1252.
57. Grabowski K, Schneider G. Properties and architecture of drugs and natural products revisited. Curr Chem Biol 2007;1(1): 115–127.
58. Quinn RJ, Carroll AR, Pham NB, Baron P, Palframan ME, Suraweera L, Pierens GK, Muresan S. Developing a drug-like natural product library. J Nat Prod 2008;7(3): 464–468.
59. Ertl P, Roggo S, Schuffenhauer A. Natural product-likeness score and its application for prioritization of compound libraries. J Chem Inf Model 2008;48(1): 68–74.
60. Bugni TS, Richards B, Bhoite L, Cimbora D, Harper MK, Ireland CM. Marine natural product libraries for high-throughput screening and rapid drug discovery. J Nat Prod 2008; 71(6): 1095–1098.
61. Higgs RE, Zahn JA, Gygi JD, Hilton MD. Rapid method to estimate the presence of secondary metabolites in microbial extracts. Appl Environ Microbiol 2001;67(1): 371–376.
62. Nielsen N-PV Carstensen JM, Smedsgaard J. Aligning of single and multiple wavelength

chromatographic profiles for chemometric data analysis using correlation optimized warping. J Chromatogr A 1998;805(1 + 2): 17–35.

63. Wolf D, Siems K. Burning the hay to find the needle: data mining strategies in natural product dereplication. Chimia 2007;61(60): 339–345.
64. Larsen TO, Smedsgaard J, Nielsen KF, Hansen ME, Frisvad JC. Phenotypic taxonomy and metabolite profiling in microbial drug discovery. Nat Prod Rep 2005;22(6): 672–695.
65. Tormo JR, Garcia JB. Automated analyses of HPLC profiles of microbial extracts: a new tool for drug discovery screening. In: Zhang L, Demain A, editors. Natural Products: Drug Discovery and Therapeutic Medicine. Totowa, NJ: Humana Press; 2005. p 57–75.
66. Peake DA, Duckworth DC, Perun TJ, Scott WL, Kulanthaivel P, Strege MA. Analytical and biological evaluation of high throughput screen actives using evaporative light scattering, chemiluminescent nitrogen detection, and accurate mass LC–MS–MS. Comb Chem High Throughput Screening 2005;8(6): 477–487.
67. Fuellbeck M, Michalsky E, Dunkel M, Preissner R. Natural products: sources and databases. Nat Prod Rep 2006;23(3): 347–356.
68. Cremin PA, Zeng L. High-throughput analysis of natural product compound libraries by parallel LC–MS evaporative light scattering detection. Anal Chem 2002;74(21): 5492–5500.
69. Fredenhagen A, Derrien C, Gassmann E. An MS/MS library on an ion-trap Instrument for efficient dereplication of natural products. Different fragmentation patterns for [M + H]+ and [M + Na]+ ions. J Nat Prod 2005;68(3): 385–391.
70. Osada H, Magae J, Watanabe C, Isono K. Rapid screening method for inhibitors of protein kinase C. J Antibiot 1988;41(7): 925–31.
71. Mewshaw RE, Edsall RJ, Yang C, Manas ES, Xu ZB, Henderson RA, Keith JC, Harris HA. ERbeta ligands. 3. Exploiting two binding orientations of the 2-phenylnaphthalene scaffold to achieve ERbeta selectivity. J Med Chem 2005;48(12): 3953–3979.
72. Pauli GF, Jaki BU, Lankin DC. Quantitative ^{1}H NMR: development and potential of a method for natural products analysis. J Nat Prod 2005;68(1): 133–149.
73. Bollard ME, Stanley EG, Lindon JC, Nicholson JK, Holmes E. NMR-based metabonomic approaches for evaluating physiological influences on biofluid composition. NMR Biomed 2005;18(3): 143–162.
74. Ward JL, Beale MH. NMR spectroscopy in plant metabolomics. In: Plant Metabolomics. Series: *Biotechnology in Agriculture and Forestry*. 2006;57:81–91.
75. Sehgal R, Nandave M, Ojha SK. Natural products: leads for drug discovery and development. The Pharma Review 2006;4(21): 40–44.
76. Pierens GK, Palframan ME, Tranter CJ, Carroll AR, Quinn RJ. A robust clustering approach for NMR spectra of natural product extracts. Magn Reson Chem 2005;43(5): 359–365.
77. Hu J-F, Garo E, Yoo H-D, Cremin PA, Zeng L, Goering MG, O'Neil-Johnson M, Eldridge GR. Application of capillary-scale NMR for the structure determination of phytochemicals. Phytochem Anal 2005;16(2): 127–133.
78. Jansma A, Chuan T, Albrecht RW, Olson DL, Peck TL, Geierstanger BH. Automated microflow NMR: routine analysis of five-microliter samples. Anal Chem 2005;77(19): 6509–6515.
79. Taggi AE, Meinwald J, Schroeder FC. A new approach to natural products discovery exemplified by the identification of sulfated nucleosides in spider venom. J Am Chem Soc 2004; 126(33): 10364–10369.
80. Bringmann G, Wohlfarth M, Rischer H, Schlauer J, Brun R. Extract screening by HPLC coupled to MS–MS, NMR, and CD: a dimeric and three monomeric naphthylisoquinoline alkaloids from *Ancistrocladus griffithii*. Phytochemistry 2002;61(2): 195–204.
81. Jaroszewski JW. Hyphenated NMR methods in natural products research. Part 2. HPLC–SPE–NMR and other new trends in NMR hyphenation. Planta Med 2005;71(9): 795–802.
82. Jaroszewski JW. Hyphenated NMR methods in natural products research. Part 1. Direct hyphenation. Planta Med 2005;71(8): 691–700.
83. Bieri S, Varesio E, Veuthey J-L, Munoz O, Tseng L-H, Braumann U, Spraul M, Christen P. Identification of isomeric tropane alkaloids from *Schizanthus grahamii* by HPLC–NMR with loop storage and HPLC–UV–MS/SPE–NMR using a cryogenic flow probe. Phytochem Anal 2006;17(2): 78–86.
84. Spraul M, Freund AS, Nast RE, Withers RS, Maas WE, Corcoran O. Advancing NMR sensitivity for LC–NMR–MS using a cryoflow probe: application to the analysis of acetaminophen metabolites in urine. Anal Chem 2003;75(6): 1536–1541.

85. Exarchou V, Krucker M, van Beek TA, Vervoort J, Gerothanassis IP, Albert K. LC–NMR coupling technology: recent advancements and applications in natural products analysis. Magn Reson Chem 2005;43(9): 681–687.
86. Lang G, Mayhudin NA, Mitova MI, Sun L, van der Sar S, Blunt JW, Cole ALJ, Ellis G, Laatsch H, Munro MHG. Evolving trends in the dereplication of natural product extracts: new methodology for rapid, small-scale investigation of natural product extracts. J Nat Prod 2008;71(9): 1595–1599.
87. Exarchou V, Godejohann M, Van Beek TA, Gerothanassis IP, Vervoort J. LC–UV–solid-phase extraction–NMR–MS combined with a cryogenic flow probe and its application to the identification of compounds present in Greek oregano. Anal Chem 2003;75(22): 6288–6294.
88. Lambert M, Staerk D, Hansen SH, Jaroszewski JW. HPLC–SPE–NMR hyphenation in natural products research: optimization of analysis of *Croton membranaceus* extract. Magn Reson Chem 2005;43(9): 771–775.
89. Newman DJ, Cragg GM, Snader KM. Natural products as sources of new drugs over the period 1981–2002. J Nat Prod 2003;66(7): 1022–1037.
90. Silverman L, Campbell R, Broach JR. New assay technologies for high-throughput screening. Curr Opin Chem Biol 1998;2(3): 397–403.
91. Zaman GJR, Garritsen A, de Boer T, van Boeckel CAA. Fluorescence assays for high-throughput screening of protein kinases. Comb Chem High Throughput Screening 2003;6(4): 313–320.
92. Croston GE. Functional cell-based uHTS in chemical genomic drug discovery. Trends Biotechnol 2002;20(3): 110–115.
93. Kumar RA, Clark DS. High-throughput screening of biocatalytic activity: applications in drug discovery. Curr Opin Chem Biol 2006; 10(2): 162–168.
94. Heilker R, Zemanova L, Valler MJ, Nienhaus GU. Confocal fluorescence microscopy for high-throughput screening of G-protein coupled receptors. Curr Med Chem 2005;12(22): 2551–2559.
95. Terstappen GC. Ion channel screening technologies today. Drug Discov Today Technol 2005;2(2): 133–140.
96. Zheng X, Hu Y, Liu J, Ouyang K. Screening of active compounds as neuromedin U2 receptor agonist from natural products. Bioorg Med Chem Lett 2005;15(20): 4531–4535.
97. Henrich CJ, Bokesch HR, Dean M, Bates SE, Robey RW, Goncharova EI, Wilson JA, McMahon JB. A high-throughput cell-based assay for inhibitors of ABCG2 activity. J Biomol Screen 2006;11(2): 176–183.
98. Marion F, Williams DE, Patrick BO, Hollander I, Mallon R, Kim SC, Roll DM, Feldberg L, Van Soest R, Andersen RJ. Liphagal, a selective inhibitor of PI3 kinase α isolated from the sponge Aka coralliphaga: structure elucidation and biomimetic synthesis. Org Lett 2006;8(2): 321–324.
99. Fowler A, Swift D, Longman E, Acornley A, Hemsley P, Murray D, Unitt J, Dale I, Sullivan E, Coldwell M. An evaluation of fluorescence polarization and lifetime discriminated polarization for high throughput screening of serine/threonine kinases. Anal Biochem 2002;308(2): 223–231.
100. Turek-Etienne TC, Lei M, Terracciano JS, Langsdorf EF, Bryant RW, Hart RF, Horan AC. Use of red-shifted dyes in a fluorescence polarization AKT kinase assay for detection of biological activity in natural product extracts. J Biomol Screen 2004;9(1): 52–61.
101. Inglese J, Auld DS, Jadhav A, Johnson RL, Simeonov A, Yasgar A, Zheng W, Austin CP. Quantitative high-throughput screening: a titration-based approach that efficiently identifies biological activities in large chemical libraries. Proc Natl Acad Sci USA 2006;103 (31): 11473–11478.
102. Zheng W, Padia J, Urban DJ, Jadhav A, Goker-Alpan O, Simeonov A, Goldin E, Auld D, LaMarca ME, Inglese J, Austin CP, Sidransky E. Three classes of glucocerebrosidase inhibitors identified by quantitative high-throughput screening are chaperone leads for Gaucher disease. Proc Natl Acad Sci USA 2007;104 (32): 13192–13197, S13192/1–S13192/5.
103. Simeonov A, Yasgar A, Jadhav A, Lokesh GL, Klumpp C, Michael S, Austin CP, Natarajan A, Inglese J. Dual-fluorophore quantitative high-throughput screen for inhibitors of BRCT-phosphoprotein interaction. Anal Biochem 2008;375(1): 60–70.
104. Titus S, Neumann S, Zheng W, Southall N, Michael S, Klumpp C, Yasgar A, Shinn P, Thomas CJ, Inglese J, Gershengorn MC, Austin CP. Quantitative high-throughput screening using a live-cell cAMP assay identifies small-molecule agonists of the TSH receptor. J Biomol Screen 2008;13(2): 120–127.
105. Johnson RL, Huang W, Jadhav A, Austin CP, Inglese J, Martinez ED. A quantitative high-

throughput screen identifies potential epigenetic modulators of gene expression. Anal Biochem 2008;375(2): 237–248.

106. Payne DJ, Gwynn MN, Holmes DJ, Pompliano DL. Drugs for bad bugs: confronting the challenges of antibacterial discovery. Nat Rev Drug Discov 2007;6(1): 29–40.

107. Malo N, Hanley JA, Cerquozzi S, Pelletier J, Nadon R. Statistical practice in high-throughput screening data analysis. Nat Biotechnol 2006;24(2): 167–175.

108. Lipinski CA. Drug-like properties and the causes of poor solubility and poor permeability. J Pharmacol Toxicol Methods 2001;44(1): 235–249.

109. Yoo H-D, Cremin PA, Zeng L, Garo E, Williams CT, Lee CM, Goering MG, O'Neil-Johnson M, Eldridge GR, Hu J-F. Suaveolindole, a new mass-limited antibacterial indolosesquiterpene from *Greenwayodendron suaveolens* obtained via high-throughput natural products chemistry methods. J Nat Prod 2005;68(1): 122–124.

110. Vogt A, Tamewitz A, Skoko J, Sikorski RP, Giuliano KA, Lazo JS. The benzo[*c*]phenanthridine alkaloid, sanguinarine, is a selective, cell-active inhibitor of mitogen-activated protein kinase phosphatase-1. J Biol Chem 2005; 280(19): 19078–19086.

111. Summers MY, Leighton M, Liu D, Pong K, Graziani EI. 3-normeridamycin: a potent non-immunosuppressive immunophilin ligand is neuroprotective in dopaminergic neurons. J Antibiot 2006;59(3): 184–189.

112. Wang J, Soisson SM, Young K, Shoop W, Kodali S, Galgoci A, Painter R, Parthasarathy G, Tang YS, Cummings R, Ha S, Dorso K, Motyl M, Jayasuriya H, Ondeyka J, Herath K, Zhang C, Hernandez L, Allocco J, Basilio A, Tormo JR, Genilloud O, Vicente F, Pelaez F, Colwell L, Lee SH, Michael B, Felcetto T, Gill C, Silver LL, Hermes JD, Bartizal K, Barrett J, Schmatz D, Becker JW, Cully D, Singh SB. Platensimycin is a selective FabF inhibitor with potent antibiotic properties. Nature 2006;441(7091): 358–361.

113. Ondeyka JG, Zink DL, Young K, Painter R, Kodali S, Galgoci A, Collado J, Tormo JR, Basilio A, Vicente F, Wang J, Singh SB. Discovery of bacterial fatty acid synthase inhibitors from a *Phoma* species as antimicrobial agents using a new antisense-based strategy. J Nat Prod 2006;69(3): 377–380.

114. Wang J, Kodali S, Lee SH, Galgoci A, Painter R, Dorso K, Racine F, Motyl M, Hernandez L, Tinney E, Colletti SL, Herath K, Cummings R, Salazar O, Gonzalez I, Basilio A, Vicente F, Genilloud O, Pelaez F, Jayasuriya H, Young K, Cully DF, Singh SB. Discovery of platencin, a dual FabF and FabH inhibitor with *in vivo* antibiotic properties. Proc Natl Acad Sci USA 2007;104(18): 7612–7616.

115. Parish CA, de la Cruz M, Smith SK, Zink D, Baxter J, Tucker-Samaras S, Collado J, Platas G, Bills G, Diez MT, Vicente F, Pelaez F, Wilson K. Antisense-guided isolation and structure elucidation of pannomycin, a substituted *cis*-decalin from *Geomyces pannorum*. J Nat Prod 2009;72(1): 59–62.

116. Koehn FE, Kirsch DR, Feng X, Janso J, Young M. A cell wall-active lipopeptide from the fungus *Pochonia bulbillosa*. J Nat Prod 2008;71 (12): 2045–2048.

117. Lum PY, Armour CD, Stepaniants SB, Cavet G, Wolf MK, Butler JS, Hinshaw JC, Garnier P, Prestwich GD, Leonardson A, Garrett-Engele P, Rush CM, Bard M, Schimmack G, Phillips JW, Roberts CJ, Shoemaker DD. Discovering modes of action for therapeutic compounds using a genome-wide screen of yeast heterozygotes. Cell 2004;116(1): 121–137.

118. Jiang N, Ou-Yang K-q, Cai S-x, Hu Y-h, Xu Z-l. Identification of human dopamine D1-like receptor agonist using a cell-based functional assay. Acta Pharmacol Sin 2005;26(10): 1181–1186.

119. Ruocco KM, Goncharova EI, Young MR, Colburn NH, McMahon JB, Henrich CJ. A high-throughput cell-based assay to identify specific inhibitors of transcription factor AP-1. J Biomol Screen 2007;12(1): 133–139.

120. Andrianasolo EH, Haramaty L, Vardi A, White E, Lutz R, Falkowski P. Apoptosis-inducing galactolipids from a cultured marine diatom, *Phaeodactylum tricornutum*. J Nat Prod 2008; 71(7): 1197–1201.

121. Andrianasolo EH, Haramaty L, Degenhardt K, Mathew R, White E, Lutz R, Falkowski P. Induction of apoptosis by diterpenes from the soft coral *Xenia elongata*. J Nat Prod 2007;70 (10): 1551–1557.

122. Oliynyk M, Samborskyy M, Lester JB, Mironenko T, Scott N, Dickens Haydock SF, Leadlay PF. Complete genome sequence of the erythromycin-producing bacterium *Saccharopolyspora erythraea* NRRL2338. Nat Biotechnol 2007;25(4): 447–453.

123. McAlpine JB, Bachmann BO, Piraee M, Tremblay S, Alarco A-M, Zazopoulos E, Farnet CM. Microbial genomics as a guide to drug

discovery and structural elucidation: ECO-02301, a novel antifungal agent, as an example. J Nat Prod 2005;68(4): 493–496.

124. Zazopoulos E, Huang K, Staffa A, Liu W, Bachmann BO, Nonaka K, Ahlert J, Thorson JS, Shen B, Farnet CM. A genomics: guided approach for discovering and expressing cryptic metabolic pathways. Nat Biotechnol 2003;21(2): 187–190.
125. Hornung A, Bertazzo M, Dziarnowski A, Schneider K, Welzel K, Wohlert S-E, Holzenkaempfer M, Nicholson GJ, Bechthold A, Suessmuth RD, Vente A, Pelzer S. A genomic screening approach to the structure-guided identification of drug candidates from natural sources. ChemBioChem 2007;8(7): 757–766.
126. Rollinger JM, Hornick A, Langer T, Stuppner H, Prast H. Acetylcholinesterase inhibitory activity of scopolin and scopoletin discovered by virtual screening of natural products. J Med Chem 2004;47(25): 6248–6254.
127. Rollinger JM, Stuppner H, Langer T. Virtual screening for the discovery of bioactive natural products. In: Petersen F, Amstutz A, editors. Natural Products as Drugs. Vol. 1,Basel: Birkhauser Verlag AG; 2008. p 211–249.
128. Shen J, Xu X, Cheng F, Liu H, Luo X, Shen J, Chen K, Zhao W, Shen X, Jiang H. Virtual screening on natural products for discovering active compounds and target information. Curr Med Chem 2003;10(21): 327–2342.

MICROBES TO MAN: FROM SOILS AND THE DEPTHS TO DRUGS

David J. Newman
Gordon M. Cragg
Natural Products Branch,
Developmental
Therapeutics
Program, NCI-Frederick,
Frederick, MD

The opinions expressed in this article are those of the authors and are not necessarily those of the US Government

1. INTRODUCTION

Although the title may appear to have a slight air of flippancy about it, we hope to show, using some specific sources and compounds derived therefrom, that microbes from all three domains, the Archaea, Bacteria and Eucarya (using the nomenclature of Woese [1,2]) are major sources of novel structures of utility in the treatment of a variety of diseases. The emphases, however, will be on modern antiinfectives, anticholesteroemics and, for obvious reasons, the multiplicity of diseases that go by the name of "cancer." We also hope to demonstrate that the advent of genomic information from microbes is leading to a significant re-evaluation of the potential of microbes from possibly all domains, since many of the genomic sequences contain previously unrecognized secondary metabolic "cryptic clusters" coding for previously unknown secondary metabolites.

By the end of the chapter, we hope that we will have shown that microbes of all types are an enormous resource, not only of novel structures that may occasionally become drugs in their own right, but that have led to a series of very elegant synthetic approaches yielding molecules that, while they may not bear too much 2D-similarity to the original natural product, would never have become drugs or leads to drugs in the absence of the original compound. A theme that also will run through the chapter is that, in the case of marine invertebrates, the original source of the natural product may or may not be the actual producer. Furthermore, we will refer to reports indicating that some of the well-known antitumor agents isolated from plants may well have a fungus involved in their biosynthesis, and in some cases, that there may be an element of "cometabolism" as defined by Dalton in the late 1970s [3,4].

2. INITIAL RECOGNITION OF MICROBIAL PRODUCTS AS ANTIBACTERIAL AND ANTIFUNGAL DRUGS

In order to set the stage for the discussion of microbial secondary metabolites as drugs, if one were asked to name the single microbial-sourced natural product that has saved the most lives, directly or indirectly since its original discovery, penicillin G (**1**) would undoubtedly be the molecule of choice. In this day and age, there are few people in developed countries who can remember the preantibiotic age with any clarity, though some older residents may have hazy memories of relatives dying at young ages due to bacterial infections; however, that is not the norm.

1. Penicillin G

The first usage of natural products as true antibacterials rather than as surface sterilants can be traced to the later stage of World War II, with the use of microbial-derived secondary metabolites such as penicillin and streptomycin being the exemplars known in the West. Though there were anecdotal reports of scientists such as Tyndall, Roberts and Pasteur in the 1870s recognizing antagonism between various bacteria, this development occurred mainly as a result of the recognition by Fleming in the late 1920s of the activity of penicillin, leading ultimately to the well known and documented use of penicillins G and V [5], as well as the discovery of streptomycin by Waksman [6] in the early 1940s. However, it also appears that in the same time frame the antibiotic Gramicidin S (Soviet Gramicidin; [7–9]) was being used as a treatment for war wounded in the USSR.

The aminoglycosides such as streptomycin, neomycin, and the gentamicins have a long and storied history as treatments for antibacterial infections, particularly in the early days when streptomycin was a treatment for both infected wounds and also for tuberculosis. There were few modifications of the basic molecule(s), however, that went into clinical use, mainly due to the complexity of chemical modification of saccharidic-based structures. Thus we will not discuss this class any further, or molecules such as the rifamycins and their manifold derivatives. Instead, due to space limitations, we will show how β-lactams, tetracyclines, macrolides, glycopeptides and pleuromutilins, even though discovered many decades ago, are still being utilized as base structures upon which to build molecules.

2.1. β-Lactam Antibiotics

The number of penicillin and cephalosporin-based molecules produced by semi- and total synthesis to date is well in excess of 20,000. Most started with modification of the fermentation product, 6-amino-penicillanic acid (**2**) or the corresponding cephalosporin, 7-amino-cephalosporanic acid (**3**), both of which can be produced by simple chemical or biochemical deacylation from penicillin or cephalosporin C. The number above is only approximate since a significant proportion of structures explored in industry were never formally published, or were only mentioned in the patent literature, particularly if they had marginal or no significant activity levels over those that had been previously reported. In order to gain an idea of the multiplicity of natural product structures that have been reported through 1979, the reader should consult the excellent review by Aoki and Okuhara from Fujisawa [10].

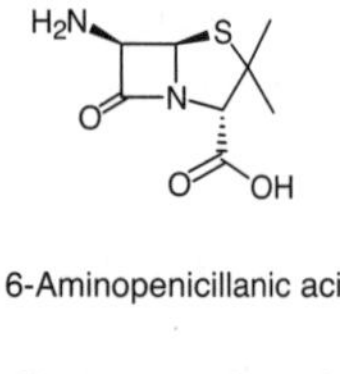

2. 6-Aminopenicillanic acid

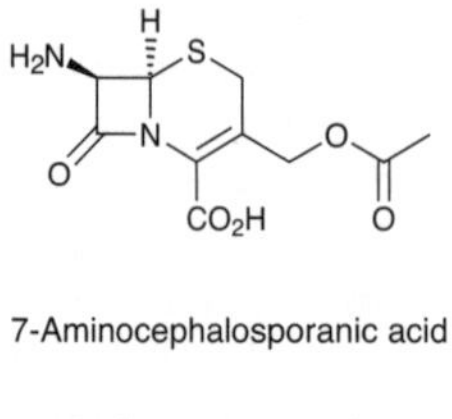

3. 7-Aminocephalosporanic acid

In 1948, the ring-expanded version of penicillin, cephalosporin C, was reported from *Cephalosporium* sp. by Brotzu [11–13]; since the original isolate was from the Mediterranean Sea off the coast of Sardinia, this could be classified as being from a marine source. The structure was reported in 1961 by the Oxford group [14,15], and in a similar manner to the chemistry around the penicillin nucleus, this ring-expanded molecule also served as the building block (as 7-aminocephalosporanic acid) for many thousands of cephalosporins. The first orally-active molecule, cephalexin (**4**) was introduced in 1970, and since that time, a multitude of cephalosporins have been synthesized with the aim of producing molecules that are more resistant to β-lactamases.

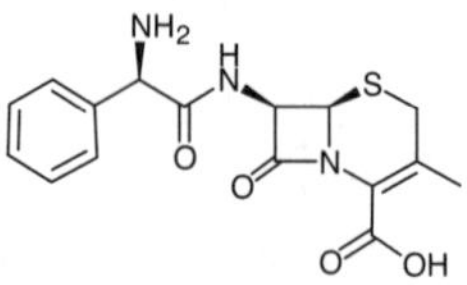

4. Cephalexin

In order to prolong the "medicinal life" of β-lactams that were no longer useful due to the presence of both constitutive and inducible β-lactamases, efforts were made in the late 1960s and early 1970s, particularly by Beecham (now part of GlaxoSmithKline (GSK)) and Pfizer, to find molecules that would have similar pharmacokinetics to the β-lactams but that would inhibit the β-lactamases that were part of the pathogenic microbe's defense systems. To this end, Beecham discovered the naturally occurring clavulanate family [16–18], and a 1:1 mixture of amoxicillin and clavulanic acid (**5**) known as Augmentin®, was launched in 1981, thus extending the franchise of this particular β-lactam well beyond its original patent date. The Pfizer [19] candidate, CP-45,899 or sulbactam (**6**), was basically penicillanic acid with a sulfoxide in place of the sulfur. Replacement of one of the gem methyl groups by a 1,2,3-triazol-1-yl-methyl substituent gave tazobactam (**7**), originally reported at a conference in 1985 by researchers from Taiho. Following publication of the synthetic details in 1987 by Micetich et al. [20], it was launched by Lederle, now Wyeth, even today, ~17 years after the last introduction, no other inhibitors have advanced to commercialization. Currently, clavulanate is combined with amoxicillin or ticarcillin, sulbactam with ampicillin, and tazobactam with piperacillin. All of these inhibit only class A serine-based β-lactamases, leaving a significant number of other β-lactamase enzymes where inhibitors are required, including the pharmacologically important zinc-containing β-lactamases [21].

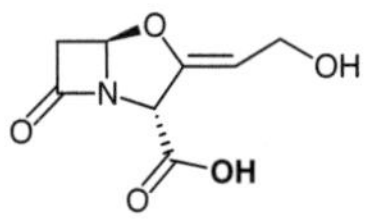

5. Clavulanic acid 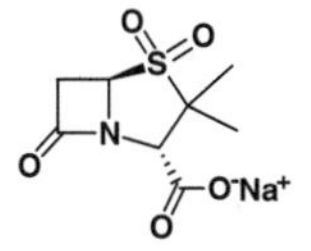 **6. Sulbactam** 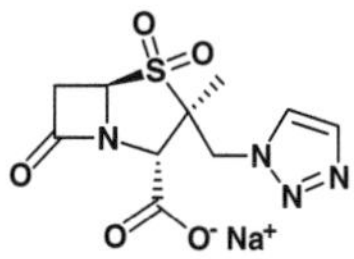 **7. Tazobactam**

In the same time frame, along with the search for β-lactamase inhibitors, efforts were underway to obtain the simplest β-lactam, a monobactam (**8**). Following many years of unsuccessful research at major pharmaceutical houses, predominately in the synthetic chemistry areas, papers from Imada et al. in 1981 [22] and a Squibb group led by Sykes [23], both reported the same basic monobactam nucleus (**8**). What is important to realize is that no molecules synthesized before the discoveries of these natural products had a sulfonyl group attached to the lactam nitrogen, which is an excellent method for stabilizing the single four-membered ring. Since that time a significant number of variations upon that theme have proceeded into clinical trials and in some cases, such as Aztreonam® (**9**) into commercial use. Recently (late 2007–early 2008), Aztreonam as the lysinate salt was submitted for approval under an Orphan drug category in both the European Union and the United States for the inhalation treatment of *P. aeruginosa* in cystic fibrosis. Currently (December 2008), the FDA is requiring further information and the status of the EU application is not yet known.

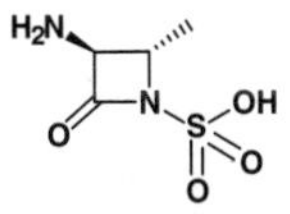

8. Monobactam nucleus 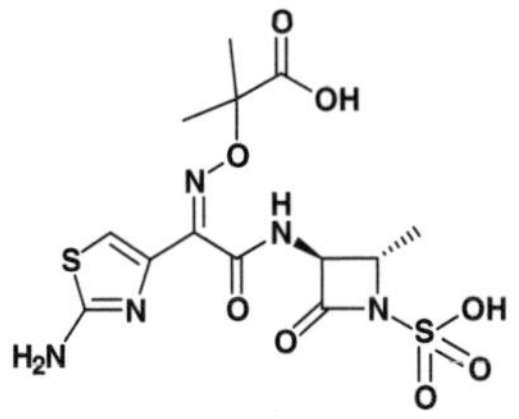 **9. Aztreonam®**

Even today, these "ancient molecular structures in drug terms" and others discovered after the early 1940s [10] are still valid as scaffolds upon which to base the synthesis of drugs. Thus, since the year 2000, three synthetic penems known as biapenem (**10**), ertapenem (**11**) and doripenem (**12**) based upon the structure of the natural product thienamycin (**13**), first reported in the patent literature in 1976, and fully described by Albers-Schoenberg et al. in 1978 [24], and two cephalosporins, cefovecin (**14**), a veterinary drug, and ceftobiprole medocaril (**15**) have been approved for commercial use. In addition to these there is one penem, tebipenem pivoxil (**16**), which is effectively a prodrug ester of the ex-Wyeth penem, tebipenem that has been preregistered in Japan with the aim of approval in early 2009.

Although we mentioned earlier that only three β-lactamase inhibitors had advanced into commercial use, there is now a potential "cephalosporinase inhibitor / cephalosporin combination" in trials. Forest Pharmaceuticals recently announced that the cephalosporin ceftaroline fosamil acetate (**17**), which is currently in Phase III clinical trials, has been combined with Novexel's synthetic β-lactamase inhibitor [25] NXL-104 (AVE-1330A; **18**) and has entered Phase II trials.

2.2. Tetracycline Antibiotics

The base molecule that was marketed from 1953 as Achromycin®, and the better known and earlier 7-chloro- and 5-hydroxy-derivatives, both marketed from 1948 as Aureomycin® (**19**) and Terramycin® (**20**), respectively, together with the dimethylamino derivative, Minocin® (**21**) marketed in 1972, are considered as members of the first (1948–1963) and second (1965–1972) generation tetracyclines. These molecules have been stalwart members of the physician's

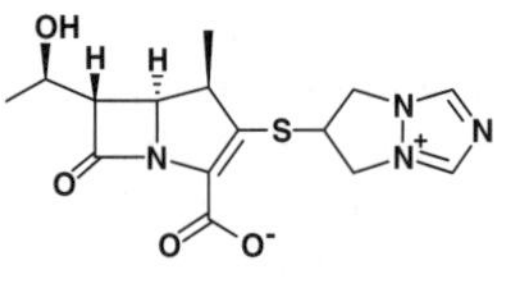

10. Biapenem

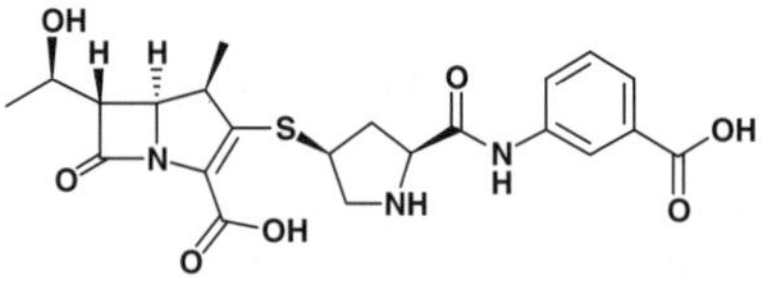

11. Ertapenem

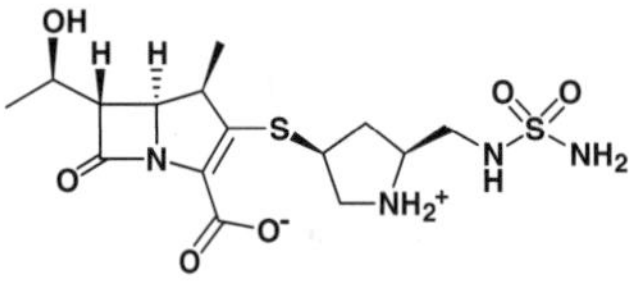

12. Doripenem

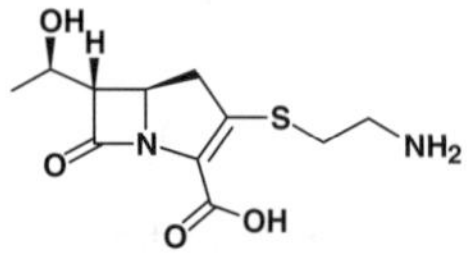

13. Thienamycin

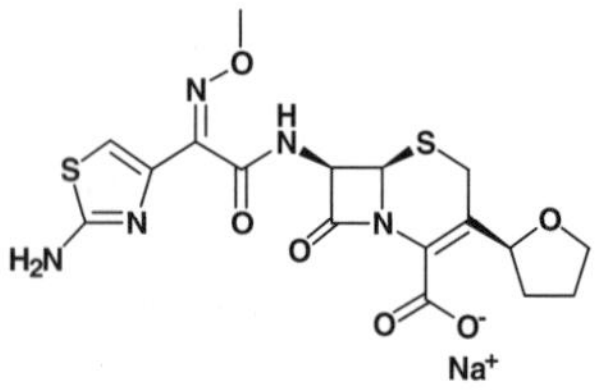

14. Cefovecin

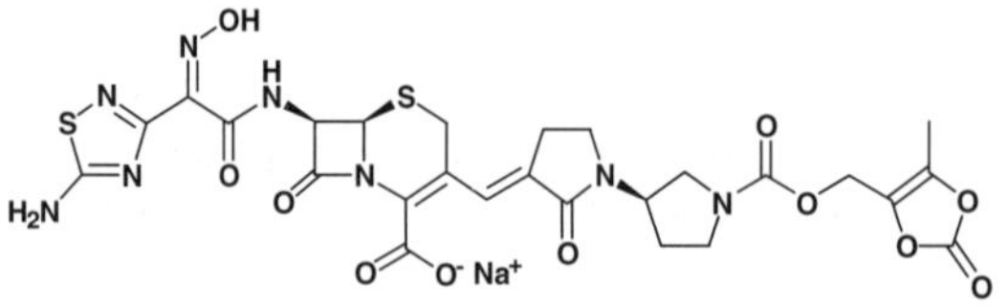

15. Ceftobiprole medocaril

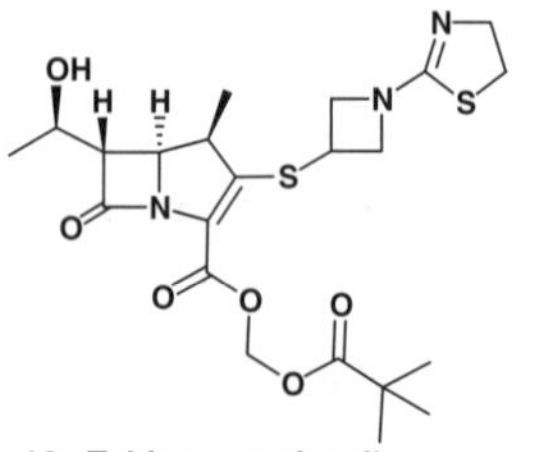

16. Tebipenem pivoxil

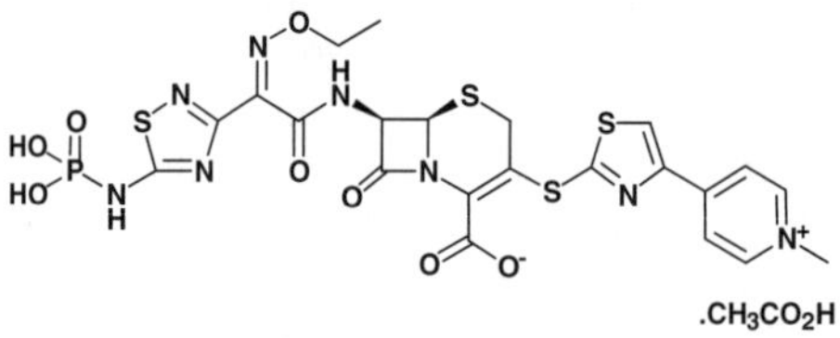

17. Ceftaroline fosamil acetate

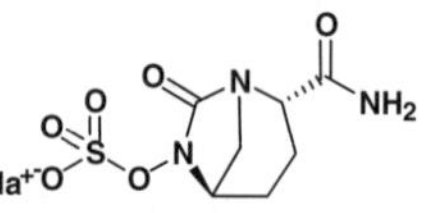

18. AVE-1330A

armamentarium for close to 60 years, and in the case of the latter three compounds, are still in use. During the earlier part of this timeframe, a large number of reports led to recognition of 6-deoxy-6-demethyltetracycline (**22**) as the minimum pharmacophore, and chemists were able to derive the optimal stereochemical and substitution patterns for antibacterial activity [26]. Clinical reports of the evolution of tetracycline resistance in *Shigella dysenteriae* in 1953 and of a multiply resistant *Shigella* in 1955 [27], led to the use of classical, and later of molecular genetics approaches, to determine the presence of multiple tetracycline efflux pumps and of protective ribosomal mechanisms. These factors are discussed in detail by Chopra and Roberts [26], and suggestive evidence of the monophyletic origin of these genes, plus the potential for cross-contamination from animal sources, was presented in 2002 by Aminov et al. [28].

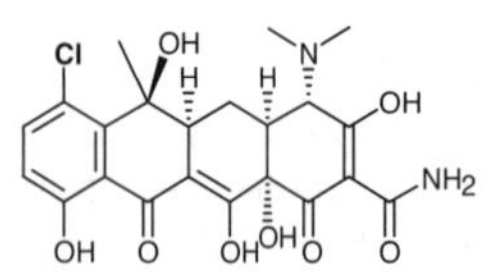

19. Aureomicin®

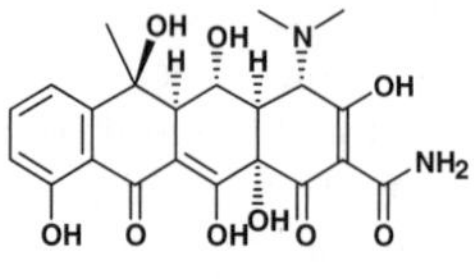

20. Terramycin®

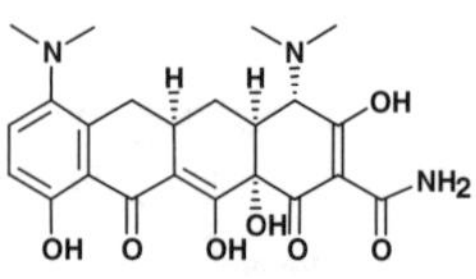

21. Minocin®

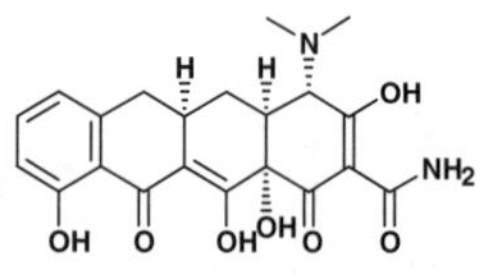

22. 6-Deoxy-6-demethytetracycline

Following on the major resistance problems with the first- and second-generation tetracyclines, a series of synthetic and semisynthetic modifications of the base pharmacophore were made, with special emphasis on position 9 of the base molecule. Previous attempts to modify at this position with 9-nitro, 9-amino or 9-hydroxy derivatives led to molecules with poor antibacterial activities; however, scientists at the then Lederle Laboratories (currently Wyeth but due to combine with Pfizer in mid-2009) discovered that 9-acylamido derivatives of minocycline (Minocin; **21**) had activities comparable to first- and second-generation molecules, but did not have activity against resistant organisms [29]. Further work at Lederle/Wyeth led to the publication in 1999 of the synthesis of GAR-936, a glycyl derivative of a modified doxycycline molecule, now known as tigecycline (**23**).

Although Woodward, together with workers at Pfizer, had identified the correct structure of the tetracycline molecule in 1952–53 through use of degradative studies on oxytetracycline, no total synthesis of a nonracemic tetracycline molecule had been reported, other than from a tetracycline derivative, until the reports from Tatsuta's group in Japan, starting from D-glucosamine [32,33]. This was followed in 2005 by a short report from the Myers group at Harvard, starting with benzoic acid [34]; their methodology was very recently expanded to give a generalized synthetic method [35] that also gives novel pentacyclic molecules never previously reported. Examples are **24**, **25**, and **26** and include the tigcycline homolog (**27**), all of which have activities *in vitro* and *in vivo* against normal and resistant microbes; more details are available in Table 4 in Sun et al. [35].

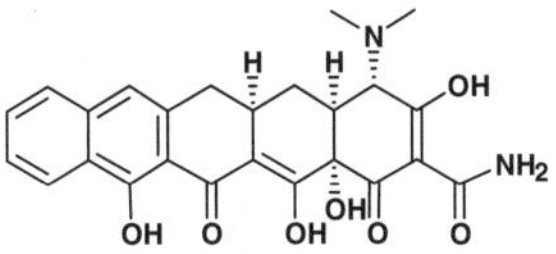

24

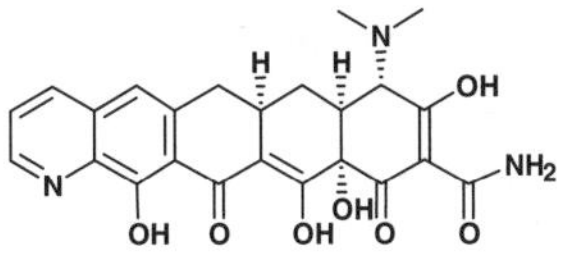

25

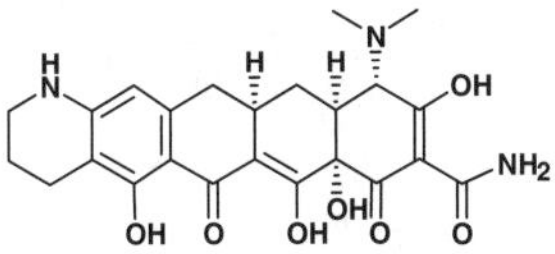

26

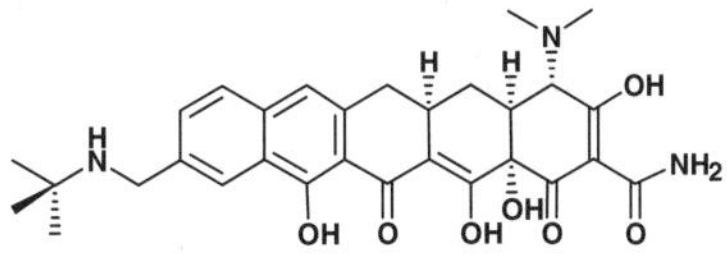

27

GAR-936 was approved in 2005 by the FDA for complicated skin and soft tissue infections, and has broad-spectrum activity including against both Gram-positive and Gram-negative bacteria and methicillin-resistant *Staphylococcus aureus* (MRSA). Thus, utilization of what are effectively relatively simple chemical modifications to an old molecule, can result in a new lease on life for these base structures, and can provide activity against clinically important infections [30,31].

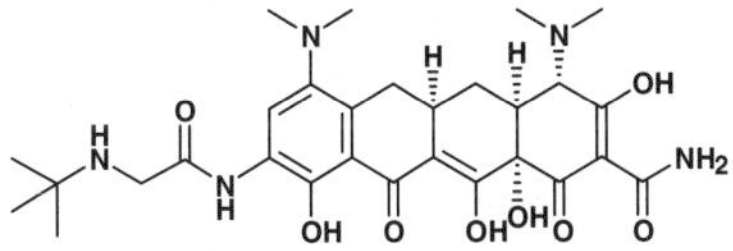

23. Tigecycline

Thus even today, 60 years after the original reports by Duggar [36], this old class of antibiotics is generating significant interest both chemically and biologically. Thus, the new knowledge from genetic analyses of tetracycline biosynthesis in bacteria, coupled to the combination of new chemical synthetic methods (see below) and advances in the biosynthetic processes as recently reported by Pickens and Tang [37], bode well for the future of this compound class.

2.3. Glycopeptide Antibiotics

Vancomycin (**28**), a natural product first approved in 1955, is still the prototype for variations around the same mechanism of action, the binding to the terminal L-Lys-DAla-DAla tripeptide in Gram-positive cell wall biosynthesis. The compounds discussed in this section are semisynthetic modifications of the same basic structural class (glycopeptides) as the prototype vancomycin, thus following in the "chemical footsteps" of the β-lactams and the tetracyclines discussed above and the macrolides discussed in a later section. Currently (December 2008) there are a number of such molecules in clinical trials, with two semisynthetic glycopeptides oritavancin (**29**) and telavancin (**30**) in preregistration status with the FDA, and dalbavancin (**31**), which though in preregistration through most of 2008, is now back in Phase III trials. In all cases their antibacterial mechanism is through inhibition of cell wall production, initially via the vancomycin target, though the exact mechanisms can vary with the individual agent. In the case of oritavancin, very recent data indicate that the agent is comparable to vancomycin in its inhibition of *trans*-glycosylation, but is more effective as a transpeptidation inhibitor [38]. As mentioned above all are semisynthetic derivatives of natural products, with oritavancin [39] being a modified chloroeremomycin (a vancomycin analog), dalbavancin [40] being based on the teicoplanin relative, B0-A40926, and telavancin (TD-6424) directly based on chemical modification of vancomycin [41].

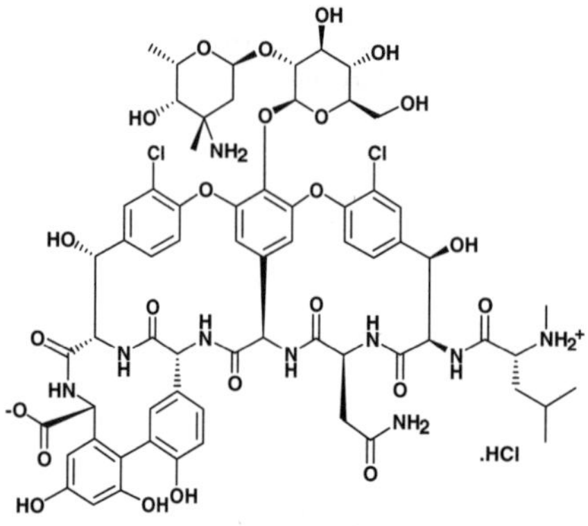

28. Vancomycin

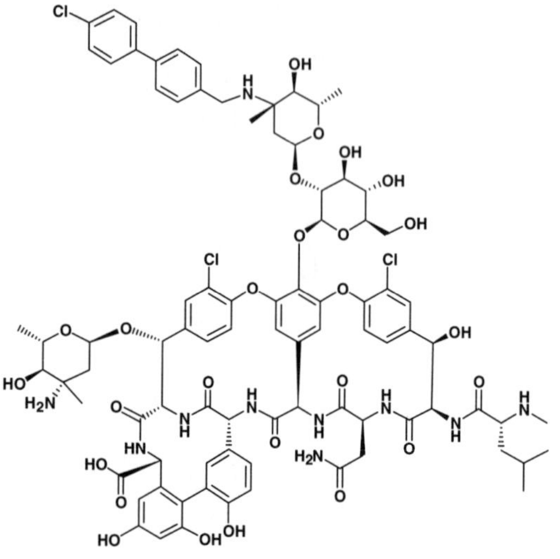

29. Oritavancin

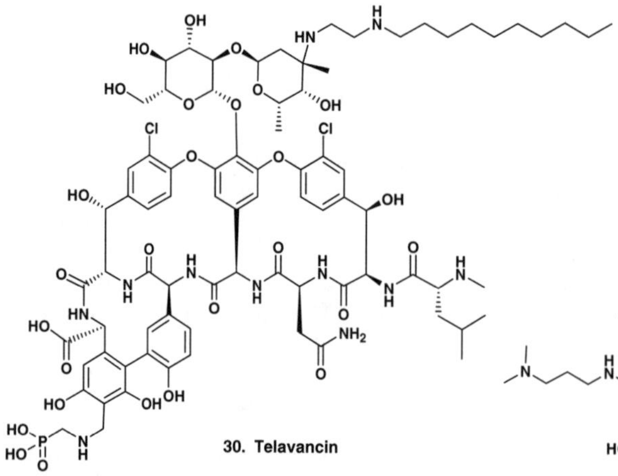

30. Telavancin

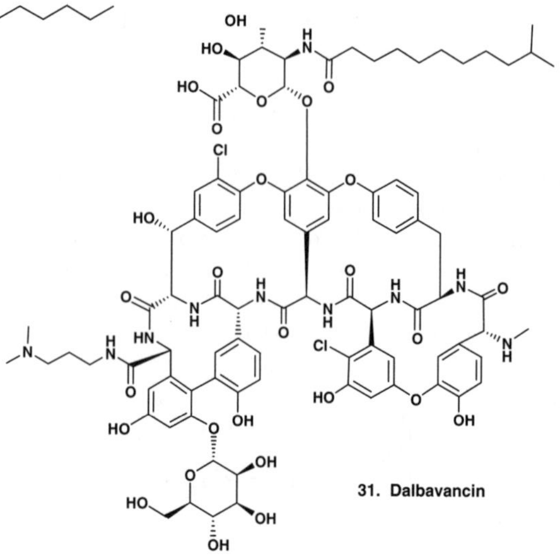

31. Dalbavancin

That one may combine the characteristics of two separate agents working at different targets within the same basic biological area is shown by the work of Theravance (also the originator of telavancin). They have successfully combined a cephalosporin with vancomycin to produce TD-1792 (**32**) which is currently in Phase II trials against complicated skin and soft tissue infections [42], and which also demonstrates potent *in vitro* activity against both sensitive and MDR Gram-positive bacteria (except for *VanA E. faecium*) [43]. Thus two old antibiotic classes can produce novel agents, again underscoring the possibilities of revisiting older structures if one understands their chemical and biological history.

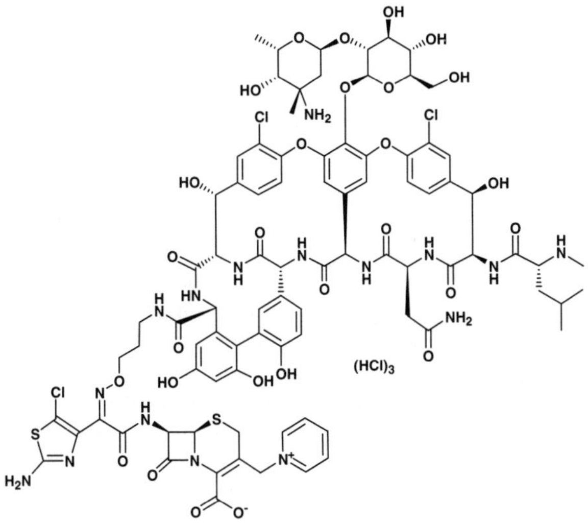

32. TD-1792

2.4. Lipopeptide Antibiotics

Although vancomycin (and possibly others in this class) does have activity against lipid II the advent of the *VanR* phenotype in pathogenic bacteria has resulted in the re-investigation of microbial secondary metabolites that have languished for a number of years, and these are now either important parts of the physician's armamentarium or are currently in clinical trials against such pathogens. The first example is ramoplanin (**33**), a lipopeptide antibiotic complex isolated from *Actinoplanes* sp. ATCC33076 consisting of factors A1, A2, and A3 [44,45], that exerts its antibacterial activity by binding to the peptidoglycan intermediate Lipid II (C_{35}-MurNAc-peptide-GlcNAc) and disrupting bacterial cell wall synthesis [46–48]. Factor A2 ("ramoplanin") (**33**) is the major component of the complex and is currently in Phase III trials with Oscient Pharmaceuticals for the treatment of *Clostridium difficile*-associated diarrhea (CDAD) [49,50]. The complex was also awarded Orphan Drug Designation in the European Union for the prevention of invasive infections due to vancomycin-resistant *Enterococci* (VRE) in colonized patients deemed at risk of infection, though currently no trials appear to be ongoing.

Another older cyclic lipopeptide that had moved around from company to company, having started at Lilly, moved to the then Lederle (now Wyeth Pfizer?) and was finally developed by Cubist as a new antibiotic against MRSA, is daptomycin (**34**), a member of a large class of complex cyclic peptides with variations in the peptidic components and the acylating fatty acids. These include the mixtures identified as the daptomycin/A21978 complex, the A54145 complex, the CDA complex, the friulimicins/amphomycins, and the

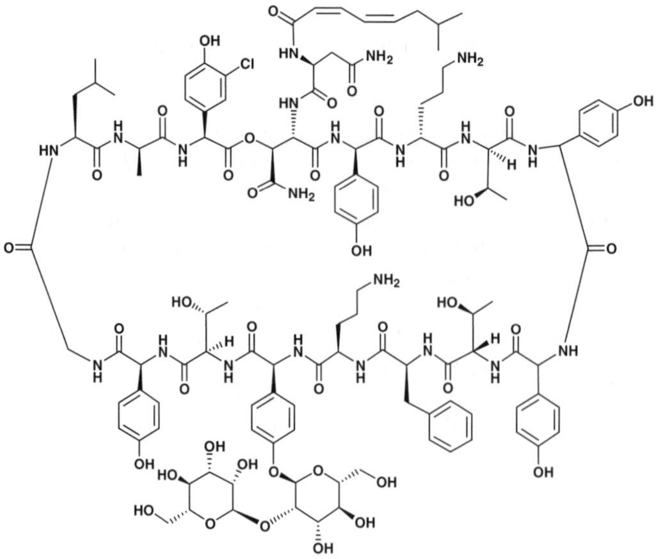

33. Ramoplanin A2

laspartomycin/glycinocins. Their base structures, as well as their biosyntheses and potential for genetic manipulation, were discussed in detail in 2005 by Baltz et al. from Cubist [51]. Further examples on the potential for such modifications were published from 2006 to late 2008, and these should be consulted as they demonstrate the potential for such "combinatorial biochemistry" to produce complex structures with modified activities [52–54].

2.5. Macrolidic Antibiotics

Following on the track of novel modifications of old structures that bind to ribosomes, thus inhibiting protein synthesis [55], there have been four molecules since 2000 formally based upon the erythromycin chemotype that have either been approved, such as telithromycin (2001; **35**), or entered clinical trials, namely cethromycin (ABT-773; Phase III; **36)** [56], EDP-420 (EP-013420, S-013420; Phase II; **37**), and the product of glyco-optimization,

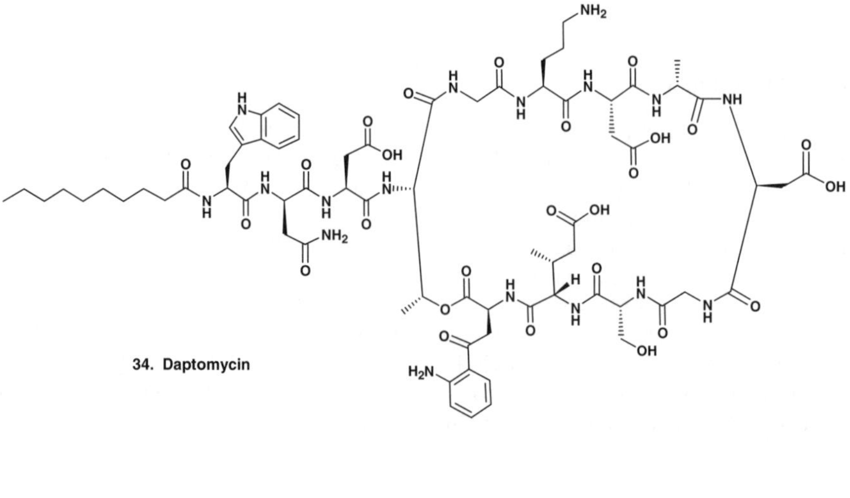

34. Daptomycin

CEM-101 (Phase I; **38**). As shown above, cethromycin (**36**) is currently in Phase III trials against community acquired pneumonia (CAP), and is being evaluated as an agent against anthrax and other biodefense targets by the National Institute of Allergy and Infectious Diseases (NIAID) and the US Army. The interesting modification of the base erythromycin structure, the "bicyclolide" EDP-420 (**37**), a novel, bridged bicyclic derivative originally designed by Enanta Pharmaceuticals [57,58], is currently in Phase II trials for treatment of CAP by both Enanta and Shionogi. Interestingly, this molecule is also quite active in a murine model of *Mycobacterium avium*, a common infection in immunosuppressed patients [59], which may well expand its usage in future trials.

From a modern chemistry perspective, Lewis et al. have recently reported the use of novel peptidic catalysts in order to acylate the unreactive secondary hydroxyl group in the 11 position in the macrolide ring of erythromycin (**39**), rather than having to proceed via the two sugar-linked primary hydroxyl groups (2′ and 4″) in the hexose substituents [60]. The earlier methods that produced the fully acylated derivatives involved difficult ester hydrolyses in order to obtain the desired products, both 11-acylated derivatives and 4″-phosphorylated compounds, that would mimic the putative method(s) of resistance by bacteria. Although the synthetic compounds were roughly 16-fold less active than erythromycin A against the organisms tested, the potential of this type of catalytic access to unreactive sites was adequately demonstrated.

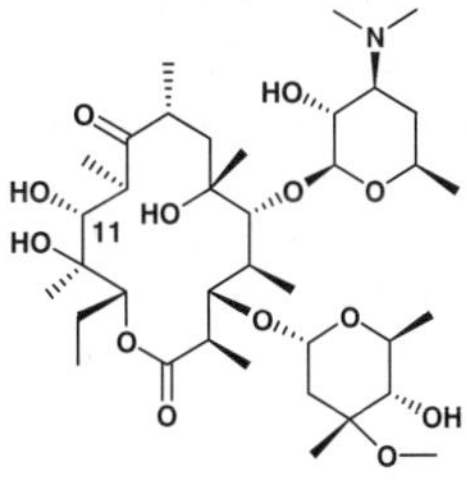

39. 11-Hydroxyl position in erythromycin

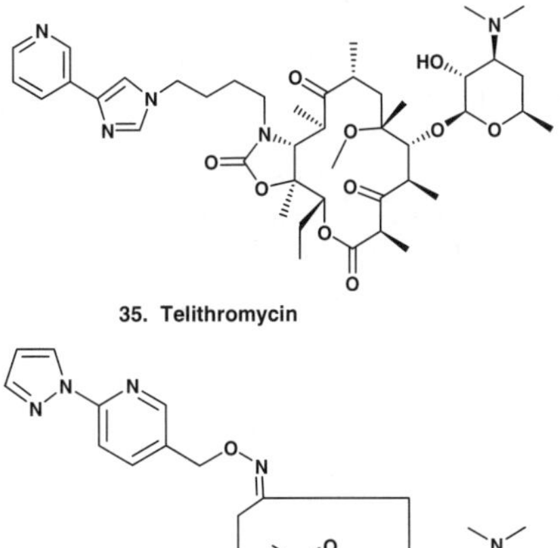

35. Telithromycin

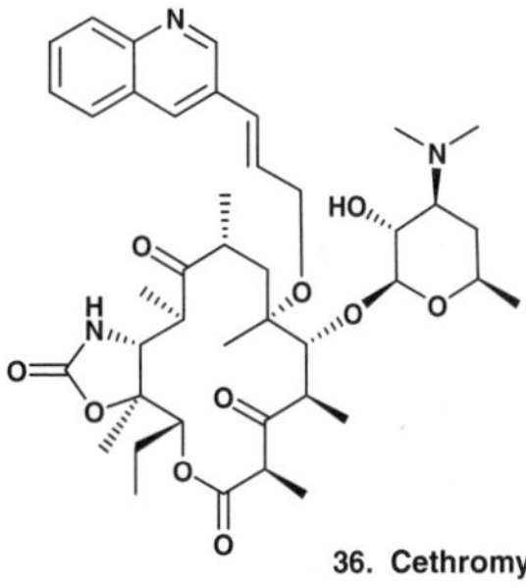

36. Cethromycin

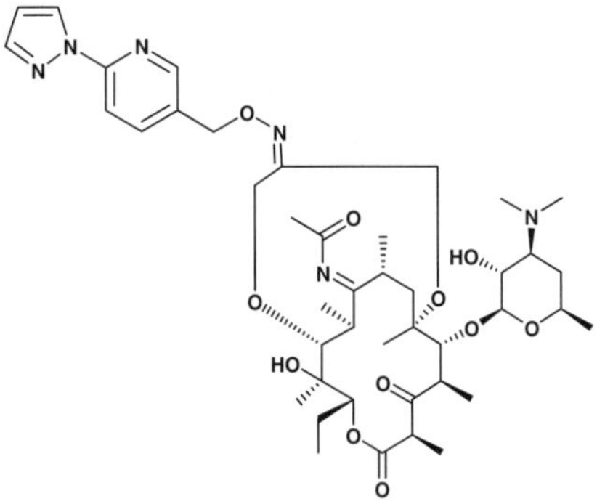

37. EDP-420

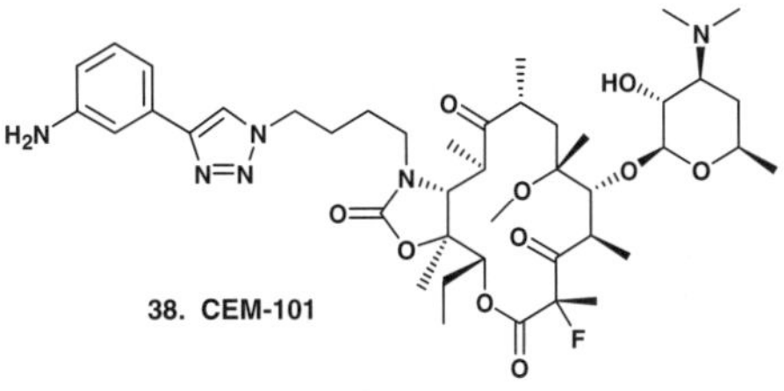

38. CEM-101

2.6. Pleuromutilin Antibiotics

Demonstrating yet again that older antibiotic structures have significant validity for today's diseases, GlaxoSmithKline received approval in 2007 for the modified pleuromutilin, retapamulin (**40**), for the treatment of impetigo in pediatric patients. The base structure, pleuromutilin (**41**) was first reported in 1951 from the basidiomycetes *Pleurotus mutilis* (FR.) Sacc and *Pleurotus passeckerianus* Pilat [61]. In the middle 1970s, a significant amount of work was reported on the use of derivatives of pleuromutilin as veterinary antibiotics [62], including approval of valnemulin (**42**) in 1999 under the trade name of Econor® by Sandoz. Thus, the subsequent utilization of the base molecule as a source of human use antibiotics is very reminiscent of the work that led to the approval of Synercid® in the late 1990s, as the base molecules in that case were also used extensively in veterinary applications.

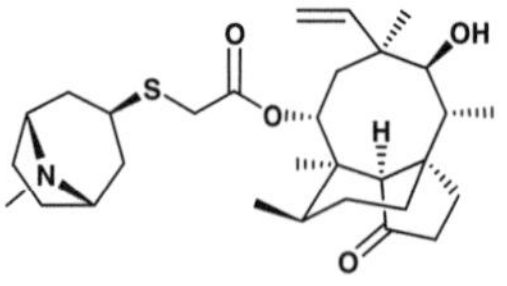

40. Retapamulin

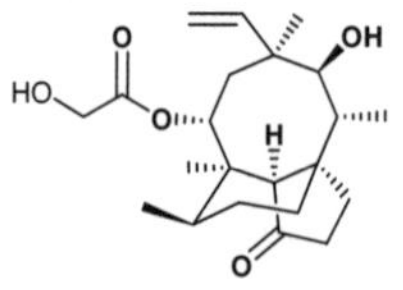

41. Pleuromutilin

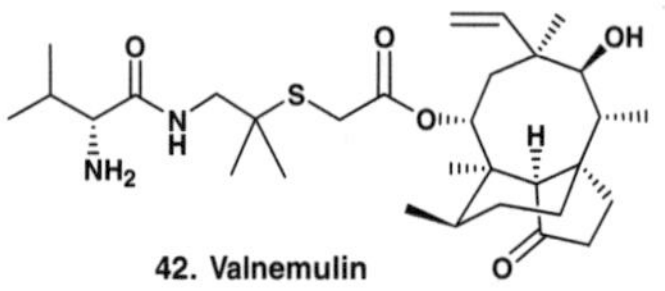

42. Valnemulin

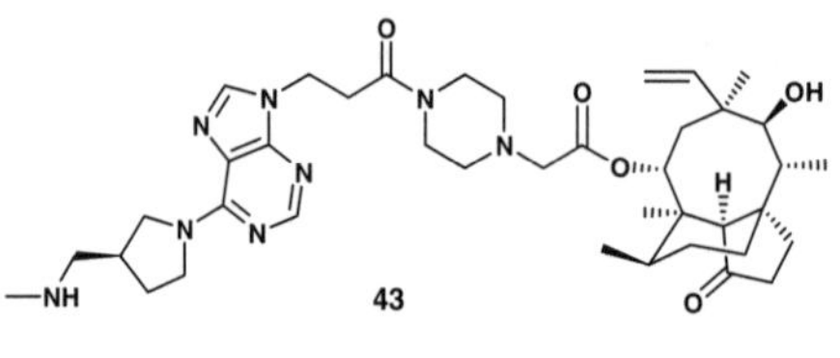

43

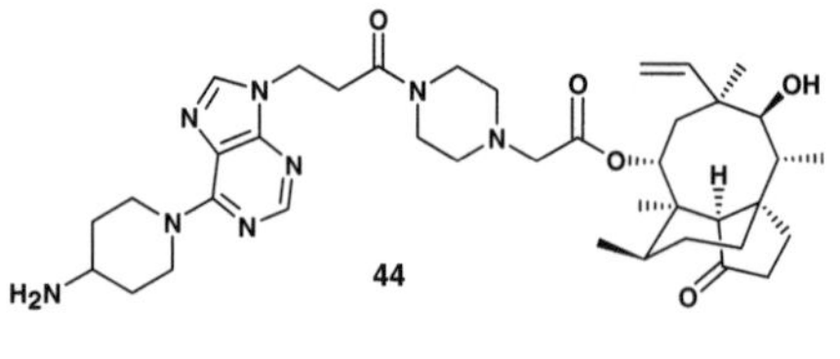

44

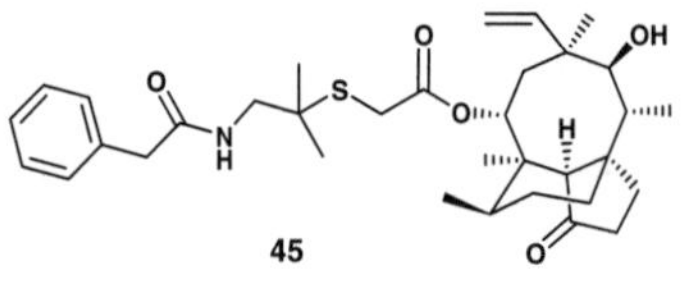

45

It is quite probable that a number of human use antibiotics based upon this "elderly" structure will enter later human trials as currently there are four "mutulins" in Phase I clinical trials against Gram-positive infections. Although structures have not yet been released, two are from GSK under the code numbers 565154 and 742510 for oral use, and the other two identified by the codes BC-3205 and BC-7013 are from Nabriva in Vienna, Austria, the former for oral use and the latter for topical use.

Continued interest in the base molecule by synthetic chemists is demonstrated by two distinctly different pleuromutilin derivatives. The first series, with a piperazine ring spacer (**43** and **44**) from Hirokawa et al. at Dainippon Sumitomo, have *in vitro* and *in vivo* activities against a plethora of resistant Gram-positive pathogens. Very recently a group from Sichuan University in China demonstrated that thiocarbamate derivatives of the base pleuromutlin structure, such as (**45**), demonstrated activity against Gram-negative *Salmonella* species, an activity not shown with the base molecule [63].

2.7. Antifungal Antibiotics

Despite considerable amounts of time and effort being expended in the early days of antibiotic discovery (middle to late 1940s), only three general use antifungal agents entered general clinical practice as a result. Since information on Russian efforts in this field at the time is effectively not available, as far as we are aware the first clinically used antifungal natural product was griseofulvin (**46**) that though launched in 1958 was originally reported in 1939. Its nonpolyene structure was defined in a series of papers in 1952 using classical techniques [64], and even today, close to 70 years after it was first described, it is still in clinical use against dermatophytes. This is the only class of fungi against which it is active, and long-term treatment is necessary due to its insolubility.

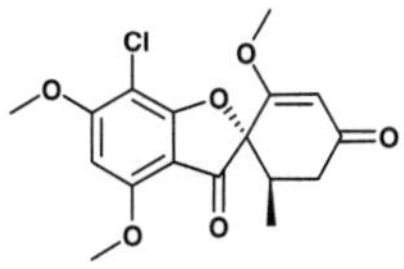

46. Griseofulvin

Perhaps the best-known clinical agent is the heptaene polyene, amphotericin B (**47**), isolated from *Streptomycetes nodosus*. It was first reported in 1956, but its full structure was not elucidated until 1970 using X-ray crystallography [65], closely followed by a description of the absolute configuration utilizing the iodo-derivative, and by mass spectroscopy [66]. Quite recently, 50 years after its initial discovery, a full review giving the highlights of the chemistry was published by Cereghetti and Carreira [67].

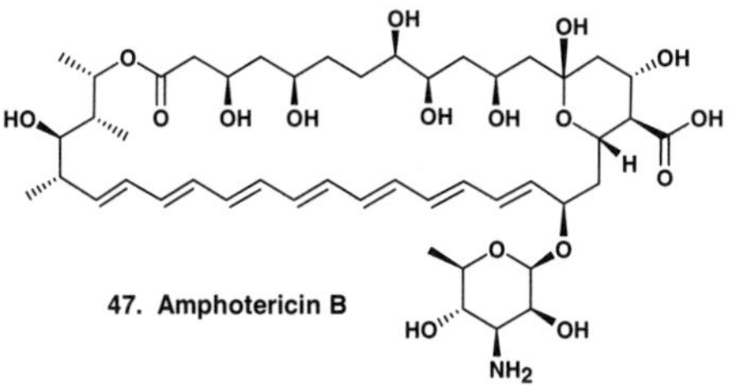

47. Amphotericin B

Though many polyenes with varying numbers of conjugated double bonds have been reported since those early days, only one other compound of this class, the tetraene nystatin (**48**), has gone into general clinical use. Like amphotericin B, its primary clinical indication is for systemic candidiasis. It was first reported from *Streptomyces noursei* and in fact was the first of this general structural class to be identified in 1950. As with amphotericin, its structure was reported in the 1970 time frame by two groups, one using classical chemical degradation plus proton NMR [68], and the other via mass spectroscopy [69]. Confirmation of the proposed hemiketal structures of both amphotericin B and nystatin was subsequently published by the Rinehart laboratory in 1976 [70].

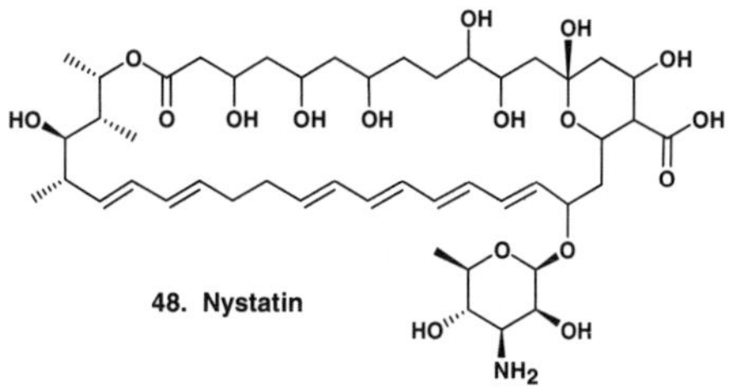

48. Nystatin

Though there have been many compounds entering preclinical trials as potential antifungal agents in the past 30 or so years, the majority of these were synthetic agents based on miconazole and its further-substituted azole derivatives. Microbial natural products of the echinocandin/pneumocandin class of glucan inhibitors have been known for many years, and sporadic attempts to develop members of this well-known but, at the time, intractable class of compounds have been pursued since their original discovery. However, it was not until the twenty-first century that three natural product-derived antifungal drugs, obtained through their chemical modification, were approved for human use as antifungal agents [71,72]. In order of FDA approval, these were: caspofungin (2001, Merck; **49**) that recently has been shown to function successfully in both invasive candidiasis and in candidaemia [73]; micafungin (2002, Astellas; **50**) that is currently in clinical trials for pediatric disease [74]; and anidulafungin (2006, Pfizer; **51**) [75,76]. Another modification of the basic echinocandin structure, aminocandin (HMR-3270; Novexel; **52**), a semisynthetic derivative of deoxymulundocandin, is currently in Phase I clinical trials, with Phase II trials reported as being scheduled [77].

3. ANTICHOLESTEROLEMICS

A major cause of elevated blood pressure is the physical blockage of the arteries by plaques of cholesterol/lipoproteins (atherosclerotic plaque). Since under normal conditions, humans usually synthesize about 50% of their cholesterol with the rest coming from diet, if the

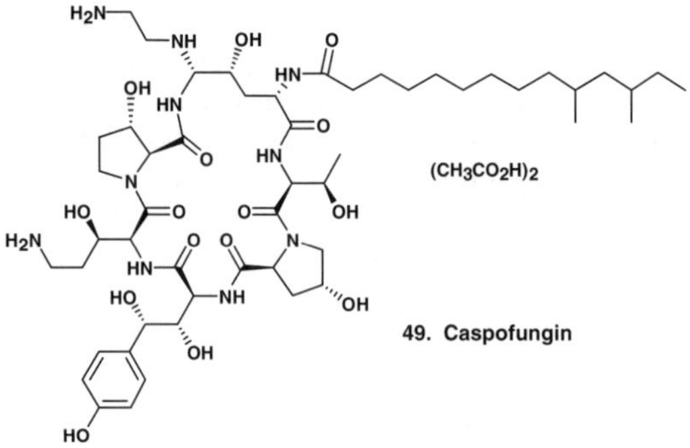

$(CH_3CO_2H)_2$

49. Caspofungin

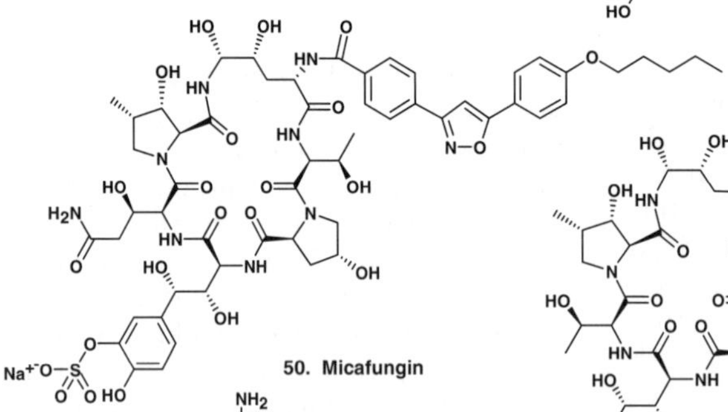

50. Micafungin

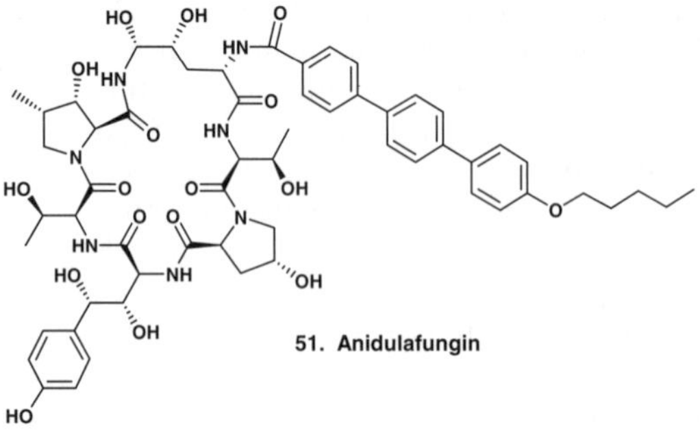

51. Anidulafungin

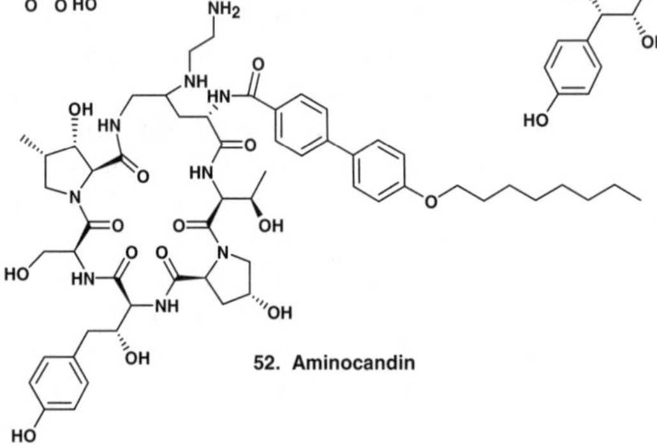

52. Aminocandin

synthesis can be inhibited, a reduction in overall cholesterol/lipoprotein levels may reduce the deleterious effects of these metabolites.

In eukaryotes (both fungi and higher organisms) the rate-limiting step in the biosynthesis of cholesterol is the reduction of hydroxymethylglutaryl coenzyme A (HMG-CoA) to produce mevalonic acid; this step was recognized by Endo in the 1970s as a potential site for inhibition of cholesterol biosynthesis [78]. By using over 6000 microbial fermentation broths as the source of compounds and measuring the inhibition of ^{14}C acetate incorporation in the production of sterols, Endo first reported the inhibitory activity of a fungal metabolite, compactin (mevastatin; **53**), in 1975 [79], followed by reports of two others in 1976 [80,81]. Mevastatin (as compactin) was reported very shortly thereafter by Brown et al. from Beecham in the United Kingdom as an antifungal agent, and was shown to be a competitive inhibitor of the HMG-CoA reductase enzyme with K_is in the nanomolar range, but was not further developed due to apparent toxicity [82]. Later work showed that material seen in sections of mouse and rat livers were deposits of cholesterol due to the strong induction of HMG-CoA reductase in rodents [78,83]. Compactin was tested in humans in 1978 when a patient with familial hypercholesterolemia (1000 mgs of cholesterol per deciliter) was treated under what would be considered "compassionate use" in the United States and her level dropped to 700 mg/dL [84]. Endo, using a similar assay, reported the isolation of the 7-methyl derivative as monacolin K (mevinolin, now known as lovastatin; **54**) from *Monascus ruber* [85,86]. This agent was patented only in Japan and concomitantly, workers at Merck in the United States discovered the same material from *Aspergillus terreus*, using an isolated HMG-CoA reductase assay and microbial broths as their source of agents, having tried their synthetic library to no avail.

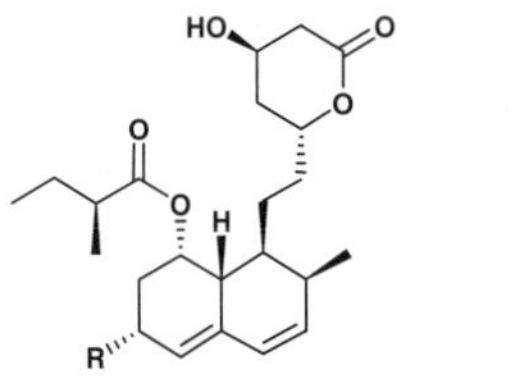

53. Compactin (Mevastatin); R = H
54. Mevinolin (Lovastatin); R = CH_3

Following submission of both structure and findings to the US Patent Office, a US patent was issued in late 1980, and lovastatin (Mevacor®; **54**) subsequently became the first commercialized HMG-CoA reductase inhibitor in 1987 [87]. Further work by Sankyo and Merck led to the entry of two more slightly modified versions. First, the 2-methylbutanoate side chain of lovastatin (**54**) was converted to 2,2-dimethyl butanoate, which was launched in 1988 as simvastatin (Zocor®; **55**) by Merck. Second, biotransformation of mevastatin led to the production of the lactone ring-opened 7-hydroxy derivative, pravastatin (**56**), which was launched in 1989 by Sankyo and subsequently licensed to Bristol-Myers Squibb.

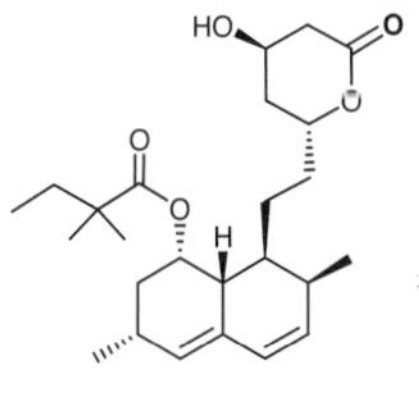

55. Simvastatin

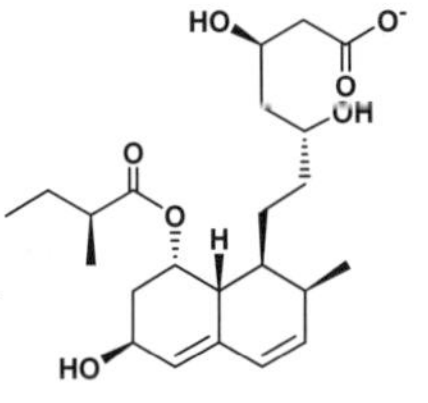

56. Pravastatin

Following the success of these agents, other small and large pharmaceutical companies used the information from mevastatin (**53**) and lovastatin (**54**) in syntheses of both the original materials, as well as potential new drugs. If one searches for the earlier examples, then the first reports outside of the patent literature are from the Merck group who identified mevinolin, demonstrating that simpler molecules where either the lactone or ring-opened "warhead" coupled to simple dichlorobenzene or fluorinated phenyl derivatives could produce compounds with activities comparable to or better than compactin [88–90].

These reports were followed up by scientists at the then Parke-Davis division of Warner-Lambert, well before the takeover by Pfizer, demonstrating that the same "natural product warhead" coupled to either fluorophenyl-substituted pyrroles [91] or 1,3,5-trisubstituted triazoles [92] were also active entities, basically comparable to compactin, though reference to the patent literature shows that a Sandoz scientist had filed a US Patent in 1985 (issued in 1986) covering the pyrazole nucleus [93]. Contemporaneously, Rosen and Heathcock published an excellent review covering the synthesis of mevinic acids that also shows some of the early "synthetic derivatives of mevinolin" [94]. From an historical viewpoint, the 2002 review by Roth [95] makes very interesting reading on the internal debate at Parke-Davis/Warner-Lambert over the development of atorvastatin (**57**) prior to their purchase by Pfizer.

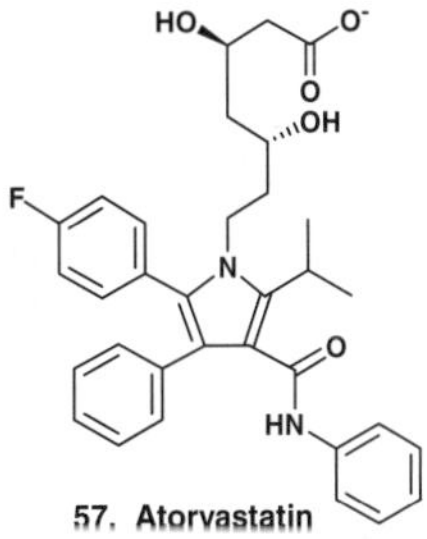

57. Atorvastatin

Thus as stated by Wilson and Danishefsky in 2006; "Atorvastatin (Lipitor®), certainly among the most commercially successful drugs ever used, differs substantially from both natural products, and its structure can be considered to have been creatively inspired by the parent compounds lovastatin and mevastatin. Clearly, though the presumed central statin pharmacophore is retained in the mega blockbuster Lipitor, the extensive periphery of the molecule has been completely reconfigured" [96]. One can therefore argue quite successfully that all of the "synthetic statins" fall under the same rubric, that is, effectively the natural product "warhead" plus varying lipophilic attachments designed to improve pharmacokinetics and/or to also avoid patent issues related to structures.

These "natural product-inspired synthetic agents" thus include the best-selling drug of all time, atorvastatin (Lipitor; **57**), whose sales in 2005/2006/2007/2008 were US $12.2/12.9/12.7/13.4 Billion worldwide, respectively, and the three further agents similar in concept to atorvastin in clinical use today that either have the lactonized or ring-opened form of the "warhead from Nature" coupled to different lipophilic entities.

3.1. Current Status of NP-Derived Anticholesterolemics

3.1.1. HMG-CoA Reductase Inhibitors Two compounds containing the natural product warhead are currently in Phase I clinical trials, BMS-644950 (**58**) from Bristol-Myers Squibb that has a lower reported myotoxicity than other current compounds [97], and RBx-10558 (**59**) from Ranbaxy.

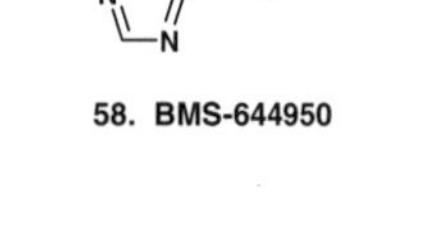

58. BMS-644950

59. RBx-10558

3.1.2. Combinations with Other Agents A very interesting combination (from at least two aspects) of compounds was approved in 2004 for the treatment of this disease complex. It was interesting both commercially and scientifically as it was a collaboration between Schering-Plough with Merck in the development and then commercialization of the Niemann-Pick C1-like Protein 1 inhibitor, ezetimibe (**60**), itself based upon the monobactam nucleus though not designed as an antibiotic, and the now generic Merck compound simvastatin (**55**) in a fixed ratio medication.

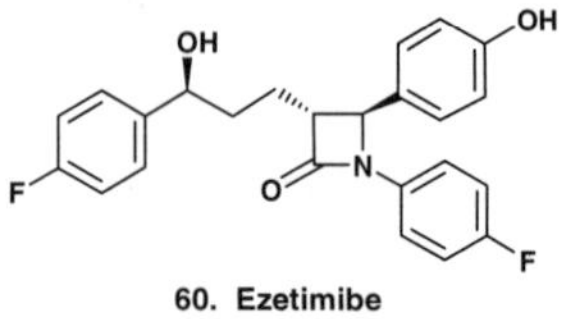

60. Ezetimibe

Inspection of the Prous Integrity® database (as of late December 2008) shows that the combination of a cholesterol synthesis inhibitor and another agent(s) with a different mechanism of action but related to lipid metabolism, is definitely "alive and well." Thus, currently the following combinations of different drugs are at various stages from approval to clinical trials. In 2002, a combination of extended release niacin plus lovastatin was approved as Advicor®, and is in Phase III trials for intermittant claudication. Merck has MK-524A (laropiprant; **61**) [98] as part of a triple combination with niacin and simvastatin. MK-524A serves as an inhibitor of the niacin-induced increase in plasma levels of prostaglandin D2 (PGD2) due to "flushing from the skin of the vasodilatory prostanoid after niacin treatment," with the underlying principle being that use of an inhibitor of either of the PGD2 receptors DP1 or DP2, plus niacin and a HMG-CoA reductase inhibitor might well be efficacious in raising HDL-C and lowering triglycerides. Merck also have the combination of atorvastin (**57**) and ezetimibe (**60**) in Phase II clinical trials, an interesting adjunct to Vytorin®.

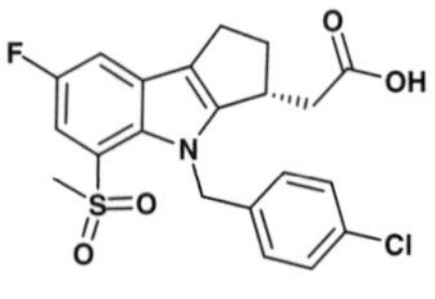

61. Laropiprant

In addition to these newer agents, there are currently three "fibrate-based" combinations with HMG-CoA reductase inhibitors in clinical trials. Sciele has fenofibrate/pravastatin combination and Astra-Zeneca and Abbott have rosuvastatin/choline fenofibrate (ABT-335), both being in Phase III, while in Phase II/III, Life Cycle Pharma have atorvastatin/fenofibrate.

Thus, 30 plus years after the first identification of natural products with HMG-CoA reductase inhibitory activity and their use as lead scaffolds in the preparation of semisynthetic variants, these agents are still being investigated as compounds in their own right, as leads to novel compounds with the same warhead, and as partners with other agents with different pharmacologic functions. In addition, they are being investigated as experimental treatments in quite different disease areas, including cancer and neuropharmacology.

4. ANTICANCER AGENTS

Since the early 1940s, the search for agents that may treat or ameliorate the scourge of cancer has involved all aspects of chemistry and pharmacology, and throughout all these years, natural products from microbes or derivatives of the original agents have played extremely important roles. Initially, these included: being the major source of drugs used for direct treatment; secondly serving as scaffolds upon which chemists would practice their medicinal chemistry skills; and currently, as modulators of specific cellular pathways in the tumor cell. Currently, the 162 (as of October 2008) agents available can be categorized as follows: N (25; 17%), ND (50; 31%), S (42; 26%), S/NM (16; 10%), S* (20; 12%), and S*/NM (7; 4%), using the nomenclature of Newman et al. [99,100].

As shown in Table 1, of the 46 commercially available compounds listed that have a natural product structure in their background, 25 (or 54%) are either directly from microbes or are derivatives of microbial secondary metabolites. In the case of trabectidin (**62**), circumstantial evidence is that it is produced by commensal microbes in the marine invertebrate.

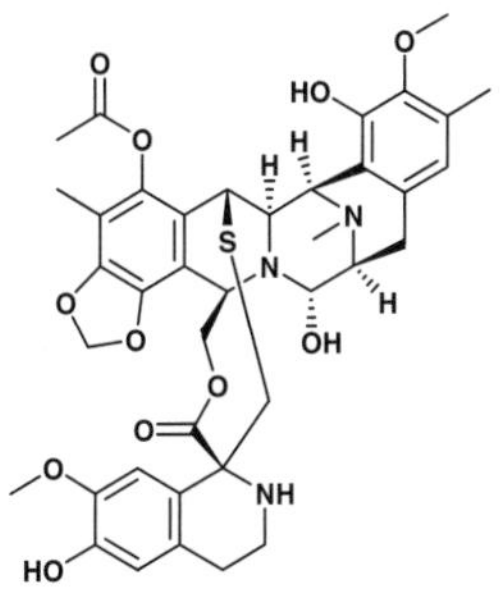

62. Trabectedin (Et743)

4.1. Actinomycins, Anthracyclines, Bleomycins, and Enediynes

It can be argued quite successfully that the discovery of actinomycins by Waksman [101] in 1940 led to two firsts; the first crystalline antibiotic and the first to demonstrate antitumor activity (actinomycin C) *in vitro* [102]. This report was then followed by a report later the same year by Schulte demonstrating the first clinical studies with these agents [103]. Over the next 60 plus years, actinomycins, usually actinomycin D (**63**), have been used as treatment for a variety of tumor types, but currently it is used primarily for treatment of rhabdomyosarcoma and Wilms' tumor in children, and in a nontumor activity, but related to its DNA-intercalating ability, actinomycin D may also significantly affect transcription factors at levels well below any antitumor activity [104]. For a more detailed history of actinomycins, their significant chemical synthetic and semisynthetic programs, and their activities, one should consult the 2005 review by Mauger and Lackner [105].

Over the two to three decades following the discovery of the actinomycins, a large variety of antibiotics from terrestrial microbial sources were tested for their abilities to act as antitumor agents. A number of basic structures were subsequently developed for use from these early studies and included the bleomycins, the mitomycins and the anthracyclines. Of these, perhaps one of the most utilized chemotypes, either directly as isolated from Nature or as semisynthetic modifications, is the chemical class known gener-

Table 1. Microbial-Sourced Antitumor Agents

Generic Name	Year Introduced	Source	Microbe
Carzinophilin	1954	N	Yes
Sarkomycin	1954	N	Yes
Mitomycin C	1956	N	Yes
Chromomycin A3	1961	N	Yes
Mithramycin	1961	N	Yes
Actinomycin D	1964	N	Yes
Bleomycin	1966	N	Yes
Doxorubicin	1966	N	Yes
Daunomycin	1967	N	Yes
Neocarzinostatin	1976	N	Yes
Aclarubicin	1981	N	Yes
Peplomycin	1981	N	Yes
Pentostatin	1992	N	Yes
Trabectedin	2007	N	Yes
Streptozocin	Pre-1977	N	Yes
Epirubicin HCl	1984	ND	Yes
Pirarubicin	1988	ND	Yes
Idarubicin hydrochloride	1990	ND	Yes
Zinostatin stimalamer	1994	ND	Yes
Valrubicin	1999	ND	Yes
Gemtuzumab ozogamicin	2000	ND	Yes
Amrubicin HCl	2002	ND	Yes
Hexyl aminolevulinate	2004	ND	Yes
Ixabepilone	2007	ND	Yes
Temirolimus	2007	ND	Yes
Imatinib mesilate	2001	S/NM	
Gefitinib	2002	S/NM	
Bortezomib	2003	S/NM	
Erlotinib hydrochloride	2004	S/NM	
Dasatinib	2006	S/NM	
Sunitinib maleate	2006	S/NM	
Lapatinib ditosylate	2007	S/NM	
Nilotinib hydrochloride	2007	S/NM	
Cytosine arabinoside	1969	S*	
Floxuridine	1971	S*	
Enocitabine	1983	S*	
Doxifluridine	1987	S*	
Fludarabine phosphate	1991	S*	
Gemcitabine HCl	1995	S*	
Capecitabine	1998	S*	
Clofarabine	2005	S*	
Nelarabine	2005	S*	
Decitabine	2006	S*	
Azacytidine	Pre-1977	S*	
Vorinostat	2006	S*/NM	

ically as the "anthracyclines," isolated from bacteria of the order *Actinomycetales*, with two of the most useful examples being daunorubicin (**64**) and its derivative doxorubicin (adriamycin) (**65**). Even today, doxorubicin (**65**) is still a major component of the treatment regimen for breast cancer. Although there have been many similar molecules isolated and described in the literature, it is doxorubicin and its more modern derivatives

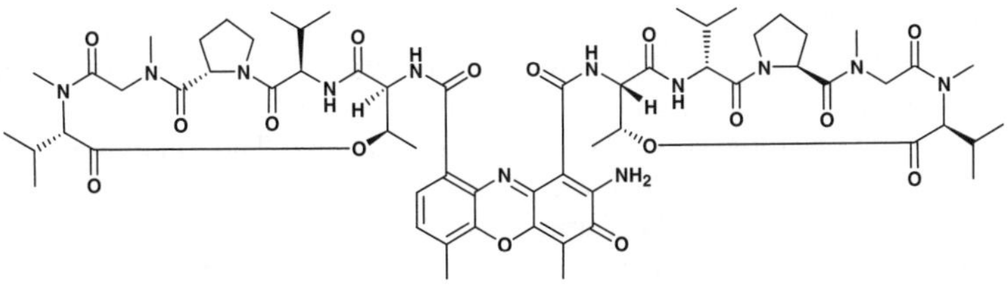

63. Actinomycin D

such as epirubicin (**66**), pirirubicin (**67**), idarubicin (**68**) and, more recently, valrubicin (**69**) and amrubicin (**70**) that have been approved for cancer treatment. The 2005 review by Arcamone [106] should be consulted for details of their history.

Another series of extremely important molecules, also from the *Actinomycetales*, is the family of glycopeptolide antibiotics known as the bleomycins (particularly bleomycin A_2 (**71**), Blenoxane®) and initially, the closely related structural class, the phleomycins.

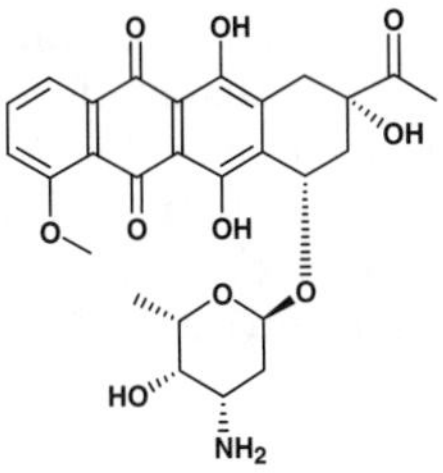

64. Daunorubicin

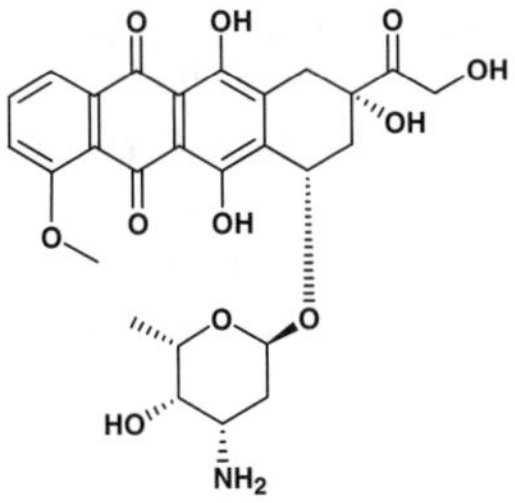

65. Adriamycin (Doxorubicin)

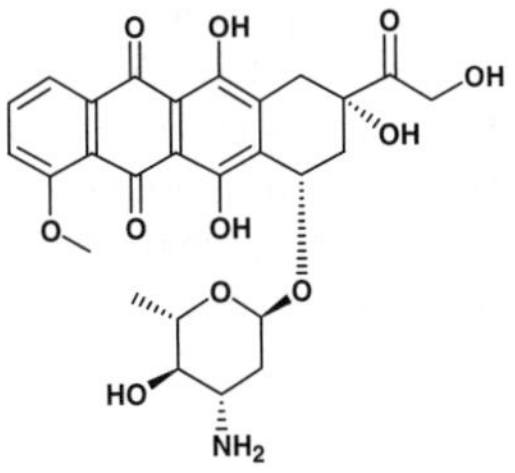

66. Epirubicin

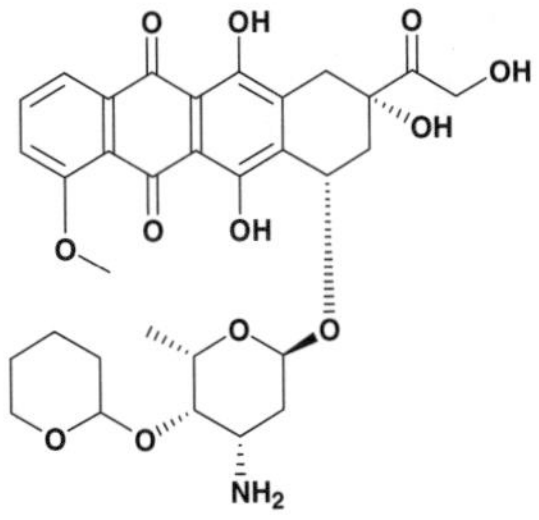

67. Pirarubicin

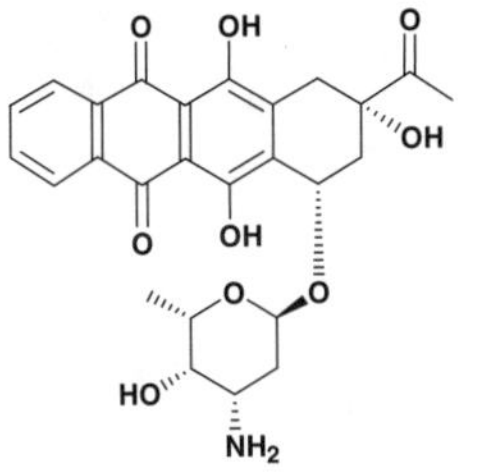

68. Idarubicin

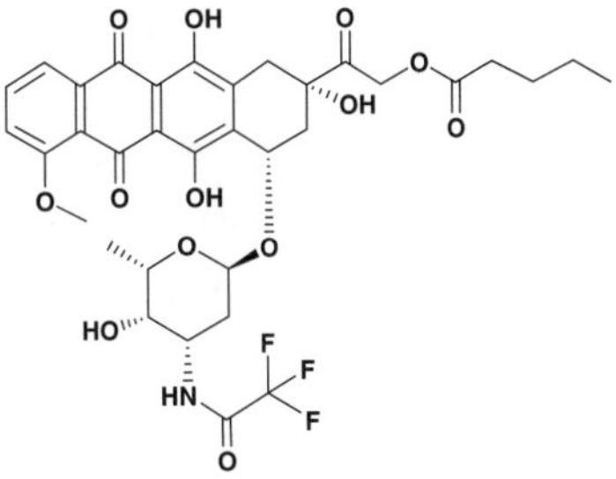

69. Valrubicin

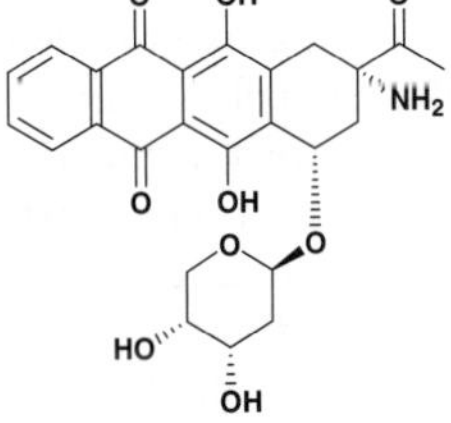

70. Amrubicin

These molecules were originally reported by Umezawa's group at the Institute of Microbial Chemistry in Tokyo, and were developed as antitumor agents by Bristol-Myers. Their "original" mechanism of action (MOA) was elucidated by Hecht and coworkers (MIT and the University of Virginia) who demonstrated that a metal ion (Cu^{2+} or Fe^{2+}) was required to activate the oxidative breakage of the DNA helix once binding to the helices occurred, and there is now evidence that they may also interact with RNA at lower concentrations. The early work in syntheses, mechanism and DNA/RNA interactions are covered by Hecht in a 2005 review [107]; later details covering solid phase syntheses of deglycobleomycin and bleomycin [108], and their biochemical evaluation [109] were published by the Hecht group. Following these publications, two recent papers have appeared, the first in 2006 identifying a potential RNA target for bleomycin [110], and the second in 2008, identifying strong DNA-binding motifs for metal-free bleomycin (since the clinical relevance of zinc and iron-bound bleomycin is unknown) [111].

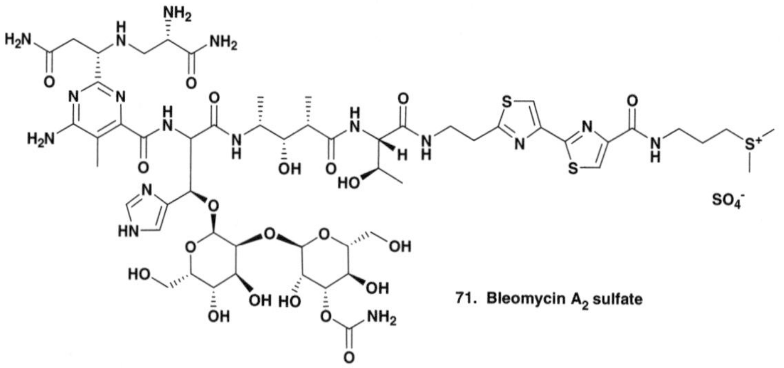

71. Bleomycin A_2 sulfate

One of the most important recently approved microbial compounds is probably the enediyne, calicheamicin γ_1^I (**72**). Calicheamicin γ_1^I (**72**) has *in vitro* cytotoxic activity at the subpicomolar level, but for a significant number of years it was not developed as it was just too toxic to pursue in spite of its exquisite activity; however, its importance lies in the fact that it was the progenitor of a new chemical class, the enediynes. These agents, which now number 13, include the much older neocarzinostatin chromophore, though two structures, sporolides A and B (**73** and **74**) and cyanosporasides A and B (**75** and **76**), are rearrangement products of putative enediynes, which will be covered later under marine products. The enediynes have been well investigated particularly by two groups. Thus, the calicheamicins from *Micromonospora echinospora* ssp. *calichensis*, representing the 10-membered category, have been extensively studied by the Wyeth (originally Lederle) discoverers [112], while from a biosynthetic perspective, a number of workers have contributed, but predominately Shen's group at Wisconsin [113].

On activation, these agents undergo an unprecedented rearrangement and interaction with DNA to produce double-stranded DNA cleavage leading to death of the cell. In 2000, Wyeth gained FDA approval for gemtuzumab ozogamicin (Mylotarg®), an antibody-warhead construct of calicheamicin and an anti-CD33 monoclonal antibody for use against acute myeloid leukemia (AML), and this is possibly the most potent antitumor compound yet approved for clinical use. The base molecule is also being investigated when linked to other monoclonal antibodies. Currently CMD-193, where calicheamicin is linked to an anti-Lewis Y antigen is in Phase I trials [114], and CMC-544 or inotuzumab ozogamicin, which is a conjugate of calicheamicin and an anti-CD22 monoclonal antibody is in Phase I/III trials against a variety of lymphomas [115]. A recent review article gives details of the current impact of monoclonal antibodies, with or without microbial-sourced attachments on cancer treatment and should be consulted for further information [116].

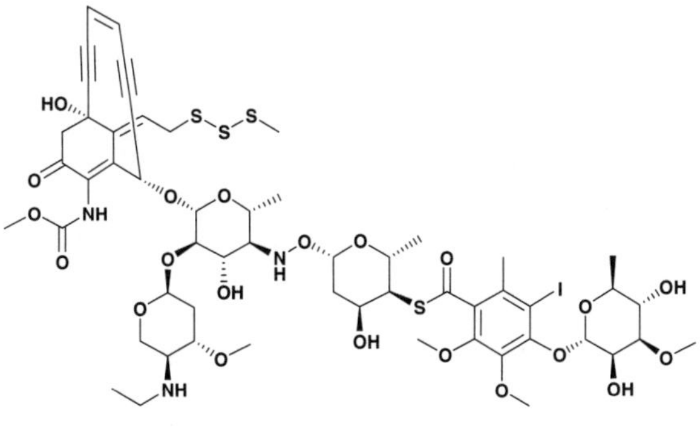

72. Calicheamicin γ1-I

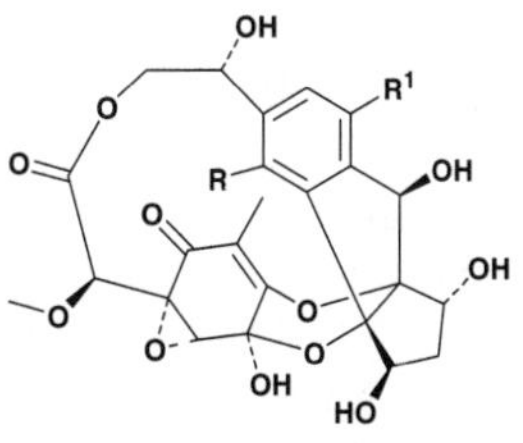

73. Sporolide A; R = Cl, R^1 = H
74. Sporolide B; R = H, R^1 = Cl

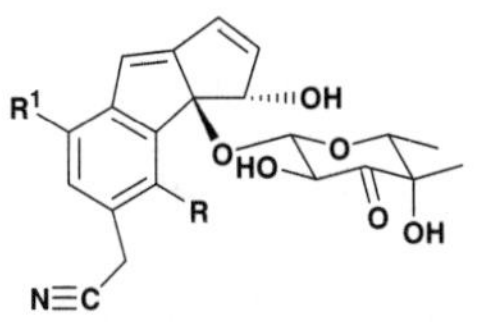

75. Cyanosporaside A; R = Cl, R^1 = H
76. Cyanosporaside B; R = H, R^1 = Cl

In addition to the isolation, development and biosynthetic work mentioned above, a significant number of papers have been published on the syntheses of various enediyne members, including papers that have shown a reassessment of the original structures by synthesis of possible isomers. Perhaps the current best example is the synthetic effort around uncialamycin (**77**) whose original structure was determined on less than 500 μg of material by Davies et al. [117], following isolation and purification from an unidentified streptomycete related to *Streptomyces cyanogenus* extracted from the lichen *Cladonia uncialis*. Due to the inability to determine the stereochemistry at position 26 in the original isolate, Nicolaou's group synthesized both enantiomers of the original structure and demonstrated that the 26 (*R*) enantiomer (**77**) is the natural product, and the 26 (*S*) or *epi*-enantiomer (**78**) the unnatural one. Further biological investigations demonstrated that both epimers were very potent antibiotics (confirming the original work of Davies et al.), and also that they were potent cytotoxins in the NCI 60 cell line panel, with 5- to 10-fold less activity for the 26-*epi* enantiomer [118]. Thus even complex molecules such as these are amenable to current synthetic methods.

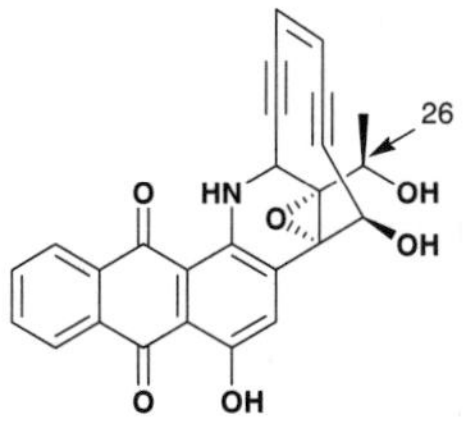

77. Uncialamycin (26 = *R*) 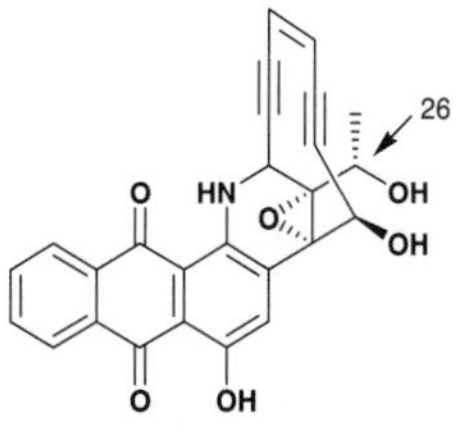   78. *epi*-Uncialamycin (26 = *S*)

4.2. Erbstatin and Lavendustin-Related Molecules, Leads to Tyrphostins

In 1986, the Umezawa group reported [119] the isolation and structure of erbstatin (**79**) and claimed that it was a protein tyrosine kinase (PTK) inhibitor, but was not competitive with ATP [120]; however later work by Posner et al. demonstrated that it was competitive with both the peptide substrate and ATP [121]. These reports led to studies by Levitski and Gazit, who argued that inhibitors directed against the peptide binding site should be less cytotoxic than those directed against the ATP site [122]. This simple natural product led to the synthesis and subsequent testing of many hundreds of compounds known collectively as the tyrphostins, which were all based upon erbstatin or the slightly simpler benzylidenemalononitriles (**80**). Inspection of the molecules produced by Sugen (see Ref. [123] and references therein) and in earlier papers from Gazit et al. [124] enables one to deduce that these structures evolved from erbstatin. The other microbial molecule that significantly aided in the derivation of PTK inhibitors was lavendustin A (**81**), first reported by Onoda et al. in 1989 [125] where the pharmacophore was considered to be an *N*-benzylaniline (**82**) rather than the *N*,*N*-dibenzylaniline moiety of lavendustin A. A subsequent revision of the kinetic properties of the natural product demonstrated that the molecule exhibited hyperbolic-mixed type inhibition against both of the substrates, the peptide and ATP [126].

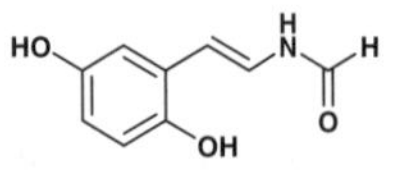

zoline system to prepare active molecules based upon the lavendustin pharmacophore, thereby generating compounds resembling Iressa® (**83**) and Tarceva® (**84**) that inhibited a variety of cell lines, but that were too unstable to develop further. However, it does show another potential relationship between these compounds and natural products. The second paper by Nussbaumer and Winiski was published in 2008 [130] and was an extension of a paper 14 years earlier [131] dealing with a series of lavendustin-based molecules that had activity in dermatologic models [132]; these compounds went into clinical trials in actinic keratoses reaching Phase II, though no current trials appear to be ongoing. What is also of interest is that in the 2008 paper, the mechanism is now reported to be tubulin inhibition following binding at the colchicine site.

79. Erbstatin

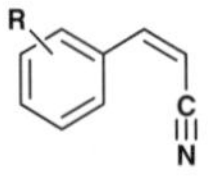

80. Benzylidene-malononitriles

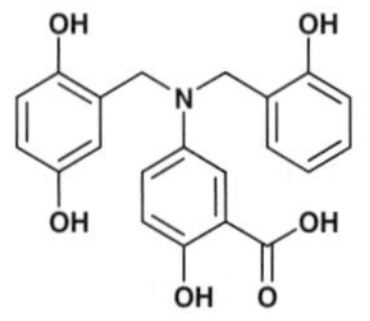

81. Lavendustin A

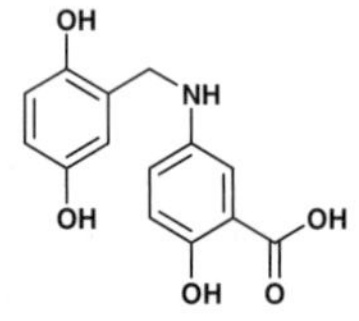

82. N-benzylaniline

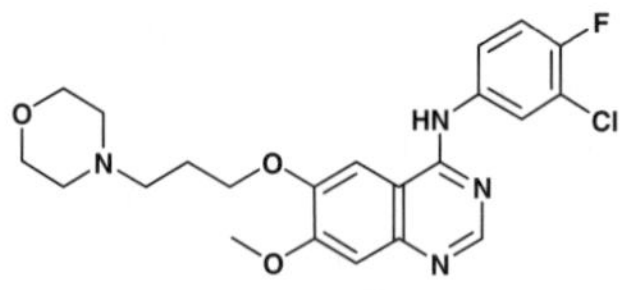

83. Iressa®

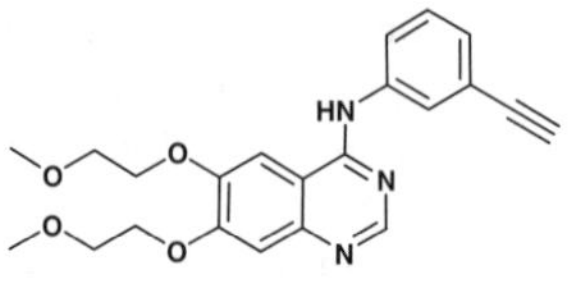

84. Tarceva®

Following a report that the amide derivative (a salicylamide-like molecule) was active against the isolated enzyme, Cushman's group proceeded to generate a series of substituted amides based on the lavendustin skeleton. On testing they were found to also inhibit tubulin polymerization, which was unexpected as prior reports had shown that polyhydroxylated *trans*-stilbenes, benzylanilines and related compounds were PTK inhibitors, whereas *cis*-stilbenes such as combretastatin had tubulin activity [127].

That lavendustin analogs are still valid starting compounds can be seen by inspection of two papers. The first in 2004 by Albuschat et al. [128] used the initial discovery by Hodge and Pierce [129] which demonstrated that a 4-aminoquinazoline is a fully functional substitute for a salicylamide, since the hydrogen bond between the hydroxyl and the amide ketone effectively generates a six-membered di-oxo ring system. The Albuschat group used this quina-

4.3. Rapamycins and Epothilones

Two classes of microbial metabolites, the rapamycins and the epothilones, currently stand out as examples of the power of modern medicinal chemistry when applied to microbial products and the production of antitumor (and other pharmaceutical area) agents that are both in use and in clinical trials.

Rapamycin (**85**) was originally reported in 1975 under the code number AY-22,989 as a potential antifungal agent produced by fermentation of a *Streptomyces hygroscopicus* bacterium collected from Easter Island [133,134]. The compound was not successful as an antifungal agent, but its potential as a possible antitumor agent was reported by the Ayerst Canada group in 1984 using syngeneic murine tumors, including some hint of oral activity [135].

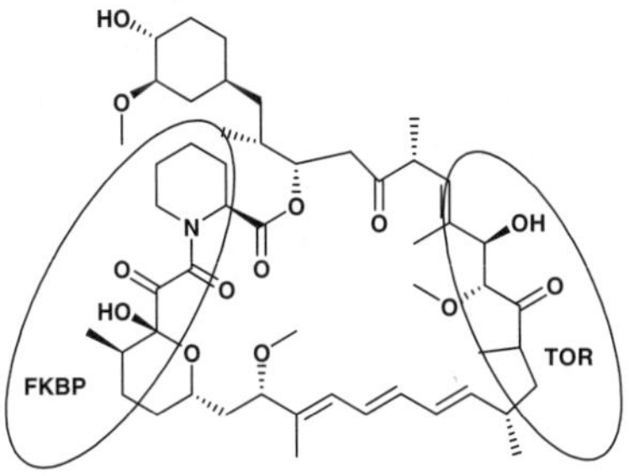

85. Rapamycin (Sirolimus)

This initial antitumor activity was not developed at that time, but the rapamycin base structure has now spawned a plethora of molecules based on the rapamycin skeleton with a variety of different pharmacologic activities including cancer. Initially, modifications were at one site (the carbon atom at C^{43}) and led to four clinical drugs, with the rapamycin base molecule being approved as sirolimus (**85**) in1999, initially as an immunosuppressive agent, and now in Phase I/II trials against various cancers. Similarly, everolimus (**86**) was launched in 2004 as an immunosuppressive agent, and it is also in Phase II/III trials against various cancers in the European Union, Japan and the United States, with approval as a renal cancer agent under the name Afinitor® being given by the FDA in March 2009. The third variation, temsirolimus (CCI-779; **87**) was approved as a treatment for renal carcinoma in the United States in 2007 and is in Phase I/II trials against various carcinomas in the United States, mainly under the auspices of the National Cancer Institute. A fourth, zotarolimus (**88**) was launched in the United States in 2005 as part of method of delivery via a drug-eluting stent for the treatment of restenosis.

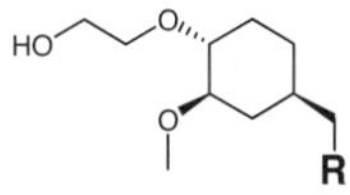

86. Everolimus, R = Macrolide Ring

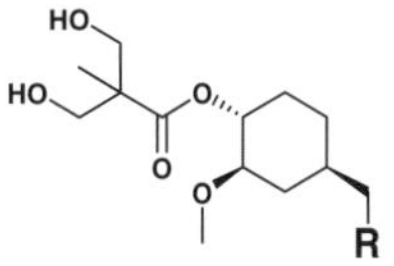

87. Temsirolimus, R = Macrolide Ring

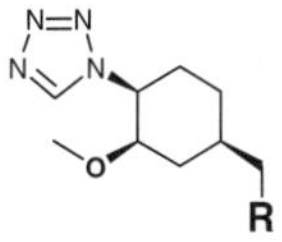

88. Zotarolimus, R = Macrolide Ring

Currently, deforolimus (A23573; **89**) is in Phase III clinical trials for cancer, and two "prodrugs" of rapamycin, Abraxis' ABI-009 (which is a nanoparticle encapsulated formulation of rapamycin) and Isotechnika's TAFA-93 (structure not yet published), are in Phase I cancer trials. All of these where the structures are known, are either the base molecule or have been modified at only the one site, the C^{43} alcoholic hydroxyl group. Such modification avoids both the FKBP-12 and the mammalian target of rapamycin (mTOR) binding sites, since modifications in other areas are thought to negate the basic biological activity of this molecule [136,137].

89. Deforolimus, R = Macrolide Ring

There is one rapamycin derivative with a modified ring structure that is currently in Phase 0 clinical trials the Wyeth compound known as ILS-920 (**90**) (with the basis for Phase 0 trials described at the following URL; http://www.cancer.gov/newscenter/pressreleases/PhaseZeroNExTQandA). ILS-920 has a modification in the triene portion of the molecule, designed to disrupt mTOR binding and appears to have a different target as it is a nonimmunosuppressive neurotrophic rapamycin analog. As such, it has demonstrated over a 200-fold higher binding affinity for FKBP52 over FKBP12, promoted neuronal survival and outgrowth *in vitro*, and bound to the β1 subunit of L-type calcium channels (CACNB1) [138]. What is of interest is that inhibition of FKBP 52 is also reported to affect tubulin interactions in cells [139] so there is a possibility that this agent may also have antitumor activity in due course, though no reports of such activity have yet been published.

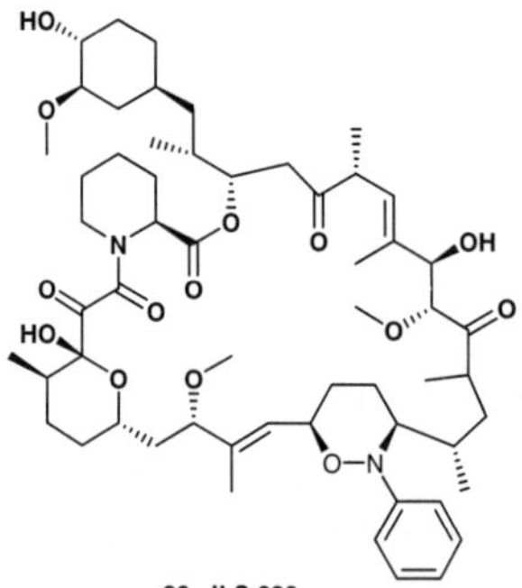

90. ILS-920

Thus from one base molecule, the original and three having slight structural modifications have been approved as drugs. In addition, there are three agents, two modified and one a molecule analogous to Abraxane® (the albumin-linked formulation of Taxol®), in cancer trials, as well as a compound showing that modifications in the macrolide ring's triene moiety produces an entirely different activity in neurology. Finally, from the patent literature other variations are in preclinical studies in a variety of pharmacologic areas.

With the identification of the myxobacterial products epothilones A and B (**91** and **92**) by Reichenbach and Hoefle in the middle to late 1980s [140] and their subsequent identification as tubulin stabilizers (similar mechanistically to paclitaxel) by Bollag et al. in 1995 [141], came an avalanche of chemical, biochemical, and even genomic modifications of the base structure in order to further explore the utility of the base skeleton. This culminated in the approval in October 2007 by the FDA of the semisynthetic epothilone 16-aza-epothilone B, known generically as ixabepilone (Ixempra®; **93**), for the treatment of breast cancer.

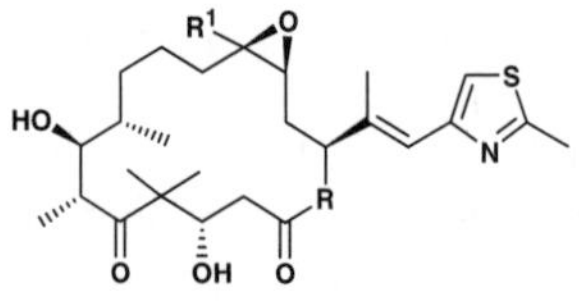

91. Epothilone A; R = O, R^1 = H
92. Epothilone B; R = O, R^1 = CH_3
93, 16-Aza-epothilone B; R = NH, R^1 = CH_3

At the present time, there are five other compounds based upon the epothilone skeleton in active development as anticancer agents listed in the Prous Integrity® database. The natural product epothilone B (patupilone; **92**) is in Phase III clinical trials in conjunction with Novartis and the original discoverers. A totally synthetic derivative, though very close to the base skeleton, sagopilone (ZK-EPO; **94**) is in Phase II trials under Bayer-Schering [142]. More details of this and the opportunities for synthesis of other agents are given in three recent reviews [143–145] and one book chapter [146], which should be consulted for the further details.

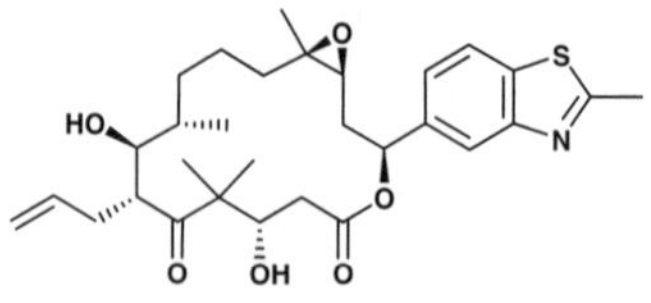

94. Sagopilone

There are two agents derived from work originating in Danishefsky's laboratory at Memorial Sloan-Kettering being developed by Kosan (now part of Bristol-Myers Squibb). The first is (*E*)-9,10-didehydroepothilone D or dehydelone (**95**) currently in Phase II clinical trials, and chemists at Kosan have performed modifications of the structure while retaining the activity [147]. The second is isoxazolefludelone [148] (**96**) currently in preclinical evaluation [149], with a recent discussion of the chemistry leading to the compound and other derivatives produced en route being reported by the same group in late 2008 [150]. The compound is scheduled for an IND submission in 2009, but no further details of such a filing are currently available.

Finally, there is a very interesting folate receptor-targeted molecule synthesized as a result of a collaboration between Endocyte and Bristol-Myers Squibb (BMS-753493; **97**), where folic acid has been linked to an aza-modified epothilone [151]. The compound is now in Phase I/II trials under the name epofolate (**97**) against folate-receptor-positive tumors.

Now that the genetic sequence of the original producing myxobacterium has been pub-

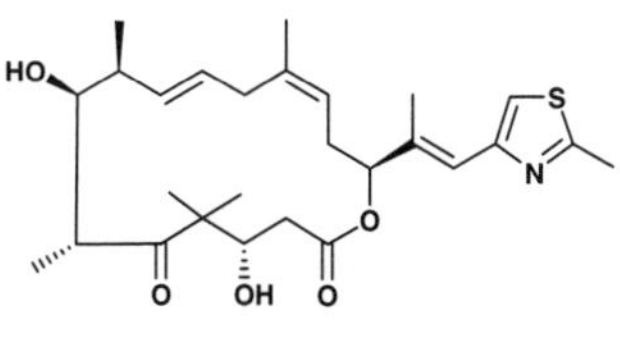

95. Dehydelone

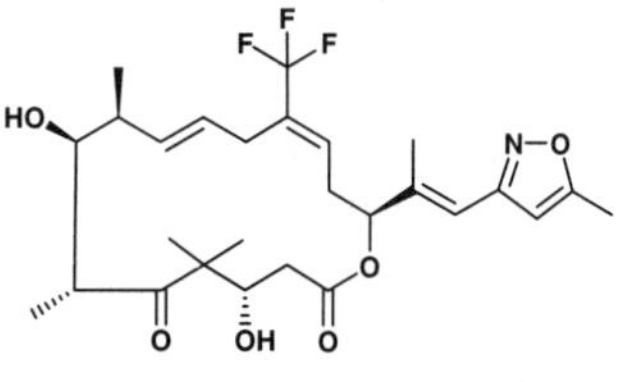

96. Isoxazolefludelone

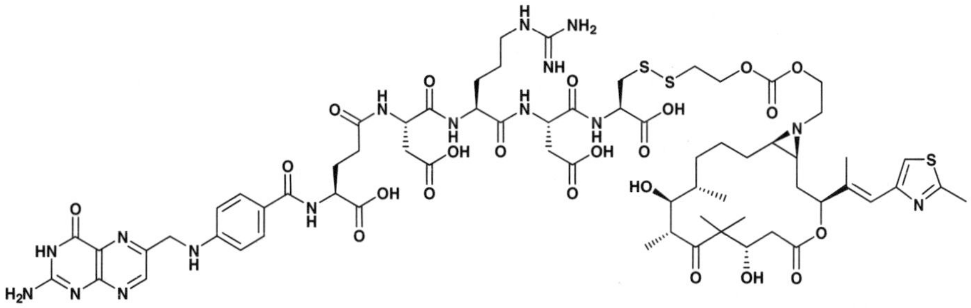

97. Epofolate

lished, we can expect further reports in the near future, not just of combinatorial biosyntheses giving rise to modifications of the base skeleton, but also of the isolation of a number of different secondary metabolites from the same microbe [152–154].

5. MARINE SOURCES

It has now become obvious as mentioned in the introduction, that a significant number of pharmacologically active agents isolated from marine invertebrates in all pharmacologic areas are produced by single-celled microbes [155,156] in, on, or around the invertebrate from which the "active agent(s)" were isolated. Thus in discussions of compounds in this section, where the actual source has been identified, it will be noted, even if the source was only identified years later.

Over the past 30 or so years, a large number of very potent compounds have been isolated, purified and assayed for their biological activities. Due to the major funding source (at least in the United States) being the NCI, a high proportion of these agents have only been tested for their antitumor activity, and some very interesting and potent agents have been discovered and are in various stages of the drug development process. Thus, we will not deal with every class of agent that has entered or is still in trials (preclinical or clinical), but will be relatively selective, with antitumor agents being the major thrust.

5.1. Bryostatins, Dolastatins, and Analogues

Although a plethora of compounds have been reported with cytotoxic activities in the past 30 or so years, only a relative few isolated from invertebrates have been proven to be microbial in origin. These include materials such as the bryostatins, now almost certainly produced by an as yet uncultured microbe that is vertically transmitted during reproduction. However, there are a significant number where there is sufficient circumstantial evidence to invoke a microbe or microbes in their biosynthesis [157]. In addition, there are now a number of compounds isolated from blue-green algae, which are prokaryotes, and from other bacteria and fungi directly isolated from both shallow and abyssal sediments, or from host invertebrates, where there is no question as to the actual producer, since when fermented, the microbe produces the compounds of interest.

Due to the relatively large number of such compounds that are now in the literature, we will only give some examples where either significant chemistry has been performed on the original material, or the compounds have significance as being the first in a chemical class, or the first to be developed by large-scale fermentation, and so on.

Perhaps the two quintessential examples would be bryostatin 1 (**98**) and dolastatin 10 (**99**). In both cases, heroic efforts were performed to obtain enough of the nominally-producing invertebrate material, *Bugula neritina* for bryostatin and the nudibranch *Dolabella auricularia* for dolastatin, to enable production of sufficient compound to permit conduct of initial clinical trials. Details of the initial work on both materials have been presented in many reviews and original papers, and the early to relatively late history of both can be found in the 2005 reviews by Newman for the bryostatins [158] and Flahive and Srirangam for the dolastatins [159].

that bryostatin-1 may have potential in amelioration of cognitive diseases [166,167].

In a similar fashion to that of bryostatin, but where the producing organism was directly isolated and identified, dolastatin 10 was shown to be produced by a cyanophyte of the genus *Symploca* [168], and subsequent work has shown the vast capabilities of cyanophytes in this respect [169]. Although dolastatin 10 entered clinical trials, as did a variety of other synthetic molecules based upon the natural product structure, at the time of writing, inspection of the NCI clinical trials database shows no ongoing clinical trials although there is some published controversy over the current status of dolastatin 10. Only auristatin PE (soblidotin; **100**) is currently in ongoing clinical trials, and there are some very interesting modifications that have been made by medicinal chemists in order to deliver this close relative of dolasta-

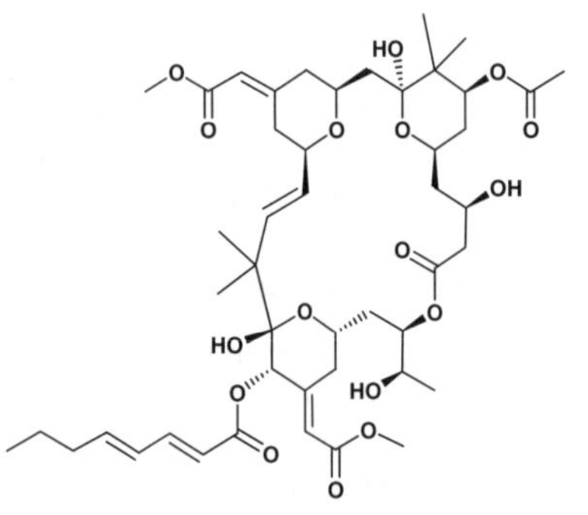

98. Bryostatin 1

99. Dolastatin 10

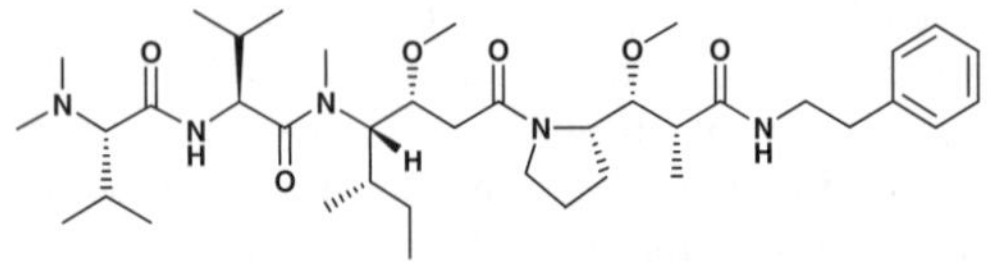

Work by Haygood and her collaborators demonstrated that in the case of the bryostatins, the probable source was an as yet uncultured microbe from which the gene cluster that would produce the base ring structure of the bryostatins was isolated and cloned, but not yet expressed [160]. Contemporaneously with this work, medicinal chemists had also been extremely busy, producing modifications of the basic structures, with some very recent publications demonstrating rapid methods of obtaining both bryostatins [161–164] and using them to identify specific protein kinase C (PKC) targets [165], and also demonstrating

tin-10 to specific tumor types by using monoclonal antibodies targeted at specific epitopes [170,171]. Currently a significant number of combinations of this base molecule with varying monoclonal antibodies are in preclinical to Phase II clinical trials, predominately against hematologic cancers.

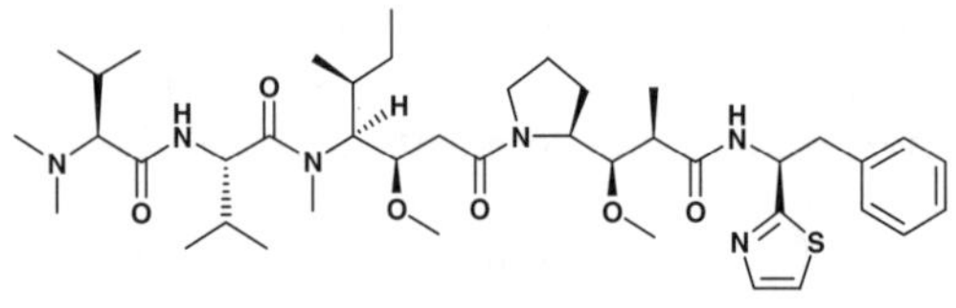

100. Auristatin PE (Soblidotin)

Thus, Seattle Genetics has SGN-35, where the antibody is an anti-CD30 linked to auristatin E, in Phase I trials and heading for Phase II [172]; CuraGen has the antibody CR011 linked to auristatin E in Phase II trials for metastatic breast cancer and melanoma [173], and Progenics has PSMA-ADC, a dimeric-specific PSMA antibody also conjugated to auristatin E in Phase I trials against prostate cancer [174]. In all of these examples, although the exact linkages between the auristatin molecule and the antibody are subtly different, all antibodies are licenced from Seattle Genetics. Hopefully these dolastatin-inspired adducts will propel this class of molecules to more advanced phases of clinical trials as the base molecule, auristatin PE (**100**), has oscillated between Phases I and II trials as it changed hands amongst a variety of small and large pharmaceutical companies.

5.2. Curacin A

Curacin A (**101**), originally isolated from the cyanophyte *Lyngbya majuscula* [175], is a potent cytotoxin acting at the colchicine site on tubulin; however, no *in vivo* activity has been reported, probably due to it being almost insoluble in any formulation. The application of combinatorial chemistry around the basic structure has enabled the synthesis of more soluble variants [176], and in 2004, the complete biosynthetic gene cluster was identified and cloned, which may permit modification via combinatorial biosynthetic techniques in due course [177]. Significantly, by application of modern modeling techniques, a common pharmacophore has been developed for colchicine-binding ligands on tubulin, including data from curacin A, and in due course this may well aid in designing new variants with improved activities [178]. Very recently, direct evidence for the production of secondary metabolites *in situ* in intact cyanophytes and invertebrates such as sponges, has been reported by Esquenazi et al. using a very clever combination of MALDI-TOF and imaging, thus proving the involvement of the microbe while in its native environment [179,180].

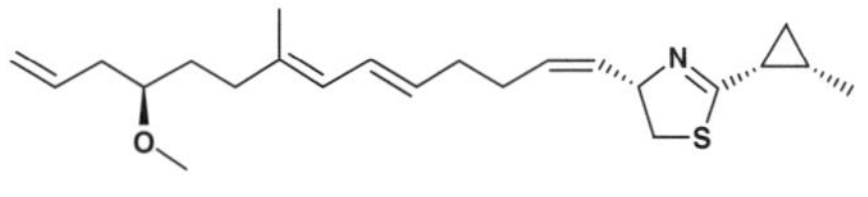

101. Curacin A

5.3. Cryptophycins

The cryptophycins are a prime example of materials isolated from cyanophytes, though in this case, the marine involvement is for one compound only, cryptophycin 24 or arenastatin (**102**) [181]. These compounds, which were originally isolated from a terrestrial *Nostoc* species [182], then rediscovered from another terrestrial strain of *Nostoc* as antitumor agents by the Moore group in Hawaii, entered clinical trials as the derivative cryptophycin 52 (**103**), in conjunction with Eli Lilly and biologists at Wayne State University [183]. The methods of large-scale syntheses and modifications to give better control of configuration were reported in the early 2000s [184,185].

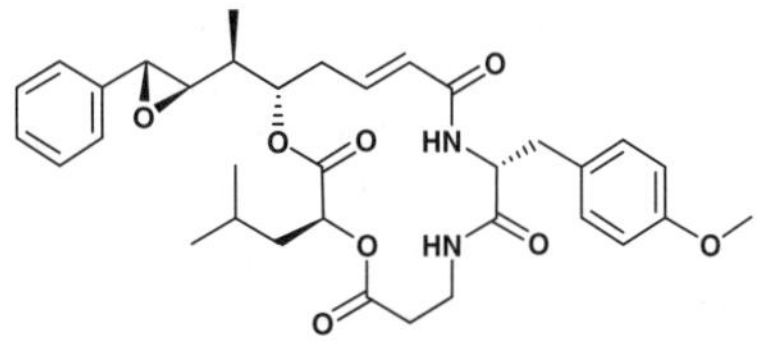

102. Arenastatin A (Cryptophycin 24)

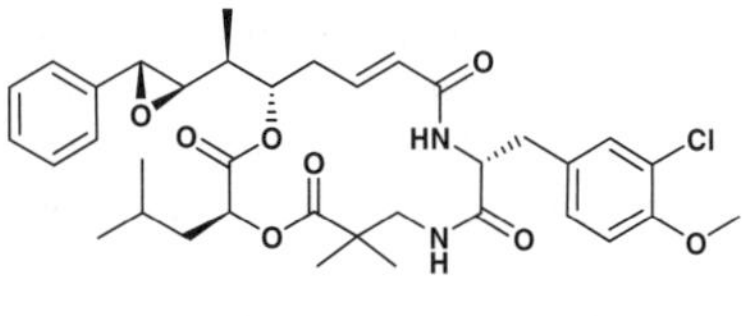

103. Cryptophycin 52 (LY-355703)

Although cryptophycin 52 was withdrawn from clinical trials by Lilly at Phase II, it was realized that the compound, arenastatin was identical to cryptophycin 24, and recently a stereoselective synthesis was reported by the Kobayashi group, the original discoverers [186] of this derivative. The identification of the gene cluster and possibilities of biosynthetic modifications were published in 2005 [187]. Currently there are possibilities of renewed interest in these molecules as there are reports [188] that further derivatives of the cryptophycins [189] may be licenced to another pharmaceutical house for development.

5.4. Peloruside A

In 2000, West and Northcote reported the isolation of the cytotoxic macrolide peloruside A (**104**) from the New Zealand marine sponge *Mycale hentscheli* [190]. This initial report was followed 2 years later by a paper from Hood et al. [191] that demonstrated that this compound was another in the series of marine-derived cytotoxins with a mechanism of action similar to that of Taxol®, and like Taxol®, peloruside A demonstrated induction of apoptosis following G_2-M arrest [191].

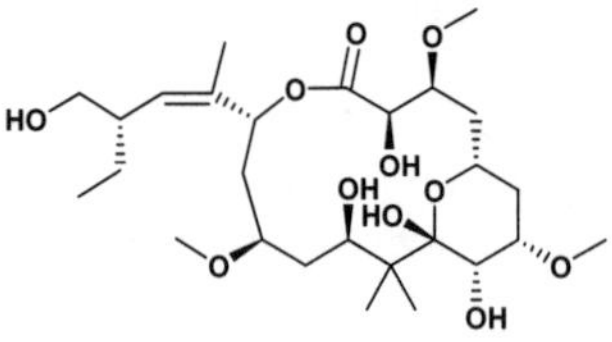

104. Peloruside A

Further biological work related to the potential mechanism has now demonstrated that peloruside A, like the epothilones is a weaker substrate for the phosphoglycoprotein efflux pump than Taxol®, and it appeared to bind at the same or a similar site to that reported for laulimalide, one significantly different from the taxoid binding site [192]. However, in a very recent paper Huzil et al. [193] consider that peloruside actually has a binding site on the exterior of the β-tubulin different from any other site so far described. This is a different binding site from that reported for the solution conformation of peloruside in the presence of tubulin by Jimenez-Barbero et al. [194] who implicated binding at a site on α-tubulin.

Peloruside A has been difficult to reisolate in any significant quantity from Nature, but by use of in-sea aquaculture techniques, successful isolation from aquacultured sponge fragments of the nominally producing sponge has been achieved [195,196]. If this is coupled to the work reported on the total syntheses [197–200] and the possibilities from the work reported by Page et al. [195] demonstrating the temporal and spatial production of peloruside by in-sea aquaculture of the producing sponge, if or when this compound enters clinical trials, these sources will permit adequate supplies to be maintained. Although the actual producer has not yet been identified, it is in all probability a microbe rather than the sponge.

5.5. Et743 (Trabectedin or Yondelis®)

The first, and currently only compound "direct from the sea" to be approved for the treatment of cancer is Et743 (trabectidin; Yondelis®; **62**), which was approved by the EMEA in September 2007 for the treatment of sarcoma, and was launched in Sweden, Germany and the United Kingdom late in 2007. Currently there are many clinical trials involving this agent listed in the NCI clinical trials list and in the Integrity® database. This compound is one of a series originally reported by two groups [201,202] in back to back papers in 1990. The molecule was recognized as a derivatized member of the saframycin class of known antibiotics and thus might have a microbial component in its production by the invertebrate.

The compound was licenced by the discoverers to the Spanish pharmaceutical company, PharmaMar, and following a very chequered development path including large-scale collections of the source tunicate *Ecteinascidia turbinata* and in-sea and on-land aquaculture, enough material was harvested to permit production of sufficient compound for the initial and early clinical trials. Though synthesized early in its study by Corey's group, the compound was ultimately produced for later clinical trials and commercialization by semisynthesis starting with cyanosafracin B (**105**). This intermediate compound was produced by fermentation of a marine-derived *Pseudomonas fluorescens*, and more details of this agent can be found in the 2009 review by Cuevas and Francesch [203].

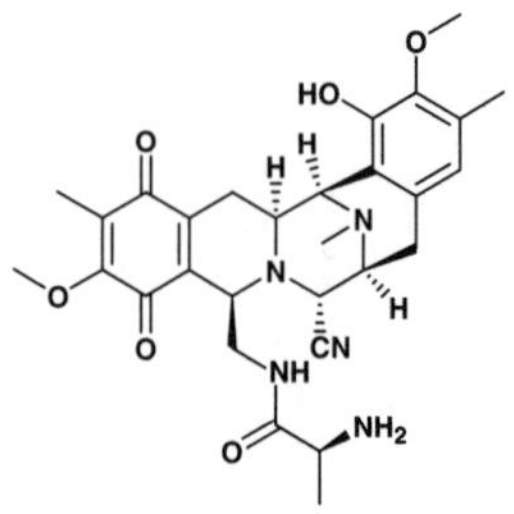

105. Cyanosafracin B

In addition to this review, Velasco et al. have also reported the identification of the producing gene cluster, thus opening up possibilities for combinatorial biosyntheses in the future [204]. A report in 2003 identified potential bacteria from Mediterranean *E. turbinata* as putative producers [205], and a recent examination of the microbial assemblages related to *E. turbinata* in both the Caribbean and Mediterranean seas, has identified five potentially persistent bacteria, with one, *Candidatus Endoecteinascidia frumentensis* occurring in *E. turbinata* in both geographic locations [206]. Though not absolute proof, this observation is indicative of microbial involvement and will need further investigation analogous to the work carried out with the putative bryostatin producer (see below).

Although a number of reports have been published in the literature over the past few years giving possibilities as to the MOA(s) of Et743 when tumor cells are treated *in vitro*, a significant problem with some of these reports is that the concentrations used in the experiments were often orders of magnitude greater than those that demonstrate activity *in vitro*. Since the "active" levels are in the low nanomolar to high picomolar range, care should be taken when evaluating published work on the MOA of this compound.

At physiologically relevant concentrations, the MOAs of Et743 have been shown to include the following: effects on the transcription-coupled nucleotide excision repair (TC-NER) process; and interaction between the Et743 DNA adduct and DNA transcription factors, in particular the NF-Y factor. Recently, pharmacogenomic analyses have identified a series of genes involved in the sensitivity of tumor cells to this agent, and the paper by Jimeno et al. [207] should be consulted in order to see the patterns identified.

In 2007, Soares et al. [208] demonstrated that the adducts formed with Et743 and DNA are stable and can be converted into double-strand breaks (DSBs) hours after initial formation. In addition, loss of homologous recombination repair function, although having no effect on the initial number of DSBs, is associated with persistence of these lesions thus giving rise to extensive chromosomal abnormalities, and hence sensitivity to the drug. In addition to this report, two other papers demonstrating further subtle aspects of its cellular interactions have recently been published, the first demonstrating that the demethylated analog Et729 has a differential affinity for the CGA DNA triplet binding site compared to Et743 [209], and the second demonstrating that in low passage sarcoma cell lines, sensitivity correlates with mutated p53 [210].

5.6. Marine Microbial Secondary Metabolites/Derivatives in Clinical Trials

Over the past 20 or so years, there have been many comments and presentations implying that a number of the agents found in marine invertebrates had "microbe(s) in their background", and this has subsequently shown to be the case using genomic information in the cases of bryostatin, Et743, and by direct isolation of microbes from the *Bryopsis* alga in the case of Kahalide F (**106**) [211]. However, it was the thesis of Fenical and Jensen that there were deep sea free-living microbes that could be cultivated, and novel agents produced by modifications of methods used for other microbial flora. That their thesis was correct is shown by extensive publications over the past 15 or so years, with a number of excellent recent reviews giving their methods/results [212–214].

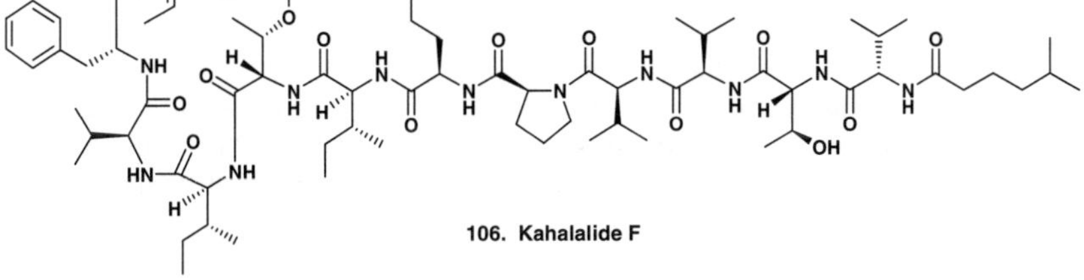

106. Kahalalide F

Salinoporamide A (**107**) was first reported [215] from a marine streptomycete of a new genus and species named initially as *Salinospora tropica*. The genus was subsequently changed to *Salinispora* but the compound retained its original name. The structure was reminiscent of the terrestrial bacterial products, lactacystin (**108**) and its rearrangement product, omuralide (**109**), a known proteasome inhibitor, and on further testing that activity was reported for the new compound in the original publication [215]. The compound had an unusual chlorine substitution and within a year or so of the publication, two academic groups had synthesized the base molecule [216,217], followed by a synthetic paper from the original developers at Nereus Pharmaceutical [218]. Subsequently, many groups have reported improved syntheses with an excellent recent review covering most of those through 2006 [219]. Since that time there have been a number of synthetic papers by Nereus Pharmaceutical scientists covering an enantiospecific synthesis [220], a chemoenzymatic technique [221], and a full description of modifications that improve the potency of this class of compounds [222].

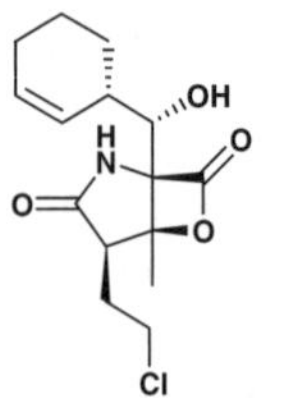

production work, the mechanism of salinosporamide as an inhibitor was worked out using a number of excellent X-ray crystallographic studies [227], together with studies on the stability of the β-lactone ring [228].

As mentioned earlier, the compound entered Phase I clinical trials in May of 2006 initially against solid tumors and leukemias, and in April of 2007, another Phase I trial against multiple myeloma was initiated. Some results from a Phase I trial have been reported demonstrating a clinical response in patients with solid tumors or lymphomas [229].

In the last year, some very interesting papers have been published on both the isolation of similar molecules from a terrestrial microbial source, the cinnabaramides A-G [230], and on the isolation and identification of other salinosporamides with varying proteasome activities from fermentation experiments [225] as mentioned earlier. Reports on the fermentation aspects of the producing organism demonstrate that, although a marine streptomycete, salinosporamide production is dependent upon sodium not halogen levels [224] and can occur in low chloride levels [231] or even with substitution of potassium for sodium [232].

107. Salinosporamide A

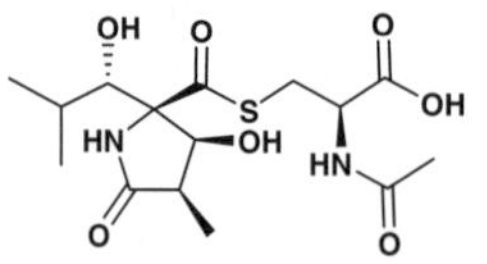

108. Lactacystin

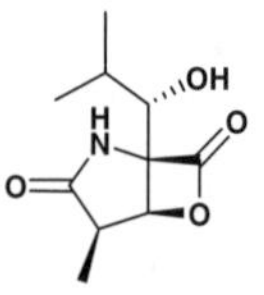

109. Omuralide

However, in addition to providing synthetic methods, Nereus Pharmaceutical scientists in concert with the now defunct fermentation group at IRL in New Zealand, were able to produce the necessary cGMP product for clinical trials by fermentation in a saline environment, the first time that this task had been successfully performed on any scale with a marine-sourced microbe. During these runs, a significant number of other salinosporamide derivatives were also isolated, and other secondary metabolites were further explored [223–226]. Concomitantly with this

These conditions are close to those normally used for terrestrial fermentations, thus giving further credence to the isolation of the cinnabaramides mentioned above.

6. GENOMIC INFLUENCE ON METABOLITES

Although the production of secondary metabolites in microbial systems is obviously controlled genomically, only in the past few years it has become obvious that to date, researchers have only observed the very "tip of the iceberg."

For over 40 years, multiple companies, both large and small, had fermented millions of soil isolates in order to find novel agents from microbes, and a constant theme was that "we keep discovering the same compounds." Thus, for this and other reasons, almost all companies ceased their fermentation-based discovery programs over the time frame from roughly the middle of the 1980s to early 2000s.

However, with the advent of two interrelated techniques, the search for new microbial agents has now entered a new phase. The first technique was the invention of the automated DNA sequencer, and the second was the discovery of the PCR technique. When these two were coupled with the advances in computer analyses of DNA and later RNA sequence data, the immense potential of the microbial universe from a secondary metabolite perspective began to be realized. In addition, one must also mention the advent of what might best be described as "combinatorial biosynthesis" or production of "unnatural natural products," whereby biosynthetic genes and/or parts of genes could be "mixed and matched" with the aim of producing novel structures, including the use of unnatural starting substrates in order to alter the final product.

In addition, and most importantly for the discovery of the putative microbial producers, is the concept of the metagenome and its analysis, whereby gross DNA samples could be extracted from soils, sea water and invertebrate organisms, then probed for genes by PCR, followed by amplification, sequencing and analyses. The first example of this was the work from Handelsman on soil antibiotics [233,234], with perhaps the most adventurous being the work of Venter's group on the marine metagenome [235]. A current example of where this has led is the recent paper from Piel and colleagues [157] demonstrating the importance of trans acyltransferases in the biosyntheses of a number of potentially important microbial secondary metabolites from as yet uncultured organisms.

Since the interplay of these discoveries and their current status would fill another book, let alone a single chapter, we are just going to give a few examples relevant to drug discovery. Perhaps the most important from a drug discovery and development perspective was the recognition by the Omura group [236] in 2001 that the avermectin producer *S. avermitilis* contained at least 25 identifiable secondary metabolite gene clusters covering roughly 6% of the total DNA; they then revised the figure to over 30 when a more thorough analysis was performed in 2003 [237]. Contemporaneously with this work, the Hopwood group also reported the sequence of the quintessential streptomycete, *S. coelicolor* in 2002 [238], again demonstrating that more than 20 secondary metabolite clusters could be identified.

Where this has led is to the interplay of microbiology, chemistry, biosynthesis, and genomics where it is now possible to use the information from the sequence of a microbe to identify the potential producing clusters (cryptic clusters) that have never been expressed under the "normal" fermentation parameters used to date. A prime recent example of this is the work on the salinosporamide producer, *S. tropica*. In 2007, Udwary et al. reported [239], the sequence of this organism, identifying 17 potential clusters including the salinosporamide locus. Over the past few years, Fenical and coworkers at Scripps Institution of Oceanography have been identifying some of these agents from both this and other isolated marine bacteria by using specific fermentation conditions. Now that the positions of the potential producing clusters have been identified, a combination of cloning, expression and fermentation is unlocking these metabolites. As mentioned earlier the sporolides (**73** and **74**) and cyanosporasides (**75** and **76**) have been isolated from this organism and now their clusters have been identified.

In addition, the genomic information permits one to produce unknown molecules by performing the "mixing and matching" referred to earlier. Thus, using information from the sequence and knowledge of the mechanism [240] of the unique chlorinase in *S. tropica*, Eustaquio and Moore produced the fluoro analog of salinosporamide by mutasynthesis [241], followed by another example of production of the unnatural antiproteolide (**110**), a novel proteasome inhibitor [242].

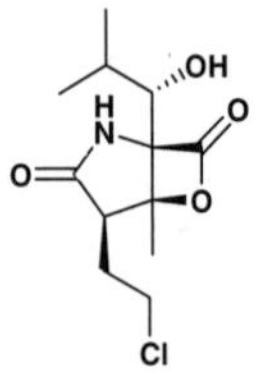

110. Antiprotealide

These are just some of the very many examples of the potential for this combination of resources, and for an up to date review of the potential of such experimental designs, the review by Watanabe should be consulted for the biosynthetic potential [243]. For the potential of the world of marine microbes, the papers from the Fenical and Jensen groups should be consulted, particularly the recent ones demonstrating the multiplicity of microbes that may be observed [244,245].

7. PLANT METABOLITES AND THE FUNGAL CONNECTION

Until a few years ago, the question as to whether plants were the actual source of some of the very important antitumor-active materials isolated from them by chemists, would probably have led to a series of somewhat derisory comments from most botanists and plant natural product chemists, but from work over the past few years, this question is now reminiscent of similar comments/questions made by marine natural products chemists 20 or so years ago with respect to sponge-derived molecules.

Over the past ~15 years, and particularly in the past 5 years, with easier access to work performed in the People's Republic of China, a significant number of reports have shown that the four major classes of "plant-derived" molecules, the taxanes, camptothecins, podophyllotoxins and even the vinca alkaloids are produced by endophytic fungi. A significant number of endophytic fungi isolated from the source plants and trees, when purified and fermented on substantial scales (up to 10 L), produce the same molecules that were originally isolated from the plants themselves.

Strobel at Montana State had demonstrated over 15 years ago that a new genus of endophytic fungus isolated from *Taxus* trees (that "produced/contained" paclitaxel) would produce very small amounts of the compound when fermented in the laboratory. He later extended this finding to identify the sources as the new genus *Taxomyces* [246] and many *Pestalotiopsis* species [247]. Following on from these reports came realization that this was not an isolated phenomenon, as over the years camptothecin [248,249], podophyllotoxin [250,251], vinblastine [252], and vincristine [253,254] have been produced by fermentation of endophytic fungi isolated from the original source plants.

Arguments have been made orally to the authors that particularly in the case of paclitaxel, it is the plant that makes the molecules as one of the sources of paclitaxel today is via large-scale plant tissue culture. Though this is a valid production method, unless the tissue culture is axenic (meaning the absence of any microbe in, on, or around the tissues), one cannot rule out the involvement of endophytes and/or other microbes. To our knowledge, the plant tissue cultures in the paclitaxel production process are not axenic.

That very specific control mechanisms are in play in these fungi can be demonstrated by the work of Keller's group on the common soil fungus, *Aspergillus nidulans* and its control of expression of the 40 plus "cryptic secondary metabolite clusters" that her group has identified from sequence studies [255], where they also demonstrated the presence of the control protein *Lae*A. Over the past few years, a significant number of related investigations have been reported from soil fungi. Five recent papers, one a primary publication demonstrating the discovery of the emericellamide (**111**) biosynthetic pathway [256], and four review papers on the activation of fungal secondary metabolite genes [257–260] have confirmed such results. The identification of the gene/gene product controlling metabolite production by microbes such as these, could provide an entry into greatly increased production of key bioactive natural products, and also lead to the discovery of a large number of previously unknown active materials, not necessarily just antitumor agents, that will challenge the skills of synthetic and medicinal chemists in the near future.

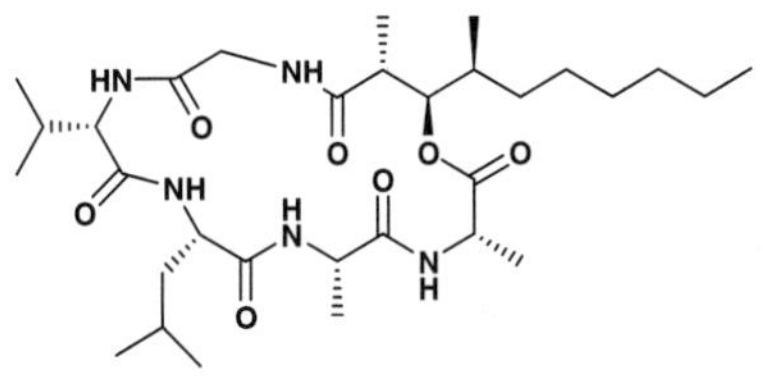

111. Emericellamide A

In addition, these discoveries also bring up the intriguing possibility that there exists some form of signaling analogous to "bacterial quorum sensing" involving these fungal endophytes, the plant itself, and the production of toxic secondary metabolites when attacked by a predator. Such a system would make more "economic sense" from an energy balance perspective rather than postulating production centrally, followed by transport to the region under attack and then removal of a material that is toxic to the host as well as the predator. However, this suggestion is presented simply as an intriguing hypothesis generated through arguing by analogy.

8. CONCLUSION

The microbial world, meaning in this case, single celled organisms from the three domains of life, are clearly important producers of known secondary metabolites with significant activity as both drugs and leads thereto, thus proving the validity of the chapter's title. Of considerable added importance, however, has been the research performed in the past few years, where the investigation of the genomic sequences of representatives of these organisms, including very well studied examples, has shown that the number of metabolites known are but a small proportion of the biosynthetic gene clusters (cryptic clusters) present. These clusters are now beginning to yield their secrets, using both modified fermentation conditions and activation of specific clusters by expression in other hosts [261].

Due to space limitations, we have only given some of the more important highlights, and have not included much commentary on what may be one of the most important microbial families from a secondary metabolite aspect, the *Myxobacteriales*. This family has yielded the epothilones, and together with fungal sources, it may well be the source of significant numbers of novel agents in the relatively near future; three recent review papers may be consulted by readers who are interested in their potential [262–264]. Thus, the number of potential bioactive agents yet to be discovered from the microbial world is uncalculable, but the number is certainly vast.

REFERENCES

1. Woese CR, Kandler O, Wheelis ML. Proc Natl Acad Sci USA 1990;87:4576–4579.
2. Winker S, Woese CR. Syst Appl Microbiol 1991;14:305–310.
3. Dalton H, Stirling DI. Philos Trans Roy Soc London, Ser B Biol Sci 1982;297:481–496.
4. Stirling DI, Dalton H. FEMS Microbiol Lett 1979;5:315–318.
5. Mateles RI, editor. Penicillin: A Paradigm for Biotechnology. Chicago: Candida Corporation; 1998.
6. Schatz A, Bugie E, Waksman SA. Proc Soc Exp Biol Med 1944;55:66–69.
7. Dudnik YV, Gause GG. SIM News 2002;52:230–233.
8. Gauze GF, Brazhnikova MG. Am Rev Sov Med 1944;2:134–138.
9. Gauze GF, Brazhnikova MG, Belozerskii AN, Paskhina TS. Byulleten Eksperimental'noi Biologii i Meditsiny 1944;18:3–6.
10. Aoki H, Okuhara M. Ann Rev Microbiol 1980;34:159–181.
11. Brotzu G. Lav Ist Ig Cagliari 1948; 1–11.
12. Crawford K, Heatley NG, Boyd PF, Hale CW, Kelly BK, Miller GA, Smith N. J Gen Microbiol 1952;6:47–59.
13. Nakajima S. Yakushigaku Zasshi (J Jpn Hist Pharm) 2003;37:119–127.
14. Newton GGF, Abraham EP. Biochem J 1956;62:651–158.
15. Abraham EP, Newton GGF. Biochem J 1961;79:377–393.
16. Brown AG, Butterworth D, Cole M, Hanscomb G, Hood JD, Reading C, Rolinson GN. J Antibiot 1976;29:668–669.
17. Howarth TT, Brown AG, King TJ. J Chem Soc Commun 1976; 266–267.

18. Reading C, Cole M. Antimicrob Agents Chemother 1977;11:852–857.
19. English AR, Retsema JA, Girard AE, Lynch JE, Barth WE. Antimicrob Agents Chemother 1978;14:414–419.
20. Micetich RG, Maiti SN, Spevak P, Hall TW, Yamabe S, Ishida N, Tanaka M, Yamazaki T, Nakai A, Ogawa K. J Med Chem 1987;30: 1469–1474.
21. Mansour TS, Bradford PA, Venkatesan AM. In: Macor JE, editor. Annual Review of Medicinal Chemistry. Vol. 43. San Diego: Academic Press; 2008. p 247–267.
22. Imada A, Kitano K, Kintaka K, Muroi M, Asai M. Nature 1981;289:590–591.
23. Sykes RB, Cimarusti CM, Bonner DP, Bush K, Floyd DM, Georgopapadakou NH, Koster WH, Liu WC, Parker WL, Principe PA, Rathnum ML, et al. Nature 1981;291:489–491.
24. Albers-Schoenberg G, Arison BH, Hensens OD, Hirshfield J, Hoogsteen K, Kaczka EA, Rhodes RE, Kahan JS, Kahan FM, Ratcliffe RW, Walton E, et al. J Am Chem Soc 1978;100:6491–6499.
25. Bonnefoy A, Dupuis-Hamelin C, Steier V, Delachaume C, Seys C, Stchyra T, Fairley M, Guitton M, Lampilas M. J Antimicrob Chemother 2004;54:410–417.
26. Chopra I, Roberts M. Microbiol Mol Biol Rev 2001;65:232–260.
27. Akiba T, Koyama K, Ishiki Y, Kimura S, Fukushima T. Jpn J Microbiol 1960;4:219–227.
28. Aminov RI, Chee-Sanford JC, Garrigues N, Teferedegne B, Krapac IJ, White BA, Mackie RI. Appl Environ Microbiol 68:2002; 1786–1793.
29. Barden TC, Buckwalter BL, Testa RJ, Petersen PJ, Lee VJ. J Med Chem 1994;37:3205–3211.
30. Kelesidis T, Karageorgopoulos DE, Kelesidis I, Falagas ME. J Antimicrob Chemother 2008;62:895–904.
31. Swoboda S, Ober M, Hainer C, Lichtenstern C, Seiler C, Wendt C, Hoppe-Tichy T, Büchler M, Weigand MA. J Antimicrob Chemother 2008;61:729–733.
32. Tatsuta K, Yoshimoto T, Gunji H, Okado Y, Takahashi M. Chem Lett 2000; 646–647.
33. Tatsuta K, Hosokawa S. Chem Rev 2005;105: 4707–4729.
34. Charest MG, Siegel DR, Myers AG. J Am Chem Soc 2005;127:8292–8293.
35. Sun C, Wang Q, Brubaker JD, Wright PM, Lerner CD, Noson K, Charest MG, Siegel DR, Wang Y-M, Myers AG. J Am Chem Soc 2008;130:17913–17927.
36. Duggar BM. Ann N Y Acad Sci 1948;51: 177–181.
37. Pickens LB, Tang Y. Metab Eng 2009;11:69–75.
38. Kim SJ, Cegelski L, Stueber D, Singh M, Dietrich E, Tanaka KS, Parr TR Jr, Far AR, Schaefer J. J Mol Biol 2008;377: 281–293.
39. Cooper RD, Snyder NJ, Zweifel MJ, Staszak MA, Wilkie SC, Nicas TI, Mullen DL, Butler TF, Rodriguez MJ, Huff BE, Thompson RC. J Antibiot 1996;49:575–581.
40. Malabarba A, Ciabatti R, Scotti R, Goldstein BP, Ferrari P, Kurz M, Andreini BP, Denaro M. J Antibiot 1995;48:869–883.
41. Leadbetter MR, Adama SM, Bazzini B, Fatheree PR, Karr DE, Krause KM, Lam M, Linsell MS, Nodwell MB, Pace JL, Quast K, et al. J Antibiot 2004;57:326–336.
42. Pace JL, Yang G. Biochem Pharmacol 2006;71: 968–980.
43. Barbachyn MR. In: Macor JE, editor. Annual Reports in Medicinal Chemistry. Vol. 43. San Diego: Academic Press; 2008. p 281–290.
44. Ciabatti R, Kettenring JK, Winters G, Tuan G, Zerilli L, Cavalleri B. J Antibiot 1989;42:254–267.
45. Walker S, Chen L, Hu Y, Rew Y, Shin D, Boger DL. Chem Rev 2005;105:449–476.
46. Breukink E, de Kruijff B. Nat Rev Drug Discov 2006;5:321–332.
47. Fang X, Tiyanont K, Zhang Y, Wanner J, Boger D, Walker S. Mol BioSyst 2006;2:69–76.
48. Tiyanont K, Doan T, Lazarus MB, Fang X, Rudner DZ, Walker S. Proc Natl Acad Sci USA 2006;103:11033–11038.
49. Freeman J, Baines SD, Jabes D, Wilcox MH. J Antimicrob Chemother 2005;56:717–725.
50. Sunenshine RH, McDonald LC. Cleve Clin J Med Chem 2006;73:187–197.
51. Baltz RH, Miao V, Wrigley ST. Nat Prod Rep 2005;22:717–741.
52. Nguyen KT, Ritz D, Gu JQ, Alexander D, Chu M, Miao V, Brian P, Baltz RH. Proc Natl Acad Sci USA 2006;103:17462–17467.
53. Doekel S, Coeffet-Le Gal M-F, Gu JQ, Chu M, Baltz RH, Brian P. Microbiology 2008; 154:2872–2880.
54. Zhou H, Xie X, Tang Y. Curr Opin Biotechnol 2008;19:590–596.
55. Poehlsgaard J, Douthwaite S. Nat Rev Microbiol 2005;3:870–881.

56. Schlünzen F, Harms J, Franceschi F, Hansen H, Bartels H, Zarivach R, Yonath A. Structure 2003;11:329–338.
57. Wang G, Niu D, Qiu YL, Phan LT, Chen Z, Polemeropoulos A, Or YS. Org Lett 2004;6:4455–4458.
58. Xiong L, Korkhin Y, Mankin AS. Antimicrob Agents Chemother 2005;49:281–288.
59. Bermudez LE, Motamedi N, Chee C, Baimukanova G, Kolonoski P, Inderlied C, Aralar P, Wang G, Phan LT, Young LS. Antimicrob Agents Chemother 2007;51:1666–1670.
60. Lewis CA, Merkel J, Miller SJ. Bioorg Med Chem Lett 2008;18:6007–6011.
61. Kavanagh F, Hervey A, Robbins WJ. Proc Natl Acad Sci USA 1951;37:570–574.
62. Meingassner JG, Schmook FP, Czok R, Mieth H. Poultry Sci 1979;58:308–313.
63. Zhang YY, Xu KP, Ren D, Ge SH, Wang YL, Wang YZ. Chinese Chem Lett 2009;20:29–31.
64. Grove JF, MacMillan J, Mulholland TPC, Rogers MAT. J Chem Soc 1952; 3977–3987.
65. Mechlinski W, Schaffner CP, Ganis P, Avitabile G. Tetrahedron Lett 1970;11: 3873–3876.
66. Borowski E, Zielinski J, Ziminski T, Falowski L, Kolodziejczyk P, Golik J, Jereczek E. Tetrahedron Lett 1970;11:3909–3914.
67. Cereghetti DM, Carreira EM. Synthesis 2006;6:914–942.
68. Chong CN, Rickards RW. Tetrahedron Lett 1970;11:5145–5148.
69. Borowski E, Zielinski J, Falowski L, Ziminski T, Golik J, Kolodziejczyk P, Jereczek E, Gdulewicz M, Shenin Y, Kotienko T. Tetrahedron Lett 1971;12:685–690.
70. Pandey RC, Rinehart KL. J Antibiot 1976;29: 1035–1042.
71. Morrison VA. Exp Rev Anti Infect Ther 2006;4:325–342.
72. Turner MS, Drew RH, Perfect JR. Exp Opin Emerg Drugs 2006;11:231–250.
73. Cornely OA, Lasso M, Betts R, Klimko N, Vazquez J, Dobb G, Velez J, Williams-Diaz A, Lipka J, Taylor A, Sable C, et al. J Antimicrob Chemother 2007;60:363–369.
74. Hope WW, Mickiene D, Petraitis V, Petraitiene R, Kelaher AM, Hughes JE, Cotton MP, Bacher J, Keirns JJ, Buell D, Heresi G, et al. J Infect Dis 2008;197:163–171.
75. Aperis G, Myriounis N, Spanakis EK, Mylonakis E. Exp Opin Investig Drugs 2006;15: 1319–1336.
76. Cappelletty D, Eiselstein-McKitrick K. Pharmacother 2007;27:369–388.
77. Pasqualotto AC, Denning DW. J Antimicrob Chemother 2008;61(Suppl1): i19–i30.
78. Endo A. Nature Med 2008;14:1050–1052.
79. Endo A. J Med Chem 1975;28:401–405.
80. Endo A, Kuroda M, Tanzawa K. FEBS Lett 1976;72:323–326.
81. Endo A, Kuroda M, Tsujita Y. J Antibiot 1976;29:1346–1348.
82. Brown AG, Smale TC, King TJ, Hasenkamp R, Thompson RH. J Chem Soc, Perkin Trans 1976;1:1165–1170.
83. Steinberg D. J Lipid Res 2006;47:1339–1351.
84. Yamamoto A, Sudo H, Endo A. Atherosclerosis 1980;35:259–266.
85. Endo A. J Antibiot 1979;32:852–854.
86. Endo A. J Antibiot 1980;33:334–336.
87. Vagelos RP. Science 1991;252:1080–1084.
88. Stokker GE, Hoffman WF, Alberts AA, Cragoe EJ Jr, Deana AA, Gilfillan JL, Huff JW, Novello FC, Prugh JD, Smith RL, Willard AK. J Med Chem 1985;28:347–358.
89. Hoffman WF, Alberts AA, Cragoe EJ Jr, Deana AA, Evans BE, Gilfillan JL, Gould NP, Huff BE, Novello FC, Prugh JD, Rittle KE, et al. J Med Chem 1986;29:159–169.
90. Stokker GE, Alberts AA, Anderson PS, Cragoe EJ Jr, Deana AA, Gilfillan JL, Hirshfield J, Holtz WJ, Hoffman WF, Huff JW, Lee TJ, et al. J Med Chem 1986;29:170–181.
91. Roth BD, Ortwine DF, Hoefle ML, Stratton CD, Sliskovic DR, Wilson MW, Newton RS. J Med Chem 1990;33:21–31.
92. Sliskovic DR, Roth BD, Wilson MW, Hoefle ML, Newton RS. J Med Chem 1990;33:31–38.
93. Wareing JR,USP 4,613,610. 1986.
94. Rosen T, Heathcock CH. Tetrahedron 1986;42:4909–4951.
95. Roth BD. Prog Med Chem 2002;40:1–22.
96. Wilson RM, Danishefsky SJ. J Org Chem 2006;71:8329–8351.
97. Ahmad S, Madsen CS, Stein PD, Janovitz E, Huang C, Ngu K, Bisaha S, Kennedy LJ, Chen BC, Zhao R, Sitkoff D, et al. J Med Chem 2008;51:2722–2733.
98. Sturino CF, O'Neill G, Lachance N, Boyd M, Berthelette C, Labelle M, Li L, Roy B, Scheigetz J, Tsou N, Aubin Y, et al. J Med Chem 2007;50:794–806.
99. Newman DJ, Cragg GM, Snader KM. J Nat Prod 2003;66:1022–1037.

100. Newman DJ, Cragg GM. J Nat Prod 2007;70:461–477.
101. Waksman SA, Woodruff HB. Proc Soc Exp Biol Med 1940;45:609–614.
102. Hackmann C. Z Krebsforsch 1952;58:607–613.
103. Schulte G. Z Krebsforsch 1952;58:500–503.
104. Gniazdowski M, Denny WA, Nelson SM, Czyz M. Curr Med Chem 2003;10:909–924.
105. Mauger AB, Lackner H. In: Cragg GM, Kingston DGI, Newman DJ, editors. Anticancer Agents from Natural Products. Boca Raton, FL: Taylor and Francis; 2005. p 281–297.
106. Arcamone FM. In: Cragg GM, Kingston DGI, Newman DJ, editors. Anticancer Agents from Natural Products. Boca Raton, FL: Taylor and Francis; 2005; p 299–320.
107. Hecht SM. In: Cragg GM, Kingston DGI, Newman DJ, editors. Anticancer Agents from Natural Products. Boca Raton, FL: Taylor and Francis; 2005. p 357–381.
108. Leitheiser CJ, Hecht SM. Curr Opin Drug Discov Dev 2003;6:827–837.
109. Ma Q, Xu Z, Schroeder BR, Sun W, Wei F, Hashimoto S, Konishi K, Leitheiser CJ, Hecht SM. J Am Chem Soc 2007;129: 12439–12452.
110. Tao Z-F, Konishi K, Keith G, Hecht SM. J Am Chem Soc 2006;128:14806–14807.
111. Akiyama Y, Ma Q, Edgar E, Laikhter A, Hecht SM. J Am Chem Soc 2008;130:9650–9651.
112. Hamann PR, Upeslacis J, Borders DB. In: Cragg GM, Kingston DGI, Newman DJ, editors. Anticancer Agents from Natural Products. Boca Raton, FL: Taylor and Francis; 2005. p 451–474.
113. Van Lanen SG, Shen B. Curr Top Med Chem 2008;8:448–459.
114. Scott AM, Herbertson RA, Lee FT, Chappell B, Micallef N, Lee ST, Saunder T, Hopkins W, Smyth FE, Tebutt NC. J Clin Oncol 2008;26: Abstract 3025.
115. DiJoseph JF, Dougher MM, Kalyandrug LB, Armellino DC, Boghaert ER, Hamann PR, Moran JK, Damle NK. Clin Cancer Res 2006;12:242–249.
116. Castillo J, Winer E, Quesenberry P. Exp Hematol 2008;36:755–768.
117. Davies J, Wang H, Taylor T, Warabi K, Huang X-H, Andersen RJ. Org Lett 2005; 7:5233–5236.
118. Nicolaou KC, Chen JS, Zhang H, Montero A. Angew Chem Int Ed 2008;47:185–189.
119. Umezawa H, Imoto M, Sawa T, Isshiki K, Matsuda N, Uchida T, Iinuma H, Hamada M, Takeuchi T. J Antibiot 1986;39:170–173.
120. Imoto M, Umezawa K, Isshiki K, Kunimoto S, Sawa T, Takeuchi T, Umezawa H. J Antibiot 1987;40:1471–1473.
121. Posner I, Engel M, Gazit A, Levitzki A. Mol Pharmacol 1994;45:673–683.
122. Levitzki A, Gazit A. Science 1995;267: 1782–1788.
123. Krystal GW, Honsawek S, Kiewlich D, Liang C, Vasile S, Sun L, McMahon G, Lipson KE. Cancer Res 2001;61:3660–3668.
124. Gazit A, App H, McMahon G, Chen J, Levitzki A, Bohmer FD. J Med Chem 1996;39: 2170–2177.
125. Onoda T, Iinuma H, Sasaki Y, Hamada M, Isshiki K, Naganawa H, Takeuchi T. J Nat Prod 1989;52:1252–1257.
126. Hsu C-Y, Persons PE, Spada AP, Bednar RA, Levitski A, Zilberstein A. J Biol Chem 1991;266:21105–21112.
127. Mu F, Coffing SL, Riese DJ II, Geahlen RL, Verdier-Pinard P, Hamel E, Johnson J, Cushman M. J Med Chem 2001;44:441–452.
128. Albuschat R, Lowe W, Weer M, Luger P, Jendrossek V. Eur J Med Chem 2004;39: 1001–1011.
129. Hodge CN, Pierce J. Bioorg Med Chem Lett 1993;3:1605–1608.
130. Nussbaumer P, Winiski AP. Bioorg Med Chem 2008;16:7552–7560.
131. Nussbaumer P, Winiski AP, Cammisuli S, Hiestand P, Weckbecker G, Stütz A. J Med Chem 1994;37:4079–4084.
132. Cammisuli S, Winiski A, Nussbaumer P, Hiestand P, Stütz A, Weckbecker G. Int J Cancer 1996;65:351–359.
133. Sehgal SN, Baker H, Vezina C. J Antibiot 1975;28:727–732.
134. Baker H, Sidorowicz A, Sehgal SN, Vezina C. J Antibiot 1978;31:539–545.
135. Eng CP, Sehgal SN, Vezina C. J Antibiot 1984;37:1231–1237.
136. Koehn FE. Curr Opin Biotechnol 2006;17: 631–637.
137. Tsang CK, Qi H, Liu LF, Zheng XFS. Drug Discov Today 2007;12:112–124.
138. Ruan B, Pong K, Jow F, Bowlby M, Crozier RA, Liu D, Liang S, Chen Y, Mercado ML, Feng X, Bennett F, et al. Proc Natl Acad Sci USA 2008;105:33–38.
139. Chambraud B, Belabes H, Fontaine-Lenoir V, Fellous A, Baulieu EE. FASEB J 2007;21: 2787–2797.
140. Hoefle G, Reichenbach H. In: Cragg GM, Kingston DGI, Newman DJ, editors. Anticancer

Agents from Natural Products. Boca Raton, FL: Taylor and Francis; 2005. p 413–450.

141. Bollag DM, McQueney PA, Zhu J, Hensens O, Koupal L, Liesch J, Goetz M, Lazarides E, Woods CM. Cancer Res 1995;55: 2325–2333.

142. Klar U, Buchmann B, Schwede W, Skuballa W, Hoffmann J, Lichtner RB. Angew Chem Int Ed 2006;45:7942–7948.

143. Altmann K-H, Gertsch J. Nat Prod Rep 2007;24:327–357.

144. Altmann K-H, Pfeiffer B, Arseniyadis S, Pratt BA, Nicolaou KC. ChemMedChem 2007;2: 396–423.

145. Feyen F, Cachoux F, Gertsch J, Wartmann M, Altmann K-H. Acc Chem Res 2008;41:21–31.

146. Altmann KH, Memmert K. In: Petersen F, Amstutz R, editors. Progress in Drug Research. Basel: Birkhauser Verlag; 2008. p 275–334.

147. Hearn BR, Zhang D, Li Y, Myles DC. Org Lett 2006;8:3057–3059.

148. Cho YS, Wu K-D, Moore MAS, Chou T-C, Danishefsky SJ. Drugs Fut 2005;30: 737–745.

149. Chou T-C, Zhang X-G, Dong H, Zhong Z, Feng L, Li Y, Sherrill M, Timmermanns P, Johnson RG Jr, Danishefsky SJ. Proc Am Assoc Cancer Res 2008; Abstract 1402.

150. Chou T-C, Zhang X, Zhong Z, Li Y, Feng L, Eng S, Myles DC, Johnson RG Jr, Wu N, Yin YI, Wilson MW, et al. Proc Natl Acad Sci USA 2008;105:13157–13162.

151. Leamon CP. Curr Opin Invest Drugs 2008;9: 1277–1286.

152. Frank B, Müller R. Pharmaz Zeit 2007;152: 64–67.

153. Schneiker S, Perlova O, Kaiser O, Gerth K, Alici A, Altmeyer MO, Bartels D, Bekel T, Beyer S, Bode E, Bode HB, et al. Nat Biotechnol 2007;25:1281–1289.

154. Wenzel SC, Müller R. Nat Prod Rep 2007;24: 1211–1224.

155. Williams PG. Trends Biotechnol 2009;27: 45–52.

156. Taylor MW, Hill RT, Piel J, Thacker RW, Hentschel U. ISME J 2007;1:187–190.

157. Nguyen T-A, Ishida K, Jenke-Kodama H, Dittmann E, Gurgui C, Hochmuth T, Taudien S, Platzer M, Hertweck C, Piel J. Nat Biotechnol 2008;26:225–233.

158. Newman DJ. In: Cragg GM, Kingston DGI, Newman DJ, editors. Anticancer Agents from Natural Products. Boca Raton, FL: Taylor and Francis; 2005. p 137–150.

159. Flahive E, Srirangam J. In: Cragg GM, Kingston DGI, Newman DJ, editors. Anticancer Agents from Natural Products. Boca Raton, FL: Taylor and Francis; 2005. p 191–214.

160. Lopanik NB, Shields JA, Buchholz TJ, Rath CM, Hothersall J, Haygood MG, Hakansson K, Thomas CM, Sherman DH. Chem Biol 2008;15:1175–1186.

161. Wender PA, Verma VA, Paxton TJ, Pillow TH. Acc Chem Res 2008;41:40–49.

162. Wender PA, DeChristopher BA, Schrier AJ. J Am Chem Soc 2008;130:6658–6659.

163. Charette A. Nature 2008;456:451–453.

164. Trost BM, Dong G. Nature 2008;456:485–488.

165. Keck GE, Kraft MB, Truong AP, Li W, Sanchez CC, Kedei N, Lewin NE, Blumberg PM. J Am Chem Soc 2008;130:6660–6661.

166. Wang D, Darwish DS, Schreurs BG, Alkon DL. Behav Pharmacol 2008;19:245–256.

167. Sun M-K, Hongpaisan J, Nelson TJ, Alkon DL. Proc Natl Acad Sci USA 2008;105: 13620–13625.

168. Luesch H, Moore RE, Paul VJ, Mooberry SL, Corbett TH. J Nat Prod 2001;64:907–910.

169. Ramaswamy AV, Flatt PM, Edwards DJ, Simmons TL, Han B, Gerwick WH. In: Proksch P, Mueller WEG, editors. Frontiers in Marine Biotechnology. Wymondham, UK: Horizon Bioscience; 2006. p 175–224.

170. Law C-L, Gordon KA, Toki BE, Yamane AK, Hering MA, Cerveny CG, Petroziello JM, Ryan MC, Smith L, Simon R, Sauter G, et al Can Res 2006;66:2328–2337.

171. Sutherland MSK, Sanderson RJ, Gordon KA, Andreyka J, Cerveny CG, Yu C, Lewis TS, Meyer DL, Zabinski RF, Doronina SO, Senter PD, et al. J Biol Chem 2006;281: 10540–10547.

172. Oflazoglu E, Kissler KM, Sievers EL, Grewal IS, Gerber H-P. Br J Haematol 2008;142: 69–73.

173. Pollack VA, Alvarez E, Tse KF, Torgov MY, Xie S, Shenoy SG, MacDougall JR, Arrol S, Zhong Z, Gerwein RW, Hahne WF, et al. Cancer Chemother Pharmacol 2007;60: 423–435.

174. Ma D, Hopf CE, Malewicz AD, Donovan GP, Senter PD, Goeckeler WF, Maddon PJ, Olson WC. Clin Cancer Res 2006;12:2591–2596.

175. Gerwick WH, Proteau PJ, Nagle DG, Hamel E, Blokhin A, Slate DL. J Org Chem 1994;59: 1243–1245.

176. Wipf P, Reeves JT, Day BW. Curr Pharm Des 2004;10:1417–1437.
177. Chang Z, Sitachitta N, Rossi JV, Roberts MA, Flatt PM, Jia J, Sherman DH, Gerwick WH. J Nat Prod 2004;67:1356–1367.
178. Nguyen TL, McGrath C, Hermone AR, Burnett JC, Zaharevitz DW, Day BW, Wipf P, Hamel E, Gussio R. J Med Chem 2005;48: 6107–6116.
179. Simmons TL, Coates RC, Clark BR, Engene N, Gonzalez D, Esquenazi E, Dorrestein PC, Gerwick WH. Proc Natl Acad Sci USA 2008; 105:4587–4594.
180. Esquenazi E, Coates C, Simmons TL, Gonzalez D, Gerwick WH, Dorrestein PC. Mol BioSyst 2008;4:562–570.
181. Kobayashi M, Aoki S, Ohyabu N, Kurosu M, Wang W, Kitagawa I. Tetrahedron Lett 1994;35: 7969–7972.
182. Schwartz RE, Hirsch CF, Sesin DF, Flor JE, Chartrain M, Fromtling RE, Harris GH, Salvatore MJ, Liesch JM, Yudin K. J Ind Microbiol 1990;5:112–124.
183. Golakoti T, Ohtani I, Patterson GML, Moore RE, Corbett TH, Valeriote FA, Demchik L. J Am Chem Soc 1994;116:4729–4737.
184. Liang J, Moher ED, Moore RE, Hoard DW. J Org Chem 2000;65:3143–3147.
185. Hoard DW, Moher ED, Martinelli MJ, Norman BH. Org Lett 2002;4:1813–1815.
186. Kotoku N, Narumi F, Kato T, Yamaguchi M, Kobayashi M. Tetrahedron Lett 2007;48: 7147–7150.
187. Beck ZQ, Aldrich CC, Magarvey NA, Georg GI, Sherman DH. Biochemistry 2005;44: 13457–13466.
188. Patterson GML.Personal Communication 2008.
189. Liang J, Moore RE, Moher ED, Munroe JE, Al-awar RS, Hay DA, Varie DL, Zhang TY, Aikins JA, Martinelli MJ, Shih C, et al. Investig New Drugs 2005;23:213–224.
190. West LM, Northcote PT. J Org Chem 2000;65: 445–449.
191. Hood KA, West LM, Rouwe B, Northcote PT, Berridge MV, Wakefield S, Miller JH. Cancer Res 2002;62:3356–3360.
192. Gaitanos TN, Buey RM, Diaz JF, Northcote PT, Teesdale-Spittle P, Andreu JM, Miller JH. Cancer Res 2004;64:5063–5067.
193. Huzil JT, Chik JK, Slysz GW, Freedman H, Tuszynski J, Taylor RE, Sackett DL, Schriemer DC. J Mol Biol 2008;378:1016–1030.
194. Jimenez-Barbero J, Canales A, Northcote PT, Buey RM, Andreu JM, Diaz JF. J Am Chem Soc 2006;128:8757–8765.
195. Page M, West L, Northcote P, Battershill C, Kelly M. J Chem Ecol 2005;31:1161–1174.
196. Page MJ, Northcote PT, Webb VL, Mackey S, Handley SJ. Aquaculture 2005;250:256–269.
197. Liao X, Wu Y, De Brabander JK. Angew Chem Int Ed 2003;42:1648–1652.
198. Jin M, Taylor RE. Org Lett 2005;7:1303–1305.
199. Ghosh AK, Xu X, Kim J-H, Xu C-X. Org Lett 2008;10:1001–1004.
200. Williams DR, Nag PP, Zorn N. Curr Opin Drug Discov Dev 2008;11:251–271.
201. Rinehart K, Holt TG, Fregeau NL, Stroh JG, Kiefer PA, Sun F, Li LH, Martin DG. J Org Chem 1990;55:4512–4515.
202. Wright AE, Forleo DA, Gunawardana GP, Gunasekera SP, Koehn FE, McConnell OJ. J Org Chem 1990;55:4508–4512.
203. Cuevas C, Francesch A. Nat Prod Rep 2009;26:322–337.
204. Velasco A, Acebo P, Gomez A, Schleissner C, Rodriguez P, Aparicio T, Conde S, Munoz R, de la Calle F, Garcia JL, Sanchez-Puelles JM. Mol Microbiol 2005;56:144–154.
205. Moss C, Green DH, Perez B, Velasco A, Henriquez R, McKenzie JD. Mar Biol 2003;143: 99–110.
206. Perez-Matos AE, Rosado W, Govind NS. Anton van Leeuwen 2007;92:155–164.
207. Jimeno J, Aracil M, Tercero JC. J Trans Med 2006;4:3.
208. Soares DG, Escargueil AE, Poindessous V, Sarasin A, de Gramont A, Bonatto D, Henriques JAP, Larsen AK. Proc Natl Acad Sci USA 2007;104:13062–13067.
209. Marco E, David-Cordonnier M-H, Bailly C, Cuevas C, Gago F. J Med Chem 2006;49:6925–6929.
210. Moneo V, Serelde BG, Fominaya J, Leal JFM, Blanco-Aparicio C, Romero L, Sanchez-Beato M, Cigudosa JC, Tercero JC, Piris MA, Jimeno J, et al. J Cell Biochem 2007;100:339–348.
211. Newman DJ, Hill RT. J Ind Microbiol Biotechnol 2006;33:539–544.
212. Fenical W, Jensen PR. Nat Chem Biol 2006;2:666–673.
213. Jensen PR, Fenical W. In: Zhang L, Demain AL, editors. Natural Products. Totowa, NJ: Humana Press, Ltd; 2005. p 315–328.
214. Jensen PR, Mincer TJ, Williams PG, Fenical W. Anton van Leeuwen 2005;87:43–48.

215. Feling RH, Buchanan GO, Mincer TJ, Kauffman CA, Jensen PR, Fenical W. Angew Chem Int Ed 2003;42:355–357.
216. Reddy LR, Saravanan P, Corey EJ. J Am Chem Soc 2004;126:6230–6231.
217. Endo A, Danishefsky SJ. J Am Chem Soc 2005;127:8298–8299.
218. Macherla VR, Mitchell SS, Manam RR, Reed KA, Chao T-H, Nicholson B, Deyanat-Yazdi G, Mai B, Jensen PR, Fenical WF, Neuteboom STC, et al. J Med Chem 2005;48:3684–3687.
219. Shibasaki M, Kanai M, Fukuda N. Chem Asian J 2007;2:20–38.
220. Ling T, Macherla VR, Manam RR, McArthur KA, Potts BCM. Org Lett 2007;9:2289–2292.
221. Fukuda T, Sugiyama K, Arima S, Harigaya Y, Nagamitsu T, Omura S. Org Lett 2008;10: 4239–4242.
222. Manam RR, McArthur KA, Chao T-H, Weiss J, Ali JA, Palombella VJ, Groll M, Lloyd GK, Palladino MA, Neuteboom STC, Macherla VR, et al. J Med Chem 2008;51:6711–6724.
223. Jensen PR, Williams PG, Oh D-C, Zeigler L, Fenical W. Appl Environ Microbiol 2007;73: 1146–1152.
224. Lam KS, Tsueng G, McArthur KA, Mitchell SS, Potts BCM, Xu J. J Antibiot 2007;60:13–19.
225. Reed KA, Manam RR, Mitchell SS, Xu J, Teisan S, Chao T-H, Deyanat-Yazdi G, Neuteboom STC, Lam KS, Potts BCM. J Nat Prod 2007;70:269–276.
226. Tsueng G, McArthur KA, Potts BCM, Lam KS. Appl Microbiol Biotechnol 2007;75: 999–1005.
227. Groll M, Huber R, Potts BCM. J Am Chem Soc 2006;128:5136–5141.
228. Denora N, Potts BCM, Stella VJ. J Pharm Sci 2007;96:2037–2047.
229. Kurzrock R, Hamlin P, Younes A, Hong D, Gordon M, Spear MA, Palladino MA, Lloyd GK, Longenecker AM, Neuteboom STC, Cropp GF, et al. Blood 2007;110: Abstract 4504.
230. Stadler M, Bitzer J, Mayer-Bartschmid A, Mueller H, Benet-Buchholz J, Gantner F, Tichy H-V, Reinemer P, Bacon KB. J Nat Prod 2007;70:246–252.
231. Tsueng G, Teisan S, Lam KS. Appl Microbiol Biotechnol 2008;78:827–832.
232. Tsueng G, Lam KS. Appl Microbiol Biotechnol 2008;78:821–826.
233. Handelsman J, Rondon MR, Brady SF, Clardy J, Goodman RM. Chem Biol 1998;5: R245–R249.
234. Rondon MR, August PR, Bettermann AD, Brady SF, Grossman TH, Liles MR, Loiacono KA, Lynch BA, MacNeil IA, Minor C, Tiong CL, et al. Appl Environ Microbiol 2000;66: 2541–2547.
235. Rusch DB, Halpern AL, Sutton G, Heidelberg KB, Williamson S, Yooseph S, Wu D, Eisen JA, Hoffman JM, Remington K, Beeson K, et al. PLoS Biol. 2007;5:398–431.
236. Omura S, Ikeda H, Ishikawa J, Hanamoto A, Takahashi C, Shinose M, Takahashi Y, Horikawa H, Nakazawa H, Osonoe T, Kikuchi H, et al. Proc Natl Acad Sci USA 2001; 98:12215–12220.
237. Ikeda H, Ishikawa J, Hanamoto A, Shinose M, Kikuchi H, Shiba T, Sakaki Y, Hattori M, Omura S. Nat Biotechnol 2003;21:526–531.
238. Bentley SD, Chater KF, Cerdeno-Tarraga A-M, Challis GL, Thomson NR, James KD, Harris DE, Quail MA, Kieser H, Harper D, Bateman A, et al. Nature 2002;417:141–147.
239. Udwary DW, Zeigler L, Asolkar RN, Singan V, Lapidus A, Fenical W, Jensen PR, Moore BS. Proc Natl Acad Sci USA 2007;104: 10376–10381.
240. Eustaquio AS, Pojer F, Noel JP, Moore BS. Nat Chem Biol 2008;4:69–74.
241. Eustaquio AS, Moore BS. Angew Chem Int Ed 2008;47:3936–3938.
242. McGlinchey RP, Nett M, Eustaquio AS, Asolkar RN, Fenical W, Moore BS. J Am Chem Soc 2008;130:7822–7823.
243. Watanabe K. Biosci Biotechnol Biochem 2008;72:2481–2506.
244. Fenical W, Jensen PR. Nat Chem Biol 2006;2: 666–673.
245. Gontang EA, Fenical W, Jensen PR. Appl Environ Microbiol 2007;73:3272–3282.
246. Stierle A, Strobel G, Stierle D. Science 1993;260:214–216.
247. Li J-Y, Sidhu RS, Bollon A, Strobel GA. Mycolog Res 1998;102:461–464.
248. Puri SC, Verma V, Amna T, Qazi GN, Spiteller M. J Nat Prod 2005;68:1717–1719.
249. Amna T, Puri SC, Verma V, Sharma JP, Khajuria RK, Musarrat J, Spiteller M, Qazi GN. Can J Microbiol 2006;52:189–196.
250. Eyberger AL, Dondapati R, Porter JR. J Nat Prod 2006;69:1121–1124.
251. Puri SC, Nazir A, Chawla R, Arora R, Riyaz-ul-Hasan S, Amna T, Ahmed B, Verma V, Singh S, Sagar R, Sharma A, et al. J Biotechnol 2006; 122:494–510.

252. Guo B, Li H, Zhang L. Yunnam univ 1998; 20:214–215.
253. Zhang LQ, Guo B, Li H, Zeng S, Shao H, Gu S, Wei R. Zhong Cao Yao (Chinese Tradit Herb Drugs) 2000;31:805–807.
254. Yang X, Zhang L, Guo B, Guo S. Zhong Cao Yao 2004;35:79–81.
255. Bok JW, Hoffmeister D, Maggio-Hall LA, Murillo R, Glasner JD, Keller NP. Chem Biol 2006;13:31–37.
256. Chiang Y-M, Szewczyk E, Nayak T, Davidson AD, Sanchez JF, Lo H-C, Ho W-Y, Simityan H, Kuo E, Praseuth A, Watanabe K, et al. Chem Biol 2008;15:527–532.
257. Hoffmeister D, Keller NP. Nat Prod Rep 2007;24:393–416.
258. Brakhage AA, Schuemann J, Bergmann S, Scherlach K, Schroeckh V, Hertweck C. In: Petersen F, Amstutz R, editors. Progress in Drug Research. Vol. 66.Basel: Birkhauser Verlag; 2008. p 3–12.
259. Fox EM, Howlett BJ. Curr Opin Microbiol 2008;11:481–487.
260. Stadler M, Keller NP. Mycolog Res 2008;112:127–130.
261. Lefevre F, Robe P, Jarrin C, Ginolhac A, Zago C, Auriol D, Vogel TM, Simonet P, Nalin R. Res Microbiol 2008;159:153–161.
262. Krug D, Zurek G, Schneider B, Garcia R, Muller R. Anal Chim Acta 2008;624:97–106.
263. Krug D, Zurek G, Revermann O, Vos M, Velicer GJ, Muller R. Appl Environ Microbiol 2008;74:3058–3068.
264. Meiser P, Weissman KJ, Bode HB, Krug D, Dickschat JS, Sandmann A, Muller R. Chem Biol 2008;15:771–781.

MULTIOBJECTIVE OPTIMIZATION FOR DRUG DISCOVERY

SEAN EKINS[1,2,3,4]
J. DANA HONEYCUTT[5]
JAMES T. METZ[6]
[1]Collaborations in Chemistry, Jenkintown, PA
[2]Department of Pharmaceutical Sciences, University of Maryland, Baltimore, MD
[3]Department of Pharmacology, Robert Wood Johnson Medical School, University of Medicine and Dentistry, Piscataway, NJ
[4]Collaborative Drug Discovery, Burlingame, CA
[5]Accelrys, San Diego, CA
[6]GPRD R4DG, Department of Fragment Screening and Lead Characterization, Abbott Laboratories, Abbott Park, IL

1. INTRODUCTION

Practitioners involved in modern drug discovery have increasingly realized that they cannot afford to focus on the biological activity at their target of interest in isolation if they are to ultimately discover a successful drug-like molecule. Simply, it takes more than a nanomolar lead to become a drug. That molecule generally has to be orally available to be absorbed, and it needs to reach its target (and then in many cases be metabolized) before being cleared from the body (but not too quickly). At each step, the molecule normally has to cross membranes whether passively, actively, or via other routes. Each of these physical processes requires different physicochemical properties—for example, moderate hydrophobicity is good for membrane penetration [1] but not for solubility. Nowadays, a pharmaceutical scientist is bombarded with information that may be real or virtual data on their molecule of interest or on a series of molecules (e.g., predicted physicochemical properties alongside measured metabolic stability). This is in addition to any data on known published molecules that are structurally similar or active at the same target. We must therefore consider the many requirements for a molecule alongside its bioactivity, which may be conflicting. This process has been termed multiobjective or multidimensional optimization. Since we and others suggested the need for a simultaneous, multiobjective optimization of various molecular properties with efficacy data [2–5], we have seen the amount of absorption, distribution, metabolism, and excretion (ADME) and toxicity data generated dramatically increase over the past decade [6]. The production of these data has also moved earlier in the process, as larger numbers of compounds were also synthesized (via combinatorial or parallel methods) and tested against the target using high-throughput screening (HTS).

Increased compound throughput in drug discovery has come at a cost, with lead compounds derived from HTS hits frequently having undesirable properties [7–9] such as increased hydrophobicity ($\log P$) and decreased solubility. This has driven some to suggest that leads should be more stringently selected in terms of their molecular properties (MW $<$ 350, $\log P < 3$, and affinity $\sim$0.1 μM [10] or other properties [11]). It has also pushed many to consider fragment-based drug discovery [12,13] or the analysis of large biological data sets to understand fragment occurrence ratios, fragment similarities, and the relationship to bioactivity or target preferences [14–16]. Others have used the rule of five [17] as a starting point for deriving rules for specific classes of drugs. For example, those marketed drugs that are inhaled tend to possess more hydrogen bonds and have generally lower $c\log P$ [18]. Similarly, large scale data set analyses enables the generation of rules such as those for predicting compound reactivity toward protein thiol groups [19]. One study used 8800 compounds with a Bayesian classifier model and an extended connectivity fingerprint (ECFP_6) to result in a model with good classification accuracy to predict reactivity [19]. Understanding the importance of fragments or transformations that can be made can also facilitate making modifications to molecules that may increase activity [20] or enable the optimization of observed side activities [21,22]. For example, subtle chemical modifications can dramatically alter the pharmacological profile, and similarly one would expect the ADME/Tox and physicochemical properties to be driving these differences. The recent slight modification of the late stage bioavailable clinical candidate for cancer, tipifarnib, resulted in a potent inhibitor for

Chagas disease active *in vivo* with more predictable drug-like qualities [23]. The trend toward understanding which parts of the molecule are important for activity has also led others to pursue measures of ligand efficiency or fit quality that balance potency with size or other properties (e.g., molecular weight, number of heavy atoms or polar surface area), such that smaller yet more efficient molecules have a better chance of having good drug-like properties [12,24–26].

We have recently used computational pharmacophores for database searching to discover seven new human pregnane X receptor antagonists [27]. We were able to compare our various antagonists to those that had already been described such as ketoconazole and sulforaphane. When we used three different molecule efficiency indices, the smaller molecules generally performed the best when measured by any of the three indices (Table 1). When we looked at the ligand efficiency versus heavy atom count, there was an exponential decrease in efficiency between 10 and 20 heavy atoms, similar to observations with much larger data sets across different targets [26].

It is now accepted that simultaneous multiobjective drug discovery is necessary [28] if we are to significantly improve the output of drug discovery and increase efficiency. We are now also seeing systems biology methods applied in drug discovery [29,30] and increasingly in toxicology in an attempt to unravel the underlying complexity of the biology. The FDA has recognized (albeit later than most) the cost and challenge of innovation [31] seen in the high failure rates [32] that the pharmaceutical industry faces, representing a major challenge to improve the identification of drug candidates and avoid toxicity or unfavorable drug–drug interactions such as induction or inhibition of drug metabolizing enzymes [6,33,34].

Clearly the most cost-effective approach for drug discovery is to simulate as much as possible (ideally all possible aspects of drug discovery and development [35,36], although this may not currently be practical) using computational methods. These have a greater throughput than *in vitro*, and could improve the quality of the molecules generated by helping to address potential liabilities that lead to latter stage failures. A recent study even proposed using a self-avoiding walk to model research projects in a chemistry department and concluded that target and process controlled success [37]. *In vitro* data can be used as inputs into computational models for oral bioavailability [38] or other ADME properties using an array of different types of machine learning algorithms such as support vector machines [39–41], Bayesian modeling [42],

Table 1. Calculated Physicochemical Properties and Ligand Efficiency Indices for PXR Antagonist Data [27]

Molecule	pIC_{50}	Molecular Weight[a] (Da)	HA	LE	FQ	ADMET Sol	ADMET Abs
SPB06257	4.78	339.41	24	0.20	0.495	2	0
SPB02372	5.24	398.69	24	0.22	0.545	1	1
Leflunomide	5.17	270.21	19	0.27	0.555	3	0
Coumestrol	4.92	270.24	20	0.25	0.535	3	0
Itraconazole	5.05	705.63	49	0.10	0.429	2	2
SPB00574	4.61	386.68	23	0.20	0.479	1	1
Ketoconazole	4.78	531.43	36	0.13	0.444	2	0
Sulforaphane	5.25	177.29	10	0.52	0.722	4	0
SPB03255	5.20	314.36	22	0.24	0.554	2	0
SPB03256	5.21	368.33	25	0.21	0.537	2	0
SPB06061	5.28	300.33	21	0.25	0.556	2	0

[a] Properties calculated using Accelrys Discovery Studio 2.0.

HA: heavy atom count; LE: ligand efficiency = pIC_{50}/HA [26]; FQ: fit quality [12]; ADMET Sol: predicted solubility level from Discovery Studio, higher is better; ADME Abs: predicted human intestinal absorption level from Discovery Studio, lower is better.

Gaussian processes, [43] or others [44]. A particular example of where computational methods can have an immediate impact is in the prediction of metabolic transformations that can guide synthesis of more metabolically stable compounds [45] that may involve many different enzymes and potential sites for metabolism. For example, metabolism databases can be used to generate transformation rules [46,47], calculate a metabolic occurrence ratio [48,49], generate machine learning models [50], and predict metabolite spectra [51]. Another method uses multiobjective recursive partitioning analysis with multiple drug metabolizing enzymes that aids the visualization of such data [52]. It would therefore be important to block potentially labile sites but at the same time consider retaining the bioactivity, solubility, and other properties of a molecule.

We and others have indicated the need for integrated simulation tools [35,53] that bring together different models to improve the decision making process. We have initially suggested a system of multidimensional scoring using many ADME/Tox filters in decision making (e.g., blood–brain barrier (BBB); human ether-á-go-go (hERG) potassium channel; CYP inhibition; human fraction absorbed; PXR; P-glycoprotein; metabolic stability; Ames mutagenicity as well as molecular properties such as solubility, rule of five, etc. [36]), which appears to have also been followed by others [54]. ADME filters have also led to the derivation of rules of thumb for most of these properties that are heavily influenced by molecular weight and $c\log P$ [55].

We can visualize relatively complex data, yet in many ways we really rely upon maps of one sort or another to visualize very complex relationships. For example, we can map compound space [56,57], cancer medicinal chemistry space [58], and natural product space [59] that would help us visualize large scale drug discovery data sets. But, we can also map physicochemical properties. One study had more than 5000 measured ADME and computational properties for 358 compounds that were visualized using Spotfire to produced binned plots, profile charts, and heatmaps [60]. Alternative methods that could be used include Infvis [61], which is a Java based tool that uses glyph-based visualization. Hierarchical generative topographic mapping [62], and Kohonen and Sammon maps [63,64] represent additional methods to identify patterns in complex data such as those generated in preclinical drug discovery.

However, when there are multiple endpoints, whether these are experimental data or computational predictions (or both) trade-offs have to occur and we really need to use a method that can consider each variable and enable the selection of the best compound(s). One could use multicriteria decision methods (MCDMs) [65], also frequently called multiobjective optimization [66], to simultaneously optimize several variables [67] using desirability functions [68] or indexes [69]. Rather than focusing on simple rules to filter or select compounds, we need to consider many more variables. We have previously indicated how such multiobjective optimization approaches may work in drug discovery simultaneously rather than optimizing single properties sequentially [3] to derive a set of Pareto-optimal compounds.

2. APPROACHES TO MULTIOBJECTIVE OPTIMIZATION

Broadly speaking, there are two approaches to numerically solving a multiobjective optimization problem. For simplicity, in illustrating the different approaches, we will assume that there are just two objectives, but all of the techniques can be generalized to any number of objectives (with some caveats). We also assume for now that we seek to optimize the properties of individual compounds, as opposed to properties of compound libraries (such as structural diversity). We discuss library optimization in next section.

The first approach to multiobjective optimization is to somehow combine all of the objectives into a single objective function. This then allows classical single-objective optimization techniques to be applied to the problem.

One variant of this approach is weighted-sum optimization, in which the overall objective is a weighted arithmetic mean of values representing the individual objectives [70]. For example, suppose we have developed models to predict a specific activity and the toxicity of chemical compounds. In order to find the compounds that have the greatest activity and the lowest toxicity, we construct the following objective function:

$$O = \frac{w_a A}{s_a} - \frac{w_t T}{s_t}$$

where A is the predicted activity of a compound; T is its predicted toxicity; w_a and w_t are positive weighting coefficients whose values we assign according to the relative importance we place on maximizing activity versus minimizing toxicity; and s_a and s_t are scaling factors to correct for possible differences in range for the A and T values. (A common approach is to use as the scaling parameter the standard deviation of the variable value over the set of compounds we are considering.)

Another variant of the single-objective approach to multiobjective optimization is to use as our objective function a weighted geometric mean of desirability functions based on the individual objectives. This approach is known as desirability optimization methodology (DOM) and has certain advantages over the weighted sum approach [67].

Having established an objective function by either means, we then seek to maximize it (or minimize it, depending on how the objective is defined). If our optimization is over a fixed set of compounds, this simply requires sorting the compounds according the value of O. If instead we wish to explore a space of compounds whose structures are not predefined, we need to iteratively "mutate" the structure, beginning from one or more seed compounds, until the objective has been maximized. At the end, we are left with a single compound that is "best" according to our criteria.

The single-objective approach to multiobjective optimization has two major drawbacks: First, a single-objective optimization yields only a single "best" solution as suboptimal solutions are not retained. It provides no information on near-optimal solutions. Nor does it provide any way, short of running multiple optimizations with differing weights, of answering such questions as, "In return for a slight decrease in activity, is it possible to greatly reduce the toxicity?"

Second, the meaning of the weights is vague. While their effect is mathematically well defined, their meaning is hard to grasp intuitively. For example, what exactly does it mean to say, "Low toxicity is twice as important as high activity"? The only way to see how the weights affect the results is to vary them and run multiple optimizations.

The second broad approach to multiobjective optimization does not require prioritization or weighting of the individual objectives. This approach is known as Pareto or trade-off optimization [70,71] and is an alternative method to the desirability function approach. One of the advantages of the Pareto optimization is that all objectives are put on an equal footing [72]. Figure 1 illustrates the basic idea.

The data in this graph are hypothetical, but show the typical pattern found in multiobjective optimization problems. We seek to maximize both the activity and the nontoxicity of our compounds. (We define nontoxicity here as simply the reciprocal of toxicity, assuming the latter is always greater than zero.) Each point on the graph represents a different compound. Ideally, we would like an activity/toxicity profile corresponding to the red point in the upper right. But according to the displayed data, the most-active compounds tend to be the most toxic, while the least-toxic compounds tend to have low activity. In short, there's no free lunch.

However, observe that many compounds (those represented by the magenta points) have relatively high toxicity, yet low activity compared to other compounds with an equal or even lower toxicity. Inspection of the graph should convince one that the only compounds of even tentative interest for this optimization problem are those represented by the blue points. To be precise, for every magenta point, at least one blue point is better with regard to toxicity or activity, while being at least as good with regard to the other property. (In the jargon of Pareto optimization, we say that the blue point *dominates* the magenta point.)

The blue points define the *Pareto-optimal curve*. (With greater than two objectives, this curve becomes a surface or hypersurface.)

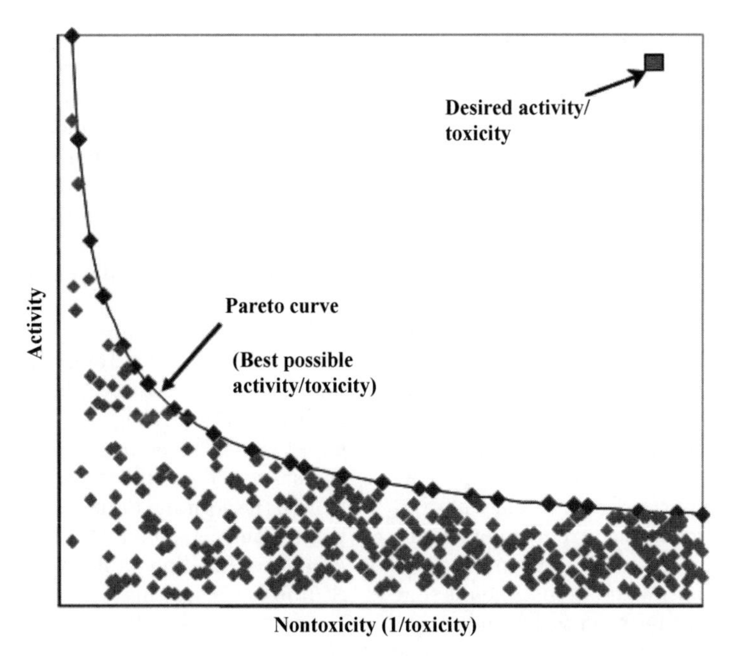

Figure 1. Illustration of Pareto optimization with hypothetical data. (This figure is available in full color at http://mrw.interscience.wiley.com/emrw/9780471266945/home.)

These are the points with the best possible trade-off between the two objectives. Note that we have not yet indicated any weights for the objectives—that is, any preference for lower toxicity over greater activity or vice versa. A preference for greater activity would have us focus on points toward the upper left of the curve. A preference for lower toxicity would have us focus on points toward the lower right of the curve.

Pareto optimization is the process of generating a set of points on the Pareto-optimal curve or surface. This accomplishes two things. First, it eliminates from consideration the vast number of compounds that are not of even tentative interest, irrespective of any weighting or priority of one objective over another. Second, it retains multiple compounds that might be of interest, allowing one to visually inspect the trade-offs involved in improving some properties at the expense of others, and to choose the best Pareto-optimal compound accordingly.

In essence, running a single Pareto optimization is equivalent to running multiple weighted sum optimizations with varying weights. Under certain conditions, the equivalence is exact. That is, under certain conditions, every Pareto-optimal point is equivalent to the solution of a weighted-sum optimization with a different (unspecified) set of weights [73].

3. MULTIOBJECTIVE OPTIMIZATION IN PRACTICE

Pareto optimization has been applied to numerous problems, including those in the area of compound and library optimization [74]. In library optimization, we seek not only to optimize properties of individual compounds in the library (though that may be one component of the problem), but properties of the library as a whole, such as structural diversity and scaffold coverage to ensure that different chemotypes are represented [75].

For example, we may wish to find a subset of 100 compounds, taken from a library of 1 million, which maximizes both the diversity and the degree to which the compounds are drug-like on average. Here, the unit being optimized is the subset rather than the individual compound, and the Pareto curve is defined by a group of these subsets. Some subsets along the Pareto curve will be more diverse and less drug-like on average, while others will be less diverse and more drug-like on average. The Pareto approach has been used by others to solve this and similar problems [74,76,77].

One drawback of the Pareto approach is that as the number of properties to be optimized increases, the number of samples on the Pareto-optimal surface tends to increase exponentially. The reason for this can be seen by considering that this "surface" is a one-dimensional curve for two optimization properties, a two-dimensional surface for three properties, a three-dimensional hypersurface for four properties, and so on. This implies a great increase in the required memory and computation time with the number of properties to be optimized. It also means that the researcher may be faced with hundreds or thousands of optimal samples to choose among at the end. To mitigate these problems and reduce the number of properties to be optimized, it may make sense to combine correlated properties into composite properties and then Pareto-optimize the composite properties (e.g., multiple measures of toxicity might be combined into an overall toxicity property.) The approach then becomes a hybrid of the single-objective and Pareto methods. Such an approach has been recently described with a case study for automated drug design for estrogen receptor antagonists using desirability indexes to reduce the number of objectives [78].

A recent review has summarized many computational multiobjective methods for molecule optimization [79], some of which are discussed below. Desirability-based multiobjective optimization (MOOP-DESIRE) was proposed for filtering combinatorial libraries [80] and global QSAR studies studying NSAIDS with analgesic, anti-inflammatory and ulcerogenic properties which all needed optimization [81]. The multiobjective evolutionary graph algorithm (MEGA) is a new method for *de novo* design of molecules that bind a target, which uses multiobjective optimization to trade off between conflicting objectives—for example, selectivity of one protein versus another, such as estrogen receptor beta over alpha selectivity [82]. Multiobjective genetic QSAR uses the Pareto ranking to produce a family of models representing a different compromise in the objectives [83]. Multiobjective optimization has also been used in pharmacophore identification to explore conformational space for multiple ligands simultaneously and align them using a genetic algorithm [84]. Another pharmacophore method uses hierarchical multiple objective ranking, which trades off internal strain, pharmacophoric overlap, and steric overlap [85]. Variants of Pareto optimization have also been used in protein design [86] and in docking-based virtual screening [87].

One commercially available set of software tools for performing Pareto optimization of compounds and compound libraries is found in the Accelrys Pipeline Pilot™ and Discovery Studio® programs (www.accelrys.com). Figure 2 shows a Pipeline Pilot protocol for optimizing a set of compound libraries to be both maximally drug-like and maximally diverse.

As indicated in the above parameter list, the first objective is to maximize the mean value of the "drug-like" property, which is the prediction of a Bayesian model trained to distinguish drug-like from baseline compounds. The second objective is to maximize structural diversity. As a measure of diversity, we use the number of distinct structural features found within a subset, based on the FCFP_4 molecular fingerprint [88].

To begin, the optimizer randomly assigns 100 compounds to each of 40 subsets. Then, using the NSGA-II algorithm for Pareto optimization [89], the population of subsets evolves over several hundred generations. The subsets that are most diverse and most drug-like are the ones that tend to survive. Figure 3 shows the progress of the optimization.

The graph displays the population of subsets every 25 generations, with a different color for each generation. The initial random population at the lower left has both low diversity and low drug-like character. As the optimization proceeds, both quantities increase, until the optimized subsets begin to converge at the upper right of the figure. Once convergence has occurred, one can choose one or more subsets considered best from the final optimal population.

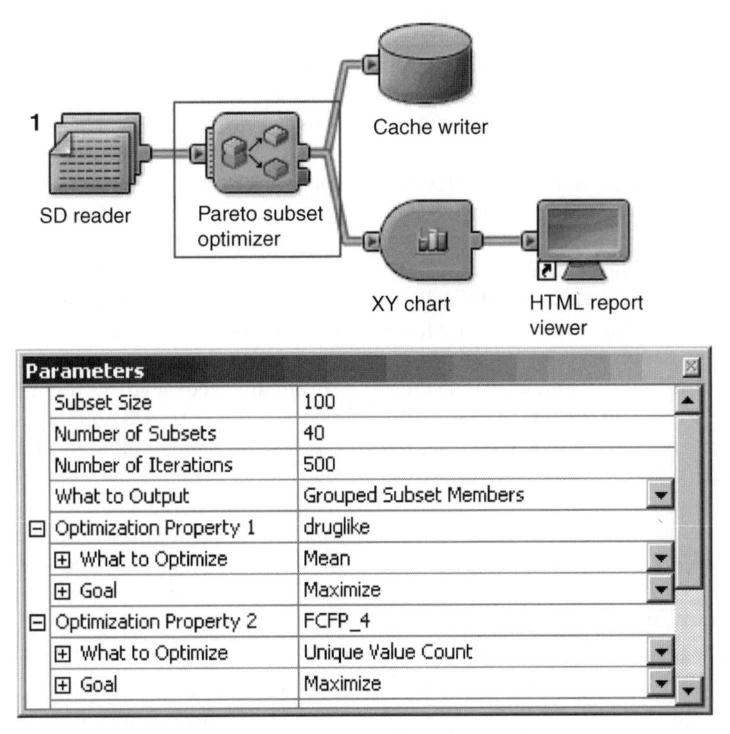

Figure 2. Pipeline Pilot protocol for optimizing a set of compound libraries to be both maximally drug-like and maximally diverse. (This figure is available in full color at http://mrw.interscience.wiley.com/emrw/9780471266945/home.)

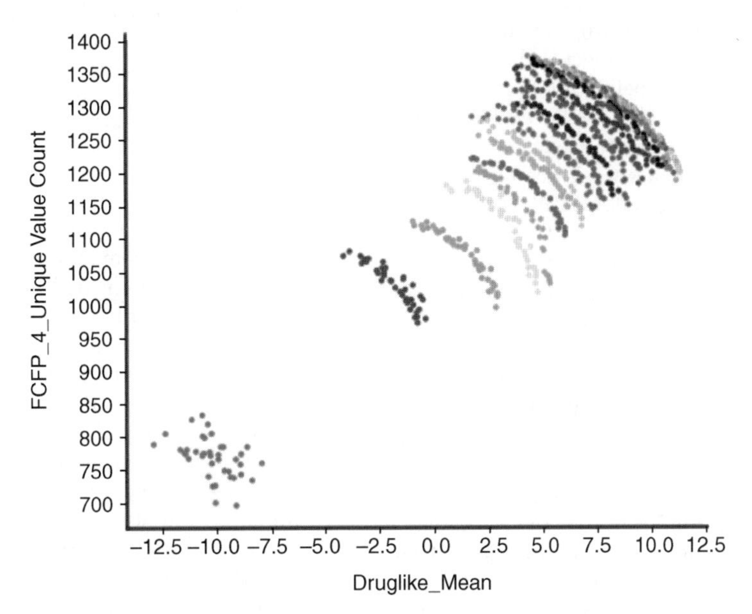

Figure 3. A graph showing the progress of Pareto optimization. The graph displays the population of subsets every 25 generations, with a different color for each generation. The initial random population at the lower left has both low diversity and low drug-like character. (This figure is available in full color at http://mrw.interscience.wiley.com/emrw/9780471266945/home.)

Another commercial implementation of Pareto optimization can be found in SAS (http://support.sas.com/documentation/cdl/en/orlsoug/59688/HTML/default/ga_sect65.htm). Desirability-based multiobjective optimization is also implemented in JMP (www.jmp.com), Minitab (www.minitab.com), STATISTICA (www.statsoft.com), and Stat-Ease (www.statease.com). The R statistics package also has "the desirability package" [90], which is readily available. Some other tools include the Pareto optimization as part of their functionality including the incremental molecule construction method OptDesign [91] and the pharmacophore method GALAHAD [85,92] (www.tripos.com).

4. PARETO LIGAND DESIGNER

There have been numerous methods published for *de novo* molecule design including genetic algorithms, which mimic to a great extent Darwinian evolution [93–96] and particle swarm optimization methods [97] with various fitness functions to direct the design of further molecules (e.g., physicochemical properties, docking, similarity, QSAR etc).

To illustrate how Pareto optimization could be applied in *de novo* molecule design, as a "proof of concept," a Pipeline Pilot protocol has been constructed utilizing the Pareto Sort component to perform simultaneous, multiobjective optimizations of a known, active CCK antagonist, **1** (Fig. 4a), reported by Evans and coworkers [98]. Although the CCK antagonist has measured biological activity (IC_{50} = 0.30 μM), the compound is predicted to have poor BBB penetration, poor aqueous solubility, medium CYP 2D6 binding probability, and a high hepatotoxicity probability using the models available in the Pipeline Pilot ADMET component package [99–102]. It should be noted that although the Accelrys Pipeline Pilot ADMET BBB model has not yet been published, the product notes indicate that the regression model was derived from a training set of 102 compounds and applied to a test set of 86 compounds (RMSE for the training set was 0.36, RMSE for the test set was 0.31). The optimizations were performed as three separate "scenarios" in which the goals were to simultaneously improve the predicted values of 2, 3, and 4 variables, while maintaining biological activity. Tanimoto similarity calculated using Accelrys ECFP_6 fingerprints calibrated using Belief theory was used as a surrogate predictor for maintaining biological activity [103].

It should be emphasized that only the initial compound had measured activity. The activities and properties of all molecules generated by the algorithm were not experimentally measured. It is assumed that the various models employed during the optimization had sufficient accuracy and domain applicability to drive the optimization toward sets of molecules with a reasonable probability of having the desired composite set of improved activities and properties.

In each of the three optimization scenarios described next, a Pipeline Pilot protocol was created, which performs the following operations:

Initialization:
i = 0 An initial set of one or more seed molecules are read in. The initial molecules may be assigned as reference molecules for the computation of Tanimoto similarity (Belief Theory). The ligand properties for the set are computed and the values are written to an overall statistics file that is used to keep track of the progress of the optimization.

Main Loop:
i = i + 1 The ligands are then passed into a Pipeline Pareto Sort component where optimal compounds along the Pareto front are identified and stored in a file. The optimal and nonoptimal compounds are then subjected to an extensive set of molecular transformations, some of which are included in the Drug Guru program [20]. Compounds resulting from the molecular transformations are then passed through a number of property and structure filters including "orange alerts" [19]. Molecules that survive the filters are then assigned to the next iteration number, passed into the Pareto Sort component, statistics are computed, and the cycle begins again (top of Main Loop).

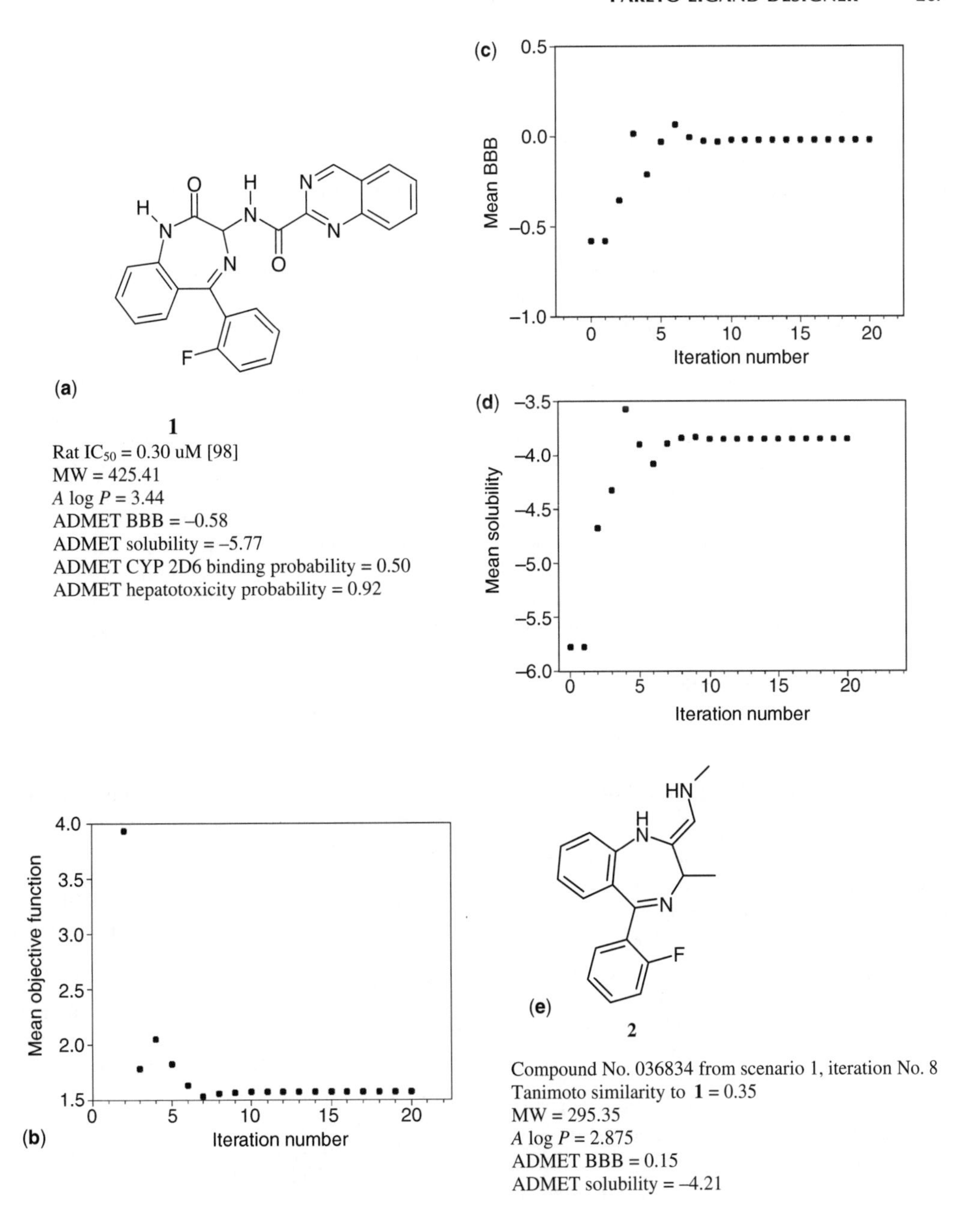

Figure 4. (a) Molecule **1** used in scenario 1 and associated properties. (b) Iteration number versus the mean value of the objective function for scenario 1. (c) Iteration number versus the mean value of BBB for scenario 1. (d) Iteration number versus the mean value of solubility for scenario 1. (e) Compound **2** from scenario 1, iteration number 8.

Optimal Pareto compounds are recycled back into the transformation/optimization selection process. They are not completely removed from the optimization loop. If a previously optimized molecule generates a new molecule with even better properties, the molecule will likely be retained in the next generation of optimized molecules. If the previously optimized molecule generates a new molecule with poorer properties, it will likely be removed, since it will not be along the Pareto front. If the previously optimized molecule generates a new molecule that has already been generated (a duplicate), the new molecule will be removed.

Table 1 in Stewart et al., [20] lists 10 examples of transformations from Drug Guru. Some of the transformations from Drug Guru have been included in Pareto Ligand Designer (kindly provided by Dr. Kent Stewart). It should be noted that some of the Drug Guru transformations implemented in Pareto Ligand Designer (PLD) give identical results as Drug Guru, while other transformations give different results. However, PLD additionally incorporates several hundred additional transformation rules that generate novel molecules beyond the capabilities of Drug Guru. Additional transformations are added as optimized molecules are found in the literature, which cannot be generated in a few iterations by PLD. This is discussed in the "Current Practical Issues and Challenges" section and is part of on-going research with PLD.

4.1. BBB and Solubility Optimization

In the first scenario, the optimization goals are to begin with the known, active CCK antagonist, **1**, and generate a set of optimized molecules with the following properties: (1) maintain or improve the biological activity, (2) maintain or decrease the molecular weight, (3) maintain $A\log P$ within a reasonable range, (4) improve (increase) the BBB partitioning, and (5) improve (increase) the aqueous solubility.

Biological activity was maintained by utilizing a minimum ECFP_6 fingerprint Tanimoto similarity filter of 0.35, corresponding to an activity belief of 16.6% [103]. The molecular weight was maintained or decreased using a filter set to 500 Da. $\log P$ was maintained in the range of 0.00–5.00 using minimum and maximum $A\log P$ filters. BBB partitioning was calculated using the Accelrys ADMET BBB component. The component calculates the value of $\log 10 \times$ ([brain concentration]/[blood concentration]). Aqueous solubility was calculated using the Accelrys ADMET solubility component. The component calculates the value of $\log 10 \times$ (molar solubility). Compounds with Pareto optimal maximum values of BBB and solubility were saved and written to files at each iteration.

Fig. 4b shows the iteration number versus the overall objective function. At each iteration, the Pareto optimizer generates five sets of compounds. The number of Pareto sets is a user-defined parameter. We then compute the mean value of the objective function over all the compounds in all five Pareto sets. Structure (e.g., "orange alerts," [19]) and property filters are turned on at various iterations during the optimization to guide the algorithm toward the creation of new chemical matter, which is likely to be of interest to practicing organic chemists. This may cause an abrupt increase in the objective function, which then typically starts to decrease again after a few additional iterations. It should be noted that the objective function tends toward zero as the properties become optimized. Clearly, the overall objective function improves significantly within the first 5 iterations and then begins to level off at about 10 iterations. Figure 4c shows the iteration number versus the mean BBB value. The BBB improves for the first 5 iterations and then shows no further improvement after about 10 iterations. Figure 4d shows the iteration number versus the mean solubility value. The solubility increases and then begins to level off at about 10 iterations.

Structure **2** (Fig. 4e) is an example of a compound generated at iteration number 8. The values of the properties including the Pareto optimized BBB and solubility are listed below the structure. For the sake of consistency, an example structure has been taken from iteration number 8 from each optimization scenario. Note that comparisons between scenarios are not entirely valid, because the optimization conditions were not identical for each scenario.

4.2. BBB, Solubility, and ADMET CYP 2D6 Binding Optimization

In the second scenario, the optimization goals are to begin with known, active CCK antagonist, **1** (Fig. 4a), and generate a set of optimized molecules with the following properties: (1) maintain or improve the biological activity, (2) maintain or decrease the molecular weight, (3) maintain $A \log P$ within a reasonable range, (4) improve (increase) the BBB partitioning, (5) improve (increase) the aqueous solubility, and (6) decrease the CYP 2D6 binding probability. The CYP 2D6 binding probability was calculated using the CYP 2D6 binding model in the Pipeline Pilot ADMET component. The component calculates the probability that a compound will be either a noninhibitor (probability = 0.00) or an inhibitor (probability = 1.00). The CYP 2D6 binding probability of compound **1** was 0.50 suggesting an intermediate probability of CYP 2D6 binding.

Figure 5a shows the iteration number versus the mean value of the objective function. The objective function has sharply decreased after 5 iterations and continues a more gradual decrease beyond 10 iterations implying slower, gradual improvements in the desired qualities of the Pareto optimized molecules. Figure 5b shows the iteration number versus the mean value of BBB. The BBB has increased at about 5 iterations, but does not show a consistent improvement afterward. This possibly suggests that BBB optimization in combination with solubility and reduction of CYP 2D6 binding may be difficult to achieve. Figure 5c shows the iteration number versus the mean value of the solubility. There is a sharp increase in solubility near iteration 5, followed by another increase of solubility, but at a slower rate. Figure 5d shows the CYP 2D6 binding probability that begins near 0.5 and slowly decreases to 0.2, but does not decrease much further. Compound 3 (Fig. 5e) is an example of a Pareto optimized ligand from iteration number 8.

4.3. BBB, Solubility, ADMET CYP 2D6 Binding, and Hepatotoxicity Optimization

In the third scenario, the optimization goals are to begin with the known, active CCK antagonist (Fig. 4a) and generate a set of optimized molecules with the following properties: (1) maintain or improve the biological activity, (2) maintain or decrease the molecular weight, (3) maintain $A \log P$ within a reasonable range, (4) improve (increase) the BBB partitioning, (5) improve (increase) the aqueous solubility, (6) decrease the probability of CYP 2D6 binding, and (7) decrease the probability of hepatotoxicity. Hepatotoxicity probability was calculated using the hepatotoxicity probability model in the Pipeline Pilot ADMET component. The component calculates the probability that a compound will be either hepatotoxic (probability = 1.00) or nonhepatotoxic (probability = 0.00). The hepatotoxicity probability for compound **1** was 0.92 suggesting a high probability of hepatotoxicity.

The results in this scenario (Fig. 6) are similar to the results in scenarios 1 and 2. It is striking that all four properties are being Pareto optimized to various extents in the first five iterations with some continued improvement, for example, solubility in additional iterations. Compound **4** (Fig. 6f), is an example of a ligand from iteration number 8. The calculated and predicted ligand properties are listed below the structure.

4.4. Current Practical Issues and Challenges

The technology presented in this section demonstrates that multiobjective optimization of 2, 3, and 4 ligand properties is clearly possible and that the resulting ligands are of reasonable quality based on predicted properties. There are a number of remaining challenges that will need to be addressed.

(1) There is a need for additional transformation reactions that can potentially get around problems of either slow convergence or poor optimization. It will also be necessary to continually survey the medicinal chemistry literature to ensure that structures that have undergone lead optimization as well as "proven" optimized structures can be readily generated by the Pipeline Pilot—Pareto program. This may be possible using a test set of known drugs with known ADME/Tox properties, for example.

(2) As has been suggested by others [97], it will be desirable to replace the chemical transformations with high yielding

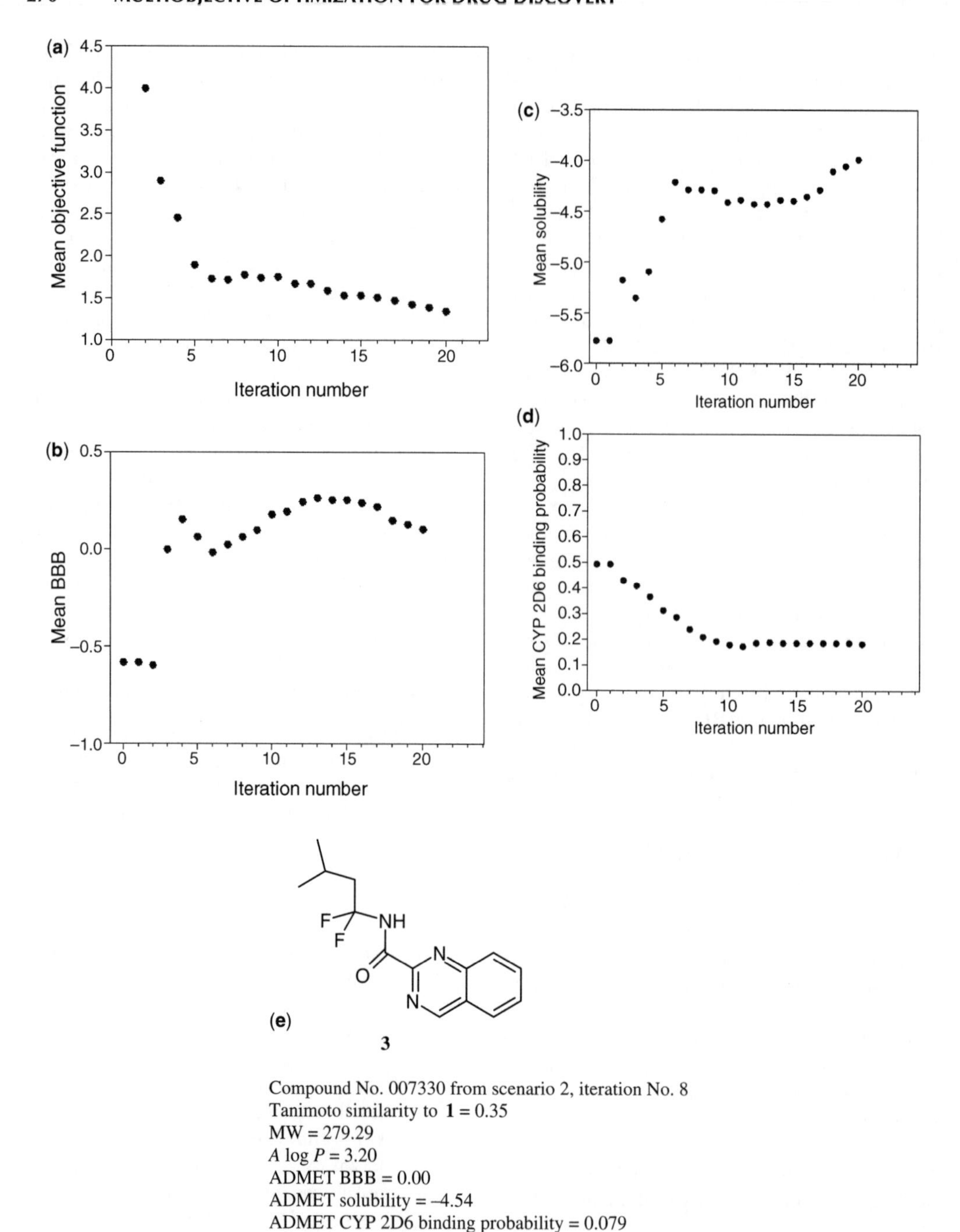

Compound No. 007330 from scenario 2, iteration No. 8
Tanimoto similarity to **1** = 0.35
MW = 279.29
$A \log P$ = 3.20
ADMET BBB = 0.00
ADMET solubility = −4.54
ADMET CYP 2D6 binding probability = 0.079

Figure 5. (a) Iteration number versus the mean value of the objective function for scenario 2. (b) Iteration number versus the mean value of BBB for scenario 2. (c) Iteration number versus the mean value solubility for scenario 2. (d) Iteration number versus the mean CYP 2D6 binding probability for scenario 2. (e) Compound **3** from scenario 2, iteration number 8.

"real" chemical reactions and keep careful track of the complete synthetic pathway or the least costly reactions. We are aware of several approaches that can be used to evaluate synthetic accessibility including methods that use a combined scoring method incorporating structural complexity, similarity to available starting materials etc (e.g., SYLVIA [104]) or relative atomic electronegativity and bond parameters (SMCM [105]).

(3) Due to facile coarse-grained parallel computation capabilities of Pipeline Pilot, it is straightforward and advantageous to run the (slow) transformation step on multiple processors. Conditions for parallel processing would also need to be optimized for maximum throughput.

(4) It will be important to continue to develop and test structure filters to remove compounds that are unstable, reactive, or of little interest to practicing organic chemists for a variety of reasons. It may also be advantageous to use "personalized" structure filters to satisfy the requirements of individual chemists. This is easy to accomplish since structure filter files can

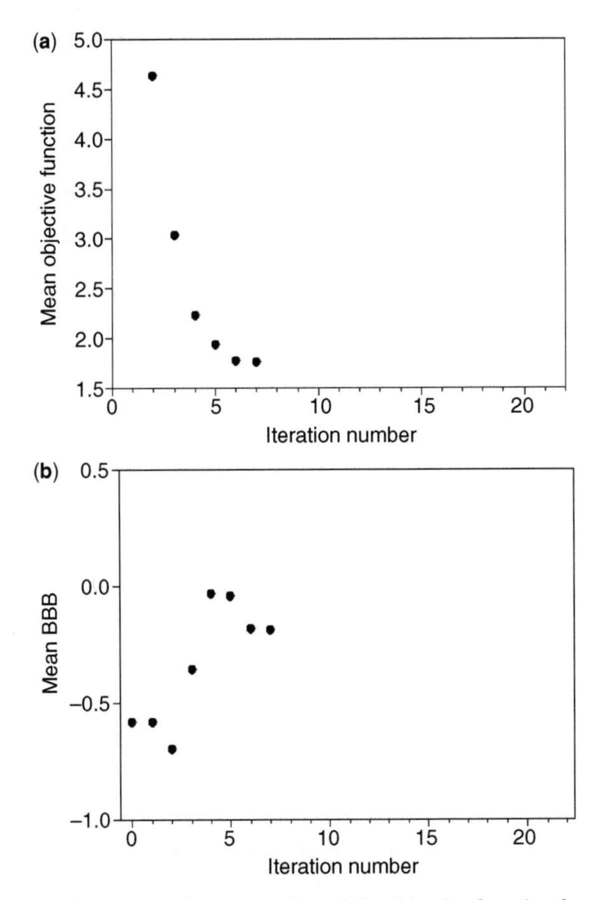

Figure 6. (a) Iteration number versus the mean value of the objective function for scenario 3. (b) Iteration number versus the mean value of BBB value for scenario 3. (c) Iteration number versus the mean aqueous solubility for scenario 3. (d) Iteration number versus the mean CYP 2D6 binding probability for scenario 3. (e) Iteration number versus the mean hepatotoxicity probability for scenario 3. (f) Compound **4** from scenario 3, iteration number 8.

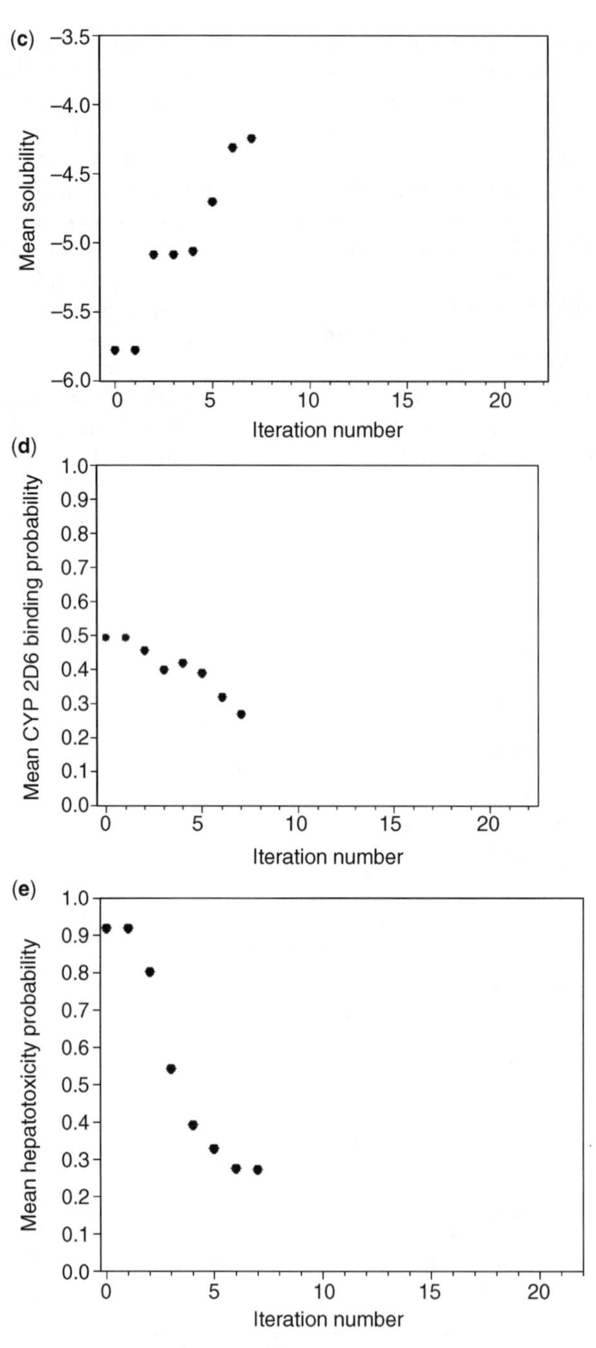

Figure 6. *(Continued)*

(f)

4

Compound No. 631614 from scenario 3, iteration No. 8
Tanimoto similarity to 1 = 0.35
MW = 380.41
$A \log P$ = 3.53
ADMET BBB = 0.052
ADMET solubility = −4.77
ADMET CYP 2D6 binding probability = 0.11
ADMET hepatotoxicity probability = 0.19

Figure 6. *(Continued)*

reside in one or more separate directories read by Pipeline Pilot.

(5) The optimization process relies on the existence of accurate predictive models with a sufficient applicability domain to cover the structures that are generated in the transformation reactions. Hence, there is an on-going need to continually develop and utilize predictive models of the highest predictive accuracy with broad applicability domains.

5. SUMMARY

Drug discovery is a costly enterprise, which requires the simultaneous optimization of many properties. As we suggest in this chapter (and many others are also showing with their various studies), it is possible to achieve such optimizations in principle using methods such as Pareto optimization. One challenge is that as the number of properties to be optimized increases the required calculation time greatly increases, meaning that efforts to combine properties into composites (e.g., as desirability functions) [68,69] should be made in order to decrease the number of properties undergoing Pareto optimization.

Tools like the Pareto Ligand Designer suggested here, MEGA [82] and other methods [97] represent new approaches to *de novo* design that have the potential to consider many properties (e.g., ADME/Tox) besides bioactivity, while at the same time rapidly generating ideas for potentially synthesizable molecules. Although these are initial steps, they represent an example of how simultaneous, multiobjective optimization can be applied in drug discovery that builds on the research in predicting ADME/Tox properties [3] and other areas. Ultimately, such approaches may represent an additional source of molecules for lead discovery or as we have indicated, they may assist in lead optimization.

ACKNOWLEDGMENTS

S.E. gratefully acknowledges Dr. Maggie A.Z. Hupcey and Dr. Peter W. Swaan for earlier discussions over several years and Dr Chris Lipinski for pointing me to many papers on property analyses. S.E. also gratefully acknowledges Accelrys for providing Discovery Studio. J.T.M would like to thank the support staff at Accelrys for many helpful discussions, suggestions, and corrections to the Pipeline Pilot protocols.

REFERENCES

1. Egan WJ, Merz KMJ, Baldwin JJ. Prediction of drug absorption using multivariate statistics. J Med Chem 2000;43:3867–3877.
2. Ekins S, et al. Progress in predicting human ADME parameters *in silico*. J Pharmacol Toxicol Methods 2000;44(1):251–272.
3. Ekins S, et al. Towards a new age of virtual ADME/TOX and multidimensional drug discovery. J Comput Aided Mol Des 2002;16: 381–401.
4. Ekins S, et al. Present and future *in vitro* approaches for drug metabolism. J Pharmacol Toxicol Methods 2000;44:313–324.
5. van De Waterbeemd H, et al. Property-based design: optimization of drug absorption and pharmacokinetics. J Med Chem 2001;44 (9):1313–1333.
6. Balani SK, et al. Strategy of utilizing *in vitro* and *in vivo* ADME tools for lead optimization

and drug candidate selection. Curr Top Med Chem 2005;5(1):1033–1038.

7. Oprea TI. Current trends in lead discovery: are we looking for the appropriate properties? J Comput Aided Mol Des 2002;16:325–334.
8. Oprea TI, et al. Is there a difference between leads and drugs? A historical perspective. J Chem Inf Comput Sci 2001;41:1308–1315.
9. Keseru GM, Makara GM. The influence of lead discovery strategies on the properties of drug candidates. Nat Rev Drug Discov. 2009;8 (3):203–212.
10. Teague SJ, et al. The design of leadlike combinatorial libraries. Angew Chem Int Ed Engl 1999;38(24):3743–3748.
11. Rishton GM. Molecular diversity in the context of leadlikeness: compound properties that enable effective biochemical screening. Curr Opin Chem Biol 2008;12(3):340–351.
12. Bembenek SD, Tounge BA, Reynolds CH. Ligand efficiency and fragment-based drug discovery. Drug Discov Today 2009;14(5–6):278–283.
13. Taylor JD, et al. Identification of novel fragment compounds targeted against the pY pocket of v-Src SH2 by computational and NMR screening and thermodynamic evaluation. Proteins 2007;67(4):981–990.
14. Sutherland JJ, et al. Chemical fragments as foundations for understanding target space and activity prediction. J Med Chem 2008;51 (9):2689–2700.
15. Hajduk PJ, Sauer DR. Statistical analysis of the effects of common chemical substituents on ligand potency. J Med Chem 2008;51(3): 553–564.
16. Aronov AM, et al. Kinase-likeness and kinase-privileged fragments: toward virtual polypharmacology. J Med Chem 2008;51(5):1214–1222.
17. Lipinski CA, et al. Experimental and computational approaches to estimate solubility and permeability in drug discovery and development settings. Adv Drug Del Rev 1997; 23:3–25.
18. Ritchie TJ, Luscombe CN, Macdonald SJ. Analysis of the calculated physicochemical properties of respiratory drugs: can we design for inhaled drugs yet? J Chem Inf Model 2009;49(4):1025–1032.
19. Metz JT, Huth JR, Hajduk PJ. Enhancement of chemical rules for predicting compound reactivity towards protein thiol groups. J Comput Aided Mol Des 2007;21(1-3):139–144.
20. Stewart KD, Shiroda M, James CA, Drug Guru: a computer software program for drug design using medicinal chemistry rules. Bioorg Med Chem 2006;14(20):7011–7022.
21. Wermuth CG. Selective optimization of side activities: another way for drug discovery. J Med Chem 2004;47(6):1303–1314.
22. Wermuth CG. Selective optimization of side activities: the SOSA approach. Drug Discov Today 2006;11(3-4):160–164.
23. Kraus JM, et al. Rational modification of a candidate cancer drug for use against Chagas disease. J Med Chem 2009;52(6):1639–1647.
24. Hopkins AL, Groom CR, Alex A. Ligand efficiency: a useful metric for lead selection. Drug Discov Today 2004;9(10):430–431.
25. Abad-Zapatero C, Metz JT. Ligand efficiency indices as guideposts for drug discovery. Drug Discov Today 2005;10(7):464–469.
26. Reynolds CH, Tounge BA, Bembenek SD. Ligand binding efficiency: trends, physical basis, and implications. J Med Chem 2008;51(8): 2432–2438.
27. Ekins S, et al. Computational discovery of novel low micromolar human pregnane X receptor antagonists. Mol Pharmacol 2008;74 (3):662–672.
28. Abou-Gharbia M. Discovery of innovative small molecule therapeutics. J Med Chem 2009;52(1):2–9.
29. Ekins S. Systems-ADME/Tox: resources and network approaches. J Pharmacol Toxicol Methods 2006;53(1):38–66.
30. Ekins S, Nikolsky Y, Nikolskaya T. Techniques: application of systems biology to absorption, distribution, metabolism, excretion, and toxicity. Trends Pharmacol Sci 2005;26: 202–209.
31. DiMasi JA, Hansen RW, Grabowski HG. The price of innovation: new estimates of drug development costs. J Health Econ 2003;22 (2):151–185.
32. Rishton GM. Failure and success in modern drug discovery: guiding principles in the establishment of high probability of success drug discovery organizations. Med Chem 2005;1 (5):519–527.
33. U.S. Food and Drug Administration. Innovation stagnation: challenge and opportunity on the critical path to new medicinal products. 2004.
34. Zientek M, et al. Development of an *in vitro* drug–drug interaction assay to simultaneously monitor five cytochrome P450 isoforms and performance assessment using drug library compounds. J Pharmacol Toxicol Methods 2008;58(3):206–214.

35. Swaan PW, Ekins S. Reengineering the pharmaceutical industry by crash-testing molecules. Drug Disc Today 2005;10:1191–1200.
36. Shimada J, et al. Integrating computer-based *de novo* drug design and multidimensional filtering for desirable drugs. Targets 2002;1: 196–205.
37. Delaney J. Modelling iterative compound optimisation using a self-avoiding walk. Drug Discov Today 2009;14(3-4):198–207.
38. Stoner CL, et al. Integrated oral bioavailability projection using *in vitro* screening data as a selection tool in drug discovery. Int J Pharm 2004;269(1):241–249.
39. Chekmarev DS, et al. Shape signatures: new descriptors for predicting cardiotoxicity *in silico*. Chem Res Toxicol 2008;21:1304–1314.
40. Kortagere S, et al. Hybrid scoring and classification approaches to predict human pregane X receptor activiators. Pharm Res 2009;26: 1001–1011.
41. Kortagere S, et al. New predictive models for blood brain barrier permeability of drug-like molecules. Pharm Res 2008;25:1836–1845.
42. Klon AE, Lowrie JF, Diller DJ. Improved naive Bayesian modeling of numerical data for absorption, distribution, metabolism and excretion (ADME) property prediction. J Chem Inf Model 2006;46(5):1945–1956.
43. Obrezanova O, et al. Gaussian processes: a method for automatic QSAR modeling of ADME properties. J Chem Inf Model 2007;47 (5):1847–1857.
44. Zhang L, et al. QSAR modeling of the blood–brain barrier permeability for diverse organic compounds. Pharm Res 2008;25(8):1902–1914.
45. Trunzer M, Faller B, Zimmerlin A. Metabolic soft spot identification and compound optimization in early discovery phases using MetaSite and LC-MS/MS validation. J Med Chem 2009;52(2):329–335.
46. Darvas F, Dorman G, Papp A. Diversity measures for enhancing ADME admissibility of combinatorial libraries. J Chem Inf Comput Sci 2000;40:314–322.
47. Darvas F, et al. MetabolExpert: its use in metabolism research and in combinatorial chemistry. In: Erhardt PW, editor. Drug Metabolism. Databases and High-Throughput Testing During Drug Design and Development. London: International Union of Pure and Applied Chemistry and Blackwell Science; 1999.
48. Boyer S, et al. Reaction site mapping of xenobiotic biotransformations. J Chem Inf Model 2007;47(2):583–590.
49. Boyer S, Zamora I. New methods in predictive metabolism. J Comp Aided Mol Des 2002;16: 403–413.
50. Embrechts MJ, Ekins S. Classification of metabolites with Kernel-partial least squares (K-PLS). Drug Metab Dispos 2007;35(3): 325–327.
51. Stranz DD, et al. Combined computational metabolite prediction and automated structure-based analysis of mass spectrometric data. Toxicol Mech Methods 2008;18:1–8.
52. Yamashita F, et al. Novel hierarchical classification and visualization method for multiobjective optimization of drug properties: application to structure–activity relationship analysis of cytochrome P450 metabolism. J Chem Inf Model 2008;48(2): 364–369.
53. Chadwick A, et al. Improving the pharmaceutical R&D Process: how simulation can support management decision making. In: Ekins S, editor. Computer Applications in Pharmaceutical Research and Development. Hoboken: John Wiley & Sons; 2006. p 247–273.
54. Wunberg T, et al. Improving the hit-to-lead process: data-driven assessment of drug-like and lead-like screening hits. Drug Discov Today 2006;11(3-4):175–180.
55. Gleeson MP. Generation of a set of simple, interpretable ADMET rules of thumb. J Med Chem 2008;51(4):817–834.
56. Oprea TI. Chemical space navigation in lead discovery. Curr Opin Chem Biol. 2002;6(3): 384–389.
57. Oprea TI, Gottfries J. Chemography: the art of navigating in chemical space. J Comb Chem. 2001;3(2):157–166.
58. Lloyd DG, et al. Oncology exploration: chartering cancer medicinal chemistry space. DDT 2006;11:149–159.
59. Rosen J, et al. Novel chemical space exploration via natural products. J Med Chem 2009; 52:1953–1962.
60. Stoner CL, et al. Implementation of an ADME enabling selection and visualization tool for drug discovery. J Pharm Sci 2004;93 (5):1131–1141
61. Oellien F, Ihlenfeldt WD, Gasteiger J. InfVis: platform-independent visual data mining of multidimensional chemical data sets. J Chem Inf Model 2005;45(5):1456–1467.
62. Maniyar DM, et al. Data visualization during the early stages of drug discovery. J Chem Inf Model 2006;46(4):1806–1818.

63. Balakin KV, et al. Comprehensive computational assessment of ADME properties using mapping techniques. Curr Drug Disc Tech 2005;2:99–113.
64. Xiao YD, et al. Supervised self-organizing maps in drug discovery. 1. Robust behavior with overdetermined data sets. J Chem Inf Model 2005;45(6):1749–1758.
65. Derringer G, Suich R. Simultaneous optimization of several response variables. J Qual Technol 1980;12:214–219.
66. Rassokin DN, Agrafiotis DK. Kolmogorov–Smirnov statistic and its application in library design. J Mol Graph Model 2000;18: 368–382.
67. Derringer GC. A balancing act: optimizing a product's properties. Qual Prog 1994;27: 51–58.
68. Govaerts B, Le Bailly de Tilleghem C. Distribution of desirability index in multicriteria optimization using desirability functions based on the cumulative distribution function of the standard normal. 2005.
69. Govaerts B, Le Bailly de Tilleghem C. Uncertainty propagation in multiresponse optimization using a desirability index. 2005.
70. Andersson J.A survey of multiobjective optimization in engineering design. 2000. Technical Report. LiTH-IKP-R-1097. Department of Mechanical Engineering, Linköping University, Linköping, Sweden.
71. Fonseca CM, Fleming PJ. An overview of evolutionary algorithms in multiobjective optimization. Evol Comput 1995;3:1.
72. Ortiz MC, et al. Vectorial optimization as a methodogical alternative to desirability function. Chemometr Intell Lab Sys 2006;83: 157–168.
73. Geoffrion AM. Proper efficiency and the theory of vector optimization. J Math Anal Appl 1968;41:491.
74. Agrafiotis DK. Multiobjective optimization of combinatorial libraries. J Comp Aid Mol Des 2002;16:335–356.
75. Gillet VJ. New directions in library design and analysis. Curr Opin Chem Biol. 2008;12 (3):372–378.
76. Gillet VJ, et al. Combinatorial library design using a multiobjective genetic algorithm. J Chem Inf Comput Sci 2002;42(2):375–385.
77. Chen G, et al. Focused combinatorial library design based on structural diversity, druglikeness and binding affinity score. J Comb Chem 2005;7(3):398–406.
78. Kruisselbrink JW, et al. Combining aggregation with Pareto optimization: a case study in evolutionary molecular design. Proceedings of EMO '09 Lecture Notes Comput Sci 2009, in press.
79. Nicolaou CA, Brown N, Pattichis CS. Molecular optimization using computational multiobjective methods. Curr Opin Drug Discov Dev 2007;10(3):316–324.
80. Cruz-Monteagudo M, et al. Desirability-based methods of multiobjective optimization and ranking for global QSAR studies. Filtering safe and potent drug candidates from combinatorial libraries. J Comb Chem 2008;10(6): 897–913.
81. Cruz-Monteagudo M, Borges F, Cordeiro MN. Desirability-based multiobjective optimization for global QSAR studies: application to the design of novel NSAIDs with improved analgesic, antiinflammatory, and ulcerogenic profiles. J Comput Chem 2008;29(14): 2445–2459.
82. Nicolaou CA, Apostolakis J, Pattichis CS. *De novo drug* design using multiobjective evolutionary graphs. J Chem Inf Model 2009;49: 295–307.
83. Nicolotti O, et al. Multiobjective optimization in quantitative structure–activity relationships: deriving accurate and interpretable QSARs. J Med Chem 2002;45:5069–5080.
84. Cottrell SJ, Gillet VJ, Taylor R. Incorporating partial matches within multi-objective pharmacophore identification. J Comput Aided Mol Des 2006;20(12):735–749.
85. Clark RD, Abrahamian E. Using a staged multi-objective optimization approach to find selective pharmacophore models. J Comput Aided Mol Des 2008.
86. Suarez M, et al. Pareto optimization in computational protein design with multiple objectives. J Comput Chem 2008;29(16):2704–2711.
87. Li H, et al. An effective docking strategy for virtual screening based on multi-objective optimization algorithm. BMC Bioinformatics 2009;10(1):58.
88. Rogers D, Brown RD, Hahn M. Using extended-connectivity fingerprints with Laplacian-modified Bayesian analysis in high-throughput screening follow-up. J Biomol Screen 2005;10(7):682–686.
89. Deb K, et al. A fast elitist non-dominated sorting genetic algorithm for multi-objective optimization: NSGA-II. 2000. Kanpur Genetic Algorithm Laboratory (KanGAL) Report 200001, Indian Institute of Technology, Kanpur, India.

90. Kuhn M, Desirability function optimization and ranking. 2008; http://neacm.fe.up.pt/pub/CRAN/web/packages/desirability/desirability.pdf.
91. Soltanshahi F, et al. Balancing focused combinatorial libraries based on multiple GPCR ligands. J Comput Aided Mol Des 2006;20(7-8):529–538.
92. Richmond NJ, et al. GALAHAD: 1. Pharmacophore identification by hypermolecular alignment of ligands in 3D. J Comput Aided Mol Des 2006;20(9):567–587.
93. Douguet D, Thoreau E, Grassy G. A genetic algorithm for the automated generation of small organic molecules: drug design using an evolutionary algorithm. J Comput Aided Mol Des 2000;14(5):449–466.
94. Pegg SC-H, Haresco JJ, Kuntz ID. A genetic algorithm for structure-based *de novo* design. J Comp Aided Mol Des 2001;15:911–933.
95. Lameijer EW, et al. The molecule evoluator. An interactive evolutionary algorithm for the design of drug-like molecules. J Chem Inf Model 2006;46(2):545–552.
96. Lameijer EW, et al. Evolutionary algorithms in drug design. Nat Comput 2005;4:177–243.
97. Schneider G, et al. Voyages to the (un)known: adaptive design of bioactive compounds. Trends Biotechnol 2009;27(1):18–26.
98. Evans BE, et al. Methods for drug discovery: development of potent, selective, orally effective cholecystokinin antagonists. J Med Chem 1988;31(12):2235–2246.
99. Cheng A, Merz KM Jr. Prediction of aqueous solubility of a diverse set of compounds using quantitative structure–property relationships. J Med Chem 2003;46(17):3572–3580.
100. Susnow RG, Dixon SL. Use of robust classification techniques for the prediction of human cytochrome P450 2D6 inhibition. J Chem Inf Comput Sci 2003;43(4):1308–1315.
101. Dixon SL, Villar HO. Bioactive diversity and screening library selection via affinity fingerprinting. J Chem Inf Comput Sci 1998;38:1192–1203.
102. Cheng A, Dixon SL. *In silico* models for the prediction of dose-dependent human hepatotoxicity. J Comput Aided Mol Des 2003;17(12):811–823.
103. Muchmore SW, et al. Application of belief theory to similarity data fusion for use in analog searching and lead hopping. J Chem Inf Model 2008;48(5):941–948.
104. Boda K, Seidel T, Gasteiger J. Structure and reaction based evaluation of synthetic accessibility. J Comput Aided Mol Des 2007;21(6):311–325.
105. Allu TK, Oprea TI. Rapid evaluation of synthetic and molecular complexity for *in silico* chemistry. J Chem Inf Model 2005;45(5):1237–1243.

GPCR HOMOLOGY MODEL DEVELOPMENT AND APPLICATION

STEPHEN L. GARLAND
FRANK E. BLANEY
Computational and Structural Sciences, GlaxoSmithKline plc, Harlow, Essex, UK

1. INTRODUCTION

G-protein-coupled receptors (GPCRs) represent one of the most important classes of proteins for drug discovery. The receptors respond to a vast range of stimuli, including light, taste, ions, amino acids, odorant molecules, monoaminergic neurotransmitters, peptide hormones, glycoproteins, chemokines, nucleotides, lipids, and prostaglandins. Intervention in the natural signaling of these molecules, or correction of aberrant signaling, provides a wealth of opportunity to develop novel therapeutic agents [1]. Well-known marketed examples include beta-2 agonists for asthma (e.g., Serevent™/salmeterol), 5HT1b/d agonists for migraine (e.g., Imitrex™/sumatriptan), H2 antagonists for peptic ulcers (e.g., Zantac™/ranitidine), and AT-II antagonists for hypertension (e.g., Cozaar™/losartan). There are around 900 GPCR sequences in the human genome, although a considerable number are associated with perception of smell or taste. These are fairly unlikely to be targets for drug discovery and are usually omitted. Even without these, there remain 365 human receptors of interest (Table 1, [2,3]). Estimates for the proportion of marketed drugs targeting these receptors varies between 25% and 50% depending in part on whether they are assessed based on number of chemical entities or total sales [4,5]. However, current drugs only target around 30 of the receptors [6], suggesting enormous further opportunity in the area.

The GPCR structure traverses the cell membrane and provides a general mechanism by which an extracellular ligand can convey a signal into the cell. The receptor is usually quite conformationally mobile and can spontaneously move from an inactive to active state ("constitutive activity"). Binding of the endogenous ligand is thought to shift the equilibrium such that the receptor spends a greater proportion of time in the active state ("agonism"). Synthetic agonists can fully or partially activate the receptor ("full agonist"/"partial agonist"), while synthetic antagonists can block agonist binding but have no effect on the conformational equilibrium ("antagonism") or both block the orthosteric site and shift the equilibrium more toward the inactive state ("inverse agonism"). The conformational change associated with agonist binding is traditionally viewed as causing activation of a G-protein complex, giving rise to intracellular signaling cascades and ultimately a physiological response. Increasingly, however, it is being found that receptor signaling is considerably more complex, potentially involving other signaling moieties and/or multiple G-protein signaling pathways and that ligands can be pathway-specific in their activity ("functional selectivity") [7]. As a result, the literature is increasingly referring to the receptors as 7-transmembrane receptors (7TMRs or 7TMs), which reflects their common structure (below) rather than being too specific about the signaling pathway. There is also the potential that GPCRs operate as homo- or heterodimers or possibly even higher order oligomers, which could introduce further complexity in their signaling properties [8]. Furthermore, ligands may bind in allosteric sites and modulate the activity of the receptor either independently of or in concert with the endogenous ligand. A positive allosteric modulator, for example, is a compound that is capable of sensitizing the receptor to the endogenous agonist but may not have any activity in its own right. Effectively, the allosteric compound amplifies the signal of the natural ligand, which potentially has substantial advantage relative to a conventional agonist in terms of controlling receptor signaling; the temporal and spatial control afforded the endogenous ligand is "borrowed" by the allosteric agent that could avoid side effects of prolonged receptor activation or activation in additional undesired tissue [9].

GPCRs share a common architecture of 7-transmembrane α-helices (TM1 through TM7) linked by three intracellular (ICL1–3) and three extracellular loops (ECL1–3). There is an extracellular N-terminal segment that

Table 1. GPCR Families and Their Sizes in Human

	A (rhodopsin)[a]	B (secretin)	B (adhesion)	C (glutamate)	F (frizzled)
Liganded	189–217	15	3	15	11
Orphan	67–95	0	30	7	0
Total	284	15	33	22	11

[a]Excluding olfactory receptors and taste type 2 receptors.
The numbers in this table were obtained from Gloriam et al. (2007) [2] and the International Union of Pharmacology (IUPHAR) list (Foord et al., 2005 [3] and http://www.iuphar-db.org/GPCR/ReceptorListForward). The number of orphan receptors in the family A/rhodopsin group will vary depending on the criteria used, specifically whether repeated evidence from unique sources is required. Reproduced from Ref. [1].

can vary considerably in length (N-terminal domain or NTD) and an intracellular C-terminus (C-term). Available structural data also indicate a helix (HLX8) running along the intracellular face of the membrane between the end of TM7 and the C-term (Fig. 1, [10]). Sequence analysis has categorized GPCRs into six subfamilies named A through F (or numbered 1–6) [11]. Families A, B, and C are the most relevant from a drug discovery perspective, with A generally being considered the most important due to its size. Family D (fungal pheromone) and family E (slime mold cAMP receptors) are not found in higher order species and, while they may well prove to be interesting in the future, very little is currently known about the *frizzled/smoothened* group of family F. There is an alternative classification scheme for GPCRs called the GRAFS system [12]. This places the receptors in five main subfamilies: *glutamate*, *rhodopsin*, *adhesion*, *frizzled/taste2*, and *secretin*. This is essentially equivalent to the A–F classification system except that the nonhuman classes D and E are omitted and family B is divided into the peptide hormone receptors (*secretin*) and the *adhesion* class. This partitioning appears justified as the two groups have major differences in their N-terminal domains (NTDs), although there are similarities in the TM regions.

Given the importance of GPCRs as drug targets there has been a strong desire to generate structural data to support the drug

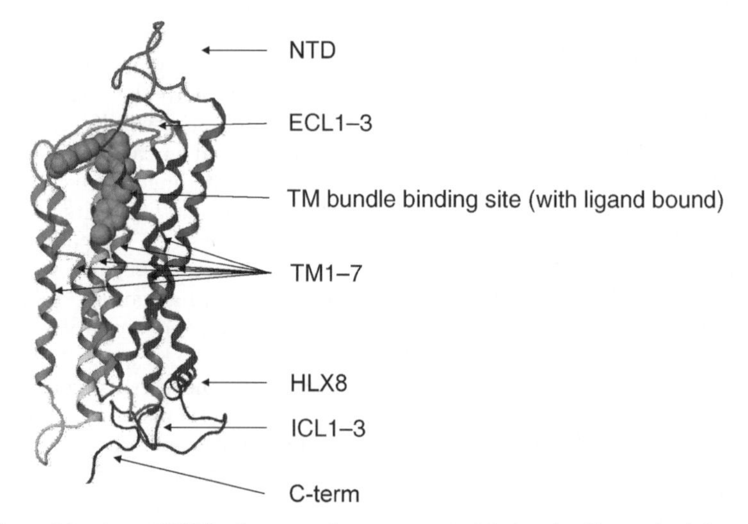

Figure 1. The architecture of 7TMs. An example receptor model showing N-terminal domain (NTD), the three extracellular loops (ECLs), the TM bundle binding site (with ligand bound), the seven transmembrane helices (TM1–7), helix 8 (HLX8), the three intracellular loops (ICLs), and the C-terminus (C-term). (Modified from Ref. [10].) (See color insert.)

discovery process. While some useful information can be elucidated via techniques such as NMR or various biophysical techniques, the mainstay for the pharmaceutical industry is X-ray crystallographic structural determination. Application to the GPCR field has, however, proven quite difficult for a number of reasons. The receptor normally sits in a lipophilic membrane environment, which is not generally amenable to crystallization. The receptors are also quite flexible which serves to destabilize crystal formation. In addition, there is a need to identify a receptor construct that will express sufficiently well and enable adequate purification for crystallization trials and the protein must remain correctly folded and retain its functional activity throughout. The first X-ray crystal structure of a GPCR was that of bovine rhodopsin [13]. The receptor has its endogenous ligand, retinal, covalently linked, and due to its role in light perception has an unusually low constitutive activity and hence limited conformational flexibility. The receptor is also highly expressed in the eye, which enabled sufficient quantity of purified material to be obtained initially without recombinant overexpression. These factors assisted the structural determination in this case. Methodology was subsequently developed enabling the crystal structure of rhodopsin to be determined via recombinant expression [14] and further developments, particularly around stabilization of receptor conformation and use of microdiffraction techniques [15], have enabled structures to be determined for GPCRs with diffusible ligands, notably the beta-2 [16,17] and beta-1 adrenoceptors [18] and the A2a adenosine receptor [19]. The stabilization of receptor conformation through point mutations [18] is particularly noteworthy since it provides a systematic way to stabilize GPCRs in a desired conformational state. However, the number of available crystal structures is currently very modest (Table 2, [13,14,16–30]) and we have yet to see robust crystallographic systems developed in which multiple structures can be solved with different ligands bound. As such, the ability to build homology models of the receptors is invaluable in support of drug design.

In addition to crystal structures of the TM bundle of GPCRs, there are a number of other structures that are useful as templates for homology model building. In particular, there are crystal structures for the NTDs of certain

Table 2. Available X-Ray Crystal Structures of Family A/*Rhodopsin* GPCR TM Regions

Structure	Reference	Resolution (Å)	PDB Code
Bovine rhodopsin	[13]	2.8	1f88
Bovine rhodopsin	[20]	2.8	1hzx
Bovine rhodopsin	[21]	2.6	1l9h
Bovine rhodopsin	[22]	2.65	1gzm
Bovine rhodopsin	[23]	2.2	1u19
Bovine rhodopsin	[24]	3.8	2i35
Bovine rhodopsin	[24]	4.1	2i36
Bovine rhodopsin (photoactivated)	[24]	4.15	2i37
Bovine rhodopsin (recombinant)	[14]	3.4	2j4y
Bovine rhodopsin (ligand free)	[29]	2.9	3cap
Bovine rhodopsin (ground state)	[25]	2.65	3c9l
Bovine rhodopsin	[25]	3.4	3c9m
Bovine rhodopsin (G_α peptide complex)	[30]	3.2	3dqb
Squid rhodopsin	[26]	3.7	2ziy
Squid rhodopsin	[27]	2.5	2z73
Human $\beta 2$ adrenergic receptor (T4)	[16]	2.4	2rh1
Human $\beta 2$ adrenergic receptor (Fab)	[17]	3.4	2r4r
Human $\beta 2$ adrenergic receptor (T4/cholesterol)	[28]	2.8	3d4s
Turkey $\beta 1$ adrenergic receptor	[18]	2.7	2vt4
Human A2a adenosine receptor (T4)	[19]	2.6	3eml

receptors. These sections of the protein are extracellular, form a discrete fold and have proven amenable to crystallization. The NTD of family A/*rhodopsin* receptors is generally quite short and not thought to form a notable structural element. For family B/*secretin* receptors, the NTD is around 120–150 residues in length and forms the bulk of the orthosteric site. The endogenous ligands in this case are all peptide hormones. They are believed to have a common binding mode where the majority of the peptide binds to the NTD, with the N-terminal section of the peptide dipping down into the receptor TM bundle serving to activate the receptor. There are currently X-ray crystal structures for the NTD of GIP receptor with GIP(1–42) bound [31], the PTH1 receptor co-crystallized with PTH(15–34) [32], GLP1 with Exendin-4(9–39) [33], and the CRF1 receptor both with and without corticoliberin bound [34]. There are also NMR structures for the NTD of CRF2b receptor [35,36] and PAC1 receptor with PACAP(6–38) bound [37]. Certain family B receptors including amylin, adrenomedullin, and CGRP are formed through heterodimerization with one of the three receptor activity modifying proteins (RAMPs). The crystal structure of the NTD of RAMP1 has been published recently [38]. The NTD of family B/*adhesion* receptors can be extremely large and often contain sequence motifs suggestive of repeating structural elements [39] but there are currently no crystal structures of these. For family C/*glutamate* receptors, the NTD is very large, comprising 900–1200 residues and includes the entire of the orthosteric binding site. The endogenous ligands are generally small amino acids such as glutamate or ions such as calcium. Their binding causes the two lobes of the NTD to close in what is called the "venus flytrap" mechanism. The change in conformation of the NTD is thought in turn to induce a conformational change in the TM bundle and give rise to receptor activation. There are currently crystal structures of mGluR1, mGluR3, and mGluR7 NTDs [40,41], including examples with orthosteric agonists and antagonists bound showing the conformational change in the NTD.

The available structural information is summarized schematically in Fig. 2, with the literature references in Tables 2, 3, and 4. While the amount of data is fairly limited compared to other target classes, it provides a good basis for building homology models for family A/*rhodopsin* receptors in the inactive (antagonist) state and for the NTDs of family

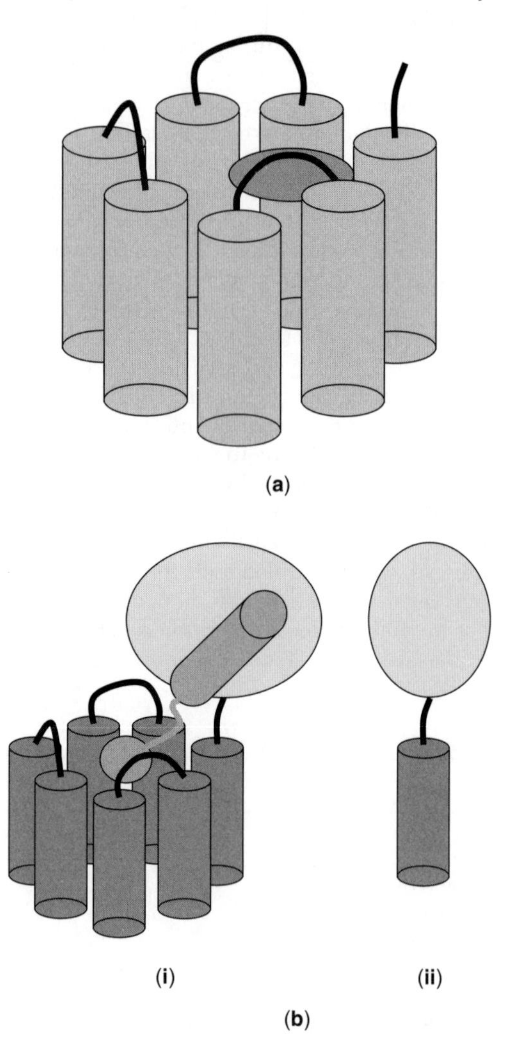

Figure 2. GPCR families and crystallographic data available for homology modeling: (a) family A/*rhodopsin*, (b) (i) family B/*secretin* and (ii) RAMP, (c) family B/*adhesion*, and (d) family C/*glutamate*. Sections with X-ray crystallographic data on which to base the models are shown in light grey with approximate location of orthosteric binding sites shown in mid grey. Regions without X-ray crystallographic data are shown in dark grey. (Modified from Ref. [1].) (This figure is available in full color at http://mrw.interscience.wiley.com/emrw/9780471266945/home.)

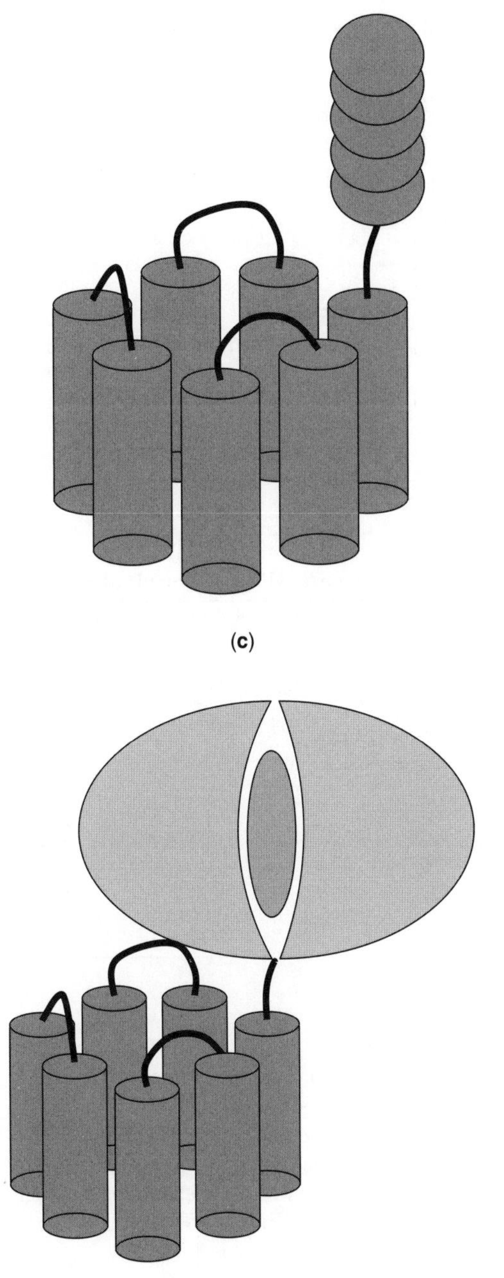

Figure 2. *(Continued)*

B/*secretin* and family C/*glutamate* receptors. The main gaps are lack of crystallographic data for the TM bundles of families B and C receptors and relatively little information about how the conformation of the TM portion of the receptor changes upon activation. For families B and C, we currently assume that the TM bundles are packed in the same way as family A but the sequence identity is low so this may not be true and as a minimum it is quite difficult to align the sequences for model building. It can also be difficult to fuse the NTD homology model onto the TM bundle homology model given the connecting region is poorly defined in the available structures. This is potentially very important for family B receptors given that the endogenous ligand spans both regions. Despite the challenges, a number of homology models have appeared in the literature for both family B [42] and family C [43–45]. For receptor activation, the "toggle switch model" has been developed based on accumulated biophysical data such as site-directed spin labeling, various fluorescent technologies, and use of modified ligands anchored at specific sites in the main ligand binding site [46]. Data of this sort have been used as the basis for generating theoretical models of the activated state of GPCRs [47–49]. More recently, crystal structures have been published for ligand-free opsin [29] and bovine rhodopsin in complex with a G-protein peptide fragment [30]. These may show the receptor in its activated state, although differences between them and the inactive structures appear to be quite modest.

2. EVOLUTION OF MODELS

GPCR homology modeling for family A/*rhodopsin* targets, particularly in the inactive (antagonist state), is now routinely performed based on available X-ray crystallographic data, but it is worth being aware of how models were built prior to publication of the bovine rhodopsin crystal structure as a great many such models are present in the literature, and it has a direct bearing on their potential reliability and utility. GPCR homology models have been in use for nearly 20 years and during that time they have effectively developed through five generations [50]: (1) *de novo* protein modeling techniques were used to try and predict the packing of the helices; (2) the

Table 3. Available X-Ray Crystal Structures and NMR Structures of Family B/*Secretin* GPCR N-Terminal Domains and Receptor Activity Modifying Proteins

Structure	Reference	Resolution (Å)	PDB Code
Human GIP NTD (GIP(1–42) complex)	[31]	1.9	2qkh
Human PTH1 NTD (PTH(15–34) complex)	[32]	1.95	3c4m
Human GLP1 NTD (Exendin-4(9–39) complex)	[33]	2.1	3c5t
Human GLP1 NTD (Exendin-4(9–39) complex)	[33]	2.3	3c59
Human CRF1 NTD (MBP fusion, ligand free)	[34]	2.76	3ehs
Human CRF1 NTD (corticoliberin complex)	[34]	3.4	3eht
Human CRF1 NTD (corticoliberin complex)	[34]	1.96	3ehu
Mouse CRF2b NTD	[35]	NMR	1u34
Mouse CRF2b NTD (astressin complex)	[36]	NMR	2jnd
Human PAC1 NTD (PACAP(6–38) complex)	[37]	NMR	2jod
Human RAMP1 NTD	[38]	2.4	2yx8

Table 4. Available X-Ray Crystal Structures of Family C/*Glutamate* GPCR N-Terminal Domains

Structure	Reference	Resolution (Å)	PDB Code
Rat mGluR1 NTD (glutamate/Gd^{2+})	[40]	4.0	1isr
Rat mGluR1 NTD (MCG (antagonist))	[40]	3.3	1iss
Rat mGluR1 NTD (glutamate)	[40]	2.2	1ewk
Rat mGluR1 NTD (ligand free)	[40]	3.7	1ewt
Rat mGluR1 NTD (ligand free)	[40]	4.0	1ewv
Rat mGluR3 NTD (glutamate)	[41]	2.35	2e4u
Rat mGluR3 NTD (DCG-IV)	[41]	2.4	2e4v
Rat mGluR3 NTD (1*S*,3*S*-ACPD)	[41]	2.4	2e4w
Rat mGluR3 NTD (1*S*,3*R*-ACPD)	[41]	2.75	2e4x
Rat mGluR3 NTD (2*R*,4*R*-APDC)	[41]	3.4	2e4y
Rat mGluR7 NTD (MES)	[41]	3.3	2e4z

bacteriorhodopsin crystal structure was used as a template. While this had low homology to GPCRs it shared the 7-transmembrane helical arrangement, so was a reasonable starting point but has subsequently been shown to have quite a different arrangement of helical packing; (3) models were built based on the electron diffraction map of rhodopsin [51]; (4) models were refined based on the 6 Å cryoelectron diffraction map of frog rhodopsin shown in 4 Å slices through the lipid bilayer [52]; and finally, (5) models were built based on the first high-resolution crystal structure of a GPCR, namely, that of bovine rhodopsin [13]. As shall be seen in the applications below, useful results could be obtained with some of the earlier models, but the quality has improved substantially following use of the bovine rhodopsin and subsequent crystal structures. While they are not discussed further here, it is worth noting that there are a few relatively recent techniques in use which do not employ homology to available crystal structures to build the receptor model, including PREDICT [53] and MembStruk [54].

3. HOMOLOGY MODEL BUILDING

The process of GPCR homology model building has been reviewed in detail [55]. It proceeds through six steps: (1) template selection, (2) sequence alignment, (3) modeling of structurally conserved regions, (4) modeling of structurally variable regions, (5) model refinement, and (6) validation.

Template selection was quite straightforward between 2000 and 2007 as there was only really one choice, that of bovine rhodopsin. Now the template could be any of those listed in Table 2. The differences between the structures are relatively modest in the TM regions so a single template could be taken as representative or a number of models could reasonably be generated based on multiple templates.

The determination of the sequence alignment might in principle be confounded by the relatively low homology between the target protein and the available template(s). However, the presence of key "functionally significant" residues means a lower identity, typically in the range of 20–30%, can yield more accurate models than for other types of proteins [56]. The use of hydropathy plots coupled with the known highly conserved sequence motifs such as the Asp-Arg-Tyr (DRY) at the bottom of TM3, a Trp in TM4, Pros in TMs 5 and 6, and the Asn-Pro-Xxx-Xxx-Tyr (NPxxY) motif in TM7 serves to generate reliable alignments of the TM regions even though the sequence identity is low [57]. This process can be further supported through the use of sequences from multiple species, if required.

The structurally conserved TM regions can then be modeled. The highly conserved residue positions are also used in the Ballesteros–Weinstein numbering convention to define the residue positions within the helices [58]. In this system, the most conserved residue in a given TM is assigned the index X.50, where X is the TM number, with remaining residues coded relative to this position. Use of this system greatly facilitates sequence comparison between receptors.

Modeling the more variable regions requires special treatment. The ICLs, HLX8, and C-term are often omitted as they are not generally thought to form part of the ligand binding site, although recent literature has emerged regarding an intracellular site on GPCRs that includes some of these elements [59–61]. The ECLs, however, frequently form part of the ligand binding site and it is here that we see some rather more substantial differences between the available crystal structures. For example, bovine rhodopsin has a β-hairpin arrangement of ECL2 that sits low down into the receptor bundle whereas beta-2 and beta-1 structures show part of ECL2 in an α-helical conformation that also sits higher and creates more space for ligand binding (Fig. 3). Hence, if the homology is low between ECL2 in the receptor of interest and any of the available crystal structures, it is potentially better to look at alternative means of modeling such as the use of loop libraries or *de novo* modeling techniques [62–65]. The

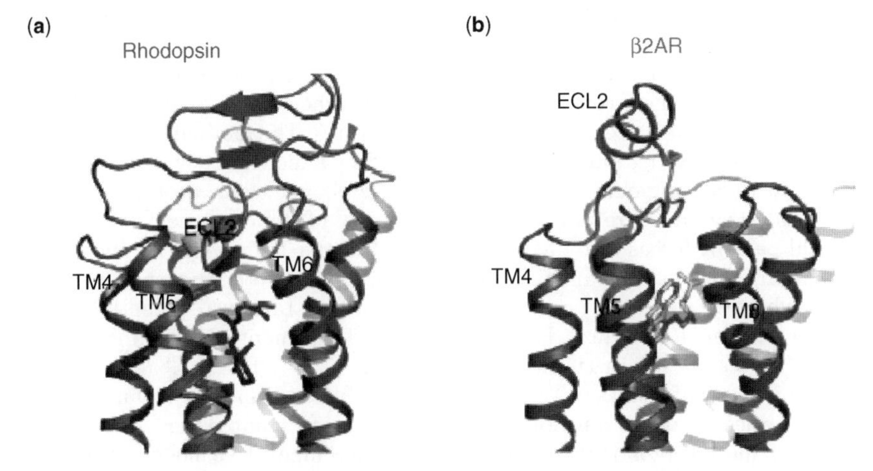

Figure 3. Differences in ECL2 conformation between (a) rhodopsin and (b) b2-AR crystal structures. (Reproduced from Ref. [15].) (See color insert.)

issue has been explored in some detail. Two *in silico* models of the beta-2 adrenoceptor were built using bovine rhodopsin as a template and compared to the available crystal structure [66]. In one, a rhodopsin-like ECL2 was used that resulted in several residues projecting into the binding site and these were found to interfere with ligand docking. For the second model, the loop regions were built completely *de novo* and this was found to yield a more accurate model that could be improved through incorporation of additional biochemical and computational data. Docking carazolol into this model using the *InducedFit* procedure implemented in the Schrodinger package [67] reproduced the mode observed in the crystal structure. An alternative approach is to omit ECL2 from the initial model. This proved effective for virtual screening in two out of three test cases [68]. In a hybrid approach, ligands of interest can be docked and then ECL2 modeled around the receptor–ligand complex—this has been applied, for example, in modeling the free fatty acid receptor FFAR1 (GPR40) [69].

Once the initial receptor model has been built, molecular dynamics (MD) refinement is generally used to resolve any anomalies that may have been introduced by the homology modeling process, for example, through changing small residues to large or vice versa. The quality of the model should then be checked, through examination of a Ramachandran plot or similar. While it is far from routine, the receptor model may also be simulated in the presence of lipid bilayer. Applying this to the mu-opioid receptor [70], for example, shows the TM bundle to become more compact, with implications for the binding site.

In the course of building these models, however, it is worth bearing in mind that the available crystal structures are not without issues. Crystal packing forces can affect protein conformation, which in turn can affect the accuracy with which the contact regions can be modeled. For a number of structures, it is also the case that the receptor construct has been altered to stabilize its conformation. This may be through a series of point mutations in varying parts of the receptor [18], in which case the conformation of the construct is biased toward binding the probe ligand. Alternatively, replacement of ICL3 with T4 lysozyme [16,19] or cocrystallization with Fab [17] has been found to assist structure solving. These modifications can clearly have an impact on the conformation of the receptor in those regions, although this is remote from the traditional TM bundle binding site but potentially critical to the intracellular site.

As the process by which GPCR homology models are built has become fairly well defined, there have been attempts to automate or semiautomate the modeling process and provide models of the whole family of receptors. GPCRmod [71] yielded 277 models of human GPCRs at the rate of approximately 3 s per model via a comparative modeling approach. The models were used for virtual screening and showed successful recovery of a given receptor for a selective antagonist probe and also that the receptor targets of a promiscuous antagonist were among the top scorers. The TASSER method [72] uses a threading assembly refinement procedure. This has been applied to 907 putative GPCR sequences, 45 of which are predicted not to be 7TMs but rather misclassified. Due to the low sequence identity between rhodopsin and most GPCRs, (only four sequences being more than 30% and 99% outside the traditional comparative modeling range, with average identity only 19.5%), the authors argue that use of the bovine rhodopsin structure alone is "highly unlikely to capture the nature of the structural differences among GPCRs, not only within the highly diverse loop regions but also within the core TM regions." As a result, they use a mixture of sensory rhodopsin, halorhodopsin, bacteriorhodopsin, and/or bovine rhodopsin structures. The method has been validated by building a model of bovine rhodopsin without use of homologous proteins with sequence identity of 30% or more; the deviation was 2.1 Å in the TM region. As such, it compares favorably with other nonrhodopsin techniques such as a statistical potential coupled with 27 experimental distance constraints (3.2 Å, [73]), PREDICT (2.9 Å, [53,74]) and MembStruk (3.1 Å, [75]). Our own experience applying automated modeling of GPCRs suggests a comparative modeling approach based on the available GPCR X-ray crystallographic data generally performs well, at least for the

TM regions. The key functionally significant residues do indeed appear to provide greater confidence in the models than might be expected given the low homology. The main issues appear to be in the more variable ECL regions, which often require manual intervention and use of alternative techniques for refinement. The large conformational space potentially available to longer loops leads to uncertainty and this can be compounded by scope for interaction between the three ECLs, meaning that the loops cannot be modeled in isolation. Ligands may need to be docked prior to modeling of ECL2, in particular, to help define the conformation and identify key residues for interaction. There are also receptors where some of the loops are very short; so short, in fact, that they cannot span the distance between the relevant helices in the available crystal structures. In these cases, it may suggest that the top of one or both of the helices in question needs to uncurl or that the helices pack differently. In either event, this will have a significant effect on the TM bundle binding site. Aside from these more variable regions, there are potential issues to be addressed in the TM regions. Bovine rhodopsin has a double Gly motif in TM2, for example, which leads to a β-bulge in the helix. This sequence is not present in all receptors, so it may suggest structural variation in this region with tightening of the helix and some implications for the TM bundle binding site. While the key functionally significant Pro residues of TMs 5, 6, and 7 are highly conserved, they are not universally conserved. Equally, some receptors have additional Pro residues in the TM regions of the sequence. Given the propensity for Pro to distort a helical conformation, these changes in sequence may suggest larger changes in the way the TM regions pack for certain receptors.

4. FUTURE

The current state of the art for GPCR homology modeling generally allows good quality models to be built for family A/*rhodopsin* receptors, particularly in the inactive state. For several years, it has also been possible to build really very good quality homology models of the NTD for family C/*glutamate* receptors in both the active and inactive states and, more recently, a considerable amount of structural data has appeared for the NTD of family B/*secretin* receptors, in some cases with the endogenous peptide agonist bound. However, there remain significant gaps in our understanding and models will continue to evolve. There may be some structural variation in the packing of TM helices in GPCRs, which will hopefully be answered as more structures become available. There is certainly scope for the TM regions of families B and C receptors to be a little different and the first crystal structures of these regions will be highly informative; at the very least, they would aid considerably in identifying the correct sequence alignment between families B/C and A. The variability of the ECLs requires careful treatment and we are already seeing the benefits of multiple crystal structures in improving our models in this respect. As well as forming a part of the binding site, the loops may play a role in gating access to the TM bundle and have a role in the kinetics of drug action. It also appears as though there is a fair degree of structural plasticity within the binding site [76,77] that relates to well-established concepts around induced fit for ligands and would benefit from further structural data. The majority of available structural data relates to the inactive state of the receptor and, even where the receptor may be in the active state, the differences appear to be modest. Further structural data would certainly be helpful in improving our understanding of the receptor activation process and hopefully allow designed efficacy. In concert with this, we will ideally see improved definition of the interaction between the receptor and G-protein or other signaling moiety. We are already seeing some developments in that respect, such as the squid rhodopsin structures [26,27], which show an intracellular extension to TMs 5 and 6 and the opsin structure cocrystallized with a G-protein fragment [30]. This may in turn lead to a structural basis for understanding functional selectivity and/or the role of receptor dimerization. There are already models for certain allosteric sites on GPCRs, but again these could be improved with supporting structural data. Finally, we currently know really very little about the structures of the NTDs of the family B/*adhesion* or any part of the family F/*frizzled* classes of receptors.

5. APPLICATIONS

While GPCR homology models are not likely to be as accurate as crystal structures, they can be of substantial use in the support of drug design, notably for antagonists of family A receptors and orthosteric ligands for family C receptors. Particular applications include aiding understanding of receptor structure/function; lead generation through understanding of the character of the binding site, formal modeling of the active site followed by 3D virtual screening and/or detailed modeling of the binding mode for a ligand at the receptor; and lead optimization through, for example, docking of lead series to support SAR interpretation and design of more potent analogs, identification of residue changes giving rise to species differences or selectivity issues and identification of regions that will tolerate groups that will improve the overall developability (ADME) profile of the ligand. While structure-based design can be applied to a reasonable degree, full *de novo* drug design is challenging even for targets with high-resolution crystal structures, so it is perhaps unsurprising that there appears to be only one successful example described thus far in the literature based on GPCR homology models [78]. An emerging technique with great potential for GPCRs is that of chemogenomics. Ligand information is "borrowed" from targets with similar binding sites and the process is greatly facilitated by the ongoing improvements in the availability of structural data and quality/coverage of homology models [79].

6. RECEPTOR STRUCTURE/FUNCTION

GPCR homology models greatly facilitate the use of biochemical experiments designed to improve our understanding of receptor structure and/or function. Techniques such as site-directed mutagenesis (SDM) [80,81] or construction of receptor chimeras [82] can be applied independently to help pinpoint regions affecting ligand binding or altering receptor function but mapping these changes onto a homology model provides much greater detail. Particular residues may be proposed as contacting the ligand or interacting with one another as part of the functional machinery of the receptor. These hypotheses can then be tested by subsequent rounds of experiment and models refined. Alternatively, receptor homology models can be used as the basis upon which to identify the endogenous ligand or surrogate ligand tool molecules for orphan receptors.

The MT2 melatonin receptor provides an elegant example of how to obtain a detailed model of the endogenous ligand binding site at the receptor in the absence of X-ray crystallographic data [83] (Fig. 4). A combination of receptor modeling and SDM was used to identify new residues critical for melatonin binding. Based on the data, V204 (V5.42), G271 (G6.55), and L272 (L6.56) were thought to participate in hydrophobic interactions with the indole moiety of melatonin, whereas Y298 (Y7.43) may specifically interact with the 5-methoxy group. The importance of H208 (H5.46) had already been established. Further refinement of the model and selection of further residues on TMs 3, 6, and 7 followed by mutation and study in a binding assay showed N268 (N6.52), A275 (A6.59), V291 (V7.36), and L295 (L7.40) to be essential for ligand binding to MT2. It was also shown that there may be a specific interaction between Y188 in ECL2 and the *N*-acetyl group of 2-iodomelatonin.

As can be seen in Table 1, depending on definition, between a quarter and a third of family A/*rhodopsin* receptors are orphans; that is their endogenous ligand is unknown. This severely hampers their prosecution as potential drug targets, not least if an antagonist is desired, in which case an agonist will be required to configure the assay. Homology models can be used to support identification of the endogenous ligand, through characterization of TM bundle binding site similarity [79] or through ligand docking. The latter has been examined for purinergic receptors [84]. There are 10 subtypes of purinergic receptors (P2Y1, 2, 3, 4, 6, 8, 11, 12, 13, and 14) and it has proven difficult to probe their pharmacology due to lack of specific tool

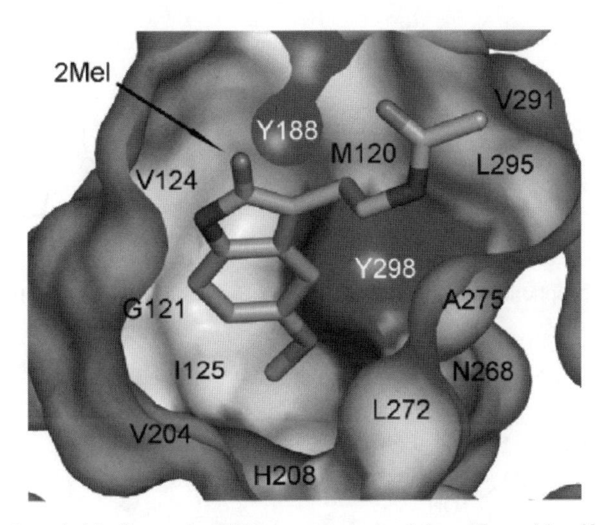

Figure 4. Model of melatonin binding to the MT2 receptor derived through combined homology modeling and site-directed mutagenesis studies. (Reproduced from Ref. [83].)

compounds. AutoDock 3.0 [85–87] was applied to a P2Y1 receptor model. The model was built using a hybrid of a rhodopsin homology-based approach with Fourier transform analysis for α-helix prediction. A virtual screen was performed of the in-house database that included animal metabolites. The top 10 solutions for each compound were energy-minimized with CHARMm [88] and 500 compounds selected as possible ligands. These included carbohydrates, fatty acids, steroids, amino acids, amines, nucleic acids, and vitamins. The known ligands all appeared in the top 30, so these were progressed, with 21 found to be available from commercial sources. Two negative controls, which scored just below the threshold, were also included. The 23 compounds were tested in the P2Y1 assay and 3 were found to be specific agonists relative to the host cell line (PRPP $ED_{50} = 15 \pm 5$ nM; CDP $ED_{50} = 6500 \pm 2400$ nM; and D-fructose 1,6-bisphosphate $ED_{50} = 14{,}000 \pm 2100$ nM) (Fig. 5). The hits gave no specific activity versus P2Y2 or P2X2 receptors. The authors found a modest correlation of 0.63 ($p < 0.05$) between the AutoDock scores and P2Y1 agonist activity for four known compounds and the three novel hits. The docking study had the advantage, of course, of not only predicting endogenous or surrogate ligands but also proposing a binding mode for the ligand at the receptor.

7. LEAD GENERATION: BINDING SITE CHARACTER, VIRTUAL SCREENING, AND LIGAND DOCKING

Homology models coupled with a chemogenomic-style sequence analysis can provide a broad picture of binding site similarity [79,89]. Similar receptors can be identified and their ligands selected for counterscreening at the target of interest to identify potential lead structures. As subsites within the TM bundle binding pocket are selected, the conventional phylogenetic relationship between targets breaks down and unexpected receptor similarities can be observed. This has already been used to effect in the discovery of ligands for peptide receptors from aminergic ligand starting points [90–92] and for a prostaglandin receptor from peptide receptor ligands [93].

More routinely, however, there is a substantial literature on the use of homology models for docking as a virtual screening method by which to identify novel hits for the target of interest [71,74,84,94–102]. This has been reviewed [103]. The main issue for GPCRs is the inherent flexibility of the protein and, even with the helices fixed in place, the potential for fairly sizeable movements of amino acid side chains within the TM bundle binding site. Many docking algorithms treat the protein as being rigid and will fail to take this into account, with consequences for their

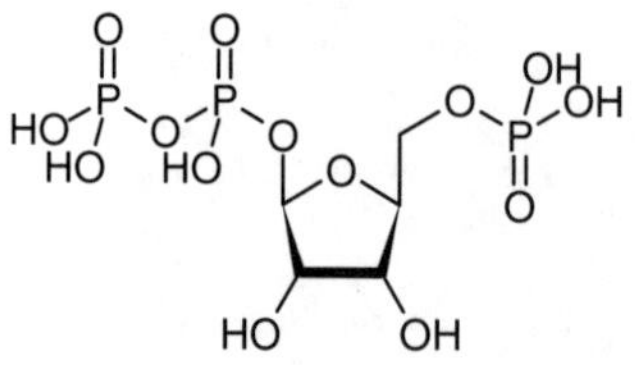

PRPP (5-phosphoribosyl-1-pyrophosphate)

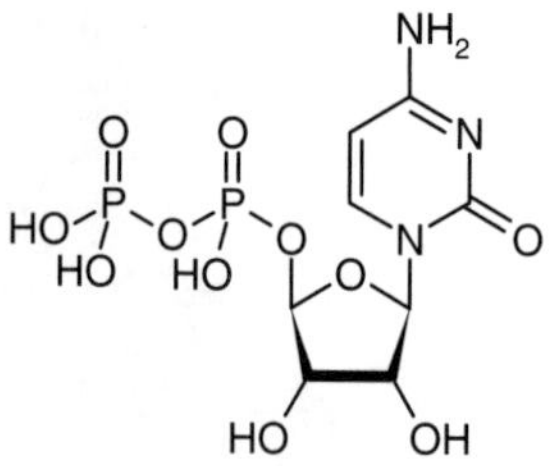

CDP (cytidine diphosphate)

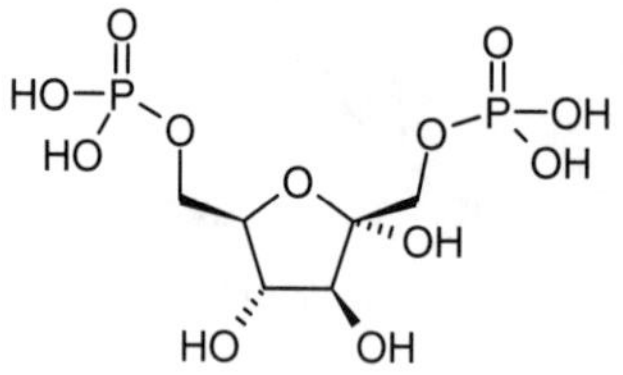

D-Fructose 1,6 biphosphate

Figure 5. Putative endogenous or surrogate agonists for the P2Y1 receptor identified through virtual screening against a homology model.

predictive accuracy. It is better to use a docking program that allows at least some protein flexibility (e.g., QXP/FLO [104]) or perhaps to use an ensemble of possible protein conformations. A number of groups have found the accuracy of the docking algorithms improves if the model is built around a known ligand ("ligand-supported" or "ligand-based" homology modeling) [76,94,105–107]. This may be partially due to improved modeling of ECL2, which might otherwise tend to encroach too deeply into the TM bundle if modeled based on rhodopsin. It also affords the opportunity to incorporate additional data such as SDM into the design process. However, while building the model this way steers the positioning of the side chains toward a good solution, it will serve to bias the compound selection toward the known chemotype used in the model building or at least toward compounds that bind to the same receptor binding site conformation. This approach is also not general insofar as it cannot be applied to orphan receptors where we lack any known ligand information. Another feature of docking to GPCRs, is that the process is greatly facilitated when there is a clear anchor point within the binding site: for example, when it is an aminergic receptor with an Asp on TM3 and the ligand has a basic center that most likely binds to this residue. Under these circumstances, the degrees of freedom are reduced and it is more likely the algorithm will arrive at appropriate solutions but the enrichment factors/success rates seen will probably not translate to other targets where we have less information about ligand binding.

Aside from these GPCR-specific issues, more general docking-type issues need to be considered. If necessary, different isomers of the compound need to be generated and docked separately. Unknown chirality or E/Z isomers of olefins would be a fairly clear example of such a problem but there may be less obvious geometric isomers such as enamines, enol ethers, or nonsymmetric *N,N*-disubstituted amides [108]. It is also important to take into account electronic factors such as different tautomeric or ionization states with different pK_a values. These can be difficult to treat correctly in a fully automated high-throughput docking (HTD) environment. The

HTD algorithm, particularly if allowing some protein flexibility, may be too slow to allow evaluation of large corporate collections and hence require some prefiltering first. This can be enabled by the homology models—factors such as the shape of the binding site or broad property constraints (presence/absence of positively or negatively ionisable side chains) could be used as a quick filter to bring the numbers down. It can also be difficult to decide *a priori* which scoring function will provide the best enrichment for a given problem.

An example of the successful application of HTD to GPCR homology models includes the alpha1a adrenergic receptor [97]. A ligand-supported homology model was built based on bovine rhodopsin with mutational and ligand SAR data incorporated to further guide the protein modeling. 2D filters and a 3D pharmacophore model were used initially to focus the selection to ca. 23 k compounds that were then docked and ranked using GOLD [110] and scored with PMF [111]. Eighty diverse compounds were selected, of which 37 had K_i values better than 10 uM, with the most active at 1.4 nM. The results are potentially comparable to those obtained when docking into a crystal structure but it is difficult to separate out the enrichment provided by the docking/homology model from the 2D and 3D prefiltering. However, while it appears possible to enrich for biologically active molecules, it seems clear that we cannot currently predict ligand affinity. In addition to searching for novel ligands, such models might prove useful for predicting off-target activities (antitargets) through a virtual "liability panel"; alpha1a adrenergic ligands are of interest for hypertension but this can be regarded as a side effect for other drug discovery programs. Predicted off-target activity would need to be confirmed in an appropriate assay given the relatively high likelihood of false positives, but it would be a useful alert to be examined in more detail.

Protein and ligand-based virtual screening methods have been compared for four receptors (alpha1a, 5HT2a, D2, and M1) using molecular docking to homology models, pharmacophore and feature tree models, 3D similarity searches, and statistical methods (PLS and PLS-DA) based on 2D molecular descriptors [109]. The ligand-based techniques show a higher enrichment than docking to a homology model, although the latter is still reasonable. Docking has potentially significant advantages, however. It is still useful when there is little or no information about active ligands available, it can probe new ways of interacting within a known site or investigate alternative interaction sites on the protein and it should also tell you how a compound binds at the target—not just that it does. The dynamic nature of GPCRs present particular challenges in this respect, though, since protein flexibility must be adequately addressed to search the full range of binding possibilities.

Use of docking to homology models can also highlight limitations in ligand-based approaches such as pharmacophore modeling. In docking a series of compounds to the CCR2 receptor model, for example, it has been found that the key basic nitrogen can be up to 5.4 Å apart [112] (Fig. 6). In each case, the ligands are still forming a strong interaction with the same Glu residue in the binding site but the side chain is conformationally mobile and the basic center may sit to either side, creating quite a large radius of potentially acceptable positions. If an equivalent superposition of ligands were to be generated without use of the homology model, the pharmacophore software would seek to force the alignment of the basic centers as a priority, which may well be incorrect. Perhaps the best approach is to create a hybrid pharmacophore and homology modeling based search, such as has been applied to the discovery of novel potent dopamine D3 receptor ligands [113,114].

Similarly, a combination of ligand-supported homology modeling with known interaction patterns from ligand to receptor has been used to score and rank order compounds from a virtual library [115]. The method utilized ligand–receptor interaction fingerprint-based similarity (IFS) [116] to encode only the patterns of interactions and not any structural features of the ligand itself. The method was evaluated in a retrospective virtual screening experiment for antagonists of the metabotropic glutamate receptor subtype 5 (mGluR5) where the authors found significantly higher enrichment than conventional scoring functions

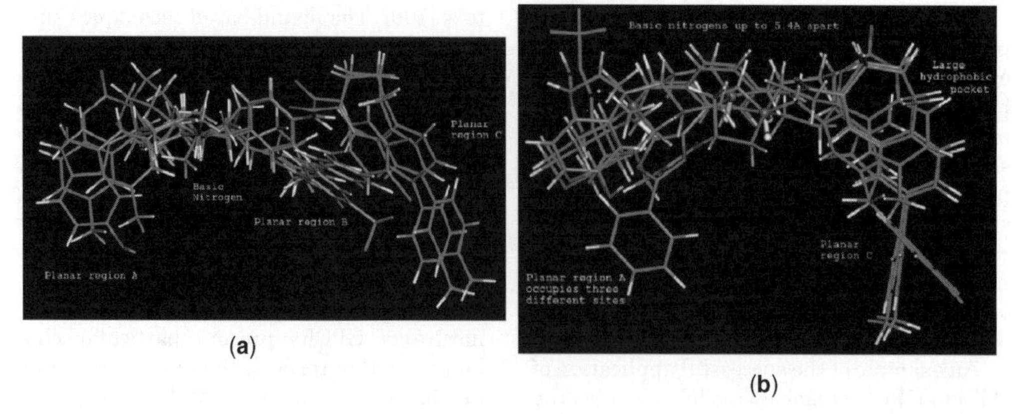

Figure 6. Comparison of ligand superpositions obtained by (a) a pharmacophore match and (b) a detailed homology modeling/SDM study of the CCR2 receptor. (Reproduced from Ref. [112].) (See color insert.)

such as Dock-Score, PMF-Score, Gold-Score, ChemScore, and FlexX-Score.

8. LEAD OPTIMIZATION—SAR INTERPRETATION, POTENCY, SELECTIVITY, QSAR, AND ADME

Homology models may be used to support SAR interpretation and the design of compounds with improved potency and/or selectivity. In principle, this is readily achieved through docking of ligands to the target of interest and evaluation of the fit through some sort of scoring function. However, for receptors that lack a key anchor point such as an acidic side chain for a basic center in the ligand, and even for some that have such a feature, multiple binding modes can often be found and it can be difficult to decide between them based solely on docking scores. When examining small numbers of compounds or supporting the optimization of a congeneric series, we have found it preferable to dock exemplars from the series manually rather than rely on an automated protocol and then to evaluate the fit through a combination of scoring functions and by eye, factoring in knowledge about the SAR/selectivities. However, sometimes it remains difficult to choose between different binding modes and the models need to be challenged directly through the synthesis of specific compounds and/or through the generation of point mutation constructs for screening. Once this is done, it is usually possible to generate a binding mode that stands up to scrutiny and is reasonably predictive for subsequent optimization. Even early models that were rather less accurate than current standards have proven successful in this respect [108].

A limiting factor is the scoring functions used in docking and our inability to predict relative potency. To address this, a number of groups have sought to develop docking-based QSAR models [98,109,117–124]. A combination of ligand-supported homology modeling and 3D-QSAR models has been applied to the discovery of A2a adenosine receptor antagonists, for example [124]. The authors used autocorrelation molecular electrostatic potential (autoMEP) vectors in combination with partial least squares (PLS) analysis as an alternative linear 3D-QSAR tool to CoMFA [125]. The authors also examined a nonlinear method based on response surface analysis (RSA) in tandem with autoMEPs (autoMEP/RSA). There were 127 known A2a antagonists used to derive two 3D-QSAR models, one based on autoMEP/PLS and the other based on autoMEP/RSA. A ligand-supported homology model was built in parallel. The two approaches were combined to predict the binding affinity of five new human A2a pyrazolo-triazolo-pyrimidine antagonists, with the conformations of newly synthesized compounds being selected from the molecular docking studies. The procedure allowed the quantitative binding affinity predictions coming from the two QSAR analyses to be merged with the qualitative information derived from the analysis of the modeled receptor–ligand complexes. There were only a limited number of compounds in the test set (five) but it provided an initial proof of concept for the tandem approach. All compounds were predicted to be 150 nM or better when, in fact, both methods overestimated the potency, with compounds active in the 230–750 nM range. However, it suggests that it is possible to correlate ligand affinities via a combined receptor modeling/QSAR approach. Importantly, the use of ligand-based homology modeling allows substantial variation in the TM bundle binding site cavity without altering the conventional rhodopsin-like receptor topology. The authors noted that the change in residue conformations could yield a change in binding site volume from 660 to 1330 $Å^3$ in A2a. This was seen to be very important; the new compounds have poor steric complementarity without it. The ligands are predicted to bind in the upper portion of the receptor, between TMs 3, 5, 6, and 7, in a region with some key differences in the A3 receptor that may account for the observed selectivities of the compounds. The model also predicts an empty pocket between TMs 5 and 6 that may be targeted by further substituents to further increase potency and/or address ADME issues.

GPCR homology modeling has been used widely to examine questions of selectivity, either between potentially closely related receptor subtypes or between species orthologs. In the former, it may be desirable to introduce multiple activities ("polypharmacology") (for example, see Ref. [126]) but usually the aim is to remove an off-target activity to avoid a potential side effect liability. For the latter, the aim is usually to introduce broadly equivalent species activity since progression of a compound into the clinic is considerably more difficult without it.

Cannabinoid receptor agonists have potential utility for treating chemotherapy-induced nausea and vomiting (CINV), glaucoma, muscle spasticity/spasm in multiple sclerosis (MS)/spinal cord injury, Alzheimer's disease (AD), and the therapeutic effects of analgesia. CB2 antagonists are also potentially useful for appetite suppression and schizophrenia. However, CB1 agonism yields the well-known side effects of altered cognition/memory, euphoria, and sedation associated with cannabis use. Knowledge of the binding mode of ligands at the CB1 and CB2 receptors has been used to help direct the design of selective compounds [127,128]. Docking and SDM studies indicate a key binding region between TMs 3, 5, 6, and 7 for three different chemotypes.

Histamine H3 receptor antagonists have potential utility in the treatment of diseases related to cognition, attention, and mood. While the affinity of histamine is broadly equivalent at both the human and rat H3 receptors, some ligands show a substantial difference in ortholog affinity [129]. Ciproxifan, for example, is ~300-fold less potent at the human receptor than rat. The source of this species difference has been identified as two relatively minor mutations on TM6. The single point mutants of the human receptor, T119A and A122V, convey more rat-like pharmacology and the double mutant T119A/A112V shows a very similar activity profile for a range of compounds. Receptor modeling has been used to rationalize this in terms of ligand binding and can suggest means of altering the ligands so as to introduce more similar species activity.

9. CONCLUSION

Structural biology of GPCRs has advanced enormously since 2007 but there are still relatively few crystal structures. Consequently, homology model building continues to play an important role in the support of drug discovery programs. While early models were useful and had some notable successes, their accuracy improved significantly with the advent of the bovine rhodopsin structure in 2000 and again in 2007 following the breakthrough of high-resolution structures for a GPCR drug target, the beta-2 adrenoceptor. These developments are not "one-offs," having enabled another structure to be produced already in the case of the adenosine A2a receptor. We can also anticipate that the technique developed for solving the beta-1 adrenoceptor structure should be general and may, importantly, enable further groundbreaking receptor structures to be solved in specific conformational states. As more structures have become available, it has been possible to probe more subtle questions such as structural variation in ECL2 and conformational changes in the TM region upon ligand binding. While some data have become available, it remains important to develop a structural understanding of the receptor activation process in order that modeling can be directed toward compounds of defined efficacy. The rather different sequence characteristics of family B/*secretin*, family B/*adhesion*, family C/*glutamate*, and family F/*frizzled* receptors also suggest that there will be a key role for structural data in understanding the sequence alignment between families and whether there are any differences in helical packing.

Despite the comparatively low sequence identity, carefully generated and validated GPCR homology models are thought to be of high quality, enabled by certain strongly conserved sequence motifs. The models can be further improved through use of known ligand/SDM information in the building process. Application of full structure-based design, particularly *de novo* design, remains challenging, but that is arguably true of any homology model. The resolution of the models is probably not high enough to predict all receptor–ligand interactions, but should be sufficient to design SDM experiments to probe specific contacts and predict the conformation/orientation of ligands, which can then be applied to a wide range of problems in the generation/optimization of drug candidates.

ABBREVIATIONS

1*S*,3*R*-ACPD	(1*S*,3*R*)-1-aminocyclopentane-1,3-dicarboxylic acid
1*S*,3*S*-ACPD	(1*S*,3*S*)-1-aminocyclopentane-1,3-dicarboxylic acid
2*R*,4*R*-APDC	(2*R*,4*R*)-4-aminopyrrolidine-2,4-dicarboxylic acid
5HT1b/d	serotonin 1b and 1d receptors
7TM or 7TMR	7-transmembrane receptor
ADME	absorption, distribution, metabolism, and excretion
AT-II	angiotensin-II receptor
Beta-2	beta-2 adrenergic receptor
CGRP	calcitonin gene-related peptide receptor
CRF2	corticotropin releasing factor receptor 2
DCG-IV	(1*R*,2*R*)-3-[(*S*)-amino(carboxy)methyl]cyclopropane-1,2-dicarboxylic acid
ECL	extracellular loop
GIP	glucose-dependent insulinotropic polypeptide receptor
GPCR	G-protein-coupled receptor
H2	histamine H2 receptor
HLX	helix
HTD	high-throughput docking
ICL	intracellular loop
MCG	(*S*)-alpha-methyl-4-carboxyphenyl glycine
MD	molecular dynamics
MES	2-(*N*-morpholino)-ethane sulfonic acid
mGluR	metabotropic glutamate receptor
NMR	nuclear magnetic resonance
NTD	N-terminal domain
PACAP	pituitary adenylate cyclase-activating polypeptide
PDB	Protein Data Bank
PTH	parathyroid hormone
PTH1	parathyroid hormone receptor 1
QSAR	quantitative structure–activity relationship
RAMP	receptor activity modifying protein
SDM	site-directed mutagenesis
TM	transmembrane

REFERENCES

1. Garland SL, Heightman TD. G-protein-coupled receptors. In: Lackey KE, editor. Gene Family Focused Molecular Design. New York: John Wiley & Sons; 2009. p. 15–52.

2. Gloriam DE, Fredriksson R, Schioth HB. The G-protein-coupled receptor subset of the rat genome. BMC Genomics 2007; 8:338.
3. Foord SM, Bonner TI, Neubig RR, Rosser EM, Pin J-P, Davenport AP, Spedding M, Harmar AJ. International Union of Pharmacology. XLVI. G-protein-coupled receptor list. Pharmacol Rev 2005;57(2):279–288.
4. Drews J. Drug discovery: a historical perspective. Science 2000;287(5460):1960–1964.
5. Esbenshade TA. G-protein-coupled receptors as targets for drug discovery. G-Protein-Coupled Receptors in Drug Discovery. *Drug Discovery Series*, Vol. 4. CRC Press LLC, Boca Raton, Fla; 2006. p. 15–36.
6. Klabunde T, Hessler G. Drug design strategies for targeting G-protein-coupled receptors. ChemBioChem 2002;3:929–944.
7. Kenakin T. Functional selectivity through protean and biased agonism: who steers the ship? Mol Pharmacol 2007;72(6):1393–1401.
8. Milligan G, Canals M, Pediani JD, Ellis J, Lopez-Gimenez JF. The role of GPCR dimerization/oligomerisation in receptor signalling. Ernst Schering Found Symp Proc 2007;2: 145–161.
9. Bridges TM, Lindsley CW. G-protein-coupled receptors: from classical modes of modulation to allosteric mechanisms. ACS Chem Biol 2008;3(9):530–541.
10. Wroblowski B, Wigglesworth MJ, Szekeres PG, Smith GD, Rahman SS, Nicholson NH, Muir AI, Hall A, Heer JP, Garland SL, Coates WJ. The discovery of a selective, small molecule agonist for the mas-related gene X1 receptor. J Med Chem 2009;52(3):818–825.
11. Attwood TK, Findlay JBC. Fingerprinting G-protein-coupled receptors. Protein Eng 1994; 7(2):195–203.
12. Schioeth HB, Fredriksson R. The GRAFS classification system of G-protein coupled receptors in comparative perspective. Gen Comp Endocrinol 2005;142(1–2):94–101.
13. Palczewski K, Kumasaka T, Hori T, Behnke CA, Motoshima H, Fox BA, Le Trong I, Teller DC, Okada T, Stenkamp RE, Yamamoto M, Miyano M. Crystal structure of rhodopsin: a G-protein-coupled receptor. Science 2000;289 (5480):739–745.
14. Standfuss J, Xie G, Edwards PC, Burghammer M, Oprian DD, Schertler GFX. Crystal structure of a thermally stable rhodopsin mutant. J Mol Biol 2007;372(5):1179–1188.
15. Kobilka B, Schertler GFX. New G-protein-coupled receptor crystal structures: insights and limitations. Trends Pharmacol Sci 2008;29(2):79–83.
16. Cherezov V, Rosenbaum DM, Hanson MA, Rasmussen SGF, Thian FS, Kobilka TS, Choi H-J, Kuhn P, Weis WI, Kobilka BK, Stevens RC, Takeda S, Kadowaki S, Haga T, Takaesu H, Mitaku S, Fredriksson R, Lagerstrom MC, Lundin LG, Schioth HB, Pierce KL, Premont RT, Lefkowitz RJ, Shenoy SK, Rosenbaum DM. High-resolution crystal structure of an engineered human β2-adrenergic G-protein-coupled receptor. Science 2007;318(5854): 1258–1265.
17. Rasmussen SGF, Choi H-J, Rosenbaum DM, Kobilka TS, Thian FS, Edwards PC, Burghammer M, Ratnala VRP, Sanishvili R, Fischetti RF, Schertler GFX, Weis WI, Kobilka BK. Crystal structure of the human β2 adrenergic G-protein-coupled receptor. Nature 2007;450 (7168):383–387.
18. Warne T, Serrano-Vega MJ, Baker JG, Moukhametzianov R, Edwards PC, Henderson R, Leslie AGW, Tate CG, Schertler GFX. Structure of a β1-adrenergic G-protein-coupled receptor. Nature 2008;454(7203):486–491.
19. Jaakola V-P, Griffith MT, Hanson MA, Cherezov V, Chien YET, Lane JR, IJzerman AP, Stevens RC. The 2.6 Angstrom crystal structure of a human A2A adenosine receptor bound to an antagonist. Science 2008;322(5905): 1211–1217.
20. Teller DC, Okada T, Behnke CA, Palczewski K, Stenkamp RE. Advances in determination of a high-resolution three-dimensional structure of rhodopsin, a model of G-protein-coupled receptors (GPCRs). Biochemistry 2001;40(26): 7761–7772.
21. Okada T, Fujiyoshi Y, Silow M, Navarro J, Landau EM, Shichida Y. Functional role of internal water molecules in rhodopsin revealed by X-ray crystallography. Proc Natl Acad Sci USA 2002;99(9):5982–5987.
22. Li J, Edwards PC, Burghammer M, Villa C, Schertler GFX. Structure of bovine rhodopsin in a trigonal crystal form. J Mol Biol 2004;343 (5):1409 1438.
23. Okada T, Sugihara M, Bondar A-N, Elstner M, Entel P, Buss V. The retinal conformation and its environment in rhodopsin in light of a new 2.2 Å crystal structure. J Mol Biol 2004;342 (2):571–583.
24. Salom D, Le Trong I, Pohl E, Ballesteros JA, Stenkamp RE, Palczewski K, Lodowski DT.

Improvements in G-protein-coupled receptor purification yield light stable rhodopsin crystals. J Struct Biol 2006;156(3):497–504.

25. Stenkamp RE. Alternative models for two crystal structures of bovine rhodopsin. Acta Crystallogr D Biol Crystallogr 2008;D64(8): 902–904.
26. Shimamura T, Hiraki K, Takahashi N, Hori T, Ago H, Masuda K, Takio K, Ishiguro M, Miyano M. Crystal structure of squid rhodopsin with intracellularly extended cytoplasmic region. J Biol Chem 2008;283(26):17753–17756.
27. Murakami M, Kouyama T. Crystal structure of squid rhodopsin. Nature 2008;453(7193): 363–367.
28. Hanson MA, Cherezov V, Griffith MT, Roth CB, Jaakola V-P, Chien YET, Velasquez J, Kuhn P, Stevens RC. A specific cholesterol binding site is established by the 2.8 Å structure of the human β2-adrenergic receptor. Structure 2008;16(6):897–905.
29. Park JH, Scheerer P, Hofmann KP, Choe H-W, Ernst OP. Crystal structure of the ligand-free G-protein-coupled receptor opsin. Nature 2008;454(7201):183–187.
30. Scheerer P, Park JH, Hildebrand PW, Kim YJ, Krauss N, Choe H-W, Hofmann KP, Ernst OP. Crystal structure of opsin in its G-protein-interacting conformation. Nature 2008;455 (7212):497–502.
31. Parthier C, Kleinschmidt M, Neumann P, Rudolph R, Manhart S, Schlenzig D, Fanghanel J, Rahfield H-U, Stubbs MT. Crystal structure of the incretin-bound extracellular domain of a G-protein-coupled receptor. Proc Natl Acad Sci USA 2007;104(35):13942–13947,S13942/ 1–S13942/11.
32. Pioszak AA, Xu HE. Molecular recognition of parathyroid hormone by its G-protein-coupled receptor. Proc Natl Acad Sci USA 2008;105 (13):5034–5039.
33. Runge S, Thogersen H, Madsen K, Lau J, Rudolph R. Crystal structure of the ligand-bound glucagon-like peptide-1 receptor extracellular domain. J Biol Chem 2008;283 (17):11340–11347.
34. Pioszak AA, Parker NR, Suino-Powell K, Xu HE. Molecular recognition of corticotropin-releasing factor by its G-protein-coupled receptor CRFR1. J Biol Chem 2008;283(47): 32900–32912.
35. Grace CRR, Perrin MH, DiGruccio MR, Miller CL, Rivier JE, Vale WW, Riek R. NMR structure and peptide hormone binding site of the first extracellular domain of a type B1 G-protein-coupled receptor. Proc Natl Acad Sci USA 2004;101(35):12836–12841.
36. Grace CRR, Perrin MH, Gulyas J, DiGruccio MR, Cantle JP, Rivier JE, Vale WW, Riek R. Structure of the N-terminal domain of a type B1 G protein-coupled receptor in complex with a peptide ligand. Proc Natl Acad Sci USA 2007;104(12):4858–4863.
37. Sun C, Song D, Davis-Taber RA, Barrett LW, Scott VE, Richardson PL, Pereda-Lopez A, Uchic ME, Solomon LR, Lake MR, Walter KA, Hajduk PJ, Olejniczak ET. Solution structure and mutational analysis of pituitary adenylate cyclase-activating polypeptide binding to the extracellular domain of PAC1-Rs. Proc Natl Acad Sci USA 2007;104(19):7875–7880.
38. Kusano S, Kukimoto-Niino M, Akasaka R, Toyama M, Terada T, Shirouzu M, Shindo T, Yokoyama S. Crystal structure of the human receptor activity-modifying protein 1 extracellular domain. Protein Sci 2008;17(11): 1907–1914.
39. Foord SM, Jupe S, Holbrook J, Bioinformatics type II G-protein-coupled receptors. Biochem Soc Trans 2002;30(4):473–479.
40. Kunishima N, Shimadat Y, Tsuji Y, Sato T, Yamamoto M, Kumasaka T, Nakanishi S, Jingami H, Morikawa K. Structural basis of glutamate recognition by a dimeric metabotropic glutamate receptor. Nature 2000;407 (6807): 971–977.
41. Muto T, Tsuchiya D, Morikawa K, Jingami H. Structures of the extracellular regions of the group II/III metabotropic glutamate receptors. Proc Natl Acad Sci USA 2007;104(10): 3759–3764.
42. Dong M, Lam PC-H, Gao F, Hosohata K, Pinon DI, Sexton PM, Abagyan R, Miller LJ. Molecular approximations between residues 21 and 23 of secretin and its receptor: development of a model for peptide docking with the amino terminus of the secretin receptor. Mol Pharmacol 2007;72(2):280–290.
43. Petrel C, Kessler A, Maslah F, Dauban P, Dodd RH, Rognan D, Ruat M. Modeling and mutagenesis of the binding site of calhex 231, a novel negative allosteric modulator of the extracellular Ca^{2+}-sensing receptor. J Biol Chem 2003;278(49):49487–49494.
44. Noeske T, Jirgensons A, Starchenkovs I, Renner S, Jaunzeme I, Trifanova D, Hechenberger M, Bauer T, Kauss V, Parsons CG, Schneider G, Weil T. Virtual screening for selective allosteric mGluR1 antagonists and structure–activity relationship investigations for cou-

marine derivatives. ChemMedChem 2007;2 (12):1763–1773.

45. Vanejevs M, Jatzke C, Renner S, Mueller S, Hechenberger M, Bauer T, Klochkova A, Pyatkin I, Kazyulkin D, Aksenova E, Shulepin S, Timonina O, Haasis A, Gutcaits A, Parsons CG, Kauss V, Weil T. Positive and negative modulation of group I metabotropic glutamate receptors. J Med Chem 2008;51(3):634–647.
46. Schwartz TW, Frimurer TM, Holst B, Rosenkilde MM, Elling CE. Molecular mechanism of 7TM receptor activation: a global toggle switch model. Ann Rev Pharmacol Toxicol 2006;46: 481–519.
47. Niv MY, Skrabanek L, Filizola M, Weinstein H. Modeling activated states of GPCRs: the rhodopsin template. J Comput Aided Mol Des 2006;20(7–8):437–448.
48. Nikiforovich GV, Marshall GR. 3D modelling of the activated states of constitutively active mutants of rhodopsin. Biochem Biophys Res Commun 2006;345(1):430–437.
49. Gouldson PR, Kidley NJ, Bywater RP, Psaroudakis G, Brooks HD, Diaz C, Shire D, Reynolds CA. Toward the active conformations of rhodopsin and the β2-adrenergic receptor. Proteins 2004;56(1):67–84.
50. Barton NP, Blaney FE, Garland SL, Tehan BG, Wall ID. 7-Transmembrane G-protein-coupled receptors: insights for drug design from structure and modelling. In: Triggle DJ, Taylor JB, editors. Comprehensive Medicinal Chemistry II. Oxford: Elsevier; 2007. p. 669–701.
51. Schertler GFX, Villa C, Henderson R. Projection structure of rhodopsin. Nature 1993;362 (6422):770–772.
52. Unger VM, Hargrave PA, Baldwin JM, Schertler GFX. Arrangement of rhodopsin transmembrane α-helixes. Nature 1997;389(6647): 203–206.
53. Shacham S, Marantz Y, Bar-Haim S, Kalid O, Warshaviak D, Avisar N, Inbal B, Heifetz A, Fichman M, Topf M, Naor Z, Noiman S, Becker OM. PREDICT modeling and *in-silico* screening for G-protein-coupled receptors. Proteins 2004;57(1):51–86.
54. Goddard WA, Abrol R. 3-Dimensional structures of G-protein-coupled receptors and binding sites of agonists and antagonists. J Nutrit 2007;137(6S-1):1528S–1538S.
55. Patny A, Desai PV, Avery MA. Homology modelling of G-protein-coupled receptors and implications in drug design. Curr Med Chem 2006;13:1667–1691.
56. Bissantz C, Bernard P, Hibert M, Rognan D. Protein-based virtual screening of chemical databases. II. Are homology models of G-protein-coupled receptors suitable targets? Proteins 2003;50(1):5–25.
57. Oliveira L, Paiva ACM, Vriend G. A common motif in G-protein-coupled seven transmembrane helix receptors. J Comput Aided Mol Des 1993;7(6):649–658.
58. Ballesteros JA, Weinstein H. Integrated methods for the construction of three-dimensional models and computational probing of structure–function relations in G protein-coupled receptors. Methods Neurosci 1995;25: 366–428.
59. Andrews G, Jones C, Wreggett KA. An intracellular allosteric site for a specific class of antagonists of the CC chemokine G protein-coupled receptors CCR4 and CCR5. Mol Pharmacol 2008;73(3):855–867.
60. Valant C, Gregory KJ, Hall NE, Scammells PJ, Lew MJ, Sexton PM, Christopoulos A. A novel mechanism of G-protein-coupled receptor functional selectivity: muscarinic partial agonist McN-A-343 as a bitopic orthosteric/allosteric ligand. J Biol Chem 2008;283(43): 29312–29321.
61. Nicholls DJ, Tomkinson NP, Wiley KE, Brammall A, Bowers L, Grahames C, Gaw A, Meghani P, Shelton P, Wright TJ, Mallinder PR. Identification of a putative intracellular allosteric antagonist binding-site in the CXC chemokine receptors 1 and 2. Mol Pharmacol 2008;74(5):1193–1202.
62. Felts AK, Gallicchio E, Chekmarev D, Paris KA, Friesner RA, Levy RM. Prediction of protein loop conformations using the AGBNP implicit solvent model and torsion angle sampling. J Chem Theory Comput 2008;4(5): 855–868.
63. Spassov VZ, Flook PK, Yan L. LOOPER: a molecular mechanics-based algorithm for protein loop prediction. Protein Eng Des Select 2008;21(2):91–100.
64. Soto CS, Fasnacht M, Zhu J, Forrest L, Honig B. Loop modeling: sampling, filtering, and scoring. Proteins 2008;70(3):834–843.
65. Peng H-P, Yang A-S. Modeling protein loops with knowledge-based prediction of sequence-structure alignment. Bioinformatics 2007;23 (21):2836–2842.
66. Costanzi S. On the applicability of GPCR homology models to computer-aided drug discovery: a comparison between *in silico* and crystal structures of the β2-adrenergic receptor. J Med Chem 2008;51(10):2907–2914.

67. Sherman W, Day T, Jacobson MP, Friesner RA, Farid R. Novel procedure for modeling ligand/receptor induced fit effects. J Med Chem 2006;49(2):534–553.
68. de Graaf C, Foata N, Engkvist O, Rognan D. Molecular modeling of the second extracellular loop of G-protein coupled receptors and its implication on structure-based virtual screening. Proteins 2008;71(2):599–620.
69. Tikhonova IG, Sum CS, Neumann S, Thomas CJ, Raaka BM, Costanzi S, Gershengorn MC. Bidirectional, iterative approach to the structural delineation of the functional "chemoprint" in GPR40 for agonist recognition. J Med Chem 2007;50(13):2981–2989.
70. Zhang Y, Sham YY, Rajamani R, Gao J, Portoghese PS. Homology modeling and molecular dynamics simulations of the mu opioid receptor in a membrane–aqueous system. ChemBioChem 2005;6(5):853–859.
71. Bissantz C, Logean A, Rognan D. High-throughput modeling of human G-protein-coupled receptors: amino acid sequence alignment, three-dimensional model building, and receptor library screening. J Chem Inf Comput Sci 2004;44(3):1162–1176.
72. Zhang Y, DeVries ME, Skolnick J. Structure modelling of all identified G-protein-coupled receptors in the human genome. PLoS Comp Biol 2006;2(2):88–99.
73. Sale K, Faulon J-L, Gray GA, Schoeniger JS, Young MM. Optimal bundling of transmembrane helices using sparse distance constraints. Protein Sci 2004;13(10):2613–2627.
74. Becker OM, Marantz Y, Shacham S, Inbal B, Heifetz A, Kalid O, Bar-Haim S, Warshaviak D, Fichman M, Noiman S. G-protein-coupled receptors: *in silico* drug discovery in 3D. Proc Natl Acad Sci USA 2004;101(31): 11304–11309.
75. Vaidehi N, Floriano WB, Trabanino R, Hall SE, Freddolino P, Choi EJ, Zamanakos G, Goddard WA. Prediction of structure and function of G-protein-coupled receptors. Proc Natl Acad Sci USA 2002;99(20):12622–12627.
76. Moro S, Deflorian F, Bacilieri M, Spalluto G. Ligand-based homology modelling as attractive tool to inspect GPCR structural plasticity. Curr Pharmaceutical Des 2006;12:2175–2185.
77. Deupi X, Dolker N, Lopez-Rodriguez ML, Campillo M, Ballesteros JA, Pardo L. Structural models of class A G-protein-coupled receptors as a tool for drug design: insights on transmembrane bundle plasticity. Curr Topics Med Chem 2007;7(10):991–998.
78. Ali MA, Bhogal N, Findlay JBC, Fishwick CWG. The first *de novo*-designed antagonists of the human NK2 receptor. J Med Chem 2005;48(18):5655–5658.
79. Gloriam DE, Foord SM, Blaney FE, Garland SL. Definition of the G-Protin-Coupled Receptor Transmembrane bundle binding pocket and calculation of receptor similarities for drug design, J. Med. Chem 2009;52(14): 4429–4442.
80. Kristiansen K. Molecular mechanisms of ligand binding, signaling, and regulation within the superfamily of G-protein-coupled receptors: molecular modeling and mutagenesis approaches to receptor structure and function. Pharmacol Ther 2004;103(1):21–80.
81. Blomenroehr M, Vischer HF, Bogerd J. Receptor mutagenesis strategies for examination of structure–function relationships. Methods Mol Biol 2004;259:307–322.
82. Gearing KL, Barnes A, Barnett J, Brown A, Cousens D, Dowell S, Green A, Patel K, Thomas P, Volpe F, Marshall F. Complex chimeras to map ligand binding sites of GPCRs. Protein Eng 2003;16(5):365–372.
83. Mazna P, Berka K, Jelinkova I, Balik A, Svoboda P, Obsilova V, Obsil T, Teisinger J. Ligand binding to the human MT2 melatonin receptor: the role of residues in transmembrane domains 3.6 and 7. Biochem Biophys Res Commun 2005;332:726–734.
84. Hiramoto T, Nokana Y, Inoue K, Yamamoto T, Omatsu-Kanbe M, Matsuura H, Gohda K, Fujita N. Identification of endogenous surrogate ligands for human P2Y receptors through an *in silico* search. J Pharmacol Sci 2004;95:81–93.
85. Goodsell DS, Olson AJ. Automated docking of substrates to proteins by simulated annealing. Proteins 8(3): 1990; 195–202.
86. Morris GM, Goodsell DS, Huey R, Olson AJ. Distributed automated docking of flexible ligands to proteins: parallel applications of AutoDock 2.4. J Comput Aided Mol Des 1996;10(4):293–304.
87. Morris GM, Goodsell DS, Halliday RS, Huey R, Hart WE, Belew RK, Olson AJ. Automated docking using a Lamarckian genetic algorithm and an empirical binding free energy function. J Comp Chem 1998;19(14):1639–1662.
88. www.charmm.org.
89. Surgand J-S, Rodrigo J, Kellenberger E, Rognan D. A chemogenomic analysis of the transmembrane binding cavity of human G-protein-coupled receptors. Proteins 2006;62(2):509–538.

90. Receveur JM, Bjurling E, Ulven T, Little PB, Norregaard PK, Hogberg T. 4-Acylamino-and 4-ureidobenzamides as melanin-concentrating hormone (MCH) receptor 1 antagonists. Bioorg Med Chem Lett 2004;14:5075–5080.
91. Martin RE, Green LG, Guba W, Kratochwil N, Christ A. Discovery of the first nonpeptidic, small-molecule, highly selective somatostatin receptor subtype 5 antagonists: a chemogenomics approach. J Med Chem 2007;50: 6291–6294.
92. Guba W, Green LG, Martin RE, Roche O, Kratochwil N, Mauser H, Bissantz C, Christ A, Stahl M. From astemizole to a novel hit series of small-molecule somatostatin 5 receptor antagonists via GPCR affinity profiling. J Med Chem 2007;50(25):6295–6298.
93. Frimurer TM, Ulven T, Elling CE, Gerlach L-O, Kostenis E, Hoegberg T. A physicogenetic method to assign ligand-binding relationships between 7TM receptors. Bioorg Med Chem Lett 2005;15(16):3707–3712.
94. Evers A, Gohlke H, Klebe G. Ligand-supported homology modelling of protein binding-sites using knowledge-based potentials. J Mol Biol 2003;334(2):327–345.
95. Becker OM, Shacham S, Marantz Y, Noiman S. Modeling the 3D structure of GPCRs: advances and application to drug discovery. Curr Opin Drug Disc Devel 2003;6(3):353–361.
96. Evers A, Klebe G. Successful virtual screening for a submicromolar antagonist of the neurokinin-1 receptor based on a ligand-supported homology model. J Med Chem 2004;47(22): 5381–5392.
97. Evers A, Klabunde T. Structure-based drug discovery using GPCR homology modeling: successful virtual screening for antagonists of the alpha1A adrenergic receptor. J Med Chem 2005;48(4):1088–1097.
98. Bissantz C, Schalon C, Guba W, Stahl M. Focused library design in GPCR projects on the example of 5-HT2c agonists: comparison of structure-based virtual screening with ligand-based search methods. Proteins 2005;61(4): 938–952.
99. Chen J-Z, Wang J, Xie X-Q. GPCR structure-based virtual screening approach for CB2 antagonist search. J Chem Inf Model 2007;47 (4):1626–1637.
100. Tikhonova IG, Sum CS, Neumann S, Engel S, Raaka BM, Costanzi S, Gershengorn MC. Discovery of novel agonists and antagonists of the free fatty acid receptor 1 (FFAR1) using virtual screening. J Med Chem 2008;51(3):625–633.
101. Cavasotto CN, Orry AJW, Murgolo NJ, Czarniecki MF, Kocsi SA, Hawes BE, O'Neill KA, Hine H, Burton MS, Voigt JH, Abagyan RA, Bayne ML, Monsma FJ. Discovery of novel chemotypes to a G-protein-coupled receptor through ligand-steered homology modeling and structure-based virtual screening. J Med Chem 2008;51(3):581–588.
102. Engel S, Skoumbourdis AP, Childress J, Neumann S, Deschamps JR, Thomas CJ, Colson A-O, Costanzi S, Gershengorn MC. A virtual screen for diverse ligands: discovery of selective G-protein-coupled receptor antagonists. J Am Chem Soc 2008;130(15):5115–5123.
103. Rognan D. Receptor-based rational design: virtual screening. Methods Principles Med Chem 2006;30:241–260. (Ligand Design for G Protein-Coupled Receptors).
104. McMartin C, Bohacek RS. QXP: powerful, rapid computer algorithms for structure-based drug design. J Comput Aided Mol Des 1997; 11(4):333–344.
105. Moro S, Bacilieri M, Deflorian F, Spalluto G. G-protein-coupled receptors as challenging druggable targets: insights from *in silico* studies. New J Chem 2006;30:301–308.
106. Colotta V, Catarzi D, Varano F, Capella F, Lenzi O, Filacchioni G, Martín C, Trincavelli L, Ciampi O, Pugliese AM, Pedata F, Schiesaro A, Morizzo E, Moro S. New 2-arylpyrazolo[3,4-*c*]-quinoline derivatives as potent and selective human A3 adenosine receptor antagonists. Synthesis, pharmacological evaluation and ligand–receptor modelling studies. J Med Chem 2007;50:4061–4074.
107. Bolcato C, Cusan C, Pastorin G, Spalluto G, Cacciari B, Klotz KN, Morizzo E, Moro S. Pyrazolo-triazolo-pyrimidines as adenosine receptor antagonists: effect of the N-5 bond type on the affinity and selectivity at the tour adenosine receptor subtypes. Purinergic Signal 2008;4(1):39–46.
108. Blaney FE, Capelli A-M, Tedesco G. 7TM models in structure-based design. In: Rognan D, editor. Ligand Design for G Protein-Coupled Receptors. Wienheim: Wiley; 2006. p. 205–239.
109. Evers A, Hessler G, Matter H, Klabunde T. Virtual screening of biogenic amine-binding G-protein coupled receptors: comparative evaluation of protein- and ligand-based virtual screening protocols. J Med Chem 2005;48 (17):5448–5465.
110. Jones G, Willett P, Glen RC, Leach AR, Taylor R. Development and validation of a genetic

algorithm for flexible docking. J Mol Biol 1997;267(3):727–748.

111. Muegge I, Martin YC, Hajduk PJ, Fesik SW. Evaluation of PMF Scoring in docking weak ligands to the FK506 binding protein. J Med Chem 1999;42(14):2498–2503.

112. Berkhout TA, Blaney FE, Bridges AM, Cooper DG, Forbes IT, Gribble AD, Groot PHE, Hardy A, Ife RJ, Kaur R, Moores KE, Shillito H, Willetts J, Witherington J. CCR2: Characterization of the antagonist binding site from a combined receptor modelling/mutagenesis approach. J Med Chem 2003;46(19):4070–4086.

113. Varady J, Wu X, Fang X, Min J, Hu Z, Levant B, Wang S. Molecular modeling of the three-dimensional structure of dopamine 3 (D3) subtype receptor: discovery of novel and potent D3 ligands through a hybrid pharmacophore- and structure-based database searching approach. J Med Chem 2003;46(21):4377–4392.

114. Hobrath JV, Wang S. Computational elucidation of the structural basis of ligand binding to the dopamine 3 receptor through docking and homology modeling. J Med Chem 2006;49(15): 4470–4476.

115. Radestock S, Weil T, Renner S. Homology model-based virtual screening for GPCR ligands using docking and target-biased scoring. J Chem Inf Model 2008;48:1104–1117.

116. Renner S, Derksen S, Radestock S, Mörchen F. Maximum common binding modes (MCBM): consensus docking scoring using multiple ligand information and interaction fingerprints. J Chem Inf Model 482008; 319–332.

117. Schafferhans A, Klebe G. Docking ligands onto binding site representations derived from proteins built by homology modelling. J Mol Biol 2001;307(1):407–427.

118. Broer BM, Gurrath M, Holtje H-D. Molecular modelling studies on the ORL1-receptor and ORL1-agonists. J Comput Aided Mol Des 2003;17(11):739–754.

119. Evers A, Klebe G. Ligand-supported homology modeling of G-protein-coupled receptor sites: models sufficient for successful virtual screening. Angew Chem Int Ed 2004;43(2):248–251.

120. Costanzi S, Joshi BV, Maddileti S, Mamedova L, Gonzalez-Moa MJ, Marquez VE, Harden TK, Jacobson KA. Human P2Y6 receptor: molecular modeling leads to the rational design of a novel agonist based on a unique conformational preference. J Med Chem 2005;48(26): 8108–8111.

121. Moro S, Braiuca P, Deflorian F, Ferrari C, Pastorin G, Cacciari B, Baraldi PG, Varani K, Borea PA, Spalluto G. Combined target-based and ligand-based drug design approach as a tool to define a novel 3D-pharmacophore model of human A3 adenosine receptor antagonists: pyrazolo[4,3-*e*]1,2,4-triazolo[1,5-*c*]pyrimidine derivatives as a key study. J Med Chem 2005;48(1):152–162.

122. Costanzi S, Tikhonova IG, Ohno M, Roh EJ, Joshi BV, Colson A-O, Houston D, Maddileti S, Harden TK, Jacobson KA. P2Y1 antagonists: combining receptor-based modeling and QSAR for a quantitative prediction of the biological activity based on consensus scoring. J Med Chem 2007;50(14):3229–3241.

123. Kellenberger E, Springael J-Y, Parmentier M, Hachet-Haas M, Galzi J-L, Rognan D. Identification of nonpeptide CCR5 receptor agonists by structure-based virtual screening. J Med Chem 2007;50(6):1294–1303.

124. Michielan L, Bacilieri M, Schiesaro A, Bolcato C, Pastorin G, Spalluto G, Cacciari B, Klotz KN, Kaseda C, Moro S Linear and non-linear 3D-QSAR approaches in tandem with ligand-based homology modelling as a computational strategy to depict the pyrazolo-triazolo-pyrimidine antagonists binding site of the human adenosine A receptor. J Chem Inf Model 2008;48:350–363.

125. Cramer RDI, Patterson DE, Bunce JD. Comparative molecular field analysis (CoMFA). 1. Effect of shape on binding of steroids to carrier proteins. J Am Chem Soc 1988;110: 5969–5967.

126. Garzya V, Forbes IT, Gribble AD, Hadley MS, Lightfoot AP, Payne AH, Smith AB, Douglas SE, Cooper DG, Stansfield IG, Meeson M, Dodds EE, Jones DNC, Wood M, Reavill C, Scorer CA, Worby A, Riley G, Eddershaw P, Ioannou C, Donati D, Hagan JJ, Ratti EA. Studies towards the identification of a new generation of atypical antipsychotic agents. Bioorg Med Chem Lett 2007;17(2):400–405.

127. Xie X-Q, Chen J-Z, Billings EM. 3D structural model of the G-protein-coupled cannabinoid CB2 receptor. Proteins 2003;53:307–319.

128. Montero C, Campillo NE, Goya P, Paez JA. Homology models of the cannabinoid CB1 and CB2 receptors. A docking analysis study. Eur J Med Chem 2005;40:75–83.

129. Yao BB, Hutchins CW, Carr TL, Cassar S, Masters JN, Bennani YL, Esbenshade TA, Hancock AA. Molecular modeling and pharmacological analysis of species-related histamine H3 receptor heterogeneity. Neuropharmacol 2003;44(6):773–786.

STRUCTURAL ALERTS FOR TOXICITY

JULIAN BLAGG
Cancer Research UK Centre for Cancer Therapeutics, The Institute of Cancer Research, Haddow Laboratories, Sutton, Surrey, UK

1. INTRODUCTION

Structural alerts, or toxicophores, are well-defined structural motifs where a significant weight of evidence exists linking the motif with an increased incidence of covalent protein/DNA modification and subsequent downstream adverse outcome (for example, irreversible CYP450 inhibition, *in vitro* genotoxicity, carcinogenicity, *in vivo* hepatotoxicity, or antibody-mediated immune response). The covalent modification of endogenous biomolecules may arise from the inherent chemical reactivity of a structural alert (see Section 2); alternatively, it can arise through the generation of reactive metabolites as a consequence of metabolism (usually oxidative, see Section 3) or photoactivation. The *in vivo* formation of covalently modified endogenous biomolecules is the primary issue with all structural alerts. This process may subsequently lead to *in vivo* toxicity outcomes and/or severe immune reactions in subsets of patients exposed to the drug (idiosyncratic drug reactions or IDRs). Rare IDRs often remain undetected until late-stage clinical trials or postmarketing when a sufficiently large patient population is exposed to the drug. Patient genetic background, presensitization, and differences in drug exposure are the key factors in predisposing certain individuals [1,2]. A well-characterized exemplar IDR is produced by molecules containing the penicillin core structure (**1**). The β-lactam core of penicillin is inherently chemically reactive: Ring opening of the strained four membered β-lactam ring (**2**) is not only inherent to the therapeutic activity (inhibition of bacterial cell wall transpeptidases) but also results in the covalent modification of other proteins bearing nucleophilic centers and the inactivation of β-lactam antibiotics by β-lactamases (Fig. 1) [3].

Idiosyncratic drug reactions, by definition, are difficult to screen for and have resulted in the withdrawal of many marketed clinical agents [4,5]. Compounds requiring low exposure to elicit their desired therapeutic effect often have a significantly lower incidence of structure derived adverse events because endogenous protective mechanisms are able to cope with low exposures to reactive metabolites (see Section 6.1). However, predicting clinical dose and exposure is currently an uncertain science. Reliance upon predicted low exposure of a clinical candidate to mitigate the presence of a structural alert shifts the attrition risk later into expensive clinical development. It is, therefore, important for the medicinal chemist to be aware of structural features that have a higher than normal propensity for covalent modification of endogenous proteins and subsequent adverse outcomes. In particular, an understanding of the metabolic pathways that can generate reactive metabolites is important in medicinal chemistry target design and selection.

This chapter outlines those structural alerts for which there is significant weight of evidence that their incorporation into drug candidates is likely to be associated with a higher incidence of the covalent modification of biomolecules and a heightened risk of downstream adverse safety or toxicity outcomes. The review highlights the pathways by which reactive metabolites may be generated from these structural alerts. In addition, key illustrative examples and a summary of the structure–toxicity relationships associated with the structural alert are included to help clarify the boundaries of each structural alert where sufficient information is available.

2. STRUCTURAL ALERTS WITH INHERENT CHEMICAL REACTIVITY

Structural moieties that are inherently reactive are an essential part of the armory by which synthetic chemists build new molecular structures. However, their presence in drug candidate molecules requires a detailed understanding of their *in vivo* fate and very

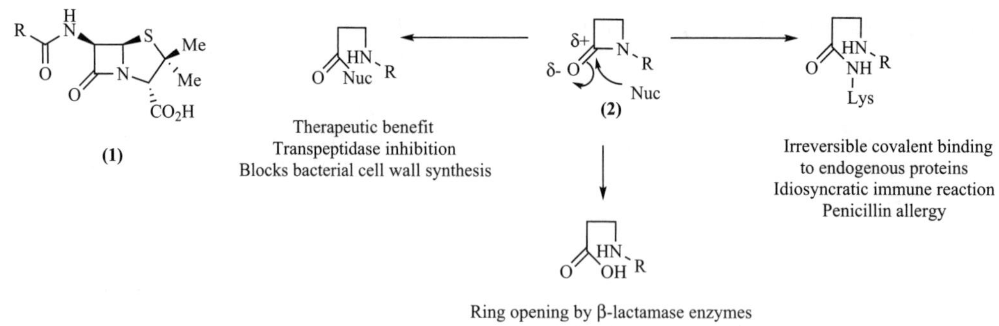

Figure 1. β-Lactam ring opening.

careful risk management. An inherently chemically reactive moiety is an integral part of the medicinal chemistry strategy for programs seeking a truly irreversible inhibitor through covalent linkage to the target protein. In this case, exquisite selectivity for the primary pharmacology target is highly desirable since irreversible binding to off-target biomolecules is likely to be significantly more deleterious than reversible off-target binding; for example, the likelihood of immunogenic responses is heightened. There are examples of irreversible or very slow offset drugs with a good long-term safety record; for example, finasteride (**3**) exhibits very slow offset kinetics from its target, the 5α-reductase enzyme, and contains an α,β-unsaturated amide or Michael acceptor structural alert (Fig. 2) [6]. However, finasteride is selectively recognized by the two 5α-reductase isoenzymes and off-target modification of other biomolecules is minimal.

The inherent chemical reactivity of a particular functional group is a key parameter in understanding the rate of reaction with biomolecules *in vivo*. Knowledge of organic chemistry principles can educate design teams to the risks involved in incorporating such functionality into target molecules, alternatively quantum mechanical calculations have been used in this setting to predict chemical reactivity [7]. The functional groups depicted in Figure 3 are common examples where the chemical reactivity of the functional group leads to the covalent modification of biomolecules *in vivo*. Structural motifs that can easily be metabolized to structural alerts depicted in Figure 3 are also potential risks, for example, the CYP450-mediated oxidation of an isolated double bond to an epoxide or the metabolic oxidation of a primary alcohol to an aldehyde. An early understanding of routes of metabolism in lead series is, therefore, an important factor in avoiding the risk of reactive metabolites in medicinal chemistry design.

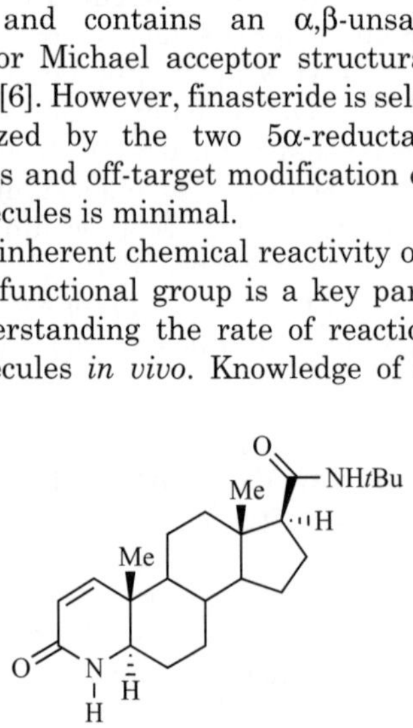

Figure 2. Finasteride.

3. STRUCTURAL ALERTS REQUIRING METABOLISM TO GENERATE A REACTIVE METABOLITE

Structural alerts that require metabolism to generate a reactive metabolite have been extensively reviewed [8–11]. This section covers a brief description of each class of structural alert and, where possible, the postulated mechanism by which the structural alert is generated. Awareness of the proposed mechanism of activation is important for medicinal chemists in deciding whether the environment of a structural alert increases or decreases the risk of reactive metabolite formation. Figure 4 summarizes the major classes of structural alert that require metabolism. However, it is important to understand that each structural

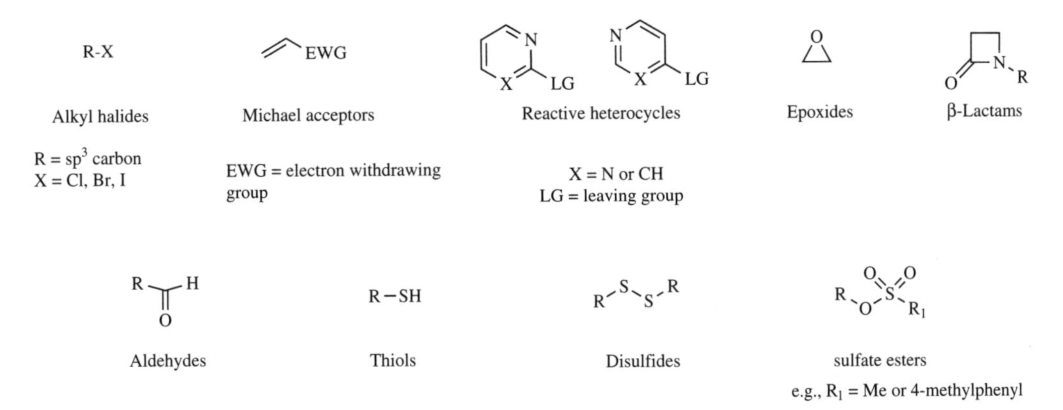

Figure 3. Structural alerts with inherent chemical reactivity.

alert has its individual structure–toxicity relationship; for example, not all anilines are equally likely to generate reactive metabolites and knowledge of the metabolic pathways involved in reactive metabolite formation highlights those anilines particularly at risk and those where the propensity to reactive metabolite formation is reduced. This section aims to provide an understanding of the factors driving the propensity to reactive metabolite formation for the structural alerts included in Figure 4.

3.1. Anilines and Hydroxyanilines

Anilines are the structural alert with the richest and most diverse structure–toxicity relationship. An understanding of these relationships is important in target design and selection in medicinal chemistry due to the relatively diverse and wide chemical space encompassed by the broadest definition of an aniline (see Fig. 4) and a significant amount of detail will be devoted to the aniline class of structural alert.

There is a considerable weight of evidence that the formation of reactive metabolites *via* the metabolic oxidation of aniline moieties is linked to adverse outcome in humans. The prototype cardioselective β-blocker practolol (**4**), the oral antiarrhythmic drug tocainide (**5**) and the first oral diabetes treatment carbutamide (**6**) have all received a black box label warning of possible serious adverse events or have been withdrawn through incidence of serious adverse events in the patient population undergoing treatment [12–14]. The closely related compounds with the same pharmacological mechanism of action where the aniline nitrogen has been replaced: atenolol (**7**), mexiletine (**8**), and tolbutamide (**9**), respectively, are notable for the significantly reduced rate or absence of the serious adverse event that led to the withdrawal or restriction on use of the precursor agent (Fig. 5).

3.1.1. Aniline Metabolism *via* Oxidation of the Aromatic Ring The formation of reactive metabolites from anilines (**10**) is mediated by oxidative metabolism through two independent pathways. The first pathway involves CYP450-mediated oxidation of the aromatic ring *ortho* or *para* to the anilinic nitrogen to generate *ortho*- and *para*-hydroxyanilines (**11**) and (**12**), respectively. These species can be further oxidized to the *ortho*- or *para*-iminoquinone reactive intermediates (**13**) and (**14**) respectively, which can be trapped by endogenous nucleophiles at the electrophilic carbon atoms to regenerate substituted aromatic hydroxyaniline systems exemplified by compounds (**15**) and (**16**) (Fig. 6).

A classic example of aniline oxidation *via* this pathway occurs in the CYP2C9-mediated transformation of diclofenac (**17**) to 4′-hydroxydiclofenac (**18**), the major metabolite in human urine. This metabolite has been implicated as the precursor of the iminoquinone reactive metabolite (**19**) and the associated elevation of liver enzymes characteristic of hepatotoxicity in humans (Fig. 7) [15].

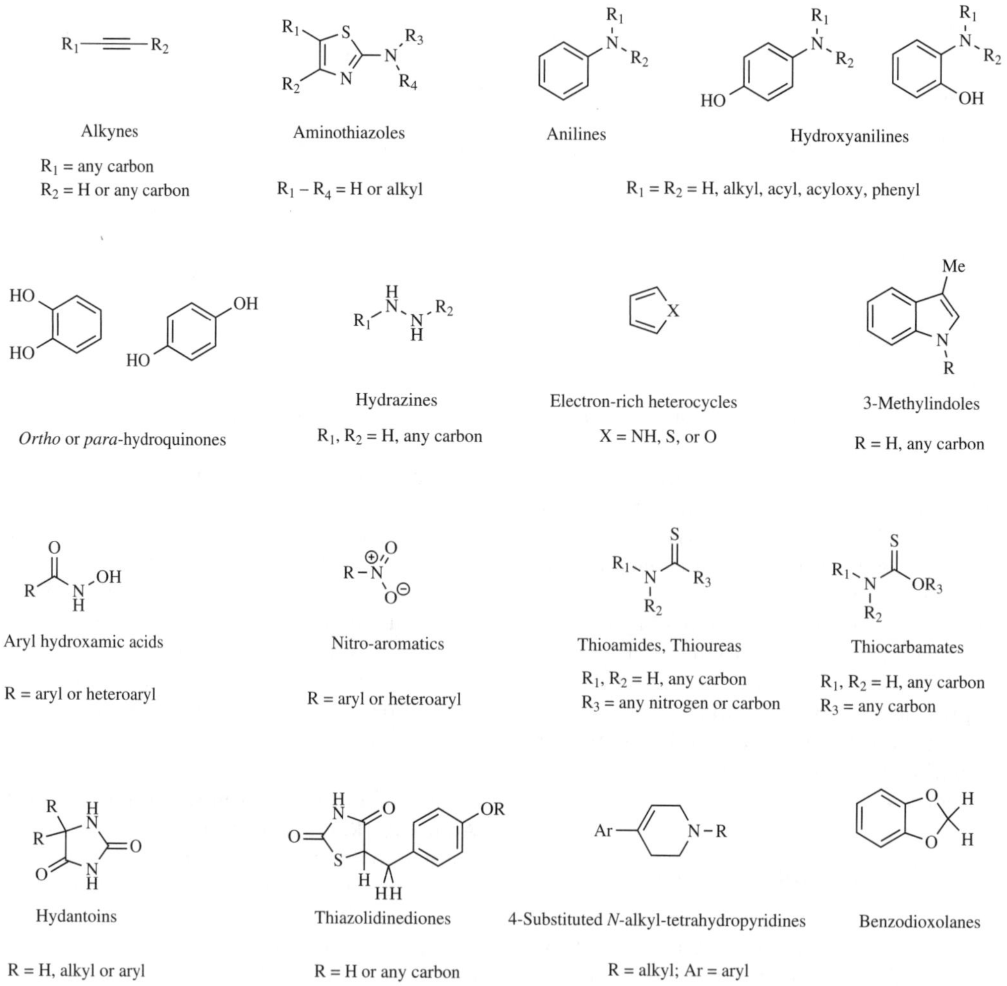

Alkynes
R_1 = any carbon
R_2 = H or any carbon

Aminothiazoles
$R_1 - R_4$ = H or alkyl

Anilines

Hydroxyanilines

$R_1 = R_2$ = H, alkyl, acyl, acyloxy, phenyl

Ortho or *para*-hydroquinones

Hydrazines
R_1, R_2 = H, any carbon

Electron-rich heterocycles
X = NH, S, or O

3-Methylindoles
R = H, any carbon

Aryl hydroxamic acids
R = aryl or heteroaryl

Nitro-aromatics
R = aryl or heteroaryl

Thioamides, Thioureas
R_1, R_2 = H, any carbon
R_3 = any nitrogen or carbon

Thiocarbamates
R_1, R_2 = H, any carbon
R_3 = any carbon

Hydantoins
R = H, alkyl or aryl

Thiazolidinediones
R = H or any carbon

4-Substituted *N*-alkyl-tetrahydropyridines
R = alkyl; Ar = aryl

Benzodioxolanes

Figure 4. Structural alerts requiring metabolism.

Compounds that already possess an *ortho*- or *para*-hydroxyaniline motif (**11** or **12**) are one step closer to the generation of an iminoquinone and therefore inherently more likely to generate reactive metabolites and irreversibly alkylate endogenous proteins or DNA. Thus, *ortho*- or *para*-hydroxyanilines are high-risk structural alerts where facile oxidative metabolism leads to precedented reactive intermediates. Acetaminophen (**20**) is the most widely studied example of this class (Fig.8). It is hepatotoxic and the role of oxidative metabolism in driving the adverse outcomes is well precedented [16]. The *para*-benzoquinone-imine reactive metabolite (**21**) is formed by oxidation of the parent (**20**) *via* a number of possible pathways and may bear an acetyl substituent on the nitrogen or a hydrogen if the acetyl group has been metabolically removed. These reactive species may be trapped by endogenous biomolecules leading to hepatotoxicity, oxidative stress and immune reactions, or by glutathione and excreted as part of an endogenous protective mechanism against reactive metabolites (Fig. 8) (see Section 6.1). Overdose of acetaminophen in human and animal species leads to depletion of glutathione and alkylation of endogenous biomolecules, predominantly through cysteine residues. The downstream

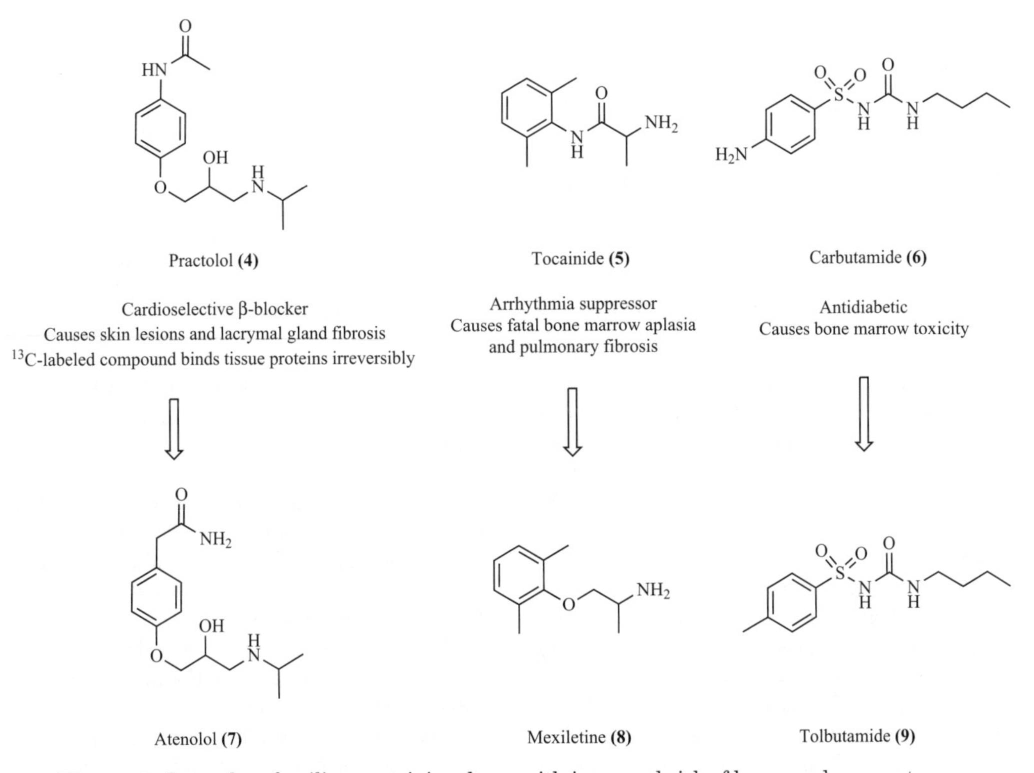

Figure 5. Examples of aniline containing drugs with increased risk of human adverse outcome.

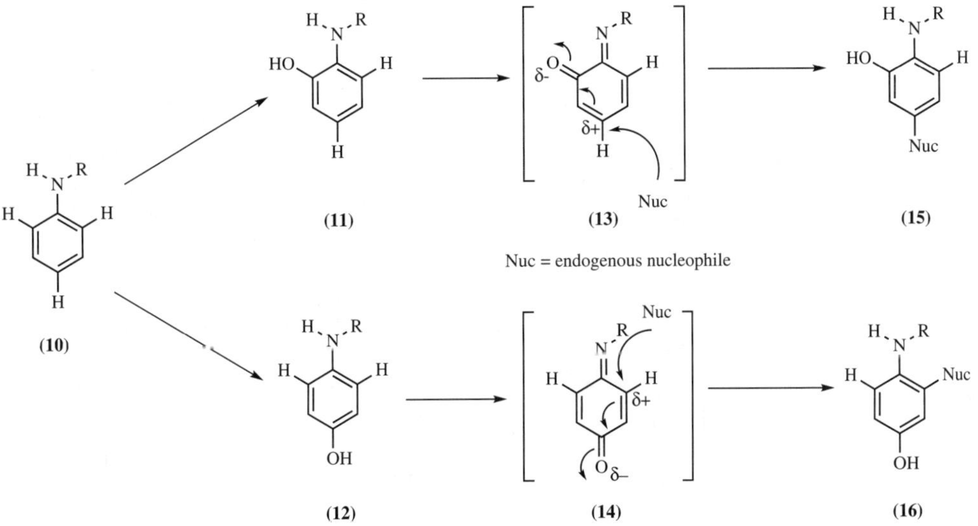

(10) (11) (12) (13) (14) (15) (16)

Nuc = endogenous nucleophile

Figure 6. Aniline aromatic ring oxidation.

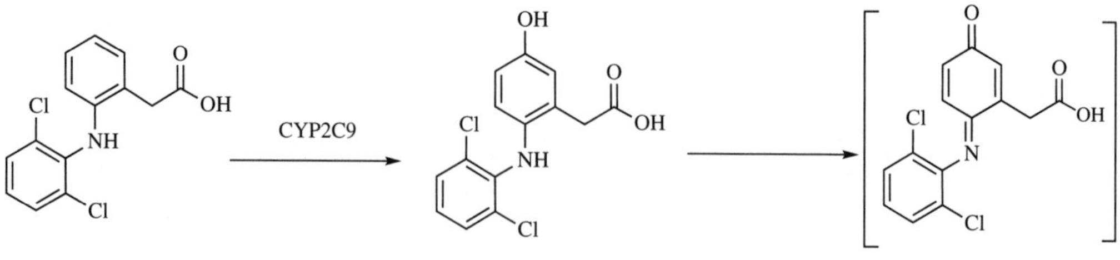

(17) (18) (19)

Figure 7. Aniline aromatic ring oxidation in diclofenac.

consequences of adduct formation are the subject of intense study. Increased levels of peroxynitrite and superoxide are consistent with the induction of downstream oxidative stress [17]. Covalent modification of *N*-terminal proline residues has also been reported [18].

The antimalarial drug amodiaquine (**22**) is a good example of the facile oxidation of the hydroxylaniline class. Amodiaquine (**22**) causes agranulocytosis and liver damage in human *via* the formation of the postulated iminoquinone metabolite (**23**) (Fig. 9) [19]. The close analog (**24**) is an equipotent antimalarial with reactive metabolite formation at undetectable levels. The *meta* relationship of the anilinic nitrogen and the phenol render conjugated iminoquinone formation impossible and nonoxidative glucuronidation of the phenol is the predominant metabolic clearance pathway [20].

3.1.2. Aniline Metabolism *via* Oxidation of the Aniline Nitrogen Atom The second pathway of aniline metabolism involves oxidation on the aniline nitrogen to generate the hydroxylamine (**25**), nitroso intermediate (**26**), and ultimately the fully oxidized nitro compounds (**27**). Both the hydroxylamine (**25**) and the nitroso species (**26**) can behave as reactive intermediates. The hydroxylamine can undergo enzyme-mediated acetylation (acetyl transferase) or sulfation (sulfotransferase) transforming the oxygen atom into a good leaving group and thereby generating a highly reactive nitrenium ion (**28**) that is readily trapped by endogenous

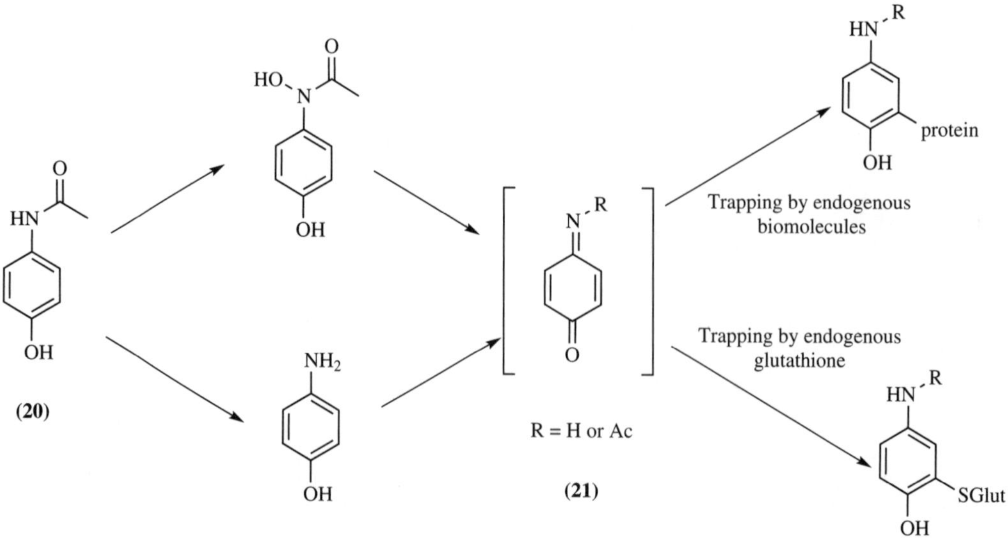

Trapping by endogenous biomolecules

Trapping by endogenous glutathione

R = H or Ac

(20) (21)

Figure 8. Acetaminophen reactive metabolite formation.

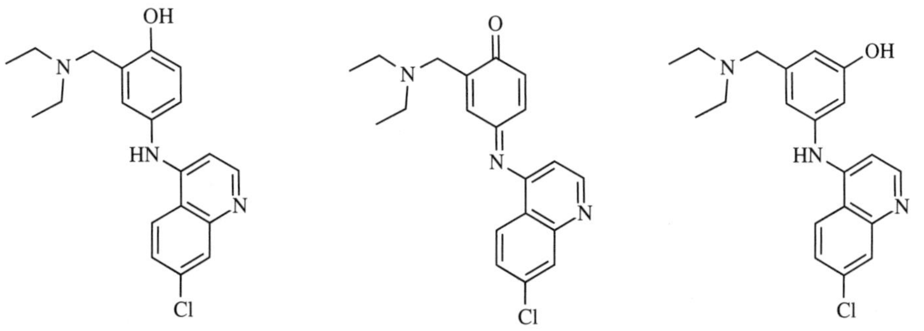

Figure 9. Amodiaquine.

biomolecules, for example, DNA, to form adducts (**29**) [21]. The nitroso species (**26**) can be directly trapped by a sulfur nucleophile, for example, the cysteine residues in hemoglobin, to generate hemoglobin adducts (**30**) [22]. Redox cycling, for example, between the nitroso species (**26**) and the nitro species (**27**) can drive the formation of reactive oxygen species (e.g., superoxide) and downstream oxidative stress responses [23] (Fig. 10).

A prerequisite for oxidative metabolism as depicted in Fig. 10 is the presence of at least one free NH on the anilinic nitrogen. For the formation of the nitroso species (**26**), then the presence of two free NHs is required either in the parent molecule or by prior oxidative metabolism to the free aniline. Thus, the propensity for anilinic nitrogens to undergo *N*-oxidation is directly related to its substitution pattern and its electronic environment. Arguments that the *N*-oxidation of anilines can be mitigated by the presence of electron withdrawing groups on the aromatic ring are too simplistic to account for the multiple pathways available in aniline metabolism [24]. There is evidence that the electronic character of the aniline merely influences the pathway of oxidation, with electron-rich anilines more likely to form DNA adducts *via* the nitrenium ion (**28**) and electron poor anilines more likely to form hemoglobin adducts *via* the nitroso species (**26**) [24,25]. The steric effect of the aniline substitution pattern also plays a significant role; for example, genotoxic response is influenced by the conformational properties of the downstream aniline–DNA adduct. [26]

A comparison of the metabolic *N*-oxidation rates of aromatic and aliphatic amines has demonstrated a > 10-fold rate enhancement for aromatic *N*-oxidation versus corresponding aliphatic amine *N*-oxidation. Rates of *N*-oxidation of neutral aromatic amines are fast and cLogP dependent whereas basic, highly protonated aliphatic amines are *N*-oxidized relatively slowly in a non-cLogP dependent manner (Fig. 11) [27].

The oxidation of anilinic nitrogen atoms is not limited to unsubstituted or alkyl-substituted anilines. For example, the oxidation to *N*-hydroxy species has also been observed for *N*-acetylanilines in rat and human [28]. Formation of *N*-hydroxy species from diphenyl anilines has also been observed in human hepatocytes [29].

There are multiple examples of compounds where *N*-oxidation of the anilinic nitrogen is associated with a higher than average incidence of adverse safety or toxicity outcomes. A well-characterized example is the oxidation of 4-aminobiphenyl (**31**), a common constituent of cigarette smoke, to the nitroso species (**33**) that is trapped by hemoglobin to form the sulfinamide adduct (**34**) (Fig. 12). This adduct has been isolated from smokers' lung tissue (154 pg/g of tissue) at levels sixfold higher than nonsmoker controls [22]. Oxidation to the *N*-hydroxylamine (**32**) and subsequent formation of DNA adducts *via* the postulated nitrenium ion (**35**) has also been reported, leading to the formation of the *N*-(deoxyguanosin-8-yl)-4-aminobiphenyl adduct (**36**) that has been isolated as a

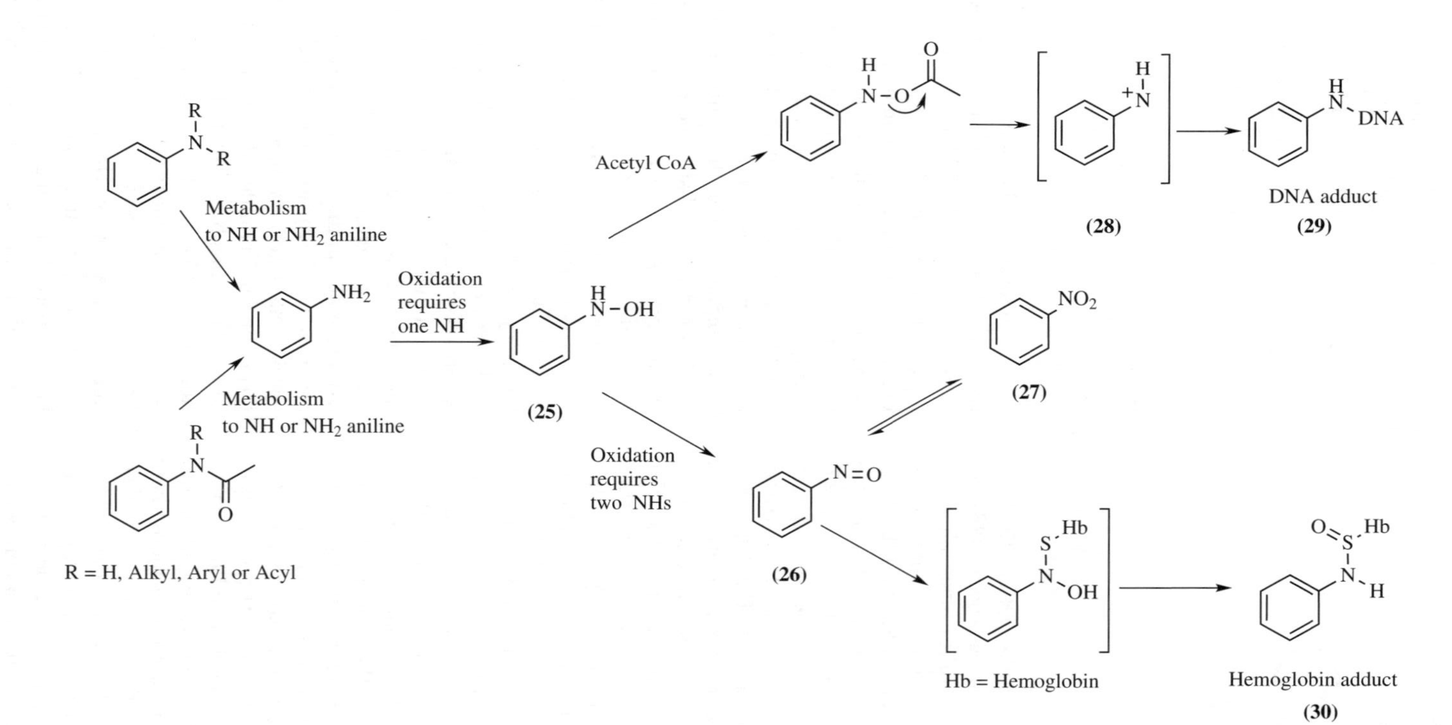

Figure 10. Aniline nitrogen atom oxidation.

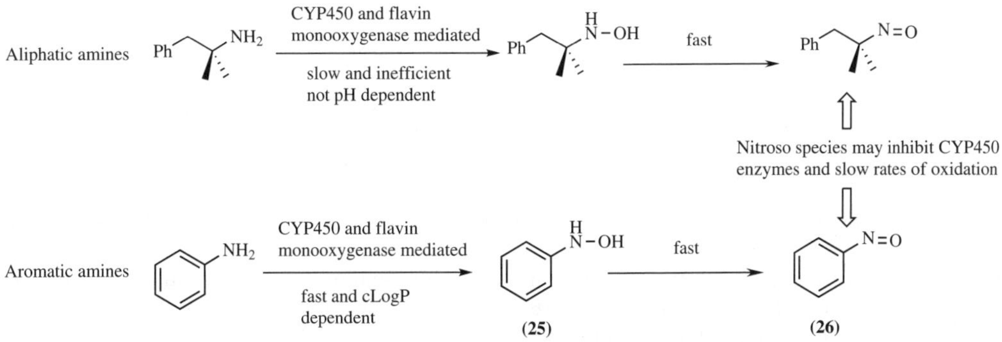

Figure 11. Aliphatic versus aromatic *N*-oxidation.

major adduct in bladder tissue taken from smokers [30] (Fig. 12).

There are examples of aniline nitrogens in structural environments where the metabolic pathways outlined above are not possible or proceed at a significantly lower rate. For these cases, the incidence of adverse outcome linked to the presence of the aniline is significantly lower and the use of these anilines in medicinal chemistry target design is a lower risk. Examples include the so-called masked anilines where metabolic or hydrolytic liberation of the free aniline or direct aniline *N*-oxidation is very slow. Anilines incorporated into sulfonamides (**37**), ureas (**38**), lactams (**39**), and cyclic carbamates (**40**) are much less susceptible to *N*-oxidation and cleavage of these functionalities to the free aniline is usually very slow. Exceptions have, however, been noted in rare cases such as lidocaine (**41**) where carboxylesterase cleavage of the urea to the free 2,6-dimethylaniline has been observed *in vitro* [31] (Fig. 13).

Heterocyclic anilines (for example, species **42** and **43**) are also less susceptible to *N*-oxidation by virtue of the lower electron density on the anilinic nitrogen and the higher propensity for other oxidative pathways in such systems, for example, heterocyclic ring *N*-oxidation (Fig. 14). One notable exception is the case of 2-azabenzimidazoles found in some cooked food, that form DNA adducts in *in vitro* preparations *via* *N*-oxidation of the exocyclic anilinic nitrogen [32]. Anilinic nitrogens incorporated into an aromatic ring (for example, species **44** and

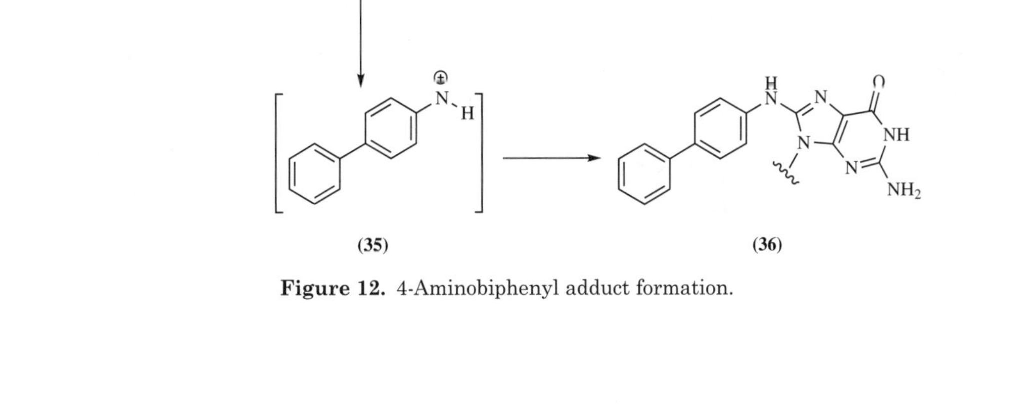

Figure 12. 4-Aminobiphenyl adduct formation.

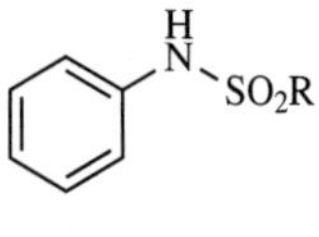

H N SO_2R H N H N R O H N O H N O O Me H N N O Me

(37) 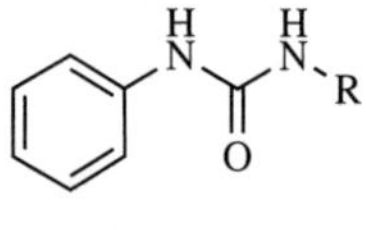 (38) 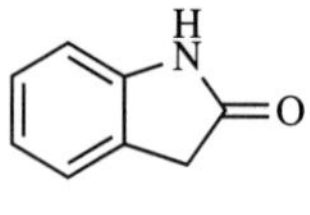 (39) 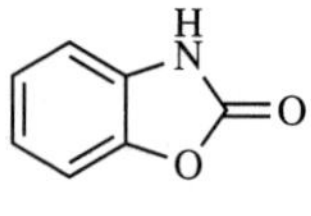 (40) 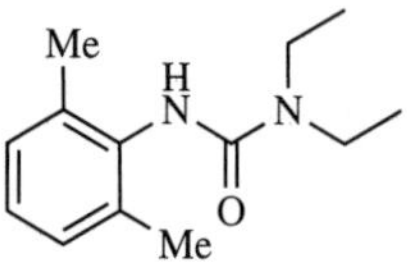 (41)

Figure 13. Masked anilines where the rate of oxidative metabolism *via* *N*-oxidation is significantly lower.

45) or where the aniline nitrogen is part of a fused aromatic ring (for example, species **46** and **47**) are also less susceptible to the known aniline metabolic pathways outlined above (Fig. 14).

One example of particular interest is the fused lactam system (**39**) (Fig. 15) where *N*-oxidation of the anilinic nitrogen is slow compared with the acyclic case (**48**) and where enzymatic cleavage of the lactam to generate

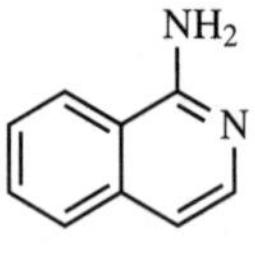

NH_2 N N NH_2 N N O N N N N H

(42) 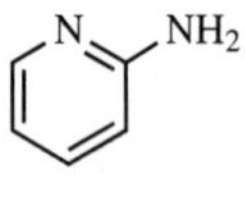 (43) 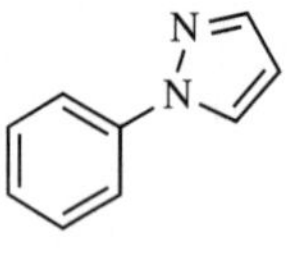 (44) 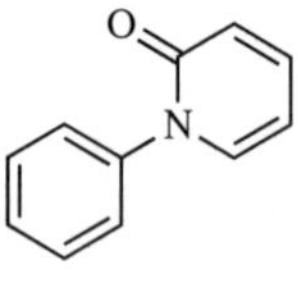 (45) 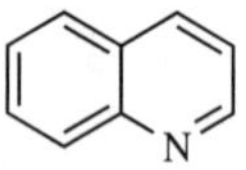 (46) 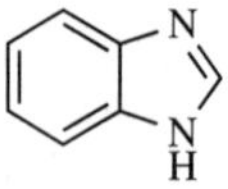 (47)

Figure 14. Anilines less susceptible to oxidative metabolism *via* *N*-oxidation.

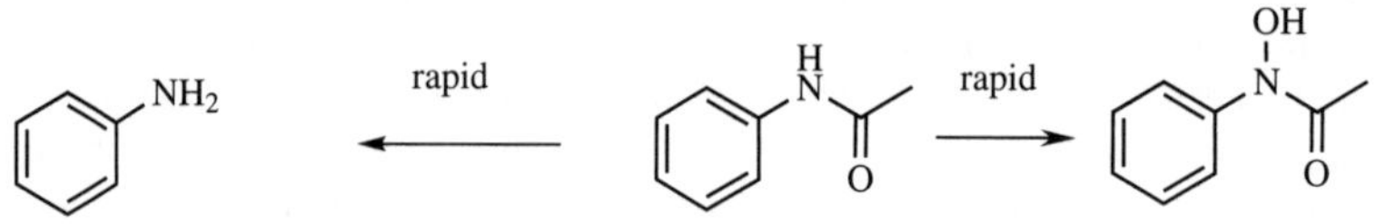

NH_2 rapid H N O rapid OH N O

(48)

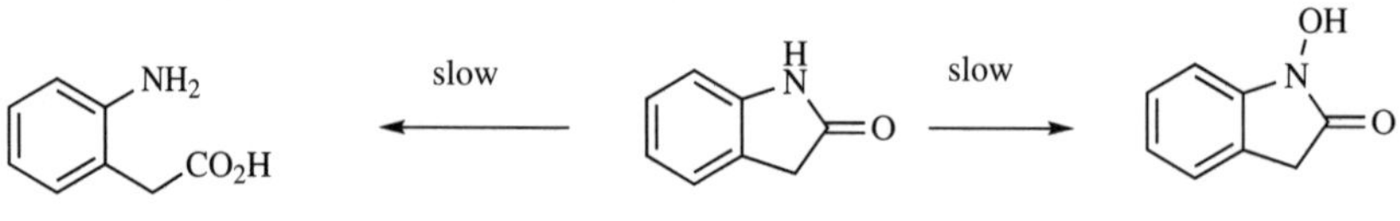

NH_2 CO_2H slow H N O slow OH N O

(39)

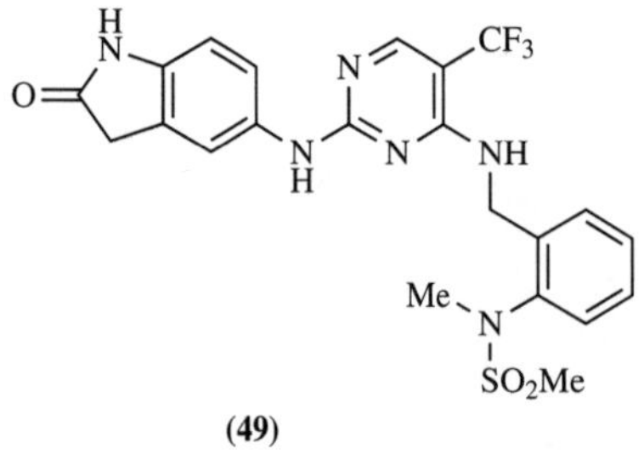

H N O N N CF_3 N H NH Me N SO_2Me N O N N CF_3 N NH Me N SO_2Me

(49) 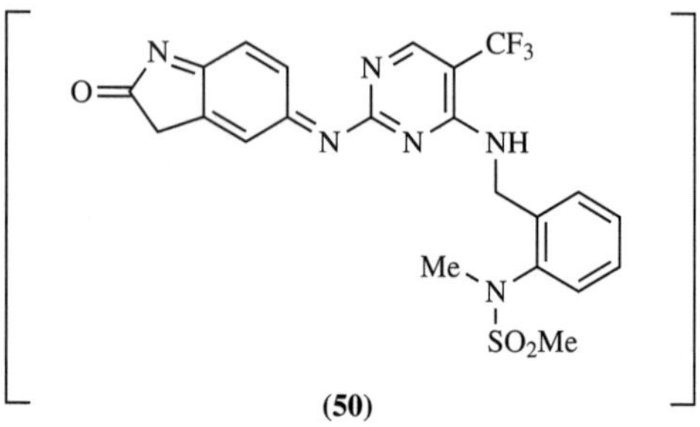 (50)

Figure 15. Masked anilines in fused lactam systems.

the free aniline is very slow compared to the acyclic case (**48**). However, oxidation *para* to the anilinic nitrogen is still possible and has been observed recently in the case of the tyrosine kinase inhibitor (**49**) leading to *in vitro* trapping of the putative quinoid reactive metabolite (**50**) (Fig. 15) [33].

Knowledge of the main metabolic pathways of anilines and the structural features that render particular classes of anilines susceptible to these pathways should help to manage the risk associated with medicinal chemistry target design involving anilines. Anilines can be categorized as high or low risk dependent upon their structural and electronic environment. An example of such a classification for two exemplar aniline motifs in differing environments is given for (**51**) and (**39**) (Fig. 16).

There are a number of common isosteres that can be used to replace an aniline to mitigate the risk of reactive metabolite formation. Common isosteres for *N*-acetyl aniline (**48**) include the indazole (**52**) sulfonamide (**53**), lactam (**39**), and 2-aminopyridine (**54**) all of which maintain a hydrogen-bond donor motif. The heterocyclic isosteres (**44**) and (**45**) maintain some of the lipophilic and electronic characteristics of the anilinic nitrogen but lose the H-bond donor functionality in *N*-acetyl aniline (**48**) (Fig. 17).

High risk aniline

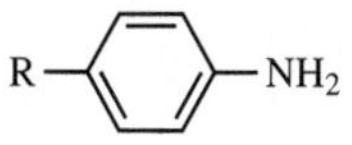

R–C₆H₄–NH₂ replaced: R—(benzene)—NH_2

(51)

All oxidative metabolic pathways are possible
Evidence for all pathways exists in similar molecules
No metabolism to the free aniline is required
Hemoglobin and DNA adducts characterized in human tissue
Multiple examples of *in vivo* adverse outcomes

Low risk aniline

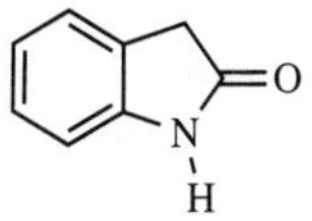

(39)

All oxidative metabolic pathways possible
Lactam ring opening slow or not observed
N-oxidation slow or not observed
Metabolic oxidation in the 5- and 7-positions is observed
Few examples of *in vivo* adverse outcomes

Figure 16. Exemplar high-risk and low-risk anilines.

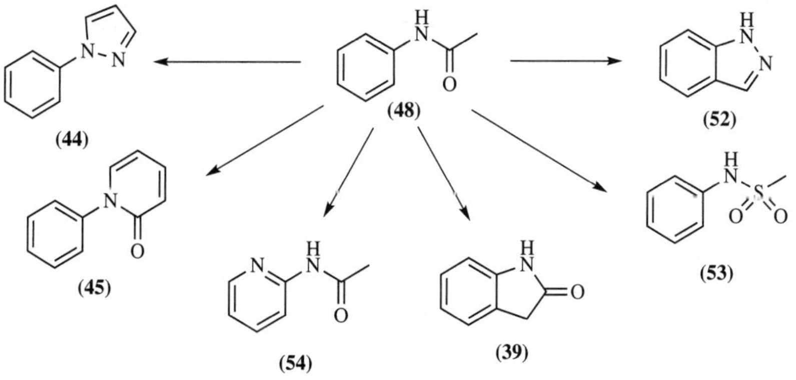

(44) **(45)** **(48)** **(52)** **(53)** **(54)** **(39)**

Figure 17. Aniline isosteric replacements.

3.2. Nitro-Aromatics and Nitro-Alkyls

The well-precedented toxicity associated with aromatic nitro groups [34] is closely associated with the reactive metabolites generated in the pathway of *N*-oxidation described above for anilines (see Section 3.1.2 and Fig. 10). The nitro group can be reduced to the same nitroso species (**26**) that is formed by the oxidation of the hydroxylamine (**25**) and the fully reduced aniline (**55**) (Fig. 18). The nitroso and hydroxylamine intermediates (**26** and **25**) can form adducts with endogenous biomolecules through the same pathways as described above in the aniline structural alert (see Section 3.1.2 and Fig. 10). The nitro-reductase activity is carried out by a number of enzymes including xanthine oxidase, aldehyde oxidase, and CYP450s [35]. Significant reduction of nitro species also occurs in the gut by anaerobic bacteria [36].

The antibacterial agent chloramphenicol (**56**) causes aplastic anemia in humans and one hypothesis is that this adverse reaction is driven by the generation of the reactive metabolite nitroso-chloramphenicol (**57**). In this case, chloramphenicol is reduced to the aniline (**58**) by intestinal bacteria, which subsequently undergoes metabolic aniline *N*-oxidation as discussed in Section 3.1.2 (Fig. 18) [37,38]. A study on the nitro-aromatic anti-Parkinson's agent tolcapone (**59**) suggests that toxicity is due to reduction of the nitro group to the aniline (**60**) that readily forms the iminoquinone reactive metabolite (**61**) (Fig. 18) [39].

As well as nitro-aromatics, simple aliphatic secondary nitroalkanes have been associated with a higher incidence of adverse toxicological outcomes, in particular carcinogenesis. For example, 2-nitro-propane (**62**) is hepatotoxic in rats and rabbits and carcinogenic in rats [40]. In male Sprague-Dawley rats it induces modifications to hepatic DNA and RNA bases [41,42]. Several analogous secondary nitro-alkanes (for example, 3-nitro-pentane) also induce genotoxicity. This genotoxic outcome is catalyzed by a sulfotransferase enzyme and a postulated mechanism involves liberation of the highly reactive aminating agent hydroxylamine-*O*-sulfonic acid (**65**) from the sulfated intermediate (**63**) and (**64**) (Fig. 19) [43,44]. This mechanism is disfavoured for primary and tertiary nitro-alkanes consistent with the lower incidence of adverse outcomes reported for these classes [42–44]. However, more data is needed to validate this hypothesis.

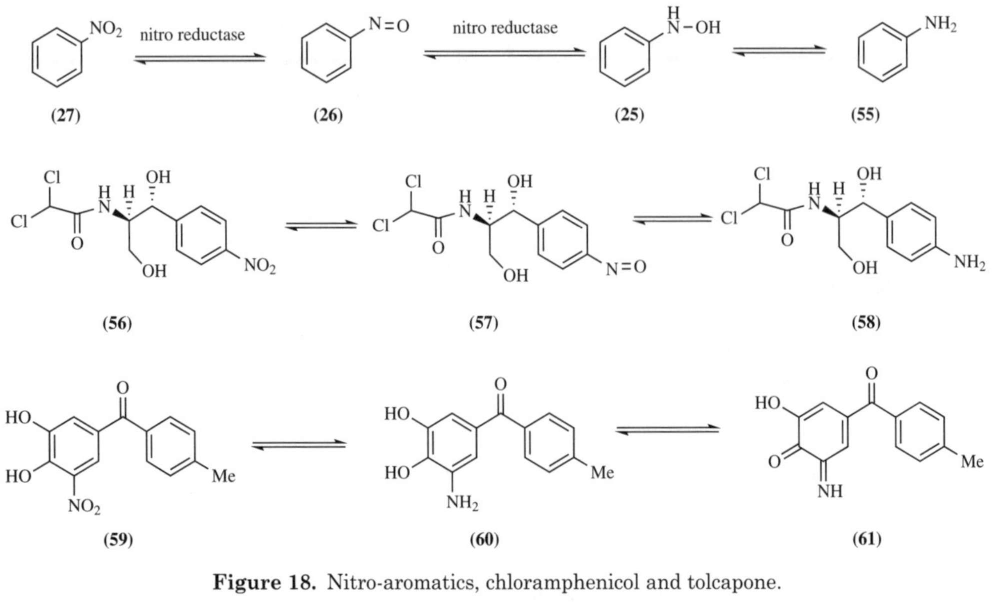

NO_2 nitro reductase N=O nitro reductase H N-OH NH_2

(27) (26) (25) (55)

(56) (57) (58)

(59) (60) (61)

Figure 18. Nitro-aromatics, chloramphenicol and tolcapone.

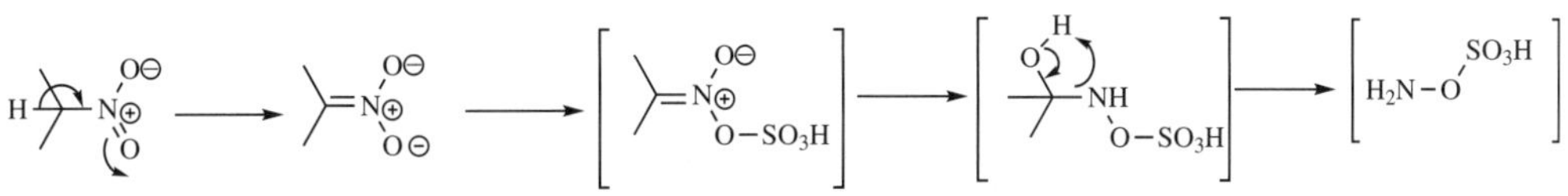

(62) (63) (64) (65)

Figure 19. Secondary nitro-alkanes as structural alerts.

3.3. Arylhydroxamic Acids

Hydroxamic acids have been of significant interest to the medicinal chemistry community as a result of their propensity to chelate metal ions, particularly zinc, and their consequent application as inhibitors of zinc-dependent proteases, for example, the matrix metalloproteinases. There are multiple examples of attrition in the arylhydroxamic acid class due to musculoskeletal syndrome (MSS) [45]; however, it is likely that this common adverse event is driven by the secondary pharmacology of the compounds rather than inherent features of the hydroxamic acid chemical structure. It is clear, however, that arylhydroxamic acids (**66**) have a significantly increased risk of mutagenic outcomes through the generation of isocyanate reactive intermediates (**68**) generated *via* Lossen rearrangement of the activated intermediate (**67**) (Fig. 20) [46,47]. The isocyanate reactive intermediates are highly electrophilic and readily trapped by endogenous nucleophiles to generate, for example, DNA adducts. A comparative study of mutagenic potential has been published with naphthalene-2-hydroxamic acid (**69**) being the most potent mutagen of the compounds tested [48]. *N*-Methylation of arylhydroxamic acids, for example, in compound (**70**) would be predicted to reduce the risk of Lossen rearrangement and reactive metabolite formation *via* this route. Similarly, it would be expected that alkyl hydroxamic acids would be less susceptible to reactive metabolite formation *via* mechanisms involving the Lossen rearrangement owing to the lower migratory propensity for alkyl groups in this mechanistic pathway. Instances of *in vitro* genotoxic and *in vivo* reproductive toxicity outcomes for alkyl hydroxamic acids may be linked to their metal chelating effects [49], which can lead to artifactual false-positive results in genotoxicity *in vitro* screens [50] by virtue of metal sequestration.

3.4. *Ortho*- or *Para*-Dihydroxyaromatics

Ortho- or *para*-dihydroxyaromatic species (**71** and **72**, respectively) are readily susceptible to oxidative formation of the corresponding *ortho*- or *para*-quinones (**73** and **74**). These reactive intermediates behave as electrophiles in a similar manner to the iminoquinones generated *via* metabolic oxidation of *ortho*- and *para*-hydroxyanilines (see Section 3.1.1). In the presence of endogenous nucleophiles, adducts (**75**) and (**76**) are formed.

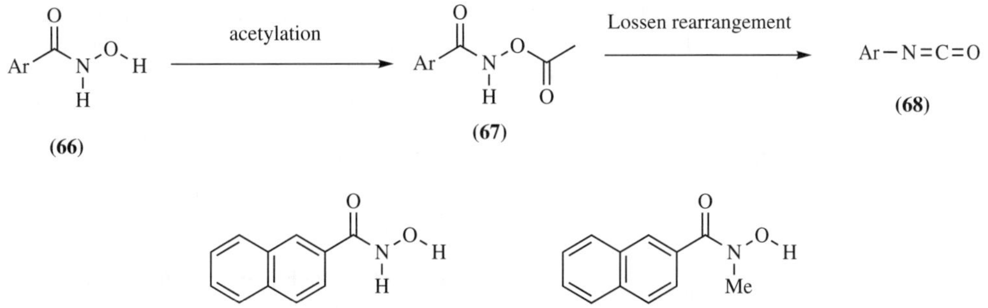

acetylation Lossen rearrangement Ar−N=C=O

(66) (67) (68) (69) (70)

Figure 20. Arylhydroxamic acid metabolism *via* the Lossen rearrangement.

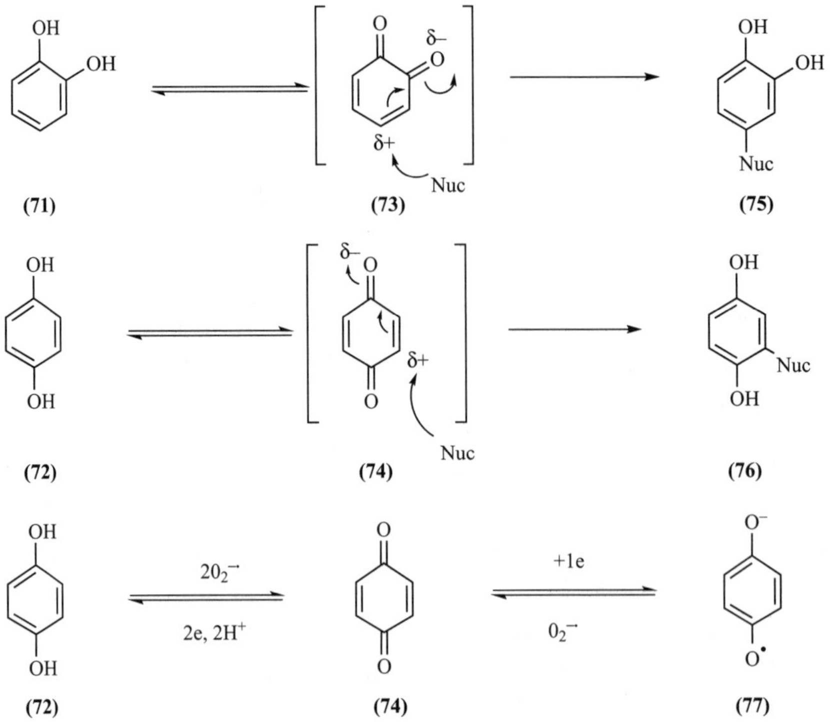

Figure 21. *Ortho* -or *para*-dihydroxyaromatics.

Alternatively, shuttling of electrons through the semiquinone species (**77**) provides a vehicle for the generation of toxic redox cycling in, for example, mitochondrial cells (Fig. 21). In addition to the catalysis of redox cycling, this pathway can also drive the formation of reactive oxygen species including superoxide, hydrogen peroxide and hydroxy radicals [51]. An example of this structural alert is hydroquinone (**72**) that has been shown to form well-characterized DNA adducts and exhibits cytotoxicity in HL-60 cells [52]. Compounds that can readily be metabolized to species (**71**) and (**72**) should also be considered as precursors to this structural alert. Examples include the tetracyclin antibiotic minocycline where oxidative metabolic generation of the *para*-quinone has been observed *in vitro* [53].

3.5. Benzodioxolanes

The benzodioxolane or methylenedioxyphenyl ring system (**78**) is commonly used in target design as an isostere for *ortho*-dimethoxy aryl ring systems. However, the methylenedioxy ring is susceptible to CYP450-mediated oxidation on the activated methylene resulting in formation of the oxidized species (**79**), the highly reactive carbene intermediate (**80**) and/or the *ortho*-dihydroxyaromatic species (**71**), an *ortho*-quinone precursor as described in Section 3.4 [54] (Fig. 22). The antimuscarinic compound zamifenacin (**81**) is an example where the formation of reactive metabolites *via* this oxidative pathway has been postulated *in vitro* [55]. Darifenacin (**82**) is a dihydrobenzofuran isostere that avoids reactive metabolite formation *via* this pathway [56] (Fig. 22). Mechanism-based inhibition of the CYP2D6 enzyme by *in situ* generated reactive intermediates of the benzodioxolane ring has been implicated in the metabolic pathway of the serotonin reuptake inhibitor paroxetine (**83**) [57].

3.6. Electron-Rich Heterocycles: Furan, Thiophene, and Pyrrole

Furan, thiophene, and pyrrole are electron-rich five-membered heterocycles widely used

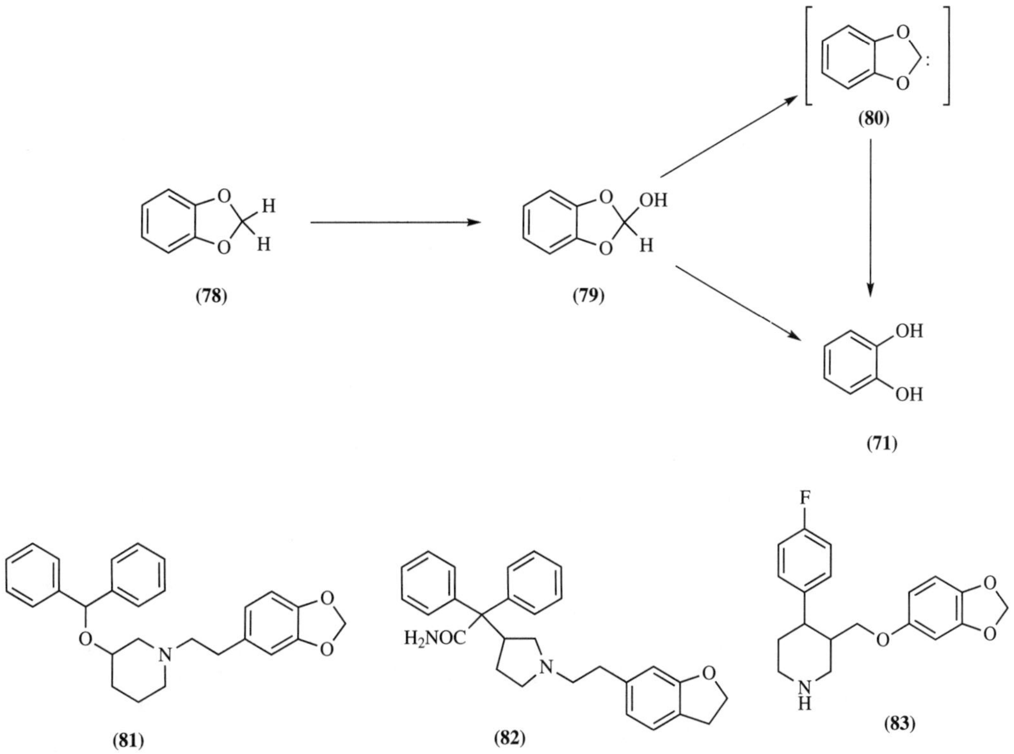

(78) (79) (80) (71)

(81) (82) (83)

Figure 22. Benzodioxolane metabolism.

as phenyl isosteres but particularly prone to oxidative metabolism. In the case of furan (**84**), epoxidation of an electron-rich 2,3-double bond produces an intermediate epoxide (**85**) that is rarely characterized. Epoxide ring opening leads to the β-keto aldehyde reactive intermediate (**86**) that can be trapped by endogenous biomolecules. Alternatively, the epoxide can form the lactone (**87**) and drive subsequent adduct formation (Fig. 23) [58]. The epoxide (**85**) is a common intermediate in all the oxidative pathways of furan metabolism; therefore, the detection of (**85**) or downstream products thereof is a signature of reactive metabolite formation. The diuretic agent furosemide (**88**) is an interesting example where a number of reactive species on the metabolic pathway have been characterized *in vivo* in rat and mouse; these may account for the low but significant incidence of human hepatic toxicity [58].

Thiophenes (**89**) undergo a similar metabolic epoxidation of the 2,3-double bond leading to the reactive epoxide (**90**); this is followed either by epoxide ring opening to give the γ-thiono-enal (**91**) or by formation of the thiolactone (**92**) both of which can also form adducts. In the case of thiophenes, there is a second possible oxidative pathway, namely, direct oxidation of the sulfur heteroatom to form the reactive intermediate (**93**) where the aromatic ring is now susceptible to nucleophilic attack by endogenous biomolecules [59,60]. Oxidation of the thiophene sulfur to the sulfenyl chloride (**94**) by HOCl generated in neutrophils similarly activates the thiophene ring to nucleophilic attack, a pathway precedented for the platelet aggregation inhibitor ticlodipine (**95**) [61] (Fig. 24). Other examples of reactive metabolite generation through thiophene oxidation include the diuretic tienilic acid (**96**) that forms the *S*-oxide *in vitro* [62], the nonsteroidal anti-inflammatory suprofen (**97**) where evidence for epoxide formation has been observed *in vitro* [63] and the anti-inflammatory compound tenidap (**98**)

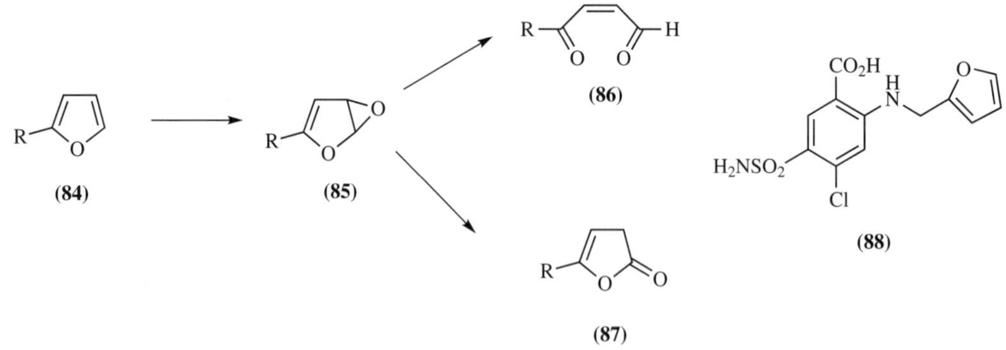

Figure 23. Furan metabolism.

where oxidation of the thiophene ring has been observed *in vivo* in the rat [64] (Fig. 24).

Pyrrole rings are susceptible to oxidation of the 2,3-double bond as illustrated above for furan and thiophene; however, this pathway has less precedent for pyrrole ring systems. Pyrrole ring nitrogen oxidation is not precedented. Examples where generation of the 2,3-epoxide reactive intermediate has been implicated in adverse safety/toxicity outcomes include the nonsteroidal anti-inflammatory agents zomepirac (**99**) and tolmetin (**100**) (Fig. 25), where generation of adducts derived from pyrrole oxidation has been reported in human liver microsomal preparations [65]. However, the evidence for significant adduct formation and associated downstream adverse events from pyrroles is less compelling than for thiophene and furan. In the case of zomepirac and tolmetin, for example, the primary route of metabolism is *via* glucuronidation of the carboxylic acid that can also lead to reactive metabolite formation (see Section 3.15).

The risk of oxidative metabolism of electron-rich five-membered heterocycles can be reduced by substitution of the ring; in particular, substitution with electron withdrawing or bulky substituents. The presence of other

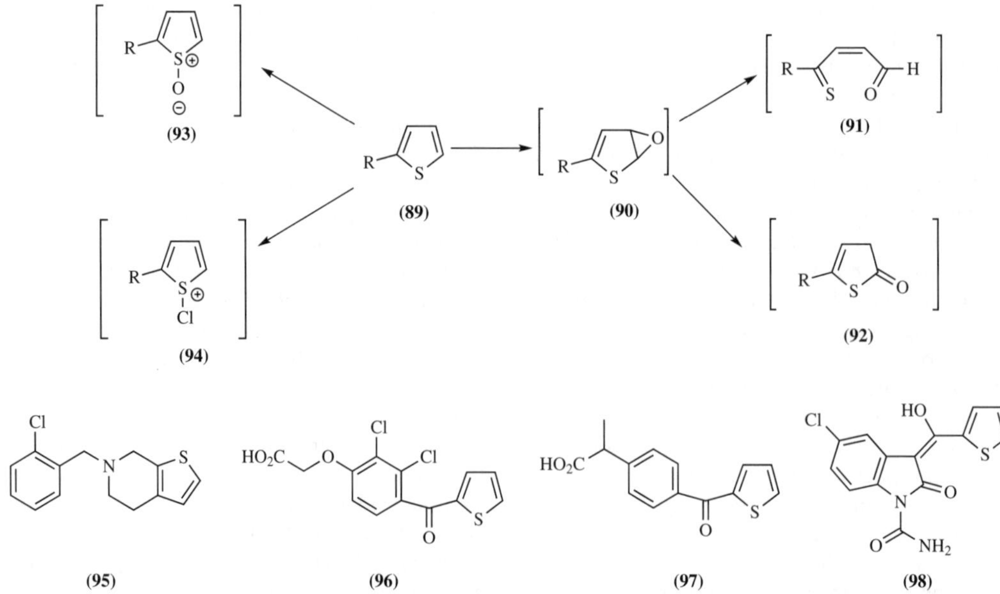

Figure 24. Thiophene metabolism.

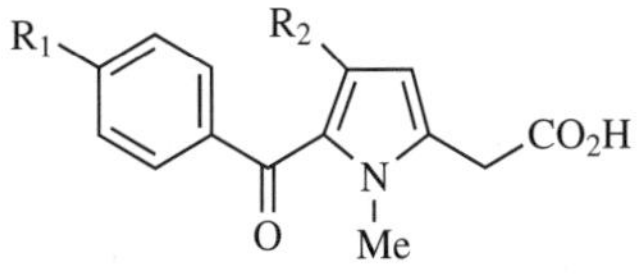

(99) Zomepirac: R_1 = Cl, R_2 = Me
(100) Tolmetin: R_1 = Me, R_2 = H

Figure 25. Pyrrole metabolism.

preferential sites of metabolism may also reduce metabolism of the five-membered ring. For example, in the case of the HIV protease inhibitor L-754394 (**101**), it has been shown that attachment of benzofuran through the 2-position in the analog L-756423 (**102**) results in reduction of the furan-associated animal toxicity observed for L-754394 [66] (Fig. 26). Similarly, the fully substituted pyrrole moiety in the lipid-lowering agent lipitor (**103**) is a protective feature against oxidative metabolism of the pyrrole ring [67] (Fig. 26).

Fused electron-rich five-membered heterocycles are also potential substrates for epoxidation for the 2,3-double bond leading to reactive metabolite formation, however, the incidence of adverse *in vivo* outcomes associated with this potential metabolic pathway is lower than that of the isolated five-membered heterocycles. The risks in these fused electron-rich five-membered heterocyclic systems can be difficult to assess and identification of the metabolic pathways in molecules containing these ring systems is recommended to build an understanding of the risks of reactive metabolite formation.

3.7. 3-Methylindoles

Although metabolic 2,3-oxidation of the indole ring system can be observed, the incidence of links to *in vivo* adverse outcomes is relatively rare and the presence of the indole ring system in endogenous amino acids, for example, tryptophan, and the aminergic agonists, for example, serotonin, reduces the

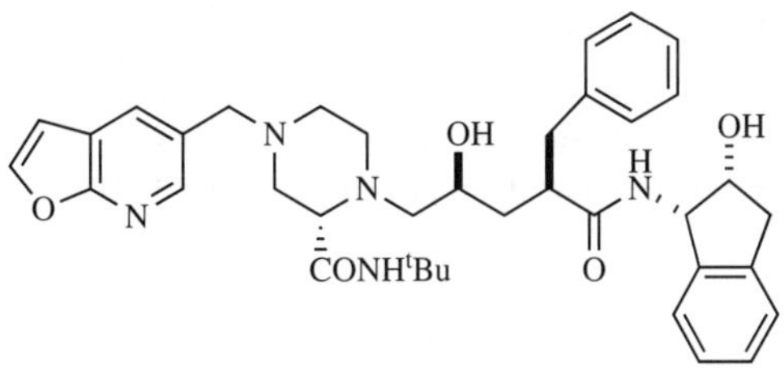

(**101**)

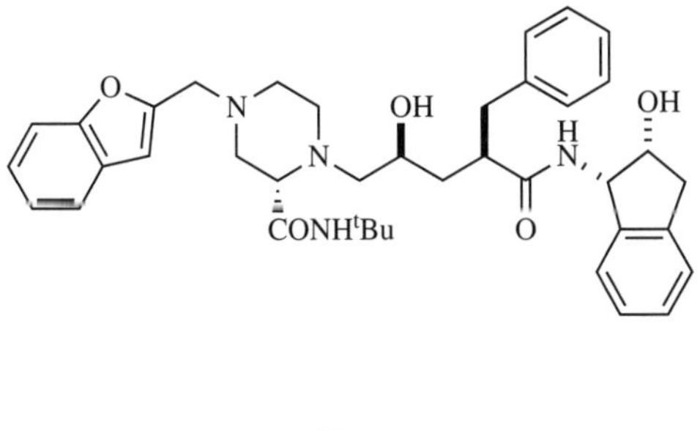

(**102**)

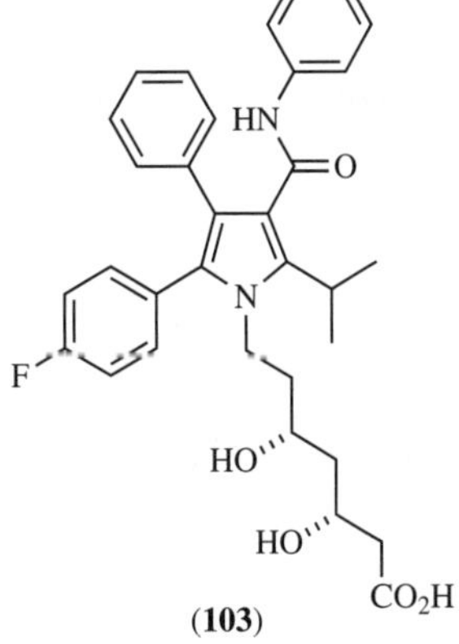

(**103**)

Figure 26. Steric hindrance of oxidation in five-membered ring heterocycles.

potential risk associated with this ring system. However, there is one specific indole ring system that has a precedented higher risk of adverse outcomes, namely, 3-methylindole (**104**). This compound is an animal pneumotoxin although sensitivity varies across species. 3-Methylindole has been shown to form adducts with glutathione, proteins, and DNA using *in vitro* preparations [68]. Oxidation of the 3-methyl group occurs either directly (**106**) or *via* epoxidation of the 2,3-double bond leading to the reactive intermediate (**105**) that can be trapped by endogenous nucleophiles (Fig. 27). The presence of a leaving group on the C3-methyl increases the likelihood of formation of electrophilic reactive intermediates (**107**). The observation of *in vitro* metabolic activation of the 3-benzylindole moiety in zafirlukast (**108**) to give the glutathione adduct (**109**) is an indication that the 3-methyl-indole activation pathway applies to other activated 3-alkyl indoles [69] (Fig 27).

3.8. Thioamides, Thioureas, and Thiocarbamates

The relatively facile metabolic oxidation of sulfur is a key step in the generation of reactive intermediates from thioamides (**110**), thioureas (**111**), and thiocarbamates (**112**) (Fig. 28). This common pathway generates electrophilic iminium ions (**113**) and (**114**) from thioamides (**110**) and thioureas (**111**), respectively. In the case of thiocarbamates (**112**) oxidation at sulfur generates the reactive intermediate (**115**) where nucleophilic attack and subsequent loss of the oxidized sulfur atom may lead to adduct formation.

There are multiple examples of compounds with documented adverse outcomes where this oxidation pathway and the associated downstream reactive intermediates have been observed. The classic example is the prototype H_2 antagonist metiamide (**116**) that caused human blood dyscrasias [70]. Replacement of the thiourea with the isosteric cyanoguanidine led to cimetidine (**117**) a very successful therapy [71]. The thiourea analog (**118**) undergoes metabolic oxidation at sulfur to form a reactive metabolite and glutathione conjugate [72]. Many simple thioureas have also demonstrated genotoxicity *in vitro* and *in vivo*. For example, thiourea itself (**111**, R = H) is tumorigenic in rats following oral administration [73] and ethylenethiourea (**119**) is classed as a human carcinogen and teratogen [74]. Tolrestat (**120**), an aldose reductase

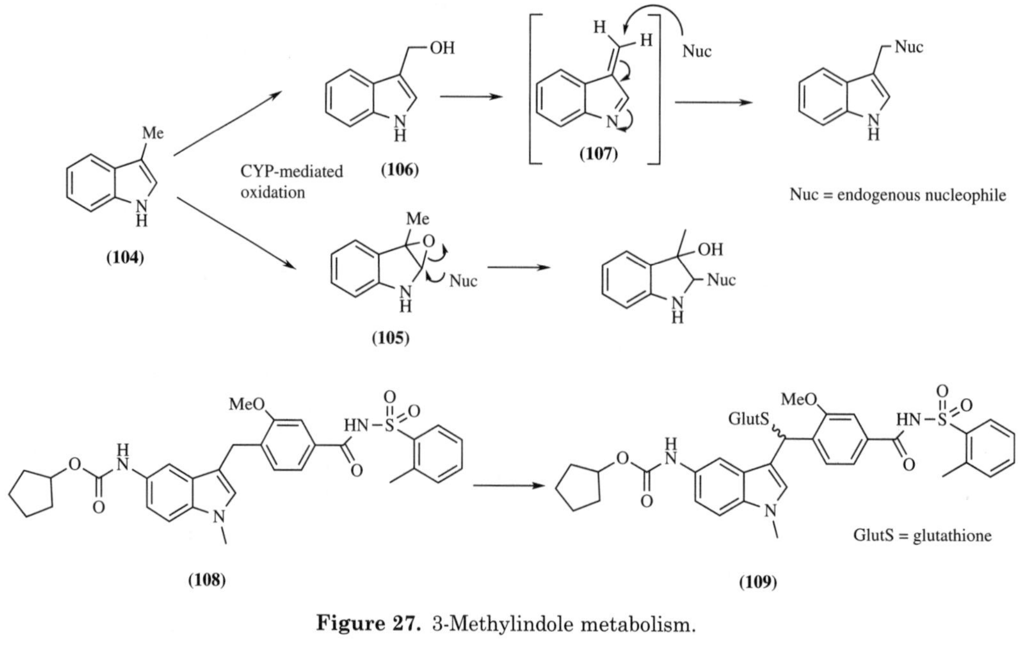

Me
(104)
CYP-mediated oxidation
OH
(106)
H H Nuc
(107)
Nuc
Nuc = endogenous nucleophile
Me O Nuc
(105)
OH Nuc
MeO HN-S O O
(108)
GlutS MeO HN-S O O
(109)
GlutS = glutathione

Figure 27. 3-Methylindole metabolism.

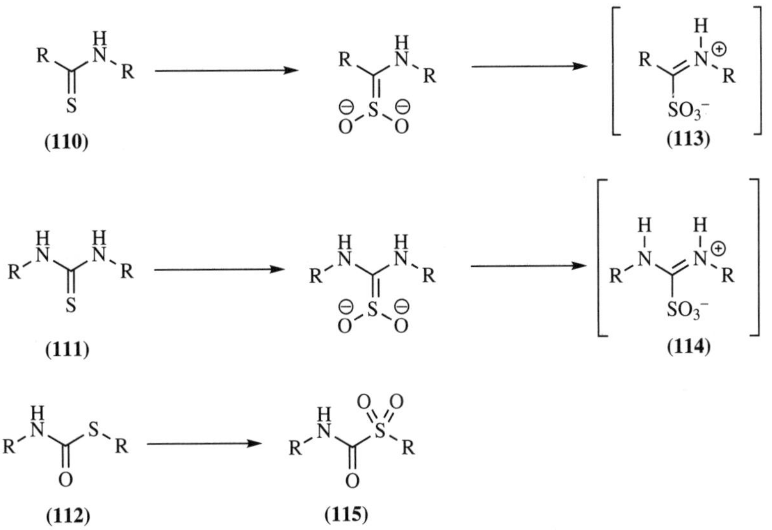

(110) (113) (111) (114) (112) (115)

Figure 28. Thioamide, thiourea, and thiocarbamate oxidative pathways.

inhibitor marketed for diabetic nephropathy, is an exemplar thioamide where oxidative metabolism on sulfur has been implicated in the severe human liver toxicity that led to its withdrawal from the market [75,76]. Some herbicidal agents are simple thiocarbamates, for example, molinate (**121**) has been found to produce male reproductive toxicity in rats through oxidation on sulfur and covalent binding to testicular carboxylesterases [77] (Fig. 29).

3.9. 2-Aminothiazoles

2-Aminothiazoles are a common motif present in commercially sourced screening libraries and an understanding of their metabolic fate and propensity for reactive metabolite formation is important in hit triage and hit follow-up tactics. Oxidative metabolism of 2-aminothiazoles (**122**) occurs predominantly at the electrophilic 4,5-double bond and results in the formation of the α-dicarbonyl metabolites (**123**) and thiourea derivatives (**124**) that can

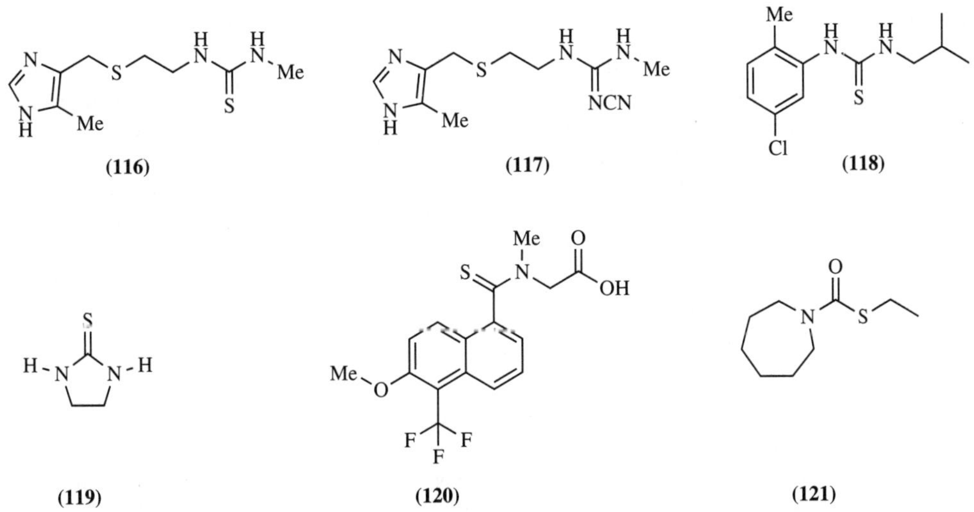

(116) (117) (118) (119) (120) (121)

Figure 29. Examples of thioamides, thioureas, and thiocarbamates.

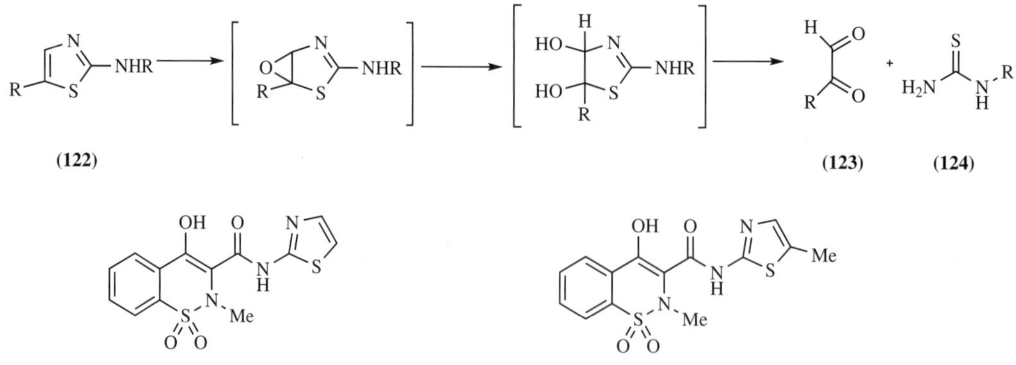

(122) (123) (124)

(125) (126)

Figure 30. 2-Aminothiazole metabolism.

be further metabolized to reactive intermediates (see Section 3.8) [78] (Fig. 30). Sudoxicam (**125**) is an example where oxidative ring opening of the aminothiazole has been characterized *in vivo* in rat, dog, and monkey [79]. In common with the five-membered electron-rich heterocyclic class, the presence of substituents at the 4- and 5-position is likely to slow the rate of oxidation and diminish metabolism *via* this pathway. For example, incorporation of a methyl group at the 5-position in the close analog meloxicam (**126**) switches metabolism to oxidation of the methyl group and little or no oxidative ring opening is observed [80,81] (Fig. 30).

3.10. Alkynes

Alkynes have been used as a rigid, vector-specific isostere for a phenyl ring. However, CYP450-mediated oxidation of alkynes (**127**) is common *via* the formation of the heme-catalyzed oxidation product (**128**) that subsequently generates reactive intermediates (**129**) and (**130**) (Fig. 31). Involvement of the active site of the CYP450 enzyme in the generation of alkyne reactive metabolites often results in the formation of CYP450 adducts through direct *in situ* metabolite trapping by amino acid side chains in the active site. Thus, testing for CYP450 inhibition is a particularly worthwhile screen for compounds containing an alkyne functionality. For example, the oral contraceptive 17α-ethynylestradiol (**131**) is a mechanism-based inhibitor of CYP3A4 *in vitro* [82]. However, the administration of 17α-ethynylestradiol to women has no impact on CYP3A4 activity; the low clinical dose is the most likely reason for the absence of drug–drug interactions [83]. Alkyne metabolism can also lead to adduct formation outside the active site of the CYP metabolizing enzyme. The HIV reverse transcriptase inhibitor efavirenz (**132**) is extensively metabolized in rat, monkey, and human; however, the metabolic fate of the compound is not the same in all species. In rats, a glutathione adduct of the alkyne was observed. However, this adduct was not detected in monkeys where nephrotoxicity is not observed [84] (Fig. 31).

3.11. Hydrazines

Hydrazines (**133**) are a well-characterized high-risk structural alert with multiple examples linking the presence of a hydrazine motif to an adverse safety or toxicity outcome. The primary route of metabolism of hydrazines is oxidation to the diazines (**134**) that rapidly fragment with loss of nitrogen to yield highly reactive radical species (Fig. 32). The antituberculosis agent iproniazid (**135**) is a good example of reactive metabolite formation *via* oxidative metabolism of a hydrazine. Formation of isopropyl radicals is implicated in the severe hepatotoxicity observed clinically with this compound [85]. Hydralazine (**136**) also forms covalent adducts, most likely *via* formation of free radical intermediate (**137**) [86]. Clinical use of hydralazine has resulted in an

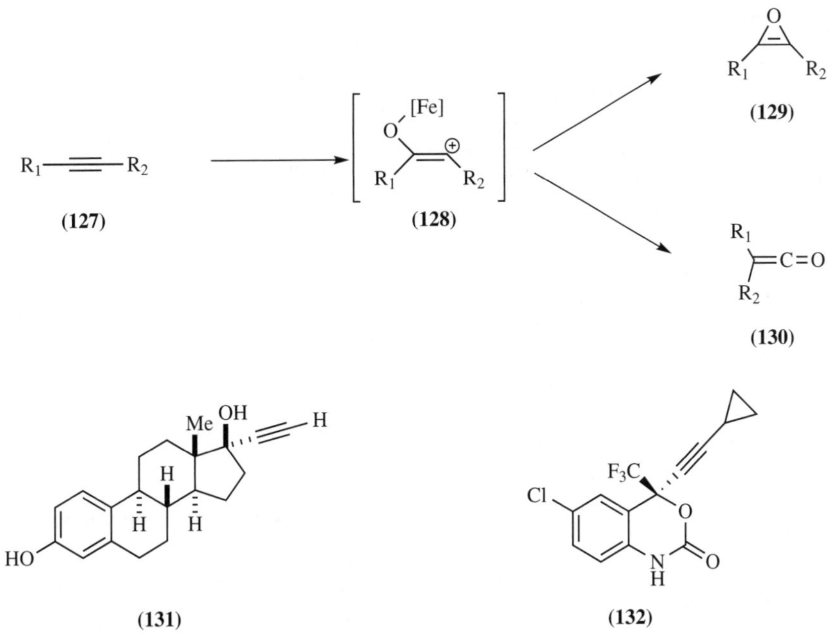

R_1—≡—R_2 (127) → [O-[Fe], ⊕, R_1, R_2] (128) → O, R_1, R_2 (129); R_1, R_2, =C=O (130)

Me, OH, H, H, H, H, HO (131)

F_3C, Cl, O, O, N, H (132)

Figure 31. Alkyne metabolism.

immunological syndrome similar to systemic lupus erythematosus (SLE). Hydralazine is polymorphically acetylated in man, and the development of the SLE-like syndrome occurs almost exclusively in slow acetylators that produce lesser amounts of acetylated metabolites and larger amounts of oxidized metabolites [87] (Fig. 32).

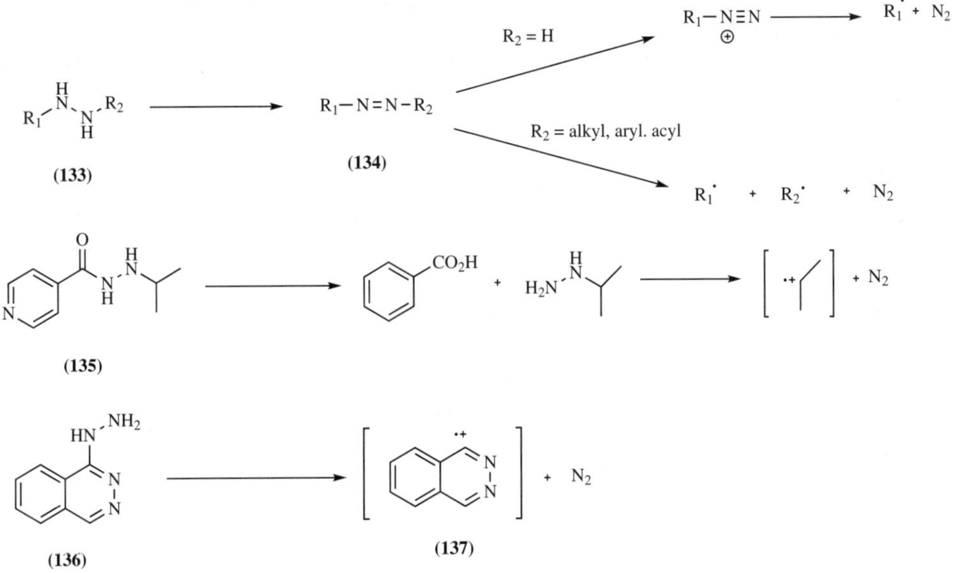

R_1-NH-NH-R_2 (133) → R_1−N=N−R_2 (134)

R_2 = H → R_1−N≡N ⊕ → $R_1^{\cdot}$ + N_2

R_2 = alkyl, aryl. acyl → $R_1^{\cdot}$ + $R_2^{\cdot}$ + N_2

O, N, H, N, H, N (135) → CO_2H + H_2N, H, N → [•+] + N_2

HN, NH_2, N, N (136) → [•+ N, N] (137) + N_2

Figure 32. Hydrazine metabolism: iproniazid and hydralazine.

3.12. Hydantoins

A hydantoin-specific oxidative metabolic pathway has been proposed to account for the human hypersensitivity responses observed with the anticonvulsant agents, phenytoin (**138**) and mephenytoin (**139**), and with the aldose reductase inhibitor sorbinil (**140**). The proposed mechanism involves hydroperoxide-mediated imide radical formation (**142**) from the parent hydantoin (**141**) and *in vitro* evidence for catalysis of the transformation by prostaglandin H synthase (cyclooxygenase) has been published [88]. Methyl hydantoins may undergo metabolic demethylation before generation of the iminium radical species (Fig. 33). Phenytoin has been shown to bind covalently to proteins *in vivo* in the mouse [89].

3.13. Thiazolidinediones

Thiazolidinedione-specific pathways of metabolism have been proposed to account for the hepatotoxicity and CYP3A4 enzyme induction observed with the withdrawn antidiabetes agent troglitazone (rezulin) (**143**) [90]. One proposed pathway involves CYP450-mediated oxidation of the activated methylene to give a quinone-like reactive intermediate (**145**) that can be trapped by endogenous biomolecules (Fig. 34). This pathway of metabolism is possible in thiazolidinedione substructures (**144**) bearing a C5-methylene substituent and a group on the pendant phenyl ring capable of stabilizing the quinone intermediate. Another proposed thiazolidinedione-specific pathway involves oxidation on the sulfur atom and ring opening to generate a reactive isocyanate and sulfenic acid (**146**) [91,92]. Rosiglitazone (**147**) has the same structural alert, but the incidence of adverse outcomes is significantly lower and not drug related. The lower clinical dose of rosiglitazone (4 mg/day) *versus* troglitazone (200 mg/day) is a likely reason for the difference in outcome. Endogenous protective mechanisms are much more likely to be able to cope with the significantly lower exposures of rosiglitazone reactive metabolites (see Section 6.1).

3.14. *N*-Alkyl-Tetrahydropyridines

1-Methyl-4-phenyl-1,2,3,6-tetrahydropyridine (MPTP) (**148**) (Fig. 35) is linked to the development of a syndrome similar to Parkinson's disease. A major metabolite of MPTP namely, MPP+ (**149**) is formed *via* monoamine oxidase B (MAO-B) mediated oxidation. MPP+ causes neuronal degeneration *via* inhibition of neuronal mitochondrial function [93]. Therefore, any compounds that could be expected to form MPP+ upon metabolism should be considered as potential structural alerts. Haloperidol (**150**) is an example where the neurotoxic pyridinium metabolite (**151**) has been detected during postmortem in the brains of patients treated

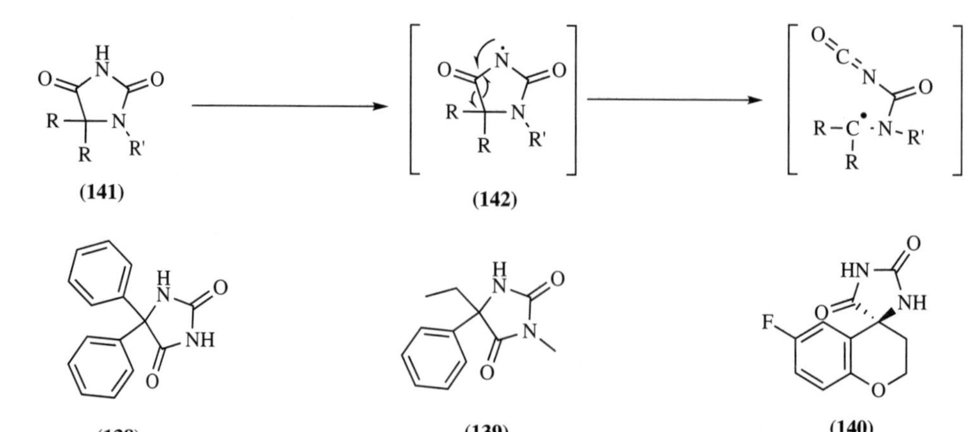

(141) (142) (138) (139) (140)

Figure 33. Hydantoin metabolism.

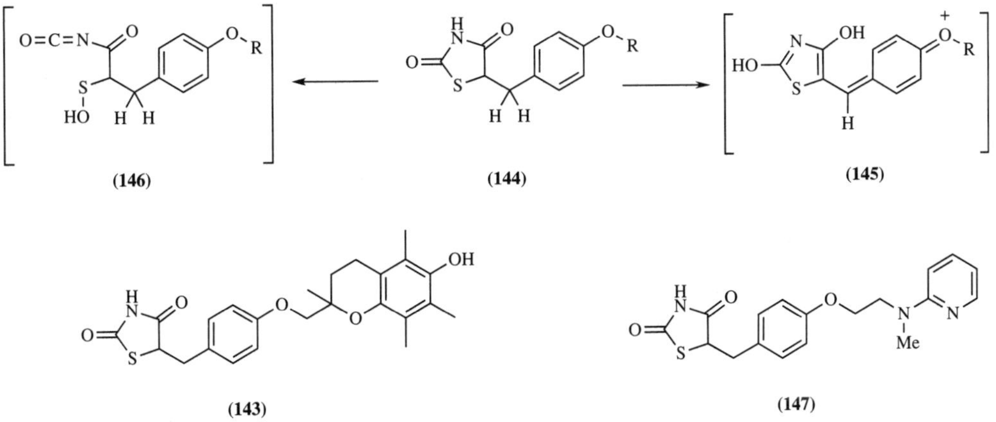

(146) (144) (145)

(143) (147)

Figure 34. Thiazolidinedione metabolism.

chronically with haloperidol. Formation of the pyridinium species (**151**) has been linked to the irreversible Parkinson's disease-like symptoms observed clinically with halopyridol [94].

3.15. Carboxylic Acid Acyl Glucuronides

Carboxylic acids rarely appear among the classical lists of structural alerts; they are prevalent in drug molecules and are an essential component in the armory of the medicinal chemist by virtue of their efficient reversible interaction with basic residues and metals in proteins of therapeutic interest. However, there is a building weight of evidence that the primary route of metabolism of carboxylic acids, namely, the formation of acyl glucuronides (**152**), may lead to reactive metabolites and irreversible covalent modification of proteins [95]. The glucuronyl transferase-mediated formation of acyl glucuronides (**152**) is followed by acid or base catalyzed migration of the acyl group on the sugar ring. Modification of endogenous biomolecules may occur *via* nucleophilic trapping of the initially formed glucuronide (**152**) or *via* Amadori rearrangement of the migrated isomers (**153**) [95–97] (Fig. 36). A correlation of the stability of the initially formed acyl glucuronide (**152**) and the propensity for adduct formation has been proposed [98]. Acyl glucuronides from acetic acids were more unstable than those from propionic acids with acyl

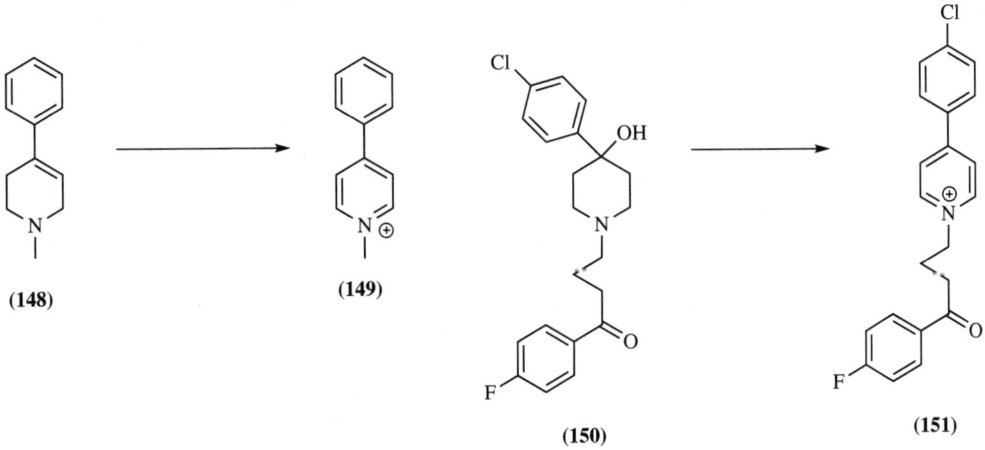

(148) (149) (150) (151)

Figure 35. *N*-Alkyl-tetrahydropyridines.

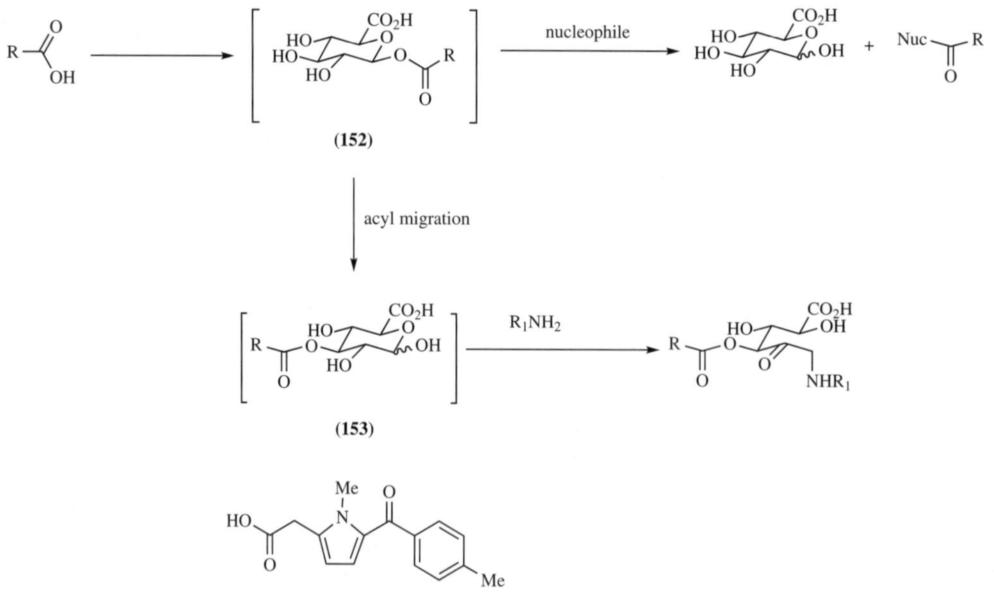

Figure 36. Carboxylic acid acyl glucuronides.

glucuronides from benzoic acids being most stable. It is notable that the most unstable glucuronide derived from tolmetin (**154**) (half-life = 16 min) irreversibly modifies plasma proteins *in vivo* [99]. A novel intramolecular trapping of a carboxylic acid acyl glucuronide leading to a reactive intermediate has also been reported [100].

4. STRUCTURAL ALERTS EXHIBITING TIGHT BINDING TO CYP450 ENZYMES

Potent inhibition of CYP450 enzymes can be a serious development risk for potential drug compounds. For example, potent CYP450 inhibition can lead to abrogation of the metabolism of coadministered medications resulting in higher plasma levels that thereby erode an otherwise safe therapeutic window [101]. Compounds that interact with CYP enzymes may do so irreversibly through covalent modification of the enzyme either directly (see Section 2) or *via in situ* reactive metabolite formation (see Section 3) [102,103]. Many of the structural alerts discussed in Section 3 rely upon oxidative metabolism for the formation of the reactive intermediate. In many cases, oxidative metabolism is performed by CYP450 enzymes and the reactive metabolite is first generated in high local concentration within the active site of the CYP enzyme. Therefore, the first protein exposed to high concentrations of the reactive metabolite is the CYP450 responsible for its generation and irreversible covalent binding to CYP enzymes is a common outcome and reliable signpost for the formation of reactive metabolites. Early testing for the irreversible inhibition of CYP enzymes is a common practice to provide early feedback on the *in situ* formation of reactive metabolites [104].

In addition to the formation of reactive intermediates, there is a subset of nitrogen heterocycles that can bind tightly *via* chelation to the heme iron of CYP450 enzymes and display slow-offset kinetics [105]. Notable examples include 2,6-unsubstituted pyridines (**155**) and imidazoles or triazoles (**156**) bearing a sterically unencumbered ring nitrogen (Fig. 37). For these reasons, binding assays for representative CYP450 enzymes are often included in secondary pharmacology screen-

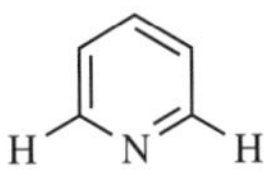

Figure 37. Chelators of the heme iron in CYP450 enzyme active sites.

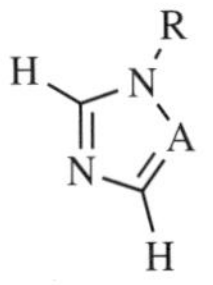

ing panels. In addition, assays to signpost the likely interaction of a compound with CYP substrates are included early in many screen cascades.

5. STRUCTURAL ALERTS WITH A PROPENSITY FOR DNA BINDING

Some moieties have a well-characterized propensity to display reversible tight binding to DNA and therefore have a higher risk of eliciting a genotoxic response [9]. Compounds may intercalate between DNA base pairs and induce a DNA conformational change, or bind to the minor groove of DNA without significant conformational change in DNA. The general class of DNA binding molecules are characterized by polycyclic aromatic compounds that intercalate. For example, 9-aminoacridines (**157**), furocoumarins (**158**), 2-phenylbenzimidazoles (**159**), and ellipticines (**160**) (Fig. 38). Not all intercalators are genotoxic, the presence of a basic center or an additional electrophilic reactive species adds to the risk [106,107].

Aristolochic acid I (**161**) serves as a well-characterized example where the presence of two structural alerts (nitro and benzodioxolane moieties) and a planar polycyclic aromatic structure has been clearly linked to human renal toxicity [108] and carcinogenic outcomes [109]. DNA adducts characterized in the patient population are linked to mutations in the oncogenes *Hras* (rodent) and *p53* (human) [109]. The nitro group of aristolochic acid is reduced to the arylhydroxamic acid (**162**) (see Section 3.2) and is a precursor to the nitrenium ion (**163**). The planar aromatic scaffold is ideally suited to DNA binding thereby facilitating adduct formation [110] (Fig. 39).

6. MANAGING THE RISKS ASSOCIATED WITH STRUCTURAL ALERTS

6.1. The Benefits of Low Exposure

Each of the structural alerts discussed in Sections 2–5 above is associated with its own structure–toxicity relationship and not all molecules containing the toxicophore will elicit an adverse reaction. There are multiple examples of marketed drugs that contain exemplars of structural alerts. However, many of these compounds are characterized by low dose and low exposure to drive the clinical benefit. At these low exposures, the endogenous protective mechanisms (glutathione trapping) are able to remove the reactive metabolites or dampen the oxidative stress resulting from their generation. There are very few clinical compounds with a total daily dose below 50 mg that are associated with a significant incidence of structure derived adverse events. A total daily dose of 10 mg or below is significantly less likely to result in an adverse outcome due to toxicity driven by the formation of a reactive metabolite [111]. The often-cited case of the successful drug lipitor (**103**) [67] is an example where the recom-

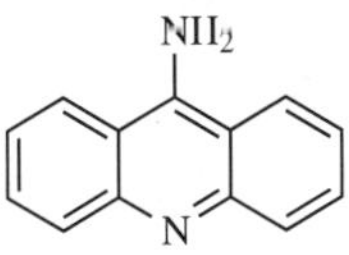

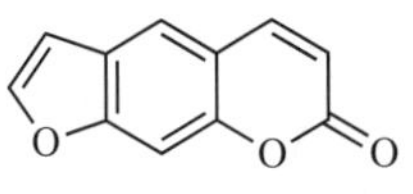

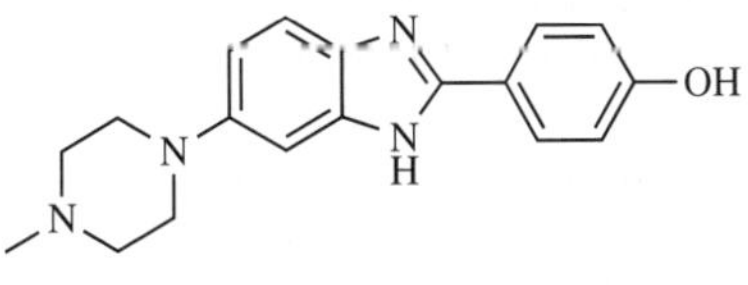

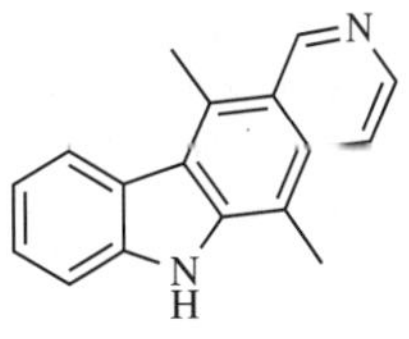

Figure 38. Structural alerts with a propensity for DNA binding.

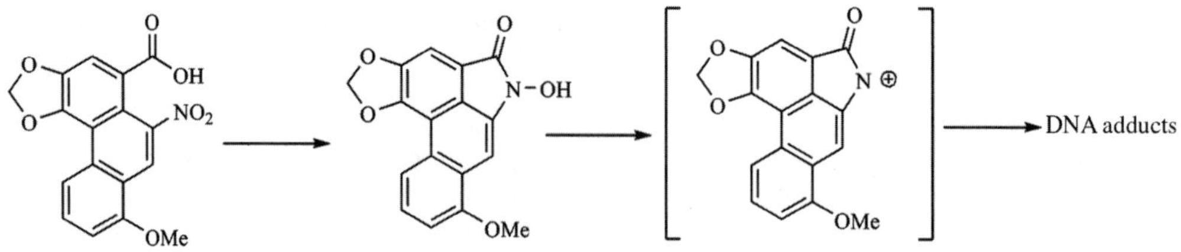

(161) (162) (163)

Figure 39. Aristolochic acid.

mended clinical dose of 10–40 mg once daily is a likely mitigating factor against the presence of the exposed aniline moiety. The comparison of troglitazone (total daily dose 200 mg) and rosiglitazone (total daily dose 4 mg) is another good example where the presence of the same structural alert in both molecules is mitigated by the low efficacious clinical exposure with the latter agent (see Section 3.13). A key issue for medicinal chemists is that accurate knowledge of the predicted dose is rarely available at the stage at which molecules are being designed. Secondly, idiosyncratic reactions as a result of reactive metabolite generation only manifest in large patient populations at the later stages of clinical evaluation when significant clinical development investment has been made. There is, therefore, an economic benefit in avoiding the incorporation of all toxicophores at the design stage; especially as toxicophore chemical space is a tiny fraction of the overall chemical space available to the medicinal chemist in target design.

6.2. Alternative Pathways of Metabolism

The presence of a structural alert may also be mitigated by the availability of an alternative metabolic clearance pathway that predominates *in vivo*. The presence of a functional group that leads to predominant Phase II clearance *via* sulfation or glucuronidation may often provide such an alternative pathway. However, as discussed in Section 3.15, glucuronidative pathways can also lead to reactive metabolite formation such that monitoring of downstream metabolic outcomes of Phase II metabolism is still important. Ketorolac (**164**) contains a pyrrole structural alert, however, Phase II conjugation and elimination predominates over oxidation of the electron-rich five-membered ring [112] (Fig. 40). Talviraline (HBY097) (**165**) is an example where oxidative metabolism of the thioether predominates over metabolism of the thioamide moiety to reactive metabolites [113] (Fig. 40).

6.3. Screening for Reactive Metabolite Formation

Should avoidance of toxicophores be impractical then early screens are available to quantify the risk associated with their presence [114]. A widely used qualitative indicator of the presence or absence of reactive metabolites is the *in vitro* reactive metabolite screen in which compounds are incubated with the endogenous trapping agent glutathione in the presence of a metabolic activation medium [115–117]. This screen does, however, have several drawbacks. For example, the enzymes responsible for activation *in vivo* may not be present in the *in vitro* activation system [118], or the reactive metabolite may abstract a hydrogen atom from the local environment faster than the rate of trapping by glutathione such that very reactive short-lived metabolites are not detected. In addition, the detection of glutathione

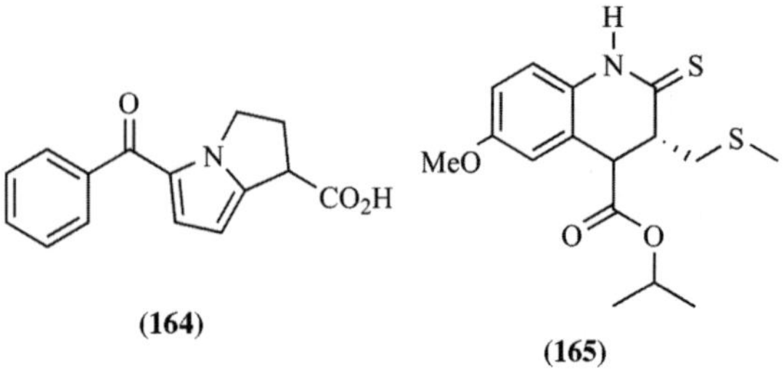

(164) (165)

Figure 40. Alternative pathways of metabolism.

adducts depends upon their stability and the power of the detection system, thus quantitation of the amount of adduct formation and the overall relevance to the *in vivo* setting is difficult [119]. The use of dansyl glutathione [120] and quaternary ammonium glutathione [121] as trapping agents has been described to facilitate a more quantitative estimation of reactive metabolite formation. The use of gene expression profiles as markers of downstream events due to reactive metabolite formation is a developing field [122]. The gene expression profiles produced by the human carcinogens 4-aminobiphenyl and benzo[a]pyrene have been published. Gene expression consistent with generalized stress response, antioxidant, glutathione transferase, and DNA repair pathways were found; however, a statistically significant dose response was not uncovered [123].

The use of radio labeled compound to detect covalent binding to endogenous proteins is an alternative method for detection of reactive metabolite formation; however, the investment required to prepare radio labeled material prevents the use of this method as an early screening tool [124]. Importantly, advanced mass spectral techniques have greatly enhanced the ability to detect adducted proteins; for example, in the case of detection of ^{14}C-naphthalene adducts in the lung [125]. The extent of correlation between covalent binding using *in vitro* human preparations and human *in vivo* hepatotoxicity has been investigated using a set of 18 radiolabeled drugs (9 hepatotoxins and 9 nonhepatotoxins). Use of human liver microsomes [126], human liver S9 preparations, or human hepatocytes [127] did not clearly distinguish hepatotoxins from nonhepatotoxins. Thus, *in vitro* covalent binding in human liver preparations is not a highly correlated predictor of human clinical hepatotoxicity.

6.4. In Silico Filtering of Structural Alerts

There are a number of well-established expert knowledge systems for the *in silico* filtering of structural alerts. These include the DEREK (deductive estimation of risk from existing knowledge) alerting system [128] and the more probabilistic method exemplified by the TOPKAT (toxicity prediction by Komputer-assisted technology) system [129]. Such systems will accurately filter a list of proposed chemical structures in high-throughput mode for the presence or absence of structural alerts. Companies and academic groups have developed proprietary structural alert computational filters that can be used to *in silico* screen large libraries of diverse compounds, for example, prior to purchase [130,131]. However, the power of these filters depends upon the electronic coding of the individual alerts and it is only with a detailed knowledge of the structure–toxicity relationships for each structural alert and expert coding systems that an accurate filter can be produced. In addition, such *in silico* filters are less useful in highlighting compounds where the generation of reactive metabolites is a function of the metabolic route and is not well signposted by the parent chemical structure, although such systems are under development [132]. Felbamate (**166**) is one such example where formation of the reactive metabolite (**167**) is facile despite the lack of recognized structural alerts in the parent structure (**133**) (Fig. 41).

7. CONCLUSION

This chapter highlights those structural alerts where there is a significant weight of evidence that their incorporation into a drug candidate molecule is likely to result in a higher than average incidence of attrition due to adverse safety/toxicity findings. The significant asso-

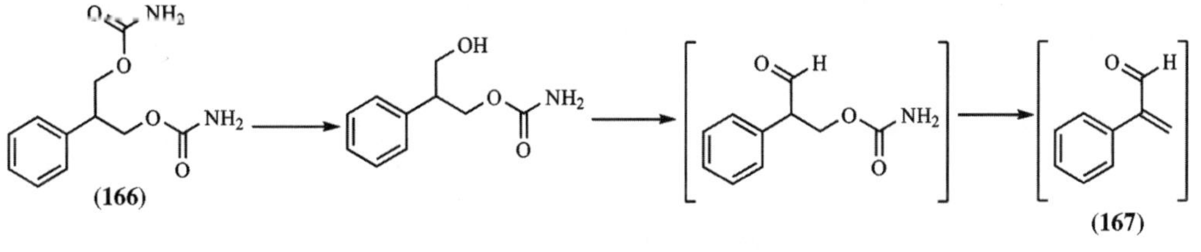

(166) (167)

Figure 41. Felbamate.

ciation between *in vivo* reactive metabolite formation from these structural alerts and idiosyncratic drug reactions in late-stage clinical development is particularly pertinent in the battle to reduce late-stage attrition of potential therapeutics. An understanding of the mechanism of formation of reactive metabolites from each of structural alert is important in understanding the structure–toxicity relationship and the boundary definition for each alert. In addition, knowledge of the common isosteres for each structural alert is important to their avoidance.

Folklore often builds around a particular structural motif based upon circumstantial evidence. This can unnecessarily restrict the diversity of chemical space in which a medicinal chemist can operate. For example, the sulfonamide group has been falsely associated with the high incidence of adverse drug reactions observed with the sulfonamide antibacterial and other sulfonamide containing drugs; the so-called "sulfa allergy." Arguments to refute this association have now been presented [134].

The main mitigating factors that can reduce the risk of incorporating a structural alert into a drug candidate include the presence of an alternative metabolic pathway and low clinical exposure. Neither of these can accurately be predicted to the human clinical situation. Therefore, building confidence that the presence of a structural alert is not going to deliver a significant incidence of an adverse drug reaction is difficult. The use of early *in vitro* or *in vivo* screens to manage the risk associated with a structural alert is fraught with uncertainty and often adds significant cost and time to a drug development program. It is often only in late-stage clinical trials with large and diverse patient populations exposed to the drug that idiosyncratic drug reactions emerge. Therefore, the avoidance of well-defined structural alerts is the lowest risk medicinal chemistry strategy.

8. SUGGESTED READING

- Lee JS, Obach RS. Fisher MB. *Drug Metabolising Enzymes: Cytochrome P450 and other Enzymes in Drug Discovery and Development*. Fontis Media: Switzerland; 2003.
- Smith DA, Van de Waterbeemd H, Walker DK. *Pharmacokinetics and Metabolism in Drug Design*. Wiley-Verlag: Weinheim; 2006.
- Kerns EH, Di L. *Drug-Like Properties: Concepts, Structure Design and Methods: From ADME to Toxicity Optimisation*. Elsevier: Burlington, MA; 2008.
- Kalgutkar AS. Role of bioactivation in idiosyncratic drug toxicity: structure–toxicity relationships. *Biotechnology: Pharmaceutical Aspects*. Springer Verlag, 2008, Vol. 9 (*Advances in Bioactivation Research*) 25–55.
- Antoine DJ, Williams DP, Park BK. Understanding the role of reactive metabolites in drug-induced hepatotoxicity: state of the science. *Exp Opin Drug Met Toxicol* 2008; 4(11), 1415–1427.

REFERENCES

1. Alfirevic A, Park BK, Pirmohamed M. Pharmacogenetics of adverse drug reactions. Pharmacogenetics 2006; 65–89.
2. Uetrecht JP. New concepts in immunology relevant to idiosyncratic drug reactions: the "danger hypothesis" and innate immune system. Chem Res Toxicol 1999;12(5):387–395.
3. Weltzien H, Padovan E. Molecular features of penicillin allergy. J Invest Dermatol 1998;110 (3):203–206.
4. Smith DA, Schmid EF. Drug withdrawals and the lessons within. Curr Opin Drug Discov Devel 2006;9:38–46.
5. Waring JF, Anderson MG. Idiosyncratic toxicity: mechanistic insights gained from analysis of prior compounds. Curr Opin Drug Discov Devel 2005;8(1):59–65.
6. Bull HG, Garcia-Calvo M, Andersson S, Baginsky WF, Chan HK, Ellsworth DE, Miller RR, Stearns RA, Bakshi RK. Mechanism-based inhibition of human steroid 5α-reductase by finasteride: enzyme-catalyzed formation of NADP-dihydrofinasteride, a potent bisubstrate analog inhibitor. J Am Chem Soc 1996;118(10):2359–2365.
7. White W. Alternative Toxicological Methods. In: Salem H, Katz SA, editors. Chapter 41. Baton Rouge: CRC Press; 2003. p. 533.

8. Blagg J. Structure activity relationships for *in vitro* and *in vivo* toxicity. Annu Rep Med Chem 2006;41:353–368.
9. Kazius J, McGuire R, Bursi R. Derivation and validation of toxicophores for mutagenicity prediction. J Med Chem 2005;48(1):312–320.
10. Kalgutkar AS, Soglia JR. Minimising the potential for metabolic activation in drug discovery. Expert Opin Drug Met Toxicol 2005;1 (1):91–142.
11. Kalgutkar AS, Gardner I, Obach RS, Shaffer CL, Callegari E, Henne KR, Mutlib AE, Dalvie DK, Lee JS, Nakai Y, O'Donnell JP, Boer J, Harriman SP. A comprehensive listing of bioactivation pathways of organic functional groups. Curr Drug Metab 2005;6(3):161–225.
12. Nelson SD. Metabolic activation and drug toxicity. J Med Chem 1982;25(7):753–761.
13. Alpert JS, Haffajee CI, Young MD. Chemistry, pharmacology, antiarrhythmic efficacy and adverse effects of tocainide hydrochloride, an orally active structural analog of lidocaine. Pharmacotherapy 1983;3(6):316–322.
14. Hare RL, Holcomb B, Page OC, Stephens JW. Toxicity of carbutamide; report of a fatal case of bone-marrow depression and anuria. N Engl J Med 1957;256(2):74–76.
15. Miyamoto G, Zahid N, Uetrecht JP. Oxidation of diclofenac to reactive intermediates by neutrophils, myeloperoxidase, and hypochlorous acid. Chem Res Toxicol 1997;10(4):414–419.
16. James LP, Mayeux PR, Hinson JA. Acetaminophen-induced hepatotoxicity. Drug Met Disp 2003;31(12):1499–1506.
17. Hinson JA, Reid AB, McCullough SS, James LP. Acetaminophen-induced hepatotoxicity: role of metabolic activation, reactive oxygen/nitrogen species, and mitochondrial permeability transition. Drug Metab Rev 36(3–4): 2004; 805–822.
18. Senter PD. Al-Abed Y, Metz CN. Benigni F, Mitchell RA. Chesney J, Han J, Gartner CG. Nelson SD. Todaro GJ. Bucala R. Inhibition of macrophage migration inhibitory factor (MIF) tautomerase and biological activities by acetaminophen metabolites. Proc Natl Acad Sci USA 2002;99:144–149.
19. Tingle MD, Jewell II, Maggs JL, O'Neill PM, Park BK. The bioactivation of amodiaquine by human polymorphonuclear leukocytes *in vitro*: chemical mechanisms and the effects of fluorine substitution. Biochem Pharmacol 1995;50 (7):1113–1119.
20. O'Neill PM, Mukhtar A, Stocks PA, Randle LE, Hindley S, Ward SA, Storr RC, Bickley JF, O'Neil IA, Maggs JL, Hughes RH, Winstanley PA, Bray PG, Park BK, Isoquine Related amodiaquine analogues: a new generation of improved 4-aminoquinoline antimalarials. J Med Chem 2003;46(23):4933–4945.
21. Famulok M, Boche G. Formation of *N*-(deoxyguanosin-8-yl) aniline in the *in vitro* reaction of *N*-acetoxyaniline with deoxyguanosine and DNA. Angew Chem Int Ed Engl 1989;28 (4):468–469.
22. Bryant MS, Skipper PL, Tannenbaum SR, Maclure M. Hemoglobin adducts of 4-aminobiphenyl in smokers and nonsmokers. Cancer Res 1987;47(2):602–608.
23. Kloehn P.-C. Massalha H, Neumann H.-G. A metabolite of carcinogenic 2-acetylaminofluorene, 2-nitrosofluorene, induces redox cycling in mitochondria. Biochim Biophys Acta 1995;1229(3):363–372.
24. Mahmud R, Tingle MD, Maggs JL, Cronin MTD. Dearden JC, Park BK. Structural basis for the hemotoxicity of dapsone: the importance of the sulfonyl group. Toxicology 1997;117(1):1–11.
25. Sabbioni, G, Sepai, O. Comparison of hemoglobin binding, mutagenicity, and carcinogenicity of arylamines and nitroarenes. Chimia 1995;49(10):374–380.
26. Marques MM, Mourato LL,G. Amorim MT, Santos MA, Melchior WB, Beland FA. Effect of substitution site upon the oxidation potentials of alkylanilines, the mutagenicities of *N*-hydroxyalkylanilines, and the conformations of alkylaniline–DNA adducts. Chem Res Toxicol 1997;10(11):1266–1274.
27. Burstyn JN, Iskandar M, Brady JF, Fukuto JM, Cho AK. Comparative studies of *N*-hydroxylation and *N*-demethylation by microsomal cytochrome P-450. Chem Res Toxicol 1991;4 (1):70–76.
28. Veronese ME. McLean S, D'Souza CA. Davies NW. Formation of reactive metabolites of phenacetin in humans and rats. Xenobiotica 1985;15(11):929–40.
29. Bort R, Ponsoda X, Jover R, Gomez-Lechon J, Castell JV. Diclofenac toxicity to hepatocytes: a role for drug metabolism in cell toxicity. J Pharmacol Exp Ther 1999;288 (1):65–72.
30. Talaska G, Al-Juburi AZ, Kadlubar FF. Smoking related carcinogen–DNA adducts in biopsy samples of human urinary bladder: identification of *N*-(deoxyguanosin-8-yl)-4-aminobiphenyl as a major adduct. Proc Natl Acad Sci USA 1991;88(12):5350–5354.

31. Alexson SE, Diczfalusy M, Halldin M, Swedmark S. Involvement of liver carboxylesterases in the *in vitro* metabolism of lidocaine. Drug Met Disp 2002;30(6):643–647.

32. Turesky RJ. Heterocyclic aromatic amine metabolism, DNA adduct formation, mutagenesis, and carcinogenesis. Drug Metab Rev 2002;34(3):625–650.

33. Walker DP, Bi FC, Kalgutkar AS, Bauman JN, Zhao SX, Soglia JR, Aspnes GE, Kung DW, Klug-McLeod J, Zawistoski MP, McGlynn MA, Oliver R, Dunn M, Li J.-C. Richter DT, Cooper BA, Kath JC, Hulford CA, Autry CL, Luzzio MJ, Ung EJ, Roberts WG, Bonnette PC, Buckbinder L, Mistry A, Griffor MC, Han S, Guzman-Perez A. Trifluoromethylpyrimidine-based inhibitors of proline-rich tyrosine kinase 2 (PYK2): structure–activity relationships and strategies for the elimination of reactive metabolite formation. Bioorg Med Chem Lett 2008;18(23):6071–6077.

34. Purohit V, Basu AK. Mutagenicity of nitroaromatic compounds. Chem Res Toxicol 2000;13 (8):673–692.

35. Zbaida S. Nitroreductases and azidoreductases. In: Ioannides C, editor. Enzyme Systems that Metabolise Drugs and Other Xenobiotics. Chapter 16. New York: John Wiley & Sons; 2002.

36. Scheline RR. Metabolism of foreign compounds by gastrointestinal microorganisms. Pharmacol Rev 1973;25(4):451–523.

37. Ascherl M, Eyer P, Kampffmeyer H. Formation and disposition of nitrosochloramphenicol in rat liver. Biochem Pharmacol 1985;34 (20):3755–3763.

38. Yunis AA. Chloramphenicol: relation of structure to activity and toxicity. Annu Pharmacol Toxicol 1988;28:83–100.

39. Smith KS, Smith PL, Heady TN, Trugman JM, Harman WD, Macdonald TL. *In vitro* metabolism of tolcapone to reactive intermediates: relevance to tolcapone liver toxicity. Chem Res Toxicol 2003;16(2):123–128.

40. Zitting A, Savolainen H, Nickels J. Acute effects of 2-nitropropane on rat liver and brain. Toxicol Lett 1981;9(3):237–246.

41. Fiala ES, Conaway CC, Mathis JE. Oxidative DNA and RNA damage in the livers of Sprague-Dawley rats treated with the hepatocarcinogen 2-nitropropane. Cancer Res 1989;49 (20):5518–5522.

42. Conaway CC, Nie G, Hussain NS, Fiala ES. Comparison of oxidative damage to rat liver DNA and RNA by primary nitroalkanes, secondary nitroalkanes, cyclopentanone oxime, and related compounds. Cancer Res 1991;51 (12):3143–3147.

43. Sodum RS, Sohn OS, Nie G, Fiala ES. Activation of the liver carcinogen 2-nitropropane by aryl sulfotransferase. Chem Res Toxicol 1994;7 (3):344–351.

44. Sodum RS, Fiala ES. *N2*-Amination of guanine to 2-hydrazino-hypoxanthine, a novel *in vivo* nucleic acid modification produced by the hepatocarcinogen 2-nitropropane. Chem Res Toxicol 1998;11(12):1453–1459.

45. Peterson JT. The importance of estimating the therapeutic index in the development of matrix metalloproteinase inhibitors. Cardiovasc Res 2006;69(3):677–687.

46. Skipper PL, Tannenbaum SR, Thilly WG, Furth EE, Bishop WW. Mutagenicity of hydroxamic acids and the probable involvement of carbamoylation. Cancer Res 1980;40(12): 4704–4708.

47. Lee MS, Isobe M. Metabolic activation of the potent mutagen, 2-naphthohydroxamic acid, in *Salmonella typhimurium* TA98. Cancer Res 1990;50(14):4300–4307.

48. Lipczynska-Kochany E, Iwamura H, Takahashi K, Hakura A, Kawazoe Y. Mutagenicity of pyridine- and quinoline-carbohydroxamic acid derivatives. Mutat Res 1984;135(3):139–148.

49. Codd R. Traversing the coordination chemistry and chemical biology of hydroxamic acids. Coord Chem Rev 252(12–14): 2008; 1387–1408.

50. Kirkland D, Pfuhler S, Tweats D, Aardema M, Corvi R, Darroudi F, Elhajouji A, Glatt H, Hastwell P, Hayashi M, Kasper P, Kirchner S, Lynch A, Marzin D, Maurici D, Meunier J.-R. Mueller L, Nohynek G, Parry J, Parry E, Thybaud V, Tice R, van Benthem J, Vanparys P, White P. How to reduce false positive results when undertaking in vitro genotoxicity testing and thus avoid unnecessary follow-up animal tests: Report of an ECVAM Workshop. Mutat Res 2007;628(1):31–55.

51. Monks TJ, Jones DC. The metabolism and toxicity of quinones, quinone-imines, quinone methides and quinone–thioethers. Curr Drug Metab 2002;3(4):425–438.

52. Levay G, Pongracz K, Bodell WJ. Detection of DNA adducts in HL-60 cells treated with hydroquinone and *p*-benzoquinone by phosphorus-32-postlabeling. Carcinogenesis 1991; 12(7):1181–1186.

53. Doerge DR, Divi RL, Deck J, Taurog A. Mechanism for the anti-thyroid action of minocycline. Chem Res Toxicol 1997;10(1):49–58.

54. Murray M. Mechanisms of inhibitory and regulatory effects of methylenedioxyphenyl compounds on cytochrome P450-dependent drug oxidation. Curr Drug Metab 2000;1(1):67–84.
55. Amacher DE, Fasulo LM, Charuel C, Comby P, Beaumont K. *In vitro* toxicity of zamifenacin (UK-76654) and metabolites in primary hepatocyte cultures. Xenobiotica 1998;28 (9):895–908.
56. Beaumont KC, Cussans NJ, Nichols DJ, Smith DA. Pharmacokinetics and metabolism of darifenacin in the mouse, rat, dog and man. Xenobiotica 1998;28(1):63–75.
57. Zhao SX, Dalvie DK, Kelly JM, Soglia JR, Frederick KS, Smith EB, Obach RS, Kalgutkar AS. NADPH-dependent covalent binding of [3*H*]paroxetine to human liver microsomes and S-9 fractions: identification of an electrophilic quinone metabolite of paroxetine. Chem Res Toxicol 2007;20(11):1649–1657.
58. Williams DP, Antoine DJ, Butler PJ, Jones R, Randle L, Payne A, Howard M, Gardner I, Blagg J, Park BK. The metabolism and toxicity of furosemide in the Wistar rat and CD-1 mouse: a chemical and biochemical definition of the toxicophore. J Pharmacol Exp Ther 2007;322(3):1208–1220.
59. Dansette PM, Bertho G, Mansuy D. First evidence that cytochrome P450 may catalyze both *S*-oxidation and epoxidation of thiophene derivatives. Biochem Biophys Res Commun 2005;338(1):450–455.
60. Treiber A, Dansette PM, El Amri H, Girault J.-P. Ginderow D, Mornon J.-P. Mansuy D. chemical and biological oxidation of thiophene: preparation and complete characterization of thiophene *S*-oxide dimers and evidence for thiophene *S*-oxide as an intermediate in thiophene metabolism *in vivo* and *in vitro*. J Am Chem Soc 1997;119(7):1565–1571.
61. Liu ZC, Uetrecht JP. Metabolism of ticlopidine by activated neutrophils: implications for ticlopidine-induced agranulocytosis. Drug Met Disp 2000;28(7):726–730.
62. Mansuy D. Molecular structure and hepatotoxicity: compared data about two closely related thiophene compounds. J Hepatol 1997;26 (Suppl 2): 22–25.
63. Neau E, Dansette PM, Andronik V, Mansuy D. Hydroxylation of the thiophene ring by hepatic monooxygenases. Evidence for 5-hydroxylation of 2-aroylthiophenes as a general metabolic pathway using a simple UV-visible assay. Biochem Pharmacol 1990;39 (6):1101–1107.
64. Fouda HG, Avery MJ, Dalvie D, Falkner FC, Melvin LS, Ronfeld RA. Disposition and metabolism of tenidap in the rat. Drug Met Disp 1997;25(2):140–148.
65. Chen Q, Doss GA, Tung EC, Liu W, Tang YS, Braun MP, Didolkar V, Strauss JR, Wang RW, Stearns RA, Evans DC, Baillie TA, Tang W. Evidence for the bioactivation of zomepirac and tolmetin by an oxidative pathway: Identification of glutathione adducts *in vitro* in human liver microsomes and *in vivo* in rats. Drug Met Disp 2006;34(1):145–151.
66. Dorsey BD, McDonough C, McDaniel SL, Levin RB, Newton CL, Hoffman JM, Darke PL, Zugay-Murphy JA, Emini EA, Schleif WA, Olsen DB, Stahlhut MW, Rutkowski CA, Kuo LC, Lin JH, Chen I.-W. Michelson SR, Holloway MK, Huff JR, Vacca JP. Identification of MK-944a: a second clinical candidate from the hydroxylamine-pentanamide isostere series of HIV protease inhibitors. J Med Chem 2000;43 (18):3386–3399.
67. Black AE, Hayes RN, Roth BD, Woo P, Woolf TF. Metabolism and excretion of atorvastatin in rats and dogs. Drug Met Disp 1999;27 (8):916–923.
68. Regal KA, Laws GM, Yuan C, Yost GS, Skiles GL. Detection and characterization of DNA adducts of 3-methylindole. Chem Res Toxicol 2001;14(8):1014–1024.
69. Kassahun K, Skordos K, McIntosh I, Slaughter D, Doss GA, Baillie TA, Yost GS. Chem Res Toxicol 2005;18:1427.
70. Taylor DC, Cresswell PR, Pepper ES. The excretion and metabolism of metiamide in the rat, dog, and man. Drug Met Disp 1979;7 (6):393–398.
71. Durant GJ, Emmett JC, Ganellin CR, Miles PD, Parsons ME, Prain HD, White GR. Cyanoguanidine–thiourea equivalence in the development of the histamine H2-receptor antagonist Cimetidine. J Med Chem 1977;20 (7):901–906.
72. Stevens GJ, Hitchcock K, Wang YK, Coppola GM, Versace RW, Chin JA, Shapiro M, Suwanrumpha S, Mangold BL. *In vitro* metabolism of *N*-(5-chloro-2-methylphenyl)-*N*′-(2-methylpropyl)thiourea: species comparison and identification of a novel thiocarbamide–glutathione adduct. Chem Res Toxicol 1997;10(7):733–741.
73. Radomski JL, Deichmann WB, MacDonald WE, Glass EM. Synergism among oral carcinogens. I. Results of the simultaneous feeding of four tumorigens to rats. Toxicol Appl Pharmacol 1965;7(5):652–656.

74. Grisolia CK. Ethylene dithiocarbamate fungicides: genotoxicity, carcinogenicity and teratogenicity. Pesticidas 1995;5:19–32.

75. Malamas MS, Sestanj K, Millen J. Synthesis and biological evaluation of tolrestat metabolites. Eur J Med Chem 1991;26(2):197–200.

76. Foppiano M, Lombardo G. Worldwide pharmacovigilance systems and tolrestat withdrawal. Lancet 1997;349(9049):399–400.

77. Jewell WT, Miller MG. Identification of carboxylesterase as the major protein bound by molinate. Toxicol Appl Pharmacol 1998;149 (2):226–234.

78. Kalgutkar AS, Driscoll J, Zhao SX, Walker GS, Shepard RM, Soglia JR, Atherton J, Yu L, Mutlib AE, Munchhof MJ, Reiter LA, Jones CS, Doty JL, Trevena KA, Shaffer CL, Ripp SL. A rational chemical intervention strategy to circumvent bioactivation liabilities associated with a nonpeptidyl thrombopoietin receptor agonist containing a 2-amino-4-arylthiazole motif. Chem Res Toxicol 2007;20 (12):1954–1965.

79. Hobbs DC, Twomey TM. Metabolism of sudoxicam by the rat, dog, and monkey. Drug Met Disp 1977;5(1):75–81.

80. Schmid J, Busch U, Trummlitz G, Prox A, Kaschke S, Wachsmuth H. Meloxicam: metabolic profile and biotransformation products in the rat. Xenobiotica 1995;25 (11):1219–1236.

81. Obach RS, Kalgutkar AS, Ryder TF, Walker GS. *In vitro* metabolism and covalent binding of enol-carboxamide derivatives and anti-inflammatory agents sudoxicam and meloxicam: insights into the hepatotoxicity of sudoxicam. Chem Res Toxicol 2008;21(9):1890–1899.

82. Lin H.-L. Kent UM, Hollenberg PF. Mechanism-based inactivation of cytochrome P450 3A4 by 17α-ethynylestradiol: evidence for heme destruction and covalent binding to protein. J Pharmacol Exp Ther 2002;301 (1):160–167.

83. Belle DJ, Callaghan JT, Gorski JC, Maya JF, Mousa O, Wrighton SA, Hall SD. The effects of an oral contraceptive containing ethinyloestradiol and norgestrel on CYP3A activity. Brit J Clin Pharm 2002;53(1):67–74.

84. Mutlib AE, Chen H, Nemeth GA, Markwalder JA, Seitz SP, Gan LS, Christ DD. Identification and characterization of efavirenz metabolites by liquid chromatography/mass spectrometry and high field NMR: species differences in the metabolism of efavirenz. Drug Met Disp 1999;27(11):1319–1333.

85. Timbrell JA. The role of metabolism in the hepatotoxicity of isoniazid and iproniazid. Drug Metab Rev 1979;10:125–147.

86. Streeter, AJ. and Timbrell, JA. Enzyme-mediated covalent binding of hydralazine to rat liver microsomes. Drug Met Disp 1983;11:179–183.

87. Nelson SD. Metabolic activation and drug toxicity. J Med Chem 1982;25:753–765.

88. Parman T, Chen G, Wells PG. Free radical intermediates of phenytoin and related teratogens. Prostaglandin H synthase-catalyzed bioactivation, electron paramagnetic resonance spectrometry, and photochemical product analysis. J Biol Chem 1998;273 (39):25079–25088.

89. Ozolins R, Terence RS, Wiley MJ, Wells PG. Phenytoin covalent binding and embryopathy in mouse embryos co-cultured with maternal hepatocytes from mouse, rat, and rabbit. Biochem Pharmacol 1995;50(11):1831–1840.

90. Menon KV,N. Angulo P, Lindor KD. Severe cholestatic hepatitis from troglitazone in a patient with nonalcoholic steatohepatitis and diabetes mellitus. Am J Gastroenterol 2001;96 (5):1631–1634.

91. Kassahun K, Pearson PG, Tang W, McIntosh I, Leung K, Elmore C, Dean D, Wang R, Doss G, Baillie TA. Studies on the metabolism of troglitazone to reactive intermediates *in vitro* and *in vivo*. Evidence for novel biotransformation pathways involving quinone methide formation and thiazolidinedione ring scission. Chem Res Toxicol 2001;14(1):62–70.

92. Tettey JN, Maggs JL, Rapeport WG, Pirmohamed M, Park BK. Enzyme-induction dependent bioactivation of troglitazone and troglitazone quinone *in vivo*. Chem Res Toxicol 2001;14(8):965–974.

93. Maret G, Testa B, Jenner P, el Tayar N, Carrupt PA. The MPTP story: MAO activates tetrahydropyridine derivatives to toxins causing parkinsonism. Drug Metab Rev 1990;22 (4):291–332.

94. Usuki E Van Der Schyf CJ, Castagnoli N, Jr. Metabolism of haloperidol and its tetrahydropyridine dehydration product HPTP. Drug Metab Rev 1998;30(4):809–826.

95. Stachulski AV, Harding JR, Lindon JC, Maggs JL, Park BK, Wilson ID. Acyl glucuronides: biological activity, chemical reactivity, and chemical synthesis. J Med Chem 2006;49 (24):6931–6945.

96. Bailey MJ, Dickinson RG. Acyl glucuronide reactivity in perspective: biological conse-

quences. Chem Biol Interact 2003;145 (2):117–137.

97. Boelsterli UA. Xenobiotic acyl glucuronides and acyl CoA thioesters as protein-reactive metabolites with the potential to cause idiosyncratic drug reactions. Curr Drug Metab 2002;3(4):439–450.
98. Wang J, Davis M, Li F, Azam F, Scatina J, Talaat R. A novel approach for predicting acyl glucuronide reactivity via Schiff base formation: development of rapidly formed peptide adducts for LC/MS/MS measurements. Chem Res Toxicol 2004;17(9):1206–1216.
99. Munafo A, Hyneck M, Benet LZ. Pharmacokinetics and irreversible binding of tolmetin and its glucuronic acid esters in the elderly. Pharmacology 1993;47(5):309–317.
100. Meng X, Maggs JL, Pryde DC, Planken S, Jenkins RE, Peakman TM, Beaumont K, Kohl C, Park BK, Stachulski AV. Cyclization of the acyl glucuronide metabolite of a neutral endopeptidase inhibitor to an electrophilic glutarimide: synthesis, reactivity, and mechanistic analysis. J Med Chem 2007;50 (24):6165–6176.
101. Rodrigues AD. Drug–Drug Interactions. New York: Marcel Dekker; 2002.
102. Fontana E, Dansette PM, Poli SM. Cytochrome P 450 enzymes mechanism based inhibitors: common sub-structures and reactivity. Curr Drug Metab 2005;6(5):413–454.
103. Kalgutkar AS, Obach RS, Maurer TS. Mechanism-based inactivation of cytochrome P450 enzymes: chemical mechanisms, structure-activity relationships and relationship to clinical drug–drug interactions and idiosyncratic adverse drug reactions. Curr Drug Metab 2007;8 (5):407–447.
104. Rodrigues AD, Lin JH. Screening of drug candidates for their drug–drug interaction potential. Curr Opin Chem Biol 2001;5(4):396–401.
105. Riley RJ, Parker AJ, Trigg S, Manners CN. Development of a generalized, quantitative physicochemical model of CYP3A4 inhibition for use in early drug discovery. Pharmaceutical Res 2001;18(5):652–655.
106. Palchaudhuri R, Hergenrother PJ. DNA as a target for anticancer compounds: methods to determine the mode of binding and the mechanism of action. Curr Opin Biotechnol 2007;18(6):497–503.
107. Kubota Y, Iwamoto T, Seki T. The interaction of benzimidazole compounds with DNA: intercalation and groove binding modes. Nucleic Acids Symp Ser 1999;42:53–54.
108. Debelle FD, Vanherweghem JL, Nortier JL. Aristolochic acid nephropathy: a worldwide problem. Kidney Int 2008;74158 169.
109. Nortier JL, Martinez MCM. Schmeiser HH, Arlt VM, Bieler CA, Petein M, Depierreux MF, De Pauw L, Abramowicz D, Vereerstraeten P, Vanherweghem JL. Urothelial carcinoma associated with the use of a Chinese herb (*Aristolochia fangchi*). N Engl J Med 2000;23:1686–1691.
110. Pfau W, Pool-Zobel BL, Von der Liethb VW, Wiessler M. The structural basis for the mutagenicity of aristolochic acid. Cancer Lett 1990;55:7–11.
111. Uetrecht J. Prediction of a new drug's potential to cause idiosyncratic reactions. Curr Opin Drug Discov Devel 2001;4(1):55–59.
112. Brocks DR, Jamali F. Clinical pharmacokinetics of ketorolac tromethamine. Clin pharmacokinet 1992;23(6):415–427.
113. Agarwal VK, Krol GJ, Krone V, Roberts D. Quantitative analysis of HBY 097 and its metabolites in human serum and urine by HPLC. J Pharm Biomed Anal 1998;16 (7):1195–1203.
114. Kalgutkar AS, Fate G, Didiuk MT, Bauman J. Toxicophores, reactive metabolites and drug safety: when is it a cause for concern? Exp Rev Clin Pharmacol 2008;1(4):515–531.
115. Hop CE, Kalgutkar AS, Soglia JR. Importance of early assessment of bioactivation in drug discovery. Ann Rep Med Chem 2006;41:369–381.
116. Caldwell GW, Yan Z. Screening for reactive intermediates and toxicity assessment in drug discovery. Curr Opin Drug Discov Devel 2006;9(1):47–60.
117. Masubuchi N, Makino C, Murayama N. Prediction of *in vivo* potential for metabolic activation of drugs into chemically reactive intermediate: correlation of *in vitro* and *in vivo* generation of reactive intermediates and *in vitro* glutathione conjugate formation in rats and humans. Chem Res Toxicol 2007; 20 (3):455–464.
118. Mitchell MD, Elrick MM, Walgren JL, Mueller RA, Morris DL, Thompson DC. Peptide-based *in vitro* assay for the detection of reactive metabolites. Chem Res Toxicol 2008;21 (4):859–868.
119. Obach RS, Kalgutkar AS, Soglia JR, Zhao SX. Can *in vitro* metabolism-dependent covalent binding data in liver microsomes distinguish hepatotoxic from nonhepatotoxic drugs? An analysis of 18 drugs with consideration of

intrinsic clearance and daily dose. Chem Res Toxicol 2008;21(9):1814–1822.

120. Gan J, Harper TW, Hsueh MM, Qu Qinling, Humphreys WG, Dansyl glutathione as a trapping agent for the quantitative estimation and identification of reactive metabolites. Chem Res Toxicol 2005;18 (5):896–903.

121. Soglia JR, Contillo LG, Kalgutkar AS, Zhao S, Hop CE, Boyd JG, Cole MA. semiquantitative method for the determination of reactive metabolite conjugate levels *in vitro* utilizing liquid chromatography–tandem mass spectrometry and novel quaternary ammonium glutathione analogues. Chem Res Toxicol 2006;19 (3):480–490.

122. Williams DP, O'Donnell C. Adverse drug reactions, transcription factors and transcriptomics. Comp Clin Pathol 2003;12(1):1–10.

123. Luo W, Fan W, Xie H, Jing L, Ricicki E, Vouros P, Zhao LP, Zarbl H. Phenotypic anchoring of global gene expression profiles induced by *N*-hydroxy-4-acetylaminobiphenyl and benzo[a]pyrene diol epoxide reveals correlations between expression profiles and mechanism of toxicity. Chem Res Toxicol 2005;18 (4):619–629.

124. Evans DC, Watt AP, Nicoll-Griffith DA, Baillie TA. Drug–protein adducts: an industry perspective on minimizing the potential for drug bioactivation in drug discovery and development. Chem Res Toxicol 2004;17 (1):3–16.

125. Lin CY, Isbell MA, Morin D, Boland BC, Salemi MR, Jewell WT, Weir AJ, Fanucchi MV, Baker GL, Plopper CG, Buckpitt AR. Characterization of a structurally intact *in situ* lung model and comparison of naphthalene protein adducts generated in this model vs. lung microsomes. Chem Res Toxicol 2005;18 (5):802–813.

126. Obach RS, Kalgutkar AS, Soglia JR, Zhao SX. Can *in vitro* metabolism-dependent covalent binding data in liver microsomes distinguish hepatotoxic from nonhepatotoxic drugs? An analysis of 18 drugs with consideration of intrinsic clearance and daily dose. Chem Res Toxicol 2008;21:1814–1822.

127. Bauman JN, Kelly JM, Tripathy S, Zhao SX, Lam WW, Kalgutkar AS, Obach RS. Can *in vitro* metabolism-dependent covalent binding data distinguish hepatotoxic from nonhepatotoxic drugs? An analysis using human hepatocytes and liver S-9 fraction. Chem Res Toxicol 2009;22:332–340.

128. http://www.chem.leeds.ac.uk/luk/derek/index.html.

129. http://www.accelrys.com/products/topkat/index.html.

130. von Korff M, Sander T. Toxicity-indicating structural patterns. J Chem Inf Model 2006;46(2):536–544.

131. Estrada E, Molina E. Automatic extraction of structural alerts for predicting chromosome aberrations of organic compounds. J Mol Graph Model 2006;25(3):275–288.

132. Serafimova R, Todorov M, Pavlov T, Kotov S, Jacob E, Aptula A, Mekenyan O. Identification of the structural requirements for mutagenicity, by incorporating molecular flexibility and metabolic activation of chemicals. II. General Ames mutagenicity model. Chem Res Toxicol 2007;20(4):662–676.

133. Thompson CD, Kinter MT, Macdonald TL. Synthesis and *in vitro* reactivity of 3-carbamoyl-2-phenylpropionaldehyde and 2-phenylpropenal: putative reactive metabolites of felbamate. Chem Res Toxicol 1996;9(8):1225–1229.

134. Smith DA, Jones RM. The sulfonamide group as a structural alert: a distorted story? Curr Opin Drug Discov Devel 2008;11(1):72–79.

PROTEIN–PROTEIN INTERACTIONS AS DRUG DISCOVERY TARGETS

Alexander Dömling
Department of Pharmaceutical Sciences, University of Pittsburgh, Pittsburgh, PA

1. INTRODUCTION

Protein–protein interactions (PPIs) are a rather new, large, but difficult to address subgroup of drug targets. Classical targets such as enzymes offer commonalities, for example, regarding the mechanism, and thus offer good and in the past successful starting points for the design of inhibitors and drugs. For example, kinases inhibitors in most of the cases were derived from mimicking the essential ATP substrate needed to phosphorylate their targets. Thus, the core chemotype is predefined by flat (hetero)aromatic rings with suitable adjacent hydrogen-bond donor and acceptor functions to mimic the corresponding ATP adenine ring system and its interaction with the kinase. PPIs, however, are very diverse in form, shape, and function and currently there is no generally used classification system. Although PPIs can be classified according to protein architectures, lifetime of interaction, affinity, and function, none of these classifications has been leveraged for drug design and discovery [1,2].

PPI interfaces are often hydrophobic and large nonpolar surface areas are buried [3]. Not uncommonly the interfaces have areas of 1200–2000 Å^2, large ones up to 4600 Å^2, and smaller ones starting from 600 Å^2 [4–6]. Hydrophobicity and van der Waals energies are the leading forces of PPIs. Large energies can result from the combination of two hydrophobic surfaces from the aqueous environment. Energy components having to take into account are desolvation, van der Waals interactions, entropy reduction, and electrostatic forces [7]. Electrostatic complementarity increases energy of PPI and the polar amino acids are often found at the rim of PPIs. They are surrounding the center of the hydrophobic PPI and shelter the PPI center from the solvent, and the comparison to an O-ring was made [8]. The dynamics of the formation of PPI has also been investigated and models proposed [9]. In these models, the specific anchor side chains are found in conformations similar to those observed in the bound complex. Once the anchors are docked, an induced fit process further contributes to forming the final high-affinity complex. The on- and off-kinetics of PPIs are of high importance for small molecules to compete with the interaction. Not all amino acids buried in the interface of interacting protein contribute equally. Instead, mutational Ala scan studies show that often a small subset of residues contributes the majority of free energy of binding [3,10–13]. These clusters of energetically important amino acid residues are usually found in the center of a PPI interface and have been termed "hotspot." These hotspots of PPIs are important to understand the physicochemical basis of these interactions, but they offer also a starting point for drug discovery and design.

PPIs are an emerging class of drug targets in terms of importance and certainly in terms of numbers. The number of classical targets, for example, proteases or channels, however, is rather limited. Druggable sites are the active sites and in certain cases allosteric sites. The number of proteases in the human genomes is ~600 and even generally assuming druggable allosteric sites only slightly increases the total number of drug targets. Similar applies for other classical targets including kinases, phosphatases, nuclear receptors, channels, and so on. The number of PPIs, however, is estimated to be several hundred thousand and thus exceeds classical targets by several orders of magnitude [14]. The number of PPIs with structural characterization and Å resolution exceeds 35,000 entries in the protein data bank (PDB) [2,15]. Additionally, it has to be noted that the number of structures of PPIs in the PDB is rapidly growing. The knowledge of the 3D structure of PPIs will be of importance for its understanding, its biological role, and the discovery of antagonists. Clearly, PPIs are involved in all basic and disease-related biochemical events and thus can comprise meaningful drug targets. This has been recently generally recognized and more and more PPI targets are in pursuit and first success stories moved into clinical trials

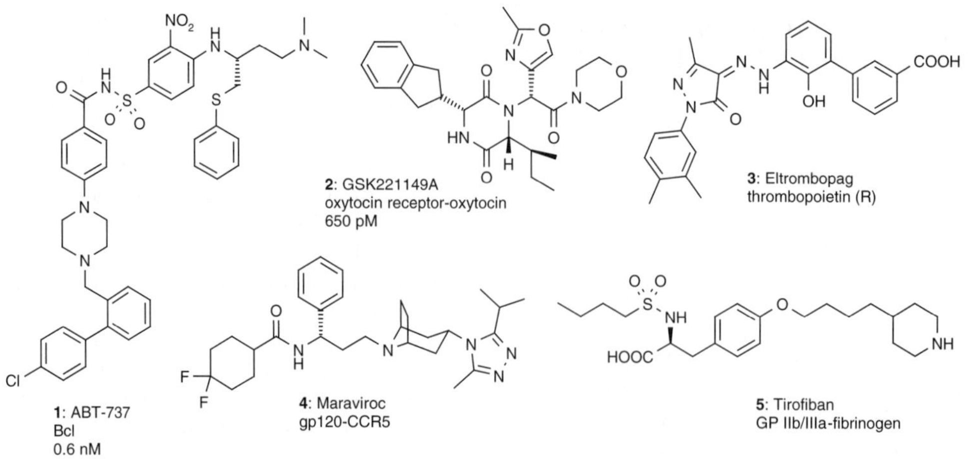

Figure 1. Small molecular weight PPI (ant)agonists have to compete with the natural interaction proteins: ABT-737 (**1**) with Bcl family proteins, GSK221149A (**2**) with oxytocin, eltrombopag (**3**) with thrombopoietin, maraviroc (**4**) with gp120, and tirofiban (**5**) with fibrinogen.

and to the market. Well-known examples are the Bcl-2 inhibitor ABT-737 (**1**) or the oxytocin antagonist GSK221149A (**2**), which all currently undergo clinical trials. PPI (ant)agonists recently receiving market approval include eltrombopag (**3**), maraviroc (**4**), and tirofiban (**5**) (Fig. 1).

It is difficult to imagine how small molecular weight inhibitors can effectively compete with a large and evolutionary optimized PPI interaction. Impressively, however, optimized compounds can have higher affinities to a protein than their natural protein interaction partner (Fig. 1). For example, ABT-737 (**1**), an antagonist of Bcl-2, Bcl-X(L), and Bcl-w with a molecular weight of 813 Da inhibits members of Bcl-2 including Mcl-1 (K_i of 0.6 nM for binding cytosolic Bcl). ABT-737 has been shown to effectively reduce the tumor burden of model species in preclinical investigations [16]. A follow-up compound ABT-263 currently undergoes clinical trials against cancer. GSK221149A (**2**) is a very potent (K_i 650 pM) and selective oxytocin antagonist and has been shown to inhibit oxytocin-induced uterine contractions in anesthetized rat. Interestingly, the compound (MW 495 Da) is a 20-fold more potent receptor antagonist than the current clinically used peptide derivative atosiban (MW 994 Da). Additionally, GSK221149A is 10-fold more potent than atosiban with a superior selectivity profile with respect to the related vasopressin receptors [17]. In contrast to the peptide derivative, however, GSK221149A is orally bioavailable [18]. GSK221149A currently undergoes extended clinical trials for preterm labor. Preterm labor (prior to 37 weeks gestation) is the largest single cause of infant morbidity and mortality and is frequently associated with long-term disability. Another success story is eltrombopag (**3**) (564 Da), the only approved PPI small-molecule agonist and targeting the thrombopoietin receptor. Eltrombopag recently received FDA and EMEA approval for the orphan indications idiopathic thrombocytopenic purpura and cirrhosis due to hepatitis C. The small molecular weight compound thus competes with the approved biotechnology product romiplostim (29,542 Da), a fusion protein analog of thrombopoietin [19]. The recently approved HIV entry inhibitor maraviroc is another example of a marketed PPI antagonist. Maraviroc (**4**) binds to the cell surface receptor CCR5, preventing an interaction with gp120 thus blocking viral cell entry. It is competing with the also approved 36-amino-acid peptide enfuvirtide (T-20). Again, the small-molecule maraviroc is orally bioavailable whereas enfuvirtide is administered twice daily by subcutaneous injection. Tirofiban (**5**) is an antiplatelet drug that is acting by blocking the interaction of fibrinogen and glycoprotein (GP) IIb/IIIa receptors in human platelets [20]. It is a mimic of the RGD sequence. Thus, in all the above cases, the optimized small molecules bind with an affinity comparable to, or better than, the native interacting protein partner. Furthermore, these examples demonstrate not only the feasibility to discover and develop small molecular weight PPI (ant)agonist into clinical trials but also the emerging interest of industry in this large and diverse class of targets. Different approaches have been described to discovery antagonists of PPIs and will be discussed in the following section.

2. TECHNIQUES TO DISCOVER PPI (ANT) AGONISTS

2.1. HTS

Currently, high-throughput screening (HTS) is the preferred method of primary screening in industry leading to hits and medicinal chemistry starting points. During HTS, large collections of synthetic and/or natural occurring compounds are screened in a functional assay. The assumption that screening a large library of sufficiently diverse compounds has payed off in numerous drug discovery projects. The resulting hits have than to be verified in secondary assays and then can eventually form the starting point for a medicinal chemistry program to receive optimized and selective leads with favorable attached physicochemical properties. The HTS approach is quite successful for a range of traditional targets. For PPIs, however, HTS approaches have a much lower success rate than for other target classes. Often, HIT rates are below 1% [21–24]. For example, in a recent report on a HTS of a corporate library of $\sim$1.2 million compounds of the cancer relevant targets p53/Mdm2 and p53/Mdm4, hit rates of 0.044% and 0.029%, respectively, were dislosed [25]. The hit rate is defined by the number of hits ($IC_{50}<$ 20 μM in FRET assay and passing LC-MS analysis checking structural integrity and purity) divided by the total number of compounds screened. The initial hit rate is of importance since only a small fraction of HTS hits can be further elaborated into suitable projects. Many primary hits will be discarded rapidly due to adverse properties such as compound aggregation, covalent binding, "frequent hitter," or recognized as false positive (promiscuous binder). It has been argued that the compound collections currently used are not particularly suitable for screening of PPIs [26]. Screening collections often are built up over several decades and are biased toward the preferred targets and drug chemotypes of the particular organization. New classes of target often require new chemotypes. Due to the diversity of structures of PPIs, it is likely that each protein–protein interface will require a new chemotype [26]. The similarity of small-molecule PPI antagonist with known drugs was investigated and the compounds generally did not show high similarity to any set of compounds against other known targets [27]. An important question arises if there are fundamental differences between drug chemotypes targeting classical drug targets as opposed to PPIs. Clearly, there are certain preferred chemotypes for certain target classes, for example, asp-proteases inhibitors mostly comprise statine-type motifs to mimic the active site water. Kinase inhibitors, however, mimic the adenine and therefore often comprise flat heteroaromatic chemotypes with adjacent hydrogen donor–acceptor moieties to address the hinge region of the kinase. Phosphatase inhibitors need acidic warhead moieties that make their development cumbersome. Certain classes of GPCRs, however, like to see the other privileged structures such as benzodiazepines, *N*-phenyl-piperazines or the 1,3-bistrifluorphenyl fragment [28]. Thus, very often, mechanism-based protein inhibitors need certain chemotypes. PPIs, however, are so diverse, functionally and structurally, that it will be rather unlikely to find certain chemotypes or "privileged backbones" with a strong preference for PPIs. Many natural products are known to antagonize PPIs, examples are shown in Fig. 2. The fungal macrolide brefeldin A (**6**) at the interface of the Arf–GEF complex with potential as antiviral or anticancer agent [29]. Camptothecin (**7**) derived from the bark of *Camptotheca acuminate* and derivatives of which are used as clinical anticancer agents binds to topoisomerase and inhibits the I-DNA complex [30]. It stabilizes a kinetic

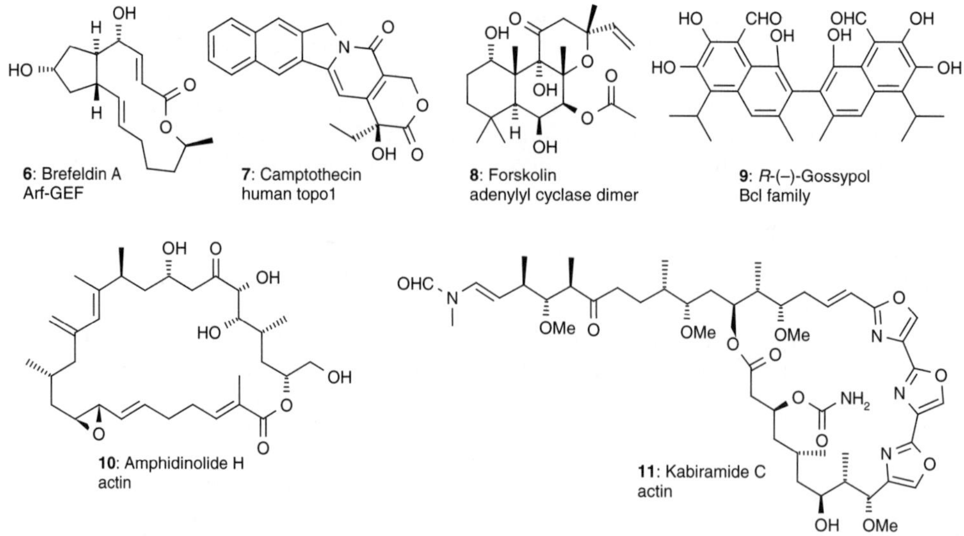

Figure 2. Representative examples of natural product PPI antagonists.

intermediate of the DNA cleavage enzyme reaction and thus induces cell death. Herbal forskolin (**8**) binds at the interface of the adenylyl cyclase dimer thereby activating the synthesis of cAMP [31]. Gossypol (**9**), a polyphenol isolated from cotton plant, has been shown to bind to Bcl-2 and antagonize Bcl family interactions [32]. The cytotoxic macrolide amphidinolide H (**10**) belongs to a large class of natural products binding to actin [33]. An example of a high resolution structure of a PPI and a natural product antagonist is shown in Fig. 3. In the structure, the marine toxin kabiramide C (**11**) is shown binding to actin. Interestingly, the same hydrophobic binding site is used by a diversity of actin binding proteins too thus forming a major interaction hub [34]. Comparing the structures of the actin–protein complexes and the kabiramide C complex show that they cover not only the same site but also the number of interacting residues is comparable. Although noteworthy, almost all heavy atoms of kabiramide C are involved in direct binding to the actin surface (Fig. 3). Another interesting observation is that most of the polar moieties of the binder are pointing to the water, whereas the hydrophobic parts are directly involved in the actin surface binding.

Another group of compounds densely functionalized compounds are based on multicomponent reactions (MCRs) [35]. MCR molecules are densely functionalized since they are assembled from several reacting starting materials in one step. It is noteworthy that many of the described (ant)agonists of PPIs are in fact derived from MCR chemistry, for example, the previously described GSK221149A (**2**). Densely functionalization of molecules, although difficult to capture, could be defined as the ratio of atoms to surface area.

Representative structures of examples of successful HTS campaigns leading to PPI antagonists are shown in figure 4. Screening 20,000 compound of an MCR scaffold biased library in an ELISA directed against vascular endothelial growth factor (VEGF) interacting with neuropilin-1 (NRP1), yielded 3-amino-1-cyano-indolizine (**12**) as hits [38]. Efficient MCR chemistry allowed for fast optimization toward **13**. HTS yeast two hybrid (YTH) technology was used to identify CL-888 (**14**) as a modulator of potassium channel KChIP/Kv4 interaction/function [39]. A new class of AICAR Tfase inhibitors were discovered through screening combinatorial libraries (>40,000 compounds) that act by inhibiting requisite enzyme dimerization. The most active compound Cappsin 2 (**15**) has a $K_i = 3\,\mu M$. Ras/Raf protein interaction inhibitors initially discovered by large-scale Y2H screening (>100,000 compounds) lead, after medicinal chemistry optimization, to (**16**) inhibiting anchorage-independent growth of HCT-116 cell line [40,41]. Moreover,

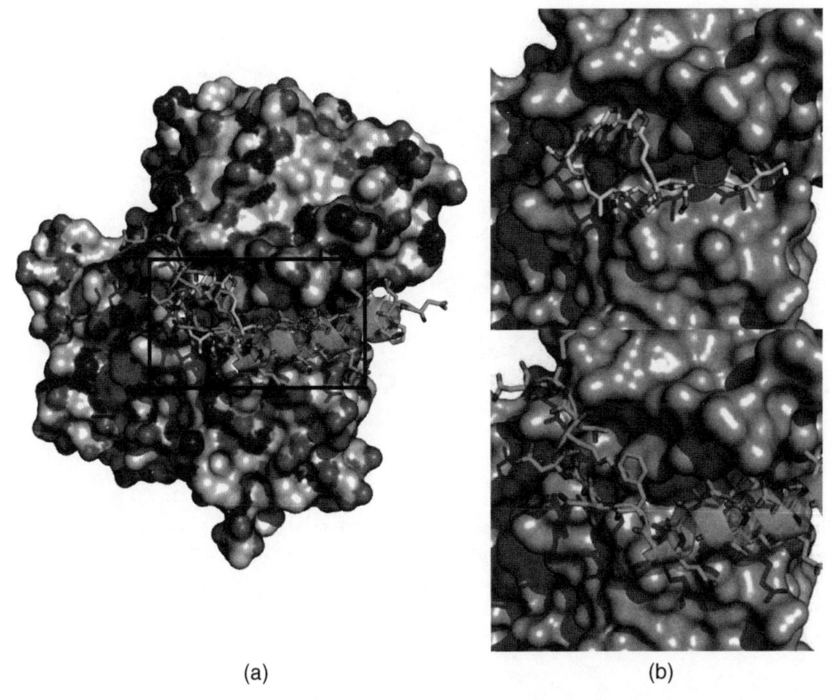

Figure 3. (a) Comparison of the high-resolution binding mode of the actin binding protein ciboulot (cyan sticks and cartoon, PDB ID: 1SQK) [36] and the natural product marine toxin kabiramide C (yellow sticks) with actin (red-blue-gray surface, PDB ID: 1QZ5) [37]. (b) closeup views of the inhibitors binding into the actin groove. Above: closeup view of the binding pocket (blue surface) and the natural product (yellow sticks). Below: The binding site covered by ciboulot (cyan sticks and cartoon) is highly overlapping with the hydrophobic binding site of kabiramide C, however, also extending into the groove via an ordered amphiphatic α-helical structure. This binding site is used by many actin binding proteins and other natural products and thus comprise a general interaction hub. (See color insert.)

the orally bioavailable compound showed a strong synergy with microtubule-targeting inhibitors, for example, taxol and showed tumor reduction in several xenograft models [42]. The spirocyclic compound (**17**) antagonizes two papillomavirus encoded proteins, E1 and E2, that are required for replication of the viral genome. Initial hits of 1,200 initial hits were found

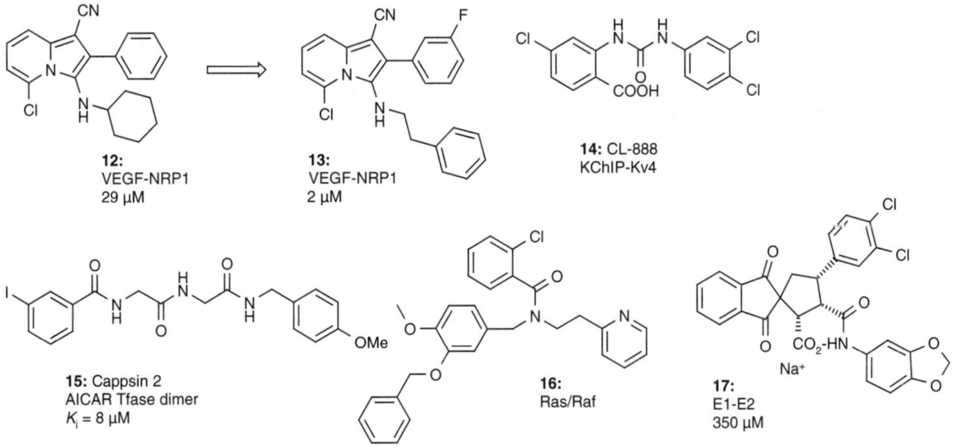

12: VEGF-NRP1 29 μM

13: VEGF-NRP1 2 μM

14: CL-888 KChIP-Kv4

15: Cappsin 2 AICAR Tfase dimer K_i = 8 μM

16: Ras/Raf

17: E1-E2 350 μM

Figure 4. Representative synthetic PPI (ant)agonists discovered by HTS.

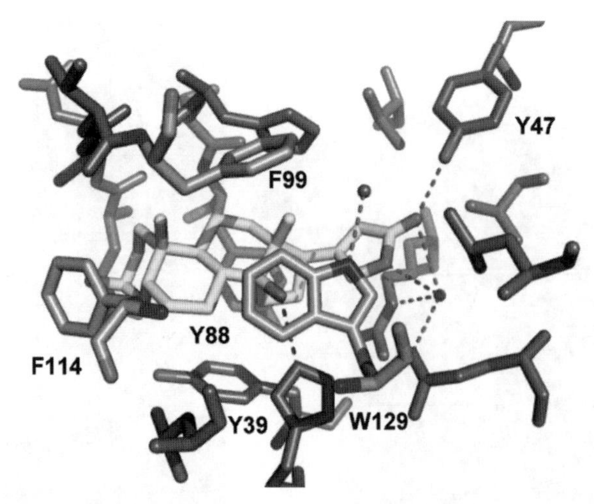

Figure 5. Example of aromatic amino acid residues and their role in the recognition process: interaction of the lipocalin binding to the small-molecule digoxigenin (PDB ID: 1LNM) [8]. A highly hydrophobic binding site comprising the aliphatic and aromatic amino acids Val, Leu, Ile (gray sticks) and Phe, Trp (pink sticks) and Tyr (blue sticks) recognizing the digoxigenin molecule (yellow sticks). The aromatic amino acid form a nest by aligning their π-aromatic rings in parallel to the hydrophobic guest. Additionally, Tyr47 forms a hydrogen bond to the lactam-carbonyl of digoxigenin. (See color insert.)

during a HTS campaign of 140,000 compounds of a corporate library including combinatorial libraries, compounds from medicinal chemistry programs, as well as externally acquired compounds. Medicinal chemistry optimization led to **17** with an IC_{50} of 350 nM in an E2-dependent E1/DNA binding assay [43].

It is well established that there is a preference for certain amino acid in the interface of interacting proteins. Systematic analysis of PDB structures of PPIs revealed area-based amino acid compositions [44–46]. The amino acid composition of the core of the interface differs considerably from the rim [8,47,48]. Also, it has been found that certain clusters of amino acids are prevalent in PPIs. For example, the three amino acids Trp/Met/Phe are preferred in functional PPIs [49]. The role of Tyr is special for the recognition in antigen–antibody structures has been investigated and the dominant role of large Tyr residues for mediating molecular contacts and of small serine/glycine residues for providing space and flexibility was revealed [50]. The nature of Tyr as an amphiphatic, large, and capable of forming nonpolar, hydrogen-bonding, and cation–π interactions make it the amino acid that is most effective for mediating molecular recognition (Fig. 5) [51]. Thus, a fruitful approach to enrich synthetic compound libraries directed to PPI hits could be to incorporate those preferred amino acids or fragments thereof into the screening compounds.

2.2. Fragment-Based Drug Discovery

Drug discovery starting from small chemical fragments with millimolar or micromolar affinity to the target and discovered by NMR, X-ray, and biochemical methods or computational screening is a relatively new technique termed fragment-based drug discovery (FBDD) [23,52]. The recent success stories of FBA have made this approach very popular [53]. A classical example of FBDD is the discovery of the two low molecular weight fragments 4′-fluoro-diphenylcarboxylicacid and tetrahydronaphtol by a 2D NMR-based screening of a 10,000-member library [54]. Subsequent structure determination by NMR spectroscopy showed the two fragments bound in proximal pockets (Fig. 6) [55]. Extensive medicinal chemistry was undertaken to link the fragments and to develop compounds with reasonable pharmacological characteristics. The final candidate ABT-263 is orally bioavailable despite a high

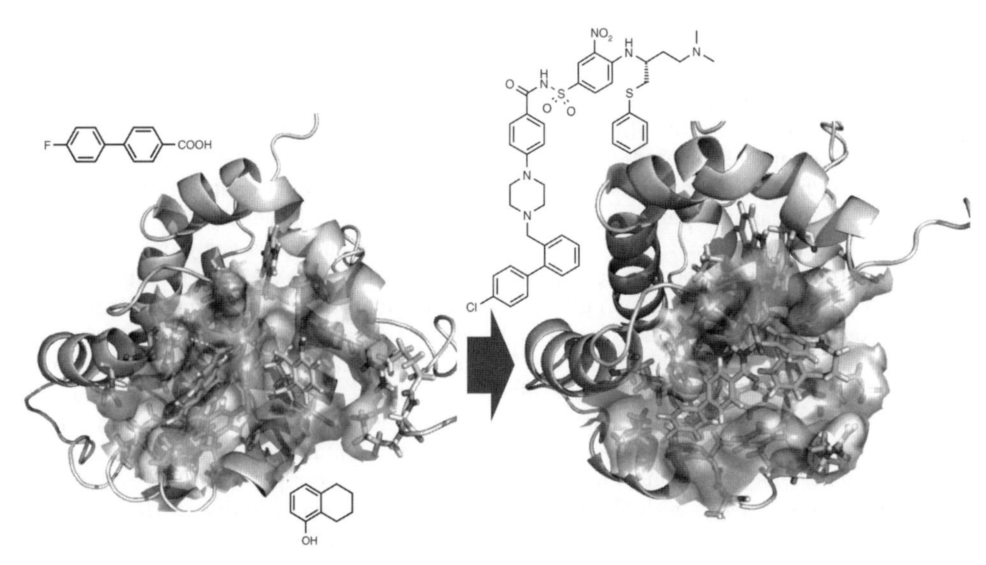

Figure 6. Fragments bound into the Bcl-x groove and the clinical candidate after lengthy optimization. Left: The 4′-fluorobiphenylcarboxylic acid in cyan sticks, the adjacent tetrahydronaphtol in yellow sticks (PDB ID: 1YSG). The carboxylic acid is making a charge–charge interaction with Arg1432 (yellow dotted lines) that is retained in the final ABT-737 **1** (cyan sticks, Right) in the acidic acylsulfonamide (PDB ID: 1YSI). Additionally, a hydrogen-bond contact is formed to the backbone amide of Gly142 and an oxygen of the sulfonamide portion. (See color insert.)

molecular weight and is currently undergoing clinical trials against cancer [56].

Another classical example of FBDD is the discovery of a potent IL-2r antagonist **18** (Fig. 7) [57]. These molecules were assembled in a fragment-based approach guided by X-ray structures and medicinal chemistry, and inspired by the previous drug discovery efforts of

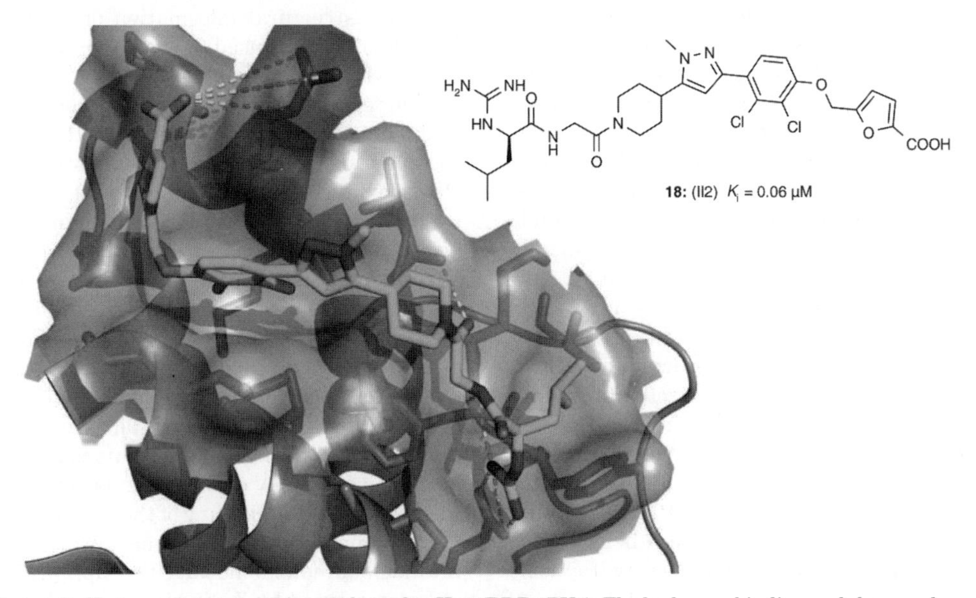

Figure 7. IL-2 receptor antagonists **18** bound to IL-2 (PDB 1PY2). The hydrogen binding and charge–charge interactions are indicated by yellow dashed lines. (This figure is available in full color at http://mrw.interscience.wiley.com/emrw/9780471266945/home.)

a group at Hoffmann-La Roche [58]. The compound binds close to the receptor contact region on IL-2 and closely mimics the electrostatic surface of the IL-2r [13]. The binding surface on IL-2 is adaptive and binds the small molecule with high-affinity and higher ligand efficacy (free binding energy per heavy atom) then IL-2r.

FBDD approaches aim to be more efficient at producing leads with better ligand efficiencies and better physical properties (relative to traditional methods, for example, HTS and subsequent medicinal chemistry). Several successful cases have shown their relevance also in PPIs. The greatest challenge in this area is the nontrivial fragment growth or linkage of fragment to drugs with acceptable PKPD profile.

2.3. Structure-Based Drug Design

The abundance and general availability of high-resolution Å information of PPIs in the PDB makes structure-based drug design (SBDD) very attractive. Intuitively, certain PPI shapes and geometries appear more suitable for structure-based PPI (ant)agonists discovery than others. An analysis of X-ray/NMR structure associated PPIs has been presented to evaluate their attractiveness in terms of drug discovery [59]. This analysis focuses primarily on the structural and physicochemical characteristics of the participating binding sites, particularly, the dimensions and hydrophobicity of the sites. Thus, it was distinguished between the large and complex protein–protein interfaces with unsuitable geometries, interactions that are to polar to allow efficient transport of drug into cells, binding pockets that are too shallow and others that appear suitable for targeting. Later targets include the interaction of HIF-1a with the transcription coactivator CBP/p300 [60], complex between EphB2 and ephrin-B2 [61], and the later discussed p53/MDM2.

It has been found that PPIs are often mediated by an amphiphatic α-helix of one protein partner sticking into a hydrophobic binding grove in the acceptor protein. The α-helix is a structurally very conservative secondary structure element with repeating hydrogen-bonding pattern and a tight geometric relationship between residues. In addition, the partners of PPIs often use other motifs or secondary structural elements than α-helix, for example, β-turn, β-sheet to interact. The design of scaffolds that can mimic the 3D substitution pattern of such secondary elements was therefore early on recognized as a rewarding task. 1,1,6-Trisubstituted indanes have been designed as conformationally constrained, nonpeptide templates that allow the incorporation of two adjacent amino acid side chains, plus a third binding group in an orientation similar to that found in α-helices (Fig. 8) [62]. Indeed, several of these compounds, for example, **19** were found to bind with micromolar affinity to the NK1 and/or NK1 receptor. Terphenyles, for example, **20** have been designed as a scaffold to mimic one face of α-helical peptide (Fig. 8) [63–65]. In several cases, nanomolar antagonists of complex PPIs have been obtained based on this backbone. A disadvantage of this backbone is the rather tedious synthetic accessibility and therefore several short and more convergent similar backbones with better accessibility have been proposed [66–68].

Numerous examples have been described using SBDD to discover PPI antagonists. For example, structural information of different PDZ domains was used to design antagonists of the NHERF1/EBP50 PDZ domain interaction, for example, **21** [69]. The proposed indole scaffold has been designed to mimic the acidic tetrapeptide sequence of the betastrand of PDZ domain ligands. Compound **21** showed an IC_{50} of 15 μM for the PDZ1 domain and a selectivity of greater than factor 10 over the PDZ2 domain using an AlphaScreen. Polycyclic trans-fused ethers, for example, **22**, reminiscent in many marine natural products have been used as a scaffold to orient substituents in a way similar to the amino acid side chains of an amphiphatic α-helices [70]. Circular dichroism analysis revealed that this compound promotes α-helicity of the recognized peptide in aqueous media. Binding of the surface plasmon resonance and fluorescence polarization hit indoloquinolizinone **23** to the ZipA protein was determined by X-ray structure analysis. The high-resolution structure severed as starting point for SBDD to discover improved

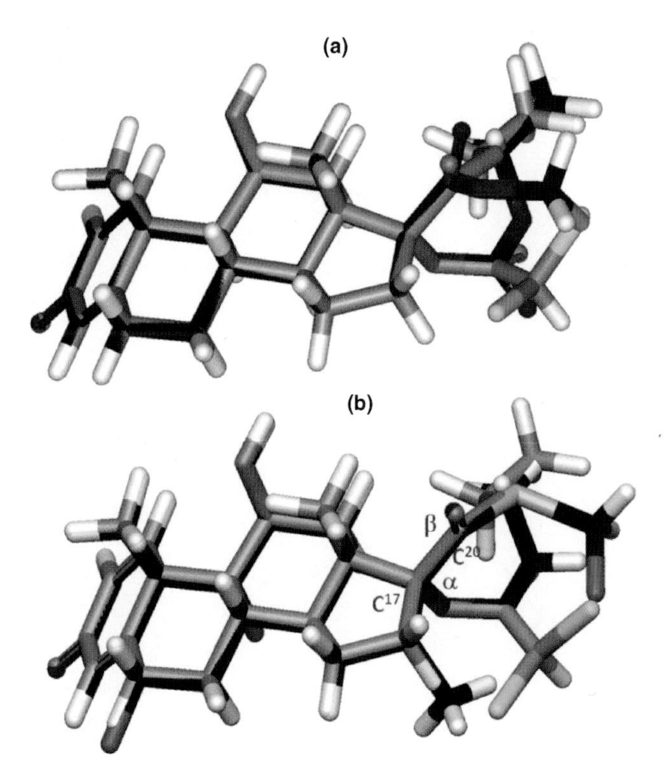

Figure 13 (Chapter 3). Overlapping pharmacophore structures of etiprednol dicloacetate (**45**) (in lighter colors) with loteprednol etabonate (**41**) (a) and fluticasone propionate (**59**) (b). Both views are from the β side, from slightly above the steroid ring system.

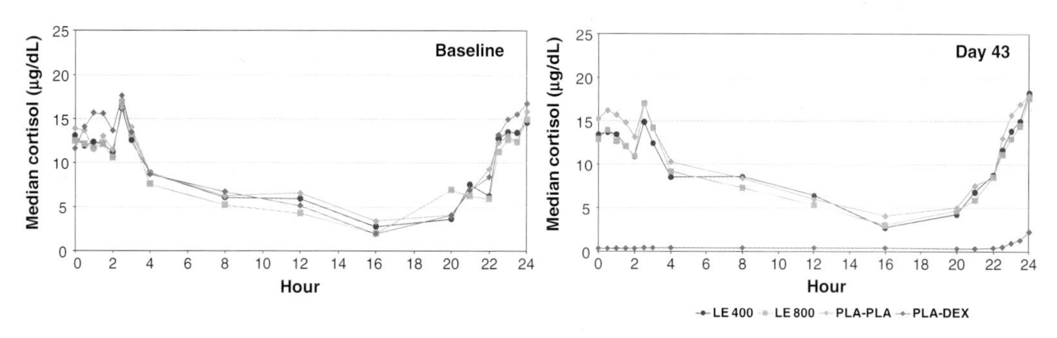

Figure 15 (Chapter 3). Median cortisol profiles obtained in a randomized, double-blind, placebo- and active-controlled, single center trial with four parallel groups designed to investigate the effects of glucocorticoid nasal sprays on the HPA axis function, as well as safety, general and local tolerability, and PK in patients ($n = 80$) with perennial allergic rhinitis (PAR) in a 6-week study.

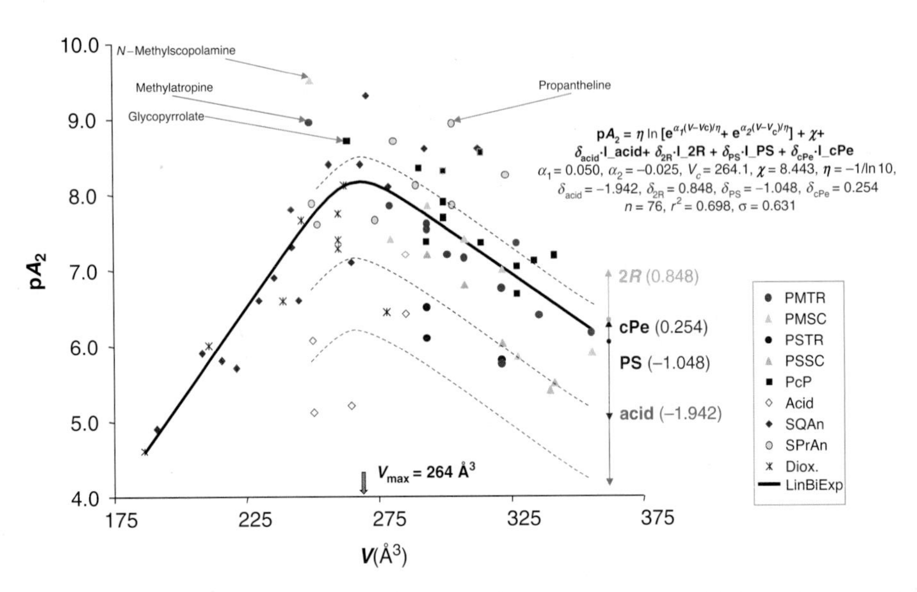

Figure 25 (Chapter 3). QSAR for the pA_2 data of all quaternary soft anticholinergics ($n = 76$) using molecular volume V as a main descriptor and the bilinear LinBiExp model with additional structural descriptors [I_acid for the presence of a carboxylic acid ($-COOH$), I_2R for enantiomerically pure $2R$ isomers, I_PS for succinic analogs where the carboxylic ester is one position away from the substitution center ($R_3 = -CH_2COOR$ versus $R_3 = -COOR$ in the malonic series), and I_cPe for the presence of a cyclopentyl substitution at the 2-position (as in glycopyrrolate)].

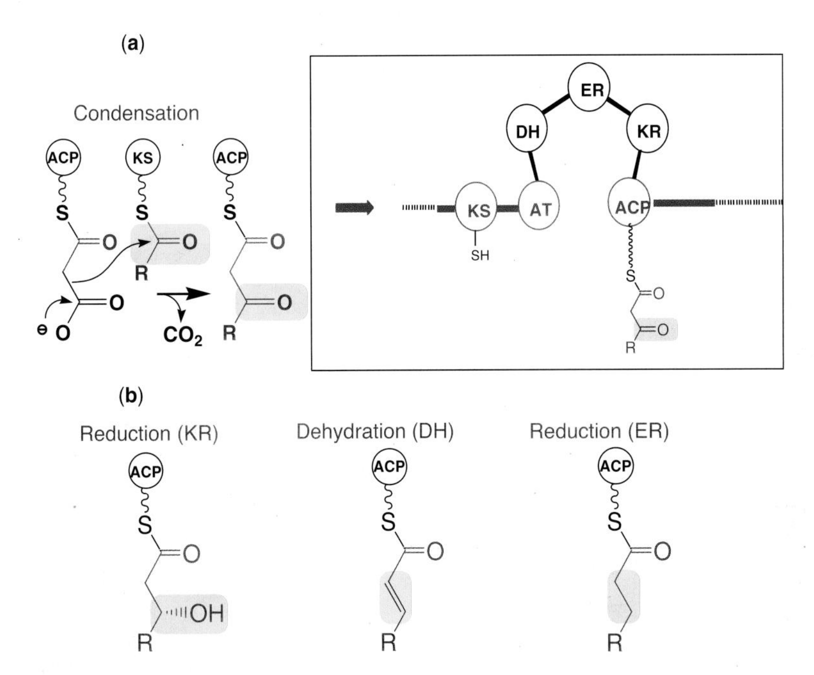

Figure 11 (Chapter 4). (a) Depiction of a single PKS module. The growing chain held by the ACP is condensed with the next acyl precursor as selected by the AT through the action of the KS, with the concomitant release of CO_2. (b) The resulting keto group is then optionally processed by the domains of the reductive loop, resulting in hydroxyl-bearing, olefinic, or saturated chains.

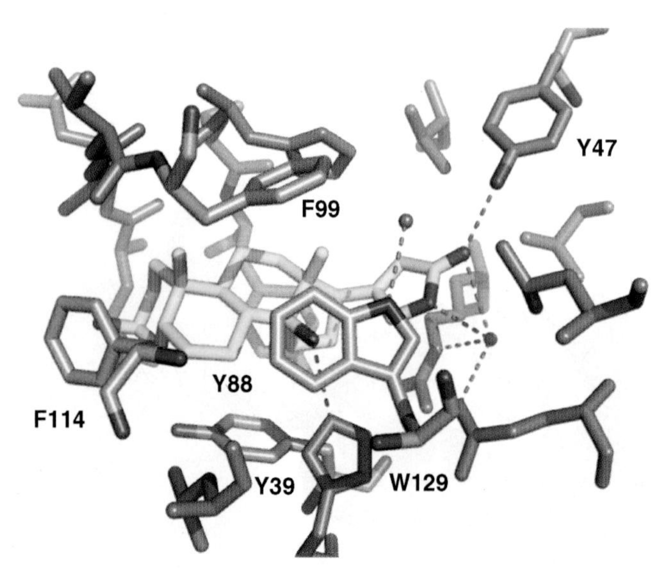

Figure 5 (Chapter 9). Example of aromatic amino acid residues and their role in the recognition process: interaction of the lipocalin binding to the small-molecule digoxigenin (PDB ID: 1LNM) [8]. A highly hydrophobic binding site comprising the aliphatic and aromatic amino acids Val, Leu, Ile (gray sticks) and Phe, Trp (pink sticks) and Tyr (blue sticks) recognizing the digoxigenin molecule (yellow sticks). The aromatic amino acid form a nest by aligning their π-aromatic rings in parallel to the hydrophobic guest. Additionally, Tyr47 forms a hydrogen bond to the lactam-carbonyl of digoxigenin.

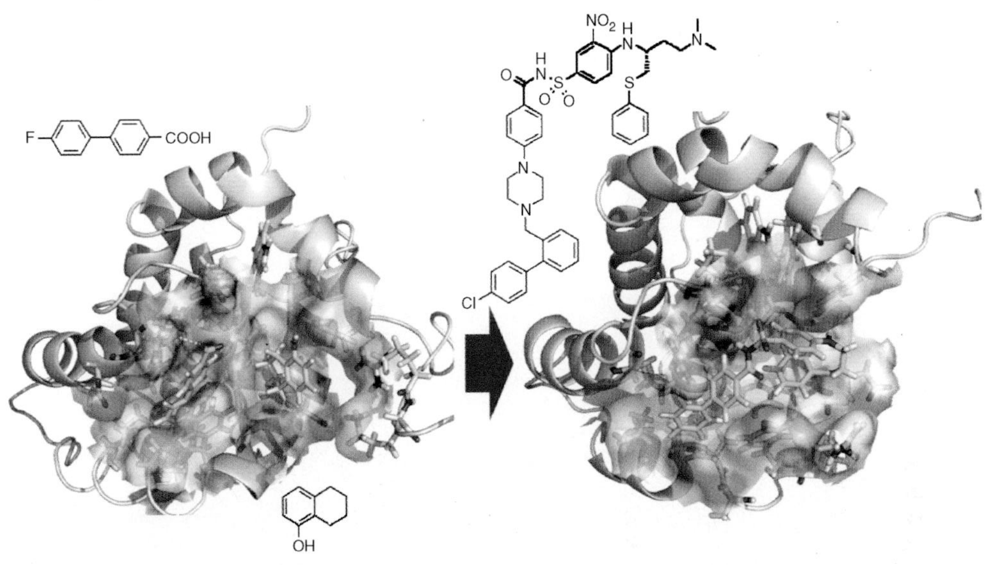

Figure 6 (Chapter 9). Fragments bound into the Bcl-x groove and the clinical candidate after lengthy optimization. Left: The 4′-fluorobiphenylcarboxylic acid in cyan sticks, the adjacent tetrahydronaphtol in yellow sticks (PDB ID: 1YSG). The carboxylic acid is making a charge–charge interaction with Arg1432 (yellow dotted lines) that is retained in the final ABT-737 **1** (cyan sticks, Right) in the acidic acylsulfonamide (PDB ID: 1YSI). Additionally, a hydrogen-bond contact is formed to the backbone amide of Gly142 and an oxygen of the sulfonamide portion.

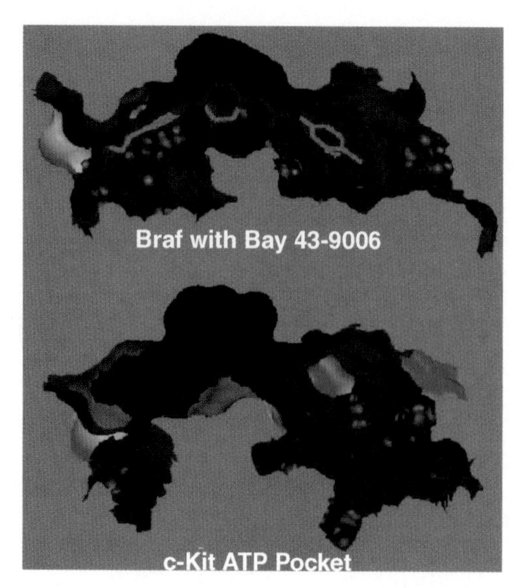

Figure 6 (Chapter 14). Target hopping with the Braf kinase inhibitor BAY 43-9006. According to SITESORTER, Braf kinase is one of the 10 most similar kinases to c-Kit. The binding site residues of 60% are conserved and colored blue; nonconserved positions are colored yellow; BAY 43-9006 is indicated in red [89].

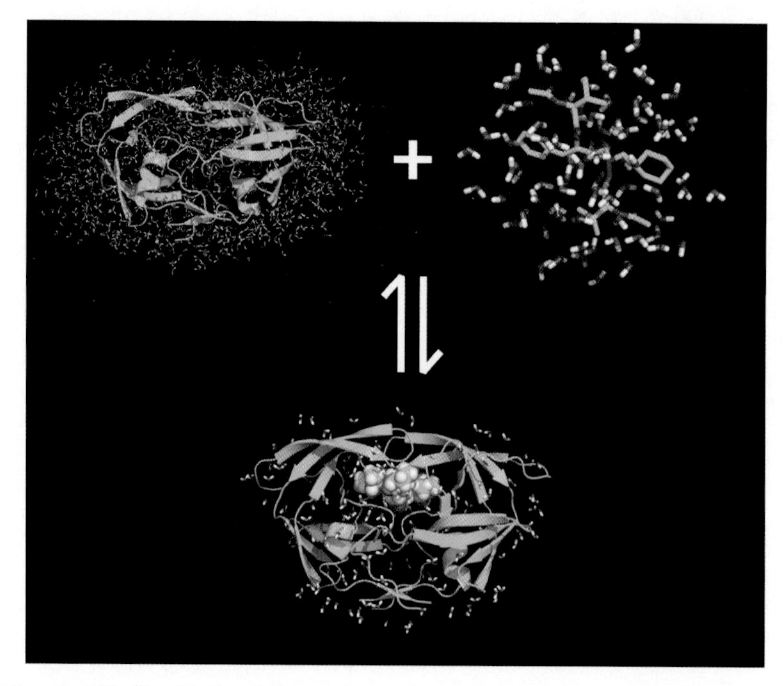

Figure 1 (Chapter 15). The biochemical and thermodynamic interpretation of docking. An uncomplexed protein (upper left) with water, ions, cofactors, etc. encounters a small molecule in solution (upper right). If the "sum" of an enormous number of things, for example, size, shape, enthalpy, entropy, etc. is (overall) favorable, the ligand may bind *reversibly* to form a protein–ligand complex (lower center).

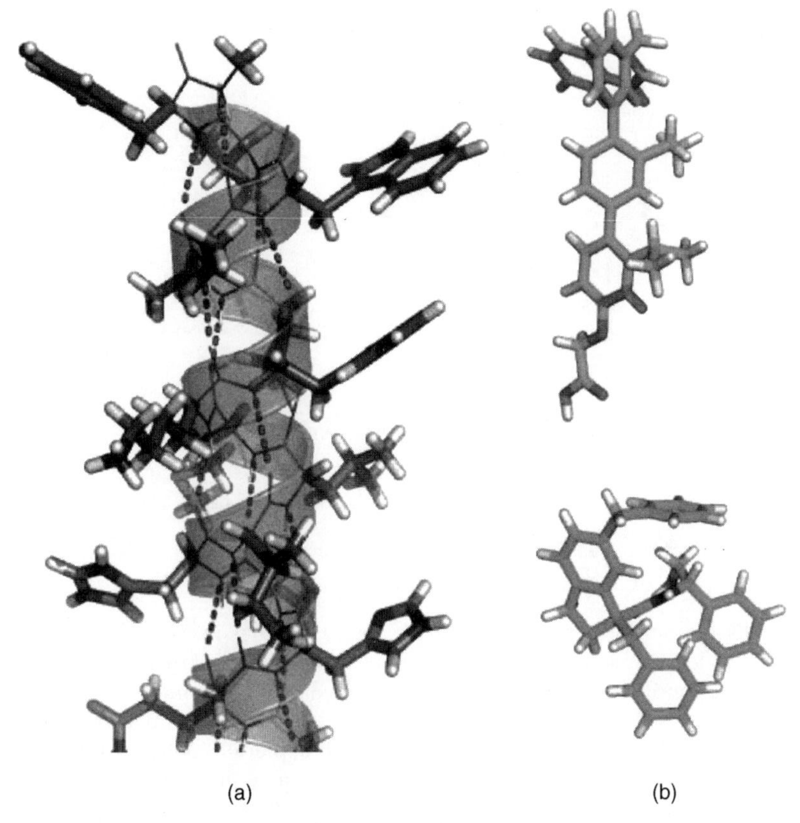

Figure 8. The structure of the α-helix ((a) light blue side chain sticks on a cartoon) is characterized by backbone NH group donates a hydrogen bond to the backbone C=O group of the amino acid four residues earlier ($i => i + 4$ hydrogen bonding). (b) Two scaffolds designed to mimic its side chains. Above: the terphenyl backbone and below the 1,1,6-trisubstituted indanes. The side chains of each backbone occupy a low energy conformational space resembling the side chains of the α-helix. (This figure is available in full color at http://mrw.interscience.wiley.com/emrw/9780471266945/home.)

compounds that also underwent structure determination [71]. 2-Aminoquinolines as ligands for the Src homology 3 (SH3) domains, for example, **24** have been designed based on available X-ray structure data [72] (Fig. 9).

Recently, a Web server and database called ANCHOR has been made available for the analysis of protein–protein interactions for small-molecule drug discovery [73]. ANCHOR exploits the so-called anchor residues, that is, amino acid side chains deeply buried at protein–protein interfaces, to indicate possible druggable pockets to be targeted by small molecules.

2.4. Computational Docking

Computational approaches that "dock" small molecules into the structures of macromolecular targets and "score" their potential complementarity to binding sites are widely used in hit identification and lead optimization [74]. Virtual docking of a small-molecule library (1900 compounds) to a site on Gβγ subunits of G-protein-coupled receptors that mediates protein interactions revealed several structurally diverse HITs that could be verified by binding assays and also selectively modulated functional Gbg–protein–protein interactions *in vitro*, chemotactic peptide signaling pathways in HL-60 leukocytes, and opioid receptor-dependent analgesia *in vivo*, compound M119 **25** [75]. The crystal structure of the Arf1-GDP/ARNO complex, which initiates the exchange reaction, was used to discover an inhibitor, LM11 **26**, using *in silico* screening of a flexible pocket near the Arf1/ARNO interface [76].

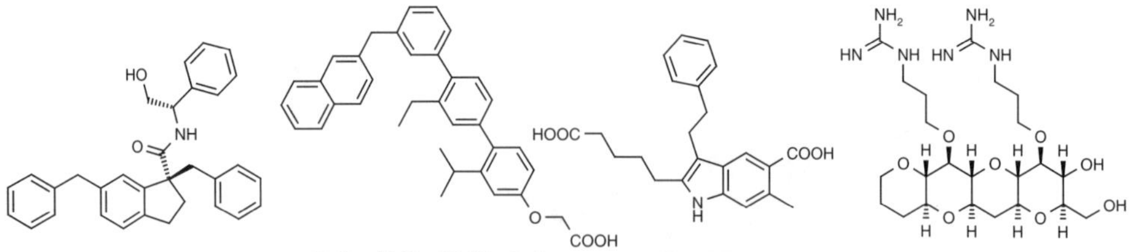

19: IC_{50} (NK1) = 3.6 μM **20:** IC_{50} (CaM-smMLCK) = 9 μM **21:** IC_{50} (PDZ1) = 15 μM **22**

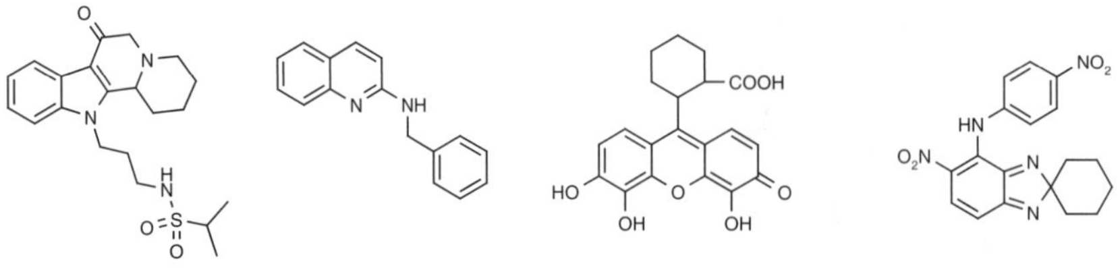

23: IC_{50} (ZIP1a) = 193 μM **24:** K_D (SH3 domain) = 193 μM **25:** IC_{50} (GrK2-bGb1g2) = 5 μM **26:**IC_{50} (Arf1-ARNO) = 49 μM

Figure 9. Structure-based designs and computational discoveries of PPI antagonists.

Recently, a detailed study on the use of chemoinformatic knowledge-based virtual screening (VS)of a large (1.2 million) corporate library to discovery p53–MDM2–MDM4 protein interaction antagonists was performed [25]. A combinations of VS methods, including homology-based similarity searching, QSAR, high-throughput docking, and UNITY pharmacophore searching provided a successful approach to the discovery of hits for this PPIs (activities with IC_{50} values between 1 and 60 μM). Depending on the activity threshold, the hit rates for the VS experiment are between 0.084 ($\leq 20\,\mu M$) and 1.836% ($\leq 64\,\mu M$) and for the HTS experiment between 0.029 and 0.366%, corresponding to an enrichment of 3 and 5 of the virtual approach over the HTS approach.

In the following, the PPI of the transcription factor p53 and its negative regulators MDM2 and MDM4 that are of particular relevance for cancer will be discussed in more detail, as a representative example of recent PPI of interest.

3. P53/MDM2/MDM4

The p53 tumor suppressor protein plays an overarching role in maintaining the genetic stability of mammalian somatic cells. It can transactivate and transrepress many hundreds of genes in response to a large number of stress signals, including those caused by DNA damage, telomere erosion, hypoxia, temperature change, nuclear disruption, and dNTP depletion. p53 has also been one of the most studied proteins in cancer research, primary because the loss of p53 function is involved in almost all human cancers. In about 50% of human cancers, p53 is inactivated as a result of missense mutation in the p53 gene. In the remainder of cancers, the tumor-suppression function of p53 can be impaired by the elevated activities of its proteinous inhibitors, such as MDM2, MDM4, and iASPP, or by the reduced activities of its activators, such as Arf, ASPP1, ASPP2, ATM, and p300.

In tumors that retain the wild-type p53, principal cellular antagonists of p53 are the MDM2 and MDM4 proteins (Fig. 10). MDM2 is transcriptionally activated by p53, which can in turn block the function of p53 by several mechanisms. First, MDM2 binds to the N-terminal transactivation domain of p53, resulting in an inhibition of p53-mediated transactivation; second, MDM2 has a nuclear export signal and shuttles the MDM2–p53 complex out of the nucleus into the cytosol. There, it serves as an ubiquitin ligase and targets p53 for degradation through the proteasome pathway. MDM2 interacts through its about 118-residue amino terminal domain with the N-terminal transac-

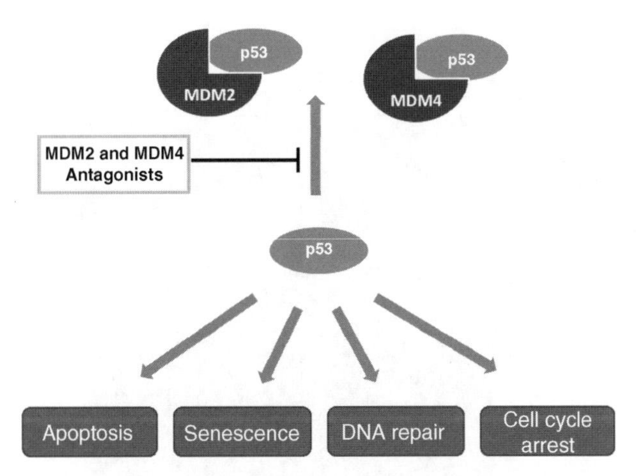

Figure 10. Transcription factor p53 plays a crucial role in tumor suppression by induction of cell cycle arrest and apoptosis. MDM2 and MDM4 are the principal negative regulators of p53 by forming tight interactions with p53. Antagonists of those interactions can free wild-type p53 and induce apoptosis in cancer cells. (This figure is available in full color at http://mrw.interscience.wiley.com/emrw/9780471266945/home.)

tivation domain of p53. In numerous studies, it was shown that the disruption of the p53–MDM2 interaction or the suppression of the MDM2 expression can activate the p53 pathway and inhibit tumor growth.

MDM4 is a close relative to MDM2 and their p53 binding domains are highly conserved, however, it is believed to regulate p53 by distinct mechanisms. Whereas MDM2 primarily regulates p53 stability and subcellular localization, MDM4 directly regulates p53 transcription. The detailed molecular mechanisms, however, are still under investigation. MDM4 is genetically amplified in a variety of tumors, including retinoblastomas; breast, colon, and lung carcinomas; and a smaller percentage of gliomas.

Recent biological investigations suggest that the combined use of MDM2 and MDM4 antagonists in cancer cells expressing wild-type p53 should activate p53 superior than agents that only antagonize MDM2, resulting in more effective antitumor activity [77–79]. Thus, the restoration of the impaired function of a single gene, p53, by disrupting the p53–MDM2 and p53–MDM4 interactions in cancers that still are equipped with the wild-type p53, offers a profoundly new avenue for anticancer therapy across a broad spectrum of cancers. The aim of current efforts is thus the identification of selective, potent, and dual action inhibitors.

The cancer biology of p53, MDM2, and MDM4 has been extensively reviewed [80–85].

Gratifyingly a wealth of Å resolution information is available on the interface of p53 and MDM2 (Table 1). MDM2 can bind to a 15-residue α-helical region of p53 ($K_d \approx 600$ nM),

Table 1. High-Resolution Structural Information on the p53/MDM2/MDM4 Interactions in the PDB

Interaction	PBD ID
MDM2–p53 peptide structures	3GO3 [86], 3EQS [87], 3JZS [88], 3JZR [88], 3LNJ [89], 3LNZ [90], 3IUX [91], 2GV2 [92], 1T4F [93], 1YCQ [94], 1YCR [94], 3IWY, 3IVJ
MDM2–small-molecule structures	3JZK [95], 1T4E [96], 1TTV [96], 1RV1 [97], 3LBK, 3LBL [98]
MDM4–p53 peptide structures	3JZO [88], 3JZP [88], 3JZQ [88], 3EQY [99], 3FE7 [99], 3FEA [99], 3IVJ, 3FDO, 3DAB [100], 3DAC [100], 2Z5S [101], 2Z5T [101]
MDM4–small-molecule structure	3LBJ [101]

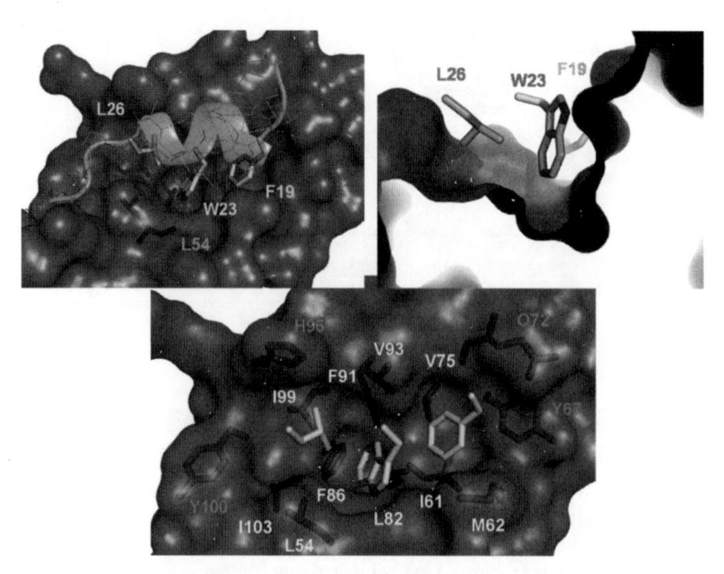

Figure 11. Interaction of p53 with MDM2 (PDB ID: 1YCR). Left above: Overview on the p53 binding to MDM2 (gray surface). The p53 binding epitope is shown as blue cartoon with yellow lines and the hotspot triad amino acid side chains (triad Leu26, Trp23, and Phe19) are marked as yellow sticks. Trp23 indole-NH is forming a hydrogen bond (cyan doted lines) to MDM2 Leu54 carbonyl (green sticks). Additionally, p53 forms several hydrogen bonds to MDM2 in the rim (cyan doted lines). Right: Above: Cut away view into the hotspot binding site with the hotspot side chains shown as yellow sticks and the MDM2 as gray surface. Note the deep triad binding site and the substantial void hydrophobic space below the Trp23. Below: Closeup of the p53 binding site on MDM2 (gray surface and highlighted hydrophobic and amphiphatic amino acids in gray and blue sticks, respectively). The binding site is flanked by the two rather flexible Tyr100 and Tyr67 located in the west and east side of the rim. Additional amphiphatic amino acids of the binding site are northwest His96 and northeast Gln72 again at the rim. The buried bottom and sides make a very hydrophobic binding pocket comprising the amino acids Phe91, Phe86, Ile99, Leu82, Ile61, and Val75, whereby the floor below the p51-Trp23 is made up from Phe86 and Leu82. The more water accessible sidewall of the binding groove is comprised of the hydrophobic amino acids Ile103, Leu54, Met62, and Val93. (This figure is available in full color at http://mrw.interscience.wiley.com/emrw/9780471266945/home.)

and the structure of the complex shows an interface that is largely hydrophobic. Physical and computational Ala-scanning mutational analysis of the 15-residue peptide identified three dominant amino acids in the center of the interface (= hotspot): Phe19, Trp23, and Leu26 (Fig. 11) [102,103]. The binding site is small and highly structured. The calculated buried accessible surface area of MDM2 in the interface is only 660 Å^2. The diameter of the MDM2-binding groove with 18 Å (from Phe100CH2 to Phe67CH2) is comparable to a small molecule. Thus, the hydrophobic triad of amino acids is at the center of design approaches for p53 MDM2 antagonists. The MDM2 target is intracellular and intranuclear. The compounds mimicking the p53 triad most likely are also hydrophobic that can facilitate the membrane transport. In the following, the different small molecular weight hits discovered and developed for this interface and their implications will be discussed. Peptide, peptoide, and oligomer approaches [92,104–107] are not considered here.

Undoubtedly, the best-documented compound in the p53 MDM2 area is Nutlin-3. It is a *cis*-imidazolidine derivative and has been discovered by HTS followed by medicinal chemistry optimization of the initial hit [97]. The binding mode of a Nultin-3 derivative **27** has been elucidated in atomic detail (Fig. 12) [96]. The 4-chlorophenyl substituent at the 4-position of the imidazoline mimics p53-Leu26, whereas the 4-chlorophenyl substituent at the 5-position corresponds to the

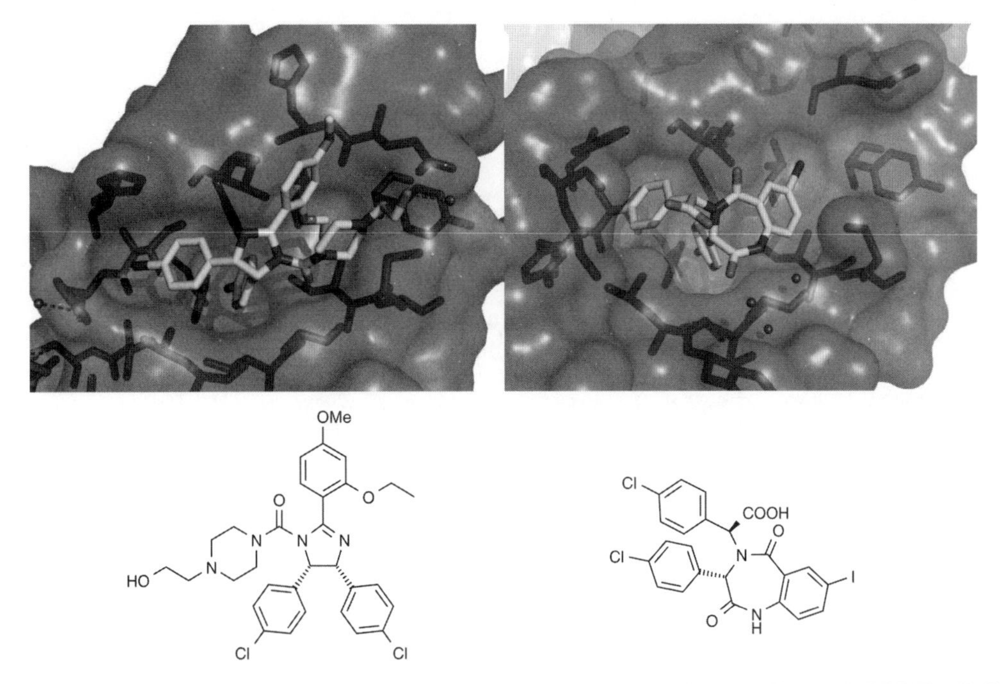

Figure 12. Binding of Nutlin-3 derivative **27** (PDB ID: 1RV1) and benzodiazepinedione (**28**) (PDB ID: 1T4E) inhibitor into MDM2 and comparison with the p53 binding mode. The binding MDM2 pockets are shown in a similar orientation than in Fig. 11. (This figure is available in full color at http://mrw.interscience.wiley.com/emrw/9780471266945/home.)

p53-Trp23 indole. The *o*-ethoxy side chain of the phenol in 2-position sticks deep into the p53-Phe19 pocket. Nutlin-3 binds very tightly to MDM2 with an IC_{50} of 90 nM. Additionally, it was also shown that Nutlin-3 is able to disrupt a preformed p53–MDM2 complex in aqueous solution [108]. An NMR technique called AIDA (antagonist-induced dissociation assay) was developed for this purpose [109]. Thus, it is comparable to or better than the natural p53 peptide (the apparent dissociation constant of natural p53 peptides ranges from 60 to 700 nM depending on the peptide length). More than hundred reports have been published on the cancer cell and xenograft biology of Nutlin-3 [110,111]. The usefulness of this compound and generally of compounds directed against the PPI p53/MDM2 has been shown in numerous studies [26]. Nutlin-3 is the prototype of a chemical biology tool: It is a commercially available, selective, highly active, and cell viable small molecular weight compound widely used in hundreds of cancer biology laboratories worldwide to decipher the p53 and related cell biology pathways.

A series of benzodiazepinedione inhibitors was described to antagonize p53–MDM2. The first hit was found by a HTS using a ThermoFluor screening assay [93]. An optimized structure **28** has a $K_i = 67$ nM (Thermofluor) and has been characterized in its binding to MDM2 in full atomic detail (Fig. 11). The two 4-chlorophenyl aromates again stick into the Leu26 and Trp23 pockets, whereas the 7-iodophenyl moiety mimics the p53–Phe19. Extensive medicinal chemistry optimization of the compound series finally leads to compounds with decent water solubility, however, retaining their affinity [112]. The chemistry used to access this compound series is very effective. Indeed, these benzodiazepinediones can be convergently synthesized in only two synthetic steps from commercially available building blocks involving a key Ugi reaction. This is in contrast to the similarly complex nultins that have to be synthesized in a lengthy sequential more than seven-step synthesis (Fig. 13) [112]. Xenograft data on a compound out of this series were published indicating

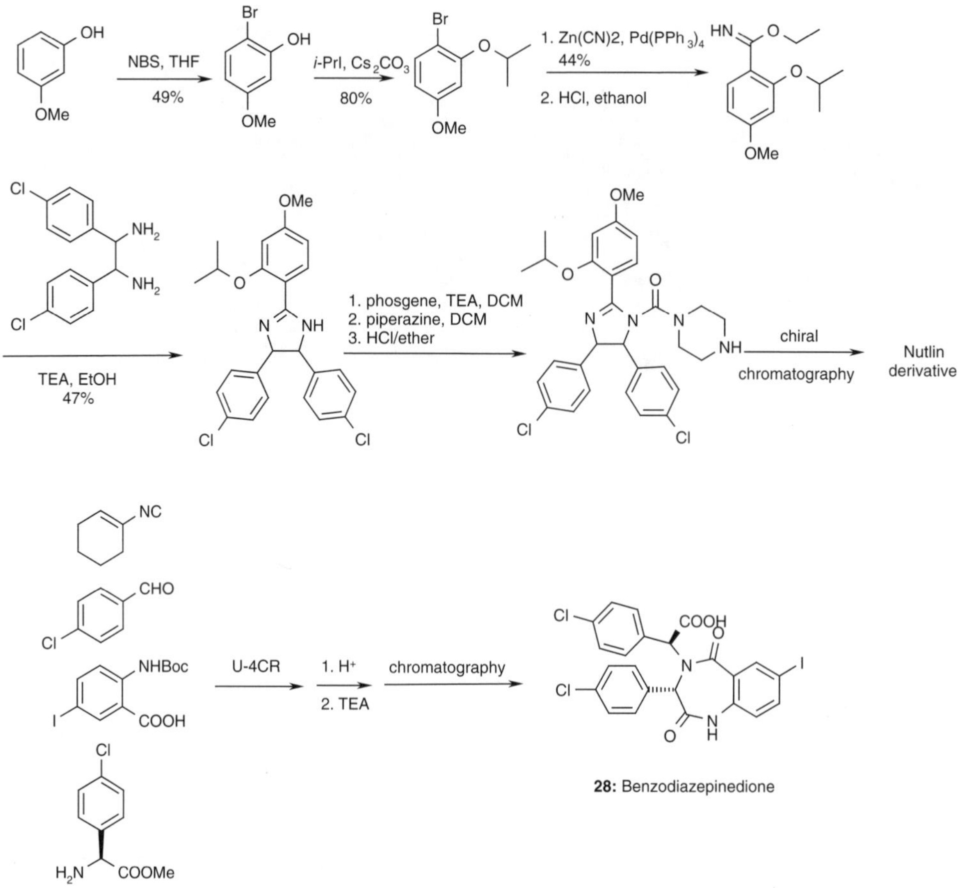

Figure 13. Comparison of the sequential nutlin and convergent benzodiazepinedione synthesis (no yields given).

potent anticancer activity in synergy with doxorubicin [113].

The first compound class antagonizing p53/MDM2 constitutes the chalcone and was discovered by multidimensional NMR-based screening (Fig. 14) [114]. Based on the HSQC signal shifts upon chalcone addition to labeled MDM2 the chlorophenyl moiety binds into the p53-Trp23 pocket and the remaining molecule fragment binds outside the hotspot. Biochemical experiments (electrophoretic gel mobility shift assay, ELISA) showed that these compounds can disrupt the MDM2/p53 protein complex, releasing p53 from both the p53/

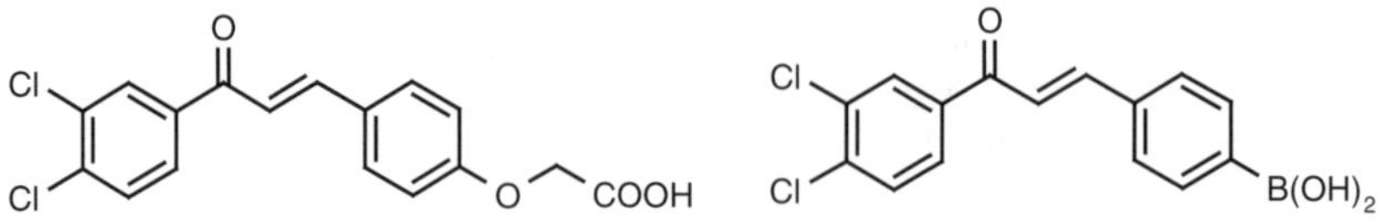

29: K_i = 49 μM (ELISA)
K_i = 90 μM (NMR)

30: GI_{50} (MDA-MB-435) = 4 μM
GI_{50} (MCF-10A) = 18 μM

Figure 14. Chalcones as p53/MDM2 antagonists.

MDM2 and the DNA-bound p53/MDM2 complexes. The authors also reported that the chalcones generally enhances the intrinsic tendency of MDM2 to aggregate at higher concentrations. Therefore, additional experiments were performed to test their specificity and to rule out a property as a general protein precipitant. Secondary assays to rule out unspecific binders (promiscuous inhibitors) are very important to perform since many lipophilic small molecules unspecifically bind to many proteins and precipitate them, a leading source of false positives in early drug discovery [115]. Improved boronic acid derived derivatives have been reported subsequently, which showed a 5–10-fold higher toxicity to human breast cancer cell lines compared to normal breast epithelial cell lines [116].

The bisthiophenefurane compound RITA (**31**) was reported to bind to p53 and thus antagonize the p53/MDM2 interaction on the basis of ELISA and fluorescence correlation spectroscopy (Fig. 15) [117]. Subsequent NMR-based analysis, however, showed no affinity of RITA toward neither p53 nor MDM2 [118]. Also was the compound unable to disrupt the preformed p53/MDM2 complex.

Guided by a rational library design of α-helic mimics, compound **32** has been discovered by FP-screening of 173 compounds (Fig. 16). As verified by NMR perturbation experiments, **32** binds into the p53 pocket of MDM2 [119].

Terphenyl derivative **33** is designed to orient the substituents in a similar 3D conformation that the amino acid side chains of the amphiphatic side of an α-helix (*i*, *i* + *3*, *i* + *7* cf. Phe19, Trp23, Leu26, Fig. 17) [120]. The compound shows a K_i of 4 μM in addition to inducing apoptosis in cancer cells at a GI_{50} of 20 μM. These terphenyl derivatives can be made by a sequential and rather lengthy synthesis.

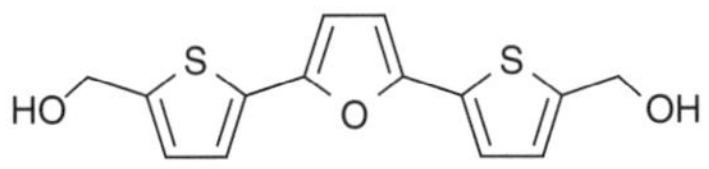

31: RITA

Figure 15. RITA, initially believed to bind p53/MDM2, has a different mode of action.

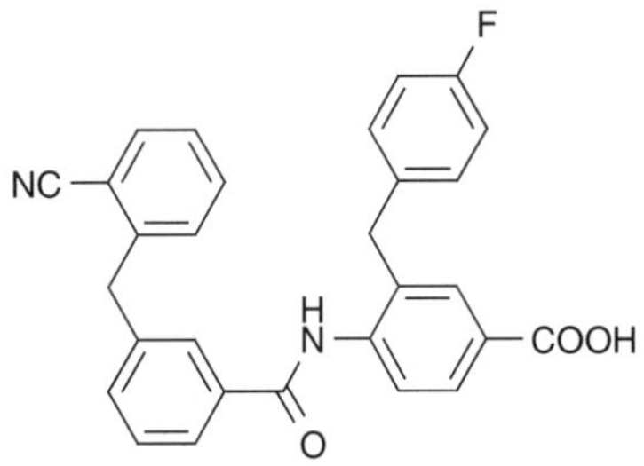

32: K_i = 12 μM

Figure 16. Proteomimetic α-helix mimic antagonizing p53/MDM2.

Isoindolone derivatives have been identified from *in silico* screening [121]. An optimized compound **34** showed a $K_i = 5$ μM in an ELISA based assay (Fig. 18). Chemical shift perturbations together with modeling was used to predict the binding mode of these compounds. For a set of 12 structurally related isoindolinones, the data suggest several orientations of binding, caused by subtle changes in the chemical structure of the inhibitors [122].

Cell-based screenings of compounds and libraries resulted in different compound classes, including norbornane derivative **35** (Fig. 19). The affinity of the compound to MDM2 has not been determined, but cellular activities showed apoptosis induction in cancer cells [123].

HTS of a natural product library of extracts (>50,000) identified the complex natural pro-

33: K_i – 4 μM

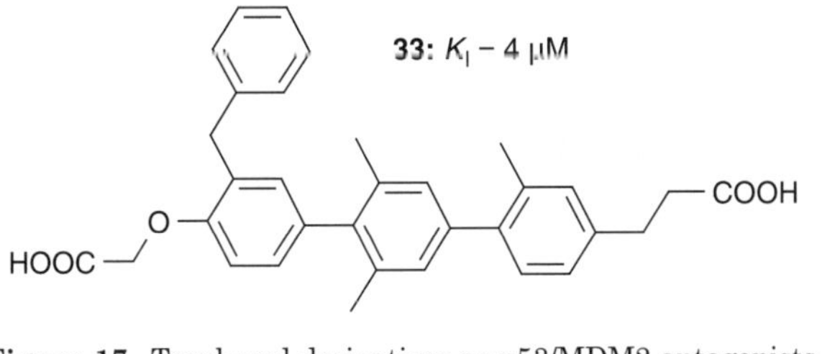

Figure 17. Terphenyl derivatives as p53/MDM2 antagonists.

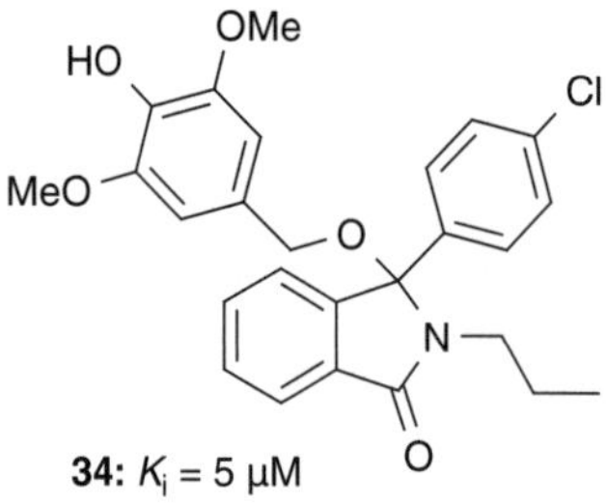

34: K_i = 5 µM

Figure 18. Isoindolone as p53/MDM2 antagonists.

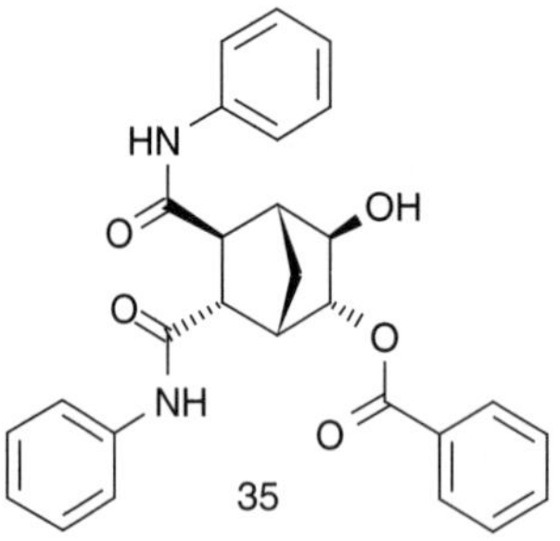

35

Figure 19. Norbornane derivatives as p53/MDM2 antagonists.

duct chlorofusin (**36**) isolated from a fungus as a low micromolar p53/MDM2 antagonist (Fig. 20) [124]. Using surface plasmon resonance, the $K_d = 4.7\,\mu M$ was determined [125]. The binding mode of the compound to MDM2 is unknown. Another natural product, hexylitaconic acid (**37**), isolated from a fermentation of a marine fungus inhibits the p53/MDM2 interaction, however, at rather high concentrations [126].

Computational screening of the NCI library afforded sulfonamide derivative **38** (Fig. 21). The K_i of 31.8 µM was determined using a peptide displacement assay [127]. Subsequent NMR-based screening, however, unequivocally showed that the compound precipitated the MDM2 protein at relevant concentrations in the NMR tube (promiscuous inhibitor?) [66]. The identification of NSC-333003 (**39**), a compound from the freely and public available NCI Diversity Set collection as a disruptor of the MDM2–p53 protein–protein interaction was facilitated by high-throughput *in silico* docking [128]. Virtual screening of 1990 compounds (3-D NCI Diversity Set) was performed and the 100 highest ranking compounds were physically screened. Amongt the compound found NSC-333003 had an IC_{50} of 20 µM in an ELISA and AlphaScreen assay. The docking hypothesis is that the benzothiazole, one phenyl and the other phenyl moieties mimics Phe19, Trp23, and Ile26, respectively. Computational screening of a 3500 compound virtual library using a highly sophisticated ensemble-based receptor model

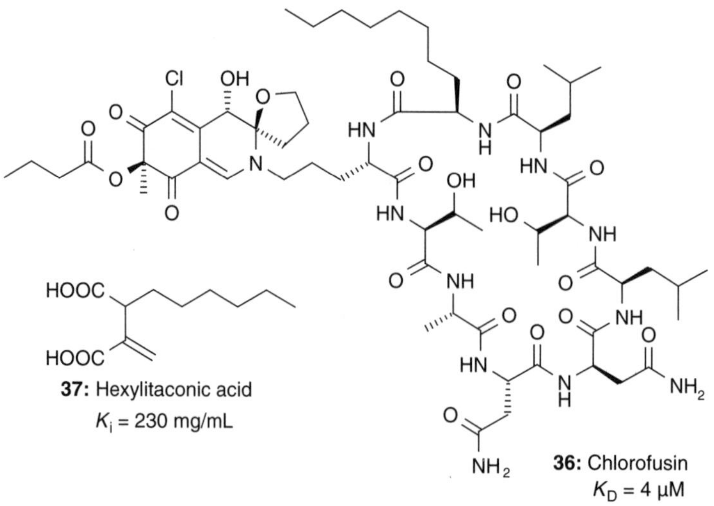

37: Hexylitaconic acid
K_i = 230 mg/mL

36: Chlorofusin
K_D = 4 µM

Figure 20. Natural product p53/MDM2 antagonists.

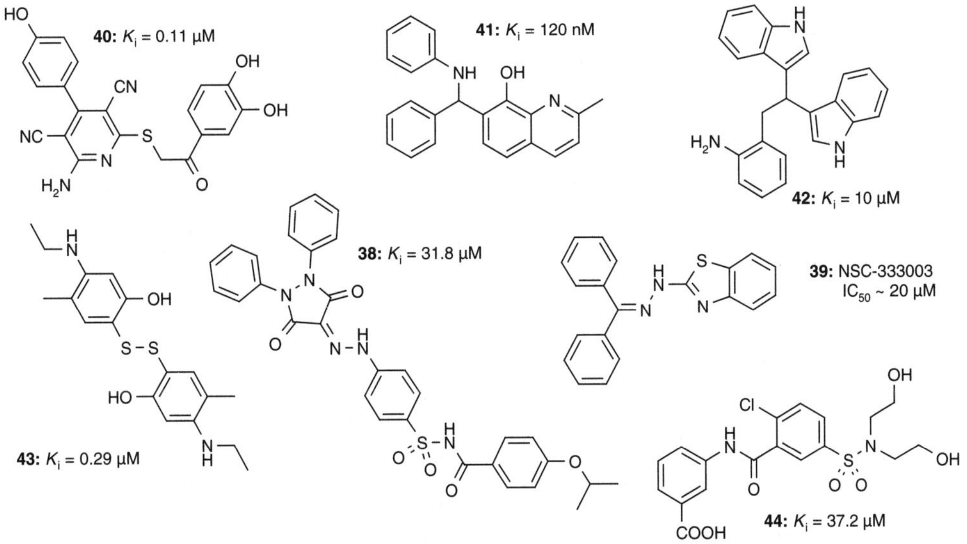

Figure 21. Different p53/MDM2 antagonists found by virtual screening.

yielded several compounds belonging to different chemotypes, including compounds **40–44** [129]. Several of the compounds showed impressive nanomolar affinity to MDM2. The affinities were determined by a fluorescence polarization assay.

Isoquinolinones and tetramic acids have been discovered by virtual screening using a ligand-based scaffold-hopping compound selection [130]. Herein, the three high-potency MDM2 binders nutlin, benzodiazepindione, and MI-43 were used as templates to search a library of commercially available compounds. As a substructure search method, a topological torsion (TT) descriptor was used and compounds were found that otherwise were not found using conventional similarity screening. The optimized compound NXN-561 (**45**) had a K_d of 5 µM whereas NXN-11 (**46**) showed some initial activity of approximately >200 µM (Fig. 22). NXN-561 separated in their enantiomers, surprisingly showed the

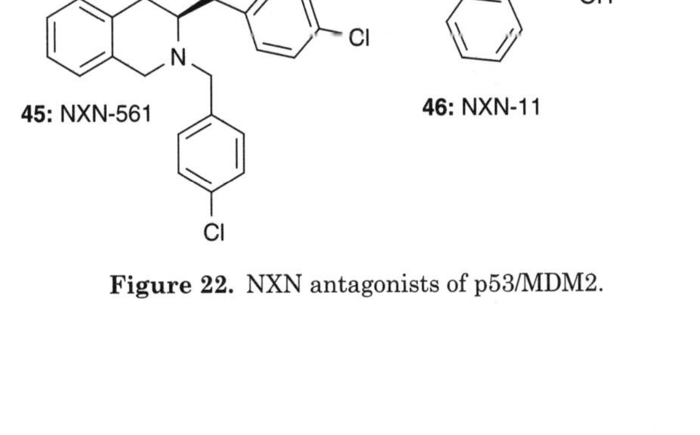

Figure 22. NXN antagonists of p53/MDM2.

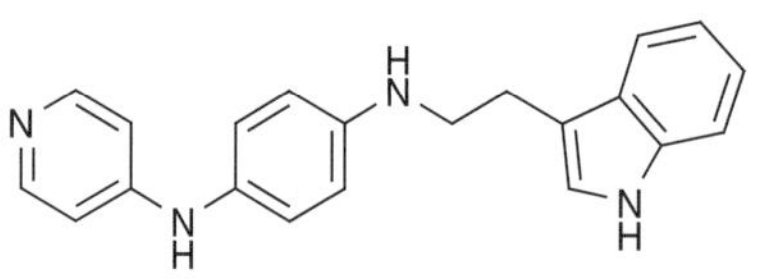

47: JNJ-26854165

Figure 23. JNJ-26854165 binds MDM2 and is currently undergoing clinical trials.

same K_d as measured by ITC- and NMR-based assays. This interesting observation can be explained by the almost planar symmetrical shape and electrostatic of the p53 binding site in MDM2. NXN-561 is able to dissociate the preformed p53/MDM2 complex as measured by the NMR AIDA antagonist-induced dissociation assay. Moreover, the compounds were shown to be able to induce specific genes downstream of p53, and induce early apoptosis. Both compound classes are again examples of molecules convergent accessible by MCR chemistry.

An indole derivative JNJ-26854165 (**47**) was reported to undergo clinical trials and seem to induce conformational changes in MDM2 upon binding (Fig. 23). Impressive activities in glioblastoma and nonsmall cell lung cancer xenografts showed a broad-spectrum tumor activity equipotent to taxol and outperforming Tarceva [131].

A novel class of spirocyclic compounds has been discovered based on the abundance of indole containing natural products with anticancer properties [132]. The series has been extensively optimized and the resulting MI-219 **48** not only is very potent (MW = 552.5; K_i (MDM2) = 5 nM) but also stimulates rapid but transient p53 activation in established tumor xenograft tissues, resulting in inhibition of cell proliferation, induction of apoptosis, and complete tumor growth inhibition [133]. To understand determinants of sensitivity or resistance to MDM2 inhibitor therapy in CLL (chronic lymphocytic leukemia), a large cohort of CLL patient-derived samples was comprehensively analyzed for response to MI inhibitors and correlated with clinically important biomarkers [134]. In this study, the wild-type p53 and several mutated forms were elaborated as potential clinical markers. A cocrystal structure of a MI-66 derivative **50** in MDM2 has been recently solved (Fig. 24). The stereochemistry of the derivative **50** differs only with respect to the phenyl group; nevertheless, the MDM2 affinities of **50** and **49** are almost identical. The cocrystal structure reveals that the oxindol part mimics the Trp23 indol side

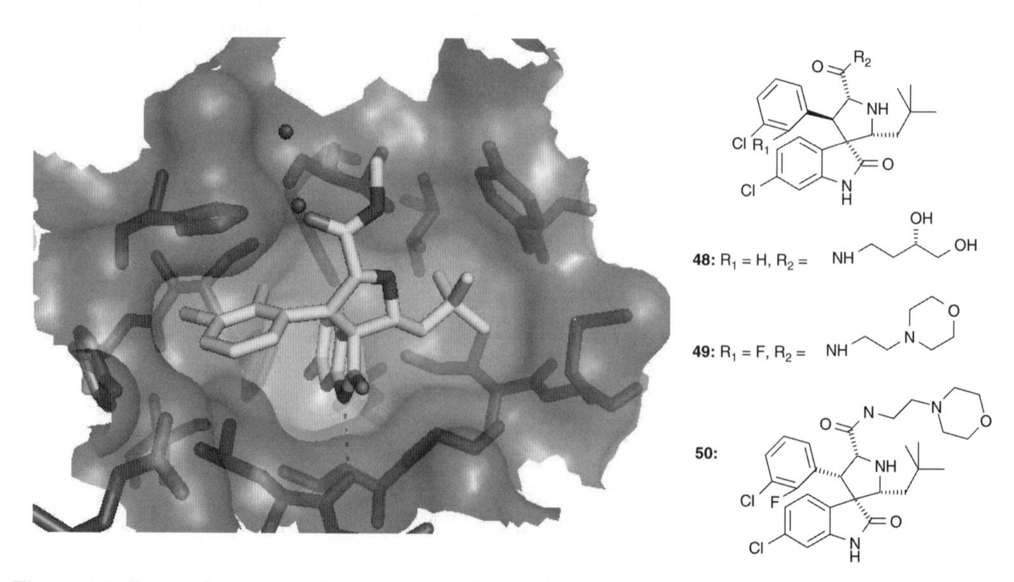

Figure 24. Cocrystal structure of structure of **50** (PDB ID: 3LBK) and different MI-219 compounds. There is no electron density to locate the morpholinoethyl side chain. (This figure is available in full color at http://mrw.interscience.wiley.com/emrw/9780471266945/home.)

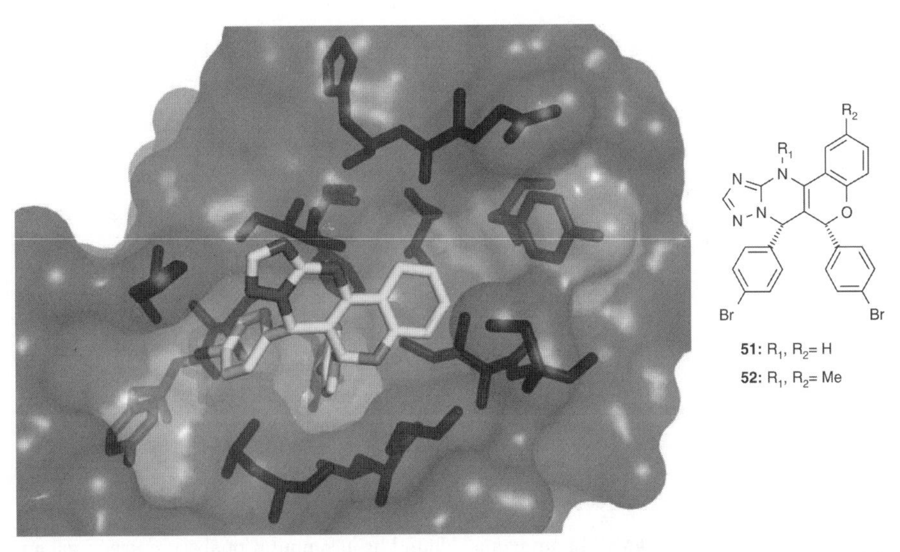

Figure 25. X-ray cocrystal structure of chromenotriazolopyridine **51** in MDM2 (PDB ID: 3JZK). (This figure is available in full color at http://mrw.interscience.wiley.com/emrw/9780471266945/home.)

chain including the hydrogen bond to the Leu54 amide. The Leu26 subpocket is filled by the 2-fluoro-3-chlorophenyl ring and the neopentyl moiety fills the Phe19 pocket. The later causes a substantial induced-fit reshaping of the binding cleft by Tyr67.

A series of chromenotriazolopyridines was discovered and optimized using HTS of ~1.4 million compounds, medicinal chemistry, and SBDD [95]. The initial screening hit **51** showed an $IC_{50} = 3.9\,\mu M$ (FRET, Förster resonanceenergytransfer) (Fig. 25). An optimized compound of the series was *N*-methylated to enhance hydrolysis stability and affinity (**52** IC_{50} $0.39\,\mu M$). Analysis of the cocrystal structure reveals the phenyl substructure of the chromene part to insert into the Ph219 pocket. The two 4-bromophenyl groups occupy the Trp23 and the Leu26 pockets, respectively. The triazole part resides on top of the hydrophobic Val93. Interestingly, in all the small-molecule MDM2 cocrystal structures, a halogen (bromo or chloro) aromate occupies the Leu26 pocket and the halogen makes a close contact with the methylene group of the Tyr100.

An new approach using an interwoven mix of different techniques delivered seven novel scaffolds (Fig. 26). For all these scaffolds, representative compounds were synthesized and they showed K_d values of 1–60 μM by NMR screening [23,135].

This method makes use of the concept of anchors in PPIs. A particular amino acid side chain deeply buried in the receptor protein (= anchor amino acid) is picked and virtual chemistry is performed involving this anchor. The resulting compound libraries based on different scaffolds and all containing the essential anchor are docked into the interface in a way that the amino acid side chain and the anchor of the small molecules overlap (constrained docking). These poses serve as starting point for energy minimization. From the resulting docking lists compounds are selected for synthesis and screening-based additional interactions the small molecules are able to capture. The chemistry used is the efficient and fast multicomponent reaction chemistry [135]. Also, this method has been currently used only to find medicinal chemistry starting points in p53/MDM2; it results in a very high hit rate (as compared to other methods) and could be a promising new approach also for other PPIs (Fig. 27).

All the scaffolds resulting from this approach are drug-like and straightforward to optimize since they come from multicomponent reactions. As an example, the imidazoline scaffold derived from the Orru-3CR with initial double digit micromolar K_i could be optimized to 1 μM compounds [23] with high water solubility [136]. One of these scaffolds, imidazolin-

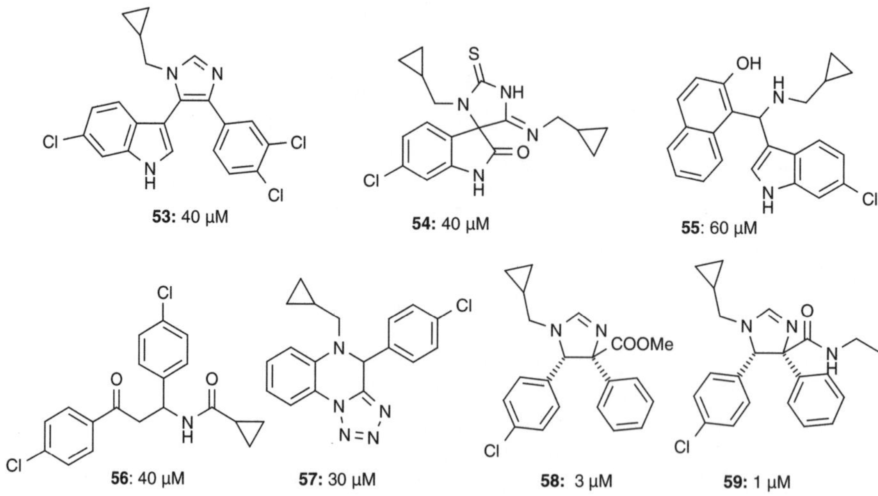

53: 40 μM 54: 40 μM 55: 60 μM

56: 40 μM 57: 30 μM 58: 3 μM 59: 1 μM

Figure 26. New MCR-based p53/MDM2 antagonists found by a computational structure-based approach.

doles, has been previous described as anticancer compounds and some derivatives show very high affinity to MDM2 [137,138]. As an encouraging evidence for the anchor approach, the crystal structure of the MCR compound **60** bound to MDM2 was solved (Fig. 28) [98]. This structure shows an almost perfect alignment of the indole form Trp23 with the indole moiety of

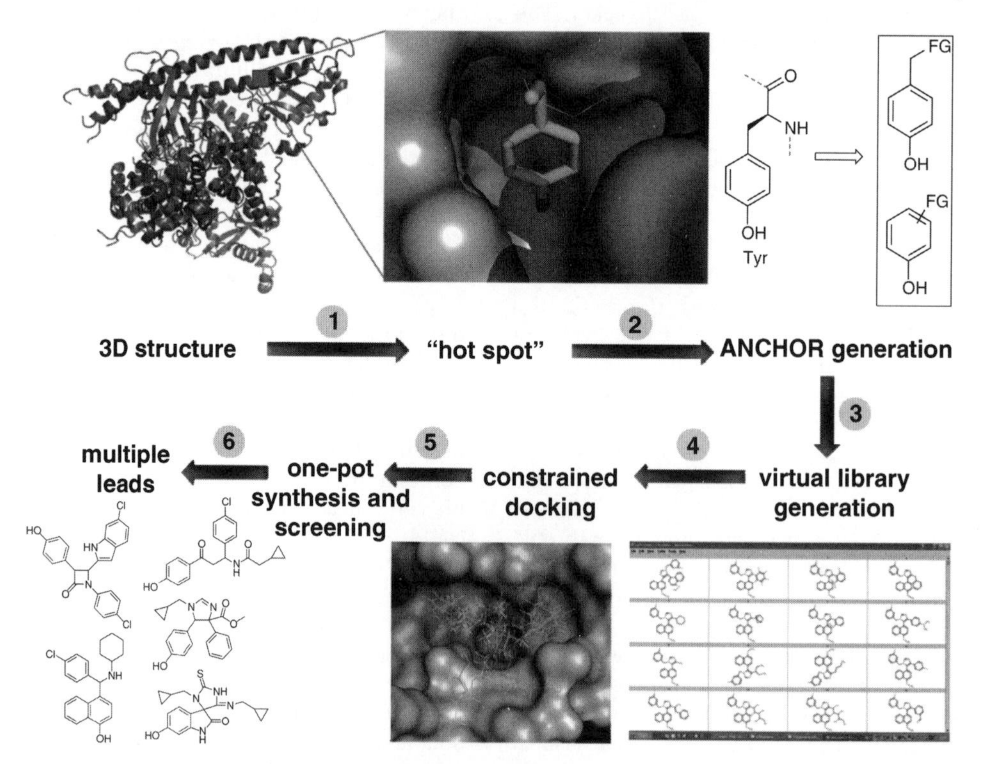

Figure 27. Schematic process of discovery of PPI antagonists, based on structural information, hotspot anchors and rapid MCR chemistry. (This figure is available in full color at http://mrw.interscience.wiley.com/emrw/9780471266945/home.)

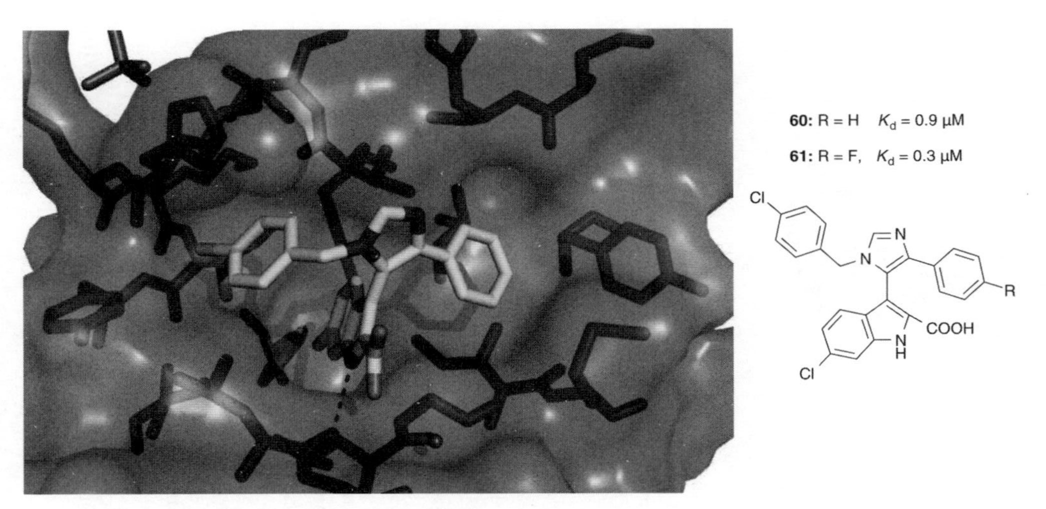

Figure 28. Cocrystal structure of imidazole **60** with MDM2 (PDB ID: 3LBK). (This figure is available in full color at http://mrw.interscience.wiley.com/emrw/9780471266945/home.)

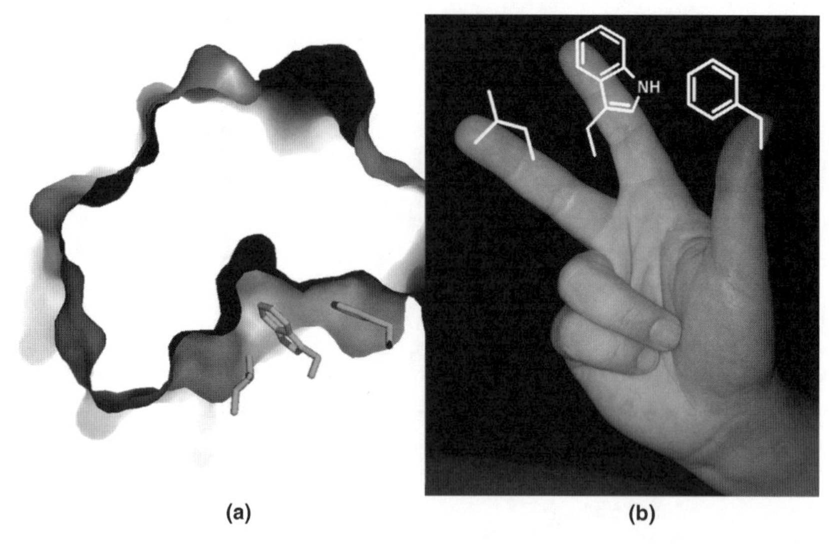

Figure 29. The "thumb–index–middle"-finger pharmacophore model for p53/MDM2 antagonists. (a) Cut-off view of the p53 binding site in MDM2 showing the hotspot side chains of Leu26, Trp23, Phe19. (b) The special orientation of the three hotspot amino acid side chains aligns well with the three fingers—thumb, index, and middle. In order to obtain tight binding, a molecule has to reproduce the three-finger shape and retain the hydrophobicity of the side chains. (This figure is available in full color at http://mrw.interscience.wiley.com/emrw/9780471266945/home.)

60. Additionally, the 4-chlorobenzyl group inserts into the Leu26 pocket and the phenyl moiety into the Phe19 pocket. An otherwise identical derivative **61** with a 4-fluorophenyl group results in a factor 3 affinity improvement.

Many different p53–MDM2 antagonists have been discovered recently mostly in the micromolar range of affinity. Binding affinities for useful therapeutic intervention, however, should be in the low nanomolar or even picomolar range. In order to achieve high affinity to the p53 binding site in MDM2, compounds with shape complementarity and sufficient lipophilicity are needed. Interestingly, the replication of the hydrogen bond of p53-Tyr23 indole-H to MDM2-Leu54 carbonyl is not necessary to obtain highly potent inhibitors. In fact, several high-affinity inhibitors are known with indole or the bioisosteric phenol instead (e.g., nultin and chromene derivatives). Only few MDM2 antagonists, however, show potent low nanomolar affinity to MDM2. For p53 hotspot mimicking, MDM2 binding compounds very sophisticated pharmacophore models have been developed, however, most described compounds also fit nicely into a very simple qualitative pharmacophore model, the three-finger model [139]. According to this model, a basic central scaffold has to suitably orient three lipophilic substituents into the three MDM2 subpockets as indicated in figure 29.

Despite abundant descriptions of MDM2 antagonists, small molecular weight MDM4 antagonists until recently have been unknown. Typically, there is a difference of a factor of 1000 or more in binding of the MDM2 antagonists and their binding to MDM4, for example, nultin-3 (MDM2, MDM4). This is surprising regarding the high-sequence homology of the MDM2 and MDM4 proteins especially in the p53 binding pocket and the very similar shape and dimension of the binding pockets. The wealth of crystal structures is a good starting point to understand the commonalities and differences of p53 binding to MDM2 versus MDM4 (Table 1, Fig. 30). The MDM2–p53 and MDM4–p53 binding pockets are most similar in the Phe19 subpockets and most different in the opposing Leu26 pockets. On the bottom of the Trp23 binding pocket, Phe86 (MDM2) is mutated to the smaller L86 (MDM4) and Ile99 (MDM2) is mutated to the Leu98 (MDM4). A shift of the two helices toward each other renders the Leu26 and Trp23 binding sites much smaller in MDM4

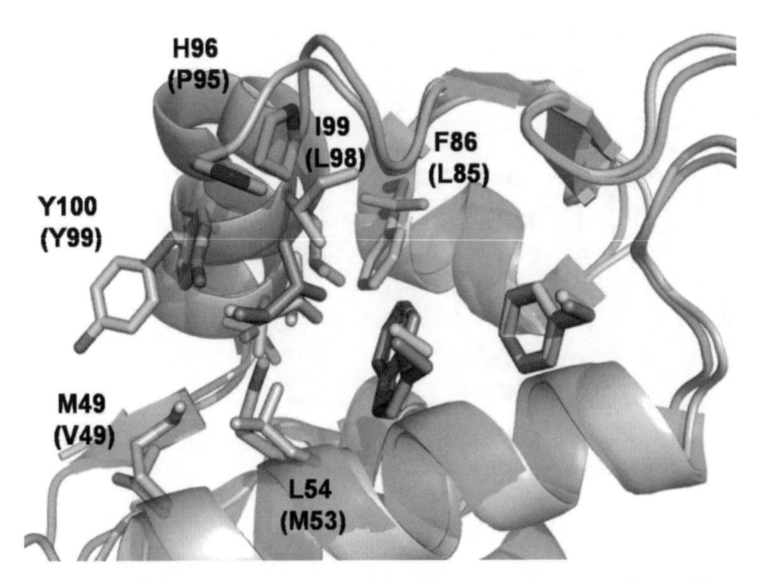

Figure 30. Key structural differences of the p53–MDM2 and p53–MDM4 complexes. Alignment of the p53 binding sites of MDM2 (PDB ID: 1YCR) and MDM4 (PDB ID: 3DAB) with their p53 peptides. MDM2 is shown in gray cartoons with selected amino acid side chains highlighted as sticks and the corresponding hotspot p53 amino acid side chains $F^{19}W^{23}L^{26}$ as yellow sticks. Selected amino acid naming is given for MDM2 and MDM4 in brackets. The corresponding colors in the MDM4–p53 complex are cyan and pink, respectively. The two complexes align with an RMSD of 1.014 Å. (This figure is available in full color at http://mrw.interscience.wiley.com/emrw/9780471266945/home.)

as compared to MDM2. Additionally, the shape and size of the Leu26 binding site in MDM4 is considerably downsized by the following mutations: His96-Pro95 and Leu54-Met53. Overall, these amino acid differences lead to a much smaller Leu26 and slightly more shallow Trp23 pockets. These subtle differences in binding site, shape and volume explain the high selectivity of current inhibitors for MDM2 over MDM4.

The class of indolo-imidazoles was recently independently discovered by two groups as potent p53–MDM2 antagonists [98,137,138,140]. Some of the compounds also exhibit micromolar MDM4 affinity. For example, compound **62** is a very potent MDM2 antagonist ($K_i = 100$ nM) but also shows weak MDM4 affinity ($K_i = 11\,\mu$M) and was therefore subjected to crystallization experiments [98]. The cocrystal structure of **62** bound to MDM4 is shown in Fig. 31. Analysis of the structure reveals that the 4-chlorobenzyl substituent that binds the Leu26 pocket does not optimally fit whereas the indole and the phenyl substituents seem to be nicely accommodated in the Trp23 and Phe19 pockets, respectively.

A high-throughput exercise of approximately 300.000 unique compounds revealed a series of compounds binding covalently to MDM4 (Fig. 32) [141]. Although one compound **63** was deemed to bind reversibly, careful NMR-based experiments revealed covalent binding (Fig. 32) [142]. The compounds described throughout this in this work comprise either Michael acceptor substructures, redox active compounds, or known promiscuous inhibitors. Such compound classes that appear as frequent hits in many biochemical HTS screens have long been known, and have been appropriately called PAINS (for pan assay interference compounds) [143]. PAINS in most cases are poor drug development choices and should be rigorously excluded from HTS libraries. Protein–protein interaction HTS hits are especially prone to be promiscuous inhibitors due to the very hydrophobic nature of most PPIs. Care must be taken to employ a battery of both primary and secondary assays, the results of which are best based on independent physicochemical assay principles.

The ligand efficiency (LE) has recently become an important measure and a useful me-

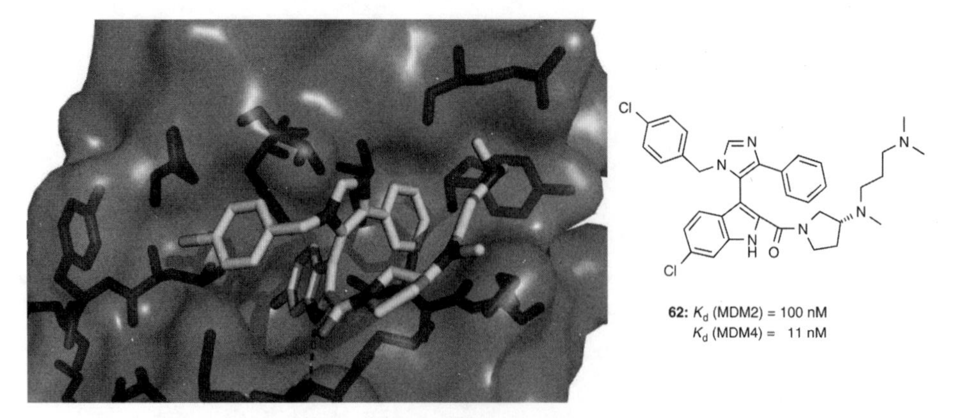

Figure 31. Cocrystal structure of imidazole **62** in MDM4 (PDB 3LBJ). (This figure is available in full color at http://mrw.interscience.wiley.com/emrw/9780471266945/home.)

Figure 32. Covalent MDM4 binder **63** and a possible mechanism of covalent binding.

tric for lead selection [144,145]. LE is defined as the binding atom per nonhydrogen heavy atom in the ligand. In order to balance potency and molecular weight of lead compounds ligand efficiency has become an important metric. Lead compounds should exhibit high target affinity but at the same time should be low in molecular weight to increase the chance of biological membrane permeation. LE of different MDM2 and MDM4 antagonists are summarized in Table 2. Interestingly, LE in all cases is higher for the much smaller synthetic

Table 2. Ligand Efficiencies of Other Small Molecules that Inhibit p53–MDM2 and p53–MDM4 Interactions

Target	Compound	PDB ID	Affinity K_i (μM)	Ligand Efficiency	Molecular Mass (Da)	Reference
MDM2	p53 peptide	1YCR	0.6	0.12	1,808	[94]
MDM2 (MDM4)	Nultin-3 (**27**)	1RV1	0.09	0.24	581	[94]
MDM2	BDA (**28**)	1T4E	0.067	0.31	566	[93]
MDM2 (MDM4)	MI-63 der. **50**	3LBL	0.036 (55)	0.26 (0.15)	577	[98]
MDM2	Chromene **51**	3JZK	11	0.26	536	[95]
MDM2 (MDM4)	Imidazole **60**	3LBK	0.916 (36)	0.26 (0.19)	462	[98]
MDM4	p53 peptide	3DAB	0.21	0.091	1,478	[100]
MDM4 (MDM2)	Imidazole **62**	3LBJ	0.11 (11)	0.21 (0.15)	630	[98]

ligand as compared to the native p53 peptide. Analysis of a stet of PPIs revealed a LE of 0.24 kcal/mol that is smaller than kinase inhibitors (0.3–0.4 kcal/mol) and the same orders of magnitude for protease inhibitors (0.25–0.35 kcal/mol) [26].

Insights into the requirements for small molecules to bind to MDM4 have been recently made possible by the high-resolution cocrystal structures of the interaction of p53 with MDM4 and the first small-molecule binding to MDM4. These structures allow to identify the details of the similarities and differences between the p53 binding sites of MDM2 and MDM4 and can thus form the basis for the discovery of potent and selective MDM4 and dual active MDM2/4 antagonists.

4. OUTLOOK

Several years ago, small molecular weight PPI antagonists were considered as the Mount Everest in drug discovery and generally regarded as too difficult to be targeted. The number of marketed products, clinical compounds, and preclinical projects clearly shows that small molecular weight (ant)agonists of PPIs can be approached. Many techniques have been described that can be used to efficiently find starting points for medicinal chemistry projects to optimize initial hits. Still, PPIs are more challenging than traditional targets based on their diversity, shape, form, and function characteristics.

REFERENCES

1. Nooren IMA, Thornton JM. Diversity of protein–protein interactions. EMBO J 2003; 22(14): 3486–3492.
2. Tuncbag N, Gursoy A, Guney E, Nussinov R, Keskin O. Architectures and functional coverage of protein–protein interfaces. J Mol Biol 2008;381(3): 785–802.
3. Moreira IS, Fernandes PA, Ramos MJ. Hot spots: a review of the protein–protein interface determinant amino-acid residues. Proteins 2007;68(4): 803–812.
4. Conte LL, Chothia C, Janin J. The atomic structure of protein–protein recognition sites. J Mol Biol 1999;285(5): 2177–2198.
5. Janin J, Chothia C. The structure of protein–protein recognition sites. J Biol Chem 1990;265 (27): 16027–16030.
6. Horton N, Lewis M. Calculation of the free energy of association for protein complexes. Protein Sci 1992;1(1): 169–181.
7. Fernández A, Scheraga HA. Insufficiently dehydrated hydrogen bonds as determinants of protein interactions. Proc Natl Acad Sci USA 2003;100(1): 113–118.
8. Bogan AA, Thorn KS. Anatomy of hot spots in protein interfaces. J Mol Biol 1998;280(1): 1–9.
9. Rajamani D, Thiel S, Vajda S, Camacho CJ. Anchor residues in protein–protein interactions. Proc Natl. Acad. Sci. 2004;101(31): 11287–11292.
10. Clackson T, Wells J. A hot spot of binding energy in a hormone–receptor interface. Science 1995;267(5196): 383–386.
11. Clackson T, Ultsch MH, Wells JA, de Vos AM. Structural and functional analysis of the 1:1 growth hormone:receptor complex reveals the molecular basis for receptor affinity. J Mol Biol 1998;277(5): 1111–1128.
12. Muller YA, Li B, Christinger HW, Wells JA, Cunningham BC, de Vos AM. Vascular endothelial growth factor: crystal structure and functional mapping of the kinase domain receptor binding site. Proc Natl Acad Sci USA 1997;94(14): 7192–7197.
13. Thanos CD, DeLano WL, Wells JA. Hot-spot mimicry of a cytokine receptor by a small molecule. Proc Natl Acad Sci USA 2006;103(42): 15422–15427.
14. von Mering C, Jensen LJ, Snel B, Hooper SD, Krupp M, Foglierini M, Jouffre N, Huynen MA, Bork P. String: known and predicted protein–protein associations, integrated and transferred across organisms: Nucl Acids Res. 2005;33(Suppl. 1); D433–D437.
15. Berman HM, Westbrook J, Feng Z, Gilliland G, Bhat TN, Weissig H, Shindyalov IN, Bourne PE. The protein data bank. Nucl Acids Res 2000;28(1): 235–242.
16. Oltersdorf T, Elmore SW, Shoemaker AR, Armstrong RC, Augeri DJ, Belli BA, Bruncko M, Dockworth TL, Dinges J, Hajduk PJ, Joseph MK, Kitada S, Korsmeyer SJ, Kunzer AR, Letai A, Li C, Mitten MJ, Nettesheim DG, Ng S, Nimmer PM, O'Connor JM, Oleksijew A, Petros AM, Reed JC, Shen W, Tahir SK, Thompson CB, Tomaselli KJ, Wang B, Wendt MD, Zhang H, Fesik SW, Rosenberg SH. An inhibitor of bcl-2 family proteins induces re-

gression of solid tumours. Nature 2005;435 (7042): 677–681.

17. McCafferty GP, Pullen MA, Wu C, Edwards RM, Allen MJ, Woollard PM, Borthwick AD, Liddle J, Hickey DMB, Brooks DP, Westfall TD. Use of a novel and highly selective oxytocin receptor antagonist to characterize uterine contractions in the rat. Am J Physiol Regul Integr Comp Physiol 2007;293(1): R299–R305.
18. Liddle J, Allen MJ, Borthwick AD, Brooks DP, Davies DE, Edwards RM, Exall AM, Hamlett C, Irving WR, Mason AM, McCafferty GP, Nerozzi F, Peace S, Philp J, Pollard D, Pullen MA, Shabbir SS, Sollis SL, Westfall TD, Woollard PM, Wu C, Hickey DMB. The discovery of gsk221149a: a potent and selective oxytocin antagonist. Bioorg Med Chem Lett 2008; 18(1): 90–94.
19. Stasi R, Evangelista ML, Amadori S. Novel thrombopoietic agents: a review of their use in idiopathic thrombocytopenic purpura. Drugs 682008; 901–912.
20. Hartman GD, Egbertson MS, Halczenko W, Laswell WL, Duggan ME, Smith RL, Naylor AM, Manno PD, Lynch RJ. Non-peptide fibrinogen receptor antagonists 1. Discovery and design of exosite inhibitors. J Med Chem 1992;35(24): 4640–4642.
21. Spencer RW. High-throughput screening of historic collections: observations on file size, biological targets, and file diversity. Biotechnol Bioeng. 1998;61(1): 61–67.
22. Robinson JA, DeMarco S, Gombert F, Moehle K, Obrecht D. The design, structures and therapeutic potential of protein epitope mimetics: Drug Discov Today. 2008;13(21–22): 944–951.
23. Murray CW, Rees DC. The rise of fragment-based drug discovery. Nature Chem 2009;1(3): 187–192.
24. Cochran AG. Antagonists of protein–protein interactions. Chem Biol 2000;7(4): R85–R94.
25. Jacoby E, Boettcher A, Mayr LM, Brown N, Jenkins JL, Kallen J, Engeloch C, Schopfer U, Furet P, Masuya K, Lisztwan J, Knowledge-based virtual screening: application to the mdm4/p53 protein–protein interaction. In: Jacoby E, editor. Chemogenomics: Methods and applications. Vol. 575, Springrs. 2009. p. 173–194.
26. Wells JA, McClendon CL. Reaching for high-hanging fruit in drug discovery at protein–protein interfaces. Nature 2007;450(7172): 1001–1009.
27. Keiser MJ, Roth BL, Armbruster BN, Ernsberger P, Irwin JJ, Shoichet BK. Relating protein pharmacology by ligand chemistry. Nat Biotech 2007;25(2): 197–206.
28. Jacoby E, Bouhelal R, Gerspacher M, Seuwen K. The 7TM G-protein-coupled receptor target family13. ChemMedChem 2006;1(8): 760–782.
29. Renault L, Guibert B, Cherfils J. Structural snapshots of the mechanism and inhibition of a guanine nucleotide exchange factor. Nature 2003;426(6966): 525–530.
30. Staker BL, Feese MD, Cushman M, Pommier Y, Zembower D, Stewart L, Burgin AB. Structures of three classes of anticancer agents bound to the human topoisomerase I–DNA covalent complex. J Med Chem 2005;48(7): 2336–2345.
31. Tesmer JJG, Sunahara RK, Gilman AG, Sprang SR. Crystal structure of the catalytic domains of adenylyl cyclase in a complex with Gsα GTPγs. Science 1997;278(5345): 1907–1916.
32. Oliver CL, Miranda MB, Shangary S, Land S, Wang S, Johnson DE. (–)-Gossypol acts directly on the mitochondria to overcome Bcl-2- and Bcl-XL-mediated apoptosis resistance. Mol Cancer Ther 2005;4(1): 23–31.
33. Yeung K-S, Paterson I. Actin-binding marine macrolides: total synthesis and biological importance. Angew Chem Int Ed 2002;41(24): 4632–4653.
34. Dominguez R. Actin-binding proteins: a unifying hypothesis. Trends Biochem Sci 2004;29 (11): 572–578.
35. Dömling A. Recent developments in isocyanide based multicomponent reactions in applied chemistry. Chem Rev 2006;106(1): 17–89.
36. Hertzog M, van Heijenoort C, Didry D, Gaudier M, Coutant J, Gigant B, Didelot G, Préat T, Knossow M, Guittet E, Carlier M-F. The [beta]-thymosin/wh2 domain: structural basis for the switch from inhibition to promotion of actin assembly. Cell 2004;117(5): 611–623.
37. Klenchin VA, Allingham JS, King R, Tanaka J, Marriott G, Rayment I. Trisoxazole macrolide toxins mimic the binding of actin-capping proteins to actin: Nat Struct. Mol Biol 2003;10(12): 1058–1063.
38. Bedjeguelal K, Bienaymé H, Dumoulin A, Poigny S, Schmitt P, Tam E. Discovery of protein–protein binding disruptors using multicomponent condensations small molecules. Bioorg Med Chem Lett 2006;16(15): 3998–4001.
39. Bowlby MR, Chanda P, Edris W, Hinson J, Jow F, Katz AH, Kennedy J, Krishnamurthy G, Pitts K, Ryan K, Zhang H, Greenblatt L. Identification and characterization of small mole-

cule modulators of kchip/kv4 function. Bioorg Med Chem 2005;13(22): 6112–6119.

40. Lu Y, Sakamuri S, Chen Q-Z, Keng Y-F, Khazak V, Illgen K, Schabbert S, Weber L, Menon SR. Solution phase parallel synthesis and evaluation of mapk inhibitory activities of close structural analogues of a ras pathway modulator. Bioorg Med Chem Lett 2004;14(15): 3957–3962.
41. Kato-Stankiewicz J, Hakimi I, Zhi G, Zhang J, Serebriiskii I, Guo L, Edamatsu H, Koide H, Menon S, Eckl R, Sakamuri S, Lu Y, Chen Q-Z, Agarwal S, Baumbach WR, Golemis EA, Tamanoi F, Khazak V. Inhibitors of Ras/Raf-1 interaction identified by two-hybrid screening revert Ras-dependent transformation phenotypes in human cancer cells. Proc Natl Acad Sci 2002;99(22): 14398–14403.
42. Skobeleva N, Menon S, Weber L, Golemis EA, Khazak V. In vitro and in vivo synergy of MCP compounds with mitogen-activated protein kinase pathway- and microtubule-targeting inhibitors. Mol Cancer Ther 2007;6(3): 898–906.
43. Yoakim C, Ogilvie WW, Goudreau N, Naud J, Haché B, O'Meara JA, Cordingley MG, Archambault J, White PW. Discovery of the first series of inhibitors of human papillomavirus type 11: inhibition of the assembly of the E1-E2-origin DNA complex. Bioorg Med Chem Lett 2003;13(15): 2539–2541.
44. Henrick K, Thornton JM. PQS: a protein quaternary structure file server. Trends Biochem Sci 1998;23(9): 358–361.
45. Zhu H, Domingues F, Sommer I, Lengauer T. Noxclass: prediction of protein–protein interaction types. BMC Bioinformatics 2006;7(1): 27.
46. Valdar WSJ, Thornton JM. Conservation helps to identify biologically relevant crystal contacts. J Mol Biol 2001;313(2): 399–416.
47. Jones S, Thornton JM. Analysis of protein–protein interaction sites using surface patches. J Mol Biol 1997;272(1): 121–132.
48. Chakrabarti P, Janin J. Dissecting protein–protein recognition sites. Proteins 2002;47(3): 334–343.
49. Ma B, Nussinov R. Trp/Met/Phe hot spots in protein–protein interactions: potential targets in drug design. Curr Top Med Chem 72007; 999–1005.
50. Koide S, Sidhu SS. The importance of being tyrosine: lessons in molecular recognition from minimalist synthetic binding proteins. ACS Chem Biol 2009;4(5): 325–334.
51. Korndörfer IP, Schlehuber S, Skerra A. Structural mechanism of specific ligand recognition by a lipocalin tailored for the complexation of digoxigenin. J Mol Biol 2003;330(2): 385–396.
52. Leach AR, Hann MM, Burrows JN, Griffen EJ. Fragment screening: an introduction. Mol Biosyst 2006;2(9): 429–446.
53. de Kloe GE, Bailey D, Leurs R, de Esch IJP. Transforming fragments into candidates: small becomes big in medicinal chemistry. Drug Discov Today 2009;14(13–14): 630–646.
54. Bruncko M, Oost TK, Belli BA, Ding H, Joseph MK, Kunzer A, Martineau D, McClellan WJ, Mitten M, Ng S-C, Nimmer PM, Oltersdorf T, Park C-M, Petros AM, Shoemaker AR, Song X, Wang X, Wendt MD, Zhang H, Fesik SW, Rosenberg SH, Elmore SW. Studies leading to potent, dual inhibitors of Bcl-2 and Bcl-XL: J Med Chem. 2007;50(4): 641–662.
55. Petros AM, Dinges J, Augeri DJ, Baumeister SA, Betebenner DA, Bures MG, Elmore SW, Hajduk PJ, Joseph MK, Landis SK, Nettesheim DG, Rosenberg SH, Shen W, Thomas S, Wang X, Zanze I, Zhang H, Fesik SW. Discovery of a potent inhibitor of the antiapoptotic protein Bcl-XL from nmr and parallel synthesis. J Med Chem 2005;49(2): 656–663.
56. Park C-M, Bruncko M, Adickes J, Bauch J, Ding H, Kunzer A, Marsh KC, Nimmer P, Shoemaker AR, Song X, Tahir SK, Tse C, Wang X, Wendt MD, Yang X, Zhang H, Fesik SW, Rosenberg SH, Elmore SW. Discovery of an orally bioavailable small molecule inhibitor of prosurvival b-cell lymphoma 2 proteins. J Med Chem 2008;51(21): 6902–6915.
57. Raimundo BC, Oslob JD, Braisted AC, Hyde J, McDowell RS, Randal M, Waal ND, Wilkinson J, Yu CH, Arkin MR. Integrating fragment assembly and biophysical methods in the chemical advancement of small-molecule antagonists of IL-2: an approach for inhibiting protein−protein interactions†. J Med Chem 2004;47(12): 3111–3130.
58. Tilley JW, Chen L, Fry DC, Emerson SD, Powers GD, Biondi D, Varnell T, Trilles R, Guthrie R, Mennona F, Kaplan G, LeMahieu RA, Carson M, Han R-J, Liu CM, Palermo R, Ju G. Identification of a small molecule inhibitor of the IL-2/IL-2rα receptor interaction which binds to IL-2. J Am Chem Soc 1997,119(32). 7589–7590.
59. Fry D, Vassilev L. Targeting protein–protein interactions for cancer therapy. J Mol Med 2005;83(12): 955–963.
60. Dames SA, Martinez-Yamout M, De Guzman RN, Dyson HJ, Wright PE. Structural basis for Hif-1α/CBP recognition in the cellular hypoxic

response. Proc Natl Acad Sci 2002;99(8): 5271–5276.

61. Himanen J-P, Rajashankar KR, Lackmann M, Cowan CA, Henkemeyer M, Nikolov DB. Crystal structure of an eph receptor–ephrin complex. Nature 2001;414(6866): 933–938.

62. Horwell DC, Howson W, Ratcliffe GS, Willems HMG. The design of dipeptide helical mimetics: the synthesis, tachykinin receptor affinity and conformational analysis of 1,1 6-trisubstituted indanes. Bioorg Med Chem 1996;4(1): 33–42.

63. Kutzki O, Park HS, Ernst JT, Orner BP, Yin H, Hamilton AD. Development of a potent Bcl-XL antagonist based on α-helix mimicry. J Am Chem Soc 2002;124(40): 11838–11839.

64. Ernst JT, Kutzki O, Debnath AK, Jiang S, Lu H, Hamilton AD. Design of a protein surface antagonist based on α-helix mimicry: inhibition of gp41 assembly and viral fusion. Angew Chem Int Ed 2002;41(2): 278–281.

65. Orner BP, Ernst JT, Hamilton AD. Toward proteomimetics: terphenyl derivatives as structural and functional mimics of extended regions of an α-helix. J Am Chem Soc 2001;123 (22): 5382–5383.

66. Antuch W, Menon S, Chen QZ, Lu YC, Sakamuri S, Beck B, Schauer-Vukasinovic V, Agarwal S, Hess S, Dömling A. Design and modular parallel synthesis of a MCR derived alpha-helix mimetic protein–protein interaction inhibitor scaffold. Bioorg Med Chem Lett 2006;16(6): 1740–1743.

67. Moisan L, Odermatt S, Gombosuren N, Carella A. Jr Synthesis of an oxazole-pyrrole-piperazine scaffold as an alpha-helix mimetic. Eur J Org Chem (10): 2008; 1673–1676.

68. Shahian T, Lee GM, Lazic A, Arnold LA, Velusamy P, Roels CM, Guy RK, Craik CS. Inhibition of a viral enzyme by a small-molecule dimer disruptor. Nat Chem Biol 2009;5(9): 640–646.

69. Mayasundari A, Ferreira AM, He L, Mahindroo N, Bashford D, Fujii N. Rational design of the first small-molecule antagonists of NHERF1/EBP50 PDZ domains. Bioorg Med. Chem. Lett. 2008;18(3): 942–945.

70. Oguri H, Tanabe S, Oomura A, Umetsu M, Hirama M. Synthesis and evaluation of α-helix mimetics based on a trans-fused polycyclic ether: Sequence-selective binding to aspartate pairs in α-helical peptides. Tetrahedron Lett 2006;47(32): 5801–5805.

71. Jennings LD, Foreman KW, Rush Iii TS, Tsao DHH, Mosyak L, Li Y, Sukhdeo MN, Ding W, Dushin EG, Kenny CH, Moghazeh SL, Petersen PJ, Ruzin AV, Tuckman M, Sutherland AG. Design and synthesis of indolo[2, 3-a]quinolizin-7-one inhibitors of the ZipA-FtsZ interaction. Bioorg Med Chem Lett 2004;14(6): 1427–1431.

72. Inglis SR, Jones RK, Booker GW, Pyke SM. Synthesis of n-benzylated-2-aminoquinolines as ligands for the tec SH3 domain. Bioorg Med Chem Lett 2006;16(2): 387–390.

73. Meireles LMC, Doemling AS, Camacho CJ. Anchor: A Web server and database for analysis of protein–protein interaction binding pockets for drug discovery. Nucl Acids Res 2010, in press.

74. Kitchen DB, Decornez H, Furr JR, Bajorath J. Docking and scoring in virtual screening for drug discovery: methods and applications. Nat Rev Drug Discov 2004;3(11): 935–949.

75. Bonacci TM, Mathews JL, Yuan C, Lehmann DM, Malik S, Wu D, Font JL, Bidlack JM, Smrcka AV. Differential targeting of Gβγ-subunit signaling with small molecules. Science 2006;312(5772): 443–446.

76. Viaud J, Zeghouf M, Barelli H, Zeeh J-C, Padilla A, Guibert B, Chardin P, Royer CA, Cherfils J, Chavanieu A. Structure-based discovery of an inhibitor of arf activation by sec7 domains through targeting of protein–protein complexes. Proc Natl Acad Sci 2007;104(25): 10370–10375.

77. Toledo F, Wahl GM. Regulating the p53 pathway: *in vitro* hypotheses, *in vivo* veritas: Nat Rev Cancer. 2006;6(12): 909–923.

78. Toledo F, Wahl GM. MDM2 and MDM4: P53 regulators as targets in anticancer therapy. Int J.Biochem Cell B 39(7–8): 1476–1482.

79. Hu B, Gilkes DM, Chen J. Efficient p53 activation and apoptosis by simultaneous disruption of binding to MDM2 and MDMX. Cancer Res 2007;67(18): 8810–8817.

80. Lane DP. P53, guardian of the genome: Nature. 1992;358(6381): 15–16.

81. Vogelstein B, Lane D, Levine AJ. Surfing the p53 network. Nature 2000;408(6810): 307–310.

82. Oren M. Decision making by p53: Life, death and cancer: Cell Death Differ. 2003;10(4): 431–442.

83. Vousden KH, Lane DP. P53 in health and disease. Nat Rev Mol Cell Biol 2007;8(4): 275–283.

84. Kruse J-P, Gu W. Modes of p53 regulation. Cell 2009;137(4): 609–622.

85. Wade M, Wahl GM. Targeting MDM2 AND MDMX in cancer therapy: better living

through medicinal chemistry?. Mol Cancer Res 712009; 1–11.

86. Czarna A, Popowicz GM, Pecak A, Wolf S, Dubin G, Holak TA. High affinity interaction of the p53 peptide-analogue with human MDM2 and MDMX. Cell Cycle 2009;8(8): 1176–84.
87. Pazgier M, Liu M, Zou G, Yuan W, Li C, Li C, Li J, Monbo J, Zella D, Tarasov SG, Lu W. Structural basis for high-affinity peptide inhibition of p53 interactions with MDM2 and MDMX. Proc Natl Acad Sci 2009;106(12): 4665–4670.
88. Phan J, Li Z, Kasprzak A, Li B, Sebti S, Guida W, Schönbrunn E, Chen J. Structure-based design of high affinity peptides inhibiting the interaction of p53 with MDM2 and mdmx. J Biol Chem 2010;285(3): 2174–2183.
89. Li C, Pazgier M, Liu M, Yuan W, Lu W. A left handed solution to peptide inhibition of the p53–MDM2 interaction Available at. http://dx.doi.org/10.2210/pdb3lnj/pdb 2010.
90. Li C, Pazgier M, Li C, Yuan W, Liu M, Wei G, Lu W-Y, Lu W,Systematic mutational analysis of peptide inhibition of the p53-MDM2/MDMX interactions Available at http://dx.doi.org/10.2210/pdb3lnz/pdb 2010.
91. Li C, Pazgier M, Liu M, Lu W-Y, Lu W. Apamin as a template for structure-based rational design of potent peptide activators of p5313. Angew Chem Int Ed 2009;48(46): 8712–8715.
92. Sakurai K, Schubert C, Kahne D. Crystallographic analysis of an 8-mer p53 peptide analogue complexed with MDM2. J Am Chem Soc 2006;128(34): 11000–11001.
93. Grasberger BL, Lu T, Schubert C, Parks DJ, Carver TE, Koblish HK, Cummings MD, LaFrance LV, Milkiewicz KL, Calvo RR, Maguire D, Lattanze J, Franks CF, Zhao S, Ramachandren K, Bylebyl GR, Zhang M, Manthey CL, Petrella EC, Pantoliano MW, Deckman IC, Spurlino JC, Maroney AC, Tomczuk BE, Molloy CJ, Bone RF. Discovery and cocrystal structure of benzodiazepinedione hdm2 antagonists that activate p53 in cells. J Med Chem 2005; 48(4): 909–912.
94. Kussie PH, Gorina S, Marechal V, Elenbaas B, Moroau J, Levine AJ, Pavletich NP. Structure of the MDM2 oncoprotein bound to the p53 tumor suppressor transactivation domain. Science 1996;274(5289): 948–953.
95. Allen JG, Bourbeau MP, Wohlhieter GE, Bartberger MD, Michelsen K, Hungate R, Gadwood RC, Gaston RD, Evans B, Mann LW, Matison ME, Schneider S, Huang X, Yu D, Andrews PS, Reichelt A, Long AM, Yakowec P, Yang EY, Lee TA, Oliner JD. Discovery and optimization of chromenotriazolopyrimidines as potent inhibitors of the mouse double minute 2-tumor protein 53 protein–protein interaction. J Med. Chem. 2009;52(22): 7044–7053.
96. Fry DC, Emerson SD, Palme S, Vu BT, Liu C-M, Podlaski F. NMR structure of a complex between MDM2 and a small molecule inhibitor. J Biomol NMR 2004;30(2): 163–173.
97. Vassilev LT, Vu BT, Graves B, Carvajal D, Podlaski F, Filipovic Z, Kong N, Kammlott U, Lukacs C, Klein C, Fotouhi N, Liu EA. In vivo activation of the p53 pathway by small-molecule antagonists of MDM2. Science 2004;303 (5659): 844–848.
98. Popowicz GM, Czarna A, Wolf S, Wang K, Wang W, Domling A, Holak TA. Structures of low molecular weight inhibitors bound to MDMX and MDM2 reveal new approaches for p53-MDMX/MDM2 antagonist drug discovery. Cell Cycle 2010;9(6): 1104–1111.
99. Kallen J, Goepfert A, Blechschmidt A, Izaac A, Geiser M, Tavares G, Ramage P, Furet P, Masuya K, Lisztwan J. Crystal structures of human MDMX (HDMX) in complex with p53 peptide analogues reveal surprising conformational changes. J Biol Chem 2009;284(13): 8812–8821.
100. Popowicz GM, Czarna A, Holak TA. Structure of the human MDMX protein bound to the p53 tumor suppressor transactivation domain. Cell Cycle 2008;7(15): 2441–2443.
101. Popowicz GM, Czarna A, Rothweiler U, Szwagierczak A, Krajewski M, Weber L, Holak TA. Molecular basis for the inhibition of p53 by MDMX. Cell Cycle 2007;6(19): 2386–2392.
102. Lin J, Chen J, Elenbaas B, Levine AJ. Several hydrophobic amino acids in the p53 amino-terminal domain are required for transcriptional activation, binding to MDM-2 and the adenovirus 5 E1B 55-kD protein. Genes Dev. 1994;8(10): 1235–1246.
103. Massova I, Kollman PA. Computational alanine scanning to probe protein–protein interactions: a novel approach to evaluate binding free energies. J Am Chem Soc 1999;121(36): 8133–8143.
104. Sakurai K, Chung HS, Kahne D. Use of a retro-inverso p53 peptide as an inhibitor of MDM2. J Am Chem Soc 2004;126(50): 16288–16289.
105. Murray JK, Farooqi B, Sadowsky JD, Scalf M, Freund WA, Smith LM, Chen J, Gellman SH. Efficient synthesis of a β-peptide combinatorial library with microwave irradiation. J Am Chem Soc 2005;127(38): 13271–13280.

106. Li C, Liu M, Monbo J, Zou G, Li C, Yuan W, Zella D, Lu W-Y, Lu W. Turning a scorpion toxin into an antitumor miniprotein. J Am Chem Soc 2008;130(41): 13546–13548.
107. Kritzer JA, Lear JD, Hodsdon ME, Schepartz A, Helical β-peptide inhibitors of the p53-HDM2 interaction. J Am Chem Soc 2004;126 (31): 9468–9469.
108. D'Silva L, Ozdowy P, Krajewski M, Rothweiler U, Singh M, Holak TA. Monitoring the effects of antagonists on protein–protein interactions with nmr spectroscopy. J Am Chem Soc 2005;127(38): 13220–13226.
109. Krajewski M, Rothweiler U, D'Silva L, Majumdar S, Klein C, Holak TA, An NMR-based antagonist induced dissociation assay for targeting the ligand–protein and protein–protein interactions in competition binding experiments. J Med Chem 2007;50(18): 4382–4387.
110. Vassilev LT. MDM2 inhibitors for cancer therapy. Trends Mol Med 2007;13(1): 23–31.
111. Secchiero P, di Iasio MG, Gonelli A, Zauli G. The MDM2 inhibitor nutlins as an innovative therapeutic tool for the treatment of haematological malignancies. Curr Pharm Des 142008; 2100–2110.
112. Parks DJ, LaFrance LV, Calvo RR, Milkiewicz KL, José Marugán J, Raboisson P, Schubert C, Koblish HK, Zhao S, Franks CF, Lattanze J, Carver TE, Cummings MD, Maguire D, Grasberger BL, Maroney AC, Lu T. Enhanced pharmacokinetic properties of 1,4-benzodiazepine-2,5-dione antagonists of the HDM2-p53 protein–protein interaction through structure-based drug design. Bioorg Med Chem Lett 2006;16(12): 3310–3314.
113. Koblish HK, Zhao S, Franks CF, Donatelli RR, Tominovich RM, LaFrance LV, Leonard KA, Gushue JM, Parks DJ, Calvo RR, Milkiewicz KL, Marugán JJ, Raboisson P, Cummings MD, Grasberger BL, Johnson DL, Lu T, Molloy CJ, Maroney AC. Benzodiazepinedione inhibitors of the HDM2:p53 complex suppress human tumor cell proliferation in vitro and sensitize tumors to doxorubicin in vivo. Mol Cancer Ther 2006;5(1): 160–169.
114. Stoll R, Renner C, Hansen S, Palme S, Klein C, Belling A, Zeslawski W, Kamionka M, Rehm T, Muhlhahn P, Schumacher R, Hesse F, Kaluza B, Voelter W, Engh RA, Holak TA. Chalcone derivatives antagonize interactions between the human oncoprotein MDM2 and p53. Biochemistry 2000;40(2): 336–344.
115. Coan KED, Maltby DA, Burlingame AL, Shoichet BK. Promiscuous aggregate-based inhibitors promote enzyme unfolding: J. Med. Chem. 2009;52(7): 2067–2075.
116. Kumar SK, Hager E, Pettit C, Gurulingappa H, Davidson NE, Khan SR. Design, synthesis, and evaluation of novel boronic-chalcone derivatives as antitumor agents: J Med Chem. 2003;46(14): 2813–2815.
117. Issaeva N, Bozko P, Enge M, Protopopova M, Verhoef LGGC, Masucci M, Pramanik A, Selivanova G. Small molecule rita binds to p53, blocks p53-HDM-2 interaction and activates p53 function in tumors. Nat Med. 2004;10(12): 1321–1328.
118. Krajewski M, Ozdowy P, D'Silva L, Rothweiler U, Holak TA. NMR indicates that the small molecule rita does not block p53-MDM2 binding in vitro. Nat Med 2005;11(11): 1135–1136.
119. Lu F, Chi S-W, Kim D-H, Han K-H, Kuntz ID, Guy RK. Proteomimetic libraries: Design, synthesis, and evaluation of p53–MDM2 interaction inhibitors. J Com Chem. 2006;8(3): 315–325.
120. Yin H, Lee G-i Park HS, Payne GA, Rodriguez JM, Sebti SM, Hamilton AD. Terphenyl-based helical mimetics that disrupt the p53/HDM2 interaction. Angew Chem Int Ed 2005;44(18): 2704–2707.
121. Hardcastle IR, Ahmed SU, Atkins H, Calvert AH, Curtin NJ, Farnie G, Golding BT, Griffin RJ, Guyenne S, Hutton C, Källblad P, Kemp SJ, Kitching MS, Newell DR, Norbedo S, Northen JS, Reid RJ, Saravanan K, Willems HMG, Lunec J. Isoindolinone-based inhibitors of the MDM2-p53 protein–protein interaction: Bioorg. Med. Chem. Lett. 2005;15(5): 1515–1520.
122. Riedinger C, Endicott JA, Kemp SJ, Smyth LA, Watson A, Valeur E, Golding BT, Griffin RJ, Hardcastle IR, Noble ME, McDonnell JM. Analysis of chemical shift changes reveals the binding modes of isoindolinone inhibitors of the mdm2-p53 interaction. J Am Chem Soc 2008;130(47): 16038–16044.
123. Zhao J, Wang M, Chen J, Luo A, Wang X, Wu M, Yin D, Liu Z. The initial evaluation of non-peptidic small-molecule HDM2 inhibitors based on p53-HDM2 complex structure. Cancer Lett 2002;183(1): 69–77.
124. Duncan SJ, Gruschow S, Williams DH, McNicholas C, Purewal R, Hajek M, Gerlitz M, Martin S, Wrigley SK, Moore M. Isolation and structure elucidation of chlorofusin, a novel p53-MDM2 antagonist from a *Fusarium* sp. J Am Chem Soc. 2001;123(4): 554–560.
125. Duncan SJ, Cooper MA, Williams DH. Binding of an inhibitor of the p53/MDM2 interaction to MDM2. Chem Commun (3): 2003; 316–317.

126. Tsukamoto S, Yoshida T, Hosono H, Ohta T, Yokosawa H. Hexylitaconic acid: a new inhibitor of p53-HDM2 interaction isolated from a marine-derived fungus Arthrinium sp. Bioorg Med Chem Lett 2006;16(1): 69–71.

127. Galatin PS, Abraham DJ. A nonpeptidic sulfonamide inhibits the p53–MDM2 interaction and activates p53-dependent transcription in MDM2-overexpressing cells. J Med Chem 2004;47(17): 4163–4165.

128. Lawrence HR, Li Z, Richard Yip ML, Sung S-S, Lawrence NJ, McLaughlin ML, McManus GJ, Zaworotko MJ, Sebti SM, Chen J, Guida WC. Identification of a disruptor of the MDM2-p53 protein–protein interaction facilitated by high-throughput in silico docking. Bioorg Med Chem Lett 2009;19(14): 3756–3759.

129. Bowman AL, Nikolovska-Coleska Z, Zhong H, Wang S, Carlson HA. Small molecule inhibitors of the MDM2-p53 interaction discovered by ensemble-based receptor models. J Am Chem Soc 2007;129(42): 12809–12814.

130. Rothweiler U, Czarna A, Krajewski M, Ciombor J, Kalinski C, Khazak V, Ross G, Skobeleva N, Weber L, Holak TA. Isoquinolin-1-one inhibitors of the MDM2-p53 interaction. ChemMedChem 2008;3(7): 1118–1128.

131. Arts J.,Preclinical anti-tumor studies with JNJ-26854165: a novel class of MDM2 antagonists in clinical development. AACR-NCIEORTC International Conference on Molecular Targets and Cancer Therapeutics, San Francisco, CA, 2007.

132. Ding K, Lu Y, Nikolovska-Coleska Z, Qiu S, Ding Y, Gao W, Stuckey J, Krajewski K, Roller PP, Tomita Y, Parrish DA, Deschamps JR, Wang S. Structure-based design of potent non-peptide MDM2 inhibitors. J Am Chem Soc 2005;127(29): 10130–10131.

133. Shangary S, Qin D, McEachern D, Liu M, Miller RS, Qiu S, Nikolovska-Coleska Z, Ding K, Wang G, Chen J, Bernard D, Zhang J, Lu Y, Gu Q, Shah RB, Pienta KJ, Ling X, Kang S, Guo M, Sun Y, Yang D, Wang S. Temporal activation of p53 by a specific MDM2 inhibitor is selectively toxic to tumors and leads to complete tumor growth inhibition. Proc Natl Acad Sci 2008;105(10): 3933–3938.

134. Malek SN. MDM2-SNP 309 allele status does not affect sensitivity to MDM2 inhibitors in CLL. Blood 2008;112(5): 2169.

135. Czarna A, Beck B, Srivastava S, Popowicz GM, Wolf S, Huang Y, Bista M, Holak TA, Domling A. Robust Generation of Lead Compounds for Protein-Protein Interactions by Computational and MCR Chemistry: p53-Hdm2 Antagonists. Angew. Chem. Intl. Ed. Engl. 2010, in press.

136. Srivastava S, Beck B, Wang W, Czarna A, Holak TA, Dömling A. Rapid and efficient hydrophilicity tuning of p53/MDM2 antagonists. J Comb Chem 2009;11(4): 631–639.

137. Dömling A, Beck B,Preparation of pyrrolylimidazoles as antibiotics and antitumor agents. Application: WO 2001025213, 20001009, 2001.

138. Boettcher A, Buschmann N, Furet P, Groell J-M, Kallen J, Hergovich Lisztwan J, Masuya K, Mayr L, Vaupel A,3-imidazolylindoles for treatment of proliferative diseases and their preparation. Application: WO 2008119741, 20080327, 2008.

139. Dömling A. Small molecular weight protein–protein interaction antagonists: an insurmountable challenge? Curr Opin Chem Biol 1232008; 281–291.

140. Beck B, Leppert CA, Mueller BK, Dömling A. Discovery of pyrroloimidazoles as agents stimulating neurite outgrowth. QSAR Comb Sci 2006;25(5–6): 527–535.

141. Reed D, Shen Y, Shelat A, Arnold A, Ferreira A, Zhu F, Mills N, Smithson D, Regni C, Bashford D, Cicero S, Schulman B, Jochemsen AG, Guy K, Dyer MA. Identification and characterization of the first small-molecule inhibitor of MDMX. J Biol Chem 2010.

142. Holak T,manuscript in preparation..

143. Baell JB, Holloway GA. New substructure filters for removal of pan assay interference compounds (pains) from screening libraries and for their exclusion in bioassays. J Med Chem 2010.

144. Hopkins AL, Groom CR, Alex A. Ligand efficiency: a useful metric for lead selection. Drug Discov Today 2004;9(10): 430–431.

145. Abad-Zapatero C, Metz JT. Ligand efficiency indices as guideposts for drug discovery. Drug Discov Today 2005;10(7): 464–469.

NMR-BASED SCREENING AND DRUG DISCOVERY

David J. Craik
Phillippa A. Smith
Richard J. Clark

The University of Queensland, Institute for Molecular Bioscience, Brisbane, Australia

1. INTRODUCTION

NMR spectroscopy has been used widely as a front-line tool in the pharmaceutical industry for several decades. Initially, the main use of NMR within this industry was in the structural characterization of molecules synthesized in the course of medicinal chemistry programs. Indeed, medicinal chemists have long regarded NMR as the premier tool for structure characterization, to confirm the identity of intermediates or to determine the conformation of lead molecules. Over the past decade, major developments in both instrumentation and methods have resulted in this traditional use of NMR in the pharmaceutical industry being augmented by a range of exciting new applications. Two of the most important of these are the use of NMR in structure-based drug design and in screening for drug discovery. Both applications differ from the traditional use of NMR in that now the macromolecular binding partner of the medicinal compound is included in the sample to be analyzed; that is, current applications of NMR in drug discovery are predominantly focused on the *interactions* between drug molecules and their macromolecular targets.

The aim of this chapter is to describe how NMR spectroscopy is used in modern drug discovery. The term *discovery* is used generically to include processes that involve rational drug design as well as those that involve discovery through NMR screening. The latter is a relatively recent development and refers to the use of NMR as a tool to screen a compound library to identify molecules that bind to a chosen macromolecular target. Of course, the distinction between "design" and "discovery" is often quite blurred. This is nowhere more evident than in the recently developed SAR-by-NMR approach [1], in which the discovery of several weakly bound ligands from a screening program is intimately linked to a design process to join them chemically. SAR-by-NMR, or its broader discipline of fragment-based screening represents an exciting new technique for lead generation or optimization and is described in more detail later in this chapter.

Drug design/discovery represents only the first stage in the drug development process. As is clear from other chapters in this volume, there are many other steps that need to be made once a lead molecule has been designed or discovered. Although other stages of the process, including toxicity studies, preclinical investigations, and clinical monitoring, do not fall within the scope of this chapter, it is worth mentioning that NMR spectroscopy contributes significantly across the whole spectrum of drug development, all of the way through into the clinical domain. For example, NMR spectroscopy has been applied for the detection of drug metabolites in biological fluids and the rapidly expanding field of metabonomics makes extensive use of NMR to detect and quantify changes in endogenous metabolites in response to a variety of stimuli, including diseases and drug therapy. Magnetic resonance imaging (MRI), likewise, has an important role in clinical investigations and is increasingly used to monitor the functional outcomes of drug therapy. We briefly address these broader applications of NMR before returning to the main topic of NMR in drug discovery.

1.1. Overview of Drug Development

Figure 1 summarizes the drug development process and indicates the role of NMR at various stages. Drug development is an iterative process and can be represented by two interconnected cycles of activity. Cycle A involves the design or discovery of an initial lead followed by its synthesis and bioassay. Based on the initial assay results, there might be several loops around this cycle before commencing the *in vivo* studies represented in Cycle B. At this stage, consideration of bioavailability, metabolism, and pharmacokinetic profiles must be made and this usually involves synthetic modifications of the lead molecules to improve their drug-like properties.

Several loops around Cycle B could be necessary before one or more development candidates are identified. Ultimately, one or more of these development candidates are identified for progression through clinical trials.

As indicated in Fig. 1, it is convenient to envisage five broad categories of NMR experiments that might contribute to this overall drug development process. These categories are listed below:

- *Small-Molecule, or Ligand-Based, NMR*: Studies drugs and drug leads, typically organic molecules with molecular weights <500 Da, to structurally characterize natural products or synthetic drug leads, and to determine their conformation.
- *Macromolecular NMR*: Studies the macromolecular targets of drugs, typically to determine their 3D structure and/or the nature of their complexes with ligands.
- *NMR Screening*: Here, NMR is used to identify lead molecules that bind to macromolecular targets. Typically, these studies seek to detect the presence of binding interactions between small molecules and macromolecules.
- *Metabolic NMR*: Analysis of endogenous molecules in biological fluids (metabonomics) and tissues with particular emphasis on how their concentrations may be modified by drug treatment. The field also includes studies of the metabolites of drugs themselves.
- *NMR Imaging*: Provides anatomical information in an animal model or human patient. This includes, for example, mon-

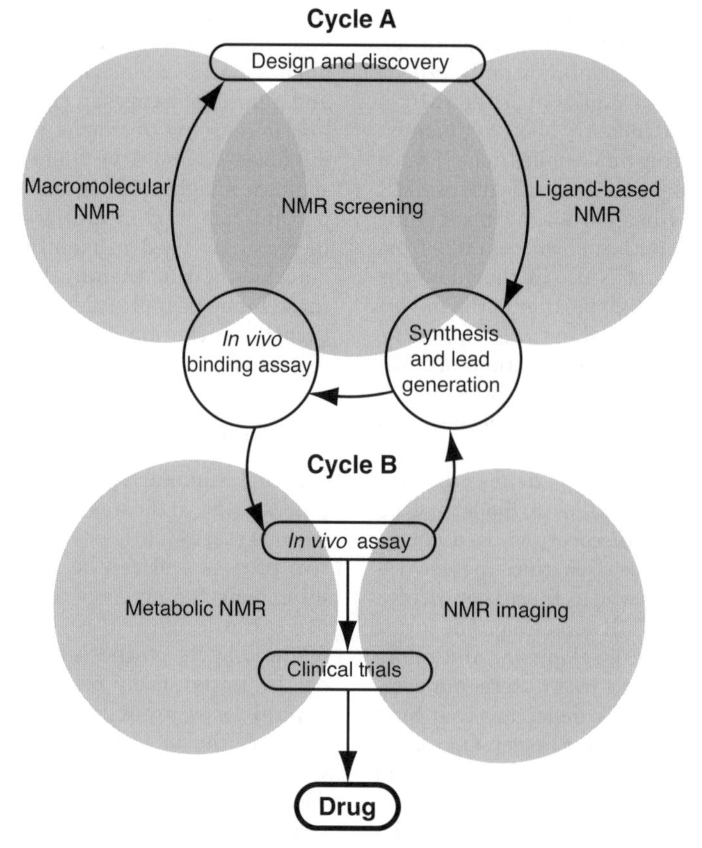

Figure 1. Overview of the drug development process and summary of various types of NMR experiments that contribute at different stages.

itoring the size of plaques in the brains of Alzheimer's disease patients or tumors in cancer patients, during drug therapy.

It is clear from these descriptions that NMR covers a wide range of applications in the pharmaceutical industry, although for the remainder of this chapter we will focus on NMR in the drug design/discovery phase of drug development, that is, on the first three categories of the above list.

1.2. Scope of Chapter

Our aim is to give a broad overview of the use of NMR as a tool in structure-based design and in screening approaches to drug discovery. The chapter also contains a description of relevant NMR methods, which are highlighted by illustrative examples. We briefly describe the instrumentation required for such studies and discuss emerging trends in the drug discovery field. This includes developments in the field of drug discovery in the postgenomic era that are likely to have an impact on the way in which NMR is used, as seen, for example, by the recent interest in structural genomics programs. NMR instrument developments are also described; for example, advances in cryoprobe technology have dramatically increased the sensitivity of NMR spectroscopy and have increased its application across the pharmaceutical industry. Finally, a section outlining some of the practical considerations in structure-based design and screening is included. Future directions for the field are mentioned throughout the discussion.

A number of reviews have described applications of NMR in drug discovery or screening and the reader is referred to these for additional information [2–28]. Books covering aspects of NMR in drug design are also available [29,30].

It is assumed that many readers will be familiar with the basic principles of NMR; however, for completeness and to define some of the terms that will be used in this chapter, it is useful to give a brief overview of these principles [31,32].

1.3. Principles of NMR Spectroscopy

The foundation of NMR is that when nuclei with a nonzero spin quantum number are placed in a magnetic field they take up one of a discrete number of quantized states. The application of radiofrequency (rf) energy produces transitions between these states. The energy changes associated with these transitions are detected as small voltages induced in a receiver coil that are subsequently amplified, digitized, and processed to yield spectra, as illustrated in Fig. 2. The most commonly studied NMR-active nucleus is the proton, ^{1}H, but in modern NMR experiments ^{2}H, ^{13}C, and ^{15}N nuclei are also very important, particularly for studies of proteins, which will become apparent later in this chapter. For these heteronuclei, it is common to isotopically enrich the sample because of their low natural abundance. Occasionally, other nuclei find

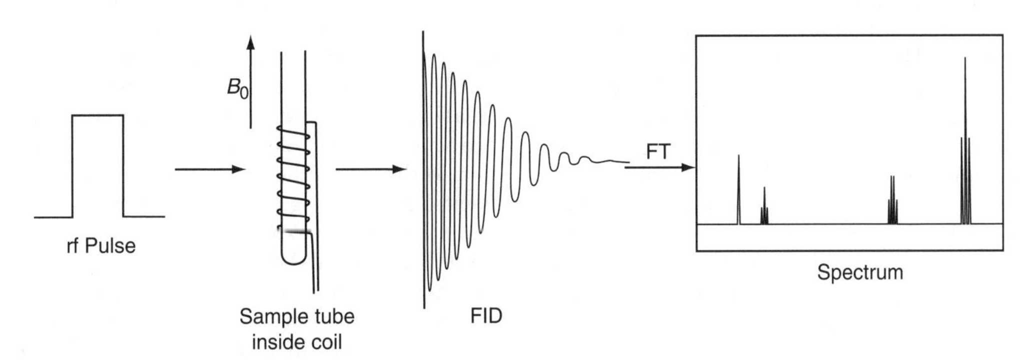

Figure 2. Overview of the principles of NMR spectroscopy. Polarization of nuclear spins by a magnetic field is perturbed by application of a rf pulse. The resultant FID signal is Fourier transformed (FT), to yield a spectrum reflecting the number and environments of nuclei in the sample.

specialist applications. For example, in fluorine-containing drugs it is possible to use ^{19}F NMR signals to monitor interaction with target proteins.

In modern spectrometers, the rf energy is supplied in the form of short pulses (typically, $\sim 10\,\mu s$) that simultaneously excite all nuclei of a given isotope type (e.g., all protons or all ^{13}C nuclei). Nuclei of a given isotope that are in different chemical environments by virtue of their atomic locations in the molecule have slightly different resonance frequencies and lead to different oscillating voltages in the receiver coil. The resultant combined signal, termed a *free induction decay* (FID), is Fourier transformed to give a spectrum that is a plot of peak intensity versus frequency, with one peak for each chemically distinct nucleus. These features are schematically illustrated in Fig. 2. The frequency axis is termed the *chemical shift* because it reflects the local chemical environment of each nucleus. The range of chemical environments of nuclei in a molecule is such that chemical shifts range up to only a few hundred parts per million (ppm) of the base resonance frequency for ^{13}C and ^{15}N. For ^{1}H the range is smaller, covering only approximately 10 ppm. Despite this small range, chemical shifts provide valuable diagnostic information on the environment of the nucleus that gives rise to the signal.

The chemical shift is an extremely important NMR parameter, but there are many other parameters that can be discerned from NMR spectra. Indeed, NMR is unique among many forms of spectroscopy in that there are so many parameters associated with a spectrum other than just peak intensity and frequency. These include coupling constants, which provide information on local conformations and molecular connectivities; nuclear Overhauser effects (NOEs), which give information on internuclear distances; and relaxation parameters, which provide information on molecular dynamics. Table 1 summarizes the main NMR parameters that may be measured and highlights their applications in the drug discovery process.

The following sections of this chapter provide specific examples of how these various parameters are useful in the drug discovery process. Before describing these examples though, it is useful to consider some of the limitations of one-dimensional NMR spectroscopy, particularly when the detected nucleus is ^{1}H, as is most commonly the case. With one signal coming from each chemically distinct proton and with those signals spread over only 10 ppm, it is clear that spectral overlap is potentially a major problem for anything but the simplest of molecules. The development of higher field NMR spectrometers, which effectively provide greater dispersion in the frequency dimension, has contributed significantly to overcoming this limitation and increasing the application of NMR for studying pharmaceutically relevant molecules. In addition to this and other instrumental developments, methodological advances have also had a key role in extending the use of NMR. Multidimensional NMR methods have revolutionized biomolecular NMR spectroscopy by removing the limitations of a single frequency dimension, and include 2D, 3D, and 4D spectra.

A simple way of illustrating multidimensional NMR is through reference to hetero-

Table 1. NMR Parameters and Their Applications in Drug Design/Discovery

Parameter	Information Relevant to Drug Design
Chemical shift	Reflects local chemical environment; provides a fingerprint marker of structure (particularly in HSQC spectra)
Coupling constants	Conformational analysis, establishing molecular connectivity
Nuclear Overhauser effect	Determining interproton distances, three-dimensional structures
Relaxation times	Molecular dynamics
Line-shape	Detecting and quantifying chemical exchange processes
Peak intensities	Reflect relative number of nuclei, molecular symmetry
Amide exchange rates/temperature coefficients	Hydrogen bonding or solvent exposure of amide protons

nuclear correlation spectroscopy, in which two or more separate frequency dimensions are correlated with one another. For example, a particularly valuable 2D experiment is $^{1}H/^{15}N$ heteronuclear single quantum coherence (HSQC) spectroscopy, in which the spectrum has two frequency axes, corresponding to ^{1}H and ^{15}N frequency dimensions, and one intensity axis. Analogous $^{1}H/^{13}C$ HSQC spectra are also used widely. Such spectra are normally represented with the intensity axis in contour form so that they may be plotted in two dimensions as a set of contour peaks. Spectral peaks occur for pairs of $^{15}N/^{1}H$ or $^{13}C/^{1}H$ nuclei that are directly bonded to one another, and with each frequency being sensitive to the local chemical environment, HSQC spectra represent a relatively simple, but highly characteristic fingerprint of the sample. Figure 3 shows the relationship between 1D and 2D spectra for the immunosuppressive drug cyclosporin, and includes a region of both the $^{1}H/^{15}N$ and $^{1}H/^{13}C$ HSQC spectra. In HSQC spectra overlap problems are alleviated because, even if two protons have the same chemical shift and would hence be overlapped in a 1D spectrum, chances are that the respective heteronuclear signals will not be overlapped, allowing the signals to be resolved in the 2D spectrum. HSQC spectra are used widely in NMR-based drug screening and we will return to them later.

Multidimensional NMR spectra are not restricted to cases where the separate frequency axes encode signals from different nuclear types. Indeed, much of the early work on the development of 2D NMR was performed on cases where both axes involved ^{1}H chemical shifts. The main value in such spectra comes from the information content in cross-peaks between pairs of protons. In COSY-type spectra (correlation spectroscopy-type spectra) cross-peaks occur only between protons that are scalar coupled (i.e., within two or three bonds) to each other, whereas in NOESY (NOE spectroscopy) spectra, cross-peaks occur for protons that are physically close in space (<5 Å apart) [33]. A combination of these two types of 2D spectra is used to assign the NMR signals of small proteins and provides sufficient information on internuclear distances to calculate 3D structures. Figure 3 includes a panel showing the COSY spectrum of cyclosporin and highlights the relationships between 1D ^{1}H NMR spectra and corresponding 2D homonuclear (COSY) and heteronuclear (HSQC) spectra.

Homonuclear 2D spectra are generally applicable for the study of proteins up to approximately 80 amino acids in size. For larger proteins, the increased number of signals leads to overlap problems and, in addition, COSY-type spectra suffer from poor sensitivity when the signal linewidths are of the same order as, or larger than, ^{1}H, ^{1}H scalar coupling constants. These limitations are reduced by use of spectra of higher dimensionality (i.e., 3D or 4D spectra) based on correlations involving heteronuclear, rather than homonuclear, coupling constants. Such spectra are important in the structure determination process for larger proteins and are typically recorded for samples that incorporate uniform labeling with ^{15}N, or both ^{13}C and ^{15}N nuclei. Multidimensional spectra that involve irradiation of ^{1}H, ^{13}C, and ^{15}N nuclei are referred to as *triple resonance spectra*. The details of how multidimensional spectra are obtained is beyond the scope of this chapter, but it suffices to say that, like most other modern NMR experiments, they involve irradiation of the sample with a set of rf pulses of defined length, frequency, and phase, with specific interpulse delays.

The above discussion provides a basic overview of some of the important methods in modern NMR spectroscopy. Before examining specific applications in drug discovery, it is useful to describe the instrumental requirements for such studies.

1.4. Instrumentation

NMR spectrometers comprise a powerful and homogeneous magnet, a radiofrequency console for generating appropriate rf pulses, a probe for applying this rf energy to the sample and receiving the resultant signals, and a computer console for controlling the experiments and acquiring the resultant data. These features are summarized in Fig. 4. Spectrometers are normally specified in terms of the resonant frequency of protons at the given magnetic field (e.g., 500 MHz corresponds to

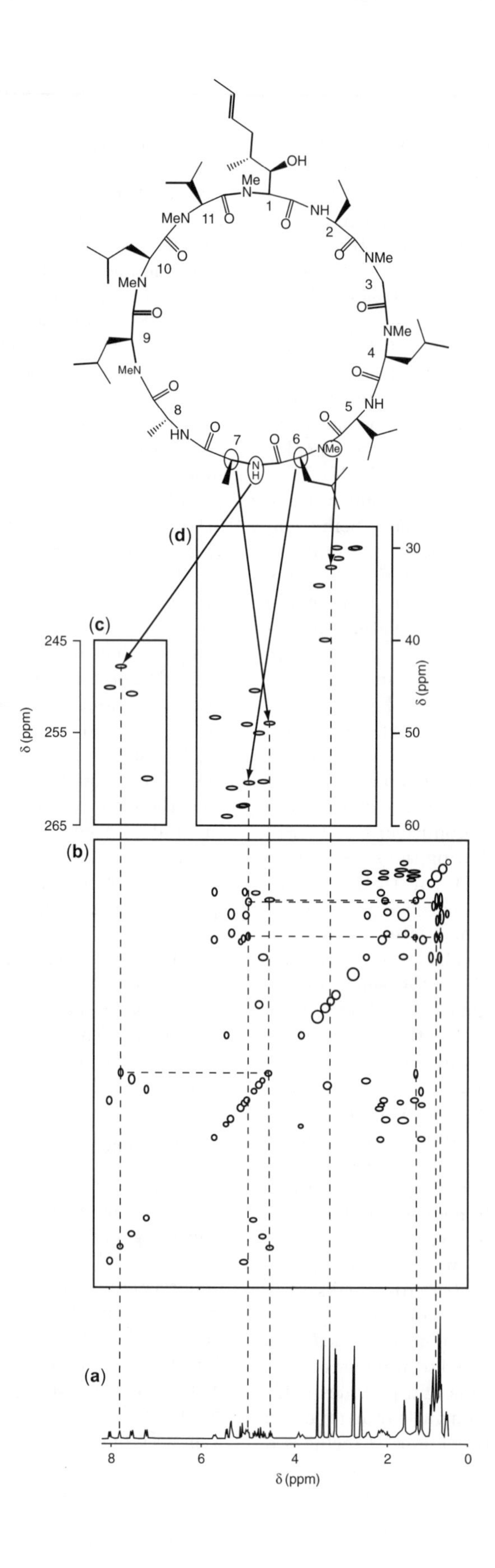

OH
Me
N
1
NH
2
NMe
3
NMe
4
NH
5
NMe
6
NH
7
HN
8
MeN
9
MeN
10
MeN
11
(d)
(c)
(b)
(a)
δ (ppm)

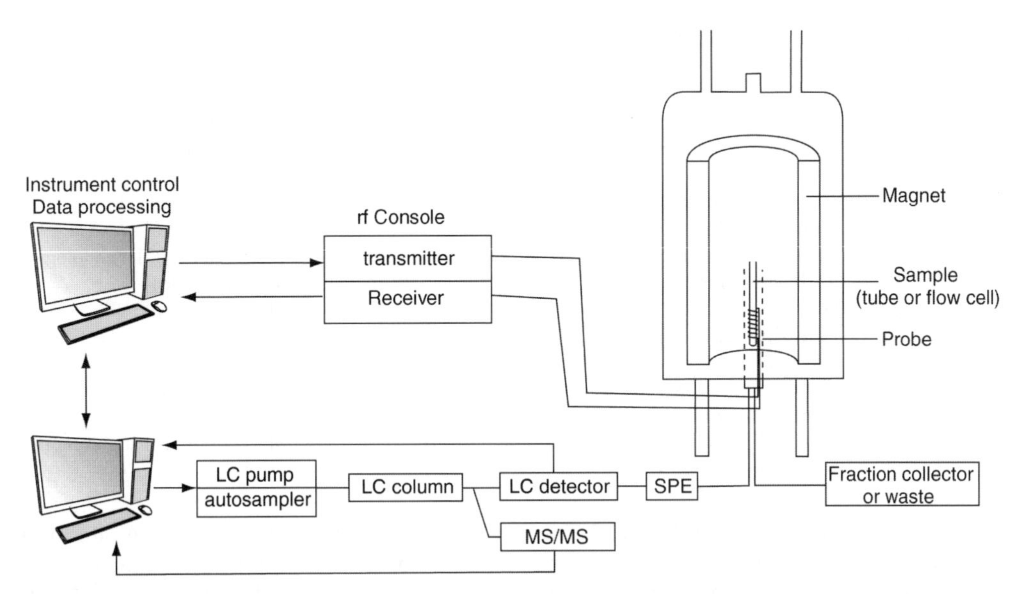

Figure 4. Schematic diagram of a modern NMR spectrometer. The extremely low temperature of the superconducting magnet (4.2 K) is well insulated from the sample chamber in the center of the magnet bore. The probe in which the sample is housed usually incorporates accurate temperature control over the range typically of 4 to 40°C for biological samples. The rf coil in the probe is connected in turn to a preamplifier, receiver circuitry, analog-to-digital converter (ADC), and a computer for instrument control and data collection/processing. Optionally the NMR instrument might be interfaced to LC and MS systems.

a magnetic field of 11.7 T). Both sensitivity and dispersion of signals increase with increasing magnetic field.

There have been some major breakthroughs in both NMR instrumentation and methodology that have greatly increased the utility of NMR for drug discovery applications. These are summarized in Table 2, which also includes some of the earlier milestones in the development of NMR. Notable among recent innovations are the use of pulsed-field gradient methods for improving spectral quality and allowing new types of experiments to be performed, transverse relaxation-optimized spectroscopy (TROSY) methods [34] for increasing the size of macromolecules that can be examined, and cryoprobes for enhancing sensitivity. The development of cryoprobes has resulted in the biggest single gain in sensitivity over recent years, effectively giving 500 MHz spectrometers the sensitivity of 800 MHz spectrometers; however, this is without a gain in resolution! The enhanced sensitivity is obtained by cooling the radiofrequency receiver coil and associated circuitry to near liquid helium temperatures, which reduces the amount of thermal noise in the NMR signal. Considerable technical barriers were overcome in developing such probes due to the large difference in temperature between the receiver coils and the sample, which are only a few millimeters apart.

Although the basic configurations of instruments tailored for structure-based design or for NMR drug screening are similar, there are some minor differences. For structure-based design applications, a relatively high-field spectrometer is required (>500 MHz), usually equipped with three or four radiofrequency channels for the simultaneous irradiation of

Figure 3. A schematic representation of the (a) 1D ^{1}H; (b) 2D DQF-COSY; (c) $^{15}N/^{1}H$-HSQC; and (d) $^{13}C/^{1}H$-HSQC spectra of the immunosuppressive agent cyclosporine illustrated at the top of the figure. Example resonances/correlations from residues 6 and 7 have been highlighted to illustrate the assignment process.

Table 2. Milestones in the Development of NMR Spectroscopy

Year	Development	Nature
1970	FT NMR	Instrumental
1975	Superconducting magnets	Instrumental
1980	2D NMR	Methodological
1985	Protein structure determination	Methodological
1990	Isotope labeling/multidimensional NMR	Methodological
1990	Pulsed field gradients	Instrumental/Methodological
1996	NMR screening	Methodological
1997	TROSY	Methodological
1998	LC-NMR/LCMS-NMR	Instrumental
2000	Cryoprobes	Instrumental
2001	Ultrahigh field (900 MHz) NMR	Instrumental
2003	Novel isotope labeling strategies	Methodological
2006	Advances in solid state NMR of biomolecules	Methodological/Instrumental

^{1}H, ^{13}C, ^{15}N, and in some cases ^{2}H nuclei. The greatest sensitivity and dispersion are obtained with the highest possible magnetic field, and currently instruments of up to 950 MHz are available. The high-field instruments provide another advantage, in that TROSY experiments [34] can be used to produce a marked improvement in spectral quality for larger proteins. Such developments have increased the size of proteins that can be structurally determined by NMR, such that sophisticated 3D and 4D experiments can now lead, in principle, to the determination of structures of proteins with molecular weights approaching 100 kDa [35].

For NMR drug screening programs, the basic requirement of a spectrometer of 500 MHz or higher remains, but in addition, an interface that allows the spectrometer to sample a library of compounds of potential binding ligands is needed. This may be done by use of either a discrete sample changer or a flow-type system. Flow systems have the potential advantage of increased throughput but have the potential disadvantage of precipitation of protein samples clogging the flow lines. In practice, this appears not to have been a major problem and both types of systems are in use in the pharmaceutical industry. The enhanced sensitivity afforded by the use of cryoprobes has also been advantageous to the study of macromolecule–ligand interactions used in screening programs [36].

Pulsed-field gradients have become integral to most modern NMR spectrometers and are routinely used both for structure determination and screening experiments. Other developments, including the interfacing of NMR spectrometers with other instruments such as liquid chromatography (LC) and/or mass spectrometry (MS) have also increased the utility of NMR in drug discovery [17,24,37,38].

LC-NMR involves the coupling of high-pressure liquid chromatography (HPLC) with NMR to enable the structural characterization of individual components in a complex mixture after on-line separation via HPLC. Mass spectrometry can also be interfaced with the LC-NMR instrument to enable the identification of analyte peaks of interest before, or in parallel with NMR analysis. This technique has been particularly useful in the study of plant natural products and metabolites. The application of these developments has been recently reviewed [38–40].

The coupling of LC-NMR with solid phase extraction (SPE) has enabled important advances for drug design, specifically for the study of natural products as potential drug candidates. LC-SPE-NMR systems involve the use of SPE cartridges to trap eluted HPLC analytes prior to NMR analysis. This additional "peak-trapping" step has led to sensitivity gains as the eluted analyte becomes highly concentrated on the SPE cartridge, thereby increasing the signal-to-noise ratio [41].

1.5. Applications of NMR in Drug Design and Discovery

Our focus here is on the use of NMR in the discovery and design phase of drug development. The major role of NMR in the *design* process comes about by its exquisite ability to provide structural information, whereas the major role of NMR in *discovery* comes through its use as a screening tool to detect the binding of novel ligands to macromolecular targets. The latter application in particular has created much recent interest in the use of NMR in the pharmaceutical industry. There are several drug leads, currently in clinical development, that have been derived from NMR strategies such as SAR-by-NMR [4,42]. As noted already, though, the discovery and design phases are often intimately connected, with lead molecules discovered in screening programs routinely being optimized by use of structure-based design approaches (Fig. 5).

The process of designing new drugs via structure-based design refers to the determination and use of the 3D structure of a lead molecule, macromolecular target, or complex to design new drugs. The questions that may be asked when embarking on structure-based design projects are

- What are the solution and bound conformations of the ligand?
- What is its charge/tautomeric state?
- Which functional groups bind to the receptor and what charge state are they in?
- What is the structure of the receptor?
- Which parts interact with the ligand?
- What is the geometry of the ligand–receptor complex?
- What are the kinetics of binding and are there dynamic motions of ligand, receptor, or the complex?

Table 3 summarizes these and other questions, and indicates the type of NMR approaches that provide answers. The remaining sections of this chapter are organized around the headings identified in Table 3.

Drug design can be categorized into ligand-based design, where the structural focus is on the small lead molecule, and receptor-based design, where the aim is to determine the structure of the macromolecular target. The NMR methods used in ligand-based design have been well established for many years, and are based on the traditional NMR techniques of organic and natural product chemists. The use of NMR to determine the 3D structures of macromolecules is a newer field, commencing in approximately 1985, and not becoming routine until the 1990s. NMR screening is a newer approach, arising in the mid-1990s. Ligand-based and receptor-based design approaches are examined in Sections 11.2 and 11.3, respectively, and screening-based approaches are examined in Section 11.4.

2. LIGAND-BASED DESIGN

Many naturally occurring molecules have potent bioactivity that renders them useful leads in the drug design process. These leads may be naturally occurring hormones, neurotransmitters,

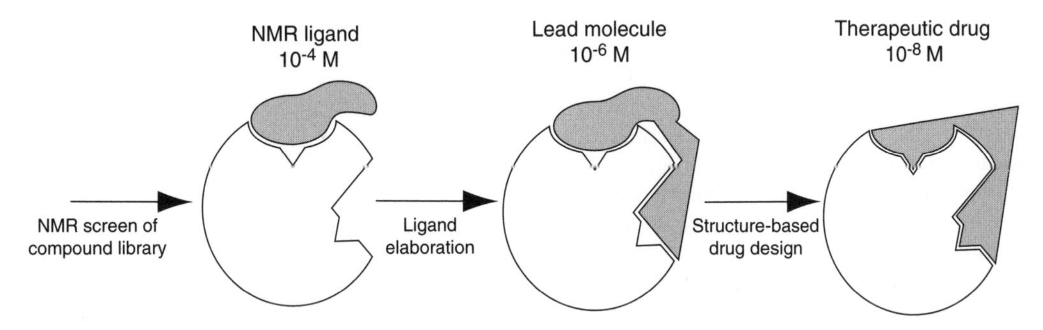

Figure 5. A summary of the relationship between NMR screening and structure-based design.

Table 3. Information on Ligands, Macromolecules and Their Complexes Sought in Structure-Based Design and Relevant NMR Technologies Used to Derive This Information

Target	Information	NMR Technology
Ligand	Solution conformation	1D/2D NMR
	Charge/tautomeric state	Chemical shift/titrations
	Solution dynamics	Line-shape/relaxation analysis
	Pharmacophore models	All of the above, and TrNOE, of multiple ligands
	Bound ligand conformation	TrNOE
Macromolecule	3D structure	2D/3D/4D NMR
	Macromolecular dynamics	Relaxation time measurements
	Structure of large macromolecules and complexes (e.g., multimeric or membrane-bound receptors)	TROSY
Ligand–macromolecular complex	Stoichiometry of complex	Chemical shift titration
	Kinetics of binding	Line width, titration analysis
	Location of interacting sites	HSQC, isotope editing
	Orientation of bound ligand	NOE docking
	Bound ligand conformation	TrNOE
	Structure of complex	3D/4D NMR
	Dynamics of complex	Relaxation time measurements

or other endogenous molecules, or they may be bioactive molecules from plants or microorganisms. Furthermore, screening programs on synthetic compound libraries frequently result in the discovery of bioactive molecules that then become starting points in drug design. The general aim of ligand- or analog-based design is to determine the structure and conformation of a known bioactive molecule and then mimic this conformation in a designed lead compound, with the aim of improving the activity or drug-like properties. The following sections examine various aspects of ligand-based design and illustrate them with examples.

2.1. Structure Elucidation

If the bioactive molecule is a synthetic product, its structure may be deduced rapidly by a simple comparison of NMR parameters (often combined with MS) of the product relative to those of the known precursor, to see whether the desired chemical transformation has taken place. If the bioactive compound is an unknown molecule discovered in an active fraction in bioassay-guided screening, then the first step is to elucidate its structure. Typical molecules that form the basis of such natural products-based drug discovery studies include "organic" natural products as well as small peptides and proteins. The approaches to structure elucidation for natural products and peptides/proteins are a little different from each other and are described in turn.

2.1.1. Structure Elucidation of Natural Products

In the case of nonpeptidic natural products, initially, the main structural focus is to elucidate the carbon framework. This normally involves a combination of 1D ^{1}H and ^{13}C NMR, followed by homonuclear (DQF-COSY, TOCSY, ROESY, or NOESY) and heteronuclear (HSQC, HMBC) 2D experiments. Heteronuclear multiple-bond correlation (HMBC) spectra are particularly valuable because they assist in tracing the carbon backbone of a molecule. Such spectra display cross-peaks between a ^{13}C nucleus and protons connected within two or three bonds and, combined with one-bond correlations from HSQC data, provide valuable information on molecular connectivity. Figure 6 shows some typical HMBC correlations seen for selected regions of taxol, a plant-derived natural product that used for the treatment for breast, lung, and ovarian cancers. Although the structure of taxol itself

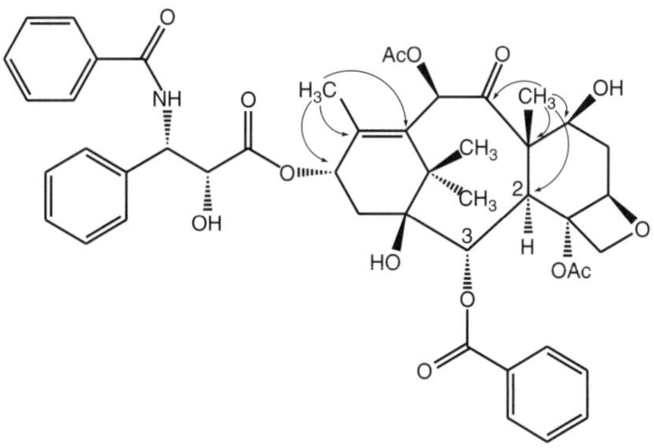

Figure 6. Illustration of the HMBC correlations (arrows) used to assign the positions of two of the methyl quaternary methyl groups in taxol.

was originally deduced from a combination of X-ray crystallography on a degradation product and a range of ^{1}H and ^{13}C spectra in the 1970s [43], before HMBC spectra had been invented, HMBC spectra have been used widely for studies of the many taxol derivatives that have been examined in the last decade [44,45].

Elucidation of the carbon framework of natural products often yields substantial information about the 3D structure at the same time, but if there are remaining questions on the stereochemistry of chiral centers or other factors affecting the 3D structure, these can usually be resolved from NOESY spectra and/or an analysis of coupling constants. We will return to the taxol example later in Section 2.2, when describing conformational analysis.

2.1.2. Structure Determination of Bioactive Peptides In contrast to the process described for organic molecules, the structure elucidation of peptide-based natural products involves two distinct steps: (1) the elucidation of the primary structure (amino acid sequence) followed by (2) a determination of secondary and tertiary structure. The primary structure determination is routinely done through Edman sequencing or by MS–MS sequencing methods. NMR has a key role in the elucidation of the secondary and tertiary structure of peptides, mainly based on 2D homonuclear NMR spectroscopy. A combination of DQF-COSY and TOCSY (total correlation spectroscopy) spectra are used to assign spin systems to general amino acid types and then NOESY spectra are used to assign the resonances to individual protons in specific amino acids in the sequence. The 3D structure is then determined by deriving a series of internuclear distance restraints and angular restraints from various NMR spectra and using them in a simulated annealing algorithm to calculate a family of structures consistent with them.

Because the structure determination of peptides and proteins represents a very important contribution of NMR to the drug development process, it is informative to describe the process in more detail. We use the peptide-based drug MVIIA as an example.

NMR Structure of Prialt® (MVIIA): A Novel Treatment for Pain MVIIA, marketed as Prialt, is a 25-amino acid peptide originally discovered from the venom of the marine cone snail, *Conus magus*. Like other ω-conotoxins it is a potent blocker of N-type calcium channels, giving it a wide range of potential therapeutic applications. When delivered intrathecally (i.e., through spinal infusion), it is an analgesic more potent than morphine that is used clinically for the treatment of intractable cancer pain [46]. Figure 7 shows the peptide sequence and illustrates selected regions of TOCSY and NOESY spectra.

As seen in Fig. 7, the TOCSY spectrum is useful for classifying spin systems to amino acid type, with typically the most useful region being the "skewers" emanating from individual NH shifts (~7–10 ppm). For each NH proton in the peptide, a series of cross-peaks to the α, β, and other side-chain protons is

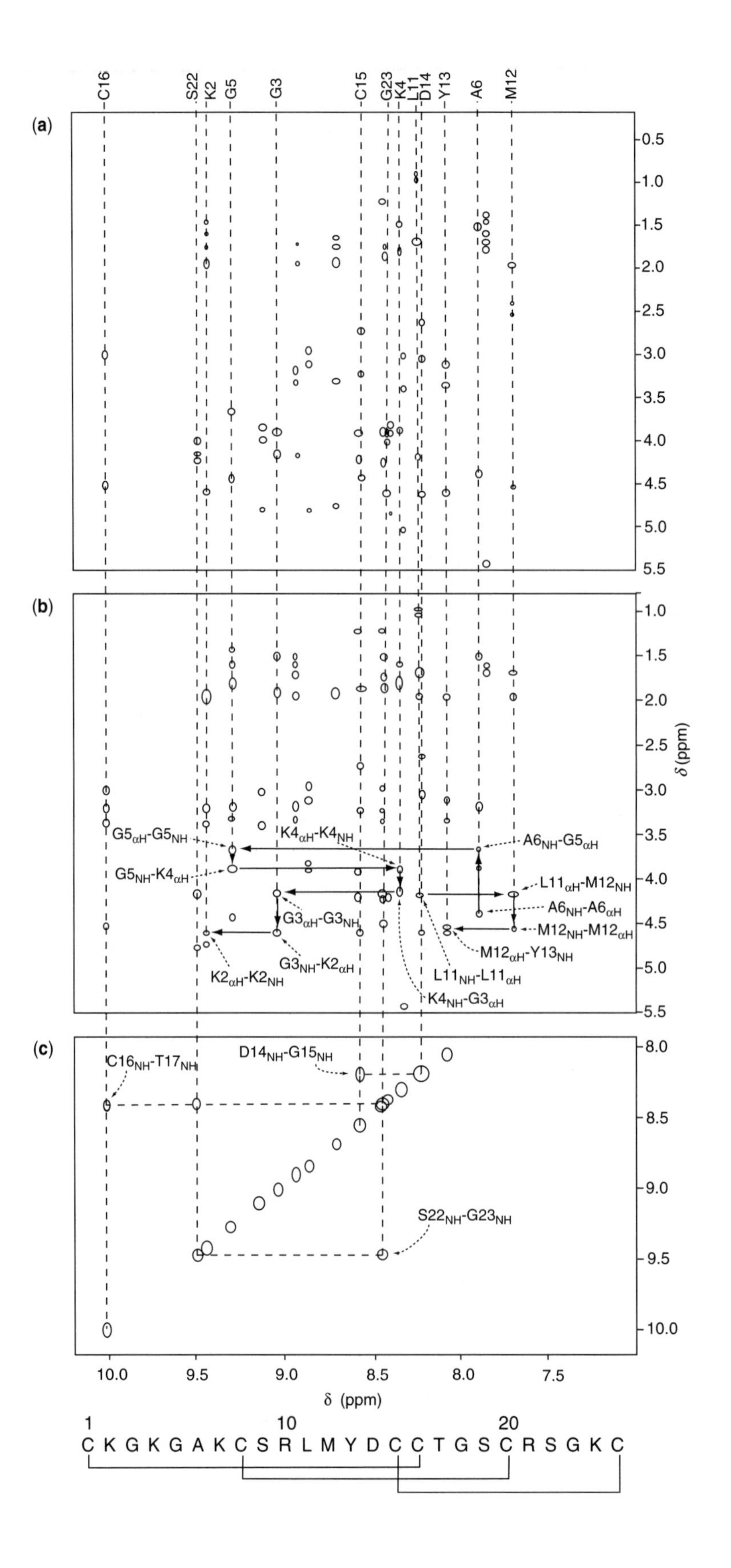

(a)
(b)
(c)
C16
S22
K2
G5
G3
C15
G23
K4
L11
D14
Y13
A6
M12
δ (ppm)
G5αH-G5NH
K4αH-K4NH
A6NH-G5αH
G5NH-K4αH
L11αH-M12NH
A6NH-A6αH
G3αH-G3NH
M12NH-M12αH
M12αH-Y13NH
G3NH-K2αH
K2αH-K2NH
L11NH-L11αH
K4NH-G3αH
C16NH-T17NH
D14NH-G15NH
S22NH-G23NH
1
10
20
C K G K G A K C S R L M Y D C C T G S C R S G K C

observed and these patterns define the spin system as belonging to a particular type of amino acid. However, note that there is some degeneracy in the resultant patterns. The NH side-chain pattern is truncated if there is a break of more than three bonds between protons within the spin system. This means, for example, that the skewers for aromatic residues such as Phe or Tyr extend only as far as the β-protons and they therefore appear similar to other "AMX" residues such as Cys, Ser, Asp, or Asn. Nevertheless, the ability to assign signals either to the individual amino acid types or to the AMX group is a useful starting point in the assignment. However, such spectra provide no information about the sequential location of an amino acid if that amino acid type is not unique in the sequence. Information regarding the sequential assignment of amino acids can be obtained from the NOESY spectrum, as illustrated in Fig. 7. The aim of the sequential assignment process is to locate adjacent amino acid spin systems, principally through a cross-peak between the αH proton of one residue (i) and the NH of the following residue ($i + 1$), often denoted as dαN(i, $i + 1$). Additional support for the assignment is usually also sought in dβN(i, $i + 1$) and dNN(i, $i + 1$) correlations. This process of connecting the i and $i + 1$ residues through analysis of the NOESY spectrum is often called a "sequential walk." At the early stages of an assignment it is impossible to be certain whether a particular cross-peak is sequential or longer range; however, as the assignment procedure progresses, ambiguities are resolved. Generally, the assignment process is highly convergent, in that once a series of correct assignments is made, the number of choices for remaining cross-peaks diminishes, in principle making their assignment easier.

Because peptides are polymers of amino acids units, the repeated NH, Hα, and side-chain protons tend to fall in characteristic chemical-shift ranges that can be useful in looking for patterns to identify amino acid types. Table 4 shows typical chemical shifts for each of the 20 common amino acids when located in a "random-coil" environment, where no secondary structure is apparent [33,47,48]. It is important to stress that these shifts can vary quite considerably in structured proteins (by up to several parts per million) and are more useful for pattern recognition purposes than for exact identification of a particular residue. In the case of the Hα protons, the differences between the actual shifts in a structured protein and these random-coil values have an additional important use because they provide an indication of the local secondary structure. Intuitively, the further a chemical shift is from a random-coil value, the more probable that it is attributed to the corresponding atom being in a structured environment.

After the assignment is complete, it is possible to derive substantial information about the secondary structure from an analysis of chemical shifts, coupling constants, and NOEs, even before the commencement of 3D structure calculations. Figure 8 shows a typical summary of the relevant NMR information, again using the data for MVIIA as an example [49,50]. Trends in these data provide a general indication of major elements of secondary structure. For example, a series of strong dαN(i, $i + 1$), relative to dNN(i, $i + 1$) NOEs, often indicates an extended or β-type structure, whereas strong dNN(i, $i + 1$) NOEs indicate local helical structure or turns. Large JαN coupling constants (>8.5 Hz) are associated with extended structure and small ones (<5 Hz) with helical structure. Similarly, deviations of chemical shifts from random-coil values, often represented in terms of "chemical-shift indices (CSI)" [51], indicate extended (positive values) or helical structure (negative values).

An additional useful parameter is the exchange rate of amide protons after dissolution of the sample in D_2O. Slowly exchanging amide protons indicate protection from solvent and possible involvement in intramolecular hydrogen bonds associated with ele-

Figure 7. Schematic representations of 2D NMR spectra of the conotoxin MVIIA. (a) The fingerprint region of the TOCSY spectrum with selected spin systems marked. (b) Fingerprint region of the NOESY spectrum showing two (K2-A6 and L11-Y13) sequential walks. (c) NH–NH region of the NOESY spectrum showing correlations between the NH protons of D14 and G15; C16 and T17; and S22 and G23.

Table 4. ^{1}H Chemical Shifts for the 20 Common Amino Acid Residues[a] in Random Coil Peptides

Residue	NH	αH	βH	Others
Ala	8.24	4.32	1.39	
Arg	8.23	4.34	1.89, 1.79	γCH_2 1.70, 1.70 δCH_2 3.32, 3.32 NH 7.17, 6.62
Asn	8.40	4.74	2.83, 2.75	γNH_2 7.59, 6.91
Asp	8.34	4.64	2.84, 2.75	
Cys	8.43	4.71	3.28, 2.96	
Gln	8.32	4.34	2.13, 2.01	γCH_2 2.38, 2.38 δNH_2 6.87, 7.59
Glu	8.42	4.35	2.09, 1.97	γCH_2 2.31, 2.28
Gly	8.33	3.96		
His	8.42	4.73	3.26, 3.20	2H 8.12 4H 7.14
Ile	8.00	4.17	1.90	γCH_2 1.48, 1.19 γCH_3 0.95 δCH_3 0.89
Leu	8.16	4.34	1.65, 1.65	γH 1.64 δCH_3 0.94, 0.90
Lys	8.29	4.32	1.85, 1.76	γCH_2 1.45, 1.45 δCH_2 1.70, 1.70 εCH_2 3.02, 3.02 εNH_3^+ 7.52
Met	8.28	4.48	2.15, 2.01	γCH_2 2.64, 2.64 εCH_3 2.13
Phe	8.30	4.62	3.22, 2.99	2,6H 7.30 3,5H 7.39 4H 7.34
Pro		4.42	2.28, 2.02	γCH_2 2.03, 2.03 δCH_2 3.68, 3.65
Ser	8.31	4.47	3.88, 3.88	
Thr	8.15	4.35	4.22	γCH3 1.23
Trp	8.25	4.66	3.32, 3.19	2H 7.24 4H 7.65 5H 7.17 6H 7.24 7H 7.50 NH 10.22
Tyr	8.12	4.55	3.13, 2.92	2,6H 7.15 3,5H 6.86
Val	8.03	4.12	2.13	γCH3 0.97, 0.94

[a] The backbone shifts (αH and NH, ppm) are from Ref. [48] The remaining shifts are from Ref. [33].

ments of secondary structure. All of the NMR and slow exchange data can be consolidated to give an accurate representation of secondary structure, as indicated in Fig. 8. In the case of MVIIA, a triple-stranded β-sheet may be deduced based on the local NOE, coupling, chemical shift, and amide-exchange NMR data.

Once all peaks in the 2D spectra have been assigned, cross-peaks in the NOESY spectrum are used to derive a series of interproton distance restraints. Dihedral angle and hydrogen bonding restraints are also routinely derived from other NMR experiments. The combined restraints are then used in a simulated annealing algorithm to calculate a family of 3D structures consistent with the input restraints [52,53]. Figure 9 shows two commonly used methods of representing such NMR-derived structures, either as a stereoview of the superimposed family of structures or as a

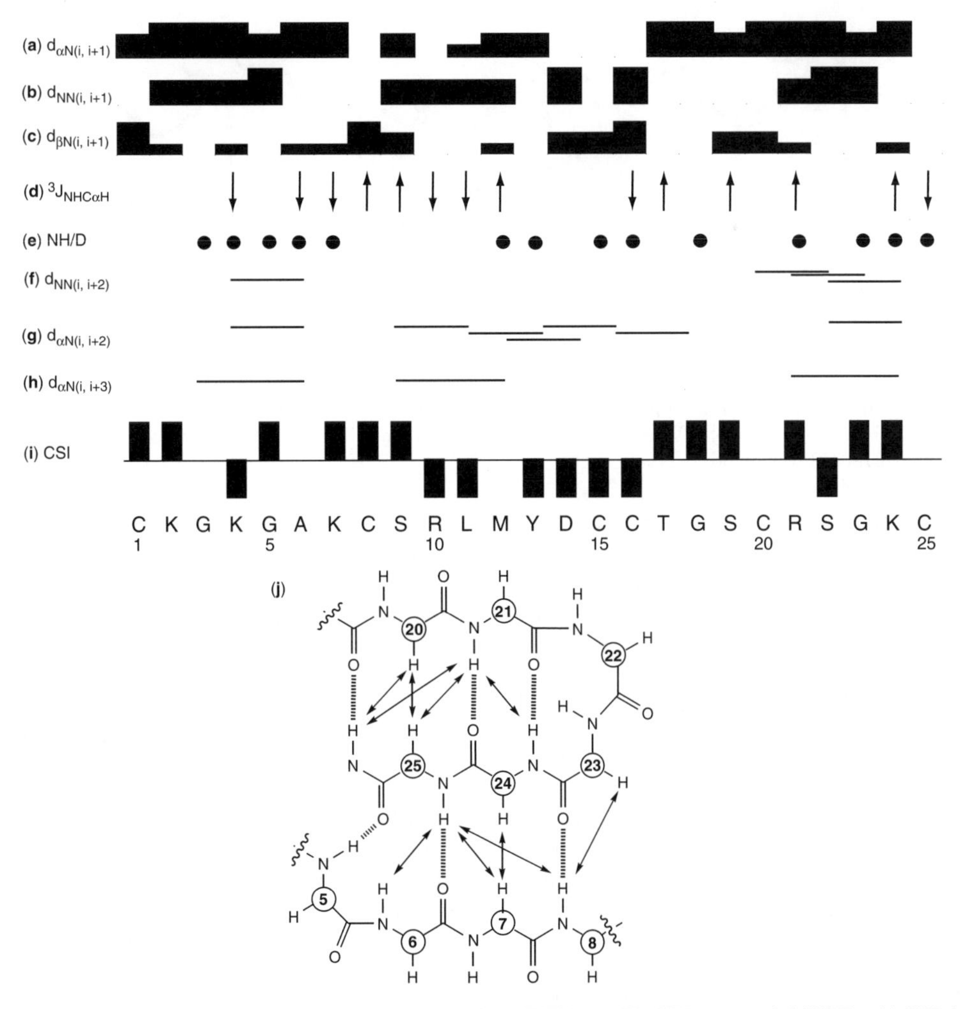

Figure 8. A summary of the NMR data observed for MVIIA. (a) Hα-NH sequential NOEs. (b) NH–NH sequential NOEs. (c) Hβ-NH sequential NOEs. (f–h) Other short-range NOEs. The thickness of the bar indicates the strength of the observed NOE (weak, medium, or strong). (d) Three-bond NH-Hα coupling data, where upward-pointing arrows indicate a large coupling (>8 Hz) and downward-pointing arrows indicate a small coupling (<5 Hz). (e) H/D exchange data, where a filled circle represents a slow exchanging NH. (i) Chemical-shift index (CSI) data. The CSI uses a scoring system that compares Hα shifts to random-coil chemical shifts. A sequence of consecutive $+1$ scores is indicative of β-structure, whereas a sequence of consecutive -1 scores suggests helical structure. (j) The β-sheet of MVIIA. Double-headed arrows indicate observed NOEs and broken lines indicate proposed H-bonds.

ribbon diagram, in which elements of second ary structure are highlighted. For the latter representation, the lowest energy or average member of the ensemble is often chosen as representative of the structure. However, it is important to examine the full ensemble to gain a complete understanding of the structure. Regions of disorder in the ensemble can be indicative of a lack of sufficient distance restraints, perhaps attributable to overlap or assignment errors, or the disorder might be related to local flexibility in the peptide structure.

In the case of MVIIA, a synthetic peptide corresponding to the natural peptide sequence was approved in the United States in 2004

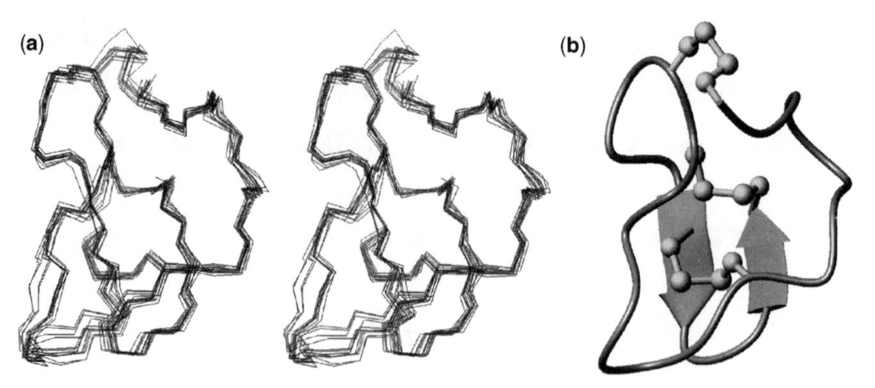

Figure 9. (a) A stereoview of the superimposed backbone structures of the 20 lowest energy conformations for MVIIA. (b) Ribbon diagram of MVIIA. (This figure is available in full color at http://mrw.interscience.wiley.com/emrw/9780471266945/home.)

under the trade name Prialt for the treatment of chronic neuropathic pain through the intrathecal (spinal infusion) route. This was the first marine-derived drug approved by the US FDA and represents a new trend in the pharmaceutical industry for peptides to be considered as useful drugs. Nevertheless in general, peptides have a range of potential disadvantages as drugs, including poor bioavailability and susceptibility to proteolytic breakdown. Thus, for many cases of drug development involving peptide-based leads, the structural information of the type described above might be used as a starting point to design smaller constrained peptides or nonpeptidic mimics. For example, this is the case in the development of the endothelin antagonists described below.

Endothelin as a Lead in Ligand-Based Design Endothelin (ET), shown in Fig. 10, is a 21-amino acid endothelial-derived constricting factor that has gained prominence as a pharmacological lead molecule. Interest in it arose because of its potent renal, pulmonary, and neuroendocrine activities. Endothelin and its isoforms have been implicated in a wide variety of disease states, including ischemia, cerebral vasospasm, stroke, renal failure, hypertension, and heart failure [54]. It exerts its pharmacological effect by acting on specific G-protein-coupled receptors. In mammalian species, two receptors, ET_A and ET_B are distributed widely in Human tissue and are distinguished by different responses to various ET isoforms. The dual endothelin receptor antagonist bosentan (trade name Tracleer®) is used for the treatment of pulmonary hypertension [55]. Research in this field is continuing for the development of drugs for other endothelin related diseases, including cancer and renal disease [56].

The NMR-derived 3D structure of ET-1 consists of several distinct regions, including a random-coil *N*-terminus, a β-turn involving residues 5–8, followed by a short helical region and a flexible C-terminal tail [57]. Inspection of the primary sequence, shown in Fig. 10, suggests that the presence of the flexible tail in solution is not surprising. Although solution structures of ET and its analogs [58–72], determined by NMR, have been valuable in defining the gross conformation of these molecules, the flexibility of the tail in solution makes it difficult to extrapolate to the bound state. Indeed, an X-ray structure of ET-1 has quite a different organization for the C-terminal tail than for the random-coil arrangement in solution [72], and the bound conformation may be different again.

There is clearly an advantage to having lead molecules with reduced flexibility, given that their solution conformation will intrinsically provide a better reflection of the bound conformation. In addition, the development of a more rigid drug will reduce unfavorable entropic contributions to binding energy. Indeed, a range of small cyclic peptides that are ET_A- or ET_B-selective antagonists, have been discovered, and provide valuable leads for the development of potential therapeutics [73]. NMR studies have been instrumental in determining their solution conformations. For

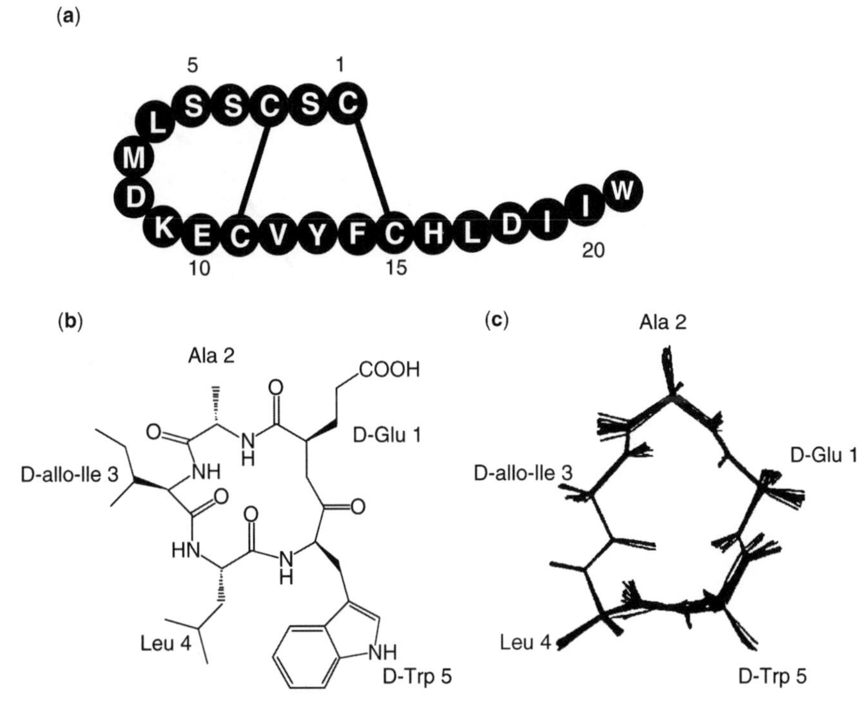

Figure 10. (a) Primary sequence and disulfide connectivities of endothelin-1 (ET-1). (b) Primary structure of the cyclic endothelin antagonist BE18257B, and (c) a family of 36 NMR structures, which demonstrate the well-defined nature of the cyclic peptide backbone.

example, the well-defined solution conformation [74] of the ET_A-selective antagonist BE18257B (shown in Fig. 10) contrasts with the flexibility of the tail region of ET that this peptide is thought to mimic. The discovery and development of these molecules illustrate the principle that cyclic peptides are often more suitable than linear peptides as lead ligands in drug design. In addition to their better-defined and less-flexible conformations than those of their linear counterparts, they generally have improved bioavailability and resistance to protease attack.

We shall return to endothelin as a lead in drug design, in relation to a nonpeptidic antagonist. The underlying theme illustrated by the endothelin example is that ligand-based design often proceeds from initial studies of flexible endogenous molecules (particularly peptides) to constrained mimics (e.g., cyclic peptides), and often culminates in the development of nonpeptidic drug leads. NMR assists by defining the structures of the lead and subsequent molecules. Although the approved endothelin antagonist, bosentan, is nonpeptidic, the cyclic peptide analogs of endothelin, including BE18257B, and a related cyclic peptide antagonist BQ-123, remain useful tools for researching the pathophysiology of the endothelin pathway [56].

2.2. Conformational Analysis

Usually only 1D or 2D NMR methods, rather than higher dimensional methods are required to determine the solution conformation of bioactive ligands. Useful tools include analysis of chemical shifts, coupling constants, and NOEs. An assumption inherent in the application of such studies to drug design is that the solution conformation will be maintained on binding to the receptor. This is justified in the case of relatively rigid ligands. However, for potentially flexible ligands the possibility of changes in conformation on binding must be considered, as noted above for the case of endothelin.

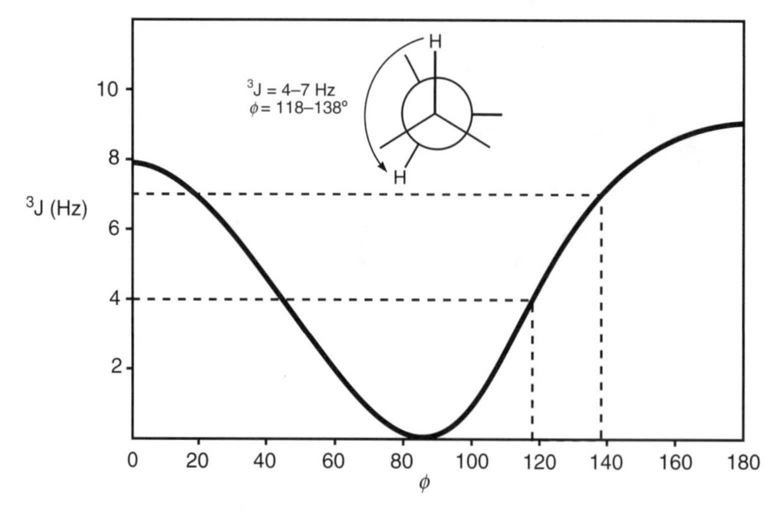

Figure 11. Illustration of the Karplus relationship between three-bond scalar coupling constants and the dihedral angle of the intervening bond. The relationship is indicated for the ϕ torsion angle of the H2 and H3 protons within the rigid core of taxol and related derivatives. See Fig. 6 for the structure of taxol.

Coupling constants and NOEs are the main NMR parameters used in determining the solution conformations of drug leads. NOEs provide information about through-space proximity. Three-bond vicinal-coupling constants are particularly valuable because their dependency on the intervening dihedral angle through the Karplus relationship allows local geometry to be determined. This is illustrated in Fig. 11 for taxol, which was also discussed in Section 2.1.1. Although there are several vicinal-coupling constants in this molecule (Fig. 6), only one $^3J_{H2H3}$ occurs in a region of the molecule that is expected to be conformationally rigid and thus suitable for conformational determination by use of coupling constants. In taxol and a range of analogs, this coupling is in the range 4–7 Hz, consistent with partially eclipsed dihedral angles of approximately 120–140° for this ring-constrained structure. This is in agreement with the X-ray structure of a taxol analog, where the angle is 120°. Note that, in general, such a Karplus analysis does not give a unique solution unless several coupling constants sampling the same dihedral angle are present and is reliant on the assumption that the molecule exists only in a single conformation in solution, which is generally believed to be the case for the core of taxol [44].

In addition to studies of the taxol core, there have been a large number of studies of the conformations of the side chains of taxol and it appears that these are flexible and that the molecule may adopt both extended and folded conformations of the side chains [44,75]. In a case like this, the observed vicinal-coupling constants are a weighted average of those from the participating conformers.

2.3. Charge State

An advantage of NMR over other structural techniques, such as X-ray crystallography, is that it has the potential to provide information not only on structure but also on the electronic properties of molecules. Many drug leads contain ionizable groups and a determination of their charge state in solution and/or at the bound site is important in the design of analogs. Simple plots of chemical shifts as a function of pH for nuclei near these ionizable groups provide a convenient way of determining the pK_a value and hence charge state. This is illustrated for Prialt in Fig. 12, where it was suspected that one of the ionizable groups in the molecule, Asp^{14}, might be involved in a stabilizing salt-bridge interaction [50]. This was confirmed by noting that the pK_a value for this residue is lowered considerably

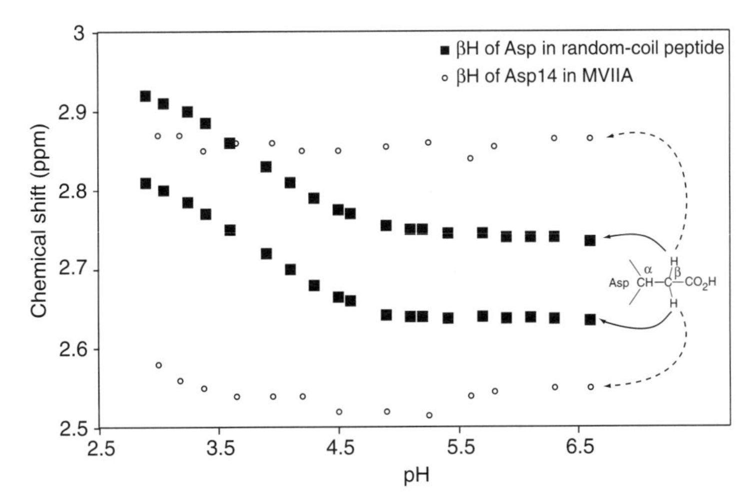

Figure 12. Chemical-shift changes of the β-protons of Asp14 in MVIIA illustrating the lack of titration of the adjacent carboxyl group, indicating its involvement in salt bridge. By contrast, the shift of a control random-coil peptide varies with an apparent pK_a value of 3.7, as expected for an uncomplexed carboxyl moiety in peptides.

relative to the usual value for Asp in a random-coil peptide. The β-proton chemical shifts were essentially independent of pH over the range 3–7 (indicating a p$K_a < 3$), whereas those of a control, random-coil peptide, titrated as expected over this range, with an apparent pK_a value of 3.7, typical of that expected for an unperturbed Asp residue.

2.4. Tautomeric Equilibria

Tautomerization is a relatively common feature of drug molecules that is amenable to analysis using chemical shifts or coupling constants as probes. This phenomenon was demonstrated, for example, in an investigation of nonpeptide endothelin analogs [76]. Starting from the modestly active compound (**1**) (Table 5), derived by screening a compound library for ET_A antagonists, the nanomolar inhibitor (**2**) was developed. Further optimization through examination of electronic and structural requirements led to the subnanomolar inhibitor (**3**), which was subsequently selected for evaluation in a number of preclinical disease models for stroke.

These molecules display keto-enol tautomerization, as illustrated in the following structures. The open form keto-acid salts and the closed form butenolides exist in a pH-dependent equilibrium in solution, and both forms exist at physiological pH. In principle, the biological activity could reside in either or both forms.

It is important to establish the position of the tautomerization equilibrium of the keto-enol forms under physiological conditions as this

Table 5. Substitution Pattern and Receptor Binding Affinity of Nonpeptidic Endothelin Antagonists[a]

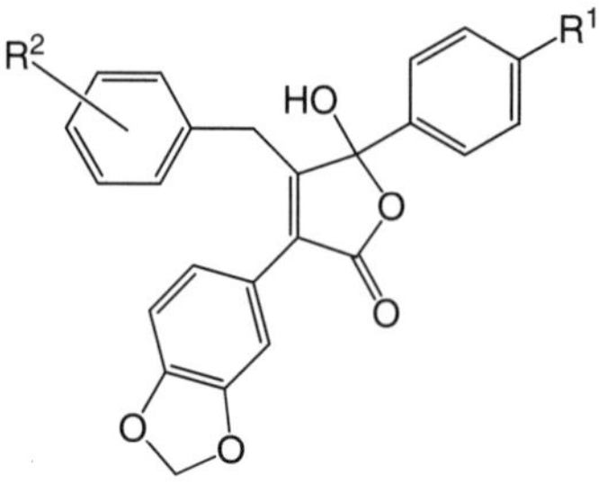

Compound	R_1	R_2	IC_{50} (nM)	
			ET_A	ET_B
(1) PD012527	Cl	H	430	27000
(2) PD 155080	OCH_3	H	>0.4	4550
(3) PD156707	OCH_3	3,4,5-OCH_3	0.3	780
(4)	OCH_3	3,5-OCH_3,4-$O(CH_2)_3SO_3Na$	0.38	1600

[a] From Refs [76,77].

can influence solubility and therefore has profound implications for formulation of drug candidates. The extent of tautomerization was established by evaluation of NMR spectra as a function of pH, from 2.65–9.05. At acidic pH, compound (2) exists essentially in the closed butenolide form. As the pH is slowly raised by addition of NaOD, the spectrum increasingly exhibits properties associated with the open form keto acid, until, at basic pH, the compound is essentially all in the open form. This tautomeric process is shown by the characteristic coupling pattern of the benzylic protons. At acidic pH the benzylic protons exhibit an AB quartet pattern consistent with the ring-closed structure. As the pH is raised, this pattern coalesces to a singlet, broad at neutral pH and sharp at basic pH, as would be expected with the open form keto-acid structure. The spectrum then returns to its original appearance after the addition of DCl to reacidify the solution, which is consistent with a reversible tautomerization process.

Pharmacological assays using the salt and closed butenolide form produced identical biological results, reflecting equilibration of the keto and enol forms at physiological pH, which made it difficult to identify the biologically active form. However, methylation of the OH group in compounds **(1)**–**(3)** resulted in a loss of activity. These analogs cannot tautomerize to form open keto acids, so it seems probable that the open form is responsible for activity.

Although it is easy to synthesize and isolate water-soluble salts of the keto acids, once they are placed in aqueous solution the tautomeric equilibrium determines the percentage of each form present. Indeed, if the closed butenolide tautomer is sufficiently water insoluble, it can precipitate out of solution and the equilibrium can drive the complete precipitation of the compound. Further development with compound **(3)** has shown that although it has good oral activity, its intravenous use is limited by the insolubility of the closed-form butenolide tautomer without the use of a specific and complex buffered formulation [77]. Thus, a series of water-soluble butenolides was developed to overcome this limitation for parenteral uses. This culminated in the development of **(4)** (Table 5). Analogs of the bu-tenolide compounds conjugated with fluorescent labels are also under investigation for use as photoprobes for imaging ET_A receptors *in vivo* [78,79].

This description of the development of **(4)** provides a good illustration of the fact that the synthesis of an active molecule is not the end of the drug development pathway, and that formulation considerations can be critical. In this case, NMR played a significant role in understanding tautomeric processes with a direct bearing on solubility and hence formulation.

2.5. Ligand Dynamics: Line-Shape and Relaxation Data

It is increasingly being recognized that the molecular dynamics of drugs in solution may have an important role in modulating biological activity [75,80–83]. For example, dynamics can influence entropic contributions to the free energy of binding. The flexibility of a ligand can be a favorable or unfavorable factor for binding, depending on the specifics of ligand binding. In general, for a ligand, the loss in entropy on binding is more unfavorable the more flexible the ligand, assuming a relatively rigid bound state of the ligand. However, flexibility of a ligand may be a favorable factor, if a degree of flexibility is required to allow access to a buried active site, or if activation of a receptor requires a conformational change mediated by ligand binding [19]. Therefore, knowledge of the flexibility of lead molecules is an important adjunct to the structural information available from NMR.

Two important NMR methods for obtaining information on ligand flexibility are line-shape analysis and relaxation measurements (usually ^{13}C or ^{15}N T_1, T_2, or heteronuclear NOE measurements). In general terms, the former is sensitive to motions on the milli- to microsecond timescale, whereas the latter covers a broader range of timescales from milli- to picosecond. To some extent, regions of flexibility can also be deduced from an examination of local regions of disorder (i.e., to lowered precision), among a family of calculated structures. Caution must be exercised because other factors can contribute to disorder, although in many cases there is a connection between disorder in a structural ensemble and molecular flexibility [84–86].

The thyroid hormones, exemplified by thyroxine (**5**), provide an example of the use of both line-shape analysis and NMR relaxation time measurements to give an insight into the internal flexibility, and the mode of action of pharmaceutically important molecules [81,82,87]. In this example, the flexibility of the thyroxine ligand is a favorable factor for receptor binding. The thyroid hormones act by binding to a nuclear receptor and modulate receptor function by apparently inducing a conformational change that directs the alignment of functionally critical secondary-structure elements of the receptor [88]. Synthetic thyroxine has been used widely for the treatment of thyroid disorders, such as hypothyroidism, since the 1950s.

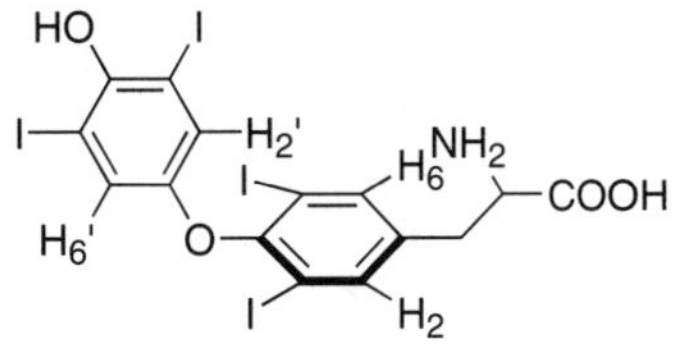

(**5**)

Table 6 shows experimental ^{13}C T_1 and heteronuclear NOE data for thyroxine, together

Table 6. ^{13}C NMR Chemical Shift and Relaxation Data for Thyroxine

				Theoretical[b]			
		Experimental[a]		Isotropic Motion		Two-State Internal Motion	
Position	Chemical Shift (ppm)	T_1 (s)	NOE	T_1 (s)	NOE	T_1 (s)	NOE
C2′,6′	127.3	0.63	2.53	0.63	2.96	0.63	2.53
C2,6	142.6	0.63	2.63	0.63	2.96	0.63	2.58
C-α	57.1	0.51	2.37	0.51	2.94	0.51	2.57
C-β	36.4	0.64	2.29	0.64	2.96	0.64	2.51

[a] Measured relaxation data at 75 MHz and 305 K (52).
[b] Theoretical best fit values based on the indicated models for molecular motion.

with best-fit theoretical values for these parameters based on two different models of the motion of thyroxine in solution. Comparison of the theoretical and experimental data indicates that a simple isotropic model, in which the drug is regarded as a rigid body tumbling in solution, is clearly unable to fit the T_1 and NOE data simultaneously. However, the two-state jump model, which incorporates a degree of internal mobility based on rapid flipping between conformational states, is better able to account for the experimental data, and this indicates that rapid (nanosecond) internal motion is present. Although it is difficult to define the nature of the internal motion precisely, it appears to involve small-amplitude torsions about the aromatic ring axes. The correlation time for overall tumbling of thyroxine was deduced to be approximately 0.35 ns, with the internal motion approximately 30-fold faster [81].

Line-shape analysis of the ^{1}H NMR spectrum of thyroxine as a function of temperature has demonstrated the presence of additional larger amplitude, but slower ring flips [89]. The two signals seen for the H2′ and H6′ protons at low temperature broadened with increasing temperature, then coalesced and sharpened as the temperature was further increased. This was attributed to exchange of the environments of the two protons brought about by 180° rotation of the "outer" ring of thyroxine. Substitution of the observed coalescence temperature (T_c) and the chemical-shift difference of the two signals at low temperature ($\delta\nu$) allowed the free energy of activation for this slow ring flip process to be established from Equation 1 [82,89].

$$\Delta G^{\neq} = 19.14 \times T_c \times \left[9.97 + \log\left(\frac{T_c}{\delta\nu}\right)\right] \quad (1)$$

The derived barriers for several thyroid hormones are in the range 36–38 kJ/mol, which corresponds to large-amplitude ring flips on the milli- to microsecond timescale. The combination of the relaxation data for the fast internal motions, and the dynamic line-shape analysis data of the slower internal motions allowed for the proposal of a unified model that accounts for both the fast and slow internal motions, as summarized in Fig. 13.

In this model, both aromatic rings of the thyroid hormones jump rapidly between two energetically equivalent conformations on a nanosecond timescale ((a) ↔ (b) and (c) ↔ (d) in Fig. 13). The magnitude of the half-angle jump is solvent dependent and corresponds to an average displacement of about 90° between the two extreme jump positions. These separate states are not detectable on the chemical-shift timescale, but lead to an average proximal environment for Ha and an average distal environment for Hb (attributed to rapid interchange between (a) and (b) in Fig. 13), which are seen in the low temperature spectra. However, these fast motions can be detected by relaxation studies. Although the rate of this motion is rapid, its amplitude is not sufficient to average the environment of proximal and distal protons. Occasionally (approximately once every 1000 jumps), the outer ring jumps further than the nominal 90° range, exchanging the environments of the proximal and distal protons ((a) ↔ (c) and (b) ↔ (d) in Fig. 13). Although the actual rate of an individual ring flip is rapid, the effective rate of the overall process is on the microsecond timescale, because, on average, a large number of small-amplitude jumps occur for every large amplitude ring flip. It is the exchanging of proximal and distal protons on the microsecond timescale that is detected by the variable-temperature line-shape studies.

The fact that thyroxine is apparently able to move so freely over a moderately large region of conformational space has implications for receptor binding. The crystal structure of the thyroid receptor ligand binding domain complexed with the thyroid agonist 3,5-dimethyl-3′-isopropylthyronine [88] shows that the thyroid hormones bind at the center of the hydrophobic core of the ligand binding domain and may play a structural role in the conformational changes that activate the receptor [90,91]. The conformational flexibility exhibited by the thyroid hormone ligands may also be required for binding. It has been suggested that the rapid "wiggling" of the aromatic rings could enable the hormone to work its way to the center of the ligand binding domain as the protein reorders itself about the ligand and may in fact trigger receptor conformational changes [81].

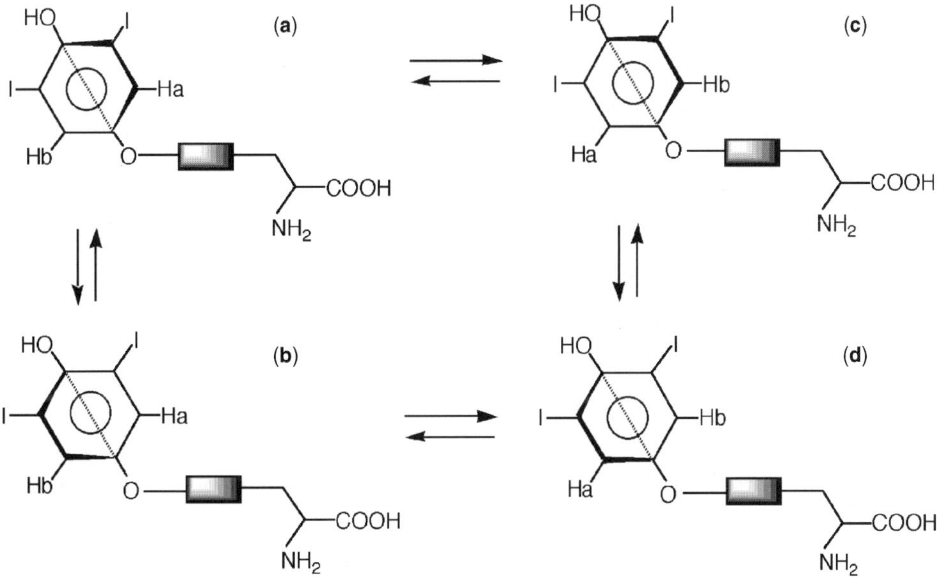

Figure 13. Schematic illustrations of motions of the outer ring of thyroxine. The dotted line through the outer ring shows the jump axis about which the ring rotates. (a) Ha is shown in the proximal position and is closer to the viewer than Hb because the torsion angle ϕ' is greater than 0°. This conformation corresponds to one of the two states of the two-state jump model and agrees with the "twist" of the outer ring observed in the crystal structure. (b) Rotation about the dotted line through the center of the outer ring moves Ha away from the viewer and brings Hb toward the viewer. This corresponds to the second state of the outer ring in the two-state jump model. (c) Hb is now in the proximal position and closer to the viewer than Ha. (d) Hb is in the proximal position and is now further from the viewer than Ha. Transition from (a) to (b) and from (c) to (d) involves small-amplitude jumps on the nanosecond timescale and is detected by NMR relaxation measurements. Although not illustrated in the figure, the inner ring also exhibits this type of motion. Transitions (a) to (c) and (b) to (d) result in 180° flips of the outer ring and exchange of the environments of Ha and Hb. This ring flip occurs on a microsecond timescale and is detected by variable temperature line-shape studies. (Adapted with permission from Ref. [81]. 1996 American Chemical Society.)

Another example of the application of line-shape analysis to ligand dynamics is described in Section 3.2 for the drug trimetrexate when bound to dihydrofolate reductase (DHFR). From that example and earlier studies on DHFR [92–94], it is clear that the techniques described above can equally be applied to ligands when bound to their receptor. In some cases, significant but highly specific mobility appears to be present at the bound site.

2.6. Pharmocophore Modeling: Conformations of a Set of Ligands

Determination of the conformations of a range of ligands that act at the same receptor site can provide significantly more information than that provided by a single ligand structure. With a sufficiently broad range of ligands, it is often possible to generate a pharmocophore model of the receptor site, which is deduced based on conserved structural features and the conformations of the ligands. This has been achieved, for example, for the ω-conotoxins, the broad class of conotoxins to which Prialt belongs [95]. It was determined, from structural studies of a range of ω-conotoxins and from literature data on various mutants with altered binding affinities, that only a localized region of the surface of these molecules is involved in receptor binding [95]. This allowed a pharmocophore model of putative receptor binding pockets to be developed. Such a pharmacophore model in principle facilitates the design of smaller, nonpeptide molecules that might have improved stability and bioavailability over their peptidic counterparts. Additionally, a pharmacophore model can be used to predict the structure–function relationships of novel molecules where there is little sequence similarity to known molecules. The NMR approach used in such pharmacophore modeling often involves a combination of many of the techniques already described. By determining information about structure and electronic properties for a range of different ligands, all acting at the same receptor site, it is often possible to infer information about the binding site, even if direct structural studies are not possible.

2.7. Limitations of Analog-Based Design

Although determination of the structure of bioactive molecules is very important, there are distinct limitations on the use of solution structures for drug design. In particular, unless the molecule is rigid, there is no certainty that the solution conformation is the same as the bioactive bound conformation. For this reason there has been a shift to approaches in which information about the bound state is obtained. Another approach has been to probe the bound conformation by making a range of constrained analogs of a flexible lead molecule, as illustrated earlier for endothelin.

The most direct way of determining the conformation of a drug lead is to determine the full 3D structure of its receptor complex. This has now been achieved in a significant number of cases but represents a substantial undertaking, as described in later sections of this chapter. A simpler approach is to use transferred NOE methods. This approach fits at the interface of ligand-based design and receptor-based design: it fits with the former because no knowledge of the receptor structure is required; however, it also fits with the latter because it requires the macromolecule of interest to be included in the sample to be analyzed. Therefore, it is appropriate to introduce the topic here and also to discuss it further in Section 3.

2.8. Conformation of Bound Ligands: Transferred NOEs

In ligand-based drug design it is not necessary to know the structure of the receptor, or even the location of the binding site, although the conformation of the ligand bound to the receptor is crucial. It is clearly better if this can be measured directly rather then be inferred from the conformation of the free ligand. In certain circumstances information on the bound conformation can be obtained from the transferred NOE (TrNOE) technique [96,97]. This method takes advantage of the fact that NOEs build up more rapidly in a ligand–macromolecule complex than they do in a free ligand, and given appropriate exchange conditions for a mixture of ligand and macromolecule (typically satisfied for $K_D \geq 10^{-7}\,M^{-1}$), signals from a free ligand can be used to determine the bound conformation.

In TrNOE experiments [98], it is not necessary to monitor signals from the macromolecule and so it is usually present in substoichiometric amounts, thus requiring only minimal amounts of what is often the more expensive component of ligand–macromolecule complexes. In addition, the molecular weight restrictions inherent in full 3D-structure determinations of complexes are ameliorated and the conformations of ligands bound to very large macromolecules may be determined. For example, TrNOE has been used to determine the structure of an antibiotic bound to the ribosome [97] and a range of other applications, including enzyme–substrate, protein–-carbohydrate, and protein–peptide interactions have been reported [22].

In addition to its application as a tool for determining bound conformations of ligands, the TrNOE method has also been used as a screening aid for the identification of ligands from mixtures that bind to a protein of interest. This application is addressed in more detail later in this chapter.

3. RECEPTOR-BASED DESIGN

Receptor-based design refers to the process of determining the 3D structure of a macromolecular target and using this information to design ligands with which it can interact. In general, there have been few cases where the structure of a macromolecule or receptor alone has been successfully used to design, *de novo*, a ligand to interact with that receptor. The most common approach is to study a ligand–

macromolecule complex and to initiate the design process based on the interaction of the lead ligand with the macromolecule. Although the structure of the complex is of more interest than the macromolecule alone, in many cases a determination of the structure of the complex follows from earlier studies on the unbound macromolecule. It is thus useful to describe the approaches to structure determination of macromolecular targets; this is followed by a discussion of the dynamic aspects of protein structures in Section 3.2, before addressing macromolecule–ligand interactions in Section 3.3.

3.1. Macromolecular Structure Determination

The two major techniques for determining 3D structures of proteins or nucleic acids are X-ray crystallography and NMR spectroscopy. X-ray crystallography is generally regarded as faster and more accurate, but NMR has some advantages over X-ray crystallography, including avoidance of the need for protein crystallization, and the complementary dynamic and structural information available from NMR studies. A disadvantage of NMR is that it is generally limited to proteins smaller than ~40 kDa. However, with the development of new labeling technologies and NMR techniques, such as TROSY [34], the ability to determine structures of larger proteins has increased significantly over recent years. Nevertheless, among all structures currently deposited in the protein database, the average size of NMR structures is about 8 kDa [99], substantially smaller than the average size of protein structures determined by X-ray crystallography. Currently, approximately 14% of structures deposited in the protein database have been determined by NMR spectroscopy.

3.1.1. Overview of Approach The basis for structure determination by NMR for proteins is similar to that described earlier for peptides whereby determining a large number of distance restraints between pairs of protons, it is possible to reconstruct a 3D image of the molecule. These distance restraints are derived primarily from NOE measurements, which detect distances up to approximately 5 Å. Such distance restraints can be supplemented by a range of other restraints, including dihedral angle restraints derived from coupling constant measurements and orientation restraints derived from residual dipolar couplings. These restraints are input into a simulated annealing algorithm, which is used to calculate a family of structures consistent with the restraints.

NMR is unique in that it can provide detailed and specific information on molecular dynamics in addition to structural information. The use of relaxation time measurements allows determination of the relative mobility of individual atomic positions within a macromolecule. The dynamic information obtained includes amplitudes as well as the rates or frequencies of internal motions. Such amplitudes are often expressed by order parameters. Order parameters are defined by the symbol S^2, which is a measure of the degree of spatial restriction of motion. S^2 can take values from 0 to 1; where $S^2 = 0$ when the internal motions are isotropic and all orientations are possible, and $S^2 = 1$ if motion is completely restricted [100]. Typically, the rates and frequencies of motion extracted from relaxation data are done via a process of modeling rather than direct determination [101].

Not surprisingly, in many cases it is observed that the termini of proteins are more flexible than internal regions. More interestingly, NMR has provided a number of examples where internal loops in proteins have been shown to have flexibility that may be associated with their function. A good example of this is HIV protease, where NMR studies have identified lowered order parameters in the flap region of the active dimer protein, which reflects the flexibility of the protein to allow entry of substrates or inhibitors into the active site. The HIV protease structure and drug targeting of the protein is discussed further in the section "HIV Protease."

In summary, a major strength of NMR is that a global picture of both the structure and the dynamics of the macromolecular target is obtained. Further, NMR provides information on ionization states of titratable groups and other electronic features within macromolecules that may have an impact on ligand binding and function.

3.1.2. Sample Requirements and Assignment Protocols Structure determination by NMR typically requires ~500 μL of a 0.5–1 mM solution of the protein of interest. It is important that the protein does not aggregate because this causes spectral broadening and may preclude assignment. The sample should be stable in solution over the time required to collect the range of NMR experiments needed for assignment and structure determination. Individual experiments may last from a few hours to several days, with several weeks of data acquisition required for larger proteins.

The particular set of NMR experiments required for NMR structure determination depends on the size of the protein. For smaller proteins ($\leq$7 kDa) it is usually possible to determine the structures using 2D NMR, without the need for isotopic labeling, by use of procedures described earlier for Prialt. For proteins in the range 7–14 kDa, ^{15}N-labeling and a combination of 2D/3D NMR experiments are usually sufficient, whereas for larger proteins ^{13}C/^{15}N labeling and 3D or 4D NMR are more or less mandatory. For proteins at the top end of the currently accessible range (35–50 kDa), there are additional advantages associated with partial deuteration of the protein and/or the labeling of specific amino acids with ^{13}C or ^{15}N, with a range of different labeling strategies used in different samples of the same protein.

Random fractional ^{2}H labeling dramatically enhances the resolution of NMR spectra via the dilution of ^{1}H/^{1}H dipolar interactions that lead to unfavorable relaxation properties [5,102]. In this strategy, all of the protons in a protein sample are randomly replaced by deuterons, resulting in each protein molecule in the NMR sample having a different ^{2}H-labeling pattern. The sum of signals from the protein molecules gives a spectrum that is nearly identical to that from a fully protonated sample, but with much sharper signals. The enhanced resolution afforded by the use of ^{2}H labeling allows for the study of larger proteins [5].

^{13}C and ^{15}N isotopes can readily be incorporated into proteins through their recombinant production in growth media enriched for the respective isotopes. Uniform labeling of proteins with ^{13}C and ^{15}N facilitates the investigation of protein structure via enabling a range of multinuclear and multidimensional NMR experiments. For example, it allows for qualitative analysis of the overall fold of a protein using ^{1}H/^{15}N HSQC and/or ^{1}H/^{13}C HSQC experiments. The unambiguous assignment of the backbone signals can be determined from the through bond correlations of the ^{1}H, ^{13}C and ^{15}N spins using a variety of 3D experiments.

3.1.3. Dynamics Proteins exhibit a range of internal motions, from the picosecond to second timescale, and a full understanding of how small drugs might interact with such a "moving target" requires more than just the time-averaged macromolecular structure. Thus, over recent years, much effort has been directed toward defining motions within proteins to investigate how motion relates to function.

The most commonly applied approach has been to use ^{13}C or ^{15}N relaxation parameters such as T_1, T_2, and the heteronuclear NOE to derive correlation times for overall motion, together with rates and amplitudes of internal motions [103]. Although the precise interpretation of NMR relaxation data in terms of motional parameters remains dependent on the appropriateness of the motional model chosen, the results from many studies on the dynamics of proteins are sufficiently clear to confirm that picosecond–nanosecond timescale motions are common [104].

NMR is also able to monitor global protein tumbling and movement of flexible loops on the microsecond–millisecond timescale as well as detecting fast molecular motions on the picosecond–nanosecond timescale. Slower motions have been correlated with function in a number of proteins, with a good example being HIV protease, described in more detail in Section 4.2.

3.1.4. Nucleic Acid Structures Most of the discussion on macromolecular targets so far has focused on proteins. DNA represents another valuable target in drug design. Most studies, in which DNA is the target, use short oligonucleotides to mimic the binding region of DNA. The regular repeating nature of DNA structures makes this a more successful approach than similar attempts to dissect out binding regions of receptor proteins, where often the whole protein must be present to maintain a viable binding site. Similar comments apply for RNA, where improvements in synthetic methods have led to an increasing number of structure determinations [105]. The principles involved in structure determination of nucleic acid targets are similar to those of proteins, but in practice nucleic acid structures are somewhat more difficult to solve [106].

3.1.5. Recent Developments A number of recently developed methods offer the potential for improving the quality of NMR structures and for increasing the size of proteins that can be examined. In particular, the use of residual dipolar couplings and anisotropic contributions to relaxation provide new kinds of restraints that promise to lead to more accurate NMR structures [46,107,108]. Advances in isotopic labeling strategies have also facilitated the study of large proteins by NMR.

The TROSY method exploits transverse relaxation phenomena to produce spectra with narrow lines and has expanded the size of protein targets that can be examined by NMR significantly. Additionally, the use of cross-relaxation-enhanced polarization transfer (CRINEPT) and cross-relaxation-induced polarization transfer (CRIPT) experiments has produced enhancements in signal to noise and has been applied to a 110 kDa protein [109] and a 900 kDa protein complex [110]. CRINEPT and CRIPT experiments utilize NOE cross-relaxation pathways to increase the efficiency of magnetization transfer in NMR experiments, thus allowing for the analysis of larger proteins, membrane proteins, and proteins that oligomerize in solution [3].

Pulse sequences such as TROSY, which reduce the undesirable effects of relaxation in larger molecules, are effective at increasing the sensitivity and resolution of NMR spectra for larger proteins, but they do not address the problem of spectral overlap and crowding. However, this difficulty can be ameliorated using isotopic labeling strategies. Uniform isotopic labeling of proteins with ^{13}C and ^{15}N has been in use for many years. Recently, complementary labeling strategies have been developed that include the selective labeling of protein domains or specific amino acids, which allows for the analysis of proteins larger than those previously accessible to NMR. These techniques have recently been reviewed [5].

Segmental labeling involves the specific labeling of a domain of a larger protein, thus allowing for the study of the behavior of that domain of interest. This is achieved using inteins, which perform splicing reactions that ligate two separately expressed regions of a protein, one labeled and one not labeled [46,111]. There are many techniques available for the selective labeling of specific amino acid types, such as methyl group selective protonation to label Ala, Val, Leu, and Ile residues in a deuterated protein [112–114]. One particularly elegant approach is the SAIL (stereoarray isotope labeling) strategy in which stereo- and regio-specific amino acid isotope patterns are utilized [115]. This strategy involves the chemical and/or enzymatic synthesis of particular amino acids labeled with ^{13}C and ^{2}H in specific positions for use in cell-free protein expression. The SAIL labeling technique, along with other labeling methods has increased the number of proteins that are accessible to NMR analysis, and the further development of these techniques will allow for the study of even more challenging proteins.

Structural genomics programs have also had a significant impact in the utilization of NMR. The demand that has arisen from such programs has stimulated new methods for the large-scale production of labeled proteins [116,117]. Now that it is easier to produce proteins, the rate of structure determination by both NMR and crystallography has increased.

3.1.6. Challenges for the Future: Membrane-Bound Proteins The majority of targets for currently known drugs are membrane-bound receptors, yet they represent a class of proteins for which least structural information is known. Membrane proteins are notoriously difficult to characterize at a structural level because they are difficult to crystallize, thus inhibiting X-ray crystallographic studies, and are both too large and too difficult to reconstitute in suitable media for NMR studies.

Nevertheless, solid-state NMR methods are beginning to show promise as a powerful tool for the structural characterization of membrane proteins [12,118]. Solid-state NMR of membrane proteins involves the preparation of the protein in a microcrystalline or other precipitated form using precipitants or lyoprotectants such as polyethyleneglycol. The preparation of the protein is important, as spectral resolution is dependent on the molecular order that can be attained. As with solution state studies, isotopically labeled proteins are required for solid-state NMR studies [12,119].

3.2. Macromolecule–Ligand Interactions

Macromolecule–ligand interactions are integral to a wide range of biological processes, including hormone, neurotransmitter or drug binding, antigen recognition, and enzyme–substrate interactions. Fundamental to each of these interactions is the recognition of a unique binding site on the macromolecule by a ligand. Through an understanding of the specific interactions involved, it may be possible to design or discover analogous ligands with altered binding properties that might inhibit the biochemical function of the macromolecule in a highly specific manner. The study of macromolecule–ligand interactions thus forms the cornerstone of most structure-based drug design applications. The macromolecule of interest may be a protein or a nucleic acid, although the majority of drug design applications have focused on protein–ligand interactions. For this reason, we will mainly refer to protein–ligand interactions in the following discussion, but will include some examples of drug-DNA interactions.

3.2.1. Overview There are several important aspects of macromolecule–ligand interactions that have a bearing on structure-based design. The simplest question that might be asked is, "What is the strength of the binding interaction?," whereas the most detailed task would be to define the atomic coordinates of the complete protein–ligand complex precisely. In between these extremes, there are many other questions important to the drug design process; these include questions about the binding stoichiometry and kinetics, the conformation of the bound ligand, and about the nature of functional group interactions between the protein and the bound ligand. These and other important questions were introduced briefly in Table 3 and are examined in more detail in this section. Before doing this, it is necessary to consider NMR timescales because the ability of NMR methods to answer questions about macromolecule–ligand complexes depends critically on the kinetics of the binding interaction. Section 3.2.2 describes how various NMR parameters depend on binding kinetics and, in particular, how fast- and slow-exchange conditions affect the interpretation of NMR data.

Once the exchange regime has been identified, the task then becomes to decide which NMR parameters can be used to answer the questions posed about the complex. Many of the NMR parameters that were described earlier for deriving information about ligands are also applicable to studies of complexes. These include chemical shifts, NOEs, and relaxation parameters. However, the presence of two interacting partners means that there are some differences in the way such parameters are measured and this has led to the development of several techniques that are particularly important for the study of macromolecule–ligand interactions, including chemical-shift mapping, isotope editing, and various NMR titrations. Section 3.2.3 describes these techniques. Illustrative examples of the application of these techniques to specific drug design problems are given in Section 3.2.4.

3.2.2. Influence of Kinetics and NMR Timescales Macromolecule–ligand interactions are characterized by an equilibrium reaction that potentially has a wide range of affinities and rates:

$$M + L \leftrightarrow ML$$

The rate constant for the forward reaction is referred to as the on rate (k_{on}), whereas dissociation of the complex is characterized by the reverse rate constant, k_{off}. The equilibrium constant for this interaction, represented in terms of the dissociation constant of the complex K_D, reflects a balance of the on and off-rates, as shown in Equation 2:

$$K_D = \frac{[M][L]}{[ML]} = \frac{k_{off}}{k_{on}} \quad (2)$$

For many protein–ligand interactions, k_{on} is of the order of $10^8\ M^{-1}\ s^{-1}$, and is typically quite similar for different ligands. The observation that K_D values may vary over a wide range, typically from millimolar to nanomolar (i.e., $K_D = 10^{-3}$–10^{-9} M) for cases of interest, is a reflection of a variation in k_{off} for different ligands. Consideration of the k_{on} value above and the range of K_D values noted suggest a range in k_{off} from 10^{-1} to $10^5\ s^{-1}$. The lifetime of the bound complex ($\tau_{ML} = 1/k_{off}$) may thus vary from much less than a millisecond to tens of seconds (10^{-5} – 10 s based on the above off-rates). The exchange rate for the second-order binding process is given by [120]

$$k = \frac{1}{\tau} = \frac{1}{\tau_{ML}} + \frac{1}{\tau_L} = k_{off}\left(1 + \frac{p_{ML}}{p_L}\right) \quad (3)$$

where p_{ML} and p_L are the mole fractions of bound and free ligand, respectively.

The appearance of an NMR spectrum of a protein–ligand complex is dependent on the rate of chemical interchange between free and bound states. In particular, the effects of exchange on an individual NMR parameter (e.g., chemical shift, coupling constant, or relaxation rate) depend on the relative magnitude of the exchange rate and the difference in the NMR parameter between the two states. Table 7 indicates the typical magnitudes of these parameters. The cases where the rate of interchange is greater than, about equal to, or less than, the parameter difference are referred to as fast, intermediate, and slow exchange, respectively.

Table 7. NMR Parameters and Their Changes on Binding

Parameter	Difference	Typical Magnitude (s^{-1})[a]
Chemical shift	$\nu_L - \nu_{ML}$	0–1000
	$\nu_M - \nu_{ML}$	0–500
Coupling constant	$J_L - J_{ML}$	0–12
	$J_M - J_{ML}$	0–12
Relaxation rate[b]	$1/T_L - 1/T_{ML}$	0–50 (for T_1, larger for T_2)
	$1/T_M - 1/T_{ML}$	0–10

[a] Ranges are approximate only and larger effects may be seen in some cases.
[b] $1/T$ refers to either $1/T_1$ or $1/T_2$.

Table 7 shows that the changes in chemical shifts on ligand binding (for signals either from the ligand or from the macromolecule) are, in general, greater than those for coupling constants or relaxation rates. Given that $100\ s^{-1}$ might represent a typical exchange rate between free and bound states, it is clear that individual NMR signals may be found in either slow, fast, or intermediate exchange on the chemical-shift timescale, but it is more probable that couplings or relaxation parameters will be in fast exchange. Thus, in most cases where the term "NMR timescale" is used in the literature, it refers to the chemical-shift timescale. The table also emphasizes that there are two types of signals that can be monitored, those from the ligand and those from the macromolecule. In general, the typical magnitude of changes to chemical shifts or couplings of either type of signal on binding are similar, although the changes to ligand signals may be larger than those from the macromolecule. However, changes to relaxation parameters for signals from ligands are more likely to be greater than protein signals. This reflects the sensitivity of relaxation parameters to molecular mobility: a ligand undergoes a greater relative change in mobility on binding than a protein, given that the relative increase in molecular weight in the complex is much greater for the ligand than for the protein.

The exchange regime (slow, intermediate, or fast) determines how a spectrum of a protein–ligand mixture changes during a titration, or as a function of temperature. Figure 14 schematically illustrates the various exchange regimes for macromolecule–ligand binding interactions. Slow exchange, corresponding to tight binding, is potentially the most useful regime, given that, in this case, much detailed information on the nature of a complex can be deduced. Nevertheless, fast exchange also allows the derivation of valuable kinetic and thermodynamic parameters. The analysis is more complex for intermediate exchange and it is uncommon for quantitative studies to be attempted for this situation.

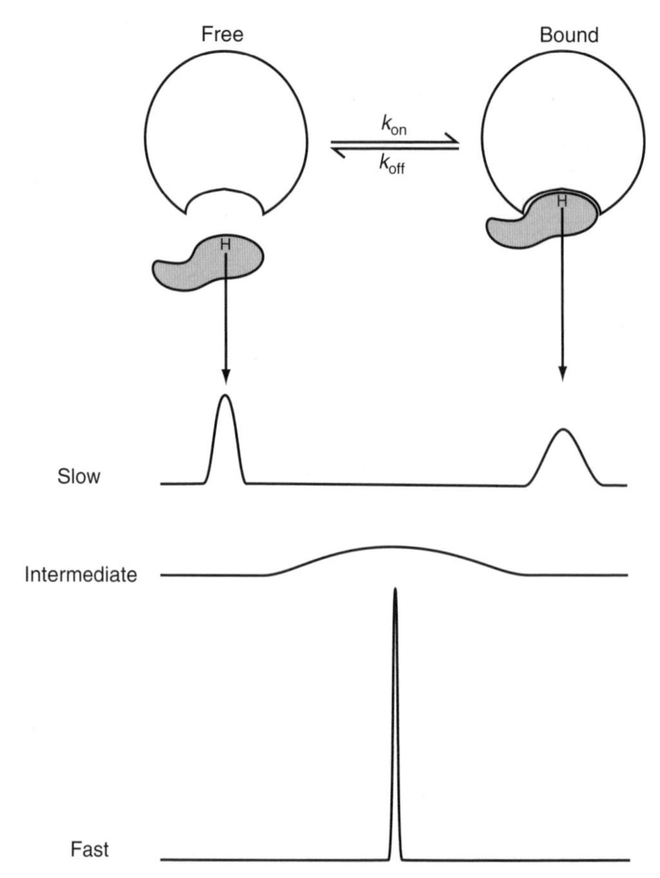

Figure 14. Schematic illustration of the effects of slow, intermediate, and fast exchange on the appearance of peaks in NMR spectra of macromolecule–ligand complexes. In the slow-exchange case separate peaks are seen for free and bound forms. Note the broader peak for the bound ligand because it now adopts the correlation time of the macromolecule. In the fast-exchange case only an averaged peak is observed.

Slow Exchange This situation applies when the rate of exchange is much slower than the difference in chemical shifts between the two states (i.e., $k \ll \nu_B - \nu_F$), where we now change to a nomenclature using subscripts to refer to the bound (B) and free (F) states. It should be understood that the signals may derive from either ligand or macromolecule, so although B is always related to the ML complex, F might refer to either free ligand (L) or free macromolecule (M). In this situation, separate peaks are potentially observable for both free and bound states at their respective chemical shifts. Whether such signals are *actually* observed depends on both the mole ratio of the individual species in a titration series and whether signals are obscured by overlap or broadening.

Addition of a ligand to a solution of a protein results in the appearance of new signals attributed to bound protein resonances, with a concurrent decrease in the intensity of the free protein resonances, reflecting the decreased proportion of free protein during the titration. Once a stoichiometric mole ratio is achieved (usually 1 : 1, but sometimes 2 : 1 or higher if multiple ligand binding sites are present), peaks from free ligand appear with increasing intensity as the excess of free ligand increases. Such a titration allows determination of the stoichiometry of the complex together with the chemical shifts of the bound states of the ligand and protein. In 1D NMR spectra, overlap of peaks makes it difficult to monitor more than a few resonances from either species and such studies are most readily done when there is a well-resolved signal on one of the interacting species. Selective isotope labels have been used in the past, but it is more common to use uniform ^{15}N- or ^{13}C-labeling of the protein and detect the chemical shifts in 2D HSQC spectra. It is often more difficult to label the ligand, but in some cases the presence of nuclei such as ^{19}F can be used to advantage.

The advantages of using ^{19}F NMR include the relatively large chemical shifts that result from minor changes to the chemical environment of the ^{19}F nucleus. A good example is the binding of the inhibitor 4-fluorobenzenesulfonamide to carbonic anhydrase [121]. Figure 15 shows ^{19}F spectra of the enzyme–inhibitor complex at various mole ratios. The broadened peak for the bound ligand has a chemical shift of approximately 6 ppm and is in slow exchange with the peak from free ligand at 0 ppm. The stoichiometry of the complex in this case is 2 : 1, so that no signal from free ligand is visible until more than 2 mol is present. The addition of increasing amounts of ligand results in an increase in the free ligand signal, but no change in the bound ligand signal.

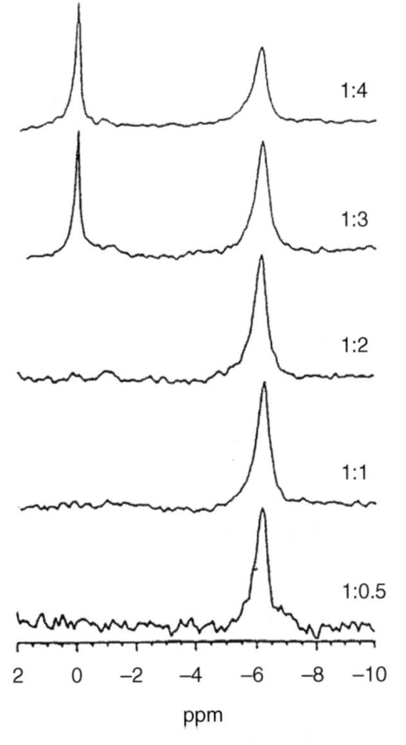

Figure 15. ^{19}F NMR spectra at 282 MHz of the 4-fluorobenzenesulfonamide-carbonic-anhydrase-1 system at various ratios of enzyme to inhibitor, as indicated on the traces. The peak at −6 ppm is caused by bound inhibitor. The enzyme concentration was 1 mM at pH 7.2 in D_2O at 25°C. (Reprinted with permission from Ref. [121]. 1988 American Chemical Society.)

Determination of the binding constant from slow exchange spectra is not usually attempted. Generally, for slow-exchange conditions to exist in the first place, the binding is submicromolar in affinity and non-NMR methods are more suitable for determining affinities in these cases. NMR studies are done at millimolar concentrations, making it difficult to determine K_D with any accuracy for tight binding systems.

In principle, kinetic information on the complex can be obtained from slow-exchange spectra, as seen from the expressions for T_2 for free and bound ligand signals:

$$\frac{1}{T_{2F,obs}} = \frac{1}{T_{2F}} + k_{off}\left(\frac{p_B}{p_F}\right) \tag{4}$$

$$\frac{1}{T_{2B,obs}} = \frac{1}{T_{2B}} + k_{off} \tag{5}$$

Because the linewidth of a peak is related to T_2 by

$$\mathrm{LW} = \frac{1}{\pi T_2} \tag{6}$$

measurements of linewidth during a titration can be used to derive k_{off}. Equations 4 and 5 show that, although the signal from bound ligand is independent of concentration, the free ligand decreases in linewidth as more ligand is added. A plot of linewidth versus ligand/macromolecule mole ratio allows k_{off} to be determined [122].

Although the determination of off-rates is of significance in assessing the stability of the complex, the major interest of complexes in the slow-exchange limit is the determination of the geometry of the complex through the use of intra- and intermolecular NOEs. A classic example is illustrated by the binding of immunosuppressant peptides, such as cyclosporin and FK506, to their receptors. These examples are discussed in more detail in the section "Immunophilins: Studies of FK506 Analog Binding to FKBP."

Fast Exchange When the exchange between free and bound states is very fast, observed NMR parameters are a simple weighted average of the two contributing states, illustrated by Equation 7 for chemical shifts and Equation 8 for linewidths.

$$\nu_{obs} = p_F \nu_F + p_B \nu_B \qquad (7)$$

$$\frac{1}{T_{2,obs}} = \frac{p_F}{T_{2,F}} + \frac{p_B}{T_{2,B}} \qquad (8)$$

These equations show that, in the fast-exchange limit, addition of a ligand to a protein solution will cause a progressive change in chemical shift. Signals from the protein initially reflect the free state, but as ligand is added the population of bound protein increases and the observed signals move toward those of the bound state. Similarly, when ligand signals are first detected they predominantly reflect the bound state, but with increasing amounts of ligand they move toward the free state. By regression analysis to Equation 7, taking into account the dependency of the mole fractions on K_D by the standard quadratic binding equation, it is possible to obtain estimates of both K_D and the bound shift. The procedure works best for ligands that bind rather weakly (e.g., millimolar dissociation constants) [123].

When exchange is somewhat slower, but still within the fast-exchange limit, there is an exchange contribution to linewidth, as shown in Equation 9:

$$\frac{1}{T_{2,obs}} = \frac{p_F}{T_{2,F}} + \frac{p_B}{T_{2,B}} + \frac{p_B p_F{}^2 4\pi^2 (\nu_B - \nu_F)^2}{k_{off}} \qquad (9)$$

In this case, a maximum in the broadening of ligand or protein peaks occurs during the titration at a mole ratio of approximately 0.3, as illustrated in Fig. 16.

The spectral changes that occur in the fast-exchange regime can conveniently be illustrated by studies on the binding of a series of terephthalamide ligands to an oligonucleotide model of DNA. The ligands, referred to as $L(NO_2)$, $L(NH_2)$, and L(Gly) (**6–8**), were synthesized as precursors for potential anticancer agents [124]. To establish whether they bind in the minor groove of AT-rich DNA, a series of NMR titration experiments was undertaken.

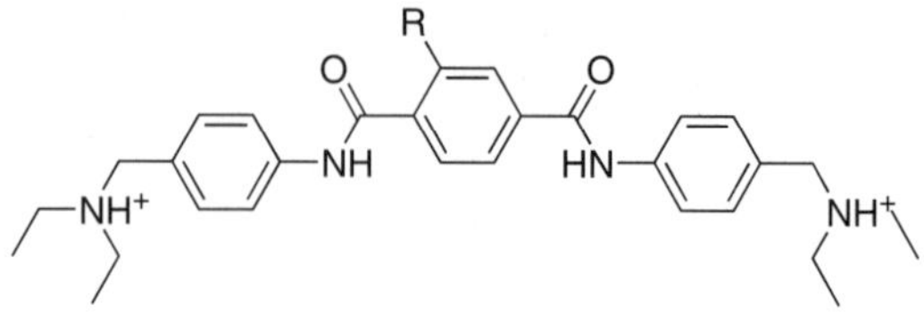

(**6**) $L(NH_2)$ R = NH_2
(**7**) $L(NO_2)$ R = NO_2
(**8**) L(Gly) R = $NHCOCH_2NH_2$

Figure 16 shows an expansion of the aliphatic region of a series of ^1H NMR spectra of 0.5 mM of the oligonucleotide $d(GGTAATTACC)_2$, to which increasing amounts of $L(NH_2)$ were added [124]; the spectra cover mole ratios of ligand to DNA duplex ranging from 0 : 1 to 2.6 : 1. Although the spectra are complicated by overlap in some regions, it is clear that addition of the ligand causes significant changes to the DNA peaks. A typical example is seen for the T6 (the thymine base in position 6) methyl peak, where addition of ligand causes both an upfield shift and broadening of the peak at certain stages of the titration. The chemical shift moves monotonically with ligand concentration up to a mole ratio of 1:1 and then reaches a plateau, remaining constant as larger amounts of ligand are added. Broadening of the peak reaches a maximum at a ligand:DNA mole ratio of approximately 0.3. Both observations are consistent with moderately fast exchange on the chemical-shift timescale between the free and ligand-bound forms of the DNA in solution. In this case, the observed spectral peaks reflect neither the free nor the bound form of DNA, but are averaged signals.

Ligand peaks are also in fast exchange, as seen with the $L(NH_2)$ methyl peak, which first appears at a ligand:DNA ratio of 1.36 : 1 as a shoulder on the overlapped T3 and T7 methyl peaks at approximately 1.27 ppm. This peak is not visible in spectra at low ligand:DNA mole

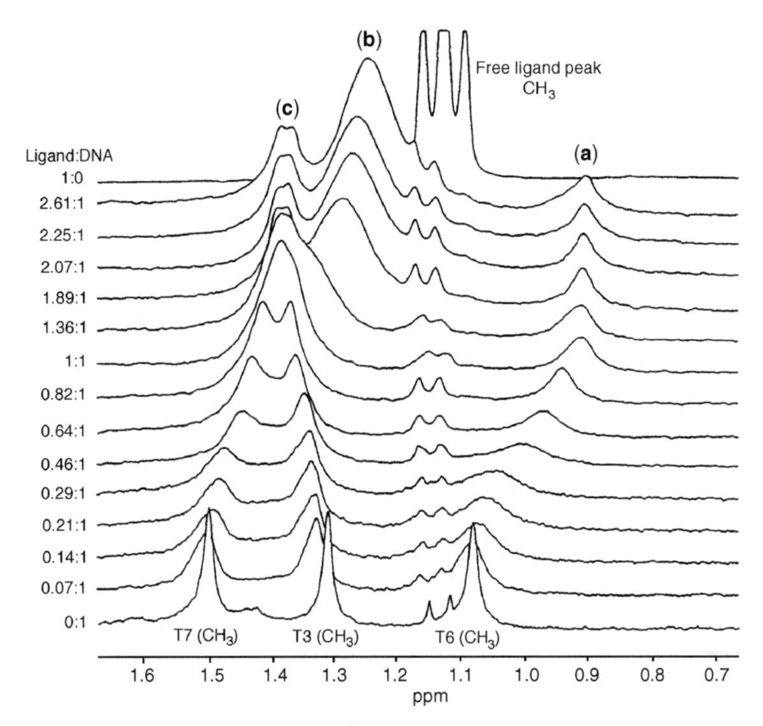

Figure 16. Expanded regions from 300-MHz ^{1}H NMR spectra for complexes between $L(NH_2)$ and $d(GGTAATTACC)_2$ recorded at 10°C. The two small peaks at 1.12 and 1.14 ppm arise from an impurity. Increasing ligand concentration causes an upfield shift of the T6 methyl resonance (a), and causes the T7 and T3 resonances to become overlapped at later stages of the titration (c). Peak (b) is an averaged resonance from the ligand methyl groups intermediate in shift between the bound and free forms of the ligand. (Reprinted with permission from Ref. [124].)

ratios due to the small population of bound species and the overlapping DNA peaks. It moves upfield with increasing ligand concentration and, again, represents an averaged peak intermediate in chemical shift between free and bound forms, reflecting fast-exchange kinetics. Eventually, the chemical shift of this signal approaches that of the free ligand at 1.1 ppm, measured in a separate experiment with a solution of ligand alone.

In the fast-exchange cases such as this, it is possible to obtain an estimate of the dissociation constant for the complex (K_D) and the bound chemical shift (ν_B) of DNA resonances by fitting the observed chemical shift changes as a function of ligand concentration to Equation 7 [123]. The parameters that best fit the experimental data for the T6 methyl peak were $K_D = 1.2 \times 10^{-6}$ M and $(\nu_B - \nu_F) = 46$ Hz. To further define the thermodynamic constants associated with binding, the linewidth data were also quantitatively examined by use of Equations 6 and 9. In the case of moderately fast exchange, a maximum linewidth was predicted at a ligand:DNA mole ratio of 0.33, and this was indeed observed. Derived binding parameters were $K_D \leq 1.0 \times 10^{-6}$ M, k_{off} 250 s^{-1}, $(\nu_B - \nu_F) = 49$ Hz, and $LW_B = 12$ Hz, consistent with the values derived from the analysis of chemical shifts. Subsequent studies with the related ligand $L(NO_2)$ showed similar binding to $L(NH_2)$. However, a third ligand, L(Gly), was found to bind somewhat more tightly, with some signals in the intermediate exchange regime.

Intermediate Exchange In this regime, the rate of exchange between bound and free states is comparable to the differences in NMR parameters associated with the exchange. In general, the spectral peaks become very broad during titrations in this exchange regime and thus analysis is difficult. This is the case, for example, for L(Gly). In the methyl region of

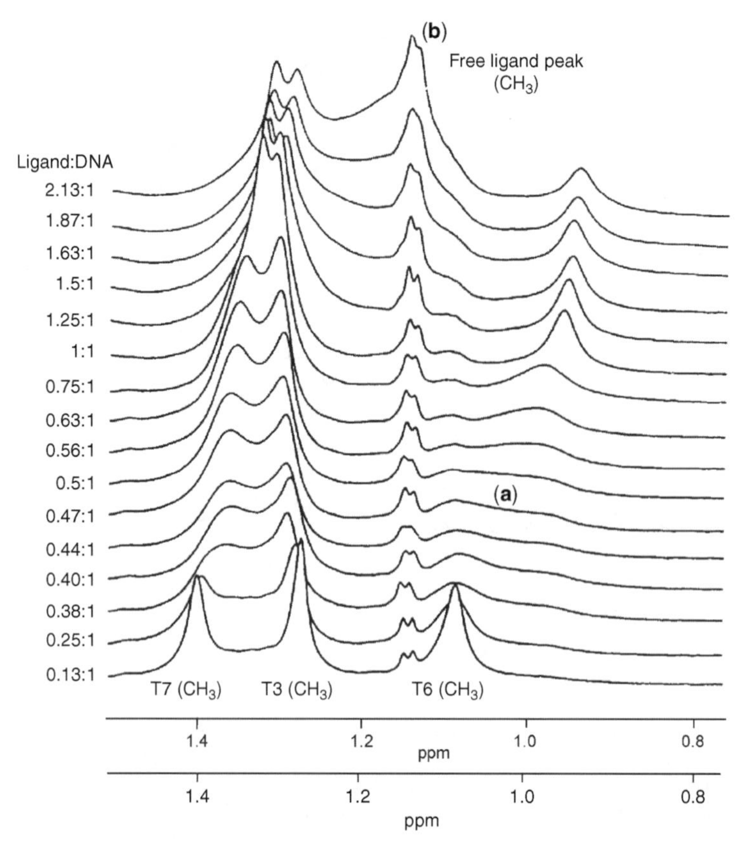

Figure 17. Expansions from the 600-MHz ^{1}H NMR spectra for complexes formed between L(Gly) and d(GGTAATTACC)$_2$ showing the methyl resonances. The two small peaks at 1.12 and 1.14 ppm are attributed to an impurity. The complex nature of the T6 methyl resonance at ligand:DNA ratios less than 1:1 (a) and the manner in which signal intensity increases at approximately 1.15 ppm at DNA:ligand ratios greater than 1:1 (b) are indicative of intermediate exchange. (Reprinted with permission from Ref. [124].)

the spectra shown in Fig. 17, the T7 CH_3 signal moves upfield and the T3 CH_3 signal moves slightly downfield with increasing ligand concentration, as seen previously for L(NH_2) and L(NO_2). However, in contrast to the other ligands, the characteristic broadening of peaks at intermediate ratios gives rise to spectral peaks that do not have a Lorentzian lineshape, suggesting kinetics in the intermediate exchange regime. The T6 CH_3 peak does not shift in the characteristic fast-exchange manner but, instead, a new broad resonance appears close to the expected position of the bound T6 CH_3 chemical shift on the first addition of ligand, and increases in intensity with increasing ligand concentration. This observation is consistent with the ligand being in slow to intermediate exchange between the free and bound forms, with $k_{\text{off}} \approx (\nu_B - \nu_F)$. Based on the magnitude of $\nu_B - \nu_F$ for this resonance, k_{off} for L(Gly) is estimated to be 50 s^{-1}, which is significantly slower than that for L(NO_2) and L(NH_2).

At a ligand:DNA ratio of approximately 1 : 1, the ratio of the integrals of the T6 methyl peak and the overlapped T3 and T7 methyl peaks is about 1 : 6. The expected value is 1 : 2, which indicates that the bound ligand methyl peak ($4 \times CH_3$) is overlapped with the T7 and T3 methyl peaks, as observed with L(NH_2) and L(NO_2). When the ligand:DNA ratio is increased beyond a 1 : 1 ratio, a new peak appears at approximately 1.15 ppm and increases in intensity as the ligand concentration is increased. This new peak corresponds to the methyl peak of the free ligand and its appearance in this manner is consistent with slow exchange on the chemical-shift timescale. To confirm this, spectra of a 2:1 mixture of L(Gly) and d(GGTAATTACC)$_2$ were acquired at different temperatures [124], as illustrated in Fig. 18.

At low temperatures, signals at 1.15 and 1.30 ppm (overlapped with the T7 CH_3 and T3 CH_3 peaks) attributable to the methyl groups from the free ligand and bound ligand, respectively, are distinguishable. As the temperature is increased, a broad peak appears between these two signals (at ~1.22 ppm). At the lower temperatures $k_{off} \leq (\nu_B - \nu_F)$, so that methyl resonances of the ligand have complex characteristics reflecting slow-intermediate exchange. At higher temperatures, $k_{off} \geq (\nu_B - \nu_F)$, so the signal appears as a fast-exchanged average between the free and bound resonances. From a qualitative analysis of the spectra, k_{off} for L(Gly) was estimated to be 50–60 s^{-1} at 283 K.

The fact that some peaks (e.g., the oligonucleotide T7 and T3 methyl signals) exhibit fast exchange, whereas others in the same spectrum of the same complex exhibit slow-intermediate characteristics, reflects the different $(\nu_B - \nu_F)$ values for different peaks. This emphasizes the point that the "exchange regime" is a relative expression and depends on both the rate of exchange and the size of the chemical-shift differences involved. In summary, the observations suggest that k_{off} for binding of L(Gly) to the oligonucleotide duplex is much slower than that for the other two derivatives. This illustrates the value of NMR as a quick method for comparing the binding of different ligands and for confirming ligand binding hypotheses.

The change in binding kinetics may be rationalized by considering the different structure of the L(Gly) ligand relative to the other ligands. It was anticipated that, upon binding to the minor groove, the terephthalamides would adopt a conformation in which the substituent on the central ring would form part of the convex edge of the ligands and therefore be directed toward the "mouth" of the groove. Given this binding arrangement, the ligand L(Gly) would have a positively charged alkylamine group positioned to interact with the negatively

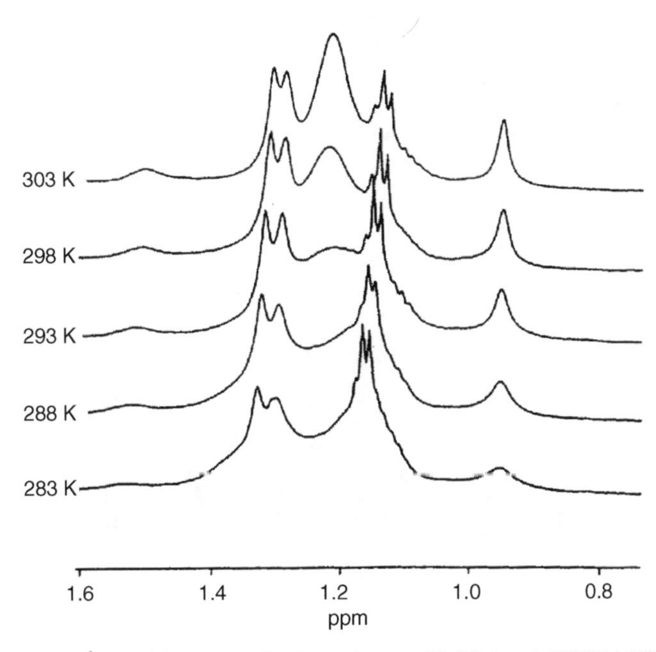

Figure 18. Expansions of ^{1}H NMR spectra of a 2:1 mixture of L(Gly) and d(GGTAATTACC)$_2$ acquired at different temperatures. (Reprinted with permission from Ref. [124].)

charged phosphate groups of the DNA backbone. The L(Gly) derivative also has a bulkier substituent than that of the other ligands and this is also consistent with some differences in its binding.

3.2.3. NMR Techniques NMR is a particularly versatile tool for the analysis of protein–ligand interactions. As well as being able to observe different nuclei, measurements may be made of a range of different NMR parameters, including chemical shifts, linewidths, coupling constants, and relaxation parameters. In addition, there are several specific NMR techniques that have been applied for the measurement of these parameters. The techniques that are particularly valuable for the study of macromolecule–ligand interactions are described in the following sections.

Chemical-Shift Mapping Chemical shifts are exquisitely sensitive markers of the local charge state and environment. Although it is not possible to construct an accurate model of a binding site from knowledge of the chemical shifts of a bound ligand, a qualitative interpretation of *changes* in chemical shifts of the macromolecule on binding provides significant insight into the location of the binding site. Initially, such studies were done using 1D NMR but are now increasingly done by 2D HSQC spectra. By simultaneously obtaining information on chemical shifts for a large number of sites in a macromolecule and seeing which ones do or do not change when a ligand binds, it is possible to deduce the location of the binding site. This procedure is referred to as chemical-shift mapping. A prerequisite of the approach is that the chemical shifts have been assigned. Chemical-shift mapping by use of HSQC spectra is used widely in NMR screening approaches and we will defer a more detailed discussion until Section 4.

The relative ease with which chemical-shift information localizes binding sites is illustrated by continuing with the example of terephthalamide binding to DNA. Figure 19 shows that, upon binding of the terephthalamides L(NH_2) and L(NO_2) to d(GGTAATTACC)$_2$, the DNA protons on the four base pairs between A5 and A8 are perturbed to a much larger degree than protons in the rest of the sequence. It is thus probable that these four residues form the binding site.

A more detailed analysis allows the binding site to be further localized to the minor, rather than major groove in the region of these bases. A4, A5, and A8 are the only residues containing easily detectable minor groove protons (H2). These resonances, which originate from the floor of the minor groove, are shifted downfield with ligand binding, whereas most other resonances are shifted upfield. This observation is consistent with the ligands binding in the minor groove [125], where adenine H2 protons on the floor of the groove experience deshielding

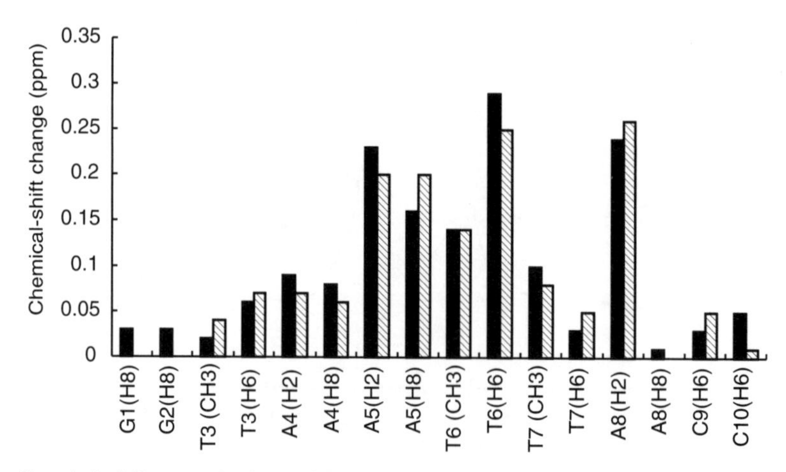

Figure 19. Chemical-shift perturbations of DNA protons upon ligand binding. The lighter and darker columns represent shifts attributed to L(NO_2) and L(NH_2) derivatives, respectively.

ring current effects and are therefore shifted downfield in the spectrum. However, significant chemical-shift changes were also observed for some major groove protons. This illustrates the general point that, sometimes, allosteric effects can cause changes at sites not directly involved in binding. Binding perturbations in the major groove have also been observed for other established minor groove binders such as distamycin [126], netropsin [127], and Hoechst 33258 [128]. Based on NOE and crystallographic data, it was concluded that the effects were caused by distortions of the B-DNA duplex upon complexation.

A comparison of the minor and major groove perturbations upon binding of the terephthalamides shows that the minor groove protons are affected to a much greater extent. This is particularly evident for A8, where the H2 proton changes in chemical shift by approximately 0.25 ppm, whereas the H8 proton is unaffected (Fig. 19). It is difficult to identify a binding mode in the major groove that would account for such a large effect on the minor groove A8 H2 resonance without a simultaneous effect on the major groove protons of T7 and A8. The observed 1 : 1 stoichiometry of the complex excludes the possibility that the ligand binds to the major and minor groove simultaneously. It is therefore more probable that binding in the minor groove causes distortion of the DNA structure so that perturbations are observed for the major groove protons of A5 and T6, but not neighboring nucleotides.

Other examples of the use of chemical-shift mapping to locate binding sites have been made for ligands binding to a range of drug targets, including immunophilins, matrix metalloproteases, and DHFR. Some of these examples are described in more detail in Section 3.2.5.

NMR Titrations There are a number of advantages to undertaking a titration of ligand against macromolecule, or vice versa, rather than just examining the final complex. These include, in the case of slow exchange, introducing the possibility of distinguishing signals from the individual components on the basis of intensities at intermediate stages of the titration, and obtaining kinetic and thermodynamic parameters associated with the interaction in the fast-exchange case. Such titrations may be done using either 1D or 2D spectra and are very useful for establishing the exchange regime of the complex, as described in Section 2.1. A variety of parameters may be monitored in these titrations, although the most common are chemical shifts and linewidths. Examples of such titrations are given in Figs 16 and 17.

Isotope Editing and Filtering Isotope editing provides a powerful way of distinguishing between the components in a complex without the need for a titration. It is one of the most useful tools for the study of macromolecule–ligand complexes, and indeed the background NMR technology that underpins isotope editing was developed specifically for the study of complexes. The principle of the approach is illustrated in Fig. 20 and is based on the use of isotopes to select for signals from either the ligand or macromolecule, or signals exclusively linking both of them.

Conceptually, the simplest approach is to deuterate the macromolecule uniformly, thereby removing its signals from ^{1}H-detected NMR spectra, and allowing signals from only the ligand to be observed. This simplifies the spectrum substantially and allows, for example, the bound conformation of a ligand to be determined from NOESY data recorded in D_2O. By rerunning the spectrum in H_2O, additional NOEs to exchangeable amide protons on the protein may be detected, thereby providing information on contacts between ligand and protein. Alternatively, ^{15}N or ^{13}C signals may be introduced selectively into either the ligand or protein and NMR editing techniques used to select only signals attached to these labels and their proximate protons [129].

Potentially, the most useful approach involves uniform labeling of one of the components with either ^{15}N or ^{13}C and leaving the other component unlabeled. It is then possible to edit the spectrum by selecting for interactions (either through bond or through space) that connect protons that are both one-bond coupled to ^{15}N or ^{13}C. Alternatively, the spectrum may be filtered to specifically remove such signals, thereby selecting only signals involving protons coupled to ^{14}N or ^{12}C

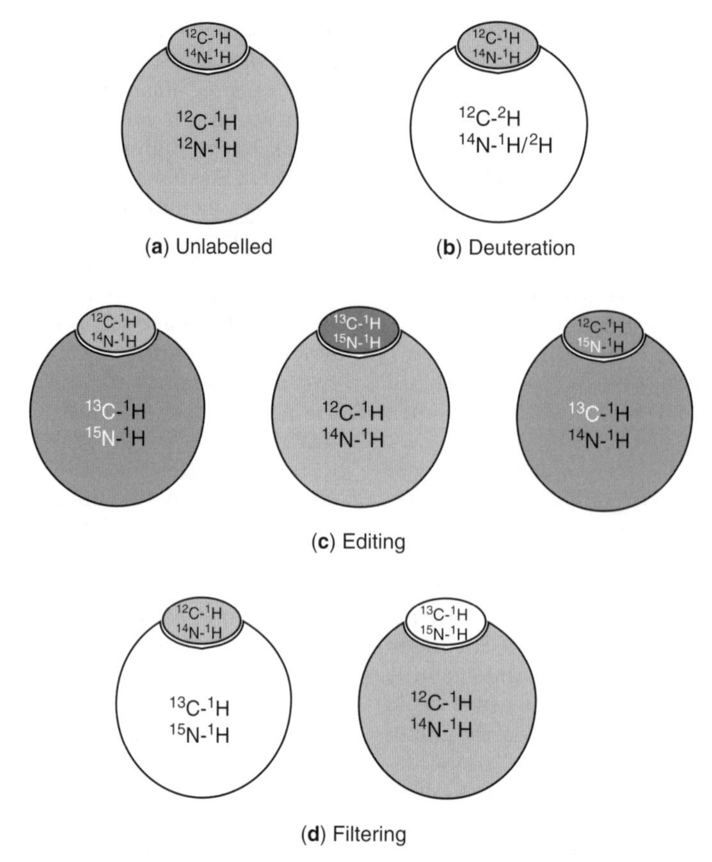

Figure 20. Isotope editing and filtering can be used to select signals from either the ligand or the protein. (a) Normal protein and ligand with no filtering or editing. (b) Selection of the ligand signals by 2H labeling of the protein. (c) Selection of protein or ligand signals by ^{13}C and/or ^{15}N labeling/editing. (d) Removal of protein or ligand signals by ^{13}C or ^{15}N filtering.

(i.e., on the unlabeled component). It is generally easier to label the protein uniformly, rather than the ligand, and editing methods are highly efficient, thus making it easy to visualize just the protein. However, visualizing ligand signals are often of interest, so filtering experiments play a valuable role [130].

Another possibility is to use half-edited/half-filtered 2D experiments to detect NOEs that specifically involve interactions between protons attached to ^{15}N or ^{13}C and those that are not. This approach is used, for example, to detect intermolecular NOEs between a labeled protein and an unlabeled ligand. However, these experiments often suffer from poor sensitivity. Examples of isotope editing/filtering are given in Section 3.2.4.

NOE Docking In many cases, the study of a complex may follow a previous structure determination of the isolated macromolecule and, in that case, it may be possible to determine much information about a complex by obtaining a relatively small number of NOEs linking the ligand and macromolecule. Gradwell and Feeney [131] have analyzed the important factors in such NOE docking experiments. In their analysis, a high-resolution X-ray structure of a protein–ligand complex was used to simulate loose distance restraints of varying degrees of quality that might typically be estimated from experimental NOE intensities. These simulated data were used to examine the effect of the number, distribution, and representation of the experimental constraints on the precision and accuracy of the calculated structures. The results showed some parallels with those from similar studies on complete protein NMR structure determinations, but it was found that more constraints per torsion angle are required to define docked structures of similar quality. This is because the conformation and orientation of the ligand are defined only by NOEs and not by covalent attachment, as is the case for amino acid side chains in a protein structure. With these considerations in mind, it appears that NOE docking can be a very cost-efficient procedure for defining the environment, orientation, and conformation of ligands.

Recently, the NOE docking method has been extended to the study of protein–protein complexes, which allows for the unambiguous assignment of sparse intermolecular NOEs [132]. This method uses the intermolecular NOEs that are available to reliably dock the protein-protein complex and allow for unambiguous assignment of the NOEs available. This gives a more accurate structure of the complex and the interaction surfaces than would otherwise be obtained. This method is applicable to the interpretation of intermolecular NOEs in conventionally labeled samples.

3.2.4. Selected Examples Applications of various NMR techniques are now illustrated with selected examples. The examples have been chosen to give a broad perspective on the types of NMR experiments that can be done and the types of information they provide. The first example covers the case of drug–nucleic acid binding and focuses on traditional NMR experiments, involving relatively standard homonuclear methods. The second example covers binding of moderately large ligands to immunophilins and highlights isotope editing techniques. The third example, covering ligand binding to a matrix metalloproteinase, also highlights the importance of NMR editing techniques and shows how relatively simple spectra involving ^{19}F-containing ligands can be very informative. The fourth example describes ligand binding to DHFR, one of the most extensively studied macromolecules by NMR, and illustrates the derivation of a range of kinetic and geometric information on intermolecular complexes. The final example, on HIV protease, describes how NMR complements X-ray studies and provides information on dynamic motions within complexes.

DNA Binding Drugs The NMR approaches that have been used to examine the interactions of minor groove binding drugs with DNA can be illustrated with studies on the bisbenzimidazole-based compound, Hoechst 33258 (**9**). It has been used widely as a fluorescent cytological DNA stain and is active as an anthelmintic agent. It also has activity against intraperitoneally implanted L1210 and P388 leukemias in mice [133].

Footprinting studies [133] have shown that sequences of four AT base pairs are a prerequisite for strong binding to DNA, consistent with similar observations for structurally related molecules such as distamycin and netropsin [125–127,134–136]. The first structural studies of Hoechst 33258 complexed to short sequences of synthetic oligonucleotides were done using X-ray crystallographic methods [137–139], but NMR and further X-ray studies followed [128,140–144]. Three of the X-ray studies [137,138,140] used the *Eco*RI sequence d(CGCGAATTCGCG)$_2$ and another [139] used the sequence d(CGCGATATCGCG)$_2$. Both sequences fulfill the requirement of at least four consecutive AT base pairs, and the resulting complexes showed similar modes of binding. In all of the X-ray studies, the Hoechst ligand was found to bind to the minor groove.

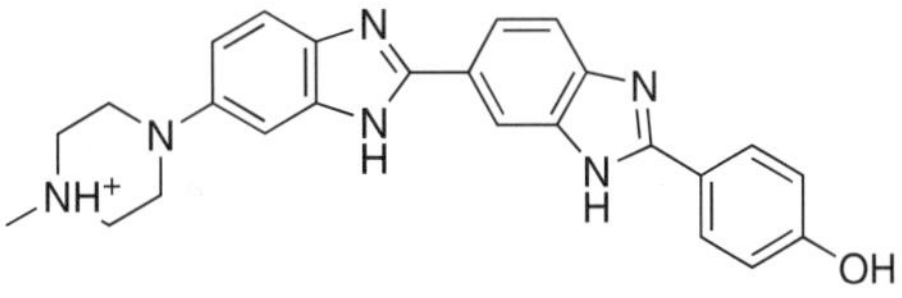

(**9**)

The NMR studies of complexes between Hoechst 33258 and oligonucleotide sequences provided complementary information to the crystal structure data [128,140–143]. Because the binding is reversible, the NMR data offer the opportunity to derive information about the kinetics of the interaction. Some NMR studies were performed with dodecanucleotide sequences used in crystallographic studies, including d(CGCGAATTCGCG)$_2$, which allowed a direct comparison with the crystallographic data. Experiments were also performed with sequences specifically designed to investigate different aspects of the interaction. The sequence d(CTTTTGCAAAAG)$_2$ was designed to offer two binding sites, and it was found that two Hoechst molecules interacted with the DNA duplex in symmetry-related orientations at the 5′-TTTT-3′ and 5′-AAAA-3′ sites [128].

Stoichiometry and Kinetics. The starting point in studies of ligand-DNA complexes is usually a titration experiment to establish the nature and stoichiometry of the complex. Complexes between the ligand and DNA duplex are obtained by adding small aliquots of ligand solution to a sample of the DNA duplex with 1D ^{1}H NMR spectra acquired after each addition. The effects observed on the NMR spectrum after each addition reveal whether an interaction is taking place and allow the interaction to be characterized as fast or slow exchange on the NMR timescale. The stoichiometry of the interaction can also be determined from the titration.

In general, the addition of Hoechst 33258 to the oligonucleotide duplexes causes a decrease in the intensity of free DNA resonances and a concomitant increase in the intensity of new resonances, which appear in previously unoccupied spectral regions. This is consistent with the free and bound forms of the DNA duplex being in slow exchange with each other. For example, when Hoechst 33258 is added to d(GGTAATTACC)$_2$, the free DNA signals completely disappear at a DNA:drug ratio of 1 : 1, and the number of new resonances is twice the number of previously observed free DNA resonances (Fig. 21). This is a common feature of complexes with 1 : 1 stoichiometry and reflects a loss of the dyad symmetry of the duplex attributed to ligand binding.

Upon addition of Hoechst 33258 to d(CTTTTCGAAAAG)$_2$, the free DNA signals completely disappeared at a ratio of 2 : 1 drug:

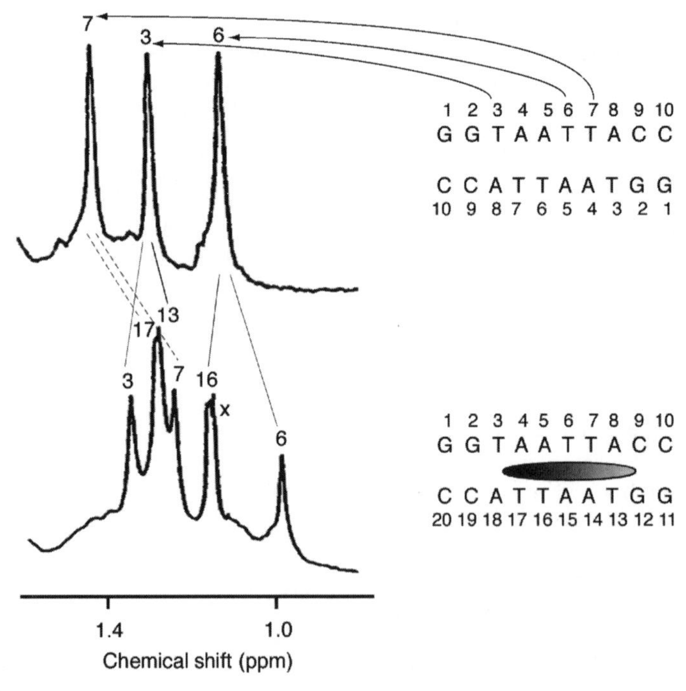

Figure 21. 1D ^{1}H NMR spectra (recorded at 20°C) illustrating the thymine methyl region for the symmetrical ligand-free duplex and for the 1:1 Hoechst:d(GGTAATTACC)$_2$ complex, which is no longer symmetrical because of the ligand binding. x corresponds to a small impurity peak. The DNA strands are numbered to the right of the spectra and the approximate location of the ligand is indicated by a black bar. (Adapted from Ref. [142]. 1993 Wiley Publishing.)

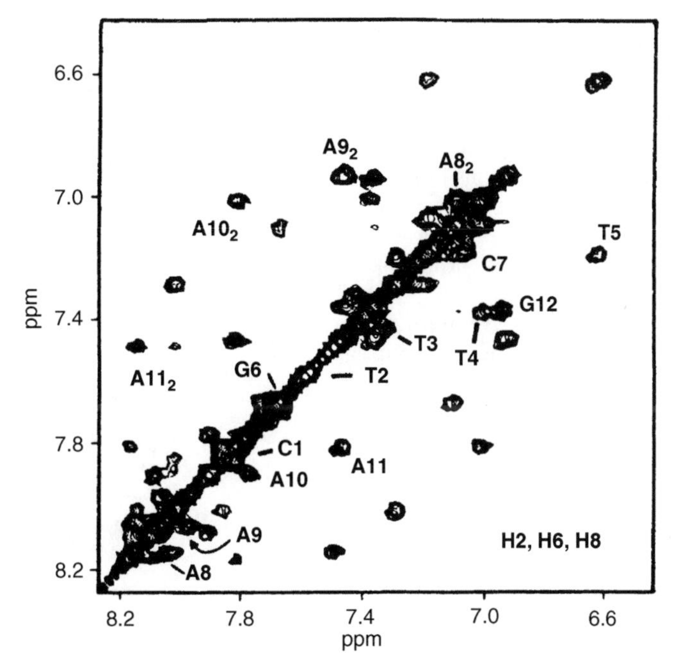

Figure 22. Aromatic region of NOESY spectrum of a 1 : 1 mixture of Hoechst versus d(CTTTTCGAAAAG)$_2$ recorded with a 200-ms mixing time. Chemical-exchange cross-peaks between protons of the free DNA and the 2 : 1 Hoechst:DNA complex are labeled with their identifying base pair. Below the diagonal the H6 and H8 cross-peaks are shown, whereas those of the adenine H2 resonances are highlighted in the upper portion of the figure and labeled with a subscript 2. (Adapted from Ref. [128] by permission of Oxford University Press.)

DNA and there was no doubling of the number of DNA resonances in the spectrum [128]. From this, it was concluded that two molecules were bound per duplex in a manner that retained the dyad symmetry of the DNA duplex. In the 2 : 1 complex, only four thymine methyl resonances were detected, as expected for a symmetrical DNA duplex. These resonances are all overlapped in the free DNA spectrum. The binding was also determined to be cooperative, in that no intermediate 1 : 1 complex was detected [128], and at a ligand: oligonucleotide ratio of 1 : 1 only signals from free DNA and from the 2 : 1 complex were detected.

The reversible nature of the Hoechst:DNA interaction is illustrated by the observation of chemical-exchange cross-peaks in NOESY spectra of mixtures of free and complexed oligonucleotides [128,141]. This is seen in the NOESY spectrum of a mixture of free and complexed d(CTTTTCGAAAAG)$_2$, shown in Fig. 22, in which many chemical-exchange cross-peaks are observed between resonances arising from the free and bound oligonucleotide. In a NOESY spectrum acquired at lower temperature, the intensity of these chemical-exchange cross-peaks is significantly reduced, indicating that the exchange is slowed at lower temperatures [128].

Overall, the ability to observe such dynamic exchange phenomena is one of the strengths of NMR relative to X-ray crystallography and several examples of these phenomena are described later in this chapter.

Binding Site. A combination of chemical-shift and NOE information can be used to locate and characterize binding sites. Chemical-shift differences between resonances arising from free and bound forms of DNA are indicative of the nature of the interaction. In all studies of the Hoechst complexes described above [128,141–144], significant changes to the chemical shifts of thymine H1′ protons and adenine H2 protons were observed, in contrast to the generally small perturbations observed for the base H8/H6 and CH_3 resonances located in the major groove. These perturbations are consistent with binding to the minor groove. In some instances, significant perturbations were observed to major groove protons located well within the binding site, reflecting changes in the conformation of the DNA duplex (e.g., base roll, propeller twisting) [128,143].

Further evidence of minor groove binding is provided by the fact that resonances arising from protons on the floor of the groove, such as the adenine H2 and imino resonances, are shifted downfield, whereas resonances from protons on the minor groove walls, such as the H1′ protons, are shifted upfield. This is a consequence of the ligands being inserted edge-on into the minor groove. The deoxyribose protons that form the walls of the minor groove are positioned above the π-plane of the aromatic rings and consequently receive upfield perturbations to their chemical shifts. However, protons positioned on the floor of the groove generally lie in the plane of the aromatic rings and experience downfield perturbations to their chemical shifts, as illustrated in Fig. 23.

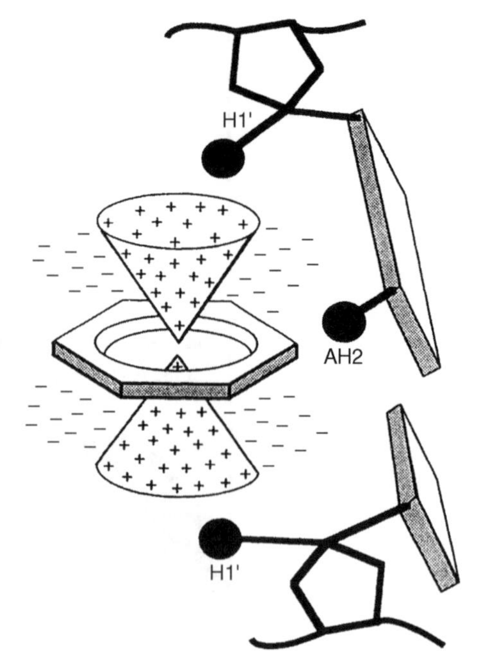

Figure 23. Schematic representation of ligand-induced ring-current effects on nucleotide protons that form the walls (deoxyribose H1′) and floor (adenine H2) of the minor groove. (+), shielding effects; (−), deshielding effects. (Reprinted from Ref. [142]. 1993 Wiley Publishing.)

The magnitude of chemical-shift changes is a strong indicator of the location of the binding site. In the case of the 2 : 1 complex with $d(CTTTTCGAAAAG)_2$, the largest chemical-shift changes occur over the 5′-TTTT-3′ and 5′-AAAA-3′ regions of the duplex. In the case of 1 : 1 complexes, where the DNA duplex contains an AT base pair segment located at the center of the sequence, greater chemical-shift perturbations are observed for resonances in that region [141–143], consistent with the binding-site location.

Assignment of the bound ligand and DNA resonances enables the identification of intermolecular NOEs, which are required for a precise determination of the binding site. The interaction of Hoechst 33258 with the oligonucleotides produced a large number of intermolecular NOEs (~30), enabling the orientation of the ligand within the binding site to be determined. The NOE contacts observed for different complexes have a few common features. The contacts generally involve DNA protons associated with the minor groove, such as ribose H1′ and adenine H2, clearly locating Hoechst in the minor groove. Protons of all four-spin systems of the ligand show NOEs to protons of the DNA, demonstrating that the interaction occurs along the entire length of the drug. Typically, protons along one edge of the ligand (e.g., NH and H4′/H4″″) exhibit close contacts to protons on the floor of the minor groove, showing that the bound drug is crescent shaped [128,141–143].

Models of the interaction of Hoechst 33258 with the oligonucleotides studied were generated based on observed intermolecular NOEs. The models of the 1:1 complexes indicate that the ligand interacts with the four AT base pairs located at the center of the sequence. In the 2 : 1 complex, the array of contacts observed indicates that the ligand is located in the minor groove at the center of the 5′-TTTT-3′ and 5′-AAAA-3′ sites [128], as illustrated in Fig. 24.

As well as defining the *location* of the binding site, intermolecular NOEs can be used to

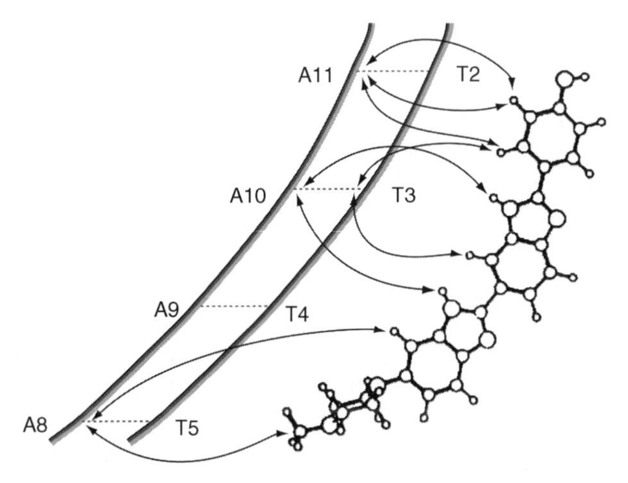

Figure 24. Schematic representation of Hoechst 33258 bound to the minor groove of the 5′-TTTT sequence Highlights of some of the NOEs that determine the position and orientation of the Hoechst molecule within the minor groove. (Adapted from Ref. [128] by permission of Oxford University Press.)

determine the *orientation* of the ligand at that site. In the case of the 2 : 1 complex, the *N*-methylpiperizine moieties were found to point toward the center of the duplex, as indicated by NOEs between the protons from the piperizine ring and the 5′-terminus of the adenine tract (Fig. 24). Corresponding NOEs were also observed between the drug phenolic protons and the 5′-terminus of the thymine tract, as well as the 3′-terminus of the adenine tract of the complementary strand.

Dynamic Processes. The binding of the Hoechst molecule to the self-complementary oligonucleotide duplexes in a 1 : 1 ratio lifts the dyad symmetry of the duplexes so that two sets of DNA resonances are observed. This indicates that the drug is in slow exchange between the free and the bound forms. However, close examination of the 2D NOE data reveals the presence of chemical-exchange cross-peaks between symmetry-related protons on opposite sides of the dyad axis of the DNA duplex. These result from dissociation of the Hoechst molecule from the duplex followed by a 180° reorientation and rebinding [142,143]. The self-complementary nature of the sequences ensures that the same complex is formed for either ligand orientation but with the net effect of interchanging the two strands with respect to the orientation of the Hoechst molecule. The rate at which this process occurs was estimated using cross-peak intensities in the NOESY spectrum [143]. The lifetime of the complex ($1/k_{ex}$) was 0.8 and 0.45 s, respectively [142,143]. These values indicate a small but significant difference in the affinity of Hoechst for TAATTA and GAATTC sites.

Intramolecular dynamic processes that are fast on the NMR timescale are also observable in the ^{1}H NMR spectrum of the bound Hoechst molecule. Peak averaging is observed for the H2/H6 and H3/H5 protons of the phenol group, which is consistent with the environments on either side of the ring being averaged by rapid ring-flipping motions about the C4–C2′ axis. This occurs despite the apparent tight fit between the phenyl ring and the walls of the minor groove, which, in a static model of the complex, must present a large barrier for rotation. The rate for this process is as high as 1000 s^{-1} [142], which is much higher than the rate of interconversion between free and bound forms of the duplex; thus, dissociation of the drug from the complex cannot be the rate-limiting factor for phenol ring flipping. Dynamic fluctuations of the DNA conformation are more likely to provide the rate-limiting step.

Summary of Solution Studies. The data obtained from these NMR studies are consistent with the bound ligand fitting tightly within the minor groove of AT tetramers, with the aromatic rings of the ligand being roughly coplanar. The AT tract provides the key recognition features required for binding, including the narrowness of the minor groove. The importance of van der Waals interactions is evident, given the large number of NOE contacts between the ligand and the walls and floor of the groove. Hydrogen bonding also plays a significant role in stabilizing the interaction, as do electrostatic interactions between the positively charged piperizine ring and the minor groove. Electrostatic interactions are also likely to play a significant role in orienting the ligand within the binding site, as shown in the 2 : 1 complex, where the piperizine rings point toward the center of the duplex where the positive charge is best stabilized [128]. The information derived from these studies, as well as from NMR studies of the interactions of other minor groove binders with DNA, is useful for the design of ligands with altered specificity or increased binding affinity, with the overall goal being the development of novel drugs.

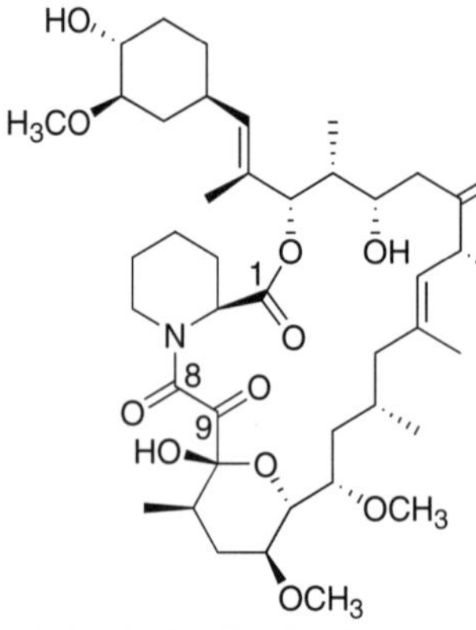

(10) FK506 R = CH_2CHCH_2
(12) Ascomycin R = CH_2CH_3

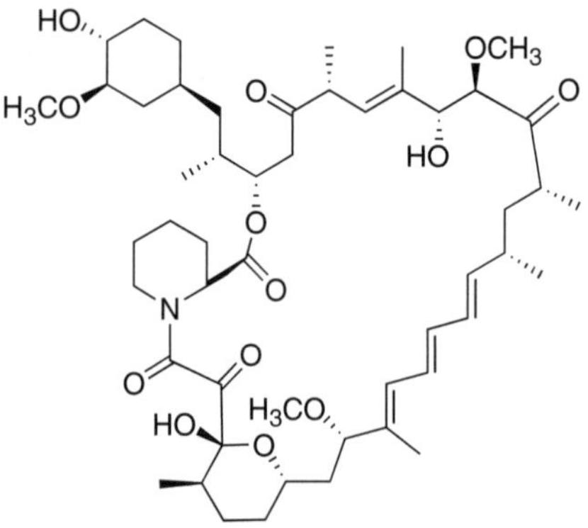

(11) Rapamycin

Immunophilins: Studies of FK506 Analog Binding to FKBP Some of the most detailed NMR investigations of the interaction between ligands and their target proteins have been made for the immunophilin class of proteins. The FK506 binding protein (FKBP) has a molecular mass of 11.8 kDa, whereas that of cyclophilin is 17 kDa. These proteins are unrelated in amino acid sequence but both have peptidyl-prolyl *cis–trans* activities that are inhibited by immunosuppressants that block signal transduction pathways leading to T-lymphocyte activation. FK506 (**10**) and rapamycin (**11**) inhibit the *cis–trans* isomerase activity of FKBP, whereas cyclosporin A (structure shown in Fig. 3) inhibits that of cyclophilin. NMR has contributed significantly to the understanding of binding interactions of both proteins.

Initial studies on FK506 focused on the structure of the free ligand to aid in the design of new analogs [145–147]. However, it was established, from studies of the cyclosporin A-cyclophilin complex, that the conformation of a molecule bound to its target site could be very different from that in the free state [148–150]. In addition, analog design is assisted by knowing the location of the binding region of the ligand. Studies were therefore undertaken to determine the bound state of the ligand as well as to identify those portions of the drug interacting with the binding protein.

The first investigations involved the analysis of ^{13}C carbonyl chemical shifts of C8 and C9 and the ^{1}H chemical shifts of the piperidine ring of FK506 bound to FKBP [151,152]. The upfield shifts of the piperidine ring protons, as well as NOEs observed between these protons and aromatic protons of FKBP, suggested that the bound site on FKBP resided in an aromatic-rich domain, which allowed a putative binding site on FKBP to be proposed. It was also evident that the pipecolinyl functionality of FK506 and analogs was involved in binding.

In another study [153], a uniformly ^{13}C-labeled ascomycin, (**12**), was prepared, allowing the bound conformation of this compound to be determined in the presence of FKBP. Not only were the assignments of side-chain methyls made possible by the ^{13}C enrichment, but ligand resonances could be distinguished readily from those of the protein. The conformation of the ligand was determined from NOEs observed in a 3D HMQC-NOESY spectrum. The resulting ascomycin structure (Fig. 25) differed considerably from that of the uncomplexed FK506 obtained by X-ray crystallography, but was similar to that of rapamycin. Specifically, the bound ascomycin displayed a *trans* orientation of the 7,8-amide bond, whereas this bond is *cis* in free FK506 and *trans* in rapamycin. This study also showed that both the piperidine ring and the pyranoyl moiety of ascomycin are involved in the binding interface of the complex with FKBP. Ligand protons that show NOEs to the protein, and are therefore likely to be in the binding interface, are indicated in Fig. 25. X-ray studies subsequently confirmed these results for both the FK506-FKBP and rapamycin-FKBP complexes, showing the *trans* orientation of the ligand amide bond in the bound conformation, and verifying the involvement of the piperidine and pyranoyl regions of the ligand in the binding interface [154].

In early studies the structure of uncomplexed FKBP was solved using ^1H NMR methods [155,156]. Although spectral overlap did not allow every structural constraint to be identified unambiguously, convergent structures defining the global fold of the 107-residue FKBP protein were obtained. Previous biochemical data allowed an aromatic cluster within the core of the structure to be identified as the ligand binding pocket. The loop regions of the protein between residues 37–43 and 83–90, situated at the open end of the binding pocket, were also of interest. The loops were the least well-defined regions of FKBP and were thought to be flexible, and perhaps involved in the binding interaction.

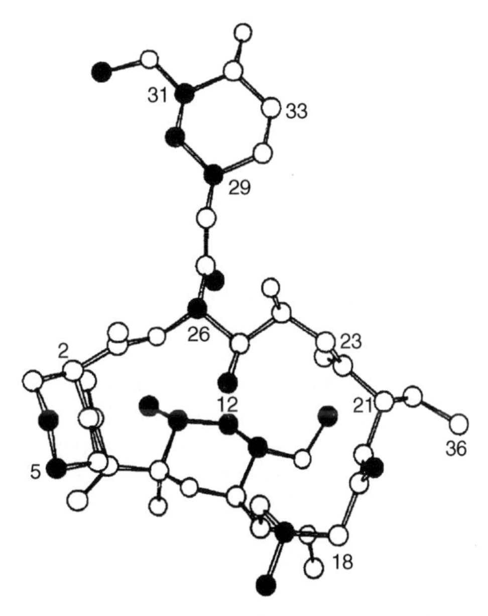

Figure 25. Three-dimensional structure of ascomycin bound to FKBP. Protons on the ligand that showed NOEs to the protein are denoted by a black shading of the carbons to which they are attached. Although no NOEs were observed from the protons at position 3 to the protein, the upfield shift of their resonances, −1.09 and 0.25 ppm, suggests that they are in close proximity to an aromatic region of FKBP. (Reprinted with permission from Ref. [153]. 1991 American Chemical Society.)

Examination of ^1H and ^{15}N chemical-shift changes on addition of ligand supported this and suggested that significant structural changes in these loop regions occurred upon ligand binding [155].

In a later study, a high-resolution structure of the complete ascomycin-FKBP complex was calculated by heteronuclear 3D and 4D NMR [157]. The extra detail afforded by the multidimensional NMR approach allowed the ligand-protein contact area to be located unambiguously and intermolecular hydrogen bonds identified. The structure of the complexed FKBP was essentially similar to that of the uncomplexed structure, except that the "ill-defined" loop regions were found to adopt well-defined conformations in the complexed proteins, as predicted by the previous studies. Although this difference may partially be a result of the differences in resolution achieved in the complexed and uncomplexed FKBP

NMR studies, it was thought that binding involved some rearrangement of the loops. This provides a good example of the dynamic nature of protein binding as revealed by NMR spectroscopy.

The dynamic aspects of the ligand-FKBP complex formation have been pursued through analysis of ^{15}N NMR relaxation data [158]. In particular, the increased backbone mobility for several residues within the flexible loops compared with that of the rest of the protein has been noted. From analysis of the ^{15}N relaxation rates of FKBP complexed with FK506, it was found that flexibility was restricted along the entire polypeptide chain [159]. This confirmed the proposition that the binding interaction of FKBP with ligand involves stabilization and structuring of the protein loops adjacent to the binding site.

In summary, it was possible to define the free and bound conformations of the ligand and, in addition, to identify the two binding interfaces involved in the interaction and demonstrate a reduction in protein mobility in a defined region of the protein upon binding. This level of analysis was possible because of the tight binding of the FKBP–ligand complex, its small size, and the availability of labeled species. The information proved to be complementary to X-ray crystallographic studies and helped to clarify the role of FKBP complex formation in immunoregulation.

Matrix Metalloproteinases Matrix metalloproteinases (MMPs), including stromelysin, collagenase, and gelatinase, are involved in tissue remodeling associated with embryonic development, growth, and wound healing. Unregulated or overexpressed MMPs have been implicated in several pathological conditions, including arthritis and cancer, and accordingly inhibitors of stromelysin and other MMPs have attracted much interest because of their potential for the treatment of these diseases.

Several NMR structural studies of stromelysin [160–164] and collagenase [165,166] complexes have been reported. The secondary structure and global fold were found to be quite similar for the catalytic domains of both enzymes and their various complexes with ligands. The active site in each enzyme is a cleft, spanning the width of the enzyme, with a catalytic zinc atom coordinated by three histidine residues located in the center. Different dynamic properties of active-site residues in stromelysin/ligand complexes [27] and of collagenase with and without bound inhibitor [165,166] have been reported. It has been proposed that structural and dynamic differences can be exploited in structure-based drug design, either to achieve broad inhibitor activity against several MMPs or to obtain inhibition that is more selective [27].

Structural data on a novel class of MMP binding inhibitors, represented by PNU-107859 (**13**) and PNU-142372 (**14**), which contain a thiadiazole moiety that coordinates the catalytic zinc atom through its exocyclic sulfur atom [167] provide useful illustrations of various NMR approaches.

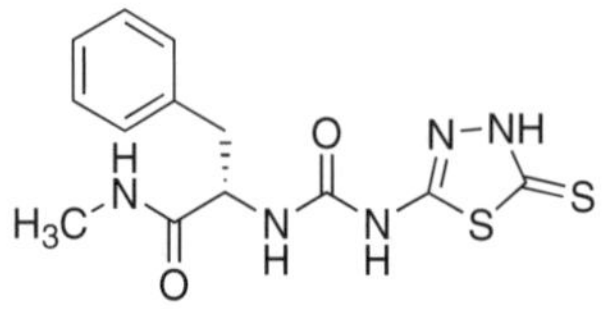

(**13**)

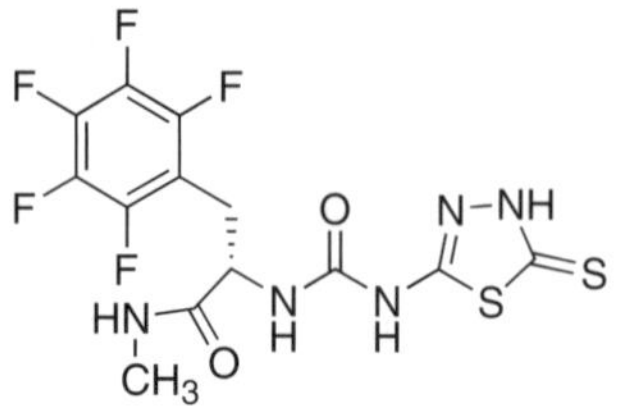

(**14**)

Isotope editing/filtering studies played an important role in defining interactions between the ligands and the stromelysin. For example, for the stromelysin/PNU-107859 complex, a 3D ^{12}C-filtered, ^{13}C-edited NOESY spectrum recorded on the [^{12}C,^{14}N]PNU-107859/[^{13}C,^{15}N]-stromelysin complex was used to assign protein/ligand NOEs. Of the 11 observed NOEs between the ligand and the protein aliphatic protons, nine involved the aromatic ring of (**13**) and one involved the terminal methyl group. NOEs were observed between (**13**) and protons of Tyr^{155}, His^{166}, Try^{168}, and Ala^{169}. All four of these residues are located in the S_1–S_3 binding sites on one side of the active site. Comparison of 2D ^{1}H-^{15}N HSQC spectra showed that differences between the ^{1}H and ^{15}N chemical shifts for the stromelysin/**13** and stromelysin/**14** complexes are concentrated in the active site. This observation indicated that no gross conformational differences in protein structure exist in the two complexes. The aromatic rings of (**13**) and (**14**) bind in the same region of the protein.

A region of the 1D ^{19}F spectrum of the stromelysin/**14** complex is shown in Fig. 26. Two separate resonances were observed for the two *ortho* fluorine atoms of the bound ligand, in contrast to the single resonance observed for both *ortho* protons of stromelysin-bound (**13**), indicating that the ring flip rate (rotation about the $C^{\beta}C^{\beta}$ bond) is reduced for stromelysin bound (**14**) compared to stromelysin bound (**13**). A ring flip rate of approximately 100 s^{-1} was estimated from the difference in linewidths for the bound *ortho* and *para* fluorine atom resonances of (**14**), which is more than two orders of magnitude slower than the ring flip rate for (**13**). Figure 26 illustrates several general principles of NMR studies of ligand macromolecule complexes. Firstly, it is important to note that the use of a ^{19}F probe nucleus produces spectra of elegant simplicity, and because there is no naturally occurring ^{19}F in the macromolecule there are no interfering signals. Secondly, the offset in chemical shift between bound and free signals reflects the different environment of the bound and free states. Thirdly, signals from the bound ligand are broader than those from the free ligand because of the higher molecular weight of the complex; however, they are still clearly visible for a complex of this size.

NMR studies have also been reported for ligands bound to collagenase. Interest has

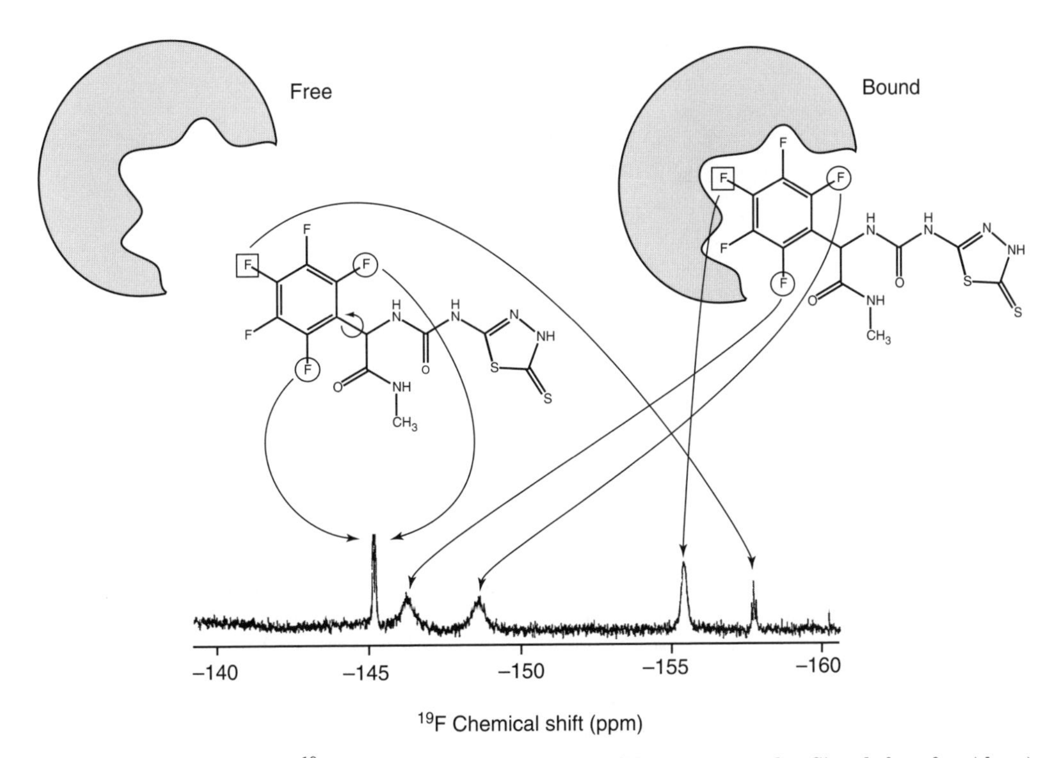

Figure 26. Region of the 1D ^{19}F spectrum of the stromelysin/PNU-142372 complex. Signals from free (sharp) and bound (broad) PNU-142372 are observed. (Adapted from Ref. [27]. Copyright 1998, with permission from Elsevier.)

focused on hydroxamate-containing ligands, where it has been shown that binding causes a decrease in mobility of some, but not all, active-site residues [165,166]. Interestingly, some active-site residues adjacent to residues that interact directly with inhibitor were found to have high mobility in both the presence and the absence of inhibitor [166]. This contrasts with what is observed for stromelysin complexed to hydroxamate ligands and a more complete understanding of the dynamics of the respective interactions may provide critical information for drug design [27].

Hydroxamate-containing ligands have also featured in transferred NOE studies to determine the bioactive conformations [168]. The NOE-derived structures of bound inhibitors were used as templates to screen a database of 260,000 compounds. More than a dozen high-affinity binders were discovered, demonstrating the value of deriving a conformationally restricted template for structure-based drug design [168]. This study also demonstrates the close synergy that exists between structure-based design and screening approaches.

Dihydrofolate Reductase Dihydrofolate reductase is an important intracellular enzyme that is the target of several clinically used drugs, including methotrexate (**15**), an anticancer compound, and trimethoprim (**16**), an antibacterial. These drugs act by inhibiting the enzyme in malignant cells and parasites, respectively. The small size of DHFR (~18 kDa) makes it amenable to structural studies and there have been numerous complexes determined using both X-ray and NMR methods. The focus here is on an illustrative example of the structure of a complex of DHFR with trimetrexate (**17**). Trimetrexate was initially investigated as an antimalarial agent but has subsequently been found to have antineoplastic activity against cancers of the breast, neck, and head. It has also been used as an antibacterial for the treatment of *Pneumocystis carinii* pneumonia in AIDS patients. As seen from the following structures, trimetrexate combines some of the features of trimethoprim and methotrexate:

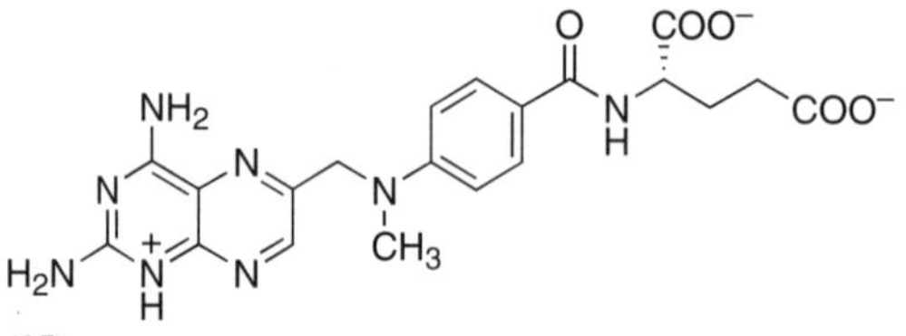

(**15**)

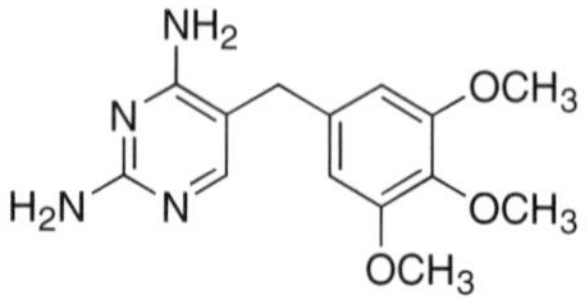

(**16**)

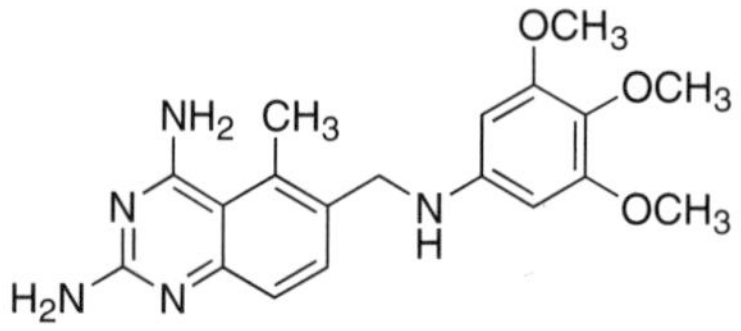

(**17**)

The 3D structure of the complex of DHFR with trimetrexate was determined using about 2000 distance restraints, 300 angle restraints, and 100 hydrogen-bonding restraints [169]. Several intermolecular protein-ligand NOEs were obtained by using an approach that monitored temperature effects of NOE signals resulting from dynamic processes in the bound ligand. At low temperature (5°C) the trimethoxy ring of bound trimetrexate flips sufficiently slowly to give narrow signals in slow exchange, which give good NOE cross-peaks. At higher temperature these broaden and their NOE cross-peaks disappear, thus allowing the signals in the lower temperature spectrum to be identified as NOEs involving ligand protons. Figure 27 shows the structure of the complex, including the orientation of the ligand in the binding site.

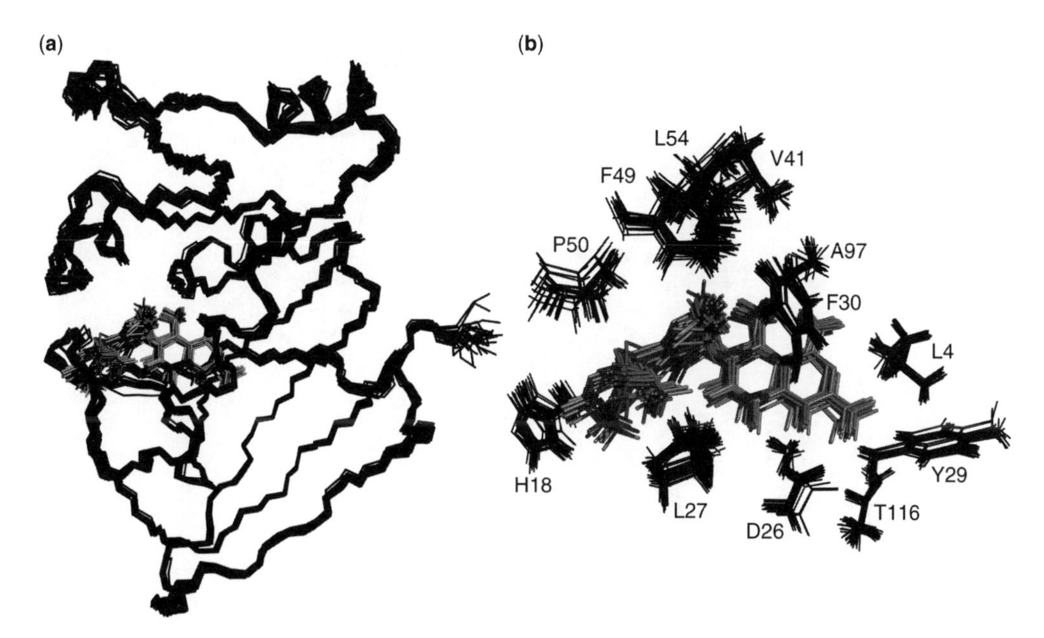

Figure 27. Stereoview of a superposition over the backbone atoms (N, Cα, and C) of residues 1–162 of the final 22 structures of the DHFR-trimetrexate complex [169]. (a) View of the protein backbone and the trimetrexate heavy atoms. (b) View of trimetrexate in the binding site of enzyme.

The binding site for trimetrexate is well defined, and was compared with the binding sites in related complexes formed with methotrexate and trimethoprim. No major conformational differences were detected between the different complexes. The 2,4-diaminopyrimidine-containing moieties in the three drugs bind essentially in the same binding pocket and the remaining parts of their molecules adapt their conformations such that they can make effective van der Waals interactions with the same set of hydrophobic amino acids.

The ring flipping of the trimethoxy aromatic ring mentioned above was detected by variable-temperature studies of the spectral line shape. The presence of such dynamic processes involving the ligand is common in macromolecule–ligand complexes and the ability of NMR methods to detect such phenomena represents one advantage of NMR over X-ray methods of structure determination. Relaxation measurements were also used to probe dynamics of the protein and no large amplitude motions were found, apart from that at the C-terminus [169]. The power of NMR methods for studying dynamics of complexes is further illustrated by an earlier study of the complex of DHFR with methotrexate [170]. In this case, an interesting correlated rotation of a carboxylate group on the ligand and Arg^{57} of the protein was detected, as illustrated in Fig. 28.

HIV Protease HIV protease is a major target for structure-based design of anti-AIDS drugs due to its essential role in the HIV life cycle. There are now more than 170 structures of HIV protease and protease inhibitor complexes in the RSCB Protein structure database and this wealth of high-resolution structural information has been the driving force behind numerous structure-based design programs [171–173]. Most of the structural information on HIV protease has been obtained from X-ray crystallography data [174]. Although there are relatively few examples of HIV protease/inhibitor complexes that have been determined by use of NMR spectroscopy, the NMR data, taken together with the structural data from X-ray experiments, have contributed to an understanding of protease-inhibitor recognition and dynamics. Indeed, studies of HIV protease/inhibitor complexes are a powerful example of the way in which complementary information

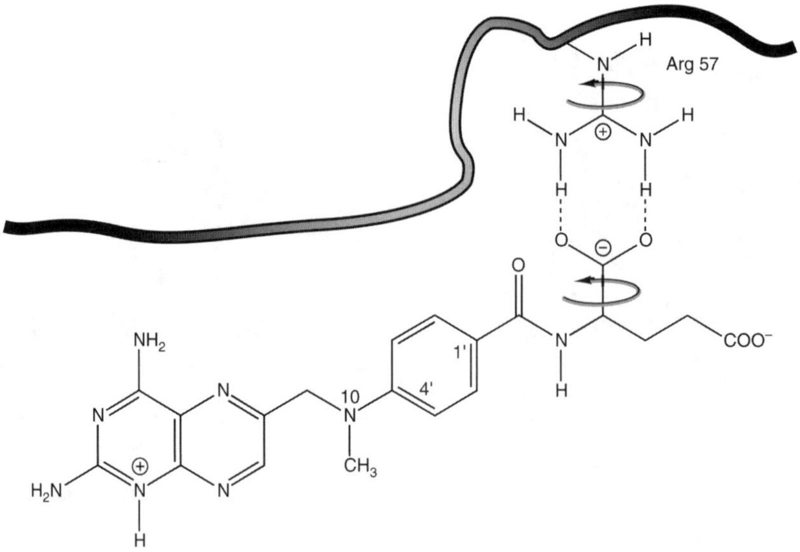

Figure 28. Correlated motions of a carboxylate group from methotrexate and Arg57 of DHFR detected by NMR. (Adapted from Ref. [170] Copyright 1997, with permission from Elsevier.)

obtained from X-ray crystallography and NMR spectroscopy can be used to facilitate structure-based drug design.

HIV protease/inhibitor complexes have a molecular mass of ~22 kDa. Although NMR spectroscopy is well suited to determination of the structure of molecules in this size range, efforts to determine the solution structure of the complex were initially hampered by the fact that the protease undergoes rapid auto-catalysis in solution. The development of potent inhibitors was required before NMR studies of the complex became feasible. The first solution structure (Fig. 29) of HIV protease bound to the cyclic urea inhibitor DMP-323 (**18**) was reported in 1996 [175].

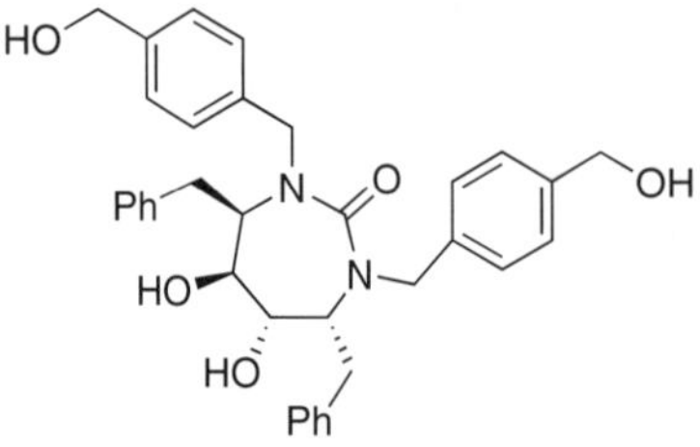

(**18**)

The protease exists as a homodimer. Each 99-residue monomer contains 10 β-strands and the dimer is stabilized by a four-stranded antiparallel β-sheet formed by the N- and C-terminal strands of each monomer. The active site of the enzyme is formed at the interface, where each monomer contributes a catalytic triad (Asp^{25}-Thr^{26}-Gly^{27}) that is responsible for cleavage of the protease substrates. The "flap region" is located above the reactive site and is formed by a hairpin from each monomer of two antiparallel β-strands joined by a β-turn. There is little difference between the solution and crystal structures of protease-inhibitor complexes, except in the regions where the polypeptide chain is disordered. However, NMR experiments in solution have allowed access to parameters that are not directly accessible from crystal data. These parameters, such as the amplitude and frequency of backbone dynamics, the protonation states of the catalytic aspartate residues, and the rate of monomer interchange, are important in understanding the interaction of HIV protease with potent inhibitors.

The cyclic urea inhibitor DMP-323 was designed by analysis of crystal structures of HIV protease/inhibitor complexes. A feature common to many of the complexes of HIV protease is a buried water molecule that bridges the inhibitor and Ile^{50} in the flaps.

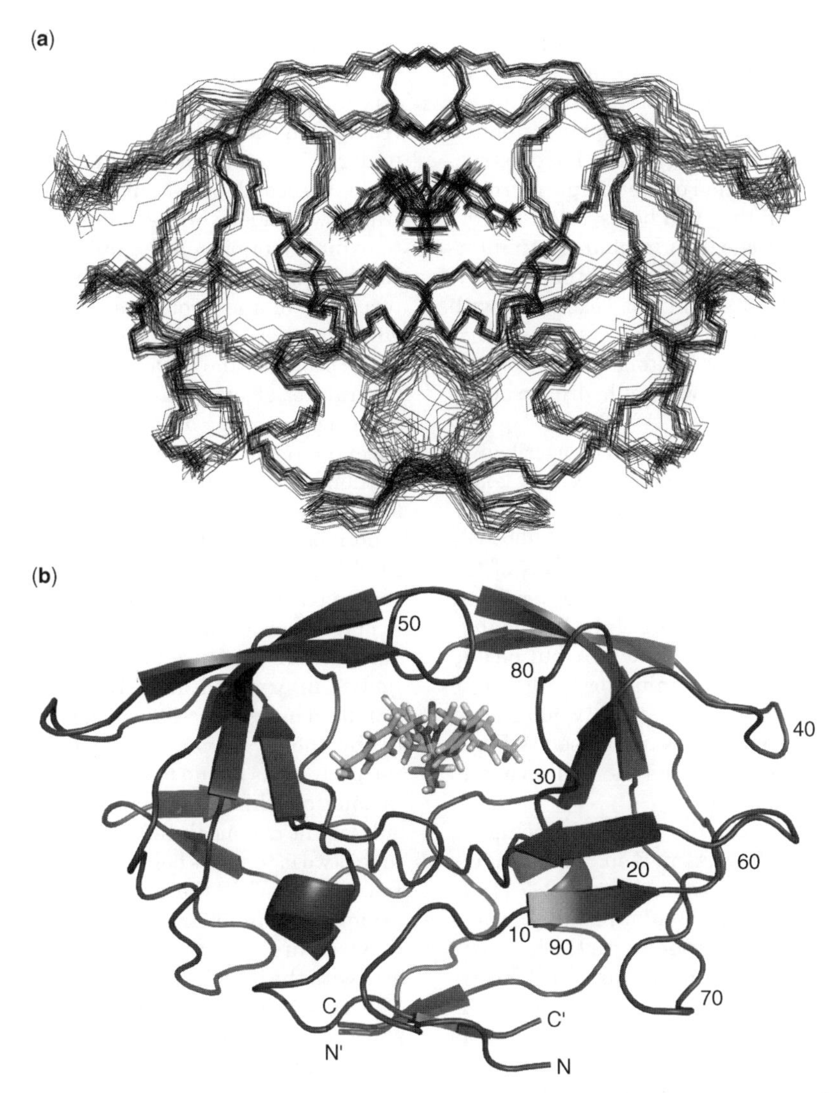

Figure 29. (a) View of the superimposed heavy atom (N, Ca, C) of the ensemble of structures of the HIV-1 proteases/DMP323 complex [168]. (b) Ribbon diagram of the minimized average structure of the complex. (This figure is available in full color at http://mrw.interscience.wiley.com/emrw/9780471266945/home.)

Interactions with this water molecule are thought to induce the fit of the flaps over the inhibitor [176]. In contrast, mammalian aspartic-protease/inhibitor complexes are unable to accommodate an equivalent water molecule [172]. This observation led to the design of a series of cyclic urea-based inhibitors that are capable of displacing the buried water molecule [176]. As well as improving the specificity of inhibitors to the viral protease, displacement of the water molecule was expected to increase the entropic contribution to inhibitor binding and thus enhance the affinity of complex formation. The cyclic urea inhibitors are highly potent and specific inhibitors of HIV protease [176] and for DMP-323 it has been shown in both the crystal structure [176] and in solution [177] that the urea moiety does indeed replace the buried water molecule.

Although DMP-323 replaces one buried water molecule, several others are observed in the crystal structure of the complex [178]. In favorable cases, NMR can be used to estimate the residence times of hydration water molecules [179], thus providing information about the timescale of the interaction of buried water with the bulk solvent. Such an analysis led to the identification of a symmetry-related pair of water molecules that may have a structural role in formation of the complex, providing information that will be useful in the design of future cyclic urea inhibitors. An interesting finding was that each of the hydroxyl protons of DMP-323 is in rapid exchange with solvent. This is a surprising result, given that two of these hydroxyl protons are completely buried and form a network of hydrogen bonds with the catalytic Asp^{25}/Asp^{125} side chains [180]. The observation is ascribed to local fluctuations in the complex that allow solvent molecules to penetrate into the binding site, a suggestion that is supported by the observation that the catalytic protons of the Asp^{25}/Asp^{125} side chains in the protease/DMP-323 complex undergo H–D exchange with solvent, even though they are buried and hydrogen bonded to the inhibitor [180]. These studies highlight that even well-ordered structures such as the protease/DMP-323 complex may be flexible on the millisecond to microsecond timescale.

Interestingly, in the DMP-323 complex, both of the catalytic Asp^{25}/Asp^{125} side chains are protonated over the pH range 2–7 [180]. The protonated Asp^{25}/Asp^{125} residues form a network of hydrogen bonds with the hydroxyl groups of DMP-323. In contrast, in the complex with the asymmetric inhibitor KNI-272, the side chain of Asp^{25} is protonated, whereas that of Asp^{125} is not. A suggested explanation is that both oxygen atoms of the Asp^{125} side chain are deprotonated to accept two hydrogen bonds, one from a bound water molecule and one from the inhibitor. In contrast, the side chain of Asp^{25} is protonated so that it can donate a hydrogen bond to the inhibitor [181]. Consequently, the protonation state of the enzyme is influenced strongly by interaction with specific inhibitors and this knowledge is essential for a detailed understanding of the protease/drug interactions.

NMR has been used to study the relationship between flexibility and enzymatic function for HIV protease. For the protease/DMP-323 complex, ^{15}N spin-relaxation studies determined that flexible residues correlate well with residues that are disordered in the NMR structure of the complex [182]. For example, residues in poorly defined loops were found to undergo large-amplitude internal motions on the nanosecond–picosecond timescale. In contrast, two regions of the molecule were found to exhibit motions on the millisecond–microsecond timescale. The first of these is at the N-terminus of the protein around Thr^4-Leu^5, which is adjacent to the major site of autolysis of the protease and it has been suggested that the rate of cleavage may regulate HIV protease activity *in vivo* [183]. Consequently, the observed flexibility may be important for regulation of protein function. The second region found to be undergoing millisecond–microsecond motion was the tips of the flaps around Ile^{50}-Gly^{51}. In crystal structures, this region of the protease is well ordered and not involved in crystal contacts, although its conformation varies from structure to structure. This motion is interpreted as a dynamic conformational exchange process, which is fast, relative to the chemical-shift timescale. Thus, when the protease is bound to a symmetric inhibitor in solution, this conformational exchange results in the chemical shifts of the flap residues in the two monomers being identical [175,182]. In contrast, when the protease is bound to an asymmetric inhibitor, such as KNI-272, crystal structures show that each monomer interacts with the inhibitor in a different way [181]. This is reflected in the fact that the chemical shifts of the monomers are different when asymmetric inhibitors are bound [178,184]. Analysis of spectra from such an asymmetric complex has revealed that the inhibitor is capable of "flipping" its orientation with respect to the two monomers without dissociating from the complex [185]. These data again highlight the importance of defining both the structural and dynamic aspects of binding to understand the requirements for potent interactions between HIV protease and its inhibitors.

The development of inhibitors of HIV protease represents a major success for structure-based drug design. When HIV was first identified in the early 1980s there were no known drugs effective for treatment of infection. A combination of X-ray crystallography, NMR spectroscopy, computer modeling, and chemical synthesis has resulted in the development of numerous effective HIV protease inhibitors. However, like other retroviruses, HIV has a high transcription error rate that results in a rapid mutational rate. One of the results of this is the production of a divergent population of viruses in which the sequence of the HIV protease produced may differ substantially [186,187]. As a consequence, drug-resistant strains of the virus emerge. Clearly, knowledge of the structural principles that govern inhibition of the protease and the mechanism by which the virus develops resistance will continue to be important in the development of effective new drugs.

4. NMR SCREENING

In the past, NMR was predominantly used in the design stage, rather than the screening stage, of drug discovery. However, new methods that make use of NMR to screen ligands for binding to a protein target have been developed and are proving to be a powerful tool in the discovery of new drug leads.

In the late 1980s, the concept of high-throughput screening (HTS) became very popular, when cloning and protein production techniques were dramatically improved, and the power and accessibility of combinatorial chemistry was realized. Classical HTS methods involve the screening of chemical libraries of up to 1 million compounds of 250–600 Da molecular weight to detect relatively potent (submicromolar) ligands (termed "hits") for a protein target. The traditional methods used for HTS include cell-based assays, surrogate systems such as yeast and phage display, and reporter assays such as those that use fluorescence-based reporting systems [188]. However, one pitfall of these methods is that they do not indicate the mode of binding or mechanism of activity against the target. Thus, the use of NMR for HTS programs gained popularity due to the ability of NMR to provide direct evidence and measurement of a binding interaction [189].

Initially, NMR was used only as a checkpoint to confirm the binding of HTS leads identified from other read-out techniques, due to the relatively high amounts of protein and long acquisition times that would be needed for NMR to be used as the primary read-out across a large library of compounds. However, over the last decade NMR has played an increasing role as the primary screening technology, mainly through the development of smaller, more directed compound libraries and the advent of fragment-based drug design (FBDD).

In contrast to HTS, FBDD involves the screening of only $\sim$100 to a few thousand fragment compounds of typically 120–350 Da molecular weight. This allows for the identification of any compounds that bind (termed "hits") to that protein target with modest affinity (millimolar to micromolar), but with high ligand binding efficiency—that is binding affinity divided by molecular weight [42]. Once an interacting fragment is identified from the initial screen, the fragment is optimized and elaborated such that favorable interactions with the binding site are maximized and drug like properties are checked. Often two fragments are found that bind to adjacent binding sites, and catenation of these fragments into one chemical entity produces a compound with a greatly enhanced binding affinity [1]. The advantage of using small fragments instead of larger molecules is that the fragment universe (compounds of 120–350 Da) is much smaller than the total chemical universe, and therefore fragment-based screening captures greater screening diversity with a smaller number of compounds [190]. This diversity results in a higher hit rate compared to HTS, and increases the likelihood of identifying novel hits for targets.

There is a wide range of experimental NMR methods which are used to screen mixtures of ligands or fragments for binding to drug targets, as summarized in Table 8. In the following sections we describe examples of these methods and discuss some of the practical considerations that need to be made when designing an NMR screening program.

4.1. Methods

4.1.1. Chemical-Shift Perturbation Perturbation of chemical shifts is a powerful indicator

Table 8. A Summary of the Methods Available for NMR Screening and Their Respective Characteristics

Screening Methodology	Signals Observed	Protein Size Limit	Labeling	Binding Information Obtained	K_D Limit (M)	K_D Determined	Suitable for HTS/FBS	Mixture Deconvolution Required?
Chemical shift perturbation, e.g., SAR by NMR	Protein	<30 kDa	^{15}N protein	Binding-site location	10^{-3}–10^{-9}	Yes	Yes	Yes
Magnetisation transfer e.g., STD NMR, WaterLOGSY	Ligand	None	None	Orientation of bound ligand	10^{-3}–10^{-8}	No	Yes	No
Diffusion based e.g., affinity NMR	Ligand	None	Not required, but occasionally used	None	10^{-3} – 10^{-7}	Yes	Yes	No
Relaxation based	Ligand	None	None	None	10^{-3}–10^{-7}	Yes	Yes	No
TrNOE	Ligand	None	None	Bound conformation	10^{-3}–10^{-7}	No	Yes	No
NOE pumping	Ligand or protein (reverse)	None	None	Bound conformation	10^{-3}–10^{-7}	No	Yes	No Yes[a]
Spin labeling e.g., slapstic	Ligand or protein	None	Spin label for either ligand or protein	Orientation, simultaneous binding	10^{-3}–10^{-6}	No	Yes[b]	Yes[b]

[a] For reverse NOE pumping.

[b] For primary screening if the protein is spin-labeled or for second-site screening if the first-site ligand is spin-labeled.

of binding of a ligand to a protein target. When a ligand binds to a protein the local chemical environment of both ligand and protein is changed, and this is reflected by a change in the chemical shifts of nuclei in close proximity to the binding site. The most common experiment used in this screening methodology is the $^{1}H/^{15}N$ HSQC, which generates a discrete signal for each amide group within the protein, effectively producing a "fingerprint" map. A reference $^{1}H/^{15}N$ HSQC spectrum, which is acquired in the absence of ligands, is compared to a spectrum recorded in the presence of ligands and any changes in the amide chemical shifts are generally indicative of ligand binding close to the corresponding amide groups. The major advantage of this technique is that, if the NMR assignment of the amide resonances is known, then the site of binding for each ligand can be determined. Care must be taken to ensure that any perturbations in chemical shifts represent direct binding interactions and not indirect conformational changes. However, direct interactions can be characterized by shifts that are obviously clustered and consistent across a range of binders. The location of the binding site is valuable information for the development of second-generation drug leads and for the optimization of binding models in computational library screening procedures. Binding affinities can also be determined by measuring the change in chemical shift as a function of ligand concentration. Chemical-shift perturbation studies are used widely for NMR screening, but their application is perhaps best exemplified by the method of structure–activity relationships by NMR (SAR by NMR)

4.1.2. Structure Activity Relationships by NMR (SAR by NMR) The technique that effectively established the broader field of FBDD screening is "SAR-by-NMR" [1,25,191–195]. SAR-by-NMR was originally developed as a method in which a potent drug candidate is derived by chemically linking two small low affinity ligands (fragments) for a target. In theory, the binding energy of the linked fragments is the sum of the binding energies of the two individual fragment compounds plus contributions to binding energy attributed to linkage. Therefore, it is possible to generate a drug lead with a nanomolar dissociation constant (K_D) from two milli- to micromolar fragments.

The first step in this process (Fig. 30) involves screening a library of fragment ligands (typically with a MW < 350) in mixtures (typically up to 10 compounds per mixture) for binding to a protein target by comparing the $^{1}H/^{15}N$ HSQC spectrum of the ^{15}N-enriched protein in the presence and the absence of ligands. Any ligand-induced changes in the chemical shifts of the protein indicate binding of one or more ligands in the mixture to the target. The mixture containing the binding ligand(s) is deconvoluted and each individual compound screened to identify the individual ligand(s) responsible for the observed chemical-shift perturbations. Once a binding ligand is identified analogs can be screened to optimize binding.

A second ligand, which binds at a proximal site, is then identified either from the original screen or by repeating the library screening with the first ligand bound to the protein. This ligand is then optimized and the structure of the ternary complex determined by use of either NMR or X-ray crystallography. The ternary complex structure provides information on the conformation and orientation of the bound ligands, which facilitates the synthesis of hybrid molecules where the two ligands are joined by a suitable linking moiety.

One of the first examples of SAR by NMR was aimed at the development of immunosuppressive agents. As noted earlier, FK506 binding protein (FKBP) inhibits calcineurin and blocks T-cell activation when complexed to the immunosuppressant FK506. This protein was used as a target for SAR by NMR screening and subsequently two ligands, (**19**) and (**20**), were identified with K_D values of 2 μM and 0.1 mM, respectively. A model of the ternary complex between the protein and both ligands was produced, which indicated that the methyl ester of (**19**) was close to the benzoyl hydroxyl group in (**20**). These two groups were linked with alkyl chains of various lengths, with the most active compound (**21**) having a three-carbon linker and a K_D of 19 nM [1].

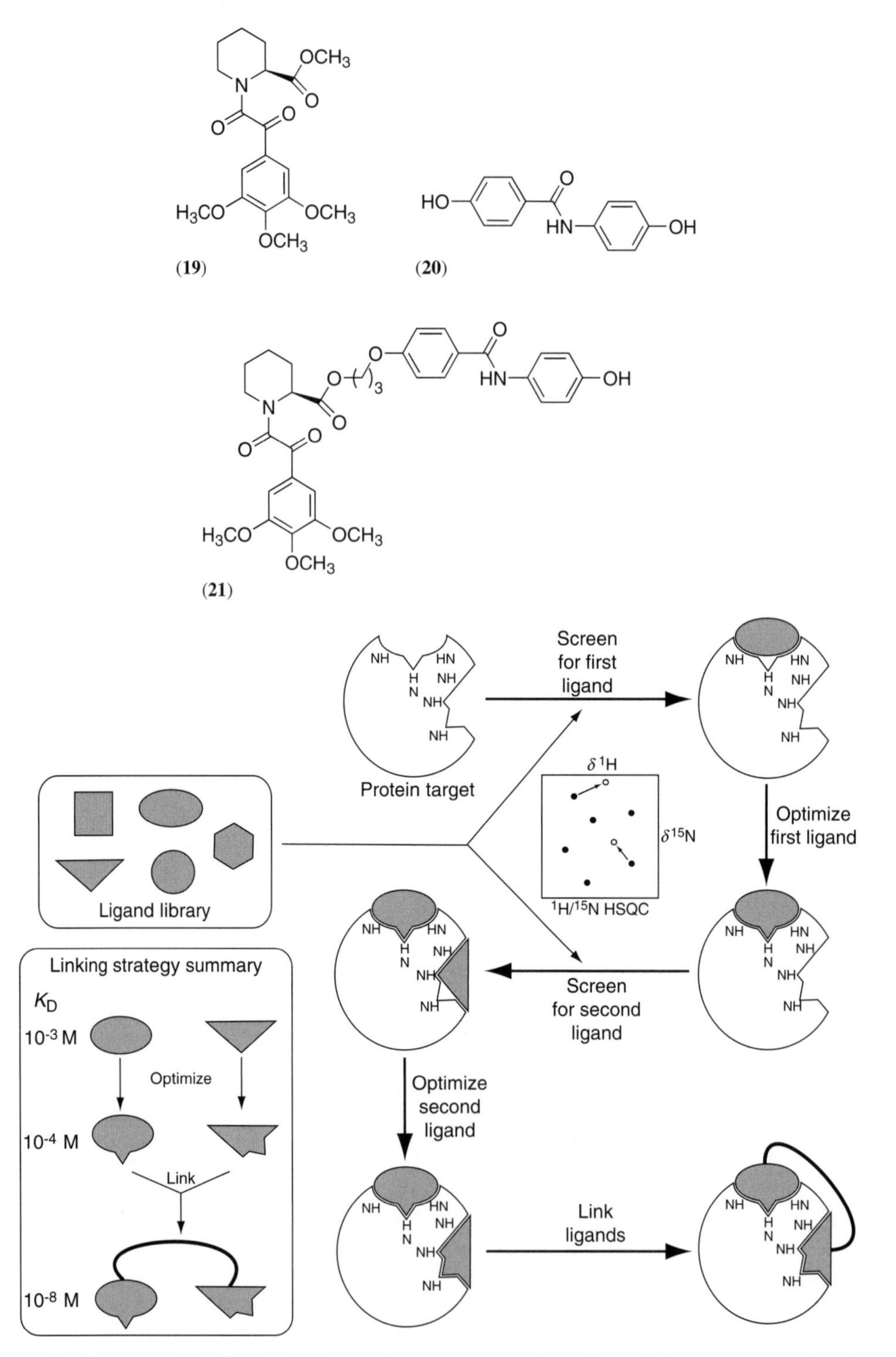

Figure 30. Summary of the SAR by NMR drug discovery methodology. A protein target is screened against a library consisting of small organic molecules by use of the $^{1}H/^{15}N$ HSQC experiment. When two ligands that bind in close proximity are identified, they are linked to form a composite ligand with an increased affinity for the target.

Inhibitors of the matrix metalloproteinase stromelysin have also been designed through the use of the SAR-by-NMR screening methodology. As mentioned in Section 3.2.4, MMPs are involved in matrix degradation and tissue remodeling, with overexpression of these enzymes being associated with arthritis and tumor metastasis. Acetoxyhydroxamate (**22**) was used as one ligand and its K_D of was determined to be 17 mM. To identify a second ligand the protein was screened against a ligand library in the presence of saturating amounts of (**22**). The library was biased for hydrophobic compounds, given that stromelysin demonstrates a substrate preference for hydrophobic amino acids and structural studies had identified a hydrophobic binding pocket supporting this observation. From the library screen, a series of biphenyl compounds was identified and analogs of these compounds were synthesized. A biphenyl derivative (**23**) was produced with a K_D of 0.02 mM. The NMR structure of a ternary complex, consisting of stromelysin, acetoxyhydroxamate (**22**), and the biaryl derivative (**24**) (chosen for its superior aqueous solubility), was determined and indicated that the methyl group of (**22**) was in close proximity to the pyrimidine ring of (**24**). With this information (**22**) and (**23**) were subsequently linked by different length linkers and the most active compound produced, (**25**), had a K_D of 15 nM [194].

A variation of SAR-by-NMR is to optimize binding or improve the pharmacological properties of known drug leads generated by other methods (e.g., natural products isolation or combinatorial chemistry). A compound can be fragmented into individual subunits and then alternative fragments identified through use of $^1H/^{15}N$ HSQC screening. These fragments can then be incorporated back into the structure in the hope of improving the binding and/or pharmacological properties of the parent compound (Fig. 31). The alternative fragment must bind in the same location as the corresponding section of the original molecule, making the $^1H/^{15}N$ HSQC screening method ideal as it provides information on the binding site of ligands.

In a demonstration of this fragmentation method, an antagonist of the interaction between leukocyte function-associated antigen 1 (LFA-1) and intracellular adhesion molecule 1 (ICAM-1) was used as a starting molecule. This interaction plays a role in the inflammatory response and specific T-cell immune responses, and inhibitors have applications in the treatment of inflammation and organ transplant rejection. The *p*-arylthiocinnamide antagonist (**26**) had an IC_{50} of 44 nM. However, it was envisaged that the activity and physical properties of the molecule could be improved by replacing the isopropyl phenyl group with a more hydrophilic moiety. Screening of a 2500-compound library provided several hits, and analogs of (**26**) were made that incorporated these ligands in place of the isopropyl phenyl group.

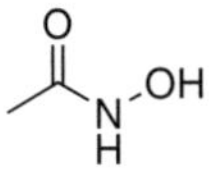

(22) 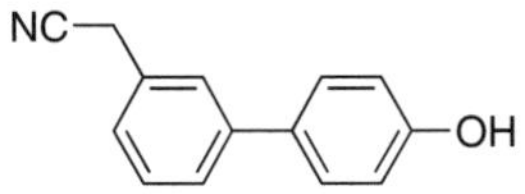 **(23)** 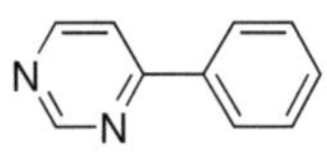 **(24)**

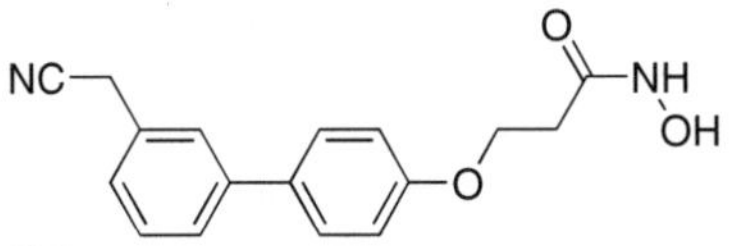

(25)

Compounds (**27**) and (**28**) had both improved aqueous solubility and pharmacokinetic profiles, with similar or improved activity (IC_{50} values of 20 and 40 nM, respectively) when

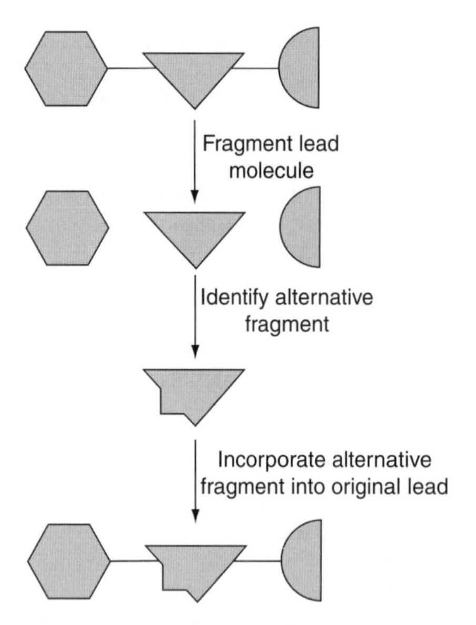

Figure 31. The fragment optimization approach developed from SAR by NMR. A known ligand of a protein is broken into fragments and small molecules based on the fragments are screened for binding. Any molecules that are found to bind can then be incorporated into the original lead compound with the hope of improving its binding and/or physicochemical properties.

compared to that of the parent compound (**26**) [196].

(**26**)

(**27**)

(**28**)

Overall, SAR-by-NMR is showing great promise as a major tool in drug design, and has led to the development of several compounds currently in clinical trials, mainly directed at cancer targets, involving the inhibition of protein:protein interactions. Nevertheless, it has a few limitations. First, the screening of large ligand libraries and subsequent deconvolution steps can be expensive, in that a large amount of ^{15}N-labeled protein is needed. This is less of a problem now than it was in the early implementations of the technique where larger libraries tended to be used. The use of smaller, drug-like libraries such as that described in the SHAPES ideology [26] also reduces the amount of protein required for screening and is an alternative approach to SAR by NMR. Furthermore, the widespread availability of cryoprobes has led to sensitivity gains that reduce the amount of protein required for screening studies. The second limitation is that the NMR assignments for the protein must be known or determined. Overall though, despite these limitations, SAR by NMR is making a major impact in the drug discovery process.

4.1.3. Magnetization Transfer Experiments

Proteins can be considered to be a large network of dipole–dipole interactions, with the potential for the efficient transfer of magnetization within them. The saturation transfer difference (STD) experiment (Fig. 32) uses this magnetization transfer to detect the binding of ligands to proteins. It relies on the fact that saturation of a single protein resonance results in saturation of all protein resonances and any ligands that bind to the protein. The STD experiment is able to detect the binding of ligands with K_D values between 10^{-3} and 10^{-8} M.

STD experiments involve irradiating an isolated protein resonance (either at low or at high field) to cause saturation of the entire protein and any binding ligands. This results in a spectrum containing reduced signal intensities from both the protein and any ligands that bind to it, but does not affect nonbinding ligands. A second spectrum of the

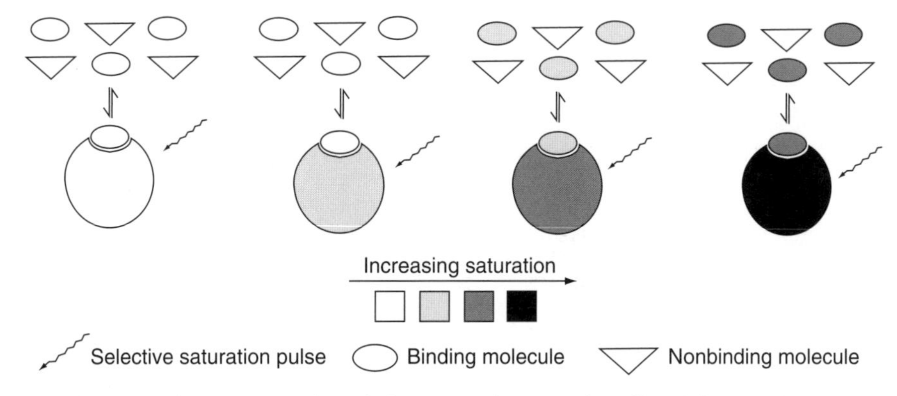

Figure 32. A schematic representation of the saturation transfer effect. The protein resonances are saturated (indicated by shading) by a selective pulse by spin diffusion. Resonances of nonbinding ligands (triangles) are not affected by this pulse but ligands that are interacting with the protein (ellipses) will also become saturated. These interacting ligands are transferred to solution through chemical exchange where they are detected.

protein and ligand library is then recorded with the saturation applied off-resonance with respect to the protein. Subtraction of the two spectra results in the STD spectrum that shows only those ligands within a mixture that bind to the protein. This subtraction is done through phase cycling to reduce artifacts from temperature or magnetic field variations [197,198].

An STD step can be added to many NMR experiments, including COSY, TOCSY, NOESY, and inversely detected ^{13}C or ^{15}N spectra. A high-resolution magic angle spinning (HR-MAS) STD experiment has been developed to study the binding of ligands to a protein immobilized on a solid support. HR-MAS STD NMR provides a way of obtaining ligand binding information for proteins, such as membrane proteins, that are difficult to work with in solution attributed to either poor solubility or conformational changes [198].

In addition to screening ligands for binding to a protein [197,199,200], the binding epitope of a molecule can also be determined by examining the intensities of ligand resonances [201–203]. The proton signals having the strongest signals will generally correspond to those that are part of the binding epitope of the ligand. For example, when methyl β-D-galactoside (**29**) binds to a target protein the H2, H3, and H4 protons are saturated to the highest degree, reflecting their proximity to the protein. This analysis was extended to the decasaccharide NA_2 (**30**) and showed that the Gal-6′ and GlcNAc-5′ residues bind edge-on to the protein, with the binding contribution of the terminal galactose residue being the greatest [202]. More recently, STD NMR has been used to investigate the glycan binding determinants of rotaviruses in host cell recognition [204]. It is important to note that quantitative interpretation of STD experiments can be made difficult by varying pathways for magnetization transfer and caution must be taken in order to avoid incorrect interpretation regarding ligand epitopes.

(**29**)

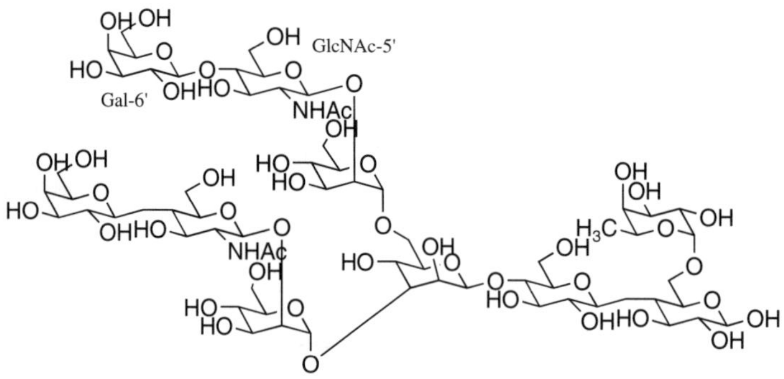

(30)

There are a number of advantages in using STD experiments for the detection of ligand binding in screening studies. Saturation transfer is an efficient process, which results in high sensitivity, and hence only small quantities of protein are required (nanomolar concentrations of a protein with MW > 10 kDa) [197,202]. In addition, protein size is not critical; in fact, as the protein becomes larger, the saturation transfer effect becomes more efficient. The acquisition time for each experiment is short and, because the experiment is ligand detected, no deconvolution of mixtures is required, making it a suitable technique for high-throughput screening of large ligand libraries. However, unlike chemical-shift perturbation techniques, STD experiments provide no information on the site of ligand binding.

A variation of saturation transfer experiments has been devised, referred to as WaterLOGSY, which uses the transfer of magnetization from water molecules to proteins and in turn to bound ligands [205]. Water is intimately associated with proteins, being bound either within their structures or on their surface. Saturation of the water resonance will lead to protein saturation through a variety of mechanisms, including concomitant saturation of αH resonances, saturation of exchanging protein resonances, and NOE interactions between water and protein. If a compound is bound to a protein it will also become saturated, and this effect can be used as an indication of ligand binding [205]. The WaterLOGSY method has the advantage of high sensitivity relative to other ligand-based screening approaches, attributable to the large excess of solvent water in biological samples and the large number of exchangeable NH protons in protein:ligand complexes. This sensitivity allows screening studies to be done with nanomolar concentrations of target proteins.

It is worth mentioning that for most compound libraries the proportion of signals occurring around the frequency of water is probably higher than it is for the STD experiment where saturation usually occurs below 0 ppm. The consequence of this is that there is the potential for more direct transfer to library compounds and this can result in more false positives. Thus, it is valuable to analyze initial hits from ligand screening by ^{15}N-based chemical-shift perturbation experiments.

4.1.4. Molecular Diffusion Molecules can be distinguished based on their diffusion coefficients, which are related to molecular size. Large macromolecules diffuse more slowly than small molecules and this diffusion difference can be exploited to screen for ligand binding. If a small molecule binds to a protein target its diffusion coefficient is altered to a value more like that of the protein. Therefore, by utilizing a diffusion filter, resonances generated by small molecules that do not bind to the protein can be removed from the spectrum.

Diffusion editing is achieved with the use of a pair of gradient pulses [206]. If field inhomogeneity is ignored, then all spins experience an identical magnetic field despite having different positions throughout the sample. The application of a field gradient has the effect of making field strength dependent on position. Under the influence of a gradient pulse, the phase of individual spins becomes dependent on their position within the sample and hence the spins are spatially "encoded." If diffusion does not occur, this spatial encoding is reversible by a second gradient of inverse polarity and no loss of NMR signal will occur. However, the second gradient pulse will be unable to "decode" the spins that have undergone diffusion and the resulting NMR signal will be reduced. Acquiring spectra of a sample with and without a diffusion filter and then subtracting them allows ligands binding to the protein to be identified. This filtering method can be "tuned" by altering the strength and duration of the gradients.

Because ligand signals are observed in this screening method, no deconvolution of the ligand mixture is required, and detected signals can be assigned directly to individual compounds within the mixture. However, signals from the protein are always present, and can pose a problem in interpreting spectra. An isotope-edited version of the diffusion experiment avoids this problem, although labeled protein is required [207]. In general though, there is no requirement for labeling of the protein or for its resonances to be assigned and thus, in theory, there is no size limit on the proteins that can be screened by use of this method. Only one sample, containing protein and ligands, is used to obtain both reference and screening data and therefore differences between the sample and the reference spectra caused by addition of the ligands (pH, salt concentration, etc.) are avoided. However, diffusion-edited experiments are potentially complicated by chemical exchange and NOE processes [208–210].

In an illustration of diffusion-edited NMR experiments referred to as "Affinity NMR" [211–214], two binding tetrapeptide ligands of vancomycin were identified from a mixture of ten peptides [215]. Similarly, the application of diffusion-editing experiments has been demonstrated by differentiating ligands of stromelysin from a mixture containing nonbinding compounds [216].

4.1.5. Relaxation Like diffusion, the transverse relaxation time (T_2) is dependent on molecular size, differing to the extent that it is sensitive to rotational motions rather than translational motions. Large molecules, such as proteins, have a short T_2 and hence exhibit broad NMR signals, whereas small molecules have a longer T_2 and hence narrower linewidths. Therefore, if a small molecule ligand binds to a protein, its T_2 value will decrease and a line-broadening effect of bound ligand signals can be observed. Alternatively, a relaxation filter can be used to remove signals from molecules with a short T_2 value. Subtraction from a reference spectrum will result in a spectrum containing only those ligands that bind to the protein. The ability to identify binding ligands using relaxation filters has been demonstrated using FKBP. In an illustration of this approach to screening, a mixture of nine compounds consisting of one known ligand of FKBP, 2-phenylimidazole (**31**), and eight nonbinding compounds (e.g., **32–34** were screened and only signals from (**31**) were observed [216].

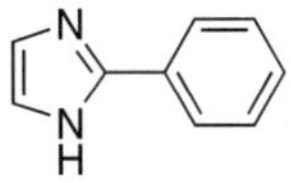

N N H (**31**) N CO_2H (**32**) O O (**33**) H N N NH_2 (**34**)

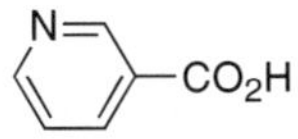

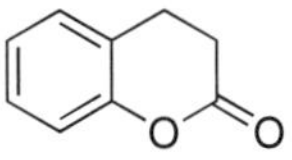

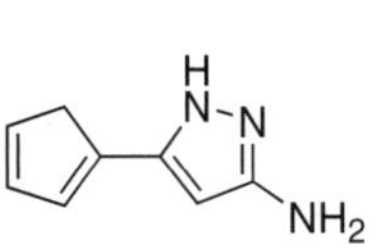

4.1.6. NOE NOE experiments can also be used to identify ligands that bind to protein targets [217–219]. Small molecules have a fast tumbling rate and, as a consequence, generally exhibit small positive NOEs. In contrast, large molecules such as proteins generate strong negative NOEs because of their slow tumbling time. When a small molecule binds to a protein, its tumbling rate is slowed to that of the protein and it exhibits strong negative NOEs. On dissociation, these are transiently retained and are known as transferred NOEs (TrNOEs) (Fig. 33). TrNOEs and those arising directly from the free ligand can be distinguished by the rate of signal build up. Transferred NOEs accumulate significantly faster and therefore can be selected for by use of shorter mixing times in the NOE experiment [218].

In practice, a 2D NOESY spectrum of the mixture of potential ligands in the absence of protein is recorded and all molecules exhibit small positive NOEs. The experiment is then repeated in the presence of protein, and ligands that bind display negative TrNOEs. Subtraction of the two spectra provides signals arising from only those compounds that bind. These TrNOEs can be interpreted to provide information regarding the bound conformation of the active ligands. However, when analyzing the conformational data care must be taken to ensure that the ligands are in fast exchange and that the observed TrNOEs are not affected by contributions from spin-diffusion [218]. Relative binding affinities between ligands can also be determined by comparison of TrNOE signal strength but, again, the fast-exchange regime and spin-diffusion effects need to be taken into account [217,218]. If all ligands are in fast exchange, the stronger binding ligands occupy more binding sites and thus give larger TrNOE intensities. Because of the need for an averaging effect, brought about by fast chemical exchange, TrNOE experiments are limited to those ligands with a K_D value from 10^{-3} to 10^{-7} M.

Transfer NOE experiments, for example, have been used to identify a bioactive disaccharide from a library of 15 mono- and disaccharides that bound to *Aleuria aurantia* agglutinin [218]. Another study described the identification of a silalyl Lewis mimetic (**35**) that binds to E-selectin from a library of ten compounds [217]. As well as being used to detect binding, TrNOEs are also used to determine bound ligand conformations, as described earlier in this chapter.

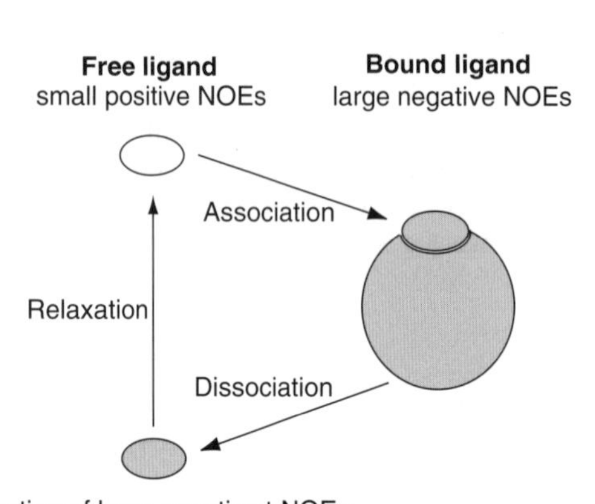

Figure 33. A schematic representation of the TrNOE experiment used to detect ligand binding (as described in Ref. [219]). The free ligand (white ellipse) exhibits only small positive NOEs, although binding to the large protein target results in the generation of large negative TrNOEs. The appearance of these large negative TrNOE signals can be used to identify ligands within a mixture that are binding to the protein and also provide some information on the bound conformation of the ligand.

(**35**)

A second technique that uses NOEs to detect binding is NOE pumping. This method was designed to alleviate some of the problems associated with diffusion-edited screening methods [208]. Signals from ligand molecules are removed using a diffusion filter and then transfer of signal from the protein to bound ligands by NOE occurs. The inverse of this is possible (known as reverse NOE pumping), which uses a relaxation filter to attenuate the protein resonances, after which the signal is transferred to the protein by NOE. Ligands may lose signal either by relaxation (for a free ligand) or through relaxation and NOE transfer (for a bound ligand). Therefore by subtracting spectra (which is done internally to reduce subtraction artifacts) from experiments with and without NOE pumping to the protein, the binding ligands can be detected [220].

NOEs between ligands bound to the same protein have also been used in screening approaches. Such NOEs are referred to as inter-ligand NOEs (ILOEs) and have been used to detect protein-mediated ligand–ligand interactions for compounds occupying adjacent sites [22] as well as identifying ligands that compete for the same binding site. The latter approach is referred to as the INPHARMA method [221] and has been used for pharmacophore mapping of binding sites.

4.1.7. Spin Labels Spin-spin relaxation rates are proportional to the product of the squares of the gyromagnetic ratios of the interacting spins. The gyromagnetic ratio of an unpaired electron is significantly larger than that of a proton and therefore any nuclear spins influenced by an unpaired electron will have substantially shortened relaxation times. The resonances of protons that are within 15–20 Å from the unpaired electron will experience this effect and be significantly broadened.

Spin labels can be used either in a primary screening method or to identify a second ligand binding site. The primary screening method requires residues around the binding pocket of the target to be spin labeled. Residues suitable for this labeling include lysine, cysteine, histidine, glutamate, aspartate, tyrosine, and methionine. Any ligands that bind to the protein in close proximity to the spin-labeled residues will be able to be identified by the induced paramagnetic relaxation enhancement from the spin label, which broadens and weakens the ligand resonance signals. The approach has been given the acronym SLAPSTIC (spin labels attached to protein side chains to identify interacting compounds) [222]. To screen for second-site ligand binding, the known first-site binding ligand is spin labeled. A reduced signal will be observed for any ligands that bind simultaneously and in close proximity to the first ligand binding site. In addition, the degree of reduction in signal intensity gives an indication of the orientation of the second ligand in relation to the first, given that the effect of the spin label is inversely proportional to the distance separating the electron and proton. This information is valuable in the design of linkers to join the two ligands.

There are several advantages to using spin-label screening. It can detect ligands that bind to the protein simultaneously, unlike other methods that can produce false positives if the first ligand binding site is not fully saturated. The concentration of protein required for screening is relatively small (~10 μM) because of the substantial enhancement of the relaxation rate by the spin label. The protein can also be unlabeled and partially purified and there is no molecular weight limit. The spin labels also quench protein signals, making interpretation of spectra easier. The experiment is easy to set up and analyze, making it amenable to automation. It is also insensitive to small changes in solvent conditions that can generate false positives in other methods. The information obtained on the orientation of ligands is also valuable and makes it an alternative to the chemical-shift perturbation methods when the proteins are large and NMR assignments have not been made.

A disadvantage of spin-label approaches is the requirement for spin-labeled proteins and ligands. Furthermore, any ligands with slow dissociation rates will show no averaging of relaxation rates and therefore tightly binding compounds ($K_D < 10^{-6}$ M) will produce false negatives. Protein spin labeling must occur adjacent but not within the binding site to minimize alteration of its binding properties.

A good example to illustrate spin-label-based approaches concerns the antiapoptotic protein Bcl-xL, which is responsible for the reduced susceptibility of cancer cells to undergo apoptosis and is therefore a target for the development of new anticancer agents. The structure of a previously identified ligand for Bcl-xL (**36**) was modified to incorporate a TEMPO spin label (**37**). By use of spin-labeled (**37**), an eight-compound library was screened for simultaneous binding to Bcl-xL. From this library an aromatic ketoxime (**38**) was

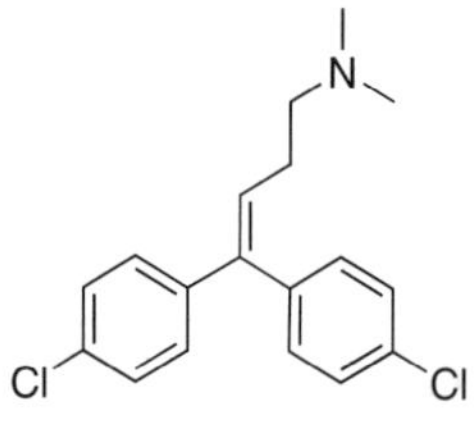

(36) 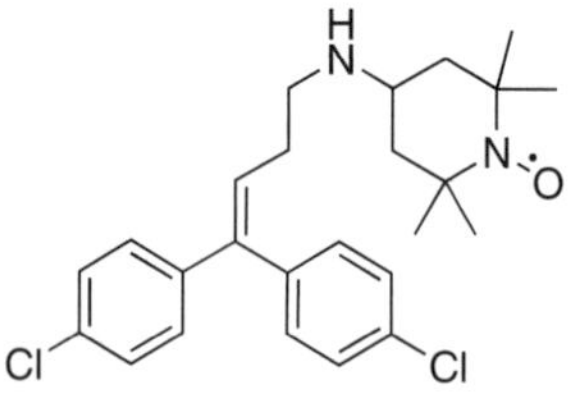 (37) 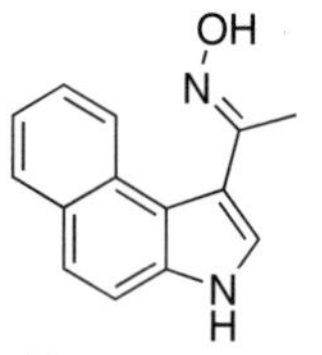 (38)

identified as binding simultaneously with and in the vicinity of (**37**). Analysis of relaxation enhancements revealed that the protons around the indole ring were closest to the spin label [221].

4.1.8. ^{19}F-Based Methods The ^{19}F nucleus has a number of useful properties that make it a valuable probe in screening studies [11].It has a particularly large chemical-shift range, making it a very sensitive probe of binding interactions. It has a large gyromagnetic ratio and a natural isotopic abundance of 100% and hence is a very sensitive to detection via NMR. ^{19}F tends to be sparsely present in naturally occurring proteins and hence there are few background signals. Fortuitously, fluorine is a favored atomic substitution by medical chemists owing to its electronic properties and hence is frequently present in pharmaceutical lead molecules. Thus ^{19}F NMR studies can readily be done on these leads, alone and in competition with nonfluorinated ligands.

Many applications of ^{19}F NMR in screening make use of relaxation phenomena. Since many compounds in screening libraries will not have fluorine present one widely used approach has been to make use of the relaxation properties of a small set of ^{19}F "spy" compounds to report on the binding of a larger set of nonfluorinated ligands via competitive displacement. This approach, developed by Dalvit and coworkers is referred to as FAXS (fluorine chemical-shift anisotropy and exchange for screening) [223].

A complementary technique that uses ^{19}F NMR for functional screening is generically referred to as FABS (fluorine atoms for biochemical screening) [224,225]. This approach requires the tagging of a substrate or cofactor of an enzyme reaction with a fluorine moiety, usually a CF_3 group to maximize sensitivity, that is used as a probe to detect the course of the reaction. Modification of the substrate via the enzyme reaction results in changes in ^{19}F chemical shifts and integration of the signals from the fluorinated substrate and product can be used to gauge the extent of inhibition of the enzyme by screened ligands. The principle of the method is illustrated in Fig. 34, with an example of its use to determine the presence of an inhibitor of HIV-1 protease in extracts from a traditional Chinese medicine [226]. FABS can in principle be used as a primary screening tool, or in the hit validation process, or in the hit to lead optimization phase of drug discovery [11].

4.2. Practical Considerations

4.2.1. Screening Approach The choice of screening method for a particular protein is determined by the characteristics of the protein target and the information that is desired from the screen. For example, the SAR-by-NMR method is suitable for small, easily expressed proteins because the NMR assignments for the target need to be known so the location of binding can be determined, and a large amount of ^{15}N-enriched protein is required. If a simple "yes/no" answer on ligand binding is wanted, then the shorter, less resource intensive ligand-observed experiments (e.g., STD, diffusion-edited, or TrNOE) may suffice.

It is also important to determine the correct NMR solvent conditions for the screening procedure. These should facilitate good solubility, with little precipitation or aggregation, and acquisition of good quality data; maintain protein structure and activity; and provide a sufficient buffering effect to allow for ligands to be added. Two methods that permit the

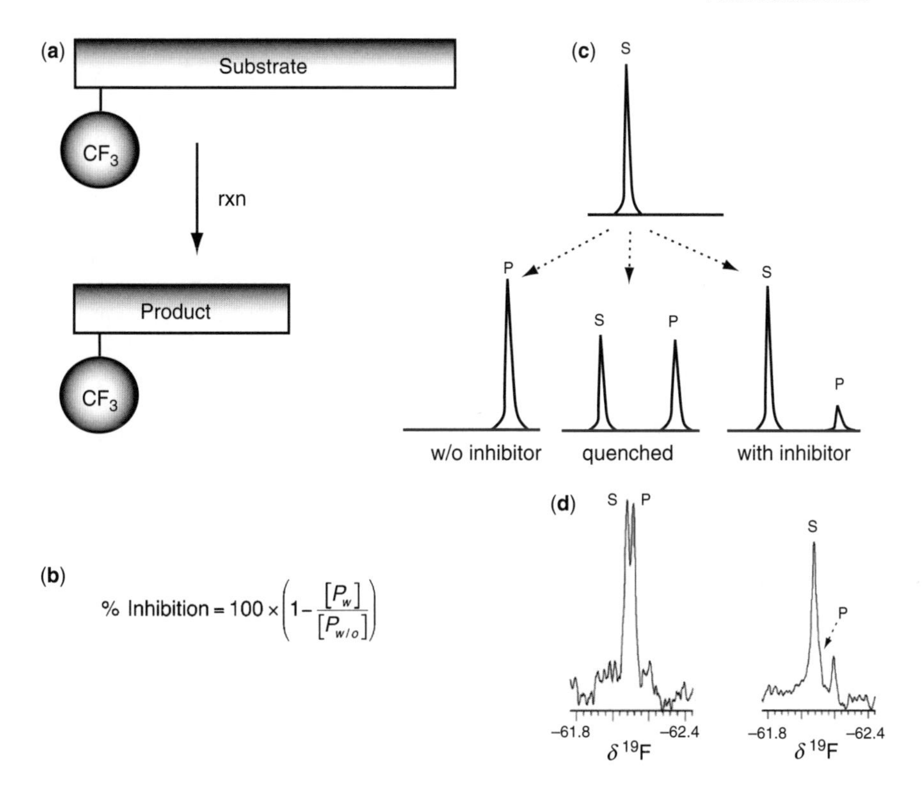

Figure 34. Schematic overview of the concept of FABS.

screening of a range of solvent conditions without the need for a large amount of protein are the microdialysis button test [227] and the microdrop-screening method [228,229].

4.2.2. Library Design Effective design and management of the ligand library to be used for screening is essential if successful results are to be obtained [16,230]. In particular, there are a number of differences in library design for HTS compared to FBDD.

Ligand Properties for HTS Diversity of ligands is an important factor to consider in the design of a HTS library for NMR screening and there are a number of factors to take into consideration. Although it would seem logical to maximize diversity, this may not always be the most efficient approach. If the system being studied exhibits neighborhood behavior, then maximizing diversity is a good option. Neighborhoods are regions of multidimensional molecular space defined by a set of molecular descriptors. By the choice of a molecule that is in the center of a neighborhood, it is possible, in theory, to represent all molecules within that molecular space. By spreading out the molecules that are selected for the library so that each neighborhood does not overlap, diversity is maximized.

However, if neighborhoods are only small then compound libraries must be very large so that the neighborhoods overlap and hence all molecular space is covered. In addition, some systems do not exhibit neighborhood behavior and relatively small changes to the structure of a compound may lead to large changes in its binding affinity for the target. Maximizing diversity can also be inefficient because many molecules do not possess physicochemical characteristics that are suitable as the basis for a drug. In practice, the more that is known about the drug target, the less diverse and more focused the library can be. However, if the library is too focused then some outlying "new" ligand type for the target being screened may be missed.

A very important aspect of library design is to select compounds that have drug-like characteristics. A simple set of rules, determined by Lipinski and coworkers, for determining whether a compound is drug-like is "the rule of five." According to this rule, the majority of orally available drugs have five or fewer hydrogen-bond donors, 10 or fewer hydrogen-bond acceptors, a $\log P$ of less than 5, and a molecular weight less than 500 [231]. Additional factors that can be taken into consideration include the number of heavy atoms, rotatable bonds, and ring systems [232–234]. Furthermore, there are a limited number of frameworks and side chains that commonly occur in many drugs. Analysis of these two structural features reveals that approximately 50% of all known drugs are represented by only 32 different frameworks, some of which are shown in Fig. 35. When atom type and bond order are included in the analysis, 41 frameworks are found to describe 24% of all drugs [235]. A similar analysis of side-chain frequency indicated that approximately 70% of all side chains present in the compound database analyzed were from the top 20 occurring side chains [236].

The presence of these common frameworks and side chains has been exploited in the SHAPES methodology [26] for NMR screening. This strategy employs a small focused library based on common frameworks and side chains to screen against protein targets through the use of relaxation and NOE experiments. The advantages of this approach are that the library is small and hence only relatively small amounts of protein are required and any hits from the library will possess drug-like characteristics. However, a disadvantage of the method is that it is unlikely to yield new drug types, given that the library is based on known drug frameworks.

Ligand Properties for FBDD Good lead fragments can have different properties to those of a good drug. In FBDD, although the affinity of any initial hit is typically later increased through synthetic manipulations, the fragment library must be of high quality to begin with to ensure that the most ideal binding epitopes are identified. These properties are desirable as they allow for favorable optimization in terms of the Lipinski criteria for a good drug.

The key properties of a fragment library include diversity, minimal structural complexity, aqueous solubility, chemically "clean" (nontoxic), and synthetically accessible. As opposed to Lipinski's "rule of five" for drug-like compounds, there is a consensus in the drug discovery community that fragments should be "rule of three" compliant [237], that is,

- Molecular weight < 300 Da
- $c\,\mathrm{Log}\,P < 3$
- H-bond donors < 3
- H-bond acceptors < 3
- Number of rotatable bonds < 3
- Total polar surface area $< 60\,\text{Å}^2$

In FBDD, it is very important to choose ligands that are synthetically accessible and/or possess suitable moieties to build upon or link to other fragments. The elaboration of chemical fragments is achieved generally either through fragment evolution, where chemical functionalities are added to the fragment systematically, or fragment linking,

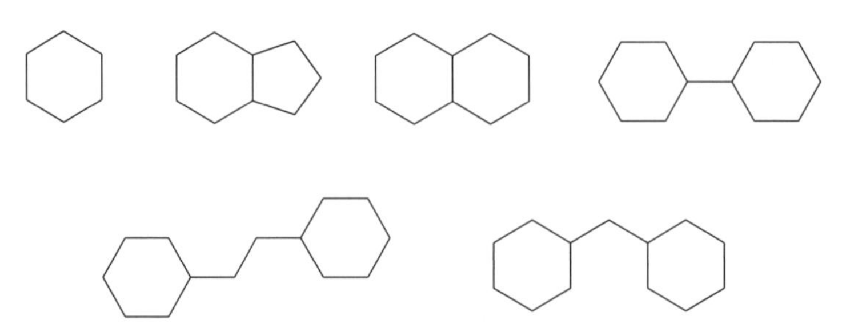

Figure 35. Examples of molecular frameworks from the SHAPES library [26].

where two or more adjacently binding hits are linked to produce a single molecule with much higher binding affinity. Alternative strategies include fragment optimization, where the fragment is reengineered to improve selectivity or efficacy, or fragment self-assembly where adjacently binding fragments with complementary reacting groups react together to form a single lead molecule at the protein surface [42].

In addition to the properties listed above, it has also been found that there is an optimal complexity for fragments that should be considered at the screening phase. Generally, the smaller the compound, the higher the hit rate for detecting an interaction. Although this might be thought of as being low selectivity, only a small portion of a ligand is responsible for the binding-site recognition and the rest of the molecule acts only to increase the overall binding energy of interaction. The relationship between molecular weight and affinity is known as the binding efficiency index and is related to the ligand efficiency.

Ligand Efficiency The concept of ligand efficiency was introduced only recently but it has already become an important consideration for the design of drugs through both the FBDD and HTS methods. Ligand efficiency (Δg) is defined as the binding energy per "heavy" (nonhydrogen) atom of a ligand hit, and is calculated by converting the K_d into the free energy of binding (Eq. 10) at 300 K;

$$\Delta G = -RT \ln K_d \qquad (10)$$

and then dividing by the number of heavy (nonhydrogen) atoms (Eq. 11) [238]:

$$\Delta g = \frac{\Delta G}{N_{\text{nonhydrogen atoms}}} \qquad (11)$$

The use of the ligand efficiency concept allows for the comparison of fragment hits from an initial screen in order to set fragment priority for further elaboration.

The binding efficiency index is calculated by dividing the percentage inhibition at a given concentration by the molecular weight, and is an alternative way of expressing the ligand efficiency. An advantage of using the binding efficiency index over ligand efficiency is that the absolute affinity (K_d) does not need to be known, as is often the case in the early stages of drug development.

Several studies have investigated how the ligand efficiency of initial fragment hits translates to success of the development of higher molecular weight inhibitors. In one study, 18 highly optimized lead compounds were deconstructed into their minimal binding elements, and the potency of the resulting compounds was compared [239]. There was a remarkably linear relationship between molecular weight and potency as the leads were optimized from the initial fragment hit, while the ligand efficiency remained constant. This relationship indicates that predictions can be made about the success of any lead optimization and any suboptimal leads can be discounted in the early stages of discovery.

Mixture Design The optimal number of compounds per mixture is dependent on the screening method. For ligand-observed experiments the limiting factor for the number of compounds in a mixture is spectral overlap. Ligands need to be chosen so that spectral overlap is minimized, making interpretation of the data manageable. In theory, protein-observed experiments could have a large number of compounds per mixture that would both minimize screening time and the requirement for large amounts of protein. However, because the experiments are protein observed then deconvolution of the mixtures and rescreening of each individual compound are required to identify hits. Therefore, the number of compounds per mixture is dependent on the hit rate in the screening procedure, given that the greater the hit rate, the more deconvolution steps required and consequently more protein and spectrometer time are needed. The number of experiments required is at a minimum when the number of compounds is equal to $1/(\text{hit rate})^{1/2}$. Thus, with a hit rate of 10% the optimal number of compounds per mixture is three [230]. In addition to these factors, if the hit rate is high then it is probable that several compounds within a mixture containing a large number of compounds may compete for the same binding pocket, which may lead to false negatives.

In mixtures of organic compounds the possibility of interactions between compounds, such as reactions or ion pairing, is also possible and should be taken into consideration, especially when using large numbers of compounds per mixture. It has been demonstrated that in random mixtures of 10 compounds in DMSO, the probability of a reaction occurring between two of the components in a mixture is 0.26 (26%). This surprisingly high value can be reduced by careful selection of mixture components (e.g., separating acids from bases) to approximately 9% [240].

4.2.3. Hardware and Automation Automation is required if libraries containing a large number of compounds are to be screened. Indeed, almost all steps of the NMR screening process, from sample preparation through to data analysis, are now automated [241]. The general setup for NMR screening consists of a robot for just-in-time preparation of each sample, which is then transferred to the magnet either through a flow system or as discrete sample tubes. There are some disadvantages in using a flow system, including the possibility of contamination of samples by previously screened compounds, the capillary line can be blocked if the protein or ligands precipitate or form aggregates, and recovery of the sample is more laborious because it has been diluted. Many of these problems can be overcome by using discrete samples, which tend to be more common.

In a medium capacity tube-based sampling set-up, approximately 50–100 samples can be screened per day and if mixtures contain 10 compounds each then libraries of several thousand compounds can be screened in a matter of days. Cryogenic probes significantly increase the signal-to-noise ratio of NMR spectra, allowing data to be obtained faster or at lower protein concentrations. For example, in one study stromelysin (50 μM) was screened against mixtures of 100 compounds (50 μM each), facilitating the screening of more than 10,000 compounds in one day [36]. The use of lower concentrations of both protein and ligands increases the stringency levels for the binding strength of ligands. At a protein/ligand concentration of 0.5 mM, ligands with dissociation constants in the millimolar range can be detected, although at a protein/ligand concentration of 50 μM the dissociation constant limit is reduced to approximately 0.15 mM. Although using higher protein/ligand concentrations can be advantageous when screening libraries containing small low affinity ligands, a higher stringency is required when screening large libraries, to reduce the number of hits obtained to a manageable number [36].

5. CONCLUSIONS

In this chapter we have given an overview of the two major approaches used in NMR and drug discovery, structure-based design and NMR-based screening. Both areas are flourishing and, together with more traditional uses of NMR, they demonstrate the versatility of NMR as a tool in medicinal chemistry and drug design. The power of NMR has been dramatically enhanced over the last decade by developments in both instruments and methodology. On the instrumental side, increases in magnetic field strengths and the development of cryoprobes have greatly increased sensitivity. Linkages of NMR to LC and MS have increased versatility. On the methods front a range of new approaches have been discovered that have enhanced the study of large molecular complexes. Advances in protein expression and labeling have played a major role in stimulating the development of new NMR methods to extract information from such complexes.

ACKNOWLEDGMENTS

Work in our laboratory on NMR in drug design and development is supported by grants from the Australian Research Council (ARC) and the National Health and Medical Research Council (NHMRC). D.J.C. is an NHMRC Professorial fellow.

GLOSSARY

Chemical Shift: A very important NMR parameter that provides information on the local electronic environment of nuclei in molecules. Chemical shifts are measured on the frequency axes of NMR spectra.

Chemical-Shift Mapping: An approach to characterizing interactions between molecules by determining which sites have their chemical shifts perturbed when a ligand binds to a macromolecule. Since chemical shifts reflect the local environment, only signals close to the binding site are perturbed by ligand binding or conformational changes.

HSQC Spectroscopy: Short for Heteronuclear Single Quantum Coherence spectroscopy. This is a powerful two-dimensional NMR method that provides a useful "fingerprint" spectrum that can be used in binding studies.

Isotope Labeling: The use of stable isotopes (e.g., ^{13}C, ^{15}N) to enhance the prevalence of NMR active nuclei in a protein to allow a sophisticated range of multinuclear NMR experiments to be carried out. Labeled proteins are easily produced using recombinant methods.

Multidimensional NMR: NMR experiments that result in spectra having multiple frequency axes so that spectral overlap is reduced. One-dimensional NMR spectra have one frequency axis (e.g., ^{1}H chemical shift) and one intensity axis, while in 2D NMR there are two frequency axes, and peak intensities are normally represented in contour form. For example, in 2D HSQC spectra the frequency axes often correspond to ^{1}H and ^{15}N chemical shifts.

NMR Screening: A generic term for a range of NMR techniques that identify ligands (usually as part of a library of compounds) that bind to a target macromolecule.

NOE (Nuclear Overhauser Effect): An NMR parameter that provides information on distances between nuclei. This distance information, in turn, can be used to determine the three-dimensional structures of molecules, particularly peptides and proteins.

Pulse Sequence: An array of radio frequency pulses that is used to perturb the nuclei in a sample in defined way, to produce a particular type of NMR spectrum. For example, different pulse sequences are used to generate NOESY or HSQC spectra.

Structure-based design: An approach to designing drugs based on knowledge of the structure of a target macromolecule and/or macromolecule–ligand complexes.

REFERENCES

1. Shuker SB, Hajduk PJ, Meadows RP, Fesik SW. Discovering high-affinity ligands for proteins: SAR by NMR. Science 1996;274: 1531–1534.
2. Orita M, Ohno K, Niimi T. Two 'Golden Ratio' indices in fragment-based drug discovery. Drug Discov Today 2009;14:321–328.
3. Skinner AL, Laurence JS. High-field solution NMR spectroscopy as a tool for assessing protein interactions with small molecule ligands. J Pharm Sci 2008;97:4670–4695.
4. Pellecchia M, Bertini I, Cowburn D, Dalvit C, Giralt E, Jahnke W, James TL, Homans SW, Kessler H, Luchinat C, Meyer B, Oschkinat H, Peng J, Schwalbe H, Siegal G. Perspectives on NMR in drug discovery: a technique comes of age. Nat Rev Drug Discov 2008;7:738–745.
5. Ohki SY, Kainosho M. Stable isotope labeling methods for protein NMR spectroscopy. Prog Nuc Mag Res Spec 2008;53:208–226.
6. Fattori D, Squarcia A, Bartoli S. Fragment-based approach to drug lead discovery: overview and advances in various techniques. Drugs R D 2008;9:217–227.
7. Zartler ER, Mo H. Practical aspects of NMR-based fragment discovery. Curr Top Med Chem 2007;7:1592–1599.
8. Middleton DA. NMR methods for characterising ligand–receptor and drug-membrane interactions in pharmaceutical research. Ann Rep NMR Spectrosc 2007;60:39–75.
9. Lepre CA. Fragment-based drug discovery using the SHAPES method. Expert Opin Drug Discov 2007;2:1555–1566.
10. Fielding L. NMR methods for the determination of protein–ligand dissociation constants. Prog Nucl Magn Reson Spectrosc 2007;51:219–242.
11. Dalvit C. Ligand- and substrate-based 19F NMR screening: principles and applications to drug discovery. Prog Nucl Magn Reson Spectrosc 2007;51:243–271.
12. Watts A. Solid-state NMR in drug design and discovery for membrane-embedded targets. Nat Rev Drug Discov 2005;4:555–568.
13. Lepre CA, Moore JM, Peng JW. Theory and applications of NMR-based screening in

pharmaceutical research. Chem Rev 2004;104: 3641–3676.

14. Pellecchia M, Sem DS, Wuthrich K. NMR in drug discovery. Nat Rev Drug Discov 2002;1:211–219.
15. Shapiro M. Applications of NMR screening in the pharmaceutical industry. Farmaco 2001; 56:141–143.
16. Peng JW, Lepre CA, Fejzo J, Abdul-Manan N, Moore JM. Nuclear magnetic resonance-based approaches for lead generation in drug discovery. Methods Enzymol. 2001;338:202–230.
17. Hicks RP. Recent advances in NMR: expanding its role in rational drug design. Curr Med Chem 2001;8:627–650.
18. Diercks T, Coles M, Kessler H. Applications of NMR in drug discovery. Curr Opin Chem Biol 2001;5:285–291.
19. Roberts GC. Applications of NMR in drug discovery. Drug Discov Today 2000;5:230–240.
20. Fry DC, Emerson SD. Applications of biomolecular NMR to drug discovery. Drug Des Discov 2000;17:13–33.
21. Craik DJ, Scanlon MJ. Pharmaceutical applications of NMR. In: Webb GA, editor. *Annual Reports on NMR Spectroscopy*. San Diego: Academic Press; 2000. p 115–173.
22. Roberts GC. NMR spectroscopy in structure-based drug design. Curr Opin Biotechnol 1999;10:42–47.
23. Moore JM. NMR screening in drug discovery. Curr Opin Biotechnol 1999;10:54–58.
24. Keifer PA. NMR tools for biotechnology. Curr Opin Biotechnol 1999;10:34–41.
25. Hajduk PJ, Meadows RP, Fesik SW. NMR-based screening in drug discovery. Q Rev Biophys 1999;32:211–240.
26. Fejzo J, Lepre CA, Peng JW, Bemis GW, Ajay Murcko MA, Moore JM. The SHAPES strategy: an NMR-based approach for lead generation in drug discovery. Chem Biol 1999;6: 755–769.
27. Stockman B. NMR spectroscopy as a tool for structure-based drug design. Prog Nucl Magn Reson 1998;33:109–151.
28. Fesik SW. NMR structure-based drug design. J Biomol NMR 1993;3:261–269.
29. Holzgrabe U, Wawer I, Diehl B. *NMR Spectroscopy in Drug Development and Analysis*. Weinheim, Germany: Wiley-VCH; 1999.
30. Craik DJ, editor. *NMR in Drug Design*. Boca Raton: CRC Press; 1996.
31. Derome AE. *Modern NMR Techniques for Chemistry Research*. New York: Pergamon Press; 1987.
32. Gunther H. *NMR Spectroscopy: An Introduction*. Chichester: John Wiley & Sons Ltd; 1980.
33. Wuthrich K. *NMR of Proteins and Nucleic Acids*. New York: Wiley; 1986.
34. Pervushin K, Riek R, Wider G, Wuthrich K. Attenuated T2 relaxation by mutual cancellation of dipole-dipole coupling and chemical shift anisotropy indicates an avenue to NMR structures of very large biological macromolecules in solution. Proc Natl Acad Sci USA 1997;94:12366–12371.
35. Tugarinov V, Muhandiram R, Ayed A, Kay LE. Four-dimensional NMR spectroscopy of a 723-residue protein: chemical shift assignments secondary structure of malate synthase. G J Am Chem Soc 2002;124:10025–10035.
36. Hajduk P, Gerfin T, Boehlen J, Haberli M, Marek D, Fesik S. High-throughput nuclear magnetic resonance-based screening. J Med Chem 1999;42:2315–2317.
37. Keifer PA. NMR spectroscopy in drug discovery: tools for combinatorial chemistry, natural products, and metabolism research. Prog Drug Res 2000;55:137–211.
38. Exarchou V, Krucker M, van Beek TA, Vervoort J, Gerothanassis IP, Albert K. LC-NMR coupling technology: recent advancements and applications in natural products analysis. Magn Reson Chem 2005;43:681–687.
39. Yang Z. Online hyphenated liquid chromatography-nuclear magnetic resonance spectroscopy-mass spectrometry for drug metabolite and nature product analysis. J Pharm Biomed Anal 2006;40:516–527.
40. Walker GS, O'Connell TN. Comparison of LC-NMR and conventional NMR for structure elucidation in drug metabolism studies. Expert Opin Drug Metab Toxicol 2008;4:1295–1305.
41. Sandvoss M, Bardsley B, Beck TL, Lee-Smith E, North SE, Moore PJ, Edwards AJ, Smith RJ. HPLC-SPE-NMR in pharmaceutical development: capabilities and applications. Magn Reson Chem 2005;43:762–770.
42. Rees DC, Congreve M, Murray CW, Carr R. Fragment-based lead discovery. Nat Rev Drug Discov 2004;3:660–672.
43. Wani MC, Taylor HL, Wall ME, Coggon P, McPhail AT. Plant antitumor agents. VI. The isolation and structure of taxol, a novel antileukemic and antitumor agent from *Taxus brevifolia*. J Am Chem Soc. 1971;93: 2325–2327.
44. Fang WS, Wang SR. Structural studies of taxol analogs for drug discovery. Expert Opin Drug Discov 2008;3:1109–1122.

45. Kingston DG. Recent advances in the chemistry of taxol. J Nat Prod 2000;63:726–734.
46. Miljanich GP. Ziconotide: neuronal calcium channel blocker for treating severe chronic pain. Curr Med Chem 2004;11:3029–3040.
47. Wishart DS, Sykes BD, Richards FM. Relationship between nuclear magnetic resonance chemical shift and protein secondary structure. J Mol Biol 1991;222:311–333.
48. Wishart DS, Bigam CG, Holm A, Hodges RS, Sykes BD. ^{1}H, ^{13}C and ^{15}N random coil NMR chemical shifts of the common amino acids. I. Investigations of nearest-neighbor effects. J Biomol NMR 1995;5:67–81.
49. Nielsen KJ, Thomas L, Lewis RJ, Alewood PF, Craik DJ. A consensus structure for omega-conotoxins with different selectivities for voltage-sensitive calcium channel subtypes: comparison of MVIIA, SVIB and SNX-202. J Mol Biol 1996;263:297–310.
50. MacLachlan LK, Middleton DA, Edwards AJ, Reid DG. A case history: NMR studies of the structure of a small protein, ω-conotoxin MVIIA. In: Reid DG, editor. *Protein NMR Techniques*. Totowa: Humana Press; 1997. p 337–362.
51. Wishart DS, Sykes BD, Richards FM. The chemical shift index: a fast and simple method for the assignment of protein secondary structure through NMR spectroscopy. Biochemistry 1992;31:1647–1651.
52. Brunger AT, Adams PD, Clore GM, DeLano WL, Gros P, Grosse-Kunstleve RW, Jiang JS, Kuszewski J, Nilges M, Pannu NS, Read RJ, Rice LM, Simonson T, Warren GL. Crystallography & NMR system: a new software suite for macromolecular structure determination. Acta Crystallogr D Biol Crystallogr 1998;54: 905–921.
53. Stein EG, Rice LM, Brunger AT. Torsion-angle molecular dynamics as a new efficient tool for NMR structure calculation. J Magn Reson 1997;124:154–164.
54. Dasgupta F, Mukherjee AK, Gangadhar N. Endothelin receptor antagonists: an overview. Curr Med Chem. 2002;9:549–575.
55. Clozel M, Brue V, Gray G, Kalina B, Loffler BM, Burri K, Cassal JM, Hirth G, Muller M, Neidhart W, Ramuz H. Pharmacological characterization of bosetan, a new potent orally-active nonpeptide endothelin receptor antagonist. J Pharmacol Exp Ther 1994;270:228–235.
56. Battistini B, Berthiaume N, Kelland NF, Webb DJ, Kohan DE. Profile of past and current clinical trials involving endothelin receptor antagonists: the novel "-sentan" class of drug. Exp Biol Med 2006;231:653–695.
57. Craik DJ, Nielsen KJ, Higgins KA. Pharmaceutical applications of NMR. In: Webb GA, editor. *Annual Reports on NMR Spectroscopy*. San Diego: Academic Press; 1995. p 143–213.
58. Krystek SR Jr, Bassolino DA, Novotny J, Chen C, Marschner TM, Andersen NH. Conformation of endothelin in aqueous ethylene glycol determined by ^{1}H-NMR and molecular dynamics simulations. FEBS Lett 1991;281:212–218.
59. Andersen NH, Chen CP, Marschner TM, Krystek SR Jr, Bassolino DA. Conformational isomerism of endothelin in acidic aqueous media: a quantitative NOESY analysis. Biochemistry 1992;31:1280–1295.
60. Saudek V, Hoflack J, Pelton JT. ^{1}H-NMR study of endothelin, sequence-specific assignment of the spectrum and a solution structure. FEBS Lett 1989;257:145–148.
61. Endo S, Inooka H, Ishibashi Y, Kitada C, Mizuta E, Fujino M. Solution conformation of endothelin determined by nuclear magnetic resonance and distance geometry. FEBS Lett 1989;257:149–154.
62. Mills RG, O'Donoghue SI, Smith R, King GF. Solution structure of endothelin-3 determined using NMR spectroscopy. Biochemistry 1992;31:5640–5645.
63. Munro S, Craik D, McConville C, Hall J, Searle M, Bicknell W, Scanlon D, Chandler C. Solution conformation of endothelin, a potent vasoconstricting bicyclic peptide. A combined use of ^{1}H NMR spectroscopy and distance geometry calculations. FEBS Lett 1991;278:9–13.
64. Reily MD, Dunbar JB Jr. The conformation of endothelin-1 in aqueous solution: NMR-derived constraints combined with distance geometry and molecular dynamics calculations. Biochem Biophys Res Commun 1991;178: 570–577.
65. Tamaoki H, Kobayashi Y, Nishimura S, Ohkubo T, Kyogoku Y, Nakajima K, Kumagaye S, Kimura T, Sakakibara S. Solution conformation of endothelin determined by means of ^{1}H-NMR spectroscopy and distance geometry calculations. Protein Eng 1991;4: 509–518.
66. Aumelas A, Chiche L, Kubo S, Chino N, Tamaoki H, Kobayashi Y. [Lys(-2)-Arg(-1)]endothelin-1 solution structure by two-dimensional ^{1}H-NMR: possible involvement of electrostatic interactions in native disulfide bridge formation and in biological activity decrease. Biochemistry 1995;34:4546–4561.

67. Aumelas A, Chiche L, Mahe E, Le-Nguyen D, Sizun P, Berthault P, Perly B. Determination of the structure of [Nle7]-endothelin by ^{1}H NMR. Int J Pept Protein Res 1991;37:315–324.

68. Dalgarno DC, Slater L, Chackalamannil S, Senior MM. Solution conformation of endothelin and point mutants by nuclear magnetic resonance spectroscopy. Int J Pept Protein Res 1992;40:515–523.

69. Boulanger Y, Biron E, Khiat A, Fournier A. Conformational analysis of biologically active truncated linear analogs of endothelin-1 using NMR and molecular modeling. J Pept Res 1999;53:214–222.

70. Arvidsson K, Nemoto T, Mitsui Y, Ohashi S, Nakanishi H. The solution structure of human endothelin-2 a ^{1}H-NMR and CD study. Eur J Biochem 1998;257:380–388.

71. Hewage CM, Jiang L, Parkinson JA, Ramage R, Sadler IH. Solution structure determination of endothelin-1 in methanol/water by NMR and molecular modelling methods. J Pept Sci 1997;3:415–428.

72. Wallace BA, Janes RW, Bassolino DA, Krystek SR Jr. A comparison of X-ray and NMR structures for human endothelin-1. Protein Sci 1995;4:75–83.

73. Iqbal J, Sanghia R, Das SK. Endothelin receptor antagonists: an overview of their synthesis and structure–activity relationship. Mini Rev Med Chem 2005;5:381–408.

74. Coles M, Sowemimo V, Scanlon D, Munro SL, Craik DJ. A conformational study by ^{1}H NMR of a cyclic pentapeptide antagonist of endothelin. J Med Chem 1993;36:2658–2665.

75. Detlefsen DJ, Hill SE, Day SH, Lee MS. Molecular flexibility profiling using NMR spectroscopy. Curr Med Chem 1999;6:353–358.

76. Patt WC, Edmunds JJ, Repine JT, Berryman KA, Reisdorph BR, Lee C, Plummer MS, Shahripour A, Haleen SJ, Keiser JA, Flynn MA, Welch KM, Reynolds EE, Rubin R, Tobias B, Hallak H, Doherty AM. Structure–activity relationships in a series of orally active gamma-hydroxy butenolide endothelin antagonists. J Med Chem 1997;40:1063–1074.

77. Patt WC, Cheng XM, Repine JT, Lee C, Reisdorph BR, Massa MA, Doherty AM, Welch KM, Bryant JW, Flynn MA, Walker DM, Schroeder RL, Haleen SJ, Keiser JA. Butenolide endothelin antagonists with improved aqueous solubility. J Med Chem 1999;42: 2162–2168.

78. Holtke C, Law MP, Wagner S, Breyholz HJ, Kopka K, Bremer C, Levkau B, Schober O, Schafers M. Synthesis, *in vitro* pharmacology and biodistribution studies of new PD 156707-derived ET(A) receptor radioligands. Bioorg Med Chem 2006;14:1910–1917.

79. Holtke C, von Wallbrunn A, Kopka K, Schober O, Heindel W, Schafers M, Bremer C. A fluorescent photoprobe for the imaging of endothelin receptors. Bioconjug Chem 2007;18: 685–694.

80. Munro SL, Andrews PR, Craik DJ, Gale DJ, ^{13}C NMR studies of the molecular flexibility of antidepressants. J Pharm Sci 1986;75:133–141.

81. Duggan BM, Craik DJ. ^{1}H and ^{13}C NMR relaxation studies of molecular dynamics of the thyroid hormones thyroxine, 3,5,3′-triiodothyronine, and 3,5-diiodothyronine. J Med Chem 1996;39:4007–4016.

82. Duggan BM, Craik DJ. Conformational dynamics of thyroid hormones by variable temperature nuclear magnetic resonance: the role of side chain rotations and cisoid/transoid interconversions. J Med Chem 1997;40:2259–2265.

83. Casarotto MG, Craik DJ. Ring flexibility within tricyclic antidepressant drugs. J Pharm Sci 2001;90:713–721.

84. Abseher R, Horstink L, Hilbers CW, Nilges M. Essential spaces defined by NMR structure ensembles and molecular dynamics simulation show significant overlap. Proteins Struc Func Genetics 1998;31:370–382.

85. Gehrmann J, Alewood PF, Craik DJ. Structure determination of the three disulfide bond isomers of alpha- conotoxin GI: a model for the role of disulfide bonds in structural stability. J Mol Biol 1998;278:401–415.

86. Balbach J, Seip S, Kessler H, Scharf M, Kashani-Poor N, Engels JW. Structure and dynamic properties of the single disulfide-deficient alpha-amylase inhibitor [C45A/C73A] tendamistat: an NMR study. Proteins 1998;33:285–294.

87. Craik DJ, Duggan BM, Munro SLA. Conformations and binding interactions of thyroid hormones and analogs. In: Choudary MI, editor. *Biological Inhibitors*. Amsterdam: Harwood Academic Publishers; 1996. p 255–302.

88. Wagner RL, Apriletti JW, McGrath ME, West BL, Baxter JD, Fletterick RJ. A structural role for hormone in the thyroid hormone receptor. Nature 1995;378:690–697.

89. Gale DJ, Craik DJ, Brownlee RTC. Variable-temperature NMR studies of thyroid hormone conformations. Magn Reson Chem 1988;26: 275–280.

90. Bourguet W, Ruff M, Chambon P, Gronemeyer H, Moras D. Crystal structure of the ligand-binding domain of the human nuclear receptor RXR-alpha. Nature 1995;375:377–382.

91. Renaud JP, Rochel N, Ruff M, Vivat V, Chambon P, Gronemeyer H, Moras D. Crystal structure of the RAR-gamma ligand-binding domain bound to all-*trans* retinoic acid. Nature 1995;378:681–689.

92. Feeney J, Birdsall B. NMR studies of prtein–ligand interactions. In: Roberts GCK, editor. *NMR of Macromolecules: A Practical Approach*. Oxford: Oxford University Press; 1993. p 181–215.

93. Feeney J, NMR studies of dynamic processes and multiple conformations in protein–ligand complexes. In: Bertini I, Molinari H, Niccolai N, editors. *NMR and Biomolecular Strucure*. New York: VCH Publishers; 1991. p 189–205.

94. Feeney J. NMR studies of interactions of ligands with dihydrofolate reductase. Biochem Pharmacol 1990;40:141–152.

95. Nielsen KJ, Adams D, Thomas L, Bond T, Alewood PF, Craik DJ, Lewis RJ. Structure–-activity relationships of omega-conotoxins MVIIA, MVIIC and 14 loop splice hybrids at N and P/Q-type calcium channels. J Mol Biol 1999;289:1405–1421.

96. Campbell AP, Sykes BD. The two-dimensional transferred nuclear Overhauser effect: theory and practice. Ann Rev Biophys Biomol Struct 1993;22:99–122.

97. Sykes BD. Determination of the conformations of bound peptides using NMR- transferred NOE techniques. Curr Opin Biotechnol 1993;4:392–396.

98. Craik DJ, Higgins KA. NMR studies of ligand macromolecule interactions. In: Webb GA, editor. *Annual Reports on NMR Spectroscopy*. London: Academic Press; 1990. p 61–138.

99. Wider G, Wuthrich K. NMR spectroscopy of large molecules and multimolecular assemblies in solution. Curr Opin Struct Biol 1999;9:594–601.

100. Lipari G, Szabo A. Model-free approach to the interpretation of nuclear magnetic-resonance relaxation in macromolecules. 1. theory and range of validity. J Am Chem Soc 1982;104: 4546–4559.

101. Craik DJ, Kumar A, Levy GC. Moldyn: a generalized program for the evaluation of molecular-dynamics models using nuclear magnetic-resonance spin-relaxation data. J Chem Inf Comput Sci 1983;23:30–38.

102. Sprangers R, Velyvis A, Kay LE. Solution NMR of supramolecular complexes: providing new insights into function. Nat Methods 2007;4:697–703.

103. Kay LE. Protein dynamics from NMR. Nat Struct Biol 1998;5:513–517.

104. Henzler-Wildman KA, Lei M, Thai V, Kerns SJ, Karplus M, Kern D. A hierarchy of timescales in protein dynamics is linked to enzyme catalysis. Nature 2007;450:913–916.

105. Bailor MH, Musselman C, Hansen AL, Gulati K, Patel DJ, Al-Hashimi HM. Characterizing the relative orientation and dynamics of RNA A-form helices using NMR residual dipolar couplings. Nat Protoc 2007;2:1536–1546.

106. Ying JF, Grishaev A, Bryce DL, Bax A. Chemical shift tensors of protonated base carbons in helical RNA and DNA from NMR relaxation and liquid crystal measurements. J Am Chem Soc 2006;128:11443–11454.

107. Tjandra N, Bax A. Direct measurement of distances and angles in biomolecules by NMR in a dilute liquid crystalline medium. Science 1997;278:1111–1114.

108. Tjandra N, Garrett AM, Gronenborn AM, Bax A, Clore GM. Defining long range order in NMR structure determination from the dependence of heteronuclear relaxation times on rotational diffusion anisotropy. Nat Struct Biol 1997;4:443–449.

109. Salzmann M, Pervushin K, Wider G, Senn H, Wuthrich K. NMR assignment and secondary structure determination of an octameric 110 kDa protein using TROSY in triple resonance experiments. J Am Chem Soc 2000;122: 7543–7548.

110. Fiaux J, Bertelsen EB, Horwich AL, Wuthrich K. NMR analysis of a 900K GroEL GroES complex. Nature 2002;418:207–211.

111. Yamazaki T, Otomo T, Oda N, Kyogoku Y, Uegaki K, Ito N, Ishino Y, Nakamura H. Segmental isotope labeling for protein NMR using peptide splicing. J Am Chem Soc 1998;120: 5591–5592.

112. Gardner KH, Kay LE. Production and incorporation of N-15, C-13, H-2 (H-1-delta 1 methyl) isoleucine into proteins for multidimensional NMR studies. Am Chem Soc 1997;119:7599–7600.

113. Goto NK, Gardner KH, Mueller GA, Willis RC, Kay LE. A robust and cost-effective method for the production of Val, Leu, Ile (delta 1) methyl-protonated ^{15}N-, ^{13}C-, ^{2}H-labeled proteins. J Biomol NMR 1999;13:369–374.

114. Rosen MK, Gardner KH, Willis RC, Parris WE, Pawson T, Kay LE. Selective methyl group protonation of perdeuterated proteins. J Mol Biol 1996;263:627–636.

115. Kainosho M, Torizawa T, Iwashita Y, Terauchi T, Ono AM, Guntert P. Optimal isotope labelling for NMR protein structure determinations. Nature 2006;440:52–57.

116. Edwards AM, Arrowsmith CH, Christendat D, Dharamsi A, Friesen JD, Greenblatt JF, Vedadi M. Protein production: feeding the crystallographers and NMR spectroscopists. Nat Struct Biol 2000;7(Suppl):970–972.

117. Christendat D, Yee A, Dharamsi A, Kluger Y, Gerstein M, Arrowsmith CH, Edwards AM. Structural proteomics: prospects for high throughput sample preparation. Prog Biophys Mol Biol 2000;73:339–345.

118. Marassi FM, Opella SJ. NMR structural studies of membrane proteins. Curr Opin Struct Biol 1998;8:640–648.

119. Watts A. NMR of drugs and ligands bound to membrane receptors. Curr Opin Biotechnol 1999;10:48–53.

120. Lian L-Y, Roberts GCK. Effects of chemical exchangeon NMR spctra. In: Roberts GCK, editor. *NMR of Macro-molecules: A Practical Approach*. Oxford: Oxford University Press; 1993.

121. Dugad L, Gerig JT. NMR-studies of carbonic anhydrase-4-fluorobenzenesulfonamide complexes. Biochemistry 1988;27:4310–4316.

122. Hyde EI, Birdsall B, Roberts GC, Feeney J, Burgen ASV. Phosphorous-31 nuclear magnetic resonance studies of the binding of oxidised coenzymes to *Lactobacillus casei* dihydrofolate reductase. Biochemistry 1980;19:3746–3754.

123. Feeney J, Batchelor JG, Albrand JP, Roberts GCK. The effects of intermedate exchange processes on the estimation of equilibrium constants by NMR. J Magn Reson 1979;33: 519–529.

124. Pavlopoulos S, Rose M, Wickham G, Craik DJ. A ^{1}H NMR analysis of the interaction between terephthalamide derivatives and the oligonucleotide duplex d(GGTAATTACC)2. Anticancer Drug Des 1995;10:623–639.

125. Pelton GJ, Wemmer DE. Structural characterization of a 2:1 distamycin A d(CGCAAATTGGC)2 determined by two-dimensional NMR. Proc Natl Acad Sci USA 1990;86:5723–5727.

126. Klevit RE, Wemmer DE, Reid BR, ^{1}H NMR studies on the interaction between distamycin A and a symmetrical DNA dodecamer. Biochemistry 1986;25:3296–3303.

127. Patel DJ, Shapiro L. Sequence-dependent recognition of DNA duplexes. Netropsin complexation to the AATT site of the d(G-G-A-A-T-T-C-C) duplex in aqueous solution. J Biol Chem 1986;261:1230–1240.

128. Searle MS, Embrey KJ. Sequence-specific interaction of Hoechst 33258 with the minor groove of an adenine-tract DNA duplex studied in solution by ^{1}H NMR spectroscopy. Nucleic Acids Res 1990;18:3753–3762.

129. Fesik SW, Luly JR, Erickson JW, Abad-Zapatero C. Isotope-edited proton NMR study on the structure of a pepsin/inhibitor complex. Biochemistry 1988;27:8297–8301.

130. Zwahlen C, Legault P, Vincent SJF, Greenblatt J, Konrat R, Kay LE. Methods for measurement of intermolecular NOEs by multinuclear NMR spectroscopy: application to a bacteriophage lambda *N*-peptide/boxB RNA complex. J Am Chem Soc 1997;119:6711–6721.

131. Gradwell MJ, Feeney J. Validation of the use of intermolecular NOE constraints for obtaining docked structures of protein–ligand complexes. J Biomol NMR 1996;7:48–58.

132. Tang C, Clore GM. A simple and reliable approach to docking protein–protein complexes from very sparse NOE-derived intermolecular distance restraints. J Biomol NMR 2006;36: 37–44.

133. Harshman KD, Dervan PB. Molecular recognition of B-DNA by Hoechst 33258. Nucleic Acids Res 1985;13:4825–4835.

134. Pelton JG, Wemmer DE. Structural modeling of the distamycin A-d(CGCGAATTCGCG)2 complex using 2D NMR and molecular mechanics. Biochemistry 1988;27:8088–8096.

135. Pelton JG, Wemmer DE. Structural characterization of a 2:1 distamycin A.d (CGCAAATTGGC) complex by two-dimensional NMR. Proc Natl Acad Sci USA 1989; 86:5723–5727.

136. Coll M, Aymami J, van der Marel GA, van Boom JH, Rich A, Wang AH. Molecular structure of the netropsin-d(CGCGATATCGCG) complex: DNA conformation in an alternating AT segment. Biochemistry 1989;28:310–320.

137. Pjura PE, Grzeskowiak K, Dickerson RE. Binding of Hoechst 33258 to the minor groove of B-DNA. J Mol Biol 1987;197:257–271.

138. Teng MK, Usman N, Frederick CA, Wang AH. The molecular structure of the complex of Hoechst 33258 and the DNA dodecamer d

(CGCGAATTCGCG). Nucleic Acids Res 1988;16:2671–2690.

139. Carrondo MA, Coll M, Aymami J, Wang AH, van der Marel GA, van Boom JH, Rich A. Binding of a Hoechst dye to d(CGCGA-TATCGCG) and its influence on the conformation of the DNA fragment. Biochemistry 1989;28:7849–7859.

140. Quintana JR, Lipanov AA, Dickerson RE. Low-temperature crystallographic analyses of the binding of Hoechst 33258 to the double-helical DNA dodecamer C-G-C-G-A-A-T-T-C-G-C-G. Biochemistry 1991;30:10294–10306.

141. Parkinson JA, Barber J, Douglas KT, Rosamond J, Sharples D. Minor-groove recognition of the self-complementary duplex d (CGCGAATTCGCG)2 by Hoechst 33258: a high-field NMR study. Biochemistry 1990;29: 10181–10190.

142. Embrey KJ, Searle MS, Craik DJ. Interaction of Hoechst 33258 with the minor groove of the A + T-rich DNA duplex d(GGTAATTACC)2 studied in solution by NMR spectroscopy. Eur J Biochem 1993;211:437–447.

143. Fede A, Labhardt A, Bannwarth W, Leupin W. Dynamics and binding mode of Hoechst 33258 to d(GTGGAATTCCAC)2 in the 1:1 solution complex as determined by two-dimensional ^{1}H NMR. Biochemistry 1991;30:11377–11388.

144. Fede A, Billeter M, Leupin W, Wuthrich K. Determination of the NMR solution structure of the Hoechst 33258-d(GTGGAATTCCAC)2 complex and comparison with the X-ray crystal structure. Structure 1993;1:177–186.

145. Taga T, Tanaka H, Goto T, Tada S. Structure of a new macrocyclic antibiotic. Acta Cryst 1987; C43:751–753.

146. Bierer BE, Somers PK, Wandless TJ, Burakoff SJ, Schreiber SL. Probing immunosuppressant action with a non-natural immunophilin ligand. Science 1990;250:556–559.

147. Karuso P, Kessler H, Mierke DF. Solution structure of FK506 from nuclear magnetic resonance and molecular dynamics. J Am Chem Soc 1990;112:9434–9437.

148. Fesik SW, Gampe RT Jr, Holzman TF, Egan DA, Edalji R, Luly JR, Simmer R, Helfrich R, Klahore V, Rich DH. Isotope-edited NMR of cyclosporin A bound to cycloshilin: evidence for a *trans* 9,10-amide bond. Science 1990;250: 1406–1409.

149. Fesik SW, Gampe RT Jr, Eaton HL, Gemmecker G, Olejniczak ET, Neri P, Holzman TF, Egan DA, Edalji R, Simmer R, Helfrich R, Hochlowski J, Jackson M. NMR studies of [U-^{13}C]cyclosporin A bound to cyclophilin: bound conformation and portions of cyclosporin involved in binding. Biochemistry 1991;31:6574–6583.

150. Weber C, Wider G, von Freyberg K, Truber R, Braun W, Widner H, Wuetrich K. The NMR structure of cyclosporin A bound to cyclophilin in aqueous solution. Biochemistry 1991;30: 6564–6574.

151. Rosen MK, Standaert RF, Galat A, Nakatsuka M, Schreiber SL. Inhibition of FKBP rotamase activity by immunosuppressant FK506: twisted amide surrogate. Science 1990;248:863–866.

152. Wandless TJ, Michnick SW, Rosen MK, Karplus M, Schreiber SL. FK506 and rapamycin binding to FKBP: common elements in immunophilin-ligand complexation. J Am Chem Soc 1991;113:2339–2341.

153. Petros AM, Gampe RT Jr, Gemmecker G, Neri P, Holzman TF, Edalji R, Hochlowski J, Jackson M, McAlpine J, Luly JR, et al. NMR studies of an FK-506 analog, [U-^{13}C]ascomycin, bound to FKBP: conformation and regions of ascomycin involved in binding. J Med Chem 1991;34:2925–2928.

154. van Duyne GD, Standaert RF, Karplus M, Schreiber SL, Clardy J. Atomic structure of FKBP-FK506, an immunophilin–immunosuppressant complex. Science 1991;252:839–842.

155. Michnick SW, Rosen MK, Wandless TJ, Karplus M, Schreiber SL. Solution structure of FKBP, a rotamase enzyme and receptor for FK506 and rapamycin. Science 1991;252: 836–839.

156. Moore JM, Peattie DA, Fitzgibbon MJ, Thomson JA. Solution structure of the major binding protein for the immunosuppressant FK506. Nature 1991;351:248–250.

157. Meadows RP, Nettesheim DG, Xu RX, Olejniczak ET, Petros AM, Holzman TF, Severin J, Gubbins E, Smith H, Fesik SW. Three dimensional structure of the FK506 binding protein/ ascomycin complex in solution by heteronuclear three- and four-dimensional NMR. Biochemistry 1993;32:754–765.

158. Cheng JW, Lepre CA, Chambers SP, Fulghum JR, Thomson JA, Moore JM. ^{15}N NMR relaxation studies of the FK506 binding protein: backbone dynamics of the uncomplexed receptor. Biochemistry 1993;32: 9000–9010.

159. Cheng JW, Lepre CA, Moore JM. ^{15}N NMR relaxation studies of the FK506 binding protein: dynamic effects of ligand binding and implications for calcineurin recognition. Biochemistry 1994;33:4093–4100.

160. Gooley PR, Johnson BA, Marcy AI, Cuca GC, Salowe SP, Hagmann WK, Esser CK, Springer JP. Secondary structure and zinc ligation of human recombinant short-form stromelysin by multidimensional heteronuclear NMR. Biochemistry 1993;32:13098–13108.
161. Gooley PR, O'Connell JF, Marcy AI, Cuca GC, Salowe SP, Bush BL, Hermes JD, Esser CK, Hagmann WK, Springer JP, Johnson BA. The NMR structure of the inhibited catalytic domain of human stromelysin-1. Nat Struct Biol 1994;1:111–118.
162. Gooley PR, O'Connell JF, Marcy AI, Cuca GC, Axel MG, Caldwell CG, Hagmann WK, Becker JW. Comparison of the structure of human recombinant short form stromelysin by multidimensional heteronuclear NMR and X-ray crystallography. J Biomol NMR 1996;7:8–28.
163. Van Doren SR, Kurochkin AV, Ye Q-Z, Johnson LL, Hupe DJ, Zuiderweg ERP. Assignments for the main-chain nuclear magnetic resonances and delineation of the secondary structure of the catalytic domain of human stromelysin-1 as obtained from triple-resonance 3D NMR experiments. Biochemistry 1993;32:13109–13122.
164. Van Doren SR, Kurochkin AV, Hu W, Ye Q-Z, Johnson LL, Hupe DJ, Zuiderweg ER. Solution structure of the catalytic domain of human stromelysin complexed with a hydrophobic inhibitor. Prot Sci 1995;4:2487–2498.
165. McCoy MA, Dellwo MJ, Schneider DM, Banks TM, Falvo J, Vavra KJ, Mathiowetz AM, Qoronfleh MW, Ciccarelli R, Cook ER, Pulvino TA, Wahl RC, Wang H, Assignments and structure determination of the catalytic domain of human fibroblast collagenase using 3D double and triple resonance NMR spectroscopy. J Biomol NMR 1997;9:11–24.
166. Moy FJ, Pisano MR, Chanda PK, Urbano C, Killar LM, Sung ML, Powers R. Assignments, secondary structure and dynamics of the inhibitor-free catalytic fragment of human fibroblast collagenase. J Biomol NMR 1997;10:9–19.
167. Jacobsen EJ, Mitchell MA, Hendges SK, Belonga KL, Skaletzky LL, Stelzer LS, Lindberg TJ, Fritzen EL, Schostarez HJ, O'Sullivan TJ, Maggiora LL, Stuchly CW, Laborde AL, Kubicek MF, Poorman RA, Beck JM, Miller HR, Petzold GL, Scott PS, Truesdell SE, Wallace TL, Wilks JW, Fisher C, Goodman LV, Kaytes PS,et al. Synthesis of a series of stromelysin-selective thiadiazole urea matrix metalloproteinase inhibitors. J Med Chem 1999;42: 1525–1536.
168. Gonnella NC, Bohacek R, Zhang X, Kolossvary I, Paris CG, Melton R, Winter C, Hu SI, Ganu V, Bioactive conformation of stromelysin inhibitors determined by transferred nuclear Overhauser effects. Proc Natl Acad Sci USA 1995;92:462–466.
169. Polshakov VI, Birdsall B, Frenkiel TA, Gargaro AR, Feeney J. Structure and dynamics in solution of the complex of *Lactobacillus casei* dihydrofolate reductase with the new lipophilic antifolate drug trimetrexate. Protein Sci 1999;8:467–481.
170. Nieto PM, Birdsall B, Morgan WD, Frenkiel TA, Gargaro AR, Feeney J. Correlated bond rotations in interactions of arginine residues with ligand carboxylate groups in protein ligand complexes. FEBS Lett 1997;405:16–20.
171. Wlodawer A, Miller M, Jaskolski M, Sathyanarayana BK, Baldwin E, Weber IT, Selk LM, Clawson L, Schneider J, Kent SB. Conserved folding in retroviral proteases: crystal structure of a synthetic HIV-1 protease. Science 1989;245:616–621.
172. Wlodawer A, Erickson JW. Structure-based inhibitors of HIV-1 protease. Annu Rev Biochem 1993;62:543–585.
173. Kempf DJ, Sham HL. HIV Protease Inhibitors. Curr Pharm Des 1996;2:225–246.
174. Wlodawer A, Vondrasek J. Inhibitors of HIV-1 protease: a major success of structure-assisted drug design. Annu Rev Biophys Biomol Struct 1998;27:249–284.
175. Yamazaki T, Hinck AP, Wang YX, Nicholson LK, Torchia DA, Wingfield P, Stahl SJ, Kaufman JD, Chang CH, Domaille PJ, Lam PY. Three-dimensional solution structure of the HIV-1 protease complexed with DMP323, a novel cyclic urea-type inhibitor, determined by nuclear magnetic resonance spectroscopy. Protein Sci 1996;5:495–506.
176. Lam PY, Jadhav PK, Eyermann CJ, Hodge CN, Ru Y, Bacheler LT, Meek JL, Otto MJ, Rayner MM, Wong YN, et al. Rational design of potent, bioavailable, nonpeptide cyclic ureas as HIV protease inhibitors. Science 1994;263:380–384.
177. Grzesiek S, Bax A, Nicholson LK, Yamazaki T, Wingfield PT, Stahl SJ, Eyermann CJ, Torchia DA, Hodge CN, Lam PY, Jadhav PK, Chang CH. NMR evidence for the displacement of a conserved interior water molecule in HIV protease by a non-peptide cyclic urea-based inhibitor. J Am Chem Soc 1994;116:1581–1583.
178. Wang YX, Freedberg DI, Grzesiek S, Torchia DA, Wingfield PT, Kaufman JD, Stahl SJ, Chang CH, Hodge CN. Mapping hydration

water molecules in the HIV-1 protease/ DMP323 complex in solution by NMR spectroscopy. Biochemistry 1996;35:12694–12704.

179. Otting G, Liepinsh E, Wuthrich K. Protein hydration in aqueous solution. Science 1991;254:974.

180. Yamazaki T, Nicholson LK, Wingfield P, Stahl SJ, Kaufman JD, Domaille PJ, Torchia DA. NMR and X-ray evidence that the HIV protease catalytic aspartyl groups are protonated in the complex formed by the protease and a non-peptide cyclic urea-based inhibitor. J Am Chem Soc 1994;116:10791–10792.

181. Baldwin ET, Bhat TN, Gulnik S, Liu B, Topol IA, Kiso Y, Mimoto T, Mitsuya H, Erickson JW. Structure of HIV-1 protease with KNI-272, a tight-binding transition-state analog containing allophenylnorstatine. Structure 1995;3: 581–590.

182. Nicholson LK, Yamazaki T, Torchia DA, Grzesiek S, Bax A, Stahl SJ, Kaufman JD, Wingfield PT, Lam PY, Jadhav PK, et al. Flexibility and function in HIV-1 protease. Nat Struct Biol 1995;2:274–280.

183. Rose JR, Salto R, Craik CS. Regulation of autoproteolysis of the HIV-1 and HIV-2 proteases with engineered amino acid substitutions. J Biol Chem 1993;268:11939–11945.

184. Freedberg DI, Wang YX, Stahl SJ, Kaufman JD, Wingfield P, Kiso Y, Torchia DA. Flexibility and function in HIV protease: dynamics of the HIV-1 protease bound to the asymmetric inhibitor kynostatin 272 (KNI-272). J Am Chem Soc 1998;120:7916–7923.

185. Katoh E, Yamazaki T, Kiso Y, Wingfield P, Stahl SJ, Kaufman JD, Torchia DA. Determination of the rate of monomer interchange in a ligand-bound homodimeric protein from NOESY cross peaks: application to the HIV protease/KNI-529 complex. J Am Chem Soc 1999;121:2607–2608.

186. Winslow DL, Stack S, King R, Scarnati H, Bincsik A, Otto MJ. Limited sequence diversity of the HIV type 1 protease gene from clinical isolates and *in vitro* susceptibility to HIV protease inhibitors. AIDS Res Hum Retroviruses 1995;11:107–113.

187. Condra JH, Schleif WA, Blahy OM, Gabryelski LJ, Graham DJ, Quintero JC, Rhodes A, Robbins HL, Roth E, Shivaprakash M, et al. *In vivo* emergence of HIV-1 variants resistant to multiple protease inhibitors. Nature 1995;374:569–571.

188. Fernandes PB. Technological advances in high-throughput screening. Curr Opin Chem Biol 1998;2:597–603.

189. Mercier KA, Powers R, Determining the optimal size of small molecule mixtures for high throughput NMR screening. J Biomol NMR 2005;31:243–258.

190. Hajduk PJ, Greer J. A decade of fragment-based drug design: strategic advances and lessons learned. Nat Rev Drug Discov 2007;6:211–219.

191. Hajduk PJ, Meadows RP, Fesik SW. Discovering high-affinity ligands for proteins. Science 1997;278:497–499.

192. Olejniczak E, Hajduk P, Marcotte P, Nettesheim D, Meadows R, Edalji R, Holzman T, Fesik S. Stromelysin inhibitors designed from weakly bound fragments: effects of linking and cooperativity. J Am Chem Soc 1997;119:5828–5832.

193. Hajduk P, Dinges J, Schkeryantz J, Janowick D, Kaminski M, Tufano M, Augeri D, Petros A, Nienbar V, Zhong P, Hammond R, Coen M, Beutel B, Katz L, Fesik S. Novel inhibtors of Erm methyltransferases from NMR and parallel synthesis. J Med Chem 1999;42: 3852–3859.

194. Hajduk P, Sheppard G, Nettesheim D, Olejniczak E, Shuker S, Meadows R, Steinman D, Carrera G, Marcotte P, Severin J, Walter K, Smith H, Gubbins E, Simmer R, Holzman T, Morgan D, Davidsen S, Summers J, Fesik S. Discovery of potent nonpeptide inhibitors of stromelysin using SAR by NMR. J Am Chem Soc 1997;119:5818–5827.

195. Hajduk P, Zhou M, Fesik S. NMR-based discovery of phosphotyrosine mimetics that bind to the LCK SH2 domain. Bioorg Med Chem Lett 1999;9:2403–2406.

196. Liu G, Huth JR, Olejniczak ET, Mendoza R, DeVries P, Leitza S, Reilly EB, Okasinski GF, Fesik SW, von Geldern TW. Novel p-arylthio cinnamides as antagonists of leukocyte function- associated antigen-1/intracellular adhesion molecule-1 interaction. 2. Mechanism of inhibition and structure-based improvement of pharmaceutical properties. J Med Chem 2001;44:1202–1210.

197. Mayer M, Meyer B. Characterization of ligand binding by saturation transfer difference NMR spectroscopy. Angew Chem Int Ed 1999;38:1784–1788.

198. Klein J, Meinecke R, Mayer M, Meyer B. Detecting binding affinity to immobilized receptor proteins in compound libraries by HR-MAS STD NMR. J Am Chem Soc 1999;121:5336–5337.

199. Hellebrandt W, Haselhorst T, Köhli T, Bäuml E, Peters T. Deuterated disaccharides for the investigation of protein-carbohydrate interac-

tions: application of bio-affinity- and STD-NMR. J Carbo Chem 2000;19:769–782.

200. Vogtherr M, Peters T. Application of NMR based binding assays to identify key hydroxy groups for intermolecular recognition. J Am Chem Soc 2000;122:6093–6099.
201. Maaheimo H, Kosma P, Brade L, Brade H, Peters T. Mapping the binding of synthetic disaccharides representing epitopes of chlamydial lipopolysaccharide to antibodies with NMR. Biochemistry 2000;39:12778–12788.
202. Mayer M, Meyer B. Group epitope mapping by saturation transfer difference NMR to identify segments of a ligand in direct contact with a protein receptor. J Am Chem Soc 2001;123: 6108–6117.
203. Meinecke R, Meyer B. Determination of the binding specificity of an integral membrane protein by saturation transfer difference NMR: RGD peptide ligands binding to integrin alpha (IIb)beta(3). J Med Chem 2001;44:3059–3065.
204. Haselhorst T, Fleming FE, Dyason JC, Hartnell RD, Yu X, Holloway G, Santegoets K, Kiefel MJ, Blanchard H, Coulson BS, von Itzstein M. Sialic acid dependence in rotavirus host cell invasion. Nat Chem Biol 2009;5: 91–93.
205. Dalvit C, Pevarello P, Tato M, Veronesi M, Vulpetti A, Sundstrom M, Identification of compounds with binding affinity to proteins via magnetization transfer from bulk water. J Biomol NMR 2000;18:65–68.
206. Hurd RE. Gradient-enhanced spectroscopy. J Magn Reson 1990;87:422–428.
207. Gonnella N, Lin M, Shapiro MJ, Wareing JR, Zhang X. Isotope-filtered affinity. NMR J Magn Reson 1998;131:336–338.
208. Chen A, Shapiro MJ. NOE Pumping: a novel NMR technique for identification of compounds with binding affinity to macromolecules. J Am Chem Soc 1998;120:10258–10259.
209. Chen A, Johnson CS Jr, Lin M, Shapiro MJ. Chemical exchange in diffusion NMR experiments. J Am Chem Soc 1998;120:9094–9095.
210. Chen A, Shapiro MJ. Nuclear Overhauser effect on diffusion measurements. J Am Chem Soc 1999;121:5338–5339.
211. Chen A, Shapiro MJ. Affinity NMR. Anal Chem 1999;71:669A–675A.
212. Lin M, Shapiro MJ. Mixture analysis in combinatorial chemistry. Applications of diffusion-resolved NMR spectroscopy. J Org Chem 1996;61:7617–7619.
213. Lin M, Shapiro MJ, Wareing JR. Diffusion-edited NMR-affinity NMR for direct observation of molecular interactions. J Am Chem Soc 1997;119:5249–5250.
214. Lin M, Shapiro MJ, Wareing JR. Screening mixtures by affinity NMR. J Org Chem 1997;62:8930–8931.
215. Bleicher K, Lin M, Shapiro MJ, Wareing JR. Diffusion edited NMR: screening compound mixtures by affinity NMR to detect binding ligands to Vancomycin. J Org Chem 1998;63: 8486–8490.
216. Hajduk PJ, Olejniczak ET, Fesik SW. One-dimensional relaxation- and diffusion-edited NMR methods for screening compounds that bind to macromolecules. J Am Chem Soc 1997;119:12257–12261.
217. Henrichson D, Ernst B, Magnani JL, Wang W, Meyer B, Peters T. Bioaffinity NMR spectroscopy: identification of an E-Sepectin antagonist in a substance mixture by transfer NOE. Angew Chem Int Ed 1999;38:98–102.
218. Meyer B, Weimar T, Peters T. Screening mixtures for biological activity by NMR. Eur J Biochem 1997;246:705–709.
219. Mayer M, Meyer B. Mapping the active site of angiotensin-converting enzyme by transferred NOE spectroscopy. J Med Chem 2000;43: 2093–2099.
220. Chen A, Shapiro MJ. NOE pumping. 2. A high-throughput method to determine compounds with binding affinity to macromolecules by NMR. J Am Chem Soc 2000;122:414–415.
221. Sanchez-Pedregal VM, Reese M, Meiter J, Blommers MJ, Griesinger C, Carlomango T. The INPHARMA Method: Protein-mediated interligand NOEs for pharmacophore mapping. Angew Chem Int Ed 2005;44: 4172–4175.
222. Jahnke W. Spin labels as a tool to identify and characterize protein–ligand interactions by NMR spectroscopy. Chembiochem 2002;3: 167–173.
223. Dalvit C, Fagerness PE, Hadden DTA, Sarver RW, Stockman BJ. Fluorine-NMR experiments for high-throughput screening: theoretical aspects, practical considerations, and range of applicability. J Am Chem Soc 2003; 125:7696–7703.
224. Dalvit C, Ardini E, Flocco M, Fogliatto GP, Mongelli N, Veronesi M. A general NMR method for rapid, efficient, and reliable biochemical screening. J Am Chem Soc 2003;125: 14620–14625.
225. Dalvit C, Ardini E, Fogliatto GP, Mongelli N, Veronesi M. Reliable high-throughput functional screening with 3-FABS. Drug Discov Today 2004;9:595–602.

226. Frutos S, Tarrago T, Giralt E. A fast and robust F-19 NMR-based method for finding new HIV-1 protease inhibitors. Bioorg Med Chem Lett 2006;16:2677–2681.

227. Bagby S, Tong KI, Liu D, Alattia JR, Ikura M. The button test: a small scale method using microdialysis cells for assessing protein solubility at concentrations suitable for NMR. J Biomol NMR 1997;10:279–282.

228. Lepre C, Moore J. Microdrop screening: a rapid method to optimize solvent conditions for NMR spectroscopy of proteins. J Biomol NMR. 1998;12:493–499.

229. Bagby S, Tong KI, Ikura M. Optimization of protein solubility and stability for protein nuclear magnetic resonance. Methods Enzymol 2001;339:20–41.

230. Lepre CA. Library design for NMR-based screening. Drug Discov Today 2001;6:133–140.

231. Lipinski CA, Lombardo F, Dominy BW, Feeney PJ. Experimental and computational approaches to estimate solubility and permeability in drug discovery and development settings. Adv Drug Deliv Rev 2001;46:3–26.

232. Ghose AK, Viswanadhan VN, Wendoloski JJ. A knowledge-based approach in designing combinatorial or medicinal chemistry libraries for drug discovery. 1. A qualitative and quantitative characterization of known drug databases. J Comb Chem 1999;1:55–68.

233. Oprea TI, Gottfries J, Sherbukhin V, Svensson P, Kuhler TC. Chemical information management in drug discovery: optimizing the computational and combinatorial chemistry interfaces. J Mol Graph Model 2000;18,512–524, 541.

234. Xu J, Stevenson J. Drug-like index: a new approach to measure drug-like compounds and their diversity. J Chem Inf Comput Sci 2000;40:1177–1187.

235. Bemis G, Murcko M. The properties of known drugs. 1. Molecular frameworks. J Med Chem 1996;39:2887–2893.

236. Bemis G, Murcko M. Properties of known drugs. 2. Side chains. J Med Chem 1999;42: 5095–5099.

237. Siegal G, Ab E, Schultz J. Integration of fragment screening and library design. Drug Discov Today 2007;12:1032–1039.

238. Hopkins AL, Groom CR, Alex A. Ligand efficiency: a useful metric for lead selection. Drug Discov Today 2004;9:430–431.

239. Hajduk PJ. Fragment-based drug design: how big is too big? J Med Chem 2006;49:6972–6976.

240. Hann M, Hudson B, Lewell X, Lifely R, Miller L, Ramsden N. Strategic pooling of compounds for high-throughput screening. J Chem Inf Comput Sci 1999;39:897–902.

241. Ross A, Senn H. Automation of measurements and data evaluation in biomolecular NMR screening. Drug Discov Today 2001;6: 583–593.

DRUG TRANSPORT AND MEMBRANE TRANSPORT PROTEINS

Peter W. Swaan
Department of Pharmaceutical Sciences, University of Maryland, Baltimore, MD

1. INTRODUCTION

To exert their desired pharmacological activity, drugs must reach their sites of action with certain minimal effective concentration. With the exception of a few drug classes, such as general anesthetics and osmotic diuretics, most therapeutic agents produce their effects by acting on specific membrane proteins or intracellular enzymes. To gain access to these cellular targets, drugs must first reach the systemic circulation by penetrating the epithelial barriers covering the absorptive surfaces of the body, such as skin, intestine, lung, and so on.

In most biological epithelia, drug molecules confront two obstacles in reaching the systemic circulation: [1] a biochemical barrier resulting from enzymatic degradation and [2] a physical barrier originating from the lipid bilayer. The first obstacle can be overcome by changing the route of administration or drug formulation, for example, by encapsulating drugs in vehicles impenetrable to metabolic enzymes. However, for hydrophilic drugs and drugs with high molecular weight, especially macromolecules, epithelial membranes still impose a formidable barrier to drug entry.

Epithelia may vary in thickness or functions among different tissues, but the general transepithelial transport mechanisms for drug molecules are similar (Fig. 1). Based on the route that drug molecules penetrate, epithelial transport can be classified into two pathways: paracellular and transcellular. In paracellular transport molecules move across the epithelium via the intercellular junctions *between* adjacent cells, whereas molecules cross the epithelium *through* the cells when they use the transcellular pathway. Depending on the nature of the driving force, transcellular transport can be further categorized into passive diffusion, receptor- and active carrier-mediated translocation. In passive diffusion, movement of drug molecules is derived by its concentration gradient. In active carrier-mediated transport, a membrane-inbedded transport protein transports molecules against a concentration gradient using an energy supply provided by either ATP hydrolysis or cotransport of ions moving down their concentration gradient, often Na^+ or H^+. Originally thought to play only a minor role in the overall drug absorption process, transport proteins have been appreciated recently to be involved in all aspects of drug absorption, distribution, and elimination. The significance of their role in drug transport will be discussed later in this chapter.

2. STRATEGIES TO ENHANCE DRUG PERMEABILITY

The permeability of drugs that are poorly absorbed because of their hydrophilic character can be influenced either by manipulating the drug or the membrane. Therefore, most strategies for absorption enhancement either change the permeability properties of the epithelial cells or alter the physicochemical properties of the compound itself. Currently, two general strategies can be employed to increase the transport of a solute across an epithelial membrane:

(1) opening of the tight junctions or changing the epithelial lipid bilayer membrane (penetration enhancement);
(2) oral lipid-based formulations;
(3) design of prodrugs with increased membrane permeability either by
  - lipophilization;
  - targeting to a carrier-mediated transport system.

Changing the permeability of the membrane by penetration enhancers is an aspecific approach that has met limited success in the drug development process and will only be discussed briefly here.

2.1. Penetration Enhancers

Permeation-enhancing agents including compounds such as surfactants, bile salts, salicy-

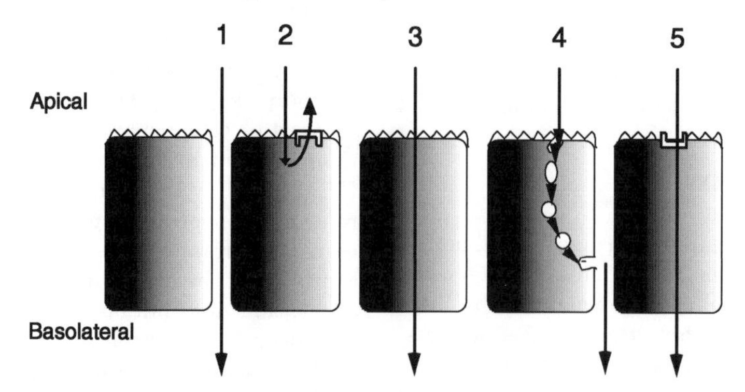

Figure 1. Routes and mechanisms of solute transport across epithelial membranes. In general, routes 2–5 are transcellular pathways, that is, compounds move *through* the cells, whereas route 1 is considered a paracellular pathway, that is, a compound moves *between* the cells. (**1**) Tight junctional pathway; (**2**) drug efflux pathway, for example, P-glycoprotein-mediated; (**3**) passive diffusion; (**4**) receptor-mediated endocytosis and/or transcytosis pathways; (**5**) carrier-mediated route. Please note that receptor and carrier proteins in epithelial cells are expressed on both the apical and basolateral surfaces.

lates, chelating agents (e.g., EDTA), or short-chained fatty acids, have been used to improve transport of poorly absorbed drugs. Most agents enhance uptake of drugs by compromising the integrity of the cell membrane. Generally, the increased absorption results from either disrupting the tight junctions or altering the membrane fluidity or both. The mechanism behind the increased permeability observed for various compounds after the coadministration of salicylates is thought to be an opening of the tight junctions. The aspecific nature of this opening of the tight junction could possibly lead to severe side effects when applied *in vivo*. The intestinal barrier becomes permeable not only for the drugs studied, but it can also become more permeable for toxic xenobiotics or even antigens, the latter leading to severe immunological side reactions. Furthermore, morphological studies indicate that salicylates cause epithelial cell damage and widen both the intercellular junctional spaces and the pores of the epithelial cell [1]. Although the basement membrane and subepithelial structures are not damaged, it is not yet completely understood whether this tissue damage is reversible or not.

Notwithstanding its obvious disadvantages, the opening of tight junctions between enterocytes by chemical modifiers is still a promising technique to increase epithelial permeability of macromolecules, including insulin. Despite a surge in research in the late 1980s and early 1990s, this approach has found rather limited clinical application due to the nonspecific behavior of many tight junctional perturbants. A greater understanding of the factors that govern junction control, has led to the discovery and development of more specific and potent permeation enhancers.

Vibrio cholera, which infects the intestinal tract and causes severe diarrhea, produces a protein known as the zonula occludens toxin (ZOT), which is able to increase the permeability of tight junctions. ZOT specifically targets the actin filaments associated with the tight junction without compromising the overall intestinal integrity or function. Interestingly, ZOT protein only seems to be effective at receptors in the jejunum and ileum but not the colon. As a result, this regulation of the paracellular pathway has been shown to be a safe, reversible, time- and dose-dependent-strategy, and limited to intestinal tissue. This controlled permeation enhancement would be preferred to the nonspecific disruptions caused by fatty acids, bile salts, and chelators. *In vivo*, ZOT has been shown to increase the absorption of insulin by 10-fold in rabbit ileum and jejunum without affecting colonic absorption [2]. In diabetic animals, those treated with ZOT and insulin orally showed comparable survival and decreases in blood glucose to those diabetic animals treated with insulin parenterally. These early findings are very promising in that ZOT may be used to enhance the intestinal absorption of proteins by safely modulating the paracellular pathway. Furthermore, ZOT in combination with other targeted delivery techniques such as nanosystems could become a highly versatile approach.

Another promising class of permeation enhancers is the biocompatible polymer chitosan and its analogs [3], which were shown to substantially increase the bioavailability of several macromolecules across the intestinal epithelium.

2.2. Lipid-Based Oral Drug Delivery

In recent years there has been an increased interest in the utility of lipid-based delivery systems to enhance oral bioavailability [4]. It is generally known that membrane permeability is directly correlated to a drug's water–lipid partition coefficient; however, the systemic availability of highly lipophilic drugs is impeded by their low aqueous solubility. In an effort to improve this solubility-limited bioavailabiliy, formulators have turned to the use of lipid excipients to solubilize the compounds prior to oral administration. Several formulations are currently on the market, for example, Sandimmun®/Neoral® (cyclosporin microemulsion), Norvir® (ritonavir), and Fortovase® (saquinavir).

From a mechanistic point of view, the interaction of lipid-based formulations with the gastrointestinal system and associated digestive processes is not completely understood and appears to be more complex than mere solubility enhancement. For example, an increasing body of evidence has shown that certain lipid excipients can inhibit both presystemic drug metabolism and intestinal drug efflux mediated by P-glycoprotein (P-gp) (see Section 5.5). Furthermore, it is well known that lipids are capable of enhancing lymphatic transport of hydrophobic drugs, thereby reducing drug clearance resulting from hepatic first-pass metabolism. As more mechanistic studies emerge we can expect more extensive application of this flexible oral drug delivery approach.

2.3. Prodrugs

The dominant factor governing passive drug transport is the lipophilicity of the compound, generally described by the oil/water partition coefficient (P, frequently expressed as $\log P$) or related parameters (e.g., distribution coefficient, D (or $\log D$), which takes into account partitioning with respect to aqueous pH). Transport across a lipid bilayer membrane involves a number of steps: diffusion across a stagnant aqueous boundary layer, interfacial transfer into the membrane, passage across the membrane, interfacial transfer out of the membrane, and diffusion across the second stagnant aqueous layer. It is now generally accepted that the overall transfer rate across the membrane increases initially with lipophilicity, but eventually reaches a maximum as diffusion across the aqueous boundary layer becomes a rate-limiting step (see below). Whereas many nutrients and drugs are readily transported across the intestinal membrane, there are a large number of highly water soluble compounds whose transfer across the intestinal membrane is limited due to extremely low P values.

An approach to increase epithelial permeation of these compounds is to alter the lipophilicity of drugs through chemical modifications. In general, a prodrug is synthesized from the parent compound by converting its hydrophilic residues into less polar moieties. During or after absorption, the parent drugs are then released from their lipophilic derivatives by hydrolysis or specific enzymatic actions. This strategy has successfully improved absorption of numerous drugs, especially for compounds with relatively small molecular weight. However, not all poorly absorbed drugs can be subjected to structural modifications. For example, macromolecules such as polypeptides and proteins, whose structures are closely related to their biological activities can potentially aggregate upon chemical modification. Upon hydrolysis, the aggregated parent molecule may no longer be biologically active.

2.3.1. Definition of Prodrugs Historically, the term prodrug was first introduced by Albert [5] who used the word "prodrug" or "proagent" to describe compounds that undergo biotransformation prior to exhibiting their pharmacological effects. Consequently, he suggested that this concept could be used for many different purposes.

Presently, the term prodrug is used to describe a compound that is converted to the pharmacologically active substance *after* administration [6]. Although many drugs are known to be inactive until biotransformed, their utility as prodrugs was based on serendipity. Today, however, rational prodrug design is widely used to overcome problems of absorption, distribution, and biotransformation associated with certain drug molecules. The prodrug concept has been most successfully applied to

- facilitate absorption and distribution of drugs with poor lipid solubility [7];
- stabilize against metabolism during oral absorption [8];
- bypass efflux transporters [9]
- increase the duration of action of drugs that are rapidly eliminated [10];
- overcome problems of poor product acceptance by patients [11];
- eliminate stability and other formulation problems [10,12];
- promote site-specific delivery of a drug [13];
- increase the aqueous solubility [10,14]; and
- lower the toxicity of a drug [15].

In fact, the vast majority of prodrugs are designed to increase the intestinal absorption of polar drugs, that is, to increase lipophilicity.

2.3.2. Prodrug Design by Increasing Lipophilicity An example of highly polar, nonlipophilic molecules with resulting poor permeability characteristics and therefore low bioavailability is the structural analogs of the natural purine and pyrimidine nucleosides. The regular use of the nucleoside analogue, 6-azauridine (Fig. 2), in the treatment of psoriasis and neoplastic diseases was impractical because of its poor oral bioavailability. This low bioavailability can be mainly attributed to poor permeability characteristics.

The synthesis of various ester prodrugs of 6-azauridine (6-AZA), such as 2′,3′,5′-triacetyl-6-AZA and 2′,3′,5′-tribenzoyl-6-AZA as well as other mono- and polyacyl derivatives was carried out in an effort to obtain orally bioavailable 6-AZA. It was shown that the triacetyl ester can be administered every 8 h and is absorbed completely. On oral dosing, the triacetyl ester is converted for 80% into 6-AZA and 17% as its 5′-monoacetyl derivative. Orally administered 2′,3′,5′-triacetyl-6-AZA caused the same clinical effects as an equivalent intravenously administered dose of 6-AZA [16].

Numerous other examples of successful ester prodrug design are presented by Yalkowski and Morozowich [17]. The pivaloyl ester prodrug of ampicillin, pivampicillin, shows about 90% bioavailability upon oral administration in humans, whereas ampicillin shows about 33%; the ester has a log P value about 2.7 log units higher than that of the parent compound. Psicofuranine is not absorbed upon oral administration in humans,

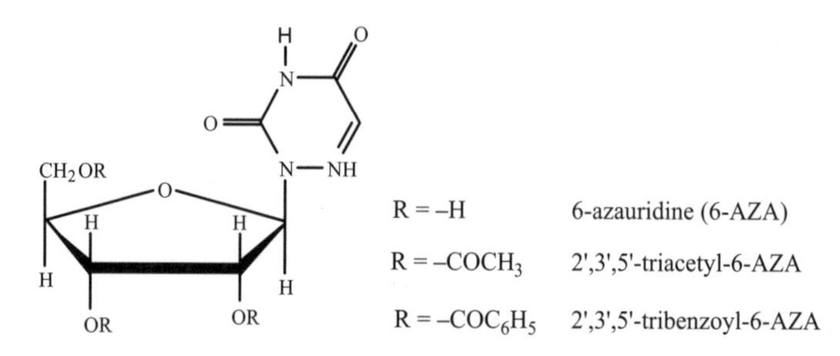

Figure 2. Structural formulas of 6-azauridine and its prodrug analogs.

having a $\log P$ of −1.95; the triacetate ester ($\log P = 0.72$) is well absorbed.

Evidently, lipophilization seems straightforward enough to create orally absorbable prodrugs of many polar substances.

2.3.3. Rationale and Considerations for the Use of Prodrugs In the rational design and synthesis of the ideal prodrug derivative, several factors should also be considered, such as

(a) *Prodrugs should be easily synthesised and purified*: elaborate synthetic schemes should be avoided due to increased costs. Multistep syntheses increase operator time and decrease yield.

(b) Prodrugs must be stable in bulk form and dosage form.

(c) *Neither the prodrug nor its metabolic derivatives should be toxic*: not only the newly created prodrug but also the derivatives formed after bioconversion should be nontoxic. Relatively safe moieties include amino acids, short to medium length alkyl esters and many inorganic and organic acid and base salt combinations. Nowadays, pivaloyl (e.g., pivampicillin, pivcephalexin), palmitoyl, propyl, and ethyl ester groups are frequently used in prodrug design.

As will be discussed below, the lipophilicity of a drug cannot be increased indefinitely to improve epithelial absorption. Because of the low permeation at both low and high $\log P$ values, an optimal lipophilicity (e.g., $-1 < \log P < 2$) has to be considered in prodrug design (see Section 3.1 and Fig. 3).

2.4. Absorption Enhancement via Targeting to Membrane Transporters

Several strategies have been developed to enhance the bioavailability of drugs with limited membrane permeability. These approaches can be categorized into methods that manipulate the membrane barrier properties or increase drug solubility.

Whereas charged, hydrophilic compounds and pharmaceutical macromolecules encounter difficulties in permeating the cell membrane, the systemic absorption of many water-soluble nutrients (e.g., sugars, vitamins), endogenous proteins (e.g., insulin, growth factors), and toxins (e.g., ricin, cholera toxin) appears to be highly efficient. The effective transcellular movement of these molecules is facilitated by specialized transport processes

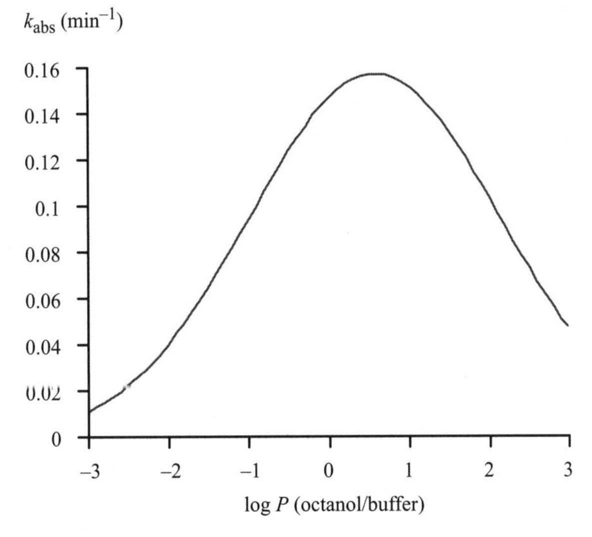

Figure 3. Empirical relationship between *in situ* intestinal absorption rate constants (k_{abs}) and apparent partition coefficients. The curve is described by $\log k_{abs} = 0.103 \log P - 0.09(\log P)^2 - 0.833$ [27]. (This figure is available in full color at http://mrw.interscience.wiley.com/emrw/9780471266945/home.)

in the epithelia, namely carrier-mediated transport and receptor-mediated endocytosis/transcytosis (discussed in more detail in subsequent sections). Both processes are operated by specific membrane associated proteins and share common features of active transport mechanisms, that is, concentration-, energy-, temperature-dependent, and subjective to structural analog inhibition. On the other hand, they also differ significantly in ways that transporter proteins are anchored in the membrane and in their ligand internalization mechanisms.

Apart from naturally occurring substrates, it is now well recognized that many drugs can be selectively taken up by active transport processes. For pharmaceutical scientists, these membrane transporters provide alternative routes for the delivery of drugs that would normally be impermeable to the biological barriers. Utilizing a method similar to the conventional prodrug approach, absorption enhancement is pursed by formation of conjugates between drugs and the endogenous ligands of the membrane transporters. Consequently, via the specific interaction ligands between the moiety and its transporter, drug candidates can be shuttled across or into the cells and eventually be released from the ligands.

Taking advantages of recent advances in molecular biology and computer modeling, scientists are now starting to design prodrugs based on the structural requirements of the transporter systems. In general, prodrug strategies involving carrier-mediated pathways have the advantage of high uptake capacity. However, the size of drug conjugates is relatively limited (~1000 Da) probably because larger conjugates fail to be shuttled through the restricted space within the carrier protein. For peptide and protein delivery, carrier-mediated pathways could only facilitate peptides up to four amino acids.

Compared to active carriers, receptor-mediated endocytosis (RME) systems have a rather limited uptake capacity, which in some cases is insufficient to elicit pharmacological activities. Yet, because of the endocytic pit formation (up to several hundred nanometers) and vesicular internalization mechanism, RME pathways are perfectly suited to accommodate large molecular weight peptide and/or protein conjugates. More importantly, recent success in transport of RME ligand–drug vehicle conjugates (e.g., nanoparticles, liposomes) via RME pathways opens new possibilities for macromolecular delivery across biological barriers. First of all, formulating pharmaceuticals in drug vehicle systems compensates the limited capacity of RME systems, resulting in 10^3- to 10^6-fold increase in uptake [18,19]. Second, drug vehicle systems also protect drug molecules from possible enzymatic degradation in the biological membrane. Furthermore, this type of conjugation avoids direct chemical reaction between drug molecules and ligands, allowing incorporation of drugs with more diverse structural properties.

3. PASSIVE DIFFUSION

In general, drug transport across any epithelium is dictated by the characteristics of the cell membrane and the physicochemical properties of drugs. In absorptive epithelia such as enterocytes, the intercellular space is sealed by tight junctions or *zonula occludens* (ZO). The junctional proteins ZO-1 and ZO-2 play essential roles in epithelial barrier function; they not only maintain cell polarity by confining surface proteins to their appropriate membrane domains, but also prevent diffusion of water-soluble molecules and backflow of the absorbed nutrients.

Studies in rat small intestine have shown that only water and small hydrophilic solutes with molecular radius smaller than 25 Å can move across the paracellular pathway [20]. Despite intense interest in the structure and regulation of the *human* intestinal tight junction, its functional dimensions have remained poorly defined. Madara and Dharmsathaphorn [21] suggested a pore size for human colonic T84 cells in the range 3.6–15 Å based on the ability of two probes of widely different size, mannitol and inulin, to cross T84 monolayers. Similarly, Ma and colleagues [22] proposed that the paracellular pore in rat colon was accessible to molecules with a radius >11 Å on the basis of significant permeation of inulin. Knipp et al. [23] used a group of structurally unrelated compounds of known hydrodynamic radii and estimated the pore radius of Caco-2 to be 5.2 Å, although the fact that all of the probes used were smaller than the estimated pore size was an acknowledged limitation. Using a series of polyethyleneglycol (PEG) beads of known radii that span that of the restrictive paracellular pore, Watson et al. [24] calculated a pore radius of 4.5 and 4.3 Å for Caco-2 and T84 cell monolayers, respectively. *In vivo* PEG absorption profiles in rat, dog, and human [25] show a significant molecular mass cut-off at ~600 Da corresponding to a hydrodynamic radius of ~5.3 Å, which compares well with data obtained in cell culture.

The overall surface area available for transcellular transport is significantly larger than that of the paracellular pathway. Therefore, the transcellular pathway is naturally the preferred route of transport for most molecules. Lipophilic and small amphiphilic compounds can traverse the epithelium efficiently by partitioning into and out of the lipid bilayers. In contrast, large hydrophilic molecules cannot diffuse freely through the cells even when thermodynamic conditions, for example, concentration gradient, favors such action. Factors influencing the transcellular passive diffusion of drugs have been thoroughly characterized. Analyzing the physicochemical properties and permeability characteristics of several thousand drug molecules, Lipinski [26] deduced that only compounds with a molecular weight lower than 500, a $\log P$ less than 5, and less than 5 hydrogen bond donors and 10 hydrogen bond acceptors are likely to permeate efficiently across the cell membrane via passive diffusion. This set of characteristics for well-permeating molecules is now popularly known as Lipinski's "rule of five" and has served in the drug industry as an extremely helpful screening mechanism for recognizing drug permeability issues early on in the drug discovery process [26].

3.1. Kinetics of Passive Diffusion

Passive diffusion refers to movement of a solute along its concentration gradient. As long as the diffusing molecule does not interact with elements of the membrane, the driving force behind the diffusion of a molecule through the lipid bilayer is the electrochemical potential difference of the compound on both sides of the membrane. The change in mass (M) of a solute as a function of time (t) during its diffusion through a membrane barrier with area S is known as the flux, J

$$J = \frac{\mathrm{d}M}{S \cdot \mathrm{d}t} \tag{1}$$

The flux, in turn, is proportional to the concentration gradient, $\mathrm{d}C/\mathrm{d}x$

$$J = -D\frac{\mathrm{d}C}{\mathrm{d}x} \tag{2}$$

where D is the diffusion coefficient of the solute in cm^2/s, C its concentration in mol/cm^3, and x is the distance in cm of movement perpendicular to the surface of the barrier. The diffusion constant, D, or diffusivity does not ordinarily remain constant and may change at higher concentrations, and is affected by temperature, pressure, solvent properties, and the chemical nature of the solute. Equation 2 is known commonly as Fick's First law. Important boundary conditions to the First Law are [1] steady state, that is, a constant rate of diffusion and [2] sink conditions, that is, homogeneously mixed compartments on both sides of the barrier. Without these boundary conditions, Fick's Second Law applies

$$\frac{\partial C}{\partial t} = -\frac{\partial J}{\partial x} \tag{3}$$

which is usually differentiated to express changes in concentration in three dimensions (x, y, z) in the general form

$$\frac{\partial C}{\partial t} = D\left(\frac{\partial^2 C}{\partial x^2} + \frac{\partial^2 C}{\partial y^2} + \frac{\partial^2 C}{\partial z^2}\right) \tag{4}$$

The reader should appreciate that this form of Fick's law does not have to be taken into consideration if the aforementioned boundary conditions are met (steady state, i.e., $\mathrm{d}C/\mathrm{d}t = 0$).

In experimental situations, a barrier (e.g., epithelial tissue) usually separates two compartments of a diffusion cell of cross-sectional area S and thickness h. If the concentrations in the membrane on the donor and receptor sides are C_1 and C_2, respectively, Fick's First Law may be written as

$$J = \frac{dM}{S \cdot dt} = D\left(\frac{C_1 - C_2}{h}\right) \quad (5)$$

where $(C_1 - C_2)/h$ is an approximation of dC/dx. The concentrations C_1 and C_2 within the membrane generally are not known but can be replaced by the partition coefficient, K, multiplied by the concentration C_d on the donor side or C_r on the receiver side. The distribution or partition coefficient, K, is expressed by

$$K = \frac{C_1}{C_d} = \frac{C_2}{C_r} \quad (6)$$

Thus,

$$J = DK\left(\frac{C_d - C_r}{h}\right) \quad (7)$$

In general, under physiological conditions, sink conditions will apply in the receptor compartment, $C_r \approx 0$,

$$J = \frac{DKC_d}{h} = P \cdot C_d \quad (8)$$

The variables D, K, and h cannot always be determined independently and commonly are lumped together to provide the permeability coefficient, P, with units of linear velocity (cm/s). P can be assessed in experimental systems and provides a relative parameter to classify solute penetration through a lipid bilayer.

Numerous studies support the idea that drug molecules are absorbed through lipid bilayer membranes in the unionized state by the process of passive diffusion. The rate of absorption, the pK_a of the diffusing solute, and the pH at the absorption site are interrelated. Equation 8 demonstrates that the absorption rate of solutes through biological membranes is directly proportional to the value of the oil/water partition coefficient. Houston et al. [27] studied the absorption of a series of carbamate esters through rat everted intestine and observed a bell-shaped relationship between the absorption rate and the partition coefficient (Fig. 3). The reader should be aware that P in this case is used to denote the partition coefficient and not permeability. It is common to express the partition coefficient in terms of $\log P$. At low P values ($\log P < -2$), the compound cannot penetrate the lipid membrane due to an excessive thermodynamic barrier. Conversely, at high P values ($\log P > 3$), the compound becomes so lipid soluble that the diffusion through the unstirred waterlayers flanking both sides of the membrane becomes the rate-limiting step in the overall absorption process. Moreover, a decrease in water solubility takes place that would make the compound highly unsoluble; this, in turn, would prevent the compound from reaching the membrane surface. Because of the low intestinal permeation at both low and high P values, an optimal P value (e.g., $-1 < \log P < 2$) has to be considered in drug design.

4. FACILITATED AND ACTIVE TRANSPORT PATHWAYS

4.1. Receptor-Mediated Transport

A distinct difference between carrier- and receptor-mediated systems should be pointed out. Carrier-mediated systems involve transport proteins that are anchored to the membrane by multiple membrane-spanning fragments or protein loops, whereas receptor-mediated systems utilize receptor proteins that span the membrane only once. Carriers operate by shuttling their substrates across the membrane via an energy-dependent (ATP or cotransport) flip-flop mechanism and receptors are internalized in vesicles after binding to their substrate, as explained in more detail below.

4.1.1. Receptor-Mediated Endocytosis Mammalian cells have developed an assortment of mechanisms to facilitate the internalization of specific substrates and target these to defined locations inside the cytoplasm. Collectively, these processes of membrane deformations are termed "endocytosis" and comprise phagocytosis, pinocytosis, receptor-mediated endocytosis (clathrin-mediated), and potocytosis (nonclathrin (caveolin)-mediated RME). The emphasis of this section is receptor-mediated endocytosis in the intestinal tract, but the interested student may consult alternative reviews covering the complete spectrum of endocytotic processes in other cell types [28,29].

RME is a highly specific cellular biologic process by which, as its name implies, various ligands bind to cell surface receptors and are subsequently internalized and trafficked within the cell. In many cells the process of endocytosis is so active that the entire membrane surface is internalized and replaced in less than a half hour [30].

RME can be dissected into several distinct events. Initially, exogenous ligands bind to specific externally oriented membrane receptors. Binding occurs within 2 min and is followed by membrane invagination until an internal vesicle forms within the cell (the early endosome, "receptosome," or CURL (compartment of uncoupling receptor and ligand) [31]. Localized membrane proteins, lipids and extracellular solutes are also internalized during this process. When the ligand binds to its specific receptor, the ligand–receptor complex accumulates in coated pits (CP). Coated pits are areas of the membrane with high concentration of endocellular clathrin subunits. The assembly of clathrin molecules on the coated pit is believed to aid the invagination process. Specialized coat proteins, which are actually a multisubunit complex, called adaptins, trap-specific membrane receptors—which move laterally through the membrane—in the coated pit by binding to a signal sequence (Tyr-X-Arg-Phe, where X = any amino acid) at the endocellular carboxy terminus of the receptor. This process ensures that the correct receptors are concentrated in the coated pit areas and minimizes the amount of extracellular fluid that is taken up in the cell. RME appears to require the GTP-binding protein dynamin, but the process by which dynamin is recruited to clathrin-coated pits remains unclear [32].

Following the internalization process, the clathrin coat is lost through the help of chaperone proteins and proton pumps lower the endosomal pH to approximately 5.5, which causes dissociation of the receptor–ligand complex [33]. CURL serves as a compartment to segregate the recycling receptor (e.g., transferrin) from receptor involved in transcytosis (e.g., transcobalamin) [34]. Endosomes may then move randomly or by saltatory motion along the microtubules [35] until they reach the trans-Golgi reticulum where they are believed to fuse with Golgi components or other membranous compartments and convert into tubulovesicular complexes and late endosomes (LE) or multivesicular bodies (MVB). The fate of the receptor and ligand are determined in these sorting vesicles. Some ligands and receptors are returned to the cell surface where the ligand is released into the extracellular milieu and the receptor is recycled. Alternatively, the ligand is directed to lysosomes (LY) for destruction while the receptor is recycled to the cell membrane. Figure 4 presents an overview of the existing possibilities in the fate of ligands and receptors.

The endocytotic recycling pathways of polarized epithelial cells (Fig. 5), such as enterocytes, are generally more complex than in nonpolarized cells. In these enterocytes a common recycling compartment exists that receives molecules from both apical and basolateral membranes and is able to correctly return them to the appropriate membrane or membrane recycling compartment (apical or basolateral recycling compartment) [36]. The signals required for this sorting step have not been defined as of yet, but are presumably similar to the peptide sequences required for proper sorting in the trans-Golgi network (TGN).

4.1.2. Structure of Cell Surface Receptors Our general understanding of RME receptor structure and related structure–function relationships has been significantly enhanced by ongoing efforts to clone mRNA sequences coding for endocytotic receptors. It appears that most RME receptors share several structural features, such as an extracellular ligand binding site, a single hydrophobic transmembrane domain (TMD) (unless the receptor is expressed as a dimer), and a cytoplasmic tail encoding endocytosis and other functional signals [37]. Two classes of receptors are proposed based on their orientation in the cell membrane: the amino terminus of Type I receptors is located on the extracellular side of the membrane, whereas Type II receptors have this same

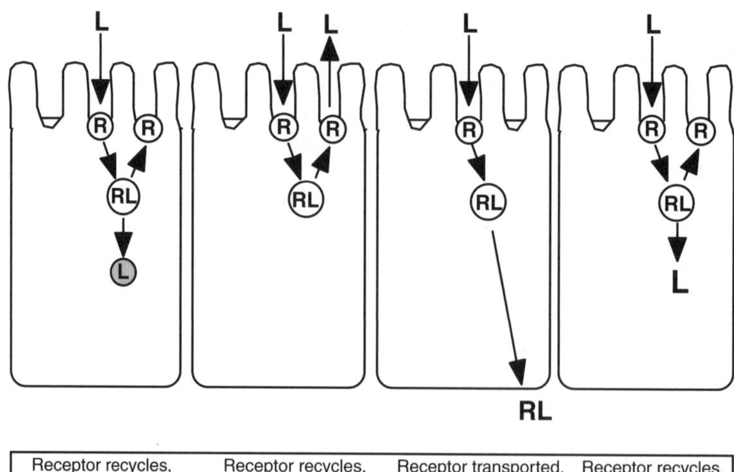

Figure 4. Schematic representation of endocytotic pathways in polarized cells. Question marks along an arrow indicate that only circumstantial evidence exists for those pathways at present time. Double-headed arrows indicate that similar magnitudes of transfer occur in either direction. Ligand and receptor complex is initially taken up from coated pits (CP) into coated vesicles (CV). The CV loses its clathrin coat as described in the text and transforms into an apical sorting endosome (ASE). The fate of receptor and ligand can be: a degradative pathway where the material is passaged to multivesicular bodies (MVB) and eventually late endosomes (LE), which fuse with lysosomes (LY); a recycling pathway to the central recycling compartment (CRC) and/or the apical recycling compartment (ARC) that eventually exocytose the contents of the vesicle; or direct exocytosis of the ASE to apical membrane. In polarized cells, similar pathways occur at the basolateral membrane. The CRC, in combination with the trans-Golgi network (TGN) interconnects the apical and basolateral sorting pathways and enables the potential for transcytosis. BLRC, basolateral recycling compartment; BLSE, basolateral sorting endosome; G, Golgi apparatus; N, cell nucleus; TJ, tight junction.

protein tail in the intracellular milieu. Although protein orientation may appear trivial, it strongly influences the eventual endocytotic mechanism [37].

4.1.3. Transcytosis One of the least understood aspects in vesicular trafficking and sorting, and possibly one of the most important aspects for successful oral drug delivery via RME, is the transport of endocytotic vesicles to the opposite membrane surface, more commonly referred to as transcytosis. Recent studies in the area of cellular biology have reported specific proteins, named TAPs (transcytosis-associated proteins), that are particularly found on transcytotic vesicles and are believed to be required for fusion with target membrane [38,39]. Other methods to stimulate transcytosis have been explored. Transcytosis of transferrin (Tf) was found to be stimulated in the presence of brefeldin A (BFA), a fungal metabolite that has profound effects on the structure and function of the Golgi apparatus. Shah and coworkers [40] showed in Caco-2 cell monolayers that BFA causes a marked decrease in the number of basolateral Tf receptors (TfR) along with a slight increase in the number of apical TfR. BFA enhanced the TfR-mediated transcytosis of both 125I-Tf and the horseradish peroxidase-Tf conjugate across Caco-2 cells in both apical-to-basolateral and basolateral-to-apical directions. Prydz and colleagues found that BFA treatment rapidly increased apical endocytosis of both ricin and horseradish peroxidase in MDCK cells, whereas basolateral endocytosis was unaffected [41].

4.1.4. Potocytosis Potocytosis has been accepted as a distinct RME pathway [42–44].

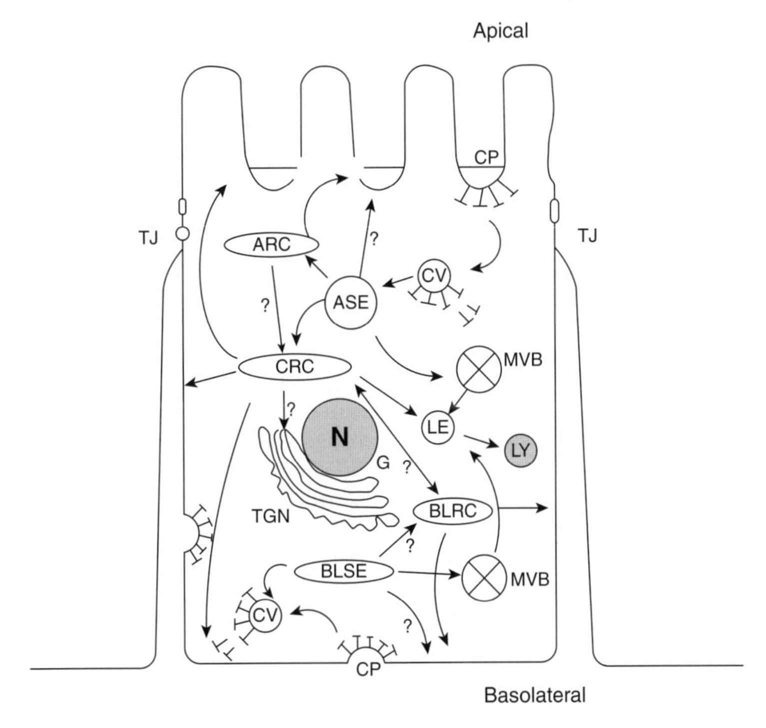

Figure 5. Intracellular sorting pathways of RME. The initial binding and uptake steps (including receptor clustering in coated or noncoated pits, internalization of the receptor–ligand complex into coated vesicles (noncoated in case of potocytosis), and fusion of vesicles to form endosomes) are common to all pathways. After entry into acidic endosomes, ligand, and receptors are sorted and trafficked independently that may result in degradation, recycling, or transcytosis of either molecule (see text). L, ligand; R, receptor; lysosomes are depicted as shaded circles. After Ref. [10].

Potocytosis, or nonclathrin coated endocytosis takes place through caveolae, which are uniform omega- or flask-shaped membrane invaginations (50–80 nm diameter) [28] and was first described as the internalization mechanism of the vitamin folic acid in a mouse keratinocyte cell line [45]. Years before the name "potocytosis" was coined, various ligands had been reported to localize in nonclathrin coated membrane regions, including cholera and tetanus toxins [46].

Morphological studies have implicated caveolae in (i) the transcytosis of macromolecules across endothelial cells; (ii) the uptake of small molecules via potocytosis involving glycosylphosphatidylinositol (GPI)-linked receptor molecules and an unknown anion transport protein; (iii) interactions with the actin-based cytoskeleton; and (iv) the compartmentalization of certain signaling molecules involved in signal transduction, including G-protein coupled receptors. Caveolae are characterized by the presence of an integral 22 kDa membrane protein termed VIP21-caveolin, which coats the cytoplasmic surface of the membrane [47,48].

From a drug delivery standpoint, the advantage of potocytosis pathways over clathrin-coated RME pathways lies in the absence of the pH lowering step, thereby circumventing the classical endosomal/lysosomal pathway [43]. This may be of invaluable importance to the effective delivery of pH-sensitive macromolecules.

4.1.5. Lipid Rafts: Clathrin- and Caveolin-Independent Endocytosis

The term *lipid raft* originated to explain flat areas of the plasma membrane that are highly enriched with the similar lipid constituents found in caveolae, including cholesterol, phospholipids, glycophospholipids, sphingomyelin, and so on.

These lipid-rich, detergent-resistant areas of the membrane serve as a meeting place for interacting molecules destined to the same intracellular compartment [49]. A clear distinction between caveolae and lipid raft endocytic internalization cannot be made at this point. It is also important to note that certain types of raft receptors will migrate to clathrin-coated pits for invagination, leading to an alternative intracellular trafficking pattern. Currently, cholera toxin B serves as a prototypical lipid raft marker, and has been utilized in targeted delivery of antigens through its conjugation to liposome carriers [50]. Coating macromolecules with lipid raft-associated ligands has allowed for cellular internalization and vesicular trafficking to nonlysosomal subcellular compartments, making this trafficking mechanism attractive to nondegradative intracellular drug delivery.

4.2. Receptor-Mediated Oral Absorption Systems

RME in Enterocytes Versus M-Cells According to Walker and Sanderson [51] the preferred route of intestinal uptake of low concentrations of antigens is through the M-cells (microfold or membranous), located in Peyer's patches, but at higher concentrations the regular enterocytes are also involved. M-cells are specialized epithelial cells of the gut-associated lymphoid tissues (GALT) that transport antigens from the lumen to cells of the immune system, thereby initiating an immune response or tolerance. Soluble macromolecules, small particles [52], and also entire microorganisms are transported by M-cells. The importance of M-cells in the uptake of particles is still a point of discussion. Hussain and colleagues [53] deduced from their own work and the work of others that "the importance of Peyer's patches as the principal site of particulate absorption may have been overemphasized, and that normal epithelial cells can also be induced, with appropriate ligands such as plant lectins and bacterial adhesins, to absorb particulate matter." In this light, it should be stressed that the surface area of M-cells is only 10% compared to the surface area of normal epithelial cells.

4.2.1. Immunoglobulin Transport

Maternal and Neonatal IgG Transport Receptor-mediated transcytosis of immunoglobulin G (IgG) across the neonatal small intestine serves to convey passive immunity to many newborn mammals [54]. In rats, IgG in milk selectively binds to neonatal Fc receptors (FcRn) expressed on the surface of the proximal small intestinal enterocytes during the first 3 weeks after birth. FcRn binds IgG in a pH-dependent manner, with binding occurring at the luminal pH (6–6.5) of the jejunum and release at the pH of plasma (7.4). The Fc receptor resembles the major histocompatibility complex (MHC) class I antigens in that it consists of two subunits: a transmembrane glycoprotein (gp50) in association with β2-microglobulin [54]. In mature absorptive cells both subunits are colocalized in each of the membrane compartments that mediate transcytosis of IgG. IgG administered *in situ* apparently causes both subunits to concentrate within endocytic pits of the apical plasma membrane, suggesting that the ligand causes redistribution of receptors at this site. These results support a model for transport in which IgG is transferred across the cell as a complex with both subunits. Interestingly, Benlounes et al. [55] showed that IgG is effectively transcytosed at lower concentrations (<300 μg/mL), whereas a degradative pathway dominates at higher mucosal IgG concentrations.

Site-directed mutagenesis of a recombinant Fc hinge fragment has been used to localize the site of the mouse IgG1 molecule that is involved in the intestinal transfer of recombinant Fc hinge fragments in neonatal mice. These studies definitively indicate that the neonatal Fc receptor, FcRn, is involved in transcytosis across both yolk sac and neonatal intestine, in addition to the regulation of IgG catabolism [56,57]. Continued binding to vesicle membranes appears to be required for successful transfer since unbound proteins are removed from the transport pathway before exocytosis. These results favor the proposal that IgG is transferred across cells as an IgG-receptor complex [58].

Drug carriers such as liposomes are not readily transported intact across epithelial barriers. Patel and Wild [59] showed that coating liposomes with appropriate IgG enhances their transport across rabbit yolk sac endoderm and enterocytes of suckling rat gut proximal small intestine. They measured the effect of liposomal transcytosis both by radiolabel assay of entrapped ^{125}I-poly(vinylpyrrolidone) and [^{3}H]-inulin, and by the hypoglycemic effect of entrapped insulin. These results suggested that transported liposomes followed a pathway of transcytosis in clathrin-coated vesicles, thus escaping lysosomal degradation.

Polymeric IgA and IgM Transport Polymeric IgA is produced by plasma cells and found in all external excretions, including bile and saliva [60,61]. In the small intestine, polymeric IgA and IgM bind to the polymeric immunoglobulin receptor (pIgR) that is located on the basolateral surface of the cell. pIgR expression can be upregulated by cytokines [62]. The pIgR–IgA complex is internalized into endosomes where it is sorted into vesicles that transcytose it to the apical surface. At the apical surface the pIgR is proteolytically cleaved, and the large extracellular fragment (known as secretory component) is released together with the ligand. The pIgR contains a cytoplasmic domain of 103 amino acids that contains several sorting signals. Targeting from the trans-Golgi network to the basolateral surface is determined by the membrane-proximal 17 residues of this domain. For endocytosis there are two signals, both of which contain tyrosines. Transcytosis of the pIgR is signaled by serine phosphorylation and may be regulated by the heterotrimeric Gs protein, protein kinase C, and calmodulin. IgG is transcytosed from the apical to basolateral surface in several epithelial tissues such as the placenta and the small intestine of newborn rats. The receptor for intestinal transport of IgG is structurally similar to class I MHC molecules [63,64].

4.2.2. Bacterial Adhesins and Invasins For many bacterial species, adherence to host cells is the initial key step toward colonization and establishing an infectious disease. Two components are necessary for the adherence process: a bacterial "adhesin" (adherence or colonization factor) and a "receptor" on the host (eukaryotic) cell surface. Bacteria usually express various cell adherence mechanisms depending on the environmental conditions and nature of the adhesins as well as receptors. In a study on the colonization mechanism of *Klebsiella*, *Enterobacter*, and *Serratia* strains, Livrelli et al. [65] found no relationship between the adhesive pattern and the production of specific fimbriae, suggesting that several unrecognized adhesive factors are involved that remain to be identified.

Bacteria causing gastrointestinal infection need to penetrate the mucus layer before attaching themselves to the epithelial surface. This attachment is usually mediated by bacterial fimbriae or pilus structures, although other cell surface components may also take part in the process. Adherent bacteria colonize intestinal epithelium by multiplication and initiation of a series of biochemical reactions inside the target cell through signal transduction mechanisms (with or without the help of toxins) [66].

Several adhesin and invasin molecules have been identified, such as a mannose-specific adhesin in *Lactobacillus plantarum* [67] and *Vibrio cholerae* [68]. Metcalfe et al. [69] found that adherence of *Escherichia coli* K-12 (K88ab) to immobilized porcine small intestine mucus was caused by a 40–42 kDa glycoprotein K88-specific receptor. Using monoclonal antibodies against fimbrial adhesins of porcine enterotoxigenic *E. coli*, K99 and K88 adhesin were detected, but not F41 and 987P adhesins [70].

The colonization mechanism of the enteropathogenic bacterium *Yersinia pseudotuberculosis* has been studied in most detail. In contrast to other infective agents, such as Salmonella strains or enteroinvasive *E. coli*, invasion and transcytosis of *Y. pseudotuberculosis* is mediated by a single 986 amino acid protein, invasin, on the bacterial surface that binds to $\alpha5\beta1$ integrin [71]. This single factor is sufficient to promote entry of inert particles by binding multiple integrin receptors during cellular uptake [72]. This phenomenon has been confirmed by Mengaud and coworkers, who identified E-cadherin as the ligand for internalin, a *Listeria monocytogenes* protein essential for entry into epithelial cells. The internalization process of many microorganism is impressively fast: within 45 min after introduction of *Y. pseudotuberculosis* into the lumen of BALB/C mice, wild-type bacteria can be found in the Peyer's patch [73]. Mutants expressing defective invasin derivatives were unable to promote efficient translocation into the Peyer's patch and instead colonized on the luminal surface of the intestinal epithelium.

The study of bacterial adhesins and invasins for the application in drug delivery strategies has become the focus of much attention. Paul and colleagues used an invasin fusion protein system for gene delivery strategies [74] and Easson et al. [75,76] used a similar approach for intestinal delivery of nanoparticles. The latter group found that latex microspheres up to 1 μm coupled to maltose-binding protein that was fused with invasin can be internalized by MDCK cell monolayers [75,76].

4.2.3. Bacterial and Plant Toxins After reaching early endosomes by RME, diphtheria toxin molecules have two possible fates. A large pool enters the degradative pathway whereas a few molecules become cytotoxic by translocating their catalytic fragment A into the cytosol [77].

The B subunit of the *E. coli* heat labile toxin binds to the brush-border of intestinal epithelial cells in a highly specific, lectin-like manner. Uptake of this toxin and transcytosis to the basolateral side of the enterocytes was observed *in vivo* [78] and *in vitro* [79].

Fisher and Wilson expressed the transmembrane domain of diphtheria toxin in *E. coli* as a maltose-binding fusion protein and coupled it chemically to high-MW poly-L-lysine. The resulting complex was successfully used to mediate the internalization of a reporter gene *in vitro* [80].

Staphylococcus aureus produces a set of proteins (e.g., staphylococcal enterotoxin A (SEA), SEB, toxic shock syndrome toxin 1 (TSST-1)), which act both as superantigens and toxins. Hamad et al. [81] found dose-dependent, facilitated transcytosis of SEB and TSST-1 in Caco-2 cells, but not SEA. They extended their studies in mice *in vivo* by showing that ingested SEB appears in the blood more efficiently than SEA.

Various plant toxins, mostly ribosome-inactivating proteins, have been identified that bind to any mammalian cell surface expressing galactose units and are subsequently internalized by RME [82]. Toxins such as nigrin b [83], α-sarcin [84], ricin and saporin [85], viscumin [86], and modeccin [87] are highly toxic upon oral administration (i.e., are rapidly internalized). The possibility exists, therefore, that modified and, most importantly, less toxic subunits of these compound can be used to facilitate the uptake of macromolecular compounds or microparticulates.

4.2.4. Viral Hemagglutinins The initial step in many viral infections is the binding of surface proteins (hemagglutinins) to mucosal cells. These binding proteins have been identified for most viruses, including rotaviruses [88], varicella zoster virus [89], semliki forest virus [90], adenoviruses [91], potato leafroll virus [92], and reovirus [93].

Etchart and colleagues [94] compared the immune response to a vaccinia virus recombinant, expressing the measles virus hemagglutinin (VV-HA), after parenteral or mucosal immunizations in mice. Oral immunizations with 10^8 pfu (plaque-forming units) of VV-HA generated low numbers of HA-specific IgA-producing cells in the lamina propria of the gut, whereas oral coimmunization with VV-HA and cholera toxin greatly enhanced the level of HA-specific spot-forming cells (IgA > IgG). Interestingly, intrajejunal immunizations with 10^8 pfu. VV-HA alone induced high levels of anti-HA IgG-producing cells in the spleen and anti-HA IgA-secreting cells in the lamina propria of the gut. This study shows that VV-A can induce measles-specific immunity in the intestine provided that it is protected from degradation in the gastrointestinal tract, or that cholera toxin is used as an adjuvant.

4.2.5. Lectins (Phytohemagglutinins) Lectins are plant proteins that bind to specific sugars, which are found on the surface of glycoproteins and glycolipids of eukaryotic cells. Such binding may result in specific hemagglutinating activity. Since lectins are relatively heat stable, they are abundant in the human diet (e.g., cereals, beans, and other seeds). Concentrated solutions of lectins have a "mucotractive" effect due to irritation of the gut wall, which explains why so-called "high fiber foods" (rich in lectins) are thought to be responsible for stimulating bowel motility [95,96].

In another study demonstrating the rapid RME-uptake of lectins, Weaver and colleagues [97] directly infused concanavalin A, conjugated with 10 nm colloidal gold particles, into the lumen of the jejunum in neonatal guinea pigs. Within 60 min, both villous and crypt epithelial cells contained gold particles, demonstrating the rapid accessibility of crypt cells to the lectin.

Hussain et al. showed that the uptake mechanism for lectins can be utilized for intestinal drug targeting *in vivo* [53]. They covalently coupled polystyrene nanoparticles (500 nm) to tomato lectin and observed 23% systemic uptake after oral administration to rats. Control animals exerted a systemic uptake of <0.5%, indicating a 50-fold increase in oral absorption. Interestingly, they showed the intestinal uptake of tomato lectin-conjugated nanoparticles via the villous tissue to be 15 times higher than uptake by GALT [53].

Although lectins are generally believed to be transported via an RME mechanism, there is substantial evidence that these compounds have significant affinity to intestinal M-cells [98,99]. Binding studies have revealed that M-cells exhibit pronounced regional and species variation in glycoconjugate expression. Sharma and colleagues [100] studied the nature of cell-associated carbohydrates in the human intestine that may mediate transepithelial transport of bacterial and dietary lectins and their processing by the lymphoid cells of Peyer's patches. Upon comparison of human and mouse glycoconjugates of follicle-associated epithelium and GALT, they found a distinct difference in glycosylation between mouse and human Peyer's patches and their associated lymphoid cells. Thus, choosing the appropriate lectin is apparently important when considering cell surface glycoconjugates as target molecules for intestinal drug delivery strategies. In the future, knowledge of the site and species related variations in M-cell surface glycoconjugate expression may allow lectins to be utilized to selectively target antigenic material and oral vaccines to the mucosal immune system at specific locations [99]. Again, it should be pointed out that the overall contribution of M-cells to the absorptive surface area of the gastrointestinal tract is minimal, which could jeopardize the widespread application of drug targeting to these specialized cells.

4.3. RME of Vitamins and Metal Ions

4.3.1. Folate The cellular uptake of free folic acid is mediated by the folate receptor and/or the reduced folate carrier. The folate receptor is a GPI-anchored 38 kDa glycoprotein clustered in caveolae mediating cell transport by potocytosis [43]. While the expression of the reduced folate carrier is ubiquitously distributed in eukaryotic cells, the folate receptor is principally overexpressed in human tumors. Two homologous isoforms (α and β) of the receptor have been identified in humans. The α-isoform is found to be frequently overexpressed in epithelial tumors, whereas the β-form is often found in nonepithelial lineage tumors [101]. Consequently, this receptor system has been used in drug-targeting approaches to cancer cells [102], but also in protein delivery [103], gene delivery [104], and targeting of antisense oligonucleotides [105] to a variety of cell types. Although considerable success has been met in other areas of drug targeting, there are currently—to our knowledge—no reports in the literature describing the use, or attempt, of this system for intestinal drug delivery purposes. This may, in part, be attributable to the low expression level of the receptor in (healthy) enterocytes. However, the fact that the α-isoform of the folate receptor is overexpressed in epithelial cell lines, local targeting to intestinal cancer cells (e.g., colon carcinoma) appears to be a fertile approach.

4.3.2. Riboflavin Although not as extensively investigated as the transferrin and folate pathways, Low and coworkers [106] demonstrated that serum albumin coupled to riboflavin showed RME-mediated uptake in distal lung epithelium. In this paper, the authors speculate that a similar uptake process exists in the small intestine. Indeed, Huang and Swaan [107,108] showed that riboflavin is taken up in human small intestinal and placental epithelial cells by a riboflavin-specific RME process.

4.3.3. Vitamin B_{12} Vitamin B_{12}, the colloquial name for cobalamin (Cbl), is a large polar molecule that must be bound to specialized transport proteins to gain entry into cells. After oral administration it is bound to intrinsic factor (IF), a protein released from the parietal cells in the stomach and proximal cells in the duodenum. The Cbl–IF complex binds to an IF-receptor located on the surface of the ileum, which triggers a yet undefined endocytotic process. After internalization, the fate of the IF–Cbl complex has yet to be clarified. It was reported that IF–Cbl complex dissociates at acidic pH, and Cbl is transferred to transcobalamin II by Ramasamy et al. [109], whereas Dan and Cutler [110] found evidence of free Cbl in endosomes and the basolateral side of the membrane after administration to the apical surface of Caco-2 cell monolayers. It is clear, however, that Cbl is transported into all other cells only when bound to transcobalamin II.

The vitamin B_{12} RME system is probably the most extensively studied system for the oral delivery of peptides and proteins. In humans, the uptake of cobalamin is approximately 1 nmol per intestinal passage, with a potential for multiple dosing (2–3 times/h) [18]. Russell-Jones and coworkers have shown that this particular system can be employed for the intestinal uptake of luteinizing hormone releasing factor (LHRH)-analogs [111], granulocyte colony stimulating factor (18.8 kDa), erythropoietin (29.5 kDa), α-interferon [19,112], and the LHRH-antagonist ANTIDE [18]. More recently, they showed (*in vitro* as well as *in vivo*) the intestinal uptake of biodegradable polymeric nanoparticles coupled to cobalamin to be two- to threefold higher compared to control (nonspecific uptake of nanoparticles) [112,113]. Thus far, the universal application of this transport system for the oral delivery of peptides and proteins seems only to be hampered by its limited uptake capacity: 1 nmol per dose. Even though this amount of uptake may be adequate for molecules such as LHRH or erythropoietin, it is clearly not sufficient for the delivery of insulin or granulocyte colony stimulating factor. However, the successes with cobalamin-conjugated nanoparticles *in vivo* [113,114] are promising and eliminate the requirement of covalently coupling cobalamin and the substrate to be delivered. This, in turn, would permit the delivery of any macromolecule via the vitamin B_{12} uptake mechanism.

4.3.4. Transferrin Transferrin, an 80 kDa iron-transporting glycoprotein, is efficiently taken up into cells by the process of carrier-mediated endocytosis. Transferrin receptors are found on the surface of most proliferating cells, in elevated numbers on erythroblasts and on many kinds of tumors. According to current knowledge of intestinal iron absorption, transferrin is excreted into the intestinal lumen in the form of apotransferrin and is highly stable to attacks from intestinal peptidases. In most cells, diferric transferrin binds to transferrin receptor, a dimeric transmembrane glycoprotein of 180 kDa [115], and the ligand–receptor complex is endocytosed within clathrin-coated vesicles. After acidification of these vesicles, iron dissociates from the transferrin/TfR complex and enters the cytoplasm, where it is bound by ferritin (Fn). The role that transferrin, TfR, and Fn have in regulating dietary iron uptake and maintaining total body iron stores is unknown. Both the TfR and Fn genes can be detected on small intestinal mRNA [115], and both proteins have been isolated from intestinal enterocytes [116,117]. The uptake of iron in the intestinal tract amounts up to 20 mg/day and is primarily mediated by transferrin. Shah and Shen [12] showed that insulin covalently coupled to transferrin was transported across Caco-2 cell monolayers by RME. Furthermore, oral administration of this complex to streptozotocin-induced diabetic mice significantly reduced plasma glucose levels (−28%), which was further potentiated by brefeldin A pretreatment (−41%) [118].

5. TRANSPORT PROTEINS

Membrane transporters in general are a large group of membrane proteins that have one (bitopic) or more (polytopic) hydrophobic transmembrane segments. These transporters are involved in almost all facets of biological processes in the cell. Their involvement in cellular function can be classified as follows (after Saier [119]):

1. Mediate entry of all essential nutrients into the cytoplasmic compartment and subsequently into organelles, thus facilitating the metabolism of exogenous sources of carbon, nitrogen, sulfur, and phosphorus.
2. Provide a means for regulation of metabolite concentrations by catalyzing the efflux of end products of metabolic pathways from organelles and cells.
3. Mediate the active extrusion of drugs and other toxic substances from either the cytoplasm or the plasma membrane.
4. Mediate uptake and efflux of ion species that must be maintained at concentrations dramatically different from those in the external milieu.
5. Participate in the secretion of proteins, complex carbohydrates, and lipids into and beyond the cytoplasmic membrane.
6. Transfer of nucleic acids across cell membranes, allowing genetic exchange between organisms and thereby promoting species diversification.
7. Facilitate the uptake and release of pheromones, hormones, neurotransmitters, and a variety of other signaling molecules that allow a cell to participate in the biological experience of multicellularity.
8. Transporters allow living organisms to conduct biological warfare, secreting, for example, antibiotics, antiviral agents, antifungal agents, and toxins of humans and other animals that may confer upon the organisms producing such an agent a selective advantage for survival purposes. Many of these toxins are themselves channels-forming proteins or peptides that serve a cell-disruptive transport function.

Polytopic membrane proteins are indispensable to the cellular uptake and homeostasis of many essential nutrients. During the past decade it has become clear that a vast number of drugs share transport pathways with nutrients. Moreover, a critical role has been recognized for transport proteins in the absorption, excretion, and toxicity of drug molecules, as well as in their pharmacokinetic and pharmacodynamic (PK/PD) profiles. Because cellular transporter expression is often regulated by nuclear orphan receptors that simultaneously regulate the translation and expression of metabolic enzymes in the cell (e.g., P-glycoprotein and cytochrome P450 regulation by the pregnane X receptor), they indirectly control drug metabolism. Thus, transport proteins are involved in all facets of drug ADME and ADMET (absorption, distribution, metabolism, excretion, and toxicology), conferring an important field of study for pharmaceutical scientists involved in these areas. As a result, in-depth knowledge of membrane transport systems may be extremely useful in the design of new chemical entities (NCE). After all, it is now well appreciated that the most critical parameter for a new drug to survive the drug development pipeline on its way to the market is its ADMET profile.

Despite the involvement of solute transporters in fundamental cellular processes, most are poorly characterized at the molecular level. As a result, we are unable to predict the interaction of drugs with this important class of membrane proteins *a priori*, and detection of drug-transporter interactions remains unacceptably serendipitous.

This chapter aims to give an overview of current strategies for modeling transporter systems illustrated by three well-characterized transport systems: the P-glycoprotein efflux pump, a prototypical ABC-transporter, and a product of the multidrug resistance (MDR-1, ABC-B1) gene that exports metabolites as well as drugs from various cell types; the small peptide transporter (PepT1, SLC15A1), which transports di- and tripeptide as well as numerous therapeutic compounds; and the apical sodium-dependent bile acid transporter (ASBT, SLC10A2), which plays a key role in intestinal reabsorption and enterohepatic recycling of bile salts, cholesterol homeostasis, and as a therapeutic target for hypocholesterolemic agents [120–122].

5.1. The ATP-Binding Cassette and Solute Carrier Genetic Superfamilies

Organic solutes such as nutrients (amino acids, sugars, vitamins, and bile acids), neurotransmitters and drugs are transferred across cellular membranes by specialized transport systems. These systems encompass integral membrane proteins that shuttle substrates across the membrane by either a passive process (channels, facilitated transporters) or an active process (carriers), the latter energized directly by the hydrolysis of ATP or indirectly by coupling to the cotransport of a counter-ion down its electrochemical gradient (e.g., Na^+, H^+, Cl^-).

Our understanding of the biochemistry and molecular biology of mammalian transport proteins has significantly advanced since the development of expression cloning techniques. Initial studies in *Xenopus leavis* oocytes by Hediger and colleagues resulted in the isolation of the intestinal sodium-dependent glucose transporter, SGLT1 [123,124]. To date, sequence information and functional data derived from numerous transporters have revealed unifying designs, similar energy-coupling mechanisms and common evolutionary origins [119]. The plethora of isolated transporters motivated the human gene nomenclature committee to classify these proteins into a distinct genetic superfamily named SLC (solute carrier). Currently, the SLC class contains 48 families with >500 members (http://www.genenames.org). The ABC superfamily contains 7 families with 55 members [125]; class B contains the well-known multidrug resistance gene (MDR1) derived P-glycoprotein (ABCB1), while class C comprises members of the multidrug resistance protein (MRP) subfamily. With the completion of the human genome project, it can be anticipated that a vast number of membrane transport proteins will be identified without known physiological function. Paulsen and colleagues [126,127] determined the distribution of membrane transport proteins for all organisms with completely sequenced genomes and identified 81 distinct families. Two superfamilies, the ATP-binding cassette (ABC) and major facilitator (MFS) superfamilies account for nearly 50% of all transporters in each organism. The other half of these genes will be members of the SLC superfamily. Furthermore, Paulsen predicts that 15% of all genes in the human genome will code for transport proteins. With a current number of estimated sequence tagged sites (STS) of 30,000 [128], we can expect an additional 4500 membrane transporters to emerge. Thus, the SLC superfamily is anticipated to consist of at least 2300 members; at this moment, only a fraction (10%) has been characterized in certain detail, that is, membrane topology, substrate specificity, organ expression pattern. Table 1 presents a concise overview of ABC and SLC members that have been classified and characterized. It should be pointed out that the family formerly known as SLC21 (OATP) has been reassigned with a distinct gene symbol (SLCO), comprising solute carrier organic anion transporters in six unique subfamilies (Table 2). These transporters play an important role in hepatic drug disposition.

Transporter nomenclature and classification can be somewhat bewildering to the uninitiated; however, the human gene nomenclature committee (HUGO) has taken on the task of directing gene classification and defining distinct subclasses in the ABC and SLC/SLCO families. Although still widely used, clear nomenclature guidelines may eradicate the persistant use of trivial names for these classes of proteins.

5.2. Therapeutic Implications of Membrane Transporters

It has been generally acknowledged that transporters play an important role in clinical pharmacology (Table 3). Several classes of pharmacologically active compounds share transport pathways with nutrients [129]. A substantial role has been recognized for transport proteins in oral absorption and drug bioavailability [130]; drug resistance, for example, efflux of antineoplastic compounds from tumor cells mediated by multidrug resistance gene products [131,132]; excretion of drugs and their metabolites, mediated by transporters in the kidney and liver; drug toxicity [133]; and drug pharmacokinetics and pharmacodynamics [134–136]. Furthermore,

Table 1. Overview of the ABC and SLC Genetic Superfamilies: Nomenclature and Expression

Gene Symbol[a]	Name/Substrate	Trivial Names	Organ Expression
ATP-binding cassette family			
ABC-A [14]	Cholesterol efflux regulatory proteins, photoreceptor proteins	CERP, RMP	
ABC-B [9]	Efflux transporters, multiple drug resistance	MDR, TAP, P-gp	Epithelial cells, tumor cells
ABC-C [14]	Cytic fibrosis transconductance regulators, multiple drug resistance associated proteins	CFTR, MRP	Lung (CFTR), liver (MRP)
ABC-D [8]	Cholesterol, fatty acid transporters, adrenoleukodystrophy-associated	ALD, ALDL	Brain
ABC-E [1]	RNase L Inhibitor	OABP, RNS4I	Ubiquitous
ABC-F [3]	TNFα-stimulated ABC-member, non-membrane bound	ABC50, GCN20	Synoviocytes
ABC-G [5]	Eye pigment transmembrane permeases	WHITE	Eye
Solute carrier family			
SLC1 [7]	High affinity glutamate transporter	EAAT	Neurons, kidney, intestine, brain, retina
SLC2 [17]	Facilitated glucose transporters 1	GLUT	Most cells
SLC3 [2]	Cystine, dibasic and neutral amino acid transporters	ATR, RBAT	Kidney and intestine
SLC4 [9]	Anion exchangers [1–3], sodium bicarbonate cotransporter [4–10]	EPB, AE1, NBC	Erythrocytes, brain, GI tract, kidney, reproductive organs
SLC5 [7]	Sodium/glucose cotransporter, iodide transporter (A5), vitamin transporters (A6), choline (A7)	SGLT, SMVT, NIS	Intestine, kidney, brain, thyroid
SLC6 [22]	Neurotransmitter transporters, GABA	GAT, GLYT	Brain
SLC7 [16]	Cationic amino acid transporter, y +	LAT, CAT	Ubiquitous (intestine, kidney)
SLC8 [3]	Sodium/calcium exchangers	NCX	Heart, brain, retina skeletal and smooth muscles
SLC9 [17]	Sodium/hydrogen exchangers	NHE	Ubiquitous (intestine, kidney)
SLC10 [7]	Sodium/bile acid cotransporters	Ntcp, ASBT	Liver, ileum
SLC11 [2]	Proton-coupled divalent metal ion transporters	NRAMP, DCT	Ubiquitous
SLC12 [9]	Sodium/potassium/chloride transporters	NKCC	Ubiquitous (kidney)
SLC13 [5]	Sodium/sulfate and dicarboxylate symporters	NaSi, NADC	
SLC14 [2]	Urea transporters	UT	Kidney and red cells
SLC15 [5]	Oligopeptide transporters	PepT	Intestine, kidney
SLC16 [14]	Monocarboxylic acid transporters	MCT	Erythrocytes, muscle, intestine, and kidney
SLC17 [9]	Sodium phosphate transporters	NPT	Kidney, intestine
SLC18 [3]	Vesicular monoamine transporters	VAT	Brain
SLC19 [3]	Folate/thiamine transporters	FOLT, THTR	Placenta, small intestine and other tissues
SLC20 [2]	Phosphate transporters	GLVR	
SLC21 [14]	Organic anion transporting polypeptides (obsolete, see Table 3)	OAT, OATP	Liver, kidney, intestine
SLC22 [25]	Organic cation transporters	OCT	Kidney, liver, intestine

(continued)

Table 1. (*Continued*)

Gene Symbol[a]	Name/Substrate	Trivial Names	Organ Expression
SLC23 [4]	Nucleobase transporters	SVCT	Brain, eye, intestine, kidney, liver
SLC24 [6]	Sodium/potassium/calcium exchangers	NCKX	Ubiquitous
SLC25 [46]	Mitochondrial carriers (citrate, adenine, carnitine)	CTP, ANT	Liver, gut, heart,
SLC26 [11]	Solute carrier family 26 (sulfate, mostly undefined)	SAT, DTD	Cartilage and intestine
SLC27 [6]	Fatty acid transporters	FATP	Adipocytes, skeletal, muscle, heart, and fat
SLC28 [3]	Sodium-coupled nucleoside transporters	CNT	
SLC29 [4]	Nucleoside transporters	ENT	Ubiquitous
SLC30 [10]	Zinc transporters	ZNT	Ubiquitous
SLC31 [3]	Copper transporters	COPT	Most organs
SLC32 [1]	GABA vesicular transporter	VGAT	
SLC33 [1]	Acetyl-CoA transporter		
SLC34 [3]	Sodium phosphate	NaPi3, NPT	Kidney, lung, pancreas
SLC35 [23]	CMP-sialic acid and UDP-galactose transporter	CST, UGALT	Ubiquitous
SLC36 [4]	Proton/amino acid symporter		
SLC37 [4]	Glycerol-3/glucose-6-phosphate transporter	G6PT1	
SLC38 [11]	Unknown substrates		
SLC39 [14]	Zinc transporter	ZIRTL	
SLC40–48 [23]	Unknown substrates/function		

[a] Number in parentheses indicate the number of presently known family members.

the pathophysiology of several hereditary diseases (i.e., clearly defined phenotypes shown to be inherited as a monogenic Mendelian traits) has been attributed to mutations in transport proteins. Most of these mutations in human genes and genetic disorders have been recorded and can be found online at the National Institute for Biotechnology Information (http://www.ncbi.nlm.nih.gov) OMIM database (Online Mendelian Inheritance in Man). For example, congenital glucose-galactose malabsorption syndrome is caused by a defect (D28N, D28G) in SGLT1 [137] and bile acid malabsorption syndrome can be attributed to a single C to T transition in the ASBT gene, resulting in a P290S mutation that abolishes bile acid reuptake [138].

In summary, solute transporters play an invaluable role in fundamental cellular processes in health and disease, and function as important mediators governing all aspects of

Table 2. Solute Carrier Organic Anion Transporter Family

SLCO1A2	OATP, OATP1A2, OATP-A
SLCO1B1	OATP-C, LST-1, OATP1B1
SLCO1B3	OATP8, OATP1B3
SLCO1C1	OATP-F, OATP1C1, OATP1
SLCO2A1	PGT, OATP2A1
SLCO2B1	OATP-B, OATP2B1
SLCO3A1	OATP-D, OATP3A1
SLCO4A1	OATP-E, OATP4A1
SLCO4C1	SLC21A20, OATP4C1, OATPX, OATP-H
SLCO5A1	OATPRP4, OATP-J, OATP5A1
SLCO6A1	OATP6A1, OATPY, MGC26949, CT48

Table 3. Top-Selling 100 FDA Approved Drugs

Molecular Target	Market %
Transporters/channels	30
Membrane receptors	25
Enzymes	20
Nuclear receptors	15
Foreign molecules (pathogens)	5

drug therapy. Despite the apparent clinical importance of SLC proteins, the knowledge of their structure and mechanism of action has lagged far behind the knowledge of these properties of proteins in general, yet this might change in the future if more research and technologies are applied to this area.

5.3. Structural Models of Transport Proteins and Methods to Design Substrates

The primary and—to a lesser extent—secondary structures of many transport systems are known. To date only 152 membrane proteins, primarily from bacterial origin, have yielded to crystallization. The resulting high-resolution structures deposited in the Protein DataBank (PDB) comprise mostly ion channels, bacterial porins, aquaporins, membrane-associated enzymes, photosystems and ATPases; MFS members are represented by several ABC transporters (BtuCD, Sav1866, EmrD, MalF) and recently obtained structures for bacterial SLC transporters of lactose (LacY), glycerol-3-phosphate (GlpT) [139], zinc (YiiP) [140], Leu (LeuT) [141], and Asp/Glu (Glt_{Ph}) [142,143]. Obtaining membrane protein structures has significantly accelerated due to strong support by programmatic initiatives from the NIH; however, overall progress is still slow and in need of alternative methods. *In silico* techniques are a highly useful tool to circumvent the difficulties associated with traditional crystallization techniques and fill the gap between our knowledge of transporter structure and transporter protein properties [144]. This trend is confirmed by an increasing number of publications that project transporter structure and mechanism using homology models as well as combinatorial *in silico* approaches that are consecutively validated by empirical methods (reviewed in Ref. [145]). It is well recognized that any two proteins that show sequence homology (i.e., share sufficient primary structural similarity to have evolved from a common ancestor) will prove to exhibit strikingly similar 3D structures [146]. Furthermore, the degree of tertiary structural similarity correlates well with the degree of primary structural similarity. Phylogenetic analyses allow application of modeling techniques to a large number of related proteins and additionally allow reliable extrapolation from one protein family member of known structure to others of unknown structure. As postulated by Kaback and coworkers [147,148], once 3D structural data are available for any one family member, these data can be applied to all other members within limits dictated by their degrees of sequence similarity.

5.4. Techniques for Studying Integral Membrane Protein Structure

There are various approaches to study the topography of integral membrane proteins in the absence of a resolved crystal structure. *In vitro* and *in vivo* translation of constructs containing one or more transmembrane sequences, epitope localization, reaction with sided reagents, and proteolysis of the purified membrane-inserted protein have all been used individually or in combination. Each has advantages and disadvantages, and only rarely does one method enable conclusive topographic analysis of the membrane-embedded segments of a polytopic integral membrane protein.

Considerable progress has been made over the past 10 years by the group of Kaback at UCLA, who have taken numerous biophysical approaches toward studying the structure of the bacterial transporter lactose permease [149]. Table 4 lists the techniques that this and other groups have used to solve the structure and topology of membrane transporters in the absence of a crystal structure.

Table 4. Techniques for Studying Polytopic Membrane Protein Structure

1.	Site-directed Ala scanning
2.	Membrane insertion scanning
3.	Site-directed thiol crosslinkers
4.	Excimer fluorescence
5.	Engineered divalent metal binding sites
6.	EPR
7.	Metal spin-labeling interactions
8.	Site-directed chemical cleavage
9.	Identification of discontinuous monoclonal antibody probes
10.	N-Glycosylation site engineering

5.4.1. Membrane Insertion Scanning Membrane topology of transporter proteins can be predicted by various computer algorithms; however, none of the existing methods can definitively define these regions and often produce a "sliding window" of groups of hydrophobic amino acid residues that are suitable candidates for the formation of a membrane spanning segment. Examples of these software programs are TopPred II [150] (http://www.biokemi.su.se), the PHD Topography neural network system developed by Rost et al. [151] (http://www.embl-heidelberg.de/predictprotein), or the hidden Markov model described by Tusnády and Simon [152] (http://www.enzim.hu/hmmtop). Sachs and coworkers at UCLA pioneered a technique aptly named "membrane insertion scanning" to determine whether a sequence of amino acids is capable of spanning a membrane. They tested the ability of individual protein segments to function as signal anchors for membrane insertion by placement between a cytoplasmic anchor encompassing the first 101 amino acids of the rabbit H^+, K^+-ATPase (HK M0)-subunit and a glycosylation flag sequence consisting of five N-linked glycosylation sites located in the C-terminal 177 amino acids of the rabbit HK M0-subunit [153]. Stop transfer properties of hydrophobic sequences can be examined in a similar manner using the first 139 N-terminal amino acids of the rabbit H^+, K^+-ATPase-subunit (HK M1) containing the first membrane sequence of the H^+, K^+-ATPase as a signal anchor upstream of individual predicted transport transmembrane sequences linked to the N-glycosylation flag. The membrane insertion scanning technique has been used to determine the membrane topology of various transporters, channels and receptors. A common criticism of membrane insertion studies is the placement of a hydrophobic amino acid sequence out of its physiological or environmental context, thus "forcing" an abbreviated sequence through the membrane. Transmembrane regions in polytopic membrane proteins may require flanking topogenic information to become integrated into the lipid bilayer and thus, topology determination by this technique alone may not be definitive.

5.4.2. Cys-Scanning Mutagenesis Studying the lactose permease of *E. coli*, a polytopic membrane transport protein that catalyzes beta-galactoside/H^+ symport, Frillingos and colleagues [154] used Cys-scanning mutagenesis in order to determine which residues play an obligatory role in the mechanism and to create a library of mutants with a single-Cys residue at each position of the molecule for structure/function studies. In general, this type of study will define amino acid side chains that play an irreplaceable role in the transport mechanism and positions where the reactivity of the Cys replacement is altered upon ligand binding. Furthermore, helix packing, helix tilt, and ligand-induced conformational changes can be determined by using the library of mutants in conjunction with a battery of site-directed techniques.

5.4.3. N-Glycosylation and Epitope Scanning Mutagenesis A technique that can be used in addition to membrane insertion scanning is based on the fact that N-glycosylation occurs only on the luminal side of the endoplasmic reticulum. This method has been successfully used to determine which domains of the receptors/channels are located extracellularly. In general, a glycosylation-free mutant is first designed and subsequently, N-glycosylation consensus sequences (*NXS/T*) are engineered into hydrophilic regions of an aglyco-mutant. Based on the positioning of the glycosylation groups within the extracellular membrane, a molecular weight shift may indicate successful glycosylation and reveals definitive information on protein topology. A drawback of this technique is the dependence of glycosylation on efficiency and accessibility of the consensus sequence to the glycosylation machinery; thus, nonglycosylated mutants do not provide conclusive information about the topology and these data should be interpreted carefully. Regardless, the topology of various proteins has been successfully solved using this technique, including the sodium dependent glucose transporter (SGLT1) [155] and the γ-aminobutyric acid transporter, GAT-1 [156].

Alternatively, small peptide epitopes can be inserted into the extramembranous parts of an SLC protein that are recognized by a well-characterized monoclonal antibody. Covitz and colleagues [157] used this technique to determine the topology of the peptide transporter, PepT1. An epitope tag, EYMPME, was inserted into different extramembranous locations of hPEPT1 by site-directed mutagenesis. The membrane topology was solved by labeling reconstituted, functionally active, EYMPME-tagged hPEPT1 mutants with an anti-EYMPME-monoclonal antibody in nonpermeabilized and permeabilized cells.

5.4.4. Excimer Fluorescence Site-directed excimer fluorescence and site-directed spin-labeling are two particularly useful techniques to study proximity relationships in membrane helices. The experiments are based upon site-directed pyrene labeling of combinations of paired Cys replacements in a mutant devoid of Cys residues. Since pyrene exhibits excimer fluorescence if two molecules are within about 3.5 Å, the proximity between paired labeled residues can be determined. Moreover, interspin distances in the range of 8–25 Å between two spin-labeled Cys residues can be measured in the frozen state. Using this technique, Kaback and coworkers showed that ligands of the lac permease cause a dramatic increase in reactivity that is consistent with the notion that the mutated amino acid positions are transferred into a more hydrophobic environment [158,159].

5.4.5. Site-Directed Chemical Cleavage The insertion of a short reporter sequences, for example, factor Xa protease cleavage sites, into hydrophilic loops has proven to be a useful alternative to N-glycosylation scanning mutagenesis. However, this approach requires isolation of homogeneous preparations of intact membranes, many tedious control experiments, and experimental difficulties associated with protease accessibility are well documented [160]. In general, in-frame factor Xa protease sites are inserted into a target sequence at positions within the NH_2- and COOH-terminal domains, and into hydrophilic loops. The factor Xa protease recognizes the tetra-peptide motif IEGR and specifically cleaves the protein sequence COOH-terminal of the arginine residue [161]. Generally, the recognition motif is tandomly (IEGRIEGR) inserted to increase the probability of cleavage [162]. After digestion of purified protein vesicles with the factor Xa enzyme, fragments are isolated on SDS-PAGE and can be analyzed to further determine membrane topology.

5.5. Case Studies

5.5.1. P-Glycoprotein: Understanding the Defining Features of Regulators, Substrates and Inhibitors The ABC efflux transporter, P-glycoprotein is a large 12 transmembrane-domain bound protein initially noted to be present in certain malignant cells associated with the multidrug resistance phenomenon that results from the P-gp-mediated active transport of anticancer drugs from the intracellular to the extracellular compartment [163]. However, P-gp is normally expressed at many physiological barriers including the intestinal epithelium, canalicular domain of hepatocytes, brush-border of proximal tubule cells and capillary endothelial cells in the central nervous system (CNS) [163]. Expression of P-gp in such locations results in reduced oral drug absorption and enhanced renal and biliary excretion of substrate drugs [164]. Moreover P-gp expression at the blood-brain-barrier is a key factor in the limited central nervous system entry of many drugs. The expressed level of P-gp as well as altered functional activity of the protein due to genetic variability in the MDR1 gene appears also to impact the ability of this transporter to influence the disposition of drug substrates [165].

Interestingly, cytochrome P450 3A4 (CYP3A4), a drug-metabolizing enzyme with broad substrate specificity, appears to coexist with P-gp in organs such as the intestine and liver. These observations lead to the hypothesis that there may be a relationship between these two proteins in the drug disposition process. Wacher et al. have described the overlapping substrate specificity and tissue distribution of CYP3A4 [166] and P-gp. Schuetz et al. found that modulators and substrates coordinately upregulate both proteins in human cell lines [167]. Similarly, P-gp mediated transport was found to be important in influencing the extent of CYP3A induction in the same cell lines and also in mice [168]. More recent data have suggested that there may be a dissociation of inhibitory potencies for molecules against these proteins. Although some molecules can interact with CYP3A4 and P-gp to a similar extent, for the most part the potency of inhibition for CYP3A4 did not predict the potency of inhibition for P-gp, and vice versa [58]. Moreover, not all CYP3A substrates such as midazolam and nifedipine are P-gp substrates [169]. The key to this linkage between P-gp and CYP3A4 appears to be their coregulation by the pregnane-X-receptor at the level of transcription [170,171]. With the description of the X-ray crystal structure of this protein with SR12813 bound, we may be closer to understanding the structural features necessary for PXR ligands. In addition, a pharmacophore for PXR ligands may also enable us to select new drugs that are less likely to induce P-gp and CYP3A4 [172].

In order to account for the observed broad substrate specificities for both CYP3A4 and P-gp, the presence of multiple drug binding sites has been proposed [173–176]. The first elegant experimentally determined signs of a complex behavior for P-gp appeared in 1996 when co-operative, competitive, and noncompetitive interactions between modulators were found to interact with at least two binding sites in P-gp [177]. The multiple site hypothesis was confirmed by other groups [178–180]. Subsequent results have indicated there may be three or more binding sites [181]. Steady-state kinetic analyses of P-gp mediated ATPase activity using different substrates indicates that these sites can show mixed-type or noncompetitive inhibition indicative of overlapping substrate specificities [182]. Other researchers have determined that immobilized P-gp demonstrates competitive behavior between vinblastine and doxorubicin, cooperative allosteric interactions between cyclosporin and vinblastine or ATP, and anticooperative allosteric interactions between ATP, vinblastine, and verapamil [183]. Clearly allosteric behavior by multiple substrates, inhibitors or modulators of CYP3A4 or P-gp complicates predicting the behavior and drug–drug interactions of new molecules *in vivo* and has important implications for drug discovery.

In terms of understanding P-gp structure–activity relationships, photoaffinity experiments have been valuable in defining the cyclosporin binding site in hamster P-gp [184] and indicating that trimethoxybenzoylyohimbine (TMBY) and verapamil bind to a single or overlapping sites in a human leukemic cell line [185]. Additional studies have shown that TMBY is a competitive inhibitor of vinblastine binding to P-gp [186]. The P-gp modulator LY 335979 has been shown to competitively block vinblastine binding [186] while vinblastine itself can competitively inhibit verapamil stimulation of P-gp-ATPase [185]. With the growth in knowledge derived from these and other studies it would be valuable to use structural information to define whether unrelated molecules are likely to interact with P-gp.

Interestingly, there has been over a decade's worth of studies that have identified pharmacophores for this efflux pump that could have been used to help predict P-gp related bioavailability issues. Early computational studies using P-gp modulators such as verapamil, reserpine, 18-epireserpine, and TMBY showed that they could be aligned, suggesting the importance of aromatic rings and a basic nitrogen atom in P-gp modulation [187,188]. A subsequent, more extensive study with 232 phenothiazines and structurally related compounds indicated that molecules with a carbonyl group that is part of an amide bond plus a tertiary amine, were active P-gp inhibitors [189]. A model built with 21 molecules of various structural classes that modulate P-gp ATPase activity suggested these molecules competed for a single binding site [190]. Similarly, 19 propafenone type P-gp inhibitors were then used to confirm the requirement for a carbonyl oxygen, suggested to form a hydrogen bond with P-gp [191]. Others have used MULTICASE to determine important substructural features like CH_2-CH_2-N-CH_2-CH_2 [192], and linear discriminant analysis with topological descriptors [193]. In 1997, the first 3D-QSAR analysis of phenothiazines and related drugs known to be P-gp inhibitors was described [194]. This was followed by Hansch-type QSAR studies with propafenone analogs [195], comparative molecular field analysis (CoMFA) studies of phenothiazines and related drugs [196], and simple regression models of propafenone analogs [197,198]. These latter models confirmed the relevance of hydrogen bond acceptors and the basic nitrogen for inhibitors [197,198] and multiple hydrogen bond donors in substrates 2.5–4.6 Å apart [199]. One study utilizing a diverse array of inhibitors with P-gp ATPase activity noted that size of the molecular surface, polarizability and hydrogen bonding had the largest impact on the ATPase activity [200]. A number of computational approaches and models of P-gp have yielded useful information that is usually derived from a series of structurally related molecules. This is not surprising as *in vitro* studies with P-gp inhibitors frequently take this approach. Neuhoff et al. suggested P-gp inhibitors with high lipophillicity and polarizability were found to be more likely to be high affinity ligands for the verapamil-binding site [201]. However, some complexity arises if one considers more structurally diverse molecules as they may bind to different sites within P-gp. This hypothesis derives from experimental results describing a complex behavior for P-gp such that co-operative, competitive, and noncompetitive interactions between modulators may occur [177], indicative of multiple binding sites within P-gp [178–180].

Specific models addressing the individual P-gp binding sites using a diverse array of inhibitors have been developed by Ekins and Erickson [172]. Utilizing a computational approach to model *in vitro* data derived from structurally diverse inhibitors of digoxin transport in Caco-2 cells, vinblastine, and calcein accumulation in P-gp expressing LLC-PK1 (L-MDR1) cells, vinblastine binding in vesicles derived from CEM/VLB100 cells and, verapamil binding in Caco-2 cells [172] suggested there may be some overlap in the binding sites for these four substrates. The inhibitor pharmacophores were suggested to be mainly large with hydrophobic and hydrogen bonding features. It is hypothesized that vinblastine, digoxin, and verapamil are likely to bind a single site as strong correlations between these models were observed. A simple P-gp substrate pharmacophore was generated using the alignment of verapamil and digoxin, onto which vinblastine was aligned [172]. This model also contained many hydrophobic and hydrogen bond acceptor features. As models are generated for the other binding sites and substrates, they may help in elucidating important structural differences between ligands for each site. Such models will be of value for modulating the bioavailability of drugs and understanding the locations of the P-gp binding site(s). The latter may be of particular importance as we are now in an era in which membrane proteins are being crystallized. The 22 Å resolution X-ray structure of the monomer for the related ABC family member (MRP1/ABCC1) is suggested to be structurally similar to P-gp as it appears to possess a pore region ringed by protein [202]. The determination of bacterial ABC family members, for example, Sav1866 [203–205], provided further structures on which to build homology models for P-gp based on amino acid sequence similarity [206] and has resulted in the construction of models for various other ABC transporters [207–210]. It is suggested that P-gp pharmacophores could be docked into these structures and the amino acids likely to correspond on both monomers defined. The question of whether P-gp behaves like this flippase or like more traditional pores still remains, but computational techniques will be at the forefront of continuing to provide some enlightenment and aid in drug discovery.

5.5.2. The Intestinal Peptide Transporter (PepT1)

Introduction Over the past two decades, targeting delivery to nutrient transport systems in the gastrointestinal tract has emerged as a strategy to increase oral availability of poorly absorbed drugs and it has met with several successes. In this approach, drug penetration is enhanced by coupling a drug molecule to natural ligands for nutrient transporters. The toxicity and immunogenicity of drug–ligand conjugates is expected to be inherently low due to the use of endogenous substrates. Alternatively, an approach named "substrate mimicry" can be used; this approach aims to create novel drug entities that mimic the three-dimensional features of natural ligands. Both approaches should result in compounds that are recognized by a specific transport protein embedded in the enterocyte brush-border membrane thereby facilitating transport across the intestinal wall.

Physiology of PEPT1 In general, the intestinal small peptide carrier (PepT1) is a proton-coupled, low affinity, active, oligopeptide transport system with a broad substrate specificity. In addition to transporting its natural substrates, di- and tripeptides occurring in food products [211–215], it shows affinity toward a broad range of peptide-like pharmaceutically relevant compounds, such as β-lactam antibiotics [216,217] and angiotensin-converting enzyme (ACE) inhibitors [218–228]. In fact, these molecules can oftentimes be viewed as "peptide-like" in their molecular composition (Fig. 6). For this reason, the transporter has been recognized as an important intermediate in the oral bioavailability of peptidomimetic compounds [220]. However, the lack of knowledge regarding structural specificity toward its substrates has prevented the use of this transporter on a more rational basis. In addition, its cellular localization to the apical membrane does not provide a mechanism for cellular exit into the basal compartment [229,230]. Currently, there exists a keen interest in understanding the structural determinants for substrates and inhibitors of this transport protein.

PepT1 belongs to a larger family of oligopeptide transporters, the proton/oligopeptide transporter (POT) family, where currently only two members have been identified in humans, hPepT1 and hPepT2. POT family members have two characteristic protein signatures assigned by Steiner and Becker, known as the PTR2 family signatures [231,232]. The first of these signature sequences include the end of transmembrane span 2 (TMD2), the intracellular loop, and TMD3 (Fig. 7). The second PTR2 signature corresponds to the core of TMD5. A third consensus sequence (GTGGIKPXV) has been proposed by Fei et al. [233], this sequence is well conserved between mammalian and *C. elegans* peptide transporters CPTA and CPTB, but is not specific to or well conserved in other members of the POT superfamily.

Structure–Transport Relationship of PEPT1 Structural information on PepT1 has been limited to its primary sequence and predicted structural membrane topology. Hydropathy

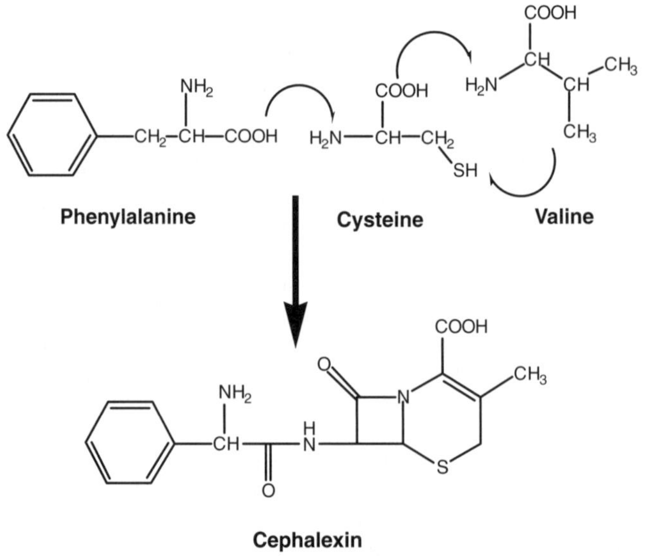

Figure 6. Molecular building blocks of the β-lactam antibiotic cephalexin. The three amino acids phenylalanine, valine, and cysteine are shown to merge via (pseudo)peptide bonds to form a peptidomimetic compound with the molecular features of a tripeptide.

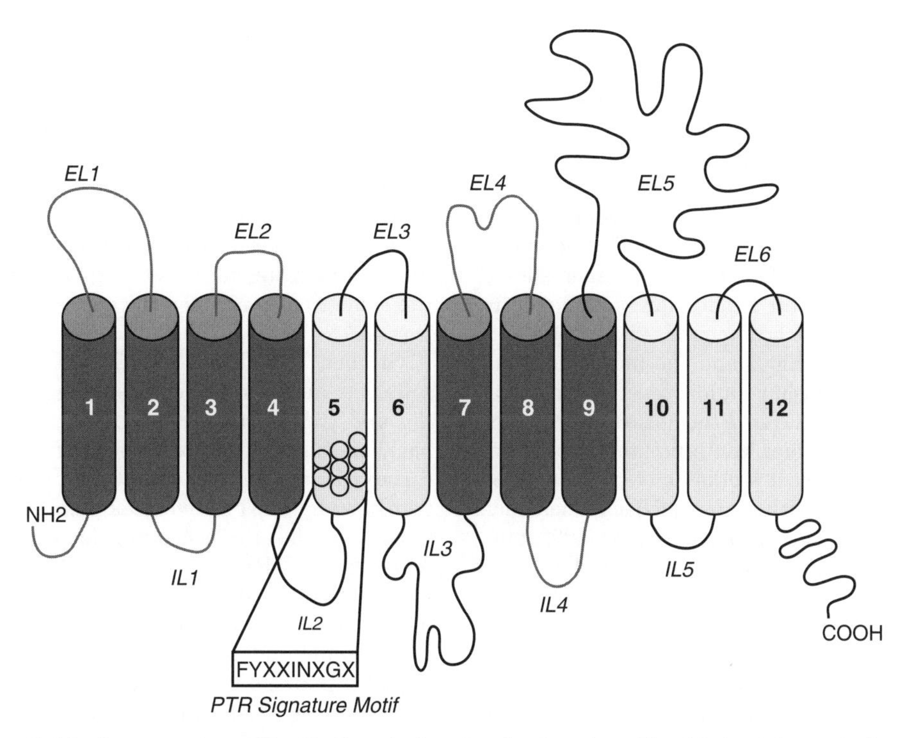

Figure 7. Membrane topology of PepT1. Protein domains that have been identified as relevant in determining the functional characteristics of the protein are indicated in red. IL, intracellular loop; EL, extracellular loop. (This figure is available in full color at http://mrw.interscience.wiley.com/emrw/9780471266945/home.)

analysis of the human, rabbit, and rat PepT1 isoforms have predicted the presence of 12 transmembrane domains in each isoform [233]. This model has been partially proven by other investigators [157,234].

The human PepT1 sequence has been predicted to contain a 2127 base pair (bp) open reading frame, that encodes a 708 amino acid, 79 kDa protein, with an estimated pI of 8.58 [233]. Hydropathic analysis also predicts that the N- and C-termini are located on the cytoplasmic side of the membrane. Site-directed mutagenesis, utilizing EYMPME (EE)-epitope tag insertion at different locations of the human PepT1 transmembrane regions demonstrated that the COOH-terminal remains intracellular, while epitope-tags at positions 106 and 412 showed that the predicted loops between 3TM-4TM, and 9TM-10TM, respectively, were localized extracellularly [157]. These results support the predicted loop and TMD numbers and orientations from TMD4 through the C-terminus [157]. The results utilizing EE epitope insertions in the amino terminal region were inconclusive, possibly due to EE effects on function, leaving ambiguity to the predicted structure from the N-terminus to TMD3.

Since no three-dimensional structure of the transporter is available, design of new substrates for PepT1 has relied on indirect structure–affinity relationships. Early studies were directed to define a pharmacophoric pattern for this transporter [235–237]. A pharmacophore or "recognition site" is defined based on a common arrangement of essential atoms or groups of atoms appearing in each active molecule. In many published papers the active analog approach (AAA), originally developed by Marshall and coworkers [238], has proven to be useful in rationalizing and predicting pharmacological data of active and inactive substrates. In the AAA, the structural requirements common to a set of compounds showing affinity for the same transporter may be used to define a pharmacophoric pattern of atoms or groups of atoms mutually oriented in space, which are necessary for binding to this transporter. Two factors have to be considered for the successful application of this technique: the included compounds should be structurally homogenous and the individual compounds should have a low level of conformational flexibility. Since the natural substrates, di- and tripeptides, contain many freely rotatable bonds, most studies have concentrated on substrates with more rigid backbones, such as a β-lactam nucleus [235,236,239,240]. Therefore, the AAA is limited in that it requires structurally similar rigid molecules. One study used β-lactams and showed that a carboxylic carbon (likely to position in a positively charged pocket), two carbonyl oxygen atoms (hydrogen bond acceptors), a hydrophobic site and finally an amine nitrogen atom (hydrogen bonding region) were important features of substrates [239].

Another study examined the three-dimensional structural features of three structurally closely related PepT1 substrates: enalapril, enalaprilat, and lisinopril (Fig. 7) [241]. Enalapril is an ester prodrug of the pharmacologically active enalaprilat. Following oral administration of enalapril, the active compound (enalaprilat) is formed by bioconversion of enalapril. Enalaprilat, a diacid, binds slowly and tightly to ACE, producing well-defined clinical effects, but is poorly absorbed from the gastrointestinal tract (3–12% bioavailability) [242]. The prodrug approach of esterifying enalaprilat to enalapril is required in order to enhance the oral bioavailability to 60–70%. Using a combined *in vitro* and molecular modeling approach we showed that intramolecular hydrogen bond formation between the lysyl side chain in enalaprilat and its carboxylic acid groups may explain decreased affinity for PepT1 [241].

Structural studies of PepT1 substrates were extended using comparative molecular field analysis of 10 known substrates for the peptide transporter with data derived from an *in situ* rat model [243]. The CoMFA approach required manual alignment of relatively rigid molecules however it allowed explanation of the variation in the permeability with respect to steric and electrostatic interaction energies of the molecules [243]. This 3D-QSAR produced by our laboratory provided a valuable starting point for future prediction of the affinity of substrates for this transporter and at the same time confirmed the earlier findings from AAA studies.

Studies by other research groups have sought to address the requirements for binding and transport by PepT1. The allowable backbone distance between the amine and the carboxylic acid terminals for binding to and transport by PepT1 have been demonstrated in part with a series of ω-amino fatty acids (ω-AFA) [244]. These studies clearly demonstrated that 4–10 methylene carbon groups (CH_2 units) could be accommodated between the termini and allow direct binding with PepT1. Binding of ω-AFAs was demonstrated to inhibit D-Phe-Ala uptake competitively in transiently transfected hPepT1 cells, with the competitive inhibition (EC_{50} values) appearing not to change substantially with chain lengths from 5 to 11 CH_2 units between the terminal, functional groups [244]. The conclusions of this study indicate that a peptide bond is not essential for substrates of PepT1, and that two ionized amino or carboxyl groups with at least four CH_2 units, providing a distance of 500–635 pm, between them are required for transport by PepT1 [244]. This study also raised considerable concerns about conformation during presentation since the $CH_2{-}CH_2$ midsection has considerable flexibility in contrast to a peptide bond.

The distance between the charged functionalities for PepT1 recognition has been demonstrated to be critical [244]. Insights into the affinity of β-amino groups were also elucidated in these studies, where 8-amino-octanoic acid effectively inhibited D-Phe-Ala uptake, whereas 2-amino-octanoic acid had no inhibitory effect [244]. Interestingly, further studies have demonstrated that the α- or β-amino carbonyl functionality may affect PepT1 substrate affinity, and serve as the key determinant in PepT2 affinity preference over PepT1, where coexpressed [245]. For example, the presence of a β-amino group resulted in the substrate having a higher affinity for transport by PepT2 compared to PepT1 [245]. However, anionic β-lactam antibiotics, lacking an α-amino group, demonstrated a preference for PepT1.

Bailey and colleagues further explored PepT1 substrate specificity by using a meta-analysis of K_i data for 42 substrates across eight classes of molecules [246]. The findings by these authors were in agreement with previous studies and provided a template consisting of an N-terminal NH_3 site, a hydrogen bond to the carbonyl group of the first peptide bond, a hydrophobic pocket and a carboxylate binding site [246]. This model was also used to classify the compounds as possessing high, medium, or low affinity for PepT1. Another group has generated K_i values for 23 β-lactams and suggested their data supported the need for all of the structural features previously described above for recognition by the transporter [247]. Moore et al. used Caco-2 cells and assessed the inhibition (K_i) of Gly-[^{3}H]-L-Pro transport by ACE inhibitors as a means to understand dipeptide transport. A two-dimensional model was constructed that indicated the types of functional groups favored for inhibition [218].

Other studies have shown the importance of certain amino acid residues in determining PepT1 peptide transport activity [248,249]. Site-directed mutations of single amino acids located within the PepT1 transmembrane domains showed that Trp at position 294 and Glu at position 595 reduced significantly the glycylsarcosine uptake by human embryonic kidney cells (HEK293), while Tyr at position 167 (TMD5) inactivated the transporter completely [248]. PepT1 has shown to present another important structural characteristic: the presence of conserved histidyl residues (H) in several transmembrane regions. Sequence alignment demonstrated the presence of H57, H121, and H260 residues (corresponding positions in the PepT1 sequence) in PepT1 and PepT2. This group is speculated to be involved in the H^+-binding, and thus being fundamental for transport function. Histidyl residues at position 57 and 121 have been demonstrated to be essential to maintain PepT1 function, while residue at position 260 remains to be completely defined [249]. The molecular mechanism(s) by which these residues may interact with H^+ remains to be elucidated.

5.5.3. The Apical Sodium-Dependent Bile Acid Transporter

Introduction The "apical sodium-dependent bile acid transporter" (ASBT) has been targeted for oral drug delivery as well. The high efficacy of ASBT combined with its high capacity makes this system an interesting target for drug delivery purposes, including local drug targeting to the intestine and improving the intestinal absorption of poorly absorbable drugs. ASBT has recently received much attention because of its pivotal role in cholesterol metabolism. Inhibition of intestinal bile acid re-absorption elicits increased hepatic bile acid synthesis from its precursor cholesterol, thereby lowering plasma cholesterol levels [250]. Application of this approach has met considerable clinical success using unspecific bile acid sequestrants, such as cholestyramine and colestipol [251]. Currently, several novel ASBT inhibitors are in clinical trials for the treatment of hypercholesterolemia [252].

Physiology of ASBT Bile acids mediate the digestion and absorption of fat and fat-soluble vitamins [253]. The total bile acid pool in humans, 3–5 g, circulates 6–10 times a day, giving rise to a daily bile acid turnover of 20–30 g in humans. Only 0.2–0.5 g of bile acids is lost in feces per day and this amount is repleted by the *de novo* synthesis of bile acids [254]. A detailed description of the enterohepatic circulation of bile acids may be found in several reviews [255–260]. The primary bile acids cholate and chenodeoxycholate are synthesized from cholesterol in the liver mediated by the enzyme cholesterol 7α-hydroxylase (CYP7A) (Fig. 8). The activity of this enzyme is under inhibitory feedback control by bile acids. In light of the importance of this pathway for the removal of cholesterol, the mechanisms for transcriptional regulation of the CYP7A gene have been extensively studied. In the ileum, conjugated bile acids are reabsorbed by ASBT (previously named IBAT, for "ileal bile acid transporter") gene SLC10A2.

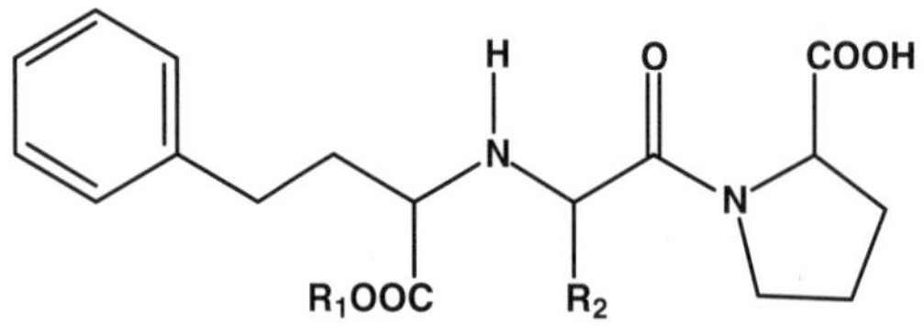

	R_1	R_2
Enalapril	N C_2H_5	N CH_3
Enalaprilat	N H	N CH_3
Lisinopril	N H	$N(CH_2)_4NH_2$

Figure 8. Molecular structure of the ACE-inhibitors enalapril, enalaprilat, and lisinopril. Enalapril and lisinopril are substrates for PepT1.

Within the cytoplasm of the enterocyte, a second 14–15 kDa soluble protein binds the absorbed bile acid and mediates its transfer across the cell. This protein is generally known as ileal bile acid binding protein (iBABP) [261], although it was originally named gastrotropin [262, 263]. The exact interaction between ASBT and iBABP is currently unknown, but there has been speculation that iBABP may bind to ASBT at the cytosolic surface and may play a role in mediating bile acid affinity for ASBT [264]. In addition to iBABP, 20, 35, and 43 kDa proteins that have remained unidentified appear to bind bile acids in the cytosol [265], although the specific affinity of these proteins for bile acids is probably low. For example, the putative 43 kDa protein was shown to share high homology with actin, which plays a role in many transport processes in the cell [266].

Expression of iBABP and ASBT is regulated by the farnesoid X receptor (FXR), a nuclear orphan receptor [267,268]. A specific binding site for FXR was found on the promotor region of the iBABP gene, named bile acid response element (BARE) [269]. The response was greatest in the presence of chenodeoxycholic acid (CDCA), whereas cholic acid was much less effective and the secondary bile acids deoxycholic and lithocholic acid had variable responses.

Molecular regulation of CYP7A appears to be regulated by the liver X receptor (LXR), a member of the steroid receptor superfamily that, in turn, appears to be regulated by oxysterols [270]. Two isoforms of LXR, α and β, have been identified; the α isoform is responsible for regulating CYP7A and some QSAR and pharmacophore studies suggest the 24-oxo ligands acting as hydrogen bond acceptors bind tightly to LXRα and may be candidates for the natural ligand [271,272]. ASBT is primarily expressed on the apical membranes of ileal enterocytes in mammals, although weak translation has been observed in the jejunum; however, these observations remain controversial and may have resulted from (patho) physiologic induction. Furthermore, ASBT has been detected in rat bile duct epithelial cells (cholangiocytes) [273], where it probably functions as part of a bile acid secretion feedback mechanism. Microinjection of guinea pig and rabbit ileal mucosal Poly(A +) mRNA into *Xenopus laevis* oocytes resulted in translation of a functional Na + /bile acid carrier that is expressed in the surface membrane [274]. As expected, incubation with rabbit jejunal-mucosa Poly(A +) mRNA did not result in expression of the bile acid transporter. Using timed photoaffinity labeling techniques the brush-border [275], cytosolic [265] and basolateral proteins [276] involved in the transfer of bile acids across the rat ileal enterocyte were revealed. A putative transport protein involved in coupled Na + -bile acid transfer across the ileal brush-border membrane had a molecular weight of 99 kDa [275,277]. Dawson and coworkers [278,279] were the first to clone and characterize a 43 kDa protein as the rat ileal bile acid transporter protein. The 99 kDa protein that was detected previously can be explained as a dimeric form of the glycosylated transport protein (Fig. 9).

Subsequent electrophsyiological characterization revealed that ASBT is an electrogenic, active cotransporter driven by a Na + -gradient across the apical membrane of ileocytes with a 2:1 Na^+:bile acid coupling stoichiometry [280]. To date, structural information on ASBT has been limited to its primary (sequence) and secondary (membrane topology) structure. Topography models of ASBT predict seven transmembrane (7TM) re-

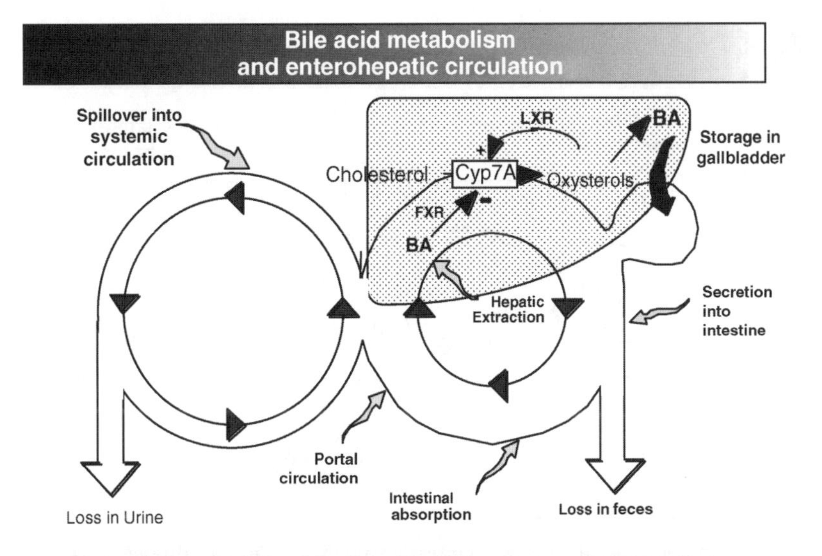

- Secretion of 600–800 mL of bile per day.
- Pool of 3–5 g of bile acids, turnover of 20–30 g per day.
- Efficient recycling by transport proteins in liver and intestine.
- 0.2–0.5 g loss in feces per day.
- Synthesized in the liver from cholesterol.
- Feedback regulation by bile acids through FXR (–) and LXR (+).

Figure 9. Biosynthesis of bile acids and the enterohepatic circulation. Bile acids are synthesized from cholesterol in the liver under feedback regulation of the nuclear orphan receptors farnesoid X receptor (FXR) and lignane X receptor (LXR). They are stored in the gallbladder and released via the bile duct into the duodenum, where they aid in the digestion of dietary fats. Intestinal uptake of bile acids takes place along the entire length of the small intestine, but active reabsorption is confined to the distal ileum to minimize loss off bile salts in the feces. The portal circulation carries bile acids from the intestine to the liver where they are actively absorbed by hepatocytes and secreted into bile. (This figure is available in full color at http://mrw.interscience.wiley.com/emrw/9780471266945/home.)

gions [279], but eight or nine membrane spanning regions have been suggested by topology/hydropathy analysis [281]. Experimental evidence indicates a *trans* position of the N- and C-terminal protein domains (N_{exo}/C_{cyt}) [138,279,282], which effectively eliminates the putative 8TM model. Based on membrane insertion scanning techniques, a 9TM model was suggested by Hallén et al. [281]; however, this model suggests the presence of two very short TM domains, presumably in the β-helix conformation. Overall, the 7TM model appears to be the most feasible topology for ASBT to date (Fig. 10). Studies with antibodies raised against the terminal amino acid sequences verify that the amino terminal is located extracellularly whereas the carboxy terminal is located at the cytoplasmic side of the membrane [283]. This topology makes ASBT unique from other members of the SLC family in that it does not adhere to the "positive-inside" rule [284]. Human ASBT contains 13 cysteine residues, of which 12 are conserved in the hamster, rat, rabbit, and mouse. Only 4 of these 12 cysteines are conserved in the human Hepatic Bile Acid Transporter (Ntcp), which has 37% homology and 48% similarity with ASBT, and only two cysteines are conserved in P3-protein, a related orphan transporter.

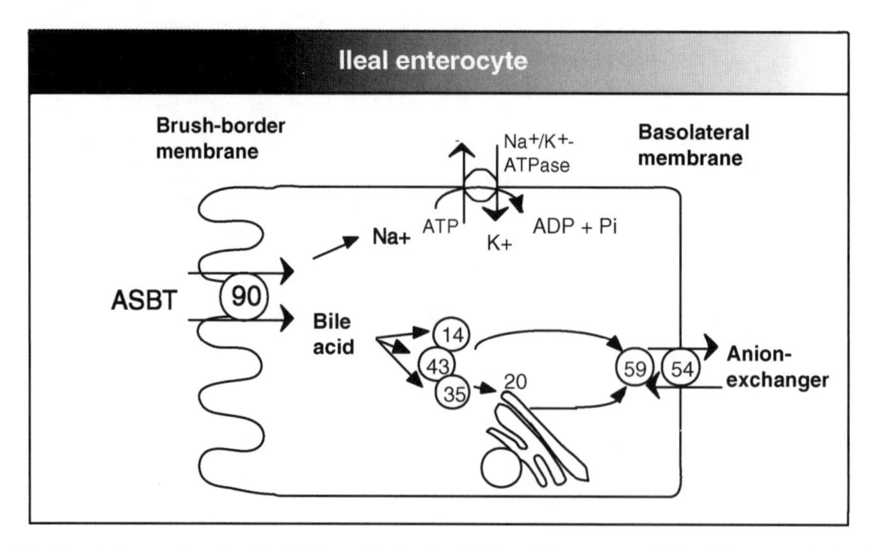

Figure 10. Physiology and molecular biology of intestinal bile acid transport. Bile acids are actively absorbed in enterocytes via a sodium-dependent cotransporter, ASBT. The sodium gradient is maintained by the sodium-potassium ATPase, located at the basolateral membrane. In the cytosol, bile acids are shuttled through the cell by the aid of various proteins, most importantly the ileal bile acid binding protein, iBABP. An anion exchanger transports bile acids across the basolateral membrane into the portal circulation.

Since bile acids are biosynthesized from cholesterol and are the only forms for cholesterol to be excreted *in vivo*, a specific nonabsorbable inhibitor of the ileal bile acid transporter system would lower plasma cholesterol level by blocking the intestinal reabsorption of bile acids and consequently raising the conversion of cholesterol to bile acids. Several compounds have been shown to be able to lower serum cholesterol in animal studies by specifically inhibiting ASBT [285–288]. These findings substantiate the feasibility of using ASBT as the new pharmacological target for cholesterol-lowering therapy.

Structure–Activity Relationships for ASBT The studies with endogenous bile acids have led to physiological understanding of the function of bile acids and the enterohepatic circulation. The search for a deeper understanding of the molecular mechanism behind the affinity and recognition of molecules by the bile acids carriers in both ileum and liver, has led researchers to modify bile acids and study the carrier affinity of these compounds. These modifications typically entail either the substitution of the hydroxyl groups at the 3, 7, or 12 positions by other functionalities, or the addition to or alterations at the C-17 side chain (Fig. 11).

In a series of seminal experiments, Lack and Weiner were the first to establish a basic structure–activity relationship for intestinal bile acid transport using the rat everted sac model [289]. Important additional findings were recorded by the group of Kramer. The following general observations have been reported:

(1) The K_m is related to whether the bile acid is conjugated or not [290,291]. In general, conjugated bile acids have a twofold higher K_m value than their unconjugated parent compounds [292].
(2) The *V*max is independent of conjugation but appears to be related to the number of hydroxyl groups attached to the sterol nucleus: trihydroxy (dihydroxy > monohydroxy [292,293]. However, no single hydroxy group is essential for transport. Triketo bile acids, such as taurodehydrocholic

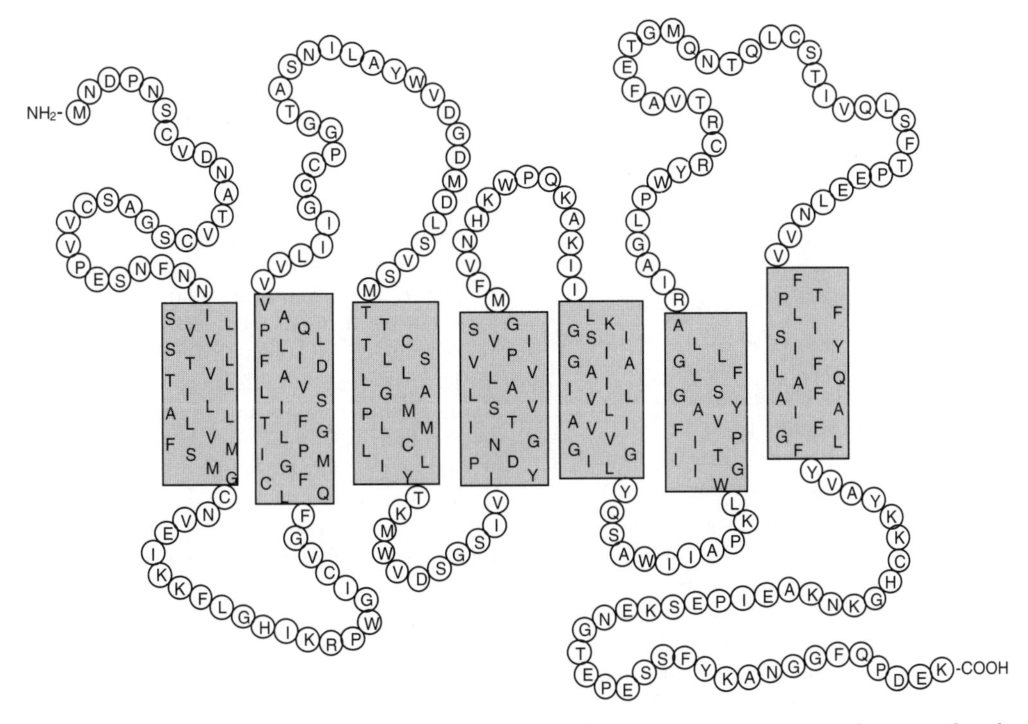

Figure 11. Membrane topology of ASBT. Hydropathy analysis proposes a seven transmembrane topology for ASBT. Studies with antibodies to terminal epitopes have confirmed the inside-out orientation of this membrane protein.

acid, are devoid of all hydroxyl groups and show considerably less transport capacity [294,295]. Krag and Phillips [296] found similar structure–*V*max relationships in humans; however, no correlation was observed between *K*T values and conjugation.

(3) For efficient transport, a bile acid molecule must possess a single negative charge [297] that should be located on the C-17 side chain of the sterol nucleus [298].

(4) Although bile acids with two negative charges around the C-17 position give minimal active transport, the addition of an extra negative charge in the form of sulfonation at the C-3 position does not preclude active transport [299]. Replacing the anionic moiety from the C-17 side chain to position 3 results in a complete loss of affinity [298].

(5) Stereospecificity of the hydroxyl groups on the sterol nucleus was shown by substitution of this group at the 3-position with a hydroxyethoxy moiety [300]. The 3α-isomer was able to inhibit transport of [^{3}H]-taurocholate in rabbit ileal brush-border membrane vesicles, whereas the 3β-isomer showed very weak affinity and was unable to inhibit taurocholic acid transport.

Lack et al. have proposed that the recognition site for carrier-mediated bile acid transport is a hydrophobic pocket on the membrane surface that consists of three components: a recognition site for interaction with the steroid nucleus; a cationic site for Coulombic interaction with the negatively charged side chain; and an anionic site for interaction with Na+. Supposedly, this anionic site could be responsible for the reduced affinity of bile acid derivatives with a dianionic side chain [301]. Generalizations on the structural requirements for ASBT affinity include

(1) the presence of at least one hydroxyl group on the steroid nucleus at position 3, 7, or 12;
(2) a single negative charge in the general vicinity of the C-17 side chain; however, an additional negative charge at the C-3 position does not prevent active transport;
(3) substitutions at the C-3 position do not interfere with active transport;
(4) substitutions at the C-17 position do not interfere with transport as long as a negative charge around the C-24 position remains present;
(5) a *cis* configuration of rings A and B within the sterol nucleus.

More recently, our group and the joint laboratories of Baringhaus and Kramer have reported on the three-dimensional structural requirements of ASBT that will be of specific use in the development of novel substrates for this transport protein. Using a training set of 17 chemically diverse inhibitors of ASBT, Baringhaus and colleagues [302] developed an enantiospecific Catalyst pharmacophore that mapped the molecular features essential for ASBT affinity: one hydrogen bond donor, one hydrogen bond acceptor, and three hydrophobic features. For natural bile acids they found that (a) ring D in combination with methyl-18 map one hydrophobic site and methyl-21 maps a second; (b) a α-OH group at position 7 or 12 constitutes a hydrogen donor; (c) the negatively charged side chain comprises the hydrogen bond acceptor; and (d) the 3α-OH group does not necessarily map a hydrogen bond functionality. The ASBT pharmacophore model is in good agreement with the 3D-QSAR model we previously developed using a series of 30 ASBT inhibitors and substrates [303]. In this study, the electrostatic and steric fields around bile acids were mapped using comparative molecular field analysis to identify regions of putative interaction with ASBT. This model enabled the *in silico* design of substrates for ASBT, especially for conjugation at the C-17 position. It should be pointed out that our model was silent on biological diversity around the C-3 position because of limited molecular variability in this region; however, alterations at that position have been described in detail by Kramer et al. [304]. The two indirect models outlined above should facilitate the rational design of (pro)drugs for targeting to ASBT. There may also be the possibility of combining models from different groups although this may present its own problems.

One serious limitation of indirect structure–activity models is the inherent lack of information on substrate–transport protein interactions at the molecular level. Currently, there is no crystal structure for ASBT and it is anticipated that suitable methods for crystallizing intrinsic membrane proteins will not be available for several years. As an alternative to high-resolution molecular information, we have recently constructed a model for the ligand binding of the ASBT transport protein using knowledge-based homology modeling [305]. Using the transmembrane domains of bacteriorhodopsin as a scaffold, the extracellular loop regions were superposed and optimized with molecular dynamics simulations. By probing the protein surface with cholic acid we identified five binding domains, three of which are located on the outer surface of the protein (Fig. 12). Another binding domain was located between two extracellular loops in close enough proximity to the first binding region to accommodate di-cholic acid conjugates; this observation explains the extreme inhibitory capacity of this class of compounds [122]. In addition to the previously described pharmacophore and 3D-QSAR models, the molecular representation of ASBT may provide a powerful tool in the design of novel substrates or inhibitors for this transporter.

Use of ASBT for Enhancing Intestinal Absorption Bile acids display a physiologic organotropism for the liver and the small intestine, which suggests that bile acid transport pathways can be exploited in two ways: to shuttle compounds across the intestinal epithelium or target them to the liver and the hepatobiliary system. This approach was first envisioned in 1948 by Berczeller in a US patent application (#2,441,129), describing the synthesis of sulphonylcholylamide, which was claimed to be particularly suitable for the treatment of "germ and virus diseases that attack the liver." It was not until the

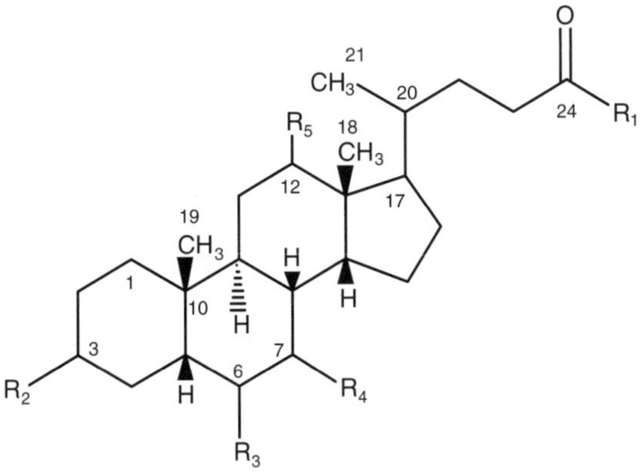

R_1	R_2	R_3	R_4	R_5	Prefix
–OH	–OH	–H	–OH	–OH	–
–OH	–OH	–H	–OH	–H	Chenodeoxy-
–OH	–OH	–H	–H	–OH	Deoxy-
–OH	–OH	–H	–H	–H	Litho-
–OH	–OH	–H	–OH (β)	–OH	Urso-
–OH	–OH	–H	–OH (β)	–H	Ursodeoxy-
–OH	–OH (β)	–H	–OH (β)	–H	Isoursodeoxy-
–OH	–OH	–H	–H	–OH (β)	Lagodeoxy-
–OH	–OH	–OH	–OH	–H	Hyo-
$–NHCH_2COOH$	–		–	–	Glyco-[1]
$–NH(CH_2)_2SO_3H$	–		–	–	Tauro-[1]

[1]The prefixes glyco- and tauro- prevail all others, that is, glyco-hyocholic acid.

Figure 12. Structure and nomenclature of bile acids.

mid-1980s when Ho [306] suggested several potential therapeutic applications of using the bile acid transport systems for (a) the improvement of the oral absorption of an intrinsically, biologically active, but poorly absorbed hydrophilic drug; (b) liver site-directed delivery of a drug to bring about high therapeutic concentrations in the diseased liver with the minimization of general toxic reactions elsewhere in the body; and (c) gallbladder-site delivery systems of cholecystographic agents and cholesterol gallstone dissolution accelerators. His experimental evidence was based on bile acid analogs with minor structural modifications (i.e., 3-iodo, 3-tosyl, and 3-benzoylcholate) that were handled like natural bile acids by the liver and intestine during *in situ* perfusion experiments. The feasibility and viability of using the bile acid transport pathway for targeting of drug–bile acid derivatives was more definitively demonstrated by the laboratory of Kramer, who primarily used the 3-position for drug conjugation. Our recent studies with small peptides attached to the C-24 position further demonstrate the general applicability of this transport system for drug targeting and delivery purposes [307].

From a structural perspective, largely based on the above outlined structure–activity relationships, the following drug-targeting strategies are feasible with bile acids:

(1) Attachment of drugs to the bile acid side chain (C-17) with positional retention of the negatively charged group.
(2) Attachment of drugs to the steroid nucleus at hydroxyl positions 3, 7, or 12 with conservation of the C-17 side chain.

Therapeutic Applications of ASBT The uses of carrier-mediated bile acid for drug delivery purposes can be divided in three groups: liver and gallbladder directed delivery, oral absorption enhancement, and lowering serum cholesterol. Oral absorption enhancement can be directed to either the liver or gallbladder or systemic delivery.

Liver and Gallbladder Delivery So far, most studies have focused on using the bile acid transport system for liver-targeting. The research groups of Kramer [308–310] and Stephan [311] have successfully shown specific hepatic delivery of chlorambucil, HMG-CoA reductase inhibitors, and L-T3, respectively. These studies prove that the coupling of drug entities to bile acids does not cause a loss of affinity for the hepatic bile acid carrier. Apart from the necessity of a negatively charged group around the C-24 position, Kim and colleagues [312] have shown that some size restrictions apply when compounds are coupled to the C-3 position. Both the 3- and the 24-positions appear to be usable coupling points for a prodrug strategem. The 24-position appears to be an attractive site in the bile acid molecule for coupling purposes. The carboxylic acid moiety is easily linked to an amine using conventional peptide synthesizing techniques [313–315], making the synthesis of these compounds relatively easy. It should be stressed, however, that the need for a negatively charged group around the C-24 position is required.

Systemic Delivery When one compares the maximal transport flux (*J*max) values of taurine conjugated bile acids measured in rat liver and distal ileum, a trend for relatively higher hepatic maximal transport rates can be observed. This observation can have important consequences when using the bile acid transporter for oral drug delivery. Using a prodrug approach with a bile acid molecule as a shuttle, liver-targeting is easily accomplished, whereas systemic drug delivery needs to address the problem of rapid biliary excretion. Thus far, no single study has unequivocally shown the release of the parent compound from the conjugate after passage across the intestinal wall. It has to be mentioned, however, that no studies so far have attempted to develop a prodrug approach in which the drug will be released prior to arrival in the liver. In that case, the drug moiety must be released from the bile acid it is coupled to either within the enterocyte or the portal vein. Only if these conditions are met, will systemic delivery using a prodrug approach be successful. Although promising, the suitability of this transport system for systemic drug delivery remains to be demonstrated.

Cholesterol-Reducing Agents Hypercholesterolemia is well known as a major risk factor for

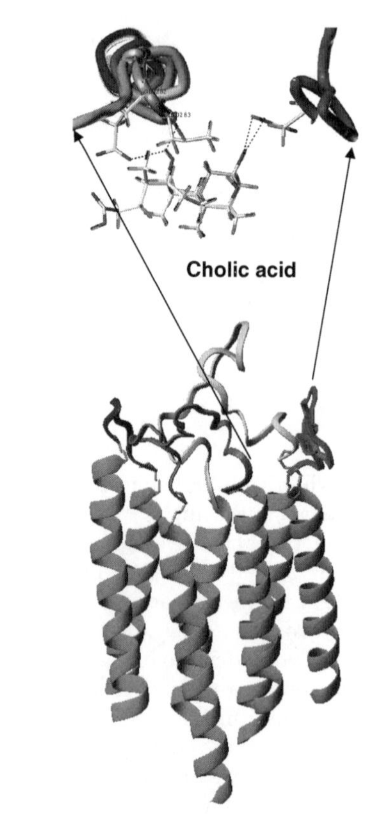

Figure 13. Putative three-dimensional structure of ASBT and its binding sites. (This figure is available in full color at http://mrw.interscience.wiley.com/emrw/9780471266945/home.)

coronary heart disease. In clinical practice, two main hypocholestrolemic agents are commonly used. One is the 3-hydroxy-3-methylglutaryl coenzyme A (HMG-CoA) reductase inhibitors (such as Lipitor®), another category are the bile acid sequestrants, such as cholestyramine and colestipol [97], which bind bile acid in the intestinal lumen and thus increase their excretion. The main drawback of these latter agents is poor compliance to patients due to adverse side effects, such as high dosages of 10–30 g/day, constipation, maldigestion and malabsorption syndromes. As an alternative method to bile acid sequestrants, any reagent that can inhibit the bile acid active transport system could block the reabsorption of bile acids and consequently reduce the serum cholesterol level. So far, several molecules have been found to possess this effect in animal studies [120–122,316].

The first such inhibitors comprised the coupling of two bile acid molecules via a spacer to allow simultaneous interaction with more than one transporter site, resulting in an efficient inhibition of bile acid reabsorption without or with only low absorption of the inhibitor itself [122] (Fig. 13). Recently it was shown that a benzothiazepine derivative, 2164U90, was able to selectively inhibit active ileal bile acid absorption in rats, mice, monkeys, and humans [317,318]. Similarly, another compound S-8921 (Fig. 14), a lignan derivative, was able to reduce serum cholesterol in hamster, mice and rabbit. The inhibition of the intestinal bile acid transport system is thought to be the underlying mechanism for an increased fecal bile acid excretion and lower plasma LDL cholesterol levels after oral administration of these drugs.

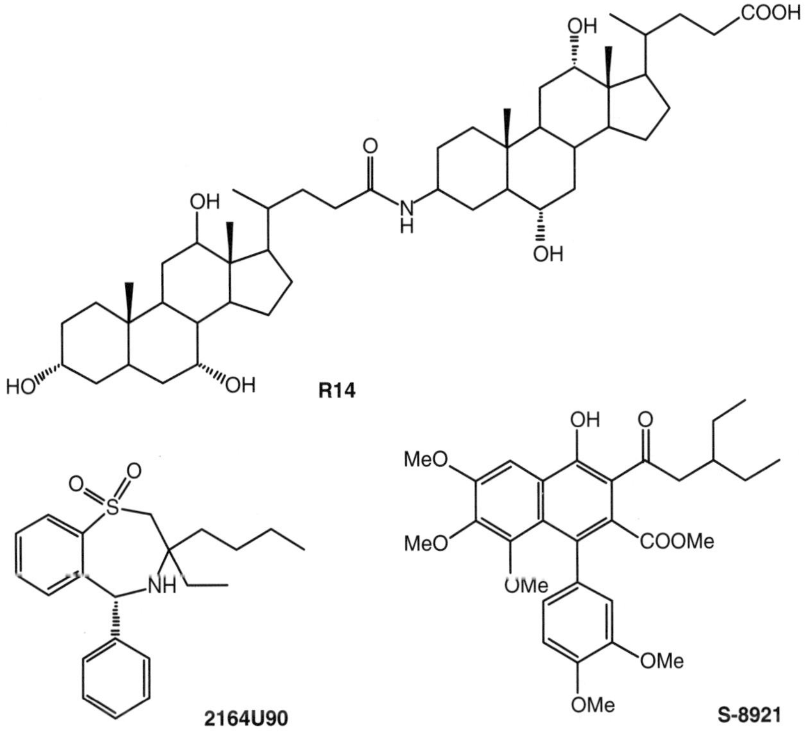

Figure 14. Structure of specific bile acid transporter inhibitors. See narrative for additional details.

6. CONCLUSIONS AND FUTURE DIRECTIONS

It is clear that over the past decade we have gained a great deal of knowledge around the functioning of SLC proteins and how they can be manipulated as drug targets for maximizing absorption and bioactivity through prodrug approaches. Although we are still without a crystal structure for the SLC proteins there has been an explosion in interest in structure–activity relationships. This has been seen largely as part of a wider growth in using computational approaches for understanding *in vitro* data and in particular in the fields of drug design, optimization and ADME properties [319]. The resolution of the three-dimensional structures of solute transporter protein models is presumably low, however, they can be justified by their ability to confirm biologically relevant phenomena. As with all models, the continued input of novel experimental data and revalidation of the model may eventually lead to highly predictive systems capable of *in silico* detection and design of novel solute transporter substrates. Furthermore, we can expect that a combination of indirect (3D-QSAR) and direct (homology models) techniques may lead to newer, higher-resolution screening systems. It is likely that the integration of bioinformatics, computational and *in vitro* models will drive our understanding of SLC protein to new levels and afford an opportunity for further possible therapeutic targets and drug design opportunities.

ACKNOWLEDGMENT

Dr. Sean Ekins (Collaborations in Chemistry, Inc., Jenkintown, PA) contributed to the P-glycoprotein section in this chapter.

ABBREVIATIONS:

AAA active analog approach
ABC ATP-binding cassette
ACE angiotensin-converting enzyme
ADME absorption, distribution, metabolism, excretion
ω-AFA ω-amino fatty acids
ASBT apical sodium-dependent bile acid transporter
6-AZA 6-azauridine; Cbl, cobalamine
CDCA chenodeoxycholic acid
CoMFA comparative molecular field analysis
CURL compartment of uncoupling receptor and ligand
CYP cytochrome P450
CYP7A cholesterol 7α-hydroxylase
FcRn neonatal Fc receptors
FXR farnesoid X receptor
GALT gut-associated lymphoid tissues
GPI glycosylphosphatidylinositol
HMG-CoA 3-hydroxy-3-methylglutaryl coenzyme A
iBABP ileal bile acid binding protein; IF, intrinsic factor
LHRH luteinizing hormone releasing factor
LXR liver X receptor
MDR multidrug resistance
MFS major facilitator superfamily
MHC major histocompatibility complex
MRP multidrug resistance-associated protein
NCE new chemical entity
NTCP Na^+-taurocholate cotransporting polypeptide
OATP organic anion transporting polypeptide
PepT1 small peptide transporter 1
P-gp P-glycoprotein
pIgR polymeric immunoglobulin receptor
POT proton/oligopeptide transporter
QSAR quantitative structure–activity relationship
RME receptor-mediated endocytosis
SEA/SEB staphylococcal enterotoxin A/B
SLC solute carrier
TAP transcytosis-associated protein
Tf transferrin

TfR	transferrin receptor
TMBY	trimethoxybenzoylyohimbine
TMD	transmembrane domain
VV-HA	measles virus hemagglutinin
ZO	zonula occludens
ZOT	toxin

REFERENCES

1. Kingham JG, Whorwell PJ, Loehry CA. Small intestinal permeability. 1. Effects of ischaemia and exposure to acetyl salicylate. Gut 1976;17:354–361.
2. Fasano A, Uzzau S. Modulation of intestinal tight junctions by Zonula occludens toxin permits enteral administration of insulin and other macromolecules in an animal model. J Clin Invest 1997;99:1158–1164.
3. van der Lubben IM, Verhoef JC, Borchard G, Junginger HE. Chitosan for mucosal vaccination. Adv Drug Deliv Rev 2001;52:139–144.
4. Porter CJ, Charman WN. *In vitro* assessment of oral lipid based formulations. Adv Drug Deliv Rev 2001;50(Suppl 1):S127–S147.
5. Albert A. Chemical aspects of selective toxicity. Nature 1958;182:421–423.
6. Stella V. Prodrugs: an overview and definition, In: Higuchiand T, Stella V, editors. Prodrugs as Novel Drug Delivery Systems. Vol. 14. Washington, DC: ACS Symposium Series; 1975. p 1–115.
7. Kamm W, Raddatz P, Gante J, Kissel T. Prodrug approach for alphaIIbbeta3-peptidomimetic antagonists to enhance their transport in monolayers of a human intestinal cell line (Caco-2): comparison of *in vitro* and *in vivo* data. Pharm Res 1999;16:1527–1533.
8. Delie F, Couvreur P, Nisato D, Michel JB, Puisieux F, Letourneux Y. Synthesis and *in vitro* study of a diglyceride prodrug of a peptide. Pharm Res 1994;11:1082–1087.
9. D'Emanuele A, Jevprasesphant R, Penny J, Attwood D. The use of a dendrimer-propranolol prodrug to bypass efflux transporters and enhance oral bioavailability. J Control Release 2004;95:447–453.
10. Gershonov E, Goldwaser I, Fridkin M, Shechter Y. A novel approach for a water-soluble long-acting insulin prodrug: design, preparation, and analysis of [(2-sulfo)-9-fluorenylmethoxycarbonyl](3)-insulin. J Med Chem 2000;43:2530–2537.
11. McCaughan B, Kay G, Knott RM, Cairns D. A potential new prodrug for the treatment of cystinosis: design, synthesis and *in-vitro* evaluation. Bioorg Med Chem Lett 2008;18: 1716–1719.
12. Shah D, Shen WC. Transcellular delivery of an insulin-transferrin conjugate in enterocyte-like Caco-2 cells. J Pharm Sci 1996;85: 1306–1311.
13. Stevens PJ, Sekido M, Lee RJ, A folate receptor-targeted lipid nanoparticle formulation for a lipophilic paclitaxel prodrug. Pharm Res 2004;21:2153–2157.
14. Dhanikula AB, Panchagnula R. Preparation and characterization of water-soluble prodrug, liposomes and micelles of Paclitaxel. Curr Drug Deliv 2005;2:75–91.
15. Takata J, Ito S, Karube Y, Nagata Y, Matsushima Y. Water-soluble prodrug of vitamin E for parenteral use and its effect on endotoxin-induced liver toxicity. Biol Pharm Bull 1997;20: 204–209.
16. Welch AD. Cancer Res 1961;21:1475.
17. Yalkowski SH, Morozowich W,In: Ariens EJ, editor. Drug Design. Vol. 9.New York, NY: Academic Press; 1980. p 121.
18. Russell-Jones GJ. Oral delivery of therapeutic proteins and peptides by the vitamin B12 uptake system. In: Taylor MD, Amidon GL, editors. Peptide-Based Drug Design: Controling Transport and Metabolism. Washington, DC: ACS; 1995. p 181–198.
19. Russell-Jones GJ, Westwood SW, Habberfield AD. Vitamin B12 mediated oral delivery systems for granulocyte-colony stimulating factor and erythropoietin. Bioconjug Chem 1995; 6:459–465.
20. Madara JL, Pappenheimer JR. Structural basis for physiological regulation of paracellular pathways in intestinal epithelia. J Membr Biol 1987;100:149–164.
21. Madara JL, Dharmsathaphorn K. Occluding junction structure–function relationships in a cultured epithelial monolayer. J Cell Biol 1985;101:2124–2133.
22. Ma TY, Hollander D, Erickson RA, Truong H, Nguyen H, Krugliak P. Mechanism of colonic permeation of inulin: is rat colon more permeable than small intestine? Gastroenterology 1995;108:12–20.
23. Knipp GT, Ho NF, Barsuhn CL, Borchardt RT. Paracellular diffusion in Caco-2 cell monolayers: effect of perturbation on the transport of hydrophilic compounds that vary

in charge and size. J Pharm Sci. 1997;86: 1105–1110.

24. Watson CJ, Rowland M, Warhurst G. Functional modeling of tight junctions in intestinal cell monolayers using polyethylene glycol oligomers. Am J Physiol Cell Physiol 2001;281: C388–C397.
25. He YL, Murby S, Warhurst G, Gifford L, Walker D, Ayrton J, Eastmond R, Rowland M. Species differences in size discrimination in the paracellular pathway reflected by oral bioavailability of poly(ethylene glycol) and D-peptides. J Pharm Sci 1998;87:626–633.
26. Lipinski CA. Drug-like properties and the causes of poor solubility and poor permeability. J Pharmacol Toxicol Methods 2000;44: 235–249.
27. Houston JB, Upshall DG, Bridges JW. Further studies using carbamate esters as model compounds to investigate the role of lipophilicity in the gastrointestinal absorption of foreign compounds. J Pharmacol Exp Ther 1975;195: 67–72.
28. Mukherjee S, Ghosh RN, Maxfield FR. Endocytosis Physiol Rev 1997;77:759–803.
29. Lehr C-M. The transcytosis approach, In: de Boer AG, editor. Drug Absorption Enhancement: Concepts, Possibilities, Limitations and Trends. Switzerland: Harwood; 1994. p 325–365.
30. Marsh M, Helenius A. Adsorptive endocytosis of Semliki Forest virus. J Mol Biol 1980;142: 439–454.
31. Pol A, Ortega D, Enrich C. Identification and distribution of proteins in isolated endosomal fractions of rat liver: involvement in endocytosis, recycling and transcytosis. Biochem J 1997;323:435–443.
32. Wigge P, Vallis Y, McMahon HT. Inhibition of receptor-mediated endocytosis by the amphiphysin SH3 domain. Curr Biol 1997;7: 554–560.
33. Yamashiro DJ, Maxfield FR. Acidification of endocytic compartments and the intracellular pathways of ligands and receptors. J Cell Biochem 1984;26:231–246.
34. Geuze HJ, Slot JW, Strous GJ, Peppard J, von Figura K, Hasilik A, Schwartz AL. Intracellular receptor sorting during endocytosis: comparative immunoelectron microscopy of multiple receptors in rat liver. Cell 1984;37: 195–204.
35. Herman B, Albertini DF. Ligand-induced rapid redistribution of lysosomes is temporally distinct from endosome translocation. Nature 1983;304:738–740.
36. Hughson EJ, Hopkins CR. Endocytic pathways in polarized Caco-2 cells: identification of an endosomal compartment accessible from both apical and basolateral surfaces. J Cell Biol 1990;110:337–348.
37. Schwartz AL. Receptor cell biology: receptor-mediated endocytosis. Pediatr Res 1995; 38:835–843.
38. Barroso M, Nelson DS, Sztul E. Transcytosis-associated protein (TAP)/p115 is a general fusion factor required for binding of vesicles to acceptor membranes. Proc Natl Acad Sci USA 1995;92:527–531.
39. Lemons PP, Chen D, Bernstein AM, Bennett MK, Whiteheart SW. Regulated secretion in platelets: identification of elements of the platelet exocytosis machinery. Blood 1997;90: 1490–1500.
40. Shah D, Shen WC. The establishment of polarity and enhanced transcytosis of transferrin receptors in enterocyte-like Caco-2 cells. J Drug Target 1994;2:93–99.
41. Prydz K, Hansen SH, Sandvig K, van Deurs B. Effects of brefeldin A on endocytosis, transcytosis and transport to the Golgi complex in polarized MDCK cells. J Cell Biol 1992;119: 259–272.
42. Severs NJ. Caveolae: static inpocketings of the plasma membrane, dynamic vesicles or plain artifact? J Cell Sci 1988;90:341–348.
43. Anderson RG, Kamen BA, Rothberg KG, Lacey SW. Potocytosis: sequestration and transport of small molecules by caveolae. Science 1992;255:410–411.
44. Anderson RG. Caveolae: where incoming and outgoing messengers meet. Proc Natl Acad Sci USA 1993;90:10909–10913.
45. Matsue H, Rothberg KG, Takashima A, Kamen BA, Anderson RG, Lacey SW. Folate receptor allows cells to grow in low concentrations of 5-methyltetrahydrofolate. Proc Natl Acad Sci USA 1992;89:6006–6009.
46. Montesano R, Roth J, Robert A, Orci L. Non-coated membrane invaginations are involved in binding and internalization of cholera and tetanus toxins. Nature 1982;296: 651–653.
47. Kurzchalia TV, Dupree P, Monier S. VIP21-Caveolin, a protein of the trans-Golgi network and caveolae. FEBS Lett 1994;346:88–91.
48. Rothberg KG, Heuser JE, Donzell WC, Ying YS, Glenney JR, Anderson RG. Caveolin, a

protein component of caveolae membrane coats. Cell 1992;68:673–682.

49. Simons K, Ikonen E. Functional rafts in cell membranes. Nature 1997;387:569–572.
50. Harokopakis E, Childers NK, Michalek SM, Zhang SS, Tomasi M. Conjugation of cholera toxin or its B subunit to liposomes for targeted delivery of antigens. J Immunol Methods 1995;185:31–42.
51. Walker WA, Sanderson IR. Epithelial barrier function to antigens. An overview Ann N Y Acad Sci 1992;664:10–17.
52. Desai MP, Labhasetwar V, Amidon GL, Levy RJ. Gastrointestinal uptake of biodegradable microparticles: effect of particle size. Pharm Res 1996;13:1838–1845.
53. Hussain N, Jani PU, Florence AT. Enhanced oral uptake of tomato lectin-conjugated nanoparticles in the rat. Pharm Res 1997;14: 613–618.
54. Berryman M, Rodewald R. Beta 2-microglobulin co-distributes with the heavy chain of the intestinal IgG-Fc receptor throughout the transepithelial transport pathway of the neonatal rat. J Cell Sci 1995;108:2347–2360.
55. Benlounes N, Chedid R, Thuillier F, Desjeux JF, Rousselet F, Heyman M. Intestinal transport and processing of immunoglobulin G in the neonatal and adult rat. Biol Neonate 1995;67:254–263.
56. Medesan C, Radu C, Kim JK, Ghetie V, Ward ES. Localization of the site of the IgG molecule that regulates maternofetal transmission in mice. Eur J Immunol 1996;26:2533–2536.
57. Medesan C, Matesoi D, Radu C, Ghetie V, Ward. ES. Delineation of the amino acid residues involved in transcytosis and catabolism of mouse IgG1. J Immunol 1997;158:2211–2217.
58. Abrahamson DR, Rodewald R. Evidence for the sorting of endocytic vesicle contents during the receptor-mediated transport of IgG across the newborn rat intestine. J Cell Biol 1981;91:270–280.
59. Patel HM, Wild AE. Fc receptor-mediated transcytosis of IgG-coated liposomes across epithelial barriers. FEBS Lett 1988;234: 321 325.
60. Mestecky J. The common mucosal immune system and current strategies for induction of immune responses in external secretions. J Clin Immunol 1987;7:265–276.
61. Hirt RP, Hughes GJ, Frutiger S, Michetti P, Perregaux C, Poulain-Godefroy O, Jeanguenat N, Neutra MR, Kraehenbuhl JP. Transcytosis of the polymeric Ig receptor requires phosphorylation of serine 664 in the absence but not the presence of dimeric IgA. Cell 1993;74:245–255.
62. Youngman KR, Fiocchi C, Kaetzel CS. Inhibition of IFN-gamma activity in supernatants from stimulated human intestinal mononuclear cells prevents up-regulation of the polymeric Ig receptor in an intestinal epithelial cell line. J Immunol 1994;153:675–681.
63. Mostov K. Protein traffic in ploarized epithelial cells: the polymeric immunoglobulin receptor as a model system. J Cell Sci Suppl 1993;17:21–26.
64. Mostov KE. Transepithelial transport of immunoglobulins. Annu Rev Immunol 1994;12: 63–84.
65. Livrelli V, De Champs C, Di Martino P, Darfeuille-Michaud A, Forestier C, Joly B. Adhesive properties and antibiotic resistance of *Klebsiella*, *Enterobacter*, and *Serratia* clinical isolates involved in nosocomial infections. J Clin Microbiol 1996;34:1963–1969.
66. Ghose AC. Adherence & colonization properties of *Vibrio cholerae* & diarrhoeagenic *Escherichia coli.* Indian J Med Res 1996;104: 38–51.
67. Adlerberth I, Ahrne S, Johansson ML, Molin G, Hanson LA, Wold AE. A mannose-specific adherence mechanism in *Lactobacillus plantarum* conferring binding to the human colonic cell line HT-29. Appl Environ Microbiol 1996;62:2244–2251.
68. Crennell S, Garman E, Laver G, Vimr E, Taylor G. Crystal structure of *Vibrio cholerae* neuraminidase reveals dual lectin-like domains in addition to the catalytic domain. Structure 1994;2:535–544.
69. Metcalfe JW, Krogfelt KA, Krivan HC, Cohen PS, Laux DC, Characterization and identification of a porcine small intestine mucus receptor for the K88ab fimbrial adhesin. Infect Immun 1991;59:91–96.
70. Thorns CJ, Wells GA, Morris JA, Bridges A, Higgins R. Evaluation of monoclonal antibodies to K88, K99, F41 and 987P fimbrial adhesins for the detection of porcine enterotoxigenic *Escherichia coli* in paraffin-wax tissue sections. Vet Microbiol 1989;20:377–381.
71. Isberg RR. Pathways for the penetration of enteroinvasive *Yersinia* into mammalian cells. Mol Biol Med 1990;7:73–82.
72. Isberg RR, Leong JM. Multiple beta 1 chain integrins are receptors for invasin, a protein that promotes bacterial penetration into mammalian cells. Cell 1990;60:861–871.

73. Marra A, Isberg RR. Invasin-dependent and invasin-independent pathways for translocation of *Yersinia pseudotuberculosis* across the Peyer's patch intestinal epithelium. Infect Immun 1997;65:3412–3421.
74. Paul RW, Weisser KE, Loomis A, Sloane DL, LaFoe D, Atkinson EM, Overell RW. Gene transfer using a novel fusion protein, GAL4/invasin. Hum Gene Ther 1997;8:1253–1262.
75. Easson JH, Jahn D, Lehr C-M, Utilizing bacterial mechanisms of cell entry for the design of bioinvasive drug carriers. Proc Int'l Symp Control Rel Bioact Mater 1997;24:101–102.
76. Easson JH, Jahn D, Lehr C-M, Invasin-modified latex beads display bioinvasive properties to cultured epithelial cells. Pharm Res 1997;14:S21.
77. Lemichez E, Bomsel M, Devilliers G, vanderSpek J, Murphy JR, Lukianov EV, Olsnes S, Boquet P. Membrane translocation of diphtheria toxin fragment A exploits early to late endosome trafficking machinery. Mol Microbiol 1997;23:445–457.
78. Lindner J, Geczy AF, Russell-Jones GJ. Identification of the site of uptake of the *E coli* heat-labile enterotoxin, LTB. Scand J Immunol 1994;40:564–572.
79. Lazorova L, Sjolander A, Russell-Jones GJ, Linder J, Artursson P. Intestinal tissue distribution and epithelial transport of the oral immunogen LTB, the B subunit of *E coli* heat-labile enterotoxin. J Drug Target 1993;1: 331–340.
80. Fisher KJ, Wilson JM. The transmembrane domain of diphtheria toxin improves molecular conjugate gene transfer. Biochem J 1997;321: 49–58.
81. Hamad AR, Marrack P, Kappler JW. Transcytosis of staphylococcal superantigen toxins. J Exp Med 1997;185:1447–1454.
82. Stirpe F, Barbieri L. Ribosome-inactivating proteins up to date. FEBS Lett 1986;195:1–8.
83. Battelli MG, Citores L, Buonamici L, Ferreras JM, de Benito FM, Stirpe F, Girbes T. Toxicity and cytotoxicity of nigrin b, a two-chain ribosome-inactivating protein from *Sambucus nigra*: comparison with ricin. Arch Toxicol 1997;71:360–364.
84. Sylvester ID, Roberts LM, Lord JM, Characterization of prokaryotic recombinant *Aspergillus ribotoxin* alpha-sarcin. Biochim Biophys Acta 1997;1358:53–60.
85. Battelli MG, Buonamici L, Polito L, Bolognesi A, Stirpe F. Hepatoxicity of ricin, saporin or a saporin immunotoxin: xanthine oxidase activity in rat liver and blood serum. Virchows Arch 1996;427:529–535.
86. Stirpe F, Sandvig K, Olsnes S, Pihl A, Action of viscumin, a toxic lectin from mistletoe, on cells in culture. J Biol Chem 1982;257: 13271–13277.
87. Sperti S, Montanaro L, Derenzini M, Gasperi-Campani A, Stirpe F. Effect of modeccin on rat liver ribosomes *in vivo*. Biochim Biophys Acta 1979;562:495–503.
88. Ludert JE, Michelangeli F, Gil F, Liprandi F, Esparza J. Penetration and uncoating of rotaviruses in cultured cells. Intervirology 1987;27:95–101.
89. Olson JK, Grose C. Endocytosis and recycling of varicella-zoster virus Fc receptor glycoprotein gE: internalization mediated by a YXXL motif in the cytoplasmic tail. J Virol 1997;71: 4042–4054.
90. Irurzun A, Nieva JL, Carrasco L. Entry of Semliki forest virus into cells: effects of concanamycin A and nigericin on viral membrane fusion and infection. Virology 1997;227: 488–492.
91. Worthington BS, Syrotuck J. Intestinal permeability to large particles in normal and protein-deficient adult rats. J Nutr 1976;106:20–32.
92. Garret A, Kerlan C, Thomas D. The intestine is a site of passage for potato leafroll virus from the gut lumen into the haemocoel in the aphid vector, *Myzus persicae* Sulz. Arch Virol 1993; 131:377–392.
93. Wolf JL, Kauffman RS, Finberg R, Dambrauskas R, Fields BN, Trier JS. Determinants of reovirus interaction with the intestinal M cells and absorptive cells of murine intestine. Gastroenterology 1983;85:291–300.
94. Etchart N, Wild F, Kaiserlian D. Mucosal and systemic immune responses to measles virus haemagglutinin in mice immunized with a recombinant vaccinia virus. J Gen Virol 1996;77:2471–2478.
95. Bardocz S, Grant G, Pasztai A. Lectins as growth factors for the small intestine and the gut. In: Bardocz S, Pusztai A, editors. Lectins—Biomedical Perspectives, London: Taylor and Francis; 1995. p 103–117.
96. Lehr C-M, Pusztai A. The potential of bioadhesive lectins for the delivery of peptide and protein drugs to the gastrointestinal tract. In: Bardocz S, Pusztai A, editors. Lectins—Biomedical Perspectives. London: Taylor and Francis; 1995. p 117–141.

97. Weaver LT, Bailey DS. Effect of the lectin concanavalin A on the neonatal guinea pig gastrointestinal mucosa *in vivo*. J Pediatr Gastroenterol Nutr 1987;6:445–453.
98. Gebert A, Rothkotter HJ, Pabst R. M cells in Peyer's patches of the intestine. Int Rev Cytol 1996;167:91–159.
99. Jepson MA, Clark MA, Foster N, Mason CM, Bennett MK, Simmons NL, Hirst BH. Targeting to intestinal M cells. J Anat 1996;189: 507–516.
100. Sharma R, van Damme EJ, Peumans WJ, Sarsfield P, Schumacher U. Lectin binding reveals divergent carbohydrate expression in human and mouse Peyer's patches. Histochem Cell Biol 1996;105:459–465.
101. Ross JF, Chaudhuri PK, Ratnam M. Differential regulation of folate receptor isoforms in normal and malignant tissues *in vivo* and in established cell lines. Physiologic and clinical implications. Cancer 1994;73:2432–2443.
102. Fan J, Kureshy N, Vitols KS, Huennekens FM. Novel substrate analogs delineate an endocytotic mechanism for uptake of folate via the high-affinity, glycosylphosphatidylinositol-linked transport protein in L1210 mouse leukemia cells. Oncol Res 1995;7:511–516.
103. Leamon CP, Low PS, Membrane folate-binding proteins are responsible for folate-protein conjugate endocytosis into cultured cells. Biochem J 1993;291:855–860.
104. Mislick KA, Baldeschwieler JD, Kayyem JF, Meade TJ. Transfection of folate-polylysine DNA complexes: evidence for lysosomal delivery. Bioconjug Chem 1995;6:512–515.
105. Wang S, Lee RJ, Cauchon G, Gorenstein DG, Low PS, Delivery of antisense oligodeoxyribonucleotides against the human epidermal growth factor receptor into cultured KB cells with liposomes conjugated to folate via polyethylene glycol. Proc Natl Acad Sci USA 1995;92:3318–3322.
106. Wangensteen OD, Bartlett MM, James JK, Yang ZF, Low PS, Riboflavin-enhanced transport of serum albumin across the distal pulmonary epithelium. Pharm Res 1996;13: 1861–1864.
107. Huang SN, Swaan PW. Riboflavin uptake in human trophoblast-derived BeWo cell monolayers: cellular translocation and regulatory mechanisms. J Pharmacol Exp Ther 2001;298: 264–271.
108. Huang SN, Swaan PW. Involvement of a receptor-mediated component in cellular translocation of riboflavin. J Pharmacol Exp Ther 2000;294:117–125.
109. Ramasamy M, Alpers DH, Tiruppathi C, Seetharam B. Cobalamin release from intrinsic factor and transfer to transcobalamin II within the rat enterocyte. Am J Physiol 1989;257:G791–G797.
110. Dan N, Cutler DF. Transcytosis and processing of intrinsic factor-cobalamin in Caco-2 cells. J Biol Chem 1994;269:18849–18855.
111. Russell-Jones GJ, Westwood SW, Farnworth PG, Findlay JK, Burger HG. Synthesis of LHRH antagonists suitable for oral administration via the vitamin B12 uptake system. Bioconjug Chem 1995;6:34–42.
112. Habberfield A, Jensen-Pippo K, Ralph L, Westwood SW, Russell-Jones GJ. Vitamin B12-mediated uptake of recombinant therapeutic proteins from the gut. Int J Pharm 1996; 145:1–8.
113. Russell-Jones GJ, Arthur L, Killinger S, Westwood SW. Vitamin B12-mediated transport of nanoparticles. Proc Intl Symp Control Rel Bioact Mater 1997;24:11–12.
114. Russell-Jones GJ, Westwood SW, Habberfield AD. The use of vitamin B12 transport system as a carrier for the oral delivery of peptides, proteins and nanoparticles. Proc Intl Symp Control Rel Bioact Mater 1996; 23:49–50.
115. Pietrangelo A, Rocchi E, Casalgrandi G, Rigo G, Ferrari A, Perini M, Ventura E, Cairo G. Regulation of transferrin, transferrin receptor, and ferritin genes in human duodenum. Gastroenterology 1992;102:802–809.
116. Anderson GJ, Powell LW, Halliday JW. Transferrin receptor distribution and regulation in the rat small intestine. Effect of iron stores and erythropoiesis. Gastroenterology 1990;98: 576–585.
117. Jeffrey GP, Basclain KA, Allen TL. Molecular regulation of transferrin receptor and ferritin expression in the rat gastrointestinal tract. Gastroenterology 1996;110:790–800.
118. Wang JS, Shen D, Shen W-C, Oral delivery of an insulin-transferrin conjugate in Streptozotocin-treated cf/1 mice. Pharm Res 1997;14: S469.
119. Saier MH Jr. Eukaryotic transmembrane solute transport systems. Int Rev Cytol 1999;190:61–136.
120. Kramer W, Wess G, Bile acid transport systems as pharmaceutical targets. Eur J Clin Invest 1996;26:715–732.

121. Tollefson MB, Vernier WF, Huang HC, Chen FP, Reinhard EJ, Beaudry J, Keller BT, Reitz DB. A novel class of apical sodium co-dependent bile acid transporter inhibitors: the 2,3-disubstituted-4-phenylquinolines. Bioorg Med Chem Lett 2000;10:277–279.
122. Wess G, Kramer W, Enhsen A, Glombik H, Baringhaus KH, Boger G, Urmann M, Bock K, Kleine H, Neckermann G, Hoffmann A, Pittius C, Falk E, Fehlhaber HW, Kogler H, Friedrich M. Specific inhibitors of ileal bile acid transport. J Med Chem 1994;37:873–875.
123. Hediger MA, Coady MJ, Ikeda TS, Wright EM. Expression cloning and cDNA sequencing of the Na^+/glucose co-transporter. Nature 1987; 330:379–381.
124. Hediger MA, Ikeda T, Coady M, Gundersen CB, Wright EM. Expression of size-selected mRNA encoding the intestinal Na/glucose co-transporter in *Xenopus laevis* oocytes. Proc Natl Acad Sci USA 1987;84:2634–2637.
125. Dean M, Rzhetsky A, Allikmets R. The human ATP-binding cassette (ABC) transporter superfamily. Genome Res 2001;11:1156–1166.
126. Paulsen IT, Sliwinski MK, Nelissen B, Goffeau A, Saier MH Jr. Unified inventory of established and putative transporters encoded within the complete genome of *Saccharomyces cerevisiae*. FEBS Lett 1998;430:116–125.
127. Paulsen IT, Sliwinski MK, Saier MH Jr. Microbial genome analyses: global comparisons of transport capabilities based on phylogenies, bioenergetics and substrate specificities. J Mol Biol 1998;277:573–592.
128. Schuler GD, Boguski MS, Stewart EA, Stein LD, Gyapay G, Rice K, White RE, Rodriguez-Tome P, Aggarwal A, Bajorek E, Bentolila S, Birren BB, Butler A, Castle AB, Chiannilkulchai N, Chu A, Clee C, Cowles S, Day PJ, Dibling T, Drouot N, Dunham I, Duprat S, East C, Hudson TJ, et al. A gene map of the human genome. Science 1996;274:540–546.
129. Swaan PW, Oie S, Szoka FC Jr. Carrier-mediated oral drug delivery. Adv Drug Del Rev 1996;20:1–4.
130. Adibi SA. The oligopeptide transporter (Pept-1) in human intestine: biology and function. Gastroenterology 1997;113:332–340.
131. Bradshaw DM, Arceci RJ. Clinical relevance of transmembrane drug efflux as a mechanism of multidrug resistance. J Clin Oncol 1998;16: 3674–3690.
132. Krishan A, Fitz CM, Andritsch I. Drug retention, efflux, and resistance in tumor cells. Cytometry 1997;29:279–285.
133. Berndt WO. The role of transport in chemical nephrotoxicity. Toxicol Pathol 1998;26:52–57.
134. Owens MJ, Nemeroff CB. The serotonin transporter and depression. Depress Anxiety 1998; 8:5–12.
135. Reith ME, Xu C, Chen NH, Pharmacology and regulation of the neuronal dopamine transporter. Eur J Pharmacol 1997;324:1–10.
136. Baly DL, Horuk R. The biology and biochemistry of the glucose transporter. Biochim Biophys Acta 1988;947:571–590.
137. Wright EM, Turk E, Zabel B, Mundlos S, Dyer J. Molecular genetics of intestinal glucose transport. J Clin Invest 1991;88:1435–1440.
138. Wong MH, Oelkers P, Dawson PA. Identification of a mutation in the ileal sodium-dependent bile acid transporter gene that abolishes transport activity. J Biol Chem 1995;270: 27228–27234.
139. Lemieux MJ, Huang Y, Wang DN. Glycerol-3-phosphate transporter of *Escherichia coli*: structure, function and regulation. Res Microbiol 2004;155:623–629.
140. Lu M, Fu D. Structure of the zinc transporter YiiP. Science 2007;317:1746–1748.
141. Yamashita A, Singh SK, Kawate T, Jin Y, Gouaux E. Crystal structure of a bacterial homologue of Na^+/Cl^--dependent neurotransmitter transporters. Nature 2005;437: 215–223.
142. Boudker O, Ryan RM, Yernool D, Shimamoto K, Gouaux E. Coupling substrate and ion binding to extracellular gate of a sodium-dependent aspartate transporter. Nature 2007;445: 387–393.
143. Yernool D, Boudker O, Jin Y, Gouaux E. Structure of a glutamate transporter homologue from *Pyrococcus horikoshii*. Nature 2004; 431:811–818.
144. Sorgen PL, Hu Y, Guan L, Kaback HR, Girvin ME. An approach to membrane protein structure without crystals. Proc Natl Acad Sci USA 2002;99:14037–14040.
145. Chang C, Ray A, Swaan P. *In silico* strategies for modeling membrane transporter function. Drug Discov Today 2005;10:663–671.
146. Doolittle RF, Johnson MS, Husain I, Van Houten B, Thomas DC, Sancar A. Domainal evolution of a prokaryotic DNA repair protein and its relationship to active-transport proteins. Nature 1986;323:451–453.
147. Kasho VN, Smirnova IN, Kaback HR. Sequence alignment and homology threading reveals prokaryotic and eukaryotic proteins si-

milar to lactose permease. J Mol Biol 2006;358:1060–1070.

148. Vardy E, Arkin IT, Gottschalk KE, Kaback HR, Schuldiner S. Structural conservation in the major facilitator superfamily as revealed by comparative modeling. Protein Sci 2004;13: 1832–1840.

149. Voss J, Wu J, Hubbell WL, Jacques V, Meares CF, Kaback HR. Helix packing in the lactose permease of *Escherichia coli*: distances between site-directed nitroxides and a lanthanide. Biochemistry 2001;40:3184–3188.

150. Claros MG, von Heijne G, TopPred II: an improved software for membrane protein structure predictions. Comput Appl Biosci 1994; 10:685–686.

151. Rost B, Fariselli P, Casadio R. Topology prediction for helical transmembrane proteins at 86% accuracy. Protein Sci 1996;5:1704–1718.

152. Tusnády GE, Simon I. Principles governing amino acid composition of integral membrane proteins: application to topology prediction. J Mol Biol 1998;283:489–506.

153. Bamberg K, Sachs G. Topological analysis of H+, K(+)-ATPase using *in vitro* translation. J Biol Chem 1994;269:16909–16919.

154. Frillingos S, Sahin-Toth M, Wu J, Kaback HR. Cys-scanning mutagenesis: a novel approach to structure function relationships in polytopic membrane proteins. FASEB J 1998;12: 1281–1299.

155. Turk E, Kerner CJ, Lostao MP, Wright EM. Membrane topology of the human Na^+/glucose cotransporter SGLT1. J Biol Chem 1996;271:1925–1934.

156. Bennett ER, Kanner BI. The membrane topology of GAT-1, a (Na^+ + Cl^-)-coupled gamma-aminobutyric acid transporter from rat brain. J Biol Chem 1997;272:1203–1210.

157. Covitz KM, Amidon GL, Sadee W. Membrane topology of the human dipeptide transporter, hPEPT1, determined by epitope insertions. Biochemistry 1998;37:15214–15221.

158. Wang Q, Voss J, Hubbell WL, Kaback HR. Proximity of helices VIII (Ala273) and IX (Met299) in the lactose permease of *Escherichia coli*. Biochemistry 1998;37:4910–4915.

159. Zhao M, Zen KC, Hubbell WL, Kaback HR. Proximity between Glu126 and Arg144 in the lactose permease of *Escherichia coli*. Biochemistry 1999;38:7407–7412.

160. Hresko RC, Murata H, Marshall BA, Mueckler M. Discrete structural domains determine differential endoplasmic reticulum to Golgi transit times for glucose transporter isoforms. J Biol Chem 1994;269:32110–32119.

161. Nagai K, Thogersen HC. Generation of beta-globin by sequence-specific proteolysis of a hybrid protein produced in *Escherichia coli*. Nature 1984;309:810–812.

162. Sahin-Toth M, Dunten RL, Kaback HR. Design of a membrane protein for site-specific proteolysis: properties of engineered factor Xa protease sites in the lactose permease of *Escherichia coli*. Biochemistry 1995;34:1107–1112.

163. Wandel C, Kim RB, Kajiji S, Guengerich P, Wilkinson GR, Wood AJ, P-glycoprotein and cytochrome P-450 3A inhibition: dissociation of inhibitory potencies. Cancer Res 1999;59: 3944–3948.

164. George A. The design and molecular modeling of CNS drugs. Curr Opin Drug Discov Dev 1999;2:286.

165. Hoffmeyer S, Burk O, von Richter O, Arnold HP, Brockmoller J, Johne A, Cascorbi I, Gerloff T, Roots I, Eichelbaum M, Brinkmann U. Functional polymorphisms of the human multidrug-resistance gene: multiple sequence variations and correlation of one allele with P-glycoprotein expression and activity *in vivo*. Proc Natl Acad Sci USA 2000;97:3473–3478.

166. Wacher VJ, Wu CY, Benet LZ. Overlapping substrate specificities and tissue distribution of cytochrome P450 3A and P-glycoprotein: implications for drug delivery and activity in cancer chemotherapy. Mol Carcinog 1995; 13:129–134.

167. Schuetz EG, Beck WT, Schuetz JD. Modulators and substrates of P-glycoprotein and cytochrome P4503A coordinately up-regulate these proteins in human colon carcinoma cells. Mol Pharmacol 1996;49:311–318.

168. Schuetz EG, Schinkel AH, Relling MV, Schuetz JD. P-glycoprotein: a major determinant of rifampicin-inducible expression of cytochrome P4503A in mice and humans. Proc Natl Acad Sci USA 1996;93:4001–4005.

169. Kim RB, Wandel C, Leake B, Cvetkovic M, Fromm MF, Dempsey PJ, Roden MM, Belas F, Chaudhary AK, Roden DM, Wood AJ, Wilkinson GR. Interrelationship between substrates and inhibitors of human CYP3A and P-glycoprotein. Pharm Res 1999;16:408–414.

170. Schuetz EG, Strom S, Yasuda K, Lecureur V, Assem M, Brimer C, Lamba J, Kim RB, Ramachandran V, Komoroski BJ, Venktaramanan R, Cai H, Sinal CJ, Gonzalez FJ, Schuetz JD. Disrupted bile acid homeostasis reveals an unexpected interaction among nuclear hormone

receptors, transporters and cytochrome P450. J Biol Chem 2001;276:(42): 39411–39418.

171. Synold TW, Dussault I, Forman BM. The orphan nuclear receptor SXR coordinately regulates drug metabolism and efflux. Nat Med 2001;7:584–590.

172. Ekins S, Erickson JA. A Pharmacophore for human pregnane-X-receptor ligands. Drug Metab Dispos 2002;30:96–99.

173. Ekins S, Ring BJ, Binkley SN, Hall SD, Wrighton SA. Autoactivation and activation of the cytochrome P450s. Int J Clin Pharmacol Ther 1998;36:642–651.

174. Korzekwa KR, Krishnamachary N, Shou M, Ogai A, Parise RA, Rettie AE, Gonzalez FJ, Tracy TS. Evaluation of atypical cytochrome P450 kinetics with two-substrate models: evidence that multiple substrates can simultaneously bind to cytochrome P450 active sites. Biochemistry 1998;37:4137–4147.

175. Houston JB, Kenworthy KE. *In vitro–in vivo* scaling of CYP kinetic data not consistent with the classical Michaelis–Menten model. Drug Metab Dispos 2000;28:246–254.

176. Greenberger LM, Yang CP, Gindin E, Horwitz SB. Photoaffinity probes for the alpha 1-adrenergic receptor and the calcium channel bind to a common domain in P-glycoprotein. J Biol Chem 1990;265:4394–4401.

177. Ayesh S, Shao YM, Stein WD. Co-operative, competitive and non-competitive interactions between modulators of P-glycoprotein. Biochim Biophys Acta 1996;1316:8–18.

178. Shapiro AB, Ling V, Positively cooperative sites for drug transport by P-glycoprotein with distinct drug specificities. Eur J Biochem 1997;250:130–137.

179. Scala S, Akhmed N, Rao US, Paull K, Lan LB, Dickstein B, Lee JS, Elgemeie GH, Stein WD, Bates SE. P-glycoprotein substrates and antagonists cluster into two distinct groups. Mol Pharmacol 1997;51:1024–1033.

180. Dey S, Ramachandra M, Pastan I, Gottesman MM, Ambudkar SV. Evidence for two nonidentical drug-interaction sites in the human P-glycoprotein. Proc Natl Acad Sci USA 1997;94:10594–10599.

181. Shapiro AB, Fox K, Lam P, Ling V. Stimulation of P-glycoprotein-mediated drug transport by prazosin and progesterone. Evidence for a third drug-binding site. Eur J Biochem 1999;259:841–850.

182. Wang EJ, Casciano CN, Clement RP, Johnson WW. Two transport binding sites of P-glycoprotein are unequal yet contingent: initial rate kinetic analysis by ATP hydrolysis demonstrates intersite dependence. Biochim Biophys Acta 2000;1481:63–74.

183. Lu L, Leonessa F, Clarke R, Wainer IW. Competitive and allosteric interactions in ligand binding to P-glycoprotein as observed on an immobilized P-glycoprotein liquid chromatographic stationary phase. Mol Pharmacol 2001;59:62–68.

184. Demeule M, Laplante A, Murphy GF, Wenger RM, Beliveau R. Identification of the cyclosporin-binding site in P-glycoprotein. Biochemistry 1998;37:18110–18118.

185. Shepard RL, Winter MA, Hsaio SC, Pearce HL, Beck WT, Dantzig AH. Effect of modulators on the ATPase activity and vanadate nucleotide trapping of human P-glycoprotein. Biochem Pharmacol 1998;56:719–727.

186. Dantzig AH, Shepard RL, Cao J, Law KL, Ehlhardt WJ, Baughman TM, Bumol TF, Starling JJ. Reversal of P-glycoprotein-mediated multidrug resistance by a potent cyclopropyldibenzosuberane modulator, LY335979. Cancer Res 1996;56:4171–4179.

187. Pearce HL, Winter MA, Beck WT. Structural characteristics of compounds that modulate P-glycoprotein-associated multidrug resistance. Adv Enzyme Regul 1990;30:357–373.

188. Pearce HL, Safa AR, Bach NJ, Winter MA, Cirtain MC, Beck WT. Essential features of the P-glycoprotein pharmacophore as defined by a series of reserpine analogs that modulate multidrug resistance. Proc Natl Acad Sci USA 1989;86:5128–5132.

189. Ramu A, Ramu N. Reversal of multidrug resistance by phenothiazines and structurally related compounds. Cancer Chemother Pharmacol 1992;30:165–173.

190. Borgnia MJ, Eytan GD, Assaraf YG. Competition of hydrophobic peptides, cytotoxic drugs, and chemosensitizers on a common P-glycoprotein pharmacophore as revealed by its ATPase activity. J Biol Chem 1996;271: 3163–3171.

191. Chiba P, Ecker G, Schmid D, Drach J, Tell B, Goldenberg S, Gekeler V. Structural requirements for activity of propafenone-type modulators in P-glycoprotein-mediated multidrug resistance. Mol Pharmacol 1996;49: 1122–1130.

192. Klopman G, Shi LM, Ramu A. Quantitative structure–activity relationship of multidrug resistance reversal agents. Mol Pharmacol 1997;52:323–334.

193. Bakken GA, Jurs PC. Classification of multidrug-resistance reversal agents using structure-based descriptors and linear discriminant analysis. J Med Chem 2000;43:4534–4541.

194. Wiese M, Pajeva IK. Molecular modeling study of the multidrug resistance modifiers *cis*- and *trans*-flupentixol. Pharmazie 1997;52: 679–685.

195. Tmej C, Chiba P, Huber M, Richter E, Hitzler M, Schaper KJ, Ecker G. A combined Hansch/ Free-Wilson approach as predictive tool in QSAR studies on propafenone-type modulators of multidrug resistance. Arch Pharm (Weinheim) 1998;331:233–240.

196. Pajeva I, Wiese M. Molecular modeling of phenothiazines and related drugs as multidrug resistance modifiers: a comparative molecular field analysis study. J Med Chem 1998;41: 1815–1826.

197. Schmid D, Ecker G, Kopp S, Hitzler M, Chiba P. Structure–activity relationship studies of propafenone analogs based on P-glycoprotein ATPase activity measurements. Biochem Pharmacol 1999;58:1447–1456.

198. Ecker G, Huber M, Schmid D, Chiba P. The importance of a nitrogen atom in modulators of multidrug resistance. Mol Pharmacol 1999;56:791–796.

199. Seelig A. How does P-glycoprotein recognize its substrates? Int J Clin Pharmacol Ther 1998;36:50–54.

200. Osterberg T, Norinder U. Theoretical calculation and prediction of P-glycoprotein-interacting drugs using MolSurf parametrization and PLS statistics. Eur J Pharm Sci 2000; 10:295–303.

201. Neuhoff S, Langguth P, Dressler C, Andersson TB, Regardh CG, Spahn-Langguth H. Affinities at the verapamil binding site of MDR1-encoded P-glycoprotein: drugs and analogs, stereoisomers and metabolites. Int J Clin Pharmacol Ther 2000;38:168–179.

202. Rosenberg MF, Mao Q, Holzenburg A, Ford RC, Deeley RG, Cole SP. The structure of the multidrug resistance protein 1 (MRP1/ ABCC1) crystallization and single-particle analysis. J Biol Chem 2001;276:16076–16082.

203. Dawson RJ, Locher KP. Structure of a bacterial multidrug ABC transporter. Nature 2006;443:180–185.

204. Hollenstein K, Frei DC, Locher KP. Structure of an ABC transporter in complex with its binding protein. Nature 2007;446:213–216.

205. Velamakanni S, Yao Y, Gutmann DA, van Veen HW, Multidrug transport by the ABC transporter Sav1866 from *Staphylococcus aureus*. Biochemistry 2008;47:9300–9308.

206. Stockner T, de Vries SJ, Bonvin AM, Ecker GF, Chiba P. Data-driven homology modelling of P-glycoprotein in the ATP-bound state indicates flexibility of the transmembrane domains. FEBS J 2009;276:964–972.

207. DeGorter MK, Conseil G, Deeley RG, Campbell RL, Cole SP. Molecular modeling of the human multidrug resistance protein 1 (MRP1/ ABCC1). Biochem Biophys Res Commun 2008;365:29–34.

208. Hazai E, Bikadi Z. Homology modeling of breast cancer resistance protein (ABCG2). J Struct Biol 2008;162:63–74.

209. Ravna AW, Sager G. Molecular model of the outward facing state of the human multidrug resistance protein 4 (MRP4/ABCC4). Bioorg Med Chem Lett 2008;18:3481–3483.

210. Ravna AW, Sylte I, Sager G. A molecular model of a putative substrate releasing conformation of multidrug resistance protein 5 (MRP5). Eur J Med Chem 2008;43:2557–2567.

211. Grimble GK, Silk DBA. Peptides in human nutrition. Nutr Res Rev 1989;2:87–108.

212. Mathews DM, Adibi SA. Peptide absorption. Gastroenterology 1976;71:151–161.

213. Adibi SA. Experimental basis for use of peptides as substrates for parenteral nutrition: a review. Metabolism 1987;36:1001–1011.

214. Furst P, Albers S, Stehle P. Dipeptides in clinical nutrition. Proc Nutr Soc 1990;49: 343–359.

215. Grimble GK. The significance of peptides in clinical nutrition. Annu Rev Nutr 1994;14: 419–447.

216. Snyder NJ, Tabas LB, Berry DM, Duckworth DC, Spry DO, Dantzig AH. Structure–activity relationship of carbacephalosporins and cephalosporins: antibacterial activity and interaction with the intestinal proton-dependent dipeptide transport carrier of Caco-2 cells. Antimicrob Agents Chemother 1997;41: 1649–1657.

217. Jiang H, Tabas LB, Dantzig AH. Interaction of β-lactam antibiotics with the human intestinal peptide transporter, hPEPT1, submitted.

218. Moore VA, Irwin WJ, Timmins P, Lambert PA, Chong S, Dando SA, Morrison RA. A rapid screening system to determine drug affinities for the intestinal dipeptide transporter 2: affinities of ACE inhibitors. Int J Pharm 2000;210:29–44.

219. Leibach FH, Ganapathy V. Peptide transporters in the intestine and the kidney. Annu Rev Nutr 1996;16:99–119.

220. Han HK, Rhie JK, Oh DM, Saito G, Hsu CP, Stewart BH, Amidon GL. CHO/hPEPT1 cells overexpressing the human peptide transporter (hPEPT1) as an alternative *in vitro* model for peptidomimetic drugs. J Pharm Sci 1999;88:347–350.

221. Covitz KM, Amidon GL, Sadee W. Human dipeptide transporter, hPEPT1, stably transfected into Chinese hamster ovary cells. Pharm Res 1996;13:1631–1634.

222. Kim JS, Oberle RL, Krummel DA, Dressman JB, Fleisher D. Absorption of ACE inhibitors from small intestine and colon. J Pharm Sci 1994;83:1350–1356.

223. Bai JP, Amidon GL. Structural specificity of mucosal-cell transport and metabolism of peptide drugs: implication for oral peptide drug delivery. Pharm Res 1992;9:969–978.

224. Hu M, Subramanian P, Mosberg HI, Amidon GL. Use of the peptide carrier system to improve the intestinal absorption of L-alpha-methyldopa: carrier kinetics, intestinal permeabilities, and *in vitro* hydrolysis of dipeptidyl derivatives of L-alpha-methyldopa. Pharm Res 1989;6:66–70.

225. Oh DM, Han HK, Amidon GL. Drug transport and targeting. Intestinal transport. Pharm Biotechnol 1999;12:59–88.

226. Boll M, Markovich D, Weber WM, Korte H, Daniel H, Murer H. Expression cloning of a cDNA from rabbit small intestine related to proton-coupled transport of peptides, beta-lactam antibiotics and ACE-inhibitors. Pflugers Arch 1994;429:146–149.

227. Ganapathy ME, Brandsch M, Prasad PD, Ganapathy V, Leibach FH. Differential recognition of beta-lactam antibiotics by intestinal and renal peptide transporters, PEPT 1 and PEPT 2. J Biol Chem 1995;270:25672–25677.

228. Ganapathy ME, Huang W, Wang H, Ganapathy V, Leibach FH. Valacyclovir: a substrate for the intestinal and renal peptide transporters PEPT1 and PEPT2. Biochem Biophys Res Commun 1998;246:470–475.

229. Shen H, Smith DE, Yang T, Huang YG, Schnermann JB, Brosius FC 3rd. Localization of PEPT1 and PEPT2 proton-coupled oligopeptide transporter mRNA and protein in rat kidney. Am J Physiol 1999;276:F658–F665,.

230. Ogihara H, Suzuki T, Nagamachi Y, Inui K, Takata K. Peptide transporter in the rat small intestine: ultrastructural localization and the effect of starvation and administration of amino acids. Histochem J 1999;31:169–174.

231. Steiner HY, Naider F, Becker JM. The PTR family: a new group of peptide transporters. Mol Microbiol 1995;16:825–834.

232. Meredith D, Boyd CA. Structure and function of eukaryotic peptide transporters. Cell Mol Life Sci 2000;57:754–778.

233. Fei YJ, Ganapathy V, Leibach FH. Molecular and structural features of the proton-coupled oligopeptide transporter superfamily. Prog Nucleic Acid Res Mol Biol 1998;58:239–261.

234. Nussberger S, Steel A, Trotti D, Romero MF, Boron WF, Hediger MA. Symmetry of H^+ binding to the intra- and extracellular side of the H^+-coupled oligopeptide cotransporter PepT1. J Biol Chem 1997;272:7777–7785.

235. Swaan PW, Tukker JJ. Essential molecular requirements for carrier-mediated peptide transport. Pharm Weekbl Sci Ed 1992;14F:62.

236. Swaan PW, Tukker JJ. Binding site mapping of the intestinal peptide carrier. Pharm Weekbl Sci Ed 1992;14M:4.

237. Tukker JJ, Swaan PW. Molecular features essential for active peptide transport. Pharm Res 1992;9:S-180.

238. Humblet C, Marshall GR. Pharmacophore identification and receptor mapping. Ann Rep Med Chem 1980;15:267–276.

239. Li J, Hidalgo IJ. Molecular modeling study of structural requirements for the oligopeptide transporter. J Drug Target 1996;4:9–17.

240. Swaan PW, Tukker JJ. Molecular determinants of recognition for the intestinal peptide carrier. J Pharm Sci 1997;86:596–602.

241. Swaan PW, Stehouwer MC, Tukker JJ. Molecular mechanism for the relative binding affinity to the intestinal peptide carrier. Comparison of three ACE-inhibitors: enalapril, enalaprilat, and lisinopril. Biochim Biophys Acta 1995;1236:31–38.

242. Kubo SH, Cody RJ. Clinical pharmacokinetics of the angiotensin converting enzyme inhibitors. A review Clin Pharmacokinet 1985;10: 377–391.

243. Swaan PW, Koops BC, Moret EE, Tukker JJ. Mapping the binding site of the small intestinal peptide carrier (PepT1) using comparative molecular field analysis. Receptors Channels 1998;6:189–200.

244. Doring F, Will J, Amasheh S, Clauss W, Ahlbrecht H, Daniel H. Minimal molecular determinants of substrates for recognition by the intestinal peptide transporter. J Biol Chem 1998;273:23211–23218.

245. Terada T, Sawada K, Irie M, Saito H, Hashimoto Y, Inui K. Structural requirements for determining the substrate affinity of peptide transporters PEPT1 and PEPT2. Pflugers Arch 2000;440:679–684.

246. Bailey PD, Boyd CAR, Bronk JR, Collier ID, Meredith D, Morgan KM, Temple CS. How to make drugs orally active: a substrate template for peptide transporter PepT1. Angew Chem Int Ed 2000;39:506–508.

247. Bretschneider B, Brandsch M, Neubert R. Intestinal transport of beta-lactam antibiotics: analysis of the affinity at the H^+/peptide symporter (PEPT1), the uptake into Caco-2 cell monolayers and the transepithelial flux. Pharm Res 1999;16:55–61.

248. Bolger MB, Haworth IS, Yeung AK, Ann D, von Grafenstein H, Hamm-Alvarez S, Okamoto CT, Kim KJ, Basu SK, Wu S, Lee VH. Structure, function, and molecular modeling approaches to the study of the intestinal dipeptide transporter PepT1. J Pharm Sci 1998;87:1286–1291.

249. Chen XZ, Steel A, Hediger MA. Functional roles of histidine and tyrosine residues in the H(+)-peptide transporter PepT1. Biochem Biophys Res Commun 2000;272:726–730.

250. Buchwald H, Stoller DK, Campos CT, Matts JP, Varco RL. Partial ileal bypass for hypercholesterolemia. 20- to 26-year follow-up of the first 57 consecutive cases. Ann Surg 1990;212: 318–329; discussion 329–331.

251. Ast M, Frishman WH. Bile acid sequestrants. J Clin Pharmacol 1990;30:99–106.

252. Takashima K, Kohno T, Mori T, Ohtani A, Hirakoso K, Takeyama S. The hypocholesterolemic action of TA-7552 and its effects on cholesterol metabolism in the rat. Atherosclerosis 1994;107:247–257.

253. Carey MD, Small DM, Bliss CM. Lipid digestion and absorption. Annu Rev Physiol 1983; 45:651–677.

254. Wilson FA. Intestinal transport of bile acids. In: Schultz SG, Field M, Frizzell RA, editors. The Gastrointestinal System IV. Bethesda, MD: American Physiological Society; 1991. p 389–404.

255. Dowling RH. The enterohepatic circulation. Gastroenterology 1972;62:122–140.

256. Hofmann AF. The enterohepatic circulation of bile acids in man. Adv Intern Med 1976;21:501–534.

257. Ewerth S. On the enterohepatic circulation of bile acids in man. Acta Chir Scand Suppl 1982;513:1–38.

258. Erlinger RH. Physiology of bile secretion and enterohepatic circulation, In: Johnson LR, editor. Physiology of the Gastrointestinal Tract. 2nd ed. New York: Raven Press; 1987. p 1557–1580.

259. Klaassen CD. Intestinal and hepatobiliary disposition of drugs. Toxicol Pathol 1988;16: 130–137.

260. Hofmann AF. Enterohepatic circulation of bile acids, In: Schultzand S, Forte J, editors. The Gastrointestinal System III. Bethesda, MD: American Physiological Society; 1989. p 567–596.

261. Stengelin S, Apel S, Becker W, Maier M, Rosenberger J, Bewersdorf U, Girbig F, Weyland C, Wess G, Kramer W. The rabbit ileal lipid-binding protein. Gene cloning and functional expression of the recombinant protein. Eur J Biochem 1996;239:887–896.

262. Vodenlich AD, Gong Y-Z, Geoghegan KF, Lin MC, Lanzetti AJ, Wilson FA. Identification of the 14 kDa bile acid transport protein of rat ileal cytosol as gastrotropin. Biochem Biophys Res Commun 1991;177:1147–1154.

263. Iseki S, Amano O, Kanda T, Fujii H, Ono T. Expression and localization of intestinal 15 kDa protein in the rat. Mol Cell Biochem 1993;123:113–120.

264. Kramer W, Corsiero D, Friedrich M, Girbig F, Stengelin S, Weyland C. Intestinal absorption of bile acids: paradoxical behaviour of the 14 kDa ileal lipid-binding protein in differential photoaffinity labelling. Biochem J 1998; 333:335–341.

265. Lin MC, Kramer W, Wilson FA. Identification of cytosolic and microsomal bile acid-binding proteins in rat ileal enterocytes. J Biol Chem 1990;265:14986–14995.

266. Lin MC, Mullady E, Wilson FA. Timed photoaffinity labeling and characterization of bile acid binding and transport proteins in rat ileum. Am J Physiol 1993;265:G56–G62,.

267. Grober J, Zaghini I, Fujii H, Jones SA, Kliewer SA, Willson TM, Ono T, Besnard P. Identification of a bile acid-responsive element in the human ileal bile acid-binding protein gene. Involvement of the farnesoid X receptor/9-*cis*-retinoic acid receptor heterodimer. J Biol Chem 1999;274:29749–29754.

268. Chiang JY, Kimmel R, Weinberger C, Stroup D. Farnesoid X receptor responds to bile acids and represses cholesterol 7alpha-hydroxylase gene (CYP7A1) transcription. J Biol Chem 2000;275:10918–10924.

269. Crestani M, Sadeghpour A, Stroup D, Galli G, Chiang JY. Transcriptional activation of the

cholesterol 7alpha-hydroxylase gene (CYP7A) by nuclear hormone receptors. J Lipid Res 1998;39:2192–2200.

270. Lehmann JM, Kliewer SA, Moore LB, Smith-Oliver TA, Oliver BB, Su JL, Sundseth SS, Winegar DA, Blanchard DE, Spencer TA, Willson TM. Activation of the nuclear receptor LXR by oxysterols defines a new hormone response pathway. J Biol Chem 1997;272: 3137–3140.
271. Janowski BA, Grogan MJ, Jones SA, Wisely GB, Kliewer SA, Corey EJ, Mangelsdorf DJ. Structural requirements of ligands for the oxysterol liver X receptors LXRalpha and LXRbeta. Proc Natl Acad Sci USA 1999; 96:266–271.
272. Spencer TA, Li D, Russel JS, Collins JL, Bledsoe RK, Consler TG, Moore LB, Galardi CM, McKee DD, Moore JT, Watson MA, Parks DJ, Lambert MH, Willson TM. Pharmacophore analysis of the nuclear oxysterol receptor LXRalpha. J Med Chem 2001;44:886–897.
273. Lazaridis KN, Pham L, Tietz P, Marinelli RA, deGroen PC, Levine S, Dawson PA, LaRusso NF. Rat cholangiocytes absorb bile acids at their apical domain via the ileal sodium-dependent bile acid transporter. J Clin Invest 1997;100:2714–2721.
274. Sorscher S, Lillienau J, Meinkoth JL, Steinbach JH, Schteingart CD, Feramisco J, Hofmann AF. Conjugated bile acid uptake by *Xenopus laevis* oocytes induced by microinjection with ileal Poly A+ mRNA. Biochem Biophys Res Commun 1992;186:1455–1462.
275. Kramer W, Burckhardt G, Wilson FA, Kurz G. Bile salt-binding polypeptides in brush-border membrane vesicles from rat small intestine revealed by photoaffinity labeling. J Biol Chem 1983;258:3623–3627.
276. Lin MC, Weinberg SL, Kramer W, Burckhardt G, Wilson FA. Identification and comparison of bile acid-binding polypeptides in ileal basolateral membrane. J Membr Biol 1988;106:1–11.
277. Burckhardt G, Kramer W, Kurz G, Wilson FA. Inhibition of bile salt transport in brush-border membrane vesicles from rat small intestine by photoaffinity labeling. J Biol Chem 1983;258:3618–3622.
278. Shneider BL, Moyer MS. Characterization of endogenous carrier-mediated taurocholate efflux from *Xenopus laevis* oocytes. J Biol Chem 1993;268:6985–6988.
279. Shneider BL, Dawson PA, Christie D-M, Hardikar W, Wong MH, Suchy FJ. Cloning and molecular characterization of the ontogeny of a rat ileal sodium-dependent bile acid transporter. J Clin Invest 1995;95:745–754.
280. Weinman SA, Carruth MW, Dawson PA. Bile acid uptake via the human apical sodium-bile acid cotransporter is electrogenic. J Biol Chem 1998;273:34691–34695.
281. Hallén S, Branden M, Dawson PA, Sachs G. Membrane insertion scanning of the human ileal sodium/bile acid co-transporter. Biochemistry 1999;38:11379–11388.
282. Sippel CJ, Dawson PA, Shen T, Perlmutter DH. Reconstitution of bile acid transport in a heterologous cell by cotransfection of transporters for bile acid uptake and efflux. J Biol Chem 1997;272:18290–18297.
283. Wong MH, Oelkers P, Craddock AL, Dawson PA. Expression cloning and characterization of the hamster ileal sodium-dependent bile acid transporter. J Biol Chem 1994;269: 1340–1347.
284. von Heijne G, Gavel Y. Topogenic signals in integral membrane proteins. Eur J Biochem 1988;174:671–678.
285. Higaki J, Hara S, Takasu N, Tonda K, Miyata K, Shike T, Nagata K, Mizui T. Inhibition of ileal Na^+/bile acid cotransporter by S-8921 reduces serum cholesterol and prevents atherosclerosis in rabbits. Arterioscler Thromb Vasc Biol 1998;18:1304–1311.
286. Ichihashi T, Izawa M, Miyata K, Mizui T, Hirano K, Takagishi Y. Mechanism of hypocholesterolemic action of S-8921 in rats: S-8921 inhibits ileal bile acid absorption. J Pharmacol Exp Ther 1998;284:43–50.
287. Love MW, Dawson PA. New insights into bile acid transport. Curr Opin Lipidol 1998;9: 225–229.
288. Hofmann AF. The continuing importance of bile acids in liver and intestinal disease. Arch Intern Med 1999;159:2647–2658.
289. Lack L. Properties and biological significance of the ileal bile salt transport system. Environ Health Perspect 1979;33:79–90.
290. Tyor MP, Garbutt JT, Lack L, Metabolism and transport of bile salts in the intestine. Am J Med 1971;51:614–626.
291. Walker S, Stiehl A, Raedsch R, Kloters P, Kommerell B. Absorption of urso- and chenodeoxycholic acid and their taurine and glycine conjugates in rat jejunum, ileum, and colon. Digestion 1985;32:47–52.
292. Aldini R, Roda A, Lenzi PL, Ussia G, Vaccari MC, Mazzella G, Festi D, Bazzoli F, Galletti G, Casanova S, et al. Bile acid active and passive ileal transport in the rabbit: effect of luminal stirring. Eur J Clin Invest 1992;22:744–750.

293. Schiff ER, Small NC, Dietschy JM. Characterization of the kinetics of the passive and active transport mechanisms for bile acid absorption in the small intestine and colon of the rat. J Clin Invest 1972;51:1351–1362.

294. Heaton KW, Lack L. Ileal bile salt transport: mutual inhibition in an *in vivo* system. Am J Physiol 1968;214:585–590.

295. Lack L, Weiner IM. Intestinal bile salt transport: structure–activity relationships and other properties. Am J Physiol 1966;210: 1142–1152.

296. Krag E, Phillips SF. Active and passive bile acid absorption in man. Perfusion studies of the ileum and jejunum. J Clin Invest 1974;53:1686–1694.

297. Gallagher K, Mauskopf J, Walker JT, Lack L. Ionic requirements for the active ileal bile salt transport system. J Lipid Res 1976; 17:572–577.

298. Lack L, Tantawi A, Halevy C, Rockett D. Positional requirements for anionic charge for ileal absorption of bile salt analogues. Am J Physiol 1984;246:G745–G749.

299. Low-Beer TS, Tyor MP, Lack L. Effects of sulfation of taurolithocholic and glycolithocholic acids on their intestinal transport. Gastroenterology 1969;56:721–726.

300. Wess G, Kramer W, Bartmann W, Enhsen A, Glombik H, Müllner S, Bock K, Dries A, Kleine H, Schmitt W. Modified bile acids: preparation of $7\alpha,12\alpha$-dihydroxy-3β- and $7\alpha,12\alpha$-dihydroxy-3α-(2-hydroxyethoxy)-5β-cholanic acid and their biological activity. Tetrahedron Lett 1992;33:195–198.

301. Lack L, Walker JT, Singletary GD. Ileal bile salt transport: *in vivo* studies of effect of substrate ionization on activity. Am J Physiol 1970;219:487–490.

302. Baringhaus KH, Matter H, Stengelin S, Kramer W. Substrate specificity of the ileal and the hepatic Na(+)/bile acid cotransporters of the rabbit. II. A reliable 3D QSAR pharmacophore model for the ileal Na(+)/bile acid cotransporter. J Lipid Res 1999;40:2158–2168.

303. Swaan PW, Szoka FC Jr, Oie S. Molecular modeling of the intestinal bile acid carrier: a comparative molecular field analysis study. J Comput Aided Mol Des 1997;11:581–588.

304. Kramer W, Stengelin S, Baringhaus KH, Enhsen A, Heuer H, Becker W, Corsiero D, Girbig F, Noll R, Weyland C. Substrate specificity of the ileal and the hepatic Na(+)/bile acid cotransporters of the rabbit. I. Transport studies with membrane vesicles and cell lines expressing the cloned transporters. J Lipid Res 1999;40:1604–1617.

305. Helsper F, Swaan PW. Knowledge-based modeling of solute carrier proteins: application to the apical sodium-dependent bile acid transporter, submitted for publication.

306. Ho NFH. Utilizing bile acid carrier mechanisms to enhance liver and small intestine absorption. Ann NY Acad Sci 1987;507:315–329.

307. Swaan PW, Hillgren KM, Szoka FC Jr, Oie S. Enhanced transepithelial transport of peptides by conjugation to cholic acid. Bioconjug Chem 1997;8:520–525.

308. Kramer W, Wess G, Schubert G, Bickel M, Girbig F, Gutjahr U, Kowalewski S, Baringhaus K-H, Enhsen A, Glombik H, Müllner S, Neckermann G, Schulz S, Petzinger E. Liver-specific drug targeting by coupling to bile acids. J Biol Chem 1992;267:18598–18604.

309. Kramer W, Wess G, Enhsen A, Bock K, Falk E, Hoffmann A, Neckermann G, Gantz D, Schulz S, Nickau L, et al. Bile acid derived HMG-CoA reductase inhibitors. Biochim Biophys Acta 1994;1227:137–154.

310. Wess G, Kramer W, Han XB, Bock K, Enhsen A, Glombik H, Baringhaus KH, Boger G, Urmann M, Hoffmann A, Falk E. Synthesis and biological activity of bile acid-derived HMG-CoA reductase inhibitors. The role of 21-methyl in recognition of HMG-CoA reductase and the ileal bile acid transport system. J Med Chem 1994;37:3240–3246.

311. Stephan ZF, Yurachek EC, Sharif R, Wasvary JM, Steele R, Howes C. Reduction of cardiovascular and thyroxine-surpressing activities of L-T_3 by liver targeting with cholic acid. Biochem Pharmacol 1992;43:1969–1974.

312. Kim D-C, Harrison AW, Ruwart MJ, Wilkinson KF, Fisher JF, Hidalgo IJ, Borchardt RT. Evaluation of the bile acid transporter in enhancing intestinal permeability to renin-inhibitory peptides. J Drug Target 1993;1: 347–359.

313. Mills CO, Martin GH, Elias E. The effect of tyrosine conjugation on the critical micellar concentration of free and glycine-conjugated bile salts. Biochim Biophys Acta 1986;876: 677–683.

314. Anwer MS, O'Maille ER, Hofmann AF, DiPietro RA, Michelotti E. Influence of side-chain charge on hepatic transport of bile acids and bile acid analogues. Am J Physiol 1985;249: G479–G488,.

315. Mills CO, Iqbal S, Elias E. Selectively reduced biliary excretion of cholyldiglycylhistamine but not of cholyltetraglycylhistamine in

ethinyl estradiol-treated rats. A possible indicator of increased bile canalicular permeability. J Hepatol 1985;1:199–210.

316. Caspary WF, Creutzfeldt W. Inhibition of bile salt absorption by blood-sugar lowering biguanides. Diabetologia 1975;11:113–117.
317. Root C, Smitth CD, Winegar DA, Brieaddy LE, Lewis MC. Inhibition of ileal sodium-dependent bile acid transport by 2164U90. J Lipid Res 1995;36:1106–1115.
318. Lewis MC, Brieaddy LE, Root C. Effects of 2164U90 on ileal bile acid absorption and serum cholesterol in rats and mice. J Lipid Res 1995;36:1098–1105.
319. Ekins S, Waller CL, Swaan PW, Cruciani G, Wrighton SA, Wikel JH. Progress in predicting human ADME parameters *in silico*. J Pharmacol Toxicol Methods 2000;44:251–272.

RECEPTOR TARGETS IN DRUG DISCOVERY

EMIR DUZIC
MICHAEL J. MARINO
MICHAEL WILLIAMS
Discovery Research Cephalon, Inc., West Chester, PA

1. INTRODUCTION

For more than a century and a half, while the applied technologies used to study receptors have undergone quantal advances, the basic conceptualization of the receptor has remained essentially unchanged as the basis for understanding disease causality and drug actions.

The receptor concept was initially proposed in the mid-nineteenth century to explain the physiological effects of transmitters/hormones in bioassays or animal models [1]. This occurred in the absence of any physical evidence for the existence of receptors [2]—the first receptor the nicotinic cholinergic only being isolated in the 1970s [1]—with the receptor concept not being formalized until the beginning of the twentieth century in the seminal receptor theories of Ehrlich and Langley [3].

The initial receptor concept was more qualitative than quantitative and met with considerable resistance from influential pharmacologists like Magnus and Dale [4,5], the latter actually describing receptor theory as being overly speculative and "unnecessary" [1]. It was not until the 1930s that Clark [6] and Gaddum [7] undertook the necessary quantitative analysis of drug action to provide crucial support for receptor theory. Subsequent work by Schild [8], Ariens [9], and Stephenson [10] elaborated on receptor theory but it was the work of Ahlquist in 1948 [11] on α- and β-adrenoceptors that ultimately established receptor theory as the basis for drug action [5,12]. Subsequent work by Black [13] on adrenergic and histamine receptor subtypes and Humphry [14] on serotonin (5-HT) receptors both of which entailed a close working relationship between medicinal chemists and pharmacologists in defining structure–activity relationships (SARs), led to the successful development of propranolol, a β-adrenoceptor antagonist, and cimetidine, the first of the histamine H_2 receptor blockers [13] as well as the $5\text{-}HT_{1B/D}$ agonist, sumitriptan [14], respectively. These agents represented novel additions to the pharmacopeia as first-in-class treatments for hypertension, gastric ulcers and migraine, respectively.

Receptors are macromolecules usually viewed as being present on the cell surface and physically associated with a number of signaling and scaffold proteins that are necessary for the expression of the physiological function of the receptor. Over time, the term receptor has been used to describe every conceivable drug target, both cellular and intracellular, and includes enzymes, DNA binding motifs, the various RNAs, and protein–protein interactions. This has further formalized the initial concept of receptors as the molecular targets at which drugs act as "magic bullets" to produce their effects, both beneficial and detrimental [3].

Ligands for traditional receptors bind to their target to mimic the effect of the natural transmitter/hormone—acting as agonists or block the effects of agonists functioning as antagonists. This simple binary view of ligand function has been modified by nuances of ligand function including partial agonists, allosteric modulators including protean and inverse agonism and constitutive activity [2,3,15] and more recently, ligand residence time [16,17] and target flexibility [18]. For other drug targets, ligands may act to block key enzymes, nucleic acid motifs and protein–protein interactions in signaling pathways or alter the cellular localization of key signaling elements.

1.1. Receptors and Drug Discovery

Drug discovery is the complex process of identifying and developing novel chemical entities (NCEs) that interact with the various receptors to treat human disease states [19]. As drug discovery efforts enter the twenty-first century, biomedical researchers have at their disposal a wealth of sophisticated technologies and information generation platforms driven by diverse scientific disciplines that have the potential to allow the more rapid development of new drugs with improved selectivity and safety profiles. These, together with the map of the human genome [20], have led to expectations (as yet to be realized) that an improved understanding of human disease states based on gene associations will result in novel targets that will lead to a level of precision in efficacy and safety in the discovery of new drugs never before possible. Thus, in the last decade, there has been an exponential increase in the mechanistic understanding of the basic pathophysiology of disorders such as pain, diabetes, some forms of cancer, and neurodegenerative diseases (Alzheimer's and Parkinson's diseases). This has, in some instances, led to major changes in disease treatment, for example, kinase inhibitors for cancer [21,22], together with a realization that molecular target selectivity, as contrasted to polypharmic activity, does not necessarily make for the most efficacious new drugs [23,24].

There are also major challenges in effectively using the copious amounts of information now available for disease-related gene association and the putative molecular targets involved in disease causality. A bewildering amount of data, much of it archival and/or overtly reductionistic (and thus of limited value), has confounded understanding of disease pathophysiology and thus the cost effective discovery and development of new drugs [25]. The effective implementation of integrative platforms such as target validation [26] and translational medicine [27] have the potential to improve productivity and reduce attrition.

Irrespective of compound or drug target sources, the use of ever more productive high-throughput screening (HTS) processes and increasingly detailed structural information on putative drug targets, it is clear that improving the rather disappointing success rate in the current drug discovery environment [28–30] is absolutely dependent on the iteration and integration of lead compound identification and optimization through the hierarchical complexity of *in vitro* and *in vivo* assays that measure efficacy, selectivity, side-effect liability, absorption, distribution, metabolism, and excretion (ADME), and potential toxicity [31,32]. It is thus critical to understand how existing knowledge of drug targets has evolved and what the realistic potential is for identifying, validating, and prioritizing new ones especially given that many target families that have known for decades, for example, neurokinin, adenosine, and muscarinic receptors, secretases, and so on, have yet to yield *bona fide* drug candidates to provide proof of concept in clinical trials.

2. CELLULAR INFORMATION PROCESSING

All cells acquire information from their extracellular milieu through the process of transmembrane signaling. The integration of the multiple inputs that are necessary to maintain cellular homeostasis and tissue viability—under both normal and adverse, disease- and trauma-related conditions—involves a variety of different extracellular signaling modalities. These include local concentrations of autocrine and paracrine hormones, pheromones and odorants, traditional neurotransmitters, neuromodulators, temperature, mechanical distention, and stress and alterations in ion (H^+/K^+) concentrations. These physical stimuli and endogenous chemicals elicit their effects through interactions with cell surface receptors (Table 1). Once receptors are activated by an *agonist* ligand, a compound capable of producing a cellular response, they transduce or couple the energy associated with the binding event to a cellular effect via a signal transduction process(es) involving alterations in protein–protein interactions or second messenger signaling systems, for example, G-protein, protein kinase or phosphatase modulation, alterations in lipid metabolism, intracellular calcium concentrations, protein–gene element interactions, and so on, to produce acute or more long-term effects on cellular and tissue function. Thus, neurotransmitters, neuromodulators, and hormones can produce transient increases in second messengers such as cyclic AMP or inositol triphosphate (IP_3) or more long-term changes that involve alterations in gene expression through modulation of transcription factors and changes in RNA function and stability.

Table 1. Drug Recognition Sites Classifiable as Receptors

Type	Class	Examples
Receptor	GPCR	Dopaminergic, adrenergic, $GABA_B$, mGluR
	Ion channel	
	LGIC	NMDA, nicotinic, $GABA_A$
	VGIC	Na_v, Ca_v
	Others	ASIC, TRPV
	Enzyme associated R	
	Tyrosine kinase	PDGF, EGF, Trk
	Others	CNTF, cytokine
	Nuclear hormone R	
	Steroid R	GR, PR
	Others	PPAR, RAR
Neurotransmitter transporter	SCDNT/SLC6	SERT, NET, GAT, DAT
Enzyme	Cell surface/extracellular	ACE, NOS, HMG CoA reductase, acetylcholinesterase
	Intracellular	Caspase, CDK, ROCK DNA polymerase

GABA: G-aminobutyric acid; mGluR: metabotropic glutamate receptor; LGIC: ligand-gated ion channel; VGIC: voltage-gated ion channel; NMDA: N-methyl-D-aspartic acid; ASIC: acid-sensing ion channel; TRPV: transient receptor potential vanilloid; PDGF: platelet-derived growth factor; EGF: epidermal growth factor; CNTF: ciliary neurotrphic factor; GR: glucocorticoid receptor; PR: progesterone receptor; PPAR: peroxisome proliferator-activated receptor; RAR: retinoic acid receptor; SERT: serotonin transporter; NET: norepinephrine transporter; GAT: GABA transporter; DAT: dopamine transporter; ACE: angiotensin-converting enzyme; NOS: nitric oxide synthase; CDK: cyclin-dependent kinase; ROCK: rho-dependent kinase.

Disease-related alterations in receptor function can occur by a functional overstimulation of receptors with their consequent desensitization, a phenomenon resulting from excess ligand availability and/or an enhanced coupling of the ligand-activated receptor to second messenger systems, or a reduction in stimulation resulting from decreased ligand availability or dysfunctional receptor coupling processes.

Thus drugs that effectively treat human disease states by restoring disease-associated defects in signal transduction can act either by replacing endogenous transmitters, for example, l-dopa treatment to replace the dopamine (DA) lost in Parkinson's disease, or blockade of excess agonist stimulation, for example, the histamine H_2 receptor antagonist, cimetidine, that blocks histamine-induced gastric acid secretion and thus reduces ulcer formation. Of the approximately 500 targets through which known drugs act, 71% of these are receptors in the GPCR, ion channel, or neurotransmitter transporter (SCDNT/*SLC6*) families. Of these drugs, approximately 85% produce their effects by antagonizing the actions of endogenous agonists [33]. Enzymes, by producing products that regulate second messenger availability, for example, adenylyl cyclase-catalyzed production of cyclic AMP or phosphodiesterase (PDE)-mediated degradation of cAMP, or by modification of protein targets, for example, by adding (protein kinases) or removing (protein phosphatases) phosphates on serine, threonine, or tyrosine residues, serve a critical role in cellular homeostasis, and as such, also represent key drug targets.

3. RECEPTOR CONCEPTS

Ligand function at a discrete target can be viewed in a binary manner with agonists mimicking the actions of endogenous transmitters/hormones and antagonists blocking the actions of endogenous agonists. This simplistic view has been modified by nuances such as partial agonists, inverse agonists, protein ligands and allosteric modulators [2,3,15], residence time [16,17] and target flexibility [18], the latter a concept with its origins in the Koshland's induced fit model [34].

The seminal concept that therapeutic agents produce their effects by acting as "magic bullets" at discrete molecular targets is attributed to Ehrlich and Langley who independently generated the experimental data that led to the evolution of the "lock and key" hypothesis [3]. A ligand (L) thus acted as the unique "key" to selectively modulate receptor (R) activity, the latter functioning as the "lock" for the "entry" of external signals into the cell. In this model, an agonist ligand forms an RL complex and has the ability to "turn" the lock (Eq. 1), whereas a receptor antagonist would occupy the lock and prevent agonist access.

$$\text{Receptor} + \text{ligand} \underset{k-1}{\overset{k+1}{\rightleftharpoons}} \text{RL} \rightarrow \text{Signaling event} \quad (1)$$

Despite the quantal advances in technologies used to study receptor interactions (e.g., FRET, BRET, etc.), the receptor–ligand (RL) concept and the related enzyme–substrate (ES) complex remain the conceptual foundations for understanding receptor and enzyme function, disease pathophysiology, and medicinal chemistry-driven approaches to drug discovery. For the past two decades, however, with the explosion in the number and diversity of receptors driven by cloning approaches and the mapping of the human genome [20] there has been an increased appreciation of the inherent complexity of receptor/drug target function.

Receptor theory is based on the classical law of mass action (LMA) as developed by Michaelis and Menten for the study of enzyme catalysis [35]. The extrapolation of classical enzyme theory to receptors is, however, an approximation. In an enzyme–substrate (ES) interaction, the substrate S undergoes an enzyme-catalyzed conversion to a product or products. Because of the equilibrium established, product accumulation has the ability to reverse the reaction process. Alternatively, the latter can be used as a substrate in other cellular pathways and is thus removed from the equilibrium situation or can act as a feedback modulator [36] to alter the ES reaction either positively or negatively (Eq. 2).

$$\text{Enzyme} + \text{Substrate} \rightleftarrows \text{ES} \rightleftarrows \text{E} + \text{Product} \quad (2)$$

For the receptor–ligand interaction, binding of the ligand to the receptor to form the RL complex results in a response driven by the thermodynamics of the binding reaction that leads to functional changes in the target cell (Eq. 1). While conformational changes occur in either the ligand, the receptor, or in both, there is no chemical change in the ligand and no chemical product derived from the ligand. Despite initiating events such as receptor internalization, receptor phosphorylation, second messenger system activation, and so on, the bound ligand is chemically unchanged by the binding event, and thus there is no equilibrium established between the RL complex and the consequences of receptor activation. After the formation of an RL complex, a functional response to receptor activation can therefore be related to the concentration of the ligand present (Fig. 1a).

3.1. Occupancy Theory

The basic premise of Clark's occupancy theory [3,6], based on Michaelis–Menten theory [35], was that the effect produced by an agonist was dependent on the number of receptors occupied by the agonist, a reflection of the agonist concentration. This theory derived from Clark's observations [6] that acetylcholine (ACh) receptor antagonists such as atropine caused a rightward shift in the ACh dose–response (DR) curve in muscle preparations when plotted logarithmically. The basic tenets of occupancy theory were (1) the RL complex formation was assumed to be reversible; (2) the association of the receptor with the ligand to form the RL complex was defined as a bimolecular process with dissociation being a monomolecular process; (3) all receptors in a given system were assumed to be equivalent to one another and able to bind

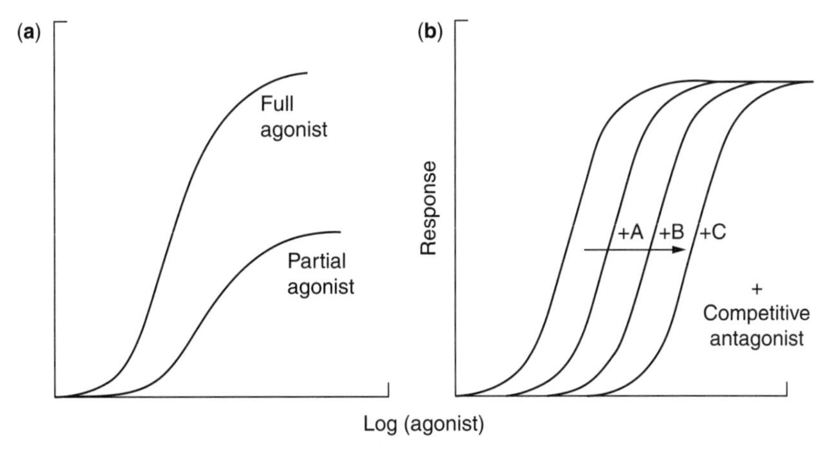

Figure 1. Dose–response curve. The addition of increasing concentrations of an agonist ligand causes an increase in a biological response. Plotted on a logarithmic scale, a sigmoidal curve is obtained. (a) A full agonist produced a maximal response, whereas a partial agonist reaches a plateau that is only part of the response seen with a full agonist. (b) In the presence of antagonist concentrations A, B, C, the dose–response curve is progressively moved to the right. Increasing agonist concentrations overcome the effects of the antagonist.

ligand independently of one another; (4) formation of the RL complex did not alter the free (F) concentration of the ligand or the affinity of the receptor for the ligand; (5) the response elicited by receptor occupancy was directly proportional to the number of receptors occupied; and (6) the biological response was dependent on an equilibrium being obtained between R and L according to Equation 1.

Although it is not always possible to determine the concentration of free ligand (F) or that of the RL complex, rearranging the latter, the equilibrium dissociation constant, K_d, equaling $k - 1/k + 1$, can be derived from Equation 3.

$$Kd = \frac{[R][L]}{[RL]} \tag{3}$$

and is equal to the concentration of L that occupies 50% of the available receptors.

Antagonist interactions with the receptor were defined by Gaddum [7] as being the result of receptor occupancy with the antagonist ligand being unable to elicit a functional response. Antagonists thus block agonist actions. Agonists can overcome the effects of a competitive (e.g., reversible/surmountable) antagonist when their concentration is progressively increased (Fig. 1b) such that in the presence of increasing fixed concentrations of a competitive antagonist, a series of parallel agonist DR curves can be generated that shift progressively to the right (Fig. 1b). A Schild regression relationship [8], a plot of log (DR-1) versus the log (antagonist) concentration (Fig. 2), can be used to derive the pA_2 value for an antagonist from the intercept of the abscissa. The pA_2 value is the negative logarithm of the affinity of an antagonist for a given receptor in a defined biological system and is equal to $-\log_{10} K_B$ where K_B is the dissociation constant for a competitive antagonist with a slope of near unity. Not all antagonists are

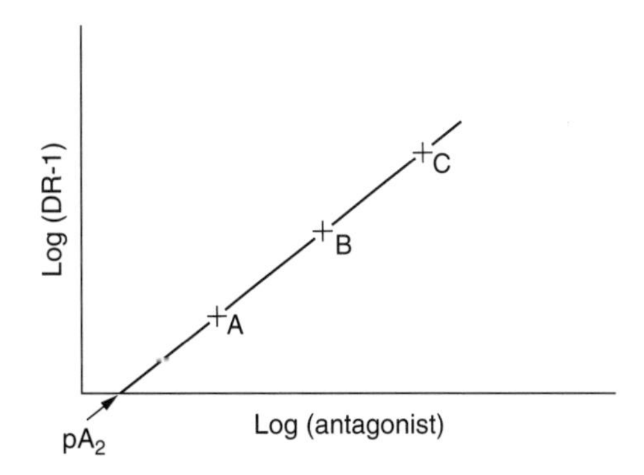

Figure 2. Schild plot regression. Data from Fig. 1b for antagonist concentrations A, B, and C can be plotted by the method of Schild [8] to yield a pA_2 value, a measure of antagonist activity. A slope of unity indicates a competitive antagonist.

competitive. Non- or uncompetitive antagonists that act at allosteric sites or that bind irreversibly to the agonist site have slopes that are significantly less than unity. The Schild plot can thus be used to determine the mechanism by which an antagonist produces its effects.

Occupancy theory was modified by Ariens [9] based on data where not all cholinergic agonists were able to elicit a maximal response in a skeletal muscle preparation, even when administered at supramaximal concentrations. This led to the introduction of the concept of the *intrinsic activity* of a ligand. A full agonist was defined as having a value of 1.0 with the value for an antagonist being zero. However, many compounds were subsequently identified that bound to the receptor but were only able to produce a portion of the response seen with a full agonist. These were defined as *partial agonists* (Fig. 1a). By definition, these compounds were also *partial antagonists.*

Other agonists have also been identified that produce a response greater than the previously defined full agonist. These are termed "super agonists." One example from the 1980s is the muscarinic cholinergic receptor agonist, L 670,207 (Fig. 3), an arecoline bioisostere that was 70% more active than arecoline and thus had an intrinsic activity of 1.7 [37]. From the activity seen in response to L 670,207, the system used to characterize these muscarinic agonists was obviously capable of a greater response than that seen with arecoline, making the latter compound a partial agonist. A more recent example of a "super agonist" was TGN1412, a CD28 antibody, Phase I clinical trials of which were nearly fatal to the human subjects [38].

The partial agonist concept was additionally refined by Stephenson [10] in introducing the concept of *efficacy*, ε, which differed from intrinsic activity in that the latter was defined as a proportion of the maximal response. This

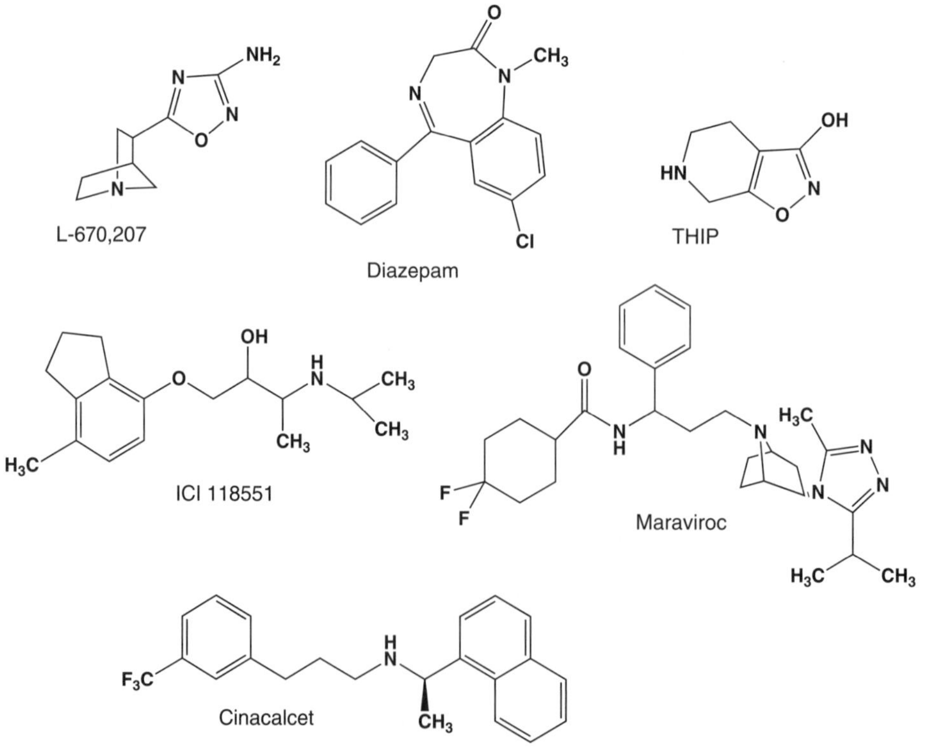

L-670,207

Diazepam

THIP

ICI 118551

Maraviroc

Cinacalcet

Figure 3. Structures of superagonist, GABA receptor modulators and allosteric modulators.

concept was extended to situations where a maximal response to an agonist could occur when only a small proportion of the total number of receptors on a tissue were occupied (a condition termed receptor reserve), as in the situations when receptors were inactivated by alkylating agents [39]. This resulted in a nonlinear occupancy relationship with the response then being defined as the stimulus, S, a product of the fraction of receptors occupied and the ligand efficacy. A nonlinear functional response clearly complicates data interpretation, especially when spare receptor or receptor reserve concepts are introduced to rationalize individual data sets. An additional issue in defining efficacy was the degree to which the receptor activation event and its blockade by antagonists was measured through events that were spatially and temporally removed from the receptor activation event and also the degree to which the response could be amplified through cofactor and signal transduction cascades. Kenakin [3] described ligand-mediated responses in a given tissue in terms of four parameters: (1) receptor density; (2) the efficiency of the transductional process; (3) the equilibrium dissociation constant of the RL complex; and (4) the intrinsic efficacy of the ligand at the receptor. *In vivo*, receptor occupancy for exogenous ligands is primarily dependent on pharmacokinetic parameters, whereas that for native endogenous ligands is most probably under intrinsic homeostatic controls including rates of production, release, metabolism and reuptake of ligands.

3.2. Rate Theory

Based on the experimental data showing the persistence of antagonist-mediated responses and agonist "fade" where maximal responses occur transiently to be followed by lesser responses of longer duration and agonist-mediated blockade of agonist effects, Paton [40] modified the concept of occupancy to include a chemically based rate term. According to rate theory, it was not only the number of receptors occupied by a ligand that determined the tissue response but also the *rate* of RL formation. The resultant effect, E, was considered equal to a proportionality factor, ϕ, that included an efficacy component and the velocity of the RL interaction, V. Thus,

$$E = \phi V \mathrm{eq} \tag{4}$$

The rate of RL formation, like that of neurotransmitter release, was measured in quantal terms with discrete "all or none" changes in receptor-mediated events. Pharmacokinetic considerations can also play a major role determining the rate of RL formation with the primary factor delineating occupancy and rate theory being the dissociation rate constant. Thus, if this factor was large, the ligand was an agonist; if the factor was small, reducing the quantal response to receptor occupancy, the ligand functioned as an antagonist. The kinetic aspects of rate theory did not, however, take into account the efficacy of transductional coupling and the potential for amplification after the initial binding event leading to its description as "a provocative conceptualization... with limited applicability" [41] a term that could be applied to many of the newer concepts of receptor theory.

3.3. The Ternary Complex Model

The effect of guanine nucleotides on agonist binding to GPCRs led to the development of the three component, or ternary complex model (TCM; see Refs [42,43]), that has as its premise that an equilibrium exists between the ligand-bound receptor and free G-protein, and the complex of receptor, ligand, and G-Protein.

$$\mathrm{R} + \mathrm{L} \rightleftarrows \mathrm{RL} + \mathrm{G} \rightleftarrows \mathrm{RLG} \tag{5}$$

In its simplest form, the TCM can account for the effect of guanine nucleotides on ligand binding, and in particular explains the finding that only agonist binding is altered by nucleotide concentrations. However, with an increased understanding of the complexity of receptor interactions, the simple TCM was extended to allosteric interactions between orthosteric and allosteric sites present on a single protein monomer and to other two-state interactions involving sites on adjacent proteins. Christopoulos [44] proposed that allosteric interactions were reciprocal such that the effects of a ligand A on the binding properties of ligand B also imply an effect on the binding of ligand B on the properties of ligand A. Similarly, because GPCRs alter the conformation of G-proteins to elicit transductional events and an alteration in cell function, changes in G-protein conformation and interactions alter receptor function and this may be reflected in desensitization.

4. CONSTITUTIVE RECEPTOR ACTIVITY

A basic principle of receptor theory is that when a receptor is activated by a ligand, the effect produced by the ligand is proportional to the concentration of the ligand, for example, it follows the LMA. It is now apparent that receptors can spontaneously form active complexes as a result of interactions with other proteins. This is especially true when receptor cDNA is *overexpressed* in engineered cell systems such that the relative abundance of a receptor is in excess of that normally occurring in the native state or associates with proteins that reflect the host cell milieu in which the receptor is expressed, rather than an intrinsic property of the receptor in its natural environment. Thus reports of receptor X unexpectedly interacting with system Y in an engineered system are often more artifactual than real and represents a major confound in the characterization of ligand efficacy [45]. A spontaneous interaction between receptor and effector can occur more frequently in a system where the proteins are in excess and where factors normally present that control such interactions are absent. This is shown graphically in Fig. 4, where constitutive activity is represented in the range of 0–50 and where the theoretical effects of inverse agonists, full and partial, are shown. A quiescent system that is more reflective of classical receptor theory shows a full agonist, partial agonist, and what is now defined as a neutral antagonist. Constitutive receptor activation has been described in terms of *protein ensemble* theory [46] and in terms of *allosteric transition* [44], where changes in receptor conformation may occur through random thermal events [47,48]. The TCM concept has been extended [49] to allow for the spontaneous activation of nonligand-bound receptor. The extended TCM includes the addition of an active (R_a) and inactive (R_i) receptor state, both of which can bind ligand. A ligand exhibiting higher affinity for R_i will tend to drive the equilibrium toward the inactive state and

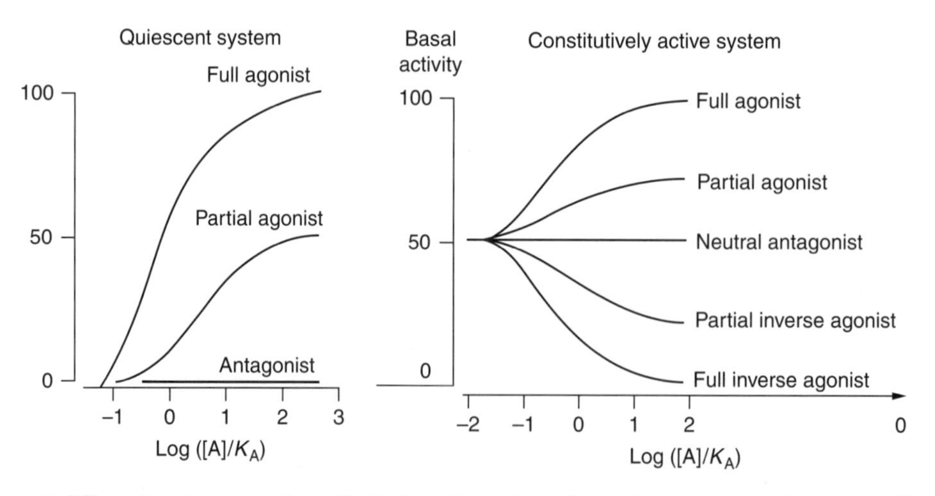

Figure 4. Efficacy in quiescent and constitutively active systems. In a quiescent system, three types of ligand can be defined, full agonist, partial agonist, and antagonist, depending on the response elicited. In a constitutively active system [45], an antagonist from a quiescent system is defined as a neutral antagonist with ligands that inhibit the activity of the constitutively active system and are defined as full and partial inverse agonists depending on the degree of inhibition.

thus decrease constitutive activity. Such ligands are termed inverse agonists and may be partial or full (Fig. 4).

5. RECEPTOR COMPLEXES AND ALLOSTERIC MODULATORS

Classical receptor theory assumed that affinity and efficacy were independent parameters [3] suggesting that no consistent relationship existed between the affinity of a ligand and its ability to elicit a full response. A ligand with relatively low affinity ($<10^{-6}$ M) could still be a full agonist when it interacted with the receptor a sufficient concentration. With the discovery of constitutive receptor activity, it appears that this lack of a consistent relationship is more reflective of an inability to measure receptor-mediated activity than a potency disconnect [46]. Thus, all ligands may have efficacy if tested in an appropriate system. The relationship between the receptor and the ligand in the classical lock and key model with the RL interaction resulting in no change in the receptor conformation essentially described a static situation. However, Hill's studies on the binding of oxygen by hemoglobin in 1909 [50] had demonstrated that identical binding sites on protein oligomers could influence one another such that the binding of the first ligand (in the case of hemoglobin in oxygen) facilitated the binding of a second, identical ligand, and so on, for sequentially bound ligands such that the saturation curve describing the interaction of the ligand with its recognition site was steeper than that which would be predicted from classical Michaelis–Menten kinetics. The process of one ligand, homologous or heterologous, interacting with the binding of another is thought to occur by a cooperative, conformational change in the binding protein for the second ligand from a site adjacent to the ligand recognition site. Koshland's induced fit model [34] extended these concepts, leading to the development of two models of cooperativity or allosterism: the *sequential or induced fit* model described by Koshland et al. [51], and the *concerted model* of Monod, Wyman, and Changeux [36], the key elements of which are outlined in Table 2.

Both models assume the existence of oligomeric protein units existing in two states that are in equilibrium with one another in the absence of ligand. Ligand binding induces a conformational change in the protein(s), moving the equilibrium of the two states to favor that with the higher affinity for the ligand. This in turn alters the kinetic and functional properties of the oligomeric complex. This model has been further refined in terms of *ligand-stabilizing conformational ensembles* [47].

The site on a receptor that defines its pharmacology and membership of a particular

Table 2. Allosteric Receptor Models

Monod, Wyman, and Changeux Concerted Model [20]
Receptor complex is a multicomponent oligomer comprised of a finite number of identical binding sites.
Subunits are symmetrically arranged each having a single ligand binding site.
Receptor complex exists in two conformational states, one of which has a preference for ligand binding.
Conformational transition state involves a simulataneous shift in the state of all subunits.
No hybrid states exist implying cooperativity.
Koshland Nemethy Filmer Sequential Model [23]
Receptor complex is a multicomponent oligomer with symmetrically arranged protomers each with a single ligand binding site.
Protomers exist in two conformational states with transition induced by ligand binding.
Receptor symmetry is lost on ligand binding.
Hybrid states of the receptor complex can be stabilized by protomers.
Stabilization is equivalent to negative cooperativity.

receptor superfamily, for example, 5-HT, nicotine, and so on, is termed the *orthosteric site*. Ligands that bind to this receptor are assumed to have a spatial overlap for the binding site such that their binding is mutually exclusive (unless an antagonist covalently binds to the orthosteric site). In contrast, the *allosteric site* (of which there may be more than one associated with a single orthosteric site and which can affect that site) is distinct from the latter in that ligands that bind to the allosteric site(s) can produce effects on ligand binding and efficacy to the orthosteric site through an indirect, conformational modulation of this site that probably involves alterations in either the association or the dissociation rates of the orthosteric ligand. The effect of an allosteric ligand on the affinity of its orthosteric ligand can be explained by the allosteric TCM model [52] that incorporates a cooperativity factor (α) that acts as a multiplier to modify the dissociation constant of the ligand at the orthosteric site. The factor, α can take on a range of positive values that determine how affinity is altered. For α values less than 1, the allosteric ligand produces a decrease in affinity for the orthosteric ligand and is thus termed an *allosteric inhibitor* or negative *allosteric modulator* (*NAM*). When α is greater than 1, the allosteric ligand produces an increase in affinity for the orthosteric ligand, and thus is termed an *allosteric potentiator or positive allosteric modulator* (PAM). If $\alpha = 1$, then there is no effect of the allosteric ligand on the affinity for the orthosteric ligand and the compound is termed a *neutral allosteric ligand*. Some forms of neutral allosteric ligand, for example, those for mGluR5 [53] neither activated nor inactivate the GPCR in the presence or absence of orthosteric agonist but block the activity of both PAMs and NAMs by occupying the allosteric site—a phenomenon known as *neutral cooperativity* or pharmacological silence.

The effect of allosteric ligands is not limited to alterations in affinity for the orthosteric ligand, since they can also produce changes in the intrinsic efficacy of the receptor–orthosteric ligand complex. An allosteric two state model has also been proposed that incorporates an additional cooperativity factor governing the transition of the receptor between active and inactive states in the presence of an allosteric ligand [54]. In this model, an allosteric modulator can alter the probability of the receptor signaling in the presence or absence of orthosteric ligand that explains allosteric ligands that exhibit intrinsic efficacy. In accord with the allosteric two-state model, a full range of positive and negative allosteric modulators as well as allosteric agonists and inverse agonists have been identified [55].

While much of the early work on allosterism derived from studies on enzymes and ligand-gated ion channels (LGICs), it is now clear that GPCRs that were once considered as monomeric proteins with only an orthosteric site also have allosteric sites and can form oligomeric complexes [15,56]. The concept of allosterism becomes more complex when considering multiple ligand sites on a receptor that have different pharmacological profiles, for example, ligand recognition sites are totally heterogenous.

The identification of allosteric ligands that can have either positive or negative effects on the function of the orthosteric site occurred in a largely serendipitous manner. The first drug identified as an allosteric modulator was the BZ, diazepam (Fig. 3), which has anxiolytic, hypnotic, and muscle relaxant activities and produces its effects by facilitating the actions of the $GABA_A$ receptor [57]. Unlike directly acting $GABA_A$ receptor agonists, for example, THIP (Fig. 3; see Ref. [58]), diazepam has a relatively modest side-effect profile.

Allosteric modulators have three putative advantages over drugs acting via orthosteric sites in that their effects are

(i) saturable such that there is a ceiling effect to their activity that can result in a good margin of safety for human use;
(ii) selective, partly as their binding sites are distinct from the orthosteric site and partly because their effects depend on the degree of cooperativity between the allosteric and the orthosteric sites.
(iii) frequently "use dependent." Thus, the actions of an allosteric modulator occur only when the endogenous orthosteric ligand is present. In the

absence of the latter, an allosteric modulator is theoretically quiescent and may thus represent an ideal prophylactic treatment for disease states associated with sporadic or chronotropic occurrence.

The "cys-loop" family of LGICs [59] that includes the $GABA_A$, glycine, $5\text{-}HT_3$, and nicotinic cholinergic receptors are the best characterized of the allosterically regulated receptors. Allsosteric GPCRs include adenosine (P1), α- and β-adrenergic, dopamine, cannabinoid, chemokine, $GABA_B$, endothelin, metabotropic glutamate, opioid, glucagon, CGRP, CRF1, neurokinin-1, P2Y, and muscarinic cholinergic receptor families, and members of the 5-HT superfamily [15]. The cone snail conotoxin, r-TIA is an allosteric modulator of the α_{1B} adrenoceptor [60]. Changes in GPCR function resulting from alterations in the ionic milieu also reflect the potential for allosteric modulation of receptor function.

Allosteric modulators of protein kinase activity distinct from ATP-competitive site inhibitors represent a new approach to drug discovery in oncology [61]. NCEs that bind to allosteric sites to not only potentiate target effects but also act as agonists in the absence of an orthosteric ligand are termed ago-allosteric modulators [62]. Two GPCR allosteric modulators have been approved, cinacalcet (Fig. 3) a first-in-class PAM calcimimetic acting at the calcium-sensing receptor and the CCR5 NAM, maraviroc (Fig. 3) that blocks HIV entry into T cells.

While it was thought that a receptor-mediated response was a predictable, linear process that involved ligand-induced activation of a protein monomer and its signal transduction pathway independently, or with very little influence, from other membrane proteins, it has become increasingly evident that receptors have the potential to physically interact both with one another and with other membrane proteins [63]. Numerous examples exist of receptor coexpression and interactions (dimerization), for example, $GABA_BR1$ and $GABA_BR2$ [64], DA with somatostatin [65] and adenosine [66], and $GABA_A$ receptors and opioid receptors with α_2-adrenoceptors [67]. These interactions are often necessary to obtain functional cell surface receptors and allow interactions with one protein partner to modulate the function of the entire signaling complex. Effects on protein subunits of receptors can be mediated through both *orthosteric* and *allosteric* sites, for example, the benzodiazepine site in the $GABA_A$ receptor [57], the glycine receptor on the NMDA receptor complex [68]. Integration of the functional effects of signal transduction pathways also affects receptor function in a cell-specific manner (receptor crosstalk). There also exist several discrete classes of receptor-associated proteins including receptor–activity-modifying proteins (RAMPs; see Ref. [69]) and trafficking chaperones [70], which number in the many hundreds and play key roles in cellular events including receptor trafficking from the endoplasmic reticulum to the cell surface and modulation of cell surface responses.

Examples of the complexity of receptor signaling at the postsynaptic level include the NMDA receptor, where proteomic analysis demonstrated more than 70 proteins other than the receptor potentially involved in the function of the receptor complex [71], and the ATP-sensitive $P2X_7$ LGIC receptor, where a signaling complex comprise some 11 proteins including laminin β-3, integrin 2,β-actin, supervillin, MAGuK, heat shock proteins, phosphatidylinositol 4-kinase and the receptor protein tyrosine phosphatase (RPTP) was identified [72]. RPTP appears to modulate $P2X_7$ receptor function via modulation of its phosphorylation state.

6. EFFICACY CONSIDERATIONS

Historical receptor theory describes a ligand efficacy continuum, with full agonism at one end and full antagonism at the other. Between the two ends of this continuum lie partial agonists that, as already noted, imply that ligands can also be partial antagonists. Antagonism *per se* implied that a ligand could bind to a receptor without producing any effect and limiting access to the native agonist to block receptor activation. Such compounds are now termed *neutral antagonists* (Fig. 4).

With the ability to measure constitutive receptor activity, some compounds, for example, the β_2-adrenoceptor antagonist, ICI 118551 (Fig. 3), inhibited constitutive activity, thus functioning as an *inverse agonist* or *negative antagonist* [73], a concept that emerged from work on the $GABA_A$/BZ receptor complex [74]. As ligands can have differing efficacies depending on the system in which they are evaluated (see Refs [45,75]; see below) making it essential that new compounds characterized in engineered cell or isolated systems be evaluated in native systems before being advanced for evaluation in more complex and sophisticated tissue/animal model systems.

The actions of a competitive antagonism can be surmounted by the addition of increasing concentrations of the agonist ligand, resulting in a functional dose–response curve that undergoes a rightward shift with approximately the same shape and maximal effect (Fig. 1). Noncompetitive or uncompetitive antagonists interact at sites distinct from the agonist recognition site and can modulate agonist binding either by proximal interactions with this site from a site adjacent to the recognition site or by allosteric modulation. The effects of noncompetitive antagonists are usually not reversible by the addition of excess agonist. This type of antagonism, whether competitive or noncompetitive, occurring at a distinct molecular target is known as *pharmacological antagonism* and involves the interactions between ligands and the receptor site (Fig. 5). In contrast, *functional antagonism* refers to a situation in which an antagonist that does not interact with a given receptor can still block the actions of an agonist of that receptor and is typically measured in intact tissue preparations or whole animal models.

In the cartoon outlined in Fig. 5, neurotransmitter A released from neuron A interacts with A-type receptors on neuron B. Antagonist α can block the effects of A on cell B by interacting with A receptors. Antagonist α is thus a pharmacological antagonist of A receptors. In the second example in Fig. 5, neuron A releases neurotransmitter A, which interacts with A-type receptors located on neuron X. In turn, neuron X releases neurotransmitter X that interacts with X-type receptors on neuron Y. Antagonist β is a competitive antagonist that interacts with X receptors to block the effect of neurotransmitter X, and in doing so, *indirectly* blocks the actions of neurotransmitter A. Antagonist β is thus a *pharmacological antagonist* of receptors for the neurotransmitter X, but a *functional antagonist* for neurotransmitter A.

In interpreting functional data in complex systems, it is always important to consider the possibility that a ligand has affinity for more than one receptor. For this reason, in advancing new ligands from *in vitro* evaluation to more complex tissue systems or animal models, it is extremely helpful to conduct an *in vitro* ligand binding profile, for example, the binding affinity of a compound at a battery of 100–150 receptors and enzymes (e.g., a Cerep or Millipore profile), to be better prepared to understand any new findings as a compound advances toward the clinic. For instance, when a ligand for a new receptor is advanced to animal models and is unexpectedly found to elicit changes in blood pressure, it is extremely helpful to know whether, in addition to its defined activity at the target

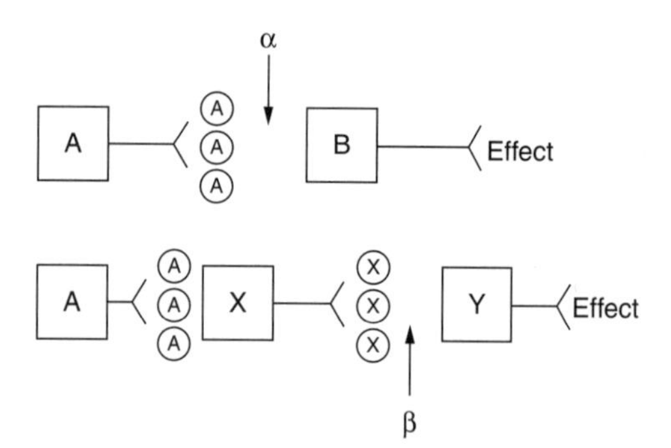

Figure 5. Pharmacological versus functional antagonism. In the top panel, neurotransmitter A is released from neuron A, directly interacting with neuron B to produce a functional response. Antagonist α blocks the effects of A, a direct pharmacological antagonism of the effects of A. In the bottom panel, neurotransmitter A is released from neuron A, directly interacting with neuron X, which in turn releases neurotransmitter X, which acts on cell Y to produce a functional response. Antagonist β blocks the effects of X on cell Y, but in the absence of other data on the actions of antagonist β, seems to block the functional effects of A because of the circuitry involved. Antagonist β thus acts as a functional antagonist.

receptor, it has some other properties that might explain the blood pressure effects. This avoids the possibly incorrect assumption that some previously unknown mechanism related to activation of the new receptor target has cardiovascular-related liabilities.

It is also important to note that potent target interactions observed in *in vitro* binding assays do not always show any functional consequence either in tissues or in *in vivo* models. This does not reduce the need to be aware of the potential for such actions in advancing a compound but also, these should not preclude advancing a compound due an overly conservative interpretation of such data especially when it is derived from systems using recombinant, rather than native receptors. The major exception to this are hERG interactions [76,77], the presence of which represents a major hurdle to compound advancement and where only human data can be considered useful in addressing false positives.

An additional concept is that of *functional selectivity* (also termed ligand-directed trafficking or biased agonism) that has been proposed as an explanation for the finding that a number of ligands for a variety of GPCRs exhibit diverse signaling properties while acting at a single receptor [78]. This suggests that the basic concept of intrinsic efficacy is, in some circumstances, an oversimplification and in fact orthosteric ligand–receptor complexes may exhibit different intrinsic efficacy for different signaling pathways, conceptually allowing for potentially novel orthosteric ligand effects. For example, a compound could theoretically act as an agonist via interactions with one signaling pathway and a neutral antagonist at another. This is a new and emerging concept in GPCR-based drug discovery that has the potential to improve precision in the targeting of NCEs.

7. RECEPTOR DYNAMICS

Receptors, present in their active form at the cell surface are not static entities. Receptors turnover as a normal consequence of cell growth, with half-lives that vary between hours and days. Ligand binding, alterations in gene function and tissue dysfunction due to disease, trauma or aging, can alter the number, half-life, and responsiveness of receptors, channels, and transporters. These processes can be modulated by the availability of the endogenous ligand. For example, in Parkinson's disease, the presynaptic nerve cells in the *substantia nigra* that normally produce DA die as the result of an as yet unknown disease etiology. This defect results in a decrease in endogenous DA levels and a consequent hypersensitivity of postsynaptic responses as cellular homeostatic events attempt to compensate for a lack of endogenous ligand, for example, modulation of DAT expression [79].

Ligand binding can also result in receptor internalization through phosphorylation-dependent events often initiated as a result of the ligand binding process, exposing serine and threonine residues in the receptor protein. Histamine H_2 antagonists acting as inverse agonists in constitutively active systems can upregulate their cognate receptors, potentially increasing cell sensitivity [80]. It is thus reasonable to assume that the target cells for endogenous effector agents, neurotransmitters, neuromodulators, and neurohormones all operate under tonic control, one example being variation with the circadian rhythm of the organism. In contrast, the effects of exogenously administered ligands, for example, drugs and NCEs, are rarely under normal homeostatic control, and as a result, their effects frequently become blunted on repeated administration or when they are administered in controlled release forms. In extreme instances, agonists that rapidly desensitize a receptor, for example, ATP at the $P2X_3$ receptor, nicotine at the $\alpha 4\beta 2$ neuronal nicotinic receptor, may appear as antagonists because their net effect is to attenuate normal receptor responses. It should not be a surprise, therefore, that the majority of effective therapeutic agents are antagonists of receptor function [33].

8. RECEPTOR RESIDENCE TIME

A basic premise of receptor theory has been that a ligand needs to be bound to its target to produce a functional effect as evidenced in Ehrlich's aphorism "*corpora non agunt nisi fixate* ("a substance will not work unless it is bound" [16]). However, as already mentioned, there are numerous exceptions to this general view where the functional effect of a ligand persists long after ligand dissociation, for example, in the absence of efficacious plasma levels. This can be due to their initiation of prolonged signaling actions that involve multiple intracellular pathways, gene activation or changes involving posttranslational modulation/modification of gene products. As a result, ligands with virtually identical *in vitro* affinity, selectivity and functional efficacy often differ in their *in vivo* actions in ways that are not readily understood.

The affinity of a ligand for its target is typically assessed *in vitro* using a binding assay or a biochemical readout under what are termed *closed system conditions* where the target is exposed to what is considered an invariant concentration of ligand throughout the assay period and an IC_{50} or K_d value is derived for the ligand/target/receptor binary complex [16]. These conditions rarely mimic the temporal and concentration (dose) relationships occurring *in vivo* that represents an open system. In the latter, association, tissue distribution, and other aspects of ADME—including metabolism—can markedly alter the relationship between the ligand and its receptor. Copeland et al., [16,17] have provided convincing evidence that the affinity of a ligand for its target is not the critical factor for determining efficacy but rather the *residence time* of the ligand at the target that has been further defined as the *binary complex residence time* [16]. The longer the residence time, the more prolonged the effect of the ligand (or drug). *In vitro*, the residence time is a function of the final equilibrium dissociation constant while in the open system situation *in vivo*, the concentration of ligand to which the target is exposed varies with the time after administration, tissue distribution (including plasma protein binding and MDR activity) and the metabolism of the ligand, making the effect of the ligand dependent on the association (k_{on}) and dissociation (k_{off}) rate constants. The off-rate can be translated into the half-life for dissociation that is a direct measure of residence time. Residence time can also affect target selectivity.

Medicinal chemistry-driven SAR activities that are focused on IC_{50} or K_d determinations can lead to optimization of the dissociative half-life but this can lead to confounds when variations in k_{on} and k_{off} result in the same K_d value even though k_{off} varies markedly. Differences in residence may explain differences in compound efficacy. For instance, in the antipsychotic field, the two classical, first-generation antipsychotic agents, haloperidol and chlorpromazine have D2 receptor residence times of approximately 1 h [17]. In contrast, the second-generation "atypical" antipsychotics, clozapine, and quietapine have residence times of less than 1 min. Similarly, examination of the IC_{50} values of the HIV protease inhibitor, saquinavir to inhibit viral replication *in vitro* at wild type and mutant forms of the protease correlated the SAR of the K_i value with differences in k_{off} [81].

9. RECEPTOR NOMENCLATURE

The availability of genome sequences from humans and nonhuman organisms has facilitated the identification of large receptor superfamilies encompassing both plasma membrane receptors (e.g., GPCRs, receptor tyrosine kinases, LGICs, integrins, toll-like receptors) and intracellular nuclear receptors (Table 1). The entire repertoire of proteins dedicated to ligand recognition has been described as the "receptorome" and encompasses approximately 5% of the human genome [82,83].

Prior to the publication of the human genome map [20], the techniques of molecular biology had already resulted in an explosion in the number of putative receptor families and subtypes within families, as well as classes of GPCRs termed *orphan receptors*. These were structurally related members of receptor classes for which the endogenous ligand and associated function was unknown.

9.1. Nomenclature Databases

There are several public domain databases available that contain information on specific families of receptors, for example, GPCRs, nuclear receptors, tyrosine kinase receptors, LGICs, and so on, and represent useful data-mining tools. These databases represent computational tools by which potential functions and molecular interactions can be derived to provide a systematic source for receptor/target nomenclature. New receptors have been (and continue to be) identified in different laboratories often simultaneously, leading to different names and considerable confusion in the literature. The International Union of Pharmacology (IUPHAR) has undertaken the development of a systematic nomenclature system, based in part on naming families of receptors for their cognate endogenous ligands and structural and genetic information. The IUPHAR database (IUPHAR-DB) has integrated peer-reviewed pharmacological, chemical, genetic, functional, and anatomical information on the 354 nonsensory GPCRs, 71 LGICs and 141 voltage-gated ion channels (VGICs on VSICs) encoded in the human, rat, and mouse genomes [84]. Other public domain databases include the GPCR database (GPRDB; see Ref. [85]), the GPCR-ligand database (GLIDA) the latter of which is primarily focused on the correlation of information between GPCR motifs, GPCRs, and their ligands [86]. Receptor tyrosine kinases (RTKs) are included in the RTKdb database [87] with the Kinomer database systematically classifying eukaryotic protein kinases [88]. The nuclear receptor database, NUREBASE contains 523 nuclear receptor protein entries from 15 metazoan species grouped into 30 nested "families" [89]. *PharmGKB: the Pharmacogenetics Knowledge Base* [90,91] links genomic, molecular, and cellular and clinical phenotype information to understand how genome variations lead to drug-related differences in clinical responses.

In addition to articles published on a regular basis in *Pharmacological Reviews*, receptor nomenclature guidelines are also found in *British Journal Pharmacology Guide to Receptors and Channels* [92] and *The Sigma-RBI Handbook of Receptor Classification and Signal Transduction* [93]. While many receptors and enzymes have been named on the basis of their ligand family, function, or homology with known drug targets, newer acronyms and names for targets have evolved from genetic studies in Drosophila where acronyms such as HERG (human ether a-go-go gene) and SOS (son of sevenless) have arisen from observed behavioral phenotypes in Drosophila associated with the function of these genes, other acronyms such as RING ("really interesting new gene"; see Ref. [94]) have tended to trivialize the nomenclature process.

10. RECEPTOR CLASSES

Receptors *per se* can be divided into four major classes (Table 1): heptahelical, 7-transmembrane (7TM) G-protein-coupled receptors (GPCRs), ion channels, transcription factor receptors and enzyme-associated receptors. Of these, the 7-TM GPCRs have historically represented the most fertile class for drug discovery, as these have been the most studied [95]. More than 50% of all prescription drugs on the market today, including 10 of the top 50 drugs, directly or indirectly target GPCRs. This receptor superfamily includes receptors for the bioamine neurotransmitters (5-HT, DA, etc.), peptide hormones and lipid signaling molecules. Ion channels may be further subdivided into ligand-gated (LGIC; see Ref. [96]), voltage-sensitive calcium and potassium (VSCC, Kir; see Refs [97–100]), and ion/pH-modulated (ASIC; acid-sensing ion channels; see Ref. [101]) subtypes, all of which have similar but distinct, multimeric structural motifs. TRPV (transient receptor potential) ion channels can be modulated by temperature, for example, vanilloid receptors. The largest group within the transcription factor receptor superfamily is the nuclear hormone receptors [102,103] that number 48 and are divided into 7 subfamilies. The enzyme-associated receptor superfamily is designated as a family of single or multi- subunit proteins that contain a subunit with a single transmembrane domain. The largest groups within this superfamily are the single subunit- receptor tyrosine kinases comprised of approximately 20 members including VEGF, PDGF, insulin and EGF receptors [104], and multimeric complexes that utilize kinases such as the nonreceptor tyrosine kinases that number approximately 58 [105] and include JAK-type kinases [106]. Also included are receptors with associated serine/threonine kinase or guanylyl cyclase activity.

Other receptor classes include the Cytochrome P450 (CYP) family comprised of approximately 17 members [107], the SMAD family of tumor suppressors [108,109], the seven members of the STAT (signal transducers and activators of transcription) receptor family [110] and the various enzymes involved in the extrinsic, intrinsic, and perforin/granzyme pathways in the apototic cascade [111]. In addition to this multitude of receptor classes, further complexity in conceptualizing receptors as distinct, classifiable entities is exemplified by recent findings related to GPCRs. By implicit definition, these receptors produce their physiological effects by coupling through the G-protein family. However, there are instances where ion channels produce their effects by coupling through G-proteins [112]. Some GPCR agonists can activate growth factor RTKs in the absence of growth factor [113]. This novel transactivation process, which places GPCR signaling downstream of RTKs, either requires the production of a ligand of the transactivated GPCR or occurs in a ligand- independent manner within an integrated signaling network. Receptor classification thus has few absolutes.

10.1. G-Protein-Coupled Receptors

The crystal structure of bovine rhodopsin [114] has driven modeling in the GPCR area although this is not a ligand-activated GPCR [115], human β_2- and β_1-adrenoceptors [116,117] and the human A_{2A} adenosine receptor [118] together with the nucleotide and amino acid sequence of other GPCRs has led to the inference of a general topological structure for GPCRs. A key feature is the conserved seven-helix transmembrane (7TM) spanning core. Based on the X-ray structure of rhodopsin, the 7TM motif is a single protein comprised of 300–500 amino acids with discrete amino acid motifs in the transmembrane regions and on the C-terminal extracellular loop that determine the ligand specificity of the receptor. Those on the amino terminal and intracellular loops designate G-protein interactions. The TM regions are linked via three extracellular and three cytoplasmatic loops. GPCRs contain extracellular N-terminus (7–600 residues), generally with N-glycosylation sites and cytoplasmatic C-termini (12–359 residues) that along with the intracellular loops serve as the principal sites for phosphorylation by second messenger kinases and GPCR kinases (GRKs). Anchoring of a palmitoylated cysteine residue to the membrane forms a fourth intracellular cytoplasmatic loop (Fig. 6). These distinct structural

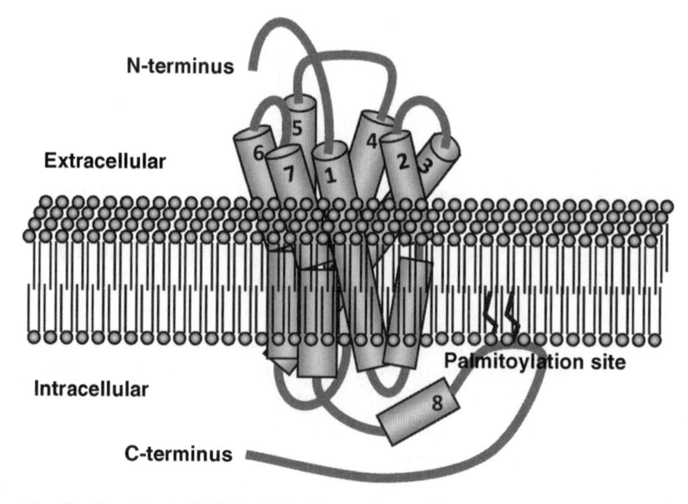

Figure 6. Schematic of a Family A GPCR. (This figure is available in full color at http://mrw.interscience.wiley.com/emrw/9780471266945/home.)

properties contribute to the specificity of each GPCR with respect to ligand binding, G-protein coupling and homologous and heterologous regulation. GPCRs interact with ligands that bind to the N-terminus and extracellular loops and/or pocket formed by the folding together of the 7TM segments. For example, RL interactions for many of the bioamine neurotransmitter receptors are thought to occur within a pocket in the 7TM motif that is generically designated to lie between TMs III, IV, and VI. Posttranscriptional alternative splicing can alter the GPCR to create isoforms that may be species, tissue, and disease-state dependent. The number of GPCRs present in the human genome, including orphan receptors, has been estimated to be in excess of 3000 with a 1000 of these coding for odorant and pheromone receptors. On the basis of sequence homology, functional domains and ligand binding mode, the GPCR superfamily has been divided into six families (A, B, C, D, E, and F). The three major families (A, B, and C) include mammalian GPCRs while Families D and E include fungal pheromone receptors and cAMP receptors and Family F consists of archaebacterial opsin [119]. More recently, mammalian GPCRs have been organized based on phylogenetic analyses into five main families under the GRAFS system (glutamate, rhodopsin, adhesion, frizzled/taste2, and secretin; see Ref. [120]). With more than 800 members, GPCRs represent by far the largest family of cell-surface molecules involved in signal transduction, accounting for approximately 2% of the human genome. These include 367 unique functional nonolfactory/nonsensory GPCR sequences [84], for which endogenous ligands are expected and which are therefore referred as liganded GPCRs [121]. From these 224 GPCRs have known ligands with the remaining 143 having no known ligand and are therefore classified as orphan receptors. Olfactory receptor genes represent the largest mammalian GPCR subgroup. These are Family A receptors encoded by single exons and are transcribed in the olfactory epithelium where they interact specifically with the G-protein, G_{olff} to transduce odorant signals. Of the liganded GPCRs, 284 are rhodopsin-like Family A receptors and include GPCRs for peptide hormones, neurotransmitters, odorants, and a large group of orphans. These are further divided on their structural similarity into subfamilies:

- 1a including rhodopsin, adrenoceptors, thrombin, and the adenosine A_{2A} receptor, with a binding site localized within the 7TM motif;
- 1b including receptors for peptides with the ligand binding site in the extracellular loops, the *N*-terminal, and the superior regions of the TM motifs;
- 1c including receptors for glycoproteins. Ligand binding to this receptor class is mostly extracellular.

Family B (secretin-like receptors) is morphologically, but not sequence, related to the Class A1c family and consists of 50 GPCRs included those activated by the hormones calcitonin, glucagon, secretin, VIP-PACAP, and MTH. Family C (metabotropic/glutamate receptors) has 17 members and includes the metabotropic glutamate receptors and three $GABA_B$ receptors. The 11 Frizzled/smoothened receptors control cell development and proliferation mediated by the secreted glycoproteins, *Wnt* and Hedgehog. The function of the *Adhesion* GPCR family [122] is unclear beyond potential roles in the CNS and immune systems. The 7TM region of this family differs from that of other GPCRs [123] with the long N-termini having multiple functional domains that are also found in TrKs, integrins, and cadhedrins. This makes them likely to interact with other membrane-bound proteins, potentially facilitating cell-to-cell communications in the absence of soluble ligands.

GPCR signaling occurs when a bound ligand stimulates conformational changes in the receptor that allow it to couple to heterotrimeric G-proteins. G-proteins are made up of three subunits: the α subunit, with a GDP/GTP binding site and β and γ subunits that form a tightly bound dimer. In the inactive state, the G-protein complex remains in a trimeric form, with the α subunit bound to GDP. In the active state, the GDP is released fro the α subunit and replaced by GTP [124]. This is thought to lead to dissociation of the α

subunit from the $\beta\gamma$ dimer. The dissociated subunits interact with a variety of effector systems to modulate cellular pathways. The existence of multiple species of α-, β-, and γ-subunits increases GPCR signaling complexity and specificity. Approximately 27 different heterotrimeric G-proteins with 23 Gα, 5 Gβ and 12 Gγ subunits have been identified. There are four α-subunit families (1) G_s that activates adenylyl cyclase; (2) $G_{i/o}$ that inhibits adenylyl cyclase and can also regulate ion channels and activation of cGMP PDE; (3) $G_{q/11}$ that activates phospholipase C; and (4) $G_{12/13}$ that is less well understood. G_{12} appears to regulate signaling via the small GTPase, *RhoA*, and by activation of PLD, while G_{13} regulates *RhoA* and Na^+/H^+ exchange [125]. Dissociated $\beta\gamma$ subunits can also modulate signaling pathways, including stimulation or inhibition of adenylyl cyclase, activation of phospholipases and regulation of potassium and calcium channels.

10.1.1. Modulators of G Protein-Coupled Receptor Signaling The investigation of the mechanisms of agonist-induced receptor signaling, desensitization, internalization, trafficking, and recycling resulted in the discovery of accessory proteins that interact with GPCRs and/or heterotrimeric G-proteins and modulated their activity [126,127]. These include cyclic nucleotide phosphodiesterases PDE, GPCR interacting proteins (GIP), second messenger-dependent kinases, for example, protein kinase C (PKC) and protein kinase A (PKA), GRKs, RAMPs [69] β-arrestins, regulators of G-protein signaling (RGS), receptor-independent activators of G-protein signaling (AGS), and various cytoskeletal proteins (e.g., tubulin, caveolin, and actin). PDEs are responsible for the hydrolytic degradation of cAMP and cGMP and comprise more than 11 families with various isoforms and splice variants [128]. The PKCs [129] are, like other members of the kinome [130], responsible for protein phosphorylation, GRKs 1-6 [131,132] phosphorylate GPCRs and the protein phosphatases [133] mediate dephosphorylation. The latter number is in excess of 300, adding to the complexity of GPCR-associated signaling processes.

Other accessory proteins include

(a) GPCR interacting proteins that are implicated in GPCR targeting to specific cellular compartments, in their assembling into functional complexes called "receptosomes," in their trafficking to and from the plasma membrane and in the fine-tuning of their signaling properties [134];
(b) RGS proteins (25 members) that modulate G-proteins by downregulating heterotrimeric G-protein signaling [135] by acting as GTPase activators;
(c) activators of G-protein signaling (see Ref. [136]) that activate heterotrimeric G-proteins independently of receptor activation and comprise three groups based on their mechanism of action [137];
(d) calmodulins that mediate calcium modulation of receptor function [138] and
(e) arrestins [139] comprised of four members that are involved in inactivation of phosphorylated receptors.

While GPCRs are generally considered to be monomeric proteins, they can, as noted [15], also form homo- and heteromeric forms (e.g., $GABA_B$, adenosine A_1 and A_{2A}, angiotensin, bradykinin, chemokine, DA, metabotropic glutamate, muscarinic, opioid, serotonin, cannabinoid, and somatostatin,) adding additional complexity to ligand-driven GPCR signaling processes and offering potentially novel targets for medicinal chemistry. Opioid receptors illustrate the potential for dimers as novel drug targets. The κ-opioid receptor (KOR) can dimerize with δ-opioid receptor (DOR) [140] to form a dimer, the pharmacology of which differs from either of the monomers. The opioid agonist, 6′-guanidinonaltrindole (6′-GNTI; Fig. 7), thought to be a KOR selective agonist, was actually selective for the dimer. It has been suggested [141] that drugs targeting dimers may have fewer side effects. Ligands interacting with μ- and δ-opioid receptors, for

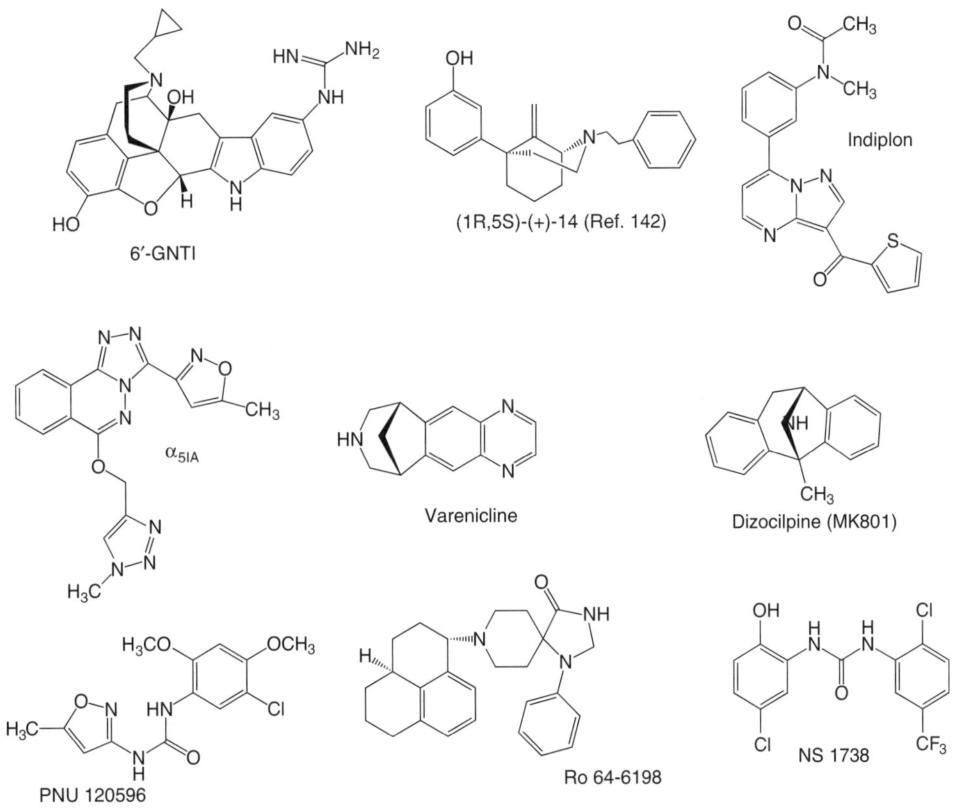

Figure 7. Novel ligands for opioid, benzodiazepine and nicotinic receptors.

example, the 5-(3-hydroxy)phenylmorphan, 1*R*,5*S*)-(+)- 14 (Fig. 7) have improved analgesic activity [142]

Applying an evolutionary trace method (ETM) to assess potential protein–protein interactions, functionally important residue clusters on transmembrane (TM) helices 5 and 6 in over 700 aligned GPCR sequences have been identified [56]. Similar clusters were found on TMs 2 and 3. TM 5 and 6 clusters were consistent with 5,6-contact and 5,6-domain swapped dimer formation. Additional application of ETM to 113 aligned G-protein sequences identified two functional sites: one associated with adenylyl cyclase, $\beta\gamma$, and RGS binding, and the other extending from the ras-like to helical domain that seems to be associated with GPCR dimer binding. Thus, GPCR dimerization and heterodimerization were concluded to be generic to all members of the GPCR superfamily and its subfamilies. From these findings, potential new approaches to ligand design include (1) antagonists that act by inhibiting dimer formation, for example, transmembrane peptide mimics; (2) bivalent compounds/binary conjugates; and (3) compounds targeting the GPCR–G-protein interface. The concept of dimerization has, however, been questioned in whether there is conclusive proof for GPCRs existing in their native membrane environments as homomers. heterodimers or oligomers [143].

10.2. Ligand-Gated Ion Channels

Ion channels consist of homo-or heteromeric complexes numbering between three (P2X) and eight (Kirs) subunits. Examples of these are the $GABA_A$/benzodiazepine and NMDA/ glycine receptor, neuronal nicotinic receptors (nAChR), and P2X receptors.

The *$GABA_A$/benzodiazepine(BZ) receptor* [57] is an LGIC that is the target site for the BZs a broad class of clinically effective anxiolytic, anticonvulsant, muscle relaxant, and hypnotic drugs that produce their therapeutic effects by enhancing the actions of the inhibitory neurotransmitter, GABA. The $GABA_A$ receptors are pentameric LGICs, the constituent subunits of which are formed from a family of six α, four β, one and two subunits, leading to the potential existence of several thousand different pentamers [144]. The functional receptor complex contains a $GABA_A$ receptor, a BZ recognition site, and by virtue of its pentameric structure, a central chloride channel. Allosteric recognition sites on this complex include those for ethanol, avermectin, barbiturates, picrotoxin, and neurosteroids such as allopregnanolone. The pharmacology and function of the allosteric sites depends on the subunit composition. The α_1 subunit, present in nearly 60% of $GABA_A$ receptors in mouse brain mediates the sedative, anticonvulsant, and amnestic effects of diazepam; the α_2, the antianxiety effects of diazepam; and the α_5, associative temporal memory. With this knowledge, rather than screening new ligands for the $GABA_A$ receptor in various animal models to derive a profile for a nonsedating anxiolytic, by understanding the structural determinants of BZ interactions with α_1 and α_2 subunits and designing compounds that preferentially interact with the latter, the process of designing novel ligands can be considerably enhanced. Unfortunately, compounds such as indiplon (Figure 7) that have α_1 subunit selectivity [145] and preferential hypnotic properties have not been approved for human use. α_{5IA} (Fig. 7) is an α_5 subunit inverse agonist with cognitive enhancing properties [146].

The *NMDA receptor* is a member of the glutamate receptor superfamily that mediates the effects of the major excitatory transmitter, glutamate and is composed of an NMDA receptor, a central ion channel that binds magnesium, the dissociative anesthetics, ketamine, and phencyclidine (PCP), the noncompetitive NMDA antagonist, dizocilpine (MK801), glycine, and polyamine binding sites, activation of which can marked alter NMDA receptor function, and some 70 other ancillary proteins, the physiological function of which remains to be determined [71]. The activation state of the receptor can define the effects of the allosteric modulators. Thus, some are termed "use dependent," reflecting modulatory actions only when the channel is opened by glutamate.

The *nicotinic cholinergic receptor, nAChR*, is another pentameric LGIC comprised of distinct functional regions and subunits [1,147]. The subunit composition of the receptor varies, imparting different functionality when the channel is activated by nicotine, or the endogenous ligand, acetylcholine ACh. Varenicline (see Ref. [148]; Fig. 7) is drug approved for smoking cessation that acts via $\alpha_2\beta_4$ nAChRs. Allosteric modulators of neuronal nAChRs include dizocilpine (MK801; Fig. 7), avermectin, galantamine, steroids, barbiturates, ancillary proteins and NCEs such as PNU-120596 (see Ref. [149]; Fig. 7) and NS-1738 (see Ref. [150]; Fig. 7).

The *P2X receptor* family is an LGIC responsive to ATP is thought to functions as a trimer that is formed from a family of seven subunits that can form both homo- and heteromers leading to the existence of at least seven distinct receptor subtypes [151,152].

Miscellaneous ion channels include the inward-rectifier potassium ion channels (Kirs or IRKs) of which 15 are known that are arranged into 6 families [98]; voltage-gated calcium (Cav; see Ref. [99]) that comprise three families Cav1-3, which have 4, 3, and 3 members, respectively. The calcium channel blockers, the dihydropyridines, for example, nifedipine (Fig. 8) interact with Cav channels [153]. There are 10 identified members of the sodium (Nav or SCN; see Ref. [100]) channel family, Nav 1.1–9 and SCN7.ASICs [101] are members of the degerin/epithelial Na^+ channel (DEG/ENaC) family that also includes BNC1 and DRASIC. The transient receptor potential (TRP) channel family with 28 members [154] are involved in temperature sensing and pain.

Far less is known regarding the structural elements involved in the ligand pharmacology

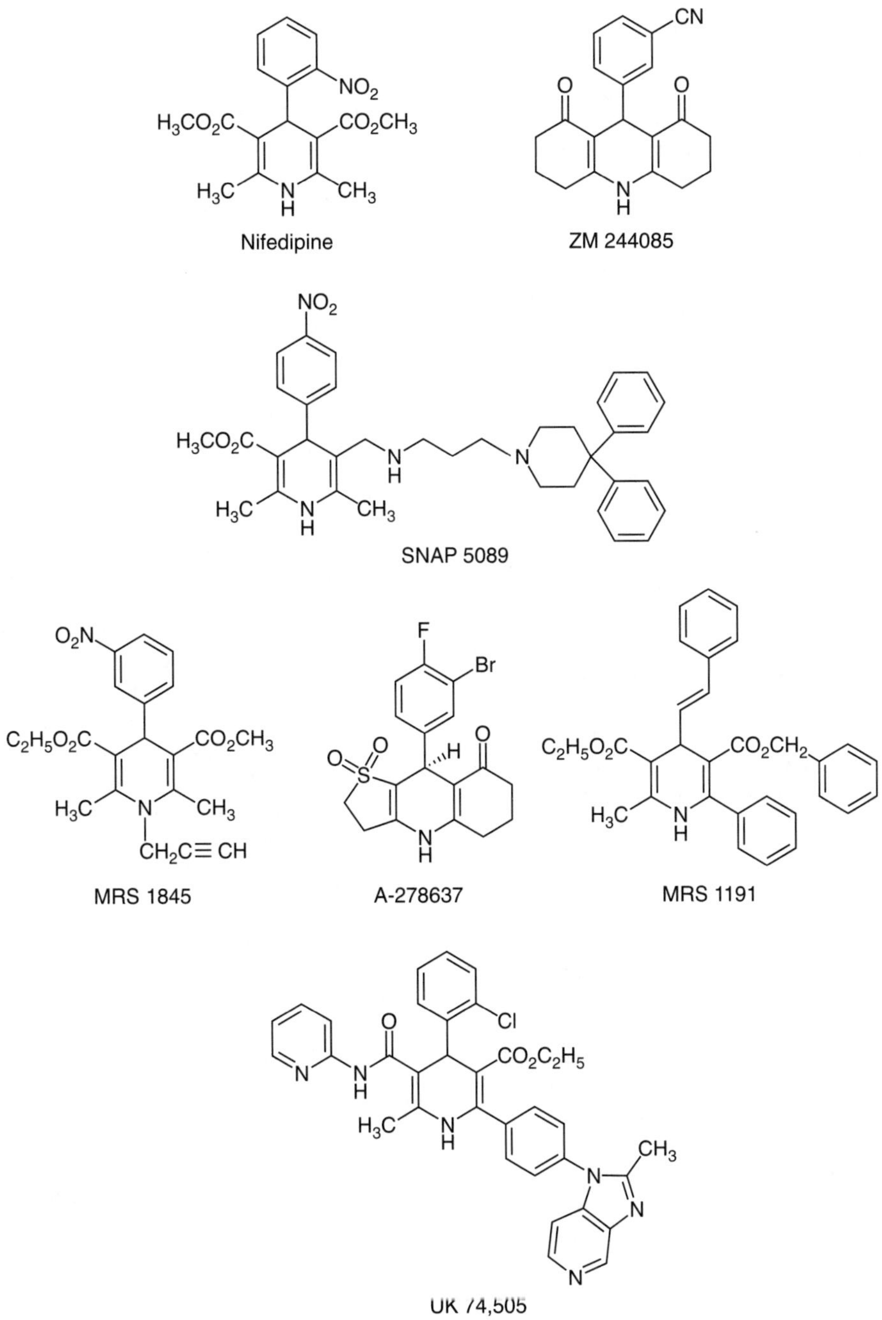

Nifedipine

ZM 244085

SNAP 5089

MRS 1845

A-278637

MRS 1191

UK 74,505

Figure 8. Dihydropyridines; calcium channel and other receptor antagonists.

and function of LGICs as compared to GPCRs, such that signaling transduction pathways are only now being extensively characterized. For example for nAChRs, the recognition site for ACh is formed between two different receptor subunits. Thus multiple orthosteric sites for ACh are possible, depending on the types of subunit forming the receptor.

10.3. Steroid Receptor Superfamily

The steroid receptor superfamily includes the glucocorticoid (GR), progesterone (PR), mineralocorticoid (MR), androgen (AR), thyroid hormone (TR), and vitamin D_3 (VDR) receptors. These receptors bind steroid hormones and are then translocated to the nucleus where they bind to hormone responsive elements (HREs) on DNA promotor regions to alter gene expression. While steroids are very effective anti-inflammatory agents, they have a multiplicity of serious side effects that limit their full use [155].

The antiestrogen, tamoxifen, [156] is the most commonly used hormonal therapy for breast cancer and has demonstrated positive effects on the cardiovascular and skeletal systems of postmenopausal women but is associated with an increased risk of uterine cancer. Tamoxifen is described as a selective estrogen receptor modulator (SERM; see Ref. [157]) with a tissue selective profile dependent on the different distribution of the α- and β-subtypes of the estrogen receptor (ERα and ERβ) that activate and inhibit transcription, respectively. These selective effects have been ascribed to differential interactions with gene promotor elements and coregulatory proteins depending on whether the ER interacts directly, or in a tethered manner with DNA [158].

10.4. Intracellular Receptors

Members of the intracellular receptor family include the cytochrome P450 (CYP) family [106], the SMAD family of tumor suppressors [108,109], intracellular kinases [128] and phosphatases, nitric oxide synthases (NOS; see Ref. [159]), caspases [159] and other proteins (e.g., XIAP, Smac) involved in apopotic pathways [110] and the transcription factor receptor superfamily that includes nuclear hormone receptors [155,161]. These include

- *Thyroid Hormone Receptor-Like* including thyroid hormone receptors, retinoic acid receptors, peroxisome proliferator-activated receptors, Rev-ErbA, RAR-related orphan receptors, liver X receptor, farnesoid X receptor, the Vitamin D receptor, Pregnane X receptor and the constitutive androstane receptor;
- *Retinoid X Receptor-Like* including hepatocyte nuclear factor-4 receptors, retinoid X receptors; testicular receptors and TLX/PNR receptors;
- *Estrogen Receptor-like* including estrogen receptors (ERs), estrogen-related receptors (ERRs), and the 3-ketosteroid receptors (glucocorticoid, mineralocorticoid, progesterone, and androgen);
- *Nerve Growth Factor IB-like* including nerve growth factor IB; nuclear receptor-related 1 (NURR1) and neuron-derived orphan receptor 1 (NOR1);
- *Steroidogenic Factor-like* including SF1/LRH1 (Steroidogenic factor 1; SF1, and liver receptor homolog-1; LRH-1 receptors;
- *Germ Cell Nuclear Factor-like*;
- *Miscellaneous* that includes the DAX (dosage-sensitive sex reversal, adrenal hypoplasia critical region, on chromosome X, gene 1), SHP (small heterodimer partner) and nuclear receptors with two DNA binding domain (2DBD-NR) families. STATs such as AP-1, NFkB, NF-AT, STAT-1, PARs, various hormone responsive elements on DNA and RNA promoters, and ribozymes are also intracellular drug targets.

10.5. RNA Targets

With the exception of the HREs on DNA, the drug targets reviewed to date are proteins and protein complexes. Emerging targets include the various RNAs that exhibit sequence-specific binding and can be selectively inhibited [162] or silenced using short 21–25 nucleotide double stranded RNAs termed small interfering RNAs (siRNAs) that effect complementary sequence-specific cleavage of mRNA. Endogenous microRNAs (miRNAS) and short hairpin RNAs (shRNAs) that resemble miRNA precursors can effect translational repression and transcript degradation and block translation [163].

10.6. Non-GPCR-Linked Cytokine Receptors

Cytokines are polypeptide mediators and are involved in the inflammatory/immune response [106]. There are three cytokine receptor families [164,165]: *hematopoietin*, which includes IL-2–IL-7, IL-9–IL-13; IL-15–IL-17; IL-19, IL-21, IL-22, GMCSF, GCSF, EPO, LIF, OSM, and CNTF, with primary signal transduction through the Jak/STAT pathway; *tumor necrosis factor*, comprising the receptors, TNFRSF1–18 that signal through NFB, TRAF, and caspases; and the *interleukin 1/ TIR* family that includes IL-1RI and IL-1RII, IL-1Rrp2, and IL1RAPL. TIGGR-1, ST2, IL-18, and Toll 1–9 also signal through NFB and TRAF.

10.7. Orphan Receptors

Orphan receptors are defined as proteins with a receptor motif that lack both a ligand and a function. Currently, 143 human GPCRs with no known ligand classified as orphan receptors. Among these, 98 belong to Family A, 34 to Family B, and 6 to Family C with no members in the Frizzled/smoothened family. Considerable efforts are currently ongoing to identify the ligands for these and their function as novel intellectual property for the drug discovery process. The most common deorphanization strategy used is known as reverse pharmacology in which the cell-surface-expressed orphan receptor is used as "bait" to bind selective ligand that include tissue extracts (usually from the same tissue region as the orphan), small molecules, peptides, proteins, and lipids [166]. The identification of an agonist ligand is dependent on the activation of an intracellular signaling cascade mediated by the expressed orphan receptor. A confound in assay design is that the signal cascade is usually unknown for a new orphan receptor. Therefore, generic assay systems amenable for HTS needed to be designed to allow the screening of large ligand collections. These include fluorescent (FLIPR-based) [167], or luminescent (aequorin-based) [168] screening technologies that can measure ligand-induced intracellular Ca^{2+} mobilization. To direct the signaling via the PLC Ca^{2+} readout, orphan receptors are often coexpressed in mammalian cells together with promiscuous G-proteins such as $G\alpha_{15/16}$, that couple the majority of GPCRs [121]. To date more than 50 orphan GPCRs have been deorphanized and ligand paired in this manner. While most of the current interest on orphan receptors is focused on GPCRs because of the considerable body of existing knowledge regarding this receptor class, it is anticipated that orphan receptors will also be discovered for other receptor classes, for example, orphan nuclear receptors.

The orphan receptor approach to drug discovery can be exemplified by the example of the orphanin/FQ receptor [169]. ORL1, a structural homolog of the opioid receptor. Identified in 1995 using a homology-based screening strategy, ORL1 had low affinity for known opioid ligands. A novel heptadecapeptide ligand for the ORL1 receptor, orphanin/ FQ, was subsequently isolated from brain regions rich in ORL1 providing the key tool to validate the target and identify a functional role for the receptor in stress-related situations in animal models. From an intensive screening program, an agonist of this receptor was identified, Ro 64–6198 (Fig. 7) that represents a novel anxiolytic/antidepressant drug candidate.

10.8. Neurotransmitter Transporters

An important factor in maintaining temporal integrity in intercellular communications is the rapid clearance of neurotransmitters from the extracellular space. Both metabolism and reuptake mechanisms are involved in limiting the effects of released neurotransmitters and provide effective drug targets for intervention in the signaling process. Neurotransmitter transporters (NTs) are members of the SLC6 family [170,171] are involved in transporting neurotransmitters, amino acids, and osmolytes. NTs are present on both plasma membrane and on intracellular vesicles where they affect vesicular packaging of neurotransmitters. Members of the plasma membrane NT family are integral membrane proteins with varying topologies whose activities are dependent on the Na^+ intracellular/extracellular gradient. Most NTs (e.g., dopamine transporters (DAT), noropinophrinc transporters (NET), serotonin transporters (SERT) also require Cl^- (Na^+/Cl^--dependent NTs, SCDNTs) while the glutamate transporter family also require K^+ (SKDGTs). These proteins bind their cognate neurotransmitters with high affinity and transport them across the plasma membrane into the cell. SERT blockers (known as selective serotonin reuptake inhibitors, SSRIs) have established an important role in treating depression- and anxiety-related disorders [172]. Orphan members of the SLC6 family have also been identified [171].

10.9. Neurotransmitter Binding Proteins

Binding proteins for corticotrophin releasing factor (CRF) [173] and acetylcholine (AChBP) [174] have been identified and exploited, to date unsuccessfully, as potential drug discovery targets. CRF binding protein is elevated under conditions of stress and may act to blunt the actions of BRF thus acting as a reservoir or "decoy receptor" thus modulating synaptic transmission. The structure of the AChBP has been used as a model system [174].

10.10. Drug Receptors

Most drugs interact with receptors (or enzymes) for which the natural ligand (or substrate) is known. There are, however, several receptors, distinct from the evolving class of orphan receptors, for which a synthesized drug is the only known ligand. The best example of this is the central benzodiazepine receptor present on the $GABA_A$ ion channel complex [57] that was originally identified using a $[^3H]$-diazepam binding assay. Because this is the site of action of the widely used BZ anxiolytic drug class, this is a *bona fide* receptor with clinical relevance. However, despite considerable efforts and a number of interesting if somewhat orthodox candidate compounds, no endogenous ligand has yet been unambiguously identified that would represent an endogenous modulator of anxiety acting through the BZ receptor. Other examples of drug receptors for which synthetic ligands were identified before the endogenous agonists were identified are the cannabinoid receptors [175], CB_1 and CB_2, through which the Δ-9-tetrahydrocannibol, the active ingredient of the psychoactive recreational drug/analgesic/antiemetic drug, marijuana acts; the vanilloid receptor, VR-1 [176], the known ligand for which is capsaicin, the ingredient in red pepper that evokes heat sensation; and the opioid receptor family [177], the site of action of morphine and other derivatives of the poppy. For each of these drug receptors, endogenous mammalian ligands, anandamide, the endovanilloid, *N*-arachidonyl-dopamine (NADA) and the enkephalins, respectively have been identified.

11. MOLECULAR BIOLOGY OF RECEPTORS

Cloning of the β_2-adrenoceptor in 1986 [116] was a key event in catalyzing molecular GPCR research. Since then, recombinant DNA methods have been extensively used to isolate and analyze the sequence of numerous receptors from different mammalian complimentary DNA (cDNA) libraries using the polymerase chain reaction (PCR) technique to clone receptors that can then be expressed in various pro- and eukaryotic cell lines and in the *Xenopus oocyte*, which agonist responses can be studied.

From the receptor/ligand modeling standpoint, the ability to specifically alter the structure of cloned receptors through the modification of a small number of nucleotides, *site-directed mutagenesis*, the removal of specific regions of the receptor gene—*deletion mutagenesis* and/or the construction of hybrid (chimeric) receptor proteins can identify the relative importance of individual amino acids in ligand recognition and receptor function. These techniques can also be used to manipulate receptors to produce recombinant expression systems with better cell surface expression of receptors, more facile signal transduction coupling and other qualities important for efficient drug discovery.

12. GENETIC VARIATION IN GPCRs

Genetic loci for GPCRs harbor numerous variants including DNA insertions or deletions and single nucleotide polymorphisms (SNPs) that can alter GPCR expression [178]. Inverse agonists would represent useful therapeutics since they would potentially provide a means to block the actions of the constitutively active receptor function, thereby contributing to inter-individual differences in disease susceptibility/progression and drug response. Genetic variations in GPCRs can influence transcription, translation, receptor folding, and expression on the cell surface or perturb receptor function contributing to disease etiology and pathophysiology [179]. Similar mutations occur in LGICs that alter function, for example, P250Δ, A251E, and T265V, in the M2 pore of the α_1-homomeric glycine receptor convert the channel conductance from being anion- to cation-selective [180].

Loss-of-function mutations of the thyroid stimulating hormone receptor (TSHR) can produce congenital hypothyroidism [181]. Loss-of-function mutations of parathyroid hormone receptors result in Blomstrand's chondrodysplasia, a disorder associated with abnormal breast and bone development [182], those in the vasopressin V_2 receptor cause X-linked nephrogenic diabetes insipidus [183] and in the CCR5 receptor lead to resistance to HIV infection [184]. Therapeutic approaches to loss-of-function mutations are challenging, as the receptor is nonfunctional. In the case of the vasopressin V_2 receptor, where loss-of-function mutations prevent appropriate sorting of the receptor to the cell surface, cell permeable V_2 receptor antagonists (e.g., SR121463B; satavaptan) rescued the mutant V_2 receptor, allowing for cell surface expression and receptor function [185].

GPCR gain-of function mutations result in the wild-type receptor being constitutively active independent of ligand. Gain-of-function constitutive activity can lead to night blindness (rhodopsin; see Ref.[186]); Jansen-type metaphyseal chondrodysphisin (parathyroid hormone-related receptor; see Ref. [187]); congenital hyperthyroidism (TSH receptor; see Ref. [188]); familial male precocious puberty (FSH receptor; see Ref. [189]) and hypocalciuric hypercalcemia and neonatal hyperparathyroidism (calcium-sensing GPCR; see Ref. [190]). Inverse agonists would represent useful therapeutics since they would potentially provide a means block the actions of the constitutively active receptor.

13. FUNCTIONAL GENOMICS: TARGET IDENTIFICATION/TARGET VALIDATION/ CONFIDENCE BUILDING

Target discovery involves the identification and early validation of targets involved in human disease state and is the essential first step in drug discovery. The various techniques applied in target identification and validation can be grouped into two broad strategies: the "molecular" and "system" approaches [191]. The molecular approach identifies new drug targets through an understanding of the cellular mechanisms thought to underlie disease phenotypes using genomics-based approaches while the systems approach involves the study of disease phenotypes in whole organisms and is driven by *in vivo* studies and complimentary clinical observations.

In recent years, the molecular approach has predominated and has led to the emergence of technologies that attempt to correlate change in gene (genomics) and protein (proteomics) expression or genetic variation (genetic association) with human disease. To date, however, genomics and proteomics have proved to be of limited utility in target identification and validation. The lack of reproducibility in genetic association studies necessitates confirmation in several independent studies using different patient cohorts. The phenotype-orientated target identification strategy can be divided into the areas of "forward genetics" (from phenotype to gene) and "reverse genetics" (from gene to phenotype).

The identification of novel receptor targets from the human genome and their subsequent use in the drug discovery process requires that the target be validated. This involves the demonstration of a functional role for the potential target in the disease phenotype. Although ultimately depend on human studies, validation in target discovery will normally require that the target is expressed in the disease-relevant cell/tissue and that target modulation in cell and/or animal models ameliorates the relevant disease phenotype. This represents a highly complex and resource intensive process that adds significantly to the cost of drug discovery.

Once a gene is associated with a disease, its protein product needs to be characterized and a biochemical function, for example, receptor, enzyme, or transporter, ascribed to it. The probable function of the protein can then be assessed and its cognate ligand or substrate identified. Not all proteins identified in this manner are obviously involved in mammalian cell function. The protein product of a novel lithium-related (NLR) gene identified in the liver of mice exposed to lithium has sequence homology to a bacterial nitrogen permease [192]. The potential role of this protein in the etiology of bipolar affective disorder remains unclear but may be more related to side effects than efficacy.

To identify what are termed "druggable" gene products as drug discovery targets, the still somewhat mystical process of functional genomics [193] can be used to construct protein–protein interaction maps [194] to identify other proteins in a pathway/network that can represent a more facile entry points to the protein cluster associated with the genetically identified target associated with the disease state. Using techniques such as yeast two-hybrid [195] and *gal*-pull down [196], the putative function of an uncharacterized protein can be assigned on the basis of the known function of its interacting partners, involving the extensive use of bioinformatics tools. Targeted gene disruption, antisense [197], and RNA silencing [163] are other techniques for assessing gene phenotypes. For instance, antisense to the rat $P2X_3$ receptor had marked hyperalgesic activity showing an unambiguous role for this receptor in chronic inflammatory and neuropathic pain states [198].

Transgenic animals, where the function of a gene is knocked out by genetic techniques, have been less useful in understanding gene and protein function [199]. In addition to being primarily limited to the mouse, this approach is high in cost and time (often taking 1–2 years to generate sufficient animals for evaluation). In many instances, the absence of a given protein has no overt effect on the phenotype of an animal because of compensatory changes that occur during the developmental phase. Alternatively, the gene knockout may lead to an animal that has a limited, if any, life span or reproductive capability.

Another target validation approach involves the use of ligands to define the function of a new target. Already discussed in regard to orphan receptors, other ligand-directed target validation models fall under the rubric of *chemical genomics* or *chemigenomics* [200]. This area of drug discovery, once known as pharmacology [201], is only now emerging from an excess of ephemeral biotech acronyms including ATLAS (any target affinity ligand screen), ALIS (automated ligand identification system, RAPTV (rapid pharmacological target validation), and so on. A key to identifying useful ligands that selectively interact with proteins of unknown function to define their function is a sufficient level of pharmacophore diversity to ensure success. Once a novel protein target is made amenable to HTS and efforts to identify its cognate ligand meet with success, the next step in target validation is to identify a "drug-like" molecule that can be used in animal models and hopefully advanced to the clinic. For ORL1, Ro 64-6198 (Fig. 7) is an excellent example of this approach [169].

This raises the key test of *target validation* or *target confidence building* [26]. Does a compound that has the appropriate potency, selectivity, and ADME properties, that is active in "predictive" animal models, and is free of other systems toxicology work in the targeted human disease state? This is the ultimate, and only, validation of a target-based drug discovery program. All events leading up to this point are more accurately defined as target confidence building. With the considerable compound attrition rates in moving from animals to diseased humans, this has not proven to be a predictable transition. A case in point is that of the NK1 receptor activated by the peptide, substance P [202]. Data from both animals and humans has implicated this receptor in a variety of human pain states including migraine and neuropathic pain. Several successful HTS campaigns run in parallel at Pfizer, Lilly, Merck, Sanofi Synthelabo (then Sanofi Aventis), and Takeda led to the identification of a number of diverse, highly efficacious and selective antagonists of the NK1 receptor that were optimized as drug candidates (Fig. 9), for example, RP-67580 CP-96345, GR-205171, lanepitant, MK-869

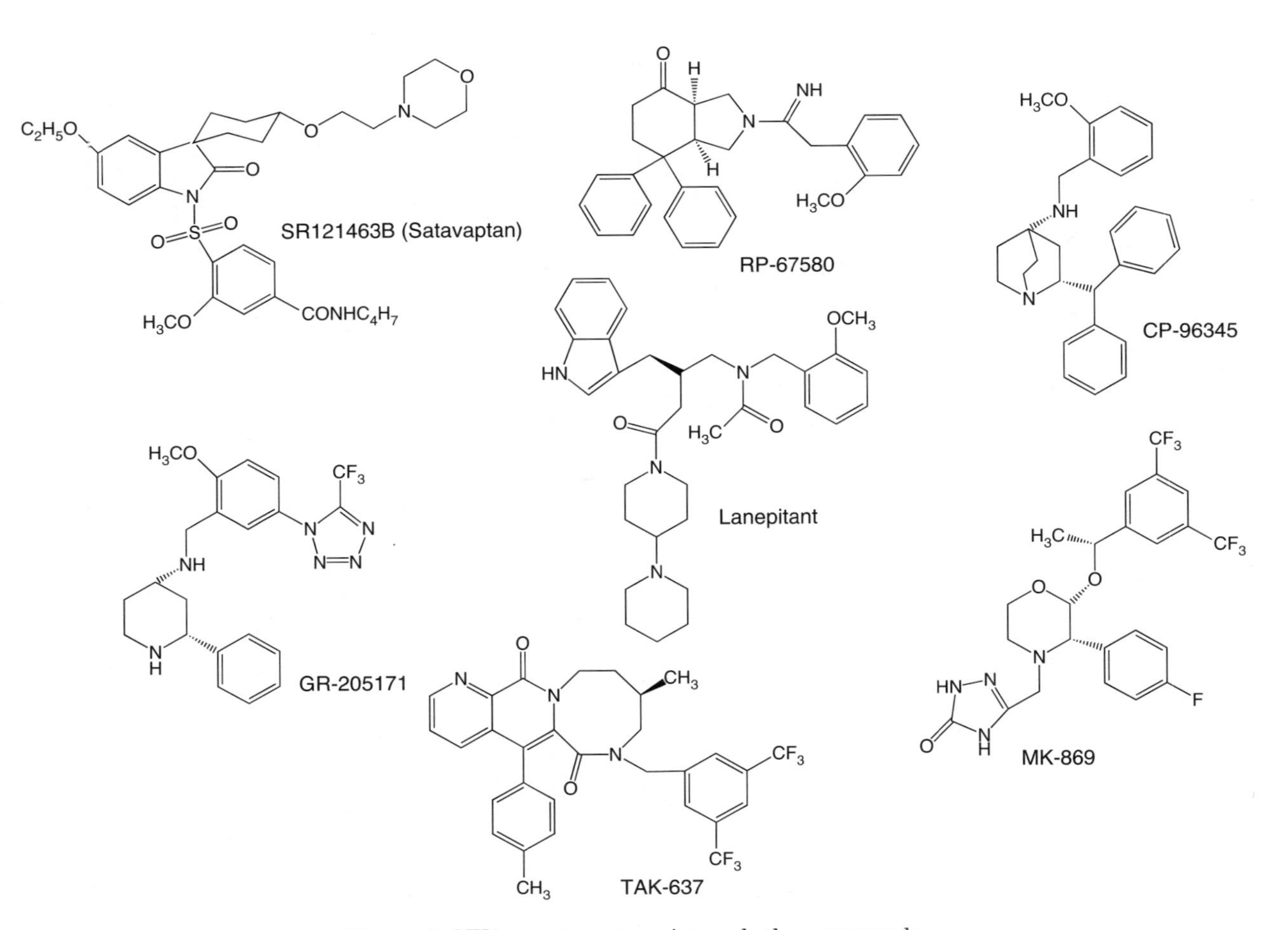

Figure 9. NK1 receptor antagonists and other compounds.

(aprepitant), and TAK-637. These compounds, with varying degrees of potency, were active in a variety of animal models of pain and were also side-effect free in Phase I human clinical trials but uniformly failed in proof of concept clinical studies as novel analgesic agents in patients with various pain conditions. The reasons for this remain unknown and have focused the relevance of animal pain models to the human condition, for example, a lack of understanding of the true human disease condition and various nuances of substance P signaling pathways. However, before the results with NK1 antagonists, drug discovery in the analgesia area was considered one of the most robust. All known analgesics, for example, aspirin and morphine, were active in one or more of the animal pain models; new receptors could be mapped in pain pathways and their function assessed using knockout and antisense procedures, and the occupancy of receptors in human brain and pain-sensing pathways could be noninvasively imaged. The only limitation was whether the side-effect profile of a putative analgesic agent acting on a novel target or suboptimal ADME characteristics would limit human exposure.

Based on the available data, the NK1 receptor has not been validated as a target for pain treatment despite an overwhelming body of robust preclinical data. This example thus serves to underline the many significant challenges of validating targets in the drug discovery process at the present time. MK-869 was also evaluated as a putative antidepressant based on data from a guinea pig vocalization model [203] but failed in late stage clinical trials, A newer NK1 antagonist, LY686017 of unknown structure has demonstrated efficacy in animal models of alcohol dependence and in the clinic [204].

Another example, more directly related to the human genome map [20], is the search for genes associated with, and by inference may be causative in, schizophrenia. This disease affects approximately 1% of the population and has a strong genetic association [205]. Numerous studies comparing schizophrenic patients with individuals lacking symptoms of the disorder have identified various targets in the genome map [206,207]. Markers between or close to D1S1653 and D1S1679 on chromosome 1q had l.o.d. scores of 2.4–6.5 [208]. A l.o.d. score of above 3 is considered to be indicative of a relevant association similar to a significant P value ($P < 0.05$). With this information, a search for the gene on chromosome 1q and the delineation of its function in humans would be a logical approach to finding novel targets that could lead to new drugs to treat schizophrenia. However, a subsequent report [209] based on data from eight separate schizophrenia populations with a sample set that was "100% powered to detect a large genetic effect under the reported recessive model," showed no evidence for a linkage between chromosome 1q and schizophrenia, a finding that could not be ascribed to ethnicity, statistical approach, or population size. Another group using prefrontal cortex tissue from two separate schizophrenic populations showed an upregulation of apolipoprotein L1 gene expression that was not seen in tissue from patients with bipolar disorder or depression [210]. The genes related to apolipoprotein L1 gene expression are clustered on the chromosome locus 22q12, providing another target for the functional genomic approach to target discovery and validation.

To an industry researcher embarking on a well-funded, science-driven approach to new targets for schizophrenia a decade ago, these later findings in 2002, as well as others showing gene associations with schizophrenia on chromosomes 6, 13, and 22 would give pause to wonder how best to proceed. Do all five locations represent *bona fide* targets for functional genomics? Should one continue work on the original chromosome 1 findings in light of the failure to replicate? Is the only real validation, in light of conflicting findings, to proceed to the identification of multiple compounds that can be tested in schizophrenic populations—to validate the genome-based approach? Given that the cost of initiating a project, finding leads, optimizing these, and running a single clinical candidate to Phase IIa clinical proof of principle is in the range of $24–28 million, how many organizations can afford the $120 million plus to undertake scientifically logical yet financially prohibitive strategies? These are difficult questions that can be applied to other psychiatric and neurological diseases, as well as any other disease state with a potential genetic causality, One answer to these questions obviously points to multifactorial genetic causes in the genesis of many diseases and thus negates the overly simplistic "one gene, one disease" mantra that heralded the age of drug discovery based on the human genome map. Strategies resulting from the answers to these questions also make the process of genome-based target and drug discovery a much more costly endeavor, with good evidence for two to three or more potential targets for each disease that need to be examined in parallel. Hopefully, the quality of the drug/disease target finally validated using NCEs in the human condition will be a magnitude of order superior to currently existing drugs and will thus justify the cost of this approach. The proof of this has yet to appear and many key opinion leaders are skeptical as to whether this will happen before 2050.

14. SPACE: CHEMICAL, BIOLOGICAL, AND DRUGGABLE

Like the limitless opportunities for targets in the human genome, it has been argued [211] that only a small fraction of available chemical space has thus far been exploited primarily due to synthetic constraints and a scaffold related bias. It has been estimated that the virtual space for small molecules with molecular weight of less than 500 is an incredible 10^{60} [212] with the CAS registry containing only some 33 million inorganic and small organic compounds with a limited number of structural motifs [213]. Further analysis using the Comprehensive Medicinal Chemistry Database [214,215] identified 143 out of 836,565 "frameworks" that described 50% of drugs in the database. Of 15,000 side chains cataloged in these 50%, 75% were from 20% of the documented side chains. The reason for this focus was ascribed to the familiarity of certain pharmacophores and their cost. Identifying validated pharmacological space for drug discovery has in instances focused on finding "new uses for old drugs" [216] by screening known drugs against new targets, the latter forming a significant part of the NIH's Molecular Libraries Initiative [217]. Efforts to increase chemical diversity have focused on fragment based [218–220] and template-guided synthetic chemistry [221,222].

15. COMPOUND PROPERTIES

Once a drug target has been identified, the challenge becomes to identify compounds that interact with the target and can be used as the basis of a lead optimization program to identify potential drug candidates.

An ideal, drug-like compound should have the appropriately unique recognition characteristics to impart affinity and selectivity for its target, have the necessary efficacy to alter the assumed deficit in cell function associated with the targeted disease state, have an established SAR (multiples of actives and inactives), be bioavailable, metabolically, and chemically stable, chirally pure, easy and cost effective to synthesize, have demonstrated *in vivo* efficacy, and a favorable intellectual property (IP) assessment.

15.1. Structure–Activity Relationships)

The term, hit-to lead describes the process of filtering and assessing initial hits from HTS and a limited chemical modifications of these to establish an initial structure–activity relationship [223]. The SAR of a compound series is a means to relate changes in chemical diversity to the biological activity of the compound *in vitro* and *in vivo,* as well as the pharmacokinetic (gut/blood–brain barrier transit, liver metabolic stability, plasma protein binding, etc.) and toxicological properties of the molecule. These SARs, when known, are frequently distinct such that changes that improve the bioavailability of a compound often decrease its activity and/or selectivity. Compound optimization is thus a highly iterative and dynamic process. Data in addition to potency and selectivity of a given hit or cluster of hit compounds are also collected to provide sufficient data to decide which class should be advanced into the lead optimization phase. Such information would include drug properties including solubility, hERG activity, stability, PK, and so on. [224]. For the purposes of the present review, however, SAR is used to describe classical compound efficacy unless otherwise stated. Quantitative SAR (QSAR) involves a more mathematical approach involving computer-assisted design. [225]. The inclusion of drug-like properties has led to the term "structure–property relationship (SPR; see Ref. [228])" Following the characterization of the SAR and the documentation of the effects of different pharmacophores and various substituents on biological activity, it is possible to theoretically model the way in which the ligand interacts with its target and thus derive a two- or three-dimensional approximation of the active site of the receptor or enzyme.

Computer-assisted molecular design (CAMD) techniques [229] provide a means to correlate biology with structure can then be used to predict—sometimes successfully—the key structural requirements for ligand binding, thus defining those regions of the receptor target that are necessary for ligand recognition and/or functional coupling to second messenger system to permit the design of new pharmacophores. The incorporation of a flexible rather than static approach to studying ligand–protein interaction [18] can add further value to CAMD efforts.

SAR can be further delineated in terms of the type of activity measured as the readout. *In vitro*, this can be displacement of a radiolabeled ligand from a receptor, receptor activation, and intrinsic activity as measured in a functional assay, blockade of receptor function by an antagonist ligand, and so on. A newer contribution in developing an SAR is the determination of the thermodynamic signatures of compounds to optimize enthalpy in the binding reaction [230]. This is not an absolute since factors that optimize enthalpy may be incompatible with other drug-like properties, for example, metabolic stability and solubility.

An additional ligand property is that of selectivity, the degree to which a ligand can interact with the target of choice as compared with related structural targets and other targets. The degree of selectivity typically determines the side-effect profile of an NCE, given that the targeted mechanism itself does not produce untoward effects when stimulated beyond the therapeutic range. As previously noted, the development of a ligand binding profile for a ligand active at a new target is very useful in assessing its effects in more complex tissue systems and underlines the need for databases. HTS screens can often find micromolar hits that are already known to be active in the nanomolar range at discrete targets.

15.2. Defining the Receptor–Ligand Interaction

The complex physiochemical interactions that describe the interaction of a small molecule with a protein despite the many sophisticated technologies used to study this interaction are still highly empirical, being implicitly defined by the SAR for a series of active and less active compounds. In an increasing number of instances, however, the ability to clone, express, and readily derive crystals of a receptor or enzyme and analyze the interaction of a ligand or substrate with these using X-ray crystallographic, NMR, and amino acid point mutation approaches has provided information on the actual topography of the selected drug target that can then be used in *de novo* ligand design.

There are a variety of approaches to deriving information on the RL interaction for use in understanding the key features required in ligand necessary to dictate a potent and selective interaction with its target. Analysis of the interactions of a series of structurally distinct pharmacophores, agonists, and antagonists with the target can be combined with point mutation changes in the target to elucidate key amino acids involved in compound recognition. This can be complimented by X-ray crystallographic and NMR-derived data to design novel pharmacophores using HTS approaches [231,232]. However, many proteins of interest, especially membrane-bound receptors and ion channels, are not available in the soluble form amenable to the use of these techniques such that the design of NCEs is based on the conceptualization of the target, for example, virtual receptors. There are also limitations to structure-based design approaches, for example, approximations of the hydration state of the isolated protein, the impact of removal from its native environment, and three-dimensional structural issues with recombinantly derived proteins. Nonetheless, these technologies have been useful in compound design. More recently, automated, high-throughput approaches have been applied to X-ray crystallography [233], providing a more rapid means to generate information on multiple ligand interactions with a given crystal.

15.2.1. Receptor Binding Assays Until the 1960s, newly synthesized compounds and compounds isolated from natural sources were assessed by a mixture of *in vivo*, whole animal screens, and classical tissue assays. While many useful therapeutic agents were identified by this approach, the cost, in terms of compound quantity as well as time and animal use were considerable. *In vivo* test paradigms also suffered from the possible elimination of interesting compounds on the basis of unknown pharmacokinetic properties as test paradigms were usually routine in terms of timing, and as a result, many potentially interesting compounds that exhibited short plasma half-lives were considered "inactive" because data on their actions was sought after their peak plasma concentration. This type of screening approach also provided little useful chemical information about discrete interaction of the drug with its target, limiting the design of analogs. The specificity of the response, ignoring caveats related to pharmacokinetics, was not ideal because the mechanism inducing the overt response and potential points of intervention to block the response were unknown. This empirical approach would predict, in the absence of any data related to specific interactions with a molecular target, that because β-adrenoceptor antagonists lower blood pressure, then any compound that lowers blood pressure is by definition a β-adrenoceptor antagonist. The 1970s saw the development of a number of *in vitro* biochemical screens that moved the measurement of the RL interaction a little closer to the molecular level. Nonetheless, the major challenge was to develop assays that measured the RL interaction independently of "downstream" events such as enzyme activation and second and third messenger systems. By such means, the ability of a compound to bind to a receptor could be determined on the basis of the SAR and thus provide the chemist with a more direct means to model the RL interaction.

Snyder [234] building on pioneering work by Rang [235], Roth [236], and Cuatrecasas [237] developed radioligand binding as a practical tool for drug discovery process in 1973 leading to a paradigm shift that led to an explosion in the identification and characterization of new receptors and their subtypes, that was enhanced by the application of tools of recombinant DNA technology to the process. Additionally, compounds could now be assessed for activity using milligram, instead of gram, quantities. While the technique of radioligand binding has been largely supplanted by activity-based assays in many high-throughput settings, it remains the gold standard for compound characterization, SAR development and selectivity determination.

It is done by measuring the RL interaction (Eq. 1) *in vitro* using a radioactive ligand, R^*, to bind with highly affinity and selectivity to receptor sites. The interaction of unlabeled ("cold") ligands with the receptor competes with the radioligand, decreasing its binding. This simple technique revolutionized compound evaluation in the 1970s, allowing SARs to be determined with milligram amounts of compound in a highly cost- and resource-effective manner. At steady state, the RL^* complex can either be separated out from free radioligand using filtration or assayed using a scintillation proximity assay.

The parameters measured in a binding assay are the dissociation constant, K_d, the reciprocal of the affinity constant, K_a. The K_d is a measure of the affinity of a radioligand for the target site: the B_{max}, usually measured in moles per milligram protein, a measure of the concentration of binding sites in a given tissue source and the IC_{50} value. The K_d and B_{max} values can be determined using a saturation curve where the concentration of radioligand is increased until all the ligand recognition sites are occupied or by measuring radioligand association and dissociation kinetics; the K_d is the ratio of the dissociation and association rate constants.

The IC_{50} value is the concentration of unlabeled ligand required to inhibit 50% of the specific binding of the radioligand. This value is determined by running a competition curve (Fig. 5) with a fixed concentration of radioligand and tissue and varying concentrations of the unlabeled ligand. To accurately determine the

IC_{50}, it is essential that sufficient data points be included. As shown in Fig. 10.5, if the data used to derive the IC_{50} value are clustered over a range that reflects 40–60% of the competition curve, much useful information is lost. Ideally, the competition curve should encompass the range of 10–90% of the competition curve and include 20 data points. Based on Michalis–-Menten kinetics, when binding is the result of the interaction of the displacer with one recognition site, 10–90% of the radioligand is inhibited over an 82-fold concentration range of the displacer. The slope of a competition curve can then be analyzed to assess the potential cooperation of the RL interactions. When binding is complex resulting in the interaction of the displacer with more than one recognition site, a greater than 82-fold concentration of displacer is required to inhibit the same 10–90% of specific radioligand binding (Fig. 10).

The IC_{50} value for a given compound is dependent on the assay conditions: the concentration of the radioligand used, the receptor density, and the affinity of the receptor, the K_d, for the radioligand. To compare the activity of a "cold" ligand across different radioligand binding assays, the Cheng–Prusoff equation [238] is used to compensate for differences in K_d and the radioligand concentration to obtain a K_i value derived by the relationship:

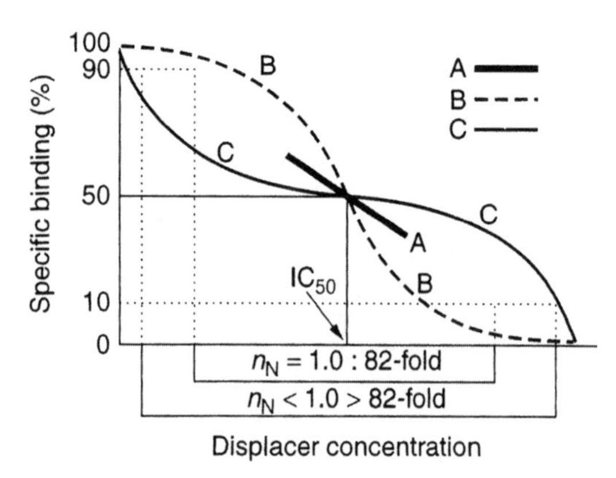

Figure 10. Measuring ligand interactions in a receptor binding assay. Binding of the radioligand is 100%. In curve A, close to the 50% point, an IC_{50} value can be obtained but ignores complexities of the displacement curve. More complete displacement, as in curves B and C, provides more information on the ligand. Displacement curve B has a Hill coefficient of unity requiring an 82-fold difference in displacer concentration to displace 10–90% of binding. Displacement curve C has a Hill coefficient of less than one, requiring a greater than 82-fold difference in displacer concentration to displace 10–90% of binding and indicating the presence of more than one site. By extending the range of concentrations used, curves B and C can assess the possibility of multiple sites being present.

$$K_i = \frac{IC_{50}}{1 + [L]/K_d} \qquad (6)$$

where [L] is the concentration of radioligand used and K_d is the dissociation constant for the radioligand at the receptor. This relationship thus corrects for inherent differences in assay conditions.

Consistent results are most often found using radioligands derived from antagonists, since the binding affinity is unaltered by changes in G-protein association or other factors that that modulate agonist binding. However, this approach is only possible once a high-affinity antagonist has been identified that is selective for the receptor of interest.

Most radioligand binding assays are easily adapted for screening using native or recombinant receptors expressed either in membrane extracts or in intact cells and have the major advantage of providing clear evidence of direct interaction of a compound and receptor, an attribute that most functional assays lack. Binding assays can be rapidly used to assess compound recognition characteristics but are generally limited in their ability to delineate agonists from antagonists, especially in a high-throughput setting and cannot be employed to study orphan GPCRs, which, by definition, lack a known ligand. However, one of the most critical factors in restricting the use of radioligand binding assay for GPCR screening is the cost or radioactive waste disposal, particularly when large amounts are used in screening campaigns. In recent years, radioactive assays have become increasingly replaced by fluorescence- or luminescence-based technologies. Fluorometric technologies include a wide range of formats including fluorescence polarization (FP; see Ref. [239]), fluorescence resonance energy transfer (FRET; see Ref. [240]), time resolved fluorescence resonance energy transfer (HTRF and LANCE™; [241]) to fluorescence correlation spectroscopy (FCS; [242]). There are also label-free binding technologies to study receptor interactions of which surface Plasmon resonance (SPR; [243]) is probably the best known.

15.2.2. Functional Assays Although binding assays have been used historically to identify novel compounds and drugs, most current technologies used for both GPCR and LGIC focused drug discovery are cell based. Cell based technologies allow the examination of receptors in a milieu that is more closely to the native state.

Biochemical assays involving the measurement of cAMP production or phosphatidylinositol turnover have given way in HTS scenarios to reporter systems where receptor activation or inhibition can be measured using a fluorescence based readout. This depends on the use of calcium-sensing dyes coupled with real time measurement using coupled charge device (CCD) cameras and data capture. While many types of plate reader are available, the fluorescence imaging plate reader (FLIPR) is widely used. Using a 96- or 384-well microtiter plate format, the throughput on a FLIPR is such that a compound libraries of 0.5 million distinct compounds can be assayed in days to weeks.

For LGICs, automated electrophysiological patch clamp systems [244] including Ionworks™ and PatchXpress™ (MDS), Port-a-Patch™ and Patchliner™ (Nanion), and Qpatch™ (Sopion) and Flyscreen (Flyion) have been developed to increase data generation. This are still at an early stage in their evolution and are dependent on the use of cell lines with stable, homogenous, and high expression of the LGIC target.

The ability to measure colocalization of two cellular components in living cells using fluorescence probes is an increasingly powerful technique in understanding ligand–receptor and protein–protein interactions [239–242]. FRET measures the proximity of two proteins through the use of a luciferase tag on one partner and green fluorescent protein (GFP) on the other [240], and FP that relies on a polarized excitation light source to illuminate a binding reaction mixture. The smaller partner (e.g., the ligand) must be fluorescently labeled, and if this is unbound then it will, by tumbling in solution, emit depolarized fluorescence providing the means to measure the amount of unbound ligand in the binding mixture and has the benefit of not requiring any washing steps.

15.3. Receptor Sources

The choice of a tissue or cell line as a receptor source has a significant impact both on the development/implementation of a HTS for a targeted receptor and on the data generated. Receptor source include primary cell types and cell lines, both "native" and engineered. The choice is driven by the availability and behavior of the cells, as well as the amplitude and reproducibility of the signal attainable in the cellular system.

The natural receptor concentration in most tissues is in the femtomole to picomole range; brain tissue has a much higher density of receptors because of more extensive nerve innervation. Expression of the drug target in a cell line can, however, lead to differences in the number of receptors expressed per clone, a factor dependent on the relative proportion of transient to stable expressing cells and the passage number of the transfectants. Thus, the number of receptors can vary affecting the apparent activity of unknown ligands that compete for binding with the radioligand. The level of recombinantly expressed receptor and coupling efficiency affect the potency of standard agonists in cellular assay. Overexpression can increase receptor reserve, leading to a leftward shift of agonist concentration response curves with partial agonists to becoming full agonists. While the assay becomes more sensitive, compound activity may be overestimated.

As the drug targets of ultimate interest are those in humans, the use of a human receptor or enzyme is ideal in defining the SAR of a potential drug series. The drawback, however, is that nearly all the toxicology and safety studies done in preparing a compound for clinical trials are conducted in rodents, dogs, and nonhuman primates. If there are no species differences between rat and human, this testing becomes a moot point. If on the other hand, the human target is substantially different from that in rat or dog or monkey, and there are many examples of this, the safety and toxicology studies may be conducted on a compound that has limited interactions with the drug target in species other than human. One approach is to incorporate human receptor orthologs into mice.

The use of transfected cell lines in NCE evaluation can lead to a number of potential artifacts. Activation of transfected receptor may result in an increase in cAMP, a second messenger effect that may already be known to be a consequence of ligand activation of the receptor in its natural state. It is also possible that the transfected receptor may activate a cell signaling pathway in the transfected cell that is not linked to the receptor in its normal tissue environment. In this instance, the second messenger readout actually functions as a "reporter," a G-protein–linked phenomenon that results from the introduction of the cDNA for a GPCR and the generic or promiscuous interaction of the receptor with the G-protein systems. The introduction of the cDNA for any GPCR may then act to elicit a similar response. It is then advisable to use caution in extrapolating events occurring in the transfected cell to the physiological milieu of the intact tissue, and there is at least one case, the dopamine D4 receptor antagonist, L-745,870 [245] identified as an antagonist in a recombinant cell system that was not only inactive but also exacerbated disease symptomatology in Phase II clinical trials for schizophrenia suggestive of partial agonist activity that was not detected in the *in vitro* recombinant cell system.

15.4. ADME

Hit selection and prioritization is based on two categories, biological profile (potency, selectivity, and specificity) and ADME-Tox physicochemical properties. In evaluating new compounds, it is only in the past 15 years that the ability of a compound to reach its putative site of action has been a priority in the discovery phase of compound identification [31]. For many years, it was naively assumed that there was a generic approach that could be used on compound with poor bioavailability that would turn them into drug candidates. With attrition rates of lead compounds in the clinical development process of 50–60% range, with one estimate of greater than 90%, this was clearly not the case. Hodgson [246] has noted, "a chemical cannot be a drug, no matter how active nor how specific its action, unless it is also taken appropriately into the body (absorption), distributed to the right parts of the body, metabolized in a way that does not instantly remove its activity and eliminated in a suitable manner—a drug must get in, move about, hang around and then get out." To effectively use ADME in the HTS and hit to lead phases of drug discovery requires that the assays are efficient and timely. This involves the increasing use of automated systems [247,248] and cassette dosing approaches where several compounds are administered to a single animal at the same time [249].

The factors involved in defining *in vivo* activity form the basis of Lipinski's "rule of five" [250]. In this widely cited, retrospective study, a number of compounds have been assessed and used to design new molecules. The physical properties that were determined as limiting bioavailability were as follows: molecular weight greater than 500; more than 5 hydrogen bond donors; more than 10 hydrogen bond acceptors; and a $C\log P$ value less than 5. Some critics have noted [251] that "the rule of five has had an oppressive effect on creative thinking in the industry" and that "nearly half the drugs ... are derived from or inspired by natural products, compounds that universally break the rules" leading to the question "When the exception is 50% of the time, how useful is a rule?" Lipinski has noted that his "rules" are actually guidelines. Veber et al. [252] undertook a retrospective evaluation of the rat oral bioavailability of 1100 NCEs having an average molecular weight of 480 from the SmithKline Beecham's drug discovery efforts and established that reduced molecular flexibility, measured by the number of rotatable bonds and low polar surface area or total hydrogen bond count (sum of acceptors and donors), were important predictors of good (>20–40%) oral bioavailability, independent of molecular weight. A molecular weight cutoff of 500 did not significantly separate compounds with acceptable oral bioavailability from those with poor oral bioavailability, the predictive value of molecular weight was more correlated with molecular flexibility than molecular weight *per se*. From this retrospective analysis, Veber et al. suggested that compounds with 10 or fewer rotatable bonds and a polar surface area equal to or less than 140 Å (representing 12 or fewer H-bond acceptors and donors) have a high probability of having good oral bioavailability in rats. Reduced polar surface area was also a better predictor of artificial membrane permeation that lipophilicity ($C\log P$), with increased rotatable bond count having a negative effect on permeation and having no correlation with *in vivo* clearance. ADME has become a routine part of early efforts of a drug discovery team with the use of a number of *in vitro* approaches, for example, Caco 2 intestinal cell lines, human liver slices, or homogenates to assess potential metabolic pathways, PgP substrate in addition to classical rat, dog, and nonhuman primate *in vivo* studies, none of which has yet shown absolute predictability to the human situation [253]. Because a hit set from an HTS campaign often comprises several hundred to several thousand compounds, experimental assessment of the ADME-Tox and physicochemical profile of the entire hit set represents a considerable challenge. For this reason, the first characterization phase usually comprises an *in silico* assessment via an *in silico* ADME-Tox scoring system [254]

In the second phase of this process, experimental *in vitro* characterization of selected screening hits is initiated. High to medium throughput assays are used to characterize the representative screening hits with respect to functional hERG activity, cytochrome p450 (CYP) 3A4 inhibition and stability in human liver microsomes.

15.5. Compound Databases

With the exponential increase in information flow resulting from the ability to make many more compounds and test these in multiple assays, the capture, analysis, and management of data is a critical success factor in drug discovery. Many databases used in industrial drug discovery are ISIS/Oracle-based using MDL structural software and a variety of data entry and analysis systems, some are PC/MAC-based, and some are on a server. They provide a management system that enables scientists to capture experimental data into a central repository for rapid access to query, collaborate, generate reports and critically analyze. Access through integrated software applications allows the linking of biological data and chemical structure [255]. The ability to capture data and then reassess its value through *in silico* approaches has the potential to be a vast improvement over the "individual memory" systems that many drug companies used for the better part of the last century. Thus, with the retirement of a key scientist, the whole history of a project or even a department disappeared, and whatever folklore existed regarding unexplained findings with compounds 10 or 20 years before was lost with the individual. There are several public databases [256] with the NIH Molecular Libraries Roadmap Initiative (MLI; see Ref. [217]) providing a new public access database, PubChem collates information on the biological activities of small molecules screened under the auspices of the MLI [244].

16. LEAD COMPOUND DISCOVERY

As evidenced by the compound code numbers used by pharmaceutical companies, many thousands of new chemical entities have been made since the industry began in the late nineteenth century. Until the advent of combinatorial chemistry, the chemical libraries at most of the major pharmaceutical companies numbered from 50,000–800,000 compounds comprised of newly synthesized compounds as well as those from fermentation and natural product sources. Approximately 2–5 million compounds were identified in the search for new drugs to treat human disease states over the past century. This number has obviously leaped to the billions with combinatorial approaches. Given decomposition and/or depletion of compounds, the 2–5 million compounds actually made could be rounded down to 1 million. The 2006 print edition of the *Merck Index* [258], a compendium of drugs and research tools, lists a total of 10,200 compounds. Thus, from a hypothetical million compounds, only 1% has proven to be of sustained interest as either therapeutic agents or research tools.

In the early 1980s, with the promise that compounds could be created and tested on a computer screen (a promise as yet unrealized), the screening of large numbers of compounds against selected targets in biological systems was viewed as irrational. Indeed, the head of research at one of the then top 20 pharmaceutical companies (a molecular biologist) told the medicinal chemists at that company in the early 1980s that the demand for their skills was becoming less and would cease by the 1990s, a viewpoint somewhat akin to the apocryphal story of the head of the U.S. Patent Office, Charles H. Duell in 1899 that "everything that can be invented has been invented." This attribution has, however, been disputed and is considered conveniently apocryphal. An increasing focus on molecular design as the single source for NCEs (SAR by computer) in the early 1980s appeared to be the death knell for serendipity in drug design and discovery. However, in 1984 Ray Chang and colleagues at Merck discovered the benzodiazepine CCK antagonist, asperlicin [259] using a screening assay. This led to a clinical candidate, MK 329. While there are many success stories of drugs being discovered from natural products [260] efforts in the pharmaceutical industry worldwide are only now remerging after the targephilic approach has proven to be more challenging than anticipated [25].

16.1. High-Throughput Screening

To identify NCEs, it is imperative that a rapid, economical, and information-rich evaluation of biological activities be available. The term high-throughput screening describes a set of techniques designed to permit rapid and automated (robotic) analysis of a library of compounds in a battery of assays that generate specific receptor- or enzyme-based signals [261]. These signals may be membrane-based (radioligand binding, enzyme catalysis) or cell-based (flux, fluorescence, and luminescence). Although *in vitro* biochemical assays, such as enzyme activity and receptor binding, have been used extensively in the past, there has been a major increase in the use of assays based on living cells. There are two major advantages of employing cell-based functional assays over ligand binding assays in primary HTS. Firstly, cell-based assays have improved sensitivity for agonism due to intracellular signal amplification and also allow identification of both agonists and antagonists. Secondly, they permit the identification of compounds with diverse pharmacological effects (e.g., allosteric modulators).

The purpose of HTS is not to identify candidate drugs (despite the wishful thinking of newcomers to the field), but rather to identify lead structures or "hits," preferably containing novel chemical features, which may serve as a guide for more tailored iterative optimization. HTS should generate as few false positive leads as possible as exploitation of leads is an expensive component of the drug discovery process. One of the main challenges of cell-based screening is to discriminate between true hits and artifacts. After the primary HTS, the activity of hits is confirmed by retesting in the same assay to remove false positives caused by the statistical variation in the assay signal. The number of false positives may comprise up to several percentages of compounds screened, depending of the cutoff set to assign a compound as a hit and on the quality of the assay. To discriminate further between true hits and artifacts, additional assays are required. Once defined as a true hit, compound selectivity can be determined by testing against related targets, preferably expressed in the same cellular background. After several rounds of chemical modification and improvements in potency, selectivity and *in vitro* drug-like properties, a lead compounds is derived. HTS are designed to give information principally about potency, and a combination of follow-up screens may provide information about selectivity, specificity, and drug-like properties. Quantitative HTS (qHTS) is a newer system developed at the NIH, the ability to generate multiple concentration–response curves (CRCs) has added significant value to the HTS process [262]. Using a fully integrated and automated qHTS screening system over 6 million CRCs were generated by the NIH from greater than 120 assays in 3 years [263] with markedly reduced volume and compound needs. As with the concerns regarding radioactivity disposal, there is a major need to limit the quantities of compound effluent from HTS assays. Interestingly, the use of zebrafish in HTS mode to screen NCEs appears in some instances to involve adding the compound to the fish tank negating the novelty of the biological target by producing unacceptable gallons of compound effluent. With success in identifying hits from HTS, a rate typically in the 0.1% range the ability to capture, store and retrieve the data for posterity is a major issue. Clearly, the quality and diversity of the compounds and the robustness of the assays play a key role in a successful screening program.

16.2. Compound Sources

New chemical entities (NCEs) are discovered or developed/optimized from the following: (1) natural products and biodiversity screening; (2) exploitation of known pharmacophores; (3) rationally planned approaches, for example, computer-assisted molecular design; (4) combinatorial or focused library chemistry approaches; and (5) evolutionary chemistry.

16.2.1. Natural Product Sources Approxi-Approximately 70% of the drugs currently in human use originate from natural sources including morphine, pilocarpine, physostigmine, theophylline, cocaine, digoxin, salicylic acid, reserpine, as well as a multitude of antibiotics [260]. Medicinal and herbal extracts form the basis for the health care of approximately 80% of the world's population; some 21,000 plant species are used worldwide. Screening of natural products led to the discovery of the immunosuppressants, cyclosporin, rapamycin, and FK 506, and there is a continued search for new compounds. Natural products because of their complexity are not facile to make and the search for novel drugs has fallen out of favor. However, in the past decade there has been a resurgence of interest as synthetic compound sources have failed to provide sufficient diversity and new drug sources. The continued destruction of habitat with the accompanying loss of animal and plant species may impede further natural product-based drug discovery. Many interesting drugs have vanished. For example, a plant called silphion by the Greeks and sylphium by the Romans grew around Cyrene in North Africa and may have been an extremely effective antifertility drug in the ancient world but was harvested to extinction [264]. While it is safe to assume that near 100% of the world's mammals are known, fewer than 1–5% of other species, bacteria, viruses, fungi, and most invertebrates, are well characterized. It has been estimated therefore that only 0.00002–0.003% of the world's estimated 3–500 × 10^6 species have been used as a source of modern drugs [265]. Exploration of environments previously assumed to be hostile to life has revealed bacterial species living at extreme depths, at extraordinary temperatures, and in the presence of high concentrations of heavy metals. The sea covers almost three quarters of the earth's surface and contains a broader genetic variation among species relative to the terrestrial environment [266]. Although a number of important molecules have been derived from marine sources, including arabinosyl nucleotides, didemnin B, and bryostatin 1 the focus on this potentially chemically productive biosphere has been limited.

The cone snails that populate Pacific and Mediterranean regions have been termed "nature's combinatorial chemistry factories," and produce a bewildering array of novel peptide conotoxins active at mammalian drug targets [267]. These toxins are typically 10–30 amino acid residues in length that contain several disulfide bridges and are rigid in structure. Some 100,000 peptides with biological activity [268] from which the potent intrathecal analgesic, N-type calcium channel blocker, ziconitide (omega-conotoxin MVIIA; see Ref. [269]) was identified. Since the utility of these peptides to the snail is survival, the side effects of these natural products make then limited in their potential for human use. Conus toxins have proven to be invaluable as molecular probes for a variety of ion channels and neuronal receptors and as templates for drug design.

Poison frogs of the *Dendrobatidae* family contain a wide variety of skin-localized poisonous alkaloids that are presumably secreted for defensive purposes. Among the chemical

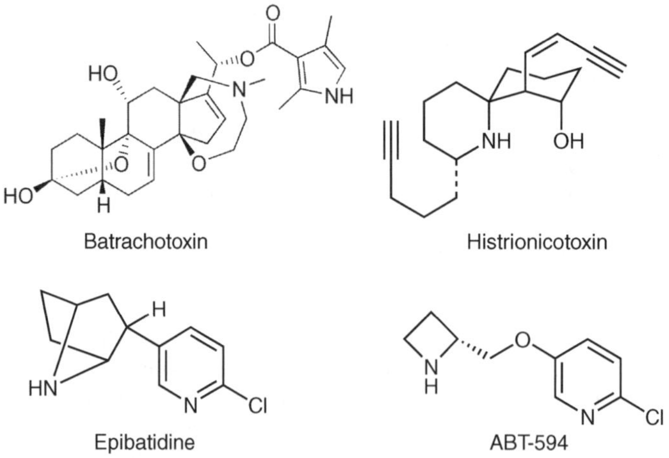

Batrachotoxin

Histrionicotoxin

Epibatidine

ABT-594

Figure 11. Alkaloids and related structures.

structures present are the batrachotoxins, pumiliotoxins, histrionicotoxins (Fig. 11), gephyrotoxins, and decahydroquinolines that target both voltage- and ligand-gated ion channels. The alkaloid epibatidine (Fig. 11), present in trace amounts in *Epipedobates tricolor,* is of particular interest because it has powerful analgesic activities, being 200 times more potent than morphine. The alkaloid is a potent neuronal nicotinic channel agonist selective for the $\alpha_4\beta_2$ subtype. Isolated by Daly and Myers in 1974 [270], the structure of epibatidine (Fig. 11) was not determined until 1992. The discovery of epibatidine as a novel and potent analgesic led to the identification of ABT-594 (see Ref. [271]; Fig. 11), which had equivalent analgesic efficacy to epibatidine but with reduced, albeit different, side-effect liabilities.

16.2.2. Pharmacophore-Based Ligand Libraries

The majority of remaining pharmaceutical companies have relatively large chemical libraries representing the cumulative synthetic efforts covering several decades of medicinal chemists within the company. Typically, the chemical diversity in these libraries is limited as the synthetic approach to drug design revolves around defined pharmacophores in lead series and the rational and systematic development of the SAR. Companies will thus have a large number of similar compounds based on the approaches and successes attendant to a therapeutic area.

The systematic modification of existing structures, both natural and synthetic, is an approach to compound optimization with improvements in potency, selectivity, efficacy, and pharmacokinetics being linked to discrete changes in molecular constituents on the basic pharmacophore.

Angiotensin converting enzyme (ACE) inhibitors are important cardiovascular drugs that block the conversion of angiotensin I, formed by the action of renin on substrate angiotensinogen, to angiotensin II that was a powerful pressor and growth factor agent. Until 1973, peptide inhibitors from the venom of the Brazilian viper were the only known inhibitors of this enzyme. The nonapeptide, teprotide, was an orally active and competitive inhibitor of ACE [272]. Benzylsuccinic acid, a potent inhibitor of carboxypeptidase A, an enzyme with structural and mechanistic similarities to ACE [273] led Ondetti to select *N*-succinyl-l-proline as a lead ($IC_{50} = 330$ pM). This was subsequently optimized (Fig. 12) based on the presence of a Zn^{2+} in the active site of both ACE and carboxypeptidase and the likely presence of hydrophobic pockets. An SH group to coordinate Zn^{2+} was incorporated into the stereoselectively active methyl analog, which eventually led to captopril and the analogs, enalapril, cilazapril, and lisinopril (Fig. 12).

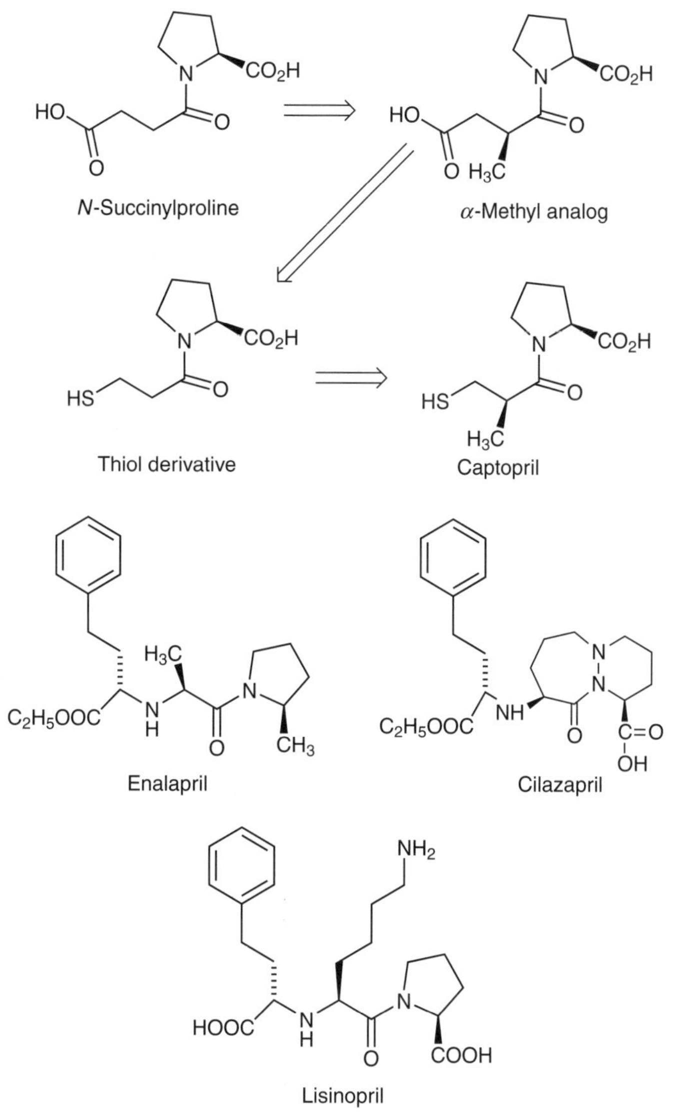

N-Succinylproline

α-Methyl analog

Thiol derivative

Captopril

Enalapril

Cilazapril

Lisinopril

Figure 12. ACE Inhibitors.

16.2.3. Privileged Pharmacophores It is increasingly apparent that a small number of common structures "basic pharmacophores," "templates," or "scaffolds"—are associated with a multiplicity of diverse biological activities at mammalian drug targets and thus represent facile starting points for synthetic approaches to ligand diversity [274].

Benzodiazepines The BZs are well established as anxiolytics, hypnotics, and muscle relaxants as represented by diazepam, clonazepam, midazolam, and triazolam (Fig. 13). The BZ nucleus also occurs in natural products. Asperlicin [259] is a naturally occurring ligand that is a weak, albeit selective, antagonist at cholecystokinin receptors and contains a BZ nucleus. A series of potent and selective BZ ligands, active at CCK_1 and CCK_2 receptors, for example, L-364,718 (Fig. 13) were derived from this lead [275]. Other BZs

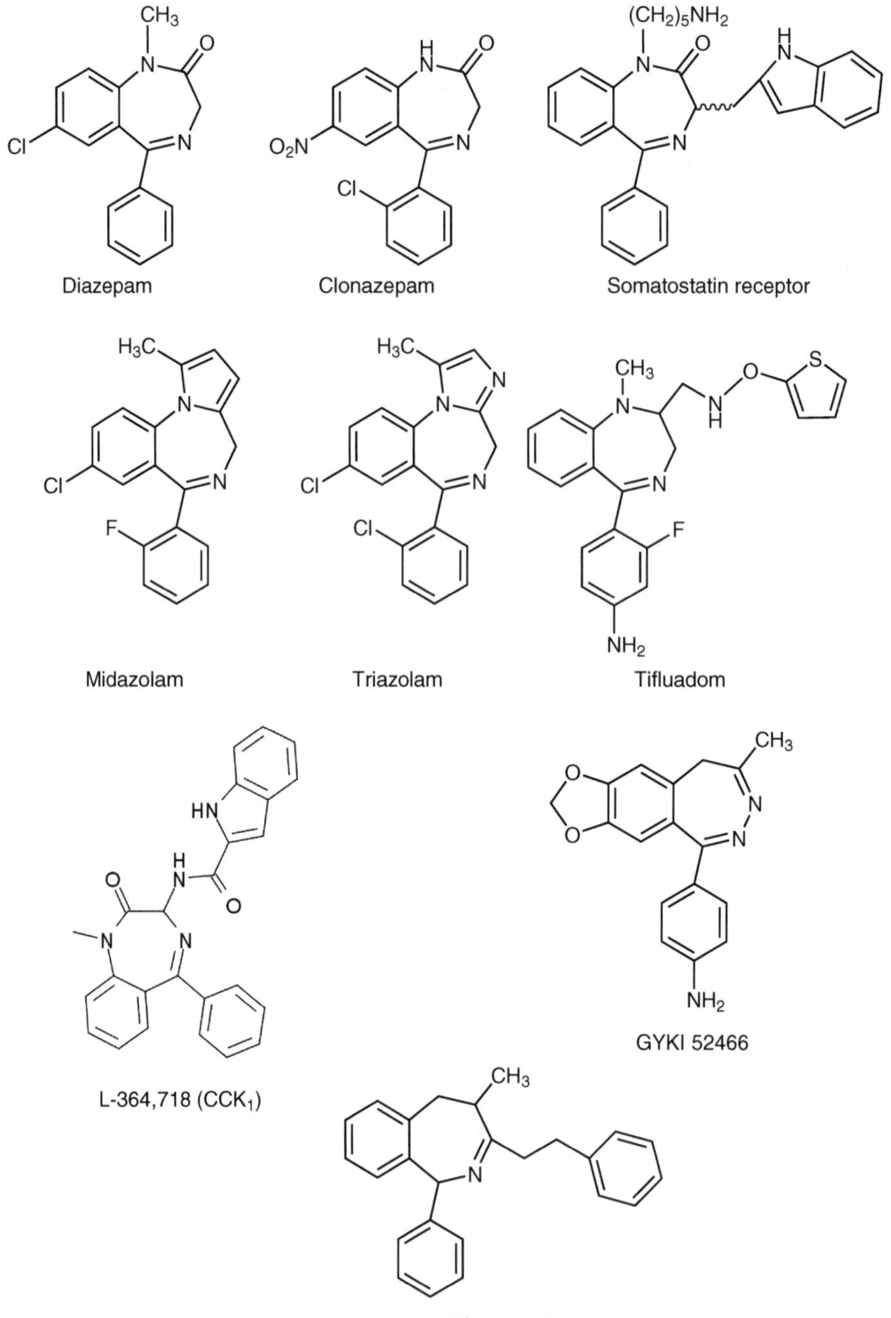

Diazepam

Clonazepam

Somatostatin receptor

Midazolam

Triazolam

Tifluadom

GYKI 52466

L-364,718 (CCK_1)

K^+ channel

Figure 13. Benzodiazepines: anxiolytics, hypnotics, and NCEs with activity at other targets.

(Fig. 13) with activity at receptors distinct from the BZ receptor are: tifluadom (opioid receptor); somatostatin; GYKI 52466 (glutamate receptor); and inward-rectifying potassium channels. This led Ben Evans to describe the BZ pharmacophore as a privileged structure [275], a term that is now in common use [274,276,277].

1,4-Dihydropyridines The DHP pharmacophore is another well-established scaffold [278]. Nifedipine (Fig. 8) and several related DHPs, including amlodipine, felodipine, nicardipine, nimodipine, and nisoldipine, are well-established antihypertensive and vasodilating agents that act through voltage-gated L-type Ca^{2+} channels in the vasculature. However, DHPs also interact with lower activity at other classes of Ca^{2+} channels, including N- and T-type channels, and "leak" channels. DHPs can block delayed rectifier K^+ channels and cardiac Na^+ channels. Other DHP analogs (Fig. 8) have activity at PAF receptors (UK 74,505), K^+ ATP channels (ZM 24405,A-278367), capacitative SOC channels (MRS 1845), and α_{1A}-adrenoceptors (SNAP 5089).

The *2-arylindole* scaffold represents a newer privileged scaffold that has resulted in antagonists for several Family A GPCRs including NPY_5, NK1, CCR_3 CCR5 5-HT_{2A}, 5-HT_6, and SST_4 (Fig. 14; see Ref. [279]).

16.2.4. Diversity-Based Ligand Libraries The issue of diversity reflects the need to enhance the scope of the library beyond ligands already well known in a company. This can be done by compound exchange with other companies, by acquiring compounds from university departments and commercial sources (Sigma, Asinex, Bader, Mayhew, Cookson, ChemBridge, etc.) that have been added to by many newer companies in China and India. From a historical perspective, many commercial sources of novel synthetic compounds included the libraries of former chemical/agricultural companies, for example, Eastman Kodak, Stauffer, FMC, and Shell) and NCEs made in the former Eastern Block countries. More recently, academic screening laboratories and the MLI represent early hit sources. Many of these compounds are, however, nonproprietary and are brokered to many companies tending to represent an IP challenging. In conjunction with the use of computerized cluster programs, the selection of compounds based on diverse structures can be considerably enhanced to provide maximum coverage of molecular space [211,280]. This can be done by generating diversity libraries that have been defined [281] as including "a degree of dissimilarity within a set of chemical structures." This definition has been disputed in the absence of criteria in additional to structure, for example, biological data such that "a relevant diversity of chemical structures, *per se*, does not exist." [281]. Nonetheless, libraries ranging in size from 2000 to 14,000 compounds can be used to rapidly identify potential leads for a new drug target using the SAR generated in an HTS assay. Such libraries can be assembled from compounds that are available in relatively large supply and may not necessarily be proprietary to the company. Their value is in rapidly eliminating unlikely structures in a systematic manner for each new target.

With the advent of combinatorial chemistry in the 1990s, the means to synthesize literally billions of molecules conceptually represented a major step forward in the exploration of "molecular space." However, combinatorial chemistry has tended to sacrifice quantity for quality [282] because what was made was what could be made in a synthetically facile manner and contributed to the "rich-get-richer" scenario of Lipkus et al. [213] where synthetic access to a pharmacophore guaranteed its use to the exclusion of more difficult synthetic structures. However, the concept of the combinatorial approach led to the approach of parallel compound synthesis [283] with exploration around lead compounds or pharmacophores to generate dedicated libraries, when coupled with HTS, clearly provides an economical and more efficient way to rapidly generate diversity in lead compounds. Such methods are likely to be more extensively used in the future to provide a highly focused combinatorial approach to generating molecular diversity.

16.3. Biologicals and Antisense

The value of naturally occurring hormones as effective drugs is not new as evidenced by the use of insulin and epinephrine. The techniques of molecular biology allow the production of an increasing number of native and modified hormones and their soluble forms. Erythropoietin (EPO) is a classic example of a successful drug with the human body as its origin [284].

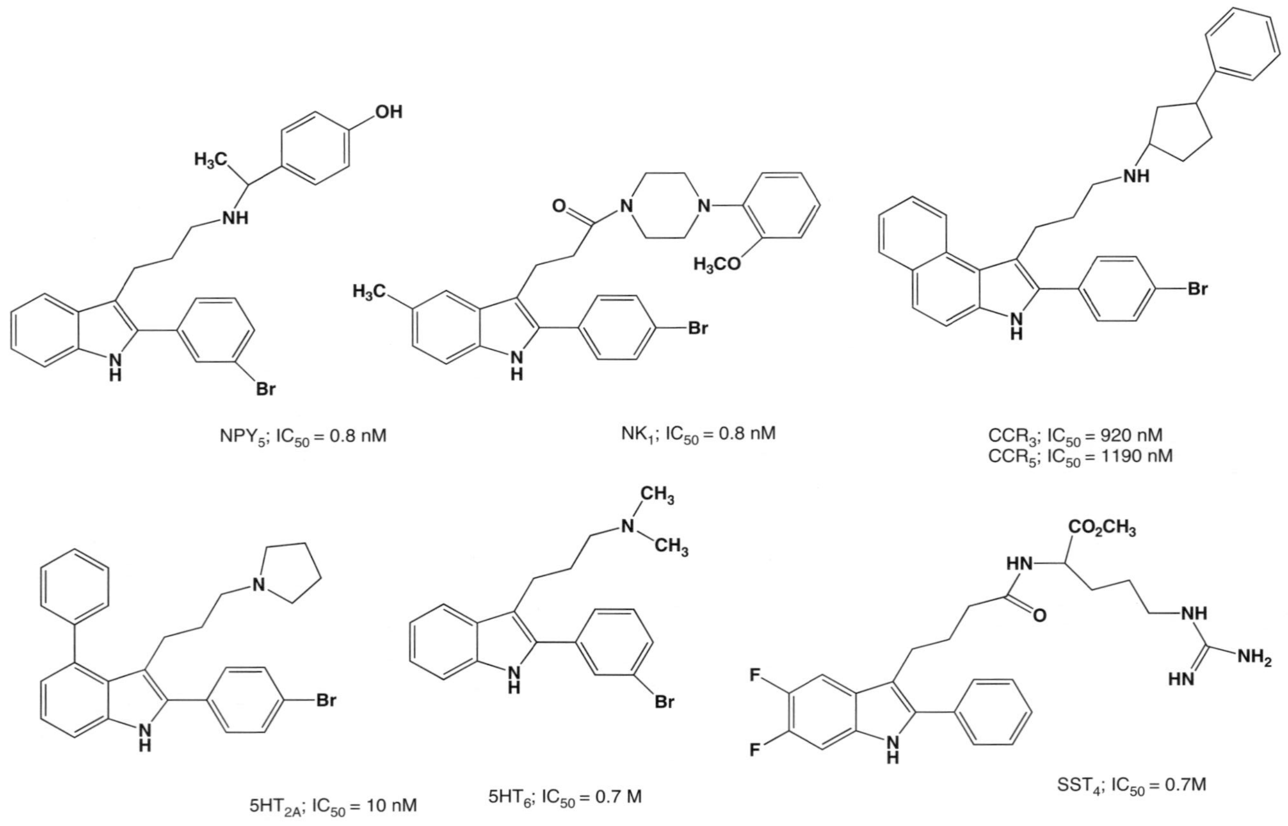

NPY$_5$; IC$_{50}$ = 0.8 nM

NK$_1$; IC$_{50}$ = 0.8 nM

CCR$_3$; IC$_{50}$ = 920 nM
CCR$_5$; IC$_{50}$ = 1190 nM

5HT$_{2A}$; IC$_{50}$ = 10 nM

5HT$_6$; IC$_{50}$ = 0.7 M

SST$_4$; IC$_{50}$ = 0.7M

Figure 14. The *2-arylindole* scaffold and activity at GPCRs.

While there was a major skepticism as to whether antibodies would become useful drugs due to their potential for immunogenicity, monoclonal therapeutic antibodies including rituximab, trastuzumab, cetuximab, and bevacizumab are highly effective treatments for cancer [285].

Similarly, chimeric monoclonal antibodies such as the soluble TNFα decoy receptors infliximab, etanercept, and adalimumab are additional examples of how cloning has altered the concept of rational drug design and what can now be considered as a drug [286]. Anticytokine therapies have been particularly amenable to this approach. However, as noted immunogenicity remains a problem and there is considerable interest in alternative approaches [287]. These include antisense oligodeoxynucleotides (ASOs; see Ref. [288]), aptamers [287] and RNA interference (RNAi; see Ref. [163]) that can modulate gene expression and that can address hitherto nondruggable protein interaction targets. Short, 21–25 nucleotide siRNAs can silence mammalian genes in a highly specific, sequence-dependent manner [289] leading to their use to generate cells, tissues, and animals with reduced targeted gene expression. siRNA is currently been investigated as a therapeutic approach to viral infection, neuropathic pain, autoimmune/inflammatory diseases targeting TNFα, apoptosis and death receptor based mechanisms, a variety of cancers, and metabolic and neurodegenerative disease disorders [163]. siRNA drugs in development include: bevasiranib, an siRNA developed to suppress expression of VEGF for the treatment of wet age-related macular degeneration and diabetic macular edema; ALN-VSP that targets two genes, kinesin spindle protein and VEGF that involve in the pathways involved in the genesis of liver cancer; and ALN-PCS that targets PCSK9 for the treatment of hypercholesterolemia. RNAi against survivin, which is markedly increased in tumors, can result in tumor suppression. Both ASOs and siRNA have challenges in their drug-like characteristics that include target delivery, off-target activity, potential immunoreactivity and stability. The latter has been addressed chemically by sugar modification and the selective introduction of phosphorothioate backbones, techniques that have been successfully used in antisense and aptamers. Aptamer-, antibody-, and cholesterol-conjugated siRNAs have also been investigated as improved delivery modes [290,291]. The long and tortuous path to developing ASOs as drugs [288] has provided a technology base that may ensure that siRNA drugs prove more amenable to use in human disease as novel therapeutics.

17. FUTURE DIRECTIONS

The gradual evolution of the receptor concept as the basis for drug discovery over the past century has led to major advances in the understanding of basic biological systems and human disease states. It is debatable, however, as whether this increased knowledge that has resulted in an exponential increase in defining the complexities of drug targets, their interactions and signaling pathways has aided or actually hindered the drug discovery process. The elegance of the lock and key concept of the receptor–ligand interaction has given way to an ever more diffuse focus on accessory proteins, alternate signaling pathways, and so on that removes the exquisite specificity imparted by the target "lock" further downstream making specificity an ever more elusive outcome. Thus rather than seeking to understand why some drug targets, for example, the histamine H_3 receptor can yield ligands that have exquisite selectivity in their interactions with target while others, for example, muscarinics and α-adrenoceptors, are notorious for their promiscuity, the strategy is to delve ever more deeply into tangential targets that lack the ligand recognition properties unique to the target proper. In many respects, this is similar to the diminishing absolutes around the human genome as the origin for all human disease that gave way, in turn to epigenetics [292] with copy number variants now appearing to be the causal agents in disease inheritance [293].

While the development of ever more sophisticated structure-based technologies and ultra, micro-HTS systems that encompass activity, ADME, toxicity, and structural data generation has resulted in an ever increasing body of knowledge, it does not appear to have added significantly to success as measured in increased numbers of quality INDs (or NDAs) but rather, together with genomics, proteomic, and other "omic" progeny, added, very significantly to the cost of the search for new medicines.

The medicinal chemist and the pharmacologist play critical roles in integrating and interpreting the data flow that constitutes much of the day-to-day workings of the drug discovery environment. Their challenge is to provide the appropriate intellectual framework to effectively use this data to find drugs rather than play a technology-driven numbers game, strong on metrics but short on outcome [25]. A situation with many similarities, as noted some 15 years ago [294] was the compact disk, a technological marvel that has replaced fragile magnetic tape and vinyl as a consumer recording media. There was/is limited evidence that this digital media (or its successor, mp3), however, much oversampled, or upsampled, or compressed has improved the quality, innovation, or longevity of the recorded music contained in the medium. Similarly, burgeoning technologies have increased the amount of information related to the receptor–ligand interaction and its downstream consequences by hundreds of fold yet, in the absence of the appropriate intellectual framework [295] and an appreciation of the human disease state as distinct from a microtiter plate [296], the results of which while well intended are disappointingly predictable.

ACKNOWLEDGMENTS

The authors would like to thank David Triggle for his contributions to previous versions of this chapter and for his continued erudite contributions to the strategy of medicinal chemistry in drug discovery, Paul Anderson for his continued advocacy of the chemistry/pharmacology interface, Bruce Dorsey for his helpful comments and the late David Horrabin for his seminal reflections on biotechnology hype.

REFERENCES

1. Changeux J-P, Edelstein SJ. Nicotinic Acetylcholine Receptors. From Molecular Biology to Cognition. New York: Odile Jacob; 2005. p 29–50.
2. Rang HP. Br J Pharmacol 2006;147:S9–S16.
3. Kenakin T. A Pharmacology Primer. 2nd ed. Amsterdam: Elsevier; 2006.
4. Parascandola J. Pharm Hist 1974;16:54–63.
5. Maehle A-H, Prüll C-R, Halliwell RF. Nat Rev Drug Discov 2002;1:637–641.
6. Clark AJ. The Mode of Action of Drugs on Cells. London: Arnold; 1933.
7. Gaddum JH. Pharmacol Rev 1957;9:211–218.
8. Arunlakshana O, Schild HO. Br J Pharmacol 1959;14:48–58.
9. Ariens EJ. Arch Int Pharmacodyn Ther 1954;99:32–49.
10. Stephenson RP. Br J Pharmacol 1956;11: 379–392.
11. Ahlquist RP. Am J Physiol 1948;155:586–600.
12. Bylund DB. Am J Physiol Endocrinol Metab 2007;293:E1479–E1481.
13. Black JW. Science 1989;245:486–493.
14. Humphrey PPA. Headache 2007;47(Suppl 1): 10–19.
15. Conn PJ, Christopoulos A, Lindsley CW. Nat Rev Drug Discov 2009;8:41–54.
16. Copeland RA, Pompilano DL, Meek TD. Nat Rev Drug Discov 2007;5:730–739.
17. Tummino PJ, Copeland RA. Biochem 2008;47:5481–5489.
18. Cozzini P, Kellogg GE, Spyrakis F, et al. J Med Chem 2008;51:6237–6255.
19. Chast F. A history of drug discovery. In: Wermuth CG, editor. The Practice of Medicinal Chemistry. 3rd ed. Burlington, MA: Academic Press; 2008. p 3–62.
20. International Human Genome Sequencing Consortium. Nature 2004;431:931–945.
21. Sawyers CL. Genes Dev 2003;17:2998–3010.
22. Cohen P. Nat Rev Drug Discov 2002;1: 309–315.
23. Roth BL, Sheffler DJ, Kroeze WK, Nat Rev Drug Discov 2004;3:353–359.
24. Hayden EC. Nature 2008;455:148.
25. Enna SJ, Williams M. J Pharmacol Exp Ther 2009; 329:404–411.
26. Kopec KK, Bozyczko-Coyne D, Williams M. Biochem Pharmacol 2005;69:1133–1139.
27. Fitzgerald GA. Nat Rev Drug Discov 2005.4:815–818.
28. Kola Landis J. Nat Rev Drug Discov 2004;3: 711–716.

29. FDA 2004; www.fda.gov/oc/initiatives/criticalpath/whitepaper.html.
30. Hughes B. Nat Rev Drug Discov 2008;7: 107–109.
31. Abdel-Rahman SM, Kauffman RE. Ann Rev Pharmacol Toxicol 2004;44:111–136.
32. Pritchard MF, Jurima-Romet M, Reimer MLJ, et al. Nat Rev Drug Discov 2003;2:542–553.
33. Sneader W. Drug Discovery: A History. Chichester, UK: Wiley; 2005.
34. Koshland DE. Proc Natl Acad Sci USA 1958; 44:98.
35. Michaelis L, Menten ML. Biochem Z 1913;49: 333–369.
36. Monod J, Wyman J, Changeux J-P. J Mol Biol 1965;12:88–118.
37. Freedman SB, Harley EA, Patel S, et al. Br J Pharmacol 1990;101:575–580.
38. Schraven B, Kalinke U. Immunity 2008;28: 591–598.
39. Furchgott RF. Ann Rev Pharmacol 1964;4:21–38.
40. Paton W. Proc R Soc Lon B Biol Sci 1961;154: 21–69.
41. Limbird L. Cell Surface Receptors. Boston, MA: Nijhoff; 1986.
42. Cuatrecasas P. Ann Rev Biochem 1974;43: 169–214.
43. DeLean A, Stadel JM, Lefkowitz RJ. J Biol Chem 1980;255:7108–7117.
44. Christopoulos A. Nat Rev Drug Discov 2002;1:198–210.
45. Kenakin TP. Pharmacol Rev 1996;48:413–463.
46. Kenakin T. Nat Rev Drug Discov 2002;1: 103–110.
47. Onaran HO, Scheer A, Cotecchia S, Costa T. Handb Exp Phamacol 2000;148:217–280.
48. Frauenfelder H, Sigar SG, Wolynes PG. Science 1991;254:1598–1603.
49. Samama P, Cotecchia S, Costa T, Lefkowitz RJ. J Biol Chem 1993;268:4625–4636.
50. Hill AV. J Physiol 1909;39:361–373.
51. Koshland DE, Nemethy G, Filmer D. Biochemistry 1966;5:365–385.
52. Ehlert FJ. Mol Pharmacol 1988;33:187–194.
53. O'Brien JA, Lemaire W, Chang TB, et al. Mol Pharmacol 2003;64:731–740.
54. Hall DA. Mol Pharmacol 2000;58:1412–1423.
55. May LT, Leach K, Sexton PM, Christopoulos A. Ann Rev Pharm Toxicol 2007;41:1–51.
56. Dean MK, Higgs C, Smith RE, et al. J Med Chem 2001;44:4595–4614.
57. Mohler H, Fristchy JM, Rudolph U. J Pharmacol Exp Ther 2002;300:2–8.
58. Krogsgaard-Larsen P, Frølund B, Liljefors T, Ebert B. Biochem Pharmacol 2004;68: 1573–1580.
59. Le Noverre N, Changeux JP. J Mol Evol 1995; 40:155–172.
60. Sharpe A, Gehrmann J, Loughnan ML, et al. Nat Neurosci 2001;4:902–907.
61. Bogoyevitch MA, Fairlie DP. Drug Disc Today 2007;12:622–633.
62. Schwartz TW, Holst B. Trends Pharmacol Sci 2007;28:366–73.
63. Milligan G. J Cell Sci 2001;114:1265–1271.
64. Margreta-Mitovic M, Jan YN, Jan LY. Proc Natl Acad Sci USA 2001;98:14643–14648.
65. Rocheville M, Lang DC, Kumar U, et al. Science 2000;288:154–157.
66. Canals M, Marcellino D, Fanelli F, et al. J Biol Chem 2003;278:46741–46749.
67. Jordan BA, Trapadize N, Gomes I, et al. Proc Natl Acad Sci USA 2001;98:343–348.
68. Coyle J, Tsai G. Psychopharmacology 2004;174: 32–38.
69. Christopoulos A, Christopoulos G, Morfi M, et al. J Biol Chem 2003;278:3293–3297.
70. Conn PM, Ulloa-Aguirre A, Ito J, Janovick JA. Pharmacol Rev 2007;59:225–250.
71. Husi H, Ward MA, Choudary JS, et al. Nat Neurosci 2000;3:661–669.
72. Kim M, Jiang L-H, Wilson HL, et al. EMBO J 2001;20:6347–6358.
73. Samama P, Pei G, Costa T, et al. Mol Pharmacol 1994;45:390–394.
74. Braestrup C, Nielsen M. Benzodiazepine receptor binding *in vivo* and efficacy. In: Olsen RW, Venter JC, editors. Benzodiazepine/GABA Receptors and Chloride Channels: Structure and Functional Properties. New York: Wiley-Liss; 1986. p 167–184.
75. Lutz M, Kenakin TP. Quantitative Molecular Pharmacology and Informatics In Drug Discovery. Chichester, UK: Wiley; 1999.
76. Redfern WS, Carlsson L, Davis AS, et al. Cardiovasc Res 2003;58:32–45.
77. Turner JR, Durham TA. Integrated Cardiac Safety. Hoboken, NJ: Wiley; 2009.
78. Urban JD, Clarke WP, von Zastrow M, et al. J Pharmacol Exp Ther 2007;320:1–13.

79. Wiesinger JA, Buwen JP, Cifelli CJ, et al. J Neurochem 2007;100:167–179.
80. Smit MJ, Leurs R, Alewijnse AE, et al. Proc Natl Acad Sci USA 1996;93:6802–6807.
81. Maschera. JBC 1996;271:33231.
82. Vassilatis DK, Hohmann JG, Zeng H, et al. Proc Natl Acad Sci USA 2003;100:4903–40908.
83. Armbruster BN, Roth BL. J Biol Chem 2005;280:5129–5132.
84. Harmar AJ, Hills RA, Rosser EM, et al. Nucleic Acids Res 2009;35:D680–D685. http://www.iuphar-db.org/index.jsp.
85. Horn F, Bettler E, Oliveira L, et al. Nucleic Acids Res 2003;31:294–297. http://www.gpcr.org/7tm/.
86. Okuno Y, Yang J, Taneishi K, et al. Nucleic Acids Res 2006;34:D673–D677. http://gdds.pharma.kyoto-pharma.kyoto-u.ac.jp:8081/glida.
87. Grassot J, Mouchiroud G, Perrière G. Nucleic Acids Res 2003;31:353–358. http://pbil.univ-lyon1.fr/RTKdb/.
88. Martin DMA, Miranda-Saavedra D, Barton GJ. Nucleic Acids Res 2009;37:D244–D250. http://www.compbio.dundee.ac.uk/kinomer/.
89. Ruau J, Duarte T, Ourjdal G, et al. Nucleic Acids Res 2004;32:D165–D167. http://www.enslyon.fr/LBMC/laudet/nurebase/nurebase.html.
90. Hernandez-Boussard T, Whirl-Carrillo M, Hebert JM, et al. Nucleic Acids Res 2008;36 (Database issue):D913–D918.
91. Klein TE, Chang JT, Cho MK, et al. Pharmacogenomics J 2001;1:167–170.
92. Alexander SPH, Mathie A. Peters JA. *Guide to Receptors and Channels (GRAC)*. 3rd ed; Br J Pharmacol 2008;153 (Suppl 2): S1-S209. http://www.nature.com/bjp/journal/vgrac/ncurrent/index.html.
93. Watling KJ. The Sigma-RBI eHandbook of Receptor Classification and Signal Transduction. 5th ed. 2006. http://www.sigmaaldrich.com/life-science/cell-biology/learning-center/ehandbook.html.
94. Rajaligam K, Dikic I. Biochem J 2009;417: e1–e3.
95. Drews J. Science 2000;287:1960–1964.
96. Hogg RC, Buisson B, Bertrand D. Biochem Pharmacol 2005;70:1267–1276.
97. Catterall WA, Gutman G. Pharmacol Rev 2005;57:385.
98. Kubo Y, Adelman JP, Clapham DE, et al. Pharmacol Rev 2005;57:4 509–526.
99. Dolphin AC. Br J Pharmacol 2006;147: S56–S62.
100. Catterall WA, Goldin AL, Waxman SG. Pharmacol Rev 2005;57(4): 397–409.
101. Lingueglia E. J Biol Chem 2007;282: 17325–17329.
102. Gronemeyer H, Gustafsson JA, Laudet V. Nat Rev Drug Discov 2004;3:950–964.
103. Zhang Z, Burch PE, Cooney AJ, et al. Genome Res 2004;14:580–590.
104. Robinson DR, Wu YM, Lin SF. Oncogene 2000;19:5548–5557.
105. Yamaoka K, Saharinen P, Pes M, et al. Genome Biol 2004;5:253. DOI: 10.1186/gb-2004-5-12-253.
106. Robinson DR, Wu YM, Lin SF. Oncogene 2000;19:5548–5557.
107. Danielson P. Curr Drug Metab 2002;3: 561–597.
108. Attisano L, Wrana JL, Science 2002;296: 1646–1647.
109. Miyazawa K, Shinozaki M, Hara T, et al. Genes Cells 2002;7:1191–1204.
110. Ihle JN, Curr Opin Cell Biol 2001;13:211–217.
111. Elmore S. Toxicol Pathol 2007;35:495–516.
112. Yevenes GE, Peoples RW, Tapia JC, et al. Nat Neurosci 2003;6:819–824.
113. Delcourt N, Bockaert J, Marin P. Trends Pharmacol Sci 2007;28:602–607.
114. Trabanino RJ, Hall SE, Vaidehi N, et al. Biophys J 2004;86:1904–1926.
115. Topiol S, Sabio M. Biochem Pharmacol 2009.
116. Cherezov V, Rosenbaum DM, Hanson MA, et al. Science 2007;318:1258–1265.
117. Warne T, Serrano-Vega MJ, Baker JG, et al. Nature 2008;454:486–491.
118. Jaakola VP, Griffith MT, Hanson MA, et al. Science 2008;322:1211–1217.
119. Fredriksson R, Lagerström MC, Lundin LG, Schiöth HB. Mol Pharmacol 2003;63: 1256–1272.
120. Schiöth HB, Fredriksson R. Gen Comp Endocrinol 2005;142:94–101.
121. Jacoby E, Bouhelal R, Gerspacher M, Seuwen K. Chem Med Chem 2006;1:760–782.
122. Yona S, Lin H-H, Gordon S, Stacey M. Trends Biochem Sci 2008;33:491–500.
123. Bjarnadóttir TK, Fredriksson R, Schiöth HB. Cell Mol Life Sci 2007;64:2104–2119.
124. Kristiansen K. Pharmacol Ther 2004;103: 21–80.

125. Worzfeld T, Wettschureck N, Offermans S. Trends Pharmacol Sci 2008;29:482–589.
126. Cooray SN, Chan L, Webb TR, Metherell L, Clark AJL. Mol Cell Endocrinol 2008. DOI: 10.1016/j.mce.2008.10.004.
127. Presland J. Biochem Soc Trans 2007;32: 888–891.
128. Jeon HY, Heo Y, Kim C, et al. Cell Mol Life Sci 2005;62:1198–220.
129. Mellor H, Parker PJ. Biochem J 1998;332(Pt II): 281–292.
130. Manning G, Whyte DB, Martinez R, et al. Science 2002;298:1912–1934.
131. Sibley DR, Benovic JL, Caron MG, Lefkowitz RJ. Cell 1987;48:913–922.
132. Hupfeld CJ, Olefsky JM. Ann Rev Physiol 2007;69:561–577.
133. Zhang Z-Y. Curr Opin Chem Biol 2001;5: 416–423.
134. Bockaert J, Fagni L, Dumuis A, Marin P. Pharmacol Ther 2004;103:203–221.
135. Neubig RR, Siderovski DP. Nat Rev Drug Discov 2002;1:187–197.
136. Cismowski MJ, Takesono A, Ma C, et al. Nat Biotechnol 1999;17:878–883.
137. Blumer JB, Cismowski MJ, Sato M, Lanier SM. Trends Pharmacol Sci 2005;26:470–476.
138. Xia Z, Storm D. Nat Rev Neurosci 2005;6:267–276.
139. Hay DL, Poyner DR, Sexton PM, Pharmacol Ther 2006;109:173–197.
140. Jordan BA, Devi LA. Nature 1999;399: 1400–1409.
141. Panetta R, Greenwood MT. Drug Discov Today 2008;13:1059–1066.
142. Hiebel A-C, Lee YS, Bilsky E, et al. J Med Chem 2007;50:3765–3776.
143. Szidonya L, Cserzo M, Hunyady L. J Endocrinol 2008;196:435–453.
144. Clayton T, Chen JL, Ernst MM, et al. Curr Med Chem 2007;14:2755–2775.
145. Petroski RE, Pomeroy JE, Das R, et al. J Pharmacol Exp Ther 2006;317:369–377.
146. Sternfeld F, Carling RW, Jelley RA, et al. J Med Chem 2004;47:2176–2179.
147. Gotti C, Moretti M, Giamarri A, et al. Biochem Pharmacol 2007;74:1102–1111.
148. Jorenby DE, Hays JT, Rigotti NA, et al. JAMA 2006;296:56–63.
149. Hurst RS, Hajos M, Raggenbass M, et al. J Neurosci 2005;25:4396–4405.
150. Timmermann DB, Grønlien JH, Kohlhaas KL, et al. J Pharmacol Exp Ther 2007;323: 294–307.
151. Jacobson KA, Jarvis MF, Williams M. J Med Chem 2002;45:4057–4093.
152. Evans RJ. Eur Biophys J 2008. DOI: 10.1007/ s00249-008-02.75-2.
153. Cheng W, Altafaj X, Ponjat M, Coronado R. Proc Natl Acad Sci USA 2005;102: 19225–19230.
154. Clapham DE, Runnels LW, Strubing C. Nat Rev Neurosci 2001;2:387–396.
155. Evans RM. Science 1988;240:889–895.
156. Jordan VC. B J Pharmacol 2006;47(Suppl 1): S269–S276.
157. Riggs BL, Hartmann LC. N Engl J Med 2003;348:618–629.
158. Shang Y, Brown M. Science 2002;295: 2465–2468.
159. Stuehr DJ. Ann Rev Pharmacol Toxicol 1997;37:339–359.
160. Riedl SJ, Shi Y. Nat Rev Mol Cell Biol 2004;5:897–907.
161. Benoit G, Cooney A, Giguere V, et al. Pharmacol Rev 2006;58:798–836.
162. Xavier KA, Eder PS, Giordano T. Trends Biotech 2000;18:349–356.
163. deFougerolles A, Vornlocher H-P, Maraganore J, Lieberman J. Nat Rev Drug Disc 2007;6:443–453.
164. Moutoussamy S, Kelly 1 PA, Finidor J. Eur J Biochem 2001;255:1–11.
165. Leung K-C. DNA Cell Biol 2004;23:463–474.
166. Civelli O, Saito Y, Wang Z, et al. Pharmacol Ther 2005;110:525–532.
167. Chambers C, Smith F, Williams C, et al. Comb Chem High Throughput Screen 2003;6: 355–362.
168. Dupriez VJ, Maes K, Le Poul E, et al. Recept Channels 2002;8:319–330.
169. Jenck W, Wichmann J, Dautzenberg FM, et al. Proc Natl Acad Sci USA 2000;97:4938–4943.
170. Hahn MK, Blakely RD. Ann Rev Pharmacol Toxicol 2007;47:401–441.
171. Broer S. Neurochem Inter 2006;48:559–567.
172. Wong DT, Perry KW, Bymaster FP, Nat Rev Drug Discov 2005;4:764–774.
173. Spiess J, Rivier J, Rivier C, Vale W. Proc Natl Acad Sci USA 1981;78:6517–6521.
174. Smit NI, Syed D, Schaap, et al. Nature 2001;411:261–268.

175. Mackie K. Ann Rev Pharmacol Toxicol 2006;46:101–122.
176. Montell C, Birnbaumer L, Flocerzi V. Cell 2002;108:595–598.
177. Waldhoer M, Bartlett SE, Whistler JL. Ann Rev Biochem 2004;73:953–990.
178. Balasubramanian S, Xia Y, Freinkman E, Gerstein M. Nucleic Acids Res 2005;33: 1710–1721.
179. Eglen RM, Bosse R, Reisine T. Assay Drug Dev Technol 2007;5:425–451.
180. Keramidas A, Moorhouse AJ, French CR, et al. Biophysical J 2000;79:247–259.
181. Jordan N, Williams N, Gregory JW, et al. J Clin Endocrinol Metab 2003;88:1002–1005.
182. Carrasquillo MM, McCallion AS, Puffenberger EG, et al. Nat Genet 2002;32:237–244.
183. Wysolmerski JJ, Cormier S, Philbrick WM, et al. J Clin Endocrinol Metab 2001;86: 1487–1488.
184. Schulz A, Sangkuhl K, Lennert T, et al. J Clin Endocrinol Metab 2002;87:5247–5257.
185. Liu R, Paxton WA, Choe S, et al. Cell 1996;86:367–377.
186. Bernier V, Lagace M, Lorengan M, et al. Mol Endocrinol 2004;18:2074–2084.
187. Dryja TP, McEvoy JA, McGee TL, Berson EL. Invest Ophthalmol Vis Sci 2000;41: 3124–3127.
188. Calvi LM, Schipani E. J Enodocrinol Invest 2000;23:545–554.
189. Corvilain B, Van Sande J, Dumont JE, Vassert G. Clin Endocrinol 2001;55:143–158.
190. Themmen AP, Verhoef-Post M. Semin Reprod Med 2002;20:199–204.
191. Hu J, Spiegel AM. Trends Endocrinol Metab 2003;14:282–288.
192. Wang JF, Chen B, Young LT. Mol Brain Res 1999;70:66–73.
193. Jones SJM. Ann Rev Genomics Hum Genetics 2006;7:315–338.
194. Futschik ME, Chaurasia G, Herzel H. Bioinformatics 2007;23:605–611.
195. Fields S, Song OK. Nature 1989;340:245–246.
196. Einarson MB, Orlinick JR. In Protein–Protein Interactions: A Molecular Cloning Manual. Cold Spring Harbor, NY: Cold Spring Harbor Laboratory Press; 2002. p 37–57.
197. Dean NM. Curr Opin Biotechnol 2001;12: 622–625.
198. Honore P, Kage K, Mikusa J, et al. Pain 2002;99:19–27.
199. Maher BA. Scientist 2002;16:22–24.
200. Chiang SL. High Throughput-Screening in Drug Discovery. In: Huser J, editor. Weinheim, Germany: Wiley-VCH; 2006. p 1–13.
201. Williams M. Biochem Pharmacol 2005;70: 1707–1716.
202. Hill RG. Trends Pharmacol Sci 2000;21: 244–246.
203. Kramer MS, Cutler N, Feighner J, et al. Science 1998;281:1640–1645.
204. George DT, Gilman J, Hersh J, et al. Science 2008;319:1536–1539.
205. Sawa A, Snyder SH. Science 2002;296: 692–695.
206. Marino MJ, Knutsen LJS, Williams M. J Med Chem 2008;51:1077–1107.
207. Lakhan SE, Kramer A. Behav Brain Funct 2009;5:2. DOI: 10.1186/1744-9081-5-2.
208. Brzustowicz LM, Hodgkinson KA, Chow WC, et al. Science 2000.288:678–682.
209. Levinson DF, Holmans PA, Laurent C, et al. Science 2002;296:739–741.
210. Mimmack ML, Ryan M, Baba H, et al. Proc Natl Acad Sci USA 2002;99:4680–4685.
211. Triggle DJ, Biochem Pharmacol 2009; 78:217–223.
212. Dobson CM. Nature 2004;432:824–198.
213. Lipkus A, Bartelt W, Funk S. J Org Chem 2008;73:4443–4451.
214. Bemis GW, Murcko MA. J Med Chem 1996;39: 2887–2893.
215. Bemis GW, Murcko MA. J Med Chem 1999;42:5095–5099.
216. Hong CR, Sullivan DJ, Jr. Nature 2007;448: 645–646.
217. Kaiser J. Science 2008;321:764–766.
218. Erlanson DA, McDowell RS, O'Brien T. J Med Chem 2004;47:3463–3482.
219. Saxty G, Woodhead SJ, Berdini V, et al. J Med Chem 2007;50:2293–2296.
220. Congreve M, Chessari G, Tisi D, Woodhead AJ. J Med Chem 2008;51:3661–3680.
221. Aronov M, Murcko MA. J Med Chem 2004;47:5616–5619.
222. Kolb HC, Sharpless BK. Drug Discov Today 2003;8:1128–1137.
223. Patani GA, LaVoie EJ. Chem Rev 1996;96: 3147–3176.
224. Li AP. Drug Discov Today 2001;6:357–366.
225. Perkins R, Fang H, Tong W, Welsh WJ. Environ Toxicol Chem 2003;22:1666–1679.

226. Van Drie JH. J Comput Aided Mol Des 2007;21:591–601.
227. Schneider G, Baringhaus K-H. Molecular Design: Concepts and Applications. Weinheim: Wiley-VCH; 2008.
228. Abou-Gharbia M. J Med Chem 2009;52:2–9.
229. Ghose AK, Herbertz T, Pippin DA, Salvino JM, Mallamo JP. J Med Chem, 2008;51: 5149–5171.
230. Freire E. Drug Discov Today 2008;13:869–874.
231. Bajorath J. Nat Rev Drug Discov 2002;1: 882–894.
232. Kerns EH, Di L. Comb Chem High Throughput Screening 2005;8:459–466.
233. Blundell T, Patel S. Curr Opin Pharmacol 2004;4:490–496.
234. Snyder SH. J Med Chem 1983;26:1667–1672
235. Rang HP. Nature 1971;231:91–96.
236. Gavln JR, III, Roth J, NevIlle DM, et al. Proc Nat Acad Sci USA 1974;71:84–88.
237. Cuatrecasas P. Drug Intell Clin Pharm 1983;17:357–366.
238. Cheng Y-C, Prusoff WC. Biochem Pharmacol 1972;22:3099–3108.
239. Owicki JC. J Biomol Screen 2000;5:297–306.
240. Hovius R, Vallottone P, Wohland T, Vogel H. Trends Pharmacol Sci 2000;21:266–273.
241. Handl HL, Gillies RJ. Life Sci 2005;77: 361–371.
242. Auer M, Moore KJ, Meyer-Almes FJ, et al. Drug Discov Today 1998;3:457–465.
243. Bieri C, Ernst OP, Heyse S, et al. Nat Biotech 1999;17:1105–1108.
244. Mathes C. Expert Opin Ther Targets 10:234–241.
245. Bristow LJ, Kramer MS, Kulagowski J, et al. Trends Pharmacol Sci 1997;18: 186–188.
246. Hodgson A. Nat Biotech 2001;19:722–726.
247. Saunders KC. Drug Discov Today Technol 2004;1:373–380.
248. Kassel DB. Curr Opin Chem Biol 2004;8: 339–345.
249. Smith NF, Raynaud FI, Workman P. Mol Cancer Ther 2007;6:428–440.
250. Lipinski CA, Lomardo F, Dominy BW, Feeney PJ. Adv Drug Deliv Rev 2001;46:3–26.
251. Jarvis LM. Chem Eng News 2007;85(45): 18.
252. Veber DF, Johnson SR, Cheng HY, et al. J Med Chem 2002;45:2615–2613.
253. Singh SS. Curr Drug Metab 2006:7:165–182.
254. Ekins S, Rose J. J Mol Graph Model 2002;20: 305–309.
255. Miller MA. Nat Rev Drug Discov 2002;1: 220–227.
256. Williams AJ. Curr Opin Drug Discov Develop 2008;11:393–404.
257. Kaiser J. Science 2005;308:774.
258. O'Neil MJ, Heckelman PE, Koch CB, Roman KJ. Merck Index. 14th ed. Whitehouse Station, NJ: Merck Research Laboratories; 2006.
259. Chang RS, Lotti VJ, Monaghan RL, et al. Science 1985;230:177–179.
260. Koehn FE, Carter GT. Nat Rev Drug Discov 2005;4:206–220.
261. Huser J. High Throughput-Screening in Drug Discovery. Wiley-VCH; Weinheim, Germany: 2006.
262. Inglese J, Auld DS, Jadhav A, et al. Proc Natl Acad Sci USA 2006;103:11473–11478.
263. Michael S, Auld D, Klumpp C, et al. ASSAY Drug Dev Technol 2008;6:637–658.
264. Riddle JM, Estes JX. Ant Sci 1992;80:226–234.
265. Bull AT, Goodfellow M, Slater JH. Ann Rev Microbiol 1992;46:219–252.
266. Fusetani N. Drugs From the Sea. Basel: S. Karger; 2000.
267. Triggle DJ. Med Chem Res 2004;13:315–324.
268. Olivera DM. J Biol Chem 2006;281: 31173–31177.
269. Wallace MS, Rauck R, Fisher R, et al. Anesth Analg 2006;106:628–637.
270. Garraffo HM, Spande TF, Williams M. Heterocycles 2009;79: in press.
271. Bannon AW, Decker MW, Holladay MW, et al. Science 1998;279:77–81.
272. Ondetti M. Ann Rev Pharmacol Toxicol 1994;34:1–16.
273. Byers LD, Wolfenden R. Biochem 1973;12: 2070–1078.
274. Langer T, Hoffmann RD, Mannhold R. Pharmacophores and Pharmacophore Searches. Wiley-VCH; 2006.
275. Evans BE, Rittle KE, Bock MG, et al. J Med Chem 1988;31:2235–2246.
276. Patchett AA, Nargund RP. Ann Rep Med Chem 2000;36:289–298.
277. Bondensgaard K, Ankersen M, Thøgersen H, et al. J Med Chem 2004;47:888–899.
278. Janis RA, Triggle DJ. J Med Chem 1983;26:775–785.
279. Willoughby CA, Hutchins SM, Rosauer KG, et al. Bioorg Med Chem Lett 2002;12:93–96.

280. Bleicher KH, Böhm H-J, Müller K, Alanine AI. Nat Rev Drug Discov 2003;2:369–378.
281. Roth HJ. Curr Opin Chem Biol 2005;9: 293–295.
282. Leeson PD, Springthorpe B. Nat Rev Drug Discov 2007;6:881–890.
283. Selway CN, Terrett NK. Biorg Med Chem 1996;4:645–654.
284. Jelkmann W. Eur J Haematol 2007;78: 183–205.
285. Carter P. Nat Rev Cancer 2001;1:118–129.
286. Breedveld FC. Lancet 2000;355:735–740.
287. De Souza EB, Cload ST, Pendergrast PS, Sah DWY. Neuropsychopharmacol Rev 2009;34: 142–158.
288. Phillips MI. Antisense Therapeutics Totowa, NJ: Humana; 2004.
289. Dillon CP, Sandy P, Nencioni A, et al. Ann Rev Physiol 2005;67:147–173.
290. Behlke MA. Oligonucleotides 2008;18: 305–320.
291. Whitehead KA, Langer R, Anderson DG. Nat Rev Drug Discov 2009;8:129–138.
292. Bird A. Nature 2007;447:396–398.
293. Cook EH, Scherer SW. Nature 2008;455: 919–923.
294. Williams M, Giordano T, Elder RA, et al. Med Res Rev 1993;13:399–448.
295. Kubinyi H. Nat Rev Drug Discov 2003;2: 665–668.
296. Horrobin DF. Nat Rev Drug Discov 2003;2: 151–154.

SELECTIVE TOXICITY

JOHN H. BLOCK
Department of Pharmaceutical Sciences, College of Pharmacy, Oregon State University, Corvallis, OR

1. OVERVIEW

Selective Toxicity was popularized by Professor Adrian Albert beginning with his lectures at University College London in 1948 and the first edition of his book in 1951. Subsequent editions appeared until the seventh edition in 1985 [1]. The latter was reprinted in Japanese as two volumes in 1993 and 1994 [2,3].

For much of the period during development of new pharmaceutical entities, the concept of selective toxicity has been limited to chemotherapy and antibiotic therapy. How does one design a cancer chemotherapeutic agent that will kill a malignant cell and not interfere with mitosis in a benign cell? How does the biochemistry of a bacterial cell differ from that of a mammalian cell so that the antibiotic is toxic to the bacterium and nontoxic to the patient's cells?

It is important to realize that nearly all types of drug therapy can be thought of as selectively toxic. With the possible exception of some forms of hormonal replacement therapy, a drug's desired pharmacological response is actually an intervention in normal biochemical processes. Consider the nonsteroidal anti-inflammatory drugs (NSAIDs) that inhibit cyclooxygenase (COX). While reducing inflammation is the desired response for these agents, the drugs are toxic to the enzyme. β-Adrenergic agonists are exogenous ligands for the same receptors as epinephrine, but are not subject to the same regulatory controls (biosynthesis, release, reuptake, metabolic disposal) as the endogenous hormone. Herbicide and pesticide use in agriculture increasingly is based on the agent being selective for a specific plant or taking advantage of some specific property of the insect which differs from those of farm animals and humans.

1.1. Categorization of Therapeutic Agents

Focusing on humans, most drugs are agonists, antagonists, or replacement agents. Discussion of partial agonists and antagonists will be found in other chapters.

1.1.1. Replacement Hormone replacement therapy is the basis for the use of insulin by a person with type 1 diabetes (formerly called insulin-dependent diabetes mellitus or IDDM) and levothyroxine for a patient with a thyroid deficiency. Insulin and thyroxine only act on specific receptors. As long as the dosing is correct, patients usually do not experience unpleasant adverse reactions. Excessive insulin causes severe, life-threatening hypoglycemia leading to insulin shock. Because of the general distribution of thyroxine receptors throughout the body, excessive thyroxine accelerates intermediate metabolism in several organs.

This concept of replacement therapy can be extended to the administration of vitamins and minerals to a patient whose body stores have been depleted, or has a medical diagnosis where increased administration of nutritional supplements is indicated (e.g., digestive disorders, malignancies, clinical depression, weight control with low calorie diets). The nutritional agents combine with specific receptors or, usually after a metabolic transformation, act as coenzymes (e.g., niacin-NAD/NADP, thiamine-thiamine pyrophosphate, folic acid-tetrahydrofolates, and cobalamin-adenosyl cobalamin). Again, when the dosing is correct, the patient does not experience adverse effects. Excessive doses of some vitamins can cause neuritis (pyridoxine/vitamin B_6), hypercalcemia (cholecalciferol/vitamin D_3 and ergocalciferol/vitamin D_2), and liver damage (retinol/vitamin A).

1.1.2. Agonists Agonists mimic the endogenous ligand when combining with the ligand's receptor. In theory, there should be excellent selectivity, but many times the receptors for the ligand are scattered widely throughout the body and located in many organs. Thus, an adrenergic agonist such as phenylephrine will constrict blood vessels while simultaneously increasing the heart rate. For some patients, the latter could lead to life-threatening tachycardia. Depending on the receptor, the active form of cholecalciferol (vitamin D_3), 1,25-dihydroxycholecalciferol (**1**) causes the synthesis of a calcium transport protein in the intestinal mucosa and regulates cell division. An analog of 1,25-dihydroxycholecalciferol, calcipotriene (**2**) is used topically for psoriasis. If administered internally, the patient could experience severe hypercalcemia leading to possible calcification of the soft tissues and blood vessel walls. In the context of selective toxicity, calcipotriene has poor selectivity.

1. 1,25-Dihydroxycholecalciferol

2. Calcipotriene

1.1.3. Antagonists Most drugs prescribed today are antagonists. The desired goal is to block responses in the cholinergic, adrenergic, rennin–angiotensin, and other integrated systems. Rarely are these biochemical systems localized in a specific organ. Inhibiting one enzyme can have unforseen metabolic consequences downstream. A classic example is the angiotensin converting enzyme (ACE) inhibitors that produce an irritating cough (to the patient and patient's family). Angiotensin converting enzyme has more than one substrate: angiotensin I and bradykinin (Fig. 1). The goal was to design drugs that inhibit the formation of angiotensin II from angiotensin I. This led to the development of the ACE inhibitors (e.g., captopril (**3**), enalapril (**4**), and lisinopril (**5**)). Angiotensin converting enzyme also degrades bradykinin, one of the many peptides involved with pain production and

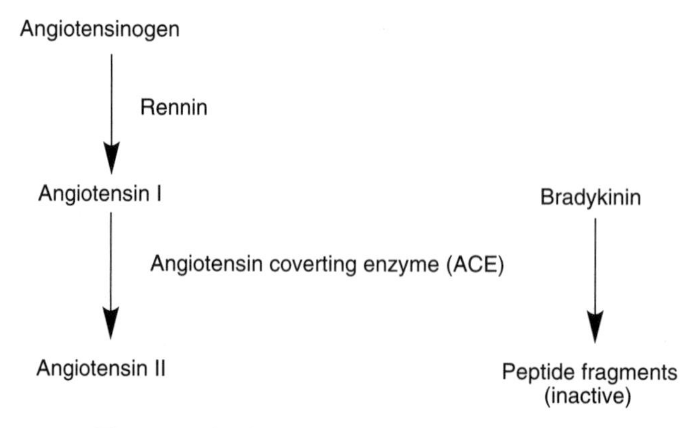

Figure 1. Angiotensin–renin catalyzed pathways.

inflammation. ACE inhibitors, by inhibiting degradation of bradykinin, cause its buildup, and this may be the cause of the annoying dry cough characteristic of these drugs.

3. Captopril

4. Enalapril

5. Lisinopril

An important approach to increasing the selectivity of a pharmacological agent is to thoroughly understand the biochemistry of the system where drug therapy is indicated. By going downstream from the angiotensin converting enzyme and employing an angiotensin II antagonist, there are fewer adverse reactions.

Asthma is a complex inflammatory disease of the respiratory system. One of the families of chemical mediators is the leukotrienes synthesized from arachidonic acid by the lipoxygenase enzyme complex (Fig. 2). The leukotrienes are one of the mediators of bronchoconstriction. In the treatments of asthma, the

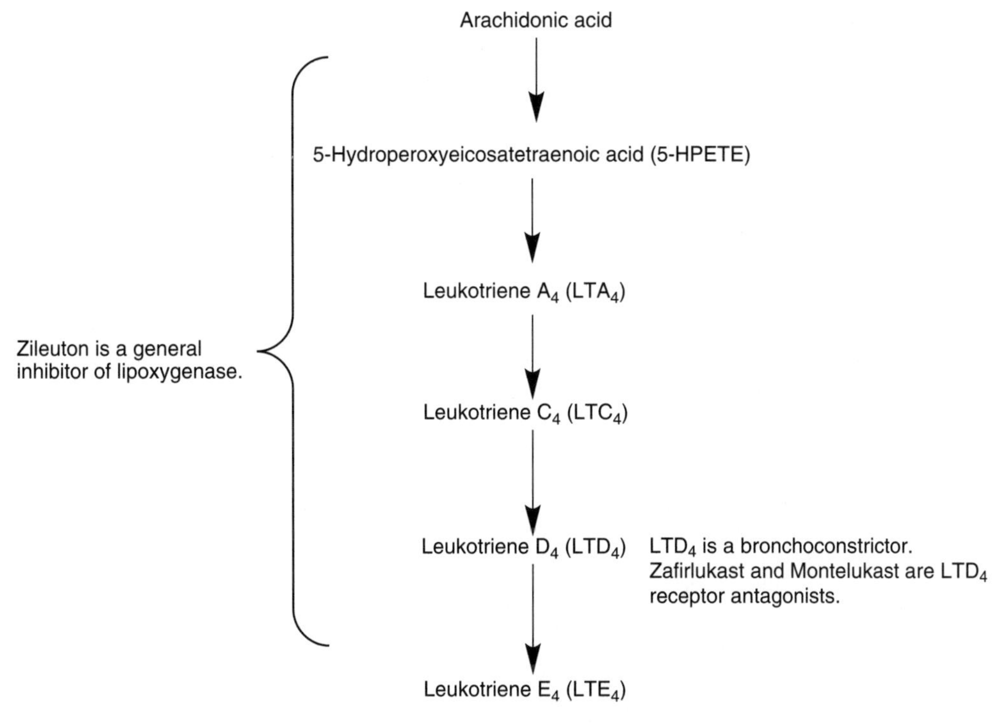

Figure 2. Outline of leukotriene biosynthesis.

leukotriene D_4 (LTD_4) receptor antagonists, montelukast sodium (**6**) and zafirlukast (**7**), exhibit less adverse responses than the 5-lipoxygenase inhibitor, zileuton (**8**), which interferes with the formation of LTA_4, LTC_4, LTD_4, and LTE_4 family of leukotrienes all of which are chemical mediators of the immune response at several sites.

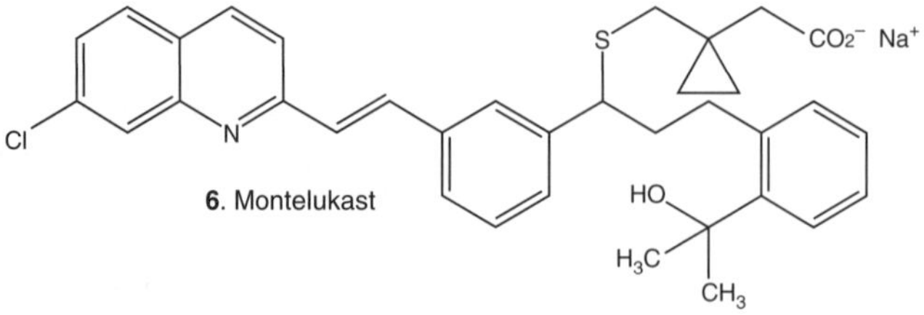

6. Montelukast

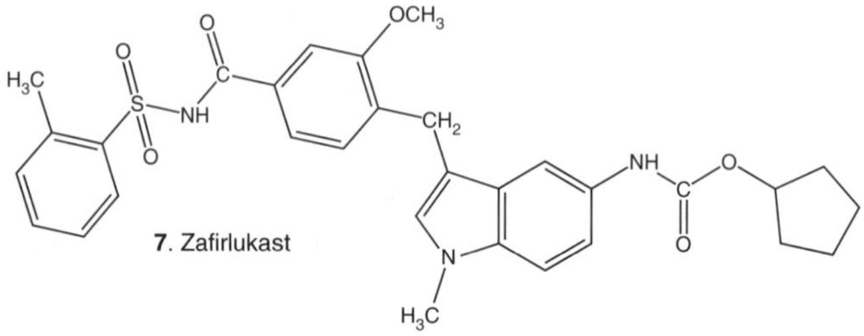

7. Zafirlukast

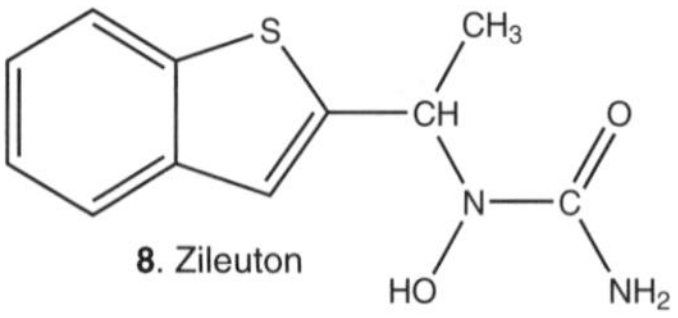

8. Zileuton

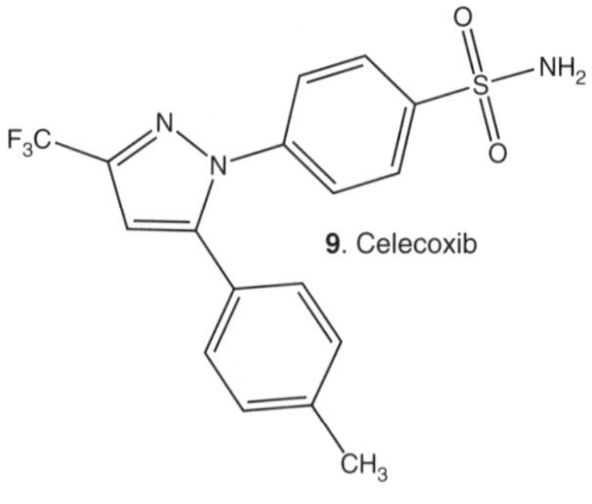

9. Celecoxib

10. Rofecoxib

Distribution of enzyme types is another way to increase the selectivity of both agonists and antagonists. The COX isozymes are distributed unevenly among the various organs. What is now known as COX-1 is a constitutive enzyme found in many organs. COX-2 is inducible by cytokines and appears at sites when there is inflammation. Sometimes, selective toxicity can produce surprising results. Two commercial COX-2 inhibitors were approved by the FDA, celecoxib (**9**) and rofecoxib (**10**), and initially appeared to exhibit fewer adverse responses in patients who must take these drugs on a chronic basis compared to the older COX-1 inhibitory NSAIDS. Then, in 2004 (5 years after the approval), rofecoxib was recalled because of increased incidence cardiovascular events including heart attack and stroke. This is not the type of selective toxicity that medicinal chemists like to see.

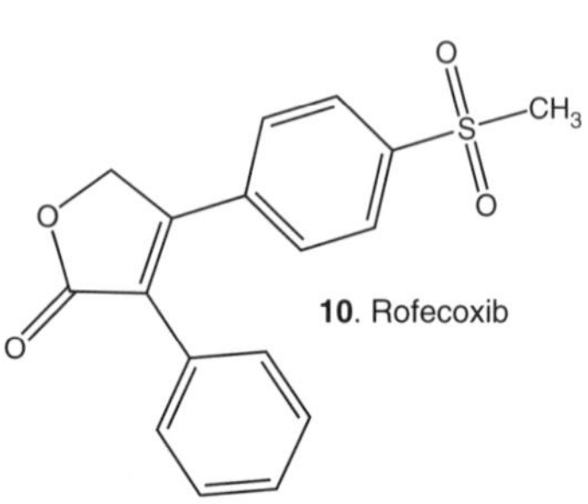

2. PRINCIPLES OF SELECTIVITY

Professor Albert, after examining the various ways that selectivity of biological response is obtained from both natural products or synthetic agents, concluded that there are three possible ways that a pharmacologically active agent exerts selectivity. These are (a) comparative distribution, (b) comparative biochemistry, and (c) comparative cytology [4].

2.1. Comparative Distribution

Comparative distribution can be caused by differences in physical area where absorption occurs or by different biodistribution of the drug. An insecticide accumulates in the insect because the insect has more exposed surface area relative to that of animals in the same environment. Radioactive iodine is used to both diagnose diseases of the thyroid and to destroy the thyroid gland where the isotope accumulates in the thyroid because of specific iodine transport proteins. Parasites that remain in the intestinal tract are easier to treat because many of the newer drugs also remain in the intestinal tract. A rapidly dividing cell line will preferentially incorporate a number of drugs including those used in the treatment of cancer.

There are several examples where drugs cause adverse responses in patients because of poor comparative distribution. Most of the drugs used in the treatment of cancers do not differentiate between benign tissues that are constantly dividing (bone marrow, intestinal mucosa, and hair follicles) and malignant cells [5]. The result is immunosuppression and/or anemia, diarrhea or constipation and loss of hair. A problem with the current antiviral drugs is that they cannot intercept free viruses, but are only effective when the virus is inside a cell and reproducing. These drugs cannot differentiate between virus-infected cells and cells free of virus.

It has been known almost from the time of its serendipitous discovery that the *cis* analogs of the platinum containing cytotoxic agents, cisplatin (**11**), and carboplatin (**12**), are much more cytotoxic relative to the *trans* isomers [6,7]. Both the *cis* and *trans* compounds enter the growing cell and form interstrand and intrastrand cross-links within the DNA double helix. The kinetics of this binding is very different between the two isomers. Further, DNA repair is more efficient with the *trans* isomers. The net result is that the *cis* isomer remains in the DNA long enough to be cytotoxic [8].

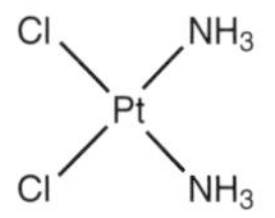

11. Cisplatin

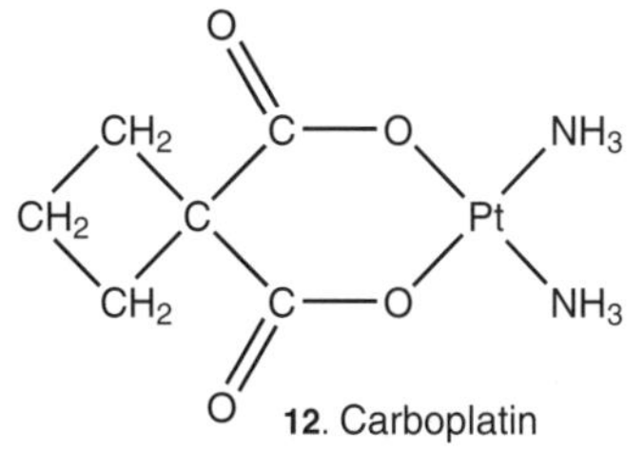

12. Carboplatin

2.2. Comparative Biochemistry

Drugs used to treat bacterial infections provide some of the best illustrations of comparative biochemistry beginning with the antibiotics. There are enough differences in the biochemistry of bacteria such that a wide variety drugs have been developed that are relatively nontoxic to the patient. This selectivity decreases in the development of antiviral agents because most viruses use the host cell's biochemistry to reproduce. Fungal metabolism is so close to mammalian biochemistry such that recently, the safety profile for the antifungal drugs has been narrow.

2.3. Comparative Cytology

The third principle, comparative cytology, refers to the comparative taxonomic structure of cells. Examples include plant versus animal cells, undifferentiated malignant cells versus fully differentiated mature benign cells, the patient's immune system recognizing nonself-cells that produces the basis for monoclonal antibody therapy, cell wall versus cell membrane, mitochondria versus chloroplasts, and the presence of mitochondria in aerobes versus their absence in anaerobes [9]. There is overlap between comparative cytology and the first two principles. It is not uncommon to see selectivity caused by two and possibly all three principles. Penicillins and cephalosporins are examples of comparative biochemistry (transpeptidase in bacteria) as well as comparative cytology (bacterial cell wall versus mammalian cell membrane).

2.4. Comparative Stereochemistry

The fourth principle of selectivity that is being used today in drug design is comparative stereochemistry. Most receptors are chiral and respond differently to drugs of different chirality. Increasingly, the U.S. Food and Drug Administration (FDA) requires new drug applications to resolve racemic mixtures to determine which stereoisomer is biologically active. Drugs with chiral centers may exhibit more complex selectivity. This can be caused by flexibility at the receptor, variations in transport across cell membranes to the site of action, and differences in metabolism for each stereoisomer. A classic example is the antitussive dextromethorphan (*d*-3-methoxy-*N*-methylmorphinan) (**13**), which is devoid of opiate activity as compared to the levo isomer (*l*-3-methoxy-*N*-methylmorphinan) (**14**).

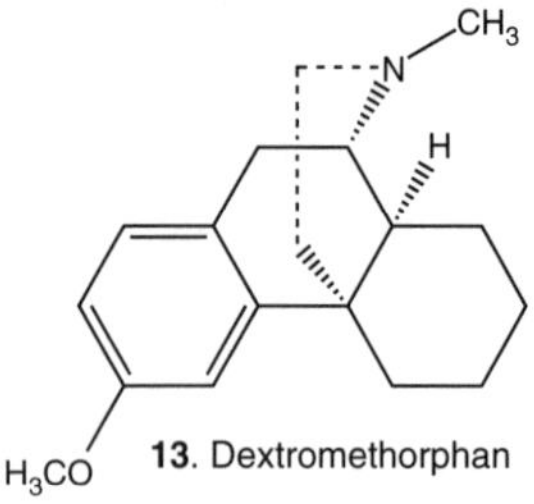

13. Dextromethorphan

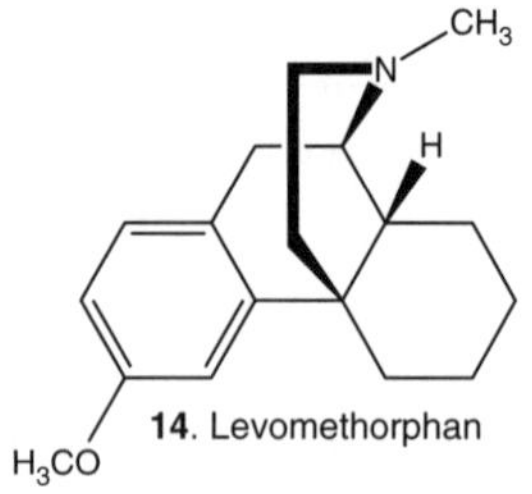

14. Levomethorphan

The routes and pharmacokinetics of a drug's metabolic degradation also may vary with the isomer. The calcium channel antagonist, verapamil (Fig. 3), illustrates why it is difficult to conclude one isomer is superior over the other. *S*-Verapamil is the more active

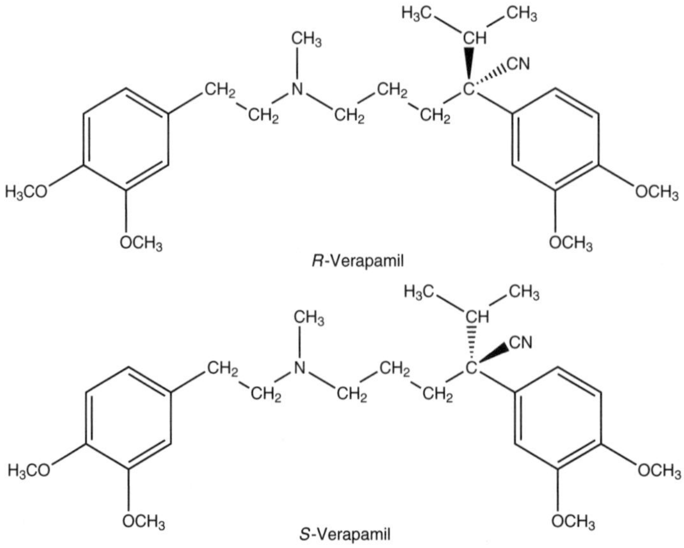

R-Verapamil

S-Verapamil

Figure 3. Verapamil isomers.

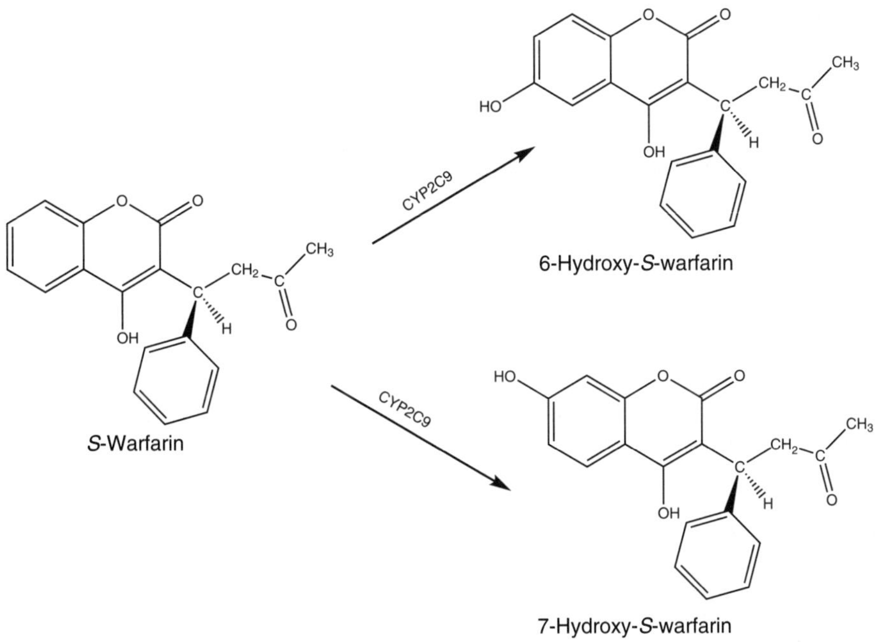

S-Warfarin

CYP2C9

6-Hydroxy-S-warfarin

CYP2C9

7-Hydroxy-S-warfarin

Figure 4. Metabolism of S-warfarin.

pharmacological stereoisomer than the less active *R*-verapamil, but the former is more rapidly metabolized by the first pass effect. (First pass refers to orally administered drugs that are extensively metabolized as they pass through the liver. It is not as significant when the drug is administered parenterally because the drug is dispersed before reaching the liver.) Therefore, intravenous administration of the racemic mixture of verapamil produces a longer duration of action than when administered orally because the more potent *S*-isomer will be metabolized more slowly.

The *S*-isomer of warfarin is more active and is metabolized by the CYP450 2C9 (CYP2C9) isozyme (Fig. 4) while the *R*-isomer is metabolized at a different position by CYP450 3A4 (CYP3A4) (Fig. 5) [10,11]. The fact that two different CYP450 isozymes are required for warfarin metabolism and polymorphism is seen with CYP2C9 increases the chances of drug–drug interactions with this potent anticoagulant. Indeed, the FDA approved package insert list 22 different pharmacological classes of drugs that can alter warfarin's pharmacological response as measured by prothrombin time/international normalized ratio (PT/INR).

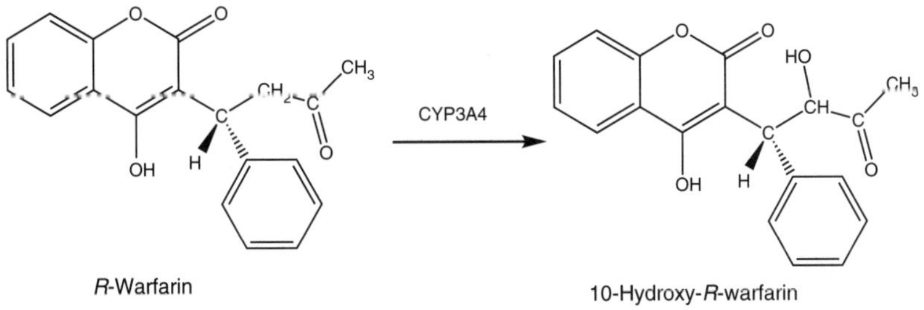

R-Warfarin

CYP3A4

10-Hydroxy-R-warfarin

Figure 5. Metabolism of *R*-warfarin.

3. ADDITIONAL EXAMPLES OF SELECTIVE TOXICITY

Several examples will be given how principles of selective toxicity can be applied to a variety of drug classes. This will not be an exhaustive list, but will serve as examples. Some will show excellent selectivity. Others will show good selectivity of one type, but fail in one of the others.

3.1. Antibiotics: Comparative Biochemistry

Comparative biochemistry is the basis for successful antibiotic therapy. The antibiotic class is one of the main reasons that human life expectancy has increased producing a significant older population. This group of drugs usually takes advantage of a microorganism's unique biochemistry. The discovery of sulfonamides (**15**) was the start of modern antibacterial chemotherapy and gave credence to the "magic bullet" ideal in drug design. This class of drugs inhibits the biosynthesis of folic acid and are bacteriocidal or bacteriostatic. Mammals cannot synthesize folic acid. (That is why folic acid is a vitamin.)

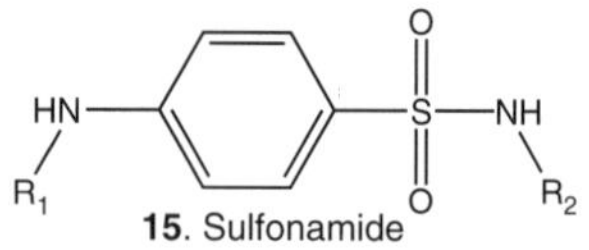

15. Sulfonamide

The sulfonamides are selective for a key reaction found only in bacteria. Most bacteria synthesize their own folic acid. In contrast, humans obtain their folic acid from food and vitamin supplements. Therefore, in bacteria sulfonamides block folic acid biosynthesis by competitive inhibition of dihydropteroate synthase which is the enzyme used by bacteria to incorporate *p*-aminobenzoic acid forming dihydropteroic acid (Fig. 6). Upon the addition of glutamic acid to the latter, the bacteria synthesize dihydrofolic acid (FAH_2, FH_2, DHF). A second antibiotic, trimethoprim (**16**), selectively inhibits bacterial dihydrofolate reductase. As long as the patient's folic acid status is adequate, there is minimal

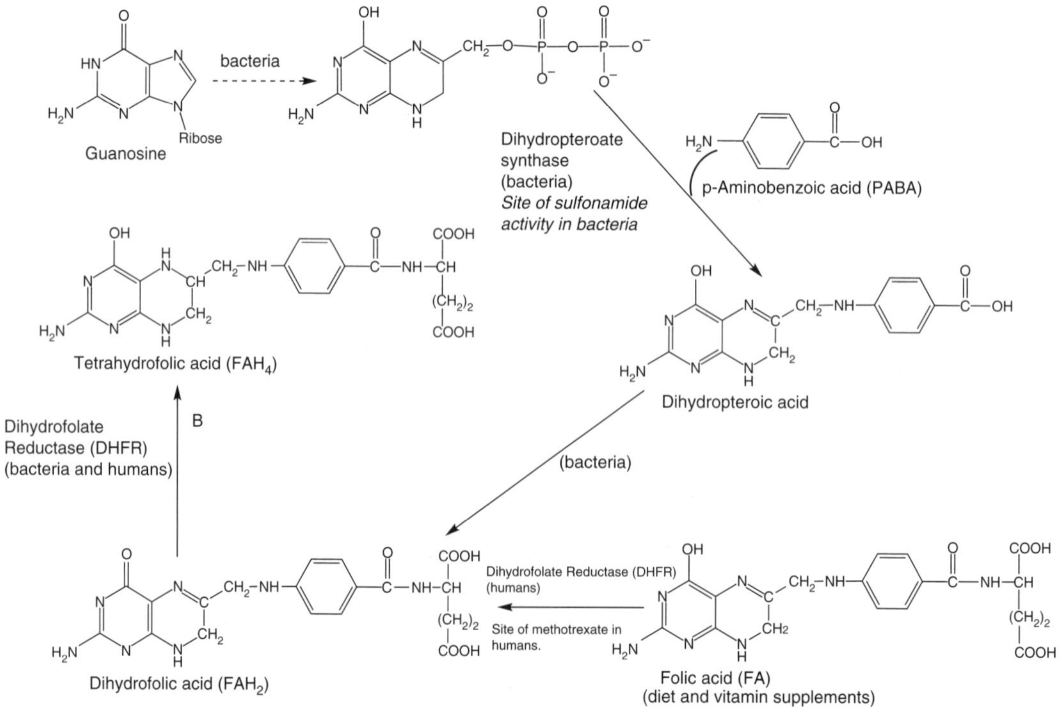

Figure 6. Folate biosynthesis.

metabolic toxicity from the sulfonamides or trimethoprim.

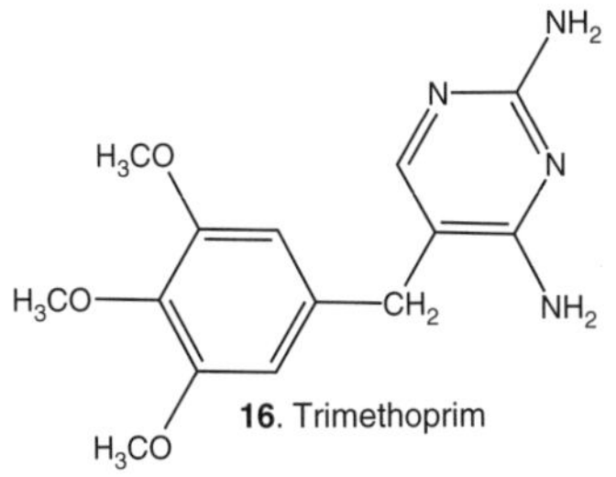

16. Trimethoprim

In contrast with these two antibacterial antibiotics, methotrexate (**17**) is one of the most used cytotoxic drugs for malignancies and in lower doses it is an immunosuppresive in autoimmune diseases (e.g., psoriasis and rheumatoid arthritis). Methotrexate inhibits mammalian dihydrofolate reductase (DHFR) an enzyme found in every cell that uses one of the coenzyme forms of folic acid. Because of its poor comparative biochemical selectivity, it is common to administer one of the tetrahydrofolates as an antidote for methotrexate toxicity.

Dihydrofolate reductase also is a potential site for antifungal antibiotics. The problem was to find a drug that is selective for the fungal version of this enzyme. The result was pyrimethamine (**18**) that shows a preference for plasmodia dihydrofolate reductase relative to the mammalian enzyme and, therefore, has found use in the treatment of malaria [12]. As explained below, developing selectively toxic antifungal drugs has been a challenge, and pyrimethamine is no exception. It can adversely affect the folate status in patients receiving this drug.

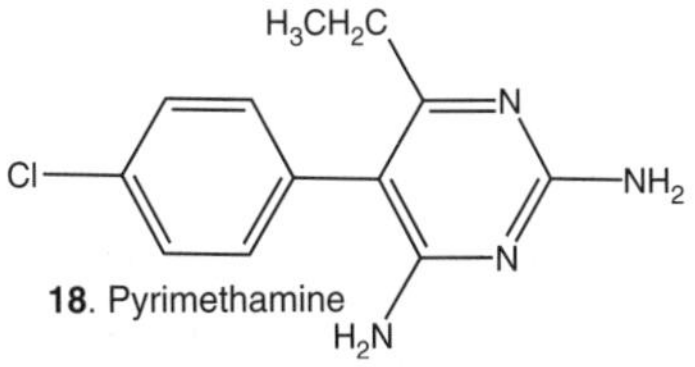

18. Pyrimethamine

Following the discovery of the antibacterial sulfonamides came the accidental discovery of the penicillins. This led to wide scale screening of microbial metabolites leading to a large number of antibiotics that are in a variety of chemical classes. Table 1 provides a summary of sites

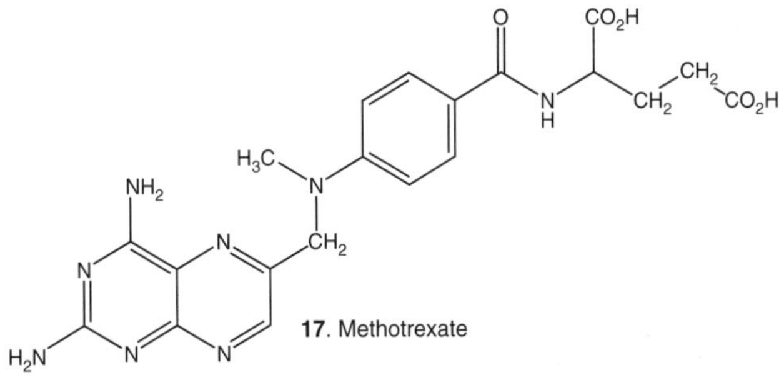

17. Methotrexate

Table 1. Summary of Sites Where Antibiotics Exert Their Antibacterial Effect

Site of Action	Antibiotic Class
Bacterial cell wall (peptidoglycan) synthesis	β-Lactams, bacitracins, vancomycin
Bacterial ribosomal units	Aminoglycosides, tetracyclines, macrolides
Organization of bacterial membrane	Polymixins, colistins, gramicidins
Bacterial DNA gyrase/topoisomerase	Quinolones
Folic acid biosynthesis	Sulfonamides
Dihydrofolate reductase	Trimethoprim

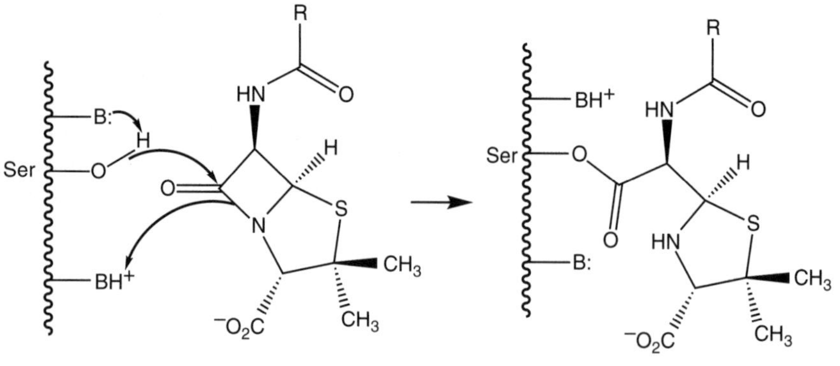

Figure 7. Mechanism of penicillin's antibacterial activity.

where antibiotics exert their antibacterial activity. The reader should refer to the anti-infectives chapters for discussions of structure–activity relationships including how to reduce the risk of antibiotic resistance and improve the drug's pharmaceutical properties.

Most of the examples in this section are based on Professor Albert's principle of comparative biochemistry. Comparative cytology also plays a role, although most of the time this principle frustrates the drug designer and clinician. Many times, antibiotics will be effective against Gram-positive bacteria, but not Gram-negative bacteria. This difference often is due to the additional proteoglycan layer on the outside of the cell wall of Gram-negative bacteria. At the macro level, the additional barrier that keeps the Gram stain from entering the bacterium also hinders the antibiotic from entering the cell.

The β-lactam antibiotics, penicillins, and cephalosporins inhibit bacterial transpeptidase preventing the microorganism from completing the synthesis of its cell wall. Mammals do not have cell walls and, thus, are not affected by β-lactam antibiotics. Mammalian lipid bilayers have completely different biochemistry from that of a bacterium.

The β-lactam antibiotics consist of two major groups, penicillins (**19**) and cephalosporins (**20**), both of which still are the most widely prescribed drugs indicated for bacterial infections. They exhibit classic selective toxicity because they inhibit a key enzyme, transpeptidase, in the biosynthesis of the bacterial cell wall, a structure not found in the lipid bilayer of mammalian cell membranes. The inhibition is complex and involves the antibiotic covalently binding to penicillin binding proteins (Fig. 7). The carbonyl carbon of the sterically strained β-lactam ring is attacked by the serine hydroxyl at the protein's active site forming a stable covalent protein-antibiotic conjugate. In general, the β-lactam antibiotics are safe. Adverse reactions usually are due to a small subset of patients who are allergic to this group of drugs.

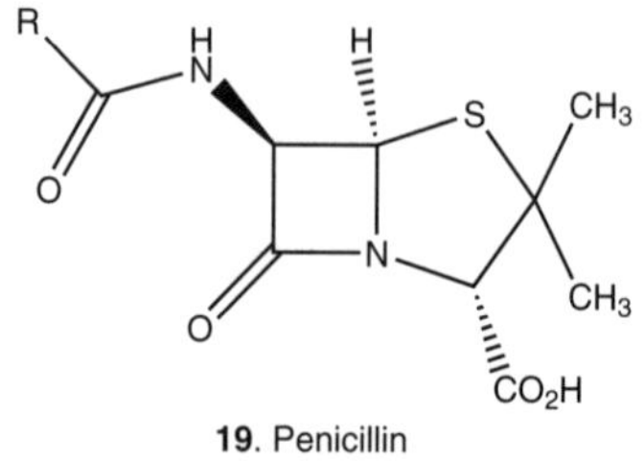

19. Penicillin

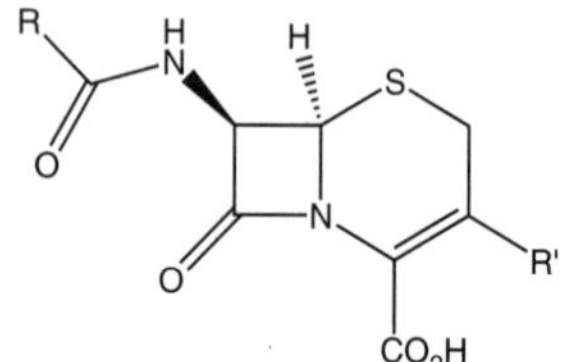

20. Cephallosporin

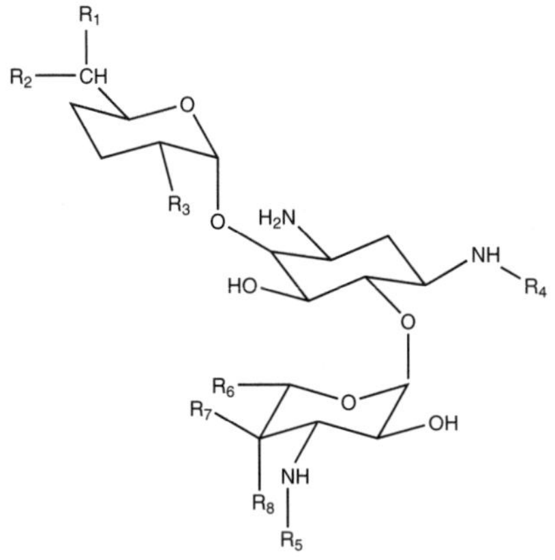

Gentamycin C_1: $R_1 = R_2 = R_5 = R_7 = CH_3$; $R_3 = NH_2$; $R_4 = H$; $R_8 = OH$
Kanamycin A: $R_1 = NH_2$; $R_2 = R_7 = H$; $R_3 = OH$; $R_4 = R_5 = R_8 = H$; $R_6 = CH_2OH$

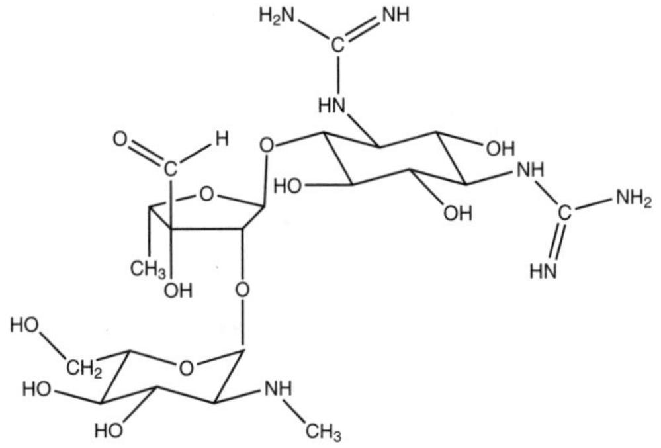

Streptomycin

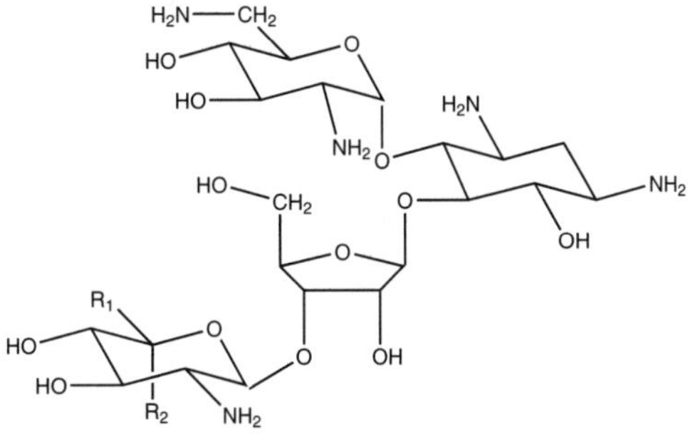

Neomycin B: $R_1 = H$; $R_2 = CH_2NH_2$
Neomycin C: $R_1 = CH_2NH_2$; $R_2 = H$

Figure 8. Aminoglycosides.

In contrast with the β-lactam antibiotics, the aminoglycoside antibiotics have more structural diversity (Fig. 8). Examples include streptomycin, the gentamycin family, the kanamycin family, and the neomycin family. The one structural characteristic they have in common is one or two amino hexoses connected to a six-membered ring substituted with alcohols and amino moieties. Sometimes, these are called aminocyclitols. Because the latter do not have an anomeric carbon, they are not sugars.

The aminoglycosides bind to the bacterial 30S ribosomal unit causing inhibition of initiation of protein synthesis and sometimes misreading of the genetic code. Fortunately, the ribosomal units in mammalian cells are sufficiently different that the aminoglycosides do not readily bind to mammalian ribosomes.

Unfortunately, the aminoglycoside antibiotics do have severe toxicities. The neomycin family has such severe nephrotoxicity that these drugs usually are only administered topically. One of the most distressing and common adverse responses is ototoxicity that can lead to permanent deafness. While the biochemical mechanism of this toxicity is poorly understood, these antibiotics concentrate in the lymphatic tissue of the inner ear.

The half-lives of the aminoglycosides are five to six times longer in the otic fluid as compared to the plasma [13]. This is an example where comparative distribution increases the toxicity of the drug.

Like the aminoglycoside antibiotics, the tetracyclines (**21**) preferentially bind to the 30S ribosomal subunit preventing elongation of peptide chains. In addition, the tetracyclines are actively transported into the bacterium causing the drug to concentrate in the susceptible cell. Mammalian cells lack this active transport system for tetracyclines. Interestingly, some resistant bacteria have an active efflux system that can pump the tetracycline back out of the cell.

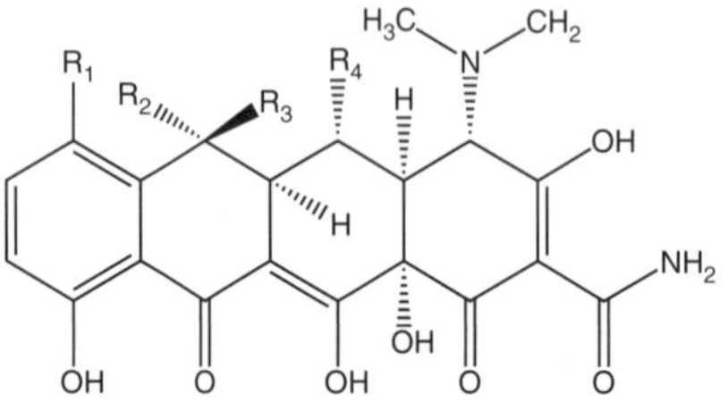

21. Tetracycline

The tetracyclines are considered relatively nontoxic. Their comparative distribution is such that adverse responses are limited. Tetracyclines ability to chelate di- and trivalent cations, particularly divalent calcium, can be a problem during certain periods of a patient's development. Tetracyclines are deposited in the bones of fetuses and growing children. They also are deposited in the developing teeth producing discoloration. Therefore, tetracyclines normally are contraindicated for infants and pregnant women.

Erythromycin (R = H) (**22**), clarithromycin (R = CH_3) (**22**) and azithromycin (**23**), produced by ring expansion of erythromycin, reversibly bind to the bacterial 50S ribosomal subunit. The result is inhibition of the growing protein chain. (Carbon #9 is labeled on erythromycin for reference.) Selective toxicity is achieved because these agents do not bind to mammalian ribosomes.

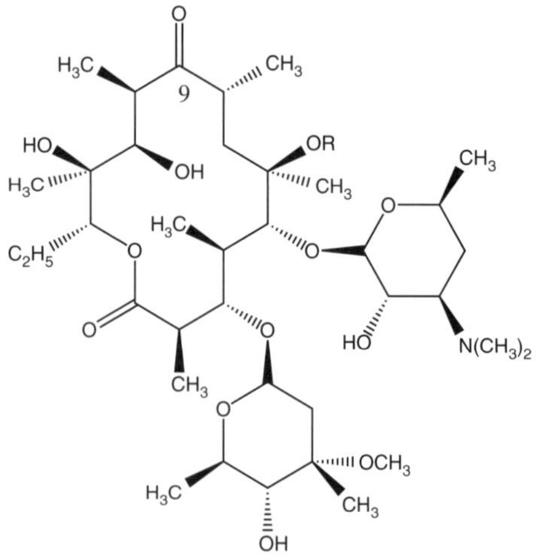

22. Erythromycin–Clarithromycin

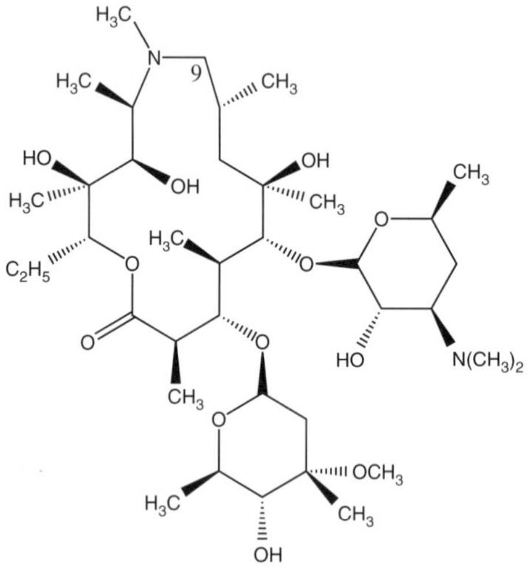

23. Azithromycin

The macrolide antibiotics are not as selective as other anti-infective drugs. They can cause severe epigastric distress, possibly functioning as a motilin receptor agonist that stimulates gastric motility. This group of macrolides also inhibits the cytochrome P450 isozymes, CYP1A2 and CYP3A4. Erythromycin can cause an increase in serum concentrations of the drugs whose metabolism is inhibited by the antibiotic.

The quinolones/fluroquinolones (**24**) are unique among the antibiotics. The quinolone group targets two bacterial DNA topoisomerase II enzymes also known as DNA gyrases. Selective toxicity arises because mammalian cells do not have DNA gyrases although they do have a topoisomerase II. The latter requires a much higher dose of quinolones for inhibition to occur. While no group of drugs administered internally is completely nontoxic, the quinolones show good selectivity.

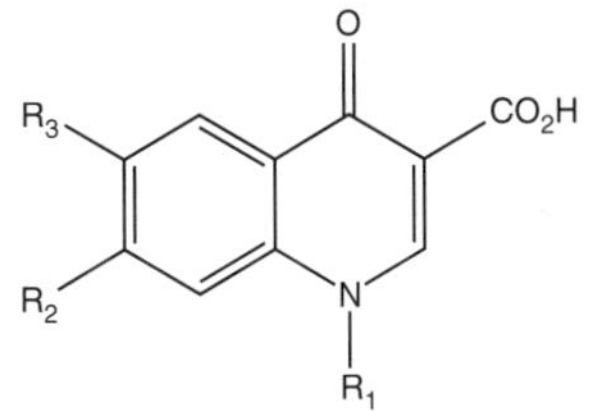

24. Quinolones/fluoroquinolones

As good as antibiotics are to illustrate principles of selective toxicity, it must be accepted that microbial antibiotic resistance is a challenge to developers of new antibiotics. This has led to searching for identifying more biochemistry unique to microbes. Examples of the latter include the QseC membrane embedded sensor histidine kinase, FtsZ bacterial protein that is the bacterial equivalent of mammalian cell-division protein β-tubulin and a biosynthetic pathway for menaquinones.

QseC is required for bacteria to respond to adrenergic signaling molecules and other signals to promote the expression of virulence factors [14]. An example of a small molecule that inhibits the binding of signals to QseC is LED209 (**25**) that prevents autophosphorylation of this sensor kinase. This results in inhibition of QseC-mediated activation of virulence gene expression. Preliminary studies inhibition of QseC does not inhibit pathogen growth. In other words, a molecule like LED209 is selectively toxic to a specific site unique to bacteria that is not found in mammals.

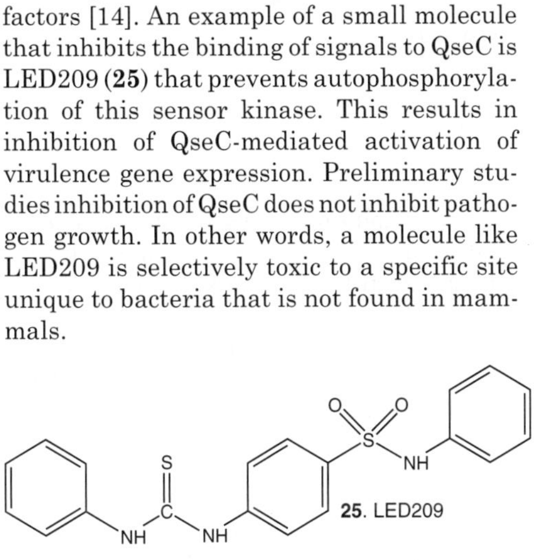

25. LED209

Another site specific to bacteria is the FtsZ quanosine triphosphatase that also is a homolog of β-tubulin. FtsZ is required for bacterial cell division. An example of a small molecule that inhibits FtsZ and, therefore, prevents cell division is PC190723 (**26**). A fascinating aspect of this site is the observation the bacterial FtsZ binding site is analogous to the taxol binding site of mammalian tubulin [15].

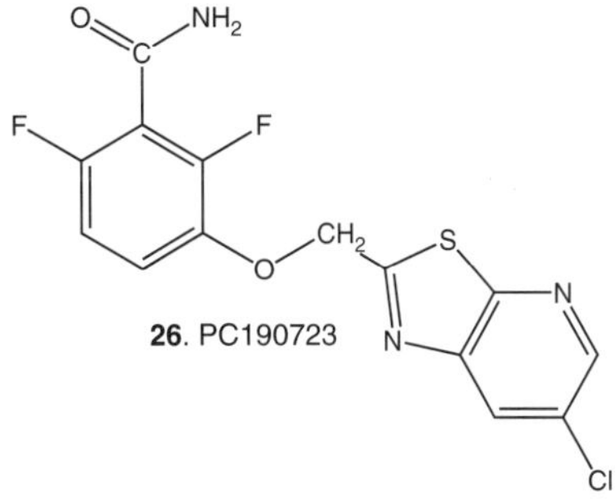

26. PC190723

While mammals require menaquinone (derived from vitamins K) in their diet, bacteria synthesize this quinone and ubiquinone and require this family of lipid-soluble molecules for the transfer of electrons within the electron transport chain. The challenge now is to develop compounds that could inhibit one or more of the steps in the bacterial menaquinone biosynthetic pathway [16].

3.2. Antivirals: Comparative Biochemistry

In contrast with antibacterial antibiotics, it has been difficult to develop antiviral and antifungal agents that are selective for the infectious agent. Viruses are complete parasites in that they can only reproduce inside the host cell. Many viruses are dependent on the host cell's polymerases, ribosomes and transfer RNAs. Most of today's antiviral drugs not only poorly differentiate between viral and human polymerases but also, usually, cannot distinguish between cells infected with the virus versus noninfected cells. Even inhibitors preferential for viral RNA-dependent DNA polymerase (reverse transcriptase), an enzyme not found in mammalian cells, can inhibit human mammalian DNA-dependent DNA polymerase leading to suppression of the patient's bone marrow.

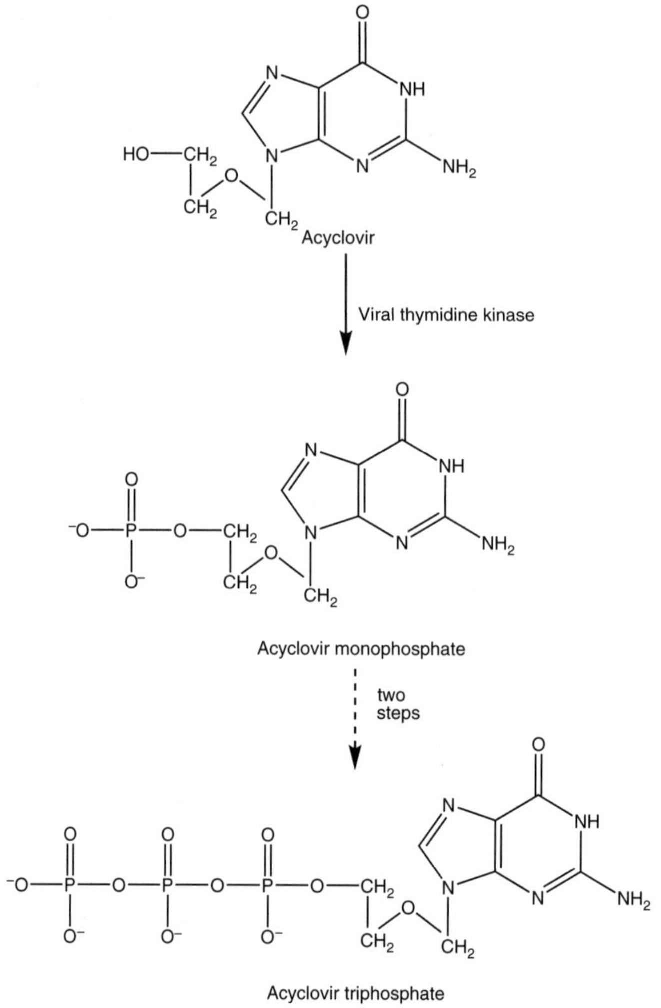

Acyclovir

Viral thymidine kinase

Acyclovir monophosphate

two steps

Acyclovir triphosphate

Figure 9. Acyclovir activation.

Acyclovir and its analogs are widely used antiviral drugs. They are indicated for herpes simplex I and II and varicella zoster (chicken pox). The herpes virus has one of the more complicated viral genomes coding for more than 160 genes. One of these structural genes codes for thymidylate kinase that is different from the mammalian kinase enzyme. The viral enzyme will phosphorylate inactive acyclovir (actually a prodrug) producing active acyclovir monophosphate (Fig. 9). The latter one is phosphorylated to the diphosphate and finally the triphosphate that is the active antiviral drug. None of the mammalian kinases can significantly phosphorylate inactive acyclovir to the active form.

With the HIV/AIDS pandemic, there has been extensive study on how this particular virus enters the host cell, reproduces within the host cell and exits the host cell. Because the human immunodeficiency virus (HIV) is a retrovirus, the initial approach was synthesis of reverse transcript inhibitors with the first group based on the nucleoside template, an example being tenofovir (**27**). These were followed by nonnucleoside reverse transcriptase inhibitors, a recent example being etravirine (**28**).

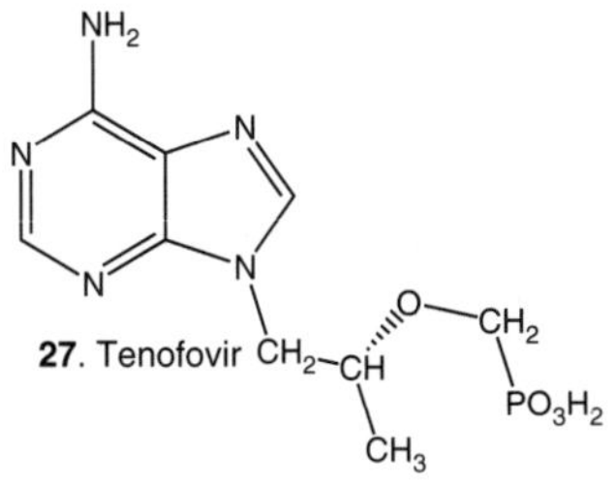

27. Tenofovir

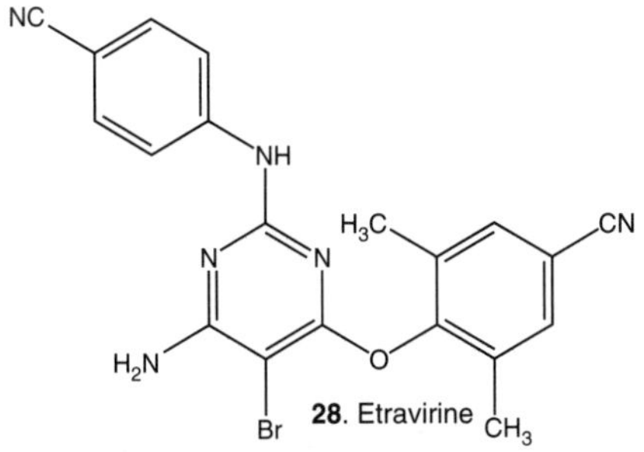

28. Etravirine

Once inside the cell, the HIV genome is read by viral reverse transcriptase producing viral DNA that must be incorporated, or integrated, into the host cell's genome. Raltegravir (**29**) inhibits strand transfer that is the final stop of provirus integration [17]. Reading the now incorporated proviron produces viral polyprotein that is cleaved by HIV-1 protease, an aspartic protease, yielding mature protein components of the HIV virion. There are a large number of HIV protease inhibitors with darunavir (**30**) being one of the more recent examples. In the case of HIV, there are several targets for antiviral therapy that are selective to the virus.

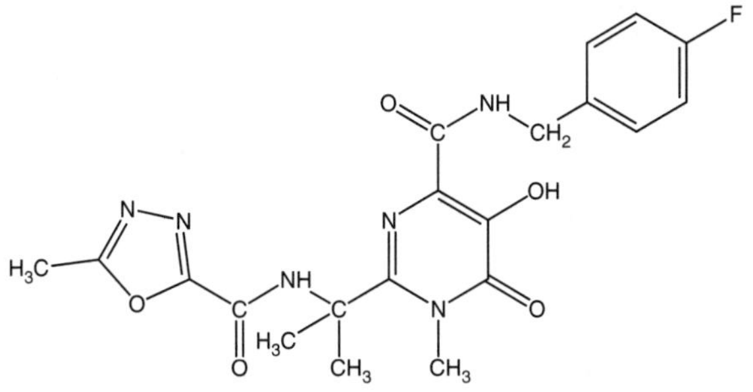

29. Raltegravir

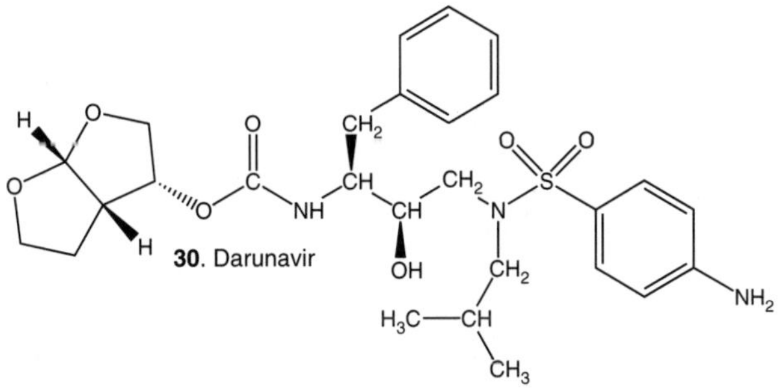

30. Darunavir

Table 2. Representative Antiviral Drugs Classified by Virus

Virus	Drug	Site of Activity	Remarks
Influenza Type A	Amantadine, Rimantadine	Blocks uncoating or entrance of the virus into the cell	Good selectivity. Restricted to influenza type A virus.
Influenza type A and type B	Oseltamivir, Zanamivir	Viral neuraminidase	Good selectivity. Effective against both influenza type A and type B virus.
Herpes simplex I and II (HSV I and II), Varicella zoster (VZV)	Acyclovir, Famciclovir	Substrate for herpes thymidine kinase	Good selectivity. This group of drugs are very poorly phosphorylated by the host cell's kinases.
Human immunodeficiency virus (HIV)	Zidovudine, Didanosine, Zalcitabine, Lamivudine, Stavudine, Abacavir Delavirdine Nevirapine Efavirenz	HIV–RNA dependent DNA polymerase (Reverse transcriptase)	Although mammals do not have reverse transcriptase, the inhibitors also inhibit the host cell's DNA-dependent DNA polymerase
	Indinavir Ritonavir Saquinavir Nelfinavir Amprenavir Lopinavir	HIV protease	This group of drugs is very effective, but they have a large number of adverse reactions. The mechanisms of these complications are poorly understood.
Cytomegalovirus (CMV)	Foscarnet, Cidofovir	Viral DNA dependent DNA polymerase	There is a preference for the viral polymerase, but these drugs also inhibit the host cell's DNA polymerase

Because the influenza virus undergoes genetic drift annually and occasionally a significant genetic shift, there has been an extensive search for drugs that that will block the virus from entering host cells. Amantadine (**31**) and rimantadine (**32**) block the uncoating of Influenza A whereas oseltamivir (**33**) and zanamivir (**34**) inhibit viral neuraminidase found in both Influenza A and B. Both sets of antiviral drugs show good selectivity for the virus, but, like the vaccines, these drugs are becoming less effective as the virus mutates [18,19]. The representative examples of antiviral drugs in Table 2 illustrate how the selectivity of this drug group varies from excellent to poor.

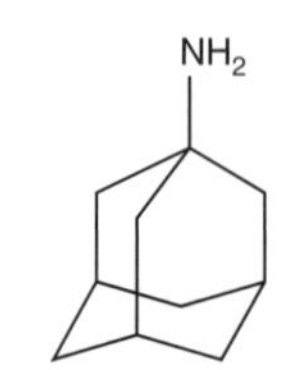

NH_2

31. Amantadine

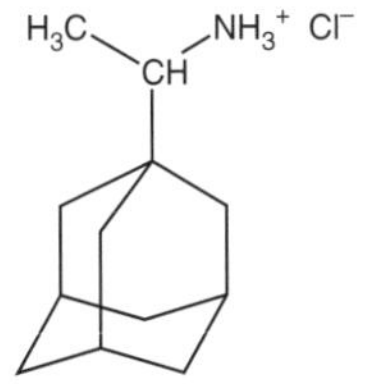

32. Rimantadine

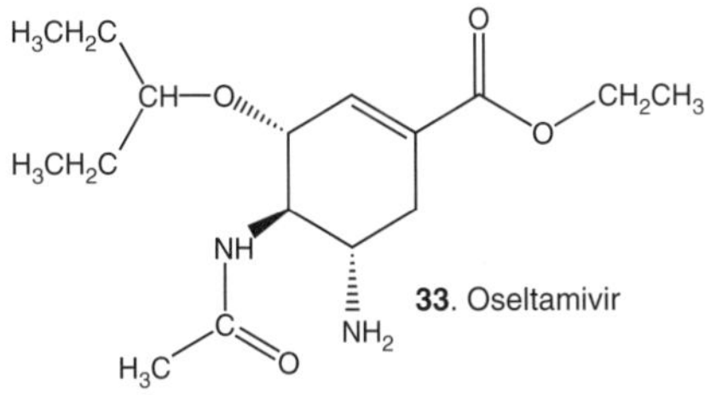

33. Oseltamivir

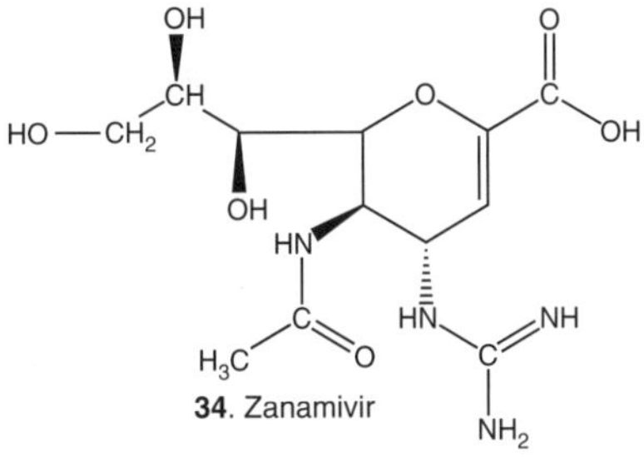

34. Zanamivir

3.3. Antimycobacterial Drugs: Comparative Biochemistry and Comparative Cytology

Mycobacteria-caused diseases are difficult to treat because of the bacterium's complex outer membrane makes it difficult to get the drug inside the microorganism. It also is very slow growing making it less susceptible to drugs that will stop the mycobacterial's reproduction. There are four drugs that are considered first line for the treatment of this difficult-to-treat disease. Three of these, isoniazid (**35**), pyrazinamide (**36**) and ethambutol (**37**) inhibit cell wall formation. Isoniazid inhibits mycolic acid biosynthesis after activation by a bacterial catalase to an acyl anion or radical [20]. Mycolic acid is an essential part of the mycobacterium's cell wall. In contrast with the β-lactam antibiotics that show excellent selectivity, isoniazid has a boxed warning regarding adverse responses involving the liver, particularly in older patients whose lifestyle may have damaged this organ. Pyrazinamide's mode of action is poorly understood but appears to disrupt the organism's membrane transport and energetics [21,22]. Like isoniazid, pyrazinamide is a prodrug and must be hydrolyzed to pyrazinoic acid. Finally, ethambutol obstructs cell wall formation by inhibiting mycobacterial arabinosyl transferase. Rifampin (**38**) shows a significant preference for DNA-dependent RNA polymerase. While the selectivity for the bacterial enzyme is very good, rifampin also induces the CYP1A2 and CYP3A4 isozymes and can be the cause of clinically significant drug–drug interactions that results in decreased serum concentrations of drugs metabolized by these isozymes.

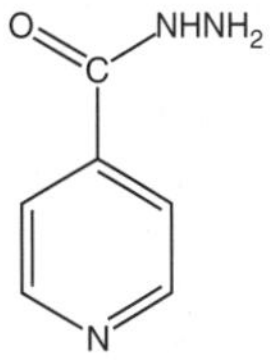

35. Isoniazid

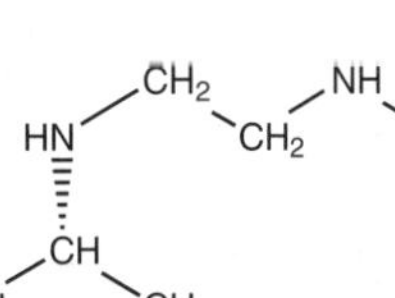

36. Pyrazinamide

37. Ethambutol

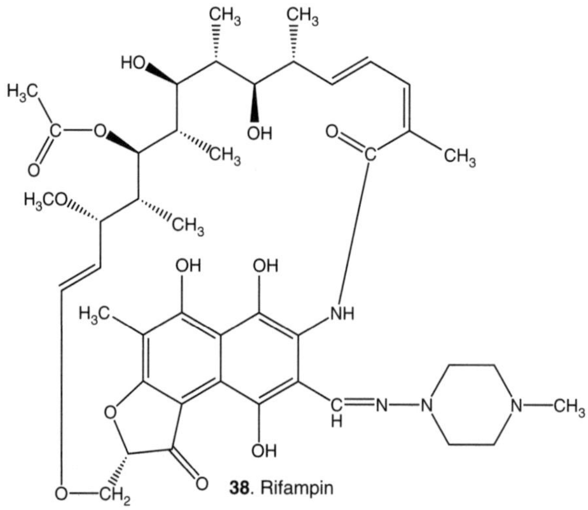

38. Rifampin

3.4. Antifungal Drugs: Comparative Biochemistry

Fungal metabolism is more similar to mammalian metabolism than to bacteria metabolism (Fig. 10). While fungi and plants produce ergosterol rather than cholesterol, fungi and mammals follow the same pathway to lanosterol from mevalonate and squalene [23]. In contrast, plants produce cycloartenol from mevalonate and squalene.

Many of today's antifungal drugs (ketoconazole (**39**) and econazole (**40**)) take advantage of the fact that fungi produce ergosterol rather than cholesterol. But ergosterol biosynthesis is very similar to that of cholesterol. The azole antifungal drugs inhibit the C-14 demethylase that removes the 14-methyl group from lanosterol thereby preventing the subsequent synthesis of ergosterol. For this reason, antifungal drugs that inhibit ergosterol's biosynthesis show poorer selectivity with the result that many antifungal drugs can only be used topically far away from the patient's organs that synthesize cholesterol. Antifungal drugs in the polyene class (amphotericin B (**41**) preferentially binds to ergosterol relative to cholesterol, possibly because fungal ergosterol has a cylindrical structure. This binding disrupts fungal membrane structure. Nevertheless, amphotericin B must be carefully dosed because of its nephrotoxity [24].

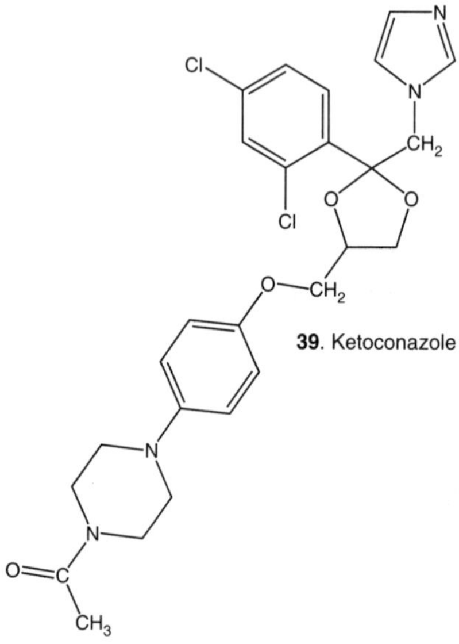

39. Ketoconazole

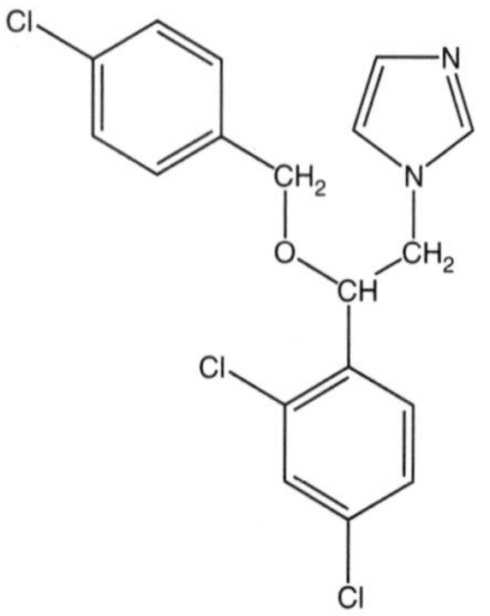

40. Econazole

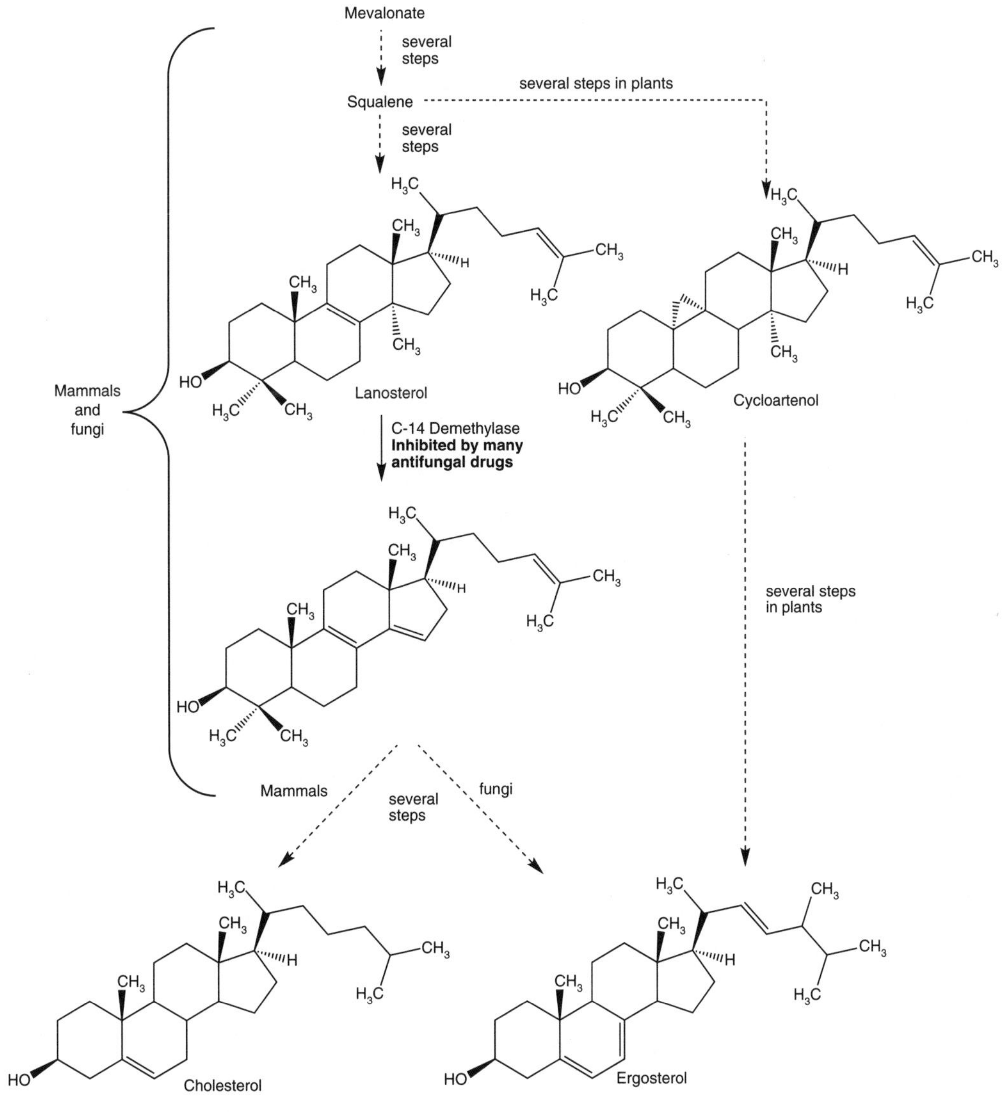

Figure 10. Outline of cholesterol and ergosterol biosynthesis.

There are unique properties of fungus for which there are drugs that take advantage of the differences between mammalian and fungal cell structure. The echinocandins (capsofungin (**42**), anidulafungin (**43**), and micafungin (**44**)) inhibit fungal 1,3-β-glucan synthase, an enzyme unique to fungus and, thereby, interfere with synthesis of the fungal cell wall polysaccharides. Relative to the antifungal agents in the preceding paragraph, the echinocandins have a limited toxicity profile [25].

3.5. Monoclonal Antibodies: Comparative Cytology

Monoclonal antibodies were going to be the fulfillment of Ehrlich's "magic bullets" particularly in the treatment of cancer where selective toxicity has been poor. They would carry the cytotoxic drug to the malignant cell where it would be released. The antibody would ignore benign cells. Alternatively, the antibody itself would attach to the malignant cell providing the initial step for further response by

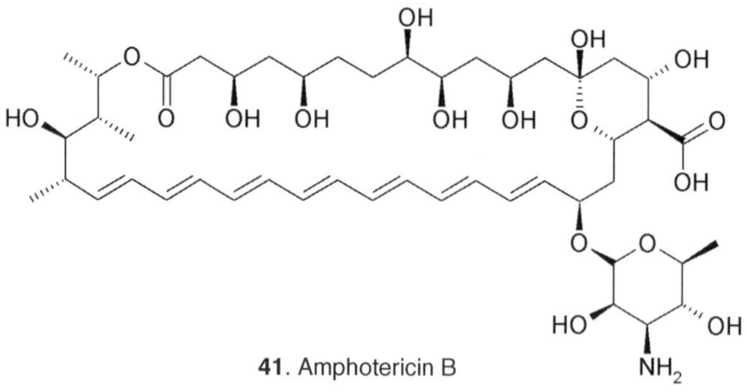

41. Amphotericin B

the patient's immune system involving complement, macrophages, and T killer cells.

The first hurdle is the main source of monoclonal antibodies (mAbs). They come from diverse nonhuman sources such as rodents. Repeated injections sensitize the patient to the animal protein. A partial solution has been "humanizing" the monoclonal antibody by using human genes to code for the constant region of the heavy and light chains and the animal genes to code for the variable regions. Depending on the relative extent of animal versus human, the results range from chimeric immunoglobulins that are 70–75% human to humanized immunoglobulins that are approximately 90–95% human. Table 3 contains a list of therapeutic monoclonal antibodies.

Their uses range from treatment of autoimmune disorder to a variety of malignancies. Their targets are just as broad. In the case of autoimmune diseases (rheumatoid arthritis, Crohn's disease, and ankylosing spondylitis), the target is tissue necrosis factor alpha (TNFα) that is part of the immune system's normal inflammation response to infection. Thus, patients taking these mAbs are at increased risk of infection. Indeed, a monoclonal antibody, efalizumab, for the treatment of psoriasis was withdrawn in 2009 because some patients developed progressive multifocal leukoencephalopathy (PML), a progressive neurlogic disease caused by a virus that infects the central nervous system. This MAB's target is a lymphocyte function-associated antigen leading to immunosuppression.

The targets for those mAbs used in the treatment of malignancies include cell surface antigens (CD52, CD20, CD33), and growth factors (vascular endothelial growth factor or VEGF, epidermal growth factor receptor or

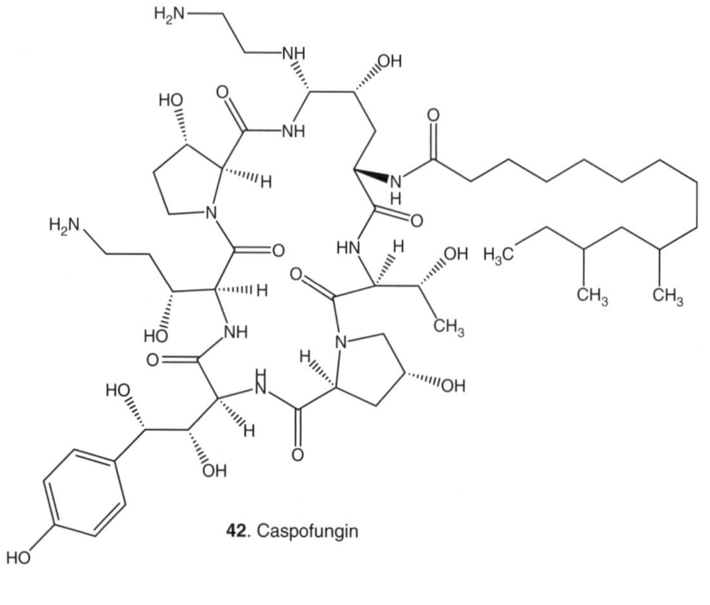

42. Caspofungin

Table 3. Therapeutic mAbs

Generic Name	mAb Type	Target	Indication
Abciximab	Chimeric	Inhibition of IIb/IIIa (an integrin on platelets)	Platelet aggregation inhibitor
Adalimumab	Human	TNFα	Rheumatoid arthritis, psoriatic arthritis, ankylosing spondylitis, Crohn's disease
Alemtuzumab	Humanized	CD52 (a glycosylphosphatidylinositol (GPI)-anchored antigen expressed on the all lymphocytes)	B-cell chronic lymphocytic leukemia
Basiliximab	Chimeric	Interleukin-2 receptor α-chain (IL-2Rα) on T cells	Acute rejection of a transplanted kidney
Bevacizumab	Humanized	Vascular endothelial growth factor (VEGF)	Nonsmall cell lung cancer, colorectal cancer; breast cancer
Cetuximab	Chimeric	Epidermal growth factor receptor (EGFR)	Colorectal cancer, head and neck cancer
Daclizumab	Humanized	IL-2Rα on T cells	Acute rejection of a transplanted kidney
Gemtuzumab ozogamicin	Humanized-conjugated	CD33 antigen, a sialic acid-dependent adhesion protein found on the surface of leukemic blasts and immature normal cells of myelomonocytic lineage	Acute myeloid leukemia
Ibritumonmab tiuxetan (with Indium-111 or yttrium-90)	Murine-conjugated	CD20 antigen found on the surface of normal and malignant B cells (but not B-cell precursors)	Low-grade or follicular CD-20 positive B-cell non-Hodgkin's lymphoma
Infliximab	Chimeric	TNFα	Rheumatoid arthritis; Psoriatic arthritis; Ankylosing spondylitis; Crohn's Disease; Ulcerative colitis
Natalizumab	Humanized	α-4 Integrin cellular adhesion molecule	Multiple sclerosis
Omalizumab	Humanized	Human immunoglobulin E (IgE)	Asthma
Palivizumab	Humanized	A antigenic site of the F protein of respiratory syncytial virus (RSV)	Lower respiratory tract disease caused by RSV
Panitumumab	Human	Epidermal growth factor receptor	Colorectal cancer
Rituximab	Chimeric	CD20 antigen found on the surface of normal and malignant B cells (but not B cell precursors)	Low-grade or follicular, CD20-positive, B-cell, non-Hodgkin's lymphoma; Rheumatoid arthritis
Tositumomab-iodine 131	Murine	CD20 antigen found on the surface of normal and malignant B cells (but not B-cell precursors)	Low-grade or follicular, CD20-positive, B-cell, non-Hodgkin's lymphoma
Trastuzumab	Humanized	Human epidermal growth factor receptor 2 (HER2)	Breast cancer
Ustekinumab	Humanized	Interleukin-12 p40 subunit, Interleukin-23	Plaque psoriasis

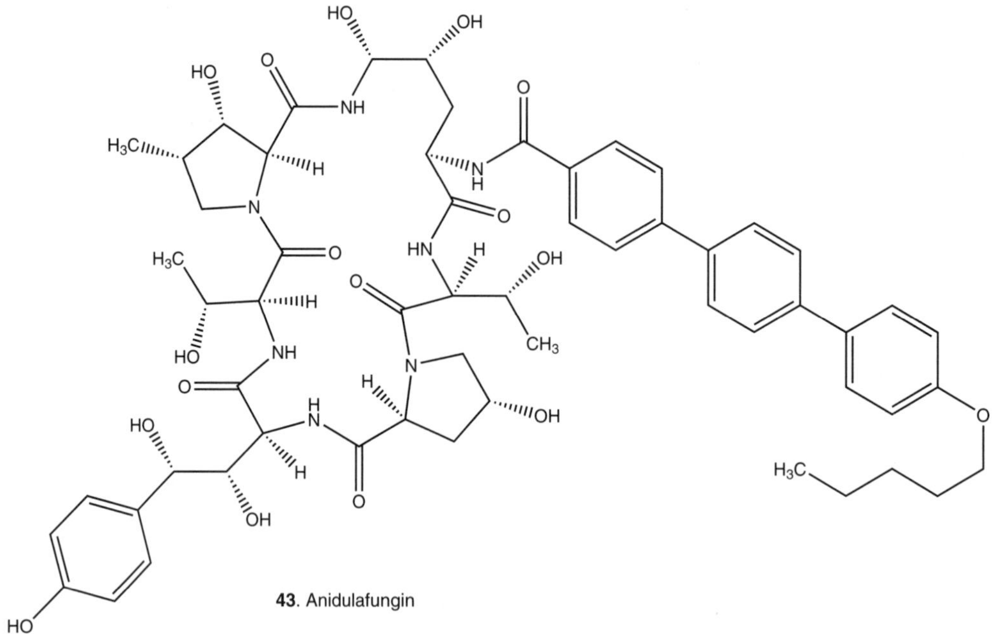

43. Anidulafungin

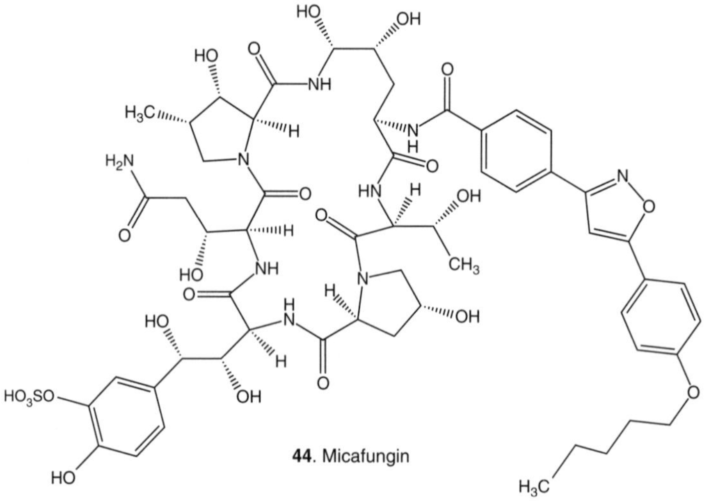

44. Micafungin

EGFR, human epidermal growth factor receptor 2 or HER2). Other targets are listed in Table 2. Note that there are only three mAbs that are carriers to toxic agents to the target cell: gemtuzumab ozogamicin, ibritumonmab tiuxetan indium-111 or yttrium-90, and tositumomab-iodine-131. All of the therapeutic mAbs carry significant warnings because the target antigens are not limited to malignant cells or tissues under attack by the patient's immune system (autoimmune diseases). In other words, the mAbs have excellent target selectivity but not distribution selectivity.

3.6. Tyrosine Kinase Inhibitors: Comparative Biochemistry

The search for selective metabolism in malignant cells has focused on unique proteins synthesized by the cell that are involved in cell division. In the case of chronic myeloid leukemia (CML), imatinib (**45**) inhibits a protein-tyrosine kinase formed by the Bcr-Abl gene. The latter, and its protein product, are not found in normal cells. While this drug does show selectivity, tyrosine kinases are common enzymes and imatinib also inhibits this tyrosine kinase receptor in platelet-derived growth factor and stem cell factor. Therefore, the patient can experience thrombocytopenia and neutropenia. A greater problem is point mutations in the Bcr-Abl gene that allosterically inhibit imatinab binding to Bcr-Abl. A second generation of Bcr-Abl inhibitors are now being prescribed in an attempt to overcome this point mutation. Dasatinib (**46**) and nilotinib (**47**) are two examples [26].

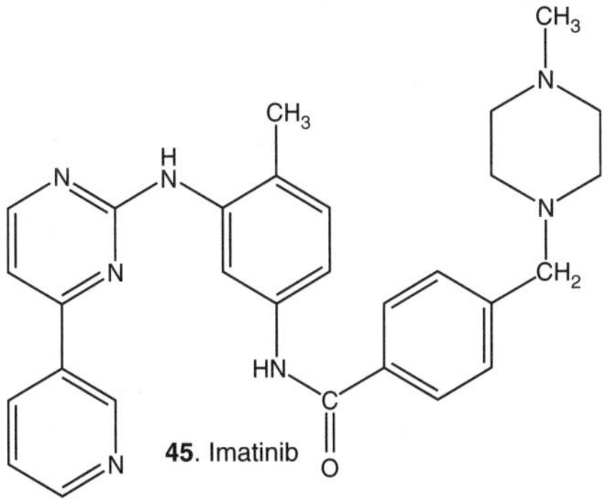

45. Imatinib

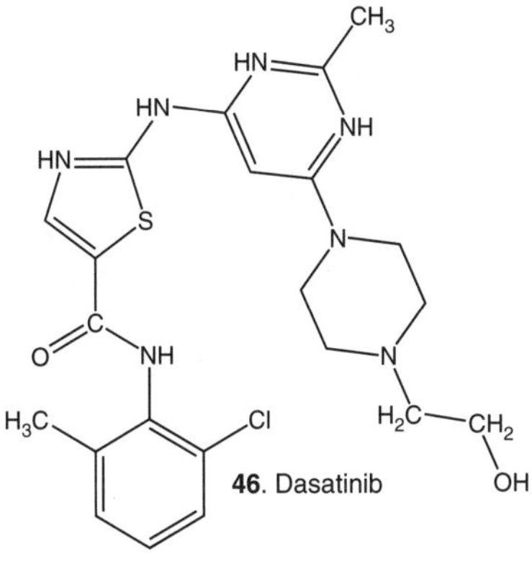

46. Dasatinib

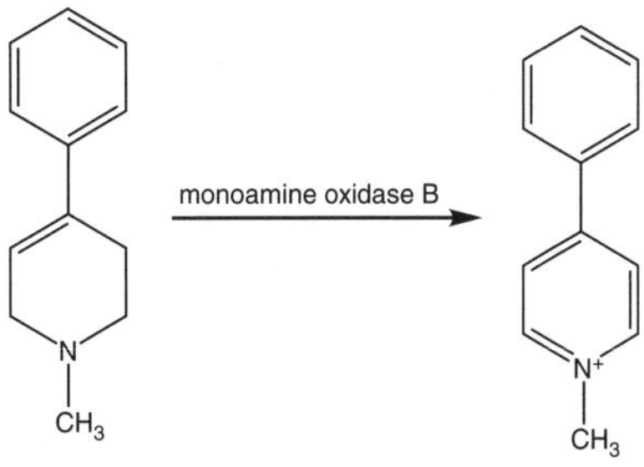

monoamine oxidase B

Figure 11. Synthesis of MPP.

explain how receptors are destroyed during the disease process. Parkinson's disease involves destruction of dopaminergic receptors. Molecular probes that are very specific to dopaminergic receptors targeted in Parkinson's disease and do not affect other receptors in the brain are useful in determining how the disease progresses. One such probe is 1-methyl-4-phenylpyridinium (MPP^+) obtained by monoamine oxidase B oxidation of 1-methyl-1,2,3,6-tetrahydropyridine (MPTP) (Fig. 11). MPP^+ inhibits mitochondrial complex I. It has been suggested that this inhibition results in less ATP for the neuron, depolarization of the mitochondria and generation of reactive oxygen species that leads to neuronal death. This model does not explain why dopaminergic neurons are more susceptible to MPP^+ relative to other aerobic cells. In a series of experiments using another mitochondrial complex I inhibitor, rotenone, it appears that MPP^+ is specific to the structure of the dopaminergic receptor [27].

One of the drug treatments indicated for Parkinson's disease is inhibition of monamine

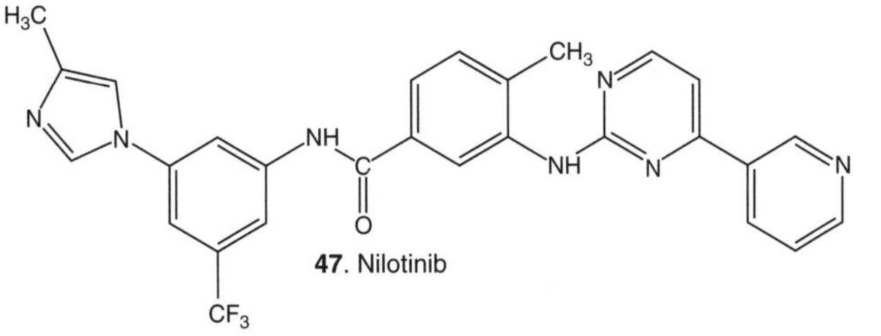

47. Nilotinib

3.7. Monoamine Oxidase Inhibitors: Comparative Biochemistry

While selective toxicity in drug design focuses on mechanisms that avoid adverse reactions, the concept can be applied to models that help

oxidase, a flavin enzyme that oxidatively deaminates neurotransmitters including serotonin, norepinephrine, epinephrine, and dopamine. These enzymes are widely distributed both centrally and peripherally. The early

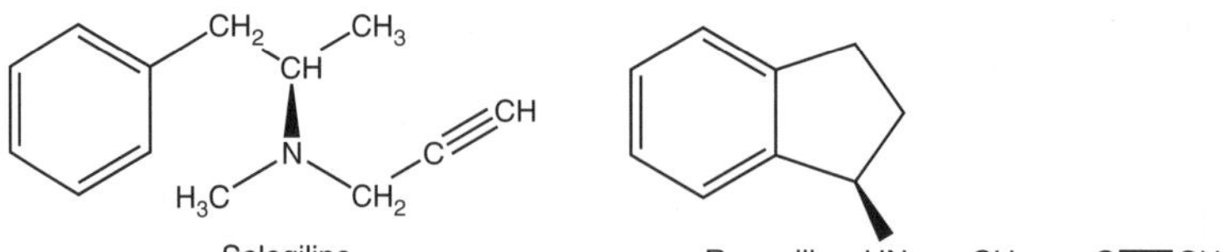

Selegiline

Rasagiline HN—CH_2—C≡CH

Figure 12. Monoamine oxidase B inhibitors.

monoamine oxidase (MAO) inhibitors were nonselective and associated with a wide variety of adverse reactions that were both physical and mental. There are two MAO classes: MAO-A that oxidatively deaminates serotonin, norepinephrine and epinephrine and MAO-B that preferentially oxidatively deaminates dopamine. One of the treatments for Parkinson's disease is to increase the brain concentration of dopamine either by administering its precursor L-DOPA or inhibit its metabolism in the brain. The latest iteration of MAO inhibitors, selegiline, and rasagiline (Fig. 12), has an acetylene pharmacophore that irreversibly inhibits MAO-B.

3.8. Levobupivacaine: Comparative Stereochemistry and Comparative Distribution

The *S*-isomer of the local anesthetic, levobupivacaine (**48**), reduces the cardiotoxicity of the racemic bupivicaine while retaining good local anesthetic activity. Many of the local anesthetics affect the neurons servicing the cardiac muscle. Sometimes, this can be useful for the treatment cardiac arrhythmias. On the other hand, particularly when administered in large amounts such that plasma levels become significant, some local anesthetics may cause depression of the myocardium, decreased cardiac output, heart block hypotension, bradycardia, and ventricular arrhythmias. Bupivacaine has been resolved with the *S*-isomer showing less cardiotoxic responses but still good local anesthetic activity [28].

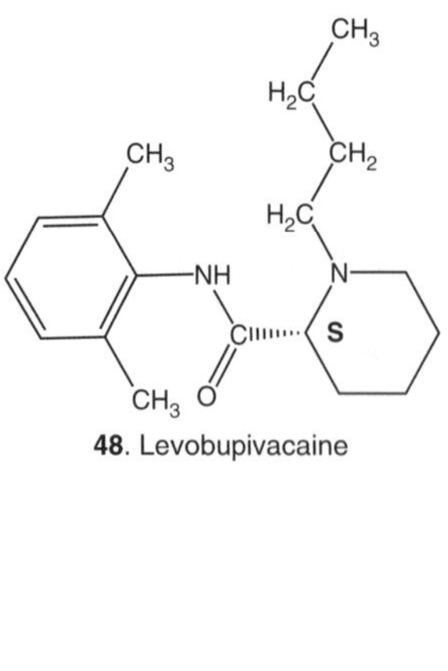

48. Levobupivacaine

3.9. Bisphosphonates: Comparative Distribution

Osteoporosis is an increasing problem in an aging population. Its mechanism is very complex because bone metabolism is intricate. Besides its structural or support role, bone is the body's calcium reservoir used to maintain calcium homeostasis. That means that there has to be a means for calcium to be stored as hydroxyapatite when not needed and removed as blood levels begin to decrease. For purposes of this discussion, consider osteoporosis resulting when the osteoblast (bone forming) and osteoclast (bone resorption) cells are not in balance.

Two examples of bisphosphonates (**49**), alendronate (R = 2-*n*-propylamine) and risedronate (R = 3-pyridyl), are used for reducing the rate of calcium loss from bone tissue. Both are administered orally, widely distribute in the tissues and then concentrate onto the hydroxyapatite where there are osteoclast cells. While the mechanism of their toxicity is not understood, this drug class can cause local irritation and, for some patients, actual damage to the gastric linings.

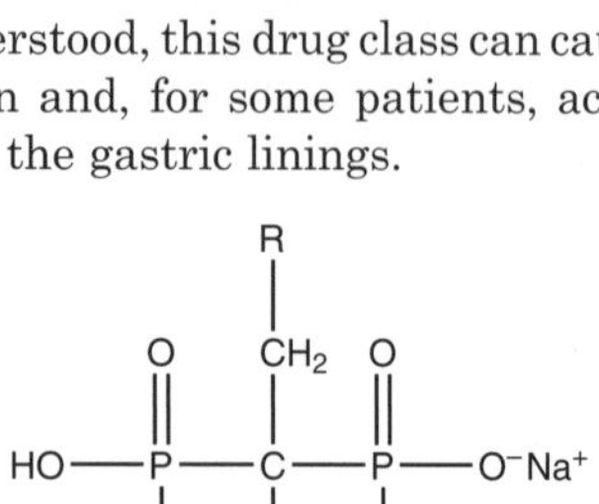

49. Bisphosphonate

3.10. Antihistamines: Comparative Biochemistry and Comparative Distribution

With two distinct classes of receptors and significant differences in their locations, antihistamines have good to excellent selectivity. First there are the H1 and H2 receptor classes.

Histamine is produced by a variety of cells. Patients taking the H1 antihistamines do so in response to the release of histamine from mast cells. These are located in the respiratory passages, skin, and gastrointestinal tract and are the cause of what patients refer to as an allergic response. The "first generation" of H1 antihistamines represented by diphenhydramine (**50**) and chlorpheniramine (**51**) are effective, but they cross the blood–brain barrier causing mild to significant sedation. Indeed, some of the early antihistamines are used as nonprescription sleep aids. They also could show anticholinergic effects. Sometimes, the combination of crossing the blood–brain barrier and anticholinergic activity was put to good use as antinausea drugs, particularly when caused by motion sickness. The familiar Dramamine® is the chlorotheophylline salt of diphenhydramine. Thus, useful as these products are, the first generation of antihistamines are examples of drugs that have "nonselective" toxicity. Remember that as used here, "toxicity" does not mean harmful to the patient. A drug's pharmacological response is an intervention in normal biochemical processes, and the first-generation H1 antihistamines are intervening not only in the normal role of histamine but also acetylcholine.

In contrast, the second generation of H1 antihistamines, as characterized by loratidine (**52**), fexofenadine (**53**) and cetirizine (**54**) show better selectivity because they are less likely to cross the blood–brain barrier, and, therefore, preferentially inhibit peripheral H1 receptors [29]. They also show less anticholinergic activity. As can be seen from their structures, they tend to be larger molecules. Nevertheless, their distribution is not an either/or situation. The second-generation H1 antihistamines still have central effects.

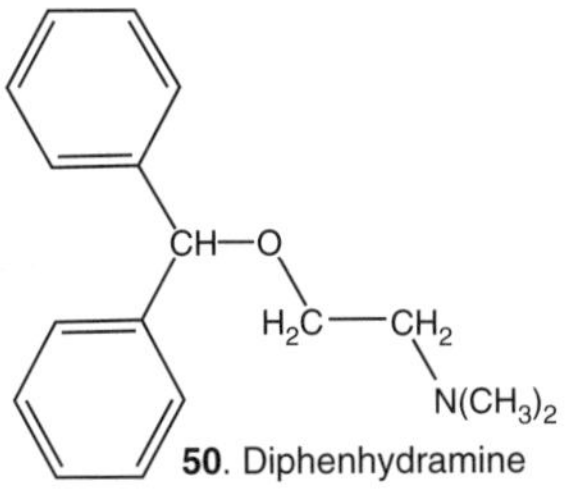

50. Diphenhydramine

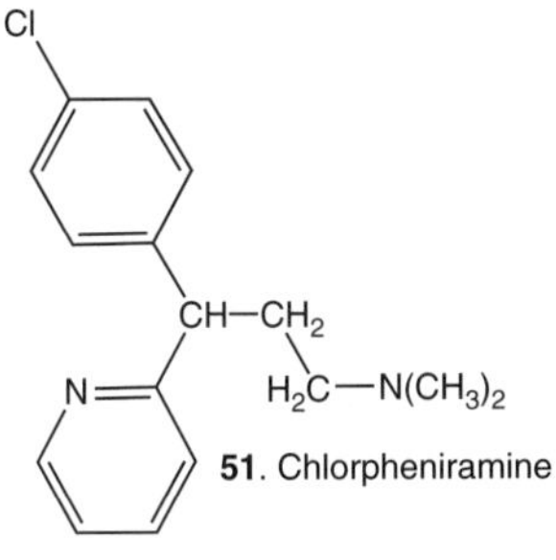

51. Chlorpheniramine

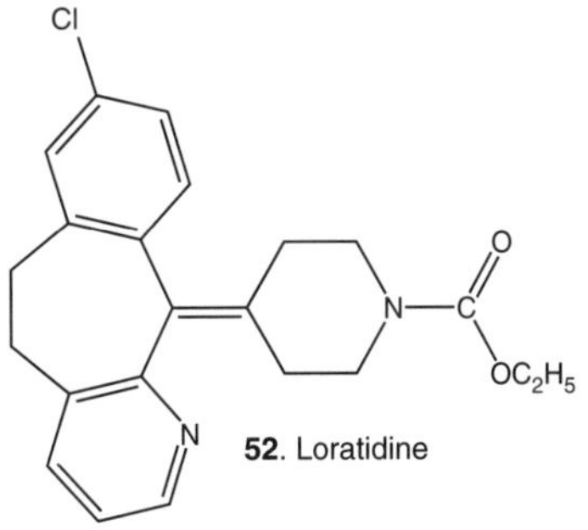

52. Loratidine

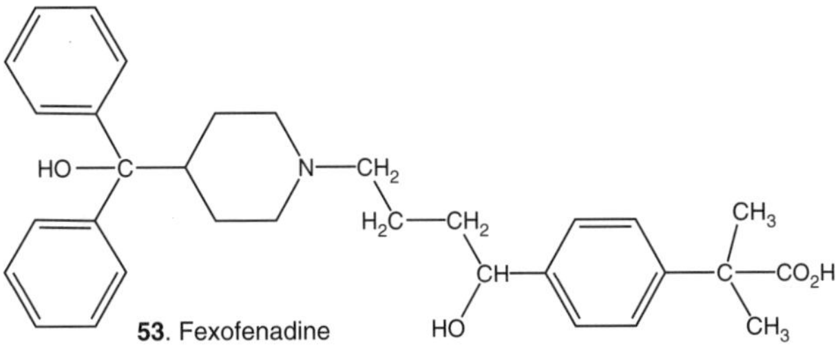

53. Fexofenadine

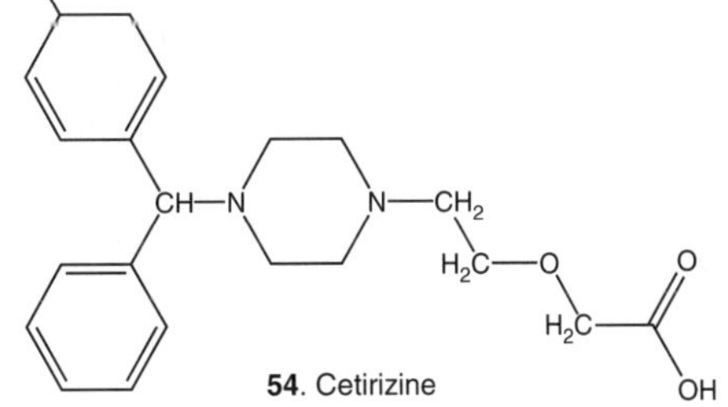

54. Cetirizine

The H2 antihistamines, or blockers, are mainly used to reduce the secretion of gastric HCl. Their structure–activity relationships are significantly different from that of the H1 antagonists. Using cimetidine (**55**) as the prototypical molecule, H2 antagonist structure–activity relationships are based on the histamine structure. The selectivity of this group of drugs for the H2 receptor in the gastric lining is remarkable when one considers how widely distributed are the H2 receptors. This is an example of how a drug's selective distribution for the H2 receptors in the gastric lining is good enough that the H2 antagonists are free enough from adverse reactions to permit their approval for nonprescription use.

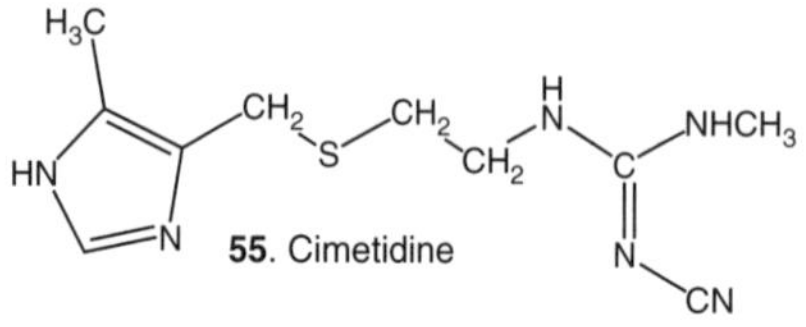

55. Cimetidine

3.11. Proton Pump Inhibitors: Comparative Distribution

More specific relative to the H2 antagonists are the proton pump inhibitors. This widely used class of drugs irreversibly blocks the hydrogen/potassium adenosine triphosphatase enzyme system (H^+/K^+ ATPase) also called the "gastric proton pump" found in the gastric lining. The latter system is the terminal stage in gastric acid secretion into the stomach, the only organ in the human body that is not damaged by low pHs [30]. The proton pump inhibitor group of drugs, lansoprazole (**56**), omeprazole (**57**) and its *S*-isomer esomeprazole (**57**), are inactive in their noncharged form and lipophilic such that they are readily distribute into the parietal cell canaliculus. The latter is acidic enough to protonate the drugs converting them into their tetracyclic planar sulphenamide active form (Fig. 13). The active sulphenamide binds covalently to cysteine residues on the H^+/K^+ ATPase [31].

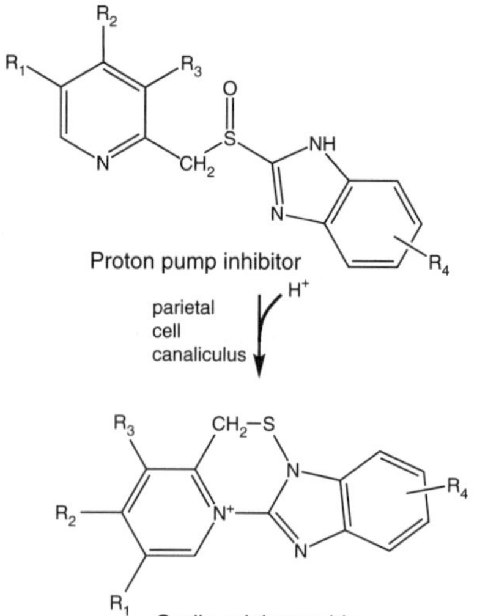

Proton pump inhibitor

parietal cell canaliculus

Cyclic sulphenamide

Figure 13. Activation of proton pump inhibitors in parietal cell canaliculus.

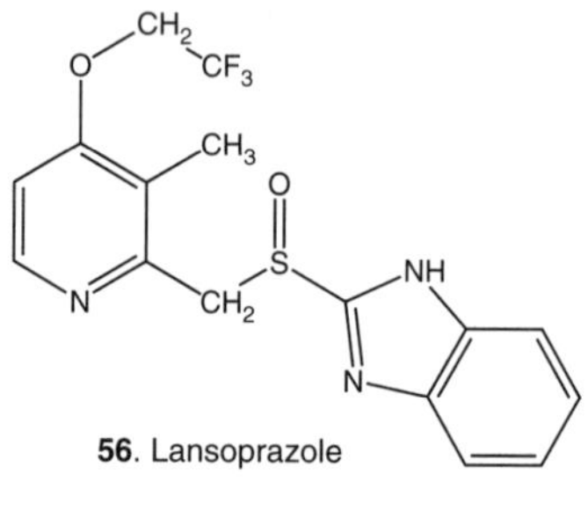

56. Lansoprazole

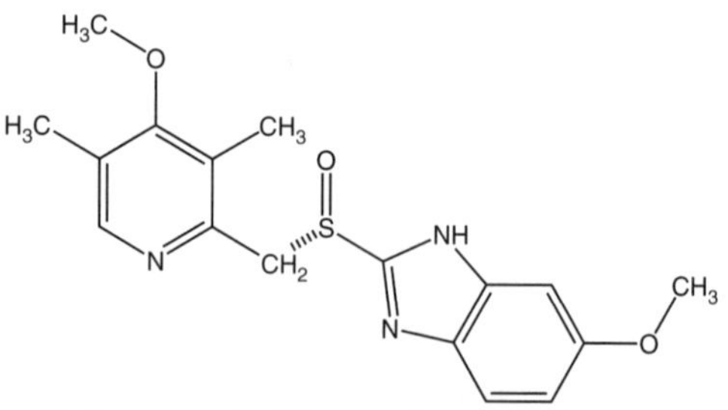

57. Esomeprazole (*S*-isomer of omeprazole)

3.12. Selective Estrogen Receptor Modulators: Comparative Biochemistry and Comparative Distribution

The pharmacology of the selective estrogen receptor modulators (SERMs) is complex. They are not simple agonists or antagonists.

Rather they can be considered variable agonists and antagonists. Their selectivity is very complex because it is be dependent on the organ where the receptor is located.

This complexity can be illustrated with tamoxifen (**58**) that is used for estrogen sensitive breast cancer and reducing bone loss from osteoporosis [32]. But prolonged treatment increases the risk of endometrial cancer. Thus, tamoxifen is an estrogen antagonist in the mammary gland and an agonist in the uterus and bone. In contrast, raloxifene (**59**) does not appear to have much agonist properties in the uterus, but, like tamoxifen, it is an antagonist in the breast and agonist in the bone.

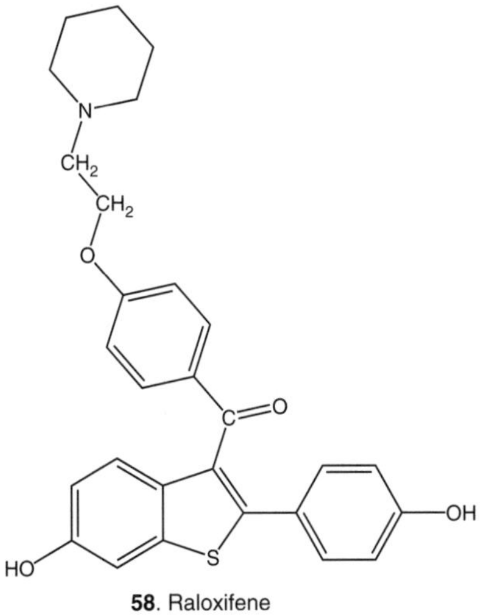

58. Raloxifene

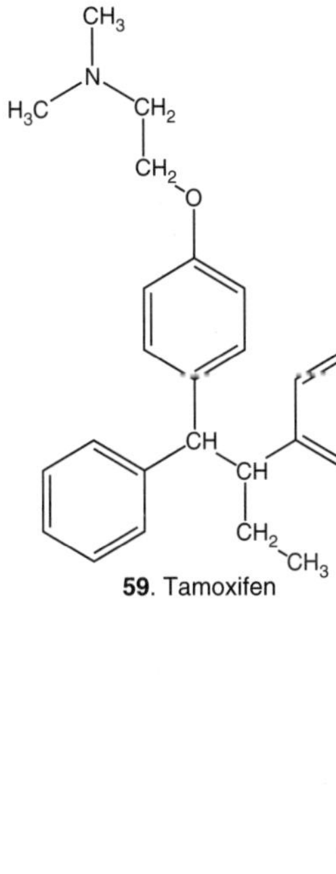

59. Tamoxifen

3.13. Phosphodiesterase Inhibitors: Comparative Distribution

There are 11 phosphodiesterase (PDE) families unevenly distributed throughout the human body [33]. They also vary in terms of their specificity for cAMP and cGMP. Phosphodiesterase type 4 (PDE4) is found in the pulmonary airways leading to the development of PDE4 inhibitors cilomilast (**60**) and roflumilast (**61**) for the possible treatment of asthma, chronic obstructive pulmonary disease and pulmonary hypertension. Phosphodiesterase 1 (PDE1) is expressed in the brain. Inhibitors of phosphodiesterase type 5 (PDE5) have an important role in maintaining a desired lifestyle: treatment of erectile dysfunction caused by a variety of conditions. They also are indicated for pulmonary arterial hypertension (PAH). Originally developed for the treatment of angina (and not effective for this purpose), male test subjects reported the ease of having an erection, and "the rest is history" [34]. A complex mechanism is involved. Nitric oxide (NO) activates guanylate cyclase forming cyclic GMP (cGMP) that is hydrolyzed by a phosphodiesterase. Sildenafil (**62**) and the newer compounds, tadalafil (**63**) and vardenafil (**64**), selectively inhibit phosphodiesterase type 5. While the selectivity is good, it must be remembered that cyclic GMP, like cyclic AMP, is ubiquitous and, therefore, the phosphodiesterases required to hydrolyze these chemical transmitters are also ubiquitous. Phosphodiesterase type 6 (PDE6) is found in the retina and explains the vision abnormalities experienced by some users of these drugs. Sildenafil shows a 10–8500 times preference for the type 5 isoform [35,36].

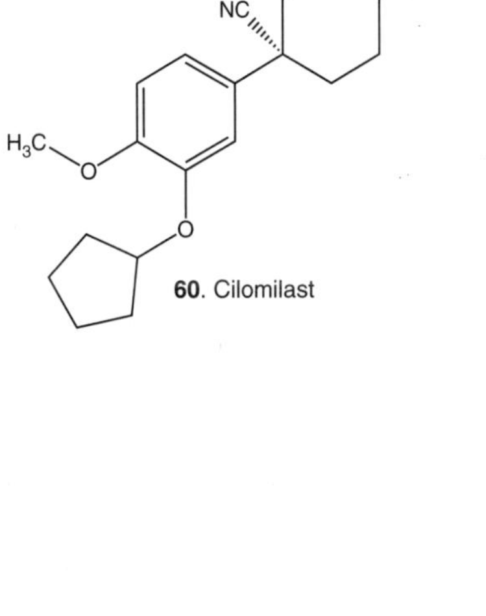

60. Cilomilast

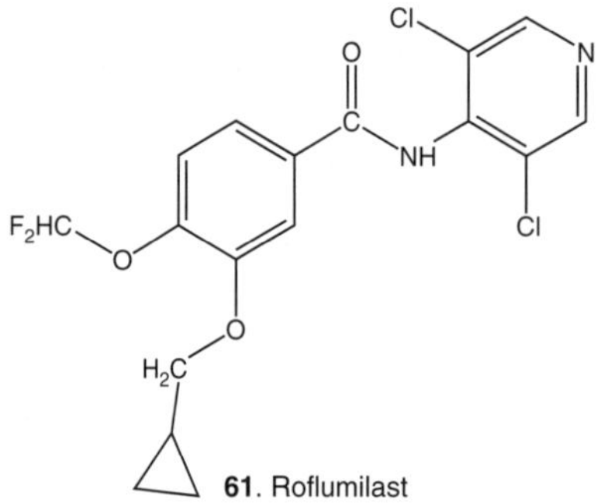

61. Roflumilast

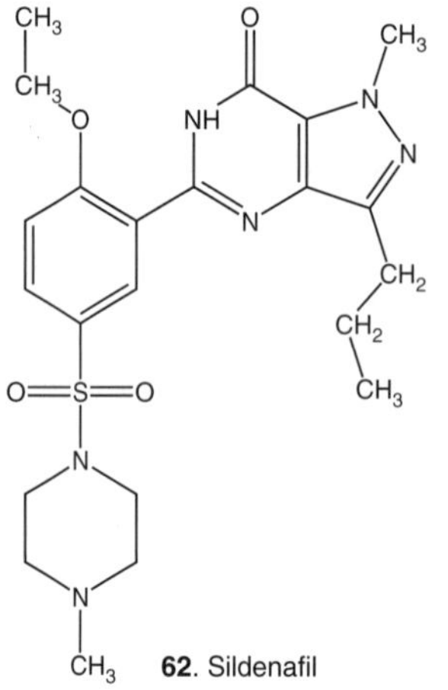

62. Sildenafil

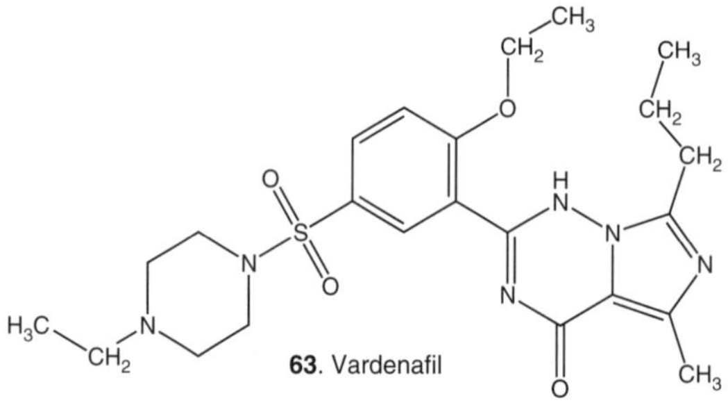

63. Vardenafil

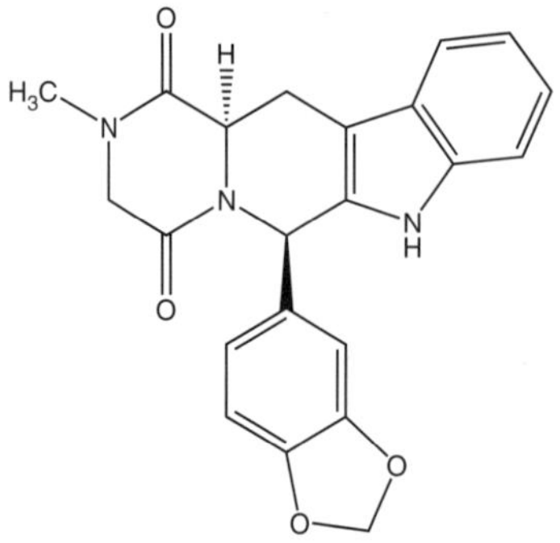

64. Tadalafil

3.14. Nonbenzodiazepine Sedatives: Comparative Distribution

The γ-aminobutyric acid ($GABA_A$) receptor is the target for a variety of drugs including anticonvulsants, muscle relaxants, anxiolytics and sedatives. An interesting type of selectivity is seen with zolpidem (**65**), eszopiclone (**66**) and zaleplon (**67**) that bind to specific receptor $GABA_A$ subunits. In the lower doses used to treat insomnia, there is sedation without any other benefits expected with $GABA_A$ agonists [37,38]. There does appear to be some potential for abuse, but it is difficult to decide if this is a distinct pharmacological syndrome [39,40].

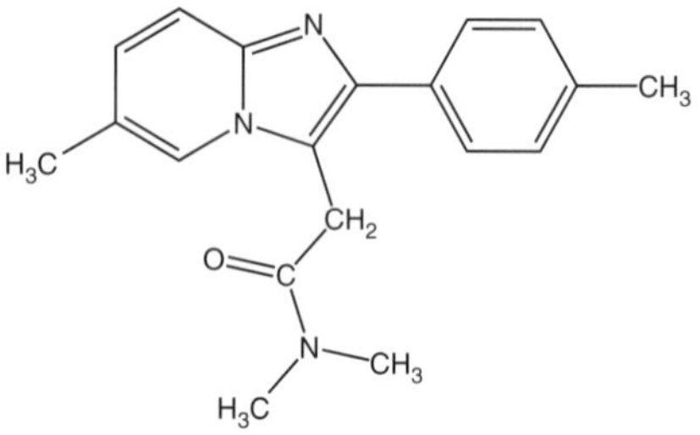

65. Zolpidem

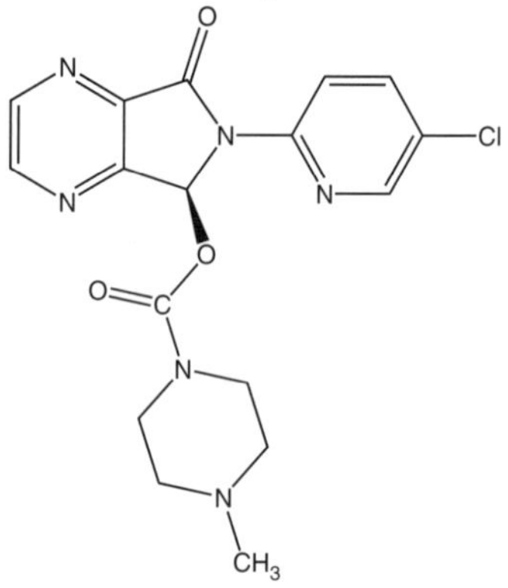

66. Eszopiclone

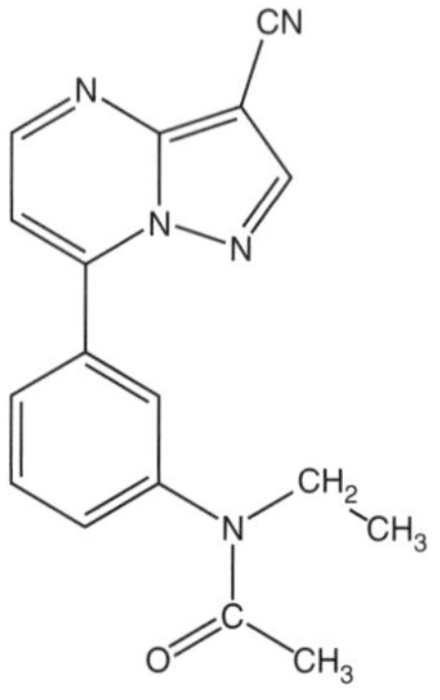

67. Zaleplon

3.15. Selective Serotonin Reuptake Inhibitors: Comparative Biochemistry and Comparative Distribution

There are several neurotransmitters that function in the human brain including serotonin, norepinephrine, epinephrine, dopamine, and acetylcholine. Some mental illnesses can be treated pharmacologically by increasing the concentration of a particular neurotransmitter. The challenge is to focus the drug therapy on one neurotransmitter (comparative biochemistry). The selective serotonin reuptake inhibitors (SSRIs) increase the extracellular level of serotonin by inhibiting its reuptake into the presynaptic cell making it available to the postsynaptic receptor. The SSRIs are indicated for depression, anxiety disorders, and some personality disorders. A challenge in the designing this class of drugs is the fact that nearly 90% of serotonin is produced in the periphery with most of the serotonin receptors expressed both within the outside of the central nervous system [41]. The goal in developing the SSRI class of antipsychotics is producing drugs that show good preference for the serotonin reuptake system (comparative biochemistry) and easily pass the blood–brain barrier (comparative distribution). With the number of warnings regarding this group of drugs (fluoxetine (**68**), paroxetine (**69**), sertraline (**70**) and citalopram and its S-isomer escitalopram (**71**)), one could argue that the SSRIs show good comparative biochemistry by being preferentially selective for the serotinergic reuptake system but poor comparative distribution by acting at the peripheral sites as well as those in the brain. When evaluating the selective toxicity of any of the antipsychotics, it must be remembered that most patients for whom this group of drugs is indicated already have shifts in their brain chemistry.

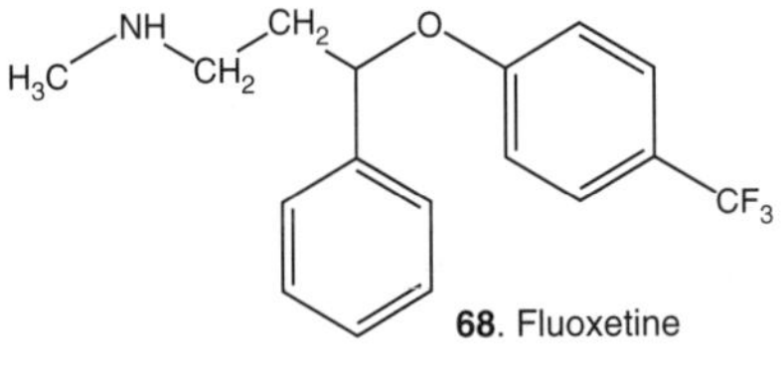

68. Fluoxetine

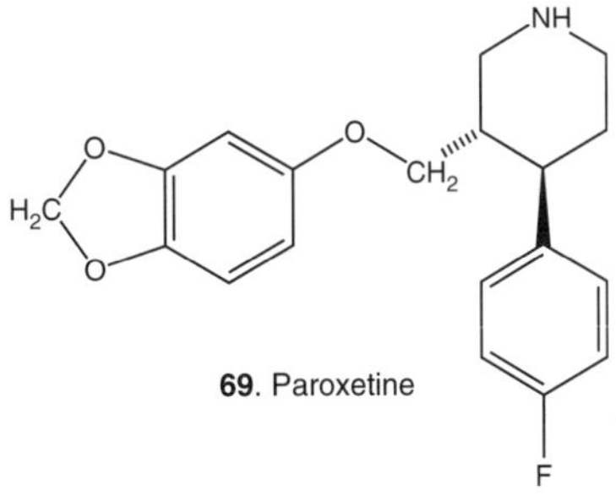

69. Paroxetine

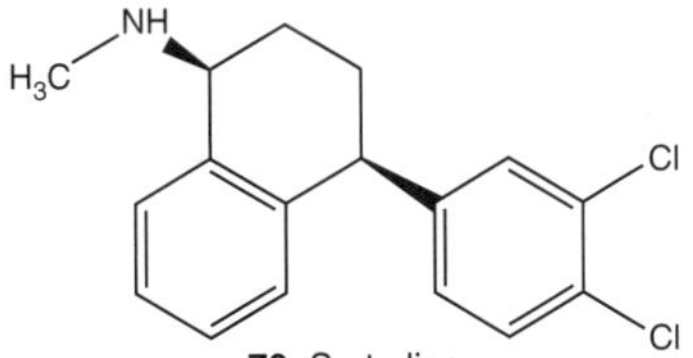

70. Sertraline

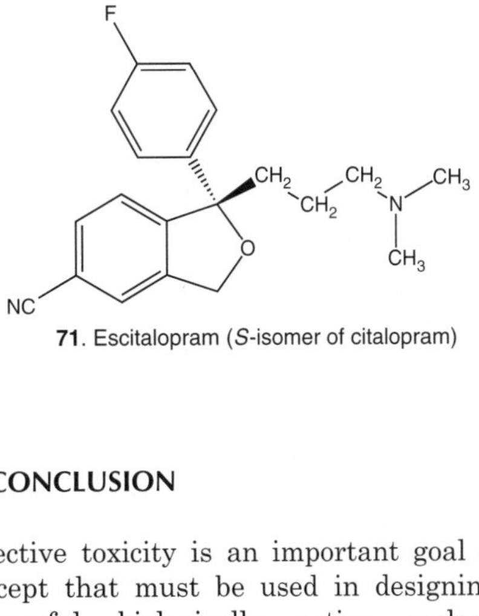

71. Escitalopram (*S*-isomer of citalopram)

4. CONCLUSION

Selective toxicity is an important goal and concept that must be used in designing a successful biologically active molecule. Today's drugs increasingly are being used to treat complex disease processes whose target receptors are found at several locations throughout the patient's body. There is selectivity within these receptors. The challenge continues to be to discover these differences and then use the principles of comparative biochemistry, distribution, cytology and/or stereochemistry in designing new and better drugs.

REFERENCES

1. Albert A. Selective Toxicity: The Physico–Chemical Basis of Therapy. 7th ed, New York: Chapman & Hall; 1985.
2. Albert A. Selective Toxicity: The Physicochemical Basis of Therapy. Vol. 1. 7th ed. (Sentaku Dokusel: Yakuzai Sayo no Seigyo to sono Bunshiteki Kiso, Jo Kan. Dai 7 Han), Tokyo, Japan: Scientific Societies Press; 1993.
3. Albert A. Selective Toxicity: The Physicochemical Basis of Therapy. Vol. 2. 7th ed. (Sentaku Dokusel: Yakuzai Sayo no Seigyo to sono Bunshiteki Kiso, Jo Kan. Dai 7 Han), Tokyo, Japan: Scientific Societies Press; 1994.
4. Albert A. Selective Toxicity: The Physico–Chemical Basis of Therapy. 7th ed, New York: Chapman & Hall; 1985. p 17–20.
5. Albert A. Selective Toxicity: The Physico–Chemical Basis of Therapy. 7th ed, New York: Chapman & Hall; 1985. p 56–117.
6. Gale GR, Rosenblum MG, Atkins LM, Walker EM, Jr, Smith AB, Meischen SJ. Antitumor action of *cis*-dichlorobis(methylamine)platinum(II). J Natl Cancer Inst 1973;51 (4):1227–1234.
7. Bitha P, Carvajal SG, Citarella RV, Child RG, Delos Santos EF, Dunne TS, Durr FE, Hlavka JJ, Lang SA, Jr, Lindsay HL, et al. Water-soluble third genration antitumor platinum complexes, [2,2-bis(aminomethyl)-1,3-propanediol-N,N′]-[1,1-cyclobutanedicarboxylato (2-)-O,O′] platinum(II). J Med Chem 1989;32 (8):2015–2020.
8. Ciccarelli RB, Solomon MJ, Varshavsky A, Lippard SJ. *In vivo* effect of *cis*- and *trans*-diamminedichloroplatinum(II) on SV40 chromosomes: differential repair, DNA–protein cross-linking, and inhibition of replication. Biochemistry 1985;24(26):7533–7540. DOI: .
9. Albert A. Selective Toxicity: The Physico–Chemical Basis of Therapy. 7th ed, New York: Chapman & Hall; 1985. p 175–205.
10. Park BK. Warfarin: metabolism and mode of action. Biochem Pharmacol 1988;37(1):19–27.
11. Redman AR. Implications of cytochrome P450 2C9 polymorphism on warfarin metabolism and dosing. Pharmacotherapy 2001;21(2):235–242.
12. Albert A. Selective Toxicity: The Physico–Chemical Basis of Therapy. 7th ed, New York: Chapman & Hall; 1985. p 118–174.
13. Huy PTB, Meulemans A, Wassef M, ManuelC, Sterkers O, Amiel C. Gentamicin persistence in rat endolymph and perilymph after a two-day constant infusion. Antimicrob Agents Chemother 1983;23(2):344–346.
14. Rasko DA, Moreira CG, Li DR, Reading NC, Ritchie JM, Waldor MK, Williams N, Taussig R, Wei S, Roth M, Hughes DT, Huntley JF, Fina MW, Falck JR, Sperandio V. Targeting QseC signaling and virulence for antibiotic development. Science 2008;321(5901):1078–1080. DOI: 10.1126/science.1160354.
15. Haydon DJ, Stokes NR, Ure R, Galbraith G, Bennett JM, Brown DR, Baker PJ, Barynin VV, Rice DW, Sedelnikova SE, Heal JR, Sheridan JM, Aiwale ST, Chauhan PK, Srivastava A, Taneja A, Collins J, Czaplewski LG. An Inhibitor of FtsZ with potent and selective anti-staphylococcal activity. Science; 321(590) 1: 1673–1675. DOI: 10.1126/science.1159961.
16. Hiratsuka T, Furihata K, Ishikawa J, Yamashita H, Itoh N, Seto H, Dairi T. An alternative menaquinone biosynthetic pathway operating in microoganisms. Science 2008;321 (5901):1670–1673. DOI: 10.1126/science.1160446.
17. Havlir DV. HIV integrase inhibitors-out of the pipeline and into the clinic. N Engl J Med 2008;359(4):416–418.
18. http://www.who.int/csr/disease/influenza/oseltamivir_faqs/en/.
19. Zűrcher Z, Yates PJ, Daly J, Sahasrabudhe A, Walters M, Dash L, Tisdale M, McKimm-Breschkin JL, Mutations conferring zanamivir resistance in human influenza virus N2 neuraminidases compromise virus fitness and are not stably maintained *in vitro*. J Antimicrob Chemother 2006;58:723–732. DOI: 10.1093/jac/dk1321.
20. Timmins GS, Deretic V, Mechanisms of action of isoniazid. Mol Microbiol 2006;62(5):1220–1227. DOI: 10.1111/j.1365-2958.2006.05467.
21. Zimhony O, Cox JS, Welch JT, Vilchèze C, Jacobs WR, Jr. Pyrazinamide inhibits the eukaryotic-like fatty acid synthetase I (FASI) of *Mycobacterium tuberculosis*. Nat Med 2000;6 (9):1043–1047. DOI: 10.1038/79558.
22. Zhang Y, Wade MM, Scorpio A, Zhang H, Sun A. Mode of action of pyrazinamide: disruption of *Mycobacterium tuberculosis* membrane transport and energetics by pyrazinoic acid. J Antimicrob. Chemother 2003;52(5):790–795.
23. Corey EJ, Matsuda SPT, Bartel B. Isolation of an Arabidopsis thaliana gene encoding cycloartenol synthase by functional expression in a yeast mutant lacking lanosterol synthase by the use of a chromatographic screen. Proc Natl Acad Sci USA 1993;90(24):11628–11637.

24. Odds FC, Brown AJP, Gow NAR. Antifungal agents: mechanisms of action. TRENDS in Microbiology. 2003;11(6):272–279. DOI: 10.1016/S0966-842X(03)00117-3.
25. Morris MI, Villmann M. Echinocandins in the management of invasive fungal infections, part 2. Am J Health Syst Pharm 2006;63 (10):1813–1820. DOI: 10.2146/ajhp050464.p2.
26. Das J, Chen P, Norris D, Padmanabha R, Lin J, Moquin RV, Shen Z, Cook LS, Doweyko AM, Pitt S, Pang S, Shen DR, Fang Q, de Fex HF, McIntyre KW, Shuster DJ, Gillooly KM, Behnia K, Schieven GL, Wityak J, Barrish JC. 2-Aminothiazole as a novel kinase Inhibitor template. Structure–activity relationship studies toward the discovery of N-(2-chloro-6-methylphenyl)-2-[[6-[4-(2-hydroxyethyl0-1-piperazinyl)]-2-methyl-4-pyrimidinyl]amino)]-1,3-thiazole-5-carboxamide (Dasatinib, BMS-354825) as a potent *pan*-Src kinase inhibitor. J Med Chem 2006;49(23):6819–6832. DOI: 10.1021/jm060727.
27. Nakamura K, Bindokas VP, Marks JD, Wright DA, Frim DM, Miller RJ, Kang UJ. The selective toxicity of 1-methyl-4-phenylpyridinium to dopaminergic neurons: the role of mitochondrial complex I and reactive oxygen species revisited. Mol Pharmacol 2000;58(2):271–278.
28. Foster RH, Markham A. Levobupivacaine: a review of its pharmacology and use as a local anesthetic. Drugs 2000;59(3):551–579.
29. Mann RD, Pearce GL, Dunn N, Shakir S. Sedation with "non-sedating" antihistamines: four prescription-event monitoring studies in general practice. Brit Med J 2000;320 (72430):1184–1187.
30. Malfertheiner P, Fass R, Quigley EMM, Modlins IM, Malagelada JR, Moss SF, Holtmann G, Goh K-L, Katelaris P, Stanghellini V, Talley NJ, Tytgat GN, Wright NA. Review article: From gastrin to gastro-oesophageal reflux disease: a century of acid suppression. Aliment Pharmacol Therap 2006;23(6):683–690; DOI: 10.1111/j.1365-2036.2006.02817.x.
31. Stedman CAM, Barclay ML. Review article: comparison of the pharmacokinetics, acid suppression and efficacy of proton pump inhibitors. 2000;14(8):963–978.
32. Dutertre M, Smith CL. Molecular mechanisms of selective estrogen receptor modulator (SERM) action. J Pharmacol Exp Ther 2000;295(2):431–437.
33. Jeon YH, Heo Y-S, Kim CM, Hyun Y-L, Lee TG, Ro S, Cho JM. Phosphodiesterase: overview of protein structures, potential therapeutic applications and recent progress in drug development. Cell Mol Life Sci 2005;62:1198–1220; DOI: 10.1007/s00018-005-4533-5.
34. Nachtsheim D. Sildenafil: a milestone in the treatment of impotence. West J.Med 1998;169 (2):112–113.
35. Ballard SA, Gingell CJ, Tang K, Turner LA, Price ME, Naylor AM. Effects of sildenafil on the relaxation of human corpus cavernosum tissue in vitro and on the activities of cyclic nucleotide phosphodiesterase isozymes. J Urol 1998;159(6):2164–2167.
36. Langtry HD, Markham A. Sildenafil: a review of its use in erectile dysfunction. Drugs 1999;57(6) :967–989.
37. Salvà P, Costa J. Clinical pharmacokinetics and pharmacodynamics of zolpidem. Therapeutic implications. Clin Pharmacokinet 1995;29 (3):142–153.
38. Pritchett DB, Seeburg PH. Gamma-aminobutyric acid A receptor alpha 5-subunit creates novel type II benzodiazepine receptor pharmacology. J Neurochem 1990;54(5):1802–1804; DOI: 10.1111/j.14714159.1990.tb01237.x.
39. Soyka M, Bottlender R, Mőller HJ. Epidemiological evidence or a low abuse potential of zolpidem. Pharmacopsychiatry 2000;33 (4):138–141.
40. Hajak G, Muller WE, Wittchen JU, Pittrow D, Kirch W. Abuse and potential for the non-benzodiazepine hypnotics zolpidem and zopiclone: a review of case reports and epidemiological data. Addiction 2003;98(10):1371–1378.
41. Berger M, Gray JA, Roth BL. The expanded biology of serotonin. Annu Rev Med 2009;60:355–366; DOI: 10.1146/annurev.med.60.042307.110802.

CHEMOGENOMICS: SYSTEMATIZATION OF DRUG DISCOVERY

EDGAR JACOBY
Novartis Institutes for BioMedical Research,
Basel, Switzerland

Chemogenomics aims toward the systematic identification of small molecules that interact with the products of the genome and modulate their biological function. The establishment and expansion of a comprehensive ligand–target SAR (structure–activity relationship) matrix is following the elucidation of the human genome a key scientific challenge for the twenty-first century. The annotation and knowledge-based exploration of the ligand–target SAR matrix is then expected to impact. Progress alongside this challenge without doubt in first place will contribute to further the fundamental understanding of the biological function of the individual proteins and ultimately provide a basis for the discovery of new and better therapies for diseases.

While historically the chemogenomics approach is based on efforts that systematically explore target gene families, today broader *in vitro* and *in silico* approaches are available to encompass wider genomes.

Herein, we review relevant aspects of chemistry, biology, and molecular informatics that are the cornerstones of chemogenomics. The focus will be on the protein world; other gene products, such as RNA [1], or whole organism phenotypic readouts [2] will not be addressed.

1. INTRODUCTION

Following the elucidation of the human and other genomes, chemogenomics, chemical genomics, chemical genetics, and chemical biology are the new front running discovery technologies that aim to unify molecular biology and chemistry [3,4]. On the long term, they aim to identify systematically new biological and chemical starting points for drug discovery projects. These disciplines—see Table 1 for definitions—integrate basic disciplines such as chemistry, biological screening and profiling, structural biology, and chemo- and bioinformatics.

The chemogenomics terminology itself, originated from the directed exploration of target gene families that was probably first emphasized in 1996 by researchers at Glaxo Wellcome, who discussed the concept of systematization of drug discovery within target families based on the analysis of gene families such as GPCRs (G-protein coupled receptors), ion channels, or proteases that had hitherto been successfully explored [5]. The Glaxo Wellcome scientists highlighted obvious advantages of system-based approaches, such as combining advances in gene cloning and expression, automation, combinatorial chemistry, and bioinformatics. The underlying molecular hypothesis for the strategy is that similar ligands should bind to similar targets and thus the gained knowledge previously from one project should be transferable to new related projects [6]. This principle was summarized and generalized first by Stephen Frye at Glaxo in 1999 as the SARAH (structure–activity relationship homology) concept, which aims to group potential drug discovery targets into families based on the relatedness of the SAR of their ligands [7]. The conservation of the binding site architecture within a target family or a subfamily thereof translates into a conservation of the architectures of ligands that bind to these targets. In 2001, Vertex described the implementation of a chemogenomics platform for kinase inhibitor discovery [8] and Novartis described chemogenomics library design principles for the biogenic amine related GPCR target family [9]. Since then, a growing number of publications appeared focusing on contributions from chemistry, biology, and molecular informatics aiming toward the central chemogenomics principle to establish and study a comprehensive genome-wide ligand–target SAR matrix. The progress and contributions to drug discovery so far achieved in these disciplines will be outlined in this chapter.

Table 1. Definition of Disciplines

Discipline	Definition
Chemical genomics Chemogenomics	The systematic identification of small molecules that interact via a specific molecular recognition mode with target proteins encoded by the genome. The term chemogenomics is historically applied more specifically to target family approaches in drug discovery.
Chemical genetics	Aims, in analogy to genetic mutations, to identify chemical compounds that induce or revert specific biological phenotypes by using cell-based or microorganism based screening of compounds.
Chemical biology	The functional and mechanistic investigation of biological systems using chemical compounds. Constitutes a more general discipline.

2. CHEMISTRY

Small chemical compounds are the first dimension of the ligand–target SAR matrix. Accordingly, the systematic expansion of the physically available and bioactive chemical space is a key objective of chemogenomics.

A few numbers help to place the chemical space discussion in perspective. The theoretical drug-like small-molecule chemical space is estimated to count more than 10^{60} compounds [10]. This astronomic number has to be placed in relation to the number of physically available chemicals and the actual number of approved drugs. The Beilstein Crossfire and CAS Registry databases, which are representative of molecules described up-to-date in the chemical literature, list, respectively, 10 and 35 million compounds. The screening collections of the major pharmaceutical companies include typically 1–4 million proprietary and nonproprietary compounds. Further, the ChemNavigator iResearch library lists 25 million commercially available compounds. Based on hit rates from target specific HTS (high-throughput screening) activities, which are typically 0.01–0.1%, the vast majority of these compounds are expected to be biologically inactive on a given target at a micromolar activity threshold. Corroboratively, a study by Pfizer identified 275,000 actives out of a total of 4.8 million nonredundant compounds in their data warehouse [11]. Finally, according to the analysis of the molecular pharmacopoeia by Overington et al., 1357 unique drugs, of which 1204 are small-molecule drugs and 153 are biological drugs, are today approved by the USFDA satisfying stringent efficacy and safety criteria [12].

The vital question is how to enlarge the physically existing chemical space into the bioactive and drug-like spaces? The main purpose of large screening collections is to supply the discovery pipeline with hit-to-lead compounds for today's and the future's portfolio of drug discovery programs and to provide tool compounds for the investigation of novel biological targets and pathways. A screening collection integrates designed focused and diversity-based compound sets from the synthetic and natural paradigms generated via corporate medicinal chemistry and combinatorial compound synthesis and external compound acquisition projects [13,14]. Effective systematic expansion of the chemical space to reach a maximum of biological binding sites appears possible when conserved molecular recognition principles are the founding hypothesis for the design of the compounds. Such principles are summarized in the following sections.

2.1. Protein Family Targeted Libraries

More than 50% of the marketed drugs target only four key gene families, including the rhodopsin-like GPCRs, nuclear receptors, ligand-gated ion channels, and voltage-gated ion channels. Historically, drug discovery has thus been focusing on a few "druggable" target families, although many therapeutic classes such as the antihistaminergics and the β-blockers were discovered before their receptors were known at the molecular basis. Protein family targeted drug design principles are thus not new. The key design principles, focusing on similarities or differences in the physicochemistry of equivalent residues lining the binding site, can also rationalize the polypharmacology of many drugs. Because protein family targeted library design requires extensive ligand-based or structure-based knowledge, it is not surprising that current design of chemical libraries directed to target classes focuses mainly on GPCRs, kinases, and nuclear receptors. In 1998, a pioneering publication by the Schultz group demonstrated the discovery of selective protein kinase inhibitors developed on the basis of trisubstituted purines to target the ATP (adenosine triphosphate) binding site of the human CDK2 (cyclin-dependent kinase 2). By iterating chemical library synthesis and biological screening, potent inhibitors of the human CDK2-cyclin A kinase complex and of the *Saccharomyces cerevisiae* Cdc28p kinase were identified [15]. One year later, researchers at Servier published a three binding site hypothesis for the interaction of ligands with biogenic amine GPCRs that had implications for the combinatorial ligand design in this family [16]. Today protein family targeted libraries with a large diversity of chemotypes are specifically designed toward subfamilies with conserved molecular recognition. Various strategies have been applied to design GPCR [17] and ion channel libraries [18], mostly based on ligand information captured in the form of molecular descriptors, pharmacophores, and substructures extracted from active reference compounds. In addition to these methods, the large amount of structure-based information available from X-ray analysis of ligand–target complexes, makes structure-based design approaches feasible in the protein kinase [19,20], protease [21], and nuclear receptor [22] classes. Recently, Vertex published the BREED approach for generating novel ligand scaffolds from the overlay of cocrystal binding sites and subsequent recombinant hybridization of the superimposed bonds of the associated ligands [23]. Using such hybridization approaches, the information content of selected chemical series can rapidly be expanded into new series that conserve the underlying interaction mode. The typical size of a protein target focused library is 100–1000 compounds if it is centered around one single chemotype; the library size can grow up to 10–20,000 compounds if it is oriented around multiple chemotypes.

The goal of a protein family targeted library is not to target a specific target exclusively, but to address by different library members different members of the target subfamilies. This coverage of a targeted library could until recently only be addressed experimentally by the analysis of hit rates. Such analyses showed that some designed libraries hit not only the primary target family but also other *a priori* not related target families. For instance, high hit rates for protein kinase targeted libraries were also observed in cell-based GPCR assays. Whether the reason for this observation resides in a direct molecular interaction, or is due to pathway interactions needs further clarification. A recent trend in protein kinase and GPCR targeted library design is to address allosteric binding sites [24]. The future will provide an answer to the question whether these binding sites can be exploited in an equally systematic manner as the classically targeted endogenous orthosteric binding sites that are expected to have a higher evolutionary conservation.

2.2. Privileged Scaffolds

The use of privileged substructures or molecular master keys is an accepted concept in medicinal chemistry. The privileged structure approach emphasizes on molecular scaffolds or selected substructures that are able to provide high-affinity ligands (agonist or antagonists) for diverse receptors and originates from the work at Merck Research Laboratories on the design of benzodiazepine-based CCK (cholecystokinin) antagonist, where the previously known κ-opioid tifluadom was identified as a lead structure [25]. A number of recent literature reviews provide impressive reference repertoires of empirically derived privileged structures for various target families [26]. The 2-aryl-indole scaffold illustrated in Fig. 1 represents a particular successful example and was shown at Merck to generate actives for diverse class A GPCRs [27].

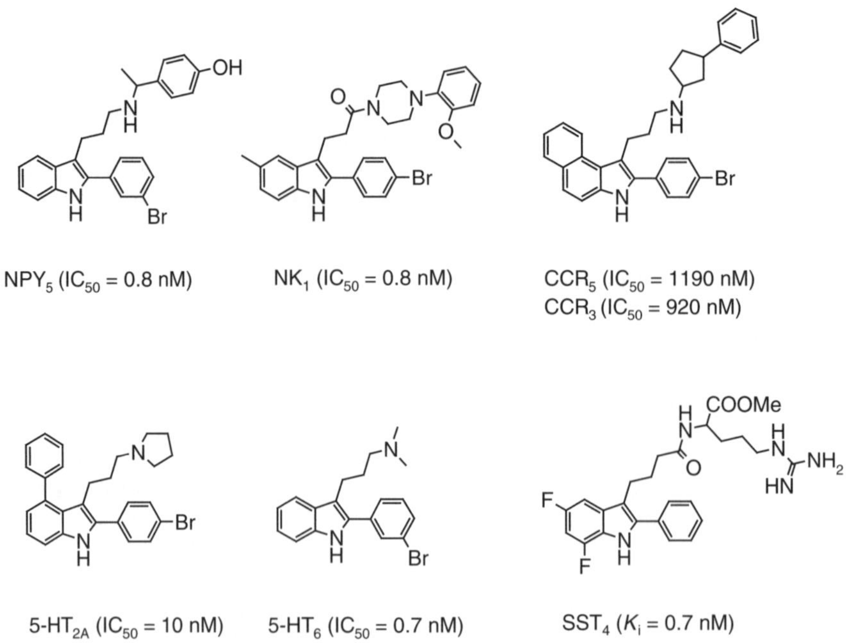

NPY_5 (IC_{50} = 0.8 nM) NK_1 (IC_{50} = 0.8 nM) CCR_5 (IC_{50} = 1190 nM) CCR_3 (IC_{50} = 920 nM)

5-HT_{2A} (IC_{50} = 10 nM) 5-HT_6 (IC_{50} = 0.7 nM) SST_4 (K_i = 0.7 nM)

Figure 1. Examples of GPCR active compounds based on the 2-aryl-indole privileged scaffold identified from a focused combinatorial library at Merck. Screening of the library against several GPCRs led to the discovery of NPY_5, NK_1, chemokine CCR_3/CCR_5, serotonin 5-HT_{2A}/5-HT_6, and SST_4 receptor antagonists [27].

The privileged structural classes will need to be analyzed further to allow a more directed use of such libraries for specific receptor subsets. To this aim, rhodopsin-based models of the 7TM domain were used at Novo Nordisk [28,29]. Modeling the interactions of three sets of privileged motif-based ligands into their receptors, including 2-aryl-indole based ligands in the serotonin 5-HT_6 and MC_4 (Melanocortin-4) receptors, spiro-piperidine-indane-based ligands in the GHS (growth hormone secretagogue) and MC_4 receptors, and 2-tetrazole-biphenyl-based ligands in the AT_1 (angiotensin 1) and GHS receptors, showed the correlation of conserved patterns of residues in the ligand binding pockets of the receptors with the recognition of specific privileged fragments. These findings imply that any one particular privileged structure can target a specific subset of receptors and that motif-based searches can be used for subsetting the receptor repertoire including the orphan receptors. The models also showed that only parts of the privileged structures are accommodated within the conserved subpocket; some contacts are between substructure elements of the full privileged motif and the nonconserved part of the pocket, which suggests the possibility for design of selective ligands based on privileged motifs.

The development of chemoinformatics methods enabling the automatic identification and extraction of privileged structures is a recognized need in the context of generating knowledge from the chemogenomics ligand–target matrix. The field was pioneered by Bemis and Murcko [30] who developed a method for decomposing molecules into frameworks, side chains, and linkers and analyzed the statistical occurrence of the frameworks within a subset of drug-like compounds listed in the CMC (Comprehensive Medicinal Chemistry) catalog. This analysis revealed that only 32 frameworks described the shapes of half the drug-like compounds in the CMC set (~5000 compounds); this number was, however, revised in a more recent study by Grabowski and Schneider to 160 frameworks [31]. A recent debate focuses on the question whether protein family privileged substructures are truly privileged. Schnur et al. analyzed the selectivity of CLASSPHARMER generated fragments for a series of protein families [32]. The conclusion from their analysis is that generating maximal common substructures and tabulating intratarget family occurrences within a drug database is insufficient; one must consider substructures with somewhat more functionality before target family privilege is observed.

2.3. Protein Secondary Structure Mimetics

Protein–protein molecular interactions are the most ubiquitous mode for controlling and modulating cellular function, intercellular communication, and signal transduction pathways. Peptide and protein mimetics libraries including β-turn and α-helix mimetics are recognized of central importance in chemogenomics [33].

A number of important hormones and protein–protein interactions make their key recognition via specific β-turn motifs. The design of drug-like active β-turn mimetics based on organic drug-like scaffolds, or based on cyclic α-peptides or β/γ-peptides advanced to a quite routine methodology. The work of Garland and Dean [34,35], defining a set of triangular distance constraints that the substitution points of a scaffold have to satisfy to mimic the specific C_α atoms of the peptide template, provided a generalized frame for the design of novel β-turn mimetic scaffolds and was in combination with database searches successfully applied for the design of CCK, SST, and MC_4 antagonists [36].

Other protein–protein interactions are via key α-helix motifs. The work of the groups of Hamilton [37] and Marshall [38] has established a solid foundation for the rational design of α-helix mimetics. 3,2′,2″-substituted terphenyl derived motifs were among the first designed motifs and were shown to be able to mimic the side-chain positions i, $i + 3$, or i 4, and $i + 7$ which are on the same face of a α-helix. Targets such as the BCL-2 (B-cell lymphoma 2) family of proteins, MDM2-p53 (mouse double minute 2 protein -p53), various calmodulin binding proteins, and HIV gp41 have been successfully targeted. Recently, the hydrophobicity and the complex syntheses of the terphenyls prompted to search for simpler and more soluble scaffolds. In this search, oligoamide foldamers and terephthalamide-based compounds were shown by the Hamilton group to be able to mimic the Bak peptide [39]. Furthermore, the design and synthesis of heterocyclic and especially pyridazine-based scaffolds were demonstrated by the Rebek group [40,41]. See Fig. 2 for representative examples of rationally designed helix mimetics.

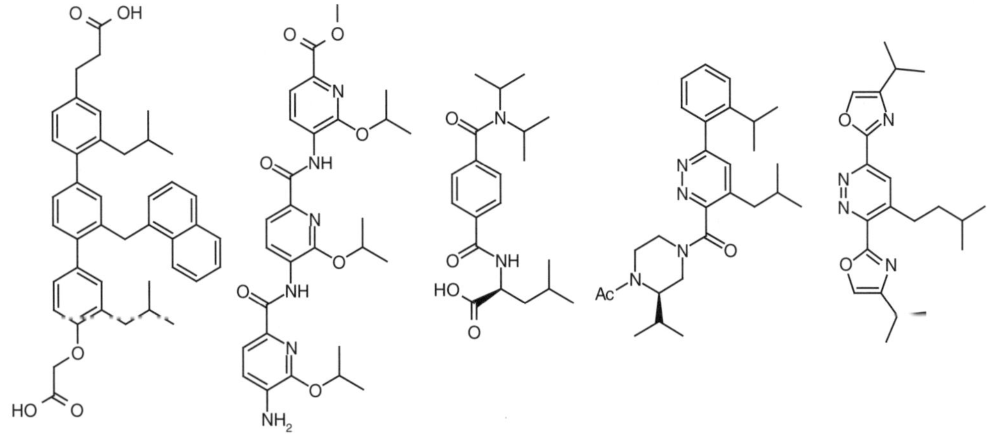

1 2 3 4 5

Figure 2. Examples of designed α-helix mimetics. **1**: terphenyl analog, **2**: oligoamide foldamer, **3**: terephthalamide-based compound, and **4** and **5**: pyridazine-based compounds [37,39–41].

Besides these rationally designed helix mimetics, diversity-based HTS and virtual screening have identified for a variety of targets a number of scaffolds which allow the correct spatial orientation of substituents for interaction with the protein target [42]. Combinatorial libraries around such scaffolds are an essential component of a chemogenomics discovery library.

2.4. Cofactor-Based Discovery

As successfully exemplified by the protein kinase ATP binding site, cofactor binding sites are evolutionary highly conserved binding sites that provide rich opportunities for enzyme based drug discovery. The detailed comparative structural analysis of cosubstrate and cofactor binding sites show indeed that cofactor analogs open a very wide target window. For instance, the common structural framework for adenine and AMP (adenosine mono-phosphate) binding is conserved in 12 unrelated protein families, including different folds, which demonstrates that ligand recognition principles have a stronger conservation than protein fold conservation; providing the basis for efficient systems-based inhibitor design strategies [43]. Denessiouk et al. analyzed in detail adenine recognition: a motif present in ATP-, CoA- (coenzyme A), NAD- (nicotinamide adenine dinucleotide), NADP- (nicotinamide adenine dinucleotide-phosphate), and FAD- (flavin adenine dinucleotide) dependent proteins [44]. Knapp et al. reviewed the challenges and successes of structure-based drug design against the human purinome by analyzing human purine nucleoside phosphorylase, human epidermal growth factor receptor kinase, and human kinesin spindle protein [45]. Given the large number of purine-dependent cellular processes—approximately 13% of the human genome is devoted to coding for purine binding proteins—purine libraries may serve as a rich source of inhibitors for many different protein targets. An additional example of cofactor based drug discovery was provided by Mozzarelli's group analyzing the vitamin B_6-derived PLP (pyridoxal 5′-phosphate) cofactor binding site, characteristic of enzymes catalyzing chemical reactions involved in amino acid metabolism [46]. Almost 1.5% of all genes in prokaryotes code for PLP-dependent enzymes, whereas the percentage is substantially lower in eukaryotes. Although approximately 4% of enzyme-catalyzed reactions cataloged by the EC (Enzyme Commission) classification are PLP dependent, only a few enzymes are targets of approved drugs and about twenty are recognized as potential targets for drugs. The work of Sem et al. on the oxidoreductase gene family provides a detailed analysis that divides this global family into structural subfamilies termed pharmacofamilies, which share pharmacophore features in their cofactor binding sites [47]. The presence of the conserved NAD(P) cofactor binding site (approximately 15% of all known enzyme functions utilize NAD(P) for catalytic function), coupled with the modular nature of this gene family, has led to the development of a highly parallel bi-ligand inhibitor design approach. In this approach, one ligand moiety is targeting the conserved cofactor site, whereas the second moiety directed into the adjacent substrate site is utilized to generate selectivity. A library designed in this manner yielded specific inhibitors for multiple oxidoreductases. The approach is in principle applicable to other cofactor-dependent enzymes.

2.5. Diversity-Oriented and Biology-Oriented Syntheses

DOS (diversity oriented synthesis), as opposed to the traditional TOS (target oriented synthesis) chemistry approach, was introduced by the Schreiber group for forward chemical genetic screening to mimic the structural complexity and the skeletal and stereochemical diversity of natural products [48]. Conversely to a convergent synthesis strategy resulting from the logic of retrosynthetic analysis of the target molecules, DOS allows the application of a diverse set of reagents and structural transformations on each synthesis intermediate. This results in diverging synthesis pathways that create a broad diversity of target molecules with different scaffolds. DOS compounds clearly share a number of characteristics with natural products including most notably the scaffold diversity and stereochemical complexity. The questions remain, however, whether these products of pure chemist imagination capture the evolutionary advantages of natural products and natural products-based compounds.

A recent successful application of the DOS principle has been provided by the Spring group (Fig. 3) with the discovery of MRSA (methicillin-resistant *Staphylococcus aureus*) antibiotics using a divergent synthetic

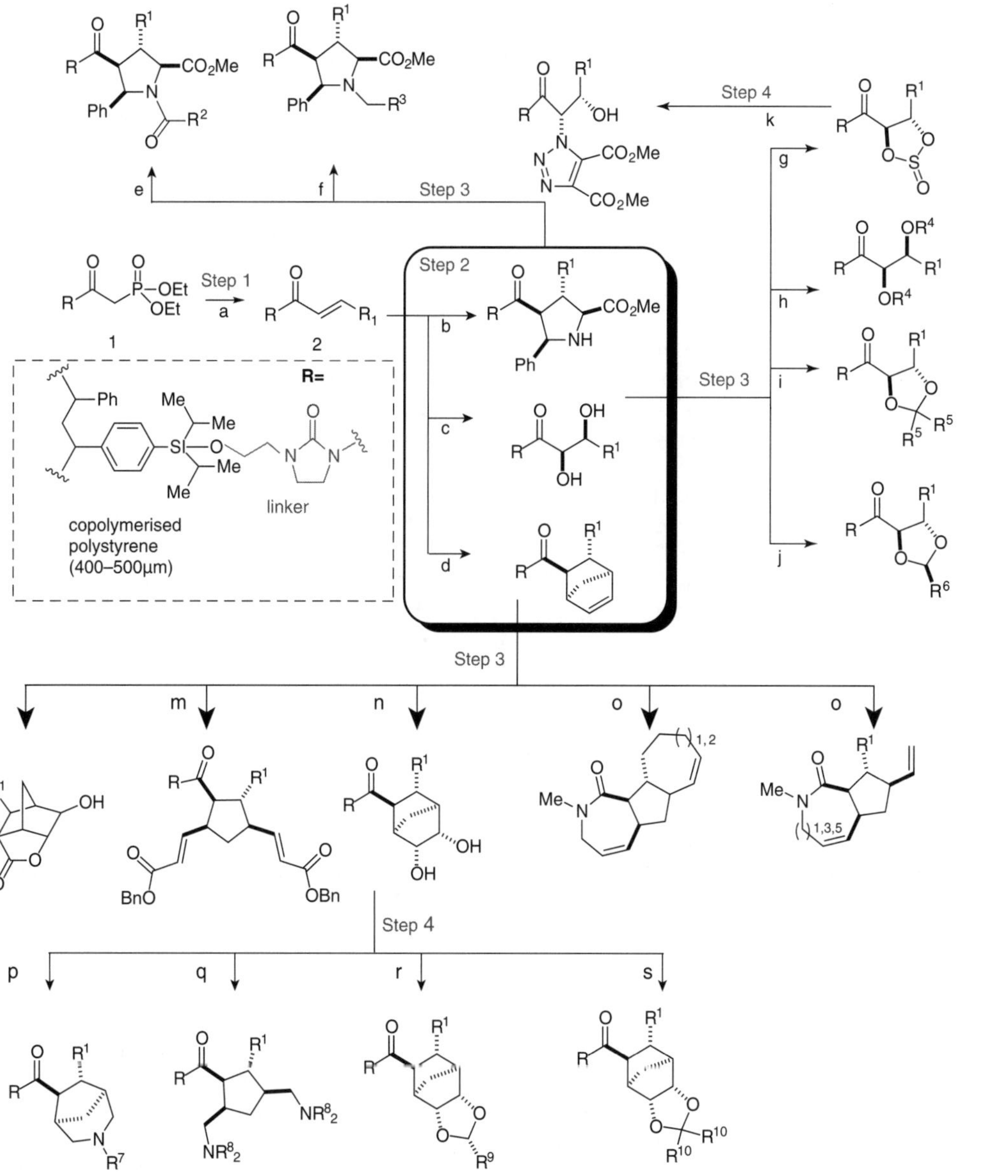

Figure 3. Example of DOS library resulting in the discovery of MRSA antibiotics. There were 242 compounds synthesized based on 18 discrete molecular frameworks using four steps of divergent reactions. The identified MRSA antibiotic hit gemmacin is based on scaffold p. See original publication for detailed reaction conditions [49].

strategy [49]. A collection of 242 small molecules was synthesized, which is claimed based on computational criteria to be more physicochemically and topologically diverse than databases of known drugs. Antibacterial screening with pathogenic strains of MRSA uncovered several hits, including gemmacin, which is likely to be a selective bacterial membrane disrupter.

BIOS (biology-oriented synthesis) was introduced by the Waldmann group [50]. BIOS centers on the generation of small compound libraries based on scaffolds of proven biological relevance. Library generation is focused on compound classes from the "biological relevant space," that is, the natural products and drugs. BIOS unifies the SCONP (structural classification of natural products) and PSSC (protein structure similarity clustering) concepts which, respectively, allow navigation in the chemical and biological spaces. The SCONP scaffold tree that is a hierarchical classification of the underlying scaffolds based on their cyclic frameworks and linkers allows the identification of biologically relevant scaffolds [51]. Correlations between different scaffold classes can be performed. Using the scaffold tree, Waldmann et al. synthesized libraries featuring among others spiroketals and α,β-unsaturated lactones that proved biological activity, respectively, in tubulin and cell cycle modulator assays. PSSC relies on bio- and chemoinformatics procedures to identify proteins in which the subfold around the ligand binding site, the so-called "ligand-sensing core," is similar [52]. Proteins that show such structural similarity are assigned to one protein structure similarity cluster. Ligands of the members of a cluster can then be expected to be prevalidated starting points for the synthesis of ligands for the other members of the same cluster. An example for the validation of the PSSC approach, was provided using the natural product dysidiolide that targets the Cdc25A phosphatase and shows cluster similarity to AChE (acetylcholine esterase) and 11β-hydroxysteroid dehydrogenase. Consequently, libraries were built around the dehydrodecalin part and around the hydroxybutenolide part. Individual members of both libraries showed, respectively, activity on the AChE and 11β-hydroxysteroid dehydrogenase targets.

The protein structure universe is estimated to contain approximately 1700 distinct protein folds and 4000 structural superfamilies, what places the PSSC approach in perspective for further systematization [53].

3. BIOLOGY

The past decade witnessed spectacular progress in the screening of small-molecular compounds against targets in both biochemical and cell-based formats. Today, modern MTS (medium-throughput screening) and HTS assays can screen 10,000–1,000,000 compounds in a turn time of a day to a week [54,55]. Impressive progress was also achieved in screening of single compounds in profiling setups against a multitude of selected target (i.e., safety pharmacology profiling) [56] or against entire genomes.

Despite this progress, the number of targets that have been evaluated up-to-date is still limited. The analysis of the global pharmacological space by Paolini et al. assigned 2876 targets to protein sequences from 55 organisms, with biologically active chemical tools for 1306 proteins [11]. Within the human genome, 836 genes were identified for which bioactive small-molecule compounds are known. More specifically, the analysis of the molecular drug targets of the 1357 FDA-approved drugs by Overington et al. showed that the known drugs act on only 324 drug targets [12]. 50% of these targets focus of four protein families and the rate of new target family innovation is 1.9 families per year. This knowledge about the biological space has to be placed in perspective to the 25–30,000 human genes of which only a fraction is expected to be disease relevant. The further growth of the biological target space is driven by the follow-up of individual genetically proven disease relevant proteins including their homologs [57] and by the systematic mechanism of action studies of specific bioactive compounds identified in phenotypic screens. The following sections address selected experimental approaches to identify the biological targets of specific compounds; also called chemogenomics target fishing.

3.1. Affinity Chromatography and Activity-Based Protein Profiling

Affinity chromatography was first described in the late 1960s as a method for selective enzyme purification. In affinity chromatography, the protein to be purified is passed through a column containing a cross-linked polymer or gel to which a specific competitive ligand of the protein of interest has been covalently linked. All proteins without substantial affinity for the bound ligand will pass directly through the column, whereas one that recognizes the ligand will be retarded in proportion to its affinity constant. Elution of the bound protein is readily achieved by changing such parameters as salt concentration or pH, or by addition of a competitive ligand in solution. Recent advances in chemical proteomics, combining LC (liquid chromatography) and tandem MS (mass spectrometry) with advanced bioinformatics analysis now permit proteome-wide studies. The technique has been used to determine all the proteins in a proteome that bind to a drug and may therefore be responsible for its efficacy. Especially in the protein kinase field, the generality of affinity chromatography has allowed a number of kinase inhibitors to be screened for target selectivity; identifying for some known kinase drugs, other kinase targets, and nonkinase targets [58].

Researchers at Cellzome refined the approach to profile the interaction of small molecules with hundreds of endogenously expressed protein kinases and purine binding proteins [59]. This subproteome is captured by immobilized nonselective kinase inhibitors (kinobeads), and the bound proteins are quantified in parallel by MS. Quantitative profiling of the drugs imatinib, dasatinib, and bosutinib in K562 cells confirmed known targets including ABL (Abelson leukemia) and SRC (Rous sarcoma) family kinases and identified the receptor tyrosine kinase DDR 1 (discoidin receptor 1) and the oxidoreductase NQO2 as novel targets of imatinib.

ABPP (activity-based protein profiling) is conceptually related to the affinity chromatography techniques [60]. The fundamental building blocks of ABPP are small-molecule probes that covalently label the active site of a subset of enzymes. Gel-based methods, where probe-labeled enzymes are visualized and quantified across proteomes by in gel-fluorescence scanning, or LC/MS-based analysis platforms exist. Several enzyme classes, including proteases and other hydrolases, kinases and nucleotide binding proteins, glycosidases, phosphatases, and cytochrome P450s were characterized at a global level in native or diseased biological systems using directed ABPP probes. ABPP has established a track record in identifying enzyme activities associated with a range of diseases, including cancer, malaria, and metabolic disorders [61].

3.2. Yeast Three-Hybrid Screens

In Y3H (yeast three-hybrid) assays, a DNA binding domain is expressed as a fusion to the DHFR (dihydrofolate reductase) or the GR (glucocorticoid receptor). These proteins bind, respectively, with high affinity to the small-molecule MTX (methotrexate) and to the steroid hormone DEX (dexamethasone). A small molecule of interest can then be displayed from the DNA binding domain if it is linked through a suitable tether to the MTX or DEX molecules. This allows the screening of cDNA libraries that are fused with an AD (activation domain) which activates reporter gene transcription. Y3H screens were first described by Licitra and Liu to screen cDNA libraries for target discovery [62]. An engineered GR protein was used as a DNA bound docking platform to display a DEX tethered FK506 hybrid ligand to target proteins that could activate reporter gene expression upon binding. FKBP12, a known target of FK506, was identified, indicating that the Y3H could be a valuable tool for the identification of novel small-molecule protein interactions.

More recently, GPC Biotech described Y3H screening to scan the proteome for targets of small-molecule kinase inhibitors—see Fig. 4 [63]. This assay uses kinase inhibitors that are linked through a flexible polyethylene glycol chain to MTX. The kinase inhibitor is displayed as a bait ligand in yeast that expresses a DHFR-LexA DNA binding domain

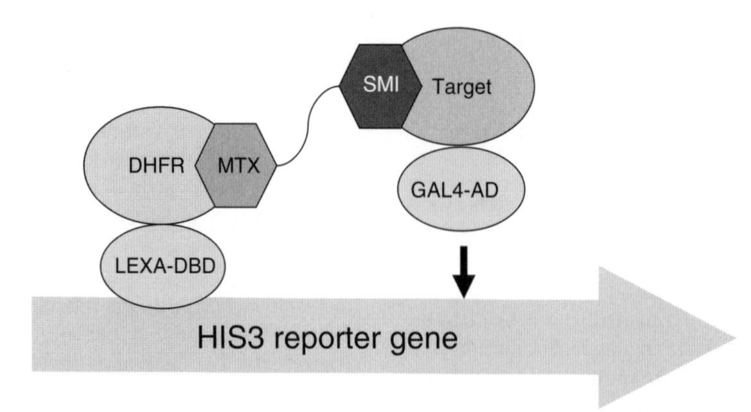

Figure 4. Example of Y3H assay to scan the proteome for targets of small-molecule kinase inhibitors. A small-molecule kinase inhibitor (SMI) is tethered to methotrexate via a flexible polyethylene glycol spacer. This bi-ligand can bind to both the DHFR-LexA-DBD (DNA binding domain) fusion protein and the kinase target-Gal4-AD (activation domain) fusion protein. The ternary complex is in turn activating the transcription of the HIS3 reporter gene [63]. (This figure is available in full color at http://mrw.interscience.wiley.com/emrw/9780471266945/home.)

fusion protein. Proteins encoded by cDNA libraries that are fused to the GAL4 AD are screened and proteins that bind to the kinase inhibitor are detected through the expression of a HIS3 auxotropic marker.

Potential targets of the CDK inhibitor purvalanol B, related purine and indenopyrazole compounds were determined. In addition to several known targets, a number of unexpected kinase targets were identified. To confirm these interactions, affinity chromatography and kinase activity assays were performed and validated most of the interactions found in the Y3H assay. The Y3H analysis appears generally most suitable for a "yes or no" detection of an interaction, with a presumed limit for detection of low-micromolar affinity type of interactions.

4. MOLECULAR INFORMATICS

Molecular informatics emerged in the last decade as a new informatics discipline that integrates aspects of chemoinformatics and bioinformatics. Given both the chemical and the biological dimensions of chemogenomics data, the two disciplines can indeed not longer be viewed as independent and adequate molecular information systems need to be designed to store and analyze the data [64–66]. The following sections address molecular information systems for chemogenomics data and, both ligand-based and structure-based data analysis and predictive modeling techniques.

4.1. Molecular Information Systems

The fast growing amount of chemogenomics data related to small-molecule interactions with biological systems generated in industrial and academic screening centers, requires the design of specific molecular information systems including systematic and comprehensive annotation of the SAR data [64,67,68]. Such annotation with molecular target, signaling pathway, and disease and therapeutic information, in combination with synonym and classification schemes result in true cross-linking of the chemical and biological knowledge spaces. Development of integrated chemical and biological ontologies turns out to be quite complex [69]. A key challenge resides in the genome-wide extension of molecular annotations and classification schemes. Given the growing detail complexity of our knowledge around biological systems, the design of data models adapts constantly to enable integration and mining of knowledge within broader dynamic systems biology and protein–protein interaction network concept spaces.

An impressive number of chemogenomics knowledge-based companies (see Table 2) have specialized in developing molecular information systems that integrate data from patents and selected literature, including 2D structures of the ligands, target sequence and classification, mechanism of action, SAR data, assay, and bibliographic information, together with chemical and biological search engines. The GeneGo MetaBase and MetaCore expert systems allow to network more than 90% of the known human proteins via physical protein–protein, protein–DNA, and protein–compound interactions.

Databases freely available include ChemBank at Harvard University and Pubchem at the NCBI. Both systems are supported by the NCI's initiative for chemical genetics and include public domain information on a fast growing number of compound–target interactions. The DrugBank database at the University of Alberta focuses on data on around 4800 drug entries including FDA-approved and experimental small-molecule drugs.

4.2. Ligand-Based Data Analysis and Predictive Modeling

The availability of large-scale ligand–target SAR matrices has triggered in recent time investigations on how to analyze and predict such data.

Table 2. Commercially and Publicly Available Compound Databases and Molecular Informatics Resources for Chemogenomics Research

Internet Resource URL/Source	Specification of Data and Information Available
http://www.aureus-pharma.com/ http://www.eidogen-sertanty.com/ http://www.evolvus.com/ http://www.gvkbio.com/ http://www.biofocus.com/ http://www.jubilantbiosys.com/ http://integrity.prous.com/	A growing number of chemogenomics knowledge-based companies, such as Aureus-Pharma, Eidogen-Sertanty, Evolvus, GVKBio, Biofocus, Jubilant Biosys, and Prous are developing molecular information systems that integrate in a comprehensive manner data from patents and selected literature together with chemical and biological search engines.
http://www.genego.com/	MetaBase is a curated database of human protein–protein and protein–DNA interactions, transcriptional factors, signaling, metabolism, and bioactive molecules. MetaCore provides intuitive tools for data visualization, mapping and exchange, multiple networking algorithms, and data mining.
http://chembank.broad.harvard.edu/	ChemBank at Harvard University provides information on the biological activities of small molecules. ChemBank houses information on more than 1.2 million unique small-molecule structures. Data from over 2500 high-throughput biological assays from 188 screening projects currently reside in ChemBank. More than 1000 proteins, 500 cell lines, and 70 species are associated with the assays.
http://pubchem.ncbi.nlm.nih.gov/	PubChem at NCBI provides information on the biological activities of small molecules. It is a component of NIH's Molecular Libraries Roadmap Initiative. PubChemCompound contains more than 19 million unique structures. PubChemBioAssay contains more than 1000 BioAssays.
http://www.sunsetmolecular.com/	The WOMBAT database contains 178,210 unique structures, totaling 416,405 biological activities on 1820 unique targets based on literature data.
http://www.drugbank.ca/	The DrugBank database combines detailed drug data with comprehensive drug target (i.e., sequence, structure, and pathway) information. The database contains nearly 4800 drug entries including FDA-approved and experimental small-molecule drugs. More than 2500 nonredundant protein (i.e., drug target) sequences are linked to these FDA-approved drug entries.

The list is not exhaustive, rather constitutes a representative compilation of selected examples in this field.

Initial analyses focused on trends of chemical properties in specific target classes and have for instance shown that protease inhibitors have in average higher molecular weight distributions or that nuclear hormone receptors have in average higher lipophilicity distributions [70]. Such distinct target class specific property profiles are of immediate relevance for property-based compound design. Target class specific chemical information can also be abstracted by machine learning techniques such as NB (Naïve Bayesian) modeling, neural networks, or others and can be used to predict target class membership of compounds. Such models were established for instance for GPCRs or protein kinases and are useful for library design [71]. Based on the SAR data available for individual targets, such modeling techniques can of course also be used to build models for the specific targets and used for the large-scale prediction of compound–target relationships [72]. Such modeling was pioneered by the PASS (prediction of activity spectra for substances) approach that allows substructure analysis-based probabilistic modeling of known biological effects based on a training data set from literature [73]. At NIBR, related NB models based on a variety of data sets are used in combination with similarity-based methods to annotate—target fishing-potential targets of black-box cell-based assays such as phenotypic assays or RGAs (reporter gene assays) [74]. The NB models trained by Nidhi et al. for 964 targets were validated by predicting the target class membership for 10 new series of active compounds; yielding 77% prediction accuracy [75]. Corroboratively, Paolini et al. built NB models for 698 protein targets based on a data set of 238,655 known active compounds; for 64% of the test compounds, at least one target was correctly predicted [11]. NB models are also used to predict the promiscuity and safety pharmacology profile of newly synthesized compounds [74,76]. These models, which need to be experimentally validated further, enable the prediction of adverse drug reactions and off-target effects from chemical structure. Based on similarity principles using the SHED molecular descriptors, the Mestres group proposed ligand-based approaches for *in silico* profiling of nuclear receptor pharmacology [77] and for the target coverage analysis of designed chemical libraries [78]. These approaches are useful for the assessment of safety pharmacology and library design, respectively. Ligand-based target classification was also investigated by Keiser et al. who computed Tanimoto similarities of 65,000 biologically active compounds divided into different target sets [79]. A statistical model was developed to rank the significance of the similarity scores, which are expressed as a minimum spanning tree to map the sets together. Although these maps are connected solely by chemical similarity, biologically sensible clusters emerged. Links among unexpected targets also emerged, in particular it was predicted that methadone, emetine, and loperamide may antagonize muscarinic M_3, α_2 adrenergic, and neurokinin NK_2 receptors, respectively. These predictions were subsequently confirmed experimentally. Altogether, these investigations highlight that many therapeutically relevant compounds interact with more targets than previously thought.

Other key motifs of chemogenomics data analyses are chemical and biological clustering and the analysis of the information content of the bioactivity fingerprints [80]. Prediction of ligand binding to proteins by affinity fingerprinting was pioneered by Kauvar et al. [81]. Affinity fingerprints, which are molecular descriptors derived from the protein binding preferences of small molecules, were reported for 122 structurally diverse compounds using a reference panel of eight proteins that collectively are able to generate unique fingerprints for approximately 75% of the small organic compounds tested. Application of multivariate regression techniques to this database enabled the creation of computational models to represent new proteins that are surprisingly effective at predicting binding potencies. This computational approach was later termed TRAP (target-related affinity profiling) and further validated by Beroza et al. demonstrating the discovery of novel inhibitors of both human intestinal carboxylesterase and cyclooxygenase-1 [82]. Using a related approach, Guba et al. discovered the first nonpeptidic SST_5 (selective somatostatin receptor subtype 5) antagonists [83]. The histamine H_1 antagonist astemizole was identified as a SST_5 receptor antagonist by a comparative sequence analysis of the consensus drug binding pocket of GPCRs. Subsequently, a similarity analysis of GPCR affinity profiles of astemizole versus a set of Roche in-house GPCR-biased combinatorial libraries revealed new chemical entry points that led to a second lead series with nanomolar binding affinity. This study illustrates nicely the similarity principle that states that structurally related ligands imply similar bioactivity principles and that affinity fingerprint diversity increases with structural dissimilarity. Hence, there is a considerable portion of diverse structures with a similar bioactivity profile that are a rich source for discovery of novel chemotypes—see Fig. 5. Conversely, the reverse of the similarity principle, that is, similar biological properties determine structural similarity, is not valid.

Similar conclusions were reached by Fliri et al. in their so termed biospectra analysis. Using a 1567-compound database profiled against 92 different assays representing a cross section of the proteome, they showed that percent inhibition values, determined at single high drug concentration, provide a precise molecular property descriptor that identifies the structure of molecules [84]. Conducting a profile search using biospectra of clotrimazole as entry point revealed three biospectra of greatest similarity in the entire 1567-compound database: tioconazole and two other related azoles. Remarkably, the biospectra similarity search identified molecules that all have antifungal activity and hence, identified molecules with similar biological response capacity. The same similarity relationships were observed by hierarchical clustering that provides an unbiased mean for establishing quantitative relationships between chemical structures and biological activity spectra. The same data set was used in a follow-up study to investigate the relationships between biospectra similarities of 24 molecules being by hierarchical clustering most closely aligned with the neurotransmitter dopamine [85]. The agonist or antagonist properties were probed and successfully distinguished. These findings indicate that biospectra analysis provides an unbiased tool for forecasting structure–response relationships and for translating broad biological effect information into chemical structure design.

A truly large-scale chemogenomics data analysis was performed by researchers at BASF [86]. Here the inhibition profile matrix of a 130,000 diverse molecule library with a set of 97 enzymes (kinases, proteases, and phosphatases) enzymes was analyzed. The profiles were used to construct a semimetric additive distance matrix. This matrix, in turn, defines a SIP (sequence-independent phylogeny) that was compared to a SBP (sequence-based phylogeny) that is a well-established tool for describing relationships between proteins. Within enzyme families, the SIP shows a good overall correlation with the SBP. More interestingly, the SIP uncovers distances within families that are not recognizable by sequence-based methods. In addition, the SIP allows the determination of distance between enzymes with no sequence homology, thus uncovering novel relationships not predicted by SBP. This chemogenomic approach, used in conjunction with SBP, should prove to be a powerful tool for choosing target combinations for drug discovery programs as well as for guiding the selection of profiling and liability targets. Similar analyses were performed previously at Eli Lilly for a subset of the kinome, reaching similar conclusions [87].

Goldstein et al. outlined recently how high-throughput kinase profiling can be used for the identification of high-quality lead compounds by providing a comprehensive assessment of the library activity from both the compound and the target perspectives [88].

4.3. Structure-Based Data Analysis and Predictive Modeling

A number of investigations have explored the ligand–target space with focus on the three-dimensional ligand–target complex structures. The data being compared and classified are binding sites and ligand binding modes. While there are many resources available for obtaining protein structure relationships, there are comparatively few resources available for understanding binding site relationships. At Eidogen-Sertanty, SITESORTER was developed as a binding site alignment and

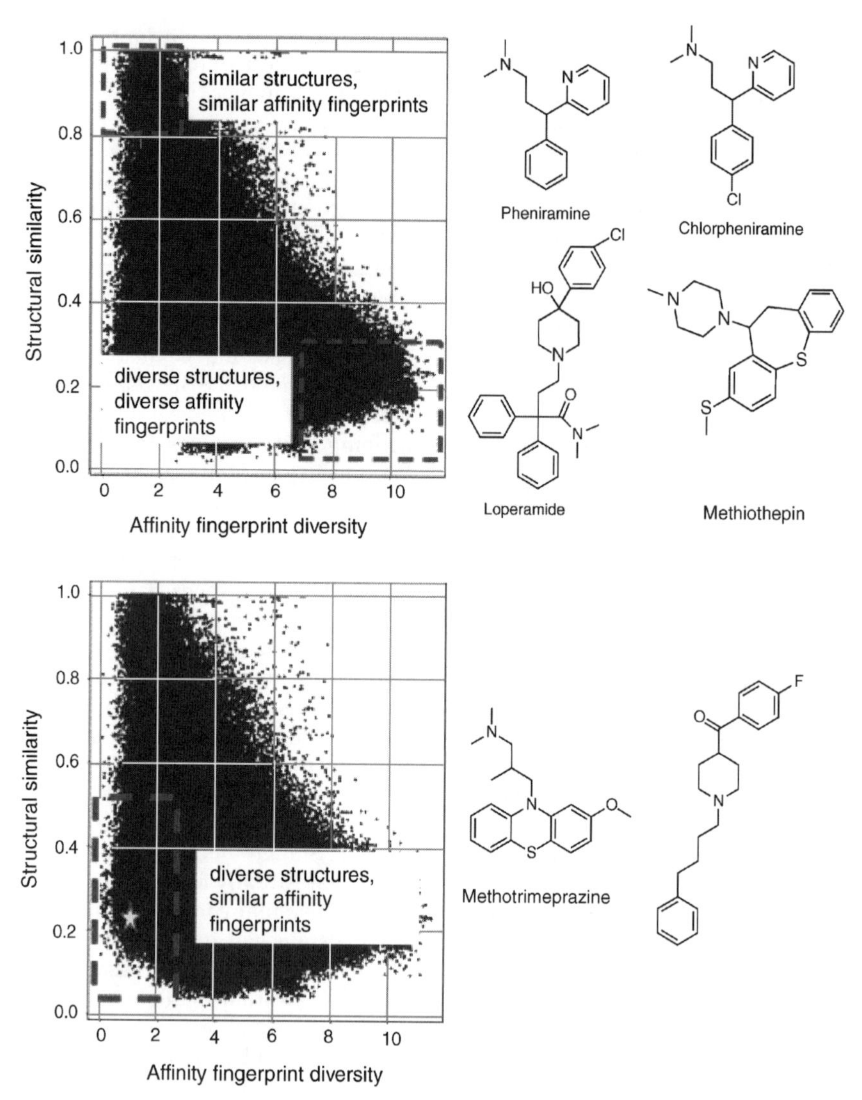

Figure 5. Illustration of the similarity principle using neighborhood plots. The underlying data are 5000 compounds containing privileged motifs from GPCR ligands tested on a panel of 15 GPCRs [83]. Structural similarity is defined by the Tanimoto coefficient of Daylight structure fingerprints, biological similarity via the Euclidean distance of the affinity profiles. Upper panel: On the upper left corner, compounds with high structural and biological similarity are highlighted. On the lower right corner, compounds with low structural and biological similarity are highlighted. Lower panel: The region encompassing biologically similar, but structurally diverse, compound pairs is highlighted.

comparison tool that uses a weighted-clique detection approach to directly overlay binding sites and avoid the requirement for structure homology [89]. By integrating SITESORTER with fully automated homology modeling and site annotation, the Eidogen-Sertanty structural informatics platform provides intra- and interfamily binding site comparisons for the entire proteome, and not just for those proteins whose structures have been resolved experimentally. Since closely related binding sites are more likely to bind to the same small molecules, binding site similarity analysis allows to infer important cross-reactivity infor-

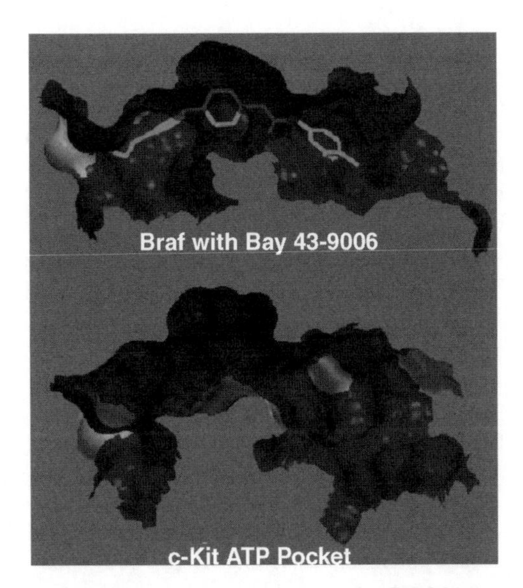

Figure 6. Target hopping with the Braf kinase inhibitor BAY 43-9006. According to SITESORTER, Braf kinase is one of the 10 most similar kinases to c-Kit. The binding site residues of 60% are conserved and colored blue; nonconserved positions are colored yellow; BAY 43-9006 is indicated in red [89]. (See color insert.)

mation. Target hopping is an important chemogenomic application and a successful example was provided by the analysis of Braf kinase, the primary target for the clinical compound BAY 43-9006— see Fig. 6. Braf is one of the 10 most similar kinases to c-Kit, which has also been shown to bind BAY 43-9006 with submicromolar affinity. This cross-reactivity could not be predicted based on the sequence similarity of the Braf and c-Kit kinase domains, since approximately one-sixth of the human kinome is more sequence similar to Braf than c-Kit.

In analogy and independently to this, the Klebe group introduced CAVBASE, a method for describing and comparing protein binding pockets on the basis of the geometrical and physicochemical properties of their active sites [90]. A cluster analysis procedure for the functional classification of binding pockets was implemented and calibrated using a diverse set of enzyme binding sites. Two relevant protein families, the α-carbonic anhydrases and the protein kinases, were used to demonstrate the scope of the cluster approach and relevant classifications of both protein families were proposed. The classification provides a new perspective on functional properties across a protein family and is able to highlight features important for potency and selectivity.

The Abagyan group developed a new computational algorithm for the accurate identification of ligand binding envelopes rather than surface binding sites [91]. The algorithm can be used to predict ligand binding pockets of uncharacterized protein structures, suggest new allosteric pockets, evaluate feasibility of protein–protein interaction inhibition, and prioritize molecular targets. The database of the known and predicted binding pockets for the human proteome structures, the human pocketome, was collected and classified. The pocketome can be used for rapid evaluation of possible binding partners of a given chemical compound.

To compare binding-sites and binding-modes within protein families, researchers at Aventis compared computed MIFs (molecular interaction fields) [92]. Starting from a structural alignment of all relevant ligand–target complexes, interaction energies are generated using different probe atoms at each point of a three-dimensional grid covering the binding sites. The MIFs are then analyzed using PCA (principle component analysis) and clustering to identify target similarities and factors that steer potency and selectivity within the given target family. The method was successfully applied to protein kinases, serine proteases, metalloproteinases, and nuclear hormone receptors.

To assess bioactivity profiles for small organic molecules, the Langer group proposed to use parallel pharmacophore-based virtual screening [93]. In a proof-of-principle study, carried out with the structure-based pharmacophore modeling tool LIGANDSCOUT and the high-performance database mining platform CATALYST, parallel pharmacophore-based virtual screening was applied on a set of 50 structure-based pharmacophore models built for various viral targets and 100 antiviral compounds. The results demonstrate that the desired enrichment, that is, a successful activity profiling, was achieved for approximately 90% of all input molecules. It was concluded that pharmacophore-based parallel screening comprises a reliable *in silico* method to predict the potential biological activities of a compound or a compound library by screening it against a series of pharmacophore queries. Subsequently a number of successful applications of parallel pharmacophore-based screening were described, including targets such as HIV protease, human histamine H_3 receptor, and PPAR (peroxisome proliferator-activated receptor). Structure-based docking is an additional promising tool to identify putative targets for a specific ligand. Instead of docking multiple ligands into a single protein cavity, a single ligand is docked in a collection of binding sites. In inverse docking hits are in fact targets that have been prioritized within the pool of the best-ranked proteins. Web-tools for inverse docking are emerging within academic groups [94]. The identification by the Rognan group of phospholipase A2 as a putative target for analogs from a 1,3,5-triazepan-2,6-dione library is a successful application of the inverse docking procedure [95]. Five selected members of the library were docked against a set of 2150 enzyme active sites and five suggested targets were selected for experimental testing.

5. CONCLUSION

Chemogenomics is a new interdisciplinary research field aiming at the genome-wide systematic identification, analysis, and prediction of ligand–protein interactions using both *in vitro* and *in silico* approaches. The knowledge-based exploration of the ligand–target SAR matrix—the chemogenomics knowledge space—will help to discover and validate an increased number of small-molecule compounds and postulated disease targets earlier and faster. While there are many obvious advantages to the herein described approaches, such as the emphasis on systematization, a key limitation to the success for chemogenomics approaches is recognized by the fact that drug discovery is not a totally structured and predictable science. Diversity and serendipity, both in chemistry and biology are drivers of many important discoveries. Conversely, chemogenomics principles such as target family discovery platforms are today part of the portfolio of the major pharmaceutical companies and important contributions to drug discovery have for instance been reached in the design of protein kinase inhibitors. A key scientific and technological challenge for the success of chemogenomics will be the informatics component that will need to render the chemogenomic data user friendly.

ACKNOWLEDGMENTS

Drs. K. Azzaoui, P. Fuerst, R. Lewis, M. Popov, B. Rohde, H.-J. Roth, U. Schopfer, and A. Schuffenhauer, (all NIBR associates) are acknowledged for various support and discussions. Figures 3, 5 and 6 were reproduced with permission from Refs [49,83,89], respectively.

REFERENCES

1. Gaither LA. Chemogenomics approaches to novel target discovery. Expert Rev. Proteomics 2007;4(3):411–419.
2. Wuster A, Madan Babu M. Chemogenomics and biotechnology. Trends Biotechnol 2008;26 (5):252–258.
3. Jacoby E. Chemogenomics: drug discovery's panacea? Mol Biosyst 2006;2(5):218–220.
4. Bredel M, Jacoby E. Chemogenomics: an emerging strategy for rapid target and drug discovery. Nat Rev Genet 2004;5(4):262–275.
5. Lehmann J. Redesigning drug discovery. Nature 1996;384(Suppl):1–5.
6. Klabunde T. Chemogenomic approaches to drug discovery: similar receptors bind similar ligands. Br J Pharmacol 2007;152(1):5–7.
7. Frye SV. Structure–activity relationship homology (SARAH): a conceptual framework for drug discovery in the genomic era. Chem Biol 1999;6 (1):R3–R7.
8. Caron PR, Mullican MD, Mashal RD, Wilson KP, Su MS, Murcko MA. Chemogenomic approaches to drug discovery. Curr Opin Chem Biol 2001;5(4):464–470.
9. Jacoby E. A novel chemogenomics knowledge-based ligand design strategy-application to G protein-coupled receptors. Quant Struct Act Relat 2001;20(2):115–123.
10. Dobson CM. Chemical space and biology. Nature 2004;432(7019):824–828.

11. Paolini GV, Shapland RH, van Hoorn WP, Mason JS, Hopkins AL. Global mapping of pharmacological space. Nat Biotechnol 2006;24 (7):805–815.

12. Overington JP, Al-Lazikani B, Hopkins AL. How many drug targets are there? Nat Rev Drug Discov 2006;5(12):993–996.

13. Jacoby E, Schuffenhauer A, Popov M, Azzaoui K, Havill B, Schopfer U, Engeloch C, Stanek J, Acklin P, Rigollier P, Stoll F, Koch G, Meier P, Orain D, Giger R, Hinrichs J, Malagu K, Zimmermann J, Roth HJ. Key aspects of the Novartis compound collection enhancement project for the compilation of a comprehensive chemogenomics drug discovery screening collection. Curr Top Med Chem 2005;5(4):397–411.

14. Jacoby E, Schuffenhauer A, Azzaoui K, Popov M, Dressler S, Glick M, Jenkins J, Davies J, Roggo S. Small molecules for chemogenomics-based drug discovery. In: Jacoby E, editor. Chemogenomics: Knowledge-Based Approaches to Drug Discovery. London: Imperial College Press; 2006. p 1–38.

15. Gray NS, Wodicka L, Thunnissen AM, Norman TC, Kwon S, Espinoza FH, Morgan DO, Barnes G, LeClerc S, Meijer L, Kim SH, Lockhart DJ, Schultz PG. Exploiting chemical libraries, structure, and genomics in the search for kinase inhibitors. Science 1998;281(5376):533–538.

16. Jacoby E, Fauchere JL, Raimbaud E, Ollivier S, Michel A, Spedding M. A three binding site hypothesis for the interaction of ligands with monoamine G protein-coupled receptors: implications for combinatorial ligand design. Quant Struct Act Relat 1999;18(6):561–572.

17. Savchuk NP, Tkachenko SE, Balakin KV. Strategies for the design of pGPCR-targeted libraries. In: Rognan D, editor. Ligand Design for G Protein-Coupled Receptors. Series: Methods and Principles in Medicinal Chemistry. Vol.30. Weinheim: Wiley-VCH; 2006. p 137– 164.

18. Baringhaus KH, Hessler G. A chemical genomics approach for ion channel modulators. In: Kubinyi H, Müller G, editors. Chemogenomics in Drug Discovery. Series: Methods and Principles in Medicinal Chemistry. Vol. 22. Weinheim: Wiley-VCH; 2004. p 221–242.

19. Harris CJ, Stevens AP. Chemogenomics: structuring the drug discovery process to gene families. Drug Discov Today 2006;11 (19–20):880–888.

20. Liao JJ. Molecular recognition of protein kinase binding pockets for design of potent and selective kinase inhibitors. J Med Chem 2007;50 (3):409–424.

21. Papp A, Szommer T, Barna L, Gyimesi G, Ferdinandy P, Spadoni C, Darvas F, Fujita T, Urge L, Dorman G. Enhanced hit-to-lead process using bioanalogous lead evolution and chemogenomics: application in designing selective matrix metalloprotease inhibitors. Expert Opin Drug Discov 2007;2(5):707–723.

22. Moore JT, Collins JL, Pearce KH. The nuclear receptor superfamily and drug discovery. In: Schreiber SL, Kapoor TM, Wess G, editors. Chemical Biology: From Small Molecules to Systems Biology and Drug Design. Weinheim: Wiley-VCH; 2007. p 891–932.

23. Pierce AC, Rao G, Bemis GW. BREED: generating novel inhibitors through hybridization of known ligands. Application to CDK2, p38, and HIV protease. J Med Chem 2004;47 (11):2768–2775.

24. Lewis JA, Lebois EP, Lindsley CW. Allosteric modulation of kinases and GPCRs: design principles and structural diversity. Curr Opin Chem Biol 2008;12(3):269–280.

25. Evans BE, Rittle KE, Bock MG, DiPardo RM, Freidinger RM, Whitter WL, Lundell GF, Veber DF, Anderson PS, Chang RS. Methods for drug discovery: development of potent, selective, orally effective cholecystokinin antagonists. J Med Chem 1988;31(12):2235–2246.

26. Müller G. Target family-directed masterkeys in chemogenomics. In: Kubinyi H, Müller G, editors. Chemogenomics in Drug Discovery. Series: Methods and Principles in Medicinal Chemistry. Vol. 22. Weinheim: Wiley-VCH; 2004. p 7–41.

27. Willoughby CA, Hutchins SM, Rosauer KG, Dhar MJ, Chapman KT, Chicchi GG, Sadowski S, Weinberg DH, Patel S, Malkowitz L, Di Salvo J, Pacholok SG, Cheng K. Combinatorial synthesis of 3-(amidoalkyl) and 3-(aminoalkyl)-2-arylindole derivatives: discovery of potent ligands for a variety of G-protein coupled receptors. Bioorg Med Chem Lett 2002;12(1):93–96.

28. Bondensgaard K, Ankersen M, Thøgersen H, Hansen BS, Wulff BS, Bywater RP. Recognition of privileged structures by G-protein coupled receptors. J Med Chem 2004;47(4):888–899.

29. Bywater RP. Privileged structures in GPCRs. Ernst Schering Found Symp Proc 2006;(2):75–91.

30. Bemis GW, Murcko MA. The properties of known drugs. 1. Molecular frameworks. J Med Chem 1996;39(15):2887–2893.

31. Grabowski K, Schneider G. Properties and architecture of drugs and natural products revisited. Curr Chem Biol 2007;1(1):115–127.

32. Schnur DM, Hermsmeier MA, Tebben AJ. Are target-family-privileged substructures truly privileged? J Med Chem 2006;49(6):2000–2009.
33. Eguchi M, McMillan M, Nguyen C, Teo JL, Chi EY, Henderson WR, Jr. Kahn M. Chemogenomics with peptide secondary structure mimetics. Comb Chem High Throughput Screen 2003;6(7):611–621.
34. Garland SL, Dean PM. Design criteria for molecular mimics of fragments of the beta-turn. 1. C alpha atom analysis. J Comput Aided Mol Des 1999;13(5):469–483.
35. Garland SL, Dean PM. Design criteria for molecular mimics of fragments of the beta-turn. 2. C alpha–C beta bond vector analysis. J Comput Aided Mol Des 1999;13(5):485–498.
36. Webb TR, Jiang L, Sviridov S, Venegas RE, Vlaskina AV, McGrath D, Tucker J, Wang J, Deschenes A, Li R. Application of a novel design paradigm to generate general nonpeptide combinatorial templates mimicking beta-turns: synthesis of ligands for melanocortin receptors. J Comb Chem 2007;9(4):704–710.
37. Fletcher S, Hamilton AD. Targeting protein–-protein interactions by rational design: mimicry of protein surfaces. J R Soc Interface 2006;3 (7):215–233.
38. Che Y, Brooks BR, Marshall GR. Protein recognition motifs: design of peptidomimetics of helix surfaces. Biopolymers 2007;86(4):288–297.
39. Rodriguez JM, Hamilton AD. Benzoylurea oligomers: synthetic foldamers that mimic extended alpha helices. Angew Chem Int Ed Engl 2007;46(45):8614–8617.
40. Volonterio A, Moisan L, Rebek J, Jr. Synthesis of pyridazine-based scaffolds as alpha-helix mimetics. Org Lett 2007;9(19):3733–3736.
41. Biros SM, Moisan L, Mann E, Carella A, Zhai D, Reed JC, Rebek J, Jr. Heterocyclic alpha-helix mimetics for targeting protein–protein interactions. Bioorg Med Chem Lett 2007;17 (16):4641–4645.
42. Parks DJ, Player MR. Alpha-helix mimetics: progress toward effective modulation of protein–protein complexes. Frontiers in Drug Design and Discovery 2007;3:5–44.
43. Denessiouk KA, Johnson MS. When fold is not important: a common structural framework for adenine and AMP binding in 12 unrelated protein families Proteins. 2000;38(3):310–326.
44. Denessiouk KA, Rantanen VV, Johnson MS. Adenine recognition: a motif present in ATP-, CoA-, NAD-, NADP-, and FAD-dependent proteins. Proteins 2001;44(3):282–291.
45. Knapp M, Bellamacina C, Murray JM, Bussiere DE. Targeting cancer: the challenges and successes of structure-based drug design against the human purinome. Curr Top Med Chem 2006;6(11):1129–1159.
46. Amadasi A, Bertoldi M, Contestabile R, Bettati S, Cellini B, di Salvo ML, Borri-Voltattorni C, Bossa F, Mozzarelli A. Pyridoxal 5′-phosphate enzymes as targets for therapeutic agents. Curr Med Chem 2007;14(12):1291–1324.
47. Sem DS, Bertolaet B, Baker B, Chang E, Costache AD, Coutts S, Dong Q, Hansen M, Hong V, Huang X, Jack RM, Kho R, Lang H, Ma CT, Meininger D, Pellecchia M, Pierre F, Villar H, Yu L. Systems-based design of bi-ligand inhibitors of oxidoreductases: filling the chemical proteomic toolbox. Chem Biol 2004;11 (2):185–194.
48. Burke MD, Schreiber SL. A planning strategy for diversity-oriented synthesis. Angew Chem Int Ed Engl 2004;43(1):46–58.
49. Thomas GL, Spandl RJ, Glansdorp FG, Welch M, Bender A, Cockfield J, Lindsay JA, Bryant C, Brown DF, Loiseleur O, Rudyk H, Ladlow M, Spring DR. Anti-MRSA agent discovery using diversity-oriented synthesis. Angew Chem Int Ed Engl 2008;47(15):2808–2812.
50. Kaiser M, Wetzel S, Kumar K, Waldmann H. Biology-inspired synthesis of compound libraries. Cell Mol Life Sci 2008;65 (7–8):1186–1201.
51. Koch MA, Schuffenhauer A, Scheck M, Wetzel S, Casaulta M, Odermatt A, Ertl P, Waldmann H. Charting biologically relevant chemical space: a structural classification of natural products (SCONP). Proc Natl Acad Sci USA 2005;102(48):17272–17277.
52. Koch MA, Wittenberg LO, Basu S, Jeyaraj DA, Gourzoulidou E, Reinecke K, Odermatt A, Waldmann H. Compound library development guided by protein structure similarity clustering and natural product structure. Proc Natl Acad Sci USA 2004;101(48):16721–16726.
53. Sadreyev RI, Grishin NV. Exploring dynamics of protein structure determination and homology-based prediction to estimate the number of superfamilies and folds BMC Struct Biol 2006;6:6.
54. Entzeroth M. Emerging trends in high-throughput screening. Curr Opin Pharmacol 2003;3 (5):522–529.
55. Schnitzer R, Sommergruber W. The use of genetically engineered cell-based assays in in- vitro drug discovery. In: Bartlett PA, Entzeroth M, editors. Exploiting Chemical Diversity

for Drug Discovery. Cambridge: The Royal Society of Chemistry; 2006. p 247–262.

56. Froloff N, Hamon V, Dupuis P, Otto-Bruc A, Mao B, Merrick S, Migeon J. Construction of a homogeneous and informative *in vitro* profiling database for anticipating the clinical effects of drugs. In: Jacoby E, editor. Chemogenomics: Knowledge-Based Approaches to Drug Discovery. London: Imperial College Press; 2006. p 175–206.
57. Hamosh A, Scott AF, Amberger JS, Bocchini CA, McKusick VA. Online Mendelian Inheritance in Man (OMIM), a knowledgebase of human genes and genetic disorders. Nucleic Acids Res 2005; 33(Database Issue): D514–D517.
58. Krishnamurty R, Maly DJ. Chemical genomic and proteomic methods for determining kinase inhibitor selectivity. Comb Chem High Throughput Screen 2007;10(8):652–666.
59. Bantscheff M, Eberhard D, Abraham Y, Bastuck S, Boesche M, Hobson S, Mathieson T, Perrin J, Raida M, Rau C, Reader V, Sweetman G, Bauer A, Bouwmeester T, Hopf C, Kruse U, Neubauer G, Ramsden N, Rick J, Kuster B, Drewes G. Quantitative chemical proteomics reveals mechanisms of action of clinical ABL kinase inhibitors. Nat Biotechnol 2007;25 (9):1035–1044.
60. Barglow KT, Cravatt BF. Activity-based protein profiling for the functional annotation of enzymes. Nat. Methods. 2007;4(10):822–827.
61. Cravatt BF, Wright AT, Kozarich JW. Activity-based protein profiling: from enzyme chemistry to proteomic chemistry. Annu Rev Biochem 2008;77:383–414.
62. Licitra EJ, Liu JO. A three-hybrid system for detecting small ligand–protein receptor interactions. Proc Natl Acad Sci USA 1996;93 (23):12817–12821.
63. Becker F, Murthi K, Smith C, Come J, Costa-Roldán N, Kaufmann C, Hanke U, Degenhart C, Baumann S, Wallner W, Huber A, Dedier S, Dill S, Kinsman D, Hediger M, Bockovich N, Meier-Ewert S, Kluge AF, Kley N. A three-hybrid approach to scanning the proteome for targets of small molecule kinase inhibitors. Chem Biol 2004,11(2):211–223.
64. Strausberg RL, Schreiber SL. From knowing to controlling: a path from genomics to drugs using small molecule probes. Science 2003;300 (5617):294–295.
65. Rognan D. Chemogenomic approaches to rational drug design. Br J Pharmacol 2007;152 (1):38–52.
66. Bajorath J. Computational analysis of ligand relationships within target families. Curr Opin Chem Biol 2008;12(3):352–358.
67. Williams AJ. Public chemical compound databases. Curr Opin Drug Discov Dev 2008;11 (3):393–404.
68. Schuffenhauer A, Zimmermann J, Stoop R, van der Vyver JJ, Lecchini S, Jacoby E. An ontology for pharmaceutical ligands and its application for *in silico* screening and library design. J Chem Inf Comput Sci 2002;42(4):947–955.
69. Schuffenhauer A, Jacoby E. Annotating and mining the ligand–target chemogenomics knowledge space. Drug Discov Today 2004;2 (5):190–200.
70. Hopkins AL, Paolini GV. Chemogenomics in drug discovery: the druggable genome and target class properties. In: Mason JS, editor. Computer-Assisted Drug Design. Comprehensive Medicinal Chemistry II. Vol. 4. Amsterdam: Elsevier; 2007. p 421–433.
71. Savchuk NP, Balakin KV, Tkachenko SE. Exploring the chemogenomic knowledge space with annotated chemical libraries. Curr Opin Chem Biol 2004;8(4):412–417.
72. Kuhn M, Campillos M, González P, Jensen LJ, Bork P. Large-scale prediction of drug–target relationships. FEBS Lett 2008;582(8): 1283–1290.
73. Stepanchikova AV, Lagunin AA, Filimonov DA, Poroikov VV. Prediction of biological activity spectra for substances: evaluation on the diverse sets of drug-like structures. Curr Med Chem 2003;10:225–233.
74. Bender A, Young DW, Jenkins JL, Serrano M, Mikhailov D, Clemons PA, Davies JW. Chemogenomic data analysis: prediction of small-molecule targets and the advent of biological fingerprint. Comb Chem High Throughput Screen 2007;10(8):719–731.
75. Nidhi, Glick M, Davies JW, Jenkins JL. Prediction of biological targets for compounds using multiple-category Bayesian models trained on chemogenomics databases. J Chem Inf Model 2006;46(3):1124–1133.
76. Azzaoui K, Hamon J, Faller B, Whitebread S, Jacoby E, Bender A, Jenkins JL, Urban L. Modeling promiscuity based on *in vitro* safety pharmacology profiling data. ChemMedChem 2007;2(6):874–880.
77. Mestres J, Martín-Couce L, Gregori-Puigjané E, Cases M, Boyer S. Ligand-based approach to *in silico* pharmacology: nuclear receptor profiling. J Chem Inf Model 2006;46(6): 2725–2736.

78. Gregori-Puigjané E, Mestres J. Coverage and bias in chemical library design. Curr Opin Chem Biol 2008;12(3):359–365.
79. Keiser MJ, Roth BL, Armbruster BN, Ernsberger P, Irwin JJ, Shoichet BK. Relating protein pharmacology by ligand chemistry. Nat Biotechnol 2007;25(2):197–206.
80. Stanton RV, Cao Q. Biological fingerprints. In: Mason JS, editor. Computer-Assisted Drug Design. Comprehensive Medicinal Chemistry II. Vol. 4. Amsterdam: Elsevier; 2007. p 807–818.
81. Kauvar LM, Higgins DL, Villar HO, Sportsman JR, Engqvist-Goldstein A, Bukar R, Bauer KE, Dilley H, Rocke DM. Predicting ligand binding to proteins by affinity fingerprinting. Chem Biol 1995;2(2):107–118.
82. Beroza P, Damodaran K, Lum RT. Target-related affinity profiling: Telik's lead discovery technology. Curr Top Med Chem 2005;5 (4):371–381.
83. Guba W, Green LG, Martin RE, Roche O, Kratochwil N, Mauser H, Bissantz C, Christ A, Stahl M. From astemizole to a novel hit series of small-molecule somatostatin 5 receptor antagonists via GPCR affinity profiling J Med Chem 2007;50(25):6295–6298.
84. Fliri AF, Loging WT, Thadeio PF, Volkmann RA. Biological spectra analysis: linking biological activity profiles to molecular structure. Proc Natl Acad Sci USA 2005;102(2):261–266.
85. Fliri AF, Loging WT, Thadeio PF, Volkmann RA. Biospectra analysis: model proteome characterizations for linking molecular structure and biological response. J Med Chem 2005; 48 (22):6918–6925.
86. Bernasconi P, Chen M, Galasinski S, Popa-Burke I, Bobasheva A, Coudurier L, Birkos S, Hallam R, Janzen WP. A chemogenomic analysis of the human proteome: application to enzyme families. J Biomol Screen 2007;12 (7):972–982.
87. Vieth M, Higgs RE, Robertson DH, Shapiro M, Gragg EA, Hemmerle H. Kinomics-structural biology and chemogenomics of kinase inhibitors and targets. Biochim Biophys Acta 2004;1697 (1–2):243–257.
88. Goldstein DM, Gray NS, Zarrinkar PP. High-throughput kinase profiling as a platform for drug discovery. Nat Rev Drug Discov 2008;7 (5):391–397.
89. Debe DA, Hambly KP, Danzer JF. Structural informatics: chemogenomics *in silico*. In: Jacoby E, editor. Chemogenomics: Knowledge-Based Approaches to Drug Discovery. London: Imperial College Press; 2006. p 157–173.
90. Kuhn D, Weskamp N, Schmitt S, Hüllermeier E, Klebe G. From the similarity analysis of protein cavities to the functional classification of protein families using cavbase. J Mol Biol. 2006;359 (4):1023–1044.
91. An J, Totrov M, Abagyan R. Pocketome via comprehensive identification and classification of ligand binding envelopes. Mol Cell Proteomics 2005;4(6):752–761.
92. Pirard B. Insight into the structural determinants for selective inhibition of matrix metalloproteinases. Drug Discov Today 2007;12 (15–16):640–646.
93. Steindl TM, Schuster D, Wolber G, Laggner C, Langer T. High-throughput structure-based pharmacophore modelling as a basis for successful parallel virtual screening. J Comput Aided Mol Des 2006;20(12):703–715.
94. Gao Z, Li H, Zhang H, Liu X, Kang L, Luo X, Zhu W, Chen K, Wang X, Jiang H. PDTD: a Web-accessible protein database for drug target identification. BMC Bioinformatics 2008;9:104.
95. Rognan D. *In silico* screening of the protein structure repertoire and of protein families. In: Jacoby E, editor. Chemogenomics: Knowledge-Based Approaches to Drug Discovery. London: Imperial College Press; 2006. p 109–131.

DOCKING AND SCORING IN DRUG DISCOVERY

FRANCESCA SPYRAKIS[1,2]
PIETRO COZZINI[1,2]
GLEN EUGENE KELLOGG[3,4]
[1] Department of General Chemistry, University of Parma, Parma, Italy
[2] Italian National Institute of Biostructures and Biosystems Rome, Italy
[3] Department of Medicinal Chemistry, Virginia Commonwealth University, Richmond, Virgina
[4] Institute for Structural Biology and Drug Discovery Virginia Commonwealth University, Richmond, Virginia

1. INTRODUCTION

1.1. Basic Principles and Concepts

The road to produce a new drug available at the local pharmacy is very long and filled with many obstacles and potholes. There are different opinions about the length of this road, and maybe even where it truly begins, but whether it is 6–12 years or 3–10 years is not very important. In any case, at the end of this road, much effort has been expended, much money has been spent and the results are far from sure and predictable. Of most interest to the manufacturers of pharmaceuticals is how broad, smooth, and long the road is *after* the drug makes it to the market because one blockbuster drug can make up for many failures.

In some cases, very small differences between quite similar molecules could be the reasons for a big success or a devastating failure. In 2005, Pfizer earned $12.2 billion for atorvastatin (Lipitor), while Bayer has had to contend with many lawsuits (2861 had been settled by 2004) for its statin drug cerivastatin (Baycol).

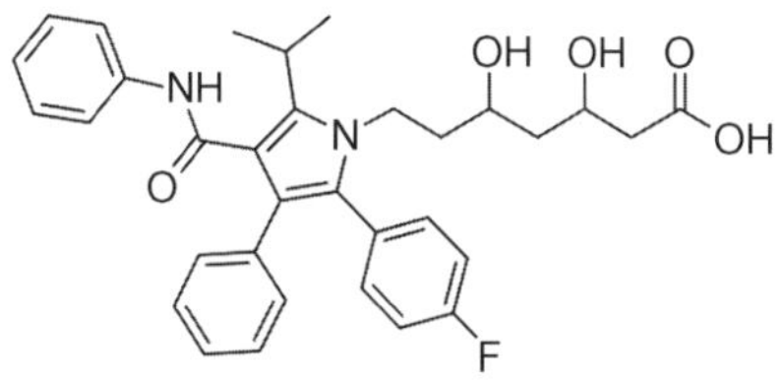

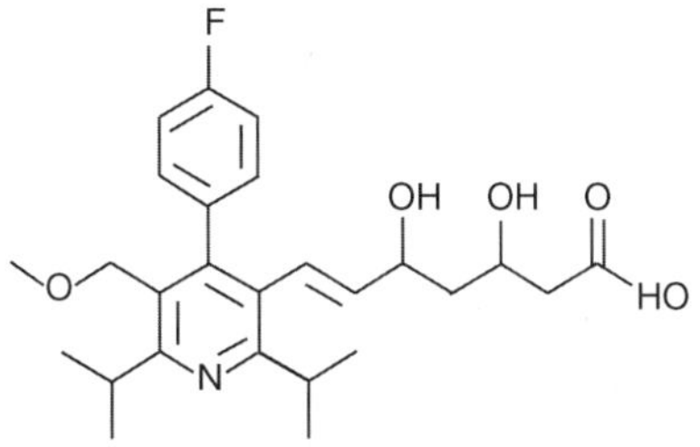

All the steps of the process to discover new drugs are exceptionally well known but many of the steps of this long road cannot be speeded up. However, with technology and ingenuity we can reduce the time needed for the first phase, the design phase. The continuous progress in *omics* technologies and structure determination techniques have resulted in a rapidly growing knowledge base of structural information about biological macromolecules and compounds. The exponential advances in computer sciences that are now offering supercomputing power at an unprecedented performance/price ratio make possible the massive (virtual) screening of huge databases of small molecules in biological receptors. In the presence of the correct data, structure-based drug discovery/design (SBDD) can provide a distinct competitive advantage for speeding the process for producing a new lead or a new drug.

The subject of this chapter is to thoroughly explore the possible features of the computational docking/scoring paradigm as a tool to predict the energetics of protein–ligand binding and the stability of protein–ligand complexes. The docking and scoring problem can be considered as the sum of two separate problems. The first is a geometric problem, that is, the positioning of a solid structure, defined in terms of triplet (Cartesian) coordinates within a cavity defined by another solid. The second is a physicochemical problem, that is, the evaluation of the energy exchange between two or more molecules during the binding event.

Many papers, probably too many, have been written to validate a particular docking/scoring method or software package and to compare the predictive quality and/or limits of a scoring function. Often the metric of success for a docking method is its ability to recreate a crystal structure pose from the separated (protein + ligand) components. Hawkins et al. (2008) summarize a large number of reasons why this is not a particularly useful or robust drill [1]. While it is not without pitfalls [1,2], an accurate evaluation of the binding-free energy between a ligand and a receptor is of paramount importance not only to define active or inactive molecules but also the closely related analogues that are particularly important for drug design.

Structure-based drug discovery/design-must be considered as an iterative process that can drive synthesis by providing feedback as to the "goodness of fit" for a ligand in the active site and potential modifications that would improve that fit. Also, of course, activity/binding assays of the ligands are part of the feedback loop. But the definition of a reliable and realistic model for the protein–ligand complex is a strong basis for a possible success of design. In modeling biological binding events we were focused on, until only a few years ago, simplifying the system as the reaction between a small molecule, the ligand (sometimes considered as a flexible object), and the fixed receptor. Other factors, notably water, were completely ignored. Now, perhaps because much of the low-hanging fruit, that is, simple ligands bound to proteins that were easy to crystallize and are mostly rigid, have been gathered, we are well aware that it is no longer possible to ignore the real properties of biological systems in SBDD. Thus, we can no longer neglect the bulk solvent properties, discrete water molecules in and around the active site, metal ions and/or cofactors if we want to predict conformations and energy more realistically. Certainly, the notion that molecules are fixed and rigid is wrong.

While it is possible to define chemical systems in terms of nuclei, electrons, and wavefunctions, that is, quantum mechanically, for the types of macromolecular systems we are dealing with, direct application of quantum theory is not practical or even possible due to their size (number of atoms) and complexity. The alternative and more accessible approach is to define the systems in terms of atoms and bonds and use classical Newtonian principles to treat them as balls and springs, respectively.

If we exclude the formation of covalent bonds, the forces acting in biomolecular associations are all of a broadly electrostatic nature. Interactions between charges (formal or partial, localized on atoms) are commonly referred to as salt bridges, dipoles, induced dipoles. Hydrogen bonds are a special case of electrostatic interactions and van der Waals interactions arise from dipole–dipole interactions. (In the largely atomistic world of molecular modeling, usually only the induced dipole-induced dipole, or in other words London force, interactions are calculated.) Even the hydrophobic effect is an emergent electrostatic property arising from the (hydrogen bond) attraction between water and polar functional groups that, in effect, "pushes" hydrophobic entities together. Thus, the "hydrophobic interaction" does not, *per se*, actually exist, but is a pragmatic and useful construct to explain the observed and apparent associations between hydrophobic (lipophilic) groups in aqueous environments. These forces are summarized in Table 1 by presenting typical values of these weak forces, thus giving the readers an idea of the magnitude of the forces involved in biological interactions. The exact value of a single specific interaction varies widely depending on the biological system and the method of determination. The data shown are composites and not necessarily specific to biological systems.

In this chapter, we will present a critical point of view of the overall state of the art in docking/scoring methods and attempt to supply the readers with all of the information

Table 1. Biomolecular Interactions and Their Energies

Interaction Type	Energy (kJ/mol)
Ion–ion [3]	150
Ion–dipole [3]	40–60
Ion–induced dipole [3]	3–15
Dipole–dipole [3]	5–25
Hydrogen bond [3,4]	10–40
Dipole-induced dipole [3]	2–10
Induced dipole–induced dipole [3]	0.05–4.0
Hydrophobic [5]	10–12
NH–π [5]	≈4
OH–π [5]	≈4
CH–π [5]	≈4
π–π [5]	≈4
Ion–π [5]	≈4

needed to make an informed and correct choice for applying an accurate and fast docking system. This chapter does not, however, rate the various algorithms and modeling programs—that is left for the readers and users. It must be made clear that the many difficulties of approximating and quantitatively estimating the great number of biological events involved in biochemical recognition and binding interaction is the main reason for the failure to obtain the "Computational Holy Grail" [6,7], that is, a "precise" and "rapid" estimation of binding free energy.

Tame well defined these limits stating: "the problem of affinity estimation lies in our limited understanding of the physics and thermodynamics of ligand binding by biomolecules." [8] However, caution must be exercised, as a purely reductionist approach to ligand binding will likely miss some of the inherent complexity driving the process (see Section 3). From a thermodynamic point of view, the formation of a protein–ligand complex can be reduced to the determination of the standard Gibb's free energy of binding (ΔG°) that is a measure of the equilibrium of the reaction $P_{aq} + L_{aq} \leftrightarrow PL_{aq}$ (see Fig. 1) in aqueous solution.

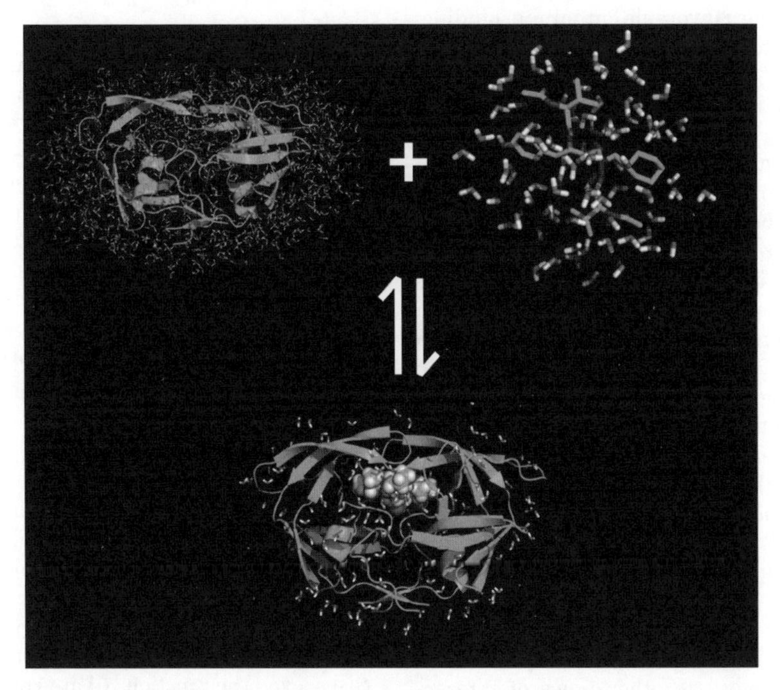

Figure 1. The biochemical and thermodynamic interpretation of docking. An uncomplexed protein (upper left) with water, ions, cofactors, etc. encounters a small molecule in solution (upper right). If the "sum" of an enormous number of things, for example, size, shape, enthalpy, entropy, etc. is (overall) favorable, the ligand may bind *reversibly* to form a protein–ligand complex (lower center).

We must start at the beginning with the "master equation,"

$$\Delta G^\circ = \Delta H^\circ - T\Delta S^\circ$$

where the enthalpic contribution (ΔH°) represents the breaking and formation of hydrogen bonds or the formation of specific polar and nonpolar contacts, and the entropic contribution ($T\Delta S^\circ$) represents the release of water from hydrophobic surfaces to solvent and the loss of conformational mobility of receptors and ligands.

The standard Gibb's free energy of binding is also related to the experimental association constant:

$$\Delta G^\circ = -RT \ln K_a$$

where K_a is the association constant for the reaction $P_{aq} + L_{aq} \leftrightarrow PL_{aq}$, but in many cases we use the reverse dissociation constant K_d, defined as $(P_{aq})(L_{aq})/(PL_{aq})$.

Even while the direct interactions between protein and ligand are of primary importance for generating a favorable enthalpy of binding, we must consider an additional and essential factor—the presence of the medium, water. In fact, the unbound components of the reaction are solvated and partial (or sometimes full) desolvation is required during the binding "reaction." The gain in enthalpy is the difference between the enthalpy of the protein–ligand association and the desolvation enthalpies for protein and ligand. There is also an entropy gain due to the release of these waters, compensated by the enthalpy gain due to the reorganization of these waters in establishing new, ordered, hydrogen bonds with the "bulk" waters. Again, there are unfavorable entropic contributions from the loss of mobility (translational and rotational) for the ligand and possibly protein side chains, while small contributions to an increase in entropy could arise from vibrations in the complex.

While hydrogen bonds are often cited as the main factor in binding energy because of their relative strength to other types of interactions, they can be capricious in biological systems where water is always near. Buried and solvent-inaccessible hydrogen bonds are considerably more robust than surface hydrogen bonds. Part of the enthusiasm for maximizing hydrogen bonds in drug design is due to their simple to understand nature and the concomitant ease in modeling and quantitating their contribution to overall binding. In contrast, hydrophobic interactions, while relatively small individually, are less influenced by proximal solvent molecules. The hydrophobic effect, as described above, is more complex to experimentally measure, understand, and/or model.

The total free energy of binding (ΔG°_{bind}) can thus be considered the sum of the interaction energy between the protein and the ligand (ΔG°_{int}), the solvation energy of the ligand, protein, and complex ($\Delta G^{\circ\,lig}_{solv}$, $\Delta G^{\circ\,prot}_{solv}$, $\Delta G^{\circ\,complex}_{solv}$, respectively), and the entropic ($T\Delta S^\circ$) and conformational ($\Delta\lambda$) changes. This can be expressed as shown in the equation below.

$$\Delta G^\circ_{bind} = \Delta G^{\circ\,complex}_{solv} - \Delta G^{\circ\,prot}_{solv} - \Delta G^{\circ\,lig}_{solv} + \Delta G^\circ_{int} - T\Delta S^\circ + \Delta\lambda$$

Overall, this equation indicates the most common and important terms that should be calculated to obtain a reliable energetic estimation of the overall binding process [9].

However, we must be reminded of the statement of Dill [10]: "Biological interactions are concerted events, not neat sum of terms where each represents an ingredient of the overall process." Thus, even the best practices in treating each of these disparate interaction types individually will not necessarily yield an accurate and reliable ΔG°_{bind} in the end.

1.2. Practical Requirements

The first requirement for successful structure-based drug design is a good knowledge of the target structure. This is not as obvious as it seems: many biomacromolecular structures (see below) are being generated and deposited in databases without context. In fact, in some cases, even the functions (if any) of the biomolecules are not known. While it is most desirable to know and understand the mechanism(s) by which the protein or enzyme can be inhibited or activated, at least knowing the location of the binding site is probably necessary.

The source of structural information employed for a docking study is usually one of the three experimental techniques, X-ray diffraction, neutron diffraction, and nuclear magnetic resonance (NMR), and one theoretical comparative technique, computational homology modeling. Probably, X-ray diffraction is the most popular method because it is relatively inexpensive to obtain a protein structure and is somewhat exempted from some of the limits that other techniques operate under and that could invalidate their predictions. The big advantage of neutron diffraction compared to X-ray diffraction is the higher resolution of neutron diffraction that often allows direct observation of hydrogen positions in structures. On the other hand, data from neutron diffraction has a much higher cost because it requires a neutron source from an accelerator. Also, somewhat larger crystals are necessary, which is not often possible with biomacromolecules available in minute quantities. NMR is limited by protein size (smaller is better) and solubility issues, but NMR allows examination of the structure in solution, absent from crystallization and packing effects, and provides some description of flexibility absent in crystallography.

Even to date there are few examples of structures derived by homology modeling studies and successfully used as targets to produce lead compounds. The uncertainty of these "model" structures compels the use of this approach only in the total absence of direct experimental structural information. As an example, only a small handful of structures are known for the massive family of G-protein-coupled receptors (GPCRs) while there are numerous drug development efforts underway to target the many diseases associated with this ubiquitous class of receptor. For these cases the only solution for docking is homology modeling using rhodopsin [11], the β1 [12], or β2 adrenergic receptors [13] as templates.

Davis et al. [14,15] published a complete review on the limits and the inaccuracy of structural data for SBDD; *Nature* published an accompanying commentary [16] discussing the extremely serious errors occurring in most important protein structures. Closer to the aim of this chapter, Mohan et al. [17] and Hawkins et al. [1] have surveyed the relationship between input structural quality and success in SBDD. In the next paragraphs, we will summarize some of the limits and the uncertainties of structural data from X-ray diffraction analysis, and define some of the terms used in evaluating structural quality, but this topic is both broad and deep.

For formal definitions of some of the structural parameters discussed here, and for more fundamentals in protein crystallography, see Giacovazzo et al. [18]. The three parameters we have to consider to understand the quality limits of structural data are resolution, *R* factor, and temperature factor. Resolution, measured in Å, is "a statement of the accuracy in data collection and not a measure of the accuracy in refinement"; with modern detectors it depends only on the intrinsic quality of the crystal. Crystallographic resolution could be defined as the width of the windows we use to analyze the crystal or, more precisely, as the "minimum interplanar spacing of the real lattice for the corresponding reciprocal lattice point that is being measured." It is directly related to Bragg's law ($d = \lambda/2\sin\theta$). In X-ray crystallography, atoms *can* be differentiated even if they are much closer together than crystallographic resolution because atoms are modeled as spheres and the entire reflection peak is analyzed and fit (not just its maximum), and because we know something about the chemistry or chemical structure of the molecule. Thus, crystallographic resolution is not the same as standard deviation in atomic positions! A resolution value under 2.0 Å is generally considered as a sign of good crystallographic data for protein structures; a resolution value near 1.0 Å is termed "atomic." In a large molecule such as a protein, differences in resolution can make the difference in determining the correct position or orientation of a residue's side chain. However, to reiterate, resolution only indicates the potential for a good quality structure as it states nothing about the quality of the *refinement*.

The second structural parameter, *R* factor (residual factor, or agreement factor), is defined as "a measure of how well the refined structure explains the observed data" or, in other words, "the percentage of difference between the crystal and the refined model." An even better, but more infrequently used, parameter is R_{free} that indicates the ability of the model to explain a small fraction of data that was not included in building the model [19].

The last parameter, *B* factor, is also called the temperature factor because one of its interpretations is "the measure of the movements of an atom along the three axes due to isotropic (or anisotropic) thermal motion," although it includes other effects such as static disorder in the crystal. In a PDB protein structure file, the mean *B* factor and the *B* factor associated with each atom of the structure are reported. A mean *B* factor value of 20 could be considered as a "normal" value but in many cases for a single atom we can find higher values. In general, $B = 20$ means that, in case of isotropic thermal motion, the atomic positions are indeterminate by about 0.25 Å. However, an important fact to remember about the *B* factors is that they are fit along with the atomic positions during refinement and are possibly subject to manipulation or overfitting and cannot be considered unbiased metrics of structure quality.

Thus, it is obvious that limitations in the quality of structural data affect the docking/scoring paradigm at different levels and, in particular, on the parameterization of scoring functions. For these reasons, several "curated" test sets for testing docking and scoring tools have become available in recent years; for example, PDBbind [20,21], BindingDB [22], Binding MOAD [23], and CCDC/Astex [24].

Another frequent problem we have to face in modeling is the making the correct interpretation of atom "type" and hybridization from structural data files. The unfortunate, but universal, "PDB format" that does not encode multiple bonds (and often no bonding at all) has led to uncounted numbers of poor geometries for starting models in SBDD.

We cannot propose a good starting model for a molecule in which we do not know the positions of the protein hydrogen atoms and, since protein structures solved using X-ray techniques usually do not reveal hydrogen positions, the tautomeric states of histidine or of bound ligands or the ionization states of ionizable functional groups on the protein or ligand cannot be directly determined. In fact, the relative positions of N and O atoms of the glutamine and asparagine side chains can also be very difficult to assign. Care has to be taken for all of these effects, especially for residues located within the protein active site, because they may *strongly* impact ligand binding. We thus must usually decide *a priori* how to proceed with proton placement, tautomerism, etc. in our models prior to docking or other computational analyses.

Finally, docking/scoring problems are, at first blush, two body interactions (e.g., protein–ligand, protein–protein, protein–DNA, organic host–guest, etc.), but this situation is seldom true in the real world. Thus, we must consider models where the interactions are between (many) more than two species: protein, ligand, and water (discrete and bulk), with the possibility of additional cofactors and/or ions, etc. Thus, during the estimation of binding energy we must consider *all* the interactions between *all* the bodies. For example, in the case of HIV-1 protease, the water (wat301) is bridging between the ligand and two amino acids of the protein and it is a key and contributes energetically to the overall stability of the complex [25].

In a recent review, Davis described the case of two different structures of the transforming growth factor-β2 (PDB codes 1TFG and 2TGI) as an enlightening example of the difficulty in correctly assigning electron density for water molecules. These two PDB files contain the structure of the same protein, solved by different laboratories at 1.8 Å (2TGI) and 1.95 Å (1TFG). The first structure reveals 58 water molecules with an average *B* factor of 31.8 $Å^2$ while the second reveals 84 water molecules with an average *B* factor of 43.3 $Å^2$ [26]. The conclusion was that "identification of water molecules in the electron-density maps can be a problem." [14] Now, think of extending this uncertainty to new ligands, putatively placed in binding pockets probably preshaped for a different ligand, surrounded and infused by an unknown number of water molecules, and you will get the idea behind docking!

2. DOCKING AND SCORING

While there is a seemingly endless array of computational techniques for lead discovery through ligand docking, the currently available approaches can be roughly divided into two classes, that is, docking algorithms and *de novo* design methods. Docking usually considers ligands as entire entities; in *de novo* design, molecules are sequentially built up by positioning a seed and extending with building blocks from a library of organic fragments, or by the simultaneous placement of several fragments that are then connected by suitable linking groups [27]. Typical *de novo* design tools are BUILDER [28], CONCEPTS [29], CONCERTS [30], DLD/MCSS [31], GenStar [32], GroupBuild [33], Grow [34], HOOK [35], Legend [36], LUDI [37], MCDNLG [38], SMOG [39], and SPROUT [40]. Despite the fact that there are significant overlaps between these two categories of modeling tools, the strategies adopted by *de novo* design approaches are not the focus of this chapter (see Refs [41,42] and references therein). Here, our interest is on docking and scoring algorithms. We will focus on ligands being docked to biomacromolecular targets. The more difficult problem of docking two macromolecules together, for example, in protein–protein associations, is beyond the scope of this chapter, and in a somewhat nascent stage of development with respect to drug discovery.

As stated by Blaney and Dixon, docking can be considered as a rational drug design approach seeking to predict the structure and the binding energy of a ligand–receptor complex given only the structures of the free ligand and receptors [43]. The first aim of a docking simulation is to correctly predict and identify the most favorable binding mode of the ligand into the target active pocket. The second is to correctly rank the different compounds in accord with the corresponding experimental binding affinities, while the third is to generate likely shortlists of potential inhibitors for experimental characterization [44]. It thus follows that in addition to a good positioning algorithm we also need a reliable scoring function able to properly predict the free energy of binding of the different generated complexes. Docking protocols can be, in fact, thought as the combination of two components, a search strategy and a scoring function [45].

Searching refers to investigating the accessible conformational space for the interaction between the target and the ligand, trying to be sufficiently accurate to not miss any valuable solutions, but also with enough computational efficiency to avoid the exploration of irrelevant regions, or of already detected local minima. Scoring refers, instead, to evaluate and rank the generated conformations, thereby discriminating between native and nonnative binding modes [46]. The ideal scoring function should be able to correctly rank the configurations generated for a single ligand, to rank different ligands with respect to one protein, and to rank different ligands with respect to their binding affinity for different proteins [46]. In a perfect simulation the highest scored complex should resemble the experimental binding mode; nevertheless this situation is rarely observed. First of all scoring functions are not often able to identify the experimental global minimum. Also, optimization algorithms are usually not fast enough to identify the global minimum for a given scoring function [47,48]. Given, in fact, the very large size of the search space, it is currently not feasible to explore all six degrees of translational and rotational freedom of the ligand along with the internal conformational degrees of freedom of both the ligand and the protein [45]. Thus, in trying to find the right balance between speed and accuracy, only a reduced region of the available conformational space and a limited number of the possible binding modes are usually explored.

Perhaps most significantly, to increase the probability of locating the global minimum, a variety of constraint strategies are adopted. Hence, the first DOCK version treated both ligand and receptor as rigid bodies, allowing the exploration of only six degrees of translational and rotational freedom. This is clearly a strong penalty since this constraint forces us to assume that the ligand conformation is similar to the experimental one, and that the protein does not experience significant adjustments upon ligand binding [45]. The numerous observations of induced fit effects and of well-known conformational adjustments occurring, for instance, in adenylate kinase [49], nuclear receptors [50,51], or GPCR [52] led, as the technology matured and developed, to incorporation of at least ligand conformational flexibility into search algorithms. Rigid body approximations for the ligand are seldom used in modern docking programs [53].

Coincident with the development of algorithms exploring ligand flexibility, algorithms and approaches for limited exploration of receptor flexibility began to be developed [48].

Similarly, rigorous scoring functions are too time demanding to be incorporated into virtual screening tools. Thus, the complexity of scoring functions is often reduced, leading, inevitably, to a loss in accuracy. Of course, it should be noted that no scoring function, at any level of complexity, is completely accurate. In trying to find the right compromise between computational time and binding prediction accuracy, some docking approaches have been implemented with a single function used for both directing the search process and ranking the resulting modified protein–ligand conformations, while others adopt a two-stage scoring approach by using a simple function to direct the search and a more rigorous one to rank the results [45].

Another major algorithm development was in the methodologies for representing molecules, in particular the macromolecular receptors. The initial explicit atoms treatment [54–56] has evolved first with molecular surface calculations [57–60] and then with regular three-dimensional grids that store the salient physicochemical receptor properties as in GRID [61], DOCKER [62], AutoDock [63–65], DOCK [66], ICM [67,68], and ProDock [69,70]. At each grid point, an interaction energy of a "probe" atom with the surrounding environment is calculated, thus providing a sort of pseudo-affinity map for each ligand atom type or interaction type occurring between the ligand and the receptor. These can then be efficiently used as a lookup table for scoring the different conformations generated by the docking algorithm.

In the early seventies, Platzer and coworkers carried out energy calculations on enzyme–ligand complexes by using molecular dynamics (MD) simulations [71,72], but the first and probably most famous docking algorithm, DOCK, that used a combinatorial approach instead of a simulation, was first described in 1982 [59]. In the following years, the original DOCK formula was significantly altered, improved, and made much more sophisticated [73–76]. As the structures of more pharmaceutical targets became available and computers became faster—fast enough to provide docking simulations within acceptable time spans, many other docking approaches were developed. Docking algorithms can be essentially classified following three criteria, that is, the allowed degrees of flexibility, the searching mechanism used, or the scoring function for ranking the different conformations. With the first classification, we could (i) place rigid ligands into rigid sites adopting the traditional "lock and key" approximation [59], (ii) place flexible ligands into rigid sites [63–65,77], or (iii) place flexible ligands into semiflexible sites. This latter is, at present, considered by many to be the most reasonable and natural approach [78–84] and will be discussed further below in Section 3.4. To date, no study considering rigid ligands with flexible proteins has been performed [85]. The most complex approach, that is, full flexibility, is under development, but the issues associated with torsion sampling, scoring, etc. make this a very difficult problem.

In the remainder of this section, we will try to provide a perspective of the state of the art in current docking tools by differentiating the diverse programs first according to their basic structure search algorithms (Section 2.1) and then with their scoring functions (Section 2.2) as used to predict the correct binding modes. As this discussion progresses, we are highlighting pitfalls and benefits of the most well-known and applied programs and identifying new and/or less utilized programs that have interesting features. The following Section 3 focuses on some of the emerging and continuing issues with docking simulations and how they are being treated by available and prototype software.

It should be emphasized at the outset that in this chapter we are neither critically evaluating the specific software tools described nor specifically commenting on the claims made by the authors regarding their algorithms, programs, and results. Generally, however, the data sets examined and the results obtained in these investigations should be viewed with some skepticism, especially if the data sets are small or lack diversity. There is voluminous literature comparing docking tools and scoring functions, but much of that is also not truly unbiased. Nevertheless, papers by Foreman et al. [86], Bissantz et al. [87], Bursulaya et al. [88], Kroemer et al. [89], Cole et al. [90], Warren et al. [91], Truchon et al. [92], and Englebienne et al. [93] give a starting point for readers interested in performance comparisons.

2.1. Structural Searching Functions

Algorithms searching the conformational flexibility of small molecules can be basically divided into three main categories, that is, systematic, stochastic, and deterministic methods [48], each in turn including different approaches.

Systematic methods, also known as combinatorial search strategies [46] are essentially based on the calculation of a grid of values for each formal degree of freedom to be explored during the search in a combinatorial way. Clearly, as the complexity of the molecules rises, and the degrees of freedom increase, the search becomes more and more complex, making the inclusion of termination criteria necessary to prevent exploring conformational space that leads to wrong solutions. This strategy was initially adopted by methods looking for shape complementarity via point complementarity or distance geometry approaches where both the ligands and binding pocket are defined by geometric and/or physicochemical descriptors and are then "matched" by various alignment procedures. Pure descriptor matching works, however, only for rigid molecule treatment. The necessity of including more flexibility led to the development of "place and join" or incremental construction approaches that are better able to investigate the conformational space without a combinatorial explosion of conformers or handling thousands of models for the ligand [46]. Sections 2.1.1–2.1.4 describe algorithms and programs that apply systematic methods for docking and model building.

Since standard minimization techniques are not only able to exhaustively sample conformational space but also progress toward the closest local minimum, minimizations are often combined with other strategies to overcome high-energy barriers [46]. Stochastic investigations of the conformational space such as Monte Carlo simulations (Section 2.1.5) and genetic algorithms (Section 2.1.6) are commonly used. These approaches search for the global minimum of the binding free energy with energy evaluations based on molecular dynamics or molecular mechanics, which should correspond to the native binding mode [94]. In contrast to systematic methods, stochastic algorithms, while more time consuming, allow better modeling of flexibility for both ligands and receptors. Given their intrinsic random nature, the major problems in stochastic docking methods are related to the difficulty of reaching reasonable convergence, that is, multiple and independent runs are usually concomitantly performed [48]. Deterministic methods, on the other hand, depend strongly on the architecture of the initial conformation, which determines the moves that can be made to generate the next state. One of the most significant drawbacks of deterministic methods is the likely probability of being trapped in local minima because of the difficulty in overcoming high energetic barriers [48]. Two divergent mathematical approaches have been explored in deterministic MD-based docking algorithms. First, approaches (Section 2.1.7) that implement simulated annealing with a variety of integrated tricks to jump over barriers have been successful in several docking programs. The second approach, heuristic tabu searches (Section 2.1.8), forces sampling of unexplored space during docking simulations by maintaining lists of already explored conformations, etc. and penalizes searches that approach these regions.

Lastly, and perhaps inevitably, a number of docking protocols that combine multiple algorithms and protocols into a single tool have recently evolved. This phenomenon of hybrid approaches and consensus docking is described in Section 2.1.9.

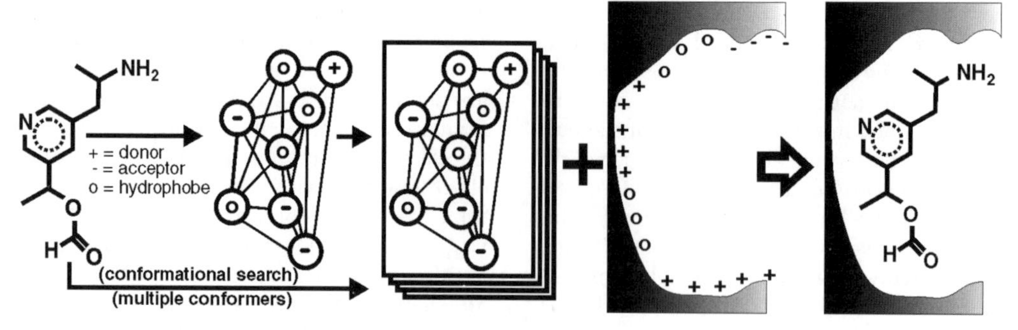

Figure 2. Point complementarity search. The molecule (or multiple conformers from a systematic search) is converted to a set of pharmacophore points. These are matched to corresponding properties of the ligand site via rigid body translations and rotations. Through match scoring the optimally docked protein–ligand model is obtained.

2.1.1. Point Complementarity Methods One of the first point complementarity approaches was proposed by Jiang and Kim, who tried to predict interaction sites by considering charge–charge, hydrogen bonding, van der Waals interaction and size and shape complementarity [58] (see Fig. 2). Molecular surfaces and volumes were described by cube representations, which allowed the effects of the conformational changes caused by complex formation to be computed as the ligand cube is translated and rotated to achieve the maximum number of matches between ligand and protein. The Jiang and Kim docking simulation used a two-step procedure, first with geometric "soft" docking to select the population, followed by screening for subpopulations with more favorable interactions between the buried surface areas. This method was tested by docking ligands extracted from their complex crystal structures and by docking ligands into uncomplexed structures (for which the complex structure was available). Comparison of these docked structures with crystal structures yielded 1.55 and 2.56 Å RMSD values, respectively.

A few years later, Miller and coworkers described FLOG (flexible ligand oriented on grid), an algorithm for matching complementary molecules in a 3D database to a given 3D structure receptor [95]. The algorithm was based on a clique search strategy [96], using a distance compatibility graph to look for compatible matches [47], that is, where each point of the graph corresponds to a possible match between protein and ligand. If pairs of corresponding nodes are connected (at an edge), then those matched features are distance compatible. Superposing these features initially orients the ligand into the target binding site. Ligand flexibility was inspected by including 25 different ligand conformations, and the possible orientations were scored with a function including explicit terms for vdW, electrostatics, hydrogen bonding, and hydrophobic interactions. While FLOG was able to select known inhibitors from a large database of drug-like compounds for dihydrofolate reductase, the highest scoring conformation did not match the crystallographic binding mode.

The LIGIN function developed by Sobolev and coworkers performs docking by maximizing a complementarity function dependent on the atomic contact surface area and the chemical properties of the interacting atoms [97]. These resulting surface contacts are weighted with respect to the positive or negative interactions they may form. Model structures are formed by maximizing the number and the quality of hydrogen bonds. While both ligand and protein are handled as rigid objects by LIGIN, the conformations of a small number of residues lining the binding cavity can be optimized. This tool was successfully applied to the CASP-2 ligand docking test set [98].

The FTDOCK (1997) program by Gabb, Jackson, and Sternberg [57] is a rigid docking program based on a point complementarity algorithm that identifies shape correspondence and favorable electrostatic interactions through Fourier correlation theory. To account for the protein's conformational adjustments due to ligand binding, the scoring function is softened to allow significant surface overlap and thus account for side-chain flexibility. The resulting score is related to the degree of surface correspondence calculated as the product of grid point values between the ligand and the protein. The predictive ability of the method was evaluated with a test set of native proteins instead of protein–ligand complexes, yielding RMSD values less than 2 Å for all but one of the 10 studied cases [57].

The SANDOCK approach by Burkhard and coworkers was first described in 1998. The aim of SANDOCK was to develop an automated method for docking of small ligands into protein pockets [99]. The target binding pocket is represented by "dots" encoding the accessibility and chemical properties of atoms potentially involved in interacting with the ligand, and the complementarity evaluation is based on both shape and chemical criteria, using a distance matching algorithm to fit the ligand onto the corresponding protein dots and then score the poses with vdW, hydrophobic, and hydrogen bond interaction terms. SANDOCK was able to properly reproduce the crystallographic structure of a thrombin–ligand complex with an RMSD value of 0.7 Å. A new shape-based polynomial time algorithm named QSDock (quadratic shape descriptors), developed by Goldman and Wipke [100], investigates the surface complementarity existing between the binding pocket and the possible ligands with a shape descriptor based on local quadratic approximations to the molecular surface. When this algorithm was evaluated on 20 different crystallographic protein–ligand structures, it successfully reproduced the experimental binding mode.

A shape complementarity-based approach was also adopted as one of the modes of FRED (fast rigid exhaustive docking), a rigid and systematic docking program developed by OpenEye [101]. Ligand poses were filtered for shape complementarity and pharmacophoric features, and scored using a consensus of different structure-based and ligand-based scoring functions. When compared to other docking algorithms in an exhaustive virtual screening study, FRED gave among the best results with the lowest variability [102].

In 2001, Diller and Merz proposed a rapid docking approach specifically designed to prioritize combinatorial libraries in virtual screening analyses [103]. The procedure first performs a ligand conformational search, then, in a recognition stage, matches ligand atoms onto precalculated hot spots in the binding site. Subsequently, a gradient-based energy minimization procedure allowing translation, orientation, and optimization of rotatable ligand bonds, while keeping the protein fixed, was performed. The initial conformational search is probably more appropriate for combinatorially derived compounds, since they usually present many rotatable bonds and clear anchor fragment are often not easily identifiable. When tested on 103 protein–ligand complexes retrieved from the PDB [104] the algorithm was able to redock these ligands with 2.0 Å or less RMSD in 90% of the cases.

Rigid body docking of fragments followed by reconstruction is used in the eHiTS algorithm as a flexible ligand docking and virtual high-throughput screening tool [105]. After the possible interaction sites in the binding pocket are identified, the ligand is decomposed into rigid fragments and flexible connecting chains. All fragments are then independently docked in all likely sites. Next, a rapid hypergraph clique detection algorithm matches the compatible fragments into pose sets. The flexible parts are fitted to construct rough binding poses that are later refined in the active site with energy minimization using a more accurate scoring function [105,106] (see Section 2.2.3).

Other, more recently documented, matching methods are: Q-fit that uses statistical thermodynamics principles for ligand atom placement [107]; the database-mining program LIDAEUS [108]; PhDOCK, a pharmacophore-based docking method for virtual screening of large 3D databases (implemented in DOCK 4.0) [109]; Ph4DOCK that exploits the pharmacophoric features in both ligands and receptor binding pockets [110]; DragHome developed by Schafferhans and Klebe for docking ligands in binding pockets in homology modeled targets [111–115]; and

PatchDOCK, developed by Schneidman-Duhovny et al., which is based on shape complementarity principles and is freely available on a web server for protein–protein and protein–ligand docking [116].

2.1.2. Distance Geometry Methods The distance geometry-based methods could be also classified as shape matching algorithms (see Fig. 3). Depending on the way that the molecular shape information is exploited, in fact, the geometric docking algorithms could be further divided into two main categories. In shape-explicit docking algorithms, the shape of both the receptor site and the ligand is investigated and the ligand orientations are generated to maximize the shape complementarity matching. In contrast, shape-implicit docking methods first generate the ligand orientations within the receptor binding pocket and then evaluate them according to shape complementarity [100], as done, for example, by Jiang and Kim [58]. The different approaches also result in a different numbers of ligand conformations, since implicit methods need to explore all degrees of freedom, while explicit techniques avoid unbiased searches and only produce conformations suitable for complementarity.

In its earliest incarnation, Kuntz's DOCK (1982), the first widely used docking simulation program, invokes a distance geometry-based approach in that rigid body algorithm places ligands into binding pocket based on distance-compatible matches [59]. The search was performed in three steps: (i) representation of the receptor site and ligand structures as spheres, (ii) matching of these representations with internal distance comparisons, and (iii) optimization of ligand position within the binding site by reducing atom overlaps and ensuring hydrogen-bonding partners. The final score in this original implementation of DOCK was based on the degree of overlap and took into account only hard sphere repulsions and hydrogen-bonding contributions. While the original DOCK was found to be well suited for generating starting conformations for refinement, this program, in particular has undergone many iterations and improvements in searching and scoring over the past 25+ years [66,117–122] (see Section 2.1.4).

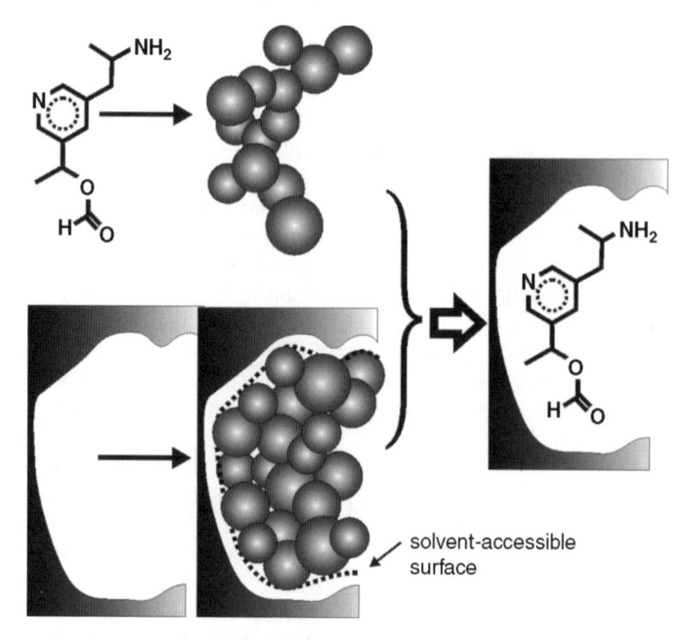

Figure 3. Distance geometry methods. Both the ligand and solvent accessible volume of the site are idealized as spheres. Optimal matches between the sphere sets are obtained by trial and error matching using distance geometry and the triangle inequality.

A similar approach was also proposed in the DockIt program developed by the Metaphorics group [123]. Again, both receptor and ligand are represented by spheres and ligand poses are generated with distance geometry and sphere docking engines. The interactions are then scored with PMF [124] and PLP [125] functions.

Other examples of distance geometry algorithms are the least squares fitting procedure proposed by Bacon and Moult [126], and the hierarchical search of geometrically compatible triplets implemented by Wallqvist and Covell in the ADAM program [127].

2.1.3. Systematic/Exhaustive Methods In exhaustive methods, such as EUDOC, a complete systematic search of the possible poses assumed by a ligand in a defined binding pocket is performed (see Fig. 4). EUDOC was developed by Perola et al. [128], starting from the docking program SYSDOC previously devised by Pang et al. [129]. The program was designed to systematically translate and rotate ligands in the putative binding pocket, looking for energetically favorable positions. The number of generated conformations is only limited by the size of the binding pocket, without accounting for any shape complementarity. EUDOC was demonstrated to be extremely efficient as a virtual screening tool and, when tested on a large data set of potential farnesyltransferase inhibitors was able to identify 21 hits, four of which experimentally inhibited the target enzyme with IC_{50} values in the range from 25 to 100 μM.

FRED, by the OpenEye group, also uses an exhaustive search method by taking a multiconformer ligand and generating a number, possibly exhaustive, of candidate poses. The ligand libraries are built with OMEGA, a tool specifically designed to rapidly and reliably produce multiconformer structure databases [130]. The ligand conformational ensemble is then generated by performing, on each conformer within the site, rigid translations and rotations, such that any single step does not produce atom displacement greater than a specified value [131]. In FRED, pose evaluation is guided by shape complementarity and pharmacophoric features [101].

2.1.4. Fragment-Based Methods Fragment-based approaches were initially developed to handle the flexibility of ligands, which had previously been considered to be virtually rigid (see Fig. 5). Molecules are divided into minor fragments, which themselves are either rigid or represented by small conformational ensembles. Then, these fragments can be treated in one of the following two ways: (i) placing the first into the receptor pocket and subsequently adding the others and (ii) simultaneously placing all fragments in the pocket and trying to link them in favorable orientations. The first approach is known as the incremental construction strategy and the second is known as "place and join."

The incremental construction approach generally performs better both in terms of the quality of conformations generated and the amount of computational time required. This

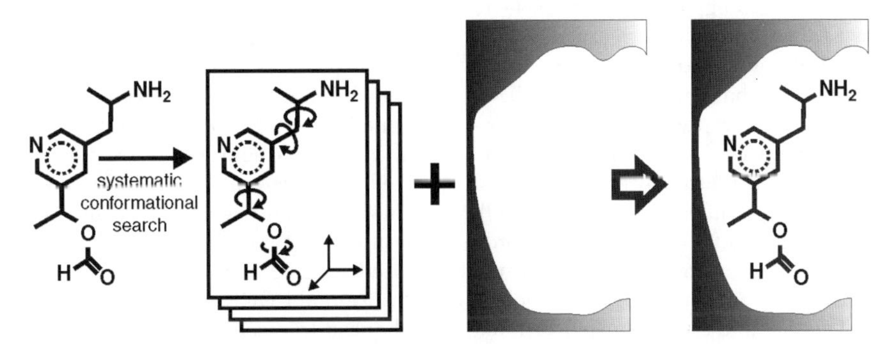

Figure 4. Exhaustive/systematic methods. Multiple, systematic or exhaustive, conformations of the ligand are optimized against the site with rigid body translations and/or rotations.

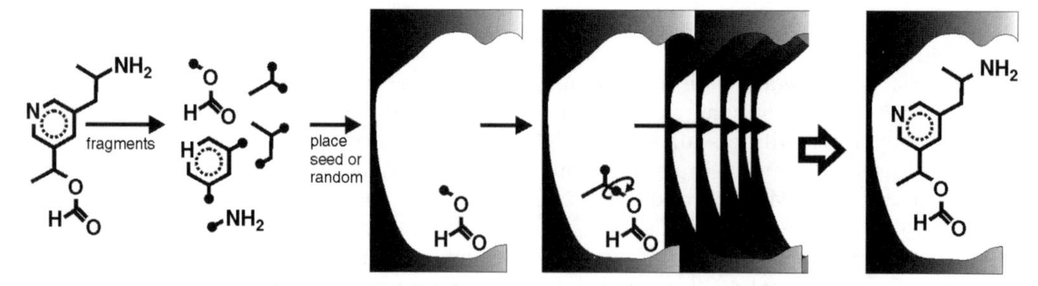

Figure 5. Fragment-based methods (incremental construction). The ligand is deconstructed into fragments that are iteratively recombined as their interactions with the binding site are scored and optimized.

is, in fact, the most widely used fragment-based molecular docking technique. Place and join can be successfully applied only when the ligands can be deconstructed into small sets of medium-sized rigid fragments that can easily be modeled into the binding pocket in an independent, but still reasonable way. Otherwise, many molecules with distorted conformations can be generated, [47] and the resulting small-molecule conformations have to be evaluated as an independent filter.

Place and join techniques have been described by DesJarlais et al. [132] and by Sandak et al. [133,134]. In particular, DesJarlais and coworkers proposed place and join as an extension of DOCK, since the shape of the protein binding pocket is reproduced by a set of overlapping spheres. Ligands are then divided into different large and rigid fragments to be docked separately and joined to, in principle, regenerate the original ligand inside the target binding pocket. This approach was validated on dihydrofolate reductase-methotrexate and prealbumin–thyroxine, and gave, in both cases, binding modes very close to those crystallographically observed [132]. Sandak and coworkers applied an extension of the geometric hashing and generalized Hough transform paradigm developed for rigid object recognition [135], and focused on significant motions induced by protein–ligand complex formation in molecular docking [133]. Initially, this technique was applied to only ligands by considering them as being formed by rigid entities connected by rotary joints (hinges). This flexible treatment is extended to whole complexes by allowing "hinge" induced motions to occur in domains, subdomains, or groups of atoms of the receptor pocket [134].

Incremental construction procedures normally have three phases: (i) selection of base or anchor fragments, (ii) placement of these fragments, and (iii) incremental construction. For long and flexible ligands, a variety of fragmentations may be investigated. The positioning of the anchor and other fragments is usually driven by descriptor matching [46]. The main weakness of the incremental construction algorithms is related to the choice of the base fragment and to the importance attributed to the functional groups forming the ligand. These weaknesses can lead to incorrect or incomplete exploration of conformational space.

The very first application of fragment-based docking was likely by Leach and Kuntz in DOCK (1992), where a method for performing conformational analysis of flexible molecules confined within a receptor binding pocket was developed [136]. The two-step procedure uniquely divided the orientation and conformational search problems. The ligand is fragmented into an anchor and different flexible portions, and the anchor is positioned in the binding pocket with the DOCK variant termed *Directed DOCK* [59,132,137]. The resulting best orientations are filtered, clustered, and fixed into the binding pocket. The conformational degrees of freedom of the remaining portion of the ligand are then explored with a systematic search algorithm. The original DOCK scoring function [138] was enhanced to resolve steric strain and to take into account hydrogen-bonding complementarity between the receptor pocket and the ligands, such that poses have both acceptable internal energy and favorable interactions with the surrounding environment of the site.

In 1996, Rarey et al. released FlexX, a fully automated approach for flexible molecular docking, that was specifically developed for virtual screening of large compound databases in reasonable computational time [77]. Again, the first step is selection of the ligand anchor fragment, first manually and subsequently automated [139]. This fragment is then positioned with a clustering algorithm driven by molecular interactions (both hydrogen bonding and hydrophobic), rather than simply by shape complementarity, that provides a better treatment of very small fragments. Each fragment added to the anchor is then evaluated in all possible conformations, and new protein–ligand interactions are improved through optimization of the whole ligand position. The best solutions are then clustered and iterations proceed until the entire ligand is reconstructed. The solutions are scored using a variant of Böhm's empirical function (see Section 2.2.3) [140]. Later improvements included better treatment of hydrophobic fragments [141], and consideration of water molecules and metal ions at the protein–ligand interface with the particle concept technique [142]. Receptor flexibility is simulated with discrete alternative protein conformations, that is, the FlexE tool that docks flexible ligands into ensembles of protein structures representing flexibility, point mutations, or alternative models [78]. The different structures are superposed and combinatorially joined to create new valid protein models. When tested with 60 ligands docked into 105 crystal structures, FlexE performed as well as cross-docking (see below) with significantly lower computational time.

Hammerhead (Welch et al.) [143], is a completely automated approach involving three steps: identification of the binding pocket with a pocket finder able to identify ideal contact geometries within the protein, docking of ligands with a fragment-based alignment and conformational search methodology, and selection of compounds for testing using a smooth nonlinear empirical scoring function that locates possible energetically favorable (hydrophobic and hydrogen-bonding) interactions [144]. As with FlexX and ADAM, interaction energies are calculated at each intermediate step to limit a possible combinatorial explosion. Hammerhead was demonstrated to rank natural binders as top scoring compounds even in cases of significant ligand flexibility. Subsequent improvements have led to SURFLEX [145] by Jain, first described in 2003, which is a fully automatic flexible molecular docking algorithm that combines the Hammerhead scoring function with a surface molecular similarity search engine method that can more rapidly generate suitable poses [146].

The incremental construction strategy implemented in DOCK version 4 combines both geometric and energy-based docking approaches [118]. The different phases involve: (i) analysis of all the possible rotatable bonds, fragmentation (with identification of the main fragment and the anchor) and organization of the remaining fragments into a sequence of layers; (ii) the anchor is then docked into the receptor binding pocket and fragments are added starting with the innermost layer; (iii) the conformations of each generated (partial) construct are optimized by minimizing both intra- and intermolecular energies, and then pruned with regard to score and novelty in orientation to avoid a combinatorial explosion of models; and (iv) the completed reconstructed ligands are again energy minimized.

The DOCK 4 method was able to reproduce the experimental pose within 1.03 Å RMSD for seven cases and within 2.00 Å RMSD for the remaining eight of the fifteen cases [118] in the initial test set. DOCK is a particularly interesting program because not only it was probably the first notable docking program but also, more importantly, it has organically evolved by incorporating and introducing new technologies and features as they became available. The list of DOCK authors, who were graduate students, postdoctorals, or otherwise associated with the Kuntz group at the University of California at San Francisco, includes many of the most prominent computational chemists currently involved in computer-aided drug discovery. From the initial simple distance geometry approach in the first version, DOCK has implemented physicochemical complementarity parameters [147,148], rigid body minimization to better refine ligand conformations [149], clique detection based on the Bron and Kerbosch approach [66,150], genetic algorithm (GA) techniques to better explore the orientational and conformational space of a flexible ligand [73], receptor flexibility based on ensembles of protein conformations [74], docking of multiple random ligand conformations generated based on the number of rotatable bonds [118,151], and a generalized Born model approximation [152] to implicitly account for water in calculations of free energy of binding for ligand–protein [122] and ligand–nucleic acid complexes [75]. The next DOCK version (5) was tested on a set of 114 protein-ligand complexes. This study found that binding poses were usually well reproduced for up to seven ligand rotatable bonds, and illustrated that more detailed modeling of the receptor prior to docking and force field improvements such as metal parameterization significantly improve docking performances and success [76]. A new pruning algorithm was then included in the anchor and grow algorithm in the 6.1 release of DOCK [153]. One consequence of the evolution of DOCK is a greater appreciation for the strength of using multiple, complementary approaches. This will be a topic later in the chapter.

Both SPECITOPE [154] and, later, SLIDE [155,156] are incremental construction algorithms specifically developed to screen large databases of compounds, incorporating both ligand and protein side-chain flexibility. The receptor pocket is analyzed for hydrophobic and hydrogen-bonding template points, after which all the ligand's hydrogen bond donor and acceptor groups are combinatorially matched within the binding pocket. The ligand is fragmented into triplets of matching points that are positioned by least squares superposition. Then the remaining part of the ligand is added and both receptor side chains and the ligand's rotatable bonds are rotated to remove intermolecular overlaps. In validation, SPECITOPE screened 140,000 different peptide fragments and identified both known inhibitors and potential new ligands [154]. SLIDE has tools to identify conserved water molecules in protein binding sites and to predict their energetic contribution to protein–ligand binding [155].

Other examples of incremental construction algorithms are given by SG-DOCK and SP-DOCK [157], MacDOCK [158], HierDOCK and HierVLS [159,160], SKELGEN [161], Propose [162], and MVP, developed in house by GlaxoSmithKline [163].

2.1.5. Monte Carlo (Simulated Annealing and Minimization Techniques) In Monte Carlo (MC) methods, the ligand pose in the binding pocket is explored by sequentially applying bond rotations, translations, and/or rigid body motions, thus sampling one degree of freedom at a time in a random way (see Fig. 6). After each modification, the energy of the new conformation is evaluated by applying selection criteria like that of the Metropolis algorithm [164]. Generally, the new generated conformation is retained if the new internal energy is lower than previously, otherwise it is rejected with a Boltzmann probability based on the effective temperature [46]. Ligands are described using internal coordinates, that is, with bond or torsion angles, which is significantly less time consuming than using 3D Cartesian coordinates. Because the most time-consuming step is energetic evaluation of the poses, precalculated grid potentials are often used [47]. Two different MC analyses can be performed: MC simulated annealing or MC minimization, which couple the Metropolis criterion with a temperature cooling schedule or with a local minimization procedure, respectively.

The starting point for the simulated annealing (SA) procedure is the input structure or one generated by docking. The temperature is modified and a new solution is generated by perturbing the current model. Then the energy is evaluated, and the Metropolis criterion is applied to decide whether to keep or reject that solution. This process continues through a cooling schedule constituted by a set of monotonically decreasing temperatures and a defined number of trials at each temperatures [165]. After a number of initial steps at a given starting temperature to equilibrate the system, the first minimization cycles are always carried out at higher temperature so that less favorable conformations are sampled

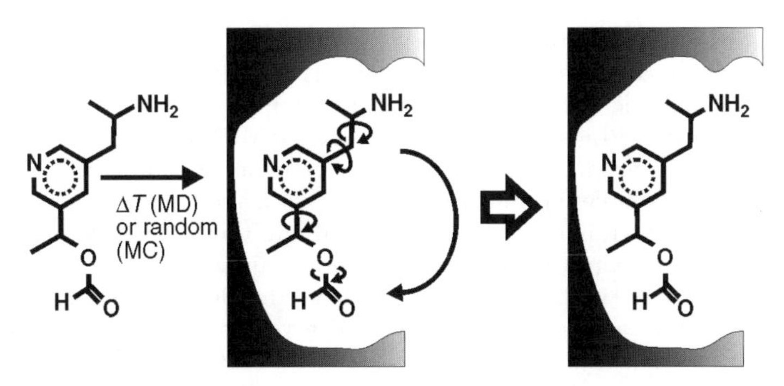

Figure 6. Monte Carlo and molecular dynamics. The ligand is placed in the active site and subjected to random (Monte Carlo) or dynamics-driven perturbations. In the case of Monte Carlo, pose iterations are evaluated with the Metropolis criterion or other algorithms. For molecular dynamics, poses are sampled and scored as the dynamics proceeds.

and high conformational energy barriers are overcome. Later cycles are at lower temperature, thus searching for low-energy conformations corresponding to local or even global minima.

Goodsell and Olson's AutoDock (1990) was one of the first docking programs based on MC/SA methods. In its earliest formulation, AutoDock used a Metropolis MC simulated annealing combined with a rapid grid-based evaluation of energy calculated with the AMBER force field. AutoDock was able to efficiently dock flexible substrates or ligands in macromolecular binding sites and reproduced the observed binding pose in many cases [63]. In 1996, AutoDock's performance was optimized by implementing parallelization and by allowing a greater number of degrees of freedom in ligands to be docked [64]. The continuing evolution of the AutoDock algorithm included, among many other features, implementation of a genetic algorithm, and will be described later.

Hart and Read (1992) developed a scheme of combining MC with simulated annealing that looks for possible binding modes of molecular fragments at a specific binding site in a target of known structure [166]. First, numerous conformations, generated by applying translational and rotational degrees of freedom to the probe fragment, are evaluated by substituting a geometric scoring function that determines the average distance of the probe to the target surface in place of the standard energy term in the Metropolis algorithm. Next, simulating annealing minimization is performed to find the most energetically favorable binding modes. On two different inhibitor–receptor systems, the algorithm reproduced the complex orientation reported in the crystal structure. The DockVision program, an integrated package that features the research and gamma docking algorithms and both Monte Carlo and genetic algorithms, developed by the same authors [167], was tested in the CASP-2 experiment and achieved the second highest success rate [168,169].

MCDOCK by Liu and Wang (1999) is a nonconventional Monte Carlo simulated annealing approach, using novel sampling techniques to find the real global minimum, and employing full ligand flexibility with a rigid receptor [170]. MCDOCK first places the ligand into the receptor pocket and then randomly modifies various degrees of freedom. Application of the Metropolis criterion, followed by a CHARMM-based scoring function, selects the most suitable conformations taking into account both protein–ligand interactions and ligand conformational energy. Testing of MCDOCK on 19 crystallographic complexes gave RMSD values ranging from 0.25 to 1.84 Å.

Bouzida and coworkers investigated the binding mode of the sb203386 and skf107457 inhibitors in HIV-1 protease with MC simulated annealing [171]. Both single protein structures and multiple conformations were used to dock the flexible ligands. While interactions were evaluated with a variety of scoring approaches, that is, simplified piecewise linear energy function (PLP), standard

AMBER and AMBER fitted with solvation and soft core smoothing component for Lennard-Jones and electrostatic terms, the best results were obtained with PLP scoring.

Two other notable docking programs that include Monte Carlo simulated annealing protocols are HADDOCK (high ambiguity driven protein–protein docking) by Dominguez and coworkers (2003) [172], a protein–protein docking program based on biochemical/biophysical information such as chemical shift perturbation data or mutagenesis data, and FDS, a flexible ligand and receptor docking tool introduced by Taylor et al. [173]. FDS uses a generalized Born/surface area (GB/SA) continuum model, a soft core energy function based on the AMBER-AA force field and a side-chain rotamer library. One interesting FDS finding was the existence of clusters of low-energy structures that are energetically indistinguishable. The consequences of this will be partially examined in Section 3.5.

MC minimization (MCM) techniques include an additional minimization step after each adjustment of the ligand conformation before application of the Metropolis algorithm. The ICM (internal coordinates mechanics) program (Totrov and Abagyan, 1994) allows both rigid and semiflexible docking [68]. ICM can be defined as a complete internal coordinate system, where all four types of variables, bond lengths, bond angles, torsion angles, and phase dihedral angles are considered. The initial step identifies a conformation with an ideal covalent geometry, that is, having small deviation from the initial structure, low-energy and optimized positions of polar hydrogens. Multistep minimization, retaining only the interactions between the polar hydrogens and the static environment, avoids trapping the structure in local minima and ameliorates strong clashes. The energy function is based on ECEPP/2 [174,175] (see below). Double-energy MC minimization optimizes the generated molecular conformations, while electrostatic and solvation contributions are separately evaluated, followed by application of the Metropolis selection. Interaction energies are calculated after any conformational changes due to MC or local minimization. The ICM program has undergone numerous developments both for docking [176–179] and as a tool for homology modeling [179–182].

A new MCM method (QXP, "quick explore") was implemented into the FLO96 package by McMartin and Bohacek in 1997. QXP is based on Monte Carlo perturbation combined with energy minimization in Cartesian space. It allows the flexible treatment of both cyclic and acyclic molecules [183]. The initial perturbation is followed by a fast search able to generate low-energy structures, which are then minimized with a conjugate gradient procedure, after which the ligand torsions are examined with the Metropolis algorithm. Ligands are positioned in the receptor binding pocket using a superposition force field that automatically assigns short range attractive forces (that is, as van der Waals interactions) to similar atoms in different molecules and then rotates and translates the ligand for the best fit with respect to the template. Then, a new MC search is initiated and the final optimized conformations are scored with the AMBER force field. For 10 of 12 analyzed cases QXP gave RMSD values less than 0.76 Å. The relative speed of QXP has made it a useful docking tool for structure-based drug design. The QXP force field is also currently operating in the MCDOCK program [170].

Caflisch, Fischer, and Karplus proposed randomly positioning the flexible ligand in the flexible binding pocket, followed by conjugate gradient minimization with CHARMM of the most promising pose candidates after each random move. A continuum model based on the Poisson–Boltzmann term and weighted solvent accessible area accounted for solvation [184]. Preliminary simulations were carried out for oligopeptide–protein systems and on a FKBP-tetrapeptide complex [185,186]. Modeled structures were demonstrated to be in agreement with crystal structures of similar complexes and consistent with the observed effects of FKBP point mutations on the enzyme activity [187]. These authors focused on the importance of taking into account not only the ligand flexibility but also the adjustments of the binding pocket upon ligand entrance, and developed a valuable tool for distinguishing between near-native and nonnative structures.

About a year later, Trosset and Scheraga released PRODOCK for protein modeling and flexible docking that defines flexibility of ligands and proteins in terms of arbitrary "levels." [69,70,188] Proteins are described with (internal coordinates) all atom models using ECEPP/3 or AMBER force fields with a new residue data dictionary. During the Monte Carlo searches, local minimizations are performed using grid-based energy evaluation based on Bezier splines to speed up the calculation and smooth the potential energy surface. Similarly, GlamDock (Tietze and Apostolakis) [189] defines the ligand search space by combining internal coordinates and a map-based description of the rigid body translation and rotation. The generated conformations are scored with a continuously differentiable empirical potential. This algorithm was validated on the scPDB database of protein–ligand complexes and gave, in particular, interesting clues about the relationship between docking accuracy and features of the protein–ligand system. Additional MCM algorithms such as Ribodock [190], YUCCA [191], and ROSETTALIGAND [192], based on Monte Carlo methodologies, have been presented and generally tested on fairly large data sets with good results.

Both MC-based techniques (MC/SA and MCM) are implemented in Affinity, the commercial program for flexible docking released by Accelrys [193]. An MC algorithm is used to rotate, translate, and perform conformational searches of the ligand within the binding pocket. Minimization or simulated annealing is then used to optimize the interactions between ligands and receptor residues. Promising results were obtained with Affinity in reproducing the crystal structures of 3-phenylpropylamine-trypsin [194], benzylsuccinate-carboxypeptidase A [195], and streptavidin-biotin [196]. Subsequently, Schrödinger developed the Glide docking and scoring program [197] that systematically searches the conformational, orientational, and positional space of the docked ligand. Candidate poses are first selected by a rough scoring function and then subjected to (flexible) torsion optimization, followed by Monte Carlo sampling of the most promising models. On a set of 282 selected crystallographic structures, Glide was demonstrated to be more accurate than other well-known docking tools [198]. Combining Glide with the prime protein structure prediction tool produced a docking method that takes into account both ligand and receptor flexibility for modeling the effects of induced fit [199].

2.1.6. Genetic Algorithms/Evolutionary Approaches

Genetic algorithms (GA) [200–204] also are classified as stochastic search methods (see Fig. 7). As defined by Jones et al., genetic algorithms are computer programs able to mimic the evolution process by manipulating a collection of structural data called chromosomes [205]. These approaches are, in fact, based on Darwin's concepts of evolution. Since 1995, GA has been successfully used in a number of docking applications [65,73,206,207]. In GA, searches do not begin with a single conformation, but from a population of conformations where each member is represented by its "chromosome." The genes making the chromosome encode for different degrees of freedom, that is, ligand rotation and translation or torsional angles.

The initial (first) generation of solutions is "evolved" by using crossover and mutation operators, respectively acting on two parent solutions or on one single solution. The resulting "children" elements are subjected to random mutation [165]. Also, a migration operator can be adopted to move individual genes among different subpopulations. Subsequently, chromosomes are decoded and the genotype is translated into the phenotype that represents the interaction of each conformation with the receptor binding pocket. A fitness function is then used to select the best-scored conformations that have the highest chance of survival and reproduction and pass to the next generation [46,47,208]. The solution is represented by the best-scored conformation(s) of the final population after a set number of generations.

The first genetic algorithm applied to molecular docking, developed in 1994 by Judson et al. [209], used a GA to search the available conformational and translational space. The competence of the basic search method was increased with a number of techniques such as allowing migration between subpopulations

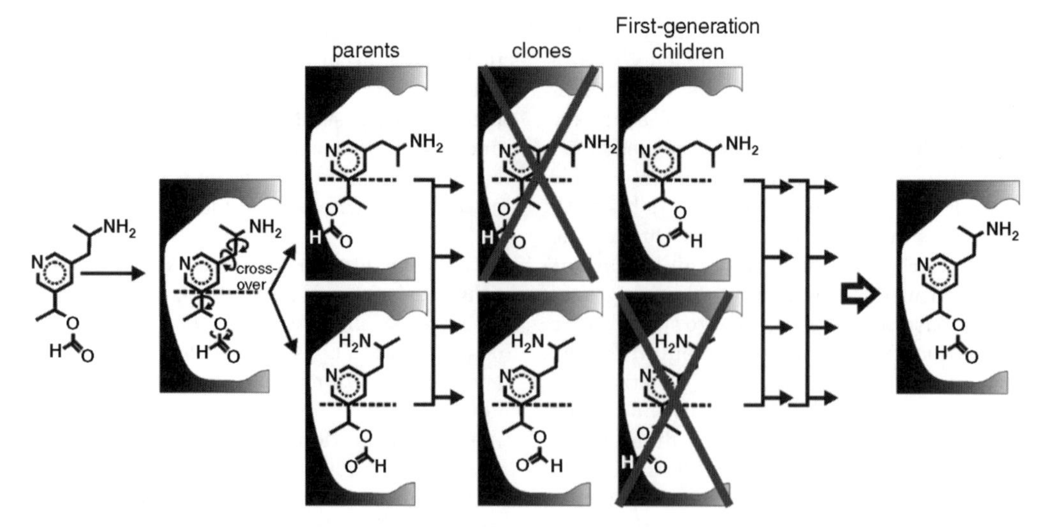

Figure 7. Genetic algorithm search. For each generation, offspring of the two parents, created by breeding the two genes (each one-half of the molecule), are evaluated. Those with the highest fitness (from scoring function), proceed to the next generation, those with low fitness are eliminated. Mutation (not shown) adds additional randomness. The final docked model is the one with the highest fitness after a predetermined number of generations.

or niches, an algorithm that first docked only part of the ligand, and gradient minimization to optimize conformation during the search. Initial tests on the known Cbz-GlyP-Leu-Leu peptide–thermolysin complex demonstrated its capability to reproduce the crystallographic pose.

GOLD (genetic optimization for ligand docking) is one of the most applied and well-known genetic algorithms for molecular docking [205,206,210]. GOLD was developed for the superposition of sets of flexible molecules, that is, to simulate the full conformational flexibility of the ligand and partial flexibility of the protein's binding pocket. Molecules are represented by chromosomes encoding not only the rotation of flexible ligand bonds but also the hydrogen bonds formed between the ligand and the protein [206]. Because the program identifies the binding pocket with a cavity detection procedure, the size and active site location defined by the user are not critical. The original GA fitness function was based on evaluation of the number and similarity of overlaid features, the volume integral of the overlay and the van der Waals energy of the molecular conformations. In fact, the first version of the scoring function was the sum of a "soft" 4–8 hydrogen-bonding potential, an intermolecular dispersion potential termed Complex_Energy, and an intramolecular potential named Internal_Energy. (See Section 2.2 for further details of the GOLD scoring function.)

The first GOLD training set of only five different protein–ligand complexes was correctly reproduced, but the significant improvements to the technique over the years have resulted in increased reliability and applicability of the method, for example, with very small ligands, metal ions, flexible cyclic ligands, and covalently bound inhibitors [205]. On a highly diverse data set, composed of different types of proteins such as metalloenzymes, with drug-like ligands or molecules able to form interesting or unusual interactions with their targets, a success rate of 71% was achieved. Only the docking of nonpolar ligands was systematically poorly reproduced, since the principle driving force of the algorithm was the identification of hydrogen-bonding contacts between proteins and ligands. While the fitness function included a term accounting for depressive forces but not hydrophobic or desolvation effects, this was resolved with the 2003 introduction of the ChemScore function [83] and, subsequently, a novel approach for scoring explicit water mediation and displacement energetic contributions [211].

Unlike GOLD, many of the other GA-based docking algorithms invoke a rigid receptor being fitted with flexible ligands. The DIVALI (docking with evolutionary algorithm) program of Clark and Ajay uses an AMBER-type potential function [207]. In an application examining four protein–ligand complexes, different binding modes other than the experimental structure were shown to be feasible and energetically accessible. Oshiro et al. developed two GA techniques for docking flexible ligands into rigid receptor binding pockets [73] that also used algorithms from the DOCK suite to define the pocket properties and AMBER-based scoring. Oshiro's two approaches differed in the ligand positioning strategy, either widely sampling the orientational and conformational space of the enzyme–inhibitor complex or exploring conformational variability in a more restricted orientational space. These methodologies were also used to simultaneously dock a large ligand database as in virtual screening.

AutoDock was implemented with a GA search in 1998 by Morris and coworkers [65]. As described above (Section 2.1.5), AutoDock combined a rapid grid-based method for energy evaluation with an MC/SA search for the generation of ligand conformations [63,64]. While good results were obtained, this implementation had significant limitations largely related to the difficulty of dealing with many degrees of freedom, that is, ligands with more than a few rotatable bonds. Thus, the 1998 AutoDock version included three new search methods along with a new empirical binding free energy force field. The combination search method of a simple genetic algorithm for global searching and local searching based on the work of Solis and Wets [212] for energy minimization, called the Lamarckian genetic algorithm (LGA), was a particularly significant advance. Here, the chromosome was composed of a string of real-valued genes including Cartesian coordinates for ligand translation, variables for ligand orientation and one value for each ligand torsion. In keeping with earlier AutoDock strategy, proteins are considered as rigid objects represented by grids. Further details concerning the AutoDock scoring functions are reported in the next section.

DARWIN is a docking program combining a genetic algorithm and a local gradient minimization using the CHARMM force field [213]. The resulting poses, again with a fixed receptor, are scored by the CHARMM-AA potential energy function. A modified version of the DelPhi program [214] is used to evaluate the solvent effects. DARWIN used parallel computing to increase the program speed. When applied to protein–carbohydrate complexes DARWIN identified conformations close to the crystallographic poses, but with a significant number of false positives. The explicit simulation of solvent effects significantly reduced the false positives, but it was noted that the presence of multiple conformations besides the crystallographic pose may reflect the flexible nature of receptors rather than a docking or scoring artifact. Thormann and Pons further developed parallelization with a docking method for virtually screening large compound libraries [215]. They reported the application of an "enhanced" genetic algorithm (EGA) that shares genetic information between subpopulations running independently on different CPUs. With this parallelization, when compared to several other algorithms, the EGA method appears to be superior in identifying lower energy solutions faster and more often. Similar GA strategies were exploited in GAMBLER (genetic algorithm multiprocessor box-oriented ligand enzyme relocator) [216] and in GAsDock [217], sufficiently accurate and efficient to be used in virtual screening applications.

Corbeil and coworkers developed and validated the FITTED 1.0 program suite to dock flexible ligands in flexible protein binding sites, considering both side-chain and main-chain dynamics [218,219]. Docking simulations also included the presence of water that with respect to the ligand's nature and conformation can be retained or displaced. FITTED 1.5, which is also applicable for virtual screening of libraries, was recently released [219]. A variety of different motions, including domain movement, rigid body, or backbone deformation, and side-chain adjustments, can be investigated by FLIP-Dock [220]. Another approach, particle swarm optimisation (PSO) with local search, is implemented in SODOCK for dealing with highly flexible ligands with genetic algorithms [221].

In contrast to genetic algorithms, in evolutionary programming (EP) methods [222] evolution depends only on mutation, that is, there is no crossover. Thus, new solutions are created from parents by a mutation operator usually based upon Gaussian or Cauchy random numbers [200]. EP methods were first used by Fogel in 1966 to simulate evolution as a learning process for generating artificial intelligence [223,224], and were later applied to molecular modeling by Gehlhaar and coworkers [125,225]. They developed a simple molecular recognition model for docking conformationally flexible ligands into rigid protein binding sites (EPDOCK). After an initial population is generated and the energy for each member of the population is calculated, comparison is made to the energies of a fixed number of randomly selected opponents. The lower energy members survive to the next generation and produce offspring by Gaussian mutation [224]. The best member of the final generation is energy minimized. In docking the AG-1343 inhibitor in the HIV-1 protease binding pocket, 34 of 100 simulations found the correct experimental structure. In this tool the simple energy function, which evaluated only steric and hydrogen-bonding interactions and intramolecular energies, was preferable to more complex functions that have more local minima [125].

In 2000, Yang and Kao proposed a new approach called family competition evolutionary algorithm (FCEA) to dock flexible ligands into rigid proteins [226]. FCEA methods were developed to solve some of the well-known problems of genetic algorithms, for example, long local searches caused by random-based mutations, and of evolutionary programming, for example, the use of self-adaptive Gaussian mutation leading to individuals that are trapped near local minima. FCEAs combine three different mutation operators, decreasing-based Gaussian mutation, self-adaptive Gaussian mutation, and self-adaptive Cauchy mutation, where their relationship is controlled by family competition and adaptive rules [227]. Yang and Kao's scoring function included only intermolecular and intramolecular terms. FCEA was tested on the crystallographic dihydropholate reductase–methotrexate complex, and in docking the trimethoprim inhibitor into the binding pocket and gave an RMSD equal to 0.67 Å. This algorithm also successfully solved global optimization [228] and was used to train neural networks [229]. The GEMDOCK method (generic evolutionary method for molecular docking), including a new empirical scoring function [230,231], is an extension of FCEA. GEMDOCK combines both discrete and continuous global and local search strategies to speed up convergence, for example, a differential evolution operator was added to reduce the disadvantages of Gaussian and Cauchy mutations and a rotamer-based mutation operator was implemented to decrease the search space of ligand structure conformations. In 100 protein–ligand complexes, the lowest energy docked ligands showed RMSD values smaller than 2.0 Å in 79% of the cases [230].

2.1.7. Molecular Dynamics Strategies Molecular dynamics simulations are, in essence, based on the calculation of solutions to Newton's equations of motions (see Fig. 6). These methods, which can be quite rigorous, have been widely used in simulating the dynamics and flexibility of proteins [232–236] and to calculate the free energy of binding of small molecules to biomacromolecules [29,237–246]. However, MD is computationally expensive, which has restricted its use in docking or virtual screening analyses. In addition, because overcoming high-energy barriers is difficult or impossible in MD, the identification of the global energy minimum is extremely difficult and the obtained conformations are often trapped in local minima. Thus, the final results depend completely on the starting conformation of the system. Nevertheless, several MD approaches have been developed despite these difficulties. The performance of MD with respect to docking problems was evaluated by Vieth et al., in a comparative study for five representative protein–ligand complexes [247]. The authors compared the efficiency of molecular dynamics, Monte Carlo, and genetic algorithm approaches and reported that MD-based docking generally provided conformations lower in energy and closer to the corresponding crystallographic structures.

CDOCKER is a molecular dynamics simulated annealing-based algorithm by Wu et al. [248]. With rigid proteins but fully flexible ligands, the final full force field minimization procedure applied to refine the docked poses yields very good docking results despite a large number of shortcuts and approximations elsewhere in the protocol.

Miranker and Karplus have developed the multiple-copy simultaneous search (MCSS) methods for determining energetically favorable positions and orientations of functional groups (ligand fragments) on protein surfaces and pockets [249]. First, 1000–5000 conformations of each functional group belonging to the protein located in the analyzed site are generated; each is subjected to energy minimization and quenched molecular dynamics. The results are plotted as maps representing the flexibility of the protein binding pocket for use in subsequent docking simulations. In the next step, the same method optimizes the positions of ligand functional groups within the active site [250]. Dock poses are then selected from overlapped maps derived from the set of functional groups found in the docked ligand.

Di Nola and coworkers released the MDD (MD docking) algorithm [251] to overcome some of the previously reported pitfalls of MD approaches by separating the motion of the ligand's center of mass from its internal and rotational motions. Also, the ligand and receptor were treated with separate thermal baths so that temperature and time constants for each of the three considered movements can be independently modified. MDD was successfully applied to the immunoglobin McPC603-phosphocoline system, where the energy barriers were overcome with a non-trapping exploratory search. Receptor flexibility was subsequently introduced by simulating a high temperature in the translational motion of the center of mass [252]. In a method with both ligand and receptor flexibility, Luty et al. sped up MD simulations by treating the solvent with an implicit model and using a grid representation of the receptor [253].

Nakajima and coworkers performed an enhanced conformational search to dock a proline-rich peptide into a Src homology 3 (SH3) domain, using a multicanonical MD simulation [254], as originally proposed by Berg and Neuhaus to improve sampling efficiency in MC calculations [255]. While promising results were obtained, the multicanonical MD approach is more time consuming than canonical MD.

Given and Gilson described a hierarchical docking protocol comprised of four different phases termed randomization, predocking, docking, and redocking. In the randomization step a variety of orientations and conformations for the isolated ligand are generated using a heating and cooling process based on a modified Langevin dynamics algorithm [256] that updates the stochastic forces less frequently. In predocking the obtained conformations are roughly docked into the binding pocket; in the docking step each pose is optimized several times from the same initial conformation with different energy functions and diverse solvent and long-range nonbonded interaction terms. The lowest scored structures proceed to the redocking phase where they are subjected to serial MD-based heat and cool cycles within the binding pocket, to obtain sets of new local minima [257].

Wang and coworkers developed a flexible ligand docking multistep approach based on a divide-and-conquer strategy [258]. The total search space is divided into several sets of subspaces, for example, conformational and orientational components, and a fine-resolution search is carried out on a selected set. Low-energy ligand conformation ensembles are obtained using a grid-based approach, while the orientation space is evaluated with geometric complementarity. Conformations are scored and ranked with AMBER [259–261] before the lowest energy poses are optimized using a three-component structure refinement. This refinement includes molecular mechanics geometry optimization to place the minimum energy structure into the binding site, a torsional-angle-driven refinement to overcome local energy barriers and short MD simulated annealing to "shake" structures around the local minima. When applied to a test set of 12 protein–ligand complexes, this method generated (lowest energy) docked conformations with RMSDs between 0.64 and

2.01 Å, compared to the crystallographic poses [258].

An interesting approach by Pak et al. (2000) used a Tsallis effective potential for docking flexible ligands into flexible receptors [262]. This algorithm was based on the infrequent smoothing of nonbonding interaction potential modified with the Tsallis mathematical transformation [263–265] as implemented in the CHARMM force field [213]. Performing MD simulations on transformed potential energy surfaces significantly improves the search and allows jumping between the original and the transformed surfaces that is normally not observed in standard MD. The authors hypothesize that the additional receptor flexibility due to the Tsallis modification may help in reducing the energy barriers when a ligand enters the receptor binding pocket [262,266], thus leading to enhanced docking performance.

2.1.8. Tabu Searches The heuristic tabu search (TS) method was originally developed by Glover [267,268] for operations research applications [165], and later applied to molecular design studies and docking simulations [269–272] (see Fig. 8). The basic idea of this approach is to avoid reinvestigating space already sampled by the search algorithm by forbidding or penalizing moves pointing in solution space already visited. Spatial restrictions are organized in a "tabu list" of solutions previously obtained. The search starts from a single solution and different conformations are generated by a mutation-like procedure using Gaussian or Cauchy random variables. Each are scored and ranked; solutions too similar to those in the tabu list are disregarded unless their energy is lower than previously recorded [165]. The remaining new solutions are added to the list. Each iteration proceeds with new moves executed starting with the last stored solution. The best-scored conformation is returned as the final solution.

Probably the best-known TS-based method is PRO_LEADS (ligand evaluation by automatic docking studies) reported in 1998 by Baxter and coworkers [273]. The algorithm exploits a "first-in, first-out" approach in that each new conformation replaces the solution with the longest residence time in a tabu list of 25 members. The scoring function originally included only a simple contact term for lipophilic and metal–ligand interactions, an explicit hydrogen-bonding term, and a penalty for ligand flexibility [274]. The scoring was later optimized for docking by adding terms to calculate the ligand internal energy and to penalize conformation clashing with the (rigid) receptor and solutions falling outside the user-defined active site. When the PRO_LEADS algorithm was applied to a set of 50 protein–ligand complexes, the lowest energy solutions had RMSD < 1.5 Å with respect to the crystallographic pose in 86% of the cases.

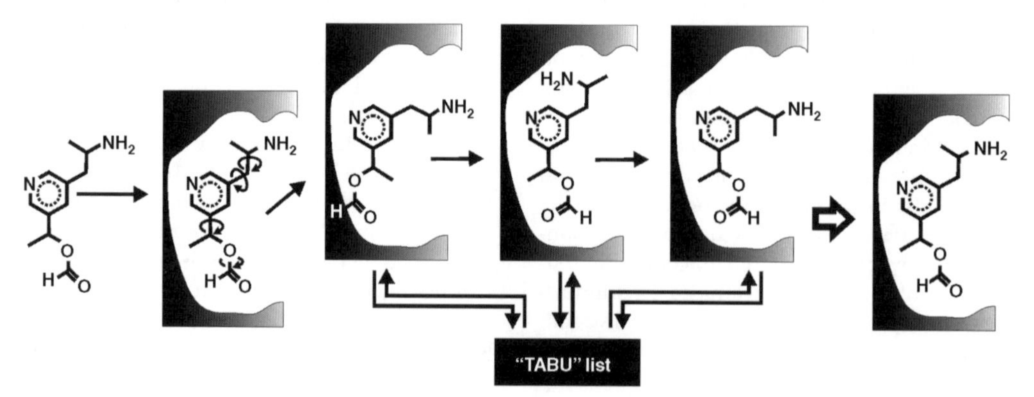

Figure 8. Tabu Search. In a tabu search, a list of previously explored conformations is maintained and each new conformation is checked against that list before scoring. Conformations with poor scores are automatically added to the list so that it is not visited again. A second list of a certain number of "good" conformations is also available—only new conformations with higher scores than the weakest member of the list can enter the list.

Promising results were obtained when PRO_LEADS was used to screen a database of 10,000 randomly chosen drug-like molecules for the thrombin, factor Xa, and estrogen receptors [275].

A tabu search approach was adopted by Tøndel and coworkers in testing the protein alpha shape (PAS) dock scoring function [276]. A mixed strategy combining a genetic algorithms and tabu search was proposed by Hou et al. [277]. During the first step, the binding pocket is investigated for surface complementarity using both GA and TS; in the following phase, hundreds of GA minimization steps are performed for each identified site. The hybrid algorithm placed the correct solutions among the best binding modes in the top conformations of the tabu list. Similarly, Pei et al. developed pose-sensitive inclined (PSI) DOCK for flexible ligand docking in 2006. This is a GA/TS hybrid implemented with a rapid shape-complementary scoring function and an improved SCORE function [278]. Good results were obtained when the program was used to reproduce the crystallographic poses and binding free energies of test sets of 194 and 64 complexes, respectively [278].

2.1.9. Hybrid Methods and Consensus Docking

The performances of docking methods and approaches have been compared in a variety of studies [86–93,165,216,247,279]. Even considering the well-known limitations of comparative analyses, as highlighted by Moitessier et al. in an outstanding review [208], these analyses demonstrated that no algorithm performs significantly better than the others and can thus be identified as a universally applicable method. Taylor et al. [45] suggest that the general strategy should be to combine a computationally inexpensive tool for initial docking with a more time consuming and accurate methodology to generate and evaluate the final conformations. Accordingly, Taylor and coworkers developed FDS [173], a two-stage approach where the ligand is initially docked in the binding pocket with a fast tool based on simultaneously satisfying sets of hydrogen bonds and a recursive distance geometry algorithm [280]. Then, after cluster analysis, the remaining conformations are minimized with an MC-based algorithm applying AMBER-AA and the GB/SA continuum model [152]. Sampling of conformational space was further increased with a rotamer library.

Leach's methodology to dock flexible ligands in proteins with side-chain flexibility [281] explores conformational degrees of freedom with dead end elimination based on MC simulated annealing and the A^* algorithm [282]. The first algorithm removes high-energy rotamer states and the second searches for minimum energy combinations of protein and ligand rotamers. The AMBER force field [259,261] with a GB/SA solvation model [152] is used to score poses. Interestingly, it was observed that the complexed structures had more rotamer combinations than the corresponding unbound proteins, suggesting that the presence of ligand is able to increase the number of accessible conformational states.

Price and Jorgensen combined simulated annealing with tabu search, applying the observation that tabu algorithms usually find crystallographic poses at lower computational cost and with higher frequency than SA protocols [165]. On each complex, a series of four MC simulations was performed with full ligand flexibility while keeping the protein rigid. The resulting lowest energy structures were submitted to an MC/FEP simulation that included effects such as internal energy differences, solvation, sterics, and intramolecular interactions. The calculated binding free energies were in excellent agreement with experimental data in all cases [283].

The two-stage method of Hoffmann and coworkers used FlexX to generate different ligand conformations followed by minimization and ranking with the CHARMM force field [284]. While this combined approach gave considerably better results than single methods, this strategy still is affected by the limitations of rigid receptor modeling and overly simplistic water treatment.

The mining minima method of the Gilson group joined concepts from genetic algorithms [200,285], the global underestimator method [286], poling [272], and tabu searches [287], adapting them to molecular docking. When compared to single approaches such as PRO_LEADS [165], AutoDock [65], FlexX [77], MCDOCK [170], GOLD [210], and

Wang's method [258] that combines systematic search and MD, mining minima is competitive in terms of speed and accuracy and generated conformations with RMSD values less than 1.5 Å in 25 docking runs. SDOCKER, by Wu and Vieth, improves docking accuracy by combining simulated annealing MD with an additional similarity force derived from existing ligand–protein X-ray structures. SDOCKER showed significantly better results compared to similarity-only approaches when the program was applied to three test systems (HIV-1 protease, thrombin, and CDK2) [288]. The MolDock program uses heuristic search and cavity prediction algorithms [289] and performed well on a set of 77 protein targets. PIPER is a protein–protein docking program based on the combination of pairwise structure-based potentials with a fast Fourier transform correlation [290], Moitessier developed and validated FITTED that combines GA and interaction site matching as reported above in Section 2.1.6 [218,219], while Grosdidier et al. released EADock, a hybrid evolutionary algorithm with two fitness functions and a sophisticated management of diversity [291]. Recently, the PLANTS program, built on a hybrid ant colony optimization (ACO) algorithm has been improved by the addition of two empirical scoring functions $\mathrm{PLANTS_{CHEMPLP}}$ and $\mathrm{PLANTS_{PLP}}$ based on PLP [125,292–294] and ChemScore [83], respectively [295].

Recently, Åqvist and coworkers proposed the combination of GOLD docking, scoring, molecular dynamics, and linear interaction energy (LIE) as a powerful tool for the prediction of the binding modes and binding affinity for a set of 43 nonnucleoside HIV-1 reverse transcriptase inhibitors [296]. The LIE method, first proposed in 1994 [297], is a semiempirical approach combining the advantages of free energy perturbation (FEP) and thermodynamic integration (TI) methodologies, with good convergence and reduced computation time. LIE estimates the free energy of binding starting from an MD or MC simulation of the bound and free state of a ligand [297]. A better correlation ($R^2 = 0.70$, average error = 0.8 kcal/mol) with experimental binding affinities was obtained with LIE predictions than with GoldScore. Several applications of the LIE method for prediction of biological interactions have been previously reported [298–302].

Consensus docking is another valuable alternative approach to obtain better and more reliable results in docking simulations [208]. In 2002, Paul and Rognan described the ConsDock procedure [303], a consensus docking method combining Dock, FlexX, and GOLD. First, the poses generated by each docking program were clustered with the definition of consensus pairs being the leaders generated by the different programs. Next, the consensus pairs are clustered in classes and, finally, the poses starting from the most populated consensus classes are ranked. When tested on a set of 100 known protein–ligand complexes, ConsDock performed better than each of the single docking tools and generated the top solution poses within 2 Å RMSD in 60% of the cases with respect to the crystallographic structures. ConsDock may also have utility as a postprocessing filter for single and multiple docking programs. Similarly, Wolf and Zimmermann created AutoxX by uniting the FlexX and AutoDock interaction models [304].

2.2. Model Scoring Functions

This section will illustrate the different approaches in designing scoring functions that have been implemented in and around docking programs. It is important to note that a scoring function maps a rather abstract concept (the measure of a binding force, which itself has many competing definitions) to a simple numeric value, providing a way to rank one placement of a ligand relative to another.

To be more precise about how scoring functions are used in docking we must quickly review one facet of the docking programs; often, but not always, scoring functions are used in two ways by the docking program: scoring after docking with complete geometric and energetic evaluation, and scoring while posing, where only a rough evaluation of the interactions may be sufficient. For example, fragment reconstruction docking programs such as FlexX must evaluate, and accept or reject, each potential fragment placement and orientation until the pose is complete. This type of scoring may not require as detailed and precise an energy evaluation as the later comparison of poses.

Divorcing the docking algorithms from the scoring functions is thus somewhat difficult. This approach was chosen for the present chapter because it is instructive to evaluate the two "phases" of docking in terms of some rather distinct algorithmic classes. There are four classes of scoring algorithms: force field-based methods, that is, using Newtonian molecular mechanics that will be described in Section 2.2.1; semiempirical approaches where molecular mechanics terms are supplemented by additional parameters or terms that are empirically derived from observation. A hydrophobic-contact surface area is such a term. These methods will be presented in Section 2.2.2; empirical scoring functions (Section 2.2.3) are mostly (or entirely) derived by analyses of training sets into QSAR-like descriptors that create a predictive equation; and knowledge-based functions that use a potentials of mean force (PMF) approach to interpret known structures into rule sets (Section 2.2.4). Lastly, consensus scoring, by using the complementary information from multiple scoring functions to make decisions on the suitability of a docking pose will be described in Section 2.2.5.

2.2.1. Force Field-Based Methods There are a number of well-known and widely applied molecular mechanics (MM) force fields: AMBER [259–261], CHARMM [213], GROMOS [305], OPLS-AA [306], DISCOVER [243], TRIPOS [307], MMFF [308], MM2 [309], MM3 [310], and MM4 [311]. Energy calculations performed by these methods are essentially the sum of electrostatic and van der Waals potentials, plus internal (i.e., intramolecular) distance, angle, and torsion contributions:

$$E_{\text{total}} = \sum_{\text{bonds}} K_r(r-r_{\text{eq}})^2 + \sum_{\text{angles}} K_e(\theta-\theta_{\text{eq}})^2 + \sum_{\text{dihedrals}} \frac{V_n}{2}[1+\cos(n\varphi-\gamma)] + \sum_{i<j}\left[\frac{A_{ij}}{R_{ij}^{12}} - \frac{B_{ij}}{R_{ij}^{6}} + \frac{q_i q_i}{\varepsilon R_{ij}}\right]$$

where, K_r is the Hook's law-like bond stretching constant, r and r_{eq} are the bond length and equilibrium bond lengths, respectively; K_e is the Hook's law-like bond stretching constant for the pseudo-bond between the two nonvertex atoms of an angle in angle bending, θ and θ_{eq} are the pseudo-bond length and equilibrium lengths, respectively; V is the torsional force constant, n represents the symmetry and periodicity, φ is the equilibrium dihedral angle, and γ is the deviation from the equilibrium and the equilibrium dihedral angle; A_{ij} and B_{ij} are the vdW repulsion and attraction parameters of the 6–12 potential, R_{ij} is the distance between ligand atoms i and protein atoms j; q is the point charge on an atom and ε is the dielectric constant.

Despite their wide application in other, more biological, environments, these methods were developed to evaluate gas-phase enthalpy of binding and not biomolecular interactions. Thus, some fundamental contributions have been largely ignored, for example, little attention has been paid to hydrophobic interactions, to solvation, and, in fact, to entropic effects in general. The advantage of force field-based scoring functions is clearly in speed. Due to their optimization for energy minimization applications where many calculation iterations are anticipated, force fields can rapidly estimate a binding energy. If a force field-based method is used to differentiate amongst docked models, its application is very similar to that of MM energy minimization, that is, the energy is calculated on an atom-by-atom basis; however, when applied to dock scoring as part of the docking algorithm, dock target molecules are often represented by grids where each grid point indicates precomputed values of the polar and nonpolar potentials, and a protein–ligand association is simply evaluated by summing up the interaction energies between the ligand atoms and the closest receptor grid points.

However, while even if extremely suitable for docking and virtual screening analyses, these simplistic, purely Newtonian, approaches quite poorly estimate the numerous entropic and solvation effects that are known to significantly affect biological interactions in aqueous medium. Thus, more accurate and computationally demanding techniques, such as the Poisson–Boltzmann equation (PBE) [214,312,313] or the generalized Born approximation (GBA) [152,314,315], have been developed and successfully applied as additional terms or layers in MM-based scoring functions.

In the earliest DOCK formulation [59], the implemented scoring function (Appendix A) was essentially based on the force field parameters of AMBER [259–261], and protein–ligand interactions were evaluated by solving a function that included a hard sphere repulsion term and a hydrogen-bonding term. Later, to optimize the scoring performance, Meng and coworkers added an electrostatic interaction energy function to DockScore based on the Poisson–Boltzmann equation, and an MM interaction energy including van der Waals and electrostatic components estimated with AMBER [66]. Only in 1994, entropic contributions were considered by applying the HINT force field [316] to four protein–ligand complexes [119], while the first attempt to directly include solvation effects in the native function was made by Shoichet, Leach, and Kuntz in 1999 [121]. This latter approach first calculated the electrostatic potential and van der Waals interactions by applying the Poisson–Boltzmann method [317] implemented in DelPhi [214] and CHEMGRID [66], respectively. Then, the standard interaction free energy was corrected by subtracting the electrostatic and nonpolar ligand solvation energies determined with HYDREN, based on Rashin's continuum electrostatic method [318,319] (Appendix A). When this DOCK implementation was used to screen the ACD (available chemical directories) in thymidylate synthase, T4 lysozyme L99A mutant, and dihydropholate reductase targets, the authors observed a remarkable change in ranking of the screened compounds [121]. In particular, the inclusion of electrostatic and nonpolar ligand desolvation penalty terms led to an improved identification of the known inhibitors—characterized by smaller size and lower formal charges. Nevertheless, accurate evaluation of the free energy of binding was still impossible since several contributions, such as entropy, receptor flexibility, and receptor desolvation, were assumed to be constant terms, and hydrophobic contributions were completely ignored.

DOCK, as noted above, has constantly evolved from its inception to the present. Thus, in terms of providing more reliable free energy predictions, the most recent, DOCK 5 and DOCK 6, versions have been significantly enhanced by including DelPhi [214] or Zap [320] electrostatics, ligand conformational entropy corrections, ligand and receptor desolvation, Hawkins–Cramer–TruhlarGB/SA and Poisson–Boltzmann/surface area (PB/SA) solvation scoring, AMBER scoring with receptor flexibility, and the full AMBER MM scoring function with implicit solvent. Moreover, the most recent releases include a distance-based movable region and a mildly performance optimized "nothing movable" region for rapid AMBER scoring (release 6.1), and a distance dielectric control for continuous score, plus metal ion and cofactor libraries to improve scoring performance (release 6.2) [153].

The GREEN approach (grid for receptor environment and energy calculation) was developed by Tomioka and Itai in 1994 [321]. The scoring function included van der Waals, electrostatic and hydrogen-bonding potentials for each 3D grid point, with energetic parameters provided by the AMBER force field (see Appendix A for further details). When GREEN was applied to the PATM-bovin trypsin system, it was shown to be a valuable tool for modeling protein–ligand complex formation, despite a rather simplistic grid-point scoring approach and the fact that the implemented approximations did not allow quantitative evaluation of binding free energy. For more accurate energetics, the authors suggested application of the PRECISE MODE tool [321], based on an AMBER calculation of intermolecular energies.

An improved function was implemented by Miller and coworkers in FLOG [95]. The scoring equation estimated docking energy by summing up van der Waals, electrostatic, hydrophobic, and acceptor, donor, and polar hydrogen-bonding contributions (see Appendix A). Thus, the score is estimated as the sum of pairwise contributions, with a well-defined method for size-normalizing the score to avoid size-dependence overestimation of the interaction energy when dealing with large and bulky molecules. Even though exhaustive ligand flexibility is allowed, the receptor is fixed and neither entropic nor solvation effects are accounted for.

As described above (see Section 2.1.5), the QXP approach developed by McMartin and coworkers was based on MC search and AMBER MM energies [183]. In particular, while the rigid body docking is guided by van der Waals and electrostatic potentials, an energy function describing superposition is used to determine the bioactive conformation of flexible ligands:

$$E_{\text{sup}} = \frac{K_{\text{sup}}(\text{dist}^2 - d_{\text{cut}}^2)^2}{d_{\text{cut}}^4}$$

where dist is the interatomic distance, d_{cut} the cutoff distance, and K_{sup} the energy constant for a perfect superposition [322]. QXP has been widely and successfully applied in a variety of studies [323–325], and has been implemented in the FLO suite [326]. More recently, QXP has been combined with a simil-consensus scoring approach by Alisarale et al., to analyze highly flexible ligands that are usually poorly predicted by docking programs [327]. The new, additional, energetic terms estimate nonbonded interactions, van der Waals energy, positive van der Waals energy, electrostatic energy, contact energy of interactions and the hydrophobic contacts count, while local MC minimization cycles are performed until the highest scored conformer corresponds to the first rank conformer. [328].

EUDOC, by Pang and coworkers [129,329], included a scoring function that was again based on the AMBER force field and the RESP charge model [261]. EUDOC calculates the interaction energy as the sum of the pairwise van der Waals and electrostatic potentials (Appendix A); that is, no grid-map point approximation is employed. Specifically designed to work with metalloprotease systems, it was validated on a set of 154 protein–ligand complexes (including 39 metalloproteases) retrieved from the PDB and extracted from the training sets of docking programs such as GOLD, FlexX, MCDOCK, and LUDI. While promising results were achieved, for example, a 97% success rate in generating the crystal pose, the authors discussed further developments to the scoring function by adding parameterization for additional metal ions and explicitly treating water molecules in the binding pocket [329].

A significant enhancement in solvent contribution treatment was made by Kollman and coworkers in developing the MM/PBSA approach, a solvent continuum method combined with a classic MM force field estimation. To the average energy, including bond, angle, torsion, van der Waals, and electrostatic terms, a Poisson–Boltzmann-based solvation free energy and a solute entropy derived from harmonic analysis of an MD trajectory were added [330]. In contrast to the more accurate, but more time consuming, FEP methods, MM/PBSA only requires MD simulations of two different states the free ligand in solution and the ligand bound to the protein. Kollman's approach was shown to be reliable and provided free energy predictions comparable to experimental values in relatively reasonable amounts of time. The method was initially tested on a biotin–avidin system [331] and on several different protein–ligand [332–334], nucleic acid–ligand, protein–protein [335–338], and protein–RNA complexes [339]. Several of these analyses suggested that the inclusion of explicit water molecules involved in binding process would further enhance the algorithm performance. In 2000, Massova et al. introduced GBA in the method, to reduce the required computational time [336].

The OWFEG (one window free energy grid) method was developed by Pearlman et al. [340] using a simplified FEP [245] that performed the analyses on a single MD trajectory. The free energy is calculated with AMBER [260] at each point of a grid approximating the protein structure built using three different probes (neutral, positively charged, and negatively charged). The final G_{score}, including Lennard-Jones, electrostatic, and hydrogen-bonding potentials, as well as bond, valence angle, and torsion contributions, is the sum of the grid score for each atom in the ligand (Appendix A). With good results on a test system, quinoline and bis-pyrimidine, on a protein–ligand system, FKBP-12·FK506, the method was further developed for docking simulations. Compared to other force field- and empirical-based approaches, OWFEG was shown to be more accurate and not significantly more time consuming [341,342], although the test set for this comparative analysis was proprietary.

Recently, the HADDOCK docking program and scoring function [172] was applied to protein–DNA docking [343]. The HADDOCK function is a weighted sum of electrostatic and van der Waals interaction terms, a desolvation contribution, relative solvent accessibility and a buried surface area term. In this study, HADDOCK considered both flexibility and explicit water contributions.

2.2.2. Semiempirical Methods Semiempirical approaches imply the use of empirical or empirically calibrated energetic terms for calculating interactions not commonly computed by molecular mechanics. Although this strategy allows the inclusion of contributions for fundamental biological interactions, for example, hydrogen-bonding or solvent effects, semiempirical models partially lose the supposedly universal applicability of many MM force fields.

As previously described above in the docking section, the first version of ICM [68], developed for protein structure prediction and protein design, included the basic energy function used in the ECEPP/2 [174,175], and in the ECEPP/3 [344] force fields. Subsequently, an alternative semiempirical scoring function, including electrostatic, hydrophobic, and entropic terms, was implemented by Schapira, Totrov, and Abagyan. In particular, the electrostatic contribution was calculated as the sum of desolvation and Coulombic ΔGs obtained by solving the Poisson equation [313,345] using a boundary element algorithm [346–348]. The hydrophobic term was derived by multiplying the solvent accessible surface area (SASA) by the surface tension, while entropy was calculated as $RT \ln(N_s)$, where N_s is the number of low-energy conformational states obtained through extensive MC simulations [349]. The final binding energy is the difference between the energies of the solvated complex and of the uncomplexed molecules, solvated. Prediction accuracy was significantly enhanced by accounting for protein flexibility by relaxing interfacial side chains. As reported in Appendix B, the scoring function was further improved with the introduction of additional terms and weights, α_1–α_5, that were validated on an extended set of complexes [350,351]. ICM has been applied in numerous protein–ligand docking and virtual screening studies [79,88,352–359].

The first GOLD version (1995) used three different energetic terms that when combined were termed the GOLD fitness function to calculate the (binding) energy of protein–ligand complex formation [210]. The hydrogen-bonding energy was estimated with a softer 4–8 (Lennard-Jones-like) potential and corrected by weighting each bond by the distance and the angle between donors and acceptors. vdW forces for the protein–ligand interaction and for the internal energy of the ligand conformation were calculated with a 6–12 Lennard-Jones potential [360]. As noted (Section 2.1.4), the original GOLD formulation was soon improved (1997) with metal parameterization and treatment for covalently bound inhibitors. Additionally, a smoother 4–8 potential for long-range protein–ligand contacts and a torsion potential term for internal energy were added [205]. Despite promising results, no estimation of entropy was made, and the GOLD fitness function was therefore not considered sufficiently accurate to provide reliable estimations of the free energy of binding, that is, comparable with experimental binding data.

A significant enhancement in this regard was the introduction of the GoldScore function, GOLD fitness minus intramolecular terms [83]. Further modifications include adapting GOLD scoring functions for virtual screening (VS) analyses [361] and implementing a variety of pharmacophoric restraints:

$$\text{Fitness}' = \text{Fitness} + C_p \cdot \text{Pharmscore}$$

where C_p is a scaling factor and Pharmscore is as defined by the following:

$$\text{Pharmscore} = \sum_i c_i p_i$$

Here p_i represents the best overlap of a matching ligand atom with pharmacophore point i and c_i is the normalized weight of pharmacophore point i.

Version 3.0 of GOLD accounted for the contribution of explicit water molecules located in the protein binding pocket and gives the user the ability to select which waters should be retained or discarded in docking simulations, and consequently, in free energy predictions. When applied to a set of 225 protein–ligand complexes the new algorithm was able to properly predict the retention of water molecules in 93% of the cases [211].

Similarly, the AMBER-based scoring function as originally implemented in AutoDock used a 6–12 Lennard-Jones potential for vdW interactions [360], a 12–10 potential for hydrogen bonding, and a traditional Coulomb potential to estimate electrostatic interactions. Both protein–ligand binding and internal ligand energies were determined [64]. AutoDock 3.0 [65] implemented not only an alternative Lamarckian genetic algorithm-based search algorithm (see Section 2.1.6) but also a new semiempirical scoring function based on the thermodynamic cycle of Wesson and Eisenberg [362], and included ligand conformational restrictions and desolvation contributions obtained with the pairwise volume-based method of Stouten [363] (Appendix B). A correlation coefficient, $R^2 = 0.96$, and standard error = 2.18 kcal/mol were obtained when tested on 30 protein–ligand complexes abstracted from Böhm's set [140]. Successful results were also achieved and reported in many other studies and applications [364–374], consistent with AutoDock's popularity. Further improvements were made in the most recent, 4.0, version by modifying the internal energy term.

The GB/SA solvation model [152,375,376], based on a continuum dielectric medium approximation [313], was implemented in the SDOCK scoring function by Zou et al. [122]. The binding free energy is calculated by summing a Lennard-Jones 6–12 potential for protein–ligand vdW interactions [360], a term for polar protein–ligand, ligand desolvation and protein desolvation electrostatic interactions [122], and two terms encoding the changes in hydrophobic and total SASA related to complex formation (Appendix B). The several adjustable parameters were obtained, when necessary, through validation on the training data set. This scoring function was then implemented in DOCK [117], as a post-DOCK processor for large database screening or for rigid/flexible docking of small databases.

Recently (2007), Naïm et al. described the solvated interaction energy (SIE) function by calibrating a combination of force field-derived terms and five solvation terms to reproduce the experimental binding free energy for a set of 99 protein–ligand complexes (Appendix B). A standard error of ±1.29 kcal/mol was obtained on the training set. In this function, it was observed that retention of water molecules in the models did not significantly improve the results and that the continuum solvation model was sufficient to provide accurate free energy predictions [377].

2.2.3. Empirical Scoring Functions In contrast to force field-based or semiempirical approaches to scoring, the empirical functions consider the free energy of binding as the sum of uncorrelated terms, each accounting for a different contribution [378]. While no physical foundation for assuming additivity has been provided to date [10,379], this paradigm has numerous pragmatic advantages and may be considered an extension of QSAR (quantitative structure–activity relationships, Chapter X). The different terms, that is, hydrogen bonds, hydrophobic interactions, ionic contacts, entropic effects, etc., are usually normalized by weighting coefficients derived from regressions over specific training sets of known and well-characterized protein–ligand complexes. Thus, the interaction energy associated with formation of a protein–ligand complex can be obtained by solving an equation of the type:

$$\Delta G_{\text{binding}} \approx \sum \Delta G_i f_i(r_1, r_p)$$

where f_i is a simple geometrical function of the ligand (r_1) and receptor (r_p) coordinates. As noted above, this is a formalism familiar to QSAR analyses. The first training sets to form the functions and derive the weighting coefficients were composed of about 50–100 complexes because of the difficulties in validating and curating the data. Both good quality crystal structures and reliable binding data are required. Wang et al. showed that good convergence could be achieved by including many more than 100 structures in training sets [380], but this requires a trade-off to more automated data curation where errors of inclusion and omission are more likely.

In recent years, empirical functions have been widely accepted and successfully applied. They are simple and fast and suitable not only for docking but also for virtual screening analyses. Nevertheless, empirical scoring functions still retain several drawbacks: (1) all empirical scoring algorithms significantly depend on the quality of the experimental binding data and of the crystallographic structures of the training sets; (2) correct interpretation of the structure, in terms of ionization states, hybridization, tautomerization, etc., both for the training set and for the complexes being analyzed, while always important, is perhaps even more critical; (3) experimental variables such as temperature, pH, or ionic strength, which can dramatically alter chemical structure, are usually not taken into account; (4) common interactions such as H-bonds are usually overestimated compared to other less frequently occurring contacts [8,381]; and (5) since molecules are generally treated as fixed objects, entropy penalty terms must be added in an attempt for more reliable energy estimations.

The first empirical scoring function, SCORE1, was developed in 1992 by Böhm, as part of the *de novo* design program LUDI [37], and was later implemented as the primary scoring function of the FlexX docking tool [382]. At first, only two contributions accounting for hydrogen-bonding and hydrophobic interactions were considered [383], but more terms (Appendix C) were soon added to account for loss of translational and rotational entropy, ideal hydrogen bond geometries, unperturbed ionic contact, or the loss of binding energy given by freezing the internal degrees of freedom of the ligand [140]. Nevertheless, SCORE1 was unable to evaluate cation–π interactions, differentiate between neutral or ionic forms of hydrogen bonds, and had no terms for protein and ligand internal conformational energy or for the contribution of water molecules. Consequently, SCORE2 was created in 1998 [384]. This function was calibrated on a training set of 82 protein–ligand structures retrieved from the PDB, and subsequently applied to several case studies [385,386]. These two SCORE functions represent a fundamental starting point for the design and development of several other empirical scoring functions [77,144,274,387].

Bohacek and McMartin's combinatorial *de novo* design program GrowMol [388] included a very simple empirical function that was similar in construct to SCORE1. It was trained on nine thermolysin-inhibitor complexes and showed particularly good correlation with the experimental binding data ($R^2 = 0.92$), but its narrow focus and limited range limited its further development or use. Analogously, the PLP1 function of Gehlhaar et al. [125,225] was developed to optimize positioning of ligands into target binding pockets, but was not intended for binding free energy predictions. The PLP1 score is calculated by summing up each (heavy) atom–atom steric and hydrogen bond interaction, adding an intramolecular term encoding a torsional potential and a nonbonded term (see Appendix C). Both PLP1 and the later PLP2 were implemented in the LigFit module of Cerius2 [389], compared to several other scoring approaches [390,391], and used in virtual screening applications [392,393].

A different strategy was adopted by Marshall's group when creating VALIDATE, a hybrid method based on MM and QSAR strategies [387]. The numerous contributions summed to compute the free energy of binding (Appendix C) include a Coulombic and a Lennard-Jones potential derived from the AMBER force field [259,260] for nonbonded electrostatic and steric interactions, a logP term calculated with the HINT force field [394,395], a term for interfacial rotatable bonds, a ligand strain energy term determined with a GB/SA solvation model [152], and two contributions for lipophilic and hydrophilic receptor–ligand complementarity [396]. Because it takes into account the effects of conformational changes, structural, and energetic complementarity, and desolvation contributions, the function was demonstrated to provide an impressively reliable estimation of binding free energy when applied to both the training and test sets. VALIDATE was further improved by introducing 15 additional descriptors for hydrogen bonds, free energy of solvation, free energy of cavity formation, ligand dipole moment and HOMO (high occupied molecular orbital) energy [397].

The scoring function implemented in FlexX [77] was a somewhat evolved SCORE1 that included a new parameter for aromatic interactions and a modified lipophilic term based on the sum of pairwise atom–atom contacts. Following, the authors introduced the particle concept that places and optimizes water molecules at the receptor–ligand interface during docking simulations [142]. Two additional terms accounting for polar-nonpolar interactions and unfavorable protein–ligand overlaps were later (2001) added [398]. FlexX has been used in many docking and VS applications [139,141,382,399–401] and still represents one of the most widely applied docking tools that is particularly well regarded because of the quality of its scoring function.

Jain developed his function in 1996 while creating a sufficiently rapid and accurate tool for virtual screening analyses and free energy predictions [144]. As with several of the approaches reported previously [140,402], this algorithm included hydrophobic and polar (hydrogen bonding) contributions, as well as terms estimating entropic and solvation effects [144,396] (Appendix C). An iterative training algorithm was used to optimize the function [403–405]. Interestingly, it was observed that hydrophobic interactions account for about 44% of the overall energy, while 26%, 25%, and 5%, respectively, are explained by polar contacts, entropic, and solvent effects. The function was implemented in the Hammerhead [143], Surflex [146], and LigandFit [406] docking programs, and was later improved by adding code for the computation of repulsive forces [407].

The ChemScore function, modeled after Böhm's SCORE1 function, included a special hydrogen bond term for estimating water-mediated contacts, but was still unable to differentiate between charged and uncharged contacts [274]. Ligand flexibility was monitored by scaling the number of rotatable bonds with their fraction of nonlipophilic character; a contact-based approach was used to compute lipophilic interactions. While good results were obtained on the calibration set of 82 PDB protein–ligand complexes, the ChemScore algorithm was still affected by many pitfalls typical of empirical-based scoring functions such as poor consideration of uncommon or negative contacts. Its estimation of the binding free energy was deemed insufficiently accurate to distinguish between very similar compounds in a lead optimization process [408].

Baxter et al. (1998) proposed an updated version of ChemScore for use in flexible docking analysis in combination with a tabu search algorithm [273]. This new version included protein–ligand clash and a ligand internal energy terms plus a modified H-bond term, all of which discourage the generation of improbable geometries in docking. More recently, a covalent energy contribution was added by Verdonk in the ChemScore version implemented in GOLD (Appendix C) [83]; numerous successful applications of GOLD with the modified ChemScore algorithm have been published [409–412].

In Horvath's approach a minimum dock energy and a free ligand energy are combined to determine "binding indexes," which correspond to the binding enthalpy and entropy [413]. The free energy comprises electrostatic, vdW, hydrophobic, and desolvation effects as functions of the position of each docked ligand in the binding pocket and is used to determine the local energy minimum after the minimization process. The following equation

$$\ln K_i = a\Delta H_i + b\Delta H_i^* + cT\Delta S_i + d$$

was calibrated on a set of 44 trypanothione reductase (TR) inhibitors with good correlation between experimental binding data and binding indexes. A later application led to the identification of new putative TR ligands with μM affinity [413].

An accurate estimation of binding free energy was made by Wang et al., who further developed the SCORE algorithm by innovatively decomposing binding affinity into several terms localized on each ligand atom [380]. The function equation (Appendix C) includes a regression constant and five terms representing vdW forces (using the AMBER vdW radius parameters [261]), metal–ligand interactions (the sum of all metal–O/N bonds), the H-bond energy (a sum of all interactions distinguishing between strong, moderate, and weak contacts), desolvation (computed with XLOGP), and deformation effects (a count of the number of ligand rotors).

The SCORE function was initially calibrated on a set of 170 protein–ligand complexes used to define the 11 adjustable parameters, and then tested on 11 endothiapesin-ligand complexes with impressively good results [380]. SCORE 3.0 [278] is an empirical algorithm based on SCORE and X-SCORE [414] that includes terms for vdW interactions, hydrogen-bonding, metal–ligand bonding, plus desolvation and entropic effects [278]. Internal ligand energy derived from the TRIPOS 5.5 force field [307] was added for docking energy evaluations (Appendix C).

A Poisson–Boltzmann electrostatics approach [415,416] was adapted for the Fresno function by Rognan et al. to include desolvation effects [417]. The algorithm estimates the overall free energy by summing desolvation and buried-polar contributions, plus terms for hydrogen-bonding, lipophilic and rotational contributions [140,274] (Appendix C). Fresno was applied to a variety of peptide–protein models, obtained through crystallographic analyses or homology modeling. It was suggested that the function may be adapted to other biological systems after appropriate recalibration. Several applications of Fresno are reported [87,418–420].

Different from the above approaches inspired by Böhm, the SLIDE scoring function implemented in SPECITOPE was developed as a fast and simple tool for screening large databases without intending to produce reliable binding free energy predictions [154]. The total energy is simply calculated by summing a term for intermolecular protein–ligand and water-mediated hydrogen bonds, HBONDS (P,L), and a term for hydrophobic complementarity, HPHOB(P,L), determined by calculating the hydrophobicity of each ligand atom and of the protein neighborhood [421]. Weighting coefficients were first derived on a set of 30 protein–ligand complexes [154] and then on a larger set of 89 complexes [155]. Further changes were introduced later to consider protein flexibility [156,422–424].

The Kasper function (Appendix C) was specifically developed for treating peptide–chaperone DnaK complexes [425]. The algorithm includes one term for conformational entropy (determined with the empirical scale of Pickett and Sternberg [426]), one for nonpolar interactions (related to the change in SAS area upon complex formation), and one for electrostatic contacts (accounting for both Coulombic interactions and solvation effects computed with a finite difference approximation of the Poisson equation [214,427]). The algorithm was trained on a set of 11 complexes including 1 crystal structure and 10 models derived form *in silico* mutation and MD simulations. A good correlation was obtained even without an entropic term. It could be concluded that, in this specific case, a reliable description of the binding process was possible by taking into account only nonpolar and electrostatic contributions. It may be, either by fortuitously or by design, that the entropic term is of similar magnitude amongst the members of the training set.

X-CSCORE is a consensus approach developed by Wang's group to create a valuable tool for virtual screening or *de novo* structure design. The method (Appendix C) is based on the combination of three different scoring functions, characterized by a diverse set of functions to estimate the hydrophobic effect [414]. In each function partial contributions accounting for vdW and H-bond interactions and for deformation and hydrophobic effects are summed up to compute the binding free energy. The algorithms adopted to predict the hydrophobic effect are Böhm's hydrophobic surface algorithm [140], Eldridge's hydrophobic contact algorithm [274], and Wang's hydrophobic matching algorithm [380]. The final energy is obtained by averaging the pK_d values given by the three functions. While the three functions were trained on a set of 200 protein–ligand complexes and then tested on a set of 30 structures, none of the three performed notably better than the other two.

Interestingly, much more reliable results were obtained by combining the three functions, yielding an average accuracy of about ±2 kcal/mol.

The empirical scoring function of the FRED docking program (McGann et al. [101]) uses a modified version of the Gaussian function [428,429] developed by Grant and Pickup [246], to which a penalty term for unfavorable atomic clashes is added. This scoring function has the unique features of adopting a simpler hypersurface and a predominance of long-range interactions—modeling well crystallographic structures, apo-proteins and homology models because of the lower sensitivity of the algorithm to conformational adjustments. Conformations similar to the crystallographic poses were obtained by applying the function on a set of 20 trypsin–ligand–protein complexes, but no reliable estimation of the free energy of binding is available. Thus, FRED was proposed as a useful prefilter tool characterized by good error tolerance and high robustness, able to reduce the search space normally explored by most common docking algorithms [101]. Later improvements to FRED were applied and compared in several docking analyses [430–432].

The eHITS docking progam's empirical scoring function (SF_e) [105] includes a distance and angle-dependent energy function for hydrogen bonding, a hydrophobicity term including contact surface area π stacking energy, an electrostatic Coulombic potential, a vdW contact energy, a metal ion interaction term (similar to that of hydrogen bonding), a penalty for incompatible contacts, an exposed surface area and an intramolecular potential (Appendix C). In the inital version of eHiTS this function was applied first for pose evaluation during docking, then for selection of the best-matching solutions, and finally for local optimization. Recently, the authors explored the impact of increasingly accurate functions on the various steps of docking. Thus, at present, a fast chemical flag-based statistical function (SF_s) is initially applied in rigid fragment-docking and pose matching phases, the more accurate empirical function (SF_e) is used in the following local minimization process, and a new, more accurate, function (SF_c) that combines statistical and empirical components with additional geometrical terms, an entropy loss estimation and scoring elements related to the coverage of the receptor surface area, is used in the final stage to predict the free energy of binding [106].

The first GEMDOCK formula combined both intermolecular and intramolecular terms for electrostatic, steric and H-bond contacts, and a penalty term to limit ligand poses placed outside the binding pocket [230,433]. The scoring function was tested on a 100 structure data set [205], and was demonstrated to identify the native binding mode in 77% of all cases and in 85% of the cases when crystallographic water molecules were maintained in the cavity. The pharmacophore-based version of the algorithm (2005) improved the energy calculation by adding two more terms (E_{pharma} and E_{ligpre}) that reduced the number of false-positive results when screening large databases [231].

The original GlideScore 2.5 SP algorithm was proposed as a ChemScore adaptation [274], where lipophilic and hydrogen-bonding terms were largely retained, while new Coulombic, vdW interaction and solvation contribution terms were added [198,434]. A new solvation model, based on a grid algorithm able to explicitly dock and score water molecules in a binding pocket, was also included. The softer scoring function (Appendix C) was suggested by the authors to be particularly suitable for docking simulations and VS analyses as it should be able to minimize false negatives [198,434]. The GlideScore 4.0 XP version added several new contributions evaluating, for example, π stacking, and π–cation interactions, ligand strain energy and hydrophobic effects, with a total of nearly 80 adjustable terms. GlideScore 4.0 XP and GlideScore 4.0 SP were evaluated and compared with a set of 198 protein–ligand complexes and indicated a better performance for the XP version that was able to predict the experimental binding free energy with an error of ±1.75 kcal/mol [1,435].

The Ligscore1 function [436] is currently implemented in the LigFit module of Cerius2 [389]. The algorithm equation in Appendix C includes a vdW term consisting of a soft Lennard-Jones 6–9 potential, an electrostatic term, and a desolvation penalty related to the quadratic polar contact surfaces between protein and ligand designed to balance favorable polar interactions. A better evaluation of desolvation effects is provided in the Ligscore2 function with a new term describing both ligand and protein contributions associated with burying of polar surface areas upon complex formation. The performance of Ligscore2 was evaluated on a set of 122 protein–ligand complexes, for which a correlation coefficient of 0.65 was found between experimentally measured binding free energies and the analogous computational estimates.

The SCFscore algorithm, recently developed by Sotriffer et al. within the "Scoring Function Consortium" project [437], was based on a training set of more than 850 protein–ligand complexes obtained from the PDB and pharmaceutical companies. More than 60 different descriptors were evaluated, including terms usually disregarded by common scoring functions, for example, interactions with aromatic ring systems. When compared with empirical functions such as SCORE1, SCORE2, ChemScore, and X-Score, SCFscore provided good correlations with experimental affinities, with significantly lower standard errors. Despite this success, the authors suggested that even more sophisticated descriptors, accounting for the contribution of water molecules, internal strain energy and imperfect steric fit, should be developed and included in next generation scoring functions.

The first description of the HINT (hydropathic interactions) scoring function was by Kellogg and Abraham in 1991 [395]. HINT is completely different from all other scoring functions being nearly completely based on hydrophobic atomic constants derived by the fragmentation of experimentally determined $\log P_{o/w}$ values, that is, the partition coefficients for 1-octanol and water [438,439]. Since $\log P_{o/w}$ is directly related to the free energy of solvent transfer, each hydrophobic atomic constant (a) corresponds to a partial δg, and encodes the atom propensity of interacting with either water or 1-octanol, or by extension, with other polar or hydrophobic atoms. The following equation is used to estimate the energy of the overall binding process:

$$\text{HINT Score} = \sum_i \sum_j (a_i S_i a_j S_j T_{ij} R_{ij} + r_{ij})$$

where i and j are interacting atoms on the two molecules, S is the solvent-accessible surface area for the atom, and R_{ij} and r_{ij} are different functions of the distance between the i and the j atoms for hydropathic and van der Waals interactions, respectively [440,441]. Good correlations were generally obtained when the algorithm was used to predict the binding free energy of protein–protein [442], DNA-ligand [443], RNA-ligand [444] and numerous protein–ligand sets [395,445–450]. The function was recently improved by including the computational titration method for predicting and optimizing the protonation state of ionizable residues at a complex interface [446,448,451], and the rank algorithm for rationalizing the role of structural water molecules in protein binding pockets [25,447,450,452]. HINT has been used to date in docking applications as a postprocessing filter [2] and has revealed the difficulties of applying out-of-the-box scoring functions on new data sets.

2.2.4. Knowledge-Based Methods

The last category of scoring function is known as the knowledge-based class of algorithms, in that they exploit the constantly increasing "knowledge" implicitly encoded in public databases. As stated by Mügge: "statistical potentials that are derived from statistical mechanics are really only statistical preferences; nevertheless due to similarity to the concept of statistically averaged forces we loosely call them potential of mean forces." [47,453] These methods, initially applied in the protein folding field [454–464], were more recently extended to the analysis of protein–ligand interactions, based on the statistical assumption that more frequent interactions are presumed to positively contribute to the overall binding process, while less common contacts are generally considered repulsive. By applying the inverse formulation of the Boltzmann law, it is possible to convert the extensive experimental information in structurally characterized protein–ligand complexes into sets of atom-pair potentials for each pair of interacting atoms:

$$E_{ij} = -kT\ln(p_{ijk}) + kT\ln(Z)$$

The variables i, j and k correspond to the protein atom types, the ligand atom types and their interatomic distances, respectively. The energy function E_{ij} represents the potential of mean force (PMF) for a state described by the variables i, j, and k, p_{ijk} is the corresponding probability density and Z is the partition function [465,466].

Differently from empirical, and particularly force field-based approaches, knowledge-based algorithms do not rely on any *physical* equation and mostly depend on statistical methods and on the quality and variety of the databases used for training [467]. Thus, unusual interactions, such as π–π or π–cation, can be poorly predicted, while more common electrostatics or H-bond contacts may be overestimated.

In Verkhivker's algorithm, one of the first knowledge-based scoring functions [468], the energy calculation is based on the approach of Sippl [455]. The distance-dependent pair potentials were derived from the statistical analysis of a set of seven HIV-1 protease-inhibitor complexes with contributions for protein–ligand, ligand–water and protein–water interactions. Terms for desolvation and conformational change upon complex formation were also added (Appendix D). The procedure for creating a PMF scoring function is generally as follows: for a given data set the different specific contacts at predetermined distances are catalogued, normalized by the frequencies observed for all types of protein and ligand atom pairs belonging to the set, and translated into mean force potentials. Because the method, as is true for other knowledge-based approaches, is calibrated on a specific training set, careful adjustment of the tunable parameters is necessary before extending the function to different protein–ligand systems.

A training set of 38 protein–ligand complexes was used by Wallqvist and coworkers to design a knowledge-based potential function based on calculated buried surface areas in protein–ligand complexes by comparing the interface area for each specific atom type in complexed and uncomplexed structures [469]. The P_{ij} preference score is then calculated from the area of the selected interacting surface elements (Appendix D); thus determining specific atom–atom and residue–residue preferences. Accordingly, the highest and lowest preference scores correspond to the most and least observed adjacent surfaces, respectively. This derived knowledge-based data set can be used to give an empirical estimation of the free energy of binding. When tested on eight protein–ligand complexes not in the training set, comparison of the experimental binding constants with predicted free energies gave a linear fit RMS deviation of 1.5 kcal/mol ($R = 0.74$) [469].

Shakhnovich and coworkers developed the SMoG96 (small-molecule growth) algorithm, based on a coarse-grained model and a potential mean force scoring function [39,470]. The probability of finding certain contacts at the protein–ligand interface was analyzed within a 5 Å cutoff range (more or less comparable to the first water coordination shell). Thus, in a sense, the average contribution of solvation entropy to the overall free energy is considered while the desolvation cost is directly included in the definition of the reference state. Although promising results were obtained when the function was tested on the Purine Nucleoside Phosphorylase, Src SH3 domain specificity pocket and HIV-1 Protease systems, the algorithm lost predictive power in differentiating ligands in the submicromolar range [470]. The new SMoG2001 version, developed and released by Ishchenko and Shakhnovich, was enhanced by extending the training and test sets to 725 and 119 PDB protein–ligand complexes, respectively. The reference state was redefined and the potential of mean forces was calculated as the logarithm of the ratio of probabilities of contact formation in the complex and in the reference state (see Appendix D for further details). Finally, to better model the loss in conformational entropy upon binding, a heuristic entropic term was further added to the original potential formula [471].

Mügge and Martin derived their landmark PMF smoothed-grained potential from the statistical analysis of 697 protein–ligand complexes retrieved from the PDB [124]. While entropic effects were only implicitly calculated, a large 12 Å cutoff was used to include solvation effects and cutoffs of 9 Å and 6 Å were used for noncarbon and carbon-carbon protein–ligand interactions, respectively. The sum of all atom pair potentials could be directly correlated to the binding free energy. A standard deviation of 1.8 $\log K_i$ units and an R^2 of 0.61 were obtained when the function was applied to predicting binding affinities in a test set of 77 protein–ligand complexes. The PMF99 potential was later implemented in DOCK4, after adding a repulsive vdW term to account for short-distance negative interactions [472]. This has been applied to several relevant biological systems [472–475]. The most recent, PMF04, version was developed on a training set of 7152 PDB protein–ligand complexes and introduced potentials for metal ions and halogens [476]; this was implemented within CScore [477] BioMedCAChe [478], DockIt [123], and the Cerius2 LigFit module [389].

The BLEEP1 and BLEEP2 mean-field scoring functions were developed by Mitchell and Thornton in 1999 [479], encoding the information derived from two sets of 351 and 188 PDB entries, respectively, with resolutions better than 2 Å. BLEEP accounts for the contribution of water molecules found at the interacting interface to better model solvent effects. As detailed in Appendix D, the overall potential corresponds to the difference between the potential of a particular atom pair and that of a reference potential given by the distance distribution for the entire data set [479] tested on a set of 90 crystallographic protein–ligand complexes, giving R^2 coefficients of 0.45 and 0.55. Most significantly, BLEEP demonstrated the importance of accounting for water-mediated effects [480].

In 2000 Gohlke, Hendlich, and Klebe created a new smooth-grained potential mean force function named DrugScore from the information encoded in a data set of 1376 protein–ligand complexes [481]. The polar or hydrophobic character of interacting atoms was expressed with distance-dependent pair potentials and solvent-accessible surface-dependent singlet potentials, and a smoothing function was applied to minimize the experimental intrinsic uncertainty. Originally, terms accounting for conformational, rotational and translational entropy or for intramolecular contacts were not included in DrugScore. In fact, all solvent-related effects were encoded within the singlet potential. In 2005, DrugScore was enhanced by Velec and coworkers [482], who extracted potentials of mean forces from the Cambridge Structural Database [483] and created DrugScore^{CSD}. Surface-dependent singlet potentials were omitted and a 15-fold increase in speed was achieved. When tested on a set of 100 protein–ligand complexes, the correct binding pose was identified in 77% of the cases (giving a correlation coefficient of 0.62) [482]. DrugScore is currently implemented in FlexX [77] and has been documented in studies of several biological systems alone and in combination with other functions [112,484–487].

The DFIRE potential uses a distance-scale finite ideal-gas reference state, rather than a canonical statistically averaged state [488]. This difference gives the potential a more physical character, and loosens the dependence on the quality and chemical nature of the database used to train the function. DFIRE was initially developed for and applied to investigations of protein folding processes [467]. When extended to docking and free energy prediction of protein–ligand, protein–protein and protein–DNA binding complexes, R coefficients of 0.63, 0.73, and 0.82, respectively, were reported [489]. Additional applications of DFIRE have been published [490–495].

Free energies predicted by the ITScore function are calculated by summing up all the interatomic interactions using the $u_{ij}^{(n)}(r)$ potential, as reported in Appendix D [496]. The pair potentials were derived from a set of 786 protein–ligand complexes with crystallographic resolution better than 2.5 Å. The reference state was obtained through an iterative approach comparing calculated and experimental potentials, with discrimination between native and decoy complex structures [496,497]. ITScore was validated on Wang's data set [390], and was shown to properly identify the correct ligand pose in 82% of the cases, and have good correlation with experimental binding energies ($R = 0.65$).

The M-Score function of Yang et al. was developed with the specific aim of addressing protein flexibility with knowledge-based scoring algorithms [498]. M-Score is a modified version of the DrugScore function to which a Gaussian distribution based on the crystallographic isotropic B factor has been added. This produced a smoother nonpolar potential, similar to those adopted by Ferrari et al. and Stahl et al. in soft docking analyses [398,499]. M-Score was trained on 896 structures, obtaining an $R = 0.49$ correlation coefficient.

2.2.5. Consensus Scoring A good scoring function should be able to discern properly positioned ligands in receptor binding pockets, and provide a reliable ranking of the generated poses by returning accurate estimations of the free energy of binding for each pose. Unfortunately, it should be quite evident that all scoring methods suffer from a variety of approximations and pitfalls, many of which necessarily arise from mediating between speed and accuracy. Also, to be sure, even the most detailed and time consuming *ab initio* approaches for free energy estimation are not completely accurate. As stated in the introduction, there is an intrinsic difficulty in translating biological phenomena to mathematical equations. Since it is currently not possible to identify a single universally performing function valid for multiple targets [500], two strategies have emerged: identifying the best-performing function for each target or merge several scoring functions in a consensus approach. The concept of consensus scoring was first introduced by Charifson in 1999 [216] based on the assumption that the combination of different functions would overcome inherent individual weaknesses and lead to better and more general scoring performance. While other strategies such as post-docking minimization [149,248,501] or topological filtering [502] have been attempted to adapt to the failings of single scoring functions, consensus scoring has been widely accepted and represents a powerful and successful approach under many conditions [503–508].

Classical consensus methods can be classified into three main categories: (1) methods that rank by "vote," quoting only conformations falling in the top n% of the list created by a scoring function; (2) methods ranking by "rank," averaging the ranking attributed by each function; and (3) methods ranking by number, averaging the actual score value calculated by the different functions used in the consensus.

A rank by vote strategy was adopted by Charifson in describing one of the first consensus scoring applications [216]. Compounds were rescored using 13 different algorithms, and the derived hit lists were compared, selecting molecules appearing in the first several positions of each list. It is important to note that this strategy has no possibility of calculating binding free energies. Nevertheless, the authors reported a significant improvement in enrichment (capability of discriminating between active and inactive compounds), and a reduction in the number of false-positive results in the hitlists filtered by consensus. An intersection approach was also adopted by Bissanz et al., who compared the screening performances of seven scoring functions, ChemScore, DOCK, FlexX, Fresno, GOLD, PMF, and SCORE, in interpreting the results provided by three different docking analyses on thymidine kinase and estrogen receptor. Interestingly, they observed that enrichments are essentially independent of the adopted *docking* tool, while more affected by the diversity and accuracy of the *scoring* methods [87]. Similar comparisons were later reported by Stahl and Rarey [398], Venkatachalam et al. [406], and by Krovat and Langer [392], who compared the performances of the seven scoring tool implemented in the Cerius2 program [389].

The rank by rank strategy was adopted by Clark and coworkers in developing the CScore program [477] (including ChemScore [274], DockScore [66], GoldScore [205], and PMF [124]), which was shown to select conformations similar to the native, and by Wang et al. [414], who applied eleven different functions to evaluate poses generated by AutoDock for 100 protein–ligand complexes. Better results were reported when three different functions were used as opposed to single or double scoring schemes. Mpamhanga and coworkers [393] proposed an average rank scoring scheme by merging force field-based, empirical, and a knowledge-based scoring functions. This rational combination of methods based on different principles was able to provide better results than random combinations of scoring functions.

As described above (Section 2.2.3) Wang et al. used a rank by number scheme in developing the X-CScore method [414]. Also described above, consensus docking, adopted by Paul and Rognan in ConsDock [303] (Section 2.1.9), was developed with the aim of predicting binding conformations rather than binding affinities. ConsDock incorporates the consensus concept throughout the docking procedure rather than in a postprocessing rescore step. Of particular note, the combination of docking tools significantly decreased the number of potential hits, while retaining almost all of the true positives [509].

A nontraditional consensus approach by Stahl and Rarey [398] merged terms from different scoring algorithms into the single scoring function ScreenScore, thus trying to exploit the strengths of each method (Appendix C). Similarly, multivariate statistical methods were exploited by Terp et al. [510] and by Jacobsson et al. [511], while a rank by median consensus strategy combined with a naïve Bayes classifier was proposed by Klon and coworkers. More recently, Betzi et al. developed GFscore (general nonlinear consensus scoring function), a method based on a nonlinear neural network, designed for the identification of true positives in large chemical libraries [512].

3. EMERGING AND CONTINUING ISSUES WITH DOCKING

One problem with docking and scoring has been, and will probably always be, related to transforming chemistry into mathematics so that ligand–biomacromolecule interactions can be "programmed" and represented on a computer. In fact, it would not be unfair to say that the more chemical (and less physical) a phenomenon is, the more difficult it is to represent in this way. This section of the chapter describes a number of these problem areas and the types of solutions that have been implemented in docking software to address them. Most of the software and algorithms described above have largely solved the core problem: producing models for ligand–biomacromolecule complexes and calculating relative scores for these models such that the user can choose one or more that are physically realistic. However, the subtleties that identify the one true and correct model, and that are more chemistry in origin, are still less well represented in these computational tools. In other words, the fairly easy to obtain *physically* realistic (i.e., Newtonian) results may not be as *chemically* realistic.

A useful concept from systems biology, *emergent properties*, which arise when individual components collectively create distinct interactive properties and functions, applies here. For example, the hydrophobic effect, where hydrophobic entities tend to self-associate as a consequence of the larger driving force inherent in polar (hydrogen-bonding) entities self-associating, cannot be inferred from a reductionist approach of considering only a single atom's physical properties. While entirely understandable from a chemist's perspective, the nuances of ionizations states, resonance structures, tautomerization, etc. are not apparent from most of the mathematical functions currently being used as docking scoring functions.

The second class of issues with docking and scoring are perhaps simpler to resolve. These are problems that can be resolved with physics, but are somewhat intractable because of their complexity. In particular, proteins and other biomacromolecules are not static under biological conditions, and, in fact, are quite flexible even at their active sites. Modeling this flexibility as part of a docking simulation is significantly more complex and time consuming compared to docking into a static structure. Also, understanding and exploiting the relationship between energetically similar models has been poorly developed as most docking protocols and energy scoring functions have been based on minimizing RMSD (root mean square difference) between docked solutions and X-ray crystal structures of the same complex. However, the existence of a crystal structure model, for which the data were collected at a very low (nonbiological) temperature, tends to bias docking experiments toward a single solution when, in fact, many others may be more than reasonable at biological temperature, while solution-phase NMR structures generally reveal a preformed family of structural models. Carlson has shown that docking to an ensemble of NMR structures is preferable to a single NMR or X-ray target in terms of inding mode prediction and virtual screening efficacy [513].

3.1. Entropy and the Hydrophobic Effect

The importance of the effect of entropy on biological binding events cannot be understated; unfortunately it is difficult to measure (see above) and perhaps even more difficult to computationally model. The underlying cause of entropy, the increase of disorder, manifests itself in a large number of ways, some of which are seemingly secondary effects and not immediately obvious. Consider the mixing of two liquids such as water and acetone. Since it is more disordered for them to be randomly distributed in the mixture, this mixing clearly results in an increase in entropy because it would be impossible to imagine them segregated. However, what about mixing water and an organic compound such as octanol? Here, because true random mixing would require the formation of *highly ordered* water clathrate cages around octanol molecules to bring them into water, the more entropically favored state is actually the two liquids in separate layers.

3.1.1. Effects Attributable to Entropy In terms of protein–ligand complex structure, there are several effects that are attributable to entropy. First, when a ligand binds the displacement of solvent (water) molecules from the binding site increases entropy as those solvent molecules are less constrained and have more degrees of freedom. Although it is somewhat counterintuitive, protein folding also is somewhat driven by entropy because more waters are disordered when the protein adopts a compact structure that shields more hydrophobic residue side chains and exposes more polar residue side chains to solvent. This is similar to the water/octanol experiment described above and suggests that accounting for hydrophobic interactions is, in part, a measure of entropy. There are other contributions to entropy including conformational effects, but recent calculations [514] suggest that the X-ray structure models of proteins are more "cleverly" packed than analogous NMR or decoy models and have higher side-chain entropy than other models with similar compactness.

3.1.2. Simulating Entropy in Docking In binding ligands to form a ligand–biomacromolecular complex structure many similar effects are observed and, again, entropy plays a key, although difficult to quantify, role. Thus, simulating the binding (or docking) process with explicit terms for entropy has proven to be an elusive goal. As discussed above, most docking scoring functions use very simplified models for even hydrophobic interactions, for example, hydrophobic surface contact areas, to recover elements of entropy in their scoring functions. Even the scoring function most identified with hydrophobic interactions, HINT, is a relatively crude instrument for simulating and quantifying all of the subtle effects contributing to entropy in the docking process. HINT inherently includes entropy and the hydrophobic effect because it is based on the partitioning of small organic molecules, fragments, and atoms between two solvents, water and 1-octanol. This is an equilibrium process—thus the partition coefficients are actually ΔGs. On the other extreme, FEP calculations using thermodynamic integration cycles, since free energy difference is path independent, can be used to estimate free energy of binding. Thus, a path is chosen between the starting (ligand unbound) and the final (ligand bound) states that includes intermediate states that are computationally accessible. Unfortunately, these calculations and similar approaches are very difficult to set up, expensive to run and nontrivial to interpret and are, thus, nonamenable at present as everyday dock scoring functions. Also, it is somewhat of a disappointment that FEP methods are actually not significantly more accurate in predicting the free energy of binding than well-calibrated empirical methods.

The computational estimation of entropy is clearly an important piece of the puzzle for dock scoring functions that currently has not been very well resolved. Much research into this problem is ongoing.

3.2. The Many Roles of Water

The water solvent plays an enormous role in biological interactions and is always present. For a protein, water is critical for achieving the correct fold, for flexibility in carrying out biological functions, and for mediating protein–protein and protein–DNA interactions. For protein–ligand complexes, in addition to their contribution to entropy described above produced when the water molecules are displaced, many waters are retained in the active site and contribute to the energetics and the process of binding in other ways. For example, waters can mediate (or bridge) interactions between the biomacromolecule and the ligand such that an otherwise unfavorable interaction is made energetically accessible. Also, in a simpler sense, water molecules can "shape" the active site by binding to residues and thereby presenting a different steric and electrostatic profile to incoming ligands than the bare biomacromolecule. While these roles have been thoroughly investigated by experimental, theoretical and computational approaches over the past several decades [515,516], water is still not well understood.

Conventionally, it is believed that there is about one water molecule for each amino acid residue in a protein. However, this number depends on the quality or resolution of the structural determination [517,518]. One water is found per residue at 2.0 Å, but around 1.6–1.7 are found at 1.0 Å resolution [519,520]. Obviously, this is a confounding factor that complicates much of what follows. Computational procedures that can predict potential sites for water molecules and their "binding" energies have been developed [61,284,521–540]. This allows "adding" waters to a biomacromolecular structure that is of poorer resolution and that has waters presumed to be "missing."

3.2.1. Predictions of Water Conservation

The first question that must be resolved in terms of docking and understanding the binding process is describing which water molecules will be displaced and which molecules will be retained, that is, their roles. This alone is a problem that has occupied much research as the structural clues available to understand water roles are few. In some ways, unraveling the binding process from the final, ligand-bound structure is much like a detective arriving on a crime scene and attempting to solve the murder from what few clues have been left behind. There is an important medicinal chemistry reason for this interest: evaluating the distribution and affinity of water molecules within enzyme active sites can provide design cues for selective inhibitors [25,161,521,541–546].

The roles of water molecules have been characterized based on a large number of different criteria: the number of hydrogen bonds formed and their interaction energies, the crystallographic thermal *B* factor, their accessible surface areas, their residence times, as well as their conservation and/or displacement resulting from ligand binding. The roles have been "named" with concepts such as "first" and "second hydration shells" [539,547], "buried," "tightly bound," "cavity," etc. The experimental structural data, as suggested above, is sometimes difficult to reconcile with simulations as it is nearly impossible to track a single water molecules experimentally. However, buried and tightly bound waters exhibit residence times typically in the hundreds of picoseconds [548–551], while surface water molecules that are more in contact with the bulk water have shorter residence times of around 5–50 ps [551–556].

Consolv [531] is a K-nearest-neighbors genetic algorithm that identifies conserved waters in biomacromolecule–ligand complexes with atomic density, atomic hydrophilicity, number of hydrogen bonds, and *B* factors. It was the first empirical calculation method for predicting conservation/nonconservation of active site waters upon ligand binding. This algorithm evaluates geometric features and is, therefore, dependent on crystallographic data quality. The microenvironment of each water molecule is characterized by Consolv and conserved or displaced waters are identified as those possessing geometrical features known to correlate with water binding. As a result, Consolv does not provide energetic estimations of the displacement or retention of water molecules and cannot discriminate between different water roles, for example, bridging, displaced waters, or cavity.

Obviously, all *conserved* waters are important molecules, and should be included in molecular modeling and docking experiments. However, the waters energetically essential are probably only those bridging or functionally displaceable. Waters deeply buried in cavities are probably too far removed from the active site. WaterScore [532], based on geometrical parameters and energetic estimations, is able to discriminate between bound and displaceable waters using *B* factor, solvent-accessible surface area, number of protein–water contacts, and the total hydrogen bond energy. WaterScore catalogues water as bound, sterically displaced and (otherwise) displaced. This method was 67% correct on a set of 46 water molecules bound to four native proteins. Bound waters usually present low *B* factors, small SASA, large atomic contacts and low hydrogen–bond energies. These factors as described by WaterScore would seem to strongly restrict water mobility, that is, conserved water molecules should be deeply buried in crevices of the binding pocket, surrounded by many protein atoms, and form multiple hydrogen bonds. This is certainly the case for water molecules locked in cavities, but are not descriptive of potential bridging waters, which should be exposed for contact with incoming ligands, and thus able to make new hydrogen bonds.

More recently a somewhat simpler model, based on HINT score and Rank has been described [450,452]. The HINT score is calculated by treating each water as an independent ligand optimized for its "site." Rank is a simple geometric parameter based on the potentiality of making hydrogen bonds for each water position. This combination, when subjected to a Bayesian-like statistical analysis [452] produced a model that predicts with up to 90% accuracy (depending on crystallographic resolution) which waters will be displaced and which are relevant. In this model, relevant waters are potentially displaceable but their locations require the presence of a functional group with equal or greater hydrogen-bonding capability.

3.2.2. Contribution of Water in Docking Simulations It is interesting that in a continuum electrostatic analysis carried out on T4 lysozyme only about half of the potential sites were actually occupied by ordered water molecules [557]. Also, it is somewhat vexing that in this case conserved (occupied) and nonconserved (unoccupied) sites had very similar free energies. The current state of the art in docking programs and their associated scoring functions is far from the ideal of: (1) *de novo* inserting a ligand into an active site that includes water, (2) displacing only the truly irrelevant and nonconserved waters, and (3) accurately estimating the free energy of binding for that ligand—including contributions from conserved and bridging waters. Nevertheless, progress has been made in these goals for several docking protocols.

One example is the FlexX particle algorithm that places and optimizes water molecules at the receptor–ligand interface during docking simulations [142]. Water molecules are placed in the binding pocket and moved during the ligand's incremental construction process to maximize the number and quality of hydrogen bonds. Then, for scoring the pose, water–ligand interactions are summed to protein–ligand interactions, with an additional penalty term for angular deviation of a definite contact from an ideal situation.

In other approaches: for example, within SLIDE the retention or displacement of water molecules upon ligand binding was predicted and a desolvation penalty term is imposed for water displacement [155]; Glide 2.5 accounts for solvation effects by explicitly docking water molecules into the target binding pocket for each generated ligand pose [198]; and another method to predict the retention or displacement of structural water molecules in binding sites and to rapidly score their energy contribution was implemented in GOLD [211] by making them individually user switchable and allowing them to rotate around their three main axes. Additionally, Moitissier et al. implemented a new potential for displaceable water molecules in AutoDock and combined water grids into a continuum grid [558], and van Dijk and Bonvin performed docking simulations from a completely solvated molecule and subsequently removed water with a

Monte Carlo procedure based on water-mediated contact propensity [559].

Despite these advancements, treatment of solvation in docking and virtual screening is a relatively underdeveloped area of computational research. Nevertheless, even imperfect approaches often yield significant improvement in binding poses and binding free energy prediction when water molecules are explicitly included [81,142,155,211,230,560,561].

3.3. Ionization and Tautomerization

Another great challenge is modeling the protonation state of the system at the active site. There are several ionizable amino acid residues whose side chains contain acids (Asp, Glu) or bases (His, Arg, Lys) in addition to the ionizable functional groups at the N- and C-termini. Also, the majority of ligands (potential drugs) contain acidic or basic functional groups. While the ionization states of these functional groups are well understood and well represented by their pK_a values when they are isolated, the solution and pH properties of microenvironments within proteins and at protein–ligand interfaces are much more complex.

3.3.1. Modeling the Positions of Protons As most crystal structures do not reveal information on the positions of hydrogen atoms, for some (ionizable) groups, both on the protein and the ligand, it is difficult to experimentally determine which protons are present, which are absent, and if they are present, to define their orientation. Because these groups are not isolated from one other and influence each other's states, their geometries and protonation states cannot be evaluated independently of each other. This is clearly of major importance for docking and scoring as simple protonation of a single carboxylate on either the ligand or an amino acid can quite significantly alter the electrostatic and hydrogen-bonding potential of a putative ligand–biomacromolecular interaction.

The solution pH or the microenvironment can also affect structure in another way: the tautomerization of specific functional groups on the ligand. This could also impact in a major way how that ligand will bind in the site. Some tautomerizations will actually force somewhat altered shapes for the ligand that would, of course, change the steric fit of the ligand in the site. Also, as shown by Czodrowski et al. [562], the pK_a of an ionizable residue at the protein–ligand interface could undergo significant electronic and structural changes upon complex formation. With this in mind, a cheap and fast charge estimation method (PEOE_PB) based on Poisson–Boltzmann calculations was developed and successfully applied to a series of ligands binding to serine proteases, trypsin and thrombin [563], and to some HIV-1 protease complexes [564].

The number of model possibilities in an ensemble grows exponentially with the number of ionizable groups in the active site. An active site can have several protonation state models that exist in equilibrium with each other.

Water molecules present in the active site further complicate the problem. Water can mediate hydrogen bonds by acting both as a Lewis base and/or as a Lewis acid to convert a weakly (or even strongly) repulsive polar interaction into a strongly favorable interaction. Thus, a water molecule can "buffer" the active site by rotating and changing its character from a donor to an acceptor when an interacting functional group is protonated or deprotonated. This was illustrated in two computational studies on a charged binding pocket [565,566], the engineered cavity in cytochrome c peroxidase, where allowing movement of a single conserved water (HOH308) produced superior docking results.

Water aside, the problem of placing and optimizing polar protons has been of great interest for crystallographers and computational chemists for a number of years. Correcting ambiguous atom placement and assigning protonation states and hydrogen orientations to ionizable residues in proteins is key for model building, for example, as a prelude to a docking or virtual screening experiment. As a start there are currently several web applications available for correcting X-ray protein crystal structures for problems with Asn, Gln, His, etc. For example, MolProbity [567] places hydrogen atoms while correcting other errors in protein structures, while NQ-Flipper [568] ignores hydrogens altogether.

3.3.2. Computational Approaches Ionization state evaluation and optimization is a much more complex problem that generally has been explored with multiple and not necessarily compatible tools. While quantum mechanics [569] and quantum mechanics– molecular mechanics (QM/MM) [570,571] approaches and methods based on molecular dynamics simulations [572–574] are available, most methods rely on solving the Poisson–Boltzmann equation, for example, with DelPhi [214], to evaluate possible protonation states and hydrogen positions [575–581]. Solving the Poisson–Boltzmann equation has also been combined with MD simulation [582–586]. None of these technologies, however, are amenable to "on the fly" evaluations that would be ideal for integration into docking programs and scoring functions. Also, these methods pursue a goal somewhat orthogonal to docking in that the results are most often described as pK_a values for protein residues. It is of considerably more interest to find the energetically accessible protonation state and hydrogen geometry for which the protein–ligand interaction is the strongest.

In terms of currently available functionality in docking programs, only a few programs attempt to define ligand tautomerization, like LigPrep in GLIDE or eHiTS, which is able to systematically evaluate all possible protonation states for both the ligand and the ionizable residues in the receptor binding pocket in a single run. For each group, all possible cases are simulated and scored, and the best state is selected independently for each. While this approach avoids the computational issues of treating the multiple states combinatorially [105,106], these ionizations are not truly independent. Most programs still require the users to provide all tautomers on input. At best, only low-level ionization state optimizations are performed by any of the current docking protocols. In general, this added dimension in dock scoring has received only marginal attention over the past few years, but most certainly will need to be enhanced before rapid and reliable docking, scoring and virtual screening of single ligands or databases can be obtained.

3.4. Flexibility

Even though only static structures are used for most structure-based drug design, that is, docking, virtual screening, etc., the importance of flexibility was recognized even with the first protein characterized by X-ray crystallography: hemoglobin [587–590]. The two different structural forms of hemoglobin, "tense" (deoxygenated) and "relaxed" (oxygenated), have markedly different quaternary structures. All proteins are inherently flexible systems and this flexibility turns out to be often essential for function—like in hemoglobin. In fact, over wide ranges in temporal and spatial scales, proteins can undergo functionally relevant conformational transitions under native state conditions [591–593]. Similar to hemoglobin, a significant degree of conformational flexibility is essential for the biological function of nuclear receptors as the molecular pharmacology of nuclear receptor ligands depends on stabilizing (or displacing) a short α-helix segment at the carboxy terminus of the receptor [50,87,594]. X-ray crystallography of farnesoid X-receptor (FXR)-ligand complexes illustrate the extent of this flexibility with the diversity of bound ligands [595,596]. To date, most successful modeling and computational drug design from structure have not considered the flexibility of the target, probably because these cases have required relatively small (induced fit) rearrangements of the targets whilst the ligand binds. The Protein Data Bank (PDB) [104] is probably overpopulated with relatively rigid proteins because these have, of course, been easier to crystallize and analyze.

3.4.1. Experimental Characterization of Flexibility As solving an X-ray crystal structure involves threading a molecular model of the structure through a map of electron density, experimental measurement techniques have focused on obtaining data that are as unambiguous as possible, that is, with well-defined electron density envelopes. Thus, to reduce disorder (and to protect valuable crystals), X-ray crystallography data are now usually obtained at extremely low, nonbiological, temperatures. One may also question whether the crystalline environment is an adequate approximation of the true biological environment for proteins and other biomacromolecules. Biological functions generally occur in aqueous or semifluid environments at or above room temperature such that functioning proteins (in solution) exist as an ensemble of energetically accessible conformations. Their three-dimensional structure may be best described when all of these states and conformations are represented (see Section 3.5).

X-Ray Crystallography While X-ray crystallography is considered the gold standard technique in obtaining experimental structural models of biological macromolecules, these models, as indicated above, are usually static, time- and space-averaged structures with little indication of their inherent flexibility or the ensemble of conformations of the protein in action. Accuracy of static structures is described by the resolution to which X-ray data have been measured. Typically, crystallographic resolution of less than 2 Å is considered good for a protein structure and resolution around 1 Å is termed "atomic." Each atom in a crystal structure will also have a *B* factor (or temperature factor) that represents the uncertainty of that atom's position; this factor is due to a mix of contributions: thermal motion, occupancy, experimental and modeling artifacts and other effects [597], so is not directly attributable to flexibility.

Recent innovations have permitted information regarding flexibility to be extracted from high-resolution crystal structure analyses, particularly from data collection on synchrotron X-ray sources: (1) time-resolved measurements using Laue diffraction for rapid data collection, or conventional diffraction on a series of crystals, can detect transient chemical states; (2) at atomic resolution anisotropic modeling of the *B* factor [598,599] allows extraction of the direction of atomic motion possibly leading to detection of both subtle and large-scale motions [600]; and (3) kinetic crystallography methodology measures a set of static structures on relevant reaction pathways. Most interestingly, transient intermediates can be observed, along with the associated structural arrangements, when biological turnover is radiation initiated in a crystal [601,602]. Also, combining kinetic crystallography with *in crystallo* UV/visible absorption and fluorescence, fluorescence lifetime, and Raman spectroscopies can provide a dynamic view of biological processes carried out by proteins [603,604].

NMR Spectroscopy NMR spectroscopy determines the three-dimensional structure of proteins in solution and thus allows direct observation of the flexibility and dynamics of biomacromolecules and their interactions with other molecules. NMR studies provide an ensemble of low-energy conformations. Each conformation of a molecule is, in effect, a static snapshot of the molecule, but a family of such snapshots provides a dynamic representation. For biological-scale molecules, multidimensional multinuclear NMR is applied: the off-diagonal cross peaks encode descriptions of the interaction between nuclei in the molecule, that is, the distances between the atoms. By applying strategies in pulse sequences, magnetization transfer and isotopic labeling (^{15}N and ^{13}C) these peaks can be assigned. Currently, the range of protein size amenable for NMR structure solution is up to 80–100 kDa [605,606]. In general, there will be a family of structures satisfying the NMR data. NMR is the only currently available technique that can monitor and discriminate protein movements over a broad range of timescales. NMR is especially appropriate in cases such as partially folded polypeptide chains that are difficult to crystallize and/or the crystals are disordered. The intrinsic experimental difficulties and limitations with NMR structure determination are more than compensated by its ability to reveal flexibility in solution at conditions similar to those where biomolecules are active.

3.4.2. Computational Molecular Dynamics

While experimental X-ray crystallography and NMR data provide a framework for understanding the structure and flexibility of biomacromolecules, these data are insufficient to understand many molecular motions as only the relatively long lived and more energetically accessible populated states will be observed. Molecular dynamics, however, can generate many protein conformations for docking and/or virtual screening studies [607–610]. Molecular dynamics approaches are often characterized by scale, depending on the nature and size of the system to be analyzed.

Coarse-Grained MD Coarse-grained approaches that smooth or average out atomistic details are appropriate for longer time-scale dynamics, that is, general protein flexibility. Brownian dynamics algorithms have several advantages over normal mode analysis because the effects of time can be incorporated and solvent effects can be introduced with residue-based potentials. Alternatively, discrete dynamics, using square wells delimited by finite or infinite energy walls instead of harmonic potentials, has the advantage of not requiring integrations of Newtonian equations because elastic collisions at each "wall" are assumed. Coarse-grained dynamics can rather accurately model general protein dynamics, for example, formation or reshaping of binding pockets. However, because of the absence of atom-level detail, these models are inadequate for structure based drug discovery and docking.

Atomic-Level MD Atomic level-of-detail molecular dynamics represents, simultaneously, both small atomic fluctuations and large protein movements, with the principal disadvantage being computational cost. The ability (or inability) to effectively sample the conformational states of the biomacromolecule or biomacromolecule-ligand complex is a second limitation. Most documented MD simulations are for only a few ns, which is too brief to ensure that the biologically significant minima are adequately sampled. By comparing crystallographic *B* factors with RMSD fluctuation per residue from the MD trajectory, it can be determined if the MD is generating new energetically accessible conformations that can be used as docking targets. In summary, while experimentally determined target conformation ensembles from X-ray or NMR are more desirable for docking, those from MD simulations are far more generally available.

3.4.3. Flexibility in Docking Simulations In considering docking simulations flexibility has been largely ignored because the conceptual and computational remedies were ill defined. In general, the goal of docking and scoring was placing ligands into static low-temperature models of biomacromolecules. Clearly, however, protein conformational changes are often initiated or stabilized by ligand binding and are, thus, essential to the protein's own function. The ability to measure or simulate dynamic changes taking place in proteins upon ligand binding is increasingly becoming a central theme in designing bioactive compounds. For drug discovery with a flexible target it is not known *a priori* how the target will adapt its conformation upon binding of a particular ligand, or how to design that ligand. Considerable effort has been expended in recent years to introduce flexibility of the active site into docking/scoring simulations. This is a challenging problem for a variety of reasons, both experimental and computational, that are steadily being surmounted with improved algorithms and increasingly inexpensive access to computer resources.

Algorithms and docking programs that fit and score flexible ligands into flexible binding sites are becoming increasingly available. The least time-consuming approaches include protein flexibility by simulating movements of side chains in the active site: (1) Version 4 of AutoDock [63–65,80,430] allows explicit protein side-chain flexibility; (2) GOLD samples alternative rotameric conformations for specified protein side chains and rotates terminal hydrogen atoms to optimize hydrogen-bond interactions [83]; (3) FlexE optimizes side-chain conformations from rotamers that have been observed in crystallographic structures [78]; and (4) SLIDE rotates protein side chains and ligand substructures with mean-field optimization to remove intermolecular van der Waals overlaps [156]. Other flexible docking tools include those that use constrained geometric simulations [611], elasticity network theory [423,612], dock ligands to ensembles of structure families either experimentally or theoretically obtained [79,81,357,613–616], evaluate complementarity of components or regions [84], or use MD to postprocess rigid docked solutions [617].

Simulating Side-chain Flexibility Including local (small scale) flexibility in docking has been shown in a few studies to greatly enhance the results [156,220]. This is especially true when docking into a site that has not been preformed for the ligand, for example, into the unliganded structure of a protein. Flexibility modeling is simplified significantly as most protein side chains move relatively little upon ligand binding because this binding often induces strain on the side chains [422], and minor rearrangements are better able to preserve intraprotein hydrogen bonds. Interestingly, water-mediated intraprotein hydrogen bonds are not well preserved and are more susceptible to rearrangement upon interaction with the ligand [618].

Induced Fit Proteins often undergo substantial rearrangement as they perform their catalytic activity or interact with other species. Gohlke's induced fit docking approach simulates significant conformational adjustments with a method derived from rigidity and elastic network theory [619,620]. First, the macromolecule is decomposed into rigid clusters using a graph-theoretical approach. Next, a rotations/translations of blocks approach [621] is applied to an elastic network model representation of the coarse-grained protein [622]. Alternatively, constrained geometric simulations are able to reproduce the motion of flexible and rigid parts of a complex with ghost template rearrangements [623]. These results strongly support the notion of "conformation selection" [591,624,625] that may be predetermined by the protein's motions in the unbound state. In other words, induced fit is a process where states with an appropriately formed binding pocket that are weakly populated in the absence of a ligand are preferentially selected by the corresponding ligand because it stabilizes those states relative to other protein conformations in the ensemble. Equilibrium then shifts to the binding-competent conformations that become predominant and, thus, experimentally observable.

Some docking programs, for example, AutoDock, use 3D grid lattices to represent atomic affinity, electrostatic potential, desolvation free energy, etc. properties around the target molecule. This has a number of advantages over atom-to-atom pairwise scoring and allows some conformational flexibility in the target by overlapping the radii of the grid points. An emerging paradigm for fully flexible protein–ligand docking [53] uses an elastic representation of the potential grids themselves.

Ensemble Docking Target structural ensembles from MD (most often), NMR, or X-ray crystallography (least often) can be used for lead docking [616]. If the scale of movement within an ensemble is consistent with the structural diversity of ligands to be docked, good models for the protein–ligand complexes can be obtained. Most basically, ensemble docking calculations are sequentially performed for each protein conformer. For example, in the relaxed complex method [607–609], MD simulations of the target is performed prior to docking and different conformations are selected, randomly, at regular time intervals, or for structural diversity for the target. Then the ligand of interest is docked into each of these structural snapshots from the MD trajectory. A more computationally efficient method is to combine or average many different conformations of a target protein into a single target grid [544]. However, the composite structure represented by the grid may not be physically "real" and the resulting ligand poses are also unrealistic. Another variant is "*in situ* cross-docking" where multiple protein structures are addressed simultaneously in a single grid target docking experiment [74,616,626].

The two key issues with ensemble docking are that (1) a complete (or even sufficiently representative) ensemble of protein conformers is seldom available and (2) as described above, the free energy scoring functions currently used in docking do not to provide acceptably accurate scores for these alternative binding modes, especially if binding is dominated by entropic contributions.

3.5. Multiple Solutions with Similar Energies

Is there more than one "correct" solution to a docking problem? We know from both experimental techniques such as NMR and through molecular dynamics and conformational analyses that there can be many conformers, even for simple molecules, that are energetically indistinct. Is it a conceit to assume that there is only one docked solution? In most training exercises, a ligand is extracted from the protein and redocked with the goal of it returning to its extracted position and conformation. In this regard, Hawkins, Warren, Skillman, and Nicholls [1] recently reiterated a long-standing point of misconception in molecular modeling. The "crystal structure" is itself a model, albeit one that (if correctly refined) fits the experimental electron density well. It is likely not unique! There may be a variety of other conformers, tautomers, ionization state variants, or even inverted molecules that also fit the electron density well (or better?) It is, in fact conceivable that the wrong molecule has been modeled by the crystallographer into the density, either by mistaken identity or because of some chemical reaction; what is thought to be the ligand may even be an artifact of crystallization such as a salt or cryprotectant. Of course the probability of these errors decreases as the resolution improves and the electron density envelopes become more defined.

Nonetheless, discounting all of the extremes listed above, multiple docking solutions, especially considering ionization state and/or tautomeric analogs, are likely to exist. The only completely reliable way to determine if such a solution is valid whilst training is to "score" it against the experimental electron density, *not* against the crystallographic ligand model [1]. In other words, RMSD is a flawed and imprecise tool for evaluating docking codes. When docking a "new" ligand into a site, where no electron density data or crystallographic ligand model exist, choosing the optimal model is even more ambiguous. The usual argument is that the docking scoring functions were trained by optimizing RMSD of docked poses, so that in a new environment, the best pose is the one with the best score, that is, most likely to match the crystallographic pose *if* a crystal structure were available. This series of assumptions clearly has problems, not the least of which is that everything relies on the initial crystal structure models, upon which the scoring function was trained, being not only "correct" but also unique.

Thus, in evaluating the results of docking experiments, choosing only the "top" candidate, and ignoring others with similar score (energy), may lead to unacceptable errors in lead selection and optimization. Another alternative is to calculate a Boltzmann-weighted energy score for the collection of poses [448] for each ligand and select candidate ligands by comparing these Boltzmann-weighted scores rather than the "best" score for each.

4. USER GUIDE

Instead of a summary at this point in the chapter, we instead have chosen to give a brief "User's Guide" that will itemize a number of helpful hints for beginning users of computational docking. However, also consider some of the comments and qualifications made in the "Conclusions and Outlook" (Section 5), following the User Guide. As can be interpreted from the discussion in Section 2 of this chapter, there are many algorithms for creating poses and many algorithms for scoring poses. It is actually true that the first part is easy: the binding pocket can be exhaustively examined with an exhaustive array of ligand conformations of which the pose or poses that you want are included. The trick, however, is to be clever about creating the poses, such that you do not have to analyze an infinite number of them to find the few of real interest. That is how the different searching algorithms vary and thus have their own strengths and weaknesses. The second part, scoring, is much more difficult—there is, to date, no universal scoring function that is both accurate and fast. However, with attention to the limitations and range of the available scoring functions, good results can be obtained.

The following are some base requirements for a beginner in docking. You need:

(1) Good "hardware," such as a PC, Macintosh, or Linux workstation. For extensive studies, such as database screening, a multi CPU cluster would be desirable.
(2) One or (preferably) more docking program packages that utilize different and complementary approaches. Although much of this is still under development, your program(s) of choice should pay some attention to protein flexibility—all docking packages of any repute have programmed ligand flexibility.
(3) One or (preferably) more internal scoring functions. At a minimum, you should have the ability to link with third-party scoring functions, perhaps in postprocessing. You will find, and it was intimated above, that scoring functions can have variable performance dependent on the data set; for example, docking ligands into a hydrophobic versus a polar cavity.
(4) Further independent scoring functions for consensus scoring. While consensus scoring is very powerful, applying it to a set of previously ranked solutions will not elevate a poorly ranked but "correct" solution above others. Thus, careful consideration must be given to false positives and false negatives—consensus scoring does not obviate that need.
(5) When choosing a final scoring function for end-stage analysis, go back to first principles—thermodynamics—and consider the scoring function in terms of the "master equation," $\Delta G = \Delta H - T\Delta S$.
(6) Define a "realistic" model and consider, as much as possible, *all* components (protein, ligand, conserved waters, metal ions, cofactors, and so on).
(7) Evaluate the quality of your input structural data, whether NMR or X-ray crystallography, particularly for the target. Examine the metrics of quality, that is, R factor, resolution, B factors, etc.
(8) If docking to a homology model, respect the process and reflect on the errors. If the target model was built with low sequence homology to the template, how much significance should the docking results be given?
(9) Have you collected all of the literature data? Even a few point mutation results on the target protein may be able to provide key information for critically evaluating and validating your docked model.
(10) Although it should be an obvious point, the fact is that experimental measures of binding affinity may have fairly large uncertainties, especially if measured by more than one laboratory (you should check), and it is meaningless to expect scoring functions to calculate binding energies with more precision.

5. CONCLUSIONS AND OUTLOOK

From the rather modest beginning of "docking" in the early 1980s to the present, this computational chemistry paradigm has emerged as one of the key technologies for drug discovery, either in one-by-one studies or in virtual screening. The success stories presented in many of the references of this chapter attest to the value of the paradigm. However, we believe that there is both promise and danger in docking as it becomes more readily available and convenient to use. Thus, we will conclude this chapter with both the "downside" and the "upside" of docking.

5.1. The Downside

If a lot of the above discussion in this chapter seems contradictory or perhaps discouraging, that was, in part, its intent. The vendors of computational chemistry software, particularly those programs designed for medicinal chemistry, are attempting to market an illusion: it is a very simple process to input a few parameters to a user friendly program and just wait a few moments for an easy to interpret, beautifully rendered, solution. The docking paradigm is particularly susceptible to this hype because of the visual impact of a seemingly well-docked structure concomitant with the "obvious" chemistry that might improve that docked ligand. Running a docking program is a computational *experiment*. As with any "wet" experiment, the assumptions and conditions behind the experiment must be evaluated, suitable controls should be run, and the obtained results must be carefully examined. In fact, it is really wise to treat any computational result with suspicion until it can be substantiated with another approach, preferably an experimental one.

It perhaps seems obvious, but running the same computational experiment with the same conditions will yield precisely the same result—each time you repeat it. This is in contrast to experimental science where repeating experiments is confidence building or, at least, provides a measure of uncertainty. To evaluate uncertainty in computational docking results, the experiment should be attempted with different methods, for example, from a different initial condition, with different parameters, or even with a different docking program. Also, consider that in a wet experiment the result is not from one, single, discrete molecule interacting with another, it is from many molecules of one type interacting with many of another; whatever numerical result that is obtained is a weighted average of many simultaneous reactions. In modeling, we *are* generally looking at one-to-one reactions and assuming that what is observed is representative. Is it? All docking algorithms use some element of randomness to drive the calculation engine and produce the result. To a very large extent, the thoroughness of random sampling directly correlates with the robustness of result. While it is easy to imagine that requesting only one docked solution from an algorithm may give a faulty answer, it is actually true that requesting 100 or 1000 or even 1 million docked poses may not produce the one "optimal" solution. This is especially true if the binding pocket is tight such that only a relatively few random starting conditions will achieve binding poses.

Although it is much simpler conceptually and perhaps more elegant to have a single "perfect" solution, the best philosophy for interpreting docking results is to not focus on one solution, but evaluate all that are reasonable. Regarding computational results as "answers" to problems is really not nearly as useful as seeing these results as clues or points to consider in designing the next "real" experiment.

5.2. The Upside

On the other hand, the outlook for docking and scoring in drug discovery is quite exciting. There is much to be optimistic about: there is an increasingly large database of crystallographic data that will not only directly provide more targets but also, ultimately, make homology-based target structure predictions more accurate; several groups are collecting and curating accurate thermodynamics data for binding, etc.—a growing necessity in calibrating and optimizing docking score functions; a growing number of resources (programs, data, etc.) are ubiquitously available to all on the internet; the availability and affordability of computational power still increases in an almost alarming manner; the field of medicinal chemistry, particularly computer-aided drug discovery, has in the past decade or so gone "mainstream" such that many bright and energetic scientists are attracted to solving problems such as docking and scoring; and our understanding of etiology and mechanisms of disease is increasingly exponentially, leading to many new and innovative approaches to treatment that will require structural data and the insight unique to computational medicinal chemists to implement.

ACKNOWLEDGMENTS

We wish to thank our colleagues for helpful discussions and inspiration: Prof. Andrea Mozzarelli (University of Parma) for enriching our understanding of biochemical phenomena, Prof. Gianluigi Ingletto (University of Parma) for insight into the fundamentals of free energy, Prof. Donald J. Abraham (Virginia Commonwealth University) for initiating our very rewarding collaboration, and Prof. J. Neel Scarsdale (Virginia Commonwealth

University) for sharing some of the more arcane knowledge of crystallographers. Our research has been sponsored by the U.S. National Institutes of Health (GM071894 to G. E.K.) and funds from the Italian Ministry of Instruction, University, and Research, within an internationalization collaborative project (to Andrea Mozzarelli).

APPENDIX A. FORCE FIELD-BASED SCORING FUNCTIONS

DOCK (1992) [66]

$$E = \sum_{i=1}^{\text{lig}} \left[\sqrt{A_{ii}} \sum_{j=1}^{\text{rec}} \frac{\sqrt{A_{jj}}}{r_{ij}^{12}} - \sqrt{B_{ii}} \sum_{j=1}^{\text{rec}} \frac{\sqrt{B_{jj}}}{r_{ij}^{6}} + 332.0 q_i \sum_{j=1}^{\text{rec}} \frac{q_j}{Dr_{ij}} \right]$$

Shoichet's approach (1999) [121]

$$E_{\text{bind}} = \sum_i q_i P_i + v_i P_v - E_{\text{solv,elec}} - E_{\text{solv,np}}$$

GREEN (1994) [321]

$$E_{\text{inter}} = \sum \sum \left(A r_{ij}^{-12} - B r_{ij}^{-6} \right) + \sum \sum \frac{q_i q_j}{\varepsilon r_{ij}} + \sum \sum \left(C r_{ij}^{-12} - D r_{ij}^{-10} \right)$$

FLOG (1994) [95]

$$\text{Score} = \sum_{j=1}^{\text{Natom}} \text{GRID}(jx, jy, jz, kj)$$

$$\begin{aligned} \text{GRID}(ix, iy, iz, k) = {} & C(k, \text{vdW}) \times V_{\text{vdW}} + C(k, \text{ESPOT}) \times V_{\text{ESPOT}} \\ & + C(k, \text{H-BondD}) \times V_{\text{H-bondD}} + C(k, \text{H-bondA}) \times V_{\text{H-bondA}} \\ & + C(k, \text{H-bondP}) \times V_{\text{H-bondP}} + C(k, \text{Hydrophobic}) \times V_{\text{Hydrophobic}} \end{aligned}$$

jx, jy, jz are the indices of the nearest grid point to the location of atom j and kj is the type of atom j;

$$V_{\text{vdW}} = - \sum_{\text{i}}^{\text{all atoms}} \text{depth}(\text{probe}, i) \left[\frac{2 \times \text{dist0}(\text{probe}, i)^9}{\text{dist}(gp, i)^9} - \frac{3 \times \text{dist0}(\text{probe}, i)^6}{\text{dist}(gp, i)^6} \right];$$

$$V_{\text{ESPOT}} = - \sum_{i}^{\text{all atoms}} \frac{q(i)}{(e(i) dist(gp, i))};$$

$$V_{\text{H-bondD}} = \sum_{i}^{\text{donors}} \text{radial term} \times \text{angle term} \times \text{dihedral term};$$

$$V_{\text{Hydrophobic}} = \begin{cases} 0.0 & \text{when } V_{\text{vdW}} < 0 \\ 0.0 & \text{when } V_{\text{H-bondD}} + V_{\text{H-bondA}} + V_{\text{H-bondP}} + V_{\text{vdW}} > F_{\min} \\ F \times V_{\text{vdW}} & \text{otherwise} \end{cases}$$

EUDOC (2001) [329]

$$E = \sum_{i<j} \varepsilon_{ij} \left(\frac{r_{ij}^{*12}}{R_{ij}^{12}} - 2 \frac{r_{ij}^{*6}}{R_{ij}^{6}} \right) + \sum_{i<j} \frac{q_i q_j}{\varepsilon_0 R_{ij}}$$

MM/PBSA (1999) [336]

$$\Delta G_{\text{bind}} = \Delta G_{\text{water}}(\text{complex}) - [\Delta G_{\text{water}}(\text{protein}) + \Delta G_{\text{water}}(\text{ligand})]$$

$$\Delta G_{\text{water}} = E_{\text{MM}} + \Delta G_{\text{solvation}} - TS;\ G_{\text{solvation}} = G_{\text{solvation-electrostatic}} + G_{\text{nonpolar}};$$

$$E_{\text{MM}} = E_{\text{internal}} + E_{\text{electrostatic}} + E_{\text{vdW}};\ E_{\text{internal}} = E_{\text{bond}} + E_{\text{angle}} - E_{\text{torsion}}$$

OWFEG (1999) [340]

$$G_{\text{score}} = X \frac{(G_{\text{grid-2}} - G_{\text{grid-1}})}{Q_{\text{probe}}} + G_{\text{grid-1}}$$

G_{score} corresponds to the grid score of each atom of the ligand;
$G_{\text{grid-1}}$ and $G_{\text{grid-2}}$ are the score on the neutral and charged probe grids;
Q_{probe} is the charge of the probe group used to generate the charged OWFEG grid.

APPENDIX B. SEMI-EMPIRICAL SCORING FUNCTIONS

GOLD (1995) [210]

$$\text{GOLD Fitness} = S_{\text{hb\_ext}} + 1.3750 \cdot S_{\text{vdw\_ext}} + S_{\text{hb\_int}} + S_{\text{vdw\_int}} + S_{\text{tors}}$$

$$S_{\text{hb}_{\text{ext}}} = \sum_{\text{prot}} \sum_{\text{lig}} [(E_{\text{da}} + E_{\text{ww}}) - (E_{\text{dw}} + E_{\text{aw}})];\ S_{\text{vdw\_ext}} = \sum_{\text{prot}} \sum_{\text{lig}} \left(\frac{A}{d_{ij}^{8}} - \frac{B}{d_{ij}^{4}} \right);$$

$$S_{\text{vdw}_{\text{int}}} = \sum_{\text{lig}} \left(\frac{C}{d_{ij}^{12}} - \frac{D}{d_{ij}^{6}} \right);\ S_{\text{tors}} = \sum \text{lig} \tfrac{1}{2} V_{ijkl} \left[1 + \frac{n_{ijkl}}{|n_{ijkl}|} \cos \left(|n_{ijkl}| \cdot \bar{\omega}_{ijkl} \right) \right]$$

GOLD (2005) [211]

$$\text{GOLD Fitness} = \sigma_o + \sum_{w} o(w)(\sigma_p + \sigma_i(w))$$

σ_{o} is the original score;
$o(w)$ is the occupancy of water w;
σ_{p} is the free energy penalty associated with the loss of rigid-body entropy;
$\sigma_{\text{i}}(w)$ represents all the interactions made by this water with protein, ligand, and other water molecules.

AutoDock (1998) [65]

$$\Delta G = \Delta G_{\text{vdW}} \sum_{i,j} \left(\frac{A_{i,j}}{r_{ij}^{12}} - \frac{B_{i,j}}{r_{ij}^{6}} \right) + \Delta G_{\text{hbond}} \sum_{i,j} E(t) \left(\frac{C_{i,j}}{r_{ij}^{12}} - \frac{D_{i,j}}{r_{ij}^{10}} + E_{\text{hbond}} \right)$$
$$+ \Delta G_{\text{elec}} \sum_{i,j} \frac{q_i q_j}{\varepsilon(r_{ij}) r_{i,j}} + \Delta G_{\text{tor}} N_{\text{tor}} + \Delta G_{\text{sol}} \sum_{i,j} S_i V_j \mathrm{e}^{(-r_{ij}^2/2\sigma^2)}$$

E_{hbond} is the estimated average energy of hydrogen bonding of water with a polar atom.

SDOCK (1999) [122]

$G_{\text{binding}} = \sigma_1 \Delta(\text{SA}_{\text{hp}}) + \beta \cdot \text{VDW} - \sigma_2 \Delta(\text{SA}) + G_{\text{POL}}$
$G_{\text{POL}} = G_{\text{screenedes}} + G_{\text{L desolve}} + G_{\text{R desolve}}$;
$\Delta(\text{SA}_{\text{hp}})$ and $\Delta(\text{SA})$ represent the change in the hydrophobic and total solvent-accessible surface area upon ligand binding;
$G_{\text{screenedes}}$ is the screened ligand–receptor electrostatic energy;
$G_{\text{L desolve}}$ is the partial desolvation energy of the ligand;
$G_{\text{R desolve}}$ is the partial desolvation energy of the receptor.

ICM scoring function (1999) [350,351]

$$E_{\text{score}} = \Delta E_{\text{IntFF}} + T\Delta S_{\text{Tor}} + \alpha_1 \Delta E_{\text{HBond}} + \alpha_2 \Delta E_{\text{HBDesol}} + \alpha_3 \Delta E_{\text{SolEl}} + \alpha_4 \Delta_{\text{HPhob}} + \alpha_5 Q_{\text{Size}}$$

ΔE_{IntFF} includes the vdW interaction and the internal force field energy of the ligand;
$T\Delta S_{\text{Tor}}$ represents the loss of ligand conformational entropy;
ΔE_{HBond} is the hydrogen-bonding term;
$\Delta E_{\text{HBDesol}}$ accounts for the disruption of hydrogen bonds with solvent upon ligand binding;
ΔE_{SolEl} is the solvation electrostatic energy change calculation determined with Poisson equation;
Δ_{HPhob} is the hydrophobic free energy proportional to the accessibile surface area;
Q_{Size} is a size correction term to avoid bias toward lager ligands.

SIE (2007) [377]

$$\Delta G_{\text{bind}}^{\text{calc}} = E_{\text{inter}}^{\text{C}} + \Delta G_{\text{bind}}^{\text{R}}\ \Delta G_{\text{bind}}^{\text{elec}} + E_{\text{inter}}^{\text{vdw}} + \Delta G_{\text{bind}}^{\text{npsol}}\ \Delta G_{\text{bind}}^{\text{np}}$$

$E_{\text{inter}}^{\text{C}}$ is the intermolecular Coulomb energy in the bound state;
$E_{\text{inter}}^{\text{vdw}}$ is the intermolecular van der Waals interaction energy in the bound state;
$\Delta G_{\text{bind}}^{\text{R}}$ electrostatic contribution of the solvation free energy to binding (change in the reaction field energy between the bound and free states);
$\Delta G_{\text{bind}}^{\text{npsol}}$ nonpolar contribution of the solvation free energy to binding (change in the nonelectrostatic solvation free energy between the bound and free states).

APPENDIX C. EMPIRICAL SCORING FUNCTIONS

SCORE1 (1994) [140]

$$\Delta G_{\text{binding}} = \Delta G_0 + \Delta G_{\text{hb}} \sum_{\text{h-bonds}} f(\Delta R, \Delta\alpha) + \Delta G_{\text{ionic}} \sum_{\text{ionic int.}} f(\Delta R, \Delta\alpha) + \Delta G_{\text{lipo}} |A_{\text{lipo}}| + \Delta G_{\text{rot}} NROT$$

$f(\Delta R, \Delta\alpha)$ is a penalty function accounting for large deviations of the hydrogen-bond geometry from ideality;
A_{lipo} is the lipophilic contact surface between the protein and the ligand;
NROT is the number of rotatable bonds.

SCORE2 (1998) [384]

$$\Delta G_{\text{binding}} = \Delta G_0 + \Delta G_{\text{hb}} \sum_{\text{h-bonds}} f(\Delta R, \Delta\alpha) \times f(N_{\text{neighb}}) \times f_{\text{pcs}} + \Delta G_{\text{ionic}} \sum_{\text{ionic int.}} f(\Delta R, \Delta\alpha) \times f(N_{\text{neighb}}) \times f_{\text{pcs}} + \Delta G_{\text{esrep}} N_{\text{repulsive contacts}} + \Delta G_{\text{lipo}} |A_{\text{lipo}}| + \Delta G_{\text{aro}} \sum_{\text{aro-int.}} f(R, \Theta) + \Delta G_{\text{lipo water}} \sum \text{unbound wate molecules} + \Delta G_{\text{rot}} NROT$$

$f(N_{\text{neighb}})$ is an empirical function which distinguish between convex and concave parts of the protein surface;
f_{pcs} is the polar contact surface per hydrogen bond.

Bohacek and McMartin's function (1994) [388]

$$\log(K_{\text{i}}) = 3.16 - 0.42(\text{PHOB}) - 0.39(\text{HBOND})$$

VALIDATE (1996) [387]

$$\Delta G_{\text{bind}}^{\text{L-R}} = \beta_1 E_{\text{vdW}}^{\text{L-R}} + \beta_2 E_{\text{coul}}^{\text{L-R}} + \beta_3 \text{SF} + \beta_4 H \log P + \beta_5 Nr_{\text{rot. bonds}} + \beta_6 \Delta H_{\text{bind}}^{\text{L}} + \beta_7 \text{CSA}_{\text{lipophilic}}^{\text{L-R}} + \beta_8 \text{CSA}_{\text{hydrophilic}}^{\text{L-R}}$$

$$E_{\text{vdW}}^{\text{L-R}} = \sum_i^L \sum_j^R \varepsilon_{ij} \left(\frac{1}{R_{12}} - \frac{2}{R_6} \right); \; E_{\text{coul}}^{\text{L-R}} = \frac{1}{4\pi\varepsilon_0} \sum_i^L \sum_j^R \frac{q_i q_j}{r_{ij}}; \; \text{SF} = \left(\sum_{ij} C_{ij}\right)/N;$$

$$Nr_{\text{rot. bonds}} = Nr_{\text{ntsb}} + \sum_i (n_i - 4); \; \Delta H_{bind}^{L} = \left| E_{\text{bind site}}^{\text{L}} - E_{\text{sol}}^{\text{L}} \right|; \; \text{CSA} = \sum_i^R \frac{(4\pi r_i^2 * \text{CP}_i)}{256}$$

β_1-β_8 are the fitted regression coefficients for the master equation;

$$C_{ij} = \begin{Bmatrix} 1, d_{ij} \leq |r_i + r_j \pm \varepsilon| \\ 0, d_{ij} > |r_i + r_j \pm \varepsilon| \end{Bmatrix}$$

CP_i is the number of contact points.

FlexX score (1996) [77]

$$\Delta G = \Delta G^{\circ} + \Delta G_{\text{rot}} \times N_{\text{rot}} + \Delta G_{\text{hb}} \sum_{\text{neutral H-bonds}} f(\Delta R, \Delta\alpha) + \Delta G_{\text{io}} \sum_{\text{ionic int.}} f(\Delta R, \Delta\alpha) + \Delta G_{\text{aro}} \sum_{\text{aro int.}} f(\Delta R, \Delta\alpha) + \Delta G_{\text{lipo}} \sum_{\text{lipo int.}} f^*(\Delta R)$$

where $f(\Delta R, \Delta\alpha)$ is scaling function penalizing deviation from the ideal geometry;

$$f^*(\Delta R) = \begin{cases} 0 & \Delta R > 0.6\,\text{Å} \\ 1 - \dfrac{\Delta R - 0.2}{0.4} & 0.2\text{Å} < \Delta R \leq 0.6\text{Å} \\ 1 & -0.2\text{Å} < \Delta R \leq 0.2\text{Å} \\ 1 - \dfrac{-\Delta R - 0.2}{0.4} & -0.6\text{Å} < \Delta R \leq -0.2\text{Å} \\ \dfrac{\Delta R + 0.6}{0.2} & \Delta R \leq -0.6\text{Å} \end{cases}$$

APPENDIX (*Continued*)

Jain's function (1996) [144]

$$F = \sum_{i,j} f_0(d(i,j)) + \sum_{i,j} f_1(d(i,j),i,j) + \sum_{i,j} f_2(d(i,j),i,j) + (l_5 \cdot \text{phbe})$$

$$+ (l_6 \cdot \text{lhbe}) + (l_7 \cdot \text{n}_{\text{rot}}) + (l_8 \cdot \log(\text{mol. weight}))$$

$$f_0(x) = l_0 g(x, n_0, n_1) + l_1 s(x, n_2 + n_1); f_1(x,i,j) = f_{1a}(x) f_{1b}(i,j)(1 + n_6 c_i)(1 + n_6 c_j)$$

$$f_2(x) = l_4 s(x, n_7) f_{1a}(x) f_{1b}(i,j)(1 + n_6 c_i)(1 + n_6 c_j);$$

$$g(x,\mu,\sigma) = e^{-(x+\mu)^2/\sigma}; \; s(x,\mu) = \frac{1}{(1 + e^{10(x+\mu)})}$$

f_0 is the hydrophobic term, f_1 the polar contribution and f_2 the repulsive term;
$(l_5 \cdot \text{phbe})$ and $(l_6 \cdot \text{lhbe})$ represent the solvation contribution;
$(l_7 \cdot \text{n\_rot})$ and $(l_8 \cdot \log(\text{mol. weight}))$ are entropic terms.

ChemScore (1997) [274]

$$\Delta G_{\text{bind}} = \Delta G_0 + \Delta G_{\text{hbond}} \sum_{il} g_1(\Delta r) g_2(\Delta \alpha) + \Delta G_{\text{metal}} \sum_{aM} f(r_{aM})$$

$$+ \Delta G_{\text{lipo}} \sum_{1L} f(r_{1L}) + \Delta G_{\text{rotor}} H_{\text{rot}}$$

$$H_{\text{rot}} = 1 + (1 - 1/N_{\text{rot}}) \sum_r (P_{\text{nl}}(r) + P'_{\text{nl}}(r))/2$$

P_{nl} and P'_{nl} are the percentages on nonlipophilic heavy atoms on either side of the rotatable bond.

ChemScore GOLD (2003) [83]

$$\Delta G'_{\text{binding}} = \Delta G_{\text{binding}} + E_{\text{clash}} + E_{\text{int}} + E_{\text{cov}}$$

$$E_{\text{clash}} = \sum \varepsilon_{\text{clash}}(r, r_{\text{clash}}); \; E_{\text{tors}} = \sum_{\text{RB}} \varepsilon_{\text{tors}}(\Theta_{\text{RB}}) + \sum_{\text{RC}} \sum_{\text{RCB}} \varepsilon_{\text{tors}}(\Theta_{\text{RCB}});$$

$$E_{\text{cov}} = \sum_{\text{CB}} \varepsilon_{\text{tors}}(\Theta_{\text{CB}}) + C_{\text{cov}} \sum_{\text{BA}} k_{\text{BA}} (\varphi_{\text{BA}} - \varphi_{0,\text{BA}})^2$$

Horvath's function (1997) [413]

$$E_{\text{S-L}} = E^{\text{rep}}_{\text{S-L}} + \omega E^{\text{att}}_{\text{S-L}} + \chi E^{\text{Coul}}_{\text{S-L}} + \lambda E^{\text{des}}_{\text{S-L}} + \eta E^{\text{hphob}}_{\text{S-L}}$$

$$E^{\text{rep}}_{\text{S-L}} = \sum_{i=1}^{\text{ligand atoms}} A_i V^{\text{rep}}_{\text{vdW}}(i); \; E^{\text{att}}_{\text{S-L}} = \sum_{i=1}^{\text{ligand atoms}} B_i V^{\text{att}}_{\text{vdW}}(i); \; E^{\text{Coul}}_{\text{S-L}} = \sum_{i=1}^{\text{ligand atoms}} Q_i V_{\text{Coul}}(i);$$

$$E^{\text{des}}_{\text{S-L}} = \sum_{i=1}^{\text{ligand atoms}} Q_i^2 V_{\text{des}}^{\text{sit}\rightarrow\text{lig}}(i) + v_i V_{\text{des}}^{\text{lig}\rightarrow\text{sit}}(i); \; E^{\text{hphob}}_{\text{S-L}} = \sum_{i=1}^{\text{ligand atoms}} \delta(i) V_{\text{hphob}}(i)$$

SCORE (1998) [380]

$$\text{p}K_{\text{d}} = K_0 + K_{\text{vdw}} + K_{\text{metal}} + K_{\text{hbond}} + K_{\text{desolvation}} + K_{\text{deformation}}$$

$$K_{\text{vdw}} = \sum_i \sum_j \text{VB}(d_{ij}); \; K_{\text{metal}} = \sum_i \sum_j \text{MB}(d_{ij});$$

$$K_{\text{hbond}} = K_{\text{SHB}} + K_{\text{MHB}} + K_{\text{WHB}} + K_{\text{SWH}} + K_{\text{MWH}} + K_{\text{WWH}}$$

$$K_{\text{HM}} = \sum_i F_i \times \text{HM}_i; \; K_{\text{RT}} = \sum_i 0.5 \times \text{RT}_i$$

VB, MB, SHB, MHB, WHB, SWH, MWH, WWH are distance function;
HM_i is an indicator of hydrophobic matching and F_i is the atomic hydrophobic scale;
RT_i is the number of rotors.

SLIDE scoring function (1998) [154]

$$\text{SCORE(P, L)} = A \cdot \text{HPHOB(P, L)} + B \cdot \text{HBONDS(P, L)}$$

$$\text{HPHOB(P, L)} = \sum_{l_i \in L, \#P_i > 0} \frac{\text{avg}\{h'(l_i), \bar{h}(P_i)\}}{\max\{\text{abs}(h'(l_i) - \bar{h}(P_i)), 32\}}; \; h'(l_i) = \max\{317 - h'(l_i), 0\};$$

$$\bar{h}(P_i) = \max\left\{(317 - \frac{1}{\#P_i} \cdot \sum_{p_j \in P_i} h(p_j)), 0\right\}$$

APPENDIX (*Continued*)

Fresno (1999) [417]	$\Delta G_{\text{binding}} = K + \alpha(\text{HB}) + \beta(\text{LIPO}) + \gamma(\text{ROT}) + \delta(\text{BP}) + \gamma(\text{DESOLV})$ $\text{HB} = \sum_{\text{HB}} g_1(\Delta r)g_2(\Delta\alpha)$; $\text{LIPO} = \sum_{1,L} f(r_{1L})$; $\text{BP} = \sum_{1,P} f(r_{1P}) + \sum_{p,L} f(r_{pL})$; $\text{ROT} = 1 + \left(\frac{1-1}{N_{\text{rot}}}\right)\sum_{\text{r}} \frac{(P_{\text{p}}(r) + P'_{\text{p}}(r))}{2}$; $\text{ROT} = N_{\text{rot}}$; $\text{DESOLV} = \Delta G^0_{\text{reac}} = \Delta G^0_{\text{reac(P-L)}} - \Delta G^0_{\text{reac(P)}} - \Delta G^0_{\text{reac(L)}}$
Kasper's scoring function (2000) [425]	$\Delta G = \Delta G_{\text{np}} + \Delta G_{\text{el}} + T\Delta S_{\text{conf}} \Delta G_{\text{np}} = \gamma_{\text{aw}} \cdot \Delta A$; $\Delta G_{\text{el}} = \Delta G_{\text{coul}} + \Delta G_{\text{solv}}$ ΔA is the change in solvent-accessible surface; γ_{aw} is the microscopic surface tension associated with the transfer from liquid alkane to water.
ScreenScore (2001) [398]	$\Delta G_{\text{bind}} = F_{\text{match}} + 0.07(F_{\text{lipo}} + F_{\text{ambig}}) + 0.3F_{\text{PLP}} + 1.6n_{\text{rot}}$
X-CSCORE (2002) [414]	$X\text{-CSCORE} = \frac{(\text{p}K_{\text{d},1} + \text{p}K_{\text{d},2} + \text{p}K_{\text{d},3})}{3}$ $\text{p}K_{\text{d},1} = C_{0,1} + C_{\text{VDW},1} \times \text{VDW} + C_{\text{H-bond},1} \times HB + C_{\text{rotor},1} \times \text{RT} + C_{\text{hydrophobic},1} \times \text{HS}$ $\text{p}K_{\text{d},2} = C_{0,2} + C_{\text{VDW},2} \times \text{VDW} + C_{\text{H-bond},2} \times \text{HB} + C_{\text{rotor},2} \times \text{RT} + C_{\text{hydrophobic},2} \times \text{HC}$ $\text{p}K_{\text{d},3} = C_{0,3} + C_{\text{VDW},3} \times \text{VDW} + C_{\text{H-bond},3} \times \text{HB} + C_{\text{rotor},3} \times \text{RT} + C_{\text{hydrophobic},3} \times \text{HM}$ $\text{VDW} = \sum_{i}^{\text{ligand}} \sum_{j}^{\text{protein}} \left[\left(\frac{d_{ij,0}}{d_{ij}}\right)^8 - 2 \times \left(\frac{d_{ij,0}}{d_{ij}}\right)^4\right]$; $\text{HB} = \sum_{i}^{\text{ligand}} \sum_{j}^{\text{protein}} \text{HB}_{ij}$ $\text{RT} = \sum_{i}^{\text{ligand}} \text{RT}_i$; $\text{HS} = \sum_{i}^{\text{ligand}} \text{SAS}_i$; $\text{HC} = \sum_{i}^{\text{ligand}} \sum_{j}^{\text{protein}} f(d_{ij})$; $\text{HM} = \sum_{i}^{\text{ligand}} \log P_i \times \text{HM}_i$
FRED (2003) [101]	$\text{GS} = \sum_{i \in A} \left[\left(\sum_{j \in B} N_{ij}\right)\left(\exp\left(-\gamma \sum_{j \in B} V_{ij}\right)\right)\right]$ $N_{ij}(d_{ij}) = \left[1 + \xi(d_{ij}^2 - D_{ij}^2)\right]\exp(+\xi(d_{ij}^2 - D_{ij}^2))$; $V_{ij}(d_{ij}) = \left(\frac{16\kappa^3}{9\pi}\right)\left(\frac{\pi R_i^2 R_j^2}{\kappa R_i^2 + \kappa R_j^2}\right)^{3/2} \exp\left(\frac{\kappa}{R_i^2 + R_j^2} d_{i,j}^2\right)$
GEMDOCK (2004) [230]	$E_{\text{tot}} = E_{\text{inter}} + E_{\text{intra}} + E_{\text{penal}}$ $E_{\text{inter}} = \sum_{i=1}^{\text{lig}} \sum_{j=1}^{\text{pro}} \left[F(r_{ij}^{Bij}) + 332.0 \frac{q_i q_j}{4r_{ij}}\right]$; $E_{\text{intra}} = \sum_{i=1}^{\text{lig}} \sum_{j=i+2}^{\text{lig}} F(r_{ij}^{Bij}) + \sum_{k=1}^{\text{dihed}} A[1 - \cos(m\theta_k - \theta_0)]$
GEMDOCK (2005) [231]	$E_{\text{tot}} = E_{\text{bind}}^{\text{MW}} + E_{\text{pharma}} + E_{\text{ligpre}}$ $E_{\text{bind}}^{\text{MW}} = \frac{E_{\text{bind}}}{(\text{NA}_l)^K}$; $E_{\text{pharma}} = \sum_{i=1}^{\text{lig}} \sum_{j=1}^{\text{hs}} \text{CW}(B_{ij})F(r_{ij}^{Bij})$; $E_{\text{ligpre}} = \text{LP}_{\text{elec}} + \text{LP}_{\text{hb}}$ NA_l is the total number of heavy atoms in a screened ligand; $\text{CW}(B_{ij})$ is a pharmacological weight function; $F(r_{ij}^{Bij})$ is a simple atomic pairwise potential function; LP_{elec} and LP_{hb} are the penalties for the electrostatic and the hydrophilic constraints.

APPENDIX (*Continued*)

GlideScore 2.5 SP (2004) [198]	$\Delta G_{\text{bind}} = C_{\text{lipo-lipo}} \sum f(r_{\text{lr}}) + C_{\text{hbond-neut-neut}} \sum g(\Delta r)h(\Delta\alpha)$ $+ C_{\text{hbond-neut-charged}} \sum g(\Delta r)h(\Delta\alpha) + C_{\text{hbond-charged-charged}} \sum g(\Delta r)h(\Delta\alpha)$ $+ C_{\text{max-metal-ion}} \sum f(r_{\text{lm}}) + C_{\text{rotb}} H_{\text{rotb}} + C_{\text{polar-phob}} V_{\text{polar-phob}}$ $+ C_{\text{coul}} E_{\text{coul}} + C_{\text{vdW}} E_{\text{vdW}} + \text{solvation terms}$
GlideScore 4.0 XP (2004) [435]	$\text{XP GlideScore} = E_{\text{coul}} + E_{\text{vdW}} + E_{\text{bind}} + E_{\text{penalty}}$ $E_{\text{bind}} = E_{\text{hyd\_enclosure}} + E_{\text{hb\_nn\_motif}} + E_{\text{hb\_cc\_motif}} + E_{\text{PI}} + E_{\text{hb\_pair}} + E_{\text{phobic\_pair}};$ $E_{\text{penalty}} = E_{\text{desolv}} + E_{\text{ligand\_strain}}$
Ligscore (2005) [436]	*Ligscore1* $\text{p}K_i = C - \beta_1 \langle E_{\text{vdW}} \rangle + \beta_2 \text{Cpos\_tot} - \beta_3 \text{TotPol}^2$ *Ligscore2* $\text{p}K_i = C - \beta_1 \langle E_{\text{vdW}} \rangle + \beta_2 \text{Cpos\_tot-}\beta_3 \left(\text{SolvPlty\_lig}^2 + \text{SolvPlty\_prot}^2\right)$ $\langle E_{\text{vdW}} \rangle = \sum_{\substack{i \in \text{ligand} \\ j \in \text{protein}}} \frac{\varepsilon_{ij} \left[2\left[\left(\frac{r_{ij}^*}{r_{ij}}\right)^9\right] - 3\left[\left(\frac{r_{ij}^*}{r_{ij}}\right)^6\right] \right]}{\sqrt[2]{\varepsilon_j} \left(\frac{\sqrt{\text{r}_j^*}}{\text{r}_{ij}}\right)^9 \alpha\left(1 + \beta \text{r}_{ij}^2\right) + 1}$ Cpos_tot is the total surface area of the ligand involved in attractive polar interactions with the protein; $\text{TotPol}^2 = \text{Cpos\_tot}^2 + \text{Cneg\_tot}^2;$ $\text{SolvPlty\_lig}^2 = (\text{Bury\_tot\_lig} - \text{Bury\_lip\_lig})^2;$ $\text{SolvPlty\_prot}^2 = (\text{Cpos\_tot} + \text{Cneg\_tot} + \text{Bury\_lip\_lig} - \text{Clip\_lip})^2$
SCORE 3.0 (2006) [278]	$\Delta G_{\text{binding}} = \Delta G_0 + \Delta G_{\text{vdw}} + \Delta G_{\text{hbond}} + \Delta G_{\text{metal}} + \Delta G_{\text{solvation}} + \Delta G_{\text{rot}}$
eHiTS (2007) [106]	The SF_E empirical function is composed by the following terms: • hydrogen-bonding term (D, H, A, and Lp represent the position of donor, hydrogen, and acceptor atoms and lone electron pair, respectively): $E_{\text{H-bond}} = E_{\max} f_{\text{dist}}(\lVert \text{H} \rightarrow \text{A} \rVert) \cos(\angle(\text{A} \rightarrow \text{Lp}, \text{D} \rightarrow \text{H}));$ • hydrophobicity term (S is the set of surface points with flags H or h): $E_{Lipo} = \sum_{p \in S_{\text{lig}}} E_{\max} f_{\text{dist}}(p, q_p), \text{ where } q_p \in S_{\text{rec}} : d(p, q_p) = \min_{q \in S_{\text{rec}}} d(p, q);$ • aromatic π-stacking (similar to E_{Lipo}, but applied on surface points with flags E,R and O with different $E_{\max}$ for each type of pair); • electrostatic potential (Coulomb formula); • van der Waals contact energy (Lennard-Jones formula); • metal ion interactions (distance and angle dependency similar to the hydrogen-bonding term); • penalty for incompatible contacts (e.g., polar–hydrophobic, same charge, evaluated similarly to E_{Lipo}, but for surface points with incompatible flags); • exposed surface atoms are scored against solvent properties; • intramolecular interactions are taken into account: ligand (conformation score) and receptor (protonation, competing interactions reduce the score toward ligand interactions).

APPENDIX D. KNOWLEDGE-BASED SCORING FUNCTIONS

Verkhivker's Potential (1995) [468]

$$\Delta G_{\text{bind}}^{\text{L-R}} = \Delta G^{\text{L-R\_interaction}} + \Delta G^{\text{L-wat\_interaction}} + \Delta G^{\text{R-wat\_interaction}} + \Delta G^{\text{nonpolar\_dedsolvation}} + \Delta G^{\text{polar\_desolvation}} + \Delta G^{\text{L\_isomerization}} + \Delta G^{\text{L-R\_rot,translation}} + \Delta G^{\text{wat\_rot,translation}} - T\Delta S^{\text{L\_conformation}} - T\Delta S^{\text{R\_conformation}}$$

$\Delta G_{\text{bind}}^{\text{L-R}}$ is the subdivided in $\Delta G^{\text{n-n}}$ (the interaction between non polar groups only) and $\Delta G^{\text{n-p,p-p,p-n}}$ (the interaction between polar and non polar groups).

Wallqvist's Potential (1995) [469]

$$\Delta G_{\text{pred}} = -\sum_{i\in A}\sum_{j\in B}\alpha_{ij}A_{ij} + \beta = -\sum_{i\in A}\sum_{j\in B}(\gamma + \delta \ln P_{ij})A_{ij} + \beta$$

$$P_{ij} = \frac{F_{ij}}{F_i F_j} = \frac{\frac{A_{ij}}{A_{\text{tot}}}}{\left(\frac{A_i}{A_{tot}}\right)\left(\frac{A_j}{A_{tot}}\right)}$$

P_{ij} is the ratio of the total interfacial area contributed by each atom pair F_{ij}, normalized by the product of the fractional contribution of each atom in the pair F_i and F_j;
α_{ij} is an effective binding parameter dependent on atom types i and j;
A_{ij} is the mutual surface jointly buried by atom types i and j in the complex;
β is a positive constant.

SMoG (1996) [471]

$$F = \sum_r\sum_p\sum_l F(r,\sigma p,\sigma l)\Delta(r,p,l)$$

$$F(r,\sigma p,\sigma l) = -\ln\left[\frac{N(r,\sigma p,\sigma l)}{C(r)\times\Omega(r,\sigma p,\sigma l)\times\sum_{\sigma p}\sum_{\sigma l}N(r,\sigma p,\sigma l)}\right]$$

$N(r,\sigma p,\sigma l)$ is the number of contacts between atoms of given atom types;
$C(r)\times\Omega(r,\sigma p,\sigma l)$ is the probability of the contacts in the reference state for a given distance interval;
$C(r)$ is a normalized constant;
$\Omega(r,\sigma p,\sigma l)$ is the term accounting for composition of the database, that is, number of atoms of the σp and σl types participating in the protein–ligand interaction.

PMF (1999) [124]

$$\text{PMF\_score} = \sum_{kl} Aij(r)$$

$$r < r^{ij}_{\text{cut-off}}$$

$$Aij(r) = -k\text{B}T\ln\left[f^{j}_{\text{Vol\_corr}}(r)\frac{\rho^{ij}_{\text{seg}}(r)}{\rho^{ij}_{\text{bulk}}}\right]$$

$k\text{B}$ is the Boltzmann constant;
$f^{j}_{\text{Vol\_corr}}(r)$ is the ligand volume correction factor;

$\frac{\rho^{ij}_{\text{seg}}(r)}{\rho^{ij}_{\text{bulk}}}$ is the pair correlation function of a protein atom of type i paired with a ligand atom of type j.

BLEEP (1999) [479]

$$\Delta E^{\text{ab}}(r) = kT\ln\left[1 + m^{\text{ab}}\sigma\right] - kT\ln\left[1 + m^{\text{ab}}\sigma\left\{\frac{g^{\text{ab}}(r)}{f(r)}\right\}\right]$$

m^{ab} is the total number of contacts between atom types a and b;
$g^{\text{ab}}(r)$ is the proportion of these contacts occurring at distance r;
$f(r)$ is the proportion of all contacts for all types occurring at distance r;
k is the Boltzmann constant.

APPENDIX (*Continued*)

DrugScore (2000) [481]

$$\Delta W_{I,J} = \sum_{ki}\sum_{kj} \Delta Wi,j(r) + (1-\gamma) \times \left[\sum_{ki} \Delta Wi(SAS, SAS0) + \sum_{kj} \Delta Wj(SAS, SAS0)\right]$$

$$\Delta Wi,j(r) = -\ln \frac{gi,j(r)}{g(r)};\ g(r) = \frac{\sum_i \sum_j gi,j(r)}{ij};\ gi,j(r) = \frac{\frac{Ni,j(r)}{4\pi r^2}}{\sum_r \frac{Ni,j(r)}{4\pi r^2}}$$

$\Delta Wi,j(r)$ are specific interactions between atoms of type i and j, located at a distance r;
ΔWi and ΔWj are nonpolar surface-dependent singlet potential for atoms of type i and j
$gi,j(r)$ is the normalized radial pair distribution function;
$g(r)$ is the normalized mean radial distribution function.

DFIRE (2003) [488]

$$\Delta G_{bind} = \frac{1}{2} \sum_{i,j}^{\text{interface}} \bar{u}(i,j,r_{ij})$$

$$\bar{u}(i,j,r) = \begin{cases} -RT \ln \dfrac{N_{\text{obs}}(i,j,r)}{\left(\dfrac{r}{r_{\text{cut}}}\right)^{\alpha} \left(\dfrac{\Delta r}{\Delta r_{\text{cut}}}\right) N_{\text{obs}}(i,j,r_{\text{cut}})} & r < r_{\text{cut}} \\ 0 & r \geq r_{\text{cut}} \end{cases}$$

$\bar{u}(i,j,r_{ij})$ is the atom–atom mean force potential;
$N_{\text{obs}}(i,j,r)$ is the number of pairs within the distance shell $r(r - \Delta r/2$ to $r + \Delta r/2)$ in a given database;
$r_{\text{cut}} = 14.5$ Å;
$\alpha = 1.61$.

ITScore (2006) [496]

$$\text{energy score} = \sum_{\text{P–L atom pair}} u_{ij}(r);\ u(r) = -k_B T \ln \left(\frac{\rho(r)}{\rho^*(r)}\right)$$

$u_{ij}(r)$ is the atom–atom pair potential;
$\rho(r)$ is the number density of the protein–ligand atom pair at distance r;
$\rho^*(r)$ is the atom pair density in a "reference" state where the interatomic interactions are zero.

M-Score (2006) [498]

$$F_{\text{SC}} = -\ln \frac{g_{i,j}(r)}{g(r)};\ g(r) = \frac{\sum_{i,j} g_{i,j}(r)}{i^* j};\ g_{i,j}(r) = \frac{\frac{N^{\text{w}}_{i,j}(r)}{4\pi r^2}}{\sum_{k,l} \frac{N^{\text{w}}_{k,j}(r)}{4\pi r^2}};$$

$$N^{\text{w}}_{i,j}(r) = \sum_{\mu}\sum_{\nu} \left[\sum_n p_{\mu}(u_n) \cdot \delta\left(\vec{r}^{\,i}_{\mu} - \vec{r}^{\,j}_{\nu} + u_n - r\right)\right]$$

$N^{\text{w}}_{i,j}$ is the weighted occurrence with a distance from r to $r + dr$ between atom pairs i and j;
r^i_{μ} is the mean position of protein atom μ (type i);
u_n is the deviation of protein atom μ from its mean position;
n is the number of atomic position of protein atom μ included;
$p_{\mu}(u_n)$ is the probabilità of observing protein atom μ at u_n;
r^j_{ν} is the mean position of ligand atom ν (type j).

REFERENCES

1. Hawkins PC, Warren GL, Skillman AG, Nicholls A. How to do an evaluation: pitfalls and traps. J Comput Aided Mol Des 2008;22: 179–190.
2. Spyrakis F, Amadasi A, Fornabaio M, Abraham DJ, Mozzarelli A, Kellogg GE, Cozzini P. The consequences of scoring docked ligand conformations using free energy correlations. Eur J Med Chem 2007;42:921–933.
3. Atkins PW. General Chemistry. Bologna: Zanichelli; 1992.
4. Jeffrey GA, Saenger W. Hydrogen Bonding in Biological Structures. Berlin: Springer-Verlag; 1994.
5. Nishio M. The CH-Pi Interactions: Evidence, Nature and Consequences. New York: Wiley-VCH; 1998.
6. Gohlke H, Klebe G. Approaches to the description and prediction of the binding affinity of small-molecule ligands to macromolecular receptors. Angew Chem Int Ed Engl 2002;41: 2644–2676.
7. Pearlman DA, Kollman PA. The calculated free energy effects of 5-methyl cytosine on the B to Z transition in DNA. Biopolymers 1990;29: 1193–1209.
8. Tame JR. Scoring functions: a view from the bench. J Comput Aided Mol Des 1999;13: 99–108.
9. Muegge I. Effect of ligand volume correction on PMF scoring. J Comput Chem 2001;22: 418–425.
10. Dill KA. Additivity principles in biochemistry. J Biol Chem 1997;272:701–704.
11. Salom D, Lodowski DT, Stenkamp RE, Le Trong I, Golczak M, Jastrzebska B, Harris T, Ballesteros JA, Palczewski K. Crystal structure of a photoactivated deprotonated intermediate of rhodopsin. Proc Natl Acad Sci USA 2006;103:16123–16128.
12. Warne T, Serrano-Vega MJ, Baker JG, Moukhametzianov R, Edwards PC, Henderson R, Leslie AG, Tate CG, Schertler GF. Structure of a beta1-adrenergic G-protein-coupled receptor. Nature 2008;454:486–491.
13. Cherezov V, Rosenbaum DM, Hanson MA, Rasmussen SG, Thian FS, Kobilka TS, Choi HJ, Kuhn P, Weis WI, Kobilka BK, et al. High-resolution crystal structure of an engineered human beta2-adrenergic G protein-coupled receptor. Science 2007;318:1258–1265.
14. Davis AM, Teague SJ, Kleywegt GJ. Application and limitations of X-ray crystallographic data in structure-based ligand and drug design. Angew Chem Int Ed Engl 2003; 42: 2718–2736.
15. DePristo MA, de Bakker PI, Blundell TL. Heterogeneity and inaccuracy in protein structures solved by X-ray crystallography. Structure 2004;12:831–838.
16. Branden CI, Jones TA. Between objectivity and subjectivity. Nature 1990;343:687–689.
17. Mohan V, Gibbs AC, Cummings MD, Jaeger EP, DesJarlais RL. Docking: successes and challenges. Curr Pharm Des 2005;11:323–333.
18. Giacovazzo C, Monaco HL, Viterbo D, Scordari F, Gilli G, Zanotti G, Catti M. Fundamentals of Crystallography. Giacovazzo C, editor. New York: Oxford University Press; 1992.
19. Brunger AT. Free *R* value: a novel statistical quantity for assessing the accuracy of crystal structures. Nature 1992;355:472–475.
20. Wang R, Fang X, Lu Y, Wang S. The PDBbind database: collection of binding affinities for protein–ligand complexes with known three-dimensional structures. J Med Chem 2004;47: 2977–2980.
21. Wang R, Fang X, Lu Y, Yang CY, Wang S. The PDBbind database: methodologies and updates. J Med Chem 2005;48:4111–4119.
22. Chen X, Lin Y, Gilson MK. The binding database: overview and user's guide. Biopolymers 2001;61:127–141.
23. Hu L, Benson ML, Smith RD, Lerner MG, Carlson HA. Binding MOAD (mother of all databases). Proteins 2005;60:333–340.
24. Nissink JW, Murray C, Hartshorn M, Verdonk ML, Cole JC, Taylor R. A new test set for validating predictions of protein–ligand interaction. Proteins 2002;49:457–471.
25. Cozzini P, Fornabaio M, Marabotti A, Abraham DJ, Kellogg GE. Free energy of ligand binding to protein: evaluation of the solvent contribution of water molecules by computational methods. Curr Med Chem 2004;11: 1345–1359.
26. Schlunegger MP, Grutter MG. An unusual feature revealed by the crystal structure at 2.2 Å resolution of human transforming growth factor-beta 2. Nature 1992;358: 430–434.
27. Böhm HJ. New computational approaches to predict protein–ligand interactions. In: Structure-Based Ligand Design. Weinheim: Wiley-VCH; 1998.

28. Roe DC, Kuntz ID. BUILDER v.2: improving the chemistry of a *de novo* design strategy. J Comput Aided Mol Des 1995;9:269–282.
29. Pearlman DA, Murcko MA. CONCEPTS: new dynamic algorithm for *de novo* drug suggestion. J Comput Chem 1993;14:1184–1193.
30. Pearlman DA, Murcko MA. CONCERTS: dynamic connection of fragments as an approach to *de novo* ligand design. J Med Chem 1996;39:1651–1663.
31. Stultz CM, Karplus M. Dynamic ligand design and combinatorial optimization: designing inhibitors to endothiapepsin. Proteins 2000;40: 258–289.
32. Rotstein SH, Murcko MA. GenStar: a method for *de novo* drug design. J Comput Aided Mol Des 1993;7:23–43.
33. Rotstein SH, Murcko MA. GroupBuild: a fragment-based method for *de novo* drug design. J Med Chem 1993;36:1700–1710.
34. Moon JB, Howe WJ. Computer design of bioactive molecules: a method for receptor-based *de novo* ligand design. Proteins 1991;11:314–328.
35. Eisen MB, Wiley DC, Karplus M, Hubbard RE. HOOK: a program for finding novel molecular architectures that satisfy the chemical and steric requirements of a macromolecule binding site. Proteins 1994;19:199–221.
36. Nishibata Y, Itai A. Confirmation of usefulness of a structure construction program based on three-dimensional receptor structure for rational lead generation. J Med Chem 1993;36: 2921–2928.
37. Böhm HJ. The computer program LUDI: a new method for the *de novo* design of enzyme inhibitors. J Comput Aided Mol Des 1992; 6:61–78.
38. Gehlhaar DK, Moerder KE, Zichi D, Sherman CJ, Ogden RC, Freer ST. *De novo* design of enzyme inhibitors by Monte Carlo ligand generation. J Med Chem 1995;38: 466–472.
39. DeWitte RS, Ishchenko AV, Shakhnovich EI. SMoG: *de novo* design method based on simple, fast, and accurate free energy estimates. 2. Case studies in molecular design. J Am Chem Soc 1997;119:4608–4617.
40. Gillet V, Johnson AP, Mata P, Sike S, Williams P. SPROUT: a program for structure generation. J Comput Aided Mol Des 1993;7:127–153.
41. Murcko MA. Recent advances in ligand design methods. In: Reviews in Computational Chemistry. New York: John Wiley & Sons, Inc.; 1997. p 1–66.
42. Clark D, Murray CW, Li J. Current issues in *de novo* molecular design. In: Reviews in Computational Chemistry. New York: John Wiley & Sons, Inc.; 1997. p 67–125.
43. Blaney J, Dixon J. A good ligand is hard to find: automated docking methods. Perspect Drug Discov Des 1993;1:301–319.
44. Bamborough P, Cohen FE. Modeling protein–-ligand complexes. Curr Opin Struct Biol 1996;6:236–241.
45. Taylor RD, Jewsbury PJ, Essex JW. A review of protein–small molecule docking methods. J Comput Aided Mol Des 2002;16:151–166.
46. Sotriffer CA, Klebe G, Stahl M, Bohm HJ. Docking and scoring functions/virtual screening. In: Burger's Medicinal Chemistry and Drug Discovery. John Wiley & Sons, Inc.; 2003. p 281–331.
47. Muegge I, Rarey I. Small molecule docking and scoring. In: Reviews Computational Chemistry. New York: John Wiley & Sons, Inc.; 2001. p 1–61.
48. Brooijmans N, Kuntz ID. Molecular recognition and docking algorithms. Annu Rev Biophys Biomol Struct 2003;32:335–373.
49. Henzler-Wildman KA, Lei M, Thai V, Kerns SJ, Karplus M, Kern D. A hierarchy of timescales in protein dynamics is linked to enzyme catalysis. Nature 2007;450:913–916.
50. Hellal-Levy C, Fagart J, Souque A, Wurtz JM, Moras D, Rafestin-Oblin ME. Crucial role of the H11–H12 loop in stabilizing the active conformation of the human mineralocorticoid receptor. Mol Endocrinol 2000;14:1210–1221.
51. Hu X, Lazar MA. Transcriptional repression by nuclear hormone receptors. Trends Endocrinol Metab 2000;11:6–10.
52. Brady AE, Limbird LE. G protein-coupled receptor interacting proteins: emerging roles in localization and signal transduction. Cell Signal 2002;14:297–309.
53. Cozzini P, Kellogg GE, Spyrakis F, Abraham DJ, Costantino G, Emerson A, Fanelli F, Gohlke H, Kuhn LA, Morris GM, et al. Target flexibility: an emerging consideration in drug discovery and design. J Med Chem 2008;51:6237–6255.
54. Levinthal C, Wodak SJ, Kahn P, Dadivanian AK. Hemoglobin interaction in sickle cell fibers. I. Theoretical approaches to the molecular contacts. Proc Natl Acad Sci USA 1975;72:1330–1334.
55. Salemme FR. A hypothetical structure for an intermolecular electron transfer complex of

cytochromes c and b5. J Mol Biol 1976;102: 563–568.

56. Wodak SJ, De Crombrugghe M, Janin J. Computer studies of interactions between macromolecules. Prog Biophys Mol Biol 1987;49:29–63.
57. Gabb HA, Jackson RM, Sternberg MJ. Modelling protein docking using shape complementarity, electrostatics and biochemical information. J Mol Biol 1997;272:106–120.
58. Jiang F, Kim SH. "Soft docking": matching of molecular surface cubes. J Mol Biol 1991;219: 79–102.
59. Kuntz ID, Blaney JM, Oatley SJ, Langridge R, Ferrin TE. A geometric approach to macromolecule–ligand interactions. J Mol Biol 1982; 161:269–288.
60. Norel R, Lin SL, Wolfson HJ, Nussinov R. Shape complementarity at protein–protein interfaces. Biopolymers 1994;34:933–940.
61. Goodford PJ. A computational procedure for determining energetically favorable binding sites on biologically important macromolecules. J Med Chem 1985;28:849–857.
62. Busetta B, Tickle I, Blundell T. DOCKER, an interactive program for simulating protein receptor and substrate interactions. J Appl Crystallogr 1983;16:432–437.
63. Goodsell DS, Olson AJ. Automated docking of substrates to proteins by simulated annealing. Proteins 1990;8:195–202.
64. Morris GM, Goodsell DS, Huey R, Olson AJ. Distributed automated docking of flexible ligands to proteins: parallel applications of AutoDock 2.4. J Comput Aided Mol Des 1996;10:293–304.
65. Morris GM, Goodsell DS, Halliday RS, Huey R, Hart WE, Belew RK, Olson AJ. Automated docking using a Lamarckian genetic algorithm and an empirical binding free energy function. J Comput Chem 1998;19:1639–1662.
66. Meng E, Shoichet B, Kuntz I. Automated docking with grid-based energy evaluation. J Comput Chem 1992;13:505–524.
67. Totrov M, Abagyan R. Flexible protein–ligand docking by global energy optimization in internal coordinates. Proteins 1997; (Suppl 1): 215–220.
68. Abagyan R, Totrov M, Kuznetsov D. ICM-A new method for protein modeling and design: applications to docking and structure prediction from the distorted native conformation. J Comput Chem 1994;15:488–506.
69. Trosset JY, Scheraga HA. Flexible docking simulations: scaled collective variable Monte Carlo minimization approach using Bezier splines, and comparison with a standard Monte Carlo algorithm. J Comput Chem 1999;20:244–252.
70. Trosset JY, Scheraga HA. Reaching the global minimum in docking simulations: a Monte Carlo energy minimization approach using Bezier splines. Proc Natl Acad Sci USA 1998;95:8011–8015.
71. Platzer KE, Momany FA, Scheraga HA. Conformational energy calculations of enzyme–substrate interactions. I. Computation of preferred conformations of some substrates of -chymotrypsin. Int J Pept Protein Res 1972; 4:187–200.
72. Platzer KE, Momany FA, Scheraga HA. Conformational energy calculations of enzyme–substrate interactions. II. Computation of the binding energy for substrates in the active site of -chymotrypsin. Int J Pept Protein Res 1972; 4:201–219.
73. Oshiro CM, Kuntz ID, Dixon JS. Flexible ligand docking using a genetic algorithm. J Comput Aided Mol Des 1995;9:113–130.
74. Knegtel RM, Kuntz ID, Oshiro CM. Molecular docking to ensembles of protein structures. J Mol Biol 1997;266:424–440.
75. Kang X, Shafer RH, Kuntz ID. Calculation of ligand–nucleic acid binding free energies with the generalized-Born model in DOCK. Biopolymers 2004;73:192–204.
76. Moustakas DT, Lang PT, Pegg S, Pettersen E, Kuntz ID, Brooijmans N, Rizzo RC. Development and validation of a modular, extensible docking program: DOCK 5. J Comput Aided Mol Des 2006;20:601–619.
77. Rarey M, Kramer B, Lengauer T, Klebe G. A fast flexible docking method using an incremental construction algorithm. J Mol Biol 1996;261:470–489.
78. Claussen H, Buning C, Rarey M, Lengauer T. FlexE: efficient molecular docking considering protein structure variations. J Mol Biol 2001; 308:377–395.
79. Cavasotto CN, Abagyan RA. Protein flexibility in ligand docking and virtual screening to protein kinases. J Mol Biol 2004;337:209–225.
80. Huey R, Morris GM, Olson AJ, Goodsell DS. A semiempirical free energy force field with charge-based desolvation. J Comput Chem 2007;28:1145–1152.
81. Osterberg F, Morris GM, Sanner MF, Olson AJ, Goodsell DS. Automated docking to multiple target structures: incorporation of protein

mobility and structural water heterogeneity in AutoDock. Proteins 2002;46:34–40.

82. Jain AN. Surflex-Dock 2.1: robust performance from ligand energetic modeling, ring flexibility, and knowledge-based search. J Comput Aided Mol Des 2007;21:281–306.
83. Verdonk ML, Cole JC, Hartshorn MJ, Murray CW, Taylor RD. Improved protein–ligand docking using GOLD. Proteins 2003; 52: 609–623.
84. Wei BQ, Weaver LH, Ferrari AM, Matthews BW, Shoichet BK. Testing a flexible-receptor docking algorithm in a model binding site. J Mol Biol 2004;337:1161–1182.
85. Jones G, Willett P. Docking small-molecule ligands into active sites. Curr Opin Biotechnol 1995;6:652–656.
86. Foreman K, Phillips A, Rosen J, Dill K. Comparing search strategies for finding global optima on energy landscapes. J Comput Chem 1999;20:1527–1532.
87. Bissantz C, Folkers G, Rognan D. Protein-based virtual screening of chemical databases. 1. Evaluation of different docking/scoring combinations. J Med Chem 2000;43:4759–4767.
88. Bursulaya BD, Totrov M, Abagyan R, Brooks CL, 3rd. Comparative study of several algorithms for flexible ligand docking. J Comput Aided Mol Des 2003;17:755–763.
89. Kroemer RT, Vulpetti A, McDonald JJ, Rohrer DC, Trosset JY, Giordanetto F, Cotesta S, McMartin C, Kihlen M, Stouten PF. Assessment of docking poses: interactions-based accuracy classification (IBAC) versus crystal structure deviations. J Chem Inf Comput Sci 2004;44:871–881.
90. Cole JC, Murray CW, Nissink JW, Taylor RD, Taylor R. Comparing protein–ligand docking programs is difficult. Proteins 2005;60: 325–332.
91. Warren GL, Andrews CW, Capelli AM, Clarke B, LaLonde J, Lambert MH, Lindvall M, Nevins N, Semus SF, Senger S, et al. A critical assessment of docking programs and scoring functions. J Med Chem 2006;49: 5912–5931.
92. Truchon JF, Bayly CI. Evaluating virtual screening methods: good and bad metrics for the "early recognition" problem. J Chem Inf Model 2007;47:488–508.
93. Englebienne P, Fiaux H, Kuntz DA, Corbeil CR, Gerber-Lemaire S, Rose DR, Moitessier N. Evaluation of docking programs for predicting binding of Golgi alpha-mannosidase II inhibitors: a comparison with crystallography. Proteins 2007;69:160–176.
94. Totrov M, Abagyan R. Drug–Receptor Thermodynamics: Introduction and Applications. Raffa R, editor. Chichester: John Wiley & Sons; 2001.
95. Miller MD, Kearsley SK, Underwood DJ, Sheridan RP. FLOG: a system to select 'quasi-flexible' ligands complementary to a receptor of known three-dimensional structure. J Comput Aided Mol Des 1994;8:153–174.
96. Rhodes N, Willett P, Calvet A, Dunbar JB, Humblet C. CLIP: similarity searching of 3D databases using clique detection. J Chem Inf Comput Sci 2003;43:443–448.
97. Sobolev V, Wade RC, Vriend G, Edelman M. Molecular docking using surface complementarity. Proteins 1996;25:120–129.
98. Sobolev V, Moallem TM, Wade RC, Vriend G, Edelman M. CASP2 molecular docking predictions with the LIGIN software. Proteins 1997; (Suppl 1): 210–214.
99. Burkhard P, Taylor P, Walkinshaw MD. An example of a protein ligand found by database mining: description of the docking method and its verification by a 2.3 Å X-ray structure of a thrombin–ligand complex. J Mol Biol 1998;277:449–466.
100. Goldman BB, Wipke WT. QSD quadratic shape descriptors. 2. Molecular docking using quadratic shape descriptors (QSDock). Proteins 2000;38:79–94.
101. McGann MR, Almond HR, Nicholls A, Grant JA, Brown FK. Gaussian docking functions. Biopolymers 2003;68:76–90.
102. McGaughey GB, Sheridan RP, Bayly CI, Culberson JC, Kreatsoulas C, Lindsley S, Maiorov V, Truchon JF, Cornell WD. Comparison of topological, shape, and docking methods in virtual screening. J Chem Inf Model 2007;47:1504–1519.
103. Diller DJ, Merz KM, Jr. High throughput docking for library design and library prioritization. Proteins 2001;43:113–124.
104. Berman HM, Westbrook J, Feng Z, Gilliland G, Bhat TN, Weissig H, Shindyalov IN, Bourne PE. The Protein Data Bank. Nucleic Acids Res 2000;28:235–242.
105. Zsoldos Z, Szabo I, Szabo Z, Johnson A. Software tools for structure based rational drug design. J Mol Struct Theochem 2003;666–667:659–665.
106. Zsoldos Z, Reid D, Simon A, Sadjad SB, Johnson AP. eHiTS: a new fast, exhaustive

flexible ligand docking system. J Mol Graph Model 2007;26:198–212.

107. Jackson RM. Q-fit: a probabilistic method for docking molecular fragments by sampling low energy conformational space. J Comput Aided Mol Des 2002;16:43–57.

108. Wu SY, McNae I, Kontopidis G, McClue SJ, McInnes C, Stewart KJ, Wang S, Zheleva DI, Marriage H, Lane DP, et al. Discovery of a novel family of CDK inhibitors with the program LIDAEUS: structural basis for ligand-induced disordering of the activation loop. Structure 2003;11:399–410.

109. Joseph-McCarthy D, Thomas BE 4th, Belmarsh M, Moustakas D, Alvarez JC. Pharmacophore-based molecular docking to account for ligand flexibility. Proteins 2003;51:172–188.

110. Goto J, Kataoka R, Hirayama N. Ph4Dock: pharmacophore-based protein–ligand docking. J Med Chem 2004;47:6804–6811.

111. Schafferhans A, Klebe G. Docking ligands onto binding site representations derived from proteins built by homology modelling. J Mol Biol 2001;307:407–427.

112. Evers A, Gohlke H, Klebe G. Ligand-supported homology modelling of protein binding-sites using knowledge-based potentials. J Mol Biol 2003;334:327–345.

113. Evers A, Klebe G. Successful virtual screening for a submicromolar antagonist of the neurokinin-1 receptor based on a ligand-supported homology model. J Med Chem 2004; 47: 5381–5392.

114. Evers A, Klebe G. Ligand-supported homology modeling of g-protein-coupled receptor sites: models sufficient for successful virtual screening. Angew Chem Int Ed Engl 2004; 43:248–251.

115. Evers A, Klabunde T. Structure-based drug discovery using GPCR homology modeling: successful virtual screening for antagonists of the alpha1A adrenergic receptor. J Med Chem 2005;48:1088–1097.

116. Schneidman-Duhovny D, Inbar Y, Nussinov R, Wolfson HJ. PatchDock and SymmDock: servers for rigid and symmetric docking. Nucleic Acids Res 2005;33:W363–W367.

117. Ewing T, Kuntz ID. Critical evaluation of search algorithms for automated molecular docking and database screening. J Comput Chem 1997;18:1175–1189.

118. Ewing TJ, Makino S, Skillman AG, Kuntz ID. DOCK 4.0: search strategies for automated molecular docking of flexible molecule databases. J Comput Aided Mol Des 2001;15: 411–428.

119. Meng EC, Kuntz ID, Abraham DJ, Kellogg GE. Evaluating docked complexes with the HINT exponential function and empirical atomic hydrophobicities. J Comput Aided Mol Des 1994;8:299–306.

120. Gschwend DA, Kuntz ID. Orientational sampling and rigid-body minimization in molecular docking revisited: on-the-fly optimization and degeneracy removal. J Comput Aided Mol Des 1996;10:123–132.

121. Shoichet BK, Leach AR, Kuntz ID. Ligand solvation in molecular docking. Proteins 1999;34:4–16.

122. Zou X, Yaxiong S, Kuntz ID. Inclusion of solvation in ligand binding free energy calculations using the generalized-Born model. J Am Chem Soc 1999;121:8033–8043.

123. http://www.metaphorics.com/products/dockit.html.

124. Muegge I, Martin YC. A general and fast scoring function for protein–ligand interactions: a simplified potential approach. J Med Chem 1999;42:791–804.

125. Gehlhaar DK, Verkhivker GM, Rejto PA, Sherman CJ, Fogel DB, Fogel LJ, Freer ST. Molecular recognition of the inhibitor AG-1343 by HIV-1 protease: conformationally flexible docking by evolutionary programming. Chem Biol 1995;2:317–324.

126. Bacon DJ, Moult J. Docking by least-squares fitting of molecular surface patterns. J Mol Biol 1992;225:849–858.

127. Wallqvist A, Covell DG. Docking enzyme–inhibitor complexes using a preference-based free-energy surface. Proteins 1996;25: 403–419.

128. Perola E, Xu K, Kollmeyer TM, Kaufmann SH, Prendergast FG, Pang YP. Successful virtual screening of a chemical database for farnesyltransferase inhibitor leads. J Med Chem 2000;43:401–408.

129. Pang YP, Kozikowski AP. Prediction of the binding sites of huperzine A in acetylcholinesterase by docking studies. J Comput Aided Mol Des 1994;8:669–681.

130. http://www.eyesopen.com/products/applications/omega.html.

131. http://www.eyesopen.com/docs/html/fred/.

132. DesJarlais RL, Sheridan RP, Dixon JS, Kuntz ID, Venkataraghavan R. Docking flexible ligands to macromolecular receptors by molecular shape. J Med Chem 1986;29:2149–2153.

133. Sandak B, Nussinov R, Wolfson HJ. An automated computer vision and robotics-based technique for 3-D flexible biomolecular docking and matching. Comput Appl Biosci 1995;11:87–99.
134. Sandak B, Nussinov R, Wolfson HJ. A method for biomolecular structural recognition and docking allowing conformational flexibility. J Comput Biol 1998;5:631–654.
135. Hecker Y, Bolle R. On geometric hashing and the generalized Hough transform. IEEE Trans Syst Man Cyber 1994;24:1328–1338.
136. Leach AR, Kuntz ID. Conformational analysis of flexible ligands in macromolecular receptor sites. J Comput Chem 1992;13: 730–748.
137. DesJarlais RL, Sheridan RP, Seibel GL, Dixon JS, Kuntz ID, Venkataraghavan R. Using shape complementarity as an initial screen in designing ligands for a receptor binding site of known three-dimensional structure. J Med Chem 1988;31:722–729.
138. Leach AR, Prout K, Dolata D. The application of artificial intelligence to the conformational analysis of strained molecules. J Comput Chem 1990;11:680–693.
139. Rarey M, Kramer B, Lengauer T. Multiple automatic base selection: protein–ligand docking based on incremental construction without manual intervention. J Comput Aided Mol Des 1997;11:369–384.
140. Bohm HJ. The development of a simple empirical scoring function to estimate the binding constant for a protein–ligand complex of known three-dimensional structure. J Comput Aided Mol Des 1994;8:243–256.
141. Rarey M, Kramer B, Lengauer T. Docking of hydrophobic ligands with interaction-based matching algorithms. Bioinformatics 1999; 15:243–250.
142. Rarey M, Kramer B, Lengauer T. The particle concept: placing discrete water molecules during protein–ligand docking predictions. Proteins 1999;34:17–28.
143. Welch W, Ruppert J, Jain AN. Hammerhead: fast, fully automated docking of flexible ligands to protein binding sites. Chem Biol 1996;3:449–462.
144. Jain AN. Scoring noncovalent protein–ligand interactions: a continuous differentiable function tuned to compute binding affinities. J Comput Aided Mol Des 1996;10:427–440.
145. Jain AN. Ligand-based structural hypotheses for virtual screening. J Med Chem 2004; 47:947–961.
146. Jain AN. Surflex: fully automatic flexible molecular docking using a molecular similarity-based search engine. J Med Chem 2003; 46:499–511.
147. DesJarlais RL, Dixon JS. A shape- and chemistry-based docking method and its use in the design of HIV-1 protease inhibitors. J Comput Aided Mol Des 1994;8:231–242.
148. Shoichet BK, Kuntz ID. Matching chemistry and shape in molecular docking. Protein Eng 1993;6:723–732.
149. Meng EC, Gschwend DA, Blaney JM, Kuntz ID. Orientational sampling and rigid-body minimization in molecular docking. Proteins 1993;17:266–278.
150. Bron C, Kerbosch J. Finding all cliques of an undirected graph. Commu ACM 1973;16: 575–576.
151. Lorber DM, Shoichet BK. Flexible ligand docking using conformational ensembles. Protein Sci 1998;7:938–950.
152. Still W, Tempczyk A, Hawley R, Hendrickson T. Treatment of solvation for molecular mechanics and dynamics. J Am Chem Soc 1990;112:6127–6129.
153. http://dock.compbio.ucsf.edu/.
154. Schnecke V, Swanson CA, Getzoff ED, Tainer JA, Kuhn LA. Screening a peptidyl database for potential ligands to proteins with side-chain flexibility. Proteins 1998;33:74–87.
155. Schnecke V, Kuhn LA. Virtual screening with solvation and ligand-induced complementarity. Perspect Drug Discov Des 2000;20: 171–190.
156. Zavodszky MI, Kuhn LA. Side-chain flexibility in protein–ligand binding: the minimal rotation hypothesis. Protein Sci 2005;14: 1104–1114.
157. Fradera X, Knegtel RM, Mestres J. Similarity-driven flexible ligand docking. Proteins 2000;40:623–636.
158. Fradera X, Kaur J, Mestres J. Unsupervised guided docking of covalently bound ligands. J Comput Aided Mol Des 2004;18:635–650.
159. Floriano WB, Vaidehi N, Zamanakos G, Goddard WA, 3rd. HierVLS hierarchical docking protocol for virtual ligand screening of large-molecule databases. J Med Chem 2004;47; 56–71.
160. Trabanino RJ, Hall SE, Vaidehi N, Floriano WB, Kam VW, Goddard WA, 3rd. First principles predictions of the structure and function of g-protein-coupled receptors: validation for bovine rhodopsin. Biophys J 2004;86: 1904–1921.

161. Alberts IL, Todorov NP, Dean PM. Receptor flexibility in *de novo* ligand design and docking. J Med Chem 2005;48:6585–6596.
162. Seifert MH. ProPose: steered virtual screening by simultaneous protein–ligand docking and ligand–ligand alignment. J Chem Inf Model 2005;45:449–460.
163. Lambert M. Docking conformationally flexible molecules with MVP. 228th ACS National Meeting, COMP-005; 2004.
164. Metropolis N, Rosenbluth A, Rosenbluth M, Teller A, Teller E. Equation of state calculations by fast computing machines. J Chem Phys 1953;21:1087.
165. Westhead DR, Clark DE, Murray CW. A comparison of heuristic search algorithms for molecular docking. J Comput Aided Mol Des 1997;11:209–228.
166. Hart TN, Read RJ. A multiple-start Monte Carlo docking method. Proteins 1992;13: 206–222.
167. Hart TN, Read RJ. DockVision. Available at http://dockvision.com.
168. Janin J. Protein–protein recognition. Prog Biophys Mol Biol 1995;64:145–166.
169. Hart TN, Ness SR, Read RJ. Critical evaluation of the research docking program for the CASP2 challenge. Proteins 1997; (Suppl 1): 205–209.
170. Liu M, Wang S. MCDOCK: a Monte Carlo simulation approach to the molecular docking problem. J Comput Aided Mol Des 1999;13: 435–451.
171. Bouzida D, Rejto PA, Arthurs S, Colson A, Freer ST, Gehlhaar DK, Larson V, Luty B, Rose P, Verkhivker GM. Computer simulations of ligand–protein binding with ensembles of protein conformations: a Monte Carlo study of HIV-1 protease binding energy landscapes. Int J Quantum Chem 1999;72:73–84.
172. Dominguez C, Boelens R, Bonvin AM. HADDOCK: a protein–protein docking approach based on biochemical or biophysical information. J Am Chem Soc 2003;125:1731–1737.
173. Taylor RD, Jewsbury PJ, Essex JW. FDS: flexible ligand and receptor docking with a continuum solvent model and soft-core energy function. J Comput Chem 2003;24:1637–1656.
174. Momany FA, McGuire R, Burgess A, Scheraga H. Energy parameters in polypeptides VII. Geometric parameters, partial atomic charges, nonbonded interactions, hydrogen bond interactions, and intrinsic torsional potentials for the naturally occurring amino acids. J Phys Chem 1975;79:2361–2381.
175. Nemethy G, Pottle MS, Scheraga HA. Energy parameters in polypeptides. 9. Updating of geometrical parameters, nonbonded interactions, and hydrogen bond interactions for the naturally occurring amino acids. J Phys Chem 1983;87:1883–1887.
176. Abagyan R, Totrov M. Biased probability Monte Carlo conformational searches and electrostatic calculations for peptides and proteins. J Mol Biol 1994;235:983–1002.
177. Cardozo T, Totrov M, Abagyan R. Homology modeling by the ICM method. Proteins 1995;23:403–414.
178. Totrov M, Abagyan R. Detailed *ab initio* prediction of lysozyme-antibody complex with 1.6 Å accuracy. Nat Struct Biol 1994;1: 259–263.
179. Abagyan R, Batalov S, Cardozo T, Totrov M, Webber J, Zhou Y. Homology modeling with internal coordinate mechanics: deformation zone mapping and improvements of models via conformational search. Proteins 1997; (Suppl 1): 29–37.
180. Maiorov V, Abagyan R. A new method for modeling large-scale rearrangements of protein domains. Proteins 1997;27:410–424.
181. Cardozo T, Batalov S, Abagyan R. Estimating local backbone structural deviation in homology models. Comput Chem 2000;24:13–31.
182. Totrov M, Abagyan R. Rapid boundary element solvation electrostatics calculations in folding simulations: successful folding of a 23-residue peptide. Biopolymers 2001;60: 124–133.
183. McMartin C, Bohacek RS. QXP: powerful, rapid computer algorithms for structure-based drug design. J Comput Aided Mol Des 1997;11: 333–344.
184. Caflisch A, Fischer S, Karplus M. Docking by Monte Carlo minimization with a solvation correction: application to an FKBP-substrate complex. J Comput Chem 1997;18:723–743.
185. Caflisch A, Niederer P, Anliker M. Monte Carlo minimization with thermalization for global optimization of polypeptide conformations in Cartesian coordinate space. Proteins 1992; 14:102–109.
186. Caflisch A, Niederer P, Anliker M. Monte Carlo docking of oligopeptides to proteins. Proteins 1992;13:223–230.
187. Apostolakis J, Pluckthun A, Caflisch A. Docking small ligands in flexible binding sites. J Comput Chem 1998;19:21–37.
188. Trosset JY, Scheraga HA. Prodock: software package for protein modeling and docking. J Comput Chem 1999;20:412–427.

189. Tietze S, Apostolakis J. GlamDock: development and validation of a new docking tool on several thousand protein–ligand complexes. J Chem Inf Model 2007;47:1657–1672.
190. Morley SD, Afshar M. Validation of an empirical RNA-ligand scoring function for fast flexible docking using Ribodock. J Comput Aided Mol Des 2004;18:189–208.
191. Choi V. Yucca: an efficient algorithm for small-molecule docking. Chem Biodivers 2005;2: 1517–1524.
192. Meiler J, Baker D. ROSETTALIGAND: protein–small molecule docking with full side-chain flexibility. Proteins 2006;65:538–548.
193. Accelrys Inc., San Diego, CA.
194. http://accelrys.com/references/case-studies/archive/studies/affinity/affinity1.pdf.
195. http://accelrys.com/references/case-studies/archive/studies/affinity/affinity2.pdf.
196. http://accelrys.com/references/case-studies/archive/studies/affinity/affinity3.pd.
197. Schrödinger Inc., San Diego, CA.
198. Friesner RA, Banks JL, Murphy RB, Halgren TA, Klicic JJ, Mainz DT, Repasky MP, Knoll EH, Shelley M, Perry JK, et al. Glide: a new approach for rapid, accurate docking and scoring. 1. Method and assessment of docking accuracy. J Med Chem 2004;47:1739–1749.
199. Sherman W, Day T, Jacobson MP, Friesner RA, Farid R. Novel procedure for modeling ligand/receptor induced fit effects. J Med Chem 2006;49:534–553.
200. Goldberg D. Genetic Algorithms in Search, Optimization, and Machine Learning. Reading, MA: Addison-Wesley; 1989.
201. Judson R. Genetic algorithms and their use in chemistry. In: Reviews in Computational Chemistry. New York: VCH Publishers; 1997. p 1–73.
202. Davis L. Handbook of Genetic Algorithms. New York: Van Nostrand Reinhold; 1991.
203. Clark DE, Westhead DR. Evolutionary algorithms in computer-aided molecular design. J Comput Aided Mol Des 1996;10:337–358.
204. Eshelman L, Schaffer J. Foundation of Genetic Algorithms 2. Morgan Kaufmann Publishers; 1993. p 187.
205. Jones G, Willett P, Glen RC, Leach AR, Taylor R. Development and validation of a genetic algorithm for flexible docking. J Mol Biol 1997;267:727–748.
206. Jones G, Willett P, Glen RC. A genetic algorithm for flexible molecular overlay and pharmacophore elucidation. J Comput Aided Mol Des 1995;9:532–549.
207. Clark K, Ajay. Flexible ligand docking without parameter adjustment across four ligand–receptors complexes. J Comput Chem 1995; 16:1210–1226.
208. Moitessier N, Englebienne P, Lee D, Lawandi J, Corbeil CR. Towards the development of universal, fast and highly accurate docking/scoring methods: a long way to go. Br J Pharmacol 2008;153(Suppl 1): S7–S26.
209. Judson R, Jaeger E, Treasurywala A. A genetic algorithm based method for docking flexible molecules. Theochem 1994;114:191–206.
210. Jones G, Willett P, Glen RC. Molecular recognition of receptor sites using a genetic algorithm with a description of desolvation. J Mol Biol 1995;245:43–53.
211. Verdonk ML, Chessari G, Cole JC, Hartshorn MJ, Murray CW, Nissink JW, Taylor RD, Taylor R. Modeling water molecules in protein–ligand docking using GOLD. J Med Chem 2005;48:6504–6515.
212. Solis F, Wets R-B. Minimization by random search techniques. Math Oper Res 1981; 6:19–30.
213. Brooks B, Bruccoleri R, Olafson B, States D, Swaminathan S, Karplus M. CHARMm: a program for macromolecular energy, minimization, and dynamics calculations. J Comput Chem 1983;4:187–217.
214. Nicholls A, Honig B. A rapid finite difference algorithm, utilizing successive over-relaxation to solve Poisson–Boltzmann equations. J Comput Chem 1991;12:435–445.
215. Thormann M, Pons M. Massive docking of flexible ligands using environmental niches in parallelized genetic algorithms. J Comput Chem 2001;22:1971–1982.
216. Charifson PS, Corkery JJ, Murcko MA, Walters WP. Consensus scoring: a method for obtaining improved hit rates from docking databases of three-dimensional structures into proteins. J Med Chem 1999; 42:5100–5109.
217. Li H, Li C, Gui C, Luo X, Chen K, Shen J, Wang X, Jiang H. GAsDock: a new approach for rapid flexible docking based on an improved multi-population genetic algorithm. Bioorg Med Chem Lett 2004;14: 4671–4676.
218. Corbeil CR, Englebienne P, Moitessier N. Docking ligands into flexible and solvated macromolecules. 1. Development and valida-

tion of FITTED 1.0. J Chem Inf Model 2007;47:435–449.

219. Corbeil CR, Englebienne P, Yannopoulos CG, Chan L, Das SK, Bilimoria D, L'Heureux L, Moitessier N. Docking ligands into flexible and solvated macromolecules. 2. Development and application of fitted 1.5 to the virtual screening of potential HCV polymerase inhibitors. J Chem Inf Model 2008;48:902–909.
220. Zhao Y, Sanner MF. FLIPDock: docking flexible ligands into flexible receptors. Proteins 2007;68:726–737.
221. Chen HM, Liu BF, Huang HL, Hwang SF, Ho SY. SODOCK: swarm optimization for highly flexible protein–ligand docking. J Comput Chem 2007;28:612–623.
222. Jones G. Encyclopedia of Computational Chemistry. Schleyer P, Allinger N, Clark T, Gasteiger J, Kollman P, Schaefer Hr, Schreiner R, editors. New York: John Wiley & Sons; 1998.
223. Fogel L, Owens A, Walsh M. Artificial Intelligence through Simulated Evolution. New York: John Wiley & Sons; 1966.
224. Fogel DB. Evolutionary Computation: Toward a New Philosophy of Machine Intelligence. Piscataway: IEEE Press; 1995.
225. Gehlhaar DK, Bouzida D, Rejto PA. Rational Drug Design: Novel Methodology and Practical Applications. Washington, DC: American Chemical Society; 1999. p 292–311.
226. Yang J, Kao C. Flexible ligand docking using a robust evolutionary algorithm. J Comput Chem 2000;21:988–998.
227. Yang J, Chen Y, Horng J, Kao C. Sixth Annual Conference on Evolutionary Programming; 1997.
228. Yang J, Kao C, Horng J. Seventh International Conference on Genetic Algorithms; 1997. p 230.
229. Yang J, Kao C, Horng J. IEEE International Conference on Robotics and Automation; 1998. p 1411.
230. Yang JM, Chen CC. GEMDOCK: a generic evolutionary method for molecular docking. Proteins 2004;55:288–304.
231. Yang JM, Shen TW. A pharmacophore-based evolutionary approach for screening selective estrogen receptor modulators. Proteins 2005;59:205–220.
232. Karplus M, McCammon JA. Dynamics of proteins: elements and function. Annu Rev Biochem 1983;52:263–300.
233. van Gunsteren WF, Mark AE. On the interpretation of biochemical data by molecular dynamics computer simulation. Eur J Biochem 1992;204:947–961.
234. Tuckerman M, Martyna G. Understanding modern molecular dynamics: techniques and applications. J Phys Chem B 2000;104: 159–178.
235. van Gunsteren WF, Hunenberger P, Mark A, Smith P, Tironi I. Computer simulation of protein motion. Comput Phys Commun 1995;91:305–319.
236. Rognan D. Molecular dynamics simulations: a tool for drug design. Perspect Drug Discov Des 1998;11:181–209.
237. Beveridge DL, DiCapua FM. Free energy via molecular simulation: applications to chemical and biomolecular systems. Annu Rev Biophys Biophys Chem 1989;18:431–492.
238. Straatsma T, McCammon J. Computational Alchemy. Annu Rev Phys Chem 1992; 43: 407–435.
239. Straatsma T. Free energy by molecular simulation. In: Reviews in Computational Chemistry. New York: VCH; 1996. p 81–127.
240. Reddy M, Erion M, Agarwal A. Reviews in Computational Chemistry. New York: VCH; 2000. p 217.
241. Lybrand T. Computer simulation of biomolecular systems using molecular dynamics and free energy pertirbation methods. In: Reviews in Computational Chemistry. New York: VCH; 1990. p 295–320.
242. Leach AR. A survey of methods for searching the conformational space of small and medium-sized molecules. In: Reviews in Computational Chemistry. New York: VCH; 1991. p 1–55.
243. Dinur U, Hagler A. New approaches to empirical force fields. In: Reviews in Computational Chemistry. New York: VCH; 1991. p 99–164.
244. Pettersson I, Liljefors T. Molecular mechanics calculated conformational energies of organic molecules: a comparison of force fields. In: Reviews in Computational Chemistry. New York: VCH; 1996. p 167–189.
245. Meirovitch H. Calculation of the free energy and the entropy of macromolecular systems by computer simulation. In: Reviews in Computational Chemistry. New York: VCH; 1998. p 1–74.
246. Kollman P. Free energy calculations-applications to chemical and biological phenomena. Chem Rev 1993;7:2395–2417.
247. Vieth M, Hirst J, Dominy B, Daigler H, Brooks C. Assessing search strategies for

flexible docking. J Comput Chem 1998;19: 1623–1631.

248. Wu G, Robertson DH, Brooks CL, 3rd, Vieth M. Detailed analysis of grid-based molecular docking: a case study of CDOCKER-A CHARMm-based MD docking algorithm. J Comput Chem 2003;24:1549–1562.

249. Miranker A, Karplus M. Functionality maps of binding sites: a multiple copy simultaneous search method. Proteins 1991;11: 29–34.

250. Caflisch A, Miranker A, Karplus M. Multiple copy simultaneous search and construction of ligands in binding sites: application to inhibitors of HIV-1 aspartic proteinase. J Med Chem 1993;36:2142–2167.

251. Di Nola A, Roccatano D, Berendsen HJ. Molecular dynamics simulation of the docking of substrates to proteins. Proteins 1994;19: 174–182.

252. Mangoni M, Roccatano D, Di Nola A. Docking of flexible ligands to flexible receptors in solution by molecular dynamics simulation. Proteins 1999;35:153–162.

253. Luty B, Wasserman Z, Stuouten P, Hodge C, Zacharias M, McCammon J. A molecular mechanics/grid method for evaluation of ligand–receptor interactions. J Comput Chem 1995;16:454–464.

254. Nakajima N, Higo J, Kidera A, Nakamura H. Flexible docking of a ligand peptide to a receptor protein by multicanonical molecular dynamics simulation. Chem Phys Lett 1997;278:297–301.

255. Berg B, Neuhaus T. Multicanonical algorithms for first order phase transitions. Phys Lett B 1991;267:249–253.

256. van Gunsteren WF, Berendsen HJ. A leap-frog algorithm for stochastic dynamics. Mol Simulat 1988;1:173–185.

257. Given JA, Gilson MK. A hierarchical method for generating low-energy conformers of a protein–ligand complex. Proteins 1998;33: 475–495.

258. Wang J, Kollman PA, Kuntz ID. Flexible ligand docking: a multistep strategy approach. Proteins 1999;36:1–19.

259. Weiner S, Kollman P, Case D, Singh U, Ghio C, Alagona G, Profeta S, Weiner P. A new force field for molecular mechanical simulation of nucleic acids and proteins. J Am Chem Soc 1984;106:765–784.

260. Weiner SJ, Kollman PA, Nguyen DT, Case DA. An all atom force field for simulations of proteins and nucleic acids. J Comput Chem 1986;7:230–252.

261. Cornell WD, Cieplak P, Bayly CI, Gould IR, Merz KM, Ferguson DM, Spellmeyer DS, Fox T, Caldwell JW, Kollman PA. A second generation force field for the simulation of proteins, nucleic acids, and organic molecules. J Am Chem Soc 1995;117:5179–5197.

262. Pak Y, Wang S. Application of a molecular dynamics simulation method with a generalized effective potential to the flexible molecular docking problems. J Phys Chem B 2000; 104:354–359.

263. Tsallis C. Possible generalization of Boltzmann–Gibbs statistics. J Stat Phys 1988;52: 479.

264. Curado E, Tsallis C. Generalized statistical mechanics: connection with thermodynamics. J Phys A 1991;24:L69–L72.

265. Pak Y, Wang S. Folding of a 16-residue helical peptide using molecular dynamics simulation with Tsallis effective potential. J Chem Phys 1999;111:4359–4361.

266. Pak Y, Enyedy IJ, Varady J, Kung JW, Lorenzo PS, Blumberg PM, Wang S. Structural basis of binding of high-affinity ligands to protein kinase C: prediction of the binding modes through a new molecular dynamics method and evaluation by site-directed mutagenesis. J Med Chem 2001;44:1690–1701.

267. Glover F, Laguna M. Modern Heuristic Techniques for Combinatorial Problems. Oxford: Blackwell; 1993. p 70–150.

268. Cvijovicacute D, Klinowski J. Taboo search: an approach to the multiple minima problem. Science 1995;267:664–666.

269. Kvasnicka V, Posphical J. Fast evaluation of chemical distance by tabu search algortihm. J Chem Inf Comput Sci 1994;34:1109–1112.

270. Goodman J, Still W. An unbounded systematic search of conformational space. J Comput Chem 1991;12:1110–1117.

271. Huber T, Torda AE, van Gunsteren WF. Local elevation: a method for improving the searching properties of molecular dynamics simulation. J Comput Aided Mol Des 1994;8:695–708.

272. Smellie A, Teig S, Towbin P. Poling: promoting conformational variation. J Comput Chem 1995;16:171–187.

273. Baxter CA, Murray CW, Clark DE, Westhead DR, Eldridge MD. Flexible docking using tabu search and an empirical estimate of binding affinity. Proteins 1998;33:367–382.

274. Eldridge MD, Murray CW, Auton TR, Paolini GV, Mee RP. Empirical scoring functions. I.

The development of a fast empirical scoring function to estimate the binding affinity of ligands in receptor complexes. J Comput Aided Mol Des 1997;11:425–445.

275. Baxter CA, Murray CW, Waszkowycz B, Li J, Sykes RA, Bone RG, Perkins TD, Wylie W. New approach to molecular docking and its application to virtual screening of chemical databases. J Chem Inf Comput Sci 2000;40: 254–262.

276. Tondel K, Anderssen E, Drablos F. Protein Alpha Shape (PAS) Dock: a new Gaussian-based score function suitable for docking in homology modelled protein structures. J Comput Aided Mol Des 2006;20:131–144.

277. Hou T, Wang J, Chen L, Xu X. Automated docking of peptides and proteins by using a genetic algorithm combined with a tabu search. Protein Eng 1999;12:639–648.

278. Pei J, Wang Q, Liu Z, Li Q, Yang K, Lai L. PSI-DOCK: towards highly efficient and accurate flexible ligand docking. Proteins 2006;62: 934–946.

279. Marcou G, Rognan D. Optimizing fragment and scaffold docking by use of molecular interaction fingerprints. J Chem Inf Model 2007;47: 195–207.

280. Leach AR, Smellie A. A combined model-building and distance–geometry approach to automated conformational analysis and search. J Chem Inform Comput Sci 1992;32:379–385.

281. Leach AR. Ligand docking to proteins with discrete side-chain flexibility. J Mol Biol 1994;235:345–356.

282. Desmer J, De Maeyer M, Hazes B, Lasters I. The dead-end elimination theorem and its use in protein side-chain positioning. Nature 1992; 356:539–542.

283. Price M, Jorgensen W. Analysis of binding affinities for celecoxib analogues with COX-1 and COX-2 from combined docking and Monte Carlo simulations and insight into the COX-2/COX-1 selectivity. J Am Chem Soc 2000;122: 9455–9466.

284. Hoffmann D, Kramer B, Washio T, Steinmetzer T, Rarey M, Lengauer T. Two-stage method for protein–ligand docking. J Med Chem 1999;42:4422–4433.

285. Holland J. Adaptation in Natural and Artificial Systems. Cambridge, MA: MIT Press; 1975.

286. Phillips A, Rosen J, Walke V. Molecular structure determination by convex global underestimation of local energy minima. DIMACS Ser Discrete Math Theoret Comput Sci 1995;23: 181–198.

287. Glover F, Laguna M. Tabu Search. Dordrecht: Kluwer Academic Publishers; 1997.

288. Wu G, Vieth M. SDOCKER: a method utilizing existing X-ray structures to improve docking accuracy. J Med Chem 2004;47:3142–3148.

289. Thomsen R, Christensen MH. MolDock: a new technique for high-accuracy molecular docking. J Med Chem 2006;49:3315–3321.

290. Kozakov D, Brenke R, Comeau SR, Vajda S. PIPER: an FFT-based protein docking program with pairwise potentials. Proteins 2006;65:392–406.

291. Grosdidier A, Zoete V, Michielin O. EADock: docking of small molecules into protein active sites with a multiobjective evolutionary optimization. Proteins 2007;67:1010–1025.

292. Verkhivker GM, Bouzida D, Gehlhaar DK, Rejto PA, Freer ST, Rose PW. Monte Carlo simulations of the peptide recognition at the consensus binding site of the constant fragment of human immunoglobulin G: the energy landscape analysis of a hot spot at the intermolecular interface. Proteins 2002;48: 539–557.

293. Verkhivker GM, Bouzida D, Gehlhaar DK, Rejto PA, Freer ST, Rose PW. Computational detection of the binding-site hot spot at the remodeled human growth hormone-receptor interface. Proteins 2003;53:201–219.

294. Verkhivker GM. Computational analysis of ligand binding dynamics at the intermolecular hot spots with the aid of simulated tempering and binding free energy calculations. J Mol Graph Model 2004;22:335–348.

295. Korb O, Stutzle T, Exner T. Empirical scoring function for advanced protein–ligand docking with PLANTS. J Chem Inf Model 2009;49: 84–96.

296. Carlsson J, Boukharta L, Åqvist J. Combining docking, molecular dynamics and the linear interaction energy method to predict binding modes and affinities for non-nucleoside inhibitors to HIV-1 reverse transcriptase. J Med Chem 2008;51:2648–2656.

297. Åqvist J. New methods for predicting binding affinity in computer-aided drug design. Protein Eng 1994;7:385–391.

298. van Lipzig MM, ter Laak AM, Jongejan A, Vermeulen NP, Wamelink M, Geerke D, Meerman JH. Prediction of ligand binding affinity and orientation of xenoestrogens to the estrogen receptor by molecular dynamics simula-

tions and the linear interaction energy method. J Med Chem 2004;47:1018–1030.

299. Almlof M, Brandsdal BO, Aqvist J. Binding affinity prediction with different force fields: examination of the linear interaction energy method. J Comput Chem 2004;25: 1242–1254.

300. Almlof M, Aqvist J, Smalas AO, Brandsdal BO. Probing the effect of point mutations at protein–protein interfaces with free energy calculations. Biophys J 2006;90:433–442.

301. Bortolato A, Moro S. *In silico* binding free energy predictability by using the linear interaction energy (LIE) method: bromobenzimidazole CK2 inhibitors as a case study. J Chem Inf Model 2007;47:572–582.

302. Nervall M, Hanspers P, Carlsson J, Boukharta L, Aqvist J. Predicting binding modes from free energy calculations. J Med Chem 2008;51:2657–2667.

303. Paul N, Rognan D. ConsDock: a new program for the consensus analysis of protein–ligand interactions. Proteins 2002;47:521–533.

304. Wolf A, Zimmermann M, Hofmann-Apitius M. Alternative to consensus scoring-a new approach toward the qualitative combination of docking algorithms. J Chem Inf Model 2007;47:1036–1044.

305. Koheler JEH, Saenger W, van Gunsteren WF. A molecular dynamics simulation of crystalline α-cyclodextrin hexahydrate. Eur Biophys J 1987;15:197–210.

306. Kaminski G, Jorgensen WL. Performance of the AMBER94, MMFF94, and OPLS-AA force fields for modeling organic liquids. J Phys Chem 1996;100:18010–18013.

307. The SYBYL software, TI http://www.tripos.com/. Tripos Inc.: St. Louis, MO. 1995.

308. Halgren TA. Merck molecular force field. I. Basis, form, scope, parameterization, and performance of MMFF94. J Comput Chem 1996;17:490–519.

309. Allinger NL. Conformational analysis. 130. MM2. A hydrocarbon force field utilizing V1 and V2 torsional terms. J Am Chem Soc 1977;99:8127–8134.

310. Allinger NL, Yuh YH, Lii JH. Molecular mechanics. The MM3 force field for hydrocarbons. 1. J Am Chem Soc 1989;111:8551–8566.

311. Bowen JP, Allinger NL. Reviews in Computational Chemistry. New York: VCH Publishers; 1991. p 81–97.

312. Sharp KA, Honig B. Electrostatic interactions in macromolecules: theory and applications. Annu Rev Biophys Biophys Chem 1990;19:301–332.

313. Honig B, Nicholls A. Classical electrostatics in biology and chemistry. Science 1995;268: 1144–1149.

314. Constanciel R, Contreras R. Self consistent field theory of solvent effects representation by continuum models: introduction of desolvation contribution. Theor Chim Acta 1986;65:1–11.

315. Bashford D, Case DA. Generalized Born models of macromolecular solvation effects. Annu Rev Phys Chem 2000;51:129–152.

316. Kellogg GE, Burnett JC, Abraham DJ. Very empirical treatment of solvation and entropy: a force field derived from $\log P_{o/w}$. J Comput Aided Mol Des 2001;15:381–393.

317. Gilson MK, Honig BH. Calculation of electrostatic potentials in an enzyme active site. Nature 1987;330:84–86.

318. Rashin AA, Namboodiri K. A simple method for the calculation of hydration enthalpies of polar molecules with arbitrary shapes. J Phys Chem 1987;91:6003–6012.

319. Rashin AA. Hydration phenomena, classical electrostatics and the boundary element method. J Phys Chem 1990;94:1725–1733.

320. Grant JA, Pickup BT, Nicholls A. A smooth permittivity function for Poisson–Boltzmann solvation methods. J Comput Chem 2001;22:608–640.

321. Tomioka N, Itai A. GREEN: a program package for docking studies in rational drug design. J Comput Aided Mol Des 1994;8:347–366.

322. McMartin C, Bohacek RS. Flexible matching of test ligands to a 3D pharmacophore using a molecular superposition force field: comparison of predicted and experimental conformations of inhibitors of three enzymes. J Comput Aided Mol Des 1995;9:237–250.

323. Bohacek RS, De Lombaert S, McMartin C, Priestle J, Gruetter M. Three-dimensional models of ACE and NEP inhibitors and their use in the design of potent dual ACE/NEP inhibitors. J Am Chem Soc 1996;118: 8231–8249.

324. Bohacek R, Boosalis MS, McMartin C, Faller DV, Perrine SP. Identification of novel small-molecule inducers of fetal hemoglobin using pharmacophore and 'PSEUDO' receptor models. Chem Biol Drug Des 2006;67:318–328.

325. Ksander GM, de Jesus R, Yuan A, Ghai RD, Trapani A, McMartin C, Bohacek R. Ortho-substituted benzofused macrocyclic lactams as

zinc metalloprotease inhibitors. J Med Chem 1997;40:495–505.

326. McMartin C. The FLO/QXP molecular modeling software. Colebrook, CT, USA: Thistlesoft. Available at cmcma@ix.netcom.com.

327. Alisaraie L, Haller LA, Fels G. A QXP-based multistep docking procedure for accurate prediction of protein–ligand complexes. J Chem Inf Model 2006;46:1174–1187.

328. Mezei M. Efficient Monte Carlo sampling for long molecular chains using local moves, tested on a solvated lipid bilayer. J Chem Phys 2003;118:3874–3879.

329. Pang YP, Perola E, Xu K, Prendergast FG. EUDOC: a computer program for identification of drug interaction sites in macromolecules and drug leads from chemical databases. J Comput Chem 2001;22:1750–1771.

330. Kollman PA, Massova I, Reyes C, Kuhn B, Huo S, Chong L, Lee M, Lee T, Duan Y, Wang W, et al. Calculating structures and free energies of complex molecules: combining molecular mechanics and continuum models. Acc Chem Res 2000;33:889–897.

331. Kuhn B, Kollman PA. A ligand that is predicted to bind better to avidin than biotin: insights from computational fluorine scanning. J Am Chem Soc 2000;122:3909–3916.

332. Kuhn B, Kollman PA. Binding of a diverse set of ligands to avidin and streptavidin: an accurate quantitative prediction of their relative affinities by a combination of molecular mechanics and continuum solvent models. J Med Chem 2000;43:3786–3791.

333. Radmer RJ, Kollman PA. The application of three approximate free energy calculations methods to structure based ligand design: trypsin and its complex with inhibitors. J Comput Aided Mol Des 1998;12:215–227.

334. Chong LT, Duan Y, Wang L, Massova I, Kollman PA. Molecular dynamics and free-energy calculations applied to affinity maturation in antibody 48G7. Proc Natl Acad Sci USA 1999;96:14330–14335.

335. Massova I, Kollman PA. A computational alanine scanning to probe protein–protein interactions: a novel approach to evaluate binding free energies. J Am Chem Soc 1999;121: 8133–8143.

336. Massova I, Kollman PA. Combined molecular mechanical and continuum solvent approach (MM-PBSA/GBSA) to predict ligand binding. Perspect Drug Discov Des 2000;18: 113–135.

337. Huo S, Massova I, Kollman PA. Computational alanine scanning of the 1:1 human growth hormone–receptor complex. J Comput Chem 2002;23:15–27.

338. Bottger A, Bottger V, Garcia-Echeverria C, Chene P, Hochkeppel HK, Sampson W, Ang K, Howard SF, Picksley SM, Lane DP. Molecular characterization of the hdm2-p53 interaction. J Mol Biol 1997;269:744–756.

339. Reyes CM, Kollman PA. Molecular dynamics studies of U1A-RNA complexes. RNA 1999;5: 235–244.

340. Pearlman DA. Free energy grids: a practical qualitative application of free energy perturbation to ligand design using the OWFEG method. J Med Chem 1999;42: 4313–4324.

341. Pearlman DA, Charifson PS. Improved scoring of ligand–protein interactions using OWFEG free energy grids. J Med Chem 2001;44: 502–511.

342. Pearlman DA, Charifson PS. Are free energy calculations useful in practice? A comparison with rapid scoring functions for the p38 MAP kinase protein system. J Med Chem 2001;44: 3417–3423.

343. van Dijk M, van Dijk AD, Hsu V, Boelens R, Bonvin AM. Information-driven protein–DNA docking using HADDOCK: it is a matter of flexibility. Nucleic Acids Res 2006;34: 3317–3325.

344. Nemethy G, Gibson KD, Palmer KA, Yoon CN, Paterlini G, Zagari A, Rumsey S, Scheraga HA. Energy parameters in polypeptides. 10. Improved geometrical parameters and nonbonded interactions for use in the ECEPP/3 algorithm, with application to proline-containing peptides. J Phys Chem 1992;96: 6472–6484.

345. Davis M, McCammon JA. Electrostatics in biomolecular structure and dynamics. Chem Rev 1990;90:509–521.

346. Zauhar RJ, Morgan RS. A new method for computing the macromolecular electric potential. J Mol Biol 1985;186:815–820.

347. Juffer AH, Botta EF, van Keilen BM, van der Ploeg A, Berendsen HC. The electric potential of a macromolecule in a solvent: a fundamental approach. J Comput Phys 1991;97:144–171.

348. Bharadwaj A, Windemuth A, Sridharan S, Honig B, Nicholls A. The fast multipole boundary element method for molecular electrostatics: an optimal approach for large system. J Comput Chem 1995;16:898–910.

349. Schapira M, Totrov M, Abagyan R. Prediction of the binding energy for small molecules,

peptides and proteins. J Mol Recognit 1999;12: 177–190.

350. Totrov M, Abagyan R. RECOMB '99: Proceedings of the Third Annual International Conference on Computational Molecular Biology. Edited by Waterman M. 1999. Association for computer machinery, New York/Lyon, France.
351. Totrov M, Abagyan R. Protein–Ligand Docking as an Energy Optimization Problem. Raffa R, editor. New York: John Wiley & Sons; 2001.
352. Strynadka NC, Eisenstein M, Katchalski-Katzir E, Shoichet BK, Kuntz ID, Abagyan R, Totrov M, Janin J, Cherfils J, Zimmerman F, et al. Molecular docking programs successfully predict the binding of a beta-lactamase inhibitory protein to TEM-1 beta-lactamase. Nat Struct Biol 1996;3:233–239.
353. Schapira M, Raaka BM, Samuels HH, Abagyan R. Rational discovery of novel nuclear hormone receptor antagonists. Proc Natl Acad Sci USA 2000;97:1008–1013.
354. Schapira M, Raaka BM, Samuels HH, Abagyan R. *In silico* discovery of novel retinoic acid receptor agonist structures. BMC Struct Biol 2001;1:1–7.
355. Fernandez-Recio J, Totrov M, Abagyan R. ICM-DISCO docking by global energy optimization with fully flexible side-chains. Proteins 2003;52:113–117.
356. Altmann SM, Muryshev A, Fossale E, Maxwell MM, Norflus FN, Fox J, Hersch SM, Young AB, MacDonald ME, Abagyan R, et al. Discovery of bioactive small-molecule inhibitor of poly adp-ribose polymerase: implications for energy-deficient cells. Chem Biol 2006;13:765–770.
357. Cavasotto CN, Kovacs JA, Abagyan RA. Representing receptor flexibility in ligand docking through relevant normal modes. J Am Chem Soc 2005;127:9632–9640.
358. Cavasotto CN, Liu G, James SY, Hobbs PD, Peterson VJ, Bhattacharya AA, Kolluri SK, Zhang X-K, Leid M, Abagyan R, et al. Determinants of retinoid X receptor transcriptional antagonism. J Med Chem 2004;47:4360–4372.
359. Cavasotto CN, Ortiz MA, Abagyan RA, Piedrafita FJ. *In silico* identification of novel EGFR inhibitors with antiproliferative activity against cancer cells. Bioorg Med Chem Lett 2006;16:1969–1974.
360. Hirschfelder JO, Curtiss CF, Bird RB. Molecular Theory of Gases and Liquids. New York: John Wiley & Sons; 1964.
361. Verdonk ML, Berdini V, Hartshorn MJ, Mooij WT, Murray CW, Taylor RD, Watson P. Virtual screening using protein–ligand docking: avoiding artificial enrichment. J Chem Inf Comput Sci 2004;44:793–806.
362. Wesson L, Eisenberg D. Atomic solvation parameters applied to molecular dynamics of proteins in solution. Protein Sci 1992; 1:227–235.
363. Stouten PF, Frommel C, Nakamura H, Sander C. An effective solvation term based on atomic occupancies for use in protein simulations. Mol simula 1993;10:97–120.
364. Goodsell DS, Lauble H, Stout CD, Olson AJ. Automated docking in crystallography: analysis of the substrates of aconitase. Proteins 1993;17:1–10.
365. Goodsell DS, Morris GM, Olson AJ. Automated docking of flexible ligands: applications of AutoDock. J Mol Recognit 1996;9: 1–5.
366. Lunney EA, Hagen SE, Domagala JM, Humblet C, Kosinski J, Tait BD, Warmus JS, Wilson M, Ferguson D, Hupe D, et al. A novel nonpeptide HIV-1 protease inhibitor: elucidation of the binding mode and its application in the design of related analogs. J Med Chem 1994;37:2664–2677.
367. Stoddard BL, Koshland DE, Jr. Prediction of the structure of a receptor–protein complex using a binary docking method. Nature 1992;358:774–776.
368. Vara Prasad JVN, Para S, Lunney EA, Ortwine DF, Dunbar JB, Ferguson D, Jr, Tummino PJ, Hupe D, Tait BD, et al. Novel series of achiral, low molecular weight, and potent HIV-1 protease inhibitors. J Am Chem Soc 1994;116:6989–6990.
369. Sotriffer CA, Flader W, Winger RH, Rode BM, Liedl KR, Varga JM. Automated docking of ligands to antibodies: methods and applications. Methods 2000;20:280–291.
370. Laederach A, Dowd MK, Coutinho PM, Reilly PJ. Automated docking of maltose, 2-deoxymaltose, and maltotetraose into the soybean beta-amylase active site. Proteins 1999;37: 166–175.
371. Rao MS, Olson AJ. Modelling of factor Xa-inhibitor complexes: a computational flexible docking approach. Proteins 1999;34:173–183.
372. Ali HI, Tomita K, Akaho E, Kambara H, Miura S, Hayakawa H, Ashida N, Kawashima Y, Yamagishi T, Ikeya H, et al. Antitumor studies. Part 1. Design, synthesis, antitumor activity, and AutoDock study of 2-deoxo-2-phe-

nyl-5-deazaflavins and 2-deoxo-2-phenylflavin-5-oxides as a new class of antitumor agents. Bioorg Med Chem 2007;15:242–256.

373. Lauria A, Ippolito M, Almerico AM. Molecular docking approach on the Topoisomerase I inhibitors series included in the NCI anti-cancer agents mechanism database. J Mol Model 2007;13:393–400.

374. Rockey WM, Elcock AH. Structure selection for protein kinase docking and virtual screening: homology models or crystal structures? Curr Protein Pept Sci 2006;7:437–457.

375. Qiu D, Shenkin PS, Hollinger FP, Still WC. The GB/SA continuum model for solvation. A Fast analytical method for the calculation of approximate Born radii. J Phys Chem 1997;101:3005–3014.

376. Hawkins GD, Cramer CJ, Truhlar DG. Pairwise solute screening of solute charges from a dielectric medium. Chem Phys Lett 1995;246: 122–129.

377. Naim M, Bhat S, Rankin KN, Dennis S, Chowdhury SF, Siddiqi I, Drabik P, Sulea T, Bayly CI, Jakalian A, et al. Solvated interaction energy (SIE) for scoring protein–-ligand binding affinities. 1. Exploring the parameter space. J Chem Inf Model 2007; 47:122–133.

378. Bohm H, Stahl M. Rapid empirical scoring functions in virtual screening applications. Med Chem Res 1999;9:445–462.

379. Mark AE, van Gunsteren WF. Decomposition of the free energy of a system in terms of specific interactions. Implications for theoretical and experimental studies. J Mol Biol 1994;240:167–176.

380. Wang R, Liu L, Lai L, Tang Y. SCORE: a new empirical method for estimating the binding affinity of a protein–ligand complex. J Mol Model 1998;4:379–394.

381. Davis AM, Teague SJ. Hydrogen bonding, hydrophobic interactions, and failure of the rigid receptor hypothesis. Angew Chem Int Ed Engl 1999;38:736–749.

382. Rarey M, Wefing S, Lengauer T. Placement of medium-sized molecular fragments into active sites of proteins. J Comput Aided Mol Des 1996;10:41–54.

383. Bohm HJ. LUDI: rule-based automatic design of new substituents for enzyme inhibitor leads. J Comput Aided Mol Des 1992;6:593–606.

384. Bohm HJ. Prediction of binding constants of protein ligands: a fast method for the prioritization of hits obtained from *de novo* design or 3D database search programs. J Comput Aided Mol Des 1998;12:309–323.

385. So SS, Karplus M. A comparative study of ligand–receptor complex binding affinity prediction methods based on glycogen phosphorylase inhibitors. J Comput Aided Mol Des 1999;13:243–258.

386. Bohm HJ, Banner DW, Weber L. Combinatorial docking and combinatorial chemistry: design of potent non-peptide thrombin inhibitors. J Comput Aided Mol Des 1999;13:51–56.

387. Head RD, Smythe ML, Oprea T, Waller CL, Green SM, Marshall G. VALIDATE: a new method for the receptor-based prediction of binding affinities of bovel ligands. J Am Chem Soc 1996;118:3959–3969.

388. Bohacek RS, McMartin C. Multiple highly diverse structures complementary to enzyme binding sites: results of extensive application of a *de novo* design method incorporating combinatorial growth. J Am Chem Soc 1994; 116:5560–5571.

389. The Cerius2 software, AI, http://www.accelrys.com/.

390. Wang R, Lu Y, Wang S. Comparative evaluation of 11 scoring functions for molecular docking. J Med Chem 2003;46:2287–2303.

391. Wang R, Lu Y, Fang X, Wang S. An extensive test of 14 scoring functions using the PDBbind refined set of 800 protein–ligand complexes. J Chem Inf Comput Sci 2004;44:2114–2125.

392. Krovat EM, Langer T. Impact of scoring functions on enrichment in docking-based virtual screening: an application study on renin inhibitors. J Chem Inf Comput Sci 2004; 44: 1123–1129.

393. Mpamhanga CP, Chen B, McLay IM, Ormsby DL, Lindvall MK. Retrospective docking study of PDE4B ligands and an analysis of the behavior of selected scoring functions. J Chem Inf Model 2005;45:1061–1074.

394. Kellogg GE, Semus SF, Abraham DJ. HINT: a new method of empirical hydrophobic field calculation for CoMFA. J Comput Aided Mol Des 1991;5:545–552.

395. Wireko FC, Kellogg GE, Abraham DJ. Allosteric modifiers of hemoglobin. 2. Crystallographically determined binding sites and hydrophobic binding/interaction analysis of novel hemoglobin oxygen effectors. J Med Chem 1991;34:758–767.

396. Oprea T, Marshall G.Receptor-based prediction of binding affinities. Perspect Drug Discov 1998; 9–11: 35–61.

397. Marshall G, Head RD, Ragno R. Thermodynamics in Biology. New York: Oxford University Press; 2000. p 87–111.
398. Stahl M, Rarey M. Detailed analysis of scoring functions for virtual screening. J Med Chem 2001;44:1035–1042.
399. Kamper A, Apostolakis J, Rarey M, Marian CM, Lengauer T. Fully automated flexible docking of ligands into flexible synthetic receptors using forward and inverse docking strategies. J Chem Inf Model 2006; 46:903–911.
400. Kramer B, Rarey M, Lengauer T. Evaluation of the FLEXX incremental construction algorithm for protein–ligand docking. Proteins 1999;37:228–241.
401. Stahl M. Modifications of the scoring function in FlexX for virtual screening applications. Perspect Drug Discov Des 2000;20:83–98.
402. Bohacek RS, McMartin C. Definition and display of steric, hydrophobic, and hydrogen-bonding properties of ligand binding sites in proteins using Lee and Richards accessible surface: validation of a high-resolution graphical tool for drug design. J Med Chem 1992;35:1671–1684.
403. Jain AN, Dietterich TG, Lathrop RH, Chapman D, Critchlow RE Jr, Bauer BE, Webster TA, Lozano-Perez T. A shape-based machine learning tool for drug design. J Comput Aided Mol Des 1994;8:635–652.
404. Jain AN, Harris NL, Park JY. Quantitative binding site model generation: compass applied to multiple chemotypes targeting the 5-HT1A receptor. J Med Chem 1995;38: 1295–1308.
405. Jain AN, Koile K, Chapman D. Compass: predicting biological activities from molecular surface properties. Performance comparisons on a steroid benchmark. J Med Chem 1994;37:2315–2327.
406. Venkatachalam CM, Jiang X, Oldfield T, Waldman M. LigandFit: a novel method for the shape-directed rapid docking of ligands to protein active sites. J Mol Graph Model 2003;21:289–307.
407. Pham TA, Jain AN. Parameter estimation for scoring protein–ligand interactions using negative training data. J Med Chem 2006;49:5856–5868.
408. Murray CW, Auton TR, Eldridge MD. Empirical scoring functions. II. The testing of an empirical scoring function for the prediction of ligand–receptor binding affinities and the use of Bayesian regression to improve the quality of the model. J Comput Aided Mol Des 1998;12:503–519.
409. Kirton SB, Murray CW, Verdonk ML, Taylor RD. Prediction of binding modes for ligands in the cytochromes P450 and other heme-containing proteins. Proteins 2005;58:836–844.
410. Cotesta S, Giordanetto F, Trosset JY, Crivori P, Kroemer RT, Stouten PF, Vulpetti A. Virtual screening to enrich a compound collection with CDK2 inhibitors using docking, scoring, and composite scoring models. Proteins 2005;60:629–643.
411. de Graaf C, Oostenbrink C, Keizers PH, van der Wijst T, Jongejan A, Vermeulen NP. Catalytic site prediction and virtual screening of cytochrome P450 2D6 substrates by consideration of water and rescoring in automated docking. J Med Chem 2006;49:2417–2430.
412. Dessalew N, Bharatam PV. Investigation of potential glycogen synthase kinase 3 inhibitors using pharmacophore mapping and virtual screening. Chem Biol Drug Des 2006; 68:154–165.
413. Horvath D. A virtual screening approach applied to the search for trypanothione reductase inhibitors. J Med Chem 1997;40:2412–2423.
414. Wang R, Lai L, Wang S. Further development and validation of empirical scoring functions for structure-based binding affinity prediction. J Comput Aided Mol Des 2002;16:11–26.
415. Warwicker J, Watson HC. Calculation of the electric potential in the active site cleft due to alpha-helix dipoles. J Mol Biol 1982; 157:671–679.
416. Klapper I, Hagstrom R, Fine R, Sharp K, Honig B. Focusing of electric fields in the active site of Cu–Zn superoxide dismutase: effects of ionic strength and amino-acid modification. Proteins 1986;1:47–59.
417. Rognan D, Lauemoller SL, Holm A, Buus S, Tschinke V. Predicting binding affinities of protein ligands from three-dimensional models: application to peptide binding to class I major histocompatibility proteins. J Med Chem 1999;42:4650–4658.
418. Bissantz C, Bernard P, Hibert M, Rognan D. Protein-based virtual screening of chemical databases. II. Are homology models of G-protein coupled receptors suitable targets? Proteins 2003;50:5–25.
419. Logean A, Sette A, Rognan D. Customized versus universal scoring functions: application to class I MHC-peptide binding free energy predictions. Bioorg Med Chem Lett 2001;11: 675–679.

420. Logean A, Rognan D. Recovery of known T-cell epitopes by computational scanning of a viral genome. J Comput Aided Mol Des 2002;16: 229–243.
421. Kuhn LA, Swanson CA, Pique ME, Tainer JA, Getzoff ED. Atomic and residue hydrophilicity in the context of folded protein structures. Proteins 1995;23:536–547.
422. Zavodszky MI, Sanschagrin PC, Korde RS, Kuhn LA. Distilling the essential features of a protein surface for improving protein–ligand docking, scoring, and virtual screening. J Comput Aided Mol Des 2002;16:883–902.
423. Zavodszky MI, Lei M, Thorpe MF, Day AR, Kuhn LA. Modeling correlated main-chain motions in proteins for flexible molecular recognition. Proteins 2004;57:243–261.
424. Sukuru SC, Crepin T, Milev Y, Marsh LC, Hill JB, Anderson RJ, Morris JC, Rohatgi A, O'Mahony G, Grotli M, et al. Discovering new classes of Brugia malayi asparaginyl-tRNA synthetase inhibitors and relating specificity to conformational change. J Comput Aided Mol Des 2006;20:159–178.
425. Kasper P, Christen P, Gehring H. Empirical calculation of the relative free energies of peptide binding to the molecular chaperone DnaK. Proteins 2000;40:185–192.
426. Pickett SD, Sternberg MJ. Empirical scale of side-chain conformational entropy in protein folding. J Mol Biol 1993;231:825–839.
427. Harvey SC. Treatment of electrostatic effects in macromolecular modeling. Proteins 1989;5: 78–92.
428. Grant JA, Pickup BT. A Gaussian description of molecular shape. J Phys Chem 1995;99: 3503–3510.
429. Grant JA, Pickup BT. Computer Simulation of Biomolecular Systems. Theoretical and Experimental Applications. Dordrecht: Kluwer/ESCOM; 1997. p 150–176.
430. Schulz-Gasch T, Stahl M. Scoring functions for protein–ligand interactions: a critical perspective. Drug Discov Today Technol 2004;1: 231–238.
431. Miteva MA, Lee WH, Montes MO, Villoutreix BO. Fast structure-based virtual ligand screening combining FRED, DOCK, and Surflex. J Med Chem 2005;48:6012–6022.
432. Kellenberger E, Rodrigo J, Muller P, Rognan D. Comparative evaluation of eight docking tools for docking and virtual screening accuracy. Proteins 2004;57:225–242.
433. Yang JM, Kao CY. Flexible ligand docking using a robust evolutionary algorithm. J Comput Chem 2000;21:988–998.
434. Halgren TA, Murphy RB, Friesner RA, Beard HS, Frye LL, Pollard WT, Banks JL. Glide: a new approach for rapid, accurate docking and scoring. 2. Enrichment factors in database screening. J Med Chem 2004;47:1750–1759.
435. Friesner RA, Murphy RB, Repasky MP, Frye LL, Greenwood JR, Halgren TA, Sanschagrin PC, Mainz DT. Extra precision glide: docking and scoring incorporating a model of hydrophobic enclosure for protein–ligand complexes. J Med Chem 2006;49:6177–6196.
436. Krammer A, Kirchhoff PD, Jiang X, Venkatachalam CM, Waldman M. LigScore: a novel scoring function for predicting binding affinities. J Mol Graph Model 2005;23:395–407.
437. Sotriffer CA, Sanschagrin P, Matter H, Klebe G. SFCscore: scoring functions for affinity prediction of protein–ligand complexes. Proteins 2008;73:395–419.
438. Hansch C, Leo AJ. Substituent Constants for Correlation Analysis in Chemistry and Biology. New York: John Wiley & Sons, Inc.; 1979.
439. Abraham DJ, Leo AJ. Extension of the fragment method to calculate amino acid zwitterion and side chain partition coefficients. Proteins 1987;2:130–152.
440. Kellogg GE, Abraham DJ. Hydrophobicity: is $\log P_{o/w}$ more than the sum of its parts? Eur J Med Chem 2000;35:651–661.
441. Kellogg GE, Joshi JS, Abraham DJ. New tools for modeling and understanding hydrophobicity and hydrophobic interactions. Med Chem Res 1992;1:444–453.
442. Burnett JC, Botti P, Abraham DJ, Kellogg GE. Computationally accessible method for estimating free energy changes resulting from site-specific mutations of biomolecules: systematic model building and structural/hydropathic analysis of deoxy and oxy hemoglobins. Proteins 2001;42:355–377.
443. Cashman DJ, Kellogg GE. A computational model for anthracycline binding to DNA tuning groove-binding intercalators for specific sequences. J Med Chem 2004;47: 1360–1374.
444. Cashman DJ, Rife JP, Kellogg GE. Which aminoglycoside ring is most important for binding? A hydropathic analysis of gentamicin, paromomycin, and analogues. Bioorg Med Chem Lett 2001;11:119–122.

445. Cozzini P, Fornabaio M, Marabotti A, Abraham DJ, Kellogg GE, Mozzarelli A. Simple, intuitive calculations of free energy of binding for protein–ligand complexes. 1. Models without explicit constrained water. J Med Chem 2002;45:2469–2483.

446. Fornabaio M, Cozzini P, Mozzarelli A, Abraham DJ, Kellogg GE. Simple, intuitive calculations of free energy of binding for protein–ligand complexes. 2. Computational titration and pH effects in molecular models of neuraminidase–inhibitor complexes. J Med Chem 2003;46:4487–4500.

447. Fornabaio M, Spyrakis F, Mozzarelli A, Cozzini P, Abraham DJ, Kellogg GE. Simple, intuitive calculations of free energy of binding for protein–ligand complexes. 3. The free energy contribution of structural water molecules in HIV-1 protease complexes. J Med Chem 2004;47:4507–4516.

448. Spyrakis F, Fornabaio M, Cozzini P, Mozzarelli A, Abraham DJ, Kellogg GE. Computational titration analysis of a multiprotic HIV-1 protease–ligand complex. J Am Chem Soc 2004;126:11764–11765.

449. Kellogg GE, Fornabaio M, Spyrakis F, Lodola A, Cozzini P, Mozzarelli A, Abraham DJ. Getting it right: modeling of pH, solvent and "nearly" everything else in virtual screening of biological targets. J Mol Graph Model 2004;22:479–486.

450. Amadasi A, Spyrakis F, Cozzini P, Abraham DJ, Kellogg GE, Mozzarelli A. Mapping the energetics of water–protein and water–ligand interactions with the "natural" HINT forcefield: predictive tools for characterizing the roles of water in biomolecules. J Mol Biol 2006;358:289–309.

451. Kellogg GE, Fornabaio M, Chen DL, Abraham DJ, Spyrakis F, Cozzini P, Mozzarelli A. Tools for building a comprehensive modeling system for virtual screening under real biological conditions: the computational titration algorithm. J Mol Graph Model 2006;24:434–439.

452. Amadasi A, Surface JA, Spyrakis F, Cozzini P, Mozzarelli A, Kellogg GE. Robust classification of "relevant" water molecules in putative protein binding sites. J Med Chem 2008;51:1063–1067.

453. Koppensteiner WA, Sippl MJ. Knowledge-based potentials-back to the roots. Biochemistry 1998;63:247–252.

454. Sippl MJ. Calculation of conformational ensembles from potentials of mean force. An approach to the knowledge-based prediction of local structures in globular proteins. J Mol Biol 1990;213:859–883.

455. Sippl MJ. Boltzmann's principle, knowledge-based mean fields and protein folding. An approach to the computational determination of protein structures. J Comput Aided Mol Des 1993;7:473–501.

456. Sippl MJ, Ortner M, Jaritz M, Lackner P, Flockner H. Helmholtz free energies of atom pair interactions in proteins. Fold Des 1996;1:289–298.

457. Bahar I, Jernigan RL. Inter-residue potentials in globular proteins and the dominance of highly specific hydrophilic interactions at close separation. J Mol Biol 1997;266:195–214.

458. Jones DT, Taylor WR, Thornton JM. A new approach to protein fold recognition. Nature 1992;358:86–89.

459. Ouzounis C, Sander C, Scharf M, Schneider R. Prediction of protein structure by evaluation of sequence–structure fitness. Aligning sequences to contact profiles derived from three-dimensional structures. J Mol Biol 1993;232:805–825.

460. Bowie JU, Luthy R, Eisenberg D. A method to identify protein sequences that fold into a known three-dimensional structure. Science 1991;253:164–170.

461. Godzik A, Kolinski A, Skolnick J. Topology fingerprint approach to the inverse protein folding problem. J Mol Biol 1992;227:227–238.

462. Godzik A, Skolnick J. Sequence–structure matching in globular proteins: application to supersecondary and tertiary structure determination. Proc Natl Acad Sci USA 1992;89:12098–12102.

463. Kocher JP, Rooman MJ, Wodak SJ. Factors influencing the ability of knowledge-based potentials to identify native sequence–structure matches. J Mol Biol 1994;235:1598–1613.

464. Wodak SJ, Rooman MJ. Generating and testing protein folds. Curr Opin Struct Biol 1993;3:247–259.

465. Bohm H, Stahl M. The use of scoring functions in drug discovery applications. In: Reviews in Computational Chemistry New York: John Wiley & Sons, Inc.; 2002. p 41–87.

466. Stahl M. Structure-based library design. In: Mannhold R, Kubinyi H, Timmerman H, editors. Virtual Screening for Bioactive Molecules. Weinheim: Wiley-VCH; 2000. p 229–264.

467. Liu S, Zhang C, Zhou H, Zhou Y. A physical reference state unifies the structure-derived

potential of mean force for protein folding and binding. Proteins 2004;56:93–101.

468. Verkhivker G, Appelt K, Freer ST, Villafranca JE. Empirical free energy calculations of ligand–protein crystallographic complexes. I. Knowledge-based ligand–protein interaction potentials applied to the prediction of human immunodeficiency virus 1 protease binding affinity. Protein Eng 1995;8:677–691.
469. Wallqvist A, Jernigan RL, Covell DG. A preference-based free-energy parameterization of enzyme–inhibitor binding. Applications to HIV-1-protease inhibitor design. Protein Sci 1995;4:1881–1903.
470. DeWitte RS, Shakhnovich EI. SMoG: *de novo* design method based on simple, fast, and accurate free energy estimates. 1. Methodology and supporting evidence. J Am Chem Soc 1996;118:11733–11744.
471. Ishchenko AV, Shakhnovich EI. Small molecule growth 2001 (SMoG2001): an improved knowledge-based scoring function for protein–ligand interactions. J Med Chem 2002; 45:2770–2780.
472. Muegge I, Martin YC, Hajduk PJ, Fesik SW. Evaluation of PMF scoring in docking weak ligands to the FK506 binding protein. J Med Chem 1999;42:2498–2503.
473. Muegge I. The effect of small changes in protein structure on predicted binding modes of known inhibitors of influenza virus neuraminidase. Med Chem Res 1999;9:490–500.
474. Sookhee H, Andreani R, Robbins A, Muegge I. Evaluation of docking/scoring approaches: a comparative study based on MMP3 inhibitors. J Comput Aided Mol Des 2000;14: 435–448.
475. Ha S, Andreani R, Robbins A, Muegge I. Evaluation of docking/scoring approaches: a comparative study based on MMP3 inhibitors. J Comput Aided Mol Des 2000;14:435–448.
476. Muegge I. PMF scoring revisited. J Med Chem 2006;49:5895–5902.
477. Clark RD, Strizhev A, Leonard JM, Blake JF, Matthew JB. Consensus scoring for ligand/ protein interactions. J Mol Graph Model 2002;20:281–295.
478. http://www.hearne.com.au/products/cache/ edition/biomedcache/.
479. Mitchell JBO, Laskowski RA, Alex A, Thornton JM. BLEEP: potential of mean force describing protein–ligand interactions. I. Generating potential. J Comput Chem 1999;20: 1165–1176.
480. Mitchell JBO, Laskowski RA, Alex A, Forster MJ, Thornton JM. BLEEP: potential of mean force describing protein–ligand interactions: II. Calculation of binding energies and comparison with experimental data. J Comput Chem 1999;20:1177–1185.
481. Gohlke H, Hendlich M, Klebe G. Knowledge-based scoring function to predict protein–ligand interactions. J Mol Biol 2000; 295:337–356.
482. Velec HF, Gohlke H, Klebe G. DrugScore (CSD)-knowledge-based scoring function derived from small molecule crystal data with superior recognition rate of near-native ligand poses and better affinity prediction. J Med Chem 2005;48:6296–6303.
483. Allen FH. The Cambridge Structural Database: a quarter of a million crystal structures and rising. Acta Crystallogr B 2002; 58: 380–388.
484. Sotriffer CA, Gohlke H, Klebe G. Docking into knowledge-based potential fields: a comparative evaluation of DrugScore. J Med Chem 2002;45:1967–1970.
485. Gohlke H, Klebe G. DrugScore meets CoMFA: adaptation of fields for molecular comparison (AFMoC) or how to tailor knowledge-based pair-potentials to a particular protein. J Med Chem 2002;45:4153–4170.
486. Oda A, Tsuchida K, Takakura T, Yamaotsu N, Hirono S. Comparison of consensus scoring strategies for evaluating computational models of protein–ligand complexes. J Chem Inf Model 2006;46:380–391.
487. Hu X, Balaz S, Shelver WH. A practical approach to docking of zinc metalloproteinase inhibitors. J Mol Graph Model 2004; 22: 293–307.
488. Zhou H, Zhou Y. Distance-scaled, finite ideal-gas reference state improves structure-derived potentials of mean force for structure selection and stability prediction. Protein Sci 2002; 11:2714–2726.
489. Zhang C, Liu S, Zhu Q, Zhou Y. A knowledge-based energy function for protein–ligand, protein–protein, and protein–DNA complexes. J Med Chem 2005;48:2325–2335.
490. Zhang C, Liu S, Zhou Y. Docking prediction using biological information, ZDOCK sampling technique, and clustering guided by the DFIRE statistical energy function. Proteins 2005;60:314–318.
491. Zhou H, Zhang C, Liu S, Zhou Y. Web-based toolkits for topology prediction of transmembrane helical proteins, fold recognition, structure and

binding scoring, folding-kinetics analysis and comparative analysis of domain combinations. Nucleic Acids Res 2005;33:W193–W197.

492. Zhou Y, Zhou H, Zhang C, Liu S. What is a desirable statistical energy function for proteins and how can it be obtained? Cell Biochem Biophys 2006;46:165–174.

493. Zhu J, Xie L, Honig B. Structural refinement of protein segments containing secondary structure elements: Local sampling, knowledge-based potentials, and clustering. Proteins 2006;65:463–479.

494. Li H, Zhou Y. Fold helical proteins by energy minimization in dihedral space and a DFIRE-based statistical energy function. J Bioinform Comput Biol 2005;3:1151–1170.

495. Li H. A model of local-minima distribution on conformational space and its application to protein structure prediction. Proteins 2006;64:985–991.

496. Huang S-Y, Zou X. An iterative knowledge-based scoring function to predict protein–ligand interactions. I. Derivation of interaction potentials. J Comput Chem 2006;27: 1866–1875.

497. Huang S-Y, Zou X. An iterative knowledge-based scoring function to predict protein–ligand interactions. II. Validation of the scoring function. J Comput Chem 2006;27: 1876–1882.

498. Yang CY, Wang R, Wang S. M-Score: a knowledge-based potential scoring function accounting for protein atom mobility. J Med Chem 2006;49:5903–5911.

499. Ferrari AM, Wei BQ, Costantino L, Shoichet BK. Soft docking and multiple receptor conformations in virtual screening. J Med Chem 2004;47:5076–5084.

500. Krovat E, Steindl T, Langer T. Recent advances in docking and scoring. Curr Comput Aided Drug Des 2005;1:93–102.

501. Perola E, Walters WP, Charifson PS. A detailed comparison of current docking and scoring methods on systems of pharmaceutical relevance. Proteins 2004;56: 235–249.

502. Stahl M, Bohm HJ. Development of filter functions for protein–ligand docking. J Mol Graph Model 1998;16:121–132.

503. Gruneberg S, Stubbs MT, Klebe G. Successful virtual screening for novel inhibitors of human carbonic anhydrase: strategy and experimental confirmation. J Med Chem 2002;45: 3588–3602.

504. Lyne PD, Kenny PW, Cosgrove DA, Deng C, Zabludoff S, Wendoloski JJ, Ashwell S. Identification of compounds with nanomolar binding affinity for checkpoint kinase-1 using knowledge-based virtual screening. J Med Chem 2004;47:1962–1968.

505. Mozziconacci JC, Arnoult E, Bernard P, Do QT, Marot C, Morin-Allory L. Optimization and validation of a docking-scoring protocol; application to virtual screening for COX-2 inhibitors. J Med Chem 2005;48:1055–1068.

506. Liu Z, Huang C, Fan K, Wei P, Chen H, Liu S, Pei J, Shi L, Li B, Yang K, et al. Virtual screening of novel noncovalent inhibitors for SARS-CoV 3C-like proteinase. J Chem Inf Model 2005;45:10–17.

507. Li J, Chen J, Gui C, Zhang L, Qin Y, Xu Q, Zhang J, Liu H, Shen X, Jiang H. Discovering novel chemical inhibitors of human cyclophilin A: virtual screening, synthesis, and bioassay. Bioorg Med Chem 2006;14:2209–2224.

508. Zhou Z, Felts AK, Friesner RA, Levy RM. Comparative performance of several flexible docking programs and scoring functions: enrichment studies for a diverse set of pharmaceutically relevant targets. J Chem Inf Model 2007;47:1599–1608.

509. Cummings MD, DesJarlais RL, Gibbs AC, Mohan V, Jaeger EP. Comparison of automated docking programs as virtual screening tools. J Med Chem 2005;48:962–976.

510. Terp GE, Johansen BN, Christensen IT, Jorgensen FS. A new concept for multidimensional selection of ligand conformations (MultiSelect) and multidimensional scoring (MultiScore) of protein–ligand binding affinities. J Med Chem 2001;44:2333–2343.

511. Jacobsson M, Liden P, Stjernschantz E, Bostrom H, Norinder U. Improving structure-based virtual screening by multivariate analysis of scoring data. J Med Chem 2003;46:5781–5789.

512. Betzi S, Suhre K, Chetrit B, Guerlesquin F, Morelli X. GFscore: a general nonlinear consensus scoring function for high-throughput docking. J Chem Inf Model 2006; 46: 1704–1712.

513. Damm KL, Carlson HA. Exploring experimental sources of multiple protein conformations in structure-based drug design. J Am Chem Soc 2007;129:8225–8235.

514. Zhang J, Liu JS. On side-chain conformational entropy of proteins. PLoS Comput Biol 2006;2: e168.

515. Rupley JA, Careri G. Protein hydration and function. Adv Protein Chem 1991;41:37–172.

516. Frauenfelder H, Fenimore PW, McMahon BH. Hydration, slaving and protein function. Biophys Chem 2002;98:35–48.
517. Karplus PA, Faerman C. Ordered water in macromolecular structure. Curr Opin Struct Biol 1994;4:770–776.
518. Thanki N, Thornton JM, Goodfellow JM. Distributions of water around amino acid residues in proteins. J Mol Biol 1988;202:637–657.
519. Levitt M, Park BH. Water: now you see it, now you don't. Structure 1993;1:223–226.
520. Carugo O, Bordo D. How many water molecules can be detected by protein crystallography? Acta Crystallogr D 1999;55:479–483.
521. de Graaf C, Vermeulen NP, Feenstra KA. Cytochrome p450 *in silico*: an integrative modeling approach. J Med Chem 2005;48: 2725–2755.
522. Li Z, Lazaridis T. Thermodynamic contributions of the ordered water molecule in HIV-1 protease. J Am Chem Soc 2003; 125:6636–6637.
523. Li Z, Lazaridis T. The effect of water displacement on binding thermodynamics: concanavalin A. J Phys Chem B 2005;109:662–670.
524. Pastor M, Cruciani G, Watson KA. A strategy for the incorporation of water molecules present in a ligand binding site into a three-dimensional quantitative structure–activity relationship analysis. J Med Chem 1997; 40: 4089–4102.
525. Kastenholz MA, Pastor M, Cruciani G, Haaksma EE, Fox T. GRID/CPCA: a new computational tool to design selective ligands. J Med Chem 2000;43:3033–3044.
526. Wade RC, Clark KJ, Goodford PJ. Further development of hydrogen bond functions for use in determining energetically favorable binding sites on molecules of known structure. 1. Ligand probe groups with the ability to form two hydrogen bonds. J Med Chem 1993;36: 140–147.
527. Wade RC, Goodford PJ. Further development of hydrogen bond functions for use in determining energetically favorable binding sites on molecules of known structure. 2. Ligand probe groups with the ability to form more than two hydrogen bonds. J Med Chem 1993; 36:148–156.
528. Boobbyer DN, Goodford PJ, McWhinnie PM, Wade RC. New hydrogen-bond potentials for use in determining energetically favorable binding sites on molecules of known structure. J Med Chem 1989;32:1083–1094.
529. Pitt WR, Murrayrust J, Goodfellow JM. AQUARIUS2-knowledge based modeling of solvent sites around proteins. J Comput Chem 1993;14:1007–1018.
530. Roe SM, Teeter MM. Patterns for prediction of hydration around polar residues in proteins. J Mol Biol 1993;229:419–427.
531. Raymer ML, Sanschagrin PC, Punch WF, Venkataraman S, Goodman ED, Kuhn LA. Predicting conserved water-mediated and polar ligand interactions in proteins using a K-nearest-neighbors genetic algorithm. J Mol Biol 1997;265:445–464.
532. Garcia-Sosa AT, Mancera RL, Dean PM. WaterScore: a novel method for distinguishing between bound and displaceable water molecules in the crystal structure of the binding site of protein–ligand complexes. J Mol Model (Online) 2003;9:172–182.
533. Jiang L, Kuhlman B, Kortemme T, Baker D. A "solvated rotamer" approach to modeling water-mediated hydrogen bonds at protein–protein interfaces. Proteins 2005;58:893–904.
534. Blundell TL. Structure-based drug design. Nature 1996;384:23–26.
535. Verlinde CL, Hol WG. Structure-based drug design: progress, results and challenges. Structure 1994;2:577–587.
536. Lloyd DG, Garcia-Sosa AT, Alberts IL, Todorov NP, Manceral RL. The effect of tightly bound water molecules on the structural interpretation of ligand-derived pharmacophore models. J Comput Aided Mol Des 2004; 18:89–100.
537. Ehrlich L, Reczko M, Bohr H, Wade RC. Prediction of protein hydration sites from sequence by modular neural networks. Protein Eng 1998;11:11–19.
538. Zhang L, Hermans J. Hydrophilicity of cavities in proteins. Proteins 1996;24:433–438.
539. Vedani A, Huhta D. An algorithm for the systematic solvation of proteins based on directionality of hydrogen-bonds. J Am Chem Soc 1991;113:5860–5862.
540. Barillari C, Taylor J, Viner R, Essex JW. Classification of water molecules in protein binding sites. J Am Chem Soc 2007;129:2577–2587.
541. Poornima CS, Dean PM. Hydration in drug design. 1. Multiple hydrogen-bonding features of water molecules in mediating protein–ligand interactions. J Comput Aided Mol Des 1995;9:500–512.
542. Poornima CS, Dean PM. Hydration in drug design. 2. Influence of local site surface shape on water binding. J Comput Aided Mol Des 1995;9:513–520.

543. Poornima CS, Dean PM. Hydration in drug design. 3. Conserved water molecules at the ligand-binding sites of homologous proteins. J Comput Aided Mol Des 1995;9:521–531.

544. Arkin MR, Randal M, DeLano WL, Hyde J, Luong TN, Oslob JD, Raphael DR, Taylor L, Wang J, McDowell RS, et al. Binding of small molecules to an adaptive protein–protein interface. Proc Natl Acad Sci USA 2003; 100:1603–1608.

545. Golke H, Klebe G. Approaches to the description and prediction of the binding affinity of small-molecule ligands to macromolecular receptors. Angew Chem Int Ed Engl 2002; 41:2644–2676.

546. Ladbury JE. Just add water! The effect of water on the specificity of protein–ligand binding sites and its potential application to drug design. Chem Biol 1996;3:973–980.

547. Eisenberg D, McLachlan AD. Solvation energy in protein folding and binding. Nature 1986;319:199–203.

548. Otting G, Liepinsh E, Wuthrich K. Protein hydration in aqueous solution. Science 1991;254:974–980.

549. Makarov VA, Andrews BK, Smith PE, Pettitt BM. Residence times of water molecules in the hydration sites of myoglobin. Biophys J 2000;79:2966–2974.

550. Likic VA, Prendergast FG. Dynamics of internal water in fatty acid binding protein: computer simulations and comparison with experiments. Proteins 2001;43:65–72.

551. Levitt M, Sharon R. Accurate simulation of protein dynamics in solution. Proc Natl Acad Sci USA 1988;85:7557–7561.

552. Lounnas V, Pettitt BM, Phillips GN, Jr. A global model of the protein–solvent interface. Biophys J 1994;66:601–614.

553. Knapp EW, Muegge I. Heterogeneous diffusion of water at protein surfaces: application to BPTI. J Phys Chem 1993;97:11339–11343.

554. Muegge I, Knapp EW. Residence times and lateral diffusion of water at protein surfaces: application to BPTI. J Phys Chem 1995;99: 1371–1374.

555. Brunne RM, Liepinsh E, Otting G, Wuthrich K, van Gunsteren WF. Hydration of proteins. A comparison of experimental residence times of water molecules solvating the bovine pancreatic trypsin inhibitor with theoretical model calculations. J Mol Biol 1993;231:1040–1048.

556. Ahlstroem P, Teleman O, Joensson B. Molecular dynamics simulation of interfacial water structure and dynamics in a parvalbumin solution. J Am Chem Soc 1988;110:4198–4203.

557. Dennis S, Camacho CJ, Vajda S. Continuum electrostatic analysis of preferred solvation sites around proteins in solution. Proteins 2000;38:176–188.

558. Moitessier N, Westhof E, Hanessian S. Docking of aminoglycosides to hydrated and flexible RNA. J Med Chem 2006;49:1023–1033.

559. van Dijk AD, Bonvin AM. Solvated docking: introducing water into the modelling of biomolecular complexes. Bioinformatics 2006;22: 2340–2347.

560. de Graaf C, Pospisil P, Pos W, Folkers G, Vermeulen NP. Binding mode prediction of cytochrome p450 and thymidine kinase protein–ligand complexes by consideration of water and rescoring in automated docking. J Med Chem 2005;48:2308–2318.

561. Pospisil P, Kuoni T, Scapozza L, Folkers G. Methodology and problems of protein–ligand docking: case study of dihydroorotate dehydrogenase, thymidine kinase, and phosphodiesterase 4. J Recept Signal Transduct Res 2002;22:141–154.

562. Czodrowski P, Dramburg I, Sotriffer CA, Klebe G. Development, validation, and application of adapted PEOE charges to estimate pK_a values of functional groups in protein–ligand complexes. Proteins 2006;65: 424–437.

563. Czodrowski P, Sotriffer CA, Klebe G. Protonation changes upon ligand binding to trypsin and thrombin: structural interpretation based on pK_a calculations and ITC experiments. J Mol Biol 2007;367:1347–1356.

564. Czodrowski P, Sotriffer CA, Klebe G. Atypical protonation states in the active site of HIV-1 protease: a computational study. J Chem Inf Model 2007;47:1590–1598.

565. Brenk R, Vetter SW, Boyce SE, Goodin DB, Shoichet BK. Probing molecular docking in a charged model binding site. J Mol Biol 2006;357:1449–1470.

566. Deng W, Verlinde CL. Evaluation of different virtual screening programs for docking in a charged binding pocket. J Chem Inf Model 2008;48:2010–2020.

567. Davis IW, Murray LW, Richardson JS, Richardson DC. MOLPROBITY: structure validation and all-atom contact analysis for nucleic acids and their complexes. Nucleic Acids Res 2004;32:W615–W619.

568. Weichenberger CX, Sippl MJ. NQ-Flipper: recognition and correction of erroneous aspar-

agine and glutamine side-chain rotamers in protein structures. Nucleic Acids Res 2007;35: W403–W406.

569. Zhong W, Gallivan JP, Zhang Y, Li L, Lester HA, Dougherty DA. From *ab initio* quantum mechanics to molecular neurobiology: a cation–pi binding site in the nicotinic receptor. Proc Natl Acad Sci USA 1998;95:12088–12093.

570. Murphy RB, Philipp D, Friesner R. A mixed quantum mechanics/molecular mechanics (QM/MM) method for large-scale modeling of chemistry in protein environments. J Comput Chem 2000;21:1442–1457.

571. Gao J, Xia X. *A priori* evaluation of aqueous polarization effects through Monte Carlo QM–MM simulations. Science 1992; 258: 631–635.

572. Kuhn B, Kollman P, Stahl M. Prediction of pK_a shifts in proteins using a combination of molecular mechanical and continuum solvent calculations. J Comput Chem 2004;25: 1865–1872.

573. Simonson T, Carlsson J, Case DA. Proton binding to proteins: pK_a calculations with explicit and implicit solvent models. J Am Chem Soc 2004;126:4167–4180.

574. Mongan J, Case DA, McCammon JA. Constant pH molecular dynamics in generalized Born implicit solvent. J Comput Chem 2004;25: 2038–2048.

575. Voges D, Karshikoff A. A model of a local dielectric constant in proteins. J Chem Phys 1998;108:2219–2227.

576. Krishtalik LI, Kuznetsov AM, Mertz EL. Electrostatics of proteins: description in terms of two dielectric constants simultaneously. Proteins 1997;28:174–182.

577. Demchuck E, Wade RC. Improving the continuum dielectric approach to calculating pK_a's of ionizable groups in proteins. J Phys Chem 1996;100:17373–17387.

578. Antosiewicz J, McCammon JA, Gilson MK. Prediction of pH-dependent properties of proteins. J Mol Biol 1994;238:415–436.

579. Yang AS, Gunner MR, Sampogna R, Sharp K, Honig B. On the calculation of pK_as in proteins. Proteins 1993;15:252–265.

580. Beroza P, Fredkin DR, Okamura MY, Feher G. Protonation of interacting residues in a protein by a Monte Carlo method: application to lysozyme and the photosynthetic reaction center of *Rhodobacter sphaeroides*. Proc Natl Acad Sci USA 1991;88:5804–5808.

581. Bashford D, Karplus M. pK_a's of ionizable groups in proteins: atomic detail from a continuum electrostatic model. Biochemistry 1990;29:10219–10225.

582. Georgescu RE, Alexov EG, Gunner MR. Combining conformational flexibility and continuum electrostatics for calculating pK_as in proteins. Biophys J 2002;83:1731–1748.

583. Bartlett GJ, Porter CT, Borkakoti N, Thornton JM. Analysis of catalytic residues in enzyme active sites. J Mol Biol 2002;324:105–121.

584. Koumanov A, Ruterjans H, Karshikoff A. Continuum electrostatic analysis of irregular ionization and proton allocation in proteins. Proteins 2002;46:85–96.

585. van Vlijmen HW, Schaefer M, Karplus M. Improving the accuracy of protein pK_a calculations: conformational averaging versus the average structure. Proteins 1998;33:145–158.

586. Nielsen JE, Vriend G. Optimizing the hydrogen-bond network in Poisson–Boltzmann equation-based pK_a calculations. Proteins 2001;43:403–412.

587. Bolton W, Perutz MF. Three dimensional Fourier synthesis of horse deoxyhaemoglobin at 2.8 Angstrom units resolution. Nature 1970;228:551–552.

588. Fermi G, Perutz MF, Shaanan B, Fourme R. The crystal structure of human deoxyhaemoglobin at 1.74 Å resolution. J Mol Biol 1984;175:159–174.

589. Kendrew JC, Bodo G, Dintzis HM, Parrish RG, Wyckoff H, Phillips DC. A three-dimensional model of the myoglobin molecule obtained by X-ray analysis. Nature 1958;181:662–666.

590. Watson H. The Stereochemistry of the Protein Myoglobin. Progr Stereochem 1969;4:299–333.

591. Tsai CJ, Kumar S, Ma B, Nussinov R. Folding funnels, binding funnels, and protein function. Protein Sci 1999;8:1181–1190.

592. Bahar I, Chennubhotla C, Tobi D. Intrinsic dynamics of enzymes in the unbound state and relation to allosteric regulation. Curr Opin Struct Biol 2007;17:633–640.

593. Henzler-Wildman K, Kern D. Dynamic personalities of proteins. Nature 2007;450:964–972.

594. Krauss G. Biochemistry of Signal Transduction and Regulation. Weinheim: Wiley-VCH; 2003.

595. Mi LZ, Devarakonda S, Harp JM, Han Q, Pellicciari R, Willson TM, Khorasanizadeh S, Rastinejad F. Structural basis for bile acid binding and activation of the nuclear receptor FXR. Mol Cell 2003;11:1093–1100.

596. Downes M, Verdecia MA, Roecker AJ, Hughes R, Hogenesch JB, Kast-Woelbern HR, Bow-

man ME, Ferrer JL, Anisfeld AM, Edwards PA, et al. A chemical, genetic, and structural analysis of the nuclear bile acid receptor FXR. Mol Cell 2003;11:1079–1092.

597. Dunitz J, Shomaker V, Trueblood K. Interpretation of atomic displacement parameters from diffraction studies of crystals. J Phys Chem 1988;92:856–867.

598. Merritt EA. Expanding the model: anisotropic displacement parameters in protein structure refinement. Acta Crystallogr D 1999;55: 1109–1117.

599. Vitkup D, Ringe D, Karplus M, Petsko GA. Why protein *R*-factors are so large: a self-consistent analysis. Proteins 2002;46:345–354.

600. Schmidt A, Lamzin VS. From atoms to proteins. Cell Mol Life Sci 2007;64:1959–1969.

601. Bourgeois D, Royant A. Advances in kinetic protein crystallography. Curr Opin Struct Biol 2005;15:538–547.

602. Schmidt M, Ihee H, Pahl R, Srajer V. Protein–ligand interaction probed by time-resolved crystallography. Methods Mol Biol 2005;305: 115–154.

603. Katona G, Carpentier P, Niviere V, Amara P, Adam V, Ohana J, Tsanov N, Bourgeois D. Raman-assisted crystallography reveals end-on peroxide intermediates in a nonheme iron enzyme. Science 2007;316:449–453.

604. http://www.sciencemag.org/content/vol316/issue5823/images/data/449/DC1/1138885s1.mov.

605. Horst R, Wider G, Fiaux J, Bertelsen EB, Horwich AL, Wuthrich K. Proton–proton Overhauser NMR spectroscopy with polypeptide chains in large structures. Proc Natl Acad Sci USA 2006;103:15445–15450.

606. Grishaev A, Tugarinov V, Kay LE, Trewhella J, Bax A. Refined solution structure of the 82-kDa enzyme malate synthase G from joint NMR and synchrotron SAXS restraints. J Biomol NMR 2008;40:95–106.

607. McCammon JA. Target flexibility in molecular recognition. Biochim Biophys Acta 2005; 1754:221–224.

608. Lin JH, Perryman AL, Schames JR, McCammon JA. The relaxed complex method: accommodating receptor flexibility for drug design with an improved scoring scheme. Biopolymers 2003;68:47–62.

609. Lin JH, Perryman AL, Schames JR, McCammon JA. Computational drug design accommodating receptor flexibility: the relaxed complex scheme. J Am Chem Soc 2002;124: 5632–5633.

610. Schames JR, Henchman RH, Siegel JS, Sotriffer CA, Ni H, McCammon JA. Discovery of a novel binding trench in HIV integrase. J Med Chem 2004;47:1879–1881.

611. Rosenfeld RJ, Goodsell DS, Musah RA, Morris GM, Goodin DB, Olson AJ. Automated docking of ligands to an artificial active site: augmenting crystallographic analysis with computer modeling. J Comput Aided Mol Des 2003; 17:525–536.

612. Zacharias M, Sklenar H. Harmonic modes as variables to approximately account for receptor flexibility in ligand–receptor docking simulations: applications to DNA minor groove ligand complex. J Comput Chem 1999;20: 287–300.

613. Nabuurs SB, Wagener M, de Vlieg J. A flexible approach to induced fit docking. J Med Chem 2007;50:6507–6518.

614. Kallblad P, Dean PM. Efficient conformational sampling of local side-chain flexibility. J Mol Biol 2003;326:1651–1665.

615. Totrov M, Abagyan R. Flexible ligand docking to multiple receptor conformations: a practical alternative. Curr Opin Struct Biol 2008;18: 178–184.

616. Zentgraf M, Fokkens J, Sotriffer CA. Addressing protein flexibility and ligand selectivity by "*in situ* cross-docking". ChemMedChem 2006;1:1355–1359.

617. Krol M, Tournier AL, Bates PA. Flexible relaxation of rigid-body docking solutions. Proteins 2007;68:159–169.

618. Kuhn LA. Strength in flexibility: modeling side-chain conformational change in docking and screening. In: Computational and Structural Approaches to Drug Discovery: Ligand–Protein Interactions. Cambridge: RSC Publishing; 2008. p 181–191.

619. Mattevi A, Rizzi M, Bolognesi M. New structures of allosteric proteins revealing remarkable conformational changes. Curr Opin Struct Biol 1996;6:824–829.

620. Ahmed A, Gohlke H. Multiscale modeling of macromolecular conformational changes combining concepts from rigidity and elastic network theory. Proteins 2006;63:1038–1051.

621. Jacobs DJ, Rader AJ, Kuhn LA, Thorpe MF. Protein flexibility predictions using graph theory. Proteins 2001;44:150–165.

622. Tama F, Gadea FX, Marques O, Sanejouand YH. Building-block approach for determining low-frequency normal modes of macromolecules. Proteins 2000;41:1–7.

623. Atilgan AR, Durell SR, Jernigan RL, Demirel MC, Keskin O, Bahar I. Anisotropy of fluctuation dynamics of proteins with an elastic network model. Biophys J 2001;80:505–515.

624. Macchiarulo A, Nobeli I, Thornton JM. Ligand selectivity and competition between enzymes *in silico*. Nat Biotechnol 2004;22:1039–1045.

625. Freire E. Statistical thermodynamic linkage between conformational and binding equilibria. Adv Protein Chem 1998;51:255–279.

626. Sotriffer CA, Dramburg I. "*In situ* cross-docking" to simultaneously address multiple targets. J Med Chem 2005;48:3122–3125.

RECENT TRENDS IN STRUCTURE-BASED DRUG DESIGN AND ENERGETICS

Adriano D. Andricopulo[1]
Rafael V.C. Guido[1]
Daniela B.B. Trivella[2]
Igor Polikarpov[2]
Andrei Leitão[3]
Carlos Montanari[4,*]
[1] Laboratório de Química Medicinal e Computacional, Centro de Biotecnologia Molecular Estrutural, Instituto de Física de São Carlos, Universidade de São Paulo, São Carlos-SP, Brazil
[2] Grupo de Cristalografia, Instituto de Física de São Carlos, Universidade de São Paulo, São Carlos-SP, Brazil
[3] Division of Biocomputing, University of New Mexico School of Medicine, Albuquerque, NM
[4] Grupo de Estudos em Química Medicinal e Produtos Naturais - NEQUIMED-PN, Departamento de Química e Física Molecular, Instituto de Química de São Carlos, Universidade de São Paulo, São Carlos-SP, Brazil

1. INTRODUCTION

The identification of promising hits and the generation of high-quality leads are crucial steps in the early stages of any drug discovery project. Recent advances in medicinal chemistry at the interface of chemistry and biology have created an important foundation in the search for new drug candidates possessing a combination of optimized pharmacodynamic and pharmacokinetic properties. In this regard, the notable impact of combinatorial chemistry and high-throughput screening (HTS) on the discovery of new lead compounds has created a large demand for small organic molecules that act on specific drug targets. Simultaneously, a variety of genomics and proteomics campaigns has fundamentally transformed key research and development (R&D) strategies in the pharmaceutical industry directed to the identification of promising new chemical entities (NCEs) [1–4]. Therefore, drug discovery has become a more rational and fully integrated multidisciplinary process.

The definition and assessment of chemical and biological space highlight the importance of exploring the intrinsic complementary nature between drug and receptor in medicinal chemistry [4,5]. In this scenario, structural and chemical determinants involved in the complex phenomena of molecular recognition, binding mode, and biological activity are elucidated on the basis of a dynamic process including biological testing, structural elucidation, thermodynamic analysis, computer-aided design and organic synthesis [6].

The evolution of molecular and structural biology combined with innovative progress in computational sciences has led to the determination of high-resolution atomic structures of several attractive molecular targets. The integration of these fields is a fundamental basis for structure-based drug design (SBDD), and has had a profound beneficial impact in drug discovery [6–8].

Current approaches in medicinal chemistry employ SBDD strategies as an integral component of hit identification, lead discovery, lead optimization, and drug candidate selection. SBDD combines the power of many scientific methods, such as X-ray crystallography; nuclear magnetic resonance (NMR); molecular modeling, synthetic organic chemistry, quantitative structure–activity relationships (QSARs), and biological evaluation. Frequently, several cycles of SBDD are needed to develop a preclinical drug candidate having acceptable pharmacodynamic and pharmacokinetic properties (Fig. 1) [1]. In this regard, X-ray crystallography and NMR are useful methods to study the 3D structure of biological receptors, as well as the geometry of ligand–receptor complexes [8,9].

The detailed information about a number of molecular targets is available in public databases (e.g., Protein Data Bank (PDB)) [10], and can be conveniently analyzed by Web-based tools such as InterPro [11–13], ExPASy [14], and Relibase [15–17]. On the basis of the receptor structure and ligand bound conformation, medicinal chemistry strategies can be applied with the aim to

*Corresponding author.

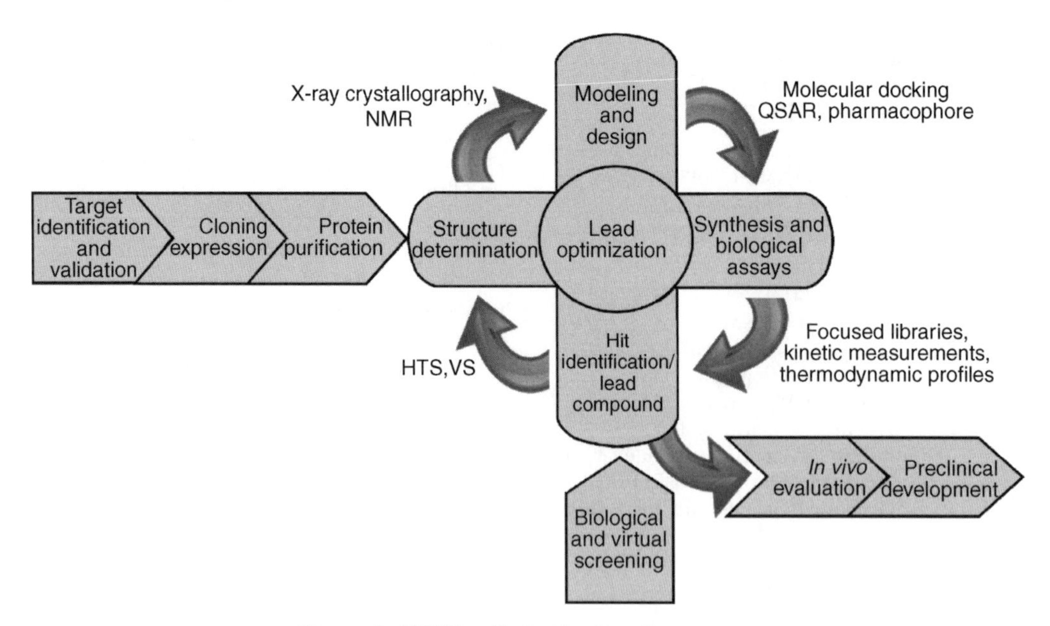

Figure 1. SBDD optimization iterative process.

design new ligands with enhanced affinity and selectivity. The wealth of information generated by these procedures can be correlated through QSAR analyses in order to investigate the fundamental aspects of chemical structure and biological activity for different data sets.

Drug discovery combines innovative strategies based on our increasing understanding of the fundamental principles of protein–ligand interactions [4,7]. Thus, besides providing invaluable information about the spatial arrangements of the target receptor, SBDD methods allow the investigation of intermolecular interactions underlying molecular recognition mechanisms and biological activity. These molecular phenomena rely on the properties and features of a binding pocket [18]. The spatial component involved in the ligand binding process is determined by the arrangement of the amino acids within the binding site, which provides structural and physicochemical constraints that must be met by any putative ligand. Concomitantly, the energetic components are driven by thermodynamics properties, such as the standard Gibbs free energy change (ΔG^0) by means of enthalpic (ΔH^0) and entropic (ΔS^0) contributions as well as the heat capacity change (ΔC_p), with ΔH^0 concerning to energetic features, ΔS^0 to configurational and ordering phenomena and ΔC_p, which relates to changes in both biomolecular and solvent hydrogen bonding. These changes are of pivotal significance in providing unique information upon binding and give rise to the forces that drive the association of molecular entities.

A detailed analysis of the stereoelectronic and thermodynamic properties of the binding pocket with a ligand bound provides useful insights into relevant interactions that can be explored to improve the biological properties of lead compounds [4]. Accordingly, an optimized combination of energetic and structural components would lead to optimal intermolecular interactions and increased receptor–ligand affinity. The knowledge of 3D protein structures and the availability of hundreds of thousands of small-molecule libraries provide compelling amount of information that leads to the use of SBDD approaches. Based on that principle, several success stories have been produced in drug discovery and development (Table 1) [19].

The focus of this chapter is to provide a perspective of the utility of SBDD approaches in drug discovery and its integration with other important computational and experi-

Table 1. Drugs Developed Employing SBDD Methods

Drug	Trade Name	Company	Molecular Target	Indication
Imitinab	Gleevec®	AstraZeneca	BCR-Abl tyrosine kinase	Cancer
Raltitrexed	Tomudex®	AstraZeneca	Thymidylate synthase	Cancer
Dorzolamide	Trusopt®	Merck & Co.	Carbonic anhydrase	Glaucoma
Captopril	Capoten®	BMS	Angiotensin I-converting enzyme	Hypertension
Oseltamivir	Tamiflu®	Roche	Neuraminidase	Influenza
Zanamivir	Relenza®	GSK	Neuraminidase	Influenza
Amprenavir	Agenerase	GSK	HIV protease	AIDS
Indinavir	Crixivan®	Merck & Co.	HIV protease	AIDS
Lopinavir	Kaletra®	Abbott	HIV protease	AIDS
Nelfinavir	Viracept®	Pfizer	HIV protease	AIDS
Ritonavir	Norvir®	Abbott	HIV protease	AIDS
Saquinavir	Invirase®	Roche	HIV protease	AIDS

mental methods, highlighting current opportunities, and future challenges in medicinal chemistry. The consistency of SBDD methods such as molecular modeling, microcalorimetry, synthesis, and biological evaluation is highlighted by several examples including (i) the combination of computational strategies with structural information to assist the drug discovery process (e.g., molecular docking, virtual screening, QSAR, and pharmacophore model), (ii) the successful development of selective agonists of thyroid hormone receptors on the basis of organic chemistry, medicinal chemistry, X-ray crystallography, and molecular modeling, (iii) the design of new transthyretin ligands with enhanced affinity, stability, and selectivity, which involved an integrated approach of *in vitro* testing, organic synthesis, pharmacophores, and X-ray crystallography, and (iv) the mutual analyses of thermodynamic and structural/modeling data in order to understand the energetic basis underlying the binding affinity phenomenon, as well as drug–macromolecule interactions.

2. COMPUTER-AIDED MOLECULAR DESIGN AND SBDD

Most drug discovery programs are driven by a combination of experimental and computational approaches [4]. The huge amount of information available has created a need for robust and reliable data processing techniques to analyze relevant molecular targets and identify privileged scaffolds for drug design. This scenario prompted the development and constant pursuit of new ways to connect chemistry, biology, and informatics, with the goal to support hit identification, lead optimization, and NCE generation [1,4,6,20].

2.1. Molecular Docking

Molecular docking is an important tool in structural molecular biology and cheminformatics, and has been a topic of intense interest for scientists in both academia and pharmaceutical industry [21]. Protein–ligand docking aims to predict and rank structures arising from the association between a given ligand and a target protein of known 3D structure [22,23]. Computer programs dedicated to docking small-molecules into protein binding sites have been receiving considerable attention in recent years due to their versatility and multiple applications in drug design. Docking methods such as DOCK [24], GOLD [25], FlexX [26], Glide [27,28], AutoDock [29], and Surflex-Dock [30] are widely used for the evaluation of intermolecular interactions of ligands into protein binding pockets, determination of binding modes for multiple ligands, as well as for predictions of relative binding affinities for structurally similar or diverse molecules. A general process of molecular docking may be seen in terms of three major steps: (i) perception of the target and ligands, (ii) generation of putative complexes (sampling), and, (iii) evaluation of the "fitness" of the complex [6,23].

Concerning the perception of the protein target and small-molecule ligands, some useful considerations include (i) evaluation of the target receptor druggability, incorporating features such as pocket size, geometry, surface complexity, and roughness, and their complementarity in shape and polarity with respect to a putative drug-like ligand, (ii) analysis and selection of the most relevant geometry of the target receptor, including macromolecular flexibility for ligand binding, that is, the accuracy and reliability of the most suitable receptor conformation for ligand recognition, (iii) assignment of the correct protonation and tautomeric states, that is, the correct definition of these crucial molecular properties for ligand binding and affinity (local dielectric conditions within the binding pocket can modulate pK_a values of functional groups, which can easily turn a hydrogen acceptor group into a donor or a charge-assisted hydrogen bond into a neutral one), (iv) definition of binding site water molecules and metals, which frequently mediate important contacts between protein and ligand (tightly bound water molecules should be considered as an integral part of the target structure and take into consideration for model development) [9].

Prediction of the binding modes for small molecules is perhaps the most straightforward step and the area where most of success has been achieved [9]. In order to sampling the ligand within the target binding site, the docking programs employ different search methods (algorithms) to treat ligand flexibility. Docking algorithms are complemented by scoring functions that are designed to predict the biological activity through the evaluation of interactions between compounds and drug targets. Thus, once the conformations of a system are sampled, the docking programs evaluate them in order to identify the most likely binding site conformations [31]. The scoring functions, in turn, make different assumptions and simplifications in the evaluation of complexes, and do not fully account for a number of physical phenomena toward understanding the molecular recognition process. For example, ligand binding events are determined by a combination of enthalpic and entropic contributions, where either one can drive specific interactions. This often presents a conceptual problem for contemporary scoring functions, because most of them are much more focused on capturing energetic than entropic effects [32]. Furthermore, scoring function programs have different parameters that rely on distinct atom type schemes and atomic partial charges calculation methods, and have been trained on diverse ligand–protein data sets. As a result, each program returns a particularly different estimate of relative binding affinity, and comparisons are nontrivial. To overcome this limitation, some approaches that assign low ranks to inactive compounds and high ranks for active compounds have been proposed. One of the most effective strategies is the combination of estimates from a variety of scoring functions into a single consensus score [33,34]. The impact of consensus scoring strategies in the enrichment of true positives (i.e., hits) can be explained by the fact that the mean of repeated samplings tends to be closer to the true value than any single sampling, thus, since useful scoring functions perform well, different methods will vote for some of the same actives.

Molecular docking is widely employed to evaluate possible binding modes of bioactive molecules. Thus, rational design can be applied to optimize both the spatial complementarity and the interaction pattern between a given ligand and the most important amino acid residues within the protein binding site. Strategic molecular modifications can be carried out to generate new analogs in order to evaluate and validate the docking model, as well as to enhance the pharmacodynamic properties of the lead compound (Fig. 1).

2.2. Structure-Based Virtual Screening (SBVS)

In general, the search for new biologically active molecules from large compound databases by means of computer-assisted techniques is a process known as virtual screening (VS) [4,9]. VS technologies have largely enhanced the impact of computational methods applied to chemistry and biology. The goal of this strategy is to reduce large chemical databases and to select a limited number of promising hits for biological test. High-performance hardware and specialized software, combined with advanced knowledge of 3D protein structure and small-molecule binding modes, have made this technology a useful complement, and in some cases, a reasonable alternative to HTS [9].

In the SBVS approach, knowledge about the 3D structure of the molecular target and its binding site are essential to perform a receptor–ligand high-throughput docking [20]. Owing to its knowledge-based feature, VS strongly depends on the amount and quality of information available about the system under investigation. Regardless of what kind of macromolecule will be employed as target, important aspects of target perception, as discussed above, must be properly considered in SBVS campaigns [9]. Another important step is the appropriate design of the databases of small-molecule ligand candidates for the screening process. Screening libraries generally contain a large number of molecules with broad chemical diversity [35]. As a general criteria, database compounds are primarily selected with respect to their drug-like (i.e., molecules that generally follow Lipinski's rule of five [36]) or lead-like (i.e., molecules that have lower molecular complexity when compared to drugs [37,38]) properties.

One of the major challenges facing SBVS programs is the selection of an appropriate docking tool. The ligands shall achieve the highest possible degree of molecular complementarity (conformational and orientational) with respect to all binding sites of the receptor active site [4]. Currently available docking tools follow slightly different concepts, which make individual programs more suitable for a specific task. In addition, some other issues, such as the treatment of ligand and receptor flexibility, and the number and quality of different protein–ligand complex structures have to be properly addressed [39–41]. Despite the inherent difficulties, there exist a number of successful stories of SBVS [42–46].

2.3. QSAR and SBDD

QSAR has a long history in the drug discovery field, and reached a tremendous impact in the optimization of promising leads that act on specific targets [47]. Medicinal chemistry studies aimed at elucidating fundamental aspects of the relationships between structural or property descriptors and biological activity are important in the understanding of the activity of interest and may enable the prediction of the biological property for new compounds. The availability of modern molecular modeling techniques and advanced 2D and 3D QSAR methods has attracted the attention of scientists for the combination of these approaches to shed light on the molecular basis of important properties such as affinity and selectivity. In this regard, QSAR techniques generate molecular descriptors (independent variable) based on the chemical structures of compounds (standard data sets), and use statistical methods (e.g., partial least squares (PLSs)) to correlate the key descriptors to the biological property value of interest (dependent variable). 3D QSAR methods such as comparative molecular field analysis (CoMFA) [48,49] and comparative molecular similarity index analysis (CoMSIA) [50,51] are widely used for the design of new small-molecule modulators with enhanced properties. A crucial step in 3D QSAR studies is the determination of the spatial molecular alignments for the data set compounds. Several strategies have been used to generate structural alignments for QSAR models. One of these approaches consists of employing molecular docking methods to align the data set compounds within the protein binding pocket (Fig. 2). On the basis of that, a grid box of suitable size is generated embedding all aligned ligands. Then, with the aid of an atom probe (e.g., positively charged sp^3 carbon atom) 3D QSAR molecular fields are calculated (e.g., steric, electrostatic, hydrophobic, hydrogen-bond donor, and hydrogen-bond acceptor). In the CoMFA and CoMSIA QSAR methods, gradual changes of the interaction properties are mapped by evaluating the potential energy at regularly spaced grid points surrounding the aligned molecules. The 3D QSAR contour maps allow easy visualization of regions in space responsible for increases or decreases in the values of a particular type of dependent variable (e.g., IC_{50}, EC_{50}, and K_i). Hence, the combination of molecular docking with 3D QSAR methods is advantageous because it allows direct visualization and interpretation of modeling results within the binding site environment, revealing essential ligand–receptor interactions (Fig. 2) [52].

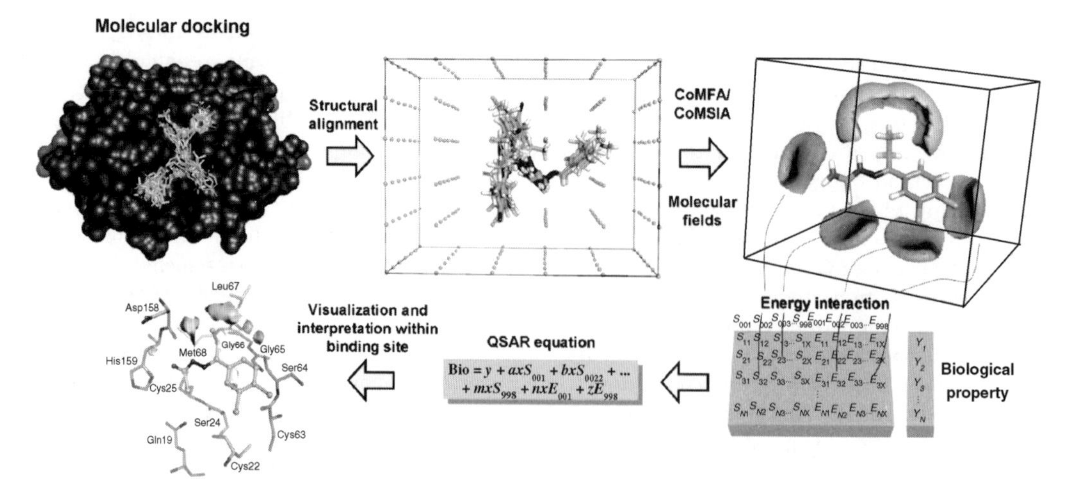

Figure 2. Integration of SBDD and QSAR in drug design.

Several examples highlight the impact of this approach in drug design [53–58].

2.4. Structure-Based Pharmacophore Modeling

Structure-based pharmacophore models are useful tools for the design of NCE candidates [59]. Pharmacophore approaches play a central role in modern medicinal chemistry and have been employed in a wide variety of drug design studies [60,61]. Ligand-based pharmacophore models (created from sets of bioactive ligands) are equally important and present also a significant number of successful applications [46,62,63]. Many methods are available for the description of ligand-based pharmacophore characteristics, including Chem-X [64], DISCO [65], Catalyst [66], GASP [67], Phase [68], and Galahad [69]. These methods have been widely employed for 3D database searching, scaffold hopping, and QSAR studies. On the other hand, the use of pharmacophore descriptors derived from receptor binding sites enables a common frame of reference to be used for structurally diverse ligands. By generating pharmacophore feature points that are complementary to the binding site, an image that represents an ideal virtual ligand (in terms of binding properties) is produced (Fig. 3). As a result, fingerprints or 3D models are obtained from these points and used for the design of new active compounds [59].

Structure-based pharmacophores obtained from molecular interaction fields (MIFs) are based on the assumption that the recognition/affinity of one molecule for the binding site is an additive function of the recognition/affinity of individual atoms that contribute to the binding process. In this approach, physicochemical properties within the binding pocket are mapped by molecular probes that simulate key ligand–receptor interactions (e.g., H-donor, H-acceptor, hydrophobic groups, acidic/basic groups, and so forth). The probes enable the identification of energetically favorable and unfavorable interactions within the binding cavity [70]. The chemical information provided by the MIF is translated into pharmacophore elements that encompasses the physicochemical properties required for binding to a given receptor pocket. A variety of methods are available for this task, nevertheless, none of them can directly derive a pharmacophore model in a fully automated fashion. For this purpose, programs such as UNITY (Tripos, Inc.), Catalyst (Accelerys, Inc.), FlexX-Pharm [71], and PhDOCK [72,73] are widely used. An important feature of the structure-based pharmacophores generated from MIF is the ability to bias the pharmacophoric complementarity selection by using the receptor shape and electronic properties.

An alternative approach to generate structure-based pharmacophore models relies on the combination of molecular interaction fin-

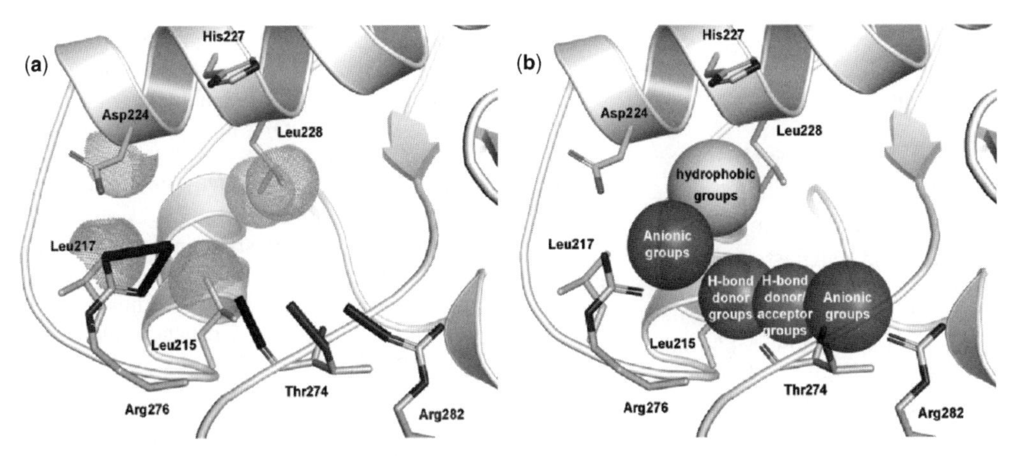

Figure 3. (a) Representation of donor-acceptor intermolecular interactions. Light dots indicate amino acid residues with hydrophobic characteristics, while dark round sticks indicate amino acid residues for specific electrostatic or hydrogen-bonding interactions. (b) Structure-based pharmacophore model. The spheres indicate regions within the binding pocket where hydrophobic, electrostatic, or hydrogen-bond groups from a putative ligand shall be placed.

gerprints (IFPs) and docking [74]. The IFP approach consists in a robust and efficient postdocking tool for processing and prioritizing the most relevant poses for either low molecular weight fragments or molecular scaffolds. These simple and robust bit strings convert 3D information about protein–ligand interactions into bit vector representations that can be quickly compared by the use of traditional metrics (e.g., Tanimoto coefficient or Euclidian distance). The rules for computing the intermolecular interactions are edited on a user-defined basis and the relevant hot spots are assigned as a protein–ligand interaction point. Subsequently, the IFP are clustered and then converted in dendrograms for visualizing the diversity of the generated poses. The poses generated by the docking procedure are scored based on the interaction fingerprint similarity [75].

Recently, a new automated method for the generation of pharmacophore models was developed combining molecular fingerprint and MIF. The fingerprints for ligands and proteins (FLAP) method exploits the significant information from crystallographic structures based on the quantification of the macromolecular fingerprints. The software enables the automated identification of the potential complementary ligand pharmacophoric features for a protein binding site. In addition, FLAP provides the automatic generation of site points for docking and automatic generation of 3D fingerprint descriptors for ligands and proteins ready for chemometric analyses and lead optimization. With the goal of identifying regions in the target binding pocket that are potentially related to pharmacophoric properties, FLAP also relies on the calculation of the MIF produced by GRID probes [76,77]. Therefore, by using GRID probes on a binding site, the locations where an energetic interaction with small molecular features should be favorable are identified and the structure-based pharmacophore is automatically produced.

In summary, structure-based pharmacophore approaches offer timely and cost-effective ways to identify new drug-like ligands for a variety of biological targets, and their utility in drug design is unquestionable. It is worth noting that each drug discovery project is different in some way, and therefore, medicinal chemists must be prepared to understand and take advantage of SBDD methods in their drug discovery and development projects.

3. STRUCTURE-BASED DRUG DESIGN: EXAMPLES

In the following sections, we will exemplify the use of SBDD approaches applied in researches carried out in the last years on the development of thyroid hormone receptor β (TRβ) selective agonists and transthyretin (TTR) amyloid inhibitors.

3.1. TRβ -Selective Agonists

The thyroid hormones, thyroxine (T4) and triiodo L-thyronine (T3), promote the reduction of plasma cholesterol levels and induce weight loss in mammals [78]. However, the administration of T3, for example, in high doses produces undesirable side effects, including tachycardia and arrhythmia [79].

These thyroid hormones interact with thyroid hormone receptors (TRs), a class of intracellular, ligand-regulated transcription factors. TRs are members of the nuclear hormone receptors (NRs) family being involved in the regulation of central metabolic functions [79]. Two major subtypes of the thyroid hormone receptors, TRα and TRβ, are expressed in humans. TRα1 is highly expressed in heart tissues and it has been argued that the thyroid hormone effects on the cardiac rate are mostly influenced by this TR isoform. On the other hand, the TRβ1 isoform is predominantly found in the liver, playing important roles in metabolism in general, and cholesterol metabolism, in particular [80]. However, as these hormones are not selective for one of the TR isoforms (IC_{50} T3: TRβ = 0.26 nM, TRα = 0.24 nM) [81,83] and since the TRβ isoform is the responsible for regulating cholesterol metabolism, systematic efforts have been done within the last years to search for TRβ selective agonists with potential use in therapeutic hypercholesterolemia treatment [82–87].

Since the crystal structures of TRα and TRβ in complex with the natural ligand T3 are available [83,84]. SBDD studies based on iterative procedures of synthesis, and/or modification, of new potential ligands have been done to search for TRβ selective agents.

3.1.1. The TRs Hormone Binding Site TRs contain three major domains: the amino-terminal domain, which influences the TR function in some cellular contexts; the DNA binding domain (DBD), that anchors the receptor to a specific DNA sequence and the carboxyl-terminal ligand binding domain (LBD), responsible for hormone binding. Connecting the DBD and LBD is a so-called "hinge" region, implicated in corepressor binding and release [78]. The LBDs of TRs, similarly to other NRs, have mostly α-helical fold, composed by 12 alpha helixes and 4 small beta strands, with a hormone binding pocket deeply buried within the interior of the protein (Fig. 4a). The LBDs can interact with a number of coregulator proteins

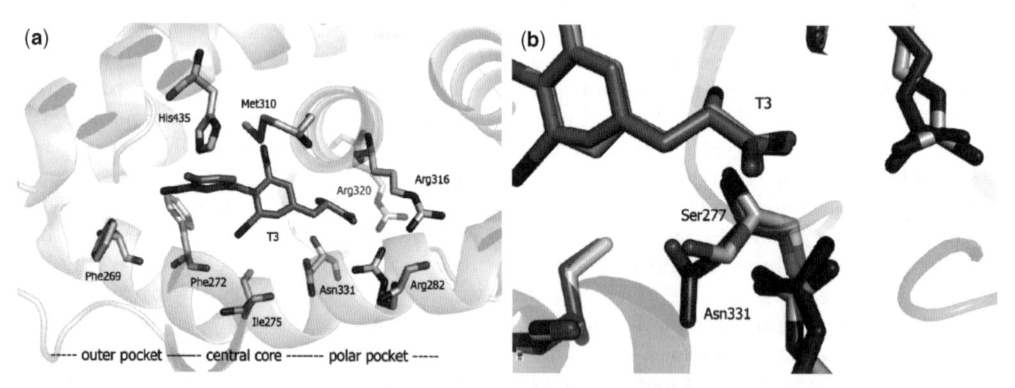

Figure 4. The TR LBD hormone binding site. (a) The TRs binding site can be divided in three main regions. The most important residues that compose each of these regions, and are involved in ligand binding, are shown as sticks. The bottom polar pocket the central hydrophobic region and the outer pocket zones are specified. The natural TRβ ligand T3 is shown in the center of the figure in gray. The secondary structure of TRβ LBD was assigned and showed as cartoon. (b) Superposition of TRβ (gray) with TRα (light gray) crystal structures and zoom at the polar pocket, where an Asn331 is found in TRβ instead of TRα Ser277. Figures were generated with the Pymol software (De Lano Scientific) and the protein coordinates were downloaded from the Protein Data Bank (TRβ LBD:T3, PDB ID: 2H6W and TRα LBD:T3, PDB ID: 2H79). Nitrogen and iodine atoms are shown in black and oxygen atoms are shown in dark gray.

such as corepressors and coactivators, and undergo hormone-induced conformational changes, which mediate transcriptional effects of the receptor [78,83]. For these reasons, the LBD is a prime target of structure-based drug design projects directed to nuclear receptors in general.

TR's hormone binding sites can be divided into three main constituent parts [84] the outer pocket, the central hydrophobic core, and the polar inner pocket (Fig. 4a). The outer pocket is largely hydrophobic but contains a single polar residue (His435 in TRβ isoform) that forms an important hydrogen-bond interaction with the 4'-hydroxyl substituent of most thyromimetics or thyroid hormones T3 and T4. The central core is mainly composed by apolar residues that interact with the thyronine rings of T3 and T4 or with halogen or methyl substituted phenyl rings of thyromimetics. The inner polar pocket contains three important arginine residues and harbors the only subtype selective residue in the TRα and TRβ hormone binding channel; TRα has a Ser277 residue, which is replaced by Asn331 in TRβ (Fig. 4b).

3.1.2. Development of TRβ-selective Agonists Based on TR:T3 Crystal Complexes In the approach to design new TRβ selective agonists based on the crystal structure of TR in complex with T3 (Fig. 4), several modifications of the T3 molecular structure were proposed. These include substitutions of the 4′-OH, the 3,5 and 3′-iodine groups attached to the T3 inner (3 and 5) and outer (4′ and 3′) phenyl rings, the linkage connecting these rings and the side chain attached to the 1 position of the T3 inner ring. The structure–activity relationships (SARs) studies followed by the synthesis and experimental analysis of the new molecules lead to the discovery of several TRβ selective candidates. The search for these TRβ selective agents were performed both by several private companies such as Karo Bio company (Sweden), Pfizer Pharmaceuticals, and by few academic groups (John Baxter's and Tom Scanlan's groups (UCSF), for example). In the following paragraphs, we will review the key published studies related to TRβ selective agonists discovery from a SBDD point of view. For this, we will focus on the compounds displayed in Fig. 5 and discuss the optimization performed at the constituent parts of these compounds that lead to more selective TRβ ligands.

The new molecules were tested experimentally using a cellular-based system with reporter genes regulated by TRα or TRβ. These experiments help to discover if the new compounds induce or suppress TR activation within a cell and also return the EC_{50} values for each test compound of each TR isoforms [81,88]. Another method used to verify TRβ selectivity promoted by a given small molecule is a T3 *in vitro* radiolabeled assay, which allows determining the K_d or IC_{50} values of each compound calculated on the basis of the radioactive T3 displacement from the receptor [81,83]. Thus, the TRβ selectivity is obtained by comparing the EC_{50} and/or the K_d values for TRβ over TRα.

T3 4′-Substituents Screening a Pfizer set of compounds, it was discovered that T3 analogs with 4'-sulfomethyl substituents displayed acceptable affinity and selectivity to TRβ ($K_d\beta$ 2 nM, $\alpha/\beta = 6x$). However, these compounds were partial TRβ agonists [82], probably due to the fact that the 4′-OH of T3 interacts with a histidine residue (His435 in TRβ or His381 in TRα) that is placed at the helix 12 of TR. Since this interaction is essential for the receptor agonist conformation, further modifications at the 4′-position of T3 were discarded.

T3,3,5-Substituents Since the synthesis of compounds containing the 3,5,3'-iodine substituted ring of T3 is technically challenging and iodines are highly susceptible to reductive deiodination, several molecules with different groups placed at 3,5,3'-positions instead of iodines were tested as TRβ selective candidates. These compounds offer a relatively simple route of chemical synthesis with higher yields and stability of the final product than the 3,5,3′-iodine compounds [81,89]. DIMIT (3,5-dimethyl-3′-isopropyl-L-thyronine) (Fig. 5), a T3 analogous, was the first reported TRβ selective agent [90].

It was experimentally demonstrated that DIMIT is twofold more selective to TRβ than the natural ligand T3. Yet, DIMIT binds with relatively low affinity to TR ($K_d TR\beta = 6.2$ nM) [91]—approximately 100-fold less than T3 ($K_d = 0.08$ nM), and the twofold β selectiv-

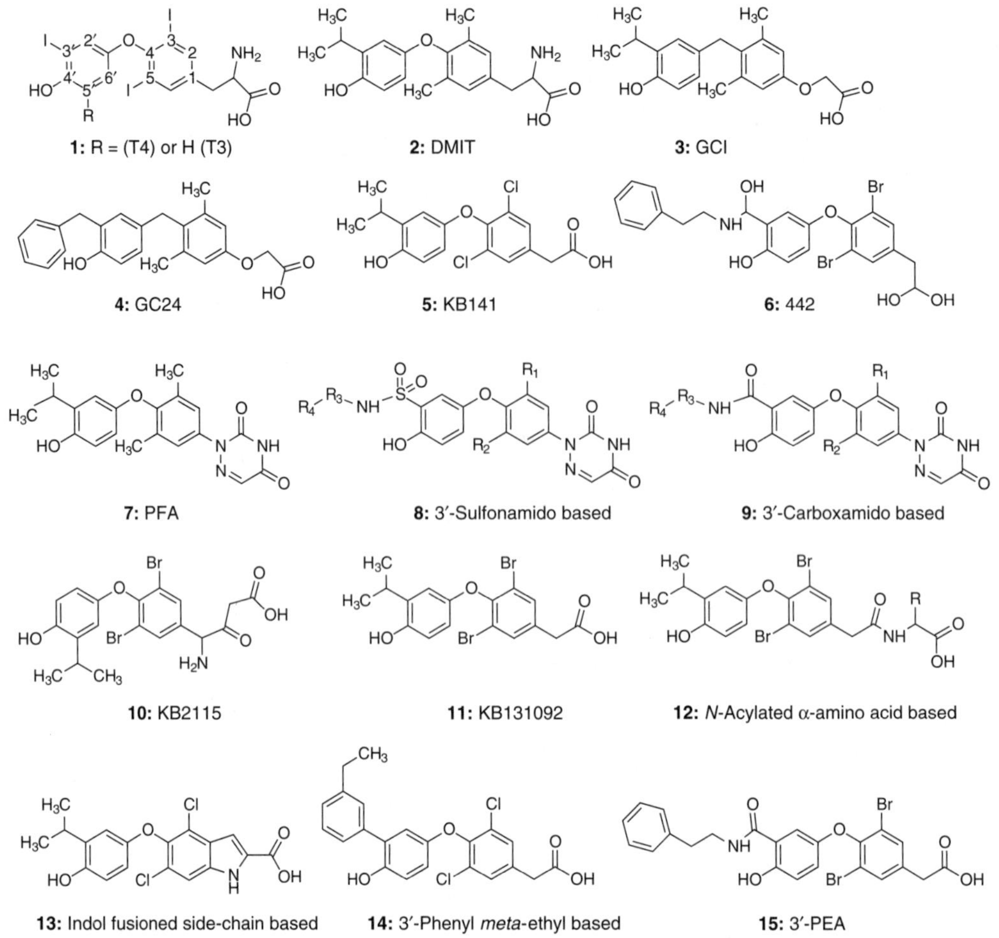

1: R = (T4) or H (T3) **2:** DMIT **3:** GCI **4:** GC24 **5:** KB141 **6:** 442 **7:** PFA **8:** 3′-Sulfonamido based **9:** 3′-Carboxamido based **10:** KB2115 **11:** KB131092 **12:** *N*-Acylated α-amino acid based **13:** Indol fusioned side-chain based **14:** 3′-Phenyl *meta*-ethyl based **15:** 3′-PEA

Figure 5. Representation of bidimensional chemical structure of TRβ agonists. **1**: T4 (thyroxine) and T3 (triiodo L-thyronine); **2**: DIMIT (3,5-dimethyl-3′-isopropylthyronine); **3**: GC1 (3,5-dimethyl-4-(4′-hydroy-3′-isopropylbenzyl)-phenoxy acetic acid); **4**: GC24 (3,5-dimethyl-4-(4-hydroxy-3-benzyl)benzylphenoxyacetic acid); **5**: KB141 (3,5-dichloro-4-[(4-hydroxy-3′-isopropylphenoxy)phenyl]acetic acid); **6**: 442 (2-(3,5dibromo-4-(4-hydroxy-3-(hydroxy(2-phenylethyl)amino(methyl)phenoxy)phenyl)ethane-1,1-; **7**: PFA (4-(4′-hydroxy-3′-isopropyl-phenoxy)-3,5-dimethyl-phenyl]-6-azauracil); **8**: 3′-sulfonamide-based compounds (R_1 and R_2 = Cl or Br); **9**: 3′-carboxamino-based compounds (R_1 and R_2 = Cl or Br); **10**: KB2115; **11**: KB-130192 (3,5-dibromo-4-[(4-hydroxy-3′-isopropylphenoxy)phenyl]acetic acid); **12**: *N*-acylated α-amino acid side-chain-based compounds; **13**: indol fusioned side-chain-based compounds; **14**: 3′-substituted KB-141 analog (3,5-dichloro-4-((4-hydroxy-3′-phenyl *meta*-ethyl phenoxy)phenyl)acetic acid and **15**: 3′-substituted KB-131092 analog (3,5-dibromo-4-((4-hydroxy-3′-phenyl ethyl amide phenoxy)phenyl)acetic acid).

ity is insufficient to guarantee the absence of cardiac side effects caused by its binding to TRα isoform. However, DIMIT was useful as a lead compound in SBDD to search for TRβ selective agonists.

Crystal structure analysis was instrumental in the comprehension of structural reasons of the low binding potency of DIMIT to TR LBD [78]. The methyl substitutions at the 3,5-position were singled out as key factors of the low affinity displayed by DIMIT in TR binding. The van der Waals radii of the methyl groups are smaller than that of the iodine substituents of T3 and were considered to negatively affect the van der Walls contacts and consequently the accommodation of biphenyl ether core into the central hydrophobic part of TR binding cavity [78].

In agreement with these observations, two compounds containing the biphenyl ether scaffold of DIMIT were designed by Karo Bio, carrying a 3,5-chlorine (KB141) or 3,5-bromide (KB130192) substituents, respectively (Fig. 5). It was demonstrated that these compounds showed 9-fold (KB130192) and 14-fold (KB141) selectivity to the TRβ isofrom. Furthermore, the compounds retained similar affinity to their analogs with 3,5-iodide atoms [81]. The compound containing the 3,5-bromide substituent (KB130192) displayed better performance to TR binding than the one with chlorines at 3,5-positions (TRβ IC_{50}: KB141 = 1.1 nM; KB130192 = 0.1 nM). The bromide, a bulkier atom, is probably better accommodated in the TR hydrophobic pocket [86]. Other compounds containing the 3,5-bromide substituents were also demonstrated to show high TRβ affinity, when compared to analogs with smaller 3,5-groups [85]. Therefore, the 3,5-bromide substitution is an interesting option in the design of high-affinity TR agonists, when a biphenyl ether scaffold (as KB130192, KB141, and DIMIT) is used. In addition, 3,5-bromide compounds can be readily prepared rendering larger yields than the 3,5-chlorine or 3,5-iodine substituted molecules [92].

T3 Linker Oppositely to the observed with compounds containing the biphenyl ether scaffold, it was demonstrated that the replacement of the ether bridge link by an ethylene linkage promotes an increase in binding affinity in compounds containing the 3,5-methyl substitutents. It is the case of the DIMIT analog containing a methylene linker, which displays fourfold higher affinity than the original compound DIMIT ($K_d\beta$ = 1.26 nM) [91].

The opposite is also true. GC1 (3,5-dimethyl-4-(4′- hydroxy-3′-isopropylbenzyl)-phenoxy acetic acid), a TRβ selective compound designed by Tom Scanlan's group, possesses the 3′-isopropyl 3,5-methyl substitutents as DIMIT; however, GC1 also carries a methylene linker and an oxyacetic acid side chain. GC1 is a TRβ selective compound (10-fold) and displays similar (K_d = 0.1 nM) TRβ affinity as T_3 (K_d = 0.087 nM) [83]. Studies of GC1 analogs revealed that changes in the methylene link of GC1, producing an ether bridged molecule, also render a TRβ selective analog with ~10-fold more selectivity than T3. However, the ether linked GC1 displays lower affinity to TRβ (K_d = 0.36 nM), further confirming the importance of the methylene linker to binding affinity in compounds with the 3,5-methyl substituents [91].

Analysis of the crystal structures of TRs in the complex with GC1 reveal that when a methylene linker is used, the distance between the two rings of the ligand is increased in comparison with compounds containing the ether linkage. The bond length is 1.53 Å for the methylene linkage, whereas in the case of the ether linkage the bond length is 1.40 Å [83]. In this case, smaller 3,5-substituents, such as the methyl substituents of GC1 are preferred because now the 3,5-methyl substituents can be better accommodated within the TR ligand binding pocket [84,91]. On the other hand, 3,5-iodine atoms are preferred (followed by bromide and later by methyl groups (I > Br > Cl-Met) [86]) when an ether linkage is present, as described above for KB141 and KB131092. Together, these observations suggest that the linker composition interfere with the preference for the chemical nature of the TR ligands 3,5-substituents, and it is based on the accommodation of these 3,5-substituents into the hydrophobic core of the TR's hormone binding site.

T3 Side Chain The side chain of TR agonists plays important roles on TRβ selectivity. It has been demonstrated that charged side chains are important in conferring TRβ selectivity. The crystal structures of TRβ selective compounds containing an oxyacetic acid (GC1), acetic acid (KB141), fused indol, azauracil, and N-acylated -α-aminoacid side chains in complex with TRα and TRβ were solved [78,81,82,84,85]. These crystal complexes show that an H-bond network is responsible for ligand selectivity. The networks are formed differently in the two TR isoforms and are dependent on the unique amino acid substitution (TRα Ser277 or TRβ Asn331) between the TRα and the TRβ binding sites [81,84].

Acetic and Oxyacetic Acid Side Chains It was demonstrated that acetic acid and oxyacetic acid side chains are responsible for TRβ selectivity of KB141 [81,91] and GC1 [84], respectively. Substitutions of the GC1 or KB141 side chains, for longer moieties, as a propionic acid, decreases TRβ selectivity and the L-alanine side chain of the above-mentioned methylene linked DIMIT analog (which differs from GC1 only in the side chain) displays better affinity to TR than DIMIT. However, the selectivity is not enhanced by this substitution [19].

The molecular basis of the ligand side chain interference with the TRβ selectivity was elucidated on the basis of the crystal structures of TRα and TRβ in complex with GC1 [84], KB-141 [81], and others TRβ selective compounds. The comparison of GC1 in complex with both TRα and TRβ [83,84] reveals that these two TR isoforms form different hydrogen-bond networks at the polar part of the binding cavity, connecting three positive charged arginine residues and the ligand GC1, which is dependent on the presence of Asn331 in the TRβ binding site. In TRβ the residue Asn331, plays an important role stabilizing the hydrogen-bond network between GC1 and one of the arginines, Arg282. Asn331 forms a hydrogen bond with Arg282, which anchors the Arg282 side chain in position to form these stable interactions with GC1 carboxylate group (Fig. 6a). However, in TRα the Asn331 residue is replaced by a Ser277 residue, which fails to engage in a hydrogen bond with Arg228 (residue that is equivalent to Arg282 in TRβ). Therefore, and opposite to TRβ, Arg228 side chain is free to sample conformational space, displaying two predominant conformations in TRα crystal structure (Fig. 6a). One of these conformations is turned outward the binding channel, lacking the possibility to form hydrogen bonds with the ligand, which explains the low affinity of GC1 to TRα, and hence the TRβ selectivity [84]. Consequently, the GC1 side chain is found in two different conformations in the TRα:GC1 crystal structure.

The direct participation of Asn331 in TRβ selectivity to GC1 is supported by mutagenesis experiments. These experiments were designed to mimic the ligand binding pocket of one TR isoform within the LBD of the other isoform (TRα Ser277 → Asn and TRβ Asn331 Ser substitutions). In this case, there is no effect on T3 binding, but the original preference for GC1 to TRβ is reversed in favor to TRα [87].

Similarly to GC1, the crystal structure of both TR isoforms in complex with KB141 reveals a comparable hydrogen-bond pattern (Fig. 6b), also implying Ser277 and Asn331 roles in TRβ selectivity [81]. In the case of KB141, the ligand binding mode is very similar in the two TRα and TRβ receptors, for this reason only one KB141 molecule is displayed in Fig. 6b. In the TRα complex, Arg228 forms two hydrogen bonds with Ser277, whereas the corresponding Arg282 in TRβ forms a bifurcated hydrogen bond with the carboxylate oxygen atom of the ligand KB141. The strong electrostatic interaction between the Arg282 and the ligand accounts for the selectivity of this ligand for TRβ [81].

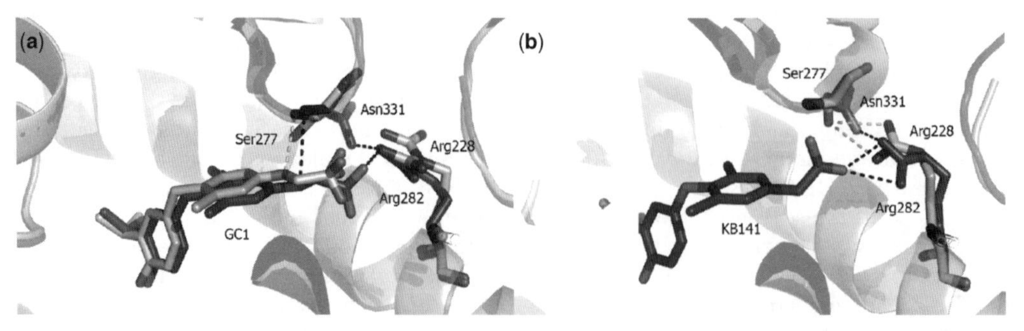

Figure 6. The side chains of (a) GC1 and (b) KB141 (PDB IDs: 1NAV (TRα) and 1NAX (TRβ)) influence TRβ selectivity by establishing different hydrogen-bond network patterns, which is dependent on the presence of Asn331 in TRβ. TRα structures are shown in light gray, whereas the TRβ structures are shown in black. Figures were generated with the Pymol software (De Lano Scientific). Nitrogen and oxygen atoms are shown in dark gray.

T3 Cyclic Side Chains: Indol and Azauracil Side Chains Compounds containing a fusion side chain to the inner ring of thyromimetics with a carboxylate attached (compound **13** in Fig. 5) were designed and tested by Karo Bio [85]. This approach was performed based on crystal structure observations that unoccupied spaces and high-flexible ligand (e.g., GC1 and KB141) and protein aminoacids (Arg, Ser/Asn) side chains may occur in the polar pocket of TRs binding site. In this direction, the rigidification of the ligand carboxylic acid side chain through the formation of indole functionality with the inner ring of thyromimetsics, and the attachment of additional substituents to this side chain, was proposed aimed at increasing TRβ selectivity [85]. In fact, a carboxylic acid attached to the fused side chain results in a 14-fold TRβ selective compound (compound **13** in Fig. 5) that also displays high potency in activating TRβ ($EC_{50} = 0.06$ nM; T3 = 0.26 nM). However, the agonist and selective performance of this compound is dependent on the resulting orientation of the terminal carboxylate group [85]. Changing the carboxylate position from *para* in compound **13** to *meta* or *ortho* results in a smaller TRβ selectivity. It is important to note that GC1 and KB130192 possesses similar affinity as **13** ($K_d \sim 0.1$ nM), but the later compound is more selective to TRβ than GC1 and KB130192, which display 5–10-fold of β selectivity.

The mechanisms, that explain TRβ selectivity in which the carboxyl substituted indol side chain is based, are very similar to the observed to the oxyacetic and acetic acid side chain of KB141. The observation of crystal structures of TRα and TRβ in complex with compound **13** reveals that the residue Arg228 forms two H-bonds with Ser277 in TRα; while the corresponding Arg282, in TRβ, forms a strong bifurcated salt bridge with the terminal carboxylate group of **13** [85], the same mechanism observed to KB141. Again, the molecular basis for selectivity is supported on the unique aminoacid difference in the binding site of the two TR isoforms, even if a fused side chain is present. In this direction, the carboxylate group appears to be the main responsible for TRβ selectivity playing similar roles as the oxyacetic and acetic acid side chains of GC1 and KB141, respectively.

Another class of TRβ selective agonists possessing a 6-azauracil-based side chain was evaluated by Pfizer Pharmaceuticals [82]. The 6-azauracil biphenyl ether having the same 3,5-methyl, 3′-isopropyl substituents as DIMIT (PFA, compound **7** in Fig. 5), displays high affinity ($K_d = 0.03$ nM, T3 = 0.08 nM) and ninefold selectivity to TRβ, suggesting that the 6-azauracil side chain is the main responsible for the significant improvement of both affinity and selectivity of PFA to TRβ. The crystal structure of PFA bound to TRβ was generated (PDB ID: 1N46) and it was verified that the 6-azauracil functionality interacts with Arg320, similarly to the observed in the TRβ in complex with T3 and GC1. However, a new interaction between the C-4 carbonyl of PFA with Arg316 was observed [82].

Minor modifications at the azauracil resulted in the reduction of binding affinity, but the selectivity was retained [82]. Nevertheless, the simplification of the six member cyclic urea—changing the nitrogen atoms for carbon—eliminates binding, showing the importance of the Arg320 interaction with the ionized imide, as well as the strict structural requirements of the 6-azauracil required for TR binding [82].

Compounds containing a side chain composed of an acetic acid condensed with α-aminoacids (*N*-acylated-α-amino acid derivatives) were more recently synthesized by Karo Bio [92]. Restrict information about the biophysical properties of these compounds are available. However, it was demonstrated that a compound based on the scaffold of compound **12** (Fig. 5, with R = isopropyl), shows better pharmacological activities than KB141. This compound reduced the lowering of TSH level—the adverse effect of KB141, in rat models [92], turning it a very promising selective thyroid hormone receptor agonist. Another compound, KB2115 (Fig. 5) is structurally similar to compound **12**, differs in the position of the oxygen and nitrogen heteroatoms in the side chain and in the 3′ → 5′ isopropyl substitution. KB2115 is now in Phase II clinical trials as a cholesterol lowering drug.

T3 3′-Substituents The observation of the crystal structures of TRα and TRβ in complex with ligands reveals that TRβ has higher flexibility at the outer part of the binding channel than TRα, as revealed by the higher B-factor values at the helix 1 and helix 3 and at the helix 11 and helix 12 loops, in TRβ structures when compared to TRα [83,89]. This opens an opportunity for the search for pharmacophores containing bulkier 3′-substituents; thus, challenging the accommodation of ligands within the TRs binding site. Since the TRβ isoform is more flexible at the outer part of the binding site, these ligands carry a promise to display high TRβ selectivity.

As mentioned above, the replacement of the 3′-iodine of T3 to an isopropyl group does not reduce agonist activity, as observed to GC1, DIMIT, and KB141, for example. Attaching a benzyl ring at the 3′-isopropyl of GC1 rendered a new potent TRβ selective agonist, named GC24. GC24 is 40-fold more selective to TRβ than T3 and displays similar affinity and activation of TRβ than the natural hormone T3 [87]. The crystal structure of TRβ in complex with GC24 (PDB ID 1Q4X) reveals that this ligand binds in a similar fashion to GC1. However, GC24 does not interact with Arg282 as GC1, evoking a new mechanism for TRβ selectivity. One of the consequences of this fact is that the mutation Asn331 → Ser in TRβ polar pocket, does not affect significantly the GC24 TRβ selectivity [87].

As GC24 has a longer 3′-substitution, the helixes 3 and 11, on the entrance of the binding channel of TRβ, suffer a displacement of 3–4 Å in relation to the observed in the structure of TRβ with GC1 [87]. It results in an enlargement of the binding channel, forming a new hydrophobic cluster with ligand and protein components. Residues Phe451 and Phe455, placed at helixes 11 and 12, respectively, make additional new contacts with 3′-benzyl of GC24, tightening ligand and protein interactions (Fig. 7b). It is important to note that the helix 12 position, relative to its agonist conformation, is maintained in TRβ after GC24 binding. Furthermore, the conformations of the residues that participate in coactivator recruitment are also preserved. On the other hand, in TRα, the outer pocket structural rearrangements are restricted by its rigidity. It hampers GC24 accommodation into the TRα binding site, decreasing significantly TRα affinity and thus resulting in TRβ selectivity [87].

Another series of 3′-substituted compounds was designed and tested by Karo Bio [86],

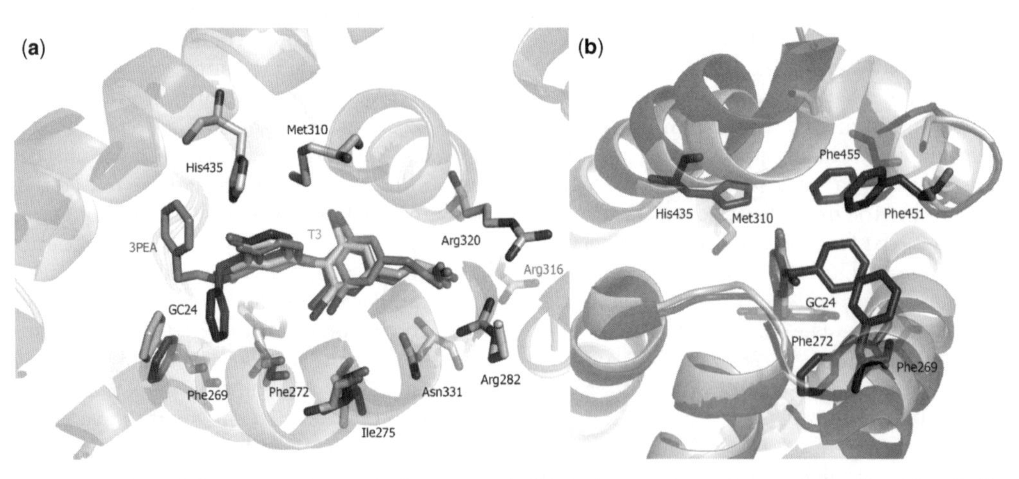

Figure 7. Thyroid hormone receptor isoforms are challenged to accommodate compounds with a bulky 3′-substitutution. (a) Comparison of GC24 (PDB ID: 1Q4X in black), 3′-PEA (PDB ID: 1R6G in dark gray), and T3 (PDB ID: 2H6W in light gray) binding TRβ. (b) The outer cavity expansion and the hydrophobic cluster formation provoked by GC24 when binding TRβ. The TRβ crystal structure in complex with T3 is shown in light gray cartoon, whereas the complex with GC24 is shown in black. The amino acid residues that composed the so-called hydrophobic cluster with GC24 are shown as sticks and named. Figures were generated with the Pymol software (De Lano Scientific). Nitrogen and iodine atoms are colored in black, oxygen and sulfur in dark gray.

starting with the KB130192 and KB141 scaffolds. Attachment of a 3′-phenyl to KB141 results in an improvement in selectivity from 14- to 31-fold. However, the 3′-phenyl substituted compound displayed threefold lower affinity to TRβ (IC_{50} = 2.9 nM) than the original analog KB141 (IC_{50} = 1.1 nM). Using a 3′-phenyl *meta*-trifluoromethyl lipophilic group further increased affinity (IC_{50} = 0.94 nM) and selectivity (25-fold) [86]. It is interesting to note that compounds with *ortho* or *para*-CF3 substituents are less selective to TRβ (three and fourfold, respectively). Testing other *meta*-susbstituents attached to the 3′-phenyl ring, leads to the observation that *meta*-ethyl (IC_{50} = 0.2 nM, 38-fold selectivity) followed by the CF3 are preferred instead *meta*-isopropyl (IC_{50} = 4 nM, 18-fold selectivity) substituents [86]. As the 3′-phenyl *meta*-ethyl KB141 analog (compound **14** in Fig. 5) shows higher affinity and selectivity to the TRβ isoform than the 3′-phenyl *meta*-isopropyl, one can argue that the selectivity to TRβ decreases when the *meta*-substitution at the 3′-phenyl ring increases. In addition, the use of ethers (methoxy, di- and trifluoromethoxy) or hydrophilic groups attached to the 3′-phenyl ring, as well as heterocyclic rings, reduces affinity and selectivity to TRβ relative to compound **14**. It indicates that the use of 3′-phenyl substituents is important to confer selectivity; however, the use of an additional group attached to this ring at the *meta*-position contributes to TRβ affinity [86].

The use of a 3′-phenoxy ring also results in TRβ selective compounds, but with somewhat limited affinity (IC_{50} TRα = 237 nM, TRβ = 6 nM, a/b = 23*x*) [86]. However, when a 3′-phenyl ethyl amide is used (3′-PEA—compound **15** in Fig. 5), the affinity is recovered preserving considerable selectivity (IC_{50} TRβ = 0.62 nM, α/β = 14*x*). Increase in the aliphatic length (propyl, butyl) decreases both affinity and agonist activity (TRα = 37% and TRβ 57% for butyl). Interestingly, a 2,2-diphenyl ethyl amide (IC_{50} TRα = 18 nM, TRβ = 0.47 nM, α/β = 23*x*) is more selective than the 3′-PEA, probably by challenging the TRα binding pocket with one additional phenyl ring.

The crystal structure of the 3′-PEA in complex with TRβ was solved (PDB ID 1R6G [86]) and compared with the KB141 structure. The side chain and biaryl ether of both compounds were accommodated in a similar fashion at the central and polar pockets of the TRβ binding site. Similarly to what was observed in the crystal complex of TRβ with GC24, the outer pocket is enlarged when 3′-PEA is bound. However, the 3′-ring of 3′-PEA is positioned differently than the 3′-ring of GC24 (Fig. 7a). The Met442 side chain moves outward the binding pocket and the ethyl linker connecting the 3′-phenyl ring to the outer ring of 3′-PEA complemented the shape of the enlarged binding site [86], being the ethyl length appropriated to fulfill the binding cavity. As the propyl or butyl substituents decreased TRβ potency, it is reasonable to assume that when a long alkyl chain is used, the steric restrictions on Met442 and other close residues will influence negatively the binding affinity and potency. The 3′-phenyl ethyl amide appears to be the limiting length of 3′-substituents, which is restricted by the protein degree of flexibility.

The 3′-substituents were also evaluated by Pfizer Pharmaceuticals, using the scaffold of PFA (compound **7** in Fig. 5) with a 3,5-Cl-3′-isopropyl substitution as the start material [82]. Aminomethyl, acylamino, sulfonyl, alkyl, aryl, alkylamino, arylamino, acyl, carboxamido, and sulfonamide 3′-substitutions were tested. The latter two, carboxamido- and sulfonamide-based substituents (compounds **8** and **9**, respectively, in Fig. 5), rendered the most potent and TRβ selective compounds. Therefore, several groups attached to the 3′-sulfonamides or carboxinamides scaffolds were tested. It was observed that the 3,5-Cl-3′-piperidinylsulfonamide derivative posses the highest affinity (K_i = 0.02 nM) and selectivity to TRβ (30-fold). In fact, it has higher affinity than T3 and is more selectivity than the original 6-azauracil 3,5-Cl-3′-isoporpyl. On the other hand, the 3′-carboxinamides, with the same substituents as the sulfonamides, showed different affinity/selectivity relationships. For example, the 3′-piperidinyl carboxamides displayed less affinity than 3′-carboxy cycloamines. The opposite occurred in the sulfonamides series. A nopinone derivative amine attached to the 3′-carboxiamide displayed a surprising 58-fold selectivity and high affinity to TRβ (K_i = 0.06 nM).

3.1.3. TRβ-Selective Agonists: Concluding Remarks Exploring the differences between the TRα and the TRβ isoforms, which include the higher flexibility of TRβ at the outer pocket and the longer Asn331 side chain at the polar part of TRβ binding site, result in new potent and β selective TR agonists.

The use of bulky 3′-groups attached to the outer ring, charged side chains, adequate linkages and 3,5-substitutions modify the ligand potential for binding and selectivity. Nonetheless, these changes in ligands affinities and TR isoform selectivities are dependent on the size of the attached groups and their orientation within the TR binding pocket. Another interesting observation is that the added chemical moieties can be dependent on each other. For example, the linkage may influences the nature of 3,5-groups attached to the inner ring, and the 3′-group may alter the position of the side chain in the other side of the binding pocket. Therefore, the rules guiding SBDD approaches should be established on the basis of a wider structural and molecular view, using adequate linkages and substitutions for each particular case in order to achieve the desired TRβ selective ligand.

3.2. Transthyretin Amyloidosis Inhibitors

The homo-tetrameric T4 carrier [93] protein TTR is involved in amyloid fibril formation [94–98]. The deposition of wild-type and mutant TTR aggregates is involved in several severe amyloid diseases, such as senile systematic amyloidosis (SSA), familial amyloid polyneuropathy (FAP), familial amyloid cardiomyopathy (FAC), and central nervous system amyloidosis (CNSA).

It was demonstrated that small molecules could bind to the TTR T4 binding sites delaying TTR aggregation via protein stabilization [100]. The first compounds shown to be capable to prevent TTR fibril formation were T4 itself and 2,4,6-triiodophenol [100]. Later, nonsteroidal anti-inflammatory drugs (NSAIDs) such as flufenamic acid (FLU), diclofenac, flurbiprofen, and diflunisal [101–103] and several classes of small molecules containing a biphenyl, dibenzofuran, stilbene, or flavone skeleton [102] were described as potential TTR fibril formation inhibitors (Fig. 8).

3.2.1. The TTR T4 Binding Site The tertiary structure of TTR monomer is composed of two

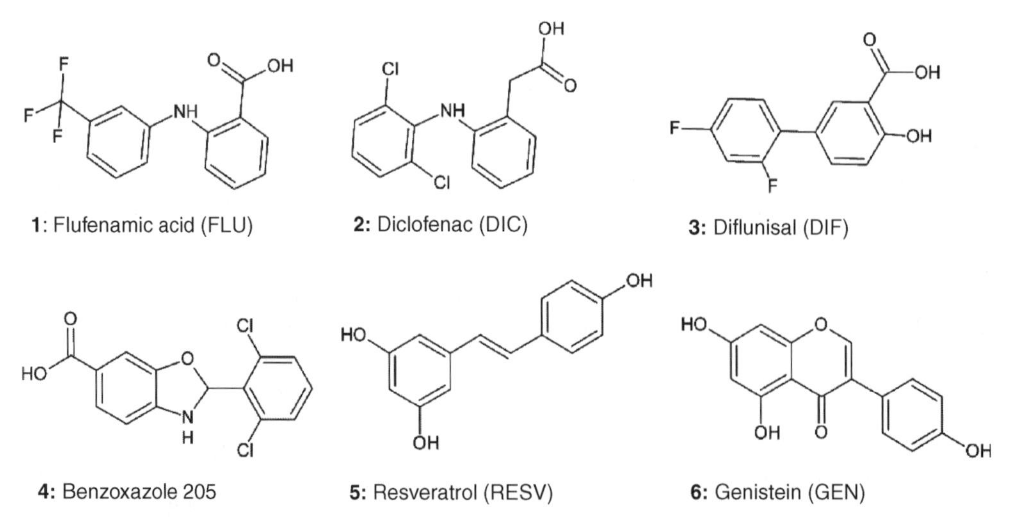

Figure 8. The principal TTR inhibitors scaffolds found in the initial TTR inhibitors screenings. NSAIDs: **1**: flufenamic acid, **2**: diclofenac, and **3**: diflunisal, **4**: benzoxazoles, such as compound 205 (2-(2,6-dichlorophenyl)-1,3-benzoxazole-6-carboxylic acid) and natural compounds **5**: stilbenes, such as resveratrol and **6**: isoflavones, such as genistein.

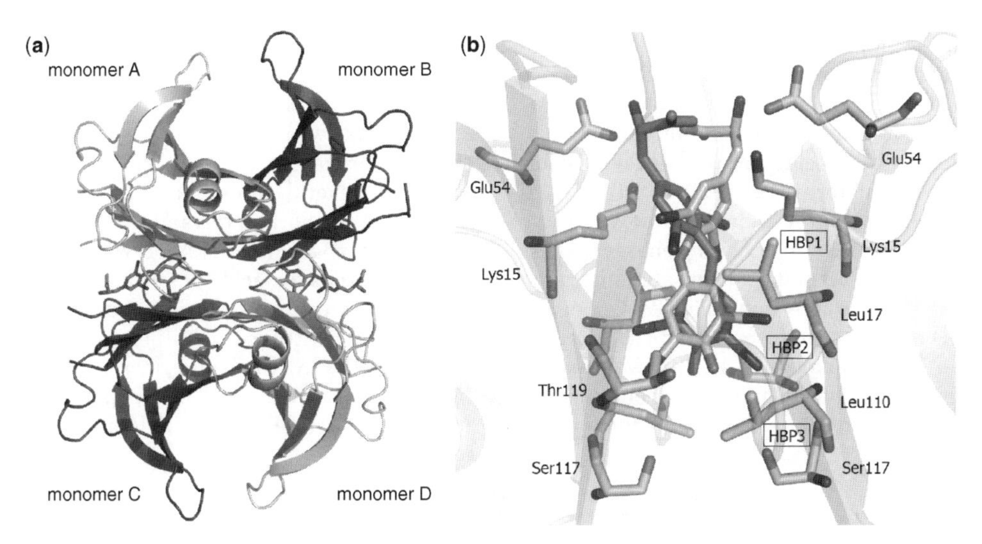

Figure 9. The TTR tetramer posses two identical T4 binding sites. (a) Crystal structure of TTR tetramer binding thyroxine (PDB ID: 1ROX). (b) A detailed view of the T4 biding site. The three hydrophobic small depressions related to the halogen binding pockets are indicated (HBP1, HBP2, and HBP3). The principal aminoacid residues involved in T4 binding to TTR are named and showed as sticks. The two symmetric T4 molecules, found in TTR:T4 crystal structure, are shown as dark and light gray. Iodine and nitrogen atoms are shown in black, whereas the oxygen atoms are shown in dark gray. Figures were generated with the Pymol software (De Lano Scientific).

β-sheets (inner and outer sheets), which are organized in eight β-strands (namely, A–H). The inner sheets of the dimer–dimer (AB–CD) interface—strands A, G, and H—form two cavities (AB and CD) of T4 binding sites (Fig. 9). These binding sites show three hydrophobic depressions, related to as halogen binding pockets (HBP1, HBP2, and HBP3). The HBPs are involved in the accommodation of T4 iodine substituents [99].

Figure 9b shows one TTR binding channel occupied by thyroxine (PDB ID 1ROX). Two symmetric related T4 molecules are generated by a twofold symmetry along the crystallographic axis Z (vertical on the figure). This particularity is current in TTR crystal structures crystallized in the $P2_12_12$ space group, what sometimes limits ligand binding interpretation. In the TTR:T4 crystal structure displayed on Fig. 9, the T4 carboxylate substituted ring is placed in the outer cavity, interacting with the carboxylic acid of the amino acid residue Glu54. Whereas the 3,5, iodine-4,hydroxyl-phenyl ring is located in the inner part, interacting with residues from the HBP2 and HBP3. This binding mode is referred as the *forward* binding mode. In contrast, some ligands were described to bind TTR in the so-called *reverse* binding mode, with its carboxylic acid or polar substituent phenyl ring placed in the inner cavity and the halogenated phenyl in the outer cavity [99].

3.2.2. Structure-Based Drug Design of TTR Fibril Formation Inhibitors Several small molecules proposed as TTR fibril inhibitors exhibit strong binding to plasma proteins and/or provoked additional undesired physiological effects, which might be a drawback in turning them into lead compounds for therapeutical development. Therefore, based on known TTR fibril formation inhibitors, structure-based drug design studies have been developed aiming at improving the TTR binding affinity and selectivity toward human plasma. The identification of molecules that have the potential to stabilize TTR tetramers and demonstrate minimum side effects under prolonged use is of considerable interest for pharmaceutical research. The NSAID and natural compounds SBDD derived molecules will be discussed in the following subsections.

NSAIDs Flufenamic acid (2-((3-(trifluoromethyl)phenyl)amino)benzoic acid), diclofenac (2-(2,6-dichlorophenyl)amino)benzeneacetic acid), and diflunisal (5-(2,4-difluorophenyl)-2-hydroxy-benzoic acid) are potent TTR fibril formation inhibitors [101,102,104,106,107,112] and are FDA-approved drugs. However, these NSAIDs act inhibiting ciclooxygenases (COXs), and therefore their long-term administration results in gastrointestinal side effects. Several SBDD studies have investigated the potential use of NSAID analogs as TTR fibril formation inhibitors. An "ideal" inhibitor should display high TTR affinity and selectivity in human plasma, as well as minimal COX inhibition and TR binding [107,123].

Flufenamic Acid The crystal structure of TTR in complex with FLU (PDB ID: 1BM7) (Fig. 10a) revealed that this small molecule mediates several intersubunit interactions, stabilizing the normal tetrameric fold of TTR, delaying fibril formation [99]. It was observed that the carboxylic acid substituent is important to allow polar interactions with residues placed at the entrance of TTR binding site, in particular the Lys15 residue, increasing TTR-ligand affinity [102,103]. The occupancy of the inner pockets (HBP3) of the TTR binding sites

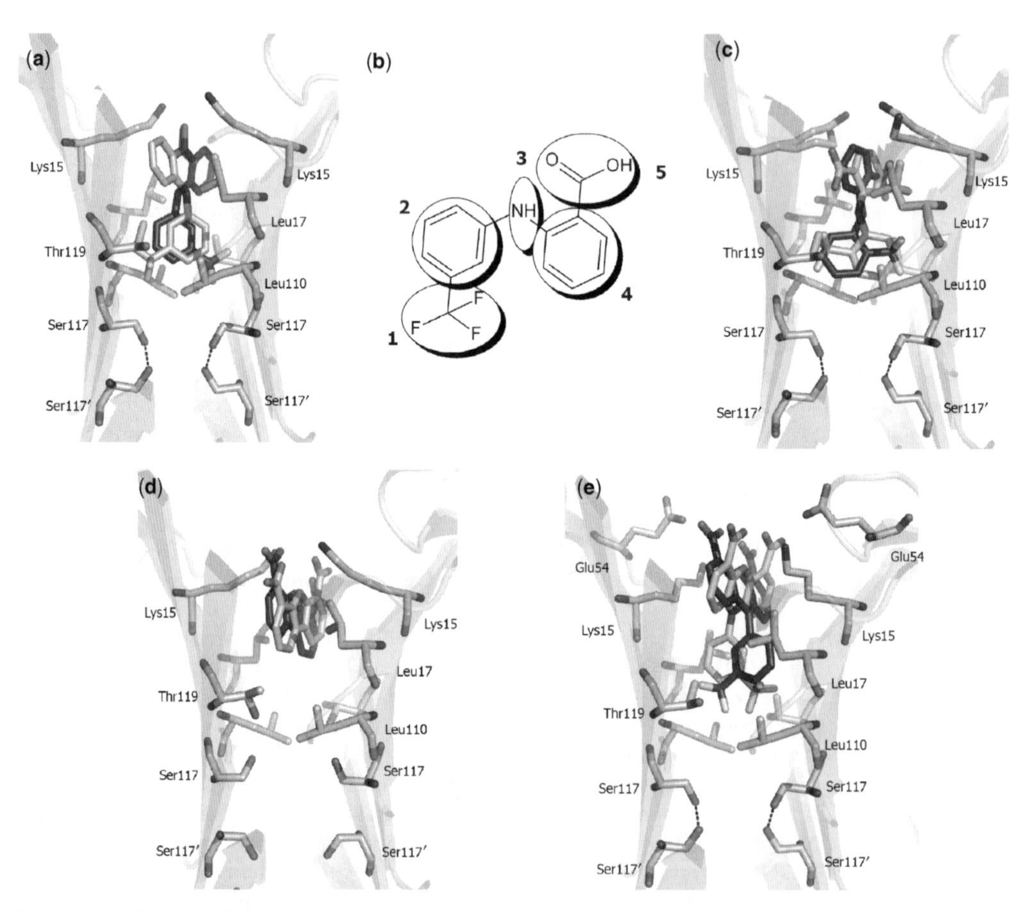

Figure 10. FLU and FLU-based molecules are potent TTR fibril formation inhibitors. (a) The TTR binding site in the crystal structure of TTR:FLU complex (PDB ID: 1BM7), showing the Ser117 side chain turned outward the cavity, in the direction of the other binding site. (b) FLU can be divided into substructures labeled **1–5**, where **1** and **5** are changeable radicals, **3** is the linker and **2** and **4** are the aromatic rings. Therefore, FLU-based TTR inhibitors (c) *ortho*-FLU (PDB ID: 1DVZ), (d) DDBF (PDB ID: 1DVU), and (e) PHENOX (PDB ID: 1DVY) were developed by employing SBDD. The color schema used was the same as in Fig. 9. Figures were generated with the Pymol software (De Lano Scientific).

was also shown as requirement for a good TTR fibril inhibitor [102,105]. The protein–ligand interactions by themselves, as well as conformational changes of residues placed at the bottom of the binding sites (mainly Ser117) are involved in new intersubunit interactions between residues from adjacent binding sites, which contribute to TTR tetramer stabilization. In particular, the side chain of Ser117 can move outward one binding channel, in the direction of the second channel, forming a new and strong interaction with an adjacent Ser117. This new interchain interaction was proposed to contribute to tetramer stability and, as a consequence, to the TTR fibril formation inhibition [103].

Starting with a 2.0 Å X-ray crystal structure of the TTR-FLU complex, a structure-based ligand design strategy was used to conceive TTR amyloid fibril inhibitors [101,107,109]. The strategy employed to design potent FLU-based TTR inhibitors had its foundation on varying the structural components of FLU. The trifluoromethyl and carboxylate substituents, the presence of one, two or more aryl rings and the linkage type and length connecting these rings were the first steps in lead optimization [101] (Fig. 10b).

The first study using FLU-based molecules resulted in the finding that aromatic-based structures might have at least two aromatic rings (bi- or tricyclic aromatic rings). Furthermore, an acidic functional group (phenol or a carboxylic acid) attached to the outer ring (substructure **5** and **4** in Fig. 10b) is also important to achieve TTR affinity and tetramer stability [101]. Based on these observations, a screening of synthetic anthranilic acid-based compounds was performed [109]. It was tested the *ortho*, *meta*, or *para* trifluoromethyl and carboxylate substitutions attached to rings **2** and **4**, respectively; and the presence of an additional methylene group composing the linker. It was observed that the presence of the methylene spacer is unfavorable to inhibiting TTR fibril formation. Furthermore, the position of the trifluoromethyl substitution attached to ring **2** influences the inhibitor potency, being the *ortho*-substitution the most favorable position for the CF3 group, in FLU-based TTR inhibitors [109].

The structure of wtTTR with *ortho*-FLU (PDB ID: 1DVZ) was solved by Klabunde et al. [107]. It was demonstrated that this molecule binds with its carboxylate-substituent phenyl ring positioned at the HBP1 and its *ortho*-CF3-substituted phenyl ring at the HBP3 and HBP3′, direct connecting different TTR monomers (Fig. 10c). This observation would suggest that *ortho*-FLU is a better TTR inhibitor than FLU. However, the direct interaction of the carboxylate group of *ortho*-FLU with the amino group of Lys15, found in the TTR:FLU crystal structure (Fig. 10a), is not observed in the TTR: *ortho*-FLU crystal complex, probably due to a minor rotation in the *ortho*-FLU carboxylate-substituted phenyl ring. The later observation can explain the lower potency of *ortho*-FLU in inhibiting TTR fibril formation related to the one verified by FLU (fibril formation inhibition at twofold ligand molar excess is 98% and 91% to FLU and *ortho*-Flu, respectively) [107].

On a subsequent approach, tricyclic compounds, such as dibenzofurans, phenoxazines, xanthenes, or dibenzo-*p*-dioxins were hypothesized to be good TTR binders. This hypothesis was formed based on the observation of the two symmetric-related binding conformations of FLU in the TTR:FLU crystal complex (Fig. 10a). The superposition of these FLU symmetric molecules leads to a tricyclic ring system, which would join the two HBP1 pockets in the binding cavity of TTR, enhancing its affinity and tetramer stability. The 4,6-dicarboxylate derivative of dibenzofuran (DDBF) was synthesized, and it was shown that DDBF indeed is a TTR fibril inhibitor, however, it has a moderate potency, showing 20% of fibril formation in the *in vitro* fibril formation assay conducted at twofold ligand molar excess [107]. The X-ray structure of TTR-DDBF (PDB ID: 1DVU) shows that the tricyclic ring system fits well into the two adjacent HBP1 of TTR, and the carboxylate groups of DDBF can form hydrogen-bond interactions with the amino groups of the two adjacent Lys15 in the outer part of the TTR binding site. This DDBF way of interaction with TTR amplifies the tetrameric contacts at the outer cavity. However, Ser117 and Thr119 are not affected by DBBF binding and do not undergo the conformational changes necessary to allow new TTR interchain interactions in the bottom of the bottom of the binding sites, as observed in most other TTR-inhibitor complexes (Fig. 10d) [101].

Therefore, a second series of compounds, the *N*-phenyl-substituted phenoxazine-4,6-dicarboxylate series [105], was designed. These compounds were synthesized based on the structure of DBBF with an additional substituted-phenyl ring, to fill the available space in the inner cavity of the T4 binding sites (see Fig. 10d). The structure of one of the two most potent phenoxazines, the *meta*-trifluoromethyl-substituent *N*-aryl phenoxazine (PHENOX), bound to TTR was solved at 1.9 Å resolution (PDB ID: 1DVY [107], Fig. 10e). Similar to the observed in the TTR:DBBF structure, the phenoxazine ring system of PHENOX occupies the outer thyroid hormone binding cavity interacting with both subunits of TTR simultaneously throughout contacts with the HBP1 and HBP1'. The *meta*-CF3 substituted phenyl ring occupies the inner TTR binding cavity, but the CF3 group of PHENOX does not bind as deep in the HBP3 as the CF3 of FLU. However, this binding mode of PHENOX could induce Ser117 and Thr119 conformational changes, resulting in the formation of additional intersubunit hydrogen-bond interaction between two Ser117 residues connecting two adjacent TTR monomers. Furthermore, the carboxylates substituents of PHENOX are in ideal distances to form electrostatic interactions with the amino group of Lys15 and hydrogen bonds with the Glu54 side chain at the entrance of the binding channel. PHENOX was the first synthetic compound described to interact with Glu54. These interactions might contribute to the high potency of PHENOX in inhibiting TTR fibril formation (3% of TTR fibril formation) [107]. Despite the similar potency of PHENOX and FLU as TTR fibril inhibitors, the mechanism of binding is different between these two compounds. isothermal titration calorimetry (ITC) experiments, conducted with FLU and PHENOX [107], showed that the binding of FLU to TTR is enthalpically driven, with a small unfavored entropic contribution. However, the binding of PHENOX is entropically favored. Since PHENOX has a large and rigid tricyclic ring system the differences in the energetic contributions of FLU and PHENOX, when binding to TTR, are probably supported in the difference of the intrinsic conformational freedom of these ligands, as well as in the displacement of water molecules caused by PHENOX binding [107].

Diclofenac The discovery that diclofenac (DIC), another NSAID, is a moderate TTR fibril formation inhibitor was important, since it has higher gastro-tolerability than FLU and other NSAIDs. The crystal structure of TTR in complex with diclofenac was solved by Klabunde et al. [107] (PDB ID: 1DVX). DIC binds in the reverse binding mode with its carboxylate substituent placed in the inner HBP3. The acetate group of DIC forms hydrogen bonds with Thr119 and induces conformation changes of Ser117 side chain, promoting formation of new intersubunit contacts. The Ser117 side chain is flexible in the TTR-DIC crystal structure, displaying two well-defined conformations. This explains, at least in part, the modest (17% of TTR fibril formation at twofold ligand molar excess) [107] DIC effect as a TTR fibril formation inhibitor. Nevertheless, DIC has low maximum therapeutic level and low selectivity to TTR in human plasma [108].

Therefore, a series of DIC analogs were designed and synthesized by Oza et al. [106] aiming at identifying the functional substituents to enhance binding affinity, selectivity and inhibition potency. They changed the position of the chlorines from 2,6 to 3,5 and changed the length of the carboxylic acid attached to the aryl ring of diclofenac. The *in vitro* studies showed that the presence of negative charged substituents is crucial for TTR fibril formation inhibition. However, the chlorine position and the length of the carboxylic acid attached to the aryl ring do not interfere in inhibitor activity. On the other hand, the selectivity to TTR in human plasma is dependent on the chlorine position, as well as on the length of the carboxylic acid substituent. 2-((3,5 Dichlorophenyl) amino) benzoic acid, displayed high TTR affinity ($K_d = 5$ nM) and selectivity (plasma stoichiometry > 1.3). The methyl instead the ethyl carboxylic acid attached to the phenyl ring in this diclofenac analog allows polar contacts with both Ser117 and Thr119 side chains in the inner part of

TTR binding site, contributing to its high affinity to TTR and probably interfering in its selectivity to TTR in human plasma.

It was verified that the diclofenac analog 2-((3,5 dichlorophenyl) amino) benzoic acid also inhibits TTR point mutants fibril formation (Val30 → Met, Val122 → Ile, Thr60 → Ala, and Ile84 → Ser) involved in FAP [110]. This diclofenac analog is more potent than diclofenac and has a smaller dissociation constant ($K_d = 5$ nM) in comparison with flufenamic acid ($K_{d1} = 30$ nM; $K_{d2} = 225$ nM), being a promising wild-type and mutant TTR inhibitor.

Diflunisal Diflunisal, other FDA-approved NSAID, also displays wild-type and FAP associated mutant TTR fibril formation inhibition [111], reducing 97% of TTR fibril formation in 1:2 (TTR:drug) concentrations [112]. Diflunisal is a high selective TTR binder in human plasma, with binding stoichiometry exceeding 1.5 [113]. Furthermore, orally administrated (2 × 250 mg/day) diflunisal concentrations in human plasma is approximately 100 μM [114] and the average TTR concentration in human blood is 3.6 μM. The properties of this NSAID include diflunisal and its analogs as promising TTR inhibitors candidates.

The crystal structure of TTR in complex with diflunisal (PDB ID: 3D2T) showed that this TTR inhibitor binds to TTR either in the forward or in the reverse binding modes [112]. The forward mode (Fig. 11b in gray) enhances diflunisal's hydrophobic interactions with the inner pockets of TTR binding sites, but the electrostatic interactions with Lys15 at the entrance are not robust, and the side chain of this aminoacid residue was found in multiple conformations. The most probable Lys15 conformation is shown in Fig. 11b. However, polar contacts with Thr119 are allowed when diflunisal binds in the reverse binding mode (Fig. 11b in dark gray).

The change of the salicylic acid substitution by a carboxylic acid at the *para*-position in one of the two phenyl rings of diflunisal and a 3,5-fluorine substituents on the other ring resulted in a diflunisal analog that binds with high affinity to TTR ($K_{d1} = 9$ nM), only in the reverse binding mode (PDB ID: 2B9A) (Fig. 11c) [112]. In this binding mode, both the aryl rings make hydrophobic interactions with the amino acid residues placed in the HBP1 and HBP2, whereas the fluorines interact simultaneously with the two HBP1 at the entrance of the ligand binding pocket. The carboxylate hydrophilic aryl ring is placed in the inner binding pocket, forming hydrogen bonds with both Ser117 and Ser117'. This binding mode preference, allows tight interactions with Ser117 and a high number of hydrophobic contacts with the HBP1 and HBP2, improving the affinity to TTR, as well as TTR selectivity in human blood plasma. This compound is a promising TTR inhibitor candidate [112].

Another diflunisal analog with an additional iodine attached to the phenyl ring of diflunisal, iododiflunisal (2′,4′-difluoro-4-hydroxy-5-iodo-(1,1′-biphenyl)-3-carboxylic acid), was also evaluated [115]. The iodine substituted phenyl ring of iododiflunisal is positioned in the outer cavity, interacting with residues from HBP1 and HBP2, as revealed by its crystal structure in complex with TTR (PDB ID: 1Y1D). The phenolic hydroxyl and carboxylate substituents are involved in direct or water mediated interactions with the amino group of Lys15, which contributes to iododiflunisal affinity to TTR [115]. The fluorine-substituted phenyl ring is positioned deep in the interior of the binding cavity, with the two fluorine atoms accommodated into the HBP3 pockets (Fig. 11d). The iodine substituent might play a crucial role in the preference for the forward binding mode observed for iododiflunisal, since in the crystal structure of TTR in complex with diflunisal both the forward and the reverse binding modes were observed [112].

Other Classes of TTR Inhibitors Derived from NSAIDs Other structure templates, based on NSAIDs, were analyzed as TTR inhibitors, by employing structure-based drug design strategy. These include oxazoles, biphenyl ethers, hydroxylated polychlorinated biphenyls (OH-PCB), and more recently, bisaryloximes and, 2-arylbenzoxazoles. These different chemical scaffolds, as well as their interaction with TTR and effects on TTR fibrilogenesis and selectivity will be discussed below.

Oxazoles and Benzoxazoles A series of benzoxazoles were synthesized based on the structure of typical TTR inhibitors. Razavi et

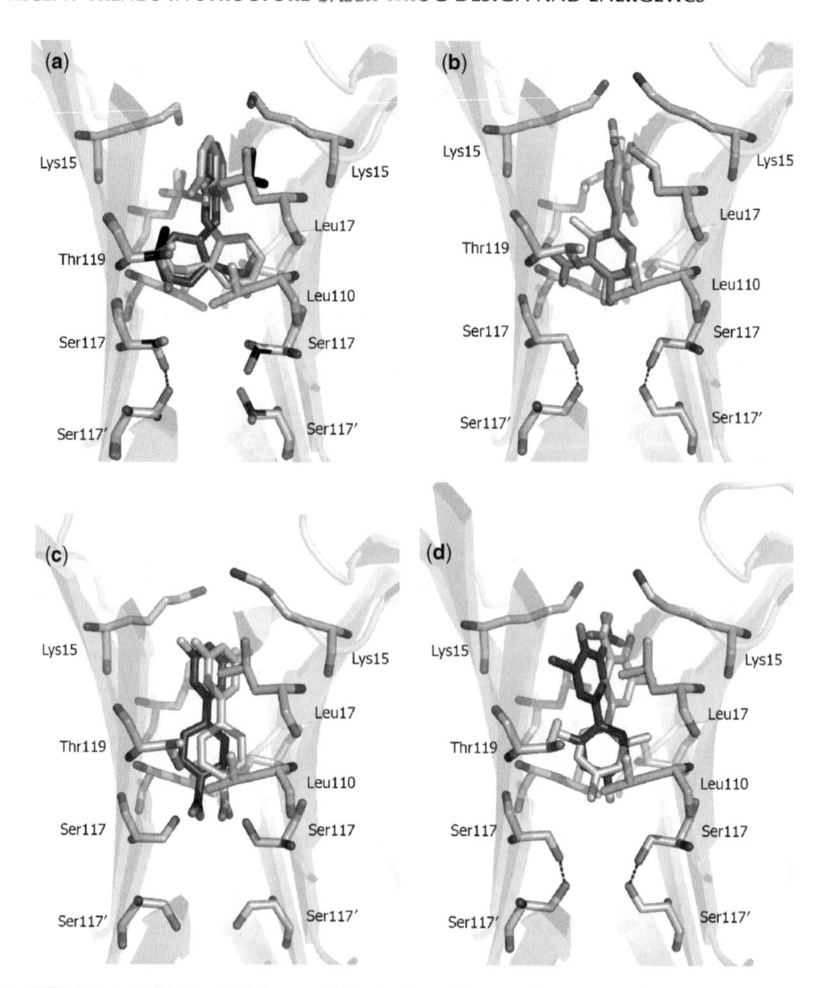

Figure 11. TTR binds NSAIDs diclofenac, diflunisal, and its analogs in the forward and/or in the reverse binding modes. The crystal structure of TTR binding (a) diclofenac (PDB ID: 1DVX), (b) diflunisal—in the forward binding mode (gray sticks) and in the reverse binding mode (dark gray sticks) (PDB ID: 3D2T)—and its analogs: (c) 3′,5′-difluorobiphenyl-4-carboxylic acid (reverse binding mode, PDB ID: 2B9A) and (d) iododiflunisal (forward binding mode, PDB ID: 1Y1D). The TTR binding site is shown in gray and the most important residues involved in ligand binding are shown as sticks. Amino acid side chains exhibiting double conformation are shown in black. The ligands are shown as black sticks, and the symmetric ligand molecules, generated on the crystallographic axis *Z*, are shown in light gray. Fluorine and chloride atoms are shown in light gray, nitrogen in black, and oxygen in dark gray. Figures were generated with the Pymol software (De Lano Scientific).

al. [116] tested benzoxazoles substituted with a carboxylic acid at C4–C7 and a halogenated phenyl ring at C2. It was observed that the position of the carboxylic acid, as well as the presence of halogen substituents are important for ligand binding and fibril formation inhibition. The crystal structure of TTR in complex with most potent compound, the 2-(2,6-dichlorophenyl)-1,3-benzoxazole-6-carboxylic acid, was solved (Fig. 12a). The 2,6-chlorine substitution on the phenyl ring caused an intramolecular readjustment of the inhibitor, placing the carboxylic acid on the benzoxazole unit in ideal distance to make polar contacts with the amine group of Lys15 side chain (Fig. 12a). Furthermore, these chlorine substituents make important van der Waals interactions in the HBP2, also contri-

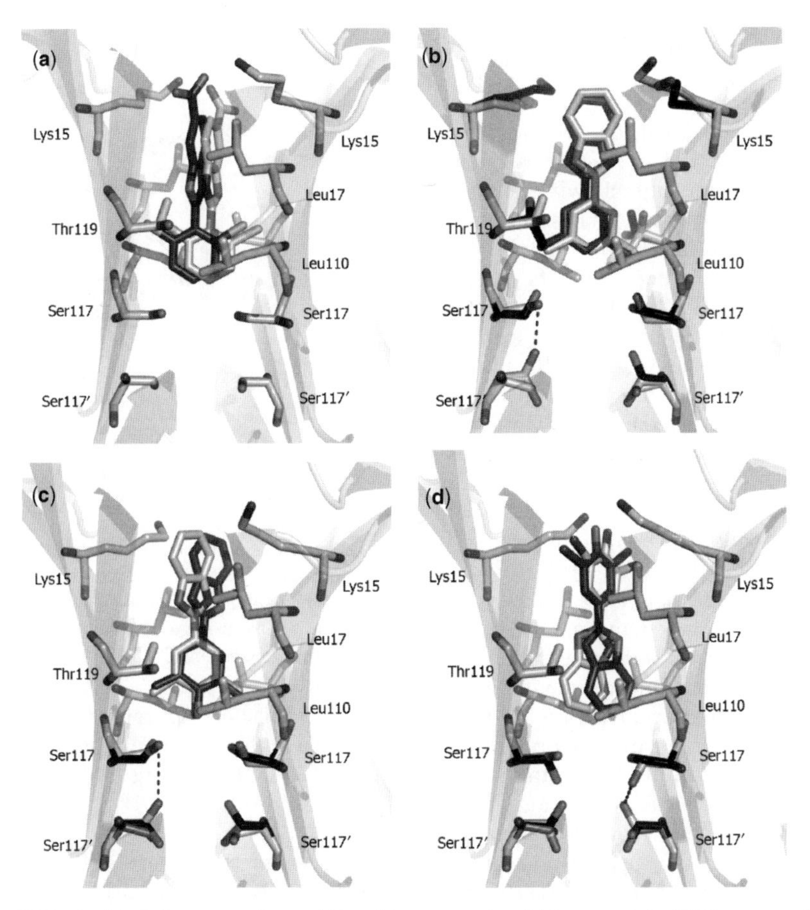

Figure 12. TTR also binds benzoxazoles and 2-aryl benzoxazoles: (a) benzoxazole 205 (compound 4 – Fig. 8) in complex with TTR (PDB ID: 2F8I). (b) TTR in complex with 3,5-methyl-2-arylbenzoxazole (PDB ID: 2QGE), (c) TTR in complex with 3,5,methyl-4-hydroxyl-2-arylbenzoxazole (PDB ID: 2QGC), and (d) 3,5-brome-4-hydroxyl 2-arylbenzoxazole (PDB ID: 2QGD). The color schema used was the same as in Fig. 11. Figures were generated with the Pymol software (De Lano Scientific).

buting to the ligand high affinity to TTR ($K_{d1} = K_{d2} = 55$ nM). In addition, this benzoxazole-based compound showed high selectivity to TTR in human plasma and similar TTR fibril formation inhibition as FLU ad PHENOX, including the benzoxazoles as potential pharmacophores to treat TTR amyloidosis [116].

Recently, Johnson et al. [117], complemented the benzoxazole series by screening different 2-arylbenzoxazole-based compounds. Several substituents attached to the 2-aryl ring of benzoxazoles were tested to evaluate the influence of thyroid hormone like aryl scaffold (3,4,5-substituted aryls) on TTR amyloidogenesis and selectivity in human plasma. Compounds with halogen or methyl 3,5-substituents were the most potent. The addition of a 4-hydroxy aryl substituent significantly increased TTR affinity. The crystal structures of TTR in complex with (3,5-methyl, PDB ID: 2QGE), (3,5-methyl-4-hydroxyl, PDB ID: 2QGC) and (3,5-brome-4-hydroxyl, PDB ID: 2QGD) 2-aryl benzoxazoles (Fig. 12b–d) revealed that the 3,5-bisaryl substituents fits well two HBPs simultaneously and the 4-hydroxyl is important to allow polar contacts with polar residues placed in the TTR binding cavity [118].

It was observed that the 2-aryl benzoxazole compounds can bind either in the forward or in the reverse binding modes, depending on the chemical nature of the 3,5-substituent used.

The compound containing the 3,5-methyl aryl substituent bound in the forward mode (Fig. 12c), whereas the compound with the 3,5-bromide aryl substituent bound in the reverse mode (Fig. 12d). The 4-hydroxyl substituent makes polar contacts with Ser117 or Lys15 side chains of both TTR subunits depending on the ligand binding mode [118]. These potential polar contacts formed between TTR and the 4-hydroxyl aryl substituent includes compounds with absence of carboxylate substituents in TTR inhibitor screenings. It is important since the carboxylate groups are involved in both COX and TR binding; therefore, increasing the potential therapeutical use of noncarboxylated TTR inhibitors. Since the 2-aryl rings of these compounds preferentially adopt conformations analogous to that observed by OH-PCBs—discussed below, and these are the most potent TTR inhibitors obtained to date [119], this inhibitor scaffold is promising candidate for further SBDD. New improvements in the other nonsubstituted aryl ring as well as the linkage connecting the rings are under development at Dr. J. W. Kelly Laboratory.

OH-PCBs It was demonstrated that hydroxylated PCBs (OH-PCBs), contrary to polychlorinated biphenyls (PCBs), display high selectivity to TTR in human plasma [119]. OH-PCBs are exceptional TTR fibril formation inhibitors, binding to TTR with high affinity ($K_d = 3.2\,\text{nM}$) in a noncoopertative process, contrasting the majority of TTR inhibitors that bind in a negative cooperative process.

The multiple chlorine substituents in the biphenyl system allow the occupation of both HBP1 and HBP3 simultaneously, enhancing TTR affinity and tetramer stability. Furthermore, an additional *para*-hydroxyl substituent mediates a hydrogen-bond network, which includes water molecules and TTR amino acid residues in the bottom of the binding cavity. This H-bond network mediates the ligand interactions with Ser117 and with the others TTR subunits, further increasing TTR-OH-PCB affinity (Fig. 13) and tetramer stability. However, the interaction of the hydroxyl group with Lys15 in the outer cavity is not observed in the 4,4′-dihydroxy-3,3,5,5′-tetrachlorobiphenyl in complex with TTR (PDB ID: 2G5U) (Fig. 13), revealing that Lys15 direct

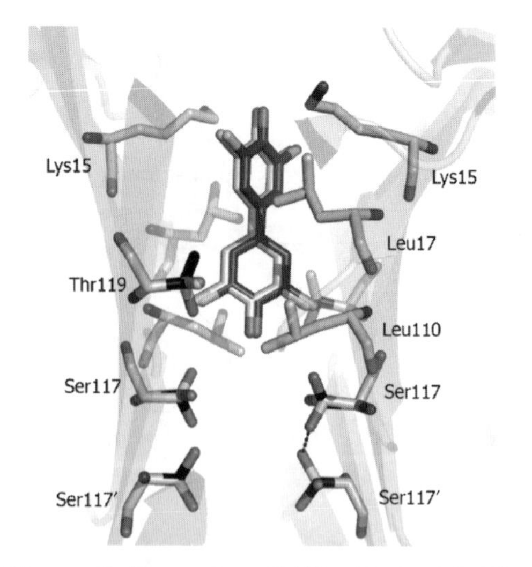

Figure 13. Interaction of TTR with the OH-PCB 4,4′-dihydroxy-3,3,5,5′-tetrachlorobiphenyl (PDB ID: 2G5U The color schema used was the same as in Fig. 11. Figure was generated with the Pymol software (De Lano Scientific).

interactions does not represent a limiting factor for a high-affinity inhibitor, and/or is compensated by the strong interactions between the chloride of these OH-PCBs and the HBPs of TTR.

Although these new TTR inhibitors scaffolds exhibit high affinity and selectivity to TTR, it was shown—in animal models and *in vitro* studies [120,121]—that PCB-OH inhibit sulfotransferases indirectly enhancing estrogenic activity and are also known to bind TR, limiting its clinical use in TTR amyloidosis treatment.

Bisaryloxime Ethers There are several FDA-approved antibacterial agents containing the oxime ether moiety, prompting active the search for TTR fibril formation inhibitors containing the oxime ether scaffold. The bisaryloxime ether series was synthesized from aryloxyamine and benzaldehyde components by Johnson et al. (2005) [118]. The resulting bisaryloxime ethers exhibit high inhibitory activity and high plasma selectivity (binding stoichiometry to TTR in human plasma >1) to TTR [118].

The crystal structure of TTR in complex with one of the bisaryloxime ethers was solved, showing a forward binding mode. The carboxylic acid substituted aromatic ring is placed in the outer cavity, interacting with the amino group of Lys15, whereas the trifluoromethyl aryl ring is positioned in the inner cavity, interacting with the HBP3. Similar to the stilbenes (discussed below), the linker heteroatoms oxygen and nitrogen do not interact with TTR. Contrarily to other class of TTR inhibitors, the bisaryloxime are unstable, limiting its use as pharmaceutical drugs. However, its high plasma selectivity can add the bisaryloxime as new scaffolds to design TTR amyloidosis inhibitors, albeit requiring further improvement of stability.

Biphenyl Ethers Following the discovery that Triclosan, a known antibacterial and antimalarial compound, inhibits TTR fibril formation, Gupta et al. [122] synthesized a number of biphenyl ethers (BPE) derivatives to test against TTR fibrillation.

It was found that most of the BPE derivatives were good TTR fibril formation inhibitors, but the 3-chloride and 1-carboxylic acid substituents are important to increase drug potency. Two new structural TTR inhibitors templates the 2-(5-mercapto-[1,3,4]oxadiazol-2-yl)-phenol and 2,3,6-trichloro-*N*-(4*H*-(1,2,4) triazol-3-yl) were also obtained, further extending the variety of structure scaffolds for TTR inhibitors.

Carborane Pharmacopores Another approach to reduce TTR inhibitors side effects consists in the use of carboranes pharmacophores [123]. It was hypothesized that both the steric bulk and the lack of π–π stacking carborane containing analogs of the promising NSAID identified would produce a poor COX inhibitor, retaining TTR dissociation inhibition enhancing its selectivity [123]. Detailed observations of the TTR:T4 crystal complex (Fig. 9b), lead to a SBDD approach to improve ligand affinity to TTR, by direct connecting a carboxylic acid functionality to a bulky carborane moiety. It was proposed that this carborane-based scaffold would improve the use of the increased volume of the outer pocket of TTR ligand binding site. Furthermore, the attachment of a phenyl ring would allow van der Waals interactions with the HBP3 of TTR. Following this structural hypothesis, Julius et al. [123] synthesized several carborane containing analogs. The carborane 1-carboxylic acid 7-(3-fluorophenyl)-1,7dicarba-*closo*-dodecaborane exhibits high inhibitory potency and low COX-1 and COX-2 inhibition. Its poor performance as COX inhibitor is probably a consequence of the short link connecting the two aromatic rings. As COX active site is known to be mainly hydrophobic, the short link present in this carborane reduces its flexibility, limiting interactions between the phenyl moiety attached to the carborane ring and COX hydrophobic ligand binding site. This presumably creates steric impediments for the ligand interactions with the COX active site.

Natural Compounds and Their Derivatives

Stilbenes Resveratrol (*trans*-3,4′,5-trihydroxystilbene), an antioxidant compound found in grape, was shown to be a good TTR fibril formation inhibitor [107]; hence, including stilbenes as a class of TTR inhibitors. The crystal structure of TTR:RESV (PDB ID: 1DVQ) reveals that the hydroxyl substituents present in RESV are important to allow polar contacts with both the inner and the outer part of TTR binding sites, contributing to TTR-RESV affinity [107]. However, the stilbene scaffold is too rigid to complement the HBPs pockets, resulting in moderated affinity to TTR.

Later, diethylbestrol (DES), a synthetic estrogen with a molecular structure similar to resveratrol (RESV), was found to be a potent competitive inhibitor for T4 binding to TTR and a potent amyloid inhibitor [124].

The crystal structure of TTR in complex with DES (PDB ID: 1TT6 and 1TZ8) showed that DES can bind to TTR both in the forward and in the reverse modes. The ethyl groups of DES play important roles in TTR ligand recognition. These hydrophobic substituents could be accommodated into the TTR HBP pockets and were pointed as the main contributors to DES effectiveness front to TTR aggregation and affinity compared to RESV [124]. However, DES and RESV exhibit estrogen and/or COX activity, limiting the use of these stilbenes as a therapeutic agent.

Flavonoids Flavonoids have been reported as good TTR inhibitors [102]. Apigenin, miricetin, and genistein were tested in acid mediated TTR fibril formation *in vitro*, showing good TTR fibril formation inhibition. However, the crystal structures of TTR in complex with those flavonoids have not been reported to date.

Genistein, a soy isoflavone, binds to TTR with high affinity in a negative cooperative process ($K_{d1} = 40$ nM; $K_{d2} = 1.6$ μM) [125]. Genistein binds with high selectivity to TTR in human plasma and cerebrospinal fluid [125,126] and displays no adverse effects in toxicity studies [127,128], which turns genistein and other flavones into a very promising candidate for TTR amyloidosis treatment. The crystal structure and the molecular mechanism explaining the allosterism displayed by genistein when biding TTR was investigated and will soon be published [129].

Other flavones containing hydroxyl substituents at different number and positions also are being currently investigated. The crystal structures, thermodynamic parameters, and the fibril formation inhibition detected in *in vitro* studies are being used to elucidate the molecular recognition of the flavone scaffold by TTR and the influence of hydroxyl substituents for binding and fibril formation inhibition [129].

3.2.3. TTR Inhibitors: Concluding Remarks

Several SBDD studies aiming at developing pharmacophores that can increase TTR affinity, stability and selectivity have been performed. Screening, structure-based design, and lead compound optimization by parallel synthesis has led to the establishment of several structurally distinct classes of selective and potent TTR amyloidogenesis inhibitors. Effective inhibitors generally have two aryls linked directly or through a spacer such as an amine, an ether, or an ethylene bridge. Optimally, one aryl is functionalized with halogens or aliphatic groups, typically occupying the inner cavity of the thyroxine binding site. The substitution of the aryl ring at the 3,5-position results in the most powerful inhibitors, due to better accommodation of the halogens and/or the alkyl groups in the HBP3 and HBP3′. The protein–drug interactions in the inner cavity can also lead to Ser117 and Thr119 side chain conformational changes, what are important for establishing new intersubunit interactions between two Ser117 from adjacent TTR subunits, leading to an enhanced protein stabilization. It has been demonstrated that a polar substitution at the aromatic ring at position 4 could conduce to the formation of a new hydrogen-bond network connecting the drug, water molecules and polar aminoacid residues placed in the bottom of the two thyroxine binding sites, also contributing to tetramer stability [119].

The other aryl ring can be substituted with a hydroxyl and/or carboxylic acid, which is able to interact electrostatically with the side chain amino group of Lys15 and/or Glu54 side chain carboxyl group at the periphery of the outer binding cavity, thus contributing to TTR-ligand affinity. A second halogen or hydrophobic substitution can enhance protein–ligand interactions at the HBP1 and/or HBP2, further improving ligand binding and bridging two TTR monomers. This results in a more stable TTR tetramer.

Ligand binding in opposite directions, with its polar substituted aryl ring placed in the inner cavity and with the other halogen substituted aryl ring in the outer cavity, is frequently observed in TTR crystal structures. The exceptional plasticity in the requirements for small molecules binding to TTR probably stems from the existence of these two different binding modes for ligand binding, the reverse and forward binding modes, each of which is able to tolerate significant structural variation. Additional opportunity for TTR-based drug design comes from the structural modifications of ligands that are neutral for their binding to TTR, but deleterious for their affinity to TR and/or COX. This strategy could contribute to the reduction of the potential side effects caused by a long-term therapeutical usage of the ligands required for TTR amyloidosis treatment.

4. STRUCTURE-BASED THERMODYNAMIC ANALYSIS OF DRUG–MACROMOLECULE INTERACTIONS

Drug discovery and development processes are intrinsically complicated and arduous, and, in the worst cases, lead to dead ends. The complexity of the steps that pave the road to a new drug comes from the multifactorial design requirements imposed by the pharmacodynamics of ligand–macromolecular interactions and pharmacokinetics (of drug absorption, distribution, metabolism, excretion and toxicology—ADME/Tox) [130]. The goal is the production of a new, safe, and efficacious drug. The processes involved in this complex pathway must be refined to be more stringent at each step of drug discovery: from the beginning, with high-throughput screenings (vHTS/HTS); through further *in vitro* and *in vivo* assays to characterize, prioritize, and propose chemical modifications of the molecules; until the final step of human testing. Here, we present a microcalorimetric assay that can improve the drug discovery process (Fig. 14).

The ligand–macromolecule interaction profile depends on many factors. Much like the typical chemical reactions, the dynamic of the cocomplex formation and disruption is governed by the kinetics and thermodynamics of the system. From the kinetics perspective, the speed of the cocomplex formation and the residence time [131] are two important points to be considered. Further, thermodynamics deals with the overall stability of the intermolecular interaction by means of the Gibbs free energy and its components: enthalpy and entropy. Both the kinetics and the thermodynamics of the system describe the pharmacodynamics behavior of the ligand–macromolecule interaction [132]. The interaction between those two entities, the ligand (or drug; a chemical structure with a small molecular weight) and the macromolecule (which is in essence the biological target—protein, DNA, etc.) is governed by energetic principles.

Thermodynamically, there is no need to understand the pathway that guides the transition from two different states, bound or unbound; only the difference in the free energy (ΔG^0) of these two states guides the cocomplex formation. Microcalorimetry is a unique technique that measures the thermodynamics of the ligand–macromolecule interaction directly [133], providing values of free energy and its components—enthalpy and entropy (Fig. 14) [134].

4.1. Binding is Driven by Thermodynamics

The interaction between the ligand (or drug) and the macromolecule (biological target, here called as receptor) is defined thermodynamically by the enthalpy, which, in turn, is composed of the energies that result from this bimolecular recognition process. The main components guiding the enthalpy change (ΔH^0) of binding—hydrogen-bond, electrostatic, and van der Waals interactions—are well known, having been widely used to describe ligand–macromolecule, in textbooks and literature [135]. Entropy change (ΔS^0) quantifies the solvation/desolvation of the ligand and the macromolecule due to the movement of water molecules and the conformational changes that come along with binding.

The stability of the ligand–receptor cocomplex can be measured by considering the enthalpy and entropy of the ligand and macromolecule unbound and bound at a given temperature, and is summarized by the Gibbs free energy change equation (Eq. 1), under defined conditions (e.g., 1 M receptor and 1 M ligand at pH 7 and 25 °C).

$$\Delta G^0 = \Delta H^0 - T\Delta S^0 \quad (1)$$

where Δ comes from two different states: bound and unbound.

The basis for interpreting and quantifying the binding affinity or equilibrium binding constant (K_B), which describes the ligand–receptor interaction via the reversible reaction between the free receptor and the ligand ($K_B = [LR]/[L][R]$), is at equilibrium under

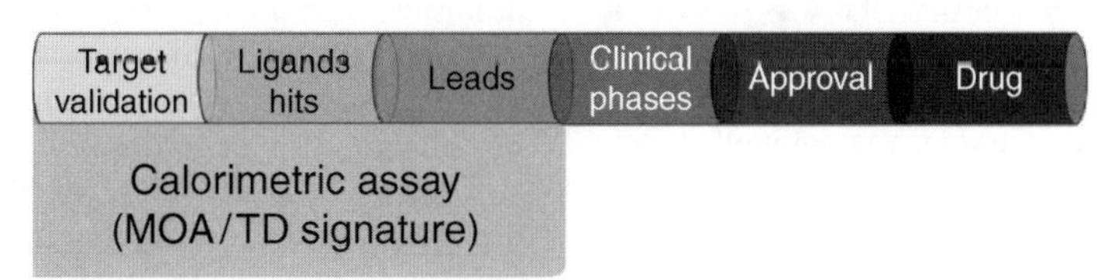

Figure 14. Main contribution of microcalorimetry in drug discovery and development.

conditions of constant pressure, related to the standard Gibbs free energy change (ΔG^0) (Eq. 2).

$$\Delta G^0 = -RT \ln K_B \quad (2)$$

where R is the gas constant (8.31 J/mol/K) and T is the temperature (in Kelvin). In order to completely describe the thermodynamics for any biomolecular interaction, the heat capacity change (ΔC_p), which is the temperature dependence of the enthalpy change ($\Delta C_p = \delta\Delta H/\delta T$), has to be determined. ITC provides a comprehensive description of the energetics for a biomolecular interaction in a single experiment; thus, allowing the thorough picture of the cocomplex thermodynamic profile.

By looking at Equations 1 and 2, one can assume that the more negative the Gibbs free energy change, the higher the affinity of the ligand–receptor cocomplex. This means that the overall balance of enthalpy and entropy should be favorable to binding. In seeking the best binding profile for the next generation of derivatives, this balance must be wisely controlled by chemical modifications of the scaffold or its branches. However, enthalpy and entropy cannot be shaped in this way without decomposing them by the fundamental types of interactions that drive cocomplex formation. For example, to prioritize a molecule (for hit or lead identification), or in the next step to chemically modify the molecule (lead optimization), it is necessary to account for the energetics that come from each type of interaction. The entropy of binding is composed of (1) the energetic contribution of the water network disruption that solvates the ligand and the macromolecule and (2) the conformational rigidity of the cocomplex compared to the unbound species. These effects usually oppose each other when considering their contributions to the free energy. This scenario is further complicated because the enthalpy of binding must also be taken into consideration.

Unlike enthalpy, entropy is not well understood because of the limitations in modeling solvent effects, that is, the nature of water and its interactions [136]. In addition, other variables, namely pH of the microenvironment, salt concentration, and the overall rotational/translational energies of the ligand and the macromolecule comprise a much more complicated picture than one might think when the entropy energy value is obtained from one experiment.

The entropic term must be carefully considered, especially when seeking new drug candidates that might contain many degrees of freedom. Awareness of the limitations regarding the entropic term is critical, and it is being typically used to optimize new chemical entities in which the rigidity of the side chains, or even the scaffold, is increased by replacing rotatable bonds with rings. This approach is useful on two fronts: (1) using rings reduces the entropy difference between the unbound and the bound states of the molecule, and (2) this modification serves as an exploratory tool regarding the biological target's flexibility in accommodating the ligand, thereby expanding knowledge of the chemical–biological space.

Enthalpy, on the other hand, is much more easily characterized and can be decomposed and modeled using computer software that gives an accurate value for ΔH, under any condition. The application of molecular modeling is out of the scope of this chapter, but it is noteworthy because of its steadily growing importance to a myriad of techniques currently employed in new drug research. Among these techniques, the computation of binding affinities [137] and the representation of the modes of binding interface and complement the microcalorimetric results.

Among the most important interactions, hydrogen bonding plays a critical role in the cocomplex stability, when the ligand is not an irreversible inhibitor that binds to the macromolecule through the use of a covalent bond (this is the mechanism of antitumoral DNA alkylators, and some painkillers that target cyclooxygenase) [138]. X-ray crystallography of many molecules reveals the hydrogen-bond network (directionality, average distance, and angle), which can also be estimated by a number of computer programs.

Charge–charge, charge–dipole, dipole–dipole, and induced dipoles interactions contribute to the electrostatics of the system. Most drugs in use are weak acids and bases, and their charge balance (ionization) is extremely dependent on the pH of the local environment. Microcalorimetric studies are also sensitive to the overall charge of the system. It is possible to correct the total energy of the system, if necessary, by running experiments with buffers of different ionization enthalpies.

Lipophilic interactions also play a role in the recognition process by means of the enthalpy value of apolar contacts between the molecular entities [139].

4.2. Structure-Based Thermodynamic Analysis

The ligand's thermodynamic signature emerges from structure-based thermodynamics analysis. The signature is critical during all phases of preclinical studies because it describes the compounds' enthalpy and entropy profiles [140]. This profile is pivotal in defining the best hit to progress to lead (hit-to-lead step) and subsequent optimization of the leads, or even for comparing drug candidates or drugs in the market.

An example of the utility of understanding a ligand's thermodynamic signature comes from a series of calorimetric studies with HIV-1 protease inhibitors (PI). Molecules in different phases of development were compared to drugs, and thermodynamic signatures were established for all of them. This study showed that entropy of the molecules has driven the discovery of the first new inhibitors, while development of the next generation of drugs relied more on enthalpy (Fig. 15) [141].

Another phenomenon currently under study is enthalpy–entropy compensation. When enthalpy is optimized in a new compound, entropy counterbalances the energy contribution that comes from enthalpy by incrementing its value. As a result, the Gibbs free energy of the new compound is similar to the precursor because the entropy loss tends to cancel out the enthalpy gain.

Enthalpy–entropy compensation can be observed for many compound classes, including protease inhibitors (Fig. 16). The two molecules share the same chemical scaffold, the only difference being the oxidation state of the sulfur atom. KNI-10075 can form a hydrogen-bond network to the target that is energetically more stable; hence, improving its enthalpy in relation to KNI-10033. However, the Gibbs energy of both compounds is almost the same, which shows that the enthalpy gain was associated with the entropy cost for KNI-10075, because of the desolvation penalty and the large loss of configurational entropy.

Although the Gibbs free energy of the two compounds might be similar after binding, the intermolecular interaction profile triggered by the enthalpy/entropy ratio might be significantly different. Freire claims that entropy-driven compounds are not as selective as enthalpy-driven compounds [142]. This means that the enthalpy–entropy balance could be

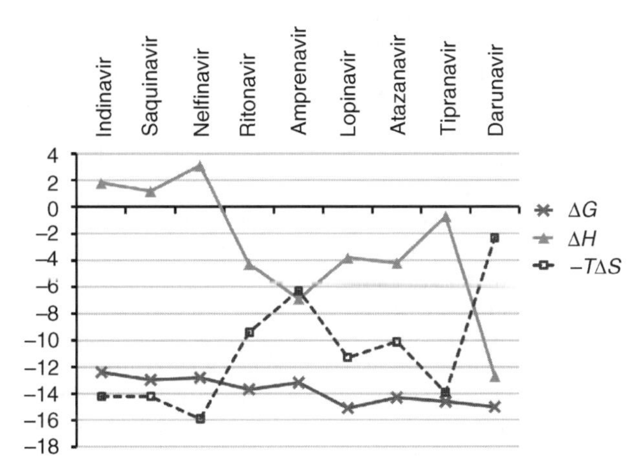

Figure 15. Thermodynamic signatures of HIV-1 protease drugs.

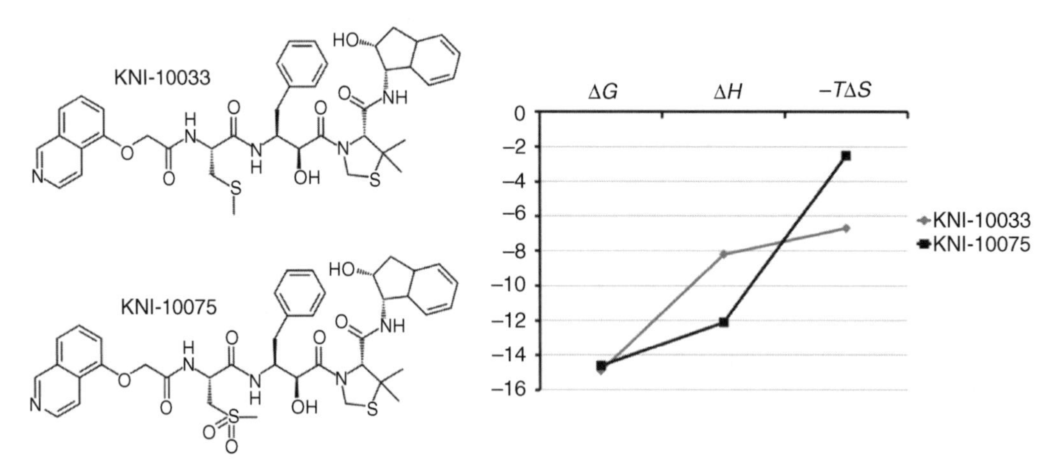

Figure 16. Enthalpy-entropy compensation for two protease inhibitors.

used in target selectivity to design new enthalpy-driven inhibitors based on a previous potent entropy-based molecule. This can be observed by looking at the protease inhibitors in Fig. 15 and the statins, which target HMG-CoA reductase enzyme.

4.3. Experimental Measurement of Thermodynamic Binding Parameters

Microcalorimetry is the only method available today for directly determining the Gibbs free energy and its components, that is, without the need of an adjuvant (such as a dye), which may introduce bias or perturb the system. The van't Hoff equation might be used to determine the thermodynamic components in an indirect way, but only as a crude estimate. Some studies have pointed out the usefulness of van't Hoff equation in estimating enthalpy, but some intrinsic deviation of values is often observed when compared with the direct calorimetric measurements [143].

Currently, three techniques are used to determine the thermodynamics of a biological system: ITC, differential scanning calorimetry (DSC), and pressure perturbation calorimetry (PPC) [143]. Among them, ITC is the most common technique used to measure ligand–macromolecule interactions.

The basic principle of ITC is the heat flow that comes from the difference in temperature (ΔT) between the reference cell, maintained at a constant temperature, and the reaction cell, where the ligand–macromolecule interaction is being studied. Usually, the ligand is titrated into the reaction cell containing a buffered solution of the macromolecule, and the temperature change is measured. Upon binding, the cocomplex formed may be exo- or endothermic; that is, it releases into or absorbs heat from the surrounding solution provided by the reference cell, depending on the Gibbs energy value (Fig. 17a). The ΔT between the reaction and the reference cells is recorded following periodical injections of the ligand, until it reaches a plateau (which means that the system is saturated). Each addition of ligand corresponds to a peak with a relative intensity and area (Fig. 17b). To avoid superimposition of the peaks, it is important to optimize the rate at which the ligand is added into the sample cell. Otherwise, it might compromise the mathematical integration of the peak essential to the quantification of the thermodynamic parameters (Fig. 17c). The heat of dilution of the ligand must be measured and then subtracted from the area under the curve of each peak.

Other binding parameters are often determined during the ITC measurements of ligand–macromolecule interaction. For example, it is possible to measure the ligand/macromolecule ratio after titration of the compound and then estimate the stoichiometry of the process (gray bar in Fig. 17c). It is also important to measure the heat of dilution of the chemicals being titrated, because this energy

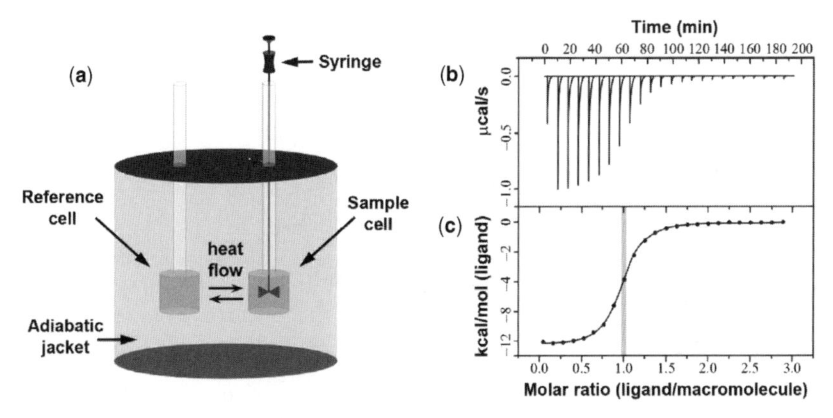

Figure 17. Isothermal titration calorimeter. Cell apparatus (a), its typical raw data plot (b) and the nonlinear mathematical transformation (c).

must be used to calculate the free energy. As an extra feature point, further experiments might be done over a range of temperatures in order to provide additional data about the stability of cocomplex formation (use of DSC is indicated in this case). In addition, it is important to stress that by fulfilling the determination of thermodynamic parameters the constant pressure heat capacity change (ΔC_p) must also be included in the whole analysis.

4.4. The Role of Isothermal Titration Calorimetry in Drug Discovery

The most common use of ITC in drug discovery is in determining the free energy of the ligand–macromolecule interaction, and involves determining the thermodynamic signature and mode of binding, as shown for protease inhibitors in Fig. 16. ITC has been applied to a variety of systems, including protein–protein, protein–carbohydrate [144], and nucleic acid [145] interactions; selectivity and resilience to mutation [146]; kinetics of the ligand–protein [147,148] interaction; and even enzyme activity.

The thermodynamic signatures of amprenavir and its derivative darunavir are compared in a retrospective study to highlight the importance of the calorimetric measurements (Fig. 16) [149]. Darunavir was originally optimized from the amprenavir at the P2/P2'-position, where the tetrahydrofurane is placed (Fig. 18). Hence, darunavir has a more significant ΔH contribution to the free energy than amprenavir. The bicycle substituent in darunavir binds to the main chain of the protease (D29 and D30) and this binding is enthalpy-driven because of the specific hydrogen-bond interactions between this moiety and the HIV-1 protease amino acids (Fig. 18) [150]. Biological tests with a set of mutant enzymes showed that darunavir had a better resilience profile, but both amprenavir and darunavir presented some cross-resistance, as expected [149,151].

Other studies claim that residence time is the key factor in darunavir resilience [152]. Most likely, not only one phenomenon is responsible for that profile but also, in fact, it is a combination of the optimized kinetics and thermodynamics properties acquired through correct fragment placement.

DNA is a major site of molecular recognition and an important target in cancer chemotherapy. Many drug classes interact with DNA to inhibit cell growth and arrest the cell cycle [153]. Using thermodynamic data from the literature [154], two classes, intercalators and minor groove binders (MGBs), were analyzed in detail. Analyses have identified different thermodynamic signatures for these classes (Fig. 19). Intercalators work as a spacer group that doubles the distance of the DNA base pairs to accommodate the compound. The process of stacking increases strain on the system, which costs energy. This is the case for actinomycin D, doxorubicin, daunorubicin, and ethidium that do not have very low Gibbs energy. MGB do not demand

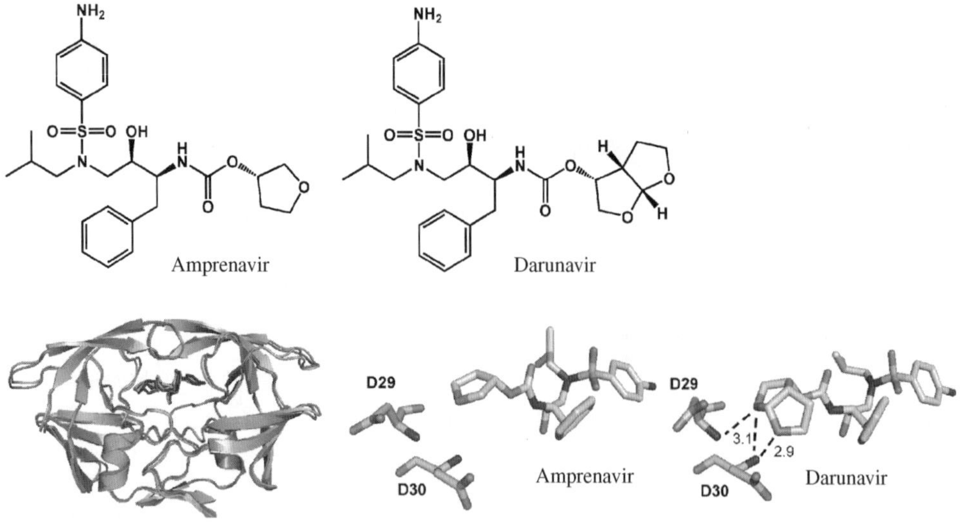

Figure 18. Chemical representations of amprenavir and darunavir and their interactions with protease superimposition of cocomplexes with amprenavir (light gray: 1HPV) and darunavir (dark gray: 1T3R). Hydrogen bonds in darunavir and D29 and D30, with distances in angstroms. No good hydrogen bond is seen in amprenavir tetrahydrofurane. Water molecules were omitted for clarity.

huge motions of the DNA structure, but they may release water molecules upon DNA binding, increasing the entropy. Berenil, Hoechst 33258, distamycin, and netropsin show the variable contribution of entropy to the Gibbs energy, but in general this contribution is much more pronounced with MGB than with intercalators.

ITC is a distinctive technique because the same equipment measures thermodynamics and kinetics. However, ITC is a low throughput assay, only useful in analyzing smaller series of molecules, despite advances that have been made to incorporate robotics. HTS can handle and process many thousands of molecules per day, while ITC still processes

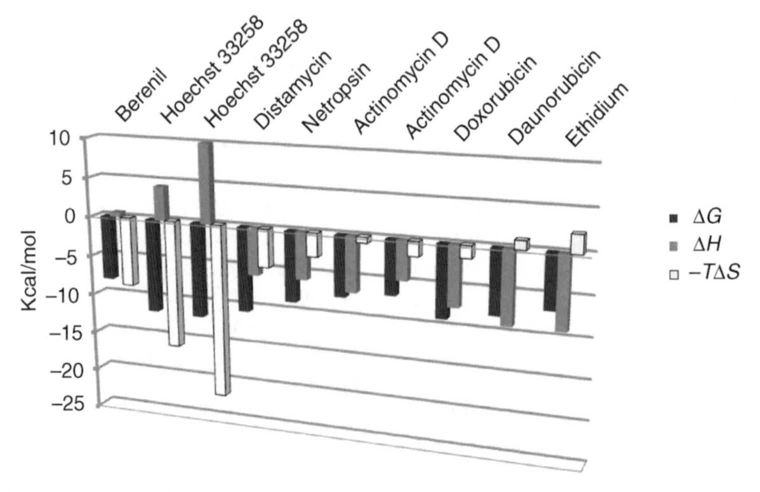

Figure 19. Thermodynamic signatures of some groove binders and intercalators. MGB: berenil, Hoechst 33258, distamycin, and netropsin. Intercalators: actinomycin D, doxorubicin, daunorubicin, and ethidium. Two ITC measurements are shown for Hoechst 33258 and actinomycin D. For the entire set, see Ref. [154].

fewer than 100. Direct quantification of ultra-high-affinity compounds (picomolar) is difficult and sometimes requires the use of competitive studies. The number of samples to be injected is huge, compared to other techniques, but new equipment with smaller sample cells is solving this problem.

Due to the inherent complexity of the drug discovery process, multiple tools are commonly used to test the compounds and to reveal their mode of binding. This includes techniques such as biochemical tests, X-ray crystallography, NMR, and molecular modeling [155]. These analyses gather information that complements the ligand–macromolecule recognition process.

4.5. Fragment-Based Drug Design Using Isothermal Titration Calorimetry

Fragment-based drug design is a research strategy that emerged in the 1990s and has become a compelling, innovative way to address drug discovery. This approach screens chemical fragments of known active compounds or fragments that will compose the ligand via spacer groups to promote their linkage (thinner bonds shown in Fig. 20) [156].

As can be seen in Fig. 20, a chemical fragment is a small molecule *per se*, binding to the target with weak affinity (usually in the micro- to milimolar range). The rule of three (Ro3) was devised to better characterize a chemical fragment. Its name comes from the numbers (multiples of three) that compose the rule. To be considered a fragment, a chemical must meet the following conditions: molecular weight (MW) < 300, hydrogen-bond donors (HBD) ≤ 3, hydrogen-bond acceptors (HBA) ≤ 3, and $\log P \approx 3$. In addition, the number of rotatable bonds ≈ 3 and a polar surface area $\approx 60\,\text{Å}^2$ [157].

One early attempt to determine the importance of chemical moieties in ligand–macromolecule stability was based on a model that linked a wide range of bioactive molecules' fragment contribution to thermodynamic data through Equation 2, leading to the generation of the Andrews equation [158]. Such a regression method is a crude way to determine the putative value of the fragment contribution, and it was used to assign the average energetic value for each compound to be compared with the experimentally measured one. The difference between those two values (average and experimental) represents how well the whole molecule interacts with the macromolecule and can be useful in defining the best approach in drug discovery.

In the fragment-based approach, new drugs can be sought by using microcalorimetry, given that the chemical groups usually have low affinity when compared to the drug itself. The chemical fragments are also useful probes to the macromolecule to select the best fragments and to help characterize the mode of binding. This is done by querying the pockets where fragments usually bind through the introduction of mutations. Those that most increase the affinity of the chemical entities are considered the "hot spots." Very often, calorimetric and molecular modeling simulations are applied in concert to compound identification and optimization through mapping the enzyme pocket with chemical fragments.

In the framework of fragment-based drug design, scaffold-hopping is useful in linking chemical fragments that bind to macromolecule hot spots to build the scaffold, and generate new and potent molecules with

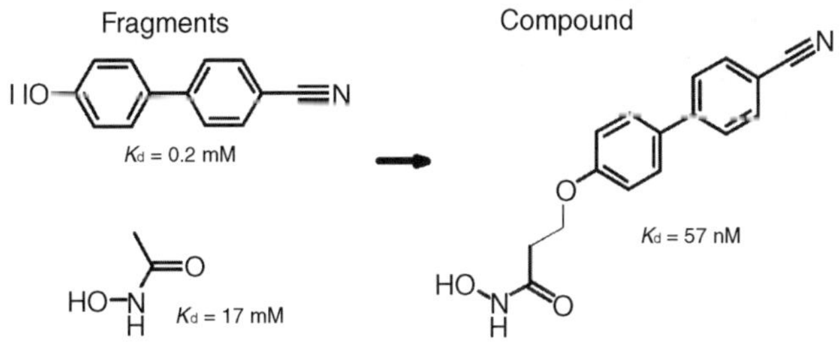

Figure 20. Application of fragment-based drug design of a matrix metalloproteinase 3 (MMP3) inhibitor [2].

improved geometric features that best match the biological target (Fig. 20) [159].

Use of substrates or cofactors is a good method for querying the macromolecule to find hot spots as well as the best fragments to start the drug discovery project. In seeking new antibiotic and herbicide inhibitors of ketopantoate reductase, fragments of the cofactor NADPH were used to query this enzyme from *Escherichia coli* (Fig. 21). Biophysical methods (ITC, NMR, X-ray crystallography, and kinetics analysis) were employed in this quest, together with site-directed mutations. Based on the ligand efficiency of the compound, the resulting inhibitor would use adenosine monophosphate (AMP) as the scaffold and the road map to the relative contributions of the moieties. Phosphate (arrow in Fig. 21) was more favorable than the reduced nicotinamide group, while β-phosphate-ribose was the least efficient (gray circle in Fig. 21) [160]. It is noteworthy that ITC with site-directed mutations was of invaluable help in revealing the incorrect crystallographic structure of the cocomplex 2′-phopho-ADP-ribose and ketopantoate reductase. The inconsistency was detected when the two X-ray crystallographic structures were superimposed and, unexpectedly, 2′-phospho-ADP-ribose bound in an inverse way in comparison to $NADP^+$ (Fig. 21, encircled rings).

Comparative results from the calorimetric measurements at two pH values (4.5 and 7.7) using mutants R31A and N98A revealed that the mismatch was caused by different pH conditions applied during the crystallization process, which led to the protonation of the acid residues in the enzyme [161]. As a reminder, the position of hydrogen atoms and the ionization states of the protein residues are defined during structure modeling because of the lack of resolution in the X-ray crystallography experiments.

Another way to determine the importance of a chemical moiety to the ligand–macromolecule interaction is by means of analogous

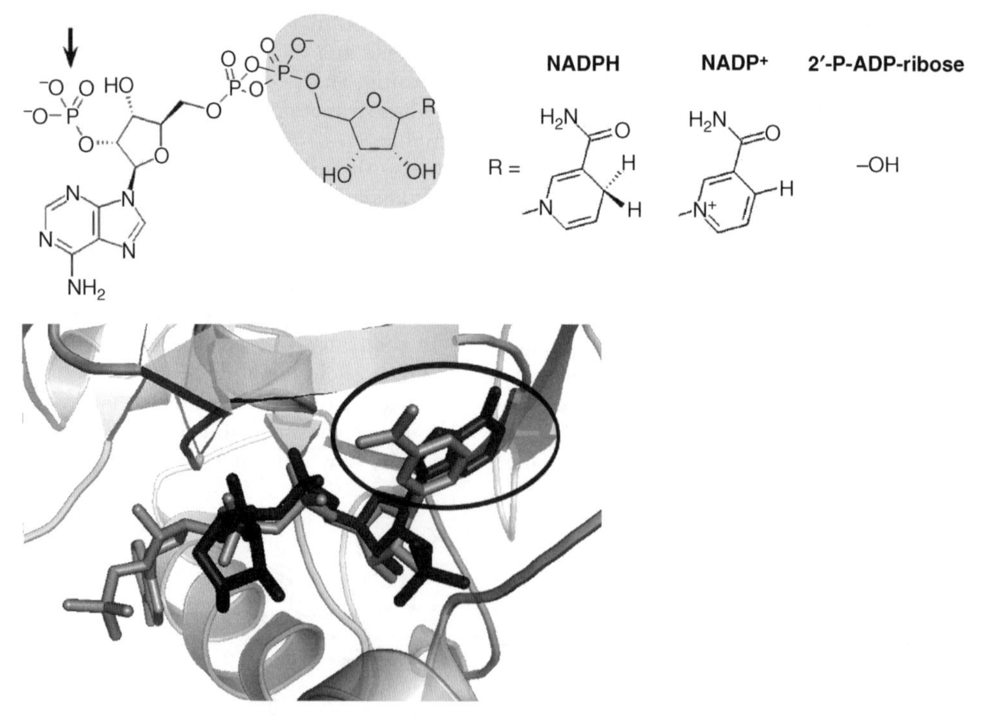

Figure 21. Redox forms of the cofactor NADPH/$NADP^+$ and the mismatch of 2′-ADP. NADPH and $NADP^+$ represent two oxidative states of nicotinamide-adenine-dinucleotide phosphate, 2′-P-ADP-ribose is 2′-monophosphoadenosine-5′-diphosphate, X-rays crystallographic structures are 2′-ADP-ribose (dark gray: 1YON) and $NADP^+$ (light gray: 2FOP).

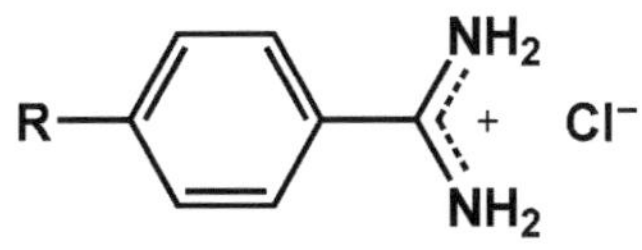

Figure 22. *p*-Alkylbenzamidinium chloride scaffold.

series. Although this cannot be considered a fragment-based approach, the contribution of chemical moieties (fragments) to the binding of a set of compounds with the same scaffold can be measured by ITC and applied in the drug discovery process. The unexpected behavior of what is considered a simple chemical modification is exemplified by a set of *p*-alkylbenzamidinium chloride inhibitors of trypsin (Fig. 22) [162]. This homologous series uses an increasing number of methyl groups at the *para*-position (from hydrogen up to hexyl, including two branched derivatives, *i*-propyl and *t*-butyl). An increased number of methyl groups were not linearly correlated with the heat capacity increment, as expected. Further analyses and decomposition of the energetic contributions, coupled with computational modeling of each ligand into the binding site, identified the importance of the dehydration process and the mobility of the hydrophobic side chains to the ligand–enzyme recognition. Molecular modeling of this series with trypsin showed that tails with an odd number of carbons tended to point out to the solvent, with higher mobility.

4.6. The Role of Solvation on Thermodynamics of Ligand–Receptor Interactions

There have been attempts to develop structure-based thermodynamic analysis of drug–macromolecule interactions, which rely on binding thermodynamics correlations with molecular structure of interest in structure-based drug design [163], addressing the role of solvation needed to fulfill the unique behavior water can have in the inhibition of a particular enzyme [136].

Ladbury et al. [163] have shown from their structure/calorimetry of reported protein interactions online (SCORPIO) [164] database analysis that the thermodynamics on enthalpy, entropy, and free-energy thermodynamic data reveal the categorized natural biological ligands from the ones originating from a medicinal chemistry program. The difference in the distributions of standard Gibbs free energy change (ΔG^0) for these groups reveals that for biological and initial leads for development the affinities are typically lower than the ones for synthetic inhibitors. Moreover, it is postulated that the products from medicinal chemistry programs have, on average, almost equal contributions to the affinity from enthalpy and entropy changes, whereas the natural biological ligands are mostly enthalpy-driven inhibitors.

Another striking observation that has arisen from their compiled peer reviewed and published ITC studies with structural information and the interactions between proteins and small-molecule ligands is the structural correlation of thermodynamic changes in free energy due to burial of apolar surfaces. Sixty-five percent of the Gibbs free energy of binding for 23 different protein–ligand interactions correlates to reduction in apolar surface area upon complex formation (including the solvent accessible surface area (SASA) for both ligands and proteins). The trend is that the increase in apolar surface area burial is accompanied by increase in affinity, but this is only of limited implication on ligand optimization, for instance, where selectivity is to play a part in affinities among functionally related proteins. Since polar interactions are the ones that impose selectivity, they have attempted to show correlations on polar interactions with affinity but found a lack of a good fitting data. Thus, what remains to be shown is the relationship between enthalpy and entropy and SASA. Nevertheless, from the ITC data on proteins interacting with small molecules for which there are accurate measured values for enthalpy and entropy the enthalpy is still seen as predominantly governing the contribution to binding.

4.7. Calorimetry in Structure-Based Drug Design: Concluding remarks

Many cellular processes like transcription, DNA replication/repair involve proteins that specifically recognize DNA sequences. Multiscale modeling is applied for searching biophysical factors, which influence the bimolecular recognition *modus operandi* system whereby a high-resolution structure of the target protein (or DNA) is used to provide information on how a potential ligand might interact with a surface. Thermodynamics governs all these processes of bimolecular recognition as well as higher complex systems. It is used for monitoring a chemical reaction initiated by adding a binding component and it is the method of choice for characterizing such interactions. ITC provides a comprehensive thermodynamic profile of molecular interaction in a single experiment whereby the energetics of association reactions in solution can be fully determined. After ligand binding, heat is absorbed or released and the measurement of this heat flow allows accurate and precise determination of (i) binding (association) constants, (ii) reaction stoichiometry, and (iii) enthalpy. From binding association constants and enthalpy, the Gibbs free energy and entropy can be obtained. If the enthalpy is measured at a range of temperatures, the change in constant pressure heat capacity can be determined and this yields correlations with burial of surface area in the binding site. The interplay of ITC and structure-based drug design in assessing the thermodynamics profile of small molecules bound to druggable targets is therefore of utmost importance to fully describe the forces that characterize the drug–target interactions.

The combination of structure- and ligand-based virtual screening techniques as a complementary approach to a biochemical screening using microcalorimetry certainly enhances the capabilities of medicinal chemists in getting better drug candidates early in the preclinical phase of the whole drug discovery and development processes. Thus far, putting all these together will shed light in the chemical–biological space that represents the most challenging effort in any strategy aimed at improving the drug-like character of hits that will end up in a new chemical entity feasible to be used as a drug.

REFERENCES

1. Andricopulo AD, Montanari CA. Mini Rev Med Chem 2005;5:585–593.
2. Hajduk PJ, Greer J. Nat Rev Drug Discov 2007;6:211–219.
3. Hopkins AL, Groom CR. Nat Rev Drug Discov 2002;1:727–730.
4. Guido RV, Oliva G, Andricopulo AD. Curr Med Chem 2008;15:37–46.
5. Overington JP, Al-Lazikani B, Hopkins AL. Nat Rev Drug Discov 2006;5:993–996.
6. Kitchen DB, Decornez H, Furr JR, Bajorath J. Nat Rev Drug Discov 2004;3:935–949.
7. Davis AM, Teague SJ, Kleywegt GJ. Angew Chem Int Ed Engl 2003;42:2718–2736.
8. Davis AM, St-Gallay SA, Kleywegt GJ. Drug Discov Today 2008;13:831–841.
9. Klebe G. Drug Discov Today 2006;11:580–594.
10. Berman HM, Westbrook J, Feng Z, Gilliland G, Bhat TN, Weissig H, Shindyalov IN, Bourne PE. Nucleic Acids Res 2000;28:235–242.
11. Mulder NJ, Apweiler R, Attwood TK, Bairoch A, Bateman A, Binns D, Bork P, Buillard V, Cerutti L, Copley R, Courcelle E, Das U, Daugherty L, Dibley M, Finn R, Fleischmann W, Gough J, Haft D, Hulo N, Hunter S, Kahn D, Kanapin A, Kejariwal A, Labarga A, Langendijk-Genevaux PS, Lonsdale D, Lopez R, Letunic I, Madera M, Maslen J, McAnulla C, McDowall J, Mistry J, Mitchell A, Nikolskaya AN, Orchard S, Orengo C, Petryszak R, Selengut JD, Sigrist CJ, Thomas PD, Valentin F, Wilson D, Wu CH, Yeats C. Nucleic Acids Res 2007;35:D224–D228.
12. Quevillon E, Silventoinen V, Pillai S, Harte N, Mulder N, Apweiler R, Lopez R. Nucleic Acids Res 2005;33:W116–W120.
13. Zdobnov EM, Apweiler R. Bioinformatics 2001;17:847–848.
14. Gasteiger E, Gattiker A, Hoogland C, Ivanyi I, Appel RD, Bairoch A. ExPASy: Nucleic Acids Res 2003;31:3784–3788.
15. Bergner A, Gunther J, Hendlich M, Klebe G, Verdonk M. Biopolymers 2001;61:99–110.
16. Hendlich M, Bergner A, Gunther J, Klebe G. J Mol Biol 2003;326:607–620.
17. Gunther J, Bergner A, Hendlich M, Klebe G, Relibase J Mol Biol 2003;326:621–636.
18. Sotriffer C, Klebe G. Farmaco 2002;57:243–251.
19. Blundell TL, Jhoti H, Abell C. Nat Rev Drug Discov 2002;1:45–54.

20. Waszkowycz B. Curr Opin Drug Discov Dev 2002;5:407–413.
21. Hartshorn MJ, Verdonk ML, Chessari G, Brewerton SC, Mooij WT, Mortenson PN, Murray CW. J Med Chem 2007;50:726–741.
22. Alvarez JC. Curr Opin Chem Biol 2004;8: 365–370.
23. Moitessier N, Englebienne P, Lee D, Lawandi J, Corbeil CR. Br J Pharmacol 2008;153 (Suppl 1):S7–S26.
24. Shoichet BK, Kuntz ID. J Mol Biol 1991;221: 327–346.
25. Jones G, Willett P, Glen RC, Leach AR, Taylor R. J Mol Biol 1997;267:727–748.
26. Rarey M, Kramer B, Lengauer T, Klebe G. J Mol Biol 1996;261:470–489.
27. Friesner RA, Banks JL, Murphy RB, Halgren TA, Klicic JJ, Mainz DT, Repasky MP, Knoll EH, Shelley M, Perry JK, Shaw DE, Francis P, Shenkin PS. J Med Chem 2004;47:1739–1749.
28. Halgren TA, Murphy RB, Friesner RA, Beard HS, Frye LL, Pollard WT, Banks JL. J Med Chem 2004;47:1750–1759.
29. Morris GM, Goodsell DS, Halliday RS, Huey R, Hart WE, Belew RK, Olson AJ. J Comput Chem 1998;19:1639–1662.
30. Jain AN. J Med Chem 2003;46:499–511.
31. Leach AR, Shoichet BK, Peishoff CE. J Med Chem 2006;49:5851–5855.
32. Salaniwal S, Manas ES, Alvarez JC, Unwalla RJ. Proteins 2007;66:422–435.
33. Clark RD, Strizhev A, Leonard JM, Blake JF, Matthew JB. J Mol Graph Model 2002;20:281–295.
34. Feher M. Drug Discov Today 2006;11:421–428.
35. Skarzynski T, Thorpe J. Acta Crystallogr D Biol Crystallogr 2006;62:102–107.
36. Lipinski CA, Lombardo F, Dominy BW, Feeney PJ. Adv Drug Deliv Rev 2001;46:3–26.
37. Hann MM, Oprea TI. Curr Opin Chem Biol 2004;8:255–263.
38. Oprea TI, Allu TK, Fara DC, Rad RF, Ostopovici L, Bologa CG. J Comput Aided Mol Des 2007;21:113–119.
39. Good AC, Cheney DL. J Mol Graph Model 2003;22:23–30.
40. Good AC, Cheney DL, Sitkoff DF, Tokarski JS, Stouch TR, Bassolino DA, Krystek SR, Li Y, Mason JS, Perkins TD. J Mol Graph Model 2003;22:31–40.
41. Smith R, Hubbard RE, Gschwend DA, Leach AR, Good AC. J Mol Graph Model 2003;22: 41–53.
42. Brenk R, Naerum L, Gradler U, Gerber HD, Garcia GA, Reuter K, Stubbs MT, Klebe G. J Med Chem 2003;46:1133–1143.
43. Evers A, Klebe G. J Med Chem 2004;47: 5381–5392.
44. Gruneberg S, Stubbs MT, Klebe G. J Med Chem 2002;45:3588–3602.
45. Kraemer O, Hazemann I, Podjarny AD, Klebe G. Proteins 2004;55:814–823.
46. Kellenberger E, Springael JY, Parmentier M, Hachet-Haas M, Galzi JL, Rognan D. J Med Chem 2007;50:1294–1303.
47. Lill MA. Drug Discov Today 2007;12: 1013–1017.
48. Cramer RD, Patterson DE, Bunce JD. J Am Chem Soc 1988;110:5959–5967.
49. Cramer RD 3rd, Patterson DE, Bunce JD. Prog Clin Biol Res 1989;291:161–165.
50. Klebe G, Abraham U. J Comput Aided Mol Des 1999;13:1–10.
51. Klebe G, Abraham U, Mietzner T. J Med Chem 1994;37:4130–4146.
52. Soderholm AA, Lehtovuori PT, Nyronen TH. J Med Chem 2006;49:4261–4268.
53. Castilho MS, Postigo MP, de Paula CB, Montanari CA, Oliva G, Andricopulo AD. Bioorg Med Chem 2006;14:516–527.
54. Sippl W. J Comput Aided Mol Des 2002;16:825–830.
55. Guido RV, Oliva G, Montanari CA, Andricopulo AD. J Chem Inf Model 2008;48:918–929.
56. Salum LB, Polikarpov I, Andricopulo AD. J Mol Graph Model 2007;26:434–442.
57. Honorio KM, Garratt RC, Polikarpov I, Andricopulo AD. J Mol Graph Model 2007;25: 921–927.
58. Borchhardt DM, Andricopulo AD. 2009;5: 66–73.
59. Guido RVC, Castilho MS, Oliva G, Andricopulo AD. Structure-based pharmacophore strategies in drug design. In: Taft CA, Silva CHTP, editors. Current Methods in Medicinal Chemistry and Biological Physics. Kerala, India: Research Signpost; 2008.
60. Mason JS, Good AC, Martin EJ. Curr Pharm Des 2001;7:567–597
61. van Drie JH. Curr Pharm Des 2003;9: 1649–1664.
62. Atlamazoglou V, Thireou T, Eliopoulos E. J Comput Aided Mol Des 2007;21:239–249.
63. Christensen HS, Boye SV, Thinggaard J, Sinning S, Wiborg O, Schiott B, Bols M. Bioorg Med Chem 2007;15:5262–5274.

64. Murrall NW, Davies EK. J Chem Inf Comp Sci 1990;30:312–316.
65. Martin Y. Distance comparisons (DISCO): a new strategy for examining 3D structure–activity relationships. In: Hansch C, Fujita T. editors. Classical and 3D QSAR in Agrochemistry. Washington, DC: American Chemical Society; 1995.
66. Barnum D, Greene J, Smellie A, Sprague P. J Chem Inf Comput Sci 1996;36:563–571.
67. Jones G, Willett P, Glen RC. J Comput Aided Mol Des 1995;9:532–549.
68. Dixon SL, Smondyrev AM, Knoll EH, Rao SN, Shaw DE, Friesner RA. J Comput Aided Mol Des 2006;20:647–671.
69. Richmond NJ, Abrams CA, Wolohan PR, Abrahamian E, Willett P, Clark RD. J Comput Aided Mol Des 2006;20:567–587.
70. Gohlke H, Hendlich M, Klebe G. Perspect Drug Discov 2000;20:115–144.
71. Rarey M, Dixon JS. J Comput Aided Mol Des 1998;12:471–490.
72. Hindle SA, Rarey M, Buning C, Lengaue T. J Comput Aided Mol Des 2002;16:129–149.
73. Joseph-McCarthy D, Alvarez JC. Proteins 2003;51:189–202.
74. Joseph-McCarthy D, Thomas BE, Belmarsh M, Moustakas D, Alvarez JC. Proteins 2003;51:172–188.
75. Marcou G, Rognan D. J Chem Inf Model 2007;47:195–207.
76. Baroni M, Cruciani G, Sciabola S, Perruccio F, Mason JS. J Chem Inf Model 2007;47:279–294.
77. Goodford PJ. J Med Chem 1985;28:849–857.
78. Wagner RL, Aprilletti JW, McGrath ME, West BL, Baxter JD, Fletterick RJ. Nature 1995;378:690–697.
79. Moreno M, de Lange P, Lombardi A, Silvestri E, Lanni A, Goglia F. Thyroid 2008;18:239–253.
80. Lazar MA. Endocr Rev 1993;14:184–193.
81. Ye L, Li YL, Mellstrom K, Mellin C, Bladh LG, Koehler K, Garg N, Collazo AMG, Litten C, Husman B, Persson K, Ljunggren J, Grover G, Sleph PG, George R, Malm J. J Med Chem 2003;46:1580–1588.
82. Dow RL, Schneider SR, Paigh ES, Hank RF, Chiang P, Cornelius P, Lee E, Newsome WP, Swick AG, Sptizer J, Hargrove DM, Patterson TA, Pandit J, Chrunyk BA, LeMotte PK, Danley DE, Rosner MH, Ammirati MJ, Simons SP, Schulte GK, tate BF, DaSilva-Jardine P. Bioorg Med Chem Lett 2003;13:379–382.
83. Wagner RL, huber BR, Shiau AK, Kelly A, Lima STC, Scanlan TS, Apriletti JW, Baxter JD, West BL, Fletterick RJ. Mol Endocr 2001;15(3):398–410.
84. Bleicher L, Aparicio R, Nunes FM, Martinez L, Dias SMG, Figueira ACM, Santos MAM, Venturelli WH, Silva R, Donate PM, Neves FAR, Simeoni LA, Baxter JD, Webb P, Skaf MS, Polikarpov I. BMC Struct Biol 2008;8(8):1–13.
85. Garcia Collazo AM, Koehler KF, Garg N, Farnegardh M, Husman B, Ye L, Ljunggren J, Mellstrom K, Sanberg J, Gryfarb M, Ahola H, Malm J. Bioorg Med Chem Lett 2006;16: 1240–1244.
86. Hangland JJ, Doweyko AM, Dejneka T, Friends TJ, Devasthale P, Mellstrom K, Sanberg J, Grynfarb M, Sack JS, Einspahr H, Farnegardh M, Husman B, Ljunggren J, Koehler K, Sheppard C, Malm J, Ryono DE. Bioorg Med Chem Lett 2004;14:3549–3553.
87. Borngraeber S, Budny MJ, Chiellini G, Cunha-Lima ST, Togashi M, Webb P, Baxter JD, Scanlan TS, Fletterick RJ. Proc Natl Acad Sci USA 2003;100(26):15358–15363.
88. Gantley DC, Bartels MJ, Gennings C, Zacharewski TR, Freshour NL, Gollapudi BB, Carney EW. Reprod Toxicol 2001;14(3):207–216.
89. Li YL, Litten C, Koehler KL, Mellstrom K, Garg N, Garcia Collazo AM, Farngard M, Grynfarb M, Husman B, Sandberg J, Malm J. Bioorg Med Chem Lett 2006;16:884–886.
90. Jorgensen EC, Wright J. J Med Chem 1970;13 (3):367–370.
91. Yoshihara HAI, Apriletti JW, Baxter JD, Scanlan TS, J Med Chem 2003;46:3152–3161.
92. Garg N, Li YL, Garcia Collazo AM, Litten C, Ryono DE, Zhang M, Caringal Y, Brigance RP, Meng W, Washburn WN, Agback P, Mellstrom K, Rehnmark S, Rahimi-Ghadim M, Norin T, Grynfarb M, Sanberg J, Grover G, Malm J. Bioorg Med Chem Lett 2007;17:4131–4134.
93. Bartalena L, Robbins J. Clin Lab Med 1993; 13(3):583–598.
94. Westermark P, Sletten K, Johansson B, Cornwell GG. Proc Natl Acad Sci USA 1990;87 (7):2843–2845.
95. Connors LH, Richardson AM, Theberg R, Costello CE. Amyloid 2000;7(1):54–56.
96. Saraiva MJM, Costa PP, Goodman DS. J Clin Inv 1984;74(1):104–119.
97. Benson MD. Am J Pathol 1996;148 (2):351–354.
98. Saraiva MJM. FEBS Lett 2001;498:201–204.

99. Wojtczak A, Cody V, Luft JR, Pangborn W. Acta Cryst 1996;D52758–765.
100. Miroy JG, Lai Z, Lashuel HA, Peterson SA, Strang C, Kelly JW. Proc Natl Acad Sci USA 1996;93:15051–15056.
101. Baures PW, Oza VB, Peterson SA, Kelly JW. Bioorg Med Chem 1999;7:1339–1347.
102. Baures PW, Peterson SA, Kelly JW. Bioorg Med Chem 1998;6:1389–1401.
103. Peterson SA, Klabunde T, Lashuel HA, Purkey H, Sacchetini J, Kelly JW. Proc Natl Acad Sci USA 1998;95:12956–12960.
104. Petrassi HM, Johnson SM, Purkey HE, Chiang KP, Walkup T, Jiang X, Powers ET, Kelly JW. J Am Chem Soc 2005;127:6662–6671.
105. Petrassi HM, Klabunde T, Sacchettini J, Kelly JW. J Am Chem Soc 2000;122(10):2178–2192.
106. Oza VB, Smith C, Raman P, Koepf EK, Lashuel HA, Petrassi HM, Chiang KP, Powers ET, Sacchettini J, Kelly JW. J Med Chem 2002;45:321–332.
107. Klabunde T, Petrassi HM, Oza VB, Raman P, Kelly JW, Sacchettini JC. Nat Struct Biol 2000;7:312–321.
108. Purkey HE, Dorrell MI, Kelly JW. Proc Natl Acad Sci USA 2001;98(10):5566–5571.
109. Oza VB, Petrassi HM, Purkey HE, Kelly JW. Bioorg Med Chem Lett 1999;9:1–6.
110. Razavi H, Powers ET, Purkey HE, Adamski-Werner SL, Chiang KP, Dendle MTA, Kelly JW. Bioorg Med Chem Lett 2005;15:1075–1078.
111. Tojo K, Sekijima Y, Kelly JW, Ikeda S. Neurosc Res 2006;56:441–449.
112. Adamski-Werner SL, Palaninathan SK, Sacchettini JC, Kelly JW. J Med Chem 2004;47: 355–374.
113. Sekijima Y, Dendle MA, Kelly JW. Amyloid 2006;13:236–249.
114. Nuernberg B, Koehler G, Brune K. Clin Pharmacokin 1991;20:81–89.
115. Gales L, Macedo-Ribeiro S, Arsequell G, Valencia G, Saraiva MJM, Damas M. Biochem J 2005;388:615–621.
116. Razavi H, Palaninathan SK, Powers ET, Wiseman RL, Purkey HE, Mohamedmohaideen NN, Deechongkit S, Chiang KP, Dendle MTA, Sacchettini JC, Kelly JW. Angew Chem Int Ed 2003;42:2758–2761.
117. Johnson SM, Connelly S, Wilson IA, Kelly JW. J Med Chem 2008;51:260–270.
118. Johnson SM, Petrassi HM, Palaninathan SK, Mohamedmohaideen NN, Purkey HE, Nichols C, Chiang KP, Walkup T, Sacchettini JC, Sharpless KB, Kelly JW. J Med Chem 2005;48:1576–1587.
119. Purkey HE, Palaninathan SK, Kent KC, Smith C, Safe SH, Sacchettini JC, Kelly JW. Chem Biol 2004;11:1719–1728.
120. Kester MH, Bulduk S, Tibboel D, Meinl W, Glatt H, Falany CN, Coughtrie MW, Bergman A, Safe SH, Kuiper GG, Schuur AG, Brouwer A, Visser TJ. Endocrinology 2000;141 (5):1897–1900.
121. Shiraishi F, Okumura T, Nomachi M, Serizawa S, Nishikawa J, Edmonds JS, Shiraishi H, Morita M. Chemosphere 2003;52(1):33–42.
122. Gupta S, Chhibber M, Sinha S, Surolia A. J Med Chem 2007;50(23):5589–5599.
123. Julius RL, Farha OK, Chiang J, Perry LJ, Hawthorne MF. Proc Natl Acad Sci USA 2007;104(12):4808–4813.
124. Morais de Sá E, Pereira PJB, Saraiva MJ, Damas AM. J Biol Chem 279:2004;51: 53483–53490.
125. Green NS, Foss T, Kelly JW. Proc Natl Acad Sci USA 2005;102:14545–14550.
126. Radović B, Mentrup B, Köhrle J. Br J Nutr 2006;95(6):1171–1176.
127. Bloedon LT, Jeffcoat AR, Lopaczynski W, Schell MJ, Black TM, Dix KJ, Thomas BF, Albright C, Busby MG, Crowell JA, Zeisel SH. Am J Clin Nut 2002;76(5):1126–1137.
128. Okazaki K, Okazaki S, Nakamura H, Kitamura Y, Hatayama K, Wakabayashi S, Tsuda T, Katsumata T, Nishikawa A, Hirose M. Arch Toxicol 2002;76(10):553–559.
129. Trivella DBB,et al., unpublished results.
130. Yu H, Adedoyin A. Drug Discov Today 2003;8:852–861.
131. Tummino PJ, Copeland RA. Biochemistry 2008;47:5481–5492.
132. Whitesides GM, Krishnamurthy VM. Q Rev Biophys 2005;38:385–395.
133. Holdgate GA, Ward WHJ. Drug Discov Today 2005;10:1543–1550.
134. Chaires JB. Annu Rev Biophys 2008;37: 135–151.
135. Böhm H-J, Schneider G, Protein–Ligand Interactions: From Molecular Recognition to Drug Design (Methods and Principles in Medicinal Chemistry). Vol. 19. Weinheim: Wiley-VCH Verlag GmbH & Co. KGaA; 2003.
136. Homans SW. Drug Discov Today 2007;12: 534–539.
137. Gilson MK, Zhou H-X. Annu Rev Biophys Biomol Struct 2007;36:21–42.

138. Rajski SR, Williams RM. Chem Rev 1998;98: 2723–2796.
139. Fernando H, Nagle GT, Rajarathnam K. FEBS J 2007;274:241–251.
140. Carbonell T, Freire E. Biochemistry 2005;44: 11741–11748.
141. Freire E. Drug Discov Today 2008;13:869–874.
142. Freire E. Drug Discov Today Technol 2004;1:295–299.
143. Weber PC, Salemme FR. Curr Opin Struct Biol 2003;13:115–121.
144. Gopalakrishnapai J, Gupta G, Karthikeyan T, Sinha S, Kandiah E, Gemma E, Oscarson S, Surolia A. Biochem Biophys Res Commun 2006;351:14–20.
145. McKnight RE, Gleason AB, Keyes JA, Sahabi S. Bioorg Med Chem Lett 2007;17:1013–1017.
146. Yanchunas J, Langley DR, Tao L, Rose RE, Friborg J, Colonno RJ, Doyle ML. Antimicrob Agents Chemother 2005;49:3825–3832.
147. Todd MJ, Gomez J. Anal Biochem 2001;296:179–187.
148. Wiggers HJ, Cheleski J, Zottis A, Oliva G, Andricopulo AD, Montanari CA. Anal Biochem 2007;370:107–114.
149. Ghosh AK, Dawson ZL, Mitsuya H. Bioorg Med Chem 2007;15:7576–7580.
150. Wang Y-F, Tie Y, Boross PI, Tozser J, Ghosh AK, Harrison RW, Weber IT. J Med Chem 2007;50:4509–4515.
151. Martinez-Cajas JL, Wainberg MA. Antivir Res 2007;76:203–221.
152. Dierynck I, De Wit M, Gustin E, Keuleers I, Vandersmissen J, Hallenberger S, Hertogs K. J Virol 2007;81:13845–13851.
153. Dervan PB. Bioorg Med Chem 2001;9: 2215–2235.
154. Chaires JB. Arch Biochem Biophys 2006;453: 26–31.
155. Morikis D, Lambris JD. Trends Immunol 2004;25:700–707.
156. Morphy R, Rankovic Z. Drug Discov Today 2007;12:156–160.
157. Rees DC, Congreve M, Murray CW, Carr R. Nat Rev Drug Discov 2004;3:660–672.
158. Andrews PR, Craik DJ, Martin JL. J Med Chem 1984;27:1648–1657.
159. Schneider G, Schneider P, Renner S. QSAR Comb Sci 2006;25:1162–1171.
160. Ciulli A, Williams G, Smith AG, Blundell TL, Abell C. J Med Chem 2006;49:4992–5000.
161. Ciulli A, Lobley CMC, Tuck KL, Smith AG, Blundell TL, Abell C. Acta Crystallogr D63:2007; 171–178.
162. Talhout R, Villa A, Mark AE, Engberts JBFN. J Am Chem Soc 2003;125:10570–10579.
163. Olsson TSG, Williams MA, Pitt WR, Ladbury JE. J Mol Biol 2008;384:1002–1017.
164. The SCORPIO database. http://www.biochem.ucl.ac.uk/scorpio/scorpio.html.

QM AND QM/MM APPROACHES TO EVALUATING BINDING AFFINITIES

Katherine E. Shaw
Christopher J. Woods
Adrian J. Mulholland
Centre for Computational Chemistry,
School of Chemistry, University of Bristol,
Bristol, UK

1. INTRODUCTION

With the average drug taking 15 years [1] to get from initial concept to market and drug approvals dropping significantly in the past decade [1], any method that can reliably aid drug discovery could be of great practical value. Ideally, drug design requires methods to calculate relative binding affinities for small molecules to their biological (e.g., protein) targets. Such methods could contribute not only in lead identification and optimization but also in development, for example, in predicting ADME or, potentially, toxicology properties. If binding of a ligand to other receptors (such as the cytochrome P450s) could be reliably predicted, it could reduce attrition rates in the later stages of drug development. For a binding affinity prediction method to be useful in the pharmaceutical industry, it should not only be accurate but also be capable of making predictions for many compounds within a reasonable timescale. In the earlier stages of drug design, this requirement precludes many of the more computationally intensive methods, as tens of thousands of compounds, or more, may need to be tested (see Volume 1 Chapter 13 and Volume 2 Chapters 1 and 15). In the later lead optimization stage, where only a handful of compounds are being considered, more thorough and demanding methods can feasibly be applied: such methods may be relatively computationally intensive. It is new methods designed to be applied at this stage in the drug discovery process that are the focus of this chapter.

There are many computational methods for predicting or calculating the binding affinity of a small molecule to a protein. Until recently, protein structure-based methods, involving detailed modeling of protein–ligand interactions, generally used a simplified, though atomically detailed, classical description of the system (using molecular mechanics (MM) force fields). There are many known limitations to such methods, for example, the conventional MM force fields cannot reliably describe charge transfer effects or metal-containing systems. The introduction of new algorithms, and ever-increasing computing power, has meant that biomolecular simulations are no longer necessarily limited to simple force fields. One approach is to develop more sophisticated empirical force fields, for example, including representation of the electronic polarizability. Computer power is still increasing rapidly, and the introduction of multicore processors and advances in parallel processing mean that more sophisticated, and potentially more accurate methods based on fundamental physics, are becoming accessible. "Quantum chemical" electronic structure methods, which are based on quantum mechanics (QM), were until recently restricted to relatively small systems. They are now increasingly applicable to protein–ligand complexes [2,3], and have the potential to improve the accuracy of binding predictions, at least in some cases. This chapter reviews some recent advances in protein–ligand binding affinity calculations, with a particular emphasis on the use of QM in methods based on molecular simulations.

1.1. Challenges in Calculating Binding Affinities

Protein–ligand binding affinities are difficult to calculate, because of the complexity of such systems, and because there are many different contributions to consider. For a drug to successfully bind, there must be good steric (the shape of the ligand is mirrored by the shape of the receptor) and electronic complementarities. The main drive to ligand binding is often hydrophobic interactions [4,5]. Hydrophobic association is a complex process, and debate continues on the nature of hydrophobic interactions. Typically, an "organic" type drug molecule is preferentially "solvated" in the lipophilic protein cavity. This releases water, from the empty receptor site, back into the bulk solvent—generally an entropically favorable process. Increasing the lipophilic contact surface between the ligand and the receptor, for example, by introducing more "fatty" groups onto the ligand in favorable locations where space allows in a binding pocket, or searching for other lipophilic pockets within the receptor, will usually increase the binding affinity.

The formation of hydrogen bonds between the ligand and the protein is also important, although they do not necessarily drive binding [6]. If hydrogen bonds that were present when the ligand was in the solvent are not replaced by interactions with the protein, the binding will be less favorable, thus reducing the binding affinity. In metalloproteins, hydrogen bonds could also be replaced by metal ligation.

Another important consideration when calculating binding affinities is entropy. If a ligand is very flexible in solution and it is then confined in a small cavity, entropy will be lost, though this can be offset by release of bound water, and potentially also changes in the receptor. Likewise, a rigid ligand may be able to bind more tightly than a flexible counterpart, as there is less of an entropic penalty [6]. There could also be loss of flexibility in the protein, particularly of the side chains, which should be taken into account.

When considering computational approaches to predicting the binding affinity, it is also useful to look at the experimental methods available, because experimental results are often used as a benchmark for computational methods. The dissociation equilibrium constant, K_d, is commonly used to quantify protein–ligand binding affinity, which is given as

$$K_d = \frac{[P][L]}{[PL]} \tag{1}$$

where [P], [L], and [PL] are the equilibrium concentrations of free protein, free ligand, and the protein–ligand complex, respectively. Note that this is not a dimensionless quantity, instead it has units of moles mol/dm^3. Compounds with a low dissociation constant are good binders. The dissociation constant can be related to the change in free energy using the following equation:

$$\Delta G = \Delta G^\circ - RT \ln K_d \tag{2}$$

where ΔG° is the free-energy change associated with binding at standard conditions (i.e., all concentrations are 1 M, the temperature is 298 K, and the pressure is 1 bar). At equilibrium, the free-energy change, ΔG, is zero, therefore Equation 2 can be used to define the standard free-energy change, ΔG°, as RT ln K_d. For this K_d should be dimensionless. This is resolved by noting that most texts on this subject ignore the ratio of the activities of each of the species in the standard state, c° [7]. In its complete form, the binding free energy associated with the process under standard states is calculated using Equation 3. In practice, as has been mentioned previously, c° is usually ignored.

$$\Delta G^\circ = RT \ln \left[\frac{K_d}{c^\circ}\right] \tag{3}$$

To measure K_d, the equilibrium concentrations of the free and the bound species need to be determined. These can be measured, for example, by equilibrium dialysis [8], fluorescence spectroscopy [9], or NMR methods (these have become increasingly useful in measuring millimolar binding affinities) [10]. Isothermal titration calorimetry (ITC) not only can measure the binding affinity but also can give the magnitude of the enthalpic and entropic contributions to the free energy [11].

In reality, experiments are rarely run to measure the dissociation constant. Instead the tendency is to measure the IC_{50} value, which dictates how much of a substance is required to inhibit the protein by 50%. It can be measured by running experimental assays and constructing a dose–response curve (showing the effect of different concentrations of inhibitor on the biological activity of the receptor). Although not a direct measurement of binding affinity, the IC_{50} value can be related to the dissociation constant [12]. When comparing experimental results to computationally predicted binding affinities, ideally the experimental values used should be from the same assay (or from assays run under the same conditions, that is, temperature and concentration need to be consistent). Often exact binding affinities are not the practical goal of a calculation; instead, the correct rank ordering of potential ligands is desired.

Calculated binding free energies are therefore often compared directly with IC_{50} values.

Unlike experimental approaches, some computational methods have the potential to analyze the different contributions to the binding affinity. This has important implications in inhibitor design, as specific interactions can be targeted. The usual starting point in all computational studies explicitly involving the protein environment is a suitable structure. The Protein Data Bank (PDB) [13] has a wealth of available crystal structures, with and without bound ligands. Structures are also available from NMR and homology modeling, although these may sometimes not be of high-enough precision or accuracy for high-quality calculations. Care must be taken as proteins are not rigid: the many degrees of freedom of these large, complex, and dynamic systems mean that extensive sampling of phase space is necessary to obtain accurate thermodynamic properties. For example, Warshel's group has shown that energy minimizations for structure generation when calculating reaction and activation free energies are insufficient and that extensive sampling (i.e., molecular dynamics (MD) or Monte Carlo (MC) simulations) is vital [14] (see Section 7.1 for more details on sampling methods). Another advantage of modeling the dynamics of the system when calculating the binding affinity is that the protein can relax with each different ligand. Therefore, a crystal structure is not necessarily required for each bound complex, as long as the conformations of the bound complexes for different ligands are similar.

To study a protein–ligand system in detail computationally, a model of the system is needed. Due to the need to sample conformations of the system, the treatment of any such model must be quick to implement. MM force fields are a good approach to this problem. Current MM methods for proteins typically describe the system as point charges joined by "springs". The "springs" account for the bonds between atoms and they are usually described using a harmonic energy term. All the bonds, angles, and dihedrals within the system are included using simple functions. The nonbonded terms are considered by including van der Waals and electrostatics terms.

1.2. Advantages of Quantum Mechanical Methods in Binding Affinity Calculations

MM simulations are based on the classical "ball and spring" description of chemical bonding. The energy of the molecule is described by a force field that describes the behavior of different types of atoms and bonds. It contains bonded terms (e.g., bond stretching, angle bending, torsion), van der Waals terms (e.g., modeled using a Lennard-Jones potential) and electrostatic terms (e.g., modeled as the Coulomb interaction between static point charges). MM force fields allow the rapid evaluation of the energy of a biomolecular system and therefore large ensembles of structures can be generated, for example, via MD or MC. To fully converge the ensemble averages required in some of the binding free energy methods discussed in this chapter, significant conformational sampling is required. However, it can be challenging to use MM models to describe electronic effects, such as polarization, as electrons (and usually electronic effects) are not modeled explicitly. Instead, atomic partial charges reflect differences in the electronegativity and hydrogen bonding potential, etc. It is increasingly being recognized that to achieve accurate binding free energies it may often be imperative to account for polarization of the ligand by the protein. This is to be expected, because the polarization of the ligand in the protein and in solvent (water) is likely to differ significantly, due to the differences in the electrostatic environment around the ligand. Experimental (e.g., spectroscopic) results show significant polarization of ligands bound to many proteins, that is, more polarization than in water [15–17]. Polarization has been shown to contribute significantly to the electrostatic interaction between proteins and ligands [18], even when the systems are not charged (calculations suggest contributions of up to 4% in dihydrofolate reductase [19] and between 32% and 39% for 3 inhibitors of HIV-1 protease [20]). In a conventional MM model, there is typically no explicit polarization. Polarization is treated implicitly by artificially increasing the atomic charges on molecular fragments within the system (by 10–20% compared to gas-phase values). These charges are typically invariant, that is, the same charges are used for the ligand regardless of whether it is bound to the protein or free in solution; hence, differences in polarization between the two environments are not accounted for. However, it has been shown in some cases that explicit treatment of polarization is necessary for accurate calculations. For example, Jiao et al. stated, based on their study of the binding of two charged ligands to trypsin [21]:

> we believe that it is critical to treat polarization explicitly to achieve chemical accuracy in predicting the binding affinity of charged systems.

A fixed charge model of the ligand lacks important physical details, which are now known to make an important contribution to binding free energies in some cases. As well as the importance of polarization effects for binding energy predictions, it has also been shown that they can play a significant role in determining the structure of protein–ligand complexes [22].

Polarizable MM force field models have been developed and can treat polarization effects successfully. They typically include polarization in one of the following two ways: by induced dipoles or by induced charges (i.e., the dipole of a molecular fragment or the charges on the atoms can fluctuate in response to changes in their surroundings). Friesner [23] and Warshel [24] have both recently reviewed polarizable MM force fields.

An alternative to polarizable MM force fields is to use QM (electronic structure/quantum chemical) methods. QM methods model polarization directly because they provide an explicit representation of (at least some of) the electrons. Work by Zhou et al., on a series of different protein–ligand complexes, led to the conclusion that QM was vital when considering significant differences in charge–charge interactions between protein–ligand complexes (i.e., ligand charges greater than $+1$) [25]. QM methods can provide an explicit model of the electrons in a system; therefore, allowing properties that rely on the electronic distribution to be calculated (see Section 7.2 for more details on QM methods). QM calculations take significantly longer than MM and so can typically only be used to study tens of atoms, unless so-called linear scaling methods are applied (see below). This computational problem can be exacerbated when accurate treatment of dispersion interactions, which may be significant in the hydrophobic contribution to binding, is required. High-level QM methods may be necessary to capture these interactions: a MM description of such interactions (e.g., in a QM/MM approach, see below) may be preferable to a low-level QM treatment.

Computer simulations that combine QM with MM are becoming increasingly important for studying biological systems [26,27] and a natural extension is to use them for binding affinity predictions [2,3,28–31]. Gleeson has studied the feasibility of using different QM/MM methods in the drug discovery process [32]. QM/MM methods combine the more detailed physical model provided by QM calculations with the speed of MM. In protein systems, in which there are typically thousands of atoms or more, the most "important" are treated using QM while the bulk of the system is treated using MM (see Fig. 1). "Important" atoms, in the case of binding affinity calculations, are those involved in protein–ligand binding (i.e., the ligand and optionally any surrounding residues). QM/MM methods and interfacing between the QM and MM regions are described in more detail in Section 7.3.

Work by Nunthaboot et al. [33] has showed that significantly different structures are sampled by MM (CHARMM22) and QM/MM (PM3/CHARMM22) MD simulations of HIV-1 integrase complexes. They found that the binding site geometry around the Mg^{2+} was very different in the two simulations. QM is required for metal-containing systems because the ill-defined atom types of metal atoms in most of the force field parameters cannot describe the complex nature of the interactions between a small molecule and a metal ion in the active site.

It can also be difficult to describe covalently bound inhibitors using MM, and QM/MM approaches have been shown to successfully model some such cases [34]. QM/MM methods have also been successfully applied to studying enzyme mechanisms [26,35,36]. This is

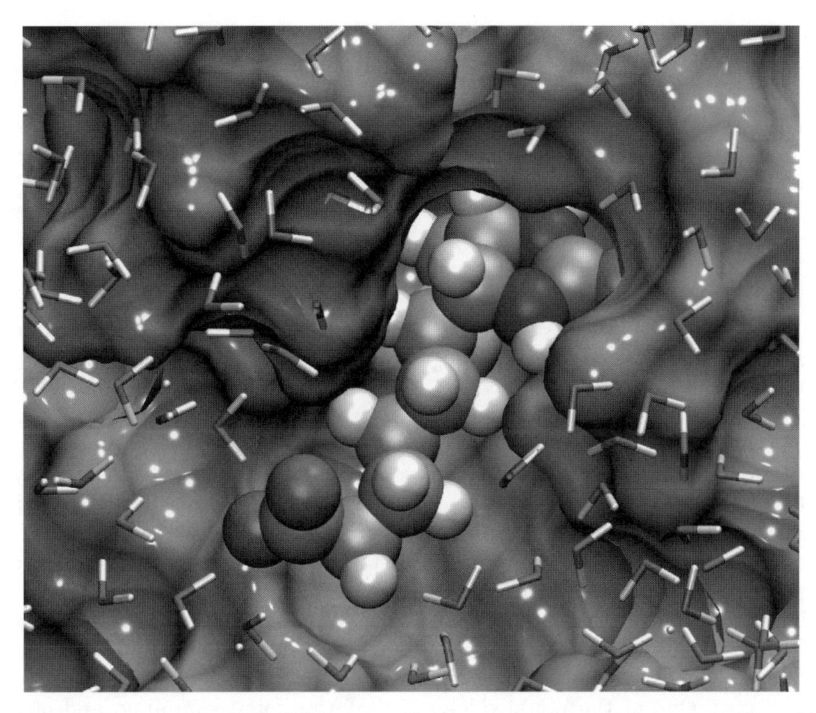

Figure 1. Streptavidin bound to biotin. This figure illustrates the division of the system into QM and MM regions for a QM/MM calculation. The ligand (shown as van der Waals spheres) is treated by QM. The protein (shown as a surface) and surrounding water (bonds shown as lines) are treated using a MM force field. (This figure is available in full color at http://mrw.interscience.wiley.com/emrw/9780471266945/home.)

another area where the application of MM can be challenging, as conventional force fields cannot model bond making and breaking.

In addition to providing a potentially more accurate and flexible physical model than MM methods, QM/MM methods are ideal for use in drug discovery as the QM model of the inhibitor requires little or no parameterization. While MM force fields have been carefully and intensively parameterized for proteins (e.g., CHARMM22 [37], OPLS/AA [38,39], and Parm99 [40] in AMBER [41]), accurate parameters for novel compounds are rarely available. Parameterization of MM methods is a time-consuming process and without thorough testing against experimental data it may be hard to judge the accuracy of the parameters. In the case of most novel compounds there is no experimental data, making an accurate parameterization process difficult. It is also possible to parameterize ligands using QM calculations [42]. Such QM calculations typically treat the ligand in the gas phase, or with a simple continuum treatment of solvent, and so do not account for any changes in polarization of the ligand upon binding.

QM/MM methods are relatively computationally efficient (compared to full QM calculations on the whole system). There are two main disadvantages with QM/MM methods: first, the computational expense of the QM calculation is significantly greater than that of the MM calculation. QM/MM calculations are typically limited by the QM part, with a trade-off needed between the size of the QM region, the accuracy of the QM method and thus the amount of time necessary for each QM energy evaluation. The second disadvantage is that the best method of describing the interactions between the QM and the MM regions is still unclear (this is discussed further in Section 7.3).

Linear scaling QM methods are another route to using QM techniques in binding affinity calculations. As mentioned above, one of the major drawbacks of using QM methods in calculations of biological systems is the high computational cost. Unlike QM/MM methods, linear scaling quantum mechanics approaches aim to treat the whole system (the ligand, the whole protein, and the solvent) using a QM method. For a system with N basis functions, the computation time for a standard semiempirical molecular orbital [43,44] or density functional theory (DFT) [45] QM calculation scales as N^3. For a traditional Hartree–Fock calculation the scaling is approximately N^4 and for MP2 [46,47] approximately N^5. Parallelization of computer code has seen some improvements in computational cost, but linear scaling methods are currently the only feasible way to treat the electronic structure of a whole protein–ligand complex quantum mechanically. Linear scaling calculations typically involve dividing a large system into smaller ones and treating each separately. This is based on the principle of locality: properties of a certain region of interest are only weakly influenced by regions that are not spatially close to that region. These methods tend to be easily parallelizable, further reducing the computational cost. One of the most widely used linear scaling QM methods is the divide and conquer (D&C) approach, which has been used for DFT [48,49] and semiempirical [50–52] QM calculations. The method divides a large system into a series of smaller subsystems, which usually consist of a core surrounded by one or two buffer regions. The Fock equations are then solved for these subsystems. A cutoff can be applied to the Fock and density matrices (i.e., only the local Fock equations are solved), greatly reducing the computational cost. An alternative linear scaling approach is to localize the molecular orbitals [53–57]. This approach has been applied in a QM/MM framework, with near quantitative results for reaction activation energies [58]. Examples of relevant applications of linear scaling QM methods include the optimization of a large portion of cytidine deaminase [59], binding enthalpy calculations on a series of protein–ligand interactions [60], DNA/RNA–ligand interactions [61] and a review of the application of linear scaling methods to study charge transfer, polarization, and solvation free energies in cold shock protein A [50].

Until recently, binding affinity calculations predominantly applied MM methods. With QM methods becoming more tractable (due to improved algorithms, such as the linear scaling approach and parallelizable code, and increased computing power) the possibility now exists to extend established binding affinity prediction methods to incorporate a QM element, and promising steps have been taken in this direction by several groups, as outlined below.

Perturbation methods, such as FEP and TI, are considered some of the most (statistically) accurate computational means of calculating *relative* binding free energies [62]. This type of approach, even with cheap MM potentials, is still not routinely practical in the drug discovery process as it can be very time consuming and is typically limited to small perturbations between the initial and final states (see Section 2 for more details on perturbation methods including QM). At the other end of the scale for predictions are QSAR (quantitative structure–activity relationship) models. They use little or no conformational sampling, and often include no structural information on the target (e.g., being entirely ligand based), and hence are not reviewed in full here. Briefly, these models involve building a mathematical model to predict a property of a molecule, in this case the *absolute* binding free energy [63]. QSAR models are fitted to experimental and/or calculated data, and use structural properties (termed molecular descriptors) to predict other properties of the molecule. The quality of QSAR results depends mainly on two factors—the type of molecular descriptor used and the method applied to extract the information. The descriptors used often give an idea of the properties that contribute to binding—and hence can aid new ligand design. The descriptors tend not to be transferable; therefore, QSAR models can have limited predictivity outside their training set. QM (and increasingly QM/MM) methods have important roles in the calculation of descriptors for QSAR. For more information on QSAR models, see Volume 1 Chapter 13.

Several methods fall between FEP and QSAR in terms of intensity, detail, and speed. These methods involve many approximations compared to the perturbation methods, which render them much quicker. Compared to the perturbative free-energy methods, it is possible to study a more diverse range of ligands and binding modes with these rapid methods, which typically aim to estimate absolute (rather than relative) binding free energies. These methods often involve fitting to experimental data (e.g., the linear interaction energy method, see Section 3.1). While this can yield accurate results for the training set, this type of method tends to be less transferable and relies on the existence of sufficient experimental data. Other methods, such as MM-GBSA and MM-PBSA (see Section 3.2) simplify the system by treating the solvent interactions implicitly.

Approximately 27 years ago, the first automated "docking" programs became available [64] and since then they have become high powered virtual high-throughput screening (vHTS) tools that are now being combined with QM calculations. Scoring functions can give a quick estimate of the relative binding affinity of ligands (for a given "pose" structure, for example, generated by docking), but involve many approximations, which can result in limited predictive power (see Section 4 for more information).

Binding affinity prediction has become a very active area of research because of its implications for drug design. However, there is not yet a "gold standard" calculation that can be used on a diverse range of systems with a high success rate [65]. Fortunately, as this review will show, the computational medicinal chemistry researcher now has an arsenal of weapons to choose from when tackling protein–ligand binding affinity calculations. Currently, the choice of method depends almost entirely on the problem at hand. One very important consideration is the prediction time—calculations must be quick for vHTS, but can be slower, and so in theory more accurate, for the lead optimization stage of drug discovery. It must be remembered that, if the calculation takes longer than synthesizing the molecule and testing it experimentally then the method is probably not practically useful. It is also important to consider the cost (including researcher time)—computation time is relatively inexpensive, but the longer the user has to spend setting up a calculation the more expensive the process becomes. In the next sections, some of the QM and QM/MM methods currently available in this field are discussed.

2. PERTURBATION BASED QM/MM BINDING AFFINITY CALCULATIONS

This section focuses on available methods that are, in principle, (classically) exact (errors arise from deficiencies in the force fields used and in statistical mechanics, rather than in approximations made in the calculations). These methods are the most computationally demanding, but equally they tend to give the best results. Perturbation methods are rooted in statistical thermodynamics. They calculate *relative* binding free-energy differences between molecules—not the absolute binding free energy of each.

Until recently, perturbation calculations were typically carried out using MM simulations. However, over the past few years, these methods have been extended to include a QM treatment of the ligands. One of the main limitations of the conventional perturbation methods is the need for validated MM force field parameters for the ligand. In drug discovery, novel compounds are usually studied; therefore, such parameters often need to be developed from the scratch for each new ligand. This development work is very time consuming, cannot be readily automated, and is potentially subject to user bias. An advantage of using QM methods to describe the ligand is that few nonstandard parameters are required. Until recently, the computational demands of incorporating QM made perturbation methods inaccessible. However, with ever-faster computers and parallelized code, such methods are now becoming practical in this setting.

2.1. Free-Energy Perturbation

2.1.1. Theoretical Background The fundamental equations behind free-energy perturbation (FEP) (Eqs 4 and 5) are based on

Zwanzig's work [66] from the 1950s. It was not until more powerful computers were available in the 1980s that they were used in this way [67], and the first calculations of relative binding free energies in proteins were by Bash and coworkers in 1987 [68].

$$G_{\mathrm{B}}-G_{\mathrm{A}} = \Delta G_{\mathrm{A}\to\mathrm{B}} = -k_{\mathrm{B}}T\ln\left[\int p_{\mathrm{A}}(q)\mathrm{e}^{-\Delta E_{\mathrm{AB}}(q)/k_{\mathrm{B}}T}\mathrm{d}q\right] \quad (4)$$

A and B are the two related systems that are in different states—in protein–ligand binding affinity calculations these states would usually be different bound ligands (or could be wild type and mutant proteins, for example). The Boltzmann probability of finding a configuration, q, in an ensemble of state A is given by $p_{\mathrm{A}}(q)$. ΔE_{AB} is the energy difference between the two states A and B. The key term in Equation 4 is $p_{\mathrm{A}}(q)$—it is this that makes these perturbation methods possible. Most configurations of state A have a very high energy, but a very low Boltzmann probability; therefore, only the low energy, thermally significant configurations of A need to be sampled to obtain the free-energy change, $\Delta G_{\mathrm{A}\to\mathrm{B}}$. Zwanzig took this and derived an equation that allows us to calculate the free-energy difference between two states, as in Equation 4, but by using ensemble averages generated during a MD or MC simulation:

$$\Delta G_{\mathrm{A}\to\mathrm{B}} = -k_{\mathrm{B}}T\ln\left\langle \mathrm{e}^{-\Delta E_{\mathrm{AB}}(q)/k_{\mathrm{B}}T}\right\rangle_{\mathrm{A}} \quad (5)$$

This is the Zwanzig equation, where the free-energy change between a reference state, A, and a perturbed state, B, can be calculated. For ΔG to converge successfully, the fluctuations in ΔE must be small, which implies good overlap between A and B. This is not true in most cases. Good overlap can be achieved by splitting the perturbation into a series of smaller steps, defined by a λ coordinate. This is used to gradually morph between the two systems. $\lambda = 0$ represents the reference system, A. The perturbed system, B, is represented at $\lambda = 1$. Intermediate λ values correspond to a nonphysical hybrid of the two systems. Now the free-energy change is calculated using Equation 6.

$$\Delta G = \sum_{i=1}^{n} -k_{\mathrm{B}}T\ln\left\langle \mathrm{e}^{-(E_{\lambda_{i+1}}-E_{\lambda_i})/k_{\mathrm{B}}T}\right\rangle_{\lambda} \quad (6)$$

where n is the number of steps over which the perturbation occurs, and λ_i is the value of λ at step i. The required ensemble averages for Equations 5 and 6 are collected during MC or MD simulations. The major advantage of this partitioning using λ is that the simulations for each λ value can be run in parallel, as the calculations are independent of one another.

The perturbation can be in either the forward (λ from 0 to 1, from system A to B) or the backward (λ from 1 to 0, from system B to A) direction. Given sufficient sampling, it is expected that the free-energy change in both directions would be identical. Many FEP simulations use double-wide sampling [69], in which the calculation is run in both the forward and the backward directions from one λ point, giving the ensembles for two consecutive windows.

When using FEP to study ligand binding affinities, the two different states, A and B, typically have different numbers of atoms. Therefore, it is usually important to account for the disappearing and appearing of atoms. In the single topology approach, the perturbation uses dummy atoms to represent atoms that are present in system A but not in system B, or vice versa. System A is "grown" into system B by extending out the dummy atoms into the relevant areas of phase space. In the dual topology approach, both ligands A and B are explicitly represented in the binding pocket throughout the calculation. They interact with the rest of the protein, but not with each other. The interactions between each ligand and the protein are scaled with λ. At $\lambda = 0$, ligand A is fully on and ligand B is fully off, that is, it does not interact with the protein. Hence, morphing along the λ coordinate can be viewed as "switching the ligands on and off". Dual topology can suffer from an "endpoint catastrophe"—it cannot easily account for appearing or disappearing atoms, as it is too computationally demanding to carry out sufficient conformational sampling [70]. This can result in the energies diverging at the endpoints (i.e., at $\lambda=0$ and $\lambda=1$). In the dual topology single coordinate approach [71], developed for use in QM/MM calculations, the two ligands are forced to adopt the same conformation during the perturbation.

2.1.2. Recent Applications One group working on QM/MM FEP is that of Reddy et al. [2,62,72]. They first introduced their method in 2004 [62], initially studying relative solvation free energies of simple systems using a semiempirical QM method (AM1) to treat the solute and MM to treat the solvent (in this case SPC/E water). The test cases were chosen because they showed relatively large structural changes (e.g., phenylalanine to isoleucine) and changes in hydrogen bond potential (e.g., acetone to acetamide), aromaticity (e.g., benzene to pyridine), or electron density (e.g., 1,1,1-trichloroethane to ethane). In 2007, they used *ab initio* QM (HF/3-21G*) to treat the same systems [72]. Their technique transforms from molecule A to B over the λ coordinate, using the thread technique (a dual topology method) to allow them to map structurally diverse compounds. In the thread technique [73], the two molecules are split into common and noncommon atoms. The common atoms are treated the same throughout the perturbation, while the noncommon atoms are switched on and off according to the λ value. Reddy et al. equilibrated their systems using 20 ps of MD at 300 K and 1 atm. They then ran 51 λ windows, each of 1.5 ps equilibration and 3 ps of data collection. To calculate the QM energies and forces they used the *Galaxy* program (AM Technologies, Inc.) [74] to separate the threaded molecule into two separate molecules (i.e., A and B) at each MD step. The QM energies and forces could then be individually calculated and combined using the λ coupling method:

$$E^{\mathrm{QM}}(\lambda) = \lambda E^{\mathrm{B}}_{\mathrm{QM}} + (1-\lambda)E^{\mathrm{A}}_{\mathrm{QM}} \tag{7}$$

The overall free-energy change is calculated using Equation 8.

$$\Delta G_{\mathrm{Tot}} = \Delta G_{\mathrm{QM}} + \Delta G_{\mathrm{MM}} \tag{8}$$

Their results show that the *ab initio* QM/MM solvation free energies are in very good agreement with experimental data, compared to the conventional MM FEP. However, this came at the cost of a significant increase in computation time (three- to fivefold more CPU time for the semiempirical QM/MM FEP than the conventional MM FEP. The *ab initio* perturbation for methanol to ethane took 169 h on a single processor at that time). The calculation time could now be trivially decreased by using parallel code. They have recently extended their work to calculate the relative binding free energies of human fructose-1,6-bisphosphatase (FBPase) inhibitors [2], a potential target for type 2 diabetes treatment. Semiempirical (AM1) QM/MM FEP calculations gave binding affinities in better agreement with experimental results than the conventional MM FEP. Calculations took 4–5 days to run on a single processor IBM RS6000.

To overcome this drastic increase in computer time, a less direct route to the free-energy change can be implemented. This involves calculating the free-energy change at a lower level of theory, and then converting this to the QM/MM level. This is best described by the Warshel cycle, shown in Fig. 2 [75,76]. Warshel and coworkers have used this cycle

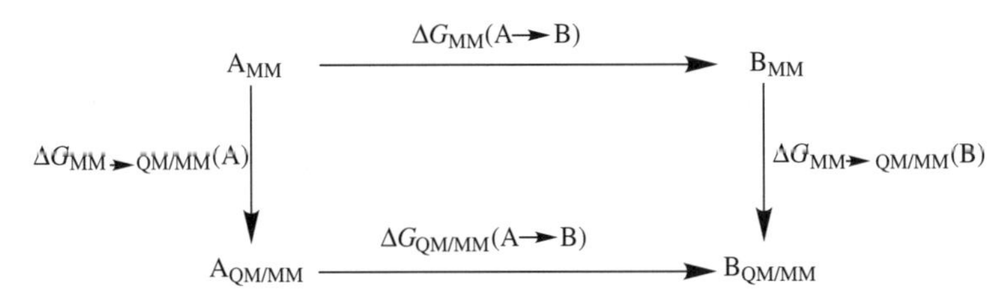

Figure 2. A thermodynamic cycle to calculate the free energy difference between A and B, $\Delta G_{\mathrm{QM/MM}}(\mathrm{A} \rightarrow \mathrm{B})$, using MM sampling of the phase space. In the implementation by Warshel [75,76], the vertical lines represent single step perturbations.

in conjunction with the empirical valence bond (EVB) method [77–81] to access high-level QM free energies within a reasonable timescale. Initially developed to compare reactions in solutions and enzymes [78], the EVB method has recently been extended to allow QM/MM simulations [75,76,82]. The EVB method describes reactions by mixing resonance structures (diabatic states) that correspond to valence-bond structures. A mapping procedure, which gradually moves between the EVB states, for example, from reactant to product, is used that also takes into account the changes in the solute (QM) charge distribution. The interactions between the solvent and solute are calculated for each resonance structure, giving the solvent contributions to the Hamiltonian. The reference EVB potential (that is ideally as close as possible to the full *ab initio* QM/MM surface) is used to accelerate the convergence of the *ab initio* FEP calculations. All the sampling is carried out on the reference potential, and a single step FEP approach is used to calculate the difference between the EVB and the *ab initio* surfaces. To improve convergence, the EVB Hamiltonian can be fitted to experimental or gas-phase data (so that the EVB and *ab initio* surfaces are more alike) [76], or a linear response approximation (LRA) can be used to calculate the free-energy change between the two systems (this method is described in full in Section 3.1) [76,83]. While no binding free-energy results have yet been reported, this method has been successful at calculating free-energy barriers in solution [75] and activation-free energies in enzymes [83].

Rod and Ryde have also developed a QM/MM FEP method that does not require full QM/MM simulations, called quantum mechanical thermodynamic cycle perturbation (QTCP) [84,85]. The method is also based on the thermodynamic cycle shown in Fig. 2. The free-energy change is calculated as

$$\Delta G^{A\to B}_{QM/MM} = -\Delta G^{A}_{MM\to QM/MM} + \Delta G^{A\to B}_{MM} + \Delta G^{B}_{MM\to QM/MM} \tag{9}$$

Each free-energy change can be calculated using FEP (so, for $\Delta G^{A\to B}_{MM}$, A and B have different structures, whereas, for $\Delta G^{A}_{MM\to QM/MM}$ and $\Delta G^{B}_{MM\to QM/MM}$, A and B are treated with different methodology). To reduce the computational cost of the MM to QM/MM perturbations, Rod and Ryde only calculate one QM/MM energy (to be included in the ensemble average) for every picosecond of simulation (for every 500 MM MD steps). They argue that MM sampling interspersed with QM/MM energy evaluations is sufficient because most of the sampling covers uninteresting phase space (i.e., it takes many simulation steps to move between two points in phase space, and so the QM/MM energy does not need to be calculated at each step). To ensure convergence they keep the QM region *fixed* during the MD simulation (the potential energy surface, for the QM internal degrees of freedom, is very different in the QM and the MM description). They treat the QM region using DFT calculations (with the PBE exchange-correlation functional) and the 6-31 + G* basis set. They have currently only used this method to calculate free-energy barriers for enzymatic reactions, gaining good agreement with experimental results, but this method could be extended to be applied to the calculation of binding free energies.

When the QM region is large, even using a reference potential to accelerate the sampling may be impractical. This would be the case, for example, when the ligand is large or if protein residues in the binding site are included in the QM region. Another approach to improving the computational efficiency of QM/MM FEP methods is to constrain or freeze the electron density of part of the system. This was originally attempted by Vaidehi et al., who replaced a MM description of the solvent with a pseudo-potential, while the solute was treated using *ab initio* techniques [86]. This allows the delocalization of electrons from the solute to the solvent to be studied, whereas if the system had been split into a QM solute and MM solvent molecules this would not be possible. The interaction between the solvent and the solute is treated using the pseudo-potential, while the interactions between solvent molecules are treated classically. The pseudo-potential is calibrated using data from *ab initio* calculations or relevant experimental results. Vaidehi et al. calculated the solvation free energy of a Li^{+} ion using a MD FEP approach, obtaining good results. This work has given rise to the constrained density functional theory (CDFT) [87] and the frozen density functional theory (FDFT) [88,89] methods. These methods are useful when the QM region is large. Initially, these approaches treated the solute and the solvent using DFT. CDFT constrained the electron density of the solvent, while in contrast it is frozen in FDFT. The electron density of the solute is then determined to minimize the total electronic energy. In larger systems, there are three regions—the QM region (e.g., the ligand), the frozen/constrained density region (which could contain, for example, active site residues), and then a normal MM region (e.g., the rest of the protein and solvent). The interactions within the frozen or constrained density region are treated classically, while the interactions between the solute and this region are treated using QM. It is this classical treatment that makes these methods tractable for large systems. The QM treatment between the two inner regions is a major advantage compared to the standard QM/MM approaches, as few approximations are required (the interaction between two QM regions is calculated, rather than attempting to combine a QM and a MM region). Wesolowski and Warshel used the FDFT method to calculate the relative QM solvation free energies of water and methane in 1994, also using a classical potential to enhance the sampling of the *ab initio* surface in MD FEP [89]. They treated the whole system using DFT, freezing the electron density of the water. The FDFT method has also been used to calculate the energy barrier for a proton-transfer reaction on a Zn^{2+} center, allowing the metal and all its ligands to be treated QM [90], and allowing charge transfer from the metal to the ligands by including an initial CDFT step. This work has been built on by Xiang and Warshel [91], calculating free-energy profiles of proton-transfer reactions in solution and in proteins. Their method combines FDFT with FEP, using a mapping analogous to EVB. Their results for enzyme-catalyzed reactions compare favorably with experimental results.

Obviously, the FDFT method cannot straightforwardly allow for charge transfer between the solute and the solvent when the solvent electron density is treated as frozen. The CDFT method was initially developed for a system where two molecules formed a weak covalent bond—hence, where different conformations of each molecule have different electron densities [87]. Different frozen densities are therefore used for different conformations, by allowing the frozen density to relax iteratively. In each successive iteration step the frozen density is set as the density of the "QM" region from the previous step, until self-consistency is achieved.

The effects of freezing the electron density of the QM region within a QM/MM calculation has been studied by Kästner et al. in calculations on *p*-hydroxybenzoate hydroxylase (PHBH). Building on the previous work of Zhang et al. [92], these workers investigated the effect of various approximations used in semiempirical AM1 QM/MM FEP on calculated free-energy barriers [93]. All the MM point charges were included in the QM calculation, to account for polarization of the QM region by the MM atoms. Initially, they ran full QM/MM optimizations of the system at varying points along a predefined reaction coordinate. These structures were used as the input for MD simulations. As with the work of Rod and Ryde, described above, the coordinates of the QM region were then kept fixed during the MD simulations, to reduce the computational cost of the calculation. In this work, Kästner et al. tested the effect of fixing the electron density of the QM region (i.e., changes in the polarization of the QM region with varying MM coordinates were not accounted for) and replacing the QM density with point charges that reproduce the electrostatic potential (ESP charges). Both these methods further reduce the computational cost of the calculations. They compared their various QM/MM FEP results (varying QM density, frozen QM density, and QM ESP charges) to the conventional MM thermodynamic integration and umbrella sampling approaches (where the whole system is simulated in the MD simulations). They found that the QM/MM FEP method produced energies in good agreement with the other methods. Freezing the density of the QM region or approximating with ESP point charges, caused an error of only approximately 0.5–0.6 kcal/mol, and resulted in significant savings in computer time. Indeed, when

ESP charges were used, the QM/MM geometry optimizations were the most time-consuming stage, meaning that higher levels of QM theory would be more feasible. In this work, they also tested different link atom approaches.

Jorgensen et al. have developed a novel QM/MM method that could be straightforwardly extended to protein–ligand binding free-energy calculations. They use a semiempirical QM method (initially AM1) to generate partial charges for the atoms in the QM region, using charge model 1 (CM1A) [94]. As their charges are calculated in the gas phase, they are scaled by a constant factor, calculated based on the simulation of 13 simple organic molecules, to account for condensed-phase polarization. These new charges can then be used to calculate the Coulomb interaction between the QM and the MM regions. Therefore, the polarization of the QM region by the MM region is included in an implicit fashion. The MM region is treated using the OPLS force field [38,39]. The van der Waals interactions are treated using MM Lennard-Jones potentials. Initially, hydration free energies were calculated, using MC to generate ensembles and double-wide FEP calculations. The results for the 13 organic molecules were in good agreement with experiment. This work has also been extended to calculate the relevant scaling factor for PDDG/PM3 calculations [95,96]. An advantage of this method is that QM calculations are only required when the conformation of the QM region changes.

2.2. Thermodynamic Integration

2.2.1. Theoretical background As with FEP, thermodynamic integration (TI) involves evaluating the ensemble average over a series of windows, defined by the λ coordinate. In the TI method, it is necessary to calculate the gradient of the free energy with respect to the scaling parameter, $\delta G/\delta\lambda$. This is in contrast to FEP, in which the difference in free energy between neighboring λ values is calculated. The TI approach avoids the requirement of having a good overlap between the λ windows, as the gradient is a property of the system at each λ value. The gradient can either be calculated analytically, by taking the ensemble average of the gradient with respect to λ, or be calculated numerically (e.g., from the endpoints). This latter numerical method, termed finite difference TI (FDTI) [97], is in many ways similar to FEP. It involves using the Zwanzig equation (Eq. 5) where the reference state A is at λ and the perturbed state B is at $\lambda \pm \Delta\lambda$ (giving the forward and backward free-energy gradients). One can imagine that when $\Delta\lambda$ equals the window width FDTI and FEP are equivalent, but in FDTI $\Delta\lambda$ above and below each window is studied and in FEP neighboring states are considered.

The free-energy gradient is then integrated with respect to λ (Eq. 10) to calculate the free-energy change, ΔG. This integral can be trivially evaluated using the trapezium rule or by quadrature.

$$\Delta G = \int_{\lambda=0}^{\lambda=1} \left\langle \frac{\partial G}{\partial \lambda} \right\rangle_{\lambda} \partial\lambda \tag{10}$$

Given the close relationship between FEP and TI, the single and dual topology algorithms, discussed in Section 2.1.1, are still applicable. This means that TI can suffer from "endpoint catastrophies", in much the same way as FEP because it also cannot easily account for the appearance and disappearance of atoms.

2.2.2. Recent Applications Cummins and Gready presented a QM/MM thermodynamic integration approach to calculating relative free energies of protonation in 2007, called the perturbed quantum mechanical atom (PQA) method [98]. They studied the relative protonation free energy in dihydrofolate (DHF) reductase, using eight different QM regions of different size. The simulations were run using semiempirical QM (at the PM3 level). Data were collected during 100 ps of QM/MM MD, run with replica exchange, at eight independent temperatures and using nine λ values. While their results showed significant variation in the total free-energy change for the different QM regions, they were in qualitative agreement with previous *ab initio* studies on smaller model complexes.

Woods et al. [99] have used a free-energy cycle, similar to the one in Fig. 2, but with the

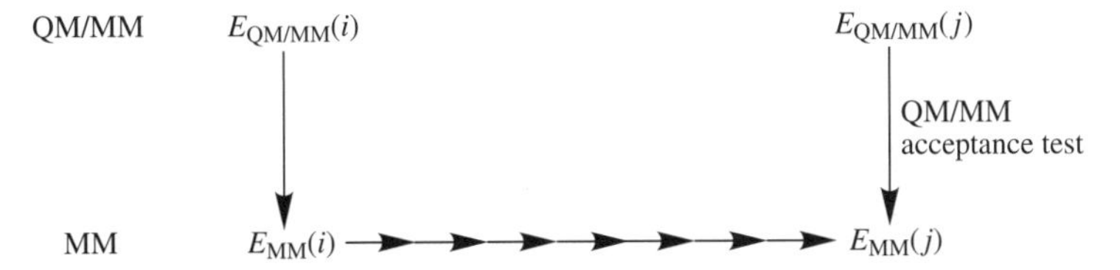

Figure 3. The Metropolis–Hastings algorithm, used to accelerate the sampling of a QM/MM system using a MM simulation [99]. MM is used to guide the MC sampling of the QM/MM system, but, due to the form of the acceptance test, a correct QM/MM ensemble is generated.

vertical steps (originally single step FEP calculations) replaced with TI calculations, where the ensembles are generated using the Metropolis–Hastings algorithm [100] (see Fig. 3). This is similar to the QTCP method of Rod and Ryde, discussed above [84,85], however, the advantage of this approach is that the QM region is not fixed at any point. Pure MM simulations are run to perturb system A into system B, and then the correction free-energy (the difference between the MM and the QM/MM) is calculated using the Hamiltonian

$$H = (1-\lambda)H_{QM/MM} + \lambda H_{MM} \tag{11}$$

Simulations are run at different values of λ, using Metropolis–Hastings MC moves to generate the necessary ensembles. The correction free energy is calculated using TI, as in Equation 12.

$$\Delta G_{QM/MM \rightarrow MM} = \int_{\lambda=0}^{\lambda=1} \left\langle \frac{\partial H}{\partial \lambda} \right\rangle_{\lambda} \partial\lambda = \int_{\lambda=0}^{\lambda=1} \langle H_{MM} - H_{QM/MM} \rangle_{\lambda} \partial\lambda \tag{12}$$

The λ coordinate is used to scale the MM system into the QM/MM system (when $\lambda = 1$ the system is purely MM, when $\lambda = 0$ the system is QM/MM). Conformational sampling can be enhanced by including replica exchange moves, as the λ coordinate is already provided by TI (the RETI method [101]). Using commodity processors, this method is able to run full-length (60 million MC steps) free-energy simulations on small molecules, using high-level *ab initio* QM (MP2/AVDZ) in a matter of days. Currently, this method has been used only to calculate relative solvation free energies, but protein–ligand binding free energy predictions are expected shortly.

3. APPROXIMATE METHODS FOR QM/MM BINDING AFFINITY PREDICTIONS

The methods in the previous section calculated, in principle, (classically) exact QM/MM relative binding free energies. In contrast, the methods discussed in this section contain many approximations, with the aim of increasing the computational efficiency, and therefore compound throughput. These approximations include a simplified treatment of entropy and/or solvent interactions and (for example) the use of pure MM calculations. The obvious advantage of these methods is that they use less computational resources, but at the cost that the results may be less accurate.

3.1. Linear Interaction Energy Method

3.1.1. Theoretical background The LIE (linear interaction energy) method, based on the LRA, was first proposed by Åqvist et al. [102], and has also been investigated by Jorgensen's group [103,104]. It involves two simulations for each ligand—first, of the free ligand in water; second, of the ligand bound to protein. Boltzmann ensembles of conformations are built using snapshots from the two simulations, and these ensembles are used to calculate interaction energies of the ligand with its environment (water or protein). The binding free energy is approximated as the difference between the ensemble averages of the ligand–solvent interaction energies for the two simulations.

In the initial formulation by Åqvist et al. [102], the binding affinity is calculated as

$$\Delta G^{\circ} \approx \beta(\langle U_{\mathrm{elec}}\rangle_{\mathrm{bound}} - \langle U_{\mathrm{elec}}\rangle_{\mathrm{free}}) + \alpha(\langle U_{\mathrm{vdW}}\rangle_{\mathrm{bound}} - \langle U_{\mathrm{vdW}}\rangle_{\mathrm{free}}) \quad (13)$$

$\langle\,\rangle$ denotes Boltzmann ensemble averages and α and β are constants designed to account for the changes in internal energy of the water and the protein (their reorganization energy upon ligand binding). The most accurate results [105] are obtained when α and β are system-specific (i.e., fitted to experimental data for a particular system). Jorgensen's group [103,104] built on this original work by adding a third term related to solvent accessible surface area (SASA) to account for cavity formation in the solvent. In some cases, further terms for hydrogen bonding have also been included.

3.1.2. Recent Applications The linear interaction energy model with continuum electrostatic solvation (LIECE) [106] has been combined with QM/MM (giving the QMLIECE model) and used to predict the binding affinities of three classes of enzyme–inhibitor complex [25]. The systems tested were 44 inhibitors of flaviviral serine protease, 24 inhibitors of retroviral aspartic protease and 73 inhibitors of CDK2. QMLIECE only showed significant improvement compared to LIECE for the 44 inhibitors of the serine protease, where the ligands had charges ranging from 0 to +3. These calculations suggested that a QM treatment is necessary when there are significant charge–charge interactions (i.e., a large variation in polarized protein charges upon binding of different ligands). The MM LIECE method is approximately two orders of magnitude faster than the explicit solvent LIE method, mainly because the MD simulations are replaced with energy optimizations. The QMLIECE calculations are slower than MM LIECE equivalents as expected, but took approximately 1 h each to run, per ligand, using the semiempirical RM1 method (a newly parameterized version of AM1) for the ligand and the CHARMM22 force field for the surrounding protein. The time required for the QM calculation scaled linearly with the size of the QM region.

Khandelwal and coworkers have used QM/MM optimized structures and single point energies for time-averaged structures, from constrained MM MD simulations, within the LIE framework, to calculate binding free energies for inhibitors of zinc-dependent matrix metalloproteinases 3 and 9 (MMP-3 and MMP-9) [29,107]. Their method involves a four-tier approach: (1) the ligands are docked into the protein using FlexX [108], with the best pose selected based on the metal binding geometry; (2) QM/MM minimizations of the complex are performed using DFT (B3LYP) combined with the OPLS-AA force field; (3) a MM MD simulation of the entire complex is run, with the metal bonds, which lie in the QM region, constrained to the optimized geometry from step 2; and (4) single point QM/MM interaction energies for the time-averaged structure from step 3 are calculated, using Equation 14.

$$\Delta\langle E_{QM/\mathrm{MM}}\rangle = \left\langle E^{\mathrm{complex}}_{\mathrm{QM/MM}}\right\rangle - \left\langle E^{\mathrm{ligand}}_{\mathrm{QM/MM}}\right\rangle - \left\langle E^{\mathrm{protein}}_{\mathrm{QM/MM}}\right\rangle \quad (14)$$

These energies are then correlated with SASA calculations in a LIE-like equation to calculate the free energy of binding. The energy terms of the time-averaged structures were used, rather than the average energies from the ensemble, after previous work suggested that similar LIE correlations were obtained for the two approaches [109,110]. Their results are in good agreement with experimental binding affinities (for example, the MMP-3 results, when compared to experiment, had a correlation coefficient of 0.9) at the expense of approximately a fourfold increase in computational time, compared to the conventional MM LIE calculations. These workers have since built on this method to study multiple ligand binding modes. This is significant in systems where the ligand undergoes large-scale motions in the binding site, and so would require extensive MD sampling. By treating these motions as different binding modes they were able to use shorter MD simulations [3]. The application of the new interaction energy equation to MMP-9 gave significant improvements in its descriptive and predictive powers (the correlation, compared to experiment, increased from an r^2 of 0.6 to 0.9).

Along the same lines as LIE is using the QM/MM interaction energy as a guide to the ligand binding energy. This approach can take the form of an unfitted LIE method, where the binding free energy is estimated using the equation

$$\Delta G_{\text{bind}} \approx E(\text{complex}) - E(\text{ligand}) - E(\text{protein}) \quad (15)$$

In an example of this approach, the binding affinities of 11 HIV-1 integrase inhibitors were calculated using 1 ns QM/MM (AM1 ligand/OPLS protein with explicit TIP3P water) MD simulations [111]. The interaction energy between the QM and the MM region (i.e., between the ligand and the protein) was averaged from 100,000 configurations from the last 100 ps of the MD simulations. The interaction energy was calculated using the following equation:

$$\begin{aligned} E_{\text{INT}} = {} & E_{\text{vdW}}(R_{\text{QM}}, R_{\text{MM}}) \\ & + E^{\text{QM}}_{\text{ELEC}}(r, R_{\text{QM}}, R_{\text{MM}}) \\ & + E^{\text{QM}}_{\text{POL}}(r, R_{\text{QM}}, R_{\text{MM}}) \end{aligned} \quad (16)$$

where E_{vdW} is the van der Waals interaction between the QM and the MM regions, E_{ELEC} is the electrostatic interaction between the QM region and the MM point charges, and E_{POL} accounts for the polarization of the QM region by the MM region, which is calculated as the gas-phase energy difference between the polarized and unpolarized wavefunctions. These workers found that the biological activity of each ligand was correlated with E_{INT}, with a correlation coefficient of 0.8.

3.2. QM/MM-PBSA and QM/MM-GBSA

3.2.1. Theoretical Background The linear interaction energy method often requires experimental data to train the coefficients used in the equation to predict the binding affinity. Other "approximate" methods exist that do not require experimental data; therefore, making them more widely applicable in the drug discovery process. The molecular mechanics, Poisson–Boltzmann model and solvent accessible surface area (MM-PBSA) [112] and the molecular mechanics, generalized-Born model and solvent accessible surface area (MM-GBSA) [113,114] methods are two such methods. They involve MM simulations of the free protein, the free ligand and the bound complex. Then, the binding free energy is estimated using implicit solvent models for the solvation free-energy contribution, greatly reducing the computational cost compared to the perturbation methods.

In the PBSA or GBSA methods (i.e., when there is no simulation involved), the solvation free energy of a system is split into two contributions, an electrostatic contribution and a SASA contribution:

$$\Delta G_{\text{solvation}} = \Delta G^{\text{electrostatic}}_{\text{solvation}} + \Delta G^{\text{SASA}}_{\text{solvation}} \quad (17)$$

The electrostatic contribution to binding is calculated by numerically solving the PB or GB equations. In the PB equation, the solvent is represented by a continuum, using an appropriate dielectric constant and ionic strength. The electrostatic contribution is calculated by solving the PB equation for the electrostatic potential throughout the system. This can be computationally demanding, and so approximations to the PB equation have been developed. An example is the GB equation, which uses Born radii to indicate how shielded an atom is from the solvent. The SASA term accounts for the nonelectrostatic contributions to the binding free energy—the solute–solvent van der Waals interactions and the cost of forming a cavity for the solute in the solvent. The binding free energy of a ligand to a protein in solution, $\Delta G^{\text{solution}}_{\text{bind}}$, can then be calculated, from the cycle shown in Fig. 4, using

$$\begin{aligned} \Delta G^{\text{solution}}_{\text{bind}} = {} & \Delta G^{\text{gas}}_{\text{bind}} + \Delta G^{\text{complex}}_{\text{solvation}} - \Delta G^{\text{protein}}_{\text{solvation}} \\ & - \Delta G^{\text{ligand}}_{\text{solvation}} \end{aligned} \quad (18)$$

This PBSA or GBSA approach is combined with MM simulations to take into account the effect of conformational change on the calculated binding affinity (giving rise to MM-PBSA and MM-GBSA, respectively). The

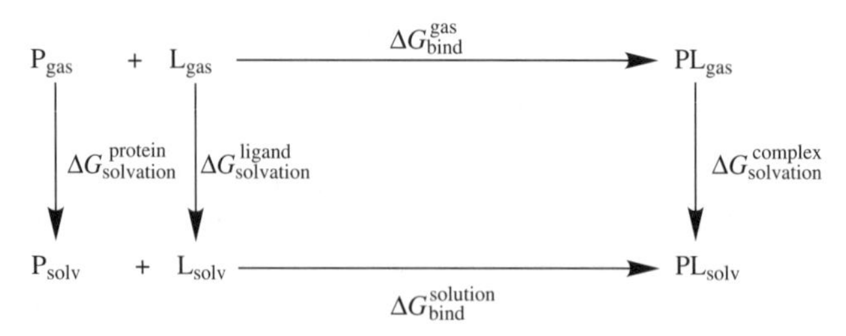

Figure 4. Thermodynamic cycle used to calculate the free energy of binding in solution in the MM-PBSA and MM-GBSA methods. P is the protein, L the ligand, and PL the bound protein–ligand complex. Each system is either in the gas phase ($_{gas}$) or in the solution phase ($_{solv}$). $\Delta G^{protein}_{solvation}$ is the solvation free energy of the protein, $\Delta G^{ligand}_{solvation}$ is the solvation free energy of the ligand, $\Delta G^{complex}_{solvation}$ is the solvation free energy of the protein–ligand complex, ΔG^{gas}_{bind} is the free energy of binding in the gas phase, and $\Delta G^{solution}_{bind}$ is the free energy of binding in solution (the quantity we are trying to calculate).

binding affinity has been shown to vary by up to 10 kcal/mol for different configurations [115]; hence, accounting for different conformations is important. A MM MD simulation for the bound complex, the free ligand and the free protein are run, in explicit solvent and the binding free energy calculated according to Equation 19.

$$\Delta G_{\text{bind}} = \langle E_{\text{MM}} \rangle + \Delta G_{\text{solvation}} - T\Delta S \qquad (19)$$

$\langle E_{\text{MM}} \rangle$ is the average MM energy of the system taken from the MD simulation (i.e., the energy of all the bonded and nonbonded interactions in the system) and the entropy change, ΔS, is often calculated using normal mode analysis. The average MM energy can be calculated either from the average of the energies of the snapshots from the ensemble, or from the single point energy of an average structure calculated over the simulation. The $\Delta G_{\text{solvation}}$ term is calculated by deleting all the explicit waters and then solving the PB or GB equations and including a SASA term (Eq. 17). The entropy term is often ignored, especially if the ligands are roughly the same size and share a common binding mode. This allows us to calculate the binding affinity using Equation 20, where the free energies of the complex, $G(C)$, the free ligand, $G(L)$, and the free protein, $G(P)$, are calculated using Equation 21.

$$\Delta G = G(C) - G(L) - G(P) \qquad (20)$$

$$G = \langle E_{\text{MM}} \rangle + G_{\text{solvation}} \qquad (21)$$

A single simulation approach has also been developed, where a simulation of the complex is run and then the MM averages for the free protein and ligand are calculated from that simulation by simply deleting the ligand and the protein, respectively. This resembles the LIE approach because it neglects changes in conformational entropy and internal energy of both the free ligand and the free protein upon binding.

These methods can be used to study more structurally diverse systems than FEP, as there are no mutations involved, because absolute rather then relative binding free energies are calculated. More recent applications have involved using a QM ligand (and in some cases also some QM protein residues) within the MM-PBSA/GBSA framework, giving rise to QM/MM-PBSA and QM/MM-GBSA, respectively.

3.2.2. Recent Applications In 2002, Pellegrini and Field [116] parameterized the GBSA model to be compatible with semiempirical AM1 QM/MM optimizations and MD simulations, using the OPLS-AA force field for the protein. The work resulted in new equations to calculate the Born radii used when implementing the GBSA approach. The parameterization involved fitting the atomic polarization energies of 19 proteins. While they recommend their model for use with protein–ligand binding affinities, no follow-up work is currently available.

Some data are available for QM/MM-PBSA implementations. Gräter et al. calculated the binding affinities of 47 trypsin inhibitors using this method [28]. They used QM/MM geometry optimizations and binding free energy calculations that include a QM component that accounted for the ligand strain and the protein–ligand electrostatic interactions, using AM1/CHARMM. They used a single conformation for the unbound ligand and protein, and compared the effect of using a rigid or a flexible complex conformation (with the binding orientation taken from docking results). The van der Waals interactions between the protein and the ligand are treated classically. They scaled their calculated van der Waals interactions by a constant, calculated by fitting to experimental data, because the ligand parameters were assigned using simple atom-element based rules. From Equation 20, the $G(C)$ and $G(L)$ terms both include QM contributions. Naturally, as the ligand was treated using QM, in the $G(L)$ term the average energy of the ligand is the QM energy. In the bound complex the electrostatic interaction between the ligand (QM) and the protein (MM) was treated using QM. Their final model gave relatively poor correlations between predicted and experimental binding affinities (correlation coefficient, r^2 of 0.2 for the flexible model and 0.3 for the rigid model). The RMSD between the predicted and the experimental values was 1.21 and 1.29 kcal/mol for the two models, respectively. They concluded that their poor results were due to the relatively rigid nature of trypsin (hence, there was no significant improvement using a flexible model) and because the experimental binding affinities fell within a small range (approximately a 3.5 kcal/mol spread). They also tested their models on FK506 binding protein (FKBP), using the same van der Waals scaling parameter as used for trypsin. They found that it was transferable (giving RMSDs of 0.66 and 1.45 kcal/mol for the flexible and rigid models respectively). Comparisons to a pure MM model also showed improved predictive power (RMSD of 1.85 kcal/mol for the MM model). Each calculation took approximately 30 min on a single processor.

Work by Wang and Wong, published in 2007 [117], used linear scaling QM to rank 5 protein kinase A inhibitors using a QM/MM-PBSA fixed conformation approach (the "backbone" of each ligand was fixed and only the conformation of the different functional groups was allowed to vary). The protein was treated using MM (CHARMM22). Each calculation took approximately 35 min on a single processor. They found that varying the dielectric constant of the protein and ligand greatly affected the predictive capabilities of the method (using a common protein dielectric constant of 2 or 4 showed no significant correlation with experiment, whereas a dielectric constant of 1.5 for the protein produced promising results).

4. QM DOCKING AND SCORING

4.1. Theoretical Background

Docking and scoring programs are high throughput, approximate methods used to predict the binding modes of a set of ligands to a target and then rank them based on their calculated binding affinity. In effect, docking and scoring occurs in two different stages—posing and ranking. A pose is the complete specification of the ligand relative to the receptor—that is, its orientation and conformation. The poses can then be scored, giving a rough estimate of the fit of the ligand in the site. The more advanced process of ranking, or more accurately evaluating the binding affinity, usually takes some of the results from the initial pose and score stage and reevaluates them using more complex calculations—for example, taking into account properties such as entropy. This review will be focusing on the ranking stage. Volume 2 Chapter 15 gives further details on the whole process. Also, many review articles have been published comparing the different methods of docking and scoring available [118–123]. Currently, there are few programs that allow for protein flexibility while creating the poses. Despite this, the initial docking and quick scoring phase can usually get a pose that is within 2 Å of the actual binding conformation. Unfortunately, the more rigorous ranking stage is not always able to distinguish this pose from other, badly docked ones when calculating the binding affinities. This is seen as a fault in the scoring function—if only the best docked poses or even crystal structures are used scoring still in many cases cannot correctly predict the binding free energy [124].

Recently, QM methods have been included into the docking and scoring process, with work concentrating not only on improving the poses but also on more accurate ranking calculations.

4.2. Recent Applications

Much recent research has concentrated on improving the quality of docked poses using QM/MM calculations. For example, Parks et al. took the 20 highest ranked FlexX [108] poses for a ligand of hepatitis C polymerase and optimized the ligand (i.e., the protein was held fixed at the same conformation in all cases) using B3LYP(6-31G*)/CHARMM22 QM/MM [125]. They then ranked the poses based on the QM/MM interaction energy (electrostatic and van der Waals contributions). From the original 20 poses, the ranking successfully picked 3 top poses that had low structural RMSDs compared to the known crystal structure, unlike the original FlexX pose that had a high structural variation.

Ferrara et al. [126] used a scoring function that uses quantum-refined force field molecular dynamics (QMFF-MD) [127,128] to improve Glide [129,130] docking results. The method involves two cycles of ligand charge refinement (calculated using DFT) followed by MM MD of the ligand (the protein is kept rigid), and has previously been used successfully on 30 inhibitors of HIV-1 protease [128]. They tested their model by attempting to find 40 CDK2 inhibitors in a database of approximately 600 nonbinders. The model was successful, resulting in a 2.8-fold enrichment compared to random screening. The correlation coefficient using their model was 0.55, compared to 0.35 for normal Glide.

The importance of accurate ligand charges in molecular docking was studied by Cho et al. [22]. They improved docking predictions by deriving new ligand charges using QM/MM calculations of the protein–ligand complex. These charges explicitly accounted for the polarization of the ligand by the protein. They ran their QM/MM calculations using the QSITE program [131], using DFT (B3LYP with the 6-31G* basis set) for the QM region. To create their docking algorithm, QM Dock, this procedure was used iteratively. The charges were recalculated and the ligand redocked until convergence, within a preset limit, was achieved. They found that the QM Dock results varied, depending on the quality of the starting structure. They reasoned that this was because their QM/MM charge calculations relied entirely upon structure. Although two docked structures may both have a RMSD compared to the crystal structure of say 2 Å, this can actually relate to quite different qualitative geometries. They hence developed a survival of the fittest (SOF) approach, where QM/MM charge evaluations and redockings were run for a series of different structures (which can be trivially run in parallel; therefore, requiring no significant increase in computational time). From this set of structures they then picked an overall best structure, based on the total Coulomb and van der Waals energy of the system. The new results showed that the average RMSD (compared to the crystal structures) is reduced from 1.77 Å for a purely MM approach to 0.88 Å for the QM/MM SOF approach.

Zhong et al. [132] used this QM/MM ligand charges approach to study the binding of duocarmycin to DNA using MM-GBSA. They compared the effect of using RESP (restrained electrostatic potential, calculated for the ligand using QM in the gas phase) and QPLD (QM polarized ligand docking, that is, the QM dock method, where the charges are calculated for the ligand using QM in the presence of the MM protein using QSITE) charges in unrestrained MD simulations of the complex. They found significant improvements in the calculated binding free energy when the QM/MM charges were used.

Work by Raha and Merz has included QM (AM1 and PM3) gas-phase interaction energies as the basis of a scoring function [30,133]. They estimate the ΔG_{bind} using

$$\Delta G_{\text{bind}} = \Delta H_{\text{I}} + \Delta \text{LJ}_6 + \Delta\Delta G_{\text{solv}} + \Delta S_{\text{solv}} + \Delta S_{\text{conf}} \quad (22)$$

where ΔH_{I} is the gas-phase interaction energy (calculated from QM heats of formation, ΔH_{f}, using the linear scaling QM method divide and conquer); ΔLJ_6 accounts for dispersion interactions, using a Lennard-Jones-like term; $\Delta\Delta G_{solv}$ is the free-energy change due to desolvation or solvation upon complexation; ΔS_{solv} is the solvent entropy (calculated using SASA approximations), and ΔS_{conf} accounts for the change in the number of freely rotatable bonds in the ligand and the interacting protein side chains. They used two approaches to predict the binding affinity. In one, applied to zinc ion-mediated ligand binding, they used Equation 22 "raw" to estimate the binding free energy (i.e., there is no fitting to experimental data [133]). For a set of 18 carbonic anhydrase inhibitors they achieve a correlation coefficient, r^2, with experiment of 0.69. They also tested this method by weighting each term in Equation 22 by fitting to experimental data for 165 noncovalent bound protein–ligand complexes [30]. They found that this method could successfully discriminate between native and decoy poses and when validated against a set of 100 diverse inhibitors out-performed the nonfitted method ($r^2 = 0.58$ when fitted, compared to 0.48 when not fitted).

5. CONCLUSIONS

This chapter has attempted to review the current state of QM and QM/MM protein–ligand binding affinity predictions. It is by no means an exhaustive review, as this is an evolving and growing field because of its significance to the drug discovery process. This review has outlined approaches that include various levels of QM methods, within a molecular simulation framework. The use of QM methods in this context is still relatively new, but we believe it has important implications for the future of accurate binding affinity calculations, because it can account for ligand polarization and requires significantly less parameterization work for novel compounds than empirical (e.g., MM) methods.

6. FURTHER READING

In addition to the references cited here, for further reading into the role of computers in drug discovery, see

- Drews F. Drug discovery: a historical perspective. Science 2000;287:1960–1964.
- Jorgenson W. The many roles of computation in drug discovery. Science 2004; 303:1813–1818.
- Raha K, Peters MB, Yu N, Wollacott AM, Westerhoff LM, Merz, KM, Jr. The role of quantum mechanics in structure-based drug design. Drug Discov Today 2007; 12:725–731.
- Kapetanovic IM. Computer-aided drug discovery and development (CADDD): *in silico*–chemico–biological approach. Chem Biol Interact 2008;171: 165–176.

For further reading on calculating protein–ligand binding affinities (not necessarily including QM), see

- Ajay A, Murcko MA. Computational methods to predict binding free energy in ligand–receptor complexes. J Med Chem 1995;38:4953–4967.
- Wong CF, McCammon JA. Protein flexibility and computer-aided drug design. Annu Rev Pharmacol Toxicol 2003;43: 31–45.
- Gilson MK, Zhou H-X. Calculation of Protein–ligand binding affinities. Annu Rev Biophys Biomol Struct 2007;36:21–42.

For further reading on biomolecular simulations, see

- Wang W, Donini O, Reyes CM, Kollman PA. Biomolecular simulations: recent developments in force fields, simulations of enzyme catalysis, protein–ligand, protein–protein, and protein–nucleic acid noncovalent interactions. Annu Rev Biophys Biomol Struct 2001;30:211–243.
- Friesner RA. Combined quantum and molecular mechanics (QM/MM). Drug Discov Today Technol 2004;1:253–260.
- Gao J, Ma S, Major DT, Nam K, Pu J, Truhlar DG. Mechanisms and free energies of enzymatic reactions. Chem Rev 2006;106:3188–3209.

- Senn HM, Thiel W. QM/MM studies of enzymes. Curr Opin Chem Biol 2007;11:182–187.
- van der Kamp MW, Shaw KE, Woods CJ, Mulholland AJ. Biomolecular simulation and modelling: status, progress and prospects. J R Soc Interface 2008;5: S173–S190.
- Senn HM, Thiel W. QM/MM methods for biomolecular systems. Angew Chem Int Ed 2009;48:1198–1229.

7. APPENDIX: METHODOLOGICAL BACKGROUND

7.1. Sampling Methods

Many of the methods mentioned above require large ensembles of structures to calculate the averages to be used in the prediction of the binding free energy. This section briefly outlines the theory behind Monte Carlo and molecular dynamics simulations—two simulation methods that are commonly used to generate these ensembles. MC and MD generally explore low-energy conformations—unless specialized techniques are used. They rarely sample the high-energy regions, which may make important contributions to the free energy and entropic properties (especially when the protein is highly flexible with many different energy minima, separated by low-energy barriers).

Molecular dynamics simulations locate and subsequently sample significant regions of the energy surface, creating a trajectory over time. Evaluating the gradient of a point on the energy surface can be used to determine a force (the negative of the gradient). This force can then be converted into an acceleration using Newton's laws of motion. Numerically integrating the forces over time yields the trajectory, a series of structures over time, which is used as the ensemble. A recent review by Adcock and McCammon gives an overview of the use of MD simulations of proteins [134].

Monte Carlo simulations randomly generate many configurations of the system being studied. The random configurations are accepted or rejected into the ensemble according to a test. Different MC methods have different acceptance tests. For example, in one of the most common MC methods, Metropolis MC, the test involves comparing the energy change, ΔE, for a random move from configuration "a" to configuration "b", to a random number between 0 and 1. The move is accepted or rejected according to Equation 23.

$$e^{(-\Delta E/k_B T)} \geq rand\,(0,1) \qquad (23)$$

7.2. Quantum Mechanical methods

Unlike MM, quantum chemical (electronic structure) methods use QM to treat (at least some of) the electrons in the system explicitly. QM methods thus can be used to study electronic effects that cannot easily be modeled by MM. In most modern quantum chemical calculations, the properties of the system are based on either the wavefunction of the system (in molecular orbital based methods) or the electron density (density functional theory).

Molecular orbital (MO) methods can be divided into two different types, *ab initio* and semiempirical. *Ab initio* MO methods, at their most sophisticated, are the most accurate calculations available to the computational chemist. There is a large variety of *ab initio* methods, using various approximations for the wavefunction, with varying accuracy and computational cost. Examples include Hartree–Fock (HF) theory, and more demanding "correlated" methods such as MP2 and CCSD that account for electron correlation (electrons interact with each other more than HF predicts). These methods are not applicable to large systems, because they have high algorithmic scaling (e.g., HF calculations scale as approximately N^4 and MP2 as approximately N^5, where N is the number of basis functions).

Semiempirical MO methods are more approximate (e.g., often only valence electrons are included and some integrals are ignored) and so have better scaling. They use parameters, derived from *ab initio* QM results or experiment, to mitigate these approximations. The approximations involved can limit the scope of these methods (e.g., accurate treatment of transition metals is challenging). Semiempirical methods are much less computationally demanding than *ab initio* calculations, and parameterizations such as AM1 and PM3 are increasingly popular for enzyme studies.

The other type of QM method, DFT calculates the electron energy from electron density of the system. Different DFT methods differ in their approaches to calculating the exchange-correlation energy. The exchange-correlation energy accounts for the Pauli exclusion principle (exchange) and for the fact that electrons interact with each other (correlation). Different functionals have been developed, and some hybrid methods exist, such as the popular B3LYP method, which uses a proportion of the HF exchange energy terms. More approximate, parameterized (semiempirical) DFT methods, such as SCC-DFTB have also been developed.

7.3. Combined Quantum Mechanics/ Molecular Mechanics Methods

Combined QM/MM methods couple the speed and efficiency of MM and with the potentially improved accuracy and broad applicability of a QM method. QM/MM methods were first developed and applied to a biomolecular system by Warshel and Levitt in 1976 [135]. A QM/ MM method was implemented in the CHARMM program by Field, Bash, and Karplus [136] and in the AMBER program by Singh and Kollman [137] using semiempirical and *ab initio* molecular orbital QM methods, respectively. An accurate description of the interactions between the QM and the MM regions is vital in any QM/MM method. Several different types of QM/MM coupling are possible [138], ranging from a simple "subtractive" approach, in which electrostatic interactions are included only by MM interactions (i.e., using MM point charges for the QM region, as in the simplest implementation of the ONIOM method [139]), through electrostatic embedding (see below), in which invariant MM atomic point charges polarize the QM region, to fully self-consistent polarization of both the QM and MM regions (e.g., using a polarizable MM force field). The second type of coupling is most common in current QM/MM modeling. In this approach, the electrostatic interactions between the two regions can be treated straightforwardly by including the MM point charges in the QM Hamiltonian, though special consideration may need to be given to such interactions in semiempirical methods (e.g., when treating the MM charges as atomic "cores" [136]). This allows for explicit polarization of the QM region by its MM surroundings. Simple MM calculations can be used to include the van der Waals contributions between the two regions, although research has shown that it may (at least in some cases) be necessary to reparameterize the Lennard-Jones parameters used to describe the interaction between the QM and the MM regions [140,141]. Other groups have suggested that further study into the representation of electrostatic interactions is required to improve the description of the interaction between the QM and the MM regions [142,143].

If the boundary between the QM and the MM regions crosses a covalent bond it must be accounted for, often with the addition of a "link" atom, which is not a part of the "real" system [136,137]. The link atom is covalently bonded to the QM side, restoring its correct valency, and the MM–link–QM bond angle and distances are often constrained. An alternative to the link atom method is to use hybrid orbitals to describe the partitioned bond. An example of this type of method is the generalized hybrid orbital (GHO) method [144], where the boundary atom is treated as both a QM and a MM atom. Fortunately, in ligand binding calculations it is often only necessary to treat just the ligand with QM; therefore, it is possible to avoid such partitioning across covalent bonds. It is still vital to treat the interactions between the two regions correctly, and research has shown that hydrogen bonds between the two regions are often too short [142,145,146], suggesting that further optimization is required. Some recent work has shown that charge leakage from QM atoms to neighboring MM point charges is negligible [147]. This work tested the accuracy of a density functional theory-based QM/ MM implementation (B3LYP/6-311 + G**/ CHARMM27) on a set of biologically relevant interactions by comparison with full QM calculations. The changes due to hydrogen bond formation in the total electron density and natural bond orbital atomic charges at the QM/MM level were compared with full QM results. The results showed that charge leakage from the QM atoms to MM atomic point charges close to the QM/MM boundary is not a serious problem with Gaussian basis sets. These results are encouraging in showing that important biomolecular interactions can be treated well at the QM/MM level employing good-quality levels of QM theory.

REFERENCES

1. PhRMA. Available at http://www.innovation.org/index.cfm/InsideDrugDiscovery. Accessed 2008 July 22.
2. Reddy MR, Erion MD. Relative binding affinities of fructose-1,6-bisphosphatase inhibitors calculated using a quantum mechanics-based free energy perturbation method. J Am Chem Soc 2007;129(30): 9296–9297.
3. Khandelwal A, Balaz S. Improved estimation of ligand–macromolecule binding affinities by linear response approach using a combination of multi-mode MD simulation and QM/MM methods. J Comp Aided Mol Des 2007; 21(1–3): 131–137.
4. Wright JS, Lyon GJ, George EA, Muir TW, Novick RP. Hydrophobic interactions drive ligand–receptor recognition for activation and inhibition of staphylococcal quorum sensing. Proc Natl Acad Sci USA 2004;101(46): 16168–16173.
5. Connelly PR, Thomson JA. Heat capacity changes and hydrophobic interactions in the binding of FK506 and rapamycin to the FK506 binding protein. Proc Natl Acad Sci USA 1992;89(11): 4781–4785.
6. Gubernator K, Bohm H-J. Structure-based ligand design. Weinheim; Wiley-VCH; 1998.
7. Ajay Murcko MA. Computational methods to predict binding free energy in ligand–receptor complexes. J Med Chem 1995;38(26): 4953–4967.
8. Harvard Apparatus. Available at http://www.nestgrp.com/pdf/Ap1/EqDialManual.pdf. Accessed 2008 July 22.
9. Nomanbhoy TK, Cerione RA. Characterization of the interaction between RhoGDI and Cdc42Hs using fluorescence spectroscopy. J Biol Chem 1996;271(17): 10004–10009.
10. Fielding L. NMR methods for the determination of protein–ligand dissociation constants. Prog Nucl Magn Reson Spectrosc 2007;51: 219–242.
11. Leavitt S, Freire E. Direct measurement of protein binding energetics by isothermal titration calorimetry. Curr Opin Struct Biol 2001;11(5): 560–566.
12. Cheng Y, Prusoff WH. Relationship between inhibition constant (K1) and concentration of inhibitor which causes 50 per cent inhibition (I50) of an enzymatic-reaction. Biochem Pharmacol 1973;22(23): 3099–3108.
13. Bioinformatics, RCfS. Available at http://www.rcsb.org/pdb. Accessed 2009 Jan 19.
14. Klahn M, Braun-Sand S, Rosta E, Warshel A. On possible pitfalls in *ab initio* quantum mechanics/molecular mechanics minimization approaches for studies of enzymatic reactions. J Phys Chem B 2005;109(32): 15645–15650.
15. Kurz LC, Ackerman JJH, Drysdale GR. Evidence from ^{13}C NMR for polarization of the carbonyl of oxaloacetate in the active-site of citrate synthase. Biochem 1985;24(2): 452–457.
16. Kurz LC, Drysdale GR. Evidence from Fourier-transform infrared-spectroscopy for polarization of the carbonyl of oxaloacetate in the active-site of citrate synthase. Biochem 1987;26(9): 2623–2627.
17. van der Kamp MW, Perruccio F, Mulholland AJ. Substrate polarization in enzyme catalysis: QM/MM analysis of the effect of oxaloacetate polarization on acetyl-CoA enolization in citrate synthase. Proteins Struct Funct Bioinf 2007;69: 521–535.
18. Gao JL, Xia XF. *A priori* evaluation of aqueous polarization effects through Monte-Carlo QM–MM simulations. Science 1992;258 (5082): 631–635.
19. Garcia-Viloca M, Truhlar DG, Gao JL. Importance of substrate and cofactor polarization in the active site of dihydrofolate reductase. J Mol Biol 2003;327(2): 549–560.
20. Hensen C, Hermann JC, Nam KH, Ma SH, Gao JL, Holtje HD. A combined QM/MM approach to protein–ligand interactions: polarization effects of the HIV-1 protease on selected high affinity inhibitors. J Med Chem 2004;47(27): 6673–6680.
21. Jiao D, Golubkov PA, Darden TA, Ren P. Calculation of protein–ligand binding free energy by using a polarizable potential. Proc Natl Acad Sci USA 2008;105(17): 6290–6295.
22. Cho AE, Guallar V, Berne BJ, Friesner R. Importance of accurate charges in molecular docking: quantum mechanical/molecular mechanical (QM/MM) approach. J Comput Chem 2005;26(9): 915–931.
23. Friesner RA. Modeling polarization in proteins and protein–ligand complexes: methods

and preliminary results. Adv Protein Chem 2005;72: 79–104.

24. Warshel A, Kato M, Pisliakov AV. Polarizable force fields: history, test cases, and prospects. J Chem Theo Comput 2007;3: 2034–2045.
25. Zhou T, Huang D, Caflisch A. Is quantum mechanics necessary for predicting binding free energy? J Med Chem 2008;51(14): 4280–4288.
26. Mulholland AJ. Computational enzymology: modelling the mechanisms of biological catalysts. Biochem Soc Trans 2008;36: 22–26.
27. Mulholland AJ. Modelling enzyme reaction mechanisms, specificity and catalysis. Drug Discovery Today 2005;10(20): 1393–1402.
28. Gräter F, Schwarzl SM, Dejaegere A, Fischer S, Smith JC. Protein/ligand binding free energies calculated with quantum mechanics/molecular mechanics. J Phys Chem B 2005;109(20): 10474–10483.
29. Khandelwal A, Lukacova V, Comez D, Kroll DM, Raha S, Balaz S. A combination of docking, QM/MM methods, and MD simulation for binding affinity estimation of metalloprotein ligands. J Med Chem 2005;48(17): 5437–5447.
30. Raha K, Merz KM. Large-scale validation of a quantum mechanics based scoring function: predicting the binding affinity and the binding mode of a diverse set of protein–ligand complexes. J Med Chem 2005;48(14): 4558–4575.
31. Alzate-Morales JH, Contreras R, Soriano A, Tunon I, Silla E. A computational study of the protein–ligand interactions in CDK2 inhibitors: using quantum mechanics/molecular mechanics interaction energy as a predictor of the biological activity. Biophys J 2007;92(2): 430–439.
32. Gleeson MP, Gleeson D. QM/MM calculations in drug discovery: a useful method for studying binding phenomena? J Chem Inf Model 2009;49(3): 670–677. DOI:10.1021/ci800419j.
33. Nunthaboot N, Pianwanit S, Parasuk V, Ebalunode JO, Briggs JM, Kokpol S. Hybrid quantum mechanical/molecular mechanical molecular dynamics simulations of HIV-1 integrase/inhibitor complexes. Biophys J 2007;93:3613–3626.
34. Lodola A, Mor M, Rivara S, Christov C, Tarzia G, Piomelli D, Mulholland AJ. Identification of productive inhibitor binding orientation in fatty acid amide hydrolase (FAAH) by QM/MM mechanistic modelling. Chem Commun 2008;(2): 214–216.
35. van der Kamp MW, Mulholland AJ. Computational enzymology: insight into biological catalysts from modelling. Nat Prod Rep 2008;25 (6): 1001–1014.
36. Mulholland AJ. Chemical accuracy in QM/MM calculations on enzyme-catalysed reactions. Chem Cent J 2007;1 (19).
37. MacKerell AD, Bashford D, Bellott M, Dunbrack RL, Evanseck JD, Field MJ, Fischer S, Gao J, Guo H, Ha S, Joseph-McCarthy D, Kuchnir L, Kuczera K, Lau FTK, Mattos C, Michnick S, Ngo T, Nguyen DT, Prodhom B, Reiher WE, Roux B, Schlenkrich M, Smith JC, Stote R, Straub J, Watanabe M, Wiorkiewicz-Kuczera J, Yin D, Karplus M. All-atom empirical potential for molecular modeling and dynamics studies of proteins. J Phys Chem B 1998;102(18): 3586–3616.
38. Jorgensen WL, Maxwell DS, TiradoRives J. Development and testing of the OPLS all-atom force field on conformational energetics and properties of organic liquids. J Am Chem Soc 1996;118(45): 11225–11236.
39. Kaminski GA, Friesner RA, Tirado-Rives J, Jorgensen WL. Evaluation and reparametrization of the OPLS-AA force field for proteins via comparison with accurate quantum chemical calculations on peptides. J Phys Chem B 2001;105(28): 6474–6487.
40. Cornell WD, Cieplak P, Bayly CI, Gould IR, Merz KM, Ferguson DM, Spellmeyer DC, Fox T, Caldwell JW, Kollman PA. A 2nd generation force-field for the simulation of proteins, nucleic-acids, and organic-molecules. J Am Chem Soc 1995;117(19): 5179–5197.
41. Case DA, Cheatham TE, Darden T, Gohlke H, Luo R, Merz KM, Onufriev A, Simmerling C, Wang B, Woods RJ. The Amber biomolecular simulation programs. J Comput Chem 2005;26 (16): 1668–1688.
42. Wang JM, Wolf RM, Caldwell JW, Kollman PA, Case DA. Development and testing of a general amber force field. J Comput Chem 2004;25(9): 1157–1174.
43. Stewart JJP. Semiempirical molecular orbital methods. In: Lipkowitz KB, Boyd DB, editors. Semiempirical Molecular Orbital Methods. New York: VCH; 1990. p 45–81.
44. Zerner MC. Semiempirical molecular orbital methods. In: Lipkowitz KB, Boyd DB, editors. Semiempirical Molecular Orbital Methods. New York: VCH; 1990. p 313–365.
45. Kohn W, Becke AD, Parr RG. Density functional theory of electronic structure. J Phys Chem 1996;100(31): 12974–12980.
46. Szabo A, Ostlund NA. Modern Quantum Chemistry. Introduction to Advanced Electro-

nic Structure Theory. Mineola, NY: Dover; 1996.

47. Friesner RA, Murphy RB, Beachy MD, Ringnalda MN, Pollard WT, Dunietz BD, Cao YX. Correlated *ab initio* electronic structure calculations for large molecules. J Phys Chem A 1999;103(13): 1913–1928.
48. Yang WT. Direct calculation of electron-density in density-functional theory. Phys Rev Lett 1991;66(11): 1438–1441.
49. Yang WT, Lee TS. A density-matrix divide-and-conquer approach for electronic-structure calculations of large molecules. J Chem Phys 1995;103(13): 5674–5678.
50. Van der Vaart A, Gogonea V, Dixon SL, Merz KM. Linear scaling molecular orbital calculations of biological systems using the semiempirical divide and conquer method. J Comput Chem 2000;21(16): 1494–1504.
51. Dixon SL, Merz KM. Semiempirical molecular orbital calculations with linear system size scaling. J Chem Phys 1996;104(17): 6643–6649.
52. Dixon SL, Merz KM. Fast, accurate semiempirical molecular orbital calculations for macromolecules. J Chem Phys 1997;107(3): 879–893.
53. Pulay P. Localizability of dynamic electron correlation. Chem Phys Lett 1983;100(2): 151–154.
54. Hampel C, Werner HJ. Local treatment of electron correlation in coupled cluster theory. J Chem Phys 1996;104(16): 6286–6297.
55. Stewart JJP. Application of localized molecular orbitals to the solution of semiempirical self-consistent field equations. Int J Quantum Chem 1996;58(2): 133–146.
56. Werner HJ, Manby FR, Knowles PJ. Fast linear scaling second-order Moller–Plesset perturbation theory (MP2) using local and density fitting approximations. J Chem Phys 2003;118 (18): 8149–8160.
57. Manby FR, Werner HJ, Adler TB, May AJ. Explicitly correlated local second-order perturbation theory with a frozen geminal correlation factor. J Chem Phys 2006;124(9): 094103.
58. Claeyssens F, Harvey JN, Manby FR, Mata RA, Mulholland AJ, Ranaghan KE, Schutz M, Thiel S, Thiel W, Werner HJ. High-accuracy computation of reaction barriers in enzymes. Angew Chem Int Ed Engl 2006;45(41): 6856–6859.
59. Lewis JP, Liu SB, Lee TS, Yang WT. A linear-scaling quantum mechanical investigation of cytidine deaminase. J Comput Phys 1999; 151(1): 242–263.
60. Nikitina E, Sulimov V, Zayets V, Zaitseva N. Semiempirical calculations of binding enthalpy for protein–ligand complexes. Int J Quantum Chem 2004;97(2): 747–763.
61. Chen XH, Zhang JZH. Theoretical method for full *ab initio* calculation of DNA/RNA–ligand interaction energy. J Chem Phys 2004;120(24): 11386–11391.
62. Reddy MR, Singh UC, Erion MD. Development of a quantum mechanics-based free-energy perturbation method: use in the calculation of relative solvation free energies. J Am Chem Soc 2004;126(20): 6224–6225.
63. Nikolova N, Jaworska J. Approaches to measure chemical similarity: a review. QSAR Comb Sci 2004;22(9–10): 1006–1026.
64. Kuntz ID, Blaney JM, Oatley SJ, Langridge R, Ferrin TE. A geometric approach to macromolecule–ligand interactions. J Mol Biol 1982; 161(2): 269–288.
65. Huang N, Kalyanaraman C, Bernacki K, Jacobson MP. Molecular mechanics methods for predicting protein–ligand binding. Phys Chem Chem Phys 2006;8(44): 5166–5177.
66. Zwanzig RW. High-temperature equation of state by a perturbation method. 1. Nonpolar gases. J Chem Phys 1954;22(8): 1420–1426.
67. Tembe BL, McCammon JA. Ligand receptor interactions. Comput Chem 1984;8(4): 281–283.
68. Bash PA, Singh UC, Brown FK, Langridge R, Kollman PA. Calculation of the relative change in binding free-energy of a protein–inhibitor complex. Science 1987;235(4788): 574–576.
69. Jorgensen WL, Ravimohan C. Monte-Carlo simulation of differences in free-energies of hydration. J Chem Phys 1985;83(6): 3050–3054.
70. Beutler TC, Mark AE, Vanschaik RC, Gerber PR, Vangunsteren WF. Avoiding singularities and numerical instabilities in free-energy calculations based on molecular simulations. Chem Phys Lett 1994;222(6): 529–539.
71. Li GH, Zhang XD, Cui Q. Free energy perturbation calculations with combined QM/MM Potentials complications, simplifications, and applications to redox potential calculations. J Phys Chem B 2003;107(33): 8643–8653.
72. Reddy MR, Singh UC, Erion MD. *Ab initio* quantum mechanics-based free energy perturbation method for calculating relative solvation free energies. J Comput Chem 2007;28(2): 491–494.
73. Singh UC, Benkovic SJ. A free-energy perturbation study of the binding of methotrexate to

mutants of dihydrofolate-reductase. Proc Natl Acad Sci USA 1988;85(24): 9519–9523.

74. Galaxy Molecular Modeling Software and AM2000 Macromolecular Simulation Package; AM Technologies, San Antonio, Texas, 1995.
75. Muller RP, Warshel A. *Ab-initio* calculations of free-energy barriers for chemical-reactions in solution. J Phys Chem 1995;99(49): 17516–17524.
76. Strajbl M, Hong GY, Warshel A. *Ab initio* QM/MM simulation with proper sampling: "first principle" calculations of the free energy of the autodissociation of water in aqueous solution. J Phys Chem B 2002;106(51): 13333–13343.
77. Warshel A. Computer modeling of chemical reactions in enzymes and solutions. New York: John Wiley & Sons; 1991.
78. Warshel A, Weiss RM. An empirical valence bond approach for comparing reactions in solutions and in enzymes. J Am Chem Soc 1980;102 (20): 6218–6226.
79. Åqvist J, Warshel A. Simulation of enzyme-reactions using valence-bond force-fields and other hybrid quantum–classical approaches. Chem Rev 1993;93(7): 2523–2544.
80. Warshel A. Computer simulations of enzyme catalysis: methods, progress, and insights. Ann Rev Biophys Biomol Struct 2003;32: 425–443.
81. Villa J, Warshel A. Energetics and dynamics of enzymatic reactions. J Phys Chem B 2001;105 (33): 7887–7907.
82. Bentzien J, Muller RP, Florian J, Warshel A. Hybrid *ab initio* quantum mechanics molecular mechanics calculations of free energy surfaces for enzymatic reactions: the nucleophilic attack in subtilisin. J Phys Chem B 1998; 102(12): 2293–2301.
83. Rosta E, Klahn M, Warshel A. Towards accurate *ab initio* QM/MM calculations of free-energy profiles of enzymatic reactions. J Phys Chem B 2006;110(6): 2934–2941.
84. Rod TH, Ryde U. Quantum mechanical free energy barrier for an enzymatic reaction. Phys Rev Lett 2005;94(13): 138302.
85. Rod TH, Ryde U. Accurate QM/MM free energy calculations of enzyme reactions: methylation by catechol *O*-methyltransferase. J Chem Theo Comput 2005;1(6): 1240–1251.
86. Vaidehi N, Wesolowski TA, Warshel A. Quantum-mechanical calculations of solvation free-energies—a combined *ab initio* pseudopotential free-energy perturbation approach. J Chem Phys 1992;97(6): 4264–4271.
87. Wesolowski TA, Weber J. Kohn–Sham equations with constrained electron density: an iterative evaluation of the ground-state electron density of interacting molecules. Chem Phys Lett 1996;248(1–2): 71–76.
88. Wesolowski TA, Warshel A. Frozen density-functional approach for *ab-initio* calculations of solvated molecules. J Phys Chem 1993;97 (30): 8050–8053.
89. Wesolowski T, Warshel A. *Ab-initio* free-energy perturbation calculations of solvation free-energy using the frozen density-functional approach. J Phys Chem 1994;98(20): 5183–5187.
90. Hong GY, Strajbl M, Wesolowski TA, Warshel A. Constraining the electron densities in DFT method as an effective way for *ab initio* studies of metal-catalyzed reactions. J Comput Chem 2000;21(16): 1554–1561.
91. Xiang Y, Warshel A. Quantifying free energy profiles of proton transfer reactions in solution and proteins by using a diabatic FDFT mapping. J Phys Chem B 2008;112(3): 1007–1015.
92. Zhang YK, Liu HY, Yang WT. Free energy calculation on enzyme reactions with an efficient iterative procedure to determine minimum energy paths on a combined *ab initio* QM/MM potential energy surface. J Chem Phys 2000;112(8): 3483–3492.
93. Kästner J, Senn HM, Thiel S, Otte N, Thiel W. QM/MM free-energy perturbation compared to thermodynamic integration and umbrella sampling: application to an enzymatic reaction. J Chem Theo Comput 2006;2(2): 452–461.
94. Kaminski GA, Jorgensen WL. A quantum mechanical and molecular mechanical method based on CM1A charges: applications to solvent effects on organic equilibria and reactions. J Phys Chem B 1998;102(10): 1787–1796.
95. Guimaraes CRW, Udier-Blagovic M, Jorgensen WL. Macrophomate synthase: QM/MM simulations address the Diels–Alder versus Michael–Aldol reaction mechanism. J Am Chem Soc 2005;127(10): 3577–3588.
96. Tubert-Brohman I, Acevedo O, Jorgensen WL. Elucidation of hydrolysis mechanisms for fatty acid amide hydrolase and its Lys142Ala variant via QM/MM simulations. J Am Chem Soc 2006;128(51): 16904–16913.
97. Mezei M. The finite-difference thermodynamic integration, tested on calculating the hydration free-energy difference between acetone and dimethylamine in water. J Chem Phys 1987;86(12): 7084–7088.

98. Cummins PL, Gready JE. Computational methods for the study of enzymic reaction mechanisms III: a perturbation plus QM/MM approach for calculating relative free energies of protonation. J Comput Chem 2005;26(6): 561–568.
99. Woods CJ, Manby FR, Mulholland AJ. An efficient method for the calculation of quantum mechanics/molecular mechanics free energies. J Chem Phys 2008;128: 014109.
100. Hastings WK. Monte-Carlo sampling methods using Markov chains and their applications. Biometrika 1970;57(1): 97–109.
101. Woods CJ, Essex JW, King MA. The development of replica-exchange-based free-energy methods. J Phys Chem B 2003;107(49): 13703–13710.
102. Åqvist J, Medina C, Samuelsson JE. New method for predicting binding-affinity in computer-aided drug design. Protein Eng 1994;7 (3): 385–391.
103. Carlson HA, Jorgensen WL. An extended linear-response method for determining free-energies of hydration. J Phys Chem 1995;99(26): 10667–10673.
104. Zhou RH, Friesner RA, Ghosh A, Rizzo RC, Jorgensen WL, Levy RM. New linear interaction method for binding affinity calculations using a continuum solvent model. J Phys Chem B 2001;105(42): 10388–10397.
105. Gilson MK, Zhou HX. Calculation of protein–ligand binding affinities. Ann Rev Biophys Biomol Struct 2007;36: 21–42.
106. Huang D, Caflisch A. Efficient evaluation of binding free energy using continuum electrostatics solvation. J Med Chem 2004;47(23): 5791–5797.
107. Khandelwal A, Balaz S. QM/MM linear response method distinguishes ligand affinities for closely related metalloproteins. Proteins Struct Funct Bioinf 2007;69: 326–339.
108. Rarey M, Kramer B, Lengauer T, Klebe G. A fast flexible docking method using an incremental construction algorithm. J Mol Biol 1996;261(3): 470–489.
109. van Vlijmen HWT, Schaefer M, Karplus M. Improving the accuracy of protein pK_a calculations: conformational averaging versus the average structure. Proteins 1998;33(2): 145–158.
110. Zoete V, Michielin O, Karplus M. Protein–ligand binding free energy estimation using molecular mechanics and continuum electrostatics. Application to HIV-1 protease inhibitors. J Comp Aided Mol Des 2003;17(12): 861–880.
111. Alves CN, Marti S, Castillo R, Andres J, Moliner V, Tunon I, Silla E. A quantum mechanics/molecular mechanics study of the protein–ligand interaction for inhibitors of HIV-1 integrase. Chem A Euro J 2008;13: 7715–7724.
112. Kollman PA, Massova I, Reyes C, Kuhn B, Huo SH, Chong L, Lee M, Lee T, Duan Y, Wang W, Donini O, Cieplak P, Srinivasan J, Case DA, Cheatham TE. Calculating structures and free energies of complex molecules: combining molecular mechanics and continuum models. Acc Chem Res 2000;33(12): 889–897.
113. Liu HY, Zou XQ. Electrostatics of ligand binding: parametrization of the generalized born model and comparison with the Poisson–Boltzmann approach. J Phys Chem B 2006;110(18): 9304–9313.
114. Still WC, Tempczyk A, Hawley RC, Hendrickson T. Semianalytical treatment of solvation for molecular mechanics and dynamics. J Am Chem Soc 1990;112(16): 6127–6129.
115. Woods CJ, King MA, Essex JW. The configurational dependence of binding free energies: a Poisson–Boltzmann study of neuraminidase inhibitors. J Comp Aided Mol Des 2001;15(2): 129–144.
116. Pellegrini E, Field MJ. A generalized-Born solvation model for macromolecular hybrid-potential calculations. J Phys Chem A 2002; 106(7): 1316–1326.
117. Wang ML, Wong CF. Rank-ordering protein–ligand binding affinity by a quantum mechanics/molecular mechanics/Poisson–Boltzmann-surface area model. J Chem Phys 2007;126(2): 026101.
118. Leach AR, Shoichet BK, Peishoff CE. Prediction of protein–ligand interactions. Docking and scoring: successes and gaps. J Med Chem 2006;49(20): 5851–5855.
119. Rester U. Dock around the clock: current status of small molecule docking and scoring. QSAR Comb Sci 2006;25(7): 605–615.
120. Ferrara P, Priestle JP, Vangrevelinghe E, Jacoby E. New developments and applications of docking and high-throughput docking for drug design and *in silico* screening. Curr Comp Aided Drug Des 2006;2(1): 83–91.
121. David L, Nielsen P, Hedstrom M, Norden B. Scope and limitation of ligand docking: methods, scoring functions and protein targets. Curr Comp Aided Drug Des 2005;1(3): 275–306.

122. Coupez B, Lewis RA. Docking and scoring: theoretically easy, practically impossible? Curr Med Chem 2006;13(25): 2995–3003.

123. Krovat EM, Steindl T, Langer T. Recent advances in docking and scoring. Curr Comp Aided Drug Des 2005;1(1): 93–102.

124. Kitchen DB, Decornez H, Furr JR, Bajorath J. Docking and scoring in virtual screening for drug discovery: methods and applications. Nat Rev Drug Discov 2004;3(11): 935–949.

125. Parks JM, Kondru RK, Hu H, Beratan DN, Yang WT. Hepatitis C virus NS5B polymerase: QM/MM calculations show the important role of the internal energy in ligand binding. J Phys Chem B 2008;112(10): 3168–3176.

126. Ferrara P, Curioni A, Vangrevelinghe E, Meyer T, Mordasini T, Andreoni W, Acklin P, Jacoby E. New scoring functions for virtual screening from molecular dynamics simulations with a quantum-refined force-field (QRFF-MD). Application to cyclin-dependent kinase 2. J Chem Inf Model 2006;46(1): 254–263.

127. Mordasini T, Curioni A, Bursi R, Andreoni W. The binding mode of progesterone to its receptor deduced from molecular dynamics simulations. Chembiochem 2003;4(2–3): 155–161.

128. Curioni A, Mordasini T, Andreoni W. Enhancing the accuracy of virtual screening: molecular dynamics with quantum-refined force fields. J Comp Aided Mol Des 2004;18(12): 773–784.

129. Friesner RA, Banks JL, Murphy RB, Halgren TA, Klicic JJ, Mainz DT, Repasky MP, Knoll EH, Shelley M, Perry JK, Shaw DE, Francis P, Shenkin PS. Glide: a new approach for rapid, accurate docking and scoring. 1. Method and assessment of docking accuracy. J Med Chem 2004;47(7): 1739–1749.

130. Halgren TA, Murphy RB, Friesner RA, Beard HS, Frye LL, Pollard WT, Banks JL. Glide: a new approach for rapid, accurate docking and scoring. 2. Enrichment factors in database screening. J Med Chem 2004;47(7): 1750–1759.

131. Schrödinger, QSITE (OR, LLC: Portland, 2000).

132. Zhong H, Kirschner KN, Lee M, Bowen JP. Binding free energy calculation for duocarmycin/DNA complex based on the QPLD-derived partial charge model. Bioorg Med Chem Lett 2008;18(2): 542–545.

133. Raha K, Merz KM. A quantum mechanics-based scoring function: study of zinc ion-mediated ligand binding. J Am Chem Soc 2004;126(4): 1020–1021.

134. Adcock SA, McCammon JA. Molecular dynamics: survey of methods for simulating the activity of proteins. Chem Rev 2006;106: 1589–1615.

135. Warshel A, Levitt M. Theoretical studies of enzymic reactions: dielectric, electrostatic and steric stabilization of carbonium-ion in reaction of lysozyme. J Mol Biol 1976;103(2): 227–249.

136. Field MJ, Bash PA, Karplus M. A combined quantum-mechanical and molecular mechanical potential for molecular-dynamics simulations. J Comput Chem 1990;11(6): 700–733.

137. Singh UC, Kollman PA. A combined *ab initio* quantum-mechanical and molecular mechanical method for carrying out simulations on complex molecular-systems: applications to the CH_3Cl + Cl^- exchange-reaction and gas-phase protonation of polyethers. J Comput Chem 1986;7(6): 718–730.

138. Bakowies D, Thiel W. Hybrid models for combined quantum mechanical and molecular mechanical approaches. J Phys Chem 1996;100 (25): 10580–10594.

139. Svensson M, Humbel S, Froese RDJ, Matsubara T, Sieber S, Morokuma K. ONIOM: a multilayered integrated MO + MM method for geometry optimizations and single point energy predictions. A test for Diels–Alder reactions and $Pt(P(t\text{-}Bu)_3)_2 + H_2$ oxidative addition. J Phys Chem 1996;100(50): 19357–19363.

140. Freindorf M, Gao JL. Optimization of the Lennard-Jones parameters for a combined *ab initio* quantum mechanical and molecular mechanical potential using the 3-21G basis set. J Comput Chem 1996;17(4): 386–395.

141. Freindorf M, Shao YH, Furlani TR, Kong J. Lennard-Jones parameters for the combined QM/MM method using the B3LYP/6-31+G*/AMBER potential. J Comput Chem 2005;26 (12): 1270–1278.

142. Riccardi D, Li GH, Cui Q. Importance of van der Waals interactions in QM/MM simulations. J Phys Chem B 2004;108(20): 6467–6478.

143. Pentikäinen U, Shaw KE, Senthilkumar K, Woods CJ, Mulholland AJ. Lennard-Jones Parameters for B3LYP/CHARMM27 QM/MM modeling of nucleic acid bases. J Chem Theo Comput 2009;5(2): 396–410. DOI:10.1021/ct800135k.

144. Gao JL, Amara P, Alhambra C, Field MJ. A generalized hybrid orbital (GHO) method for the treatment of boundary atoms in combined

QM/MM calculations. J Phys Chem A 1998;102 (24): 4714–4721.

145. Gao JL. The hydration and solvent polarization effects of nucleotide bases. Biophys Chem 1994;51(2–3): 253–261.
146. Luque FJ, Reuter N, Cartier A, Ruiz-Lopez MF. Calibration of the quantum/classical Hamiltonian in semiempirical QM/MM AM1 and PM3 methods. J Phys Chem A 2000;104(46): 10923–10931.
147. Senthilkumar K, Mujika JI, Ranaghan KE, Manby FR, Mulholland AJ, Harvey JN. Analysis of polarization in QM/MM modelling of biologically relevant hydrogen bonds. J R Soc Interface 2008;5: S207–S216.

ASSESSING THE BIOAVAILABILITY OF DRUG DELIVERY SYSTEMS: MATHEMATICAL MODELING

Jean-Maurice Vergnaud[1]
Iosif-Daniel Rosca[2,3]
[1]University of Saint-Etienne, Saint-Chamond, France
[2]University of Bucharest, Bucharest, Romania
[3]Concordia University, Montreal, Canada

1. INTRODUCTION

The purpose of this chapter is to describe a method of calculation leading to what it could be called a "quantitative pharmacokinetics," since the plasma drug profiles are reckoned.

This is the answer to the questions set by the U.S. Food and Drug Administration (FDA) in the late 1980s about a possible correlation between the *in vivo* tests and the *in vitro* tests for oral dosage forms with controlled release. This question was not only of high concern from the theoretical point of view but also of great importance from the practical applications, since the *in vitro* tests are easy to perform, while the *in vivo* tests necessitate healthy volunteers and patients as shown along the three phases of clinical trials. On their experience acquired with immediate-release dosage forms, in 1984 [1] and in 1988 [2], the FDA managed two workshops in the way of considering the issue set by the establishment of *in vitro/in vivo* correlations with sustained-release forms orally taken. But the result was deceitful as said in 1993 [3]: "Both workshops concluded that the state of science and technology at that time did not always permit meaningful correlations. Unlike the case of immediate-release dosage forms, the correlations, when possible, would have to be product specific." As a matter of fact, the results of these correlations are of little use, if not of no use, for evaluating the plasma profile of a drug associated with a dosage form with controlled release. The main mistake comes from that the straight line drawn in these correlations does not pass through the origin. At time 0, the amount of drug either released out of the dosage form or located in the plasma is 0; then, by plotting the results obtained either with *in vitro* or with *in vivo* experiments, the curve should pass through the origin in this diagram. Also, it is difficult to find any study showing this fact. However, the problem was resolved numerically rather in a simple way by building a numerical model taking all the known facts into account, for example, the kinetics of release of the drug from the dosage form (with a value of the coefficient of convection to be previously determined), followed by the kinetics of absorption into the plasma through the gastrointestinal membrane and terminated by the kinetics of elimination. The only but main difference with the process done with the immediate-release dosage form is that in this last case, the totality of the drug initially in the dosage form is instantaneously released in the gastrointestinal tract and thus available for the stage of absorption. Then, after publication of the method, the FDA invited one of the authors two times to present it, the first for dosage forms whose release is controlled by diffusion in 1996 [4] and the other when controlled by erosion [5]. It should be said that a linear correlation cannot exist when the release of the drug is not linear with time or instantaneous.

Generally, there is always some interest in using a mathematical or a numerical treatment in order to study a phenomenon, and of course, to test it by comparing the theoretical results obtained with the experimental ones. Before calculating, it is necessary to describe the process with the various stages, and thus to make some assumptions. These assumptions are either dictated by the process or needed to help the theoretical treatment. Let us note, at that time, that various problems cannot be resolved by a simple theoretical treatment and the solution cannot be expressed by an analytical solution. Generally, this is the case of the differential equations with partial derivatives when the boundary conditions are not simple: a typical example appears with the evaluation of the plasma drug profile obtained with oral sustained-release dosage forms, whatever the process of control of the drug release; another example is shown with the evaluation of the plasma drug profile generated by a patch with the transdermal drug delivery.

Various ways of drug delivery are considered, and the method is applied to all of them, by selecting a few drugs. Then, the plasma drug level is obtained with the following cases:

- The intravenous drug delivery, with bolus (single and repeated), with continuous infusion and with repeated short infusions.
- The oral drug delivery with immediate-release dosage forms and especially with sustained-release dosage forms whose release is controlled either by diffusion or by erosion.
- The transdermal drug transfer through the skin using a patch.
- Moreover, the drug profile is also evaluated in a few tissues.Following this way of calculation, applications of great concern are made, using the values of the pharmacokinetics parameters taken in the literature.
- The intervariability of the patients can be considered and the drug delivery adapted.
- The noncompliance of the patients can easily be observed through the drug profiles.
- As the drug profiles obtained with the oral dosage forms depend mainly on the nature of the drug, the characteristics of the controlled-release dosage forms are defined in terms of the half-life time obtained with them, in relation with the half-life time obtained through bolus injection.

As often as possible, dimensionless numbers are used, as they can provide the results in a general way. For instance, instead of presenting the amount of the drug released or located in the plasma, this amount M_t at time t depending on the nature of the drug and on the whole characteristics of the dosage form, the dimensionless number M_t/M_{in} expressing the amount at time t as a fraction of the amount initially in the dosage form, is used. In the same way, for the time, a dimensionless time is often used, by taking the half-life time of the drug $t_{0.5}$ as the unit time.

When it comes to remarking that a book [6] has already been written on this subject by the authors, it can be added that various changes and improvements have also been made in this chapter, taking advantage of three more recent papers of their own. Thus, finally, the position of the quantitative pharmacokinetics described here is reinforced.

2. DEFINITIONS

2.1. Active Agent (Drug) and Dosage Form (Supply Form of the Drug)

2.1.1. Drug and Dosage Form According to the World Health Organization, a "drug" is "any substance that is used to modify or explore physiological systems or pathological states for the benefit of the recipient." In a few words, the drug is the active biological substance responsible for a pharmaceutical effect.

The dosage form is the supply form of the drug for medication. It consists of the active agent and excipients. For immediate-release dosage forms, or conventional forms, excipients are soluble.

2.1.2. Excipients in Controlled (Sustained)-Release Dosage Forms In controlled-release dosage forms, or sustained-release dosage forms, the excipient plays an important role, as it should control the rate of release of the drug along the gastrointestinal tract (GIT). These monolithic devices are prepared by dispersing the drug in the polymer acting as a consistent matrix. The polymer should be biocompatible and pass along the GIT without any interference with the patient's body. Nonbiodegradable polymers are used for sustained-release dosage forms whose drug release is controlled by diffusion. Degradable polymers (called erodable polymers) are used when the process of drug release is controlled by both erosion and diffusion, but essentially by erosion.

2.2. Goals of Biopharmaceutics

According to Levy in 1958, "Biopharmaceutics is the study of factors that affect the extent and speed of absorption and liberation of the drug from its various physiological forms. It is concerned with the dependence of absorption, distribution, metabolism, retention, or excretion of the drug by the patient on the physiochemical characteristics of the dosage form. It deals essentially with the optimum availability of the drug from the dosage form employed for the target site, that is, the influence of the drug formulation on the biological activity of the drug." In other words, an effective remedy being not necessarily obtained with a pharmacologically active substance, and its efficacy depending on the way by which it is delivered to the organism, this fact confers a primary importance on Galenic studies.

2.2.1. Biopharmaceutical Stage with the Liberation of the Drug

The drug is liberated very fast with immediate-release dosage forms, and much slower with controlled-release dosage forms, as shown through *in vitro* tests.

Through *in vivo* tests made on patients, the drug is dissolved in the GI liquid, either in the stomach or the intestine, depending on the nature of the excipient or of the coating.

2.2.2. Pharmacokinetic Stage Upon i.v. Drug Delivery with Two Stages

- Distribution of the drug through the blood volume and bounding to the plasmatic proteins, followed by diffusion through the tissues to the target organs.
- Elimination of the drug, essentially through the kidney.

2.2.3. Pharmacokinetic Stage with Oral Drug Delivery, with Several Steps

- Absorption of the dissolved drug into the circulating blood, with a possible metabolism in the liver responsible for a decrease of the bioavailability of the drug.
- Distribution of the drug that is more or less bound to the plasmatic proteins; this fact leads to an apparent plasmatic volume that can be larger than the blood volume.
- Diffusion of the drug to the target organs.
- Elimination of the drug, either in its non-modified form or in the form of metabolites, through various means of excretion, and essentially through the kidneys.
- Clearance.

2.2.4. Pharmacokinetic Stage Through Transdermal Drug Delivery, with Several Steps

- Diffusion of the drug through the skin under transient and stationary conditions.
- Distribution of the drug into the blood volume with bounding to the plasmatic proteins.
- Elimination of the drug.

2.2.5. Characteristics of the Elimination Stage by the Pharmacokinetic Parameters

- Rate of elimination, expressed in terms of a first-order reaction with the rate constant k_e.
- Half-life time of residence in the blood $t_{0.5}$ for i.v. drug delivery or $T_{0.5}$ for oral dosage forms.

2.3. Pharmacokinetics

Pharmacokinetics deals with what the body does to the drug, and thus is concerned with the kinetics of absorption, distribution, metabolism, and excretion of the drug and its metabolites produced by the body itself.

The case of the drug delivery through oral dosage forms is the most complex, as there are in addition of the other two ways of drug delivery the following two steps before the drug enters the blood volume:

- The liberation of the drug from the dosage form in the stomach or the GI tract.
- The absorption of the drug in the blood volume after its transport through the membrane or gastrointestinal membrane. This diffusion of the drug through these membranes is assumed to be under stationary state, and thus can be expressed in terms of a first-reaction with a rate constant of absorption k_a.

Because of the steps of binding and debinding of the drug to the plasmatic proteins, the apparent blood volume, V_p, may be larger than the blood volume itself.

2.4. Pharmacodynamics

Pharmacodynamics being concerned with what the drug does to the body, has as objective the therapeutic effect on the patient.

2.5. Absorption of the Drug in the Blood with Oral Dosage Forms

Absorption is the transport of the drug through the GI membrane from the GIT to the bloodstream or the lymphatic system.

In terms of mathematics, this transport is expressed by a first-order kinetics with the rate constant of absorption k_a whose unit is per hour.

Application of the first-order equation to the drug remaining in the dosage form gives

$$-\frac{dM}{dt} = k_a \cdot M \tag{1}$$

whose integration from time 0 when $M'_t = M_{in}$ to infinite time when $M'_t = 0$ leads to

$$M'_t = M_{in} \cdot \exp(-k_a t) \tag{2}$$

Thus in the case of an immediate-release dosage form, the amount of drug that has passed through the gastric membrane at time t, M_t, is equal to

$$M_t = M_{in} - M'_t \tag{3}$$

where M_{in} is the amount of drug initially in the dosage form and in the GIT.

In the case of a controlled-release dosage form, the process is far more complex since the amount of drug transferred through the membrane is controlled not only by the rate of absorption through the GI membrane but also by the kinetics of drug release out of the dosage form (Section 4).

2.6. Elimination of the Drug

The drug is inactivated by various processes, based on the alteration of the molecule through metabolism, and excretion via various organs.

2.6.1. Metabolism Metabolism is the sum of all chemical reactions requiring the presence of enzymes as catalysts, involved in the biotransformation of the molecule: catabolic reaction breaking the drug into simpler substances, and anabolism by which new molecules are synthesized.

These metabolites formed are generally more water soluble than the drug itself, facilitating their excretion through the kidneys.

2.6.2. Biological Half-life Time of the Drug Biological half-life time of the drug, $t_{0.5}$, is the main pharmacokinetic parameter of the patient. This is the time necessary for the concentration of the unbound drug in the blood to be reduced by half from the initial value. The rate of elimination being expressed by a first-order kinetics with the rate constant of elimination k_e, the equation of elimination becomes

$$-\frac{dC}{dt} = k_e \cdot C \tag{4}$$

Integration of this equation from time 0, with C_0, to time t with C_t, gives

$$\ln \frac{C_t}{C_0} = -k_e \cdot t \tag{5}$$

which leads to the expression for the half-life time $t_{0.5}$:

$$\ln 2 = k_e \cdot t_{0.5} = 0.693 \tag{6}$$

Equation 5 can be rewritten as follows:

$$C_t = C_{in} \cdot \exp(-k_e \cdot t) \tag{7}$$

2.6.3. Excretion and Clearance via Various Organs Excretion of the drug and its metabolites from the body is mainly undertaken through the kidney and the intestinal tract. However, the major route of elimination is via the kidney, the rate and extent of the renal excretion being determined by glomerular filtration, tubular reabsorption and tubular secretion.

Clearance, Cl, is a pharmacokinetic parameter that is related to the proportion of the drug excreted or altered by the flow of blood Q through the organ of excretion:

$$Cl = Q \cdot E \tag{8}$$

where E is the proportion of the drug excreted or altered.

The organs of excretion being mainly the liver, with the hepatic clearance, Cl_{h}, and the kidney with the renal clearance, Cl_{r}, the total value of the clearance, Cl, is

$$Cl = Cl_{\mathrm{r}} + Cl_{\mathrm{h}} \tag{9}$$

In a few words, clearance is the blood volume circulating through the organ of secretion necessary to eliminate the drug per unit time.

The clearance can also be expressed in terms of the dose and of the area under the curve (AUC), and also in terms of the drug concentration C and apparent plasmatic volume V_{p}.

$$Cl = \frac{\mathrm{Dose}}{\mathrm{AUC}} = \frac{C \cdot V_{\mathrm{p}}}{\mathrm{AUC}} \tag{10}$$

Using the rate constant of elimination k_{e} and the apparent plasmatic volume V_{p}, it becomes

$$Cl = k_{\mathrm{e}} \cdot V_{\mathrm{p}} \tag{11}$$

Equation 11 expresses that the clearance is the apparent volume of distribution getting free of drug per unit time.

2.7. Therapeutic Index

The therapeutic index, T.I., introduced by Erlich, is defined as the ratio relating the median lethal dose LD_{50} and the median effective dose ED_{50}, the first causing the death of 50% of experimental animals, and the second being the concentration at which the drug is effective in 50% of cases.

$$\mathrm{T.I.} = \frac{LD_{50}}{ED_{50}} \tag{12}$$

The higher the therapeutic index is, the greater the safety margin.

The concept of therapeutic index comes from the threshold model, based on the idea that there is a threshold concentration for the drug to be active. With the linear model, the patient's response to the dose is zero when the dose is zero. Let us remark also the hormesis theory that suggests that there are advantageous effects using low doses of a substance toxic at higher level [7].

2.8. Bioavailability

This term refers to the amount of drug absorbed by the organism that arrives at the target site in a given time. This concept introduced by Oser [8] as the "physiological availability" is defined as the ratio of the amount of unchanged drug reaching the circulating system after administration of a test dose to that after administration of a standard dosage form [9].

The value of AUC is calculated by integrating the profile of drug concentration with respect to time, for example, from time 0 to time t, in the following equation:

$$\mathrm{AUC}_{0-t} = \int_0^t C_t \, \mathrm{d}t \tag{13}$$

Absolute bioavailability can be evaluated by comparing the results obtained after the drug administration, orally and intravenously, and measuring the drug concentration at various times in both cases. The ratio of AUC gives the value of this absolute bioavailability. The bioavailability is thus expressed by the ratio of these values obtained with the oral dose and through i.v. injection:

$$\frac{\mathrm{AUC}_{\mathrm{oral}}}{\mathrm{AUC}_{\mathrm{i.v.}}} \tag{14}$$

2.9. Determination of the Pharmacokinetic Parameters

The following pharmacokinetic parameters should be determined before calculation:

- The rate constant of elimination, k_{e}, which enables calculation of the half-life time $t_{0.5}$.

- The apparent volume of distribution, V_p, which is often larger than the volume of blood.
- The rate constant of absorption, k_a, of interest for oral administration.
- The systemic clearance Cl.

2.9.1. Determination of Rate Constant of Elimination Calculation is made using the drug profile obtained during the elimination stage following a single bolus injection (Fig. 1), where the drug concentration in the blood decreases exponentially with time. The concentration at time 0, C_0, taken as unit concentration in Fig. 1 cannot be measured for two reasons: The first is that it takes some time for the drug to be distributed throughout the body; this phase called phase of distribution is associated with the half-life time $t_{0.5\alpha}$. The other reason is that it is not experimentally possible to measure the drug concentration at the time of injection.

But there is a small difference between the values of AUC calculated either by taking into account the curve obtained with the two half-life times $t_{0.5\alpha}$ and $t_{0.5\beta}$ or by considering the phase of elimination with only the half-life time $t_{0.5\beta}$ in extrapolating the initial drug concentration C_0.

Fortunately, the drug concentration can easily be measured at any time. Thus, by applying Equation 7 to the concentrations C_1 and C_2 measured at the times t_1 and t_2:

$$C_1 = C_0 \cdot \exp(-k_e t_1)$$
$$C_2 = C_0 \cdot \exp(-k_e t_2) \tag{15}$$

The unknown concentration C_0 disappears by dividing these equations by themselves:

$$\frac{C_1}{C_2} = \exp[-k_e(t_1 - t_2)] \tag{16}$$

Thus, the rate constant k_e is obtained from Equation 16

$$k_e = \frac{1}{t_2 - t_1} \ln \frac{C_1}{C_2} \tag{17}$$

and the drug concentration at time 0 is given by

$$C_0 = C_1 \cdot \exp(k_e t_1) = C_2 \cdot \exp(k_e t_2) \tag{18}$$

Example: From the curves in Fig. 1, for the values of the ratios obtained for C_1/C_2 at times of 4 and 6 h, the values of k_e equal to 0.231 h^{-1} and of 0.115 h^{-1} are obtained for ciprofloxacin

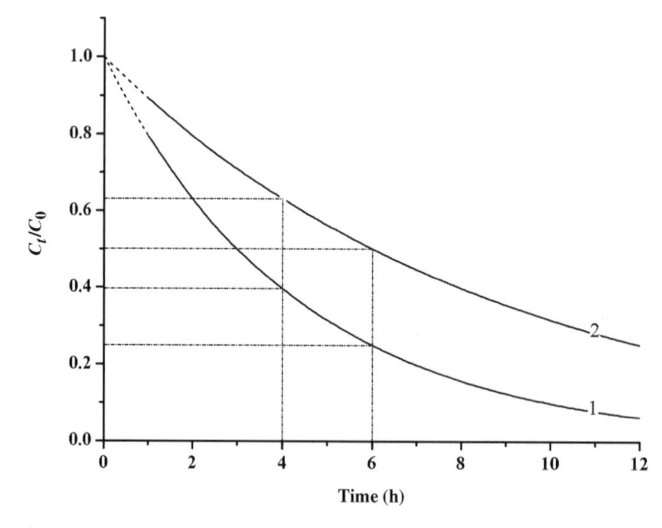

Figure 1. Decrease in concentration with time of the drug in the blood delivered with i.v. bolus injection. The drug concentration is expressed as a fraction of the initial concentration C_0 at time 0 obtained by extrapolation. The concentrations are measured at 4 and 6 h for the drug: ciprofloxacin with $k_e = 0.231\,h^{-1}$ (curve 1) and with $k_e = 0.115\,h^{-1}$ (curve 2).

by using Equation 17 as well as Equation 6. As C_0 cannot be measured, the extrapolated value for this initial concentration C_0 is obtained using Equation 18.

2.9.2. Apparent Volume of Distribution The value of the apparent volume of distribution is determined from the measurements made with i.v. injection. After calculating the initial concentration of the unbound drug in the blood at time 0, C_0, the value of the apparent volume of distribution, V_p, is evaluated using the following equation where Dose represents the amount of drug injected:

$$V_p = \frac{\text{Dose}}{C_0} \tag{19}$$

2.9.3. Rate Constant of Absorption The rate constant of absorption, k_a, is calculated by iteration using Equation 20 giving the value of the time t_{max} at which the peak of the drug concentration in the blood is attained when the oral dosage form with immediate release is taken. This equation is established in Section 3.

$$(k_a - k_e) \cdot t_{max} = \ln \frac{k_a}{k_e} \tag{20}$$

2.9.4. Systemic Clearance The total systemic clearance, Cl, can be obtained from measurements made with i.v. injection using Equation 10 or more simply with Equation 11. The rate of elimination represents the whole phenomenon of elimination of the drug from the blood, including renal and hepatic contributions, and the rate constant of elimination k_e can be considered, provided that these processes obey a first-order kinetics.

2.9.5. Linear or Nonlinear Pharmacokinetics The pharmacokinetics for a drug is linear when the drug concentration in the blood is proportional to the dose, whatever the way of administration. This linearity is obtained when the processes of absorption, metabolism, and elimination, without forgetting the process in which the plasmatic proteins bind and debind the drug, are all controlled by first-order kinetics, when the rate constant in these processes remains constant whatever the dose, and when the plasmatic volume remains constant. In fact, this linearity is observed within a limited range of the dose. In other cases, the pharmacokinetics is nonlinear.

3. INTRAVENOUS ADMINISTRATION OF DRUGS

The intravenous (i.v.) administration is an effective way to deliver the drug in the patient's body, which is mainly used at hospital. If it is not the main purpose in this chapter to deeply study these methods, they are, however, considered for various reasons:

- The administration of single dose in bolus injection is the way to attain the pharmacokinetic parameters of the patient, which are used in calculation of the plasma drug level obtained through other ways of drug delivery.
- These methods should be adapted to the patients according to their intervariability.

3.1. Administration of a Single Dose (Bolus Injection)

The drug administered intravenously into the blood circulation undergoes distribution into tissues before clearance, as shown in Fig. 2. The three following pharmacokinetic parameters necessary to characterize the process are

- The rate constant of elimination k_e expressed in h^{-1}, and the half-life time $t_{0.5}$ of the drug in the body.
- The apparent volume of distribution V_p expressed either in liters or much better in liter per kilogram.
- The systemic clearance reduced mainly to the renal one.

3.1.1. Mathematical Treatment for Evaluation of the Parameters The equations being established in Section 2.9, only the main results are presented.

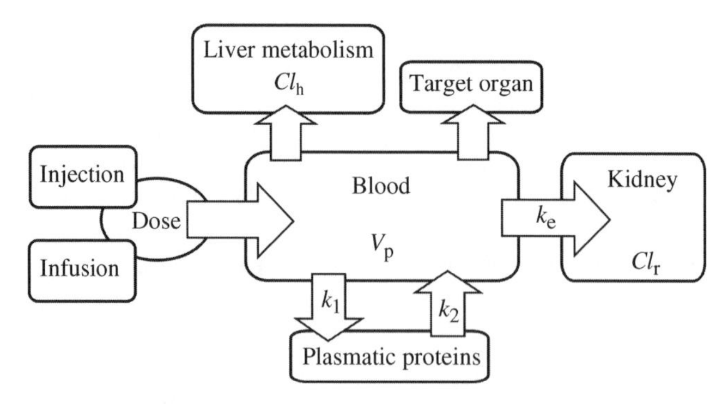

Figure 2. Path followed by the drug with intravenous delivery.

The equation that expresses the decrease in the drug concentration is the first-order kinetics with the rate constant of elimination k_e:

$$-\frac{dC}{dt} = k_e \cdot C \tag{21}$$

It becomes after integration from time 0, with C_0, to time t with C_t:

$$\ln \frac{C_t}{C_0} = -k_e \cdot t \tag{22}$$

Thus, the expression for the half-life time $t_{0.5}$ is given by the relationship

$$\ln 2 = k_e \cdot t_{0.5} = 0.693 \tag{23}$$

The Equation 22 can be rewritten as follows:

$$C_t = C_0 \cdot \exp(-k_e t) \tag{24}$$

By applying this equation to each of the two values of the drug concentration, C_1 obtained at time t_1 and C_2 at time t_2, and dividing these equations by themselves, the initial unknown concentration C_0 disappears, leading to

$$\frac{C_1}{C_2} = \exp[-k_e(t_1 - t_2)] \tag{25}$$

Thus, the rate constant k_e is obtained from Equation 25

$$k_e = \frac{1}{t_2 - t_1} \ln \frac{C_1}{C_2} \tag{26}$$

and the unknown drug concentration at time 0, C_0, is extrapolated as follows:

$$C_0 = C_1 \cdot \exp(k_e t_1) = C_2 \cdot \exp(k_e t_2) \tag{27}$$

After calculating the initial concentration of the unbound drug in the blood at time 0, C_0, the value of the apparent volume of distribution, V_p, can be evaluated using the following equation where Dose represents the amount of drug injected:

$$V_p = \frac{\text{Dose}}{C_0} \tag{28}$$

3.1.2. Results Obtained from Calculation The pharmacokinetic parameters are calculated using the experimental curves describing the exponential decrease with time of the free drug in the blood, as shown in Fig. 3 for ciprofloxacin. Only three curves are drawn for three different values selected among the values of the parameters that are extending over a rather wide range, as stated in Table 1.

Then, the main factor of the intervariability of the patients already appears with the rate constant of elimination and thus the residence time in the plasma, without speaking of the variation of those parameters resulting from a pathology.

The effect of the nature of the drug on the same patient is also of great importance, as shown in Fig. 4 where are drawn the plasma profiles of the following four drugs, three of them belonging to the quinolone family:

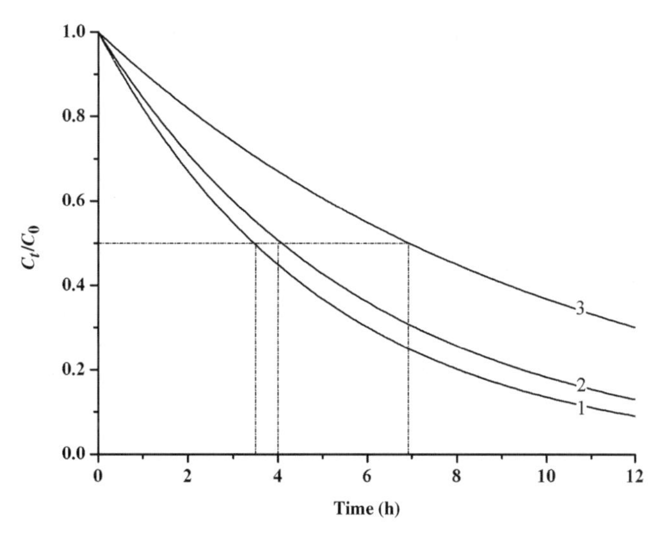

Figure 3. Kinetics of elimination of ciprofloxacin with intravenous bolus injection, showing the intervariability of the patients: $k_e = 0.231\,h^{-1}$ (curve 1); $k_e = 0.17\,h^{-1}$ (curve 2); $k_e = 0.1\,h^{-1}$ (curve 3).

Table 1. Pharmacokinetic Parameters of Ciprofloxacin

Reference(s)	[10]	[11]	[12,13]	[14]	[15]
V_p (l/kg)	2–3	1.7–2.7			
$t_{0.5}$ (h)	3–6	2.9–3.7	3.1–4.9	2.2–7.8	2.7–4.5
k_e (h^{-1})	0.17–0.12	0.24–0.19	0.22–0.14	0.31–0.09	0.26–0.15

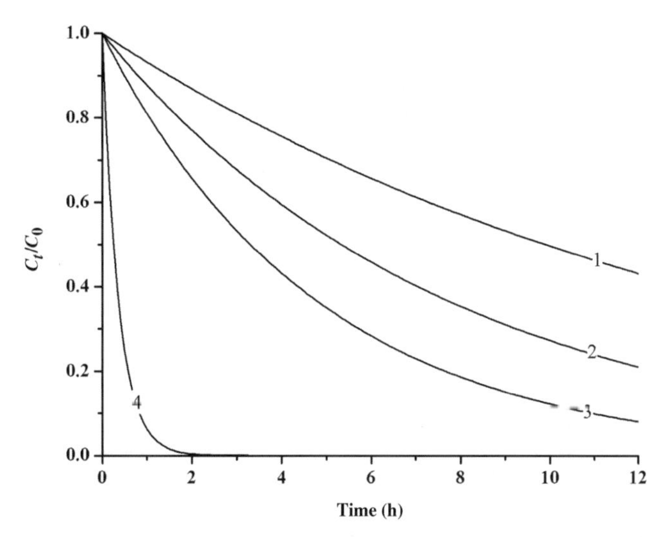

Figure 4. Kinetics of elimination of drugs (qinolones) with intravenous bolus injection: curve (1), péfloxacin with $k_e = 0.07\,h^{-1}$; curve (2), ofloxacin with $k_e = 0.13\,h^{-1}$; curve (3) ciprofloxacin with $k_e = 0.231\,h^{-1}$; curve (4) acetylsalicylic acid with $k_e = 2.77\,h^{-1}$.

péfloxacin (1) [16], ciprofloxacin (3) [15,16], ofloxacin (2) [17], whose values of the half-life time are given in the literature, as well as that obtained with (4) acetylsalicylic acid [18].

The dimensionless number C_t/C_0 appears on the ordinate of Fig. 4, where the initial drug concentration, C_0, is obtained by extrapolation for time 0; thus this concentration C_0 is taken as unit of drug concentration.

3.1.3. Remark on the Expression of the Rate of Elimination

It is surprising, if not wrong, theoretically speaking, to express the process of the drug transfer through the membranes of the kidney by a first-order kinetics. This transfer through a membrane is mathematically defined by a differential equation with partial derivatives [19]—as it will be shown in the section devoted to the transdermal drug transport—the solution of which is given by a series showing two stages in succession: a transient followed by a stationary transport. However, it could be said with a rather good approximation that when the membrane is very thin, and the diffusivity of the drug rather large, the transient stage could be reduced in time, and thus the first-order reaction may be applied.

3.2. Repeated Intravenous Bolus Injections (DIV)

When bolus i.v. doses are administered far apart in time, they behave independently in the same way as shown in Fig. 3 when they are alone. But this is not the most desirable profile of concentration, as a certain minimum concentration for the troughs is needed to maintain efficacy and the peaks should not exceed a maximum value to prevent side effects to occur. Delivery bolus i.v. should thus be sufficiently close together so that the following doses are administered prior to the full elimination of the preceding doses and some accumulation will develop. This fact is observed in Fig. 5 where the drug concentration at time t, C_t as a fraction of the initial concentration C_0, is drawn as a function of time, for repeated doses with the same interval of time equal to the half-life time $t_{0.5}$ of the drug, ciprofloxacin, when it is either 3 h (curve 1) with $k_e = 0.231\,h^{-1}$ or when it is or 6 h (curve 2) with $k_e = 0.1155\,h^{-1}$.

The following comments are worth noting from Fig. 5:

- The so-called steady state—meaning that the profiles are exactly reproduced by each following dose—is attained after a few injections.

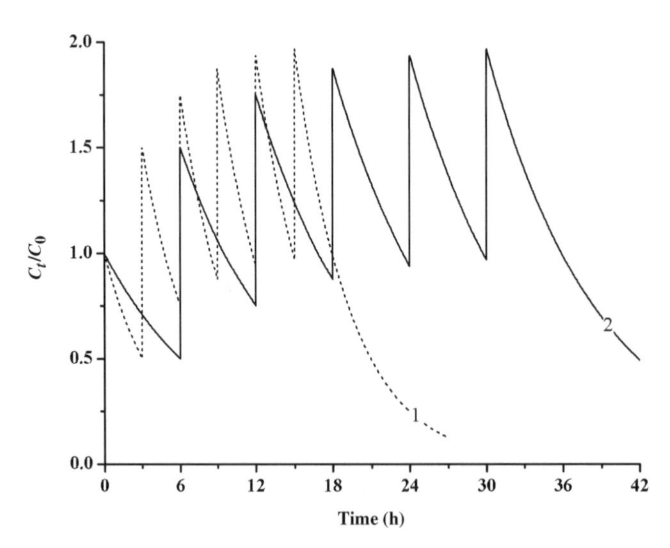

Figure 5. Profiles of drug concentration in the blood with repeated i.v. bolus injections with the constant interval of time, equal to the half-life of time: $t_{0.5} = 3$ h with $k_e = 0.231\,h^{-1}$. (curve 1); $t_{0.5} = 6$ h with $k_e = 0.1155$ h^{-1} (curve 2). Ciprofloxacin.C_0 is unit concentration.

- The effect of the intervariability of the patients toward the same drug, ciprofloxacin, appears with these two profiles that are obtained with two extreme values of the rate constant of elimination k_e and of the half-life time $t_{0.5}$.

3.2.1. Calculation of the Drug Concentration at Peaks and Troughs

Calculation is made by taking care that each dose provokes a drug concentration in the blood that comes in addition to the remaining concentration resulting from the previous doses. When the pharmacokinetics is linear, this law of addition for the doses and their associated concentrations is right, meaning that the concentration of the drug in the body is proportional to the dose injected. By introducing the interval of time T between each following doses, and by using the dimensionless number $\theta = T/t_{0.5}$, Equation 24 is written in the form

$$C_T = C_0 \cdot \exp(-k_e.T) = C_0 \cdot \exp(-0.693 \cdot \theta) \tag{29}$$

The principle of calculation is as follows, when the same dose is repeated, provoking the additional drug concentration C_0, when the process is linear:

At time θ associated with the first dose, the concentration is given by Equation 29.

The second dose injected at that time T causes the new drug concentration to reach the first peak:

$$C_T = C_0 + C_0 \cdot \exp(-0.693 \cdot \theta) = C_0 \cdot (1 + \exp(-0.693 \cdot \theta)) \tag{30}$$

The second peak is reached when the third dose is administered at time $2T$ (or 2θ):

$$C_{2T} = C_0 + C_0 \cdot (1 + \exp(-0.693 \cdot \theta)) \cdot \exp(-0.693 \cdot \theta) \tag{31}$$

which can be rewritten in terms of a series:

$$\frac{C_{2T}}{C_0} = 1 + \exp(-0.693 \cdot \theta) + \exp^2(-0.693 \cdot \theta) \tag{32}$$

For a number of doses n, the series becomes

$$\frac{C_{nT}}{C_0} = \sum_0^n \exp^n(-0.693 \cdot \theta) = \sum_0^n \exp^n\left(\ln(0.5)^\theta\right) = \sum_0^n (0.5)^{\theta \cdot n} \tag{33}$$

This is a convergent series whose terms follow a geometrical progression from 1 to 0, and the sum of this series is well known, so that Equation 33 is resolved finally when n is so large that it can be considered as infinite. Thus, after a sufficient number of repeated doses for the steady sate to be attained, the concentration is given by

$$\frac{C_{nT}}{C_0} = (1-(0.5)^\theta)^{-1} \tag{34}$$

As the values expressed in Equations 33 and 34 are the concentration of the peaks, it is easy to get the extreme values of the peaks and troughs through which the drug concentration alternates. The concentration of the peak increases with n according to

$$\frac{C_n^{Max}}{C_0} = \sum_0^n (0.5)^{\theta.n} \tag{35}$$

while the concentration of the following troughs is given by the obvious relation:

$$\frac{C_n^{\min}}{C_0} = -1 + \sum_0^n (0.5)^{\theta.n} \tag{36}$$

The values of the maximum values of the drug concentration in the body are shown in Table 2 as a function of the number of DIV, n, for three values of the interval of time T between doses, expressed in terms of the half-life time, that is: $\theta = 1$; 1.5; and 2, successively.

The values of the minimum values of the drug concentration in the body as a function of the number of DIV for the same values of the ratio of time θ are easily obtained by using the obvious relationship:

$$C_n^{\min} = C_n^{\mathrm{Max}} - C_0 \tag{37}$$

Table 2. Values of the Peaks Concentration C_n^{Max} as a Function of n and θ

θ/n	1	2	3	4	5	6	7	8	9	10	∞
1	1.5	1.75	1.87	1.94	1.97	1.98	1.992	1.996	1.997	1.998	2
1.5	1.35	1.48	1.52	1.54	1.545						1.547
2	1.25	1.31	1.328	1.332	1.333						1.333

3.2.2. Effect of the Nature of the Drug The effect of the nature of the drug was already shown in Fig. 4 with a single dose, and it is obvious that a similar effect is obtained with repeated doses. This fact is also obvious with the change in the pharmacokinetic parameters of the same drug resulting from the different answer of the patient to the drug, in the following section.

3.2.3. Effect of the Patient's Intervariability Toward a Drug The effect of the patient's intervariability appears in Fig. 5 where two curves are calculated with some extreme values of the rate constant of elimination of ciprofloxacin ranging from 0.1155–0.231 h^{-1}, and the half-life time varied from 3 to 6 h. These two curves represent the profiles of concentration obtained when the doses are administered in succession at the same interval of time equal to the half-life time $t_{0.5}$.

The following two results are worth noting.

- The one, more evident, is that the time of duration of the DIV is shorter when the value of the rate constant is larger and the half-life time shorter.
- The other result, more important perhaps, is that the concentration of the drug alternates between higher peaks and higher troughs. When the doses are delivered at the interval of time T taken as equal to the half-life time, the drug concentration at peaks and troughs are the same, whatever the rate constant of elimination. However, these peaks and troughs are obtained at shorter times when the rate constant of elimination is larger.

3.2.4. Plasma Drug Profiles with Master Curve and Dimensionless Numbers A drug being associated with its typical pharmacokinetic parameters, normally it would be necessary to calculate and draw for each drug the profile of concentration as a function of time in case of repeated injections. This drawback is avoided by using a master curve with dimensionless coordinates, as shown in Fig. 6 where the concentration of the drug, C_t, as a fraction of the initial concentration, C_0, is drawn as a function of the dimensionless time θ, this time being obtained by taking the half-life of time as unit of time. The plasma drug profiles obtained with the two different pharmacokinetic parameters of the ciprofloxacin: the one with the couple ($k_e = 0.1155\,h^{-1}$, $t_{0.5} = 6\,h$), the other with ($k_e = 0.231\,h^{-1}$, $t_{0.5} = 3\,h$), are shown in Fig. 6, leading to the following main results:

- These profiles are similar by using the dimensionless numbers for the concentration and time, provided that the followings are applied: the unit concentration C_0 is the initial concentration obtained for each drug, the period of time T is equal to the half-life time for each drug $t_{0.5}$, and the curve is calculated with the associated value of k_e.
- This master curve can be used for all drugs provided that the above conditions are met.

3.2.5. Use of Master Curve with Different Intervals of Time Between Doses The master curve can also be applied with different intervals of time between the doses.

Among the two curves drawn in Fig. 7 the dotted curve (1) obtained when the interval of time T between each injection is equal to the half-life time of the drug, while the full line curve (2) is drawn when the interval of time T is 1.5 the half-life time of the same drug. The effect of the interval of time on the profile of the drug concentration clearly appears in this

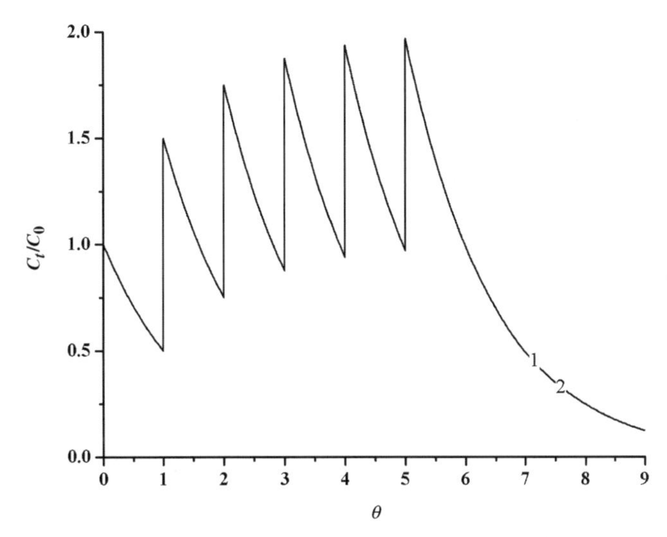

Figure 6. Master curve for the two plasma drug profiles of ciprofloxacin, drawn with dimensionless numbers for the coordinates: concentration C_t expressed in terms of C_0 taken as unit concentration; time θ expressed with the half-life time $t_{0.5}$ taken as unit of time. The curves calculated with ($t_{0.5} = 3\,h$ with $k_e = 0.231\,h^{-1}$) and ($t_{0.5} = 6\,h$ with $k_e = 0.1155\,h^{-1}$) are superimposed.

figure, which can be defined by the statement: the shorter the interval of time between every dose, the higher the concentrations of the drug, with a special mention for the peaks and troughs.

3.2.6. Method of Using the Master Curves The protocol for using the master curves shown in Fig. 7 is as follows. When the half-life time is known, for example, 6 h, θ representing the dimensionless time $t/t_{0.5}$ can be associated

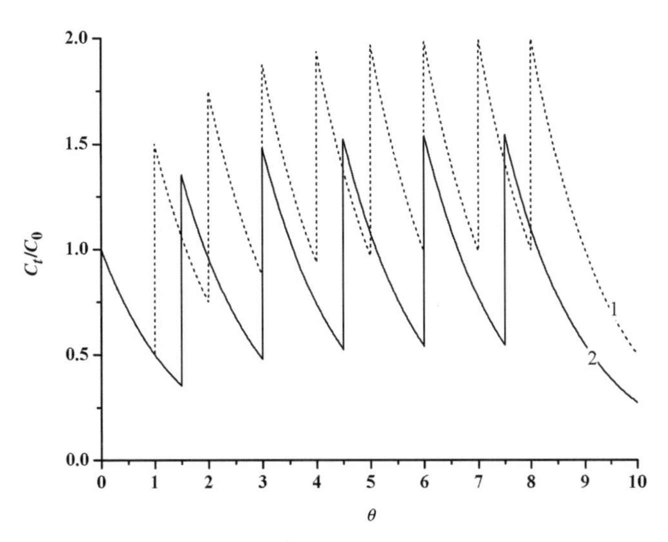

Figure 7. Master curve for the plasma drug profiles of ciprofloxacin, drawn with dimensionless numbers for the coordinates: the concentration C_t is expressed in terms of C_0 taken as unit concentration, and time θ is expressed in terms of the half-life time $t_{0.5}$ taken as unit of time. (1) Dotted line: interval of time $T = t_{0.5}$. (2) Full line: interval of $T = 1.5 \cdot t_{0.5}$

with normal time (in hours), and for $\theta = 10$ it corresponds 60 h. Thus, each dose is injected at the interval of 6 h in case of the curve 1 ($\theta = 1$) or of 9 h in case of the curve 2 ($\theta = 1.5$). The concentration of the drug in these two cases, at any time and especially for the peaks and troughs, is expressed in terms of the concentration C_0 defined by the Equation 27. The precise values at the peaks and troughs are given in Table 2 for the following doses delivered in succession (with the number n) and for various values of the interval of time T expressed in terms of the half-life time through the dimensionless time θ.

3.2.7. Calculation of AUC AUC is evaluated by using the following relations:

For a single injection, it is easily obtained:

$$\mathrm{AUC}_{0-\infty} = C_0 \cdot \int_0^\infty \exp(-k_e \cdot t) \cdot \mathrm{d}t = C_0 \cdot \frac{t_{0.5}}{0.693} \tag{38}$$

For two doses injected in succession at the interval of time T, the AUC is the sum of the integrals:

$$\mathrm{AUC}_{0-\infty} = \mathrm{AUC}_{0-T} + \mathrm{AUC}_{T-\infty} = C_0 \cdot \int_0^T \exp(-k_e \cdot t) \cdot \mathrm{d}t + C_1^{Max} \cdot \int_T^\infty \exp(-k_e \cdot t) \cdot \mathrm{d}t \tag{39}$$

and for n injections, the recurrent relation is obtained with the corresponding integrals:

$$\mathrm{AUC}_{0-\infty} = \mathrm{AUC}_{0-T} + \mathrm{AUC}_{T-2T} + \ldots \mathrm{AUC}_{(n-1)T-nT} + \mathrm{AUC}_{nT-\infty} \tag{40}$$

It is of interest to predict by calculating the value of the AUC associated with the system of n injections, when the half-life time of the drug is known, for various values of the interval of time T of the successive injections. In Figs 5–7 where the profile of the drug concentration is drawn, the successive areas corresponding to the amounts of drug delivered in the body over each period of time T can be evaluated as follows:

For the first period of time, when the initial concentration at time 0 is C_0 in the body, the amount is obtained by integrating the concentration $C_t.\mathrm{d}t$ from 0 to T:

$$\mathrm{AUC}_{0-T} = C_0 \cdot \int_0^T \exp(-k_e \cdot t) \cdot \mathrm{d}t = \frac{C_0}{k_e}[1-\exp(-0.693 \cdot \theta)] = \frac{C_0}{k_e}\left[1-(0.5)^\theta\right] \tag{41}$$

In the same way, within the interval of time between T and $2T$, the AUC is

$$\mathrm{AUC}_{T-2T} = C_1^{\mathrm{Max}} \cdot \int_0^T \exp(-k_e \cdot t) \cdot \mathrm{d}t = \frac{C_1^{\mathrm{Max}}}{k_e}[1-\exp(-1.693 \cdot \theta)] = \frac{C_1^{\mathrm{Max}}}{k_e}\left[1-(0.5)^\theta\right] \tag{42}$$

where C_1^{Max} is the peak concentration obtained at time T after the second injection.

The following recurrent relationship allows calculation of the amount of drug in the body over any period of time, for example, $(n-1)T$ and T, using the corresponding maximum value of the initial drug concentration in the body C_{n-1}^{Max}:

$$\mathrm{AUC}_{(n-1)T-nT} = C_{n-1}^{\mathrm{Max}} \cdot \int_0^T \exp(-k_e \cdot t) \cdot \mathrm{d}t = \frac{C_{n-1}^{\mathrm{Max}}}{k_e}[1-\exp(-0.693 \cdot \theta)] = \frac{C_{n-1}^{\mathrm{Max}}}{k_e}\left[1-(0.5)^\theta\right] \tag{43}$$

Finally, the amount of drug that passed in the body during n successive injections delivered with a period of time T, is given by Equation 40 that becomes

$$\mathrm{AUC}_{0-\infty} = \frac{C_n^{\mathrm{Max}}}{k_e} + \frac{1}{k_e}\left[1-(0.5)^\theta\right] \cdot \sum_0^{n-1} C_n^{\mathrm{Max}} \tag{44}$$

Table 3. Values of AUC for Each Multidose as a Fraction of AUC with a Single Dose

Dose Number	1	2	3	4	nth	Last
$\theta = 1$	0.5	0.75	0.875	0.94	1	2
$\theta = 1.5$	0.65	0.87	0.96	0.98	1	1.547

The partial AUC associated with every successive dose, given by the Equation 43, is expressed as a fraction of the total AUC obtained with a single dose taken alone, in Table 3 for the two values $\theta = 1$ and $\theta = 1.5$.

The AUC obtained with a single dose is given in Equation 41. This calculation is made by using the data shown in Table 2 for the maximum values at peaks for $\theta = 1$ and 1.5.

Similar calculation can be made for other values of the interval of time T by evaluating the concentration at the following peaks with the Equation 35 for the desired value of θ. The total AUC is the sum of each partial AUC, and when n is large, it is nth times that obtained with the single dose, with a slight change in the last one.

3.2.8. Adaptation of the Therapy to the Patient During the DIV

The inconvenience of the intravenous therapy with regard to the way based on oral dosage forms, is that it is administered generally at the hospital. Nevertheless, an advantage exists for intravenous administration, because a change in the conditions of the treatment is possible, by varying either the dose injected or the interval of time between the following injections. As shown in Table 1 for ciprofloxacin, depending on the response of each patient to the dose of the drug, which is the basis of the intervariability, the values of the two main parameters V_p and $t_{0.5}$ (apparent volume of distribution and half-life time) are ranging within a wide extent. Thus, it is necessary to cope with this problem in adapting the right operational conditions of the therapy during the treatment. The protocol is as follows with the curves in Fig. 8:

After the first injection with the predetermined dose, which gives the initial concentration C_0, regardless the interval of time T or the dimensionless time θ, two analyses are made at two times selected between 0 and T. These

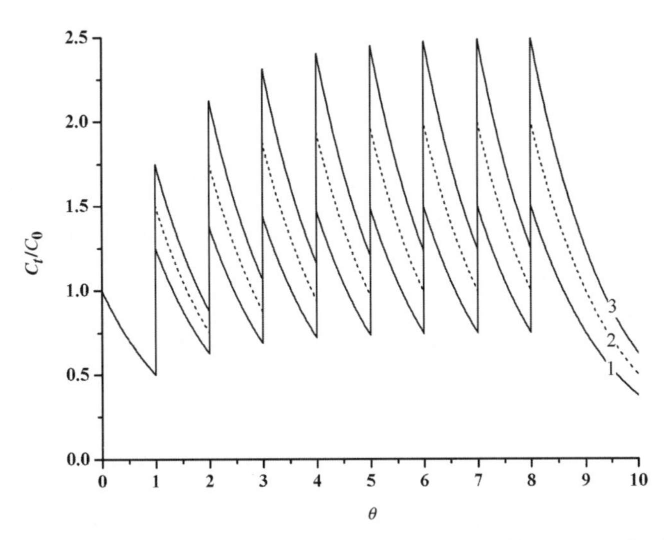

Figure 8. Use of master curves in expressing the plasma drug profiles obtained with repeated bolus injections during the adaptation of the therapy to the intervariability of the patient. C_0 is the first dose selected for the therapy. Depending on the analysis, it can be modified for the second and following injections. In all cases, it remains the new unit drug concentration. θ is the dimensionless time with the half-life time as unit. Depending on the new value of the half-life time obtained from analysis, it can be varied from the second injection.

two measurements enable the therapist to calculate the initial concentration C_0 and the half-life time of the drug for the patient by using the Equations 26 and 27. Then, two cases may occur, depending either on the apparent volume of distribution or on the half-life time:

- If the value obtained for the volume of distribution is different from the one that was estimated in selecting the first dose, the second dose and the following ones are adjusted by either increasing or decreasing their value, by using Equation 28.
- If the new value of the half-life time obtained by calculation is quite different from the one firstly expected, the curves in Fig. 8 help the therapist in selecting the new interval of time T between successive doses. For instance, when the desired therapy is done with $T = t_{0.5}$, the interval of time is adjusted to the new value of the half-life time obtained with this calculation. The effect of a change in the dose and the corresponding plasma drug profiles can be appreciated in Fig. 8 where the second dose and the following ones are either 30% more or less than the first one, when the interval of time between two successive doses is equal to the half-life time of the drug.

3.3. Continuous Intravenous Infusion

3.3.1. Mathematical Expression of the CIV The general equation for the continuous intravenous infusion (CIV) taking into account both the rate of drug delivered in the body through infusion and the rate of drug eliminated is written as follows:

$$\frac{dM}{dt} = U - k_e \cdot M \tag{45}$$

M being the mass of drug in the body and U the mass of the drug administered per unit time.

By dividing both terms by the volume of distribution, V_p, this equation is expressed in terms of concentration of the drug:

$$\frac{dC}{dt} = \frac{U}{V_p} - k_e \cdot C \tag{46}$$

By integrating this equation in terms of time, between 0 and t, the plasma drug profile is obtained, related to the pharmacokinetic parameters.

$$C_t = \frac{U}{k_e \cdot V_p}[1 - \exp(-k_e \cdot t)] \tag{47}$$

From the Equation 47, the concentration of the free drug in the blood is proportional to the rate of infusion and inversely proportional to both the rate constant of elimination and the apparent volume of distribution.

Equation 47 can also be expressed in terms of the half-life time of the drug:

$$C_\theta = \frac{U \cdot t_{0.5}}{V_p \cdot 0 \cdot 693}[1 - \exp(-0.693 \cdot \theta)] = \frac{U \cdot t_{0.5}}{V_p \cdot 0.693}\left[1 - (0.5)^\theta\right] \tag{48}$$

For infinite time, or rather for a time long enough with regard to the half-life time, the term in exponential vanishes, and the asymptotical value of the concentration becomes

$$C_\infty = \frac{U \cdot t_{0.5}}{V_p \cdot 0.693} = \frac{U}{V_p \cdot k_e} \tag{49}$$

A master curve (Fig. 9) is obtained, by plotting the concentration at any time C_θ as a fraction of the asymptotical value of this concentration C_∞, as a function of the dimensionless time θ. This curve, in the same way as for bolus injection (Fig. 6), can be used whatever be the drug and its pharmacokinetic parameters. It can be noted that the ratio of these concentrations is equal to 0.5 when $\theta = 1$, that is, when the time is equal to the half-life time of the drug.

The values given in Table 4 show that the concentration of the drug in the body increases with time up to an asymptotic value, enabling the therapist to know the time at which the process of the drug infusion may be considered as performed under steady state.

The rate of administration U can thus be evaluated in terms of the half-life time of the drug as well as of the apparent volume of

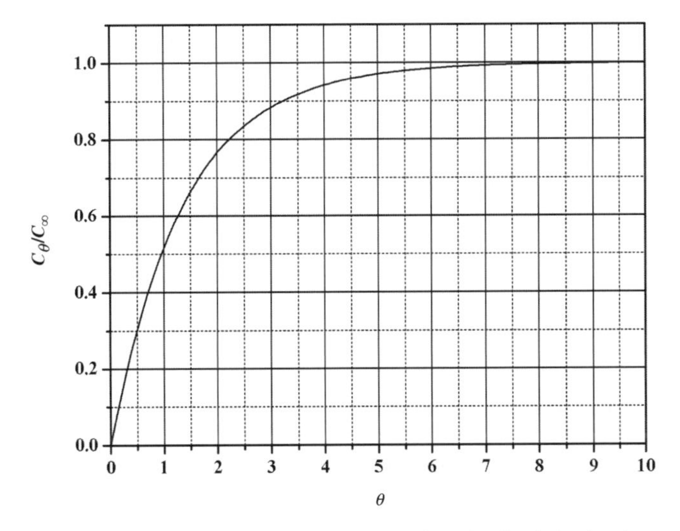

Figure 9. Master curve describing the profile of concentration obtained during intravenous infusion at constant rate. The unit concentration is C_∞ and the unit time is the half-life time of the drug $t_{0.5}$.

distribution in order to attain the optimal drug level in the patient's body, when these pharmacokinetic parameters are known.

3.3.2. Adapting the Treatment During the Course of Infusion Of course, intravenous therapy exhibits some drawbacks with regard to the method based on oral dosage forms, the essential one being that it is administered generally at the hospital. Nevertheless an advantage exists for intravenous administration, resulting from the fact that a change in the course of delivery is possible by varying the rate of infusion.

As shown in Section 3.2, the pharmacokinetic parameters vary within a large range from one patient to another, and it should be necessary to adapt the rate of infusion to each patient. When neither the apparent volume of distribution nor the half-life time are known, it seems difficult to predict the right rate of drug adapted to a patient during the course of infusion. In fact, as shown in Fig. 9 and more precisely in Table 4, it is not obligatory to deliver the drug at the first rate previously selected over a long period of time before making analyses, and thus for the therapist to decide how to vary the rate of drug in order to adapt it to the patient.

The following protocol can be used: The concentration of the free drug is measured at various times up to the time at which an insignificant change in the concentration occurs. Thus, from the values obtained, two are retained, C_1 at time t_1, and C_2 at time t_2, respectively. The concentration C_1 is expressed in terms of time (or θ) by the Equation 48, while at the longer time t_2, the concentration is nearly constant at the value given in Equation 49. By making the ratio of these two concentrations the simple relation is obtained, where appears only the time t_1 and the two concentrations:

$$\frac{C_1}{C_2} = 1-(0.5)^{\theta_1} \tag{50}$$

$$\text{with} \quad \theta_1 = \frac{t_1}{t_{0.5}} \tag{51}$$

Table 4. Time Necessary to Reach Steady State (Constant Concentration)

θ	0	1	2	3	4	5	6	7	8	9	10	∞
$\frac{C_\theta}{C_\infty}$	0	0.5	0.75	0.87	0.94	0.97	0.984	0.992	0.996	0.998	0.999	1

The value of half-life time is obtained from the above two equations as

$$t_{0.5} = t_1 \cdot 0.693 \cdot \left[\ln \frac{C_2}{C_2 - C_1}\right] \quad (52)$$

The apparent volume of distribution is given, by rewriting Equation 49, as

$$V_p = \frac{U \cdot t_{0.5}}{C_\infty \cdot 0.693} \quad (53)$$

The pharmacokinetic parameters of the patient being known, it is possible to adapt the treatment, as shown in Fig. 10 where the profiles of the drug concentration are drawn when the rate of infusion U is varied by 30% of its original value after the last time t_2 used for calculating the pharmacokinetic parameters.

3.3.3. Value of AUC with Infusion Delivery

The value of the AUC can be evaluated in the following cases of infusion:

- During the continuous intravenous infusion up to the time t at which it is administered:

$$\mathrm{AUC}_{0-t} = \frac{U}{k_e \cdot V_p} \int_0^t [1 - \exp(-k_e . t)] \cdot \mathrm{d}t = \frac{U}{k_e \cdot V_p}\left[t - \frac{1}{k_e}(1 - \exp(-k_e \cdot t))\right] \quad (54)$$

This equation can be rewritten in a more simple way when the time is long enough for the exponential term to vanish and the concentration to attain a constant value:

$$\mathrm{AUC}_{0-t} = \frac{U}{k_e \cdot V_p}\left(t - \frac{1}{k_e}\right) = \frac{U \cdot t_{0.5}}{0.693 \cdot V_p}\left(t - \frac{t_{0.5}}{0.693}\right) \quad (55)$$

- When the intravenous infusion is stopped, the drug in the body is eliminated leading to the following value of the AUC, resulting from the elimination from the body of the constant concentration of the drug expressed by the Equation 49:

$$\mathrm{AUC}_{0-\infty} = \frac{U}{k_e V_p} \int_0^\infty \exp(-k_e \cdot t) \cdot \mathrm{d}t = \frac{U}{V_p \cdot k_e^2} = \frac{U}{V_p}\left(\frac{t_{0.5}}{0.693}\right)^2 \quad (56)$$

The total value of the AUC from the beginning to the end of the process of infusion is obtained by adding the two values given in Equations 55 and 56.

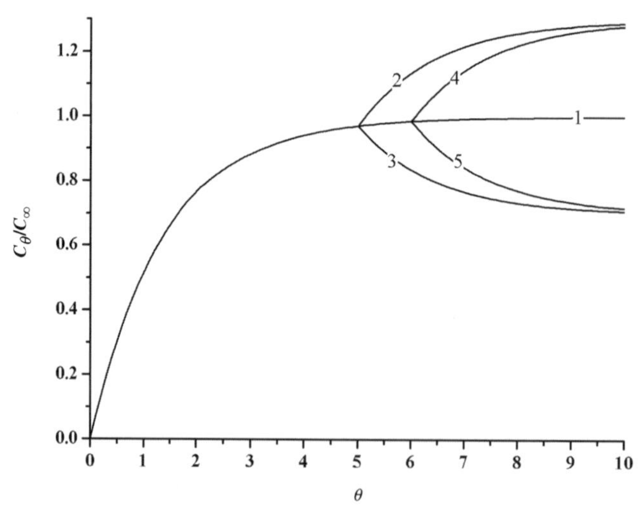

Figure 10. Master curve showing the profile of concentration obtained during intravenous infusion at constant rate, when the rate of infusion is varied by 30% of its initial value, at the dimensionless times: $\theta = 5$ and $\theta = 6$. The unit concentration is C_∞ and the unit time is the half-life time of the drug $t_{0.5}$.

3.4. Repeated Short Infusions at Constant Flow Rate Over a Finite Period of Time

3.4.1. Mathematical Treatment for the Drug Concentration [20,21] For various reasons, hospitalized patients are administered the drug dose i.v. at a constant flow rate U over a short period of time, t_i, that is, from 1 to 60 min, and after this time the plasma drug concentration is let decrease over another period of time called T (Fig. 11). The same process of drug delivery is repeated a number of times in succession (Fig. 12).

The mathematical treatment of this method of drug delivery is derived from both the previous parts, that is, the infusion, followed by the elimination:

First Infusion–Elimination Stage At the time of infusion, t_i, the plasma drug concentrations are given by Equations 47 and 49 that can be rewritten as follows:

$$C_{t_i} = C_\infty[1-\exp(-k_e t_i)] \quad (57)$$

When the drug flow rate is stopped at time, t_i, the plasma drug concentration, C_{t_i}, decreases up to time T according to Equation 24, thereby becomes

$$C_T = C_{t_i}\exp(-k_e T) \quad (58)$$

Second Infusion–Elimination Stage In the case of repeated short infusions with the same infusion time, t_i, the differential equation (Eq. 46) should be integrated from $t = 0$ when $C = C_T$, to time t_{2i} at which the concentration for the second peak, noted $C_{t_{2i}}$, becomes

$$C_{t_{2i}} = C_\infty[1-\exp(-k_e t_i)] + C_{1T}\exp(-k_e t_i) \quad (59)$$

And after the same time T of elimination, the concentration for the second trough, represented by C_{2T}, is expressed as follows:

$$C_{2T} = C_{t_{2i}}\exp(-k_e T) \quad (60)$$

Third Infusion–Elimination Stage Making similar calculation, the concentrations for the third peak and trough are defined by the following relations:

$$C_{t_{3i}} = C_\infty[1-\exp(-k_e t_i)] + C_{2T}\exp(-k_e t_i) \quad (61)$$

$$C_{3T} = C_{t_{3i}}\exp(-k_e T) \quad (62)$$

Generalization for the nth Infusion–Elimination Stage The concentrations for the peaks and troughs can be expressed in terms of a series:

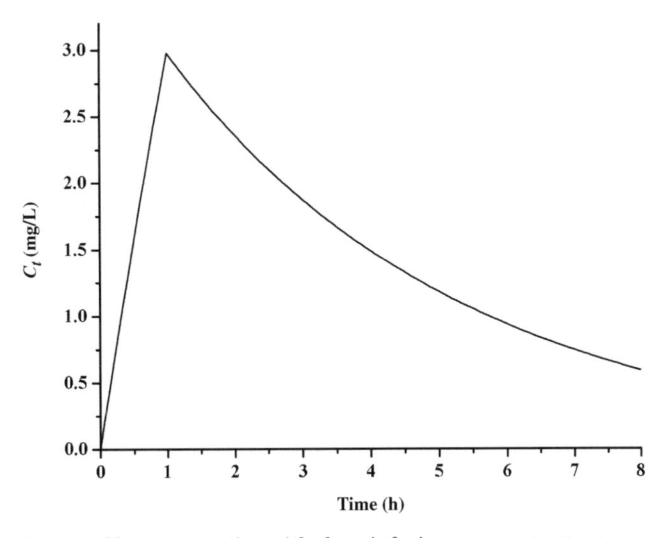

Figure 11. Plasma drug profile concentration with short infusion at a constant rate over 1 h period followed by elimination over 7 h. Rate of infusion = 500 mg/h; $k_e = 0.231\,h^{-1}$; $t_{0.5} = 3\,h$; $V_p = 150\,L$; C_{Max} is the concentration at the peak.

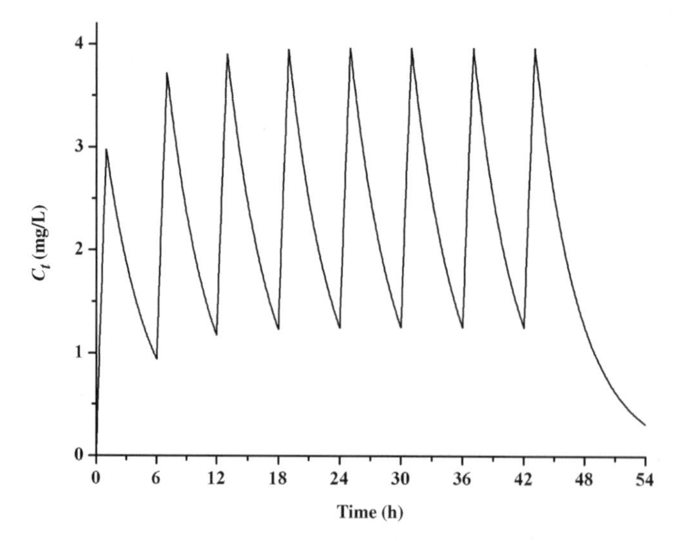

Figure 12. Plasma drug profiles obtained with repeated short infusions. $U = 500$ mg/h over 1 h period, followed by elimination over 5 h, with $K_e = 0.231\,h^{-1}$ and $t_{0.5} = 3$ h; $V_p = 150$ L; $C_{Max} = 3$ mg/L for the first injection.

$$C_{t_{ni}} = C_\infty[1-\exp(-k_e t_i)]\sum_{j=0}^{n-1}\exp^j[-k_e(t_i+T)] \tag{63}$$

$$C_{nT} = C_{t_{ni}}\exp(-k_e T) \tag{64}$$

By using the value of the series, Equation 63 can be rewritten as follows:

$$C_{t_{ni}} = C_\infty[1-\exp(-k_e t_i)]\frac{1-\exp^n[-k_e(t_i+T)]}{1-\exp[-k_e(t_i+T)]} \tag{65}$$

Case of a Large Number n of Stages When n is large, the $\exp^n$ term in the numerator tends to 0, as the exponential term is lower than 1. Thus the concentrations for highest peaks and troughs are given by

$$C_{t_{ni}} = C_\infty\frac{1-\exp(-k_e t_i)}{1-\exp[-k_e(t_i+T)]} \tag{66}$$

$$C_{nT} = C_{t_{ni}}\exp(-k_e T) \tag{67}$$

The drug concentrations for the peaks and troughs are shown in Table 5 when the value of the infusion time is: $t_i = 1$ h, and the elimination time T is equal to the half-life, t_i, of the drug, that is, 3 h for the higher value of k_e and 6 h for its lower value, respectively. In both cases, the rate of infusion U is 500 mg/h, and the apparent volume of distribution V_p is

Table 5. Drug Concentrations at the Peaks and Troughs

Peak Number Number	$k_e = 0.231\,h^{-1}$ Minimum	$t_{0.5} = 3$ h Maximum	$k_e = 0.1155\,h^{-1}$ Minimum	$t_{0.5} = 6$ h Maximum
1	0	1	0	1
2	0.5	1.3969	0.5	1.445
3	0.698	1.554	0.723	1.644
4	0.777	1.6169	0.822	1.733
5	0.8084	1.6417	0.866	1.772
6	0.8208	1.6515	0.886	1.79
7	0.8258	1.6554	0.895	1.797
8	0.8277	1.657	0.898	1.80

150 L. Two values of the elimination rate constant, k_e, are considered, $k_e = 0.231\,h^{-1}$ and $k_e = 0.1155\,h^{-1}$. The concentrations, expressed in mg/L, obtained for the first peak and the second trough (by considering that the first trough is equal to 0, and thus before the first infusion) and for the two values of the drug elimination rate constant, k_e, are, respectively:

$$k_e = 0.231\,h^{-1} \quad C_{t_i} = 2.976\,mg/L$$

$$C_T = 0.9377\,mg/L \qquad t_{0.5} = 3\,h$$

$$k_e = 0.1155\,h^{-1} \qquad C_{t_i} = 3.149\,mg/L$$

$$C_T = 1.772\,mg/L \qquad t_{0.5} = 6\,h$$

The concentrations attained at the first peak and second trough are evaluated in milligram per liter. However, in the following short infusions, the value of the concentration of this first peak is taken as the unit of concentration for the successive peaks and troughs, as shown in Table 5.

The following conclusions can be drawn from the above values:

- The drug concentrations for the peaks and troughs increase regularly with the number of the repeated short infusion-elimination stages, according to Equations 64 and 65, with an asymptotic tendency, whatever the value of the drug elimination rate constant, k_e.
- The drug concentrations at peaks and troughs increase rapidly from the first to the second stage, that is, by 40%, but only by 1.2% between the seventh and eighth stages, when $k_e = 0.231\,h^{-1}$.
- The effect of the drug elimination rate constant, k_e, is not very significant, and the increase in concentration at the peaks and troughs is slightly larger when the drug elimination rate constant, k_e, is lower, as determined by Equations 66 and 67, and shown in Table 5.
- When the drug elimination rate constant, k_e, vary by 80%, the values of the first peak, C_{t_i}, only differ by only 4.7%. This fact is clearly shown in Fig. 13.

3.4.2. AUC for the Repeated Short Infusions

The AUC is obtained by integrating the various equations expressing the drug concentration as a function of time. Thus, for the first infusion–elimination stage, the two steps are considered successively: the short infusion

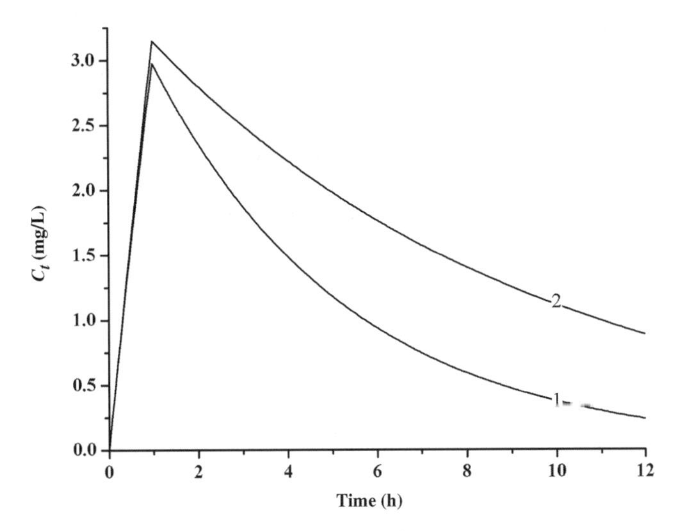

Figure 13. Effect of the drug elimination rate on the plasma drug profile with an infusion rate $U = 500$ mg/h over 1 h period, followed by elimination over 11 h. Curve 1 with $k_e = 0.231\,h^{-1}$ and $t_{0.5} = 3\,h$ and curve 2 with $k_e = 0.115\,h^{-1}$ and $t_{0.5} = 6\,h$; $V_p = 150\,L$ in both cases.

from time $t=0$ to t_i, when the concentration increases from $C_0=0$ to C_{t_i}; the elimination stage from time t_i and concentration C_{t_i} to time T when the drug concentration has decreased to C_T. During the short infusion stage, the value of AUC is given by

$$\mathrm{AUC}_{0-t_i} = C_\infty \int_0^{t_i} [1-\exp(-k_e t)]dt \qquad (68)$$

which gives

$$\mathrm{AUC}_{0-t_i} = C_\infty \left[t_i - \frac{1}{k_e}[1-\exp(-k_e t_i)] \right] \qquad (69)$$

During the elimination stage of duration T, which starts at the time t_i considered as the beginning of this stage, integration takes place from time 0 to time T, as follows:

$$\mathrm{AUC}_{t_i-T} = C_{t_i} \int_0^{T} \exp(-k_e t)dt \qquad (70)$$

which results in:

$$\mathrm{AUC}_{t_i-T} = \frac{C_{t_i}}{k_e}[1-\exp(-k_e T)] \qquad (71)$$

And the total AUC_{0-T} for the first infusion–elimination stage is expressed by

$$\mathrm{AUC}_{0-T} = C_\infty \left[t_i - \frac{1-\exp(-k_e t_i)}{k_e} \exp(-k_e T) \right] \qquad (72)$$

For instance, for the 1 h period infusion at 500 mg/h, $V_p = 150$ L, the AUC values are

- with $k_e = 0.231\,\mathrm{h}^{-1}$, $t_{0.5} = 3$ h, and $C_\infty = 14.43$, $\mathrm{AUC}_{0-T} = 6.31$ (mg/L)h
- with $k_e = 0.1155\,\mathrm{h}^{-1}$, $t_{0.5} = 6$ h, and $C_\infty = 28.86$, $\mathrm{AUC}_{0-T} = 15.23$ (mg/L)h

As shown from these two AUC_{0-T} values, for the first stage as well as for the other ones, the AUC is larger when the drug elimination rate constant k_e is lower, and this is for two reasons: first, the concentrations at peak C_{t_i} and at trough C_T are higher for the lower value of k_e; and secondly, the period of time $T = 6$ h is longer when $k_e = 0.1155\,\mathrm{h}^{-1}$, while it amounts to $T = 3$ h when $k_e = 0.231\,\mathrm{h}^{-1}$.

3.4.3. Adaptation of Therapy to the Individual Patient

Evaluation of the Intervariability Between Patients Figure 14 shows the curves expressing the decrease in concentration of the free drug in the blood circulation as a function of time, calculated for two extreme values of the drug elimination rate constant, k_e, obtained for ciprofloxacin for various patients. Some values of this drug elimination rate constant, k_e, that have been reported in the literature, are shown in Table 1. The significant difference in these drug concentration profiles makes it clear that the selection of the pharmacokinetic parameters required for determining the drug dosage for unknown patients is of major concern, and necessitates a precise determination of the value of the drug's half-life $t_{0.5}$ for each patient before initiating therapy. Furthermore, it clearly appears that interindividual behavior toward the drug differs according to whether the patient is healthy or not, as has in particular been shown for ciprofloxacin.

With the aim of adapting the drug therapy in question to the individual patient, two blood analyses are made at the beginning of the treatment period, that is, at times t_1 and t_2, during the period of decrease in drug concentrations from the first peak for the drug concentration C_{t_i}, as shown in Fig. 14. Calculation is made following the method described for bolus injection, which leads to the determination of plasma drug concentrations:

$$C_{t_1} = C_{t_i} \exp[-k_e(t_1 - t_i)] \qquad (73)$$

$$C_{t_2} = C_{t_i} \exp[-k_e(t_2 - t_i)] \qquad (74)$$

These two concentrations enable the determination of the patient's pharmacokinetic parameters of the patient, whatever the drug–protein binding:

$$k_e = \frac{1}{t_2 - t_1} \ln \frac{C_{t_1}}{C_{t_2}} \qquad (75)$$

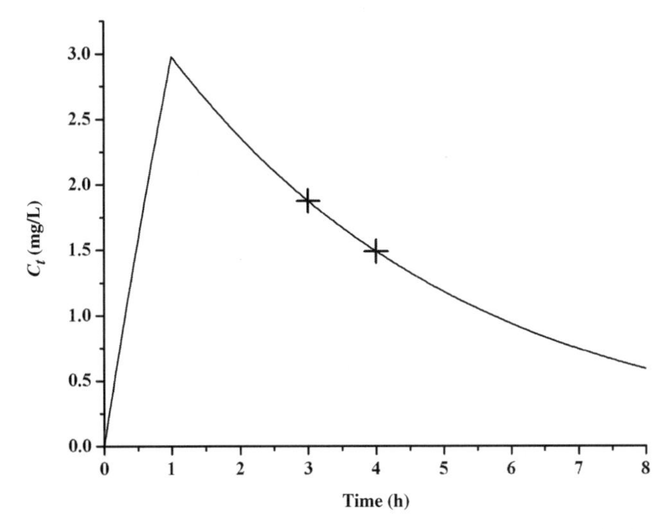

Figure 14. Scheme showing the process when two analyses are made during the elimination stage following the first short infusion. The concentrations are $C_1 = 1.875$ at time $t_1 = 3\,\text{h}$ and $C_2 = 1.488$ at time $t_2 = 4\,\text{h}$. Pk parameters: $k_e = 0.231\,\text{h}^{-1}$ and $t_{0.5} = 3\,\text{h}$; $U = 500\,\text{mg/h}$ over 1 h. $C_{Max} = 2.976$; $V_p = 150\,\text{L}$.

$$C_{t_i} = C_{t_1}\exp[-k_e(t_1 - t_i)] = C_{t_2}\exp[-k_e(t_2 - t_i)] \quad (76)$$

From Equations 49 and 57, the value of the volume of distribution V_P is obtained as

$$V_p = \frac{U}{k_e C_{t_i}}[1 - \exp(-k_e t_i)] \quad (77)$$

As shown in Fig. 14, two measurements of concentrations $C_{t_1} = 1.875\,\text{mg/L}$ and $C_{t_2} = 1.49\,\text{mg/L}$ made at time $t_1 = 3\,\text{h}$ and at time $t_2 = 4\,\text{h}$ with time $t_i = 1\,\text{h}$ considered as the start of time for calculation, the value of $k_e = 0.231\,\text{h}^{-1}$ is obtained from Equation 75, the value of $C_{t_i} = 2.976\,\text{mg/L}$ from Equation 76; and Equation 77 enables the following value $V_p = 150\,\text{L}$ to be obtained when the rate of infusion is $U = 500\,\text{mg/h}$.

Effect of the Intervariability Between Patients When the same short infusion is given to two patients with extreme values of the drug elimination rate constant, k_e, shown in Table 1, the plasma drug profiles are seen in Fig. 13 with $k_e = 0.231\,\text{h}^{-1}$ (curve 1) and with $k_e = 0.1155\,\text{h}^{-1}$ (curve 2), under similar operational conditions, that is, $U = 500\,\text{mg/h}$ and $V_p = 150\,\text{L}$.

When drug therapy is maintained similar for these two patients with the above pharmacokinetic parameters using the same repeated short infusions and under the same operational conditions, that is, time of infusion $t_i = 1\,\text{h}$ and time of drug elimination $T = 5\,\text{h}$, with the rate of infusion $U = 500\,\text{mg/h}$, the two curves expressing the plasma drug profiles are obtained (Fig. 15): curve 1 with $k_e = 0.231\,\text{h}^{-1}$ and curve 2 with $k_e = 0.1155\,\text{h}^{-1}$.

The following conclusions can be drawn from the Figs 13 and 15:

- The value of the drug concentration at the first peak C_{t_i} shown in Figs 13 and 14 are almost similar, whatever the value of the drug elimination rate constant, k_e. These values being 2.976 mg/L and 3.148 mg/L, respectively, they differ from one another by only 5.7% while the drug elimination rate constant differs by 100%.
- As shown in Fig. 15, the plasma drug profiles differ significantly for these two patients from the first peak of the drug concentration to the following peaks. This difference results from the fact that the drug elimination rate constant plays a major role during the elimination stage, with an obvious conclusion: the

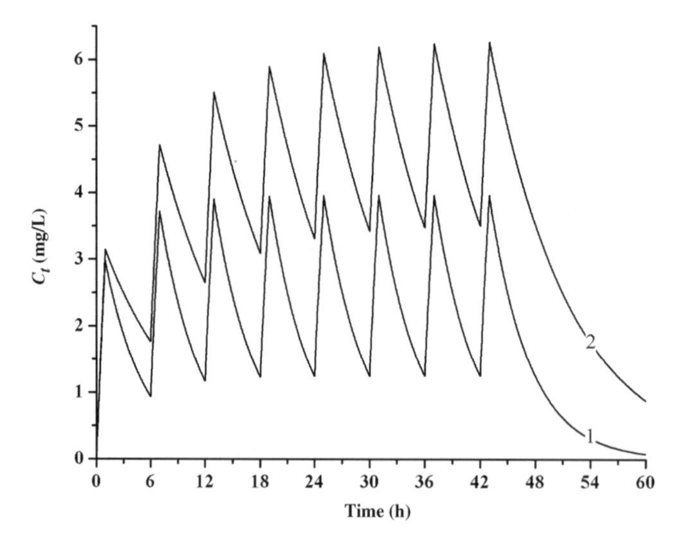

Figure 15. Plasma drug profiles obtained with repeated short infusion–elimination stages, for two patients with extreme values of the drug elimination constant. Curve 1 with $k_e = 0.231\,h^{-1}$ and $t_{0.5} = 3\,h$; curve 2 with $k_e = 0.115\,h^{-1}$ and $t_{0.5} = 6\,h$; $V_p = 150\,L$ in both cases. Infusion over 1 h period with $U = 500$ mg/h, followed by elimination over 5 h.

higher the drug elimination rate constant, the lower the value of the drug concentration.

- The so-called steady state, meaning that the drug concentration profiles are reproduced in a similar manner for the following stages, is attained within a much shorter period of time when the drug elimination rate constant is higher (curve 1).

Adaptation of Therapy to the Intervariability Between Patients At the end of the first infusion-elimination stage, that is, at time, T, depending on the critical pharmacokinetic parameters of the patient determined through Equations 75 and 77, various parameters can be varied in order to adapt therapy to the particular patient, such as the dosage and rate of drug delivery U, and time t_i over which it is delivered, as well as time T characterizing the frequency of repeated doses. The effects of these two parameters are considered successively.

Change in the Dosage As shown in Fig. 16, as soon as the first trough is reached, the dosage in the second infusion is varied by 20% with respect to that used for the first infusion. The three curves appearing in this figure may lead to some interesting conclusions:

- When the plasma volume measured during the first elimination stage is found to be different from that, which was assumed at the beginning of therapy—which determined the dose used in the first infusion—the dose must be varied during the following infusion stages.
- In Fig. 16, the dose is varied by 20% from the initial dose at the second short infusion. As a result of this change in therapy, the values of the drug concentration also varied.
- The so-called steady state is not attained exactly at the same time points for the three drug profiles: at the fifth peak in curve 1 when the dose is not varied, at the sixth peak in curve 2 when the dose is increased, and at the fourth peak (curve 3) when the dose is decreased. Thus, an increase in the dose is responsible for delaying the attainment of steady state.

Change in the Dosage and the Drug Elimination Period of Time Hours are frequently used for unit time, but some interest could be found in

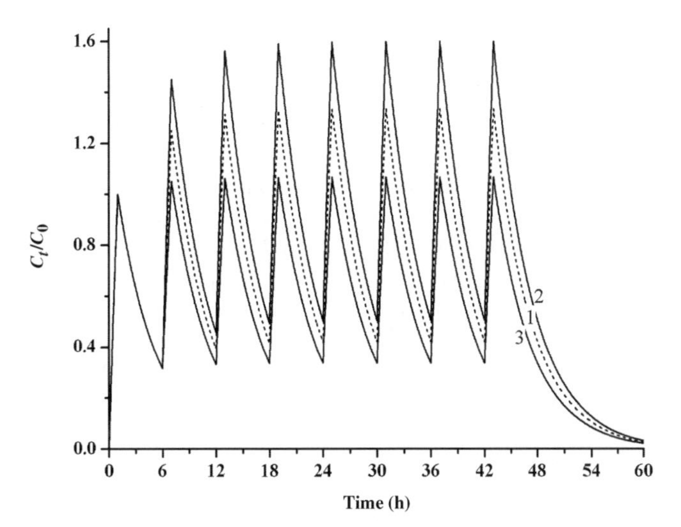

Figure 16. Plasma drug profiles with repeated short infusions delivered every 6 h. After the first infusion–elimination stage, the rate of infusion is varied by 20% (curves 2 and 3) or is constant at 500 mg/h over 1 h. $k_e = 0.231\ h^{-1}$ and $t_{0.5} = 3$ h; $V_p = 150$ L (curve 1). Dimensionless number is used for the ordinate, C_0 being the drug concentration at the first peak.

expressing time in terms of the drug half-life, $t_{0.5}$, in the patient's body. In the same manner, the unit for plasma drug concentrations is that of the value attained at the first peak. Two dimensionless numbers are obtained: C_t/C_0 for the dose and $\theta = t/t_{0.5}$ for the elimination time. The curves in Fig. 17 are built using these dimensionless numbers either for the drug concentration or for the time during the various stages of drug elimination. However, for the infusion stage, the time is expressed in hours, and in this Fig. 17 it is 1 h.

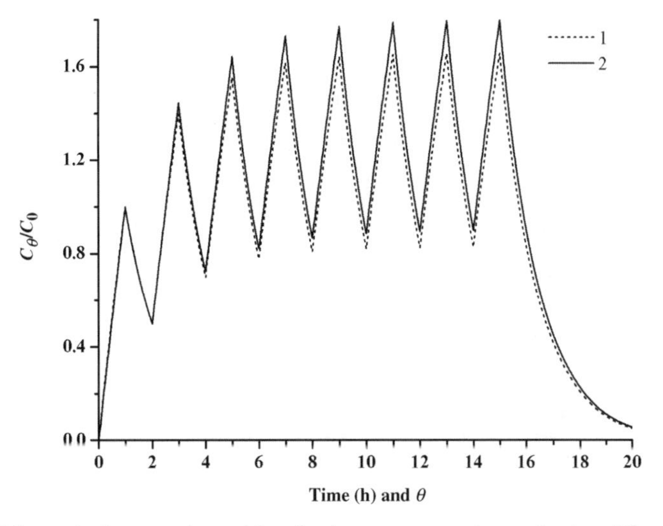

Figure 17. Use of dimensionless numbers either for drug concentration or for time. Plasma drug level with repeated short infusion–elimination, for two patients: curve 1 ($k_e = 0.231\ h^{-1}$ and $t_{0.5} = 3$ h); curve 2 ($k_e = 0.115\ h^{-1}$ and $t_{0.5} = 6$ h); $V_p = 150$ L for both. Infusion over 1 h period with $U = 500$ mg/h. $\theta = t_{0.5}$ for each patient, that is, $t_{0.5} = 3$ h (curve 1); $t_{0.5} = 6$ h (curve 2). Dimensionless number is used for the ordinate, C_0 being the drug concentration at the first peak.

The plasma drug profiles are given for the two patients with the values of the drug elimination rate constant shown in Table 1, that is, $k_e = 0.231\,h^{-1}$ (curve 1) and $k_e = 0.1155\,h^{-1}$ (curve 2).

The following conclusions can be drawn:

- As already noted, when the values of the different drug elimination rate constant, k_e, differ from one another by approximately 80%, the two plasma drug profiles are nearly similar, when the rate of infusion is 500 mg/L over a 1 h period.
- Let us recall that the concentrations, C_0, attained at the first peak are 2.976 mg/L (curve 1) and 3.148 mg/L (curve 2), respectively.

Change in the Infusion Rate Not only can the dose be varied as shown in Fig. 16 but also the rate of drug delivery during the infusion stage. The dose being kept constant at 500 mg, the infusion rate is 500 mg/h over a 1 h period in Fig. 17, 1000 mg/h over a 0.5 h period in Fig. 18, and 250 mg/h over a 2 h period in Fig. 19. Comparison between Figs 17–19 leads to a few interesting observations:

- It can be seen that the lower the rate of infusion, the lower the plasma drug level, as shown in Figs 17–19, whatever the value of the drug elimination rate constant, k_e.
- The plasma drug profiles are nearly similar when the infusion rate is equal to or larger than 500 mg/h, a significant difference appears between the curves 1 and 2 when it is lower (Fig. 19). This is due to the longer time of infusion in Fig. 19. In fact, when the time is expressed in hours, the plasma drug profiles associated with the infusion differ notably during the infusion process for patients with different rates of drug elimination.
- The infusion time is expressed in hours, while dimensionless time is used for the drug elimination stage. Thus, in Figs 17–19, the drug elimination time is 3 h for curve 1, and 6 h for curve 2.
- The drug concentration at the first peak is 3.148 mg/L when $k_e = 0.1155\,h^{-1}$, and 2.976 mg/L when $k_e = 0.231\,h^{-1}$.
- Of course, the value of AUC expressed in (mg/L)h decreases when the drug elimination rate is increased. When steady

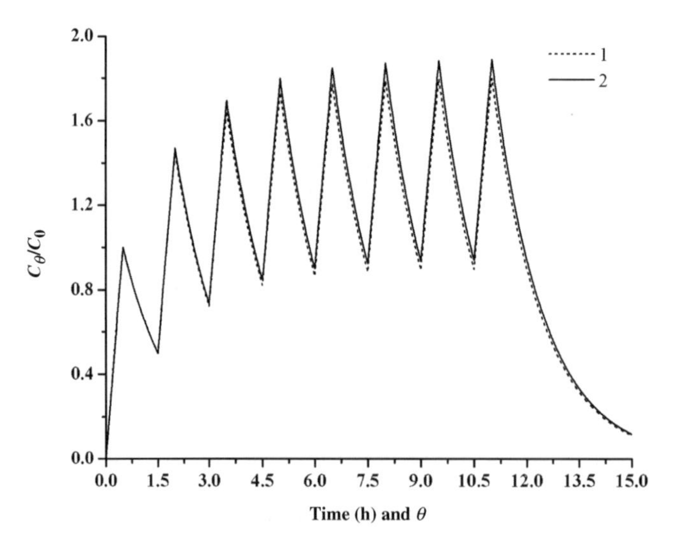

Figure 18. Dimensionless numbers either for drug concentration or for time. Plasma drug level with repeated short infusion–elimination, for two patients: curve 1 ($k_e = 0.231\,h^{-1}$ and $t_{0.5} = 3\,h$); curve 2 ($k_e = 0.115\,h^{-1}$ and $t_{0.5} = 6\,h$); $V_p = 150\,L$ for both. Infusion over half-hour period with $U = 1000$ mg/h. $\theta = t_{0.5}$ for each patient, that is, $t_{0.5} = 3\,h$ (curve 1); $t_{0.5} = 6\,h$ (curve 2). Dimensionless number is used for the ordinate, C_0 being the drug concentration at the first peak.

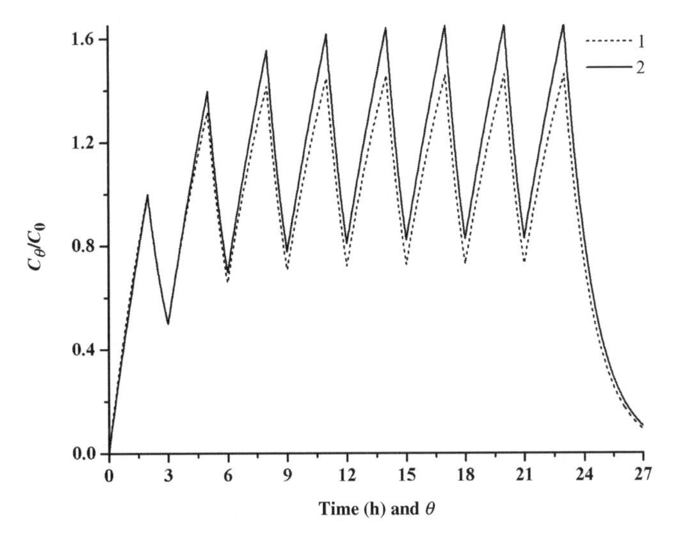

Figure 19. Dimensionless numbers either for drug concentration or for time. Plasma drug level with repeated short infusion–elimination, for two patients: curve 1 ($k_e = 0.231\,h^{-1}$ and $t_{0.5} = 3\,h$); curve 2 ($k_e = 0.115\,h^{-1}$ and $t_{0.5} = 6\,h$); $V_p = 150\,L$ for both. Infusion over 2 h period with $U = 250$ mg/h. $\theta = t_{0.5}$ for each patient, that is, $t_{0.5} = 3$ h (curve 1); $t_{0.5} = 6$ h (curve 2). Dimensionless number is used for the ordinate, C_0 being the drug concentration at the first peak.

state is attained, the value of AUC is inversely proportional to the value of the drug elimination rate.

- For the same dose injected, the time of infusion plays an important role, as shown in Figs 17–19. At the beginning of therapy, AUC is larger when the time of infusion is longer. Then the values of AUC for each stage of infusion–elimination increase slightly with the number of stages. When steady state is attained, the values of AUC for each stage of infusion–elimination are similar, whatever the rate of infusion.

3.5. Conclusions on the Methods of Intravenous Drug Delivery

A few conclusions are worth noting at the end of Section 3.

- The first is concerned with the master curves built using dimensionless numbers. The drug concentration in the blood at time t is expressed as a fraction of the initial concentration at time 0. The time, generally expressed in hours, is thus defined by considering the half-life time of the patient–drug couple as unit time. The advantage of these master curves is that there are general, of use whatever the patient and the drug.
- Great attention has been given to the intervariability of the patient. Some therapists are thinking that if the sensitivity of the patients to Vioxx had been taken into account, by taking care of the side effects, this drug would perhaps not have known this bad issue.
- In a few words, the therapy based on the intravenous drug delivery at the hospital is not only desirable but also easily possible.

4. ORAL DOSAGE FORMS WITH IMMEDIATE RELEASE

Nomenclature

a amount of drug initially in the dosage form (for calculation in Section 4.1.2);

AUC area under the curve expressing the drug concentration versus time;

AUC_{0-t} area under the curve from the time at which the dose is taken up to time t;

C_t free drug concentration in the blood at time t;

C_∞ drug concentration if all the drug in the dosage form were instantaneously in the blood;

GIT gastrointestinal tract;

i.v. intravenous;

k_a rate constant of absorption;

k_e rate constant of elimination;

M_t amount of drug in various places at time t (in Fig. 21);

M_{in} amount of drug initially in the dosage form (in Fig. 21);

t, t_{Max} time, time at which the peak of concentration occurs;

T interval of time between doses;

$t_{0.5}$ half-life time of the drug obtained with bolus i.v.;

$T_{0.5}$ half-life time of the drug obtained with oral dosage form;

V_p apparent volume of distribution;

w amount of drug eliminated out of the blood;

y, z amount of drug remaining along the GIT, lying in the blood, respectively.

Dimensionless Numbers

C_{Max}/C_∞ drug concentration at the peak in the blood as a fraction of C_∞;

M_t/M_{in} amount of drug in various places as a fraction of the initial amount in the dosage form;

$\theta = t/t_{0.5}$ dimensionless time, $t_{0.5}$ being the unit time;

$\theta' = t/T_{0.5}$ dimensionless time, $T_{0.5}$ being the unit time.

4.1. Administration of Oral Single Dose

4.1.1. Principle of the Drug Transport The scheme of drug transport is drawn in Fig. 20 showing the path followed by the drug from absorption into to excretion out of the organism with the successive stages:

- Liberation of the drug from the dosage form along the GIT, either in the stomach or in the intestine when a special coating protects the dosage form and the drug from the aggression of the acid gastric liquid. The supply form with its excipient exhibits a fast rate of dissolution, so as to provoke an immediate release of the drug.
- Diffusion of the drug through the GI membrane into the liver and plasma compartment. In the liver, a first-pass hepatic may take place, giving active or inactive metabolites, this metabolism reaction following a first-order kinetics in case of linear pharmacokinetics. A main difference between the intravenous and oral delivery appears in that place, as with oral delivery the drug passes through the liver where it can be metabolized.
- The stage of absorption, controlled by diffusion through the GI membrane, and expressed in terms of rate transport by a first-order kinetics with the rate constant of absorption k_a.
- After bounding to the plasmatic proteins, the stage of distribution into tissues of the free drug leading to the therapeutic action.
- Finally, the stage of clearance and elimination, expressed by a first-order

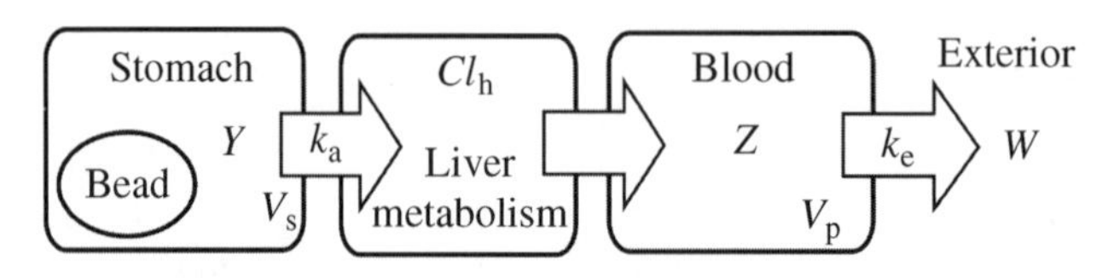

Figure 20. Path followed by the drug when orally taken.

kinetics with the rate constant of elimination k_e.

Hence, three pharmacokinetic parameters are necessary to describe the process: the rate constant of absorption, k_a, the apparent volume of distribution, V_p, taking into account the amount of drug bounding to the plasmatic proteins, and the rate constant of elimination, k_e. Thus, the fundamental differences between the intravenous and oral delivery of the drug lies in the presence of the diffusion of the drug through the GI membrane and the passage through the liver. The rate constant of absorption being larger than the rate constant of elimination, an accumulation of drug occurs in the blood, provoking the formation of a peak of concentration, as shown in Fig. 21, where the amount of drug as a fraction of the initial amount of drug located in the dosage form is expressed in terms of time in the following parts: along the GIT (curve 1) a strong decrease results from the instantaneous dissolution and fast rate of absorption; an accumulation in the blood compartment with a peak of concentration (curve 2); the amount of drug eliminated (curve 3). These three curves are obtained by calculation using ciprofloxacin as the drug.

4.1.2. Calculation of the Profiles of Drug Concentration At time t, y being the amount of drug that has left the GIT, z the amount remaining in the blood compartment, and w the drug eliminated, the three basic equations are written for the successive two stages:

	$Y \rightarrow k_a$	$Z \rightarrow k_e$	W
$t=0$	a	0	0
t	$a-y$	z	w

$$\frac{dy}{dt} = k_a(a-y) \tag{78}$$

$$y = z + w \tag{79}$$

$$\frac{dw}{dt} = k_e \cdot z = k_e(y-w) \tag{80}$$

Equation 78 expressing the process of absorption can be resolved independently as

$$y = a \cdot [1-\exp(-k_a \cdot t)] \tag{81}$$

while Equation 80, representing the amount of drug eliminated, becomes

$$\frac{dw}{dt} + k_e \cdot w = a \cdot k_e \cdot [1-\exp(-k_a \cdot t)] \tag{82}$$

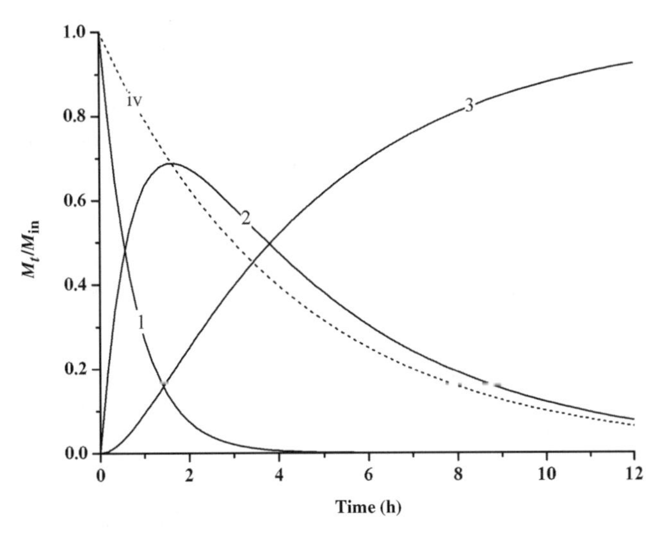

Figure 21. Kinetics of the drug in various compartments: (1) Amount of drug along the GIT. (2) Amount of drug in the blood. (3) Amount of drug eliminated. Ciprofloxacin with $k_e = 0.231\ h^{-1}$ ($t_{0.5} = 3$ h) and $k_a = 1.3\ h^{-1}$.

The treatment of Equation 82 leads to the expression of the amount of drug eliminated w:

$$\frac{w}{a}=1+\frac{1}{k_a-k_e}[k_e\cdot\exp(-k_a\cdot t)-k_a\cdot\exp(-k_e\cdot t)] \quad (83)$$

By replacing w and y in Equation 79, the amount of drug in the blood z is given by

$$\frac{z}{a}=\frac{k_a}{k_a-k_e}[\exp(-k_e\cdot t)-\exp(-k_a\cdot t)] \quad (84)$$

The characteristics of the peak in the blood are obtained by differentiating the Equation 84 with respect to time, and writing that this new expression is equal to 0, the slope of the curve at the peak being flat. The time at which the peak is attained, t_{Max}, and the amount of drug present in the blood at that time, z_{Max}, are given by the two following relations:

$$t_{\mathrm{Max}}=\frac{1}{k_a-k_e}\ln\frac{k_a}{k_e} \quad (85)$$

$$\frac{z_{\mathrm{Max}}}{a}=\left[\frac{k_e}{k_a}\right]^{\frac{k_e}{k_a-k_e}} \quad (86)$$

The amount of the free drug in the blood is expressed as a fraction of the amount of drug initially in the oral dosage form. The plasma drug concentration, and its value at the peak, is obtained by dividing the amount by the apparent plasmatic volume

$$C=\frac{z}{V_p} \text{ and } C_{\mathrm{Max}}=\frac{z_{\mathrm{Max}}}{V_p} \quad (87)$$

The three stages of the overall process of drug delivery and transport through the body can be distinctly comprehended in Fig. 21. The stage of absorption is completely ended at less than 4 h (curve 1). The stage of elimination starts after a short while less than 1 h (curve 3).

The amount of drug in the blood as a fraction of the amount initially in the dosage form varies as a function of time (curve 2), represents also the concentration of the free drug in the blood, as the pharmacokinetics is linear and the Equation 87 holds. With the values of the rate constant of absorption and elimination selected for calculation with ciprofloxacin, the peak is attained at 2 h with a maximum concentration C_{Max}/C_0 equal to 0.69, C_0 representing the concentration that would be obtained at time 0 with the bolus intravenous administration of a dose equal to the amount of drug, a, initially located in the dosage form. Thus 1.6 h after the dosage form has been taken, the drug concentration at the peak is 0.688 times the concentration that would be attained at time 0, C_0 (Section 3), with the bolus i.v. delivered with the same amount of drug.

It is also worth noting that in Fig. 21 the curves 1 and 2, as well as 2 and 3, intersect at the ordinate 0.5. The reason for the intercept of the curves 1 and 2 is that the process of elimination does not interfere during the stage of absorption, and that during the stage of elimination the process of absorption has come to the end for the curves 2 and 3.

4.1.3. Effect of the Pharmacokinetic Parameters on the Plasma Drug Profile

The apparent volume of distribution plays a role only in decreasing the concentration of the free drug in the blood; thus this parameter does not interfere in the ordinates of the various figures when the drug concentration in the blood is expressed as a fraction of the drug concentration C_∞. This value C_∞ is the drug concentration that would be in the blood if the total amount of drug initially in the dosage form were instantaneously absorbed in the blood under the same conditions. Thus, C_∞ is the same as that written C_0 in case of bolus i.v. administration (Section 3.1), by taking into account the effect of the first-pass hepatic.

The rate constants of absorption and elimination act upon the rate of drug transport, and thus their effect can be seen not only along the time abscissa but also on the ordinate and especially for the value of the peak concentration.

Effect of the Rate Constant of Absorption on the Plasma Drug Profile The effect of the value of the rate constant of absorption k_a is precisely

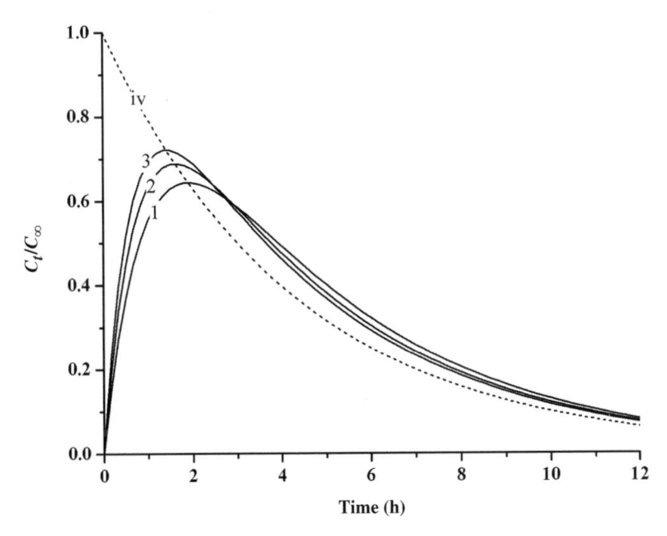

Figure 22. Effect of the value of the rate constant of absorption on the profiles of concentration of the drug (ciprofloxacin) in the blood as a function of time. $k_e = 0.231\,h^{-1}$; $k_a = 1\,h^{-1}$ (1); $k_a = 1.3\,h^{-1}$ (2); $k_a = 1.6\,h^{-1}$ (3).

determined in Fig. 22 where the ratio of the drug concentration C/C_∞ is plotted against time, when the rate constant of absorption is varied between $1\,h^{-1}$ and $1.6\,h^{-1}$, while the rate constant of elimination k_e is kept constant at $0.231\,h^{-1}$. The following three facts are worth noting:

- During the stage of absorption, the abrupt rise in the kinetics curve is sharper with the higher value of the rate constant of absorption (curve 3).
- The higher peak is reached at a shorter time when the rate constant of absorption is higher, as shown in curve 3.
- The curve obtained with the lower rate constant of absorption during the stage of elimination, exhibits a longer value for the half-life time, the difference between the extreme values of this half-life time being less than 0.5 h.

Effect of the Rate Constant of Elimination on the Plasma Drug Profile As shown in Table 6, the values of the rate constant of elimination largely vary from one patient to another [10–13,15,16,22]. Calculation made by considering the two extreme and the median values of k_e for ciprofloxacin lead to the profiles of the relative concentration C/C_∞ of the drug in the blood drawn in Fig. 23 ($k_a = 1.3\,h^{-1}$ and the rate constant of elimination k_e is varied from 0.1 to $0.3\,h^{-1}$). These curves lead to the following conclusions:

- The three curves are well superimposed during the stage of absorption, before diverging approximately 0.5 h after the uptake.
- The characteristics of the three peaks are quite different, either for the time t_{Max} or for the concentration $\frac{C_{Max}}{C_\infty}$. The lower peak at 0.64 is attained 1.47 h after the uptake,

Table 6. Pharmacokinetics Parameters of the Four Drugs

Drug	Half-Life Time (h)	k_e (h^{-1})	Reference(s)
Ciprofloxacin	7.7–3.15	0.09–0.22	[10–13,15,16,22]
Acetylsalicylic acid	0.25–0.33	2.77–2.1	[18]
Ofloxacin	4.4–6.4	0.16–0.11	[17]
Péfloxacin	7.5–12.5	0.06–0.09	[16]

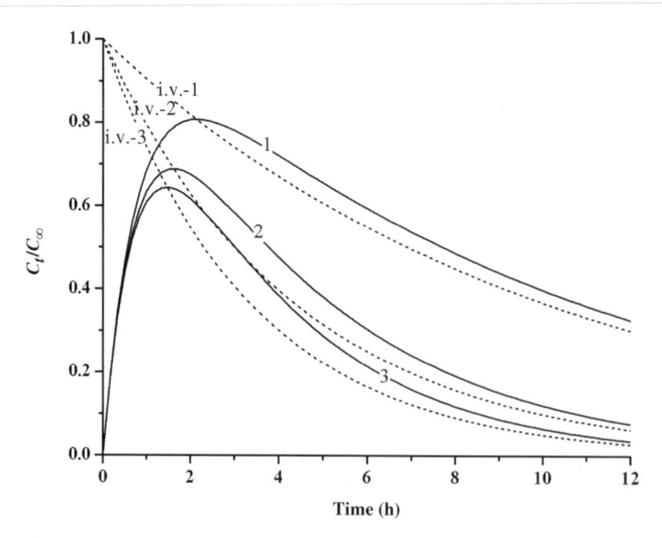

Figure 23. Effect of the value of the rate constant of elimination on the profiles of concentration of the drug (ciprofloxacin) in the blood as a function of time. $k_a = 1.3\,h^{-1}$; $k_e = 0.1\,h^{-1}$ (1); $k_e = 0.231\,h^{-1}$ (2); $k_e = 0.3\,h^{-1}$ (3).

for the higher value of the rate constant of elimination, while the higher peak at 0.82 is reached after 2 h for the lower value of this rate constant.

- The half-life time $T_{0.5}$ of the drug orally administered with immediate release, taken as the time necessary for the concentration to attain half the value of the peak, is of great concern.
- Thus the so-called half-life time of the drug $T_{0.5}$, which is quite different from the half-life time $t_{0.5}$, obtained with these three values of the rate constant of elimination varies from 10 h (curve 1 with $k_e = 0.3\,h^{-1}$) to 4.7 h (curve 3) with $k_e = 0.1\,h^{-1}$.

4.1.4. Effect of the Nature of the Drug

In this study concerning the nature of the drug, the effects of the rate constants of absorption and elimination being combined, they are thus considered simultaneously together.

The profiles of concentration of the free drug in the blood are drawn in Fig. 24 as they are calculated for various drugs: ciprofloxacin (1), acetylsalicylic acid (2), ofloxacin (3) and Péfloxacin (4). Some pharmacokinetics parameters are listed in Table 6.

As shown in Table 6, the values of the pharmacokinetics parameters are scattered within a wide range. This fact results from two reasons: the one is that the patients behave differently toward a drug, and the other that ill patients and healthy volunteers respond to the drug in a different way. Among the values presented in Table 6, some values for the half-life time are selected and used for calculation. These values are shown in Table 7 as well as the main characteristics obtained for the profiles drawn in Fig. 24, such as the time at which the peak of concentration is attained and the maximum value reached for this concentration, as well as the half-life time obtained with this dosage form which can be compared with the corresponding half-life time determined through bolus i.v. administration.

4.2. Multiple Doses Administration of Oral Dosage Forms with Immediate Release

Oral doses administered far apart in time behave independently. But as a minimum concentration is needed to maintain efficacy and a maximum concentration should not be exceeded to prevent side effects, the desired profile of concentration should be attained with repeated doses, especially for a long duration of therapy. Hence, oral dosage forms have to be taken sufficiently close together so that the following doses are administered in the

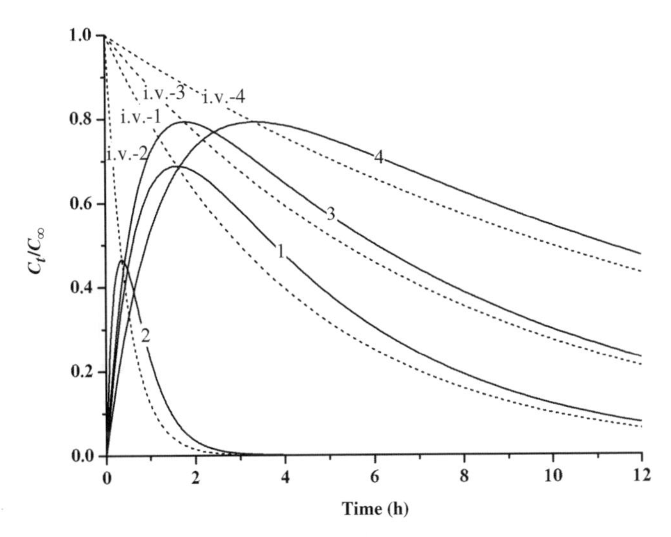

Figure 24. Effect of the nature of the drug on the profile of concentration in the blood as a function of time. Ciprofloxacin (1) with $k_a = 1.3/h$ and $k_e = 0.231\,h^{-1}$; acetylsalicylic acid (2) with $k_a = 3.5/h$ and $k_e = 2.1/h$; ofloxacin (3) with $k_a = 1.5\,h^{-1}$ and $k_e = 0.13\,h^{-1}$; péfloxacin (4) with $k_a = 0.8\,h^{-1}$ and $k_e = 0.07\,h^{-1}$.

blood prior the full elimination of the preceding dose. This system of drug delivery gives way to some accumulation of drug in the blood over a prolonged period of time and the drug concentration alternates between peaks and troughs. Good knowledge of this process is of help in optimizing the treatment so as to make the side effects smoothed out. Comparison is made between oral dosage forms taken at regular interval and repeated bolus intravenous injections. The ultimate objective is to find the right interval of time for which the quality of the oral treatment may be considered equal to that obtained with i.v. injections.

4.2.1. Principle for Repeated Oral Doses and Parameters The pharmacokinetics for the drug when orally taken is governed by three parameters: the rate constants of absorption k_a and of elimination k_e, the apparent volume of distribution V_p. Moreover, in the multiple dose administration, the interval of time between the successive doses becomes a parameter of great concern.

The plasma drug profiles are calculated by taking ciprofloxacin as the drug, and considering three values of the pharmacokinetic parameters, with two different values of the interval of time T between every dose: 6 h (Fig. 25) and 12 h (Fig. 26). These profiles alternate between peaks and troughs. The values for the peaks and troughs are listed in Table 8.

From the three profiles of drug concentration depicted in Figs 25 and 26, as well as from the data gathered in Table 8, the following conclusions can be drawn:

- The interval of time T between successive oral doses plays an important role, acting upon the values of the peaks and troughs, as well as on the comfort of the patient.

Table 7. Characteristics of the Profiles Obtained with the Four Drugs

Drug	k_e (h)	k_a (h)	t_{max} (h)	$\frac{C_{Max}}{C_\infty}$	$T_{0.5}$ (h)	$t_{0.5}$ (h)
Ciprofloxacin	0.231	1.3	1.62	0.69	5.5	3
Acetylsalicylic acid	2.1	3.5	0.367	0.465	1	0.33
Ofloxacin	0.13	1.5	1.78	0.793	7.8	5.33
Péfloxacin	0.07	0.8	3.34	0.791	14	9.9

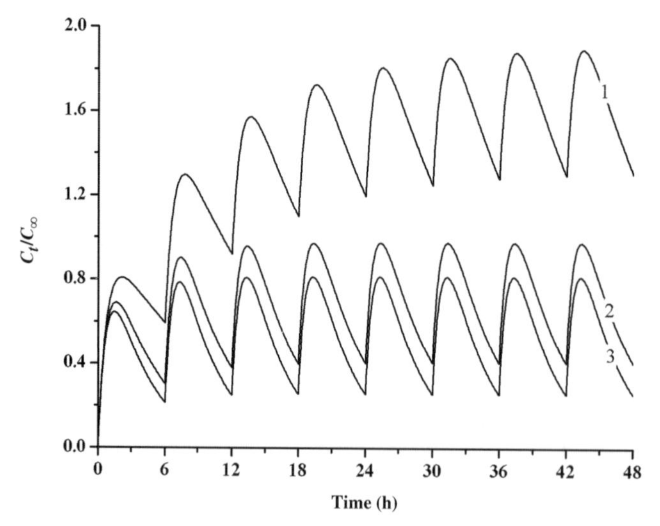

Figure 25. Profile of drug (ciprofloxacin) concentration with repeated doses, with the interval of time $T = 6\,\text{h}$; $k_a = 1.3/h$ and $k_e = 0.1\,\text{h}^{-1}$ (1); $k_e = 0.231\,\text{h}^{-1}$ (2); $k_e = 0.3\,\text{h}^{-1}$ (3).

- The time elapsed between two successive peaks slightly increases from the second dose to the following to reach finally the value selected for the interval of time between each dose. Thus, the time between two successive peaks is equal to the interval of time, after six doses for $T = 6\,\text{h}$, but after three doses for $T = 12\,\text{h}$.
- The values of the drug concentration at the peaks and troughs increase from the first to the nth dose up to constant values, when the so-called steady state is attained, the steady state meaning that the same profile of concentration is reproduced for the successive doses.
- The increase in the peaks and troughs depends on the value of the interval of time between the successive doses. As shown in Table 8, when the interval of time T is 6 h, under steady state the con-

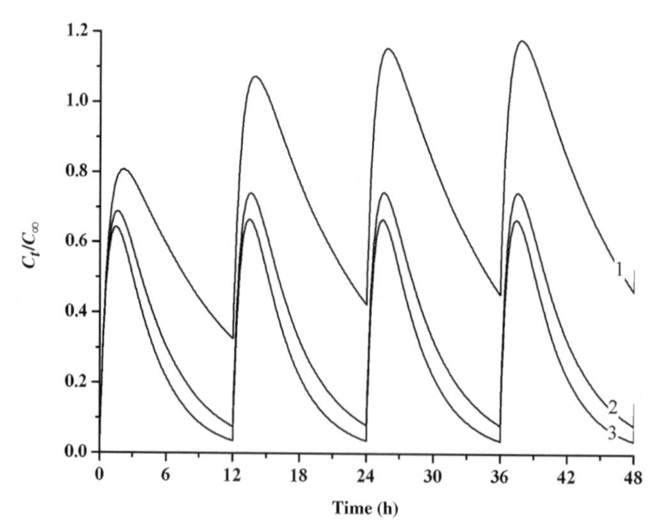

Figure 26. Profile of drug (ciprofloxacin) concentration with repeated doses, with the interval of time $T = 12\,\text{h}$; $k_a = 1.3/h$ and $k_e = 0.1\,\text{h}^{-1}$ (1); $k_e = 0.231\,\text{h}^{-1}$ (2); $k_e = 0.3\,\text{h}^{-1}$ (3).

Table 8. Effect of T on the Values of the Peaks and Troughs for Ciprofloxacin

$k_a = 1.3/h$ (1) $k_e = 0.1\,h^{-1}$; $t_{0.5} = 6.93\,h$ (2) $k_e = 0.231\,h^{-1}$; $t_{0.5} = 3\,h$ (3) $k_e = 0.3\,h^{-1}$; $t_{0.5} = 2.31\,h$

T(h)/Dose		1	2	3	4	6	7	8
6	Peak 1	0.8	1.29	1.57	1.70	1.84	1.87	1.88
	2	0.69	0.89	0.96	0.97			
	3	0.64	0.77	0.8	0.81			
	Trough 2	0.30	0.39	0.41	0.41			
12	Peak 1	0.81	1.07	1.15	1.17			
	2	0.68	0.74	0.74				
	3	0.64	0.66					
	Trough 2	0.08	0.08					

centrations at peaks and troughs are around 2.3 times the corresponding value for the first dose; this ratio is only 1.45 when the interval of time T is 12 h.

- The effect of the inter variability of the patients clearly appears with the values at the peaks and troughs, leading to a difficult conclusion for adapting the therapy.
- In terms of conclusion, the interval of time between two successive doses is of concern not only to facilitate the patient's comfort but also to obtain a smoothly steady state profile located between the effective and the lethal levels. There is also another way to adapt the profiles of concentration (within the frame between the ED and the LD levels associated with the therapeutic index), by evaluating the effect of the dose itself.

4.2.2. Effect of the Nature of the Drug in Repeated Doses It is of great interest to consider the effect of the nature of the drug taken in repeated doses, in the same way as for single dose. The drugs selected are the same as those studied with a single dose, for example, ciprofloxacin, acetylsalicylic acid, ofloxacin, and péfloxacin, whose pharmacokinetic parameters (rate constant of absorption and elimination) are shown in Table 7. Thus, comparison between the profiles of concentration obtained with a single dose (Fig. 24) is possible with the corresponding ones obtained in multidoses taken at the interval of time of 6 h (Fig. 27) and of 12 h (Fig. 28). Of course, in the case of repeated doses, the drug concentration alternates between high peaks and low troughs. The particular values measured for the successive peaks and troughs from the first dose up to the so-called steady state are collected in Table 9 for the drugs with $T = 6$ or 12 h. The drugs are noted: ciprofloxacin (1), acetyl salicylic acid (2), ofloxacin (3), and péfloxacin (4).

In Table 9, the concentrations at peaks C_{Max}/C_{∞} and troughs C_{Min}/C_{∞}, are noted as Max and Min, C_{∞} being the value attained if all the drug was delivered instantaneously (C_0 through i.v.).

From the plasma drug profiles drawn in Figs 27 and 28, as well as the data collected in Table 9, the following remarks are worth noting:

- Of course, as shown in Figs 25 and 26 with ciprofloxacin alone, the profiles of concentration for the various drugs alternate between high peaks and low troughs.
- As already discussed (Section 4.2.1), the time elapsed between two successive peaks is slightly lower than the interval of time T at the beginning of the process with multidoses. This time increases from dose to dose up to the interval of time T, which is attained after three doses for ciprofloxacin.
- For $T = 6$ h, the values of the concentrations at peaks and at troughs increase from dose to dose up to the fifth dose where the steady state is attained for ciprofloxacin, and for ofloxacin, and at the eighth dose at least for péfloxacin.

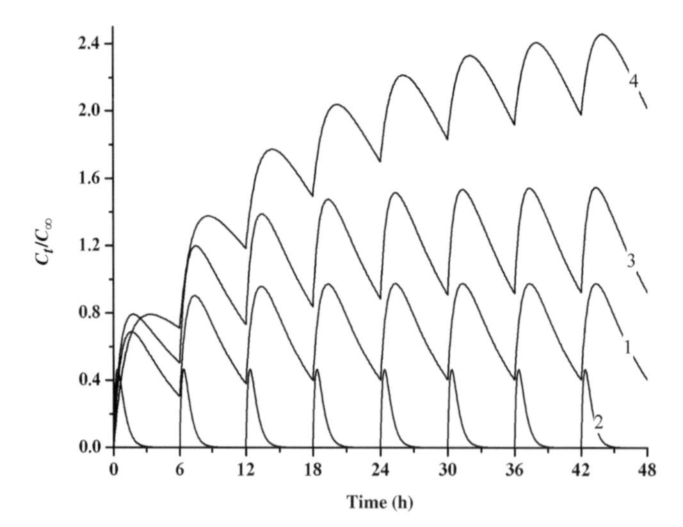

Figure 27. Profiles of drug concentrations with repeated doses for various drugs taken at the interval of time $T = 6$ h; Ciprofloxacin (1) with $k_a = 1.3/h$ and $k_e = 0.231\,h^{-1}$; acetylsalicylic acid (2) with $k_a = 3.5/h$ and $k_e = 2.1/h$; ofloxacin (3) with $k_a = 1.5\,h^{-1}$ and $k_e = 0.13\,h^{-1}$; péfloxacin (4) with $k_a = 0.8\,h^{-1}$ and $k_e = 0.07\,h^{-1}$.

- In case of acetylsalicylic acid, there is no accumulation of drug as the interval of time T between two doses is much larger than the half-life time of the drug.
- The maximum concentration at peaks for a drug orally taken is lower than that attained with i.v. injections administered in multidoses, this fact being without considering the loss in concentration resulting from the first-pass hepatic. The reasons are for a large part that the process of absorption with oral dosage forms through the GIT membrane is not instantaneous in contrast with the bolus intravenous injection, and for a minor proportion that the process of elimination takes place already during the stage of absorption.

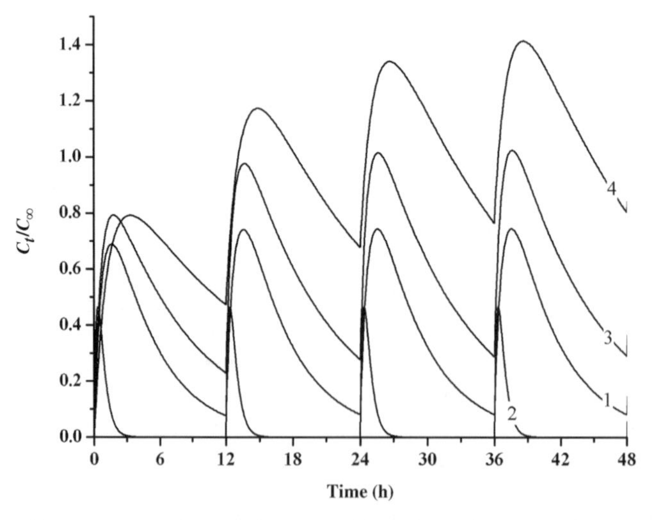

Figure 28. Profiles of drug concentrations with repeated doses for various drugs taken at the interval of time $T = 12$ h; Ciprofloxacin (1) with $k_a = 1.3/h$ and $k_e = 0.231\,h^{-1}$; acetylsalicylic acid (2) with $k_a = 3.5/h$ and $k_e = 2.1/h$; ofloxacin (3) with $k_a = 1.5\,h^{-1}$ and $k_e = 0.13\,h^{-1}$; péfloxacin (4) with $k_a = 0.8\,h^{-1}$ and $k_e = 0.07\,h^{-1}$.

Table 9. Effect of the Nature of the Drug on Peaks and Troughs Concentrations

T(h)	Drug	Dose	1	2	3	4	5	6
6	1	Max.	0.7	0.9	0.96	0.97	0.98	
		Min.	0.31	0.39	0.4	0.41		
12		Max.	0.7	0.73	0.74			
		Min.	0.08	0.08				
6	2	Max.	0.47	0.47				
		Min.	0					
12		Max.	0.47					
		Min.	0					
6	3	Max.	0.8	1.2	1.38	1.47	1.52	1.53
		Min.	0.49	0.74	0.83	0.89	0.9	
12		Max.	0.8	0.97	1.02	1.02		
		Min.	0.24	0.28	0.29			
6	4	Max.	0.8	1.37	1.77	2.03	2.22	2.32
		Min.	0.72	1.17	1.49	1.7	1.82	1.92
12		Max.	0.8	1.17	1.34	1.4		
		Min.	0.47	0.68	0.76			

- When the steady state is reached with $T = 6\,h$, the peaks are 1.4 times larger than that attained for the first dose with ciprofloxacin, while they are up to 1.9 for ofloxacin. These constant values for acetyl salicylic acid result from the fact that the interval of time T, taken at 6 h, is far much larger than the half-life time of this drug, so that the following dose is delivered when the preceding dose has been eliminated. The value of the interval of time T of 6 or 12 h may be correct for ciprofloxacin and ofloxacin whose half-life time values $T_{0.5}$ are 5.5 and 7.8 h, respectively, while even 12 h is far too short for péfloxacin whose half-life time $T_{0.5}$ is 14 h.
- From these observations, it can be said that the interval of time should be carefully selected in relation to the value of the half-life time of the drug, for a prolonged therapy necessitating a steady profile of concentration alternating between peaks and troughs not too far apart, and totally situated within the optimal therapeutic level.

4.2.3. Master Curves with Dimensionless Number for the Interval of Time θ and θ' The best solution is to build and use master curves based on a dimensionless number expressing the time as a fraction of the half-life time of each drug. Two dimensionless numbers can be used: the one θ taking as unit time the half-life time $t_{0.5}$ obtained with bolus iv injection; the other θ' taking as unit time $T_{0.5}$, the time at which the concentration is half that of the peak obtained with a dose alone. In master curves, the concentration of the drug at time t as a fraction of the concentration associated with the total amount of drug initially in the GIT and the dosage form, is expressed in terms of these dimensionless times.

Master curves are drawn for various values of the rate constant of elimination of ciprofloxacin by taking for the dimensionless time either θ (Fig. 29) or θ' (Fig. 30).

Other master curves are drawn for the four drugs: ciprofloxacin (1), acetylsalicylic acid (2), ofloxacin (3), and péfloxacin (4) using either θ (Fig. 31) or θ' (Fig. 32).

The effect of the value given to the dimensionless time is also considered: for ciprofloxacin in Fig. 29 using $\theta = 1$ and Fig. 33 with $\theta = 2$; for the four drugs selected (ciprofloxacin (1), acetylsalicylic acid (2), ofloxacin (3), and péfloxacin (4)) in Fig. 31 with $\theta - 1$ or with $\theta' = 1$ (Fig. 32) and with $\theta = 2$ (Fig. 34).

The curves in Figs 29–34 are worth pointing out, leading to the following comments:

- The advantage of these master curves is that all the drugs, whatever their

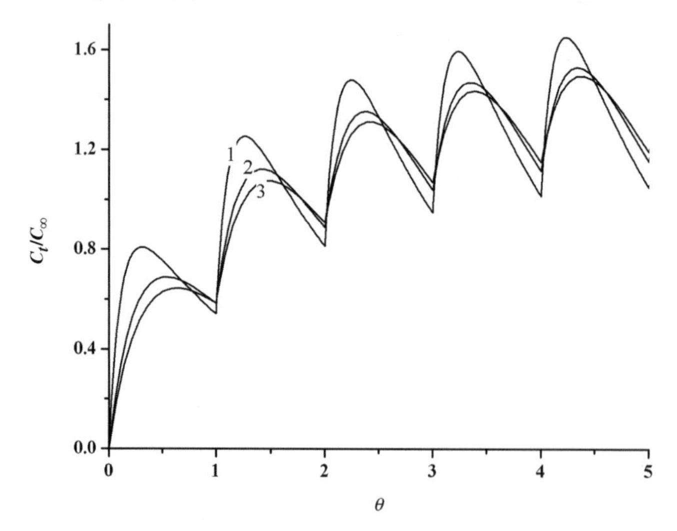

Figure 29. Master curves for the plasma drug profile (ciprofloxacin) with repeated doses, with the interval of time $T(\theta)$; $k_a = 1.3/h$ and $k_e = 0.1\,h^{-1}$ (1); $= 0.231\,h^{-1}$ (2); $= 0.3\,h^{-1}$ (3). $T = t_{0.5}$, the half-life time through bolus i.v. $\theta = 1$.

pharmacokinetic parameters, can be drawn in the same figure, as shown in Figs 29–34.

- The plasma drug profiles obtained with repeated doses slightly depend on the value taken for the time T between successive doses, as shown by comparing the curves in Figs 29 and 30 for ciprofloxacin, or in Figs 31 and 32 for the four drugs.
- The plasma drug profiles look similar for ofloxacin and péfloxacin in Figs 31 and 32 whatever the value taken for the time T between successive doses. The same similarity appears in Fig. 34 with a longer time $\theta = 2$ ($T = 2\ t_{0.5}$).
- There is a particular case with acetylsalicylic acid for which the half-life time is as short as 0.33 h, the interval of time T

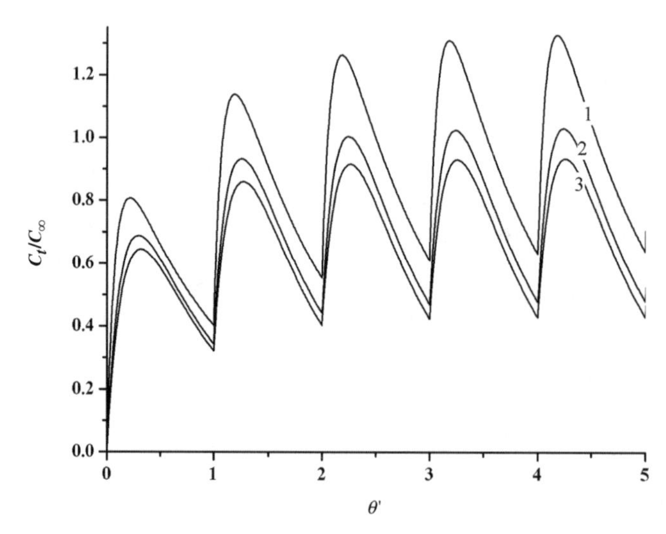

Figure 30. Master curves for the plasma drug profile (ciprofloxacin) with repeated doses, with the interval of time θ; $k_a = 1.3/h$ and $k_e = 0.1\,h^{-1}$ (1); $k_e = 0.231\,h^{-1}$ (2); $k_e = 0.3\,h^{-1}$ (3). $\theta' = 1$. $T = T_{0.5}$, the half-life time with the oral dosage form, measured from Fig 23.

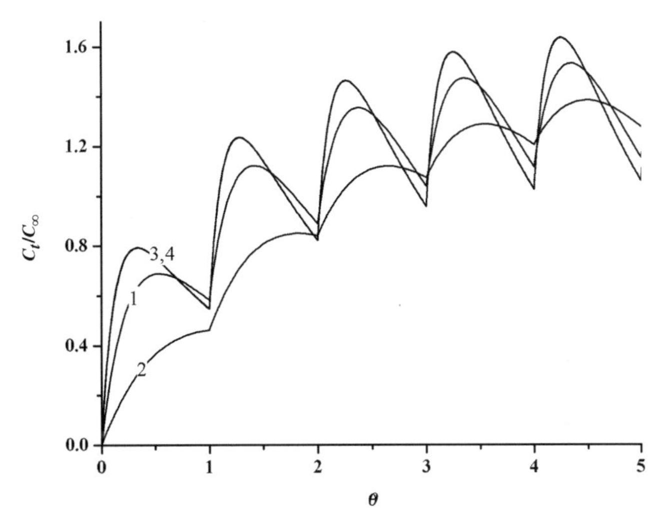

Figure 31. Master curves drawn for the plasma drug profiles, when the doses are taken in succession with the interval of time $T = t_{0.5}$, the half-life time through bolus i.v. $\theta = 1$. Ciprofloxacin (1) with $k_a = 1.3/h$, $k_e = 0.231\,h^{-1}$, $t_{0.5} = 3\,h$; acetylsalicylic acid (2) with $k_a = 3.5/h$, $k_e = 2.1/h$, $t_{0.5} = 0.33\,h$; ofloxacin (3) with $k_a = 1.5\,h^{-1}$, $k_e = 0.13\,h^{-1}$, $t_{0.5} = 5.33\,h$; péfloxacin (4) with $k_a = 0.8\,h^{-1}$, $k_e = 0.07\,h^{-1}$, $t_{0.5} = 9.9\,h$.

being also 0.33 h so as to have $\theta = 1$ for the curve 2 in Fig. 31 and being 0.66 h with $\theta = 2$ in Fig. 34.

- The values of the maximum (peaks) and minimum (troughs) concentration increase from dose to dose, reaching the values found under the steady state. When the period of time is equal to the half-life time ($\theta = 1$), the concentrations at peaks and troughs under steady state are about twice those obtained in the first dose for ciprofloxacin (Fig. 29). With a

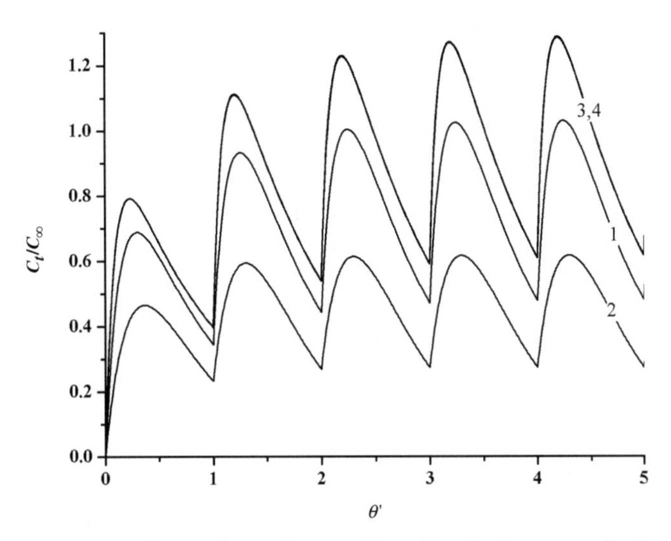

Figure 32. Master curves drawn for the plasma drug profiles, when the doses are taken in succession with the interval of time $T = T_{0.5}$, the half-life time with the oral dosage form, measured from Fig. 24, $\theta' = 1$. Ciprofloxacin (1) with $k_a = 1.3/h$, $k_e = 0.231\,h^{-1}$; acetylsalicylic acid (2) with $k_a = 3.5/h$, $k_e = 2.1/h$; ofloxacin (3) with $k_a = 1.5\,h^{-1}$, $k_e = 0.13\,h^{-1}$; péfloxacin (4) with $k_a = 0.8\,h^{-1}$, $k_e = 0.07\,h^{-1}$.

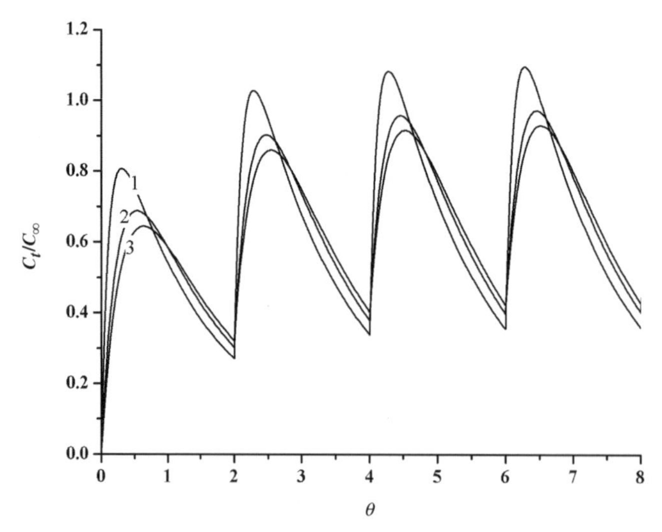

Figure 33. Master curves for the plasma drug profile (ciprofloxacin) with repeated doses, with the interval of time $T(\theta)$; $k_a = 1.3/h$ and $k_e = 0.1\,h^{-1}$ (1); $k_e = 0.231\,h^{-1}$ (2); $k_e = 0.3\,h^{-1}$ (3). $T = 2t_{0.5}$, the half-life time through bolus i.v. $\theta = 2$.

period of time twice the half-life time ($\theta = 2$), the increase in the concentration is less important, and the concentrations at peaks and troughs under steady state are 1.33–1.39 times larger than those obtained in the first dose, for the same drugs (Fig. 33).

- The steady state is attained, with steady values for the peaks and troughs, after a few doses, for example, approximately 5–7 when the interval of time is equal to the half-life time ($\theta = 1$) and approximately 3–4 when the interval of time is twice the half-life time ($\theta = 2$).

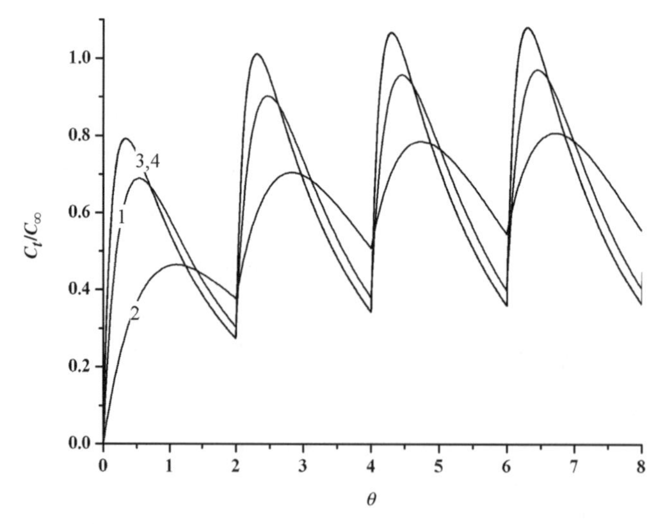

Figure 34. Master curves drawn for the plasma drug profiles, when the doses are taken in succession with the interval of time $T = t_{0.5}$, the half-life time through bolus i.v. $\theta = 2$. Ciprofloxacin (1) with $k_a = 1.3/h$, $k_e = 0.231\,h^{-1}$, $t_{0.5} = 3\,h$; acetylsalicylic acid (2) with $k_a = 3.5/h$, $k_e = 2.1/h$, $t_{0.5} = 0.33\,h$; ofloxacin (3) with $k_a = 1.5\,h^{-1}$, $k_e = 0.13\,h^{-1}$, $t_{0.5} = 5.33\,h$; péfloxacin (4) with $k_a = 0.8\,h^{-1}$, $k_e = 0.07\,h^{-1}$, $t_{0.5} = 9.9\,h$.

4.3. Area Under the Curve

AUC is calculated either with a single dose or with the repeated doses. In this calculation, the effect of the first-pass hepatic is not considered. However, the metabolism following this first-pass hepatic can be taken into account provided that some data were supplied from *in vivo* measurements either with bolus i.v. or with oral dosage form and the same amount of drug delivered.

4.3.1. Area Under the Curve with a Single Dose

The profile of the free drug concentration in the blood defined by Equation 84 is rewritten by using the ratio of the concentrations:

$$\frac{C_t}{C_\infty} = \frac{k_a}{k_a - k_e}[\exp(-k_e \cdot t - \exp(-k_a \cdot t)] \quad (88)$$

C_t being the concentration at time t, and C_∞ the concentration associated with the amount of drug in the dosage form (equal to the concentration C_0 obtained with bolus i.v. by neglecting the effect of the first-pass hepatic).

Integration of the above equation with respect to time, between time 0 and ∞, gives

$$\mathrm{AUC}_{0-\infty} = \int_0^\infty C_t \cdot \mathrm{d}t = \frac{C_\infty \cdot k_a}{k_a - k_e} \left[\frac{1}{k_e}\exp(-k_e \cdot t) - \frac{1}{k_a}\exp(-k_a \cdot t)\right]_\infty^0 \quad (89)$$

which reduces to

$$\mathrm{AUC}_{0-\infty} = \frac{C_\infty}{k_e} = \frac{C_\infty \cdot t_{0.5}}{0.693} \quad (90)$$

This value of the AUC for a single dose orally taken is equal to the value obtained by integrating Equation 22 under the same conditions of time with bolus i.v., This is true, without speaking of the first-pass hepatic, the metabolism reducing the amount of drug available in case of oral delivery. Thus, according to the value of the bioavailability found with ciprofloxacin varying within the 45–77% range, the actual value of the AUC associated with a single oral dose will be 45–77% that obtained with the bolus i.v.

4.3.2. Area Under the Curve with Repeated Doses

By dividing the AUC associated with each dose taken in succession with the same interval of time by the AUC obtained with a single dose far apart from the others, a dimensionless number is obtained:

$$\frac{\mathrm{AUC}_{nT-(n+1)T}}{\mathrm{AUC}_{0-\infty}} = k_e \cdot AUC_{nT-(n+1)T} = \frac{0.693}{t_{0.5}}\mathrm{AUC}_{nT-(n+1)T} \quad (91)$$

with $\mathrm{AUC}_{0-\infty}$ the area under the curve associated with a single dose,

$\mathrm{AUC}_{nT-(n+1)T}$ the area under the curve for the nth dose between the time nT and $(n+1)T$ when these doses are taken in succession with the constant interval of time T, and k_e is the rate constant of elimination of the drug.

The values of this ratio of the area under the curve are shown in Table 10 for four drugs: ciprofloxacin (1); acetylsalicylic acid (2); ofloxacin (3); péfloxacin (4) taken in succession with the interval of time $\theta = 1$ or $\theta = 2$.

The profiles of drug concentration obtained with the doses taken in succession, shown in Figs 31 and 34, as well as the values of the AUC collected in Table 10, enable one to draw some conclusions:

- The concentration increases from dose to dose, exhibiting peaks and troughs as already shown in Table 8 for ciprofloxacin and Table 9 for the four drugs, whatever the interval of time elapsed between each dose uptake. After a number of doses, depending on the value of the interval of time, for example, 5–7 for $\theta = 1$ and 3.4 for $\theta = 2$, the stationary state is attained. The area under the curve associated with each dose between each interval of time increases in the same way up to a constant value when the stationary state is attained.
- Under stationary state conditions, each dose is responsible for an area under the curve equal to that obtained with a single dose taken alone far apart from the others.

Table 10. Values of $AUC_{nT-(n+1)T}$ as a Fraction of $AUC_{0-\infty}$ for Each Drug

Drug	Doses	1	2	3	4	5	6	7	8
1	$\theta = 1$	0.40	0.70	0.85	0.92	0.96	0.98	0.99	0.995
Curve 1	$\theta = 2$		0.70		0.92		0.98		0.995
2	$\theta = 1$	0.22	0.52	0.73	0.86	0.93	0.96	0.98	0.99
Curve 2	$\theta = 2$		0.52		0.86		0.96		0.99
3	$\theta = 1$	0.45	0.73	0.86	0.93	0.97	0.98	0.99	0.996
Curve 3	$\theta = 2$		0.73		0.93		0.98		0.996
4	$\theta = 1$	0.45	0.73	0.86	0.93	0.97	0.98	0.99	0.996
Curve 4	$\theta = 2$		0.73		0.93		0.98		0.996

- The value of the interval of time between successive doses acts upon the profiles of concentration as well as upon the value of the AUC. It is worth noting that the area under the curve associated with each even-numbered dose taken with $\theta = 1$ is nearly the same as that obtained with $\theta = 2$. Thus under steady state, the area under the curve is proportional to the number of doses taken, whatever the interval of time between the dose uptake.

4.4. Comparison Between Oral and i.v. Drug Delivery

The dosage regimen or schedule is the plan of administration of one or several drugs. The dosage regimen is prescribed by the physician who bases his decision on the information presented to him by the manufacturer of the drug. An exact dosage schedule is of particular importance when it is necessary to maintain a constant therapeutic concentration in the blood and tissues over a long period of time.

For very serious diseases, the right way for the patient is to have the drug administered at the hospital. In this place, the drug can be delivered either through repeated i.v. injections with the interval of time precisely defined depending on his pharmacokinetics and pharmacodynamics response or through continuous infusion with the right dose.

For less serious diseases, the therapy is accomplished at home by the patient himself; and it comes to discovering the problem of the patient's compliance. Almost all conventional self-administered drugs through oral dosage forms first reach a therapeutic peak concentration and then, because of their short biological half-life time, rapidly fall below the minimum concentration for therapeutic activity. At this point intervene the patient compliance and the limited reliability of the patient. Often the physician is not aware of how frequently the patient changes the dosage schedule carefully prescribed. Investigations conducted in the United States and Europe demonstrated that not less than 50% of patients do not comply with the regimen. This percentage varies from disease to disease, and it can be said that patients have more difficulty in complying with the treatment of a disease causing no particular apparent complaint.

Therapeutic systems, and especially oral dosage forms with controlled release, certainly offer the opportunity of improving patient compliance by reducing some problems connected with their immediate-release counterparts. All compliance studies made it clear that reducing the frequency of drug administration improves compliance, and the once a day (or at least twice a day) schedule is of help in establishing a routine procedure, with the uptake either in the morning getting out of bed or in the evening just before getting into bed.

Finally, the main difference between these two ways of drug delivery comes from the fact that the drug should go through the first-pass hepatic that may metabolize the drug, these metabolites having different therapeutic effects. Moreover, instead of being distributed directly in the blood with i.v. administration, the drug should follow the GIT tract along which it could be transformed, without speaking of the other possible transformation done through the GIT membrane.

Comparison can be easily made in Figs 21–24 where the profiles of plasma drug concentration obtained either by bolus i.v. or with oral administration are drawn.

5. *IN VITRO–IN VIVO* CORRELATIONS FOR SUSTAINED-RELEASE DOSAGE FORMS TAKEN ORALLY

Nomenclature

FDA	U.S. Food and Drug Administration;
GIT	gastrointestine tract;
k_a	rate constant of absorption of the drug into the plasma;
k_e	rate constant of elimination from the plasma;
M	amount of drug released out of the dosage form at time t;
Y	amount of drug in the GIT at time t;
Z	amount of drug in the plasma at time t;
W	amount of drug eliminated out of the plasma at time t;
x, t	abscissa and time, respectively.

5.1. Interest of *in vitro/in vivo* Correlations

5.1.1. Establishment of *in vitro/in vivo* Correlations In 1984 [1] and in 1988 [2], based on their experience acquired with immediate-release dosage forms, the FDA managed two workshops in order to consider the issue of *in vitro/in vivo* correlations with oral dosage forms with controlled release. But, "both workshops concluded that the state of science and technology at that time did not always permit meaningful correlations. Unlike the case of immediate-release dosage forms, the correlations, when possible, would have to be product specific."

It is a pity, essentially from the economical point of view, as when the *in vivo* experiments are very costly, necessitating highly time consuming experiments made on several patients, the *in vitro* experiments are easy to make and quickly obtained.

Because there was no generally recognized acceptable correlations, the FDA required bioavailability testing for the regulatory issues imposed to immediate-release dosage forms, for example, site of manufacturing change, formulation changes, batch size scale-up. Thus, the FDA imposed a bioavailability requirement as a condition of approval.

In June 1992, the FDA, the Canadian Health Protection Branch, and USP cosponsored an open hearing in Toronto [23] concerning *in vitro* and *in vivo* evaluation of dosage forms, dealing essentially with controlled release, in order to answer the questions: how to establish and use *in vitro/in vivo* correlations for controlled-release dosage forms.

Three levels of correlation were defined:

- *Level A.* A unit-to-unit relationship established by comparing the *in vitro* curve to the input function resulting from deconvolution of the plasma concentration–time curve.
- *Level B.* Mean *in vitro* dissolution time compared to either the mean residence time or mean *in vivo* dissolution time.
- *Level C.* Comparison of a single pharmacokinetic parameter such as AUC, or C_{Max} to mean *in vitro* dissolution time.

From the scientific point of view, the "Level A" correlation is preferred to Levels B or C, in this order [3,24] for the following reasons:

- It uses each plasma level and dissolution time point generated and is therefore reflective of the entire curve.
- Since it is predictive of the *in vivo* performance of the dosage forms, it is an excellent quality control procedure.
- The boundaries of the *in vitro* dissolution curve can be justified on the basis of convolution and deconvolution.

The "Level A" correlation has been reported [25] to describe pretty well the results obtained with oral osmotic pump. It was also employed to design the Theodur® controlled release theophylline formulations [26].

A mapping approach [27] was said to be desirable when a Level A cannot be established.

5.1.2. Requirements for Establishing Level A Correlations Among the three levels of correlation defined, Level A has been studied from a mathematical and a biopharmaceutical point of view [24].

Two definitions on *in vitro/in vivo* correlations have been proposed by a USP working group and by the FDA [24,28,29].

> *USP Definition.* The establishment of a relationship between a biological property, or a parameter derived from a biological property produced by a dosage form, and a physiochemical characteristic of the same dosage form.
>
> *FDA Definition.* To show a relationship between two parameters. Typically a relationship is sought between *in vitro* dissolution rate and *in vivo* input rate. This initial relationship may be expanded to critical formulation parameters and *in vivo* input rate.

Several prior conditions have to be met before attempting to establish correlations [24]:

(a) The pharmacokinetic profile of the drug must be linear. If the first-pass hepatic depends on the input rate of the drug in the organism, the drug may have linear pharmacokinetics but nonlinear input kinetics.

(b) The stage of absorption through the GI wall must be limited by the release of the drug from the dosage form. Level A correlation is useful above all for sustained-release dosage forms in which the biopharmaceutical design is based on the drug release.

(c) The intra- and intersubject variability of kinetics must be ascertained beforehand since it is useless establishing a correlation if the variability of the response is very wide.

(d) Establishment of a Level A correlation is appropriate only after single administration since it is a pharmaceutical parameter. It is necessary that the phenomena observed and the relationships obtained after the single administration persist after the second dose and at the so-called steady state when the curves are reproducible for each dose.

(e) It is necessary to administer every dosage form to the same patients, particularly an intravenous injection or a solution so that convolutions and deconvolutions may be performed.

(f) Level A correlations must be established as early as possible in the dosage form development.

(g) Numerous dissolution methods, not limited to those officially prescribed, must be available to ensure proper differentiation and discrimination.

5.1.3. Mathematical Treatment for *In Vitro/In Vivo* Correlations As designated by the definitions proposed by the FDA and USP groups [29], the level A specifies one-to-one correlation. The relationship between *in vitro* dissolution (expressed by the kinetics $x=f(t)$) and *in vivo* input kinetics (expressed by the relation $y=f(t)$) would be linear, in the form

$$y = a \cdot x + b \qquad (92)$$

the intercept b being different from 0, and positive whenever possible.

This term b means that the *in vivo* input kinetics lag behind the *in vitro* kinetics; this lag time would correspond with the time necessary for the drug to pass through the GI membrane. Surely this lag time exists, but it would be rather low, not exceeding a few minutes. However, as the process of absorption is usually assumed to be described by a first-order kinetics, this fact means that the time necessary for the drug to go across the GI membrane is zero, or at least very low. In conclusion, Equation 92 does not hold well, as it stands to reason that at time 0, there is no drug either in the GI or in the plasma, whatever the *in vitro* measurement system, leading to the fact that the term b in Equation 92 should be zero.

A few examples have been given when the OROS® osmotic oral dosage form releases the drug (salbutamol sulfate) [30]. The dissolution kinetics were determined by using four types of methods: the USP rotating paddle (a), the flow-through cell (b), the index release rate tester (c), and the Bio disc (d). The percentage of drug released *in vivo* (*y*) was plotted against the percent drug released *in vitro*, in the four cases. Fitting was tempted for a linear relationship between these two kinetics $y = f(x)$. The correlation measured with the r^2 coefficient, between 0.91 and 0.986 is rather poor, meaning that the linear relation between *y* and *x* is not the best one. The shape of the curves in Fig. 35, expressing the *in vitro/in vivo* correlation with OROS form obtained by plotting the percent released *in vivo* versus the percent released *in vitro* measured using the USP rotating paddle, evidently shows that there is no effective use to try drawing a straight line while a parabolic curve would better represent the phenomenon. A fact should be considered with the OROS systems: the process of release does not start as soon as the OROS form is put in contact with the liquid; on the contrary, the rate of drug release increases very slowly and 1 h at least is necessary for the constant rate to be attained, as shown in various studies [30,31].

5.1.4. Objectives of *In Vitro* Dissolution Tests

In vitro dissolution testing stands for an important tool for characterizing the biopharmaceutical quality of a dosage form at different stages of the drug development. In the early stages of this drug development, perfect knowledge on *in vitro* dissolution properties is decisive for selecting among various alternative dosage forms the one that will enable a better further development of the drug itself. *In vitro* dissolution data are helpful in the evaluation and interpretation of possible risks, especially in the case of modified-release dosage forms, for example, regarding the dose delivery, the food effects on bioavailability that are influencing the gastrointestinal conditions. Of course, *in vitro* dissolution tests are of great importance when assessing changes in the manufacturing process. However, none of these purposes can be fulfilled by *in vitro* dissolution test without sufficient knowledge of its *in vivo* relevance, that is, by studying *in vitro/in vivo* correlations. These *in vitro/in vivo* correlations have been defined in different

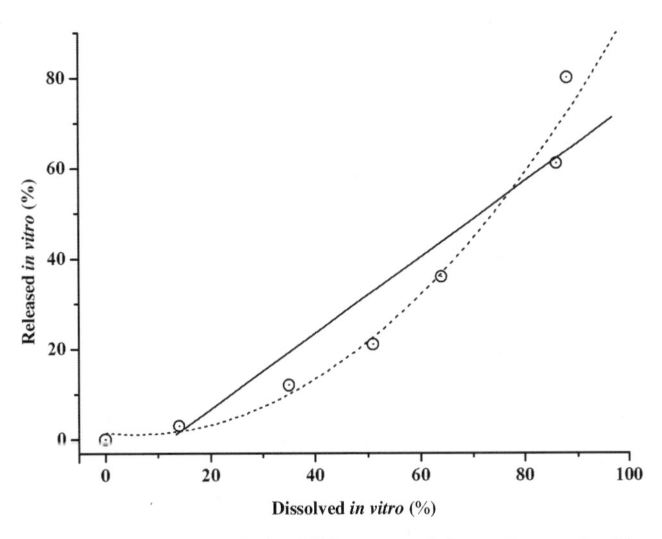

Figure 35. *In vitro/in vivo* correlations, with OROS form containing salbutamol sulfate, obtained with the USP rotating paddle. O: experiments; straight line: from Ref. [24]; dotted line: with parabolic tendency. The experimental value O for $x = y = 0$ has been added for the parabolic tendency. Retraced from Fig. 3 and Ref. [24] with permission from Medecine and Hygiene.

ways, and have been subject to much controversy [32,33]. A meaningful correlation has to be quantitative so as to allow interpolation between data and thus making *in vitro* model predictive

All these factors are of great concern and should be kept in mind, and especially the intervariability of the response of the patients toward a drug (Sections 7 and 8).

In order to increase their predictive value, dissolution tests aim to mimic *in vivo* behavior along the GIT, and thus the *in vitro* models are adjusted so as to simulate the physiological conditions as closely as possible [34]. Nevertheless, several examples demonstrate that this procedure could also lead to misinterpretations [35]. Thus, from numerous scientific data, it is concluded that there is no need of close simulation of physiological conditions. However, special care with pH, agitation, ionic strength, surface tension, should give conditions nearly similar to the physiological reality and so avoid artificial data and misinterpretations. High rates of agitation are not necessary and should therefore be avoided since it can cause a loss of sensitivity in distinguishing real differences in the dissolution rates due to stronger agitation [36,37].

5.1.5. Conclusions on the *In Vitro/In Vivo* Correlations

- Generally, good correlation is supposed to be obtained when a straight line represents the percent absorbed versus the percent dissolved, for the drug. But, it is clear that for salbutamol sulfate delivered through an OROS system [24,30], the percent released *in vitro* versus the percent absorbed *in vivo* exhibits a kind of parabolic tendency more than a linear relationship, as shown in Fig. 35.
- *In vitro/in vivo* correlations are set up for mean data acquired with healthy volunteers, usually after single administration, and from averaged data.
- Moreover, while the variability between *in vitro* results is generally good (5%), the intersubject variability is far much larger (more than 100%).

From these two facts, it seems that the safest approach would be to consider this *in vitro/in vivo* correlation as a development tool for modified release forms, with possible quality control applications, rather than an infallible monitoring method [24].

5.2. New Way of Calculating the Plasma Drug Profile by Numerical Methods

The plasma drug level has been calculated with oral dosage forms whose release is controlled by diffusion in various papers. The first one [38] was concerned with a spherical dosage form containing theophylline as a drug, and the calculation has been made by considering the change in the pH by following the pH-time history along the gastrointestinal transit described by scintigraphy [39]. The second paper done with a spherical dosage form was dealing with salicylate sodium as the drug dispersed in the polymer Eudragit® RL, and the profile of drug concentration has been calculated in various parts such as the gastrointestine, the blood volume, the drug eliminated as well as the drug released from the dosage form; thus it was found that the amount of drug in the gastrointestine was very low [40]. Another study was devoted to the prediction on *in-vivo* blood level with controlled-release dosage forms, by considering especially the gastrointestinal tract time, which is, according to some authors, not more than 10 h [39], and from this fact arises a limitation that is a drawback [41]. The effect of the dose frequency on the plasma drug profile was also regarded [42]. Original dosage forms were also considered as potential issues for delivering the drug in a sustained release way, by using lipidic Gelucire® as excipient [43]. The plasma drug level was also calculated with a more complex system based on spherical materials made of a core and a shell with a lower concentration in the shell, in order to reduce the drawback of the diffusion process leading to a rate of drug delivery very high at the beginning and which decreases exponentially with time [44]. Moreover, the results concerning the process controlled by diffusion as well as by erosion were collected in a more general paper [45] and studied more deeply later [46]. This method developed and published in these above papers was presented in 1996 in front of a panel of specialists of the FDA when the process of drug release is controlled by diffusion in 1996 [47] and by erosion in 1998 [48]. And more recently, the plasma drug level has been related to the characteristics of the dosage forms such as the dimensions, the nature of the polymer and thus the diffusivity, for various drugs [49].

5.2.1. Assumptions for Calculation The following assumptions are made in order to define the problem precisely:

(1) The process of drug transport is divided into three stages, connected with each other: transport through and out of the dosage form into the gastrointestinal liquid, transport from this liquid to the blood through the gastrointestinal membrane with the rate constant of absorption, transport from the blood to the surrounding with the rate constant of elimination.
(2) The dosage form, whatever its shape, may keep or not its dimensions constant during the process, provided that the kinetics of these changes are known.
(3) The process of drug delivery from the dosage form in the gastric liquid is simplified, by considering only the diffusion of the drug through the dosage form, with a constant diffusivity. Nevertheless this assumption is not mandatory.
(4) The two phenomena of absorption into and elimination out of the blood are described by first-order kinetics.
(5) If the rate constant of absorption k_a remains constant along the GIT, and the rate constant of elimination K_e is often supposed remaining constant during the process, the present numerical model is able to take into account a dependence of the rate of drug release with the pH and the GIT history, provided that it is known.

5.2.2. Mathematical Treatment The rate of drug delivery by the dosage form at time t, whatever its shape, is given by $\frac{dM}{dt}$:

Thus the amount of drug in the gastrointestine, Y, is expressed by the relation

$$\frac{dY}{dt} = \frac{dM}{dt} - k_a \cdot Y \tag{93}$$

The amount of drug in the plasma compartment is Z, at time t, and given by:

$$\frac{dZ}{dt} = k_a \cdot Y - k_e \cdot Z \tag{94}$$

while the amount of drug eliminated at time t, W, is expressed by:

$$\frac{dW}{dt} = k_e . Z \tag{95}$$

5.2.3. Numerical Treatment The mathematical treatment of this above system of three equations is possible only when the drug is release immediately in the GI or when the rate of delivery is constant. In all other cases, a numerical method is necessary to solve the problem. It is run step by step, with a constant increment of time Δt. Two general ways can be followed, by using either an old explicit method or an implicit method, the most well-known being the Crank-Nicolson method.

6. SUSTAINED-RELEASE DOSAGE FORMS. *IN VITRO* TESTS

Nomenclature

a, b, c	half the sides of a solid, parallelepiped, cube, sheet;
β_n	positive roots of $\beta \cdot \tan \beta = G$ in Equation 104, for a sheet;
C, C_s, C_{ext}	concentration, on the surface, in the external liquid;
D	diffusivity, or coefficient of diffusion (cm^2/s);
$\frac{\partial C}{\partial x}$, $\frac{\partial C}{\partial r}$	longitudinal and radial gradient of concentration, respectively;
F	flux of diffusing substance mass/unit area/unit time;

G — dimensionless number $G = h \cdot L/D$ in Equation 105, for a sheet;
h — coefficient of convection on the surface of the dosage form (cm/s);
H — half the length of a cylinder;
γ_n — roots of Equation 121, for a sphere;
K — dimensionless number in Equation 120, for a sphere;
L — half the thickness of a sheet;
M_t — amount of drug released up to time t;
M_∞ — amount of drug released after infinite time;
M_{in} — amount of drug initially in the dosage form;
n, m, p — integers used for calculation in series;
q_n — roots of the Bessel function in radial diffusion, in Equation 129 for a cylinder;
R — radius of the sphere, radius of a cylinder;
t, t_r — time and time of full erosion of the dosage form, respectively;
v — linear rate of erosion, in Equation 135;
x, r — longitudinal, radial abscissa along which diffusion takes place, respectively;
DBS, ATBC — plasticizer: dibutyl sebacate and acetyl tributyl citrate, respectively.

The various sustained release oral dosage forms are generally made of polymers through which the drug is dispersed. Depending on the nature of the polymer, the process of drug release is controlled either by diffusion or by erosion, and sometimes by both these processes. Thus, these two processes are considered in succession.

Experimentally, the kinetics of drug release is determined through *in vitro* measurements. Nevertheless calculation is always useful, showing the best line to be taken along the way of the research, allowing reducing the number of tedious experiments and offering logical shortcuts. Moreover, the equations expressing the kinetics of drug release are necessary for calculating the plasma drug level.

6.1. Drug Release Controlled by Diffusion

In this case, the polymer is stable in the GI and pass unchanged through the body. It is not the purpose of this book to deeply examine the process of diffusion through polymers as three books are worth noting for people interested in such a study. In Ref. [50], the results obtained in a fundamental study on "heat conduction" [51] have been applied to diffusion. In this book, the mathematical treatment is made by selecting various shapes: sheets and parallelepipeds, spheres and cylinders, when the diffusivity is constant. In Ref. [52], not only the mathematical but also the numerical treatment of diffusion is made, allowing the reader to consider more complex matter transfers, especially when the diffusivity is concentration dependent, and various practical applications are presented. Ref. [53] is devoted to the applications of diffusion for the drug liberation from oral sustained-release dosage forms, as they are determined through *in vitro* tests.

6.1.1. Principles of Diffusion and Basic Equations

Fick, in 1855, put diffusion on a quantitative basis by adopting the same equation as that established by Fourier in 1822 for heat conduction. In an isotropic substance, the rate of diffusion through unit area of a section is proportional to the gradient of concentration measured normal to this section:

$$F = -D \cdot \frac{\partial C}{\partial x} \tag{96}$$

D being the diffusivity and $\frac{\partial C}{\partial x}$ the gradient of concentration, the negative sign arising because diffusion occurs in the direction opposite to that of increasing concentration.

Boundary Conditions The boundary conditions express the value of the concentration and of the gradient of concentration of diffusing substance on the surface of the dosage form.

- When the drug does not diffuse out of the dosage form, the mathematical condition is obtained by writing that the rate of transfer through its surface is zero:

$$\frac{\partial C}{\partial x} = 0 \qquad (97)$$

- When the rate of matter transfer through the external surface is finite, a finite coefficient of matter transfer through this surface is introduced, leading to the general equation:

$$-D \cdot \left(\frac{\partial C}{\partial x}\right)_s = h \cdot (C_s - C_{ext}) \qquad (98)$$

where C_s and C_{ext} are the drug concentration on the surface of the dosage form and in the surrounding liquid, respectively. This condition on the surface expresses that the rate at which the drug leaves the surface is constantly equal to the rate at which the drug is brought to this surface by internal diffusion. This rate, per unit area, is proportional to the difference between the actual concentration on the surface C_s and the concentration required to maintain equilibrium with the surrounding liquid C_{ext}.

- When the coefficient of matter transfer is very large, tending to infinity, it clearly appears that the concentration on the surface is equal to that in the surrounding liquid as soon as the dosage form is put in contact with the liquid and the process of diffusion starts:

$$h \to \infty \cdot C_s = C_{ext} \qquad (99)$$

Initial Conditions Generally, the drug concentration is initially uniform through the dosage form.

Process of Drug Release The drug is only released from the dosage form when it is brought in contact with a liquid. The actual process of drug release is complex as proved in various studies, one with Eudragit [54] and the other with Carbopol® [55] when dosage forms made of these polymers and acetylsalicylic acid as the drug were immerged in synthetic gastric liquid, following a previous study made on plasticized PVC [56]. A double matter transfer takes place, as the liquid diffuses through the polymer, dissolves the drug, enabling the drug to diffuse through the liquid located in the polymer; and thus the diffusivity of the drug depends on the local concentration of liquid situated in the polymer. The solution of this difficult problem necessitates complex experiments, by weighing the dosage form and measuring the drug concentration in the liquid simultaneously, so as to determine the kinetics of the transfer of both the liquid and the drug, and a numerical treatment is necessary for calculation [52,53]. But, finally, pharmacists being only interested in the drug release into the liquid, the diffusion of the drug alone through the polymer is considered in the following sections.

***In vitro* Dissolution Tests** Biopharmaceutical characterization of drug formulations requires a suitable dissolution model apparatus. The paddle and the rotating basket models are described in USP XXII, European, Japanese and other Pharmacopoeias worldwide. The alternative flow-through cell was also developed if a change in pH in the test medium is required [32]. In order to increase their predictive value, dissolution tests aim to mimic *in vivo* behavior along the GIT, and thus the models are adjusted to simulate the physiological conditions as closely as possible [34]. Special care with pH, agitation with controlled stirring rate, could give conditions similar to physiological reality. It was said [36] that high rates of agitation are not necessary, but in motionless liquid the process of drug release is controlled not only by diffusion through the dosage form but also by convection in the liquid, and increasing the stirring rate reduces the convection effect.

Parameters of diffusion The diffusivity D is the essential parameter of diffusion, the coefficient of convection h at the liquid-dosage form interface may appear of concern, especially at the beginning of the process when the stirring rate is too low. The shape given to the dosage form and more particularly its dimensions represent the other parameters of interest.

6.1.2. Kinetics of Drug Release from Thin Films (Thickness 2L) Fickian diffusion is expressed by the second law of diffusion (which is established from Equation 96 called first law of diffusion, by evaluating the matter balance through a thin sheet of thickness dx during the small time dt) that can be written in the form of a partial derivative equation:

$$\frac{\partial C}{\partial t} = D \cdot \frac{\partial^2 C}{\partial x^2} \tag{100}$$

C being the concentration of the drug in the dosage form, which depends on time t and position x, and D the diffusivity of the drug through the dosage form.

The respective initial and boundary conditions are written as follows:

$$t = 0 \quad -L \leq x \leq +L \quad C = C_{\text{in}} \tag{101}$$

$$t > 0 - D \cdot \left(\frac{\partial C}{\partial x}\right)_L = h \cdot (C_s - C_{\text{ext}}) \tag{102}$$

At the beginning of the experiment ($t = 0$), the drug is uniformly distributed through the film of thickness $2L$, C_{in} being this concentration.

Equation 100 has two types of solutions, one expressed in terms of an exponential series, and the other in terms of error-function series. The general solution of the problem set up by Equations 100 and 102 with a finite coefficient of convection h at the interface is [50–53]

$$\frac{M_\infty - M_t}{M_\infty} = \sum_{n=1}^{\infty} \frac{2 \cdot G^2}{\beta_n^2(\beta_n^2 + G^2 + G)} \exp\left(-\frac{\beta_n^2}{L^2} Dt\right) \tag{103}$$

where the β_n are the positive roots of

$$\beta \cdot \tan \beta = G \tag{104}$$

and the dimensionless term G being expressed by

$$G = \frac{h \cdot L}{D} \tag{105}$$

where L is half the thickness of the sheet.

Some roots of Equation 104 are supplied for various values of G in Books 60–63, but the best way is to calculate them, as a table cannot provide all the β_n values for any value of G.

When the rate of stirring is high enough, the coefficient h becomes so large that it can be considered as infinite, and the drug concentration on the surface C_s reaches its value at equilibrium as soon as the dosage form is put in contact with the liquid. In this case, Equation 103, when $G > 50$–100, reduces to

$$\frac{M_\infty - M_t}{M_\infty} = \frac{8}{\pi^2} \sum_{n=0}^{\infty} \frac{1}{(2n+1)^2} \exp\left[-\frac{(2n+1)^2 \pi^2}{4L^2} Dt\right] \tag{106}$$

M_t and M_∞ being the amount of drug released after time t and infinite time, respectively. In fact, M_∞ is also the amount of drug initially located in the dosage form, when the volume of liquid is more than 50 times that of the dosage form, n is an integer ranging from 1 to infinity (generally, 5–7 is enough), and L is half the thickness of the sheet.

By using the error-function, the following series is obtained [50–53]:

$$\frac{M_t}{M_\infty} = \frac{2\sqrt{D \cdot t}}{L} \left[\frac{1}{\sqrt{\pi}} + 2 \cdot \sum_{n=1}^{\infty} (-1)^n \cdot \text{ierfc}\left(\frac{n \cdot L}{\sqrt{D \cdot t}}\right)\right] \tag{107}$$

where ierfc is the integral of the error-function complement, and n an integer.

Equations 106 and 107 reduce significantly either for long times or for short times, this time being related to the ratio M_t/M_∞:

$$t \text{ is large with } \frac{M_t}{M_\infty} > 0.5\text{–}0.7 \quad \frac{M_\infty - M_t}{M_\infty} = \frac{8}{\pi^2} \cdot \exp\left[-\frac{\pi^2 \cdot D \cdot t}{4 \cdot L^2}\right] \tag{108}$$

$$t \text{ is small with } \frac{M_t}{M_\infty} < 0.6 \quad \frac{M_t}{M_\infty} = \frac{2}{L} \cdot \sqrt{\frac{D \cdot t}{\pi}} \tag{109}$$

Equations 108 and 109 lead to the following simple relation:

$$\text{For } \frac{M_t}{M_\infty} = 0.5 \quad \frac{D \cdot t}{L^2} = 0.196 \tag{110}$$

Dimensionless Numbers Two dimensionless numbers are of great concern, the one related to the kinetics, expressed in terms of time, and the other with the coefficient of convection at the interface:

$$\frac{D \cdot t}{L^2} \text{ and } G = \frac{h \cdot L}{D} \tag{111}$$

The first one shows that, for a given diffusivity associated with the drug–polymer couple, the time of release is proportional to the square of the thickness of the sheet.

Moreover, for $G > 100$, the Equation 109 shows that there is a vertical tangent at the origin of the curve M_t/M_∞ versus time, and that a straight line passing through the origin of time is obtained by plotting M_t/M_∞ versus $\sqrt{t}$, whose slope is proportional to the diffusivity.

6.1.3. Kinetics of Release from the Sphere For a radial diffusion and a constant diffusivity, the diffusion equation becomes

$$\frac{\partial C}{\partial t} = \frac{D}{r^2} \cdot \frac{\partial}{\partial r}\left[r^2 \cdot \frac{\partial C}{\partial r}\right] \tag{112}$$

The solution of this equation gives the amount of drug released after time t, M_t, as a fraction of the corresponding quantity after infinite time, M_∞:

$$\frac{M_\infty - M_t}{M_\infty} = \frac{6}{\pi^2} \sum_{n=1}^{\infty} \frac{1}{n^2} \exp\left[-\frac{n^2 \cdot \pi^2}{R^2} D \cdot t\right] \tag{113}$$

When the liquid is not stirred, the equation becomes

$$\frac{M_\infty - M_t}{M_\infty} = \sum_{n=1}^{\infty} \frac{6 \cdot K^2}{\gamma_n(\gamma_n^2 + K^2 - K} \exp\left[-\frac{\gamma_n^2}{R^2} D \cdot t\right] \tag{114}$$

with the dimensionless number K in a form similar as G:

$$K = \frac{h \cdot R}{D} \tag{115}$$

where the γ_n's are the roots of

$$\gamma_n \cdot \cot \gamma_n = 1 - K \tag{116}$$

Some roots of Equation 121 are given for various values of K [50–53], but the best way is to use a program, as it is impossible to give these values for every K.

6.1.4. Calculation for the Parallelepiped of Sides 2*a*, 2*b*, 2*c* The amount of drug released from the parallelepiped after time t, M_t, as a fraction of the corresponding amount after infinite time, M_∞, is expressed in terms of time by the product of the three series corresponding with one-dimensional transport through the three sheets of thickness $2a$, $2b$, $2c$:

$$\frac{M_\infty - M_t}{M_\infty} = \frac{512}{\pi^6} \sum_{m=0}^{\infty} \frac{1}{(2m+1)^2}$$

$$\exp\left[-\frac{(2m+1)^2\pi^2}{4 \cdot a^2} D \cdot t\right] \cdot \sum_{n=0}^{\infty} \frac{1}{(2n+1)^2}$$

$$\exp\left[-\frac{(2n+1)^2\pi^2}{4 \cdot b^2} D \cdot t\right] \cdot \sum_{p=0}^{\infty} \frac{1}{(2p+1)^2}$$

$$\exp\left[-\frac{(2p+1)^2\pi^2}{4 \cdot c^2} D \cdot t\right] \tag{117}$$

For small times of release, when $\frac{M_t}{M_\infty} < 0.6$, the simple relation is obtained:

$$\frac{M_t}{M_\infty} = \left(\frac{2}{a} + \frac{2}{b} + \frac{2}{c}\right)\sqrt{\frac{D \cdot t}{\pi}} - \left(\frac{4}{ab} + \frac{4}{ac} + \frac{4}{bc}\right)$$

$$\frac{D \cdot t}{\pi} + \frac{8}{abc}\left(\frac{D \cdot t}{\pi}\right) \tag{118}$$

For long times of release, when $\frac{M_t}{M_\infty} > 0.7$, the first term in the exponential series becomes preponderant, and the amount of drug released can be expressed in terms of time by

$$\frac{M_\infty - M_t}{M_\infty} = \frac{512}{\pi^6} \exp\left[-\left(\frac{1}{a^2} + \frac{1}{b^2} + \frac{1}{c^2}\right)\frac{\pi^2}{4} D \cdot t\right] \quad (119)$$

6.1.5. Calculation for the Cube of Sides 2*a* For the cube of sides $2a$, Equation 122 with $2a = 2b = 2c$, reduces to

$$\frac{M_\infty - M_t}{M_\infty} = \frac{512}{\pi^6}\left[\sum_{m=0}^{\infty} \frac{1}{(2m+1)^2} \exp\left(-\frac{(2m+1)^2 \pi^2}{4 \cdot a^2} D \cdot t\right)\right]^3 \quad (120)$$

For the cube of sides $2a$, the kinetics of drug release for short times is easily obtained by putting $2a = 2b = 2c$ in Equation 123:

$$\frac{M_t}{M_\infty} = \frac{6}{a}\sqrt{\frac{D.t}{\pi}} - \frac{12}{a^2} \cdot \frac{D \cdot t}{\pi} + \frac{8}{a^3}\left(\frac{D \cdot t}{\pi}\right)^{1.5} \quad (121)$$

and for long times, Equation 124 or 125 reduces to

$$\frac{M_\infty - M_t}{M_\infty} = \frac{512}{\pi^6} \exp\left(\frac{3\pi^2}{4 \cdot a^2} D \cdot t\right) \quad (122)$$

6.1.6. Calculation for a Cylinder of Radius *R* and Height 2*H* In the cylinder of finite length, the diffusion is both radial and longitudinal. The solution of the problem is thus expressed by the product of the solutions obtained either for the radial diffusion only or for the longitudinal diffusion only.

For the infinite coefficient of convection at the surface, the solution is

$$\frac{M_\infty - M_t}{M_\infty} = \frac{32}{\pi^2} \sum_{n=1}^{\infty} \frac{1}{q_n^2} \exp\left(-\frac{q_n^2}{R^2} D \cdot t\right) \cdot \sum_{p=0}^{\infty} \frac{1}{(2p+1)^2} \exp\left(-\frac{(2p+1)^2 \pi^2}{4.H^2} D \cdot t\right) \quad (123)$$

where the q_ns are the roots of the Bessel function of the first kind of order zero:

$$J_0(q_n \cdot R) = 0 \quad (124)$$

these roots of the Bessel function being given in Tables 5.1–5.4.

6.1.7. Effect of the Shape of Dosage Forms of Similar Volume As it is of interest to know precisely the effect of the shape given to dosage forms of similar volume on the kinetics of release of the drug obtained under similar conditions of stirring in the same synthetic gastrointestinal liquid, the kinetics obtained with the following shapes are considered in succession: a parallelepiped, a cube, a cylinder and a sphere, this sphere having a radius of 0.182 cm, as shown in Fig. 36. These kinetics of drug release from the dosage forms of same volume and different shapes controlled by diffusion lead to a few comments of interest:

- The kinetics of drug release depends slightly on the shape given to the dosage forms; it is faster for the parallelepiped, resulting from the presence in this solid of the thinner thicknesses of two of its three sides, and it is slower for the sphere.
- The shapes of these kinetic curves of drug release are typical; starting with a high rate at the beginning of the process associated with a nearly vertical tangent, the rate of drug release decreases with time in an exponential way.
- Theoretically speaking, a drawback appears with this fact that the whole drug initially in dosage form is released after infinite time, resulting from the diffusion process.

6.1.8. Kinetics of Drug Release with Spherical Dosage Forms Spherical beads are commonly used, either in the form of a solid or in the form of small granules gathered in an envelope, and moreover the effect of the shape is small on the kinetics of release. The kinetics of drug release is determined through *in vitro* experiments by using either Equation 117 when the synthetic liquid is strongly stirred or Equations 114–116 when the coefficient of convective transfer at the interface has to be taken into account.

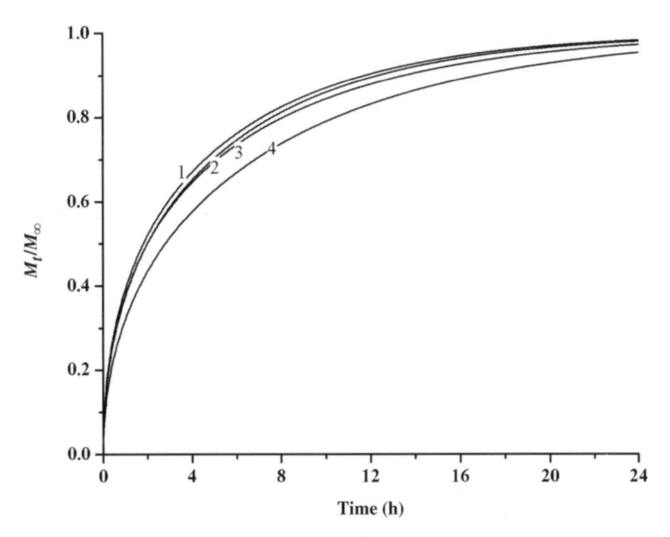

Figure 36. Curves expressing the kinetics of release of a drug from dosage forms of same volume and different shapes, controlled by diffusion. (1): Parallelepiped with $2a = 2b = c$. (2): Cube of sides $2a$. (3): Cylinder with $R = H$. (4): Sphere with $R = 0.182$ cm. $D = 2.5 \times 10^{-7} \text{cm}^2/\text{s}$.

When the data concerning the diffusion are known, for example, the diffusivity D and the coefficient of convective transfer h, the curves expressing the amount of drug released as a function of time can be obtained by calculation.

The effect of the parameters of diffusion, for example, the diffusivity D and the coefficient of convection h, as well as the radius of the sphere, is shown in Fig. 37 where the kinetics of drug release are drawn by using the following dimensionless number: the amount of drug released at time t as a fraction of the corresponding amount after infinite time (which in fact is the amount of drug initially located in the dosage form) as ordinate, and the dimensionless time $\sqrt{D \cdot t}/R$ as abscissa. The various curves depicted for various values of the dimensionless number $K = h \cdot R/D$ clearly show the effect of the coefficient of

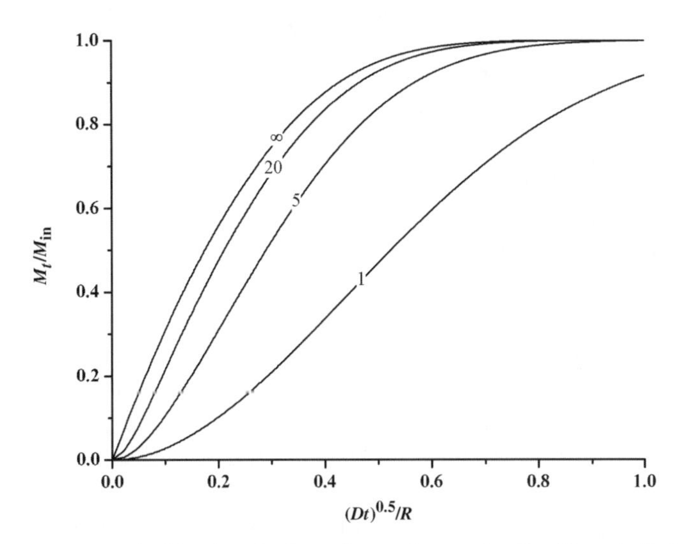

Figure 37. Master curves for the kinetics of release from spheres of radius R, controlled by diffusion, with various values of the coefficient of convection and $K = \frac{h \cdot R}{D}$

convective transport on the bead surface. For K larger than 100 or even 50, the process is controlled by diffusion and Equation 113 can be used; when K is lower than 50, the process is controlled by both diffusion and convection, necessitating the use of the group of Equations 114–116. In all cases, the amount of drug released at any time is proportional to the square of the radius of the spherical dosage form.

6.1.9. Example of Kinetics of Drug Release An example is given for the kinetics of drug release obtained either by experiments or by calculation, for thin films containing Diltiazem HCl as a drug when the release is controlled by addition of the convenient percent plasticizer [57] (Fig. 38). These curves can lead to the following comments:

- The kinetics obtained by experimentation can be expressed by Equation 103 and even Equation 106 as the liquid is strongly stirred.
- The effect of the addition of plasticizer clearly appears on the kinetics of drug release.
- Let us remark that dimensionless numbers are used either for the amount of drug release or for the time. The amount of drug released at time t is expressed as a fraction of the corresponding value at infinite time, which is also the amount of drug initially in the dosage form. The time is expressed as a fraction of the square of the thickness of each film, as shown in Equation 106.

6.1.10. Determination of the Parameters of Diffusion Evaluation of the parameters of diffusion, for example, the diffusivity D and secondary the coefficient of convection h is a problem for specialists of diffusion [50,52–54]. On the whole, the diffusivity is obtained from the slope of the straight line obtained by plotting the relative amount of drug release as a function of the square root of time, when the liquid is strongly stirred. The presence of a finite coefficient of convection appears when the kinetics or release exhibits a S-shape, as shown in Fig. 37.

6.2. Drug Release Controlled by Erosion

In this case, the dosage forms are made of the same drug dispersed in the same erodible polymer, and various shapes are selected in order to make comparisons.

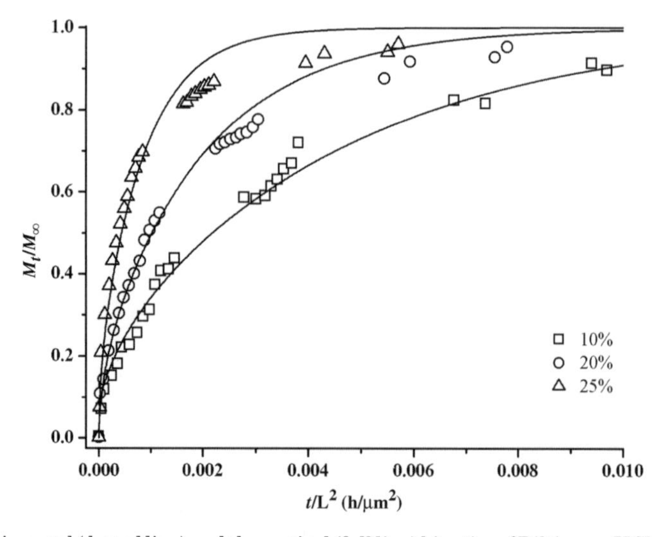

Figure 38. Experimental (dotted line) and theoretical (full line) kinetics of Diltiazem HCl release with dosage forms made of EC films with various percent of plasticizer (DBS): 25%; 20%; 10%. Reprinted from Ref. [57] with permission from Wiley-Liss.

The rate of erosion constant during the process, is expressed in the linear form

$$v = \frac{dL}{dt} \quad (125)$$

The initial concentration of the drug is uniform in the dosage form.

The kinetics of release is determined for various shapes given to the dosage forms.

6.2.1. Kinetics of Release for a Sheet of Thickness $2L_0$ The sheet of thickness $2 \cdot L_0$ at time 0 is immersed on both sides in the liquid, and at time t, half the thickness is reduced to L_t, expressed as follows:

$$L_t = L_0 - v \cdot t \quad (126)$$

and it is completely eroded after the time of erosion t_r:

$$L_t = 0 \text{ for } t_r = \frac{L_0}{v} \quad (127)$$

By introducing the time of erosion, Equation 126 becomes

$$L_t = L_0 \cdot \left[1 - \frac{t}{t_r}\right] \quad (128)$$

And finally, the amount of drug released at time t, M_t, is proportional to time

$$\frac{M_t}{M_{in}} = \frac{t}{t_r} \quad (129)$$

where M_{in} is the amount of drug initially in the sheet.

6.2.2. Kinetics of Release for a Sphere of Initial Radius R_0 At time t, the new radius is given by

$$R_t = R_0 - v \cdot t = R_0 \left[1 - \frac{t}{t_r}\right] \quad (130)$$

and the amount of drug released at time t, M_t, is expressed as a fraction of the amount initially in the sphere, in terms of time:

$$\frac{M_t}{M_{in}} = 1 - \left[1 - \frac{t}{t_r}\right]^3 \quad (131)$$

6.2.3. Kinetics of Release for a Cylinder of Radius R_0 and Height $2H_0$ The amount of drug released at time t, M_t, is given as

$$\frac{M_t}{M_{in}} = 1 - \left[1 - \frac{v \cdot t}{R}\right]^2 \cdot \left[1 - \frac{v \cdot t}{H}\right] \quad (132)$$

When $H < R$, the time of full erosion is given by

$$t_r = \frac{H}{v} \quad (133)$$

and the amount of drug released at time t is

$$\frac{M_t}{M_{in}} = 1 - \left[1 - \frac{H}{R}\frac{t}{t_r}\right]^2 \cdot \left[1 - \frac{t}{t_r}\right]$$

6.2.4. Kinetics of Release with a Parallelepiped of Sides $2a$, $2b$, $2c$ For this parallelepiped the amount of drug released at time t, M_t, is given by

$$\frac{M_t}{M_{in}} = 1 - \frac{(a - v \cdot t)(b - v \cdot t)(c - v \cdot t)}{a \cdot b \cdot c} \quad (134)$$

When the sides are such as $a < b < c$, the time of full erosion is given by

$$t_r = \frac{a}{v} \quad (135)$$

and the amount of drug released at time t is

$$\frac{M_t}{M_{in}} = 1 - \left[1 - \frac{t}{t_r}\right]\left[1 - \frac{a}{b}\frac{t}{t_r}\right]\left[1 - \frac{a}{c}\frac{t}{t_r}\right] \quad (136)$$

6.2.5. Kinetics of Release for a Cube of Sides $2a$ For the cube of sides $2a$, the amount of drug released at time t is obtained from Equation 136 when the sides are equal to $2a$:

$$\frac{M_t}{M_{in}} = 1 - \left[1 - \frac{t}{t_r}\right]^3 \quad (137)$$

Equation 137 for the cube and Equation 131 for the sphere are similar from the first look, but the time of erosion is not the same, being

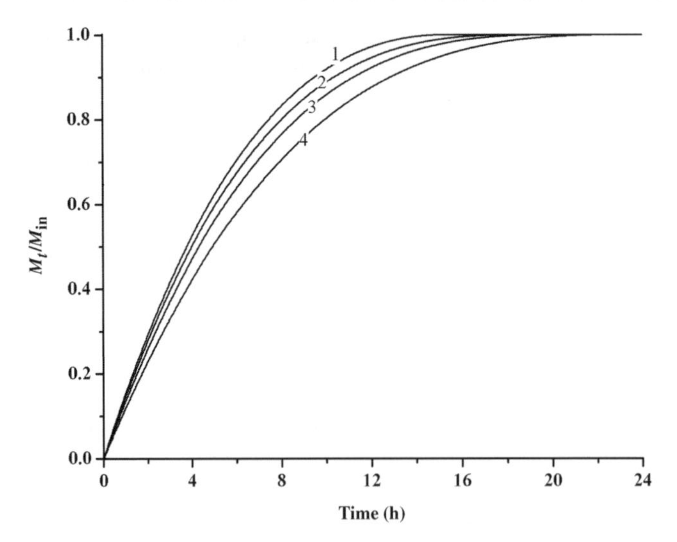

Figure 39. Kinetics of release of a drug from dosage forms of same volume and different shapes, controlled by erosion. (1): Parallelepiped with $2a = 2b = c$. (2): Cube. (3): Cylinder with $R = H$. (4): Sphere. Time of full erosion = 24 h for the sphere.

proportional to either the radius of the sphere or half the edge of the cube. As half the edge of the cube, a, is 0.806 times the value of the radius of the sphere having the same volume, thus the time of erosion is 0.806 times shorter for the cube than for the sphere. This is the reason why the kinetics of erosion is faster for the cube than for the sphere, as shown in Fig. 39. In the same way, the time of erosion is proportional to the thinner dimension of a solid, and this fact explains that the kinetics of erosion is faster for the parallelepiped than for the cube of same volume. Moreover, the time of full erosion may be longer than 24 h, leading to kinetics of release of interest as shown in Fig. 40.

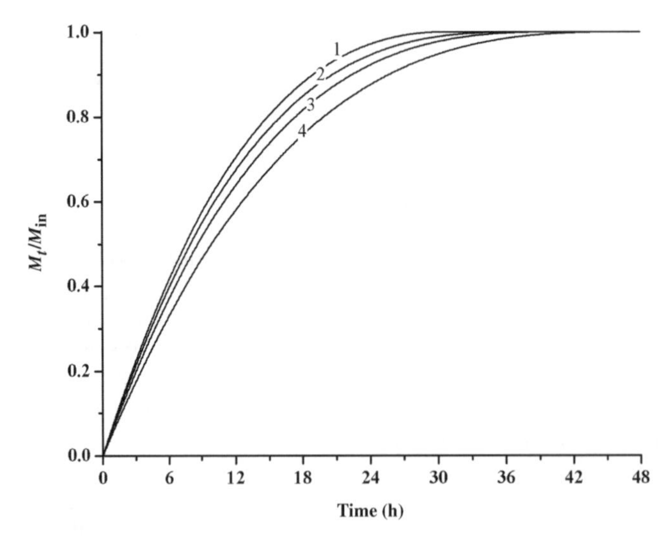

Figure 40. Kinetics of release of a drug from dosage forms of same volume and different shapes, controlled by erosion. (1): Parallelepiped with $2a = 2b = c$. (2): Cube. (3): Cylinder with $R = H$. (4): Sphere. Time of full erosion = 48 h for the sphere.

6.2.6. Comparison Between the Processes of Diffusion and Erosion Comparison between the kinetics of drug release controlled either by diffusion or by erosion is of great concern. The curves expressing the kinetics of drug release are shown either in Fig. 36 when they are controlled by diffusion or in Figs 39 and 40 when they are controlled by erosion. The following conclusions are worth noting:

- The slopes of these curves are quite different. The rate of drug release controlled by diffusion very high at the beginning of the process, decreases exponentially with time. On the other hand, the rate of drug release controlled by erosion starts rather slowly and decreases slowly with time.
- More precisely, the kinetics with process controlled by diffusion are far from being linear with time, while the erosion process leads to kinetics nearly linear with time. Thus, it can be said that for a sheet, a straight line would be obtained with erosion.
- With the process of erosion, the time of full erosion is well defined, meaning that the whole drug initially in the dosage form is totally released up to that time. In contrast with this fact, the process of drug release controlled by diffusion comes to an end after infinite time, at least mathematically speaking.
- By making sure that not only the drug but also the excipient and additives are released and dissolved away after a finite time, dosage forms whose release is controlled by erosion may be used as adhesive forms either on the stomach or the intestinal wall [58–60]. As shown in Sections 7 and 8, great interest is found to get a very long time of release of the dosage form, for example, several days, with a nearly constant rate of drug delivery. A few polymers are capable of swelling and binding with the macromolecules of the mucus. Some of them are erodible, especially in the family of polysaccharides, alginate, carraghenate, dextrn, chitosan and starch, and protein derivatives such as gelatine.

7. PLASMA DRUG PROFILE WITH DIFFUSION-CONTROLLED DOSAGE FORMS

Nomenclature

D	diffusivity of the drug through the polymeric matrix of the sustained-release dosage form, expressed in cm^2/s;
D/R^2	diffusivity as a fraction of the square of the radius of the sustained-release dosage form;
GIT	gastrointestinal tract;
i.v.	intravenous bolus injection;
H	height of a cylinder;
k_a	rate constant of absorption of the drug (per hour);
k_e	rate constant of elimination of the drug (per hour);
M_t	amount of drug in the blood at time t;
M_{in}	amount of drug initially in the dosage form;
R	radius of the spherical dosage form;
t	time;
$t_{0.5}$	half-life time of the drug in bolus i.v. (h);
$T_{0.5}$	half-life time of the drug obtained with the sustained-release dosage form (h);
$T'_{0.5}$	half-life time of the drug obtained with the immediate-release dosage form (h);
V_p	apparent plasmatic volume (L);
dM/dt	rate of drug released by the dosage form in the gastrointestinal volume (in Eq. 138);
W	amount of drug eliminated (in Eq. 140);
Y	amount of drug in the gastrointestinal volume (in Eq. 138);
Z	amount of drug in the plasmatic volume (in Eq. 139).

7.1. Method of Calculation

The plasma drug level has been calculated with oral diffusion-controlled oral dosage forms in various papers: With a spherical dosage form containing theophylline as a drug [38], by considering the change in the pH along the gastrointestinal transit described by scintigraphy [39]; with a spherical dosage form containing salicylate sodium dispersed in polymer Eudragit RL, and the profile of drug concentration was calculated in various parts such as the gastrointestine, the blood volume, the drug eliminated as well as the drug released from the dosage form, this last information proving that the amount of drug in the gastrointestine is very low [40]; prediction of *in vivo* blood level was established, by considering the gastrointestinal tract time, which is, according to some authors, not more than 10 h [39], this limitation being a drawback [41]; the effect of the dose frequency on the plasma drug profile was also regarded [42]; original dosage forms were also considered by using lipidic Gelucire^{R} as excipient [43]; the plasma drug level was calculated with a more complex system made of a core and a shell with a lower concentration in the shell, in order to reduce the drawback of the diffusion process with a very high rate of release at the beginning, which decreases exponentially with time [44]; finally, the results concerning the process controlled by diffusion as well as by erosion were collected in a more general paper [45] and studied more intensely later [46]. And recently, the plasma drug level has been related to the characteristics of the dosage forms such as the dimensions, the nature of the polymer and thus the diffusivity, for various drugs [61,62]. This method of working published in these papers was presented in 1996 in front of a panel of specialists of the FDA [63].

7.1.1. Assumptions for Calculation

A few necessary assumptions are made in order to define the problem:

(1) The process of drug transport is divided into the following stages that are connected with each other: transport through and out of the dosage form into the gastrointestinal liquid, transport from this drug to the blood through the gastrointestinal membrane with the rate constant of absorption, transport from the blood to the surrounding with the rate constant of elimination (Fig. 41).

(2) The drug release from the dosage form in the gastric liquid is controlled by diffusion with a diffusivity that can be constant or not [64].

(3) The stages of absorption in the blood and elimination out of the blood are described by first-order kinetics.

(4) The rate constant of absorption k_a remains constant along the GIT as well as the rate constant of elimination k_e. The numerical model can take into account a dependence of the rate of drug release with the pH and the GIT history when it is known.

7.1.2. Mathematical Treatment

The rate of drug delivery by the dosage form at time t, whatever its shape, is given by $\mathrm{d}M/\mathrm{d}t$, leading to the amount of drug in the gastrointestine, Y, expressed by the relation

$$\frac{\mathrm{d}Y}{\mathrm{d}t} = \frac{\mathrm{d}M}{\mathrm{d}t} - k_a \cdot Y \qquad (138)$$

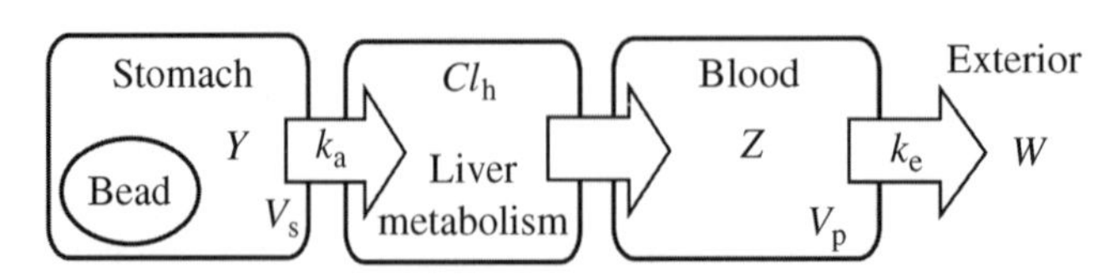

Figure 41. Scheme of the process of drug delivery with the following stages: release of the drug along the GIT controlled by diffusion, absorption into the plasma compartment, and elimination. The first-pass hepatic is shown.

The amount of drug in the plasmatic compartment Z, at time t, is given by

$$\frac{dZ}{dt} = k_a \cdot Y - k_e \cdot Z \qquad (139)$$

while the amount of drug eliminated at time t, W, is expressed by

$$\frac{dW}{dt} = k_e \cdot Z \qquad (140)$$

7.1.3. Numerical Treatment The mathematical treatment of this system of three equations can be resolved only when the drug is release immediately in the GI or when the rate of delivery is constant. In other cases, a numerical method is necessary to solve the problem. It is run step by step, with a constant increment of time Δt.

7.1.4. Expression of the Sustained Release Through the Half-Life Time $T_{0.5}$ With bolus i.v. drug delivery, the main parameter is the half-life $t_{0.5}$. In a similar way, for oral dosage form and especially for sustained-release dosage forms, the time $T_{0.5}$ at which the concentration falls down to half its peak value is the parameter of interest. This time enables one to evaluate in a quantitative way the effect of the polymeric matrix as well as the dimensions of the dosage form on the sustained release of the drug.

7.2. Results Obtained with a Single Dose

Instead of displaying the plasma drug concentration that necessitates good knowledge of the pharmacokinetic data of this drug, the amount of the drug present at each time in the blood as a fraction of the initial amount of the drug in the dosage form is used in all the figures. However, it is easy to determine the drug concentration when the apparent plasmatic volume is known, the drug amount in the blood being proportional to the concentration.

7.2.1. Calculation of the Amount of Drug in Various Places The amount of drug, expressed as a fraction of the amount of drug initially in the dosage forms, is drawn as a function of time in various places and compartments, for ciprofloxacin. In order to make comparisons, the curves resulting from the immediate-release dosage form are also given. The curves are noted as follows:

- 1; 2; 3: kinetics of drug release out of the dosage form;
- 1′; 2′; 3′: drug profiles in the plasma compartment;
- 1″; 2″; 3″: kinetics of elimination of each of the drug.

The curves (1; 1′; 1‴) refer to the immediate-release dosage forms whose half-life time $T'_{0.5}$ in the plasma is 6.6 h, while the curves (2; 2′; 2″) denote the kinetics obtained with controlled-release dosage forms whose half-life time $T_{0.5}$ in the plasma is 8.2 h, and the curves (3; 3′; 3″) represent the kinetics obtained with controlled-release dosage forms whose half-life time $T_{0.5}$ in the plasma is 12 h.

From of all the curves drawn in Fig. 42, the following conclusions are worth noting:

- The kinetics of drug released from the three dosage forms, expressed by the curves (1; 2; 3) are quite different. The kinetics of drug release from immediate release is so fast that the curve 1 is drawn along the axis mentioned for amount.
- Each dosage form (2 and 3), spherical in shape, has a different radius: 87 μm for 2, and 174 for 3. Of course, the larger radius is responsible for the longer time of release and for the more prolonged drug release.
- The characteristics of the plasma drug profiles obtained with these dosage forms are collected in Table 11, by considering the ratio D/R^2 expressed in per second unit as reference.
- In terms of conclusion, it can be seen, that, depending on the value of the main characteristic of the dosage form with controlled release, the plasma drug level is spread over time, with a definite maximum value at peak and for the half-life time $T_{0.5}$.

7.2.2. Effect of the Shape of the Dosage Form The effect of the shape given to the dosage forms is considered in Fig. 43 where the

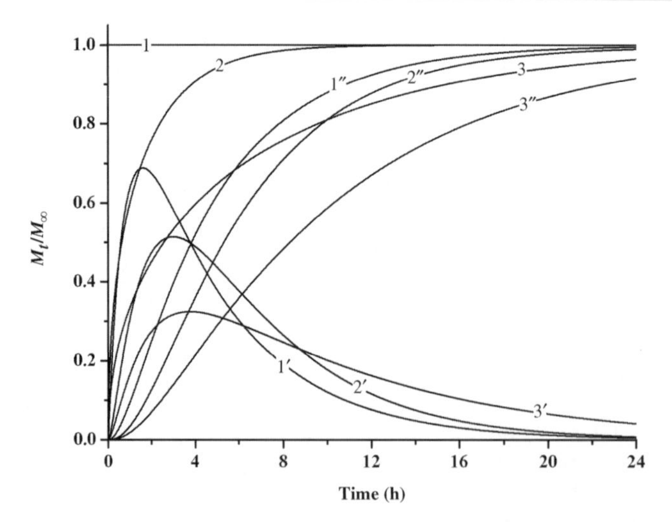

Figure 42. Relative amount of ciprofloxacin versus time obtained with different dosage forms: immediate release (1; 1′; 1″); controlled release from spherical dosage forms with $R = 87\,\mu\text{m}$ (2; 2′; 2″) and with $R = 174\,\mu\text{m}$ (3; 3′; 3″). 1; 2; 3: Kinetics of drug release from the dosage forms. 1′; 2′; 3′: Plasma drug profiles. 1"; 2"; 3": Kinetics of drug elimination out of the plasma. $D = 10^{-9}\text{cm}^2/\text{s}$; $K_a = 1.3\,\text{h}^{-1}$; $K_e = 0.231\,\text{h}^{-1}$.

Table 11. Plasma Drug Level Characteristics for Ciprofloxacin with D/R^2

D/R^2 (s^{-1})	0.57	1.12	Immediate release
$T_{0.5}$ (h)	12	8.2	6.6
M_t/M_∞ max	0.82	0.52	0.68

plasma drug level is drawn for various dosage forms of same volume and different shapes. The same volume specifies that there is the same amount of drug in every dosage form and the concentration of the drug is similar in each polymeric system. The following shapes are examined: parallelepiped with $2a = 2b = c$ (2);

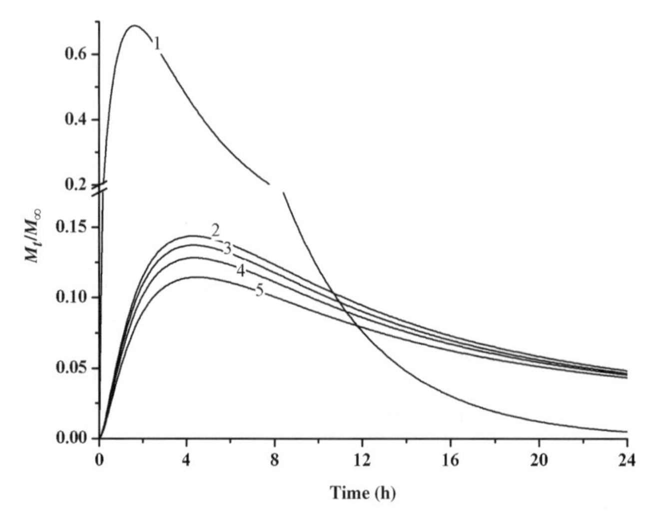

Figure 43. Plasma drug profiles with dosage forms of same volume and various shapes; 1: Immediate release; 2: Parallelepiped with $2a = 2b = c$; 3: Cube; 4: Cylinder with $R = H$; 5: Sphere with $R = 0.182\,\text{cm}$. $D = 10^{-8}\text{cm}^2/\text{s}$; Ciprofloxacin with $K_a = 1.3\,\text{h}^{-1}$; $K_e = 0.231\,\text{h}^{-1}$.

cube (3); cylinder with $R = H$ (4) and a sphere of radius $R = 0.182$ cm, while curve 1 corresponds with the immediate-release dosage form. From these curves, some marks of interest are made:

- The plasma drug levels are far more constant with the released dosage forms than with the immediate-release counterpart, with lower peaks and higher concentration after 24 h of release, which is the trough when these dosage forms are taken once a day.
- The shape given to the dosage forms of same volume plays a minor role. Nevertheless, a more constant level is acquired with the sphere (curve 5) with a lower peak and larger trough.

7.2.3. Effect of the Nature of the Drug Even in the dosage form with controlled release, the drug exhibits its own behavior. As already said (Section 4), the effect of the rate constant of elimination is very important. The plasma drug levels obtained with the four drugs: ciprofloxacin (1), acetyl salicylic acid (2), ofloxacin (3), and péfloxacin (4) are built in Fig. 44. The amount of drug at time t in the blood as a fraction of the amount of drug located in the dosage form is expressed as a function of time (hours), for dosage forms spherical in shape of same radius made of the same polymer matrix with the same diffusivity. The following obvious conclusions can be traced:

- The drug concentration in the plasma is not shown, as it depends on the nature of the drug, with its typical metabolism and volume of distribution. Nevertheless these curves are similar to the blood concentration profiles, as the plasma concentration is proportional to the amount of drug in the blood when the pharmacokinetic is linear.
- The same polymer matrix and dimensions of the dosage form selected for calculation, makes comparison easy between the four drugs ($D = 10^{-8}\,cm^2/s$ and $R = 0.182$ cm).
- The rate at which the drug emerges in the blood compartment is very high for the drugs, and much higher for acetyl salicylic acid (2) that has a higher rate constant of absorption.
- The amount of drug at the peak (or the concentration) is larger for the drug 4 that has the lower rate constant of elimination. The well-known statement

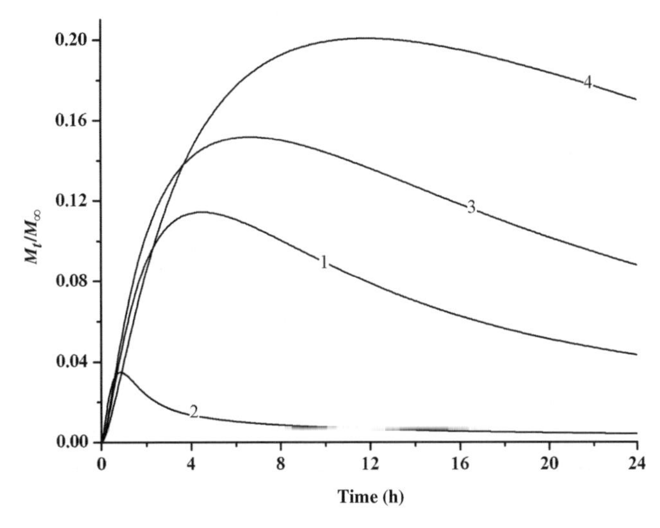

Figure 44. Plasma drug profiles with for drugs in similar sustained-release dosage forms, spherical in shape, made of the same polymer with the same diffusivity. $R = 0.182$ cm. $D = 10^{-8}\,cm^2/s$; Ciprofloxacin (1) with $K_a = 1.3\,h^{-1}$ and $K_e = 0.231\,h^{-1}$; acetylsalicylic acid (2) with $K_a = 3.5\,h^{-1}$ and $K_e = 2.1\,h^{-1}$; ofloxacin (3) with $K_a = 1.5\,h^{-1}$ and $K_e = 0.13\,h^{-1}$; péfloxacin (4) with $K_a = 0.8\,h^{-1}$ and $K_e = 0.07\,h^{-1}$.

holds: the higher the rate constant of elimination, the lower the amount of drug at the peak.

- The concentration of acetyl salicylic acid (2) is very low, because of the high rate of elimination.
- The half-life time obtained with this same polymer matrix varies with the nature of the drug and the following statement remains: the higher the half-life time of the drug in bolus i.v., the higher the half-life time with the sustained-release dosage form.

7.2.4. Effect of the Intervariability of the Patients on the Plasma Drug Profile The best way to determine the effect of the intervariability of the patients is to consider the mean and the extreme values of the pharmacokinetic parameters found in the literature for the drugs, and among them, the rate constant of elimination. Four drugs are selected: ciprofloxacin [10–15,17], acetyl salicylic acid [10,18,64], ofloxacin [16,17], and péfloxacin [16], whose values of their rate constant of elimination are given in Table 12.

The plasma drug profiles calculated with sustained-release dosage forms characterized by the value of the ratio D/R^2 equal to 3×10^{-7} s^{-1}, are shown in Fig. 45 for ciprofloxacin, Fig. 46 for acetyl salicylic acid, Fig. 47 for ofloxacin, and Fig. 48 for péfloxacin. For each drug, the same value of this ratio D/R^2 is used, as well as the three values of the rate constant of elimination shown in Table 12.

These curves and of the values in Table 12, lead to the following comments:

- The polymer matrix of the dosage form is able to prolong the time over which the drug is delivered, but the increase in the half-life time $T_{0.5}$ obtained with the sus-

Table 12. Pharmacokinetic Parameters of the Drugs

Drug	k_e (h)	k_a (h)	$T'_{0.5}$ (h)	$t_{0.5}$ (h)	$T_{0.5}$ (h)
Ciprofloxacin	0.1; 0.231; 0.3	1.3	5.5	6.93; 3; 2.31	35.4; 17.7; 14.3
Acetylsalicylic acid	2.1; 2.5; 2.77	3.5	1	0.33; 0.28; 0.25	2.9; 2.6; 2.4
Ofloxacin	0.1; 0.13; 0.16	1.5	7.8	6.93; 5.33; 4.33	32.5; 28.3; 23.8
Péfloxacin	0.06; 0.07; 0.09	0.8	14	11.5; 9.9; 7.7	54.3; 48; 39.3

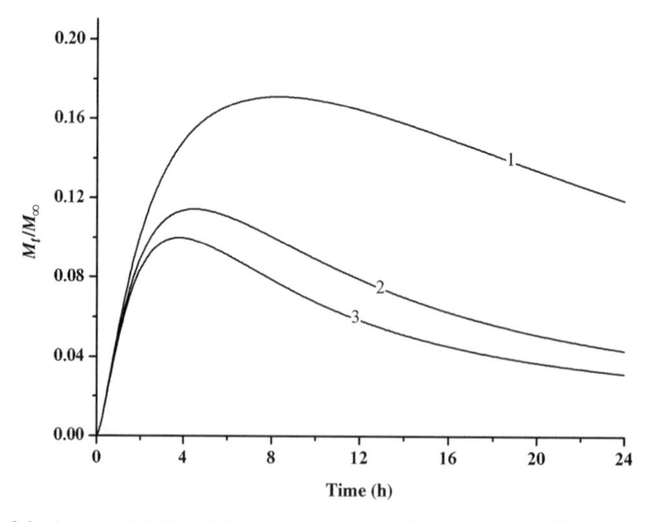

Figure 45. Effect of the intervariability of the patients on the plasma drug profiles obtained for ciprofloxacin in the sustained-release dosage forms with. $D = 10^{-8}\,cm^2/s$; $R = 0.182\,cm$. $K_a = 1.3\,h^{-1}$ and $K_e = 0.1\,h^{-1}$ (1); $K_e = 0.231\,h^{-1}$ (2); $K_e = 0.3\,h^{-1}$ (3), respectively.

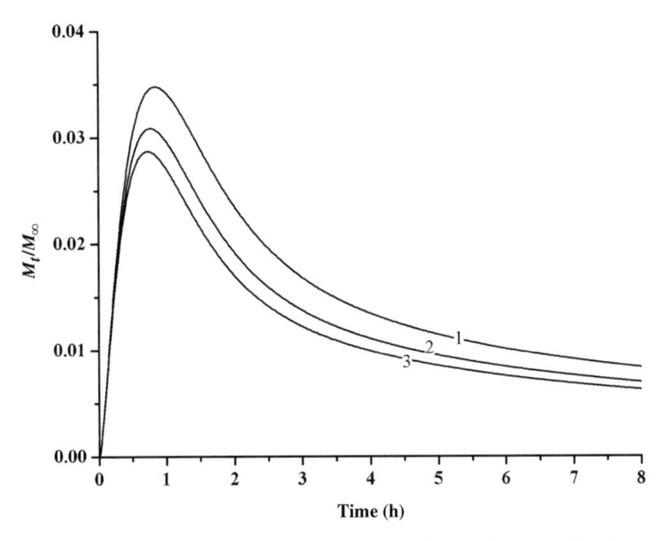

Figure 46. Effect of the intervariability of the patients on the plasma drug profiles for acetyl salicylic acid in the sustained-release dosage forms with. $D = 10^{-8} cm^2/s$; $R = 0.182$ cm. $K_a = 3.5 h^{-1}$ and $K_e = 2.1 h^{-1}$ (1); $K_e = 2.5 h^{-1}$ (2); $K_e = 2.77 h^{-1}$ (3), respectively.

tained-release dosage form also largely depends on the half-life time $t_{0.5}$ of the drug delivered through bolus i.v. Thus, in the case of ciprofloxacin, for the mean value of $t_{0.5}$, the corresponding value of $T'_{0.5}$ is multiplied by nearly 2. The half-life times $T_{0.5}$ allowed by using this sustained release system are approximately three times longer than the corresponding values $T'_{0.5}$ obtained with the immediate-release dosage forms for the same drug.

- It makes it clear that the intervariability exists not only when the drug is delivered through bolus i.v. but also in the case of oral sustained-release dosage forms.

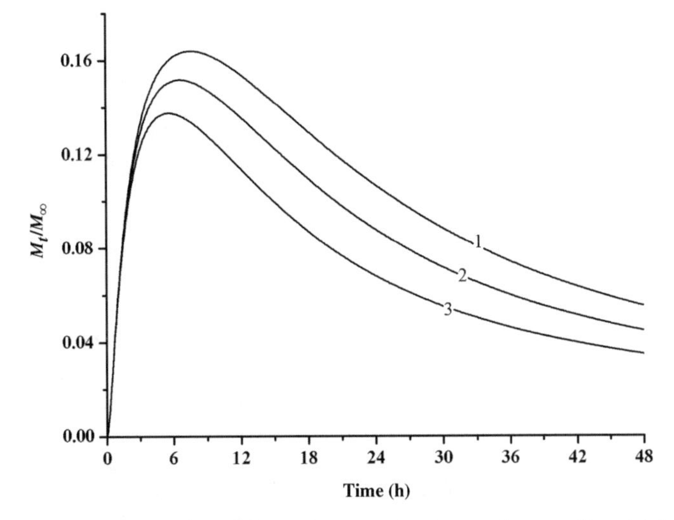

Figure 47. Effect of the intervariability of the patients on the plasma drug profiles obtained for ofloxacin in the sustained-release dosage forms with $D = 10^{-8} cm^2/s$. $R = 0.182$ cm. $K_a = 1.5 h^{-1}$, and, respectively, for $K_e = 0.11 h^{-1}$ (1); $k_e = 0.13 h^{-1}$ (2); $k_e = 0.16 h^{-1}$ (3).

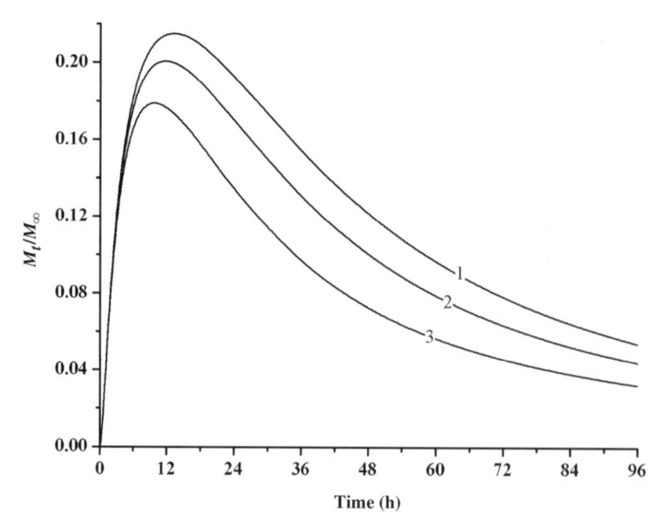

Figure 48. Effect of the intervariability of the patients on the plasma drug profiles obtained for péfloxacin in the sustained-release dosage forms with $D = 10^{-8} cm^2/s$. $R = 0.182$ cm. $K_a = 0.8 h^{-1}$, and, respectively, for $K_e = 0.06 h^{-1}$ (1); $k_e = 0.07 h^{-1}$ (2); $k_e = 0.09 h^{-1}$ (3).

Nevertheless, if we consider the ratio of the larger value for $T_{0.5}$ to the lower one with the sustained-release dosage form, it is found 2.5 for ciprofloxacin, while the same ratio is 3 for the half-life times obtained through bolus i.v. The same fact can be observed for acetyl salicylic acid, since the ratios of the half-life times are, respectively, 1.2 and 1.32, as well as for the other two drugs. Thus, more constant plasma drug profiles can be expected over a long period of time with repeated doses, when they are delivered with the appropriate sustained-release dosage forms, and that, much better than with either bolus iv or oral immediate-release dosage forms.

- The variation in the concentration at the peaks (measured at the maximum value of the ratio M_{Max}/M_{in}) could also be of interest. These curves drawn in Figs 144.45-144.48 show that the effect of the intervariability of the patients is of great concern, whatever the system of drug delivery, the values of the ratio of these maximum values measured in the extreme cases of the rate constant of elimination being 1.7, 1.2, 1.2, and 1.2 for these four drugs, respectively, when they are dispersed through a polymer matrix characterized by the ratio of $3 \times 10^{-7} s^{-1}$.

7.3. Plasma Drug Profiles with repeated Doses

7.3.1. Effect of the Nature of the Drug in the Same Polymer Matrix The four dosage forms previously studied when delivered with one dose (Section 7.2.3) are now considered in multidosing taken once a day. The plasma drug profiles are drawn in Fig. 49 for ciprofloxacin (curve 1), acetyl salicylic acid (2), ciprofloxacin (3), and péfloxacin (4), by plotting the amount of drug in the blood as a fraction of the amount of drug initially in the dosage form, as a function of time. In all these cases, the polymer matrix is the same, defined by the ration of the diffusivity by the square of the radius $D/R^2 = 3 \times 10^{-7} s^{-1}$.

The following conclusions can be drawn:

- The advantage of the controlled release systems over their immediate-release counterparts appears clearly for every drug, since the dosage forms with controlled release are responsible for more constant plasma drug levels, with lower peaks and higher troughs.

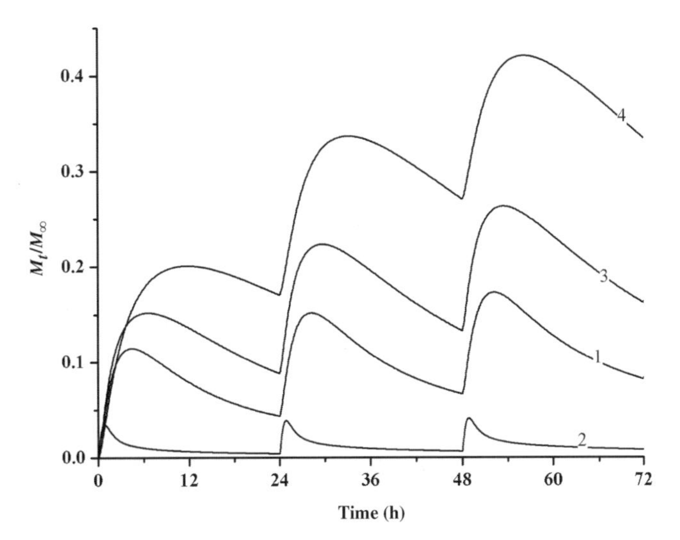

Figure 49. Plasma drug profiles with drugs in similar sustained-release dosage forms, spherical in shape, made of the same polymer with the same diffusivity. $R = 0.182\,\text{cm}$. $D = 10^{-8}\text{cm}^2/\text{s}$, taken once a day; Ciprofloxacin (1) with $K_a = 1.3\,\text{h}^{-1}$ and $k_e = 0.231\,\text{h}^{-1}$; acetylsalicylic acid (2) with $k_a = 3.5\,\text{h}^{-1}$ and $k_e = 2.1\,\text{h}^{-1}$; ofloxacin (3) with $k_a = 1.5\,\text{h}^{-1}$ and $k_e = 0.13\,\text{h}^{-1}$; péfloxacin (4) with $k_a = 0.8\,\text{h}^{-1}$ and $k_e = 0.07\,\text{h}^{-1}$.

- The value of the ratio D/R^2 in all cases is $3 \times 10^{-7}\,\text{s}^{-1}$, whatever the nature of the dosage forms. This value is convenient for ciprofloxacin (1) and for ofloxacin (3), leading to rather constant plasma drug level. Nevertheless, it is far too large for acetyl salicylic acid (2), and too small for péfloxacin (4).
- Of course, the so-called steady state (meaning that the profiles are similar for each dose taken in succession) in only attained when the trough associated with the previous dose is very low. Thus for ciprofloxacin (1) and for ofloxacin (3), it is attained at the fourth dose.
- Recall that the once a day dosage is surely the best one, for various reasons, the one resulting from the easier compliance of the patient as well as his comfort.

7.3.2. Effect of the Intervariability of the Patients The effect of the intervariability of the patients on the plasma drug profiles is shown for the four drugs: ciprofloxacin (Fig. 50), acetyl salicylic acid (Fig. 51), ofloxacin (Fig. 52), and péfloxacin (Fig. 53), when they are dispersed through the same spherical matrix characterized by the ratio $D/R^2 = 3 \times 10^{-7}\,\text{s}^{-1}$.

A few comments can be given from the observation of the curves in Figs 50–53:

- First at all, the frequency of the doses should be quite different for the four drugs. The dosage forms with ciprofloxacin (Fig. 50) and ofloxacin (Fig. 52) should be taken once a day. But because of its high rate of elimination, acetyl salicylic acid (Fig. 51) should be taken only twice a day; and, resulting from its very low rate of elimination, péfloxacin (Fig. 53) could be taken every other day.
- In all cases, the effect of the intervariability of the patients clearly appears, either after the first dose or during the course of the treatment.
- Of course, the following statement holds: the lower the rate constant of elimination, the larger the plasma drug level with a higher peak.

Let us remark the case of acetyl salicylic acid (Fig. 51). Because of the high rate of elimination, the concentration of the drug in the plasma is very low with this dosage form. To obtain a rather flat plasma drug level, the diffusivity D should be low and the value of the radius D rather high.

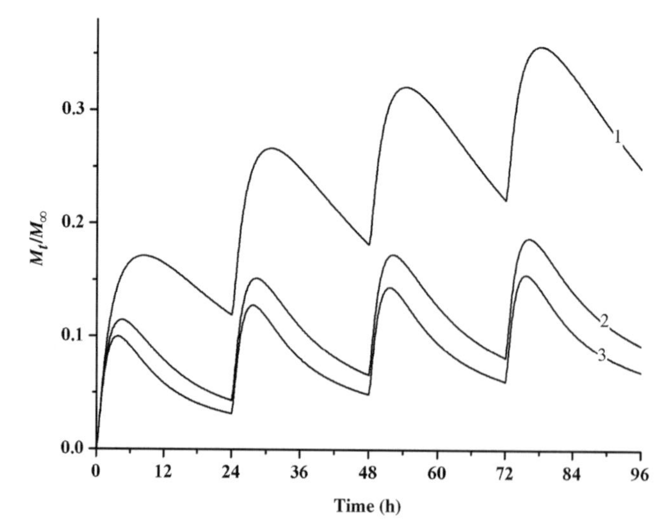

Figure 50. Effect of the intervariability of the patients on the plasma drug profiles obtained for ciprofloxacin in the sustained-release dosage forms, taken once a day; with. $D = 10^{-8}\text{cm}^2/\text{s}$; $R = 0.182\,\text{cm}$. $k_a = 1.3\,\text{h}^{-1}$ and $k_e = 0.1\,\text{h}^{-1}$ (1); k_e $0.231\,\text{h}^{-1}$ (2); k_e $0.3\,\text{h}^{-1}$ (3).

Thus the solution could be a great number of granules.

- It is also clear that there is some difficulty to adapt the treatment to the patient, for two reasons: first, the pharmacokinetic parameters of the patient should be evaluated by i.v. before starting the treatment; second, the dosage forms should be presented in such a way that they could be divided in various parts.

7.3.3. Reasons for the Patients' Noncompliance

Some precise investigations conducted in the United States prove that not less than 40–60% of patients fail to comply with the regimen [65,66]. In a few words, the compliance

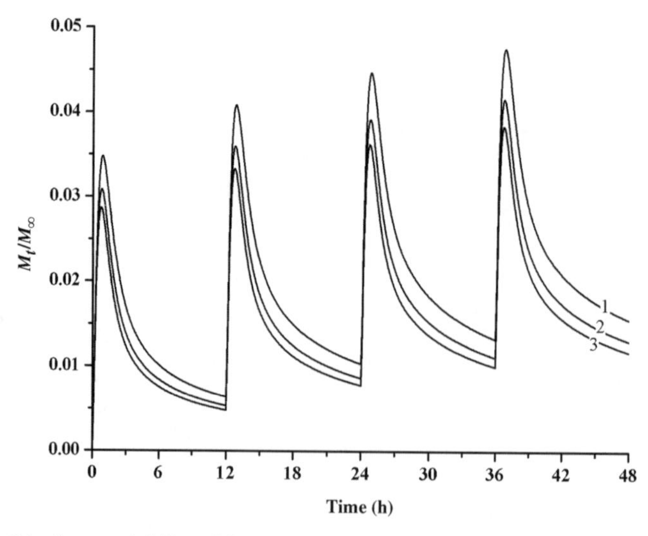

Figure 51. Effect of the intervariability of the patients on the plasma drug profiles for acetyl salicylic acid in the sustained-release dosage forms with. $D = 10^{-8}\text{cm}^2/\text{s}$; $R = 0.182\,\text{cm}$. taken twice a day; $k_a = 3.5\,\text{h}^{-1}$ and $k_e = 2.1\,\text{h}^{-1}$ (1); $k_e = 2.5\,\text{h}^{-1}$ (2); $k_e = 2.77\,\text{h}^{-1}$ (3)

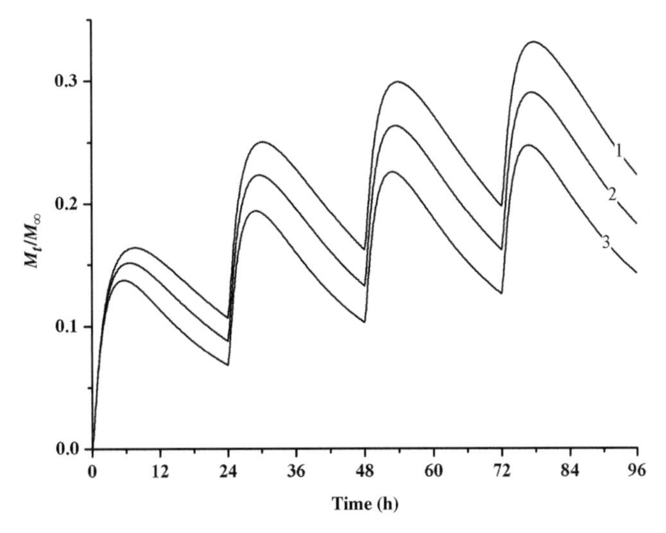

Figure 52. Effect of the intervariability of the patients on the plasma drug profiles obtained for ofloxacin in the sustained-release dosage forms with $D = 10^{-8} cm^2/s$. $R = 0.182$ cm. taken once a day; $k_a = 1.5 h^{-1}$, and for $k_e = 0.11 h^{-1}$ (1); $k_e = 0.13 h^{-1}$ (2); $k_e = 0.16 h^{-1}$ (3).

of the patient is worse than it could be believed, dependent on the disease pattern, and variable with respect to the treating physician. The main causes of noncompliance refer primarily to the following three factors: nature of the disease, type of medication, interaction of the physician or pharmacist with the patient.

Nature of the Disease Severe symptoms and obvious signs of the relentless progress of a cruel disease cause a consciousness of illness that exerts a psychological pressure on the patient. This pressure is able to stimulate a great reliability in following the prescribed treatment. On the other hand, patients have more difficulty in complying with the

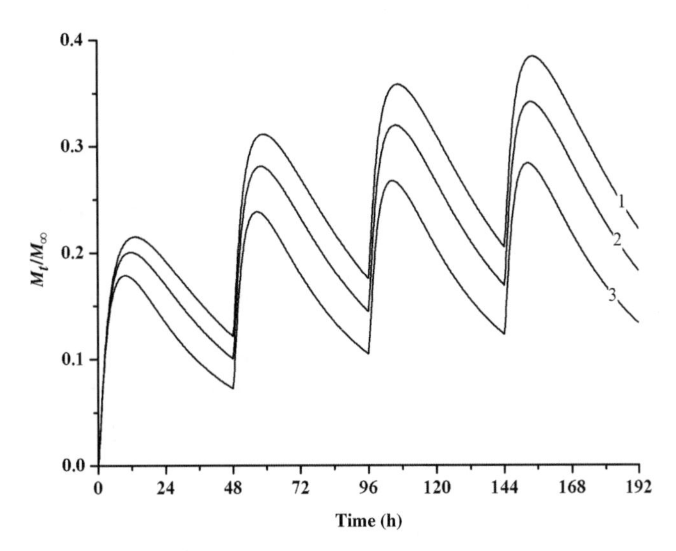

Figure 53. Effect of the intervariability of the patients on the plasma drug profiles obtained for péfloxacin in the sustained-release dosage forms with $D = 10^{-8} cm^2/s$. $R = 0.182$ cm. $k_a = 0.8 h^{-1}$, and, respectively, for $k_e = 0.06 h^{-1}$ (1); $k_e = 0.07 h^{-1}$ (2); $k_e = 0.09 h^{-1}$ (3). Taken every other day.

treatment of a disease causing no distinct complaint, such as high blood pressure. Moreover, it seems difficult to patients to appreciate the necessity of the therapy when this therapy causes disturbing side effects, while the disease gives rise to little complaint.

Type of Medication The easiness of the patient to cooperate with the physician decreases regularly with the duration of the treatment. And willingness to cooperate also decreases with increasing complexity of the regimen, the length of the treatment, the complexity of dosing prescriptions. These problems become particularly serious in chronic diseases.

Influences on the Patients It was found that the influence of the physician and the pharmacist is decisive on the patient's compliance [65,66]. The patient has to be informed precisely of "why" the drug must be taken, and "how often," "how long," and "at what dose." The treating physician is the right person responsible for providing information on the "why" and "how" of the drug administration, while the pharmacist shares this responsibility in bringing his professional support.

Various Cases of Noncompliance Forgetfulness, the first case of noncompliance, can have serious consequences in drug treatment. It should be said that if the patient omits one or two doses, this omission cannot be repaired by any subsequent regular dosage or worse, by an adapted dosage determined by simple calculation.

Therapeutic systems certainly offer the opportunity of improving the patient compliance by reducing some difficulties connected with conventional dosage forms. First, all compliance studies make it clear that reducing the frequency of drug administration improves compliance, the once a day dosing being the ideal objective. Second, the patient would understand that increasing compliance is associated with reducing his bad and painful side effects.

Moreover, a simple way to convince patients that compliance is beneficial and noncompliance may be dangerous is to show them the resulting plasma drug levels obtained with omission, wrong dosage at the wrong time. This is especially obvious and necessary with sustained-release dosage forms containing large amount of drugs.

As far as know, only a few quantitative studies have been made on the problem of noncompliance by considering the process of drug delivery, in order to evaluate the change in the plasma drug profiles resulting from the patients' noncompliance [65–67].

7.3.4. Effect of Noncompliance on the Plasma Drug Level Perfect compliance with once a day dosing is associated with regular plasma drug profiles, as depicted in Fig. 54 with ciprofloxacin taken regularly once a day over 4 days. A regular profile is obtained with peaks and troughs located between the minimal curative and the maximal tolerated level, in spite of the fact already seen that the drug concentration increases regularly up to the steady state.

In Fig. 55 appears the plasma drug level with an omission the second day. Of course, a deep decrease in the first trough is prolonged over one day.

In Fig. 56 is drawn the plasma drug profile obtained with the worst case of noncompliance, when the patient having missed the second dose compensates in taking a double dose the third day. The concentration thus reaches a very high level, which could pass through the median lethal level, provoking at least bad side effects to occur, if not worse.

7.4. Prediction of the Characteristics of the Controlled-Release Dosage Form

As an extension of the results shown in this section, it is of interest to know the main characteristics of the sustained release that are necessary, before building the dosage form able to obtain the desired plasma drug level.

7.4.1. Relationships Between the Half-Life Times $T_{0.5}$ and $t_{0.5}$ Relationships between the half-life time obtained when the drug is delivered through bolus i.v. and the other half-life times obtained with various sustained-release dosage forms would be useful to predict the right sustained-release dosage form able to comply with the desired requirement of therapy. These relationships are expressed in the way of two diagrams, the one (Fig. 57) when the half-life time $t_{0.5}$ is between 1 and 10 h, and

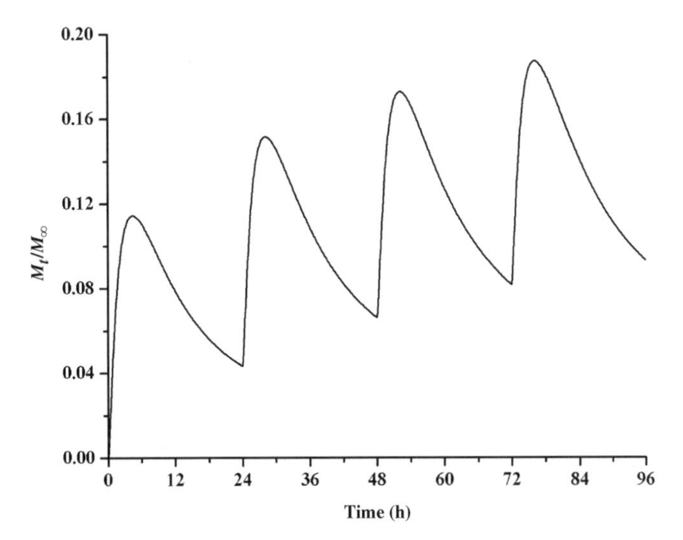

Figure 54. Effect of the patient's noncompliance. The dose is taken regularly once a day without omission regularly over 4 days. Ciprofloxacin with $D = 10^{-8} cm^2/s$; $R = 0.182$ cm. $k_a = 1.3 h^{-1}$ and: $k_e = 0.231 h^{-1}$ (1).

the other (Fig. 58) when the half-life time $t_{0.5}$ is far much lower, between 0 and 1 h.

From the observation of these two figures, the following conclusions can be drawn:

- These diagrams are built by using various values of the ratio D/R^2 expressed in terms of s^{-1}, which combines the characteristics of the sustained release system, for example, the diffusivity of the polymer matrix and the main dimension of the dosage form, which is either the radius in the case of a sphere or the smaller size taken in any other shape of the dosage forms.
- The values of the ratio D/R^2 s^{-1} should be larger than $10^{-7} s^{-1}$, because lower values of this ratio give rise to a very low

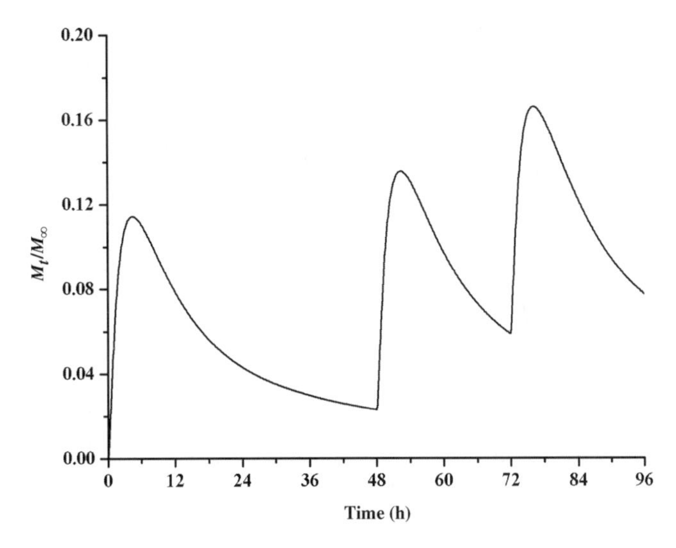

Figure 55. Effect of the patient's noncompliance. The dose is taken the first day, omitted the second day, and taken regularly the third and fourth day. Ciprofloxacin with $D = 10^{-8} cm^2/s$; $R = 0.182$ cm. $k_a = 1.3 h^{-1}$ and $k_e = 0.231 h^{-1}$ (1).

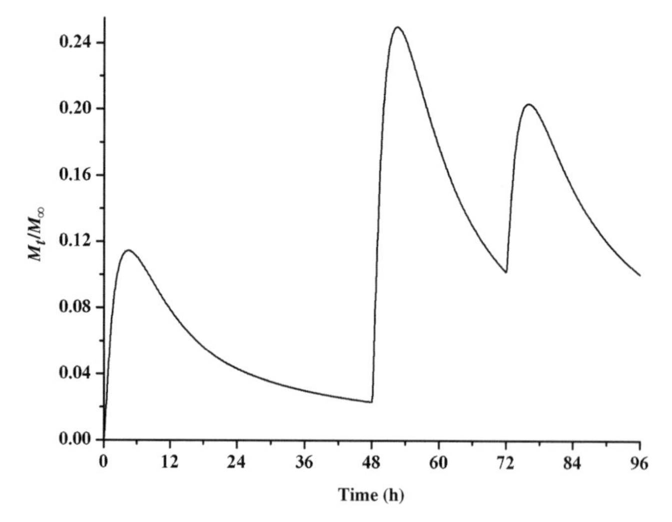

Figure 56. Effect of the patient's noncompliance. The dose is taken the first day, omitted the second day, taken with a double dose the third, and a single dose the fourth day. Ciprofloxacin with $D = 10^{-8} \mathrm{cm}^2/\mathrm{s}$; $R = 0.182\,\mathrm{cm}$. $k_a = 1.3\,\mathrm{h}^{-1}$ and $k_e = 0.231\,\mathrm{h}^{-1}$ (1).

plasma drug level, as the radius for the spherical dosage form cannot be larger than 0.5 cm., and the corresponding value of the diffusivity would be responsible for a very low rate of drug release from the dosage form.

- Two diagrams are necessary to express the results, the one with $1 < t_{0.5} < 10$ (Fig. 57), and the other with $0 < t_{0.5} < 1$ (Fig. 58) for a better accuracy, especially for the very low values of the half-life time obtained when the drug is delivered through bolus i.v.
- It clearly appears that the obvious statement holds: the higher the value of $t_{0.5}$, the higher the value of $T_{0.5}$ obtained with

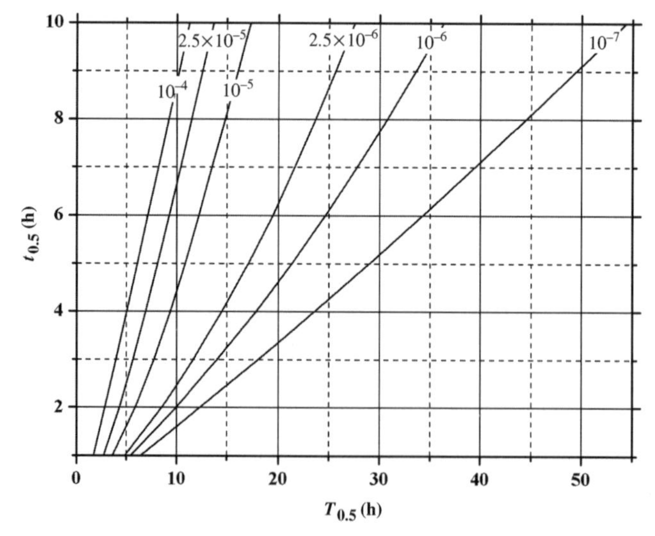

Figure 57. Diagram expressing the relation between the half-life time $t_{0.5}$(h) obtained with bolus i.v. injection and the half-life time $T_{0.5}$(h) obtained with release dosage form, with various values of the ratio $D/R^2/s$ with $1 < t_{0.5} < 10$.

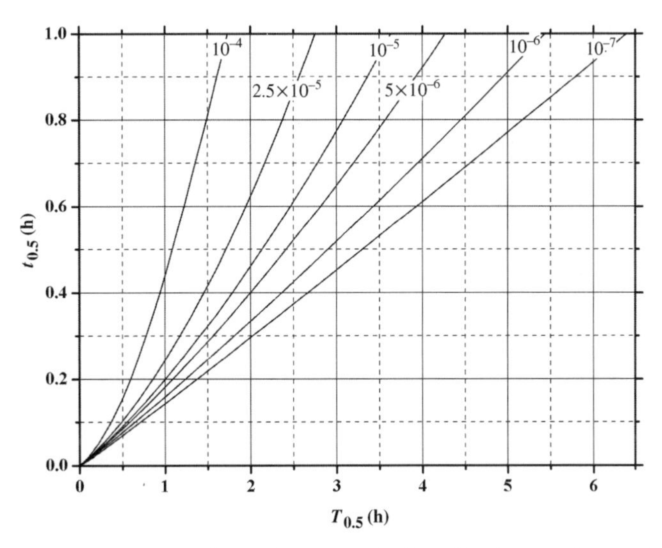

Figure 58. Diagram expressing the relation between the half-life time $t_{0.5}$(h) obtained with bolus i.v. injection and the half-life time $T_{0.5}$(h) obtained with release dosage form, with various values of the ratio $D/R^2/s$ with $0 < t_{0.5} < 1$.

the sustained release system. Thus, for a drug such as acetyl salicylic acid whose half-life time through bolus i.v. is as low as 0.33 h, the larger value of the half-life time $T_{0.5}$ obtained with a sustained-release dosage form cannot be expected much more than 2–2.5 h.

7.4.2. Plasma Drug Profile Expressed in Terms of Concentration The various plasma drug profiles have been expressed in terms of time for the abscissa, by considering the amount of drug in the plasma as a fraction of the amount initially in the dosage forms for the ordinate. The interest of this presentation is that a dimensionless number appears in the ordinate that can be used whatever the amount of drug initially in the dosage form. But physicians are more sensitive to the concentration, and the analysis leads directly to the drug concentration, it is necessary to be able to convert the above ordinate into the plasma drug concentration. The conversion is easy by using the apparent volume of distribution V_p (which can be highly different from the plasma volume) and the amount initially dispersed into the dosage form, as there is proportionality between the amount of drug in the plasma and the resulting concentration. However, besides the apparent volume of distribution, another pharmacokinetic parameter appears of interest with the rate constant of elimination or the half-life time, which is determined with bolus i.v. Nevertheless, when these two parameters are known, the convertibility of the amount of drug into the drug concentration is easy to obtain, whatever may be the values of these two pharmacokinetic parameters and the relation, which exist between them. Thus, an example is given in the case of the intervariability of the patients: the plasma drug profiles shown in Fig. 45 for ciprofloxacin (Section 7.2.4) expressed in terms of the ratio of mass of drug M/M_{in} are converted into plasma drug concentrations in Fig. 59. The six curves represent the intervariability of the patients by considering not only the change in the rate constant of elimination but also two extreme values of the apparent plasmatic volume (150 and 225 L).

7.5. Conclusions on Controlled-Release Dosage Forms

A few conclusions can be drawn on the dosage forms whose release is controlled by diffusion:

- Because of the stability of the polymer, these dosage forms pass through the GIT without any problem.

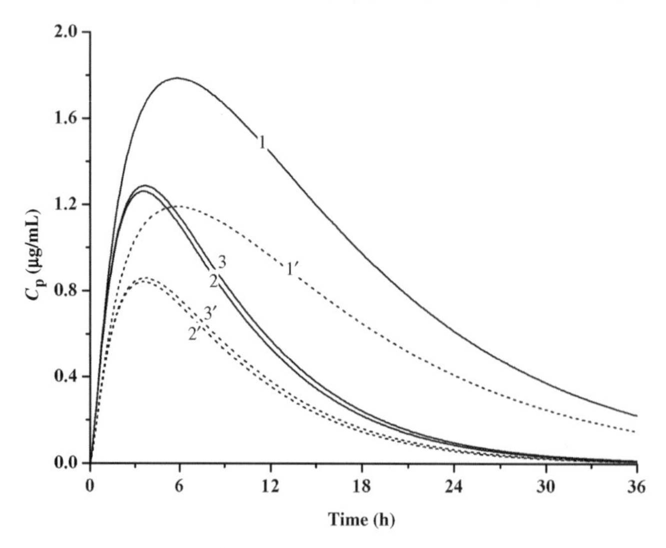

Figure 59. Plasma drug concentration of ciprofloxacin as a function of time with a single dose. Intervariability of the patients with the simultaneous effect of the apparent plasmatic volume and of the rate constant of elimination. $k_a = 1.3\,h^{-1}$ and $k_e = 0.09\,h^{-1}$ (1 and 1′); and $k_e = 0.231\,h^{-1}$ (2 and 2′); and $k_e = 0.22\,h^{-1}$ (3 and 3′) with $V_p = 150\,L$ (1; 2; 3) and $V_p = 225\,L$ (1′; 2′; 3′); $M_{in} = 500\,mg$.

- The pharmacokinetic parameters of the drug with the patient remain the main parameters.
- Without doubt, the main interest for these types of dosage forms stands for the possibility of taking them once a day. This fact not only allows the comfort of the patient but also enables him to follow perfectly well the regimen.
- A question arises either for the company in charge of preparing the dosage forms or for the therapists: how to prepare the right formulation for any drug, whatever its pharmacokinetic parameters. By using the diagrams drawn in Figs 57 and 58 where the half-life times obtained with bolus i.v. $t_{0.5}$ or with the dosage form $T_{0.5}$ are connected with each other, it is easy to select the right parameters necessary to define the dosage form for a given purpose and for any drug. Thus, examples are given for the following two drugs: With ciprofloxacin (Fig. 60) whose plasma drug profiles are drawn using three values of the ratio D/R^2 ranging from $10^{-6}\,s^{-1}$ (curve 1) to $10^{-7}\,s^{-1}$ (curve 3); the rather difficult problem set out with péfloxacin because of its very large rate of elimination ($t_{0.5} = 9.9\,h$) has found a solution with three large values of the ratio D/R^2 ranging from $1.5.10^{-5}\,s^{-1}$ (1) to $0.9.10^{-5}\,s^{-1}$ (3) in Fig. 61.

8. PLASMA DRUG PROFILE WITH EROSION-CONTROLLED DOSAGE FORMS

Nomenclature

AUC	area under the curve;
C	plasma drug concentration;
GIT	gastrointestinal (tract time);
H	half the length of a cylinder;
i.v.	intravenous bolus injection;
k_a	rate constant of absorption of the drug (per hour);
k_e	rate constant of elimination of the drug (per hour);
L	half the thickness of the sheet;
M_t	amount of drug released up to time t;
M_{in}	amount of drug initially in the dosage form;

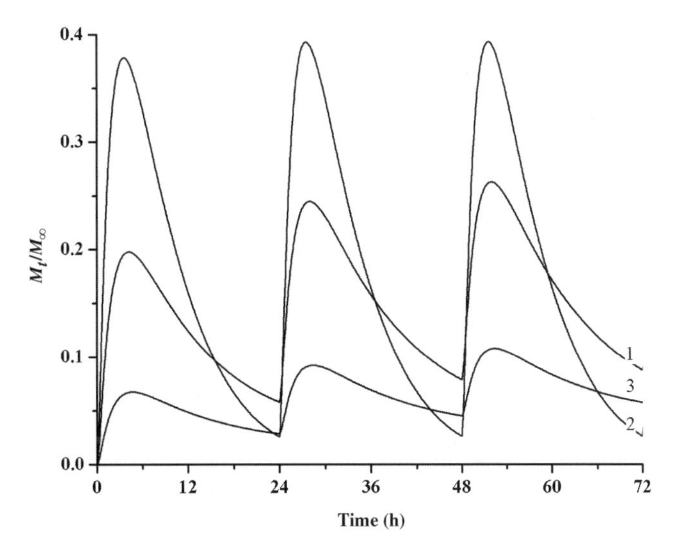

Figure 60. Drug plasmatic profiles for various oral dosage forms, taken once a day, with ciprofloxacin, and various polymers defined by $D/R^2 = 10^{-6}\,\text{s}^{-1}$ (1); $5 \times 10^{-6}\,\text{s}^{-1}$ (2); $10^{-7}\,\text{s}^{-1}$ (3). $k_e = 0.231\,\text{h}^{-1}$ and $t_{0.5} = 3\,\text{h}$. $k_a = 1.3\,\text{h}^{-1}$.

M_t/M_{in} dimensionless number expressing the drug release from the dosage form;

dM/dt rate of drug release by the dosage form in the gastrointestine at time t;

R radius of the sphere or radius of the cylinder;

t, t_r time and time of full erosion, respectively;

$t_{0.5}$ half-life time of the drug delivered in bolus i.v.;

$T_{0.5}$ half-life time of the drug delivered with the sustained-release dosage form;

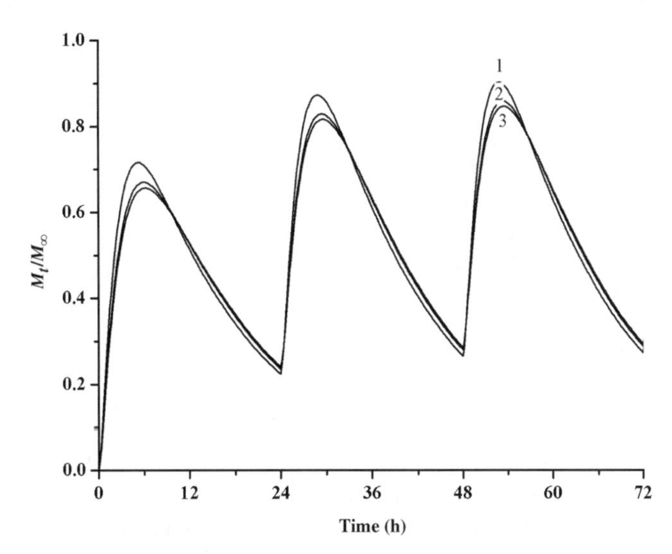

Figure 61. Drug plasmatic profiles for various oral dosage forms, taken once a day, with péfloxacin and various polymers defined by $D/R^2 = 1.5 \times 10^{-5}\,\text{s}^{-1}$ (1); 10^{-5}(2); $0.9 \times 10^{-5}\,\text{s}^{-1}$ (3). $k_e = 0.07\,\text{h}^{-1}$ and $t_{0.5} = 9.9\,\text{h}$. $k_a = 0.8\,\text{h}^{-1}$.

V_p apparent plasmatic volume (L);
v linear rate of erosion (cm/s);
W amount of drug eliminated (Eq. 143);
Y amount of drug in the gastrointestinal volume (Eq. 141);
Z amount of drug in the plasmatic volume (Eq. 142).

Oral dosage forms with release controlled by erosion are prepared by dispersing the drug through an erodible polymer. Some erodible polymers exist for which the process seems to be strictly controlled by erosion [68–71]. Biodegradable polymers can be obtained in two ways, either by using pure polymers or by introducing some additives in the polymer that dissolve by provoking a progressive disintegration. An example of the first case is observed with Na carboxymethyl cellulose, the dissolution being related to the rheological properties of the gelled polymer [72]. Another example has been pointed out with Gelucires for which the process of drug release is controlled either by diffusion with lipidic material or in the other case by erosion [68,71], depending on the hydrophilic–lipidic balance.

Moreover, the great interest of these erodible dosage forms results from the fact that they can be used in bioadhesive delivery systems, as over a given period of time it is sure that all the components of the dosage forms are eliminated. Thus, the problem of bioadhesion becomes of high concern. Various books and reviews have tempted to present an extensive review on these polymers [73,74], and only a few examples give a brief overview on the importance of the problems encountered [75–78]. On the whole, mucoadhesion has been related to the rheological properties of the polymer [59,79,80] as well as to its molecular weight by considering the poly-dispersity [81] or in insisting on the moisture absorption of hydroxypropyl cellulose [81]. Various methods and techniques have been used to study the mechanism of bioadhesion [58,83–85] and its subsequent effect with the retention of the dosage form [60]. Thus these various studies have tempted to give the answer to the fundamental questions that have been put forward concerning the bioadhesion [86] and its consequences for the dosage forms with erosion-controlled release.

8.1. Method of Calculation

The mathematical treatment of the kinetics of release of the drug from dosage forms controlled by erosion obtained has been developed in Section 6. Thus, the equations have been established for various shapes, expressing the kinetics of release of the drug as they are determined through *in vitro* measurements.

8.1.1. Assumptions for Calculating the Plasma Drug Profile

(1) The process of drug transfer is described in Fig. 62 where the following stages appear in succession: release of the drug out of the dosage form along the GI tract, kinetics of absorption into and elimination out of the plasma compartment.
(2) The process of drug release out of the dosage form is controlled by erosion, with a constant rate (in fact, this strict assumption is not necessary provided that the process of the change in the rate of erosion along the GIT is known).
(3) The phenomena of absorption and elimination are described by first-order kinetics with the two rate constants k_a and k_e expressed in h^{-1}.
(4) The rate constant of absorption remains constant along the GI tract time (in fact, the numerical model can take into account any change in this rate constant).
(5) The pharmacokinetics process is linear, meaning that the drug concentration in the plasma is proportional to the dose.

8.1.2. Mathematical and Numerical Treatment

The rate of drug delivery by the dosage form at time t, whatever the shape of the dosage form, is expressed by the derivative dM/dt.

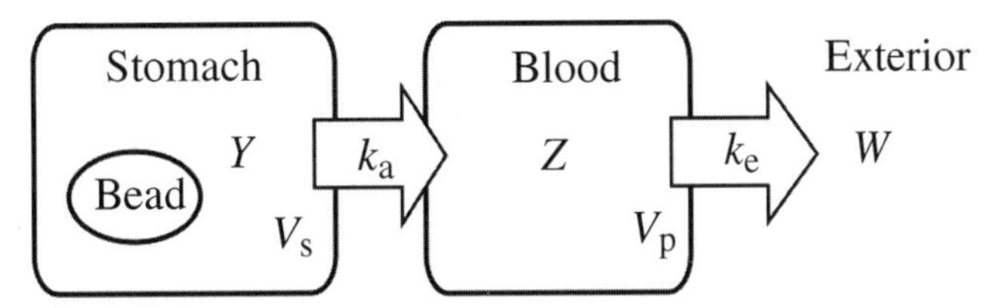

Figure 62. Scheme of the process of drug delivery in the body.

The amount of drug in the gastrointestine is Y_t, related to the rate constant of absorption k_a by the following equation:

$$\frac{dY}{dt} = \frac{dM}{dt} - k_a \cdot Y \tag{141}$$

The amount of drug in the plasma compartment Z_t is expressed by

$$\frac{dZ}{dt} = k_a \cdot Y - k_e \cdot Z \tag{142}$$

and the amount of drug eliminated at time t is W:

$$\frac{dW}{dt} = k_e \cdot Z \tag{143}$$

As the mathematical treatment is not feasible, the plasma drug level Z, as well as the kinetics of elimination, are evaluated by using a numerical method, run step by step, with constant increment of time.

8.2. Plasma Drug Profile with a Single Dose

Various factors intervene in the case of oral dosage forms whose release is controlled by erosion: the shape of the dosage form, and much more important, the time of full release t_r, this time taking into account the dimensions of the dosage form, a well as the nature of the drug with its pharmacokinetic parameters.

8.2.1. Effect of the Shape of the Dosage Forms The effect of the shape given to the dosage forms on the kinetics of release is very small, whatever the release is controlled by diffusion or erosion [46,49], provided that their volume is the same. Thus, spherical dosage forms will be considered in this section.

8.2.2. Effect of the Nature of the Drug The effect of the nature of the drug on the plasma drug profiles is studied by considering the time of full erosion of 48 h and the following drugs in Fig. 63: ciprofloxacin (curve 1), acetyl salicylic acid (curve 2), ofloxacin (3), and péfloxacin (4) whose pharmacokinetics parameters are shown in the legend of this figure. These curves expressing the variation of the amount of drug in the plasma as a fraction of the amount initially in the dosage form with time lead to a few comments:

- The plasma drug profile largely depends on the nature of the drug, and the statement holds: the larger the rate constant of elimination, the faster the rate of drug elimination and the lower the plasma drug concentration.
- The peak for the profile of péfloxacin, which exhibits the lower rate constant of elimination is not only the larger but also attained at a longer time.
- The rate constant of absorption does not intervene much in the plasma drug profile.

8.2.3. Effect of the Intervariability of the Patients The effect of the intervariability of the patients is considered in evaluating the plasma drug profile by keeping the same time of full erosion of 48 h and by varying the value of the rate constant of elimination of the drugs within the range defined by various authors. Thus, the plasma drug profiles are drawn for ciprofloxacin in Fig. 64 [11–15,22] and in Fig. 65 for acetyl salicylic acid in [18,87]. The variability of the patients toward the other drugs has been shown in the preceding chapters, with these following references: péfloxacin [16] and ofloxacin [17].

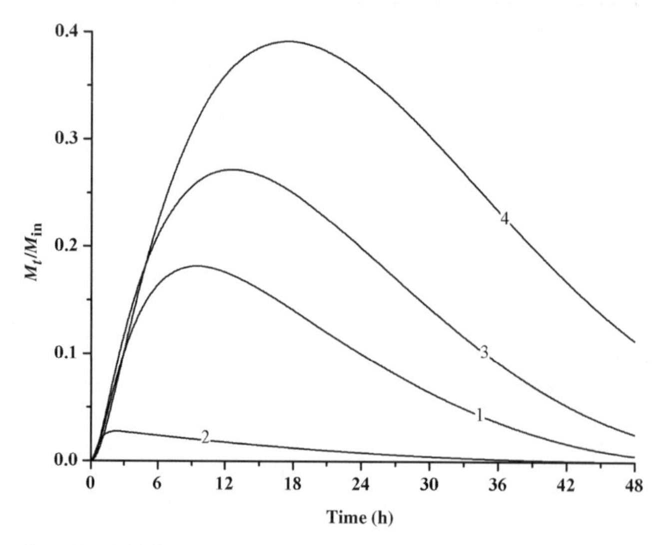

Figure 63. Plasma drug level M_t/M_{in} obtained with various drugs in spherical dosage forms and the same conditions. (1) Ciprofloxacin with $k_a = 1.3\,h^{-1}$ and $k_e = 0.231\,h^{-1}$. (2) Acetyl salicylic acid with $k_a = 3.5\,h^{-1}$ and $k_e = 2.1\,h^{-1}$. (3) Ofloxacin with $k_a = 1.5\,h^{-1}$ and $k_e = 0.13\,h^{-1}$. (4) Péfloxacin with $k_a = 0.8\,h^{-1}$ and $k_e = 0.07\,h^{-1}$. Time of erosion $t_r = 48\,h$.

Some conclusions of concern appear when it is desired to adapt the therapy to the patient:

- The effect of the intervariability of the patients toward the drugs results from the change in the rate constant of elimination. Thus, the statement holds: the larger the rate constant of elimination (or the shorter the half-life time), the lower the concentration of the drug in the plasma and the shorter the new half-life time obtained with the dosage form whose release is controlled by erosion.

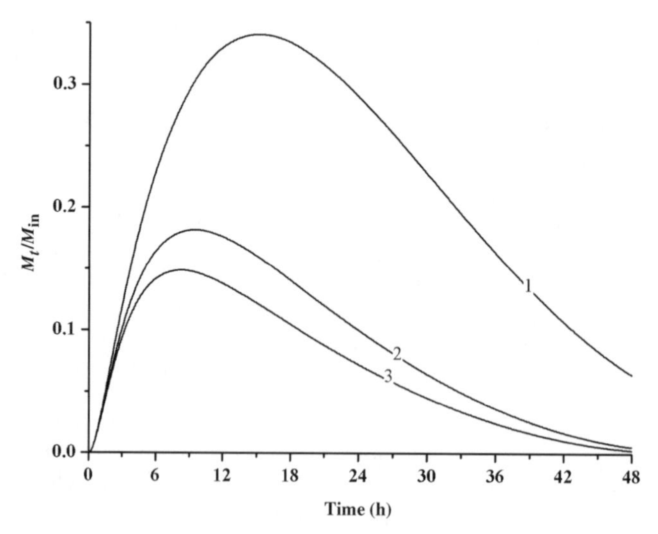

Figure 64. Plasma drug level M_t/M_{in} obtained with ciprofloxacin, in spherical dosage forms administered under the same conditions. Intervariability of the patients with $k_a = 1.3\,h^{-1}$ and various values of k_e: (1) $0.09\,h^{-1}$, (2) $0.231\,h^{-1}$, and (3) $0.3\,h^{-1}$. $t_r = 48\,h$.

Table 13. Pharmacokinetic Parameters of the Drugs

Drug	Ciprofloxacin	Acetyl Salicylic Acid	Ofloxacin	Péfloxacin
$k_a(h^{-1})$	1.3	3.5	1.5	0.8
$k_e(h^{-1})$	0.231	2.5	0.13	0.07
$t_{0.5}(h)$	3	0.28	5.33	9.9
$T'_{0.5}(h)$	5.5	1	7.8	14
$T_{0.5}(h)t_r = 24\,h$	16.6	8.8	20.3	26.9
$T_{0.5}(h)t_r = 48\,h$	25.5	16	30.8	39.4

- More precisely, these facts listed in Table 13 are also shown in Fig. 64 for ciprofloxacin and Fig. 65 for acetyl salicylic acid. In Table 13, let us remark that, in the same way as in Table 12 concerned with diffusion-controlled dosage forms, the half-life time obtained with these dosage forms, $T_{0.5}$, is the time at which the plasma drug concentration is half the value attained at the peak. The half-life time $T'_{0.5}$ is, in the same way, obtained with the immediate-release dosage form. Moreover, the time $T_{0.5}$ can be obtained in different manners depending on the value of the time of full erosion t_r.
- According to the possibility offered by the erosion-controlled dosage forms, two very different values for the time of erosion are selected, the one equal to 24 h, and the other much longer, at 48 h. The effect of this time of erosion on the half-life time of the drug is of great concern, as from 24 to 48 h, the half-life time of the drug is increased by approximately 50%, whatever the nature of the drug.
- Following this increase in the half-life time of the drug allowed by the time of erosion, it can be seen in Fig. 65 that acetyl salicylic acid could be taken once a day, in spite of its very high rate constant of elimination.

8.2.4. Comparison Between the Processes of Diffusion and Erosion

Comparison between the processes of drug release controlled either by diffusion or by erosion is of great interest.

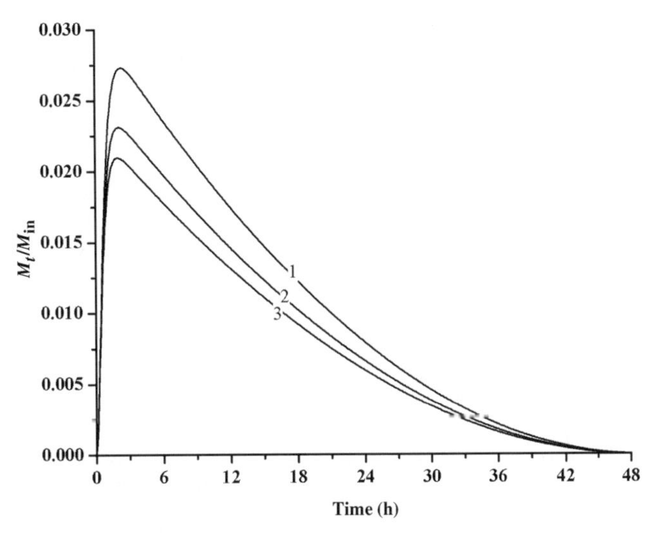

Figure 65. Plasma drug level M_t/M_{in} obtained with acetyl salicylic acid, in spherical dosage forms administered under the same conditions. Intervariability of the patients with $k_a = 3.5\,h^{-1}$ and various values of k_e: (1) $2.1\,h^{-1}$, (2) $2.5\,h^{-1}$, and (3) $2.77\,h^{-1}$. $t_r = 48\,h$.

The maximum value at the peak defined by the ratio M_{max}/M_{in}, which is proportional to the maximum drug concentration, as well as the values of the half-life time $T_{0.5}$ obtained with the controlled-release dosage forms, can give a good idea of the release efficiency provided by these two processes of release. Let us remark that the process of diffusion allows the drug to be nearly totally released within 24 h, while the time of erosion of 48 h means that the drug is totally released within 48 h.

The following conclusions in these two processes of controlled drug release are given:

- The time of erosion of 48 h can be selected only when the erosion-controlled dosage forms are made of bioadhesive polymers, meaning that these dosage forms can be kept in the GI over a period of time of 48 h, exceeding largely the usual GIT.
- The process of diffusion, for example, allows the dosage forms to release nearly all the drug over a period of time of 24 h, in the best case.
- In case of drugs having a high rate of elimination, as acetyl salicylic acid, only the erosion process makes it possible a once a day dosage, as precisely shown in Fig. 65.

8.3. Plasma Drug Profile with Repeated Doses

The following parameters should be examined in succession when the dosage forms are taken in repeated multidoses: the nature of the drug, the intervariability of the patients and the time of full erosion. As already shown, the effect of the shape given to the dosage form is of slight importance. Moreover, the effect of the noncompliance on the plasma drug profile is precisely determined.

8.3.1. Effect of the Nature of the Drug The effect of the nature of the drug clearly appears in Fig. 66, where are drawn the plasma drug profiles obtained with dosage forms, spherical in shape, taken once a day, when the drug is: ciprofloxacin (curve 1), acetyl salicylic acid (curve 2), ofloxacin (curve 3), and péfloxacin (curve 4). These dosage forms are erosion controlled with a time of full erosion $t_r = 48$ h.

The conclusions are similar to those given with a single dose in Section 8.2.2 with Fig. 63:

- The effect of the nature of the drug taken in repeated doses is so important that the profiles drawn in Fig. 66 are quite different for these three drugs.

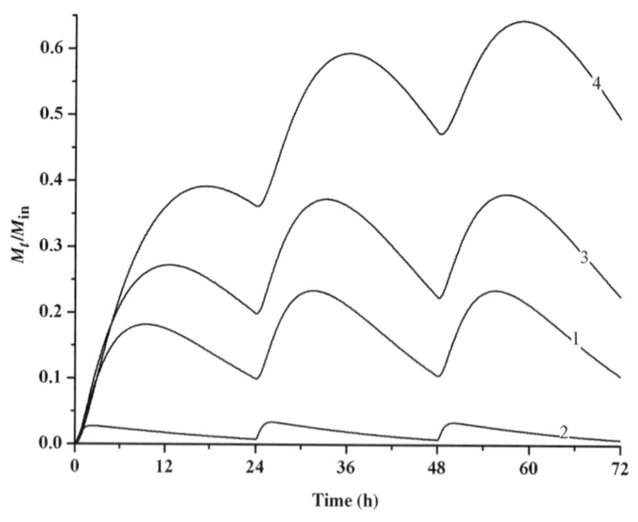

Figure 66. Plasma drug level M_t/M_{in} obtained with repeated doses taken once a day, with dosage forms spherical in shape with a time of erosion of 48 h, and various drugs: (1) Ciprofloxacin with $k_a = 1.3\,h^{-1}$ and $k_e = 0.231\,h^{-1}$. (2) Acetyl salicylic acid with $k_a = 3.5\,h^{-1}$ and $k_e = 2.1\,h^{-1}$. (3) Ofloxacin with $k_a = 1.5\,h^{-1}$ and $k_e = 0.13\,h^{-1}$. (4) Péfloxacin with $k_a = 0.8\,h^{-1}$ and $k_e = 0.07\,h^{-1}$. Time of erosion $t_r = 48$ h.

- Thus for ciprofloxacin (curve 1), the dosage forms made with a polymer such as $t_r = 48\,h$, taken once a day, provokes an interesting plasma drug profile. This profile alternates between peaks and troughs that are not far apart, their ratio being as low as 1.56, while the peak concentration increases from dose to dose up the third dose at which the steady state is attained.
- For acetyl salicylic acid (curve 2), the plasma concentration is very low, resulting from the very large rate constant of elimination, but the concentration at the troughs is not zero, when
- $t_r = 48\,h$, since the ratio of the concentrations at peaks and troughs does not exceed 3.6.
- For ofloxacin (curve 3), the plasma drug profile increases up to the third dose for which the steady state is attained.
- But, it is true that for péfloxacin (curve 4), with a very low rate of elimination, a shorter time of erosion around 24 h should be considered so as to reduce the half-life time $T_{0.5}$.

8.3.2. Effect of the Intervariability of the Patients The effect of the intervariability of the patients is shown with ciprofloxacin in Fig. 67 dispersed in erosion-controlled dosage forms with a time of full erosion, $t_r = 48\,h$, when these dosages are taken once a day. The various values of the pharmacokinetic parameters are shown either of the legend of this figure.

The results are similar as those obtained with a single dose, with in addition the effect resulting from the repeated dose:

- The effect of the intervariability of the patients is of great concern, as shown in Fig. 67. For the lower value of the rate constant of elimination k_e (curve 1), the ratio of the peak to troughs tends to 1.46 at the steady state that is attained at the third dose. With the larger value of k_e (curve 3), the same peak to trough ratio is of 2.2 at the steady state that is reached at the second dose. But the most important effect of the intervariability of the patients stands in the ratio of the plasma drug concentrations obtained with these two extreme categories of patients (curves 1 and 3) since the ratio of the peaks is 2.1, while the ratio of the troughs is 4.

8.3.3. Effect of the Time of Full Erosion t_r of the Polymer Matrix The effect of the time of full erosion on the plasma drug level can be seen in Fig. 68, with the plasma drug profile obtained

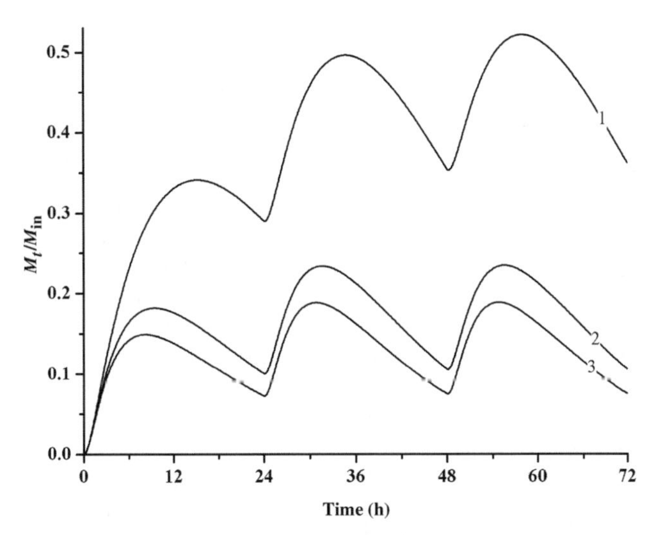

Figure 67. Plasma drug level M_t/M_{in} obtained with repeated doses taken once a day, with dosage forms spherical in shape with a time of erosion of 48 h, and ciprofloxacin. Effect of the intervariability of the patients with $k_a = 1.3\,h^{-1}$ and various values of k_e: (1) $0.09\,h^{-1}$, (2) $0.231\,h^{-1}$, and (3) $0.3\,h^{-1}$. $t_r = 48\,h$.

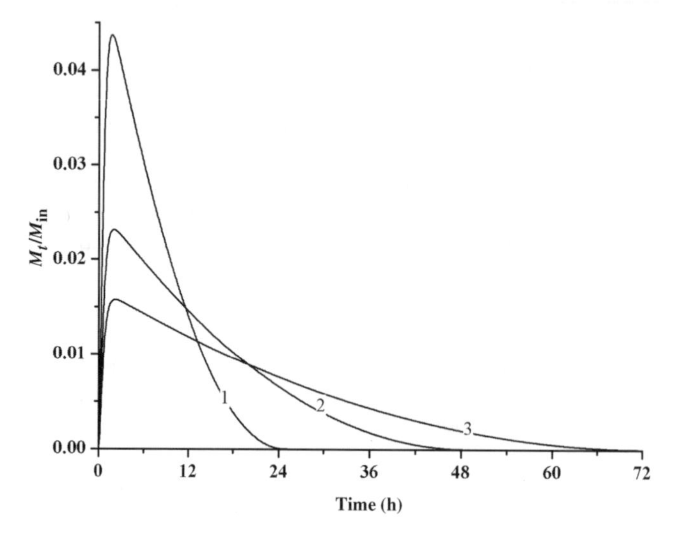

Figure 68. Effect of the time of erosion on the plasma drug profile of acetyl salicylic acid with $k_a = 3.5\,h^{-1}$ and $k_e = 2.5\,h^{-1}$; $t_r = 24\,h$ (1); $t_r = 48\,h$ (2): $t_r = 72\,h$ (3).

for different values of t_r: 24 h (curve 1); 48 h (curve 2); 72 h (curve 3). Acetyl salicylic acid is selected for the drug, in dosage forms of same volume, and spherical in shape.

The following obvious conclusions are given from these curves:

- The effect of the time of full erosion on the plasma drug profiles is of great concern, especially when it is wanted to take the dosage form once a day.
- This effect is so apparent that only dosage forms with a value of t_r much larger than 24 h, for example, 48 or 72 h, can be taken once a day without provoking a too low drug concentration at the trough.

8.3.4. Perfect Compliance and Noncompliance: Omission Followed by Double Dose Perfect compliance with once a day dosing is associated with regular plasma drug profiles, as depicted in Fig. 69 for ciprofloxacin whose pharmacokinetic parameters are shown in Table 13 and the figure caption. The dosage forms are made of a polymer whose time of erosion is 24 h, and the plasma drug profiles are expressed in terms of time by using the dimensionless number M_t/M_{in} for the ordinate. The plasma drug profile alternates between peaks and troughs at rather high levels with a peak/trough ratio of 2.

The effect of omission appears in Fig. 70, with the same drug dispersed in the same dosage forms with a time of erosion, t_r of 24 h. The omission taking place on the second day provokes a deep trough, and on the third day, the right dosing starts again providing a plasma drug profile as normal as for the first dose.

Following an omission, some patients, perhaps interested in arithmetic calculation, would believe in compensating for this omission in taking a double dose afterward. As a matter of fact, instead of compensating errors, there is a wrong dosage in addition to omission.

The plasma drug profile obtained in the case of omission the second day and compensation in taking a double dose the third day, is drawn for ciprofloxacin when they are orally delivered with the same sustained-release dosage forms.

The following comments can be made from Figs 69–71:

- From the first approach, it clearly appears that after a lack of drug in the plasma resulting from the omission on the second day, there is an excess of drug

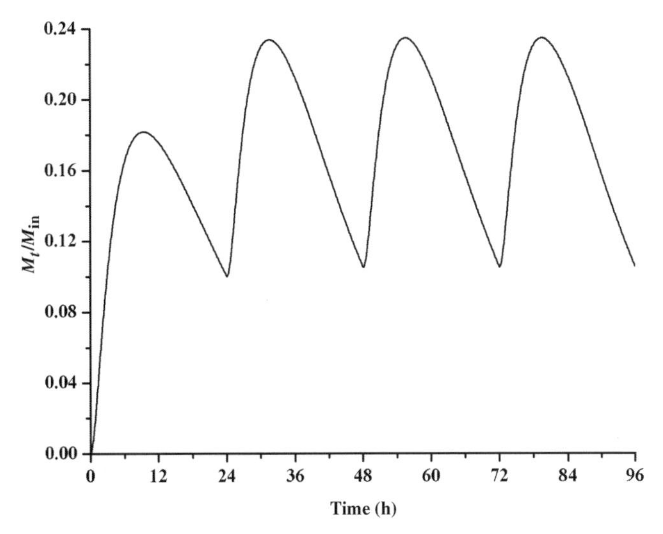

Figure 69. Effect of compliance on the plasma drug profile with ciprofloxacin with $k_a = 1.3\,h^{-1}$ and $k_e = 0.231\,h^{-1}$. Taken once a day, regularly.

the third day resulting from double dosing. On the whole, the peak on the third day is twice as high as the peak obtained under normal right conditions.

- The time of full erosion, t_r, has been found to play an important role on the process [49]. The statement holds: the longer the time of erosion, the more important the effect of the wrong dosage.
- The nature of the drug, with its rate constant of elimination, plays a role on the effect of omission and double dosing. It has been found that the higher the rate of elimination of the drug, the lower the trough resulting from omission; moreover, the lower the rate constant of elimination of the drug, the larger the peak associated with double dosing [49].

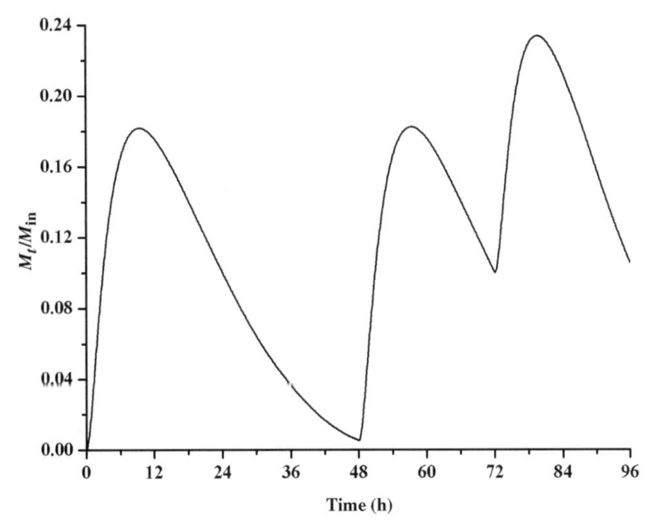

Figure 70. Effect of noncompliance on the plasma drug profile with ciprofloxacin with $k_a = 1.3\,h^{-1}$ and $k_e = 0.231\,h^{-1}$. Taken once a day, with a dose missing on the second day.

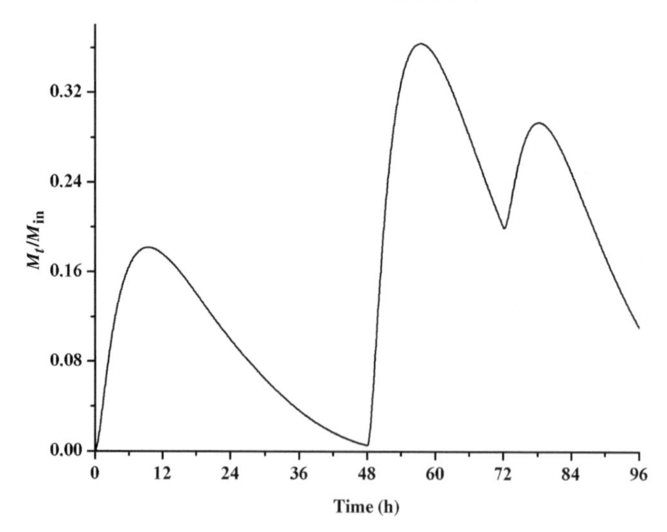

Figure 71. Effect of noncompliance on the plasma drug profile with ciprofloxacin with $k_\alpha = 1.3\,\text{h}^{-1}$ and $k_e = 0.231\,\text{h}^{-1}$. Taken once a day, with a dose missing on the second day, and a double dose on the third day.

8.4. Prediction of the Characteristics of the Dosage Forms

It is of interest to predict by calculation the best characteristics of the dosage forms able to deliver the drug for the desired therapy. Thus, on one hand, the correlation between the half-life times of the drugs is evaluated; on the other hand, plasma drug profiles are expressed in terms of drug concentration taking into account the apparent plasma volume.

8.4.1. Relationships Between the Half-Life Times $t_{0.5}$ and $T_{0.5}$ The effect of the time of full erosion t_r has been defined, not only with single dose but also in repeated doses with the important factor of the dose frequency.

Figs 72 and 73 show the characteristics of the dosage form necessary to obtain the desired value of the half-life time of the drug $T_{0.5}$ dispersed in the erodible polymer, as a function of the half-life time of the drug itself $t_{0.5}$ measured through i.v.

$$t_r = \frac{R}{v} = \frac{L}{v} \tag{144}$$

where R is the radius of the dosage form, spherical in shape, and L is the smaller dimension of dosage forms having another shape.

Thus, for a given rate of erosion, v, the main dimension of the dosage form can be adjusted by using Equation 144.

The first diagram in Fig. 72 is drawn for drugs having a low half-life time $t_{0.5}$, such as acetyl salicylic acid, while the second diagram in Fig. 73 is concerned with drugs having much larger half-life times.

From these diagrams, it can be seen that the drug plays a major role, even dispersed through a polymer, and that it is not easy to prepare dosage forms able to deliver the drug over a long period of time when this drug has a very short half-life time, as already shown in Fig. 68 for acetyl salicylic acid.

8.4.2. Plasma Drug Profiles Expressed in Terms of Concentration By introducing the amount of drug M_{in} initially in the dosage form and dividing the amount M_t by the apparent plasmatic volume, the ordinate M_t/M_{in} is expressed in terms of plasma drug concentration. The plasma drug profiles are drawn by considering the plasma drug concentration, with ciprofloxacin dispersed in dosage forms, spherical in shape, with a time of full erosion of 48 h in Fig. 74, when the amount of drug is 500 mg.

The following conclusions can be drawn from these curves:

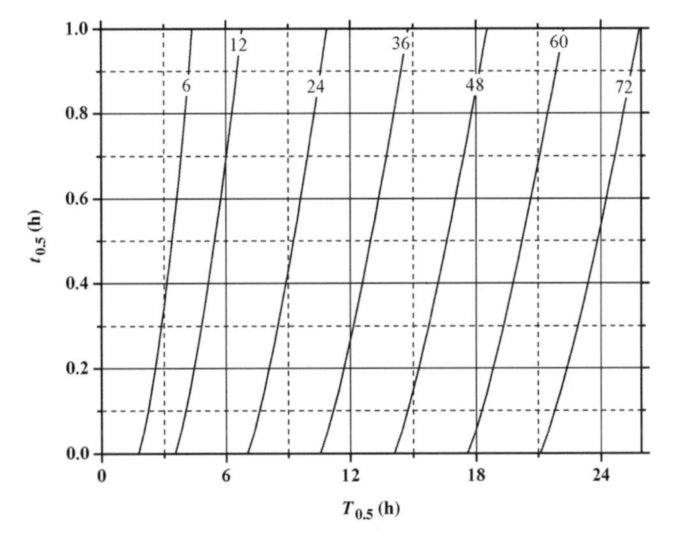

Figure 72. Diagram showing the relation between the half-life times $t_{0.5}$ and $T_{0.5}$ for various values of the time of erosion of dosage forms, spherical in shape, when $1 < t_{0.5} < 1$.

- Of course, the curves drawn in Fig. 74 calculated with the time of full erosion of 48 h are similar to those drawn in Fig. 75 obtained under the same conditions, and expressed with the ordinate M_t/M_{in}.
- The intervariability of the patients is also considered, with three values of the rate constant of elimination (curves 1; 2; 3) and two values of the apparent plasmatic volume (full lines: with 150 L; dotted lines: with 225 L).

8.5. Conclusions on Dosage Forms Controlled by Erosion

Following the failure in the attempt of the FDA to define *in vitro/in vivo* correlations, one of the authors was invited to present his

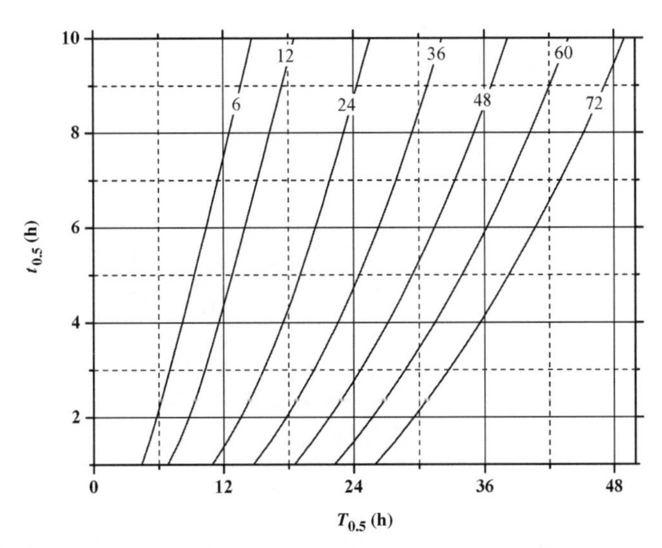

Figure 73. Diagram showing the relation between the half-life times $t_{0.5}$ and $T_{0.5}$ for various values of the time of erosion of dosage forms, spherical in shape, when $1 < t_{0.5} < 10$.

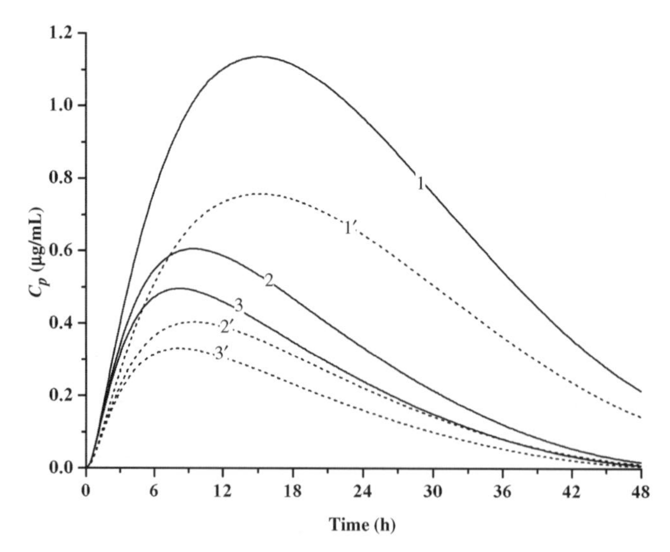

Figure 74. Plasma drug level concentration, C_t, obtained with one dose taken with dosage forms, spherical in shape, having a time of erosion of 48 h, and containing 500 mg ciprofloxacin. Effect of the intervariability of the patients with $k_a = 1.3\,h^{-1}$ and various values of k_e and of the apparent plasmatic volume: (1 and 1′) $0.09\,h^{-1}$; (2 and 2′) $0.231\,h^{-1}$; (3 and 3′) $0.3\,h^{-1}$. (1, 2, 3) 150 L; (1′, 2′, 3′) 225 L.

method [5] to resolve this problem in the case of erosion-controlled dosage forms. More recently, a paper collected a few results on this type of dosage forms [88]. The main interest of these dosage forms stands on the fact that they can adhere to the wall surfaces of the GIT. Thus the time of full release of the drug can greatly extend the length of the GIT time. This result is of great concern, as a uniform plasma drug level is obtained, leading to a better therapy avoiding side effects. Moreover, drugs with very high value of the rate constant of elimination can also been delivered by using such oral dosage forms, providing that the time of full release was long enough.

9. TRANSDERMAL THERAPEUTIC SYSTEMS

Nomenclature

C concentration of the drug;

$C_{x,t}$ concentration of the drug at position x and time t;

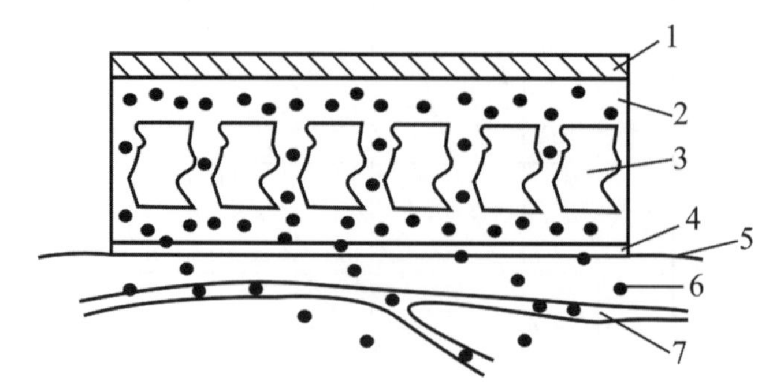

Figure 75. Scheme of a general transdermal therapeutic system applied to the skin. Covering membrane (1); drug reservoir (2); micropore membrane controlling drug delivery (3); adhesive contact surface (4); surface of skin (5); drug molecules (6); and capillary (7).

C_0	constant concentration of drug on the surface of the skin;
D	diffusivity of the drug in the skin, in the TTS in 9.1 and 9.3;
h	coefficient of drug transfer after removal of the TTS;
k_a	rate constant of absorption of the drug (h^{-1});
k_e	rate constant of elimination of the drug (h^{-1});
L	thickness of the skin (cm);
M_t	amount of drug transferred through the skin after time t;
n	integer in series;
TTS	transdermal therapeutic system, patch;
t	time;
$t_{0.5}$	half-life of the drug (h);
V_p	plasmatic volume (L);
x	abscissa taken along the thickness of the skin, for calculation;
X	amount of drug at the internal surface of the skin;
Y	amount of drug in the blood compartment;
Z	amount of drug eliminated.

The transdermal therapeutic systems deliver the drug systemically, without the need of a needle to inject it into the circulatory system of the patients. The terms defined by the FDA are as follows: patch for the common name, and patch with extended release, as well as patch with extended release electrically controlled.

9.1. General Mechanism of Drug Delivery

The structure of the TTS (Fig. 75) [89–91] consists of the following features:

- A covering membrane in contact with the drug reservoir protecting it;
- The drug reservoir containing the drug in liquid state or in dissolved state in a liquid or a gel;
- A delivery control element;
- An adhesive layer keeping the TTS in contact with the skin;
- A protective foil that should be removed from the system before application on the skin surface.

Thus, the outward appearance of a TTS is a patch. The TTS may be as thin as 150 μm and can cover an area of 5–20 cm^2 of the skin surface. This patch should be applied to the parts of the body where the skin has a constant thickness and are supplied by a relatively constant and high blood flow. Some examples are: behind the ear, on the upper arm, on the chest.

The process of drug delivery by the TTS is mainly the diffusion of the drug through the skin from the TTS-skin interface to the sub capillary plexus. The diffusion starts as soon as the TTS is applied to the skin, first under transient conditions as it takes some time for the drug to cross the skin and to saturate the binding sites of the skin, followed by a steady state that is attained with a constant rate of drug delivery associated later with a constant drug concentration in the plasma. Finally, when the TTS is removed, the plasma drug concentration falls at a rate that depends on the rate of elimination of the drug.

From this first approach, the following parameters intervene in the TTS process:

(1) The solubility of the drug in the skin, and especially on the stratum corneum surface, which depends only on the nature of the drug and of the epidermis.
(2) The diffusion of the drug through the skin, which depends on the nature of the drug and on the characteristics of the skin.
(3) The pharmacokinetic parameters of the drug, such as the rate constant of absorption into and the rate constant of elimination out of the blood.
(4) The amount of drug contained in the reservoir that is crucial to the length of time over which the TTS remains functional.
(5) The TTS itself exhibits a limited rate of release of the drug, but generally this rate is much larger than the

rate of transport of the drug through the skin.

9.2. Properties and Role of the Skin

9.2.1. Main Characteristics of the Skin The skin is the largest organ and one of the toughest tissues of the body with the following characteristics: Self-healing, water-resistant, temperature regulating, infection repelling, and highly sensitive.

Basically, it consists of two layers:

- The outer layer, epidermis, is itself divided into four layers: The innermost layer; the germinative or Malpighian layer, composed of cells constantly replicating and rising through the next two layers; the stratum granulosum; the stratum lucidum, in which their keratin content increases by degrees to become the outer layer, called stratum corneum.
- The dermis (or corium) lies underneath, under the epidermis but lying on top of the thin basal membrane. In its thick single stratum of connective tissue lie sensory nerve endings, blood capillaries, lymph vessels, tiny muscle fibers, sweat glands, sebaceous glands, and hair follicles.

The thickness of the stratus corneum is between 10 and 20 μm, while the epidermis thickness is ranging from 0.04 to 1.5 mm, and the dermis thickness varies from 1 to 4 mm depending upon the part of the body.

A part of the function of the skin, relatively impermeable to most substances, is to provide a protective barrier against invasion from microorganisms and toxic materials, as well as to prevent the loss in water or other physiological elements from the body. For that reason, it has been considered for a long time as an inappropriate site for the administration of drugs for systemic use.

The stratum corneum is impermeable to the hydrophilic molecules and slightly permeable to the lipophilic molecules. The other parts of the skin, either the living epidermis or the dermis are rather permeable to hydrophilic molecules.

The permeability of the stratum corneum can be increased either by stripping or much better by using lipophilic enhancers of diffusion. Amphiphile molecules, both hydrophilic and lipophilic, could be another solution.

9.2.2. Clinical Possibilities Offered by the Skin Thus, the skin is a way to attain the systemic circulation that offers various advantages:

(1) The phenomena affecting the GI absorption of oral dosage forms are circumvented.
(2) The liver being bypassed, as well as the first-pass hepatic, the drug metabolism is reduced to that only which could intervene through the skin.
(3) Drugs with a narrow therapeutic index can be used more easily.
(4) Drugs with a short biological half-life can be employed.
(5) The treatment can be started and terminated at any time by putting or removing the patch.
(6) The compliance of the patient is easily improved.
(7) There is no problem of time of delivery, and the steady state is attained with a constant plasma drug level.
(8) Over-dosage when the drug enters the circulation can be avoided, and so the side effects associated to it are reduced.

9.3. Calculation of the Plasma Drug Level

9.3.1. Assumptions Some assumptions are able to make the process clear and help calculation:

(1) The TTS is capable to maintain a constant drug concentration on the surface of the skin.
(2) The contact at the TTS-skin interface is perfect, so that the drug concentration on the surface of the skin reaches the constant value as soon as the patch is put in contact with the skin. Nevertheless, the effect of the pressure has been found of concern [92].

(3) The drug dissolves on the surface of the skin, and a partition factor may appear.
(4) The drug diffuses through the various parts of the skin. Considered as homogeneous, the diffusivity through it is constant.
(5) In the innermost part of the skin, the drug is absorbed in the capillaries. This stage of absorption follows a first-order kinetics with the rate of absorption, in the same way as it is absorbed in the blood compartment when it is orally taken, as these two processes are similar [91,93–96].
(6) The drug is eliminated from the blood compartment with the first-order kinetics of elimination.

9.3.2. Calculation of the Drug Transfer into the Blood (Fig. 76) Two boundary conditions for the skin of thickness L playing the role of a membrane are as follows: the drug concentration is constant onto the external surface of the skin; the drug concentration at the innermost of the skin is very low, since the drug present in this place is absorbed in the capillaries. These conditions are written as follows:

$$t > 0 \quad 0 < x < L \quad C = 0 \tag{145}$$

$$t > 0 \quad x = 0 \quad C = C_0 \tag{146}$$

$$x = L \quad C = \varepsilon$$

The one-directional transient diffusion through the skin with constant diffusivity is expressed by the partial derivative equation:

$$\frac{\partial C}{\partial t} = D \cdot \frac{\partial^2 C}{\partial x^2} \tag{147}$$

The skin playing the role of a membrane for the drug transport, with constant concentrations on each surface, the drug concentration developed through its thickness L is thus expressed by the series [97,98]

$$\frac{C_{x,t}}{C_0} = 1 - \frac{x}{L} - \frac{2}{\pi}\sum_{n=1}^{\infty}\frac{1}{n}\cdot\sin\frac{n\pi x}{L}\cdot\exp\left(-n^2\pi^2\frac{D\cdot t}{L^2}\right) \tag{148}$$

The amount of drug that emerges from the internal face of the skin is

$$\frac{M_t}{C_0} = \frac{D\cdot t}{L} - \frac{L}{6} - \frac{2L}{\pi^2}\sum_{n=1}^{\infty}\frac{(-1)^n}{n^2}\exp\left(-n^2\pi^2\frac{D\cdot t}{L^2}\right) \tag{149}$$

The drug concentration in the skin $C_{x,t}$ increases with time, and the profiles of drug concentration through the thickness of the skin tends to become linear when the series in Equation 148 vanishes.

In the same way, from Equation 149, the amount of drug leaving the skin increases slowly with time as the series in this equation vanishes with time. Thus, after a long period of time, Equation 149 tends to become linear,

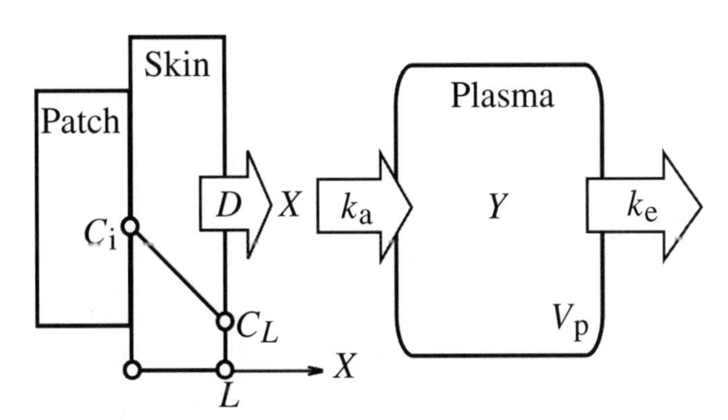

Figure 76. Scheme of the process with the patch, the skin, and the blood compartment.

and the asymptote being given by

$$M_t = \frac{D \cdot C_0}{L}\left(t - \frac{L^2}{6D}\right) \quad (150)$$

the slope of which is

$$\text{Slope} = \frac{M_t}{t} = \frac{D \cdot C_0}{L} \quad (151)$$

and the intercept on the time-axis is

$$t_i = \frac{L^2}{6D} \quad (152)$$

Transfer into the Blood Compartment The amount of drug at the internal surface of the skin, X_t is given by the relation

$$\frac{dX}{dt} = \frac{dM}{dt} - k_a \cdot X_t \quad (153)$$

and the amount of drug in the blood compartment is obtained by the relationship

$$\frac{dY}{dt} = k_a \cdot X - k_e \cdot Y \quad (154)$$

with the rate constants k_a and k_e of the first-order kinetics of absorption and elimination.

The amount of drug eliminated is also obtained, by the relation

$$\frac{dZ}{dt} = k_e \cdot Y \quad (155)$$

Calculation is made step by step, by using a numerical method with a constant increment of time, as no analytical solution of the problem can be found.

9.4. Transdermal Delivery of a Drug: Metropolol

9.4.1. Experimental Part

Choice of the Drug [99] Metoprolol, a β_1-selective adrenergic blocking agent is a first choice drug in the treatment of mild to moderate hypertension and stable angina and is beneficial in postinfraction patients. But, it is also subjected to extensive hepatic first-pass metabolism following oral administration and has a short biological half-life [100]. The transdermal route of administration is capable of avoiding the hepatic first-pass effect and achieving higher systemic bioavailability of this drug. In this chapter, [99], various experiments have been made: preparation of the multilaminate adhesive device able to deliver the drug through the transdermal way; skin permeation studies on Valia–Chien glass cells diffusion, whole bioavailability studies with the plasma level determination of the drug in hairless rats after an i.v. bolus injection and the plasma level obtained in the same rats after oral administration; determination of the plasma levels in these rats after application of an adhesive device.

Thus, from these experiments, it is possible to determine the values necessary for calculating the plasma drug level, for example, the characteristics of the skin of the rats such as its thickness and the diffusivity while the concentration of the drug on the skin surface is calculated, as well as the pharmacokinetic parameters of the rats (plasmatic volume and rate constants of absorption and elimination).

Preparation of the Patch [99] A weighed amount of metoprolol was mixed thoroughly with the specially formulated polyacrylate adhesive, and a uniform layer of a fixed thickness (1.2 mm) was made on heat-sealable backing membrane with a laboratory coating device. The whole system was cured at 20 °C, and the laminate was then covered by a release liner, cut into 10 cm^2 pieces and used in the subsequent experiments.

Skin Permeation Studies [99] The freshly excised full-thickness hairless rat skin was mounted on Valia–Chien glass diffusion cells, with the stratum corneum side in contact with the drug-releasing surface of the patch and the dermal side facing the receptor solution; this solution (pH 7.4 Sorensen's phosphate buffer) kept at 37°C and stirred conveniently was analyzed at intervals, by withdrawing samples and replacing them by an equal volume of fresh solution. The sink condition was verified in the glass diffusion cells.

Bioavailability Studies [99] Three categories of rats were considered for the following three series of experiments. In the first group (6), the rats received an i.v. administration of drug (50 mg/kg) and blood samples were analyzed at intervals. The second group of rats (6) was administered a single 75 mg/kg oral dose, and blood samples were analyzed at intervals. The third group received on the abdomen their transdermal administration with a 10 cm^2 adhesive device containing the drug dose, and blood samples analyzed. The transdermal dose and rate of delivery were calculated from the *in vitro* hairless rat skin permeation studies.

9.4.2. Results Obtained from *In Vitro* Experiments The kinetics of drug transfer through the patch-skin system is shown in Fig. 77, obtained either from experiments (dotted line) [99] or from calculation (full line) [96].

The following comments can be drawn from the kinetics calculated by using Equation 149:

- The kinetics of drug transfer through the patch-skin system corresponds with the transfer through a membrane with constant concentration on each side [97,98].
- From the beginning to approximately 5–7 h into the process, the amount of drug transferred increases slowly with time, following Equation 149, and the process is controlled by transient diffusion.
- After 5–7 h, the kinetics becomes linear, the process being controlled by stationary diffusion, as the series in Equation 149 vanishes.
- From the intercept on the time-axis, it is possible to obtain a relationship between the thickness of the skin and the diffusivity of the drug, according to Equation 152.
- The value of the slope, in the linear part of the kinetics of drug transport through the skin, gives a relationship between the same parameters shown in Equation 151. In addition, the drug concentration, C_0, is maintained on the external surface of the skin by the patch.
- As the thickness of the skin has been measured experimentally [99], the values of the other parameters are evaluated by using the Equations 151 and 152.
- The good agreement between the calculated and the experimental curves in Fig. 77 is proof of the accuracy of the values of the parameters of diffusion given in Table 14.
- It is worth noting that the patch is capable of maintaining the drug available at a constant concentration, C_0, on the surface of the skin over a period of time of at

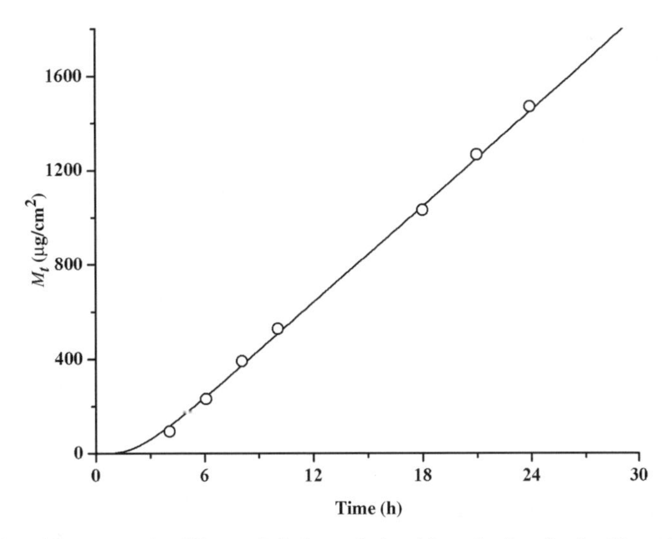

Figure 77. Kinetics of drug transfer (Metoprolol) through the skin: calculated using Equation 149 (solid line); experimental *in-vitro* data on hairless rats (dotted line). Reprinted from Ref. [96] with permission from Elsevier.

Table 14. Parameters of Diffusion Through the Skin

Slope = 68×10^{-6} g/cm^2 h;
$t_i = 2.6$ h;
$L = 0.035$ cm;
$D = 2.2 \times 10^{-8}$ cm^2/s;
$C_0 = 0.03$ g/cm^3;
$h = 6 \times 10^{-6}$ cm/s.

least 24 h, as it is illustrated by the shape of the curve in Fig. 77.

9.4.3. Results Concerned with *In Vivo* Experiments The pharmacokinetic parameters are obtained from the drug level-time history measured in the blood [99]. From the intravenous bolus injection, the rate constant of elimination, k_e, is related to the time of half-life of the drug: $\ln 2 = 0.693 = k_e \cdot t_{0.5}$.

The rate constant of absorption is determined from the oral absorption of the drug [99], either by calculation using the main characteristics of the peak or by using a numerical model using the whole plasma drug curve [96]. The volume of distribution is evaluated [99] from the plasma drug level by using the Equation 19.

The values of the pharmacokinetic parameters are shown in Table 15

9.4.4. Calculation of the Plasma Drug Level By using the numerical model described in the theoretical part and the data shown in Tables 14 and 15, the plasma drug level obtained by transdermal delivery has been calculated.

The plasma drug profiles obtained either by calculation (full line) [96] or by experiments (dotted line) [99] are drawn in Fig. 78, leading to some conclusions:

- Good agreement is observed between the experimental and the calculated curves. The mean values of the experiments are shown, as well as the upper and lower values obtained with 6 rats corresponding to their intervariability.
- In the same way as for the kinetics of the drug transfer through the skin, it takes some time for the drug to get into the blood compartment. For up to 5–7 h, under transient diffusion through the skin, the plasma drug level increases.
- After approximately 10 h, the stationary conditions are attained, and the plasma drug level remains constant up to the removal of the patch.
- When the patch is removed after 24 h of drug delivery, the plasma drug level decreases very quickly, as shown in Fig. 78 (curve 2). A problem arises in calculating the drug level after the removal of the patch by using the numerical model. Two extreme assumptions can be made for the drug located in the skin:

  (1) The drug remains in the skin, and calculation is made by writing that $k_a = 0$.
  (2) The drug enters the blood compartment entirely by diffusion, resulting from the gradient of concentration. Calculation is made by keeping k_a constant and $\partial C/\partial x = 0$ at the external surface, meaning that the drug located in the skin diffuses only and totally into the blood compartment.

- As neither the curve 1 obtained with the first assumption nor the curve 3 determined with the second assumption is able to fit the experimental curve expressing the decrease in the drug concentration after removal of the patch, another assumption is made, by considering that the drug can leave the skin on both sides. Thus the drug transfer into the blood compartment is maintained by keeping the rate constant of absorption constant and writing that there is a drug transfer outside the skin obeying the following equation:

$$D \cdot \frac{\partial C}{\partial x} = h \cdot C \qquad (156)$$

Table 15. Pharmacokinetic Parameters of Metoprolol in Rats [99]

$k_a = 1.1\,\text{h}^{-1}$	$k_e = 0.78\,\text{h}^{-1}$	$V_p = 9$ L

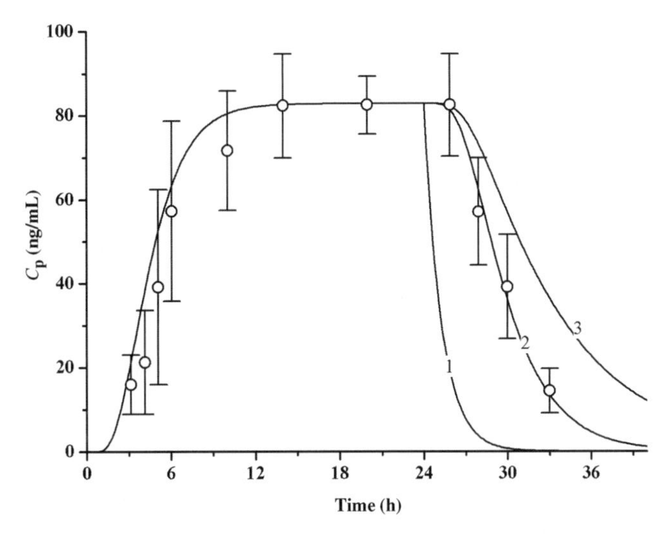

Figure 78. Plasma drug level (Metoprolol) obtained with the patch on hairless rats: calculated using the numerical model (solid line); experimental *in-vivo* data (dotted line). Curve 1 with $k_a = 0$; curve 3 with $h = 0$; curve 2 with $h = 6 \times 10^{-6}$ cm/s. Reprinted from Ref. [96] with permission from Elsevier.

Equation 156 expresses that the rate at which the drug is brought to the external surface of the skin by diffusion is constantly equal to the rate of elimination outside the skin, h being the coefficient of the transfer of the drug out of the skin. The value of h given in Table 14 enables the theoretical curve to fit the experimental one. The coefficient of convection exists when a substance diffuses out of a solid [97,98], playing an important role in the transport of a contaminant or food at the package-food interface [101].

9.5. Conclusions and Future Prospects

Various studies have been carried out on transdermal drug delivery, and even a large bibliography made of 148 references [90] cannot be exhaustive. Some other papers on the subject of diffusion are also worth noting:

In the first [102], the diffusion phenomena have been deeply explored not only through the skin but also in the tissue between the skin and the capillaries, and the profiles of drug concentration developed through the dermis have been determined experimentally. The remaining problem is that the calculated plasma drug level did not correspond with the experimental one, especially at the beginning of the process, as this model predicted that the steady state would be reached after approximately 35–50 min, whereas the experimental values for this time are far much larger.

The effect of the site to which the patch is placed has been studied by making *in vitro* experiments, showing that the upper dorsal position allows the larger rate of drug delivery [103] for humans and rats.

The effect of enhancers has also been widely studied, and only a few reports are mentioned here: The relative effect of laurocapram (Azone) and transcutol on the solubility and diffusivity of a drug, the 4-cyanophenol, in human stratum corneum [104]; A mechanistic investigation of the *in vitro* human skin permeation enhancing effect of laurocapram was deeply made, leading to this conclusion that the less the lipophilicity value of the drug, the greater the enhancer effect observed [105]; other complex models were used for predicting the penetration of monofunctional solutes from aqueous solutions through the human stratum corneum [106].

The contribution of the various skin layers to the resistance to permeation was considered, showing that not only the stratum corneum is the main barrier to permeation but also that there is a threefold difference between the fluxes through unstripped and stripped full thickness for a drug [107].

The effect of the adhesive layer, this part of the patch in contact with the skin, was deeply studied [92] showing that the drug concentration should reach saturation in this polymeric layer, in order not to play the role of a barrier.

Disposition of multiple dose transdermal system was considered with Nicotine TTS [108], and the analysis of the components of variance contributing to the variability in nicotine delivery indicated that *in vivo* transdermal permeation of that drug is rate limited by both the device and the intrinsic skin diffusivity.

Various studies have been made in order to explore the possibility of building *in vitro–in vivo* correlations [109] in the same way as those managed by the FDA for oral dosage forms, and concluding to some difficulties to accomplish this task.

If the first-pass hepatic is generally circumvented with transdermal drug delivery, a possibility of metabolism remains through the skin, and thus propanolol is partly metabolized in human skin tissues, and its metabolites are retained in the skin [110,111].

The ageing effect is also of great importance on the transdermal transfer, as the permeation properties of the skin are altered with age. If the absorption rate does not change very much for lipophilic materials such as testosterone or oestradiol, this rate decreases significantly for less lipophilic materials such as hydrocortisone. This is due to various factors: the time necessary for repairing the stratum corneum is increased, while the number of subdermis capillaries decreases [112,113].

Comparison of the pharmacokinetics of transdermal systems has been made. The first clinical study on human, has compared the pharmacokinetic parameters and the plasma concentration–time profiles for the two nicotine transdermal products Nicoderm and Habitrol [114] either in single dose or multiple dose. Thus, the observed differences in plasma profiles were attributed to the technical designs of these two systems. The nicotine concentration present in the adhesive layer acts on the rapid rise in the plasma concentration, while the rate-controlled membrane determining a constant rate of drug from the reservoir is responsible for a steady plasma nicotine concentration.

If conventional transdermal systems, by way of a patch, are easy to use, they are not able to deliver a drug having a large molecular weight such as peptides or proteins. Thus, various improvements are developed and a brief survey is given [90].

- Another way of increasing the permeability of the skin consists of using the transdermal iontophoresis by applying a potential electrical gradient between the skin site and another place on the body with two electrodes. This technique is useful for drugs that are predominantly ionized [115–118]. Nevertheless, if ions are transported faster, the transport of neutral molecules is facilitated, as the electric current increases the permeability of the stratum corneum. Interactions occur between penetration enhancers and iontophoresis [119].
- Electroporation, in making the stratum corneum porous by way of electric discharges, which increases largely the rate of drug transport through the skin.
- Transdermic administration by using ultra-sounds called sonophoresis [120].
- Using a patch containing microneedles able to perforate the stratum corneum and project the drug in the dermis. This system is able to work either in bolus or in continuous way.

9.6. Characteristics of the Transdermal Therapeutic Systems

A general study was made on the TTS and the way to improve them, the main purpose of them being to be able to bring the drug to the skin surface with a constant concentration over a long time [91]. The TT Systems were classified into three categories. The first system, very simple system is where the drug is delivered to the skin by diffusion through the polymer layers of the system. The second system, more complex, consists of these above layers with in addition a reservoir for the drug. The third system differs from the other two in the sense that the drug is transferred not by diffusion but by convection to the external surface of the skin.

The following conclusions were obtained from these theoretical studies:

- The first system is not able to apply a constant drug concentration on the skin, whatever the value of the diffusivity of the drug through the polymer.
- The addition of a reservoir to the previous TT System brings an interesting improvement.
- But the best device was found with the TTS when the drug release is controlled by convection. The main effect of the convection is that the rate of drug transfer through the TTS is so high that the drug concentration on the skin surface is maintained constant, at least for a period of time over which the mean drug concentration remains nearly constant in the TTS.
- The results are expressed in terms of kinetics of drug transfer through the skin (Fig. 79), as well as by the plasma drug profiles (Fig. 80).

These curves lead to the following conclusions:

- The rate of drug transfer through the TTS is so large that the concentration of the drug is maintained constant on the skin surface.
- The kinetics of drug transfer through the skin is drawn for various values of the constant drug concentration in the TTS ranging from 0.075 to 0.01 g/mL (Fig. 79). Of course, the higher the drug concentration in the TTS, the higher the rate of drug transfer through the skin, provided that the solubility of the drug in the skin is sufficiently high. In fact, when the process has reached the stationary conditions, the rate of drug transfer through the skin is proportional to this drug concentration. We have to note that when the solubility of the drug in the skin is not so large, an increase in the concentration in the TTS will find a limit for the concentration in the skin, and thus a partition factor will occur; but nevertheless the concentration of the drug in the skin is limited by the solubility of the drug in the skin.
- Good agreement is observed between the theoretical and the experimental drug levels when the constant concentration of the drug is 0.03 g/mL on the skin surface (Fig. 80, curve 3). As shown with the other curves, the plasma drug level largely depends on the value of this constant

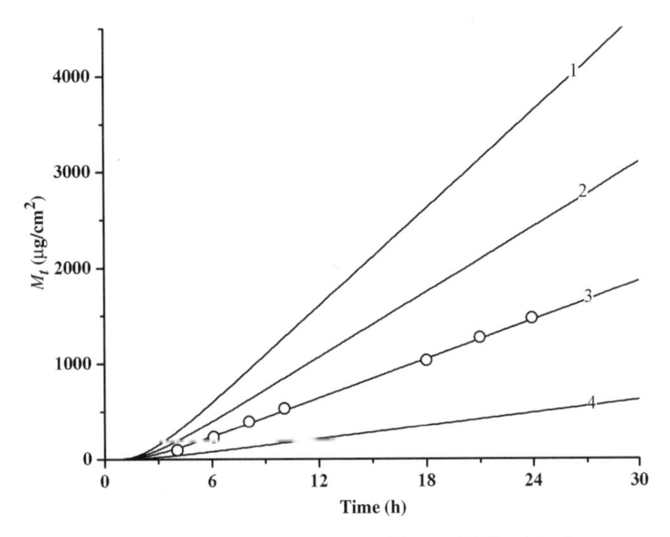

Figure 79. Kinetics of drug transfer through the skin with the TTS with drug convective transfer, and various values of the constant drug concentration in the TTS: Skin: $L = 0.035$ cm; $D = 6 \times 10^{-6}$ cm^2/s; drug concentration: (1) 0.075 g/mL; (2) 0.05 g/mL; (3) 0.03 g/mL; (4) 0.01 g/mL. Reprinted from Ref. [91] with permission from Elsevier.

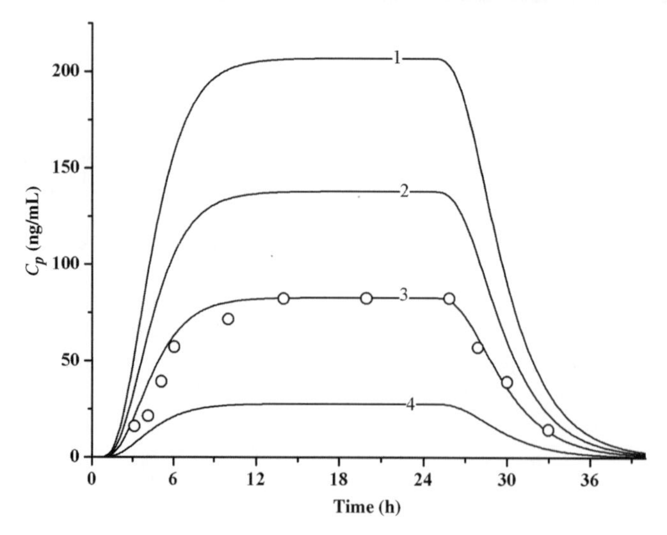

Figure 80. Plasma drug levels with the convective TTS and various values of the constant drug concentration in the TTS and the surface of the skin. Metoprolol with $k_a = 1.1\,h^{-1}$ and $k_e = 0.78\,h^{-1}$. Skin: $L = 0.035$ cm; $D = 2.2 \times 10^{-8}\,cm^2/s$; Drug concentration: (1) 0.075 g/mL; (2) 0.05 g/mL; (3) 0.03 g/mL; (4) 0.01 g/mL. Reprinted from Ref. [91] with permission from Elsevier.

concentration of the drug on the surface of the skin. In fact, the drug level at the plateau is proportional to the drug concentration in the TTS applied to the skin surface, or rather to the drug concentration on the skin surface, as this concentration on the skin surface cannot be higher than the solubility of the drug in the stratum corneum and the skin.

- The time necessary for the drug to cross the skin before reaching the blood stream is the same, whatever the drug concentration in the skin; in fact, this lag time depends only on the characteristics of the skin, e.g., its thickness and the drug diffusivity through it.
- After removal of the TTS, the plasma drug level falls in all cases, and the equation 10.12 applies, meaning that only a part of the drug located in the skin is delivered in the plasma.

10. TRANSFER OF THE DRUG IN TISSUES

Nomenclature

A_1, A_{bs}	area of the lung in contact with the blood, with the mucus, respectively;
$C_{1_{x,t}}$	drug concentration in the lung at position x and time t;
C_{bs_t}	uniform drug concentration in the mucus at time t;
C_{Pt}, C	uniform drug concentration in the blood at time t, drug concentration, respectively;
D	diffusivity of the drug in the tissue (lung, capillaries wall);
GIT	gastrointestine tract;
h	convection coefficient in the mucus;
k_a; k_e	rate constant of absorption, of elimination (h^{-1})
K	partition factor at the lung–blood interface;
K'	partition factor at the mucus–-lung interface;
L; L_1	thickness of the lung;
L'; L_{bs}	thickness of the mucus;
M_t	amount of drug transferred in the blood up to time t;
M_{bs_t}	amount of drug transferred in the mucus up to time t;
t_r	time of full erosion of the erodible dosage form;
V_p	apparent plasmatic volume (L);

x, t indicates the position, time;
Y amount of drug available along the GI;
Z amount of drug in the blood;
W amount of drug eliminated.

10.1. General Study on the Drug Transfer in the Tissues and Organs

A general study was made in a book published recently, devoted on the drug transfer from the plasma in the tissues and various organs [49]. In this chapter, only the drug transfer into a few tissues is considered: into and through the lung and bronchial mucus; in the blister fluid; into the endocarditic vegetation.

10.2. Drug Transfer into and Through Lung and Bronchial Mucus

The drug can be transferred into the lungs and the bronchial secretion through two ways: the one by inhalation when the gaseous or atomized form of the drug is inhaled [121], the other either through intravenous delivery or with oral dosage forms [12,67,122–126].

Only the second way of drug delivery is considered in this book, by using oral dosage forms either with immediate release or with sustained release, as well as i.v. delivery.

Some papers are concerned with the modeling of the transfer of the drug into the lung and the mucus from the blood compartment, and they are briefly analyzed.

The concentration–time of ciprofloxacin has been assessed in the blood compartment and in the lung tissue following oral administration of the drug. Immediate-release and erosion-controlled dosage forms have been examined. A numerical model based on finite differences and taking into account all relevant data has been built: the kinetics of drug release in the gastrointestinal tract, drug absorption in the blood compartment and elimination, and the transient diffusion of the drug throughout the lung tissue. A partition coefficient for the drug at the tissue-blood interface has been considered to express the increase in the drug concentration at the tissue surface. The effect of the dose frequency and of the erosion rate of the dosage forms on the antibiotic concentration versus time curves in the plasma and the lung tissue has been studied in detail [122].

The drug level was also calculated in the plasma, the lung tissue and bronchial secretion using ciprofloxacin as the drug with erosion-controlled dosage forms. The numerical model built and tested for this calculation, takes into account the following stages: the kinetics of drug release along the GI tract, the absorption into and elimination out of the plasma, the transient diffusion through the lung tissue and convection into the bronchial secretion [123]. Emphasis was also placed upon the effect of the patient's noncompliance [67].

This diffusion model [122] was used for evaluating the transport of ciprofloxacin into the lung tissue, and the drug concentration–-time history obtained both from calculation and experiment in the plasma and the lung were found to be in good agreement; the ciprofloxacin level in the lung tissue followed the ciprofloxacin plasma level with a lag time resulting from the time necessary for the drug to diffuse through the lung [12].

The diffusion of the drug was also considered for evaluating the degree of penetration of the antibiotic into the lung tissue with moxifloxacin at a dose of 400 mg administered intravenously or orally once daily, and the results were correlated to microbiological data, in order to estimate the clinical efficacy of the drug; this drug exhibits high penetration in lung tissue, with tissue concentrations far above the MIC 90 for most of the susceptible pathogens involved [124].

Other studies were made on the estimation of the drug concentrations either in the plasma or the lung and epithelial lining fluid for ceftazidime administered in continuous infusion to critically ill patients with severe nosocomial pneumonia [125]. In the same way, the concentrations of piperacillin/tazobactam in the plasma and lung tissue was determined in case of the steady state drugs concentrations [126].

In fact, as it was shown in a review paper [45], after the evaluation of the concentration–time profile of the drug in the blood, it is possible to calculate the drug transfer by transient diffusion throughout the lung and by convection into the mucus, and thus whatever the way of the drug delivery in the body.

10.2.1. Mathematical and Numerical Treatment Four stages are considered in succession: release of the drug out of the dosage form along the GI tract, absorption into and elimination out of the plasma, diffusion through the lung tissue and convection into the bronchial secretion.

The following assumptions are made in order to make clear the process:

(1) The kinetics of drug release along the GI tract is either very fast in the case of immediate-release dosage forms or controlled in case of sustained-release dosage forms.
(2) The kinetics of drug release does not vary along the GI tract time, as well as the rate constant of absorption [127]. (In fact, this assumption is not necessary, provided that the variation is known).
(3) The stages of absorption and elimination of the drug are expressed by first-order kinetics with the rate constants of absorption and elimination.
(4) The rate constants of absorption and elimination, as well as the apparent volume of distribution, are determined from *in vivo* data, either in the case of the lung [128,129] or in the case of the mucus [15]. These values are kept constant in the multidoses treatment [130].
(5) The typical representation of the lung is a tree, but the distribution of the area and volume as a function of its radius is not known, the lung is assumed to have a uniform thickness L. The drug transport throughout the lung tissue is controlled by one-directional transient diffusion with a constant diffusivity. A partition factor K is introduced at the plasma-lung interface, to express that the drug concentration on the lung tissue surface is K times that in the plasma [122], following the experimental data [128].
(6) The transport of the drug from the lung tissue into the bronchial secretion (mucus) is controlled by convection with a convective transfer coefficient into the secretion next to the lung tissue surface. A partition factor K' is introduced [123], meaning that the drug concentration in the secretion is K' times that at the surface of the lung tissue in contact with the secretion. Because of the convection process whose rate is much larger than that with diffusion, the drug concentration is uniform in the bronchial secretion [131].

Evaluation of the Amount of Drug in the GI Tract, the Plasma, the Lung, and Mucus The rate of drug released by the dosage form is $\mathrm{d}M/\mathrm{d}t$ whatever the type of release is

The amount of drug in the GI, Y, is obtained by the equation

$$\frac{\mathrm{d}Y}{\mathrm{d}t} = \frac{\mathrm{d}M}{\mathrm{d}t} - k_{\mathrm{a}} \cdot Y \tag{157}$$

The amount of drug in the plasma Z is given by

$$\frac{\mathrm{d}Z}{\mathrm{d}t} = k_{\mathrm{a}} \cdot Y - k_{\mathrm{e}} \cdot Z \tag{158}$$

and the amount of drug eliminated, W, is given by

$$\frac{\mathrm{d}W}{\mathrm{d}t} = k_{\mathrm{e}} \cdot Z \tag{159}$$

Transfer of the Drug into the Lung Tissue (Fig. 81) The general equation expressing the mass balance for the drug at any time t is expressed by

$$M_t = Y_t + Z_t + W_t + A_1 \cdot \int_0^L Cl_{x,t} \cdot \mathrm{d}x + A_{\mathrm{bs}} \cdot L_{\mathrm{bs}} \cdot C_{\mathrm{bs}_t} \tag{160}$$

where M_t is the amount of drug released from the dosage form up to time t; A_1 and A_{bs} are the area of the lung in contact with the plasma and

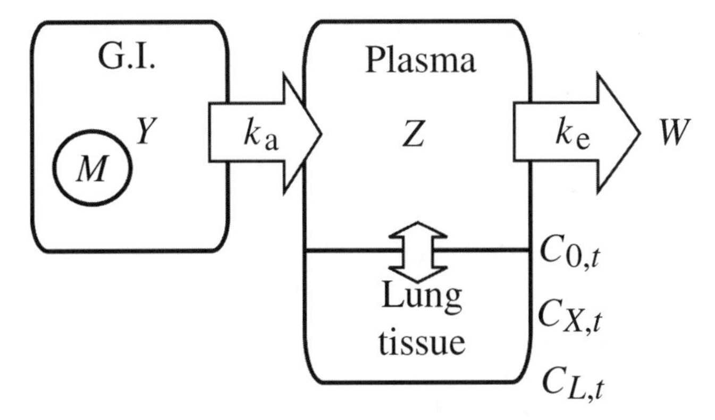

Figure 81. Scheme of the path followed by the drug from the GI tract to the lung. *Y*: The amount of drug along the GI tract at time *t*; Z: The amount of drug in the plasma at time *t*.

the bronchial secretion, respectively; L and L_{bs} are the mean thickness of the lung and of the bronchial secretion, respectively; $C_{1_{x,t}}$ and C_{bs_t} represents the concentration of the drug in the lung at time t and position x, and the uniform concentration in the bronchial secretion at time t, respectively.

The equation of diffusion with constant diffusivity through the thickness of the lung tissue is

$$\frac{\partial C}{\partial t} = D \cdot \frac{\partial^2 C}{\partial x^2} \qquad (161)$$

with the boundary conditions in contact with the plasma:

$$t > 0 \quad x = 0 \quad C_{1_{0,t}} = K \cdot C_{\mathrm{p}_t} \qquad (162)$$

where C_{p_t} represents the uniform concentration of free drug in the plasma at time t, and K is the partition factor at the lung–plasma interface.

Transfer of Drug in the Bronchial Secretion Depending on the viscosity of the material, the transfer of a substance into this material is controlled either by convection or by diffusion [131]. In the case of a liquid like the bronchial secretion, the transfer can be assumed to be controlled by convection:

$$-D \cdot \left(\frac{\partial C}{\partial x}\right)_{L,t} = h \cdot (Cl_{L,t} - K' \cdot Cbs_t) \qquad (163)$$

Thus, the rate at which the drug is brought to the interface by diffusion through the lung is constantly equal to the rate at which the drug enters the bronchial secretion [131]. The value of the coefficient of convective transfer h into the bronchial secretion depends on its viscosity. K' is the partition factor at the lung-bronchial secretion interface, meaning that the uniform concentration of the drug in the bronchial secretion is K' times that on the lung surface in contact with the liquid.

When no evaporation or excretion occurs out of the bronchial secretion, the other boundary condition stands at the external surface:

$$\left(\frac{\partial C_{bs}}{\partial x}\right)_{L'} = 0 \qquad (164)$$

and the uniform concentration in the bronchial secretion is obtained by

$$C_{bs_t} = \frac{M_{bs_t}}{L'} \qquad (165)$$

where M_{bs_t} is the amount of drug transferred into the bronchial secretion per unit area of lung-bronchial secretion interface at time t, and L' is the thickness of the bronchial secretion.

10.2.2. Experimental Part

Experiments on Lung Tissue [128] The pharmacokinetics of ciprofloxacin in the plasma and lung tissue at steady state were determined on 38 patients subjected to lung surgery, after they were given 500 mg of drug through an immediate-release dosage form.

Table 16. Pharmacokinetic Parameters of Ciprofloxacin

$k_a = 1.29\,h^{-1}$	$k_e = 0.12–0.2–0.32\,h^{-1}$	$V_p = 120\,L$	$D = 5 \times 10^{10}\,cm^2/s$
$L = 10\,\mu m$ (lung)	$K = 3.1$ (lung–serum interface)	$L' = 10\,\mu m$ (mucus)	
$K' = 0.125$ (mucus–lung interface)	$D = 2.7 \times 10^{-6}\,cm^2/s$	$\frac{h \cdot L}{D} = 5$	

Plasma samples, two for each patient, were obtained and analyzed.

Experiments on Bronchial Secretion [15] Eight adults (four females and four males) with a mean age of 24 years were given 500 mg oral ciprofloxacin every 8 h for a total of 10 days. Pharmacokinetic studies were performed on day 3 of therapy. Blood and sputum were taken at various times for analysis of the drug.

The data collected by the authors of Ref. [15,128] are in Table 16.

10.2.3. Results Obtained with Lung Tissue

Good correlation between the experimental and the calculated values is observed in Fig. 82 for the profiles of drug concentration either in the plasma (curve 1) or in the lung tissue (2). The mean theoretical value for the lung is only drawn in this Fig. 82, as the variation in the concentration is rather small, resulting from the very thin thickness of the tissue. The two values used for the rate constant of elimination of the drug in calculating these profiles show the strong effect of this parameter.

Other drug profiles are drawn with an erosion-controlled dosage form containing 500 mg of ciprofloxacin whose time of erosion is 24 h either for a single dose (Fig. 83) or for repeated doses using the same dosage forms taken once a day (Fig. 84).

Some comments are made regarding the above-mentioned figures:

- A short time lag appears in Fig. 82 between the peak of the drug concentration reached in the lung tissue and in the blood. It is due to the time required for the drug to diffuse through the lung tissue, and especially to reach to its mid-plane.
- A higher concentration is observed in the lung tissue than in the blood compartment as determined by the experiments [128]. This fact necessitates in calculation to take a partition factor of

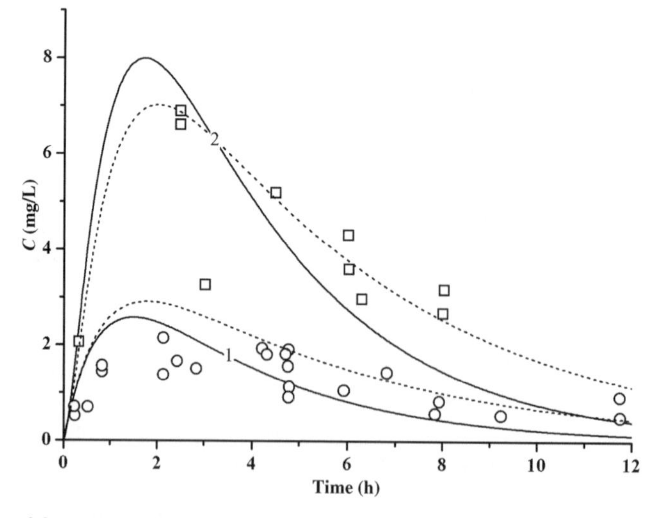

Figure 82. Profile of drug concentration in the plasma (1) and lung (2) with an immediate-release dosage form containing 500 mg ciprofloxacin. $k_a = 1.2\,h^{-1}$; $k_e = 0.32\,h^{-1}$ (full line); and $k_e = 0.2\,h^{-1}$ (dotted line). $V_p = 120\,L$; $K = 3.1$; $D = 5 \times 10^{-10}\,cm^2/s$; $L = 10\,\mu m$.

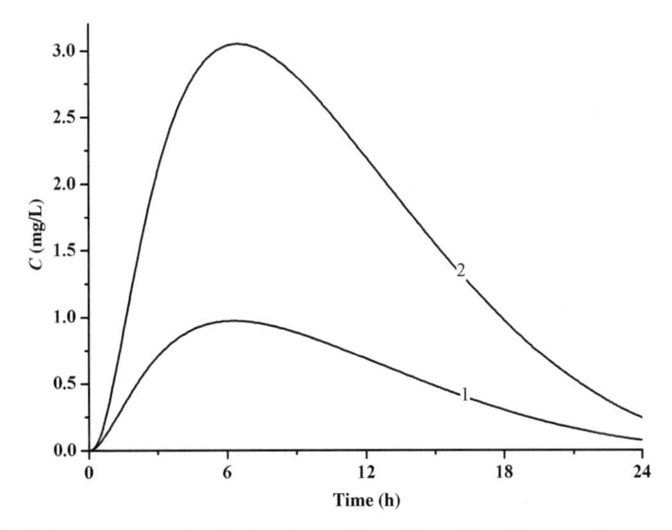

Figure 83. Profile of drug concentration in the plasma (1) and lung (2) with an erosion-controlled dosage form with a time of full erosion of 24 h, and 500 mg ciprofloxacin $k_a = 1.2\,h^{-1}$; $k_e = 0.32\,h^{-1}$. $V_p = 120\,L$; $K = 3.1$; $L = 10\,\mu m$; $t_r = 24\,h$.

3.3 for the drug at the blood-lung tissue interface.

- Of course, it is possible to extend the numerical model in calculating the drug profile in both the blood and the lung tissue when sustained-release dosage forms are taken either in single dose [Fig. 83] or with multidoses [Fig. 84].
- It can be shown that the steady state is promptly attained in Fig. 84 with sustained-release dosage forms controlled by erosion taken once a day, with a more constant drug level than that obtained with immediate-release dosage forms.
- The experimental values are widely scattered. This dispersion results from the fact that various patients have been

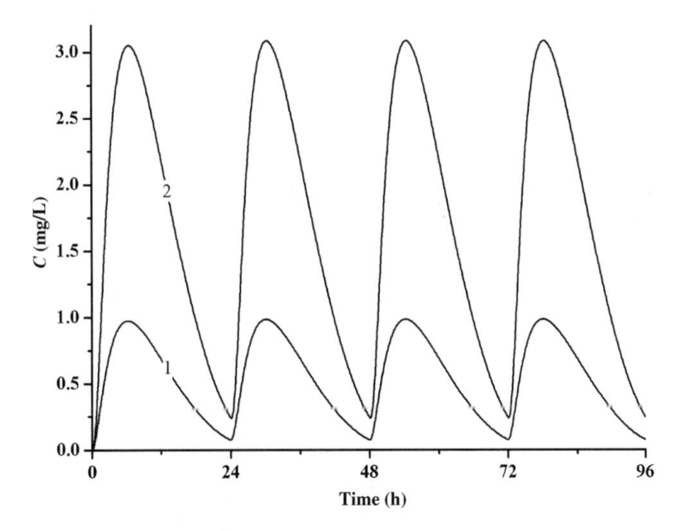

Figure 84. Profile of drug concentration in the plasma (1) and lung (2) with repeated doses taken once a day, each being an erosion-controlled dosage form with a time of full erosion of 24 h, and 500 mg ciprofloxacin $k_a = 1.2\,h^{-1}$; $k_e = 0.32\,h^{-1}$. $V_p = 120\,L$; $K = 3.1$; $L = 10\,\mu m$; $t_r = 24\,h$.

considered, as only two tissue and blood samples were taken for each patient. The intervariability of the patients plays an acute role, explaining the dispersion obtained through these tedious experiments, and in calculation two values of the rate constant of elimination are used.

10.2.4. Results Obtained with the Bronchial Secretion The scheme of the process in Fig. 85, which shows an extension of the process elaborated for the drug transfer through the lung tissue, with in addition the convection of the drug into the highly viscous liquid of the mucus.

The profiles of drug concentration in the mucus and in the blood, are drawn, as they are obtained by experiments [15] and calculation [Fig. 86].

The following conclusions can be drawn from these curves:

- Good agreement is obtained between the experimental and the calculated curves, proving the validity of the model. The maximum for each curve is also well evaluated either for the time or for the drug concentration in these two sites.
- The maximum of the drug level is attained after a longer time in the bronchial secretion (2.5 h) than in the blood (1.5 h). This lag time results from the time necessary for the drug to diffuse through the lung tissue before entering the mucus [67,123].
- The experiments made with samples taken directly in the sputum and the blood being not so tedious as those made in the lung tissue, that fact explains the better profiles obtained.
- Of course, for calculation, the drug transport has been evaluated either through the lung tissue or in the mucus, while only the drug profiles are drawn in the blood and the mucus in Fig. 86. In Fig. 87, the three profiles of the drug concentration in the blood (curve 1), in the lung tissue (curve 2) and in the mucus (curve 3) are calculated with an erosion-controlled dosage form having a time of full erosion of 24 h.
- The profiles are calculated (Fig. 88) in the blood (1), in the lung tissue (2), and in the mucus (3) with multidoses taken once a day with erosion-controlled dosage forms having a time of full erosion of 24 h.

10.3. Drug Transport in the Blister Fluid

10.3.1. Short Bibliography Survey It is of great interest to get good knowledge on the diffusion of antibiotics in the extracellular fluid, since this is the obligatory path to infections loci [132]. More specifically, various experimental studies have been especially

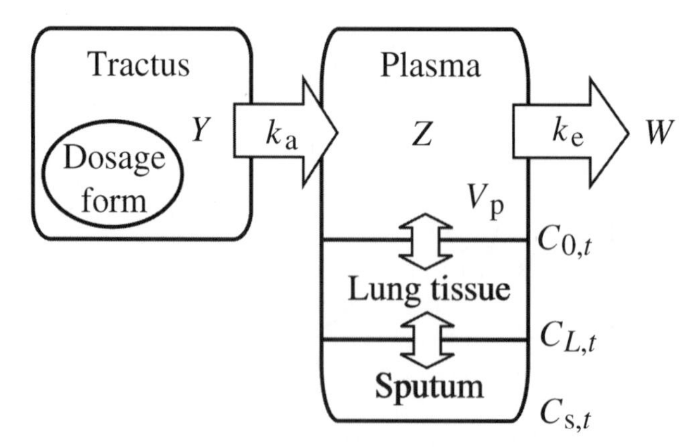

Figure 85. Scheme of the process of the drug transfer from the GI tract to the mucus, through the blood and the lung tissue.

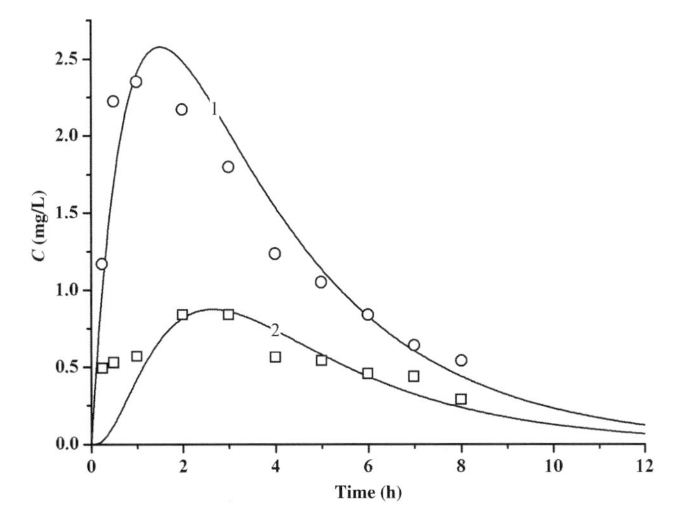

Figure 86. Profile of drug concentration in the plasma (1) and mucus (2) with an immediate-release dosage form containing 500 mg ciprofloxacin. $k_a = 1.2\,h^{-1}$ and $k_e = 0.2\,h^{-1}$. $V_p = 120\,L$; $K = 3.1$; $D = 5 \times 10^{-10}\,cm^2/s$; $L = 10\,\mu m$; $L' = 10\,\mu m$; $K' = 0.125$; $h = 2.7 \times 10^{-6}\,cm/s$; and $\frac{h \cdot L}{D} = 5$. Reprinted from Ref. [123] with permission from Kluwer.

made on the penetration of the drug into the chemically induced blister fluid that has been shown to be similar in composition to the exudates of a mild inflammatory reaction [133,134]. However, there is little information about the diffusivity of antibiotics through the tissues, it has been shown that the penetration of antibiotics into tissues is governed by Fickian diffusion [135,136]. The concentration of ciprofloxacin was measured either in the blood compartment or in the blister fluid of healthy male volunteers after they were given a single immediate-release oral dose of 500 mg [13].

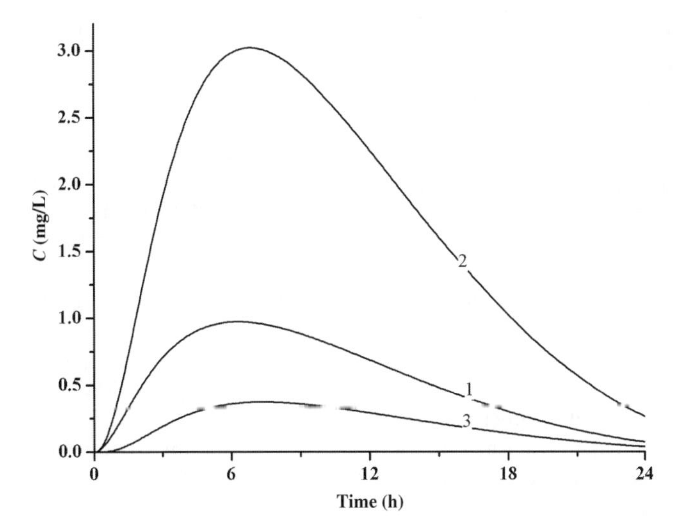

Figure 87. Profile of drug concentration in the plasma (1), in the lung (2), and in the mucus (3) with an erosion-controlled dosage form containing 500 mg ciprofloxacin. $k_a = 1.2\,h^{-1}$ and $k_e = 0.2\,h^{-1}$; $V_p = 120\,L$; $K = 3.1$; $D = 5 \times 10^{-10}\,cm^2/s$; $L = 10\,\mu m$; $L' = 10\,\mu m$; $K' = 0.125$; $t_r = 24\,h$; $h = 2.7 \times 10^{-6}\,cm/s$; and $\frac{h \cdot L}{D} = 5$.

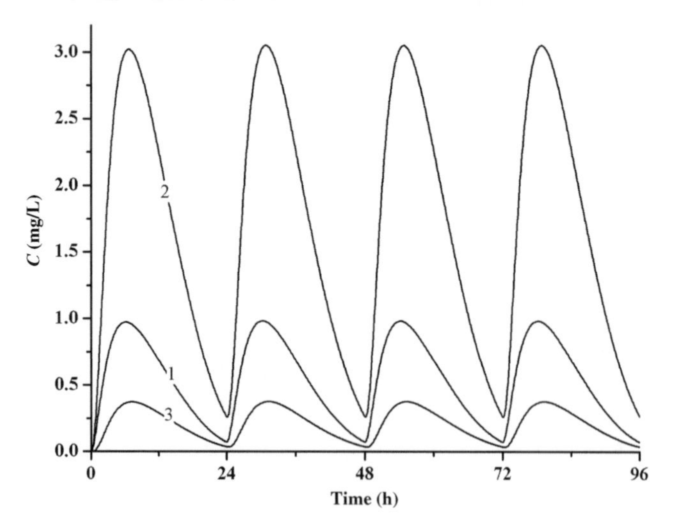

Figure 88. Profile of drug concentration in the plasma (1), in the lung (2), and in the mucus (3) with erosion-controlled dosage forms containing 500 mg ciprofloxacin taken once a day. $k_a = 1.2\,h^{-1}$ and $k_e = 0.2\,h^{-1}$; $V_p = 120\,L$; $K = 3.1$; $D = 5 \times 10^{-10}\,cm^2/s$; $L = 10\,\mu m$; $L' = 10\,\mu m$; $K' = 0.125$; $t_r = 24\,h$; $h = 2.7 \times 10^{-6}\,cm/s$; and $\frac{h \cdot L}{D} = 5$.

Finally, the process of drug transport in the blister fluid was studied by considering the transient diffusion though the capillaries wall with the drug concentration into the blood varying according to the pharmacokinetic parameters of the patient [137]. In order to simplify the problem, the transport was assumed to be unidirectional through the thickness of the wall, and the diffusivity constant. As no analytical solution exists, a numerical model based on finite differences was built and tested by comparing the theoretical with the experimental values found in the literature [13,138].

As it was not possible to evaluate the thickness of the capillaries wall, the diffusion-controlled transport through this tissue was defined by the diffusivity as a fraction of the square of this thickness, this ratio having the dimension of $(\text{time})^{-1}$, in the same way as the rate constant of a first-order reaction.

10.3.2. Mathematical and Numerical Treatment Two kinds of calculation are made in succession, the one concerned with the determination of the profile of the drug in the blood, and the other with the diffusion-controlled transport of the drug through the capillaries wall to the blister fluid.

Assumptions A few assumptions are made in order to make the process clear (Fig. 89):

(1) The drug transport is controlled by unidirectional diffusion through the thickness of the capillaries wall.
(2) The diffusion is Fickian with a constant diffusivity.
(3) The concentration of the drug on the surface of the wall in contact with the blood volume is the same as that in the plasma volume, at any time.
(4) The concentration of the drug on the surface of the wall in contact with the blister fluid is constantly equal to that of the blister fluid.

Mathematical Treatment of the Drug Transport The profile of drug concentration in the blood is calculated in the same way as made for the lung tissue, when the drug is administered through oral immediate-release dosage forms.

The amount of drug in the GI, Y, is obtained by the equation

$$\frac{dY}{dt} = \frac{dM}{dt} - k_a \cdot Y \qquad (166)$$

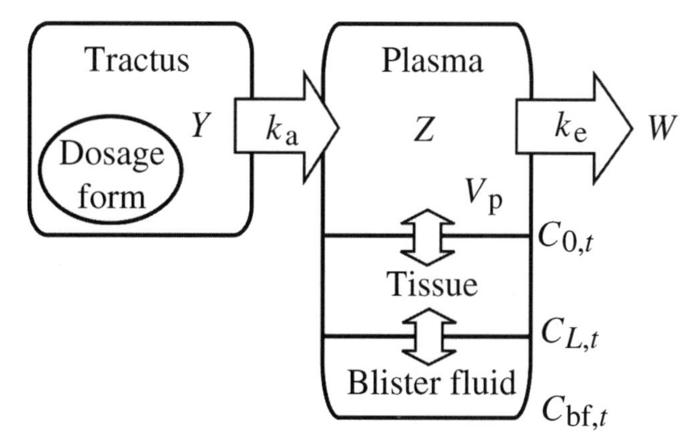

Figure 89. Scheme of the process of drug transfer in the blister fluid by diffusion through the capillaries wall.

The amount of drug in the plasma Z is given by

$$\frac{dZ}{dt} = k_a \cdot Y - k_e \cdot Z \tag{167}$$

Diffusion of the Drug Through the Capillaries Wall The equation of diffusion through the wall of thickness L is

$$\frac{\partial C}{\partial t} = D \cdot \frac{\partial^2 C}{\partial x^2} \tag{168}$$

and the initial conditions express that there is no drug in the wall

$$t = 0 \quad 0 < x < L\, C = 0 \tag{169}$$

The boundary conditions are for the surface in contact with the plasma

$$t > 0 \quad x = 0 \quad C_{0,t} = C_{\text{plasma},t} \tag{170}$$

and for the surface in contact with the blister fluid:

$$t > 0 \quad x = L\ C_{L,t} = \text{blister fluid concentration} \tag{171}$$

As no analytical solution exists for the problem with fluctuating concentrations on each surface, a numerical method with finite differences is built and used to resolve the problem.

10.3.3. Calculated and Experimental Results

The validity of the model has been tested by comparing the drug concentration either in the blood or in the blister fluid obtained by calculation and by experiments.

The drug profiles in the blood (1) and the blister fluid (2) shown in Fig. 90 for a single oral dose with immediate release, and in Fig. 91 with the same dosage form delivered in multidoses, by using the experimental values shown in Table 17 [138], lead to the following comments:

- Good agreement is obtained between the calculated and the experimental values either in the blood or in the blister fluid for the oral dosage form with immediate release, as shown in Fig. 90. This is due to the fact that the method used in sampling the blister fluid for analysis is highly convenient and efficient. There is also a little intervariability between the six healthy volunteers who were administered the drug.
- With the oral administration in Fig. 90, a maximum of the drug concentration is attained at 1.33 h in the blood and at approximately 4 h in the blister fluid. Thus the drug concentration in the blister fluid lags far behind the corresponding one in the blood, resulting from the process of diffusion through the capillaries wall.
- As shown in Fig. 91 with repeated doses at the rate of three times a day using the same oral dosage form, the drug concen-

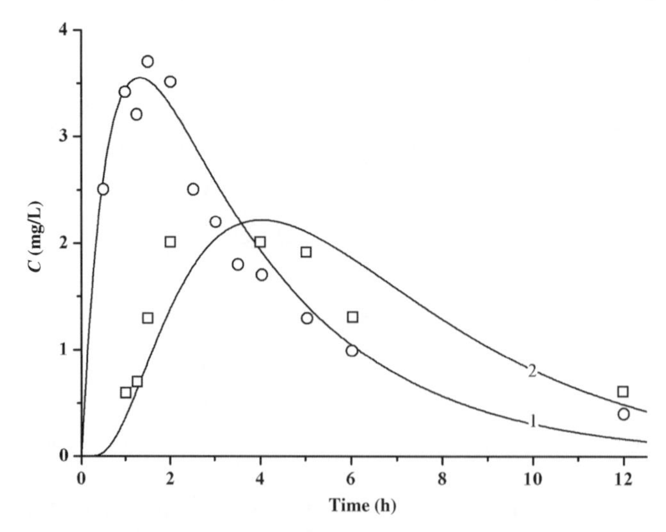

Figure 90. Profiles of drug concentration in the blood (1) and in the blister fluid (2) obtained with an oral dosage form with immediate-release containing 750 mg ciprofloxacin. $k_a = 1.5\,h^{-1}$ and $k_e = 0.3\,h^{-1}$; $V_p = 140$ L; $D/L^2 = 0.2\,h^{-1}$. Reprinted from Ref. [122] with permission from Kluwer.

tration in the blister fluid alternates between peaks and troughs in the same way as that in the blood with the time lag of approximately 2.6 h.

- Of course, resulting from the process of diffusion with no partition factor that could increase the drug concentration in the same way as in the lung, the profile of drug concentration in the blister fluid is flatter than that in the blood.
- The values of the parameters of diffusion necessary for fitting the theoretical and experimental curves are around $0.2\,h^{-1}$ for the ratio D/L^2. This ratio has the same dimension as the rate constant of absorption.

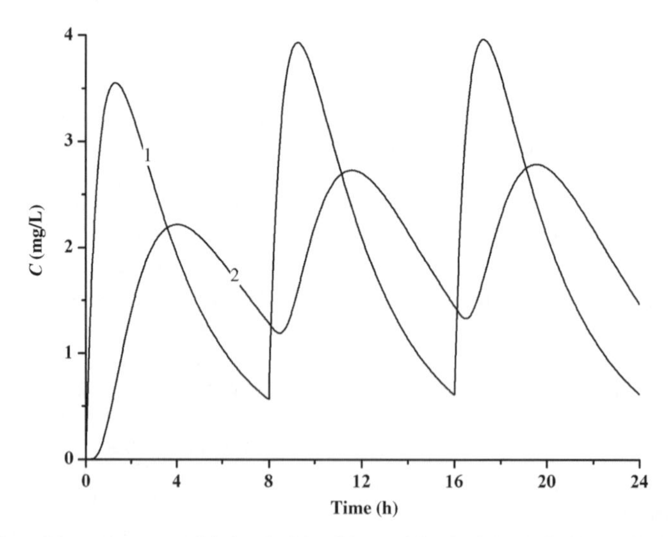

Figure 91. Profiles of drug concentration in the blood (1) and in the blister fluid (2) obtained with repeated oral dosage form with immediate-release containing 750 mg ciprofloxacin, taken three times a day. $k_a = 1.5\,h^{-1}$ and $k_e = 0.3\,h^{-1}$; $V_p = 140\,L$; $D/L^2 = 0.2\,h^{-1}$.

Table 17. Pharmacokinetic and Diffusion Parameters of Ciprofloxacin

$k_a = 1.5\,h^{-1}$	$k_e = 0.3\,h^{-1}$	$V_p = 140\,L$	$\frac{D}{L^2} = 0.2\,h^{-1}$

10.4. Transfer of Drug into Endocarditis

10.4.1. Short Bibliography Bacterial endocarditis is a difficult infection to cure, due to simultaneous reasons: poor penetration of antibiotics into infected vegetations; altered metabolic state of bacteria within the lesion; and absence of adequate host-defense cellular response that could cooperate with antibiotic action. In spite of the studies made on animal with the objective to get a better understanding of the pathophysiology of the infection [139], the definition and improvement of therapeutic regimens of endocarditis in human remains of great importance due to the difficulties encountered in clinical trials. If it is usually effective to treat a patient using antibiotics when the bacterial infection is in the blood compartment, such treatment becomes difficult in the case of a vegetation infected by bacteria [140]. The striking example of the persistence of *Streptococcus mitis* in such a situation has been observed over some 25 days after antibiotic therapy [141]. This well-known fact appears in patients with endocardial vegetations containing bacteria, and the aggressive treatment of the infectious endocarditis required doses of antibiotics over a long period of time to achieve an effective cure [142]; such treatment is needed because the antibiotics present in the blood compartment should diffuse within the vegetation before exerting their activity against those bacteria located in the vegetation. This diffusion stage of the drug is controlled by transient radial diffusion, and the time necessary for the drug to reach the center of this sphere is proportional to the square of the radius of the vegetation [131]. Thus, the treatment of endocarditis should require a prolonged antibiotic therapy, and it is of great concern to know the terms of the adequate therapy; individualized drug dosage regimens have made therapy possible even in patients with impaired renal function [142].

The knowledge brought with the typical characteristics of the diffusion through the spherical endocarditis enabled the medical doctors to define clearly the special features of the therapy, with the amount of drug and the duration of the drug delivery [142,143], and they were applied to a successful treatment of various patients at hospital by using repeated infusions with the calculated dose and duration of therapy [144]. Moreover, an *in vitro* method was developed able to evaluate precisely the diffusivity of a drug [145].

10.4.2. Mathematical and Numerical Treatment of the Drug Transport

Assumptions The following assumptions are made to clarify the problem:

(1) The drug profile in the blood is obtained with discontinuous infusions, repeated twice a day, each dose of 500 mg being delivered with a constant rate over 0.5 h.
(2) The vegetation, spherical in shape, is in contact with the blood.
(3) The drug transport through (into and out of) the sphere is controlled by radial diffusion, with a finite coefficient of convection at the blood–vegetation interface.

Mathematical Treatment The drug profile in the blood is calculated by using the Equations .

The equation of the radial diffusion is [131]

$$\frac{\partial C}{\partial t} = \frac{1}{r^2} \cdot \frac{\partial}{\partial r}\left[D \cdot r^2 \cdot \frac{\partial C}{\partial r}\right] \tag{172}$$

The initial condition expresses the fact that the vegetation is free from drug:

$$t = 0 \quad 0 < r < R \quad C = 0 \tag{173}$$

The boundary condition expresses the fact that at the blood–vegetation interface, the rate of drug transfer by convection is constantly equal to the rate transfer by diffusion into the sphere.

$$-D\frac{\partial C}{\partial r} = h(C_{\text{blood},t} - C_{R,t}) \tag{174}$$

Because of the variable concentration of the drug in the blood, the numerical treatment is not feasible, and a numerical method should be used, with a constant increment of time and by considering the spherical penetration along the increment of the radius [131].

10.4.3. Drug Profiles Calculated and Measured in the Vegetation During the Treatment The profiles of drug concentration are drawn in Fig. 92 with the ratio $D/R^2 = 10^{-6}\,\text{s}^{-1}$ and in Fig. 93 with the ratio $D/R^2 = 5.10 \times 10^{-6}\,\text{s}^{-1}$ are calculated in various places: in the blood (1), on the surface of the vegetation (2), at half its radius (3), and at the center of the sphere (4). These curves lead to some comments of interest for the therapy:

- The drug diffuses slowly along the radius of the spherical vegetation, and the concentration–time profiles at different positions in the sphere are quite different.
- The effect of the characteristics of the vegetation with its radius and the value of the diffusivity of the drug through it is of great importance on the penetration of the drug. Thus with the lower value of the ratio D/R^2, in Fig. 92, the steady state in not yet attained after 72 h of infusion, while it is attained after the second dose with the higher value of this ratio (Fig. 93).
- The drug concentration in various positions of the vegetation alternates between peaks and troughs in the same way as in the blood compartment, as shown in Fig. 93 with the larger value of the ratio D/R^2.
- The ratio D/R^2 is of great concern, as both the diffusivity and the radius of the vegetation are difficult to determine with precision, and thus instead of having two unknowns with the diffusivity and the radius, there is only one.
- The obvious statement holds: the smaller the radius of the vegetation and the larger the diffusivity of the drug through it, the faster the drug transport into the vegetation.
- It is worth noting again that, as the process is controlled by diffusion, the time of penetration in the vegetation is proportional to the square of its radius.

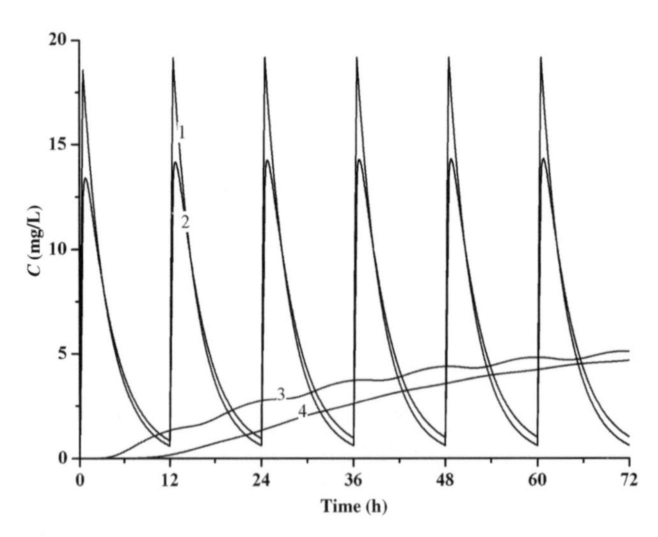

Figure 92. Profiles of drug concentration in the blood (1) and in various positions in the vegetation: on the surface (2); at the mid-radius (3); in the center (4); with the value of the ratio $D/R^2 = 10^{-6}\,\text{s}^{-1}$, with amikacin delivered through infusions of 500 mg over 30 min twice a day. $k_e = 0.30\,\text{h}^{-1}$; $V_p = 25\,\text{L}$.

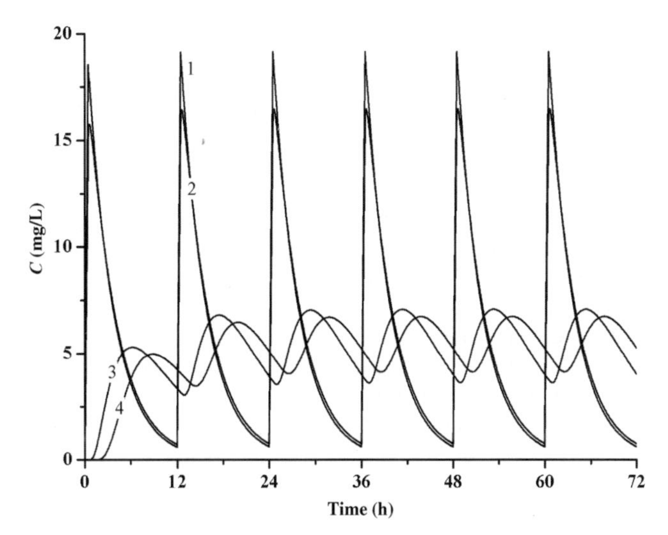

Figure 93. Profiles of drug concentration in the blood (1) and in various positions in the vegetation: on the surface (2); at the mid-radius (3); in the center (4); with the value of the ratio $D/R^2 = 5 \times 10^{-6}\,\mathrm{s}^{-1}$, with amikacin delivered through infusions of 500 mg over 30 min twice a day. $k_e = 0.30\,\mathrm{h}^{-1}$; $V_p = 25\,\mathrm{L}$.

11. CONCLUSIONS

In terms of a very short conclusion for this chapter, it should be said that the following facts of great concern have been considered and a solution has been found for them:

A good knowledge of the process of drug delivery and transport in the body and the tissues has been provided, on the ground that it makes the mathematical treatment applied possible, leading to the following advantages:

(1) The problem of the intervariability of the patients can be not only considered seriously but also resolved, either at the hospital or at home.

-At the hospital, the treatment has been deeply studied in Section 3 with the intravenous drug delivery. Some conclusions are of concern. The protocol followed by the treatment usually done has been established on patients who are in the medium position at the end of the clinical trials (phase III and IV). But it does not correspond with all the patients. A simple addition or modification at the beginning of this protocol enables the evaluation of the values of the pharmacokinetic (pk) parameters of the patient: two analyses made at 2 h at intervals and a simple calculation provides this piece of information. And, afterward, the right treatment can be promptly determined with the help of the numerical model and applied to the therapy of the patient.

-At home, the problem is not so simple and it cannot be resolved so easily. But this problem has to be considered by the three persons concerned with the treatment: the patient, the doctor who prescribes, and the pharmacist who delivers the drug. A good connection between these three partners should determine the side effects of the patient and thus should improve the therapy by reducing them. It is a pity that a drug such as Vioxx would lead to a disaster, because it was delivered without any great care. An effective drug acts on various organs and can disturb the patients who suffer from various pathologies.

All these considerations lead to use dosage forms easily breakable, so as to allow the patient to take the dosage form at the right amount necessary for reducing the side effects.

(2) The important and more often forbidden problem of noncompliance should be considered seriously. The sustained release oral dosage forms are an improvement for the therapy at home since they allow the once a day dosage. But the patient should be given a perfect piece of information on the dangers of the bad compliance. And the best way is to show him the effect on the plasma drug profile: for instance, when he sees that doubling the dose makes the drug concentration approach the lethal dose.

The preparation of these sustained-release dosage forms should be made in terms of the pk parameters of the drug. It would be a huge work if a few diagrams were not built, so as to connect the half-life time of the drug with the characteristics of the dosage form.

(3) Finally, these diagrams should be of help for the laboratories that prepare the drugs and dosage forms. Finding a new drug is too expensive: US$1 billion is often considered as the amount necessary for preparing the right drug after 7–15 years of study.

Thus, the method described in this chapter is the answer to the important question that always arises: How to use the drug in a better way to the benefit of the patient.

REFERENCES

1. Skelly JP, Barr WH, Benet LZ, et al. Report of the workshop on controlled-release dosage forms: issues and controversies. Pharm Res 1987;4:75–77.
2. Skelly JP, Amidon GL, Barr WH, et al. *In vitro* and *in vivo* testing and correlation for oral controlled/modified release dosage forms. Pharm Res 1990;7:975–982.
3. Skelly JP, Shiu GF. *In vitro/in vivo* correlations in biopharmaceutic: scientific and regulatory implications. Eur J Drug Metab. Pharmacokinet 1993;18(1):121–129.
4. Vergnaud JM. Invited speaker by the FDA. Method of calculation of the plasma drug level with oral sustained-release dosage forms controlled by diffusion. Washington, DC. 1996.
5. Vergnaud JM. Invited speaker by the FDA. Method of calculation of the plasma drug level with oral sustained-release dosage forms controlled by erosion. Washington, DC. 1998.
6. Vergnaud JM, Rosca ID, Assessing Bioavailability of Drug Delivery Systems. Boca Raton, FL: Taylor & Francis; 2006.
7. Hogue C. Low-dose effects. Chem Eng News 2004;5:50–54.
8. Oser BL, Melnik D, Hochberg M, Ind. Eng. Chem. Anal. Chem. Ed. 1945;17:405.
9. Levy G, Gibaldi M, Jusko J. Multi compartment pharmacokinetic models and pharmacologic effects. J Pharm Sci 1969;58:422–424.
10. Vidal dictionnaire. Vidal, 2 rue Béranger, 75140 Paris, cedex 03, France.
11. Campoli-Richards DM, Monk JP, Price A, Benfield P, Todd PA, Ward A. Ciprofloxacin: a review of its antibacterial activity, pharmacokinetic properties and therapeutic use. Drugs 1988;35:373–447.
12. Breilh D, Saux MC, Maire P, Vergnaud JM, Jelliffe RW. Mixed pharmacokinetic population study and diffusion model to describe ciprofloxacin lung concentrations. Comput Biol Med 2001;31(3):147–155.
13. Crump B, Wise R, Dent J. Pharmacokinetics and tissue penetration of ciprofloxacin. Antimicrob. Agents Chemother 1983;24:784–786.
14. Fabre D, Bressolle F, Gomeni R, Arich C, Lemesle F, Beziau H, Galtier M. Steady-state pharmacokinetics of ciprofloxacin in plasma from patients with nosocomial pneumonia: penetration of the bronchial mucosa. Antimicrob. Agents Chemother 1991;35(12):2521–2525.
15. Smith MJ, White IO, Bouryer H, Willis J, Hodson ME, Batten JC. Pharmacokinetics and sputum penetration of ciprofloxacin in patients with cystic fibrosis. Antimicrob Agents Chemother 1986;30:614–616.
16. Sorgel F, Jaehde U, Naber K, Stephan U. Pharmacokinetic disposition of quinolones in human body fluids and tissues. Clin Pharmacokin 1989;16(1):5–24.
17. Hoppfler D, Koeppe P. Pharmacokinetics of ofloxacin in healthy subjects and patients with impaired renal function. Drugs 1987;1:51–55.

18. Clissold SP. Aspirin and related derivatives of salicylic acid. Drugs 1986;32:8–26.
19. Vergnaud JM, Rosca ID, Case of membrane of thickness L. In: Vergnaud JM, Rosca ID,Editors. Assessing Food safety of Polymer Packaging. Shawbury, UK: Smithers Rapra Limited; 2006. p 41–49.
20. Rosca ID, Vergnaud JM. Adapting therapy with repeated short-infusions to inter individual variability between patients. Eur J Drug Metabol Pharmacokinet 2007;32:87–99.
21. Vergnaud JM, Rosca ID. Intravenous administration. In: Vergnaud JM, Rosca ID,Editors. Assessing Food Bioavailability of Drug Delivery Systems. USA: CRC Press Taylor & Francis Group; 2005. p 15–31.
22. Fabre D, Bressole F, Gomeni R. Steady-state pharmacokinetics of ciprofloxacin in plasma from patients with nosocomial pneumonia: penetration of the bronchial mucosa. Antimicrob. Agents Chemother. 1991;35: 2521–2525.
23. Skelly JP, Van Buskirk GA, Savello DR, et al Scale-up of immediate oral solid dosage forms. Pharm Res 1993;10:313–316.
24. Cardot JM, Beyssac E. *In vitro/in vivo* correlations: scientific implications and standardisation. Eur J Drug Metab Pharmacokinet 1993;18(1):113–120.
25. Leeson L.USP Open Hearing, Toronto, Canada, June 15–18, 1992.
26. Skelly JP. Bioavailability and bioequivalence. J ClinPharmacol 1976;16:539–545.
27. Shah VP, Skelly JP, Barr WH, Malinowski H, Amidon GL. Scale-up of controlled release products-preliminary considerations. Pharm Tech 1992; 35–39.
28. Subcommittee on Biopharmaceutics, *In vitro–in vivo* correlation for extended-release oral dosage forms. Pharmacopoeia Forum, Stimuli to the revision process. USP. 1988; 4160.
29. AAPS/FDA/USP. *In vitro* and *in vivo* testing and correlation for oral controlled/modified release dosage forms, Washington, DC. 1988.
30. Civiale C, Ritschel WA, Shiu GK, Aiache JM, Beyssac E. *In vitro/in vivo* correlation of salbutamol sulphate release from a controlled release osmotic pump delivery system. Methods Find Exp Clin Pharmacol 1991;13:491–498.
31. Heilmann K, Therapeutic Systems. 2nd ed.; Stuttgart: Georg Thieme Verlag Publications; 1984. p 50.
32. Siewert M. Perspectives of *in vitro* dissolution tests in establishing *in vitro/in vivo* correlations. Eur J Drug Metab Pharmacokinet 1993;18(1):7–18.
33. Süverkrüp R. *In vitro/in vivo* correlations: concepts and misconceptions. Acta Pharm Technol 1986;32:105–108.
34. Das SK, Gupta BK. Simulation of physiological pH-time profile in *in vitro* dissolution study. Relationship between dissolution rate and bioavailability of controlled release dosage forms. Drug Dev Ind Pharm 1988;14:537–544.
35. Dominguez-Gil A. Contribution of biopharmaceutics and pharmacokinetics to improve drug therapy. Eur J Drug Metab Pharmacokinet 1993;18(1):1–5.
36. Shah VP, Burbarg M, Noory A, Dighe S, Skelly JP. Influence of higher rates of agitation on release patterns of immediate-release drug products. J Pharm Sci 1992;81:500–503.
37. Hamlin WE, Nelson E, Ballard BE, Wagner JG. Loss of sensitivity in distinguishing real differences in dissolution rates due to increasing agitation. J Pharm Sci 1962;51:432–435.
38. Nia B, Ouriemchi EM, Vergnaud JM. Calculation of the blood level of a drug taken orally with a diffusion controlled dosage form. Int J Pharm 1995;119:165–171.
39. Sournac M, Beyssac E, Maublant JC, Aiache JM, Veyre A, Bourgart J. Scintigraphic study of the gastro-intestinal transit and correlations with the drug absorption kinetics of a sustained release theophylline tablet. II. Administration in non fasting state. J. Control. Release. 1988;15:113–120.
40. Ouriemchi EM, Bouzon J, Vergnaud JM. Modelling the process of controlled release of drug in *in vitro* and *in vivo* tests. Int J Pharm 1995;113:231–240.
41. Ouriemchi EM, Vergnaud JM. Prediction of *in-vivo* blood level with controlled-release dosage forms. Effect of the gastrointestinal tract time,. J. Pharm. Pharmacol. 1996;48:391–396.
42. Ouriemchi EM, Vergnaud JM. Calculation of the plasma drug level with oral controlled release dosage forms. Effect of the dose frequency. Int J Pharm 1996;127:177–184.
43. Aïnaoui A, Vergnaud JM. Modelling the plasma drug level with oral controlled release dosage forms with lipidic Gelucire. Int J Pharm 1998;169:155–162.
44. Ouriemchi EM, Vergnaud JM. Plasma drug level assessment with controlled release dosage forms with a core and shell and lower concentration in the shell. Int J Pharm 1999; 176:251–260.

45. Vergnaud JM. Use of polymers in pharmacy for oral dosage forms with controlled release. Recent Res Devel Macromol Res 1999;4:173–209.
46. Aïnaoui A, Vergnaud JM. Effect of the nature of the polymer and of the process of drug release (diffusion or erosion) for oral dosage forms. Comput Theoret Polym Sci 2000;10:383–390.
47. Vergnaud, JM. Invited speaker by the FDA. Modelling the process of drug delivery with oral dosage forms controlled by diffusion. Washington, DC. 1996.
48. Vergnaud, JM. Invited speaker by the FDA. Modelling the process of drug delivery with oral dosage forms controlled by erosion. Washington, DC. 1998.
49. Vergnaud JM, Rosca ID. Assessing Bioavailability of Drug Delivery Systems. CRC Press; Boca raton, FL, USA: 2005.
50. Crank J. The Mathematics of Diffusion. Oxford, UK: Clarendon Press; 1975; Chapters 1–6.
51. Carslaw HS, Jaeger JC. Conduction of Heat in Solids. Oxford, UK: Clarendon Press; 1959; Chapters 1–7.
52. Vergnaud JM. Liquid Transport Processes in Polymeric Materials. Englewood cliffs, NJ: Prentice Hall; 1991; Chapters 1–3.
53. Vergnaud JM. Controlled Drug Release of Oral Dosage Forms. Chichester, UK: Hellis Horwood; 1993; Chapters 1–5.
54. Droin A, Chaumat C, Rollet M, Taverdet JL, Vergnaud JM. Model of matter transfers between sodium salicylate-Eudragit matrix and gastric liquid. Int J Pharm 1985;27:233–243.
55. Malley I, Bardon J, Rollet M, Taverdet JL, Vergnaud JM. Modelling of controlled release in case of Carbopol-sodium salicylate matrix in gastric liquid. Drug Dev Ind Pharm 1987;13:67–81.
56. Messadi D, Vergnaud JM. Simultaneous diffusion of benzyl alcohol into plasticized PVC and of plasticizer from polymer into liquid. J Appl Polym Sci 1981;26:2315–2324.
57. Siepmann J, Ainaoui A, Vergnaud JM, Bodmeier R. Calculation of the dimensions of drug–polymer devices based on diffusion parameters. J Pharm Sci 1998;87(7):827–832.
58. Rossi S, Bonferoni MC, Caramella C, Ironi L, Tentoni S. Model-based interpretation of creep profiles for the assessment of polymer–mucin interaction. Pharm Res 1999;16:1456–1463.
59. Riley RG, Tsibouklis J, Dettmar PW, Hampson F, Davis JA, Kelly G, Wilber WR. An investigation of mucus/polymer rheological synergism using synthesised and characterised poly(acrylic acids). Int J Pharm 2001;217:87–100.
60. Riley RG, Smart JD, Tsibouklis J, Young SA, Hampson F, Davis A, Kelly G, Dettmar PW, Wilber WR. An *in vitro* model for investigating the gastric mucosal retention of C^{14}-labelled poly(acrylic acid) dispersions. Int. J. Pharm. 2002;236:87–96.
61. Aïnaoui A, Siepmann J, Bodmeier R, Vergnaud JM. Calculation of the dimensions of dosage forms with release controlled by diffusion for *in vivo* use. Eur J Pharm Biopharm 2001;51:17–24.
62. Vergnaud JM. Rosca ID. Assessing Bioavailability of Drug Delivery Systems CRC Press Taylor & Francis; Boca Raton, FL: 2006.
63. Vergnaud, JM. Invited speaker by the FDA. Calculation of the plasma drug profile with oral dosage forms controlled by diffusion. Washington, DC. 1966.
64. Droin A, Chaumat C, Rollet J, Taverdet JL, Vergnaud JM. Model of matter transfer between sodium salicylate-Eudragit matrix and gastric liquid. Int J Pharm 1985;27:233–243.
65. Heilmann K, Therapeutic Systems. 2nd ed.; Stuttgart: Georg Thieme Verlag; 1984. p 20.
66. Caron HS, Roth HP. Patients' cooperation with a medical regimen. J Am Med Assoc 1968;203:120.
67. Ouriemchi EM, Vergnaud JM. Assessment of the drug level in bronchial secretion with patient's non-compliance and oral dosage forms with controlled release. Inflammopharmacol. 2000;8(3): 267–283.
68. Vergnaud JM. Controlled Drug Release of Oral Dosage Forms. London: Ellis Horwood Publications; 1993; Chapter 12.
69. Heller J. Biodegradable polymers in controlled drug delivery. CRC Crit Rev Ther Drug Carrier Syst 1984;1:39–90.
70. Vergnaud JM. Liquid Transport Processes in polymeric Materials Prentice Hall Publications; 1991; Chapter 10.
71. Bidah D, Vergnaud JM. Kinetics of *in-vitro* release of Na salicylate dispersed in Gelucire. Int J Pharm 1990;58:215–220.
72. Bonferoni MC, Rossi S, Ferrari F, Bertoni M, Caramella C. Influence of medium on dissolution-erosion behaviour of Na carboxymethylcellulose and on viscoelastic properties of gels. Int J Pharm 1995;117(1):41–48.
73. Falson-Rieg F, Faivre V, Pirot F. Nouvelles formes médicamenteuses. Tec & Doc Publishers; 2004; Chapter 1.

74. Peppas NA, Sahlin JJ. Hydrogels as mucoadhesive and bioadhesive materials: a review. Biomaterials. 1996;17:1553–1561.
75. Ponchel G, Montisci M, Dembri A, Durrer C, Duchene D. Mucoadhesion of colloidal particulate systems in the gastro-intestinal tract. Eur J Pharm Biopharm 1997;44:25–31.
76. Sakuma S, Sudo R, Suzuki N, Kikuchi H, Akashi M, Hayashi M. Mucoadhesion of polystyrene nanoparticles having surface hydrophilic polymeric chains in the gastro-intestinal tract. Int J Pharm 1999;177:161–172.
77. Lavelle EC. Targeted delivery of drugs to the gastro-intestinal tract. Crit Rev Ther Drug. 2001;18:341–385.
78. Lehr CM. From sticky stuff to sweet receptors. Achievements, limits and novel approaches to bioadhesion. Eur J Drug Metab Pharmacokinet 1996;21:139–148.
79. Kocevar-Nared J, Kristl J, Smid-Korbar J. Comparative rheological investigation of crude gastric mucin and natural gastric mucus. Biomaterials 1997;18:677–681.
80. Madsen F, Eberth K, Smart JD. A rheological examination of the mucoadhesive/mucus interaction: the effect of mucoadhesive type and concentration. J Control Release 1998;50: 167–178.
81. Jabbari E, Peppas NA. Molecular weight and polydispersity effects at the interface between polystyrene and poly(vinyl methyl ether). J Mater Sci 1994;29:3969–3978.
82. Repka MA, McGinity JW. Physical-mechanical, moisture absorption and bioadhesive properties of hydroxy propyl cellulose hot-melt extruded films. Biomaterials 2000;21:1509–1517.
83. Marshall P, Snaar JE, Ng YL, Bowtell RW, Hampson FC, Dettmar PW, Melia CD. A novel application of NMR microscopy: measurement of water diffusion inside bioadhesive bonds. Magn Reson Imag 2001;19:487–488.
84. Kockisch S, Rees GP, Young SA, Tsibouklis J, Smart JD. A direct staining method to evaluate the mucoadhesion of polymers from aqueous dispersion. J Control Release 2001;77:1–6.
85. Quintanar-Guerrero D, Villalobos-Garcia R, Alvarz-Colin E, Cornejo-Bravo JM. *In vitro* evaluation of the bioadhesive properties of hydrophobic polybasic gels containing *N*-dimethylaminoethyl methacrylate-co-methylmethacrylate. Biomaterials 2001;22:957–961.
86. Duchêne D, Ponchel G. Bioadhesion of solid oral dosage forms, why and how?. Eur. J. Pharm. Biopharm 1997;44:15–23.
87. Rowland M, Riegelman S, Harris PA, Shalkoff SD, Eyring EJ. Kinetics of acetylsalicylic acid disposition in man. Nature 1967;215: 413–414.
88. Rosca ID, Vergnaud JM. Evaluation of the characteristics of oral dosage forms with release controlled by erosion. Comput Biol Med 2008;38:668–675.
89. Heilmann K. Therapeutic Systems. Stuttgart: Georg Thieme Verlag; 1984. p 37.
90. Falson-Rieg F, Pirot F. Dispositifs transdermiques. Nouvelles Formes Médicamenteuses. Tec.et Doc., Paris 2004;14:259.
91. Ouriemchi EM, Vergnaud JM. Processes of drug transfer with three different polymeric systems with transdermal drug delivery. Comput Theoret Polym Sci 2000;10:391–400.
92. Roy SD, Guttierez M, Flynn GL, Cleary GW. Controlled transdermal delivery of fentanyl: characterizations of pressure-sensitive adhesives for matrix patch design. J Pharm Sci 1996;85:491–495.
93. Ouriemchi EM, Ghosh TP, Vergnaud JM. Modelling the transdermal transfer of Metoprolol. American Chemical Society Meeting, Boston, August 23–27, 2008.
94. Ouriemchi EM, Ghosh TP, Vergnaud JM. Assessment of plasma drug level with transdermal delivery. American Chemical Society Meeting, Division of Medicinal Chemistry, Anaheim, CA, 1999.
95. Ghosh TP, Rosca ID, Vergnaud JM. Process of transdermal drug transfer from a polymer device. Third World Meet on Pharmaceutics, Berlin, Germany, April 2000.
96. Ouriemchi EM, Ghosh TP, Vergnaud JM. Transdermal drug transfer from a polymer device: study of the polymer and the process. Polym Testing 2000;19:889–897.
97. Crank J. The Mathematics of Diffusion. Oxford, UK: Clarendon Press; 1975. p 49.
98. Vergnaud JM. Controlled Drug Release of oral Dosage Forms. Chichester, UK: Ellis Horwood; 1993. p 35.
99. Ghosh TP, Adir J, Xiang JL, Onylofour S. Transdermal delivery of Metropolol: *in vitro*-skin permeation and bioavailability in hairless rats. J Pharm Sci 1995;84:158–160.
100. Regardh CG, Johnsson G. Clinical pharmacokinetics of Metropolol. Clin Pharmacokinet 1980;5:557–569.
101. Laoubi S, Vergnaud JM. Processes of chemical transfer from a packaging into liquid or solid food by diffusion–convection or by diffusion.

Plastics Rubber Compos Process Appl. 1996;25:83–91.

102. Gao K, Wientjes MG, Au JL. Use of drug kinetics in dermis to predict *in vivo* blood concentration after topical application. Pharm Res 1995;12:2012–2017.

103. Ho H, Chien YW. Kinetic evaluation of transdermal nicotine delivery systems. Drug Dev Ind Pharm 1993;19:295–313.

104. Harrison JE, Watkinson AC, Green DH, Hadgraft J, Brain K. The relative effect of Azone and Transcutol on permeant diffusivity and solubility in human stratum corneum. Pharm Res 1996;13:542–546.

105. Diez-Sales O, Watkinson AC, Herraez-Dominguez M, Javaloyes C, Hadgraft J. A mechanistic investigation of the *in vitro* human skin permeation enhancing effect of Azone. Int J Pharm 1996;129:33–40.

106. Roberts MS, Pugh WJ, Hafgraft J, Watkinson AC. Epidermal permeability-penetrant structure relationships. 1. An analysis of methods of predicting penetration of mono-functional solutes from aqueous solutions. Int J Pharm 1995;126:219–233.

107. Anigbogu ANC, Williams AC, Barry BW. Permeation characteristics of 8-Methoxypsoralen through human skin; relevance to clinical treatment. J Pharm Pharmacol 1996;48:357–366.

108. Kochak GM, Sun JX, Choi RL, Piraino AJ. Pharmacokinetic disposition of multiple-dose transdermal nicotine in healthy adult smokers. Pharm Res 1992;9:1451–1455.

109. Hadgraft J. Pharmaceutical aspects of transdermal nitroglycerin. Int J Pharm 1996;135:1–11.

110. Krishna R, Pandit JK. Carboxymethyl cellulose-sodium based transdermal drug delivery system for propanolol. J Pharm Pharmacol 1996;48:367–370.

111. Ademola JI, Chow CA, Wester RC, Maibach HI. Metabolism of propanolol during percutaneous absorption in human skin. J Pharm Sci 1993;82:767–770.

112. Mary S, Makki S, Agache P. Perméabilité cutanée en fonction de l'âge. Cosmétologie 1996;12:34–37.

113. Harwell JD, Maibach HI. Percutaneous absorption and inflammation in aged skin: a review. J. Am. Acad. Dermatol. 1994;31:1015–1021.

114. Gupta SK, Richard A, Okherolm RA, Eller M, Wey G, Rolf CN, Gorshine J. Comparison of the pharmacokinetics of two nicotine transdermal systems: Nicoderm and Habitrol. J Clin Pharmacol 1995;35:493–498.

115. Green PG, Hine RS, Kim A, Callander C, Yamane G, Szoka FC, Guy RH. Transdermal iontophoresis of amino acids and peptides *in vivo*. J Control Release 1992;21:187–190.

116. Brand RM, Guy RH. Iontophoresis of nicotine *in vitro*: Pulsatile drug delivery across the skin. J Control Release 1995;33:285–297.

117. Hirvonen J, Kalia YN, Guy RH. Transdermal delivery of peptides by iontophoresis: structure, mechanism and feasibility. Nature Biotechnol 1996;14:1710–1713.

118. Pirot F, Beradesca E, Kalia YN, Singh M, Maibach HI, Guy RH. Stratum corneum thickness and apparent water diffusivity: facile and non invasive quantitation *in vivo*. Pharm Res 1998;15:490–492.

119. Kalia YN, Guy RH. Interaction between penetration enhancers and iontophoresis. J Control Release 1997;44:33–42.

120. Bommannan D, Okuyama H, Stauffer P, Guy RH, Sonophoresis The use of high-frequency ultrasound to enhance transdermal drug delivery. Pharm Res 1992;9:559–564.

121. Falson-Rieg F, Faivre V, Pirot F. Microspheres pour administration nasale; Dispositifs d'administration pulmonaire In: Nouvelles formes médicamenteuses. Paris: Tec & Doc; 2004; Chapters 8 and 10.

122. Saidna M, Ouriemchi EM, Vergnaud JM. Assessment of antibiotic levels in lungs tissue with erosion-controlled dosage forms. Eur J Drug Metab Pharmacokinet 1997;22(3):237–243.

123. Saidna M, Ouriemchi EM, Vergnaud JM. Calculation of the antibiotic level in the lung tissue and bronchial secretion with an oral controlled-release dosage form. Inflammopharmacology 1998;6:321–337.

124. Breilh D, Jougon J, Djalarouti S, Gordier JB, Xuereb F, Velly JF, Arvis P, Landreau V, Saux MC. Diffusion of oral and intravenous 400 mg once-daily moxifloxacin into lung tissue at pharmacokinetic steady-state. J Chemother 2003;15(6):558–562.

125. Boselli E, Breilh D, Rimmelé T, Saux MC, Chassard D, Allaouchiche B. Plasma and lung concentrations of ceftazidime administered in continuous infusion to critically ill patients with severe nosocomial pneumonia. Intensive Care Med 2004;30(5):989–991.

126. Boselli E, Breilh D, Cannesson M, Xuereb F, Rimmelé T, Chassard D, Saux MC, Allaouchiche B. Steady-state plasma and intrapulmonary concentrations of piperacillin/tazobactam 4 g/0.5 g administered to critically ill

patients with severe nosocomial pneumonia. Intensive Care Med 2004;30(5):976–979.

127. Amidon GL, Lennernas H, Shah VP, Crison JR. A theoretical basis for a biopharmaceutical drug classification: the correlation of *in vitro* drug product dissolution and *in vivo* bioavailability. Pharm Res 1995;12:413–420.

128. Breilh D.Diffusion de la ciprofloxacine et de son principal métabolite la désethyl- ciprofloxacine dans le parenchyme pulmonaire humain. Etude des modélisations plasmatiques et simulation des concentrations pulmonaires de ciprofloxacine selon un modèle physiologique-pharmacocinétique. *Diplôme d'Etudes Approfondies*. Pessac, France. 1994.

129. Smith BR, Frocq JL. Bronchial tree penetration of antibiotics. Chest 1983;6:904–908.

130. Le Bel M, Vallée F, Bergeron M. Tissue penetration of ciprofloxacin after single and multiple doses. Antimicrob Agents Chemother 1986;29:501–505.

131. Vergnaud JM. The diffusion equations and basic considerations In: Controlled drug release of oral dosage forms. Chichester, UK: Ellis Horwood; 1993; Chapter 1.

132. Saux P, Martin C, Mallet MN, Papazian L, Bruguerolle B, De Micco P, Goulin F. Penetration of ciprofloxacin into bronchial secretions from mechanically ventilated patients with nosocomial bronchopneumonia. Antimicro Agents Chemother 1994;38:901–904.

133. Wise R, Gillet AP, Cadge B, Burgham SR, Baker S. The influence pf protein binding upon tissue fluid levels of six β-lactam antibiotics. J Infect Dis 1980;142:77–82.

134. Wise R, Donovan IA. Tissue penetration and metabolism of ciprofloxacin. The Am J Med 1987;82:(Suppl 4A) 103–107.

135. Bergan T. Pharmacokinetics of tissue penetration of antibiotics. Rev Infect Dis 1981;3: 45–66.

136. Bergeron M. Tissue penetration of antibiotics. Clin Biochem 1986;19:90–100.

137. Bakhouya A, Saïdna M, Vergnaud JM. Calculation of the blister-fluid history with ciprofloxacin administered orally or by infusion. Int J Pharm 1997;146:225–232.

138. Catchpole C, Andrews JM, Woodcock J, Wise R. The comparative pharmacokinetics and tissue penetration of single dose ciprofloxacin 400 mg iv and 750 mg po. J Antimicrob Chemother 1994;33:103–110.

139. Carbon C. Animal models of endocarditis. Int J Bio-Med Comput 1994;36:59–67.

140. Rosca ID, Vergnaud JM. Modelling the process of diffusion and antibacterial activity of antibiotics in endocardial vegetations. Pharm Sci 1995;1:391–394.

141. Voiriot P, Weber M, Gerard A. Persistence of Spectrococcus mitis in aortic vegetation after 25 days of penicillin-netilmicin combination therapy. N Engl J Med 1988;318:1067–1068.

142. Maire P, Barbaut X, Vergnaud JM, ElBrouzi A, Confesson M, Pivot C, Chuzeville M, Ivanoff N, Brazier JL, Jelliffe RW. Computation of drug concentrations in endocardial vegetations in patients during antibiotic therapy. Int J Biomed Comput 1994;36:77–85.

143. Confesson MA, Maire P, Barbaut X, Vergnaud JM, ElBrouzi A, Jelliffe RW. Traitement antibiotique de l'endocardite infectieuse: apport de la simulation des concentrations d'antibiotiques dans les végétations cardiaques. Soc de Pharmacie de Lyon 1992; 177.

144. Confesson MA, Barbaut X, Maire P, Vergnaud JM, ElBrouzi A, Jelliffe RW. Concentrations calculées en aminoside dans des végétations d'endocardites. Thérapie 1994;49:27–34.

145. Senoune A, Benghalem A, Erdogan K, ElBrouzi A, Vergnaud JM. Intravegetation antimicrobial distribution in endocarditis: a numerical model and establishment of the conditions for an *in-vitro* test. Int J Biomed Comput 1994;36: 69–75.

PROTEIN FLEXIBILITY IN *IN SILICO* SCREENING

Holger Gohlke
Hannes Kopitz
Department of Mathematics and Natural Sciences,
Heinrich-Heine-University,
Düsseldorf, Germany

1. INTRODUCTION

Flexibility and mobility are essential for the function of biological macromolecules. As a prominent example, adenylate kinase undergoes large conformational changes of its domains during a catalytic cycle [1,2]. These movements are coupled to small-amplitude fluctuations on the picosecond timescale of backbone atoms, such that local backbone conformational changes are required for the lid closure to occur [3]. As a another example, the ribosome behaves in a highly dynamic fashion and so acts as a molecular "machine," with the large-scale, low-frequency, ratchet-like movements of both subunits as the most dominant motions [4–10]. However, motions on a smaller scale are no less important, as demonstrated by the active role of the ribosomal exit tunnel in cotranslational processes [11]. Overall, these examples demonstrate that biological macromolecules have an intrinsic ability to switch between conformationally distinct states under native conditions and that conformational transitions occur on a wide range of scales, both in time and in space.

The ability to undergo conformational transitions becomes particularly pronounced in the case of ligand binding to several pharmacologically important proteins, for example, HIV-1 protease [12] (Fig. 1), aldose reductase [13], FK506 binding protein [14], renin [15], and DHFR [14]. The mutual conformational adaptation of binding partners is referred to as *plasticity*. Receptor plasticity is also a hallmark of DNA [16] or RNA [17–20] targets. Frequently, hydrophobic interactions between receptor and ligand lead to the observed plasticity, giving rise to the notion of a "hydrophobic collapse" of a receptor around a ligand [21]. Likewise, binding pockets that show large conformational changes have been found to be dominated by hydrophobic–hydrophobic or aromatic–aromatic residue pair interactions, whereas pockets that do not undergo conformational changes are dominated by mostly polar interactions [22]. Disorder-to-order transitions of intrinsically unfolded proteins upon complex formation are found as an extreme case of collapse [23–25].

Motion (or mobility) of the biomacromolecule is a prerequisite for receptor plasticity. Mobility describes actual movements in terms of directions and amplitudes. Flexibility (and the opposite, *rigidity*), however, is a static property that only determines the possibility of motion, whereas nothing actually moves [26]. As such, flexibility is not necessarily a prerequisite for mobility, as rigid parts of a biomolecule, for example, domains, may well move as a whole, for example, when connected by hinges. Still, knowing which parts of a biomolecule are flexible or rigid is valuable because it considerably simplifies the task of modeling biomacromolecular mobility.

The above examples of receptor plasticity demonstrate that the "rigid receptor hypothesis" [21], which is based on the "lock and key" model [27] of molecular recognition and has served as an underlying principle in structure-based ligand design (SBLD) and *in silico* screening, is no longer tenable. Instead, the ability to understand and predict receptor plasticity, and hence, to deal with protein flexibility and mobility, becomes central for a more in-depth understanding of molecular recognition processes [28] and success in SBLD and *in silico* screening. In fact, if protein–ligand docking is performed with the assumption of a rigid active site in those cases where actual protein movements have been observed, a dramatic drop of the docking accuracy is observed [29,30]. Whereas a docking success rate of 76% was reported for docking a ligand back to the protein structure derived from the ligand's cocrystal structure ("redocking"), this rate dropped to only 49% when the ligands were docked against protein structures derived from other ligand's cocrystal structures ("cross-docking") [30]. Similar drop-offs have also been reported by others [31,32]. The drop in docking accuracy

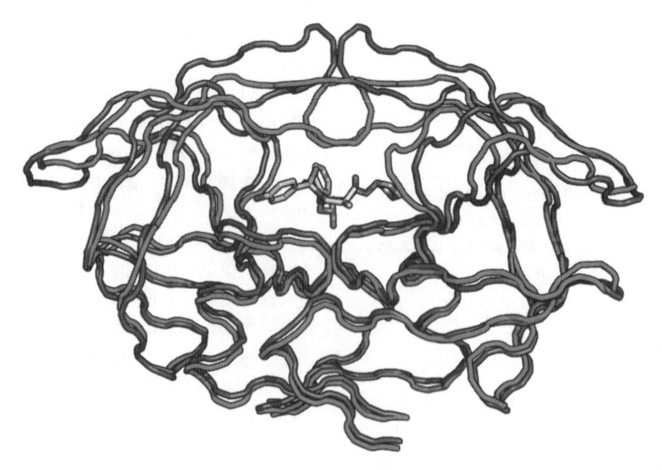

Figure 1. HIV-1 protease as an example for conformational changes between *apo* (dark grey/blue) and *holo* (light grey/green) structures. The *holo* structure is bound to amprenavir. (This figure is available in full color at http://mrw.interscience.wiley.com/emrw/9780471266945/home.)

was often found to be mirrored by the degree to which the protein moves upon ligand binding [31,33], so that docking to an *apo* form usually shows the largest deterioration [29].

Notably, biomacromolecular flexibility and mobility is not limited to influencing only steric complementarity and, hence, direct interactions between the binding partners. Rather, flexibility and mobility and their changes upon complex formation can give rise to pronounced additional, indirect energetic and entropic contributions to the binding affinity [28]. At present, these contributions are almost completely neglected in current SBLD approaches. Finally, protein flexibility and mobility provide a basis for *allosteric regulation*, which is considered to be advantageous compared to competitive regulation [34]. In contrast to the "classical view" [35,36], where allostery is explained by the occurrence of conformational changes, the "modern view of allostery" [37] considers the allosteric modulator to change the population of conformational substates that are already accessible in the global ensemble of the system [38,39]. This leads to differences in binding or signal transduction properties of the system [34,36]. Thus, this view emphasizes the role of (changes in) the protein flexibility as an entropic carrier of allosteric information [37,40–42]. Several opportunities exist to exploit the "modern view of allostery" for SBLD [43], which require a thorough understanding and accurate modeling of protein flexibility and mobility, however.

In this chapter, we will focus on the first two aspects of biomacromolecular flexibility and mobility in SBLD: modeling receptor plasticity and the energetic and entropic contributions to the binding affinity. With respect to the first, we particularly consider how protein plasticity can be incorporated into docking approaches. We feel that, from a computational point of view, these fields have seen the most progress over recent years.

2. MODELING RECEPTOR PLASTICITY IN SBLD

For incorporating protein plasticity into docking approaches, two requirements need to be met. First, one needs to detect what can move and how. Second, this knowledge needs to be transformed into a docking algorithm. As for the former, knowledge about moving protein parts can be gained from experimental information as well as established techniques such as molecular dynamics (MD) simulations, graph–theoretical approaches, or normal mode analysis. As for the latter, a multitude of approaches has been presented that range from considering protein plasticity only implicitly to modeling side-chain movements to also including backbone changes. In all cases,

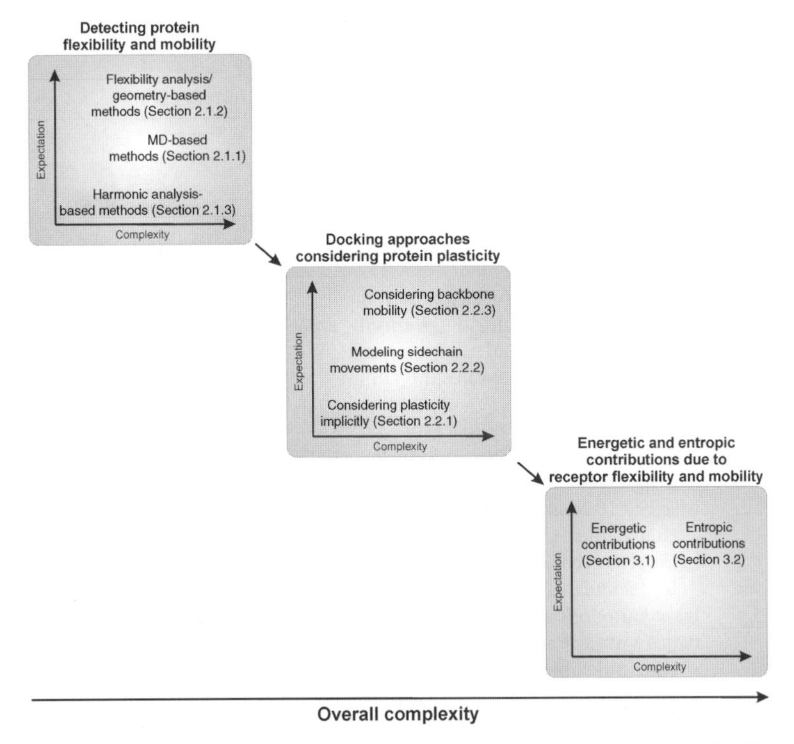

Figure 2. Flow chart showing the different approaches described in this chapter. For each subgroup, the level that we expect each method to impact the modeling of protein flexibility in *in silico* screening is related to the method's complexity. In addition, the overall complexity of the different tasks is depicted. Numbers in parentheses denote the section where each approach is described.

making a tradeoff between accuracy and efficiency is required. Figure 2 shows a flow chart summary of the different approaches described in this chapter. Table 1 lists Web site addresses of programs or servers implementing these algorithms.

2.1. Determining Protein Flexibility and Mobility

Protein plasticity comprises a range of possible movements, from single side chains to drastic structural rearrangements as seen in calmodulin [44]. Recent studies indicate, however, that with little effort much should be gained. Najmanovich et al. [45] indicated that many conformational changes can be captured by side-chain motions only, based on the finding that rotations in side-chains of up to three residues account for ≈85% of all the cases where there is a conformational change upon ligand binding. These findings were corroborated by a study by Zavodszky and Kuhn [46], according to which side-chain rotations were proven to be necessary for correctly docking about half of the protein–ligand complexes considered. However, 95% of the rotations were smaller than 45°. Regarding backbone φ/ψ changes, Gunasekaran and Nussinov [22] concluded that such dihedral angle changes upon ligand binding are in general minimal, with the most frequent large changes between the right-handed α-helical and the extended regions. On the contrary, changes between the α_R and the α_L regions were almost never seen, as they would require crossing a high-energy barrier. Apparently, these studies suggest that a protein follows a minimal energy penalty pathway to achieve required conformational changes. Considering such findings will help in developing efficient fully flexible protein–ligand docking approaches (see Section 2.2).

Table 1. Web site Addresses of Programs or Servers Implementing Algorithms Described in this Chapter

Program/Server	Web Address	Section	Reference(s)
BOSS	http://www.cemcomco.com/BOSS_and_MCPRO_Distribution125.html	2.1.1	[51]
FIRST	http://flexweb.asu.edu/software/first/	2.1.2	[56]
ProFlex	http://www.kuhnlab.bmb.msu.edu/software/proflex/		
ROCK	http://www.bch.msu.edu/~kuhn/software/rock/index.html	2.1.2	[63]
FRODA	http://flexweb.asu.edu/software/first/	2.1.2	[65]
CONCOORD	http://www.mpibpc.mpg.de/groups/de_groot/concoord/index.html	2.1.2	[69,70]
Ognm	http://ignm.ccbb.pitt.edu/GNM_Online_Calculation.htm	2.1.3	[83,84]
SOFTDOCK	http://bio.iphy.ac.cn/#Software	2.2.1	[111]
DrugScoreONLINE	http://pc1664.pharmazie.uni-marburg.de/drugscore/	2.2.1, 2.2.3	[114]
M-Score	http://sw16.im.med.umich.edu/lab/members/chaoyie/index.html#mscore	2.2.1	[116]
GOLD	http://www.ccdc.cam.ac.uk/products/life_sciences/gold/	2.2.2	[123–125]
AutoDock 4	http://autodock.scripps.edu/	2.2.2	[130]
SLIDE	http://www.bch.msu.edu/~kuhn/software/slide/index.html	2.2.2	[131]
FlexE	http://www.biosolveit.de/FlexX-Ensemble/	2.2.3	[141]
ATTRACT in PTools	http://www.ibpc.fr/chantal/www/ptools/index.html	2.2.3	[142]
FLIPDock	http://flipdock.scripps.edu/	2.2.3	[146]
Fleksy	http://yasara.org/docking.htm#fleksy	2.2.3	[150]
RosettaLigand	http://www.rosettacommons.org/software/index.html	2.2.3	[151,152]
PyRosetta	http://pyrosetta.org/		
Surflex	Available via email request to ajain@jainlab.org	2.2.3	[153]
Discovery Studio	http://accelrys.com/products/discovery-studio/	2.2.3	[154]
M2, ACCENT	http://gilsonlab.umbi.umd.edu/software1a.html	3.2	[36,185,186]

Here, we focus on techniques for the prediction of protein flexibility and mobility that are either characterized by well-established methods that are increasingly applied in SBLD or considerable developments recently. In particular, we emphasize three distinct areas of research: molecular dynamics simulation-based methods, flexibility analysis-based and geometry-based methods, and harmonic analysis-based methods.

2.1.1. Molecular Dynamics Simulations-Based Methods Molecular dynamics simulation is one of the most widely applied and accurate computational techniques currently being used in the field of macromolecular computation [47–49]. By analyzing the trajectory of a molecular structure through phase space, important information related to SBLD can be derived. This includes the analysis of flexible and mobile regions of the macromolecule and the generation of multiple conformations of a protein.

Regarding receptor plasticity, the potential of MD to enhance SBLD was demonstrated by a recent study, which predicted an intermittent opening of an unknown favorable binding trench adjacent to the catalytic site for HIV-1 integrase [14]. The prediction was experimentally validated later on [50]. Likewise, Carlson et al. [51] developed a new method for generating dynamic pharmacophore models to compensate for the inherent plasticity of an active site. This method identifies conserved binding regions over multiple configurations of the active site and uses those regions to define a complementary model.

Improvements in MD simulation techniques and increased computational power have recently allowed performing MD simulations on *unbound* protein states that clearly show a potential for generating conformations that mimic *bound* states. Using an artificially low solvent viscosity in a MD simulation [52] of unliganded HIV-1 protease, a comprehensive sampling of the protein's conformational space was achieved, which shaded light on the flap dynamics of the protein. Remarkably, multiple conversions between a "closed" and a "semiopen" form observed in crystal structures of inhibitor bound and unliganded protein, respectively, were observed, including reversal of the flap "handedness." In an MD study [53], starting from the unbound form of aldose reductase, a set of distinct conformational substates was identified, which may prove useful as input for flexible docking approaches (see Section 2.2). Similarly encouraging results were reported for the identification of transient pockets in protein–protein binding epitopes by the groups of Helms [54] and Gohlke [55]. Overall, these results are in agreement with the "conformational selection" model [38,39], according to which an appropriate receptor conformation is "picked" during binding from an ensemble of rapidly interconverting conformational species of the unbound macromolecules.

2.1.2. Methods Based on Flexibility Analysis and Geometry-Based Methods Methods based on flexibility analysis of biomacromolecules and geometric simulations are becoming valuable alternatives to force field-based MD simulations for predicting receptor flexibility and mobility due to their lower computational burden. For analyzing protein flexibility, a single, static structure is, first, represented as a so-called bond-bending network or molecular framework, where nodes represent atoms and edges represent covalent and noncovalent interactions. A fast combinatorial algorithm, the "pebble game," then identifies rigid, overrigid, and flexible regions by counting bond-rotational degrees of freedom in the network. This algorithm has been implemented into the FIRST (floppy inclusion and rigid substructure topology) software [56]. FIRST analyses have been used to accurately identify rigid regions as well as collectively and independently moving regions in a series of proteins [57,58]. Recently, the approach has also been extended to RNA structures [59,60]. FIRST results allow investigation of the internal degrees of freedom within a biomacromolecule at different levels of detail: (1) flexibility characteristics at the bond level are instructive for the analysis of binding site regions; (2) flexibility characteristics of larger regions can be related to potential global movements; and (3) rigid cluster decompositions provide hints about movements of structural parts as rigid bodies. The results also allow analysis of the long-range aspects of rigidity percolation, according to which local interactions between a ligand and a receptor may lead to a rigidification of the system at distances beyond the interaction range. Such an effect has been found in the case of a protein–protein complex formation [61], where additional interactions across the interface led to a propagation of rigidity through the binding partners.

As mentioned above, the flexibility analysis provides a natural coarse-graining of macromolecules based on rigid regions [62]. This knowledge can be further exploited for simulating protein mobility using constrained geometric simulation (CGS) [63–65] or as input for elastic network-based approaches (see Section 2.1.3) [66]. CGS-based methods either explore the rigidity-restricted conformational space by satisfying ring closure equations, as given in the program ROCK (rigidity optimized conformational kinetics) [63], or rearrange flexible and rigid parts as "ghost templates," as implemented in FRODA (framework rigidity optimized dynamic algorithm) [65]. From an application point of view, structures resulting from these approaches were used in flexible docking to cyclophilin and the estrogen receptor [67] or in docking studies of the multisubunit protein complex photosystem I [68].

An alternative geometry-based simulation method is CONCOORD (from CONstraint to

COORDinates) [69,70], which applies concepts from distance geometry [71] and generates conformations by satisfying distances and angle constraints. In an application to hyaluronate lyase [72], whose size precludes the use of MD to investigate biologically relevant timescales, flexibility (allosteric) information and functional implications were derived by CONCOORD. Furthermore, Mustard and Ritchie [73] showed that docking to multiple structures, which were obtained by an essential dynamics study following a CONCOORD run, generates better docking predictions than docking only to unbound or modeled structures.

2.1.3. Harmonic Analysis-Based Methods For a long time, normal mode analyses (NMA) have been used to study large-amplitude motions in biomacromolecules [74–76]. About a decade ago, elastic network models (ENMs) were developed, which use simplified force fields [77] and coarse-grained macromolecular models [66,78–82]: the Gaussian network model (GNM) [83,84] and the anisotropic network model (ANM) [85–87]. Since then, ENM have been applied to a vast range of problems concerning the flexibility/mobility of biomacromolecules [88–91].

Notably, conformational changes upon ligand binding are found for most of the proteins to occur along the lowest energy (frequency) modes calculated by NMA or ENM of the *unbound* protein. These modes usually involve large-amplitude and correlated motions [74,78,87,92]. Accordingly, ENMs have been mostly applied as an *a posteriori* analysis in combination with experimental studies, for example, for examining functional dynamics in *Escherichia coli* adenylate kinase, HIV-1 reverse transcriptase, and influenza virus hemagglutinin [93–95], cooperative and allosteric dynamics in tryptophan synthase [96], and binding effects in HIV-1 reverse transcriptase [94]. In agreement with the above findings, ENM have also been successfully used for efficient conformation and pathway generation, which can be exploited for docking studies [97–100] (see Section 2.2.3).

In addition to large-scale dynamical analysis, in particular, GNM has been successfully applied for residue level analyses, too [84,101–103]. At first sight, this is surprising, considering the simplicity and coarse-graining of the underlying model. Yet, high-frequency modes of GNM have been shown to be important for the identification of binding "hot spot" residues [84], catalytic residues [102], and protein-binding sites [101]. In SBLD, these studies bear the potential for efficiently identifying binding hotspot and catalytic residues. As a hybrid approach, harmonic analyses have been combined with methods that provide atomic detail such as MD. MD/NMA hybrid methods have been proposed [104–106] for amplifying collective motions along normal mode directions in a conventional MD; thus, leading to a better sampling of certain phase space regions. This method was successfully used for docking in the case of HIV-1 protease [105].

Loop motions play an important role in accommodating ligands in binding pockets, but are hard to predict. As such, for highly mobile receptors, the first few low-frequency modes identified by NMA or ENM may not describe movements of the loop region. To overcome this problem, Cavasotto et al. [100] introduced a measure of relevance of normal modes for important loop conformational changes and found that only a few low-frequency modes, usually not among the first ones, are sufficient to represent binding pocket mobility in protein kinases.

Recently, a FIRST analysis-based coarse-graining of protein structures was combined with an ENM analysis, resulting in a rigid cluster normal mode analysis (RCNMA) [66]. With this approach, predicted directions and magnitudes of receptor motions upon ligand binding agree well with experimentally determined ones (Fig. 3a), despite embracing, in extreme cases, >50% of the protein into one rigid cluster. In fact, the results of the method are in general comparable with when no coarse-graining or a uniform coarse-graining is applied; and the results are superior if the movement is dominated by loop or fragment motions. This can be explained by the fact that the appropriate coarse-graining removes irrelevant modes of the system, whereas modes related to flexible regions become more emphasized. Overall, low-frequency modes representing functional motions in flexible

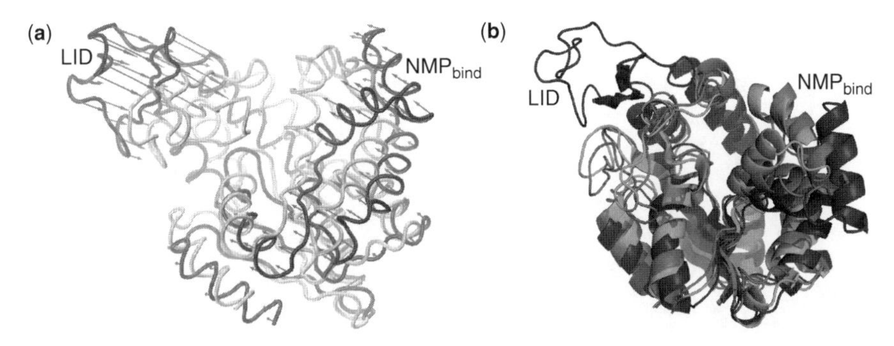

Figure 3. Superimposition of *apo* (dark grey/blue) and *holo* (light grey/green) conformations of adenylate kinase. In (a), the amplitudes and directions of motions as predicted by the mode most involved in the conformational change are depicted as red arrows. The mode was calculated using the RCNMA approach [66] and the *apo* protein conformation. The amplitudes of the motions were scaled for best graphical representation. In (b), additionally, a structure generated by NMSim calculations [107] is given in medium grey/magenta. This structure has moved by about 5 Å RMSD from the *apo* starting structure and comes as close as 2.4 Å RMSD to the *holo* conformation, although no knowledge about the bound ligand was used during simulation. Figure 3a is reproduced from Ref. [66]. (This figure is available in full color at http://mrw.interscience.wiley.com/emrw/9780471266945/home.)

regions are obtained. As a further extension of the approach, biomacromolecule conformations were then generated by deforming structures along low-energy normal mode directions predicted by RCNMA plus random direction components. Afterwards, the generated structures were iteratively corrected regarding steric clashes or stereochemical violations. The last step, termed NMsim, is similar in spirit to MD/NMA hybrid methods, but uses constrained geometric simulations instead of more time-consuming MD simulations. In total, when applied repeatedly over all three steps, the procedure efficiently generates series of conformations that lie preferentially in the low energy subspace of normal modes (Fig. 3b) [107].

2.2. Protein–Ligand Docking Approaches that Consider Protein Plasticity

Once mobility information about a protein is known, this knowledge needs to be incorporated in the protein–ligand docking algorithms. Computational efficiency is critical in this step for a thorough sampling of protein conformational space in the available computational time. Further difficulties arise when it comes to recognizing energetically accessible protein conformations among the set of decoys generated (see Section 3.1). Both findings led to the conclusion that docking including receptor plasticity is as difficult as the protein folding problem [108]. In addition to algorithmic developments, both the development of computer hardware and a better understanding of biophysical processes on the microscopic level, which lead to new models for the description of movements in biomacromolecules, have aided in considering receptor plasticity. Still, fully flexible docking requires a tradeoff between prediction quality and calculation time.

In the following sections, an overview of current fully flexible protein–ligand docking methods, which have seen a lively development recently, is given (Fig. 4). These methods can be divided into three major groups: (1) Plasticity is considered *implicitly*. In many cases, this is achieved by modifying already existing methods; (2) Side-chain conformational changes in the binding pocket are modeled; (3) Large-scale conformational changes including backbone motions are taken into account.

2.2.1. Considering Plasticity Implicitly From an algorithmic point of view, soft docking is the simplest approach for considering receptor plasticity. Here, repulsive contributions to the energy function are reduced either by capping van der Waals contributions or by reducing the van der Waals radii of the receptor atoms [109,110]. This allows a ligand atom to

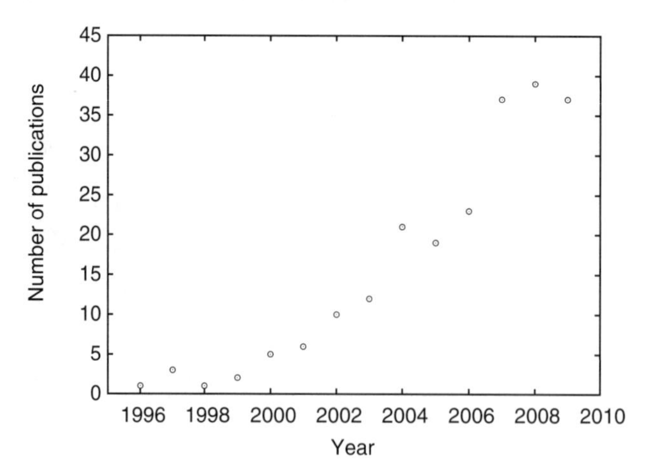

Figure 4. The number of publications dealing with protein plasticity in docking over the past 10 years. The data was derived from the ISI Web of Science database (updated Nov 30, 2009) using the query "docking AND ("flexible protein" OR "protein flexibilit*" OR "protein plasticit*")."

interpenetrate a receptor surface to a certain degree. In the case of only small side-chain rearrangements in the pocket upon binding, this should be sufficient to find a right binding pose for a ligand even when docking to an *apo* protein conformation. As a main advantage, this approach can be easily implemented into existing docking algorithms [111]. Furthermore, the approach should not be slower than docking considering ligand flexibility only, because no additional degrees of freedom are added to the search problem.

Along these lines, the reduced steepness of knowledge-based potentials has been recognized as an advantage over force field-based or empirical scoring functions [33,112,113], as the former functions are more robust to small changes in a receptor conformation. The potential softness can be amplified during the derivation process by "smearing" observed interactions over a larger distance range, as applied for the DrugScore scoring functions [114,115]. When extending this idea by more explicitly considering atom mobilities of protein atoms based on crystallographic B factors, however, only a slight improvement in the correlation between experimental and calculated binding affinities was found compared to when protein atomic mobility was not considered [116].

Finally, receptor plasticity can also be represented based on multiple structures, which have been derived from experiment or computational approaches described above (see Section 2.1). Here, we will focus on approaches where properties of several conformations are combined into one protein representation, leading to a mean-field approach, rather than considering the trivial solution to run a parallel docking against every conformation [117,118]. Knegtel et al. [119] merged interaction energy contributions of different conformations of the target protein derived from NMR or crystallography into one "average" grid and then, performed a rigid ligand–rigid receptor docking with DOCK [120]. When tested on a database of about 150 small molecules to which known HIV-1 protease ligands were added, all active ligands could be ranked within the top 17% of the database when using the merged grids. This compares favorably to recognizing at most three inhibitors within the top 25% in the case of single structure grids. Similarly, Broughton et al. [121] used combinations of several conformations to obtain merged grids. Here, the conformations were obtained by applying short MD simulations. Österberg et al. [122] pointed out that the way how grids are combined and weighted has a dramatic influence on the outcome of the docking result. They showed that naïve mean, minimum, and maximum grids performed badly, whereas a weighted averaging already applied by

Knegtel et al. as well as a Boltzmann-weighting scheme succeeded in carrying over important binding determinants of multiple structures into a single grid representation, resulting in good dockings. After all, while promising, one should note that mean-field representations bear the risk to lead to unphysical protein representations. For example, if a moving side-chain leads to two distinct interaction regions in two different receptor conformations, the combined grid may falsely represent both these regions.

2.2.2. Modeling Side-Chain Conformational Changes In proteins that undergo only small conformational changes upon ligand binding, side chains move as little as necessary to achieve a collision-free complex, according to the "minimal rotation hypothesis" [46]. This provides the basis for docking approaches in that protein plasticity is modeled by avoiding collisions of side chains with the ligand in the binding pocket. One of the early docking programs taking partial receptor plasticity into account was GOLD [123–125], which optimizes ligand placement and conformations of some terminal bonds of receptor side chains by a genetic algorithm (GA). Thereby, the program fosters the construction of a hydrogen-bond network. GOLD was evaluated on a test set of 100 protein complexes of which it could predict "good" complex configurations in 73% of the cases, albeit with more moderate results for hydrophobic ligands. Systematic problems in ranking very polar ligands and in ranking general ligands in large cavities have been reported, too [126]. Based on earlier developments [127–129], the latest AutoDock (version 4) now also supports side-chain mobility [130]. The user can chose side chains that should be considered explicitly during the docking run. Both the ligand configuration and the side-chain conformations are simultaneously optimized during docking, similar to GOLD. As a drawback, the number of side chains set flexible needs to be restricted to assure suitable calculation times.

As an alternative, the program SPECITOPE [109] uses side-chain rotations at a late stage of the docking in order to remove clashes between the ligand and the receptor. The side-chain rotations are done in an iterative manner: if a clash is detected, the bond of the side-chain next to the clashing atoms is rotated to remove the clash. If a clash free conformation cannot be obtained that way, another bond next to the ligand will be used. The approach was used to screen a library of 140,000 peptide fragments for the targets serine protease, a DNA repair enzyme, an aspartic proteinase, and a glycosyltransferase, each of which took about an hour. SPECITOPE was able to narrow down the library of ~140,000 peptides to 10–40 potential ligands for each of the investigated systems. Based on this concept, Schnecke and Kuhn have also proposed the docking algorithm SLIDE [131]. Here, an anchor fragment of the ligand is placed in a first step. Afterwards, the remaining fragments are added according to their database conformation, and clashes of the protein side chains and the ligand are corrected by rotation around the side-chain and ligand bonds in the nonanchor regions. As a disadvantage, only a restricted flexibility of the ligand and receptor is considered that way.

By combining soft docking with a postdocking optimization, Mizutani et al. [110] recently proposed an approach based on ADAM [132] that allows the consideration of side-chain movements. After a docking with ADAM into a "softened" cavity, clashes with protein atoms are detected and removed with the energy minimization program BLUTO by moving protein atoms locally. For a test case of 18 complexes, this approach performed better than the docking algorithms FlexX [133], Glide [134], and GOLD [124] with respect to ligand root mean square deviation (RMSD) of the top-ranked solutions.

Rotamer libraries that contain information about favorable discrete conformations provide a way to account for side-chain mobility without sampling the full conformational space of the side chain [135–137]. Following this idea, Frimurer et al. [135] used a rotamer library to generate an ensemble of tyrosine phosphate B1 structures by using three selected active site residues. The conformers were then used for flexible-ligand docking with FlexX [133]. Similarly, Kallblad and Dean [136] used a rotamer library to generate numerous conformers from a single MMP-1 structure. Subsequently, a core ensemble of conformers was extracted by statistical methods. A synthetic inhibitor was then rigidly docked to this core ensemble using GOLD. This resulted in a good docking pose (RMSD 1.25 Å), in contrast to docking to the experimental conformation or an energy-minimized version.

2.2.3. Considering Backbone Mobility The consideration of receptor backbone degrees of freedom for docking results in a very large search space. In addition, when large-scale macromolecular motions are modeled the issue of scoring the generated receptor conformation becomes critical: disfavorable receptor conformations need to be recognized as such, as they will contribute adversely to the overall binding affinity (see Section 3.1). Approaches that consider backbone mobility follow one of four main strategic ideas: (1) (parallel) docking into multiple conformations; (2) structurally combining multiple conformations; (3) modeling protein motions in reduced coordinates; and (4) pose optimization with MD simulations.

Regarding parallel docking, the algorithm of IFREDA (ICM-flexible receptor docking algorithm) [32] consists of three main steps. In the first step, an ensemble of receptor conformations is generated by placing the ligand at arbitrary positions and orientations into a binding pocket, followed by an energy minimization of these conformations. In this step, side-chain and essential backbone movements are taken into account, leading to an "induced-fit" of the binding pocket. As a drawback, a number of binders need to be known already. In the second step, the receptor conformations are used for a fast rigid receptor, flexible ligand docking. This is finally followed by merging the screening scores and keeping the best rank for each compound. IFREDA was tested for seven protein kinases complexes, for which it was able to generate correct conformations (RMSD < 2 Å) for 70% of the ligands in a cross-docking experiment. The dataset included structures with a backbone motion of up to 2 Å RMSD. In an attempt to improve the performance over sequential docking to multiple receptor conformations, novel ensemble docking methods have been introduced [138–140]. Here, the algorithm automatically selects a good protein structure during docking by simultaneously optimizing ligand configuration and protein conformation.

As for structurally combining multiple conformations, FlexE [141] uses experimentally determined or computed conformations as a "basis set" of receptor plasticity. By structural combination, novel adopted conformations of the receptor are generated. The search for receptor conformations is based on a so-called united protein description. This protein description prunes the search space, but at the same time allows combining alternatives to new structures that were not contained in the ensemble. That way, FlexE can account for mobility in terms of loop movements and side-chain movements. Still, the authors propose to combine FlexE with a rotamer library approach to achieve even better side-chain predictions. Likewise, FlexE's capability to consider large main-chain variations, such as domain movements, is limited.

Vibrational modes, either obtained from normal mode analysis, ENM, or MD-derived principal components (see Section 2.1.3), describe receptor movements in terms of collective motions. Using such modes as independent variables to describe receptor motions, thus, considerably reduces the complexity of the conformational search space. One of the earliest approaches along these lines was presented by Zacharias and Sklenar [142]. In docking ligands to DNA, harmonic modes derived from the energy-minimized free DNA were combined to model receptor plasticity. About 5–40 low-frequency eigenvectors were sufficient to successfully deform the DNA from the unbound structure to the ligand-bound form, sometimes coming as close as 0.5 Å to the latter. In this early approach, high-frequency movements, for example, rotations of binding pocket side chains, were not considered. This drawback has been overcome recently for protein–ligand docking by May and Zacharias by additionally modeling side-chain mobility in terms of discrete rotamers [143]. Notably, the use of modes may not be successful when it comes to modeling curvilinear receptor motions. It is important to point out, however, that with the use of harmonic modes it is possible to calculate approximate energies for the protein deformation during the docking run. This is an advantage over approaches that are purely geometry-based such as FlexE. In another approach, Cavasotto et al. [100] used relevant low-frequency modes to describe loop mobility in cAMP-dependent protein kinase (see Section 2.1.3).

A novel reduced representation of protein motions for docking has been introduced by Kazemi et al. [144], who developed an accurate representation of intermolecular interactions that makes use of the high efficiency in evaluating protein–ligand interaction energies from lookup tables even in the case of a moving protein. The new lookup table function for potential fields is based on irregular, deformable 3D grids (Fig. 5). The underlying idea is to adapt a 3D grid with precalculated potential field values, which were derived from an initial protein conformation, to another conformation by moving intersection points in space, but while keeping the potential field values constant. The approach was tested by docking into deformed potential grids for which DrugScore [114] potential values were initially calculated based on *apo* protein structures. The grids were then deformed towards *holo* conformations. For data sets of HIV-1 protease and the kinases CAPK, CDK2, and LCK, this resulted in docking success rates of 67–100%. Another sophisticated data structure for encoding the conformational space of a receptor is termed flexibility tree [145]. This data structure describes the receptor as a nested system of molecular fragments, which can be involved in a variety of movement types. Implemented into the protein–ligand docking software FLIPDock [146], the data structure provides a small number of variables with which conformational subspaces are parameterized and that can be searched during docking; thus, allowing an effective modeling of receptor conformational changes.

Finally, a number of docking approaches have been published recently that are based on a two-step protocol. In the first step, methods with low accuracy are used, which are fast and applicable to large compound databases. The second step then is more accurate and time consuming. The overall goal here is to screen many compounds without compromising the accuracy. In most of the cases, this is accomplished by docking in combination with MD simulations. The docking is used for two purposes: first, as a filter to identify compounds that can actually bind to the considered target and, second, to provide decoy complex structures as starting points for MD. In

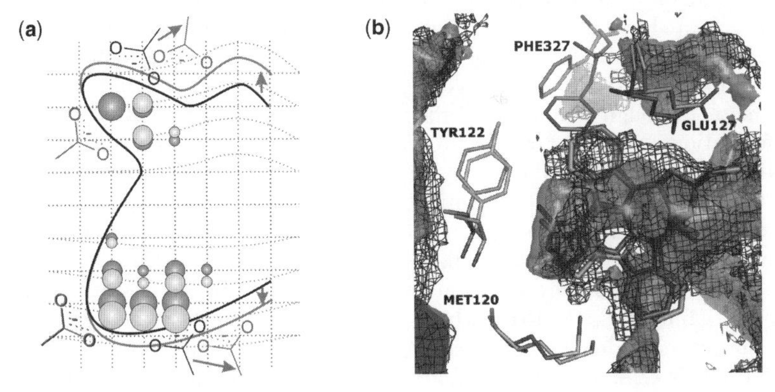

Figure 5. (a) Schematic view of the spatial deformation of potential fields inside a binding pocket according to movements of the surrounding protein. This deformation is modeled using irregular, deformable 3D grids [144]. (b) Docking of staurosporine into potential fields generated from a CAPK *apo* structure but deformed to a *holo* structure. Attractive potential fields for aromatic carbon and side-chain conformations of the *apo* structure are depicted in dark grey/blue. Deformed potential fields and side-chain conformations of the *holo* structure are depicted in light grey/orange. Staurosporine carbon atoms are displayed in light grey/green for the native structure and dark grey/magenta for the solution found for docking into the deformed grids (RMSD to the native structure: 0.64 Å). Note how the most significant movement of Phe327 "drags along" the potential field. Reproduced from Ref. [144]. (This figure is available in full color at http://mrw.interscience.wiley.com/emrw/9780471266945/home.)

the subsequent MD step, induced fit effects of the protein will be considered. During the MD step, it is also possible to incorporate explicit water molecules into the complex structure prediction, which is not provided by most of the docking approaches. An intriguing example of such an application is given by the insight gained into the preferential binding of cyclin-dependent kinase inhibitors. Here, Park et al. [147] studied the inhibition of CDK2 and CDK4 by three different selective inhibitors. The inhibitors reduced the mobility of a disordered loop of CDK4 but did not seriously affect CDK2. It was also discovered that the tighter binding of the inhibitor to CDK4 was an effect of a smaller number of water molecules moving into the binding site upon ligand binding. Such insights would not have been possible without the MD step. Furthermore, Cavalli et al. [148] pointed out that the combination of docking with subsequent MD simulations allows identifying those starting structures as being in good agreement with the experimental binding mode that lead to stable trajectories. Similar conclusions were also reported by Zoete et al. [149].

For approaches applying MD simulations, it has to be considered that even short MD runs require considerable computational time and, hence, are limited to only selected cases. Using energy minimization instead of MD simulations helps to overcome this problem, although a more thorough sampling of the protein conformational space is now required in the initial docking step. This philosophy is followed by the Fleksy approach [150], which uses a FlexE-based ensemble docking, followed by an effective Yasara-based complex optimization. Similarly, the previously published RosettaLigand method [151], which only allowed for protein side-chain degrees of freedom so far, has been extended by a stringent gradient-based minimization in the final step [152], now allowing side-chain and backbone torsions of the receptor to vary. Other approaches following the strategy of ensemble docking with subsequent receptor optimization have recently emerged [153,154]. Next, the accuracy of the docking tool used in the first step is critical for the stability of the MD trajectories and, consequently, for the success of the complex prediction. Finally, appropriate force field parameters for ligands are required that are compatible with biomolecular force fields. Those are available from, for example, the general Amber force field (GAFF) [155], which is compatible with existing Amber biomolecular force fields and has parameters for most organic and pharmaceutical small molecules.

3. ENERGETIC AND ENTROPIC CONTRIBUTIONS TO BINDING AFFINITY DUE TO RECEPTOR FLEXIBILITY AND MOBILITY

3.1. Energetic Contributions

Receptor flexibility and mobility are not limited to influencing only steric complementarity and, hence, direct interactions between the binding partners in molecular recognition. Rather, pronounced contributions to the binding affinity arise in a more indirect manner. As such, *conformational variability of the receptor* leads to a disfavorable *reorganization energy* upon binding. This contribution can be large, even for relatively well-preorganized binding sites. In binding free energy calculations of ligands binding to avidin and streptavidin, Lazaridis et al. obtained values between 9 and 36 kcal/mol [156]. Notably, these numbers are of the same order of magnitude as the binding free energies and should clearly be included to correct predicted binding affinities when it comes to flexible receptor docking [157]. In fact, in a recent assessment of 18 available scoring functions for their accuracy to predict binding affinities and to rank-order compounds by their affinity, it was found that most of the assessed scoring functions were much less accurate when a nonnative protein conformation was provided instead of the native one [158], perhaps due to missing reorganization energy contributions. Still, we note that these values are in general difficult to compute because they correspond to the small difference between large conformational energies of the bound and the unbound receptor. Further complications arise from the fact that the protein reorganization energy may not be the same even for slightly different ligands [156].

3.2. Entropic Contributions

Changes in the receptor flexibility upon complex formation, that is, changes in the internal degrees of freedom, lead to *configurational entropy* contributions. Usually, recognition sites that are in direct contact with the binding partner become less flexible, although influences on the flexibility of residues distant from the epitope are also seen. This can be explained in that perturbations at the binding site can propagate to remote locations by altering the dynamic network of interactions in proteins [159–161]. Conversely, it is increasingly recognized that complex formation may also lead to an increase in configurational entropy (which is related to an increase in flexibility of the system), compensating for the loss of translational and rotational entropy upon association [6,162–169]. For some cases, transfer of flexibility to other protein parts (within one binding partner) has been described, leading to a redistribution of protein configurational entropy, also potentially reducing the total entropy loss [38,39,170–173].

Computing configurational entropies is notoriously difficult [174–176], and the methods can be computationally demanding. Methods that belong to this category are harmonic analyses based on static structures [177–180] or covariance matrices obtained from MD simulations [181–183]. These are largely applied in the context of MM/PB(GB)SA calculations or related approaches, with varying success [174]. Other high-level methods for calculating receptor entropies are given by the "mining minima," [184] "second generation mining minima" [36,185] and "mutual information expansion approximation" [186] methods, which, so far, have been applied to host/guest systems only. Finally, entropic contributions can also be obtained by decomposing free energies computed by free energy perturbation or thermodynamic integration approaches [187,188] into energetic and entropic components. The overall entropy difference then contains changes in the solvent entropy in addition to solute contributions. Dauntingly, when investigating different ways to do so on a test system of liquid water, van Gunsteren and coworkers concluded that "none of the considered techniques seems suitable to give a perspective for the calculation of the entropy of ligand–protein binding" [189]. As a major obstacle in that respect, the limited sampling of the configurational phase space has been identified.

All of the above methods are inadequate in terms of computational requirements for applications in lead finding and virtual screening. Faster but approximate approaches have been developed for application within scoring functions. In the majority of cases, ligand entropy changes are estimated as sums of constant terms penalizing the restriction of external and internal degrees of freedom of the binding partners, as originally introduced by Böhm [190]. More sophisticated approaches to treat ligand entropy contributions have been introduced in recent years [191–195]. Receptor contributions are still ignored by approximate approaches, however, except for empirical scales considering the restriction of side-chain motions within binding pockets [178]. Clearly, much more work is required in this area, as the above rigorous techniques reveal that receptor contributions should not be neglected.

4. SUMMARY AND OUTLOOK

Computational approaches in structure-based ligand design and *in silico* screening that address issues of protein flexibility and mobility have been reviewed. Motivated by the "conformational selection" model, according to which conformational states adopted by a *holo* protein may be found already by investigating conformational fluctuations of the *apo* structure, in a first step one detects what can move and how. MD simulations, graph theoretical and geometry-based approaches, or harmonic analysis-based methods are used for this. In most of the cases, these methods describe protein motions at a certain scale. Approaches that describe motions on multiple scales are notable exceptions; they usually combine two or more of the established "single-scale" techniques. We expect that further developments along the lines of multiscale modeling will be particularly fruitful. Subsequently, the knowledge about moving protein parts needs to be included into a docking strategy. Three main routes of algorithmic development have been identified: (1) implicit consideration of plasticity; (2) modeling of side-chain motions; and (3) accounting for large-scale conformational changes including backbone motions. Only in the latter class, receptor motions at different scales can be accounted for, and we are awaiting further exciting developments here. While the impact of receptor flexibility and mobility on structure is well received by now, energetic and entropic consequences are largely neglected so far. Progress in this area is highly necessary, in particular, as considering receptor plasticity during docking yields ample possibilities for identifying false positive ligand candidates.

ACKNOWLEDGMENTS

This work was supported by funds from Heinrich-Heine-University, Düsseldorf, Germany. We are grateful to Daniel Cashman for critically reading the manuscript.

REFERENCES

1. Vonrhein C, Schlauderer GJ, Schulz GE. Movie of the structural-changes during a catalytic cycle of nucleoside monophosphate kinases. Structure 1995;3(5): 483–490.
2. Wolf-Watz M, Thai V, Henzler-Wildman K, Hadjipavlou G, Eisenmesser EZ, Kern D. Linkage between dynamics and catalysis in a thermophilic–mesophilic enzyme pair. Nat Struct Mol Biol 2004;11(10): 945–949.
3. Henzler-Wildman KA, Lei M, Thai V, Kerns SJ, Karplus M, Kern D. A hierarchy of timescales in protein dynamics is linked to enzyme catalysis. Nature 2007;450(7171): 913–916.
4. Frank J, Agrawal RK. A ratchet-like intersubunit reorganization of the ribosome during translocation. Nature 2000;406(6793): 318–322.
5. Berisio R, Schluenzen F, Harms J, Bashan A, Auerbach T, Baram D, Yonath A. Structural insight into the role of the ribosomal tunnel in cellular regulation. Nat Struct Biol 2003;10(5): 366–370.
6. Arumugam S, Gao G, Patton BL, Semenchenko V, Brew K, Van Doren SR. Increased backbone mobility in β-barrel enhances entropy gain driving binding of N-TIMP-1 to MMP-3. J Mol Biol 2003;327:719–734.
7. Tama F, Valle M, Frank J, Brooks CL 3rd. Dynamic reorganization of the functionally active ribosome explored by normal mode analysis and cryo-electron microscopy. Proc Nat Acad Sci 2003;100(16): 9319–9323.
8. Sanbonmatsu KY, Joseph S, Tung CS. Simulating movement of tRNA into the ribosome during decoding. Proc Natl Acad Sci USA 2005;102(44): 15854–15859.
9. Horan L, Noller H. Intersubunit movement is required for ribosomal translocation. Proc Natl Acad Sci USA 2007;104(12): 4881–4885.
10. Korostelev A, Trakhanov S, Laurberg M, Noller HF. Crystal structure of a 70S ribosome-tRNA complex reveals functional interactions and rearrangements. Cell 2006;126(6): 1065–1077.
11. Fulle S, Gohlke H. Statics of the ribosomal exit tunnel: implications for cotranslational peptide folding, elongation regulation, and antibiotics binding. J Mol Biol 2009;387(2): 502–517.
12. Wlodawer A, Vondrasek J. Inhibitors of HIV-1 protease: a major success of structure-assisted drug design. Annu Rev Biophys Biomol Struct 1998;27:249–284.
13. Wilson DK, Tarle I, Petrash JM, Quiocho FA. Refined 1.8 Å structure of human aldose reductase complexed with the potent inhibitor zopolrestat. Proc Natl Acad Sci USA 1993;90 (21): 9847–9851.
14. Lin JH, Perryman AL, Schames JR, McCammon JA. Computational drug design accommodating receptor flexibility: the relaxed complex scheme. J Am Chem Soc 2002;124(20): 5632–5633.
15. Vieira E, Binggeli A, Breu V, Bur D, Fischli W, Guller R, Hirth G, Marki HP, Muller M, Oefner C, Scalone M, Stadler H, Wilhelm M, Wostl W. Substituted piperidines: highly potent renin inhibitors due to induced fit adaptation of the active site. Bioorg Med Chem Lett 1999;9(10): 1397–1402.
16. Allemann RK, Egli M. DNA recognition and bending. Chem Biol 1997;4(9): 643–650.
17. Puglisi JD, Chen L, Frankel AD, Williamson JR. Role of RNA structure in arginine recognition of TAR RNA. Proc Natl Acad Sci USA 1993;90(8): 3680–3684.
18. Puglisi JD, Tan R, Calnan BJ, Frankel AD, Williamson JR. Conformation of the TAR RNA-arginine complex by NMR spectroscopy. Science 1992;257(5066): 76–80.

19. Fourmy D, Yoshizawa S, Puglisi JD. Paromomycin binding induces a local conformational change in the A-site of 16 S rRNA. J Mol Biol 1998;277(2): 333–345.
20. Fulle S, Gohlke H. Molecular recognition of RNA: challenges for modelling interactions and plasticity. J Mol Recognit 2009. DOI: 10.1002/jmr.1000.
21. Davis AM, Teague SJ. Hydrogen bonding, hydrophobic interactions, and failure of the rigid receptor hypothesis. Angew Chem Int Ed Engl 1999;38(6): 736–749.
22. Gunasekaran K, Nussinov R. How different are structurally flexible and rigid binding sites? Sequence and structural features discriminating proteins that do and do not undergo conformational change upon ligand binding. J Mol Biol 2007;365(1): 257–273.
23. Wright PE, Dyson HJ. Intrinsically unstructured proteins: re-assessing the protein structure–function paradigm. J Mol Biol 1999;293 (2): 321–331.
24. Levy Y, Cho SS, Onuchic JN, Wolynes PG. A survey of flexible protein binding mechanisms and their transition states using native topology based energy landscapes. J Mol Biol 2005;346:1121–1145.
25. Liu J, Tan H, Rost B. Loopy proteins appear conserved in evolution. J Mol Biol 2002;322: 53–64.
26. Gohlke H, Thorpe MF. A natural coarse graining for simulating large biomolecular motion. Biophys J 2006;91(6): 2115–2120.
27. Fischer E. Einfluss der konfiguration auf die wirkung der enzyme. Ber Dtsch Chem Ges 1894;27:2985–2993.
28. Gohlke H, Klebe G. Approaches to the description and prediction of the binding affinity of small-molecule ligands to macromolecular receptors. Angew Chem Int Ed 2002;41: 2644–2676.
29. Erickson JA, Jalaie M, Robertson DH, Lewis RA, Vieth M. Lessons in molecular recognition: the effects of ligand and protein flexibility on molecular docking accuracy. J Med Chem 2004;47(1): 45–55.
30. Murray CW, Baxter CA, Frenkel AD. The sensitivity of the results of molecular docking to induced fit effects: application to thrombin, thermolysin and neuraminidase. J Comput Aided Mol Des 1999;13(6): 547–562.
31. Verdonk ML, Mortenson PN, Hall RJ, Hartshorn MJ. Murray CW. protein–ligand docking against non-native protein conformers. J Chem Inf Model 2008;48(11): 2214–2225.
32. Cavasotto CN, Abagyan RA. Protein flexibility in ligand docking and virtual screening to protein kinases. J Mol Biol 2004;337(1): 209–225.
33. Ferrara P, Gohlke H, Price DJ, Klebe G, Brooks CL. Assessing scoring functions for protein–ligand interactions. J Med Chem 2004;47: 3032–3047.
34. Christopoulos A. Allosteric binding sites on cell-surface receptors: novel targets for drug discovery. Nat Rev Drug Discov 2002;1: 198–210.
35. Monod J, Wyman J, Changeux J-P. On the nature of allosteric transitions: a plausible model. J Mol Biol 1965;12:88–118.
36. Changeux J-P, Edelstein SJ. Allosteric mechanisms of signal transduction. Science 2005;308:1424–1428.
37. Kern D, Zuiderweg ER. The role of dynamics in allosteric regulation. Curr Opin Struct Biol 2003;13(6): 748–757.
38. Tsai C-J, Ma B, Nussinov R. Folding and binding cascades: shifts in energy landscapes. Proc Natl Acad Sci USA 1999;96:9970–9972.
39. Tsai C-J, Kumar S, Ma B, Nussinov R. Folding funnels, binding funnels, and protein function. Prot Sci 1999;8:1181–1190.
40. Volkman BF, Lipson D, Wemmer DE, Kern D. Two-state allosteric behavior in a single-domain signaling protein. Science 2001;291 (5512): 2429–2433.
41. Rousseau F, Schymkowitz J. A systems biology perspective on protein structural dynamics and signal transduction. Curr Opin Struct Biol 2005;15:23–30.
42. Gunasekaran K, Ma B, Nussinov R. Is allostery an intrinsic property of all dynamic proteins? Proteins 2004;57:433–443.
43. Hardy JA, Wells JA. Searching for new allosteric sites in enzymes. Curr Opin Struct Biol 2004;14(6): 706–715.
44. Barnard D, Diaz B, Hettich L, Chuang E, Zhang XF, Avruch J, Marshall M. Identification of the sites of interaction between c-Raf-1 and Ras-GTP. Oncogene 1995;10(7): 1283–1290
45. Najmanovich R, Kuttner J, Sobolev V, Edelman M. Side-chain flexibility in proteins upon ligand binding. Proteins 2000;39(3): 261–268.
46. Zavodszky MI, Kuhn LA. Side-chain flexibility in protein–ligand binding: the minimal rotation hypothesis. Protein Sci 2005;14(4): 1104–1114.

47. Hansson T, Oostenbrink C, van Gunsteren W. Molecular dynamics simulations. Curr Opin Struct Biol 2002;12(2): 190–196.
48. Karplus M, McCammon JA. Molecular dynamics simulations of biomolecules. Nat Struct Biol 2002;9(9): 646–652.
49. Adcock SA, McCammon JA. Molecular dynamics: survey of methods for simulating the activity of proteins. Chem Rev 2006;106(5): 1589–1615.
50. Hazuda DJ, Anthony NJ, Gomez RP, Jolly SM, Wai JS, Zhuang L, Fisher TE, Embrey M, Guare JP Jr, Egbertson MS, Vacca JP, Huff JR, Felock PJ, Witmer MV, Stillmock KA, Danovich R, Grobler J, Miller MD, Espeseth AS, Jin L, Chen IW, Lin JH, Kassahun K, Ellis JD, Wong BK, Xu W, Pearson PG, Schleif WA, Cortese R, Emini E, Summa V, Holloway MK, Young SD. A naphthyridine carboxamide provides evidence for discordant resistance between mechanistically identical inhibitors of HIV-1 integrase. Proc Natl Acad Sci USA 2004;101(31): 11233–11238.
51. Carlson HA, Masukawa KM, Rubins K, Bushman FD, Jorgensen WL, Lins RD, Briggs JM, McCammon JA. Developing a dynamic pharmacophore model for HIV-1 Integrase. J Med Chem 2000;43(11): 2100–2114.
52. Hornak V, Okur A, Rizzo RC, Simmerling C. HIV-1 protease flaps spontaneously open and reclose in molecular dynamics simulations. Proc Natl Acad Sci USA 2006;103(4): 915–920.
53. Sotriffer CA, Krämer O, Klebe G. Probing flexibility and "induced-fit" phenomena in aldose reductase by comparative crystal structure analysis and molecular dynamics simulations. Proteins 2004;56:52–66.
54. Eyrisch S, Helms V. Transient pockets on protein surfaces involved in protein–protein interaction. J Med Chem 2007;50(15): 3457–3464.
55. Cozzini P, Kellogg GE, Spyrakis F, Abraham DJ, Costantino G, Emerson A, Fanelli F, Gohlke H, Kuhn LA, Morris GM, Orozco M, Pertinhez TA, Rizzi M, Sotriffer CA. Target flexibility: an emerging consideration in drug discovery and design. J Med Chem 2008;51 (20): 6237–6255.
56. Jacobs D, Rader AJ, Kuhn L, Thorpe MF. Protein flexibility predictions using graph theory. Proteins 2001;44(2): 150–165.
57. Jacobs DJ, Kuhn LA, Thorpe MF. Flexible and rigid regions in proteins. In: Thorpe MF, Duxbury PM, editors. Rigidity Theory and Applications. New York: Kluwer Academic/Plenum Publishers; 1999. p 357–384.
58. Jacobs DJ, Rader AJ, Kuhn LA, Thorpe MF. Protein flexibility predictions using graph theory. Proteins 2001;44(2): 150–165.
59. Fulle S, Gohlke H. Analyzing the flexibility of RNA structures by constraint counting. Biophys J 2008;94(11): 4202–4219.
60. Fulle S, Gohlke H. Constraint counting on RNA structures: Linking flexibility and function. Methods 2009;49(2): 181–188.
61. Gohlke H, Kuhn LA, Case DA. Change in protein flexibility upon complex formation: analysis of Ras-Raf using molecular dynamics and a molecular framework approach. Proteins 2004;56:322–337.
62. Gohlke H, Thorpe M. A natural coarse graining for simulating large biomolecular motion. Biophys J 2006.
63. Lei M, Zavodszky M, Kuhn L, Thorpe MF. Sampling protein conformations and pathways. J Comput Chem 2004;25(9): 1133–1148.
64. Thorpe MF, Lei M, Rader AJ, Jacobs DJ, Kuhn LA. Protein flexibility and dynamics using constraint theory. J Mol Graph Model 2001;19(1): 60–69.
65. Wells S, Menor S, Hespenheide BM, Thorpe MF. Constrained geometric simulation of diffusive motions in proteins. Phys Biol 2005;2:1–10.
66. Ahmed A, Gohlke H. Multi-scale modeling of macromolecular conformational changes combining concepts from rigidity and elastic network theory. Proteins 2006;63:1038–1051.
67. Zavodszky M, Lei M, Thorpe MF, Day A, Kuhn L. Modeling correlated main-chain motions in proteins for flexible molecular recognition. Proteins 2004;57(2): 243–261.
68. Jolley CC, Wells SA, Hespenheide BM, Thorpe MF, Fromme P. Docking of photosystem I subunit C using a constrained geometric simulation. J Am Chem Soc 2006;128(27): 8803–8812.
69. de Groot BL, Vriend G, Berendsen HJ. Conformational changes in the chaperonin GroEL: new insights into the allosteric mechanism. J Mol Biol 1999;286(4): 1241–1249.
70. de Groot BL, van Aalten DMF, Scheek RM, Amadei A, Vriend G, Berendsen HJC. Prediction of protein conformational freedom from distance constraints. Proteins 1998;29(2): 240–251.
71. Crippen GM. Novel-approach to calculation of conformation: distance geometry. J Comput Phy 1977;24(1): 96–107.
72. Singh S, Jedrzejas MJ, Air GM, Luo M, Laver WG, Brouillette WJ. Structure-based inhibi-

tors of influenza virus sialidase. A benzoic acid lead with novel interaction. J Med Chem 1995;38(17): 3217–3225.

73. Mustard D, Ritchie D. Docking essential dynamics eigenstructures. Proteins 2005;60 (2): 269–274.
74. Go N, Noguti T, Nishikawa T. Dynamics of a small globular protein in terms of low-frequency vibrational modes. Proc Natl Acad Sci USA 1983;80(12): 3696–700.
75. Brooks B, Karplus M. Harmonic dynamics of proteins: normal modes and fluctuations in bovine pancreatic trypsin inhibitor. Proc Natl Acad Sci USA 1983;80(21): 6571–6575.
76. Case D. Normal mode analysis of protein dynamics. Curr Opin Struct Biol 1994;4(2): 285–290.
77. Tirion MM. Large amplitude elastic motions in proteins from single-parameter atomic analysis. Phys Rev Lett 1996;77:1905–1908.
78. Tama F, Gadea FX, Marques O, Sanejouand Y-H. Building-block approach for determining low-frequency normal modes of macromolecules. Proteins 2000;41:1–7.
79. Kurkcuoglu O, Jernigan RL, Doruker P. Mixed levels of coarse-graining of large proteins using elastic network model succeeds in extracting the slowest motion. Polymer 2004;45:649–657.
80. Doruker P, Jernigan RL, Bahar I. Dynamics of large proteins through hierarchical levels of coarse-grained structures. J Comp Chem 2002;23(1): 119–127.
81. Li G, Cui Q. A coarse-grained normal mode approach for macromolecules: an efficient implementation and application to Ca^{2+}-ATPase. Biophys J 2002;83(5): 2457–2474.
82. Durand P, Trinquier G, Sanejouand Y-H. A new approach for determining low-frequency normal modes in macromolecules. Biopolymers 1994;34:759–771.
83. Bahar I, Atilgan AR, Erman B. Direct evaluation of thermal fluctuations in proteins using a single-parameter harmonic potential. Fold Des 1997;2(3): 173–181.
84. Bahar I, Erman B, Haliloglu T, Jernigan RL. Efficient characterization of collective motions and interresidue correlations in proteins by low-resolution simulations. Biochemistry 1997;36(44): 13512–13523.
85. Hinsen K. Analysis of domain motions by approximate normal mode calculations. Proteins 1998;33(3): 417–429.
86. Atilgan AR, Durell SR, Jernigan RL, Demirel MC, Keskin O, Bahar I. Anisotropy of fluctuation dynamics of proteins with an elastic network model. Biophys J 2001;80(1): 505–515.
87. Tama F, Sanejouand YH. Conformational change of proteins arising from normal mode calculations. Protein Eng 2001;14(1): 1–6.
88. Tozzini V. Coarse-grained models for proteins. Curr Opin Struct Biol 2005;15:144–150.
89. Ma J. Usefulness and limitations of normal mode analysis in modeling dynamics of biomolecular complexes. Structure 2005;13(3): 373–380.
90. Ma J. New advances in normal mode analysis of supermolecular complexes and applications to structural refinement. Curr Protein Peptide Sci 2004;5(2): 119–123.
91. Bahar I, Rader AJ. Coarse-grained normal mode analysis in structural biology. Curr Opin Struct Biol 2005;15:586–592.
92. Brooks B, Karplus M. Normal modes for specific motions of macromolecules: application to the hinge-bending mode of lysozyme. Proc Natl Acad Sci USA 1985;82(15): 4995–4999.
93. Bahar I, Erman B, Jernigan R, Atilgan A, Covell D. Collective motions in HIV-1 reverse transcriptase: examination of flexibility and enzyme function. J Mol Biol 1999;285(3): 1023–1037.
94. Temiz NA, Bahar I. Inhibitor binding alters the directions of domain motions in HIV-1 reverse transcriptase. Proteins 2002;49:61–70.
95. Isin B, Doruker P, Bahar I. Functional motions of influenza virus hemagglutinin: a structure-based analytical approach. Biophys J 2002;82 (2): 569–581.
96. Bahar I, Jernigan RL. Cooperative fluctuations and subunit communication in tryptophan synthase. Biochemistry 1999;38(12): 3478–3490.
97. Kim MK, Jernigan RL, Chirikjian GS. Efficient generation of feasible pathways for protein conformational transitions. Biophys J 2002;83(3): 1620–1630.
98. Kim MK, Chirikjian GS, Jernigan RL. Elastic models of conformational transitions in macromolecules. J Mol Graph Model 2002;21(2): 151–160.
99. Miyashita O, Onuchic JN, Wolynes PG. Nonlinear elasticity, proteinquakes, and the energy landscapes of functional transitions in proteins. Proc Natl Acad Sci USA 2003;100 (22): 12570–12575.
100. Cavasotto CN, Kovacs JA, Abagyan RA. Representing receptor flexibility in ligand dock-

ing through relevant normal modes. J Am Chem Soc 2005;127(26): 9632–9640.

101. Ertekin A, Nussinov R, Haliloglu T. Association of putative concave protein-binding sites with the fluctuation behavior of residues. Protein Sci 2006;15(10): 2265–2277.

102. Yang LW, Bahar I. Coupling between catalytic site and collective dynamics: a requirement for mechanochemical activity of enzymes. Structure 2005;13(6): 893–904.

103. Bahar I, Atilgan AR, Demirel MC, Erman B. Vibrational dynamics of folded proteins: significance of slow and fast modes in relation to function and stability. Phys Rev Lett 1998;80: 2733–2736.

104. He J, Zhang Z, Shi Y, Liu H. Efficiently explore the energy landscape of proteins in molecular dynamics simulations by amplifying collective motions. J Chem Phys 2003;119 (7): 4005–4017.

105. Tatsumi R, Fukunishi Y, Nakamura H. A hybrid method of molecular dynamics and harmonic dynamics for docking of flexible ligand to flexible receptor. J Comput Chem 2004;25(16): 1995–2005.

106. Zhang Z, Shi Y, Liu H. Molecular dynamics simulations of peptides and proteins with amplified collective motions. Biophys J 2003;84 (6): 3583–3593.

107. Ahmed A, Gohlke H. Multiscale modeling of macromolecular conformational changes. In: Nithiarasu P, Löhner R, van Loon R, editors. 1st International Conference on Computational and Mathematical Biomedical Engineering (CMBE09), Swansea, 2009. p 219–222.

108. Halperin I, Ma B, Wolfson H, Nussinov R. Principles of docking: an overview of search algorithms and a guide to scoring functions. Proteins 2002;47(4): 409–443.

109. Schnecke V, Swanson C, Getzoff E, Tainer J, Kuhn LA. Screening a peptidyl database for potential ligands to proteins with side-chain flexibility. Proteins 1998;33:74–87.

110. Mizutani MY, Takamatsu Y, Ichinose T, Nakamura K, Itai A. Effective handling of induced-fit motion in flexible docking. Proteins 2006;63(4): 878–891.

111. Jiang F, Kim SH. "Soft docking": matching of molecular surface cubes. J Mol Biol 1991;219 (1): 79–102.

112. Gohlke H, Klebe G. Statistical potentials and scoring functions applied to protein–ligand binding. Curr Opin Struct Biol 2001;11: 231–235.

113. Brylinski M, Skolnick J. Q-Dock: low-resolution flexible ligand docking with pocket-specific threading restraints. J Comp Chem 2008;29 (10): 1574–1588.

114. Gohlke H, Hendlich M, Klebe G. Knowledge-based scoring function to predict protein–ligand interactions. J Mol Biol 2000;295(2): 337–356.

115. Pfeffer P, Gohlke H. DrugScoreRNA-knowledge-based scoring function to predict RNA–ligand interactions. J Chem Inf Model 2007;47 (5): 1868–1876.

116. Yang CY, Wang R, Wang S. M-score: a knowledge-based potential scoring function accounting for protein atom mobility. J Med Chem 2006;49(20): 5903–5911.

117. Carlson HA. Protein flexibility and drug design: how to hit a moving target. Curr Opin Chem Biol 2002;6:447–452.

118. Totrov M, Abagyan R. Flexible ligand docking to multiple receptor conformations: a practical alternative. Curr Opin Struct Biol 2008;18(2): 178–184.

119. Knegtel RM, Kuntz ID, Oshiro CM. Molecular docking to ensembles of protein structures. J Mol Biol 1997;266(2): 424–440.

120. Kuntz ID, Blaney JM, Oatley SJ, Langridge R, Ferrin TE. A geometric approach to macromolecule–ligand interactions. J Mol Biol 1982; 161:269–288.

121. Broughton HB. A method for including protein flexibility in protein–ligand docking: improving tools for database mining and virtual screening. J Mol Graph Model 2000;18 (3): 247–257, 302–304.

122. Österberg F, Morris GM, Sanner MF, Olson AJ, Goodsell DS. Automated docking to multiple target structures: incorporation of protein mobility and structural water heterogeneity in AutoDock. Proteins 2002;46: 34–40.

123. Jones G, Willett P, Glen RC. Molecular recognition of receptor sites using a genetic algorithm with a description of desolvation. J Mol Biol 1995;245:43–53.

124. Jones G, Willett P, Glen RC, Leach AR, Taylor R. Development and validation of a genetic algorithm for flexible docking. J Mol Biol 1997;267(3): 727–748.

125. Sousa SF, Fernandes PA, Ramos MJ. Protein–ligand docking: current status and future challenges. Proteins 2006;65(1): 15–26.

126. Kellenberger E, Rodrigo J, Muller P, Rognan D. Comparative evaluation of eight docking

tools for docking and virtual screening accuracy. Proteins 2004;57(2): 225–242.

127. Goodsell DS, Olson AJ. Automated docking of substrates to proteins by simulated annealing. Proteins 1990;8(3): 195–202.

128. Goodsell DS, Morris GM, Olson AJ. Automated docking of flexible ligands: applications of AutoDock. J Mol Recognit 1996;9(1): 1–5.

129. Morris GM, Goosell DS, Huey R, Hart WE, Belew RK, Olson AJ. Automated docking using a Lamarckian genetic algorithm and an empirical binding free energy function. J Comput Chem 1998;19:1693–1662.

130. Morris GM, Huey R, Lindstrom W, Sanner MF, Belew RK, Goodsell DS, Olson AJ. AutoDock4 and AutoDockTools4: automated docking with selective receptor flexibility. J Comp Chem 2009;30(16): 2785–2791.

131. Schnecke V, Kuhn LA. Virtual screening with solvation and ligand-induced complementarity. Persp Drug Discov Des 2000;20:171–190.

132. Yamada M, Itai A. Development of an efficient automated docking method. Chem Pharm Bull 1993;41:1200–1202.

133. Rarey M, Kramer B, Lengauer T, Klebe G. A fast flexible docking method using an incremental construction algorithm. J Mol Biol 1996;261(3): 470–489.

134. Friesner RA, Banks JL, Murphy RB, Halgren TA, Klicic JJ, Mainz DT, Repasky MP, Knoll EH, Shelley M, Perry JK, Shaw DE, Francis P, Shenkin PS. Glide: a new approach for rapid, accurate docking and scoring. 1. Method and assessment of docking accuracy. J Med Chem 2004;47:1739–1749.

135. Frimurer TM, Peters GH, Iversen LF, Andersen HS, Moller NP, Olsen OH. Ligand-induced conformational changes: improved predictions of ligand binding conformations and affinities. Biophys J 2003;84(4): 2273–2281.

136. Kallblad P, Dean PM. Efficient conformational sampling of local side-chain flexibility. J Mol Biol 2003;326(5): 1651–1665.

137. Leach A. Ligand docking to proteins with discrete side-chain flexibility. J Mol Biol 1994; 235:345–356.

138. Sotriffer CA, Zentgraf M. Addressing selectivity and protein flexibility by simultaneous docking to multiple targets. Protein Sci 2004;13:75–75.

139. Bottegoni G, Kufareva I, Totrov M, Abagyan R. Four-dimensional docking: a fast and accurate account of discrete receptor flexibility in ligand docking. J Med Chem 2009;52(2): 397–406.

140. Huang SY, Zou XQ. Ensemble docking of multiple protein structures: considering protein structural variations in molecular docking. Proteins 2007;66(2): 399–421.

141. Claussen H, Buning C, Rarey M, Lengauer T. FlexE: efficient molecular docking considering protein structure variations. J Mol Biol 2001;308(2): 377–395.

142. Zacharias M, Sklenar H. Harmonic modes as variables to approximately account for receptor flexibility in ligand–receptor docking simulations: applications to DNA minor groove ligand complex. J Comput Chem 1999;20: 287–300.

143. May A, Zacharias M. Protein–ligand docking accounting for receptor side chain and global flexibility in normal modes: evaluation on kinase inhibitor cross docking. J Med Chem 2008;51(12): 3499–3506.

144. Kazemi S, Kruger DM, Sirockin F, Gohlke H. Elastic potential grids: accurate and efficient representation of intermolecular interactions for fully flexible docking. ChemMedChem 2009;4(8): 1264–1268.

145. Zhao Y, Stoffler D, Sanner M. Hierarchical and multi-resolution representation of protein flexibility. Bioinformatics 2006;22(22): 2768–2774.

146. Zhao Y, Sanner MF. FLIPDock: docking flexible ligands into flexible receptors. Proteins 2007;68(3): 726–737.

147. Park H, Yeom MS, Lee S. Loop flexibility and solvent dynamics as determinants for the selective inhibition of cyclin-dependent kinase 4: comparative molecular dynamics simulation studies of CDK2 and CDK4. Chembiochem 2004;5(12): 1662–1672.

148. Cavalli A, Bottegoni G, Raco C, De Vivo M, Recanatini M. A computational study of the binding of propidium to the peripheral anionic site of human acetylcholinesterase. J Med Chem 2004;47(16): 3991–3999.

149. Zoete V, Meuwly M, Karplus M. Investigation of glucose binding sites on insulin. Proteins 2004;55(3): 568–581.

150. Nabuurs SB, Wagener M, De Vlieg J. A flexible approach to induced fit docking. J Med Chem 2007;50(26): 6507–6518.

151. Meiler J, Baker D. ROSETTALIGAND: protein-small molecule docking with full side-chain flexibility. Proteins 2006;65(3): 538–548.

152. Davis IW, Baker D. ROSETTALIGAND docking with full ligand and receptor flexibility. J Mol Biol 2009;385(2): 381–392.

153. Jain AN. Effects of protein conformation in docking: improved pose prediction through protein pocket adaptation. J Comput Aided Mol Des 2009;23(6): 355–374.
154. Koska J, Spassov VZ, Maynard AJ, Yan L, Austin N, Flook PK, Venkatachalam CM. Fully automated molecular mechanics based induced fit protein–ligand docking method. J Chem Inf Model 2008;48(10): 1965–1973.
155. Wang J, Wolf RM, Caldwell JW, Kollman PA, Case DA. Development and testing of a general amber force field. J Comput Chem 2004;25(9): 1157–1174.
156. Lazaridis T, Masunov A, Gandolfo F. Contributions to the binding free energy of ligands to avidin and streptavidin. Proteins 2002;47(2): 194–208.
157. Wei BQ, Weaver LH, Ferrari AM, Matthews BW, Shoichet BK. Testing a flexible-receptor docking algorithm in a model binding site. J Mol Biol 2004;337(5): 1161–1182.
158. Englebienne P, Moitessier N. Docking ligands into flexible and solvated macromolecules. 4. Are popular scoring functions accurate for this class of proteins? J Chem Inf Model 2009;49(6): 1568–1580.
159. Schymkowitz JW, Rousseau F, Wilkinson HR, Friedler A, Itzhaki LS. Observation of signal transduction in three-dimensional domain swapping. Nat Struct Biol 2001;8(10): 888–892.
160. Freire E. Can allosteric regulation be predicted from structure? Proc Natl Acad Sci USA 2000;97(22): 11680–11682.
161. Pan H, Lee JC, Hilser VJ. Binding sites in Escherichia coli dihydrofolate reductase communicate by modulating the conformational ensemble. Proc Natl Acad Sci USA 2000;97 (22): 12020–12025.
162. Yu L, Zhu CX, Tse-Dinh YC, Fesik SW. Backbone dynamics of the C-terminal domain of *Escherichia coli* topoisomerase I in the absence and presence of single-stranded DNA. Biochemistry 1996;35:9661–9666.
163. Olejniczak ET, Zhou MM, Fesik SW. Changes in the NMR-derived motional parameters of the insulin receptor substrate 1 phosphotyrosine binding domain upon binding to an interleukin 4 receptor phosphopeptide. Biochemistry 1997;36(14): 4118–4124.
164. Zidek L, Novotny MV, Stone MJ. Increased protein backbone conformational entropy upon hydrophobic ligand binding. Nat Struct Biol 1999;6:1118–1121.
165. Loh AP, Pawley N, Nicholson LK, Oswald RE. An increase in side chain entropy facilitates effector binding: NMR characterization of the side chain methyl group dynamics in Cdc42Hs. Biochemistry 2001;40(15): 4590–600.
166. Forman-Kay JD. The 'dynamics' in the thermodynamics of binding. Nat Struct Biol 1999;6:1086–1087.
167. Kay LE, Muhandiram DR, Wolf G, Shoelson SE, Forman-Kay JD. Correlation between binding and dynamics at SH2 domain interfaces. Nat Struct Biol 1998;5:156.
168. Kay LE, Muhandiram DR, Farrow NA, Aubin Y, Forman-Kay JD. Correlation between dynamics and high affinity binding in an SH2 domain interaction. Biochemistry 1996;35: 361–368.
169. Steinberg IZ, Scheraga HA. Entropy changes accompanying association reactions of proteins. J Biol Chem 1963;238:172–181.
170. Stivers JT, Abeygunawardana C, Mildvan AS, Whitman CP. ^{15}N NMR relaxation studies of free and inhibitor-bound 4-oxalocrotonate tautomerase: backbone dynamics and entropy changes of an enzyme upon inhibitor binding. Biochemistry 1996;35:16036–16047.
171. Hodsdon ME, Cistola DP. Ligand binding alters the backbone mobility of intestinal fatty acid-binding protein as monitored by ^{15}N NMR relaxation and ^{1}H exchange. Biochemistry 1997;36(8): 2278–2290.
172. Lee AL, Kinnear SA, Wand AJ. Redistribution and loss of side chain entropy upon formation of a calmodulin-peptide complex. Nat Struct Biol 2000;7(1): 72–77.
173. Canino LS, Shen T, McCammon JA. Changes in flexibility upon binding: application of the self-consistent pair contact probability method to protein–protein interactions. J Chem Phys 2002;117:9927–9933.
174. Gohlke H, Case DA. Converging free energy estimates: MM-PB(GB)SA studies on the protein–protein complex Ras-Raf. J Comput Chem 2004;25:238–250.
175. van Gunsteren WF, Bakowies D, Baron R, Chandrasekhar I, Christen M, Daura X, Gee P, Geerke DP, Glattli A, Hunenberger PH, Kastenholz MA, Oostenbrink C, Schenk M, Trzesniak D, van der Vegt NF, Yu HB. Biomolecular modeling: goals, problems, perspectives. Angew Chem Int Ed Engl 2006;45(25): 4064–4092.
176. Meirovitch H. Recent developments in methodologies for calculating the entropy and free energy of biological systems by computer si-

mulation. Curr Opin Struct Biol 2007;17(2): 181–186.

177. Case DA. Normal mode analysis of protein dynamics. Curr Opin Struct Biol 1994;4: 285–290.

178. Pickett SD, Sternberg MJ. Empirical scale of side-chain conformational entropy in protein folding. J Mol Biol 1993;231(3): 825–839.

179. Go N, Scheraga HA. Analysis of contribution of internal vibrations to statistical weights of equilibrium conformations of macromolecules. J Chem Phys 1969;51(11): 4751.

180. Go N, Scheraga HA. Use of classical statistical-mechanics in treatment of polymer-chain conformation. Macromolecules 1976;9(4): 535–542.

181. Levy RM, Karplus M, Kushick J, Perahia D. Evaluation of the configurational entropy for proteins: application to molecular dynamics simulations of an α-helix. Macromolecules 1984;1984:1370–1374.

182. Schlitter J. Estimation of absolute and relative entropies of macromolecules using the covariance matrix. Chem Phys Lett 1993;215: 617–621.

183. Karplus M, Kushick JN. Method for estimating the configurational entropy of macromolecules. Macromolecules 1981;14(2): 325–332.

184. Head MS, Given JA, Gilson MK. "Mining minima": direct computation of conformational free energy. J Phys Chemi A 1997;101(8): 1609–1618.

185. Nam KY, Chang BH, Han CK, Ahn SG, No KT. Investigation of the protonated state of HIV-1 protease active site. Bull Korean Chem Soc 2003;24(6): 817–823.

186. Killian BJ, Kravitz JY, Gilson MK. Extraction of configurational entropy from molecular simulations via an expansion approximation. J Chem Phys 2007;127 (2): 024107-1–024107-16.

187. Kirkwood JG. Statistical mechanics of fluid mixtures. J Chem Phys 1935;3:300–313.

188. Beveridge DL, DiCapua FM. Free energy via molecular simulation: applications to chemical and biomolecular systems. Annu Rev Biophys Biophys Chem 1989;18:431–492.

189. Peter C, Oostenbrink C, van Dorp A, van Gunsteren WF. Estimating entropies from molecular dynamics simulations. J Chem Phys 2004;120(6): 2652–2661.

190. Böhm HJ. Prediction of binding constants of protein ligands: a fast method for the prioritization of hits obtained from de novo design or 3D database search programs. J Comput Aided Mol Des 1998;12(4): 309–323.

191. Salaniwal S, Manas ES, Alvarez JC, Unwalla RJ. Critical evaluation of methods to incorporate entropy loss upon binding in high-throughput docking. Proteins 2007;66(2): 422–435.

192. Eldridge MD, Murray CW, Auton TR, Paolini GV, Mee RP. Empirical scoring functions. I. The development of a fast empirical scoring function to estimate the binding affinity of ligands in receptor complexes. J Comput Aided Mol Des 1997;11(5): 425–445.

193. Mammen M, Shakhnovich EI, Deutch JM, Whitesides GM. Estimating the entropic cost of self-assembly of multiparticle hydrogen-bonded aggregates based on the cyanuric acid center dot melamine lattice. J Org Chem 1998;63(12): 3821–3830.

194. Ruvinsky AM. Calculations of protein–ligand binding entropy of relative and overall molecular motions. J Comput Aided Mol Des 2007;21(7): 361–370.

195. Salaniwal S, Manas ES, Alvarez JC, Unwalla RJ. Critical evaluation of methods to incorporate entropy loss upon binding in high-throughput docking. Proteins 2007;66(2): 422–435.

INDEX